P9-DVF-707

ENCYCLOPEDIA OF NEUROSCIENCE

ENCYCLOPEDIA OF NEUROSCIENCE

EDITOR-IN-CHIEF

LARRY R. SQUIRE
Departments of Psychiatry, Neurosciences, and Psychology
University of California
San Diego and VA Medical Center
San Diego, CA
USA

AMSTERDAM • BOSTON • HEIDELBERG • LONDON • NEW YORK • OXFORD
PARIS • SAN DIEGO • SAN FRANCISCO • SINGAPORE • SYDNEY • TOKYO
Academic Press is an imprint of Elsevier

Academic Press is an imprint of Elsevier
32 Jamestown Road, London NW1 7BY, UK
30 Corporate Drive, Suite 400, Burlington, MA 01803, USA
525 B Street, Suite 1900, San Diego, CA 92101-4495, USA

The following articles are US government works in the public domain and are not subject to copyright:

Adolescent Brain Development and the Risk of Psychiatric Disorders
AMPA Receptors: Molecular Biology and Pharmacology
Basal Ganglia and Oculomotor Control
BDNF in Synaptic Plasticity and Memory
Blindsight: Residual Vision
Brain Damage: Functional Reorganization
Calcium Channels and SNARE Proteins
Cognition: Basal Ganglia Role
Consciousness: Neurophysiology and Visual Awareness in
Depression and the Brain
Drug Addiction: Neuroimaging
Electrical Self-Stimulation
Epilepsy
Eye Movement Disorders
Eyeblink Conditioning
Forebrain Development: Holoprosencephaly (HPE)
Functional Amnesia
Fusion Pore
Galanin and Receptors
Genetics of Human Anxiety and Its Disorders
Gonadotropin-Releasing Hormone: GnRH-1 System
Hair Cell Differentiation
Information Coding
Intracellular Calcium and Neuronal Death
Long-Term Depression (LTD): Endocannabinoids and Cerebellar LTD
Mood Stabilizers
Motor Skill Learning
Neural Crest Cell Diversification and Specification: Melanocytes
Neurodegeneration in Psychiatric Illness
Neuropsychology of Primate Reward Processes
Neurosecretion (Regulated Exocytosis in Neuroendocrine Cells)
NMDA Receptors, Cell Biology and Trafficking
Notch Pathway: Lateral Inhibition
Nutrient Sensing: Carbohydrates
Nutrition
Oculomotor System: Models
Panic Disorder
Perception and Eye Movements
Post-Traumatic Stress Disorder: Neurobiology
Post-Traumatic Stress Disorder: Overview
Representation of Reward
Retinal Models
Schwann Cells and Axon Relationship
Semantic Memory
Spasticity
Stress Response: Genetic Consequences
Superior Colliculus
Thyroid Hormone and Transcriptional Regulation in the CNS
TIP39 (Tuberoinfundibular Peptide of 39 Residues)
Vasopressin/Oxytocin and Receptors

British Library Cataloguing in Publication Data
A catalogue record for this book is available from the British Library

Library of Congress Catalog Number: 2009923450

ISBN: 978-0-08-044617-2

For information on all Elsevier publications
visit our website at books.elsevier.com

PRINTED AND BOUND IN SPAIN
09 10 11 12 13 10 9 8 7 6 5 4 3 2 1

SENIOR EDITORS

Thomas D. Albright

Systems Neurobiology Laboratories
Salk Institute for Biological Studies
La Jolla, CA
USA

Floyd E. Bloom

Molecular and Integrative Neuroscience Department
The Scripps Research Institute
La Jolla, CA
USA

Fred H. Gage

Laboratory of Genetics
Salk Institute for Biological Studies
La Jolla, CA
USA

Nicholas C. Spitzer

Neurobiology Section
University of California
La Jolla, CA
USA

ASSOCIATE EDITORS

HOW TO USE THE ENCYCLOPEDIA

The Encyclopedia of Neuroscience is intended for use by students, research professionals, and interested others. Articles have been chosen to reflect major disciplines in the study of neuroscience and common topics of research by academics in this domain. Each article serves as a comprehensive overview of a given area, providing both breadth of coverage for students, and depth of coverage for research professionals. We have designed the encyclopedia with the following features for maximum accessibility for all readers.

Articles in the encyclopedia are arranged alphabetically by subject in the Contents list. There are two indexes. The Subject Classification index appears at the start of Volume 1 and groups entries under subject headings that reflect the broad themes of Neuroscience. This index is useful for making quick connections between entries and locating the relevant entry for a topic that is covered in more than one article. Under some section headings, you will find sub-headings if it is appropriate for the entries in a subject area to be grouped into more specific sub areas. The Subject Index is located in Volume 10. Some topics are covered in a multitude of articles from differing perspectives, while other topics may have only one entry. We encourage use of the indexes for access to a subject area, rather than use of the Contents list alone, so that a reader has a full notion of the coverage of that topic.

Each article contains cross-references to other related encyclopedia articles, suggested further readings where applicable, and many contain relevant websites for additional information. Each article has been cross-referenced to other related articles in the encyclopedia at the close of each article. We encourage readers to use the cross-references to locate other encyclopedia articles that will provide more detailed information about a subject.

The Further Reading sections include recent secondary sources to aid the reader in locating more detailed or technical information. Review articles and research articles that are considered of primary importance to the understanding of a given subject area are also listed. These suggested further readings are not intended to provide a full reference listing of all material covered in the context of a given article, but are provided as next steps for a reader looking for additional information.

SUBJECT CLASSIFICATION

This subject classification is for use as a thematic guide to the contents of this Encyclopedia, it is divided by subject area into main sections and subsections, most of which are subdivided as appropriate. Every article in this Encyclopedia is listed under at least one section, many also receive repeat entries under other relevant sections. Taxonomies of this kind are always difficult and variations always possible. The reader is encouraged to browse in other categories than simply those which may at first seem obvious.

Speech

Emotion in Speech
Lexical Impairments Following Brain Injury
Reading
Sentence Comprehension
Sentence Production
Speech Perception: Adult
Speech Perception: Cortical Processing
Speech Perception: Development
Speech Perception: Neural Encoding
Speech Production: Adult
Speech Production: Development
Word Learning
Word Production
Word Recognition

DISEASE AND DYSFUNCTION

AMPA Receptors: Disease
$GABA_A$ Receptors and Disease
NMDA Receptors and Disease+C464
PANDAS (Pediatric Autoimmune Neuropsychiatric Disorders Associated with Streptococcal Infections)

Cognitive dysfunction (*see also* dementia)

Agnosia
Agraphia
Alexia
Amnesia: Declarative and Nondeclarative Memory
Animal Models of Amnesia
Aphasia: Sudden and Progressive
Apraxia: Disease
Cognition in Aging and Age-Related Disease
Cognitive Deficits in Schizophrenia
Cognitive Dysfunction in Psychiatric Disorders
Coma and Other Pathological Disorders of Consciousness
Coma
Dementia and Language
Diabetes Type 2 and Stress: Impact on Memory and the Hippocampus
Dyslexia: Neurodevelopmental Basis
Frontal Lobe Syndrome
Functional Amnesia
Language Following Congenital Disorders (not SLI)
Language: Aphasia
Language: Learning Impairments
Lexical Impairments Following Brain Injury
Memory Disorders
Neglect Syndrome and the Spatial Attention Network
Prosopagnosia
Sleep Deprivation and Brain Function
Stress and Cognition
Vegetative State

CNS infections

Cysticercosis: Cerebral
Encephalitis
Glial Responses to Virus Infection
Infectious Agents in Neurodegenerative Disease
Parasitic Diseases: Nervous System Effects
Post-Polio Syndrome
Prion Diseases
Prion Transport
Variant Creutzfeldt–Jakob Disease
Zoster and Postherpetic Neuralgia

Developmental disorders

Adolescent Brain Development and the Risk of Psychiatric Disorders
Angelman Syndrome
Autism
Cornelia De Lange Syndrome
Corpus Callosum: Agenesis
Double Cortex
Down Syndrome: A Disorder of Histogenesis
Dyslexia: Neurodevelopmental Basis
Folate Deficiency States
Heritable Microgyrias
Holoprosencephaly
Joubert Syndrome
Language Following Congenital Disorders (not SLI)
Lissencephaly Type I and Periventricular Heterotopia
Microcephaly Vera
Rett Syndrome
Syringomyelia

Genetic/chromosomal disorders

Angelman Syndrome
Animal Models of Huntington's Disease
Animal Models of Inherited Retinal Degenerations
Axonal Transport and Huntington's Disease
CADASIL (Cerebral Autosomal Dominant Arteriopathy with Subcortical Infarcts and Leukoencephalopathy)
Cell Replacement Therapy for Huntington's Disease
Cholinergic System Imaging in the Healthy Aging Process and Alzheimer Disease
Congenital Muscular Dystrophy
Cornelia De Lange Syndrome

Nervous system injury

Neurological disorders

Headache
Hemiplegic Migraine and Channel Disorders
Hepatic Encephalopathy
Hereditary Spastic Paraplegia
Landau Kleffner Syndrome
Neurotrophic Factor Therapy: GDNF and CNTF
Neurotrophic Factor Therapy: NGF, BDNF and NT-3
Porphyria: Acute Intermittent
Reye Syndrome
Syringomyelia

Auditory/vestibular disorders

Cochlear Prosthesis
Deafness
Motion Sickness
Tinnitus
Vertigo

Autonomic disorders

Animal Models of Stroke
Autoimmune Autonomic Neuropathy
Autonomic Disorders
Cerebrovascular Disease
Exercise in Neurodegenerative Disease and Stroke
Glia and Stroke
Map Plasticity and Recovery from Stroke
Stroke: Injury Mechanisms
Stroke: Neonate vs. Adult
Stroke

Cerebrovascular disorders

Animal Models of Stroke
Cerebrovascular Disease
Exercise in Neurodegenerative Disease and Stroke
Glia and Stroke
Map Plasticity and Recovery from Stroke
Stroke: Injury Mechanisms
Stroke: Neonate vs. Adult
Stroke

Dementia

Acetylcholinesterase Inhibitors and Alzheimer's Disease
Aging of the Brain and Alzheimer's Disease
Alzheimer's Disease: An Overview
Alzheimer's Disease: Molecular Genetics
Alzheimer's Disease: MRI Studies
Alzheimer's Disease: Neurodegeneration
Alzheimer's Disease: Transgenic Mouse Models
Animal Models of Alzheimer's Disease
Axonal Transport and Alzheimer's Disease
Brain Glucose Metabolism: Age, Alzheimer's Disease, and ApoE Allele Effects
Cholinergic System Imaging in the Healthy Aging Process and Alzheimer Disease
Dementia and Language
Dementia
Sleep and Circadian Rhythm Disorders in Human Aging and Dementia
Sleep in Aging

Demyelination

Axonal Injury in Demyelinating Disease and CNS Injury
Demyelinating Diseases
Demyelination and Demyelinating Antibodies
Transplantation of Myelin Forming Cells

Epilepsy/seizures

Epilepsy: Channelopathies
Epilepsy: Neuronal Death
Epilepsy
Induced Seizures as Therapy in Man
Neuropeptides: Epilepsy

Malignancy

Glioma
Medulloblastoma
Opsoclonus/Myoclonus and Neuroblastoma
Retinitis Pigmentosa
Tumors of the Brain and the Spinal Cord

Movement disorders

Animal Models of Huntington's Disease
Animal Models of Parkinson's Disease
Apraxia: Sensory System
Balint Syndrome
Basal Ganglia: Functional Models of Normal and Disease States
Cell Replacement Therapy: Parkinson's Disease
Cerebellum: Clinical Pathology
Deep Brain Stimulation and Movement Disorder Treatment
Deep Brain Stimulation and Parkinson's Disease
Hereditary Spastic Paraplegia
Huntington's Disease: Neurodegeneration
Huntington's Disease
Motor Gait and Falls
MPTP Parkinsonism Model
Parkinson's Disease: Alpha-Synuclein and Neurodegeneration
Parkinsonian Syndromes
Restless Leg Syndrome

Neurodegenerative disorders

Aluminum
Alzheimer's Disease: An Overview
Alzheimer's Disease: Neurodegeneration
Alzheimer's Disease: Transgenic Mouse Models
Amyotrophic Lateral Sclerosis (ALS)
Amyotrophic Lateral Sclerosis (ALS): Disease Mechanisms
Animal Models of Huntington's Disease
Animal Models of Inherited Retinal Degenerations
Animal Models of Motor and Sensory Neuron Disease
Apoptosis in Neurodegenerative Disease
Axonal Transport and ALS
Axonal Transport and Alzheimer's Disease
Axonal Transport and Huntington's Disease
Axonal Transport and Neurodegenerative Diseases
Cell Replacement Therapy for Huntington's Disease
Cell Replacement Therapy: Mechanisms of Functional Recovery
Cell Replacement Therapy: Parkinson's Disease
Congenital Muscular Dystrophy
Epilepsy: Neuronal Death
Excitotoxicity in Neurodegenerative Disease
Exercise in Neurodegenerative Disease and Stroke
Huntington's Disease: Neurodegeneration
Huntington's Disease
Infectious Agents in Neurodegenerative Disease
Inflammation in Neurodegenerative Disease and Injury
Inherited Macular Degenerations: Animal Models
Neural Stem Cells and CNS Diseases
Neuroprotection: Pharmacological Approaches
Oxidative Damage in Neurodegeneration and Injury
Parkinson's Disease: Alpha-Synuclein and Neurodegeneration
Post-Polio Syndrome
Protein Folding and the Role of Chaperone Proteins in Neurodegenerative Disease
Role of NO in Neurodegeneration
Spinocerebellar Atrophy
Sporadic Degenerative Ataxias and the Dominantly Inherited Spinocerebellar Ataxias
Transgenic Models of Neurodegenerative Disease
Triplicate Repeats: Huntington's disease
Wallerian Degeneration

Neuromuscular disorders

Babinski's Reflex/Sign
Congenital Muscular Dystrophy
Cranial–Cervical Dystonia
Dystonia: Myoclonus–Dystonia
Dystrophin, Associated Proteins, and Muscular Dystrophy
Eye Movement Disorders
Gap Junction Abnormalities and Disorders of the Nervous System
Mitochondrial Encephalomyopathies
Neuromuscular Junction (NMJ): Inherited and Acquired Disorders
Opsoclonus/Myoclonus and Neuroblastoma
Spasticity
Spinocerebellar Atrophy
Torsion Dystonia
Tourette's Disorder
Writer's Cramp

Neuropathies

Autoimmune Autonomic Neuropathy
Bell's Palsy
Diabetic Neuropathy
Neuropathy: Chemically-Induced
Neuropathy: Metabolically-Induced
Neuropathy: Peripheral
Peripheral Nerve Regeneration: An Overview
Trigeminal Neuralgia

Neurotransmitter systems

Autonomic Nervous System: General Overview
Segawa Dopa Responsive Dystonia
Torsion Dystonia
Tourette Syndrome as a Neurotransmitter Disorder

Psychiatric disorders

Adolescent Brain Development and the Risk of Psychiatric Disorders
Attention Deficit Hyperactivity Disorder (ADHD): Methylphenidate (Ritalin) and Dopamine
Attention Deficit Hyperactivity Disorder
Borderline Personality Disorder
Cognitive Dysfunction in Psychiatric Disorders
Eating Disorders
Executive Functions: Eye Movements and Neuropsychiatric Disorders
Eye Tracking and Mental Illness
Induced Seizures as Therapy in Man
Neurodegeneration in Psychiatric Illness
Neuropeptides: Mental Disease
Obsessive–Compulsive Disorder
Prepulse Inhibition of Startle in Humans and Laboratory Models
Psychiatric Disorders Associated with Disturbed Sleep and Circadian Rhythms
Psychiatric Disorders: Functional Genomics

Psychiatric Genomics and Expression Profiling
Psychotherapeutic Approaches to Psychiatric Disorders
Spines and Mental Disorders

Addiction/drug use

Addiction: Neurobiological Mechanism
Alcoholism
Drug Addiction: Behavioral Neurophysiology
Drug Addiction: Behavioral Pharmacology of Drug Addiction in Rats
Drug Addiction: Cellular Mechanisms
Drug Addiction: Neuroimaging
Drugs Addiction: Actions
Fetal Alcohol Syndrome
Substance Abuse and Dependence

Anxiety disorders

Anxiety Disorders
Anxiety: Drug Therapy
Dreams and Nightmares in PTSD
Panic Disorder
Posttraumatic Stress Disorder: Neurobiology
Posttraumatic Stress Disorder: Overview

Emotional disorders and phobias

Emotional Disorders: Treatment
Genetics of Human Anxiety and Its Disorders
Nightmares
Panic Disorder as an Emotional Disorder
Phobia and Human Evolution
Posttraumatic Stress Disorder as an Emotional Disorder

Mood (affective) disorders

Bipolar Disorder
Depression and the Brain
Mood Stabilizers
Neuroendocrinology of Affective Disorders
Premenstrual Dysphoric Disorder and Postpartum Major Depression: Chronobiology
Stress and Suicide
Stress, Cytokines and Depressive Illness
Stress, the HPA Axis and Depressive Illness
Suicide

Schizophrenia and psychoses

Antipsychotic Drugs
Cognitive Deficits in Schizophrenia
Neuroleptics
Schizophrenia: Epidemiology, Clinical Features, Course and Outcome
Schizophrenia: Genetics

Sleep disorders

Cataplexy
Genetics of Circadian Disorders in Humans
Immune Function During Sleep and Sleep Deprivation
Metabolic Syndrome and Sleep
Narcolepsy
Nightmares
Parasomnias
Psychiatric Disorders Associated with Disturbed Sleep and Circadian Rhythms
Sleep and Circadian Rhythm Disorders in Human Aging and Dementia
Sleep Apnea
Sleep Deprivation and Brain Function
Sleep in Adolescents
Sleep in Aging
Sleeping Sickness

EVOLUTION AND COMPARATIVE BIOLOGY OF NERVOUS SYSTEMS

Evolution of organisms/systems

Allometric Analysis of Brain Size
Basal Ganglia: Evolution
Bird Brain: Evolution
Bird Song Systems: Evolution
Brain Asymmetry: Evolution
Brain Connectivity and Brain Size
Brain Development: The Generation of Large Brains
Brain Evolution: Developmental Constraints and Relative Developmental Growth
Brain Evolution: The Radiator Theory
Brain Fossils: Endocasts
Brain Modules: Mosaic Evolution
Brain Scaling Laws
Brains of Primitive Chordates
Broca's Area: Evolution
Cerebellum: Evolution and Comparative Anatomy
Cerebral Fissure Patterns
Circadian Systems: Evolution
Cone Photopigment Evolution
Evolution of Sensory Receptor Specializations in the Glabrous Skin
Evolution of the Limbic System
Evolution of Vertebrate Brains
Evolution of Vertebrate Respiratory Control
Frontal Cortex Evolution in Primates
Gene Expression in the Evolution of the Human Brain

Homology and Homoplasy
Kinship Signals in Animals
Language Evolution
Microcephaly Vera
Neocortex: Origins
Neuropeptides Phylogeny and Evolution
Olfactory Receptor Genes: Human Loss During Evolution
Phobia and Human Evolution
Primate Communication: Evolution
Sexual Selection and the Evolution of Animal Signals
Social Brain: Evolution
Thalamus: Evolution in Vertebrates
Triune Brain Concept: A Comparative Evolutionary Perspective
Vertebrate Eyes: Evolution
Visual System: Adaptive Regression and Progression in Subterranean Mammals
Vomeronasal System Evolution

Comparative biology

Amygdala: Structure and Circuitry in Rodents and Felines
Animal Intelligence: The Search for Animal Intelligence
Auditory Systems in Insects
Birdsong Learning: Evolutionary, Behavioral, and Hormonal Issues
Brain Scaling Laws
Cerebellum: Evolution and Comparative Anatomy
Cetacean Brains
Circadian Organization in Non-Mammalian Vertebrates
Comparative Biology of Invertebrate Neuromuscular Junctions
Comparative Neurobiology: History
Connectivity of Primate Reward Centers
Genetic Regulation of Circadian Rhythms in *Drosophila*
Hearing and Echolocation in Dolphins
Learning and Memory in Invertebrate Models: *Tritonia*
Learning and Memory in Invertebrates: *Aplysia*
Learning and Memory in Invertebrates: *C. Elegans*
Learning and Memory in Invertebrates: *Drosophila*
Learning and Memory in Invertebrates: *Hermissenda*
Learning and Memory in Invertebrates: Honey Bee
Learning and Memory in Invertebrates: *Limax*
Learning and Memory in Invertebrates: Mollusks
Marsupial Neocortex
Neocortical Organization in Monotremes
Neuromuscular Connections: Vertebrate Patterns of
Non-Primate Models of Normal Brain Aging
Olfaction in Invertebrates: Honeybee
Olfactory Bulb Anatomy
Olfactory Cortex: Comparative Anatomy
Primate Interneurons
Referentiality and Concepts in Animal Cognition
Reward Neurophysiology and Primate Cerebral Cortex
Rodent Aging
Social Interaction Effects on Reward and Cognitive Abilities in Monkeys
Stroke
Visual System: Invertebrates

HISTORY

Cajal's Place in the History of Neuroscience
Comparative Neurobiology: History
History of Neuroscience: Early Neuroscience
Morphogens: History
Neuroeconomics: History
Neuron Doctrine: Historical Background
Neuropeptides: Discovery
Neurophysiology: Past and Present
Sleep Research and Sleep Medicine in Historical Perspective
Stress: Definition and History

INTERCELLULAR COMMUNICATION

Neuroendocrinology

Adrenal Steroids: Biphasic Effects on Neurons
Angiotensin Actions on and Within Brain
Birdsong Learning: Evolutionary, Behavioral, and Hormonal Issues
Chromaffin Cells: Model Cells for Neuronal Cell Biology
Chronic (Repeated) Stress: Consequences, Adaptations
Circumventricular Organs in Neuroendocrine Control
Clock Gene Regulation of Endocrine Function
Corticotropin-Releasing Hormone and Urocortins: Binding Proteins and Receptors
Corticotropin-Releasing Hormone: Integration of Adaptive Responses to Stress
Emotional Hormones and Memory Modulation
Endocrine Function During Sleep and Sleep Deprivation
Estrus and Menstrual Cycles: Neuroendocrine Control

Neuroimmunology

Neuromuscular and gap junctions

Neuropeptides and neurotrophic factors

Prolactin-Releasing Peptide
Proopiomelanocortin
Retrograde Neurotrophic Signaling
Somatostatin and Receptors
Substance P/Tachykinins and its/their Receptors
TIP39 (Tuberoinfundibular Peptide of 39 Residues)
Vasoactive Intestinal Peptide and Pituitary Adenylate Cyclase Activating Peptide Receptors
Vasopressin/Oxytocin and Receptors

Neuropharmacology

AMPA Receptors: Molecular Biology and Pharmacology
Cognition: Neuropharmacology
Drug Addiction: Behavioral Pharmacology of Drug Addiction in Rats
Drugs Addiction: Actions
Exercise: Optimizing Function and Survival at the Cellular Level
Metabotropic Glutamate Receptors (mGluRs): Molecular Biology, Pharmacology and Cell Biology
Neuropeptide Receptors – Drug Development
Pharmacology of Sleep: Adenosine

Neurotransmitters/receptors

Astrocyte: Neurotransmitter and Hormone Receptors
Cotransmission
Neuropeptides and Coexistence
Neurotransmitter and Hormone Receptors on Oligodendrocytes and Schwann Cells
Neurotransmitters and Growth Factors: Overview
Nitric Oxide

Amino acids and purines

Adenosine Triphosphate (ATP) as a Neurotransmitter
Adenosine Triphosphate (ATP)
Adenosine
AMPA Receptor Cell Biology/Trafficking
AMPA Receptors: Disease
AMPA Receptors: Molecular Biology and Pharmacology
Calcium Waves: Purinergic Regulation
d-Serine: From its Synthesis in Glial Cell to its Action on Synaptic Transmission and Plasticity
GABA Synthesis and Metabolism
$GABA_A$ Receptor Synaptic Functions
$GABA_A$ Receptors and Disease
$GABA_A$ Receptors: Developmental Roles
$GABA_A$ Receptors: Molecular Biology, Cell Biology and Pharmacology
$GABA_B$ Receptor Function
$GABA_B$ Receptors: Molecular Biology and Pharmacology
Gamma-Aminobutyric Acid (GABA)
Glial Glutamate Transporters: Electrophysiology
Glutamate Receptor Clusters: Narp, EphB2 Receptor, Stargazin
Glutamate Receptor Organization: Ultrastructural Insights
Glutamate Regulation of Dendritic Spine Form and Function
Glutamate
Glutamatergic and Gabaergic Systems
Glycine Receptors: Molecular and Cell Biology
Herbal Products and GABA Receptors
Kainate Receptor Functions
Kainate Receptors: Molecular and Cell Biology
Long-Term Depression (LTD): Metabotropic Glutamate Receptor (mGluR) and NMDAR-Dependent Forms
Long-Term Potentiation (LTP): NMDA Receptor Role
Long-Term Potentiation and Long-Term Depression in Experience-Dependent Plasticity
Metabotropic Glutamate Receptors (mGluRs): Functions
Metabotropic Glutamate Receptors (mGluRs): Molecular Biology, Pharmacology and Cell Biology
NMDA Receptor Function and Physiological Modulation
NMDA Receptors and Development
NMDA Receptors and Disease+C464
NMDA Receptors, Cell Biology and Trafficking
P2X Receptors
Pharmacology of Sleep: Adenosine
Postsynaptic Density/Architecture at Excitatory Synapses
Purinergic Receptors
Purines and Purinoceptors: Molecular Biology Overview
Transporter Proteins in Neurons and Glia

Amines and acetylcholine

Acetylcholine Neurotransmission in CNS
Acetylcholinesterase Inhibitors and Alzheimer's Disease
Acetylcholinesterase
Adenosine Receptor Mediated Functions
Adrenergic Receptors

Synaptic structure and organization (pre- and post-)

METHODS AND TECHNIQUES

Animal models/methods

Animal Models of Inherited Retinal Degenerations
Animal Models of Parkinson's Disease
Animal Models of Stroke
BAC Transgenesis: Cell-Type Specific Expression in the Nervous System
Drosophila Apterous Neurons: From Stem Cell to Unique Neuron
Drug Addiction: Behavioral Pharmacology of Drug Addiction in Rats
Episodic Memory: Assessment in Animals
Executive Function and Higher-Order Cognition: Assessment in Animals
Inherited Macular Degenerations: Animal Models
Invertebrate Models to Study Learning and Memory: Lymnaea
Learning and Memory in Invertebrates: *Aplysia*
Learning and Memory in Invertebrates: *C. Elegans*
Learning and Memory in Invertebrates: *Drosophila*
Learning and Memory in Invertebrates: *Hermissenda*
Learning and Memory in Invertebrates: Honey Bee
Learning and Memory in Invertebrates: *Limax*
Learning and Memory in Invertebrates: Mollusks
Learning and Memory in Invertebrate Models: *Tritonia*
Mammalian Sleep and Circadian Rhythms: Flies
Neural Induction in Chicks
Non-Primate Models of Normal Brain Aging
Procedural Learning in Animals
Rodent Aging
Spatial Memory: Assessment in Animals
Transgenic Models of Neurodegenerative Disease
VelociGene and VelociMouse: High-Throughput Approaches for Generating Targeted Mutations in Mice on a Genome-Wide Scale

Biochemistry/cell and molecular biology

Alzheimer's Disease: Transgenic Mouse Models
Atomic Force Microscopy Methodologies
BAC Transgenesis: Cell-Type Specific Expression in the Nervous System
BAC Use in the Study of the CNS
Cell Culture: Autonomic and Enteric Neurons
Cell Culture: Primary Neural Cells
Cellular Dynamics Revealed by Digital Holographic Microscopy
Chromaffin Cells: Model Cells for Neuronal Cell Biology
Decoding Neuron Transcriptome by SAGE
Engineering Viruses for CNS studies
Fluorescence Microscopy in the Neurosciences
Fluorescent Biomarkers in Neurons
Glial Ion Homeostasis: A Fluorescence Microscopy Approach
Imaging Studies Using Reporter-Gene Transgenic Rats
Mass Spectroscopy of Proteins
Memory: Genetic Approaches
Microarray use for the Analysis of the CNS
Microglia identification Methods
Monoamines: Release Studies
Neurophysiology: Past and Present
Neuroproteomics
Nucleic Acid Introduction Into Primary Neurons and Glia
Oligodendrocyte and Schwann Cell Identification Methods
Optical Monitoring of Exo- and Endocytosis
Photolysis of Caged Glutamate For Use in the CNS
RNA Binding Protein Methods
Rodent Behavior: Approaches
Single Cell Electroporation
Single Cell Genomic DNA Analysis
Single Cell Molecular Analysis Procedures
Single Cell PCR Coupled with Electrophysiology
Single-Nucleotide Polymorphism (SNP) Analysis
siRNA: Utility
Synaptosomes
Ultrastructural Analysis of Spine Plasticity
VelociGene and VelociMouse: High-Throughput Approaches for Generating Targeted Mutations in Mice on a Genome-Wide Scale
Viral Vectors in the CNS

Electrophysiological Methods

Aging: Brain Potential Measures and Reaction Time Studies
Auditory Evoked Potentials
Deep Brain Stimulation
Electroencephalography (EEG)
Electrolocation
Electromyography (EMG) and Nerve Conduction Studies
Electrophysiology: EEG and ERP Analysis
Electroretinography
Event-Related Potentials (ERPs) and Cognitive Processing
Event-Related Potentials (ERPs)
Evoked Potentials: Clinical
Evoked Potentials: Recording Methods
Executive Function and Higher-Order Cognition: EEG Studies
Glial Glutamate Transporters: Electrophysiology
Human Depth Electrodes

Human Methods

Imaging Techniques

Other Systems Neuroscience Methods

Computational neuroscience* see under *Neural basis of behaviour

MOLECULAR AND CELL BIOLOGY OF NERVOUS SYSTEMS

Cellular Excitability: axon, dendrite, soma

Calcium Channel and Calcium-Activated Potassium Channel Coupling
Calcium Channel Subtypes Involved in Neurotransmitter Release
Calcium Channels and SNARE Proteins
Calcium Channels
Calcium Homeostasis in Glia
Calcium Waves in Glia
Calcium Waves: Purinergic Regulation
Cotransmission
Cyclic Nucleotide-Gated and Hyperpolarization-Activated Channels
Dendritic Signal Integration
Dorsal Root Ganglion Neurons
Glial Influence on Synaptic Transmission
Glutamate Receptor Clusters: Narp, EphB2 Receptor, Stargazin
Glutamate Regulation of Dendritic Spine Form and Function
Hair Cells: Sensory Transduction
Hodgkin–Huxley Models
Information Coding
Inwardly Rectifying Potassium Channels
Ion Channel Localization in Axons
Ion Channel Localization in Cell Bodies and Dendrites
Ionic Channels in Glia
Large Conductance Calcium-Activated Potassium Channels
Myelin: Molecular Architecture of CNS and PNS Myelin Sheath
Neuromodulation of Calcium Channels
Neuromodulation of Sodium Channels
Neuronal Pacemaking
Neutrotransmission and Neuromodulation: Acetylcholine
Potassium Channel Regulation
Schwann Cells and Axon Relationship
Sodium Channels
Spike-Timing-Dependent Plasticity (STDP)
Spine Plasticity
Synaptic Transmission: Models
Transient Receptor Potential (TRP) Channels
Two-P-Domain (K_{2P}) Potassium Channels: Leak Conductance Regulators of Excitability
Ultrastructural Analysis of Spine Plasticity
Voltage Gated Potassium Channels: Structure and Function of Kv1 to Kv9 Subfamilies
Voltage-Gated Calcium Channels
Voltage-Gated Potassium Channels (Kv10–Kv12)

Gene expression and regulation in the nervous system

Alternative Splicing in the Nervous System
Alzheimer's Disease: Molecular Genetics
Alzheimer's Disease: Transgenic Mouse Models
Aversive Emotions: Genetic Mechanisms of Serotonin
BAC Transgenesis: Cell-Type Specific Expression in the Nervous System
Circadian Gene Expression in the Suprachiasmatic Nucleus
Clock Gene Regulation of Endocrine Function
Clock Genes and Metabolic Regulation
Decoding Neuron Transcriptome by SAGE
Dendrites: Localized Translation
Gene Expression Dysregulation in CNS Pathophysiology
Gene Expression in Normal Aging Brain
Gene Expression in the Evolution of the Human Brain
Gene Expression Regulation: Activity-Dependent
Gene Expression Regulation: Chromatin Modification in the CNS
Gene Expression Regulation: Steroid Hormone Effects
Gene Therapy: Direct Viral Delivery
Gene Therapy: Genetically Modified Cells
Genetic Influence on CNS Gene Expression: Impact on Behavior
Genetic Regulation of Circadian Rhythms in *Drosophila*
Genomic Disorder and Gene Expression in the Developing CNS
Genomics of Brain Aging: Apolipoprotein E
Genomics of Brain Aging: Nuclear and Mitochondrial Genomes
Genomics of Brain Aging: Twin Studies
Hippocampus: Molecular Anatomy
Hox Genes Expression
Microarray use for the Analysis of the CNS
Molecular Anatomy of the Mammalian Brain
Neuromuscular Junction: Neuronal Regulation of Gene Transcription at the Vertebrate
Olfactory Insights from Transcriptional Profiling
Pain and Genes
Plasticity, and Activity-Dependent Regulation of Gene Expression
Psychiatric Disorders: Functional Genomics
Psychiatric Genomics and Expression Profiling
Single Cell Genomic DNA Analysis
Sleep and Sleep States: Gene Expression
Sox Gene Expression
Stress Response: Genetic Consequences
Thyroid Hormone and Transcriptional Regulation in the CNS
Transcription and Reward Systems
Transcription Control and the Circadian Clock
Transcription Factors in Synaptic Plasticity and Learning and Memory

Transcriptional Networks and the Spinal Cord
Transcriptional Silencing
VelociGene and VelociMouse: High-Throughput Approaches for Generating Targeted Mutations in Mice on a Genome-Wide Scale

Glia: origins, diversity and functions

Activity-Dependent Metabolism in Glia and Neurons
Astrocyte: Calcium Signaling
Astrocyte: Identification Methods
Astrocyte: Neurotransmitter and Hormone Receptors
Astrocyte: Response to Injury
Axon Guidance by Glia
Axonal Regeneration: Role of the Extracellular Matrix and the Glial Scar
Bergmann Glial Cells
Calcium Homeostasis in Glia
Calcium Waves in Glia
Cytokine Receptors in Glia
Enteric Nervous System: Glial Cells and Interstitial Cells of Cajal
Ependymal Cells
Gap Junctions and Hemichannels in Glia
Glia and Stroke
Glia Control of Blood Flow
Glial Cells: Astrocytes and Oligodendrocytes During Normal Brain Aging
Glial Cells: Invertebrate
Glial Cells: Microglia During Normal Brain Aging
Glial Cells: T Cell Interactions
Glial Energy Metabolism: A NMR Spectroscopy Perspective
Glial Energy Metabolism: Overview
Glial Glutamate and GABA Metabolism
Glial Glutamate Transporters: Electrophysiology
Glial Glutamate Transporters
Glial Glycogen Metabolism
Glial Growth Factors
Glial Influence on Synaptic Transmission
Glial Ion Homeostasis: A Fluorescence Microscopy Approach
Glial Plasticity and Neuroendocrine Regulation
Glial Responses to Injury
Glial Responses to Virus Infection
Glial Steroid Metabolism
Glioma
Intermediate Filaments
Ionic Channels in Glia
Macroglial Lineages
Microglia identification Methods
Microglia Properties
Microglial Response to Injury
Myelin: Molecular Architecture of CNS and PNS Myelin Sheath
Neuron–Glia Coupling in Glutathione Metabolism
Neuron-Glia pH Regulation
Neuropeptides and Receptors in Glia
Neurotransmitter and Hormone Receptors on Oligodendrocytes and Schwann Cells
Neurotransmitter Release from Astrocytes
Oligodendrocyte and Schwann Cell Identification Methods
Oligodendrocyte Morphology
Oligodendrocyte Specification
Potassium Homeostasis in Glia
Radial Glial Cells: Brain Functions
Retinal Glia
Schwann Cell Development
Schwann Cell Morphology
Schwann Cells and Axon Relationship
Schwann Cells and Plasticity of the Neuromuscular Junction
Transplantation of Myelin Forming Cells
Transporter Proteins in Neurons and Glia

Intracellular signalling cascades and second messengers

Cyclic AMP (cAMP) Role in Learning and Memory
Calcium-Calmodulin Kinase II (CaMKII) in Learning and Memory
Eph Receptor Signaling and Spine Morphology
MAP Kinase Signaling in Learning and Memory
Neural Crest Cell Diversification and Specification: ErbB Role
Notch Pathway: Lateral Inhibition
Notch Signal Transduction: Molecular and Cellular Mechanisms
Retinoic Acid Signaling and Neural Patterning
Sonic Hedgehog and Neural Patterning
Synaptic Plasticity: Diacylglycerol Signalling role
Wnt Pathway and Neural Patterning

Intracellular transport and cytoskeleton: axon, dendrite, soma

Actin Cytoskeleton in Growth Cones, Nerve Terminals, and Dendritic Spines
AMPA Receptor Cell Biology/Trafficking
Axonal and Dendritic Identity and Structure: Control of
Axonal and Dendritic Transport by Dyneins and Kinesins in Neurons
Axonal mRNA Transport and Functions
Axonal Transport and ALS

Intracellular regulation of energy and ions

Secretion and vesicle re-cycling

PHR (Pam/Highwire/RPM-1)
Piccolo and Bassoon
Presynaptic Receptor Signaling
Rab3
Rab3A Interacting Molecules (RIMs)
RNA Granules: Functions Within Presynaptic Terminals and Postsynaptic Spines
SNAREs
Synapsins
Synaptic Plasticity: Short-Term Mechanisms
Synaptic Vesicle Protein-2 (SV2)
Synaptic Vesicles
Synaptojanin
Synaptotagmins
SynCAMs
Synucleins
Vesicle Pools
Vesicular Neurotransmitter Transporters
Vesicular Sorting to Axons and Dendrites
Voltage-Gated Calcium Channels

MOTOR AND NEUROMUSCULAR SYSTEM

Anatomy and physiology

Command Systems
Evolution of Vertebrate Respiratory Control
Fatigue
Fatigue
Functional Imaging of the Motor System
Motor Autonomic Transmission
Motor Control of Feeding and Drinking
Motor Neuron Specification in Vertebrates
Population Coding
Spasticity
Startle Response
Trigeminal Motor System

Basal ganglia

Basal Ganglia and Oculomotor Control
Basal Ganglia: Acetylcholine Interactions and Behavior
Basal Ganglia: Evolution
Basal Ganglia: Functional Models of Normal and Disease States
Basal Ganglia: Habit
Basal Ganglia: Internal Organization
Basal Ganglia: Motor Functions
Basal Ganglia: Physiological Circuits
Cognition: Basal Ganglia Role
Deep Brain Stimulation and Movement Disorder Treatment
Deep Brain Stimulation and Parkinson's Disease
Dopamine Neurons: Reward and Uncertainty
Functional Electrical Stimulation
MPTP Parkinsonism Model
Procedural Learning: Striatum
Striatum: Internal Physiology

Biomechanics

Biomechanics: Hydroskeletal
Kinematics and Dynamics
Vestibulospinal System and Eye–Head/Neck Movement

Cerebellum

Cerebellar Deep Nuclei
Cerebellar Lesions and Effects on Posture, Locomotion and Limb Movement
Cerebellar Microcircuitry
Cerebellum and Oculomotor Control
Cerebellum: Clinical Pathology
Cerebellum: Evolution and Comparative Anatomy
Cerebellum: Models
Cognition: Cerebellum Role
Eyeblink Conditioning
Olivocerebellar System
Procedural Learning: Cerebellum Models
Synaptic Plasticity: Cerebellum

Locomotion and central pattern generators

Central Pattern Generators: Sensory Feedback
Central Pattern Generators
Cerebellar Lesions and Effects on Posture, Locomotion and Limb Movement
Motor Gait and Falls
Pattern Generation
Swim Oscillator Networks
Swimming: Neural Mechanisms
Walking in Invertebrates

Motor control

Balance and Posture Control: Human
Basal Ganglia: Habit
Bayesian Models of Motor Control
Computational Approaches to Motor Control
Eyeblink Conditioning
Fatigue
Motor Primitives
Motor Psychophysics
Motor Timing
Network Control
Orofacial Motor Control
Posterior Parietal Cortex and Arm Movement
Posterior Parietal Cortex and Tool Usage and Hand Shape

Motor cortex

Motor learning

Oculomotor control and spatial cognition

Sensorimotor integration *see under* Sensory systems

Spinal cord/brainstem

NERVOUS SYSTEM DEVELOPMENT, REPAIR AND AGING

Aging of the brain: neural and cognitive changes

Anatomical/metabolic changes

Cholinergic System Imaging in the Healthy Aging Process and Alzheimer Disease
Cognition in Aging and Age-Related Disease
Functional Neuroimaging Studies of Aging
Neuromuscular Junction (NMJ): Aging
Nutrition
Sleep in Aging

Animal models

Aging and Memory in Animals
Aging: Invertebrate Models of Normal Brain Aging
Non-Primate Models of Normal Brain Aging
Rodent Aging

Cell biology

Calcium and Signal Transduction
Glial Cells: Astrocytes and Oligodendrocytes During Normal Brain Aging
Glial Cells: Microglia During Normal Brain Aging
Lipids and Membranes in Brain Aging
Lipofuscin and Lipofuscinosis
Metal Accumulation During Aging
Synaptic Plasticity: Neuronal Sprouting
Synaptic Plasticity: Neuronogenesis and Stem Cells in Normal Brain Aging

Cognitive changes (learning and memory)

Aging and Memory in Animals
Aging: Brain Potential Measures and Reaction Time Studies
Conditioned Reflex
Conditioned Taste Aversion
Language in Aged Persons
Synaptic Plasticity: Learning and Memory in Normal Aging

Genomics of brain aging

Gene Expression in Normal Aging Brain
Genomics of Brain Aging: Apolipoprotein E
Genomics of Brain Aging: Nuclear and Mitochondrial Genomes
Genomics of Brain Aging: Twin Studies

Neuroendocrine aging

Neuroendocrine Aging: Hypothalamic–Pituitary–Gonadal Axis in Women
Neuroendocrine Aging: Pituitary Metabolism
Neuroendocrine Aging: Pituitary–Gonadal Axis in Males
Neuroendocrine Aging: Pituitary-Adrenal Axis
Neuroimmune System: Aging
Thermoregulation: Autonomic, Age-Related Changes

Sensory aging

Conditioned Taste Aversion
Sensory Aging: Chemical Senses
Sensory Aging: Hearing
Sensory Aging: Vision

Cell fate determination

Pattern formation

Anterior–Posterior Spinal Cord Patterning of the Motor Pool
Floor Plate Patterning of Ventral Cell Types: Ventral Patterning
Forebrain Development: Holoprosencephaly (HPE)
Forebrain Development: Prosomere Model
Forebrain: Early Development
Hox Genes Expression
Lissencephaly Type I and Periventricular Heterotopia
Microcephaly Vera
Midbrain Patterning
Morphogens: History
Neural Patterning: Arealization of the Cortex
Neural Patterning: Eye Fields
Neural Patterning: Midbrain–Hindbrain Boundary
Retinoic Acid Signaling and Neural Patterning
Segmentation: Segmental Boundaries, Establishment and Morphogenesis (EPH)
Segmentation: Spinal Cord Segmentation and A–P Somite Patterning
Sonic Hedgehog and Neural Patterning
Sox Gene Expression
Wnt Pathway and Neural Patterning

Cellular differentiation/specification

Cerebral Cortex: Symmetric vs. Asymmetric Cell Division
Differentiation: the Cell Cycle Instead
Dopaminergic Differentiation
Drosophila Apterous Neurons: From Stem Cell to Unique Neuron
Hair Cell Differentiation
Helix–Loop–Helix (bHLH) Proteins: Hes Family

Connectivity/integration

Activity-dependent development

Pathfinding and target recognition

Synapse formation and elimination

Neural repair, regeneration and functional recovery

Peripheral Nerve Regeneration: An Overview
Protein Folding and the Role of Chaperone Proteins in Neurodegenerative Disease
Sensorimotor Plasticity and Control of Movement Following Spinal Cord Injury +D18
Sensory Re-education
Spinal Cord Regeneration after Injury: Bridging by Schwann Cells and Bioimplants
Spinal Cord Regeneration and Functional Recovery: Strategies
Stem Cells and CNS Repair

Neurogenesis, Neural induction and migration

Actin Cytoskeleton in Growth Cones, Nerve Terminals, and Dendritic Spines
Autonomic Nervous System Development
Bone Morphogenetic Protein (BMP) Signaling in the Neuroectoderm
Drosophila Apterous Neurons: From Stem Cell to Unique Neuron
Holoprosencephaly
Joubert Syndrome
Lissencephaly Type I and Periventricular Heterotopia
Lissencephaly
Neural Crest
Neural Induction in Chicks
Neural Stem Cells: Adult Neurogenesis
Neural Stem Cells: Ocular
Neurogenesis and Neural Precursors, Progenitors, and Stem Cells in the Adult Brain
Neuronal Motility and Structure: Cdk5 Pathways
Neuronal Motility and Structure: MARK and GSK Pathways
Neurulation
Sox Gene Expression
Stem Cells and CNS Repair
Synaptic Plasticity: Neuronogenesis and Stem Cells in Normal Brain Aging

Neuronal degeneration and cell death

Aluminum
Alzheimer's Disease: An Overview
Alzheimer's Disease: Neurodegeneration
Alzheimer's Disease: Transgenic Mouse Models
Amyotrophic Lateral Sclerosis (ALS)
Amyotrophic Lateral Sclerosis (ALS): Disease Mechanisms
Apoptosis in Nervous System Injury
Apoptosis in Neurodegenerative Disease
Autonomic and Enteric Nervous System: Apoptosis and Trophic Support During Development
Autophagy and Neuronal Death
Axonal Injury in Demyelinating Disease and CNS Injury
Blood–Brain Barrier and Neurovascular Mechanisms of Neurodegeneration and Injury
Brain Injury: Magnetic Resonance Studies of Metabolic Aspects
Epilepsy: Neuronal Death
Excitotoxicity in Neurodegenerative Disease
Glial Glutamate Transporters
Inflammation in Neurodegenerative Disease and Injury
Inherited Macular Degenerations: Animal Models
Intracellular Calcium and Neuronal Death
Mitochondrial Dysfunction in Nervous System Injury
Neurodegeneration in Psychiatric Illness
Neuroprotection: Endogenous Mechanisms
Neuroprotection: Pharmacological Approaches
Neurotoxins and their Neurotoxicology
Neurotoxins
Oxidative Damage in Neurodegeneration and Injury
Parkinson's Disease: Alpha-Synuclein and Neurodegeneration
Programmed Cell Death
Proteasome Role in Neurodegeneration
Role of NO in Neurodegeneration
Transport Dependent Damage Signaling
Vascular Issues in Neurodegeneration and Injury
Wallerian Degeneration

Organ/system development

Adolescent Brain Development and the Risk of Psychiatric Disorders
Autonomic Nervous System Development
Autonomic Neuroplasticity: Development
Axonal and Dendritic Identity and Structure: Control of
Brain Development: The Generation of Large Brains
Brain Evolution: Developmental Constraints and Relative Developmental Growth
Cochlear Development
Energy Homeostasis: Hypothalamic Development
Enteric Nervous System Development
Neuromuscular Junction (NMJ): Mammalian Development
Pituitary Gland (Cell Types, Mediators, Development)

Schwann Cell Development
Sexual Differentiation of the Brain
Sexual Differentiation of the Central Nervous System
Visual Development
Visual System Development: Invertebrates

NEURAL BASIS OF BEHAVIOR

Computational neuroscience and neural networks

Active Perception
Attention: Models
Attractor Network Models
Auditory Cortex: Models
Axonal Pathfinding
Bayesian Cortical Models
Bayesian Models of Motor Control
Birdsong Learning
Brain scaling laws
Brain–Computer Interface
Circadian Rhythm Models
Computational Approaches to Motor Control
Computational Methods
Computational Neuroethology
Connectionist Models of Language Processing
Connectionist Models
Consciousness: Theoretical and Computational Neuroscience
Consciousness: Theories and Models
Electrophysiology: EEG and ERP Analysis
Emotion: Computational Modeling
Executive Function and Higher-Order Cognition: Computational Models
Gain Modulation
Hippocampus: Computational Models
Hodgkin–Huxley Models
Homeostasis at Multiple Spatial and Temporal Scales
Ideal Observer Theory
Information Coding
Learning, Action, Inference and Neuromodulation
Memory: Computational Models
Neural Integrator Models
Neural Oscillators and Dynamical Systems Models
Neuromorphic Systems
Neuroplasticity: Computational Approaches
Olfactory Coding
Population Codes: Theoretic Aspects
Retinal Models
Self-Organizing Maps
Sensorimotor Integration: Models
Sleep Oscillations
Spike-Timing-Dependent Plasticity Models
Spiking Neuron Models
Statistical Analysis of Visual Perception
Statistical Learning of Language
Statistical Tests and Inferences
Stomatogastric Ganglion Models
Swim Oscillator Networks
Synaptic Transmission: Models
Synfire Chains
Visual Cortical Models of Orientation Tuning
Visual Motion Models

Consciousness

Awareness: Functional Imaging
Blindsight: Residual Vision
Cognition: Neuropharmacology
Coma and Other Pathological Disorders of Consciousness
Coma
Consciousness: Neural Basis of Conscious Experience
Consciousness: Neurophysiology and Visual Awareness in
Conciousness: Philosophy
Consciousness: Theoretical and Computational Neuroscience
Consciousness: Theories and Models
The AIM Model of Dreaming, Sleeping, and Waking Consciousness
Vegetative State

Decision-making and neuroeconomics

Animal Communication: Honesty and Deception
Decision-Making and Neuroeconomics
Decision-Making and Vision
Decision-Making in Financial Markets
Delayed Reinforcement: Economics
Delayed Reinforcement: Neuroscience
Game Theory and the Economics of Animal Communication
Games in Monkeys: Neurophysiology and Motor Decision-Making
Neuroeconomics: History
Neuroethological Perspective
Reasoning and Problem Solving: Models
Reward Decision-Making
Social Cognition
Social Interaction

Emotion

Depression and the Brain
Emotion in Speech

Emotion Systems and the Brain
Emotion: Computational Modeling
Emotional Control of the Autonomic Nervous System
Nightmares
Sexual Behavior: Neuroendocrine Control

Aversive

Amygdala: Contributions to Fear
Aversive Emotions: Genetic Mechanisms of Serotonin
Aversive Emotions: Molecular Basis of Unconditioned Fear
Emotion: Neuroimaging
Fear Conditioning and Synaptic Plasticity
Genetics of Human Anxiety and Its Disorders
Learned Flavor Aversions and Preferences
Panic Disorder
Pharmacology of Fear Extinction
Phobia and Human Evolution
Posttraumatic Stress Disorder as an Emotional Disorder

Cognition and

Appetitive Systems: Amygdala and Striatum
Emotion and Vigilance
Emotional Hormones and Memory Modulation
Emotional Influences on Memory and Attention
Social Interaction Effects on Reward and Cognitive Abilities in Monkeys

Social

Aggression: Hormonal Basis
Aggression: Neurochemical and Molecular Mechanisms
Emotional Learning in Humans
Neuroendocrinology of Social/Affiliative Behavior
Pheromones in Humans and Social Chemosignals
Social Brain: Evolution
Social Emotion: Neuroimaging
Social Interaction Effects on Reward and Cognitive Abilities in Monkeys
Social Interaction

Executive function and higher-order cognition (*see also* **Oculomotor control and spatial cognition**)

Agnosia
Amnesia: Declarative and Nondeclarative Memory
Alexia
Animal Models of Amnesia
Attention and Eye Movements
Attention: Models
Attentional Functions in Learning and Memory
Attentional Mechanisms in Ventral Pathway
Attentional Networks in the Parietal Cortex
Attentional Networks
Brain–Computer Interface
Cognition: An Overview of Neuroimaging Techniques
Cognition: Basal Ganglia Role
Cognition: Cerebellum Role
Cognition: Neuropharmacology
Cognitive Control and Development
Cognitive Neuroscience: An Overview
Emotion and Vigilance
Emotional Influences on Memory and Attention
Executive Function and Higher-Order Cognition: Assessment in Animals
Executive Function and Higher-Order Cognition: Computational Models
Executive Function and Higher-Order Cognition: Definition and Neural Substrates
Executive Function and Higher-Order Cognition: EEG Studies
Executive Function and Higher-Order Cognition: Neuroimaging
Executive Functions: Eye Movements and Neuropsychiatric Disorders
Frontal Lobe Syndrome
Functional Amnesia
Hemispheric Specialization and Cognition
Human Methods: Psychophysics
Humans
Inhibitory Control over Action and Memory
Neglect Syndrome and the Spatial Attention Network
Neural Coding of Spatial Representations
Neuropsychological Testing
Parietal Cortex and Spatial Attention
Perception and Eye Movements
Prefrontal Cortex: Structure and Anatomy
Prefrontal Cortex
Prosopagnosia
Psychophysics of Attention
Reasoning and Problem Solving: Models
Referentiality and Concepts in Animal Cognition
Sleep Deprivation and Brain Function
Social Interaction Effects on Reward and Cognitive Abilities in Monkeys
Spatial Cognition and Executive Function
Spatial Cognition
Spatial Cognitive Maps
Spatial Orientation: Our Whole-Body Motion and Orientation Sense
Strategic Control of Memory

Hemispheric specialization

Homeostasis

Energy balance

Fluid/mineral balance

Food intake

Thermoregulation

Intelligence

Learning and Memory

Animal models

Aging and Memory in Animals
Animal Models of Amnesia
Invertebrate Models to Study Learning and Memory: Lymnaea
Learning and Memory in Invertebrate Models: *Tritonia*
Learning and Memory in Invertebrates: *Aplysia*
Learning and Memory in Invertebrates: *C. Elegans*
Learning and Memory in Invertebrates: *Drosophila*
Learning and Memory in Invertebrates: *Hermissenda*
Learning and Memory in Invertebrates: Honey Bee
Learning and Memory in Invertebrates: *Limax*
Learning and Memory in Invertebrates: Mollusks
Procedural Learning in Animals

Synaptic plasticity/neuroplasticity

Adult Cortical Plasticity
BDNF in Synaptic Plasticity and Memory
Calcium-Calmodulin Kinase II (CaMKII) in Learning and Memory
Cortical Plasticity and Learning: Mechanisms and Models
Cyclic AMP (cAMP) Role in Learning and Memory
Cytoskeleton in Plasticity
Endocannabinoid Role in Synaptic Plasticity and Learning
Extracellular Matrix Molecules: Synaptic Plasticity and Learning
Fear Conditioning and Synaptic Plasticity
Fragile X Syndrome
$GABA_A$ Receptor Synaptic Functions
Hebbian Plasticity
Long-Term Depression (LTD): Endocannabinoids and Cerebellar LTD
Long-Term Depression (LTD): Metabotropic Glutamate Receptor (mGluR) and NMDAR-Dependent Forms
Long-Term Depression: Cerebellum
Long-Term Potentiation (LTP): Mossy Fiber cAMP-Dependent Presynaptic LTP
Long-Term Potentiation (LTP): NMDA Receptor Role
Long-Term Potentiation (LTP)
Long-Term Potentiation and Long-Term Depression in Experience-Dependent Plasticity
MAP Kinase Signaling in Learning and Memory
Metaplasticity
Neuroplasticity: Computational Approaches
Pain and Plasticity
Plasticity, and Activity-Dependent Regulation of Gene Expression
Plasticity of Intrinsic Excitability
Post-Tetanic Potentiation (PTP)
Presynaptic Facilitation
Presynaptic Inhibition
Spike-Timing-Dependent Plasticity (STDP)
Stress and Neuronal Plasticity
Synaptic Capture and Tagging
Synaptic Depression
Synaptic Mechanisms of Learning
Synaptic Plasticity and Place Cell Formation
Synaptic Plasticity: Cerebellum
Synaptic Plasticity: Diacylglycerol Signalling role
Synaptic Plasticity: Learning and Memory in Normal Aging
Synaptic Plasticity: Neuronogenesis and Stem Cells in Normal Brain Aging
Synaptic Plasticity: Short-Term Mechanisms
Transcription Factors in Synaptic Plasticity and Learning and Memory
Ubiquitin-Proteasome System and Plasticity

Systems/mechanisms (of learning and memory)

Aging and Memory in Animals
Aging and Memory in Humans
Aging: Brain Potential Measures and Reaction Time Studies
Amnesia: Declarative and Nondeclarative Memory
Attentional Functions in Learning and Memory
Basal Forebrain and Memory
Birdsong Learning: Evolutionary, Behavioral, and Hormonal Issues
Birdsong Learning
Birdsong: The Neurobiology of Avian Vocal Learning
Conditioned Reflex
Conditioned Taste Aversion
Conditioning: Simple Neural Circuits in the Honeybee
Conditioning: Theories
Declarative Memory System: Anatomy
Emotional Hormones and Memory Modulation
Emotional Influences on Memory and Attention
Emotional Learning in Humans
Episodic Memory: Assessment in Animals
Episodic Memory
Event-Related Potentials (ERPs)
Eyeblink Conditioning
Functional Amnesia

Hippocampus and Neural Representations
Hippocampus: Computational Models
Hormones and Memory
Humans
Learned Flavor Aversions and Preferences
Learning, Action, Inference and Neuromodulation
Memory Consolidation: Cerebral Cortex
Memory Consolidation: Systems
Memory Disorders
Memory Representation
Memory: Computational Models
Memory: Genetic Approaches
Motor Autonomic Transmission
Motor Skill Learning
Multiple Memory Systems
Operant Conditioning of Reflexes
Perceptual Learning and Sensory Plasticity
Perceptual Learning: Neural Mechanisms
Pharmacology of Fear Extinction
Prediction Errors in Neural Processing: Imaging in Humans
Priming
Procedural Learning in Animals
Procedural Learning in Humans
Procedural Learning: Cerebellum Models
Procedural Learning: Classical Conditioning
Procedural Learning: Striatum
Recognition Memory
Reinforcement Models
Respiration
Reward and Learning
Semantic Memory
Short Term and Working Memory
Sleep-Dependent Memory Processing
Spatial Memory: Assessment in Animals
Statistical Learning of Language
Strategic Control of Memory
Synaptic Plasticity: Learning and Memory in Normal Aging
Visual Associative Memory
Working Memory: Capacity Limitations

Neuroethology

Behavioral Hierarchies
Computational Neuroethology
Development of Behavior
Electrocommunication
Extinction: Anatomy
Hormones and Behavior
Magnetic Sense in Animal Navigation
Neuroendocrine Control: Maternal Behavior
Neuroethological Perspective
Neuromodulation
Visually Guided Behavior

Reward and reward systems

Addiction: Neurobiological Mechanism
Appetitive Systems: Amygdala and Striatum
Connectivity of Primate Reward Centers
Cortical Processing of the Reward Value of Food
Dopamine: Cellular Actions
Drug Addiction: Behavioral Neurophysiology
Drug Addiction: Behavioral Pharmacology of Drug Addiction in Rats
Drug Addiction: Cellular Mechanisms
Drug Addiction: Neuroimaging
Drugs Addiction: Actions
Electrical Self-Stimulation
Energy Homeostasis: Endocannabinoid System
Goal-Directed Behavior Theories
3,4-Methylenedioxymethamphetamine (MDMA, "Ecstasy")
Neuropsychology of Primate Reward Processes
Prefrontal Contributions to Reward Encoding
Psychopharmacology of Reward and Appetite in Rats
Representation of Reward
Reward and Learning
Reward Decision-Making
Reward Neurophysiology and Orbitofrontal Cortex
Reward Neurophysiology and Primate Cerebral Cortex
Reward Processing: Human Imaging
Reward Systems: Human
Serotonin-Related Psychedelic Drugs
Social Interaction Effects on Reward and Cognitive Abilities in Monkeys
Substance Abuse and Dependence
Transcription and Reward Systems

Rhythmicity and synchronization

Circadian rhythms **(see also *Sleep, dreaming and arousal*)**

Circadian Function and Therapeutic Potential of Melatonin in Humans
Circadian Gene Expression in the Suprachiasmatic Nucleus
Circadian Genes and the Sleep–Wake Cycle
Circadian Metabolic Rhythms Regulated by the Suprachiasmatic Nucleus
Circadian Organization
Circadian Organization in Non-Mammalian Vertebrates
Circadian Oscillations in the Suprachiasmatic Nucleus
Circadian Regulation by the Suprachiasmatic Nucleus
Circadian Regulation in Invertebrates

Circadian Rhythm Models
Circadian Rhythms in Sleepiness, Alertness, and Performance
Circadian Rhythms: Influence of Light in Humans
Circadian Systems: Evolution
Clock Gene Regulation of Endocrine Function
Clock Genes and Metabolic Regulation
Entrainment of Circadian Rhythms by Light
Genetic Regulation of Circadian Rhythms in *Drosophila*
Genetics of Circadian Disorders in Humans
Mammalian Sleep and Circadian Rhythms: Flies
Melatonin Regulation of Circadian Rhythmicity in Vertebrates
Non-Photoreceptor Photoreception
Peripheral Circadian Oscillators
Photoreceptors and Circadian Clocks
Psychiatric Disorders Associated with Disturbed Sleep and Circadian Rhythms
Serotonin and the Regulation of Mammalian Circadian Rhythms
Shift Work and Circadian Rhythms
Single Cell Neuronal Circadian Clocks
Sleep and Circadian Rhythm Disorders in Human Aging and Dementia
Sleep and Waking in *Drosophila*
Sleep: Development and Circadian Control
Transcription Control and the Circadian Clock

Oscillatory neural activity

Central Pattern Generators: Sensory Feedback
Central Pattern Generators
Gap Junctions and Neuronal Oscillations
Information Coding
Neural Oscillators and Dynamical Systems Models
Neuronal Pacemaking
Pulsatility in Neuroendocrine Systems
Sleep Oscillations and PGO Waves
Sleep Oscillations
Spiking Neuron Models
Swim Oscillator Networks

Seasonal rhythms

Photoperiodic Regulation of Reproductive Cycles
Seasonal Changes in Night-Length and Impact on Human Sleep
Seasonal Hormonal Changes and Behavior
Seasonal Timing: Neural Mechanisms

Sleep, dreaming and arousal

Autonomic Dysregulation During REM Sleep
Cataplexy
Circadian Function and Therapeutic Potential of Melatonin in Humans
Circadian Genes and the Sleep–Wake Cycle
Circadian Rhythms in Sleepiness, Alertness, and Performance
Coma
Dopamine Control of Arousal
Dream Function
Dreams and Dreaming: Incorporation of Waking Events
Dreams and Nightmares in PTSD
Dreams, Dreaming Theories and Correlates of Nightmares
Endocrine Function During Sleep and Sleep Deprivation
Genetics of Circadian Disorders in Humans
Hibernation
Immune Function During Sleep and Sleep Deprivation
Mammalian Sleep and Circadian Rhythms: Flies
Metabolic Syndrome and Sleep
Napping
Narcolepsy
Nightmares
Parasomnias
Pharmacology of Sleep: Adenosine
Psychiatric Disorders Associated with Disturbed Sleep and Circadian Rhythms
Reticular Activating System
Seasonal Changes in Night-Length and Impact on Human Sleep
Shift Work and Circadian Rhythms
Sleep and Circadian Rhythm Disorders in Human Aging and Dementia
Sleep and Sleep States: Gene Expression
Sleep and Sleep States: Hippocampus–Neocortex Dialog
Sleep and Sleep States: Histamine Role
Sleep and Sleep States: Hypothalamic Regulation
Sleep and Sleep States: Network Reactivation
Sleep and Sleep States: PET Activation Patterns
Sleep and Sleep States: Phylogeny and Ontogeny
Sleep and Sleep States: Thalamic Regulation
Sleep and Waking in *Drosophila*
Sleep Apnea
Sleep Architecture
Sleep Deprivation and Brain Function
Sleep Deprivation: Neurobehavioral Changes
Sleep in Adolescents
Sleep in Aging
Sleep Mentation in REM and NREM: A Neurocognitive Perspective
Sleep Oscillations and PGO Waves

Stress and the nervous system

SENSORY SYSTEMS

Auditory and vestibular sensation

Chemical senses

Somatosensation

Anatomy/physiology

Haptics/touch

Pain perception/processing

Sensorimotor integration

Vision

Early vision

Eye and retina

Subcortical stages

Early visual processing

Mid-level vision

High-level vision

Visual attention

Visual system development

Activity in Visual Development
Retinal Development: An Overview
Visual Deprivation
Visual System Development: Invertebrates

Vision for motor control

Nystagmus
Optic Tectum: Sensorimotor Integration
Vision for Action and Perception
Visual–Vestibular Interactions

Computational vision

Visual Cortical Models of Orientation Tuning
Visual Motion Models

Comparative vision

Visual System Development: Invertebrates
Visual System: Adaptive Regression and Progression in Subterranean Mammals

Eye movements see under *Oculomotor control*

STRUCTURAL ORGANISATION OF THE BRAIN

Brain–Computer Interface
Magnocellular Neurosecretory System: Organization, Plasticity, Model Peptidergic Neurons
Sexual Differentiation of the Brain
Sexual Differentiation of the Central Nervous System

Brainstem and spinal cord

Brainstem and Cranial Nerves
Reticular Activating System
Reticular Formation and the Brain Stem

Cerebral cortex

Attentional Networks in the Parietal Cortex
Auditory Cortex: Models
Barrel Cortex Circuits
Brain Damage: Functional Reorganization
Brodmann's Areas
Cerebral Cortex: Inhibitory Cells
Cerebral Cortex
Cortical Plasticity and Learning: Mechanisms and Models
Entorhinal Cortex
Frontal Cortex Evolution in Primates
Humans
Memory Consolidation: Cerebral Cortex
Microcephaly Vera
Olfactory Bulb Anatomy
Olfactory Cortex: Comparative Anatomy
Orbitofrontal Cortex: Visual Functions
Perirhinal Cortex: Neural Representations
Perirhinal Cortex
Prefrontal Cortex: Structure and Anatomy
Somatosensory Cortex: Functional Architecture
Somatosensory Cortex
Visual Cortex in Humans
Visual Cortex: Mapping of Functional Architecture Using Optical Imaging
Visual System: Functional Architecture of Area V2
Visual System: Multiple Visual Areas in Monkeys

Circuitry/integration

Amygdala: Structure and Circuitry in Primates
Amygdala: Structure and Circuitry in Rodents and Felines
Auditory Cortex Structure and Circuitry
Auditory System: Central Pathways
Barrel Cortex Circuits
Brain Connectivity and Brain Size
Brainstem and Cranial Nerves
Brainstem Respiratory Circuits
Cerebellar Microcircuitry
Cerebellum: Models
Cerebral Cortex: Inhibitory Cells
Cholinergic Pathways in CNS
Connectivity of Primate Reward Centers
Corticothalamic Connections: Structure and Function
Corticothalamic Connections: Ultrastructure
Enteric Nervous System: Neural Circuits and Chemical Coding
Hypothalamic Structure–Function Relationships
Neonatal Circuits
Neural Circuitry in the Somatosensory System
Olfactory Bulb Anatomy
Olfactory Glomeruli: Structure and Circuitry
Olivocerebellar System
Photoreceptor Mosaic
Pulvinar Structure and Circuitry in Primates
Reflex Circuits
Taste: Vertebrate Central Pathways

TABLE OF CONTENTS

VOLUME 1

A

VOLUME 2

B

VOLUME 3

C

E

VOLUME 4

E

F

G

H

VOLUME 5

H

I

J

K

L

VOLUME 6

N

VOLUME 7

O

P

VOLUME 8

R

S

VOLUME 9

S

VOLUME 10

U

V

FOREWORD

What is an encyclopedia? The term is derived from two Greek words: *enkuklios*, which means cyclical, and *paideia*, which means education. In the early sixteenth century, copyists of Latin manuscripts combined the two words into a Latin designation which comes to us in English with the same spelling and with the meaning 'general course of instruction' (The American Heritage Dictionary, 2000, Boston: Houghton Mifflin, p. 589). The original Greek provides a dynamic connotation, an effort to approach knowledge in a probing, integrated fashion. In Latin and then English, the term has generally been applied to reference volumes addressing broad areas of knowledge.

The Greek emphasis on creative approaches to information is particularly apt for the neurosciences. The term 'neuroscience' is of remarkably young vintage. Francis Schmitt, in his foreword to the first edition of this encyclopedia, related his process of concatenating scientists from disparate fields into an invisible college whose workshops attacked the brain's most recalcitrant puzzles. He dubbed the organization the Neuroscience Research Program (NRP). The early NRP 'associates' included giants such as Melvin Calvin, whose Nobel Prize honored his work on photosynthesis; the Nobel laureate physical chemist Manfred Eigen; the biochemist Albert Lehninger; and Marshall Nirenberg, then a young molecular biologist in the throes of breaking the genetic code.

When Schmitt established the NRP, neuroscience as an integrated endeavor hardly existed. The invention of the microelectrode was permitting neurophysiologists to record from single cells, but characterizing them biochemically was impossible. A major step forward was the emergence in the mid-1960s of the Falck-Hillarp fluorescence microscopic techniques, which permitted selective visualization of catecholamine and serotonin neurons. Mapping the aminergic pathways soon led to collaborative efforts of behaviorists, who could examine the consequences of selective lesions, and biochemists, who no longer were relegated to monitoring neurotransmitters in homogenates of the whole brain. Advances in the 1970s in receptor biochemistry, their localization by autoradiography, and neuropeptide immunohistochemistry further enhanced discourse among neurophysiologists, neuroanatomists, neurochemists, and neuropharmacologists. In the past two decades, the tools of molecular biology have furthered the dialogue.

The explosion of the neurosciences can also be documented through the chronicles of the Society for Neuroscience (SFN). The SFN was founded in 1970 with Vernon Mountcastle as the first elected president, and its inaugural annual meeting in Washington, DC hosted a few hundred researchers. When I served as president in 1980, SFN numbered 7000 members. One of my key tasks was to combat attacks on the *raison d'être* of the society. Some argued, "We have too many scientists in this organization. Let's split into two societies, the 'Wets' and the 'Drys.'" Instead, to emphasize the integrated nature of the field, we launched the *Journal of Neuroscience*. Also, we argued that growth might plateau and that careful meeting organization would prevent individuals from getting 'lost in a crowd.' My prediction about a plateau was off the mark. As of this writing, May 2007, SFN numbers about 38 000 active members, with up to 35 000 people attending each annual meeting, dwarfing any other biomedical research society.

Most would agree that neuroscience is the most integrated scientific discipline. As such, the concept of an encyclopedia, in the original Greek sense of a circle of learning, is notably appropriate yet immensely challenging. The current edition, like earlier ones, succeeds by careful attention to organization and, most importantly, to the selection of the finest researchers as Associate Editors for individual topics. The Associate Editors are all seasoned veterans yet active researchers whose vision remains at the forefront of their field. All areas of importance are covered, from 'soup to nuts.' Emphasis is elegantly balanced between molecular and systems neuroscience.

In this era of rapid advances, one can question whether an encyclopedia, comprising a snapshot in time, serves a meaningful function. Might not all the information in such an enterprise be obsolete soon after publication? An effective encyclopedia, exemplified in these volumes, integrates disparate areas in a lucid, reader-friendly format. Such a publication can be provocative and invigorating to the most sophisticated professionals. At the same time, the entries are presented in such an inviting fashion that the encyclopedia serves for novices as the ideal entrée into the world of the nervous system.

Solomon H. Snyder
Distinguished Service Professor,
Johns Hopkins University

PREFACE

During the second half of the twentieth century, the study of the nervous system moved from a peripheral position within the biological and psychological sciences to become an interdisciplinary field called neuroscience. The new discipline brought biochemists, cell biologists, anatomists, physiologists, psychologists, neurologists, and psychiatrists – scientists and clinicians from diverse backgrounds, all drawn to the promise and excitement of studying the brain. They aimed to discover the mechanisms of neuronal function, elucidate the neural substrates of behavior and cognition, and learn about the diseases of the nervous system. The development of the discipline was catalyzed in 1969 by the formation of the Society for Neuroscience, which now has nearly 37 000 members. The first academic training programs for neuroscience were established in medical schools (the Department of Neurosciences at the University of California, San Diego in 1965 and the Department of Neurobiology at Harvard University in 1966). The first undergraduate training programs in neuroscience were established in 1972 at Amherst College and at Oberlin College, alma mater of Nobelist Roger Sperry and three Past-Presidents of the Society for Neuroscience. Today, there are more than 300 neuroscience departments and programs around the world.

The *Encyclopedia of Neuroscience* is intended to catalog and explicate the rich, diverse subject matter of the discipline and to facilitate communication among its subspecialties. It is meant to be an authoritative source of information for all areas of neuroscience. It will hopefully make neuroscience more accessible to a wide range of readers, from students making their first acquaintance with the field to general readers seeking information about specific topics. It should also serve as a useful reference for working neuroscientists and be useful as well to undergraduate and graduate students in neuroscience training programs, teachers in the life sciences, clinicians, and science writers.

The inaugural edition of this encyclopedia, which was the first comprehensive reference work for the field, was published in 1987 under the able leadership of George Adelman. It included some 700 entries and appeared in two volumes. The second edition, edited by George Adelman and Barry Smith, appeared in 1999 in two volumes and included more than 800 entries. A CD-ROM version of this edition was also published. A third edition, published only in electronic form, appeared in 2004.

The *Encyclopedia of Neuroscience* appears in ten volumes and includes nearly 1500 entries. The full work has also been published online at Science Direct, which can be accessed with subscription at www.sciencedirect.com. To assemble the entries, the Senior Editors identified 46 major areas of the discipline and then invited 46 Associate Editors, all experts in their field, to survey the content of neuroscience within each of these areas. Each Associate Editor then invited 30 to 40 authors to prepare articles on specific topics, with the objective of obtaining complete coverage for each area. Many of the authors are the recognized leaders in their field. The result is a compendium of expert articles representing the current world of neuroscience – the most important research, the most powerful tools, and the most promising applications.

Most of the entries are self-contained reviews that can be read as independent articles. Extensive cross-listing at the conclusion of each entry directs readers to articles on related topics. The principal organization of the *Encyclopedia* lies in the alphabetically arranged list of entries. In addition, the comprehensive subject classification will help readers find related topics and appreciate the structure of the discipline.

While no single reference work in neuroscience can claim to include every notable idea and fact about the brain, the Senior Editors hope that these volumes provide a summary of contemporary neuroscience that is both comprehensive and instructive. Neuroscience is still a developing field, but the *Encyclopedia* will have succeeded if it conveys the considerable promise that neuroscience offers for conquering the diseases that affect the nervous system and for understanding the brain, the mind, and ourselves.

The Senior Editors are grateful to the Developmental Editors at Elsevier, Michael Bevan, Joanna De Souza and Richard Berryman, and the Editorial Assistants, Caroline Phipps, Afandi Mohamed and Nicky Carter, for capably managing the formidable task of assembling and organizing the contents of the Encyclopedia. Andrew Lowe and Laura Jackson, the Project Managers, diligently brought the project through its several stages of production.

Larry R. Squire
Editor-in-Chief

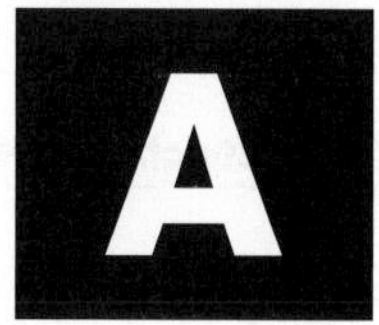

Acetylcholine Neurotransmission in CNS

M-M Mesulam, Northwestern University, Chicago, IL, USA

Introduction

Acetylcholine (ACh) was the first neurotransmitter to be discovered. It is synthesized by choline acetyltransferase (ChAT) from choline and acetate as precursors. Its action on cholinergic receptors is terminated through hydrolysis by acetylcholinesterase (AChE). Neurons and axons that contain ChAT and that use ACh for neurotransmission are designated 'cholinergic,' and the term 'cholinoceptive' is used to designate neurons that respond to ACh through muscarinic or nicotinic receptors.

Cholinergic innervation is ubiquitous at all levels of the neuraxis but also shows precise patterns of origin and distribution. The vast majority of cholinergic innervation of the thalamus arises from the pedunculopontine and laterodorsal tegmental nuclei (also known as Ch5 and Ch6); the vast majority of cholinergic innervation for the caudate and putamen arises from intrinsic cholinergic neurons within the striatum; and the vast majority of cholinergic innervation for the cerebral cortex, hippocampus, and amygdala arises from the basal forebrain.

The human cerebral cortex has no cholinergic neurons but receives massive cholinergic innervation. Nearly all of it arises from four partially overlapping basal forebrain cell groups where cholinergic and noncholinergic neurons are intermingled. The Ch1–Ch4 nomenclature was introduced to designate the cholinergic (i.e., ChAT-containing) neurons within these four cell groups. According to this nomenclature, Ch1 designates the cholinergic cells associated predominantly with the medial septal nucleus; Ch2, those associated with the vertical nucleus of the diagonal band; Ch3, those associated with the horizontal limb of the diagonal band nucleus; and Ch4, those associated with the nucleus basalis of Meynert. Tracer experiments in laboratory animals have shown that Ch1 and Ch2 provide the major cholinergic innervation for the hippocampal complex, Ch3 for the olfactory bulb, and Ch4 for the rest of the cerebral cortex and the amygdala. The term 'nucleus basalis' can be used to designate the cholinergic as well as noncholinergic components in this nucleus, whereas the more restrictive Ch4 designation is reserved for its cholinergic neurons.

ACh Neurotransmission

Anatomy of the Nucleus Basalis

The nucleus basalis is a phylogenetically progressive nucleus which displays its greatest differentiation in the cetacean and human brains. The human nucleus basalis extends from the level of the olfactory tubercle to that of the posterior amygdala, spanning a distance of 13–14 mm in the anteroposterior axis and attaining a mediolateral width of 18 mm within the substantia innominata (subcommissural gray). It contains approximately 200 000 neurons in each hemisphere. In the human brain, Ch4 neurons express ChAT, the vesicular ACh transporter, AChE, calbindin-d28k, the high affinity nerve growth factor receptor (NGFr) trkA, and the low affinity p75 NGFr. A minority of Ch4 neurons are NGFr-negative and, at least in the rat, project preferentially to the amygdala. The nucleus basalis also contains a complex mosaic of noncholinergic neurons that are nicotinamide adenine dinucleotide phosphate diaphorase (NADPHd)-positive, γ-aminobutyric acid (GABA)ergic, peptidergic, or tyrosine hydroxylase-positive. Based on its location, connectivity, and cellular morphology, the nucleus basalis has been conceptualized as a site of confluence for the limbic system and the ascending reticular activating system.

Inputs and Neurotransmitter Circuitry of the Nucleus Basalis

Although the primate nucleus basalis projects to the entire cerebral cortex, it receives major cortical projections from only the limbic and paralimbic regions of the brain, including the amygdala. The cortical inputs are mostly glutamatergic but can also be GABAergic. The primate nucleus basalis receives dopamine, norepinephrine, and serotonin inputs, presumably all arising in the brain stem.

In the monkey, cholinergic synapses onto cholinergic Ch4 neurons are quite frequent and take the form of large asymmetrical synapses. The precise source of the cholinergic input to the nucleus basalis is unknown but could include collaterals from cholinergic Ch1–Ch4 neurons of the basal forebrain or ascending projections from the Ch5–Ch6 group of pontomesencephalic cholinergic nuclei. In the rat and monkey, the m2 muscarinic receptor subtype is prominently expressed by only a third of Ch4 neurons. Therefore, the often cited assumptions that all Ch4 neurons express m2 and that m2 is a universal presynaptic marker of cortical cholinergic innervation need to be modified.

Topography and Distribution of Cortical Cholinergic Projections

The density of cholinergic axons is higher in the more superficial layers of the cerebral cortex, suggesting that the axons which enter the cortex from the underlying gray matter undergo branching as they course toward the pial surface, an interpretation which is supported by physiological evidence. The density of cholinergic innervation is lower within unimodal and heteromodal association areas than in paralimbic areas of the brain. Core limbic areas such as the amygdala and hippocampus contain the highest densities of cholinergic innervation.

Trajectory of Cholinergic Pathways from the Nucleus Basalis to the Cerebral Cortex

Two bundles of cholinergic fibers extend from the nucleus basalis to the cerebral cortex and amygdala in the human brain. A medial pathway joins the white matter of the gyrus rectus, curves around the rostrum of the corpus callosum to enter the cingulum, and merges with fibers of the lateral pathway within the occipital lobe. It supplies the parolfactory, cingulate, pericingulate, and retrosplenial cortices in the medial parts of the cerebral hemispheres. The lateral pathway is subdivided into a capsular division traveling in the white matter of the external capsule and a perisylvian division traveling within the claustrum. Branches of the perisylvian division supply the frontoparietal operculum, insula, and superior temporal gyrus. Branches of the capsular division innervate the remaining parts of the frontal, parietal, and temporal neocortex.

Cholinergic Synapses in the Cerebral Cortex

In the rat and cat cerebral cortex, ChAT-immunoreactive fibers are almost exclusively unmyelinated, display numerous varicosities, and make mostly symmetrical synapses on the perikaryon, dendritic shaft, and spines of pyramidal as well as nonpyramidal neurons. In the temporal cortex of the human brain, most ChAT-positive terminals form identifiable synaptic specializations. These are usually quite small and symmetric and located predominantly on the dendritic shafts and spines of pyramidal neurons.

Cholinergic Receptors in the Cerebral Cortex

The ACh released by cholinergic axons exerts its influence on the cerebral cortex by interacting with neurons expressing muscarinic and nicotinic receptors. In mammals, the m1 subtype is the most common species of cholinergic receptor in the cerebral cortex. Muscarinic receptor subtypes are found not only at traditional postsynaptic cholinoceptive sites but also at additional locations where they can function as postsynaptic modulators of noncholinergic transmission or presynaptic autoreceptors that influence the release of ACh.

Cholinoceptive Neurons of the Cerebral Cortex

Incoming cholinergic axons innervate at least three kinds of cortical neurons: glutamatergic neurons, GABAergic interneurons, and NADPHd-positive infracortical neurons. The one marker most closely associated with cholinoceptive neurons is AChE, the enzyme which terminates cholinergic neurotransmission through the rapid hydrolysis of ACh into acetate and choline. Although the vast majority of cholinoceptive neurons probably express some AChE, only a subset yields an AChE-rich histochemical staining pattern. In the cerebral cortex of the adult rat, the AChE-rich cytochemical pattern is limited to a few polymorphic neurons. The situation is dramatically different in the human cerebral cortex, which contains a dense network of AChE-rich cortical neurons, especially in layers III and V of premotor and sensory association cortex. These neurons are mostly pyramidal in shape and are likely to represent a glutamatergic contingent of cholinoceptive neurons. The cerebral cortex also contains GABAergic cholinoceptive interneurons that respond to ACh through muscarinic receptors. As shown by Freund and Meskenaite, these neurons may receive additional inputs from the GABAergic projection neurons of the nucleus basalis.

Functionality of Cortical Cholinergic Innervation

Although the primate Ch4 projects to all cortical areas, it receives major inputs almost exclusively from limbic and paralimbic regions of the cerebral cortex. This creates a unique circuitry through which cholinergic projections can modulate neural activity in all cortical areas in ways that reflect the prevailing influence of the limbic system. Experiments in rodents and cats have shown that the principal action of cholinergic stimulation at m1 receptor sites is to induce a relatively prolonged reduction of potassium conductance so as to make the cholinoceptive neuron more responsive to other incoming excitatory inputs. A sensory event that activates Ch4 would thus be expected to have a stronger impact on cortical circuitry and to attract more robust neural encoding. These aspects of receptor physiology and the preferential relationship to the limbic system suggest that the attentional modulation of behaviorally relevant events and their preferential encoding in memory are two cognitive faculties most likely to be under the influence of cortical cholinergic innervation.

Memory and Cortical Cholinergic Innervation

The characteristically intense cholinergic innervation of limbic areas, including the hippocampus and entorhinal cortex, had promoted the opinion that cortical cholinergic neurotransmission plays a crucial role in memory function. In rodents and monkeys, damaging the cholinergic neurons in the basal forebrain causes learning and memory impairments in some experiments but not in all. Turchi and colleagues showed that selective immunotoxic lesioning of cholinergic axons in the parahippocampal–rhinal region of the monkey brain leads to distinct memory deficits in the delayed nonmatching-to-sample task. The deficit was as severe as that obtained by lesioning the parahippocampal–rhinal cortex, suggesting that the mnemonic function of this region is dependent on its cholinergic innervation. Humans respond to the muscarinic antagonist scopolamine with a memory impairment somewhat similar to that seen in Alzheimer's disease (AD). The role of ACh in hippocampal long-term potentiation, its ability to suppress proactive interference, and its participation in memory consolidation may provide some of the cellular mechanisms that mediate its importance for memory function.

Attentional State and Cortical Cholinergic Innervation

There is considerable agreement concerning the importance of cortical cholinergic innervation for various aspects of attention, including the setting of signal-to-noise ratios during information processing and the online holding of information. The nucleus basalis projects to all cortical areas while receiving its cortical inputs almost exclusively from components of the limbic system, leading to an arrangement that promotes the selective release of cortical ACh in response to events that are of limbic relevance. Cortical cholinergic innervation is thus in a position to preferentially enhance the cortical impact of events that are of emotional and motivational significance. In keeping with this formulation, neurons of the nucleus basalis in the monkey are sensitive to novel and motivationally relevant sensory events.

Cortical ACh and Plasticity

Neuroplasticity is a life-long process that mediates the structural and functional reaction of dendrites, axons, and synapses to experience, attrition, and injury. One of the most interesting functional correlates of cortical ACh is its role in mediating neuroplasticity. In an experiment by Baskerville and colleagues, all whiskers except for D2 and D3 were trimmed in rats. This led to a pairing between the D2 and D3 barrel fields so that the D2 neurons in the cerebral cortex started to show a greater responsivity to stimulation of D3 than to stimulation of the adjacent D1, which had been trimmed. This pairing, indicative of experience-induced synaptic plasticity, could not be obtained in rats with selective immunotoxic lesions of the cholinergic neurons in the nucleus basalis. Furthermore, Kilgard and Merzenich showed that pairing auditory stimuli with the electrical stimulation of the nucleus basalis in adult rats caused a long-lasting reorganization of primary auditory cortex so that the area optimally responsive to the paired tone expanded substantially. This plasticity was also abolished following the selective immunotoxic destruction of cholinergic nucleus basalis neurons. Based on these observations, it appears that cortical cholinergic denervation can undermine the learning-dependent reorganization of cortical circuitry and perhaps also the ability of the brain to keep itself in good repair in response to attrition and injury.

Cortical Cholinergic Innervation and AD

AD is associated with a prominent loss of cortical (but not thalamic or striatal) cholinergic innervation (**Figure 1**). The modest therapeutic effect of cholinesterase inhibitors such as galantamine, donepezil, and rivastigmine in AD is attributed to their cholinomimetic effects. The initial reports of cortical cholinergic denervation and cell loss in the nucleus basalis were based on patients who had advanced AD. Recent evidence shows that neurofibrillary degeneration of the Ch4–nucleus basalis complex starts in the course of normal aging and that it becomes more pronounced even in the prodromal stages of AD (also known as mild cognitive impairment (MCI)). By early stages of AD, this cholinergic cytopathology is already quite advanced. There is a significant inverse correlation between the intensity of neurofibrillary degeneration in the nucleus basalis and performance in delayed memory tests throughout the age–MCI–AD continuum.

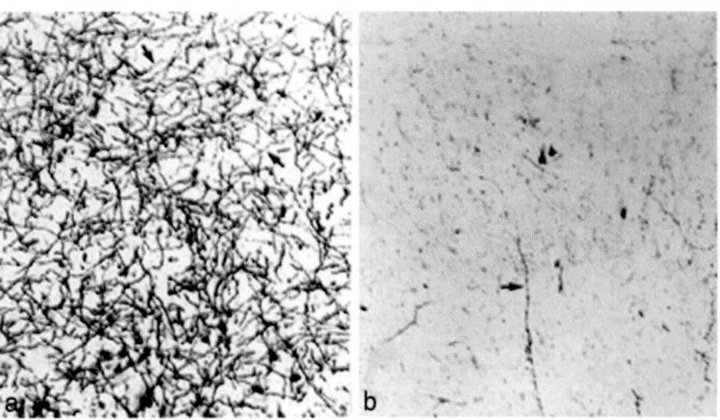

Figure 1 Cholinergic fibers (arrows) of the temporal neocortex in a cognitively normal 70-year-old (a) and in a patient of similar age with Alzheimer's disease (b). Magnification 100×.

The presence of cytopathology in the nucleus basalis of cognitively normal older people may initially raise the possibility that it may have no functional relevance. Such an inference would be based on the erroneous assumption that 'normal for age' reflects a stability of cognitive function. However, a person in the age range of 70–100 may have experienced a considerable decline of performance from a former baseline and still remain within the normal range for age. The nucleus basalis cytopathology observed in cognitively normal people may have contributed to the emergence of these age-related changes. The substantially more severe nucleus basalis lesion of AD may contribute to the emergence of the dementia by interfering with cholinergic neurotransmission and also by eroding the capacity for neuroplasticity to the point that the brain cannot keep itself in good repair.

Numerous hypotheses have been advanced to explain why the pathway from the basal forebrain to the cerebral cortex is so vulnerable to AD. Biosynthetic bottlenecks in the production of ACh, special vulnerabilities to amyloid, perturbation of intracellular calcium homeostasis, and impairments in the axonal transport of growth factors have been invoked as possible explanations, but proof is lacking. The answer may lie in the anatomical affiliations of the Ch4–nucleus basalis complex. This cell group, together with the hippocampus, amygdala, and entorhinal cortex, is part of an uninterrupted band of core limbic areas. These are the regions of the brain that attract the greatest neurofibrillary degeneration and related cell death in the course of aging and AD, perhaps because they sustain an unusually high neuroplasticity burden throughout the life span. The severe and selective loss of cortical cholinergic innervation may thus reflect the location and connectivity of the nucleus basalis rather than its relationship to cholinergic neurotransmission.

See also: Acetylcholinesterase Inhibitors and Alzheimer's Disease; Cholinergic Pathways in CNS; Cholinergic Neurotransmission in the Autonomic and Somatic Motor Nervous System; Cholinergic System Imaging in the Healthy Aging Process and Alzheimer Disease; Nicotinic Acetylcholine Receptors.

Further Reading

Baskerville KA, Schweitzer JB, and Herron P (1997) Effects of cholinergic depletion on experience-dependent plasticity in the cortex of the rat. *Neuroscience* 80: 1159–1169.

Berger-Sweeney J, Heckers S, Mesulam M-M, Wiley RG, Lappi DA, and Sharma M (1994) Differential effects upon spatial navigation of immunotoxin-induced cholinergic lesions of the medial septal area and nucleus basalis magnocellularis. *Journal of Neuroscience* 14: 4507–4519.

De Long MR (1971) Activity of pallidal neurons during movement. *Journal of Neurophysiology* 34: 414–427.

De Rosa E, Desmond JE, Anderson AK, Pfefferbaum A, and Sullivan EV (2004) The human basal forebrain integrates the old and the new. *Neuron* 41: 825–837.

Freund TF and Meskenaite V (1992) g-Amynobutyric acid-containing basal forebrain neurons innervate inhibitory interneurons in the neocortex. *Proceedings of the National Academy of Sciences of the United States of America* 89: 738–742.

Gorry JD (1963) Studies on the comparative anatomy of the ganglion basale of Meynert. *Acta Anatomica* 55: 51–104.

Heckers S, Ohtake T, Wiley RG, Lappi DA, Geula C, and Mesulam M-M (1994) Complete and selective cholinergic denervation of rat neocortex and hippocampus but not amygdala by an immunotoxin agoinst the p75 NGF receptor. *Journal of Neuroscience* 14: 1271–1289.

Kilgard MP and Merzenich MM (1998) Cortical map reorganization enabled by nucleus basalis activity. *Science* 279: 1714–1718.

McCormick DA and Prince DA (1985) Two types of muscarinic responses to acetylcholine in mammalian cortical neurons. *Proceedings of the National Academy of Sciences of the United States of America* 82: 6344–6348.

Mesulam M-M (1996) The systems-level organization of cholinergic innervation in the cerebral cortex and its alterations in Alzheimer's disease. *Progress in Brain Research* 109: 285–298.

Mesulam M-M and Geula C (1991) Acetylcholinesterase-rich neurons of the human cerebral cortex: Cytoarchitectonic and ontogenetic patterns of distribution. *Journal of Comparative Neurology* 306: 193–220.

Mesulam M-M, Shaw P, Mash D, and Weintraub S (2004) Cholinergic nucleus basalis tauopathy emerges early in the aging-MCI-AD continuum. *Annals of Neurology* 55: 815–828.

Ridley RM, Murray TK, Johnson JA, and Baker HF (1986) Learning impairment following lesion of the basal nucleus of Meynert in the marmoset: Modification by cholinergic drugs. *Brain Research* 376: 108–116.

Sarter M and Bruno JP (2000) Cortical cholinergic inputs mediating arousal, attentional processing and dreaming: Differential afferent regulation of the basal forebrain by telencephalic and brainstem afferents. *Neuroscience* 95: 933–952.

Turchi J, Saunders RC, and Mishkin M (2005) Effects of cholinergic deafferentation of the rhinal cortex on visual recognition memory in monkeys. *Proceedings of the National Academy of Sciences of the United States of America* 102: 2158–2161.

Wilson FAW and Rolls ET (1990) Neuronal responses related to novelty and familiarity of visual stimuli in the substantia innominata, diagonal band of Broca and periventricular region of the primate basal forebrain. *Experimental Brain Research* 80: 104–120.

Záborszky L, Gaykema RP, Swanson DJ, and Cullinan WE (1997) Cortical input to the basal forebrain. *Neuroscience* 79: 1051–1078.

Acetylcholinesterase

P Taylor, S Camp, and Z Radić, University of California at San Diego, La Jolla, CA, USA

Introduction

The turnover number of acetylcholinesterase (AChE) approaches $1.5 \times 10^4\ s^{-1}$, making it one of the most efficient enzymes known. Rapidity of catalysis of released acetylcholine in a submillisecond time frame is essential in the skeletal neuromuscular junction to allow the next volley of released acetylcholine to trigger a postsynaptic excitatory potential. AChE (EC 3.1.1.7) is distinguished from butyrylcholinesterase (BChE; EC 3.1.1.8) by its catalytic selectivity for acetylcholine over butyrylcholine hydrolysis. AChE is typically synthesized in nerve, muscle, and certain hematopoietic cells. In excitable tissues, AChE expression is regulated by tissue-specific development, and the enzyme is localized on the extracellular surface of both nerve and muscle. The AChE found in the neuromuscular junction of skeletal muscle is synthesized by the muscle rather than the nerve cell. AChE inhibitors are employed in the treatment of Alzheimer's disease, myasthenia gravis, glaucoma, smooth muscle atony, and assorted disorders of autonomic nervous system functions.

BChE is synthesized largely in the liver and is exported into the plasma. It displays a wider substrate range for catalysis than does AChE and is thought to play a primary role in the metabolism of dietary esters, perhaps only in selected species. Genetic polymorphisms that preclude the expression of BChE in humans yield no apparent phenotype, unless drugs in doses requiring BChE catalysis for rapid degradation are given. BChE has no apparent role in the nervous system, and its protective role only becomes manifest in AChE-deleted transgenic mice.

Genes Encoding the Cholinesterases

To date, about 150 DNA sequences encoding cholinesterases, extending from nematodes and insects to humans, have been reported. Although in mammals a single AChE subserves all cholinergic nervous system function, this is not the case in lower species, such as nematodes and certain insects, in which multiple AChE species exist and influence nervous system function. In *Drosophila*, a single cholinesterase is found; its structure shows features of both AChE and BChE.

Distinct, but single, genes in mammals encode AChE and BChE, yielding gene products with ~50% amino acid identity. AChE exists in multiple molecular species that arise from three alternative splice options in the gene, not found in BChE. The AChE gene contains three constitutive exons within the open reading frame, starting in exon 2; in the following two exons, three options exist: reading through exon 4, splicing to exon 5, and splicing to exon 6 (**Figure 1**). These options yield a soluble monomer, glycophospholipid-linked dimers, or a tetrameric form of AChE, respectively. The monomeric subunits of all of these forms, although differing in their very C-termini, contain identical sequences forming the catalytic and inhibitor binding sites. The monomeric unit is glycosylated at three or four N-linked sites and has a peptide molecular weight of ~62 000 (**Figure 2**).

Exon 6 encodes a C-terminal region that contains a proline-attachment domain, enabling it, when fully assembled as a tetramer, to associate with proline-rich sequences of two structural subunits. The disulfide-linked assembly with the first structural subunit yields a heteromeric tetramer of catalytic subunits with the single structural subunit (**Figure 2**). This structural subunit contains the requisite hydrophobicity to be attached to the outer leaflet of the cell membrane. Assembly with a second type of structural subunit yields a more complex heteromeric structure composed typically of three AChE tetramers associated with a collagen-containing unit. This subunit also has a proline-rich sequence at its N-terminus for association and attachment through a disulfide linkage. The C-terminal portion of the collagen-containing subunit appears to associate with perlecan and perhaps other extracellular synaptic proteins. Each subunit of the triple-helical collagen tail associates with a tetramer. Accordingly, the assembled species approaches a molecular mass of 10^6 Da. The membrane-associated heteromeric species is the predominant form in brain, while the collagen-containing heteromer is localized to the basal lamina of the neuromuscular junction.

Regulation of AChE gene expression occurs at multiple levels to account for the great enhancement in expression associated with muscle and nerve differentiation. Accordingly, transcriptional activation and mRNA stabilization have been described in muscle differentiation. The primary promoter region of the *AChE* gene is selectively regulated in skeletal muscle, in which an intron between exons 1 and 2 serves to interact with the promoter to enhance expression during the myoblast-to-myotube conversion. This 250-bp intronic region of the gene, containing several regulatory elements, is well conserved between mammalian species. Its importance in controlling

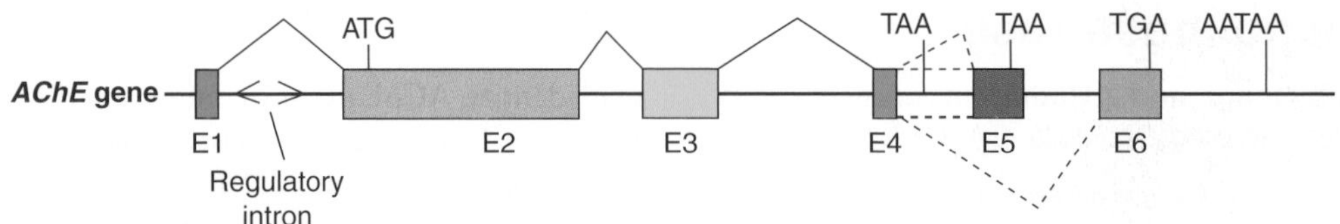

Figure 1 Acetylcholinesterase (AChE) gene structure. The gene is compacted into approximately 7.5 kbp with six closely spaced exons (E1–E6). Although activity from a more distal 5′ promoter has been reported, the primary promoter is immediately 5′ of the cap site of the start of transcription. Its activity is regulated by a 250-bp region in the intron between exons 1 and 2. Exon 2 encodes the start methionine, and the region 3′ of the start site in exon 2, exon 3, and exon 4 encodes the invariant region of the multiple spliced forms of AChE. Following exon 4, three alternatives are possible: (1) reading through exon 4 into the intron, giving rise to a soluble monomeric form of AChE; (2) splicing to exon 5, giving rise, after processing, to a glycophospholipid-linked dimer; and (3) splicing to exon 6, yielding a sequence predisposed to attaching to proline rich sequences in the structural subunits.

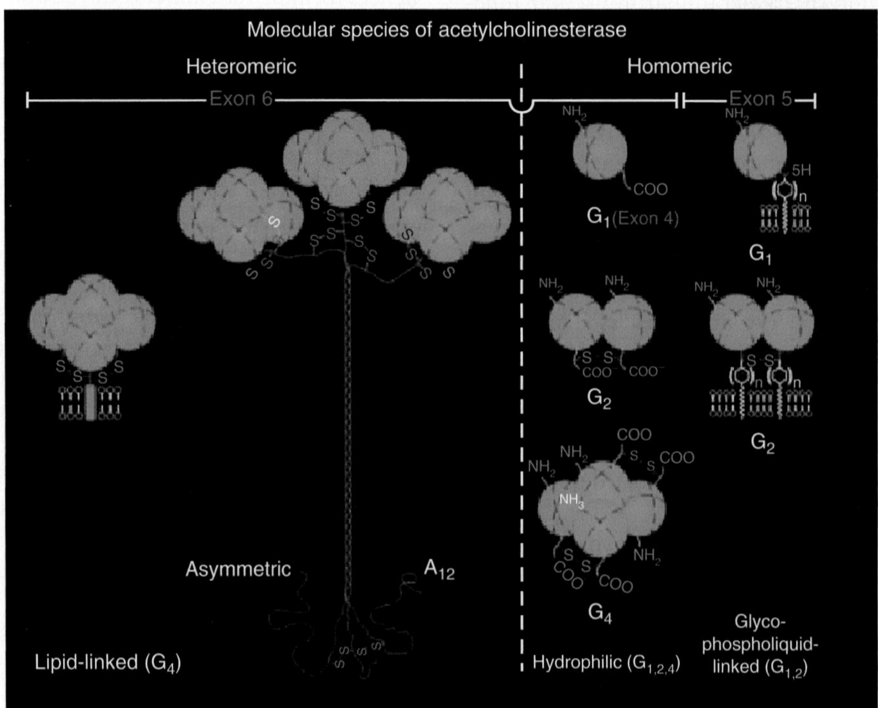

Figure 2 Molecular forms of acetylcholinesterase (AChE). The molecular species arising from alternative mRNA processing of the gene in **Figure 1** and the subunit assembly are shown. Depending on the splice alternative, AChE can exist in multiple forms of assembly (red), ranging from homomeric monomers, glycophospholipid (yellow) attached dimers and tetramers, to heteromeric species formed by disulfide attachment to a hydrophobic, lipid-linked subunit (fushia) or a subunit with a collagen-containing sequence (blue).

expression *in vivo* has been documented in transgenic mice in which selective regions of the *AChE* gene have been deleted.

Structure of the Cholinesterases

About 100 structures of cholinesterases as apo forms or complexed or conjugated with various ligands have been deposited in the Protein Data Bank. From these structures the molecular bases for efficient catalysis and inhibitor selectivity are beginning to be uncovered. The active center of the enzyme is centrosymmetric to the subunit and lies at the bottom of a narrow gorge, 18–20 Å in depth and lined with aromatic residues (**Figure 3**). Hydrogen bonding between Asp334, His447, and Ser203 (the catalytic triad) renders the serine an effective nucleophile. The side chains of Phe295 and Phe297 outline the space restricted acyl pocket in AChE, and contain smaller side chains for BChE. Aromatic residues of Trp86, Tyr337, and Try338 surround the choline subsite, forming a π orbital nest for the choline moiety. At the rim of the gorge lies a peripheral site, distinguished by an aromatic patch, where allosteric inhibitors, such as propidium, and the peptidic snake toxin fasciculin bind. Dimers of AChE are stabilized through a four-helix bundle. An amphiphilic, tryptophan-containing sequence, encoded only by exon 6, stabilizes the tetrameric form.

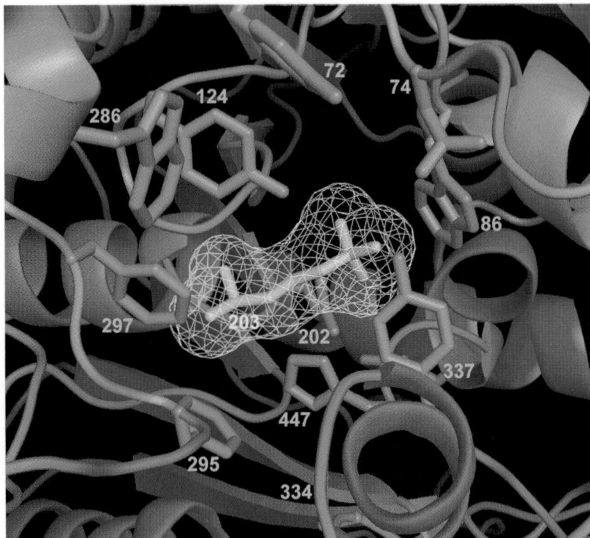

Figure 3 Structure of the active center of mouse acetylcholinesterase. Shown are residue side chains that constitute the catalytic triad (Ser203, His447, and Glu334), the acyl pocket (Phe295 and Phe297), the choline subsite (Trp86 and Try337, among other residues), and the peripheral site near the gorge entrance (Trp286, Try72, and Try124). Asp74, residing midway down the gorge, may have a gating or orienting role for entering cationic substrates.

Inhibitor Interactions

The primary AChE inhibitors used clinically or of toxicologic consequence associate with the active center. These include reversibly bound inhibitors, such as edrophonium and tacrine, that associate with the choline subsite; carbamoylating agents, such as physostigmine and neostigmine, that carbamylate transiently the active center serine; the alkyl phosphate agents, such as diisopropyl fluorophosphate (DFP), echothiophate, and diazinon, and nerve agents such as sarin and soman, that phosphorylate or phosphonylate the serine, forming a stable bond. The alkyl phosphates that form a dimethoxyphosphoryl or diethoxyphosphoryl conjugates with the serine are primarily used as pesticides, whereas the methylphosphonylating agents, such as sarin, have been insidiously employed in chemical terrorism in isolated areas of the globe.

Reactivating Agents and Antidotes

Oximes containing a pyridinium aldoxime moiety serve as nucleophiles for reactivating the alkyl phosphate conjugates with the enzyme and are used as antidotes. Their quaternary structure limits their capacity to reactivate alkyl phosphate-inhibited enzyme in the central nervous system. Atropine and benzodiazepines are also used as antidotes to control excessive muscarinic stimulation and convulsions.

See also: Acetylcholine Neurotransmission in CNS; Acetylcholinesterase Inhibitors and Alzheimer's Disease; Basal Ganglia: Acetylcholine Interactions and Behavior; Cholinergic System; Cholinergic Pathways in CNS; Cholinergic Neurotransmission in the Autonomic and Somatic Motor Nervous System; Cholinergic System Imaging in the Healthy Aging Process and Alzheimer Disease; Neuromuscular Junction (NMJ): Acetylcholinesterases; Neutrotransmission and Neuromodulation: Acetylcholine; Nicotinic Acetylcholine Receptors.

Further Reading

Bourne Y, Taylor P, and Marchot P (1995) Acetylcholinesterase inhibition by fasciculin: Crystal structure of the complex. *Cell* 83: 493–506.

Cygler M, Schrag JD, Sussman JL, et al. (1993) Relationship between sequence conservation and three-dimensional structure in a large family of esterases, lipases, and related proteins. *Protein Science* 2: 366–382.

Dvir H, Harel M, Bon S, et al. (2004) The synaptic acetylcholinesterase tetramer assembles around a polyproline II helix. *EMBO Journal* 23: 4394–4405.

Giacobini E (ed.) (2000) *Cholinesterases and Cholinesterase Inhibitors*. London: Martin Dunitz Publishers.

Giacobini E (ed.) (2003) *Butyrylcholinesterase – Its Function and Inhibitors*. London: Martin Dunitz Publishers.

Hotelier T, Renault L, Cousin X, et al. (2004) ESTHER, the database of the alpha/beta-hydrolase fold superfamily of proteins. *Nucleic Acids Research* 32: D145.

Li Y, Camp S, and Taylor P (1993) Tissue-specific expression and alternative mRNA processing of the mammalian acetylcholinesterase gene. *Journal of Biological Chemistry* 268: 5790–5797.

Massoulie J (2002) The origin of the molecular diversity and functional anchoring of cholinesterases. *Neurosignals* 11: 130–143.

Quinn DM (1987) Acetylcholinesterase: Enzyme structure, reaction dynamics, and virtual transition states. *Chemical Reviews* 87: 955–979.

Radić Z and Taylor P (2001) Interaction kinetics of reversible inhibitors and substrates with acetylcholinesterase and its fasciculin 2 complex. *Journal of Biological Chemistry* 276: 4622–4633.

Radić Z and Taylor P (2006) Structure and function of cholinesterases. In: Gupta R (ed.) *Toxicology of Organophosphate and Carbamate Compounds*, pp. 161–186. Amsterdam: Elsevier.

Sussman JL, Harel M, Frolow F, et al. (1993) Atomic structure of acetylcholinesterase from *Torpedo californica*: A prototypic acetylcholine-binding protein. *Science* 253: 872–879.

Taylor P (2006) Acetylcholinesterase agents. In: Brunton LL (ed.) *Goodman & Gilman's Pharmacological Basis of Therapeutics*, 11th edn., pp. 201–216. New York: McGraw-Hill.

Relevant Website

http://bioweb.ensam.inra.fr – ESTHER Database (analysis of protein and nucleic acid sequences homologous to cholinesterases).

Acetylcholinesterase Inhibitors and Alzheimer's Disease

H Kaduszkiewicz and H van den Bussche,
University Medical Center Hamburg-Eppendorf,
Hamburg, Germany

Introduction

The age specific prevalence of Alzheimer's disease (AD) doubles every 5 years of life, affecting 1.2% of those aged 65–69, 2.8% of those aged 70–74, 6.0% of those aged 75–79, etc. The disease causes irreversible memory loss, behavioral and cognitive decline, personality changes, and a decreasing ability to cope with everyday life. Amyloid plaques, neurofibrillary tangles, and a loss of neurons are the typical cellular pathology observed in the central nervous system of AD patients, though these changes are not pathognomonic. In the 1970s, postmortem examinations of the brains of patients with largely end-stage AD also reported markedly reduced activity of choline acetyltransferase, an acetylcholine-synthesizing enzyme. The cholinergic deficits were reported to correlate with cognitive deficits and disease severity. They were also linked to neuropsychiatric symptoms of AD, when medial frontal and limbic cholinergic deficits were identified. The discovery of widespread loss of cholinergic innervation triggered development of cholinesterase inhibitors, which aim to raise acetylcholine levels in the brain by blocking the enzymes metabolizing this molecule. Donepezil, rivastigmine, and galantamine are the three acetylcholinesterase (AChE) inhibitors licensed for the treatment of mild to moderately severe AD. Their characteristics are listed in **Table 1**.

Preclinical Testing

Originally, preclinical testing of the AChE inhibitors was promising. For example, donepezil at doses of $2.5\ mg\ kg^{-1}$ showed maximum increases of the extracellular acetylcholine concentration in the hippocampus of rats of 499% of the prelevel at about 1.5 h after administration. The time courses of brain AChE inhibition with donepezil correlated strongly with the extracellular acetylcholine-increasing action. Donepezil demonstrated efficacy in tests of reference memory in animals, but had less consistent activity in tests of working memory. AChE activity was also inhibited in samples of postmortem human brain, fresh brain cortex biopsies, and human erythrocytes. In addition, rapid, sustained, and dose-dependent inhibition of cerebral spinal fluid AChE was demonstrated in AD patients.

Some findings suggested a neuroprotective potential of AChE inhibitors. For example, donepezil was found to attenuate A-beta(25-35)-induced toxicity in rat pheochromocytoma PC12 cells in clinically relevant concentrations. Additional hope came from findings of special action of galantamine directly interacting with nicotinic acetylcholine receptors. These allosterically potentiating ligands sensitize nicotinic receptors by increasing the probability of channel opening induced by acetylcholine and nicotinic agonists and by slowing down receptor desensitization.

Clinical Efficacy

Despite the promising preclinical findings, the evaluation of clinical efficacy did not reveal striking effects. In fact, the study results are interpreted so diversely that 10 years after licensure, the clinical efficacy of AChE inhibitors is still controversial.

In this article, the manifold discussions on the study results on AChE inhibitors will be presented, as they are crucial for the estimation of clinical efficacy. There is no direct way to deduce a clinical benefit from study results. In fact, there is still a long way to go to accomplish this, namely:

1. Selection of relevant studies
2. Identification of the methodological quality of the studies
3. Assessment of the relevance of the methodological quality for the results
4. Discussion of the external validity of the trials
5. Determination of the clinical relevance of effects and side effects measured in the trials
6. Comprehension of the clinical experience of patients, caregivers, and health personnel

Not until these steps are taken is an estimation of clinical benefit possible.

Selection of Relevant Studies

In order to assess the efficacy of cholinesterase inhibitors, there is wide consensus that the analysis of randomized controlled double blind trials is the first choice. In these studies the agent should have been tested against placebo, and clinical parameters should have been used as outcome measures. To date, 18 randomized controlled trials (RCTs) on donepezil, five on rivastigmine, and seven on galantamine have been published. The majority of these studies included patients with mild to moderate stages of AD; the minority focused on patients in the severe stage.

Table 1 Characteristics of the three licensed AChE inhibitors for the treatment of AD patients

	Donepezil hydrochloride (Aricept®)	*Galantamine hydrobromide (Reminyl®)*	*Rivastigmine tartrate (Exelon®)*
Active principle	Reversible and highly centrally selective inhibitor of acetylcholinesterase (AChE) with minimal effects on butrylcholinesterase (BuChE)	Dual effect on the cholinergic system: selective competitive and reversible inhibitor of AchE and allosterical modulator of presynaptic nicotinic acetylcholine receptors (nAChR)[a]	'Pseudoirreversible'[b] centrally selective inhibitor (focus on cortex and hippocampus) of AChE and BuChE
Molecule type	Piperidine-based molecule	Tertiary alkaloid	Carbamate type
$t_{1/2}$	Approximately 70 h	7–8 h	0.6–2 h
Bioavailability	100%; 96% bound to plasma proteins	88%; 18% bound to plasma proteins	36%; 40% bound to plasma proteins
T_{max}	4 h	1 h	1 h
Steady state	After 2–3 weeks	After 2–3 days	
Administration	Once daily	Twice daily	Twice daily
Metabolism	One active metabolite (6-O-desmethyldonepezil), hepatic metabolism by CYP 450 isoenzymes 3A4 and 2D6 and glucuronidation	Hepatic metabolism by CYP 450 isoenzymes 2D6 and 3A4	95% metabolism by AChE (decarbamylation); no involvement of the CYP 450 system
Elimination	Excreted in urine and hepatic metabolism	>90% renal elimination of metabolites	Renal elimination of metabolites (>90% within 24 h)

[a]*In vitro*, galantamine binds to a site on nAChR that is different from the binding site of the natural agonist, acetylcholine (described as allosteric, meaning 'other site'). When galantamine and acetylcholine bind simultaneously to nAChR, the response of these receptors to acetylcholine is amplified.
[b]Pseudoirreversible means that AChE is carbamylated and after some hours hydrolyzed, that is, AChE regenerates without new synthesis.

Mostly, cognitive performance was assessed by means of the Alzheimer's Disease Assessment Scale – Cognitive subscale (ADAS-Cog), which ranges from 0 to 70 points, 0 indicating no impairment and 70 points indicating final dementia. On average, the mean difference between the groups receiving AChE inhibitors and placebo was 3 points, which was statistically significant in favor of AChE inhibitors. This corresponds to an effect size of circa 0.5 standard deviations. In addition to the ADAS-Cog scale, the CIBIC-plus scale (Clinician's Interview Based Impression of Change with caregiver input) was often used. This semistructured interview, performed first with the caregiver and then with the patient, provides a global clinical assessment of change. The change between baseline and endpoint assessment is marked on a 7-point Likert-like scale, where 4 represents no change compared with baseline, 1–3, different degrees of improvement, and 5–7, degrees of deterioration. On this scale, the mean difference between the active treatment and the placebo group amounted to some 0.5 points in favor of donepezil.

Concerning the activities of daily living, various instruments were used in the studies. For example, in the rivastigmine and galantamine studies, improvements in the range of 3 points on the DAD (Disability Assessment for Dementia Scale, total range 0–100) and the PDS (Progressive Deterioration Scale, total range 0–100) were seen.

Identification of the Methodological Quality of the Studies

For a comprehensive assessment of methodological quality, several questions have to be answered. As most important aspects the following are considered:

- *Randomization/allocation concealment.* Was the randomization of the patient to the treatment or placebo groups really by chance? And did neither patients nor investigators know who was randomized to which group?
- *Blinding.* Were the investigators assessing the patients' performance reliably blinded to the treatment of the patients? Did they not know about the side effects the patients experienced?
- *Handling of dropouts.* Which strategies were used to include results of those patients into the analyses who dropped out of the study?
- *Violation of the intention-to-treat-principle.* Was the intention-to-treat-principle (i.e., inclusion of all randomized patients into endpoint analyses) violated, and if yes, to what extent?

The methodological quality of the studies altogether can be judged as moderate or even partly insufficient. In the first place, this is due to the fact that in most of the studies the intention-to-treat-principle was violated. Also, in more than half of the studies the dropout rates in at least one of the treatment arms amounted to 20% or more. Finally, precise information on randomization, allocation concealment, and blinding was mostly missing in the study reports.

Assessment of the Relevance of the Methodological Quality for the Results

After having identified the methodological shortcomings of the studies, the crucial question is whether these shortcomings might have biased the results substantially. To date, this question is controversial.

Two state institutions of health technology assessment have lately extensively evaluated the trials on AChE inhibitors. Both the British National Institute for Health and Clinical Excellence (NICE) and the German Institut für Qualität und Wirtschaftlichkeit im Gesundheitswesen (Institute for Quality and Economy in Health Care) (IQWiG) confirmed the methodological shortcomings of the trials but did not assume that these shortcomings biased the results substantially. Because a definite estimation of bias caused by methodological shortcomings can only be given when original data are available, the assumption of NICE and IQWiG – who did not have the data – is not conclusive.

Another question concerns the method used to account for dropouts. In most of the studies the LOCF ('last observation carried forward') method was used: when a patient drops out of the study prematurely, the last test results available are used as if they were results of endpoint assessment. In the case of a progressively deteriorating disease like AD, the use of LOCF can lead to better results in groups with higher dropout rates and earlier dropout time points. As in the majority of the AChE inhibitor trials, more patients dropped out of the verum groups; this may have led to better results in comparison to placebo. However, this theoretical criticism can again only be substantiated on the basis of original data. Unfortunately, the pharmaceutical companies do not disclose these data.

Discussion of the External Validity of the Trials

In the majority of the RCTs, patients with other-severe 'not controlled' medical or psychiatric diseases, insulin-dependent diabetes mellitus, other endocrinological dysfunction, asthma, or obstructive pulmonary disease were excluded. A Canadian research group investigated to what extent a cohort of 6424 older adults who were newly dispensed donepezil in Ontario was represented in the RCTs on cholinesterase inhibitors. They found that 51% to 78% of the patients would not have been included in the trials. This means that, strictly speaking, the validity of the results of these trials is limited to some half of the patients.

A second point concerning the external validity of the trials is their short duration, as most of the RCTs do not exceed 6 months. In consequence, there is hardly any RCT-based evidence for treatment lasting for more than 6 months.

Determination of the Clinical Relevance of Effects and Side Effects as Measured in the Trials

What does a mean group difference of 3 points on a 70 points cognitive scale mean? The answer is unclear. A common answer given is that a US Food and Drug Administration (FDA) panel defined a change of 4 points on the ADAS-Cog as clinically relevant. But irrespective of the scientific basis of this consensus definition, these 4 points apply to changes in individuals and not in groups. The answer on the question of clinical relevance could be clearer if the distribution of changes had been published. Assuming a normal distribution, IQWiG calculated that the difference in success rates between the cholinesterase inhibitor and placebo groups amounts to 13–16% if success is defined as an individual improvement of $\geq$4 points. This means that at least six to eight patients have to be treated with the active drug in order to see success in cognition in one patient. Concerning activities of daily living, the number needed to treat amounts to 13–14. But these calculations have to be interpreted with caution, as they are based on the assumption of a normal distribution of the results.

The leading question concerning clinical relevance is the impact of treatment on life quality of patients and caregivers. Does the one out of six to eight patients who scores 4 points better on the ADAS-Cog experience any benefit in everyday life? There is a consensus that the data from the RCTs are not sufficient (either results are inconsistent or life quality was not evaluated) to ascertain a benefit concerning life quality of patients and caregivers. The same applies to institutionalization and a desirable disease-modifying effect: There is no evidence from RCTs that cholinesterase inhibitors delay institutionalization of patients or that they modify disease progression.

The question of side effects is also not controversial. AChE inhibitors have side effects: nausea, vomiting, diarrhea, weight loss, and agitation are typical. They occur in 5–10% of patients taking donepezil, 5–20% of patients taking galantamine, and 10–40% of patients taking rivastigmine. They can occur just in the titration phase of the drug, but can also persist and are the main reason for dropout from the trials and for discontinuation of therapy in clinical practice.

Comprehension of the Clinical Experience of Patients, Caregivers, and Health Personnel

Clinicians state that they see a clear benefit in 5% to 15% of their patients, but it has not been possible to identify these responders before treatment. Here again, hypotheses could be deduced from the studies that already exist, but the pharmaceutical companies do not make the original data available.

Another point is that some clinicians or caregivers point out that they see improvement in areas that are difficult to measure in standardized tests. For example, in their opinion the patient appears to be better regulated. Such an observation is in fact difficult to measure and to quantify. One of the latest studies on galantamine tried to overcome this dilemma and measured treatment outcomes by means of the Goal Attainment Scaling (GAS). The GAS is a personalized outcome measure in which people set goals according to their own needs and define improved or worsened states in their own words. After 16 weeks, at the end of double blind treatment, clinicians reported statistically significant changes on the GAS, whereas patients/caregivers did not. Even though the data do not support the assumption that the GAS is more sensitive in detecting treatment effects than other commonly used instruments in cholinesterase inhibitor trials, this instrument highlights the views of patients and caregivers as to whether treatment is seen as meaningful. Further research on this subject is needed.

Consequences of Unclear Clinical Benefit

Reflecting the controversial estimation of clinical benefit, national regulations concerning prescription and treatment documentation differ greatly. For example, in Germany anybody with mild to moderate AD can be treated with cholinesterase inhibitors as long as the treating doctor prescribes the drug, without any further restrictions. In many countries (e.g., Belgium, France, the UK, Australia), in contrast, there are clear rules concerning initiation, evaluation, and termination of therapy. In the Netherlands, donepezil and galantamine (and memantine) are not covered by the statutory health insurance – only rivastigmine is – on the basis of a European Union-wide admission procedure. In Belgium, a patient registry is to be established to follow these patients. Data from such registries could be used to analyze prescription patterns and effects in the great variety of patients treated under real life conditions.

Imprecise Cost Estimates

Costs for the drugs in Germany amount to \$1600 to \$2000 per year. Not included are all other costs incurred, for example, costs of doctor consultations to get the prescription or costs of diagnosis and treatment of side effects.

The data basis for cost estimates in relation to efficacy is extremely arguable. In their final appraisal document, the NICE reports amounts of \$39 000 to \$260 000 per quality adjusted life year gained. A huge range is also reported for the costs of gaining an additional year in a nonsevere state: \$2200 up to \$26 000. All these cost calculations are difficult to follow.

Summary

- The appropriateness of including RCTs only in assessments of efficacy of cholinesterase inhibitors is questioned, but there is a broad consensus that RCTs constitute the basis.
- The published RCTs on cholinesterase inhibitors have methodological shortcomings.
- The relevance of the methodological shortcomings is valued differently.
- Without original data it is not possible to prove whether the methodological shortcomings substantially biased the results or not.
- If the methodological shortcomings are ignored, the study results show small differences between the cholinesterase inhibitor and placebo groups. The clinical relevance of these differences is controversial.
- There is no evidence from RCTs concerning benefits on life quality of patients and caregivers and delay of institutionalization.
- Due to broad exclusion criteria and the duration of the trials mostly not exceeding 6 months, the external validity of the trials is rather low.
- Cholinesterase inhibitors have side effects.
- No disease-modifying effect has been able to be measured.
- Trial data from the industry should be accessible for research purposes.
- Further research directed at improved pharmacological therapy for patients with AD is needed.

See also: Acetylcholine Neurotransmission in CNS; Acetylcholinesterase; Aging of the Brain and Alzheimer's Disease; Alzheimer's Disease: An Overview; Basal Forebrain and Memory; Cholinergic System; Cholinergic Pathways in CNS; Cholinergic Neurotransmission in the Autonomic and Somatic Motor Nervous System; Cholinergic System Imaging in the Healthy Aging Process and Alzheimer Disease; Nicotinic Acetylcholine

Receptors; Sleep–Wake State Regulation by Acetylcholine.

Further Reading

Burns A and O'Brien J on behalf of the BAP Dementia Consensus Group (2006) Clinical practice with anti-dementia drugs: A consensus statement from British Association for Psychopharmacology. *Journal of Psychopharmacology* 20(6): 732–755.

Gill SS, Bronskill SE, Mamdani M, et al. (2004) Representation of patients with dementia in clinical trials of donepezil. *Canadian Journal of Clinical Pharmacology* 11(2): e274–e285.

IQWiG (2006) Cholinesterasehemmer bei Alzheimer Demenz [Cholinesterase inhibitors in Alzheimer's disease]. Vorbericht A05/19-A. Köln: Institut für Qualität und Wirtschaftlichkeit im Gesundheitswesen (IQWiG).

Kaduszkiewicz H, Zimmermann T, Beck-Bornholdt H-P, and van den Bussche H (2005) Cholinesterase inhibitors for patients with Alzheimer's disease: Systematic review of randomized clinical trials. *British Medical Journal* 331: 321–327.

National Institute for Health and Clinical Excellence (NICE) (2006) Alzheimer's disease – Donepezil, rivastigmine, galantamine and memantine (review) – Final appraisal document. 26 May 2006.

Rockwood K, Fay S, Song X, MacKnight C, Gorman M, and Video-Imaging Synthesis of Treating Alzheimer's Disease (VISTA) Investigators (2006) Attainment of treatment goals by people with Alzheimer's disease receiving galantamine: A randomized controlled trial. *Canadian Medical Association Journal* 174(8): 1099–1105.

Further Reading

Actin Cytoskeleton in Growth Cones, Nerve Terminals, and Dendritic Spines

S Halpain, B Calabrese, and L Dehmelt, The Scripps Research Institute, La Jolla, CA, USA

Introduction

The range and complexity of cell morphologies found in neurons are largely defined by the structure and dynamic organization of the underlying neuronal cytoskeleton. Neurons contain three major cytoskeleton types: microtubules, intermediate filaments, and filamentous actin. Here we review the properties and roles of actin in early and late neuronal development.

In the brain, neurons develop within an intricate, layered, three-dimensional matrix, which makes it difficult to analyze the detailed organization of the neuronal cytoskeleton. Therefore, many studies on actin organization during neuromorphogenesis have been performed using dissociated neuronal cultures grown on glass coverslips. In such a simplified model system, neurons first display a quasi-symmetrical shape. Actin is concentrated in the cell periphery in one or more flattened, veil-like structures, usually referred to as lamellipodia. During neurite initiation these lamellipodia divide into smaller lamellipodia, which then slowly move outward, trailed by a condensed, microtubule-rich shaft. If the shaft reaches significant length, the whole neuronal protrusion is called a neurite, and the lamellipodium at the neurite tip is called a growth cone (**Figure 1**, left).

In one of the classical neuronal model systems – primary hippocampal neurons – one of the neurites will grow significantly longer and accumulate certain markers, such as antigenicity to the tau-1 antibody. This neurite usually continues to grow rapidly and will become the future axon. During this stage, the remaining smaller neurites are called minor neurites, and these eventually mature and become the future dendrites.

As the dendritic arbor develops, many filopodia rapidly protrude and retract from the dendritic shaft, with a lifetime of minutes. Over a protracted period of time, the number of filopodia declines, as the number of synaptic contacts and dendritic spines increase. Dendritic spines are the postsynaptic receptive regions of most excitatory synapses. They consist of a bulbous head connected to the dendritic shaft by a narrow neck (**Figure 1**, right). Normally each spine is connected across the synapse to a specialized region of an axon called a presynaptic bouton or nerve terminal. It is thought that the shape of spines is critical for the normal function of such synapses and essential for the normal development of neuronal networks.

In addition to being one of the most abundant cellular proteins, actin is also one of the most versatile building blocks in cells. By itself, monomeric actin (G-actin) is a simple globular protein with intrinsic adenosine triphosphatase (ATPase) activity. If ATP is bound, G-actin can polymerize into a linear, two-stranded helical filament (F-actin). The intrinsic ATPase activity converts ATP-actin monomers to ADP-actin. The two ends of an actin filament have distinct biochemical activities: the (+) end has a higher affinity for monomeric actin and therefore elongates faster under standard conditions. The (−) end has lower monomer-binding activity and elongates slower than the (+) end does. At equilibrium, this difference of affinities leads to a process termed 'treadmilling,' which is characterized by net monomer addition at the (+) end and net depolymerization at the (−) end. Filament behavior is regulated by a variety of actin-regulating proteins that either bind to the (+) end, the (−) end, or along the sides of filaments.

The overall organization of actin filaments in cells can be controlled by regulating the local rates of *de novo* filament formation (nucleation), addition of monomers to existing filaments (polymerization), removal of monomers from filament ends (depolymerization), breaking of intact filaments (severing), sliding of individual filaments, and physical cross-linking of filaments to other cellular structures. Throughout this article, we discuss various examples of how precise tuning of such activities within neurons leads to the specific organization and dynamic properties of key neuronal structures.

Early Development

Actin Nucleation and Control of Actin Polymerization Rates in Growth Cones

The 'dendritic nucleation model' is a popular model for how actin filaments might maintain and control protrusive cell behavior at the plasma membrane. In this case the term 'dendritic' refers not to neurites but to the branched organization of actin filaments that defines this particular mode of actin assembly. The dendritic nucleation model is often thought to be generally applicable; however, most data that led to its proposal were derived from nonneuronal cells, notably fibroblasts. In a nutshell, this model proposes that rapid nucleation and ATP-dependent actin

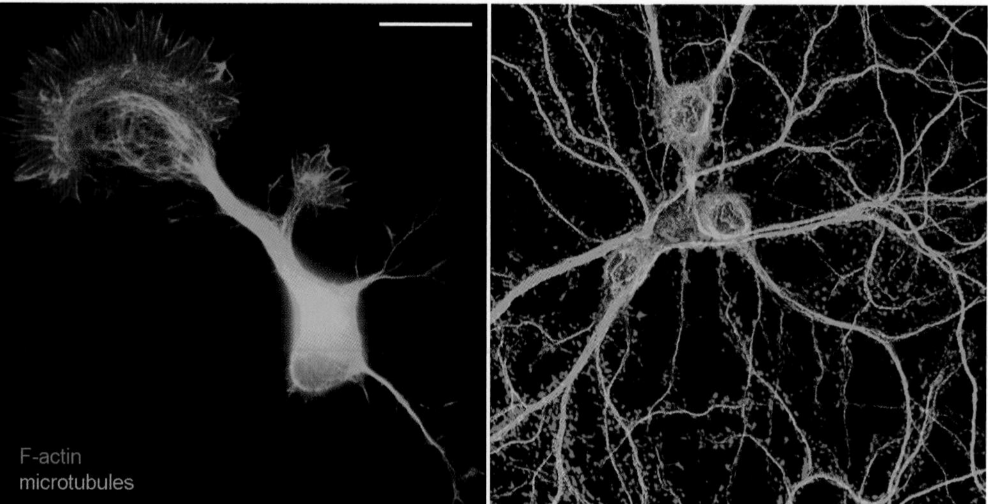

Figure 1 Localization of filamentous actin (F-actin) in rat cultured hippocampal neurons. Left: hippocampal neuron (1 day after plating) stained for F-actin (red) and neuronal tubulin (green). Right: hippocampal neurons (3 weeks after plating) stained for F-actin (red) and MAP2 (a microtubule-associated protein specific to dendrites; green). Note that F-actin is enriched in small clusters along the dendrites, which correspond to dendritic spines. Scale bar = 20 μm.

polymerization at the leading edge of motile cells are sufficient to drive membrane protrusion. To be effective, the turnover of actin must be very high, in order to maintain a constant supply of monomers to drive protrusion. Therefore, depolymerization must be ensured shortly after the initial polymerization. The zones of polymerization and depolymerization are separated by a zone of active myosin motor- and actin polymerization-driven flux. In this zone, the level of adhesion is thought to be important to regulate the level of protrusion or slippage – a process first suggested in the 'clutch hypothesis' (see below).

Some key aspects of this model are likely to apply to growth cones as well. However, there are also some very significant differences. Most notably, the dendritic nucleation model proposes that actin filament nucleation at the leading edge is driven by a actin-related protein (ARP) assembly termed the ARP2/3 complex, which binds to the sides of existing filaments and primes the nucleation of a new filament in a characteristic 60° angle off the mother filament. In fibroblasts, this mechanism leads to the formation of an intricate network of interdigitated filaments of 'dendritic' (branched) appearance. However, such dendritic networks are not observed in neuronal growth cones, suggesting that the way in which actin filaments initially form differs significantly between neuronal and nonneuronal systems.

It is believed that in addition to the ARP2/3 complex, another class of actin-binding proteins called formins plays an important role in actin nucleation. In the case of growth cones, formin-mediated actin nucleation might be dominant. In fact, whereas ARP2/3-induced nucleation is necessary for fibroblast protrusion, in growth cones, this nucleation complex is a negative regulator of neurite protrusion, suggesting a distinct role for ARP2/3 in the growth cone. Interestingly, ARP2/3-mediated cell protrusions are typically characterized by broad, flat lamellipodia, usually found at the leading edge of fibroblasts, whereas formin-based cell protrusions are characterized by actin spike structures called filopodia, which in turn are frequently formed in growth cones (**Figures 1** (**left**) and **2**(**a**)).

Apart from this obvious difference in the actin nucleation mechanism, other aspects of actin polymerization regulation appear to be similar in fibroblasts and growth cones. In both systems, for example, rapid polymerization at the leading edge, promoted by actin bound to profilin, is thought to drive membrane protrusion. β-Thymosin is present in both neuronal and nonneuronal systems and is known to sequester actin monomers, thereby limiting polymerization rates. Finally, depolymerization is thought to be the rate-limiting step in actin turnover control (see the section titled 'Control of actin depolymerization and turnover in early neuromorphogenesis') and thus is thought to be a key modulator of polymerization rates as well.

Substrate Adhesion and Generation of Traction in Axon Outgrowth and Guidance

Once actin is polymerized at the leading edge of the growth cone, it enters a zone of rapid retrograde flow usually referred to as the lamellipodium. According to the 'clutch hypothesis' proposed by Mitchison and

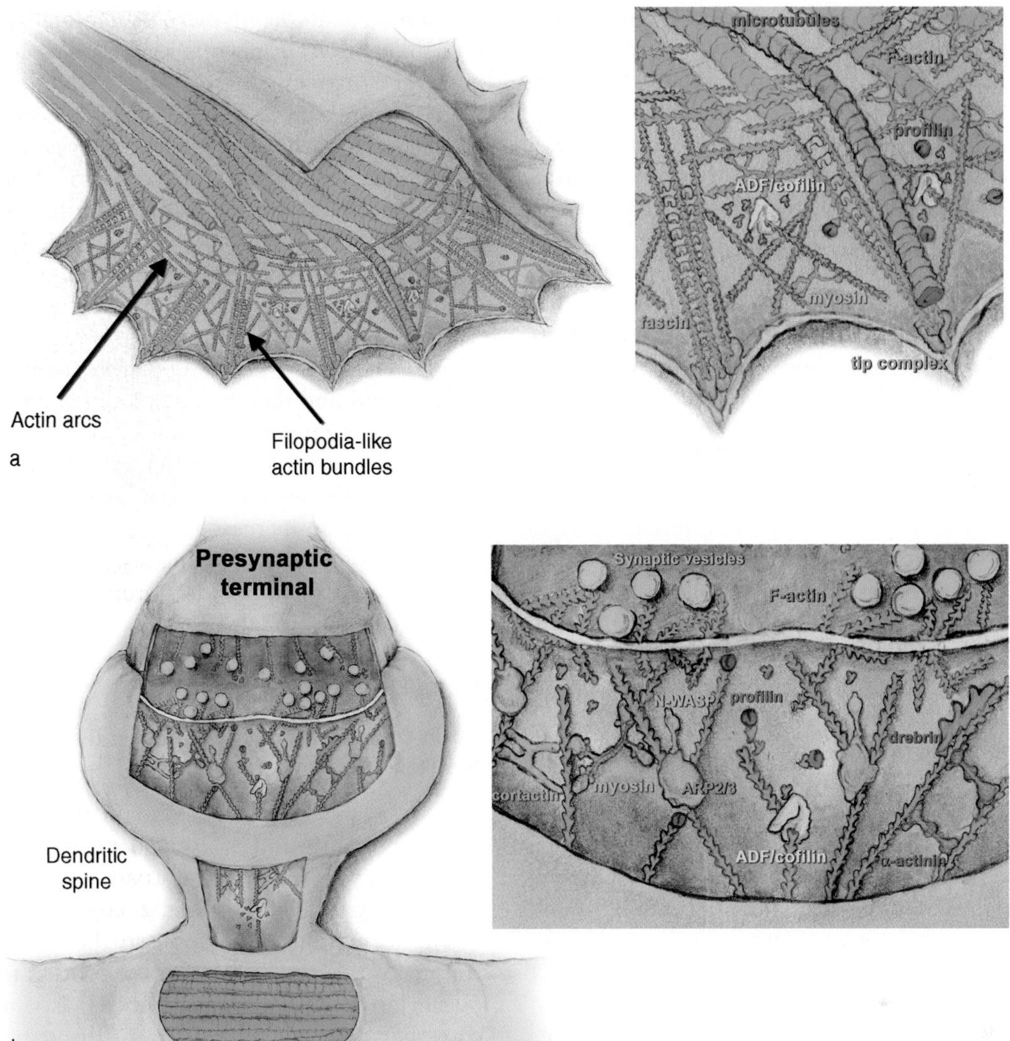

Figure 2 Cytoskeletal organization in the growth cone and the synapse. (a) Within the growth cone, actin is concentrated in the peripheral lamella. Actin-binding proteins, such as profilin, cofilin, and myosin, are thought to play a role in regulating actin dynamics and/or organizing actin into specific structures. (b) Actin filaments are distributed throughout the presynaptic nerve terminal, preferentially concentrated around synaptic vesicles. However, mature nerve terminals contain much less actin compared to either growth cones or dendritic spines. Actin regulatory molecules (such as cofilin and profilin) control the extent and rate of actin polymerization, and, ultimately, dendritic spine shape or growth cone motility. ADF; actin-depolymerizing factor; N-WASP, neural Wiskott–Aldrich syndrome protein; ARP2/3, actin-related protein complex 2/3. Drawings by James Lim; colorization by Barbara Calabrese and Leif Dehmelt.

Kirschner, this retrograde flow acts as a motor for cell protrusion. By analogy to a car engine clutch, the retrograde flow is constitutively active, even in resting growth cones. This retrograde flow does not produce any forward movement unless it is linked to the substratum by formation of cell adhesions. Thus, the temporal regulation of the extent of adhesion or slippage is thought to control the speed of growth cone progression.

The giant growth cones of the sea snail *Aplysia* have been instrumental in revealing how local manipulations of adhesion affect growth cone behavior and protrusion. Such studies revealed that adhesion to a bead coated with the signaling molecule ApCAM (*Aplysia* cell adhesion molecule) is not sufficient by itself to initiate a full response of the growth cone. However, if the coated bead is mechanically restrained to allow the growth cone to pull on it, Src kinase is locally activated at the bead interaction site. This is followed by microtubule targeting toward the bead, modulation of retrograde flow, growth cone advance, and axonal turning behavior.

Retrograde flow is driven both by forces generated distally from actin polymerization at the leading edge and proximally by myosin motor activity within the lamellipodium (**Figure 2**). There is still some controversy about the roles of myosin II and myosin I in growth cones. These motors have been suggested either to participate in the organization and maintenance of structural integrity of the lamellipodium or in the generation of the retrograde flow. It is clear, however, that myosin II is mostly localized to the proximal part of the growth cone, where it co-localizes with contractile filamentous structures called actin arcs. These structures accumulate at the lateral sides of the growth cone and are thought to compress microtubules into parallel arrays. Furthermore, actin arcs are also thought to mediate contractile responses during neurite retraction.

Roles for Actin–Microtubule Interactions in Early Neurite Development

Several laboratories have reported physical, rigid interactions between microtubules and filamentous actin within cells. This physical coupling might have several nonexclusive roles: (1) As mentioned earlier, actin arcs might shape the distribution of microtubules within growth cones and might even be involved in the formation and/or maintenance of the tight microtubule bundles typically found in neurons. (2) Similar to the clutch hypothesis, physical linkage of actin filaments to a stable scaffold of microtubules might promote protrusion of the leading cell edge. (3) Microtubules might be involved in signaling cross-talk mediated by regulatory molecules found on the two cytoskeletons. (4) Dynein-generated forces on microtubule arrays are suggested to counteract actomyosin-driven neurite retraction. Interestingly, microtubules also appear to have an instructive role in growth cone turning, as local pharmacological manipulations of microtubule dynamics are sufficient to drive growth cone turning behavior.

Several potential microtubule-actin cross-linkers or signaling mediators have been identified, including the coronin-like protein pod-1, microtubule-associated protein MAP2c, the dynein–dynactin complex, and the IQGAP1/Rac1 complex. These proteins have been implicated in several stages of early neurite development, including neurite initiation, neurite growth, and axon pathfinding.

Control of Actin Depolymerization and Turnover in Early Neuromorphogenesis

The dynamic behavior of actin filaments is not only dependent on the rates of polymerization, nucleation, and motor-dependent filament sliding, but is also dependent on a constant supply of active monomers that can be incorporated at the filament ends. Thus, depolymerization is required for rapid actin turnover, and biochemical studies suggest that this might be the rate-limiting step in the cycle of actin dynamics. Therefore, actin-depolymerizing factors (ADFs), such as cofilin and ADF, play important roles as regulators of actin dynamics.

ADF/cofilin binds filamentous actin and increases its depolymerization rate by altering the helical twist of the filament. The activity of ADF/cofilin is controlled by multiple mechanisms, including phosphorylation of serine 3 by LIM and TES protein kinases, dephosphorylation by the protein phosphatase 'slingshot' and 'chronophin' and binding of 14-3-3 proteins. Furthermore, another actin-binding protein called tropomyosin protects filaments against ADF/cofilin and thereby adds another level of depolymerization rate regulation.

Consistent with the idea that ADF/cofilin regulates a rate-limiting step during actin turnover, overexpression leads to an increase in axonal outgrowth, while inhibition of its activity using dominant-negative mutants or by overexpression of an inactivating kinase slows axon outgrowth.

Late Development

Actin Organization and Function during the Development of Synapses

The actin cytoskeleton is the main structural component of both pre- and postsynaptic terminals. Here we focus on excitatory synapses, where the function of the actin cytoskeleton is best understood.

The formation of synapses in the vertebrate central nervous system is a complex process that occurs over a protracted period of time. It is generally the case that dendritic filopodia are the structural precursors of dendritic spines during synaptogenesis, although direct emergence of new spines from the dendrite shaft has also been observed. Dendritic filopodia are long, thin protrusions characterized by a highly transient and motile behavior. In these protrusions F-actin is sparse until pre- and postsynaptic transmembrane cell adhesion proteins, such as the cadherin–catenin complex, link adjacent synapse precursors. The F-actin accumulates within early filopodial synaptic contacts, where it recruits synaptic signaling molecules. Simultaneously, the dynamic dendritic filopodia begin to morphologically change into intermediate structures sometimes called 'protospines.' These intermediate structures have also been referred to as cluster-type filopodia, or synaptic filopodia, and they can eventually mature into more stable dendritic spines. Only a fraction of the filopodia that emerge from a dendrite persist to become dendritic spines.

Mature, postsynaptic dendritic spines are highly enriched in F-actin compared to the dendritic shaft (**Figure 1**, right). Based on measurements of green fluorescent protein (GFP)–actin fluorescence recovery after photobleaching in cultured hippocampal neurons, only approximately 5% of the total actin in spines is relatively stable, while the majority of the actin almost completely turns over in a 2 min period. Unlike F-actin, G-actin is distributed homogeneously throughout axonal and dendritic processes in hippocampal neurons, which indicates that its main role at the synapse is to maintain the synaptic F-actin pool. The characteristic shape of dendritic spine heads is highly dependent on actin dynamics and the fraction of polymerized actin. Disruption of filamentous actin by the toxin latrunculin A transforms dendritic spines into filopodia-like processes.

F-actin is found within the presynaptic nerve terminal and is preferentially concentrated around synaptic vesicles (**Figure 2(b)**). In resting conditions, approximately 30% of actin is in a polymerized state, with a turnover halftime of 20 s. During electrical activity, actin is further polymerized and recruited around vesicles from nearby axonal regions, leading many groups to propose a propulsive role for actin, either in maintaining the vesicle cluster or in guiding vesicle recycling. However, the precise role for actin in nerve terminals remains controversial, and may differ depending on the nature of the synapse. Although actin plays an active role in synaptic vesicle endocytosis in some systems, recent evidence indicates that in mammalian central synapses it passively inhibits vesicle fusion at the active zone, and helps sequester synaptic vesicles in the reserve pool. Synaptic vesicle trafficking from the reserve pool to the readily releasable pool can occur independently of actin. Thus, actin may serve primarily as a scaffold for other regulatory proteins involved in vesicle trafficking within the terminal. The recruitment and retention of presynaptic matrix proteins, such as bassoon and piccolo, also are dependent on actin.

The higher content of F-actin in dendritic spines compared to nerve terminals correlates with more dynamic shape changes. Nevertheless, the pre- and postsynaptic cytomatrix components are strongly interconnected, and therefore their movements are physically coupled in established synapses. Engulfment of spinules (membrane extensions emerging from the head of big mushroom spines) by presynaptic axons is a process known as transendocytosis and suggests retrograde signaling or coordinated remodeling of pre- and postsynaptic membranes, which is likely modulated by actin.

Actin Dynamics and Spine Morphogenesis

In recent years, imaging techniques both *in vivo* and *in vitro* have revealed that a majority of mature dendritic spines are stable over months in the adult brain. They rarely form and disappear; instead, the dendritic spine head is constantly changing shape (morphing). This rapid spine motility is caused by regulated polymerization and depolymerization of actin, based on studies using drugs and cDNAs that perturb actin polymerization.

The Rho family of GTPases is one common denominator controlling spine morphology. Multiple signaling pathways converge on the Rho family of small GTPases (particularly RhoA, Rac1, and Cdc42), which ultimately converge on the actin cytoskeleton to regulate spine morphology and dynamics. Generally speaking, RhoA inhibits, whereas Rac and Cdc42 promote, the growth and/or stability of dendritic spines.

An expanding number of guanine nucleotide exchange factors (GEFs) and GTPase-activating proteins (GAPs) control these Rho family GTPases. For example, kalirin-7, a GEF for Rac1, mediates effects of the transmembrane signaling molecule EphB on spine maturation by activating Rac1 and its downstream effector p21-activated kinase (PAK). PAK1 and PAK3 are Rac effectors that promote formation and/or growth of spines.

PAK also stimulates LIM kinase (LIMK), which phosphorylates and inactivates ADF/cofilin, an actin-depolymerizing protein that mediates reorganization of the actin cytoskeleton. Recently, cofilin has been detected by immunoelectron microscopy to concentrate in the periphery of the spine head and within the postsynaptic density. Mice expressing dominant-negative PAK in the forebrain showed reduced spine density in cortical neurons, impaired long-term synaptic plasticity, and reduced memory consolidation. PAK activity is markedly reduced in Alzheimer's disease due to a reduction in phospho-PAK, which in turn results in an increased cofilin activity and downstream loss of the spine actin-regulatory protein drebrin.

Insulin receptor substrate p53 (IRSp53), an adaptor protein that connects Rac1 to the Wiskott–Aldrich syndrome protein WAVE2, is implicated in filopodia and lamellipodia formation in nonneural cells. IRSp53 is enriched in the postsynaptic density (PSD) by binding to Shank and PSD-95 family scaffolds. Overexpression of IRSp53 increases spine density, whereas RNA interference (RNAi) or dominant-negative inhibition of IRSp53 causes reduction of spine density and size. The Abl-interactor (Abi) adaptor proteins, so-named because they bind to the Abl tyrosine kinases, also play a part in Rac GTPase signaling and actin regulation. Abi2 is highly expressed in brain; and Abi2-deficient mice show reduced spine density and a decrease in the relative proportion of mushroom spines, which is thought to be associated with deficits in memory.

It is worth noting that components of Rho GTPase signaling pathways are highly represented among the genes so far identified in hereditary forms of human nonsyndromic mental retardation. These mutant genes include X-linked genes, such as PAK3 (see earlier) and oligophrenin, which is a Rho GAP. RNAi suppression of oligophrenin decreases spine length, mimicking the effect of active RhoA.

Numerous scaffold proteins and actin-binding proteins are concentrated in spines and are known to control spine morphogenesis (**Figure 2(b)**). Many of these are regulated by the small GTPases described earlier. Overexpression of α-actinin 2, a protein linking *N*-methyl-D-aspartate (NMDA) receptors to the actin cytoskeleton, increases the length and density of dendritic protrusions in cultured hippocampal neurons. Clustering of the actin-binding protein drebrin occurs early in development, supporting its involvement in spinogenesis and synaptogenesis. Cortactin is an activator of the ARP2/3 actin nucleation machinery that interacts with the postsynaptic density scaffold Shank. RNAi knockdown of cortactin results in depletion of dendritic spines, whereas overexpression of cortactin causes enlargement of spines.

Myosin II is also enriched in postsynaptic density preparations and acts as a molecular motor that produces tension on actin filaments. Inhibition of myosin IIB destabilizes mushroom spines, suggesting that the structure and function of spines are regulated by an actin-based motor in addition to actin polymerization.

Effects of Synaptic Activity on Actin Polymerization in Dendritic Spines

Synaptic plasticity is associated with changes in actin polymerization and spine morphology. Long-term potentiation (LTP) is associated with formation of new spines, enlargement of existing spines, and a shift of the actin turnover equilibrium toward polymerization of F-actin. By contrast, long-term depression (LTD) is associated with shrinkage and/or retraction of spines due to an increase in actin depolymerization. In addition, synaptic activity is thought to play an important role in determining whether a synapse will be stabilized or eliminated.

During induction of LTP, α-amino-3-hydroxy-5-methyl-4-isoxazole propionic acid (AMPA) receptors are rapidly recruited to dendritic spines through Ca^{2+}-permeable NMDA receptor activation. In response to NMDA receptor activation, profilin, a promoter of F-actin assembly, is targeted to spine heads, while cortactin redistributes from spines to the dendritic shaft. Distinct but overlapping signaling pathways are involved in LTD and spine shrinkage. During induction of LTD, NMDA receptor-mediated calcium influx triggers phosphatase 2B (calcineurin) activity. Calcineurin in turn activates a downstream phosphatase 'slingshot' and promotes actin severing and depolymerization through ADF/cofilin.

Synaptic activity can also affect the relationship between F-actin and the plasma membrane of dendritic spines. Myristoylated alanine-rich protein kinase C substrate (MARCKS), a major target of protein kinase C (PKC), might be involved in local coupling of activity-dependent calcium entry with regulation of actin assembly–disassembly cycles. During synaptic activity, when calcium levels increase and stimulate PKC, MARCKS detaches from the plasma membrane, thereby freeing phosphatidylinositol 4,5-bisphosphate to interact with actin-binding proteins such as profilin and ARP2/3, which in turn can locally alter actin dynamics and membrane trafficking.

Comparing Growth Cones, Dendritic Spines, and Nerve Terminals

One might be tempted to generalize the role of actin in growth cones, synaptic terminals, and spines, and introductions to this topic often try to point out similarities. However, the current state of our knowledge does not allow an in-depth comparison. Nonetheless, certain parellels exist (**Figures 2(a)** and **2(b)**). In growth cones, a relatively stable, less dynamic F-actin population is found toward the growth cone center, which is thought to be involved in contractile behavior. Highly dynamic F-actin is concentrated in the periphery of the growth cone, where it is thought to mediate membrane protrusion, resulting in motile filopodia and lamellipodia. Similarly, a stable pool of F-actin is present in the dendritic spine core, while a dynamic pool lies peripherally in its shell, which exhibits highly dynamic, sometime protrusive behavior.

In growth cones, the organization and dynamics of the underlying actin structures have been analyzed using both high-contrast electron microscopy and dynamic speckle microscopy. These analyses converge in a detailed working model of growth cone behavior. In spines, defining the detailed actin organization and dynamics has been more difficult. Ultrastructural analysis of filament polarity suggests that most actin filaments are oriented with their barbed end pointing toward the head of the spine protrusion, similar to the preferred orientation of actin filaments toward the growth cone leading edge. One notable difference is that spines generally lack the dense parallel organization of F-actin that characterizes growth cone filopodia. However, at times, rapidly growing, elongated filopodia-like structures are observed on spine heads. The purpose of such filopodial-like protrusions emerging from spine heads remains unclear. In growth cones, and along developing dendrites, such filopodial protrusions are thought to extend the structure's sensory radius to probe the environment for guidance cues or neurotransmitters. Speckle microscopic analysis of actin dynamics has not been performed in spines; however, indirect data on spine actin dynamics have been derived from photobleaching recovery experiments (see the section titled 'Actin organization and function during the development of synapses'). These experiments reveal a highly dynamic F-actin population, which may be roughly comparable to the rapidly turning over F-actin of growth cone lamellipodia.

While there might be some similarities between actin behavior in growth cones and spines, the role and organization of actin in nerve terminals appear to be distinct. F-Actin is distributed throughout the presynaptic nerve terminals and is preferentially concentrated around synaptic vesicles. However, mature nerve terminals do not exhibit significant protrusive behavior and contain much less actin compared to either growth cones or spines. In nerve terminals, a specific role for actin remains unclear.

See also: Axonal Regeneration: Role of Growth and Guidance Cues; Axonal and Dendritic Identity and Structure: Control of; Cytoskeletal Interactions in the Neuron; Cytoskeleton in Plasticity; Dendritic Spine History; Glutamate Regulation of Dendritic Spine Form and Function; Growth Cones; LIM Kinase and Actin Regulation of Spines; Microtubules: Organization and Function in Neurons; Spine Plasticity.

Further Reading

Calabrese B, Wilson MS, and Halpain S (2006) Development and regulation of dendritic spine synapses. *Physiology (Bethesda)* 21: 38–47.

Carlisle HJ and Kennedy MB (2005) Spine architecture and synaptic plasticity. *Trends in Neurosciences* 28: 182–187.

Dehmelt L and Halpain S (2004) Actin and microtubules in neurite initiation: Are MAPs the missing link? *Journal of Neurobiology* 58: 18–33.

Dent EW and Gertler FB (2003) Cytoskeletal dynamics and transport in growth cone motility and axon guidance. *Neuron* 40: 209–227.

Ethell IM and Pasquale EB (2005) Molecular mechanisms of dendritic spine development and remodeling. *Progress in Neurobiology* 75: 161–205.

Lippman J and Dunaevsky A (2005) Dendritic spine morphogenesis and plasticity. *Journal of Neurobiology* 64: 47–57.

Luo L (2002) Actin cytoskeleton regulation in neuronal morphogenesis and structural plasticity. *Annual Review of Cell and Developmental Biology* 18: 601–635.

Oertner TG and Matus A (2005) Calcium regulation of actin dynamics in dendritic spines. *Cell Calcium* 37: 477–482.

Pollard TD and Borisy GG (2003) Cellular motility driven by assembly and disassembly of actin filaments. *Cell* 112: 453–465.

Sarmiere PD and Bamburg JR (2004) Regulation of the neuronal actin cytoskeleton by ADF/cofilin. *Journal of Neurobiology* 58: 103–117.

Strasser GA, Rahim NA, VanderWaal KE, et al. (2004) Arp2/3 is a negative regulator of growth cone translocation. *Neuron* 43: 81–94.

Suter DM, Schaefer AW, and Forscher P (2004) Microtubule dynamics are necessary for SRC family kinase-dependent growth cone steering. *Current Biology* 14: 1194–1199.

Zhang W and Benson DL (2002) Developmentally regulated changes in cellular compartmentation and synaptic distribution of actin in hippocampal neurons. *Journal of Neuroscience Research* 69: 427–436.

Zhang XF, Schaefer AW, Burnette DT, et al. (2003) Rho-dependent contractile responses in the neuronal growth cone are independent of classical peripheral retrograde actin flow. *Neuron* 40: 931–944.

Action Potential Initiation and Conduction in Axons

J H Caldwell, University of Colorado, Aurora, CO, USA

Introduction

Transmission of electrical signals in biological systems operates under constraints imposed by the materials used to build the conducting pathways. The conduction medium in nerves is a membrane-enclosed dilute salt solution rather than a metal wire, for example, as in macroscopic physical systems. Axons have been compared for decades to undersea cables that are surrounded by salt water. However, the axon relies on a salt solution both inside and outside the axon. Since the resistivity of mammalian physiological saline is about 1 ohm-meter and that of copper is about 1.7×10^{-8} ohm-meter, the ease of current flow in 1 mm of an axon (a typical length constant in a large-diameter axon) is equivalent to that of a wire that is 170 km long. The difficulty of passive propagation of signals in an axon is compounded by two factors. First, the signal is carried by cell membranes that have a capacitance ($\sim 1\ \mu F\ cm^{-2}$) that needs to be charged to propagate a voltage change. Second, the membrane is inherently leaky due to a requirement for ion channels that are open to set the negative resting membrane potential. Thus, passive conduction of a signal in neurons is effective over distances of only a few millimeters. For this reason, it does not matter whether you are a worm or a whale: transmission of signals in axons requires a booster mechanism to replenish the decrementing electrical signal. The analogy between undersea cables and axons has other similarities and divergences. Undersea cables use repeater or booster stations that amplify the analog signal and send it on. The nervous system also uses a booster system but transforms the analog signal into a digital one (i.e., the continuously varying voltage signal becomes either 'on' or 'off'). This encoding is a trade-off that neurons accept for the sake of long-distance communication. The amplitude of the graded signal is transformed into the timing and frequency of action potentials.

Ionic Basis of the Action Potential in Axons

Sodium and Potassium Channels

The fundamental ionic mechanism of a propagating, regenerative increase in sodium conductance was described by Hodgkin and Huxley for the squid giant axon, and this process holds in both the central nervous system (CNS) and peripheral nervous system (PNS) of both vertebrates and invertebrates. The action potential is generated by the opening and subsequent inactivation of voltage-gated sodium channels and, with a slight delay, the opening of voltage-gated potassium channels. This ionic interplay of opening and closing of sodium and potassium channels is present in such diverse phyla that it must have evolved hundreds of millions of years ago. This conservation of the molecular foundation of the action potential has additional complexities that are not yet completely understood. In mammals there are 10 voltage-gated sodium channels (Nav1 and Nav2 families), and many of these are expressed in neurons and localized in axons. In addition, there are at least 20 voltage-gated potassium channels in vertebrates (Kv1 to Kv4 and KCNQ families). Thus, the molecular species and combinations of sodium and potassium channels utilized for conduction are potentially quite large. This suggests that characteristics of different axon types (e.g., frequency of firing and ability to maintain conduction) are dependent on the isoforms of sodium and potassium channels present in the axon.

Sodium–Potassium Pump

Although the immediate basis of the action potential is the activation of voltage-gated sodium and potassium channels, long-term support of the action potential requires sodium–potassium pumps to maintain the concentration gradients. It has been often emphasized that if the sodium pump is blocked (e.g., with ouabain) in a squid axon, hundreds of thousands of action potentials can still be generated because few sodium and potassium ions cross the membrane for each action potential. The squid axon illustrates the fact that the energy for ion flux (the concentration gradient of sodium and potassium between the inside and outside of the axon) has already been established by the pump. However, the squid giant axon is 1 mm in diameter, and when considering small-diameter axons that are less than a micrometer in diameter (e.g., pain and temperature fibers in the periphery or parallel fibers in the cerebellar cortex), sodium and potassium fluxes during each action potential are significant. For these small-diameter axons, the ability to maintain action potential firing is highly dependent on sodium pump activity.

Structural and Functional Differences of Vertebrate Axons

Axons fall into two major categories depending on the structure of the glial cells that envelop them. The first category is the unmyelinated axon, which describes all invertebrate axons and small axons of vertebrates, typically axons with a diameter below 1 μm. The unmyelinated axon is usually loosely surrounded by a glial cell or in some cases, such as parallel fibers in the cerebellar molecular layer, is not covered by a glial cell. The speed of conduction of an action potential in an unmyelinated axon is proportional to the square root of the axon diameter. Thus, invertebrates have large-diameter axons for signals that need to be propagated rapidly. The squid giant axon, which is part of a circuit used for rapid propulsion in the escape response, is as large as 1 mm in diameter, conducts at $25\,\mathrm{m\,s^{-1}}$ (at 25 °C), and is formed by the fusion of axons of many neurons.

Vertebrates have evolved an alternative strategy for increasing the speed of action potential conduction. Vertebrate axons larger than about 1 μm are tightly wrapped by many layers of the glial cell, creating the second category, the myelinated axon. Myelination occurs in a repeating pattern, with long wrapped regions (internodes that are up to 1–2 mm in length) interrupted by a very short bare region (the node of Ranvier, 1–2 μm in length). In the PNS the glial cell is a Schwann cell. Each Schwann cell can envelop many unmyelinated axons, but when myelination occurs, one Schwann cell is devoted to the formation of one myelinated internode. Myelin in the CNS is formed by oligodendrocytes, and one oligodendrocyte sends out tens of processes, each one forming an internode on a different axon. Functionally, the myelin acts as an insulator, by reducing the leak of current through the membrane in the internodal regions. The myelin also reduces the effective capacitance of the internodal region, which in turn reduces the capacitive current required to charge the internodal membrane. Consequently, current travels rapidly with little loss in the internodal regions to the node of Ranvier, where the current is boosted or regenerated by voltage-gated sodium channels concentrated at the node (described later). Conduction velocity in myelinated axons is proportional to the axon diameter, and the general rule of thumb is that for axons with an outside diameter greater than 11 μm, the speed of conduction, in meters per second, is about six times the axon diameter, in micrometers. For smaller axons the proportionality factor is 4.5. An axon 20 μm in diameter, which is one of the largest in the mammalian nervous system, conducts at $120\,\mathrm{m\,s^{-1}}$, about five times faster than the squid axon, even though it is 50 times smaller. Thus, myelination not only increases speed of conduction, but also does this with an economy of space. This concept of economizing the volume used for conduction is invoked to explain why we have many more unmyelinated axons than myelinated ones: more information can be carried in a given volume with small, unmyelinated axons. In the mammalian nervous system, pain and temperature information is carried by small, unmyelinated axons in the PNS, and the molecular layer of the cerebellum in the CNS is densely packed with parallel fibers that are unmyelinated axons of granule cells.

Initiation of the Action Potential

Stimulation Required for Electrogenesis

A rapid membrane depolarization is necessary to open voltage-gated sodium channels and start the action potential. This depolarization can be achieved artificially by inserting a microelectrode into an axon and injecting current, by extracellularly stimulating an axon with an electrode, or even by mechanically hitting the nerve – for example, when we hit our ‘funny bone’ (the ulnar nerve near the elbow). Naturally produced depolarizations fall into two categories. For most neurons, synaptic input to the dendrites and cell body provides the required depolarization. Sensory neurons, such as stretch receptors in muscle or cutaneous receptors in the skin, propagate action potentials centrally; a sensory signal in the periphery creates a depolarization called the receptor or generator potential that opens sodium channels to produce action potentials.

Site of Initiation

Unmyelinated axons The simplest treatment of a neuron separates it into three regions: soma, dendrite, and axon. The elementary concept of a passive dendritic tree that simply receives excitatory and inhibitory synaptic inputs and sends these inputs to the cell body, where they are summated, is now known to be an oversimplification. Many neurons have voltage-dependent sodium and calcium channels in dendrites. However, only in rare cases can dendrites initiate action potentials that are propagated orthodromically to the soma. Thus, the integration of excitatory and inhibitory potentials takes place at the soma. The soma contains a variety of voltage-gated channels, but sodium channel density is highest at the axon hillock, an enlargement of the axon at the point it leaves the cell body (**Figure 1**). For example, sodium channel density is over sevenfold higher on the initial segment of a neurite (presumptive nascent axon),

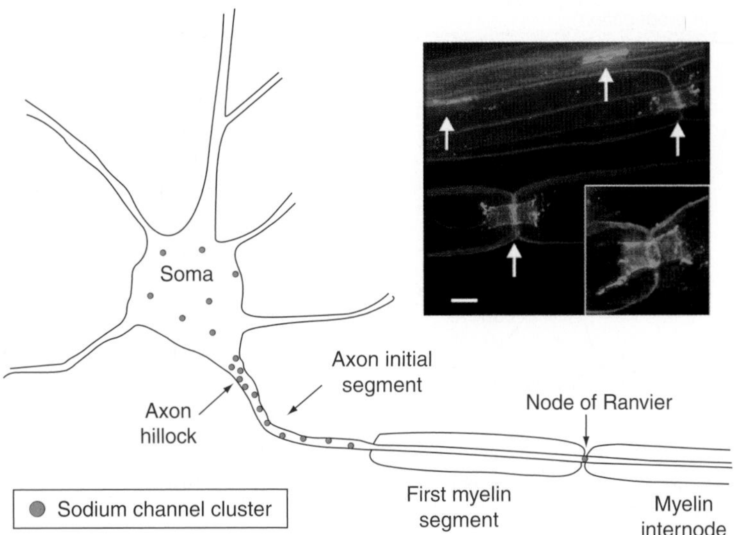

Figure 1 Sites of action potential initiation and sodium channel clustering. Action potentials begin at the point of lowest threshold, which is a function of the balance between sodium, potassium, and leak currents. In general, the action potential originates where sodium channels are clustered at high density, and for both unmyelinated and myelinated axons this is usually at the axon hillock. In myelinated neurons initiation can also occur in the initial segment or at the first node of Ranvier, where sodium channels are highly concentrated. Inset: Four nodes of Ranvier (arrows) in a teased sciatic nerve with nodal labeling by an anti-Nav1.6 antibody (red) and paranodal labeling with an anti-Caspr antibody (green). The node shown at higher power within the inset is tilted such that the labeling of the entire circumference of the nodal membrane can be seen. Scale bar = 5 μm (3.3 μm for the tilted node). (Inset) Reproduced from Caldwell JH, Schaller KL, Lasher RS, et al. (2000) Sodium channel Na(v)1.6 is localized at nodes of Ranvier, dendrites, and synapses. *Proceedings of the National Academy of Sciences of the United States of America* 97: 5616–5620.

compared to the soma in cultured spinal cord neurons. The higher density at the axon hillock has been confirmed with immunolabeling of neurons in many regions of the CNS. The main consequence of the increased sodium channel density at the axon hillock is that the threshold for action potential initiation is lowest there. Thus, for unmyelinated axons, action potential initiation takes place at the axon hillock.

Myelinated axons The site of action potential origination in myelinated axons was also shown to be the axon hillock about 50 years ago. More recently, with improvements in the ability to measure voltage changes optically and with patch clamp electrodes, the precise location of the origin of the action potential has been identified. Action potentials can originate not only at the axon hillock, but also in the axon initial segment, 30–40 μm from the soma and close to the first myelinated segment. In some neurons the action potential even originates at the first node of Ranvier, where sodium channels are highly concentrated (**Figure 1**). For both myelinated and unmyelinated axons, once the action potential begins in the axon, it not only propagates orthodromically toward the nerve terminals but also propagates antidromically, back into the soma and dendrites.

Conduction in Unmyelinated and Myelinated Axons

Unmyelinated Axons

Before considering a propagating action potential, it is useful to understand the currents that underlie a stationary action potential. It is possible to control the membrane potential experimentally along the length of an axon. In this case, a short stimulus current can be applied to bring the entire length of membrane to threshold, and the whole axon subsequently and simultaneously undergoes an action potential. Once the applied stimulus is over, total membrane current in this artificial situation is zero. Since membrane current is the sum of capacitive and ionic currents, ionic and capacitive currents are equal and opposite. Ionic currents through sodium and potassium channels simply change the membrane potential by charging membrane capacitance. For a propagating action potential, the relationship between ionic and capacitive currents is more complex (**Figure 2**). The reason for this complexity is that once the action potential is initiated, either by transduction of a sensory stimulus or by summation of postsynaptic potentials, sodium current in the active region not only depolarizes the active region further, but also provides depolarizing

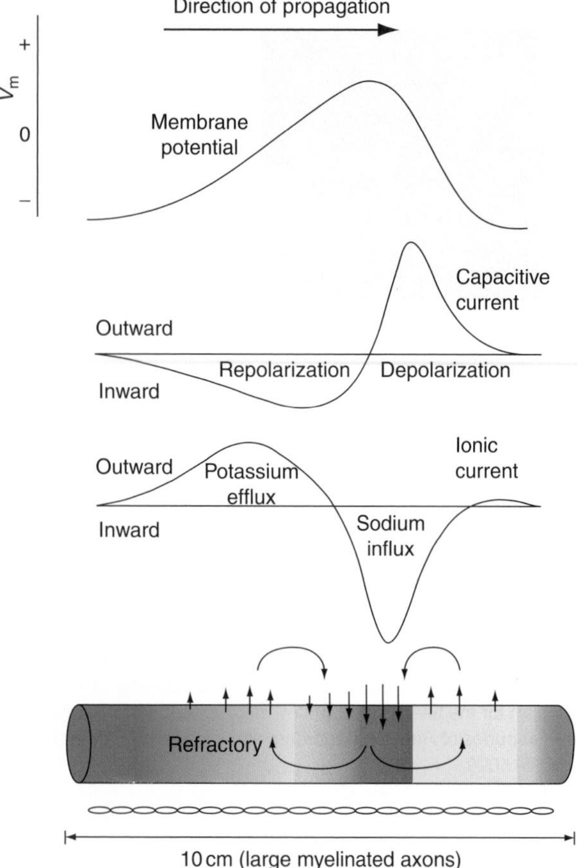

Figure 2 Ionic and capacitive currents that underlie a propagating action potential. The membrane potential change of an action potential propagating from left to right is illustrated at the top. Capacitive and ionic currents are schematically shown below the membrane potential and are drawn to illustrate the major relationships. Since capacitive current is proportional to the first derivative of the membrane potential, peak capacitive currents occur at the maximum slopes of depolarization and repolarization, and capacitive current is zero at the peak of the action potential. Total membrane current (the sum of capacitive and ionic currents) is proportional to the second derivative of the membrane potential and is zero at the maxima of the capacitive current where total current changes between inward and outward (yellow to red boundary and red to green boundary in the schematic of the axon at the bottom). For an action potential propagating at a constant velocity, the scale bar at the bottom can be thought of as time at a fixed point on the axon (typical action potential duration is 1 ms) or as distance over which the action potential is occurring at one instant of time (10 cm for an action potential with a duration of 1 ms and conduction velocity of $100\,\text{m}\,\text{s}^{-1}$). A series of myelinated nodes is shown above the scale bar to illustrate that many nodes participate at any given time. For the fastest conducting axons, there would be five times as many nodes as are illustrated here.

current to the adjacent region of the axon at the leading edge of the action potential to bring it beyond threshold (**Figure 2**). The depolarization at the leading edge of the action potential is primarily a capacitive current until threshold is reached. Current also spreads longitudinally behind the action potential, but an action potential is not created in the retrograde or backward direction because of the residual changes in the state of potassium and sodium channels. Potassium channels are still activated and are holding the membrane potential near the resting potential while sodium channels in this region are still recovering from the depolarization; they are inactivated, and this part of the axon is temporarily refractory to action potential generation.

The conduction velocity of unmyelinated axons depends on how much current is injected into the axon by the sodium channels, how far the current can spread longitudinally, and how quickly the adjacent membrane can be brought to threshold. The amount of current depends on the density of sodium channels. Since more sodium channels provide more current, one might think that an increase in channel density will always increase conduction velocity. This proportionality is valid only for low to moderate sodium channel densities because the channels act as dipoles (the source of their voltage sensitivity) and add additional capacitance to the membrane. The time required to charge the membrane is the product of the specific membrane resistance and capacitance. At very high channel density, the effect of the added capacitance outweighs the additional current provided because it takes longer to charge the membrane and conduction velocity is decreased. Typical sodium channel density of unmyelinated axons is 50–500 channels μm^{-2}, with potassium channel density about tenfold lower. An especially low density (2–3 channels μm^{-2}) has been reported in garfish olfactory nerve and neonatal rat optic nerve.

Channel subtypes In general, all neurons express multiple subtypes of sodium and potassium channels. All channel subtypes in the Nav1 sodium channel family have the basic features described by Hodgkin and Huxley over 50 years ago; they are activated by depolarization, with subsequent inactivation that is removed when the membrane is repolarized. Ion selectivity seems to be the same for all the subtypes, but the details of voltage dependence, the kinetics of opening and closing, and the modulation of these gating properties vary from one subtype to another. The multiplicity of potassium channel subtypes is much greater than that of sodium channels, and their properties are also more variable.

There is evidence that neurons use different channel subtypes in subcellular regions of the cell. This would allow the cell to fine-tune the excitability of the cell in different regions. The expression and targeting of different sodium and potassium channel proteins to the unmyelinated axon remain an active area of research. Some sodium channel subtypes in the PNS, such as Nav1.8 and Nav1.9, seem to be predominantly expressed in small dorsal root ganglion neurons and are targeted to unmyelinated axons. Parallel fibers in the cerebellar cortex utilize Nav1.6.

Myelinated Axons

The current available for depolarizing the next axonal segment to threshold is dependent on the loss of current through the membrane (its leakiness) and the decrease due to capacitive current required to charge the membrane and change the membrane potential. The number of wraps of the axon by the myelin and the length of the myelin internode have important electrical consequences for both the ionic and the capacitive currents. Since the extracellular fluid between each wrap is squeezed to a negligible volume and since the cytoplasm is also squeezed out of the glial wraps, the axonal membrane is essentially increased in thickness by the myelin membranes. The number of wraps by myelin varies from a low of about 10 to as many as 150, with each wrap consisting of a pair of membranes. For example, a large myelinated axon with 150 wraps will decrease ionic current loss through the internodal membrane by a factor of 300. Because capacitance is inversely related to the distance between the charged surfaces (in this case the thickness of a membrane), capacitance and capacitive current will also be reduced 300-fold. The current thus moves rapidly in the internode, with little loss through the membrane or in charging the membrane, essentially jumping from one node to the next. This is described as saltatory conduction. Each node acts as a booster station to ensure propagation to the next node, and to accomplish this regeneration of the signal, sodium channels are highly concentrated at each node (2000–3000 channels μm^{-2}) and are about 100-fold lower in density in the internodal membrane.

Increasing the internodal distance increases the speed of conduction because the current is jumping farther. However, there is an optimal internodal length. If internodal distances were to become very large, conduction velocity is predicted to decrease. This decrease in velocity is due to the loss of current in the internodal region, slowing the rate of rise of depolarization at the next node. Internodal distances are found to be about 100-fold greater than the axon diameter (in agreement with internodal distances predicted to optimize the conduction speed) and range from a few hundred micrometers to 1–2 mm.

The mental image of an action potential occupying a single node and hopping from one node to the next is a common misconception. Although the action potential is jumping from one node to the next at the leading edge, many nodes are simultaneously participating. The extent of axons actively involved in the action potential is dependent on the speed of conduction. The fastest conducting myelinated fibers have a speed of 100–120 m s^{-1}. If the action potential duration is 1 ms, an action potential traveling at 100 m s^{-1} will, at a given instant of time, occupy 10 cm of the axon, or approximately 100 nodes, since the internodal length is on the order of 1 mm for large-diameter axons (**Figure 2**).

A measure of the reliability of conduction is called the safety factor, which is defined as the current in excess of that required to reach threshold and maintain propagation. A safety factor of 2 means that the current generated by the sodium channels is twice the minimum needed for conduction. Axons have a safety factor of about 5, and this excess is important because it speeds conduction (allowing the membrane to reach threshold faster) and provides the extra current needed at branch points. Axons branch hundreds of times, each branch imposing an increased load on the current provided by the upstream axonal membrane. If several branches occur close together, conduction can fail in some branches, especially during high-frequency firing. For similar reasons, additional current is also needed at the synaptic terminal where additional membrane must be depolarized. The internodal distances in motor axons decrease as the synaptic terminal is approached, and in some cases are as short as 10–20 μm. The effect of decreasing internodal distance is to concentrate nodes of Ranvier near the synaptic terminal, to provide the necessary current for terminal depolarization. It is not known if the terminals have sodium channels, since immunolabeling with antibodies specific for voltage-gated sodium channels have failed to show this.

Channel subtypes Many subtypes of sodium channels can be targeted to nodes of Ranvier, and during development several subtypes are found at neonatal nodes. In the adult mammal almost all nodes of Ranvier in the PNS and CNS contain predominantly one subtype, Nav1.6. The switch between neonatal and adult subtypes at the node coincides with the formation of compact myelin. Three types of potassium channels have been identified pharmacologically (inward rectifier, slow outward rectifier, and 4-aminopyridine-sensitive channels) and attributed primarily to internodal membrane. Subtype-specific antibodies have shown that Kv1.1, Kv1.2, and Kv1.4 are present in the internodal region, with the highest density in the juxtaparanodal region at the boundary

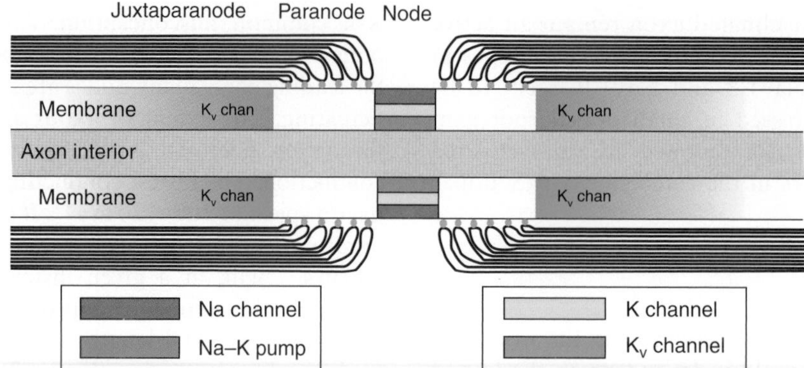

Figure 3 Ion channels and pumps concentrated in the vicinity of the node of Ranvier. Saltatory conduction in myelinated nerves is dependent on a high concentration of sodium channels at the node of Ranvier, to provide the inward current needed for depolarization of the next node. Repolarization is accomplished not only by inactivation of the sodium channels but also by a high resting potassium conductance at the node (KCNQ potassium channels) and by voltage-gated potassium channels excluded from the node and concentrated in the juxtaparanodal region near the node. Sodium–potassium pumps are concentrated at the node to maintain the concentration gradients. The thickness of the membrane (5 nm) is highly exaggerated relative to the axon diameter (>1 μm).

of the paranode. In addition, there is a high resting potassium conductance at some nodes, and a high concentration of KCNQ2 and possibly KCNQ3 potassium channels is co-extensive with the high concentration of sodium channels at the node. These separate highly aggregated clusters of channels are illustrated in **Figure 3**. As mentioned earlier, the flux of sodium and potassium ions needed to charge and discharge the membrane for each action potential is small, but maintenance of the ion gradients is dependent on sodium–potassium pumps. These pumps are highly concentrated in the nodal membrane.

Summary

The essential features of action potential initiation and propagation were determined over 50 years ago. Research into the electrical excitability of neurons is, however, far from moribund. Recent advances in molecular biology have revealed a multiplicity of sodium and potassium channel subtypes in neurons. Subtle changes in the activation, inactivation, and kinetics of voltage-gated sodium and potassium channels are predicted to have large effects on action potential threshold and rate of firing. The subcellular placement of specific isoforms and the modulation of these isoforms are also critical parameters of neuronal excitability. Many basic questions at the cellular and subcellular level remain. What is the lifetime of these channels in different regions, such as the axon hillock, initial segment, or node of Ranvier? Are there intracellular pools of the channels that can be rapidly inserted to provide plasticity at the level of conduction? What are the signals that target channels to the nodal region and adjacent paranodal and juxtaparanodal regions of myelinated axons? How does the cell achieve a balance between channel synthesis and degradation in the cell body, as well as insertion and retrieval at specific sites such as the node of Ranvier? How is the distribution of channels established during development? Maintenance of electrical excitability during adulthood is a process of continual remodeling that requires constant feedback with signals from target cells and interactions with glial cells. These molecular signals and interactions remain unknown.

See also: Demyelinating Diseases; Demyelination and Demyelinating Antibodies; Ion Channel Localization in Axons; Myelin: Molecular Architecture of CNS and PNS Myelin Sheath; Schwann Cells and Axon Relationship; Sodium Channels; Voltage Gated Potassium Channels: Structure and Function of Kv1 to Kv9 Subfamilies; Voltage-Gated Potassium Channels (Kv10–Kv12).

Further Reading

Ariyasu RG, Nichol JA, and Ellisman MH (1985) Localization of sodium/potassium adenosine triphosphatase in multiple cell types of the murine nervous system with antibodies raised against the enzyme from kidney. *Journal of Neuroscience* 5: 2581–2596.

Baker M, Bostock H, Grafe P, et al. (1987) Function and distribution of three types of rectifying channel in rat spinal root myelinated axons. *Journal of Physiology* 383: 45–67.

Boiko T, Van Wart A, Caldwell JH, et al. (2003) Functional specialization of the axon initial segment by isoform-specific sodium channel targeting. *Journal of Neuroscience* 23: 2306–2313.

Caldwell JH, Schaller KL, Lasher RS, et al. (2000) Sodium channel Na(v)1.6 is localized at nodes of Ranvier, dendrites, and synapses. *Proceedings of the National Academy of Sciences of the United States of America* 97: 5616–5620.

Catterall WA (1981) Localization of sodium channels in cultured neural cells. *Journal of Neuroscience* 1: 777–783.

Colbert CM and Johnston D (1996) Axonal action-potential initiation and Na^+ channel densities in the soma and axon initial segment of subicular pyramidal neurons. *Journal of Neuroscience* 16: 6676–6686.

Hodgkin AL (1975) The optimum density of sodium channels in an unmyelinated nerve. *Philosophical Transactions of the Royal Society of London* 270: 297–300.

Jack JJB, Noble D, and Tsien RW (1975) *Electric Current Flow in Excitable Cells*. London: Oxford University Press.

Nicholls JG, Martin AR, Wallace BG, et al. (2001) *From Neuron to Brain*, 4th edn. Sunderland, MA: Sinauer Associates.

Palmer LM and Stuart GJ (2006) Site of action potential initiation in layer 5 pyramidal neurons. *Journal of Neuroscience* 26: 1854–1863.

Quick DC, Kennedy WR, and Donaldson L (1979) Dimensions of myelinated nerve fibers near the motor and sensory terminals in cat tenuissimus muscles. *Neuroscience* 4: 1089–1096.

Active Perception

D H Ballard, The University of Texas at Austin, Austin, TX, USA

Perception Is Active Perception

What makes perception 'active'? Our everyday perception seems to require some engagement of attention on our part. We can be unaware of an ant crawling on the surface of our skin and at another time be conscious of its presence. Furthermore, this consciousness has an aim: we need to intercept the ant's path to get rid of it. Such is the nature of active perception. The awareness of these events has a neural trace in the brain's cortex. In binocular rivalry, separate images are presented to each eye, but the brain 'chooses' one to see on a momentary basis having an interval of a few seconds. Neurons in the brain's cortex respond to the perceived stimulus with the effects being more pronounced in higher cortical areas. In another example, microstimulation of the medial temporal (MT) cortical motion area affects a monkey's report of stimulus direction of motion. Thus, clearly signals related to perception are evident in the brain.

One or both of two additional constraints influence active perception. The first is the addition of a task agenda. The requirements of a specific task almost always allow shortcuts in the computation of the requisite information. In following a car on a highway, the observer only need lock the gaze on the rear of the car. Once this is done, the looming reflex provides a reliable measure of the change in following distance. The second constraint is the addition of the use of the body's mobility. Changing the body's posture or motion in a constrained trajectory can also lead to dramatic computational cost reductions, as any tennis player is acutely aware.

Early Vision

Early theoretical thinking in active perception has been inevitably influenced by our phenomenological experience of seeing a coherent three-dimensional scene. Although we are admonished not to think of the job of visual computation as producing a 'picture in the head,' all work in vision is heavily influenced by the seminal work of David Marr, who proposed that the job of vision was a form of this Cartesian idea. Marr proposed that the brain, particularly the cortex, constructs a layered series of representations that gradually progress to a more physical description of the 'outside' world. Evidence for such a progression is very solid, with the most well-known example being the MT visual cortical area, in which the representation of motion is transformed from a retinotopic representation of photometric change to a retinotopic image of local velocities, the latter termed 'optic flow.'

Active Perception Avoids Marr's 'Ill-Posed Problem'

Inspired by the discoveries of retinotopic maps of visual properties of increasing abstraction, early theoretical work focused on computing such representations. Initially, motor systems were seen as divorced from this effort, but the upshot was that the computational problems were very difficult. Marr characterized them as ill-posed problems because all the methodologies seemed to require iterating over the visual array of information. However, motor systems integration has largely revised that view because the ability to move the head, hand, and eyes provides important constraints that vastly simplify the computation required.

Consider the computation of optic flow, which is an instantaneous measurement of retinal velocity. Each point x, y in the optic flow 'image' indexes a velocity $u(x,y)$, $v(x,y)$. Fundamental conservation considerations lead to the constraint that the image velocity is the source of the spatial and temporal photometric variations. This can be codified as the flow equation that relates the two components of the optic flow velocity to the derivatives of the time-varying image $I(x, y, t)$ as shown:

$$I_x u + I_y v = I_t$$

where I_x, I_y, and I_t represent partial derivatives of the image that can be approximated with standard Gabor filter ('simple cell') measurements. Because these derivatives can be measured, the result at every point is one equation in two unknowns. This indeterminacy caused Marr to refer to this is an ill-posed problem. Without a further constraint this problem has to be solved with costly iterations over the image array.

However, consider the case of a static scene and further note that the motion of the head is constrained by the use of fixation to lock the gaze onto a specific point in the scene. In this case, active perception provides another constraint that greatly simplifies this problem. The actual direction of motion is known, providing another equation that now makes the problem well posed and easily solved for each retinotopic neighborhood. Furthermore, in this special case, the velocities are now correlated with depth so that a depth cue is obtained as a side effect. Kinetic depth is extremely fundamental because it is the first of several depth cues obtained during infant development.

Dynamic Frames of Reference

The key to understanding how these calculations are simplified is to realize that the primate visual system has a special capability with regard to its ability to fixate. Five independent stabilization systems – the vestibulo-occlular reflex, optokenetic nystagmus, saccades, vergence, and pursuit systems – allow it to lock its fixation point on a specific location in space. For humans, this ability is especially important because of the very expanded spatial resolution of photoreceptors at the fixation point. In one degree of visual area, characterized as the width of a thumb held at arm's length, the spatial resolution is approximately 100 times greater than the periphery. Thus, the visual system has evolved to rely on the point of fixation for high-precision visual calculations.

It sometimes takes more than a moment's reflection to realize that the point of fixation is not in the body. Thus, the fundamental computations of vision, as well as audition, are about a point that is not only distal in space but also dynamically changes from moment to moment. A way to remember this is to recall the coding for depth using binocular disparity. In striate and subsequent cortical areas, depth is coded in a relative system. Neurons have receptive fields that are tuned to positive, negative, and zero or near-zero disparities. Zero disparity means that the cell's best response occurs inputs from the left and right eyes match, and this is of course at the distal point of fixation. It is estimated that humans make more than 150 000 such fixations every day. Laboratory tests show that each such fixation has a specific purpose.

The interaction of increased resolution with active perception was first appreciated in studies of reading. When reading text on a computer terminal, the text outside of approximately 13 characters around the fixation point can be changed almost arbitrarily without consequence to reading speed, as long as the characters under each momentary fixation are correct. However, this result is not specialized for reading but, as subsequent work has shown, is likely to be a general property of the perceptual system.

Modeling Active Perception with Robots

The importance of active perception led to motorized camera systems as a basic input device for models. The very first such device was built by a team led by Ruzena Bajczy at the University of Pennsylvania. The first system with real-time image processing capability was built at the University of Rochester. **Figure 1** shows the combined system for studying hand–eye coordination. In order to provide the degrees of freedom of a head, the binocular camera system is mounted on a Puma industrial robot arm. It coordinates the movements of a robot hand that is mounted on a similar arm.

The development of robot active perception systems was a breakthrough because it allowed the computations in active perception to be studied directly. One of the first findings was that although the general problem of perceiving a scene at some level of detail in its entirety was very computationally expensive, there were subproblems of the general problem that could be done very cheaply. **Figures 2(a)** and **2(c)** show examples of the classical dichotomy of looking for a specific colored object in a scene when the colors are known in advance together with the complementary problem of having a close look at a colored object and trying to identify that object from a library of objects. These kinds of results can be related to the organization of cortex, which typically has separate cortical sensory maps devoted to object identity and object location. The former are located in temporal cortex and the latter are located in dorsal cortex.

Experiments with active perceiving robot systems showed that each of these problems could be done very cheaply. Work has continued along these lines and now incorporates probabilistic methods that are discussed later.

The Quantal Nature of Data Acquisition

One of the most remarkable findings of active perception is its inherently quantal nature. Since 'active' denotes a specific agenda, the perceiver almost always needs very specific information from each fixation. Thus, the larger percept of a scene may be replaced by a question: 'Is the cup I am filling with tea full yet?' This requires monitoring the rising level of liquid in a cup. This computation in turn can be modeled with simple template matching, which is vastly simpler than the general-purpose computation that would be required to create a three-dimensional model of the cup and its immediate venue.

The quantal nature of active perception is very dramatically revealed in sports in which the speed of travel of the ball places high demands on rapid perception. Interestingly, many such sports are handled identically by the perceptual system. Consider the example of cricket, in which the ball bounces on the cricket pitch at approximately 90 mph. When

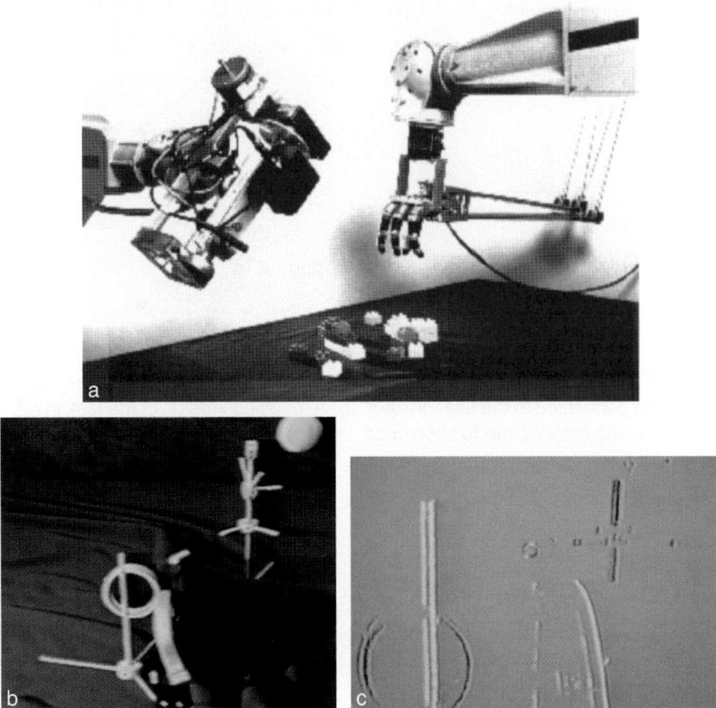

Figure 1 (a) Binocular camera vision system combined with a MIT-Utah robot hand. Motors drive the cameras in pitch and yaw at near saccadic speeds. (b) Table scene containing simple toy objects. (c) Kinetic depth image obtained in real time by moving the head horizontally in a frontoparallel plane. Points in the scene in front of the fixation point are coded as white and those behind in black.

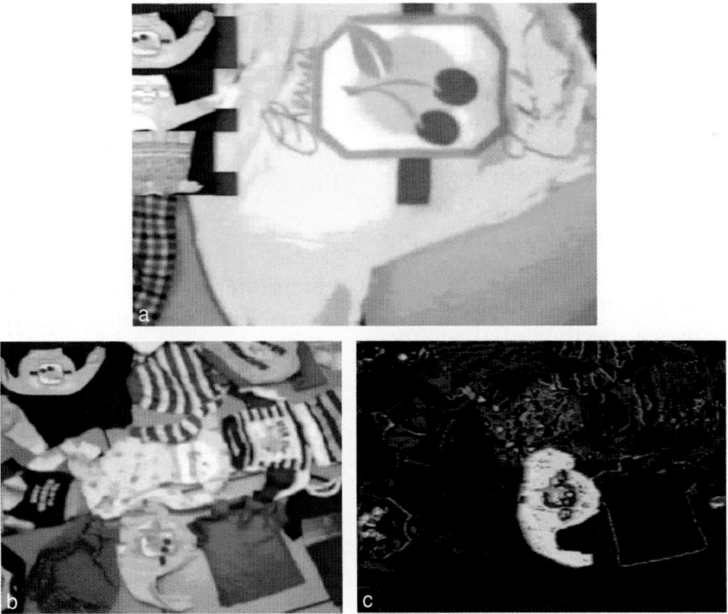

Figure 2 (a) A 'what' task solved by focusing on a single object and matching its colors against a color database. (b) A 'where' task: a floor strewn with colored shirts, one of which needs to be found. (c) The location of its colors can be established by looking for localized clusters of them in the image.

analyzing the pitch, the batsman first fixates the bowler's hand at the release point and then saccades to the estimated bounce point. At the bounce point, the computation required is to estimate the path of the ball, on which the rest of the batsman's motion is determined, based on extensive training from similar pitches. This strategy is mirrored in table tennis, in which players fixate the bounce point on the table,

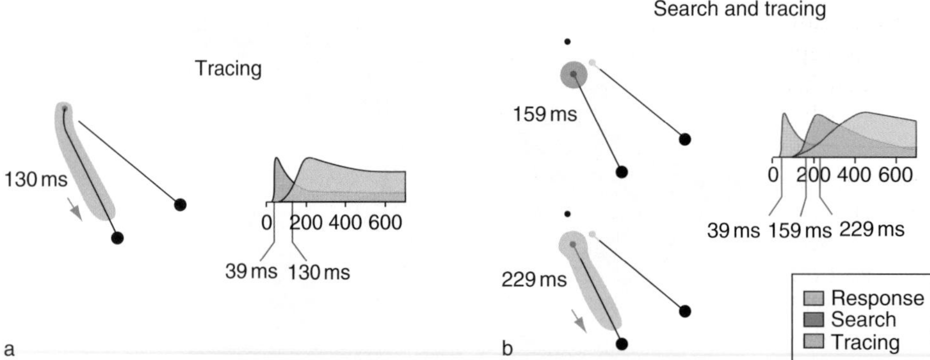

Figure 3 (a) A monkey is trained to fixate a small spot on an otherwise blank CRT screen. Subsequently, two lines appear and the monkey must fixate the end of the line connected to the fixation spot. (b) A modified version of the task in (a), where now the monkey must fixate the line that has a stub of the same color as the fixation spot.

and in squash, in which players fixate the bounce point on the front wall. However, in each of these cases, information gleaned at the fixation point keys an elaborate motor response.

Neural evidence of the quantal nature of active perception comes from experiments such as those by Pieter Roelfsema, in which monkeys fixate a small spot on a blank screen. While they hold fixation, two lines that are oriented radially with respect to the spot appear. One is connected to the fixation spot and the monkey's task is to make a saccadic eye movement to its other terminal. The experiment takes advantage of the fact that in the absence of cues, it takes approximately 250–300 ms to program the saccade. In addition to this protocol, the experimenters monitor cells in striate cortex with receptive fields that are oriented coincident to the line's orientation. The crucial finding is that the firing rate of the cell is increased at a specific point in time, approximately 130 ms after the target onset. The interpretation is that the monkey uses a specific visual routine to trace along the connected line to find its terminal. In a modification of this protocol, a monkey has to make a saccade to the end of the line that has a stub which is the same color as the fixation spot. Under the visual routines hypothesis, the tracing should now start later in time. In fact, the elevated onset in firing now starts at 229 ms (**Figure 3**). Experiments such as these provide corroborating evidence for active perception and suggest that it may consist of elemental operations that can be combined to solve more complicated tasks.

Bayes' Rule

The need for visual routines is compelling, but at the same time they introduce a modeling complexity. One is led to the same conclusions as Shimon Ullman and Steve Kosslyn, two of the original visual routine proponents: there must be some library of standard operations that can be composed into short programs that solve specific tasks. How such a library gets established through learning and development is very much an open question, but a very helpful way of characterizing the computations of a routine abstractly has emerged via Bayesian probability theory. The reader will be very familiar with the following version of Bayes' rule:

$$P(M|D) = \frac{P(D|M)P(M)}{P(D)}$$

Within a constant of proportionality we can write this as

$$P(M|D) \propto P(D|M)P(M)$$

Let us interpret this using the example of anticipating the cricket ball bounce location on the cricket pitch. The batsman needs to know exactly where the ball will bounce, and to do this he or she has to integrate two sources of information. In this case, $P(M)$ models the probability distribution of likely locations of the ball based on numerous prior pitches. The term $P(D|M)$ is data on the location of the ball estimated by the batter for a particular pitch. Multiplication of these two distributions provides the Bayesian estimate for the likely bounce point of the ball. This kind of Bayesian formulation has been repeated in numerous experiments, suggesting that it is a general mechanism that the brain uses to combine perceptual estimates with previous experience.

Virtual Environments

Virtual environments have proven to be a breakthrough for the study of active perception because they allow specific features to be present or absent during an active perception task. Typically, subjects can wear a binocular head-mounted color display to see scenes that are rich in visual cues. Magnetic and inertial-based sensors on the head-mounted display (HMD) allow for the scene to be updated at frame rates, thus providing a sense of immersion in a three-dimensional world. Because the subject can execute different tasks based on the visual display contents, the motor system provides a new and reliable arbiter of what was perceived.

For example, consider the block-copying task used by Triesch et al. Subjects perform the task of picking up blocks in a virtual environment. Subjects have a sense of touch delivered through a virtual force feedback device. The device shown in **Figure 4** provides a realistic sensation of force through opposed kinematic links that can be back-driven with cable-linked electric motors based on the position of two fingers, typically the finger and thumb used in a pinch grip.

Experimental subjects are instructed through visual cues in the virtual reality environment to pick up one of several blocks based on a feature such as color and place it on a conveyor belt based on a second, possibly the same, feature. The key manipulation is that on a small percentage of the trials the features of the block change between the pickup and put down. Subjects are told to discard the block if they notice any such changes.

The key observation is that on missed color changes, where the subjects behave as if nothing happened, the block is sorted by its original color rather than its new color. This decision is made even when the subjects have fixated the block after the color change. The implication is that, consistent with neural data, the acquisition of the color is done in a quantal manner. When the color is registered (represented) early on in the experiment, later there is no need to check it in the neural protocol.

Tasks with Sequential Steps

Simple tasks such as staying on a sidewalk require only the measurement of the sidewalk edge. The history of the traverse is not needed. However, more complicated behaviors, such as making a sandwich, require more state. If you want to put peanut butter on a slice of bread, you must be holding the knife. These causal dependencies can be represented in a graphical network. Such a network is a suitable tool for this class of problems because it uses easily observable evidence to update or infer the probabilistic distribution of the underlying random variables. A Bayesian graphical network represents the causalities with a directed acyclic graph; its nodes denote variables and edges denote causal relations that can be annotated with conditional probability distributions. In this active perception case, the state of the agent is dynamically changing, and the observations are updated throughout the task execution process, so one must specify the temporal evolution of the network. **Figure 5** illustrates the two-slice representation of a dynamic Bayesian network. Shaded nodes are observed; the others are hidden. Probability distribution matrices determine causalities, represented by straight arrows. The state of the lowest hidden node is determined by its prior distribution in the first time/slice, and thereafter it is jointly determined by its previous state and the transition matrix, as denoted by the curved arrow. The two-slice representation can be easily unrolled to address behaviors with arbitrary numbers of slices. At each moment, the observed sensory data (gray nodes), along with its history, are used to compute the probability of the hidden nodes being in certain states:

$$P(S^{t\in[1,T]}) = P(S^1)P(O^1|S^1)\prod_{t=2}^{T} P(S^t|S^{t-1})P(O^t|S^t)$$

Although this equation might look complex, it is simply an expression for the probability of being in the

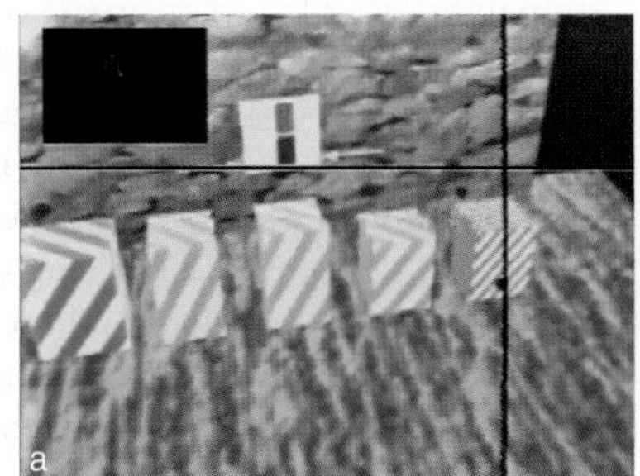

Figure 4 (a) The view from a subject's helmet in a block manipulation task. A subject is about to pick up a blue block using virtual force fingertips indicated by the red dots on the screen. (b) After moving it closer, a cue on the table instructs that a blue block goes on the conveyor belt to the left, and a red block goes on the conveyor belt to the right. If the block changes color, as it has in this case, it is to be discarded, but instead the subject follows the instructions for a blue block.

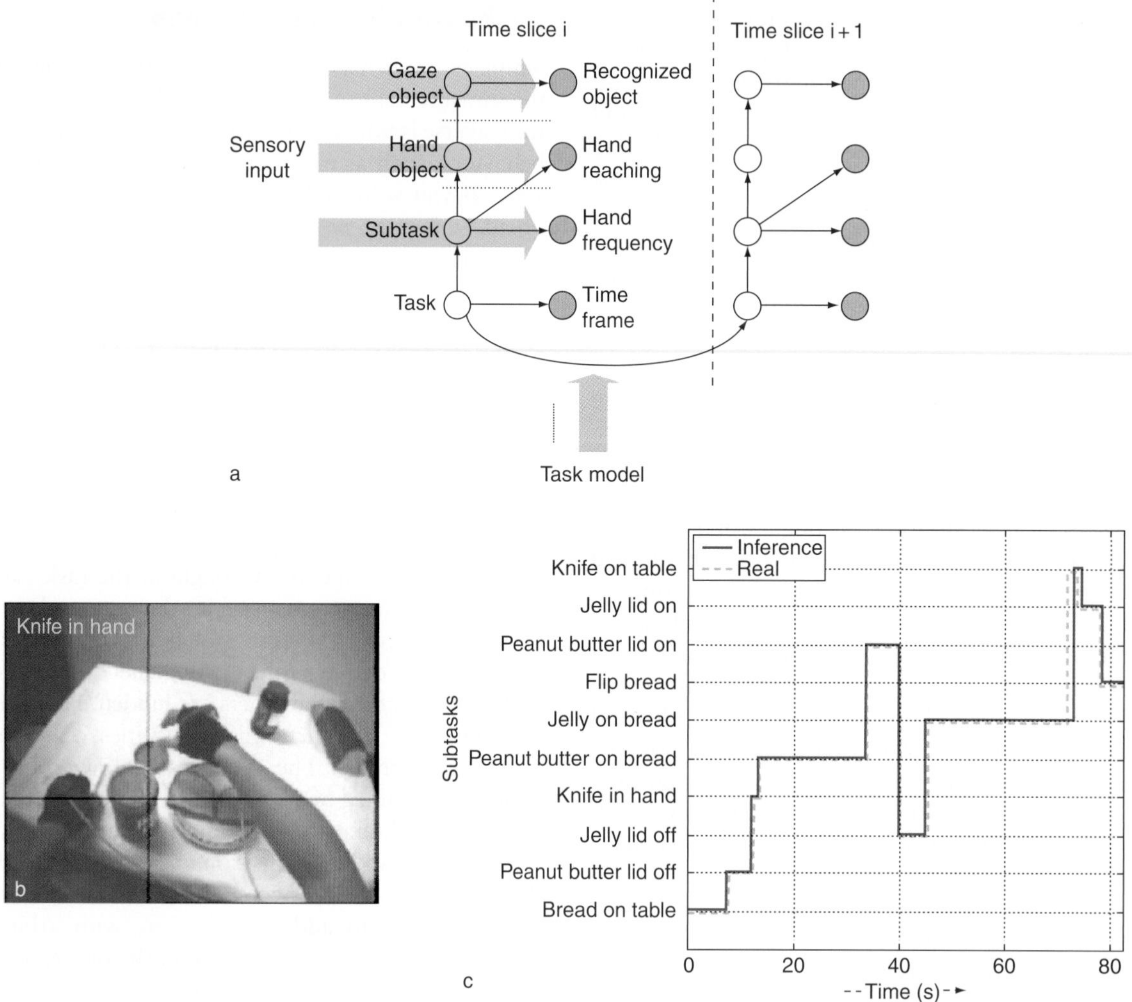

Figure 5 (a) The basic structure of the dynamic Bayesian network (DBN) used to model sandwich making. Visual and hand measurements provide input to the shaded nodes. The sequencing probabilities between subtasks are provided from a task model that in turn is based on human subject data. (b) A frame in the video of a human subject in the process of making a sandwich showing that the DBN has correctly identified the subtask as 'knife-in-hand.' (c) A trace of the entire sandwich-making process showing perfect subtask recognition by the DBN. This trace uses the entire data set in an off-line mode. In the online mode in which the classifications have to be made on the partial data sets, small errors can be made.

states $\{S^t, t = 1, \ldots, T\}$ and seeing observations $\{O^t, t = 1, \ldots, T\}$ from those states.

A behavior-recognition algorithm computes the states of each hidden node S^tat time t that maximize the probability of observing the given sensory data:

$$S^t = \arg\max_s P(S^t = S|O^{[1,T]})$$

Conclusion

This introduction to active perception has traced the development of its key ideas. The notion that physical descriptions of the world are represented in visual maps in the cortex heralded the beginning of understanding how visual information is represented, but early static models were very computationally inefficient. Active perception research, driven by advances in robotics technology, provided a more behavioral framework that led to dramatic computational simplifications. Two other contributions to active perception are the Bayesian revolution and virtual reality technology. The former has demonstrated that the perception of almost all visual information is governed by prior information stored as probability distributions. Complicated sequential tasks can use graphical structures to organize complex probabilistic dependencies. Virtual reality technology provides a way of probing these dependencies because the virtual sensory world can be manipulated on a timescale that is sufficiently fast to interact with the brain's management of such information.

See also: Bayesian Cortical Models; Bayesian Models of Motor Control; Binocular Rivalry; Contextual Interactions in Visual Perception; Ideal Observer Theory; Spatial Transformations for Eye–Hand Coordination; Statistical Analysis of Visual Perception.

Further Reading

Aloimonos J, Weiss I, and Bandyopadhyay A (1988) Active vision. *International Journal of Computer Vision* 1(4): 333–356.

Ballard DH (1991) Animate vision. *Artificial Intelligence* 48(1): 57–86.

Bishop C (2006) *Pattern Recognition and Machine Learning*. New York: Springer.

Britten KH, Shadlen MN, Newsome WT, and Movshon JA (1992) The analysis of visual motion: A comparison of neuronal and psychophysical performance. *Journal of Neuroscience* 12: 4745–4765.

Droll JA, Hayhoe MM, Triesch J, and Sullivan B (2003) Task relevance of object features modulates the content of visual working memory. *Vision Research* 3(9).

Gray WD (ed.) (2007) *Integrated Models of Cognitive Systems*. Oxford: Oxford University Press.

Jordan MI (ed.) (1998) *Learning in Graphical Models*. Cambridge, MA: MIT Press.

Land M and Hayhoe M (2003) In what ways do eye movements contribute to everyday activities? *Vision Research* 41: 3559–3566.

Landy MS, Maloney LT, and Pavel M (eds.) (1995) *Exploratory Vision: The Active Eye*. New York: Springer.

Marr D (1982) *Vision*. New York: Freeman.

Pearl J (2000) *Causality: Models, Reasoning and Inference*. Cambridge, UK: Cambridge University Press.

Roelfsema PR, Khayat PS, and Spekreijse H (2003) Subtask sequencing in the primary visual cortex. *Proceedings of the National Academy of Sciences of the United States of America* 100: 5467–5472.

Triesch J, Ballard D, Hayhoe M, and Sullivan B (2003) What you see is what you need. *Journal of Vision* 3: 86–94.

Weiss Y, Simoncelli EP, and Adelson EH (2002) Motion illusions as optimal percepts. *Nature Neuroscience* 5: 598–604.

Active Zone

P S Kaeser, University of Texas Southwestern Medical Center, Dallas, TX, USA

Introduction

Synapses are highly specialized contacts between nerve cells that transmit signals from the presynaptic neuron to the postsynaptic cell. They are composed of the presynaptic terminal, the synaptic cleft, and the postsynaptic membrane. The terminal of the presynaptic neuron translates the arriving action potential into a chemical signal, which in turn stimulates a response on the postsynaptic cell. In brief, the action potential induces opening of presynaptic voltage-gated calcium channels, and the local increase in calcium in the presynaptic terminal promotes fusion of neurotransmitter filled synaptic vesicles at the presynaptic membrane on a millisecond timescale. Fusing vesicles release their content into the synaptic cleft, and the neurotransmitters induce a response in the target cell upon binding to postsynaptic receptors.

This brief overview of synaptic transmission demonstrates the asymmetric nature of synapses, where both synaptic compartments – namely the presynaptic terminal and the postsynaptic membrane – contain highly specialized components. In principle, the presynaptic terminal is an organ of membrane trafficking and the postsynaptic compartment is a signal transduction machinery. The apparatus of postsynaptic reception is the postsynaptic membrane with the receptors and the postsynaptic density (PSD), which consists of signaling proteins, scaffolding proteins, and cytoskeletal proteins that enable the postsynaptic cell to sense the presynaptic release of neurotransmitters and respond appropriately. The fusion of synaptic vesicles at the presynaptic membrane is under tight spatial and temporal control. Synaptic vesicle fusion only occurs at highly specialized hot spots called active zones (AZs).

Definition of Active Zones and Functional Participation in the Synaptic Vesicle Cycle

Originally, AZs were described morphologically in seminal studies by Couteaux and Pecot-Dechavassine and Akert et al., and they appear as electron dense material that is tightly attached to the presynaptic membrane (**Figure 1**). Functionally, AZs are defined as sites where neurotransmitters are released into the synaptic cleft, and biochemically, AZs are composed of a network of insoluble proteins, containing AZ-specific and other proteins.

In the presynaptic nerve terminal, synaptic vesicles undergo a series of events in order to release their content into the synaptic cleft before being recycled and refilled. Investigators divide the synaptic vesicle cycle into nine steps. Synaptic vesicles are filled with neurotransmitters (step 1), and the vesicles then cluster around the AZ (step 2). They are docked to the AZ (step 3), where they gothrough an ATP-dependent priming reaction (step 4) that enables them to undergo fast, calcium-dependent fusion pore opening (step 5) which is initiated by the arriving action potential in the presynaptic nerve terminal. Upon releasing neurotransmitters into the synaptic cleft, three pathways have been proposed: Vesicles might remain in the docked stage (step 6, also referred to as 'kiss-and-stay'), or they might be undocked and locally refilled (step 7, called 'kiss-and-run'). Alternatively, they may be recycled through clathrin-dependent endocytosis of synaptic vesicles (step 8), which leads to either direct refilling or refilling after being passed through the endosome (step 9). It is important to note that only approximately 10–20% of all action potentials that arrive in a nerve terminal induce exocytosis of synaptic vesicles.

AZs participate in multiple steps within the synaptic vesicle cycle. Their core function is to bring synaptic vesicles in close proximity to the presynaptic plasma membrane and to presynaptic calcium channels in order to enable the synapse for fast, calcium-dependent fusion. In addition, AZs participate in a regulatory manner by modifying the probability of neurotransmitter release. This involvement in synaptic plasticity is mediated through AZ-specific proteins, and in principle it can be achieved by changing the number of either docked or primed vesicles. The molecular events that occur within the AZ in order to dock and prime synaptic vesicles and to change synaptic strength are under intense investigation, and aspects of these events are discussed after a more in-depth examination of the structure of AZs.

Morphology of the Active Zone

The fine structure of AZs is diverse among different species, and it is also variable among different kinds of synapses. Nevertheless, all AZs share common features. In electron microscopic images, they appear as a dense structure that is tightly attached to the presynaptic membrane (**Figure 1**). The electron dense nature of AZs reflects the very high protein content, and they consist of a network of highly

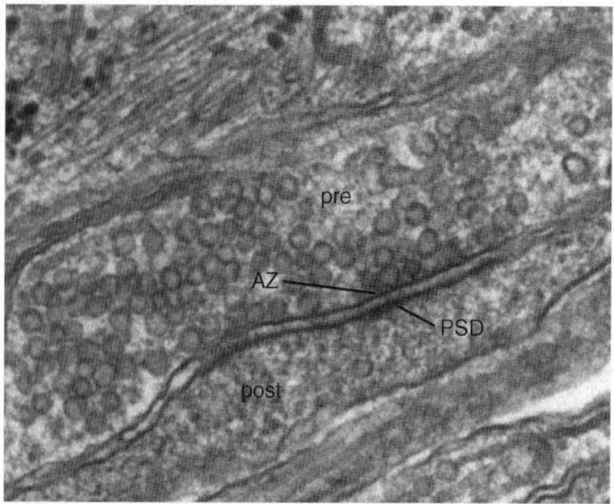

Figure 1 Electron micrograph of a central nervous synapse. Synaptic vesicles tether in the presynaptic terminal (pre) around the active zone (AZ), which consists of electron dense material that is tightly associated with the presynaptic membrane. The postsynaptic density (PSD) is precisely opposed to the AZ at the postsynaptic cell (post). Courtesy of Dr. Xinran Liu, University of Texas Southwestern Medical Center, Dallas, TX.

insoluble proteins that interact with each other and with multiple cellular proteins. AZs are precisely opposed to PSDs, and they are separated from PSDs by the presynaptic plasma membrane, the synaptic cleft, and the neurotransmitter receptor containing postsynaptic plasma membrane. This architecture of precise opposition of the area of release and sensing of neurotransmitters is crucial because it minimizes the distance that neurotransmitters have to diffuse before they bind to the postsynaptic receptors. In mammalian synapses, the variability of this geometric assembly of the AZ and the PSD correlates with the synapse type. Whereas excitatory, glutamatergic synapses are typically asymmetric with a PSD that is thicker than the AZ, inhibitory GABAergic and glycinergic synapses appear more symmetric.

In typical central nervous system mammalian synapses, electron microscopy has revealed that AZs consist of a hexagonal grid, and synaptic vesicles are embedded in depressions of the grid close to the presynaptic membrane (**Figures 2(a)** and **2(b)**). The electron

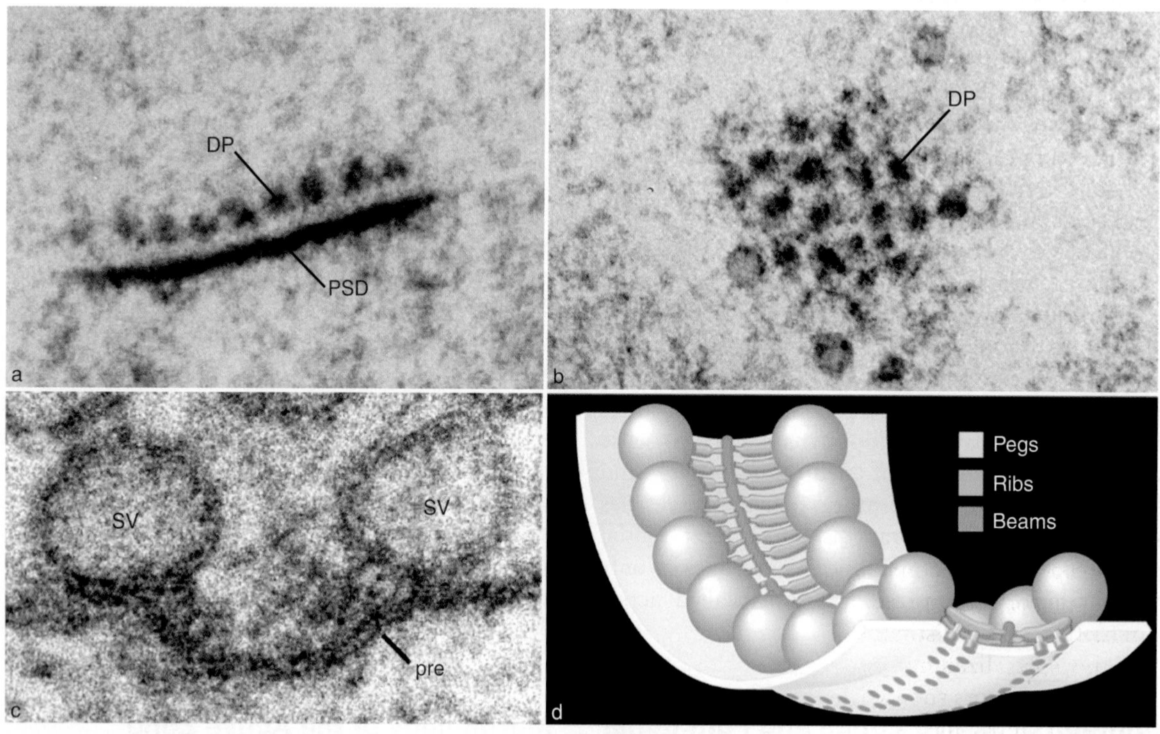

Figure 2 Fine structure of active zones (AZs) of a mouse central nervous system synapse and the frog neuromuscular junction. (a and b) Phosphotungstic acid (PTA) staining of a central nervous synapse in cross section (a) and as top view from the presynaptic terminal (b). PTA selectively visualizes the proteinacious contents of AZs and postsynaptic densities (PSDs); the synaptic plasma membrane is not stained. Whereas the PSD appears as a regular, dense band, the AZ forms dense projections (DPs) that protrude approximately 50 nm into the presynaptic terminal. The DPs are assembled in a regular, hexagonal grid and are connected with thin fibrils. This regular assortment allows the synaptic vesicles to closely approach the presynaptic plasma membrane in the depressions of the grid. Courtesy of Dr. Xinran Liu, University of Texas Southwestern Medical Center, Dallas, TX. (c) Cross section and a model (d) of a frog neuromuscular junction based on conventional two-dimensional electron microscopy. The AZ material is located between the synaptic vesicles (SVs) forming a regular structure with a central beam, ribs that connect the central beam with the SVs, and pegs that connect the ribs with the synaptic plasma membrane containing the calcium channels. pre, presynaptic membrane. Reprinted by permission from Macmillan Publishers Ltd: *Nature* (Harlow ML, Ress D, Stoschek A, Marshall RM, McMahan UJ (2001) The architecture of active zone material at the frog's neuromuscular junction. *Nature* 409: 479–484), copyright 2001.

dense protrusions, called dense projections, are relatively small compared to other synapses; they extend for approximately 50 nm from the presynaptic plasma membrane into the presynaptic terminal, and they are connected to each other by thin fibrils. Each dense projection has six immediate neighbors, and they are evenly spaced with a distance of approximately 50–100 nm. This highly organized architecture leaves slots for docking of synaptic vesicles in close proximity to the membrane.

Harlow et al. analyzed the fine structure of the AZ at frog neuromuscular junction (NMJ) in detail (**Figures 2(c)** and **2(d)**). At these synapses, the vesicles line up at both sides along an elongated ridge at the presynaptic terminals of motor neurons. Each synaptic vesicle is connected via ribs to a central beam, and each rib contains two connections (pegs) to the presynaptic plasma membrane. It is assumed that these pegs consist of macromolecules that connect the synaptic vesicle anchoring ribs with presynaptic calcium channels. Although this study leaves the molecular identities of ribs, pegs, and beams open, it clearly shows the highly specialized organization of presynaptic AZs.

The size and morphology of dense projections vary greatly between different synapses and different species. In contrast to the rib- and beamlike structure at the frog NMJ, the *Caenorhabditis elegans* NMJ forms plaque-like AZs, and *Drosophila* NMJs contain T-shaped projections called 'T-bars.' At so-called ribbon synapses in the rat retina, long, prominent ribbonlike protrusions extend into the cytoplasm and tether synaptic vesicles around them. Many molecular and morphological studies have been performed on ribbon synapses because they are relatively easy to access and visualize. Another sensory synapse with a very specialized architecture is the hair cell in the auditory pathway, where ovoid or spherical dense projections are surrounded by a halo of synaptic vesicles. Both the photoreceptor synapse and the ribbon synapse are functionally very different because they do not respond to incoming action potentials but, rather, to sensory stimuli that produce graded receptor potentials. Their distinct function presumably led to a highly specialized morphology.

Some investigators subdivide AZs morphologically into three compartments: the presynaptic plasma membrane, the cytomatrix immediately attached to the membrane (also referred to as 'CAZ' – cytomatrix of the active zone), and the electron dense protrusions into the cytoplasm, called 'dense projections' (also referred to as membrane thickenings or AZM – active zone material). Although this subdivision is morphologically correct, AZs are functional units, and many proteins involved in their architecture do not respect these morphological borders.

Molecular Components of the Active Zone

AZs consist of a dense, insoluble protein network that is tightly attached to the presynaptic plasma membrane. Although several protein/gene families have been described to be involved in the formation, maintenance, and function of AZs, the molecular characterization has only just begun. Many interactions of proteins involved in AZs have been illustrated, but most of them are poorly understood in terms of their functional importance. In addition, the fact that most AZ proteins are large and highly insoluble makes biochemical characterization difficult, and the results of binding studies that have identified new interactions *in vitro* are difficult to confirm *in vivo*. Despite these difficulties, investigators have made significant progress in identifying molecules that participate in AZs. Gene knockdown/out studies in *Drosophila*, *C. elegans*, and mice have provided fascinating insights into potential AZ functions.

In principle, AZs consist of two classes of molecules: non-AZ-specific and AZ-specific (or enriched) components. Among the non-AZ-specific members are five functional classes of proteins that are involved in different processes. First, there are proteins that are directly involved in synaptic vesicle fusion and associated with the presynaptic plasma membrane, such as the SNAREs SNAP25 and syntaxin, and also Munc18, a protein associated with the presynaptic release machinery that is required for synaptic transmission. Second, although the link between the AZ and the cytoskeleton is only poorly understood, it is clear that cytoskeletal proteins such as actin, spectrin, myosin, and tubulin participate in the structural backbone of the AZ. Third, synaptic scaffolding molecules are tightly associated with both AZ and PSD, and presynaptically CASK, Mints, Velis, and SAP97s have been identified. Fourth, as mentioned previously, the core function of the AZ is to bring together synaptic vesicles and calcium channels at the presynaptic terminal. Thus, voltage-gated calcium channels are an integral part of the AZ, but the molecular link between the intracellular portions of the channel subunits and the dense protein mesh of the AZ is only poorly understood. Finally, cell adhesion molecules play a crucial role during synaptic development and in mature synapses. Several molecules have been shown to reach from the AZ through the presynaptic plasma membrane into the synaptic cleft to form either homophilic or heterophilic interactions with proteins in the synaptic cleft or with proteins that reach through the postsynaptic membrane into the PSD. Such proteins are integrins, cadherins, neurexins, and SynCAMs. Several fascinating studies have shown how these transmembrane molecules induce synapse formation, and neurexins with their postsynaptic interaction partner neuroligin have been suggested to be crucially involved in the pathogenesis of neurodevelopmental defects such as autism.

Six families of AZ-specific proteins have been identified (**Figure 3**): Munc13s, Rab3A interacting molecules (RIMs), RIM-binding proteins (RBPs or RIM-BPs), α-Liprins, ELKS (proteins rich in E, L, K, and S), and Piccolo/Bassoon. They all consist of large multidomain proteins that are encoded by multiple genes and that are involved in protein–protein interactions. **Figure 3** demonstrates that these proteins interact with each other in multiple ways to form the dense material that can be observed in electron microscopic images. **Table 1** provides an overview of the members of these proteins, their evolutionary conservation, and their involvement in AZ function. Next, the functional and biochemical involvement of the AZ-specific proteins is discussed.

Munc13s were initially identified in *C. elegans* as phorbol ester (β-PE)/diacylglycerol (DAG)-binding proteins encoded by the unc13 gene, and the phenotype in unc13 mutant worms suggested that these proteins were involved in neurotransmitter release. Since then, data obtained from *C. elegans*, *Drosophila*, and mouse models have suggested that synaptic transmission is crucially dependent on Munc13-mediated vesicle priming. In mammals, the four Munc13 genes (Munc13-1 to -4) produce three brain-specific Munc13 isoforms (Munc13-1, bMunc13-2, and Munc13-3), the ubiquitous Munc13-2, and the nonneuronal Munc13-4 (**Table 1**). Munc13s consist of a conserved C-terminal region that contains a C1 domain and two C2 domains flanking the Mun homology domain. The Mun domain is sufficient to rescue the defective synaptic transmission in synapses lacking Munc13. In addition, Munc13-1 and ubMunc13-2 have an N-terminal domain that binds to the RIM zinc finger and calmodulin and that contains a C2 domain (**Figure 3**). It is interesting to note that different synapses have been shown to express different Munc13 isoforms, and the expression pattern of these isoforms determines parameters of short-term synaptic plasticity. The central C1 domain of Munc13 binds β-PE/DAG, and this interaction mediates β-PE/DAG-induced augmentation in the hippocampus. The conserved C-terminal region of Munc13s has been suggested to bind to the synaptic plasma membrane SNARE syntaxin, two proteins that have been implicated in vesicular trafficking/exocytosis (DOC2α and msec7-1 ARF-GEF), and a novel spectrin (spectrin β-spIIIΣ) that potentially links the AZ to the cytoskeleton. Functional experiments in *C. elegans* and *in vitro* binding studies have proposed a molecular mechanism by which Munc13s promote the open confirmation of syntaxin, which in turn results in a loose assembly of the SNARE complex, bringing synaptic vesicles into a fusion competent state. However, this model, which involves a direct interaction between Munc13 and syntaxin, remains controversial because nuclear magnetic resonance measurements

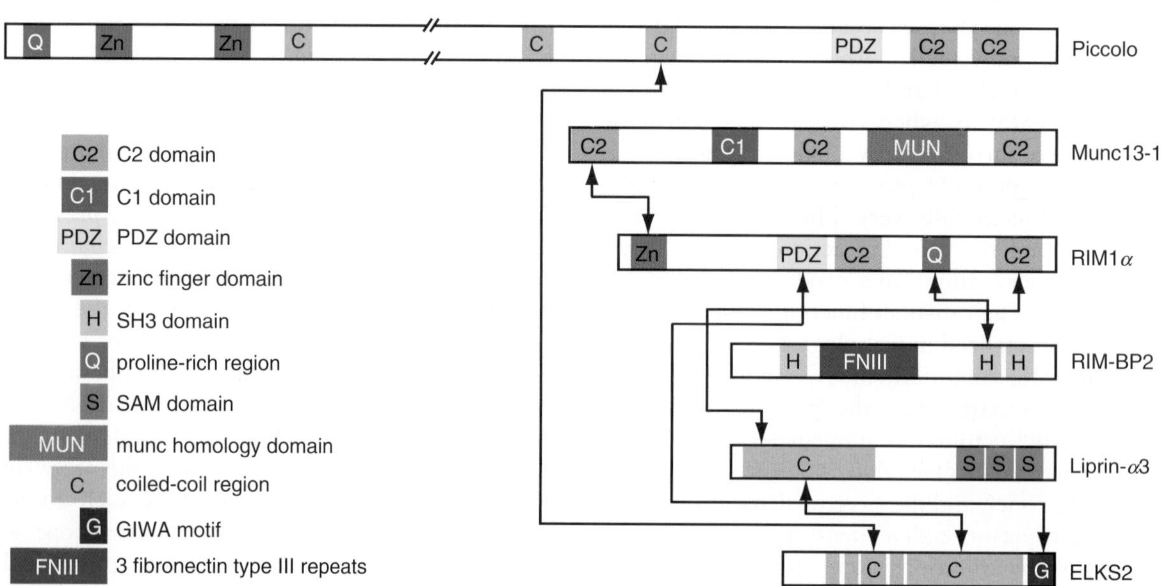

Figure 3 Schematic overview showing a representative member of each family of active zone (AZ) proteins. Interactions among AZ proteins are indicated with black arrows. All AZ-specific proteins are large proteins that contain multiple domains. Together, they form an insoluble protein scaffold that brings together synaptic vesicles, the presynaptic plasma membrane, and voltage-gated calcium channels at the presynaptic nerve terminal.

Table 1 Active zone proteins

Protein family	*Mammalian protein isoforms*[a]	*Drosophila/C. elegans homologs*	*Proposed functions*
Munc13s	Munc13-1 ubMunc13-2, bMunc13-2 Munc13-3 Munc13-4	Dunc13/unc-13	SV priming, scaffolding, synaptic plasticity, cytosceleton anchoring
RIMs	RIM1α RIM2α, RIM2β, RIM2γ RIM3γ RIM4γ	RIM-PA/unc-10	SV docking and priming, short- and long-term plasticity, scaffolding, possibly channel anchoring
ELKS	ELKS1A, ELKS1B ELKS2	*bruchpilot* (*brp*)?/ELKS	Scaffolding, NF-κB signaling
Piccolo/ Bassoon	Piccolo/Aczonin Bassoon	Not identified to date	Scaffolding, anchoring, membrane trafficking, endocytosis
α-Liprins	Liprin-α1 Liprin-α2 Liprin-α3 Liprin-α4	Dliprin/syd-2	Scaffolding, receptor anchoring, membrane trafficking
RIM-BPs	RIM-BP1 RIM-BP2	Not identified to date	Scaffolding, channel anchoring

[a]Each gene is represented on one line. For example, Munc13s are encoded by four mammalian genes, and the Munc13-2 gene produces two protein isoforms (ubMunc13-2 and bMunc13-2).

have not confirmed a direct interaction between the two proteins.

The N-terminal RIM zinc finger forms a tripartite complex with Munc13 and Rab3A, a small GTPase that is located on synaptic vesicles. This biochemical interaction is important because it links Munc13s and RIMs directly to synaptic vesicles. A detailed morphological analysis of *C. elegans* has shown that RIM is necessary for connecting synaptic vesicles to dense projections but not for bringing them in close proximity to the synaptic plasma membrane. In addition, the RIM C2B domain has been suggested to bind to the calcium sensor synaptotagmin 1 *in vitro*, providing another potential link between AZs and synaptic vesicles. RIMs contain a number of domains and interaction motifs that are involved in scaffolding at AZs by binding to multiple other AZ proteins (**Table 1**). With their central PDZ domain, RIMs interact with ELKS, and with the C-terminal C2B domain they bind to α-Liprins (**Figure 3**). RIMs are a conserved gene family, with RIM-PA in *Drosophila*, unc10 in *C. elegans*, and four genes (RIM1–4) in mammals. The RIM2 gene has additional internal promoters to produce shorter isoforms (RIM2β and RIM2γ), and multiple sites of alternative splicing enhance the variety of RIM1 and RIM2. The absence of RIM1α in a mouse model leads to a large deficit in multiple parameters of synaptic plasticity, including a reduction of release probability at excitatory synapses and a lack of presynaptic long-term potentiation. This functional deficit is also reflected in the fact that these animals show severe shortcomings in various tests for learning and memory. Electrophysiological and behavioral analysis of the RIM1α mutant mice demonstrates how AZs are critically involved in modulating synaptic transmission, being an essential component for the brain's capacity to adapt to environmental inputs. When RIM1α and RIM2α, two of the six major RIM isoforms, are absent, the mice show a drastic calcium-dependent release deficit at the NMJ, and they die immediately after birth. Thus, RIMs are critical not only for modulating the plasticity of synapses but also for normal synaptic release. RIM proteins contain a proline-rich region in the linker between the central and the C-terminal C2 domains that binds to RIM-BPs.

RBPs are produced by two genes – RIM-BP1 (also referred to as PRAX-1) and RIM-BP2 (**Table 1**) – and contain three Sarc homology 3 (SH3) domains and three fibronectin type III repeats (**Figure 3**). No homologs have been identified in either *C. elegans* or *Drosophila*. An *in vitro* study showed that RIM-BPs might bind to voltage-gated calcium channels with their SH3 domains, and thus they can potentially link RIMs and, indirectly, all other AZ-specific proteins to presynaptic calcium channels. In addition, one study suggested a direct interaction between RIMs and calcium channels. Although these data are very interesting, they have not been confirmed *in vivo*, and functional data have only been provided for the RIM-BP–calcium channel interaction using a neurosecretory PC12 cell line. The *in vivo* existence of the potential calcium channel–RIM-RIM-BP complex remains to be elucidated.

Liprins have been identified as a family of proteins that interact with LAR transmembrane tyrosine phosphatases, and they consist of two subfamilies – Liprin-α and Liprin-β. Liprin-β proteins are not brain specific, and Liprin-αs are encoded by four genes (1–4), of which Liprin-α2 and Liprin-α3 are brain specific (**Table 1**). Mammalian Liprin-α function has only been addressed *in vitro*, but it has been found that Liprin-α3 forms a complex at the AZ with ELKS and RIM through its N-terminal coiled-coil regions (**Figure 3**). In *C. elegans*, it has been convincingly shown that the absence of the Liprin homolog syd-2 disrupts the regular AZ structure, and that syd-2-deficient synapses have a defect in neurotransmitter release but a normal amount of synaptic vesicles. It has been concluded that syd-2 is involved in the structural assembly of AZs, and that it probably functions as an intracellular anchor of LAR transmembrane tyrosine phosphatase signaling in synaptic junctions. The structural involvement of Liprins was later confirmed by analysis of Dliprin-deficient flies, which showed a deficit in synaptic morphogenesis. Importantly, Liprins were the first AZ proteins to provide a molecular link between the presynaptic plasma membrane and the electron dense material at the AZ through their interaction with LAR transmembrane tyrosine phosphatases. In addition, Liprin-αs have been shown to bind to GIT (which in turn binds to Piccolo) and the motor protein KIF1a via coiled-coil domains, to CASK and Liprin-β via the SAM domain, and to the scaffolding protein GRIP with its C-terminal PDZ binding motif, supporting the function as a scaffolding protein at AZs.

ELKS proteins were originally described as proteins that bind to the IKK complex and take part in NF-κB signaling. Later, they were identified in mammals as Rab6-interacting proteins (Rab6IP2) that are involved in Golgi trafficking, and they were also described as AZ proteins called CAST. They were abbreviated as ERCs (ELKS/Rab6IP2s/CAST). A genetic analysis revealed that mammals have two genes, ELKS1 and ELKS2. The ELKS1 gene produces two main isoforms – a ubiquitously expressed ELKS1A and the AZ protein ELKS1B, which is alternatively spliced at the C-terminus and binds to the RIM PDZ domain. ELKS2 also has the PDZ binding motif at its C-terminus (**Figure 3**) and binds to RIM. ELKS is conserved in *C. elegans*, and Liprin-α's function in synapse assembly requires ELKS in the nematode NMJ. In *Drosophila*, there is only a distant homolog to ELKS, which has a partially conserved N-terminus but a nonhomologous C-terminus that resembles cytoskeletal proteins and lacks the RIM-interacting C-terminal motif. Investigators called this protein *bruchpilot* (*brp*; 'crash pilot') due to the unstable flight that these flies demonstrate when *brp* is absent. *brp* null mutant analysis suggested that T-bars, the dense projections at the neuromuscular synapse of *Drosophila*, are absent, calcium channel density is reduced, and synaptic release is deficient. Whether ELKS is equally important at mammalian synapses has to be investigated.

Piccolo (also called Aczonin) and Bassoon are the largest AZ proteins (530 and 420 kDa, respectively), and they are exceptional because they are not conserved in *C. elegans* or flies (**Table 1**). They contain multiple domains, and 10 regions of homology between Piccolo and Bassoon have been summarized as Piccolo–Bassoon homology domains. The large multidomain structure that includes many protein interaction domains make it likely that Piccolo and Bassoon act as AZ scaffolding proteins. They have the potential to act as calcium sensors with their unusual C2 domains, and they are potentially targeted to the presynaptic plasma membrane via N-terminal myristoylation. Interestingly, in a mouse mutant model that expresses a truncated Bassoon which lacks the central region anchoring Bassoon to the AZ, the dense projections at ribbon synapses are freely floating in presynaptic terminals and synaptic transmission is impaired in these synapses. The delivery of these large proteins to AZs depends on an intact Golgi apparatus, in line with the hypothesis that presynaptic AZs might be preassembled as a complex in the Golgi apparatus and then transported in so-called Piccolo–Bassoon transport vesicles to the presynaptic terminal during synaptogenesis.

Conclusion

Presynaptic AZs consist of a highly specialized network of proteins that acts in neurotransmitter release. Although there is wide variability in the morphology of AZs in different synapses and among various species, their function and molecular architecture are largely conserved. We have only begun to understand how the various proteins present in AZs work together, and crucial issues such as the tight membrane association, the link to the cytoskeleton, the docking mechanism of synaptic vesicles, and the clustering of calcium channels at these hot spots of neurotransmitter release are only partially understood. A fascinating question that has not been addressed regards the use-dependent structural plasticity of AZs and how it is involved in processes of memory formation and learning.

See also: Exocytosis: Ca^{2+}-Sensitivity; Liprins, ELKS, and RIM-BP Proteins; Munc13 and Associated Molecules; Piccolo and Bassoon; Presynaptic Development and Active Zones; Rab3A Interacting Molecules (RIMs); Synaptic Plasticity: Short-Term Mechanisms; Synaptic Vesicles.

Further Reading

Akert K, Moor H, and Pfenninger K (1971) Synaptic fine structure. *Advances in Cytopharmacology* 1: 273–290.

Augustin I, Rosenmund C, Sudhof TC, and Brose N (1999) Munc13-1 is essential for fusion competence of glutamatergic synaptic vesicles. *Nature* 400: 457–461.

Couteaux R and Pecot-Dechavassine M (1970) Synaptic vesicles and pouches at the level of 'active zones' of the neuromuscular junction. *Comptes Rendus Hebdomadaires des Seances de l'Academie des Sciences* 271: 2346–2349.

Harlow ML, Ress D, Stoschek A, Marshall RM, and McMahan UJ (2001) The architecture of active zone material at the frog's neuromuscular junction. *Nature* 409: 479–484.

Phillips GR, Huang JK, Wang Y, et al. (2001) The presynaptic particle web: Ultrastructure, composition, dissolution, and reconstitution. *Neuron* 32: 63–77.

Schoch S and Gundelfinger ED (2006) Molecular organization of the presynaptic active zone. *Cell and Tissue Research* 326: 379–391.

Sudhof TC (2004) The synaptic vesicle cycle. *Annual Review of Neuroscience* 27: 509–547.

tom Dieck S, Sanmarti-Vila L, Langnaese K, et al. (1998) Bassoon, a novel zinc-finger CAG/glutamine-repeat protein selectively localized at the active zone of presynaptic nerve terminals. *Journal of Cell Biology* 142: 499–509.

Wang Y, Liu X, Biederer T, and Sudhof TC (2002) A family of RIM-binding proteins regulated by alternative splicing: Implications for the genesis of synaptic active zones. *Proceedings of the National Academy of Sciences of the United States of America* 99: 14464–14469.

Zhai RG and Bellen HJ (2004) The architecture of the active zone in the presynaptic nerve terminal. *Physiology* 19: 262–270.

Zhen M and Jin Y (1999) The liprin protein SYD-2 regulates the differentiation of presynaptic termini in *C elegans*. *Nature* 401: 371–375.

Activity in Visual Development

M B Feller, University of California at Berkeley, Berkeley, CA, USA
E S Ruthazer, McGill University – Montreal Neurological Institute, Montréal, QC, Canada

Introduction

The development of neuronal circuits involves initial coarse wiring under the guidance of molecular cues and the refinement of connections through mechanisms that are governed by patterned spontaneous activity and sensory experience. This fine-tuning of neuronal circuits is evident in the development of sensory maps in the brain. Sensory maps are organized layouts of neurons in which cells that prefer specific stimulus features are found in close physical proximity. The term map can apply either to the central representation of the sensory periphery, as in the case of retinotopic or other topographic maps, or to the orderly representation of higher order stimulus features such as orientation or interaural time difference.

Neural activity can be either permissive or instructive for circuit formation. When serving a permissive role, the presence of activity acts as a switch to regulate other downstream signaling events. In this case, neural activity essentially converts neurons from one state to another. An example of this is the ability to delay the onset of the critical period for plasticity by dark-rearing animals. Even brief visual experience can activate critical period plasticity in such animals. In contrast, an instructive role for activity is where the specific levels or patterns of neuronal firing carry information that allows different neurons to be distinguished from one another exclusively on the basis of these activity patterns. It is often difficult to prove unambiguously that activity plays an instructive role in a system because the simplest experiments involving blocking activity cannot distinguish between instructive and permissive roles. Nonetheless, there is compelling evidence from experiments in which activity patterns but not levels are altered, such as the ocular dominance shift in strabismic animals or the requirement for retinal waves of activity in the refinement of the retinocollicular map, that activity can also play an instructive role.

Activity in the developing visual system is not limited to that driven by visual experience. Throughout the visual system, in the thalamus, colliculus, and cortex, visual experience appears to be important in the maintenance of functional maps, but early development in these structures actually precedes vision. In this latter case, internally generated patterns of spontaneous activity play a key role in circuit refinement.

Sources of Activity in the Developing Visual System

In mammals, retinal ganglion cell (RGC) axons reach their targets before mammals' eyes open, and even before they have functional photoreceptors. In these early stages of development, in the absence of light responses, retinas are spontaneously active. These early activity patterns are called 'retinal waves' because they propagate across the ganglion cell layer, correlating the firing of tens to hundreds of RGCs. The synaptic circuits that mediate retinal waves are transient, with retinal waves disappearing as light responses are first developing. In contrast, in lower vertebrates, such as turtles, frogs, and chicks, there is an extended time during which waves and visual responses overlap. The synaptic mechanisms underlying retinal waves are described elsewhere in this encyclopedia.

An interesting feature of retinal waves is that they coincide with the period of development when visual responses are first detected in the retina. In mice, light-evoked responses have been detected as early as P10, which is 3 or 4 days before eye opening. In ferrets, which are born at approximately the same developmental stage as mice but have an elongated developmental period lasting 4 weeks until eye opening, light-driven responses are detectable in the dorsal lateral geniculate nucleus of the thalamus (dLGN) and visual cortex as much as 14 days before eye opening. Several experiments in mice, rats, and ferrets indicate that both spontaneous and light-evoked activity are detected in visual cortex before eye opening, indicating that both may influence developmental events. Visual deprivation by dark rearing, even when the eyelids are closed, alters the refinement of circuits within the retina and dLGN. Similarly, dark rearing and/or pharmacological manipulations of spontaneous activity have distinct influences on the development of RGCs in turtle retina, which have an extended period of light-evoked activity and retinal waves.

Once the eyes open, vision improves quickly, as determined by several measures. There is an immediate and steady increase in acuity and contrast sensitivity and a more gradual increase in spectral sensitivity. Neurons tuned to several features of the visual scene can be detected at eye opening, including ocular dominance and orientation.

Another source of neural activity that may be critically important for development is the activity patterns that are intrinsic to local circuits. Even in the absence of sensory input, there is a tremendous amount of spontaneous activity, some of which can be highly patterned. For example, spindle waves, which are fast oscillations in the cortical field potential generated by thalamocortical circuits, persist when the eyes are removed.

Retinotopic Maps in Tectum/Superior Colliculus

The two primary targets of RGCs in the brain are the superior colliculus (SC) and the LGN of the thalamus. In these targets, RGCs establish an arrangement of connections in target fields, termed a retinotopic map, that reflects the spatial arrangement of the RGCs in the retina, and eye-specific maps with inputs from the two retinas layering in neighboring but nonoverlapping regions.

The precise retinotopic and eye-specific targeting of RGCs axons observed in the adult emerges from initially diffuse and overlapping projections, prior to visual experience. There is a clear role for both neural activity and molecular factors, such as the ephrins and their corresponding receptors, for the establishment of these maps, although the relative importance of the two throughout the process of axon targeting and refinement is the subject of ongoing research. This article reviews the evidence that activity plays a role in the establishment of retinotopic maps.

The degree of retinotopic mapping can be assayed by different techniques. Most studies have relied on small focal injections of anterograde tracers (e.g., DiI) into the retina to visualize the axonal arbors in the SC/tectum, which is referred to as the termination zone. In addition, retinotopic maps have also been assayed by the spatial distribution of RGCs that are labeled by focal injections of retrograde tracers into the SC. Third, physiological measures in *in vitro* slices containing the optic tract and SC can assay the number of functional retinal inputs onto individual SC neurons. Last, *in vivo* physiological measurements of receptive sizes of SC/tectum neurons reveal the physiological consequences of topographic refinement.

The first preparations used for establishing a role for activity in retinotopic map formation were frogs and fish. These species have two advantages. First, topographic refinement occurs throughout life. The retina is constantly adding new cells at its periphery, whereas the tectum grows from the caudal end. Consequently, the retinal projections must constantly shift in order to maintain a retinotopic map. Second, in these species, RGC axons regenerate after injury, and hence maps can be studied while reforming in this more adult stage. Blockade of activity during either development or regeneration does not affect the course topography of projections but does profoundly affect the development of fine topography – the projections that mediate the fine point-to-point connectivity between RGCs and tectal neurons. These classic experiments led to the generation of a major dogma in developmental neuroscience that molecular cues mediate the development of course maps, whereas activity is important for the establishment of fine topography.

There has been growing evidence for activity also playing a role in the refinement of maps in mammalian systems. One fundamental difference between refinement in mammals and in frogs and fish is the location of axon branches that undergo refinement. In frogs and fish, refinement is mediated by small-scale changes in higher order axon branching emerging from the tip of the RGC axon. In contrast, during the development of retinocollicular maps in mammals (and similar to retinotectal maps in chicks), RGC axons overshoot their targets in the A–P axis. Branching in the appropriate retinotopic location occurs along the RGC axonal shaft, at sites anterior to the growth cone. Then, in what appears to be a distinct process, the overshooting axon and, in some cases, entire axonal branches are eliminated.

Pharmacological blockade in mammals leads to small, although significant, effects on the final level of retinotopy. To demonstrate a role for correlated retinal activity, mice that lack β_2-containing nicotinic acetylcholine receptors (nAChRs), which exhibit a pattern of retinal activity in which RGCs spike in a seemingly random pattern with little correlation between the spike trains of neighboring RGCs, have been examined. β_2-nAChR−/− mice exhibit less retinotopic refinement than wild-type mice. The absence of retinal waves in β_2-nAChR−/− mice is correlated with the irregular refinement of retinotopic maps despite the presence of approximately normal levels of activity in individual RGCs. Similar results were obtained with intraocular injections of nAChR antagonists, indicating that disruption of retinal waves can prevent the retinotopic refinement of retinocollicular projections.

Eye-Specific Maps in the Lateral Geniculate Nucleus

RGC axons project to the dLGN of the thalamus terminating in regions that are organized topographically and are segregated into eye-specific layers (i.e., projections from one eye end in regions spatially distinct from those of the other eye). When RGC projections from the eyes first grow into the dLGN, they are partially intermixed. The eye-specific layers then emerge gradually as the termination fields of the eyes segregate into regions containing either ipsilateral or contralateral retinal projections. This process is known to be activity dependent since intracranial infusion of TTX, a blocker of voltage-activated sodium channels, prevents segregation. Moreover, experiments have revealed that the activity driving this segregation comes from the retina since prolonged desynchronization of spontaneous retinal activity by parmacological disruption of nAChR activation in the eye also prevents layer formation. Blocking spontaneous activity in a single eye also significantly alters the distribution of RGC axons, indicating that competition from the two eyes is critical for the formation of eye-specific layers.

A wide array of transgenic mice and pharmacological manipulations have been used to gain insights into the mechanisms that mediate the formation of eye-specific layers. A few studies have taken advantage of the fact that the cellular basis of retinal waves switches from one mediated by nAChR to one mediated by ionotropic glutamate receptors to transiently block retinal waves. If nAChR-mediated retinal waves are eliminated during the initial period of refinement, either by pharmacological manipulation or by using β_2-nAChr−/− mice, eye-specific layers fail to form. However, even in the absence of the initial establishment of layers, RGC axons segregate into local eye-specific regions, with ipsilaterally regions segregated into small islands within the contralateral region. Thus, axons segregate without forming distinct eye-specific layers, indicating that eye-specific segregation and layer formation are separable processes that may occur through different mechanisms.

Retinal activity is essential not only for the establishment but also for the maintenance of eye-specific layers. In ferrets, intraocular injections of APB block glutamate-mediated waves after layers have been established and it has been shown to result in desegregation. In no-b-wave (*nob*) mice, RGCs fire in very frequent synchronous bursts that desegregate after layers have been established.

Whether retinal waves provide an instructive or permissive signal for driving eye-specific segregation is controversial. β_2-nAChR−/− mice do not form eye-specific layers, but pharmacological and genetic manipulations that significantly disrupt nearest neighbor correlations by increasing the uncorrelated firing of RGCs do not prevent layers from forming. The resolution of this controversy may rely on gaining insights into what aspect of the highly correlated activity is critical or driving refinement. Information required for activity-dependent segregation might be encoded in the slow periodic firing generated in individual neurons by waves. These periodic bursts of action potentials lead to substantial increases of intracellular calcium concentration in the participating neurons. There is growing evidence that periodic changes in intracellular calcium occurring on the order of minutes can profoundly influence a variety of intracellular processes. Thus, the periodicity of circuit activation may be tuned to the periodicity of intracellular signaling required to ensure the normal maturation of neurons in the retina or the segregation of retinal inputs in the dLGN.

Similarly, important information might be encoded in the spatial pattern of the activity. Synchronous activation of cells contains no distinct spatial information regarding the relative positions of cells involved in each event. However, the propagating activity seen in the retina synchronizes the activity of subsets of cells, thereby encoding their relative positions. A single retinal wave synchronizes firing of cells along a wavefront with a particular orientation on the retina, generating an activity pattern that might be used to establish orientation selectivity in visual cortical neurons. Activity patterns averaged over a large number of waves would lose orientation information but would maintain highly correlated firing among neighboring neurons, thus providing information that might be used to establish topographic projections.

The resolution of the question of whether the retinal waves are instructive or permissive for map refinement will rely on better targeted disruptions of spontaneous retinal activity based on a deeper understanding of the cellular mechanisms underlying plasticity.

Ocular Dominance Column Formation and Plasticity

The segregation of retinal inputs from each eye into eye-specific layers in the dLGN sets the stage for further segregation of eye-specific inputs in the thalamocortical projection to primary visual cortex. Transneuronal labeling studies in which radioactive amino acid or other anterograde neuronal traces that can jump synapses such as wheat germ agglutinin–horseradish peroxidase (WGA-HRP) are injected into one eye reveal a high degree of segregation of thalamocortical afferents into ocular dominance columns (ODCs) in layer 4 of the visual cortex of carnivores and certain primates, including humans.

Note, however, that eye-specific segregation is not evident in the visual cortices of all mammalian species: mice, rats, and even highly visual animals such as squirrels lack ODCs. Regardless of whether a species has segregated ODCs in the binocular zone of its visual cortex, binocular responsiveness (along with orientation and direction selectivity) of neurons is an important emergent property of the cortex not found at earlier levels of the visual system in normal adult animals.

This binocularity has proven to be a remarkably useful tool for studying cortical developmental plasticity. The pioneering work of Hubel and Wiesel revealed that deprivation of visual information through one eye by eyelid suture or image defocusing, known as monocular deprivation (MD), results in a dramatic shift of the responsiveness of cortical neurons to favor the nondeprived eye. These changes in the response properties of individual neurons are accompanied in most cases by a corresponding loss of visual acuity through the deprived eye, or amblyopia. The ocular dominance (OD) shift in response to MD is particularly powerful during a limited critical period in development, although evidence suggests that some degree of shift is possible even in adults. During the critical period, MD for as little as 1 day leads initially to a reduction of responsiveness to the deprived eye followed by an enhancement of the response driven by the nondeprived eye.

These changes are stronger and more rapid in extragranular (outside layer 4) layers of visual cortex, suggesting that plasticity of local cortical circuitry may guide the process. Nonetheless, the MD shift does propagate to cells in layer 4 and ultimately back to the dLGN. With less than 1 week of MD in the cat, the axonal arbors of thalamocortical neurons representing the deprived eye shrink, whereas those representing the nondeprived eye expand. Bulk transneuronal labeling from the eyes, as well as physiological assays, also reveals a gross shrinkage of deprived eye ODCs and a corresponding expansion of nondeprived eye columns. This propagation of OD plasticity back to successively earlier stages of the visual processing stream suggests the existence of retrograde messengers. This is further supported by numerous experiments in which blockade of spiking activity or of synaptic transmission through *N*-methyl-D-aspartate (NMDA) receptors in cortical neurons results in a disruption of the OD plasticity of thalamic afferents.

In addition to MD, misalignment of the eyes, or strabismus, during the critical period can also lead to a shift in the distribution of OD responses of cortical neurons, including the selective loss of binocular responses accompanied by a sharpening of ODC borders. This observation, together with the fact that comparable periods of binocular deprivation do not result in significant loss of visual responsiveness, argues that the equilibrium of inputs representing the two eyes is the result of a developmental competitive process.

This raises the question of whether a competitive process such as ocular dominance plasticity could be responsible for the initial segregation of ODCs. This question remains unresolved. Computer simulations, as well as the finding that RGC axons spontaneously segregate in an activity-dependent manner into ODC-like stripes in the optic tectum of fish and amphibia, indicate that the information contained in the spontaneous firing of retinal ganglion cells should be sufficient in principle to drive segregation into ODCs without the need for a molecular scaffold. On the other hand, ODCs are evident soon after thalamocortical innervation prior to the onset of the critical period for MD effects, and they do not appear to be disrupted by monocular enucleation at this early time. The presence of spontaneous activity within the already segregated dLGN at this stage, however, does not permit activity-dependent segregation of thalamocortical inputs to be ruled out at this point.

Molecular Mechanisms of Plasticity

It has become increasingly clear that developmental plasticity in the visual system is not mediated by a single mechanism. For example, in the dLGN of the ferret, segregation of eye-specific layers does not appear to require NMDA receptor (NMDAR) activation, whereas the segregation of inputs from on-center and off-center RGCs into sublaminae in the dLGN is prevented by application of NMDAR antagonists. In the visual cortex, NMDARs appear to play a key role in ocular dominance plasticity because pharmacological blockade or genetic knock-down of cortical NMDARs prevents the shift of OD in response to MD. The fact that current through NMDARs in response to presynaptic glutamate release is blocked by Mg^{2+} ions except when relieved by concurrent depolarization of the postsynaptic neuron allows NMDARs to serve as molecular detectors of correlated pre- and postsynaptic firing. Consistent with the role of NMDARs in the induction of synaptic plasticity, mutant mice deficient in the alpha isoform of Ca^{2+}/calmodulin-dependent protein kinase (CaMKII) or the serine/threonine phosphatase calcineurin, required for NMDAR-mediated long-term potentiation and long-term depression, respectively, lack normal OD plasticity.

Long-lasting neuronal plasticity generally requires protein synthesis. This is also true for OD plasticity because application of cyclohexamide to visual cortex (but not LGN) to inhibit protein synthesis prevents the OD shift. The identities of the gene products required for OD plasticity have not been revealed. However, at least two key regulators of gene transcription, the extracellular signal-related kinase (ERK) and cyclic adenosine monophosphate response element-binding (CREB) transcription factors, which have both been implicated in long-term synaptic plasticity, are required to produce an OD shift. These protein synthesis-dependent pathways may be important for long-lasting structural plasticity, such as axonal arbor and dendritic spine remodeling in visual cortex.

Structural plasticity involves both the assembly of new connections and the dismantling of existing connections. Existing connections may be stabilized by interactions with the extracellular matrix and through cell–cell adhesion and signaling. Consistent with this model, activity of the serine protease tissue plasminogen activator has been shown to facilitate the OD shift during the critical period, leading to dendritic spine remodeling. After the critical period, a large degree of plasticity can be restored under conditions that reduce signaling by outgrowth inhibitory molecules such as chondroitin sulfate proteoglycans or the myelin inhibitor receptor NogoR.

These myriad molecular signaling cascades all nonetheless share a requirement for discriminable differences between the patterned neural activity in the two eyes. The ability of cortical neurons to detect these differences appears to rely critically on the balance between excitation and inhibition in the circuit. Evidence points to the developmental maturation of inhibitory circuitry, GABAergic basket cells in particular, as a key event for initiating the critical period for OD plasticity. The critical period in mice is opened precociously by augmenting the immature endogenous inhibitory circuitry with administration of the GABA-A receptor partial agonist diazepam. The excitatory–inhibitory balance may be regulated in part by the activity-regulated expression of brain-derived neurotrophic factor (BDNF). Such activity-dependent control of the susceptibility to undergo plastic changes, known as 'meta-plasticity,' is a prediction of the influential Bienenstock–Cooper–Munro (BCM) model for neuronal plasticity. The BCM model posits that activity levels determine a sliding threshold of input strengths required for synaptic modification, above which synaptic strengthening occurs and below which synapses are weakened. Alternatively, an important role for inhibition may be to help sharpen the temporal precision of firing of postsynaptic neurons in response to sensory inputs. Spike timing-dependent plasticity, in which synaptic changes depend critically on whether a postsynaptic neuron fires before or after its presynaptic partner, may be facilitated by maintenance of an appropriate excitatory–inhibitory balance.

The increasing availability of useful transgenic mouse models for the study of activity-dependent developmental plasticity ensures that many more candidate genes will be found to participate in this process.

See also: Adult Cortical Plasticity; Spontaneous Patterned Activity in Developing Neural Circuits; Vision: Light and Dark Adaptation; Visual Associative Memory; Visual Deprivation; Visual Development.

Further Reading

Demas J, Sagdullaev BT, Green E, et al. (2006) Failure to maintain eye-specific segregation in nob, a mutant with abnormally patterned retinal activity. *Neuron* 50: 247–259.

Grubb MS and Thompson ID (2004) The influence of early experience on the development of sensory systems. *Current Opinion in Neurobiology* 14: 503–512.

Hensch TK (2004) Critical period regulation. *Annual Review of Neuroscience* 27: 549–579.

Hofer SB, Mrsic-Flogel TD, Bonhoeffer T, and Hübener M (2006) Lifelong learning: Ocular dominance plasticity in mouse visual cortex. *Current Opinion in Neurobiology* 16(4): 451–459.

Huberman AD, Wang GY, Liets LC, Collins OA, Chapman B, and Chalupa LM (2003) Eye-specific retinogeniculate segregation independent of normal neuronal activity. *Science* 300: 994–998.

McLaughlin T, Torborg CL, Feller MB, and O'Leary DD (2003) Retinotopic map refinement requires spontaneous retinal waves during a brief critical period of development. *Neuron* 40: 1147–1160.

Pfeiffenberger C, Yamada J, and Feldheim DA (2006) Ephrin-As and patterned retinal activity act together in the development of topographic maps in the primary visual system. *Journal of Neuroscience* 26(50): 12873–12884.

Ruthazer ES and Cline HT (2004) Insights into activity-dependent map formation from the retinotectal system: A middle-of-the-brain perspective. *Journal of Neurobiology* 59: 134–146.

Torborg CL and Feller MB (2005) Spontaneous patterned retinal activity and the refinement of retinal projections. *Progress in Neurobiology* 76: 213–235.

Wiesel T (1981) *The Postnatal Development of the Visual Cortex and the Influence of Environment,* Nobel lecture.

Activity-Dependent Metabolism in Glia and Neurons

K A Kasischke, University of Rochester Medical Center, Rochester, NY, USA

Introduction

The extraordinary density and cellular diversity of the brain can possibly be recognized as a solution to an optimization problem in which the complex and energy-demanding tasks of this organ must be organized within the spatial constraints imposed by oxygen and substrate delivery through the vascular system and diffusion. The intimate spatial and functional interactions between the two most prominent cell types of the central nervous system – neurons and astrocytes – epitomize the division of metabolic and signaling pathways between fundamentally differing cells and provide both a key and a challenge for our understanding of this organ.

The historical view of energy metabolism in the human brain was centered on the neuron as the foremost energy consumer, with its energy needs exclusively met by oxidative metabolism of glucose. However, during the past few decades conclusive evidence has emerged that glial cells participate directly in neuronal energy metabolism in such a way that important metabolic pathways are strictly compartmentalized between neurons and their adjacent glial partners, particularly astrocytes. For instance, synthesis of glutamine in the central nervous system (CNS) occurs exclusively in astrocytes. Consequently, neurons depend on glutamine released by astrocytes to produce the neurotransmitters glutamate and γ-aminobutyric acid (GABA). In contrast, the enzyme glutaminase and the rate-limiting enzyme in the biosynthesis of the inhibitory neurotransmitter GABA, glutamic acid carboxylase, are present preferentially in neuronal cells.

The complex and highly dynamic metabolic interactions between neurons and astrocytes in the resting and in the activated brain have become a major and contested field in contemporary neuroscience. Our current knowledge of the metabolic compartmentalization and activity-dependent metabolism in glia and neurons of the brain relies primarily on biochemical studies on brain tissue slices and cultured neural cells using radioactively labeled substrates. This knowledge has been further expanded by the noninvasive measurement of metabolic fluxes in biochemical pathways in the intact brain. These remarkable insights have been made possible by the successful application of magnetic resonance spectroscopy (MRS) to study energy metabolism in both experimental animals and humans. However, none of the established methodologies can provide both the spatial and the temporal resolution to directly investigate cellular energy metabolism in the intact, living brain. Thus, fundamental questions relating to the spatial and temporal organization of the metabolic pathways in neurons and glial cells remain unresolved. For example, current measurements of metabolic fluxes in neurons and astrocytes depend critically on assumptions regarding the size and turnover rate of substrate pools on which there is no universal agreement. Furthermore, there is no general consensus on which substrates in addition to blood glucose the activated brain may utilize at rest and upon activation. Lactate may provide an alternative or even preferred oxidative substrate for neurons, and astrocytes readily metabolize additional substrates such as glycogen, acetate, and glutamate.

The Glutamate–Glutamine Cycle as the Axis of Metabolic Neuron–Glia Coupling

The field of metabolic interaction between astrocytes and neurons began to emerge when it was recognized in the 1950s and 1960s that the metabolism of glutamate was apparently divided between two distinct compartments within the brain. These two metabolic compartments were characterized in groundbreaking biochemical studies using radioactively labeled substrates on cortical tissue slices which identified a 'large,' presumably neuronal pool that contains the majority of glutamate and a 'small,' presumably glial pool that contains the majority of glutamine and in addition a small glutamate pool. These findings laid the foundation for the proposal of the glutamate–glutamine cycle by Benjamin and Quastel in 1975, with "flow of glutamate from neurons to glia, its conversion there into glutamine, and the return of glutamine to neurons and its conversion there to glutamate to complete the cycle." The absence of glutamine synthetase in neurons, endothelial cells, and choroid epithelium and its restricted localization in glial cells corroborated the glial origin of the small pool. The subsequent finding that glutamine synthetase is exclusively localized in astrocytes defined a specific metabolic role for these nonneuronal cells in glutamatergic neurotransmission. The unique metabolic position of astrocytes was further substantiated by the discovery that pyruvate carboxylase is another astrocyte-specific enzyme. The exclusive presence of pyruvate carboxylase implies a distinctive role of oxidative metabolism in astrocytes in CO_2 fixation and anaplerotic metabolism, specifically the replenishment of citric acid cycle intermediates as precursors for amino acid neurotransmitters.

The first step of the glutamate–glutamine cycle is the uptake of synaptically released glutamate. Glutamate uptake occurs through plasma membrane-bound glutamate transporters. Although glutamate transporters are ubiquitously expressed in the CNS, their compartmentalized and cell-specific distribution provides another remarkable example of differentiation between the neuronal and glial compartments. The glutamate transporter subtype GLAST (EAAT1) is ubiquitously expressed, GLT1 (EAAT2) is highly specific for astrocytes, and EAAC (EAAT3) is predominately neuronal. It is believed that the functional role of neuronal transporters for glutamate uptake is minor; in contrast, the glial compartment has a high capacity for glutamate uptake. Experimental evidence implies astrocytes as the principal uptake site for synaptically released glutamate, even though their intrinsic transport rate is relatively slow. Glutamate transporters are highly expressed on the glial outer cell membranes, thus providing a high-affinity, high-capacity binding site for extracellular glutamate which effectively serves as a glutamate buffer. Electrophysiological measurements have shown that extracellular glutamate remains elevated for several milliseconds in the extrasynaptic space and activates astrocytic transporters even at considerable distance from the glutamate release site. Consequently, in mice with suppressed or missing transporter activity, failure of astrocytic glutamate uptake causes elevated glutamate levels, hyperexcitability, and excitotoxicity, as well as neurodegeneration.

Glutamate uptake directly links neuronal activity and astrocytic energy metabolism because cell culture studies have demonstrated immediate activation of both glycolytic and oxidative energy metabolism in astrocytes upon exposure to glutamate. Glutamate uptake is an energy-requiring process because it depends on the Na^+/K^+-ATPase to clear the three Na^+ ions from the intracellular space that are cotransported with each glutamate molecule. Intra-astrocytic glutamate can then be converted to glutamine by glutamine synthetase or be oxidized by the Krebs cycle, in which it is converted to the intermediary metabolites malate or oxaloacetate to yield diverse products such as lactate, aspartate, glutamine, and glutamate. Glutamate oxidation in astrocytes is a conceptually important pathway in cellular energy metabolism because it provides a significant source of oxidatively generated ATP that does not utilize glucose.

Although the concept of the glutamate–glutamine cycle was formulated early, the actual fraction of the total CNS glutamate involved in this pathway was not known. It was initially assumed that two functionally distinct glutamate pools existed – one small transmitter pool and one large metabolic pool. However, early MRS measurements in humans predicted a high rate of glutamate and glutamine labeling from ^{13}C-glucose, arguing against the existence of substantial glutamate and glutamine pools which do not participate in the cycle. It is current knowledge that the glutamate pool is uniform and that the glutamine–glutamate cycle is highly active in both rodents and humans.

Nuclear Magnetic Resonance Spectroscopy of the Glutamate–Glutamine Cycle and Resulting Quantitative Models of Neuronal–Glial Energetics

Our current understanding of the glutamate–glutamine cycle provides the fundament for the quantitative measurement of neuronal and astroglial energy metabolism and of glutamatergic and GABAergic neurotransmission by MRS studies. MRS is often referred to as nuclear magnetic resonance (NMR) spectroscopy. MRS allows the measurement of concentrations and synthesis rates of certain labeled molecules, such as glucose, glutamate, aspartate, GABA, ammonia, acetate, or lactate, in defined regions in the intact, living brain with high chemical specificity. These molecules need to be labeled with an MRS observable nucleus, typically ^{1}H, ^{13}C, or ^{15}N, and are applied intravenously to the experimental subject. The appearance and disappearance of these labels over time are then observed by MRS measurements. The obtained spectra are quantitative and even sensitive to the position of the labeled nucleus within the molecule, which provides an exceptional window to measure the activity of intracellular pathways *in vivo*.

Original MRS studies in anesthetized rodents reported an approximately 250% increase in the Krebs cycle flux during sustained sensory activation of the rat brain. The important implication is that oxidative glucose metabolism provides the required energy not only under resting but also under activated conditions. Further studies using functional imaging modalities such as positron emission tomography (PET) and magnetic resonance imaging (MRI) have consistently reported increases in regional cerebral oxygen metabolism during activation, albeit with much smaller amplitudes. MRS studies in anesthetized rats have shown that cerebral glucose metabolism in the Krebs cycle and glutamatergic activity are linearly coupled with a slope close to 1 over a wide range of neural activities from deep anesthesia to shallow anesthesia closely approaching the awake, resting state. The quantitative coupling between glucose metabolism and glutamatergic neurotransmission with a 1:1 stoichiometry has also been reported in humans. The far-reaching conclusions from these MRS studies are that the majority (up to 80%) of oxidative energy production in the CNS is devoted to supporting glutamatergic neurotransmission and that oxidative glucose metabolism provides a quantitative measure of synaptic glutamate release. The localization of activity-dependent glucose uptake in the CNS to the projection zones of activated functional pathways in the neuropil provides independent support for this model.

In extension of these studies, a strictly quantitative model of neuronal–glial coupling that integrates the astrocyte–neuron lactate shuttle hypothesis was proposed. In this model, two ATPs are consumed to fuel the glutamate–glutamine cycle. The first ATP is utilized by the Na^+/K^+-ATPase for the restoration of the membrane potential after co-uptake of Na^+ with glutamate, and the second ATP is used for the synthesis of glutamine. As a consequence of the predicted 1:1 stoichiometry, one molecule of glucose would be oxidized for each molecule of synaptically secreted glutamate. The two ATPs needed to fuel the glutamate–glutamine cycle could be readily supplied by glycolytic production of two ATPs from the same glucose. Lactate as the product of astrocytic glycolysis would then be secreted in the extracellular space to be taken up and completely oxidized by neurons. An important implication of this model is the possibility of precisely quantifying glutamatergic neurotransmission as the very essence of functional brain activity in functional neuroimaging studies. Regarding the 1:1 stoichiometry coupling between glutamatergic neurotransmission and neuronal–glial glucose oxidation, a major revision has been put forward. It is still upheld that glial lactate is the major substrate for neuronal oxidation but with the fraction approximately 30% less than initially predicted. Glial cells produce at least 8% and neurons produce at least 88% of total oxidative ATP. Approximately 26% of the total glucose is taken up by neurons, leaving the majority of glucose for the glial compartment.

A caveat for the quantitative interpretation of glutamate–glutamine cycling as a measure of astrocytic energetics is that the previously mentioned oxidation of glutamate as an additional significant cellular energy source needs to be considered. As discussed later, astrocytes are likely to be highly oxidative cells, and their energetics may be underestimated if quantified by glutamate–glutamine cycling alone.

Nonlinear Coupling between Glucose and Oxygen Utilization in the Activated Brain

Further complicating the quantitative interpretation of energy metabolism in neurons and glial cells, the coupling between glucose utilization and oxygen consumption during activation is not linear. The apparent deviation from the long-held dogma that the brain is an entirely oxidative organ has been the subject of long-lasting discussions. The controversy was initiated by a landmark PET study that reported increases in glucose utilization in significant excess of oxygen consumption during focal neural activity. The proposal that the energy demands of the activated brain are fueled by glycolysis instead of oxidative metabolism marked a paradigm shift in the field of brain energy metabolism. Numerous studies reported increased lactate levels in the activated brain using differing methodologies, further substantiating this view. The fact that lactate accumulation occurs during a period of augmented oxygen delivery to the brain argued against the interpretation that increased lactate in the activated brain is the consequence of hypoxia. Moreover, the elevated oxyhemoglobin levels and decreased deoxyhemoglobin levels as a consequence of increased blood flow in the activated areas provide the basis for functional magnetic resonance imaging (fMRI) using the blood oxygen level-dependent signal.

Although it is now accepted that the ratio of the regional cerebral oxygen to glucose metabolism (oxygen–glucose index) decreases upon focal activation of the brain, a coherent and validated model of the underlying changes in neuronal and glial energy metabolism is still lacking. For example, the decrease in the oxygen–glucose index during activation is not consistent with the concept of linear coupling between glutamatergic neurotransmission and is at odds with the previously mentioned measurements of significant and sustained elevations of oxidative metabolism following sensory stimulation. A theoretical model to reconcile these contractions predicts that a fraction of the glucose taken up by astrocytes is cycled through the glial glycogen pool to supply rapid glial energy for clearance of glutamate while the remaining glucose is processed into lactate to support oxidative metabolism in neurons. In this scenario, both a transient decrease in the oxygen–glucose index and the increased production of lactate are accounted for, whereas the linear coupling between glutamate–glutamine cycling and oxidative metabolism in neurons remains preserved. An alternative explanation for the transient uncoupling of oxygen to glucose metabolism is that increased glial flux through pyruvate carboxylase removes carbons from oxidative glucose metabolism to supply anaplerotic metabolism in the glial Krebs cycle.

The Astrocyte–Neuron Lactate Shuttle Hypothesis

The astrocyte–neuron lactate shuttle hypothesis by Pellerin and Magistretti provides a possible mechanism for the quantitative coupling of glutamatergic metabolism to glucose utilization in the CNS and also an explanation for the apparent uncoupling of glucose and oxidative utilization in the activated brain. In its original form, this model predicted that uptake of synaptically released glutamate triggers glycolytic production of lactate in astrocytes. The lactate is released and immediately taken up by surrounding neurons to fuel activity-dependent oxidative metabolism. The lactate shuttle model was primarily based on the experimental observation that glutamate uptake activates glycolytic metabolism in astrocytes with significant accumulation of lactate in the culture medium. This hypothesis is indirectly supported by the cell-specific distribution of lactate dehydrogenase isoforms, which were interpreted as facilitating the proposed directed transport of lactate from astrocytes to neurons. In its extreme form (discussed previously), the model proposed a strict and linear coupling between synaptic glutamate release and astrocytic glycolysis with export of lactate for complete oxidation in neurons.

However, the astrocyte–neuron lactate shuttle hypothesis did not specifically address the role of glycolysis in neurons and of oxidative metabolism in astrocytes and was at odds with PET measurements of cerebral oxygen consumption and theoretical considerations which suggested that neurons increase their oxidative metabolism in parallel with an increase in pyruvate, implying that glycolysis in neurons, not in astrocytes, determines the kinetics of the metabolic response to activation. Furthermore, a stringent explanation of why glutamate uptake in astrocytes should be strictly coupled and exclusively fueled by astrocytic glycolysis was lacking and, most important, the possibility that glutamate oxidation may easily provide the required energy was not excluded. Numerous NMR spectroscopy studies have provided substantial evidence that a major portion of the energy for glutamine synthesis in astrocytes is indeed derived from oxidative metabolism and not glycolysis. Finally, the metabolic consequences of glutamate uptake in astrocytes are not restricted to activation of glycolysis since numerous studies have established that glutamate is capable of increasing the oxidative rate in astrocytes as well.

Based on these and other findings, the astrocyte–neuron lactate model has been the subject of considerable debate and criticism. In an amendment of their model, Pellerin and Magistretti loosened the temporal coupling between initial glycolysis in astrocytes and successive oxidative metabolism in neurons and proposed that both pathways could occur in arbitrary sequence in activated brain areas. Observations of activity-dependent fluctuations of the fluorescent coenzyme nicotinamide adenine dinucleotide (NADH) are indeed consistent with spatiotemporal partitioning of oxidative and glycolytic metabolism between astrocytes and neurons, albeit in a reversed temporal sequence with early oxidative metabolism in neurons followed by subsequent activation of glycolysis in astrocytes (**Figure 1**).

Further experimental support for the astrocyte–neuron lactate shuttle hypothesis derives from NMR spectroscopy studies which demonstrated that activity-dependent glycolytic lactate production occurs primarily in a nonneuronal compartment, strongly implicating astrocytes as the primary lactate source. Independent observations that glutamate increases membrane transport of glucose in cultured astrocytes and mediates inhibition of glucose transport in cultured hippocampal neurons further sustain this model. In the latest amendment of the lactate shuttle model, it has been proposed that astrocytic lactate is used as an oxidative substrate for neurons even under resting conditions.

In summary, there is little doubt that astrocytes can release lactate and that neurons can utilize external lactate as an oxidative substrate. However, conclusive proof for the astrocyte–neuron lactate shuttle hypothesis requires the direct experimental observation of a directed transfer of lactate from astrocytes to neighboring neurons.

Subcellular Differentiation of Energy Metabolism within Neurons and Astrocytes

The compartmentation and differentiation of energy metabolism extends even within neuronal and glial cells. Potentially highly relevant results for the understanding of activity-dependent metabolism in glial cells and neurons can be found in older histological literature on enzyme localization. For example, most of the glycolytic activity in neurons is localized within the perikarya and proximal dendrites. In contrast, the distal segments of neurons typically contain numerous mitochondria, with the highest densities in the smaller dendritic branches and axon terminals. The subcellular distribution of cytochrome oxidase activity follows an analogous pattern. An electron microscopy study found consistent and substantial staining for hexokinase in astroglia and neurons, with the remarkable

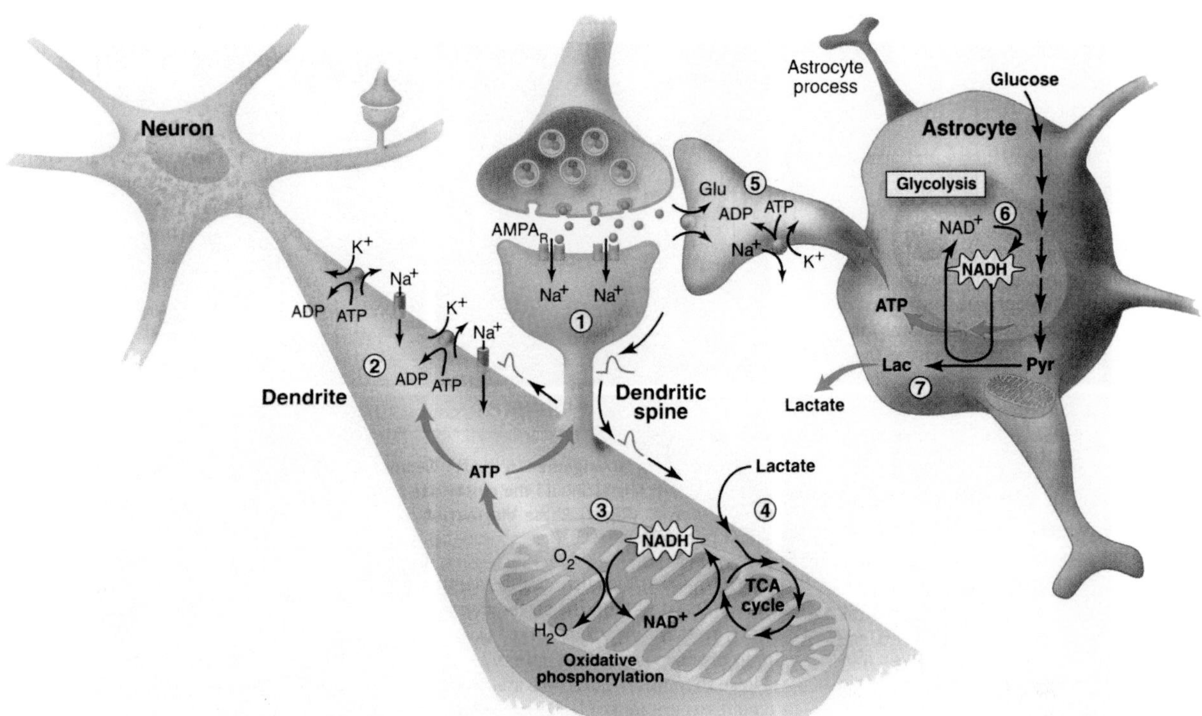

Figure 1 Brain energetics in the limelight. Separate activation of oxidative phosphorylation (respiration) in neurons (brown) and glycolysis in astrocytes (gray), as revealed by two-photon fluorescence imaging of NADH (4). (1) Stimulation of excitatory (glutamatergic) neurons activates postsynaptic AMPA receptors and induces an excitatory postsynaptic potential (EPSP) in the dendritic spine of the neuron. (2) The depolarization propagates from the dendritic spine to the dendrite, where it may cause further opening of voltage-gated sodium channels and activation of the Na^+/K^+-ATPase, leading to an increased demand for energy (ATP). (3) In response, oxidative phosphorylation is rapidly activated, causing a decrease in mitochondrial NADH content (the so-called 'dip' in the fluorescent signal). (4) Recovery of mitochondrial NADH in dendrites is accomplished by stimulation of the TCA cycle, fueled largely by lactate from the extracellular pool. (5) In parallel, but delayed in time, glutamate reuptake in astrocytes (gray) activates the glial Na^+/K^+-ATPase. (6) The increased energy demand leads to a strong enhancement of glycolysis in the cytoplasm of astrocytes, as indicated by the large increase in cytosolic NADH fluorescence (the so-called 'overshoot'). (7) To maintain the high glycolytic flux, NAD^+ must be regenerated via the conversion of pyruvate to lactate through the activity of the enzyme lactate dehydrogenase. Release of lactate into the extracellular space not only replenishes the extracellular pool but also may sustain the late phase of neuronal activation. *In vivo*, glucose is delivered from the blood to both the extracellular space and astrocytes (via astrocytic protrusions called end-feet that are in close contact with the blood vessel wall). AMPA$_R$, α-amino-3-hydroxy-5-methyl-4-isoxazole propionic acid receptors; GLU, glutamate; LAC, lactate; PYR, pyruvate; TCA, tricarboxylic acid. Reprinted from Pellerin L and Sutliff K, in Pellerin L and Magistretti PJ (2004) Let there be (NADH) light. *Science* 305: 50–52, with permission from AAAS.

exception that no hexokinase activity was found in the terminal segments of both Purkinje and granule cell dendrites. The corollary is that these dendritic terminals apparently do not depend on local metabolism of glucose to satisfy their energy needs. Another interesting finding is the localized presence of glycolytic enzymes in the postsynaptic density of the dendritic spines, with the implication that postsynaptic energy metabolism in the vicinity of the postsynaptic density may be fueled by glycolytic ATP production.

An important question concerns the number, density, and activity of mitochondria in astrocytes and neurons (**Figure 2**). A widely held misconception is that astrocytes contain substantially fewer mitochondria than neurons. However, even in the nineteenth century the presence of 'gliosomes' – large, dense granules in astrocyte processes – was noted by light microscopists. Later, these gliosomes were identified as mitochondria. An electron microscopy study reported that neurons and glia have approximately the same number of mitochondria, with the crystal volume fraction or crystal packing density being considerably higher in neurons (17.3%) than in astrocytes (11%). Numerous studies have found that the density of mitochondria is approximately equal within astrocytes and neurons, with the important implication that astrocytes are highly oxidative cells. These reports are in good agreement with MRS studies that report high glial Krebs cycle fluxes which account for approximately 30% of the total Krebs cycle activity, a percentage similar to the volume fraction of astrocytes in the CNS.

A striking subcellular differentiation between glial cells and neurons is the exclusive presence of glycogen

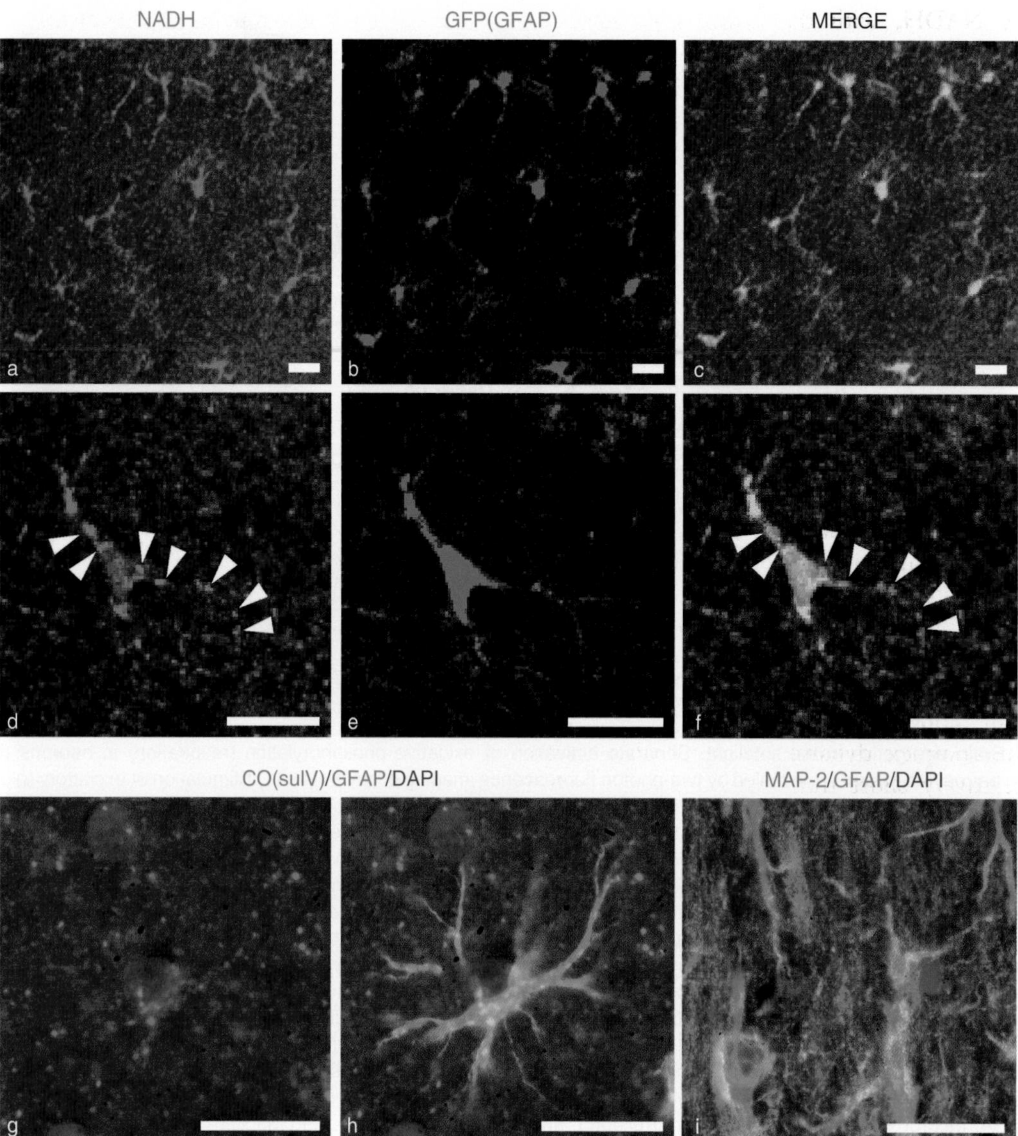

Figure 2 Astrocytes exhibit a characteristic NADH fluorescence pattern. (a–c) Astrocytes can be identified by means of their characteristic NADH fluorescence pattern as demonstrated by the 1:1 correlation of the disseminated, intrinsically bright fluorescent cells in the neuropil of CA1 (stratum radiatum) with green fluorescent protein (GFP)-expressing astrocytes. (d–f) A GFP-expressing astrocyte shown at high magnification with confluent cytoplasmic NADH fluorescence and granular mitochondrial NADH fluorescence (arrowheads) along its processes. (g) Distribution of mitochondria (cytochrome oxidase subunit IV immunolabeling) in the neuropil of CA1 (stratum radiatum). (h) Additional glial fibrillary acidic protein (GFAP) labeling of the same section shows that the concentration of mitochondria in astrocytes is not elevated in comparison with the surrounding neuronal neuropil. (i) Astrocytes (GFAP immunolabeling) and dendrites (MAP-2 immunolabeling) are the two dominant compartments in the neuropil of the stratum radiatum of CA1. Nuclei were counterstained with DAPI in (g–i). Scale bars = 20 μm. Reprinted from Kasischke KA, Vishwasrao HD, Fisher PJ, Zipfel WR, and Webb WW (2004) Neural activity triggers neuronal oxidative metabolism followed by astrocytic glycolysis. *Science* 305: 99–103, with permission from AAAS.

in astrocytes. The greatest accumulation of astrocytic glycogen is found in areas of high synaptic activity and in close proximity to neuronal perikarya. Consequently, it was proposed that astrocytic glycogen levels may be influenced by changes in neuronal activity and that astrocytic glycogen may even be considered as a store for lactate rather than for glucose. The unique morphology of astrocytes further suggests a special necessity and role for glycolysis and glycogenolysis, simply because the fine astrocyte processes and especially their filopodial and lamellipodial extensions are too narrow for mitochondria. The contribution of mitochondrial ATP provided by diffusion into this exceptionally narrow compartment is uncertain. A special role for glycolysis in astrocytes is further supported by the distribution of the reduced coenzyme NADH, with the cytoplasmic levels of NADH in astrocytes apparently substantially higher than in the surrounding neuronal neuropil, indicating the possibility of higher glycolytic capacities in astrocytes than in neurons (**Figure 2**).

Activity-Dependent Metabolic Transitions in Neurons and Astrocytes

Principal questions related to activity-dependent metabolism in neurons and glia remain unresolved. Prominent examples are the role of cerebral glycogen metabolism, conclusive evidence for the cellular localization of lactate production and utilization or efflux, and the exact extent and function of oxidative metabolism in the glial compartment. A fundamental limitation has been the lack of spatial resolution of current functional neuroimaging and spectroscopy techniques to directly resolve the intertwined neuronal and glial compartments in the intact living brain.

Another important consideration is that many reported measurements have been obtained under steady-state conditions, and not much is known about the complex dynamic changes which occur in the relatively short transitional period between the resting state and the fully activated state. The methodological challenges are significant because metabolic transitions typically occur on a millisecond to second timescale and are difficult to capture with the relatively low temporal resolution offered by MRS, PET, and fMRI. Further complicating their interpretation, many reported metabolic transitions are biphasic and actually the sum of overlapping changes in utilization and delivery of a substrate at spatially close, but functionally and anatomically distinct, locations. Examples of such biphasic transitions include intrinsic imaging and spectroscopy of blood oxygenation, the measurement of interstitial tissue oxygen, interstitial glucose fluctuations, lactate dynamics, and NADH fluorescence fluctuations in electrically activated hippocampal tissue slices.

The consideration of metabolic transitions implies that coupling between neuronal and glial metabolism in the activated brain is more complex than that under steady-state conditions. A redox switch hypothesis has been proposed which may explain how the activity state of the brain directly determines the flux in cellular energy pathways. Here, the cytoplasmic redox state of the NADH/NAD^+–lactate/pyruvate couple in neurons and astrocytes acts as a metabolic switch, favoring either glucose or lactate as the preferred oxidative substrate for neurons.

Answers to difficult experimental questions can be expected from the successful application of novel methodologies. Perhaps most prominently among them, multiphoton microscopy has emerged as a popular and powerful tool providing many opportunities for the direct cellular resolution of neurometabolic and neurovascular coupling. For example, two-photon imaging and spectroscopy of NADH fluctuations in hippocampal tissue slices has provided evidence for enhanced glycolytic capacities in astrocytes and enabled the resolution of activity-dependent oxidative and presumably glycolytic responses in astrocytic and neuronal processes. Based on the observed NADH transitions, a model for neurometabolic coupling has been proposed in which early oxidative metabolism in neurons is eventually sustained by late activation of glycolysis in astrocytes. Similar observations were obtained using two-photon calcium imaging in the mouse barrel cortex *in vivo*, with delayed astrocytic calcium responses following prolonged whisker stimulation. In both exemplary cases, the observed astrocytic calcium and NADH responses prevailed after pharmacological blockade of postsynaptic activity, strongly implying a direct activation of astrocytic signaling and metabolism by synaptically released glutamate.

It will be exciting to determine whether microscopic functional imaging of the astrocytic and neuronal compartments in intact animals will provide a useful complement to established macroscopic imaging and spectroscopy technologies, and whether microscopic and macroscopic imaging approaches in combination with molecular and physiological techniques will ultimately yield a unifying model for the exceedingly complex metabolic interactions between neurons and astrocytes.

See also: Glial Glutamate Transporters: Electrophysiology; Glial Glutamate Transporters; Glial Glycogen Metabolism; Glial Glutamate and GABA Metabolism; Glial Energy Metabolism: A NMR Spectroscopy Perspective; Glutamate Receptor Organization: Ultrastructural Insights.

Further Reading

Balazs R, Machiyama Y, Hammond VJ, Julian T, and Richter D (1970) The operation of the gamma-aminobutyrate pathway of the tricarboxylic acid cycle in brain tissue *in vitro*. *Biochemical Journal* 116: 445–461.

Garcia-Espinosa MA, Rodrigues TB, Sierra A, et al. (2004) Cerebral glucose metabolism and the glutamine cycle as detected by *in vivo* and *in vitro* ^{13}C NMR spectroscopy. *Neurochemistry International* 45: 297–303.

Gjedde A, Marrett S, and Vafee M (2002) Oxidative and nonoxidative metabolism of excited neurons and astrocytes. *Journal of Cerebral Blood Flow and Metabolism* 22: 1–14.

Gruetter R (2004) *In vivo* ^{13}C-NMR studies of compartmentalized cerebral carbohydrate metabolism. *Neurochemistry International* 41: 143–154.

Hertz L (2003) Intercellular metabolic compartmentation in the brain: Past, present, and future. *Neurochemistry International* 45: 285–296.

Hertz L, Peng L, and Dienel GA (2007) Energy metabolism in astrocytes: High rate of oxidative metabolism and spatiotemporal dependence on glycolysis/glycogenolysis. *Journal of Cerebral Blood Flow and Metabolism* 27: 219–249.

Hyder F, Patel AB, Gjedde A, Rothman DL, Behar KL, and Shulman RG (2006) Neuronal–glial glucose oxidation and glutamatergic–GABAergic function. *Journal of Cerebral Blood Flow and Metabolism* 26: 865–877.

Kasischke KA, Vishwasrao HD, Fisher PJ, Zipfel WR, and Webb WW (2004) Neural activity triggers neuronal oxidative metabolism followed by astrocytic glycolysis. *Science* 305: 99–103.

Magistretti PJ, Pellerin L, Rothman DL, and Shulman RG (1999) Energy on demand. *Science* 283: 496–497.

Norenberg MD and Martinez-Hernandez A (1979) Fine structural localization of glutamine synthetase in astrocytes of rat brain. *Brain Research* 161: 303–310.

Patel HJ and Balazs R (1970) Manifestation of metabolic compartmentation during the maturation of the rat brain. *Journal of Neurochemistry* 17: 955–971.

Pellerin L and Magistretti PJ (1994) Glutamate uptake into astrocytes stimulates aerobic glycolysis: A mechanism coupling neuronal activity to glucose utilization. *Proceedings of the National Academy of Sciences of the United States of America* 91: 10625–10629.

Pellerin L and Magistretti PJ (2004) Let there be (NADH) light. *Science* 305: 50–52.

Rothman DL, Sibson NR, Hyder F, Shen J, Behar K, and Shulman RG (1999) *In vivo* nuclear magnetic resonance spectroscopy studies of the relationship between the glutamate–glutamine cycle and functional neuroenergetics. *Philosophical Transactions of the Royal Society of London, Series B* 354: 1165–1177.

Sibson NR, Dhankhar A, Mason GF, Rothman DL, Behar KL, and Shulman RG (1998) Stoichiometric coupling of brain glucose metabolism and glutamatergic neuronal activity. *Proceedings of the National Academy of Sciences of the United States of America* 95: 316–321.

Sonnewald U, Westergaard N, and Schousboe A (1997) Glutamate transport and metabolism in astrocytes. *Glia* 21: 56–63.

Activity-Dependent Regulation of Glucose Transporters

L F Barros, Centro de Estudios Científicos, Valdivia, Chile

Introduction

The energetic needs of the mammalian brain are provided almost exclusively by glucose, which is metabolized 10 times faster in the brain than in the rest of the body. Other energy sources, such as fatty acids, ketone bodies, and lactate, play minor roles in adulthood but are important during early embryonic development and lactation and under special circumstances (e.g., during strenuous exercise). After entering the intermediary metabolism at hexokinase, glucose is oxidized to CO_2 by the successive action of glycolysis and oxidative phosphorylation. In the brain the process is 90% efficient, with the expense of nearly six O_2 per glucose molecule. The free energy extracted from the sugar is spent mostly by neuronal cation pumps to restore ion gradients dissipated by postsynaptic currents and action potentials, thus a good correlation exists between neuronal signaling and local glucose usage. Like glucose, isotope-labeled tracer glucose analogs are preferentially metabolized in active regions, an observation that has permitted functional mapping of the brain by autoradiography and PET scanning.

Temporal Fluctuations in Local Glucose Demand

Throughout day and night, during mental exertion or rest, the overall metabolic demand of the brain remains fairly constant; however, local metabolic demand varies. For example, a striatal neuron jumping from electric silence to a firing rate of 20 Hz will increase its rate of ATP hydrolysis by a factor of 15. Because neurons do not store glycogen or other high-energy molecules, the new energy demand must be provided by glucose and/or lactate, the only readily available fuels present in the neuronal cytosol. The concentrations of glucose and lactate, about 1 mM, will sustain a 15-fold increase in demand for no longer than tens of seconds, implying that sustained neuronal activity must somehow increase the delivery of glucose/lactate to active regions. To examine how this may be achieved, it is useful to consider the architecture of the neuropil, the tissue where most ATP is spent. In the neuropil, dendrites and axons are surrounded by fine astrocytic processes, each astrocyte housing about 10^4 synapses. Astrocytes are in turn fixed to a capillary by means of their end feet, thin processes interconnected by glucose-permeable gap junctions. This complement of capillary, neighboring astrocytes, and their cargo of neuronal processes and synapses is termed the gliovascular unit (**Figure 1**). Due to the slowness of diffusion over long distances relative to the rapid fluxes involved, neuronal processes are energetically isolated from their soma. Instead, the fuel must be supplied locally, which defines the gliovascular unit as the metabolic unit of the brain. At variance with neurons, which can be fueled by either glucose or lactate, the gliovascular unit's only energy substrate is glucose, for lactate is not normally present at significant concentrations in the blood.

Although the firing rate of many neuronal types is known to be highly variable, it is not equally clear how variable is the metabolic demand of the gliovascular unit. The latter will depend on an unknown factor, which is the functional homogeneity of the neuronal processes in a given unit. For example, if all processes were to increase their metabolic demand by 15-fold, the whole unit will require an increase in glucose delivery of similar magnitude, the metabolic demand of endothelium and astrocyte being relatively smaller. But if a significant fraction of the processes in the unit belongs to other circuits, which do not become activated, or perhaps decrease their electrical activity, the maximum metabolic burden on the unit will be correspondingly lower. A further mechanism of metabolic load distribution is provided by glucose and lactate present in neighboring gliovascular units, their efficacy depending on how permeable is the interphase between units. In summary, sustained neuronal activity must be accompanied by enhanced flux of fuels into neurons and enhanced glucose flux into the tissue. Thanks to the spatial buffering provided by the gliovascular unit, the large increases in fuel delivery demanded by single processes need not be fully translated into increased glucose delivery from blood to brain. Studies utilizing autoradiography and positron emission tomography (PET) scanning, techniques that measure the uptake of glucose analogs, have reported maximum activity-dependent increases in brain metabolism of around twofold. Because these techniques at best average the behavior of thousands of gliovascular units over 20 min or longer, this value should be considered as a lower estimate of the actual degree of stimulation. As already discussed, the highest estimate is provided by the increase in metabolic flux in single neurons (e.g., 15-fold). Somewhere in between these two extremes lies the increase in glucose flux into the gliovascular unit, its precise

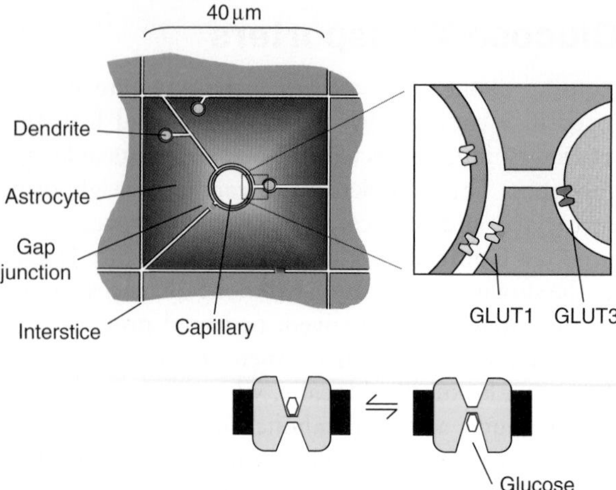

Figure 1 The gliovascular unit is the metabolic unit of the brain neuropil. Glucose diffuses into the gliovascular unit through facilitative transporters, present in endothelium and glia (GLUT1) and in neurons (GLUT3). GLUT transporters alternate their sugar binding site between the extracellular space and the cytosol, a process which is favored by the presence of the sugar.

quantification awaiting the development of new metabolic measurement techniques, capable of resolving at the level of micrometers and seconds.

Rate Limitation and Control of Glucose Flux into Brain Cells

Except for a few layers of cells lining the surface of the organ and the ventricles, the brain parenchyma is fed by blood-borne glucose. Overall glucose extraction by the organ is only 10%, which is a lower estimate of local extraction than occurs in active zones. Once inside brain cells, glucose enters the intermediary metabolic pathway at hexokinase, which is expressed in both astrocytes and neurons. By moving glucose beyond the permeability barriers imposed by the endothelial blood–brain barrier and the plasma membrane of neurons and astrocytes, hexokinase generates the concentration gradient that drives the sugar into the brain. This sole stimulation by hexokinase, by making the gradient steeper, may in principle augment flux. However, measurements of brain glucose concentration with microprobes and nuclear magnetic resonance (NMR) spectroscopy show that the gradient is nearly maximal even under resting conditions, leaving little room for further increase during activation. In an extreme case, in the striatum of the resting rat, glucose drops from 8 mM inside the capillary to 0.35 mM in the interstice, so hexokinase stimulation, however intense, may only enhance the consumption rate of glucose by less than 3%. In other regions of the brain, where glucose levels are about 1 mM, maximum flux stimulation would be around 15%. In addition to the gradient, the other factor that determines flux is permeability. But an isolated increase in permeability is not sufficient to modify flux either, because hexokinase, with its K_m of 50 μM, is nearly saturated. The result of this particular balance between permeability and metabolism is that the increase in glucose consumption demanded by neuronal activity requires commensurate stimulation of hexokinase and glucose permeability. Both glucose transporters (GLUTs) and hexokinase exert control over the rate of glucose consumption by the brain.

Mechanisms of Glucose Permeation into the Brain

Glucose is a hydrophilic molecule, thus its rate of diffusion across lipid bilayers is negligible. Glucose permeation into most mammalian cells is mediated by facilitative transporters of the GLUT family, of which 14 members have been described to date. Most of what is known about the function of GLUTs was learned from kinetic studies of the glucose transporter GLUT1, which is abundantly expressed in readily available erythrocytes. GLUT1 has a single hexose-binding site that becomes exposed alternatively to the extracellular space and the cytosol (**Figure 1**), a process that occurs 200 times per second in the absence of substrate and 10 times faster when glucose is bound to the site. The binding site is never exposed simultaneously to both sides of the plasma membrane, a key kinetic property that separates transporters from ion channels, which have a 'see-through' pore during the open state. GLUT carriers, however, do resemble channels insofar as they are unable to generate concentration gradients. The cloning of the *glut1* gene in 1985 revealed a highly hydrophobic 492-amino-acid protein, predicted to span the plasma membrane 12 times, with both ends and a large central loop oriented toward the cytosol. Thanks to biochemical work and mutagenesis, the sugar binding site has been mapped in detail, but the long-sought mechanism of catalysis remains elusive. Obtaining well-diffracting crystals of GLUT proteins has proved difficult; meanwhile a three-dimensional (3-D) structure of GLUT1 has been proposed based on the crystal structure of other transporters of the same superfamily. The transporter is visualized as having conic vestibules at each side. A large conformational change is thought to inundate the binding site with the water phase in alternating fashion, from one vestibule to the other.

The main GLUT isoform in the brain capillary endothelium is GLUT1, which according to immunogold studies is present at similar densities at both luminal and abluminal membranes, with a third pool in the cytoplasm. Endothelial GLUT1 is a kinetically symmetric carrier, displaying K_m values of zero-trans influx and efflux of about 5 mM. The concentration of glucose in the brain interstitium is 10–20% of that in the blood, indicating that 80–90% of the overall flux resistance is endothelial. In astrocytes, the main glucose transporter is also GLUT1, which is conspicuously abundant in capillary-ensheathing end feet. Astrocytic GLUT1 is not glycosylated, as is the case for endothelial and erythrocytic GLUT1, a structural difference without apparent functional consequences. In neurons the main isoform is GLUT3, similar to GLUT1 in substrate specificity and affinity, but with a higher turnover rate. In microglia, the main isoform expressed is GLUT5, which has a high affinity for fructose. The low-affinity transporter GLUT2, found in liver and pancreatic islet cells, is also present in the hypothalamus and in tanycytes, astroglial cells that line the ventricles, where they may participate in gluco-sensing by the central nervous system. GLUT4, the carrier responsible for insulin-regulatable glucose transport in muscle cells and adipocytes, is also expressed in neurons scattered across the brain cortex. Several other isoforms of the GLUT carrier, including the intraneuronal GLUT8, are expressed at low levels throughout the brain; their physiological roles have yet to be determined. Of note, GLUTs can also transport dehydroascorbate, the oxidized form of vitamin C.

Acute Regulation of Glucose Transport

Neuronal activity causes an acute increase in local energy demand, which requires stimulation of glucose phosphorylation and a concomitant stimulation of glucose permeability. A possible molecular mechanism that may contribute to the increase in permeability was suggested by the observation that GLUT1 in cultured astrocytes is stimulated by the excitatory neurotransmitter glutamate, a phenomenon that develops in seconds, is reversible, and is not present in other cell types expressing GLUT1. The effect of glutamate is mediated by the Na^+–glutamate cotransporter and explained by a modification in transport capacity (V_{max}), but it is not yet known whether it results from translocation of GLUT1 to the plasma membrane or from an increase in the intrinsic activity of surface-resident transporters. Cultured neurons also react to glutamate reversibly and in seconds but, in contrast to that observed in astrocytes, the effect is an inhibition of hexose uptake. The neuronal inhibition is mediated by α-amino-3-hydroxy-5-methyl-4-isoxazole propionic acid (AMPA) receptors, and, again, it remains to be demonstrated whether this inhibition responds to changes in surface transporter number or intrinsic activity. Both the stimulation of astrocytic GLUT1 and the inhibition of neuronal GLUT3 by glutamate are mediated by Na^+, which also commands the activation of astrocytic glycolysis by glutamate and participates in the waves of glycolytic activation that spread between adjacent gliovascular units, making Na^+ a key signal for the control of glucose distribution and usage in the brain. If these phenomena, observed in cultured cells, were to occur *in situ*, it would mean that glutamate redirects glucose toward astrocytes when fuel is badly needed in neurons. A possible solution for this paradox is provided by the astrocyte-to-neuron lactate-shuttle hypothesis (ANLSH), a model proposed by Pellerin and Magistretti that suggests astrocytic lactate, and not glucose, is the preferential substrate for active neurons. The proposed participation of glucose transporters in ANLSH is illustrated in **Figures 2** and **3**. In response to synaptic glutamate, astrocytes rapidly activate GLUT1 and the glycolytic machinery, generating lactate that diffuses to neurons through monocarboxylate transporters (MCTs). The simultaneous inhibition of neuronal GLUT3 by glutamate, mediated by AMPA receptors, slows down neuronal glycolysis, releasing NAD^+ at glyceraldehyde 3-phosphate dehydrogenase (GAPDH). NAD^+ is thus made available for lactate dehydrogenase (LDH), which drives lactate into the neuron. The energy cycle is completed when the ATP generated by neuronal mitochondria from pyruvate is used by the Na^+/K^+-ATPase to restore the ion gradients dissipated during electrical activity.

Given that the larger fraction of the total glucose gradient extends across the endothelium, endothelial GLUT1 is predicted to be acutely upregulated during neuronal activity – otherwise, glucose flux would not increase sufficiently during hexokinase stimulation. The detection of this phenomenon, which is expected to be highly local, is beyond current technology. However, a sizable increase in the V_{max} for glucose transport in whole brain is observed after only 3 min of seizures in the rat. A potential mechanism for the stimulation is suggested by the pool of GLUT1 inside endothelial cells, which may translocate to the cell surface and increase endothelial glucose permeability in response to neuronal activity, much in the way that GLUT4 translocates in muscle cells in response to insulin or contraction.

Slower Regulation of Glucose Transport

In addition to the acute response to glutamate, which is specific to astrocytes, slower stimulations of GLUT1 can be elicited in most cell types. Hormones, kinase

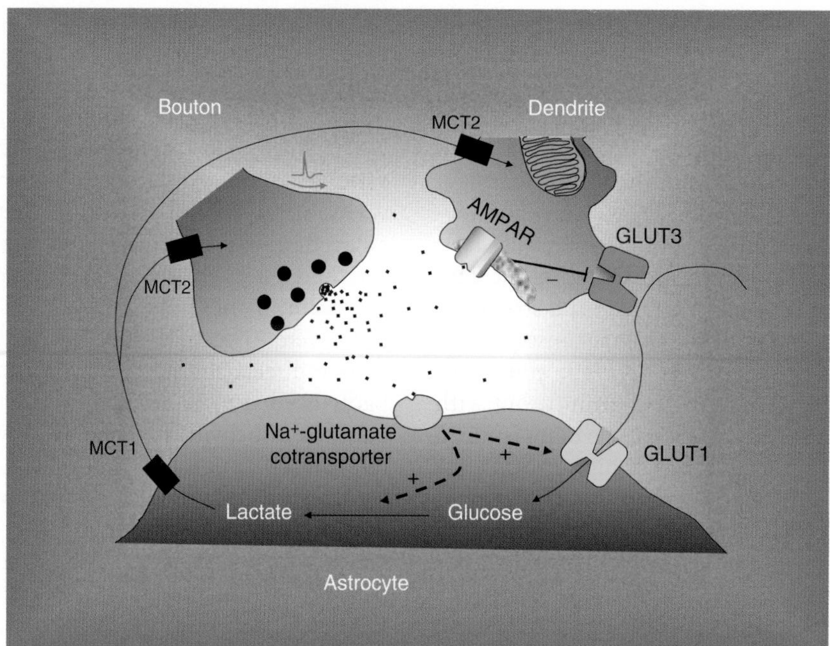

Figure 2 Glutamate released from synaptic boutons activates α-amino-3-hydroxy-5-methyl-4-isoxazole propionic acid receptors (AMPARs) in neurons and Na^+–glutamate cotransporters in astrocytes, leading to stimulation of the astrocytic glucose transporter GLUT1 and inhibition of the neuronal glucose transporter GLUT3. Glutamate also activates astrocytic glycolysis via the Na^+–glutamate cotransporter. The result is a surge in lactate, which diffuses through monocarboxylate transporters (MCTs) into neurons, where oxidation is completed.

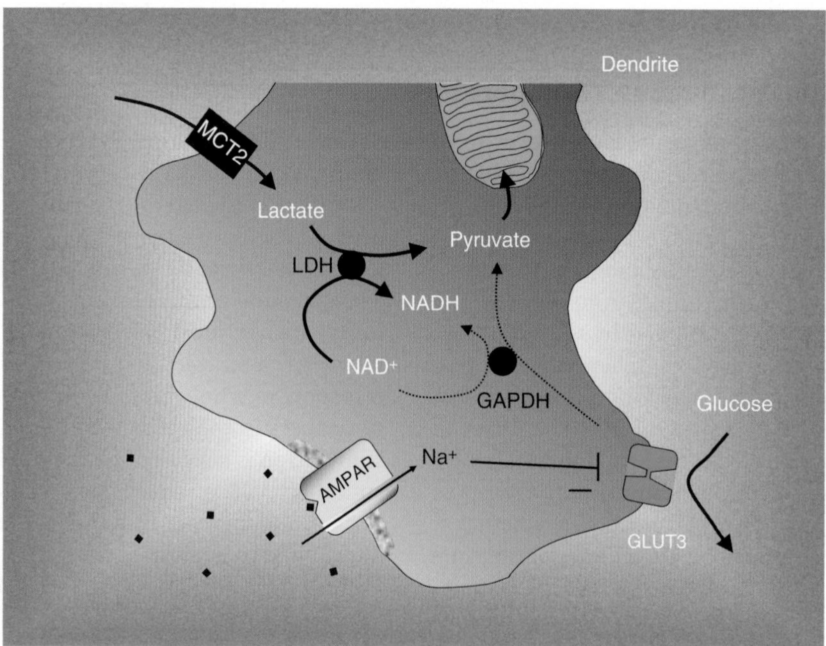

Figure 3 Glutamate opens ionotropic α-amino-3-hydroxy-5-methyl-4-isoxazole propionic acid receptors (AMPARs), allowing the influx of Na^+ into postsynaptic spines and dendrites. Na^+, by an unknown mechanism, inhibits the glucose transporter GLUT3, decreasing the rate of glycolysis in neurons. The reduced flux through glyceraldehyde 3-phosphate dehydrogenase (GAPDH) releases NAD^+, which becomes available for lactate dehydrogenase (LDH) to deal with the lactate wave from astrocytes. Pyruvate is then metabolized in the mitochondria.

activators, hypoxia, metabolic poisons, viral infection, osmotic stress, and calcium ionophores all trigger a stereotyped biphasic response. In the first hour, the V_{max} for transport increases to two- to threefold. If the stimulus continues, a second phase that lasts 12–24 h takes the V_{max} to tenfold the basal value. The first phase of the response is due to translocation or activation of preexisting carriers. The delayed phase results from an increase in the total number of carriers, brought about by transactivation of the *glut1* gene. The same kind of adaptive phenomenon appears to be present in the brain, where increased GLUT1 expression has been reported in response to seizures, hypoxia, ischemia, and hypoglycemia. In contrast, the behavior of neuronal GLUT3 expression under metabolic demands is less consistent. Reportedly, GLUT1 expression, but not GLUT3 expression, increases in the hippocampus during learning.

Choice of Functional Probes for the Study of Glucose Transporters

Glucose transport in cell populations is often studied with isotope-labeled hexoses, both in culture and *in vivo*. After entering cells, labeled glucose is rapidly metabolized to lactate and CO_2. Because both products can leave the cell very rapidly, glucose uptake is not a reliable measure of glucose transport. 3-O-Methyl-D-glucose (3MG) is a good nonmetabolized substrate for GLUTs, but in brain cells it is transported so quickly that accurate measurement of initial rates at room temperature or higher is not easy. 2-Deoxyglucose (DG) is transported with high efficiency and then phosphorylated by hexokinase but does not proceed further. Although DG is by far the most popular substrate used to characterize GLUTs in the brain and elsewhere, DG data must be interpreted with caution, for depending on the times chosen for assay and the relative number of GLUTs and hexokinase, DG uptake rates may reflect the phosphorylation step rather than the transport step. Contrary to common wisdom, a linear time course of uptake should not be considered proof of the contrary. The transport of sugars into single cells can also be studied by fluorescence microscopy. A fluorescent derivative of DG, 6-NBDG, is a hexose that binds GLUTs with apparent affinity similar to that of glucose but translocates with very low probability, a useful property which allows real-time monitoring of uptake rates. 2-NBDG behaves similarly but it is phosphorylated by hexokinase. As with DG, whether changes in 2-NBDG uptake reflect modulation of transport or phosphorylation must be determined under each experimental condition. Finally, GLUTs can be studied by following osmotically obliged water movements in calcein-loaded cells using confocal microscopy. This technique works best for cells endowed with a high-capacity, low-affinity transporter and provides direct estimates of V_{max} values.

See also: Activity-Dependent Metabolism in Glia and Neurons; Glial Energy Metabolism: Overview; Glial Glutamate Transporters: Electrophysiology; Glial Glutamate Transporters; Glial Glycogen Metabolism; Noradrenaline.

Further Reading

Attwell D and Laughlin SB (2001) An energy budget for signaling in the grey matter of the brain. *Journal of Cerebral Blood Flow & Metabolism* 21: 1133–1145.

Barros LF, Bittner CX, Loaiza A, et al. (2007) A quantitative overview of glucose dynamics in the gliovascular unit. *Glia* 55(12): 1222–1237.

Barros LF, Porras OH, and Bittner CX (2005) Why glucose transport in the brain matters for PET. *Trends in Neurosciences* 28: 117–119.

Loaiza A, Porras OH, and Barros LF (2003) Glutamate triggers rapid glucose transport stimulation in astrocytes as evidenced by real-time confocal microscopy. *Journal of Neuroscience* 23: 7337–7342.

Pellerin L and Magistretti PJ (1994) Glutamate uptake into astrocytes stimulates aerobic glycolysis: A mechanism coupling neuronal activity to glucose utilization. *Proceedings of the National Academy of Sciences of the United States of America* 91: 10625–10629.

Pellerin L and Magistretti PJ (2004) Neuroenergetics: Calling upon astrocytes to satisfy hungry neurons. *Neuroscientist* 10: 53–62.

Porras OH, Loaiza A, and Barros LF (2004) Glutamate mediates acute glucose transport inhibition in hippocampal neurons. *Journal of Neuroscience* 24: 9669–9673.

Vannucci SJ, Maher F, and Simpson IA (1997) Glucose transporter proteins in brain: Delivery of glucose to neurons and glia. *Glia* 21: 2–21.

Activity-Dependent Remodeling of Presynaptic Boutons

C H Bailey and E R Kandel, College of Physicians and Surgeons of Columbia University, New York, NY, USA

Introduction

Activity-dependent changes in the structure of the synapse accompany various forms of learning and memory. This is particularly well documented at the presynaptic connections of identified sensory neurons in the marine invertebrate *Aplysia californica*. Long-term sensitization of the gill-withdrawal reflex in *Aplysia* has been extensively studied in this respect and is associated with the growth of new synaptic connections between the sensory neurons and their postsynaptic target neurons. Recent *in vitro* studies of the sensory-to-motor neuron synapse reconstituted in dissociated cell culture have begun to define the mechanisms that underlie these presynaptic structural changes and their functional contribution to the different temporal phases of long-term memory. Insights provided by these studies in *Aplysia* suggest that the presynaptic remodeling and growth of new synapses induced by learning in the adult brain may reutilize some of the cellular and molecular mechanisms important for *de novo* synapse formation during development.

Learning-Induced Remodeling and Growth of Sensory Neuron Presynaptic Varicosities during Long-Term Sensitization in *Aplysia*

The structural mechanisms contributing to implicit memory storage have been most extensively studied for the gill-withdrawal reflex of *Aplysia.* As is the case with other types of defensive reflexes, several different forms of implicit learning can modify the gill-withdrawal reflex. We focus here on 'sensitization,' an elementary form of nonassociative learning by which an animal learns about the properties of a single noxious stimulus. The animal learns to strengthen its defensive reflexes and to respond vigorously to a variety of previously neutral stimuli after it has been exposed to a potentially threatening stimulus. In *Aplysia,* sensitization of the gill-withdrawal reflex can be induced by a strong stimulus applied to the tail. This activates facilitatory interneurons that synapse on identified sensory neurons to strengthen the synaptic connection between the sensory neurons and their target motor neurons. The behavioral memory for sensitization of the gill-withdrawal reflex is graded and retention is proportional to the number of training trials. A single stimulus to the tail gives rise to short-term sensitization lasting minutes to hours, while repetition of this stimulus produces long-term sensitization that can last for days or weeks.

Short- and long-term sensitizations lead to enhanced synaptic transmission at the monosynaptic connection between identified mechanoreceptor sensory neurons and motor neurons. This monosynaptic pathway can be reconstituted in dissociated cell culture in which serotonin (5-hydroxytryptamine (5-HT)), a modulatory neurotransmitter normally released by sensitizing stimuli, can substitute for the tail shock used during behavioral training in the intact animal. In parallel to behavioral sensitization, a single application of 5-HT produces short-term changes in synaptic effectiveness, whereas five spaced applications given over a period of 1.5 h produce long-term changes lasting 1 or more days.

Biophysical studies of this monosynaptic connection suggest that both the similarities and the differences in short- and long-term memory reflect, at least in part, intrinsic cellular mechanisms of the nerve cells participating in memory storage. Thus, studies of the connections between sensory and motor neurons in both the intact animal and in cells in culture indicate that the short-term changes are surprisingly similar to the long-term changes. A component of the increase in synaptic strength observed during both the short- and long-term changes is due, in each case, to enhanced release of transmitter by the sensory neuron, accompanied by an increase in the excitability of the sensory neuron, which is attributable to the depression of specific sets of potassium channels.

Despite this phenotypic similarity, the short-term changes differ fundamentally from the long-term changes in two important ways. First, the short-term change involves only covalent modification of preexisting proteins and an alteration of preexisting connections. Neither short-term behavioral sensitization in the animal nor short-term facilitation in dissociated cell culture requires ongoing macromolecular synthesis: the short-term change is not blocked by inhibitors of transcription or translation. By contrast, these inhibitors selectively block the induction of the long-term changes both in the semi-intact animal and in primary cell culture.

Second, the long-term but not the short-term process involves a structural change (see later). Long-term sensitization training is associated with the growth of new synaptic connections by the sensory neurons onto their follower cells. This synaptic growth can be induced in the intact ganglion by the intracellular injection of cyclic adenosine monophosphate (cAMP), a second messenger activated by 5-HT, and can be reconstituted in sensory–motor neuron co-cultures by repeated presentations of 5-HT.

In the early 1980s, studies in *Aplysia* began to explore the morphological basis of the synaptic plasticity that may underlie the transition from short-term to long-term memory. By combining selective intracellular labeling techniques with the analysis of serial thin sections and transmission electron microscopy, complete reconstructions of unequivocally identified sensory neuron synapses were quantitatively analyzed from both control and behaviorally modified animals. The storage of long-term memory for sensitization (lasting several weeks) was accompanied by a family of distinct structural changes at identified sensory neuron synapses. These changes reflected a learning-induced remodeling of the functional architecture of presynaptic sensory neuron varicosities (boutons) at two different levels of synaptic organization: (1) alterations in focal regions of membrane specialization of the synapse that mediate transmitter release (the number, size, and vesicle complement of sensory neuron active zones were larger in sensitized animals than in controls) and (2) a parallel but more pronounced and widespread effect involving modulation of the total number of presynaptic varicosities per sensory neuron. Sensory neurons from long-term sensitized animals exhibited a twofold increase in the total number of synaptic varicosities, as well as an enlargement in the size of each neuron's axonal arbor. The duration of the increases in active zone and varicosity number, which persisted unchanged for at least 1 week and were only partially reversed at the end of 3 weeks, paralleled the behavioral time course of memory, indicating that only changes in the number of sensory neuron synapses contribute to the retention of long-term sensitization.

These studies in *Aplysia* demonstrated that clear structural changes could accompany long-term behavioral modifications and indicated, for the first time, that learning-induced changes could be detected at the level of identified synaptic connections known to be critically involved in the behavior. Results from these initial studies of structural synaptic plasticity also provided evidence for an intriguing notion – that active zones are plastic rather than immutable components of the synapse and that even elementary forms of learning can alter the organization and number of transmitter release sites in the presynaptic compartment to modulate the functional expression of synaptic connections. In addition, complete reconstructions of sensory neuron varicosities in control animals revealed that approximately 60% of these presynaptic terminals lacked an active zone, suggesting the possibility of nascent or silent synapses in the adult brain. (The extent to which learning and memory can convert these immature synapses into mature and functionally competent synaptic connections is discussed in the following section.) Finally, these studies indicated that the growth of new sensory neuron synapses may represent the final and perhaps most stable phase of long-term memory storage, and suggested that the stability of the long-term process may be achieved, at least in part, because of the relative stability of synaptic structure.

This long-lasting growth of new synaptic connections between sensory neurons and their follower cells (both interneurons and motor neurons) during long-term sensitization can be reconstituted in dissociated sensory–motor neuron co-cultures by repeated applications of 5-HT. In culture, the structural change can be correlated with the long-term (24–72 h) enhancement in synaptic effectiveness and depends upon the presence of an appropriate target cell similar to the synapse formation that occurs during development.

Long-Term Facilitation Is Associated with Presynaptic Activation of Silent Varicosities and Growth of New Functional Synaptic Varicosities

These earlier studies did not examine whether the increase in synaptic strength during long-term sensitization resulted from the conversion of preexisting but nonfunctional (silent) synapses to active synapses, or from the addition of newly formed functional synapses, or both. To address these issues, more recent *in vitro* studies of the sensory-to-motor neuron synapse in culture have monitored both functional and structural changes simultaneously, so as to follow remodeling at the same specific synaptic varicosities continuously over time and to examine the functional contribution of these presynaptic structural changes to the different time-dependent phases of long-term facilitation (LTF). Toward that end, time-lapse confocal imaging of individual presynaptic varicosities of sensory neurons was combined with different fluorescent markers: the whole cell marker Alexa-594, and two presynaptic marker proteins, synaptophysin-eGFP, which monitors changes in the distribution of synaptic vesicles within individual varicosities, and synapto-PHluorin (synPH), a monitor of active transmitter release sites. Repeated pulses of 5-HT were found to induce two temporally, morphologically, and molecularly distinct classes of presynaptic changes: (1) a rapid activation of silent presynaptic terminals through the filling of preexisting empty varicosities with synaptic vesicles, which requires translation but not transcription, and (2) a generation of new synaptic varicosities which occurs more slowly and requires both transcription and translation. The enrichment of preexisting but empty varicosities with synaptophysin is completed within 3–6 h, parallels intermediate-term facilitation, and accounts for approximately 32% of the newly activated synapses evident at 24 h. By contrast, the new sensory neuron varicosities, which account for 68% of the newly activated synapses at 24 h, do not form until 12–18 h after exposure to five pulses of 5-HT. The rapid activation of silent presynaptic terminals suggests that in addition to its role in LTF, this modification of preexisting synapses may also contribute to the intermediate phase of synaptic plasticity and memory storage (**Figure 1**).

In this study, a reduced 5-HT protocol was used to induce selectively facilitation in the intermediate-term time domain without inducing LTF. Isolated intermediate-term facilitation was also accompanied by the redistribution and clustering of synaptic vesicle proteins into empty sensory neuron varicosities at 0.5 and 3 h, similar to what occurred when intermediate-term facilitation and LTF were recruited together. However, the presynaptic structural changes induced by the reduced 5-HT protocol differed from those induced by long-term training in at least two ways. First, there was no growth of new sensory neuron varicosities in the isolated intermediate phase. Second, unlike the filling of preexisting empty varicosities during the intermediate-term phase induced by the long-term protocol, the newly filled varicosities did not persist for 24 h and were unaffected by inhibitors of protein synthesis, suggesting that the structural remodeling induced by the reduced 5-HT protocol involved only a simple rearrangement of preexisting synaptic components. This may reflect a fundamental difference in the molecular mechanisms recruited by the two 5-HT protocols. Although both protocols induce intermediate-term facilitation, the long-term protocol may activate additional molecular events (including the machinery for translational activation) required to set up the long-term phase, perhaps by stabilizing the intermediate phase. At present it is not known how the covalent modifications that lead to the rearrangement of preexisting synaptic proteins at empty varicosities are converted by the long-term protocol to a more stable, protein-synthesis-dependent process.

In most model systems used to study long-lasting forms of synaptic plasticity, the functional contribution of the structural changes that accompany memory storage remains largely unknown. In particular, one would like to know if changes in the number or structure of synaptic connections induced by learning are functionally effective and capable of contributing to the storage of long-term memory. The synPH imaging of the sensory to motor neuron synapse in *Aplysia* represents an important step in addressing this issue. Previous studies have shown that specific 5-HT protocols or experimental manipulation in *Aplysia* can induce LTF at 24 h without the formation of new varicosities. How might such an increase in synaptic strength persist for 24 h in the absence of synaptic growth? Clearly, additional modifications

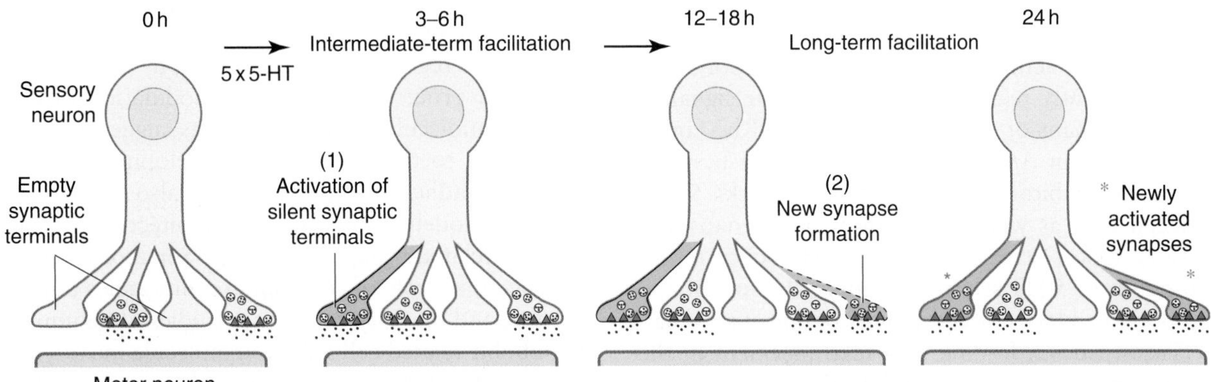

Figure 1 Time course and functional contribution of two distinct presynaptic structural changes associated with intermediate- and long-term facilitation in *Aplysia*. Repeated pulses of 5-HT in sensory to motor neuron co-cultures trigger two distinct presynaptic structural changes: (1) the rapid clustering of synaptic vesicles to preexisting silent sensory neuron varicosities (3–6 h) and (2) the slower generation of new sensory neuron synaptic varicosities (12–18 h). The resultant newly filled and newly formed varicosities are functionally competent (capable of evoked transmitter release) and contribute to the synaptic enhancement that underlies LTF. The rapid filling and activation of silent presynaptic terminals at 3 h suggest that, in addition to having a role in LTF, this modification of preexisting varicosities may also contribute to the intermediate phase of synaptic plasticity. Red triangles represent transmitter release sites (active zones).

of preexisting connections, including the activation of previously silent synapses previously outlined, may play an important role in the initial phases of synaptic maintenance and highlight the fact that there are likely to be multiple types of structural mechanisms that can contribute to LTF at 24 h.

Of the two classes of presynaptic structural plasticity induced by 5-HT in culture, synaptic growth appears to contribute more to the synaptic enhancement present at 24 h than does the activation of preexisting silent synapses. It will be of interest to see if the functional contribution by newly formed synapses increases with time when the growth process is more fully developed and memory storage is likely to be more stable. This would be consistent with the earlier studies in the intact animal outlined previously, which have shown that only the increases in the number of sensory neuron varicosities and active zones persist for several weeks in parallel with the behavioral duration of the memory, as well as more recent work in culture which has demonstrated that synaptic growth plays a more prominent role in the expression of the later phases of LTF.

5-HT-Induced Regulation of the Presynaptic Actin Network Is a Nodal Point for Learning-Related Synapse Remodeling and Growth

The 5-HT-induced enrichment of synaptic vesicle proteins and concomitant recruitment of active zone components in both preexisting and newly formed sensory neuron synapses during LTF involve an activity-dependent rearrangement of the actin cytoskeleton. How do repeated applications of 5-HT lead to reorganization of the actin cytoskeleton? The balance between actin polymerization and depolymerization is tightly regulated by extracellular signaling molecules, many of which act through the Rho family of GTPases. In *Aplysia*, the application of toxin B, a general inhibitor of the Rho family, blocks 5-HT-induced LTF, as well as growth of new synapses in sensory–motor neuron co-cultures. Moreover, repeated pulses of 5-HT selectively induce the spatial and temporal regulation of the activity of one of the small GTPases, Cdc42, leading to a rearrangement of the presynaptic actin network, followed by the assembly, insertion, and functional maturation of active transmitter release sites at sensory neuron varicosities. The 5-HT activation of ApCdc42 is dependent on signaling through the phosphatidylinositol 3-kinase (PI3K) and phospholipase C (PLC) pathways and, in turn, ApCdc42 activates the downstream effectors PAK (p21–Cdc42/Rac-activated kinase) and neuronal Wiskott–Aldrich syndrome protein (N-WASP).

The activation of ApCdc42 in sensory neurons leads to the outgrowth of filopodia from presynaptic varicosities. Interestingly, 5-HT stimulation by itself naturally induces filopodia, which is dependent on the activation of ApCdc42. Filopodia have been proposed to be a morphological precursor of dendritic spines in the mammalian nervous system, and this process may be regulated by neuronal activity. The 5-HT-induced activation of Cdc42 in *Aplysia* triggers not only the formation of filopodia but also the molecular maturation of new transmitter release sites. A major synaptic vesicle protein, synaptophysin, accumulates at the tips of 5-HT-induced filopodia, some of which then give rise to new varicosities. These observations support the following ideas: (1) filopodia represent one of the morphological precursors for the growth of new presynaptic varicosities during learning-related synaptic plasticity and (2) the formation of filopodia and initial assembly of the presynaptic compartment can be induced by the activation of Cdc42. Thus, 5-HT-induced regulation of the Cdc42 signaling pathways and the consequent reorganization of the presynaptic actin network appear to be a part of the initial molecular cascade required for the growth of new sensory neuron varicosities associated with the storage of long-term memory.

Activity-Dependent Modulation of Cell Adhesion Molecules and the Initiation of Learning-Related Presynaptic Growth

Studies in both higher invertebrates and mammals have shown that at critical developmental stages the refinement of synaptic connections, both their growth and regression, is determined by an activity-dependent process involving the modulation of cell adhesion molecules. This in turn has suggested that, in addition to their role in the development of the brain, cell adhesion molecules may also participate in the remodeling of synaptic architecture during learning-related synaptic plasticity. Some of the first evidence for a role of cell adhesion molecules during learning and memory came from studies of an immunoglobulin-related cell adhesion molecule in *Aplysia*, designated ApCAM, which is homologous to NCAM in vertebrates and Fasciclin II in *Drosophila*. 5-HT-induced synaptic growth in sensory–motor neuron co-cultures is associated with a decrease in ApCAMs on the surface membrane of the sensory neuron. This downregulation is particularly prominent at sites at which the processes of the sensory neurons contact one another and is achieved by the protein-synthesis-dependent activation of a coordinated program of clathrin-mediated endocytosis, leading to the internalization and apparent degradation of ApCAM. *Aplysia* neurons express two isoforms of ApCAM – a membrane form and a phosphoinositol-linked form. Only the transmembrane isoform is internalized following exposure to 5-HT. This internalization can be blocked by overexpression of transmembrane ApCAM with a point mutation in the two mitogen-activated protein kinase (MAPK) phosphorylation consensus sites, as well as by injection of a specific MAPK antagonist into sensory neurons, suggesting that activation of the MAPK pathway is important for the internalization of ApCAMs and may represent one of the initial and perhaps permissive stages of learning-related synaptic remodeling and growth in *Aplysia*. Furthermore, overexpression of the transmembrane isoform, but not the glycosyl phosphatidylinositol (GPI)-linked isoform of ApCAM, blocks both LTF and 5-HT-induced synaptic growth. Long-term facilitation can also be blocked by overexpression of the cytoplasmic tail portion of ApCAM, designed to bind proteins such as MAP kinase p42. Taken together, these findings confirm that the extracellular domain of transmembrane ApCAM has an inhibitory function that needs to be neutralized by internalization to induce LTF and synaptic growth and that the cytoplasmic tail provides an interactive platform for both signal transduction and the internalization machinery.

The Presynaptic Remodeling and Growth of New Synapses Induced by Learning in the Adult Brain May Reutilize Mechanisms That Govern *De Novo* Synapse Formation during Development

As already discussed, long-term facilitation of the sensory to motor neuron synapse is accompanied by two temporally and morphologically distinct classes of presynaptic structural change: the rapid activation of silent preexisting varicosities by filling with synaptic vesicles and the slower growth of new functional varicosities. These findings, the first to be made on individually identified presynaptic varicosities, suggest that the duration of the changes in synaptic effectiveness that accompany memory storage may be reflected by the differential regulation of two fundamentally disparate forms of presynaptic compartment: (1) nascent (empty) silent varicosities that can be rapidly and reversibly remodeled into active transmitter release sites and (2) mature, more stable and functionally competent varicosities that, following long-term training, may undergo a process of fission to form new, stable synaptic contacts. What are the cellular and molecular mechanisms responsible for these two distinct classes of learning-related presynaptic structural change?

Some potential insights into this question are suggested by recent live imaging studies of developing synapses in the mammalian central nervous system (CNS). In *Aplysia*, the 5-HT-induced enrichment and subsequent activation of preexisting silent varicosities occur very rapidly – initial changes in the recruitment of synaptic vesicle proteins to empty presynaptic varicosities can be detected at 30 min after 5-HT training. This short delay is similar to what has been reported for the establishment of functional transmitter release sites in cultured hippocampal neurons and suggests that organization of the presynaptic compartment could be achieved by a rapid recruitment of preassembled active zone components to the sites of cell contact.

One attractive hypothesis that has been proposed to account for this rapid differentiation of the presynaptic compartment is based on a unitary model of active zone assembly. Here proteins that comprise the cytoskeletal matrix of the active zone (CAZ) are packaged into transport vesicles for delivery and fusion with the plasma membrane at nascent synaptic contacts. These precursor vesicles contain multiple active zone components, including proteins Piccolo and Bassoon – hence the name Piccolo/Bassoon transport vesicles (PTVs) – as well as other CAZ scaffolding molecules implicated in synaptic vesicle exocytosis, such as Rim, Munc13s, Munc18–1, syntaxin, Snap25, N-type calcium channels, α-liprin, and ERC/CAST, but typically not synaptic vesicle proteins. The assembly of each new active zone appears to be preceded by the recruitment and fusion of integer multiples of PTVs – typically two to five.

This model of developing synapses in the mammalian CNS provides a molecular mechanism of active zone assembly that is consistent with the rapid remodeling and presynaptic activation of preexisting empty sensory neuron varicosities induced by 5-HT in *Aplysia* culture. Moreover, the apparent heterogeneity in the content of these mobile preassembled packets – some of which contain synaptic vesicle proteins and the others which contain components required for assembly of the active zone – could explain why more than half of the sensory neuron varicosities enriched in synaptophysin following 5-HT treatment are not functional. It is likely the maturation of transmitter release sites is not yet complete in these varicosities and that varicosities which are enriched only in synaptic vesicle proteins, but lack a fully differentiated active zone, would not be functionally competent.

How are these modular transport packets induced by learning for presynaptic assembly targeted to empty sensory neuron varicosities? In *Aplysia* culture, only a specific subset of presynaptic varicosities becomes enriched in synaptic vesicles and activated, whereas others do not. Since repeated 5-HT treatment leads to a differential enrichment only in specific sensory neuron varicosities, proteins produced locally at each varicosity might contribute to this structural alteration. Alternatively, these protein components might not be synthesized within or nearby each varicosity, but might be transported and captured at specific varicosities that have been 'tagged' following 5-HT stimulation. Thus, the 5-HT-induced recruitment of synaptic vesicle and active zone proteins at a specific subset of sensory neuron varicosities, and the subsequent functional activation of these previously silent synapses, may be one of the local structural consequences of long-term synapse-specific plasticity. It should be noted that although considerable turnover of synapses and concomitant renewal of active zone proteins characterize the mature nervous system, PTVs are exceedingly rare when compared to developing synapses. It will be of interest to see if this developmental mechanism for presynaptic differentiation can be induced by learning and memory in the adult brain and if the learning-related recruitment of active zone proteins and subsequent assembly of transmitter release sites at empty sensory neuron varicosities are associated with an increase in the frequency of these modular precursor transport vesicles.

The second general class of learning-related presynaptic structural change associated with LTF in *Aplysia* is the 5-HT-induced formation of new sensory neuron varicosities. In culture, this increase in synapse number appears to be accomplished, at least in part, by the division or splitting of preexisting varicosities. Based on time-lapse confocal imaging, the apparent sequence of structural plasticity that gives rise to new varicosities involves the 5-HT-induced recruitment of synaptic vesicle proteins and components of the active zone to a preexisting varicosity, leading to both an enrichment of these presynaptic constituents and to an overall increase in the size of that varicosity. This remodeling and growth of preexisting varicosities ultimately result in the generation of new varicosities that are also enriched in synaptic vesicles and active zone proteins.

Precisely how these new varicosities come into being is still not clear. The close spatial association of preexisting and newly formed sensory neuron varicosities in time-lapse images, including the occasional confluence of both presynaptic compartments, suggests a process involving either a physical division of the original varicosity or a budding off of some components of the transmitter release site and associated synaptic vesicle cluster of that varicosity. Aspects of this reorganization of the presynaptic compartment that precedes learning-related synaptic growth in *Aplysia* have been reported at developing synapses in mammals. For example, recent imaging studies of the early stages of synapse formation have shown that presynaptic sites formed immediately after initial contact of axonal and dendritic processes are highly unstable. Moreover, even apparent mature presynaptic sites are relatively unstable, as occasionally 'orphan release sites' break off from fully formed boutons and either migrate to adjacent presynaptic sites or participate in the formation of completely new ones.

Combined, these observations indicate that differentiation of the presynaptic compartment, either induced by learning in the mature nervous system or as a mechanistic step during development, is a highly dynamic and rapid process that can recruit both preexisting proteins as well as preassembled synaptic components. These findings also suggest that there may be an upper limit to the size of an active zone and associated synaptic vesicle cluster that a neuron can construct. Any increase beyond this limit may cause the entire transmitter release apparatus to become unstable, perhaps because of some inherent metabolic or thermodynamic constraints. The final result is that smaller units of active zone material and cognate synaptic vesicle clusters can bud from the established site and translocate along the axon, where, under appropriate conditions, they may participate in the assembly of new presynaptic sites.

The instability of nascent presynaptic compartments during the early stages of neuronal differentiation is characterized by the dispersion of mobile packets of synaptic vesicles, synaptic vesicle precursors, and active zone precursors and a renewal of their migration once the transient pre- and postsynaptic contacts break up. As the neurons mature, an increasing proportion of these initial contacts develop into more stable and functionally competent presynaptic terminals. Results in the *Aplysia* sensory-to-motor neuron culture preparation indicate that these mature synaptic contacts can be selectively destabilized in an activity-dependent fashion during learning and memory.

How does the long-term process stabilize the transformation of labile, nascent presynaptic compartments into more mature and functionally competent release sites? Conversely, how are mature and fully functional presynaptic compartments destabilized to give rise to the formation of new release sites? Both the stabilization of presynaptic assembly and the recruitment of destabilizing factors that may remove molecular inhibitory constraints and permit fission are likely to require some regulatory interaction with the postsynaptic neuron. Several studies have now suggested a potential role of the postsynaptic neuron in modulating the 5-HT-induced structural changes observed at presynaptic terminals in *Aplysia*. This interaction between postsynaptic and presynaptic neurons is critical for the formation and maturation of synapses during development, and may play a key role in both the pre- and postsynaptic expression of the structural plasticity associated with learning and memory storage. Moreover, since the two classes of 5-HT-induced changes in presynaptic structure that accompany LTF in *Aplysia* culture share their postsynaptic counterpart, there must also be transsynaptic signals, both anterograde and retrograde, to coordinate and regulate the learning-induced structural remodeling in an ongoing manner.

Obvious candidates for the molecules involved in triggering and stabilizing the presynaptic differentiation associated with long-term memory are those involved in synaptogenesis during development. With the extensive investigation in recent years of the developing neuromuscular junction and central synapses of both vertebrates and invertebrates, the list of such molecules has increased considerably. Once neuritic outgrowth and axonal pathfinding are completed and the incipient axon–target interactions occur, signaling molecules begin to engage in bidirectional communication to coordinate the differentiation of pre- and postsynaptic membrane specializations. Several cell surface molecules have been implicated in regulating differentiation of the presynaptic compartment during synaptogenesis, including WNT-7a, SynCAM, and the fibroblast growth factor (FGF) 22; these molecules might mediate this process through their adhesion and signaling capabilities. Some of these transsynaptic signaling systems employ heterophilic interactions that, in principle, could introduce an element of directionality required for the coordinated functional differentiation of the pre- and postsynaptic compartment. Among the transsynaptic molecular candidates, the β-neurexin–neuroligin interaction is particularly intriguing.

Neurexins are presynaptic transmembrane proteins present in many variants. Neurexins associate with synaptic vesicles by interaction with presynaptic scaffolding proteins such as CASK and Mints, which are present in the cytomatrix of the active zone and direct binding to synaptotagmin. Neuroligins are postsynaptic transmembrane proteins and bind to the PDZ domains of PSD-95, a scaffolding protein in the postsynaptic compartment of excitatory synapses. Thus, the β-neurexin–neuroligin interaction may act as a transsynaptic bridge, bringing synaptic vesicles into alignment with neurotransmitter receptor–ion channel complexes in the postsynaptic density. This hypothesis gained support when neuroligin expressed in nonneuronal cells was shown to cluster synaptic vesicles in contacting glutamatergic axons. Moreover, antibody-induced clustering of recombinant neurexin directly induced the co-clustering of synaptic vesicles.

Although details of neuroligin-induced presynaptic differentiation are not completely understood, aspects of the proposed consequences of this transsynaptic interaction during development might provide additional molecular insights into the mechanisms that underlie the two learning-related presynaptic changes in *Aplysia*. Neuroligin activity depends on the lateral clustering of individual neuroligin molecules that appear to induce the clustering of neurexin in the presynaptic membrane. These lateral clusters of neurexin may, in turn, activate signaling cascades that promote presynaptic assembly by recruiting scaffolding molecules such as CASK, Mint, and Veli, which interact directly with the PDZ-binding motifs in the cytoplasmic tail of neurexin.

Since CASK also interacts with proteins that regulate the actin–spectrin cytoskeleton, additional neurexin molecules and/or other synaptic components could be inserted into this presynaptic scaffold at the cytomatrix of the active zone, perhaps facilitating the 5-HT-induced growth and stabilization of the newly formed release sites at empty sensory neuron varicosities. Conversely, activity-dependent stimulation that gives rise to learning and memory could lead to a selective and synapse-specific interruption of the adhesive and/or signaling capabilities of the neuroligin–neurexin transsynaptic interaction at a subset of mature sensory neuron varicosities. This learning-induced alteration might, in turn, destabilize the cytoskeleton matrix of the presynaptic compartment – a transient and permissive step that would allow units of active zone material and associated synaptic vesicles to break off and to participate in the establishment of new presynaptic sites.

Conclusions

The morphological correspondence between studies in higher invertebrates and mammals suggests that learning resembles a process of neuronal growth and differentiation across a broad segment of the animal kingdom. Many of the mechanisms and signaling interactions utilized for experience-dependent remodeling of the synapse and the growth of new synaptic connections appear to share features in common with those that govern synaptogenesis. Recent studies of the synaptic growth that accompanies LTF in *Aplysia* have begun to characterize the sequence of cellular and molecular events responsible for both the initiation and persistence of these presynaptic structural changes. This, in turn, has revealed that molecules and mechanisms important for synapse formation during the development of the nervous system may be reutilized in the adult for the purposes of synaptic plasticity and memory storage.

See also: Active Zone; Cell Adhesion Molecules at Synapses; Endocytosis and Presynaptic Scaffolds; Learning and Memory in Invertebrates: *Aplysia*; Presynaptic Facilitation; Presynaptic Inhibition; Presynaptic Development: Functional and Morphological Organization; Transcription Factors in Synaptic Plasticity and Learning and Memory.

Further Reading

Bailey CH and Kandel ER (1993) Structural changes accompanying memory storage. *Annual Review of Physiology* 55: 397–426.

Bailey CH, Kandel ER, and Si K (2004) The persistence of long-term memory: A molecular approach to self-sustaining changes in learning-induced synaptic growth. *Neuron* 44: 49–57.

Goda Y and Davis GW (2003) Mechanisms of synapse assembly and disassembly. *Neuron* 40: 243–264.

Huntley GW, Benson DL, and Colman DR (2002) Structural remodeling of the synapse in response to physiological activity. *Cell* 108: 1–4.

Kandel ER (2001) The molecular biology of memory storage: A dialogue between genes and synapses. *Science* 294: 1030–1038.

Lamprecht R and LeDoux J (2004) Structural plasticity and memory. *Nature Reviews Neuroscience* 5: 45–54.

Scheiffele P (2003) Cell–cell signaling during synapse formation in the CNS. *Annual Review of Neuroscience* 26: 485–508.

Ziv NE and Garner CG (2004) Cellular mechanisms of presynaptic assembly. *Nature Reviews Neuroscience* 5: 385–399.

Addiction: Neurobiological Mechanism

G F Koob, The Scripps Research Institute, La Jolla, CA, USA

Neurocircuitry of Drug Reward, Dependence, and 'Craving'

Substance dependence can be defined as a chronically relapsing disorder characterized by (1) compulsion to seek and take the drug, (2) loss of control in limiting intake, and (3) the emergence of a negative emotional state (e.g., dysphoria, anxiety, and irritability) when access to the drug is prevented (defined here as withdrawal). Addiction and substance dependence (as currently defined by the *Diagnostic and Statistical Manual of Mental Disorders*, 4th edn.) are used interchangeably throughout this article to refer to a final stage of a usage process that moves from drug use to abuse to addiction. The term dependence has two meanings: (1) to describe an acute withdrawal syndrome and (2) to describe a syndrome in which a subject meets the criteria for substance dependence. Other terms that need to be defined include 'reward,' which connotes a reinforcing stimulus with positive hedonic valence, and 'anti-reward.' Anti-reward is a concept based on the hypothesis that there are brain systems in place to limit reward and is represented at the neurocircuit level by a between-system neuroadaptation to activation of the reward system.

Clinically, the occasional but limited use of a drug with the potential for abuse or dependence (the second meaning given previously) is distinct from escalated drug use and the emergence of a chronic drug-addicted state. An important goal of current neurobiological research is to understand the neuropharmacological and neuroadaptive mechanisms within specific neurocircuits that mediate the transition from occasional, controlled drug use to the loss of behavioral control over drug-seeking and drug-taking that defines chronic addiction.

Addiction has been conceptualized as a chronic relapsing disorder with roots in both impulsivity and compulsivity and neurobiological mechanisms that change as an individual moves from one domain to the other. Subjects with impulse control disorders experience an increasing sense of tension or arousal before committing an impulsive act; pleasure, gratification, or relief at the time of committing the act; and finally regret, self-reproach, or guilt following the act. In contrast, individuals with compulsive disorders experience anxiety and stress before committing a compulsive repetitive behavior and then relief from the stress by performing the compulsive behavior. In addiction, drug-taking behavior progresses from impulsivity to compulsivity in a three-stage cycle: binge/intoxication, withdrawal/negative affect, and preoccupation/anticipation. As individuals move from an impulsive to a compulsive disorder, the drive for the drug-taking behavior shifts from positive to negative reinforcement.

Much of the progress in understanding the mechanisms of addiction has derived from the study of animal models of addiction. Although no animal model of addiction fully emulates the human condition, animal models do permit investigation of specific elements of the process of drug addiction. Such elements can be defined by models of different systems, models of psychological constructs such as positive and negative reinforcement, and models of different stages of the addiction cycle. Although much focus in animal studies has been on the synaptic sites and molecular mechanisms in the nervous system on which drugs with dependence potential act initially to produce their positive reinforcing effects, new animal models of components of the negative reinforcing effects of dependence, the transition to dependence, and propensity to relapse have been developed and are beginning to be used to explore how the nervous system adapts to drug use. The neurobiological mechanisms of addiction that are involved in various stages of the addiction cycle have a specific focus on certain brain circuits and the neurochemical changes associated with those circuits during the transition from drug taking to drug addiction and how those changes persist in the vulnerability to relapse.

A key element of drug addiction is how the brain reward system changes with the development of addiction, and one must understand the neurobiological bases for acute drug reward to understand how the reward systems change with the development of addiction. A principle focus of research on the neurobiology of the positive reinforcing effects of drugs with dependence potential has been on the activation of the circuitry related to the origins and terminals of the mesocorticolimbic dopamine system. There is compelling evidence for a critical role of this system in drug reward associated with psychostimulant drugs, and there is evidence that all major drugs of abuse activate this system as measured either by increased extracellular levels of dopamine in the terminal areas (e.g., the nucleus accumbens) or by activation of the firing of neurons in the ventral tegmental area. However, although selective neurotoxin-induced lesions of the mesolimbic dopamine system do block cocaine, amphetamine, and nicotine self-administration, rats continue to self-administer heroin and alcohol in the absence of the mesocorticolimbic dopamine system, and place preference studies show robust place preferences to morphine and nicotine in the presence of major dopamine receptor blockade. Together these results suggest that activation of the mesolimbic dopamine system is a component of drug seeking in general but only critical for the rewarding effects of stimulant drugs.

Specific components of the basal forebrain associated with the amygdala also have been identified with drug reward. One hypothetical construct, the extended amygdala, includes not only the central nucleus of the amygdala (CeA) but also the bed nucleus of the stria terminalis (BNST) and a transition zone in the medial subregion of the nucleus accumbens (shell of the nucleus accumbens), and these regions share certain cytoarchitectural and circuitry similarities (**Figure 1**). As the neural circuits for the reinforcing effects of drugs with dependence potential have evolved, the role of neurotransmitters/neuromodulators also has evolved, and multiple neurotransmitter systems have been identified to have a role in the acute reinforcing effects of drugs of abuse in these basal forebrain areas: mesolimbic dopamine, opioid peptide, γ-aminobutyric acid (GABA), glutamate, endocannabinoids, and serotonin.

The neural substrates and neuropharmacological mechanisms for the negative motivational effects of

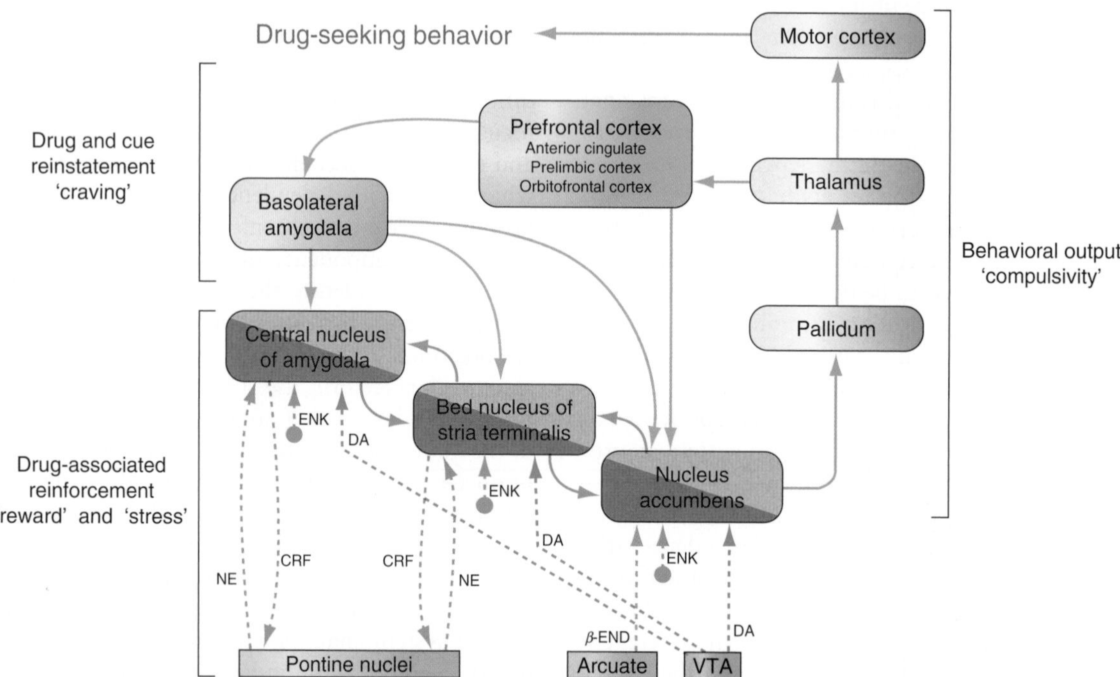

Figure 1 Key common neurocircuitry elements in drug-seeking behavior of addiction. Three major circuits that underlie addiction can be distilled from the literature. A drug-reinforcement circuit ('reward' and 'stress') is composed of the extended amygdala, including the CeA, the bed nucleus of the stria terminalis, and the transition zone in the shell of the nucleus accumbens. Multiple modulator neurotransmitters are hypothesized, including dopamine and opioid peptides for reward and corticotropin-releasing factor and norepinephrine for stress. The extended amygdala is hypothesized to mediate integration of rewarding stimuli or stimuli with positive incentive salience and aversive stimuli or stimuli with negative aversive salience. During acute intoxication, valence is weighted on processing rewarding stimuli, and during the development of dependence aversive stimuli come to dominate function. A drug- and cue-induced reinstatement ('craving') neurocircuit is composed of the prefrontal (anterior cingulate, prelimbic, and orbitofrontal) cortex and basolateral amygdala, with a primary role hypothesized for the basolateral amygdala in cue-induced craving and a primary role for the medial prefrontal cortex in drug-induced craving, based on animal studies. Human imaging studies have shown an important role for the orbitofrontal cortex in craving. A drug-seeking ('compulsive') circuit is composed of the nucleus accumbens, ventral pallidum, thalamus, and orbitofrontal cortex. The nucleus accumbens has long been hypothesized to have a role in translating motivation to action and forms an interface between the reward functions of the extended amygdala and the motor functions of the ventral striatal–ventral pallidal–thalamic–cortical loops. The striatal–pallidal–thalamic loops reciprocally move from prefrontal cortex to orbitofrontal cortex to motor cortex, ultimately leading to drug-seeking behavior. Note that for the sake of simplicity, other structures are not included, such as the hippocampus (which presumably mediates context-specific learning, including that associated with drug actions). Also note that dopamine and norepinephrine both have widespread innervation of cortical regions and may modulate function relevant to drug addiction in those structures. DA, dopamine; ENK, enkephalin; CRF, corticotropin-releasing factor; NE, norepinephrine; β-END, β-endorphin. Reproduced from Koob GF and Le Moal M (2006) *Neurobiology of Addiction*. London: Academic Press, with permission from Elsevier.

drug withdrawal may involve disruption of the same neural systems implicated in the positive reinforcing effects of drugs. Measures of brain reward function during acute abstinence from all major drugs with dependence potential have revealed increases in brain reward thresholds as measured by direct brain stimulation reward. These increases in reward thresholds may reflect decreases in the activity of reward neurotransmitter systems in the midbrain and forebrain implicated in the positive reinforcing effects of drugs and as such represent a 'within-system' neuroadaptation. Examples of such changes at the neurochemical level include decreases in dopaminergic and serotonergic transmission in the nucleus accumbens during drug withdrawal as measured by *in vivo* microdialysis, increased sensitivity of opioid receptor transduction mechanisms in the nucleus accumbens during opiate withdrawal, and decreased GABAergic transmission. One also sees differential regional changes in nicotine receptor function during nicotine withdrawal. The decreases in reward neurotransmitters have been hypothesized to contribute significantly to the negative motivational state associated with acute drug abstinence and also the long-term biochemical changes that contribute to the clinical syndrome of protracted abstinence and vulnerability to relapse.

Different neurochemical systems involved in arousal and stress modulation also may be engaged within the neurocircuitry of the brain stress systems in an attempt to overcome the chronic presence of the perturbing drug and to restore normal function despite the presence of drug. Glutamate, an excitatory neurotransmitter, has been implicated in neuroadaptation to repeated exposure to drugs of abuse in two major domains. First, glutamate hyperactivity in the basal forebrain has been linked to the hyperexcitability associated with ethanol withdrawal, and this hyperexcitability has been observed in slices of the hippocampus, nucleus accumbens, and amygdala. This hyperexcitability is linked to the protracted abstinence state in alcohol dependence and is hypothesized to be a neural substrate for the anti-relapse effects of acamprosate, a medication for the treatment of alcoholism. Second, glutamate neuroplasticity has been implicated in cocaine-induced reinstatement where increased glutamate release combined with reduced basal glutamate function in the prefrontal cortex-to-nucleus accumbens core pathway has been hypothesized to explain increased glutamate release in response to repeated cocaine administration. In both models, increased activation of glutamatergic function contributes to increased drug seeking in addiction.

Chronic administration of drugs with dependence potential also dysregulates both the hypothalamic–pituitary–adrenal axis and the brain stress system mediated by corticotropin-releasing factor (CRF). Common responses include an activated pituitary adrenal stress response and elevated adrenocorticotropic hormone and corticosteroids and an activated brain stress response with activated amygdala CRF during acute withdrawal from all major drugs of abuse. Acute withdrawal from drugs of abuse may also increase the release of norepinephrine in the BNST and decrease levels of neuropeptide Y (NPY) in the central and medial nuclei of the amygdala (**Figure 1**).

These results suggest not only a change in function of neurotransmitters associated with the acute reinforcing effects of drugs (dopamine, opioid peptides, serotonin, and GABA) during the development of dependence but also recruitment of the brain arousal and stress systems (glutamate, CRF, and norepinephrine) and dysregulation of the NPY brain antistress system. These changes would represent a 'between-system' neuroadaptation. Thus, reward mechanisms in dependence are compromised by disruption of neurochemical systems involved in processing natural rewards and by recruitment of the anti-reward systems that represent neuroadaptation to the chronic exposure of the brain reward neurocircuitry to drugs of abuse.

The neuroanatomical entity termed the extended amygdala thus may represent a common anatomical substrate for acute drug reward and a common neuroanatomical substrate for the negative effects on reward function produced by stress that help drive compulsive drug administration. The extended amygdala receives numerous afferents from limbic structures such as the basolateral amygdala and hippocampus, and it sends efferents to the medial part of the ventral pallidum and a large projection to the lateral hypothalamus, thus further defining the specific brain areas that interface classical limbic (emotional) structures with the extrapyramidal motor system.

Animal models of craving involve the use of drug-primed reinstatement, cue-induced reinstatement, or stress-induced reinstatement in animals that have acquired drug self-administration and then have been subjected to extinction of responding for the drug. Most evidence from animal studies suggests that drug-induced reinstatement is localized to a medial prefrontal cortex/nucleus accumbens/ventral pallidum circuit mediated by the neurotransmitter glutamate. In contrast, neuropharmacological and neurobiological studies using animal models for cue-induced reinstatement involve the basolateral amygdala as a critical substrate with a possible feed-forward mechanism through the prefrontal cortex system involved in drug-induced reinstatement.

Stress-induced reinstatement of drug-related responding in animal models appears to depend on the activation of both CRF and norepinephrine in elements of the extended amygdala (CeA and BNST).

In summary, three neurobiological circuits have been identified that have heuristic value for the study of the neurobiological changes associated with the development and persistence of drug dependence. The acute reinforcing effects of drugs of abuse that comprise the binge/intoxication stage of the addiction cycle most likely involve actions localized to a nucleus accumbens–amygdala reward system, dopamine inputs from the ventral tegmental area, local opioid peptide circuits, and opioid peptide inputs in the arcuate nucleus of the hypothalamus. In contrast, the symptoms of acute withdrawal important for addiction, such as dysphoria and increased anxiety associated with the withdrawal/negative affect stage, most likely involve decreases in function of the extended amygdala reward system and recruitment of brain stress neurocircuitry. The preoccupation/anticipation (or craving) stage involves key afferent projections to the extended amygdala and nucleus accumbens, specifically the prefrontal cortex (for drug-induced reinstatement) and the basolateral amygdala (for cue-induced reinstatement). Compulsive drug-seeking behavior is hypothesized to be driven by ventral striatal–ventral pallidal–thalamic–cortical loops (**Figure 1**).

Molecular and Cellular Targets within the Brain Circuits Associated with Addiction

Cellular mechanisms within the neurocircuitry of the basal forebrain have been focused on changes in patterns of neuronal firing in freely moving animals, glutamate-dependent plasticity in the nucleus accumbens, and GABAergic plasticity in the CeA. In freely moving animals, extracellular recordings have revealed a neuronal population in the nucleus accumbens that exhibits phasic excitatory responses that are time locked to drug-related events and, with repeated intravenous self-administration, decreases in background firing, suggesting a net enhancement of drug-related signals. This differential inhibition of drug reward-related firing and background firing may represent a form of filtering that could narrow the ensemble of neurons in the nucleus accumbens to those that mediate the strengthening of associations.

Long-term potentiation at excitatory synapses has also been observed in nucleus accumbens and ventral tegmental area slices. Increases in synaptic strength in ventral tegmental area slices have been reported even after a single *in vivo* administration of cocaine. These changes in synaptic strength have been hypothesized to share some of the same mechanisms with long-term potentiation, such as a dependence on glutamatergic neurotransmission.

Cellular mechanisms in the amgydala involving both glutamate and GABA have been identified as the basis for changes in reward system neuroadaptation to excessive alcohol exposure. Acute alcohol markedly decreased both *N*-methyl-D-aspartate (NMDA) and non-NMDA-mediated excitatory postsynaptic currents in CeA neurons from naive control rats. Early (on the order of hours) and long-term (weeks) withdrawal from chronic alcohol treatment significantly increase the sensitivity of CeA NMDA responses to acute alcohol. In contrast, acute ethanol significantly increased evoked $GABA_A$ receptor-mediated inhibitory postsynaptic currents in a majority of neurons in CeA slices via increased GABA release in the CeA. Also, chronic ethanol treatment did not change the effect of acute ethanol in augmenting inhibitory postsynaptic current sizes and the frequency of mini-postsynaptic currents, suggesting a lack of tolerance for the presynaptic effects of ethanol. Studies have also shown a key role for CRF in driving alcohol-induced GABA release in the CeA. Thus, CRF and GABA interactions may represent a cellular substrate underlying the neuroadaptations associated with the development of an anti-reward construct in dependence.

Acknowledging that all drugs of abuse share some common neurocircuitry actions, namely inhibition of medium spiny neurons in the ventral striatum and interneurons in the CeA, the search at the molecular level has led to examining mechanisms for these cellular changes. Repeated perturbation of intracellular signal transduction pathways may cause changes in neuronal function and/or changes in nuclear function and altered rates of transcription of particular target genes. Altered expression of such genes would lead to presumably long-term altered activity of the neurons where such changes occur and ultimately to changes in neural circuits in which those neurons operate.

Two transcription factors in particular have been implicated in the plasticity associated with addiction: cyclic adenosine monophosphate (cAMP) response element binding protein (CREB) and ΔFosB. CREB regulates the transcription of genes that contain a cAMP response element site within the regulatory regions and can be found ubiquitously in genes expressed in the central nervous system such as those encoding neuropeptides, synthetic enzymes for neurotransmitters, signaling proteins, and other transcription factors. CREB can be phosphorylated by protein kinase A and by protein kinases regulated by growth factors, placing it at a point of convergence for several intracellular messenger pathways that can regulate the expression of genes.

Much work in the addiction field has shown that activation of CREB in the nucleus accumbens, one part of the brain reward circuit, is a consequence of chronic exposure to opiates, cocaine, and alcohol and deactivation in the CeA, another part of the reward circuit. The activation of CREB in the nucleus accumbens with psychostimulant drugs is linked to the symptoms of psychostimulant withdrawal, possibly through induction of the opioid peptide dynorphin which binds to κ opioid receptors and has been hypothesized to represent a mechanism of motivational tolerance and dependence. These molecular adaptations decrease an individual's sensitivity to the rewarding effects of subsequent drug exposures (tolerance) and impair the reward pathway (dependence) so that after removal of the drug the individual is left in an amotivational, dysphoric, or depressed-like state.

In contrast, decreased CREB phosphorylation has been observed in the CeA during alcohol withdrawal and has been linked to decreased NPY function and consequently increased anxiety-like responses associated with acute alcohol withdrawal. These molecular changes may occur simultaneously and point to transduction mechanisms that could produce neurochemical changes in the neurocircuits outlined previously and are critical for breaks with reward homeostasis in addiction.

These transcription factors can change gene expression and produce long-term changes in protein expression and, as a result, neuronal function. Whereas acute administration of drugs of abuse can cause a rapid (within hours) activation of members of the Fos protein family, such as c-*fos*, FosB, Fra-1, and Fra-2 in the nucleus accumbens, other transcription factors, isoforms of ΔFosB, accumulate over longer periods of time (days) with repeated drug administration. Animals with activated ΔFosB have exaggerated sensitivity to the rewarding effects of drugs of abuse, and ΔFosB may be a sustained molecular 'switch' that helps to initiate and maintain a state of addiction.

Genetic and molecular genetic animal models have provided some convergence of data to support the neuropharmacological substrates identified in neurocircuitry studies. High alcohol-preferring rats have been bred that show high voluntary consumption of alcohol, increased anxiety-like responses, and numerous neuropharmacological phenotypes such as decreased dopaminergic activity and decreased NPY activity. In an alcohol-preferring and -nonpreferring cross, a quantitative trait locus was identified on chromosome 4, a region to which the gene for NPY has been mapped. In the inbred preferring and nonpreferring quantitative trait loci analyses, loci on chromosomes 3, 4, and 8 have been identified which correspond to loci near the genes for the dopamine D_2 and serotonin $5HT_{1B}$ receptors.

Advances in molecular biology have led to the ability to systematically inactivate the genes that control the expression of proteins that make up receptors or neurotransmitter/neuromodulators in the central nervous system using the gene knockout and transgenic knockin approaches. Although these approaches do not guarantee that these genes are the ones that convey vulnerability in the human population, they provide viable candidates for exploring the genetic basis of endophenotypes associated with addiction.

Notable positive results with opioids with gene knockout studies in mice have focused on knockout of the μ opioid receptor. Opiate (morphine) reinforcement as measured by conditioned place preference or self-administration is absent in μ knockout mice, and there is no development of somatic signs of dependence to morphine in these mice. Indeed, all morphine effects tested, including analgesia, hyperlocomotion, respiratory depression, and inhibition of gastrointestinal transit, are abolished in μ knockout mice. Knockout of the μ opioid receptor also decreases nicotine reward, cannabinoid reward, and alcohol drinking in mice.

Knockout and neuropharmacological studies have implicated numerous neurotransmitter systems in ethanol preference, including opioid, dopamine, GABA, and serotonin. Novel modulatory effects on ethanol preference have been suggested by protein kinase and G-protein channel knockout studies.

Selective deletion of the genes for expression of different dopamine receptor subtypes and the dopamine transporter has revealed significant effects to challenges with psychomotor stimulants. Dopamine D_1 receptor knockout mice show no response to D_1 agonists or antagonists and show a blunted response to the locomotor-activating effects of cocaine and amphetamine. D_1 knockout mice are also impaired in their acquisition of intravenous cocaine self-administration compared to wild-type mice. D_2 knockout mice have severe motor deficits and blunted responses to psychostimulants and opiates, but the effects on psychostimulant reward are less consistent. Dopamine transporter knockout mice are not only dramatically hyperactive but also show a blunted response to psychostimulants. Although developmental factors must be taken into account for the compensatory effect of deleting any one or a combination of genes, it is clear that D_1 and D_2 receptors and the dopamine transporter play important roles in the actions of psychomotor stimulants.

Brain Imaging Circuits Involved in Human Addiction

Brain imaging studies using positron emission tomography with ligands for measuring oxygen utilization or glucose metabolism or using magnetic resonance imaging techniques are providing dramatic insights into the neurocircuitry changes in the human brain associated with the development and maintenance and even vulnerability to addiction. These imaging results bear a striking resemblance to the neurocircuitry identified by human studies. During acute intoxication with alcohol, nicotine, and cocaine, there is an activation of the orbitofrontal cortex, prefrontal cortex, anterior cingulate, extended amygdala, and ventral striatum. This activation is often accompanied by an increase in availability of the neurotransmitter dopamine. During acute and chronic withdrawal, there is a reversal of these changes with decreases in metabolic activity, particularly in the orbitofrontal cortex, prefrontal cortex, and anterior cingulate, and decreases in basal dopamine activity as measured by decreased D_2 receptors in the ventral striatum and prefrontal cortex. With limited studies, cue-induced reinstatement appears to involve a reactivation of these circuits much like acute intoxication. Two strongly represented markers for active substance dependence in humans across drugs of different neuropharmacological actions are decreases in prefrontal cortex metabolic activity and decreases in brain dopamine D_2 receptors that are hypothesized to reflect decreases in brain dopamine function.

Conclusions

Much progress in neurobiology has provided a heuristic neurocircuitry framework with which to identify the neurobiological and neuroadaptive mechanisms involved in the development of drug addiction. The brain reward system implicated in the development of addiction is composed of key elements of the basal forebrain, including the nucleus accumbens and amygdala and their connections. Neuropharmacological studies in animal models of addiction have provided evidence for the dysregulation of specific neurochemical mechanisms in specific brain reward systems (dopamine, opioid peptides, and GABA). There is also recruitment of brain arousal and stress systems (glutamate, CRF, and norepinephrine) and dysregulation of brain anti-stress systems (NPY) that provide the negative motivational state associated with drug abstinence. Additional neurobiological and neurochemical systems have been implicated in animal models of relapse, with the prefrontal cortex and basolateral amygdala (and glutamate systems therein) being implicated in drug- and cue-induced relapse, respectively. The brain stress systems in the extended amygdala are directly implicated in stress-induced relapse. The changes in reward and stress systems are hypothesized to remain dysregulated and outside of a homeostatic state, and combined with compromised executive function they convey the vulnerability for development of dependence and relapse in addiction. Genetic studies in animals suggest roles for the genes encoding the neurochemical elements involved in the brain reward (dopamine and opioid peptide) and stress (NPY) systems in the vulnerability to addiction, and molecular studies have identified transduction and transcription factors that may mediate the dependence-induced reward dysregulation (CREB) and chronic vulnerability changes (ΔFosB) in neurocircuitry associated with the development and maintenance of addiction. Human imaging studies reveal similar neurocircuits involved in acute intoxication, chronic drug dependence, and vulnerability to relapse.

See also: Drug Addiction: Behavioral Neurophysiology; Drug Addiction: Behavioral Pharmacology of Drug Addiction in Rats; Drug Addiction: Cellular Mechanisms; Drug Addiction: Neuroimaging; Drugs Addiction: Actions; Transcription and Reward Systems.

Further Reading

Contet C, Kieffer BL, and Befort K (2004) Mu opioid receptor: A gateway to drug addiction. *Current Opinion in Neurobiology* 14: 370–378.

Crabbe JC, Phillips TJ, Harris RA, Arends MA, and Koob GF (2006) Alcohol-related genes: Contributions from studies with genetically engineered mice. *Addiction Biology* 11: 195–269.

Everitt BJ and Wolf ME (2002) Psychomotor stimulant addiction: A neural systems perspective. *Journal of Neuroscience* 22: 3312–3320. (Erratum: 22(16), 1a.)

Gardner EL and Vorel SR (1998) Cannabinoid transmission and reward-related events. *Neurobiology of Disease* 5: 502–533.

Goldstein RZ and Volkow ND (2002) Drug addiction and its underlying neurobiological basis: Neuroimaging evidence for the involvement of the frontal cortex. *American Journal of Psychiatry* 159: 1642–1652.

Heimer L and Alheid G (1991) Piecing together the puzzle of basal forebrain anatomy. *Advances in Experimental Medicine and Biology* 295: 1–42.

Heinrichs SC and Koob GF (2004) Corticotropin-releasing factor in brain: A role in activation, arousal, and affect regulation. *Journal of Pharmacology and Experimental Therapeutics* 311: 427–440.

Koob GF, Bartfai T, and Roberts AJ (2001) The use of molecular genetic approaches in the neuropharmacology of corticotropin-releasing factor. *International Journal of Comparative Psychology* 14: 90–110.

Koob GF and Le Moal M (2001) Drug addiction, dysregulation of reward, and allostasis. *Neuropsychopharmacology* 24: 97–129.

Koob GF and Le Moal M (2006) *Neurobiology of Addiction.* London: Academic Press.

McFarland K and Kalivas PW (2001) The circuitry mediating cocaine-induced reinstatement of drug-seeking behavior. *Journal of Neuroscience* 21: 8655–8663.

Nestler EJ (2004) Historical review: Molecular and cellular mechanisms of opiate and cocaine addiction. *Trends in Pharmacological Sciences* 25: 210–218.

Pandey SC (2004) The gene transcription factor cyclic AMP-responsive element binding protein: Role in positive and negative affective states of alcohol addiction. *Pharmacology and Therapeutics* 104: 47–58.

Shalev U, Grimm JW, and Shaham Y (2002) Neurobiology of relapse to heroin and cocaine seeking: A review. *Pharmacological Reviews* 54: 1–42.

Siggins GR, Roberto M, and Nie Z (2005) The tipsy terminal: Presynaptic effects of ethanol. *Pharmacology and Therapeutics* 107: 80–98.

Adenosine

K A Jacobson and Z-G Gao, National Institutes of Health, Bethesda, MD, USA

Introduction

The behavioral stimulant effects of the alkylxanthine caffeine and related xanthines are well known. The mechanism of these effects at moderate xanthine doses is now known to be through antagonism of one or more subtypes of adenosine receptors (ARs), rather than through inhibition of phosphodiesterases or stimulation of calcium release, which occur only at higher concentrations of caffeine. At very high doses, caffeine depresses locomotor activity, which does not appear to be related to ARs. Adenosine, an endogenous agonist of the ARs, acts as a local modulator of the action of various neurotransmitters, including biogenic amines and excitatory amino acids. In fact, adenosine modulates the release of many, if not most, neurotransmitters, both in the brain and in the peripheral nervous system.

The ARs activated by extracellular adenosine are classified as four subtypes: A_1, A_{2A}, A_{2B}, and A_3 (**Figure 1**). A_{2A} and A_{2B} ARs are relatively close in sequence identity (59%, for the human homologs), as are A_1 and A_3 ARs (49%, for the human homologs). Typically, adenosine has an affinity of 10–30 nM at the high-affinity binding sites of the A_1 and A_{2A} ARs. A_{2B} AR has the lowest affinity (high micromolar) of all of the subtypes for adenosine, and its affinity at the A_3 AR is intermediate. The classical second messengers associated with these subtypes are stimulation of adenylate cyclase through G_s proteins for the A_{2A} and A_{2B} subtypes and inhibition of adenylate cyclase through G_i proteins for the A_1 and A_3 subtypes. In recent years, it has become apparent that other effector mechanisms are important in the physiological actions of adenosine, such as phosphoinositide 3-kinase and mitogen-activated protein kinases (MAPKs), which may be activated through the β and γ subunits of the G-proteins.

In general, the role of adenosine in the brain and peripherally is protective: that is, when released or produced in response to organ stress or tissue damage, adenosine tends to increase the ratio of oxygen supply to oxygen demand. It also suppresses various cytotoxic processes, such as cytokine-induced apoptosis. In the brain, both neuronal and glial cell functions are regulated by adenosine. In the autonomic and enteric nervous systems, adenosine has been shown to modulate neurotransmitter release and the interaction of neurotransmitters with receptors. ARs are critical in controlling bladder function via a neurological mechanism. In the intestine, adenosine also plays an essential role in modulating the function of neurons located in the myenteric plexuses between muscle coats and submucous plexus. In the peripheral immune system, adenosine has been shown to 'put the brakes' on excessive inflammation and to promote tissue protection against ischemic damage. In the cardiovascular system, adenosine promotes vasodilation, vascular integrity, and angiogenesis, and counteracts the lethal effects of prolonged ischemia on cardiac myocytes. Nearly every cell type in the body expresses one or more of the subtypes of ARs, indicating the fundamental nature of adenosine as a cytoprotective mediator in both the peripheral and the central nervous systems.

Medicinal chemists have succeeded in designing and synthesizing for the ARs numerous selective and potent ligands that act competitively with adenosine at the orthosteric site of the receptors. Highly selective agonists and antagonists of the A_1, A_{2A}, and A_3 ARs have been reported both as research tools and as experimental therapeutic agents, but for the A_{2B} ARs only selective antagonists are currently known. Some of these selective agents are excluded from the central nervous system when administered peripherally, due to lack of penetration of the intact blood–brain barrier. Thus, these are important pharmacological probes for separating the central and peripheral actions of adenosine. It should be noted that nucleosides, such as adenosine and its analogs, by virtue of the hydrophilic 9-ribose moiety, tend to enter the brain only to a small degree, if at all. Most of the adenosine antagonists are nonnucleoside heterocyclic derivatives, which tend to be more hydrophobic than are the adenosine agonists. Therefore, a given adenosine agonist cannot be assumed to enter the brain in sufficient amounts to exogenously stimulate one of the ARs, unless there is supportive evidence for its action in the central nervous system. The half-life of adenosine in peripheral circulation can be as short as 1 s, and therefore its peripheral administration is unlikely to affect the concentration of extracellular adenosine in the brain. In contrast, a significant number of adenosine antagonists enter the brain when administered peripherally. The simple xanthines, caffeine and theophylline, readily permeate the blood–brain barrier.

Envisioned therapeutic applications of AR agonists and antagonists include treatment of inflammatory bowel diseases, arrhythmia, bladder dysfunction, ischemia, neurodegenerative diseases (such as Parkinson's

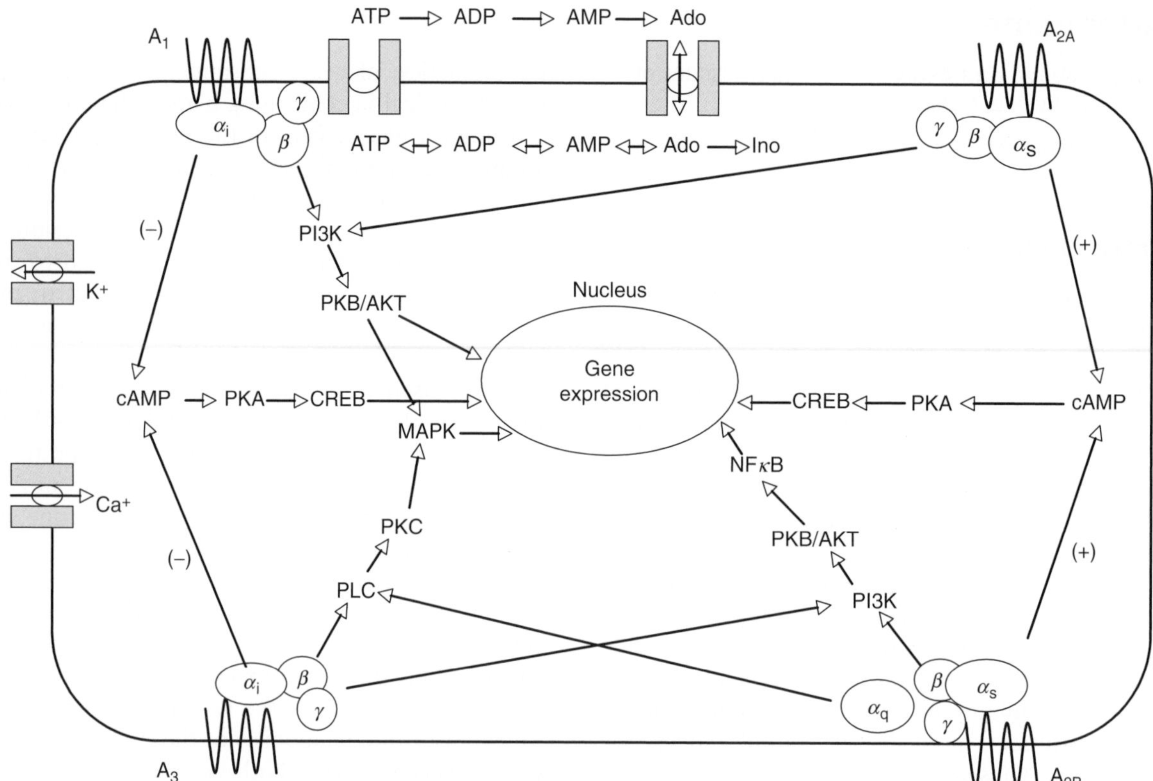

Figure 1 Signaling pathways associated with the four subtypes (A_1, A_{2A}, A_{2B}, and A_3) of adenosine receptors. ATP, adenosine triphosphate; ADP, adenosine diphosphate; AMP, adenosine monophosphate; Adp, adenosine; Ino, inosine; CREB, cyclic AMP response element-binding protein; MAPK, mitogen-activated protein kinase; NFκB, nuclear factor-kappa B; PI3K, phosphatidylinositol 3-kinase; PKA, protein kinase A; PKB, protein kinase B; PKC, protein kinase C; PLC, phospholipase C.

disease and Alzheimer's disease), pain, psychiatric disorders, and sleep disorders. The use of both genetic deletion of the receptor (now each of the subtypes has been deleted in mice strains) and selective agonists/antagonists has contributed toward these concepts.

Sources of Adenosine, Its Transport Mechanisms, and Metabolism

Adenosine production in the synapse is not through vesicular release in response to nerve firing, as is the case for classical neurotransmitters. Rather, adenosine acts as a local autacoid, the release of which increases upon stress to an organ or tissue. Most cells in culture and *in situ* produce adenosine and release it extracellularly, which tends to influence the outcome of pharmacological studies if not properly controlled. The levels of extracellular adenosine in a given tissue or organ may vary widely, depending on stress factors present, leading to highly variable basal levels of stimulation of the ARs by endogenous adenosine.

The source of extracellular adenosine may be both from inside the cell, where it is present in millimolar concentrations, and from the breakdown of extracellular adenine nucleotides, which activate signaling pathways through their own superfamily of receptors (P2Y metabotropic and P2X inotropic receptors). Several nucleotide precursors of adenosine, such as ATP and adenosine diphosphate (ADP), act at P2 receptors, but the monophosphate, AMP, which does not act at any of the P2 receptors, is weakly active as an agonist of ARs. The level of extracellular adenosine may be as low as ~20 nM in the resting brain and as high as 100 μM in severe ischemic conditions in the brain or elsewhere in the body. As a hydrophilic small molecule, adenosine does not diffuse across the plasma membrane freely; rather, it may pass through an equilibrative transporter such as the ENT1 nucleoside transporter. Levels of extracellular adenosine may also rise as a result of the enzymatic hydrolysis of extracellular adenine nucleotides by widely occurring ectonucleotidases or as a result of cell lysis. Ectonucleotidases, which are also ubiquitously expressed on the cell surface but with characteristic distribution patterns, cleave the phosphate moieties of adenine nucleoside 5′-phosphate derivatives, to culminate in the formation of adenosine. There are many classes of ectonucleotidases; however, the most relevant species in the family of ectonucleoside triphosphate dihydrolases (E-NTPDases) are apyrase (NTPDase1, which converts ATP and ADP to AMP) and NTPDase2 (which converts ATP to ADP). A separate enzyme ecto-5′-nucleotidase (CD73) converts AMP to adenosine. CD73 is characteristically found on the surface of astrocytes but not of neurons.

Unlike classical neurotransmitters, adenosine does not have a rapid synaptic uptake system (as for the biogenic amines), or a rapid chemical inactivation system (as for acetylcholine). Rather, adenosine is metabolized extracellularly by widespread enzymes adenosine kinase (AK; to produce AMP) and adenosine deaminase (to produce inosine) and therefore is inactivated with respect to the ARs. The metabolite inosine may have its own biological actions. For example, there is evidence that inosine activates the A_3 AR.

Inhibition of the metabolism of extracellular adenosine or its uptake proteins is being explored for therapeutic purposes. AK inhibitors have been proposed for the treatment of pain and seizure; however, the promising clinical development of these efficacious compounds has been discontinued due to their toxicity.

Receptor Structure, Signaling Pathways, and AR Regulation

Receptor Structure

All of the ARs are G-protein-coupled receptors (GPCRs), which structurally consist of seven transmembrane helices connected by extracellular and cytoplasmic loops. There is a strong sequence homology between A_1 and A_3 ARs and between A_{2A} and A_{2B} ARs. The neurotransmitter receptors that are closest in primary sequence to the ARs are the biogenic amine receptors. Each of the ARs has been modeled both in overall three-dimensional structure and in the nature of molecular recognition in the putative ligand binding site, based on homology to bovine rhodopsin or by other methods, such as the recently developed threading assembly refinement method. Essential residues for the binding of agonists and antagonists and for the activation of the receptors have been defined through modeling and mutagenesis. Use of the molecular models to design *de novo* new ligands has not yet met with extensive success. However, confidence in the receptor models and in their predictive abilities has increased gradually since the first models were reported in the early 1990s. The neoceptor approach to reengineering GPCRs in general, and ARs specifically, has validated binding site hypotheses for adenosine at its receptors.

The binding and activation steps of receptor action have been dissected computationally, although not yet in a global fashion, largely due to the lack of a suitable template for the activated conformation. The conformational dynamics of the activation of the A_3 AR have been approximated with respect to isolated portions of the receptor.

Signaling Pathways

Two AR subtypes, A_1 and A_3, couple through G_i to the inhibition of adenylate cyclase, while the other two subtypes, A_{2A} and A_{2B}, stimulate adenylate cyclase through G_s or G_{olf} (for A_{2A}). The A_{2B} AR is also coupled to activation of phospholipase C through G_q. Furthermore, each of these receptors may couple through the β and γ subunits of the G-proteins to other effector systems, including ion channels and phospholipases. ARs have been found to couple to MAPKs in a variety of circumstances, leading to effects on differentiation, proliferation, and cell death. The A_3 AR can reduce apoptosis through the activation of Akt. ARs can couple to ion channels; for example, activation of the A_1 AR can induce the influx of calcium ions or the efflux of potassium ions.

Cross-talk occurs between ARs and other receptors. For example, an otherwise subthreshold concentration of acetylcholine (ACh), as might be present in the Alzheimer's brain, still produces a strong calcium signal when the A_1 AR is costimulated. Cross-talk occurs with the striatal dopamine receptor system, in which a direct physical association (dimerization) occurs between A_{2A} and D_2 receptors, and between other subtypes. A_{2A} AR stimulation also transactivates the neuroprotective Trk neurotrophin receptor pathway, even in the absence of neurotrophic growth factor. Recently, heterodimers of adenosine A_1 receptors and either $P2Y_1$ or $P2Y_2$ nucleotide receptors have been characterized pharmacologically.

Regulation

Similar to the function and regulation of other GPCRs, both activation and desensitization of the ARs occur after agonist binding. Interaction of the activated ARs with the G-proteins leads to second-messenger generation and classical physiological responses. Interaction of the activated ARs with G-protein receptor-coupled kinases (GRKs) leads to their phosphorylation. Downregulation of ARs should be considered in both the basic pharmacological studies and with respect to the possible therapeutic application of agonists. AR responses desensitize rapidly, and this phenomenon is associated with receptor downregulation, internalization, and degradation. Mutagenesis has been applied to analyze the molecular basis for the differences in the kinetics of the desensitization response displayed by various AR subtypes. The most rapid downregulation among the AR subtypes is generally seen with the A_3 AR, due to phosphorylation by GRKs.

Novel and Definitive Ligands

Potent and selective AR antagonists have been prepared for all four AR subtypes, and selective agonists are known for three subtypes. Thus, numerous pharmacological tools are available for *in vitro* and *in vivo* use. Potent and selective A_{2B} AR agonists are yet to be reported, although several research groups have identified lead compounds. Molecular modeling of the ARs and ligand docking have provided insights into the putative binding sites of all of the subtypes, which has aided in ligand design.

Adenosine Agonists

Until recently, nearly all AR agonists have been purine nucleoside derivatives. With the synthesis of pyridine-3,5-dicarbonitrile derivatives, there is now an example of a nonnucleoside chemical class that fully activates ARs. Medicinal chemists have extensively explored the structure–activity relationships of adenosine derivatives as agonists of the ARs (**Figure 2**). In general, for the adenine moiety of adenosine, modifications at the N^6 position have led to selectivity for the A_1 AR and the A_3 AR, and modifications at the 2 position, especially with ethers, secondary amines, and alkynes, have led to selectivity for the A_{2A} AR. Commonly used A_1 AR agonists that are N^6-cycloalkyl derivatives are cyclopentyladenosine (CPA), its 2-chloro analog CCPA, and endo-norbornyl adenosine ($S(-)$-ENBA) (**Figure 2(a)**). At the human ARs, $S(-)$-ENBA is more highly A_1 AR-selective than are CPA and CCPA. Adenosine amine congener (ADAC) is an amine-functionalized congener designed for covalent coupling to carriers with the retention of A_1 AR activity. The A_{2A} AR-enhancing effects of substitution at the 2 position are further boosted with a uronamido substitution at the 5′ position of adenosine (**Figure 2(b)**). For example, the 5′-*N*-ethyl derivatives CGS21680 and ATL-146e are both selective in binding to the rat A_{2A} AR and less selective at the human subtypes. When the 5′-*N*-alkyluronamide group alone is present, high affinity at the A_{2A} AR, but not selectivity, is typically observed, as with the nonselective agonist 5’-*N*-ethylcarboxamide adenosine (NECA). The N^6-substituted A_1 AR agonists NNC-21-0136 and GR79236 and the doubly substituted selodenoson have been clinical candidates. A_1 AR agonists are of interest for use in treating cardiac arrhythmias (for which adenosine itself, under the name Adenocard, is in widespread use). AMP579 is a carbocyclic nucleoside that activates both A_{2A} and A_1 ARs. A_{2A} AR agonists such as ATL-146e, binodeson, CVT-3146, and MRE0094 have been clinical candidates. Such agonists are of interest for use as vasodilatory agents in cardiac imaging and in suppressing inflammation.

Numerous exceptions to the preceding generalizations concerning selectivity of specific substitution sites on adenosine have been found. For example, the mono-N^6-substituted derivative of adenosine, N^6-[2-(3,5-dimethoxyphenyl)-2-(2-methylphenyl)-ethyl] adenosine (DPMA), is potent and selective at the rat A_{2A} AR. At the human ARs, the A_{2A} AR and A_1 AR affinities of DPMA are roughly equal. N^6-Arylmethyl derivatives of adenosine, such as N^6-benzyl derivatives, tend toward selectivity for the A_3 AR. MRS3997, an adenosine derivative that is monosubstituted at the 2 position with a (6-bromotryptophol)ether moiety, is a promising lead for the A_{2B} AR. It is to be noted that MRS3997 fully activates human A_{2A} and A_{2B} ARs equipotently.

The A_3 agonists (**Figure 2(c)**) IB-MECA and Cl-IB-MECA are widely used as selective agonists of the A_3 AR, although yet more selective agents are now known (e.g., the conformationally locked (*N*) methanocarba derivative MRS3558 and the 4′-thio nucleoside LJ568). CP-608039 is a selective A_3 agonist that was developed for cardioprotection.

Adenosine Antagonists

The structures of A_3 antagonists are typically more chemically diverse than are the classical xanthine antagonists of the A_1 and A_2 receptors (**Figure 3**). The dihydropyridine derivatives MRS1191 and MRS1334 and the pyridylquinazoline derivative VUF5574 are potent, selective A_3 AR antagonists in the human, but are weak or inactive at the rat A_3 AR. Nevertheless, MRS1191 has been used successfully in murine species. MRS1220 is very potent at the human but not at the rat or the mouse receptor. The pyridine derivative MRS1523 and the nucleosides MRS1292, however, are selective A_3 AR antagonists in both rats and humans. Selective A_3 AR antagonists, such as the aforementioned compounds and the heterocyclic derivatives OT-7999 and MRE-3008-F20 and the adenine derivative MRS3777, are in general of interest for the treatment of cancer, stroke, inflammation, and glaucoma.

Variations in the relative efficacy of nucleosides, depending on structure, have been noted. This phenomenon is especially pronounced for the A_3 AR, at which changes on the adenine moiety (N^6 and 2 positions) and ribose moiety can either reduce efficacy to the point of pure antagonism (i.e., combination of 2-Cl and N^6-(3-iodobenzyl)) or can guarantee a robust, nearly full activation of the A_3 AR (i.e., 5′-uronamide).

Although the classical AR antagonists are xanthines derivatives such as caffeine and theophylline, today an enormous diversity of heterocyclic structures have been reported as AR antagonists. The micromolar affinity of the naturally occurring antagonists has been greatly exceeded with the introduction of selective

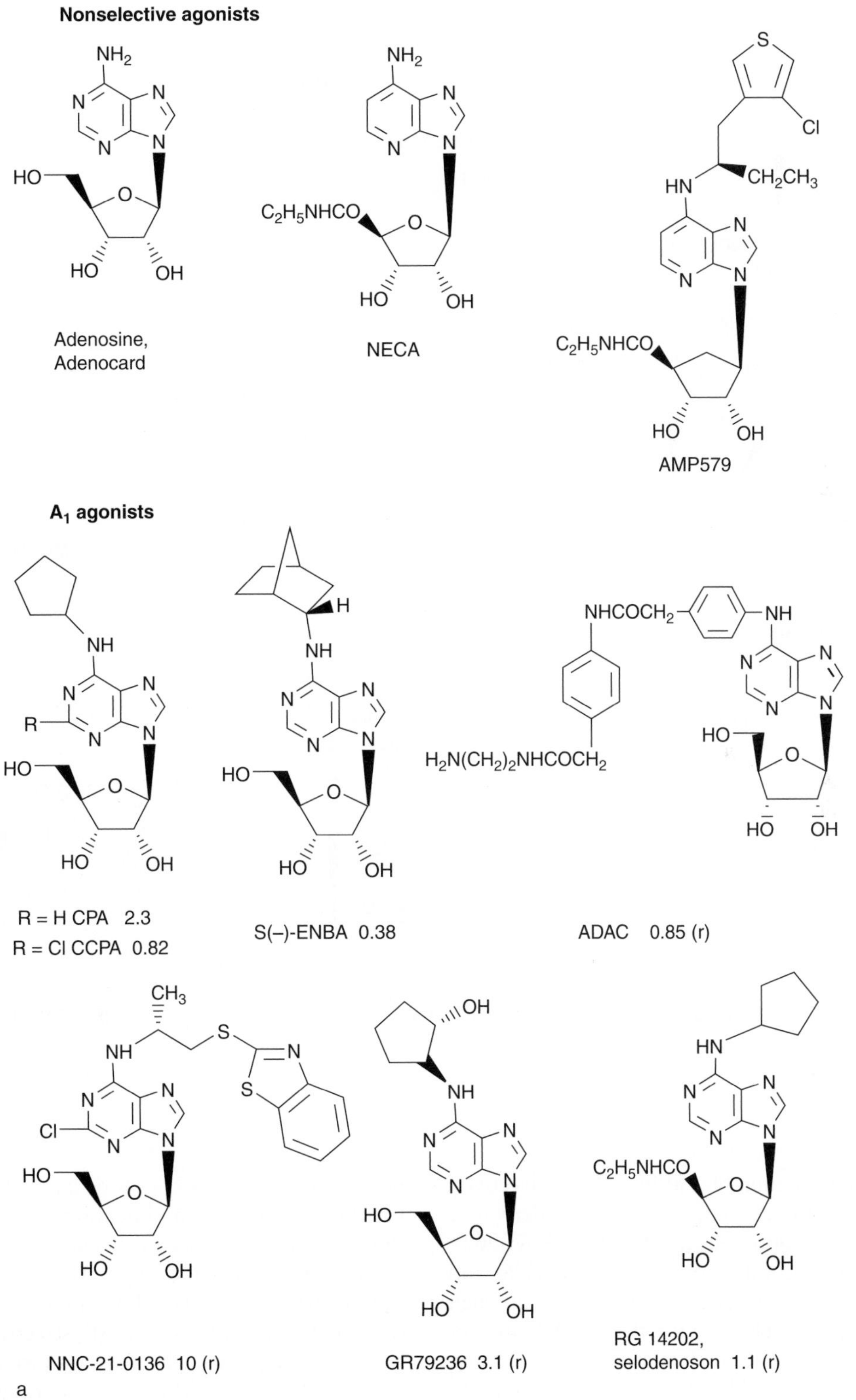

Nonselective agonists

Adenosine, Adenocard

NECA

AMP579

A_1 agonists

R = H CPA 2.3
R = Cl CCPA 0.82

S(–)-ENBA 0.38

ADAC 0.85 (r)

NNC-21-0136 10 (r)

GR79236 3.1 (r)

RG 14202, selodenoson 1.1 (r)

a

Figure 2 Continued

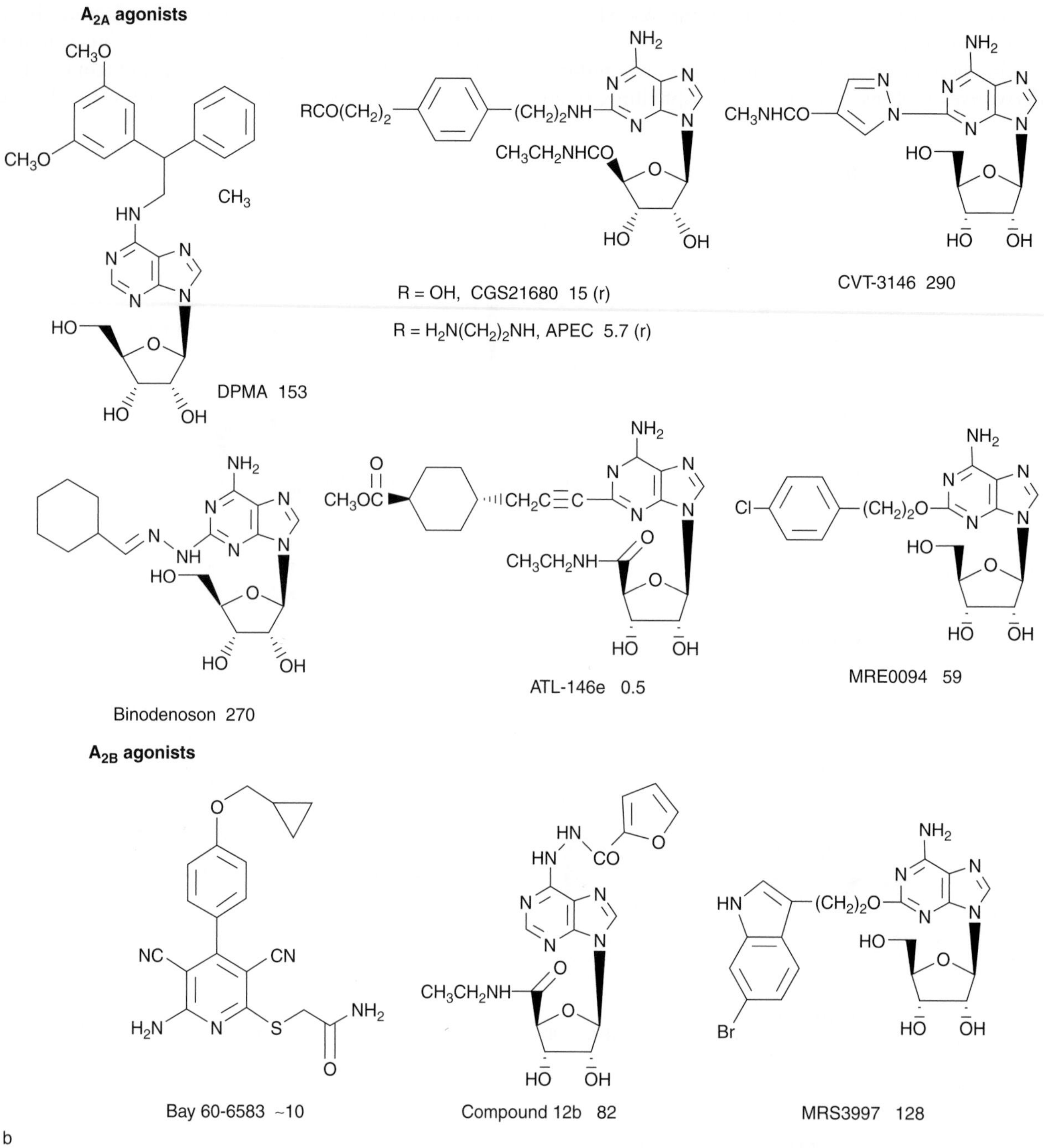

A_{2A} agonists

DPMA 153

R = OH, CGS21680 15 (r)

R = $H_2N(CH_2)_2NH$, APEC 5.7 (r)

CVT-3146 290

Binodenoson 270

ATL-146e 0.5

MRE0094 59

A_{2B} agonists

Bay 60-6583 ~10

Compound 12b 82

MRS3997 128

b

Figure 2 Continued

antagonists of subnanomolar affinity. High selectivity of xanthines at the A_1 (e.g., the epoxide derivative BG 9719 and the more selective BG9928; KW3902), A_{2A} (e.g., KW6002), and A_{2B} (e.g., MRS1706 and MRS1754) ARs has been achieved (**Figure 3**). Although xanthines that are selective for the A_3 AR are not available, certain cyclized derivatives of xanthines, such as PSB-11, are A_3 AR selective. In general, modifications of the xanthine scaffold at the 8 position with aryl or cycloalkyl groups has led to selectivity for the A_1 AR, and modifications at the same position with alkenes (specifically, styryl groups) has led to selectivity for the A_{2A} AR. 8-Cyclopentyl-1, 3-dipropylxanthine (DPCPX) is highly A_1 AR selective in the rat but less A_1 AR selective among the human AR subtypes. The 8-styrylxanthine derivatives KW6002, CSC, and MSX are moderately potent A_{2A} AR antagonists. Some 8-styrylxanthine derivatives,

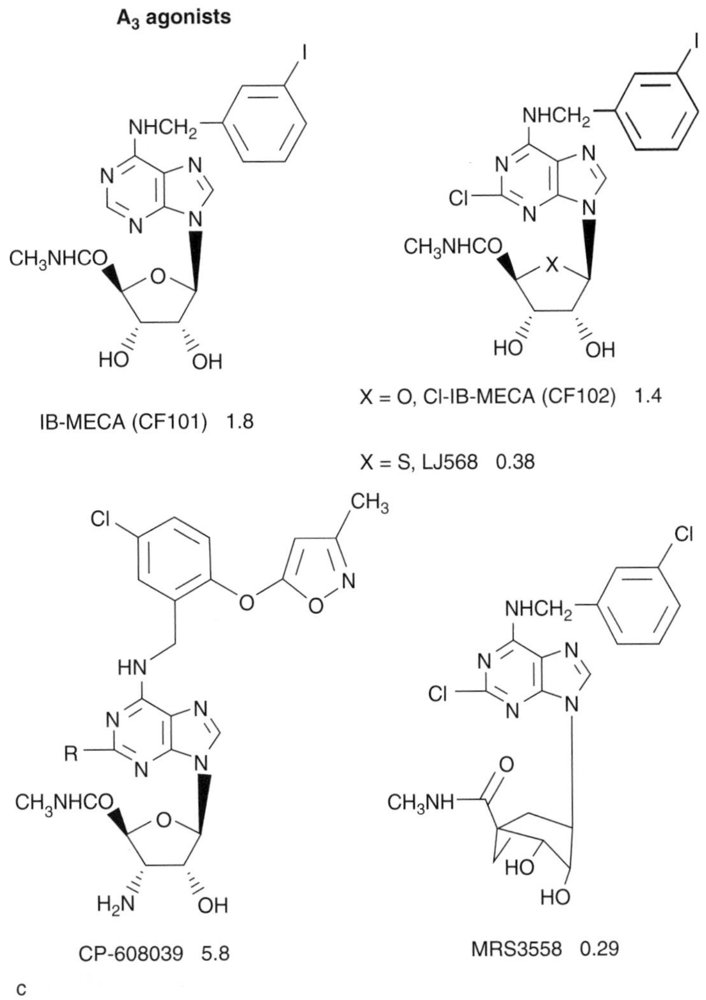

Figure 2 Structures of (a) nonselective and A_1, (b) A_{2A} and A_{2B}, and (c) A_3 adenosine receptor agonist probes used as pharmacological tools, and in some cases as clinical candidates. The K_i values (nM) in binding to the appropriate human adenosine receptor are indicated after the name (all are human adenosine receptors, unless indicated 'R' for rat). Values indicated for the A_{2B} adenosine receptor are functional EC_{50} values. Bay 60-6583 is a nonnucleoside that potently and selectively activates the A_{2B} adenosine receptor. MRS3997 is a mixed agonist for the A_{2A} and A_{2B} adenosine receptors.

especially CSC, have been discovered to inhibit monoamine oxidase-B as well. A persistent problem in the use of xanthine derivatives as AR antagonists of the A_1 AR is their interaction at the A_{2B} AR. Use of adenine derivatives, such as the inverse agonist WRC-0571, provides A_1 AR-selective antagonists having low affinity at the A_{2B} AR.

The triazolotriazine ZM241,385 and the pyrazolotriazolopyrimidines SCH-58261 and SCH442,416 are highly potent and selective A_{2A} AR antagonists. SCH442,416 displays >23 000-fold selectivity for the human A_{2A} AR (K_i, 0.048 nM), in comparison to human A_1 AR, and an $IC_{50} > 10\,\mu M$ at the A_{2B} and A_3 ARs. A_{2A} AR antagonists, such as the xanthine KW6002 and the nonxanthines SCH58261, SCH442416, VER6947, and VER7835, are of interest for use in treating Parkinson's disease.

There is a marked species dependence of antagonist affinity at the A_3 AR. Therefore, commonly used antagonists must be treated with caution in species other than humans. In general, one must be cognizant of potential species differences for both AR agonists and antagonists.

Radioligands

Radioligands commonly used for the ARs are A_1 agonist [^{3}H]CCPA, antagonist [^{3}H]DPCPX, A_{2A} agonist [^{3}H]CGS21680, antagonist [^{3}H]ZM241,385 or [^{3}H]SCH58261, A_3 agonist [^{125}I]I-AB-MECA, and antagonist [^{3}H]PSB-11. Ligands for *in vivo* positron emission tomographic (PET) imaging of A_1 and A_{2A} ARs have been developed. For example, the xanthine [^{18}F]CPFPX and the nonxanthine [^{11}C]FR194921 have been developed as centrally active PET tracers for imaging the A_1 AR in the brain. Potent fluorescent ligands have been reported for A_1, A_{2A}, and A_3 ARs.

Allosteric Modulation

In addition to directly acting AR agonists and antagonists, allosteric modulators of agonist action are under consideration for disease treatment. Such modulators, either positive enhancers or negative allosteric inhibitors, might have advantages over the directly acting (orthosteric) receptor ligands. The action of the allosteric compounds would depend on the presence of a high local concentration of

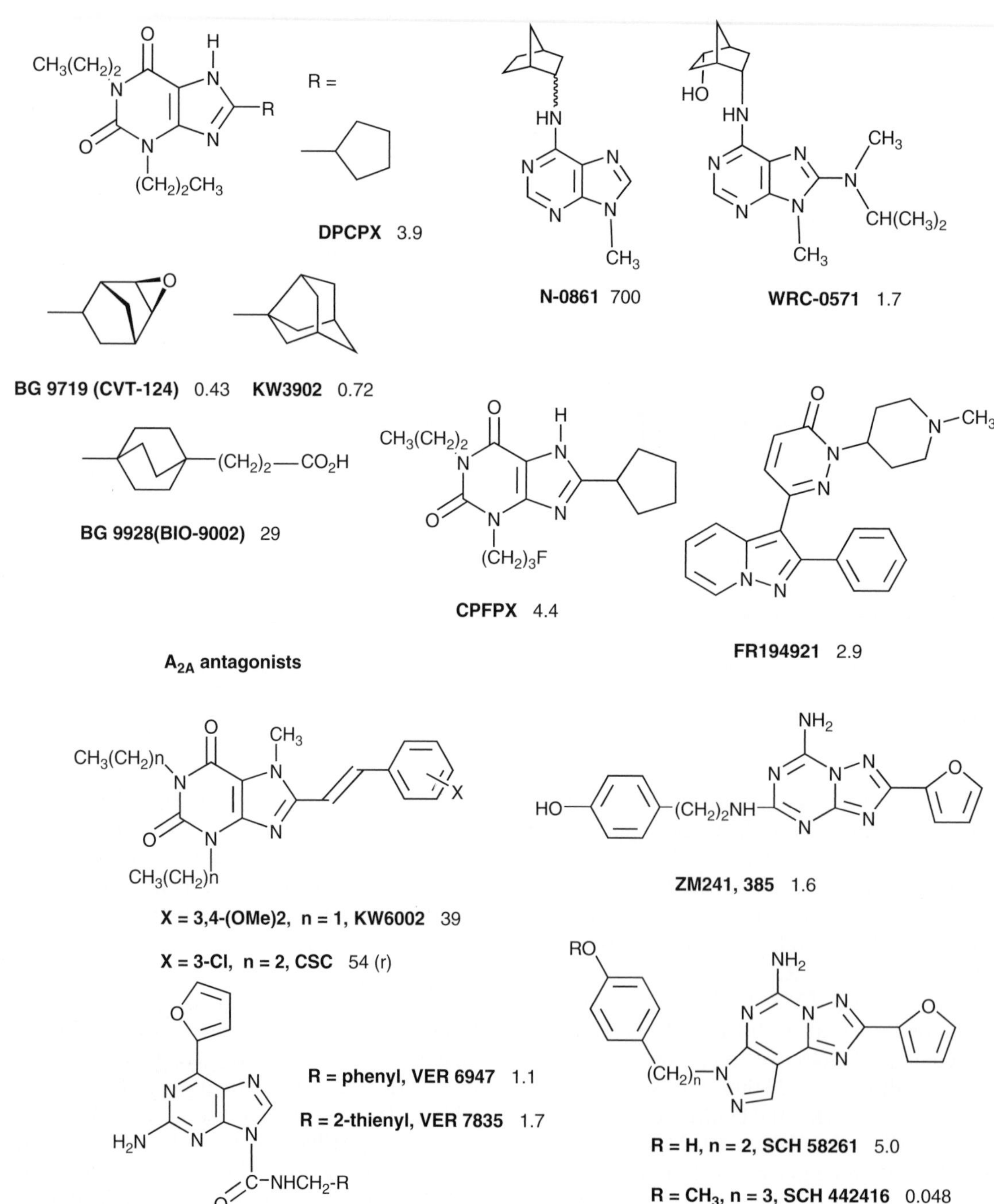

Figure 3 Continued

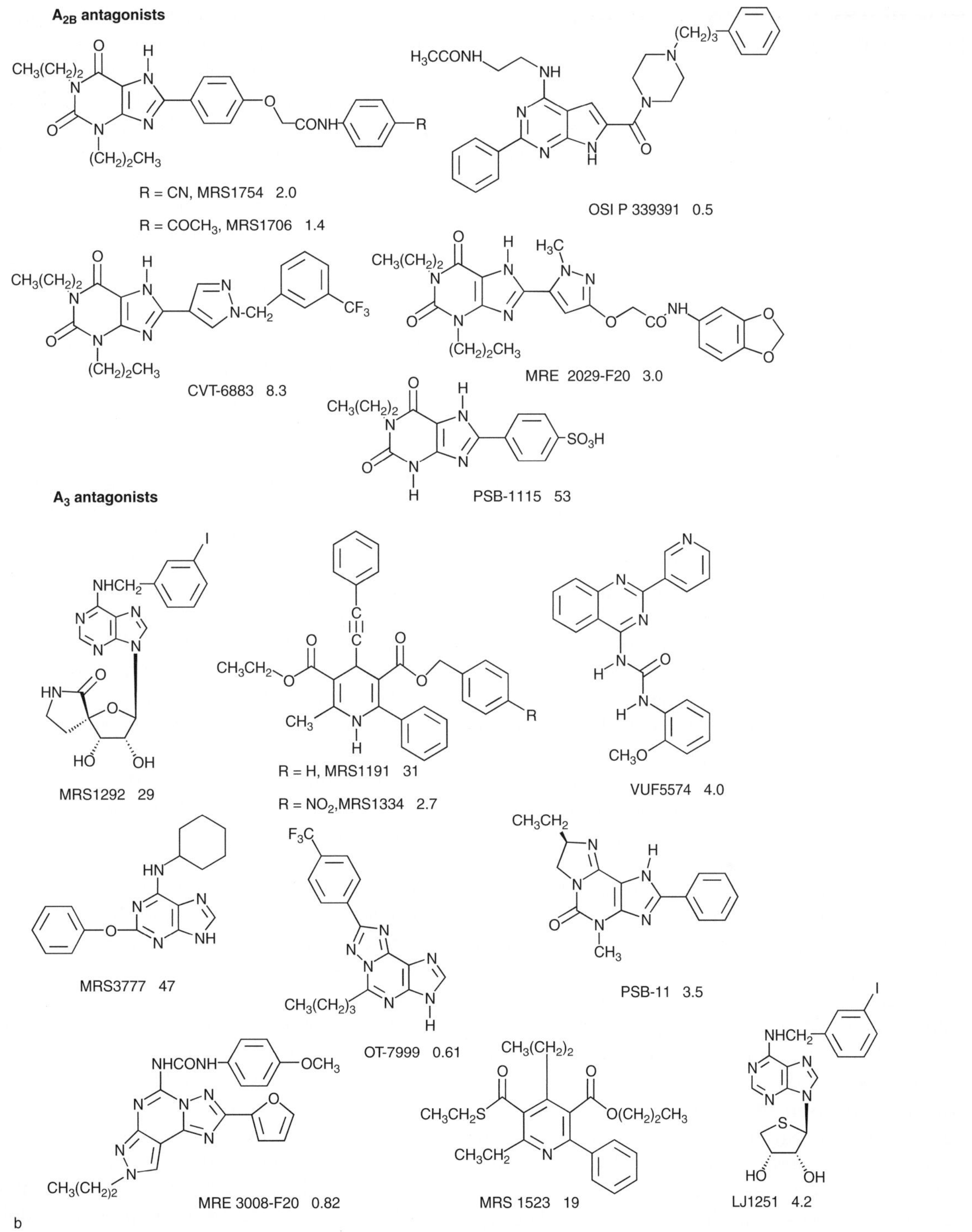

A_{2B} antagonists

R = CN, MRS1754 2.0

R = $COCH_3$, MRS1706 1.4

OSI P 339391 0.5

CVT-6883 8.3

MRE 2029-F20 3.0

PSB-1115 53

A_3 antagonists

MRS1292 29

R = H, MRS1191 31

R = NO_2, MRS1334 2.7

VUF5574 4.0

MRS3777 47

OT-7999 0.61

PSB-11 3.5

MRE 3008-F20 0.82

MRS 1523 19

LJ1251 4.2

b

Figure 3 Structures of selected (a) A_1 and A_{2A} and (b) A_{2B} and A_3 adenosine receptor antagonist probes used as pharmacological tools, and in some cases as clinical candidates. The K_i values (nM) in binding to the appropriate human adenosine receptor are indicated after the name (all are human adenosine receptors, unless indicated 'R' for rat).

adenosine, which often occurs in response to a pathological condition. In some cases (dependent on tissue, receptor subtype, and other conditions), one would wish to boost the adenosine effect, and therefore an allosteric enhancer would be useful; in other cases, the elevated adenosine may be detrimental, in which case one would want to apply a negative modulator. Positive allosteric modulators have been explored for the A_1 (benzoylthiophene derivatives) and A_3 (imidazoquinoline derivatives) AR subtypes.

Role of ARs in Autonomic Nervous System Disorders

Distribution

ARs are widely distributed in the autonomic and enteric nervous systems. Distribution of neural ARs in the human intestine has been investigated. Messenger RNAs (mRNAs) of subtypes of AR are differentially expressed in neural and nonneural layers of the jejunum, ileum, colon, and cecum. The A_1 AR is expressed in jejunal myenteric neurons and colonic submucosal neurons. The A_{2A} AR is also found in other neurons, but A_{2B} AR immunoreactivity is more prominent than that of the A_{2A} AR in myenteric neurons, nerve fibers, and glia. The A_3 AR largely occurs in substance P-positive jejunal submucosal neurons and less in vasoactive intestinal peptide (VIP) neurons.

The AR that mediates the relaxation of bladder strips induced by AR agonists such as CGS21680 and NECA has been classified as A_{2A}; however, the highest mRNA levels are found for A_{2B} transcripts in the bladder. However, reports of tissue distribution based on mRNA levels might not correspond to protein levels. Thus, quantitative determination with a potent and selective A_{2B} AR radioligand may be necessary. Western blot analysis shows that all four ARs are expressed in the uroepithelium. A_1 ARs are prominently localized to the apical membrane of the umbrella cell layer, whereas A_{2A}, A_{2B}, and A_3 ARs are localized intracellularly or on the basolateral membrane of umbrella cells and the plasma membrane of the underlying cell layers.

Various subtypes of ARs have also been detected in the heart, blood vessels, kidney, and other organs throughout the body. A_1, A_{2A}, and A_{2B} ARs are all expressed on normal human airway smooth muscle cells, and both A_1 and A_{2A} ARs are expressed in vagal pulmonary C fibers. A_1 and A_{2A} ARs are highly expressed in gastric mucosa. In the brain, the A_1 AR is widely expressed in almost all areas. Pre- and postsynaptic activation of the A_1 AR inhibits synaptic transmission, in part by suppressing the release of excitatory transmitters. The A_{2A} AR is less widely expressed, with greatest density in the striatum, nucleus accumbens, and olfactory tubercle. The A_{2B} and A_3 ARs are expressed at low density in most brain regions, and are implicated in purinergic signaling in neuronal–glial interactions.

Functions of ARs in the Autonomic and Enteric Systems

The enteric nervous system contains several hundred million neurons located in the myenteric plexuses between muscle coats and submucous plexus. ARs in the enteric nervous system are critical for the control of motor and secretomotor functions. Adenosine is known to suppress intestinal motility by activating putative neural A_1 ARs in the small intestine. A_{2A} and A_{2B} ARs in the myenteric neurons were also suggested to contribute to effects of adenosine on motility. ARs in circular muscle may contribute to the postjunctional actions of adenosine on motility. Adenosine directly modulates intestinal tone in the rat by causing relaxation via A_{2B} ARs or contraction via A_1 ARs in longitudinal muscle cells. Electrophysiological studies in rodents provided evidence for pre- and postsynaptic A_1 AR-mediated inhibition of slow synaptic transmission and presynaptic inhibition of fast synaptic transmission. A_{2B} AR gene expression products are widely expressed in the mucosa of the human intestinal tract, where they are postulated to be involved in the pathophysiology of diarrheal diseases. Both A_{2A} and A_3 ARs have been found in mucosal tissues, suggesting that they influence secretion and/or absorption in the human intestine. The ARs are involved in neuroplastic changes occurring in inflamed gut. The A_{2A} AR modulates the activity of colonic excitatory cholinergic nerves via facilitatory control on inhibitory nitrergic pathways, and such a regulatory function is enhanced in the presence of bowel inflammation.

Activation of the A_{2A} AR inhibits stress-induced gastric inflammation and damage. Thus, selective A_{2A} AR agonists may be useful for preventing ulcers and gastric inflammation. Both A_1 and A_{2A} ARs are involved in gastrin release. The modification of AR expression by changes in intraluminal acidity may represent a novel regulatory feedback mechanism to control gastric acid secretion.

The discharge of urine from bladder is controlled by the nervous system. The causes of bladder dysfunction related to the nervous system include multiple sclerosis, traumatic or developmental brain or spinal cord injury, or Parkinson's disease. Although it is generally agreed that ACh acting on smooth muscle muscarinic receptors is the primary neurologic mechanism controlling bladder emptying, neural stimulation of the bladder is only partially inhibited in many cases by the muscarinic receptor antagonist atropine. The atropine-resistant component of parasympathetic contraction was later found to be ATP sensitive. There is plenty of evidence to suggest that ARs also play an important role in bladder function. It has been shown that adenosine-evoked membrane hyperpolarization and relaxation of bladder smooth muscle is mediated by A_{2A} AR-mediated activation of K_{ATP} channels via adenylate cyclase and elevation of cAMP. High mRNA levels of A_{2B} transcript are also found in the bladder.

Adenosine was found to be released from the uroepithelium, which was potentiated tenfold by stretching the tissue. It is generally accepted that the sensation of bladder fullness is relayed through the mucosal layer by afferent nerves, which are activated by the release of neurotransmitters such as ATP from the urothelium when it is stretched as the bladder fills. It is suggested that adenosine reduces the force of nerve-mediated contractions by acting predominantly at the presynaptic sites at the nerve muscle junction via the A_1 AR.

The autonomic nervous system, a complex and self-organized entity, plays a key role in regulating cardiovascular function. In the heart, a number of intrinsic nerves in the atrial and intra-atrial septum have been shown to release ATP, ACh, 5-hydroxytryptamine, and other neurotransmitters. Adenosine may regulate both the release and the interaction with their receptors of these neurotransmitters. Adenosine-mediated myocardial protection has been suggested to be through a neurogenic pathway. The infarct-reducing effect of intravenous adenosine in intact rats was blocked with both the ganglionic blocker hexamethonium and the nitric oxide synthase inhibitor N^{ω}-nitro-L-arginine. The A_1 AR was determined to be the primary subtype involved in the modulation of norepinephrine release from cardiac nerve terminals using isolated rat hearts.

Vascular tone in blood vessels is controlled by perivascular nerves and the endothelial cells. Adenosine acts directly on smooth muscles as well as modulates the release of neurotransmitters, such as ACh and ATP. A study of neurotransmitter release following electrical depolarization of nerve endings from the rat mesenteric artery suggests that activation of the presynaptic A_{2A} AR and A_3 AR modulates neurotransmission by inhibiting the release of norepinephrine but not neuropeptide Y. The A_1 ARs, not P2 receptors, inhibit prejunctionally sympathetic neurotransmission in the hamster mesenteric arterial bed. Activation of both A_1 AR and A_{2A} AR is required to attenuate neurogenic coronary constriction due to sympathetic stimulation.

In the respiratory system, adenosine has been shown to stimulate vagal pulmonary C fiber terminals through activation of the A_1 AR, and subsequently cause

Table 1 Exploration of the role of adenosine receptors in disorders of the nervous system

Condition/system	*Model*[a]	*Subtypes implicated*	*Relationship*
Aggression	A_1 KO	A_1	Increased aggressiveness
	A_{2A} KO	A_{2A}	Increased aggressiveness
Alzheimers's disease	β-Amyloid	A_{2A}	Reduced neurotoxicity by antagonists
Anxiety	A_{2A} KO	A_{2A}	Increased anxiety
Cardiac arrhythmias	Various	A_1	Agonist protects
Depression	Swim	A_{2A}	Antagonist improves
	A_3 KO	A_3	Increased despair-like behavior
Epilepsy	Various	A_1	Agonist protects in some models
Huntington's disease	3-NP	A_1	Agonist protects striatal damage
	3-NP	A_{2A}	Antagonist protects striatal damage
Memory	A_1 KO	A_1	No change
Neurodegeneration	Ischemia	A_1	Agonist protects
	Ischemia	A_{2A}	Antagonist protects
	Ischemia	A_3	Chronic agonist protects
	A_1 KO	A_1	No effect in adults, benefits newborns
	A_{2A} KO	A_{2A}	Detrimental in newborns, benefits adults
	A_3 KO	A_3	Detrimental in adults
Pain	Constriction injury	A_1	Agonist protects
	Formalin	A_{2A}	Agonist protects
	A_1 KO	A_1	Increased thermal nociception
	A_{2A} KO	A_{2A}	Lowers thermal nociception
Parkinson's disease	6-OHDA	A_{2A}	Antagonist protects
	MPTP	A_{2A}	Antagonist protects
	Haloperidol	A_{2A}	Decreased catalepsy by antagonists
Pulmonary inflammation	ADA KO mice	A_{2B}	Antagonist protects
Sleep disorders	A_{2A} KO	A_{2A}	Agonist lost effect

[a]ADA, adenosine deaminase; KO, knockout; 6-OHDA, 6-hydroxydopamine; MPTP, 1-methyl-4-phenyl-1,2,3,6-tetrahydropyridine; 3-NP, 3-nitropropionic acid.

bronchoconstriction, which was significantly attenuated by A_1 AR antagonists. Recent evidence suggests that both A_1 AR and A_{2A} AR are involved in the activation of vagal sensory C fibers in the lung. The A_1 AR has also been suggested to be directly involved in the mobilization of calcium in human bronchial smooth muscle cells. The A_{2B} AR is known to mediate mast cell degranulation in large animals and humans, and A_{2B} AR antagonists are of potential clinical applications for asthma.

A_1 AR agonists reduce pain signaling in the spinal cord, where the receptors are highly expressed. In humans, infusion of adenosine in the spinal cord is effective in decreasing postoperative pain. Recent studies suggest that A_1 ARs might be more important in chronic pain than in acute pain. The A_1 AR agonists are being evaluated in phase II clinical trials for the treatment of pain and migraine. An A_1 AR-selective allosteric enhancer is also used in clinical trials as a treatment for neuropathic pain. Peripherally administered A_{2A} AR agonists have an antinociceptive effect.

In summary, in the peripheral nervous systems, ARs are related to gut inflammation, rheumatoid arthritis, ischemia, Crohn's disease, constipation, bladder disorders, and a variety of other conditions. In the brain, ARs appear to regulate important functions in cerebral ischemia, dementia, Parkinson's disease and other neurodegenerative diseases, pain, sleep disorders, anxiety, and schizophrenia (**Table 1**).

Genetic Deletion of ARs

Deletion of each of the four subtypes has been carried out, and the resulting single-AR knockout (KO) mice are viable and not highly impaired in function. The pharmacological profile indicates that the analgesic effect of adenosine is mediated by the A_1 AR, and analgesia is lost in mice in which the A_1 AR has been genetically eliminated. Genetic KO of the A_1 AR in mice removes the discriminative-stimulus effects, but not the arousal effect, of caffeine, and increases anxiety and hyperalgesia. Study of A_{2A} AR KO mice reveals functional interaction between the spinal opioid receptors and peripheral ARs. A_1 AR KO mice demonstrate a decreased thermal pain threshold, whereas A_{2A} AR null mice demonstrate an increased threshold to noxious heat stimulation, supporting an A_1 AR-mediated inhibitory and an A_{2A} AR-mediated excitatory effect on pain transduction pathways. KO of the A_{2A} AR eliminates the arousal effect of caffeine. Genetic KO of the A_{2A} AR also suggests a link to increased anxiety and protects against damaging effects of ischemia and the striatal toxin 3-nitropropionic acid. Genetic KO of the A_3 AR leads to increased neuronal damage in a model of carbon monoxide-induced brain injury. Neutrophils lacking A_3 ARs show correct directionality but diminished speed of chemotaxis. Although studies on A_{2B} ARs KO mice have been reported, the importance of A_{2B} ARs in the brain still awaits future investigation.

In conclusion, adenosine is a ubiquitous neuromodulator of many functions, by activating one or more of the widely distributed AR subtypes. Selective AR agonists and antagonists have been developed, some of which are in advance stages of clinical trials for therapeutic applications. The use of both knockout animals and selective drugs has contributed toward elucidation of the physiological role of adenosine and the signaling pathways involved.

See also: Adenosine Receptor Mediated Functions; Autonomic Disorders; Autonomic and Enteric Nervous System: Apoptosis and Trophic Support During Development; Autonomic Nervous System; Neuromodulation; Pharmacology of Sleep: Adenosine; Stimulant and Wake-Promoting Substances.

Further Reading

Burnstock G (2007) Physiology and pathophysiology of purinergic neurotransmission. *Physiological Reviews* 87: 659–797.

Christofi FL, Zhang H, Yu JG, et al. (2001) Differential gene expression of adenosine A1, A2a, A2b, and A3 receptors in the human enteric nervous system. *Journal of Comparative Neurology* 439: 46–64.

Costanzi S, Ivanov AA, Tikhonova IG, et al. (2007) Structure and function of G protein-coupled receptors studied using sequence analysis, molecular modelling, and receptor engineering: Adenosine receptors. In: *Frontiers in Drug Design and Discovery,* vol. 3. Hilversum, The Netherlands: Bentham Science Publishers, Inc.

Fields RD and Burnstock G (2006) Purinergic signalling in neuron-glia interactions. *Nature Reviews Neuroscience* 7: 423–436.

Franco R, Casado V, Mallol J, et al. (2006) The two-state dimer receptor model: A general model for receptor dimers. *Molecular Pharmacology* 69: 1905–1912.

Fredholm BB, IJzerman AP, Jacobson KA, et al. (2001) International Union of Pharmacology. XXV. Nomenclature and classification of adenosine receptors. *Pharmacological Reviews* 53: 527–552.

Fredholm BB, Chen JF, Masino SA, et al. (2005) Actions of adenosine at its receptors in the CNS: Insights from knockouts and drugs. *Annual Review of Pharmacology and Toxicology* 45: 385–412.

Jacobson KA and Gao ZG (2006) Adenosine receptors as therapeutic targets. *Nature Reviews Drug Discovery* 5: 247–264.

Linden J (2005) Adenosine in tissue protection and tissue regeneration. *Molecular Pharmacology* 67: 1385–1387.

McGaraughty S, Cowart M, Jarvis MF, et al. (2005) Anticonvulsant and antinociceptive actions of novel adenosine kinase inhibitors. *Current Topics in Medicinal Chemistry* 5: 43–58.

Moro S, Gao ZG, Jacobson KA, et al. (2006) Progress in pursuit of therapeutic adenosine receptor antagonists. *Medicinal Research Reviews* 26: 131–159.

Ruggieri MR Sr. (2006) Mechanisms of disease: role of purinergic signaling in the pathophysiology of bladder dysfunction. *Nature Clinical Practice Urology* 3: 206–215.

Schulte G and Fredholm BB (2003) Signalling from adenosine receptors to mitogen-activated protein kinases. *Cell Signaling* 15: 813–827.

Yan L, Burbiel JC, Maass A, et al. (2003) Adenosine receptor agonists: From basic medicinal chemistry to clinical development. *Expert Opinion in Emerging Drugs* 8: 537–576.

Yu L, Huang ZL, Mariani J, et al. (2004) Selective inactivation or reconstitution of adenosine A_{2A} receptors in bone marrow cells reveals their significant contribution to the development of ischemic brain injury. *Nature Medicine* 10: 1081–1087.

Adenosine Receptor Mediated Functions

R W Greene, University of Texas Southwestern Medical Center, Dallas, TX, USA

Introduction

The purine adenosine (Ado) forms the backbone of adenosine triphosphate (ATP), the cell's primary energy currency, and accordingly, is well situated to act as a sensor for a cell's metabolic state. Both neurons and cardiac cells express membrane receptors for adenosine as well as for equilibrative adenosine transporters, so that changes in intracellular adenosine metabolism may be reflected extracellularly, and this in turn may exert a paracrine-like effect through activation of Ado receptors.

Extracellular Ado concentration is increased by metabolic stressors that increase the ratio of metabolite demand to metabolite availability (**Figure 1**). Increased electrical activity due to electrical stimulation, increased extracellular potassium, or increased extracellular glutamate (or related agonists) all increase metabolite demand and extracellular Ado. A reduction of oxygen (hypoxia or ischemia), a reduction of glucose, or inhibition of the Kreb's cycle all reduce metabolite availability and increase Ado.

In the mammalian central nervous system, the most prevalently expressed membrane surface Ado receptor (the A1AdoR) exerts an inhibitory effect on excitability. Taken together with the extracellular Ado correlation with metabolism, the primary components of a negative feedback system to maintain the homeostasis between the metabolic state of a neuron (or cardiac cell) and its electrical excitability are in place. Thus, in response to a metabolic stress of local cells, whenever there is an increase of metabolite demand relative to metabolite availability, Ado may be released into the extracellular environment to act on A1AdoRs and inhibit the excitability of these same cells. The reduced excitability results in a reduced metabolic demand, restoring normal metabolic state. This Ado-mediated homeostasis has profound effects on synaptic function, and accordingly on circuit systems and ultimately on behavioral levels, as discussed in the following sections.

Ado Metabolism and Extracellular Ado Concentration

Extracellular Ado derives from a number of sources. ATP is released from synaptic vesicles from both presynaptic terminals and from glia in an activity-dependent manner (**Figure 2**). The ATP is rapidly broken down by ectonucleotidases to form Ado in the extracellular medium, where it has been shown to both reduce postsynaptic excitability and inhibit presynaptic release of neurotransmitter. Other sources include cyclic adenosine monophosphate (cAMP) that may be directly transported from the intracellular to extracellular space and, especially, intracellular Ado. Intracellular Ado is maintained in equilibrium with extracellular Ado by equilibrative, facilitated transport. A slowing of intracellular Ado metabolism can increase extracellular Ado by direct transport of Ado to the outside in extreme or pathological conditions, when intracellular concentration exceeds that on the outside (for either neurons or glia or both), and in less-extreme or in physiological conditions, by a reduction of the flux from the extracellular to intracellular space.

Under physiological conditions, intracellular Ado concentration is maintained primarily by the low-capacity (but high-affinity) enzyme adenosine kinase (AK). This enzyme requires the binding of Ado, AMP, ADP, and ATP as it phosphorylates Ado to AMP by the dephosphorylation of ATP to ADP. This enzyme maintains a ratio of about 10 000:1 between ATP, ADP, AMP, and Ado, consistent with an extremely small change in ATP being reflected as a large percentage change in Ado that will in turn be reflected extracellularly. In addition, this enzyme undergoes substrate inhibition, suggesting that as the intracellular concentration of Ado increases, AK will be inhibited, leading to a still more rapid increase in concentration. This mechanism provides a threshold for Ado production rate that, once exceeded, would be expected to greatly reduce the excitability of the neuronal tissue, within the locale of the production. Finally, at a high intracellular concentration of Ado, as might be expected during ischemia, hypoglycemia, or epileptic activity, the capacity of AK is exceeded and low-affinity, high-capacity pathways become active to catabolize Ado, including the enzymes adenosine deaminase and *S*-adenosine methionine hydrolase.

Adenosine Receptor Activation

Extracellular Ado modulates neuronal excitability through activation of the A1 Ado receptor (A1R) or by activation of the A2a receptor. The expression of the A2a receptor is limited primarily to the striatum, including the nucleus accumbens, while the expression of the A1R is widespread throughout the central nervous system (CNS). Two other Ado receptors – the

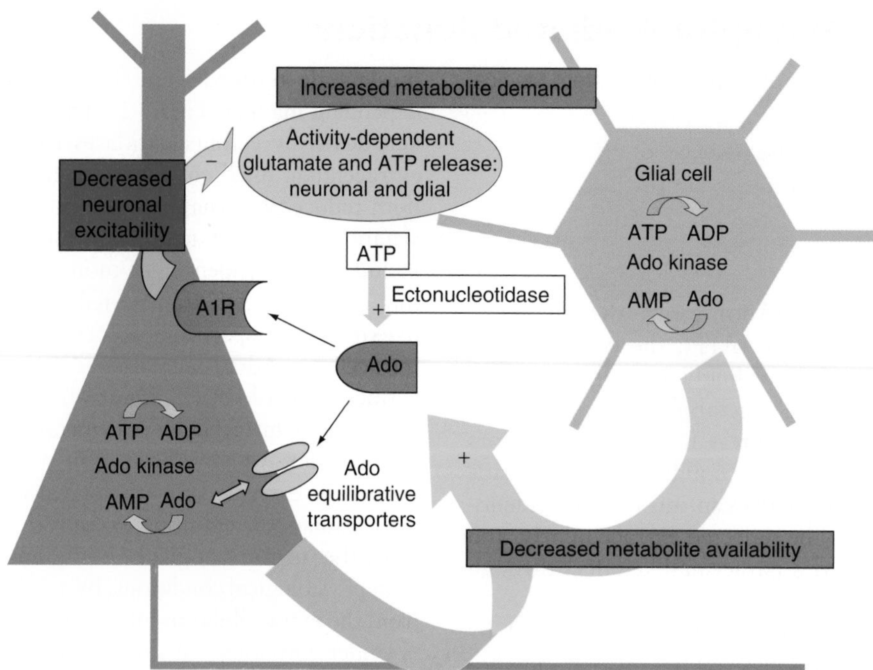

Figure 1 Metabolite demand and metabolite availability ratio, depicting extracellular adenosine (Ado) and the A1 Ado receptor (A1R).

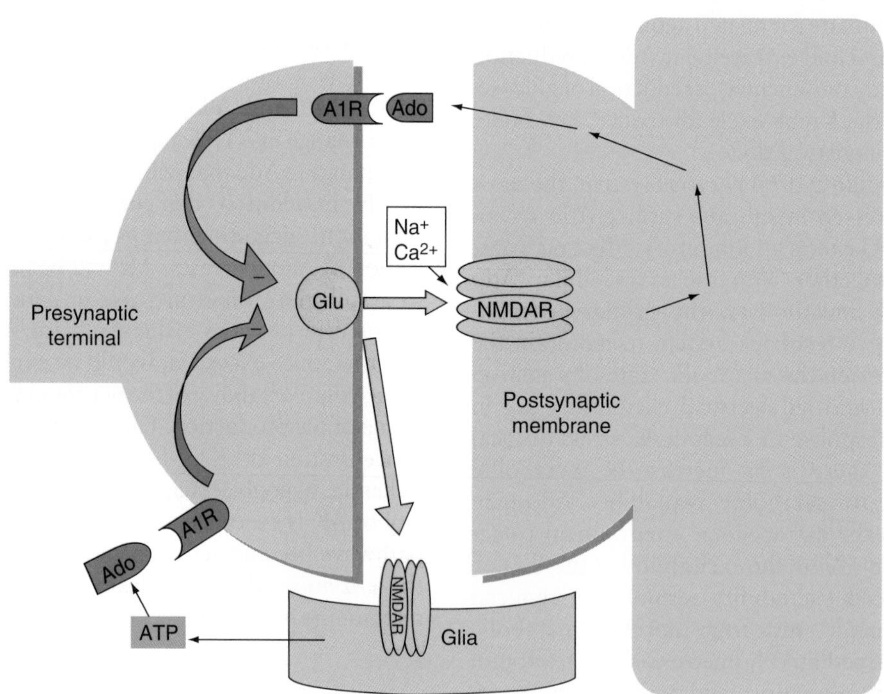

Figure 2 Adenosine (Ado) metabolism. Adenosine triphosphate (ATP) is broken down to form Ado in the extracellular medium, where it has been shown to both reduce postsynaptic excitability and inhibit presynaptic release of neurotransmitter. A1R, A1 Ado receptor; NMDAR, *N*-methyl-D-aspartate receptor.

A2b and A3 receptors – have been cloned; these are expressed in the CNS, although at considerably lower levels than the A1R is. The physiological actions of the A2b and the A3 receptors are not known. All of the receptors are metabotropic and coupled to G-proteins as follows: A1R is primarily coupled to G_i, A2a and A2b receptors are primarily coupled to G_s, and the A3 receptor is primarily coupled to G_{i3}/G_q.

The A2a receptor activation suppresses motor striatal-dependent motor activity and A2a knockout mice have increased motor activity. The electrophysiological mechanism responsible for A2a-dependent motor activity modulation is not known. Evidence from *in vitro* slice intracellular recordings suggests both enhancement and inhibition of GABAergic synaptic transmission. The enhancement, observed in the globus pallidus, is mediated by an A2a-induced cAMP activation of protein kinase A (PKA) to increase miniature synaptic current frequency, suggestive of a presynaptic effect.

A1 Receptor Electrophysiological Postsynaptic Effects

The best-characterized and most significant electrophysiological effects of Ado are mediated by A1 receptors on both the pre- and postsynaptic membranes that are found throughout the CNS. The A1Rs have the highest affinity for Ado ($<100\,\text{nM}$), compared to A2a affinity that is $\sim 2\times$ less sensitive and A2b and A3 receptors that are more than an order of magnitude less sensitive still.

The postsynaptic effects include (1) an increase in the G-protein-dependent, inwardly rectifying potassium current (GIRK); (2) an increase in the calcium-dependent, potassium-mediated, slow after-hyperpolarization; (3) a decrease in hyperpolarization-activated current (I_h); and (4) a decrease in voltage-gated calcium currents. All four effects, but especially the GIRK current increase, may reduce the responsiveness of the neuron. The GIRK current increase from A1R activation at high concentration of Ado ($>100\,\text{nM}$) can render a neuron completely unresponsive to excitatory input and it desensitizes only very slowly, if at all. The increase in GIRK current is less sensitive than are the other Ado effects and thus is more likely to manifest in conditions of extreme or pathological metabolic stress, when Ado release is particularly high.

The increase in GIRK conductance coupled with a decrease in I_h facilitates single-cell oscillation of a 'burst' and a 'pause' in the frequency range of 0.5–4.5 Hz. This is the same frequency that predominates during slow-wave sleep (this is also called the delta frequency range) and imparts the name 'slow wave' to this CNS state. The oscillation is dependent on the low-threshold calcium current (I_T) that provides the depolarizing drive engendering a burst of action potentials during the burst phase of the oscillation. The depolarization results in an inactivation of the I_T (it turns off and cannot be turned back on until the inactivation is removed by hyperpolarization) and a deactivation of I_h (I_h simply turns off and will turn back on when the membrane potential falls within the range of activation for this current). As a result, the membrane potential hyperpolarizes beginning the 'pause' phase of the oscillation. With hyperpolarization, inactivation of I_T is removed (so that with depolarization it may become active, to cause a burst) and I_h is activated, causing the cell to depolarize, until threshold for the I_T- dependent burst is reached and the oscillation is repeated. The GIRK channel facilitates this effect by enhancing the hyperpolarization and, due to its inwardly rectifying property, the conductance is less with depolarization, allowing the burst to readily occur. Ado, acting through A1Rs, increases GIRK channel conductance and decreases I_h, both effects resulting in more hyperpolarization with greater removal of I_T inactivation and a more robust ensuing burst response. If many thalamocortical neurons are influenced by the A1R activation, then it is likely that a more powerful delta oscillation will be observed on the cortical electroencephalogram (EEG) during slow-wave sleep.

A1 Receptor Presynaptic Effects

A1 presynaptic receptors cause presynaptic inhibition. The presynaptic inhibition has been observed mostly with excitatory glutamatergic terminals in the cortex. However, Ado-mediated presynaptic inhibition is well documented for GABAergic terminals as well in the hypothalamus and brain stem tegmentum and most likely in other areas. Presynaptic inhibition of both excitatory and inhibitory inputs amounts to a relative functional deafferentation, predisposing the neuron to its own voltage-gated channel activity that can (as, for example, with thalamic neurons) result in slow oscillations. This may be important for behavioral-state-related activity, as discussed later.

Presynaptic inhibition mediated by A1R activation occurs in both the periphery and the CNS. At the neuromuscular junction, there is evidence for a role of reduced calcium ion flux across the membrane in the Ado-mediated presynaptic inhibition. At CNS excitatory glutamate synapses, a decreased calcium flux may also have a role, but this has yet to be convincingly demonstrated. However, A1R activation causes a well-characterized decrease in synaptic release in the CNS, due to fewer vesicles being released from the pool of vesicles used in action potential-dependent synaptic vesicle release. This decrease in vesicular release from the readily releasable pool is independent of calcium flux across the presynaptic membrane since osmotically evoked release (i.e., independent of calcium flux) is reduced by A1R activation. This effect is very sensitive to extracellular Ado and exerts an inhibitory Ado tone under physiological conditions.

Adenosine-Mediated Homeostasis of Synaptic Glutamate Release

Activation of *N*-methyl-D-aspartate (NMDA) receptors in cortical slices results in an increase of Ado release. Of particular note, this release is sufficient to exert electrophysiological inhibitory effects on glutamate-mediated synaptic release, mediated by A1Rs. It has been recently observed that synaptically released glutamate can, through NMDA receptor activation, cause Ado release that, in turn, activates presynaptic A1Rs on glutamatergic terminals. A complete negative feedback circuit is thus formed, since these presynaptic A1Rs inhibit glutamate synaptic vesicle release of vesicles from the readily releasable pool. (Consistent with inhibition of release from the readily releasable pool, the A1R inhibition also reduces action potential-evoked glutamate excitatory postsynaptic currents.) Thus, synaptically released glutamate activates NMDA receptors that cause an increase in extracellular Ado that feeds back onto presynaptic A1Rs to inhibit the release of glutamate. This negative feedback control exerts a tonic restraint on synaptic glutamate release under conditions that are comparable to those of *in vivo* synaptic release. Following an increase in tonic synaptic activity (the experimental manipulation employed was a small hypertonic stimulus in the presence of tetrodotoxin while monitoring miniature excitatory postsynaptic current frequency as a measure of vesicular release; see **Figure 3**), the activity returns to its former level over ~25–30 min.

The return to the original level of synaptic activity requires NMDA receptor activation, an increase in extracellular Ado, and A1R activation. The source of the extracellular Ado could be neuronal, glial, or both. The mechanism of the Ado release is also unclear, as it may result from a direct release of Ado, from ectonucleotidase-mediated breakdown of extracellular ATP (for example, released from synaptic vesicles), or from slowed reuptake of Ado or some combination of these mechanisms.

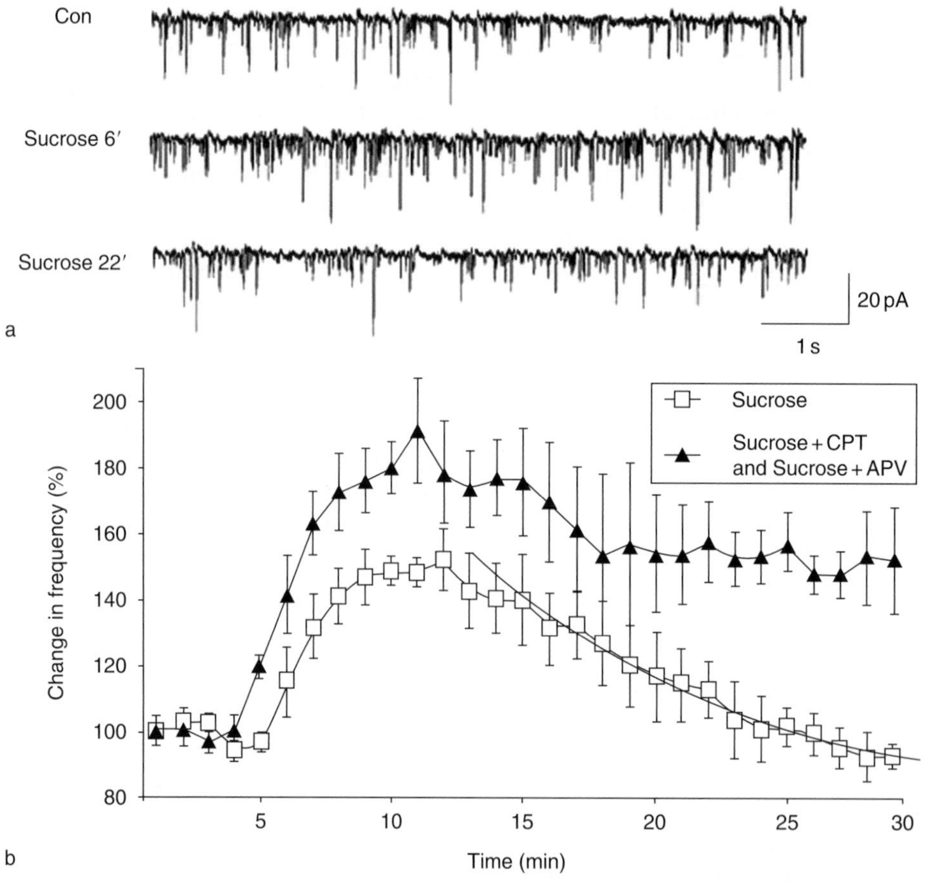

Figure 3 (a) The frequency of miniature excitatory postsynaptic currents mediated by glutamate in the presence of tetrodotoxin increases in hypertonic medium and then (over the next 25–30 min) spontaneously decreases back to the original baseline frequency. The decay occurs despite the continued presence of hypertonic medium. (b) The decay is blocked by an antagonist of the A1 receptor, by catabolism of adenosine on the addition of adenosine deaminase, or by antagonists of the *N*-methyl-D-aspartate receptor. APV, 2-amino-5-phosphonovaleric acid; CON, control; CPT, 8-cyclopenthyl-1,3-dimethylxanthine.

Adenosine and Glia

Glial cells receive glutamatergic synaptic input that can activate calcium-permeable α-amino-3-hydroxy-5-methyl-4-isoxazole propionic acid (AMPA) postsynaptic glial receptors. For example, in the hippocampus, Schaffer collateral input from the CA3 field pyramidal neurons to the CA1 field innervates not only pyramidal neurons and interneurons, but glia as well. In addition, these glial cells have vesicles that release their contents, including ATP, in a calcium-dependent manner. It was recently demonstrated that ATP released from glial vesicles can be broken down to Ado by ectonucleotidase and can reduce neuronal excitability in the hippocampus; this, in turn, reduces the likelihood of long-term potentiation of excitatory synaptic input to pyramidal cells. It is conceivable that the metabolic state of the hippocampal neural tissue determines the concentration of Ado in the extracellular medium derived from glial sources and, thus, affects hippocampal-dependent learning and memory. Indeed, when rodents explore a new environment, local Ado concentration, assessed by microdialysis, increases, suggesting a link with hippocampal activity and information processing. A similar correlation has been observed with motor activity and striatal Ado concentration.

Adenosine and Slow-Wave Sleep

A necessary condition for the transition from wake to sleep is the relative quiescence of the two cholinergic thalamocortical activating centers in the brain stem and basal forebrain. Ado, released locally, exerts an inhibitory tonus on the neurons of these two thalamocortical activating centers that is capable of reducing their excitability and action potential firing rates. The Ado exerts its inhibitory effect through activation of both presynaptic and postsynaptic A1Rs as described earlier. An increase in extracellular Ado in either or both of the two cholinergic arousal centers, resulting from either a local increase in metabolic demand (for example, increased neuronal activity) or a decrease in metabolite availability, can facilitate a transition from waking to sleep. Thus local metabolic factors may be linked through Ado to a change of behavioral state. Indeed, a local increase in extracellular Ado is sufficient to facilitate the transition from waking to sleep, as demonstrated by the sleep-inducing effects of local microdialysis of Ado into the brain stem or the basal forebrain cholinergic arousal centers.

The rebound sleep response to sleep deprivation suggests a homeostatic control of sleep, with characteristic dynamics that have been modeled as 'Process S' by Borbely and colleagues. After 6 h or more of enforced waking, a well-characterized rebound sleep ensues in mammals, including C57/BL mice. The neurophysiological mechanism(s) and substrates of this process are unknown, although recent evidence indicates that it is under genetic control. Synchronized, electrical, neuronal oscillations in the delta wave frequency range of 0.5–4.5 Hz can be quantified as delta power using Fourier analysis of the surface EEG wave form, and the delta power during the nonrapid eye movement sleep (NREM) period following sleep deprivation is proportional to the prior sleep loss. Further, delta power decreases during an NREM episode with a time course that is independent of the circadian phase. Together, these findings suggest a sleep homeostatic process that is closely associated with delta power.

Sleep deprivation elicits an increase in the extracellular level of Ado, and when sleep is allowed the Ado level returns toward baseline in the basal forebrain, where changes in extracellular Ado may be reasonably assessed by currently available microdialysis techniques. Thus, it may be predicted that a localized disruption of normal A1R-mediated inhibition would dampen the normal homeostatic response (both the rebound NREM duration and total delta power increase) to prolonged waking.

The observed increase in Ado with prolonged waking suggests a state-related control of Ado levels in the cholinergic arousal centers. Changes in Ado concentration are considered to regulate neurotransmission in the brain, especially with respect to glutamatergic neurotransmission, as described previously for the negative feedback circuit mediated by Ado at glutamatergic synapses that drive neurons in the cholinergic arousal centers. It has now been demonstrated that NMDA receptor (NMDAR) activation is sufficient to exert an Ado-mediated inhibition in the brain stem cholinergic arousal center to affect sleep/wake states in a manner that is closely related to glutamate-mediated afferent input. Glutamate-mediated input to the brain stem arousal center is greatly reduced during NREM sleep as evidenced by the decreased neuronal activity during this state, consistent with a negative feedback role (not necessarily an exclusive one) for NMDAR-induced adenosine-mediated inhibition in sleep/wake homeostasis (**Figure 4**). Accordingly, with prolonged glutamate activity and NMDAR activation of waking, extracellular Ado concentration will increase, reducing the glutamate-mediated drive onto the arousal centers, facilitating the transition from waking to sleep. Thus, an Ado-mediated negative feedback loop to maintain homeostasis of tonic glutamate-mediated transmission and homeostasis between metabolic and electrophysiological states, localized to the arousal centers, can have a global effect, as the widely divergent cholinergic (and possibly

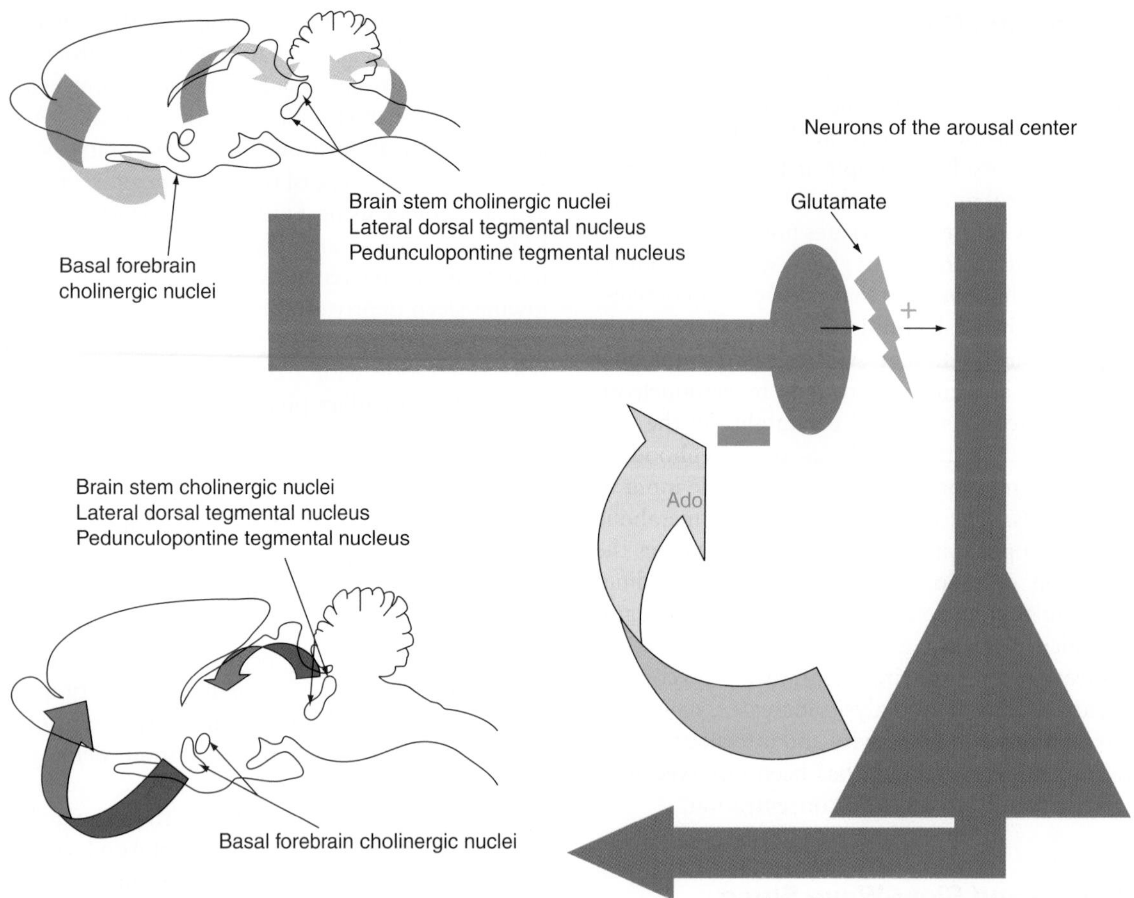

Figure 4 Neurons of the brain stem and forebrain arousal centers: from local to global effects, depicting the negative feedback circuit mediated by adenosine (Ado) at glutamatergic synapses.

gabaergic as well) tone exerted by these centers is diminished by Ado-dependent influence. Additionally at the target sites of the cholinergic centers' tone, especially the thalamocortical systems, extracellular Ado may be increased, as determined by the degree of metabolic demand placed on the particular system by the waking activity. In regions where the Ado is high, synchronous oscillatory activity in the delta frequency range (0.5–4.5 Hz) may be enhanced by Ado. During periods of prolonged wakefulness the increase in Ado may be widespread and may account, at least in part, for the observed rebound delta power observed in the surface EEG following periods of sleep deprivation. The burst–pause neuronal activity underlying the delta wave generation must be associated with an oscillation of intracellular calcium concentration, with the potential for eliciting a sleep functionally relevant change in intracellular metabolism. In summary, Ado might act at two sites to facilitate slow-wave sleep, as follows: first, at the cholinergic arousal centers, to reduce the modulatory influence of these centers on the rest of the CNS, and, second, at their target sites, to facilitate delta frequency oscillations.

See also: Adenosine Triphosphate (ATP); Adenosine Triphosphate (ATP) as a Neurotransmitter; NMDA Receptors and Development; NMDA Receptors, Cell Biology and Trafficking; NMDA Receptor Function and Physiological Modulation; Pharmacology of Sleep: Adenosine.

Further Reading

Brambilla D, Chapman D, and Greene R (2005) Adenosine mediation of presynaptic feedback inhibition of glutamate release. *Neuron* 46: 275–283.

Dunwiddie TV and Masino SA (2001) The role and regulation of adenosine in the central nervous system. *Annual Review of Neuroscience* 24: 31–55.

Greene RW and Haas HL (1991) The electrophysiology of adenosine in the mammalian central nervous system. *Progress in Neurobiology* 36: 329–341.

Manzoni OJ, Manabe T, and Nicoll RA (1994) Release of adenosine by activation of NMDA receptors in the hippocampus. *Science* 265: 2098–2101.

Pascual O, Casper KB, Kubera C, et al. (2005) Astrocytic purinergic signaling coordinates synaptic networks. *Science* 310: 113–116.

Porkka-Heiskanen T, Strecker RE, Thakkar M, et al. (1997) Adenosine: A mediator of the sleep-inducing effects of prolonged wakefulness. *Science* 276: 1265–1268.

Rainnie DG, Grunze HC, McCarley RW, et al. (1994) Adenosine inhibition of mesopontine cholinergic neurons: Implications for EEG arousal. *Science* 263: 689–692.

Adenosine Triphosphate (ATP)

G Burnstock, Royal Free and University College School of Medicine, London, UK

Early History

Drury and Szent-Györgyi, in 1929, were the first to demonstrate the potent extracellular actions of adenosine 5′-triphosphate (ATP) and adenosine on the heart and coronary blood vessels. In 1948, Emmelin and Feldberg demonstrated that intravenous injection of ATP into cats caused complex effects that affected both peripheral and central mechanisms. Injection of ATP into the lateral ventricle produced muscular weakness, ataxia, and a tendency of the cat to sleep. Application of ATP to various regions of the brain produced biochemical or electrophysiological changes. Holton presented in 1959 the first hint of a transmitter role for ATP in the nervous system by demonstrating the release of ATP during antidromic stimulation of sensory nerves supplying the ear artery. Buchthal and Folkow recognized a physiological role for ATP at the neuromuscular junction in 1948, finding that acetylcholine (ACh)-evoked contraction of skeletal muscle fibers was potentiated by exposure to ATP. Presynaptic modulation of ACh release from the neuromuscular junction by purines in the rat was also reported.

The existence of nonadrenergic, noncholinergic (NANC) neurotransmission in the gut and bladder was established in the mid-1960s. Several years later, after many experiments, Burnstock and his colleagues published a study that suggested that the NANC transmitter in the guinea pig taenia coli and stomach, rabbit ileum, frog stomach, and turkey gizzard was ATP. The experimental evidence included mimicry of the NANC nerve-mediated response by ATP; measurement of release of ATP during stimulation of NANC nerves with luciferin-luciferase luminometry; histochemical labeling of subpopulations of neurons in the gut with quinacrine, a fluorescent dye known to selectively label high levels of ATP bound to peptides; the later demonstration that the slowly degradable analogue of ATP, α,β-methylene ATP (α,β-meATP), which produces selective desensitization of the ATP receptor, blocked the responses to NANC nerve stimulation. Soon after, evidence was presented for ATP as the neurotransmitter for NANC excitatory nerves in the urinary bladder. The term 'purinergic' was proposed in a short letter to *Nature* in 1971, and the evidence for purinergic transmission in a wide variety of systems was presented in *Pharmacological Reviews* in 1972 (**Figure 1**). This concept met with considerable resistance for many years. This was partly because it was felt that ATP was established as an intracellular energy source involved in various metabolic cycles and that such a ubiquitous molecule was unlikely to be involved in extracellular signaling. However, ATP was one of the biological molecules to first appear, and therefore it is not really surprising that it should have been utilized for extracellular, as well as intracellular, purposes early in evolution. The fact that potent ectoATPases were described in most tissues in the early literature was also a strong indication for the extracellular actions of ATP. Purinergic neurotransmission is now generally accepted, and a volume of *Seminars in Neuroscience* was devoted to purinergic neurotransmission in 1996.

Purinergic Cotransmission

Another concept that has had a significant influence on our understanding of purinergic transmission was that of cotransmission. Burnstock wrote a commentary in *Neuroscience* in 1976 titled, "Do some nerves release more than one transmitter?" This position challenged the single-neurotransmitter concept, which became known as 'Dale's Principle,' even though Dale himself never defined it as such. The commentary was based on hints about cotransmission in the early literature describing both vertebrate and invertebrate neurotransmission and more specifically, with respect to purinergic cotransmission, on the surprising discovery in 1971 that ATP was released from sympathetic nerves supplying the taenia coli as well as from NANC inhibitory nerves. The excitatory junction potentials (EJPs) recorded in the vas deferens were blocked by α,β-meATP, a selective desensitizer of P2X receptors (**Figures 2(a)** and **2(b)**). This finding clearly supported the earlier demonstration of sympathetic cotransmission in the vas deferens in the laboratory of Dave Westfall, following an earlier report of sympathetic cotransmission in the cat nictitating membrane. Purinergic cotransmission was later described in the rat tail artery and in the rabbit saphenous artery. Noradrenaline (NA) and ATP are now well established as cotransmitters in sympathetic nerves (see **Figure 3 (a)**), although the proportions vary in different tissues and species, during development and aging, and in different pathophysiological conditions.

ACh and ATP are cotransmitters in parasympathetic nerves supplying the urinary bladder. Subpopulations of sensory nerves have been shown to utilize ATP in addition to substance P and calcitonin gene-related peptide; it seems likely that ATP cooperates

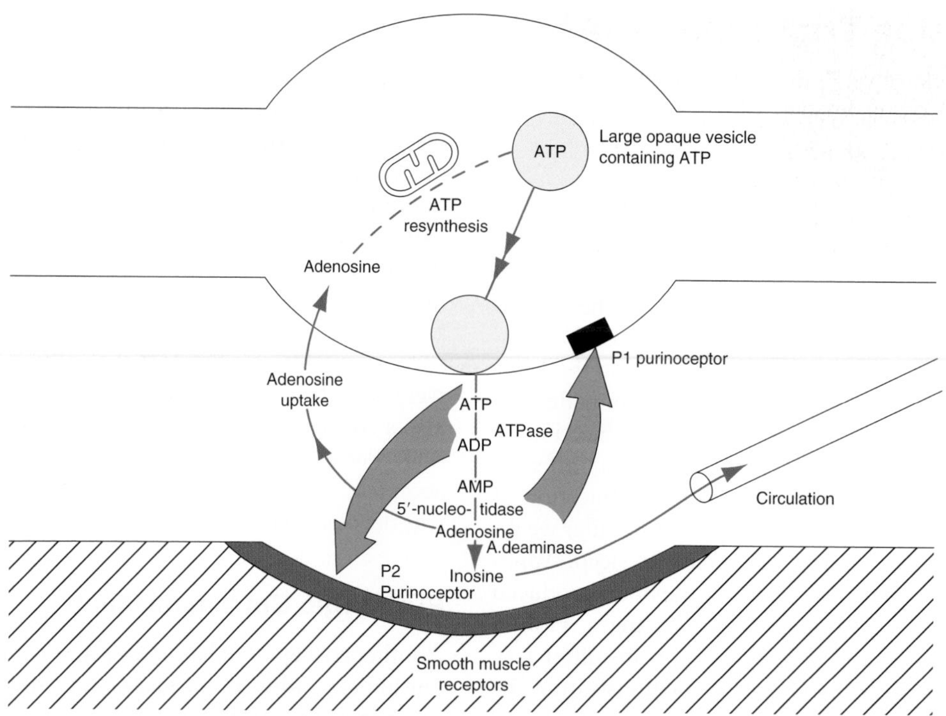

Figure 1 Purinergic neuromuscular transmission depicting the synthesis, storage, release, and inactivation of adenosine 5′-triphosphate (ATP). ATP, stored in vesicles in nerve varicosities, is released by exocytosis to act on postjunctional P2 purinoceptors on smooth muscle. ATP is broken down extracellularly by ATPases and 5′-nucleotidase to adenosine, which is taken up by varicosities to be resynthesized and restored in vesicles. Adenosine acts prejunctionally on P1 purinoceptors to modulate transmitter release. If adenosine is broken down further by adenosine deaminase to inosine, it is removed by the circulation. Adapted from Burnstock G (1972) Purinergic nerves. *Pharmacological Reviews* 24: 509–581, with permission from the American Society for Pharmacology and Experimental Therapeutics.

with these peptides in axon reflex activity. ATP, vasoactive intestinal polypeptide and nitric oxide (NO) are cotransmitters in NANC inhibitory nerves, but that they vary considerably in proportion in different regions of the gut. More recently, ATP has been shown to be a cotransmitter with NA, 5-hydroxytryptamine, glutamate, dopamine and γ-aminobutyric acid (GABA) in the central nervous system (CNS) (see **Figure 3(b)**). ATP and NA act synergistically to release vasopressin and oxytocin, which is consistent with ATP cotransmission in the hypothalamus. ATP, in addition to glutamate, is involved in long-term potentiation in hippocampal CA1 neurons associated with learning and memory.

Receptors for Purines and Pyrimidines

Implicit in the purinergic neurotransmission hypothesis was the presence of purinoceptors. A basis for distinguishing two types of purinoceptor, identified as P1 and P2 for adenosine and ATP/adenosine diphosphate (ADP), respectively, was recognized in 1978. This helped resolve some of the ambiguities in earlier reports, which were complicated by the breakdown of ATP to adenosine by ectoenzymes so that some of the actions of ATP were directly on P2 receptors, whereas others were due to indirect action via P1 receptors.

At about the same time, two subtypes of P1 (adenosine) receptor were recognized, but it was not until 1985 that a pharmacological basis for distinguishing two types of P2 receptors (P2X and P2Y) was proposed. A year later, two further P2 receptor subtypes were named, a P2T receptor selective for ADP on platelets and a P2Z receptor on macrophages. Further subtypes followed, perhaps the most important being the P2U receptor, which could recognize pyrimidines such as uridine 5′ triphosphate (UTP) in addition to ATP. However, to provide a more manageable framework for newly identified nucleotide receptors, Abbracchio and Burnstock proposed in 1994 that purinoceptors should belong to two major families: a P2X family of ligand-gated ion channel receptors and a P2Y family of G-protein-coupled receptors. This was based on studies of transduction mechanisms and the cloning of nucleotide receptors: P2Y receptors were cloned first, in 1993, and a year later P2X receptors were cloned. This nomenclature has been widely adopted, and currently seven P2X subtypes and eight P2Y

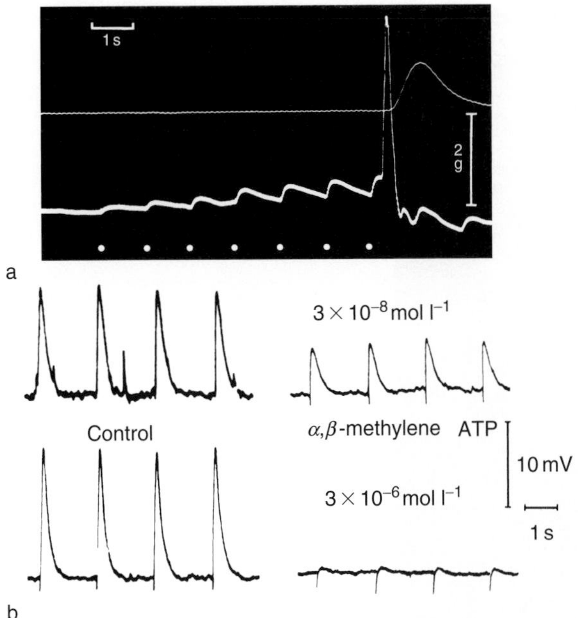

Figure 2 (a) EJPs in response to repetitive stimulation of adrenergic nerves (white dots) in the guinea pig vas deferens. The upper trace records the tension, the lower trace the electrical activity of the muscle recorded extracellularly by the sucrose gap method. Note both summation and facilitation of successive junction potentials. At a critical depolarization threshold, an action potential is initiated which results in contraction. (b) The effect of various concentrations of α,β-methylene ATP (α,β-meATP) on EJPs recorded from guinea pig vas deferens (intracellular recordings). The control responses to stimulation of the motor nerves at 0.5 Hz are shown on the left. After at least 10 min in the continuous presence of the indicated concentration of α,β-meATP, EJPs were recorded using the same stimulation parameters. (a) Reproduced from Burnstock G and Costa M (eds.) (1975) Adrenergic neuroeffector transmission. In *Adrenergic Neurones: Their Organisation, Function and Development in the Peripheral Nervous System*, pp. 51–106. London: Chapman and Hall, with permission from Springer Science and Business Media. (b) Reproduced from Sneddon P and Burnstock G (1984) Inhibition of excitatory junction potentials in guinea-pig vas deferens by α,β-methylene-ATP: Further evidence for ATP and noradrenaline as cotransmitters. *European Journal of Pharmacology* 100: 85–90, with permission from Elsevier.

receptor subtypes are recognized. Four subtypes of P1 receptor have been cloned and characterized.

P2X receptors in general mediate fast neurotransmission but are sometimes located prejunctionally to mediate increase in release of cotransmitters, for example, glutamate in terminals of primary afferent neurons in the spinal cord. $P2X_3$ and $P2X_{2/3}$ receptors are prominent in sensory neurons and are involved in nociception. $P2X_7$ receptors are involved in cell death.

P2Y receptors are particularly involved in prejunctional inhibitory modulation of transmitter release, as well as long-term (trophic) events such as cell pro liferation. $P2Y_1$ receptors are widespread in many regions of the brain, while the $P2Y_2$ receptors have been localized on pyramidal neurons in the hippocampus and prefrontal cortex, on supraoptic magnocellular neurosecretory neurons in the hypothalamus, and on neurons in the dorsal horn of the spinal cord. In addition, mRNA but not protein has been reported for $P2Y_4$ and $P2Y_6$ receptor subtypes in the cerebellum and hippocampus, while $P2Y_{12}$ receptor mRNA has also been described in the cerebellum and $P2Y_{13}$ in the cortex. In the periphery, $P2Y_{1,2,4,6}$ receptors have been described on subpopulations of sympathetic neurons, $P2Y_2$ and $P2Y_4$ receptors in intracardiac ganglia, $P2Y_1$ and $P2Y_2$ receptors on sensory neurons (although $P2Y_4$ and $P2Y_6$ mRNA have also been reported) while $P2Y_1$ receptors appear to be the dominant subtype on enteric neurons. $P2Y_{1,2,4,6}$ functional receptors have been described on astrocytes in the CNS and also on microglia, where functional $P2Y_{12}$ receptors have also been identified. $P2Y_1$ and $P2Y_2$ receptors have been located in Schwann cells and oligodendrocytes, where functional $P2Y_{12}$ receptors also appear to be present. $P2Y_2$ (and/or $P2Y_4$) receptors are expressed on enteric glial cells. There is also emerging evidence for P2Y receptors on stem cells.

ATP Release and Degradation

There is clear evidence for exocytotic vesicular release of ATP from nerves, and the concentration of nucleotides in vesicles is claimed to be up to 1000 mmol l^{-1}. It was generally assumed that the main source of ATP acting on purinoceptors was damaged or dying cells. However, it is now recognized that ATP release from many cells is a physiological or pathophysiological response to mechanical stress, hypoxia, inflammation, and some agonists. There is debate, however, about the ATP transport mechanisms involved. There is compelling evidence for exocytotic release from endothelial and urothelial cells, osteoblasts, astrocytes, mast, and chromaffin cells, but other transport mechanisms have also been proposed, including ATP binding cassette transporters, connexin hemichannels, and plasmalemmal voltage-dependent anion channels.

Much is now known about the ectonucleotidases that break down ATP released from neurons and nonneuronal cells. Several enzyme families are involved: ecto-nucleoside triphosphate diphosphohydrolases (E-NTPDases), of which NTPDase1, 2, 3, and 8 are extracellular; ectonucleotide pyrophosphatase of 3 subtypes; alkaline phosphatases; ecto-5′-nucleotidase; and ecto-nucleoside diphosphokinase. NTPDase1 hydrolyzes ATP directly to adenosine monophosphate (AMP) and UTP to uridine diphosphate (UDP), while NTPDase2 hydrolyzes ATP to ADP and 5′-nucleotidase AMP to adenosine.

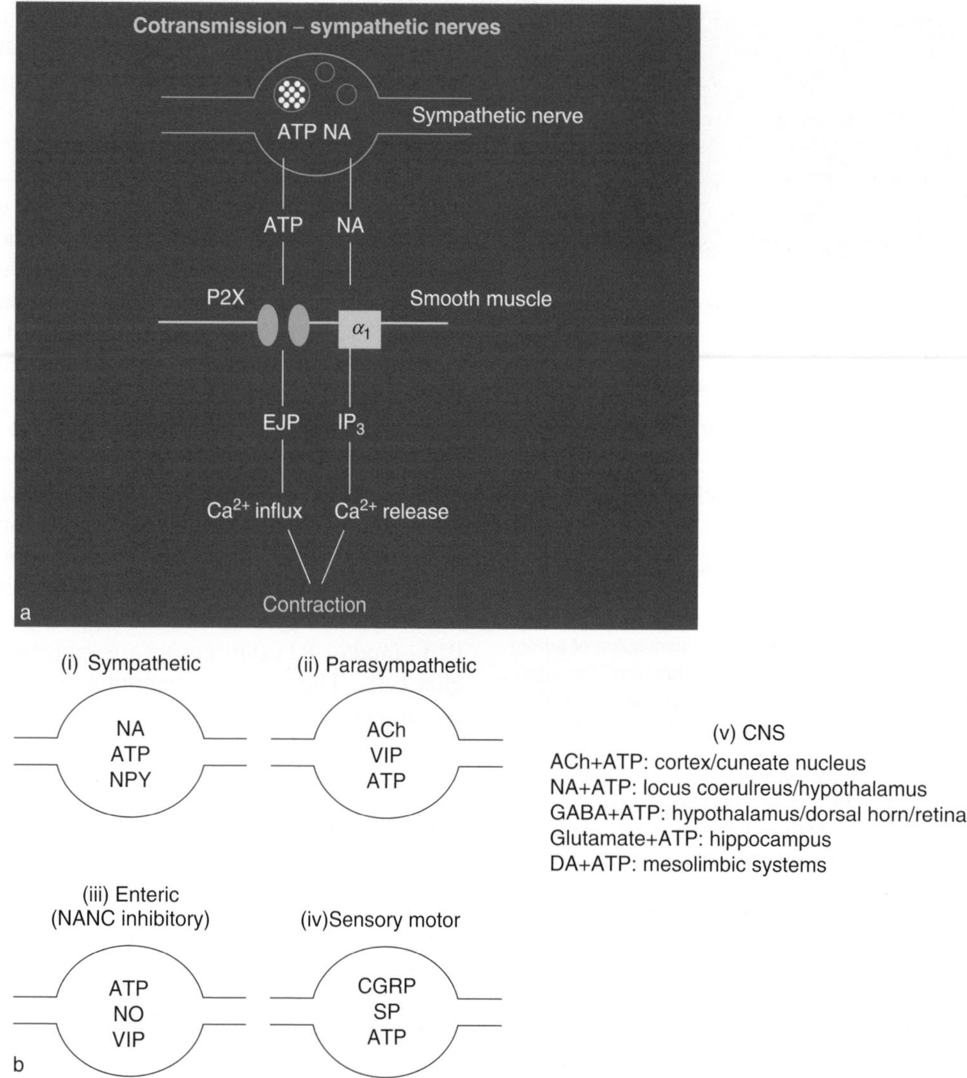

Figure 3 (a) Cotransmission in sympathetic nerves. Adenosine 5′-triphosphate (ATP) and noradrenaline (NA) from terminal varicosities of sympathetic nerves can be released together. With NA acting via the postjunctional α_1-adrenoceptor to release cytosolic Ca^{2+}, and with ATP acting via $P2X_1$-gated ion channels to elicit Ca^{2+} influx, both contribute to the subsequent response (contraction). IP_3 is inositol triphosphate, EJP is excitatory junction potential. (b) Schematic diagram of the principal cotransmitters with ATP in the nervous system. Nerve terminal varicosities of (i) sympathetic, (ii) parasympathetic, (iii) enteric (NANC inhibitory), (iv) sensory-motor neurons, and (v) central nervous system (CNS). (a) Adapted from Kennedy C, McLaren GJ, Westfall TD, and Sneddon P (1996) ATP as a co-transmitter with noradrenaline in sympathetic nerves – Function and fate. In: Chadwick DJ and Goode J (eds.) *P2 Purinoceptors: Localization, Function and Transduction Mechanisms*, pp. 223–235. Chichester: John Wiley and Sons, with permission from John Wiley & Sons. (b) Reproduced from Burnstock G (2007) Physiology and pathophysiology of purinergic neurotransmission. *Physiological Reviews* 87: 659–797, with permission from The American Physiological Society.

Physiology of Purinergic Neurotransmission

Purinergic signaling appears to be a primitive system that is involved in many nonneuronal and neuronal mechanisms, in both short-term and long-term (trophic) events, including exocrine and endocrine secretion, immune responses, inflammation, mechanosensory transduction, platelet aggregation, endothelial-mediated vasodilatation and in cell proliferation, differentiation, migration, and death in development and regeneration.

The first clear evidence for nerve–nerve purinergic synaptic transmission was published in 1992. Synaptic potentials in the coeliac ganglion and in the medial habenula in the brain were reversibly antagonized by the antitrypanosomal agent suramin. Since then, many articles have described either the distribution of various P2 receptor subtypes in the brain and spinal cord or electrophysiological studies of the effects of purines in brain slices, isolated neurons, and glial cells. Synaptic transmission has also been demonstrated in the myenteric plexus and in various sensory, sympathetic, parasympathetic, and pelvic ganglia. Adenosine produced by the ectoenzymatic breakdown of ATP acts through presynaptic P1 receptors to inhibit the release of excitatory neurotransmitters in both the peripheral and the central nervous systems. Purinergic signaling is also implicated in higher order cognitive functions, including learning and memory in the prefrontal cortex.

CNS Control of Autonomic Function

Functional interactions seem likely to occur between purinergic and nitrergic neurotransmitter systems; these interactions might be important for the regulation of hormone secretion and body temperature at the hypothalamic level and for cardiovascular and respiratory control at the level of the brain stem. The nucleus tractus solitarius (NTS) is a major integrative center of the brain stem involved in reflex control of the cardiovascular system; stimulation of P2X receptors in the NTS evokes hypotension. P2X receptors expressed in neurons in the trigeminal mesencephalic nucleus might be involved in the processing of proprioceptive information.

Neuron–Glia Interactions

ATP is an extracellular signaling molecule between neurons and glial cells. ATP released from astrocytes might be important in triggering cellular responses to trauma and ischemia by initiating and maintaining reactive astrogliosis, which involves striking changes in the proliferation and morphology of astrocytes and microglia. Some of the responses to ATP released during brain injury are neuroprotective, but at higher concentrations, ATP contributes to the pathophysiology initiated after trauma. Multiple P2X and P2Y receptor subtypes are expressed by astrocytes, oligodendrocytes, and microglia. ATP and basic fibroblast growth factor (bFGF) signals merge at the mitogen-activated protein kinase cascade, which underlies the synergistic interactions of ATP and bFGF in astrocytes. ATP can activate $P2X_7$ receptors in astrocytes to release glutamate, GABA, and ATP, which regulate the excitability of neurons.

Microglia, immune cells of the CNS, are also activated by purines and pyrimidines to release inflammatory cytokines such as interleukins 1β (IL-1β) and IL-6 and tumor necrosis factor α. Thus, although microglia might play an important role against infection in the CNS, overstimulation of this immune reaction might accelerate the neuronal damage caused by ischemia, trauma, or neurodegenerative diseases. $P2X_4$ receptors induced in spinal microglia gate tactile allodynia after nerve injury. $P2X_7$ receptors mediate superoxide production in primary microglia and are upregulated in a transgenic mouse model of Alzheimer's disease, particularly around β-amyloid plaques.

Purine Transmitter and Receptor Plasticity

The purinergic neurotransmission field is expanding rapidly; there is increasing interest in the physiology and pathophysiology of this neurosignaling system, and therapeutic interventions are being explored. The autonomic nervous system shows marked plasticity: that is, the expression of cotransmitters and receptors shows dramatic changes during development and aging, in nerves that remain after trauma or surgery, and in disease conditions. There are several examples where the purinergic component of cotransmission is increased in pathological conditions. The parasympathetic purinergic nerve-mediated component of contraction of the human bladder is increased to 40% in pathophysiological conditions such as interstitial cystitis, outflow obstruction, idiopathic instability, and also some types of neurogenic bladder. ATP also has a significantly greater cotransmitter role in sympathetic nerves supplying hypertensive compared to normotensive blood vessels. Upregulation of $P2X_1$ and $P2Y_2$ receptor mRNA in hearts of rats with congestive heart failure has been reported, and there is a dramatic increase in expression of $P2X_7$ receptors in the kidney glomerulus in diabetes and hypertension.

Neuroprotection

In the brain, purinergic signaling is involved in nervous tissue remodeling following trauma, stroke, ischemia, or neurodegenerative disorders. The hippocampus of chronic epileptic rats shows abnormal responses to ATP associated with increased expression of $P2X_7$ receptors. Neuronal injury releases fibroblast growth factor, epidermal growth factor, and platelet-derived growth factor. In combination with these growth factors, ATP can stimulate astrocyte proliferation, contributing to the process of reactive astrogliosis and to hypertrophic and hyperplasic responses. P2Y receptor antagonists have been proposed as potential neuroprotective agents in the cortex, hippocampus, and cerebellum. Blockade of

A_{2A} (P1) receptors antagonizes tremor in Parkinson's disease. ATP–$MgCl_2$ is being explored for the treatment of spinal cord injuries.

Dual Purinergic Neural and Endothelial Control of Vascular Tone and Angiogenesis

ATP and adenosine are much involved in the mechanisms underlying local control of vessel tone in addition to cell migration, proliferation, and death during angiogenesis, atherosclerosis, and restenosis following angioplasty. ATP, released as a cotransmitter from sympathetic nerves, constricts vascular smooth muscle via P2X receptors, whereas ATP released from sensory-motor nerves during 'axon reflex' activity dilates vessels via P2Y receptors. Furthermore, ATP released from endothelial cells during changes in flow (shear stress) or hypoxia acts on P2Y receptors in endothelial cells to release NO, resulting in relaxation (**Figure 4**). Adenosine, following breakdown of extracellular ATP, produces vasodilatation via smooth muscle P1 receptors.

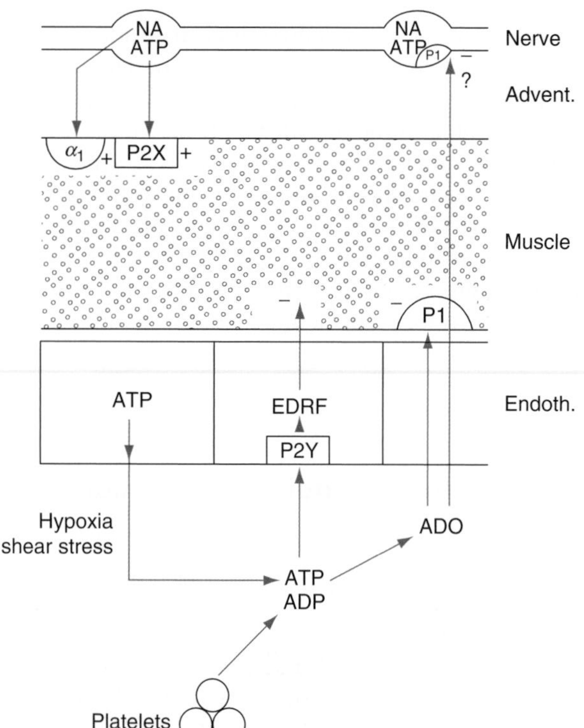

Figure 4 A schematic representation of the interactions of adenosine 5′-triphosphate (ATP) released from perivascular nerves and from the endothelium (Endoth.). ATP is released from endothelial cells during hypoxia to act on endothelial P2Y receptors, leading to the production of endothelium-derived relaxing factor (EDRF), nitric oxide (NO), and subsequent vasodilation (–). In contrast, ATP released as a cotransmitter with noradrenaline (NA) from perivascular sympathetic nerves at the adventitia (Advent.)–muscle border produces vasoconstriction (+) via P2X receptors on the muscle cells. Adenosine (ADO), resulting from rapid breakdown of ATP by ectoenzymes, produces vasodilation by direct action on the muscle via P1 receptors and acts on the perivascular nerve terminal varicosities to inhibit transmitter release. Reproduced from Burnstock G (1987) Local control of blood pressure by purines. *Blood Vessels* 24: 156–160, with permission from S. Karger AG, Basel.

Pain and Purinergic Mechanosensory Transduction

The involvement of ATP in the initiation of pain was recognized first in 1966 and later in 1977 using human skin blisters. A major advance was made when the $P2X_3$ ionotropic receptor was cloned in 1995 and shown later to be predominantly localized in the subpopulation of small nociceptive sensory nerves that label with isolectin B4 in dorsal root ganglia whose central projections terminate in inner lamina II of the dorsal horn. A unifying purinergic hypothesis for the initiation of pain was proposed in 1996 with ATP acting via $P2X_3$ and $P2X_{2/3}$ receptors associated with causalgia, reflex sympathetic dystrophy, angina, migraine, and pelvic and cancer pain. This has been followed by an increasing number of published reports expanding on this concept for acute, inflammatory, neuropathic, and visceral pain. $P2Y_1$ receptors have also been demonstrated in a subpopulation of sensory neurons that colocalized with $P2X_3$ receptors.

A hypothesis was proposed that purinergic mechanosensory transduction occurred in visceral tubes and sacs, including ureter, bladder, and gut, where ATP, released from epithelial cells during distension, acted on $P2X_3$ homomultimeric and $P2X_{2/3}$ heteromultimeric receptors on subepithelial sensory nerves, initiating impulses in sensory pathways to pain centers in the CNS (**Figure 5(a)**). Subsequent studies of bladder, ureter, gut, tongue, and tooth pulp have produced evidence in support of this hypothesis. $P2X_3$ knockout mice were used to show that ATP released from urothelial cells during distension of the bladder act on $P2X_3$ receptors on subepithelial sensory nerves to initiate both nociceptive and bladder voiding reflex activities. In the distal colon, ATP released during moderate distension acts on $P2X_3$ receptors on low-threshold intrinsic subepithelial sensory neurons to influence peristalsis, whereas high-threshold extrinsic subepithelial sensory fibers respond to severe distension to initiate pain (see **Figure 5(b)**).

ATP is also a neurotransmitter released from the spinal cord terminals of primary afferent sensory nerves to act at synapses in the central pain pathway. Using transverse spinal cord slices from postnatal rats, excitatory postsynaptic currents have been shown to be

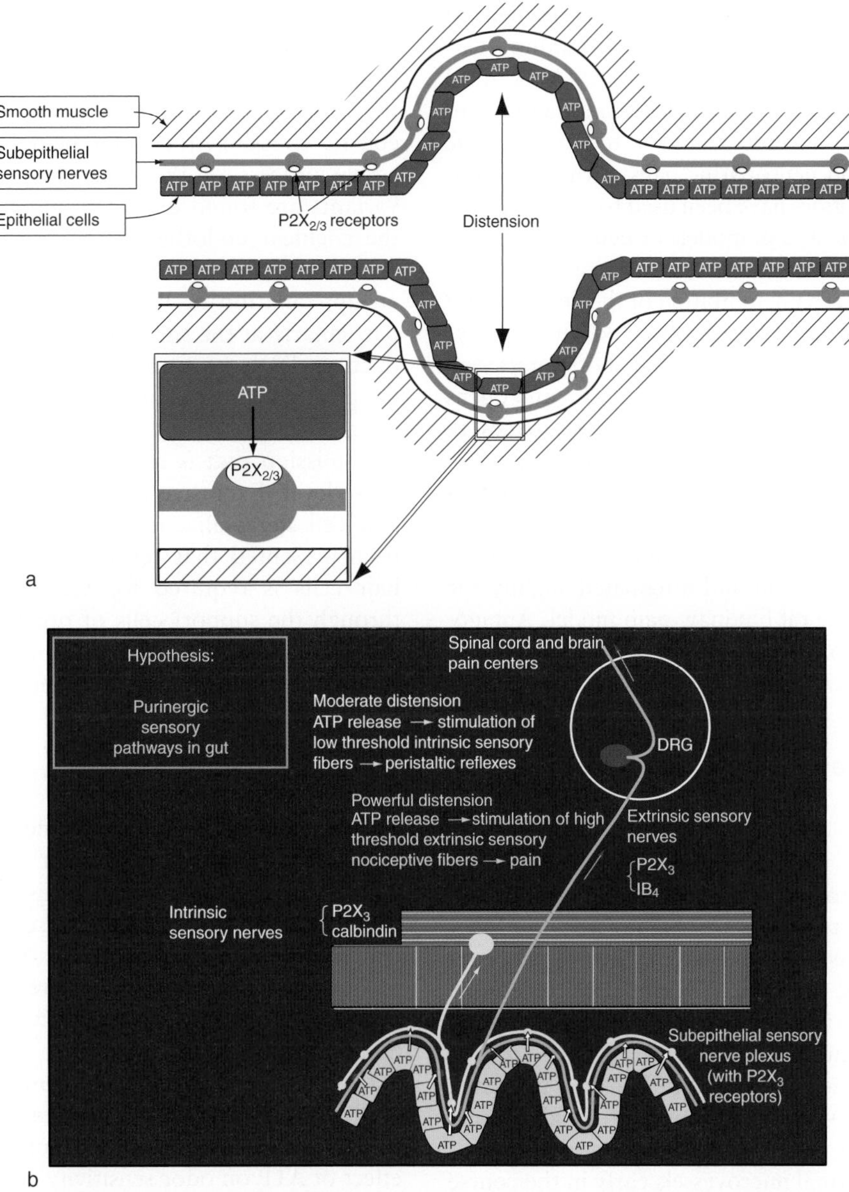

Figure 5 (a) Schematic representation of the hypothesis for purinergic mechanosensory transduction in tubes (e.g., ureter, vagina, salivary, and bile ducts and gut) and sacs (e.g., urinary and gall bladders and lung). It is proposed that distension leads to the release of adenosine 5′ triphosphate (ATP) from the epithelium lining the tube or sac, which then acts on $P2X_{2/3}$ receptors on subepithelial sensory nerves to convey sensory (nociceptive) information to the central nervous system (CNS). (b) Schematic of a novel hypothesis about purinergic mechanosensory transduction in the gut. It is proposed that ATP released from mucosal epithelial cells during moderate distension acts preferentially on $P2X_3$ receptors on low-threshold subepithelial intrinsic sensory nerve fibers (labeled with calbindin), contributing to peristaltic reflexes. ATP released during extreme distension also acts on $P2X_3$ receptors on high-threshold extrinsic sensory nerve fibers (labeled with isolectin B_4 (IB_4)) that send messages via the dorsal root ganglia (DRG) to pain centers in the CNS. (a) Adapted from Burnstock G (1999) Release of vasoactive substances from endothelial cells by shear stress and purinergic mechanosensory transduction. *Journal of Anatomy* 194: 335–342, with permission from Blackwell Publishing. (b) Adapted from Burnstock G (2001) Expanding field of purinergic signaling. *Drug Development Research* 52: 1–10, with permission of Wiley-Liss, Inc.

mediated by P2X receptors activated by synaptically released ATP, in a subpopulation of less than 5% of the neurons in lamina II, a region known to receive major input from nociceptive primary afferents.

There is an urgent need for selective $P2X_3$ and $P2X_{2/3}$ receptor antagonists that do not degrade *in vivo*. Pyridoxal-phosphate-6-azophenyl-2′, 4′-disulphonic acid is a nonselective P2 receptor antagonist but has the advantage that it dissociates about 100 to 10 000 times more slowly than other known antagonists. The trinitrophenyl-substituted nucleotide TNP-ATP is a selective and very potent antagonist at both

$P2X_3$ and $P2X_{2/3}$ receptors. 5-((3-Phenoxybenzyl)[(1*S*)-1,2,3,4-tetrahydro-1-naphthalenyl]amino carbonyl)-1,2,4-benzenetricarboxylic acid (A-317491) is a potent and selective nonnucleotide antagonist of $P2X_3$ and $P2X_{2/3}$ receptors, and it reduces chronic inflammatory and neuropathic pain in the rat. Antisense oligonucleotides have been used to downregulate the $P2X_3$ receptor, and in models of neuropathic (partial sciatic nerve ligation) and inflammatory (complete Freund's adjuvant) pain, inhibition of the development of mechanical hyperalgesia was observed within 2 days of treatment. $P2X_3$ double-stranded-short interfering RNA also relieves chronic neuropathic pain and opens up new avenues for therapeutic pain strategies in humans. Tetramethylpyrazine, a traditional Chinese medicine used as an analgesic for dysmenorrhoea, is claimed to be a P2X receptor antagonist, and it inhibited significantly the first phase of nociceptive behavior induced by 5% formalin and attenuated slightly the second phase in the rat hindpaw pain model. Antagonists to P2 receptors are also beginning to be explored in relation to cancer pain.

Special Senses

Eye

$P2X_2$ and $P2X_3$ receptor mRNA is present in the retina and receptor protein expressed in retinal ganglion cells. $P2X_3$ receptors are also present on Müller cells, which release ATP during Ca^{2+} wave propagation. ATP, acting via both P2X and P2Y receptors, modulates retinal neurotransmission, affecting retinal blood flow and intraocular pressure. Topical application of diadenosine tetraphosphate has been proposed for the lowering of intraocular pressure in glaucoma. The formation of $P2X_7$ receptor pores and apoptosis is enhanced in retinal microvessels early in the course of experimental diabetes, suggesting that purinergic vasotoxicity might have a role in microvascular cell death, a feature of diabetic retinopathy. The possibility has been raised that alterations in sympathetic nerves might underlie some of the complications observed in diabetic retinopathy; ATP is well established as a cotransmitter in sympathetic nerves, raising the potential for P2 receptor antagonists in glaucoma. $P2Y_2$ receptor activation increases salt, water, and mucus secretion and thus represents a potential treatment for dry eye conditions.

Ear

Both P2X and P2Y receptors have been identified in the vestibular system. ATP regulates fluid homeostasis, cochlear blood flow, hearing sensitivity, and development and thus might be useful in the treatment of Ménières disease, tinnitus, and sensorineural deafness. ATP, acting via P2Y receptors, depresses sound-evoked gross compound action potentials in the auditory nerve and the distortion product otoacoustic emission, the latter being a measure of the active process of the outer hair cells. P2X splice variants are found on the endolymphatic surface of the cochlear endothelium, an area associated with sound transduction. Sustained loud noise produces an upregulation of $P2X_2$ receptors in the cochlear, particularly at the site of outer hair cell sound transduction. $P2X_2$ receptor expression is also increased in spiral ganglion neurons, indicating that extracellular ATP acts as a modulator of auditory neurotransmission that is adaptive and dependent on the noise level. Excessive noise can irreversibly damage hair cell stereocilia, leading to deafness. Data have been presented that release of ATP from damaged hair cells is required for Ca^{2+} wave propagation through the support cells of organ of Corti, involving P2Y receptors, and this might constitute the fundamental mechanism to signal the occurrence of hair cell damage.

Nasal Organs

The olfactory epithelium and vomeronasal organs contain olfactory receptor neurons that express $P2X_2$, $P2X_3$, and $P2X_{2/3}$ receptors. It is suggested that the neighboring epithelial supporting cells or the olfactory neurons themselves can release ATP in response to noxious stimuli, acting on P2X receptors as an endogenous modulator of odor sensitivity. Enhanced sensitivity to odors was observed in the presence of P2 antagonists, suggesting that low-level endogenous ATP normally reduces odor responsiveness. It has been suggested that the predominantly suppressive effect of ATP on odor sensitivity could be involved in reduced odor sensitivity that occurs during acute exposure to noxious fumes and might be a novel neuroprotective mechanism.

See also: Adenosine; Adenosine Triphosphate (ATP) as a Neurotransmitter; Adenosine Receptor Mediated Functions; Cotransmission; Pharmacology of Sleep: Adenosine; Purinergic Receptors; Purines and Purinoceptors: Molecular Biology Overview.

Further Reading

Abbracchio MP and Burnstock G (1994) Purinoceptors: Are there families of P2X and P2Y purinoceptors? *Pharmacology and Therapeutics* 64: 445–475.

Abbracchio MP and Burnstock G (1998) Purinergic signalling: Pathophysiological roles. *Japanese Journal of Pharmacology* 78: 113–145.

Bodin P and Burnstock G (2001) Purinergic signalling: ATP release. *Neurochemical Research* 26: 959–969.

Burnstock G (1972) Purinergic nerves. *Pharmacological Reviews* 24: 509–581.

Burnstock G (1987) Local control of blood pressure by purines. *Blood Vessels* 24: 156–160.

Burnstock G (1999) Release of vasoactive substances from endothelial cells by shear stress and purinergic mechanosensory transduction. *Journal of Anatomy* 194: 335–342.

Burnstock G (2001) Expanding field of purinergic signaling. *Drug Development Research* 52: 1–10.

Burnstock G (2001) Purine-mediated signalling in pain and visceral perception. *Trends in Pharmacological Sciences* 22: 182–188.

Burnstock G (2002) Purinergic signalling and vascular cell proliferation and death. *Arteriosclerosis, Thrombosis and Vascular Biology* 22: 364–373.

Burnstock G (2001) Purinergic signalling in gut. In: Abbracchio MP and Williams M (eds.) *Handbook of Experimental Pharmacology, Vol. 151: Purinergic and Pyrimidinergic Signalling II: Cardiovascular, Respiratory, Immune, Metabolic and Gastrointestinal Tract Function*, pp. 141–238. Berlin: Springer.

Burnstock G (2001) Purinergic signalling in lower urinary tract. In: Abbracchio MP and Williams M (eds.) *Handbook of Experimental Pharmacology, Vol. 151: Purinergic and Pyrimidinergic Signalling I: Molecular, Nervous and Urinogenitary System Function*, pp. 423–515. Berlin: Springer.

Burnstock G (2003) Purinergic receptors in the nervous system. In: Schwiebert EM (ed.) *Current Topics in Membranes, Vol. 54: Purinergic Receptors and Signalling*, pp. 307–368. San Diego, CA: Academic Press.

Burnstock G (2004) Cotransmission. *Current Opinion in Pharmacology* 4: 47–52.

Burnstock G (2006) Pathophysiology and therapeutic potential of purinergic signalling. *Pharmacological Reviews* 58: 58–86.

Burnstock G (2007) Physiology and pathophysiology of purinergic neurotransmission. *Physiological Reviews* 87: 659–797.

Burnstock G and Costa M (eds.) (1975) Adrenergic neuroeffector transmission. In *Adrenergic Neurones: Their Organisation, Function and Development in the Peripheral Nervous System*, pp. 51–106. London: Chapman and Hall.

Burnstock G and Knight G (2004) Cellular distribution and functions of P2 receptor subtypes in different systems. *International Review of Cytology* 240: 31–304.

Dunn PM, Zhong Y, and Burnstock G (2001) P2X receptors in peripheral neurones. *Progress in Neurobiology* 65: 107–134.

James G and Butt AM (2002) P2Y and P2X purinoceptor mediated Ca^{2+} signalling in glial cell pathology in the central nervous system. *European Journal of Pharmacology* 447: 247–260.

Kennedy C, McLaren GJ, Westfall TD, and Sneddon P (1996) ATP as a co-transmitter with noradrenaline in sympathetic nerves – Function and fate. In: Chadwick DJ and Goode J (eds.) *P2 Purinoceptors: Localization, Function and Transduction Mechanisms*, pp. 223–235.

Sneddon P and Burnstock G (1984) Inhibition of excitatory junction potentials in guinea-pig vas deferens by α, β-methylene-ATP: Further evidence for ATP and noradrenaline as cotransmitters. *European Journal of Pharmacology* 100: 85–90.

Zimmermann H (2001) Ectonucleotidases: Some recent developments and a note on nomenclature. *Drug Development Research* 52: 44–56.

Adenosine Triphosphate (ATP) as a Neurotransmitter

A Verkhratsky, The University of Manchester, Manchester, UK
O Krishtal, Bogomoletz Institute of Physiology, Kiev, Ukraine

Historic Remarks

The molecule of adenosine 5′-triphosphate, ATP, was discovered in 1929 by Karl Lohman in Heidelberg and by Cyrus Hartwell Fiske and Yellapragada SubbaRow at Harvard. In the same year, the role for purines and ATP as extracellular signaling molecules was also suggested by Drury and Szent-Györgyi, who found that purines exert a potent negative chronotropic effect on the heart and trigger dilatation of coronary vessels. The signaling function of ATP in peripheral tissues was subsequently confirmed by numerous experiments.

In 1959 Pamela Horton made a seminal discovery that ATP can be released from nerves upon electrical stimulation. In the 1960s nonadrenergic, noncholinergic inhibitory transmission was described in the taenia coli and stomach, activity that subsequently was demonstrated to be mediated by ATP. These experimental observations led Geoffrey Burnstock to suggest a distinct type of 'purinergic' neurotransmission. Meanwhile, the first indications of excitatory action of extracellular ATP on central neurons were obtained in experiments on neurons from the nucleus cuneatus. Further indications that ATP may act as a neurotransmitter in neuronal–neuronal synapses were obtained by Thomas White, who noted the release of ATP from brain synaptosomes treated by veratridine or high extracellular K^+; he subsequently found depolarization-induced ATP release from synaptosomes isolated from cortex and striatum and also demonstrated that in central nervous tissue, ATP was not co-released with acetylcholine (ACh).

Investigations of purinergic nerves, primarily led by Burnstock, resulted in discovery of specific family of purinoreceptors, which are classified as two fundamentally distinct classes, the P1 (A1) receptors, sensitive to adenosine, and the P2 receptors, activated by ATP and its analogs. The P2 receptors are represented by the ionotropic P2X and metabotropic P2Y receptors. The P2X receptors are ligand-gated cationic channels, which are permeable to Na^+, K^+, and Ca^{2+}. Seven P2X receptor subunits ($P2X_1$–$P2X_7$) encoded by distinct genes have been identified. These subunits may form homo- or heteromeric receptors, with each functional receptor containing at least three monomers. The P2Y receptors are classical seven-transmembrane-domain metabotropic receptors coupled to G-proteins. These receptors are represented by at least 14 subtypes. Activation of neuronal P2Y receptors regulates K^+ channels, potentiates high-voltage-activated Ca^{2+} channels, and triggers inositol 1,4,5-trisphosphate (IP_3)-mediated release of Ca^{2+} from endoplasmic reticulum Ca^{2+} stores.

ATP-Mediated Transmission in the Peripheral Nervous System

Peripheral Nerves

ATP acts as a nerve-to-effector transmitter in many tissues. As a rule, ATP is co-released with other neurotransmitters – for example, with noradrenalin (NA) or ACh. This co-release is well documented for the sympathetic nervous system. ATP mediates neuroeffector transmission between the hypogastric nerve and vas deferens, being in fact the main mediator triggering the excitatory junction potential and initiating contraction of smooth muscle. Both ATP and NA, co-released from the same terminals, are instrumental in sympathetic control of vascular tone. The relative fraction and hence the relative functional importance of purinergic component will vary among different types of vessels, the most significant activity occurring in the arterioles of the mesentery and submucosal plexus of the intestine; here ATP acts as a main excitatory transmitter, whereas NA, which is co-released with ATP, modulates the release of the latter. In the parasympathetic system, ATP acts as an excitatory cotransmitter together with ACh in the terminal innervating the urinary bladder. Similarly, ATP is co-released with ACh from motor nerve terminals. In these terminals ATP mostly potentiates the excitatory action of ACh, acting through postjunctional metabotropic receptors. In the neuromuscular junction, ATP also appears to be a source of adenosine, which, by activating prejunctional P1(A1) receptors, regulates release of ACh. Notably, however, P2X receptors are found in immature skeletal muscle cells, and several P2X receptors subunits have been identified in developing myocytes. Therefore, P2X-mediated excitatory transmission is likely operative during yet unknown developmental steps.

Sensory Neurons

ATP-induced currents, mediated through opening of ionotropic P2X receptors, were initially recorded from neurons acutely isolated from nodose, vestibular, trigeminal, and dorsal root ganglia (DRG). These currents were concentration dependent (with K_D for ATP ~5 μM) and were carried mostly by sodium ions, as the substitution of the latter by tris(hydroxymethyl)aminomethane ($Tris^+$), tetraethylammonium (TEA^+), or $choline^+$ markedly suppressed the ATP-induced responses (**Figure 1**). This initial observation was confirmed and extended by many groups, and purinoreceptor-mediated currents in sensory neurons have been characterized in minute detail. Analysis of mRNA expression reveals the presence of transcripts for $P2X_{1-6}$ receptors, with $P2X_3$ mRNA showing the highest level. Similarly, immunohistochemical analysis of sensory neurons in DRG shows the appearance of $P2X_{1-6}$ subunits, with clear predominance of the $P2X_3$ subunit and significant expression of the $P2X_2$ subunit. Both $P2X_3$ and $P2X_2$ subunits are mostly confined to small and medium-size sensory neurons putatively executing nociceptive functions. Expression of various P2X subunits is reflected by heterogeneity of ATP-induced currents evoked in sensory neurons. Some authors suggest the participation of P2X receptors in the functioning of the synapses between sensory neurons and the neurons of dorsal horn. Sensory neurons also express metabotropic $P2Y_1$ receptors, which mediate long-lasting depolarization.

Autonomic Neurons

Sympathetic neurons express $P2X_{1-6}$ purinoreceptor subunits, with a particularly high presence of $P2X_2$ subunits. External application of ATP to cultured celiac neurons triggers depolarization and inward ion currents, which are blocked by suramin and reactive blue 2. Synaptic currents recorded from these neurons display similar properties, suggesting the involvement of P2X receptors. In parasympathetic neurons, ATP has an excitatory action mediated through ATP-induced currents, which have been recorded

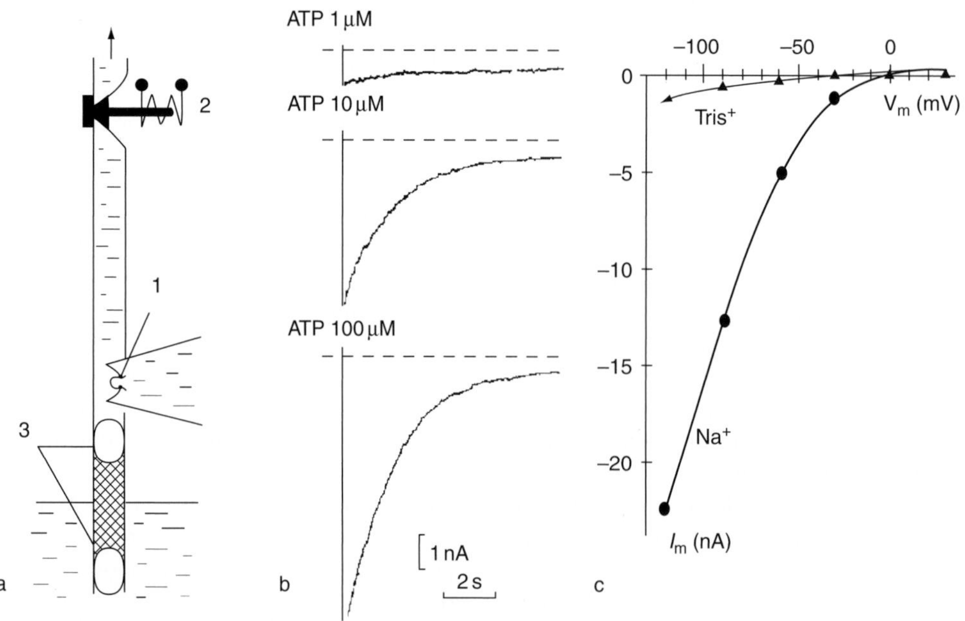

Figure 1 First recordings of ATP-induced membrane currents in acutely isolated sensory neurons. (a) The 'square pulse' application technique used for rapid application of ATP to internally dialyzed sensory neurons. The tip of the micropipette (with the cell; 1) is inserted into a plastic tube. An invagination in the tip of the micropipette prevents the cell from damage. The lower end of the tube can be exposed to different external solutions or to air. Suction applied to the upper end of the tube (arrow) is controlled by an electromagnetic valve (2). A preprogrammed sequence of current pulses applied to the valve allows a column of test solution to form in the tube; the test solution is separated from the normal solution by air bubbles (3). Another sequence of pulses exposes the cell to the test solution for the desired period of time by rapid displacement of the column along the tube. The electrical recording is unharmed, since there is a thin layer of saline between the air bubbles and the walls of the tube. (b) ATP-activated inward current (ordinate) elicited by application of different concentrations of ATP as indicated on the graph. (c) Voltage dependence of the peaks of ATP-activated current. ATP (5 μM) was applied in normal saline (•) and in an external solution in which Na was substituted with tetraethylammonium (▲). Holding potential was −90 mV throughout. Reproduced from Krishtal OA, Marchenko SM, and Pidoplichko VI (1983) Receptor for ATP in the membrane of mammalian sensory neurones. *Neuroscience Letters* 35: 41–45, with permission.

from variety of neurons isolated from different ganglia (e.g., intramural ganglia from the guinea pig urinary bladder, chick ciliary ganglia, and rabbit vesical parasympathetic ganglia). These ionotropic purinergic responses are mostly mediated through receptors constructed from $P2X_{2/3}$ subunits. Finally, a majority of myenteric neurons from various species demonstrate both P2X current responses and P2Y/IP_3-mediated Ca^{2+} signals.

ATP-Mediated Synaptic Transmission in the Central Nervous System

P2X-Mediated Synaptic Currents

All types of P2X receptors ($P2X_{1-7}$) and many P2Y receptors ($P2Y_1$, $P2Y_6$, $P2Y_{11}$, $P2Y_{12}$, $P2Y_{13}$, and $P2Y_{14}$) are expressed in the brain. Several types of P2X subunits can be co-expressed in the same neuron, and this expression can be quite heterogeneous even within the same brain region. As a consequence, ATP-mediated currents recorded from central nervous system (CNS) neurons display heterogeneous kinetics and sensitivity to pharmacological modulators (**Figure 2**).

In the CNS, the ATP-induced excitatory responses were initially demonstrated in cultured dorsal horn neurons. Ten year later, the ATP-mediated synaptic transmission was discovered in brain slices of medial habenula, where both excitatory postsynaptic currents and spontaneous 'miniature' postsynaptic currents were identified (**Figure 3(a)**). Subsequently, ATP-mediated synaptic responses were found in other brain areas, including spinal cord, hippocampus, and

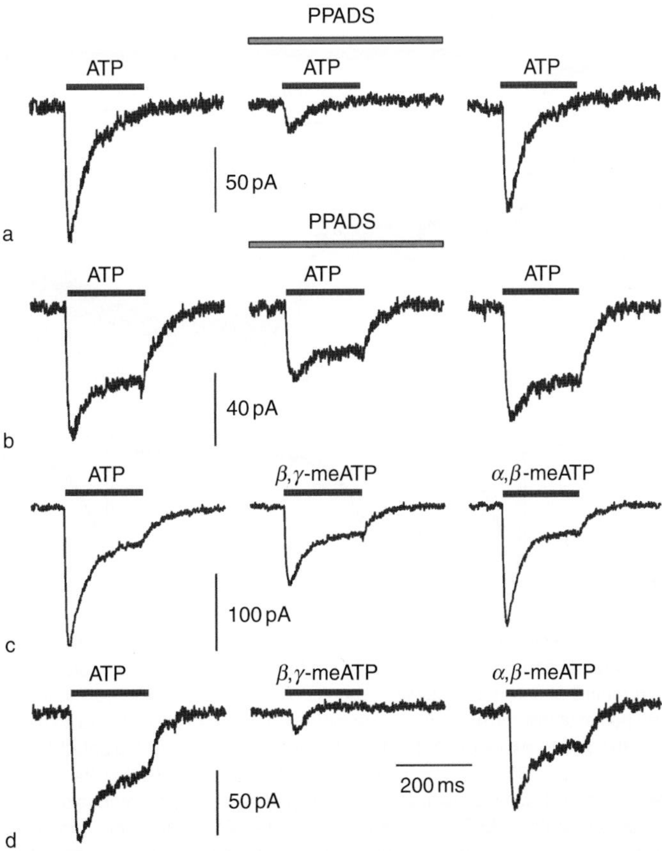

Figure 2 Heterogeneity of P2X receptor-mediated currents in the isolated pyramidal neocortical layer V neurons. (a, b) Examples of inward currents evoked by short (200 ms) applications of ATP (20 μM). The traces (from left to right) represent control, response to ATP under action of pyridoxal phosphate-6-azophenyl-2′,4′-disulfonic acid (PPADS; 30 μM), and response to ATP after washout of drug. (c, d) Examples of currents evoked by application of various agonists of the P2X receptor. The traces (from left to right) represent response to ATP (20 μM), response to β,γ-methylene-ATP (25 μM), and response to α,β-methylene-ATP (20 μM). Groups of traces (a, b; and c, d) were recorded from different cells. Recordings were made with a 5 min time interval between applications at the holding potential of –80 mV. Reproduced from Pankratov Y, Lalo U, Krishtal O, et al. (2003) P2X receptor-mediated excitatory synaptic currents in somatosensory cortex. *Molecular and Cellular Neuroscience* 24: 842–849, with permission.

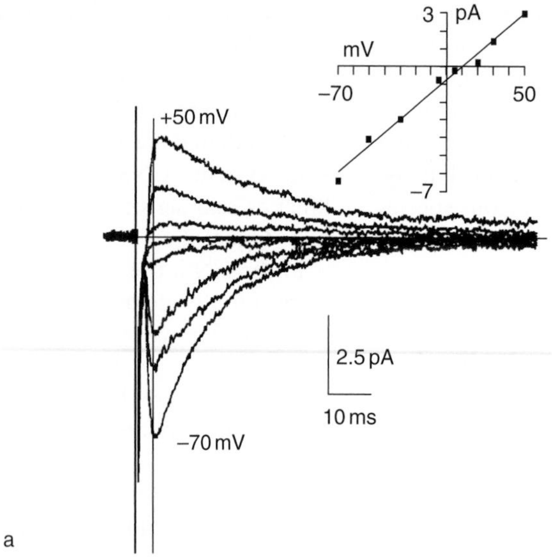

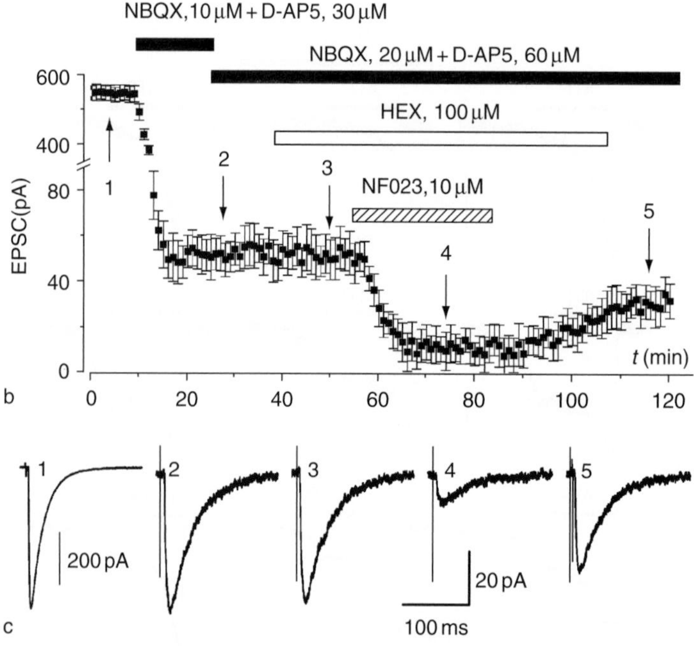

Figure 3 ATP-mediated synaptic transmission in CNS neurons. (a) Voltage dependence of ATP-mediated synaptic currents, recorded from a 24-day-old male rat medial habenula neuron in Krebs solution containing 5 mM Ca^{2+}. Stimulus pulses (13 V, 200 μs) were delivered at 2 Hz, and 53–215 currents were recorded before the holding potential was stepped to the next voltage. The traces are averages of all responses (including failures). The peaks of the currents shown are plotted in the inset and fitted to a straight line using a linear regression. Peak current was measured from the single exponential fit of each current at the time indicated by the vertical line. The membrane potentials shown in this plot are −70, −50, −30, −5, +5, +20, +30, and +50 mV. Note that the voltage dependence of peak current amplitude is well fitted by a straight line, with clear outward currents at positive holding potentials. (b) Dissection of the ATP-mediated component of excitatory postsynaptic currents (EPSCs) in pyramidal neurons of layer II of the somatosensory cortex. (Top) Changes in amplitude of the EPSC following bath application of the glutamatergic antagonists 2,3-dioxo-6-nitro-1,2,3,4-tetrahydrobenzo[*f*]quinoxaline-7-sulfonamide (NBQX) and D-(−)-2-amino-5-phosphopentanoic acid (D-AP5), the cholinergic antagonist hexamethonium (HEX), and the P_{2X} receptor antagonist NF023 at various time points (1–5) . Each point represents the mean ± SD for six sequential trials (holding potential, −80 mV; stimulation frequency, 0.1 Hz). (Bottom) Examples of the EPSCs (average of six traces) recorded at moments 1–5 as indicated on the upper graph. (a) Reproduced from Edwards FA, Robertson SJ, and Gibb AJ (1997) Properties of ATP receptor-mediated synaptic transmission in the rat medial habenula. *Neuropharmacology* 36: 1253–1268, with permission. (b) Reproduced from Pankratov Y, Lalo U, Krishtal O, et al. (2002) Ionotropic P2X purinoreceptors mediate synaptic transmission in rat pyramidal neurones of layer II/III of somato-sensory cortex. *Journal of Physiology* 542: 529–536, with permission.

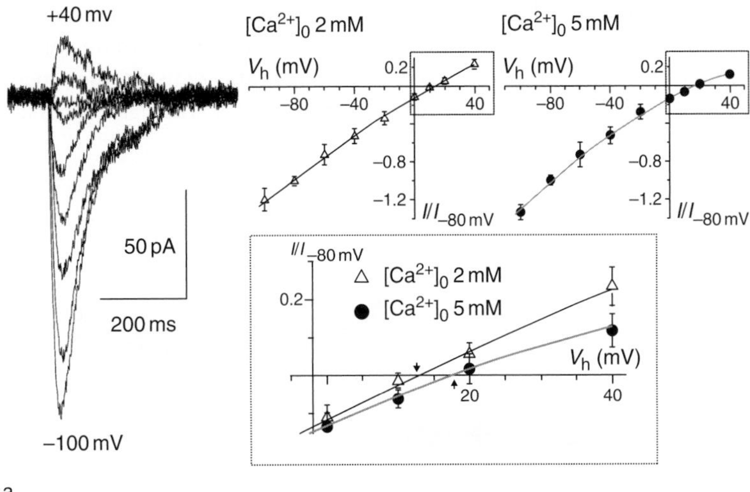

a

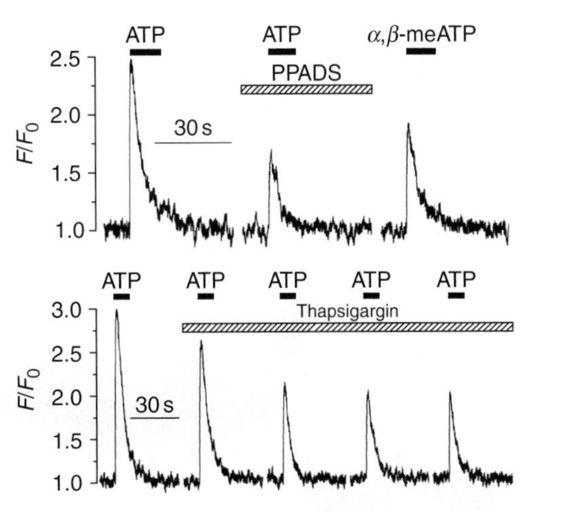

b

Figure 4 Calcium permeability of P2X receptors in acutely isolated pyramidal cortical neurons. (a) Left panel shows examples of inward currents induced in the pyramidal neurons of somatosensory cortex layer II/III by the application of ATP (10 μM) at membrane potentials ranging from −100 to +40 mV. Shown on the right are voltage–current relationships for the ATP-induced currents measured at different extracellular calcium concentrations. Amplitudes of currents were normalized to the maximal value measured at −80 mV. Each point is the mean ± SD for seven cells. Lines represent the cubic polynomial fit. Lower inset demonstrates the superposition of voltage–current relationships in the vicinity of reversal potential. (b) ATP-induced $[Ca^{2+}]_i$ transients in CA1 pyramidal neurons. Examples in the upper graph show $[Ca^{2+}]_i$ transients induced in the pyramidal cells by fast application of ATP (100 μM) and α,β-methylene-ATP (100 μM) to the hippocampal slice in the control and after bath application of 20 μM pyridoxal phosphate-6-azophenyl-2′,4′-disulfonic acid (PPADS). All traces were recorded in the same cell at 5 min intervals. Note the substantial amplitude of the response to α,β-methylene-ATP, which is attributable only to the calcium influx via ionotropic ATP receptors. Examples in the lower graph show $[Ca^{2+}]_i$ transients induced by repetitive fast application of ATP (100 μM) in the presence of 1 μM thapsigargin at 5 min time intervals. It is worth noting that the amplitudes of the third and the following transients, which are attributable to entry of the extracellular calcium via ionotropic ATP receptors, comprise ~40% of the initial response, which represents the combined activity of P2X and P2Y purinoreceptors. (a) Reproduced from Pankratov Y, Lalo U, Krishtal O, et al. (2002) Ionotropic P2X purinoreceptors mediate synaptic transmission in rat pyramidal neurones of layer II/III of somato-sensory cortex. *Journal of Physiology* 542: 529–536, with permission. (b) Reproduced from Pankratov YV, Lalo UV, and Krishtal OA (2002) Role for P2X receptors in long-term potentiation. *Journal of Neuroscience* 22: 8363–8369, with permission.

cortex. The purinergic nature of the synaptic responses in central neurons was usually confirmed based on neuronal sensitivity to P2X receptors inhibitors or by desensitization induced following incubation with α,β-methylene-ATP (α,β-meATP). An example of the analysis of evoked postsynaptic currents mediated by P2X receptors is shown in **Figure 3(b)**.

As a rule, P2X-mediated synaptic currents are not very large, rarely exceeding 50–100 pA in amplitude and representing 5–15% of the synaptic currents mediated by glutamate. Nonetheless, ATP-mediated synaptic transmission can have functional importance, especially considering that activation of postsynaptic P2X receptors can provide a route for Ca^{2+} entry at resting membrane potentials. This is in contrast to another main source for Ca^{2+} entry in the postsynaptic density, that mediated by *N*-methyl-D-aspartate (NMDA) receptors; activation of NMDA

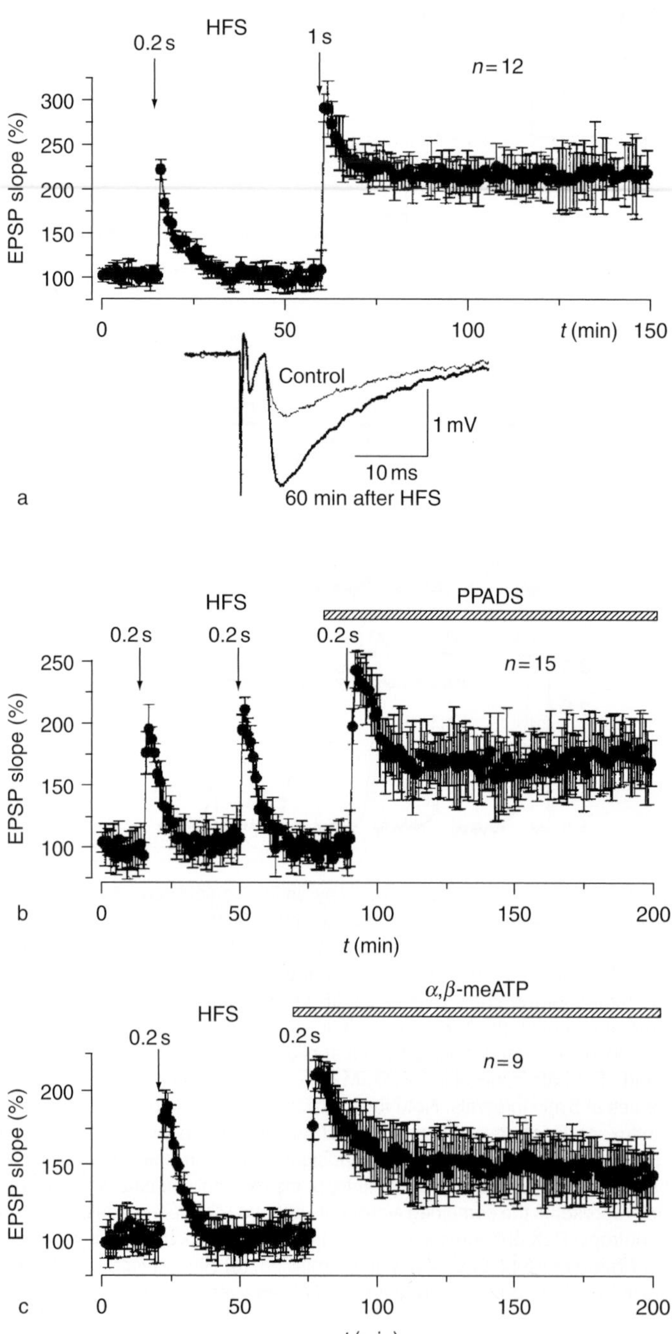

Figure 5 Inhibition of P2X receptors facilitates long-term potentiation initiation in the CA1 region of the hippocampus: changes in CA1 field potentials induced by 100 Hz high-frequency stimulation (HFS) delivered to the Schaffer collateral in the control and after inhibition of the ATP receptors. (a) The time course of the potentiation evoked by 0.2 and 1 s of 100 Hz stimulation trains in control conditions. Examples of field excitatory postsynaptic potentials (EPSPs) recorded before and 60 min after the 1-s-long HFS train are indicated in the inset. Each trace represents the average of ten EPSPs. (b, c)The time course and magnitude of the potentiation evoked by the 0.2 s HFS train in control and after bath application of 20 μM pyridoxal phosphate-6-azophenyl-2′,4′-disulfonic acid (PPADS) and α,β-methylene-ATP, respectively. Baseline stimulation frequency is 0.08 Hz. Reproduced from Pankratov YV, Lalo UV, and Krishtal OA (2002) Role for P2X receptors in long-term potentiation. *Journal of Neuroscience* 22: 8363–8369, with permission.

receptors requires cell predepolarization, which removes the Mg^{2+} block. Functional P2X receptors expressed in the CNS display substantial Ca^{2+} permeability; Ca^{2+} permeability, relative to monovalent cations, can range between 2 and 12 (**Figure 4(a)**). Application of ATP triggers substantial Ca^{2+} signals in central neurons; the signals are mediated by Ca^{2+} entry through P2X receptors/voltage-gated Ca^{2+} channels and by P2Y-mediated Ca^{2+} release from intracellular stores (**Figure 4(b)**). Ca^{2+} signals induced by stimulation of P2 receptors are functionally important; for example, these signals may modulate NMDA receptors and facilitate long-term potentiation in CA1 hippocampal neurons (**Figure 5**).

Mechanisms of ATP Release

The overall concentration of ATP in the brain is rather high, varying between 2 and 4 mM in different regions. The cellular membranes are not permeable to ATP, therefore ATP-mediated intercellular signaling requires specialized pathways for ATP release. There are several of these pathways, including both exocytotic release from presynaptic vesicles and non-exocytotic release through plasmalemmal pores.

Vesicular ATP release is widespread throughout the peripheral and CNSs. ATP can be accumulated and stored in synaptic vesicles together with other neurotransmitters and then both can be co-released. This co-release of ATP with other neurotransmitters is abundant in the peripheral nervous system (PNS); in the CNS, ATP can be co-released with γ-aminobutyric acid (GABA; as seen in cultured neurons from spinal cord and lateral hypothalamus) or with glutamate (as seen in hippocampal organotypic slices). Early studies of ATP release from synaptosomal preparations also indicated that some fraction of ATP could be co-released with ACh or NA, but this co-release is probably of minor importance. Alternatively, ATP may be released on its own from specific ATP-containing vesicles, which can be present in specific ATP-containing terminals or in shared glutamatergic terminals (**Figure 6**).

Nonvesicular ATP release may occur through gap junction hemichannels or via volume-sensitive chloride channels, or even through $P2X_7$ receptors, known to form large transmembrane pores following stimulation. These mechanisms are predominantly associated with glial cells.

ATP Mediates Neuronal–Glial Signaling

Purinergic transmission is particularly important for neuronal–glial integration in the CNS. All types of

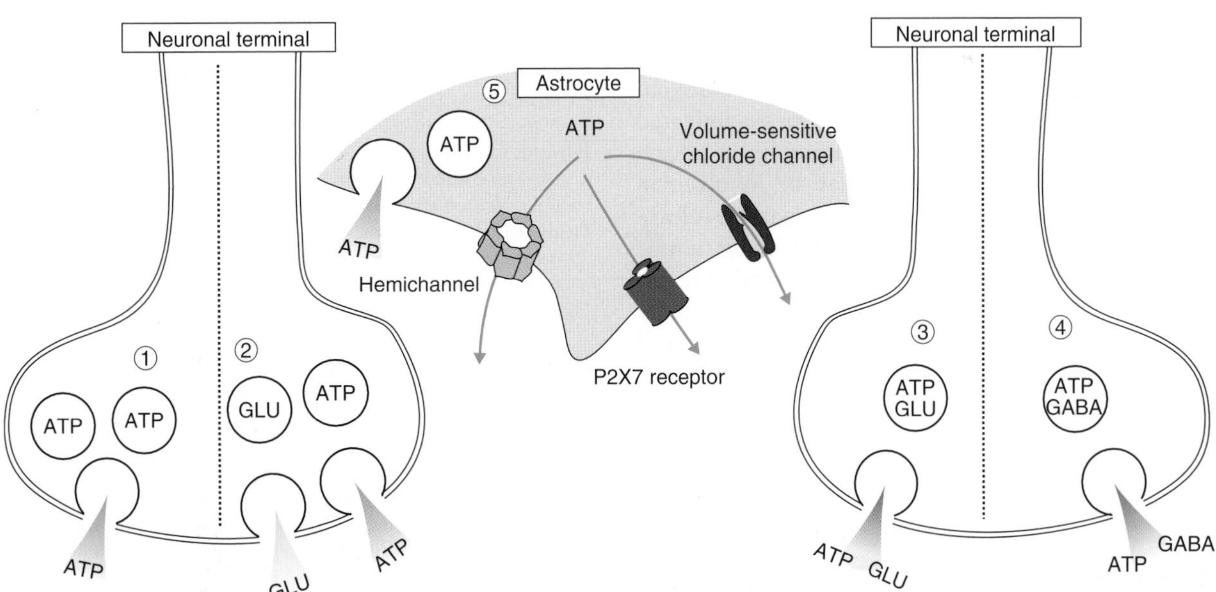

Figure 6 Adenosine triphosphate (ATP) release pathways in the CNS, where ATP can be stored and released from the synaptic terminals in several distinct fashions. (1) ATP can act as a sole transmitter (e.g., in terminals from medial habenula). (2) Vesicles containing ATP can coexist with vesicles containing glutamate (GLU) in the same terminals; release of ATP occurs simultaneously with but independently from the release of glutamate (this mechanism is, for instance, operative in the hippocampus). (3) In principle, ATP can be co-stored and co-released with glutamate or other neurotransmitters, although this is not supported by direct experimental evidence from *in situ* preparations. (4) ATP can be co-stored and co-released with γ-aminobutyric acid (GABA; e.g., in cultured spinal or lateral hypothalamic neurons). (5) ATP can be also released from glial cells via regulated exocytosis or through plasmalemmal channels.

glia (i.e., the astrocytes, oligodendrocytes, and microglia in the CNS and Schwann cells in the PNS) express functional ATP receptors. Application of exogenous ATP triggers Ca^{2+} responses in both cultured glia and glial cells in *in situ* preparations. These Ca^{2+} signals are usually driven through the metabotropic route, mediated by P2Y receptors and IP_3-induced Ca^{2+} release from intracellular stores. The functional role of P2X receptors in macroglia is much less clear: although P2X-mediated ion currents are detected in cultured astrocytes, ATP-mediated ionotropic responses *in situ* remain mostly uncharacterized. Reverse transcriptase–polymerase chain reaction (RT-PCR) revealed expression of mRNA for all types ($P2X_{1-7}$) of receptors in astroglial cells in nucleus accumbens and in all receptors save $P2X_6$ in astrocytes from cortex. Functional expression of P2X receptors was found in oligodendrocytes from the optic nerve.

ATP acts as a mediator not only in neuronal–glial communications but also in glial–glial and potentially glial–neuronal signaling. ATP is generally acknowledged as a 'glio' transmitter, and regulated release of ATP from astrocytes has been observed in several glial preparations. ATP released by the glial cells is instrumental for generation and maintenance of propagating Ca^{2+} waves, which are the substrate for glial excitability and long-range signaling. The ATP released from glial cells can also signal to neurons either by activating neuronal P2 receptors, or, following degradation to adenosine, by stimulating neuronal A1 receptors.

ATP-mediated signaling is particularly important for pathological reactions of glial cells. Astroglial cells express $P2X_7$ receptors, which can be activated following massive release of ATP upon brain injury; stimulation of P2X7 receptors may be involved in initiation of reactive gliosis. Similarly, massive release of ATP can act as a specific signal for activation of microglia; microglial cells express an extended repertoire of P2X receptors (which can be substantially modified in the course of microglial activation) and P2Y receptors linked to intracellular Ca^{2+} signaling. The P2 receptors control motility of microglial processes, and upon overstimulation they trigger rapid movement of these processes toward the site of injury.

Conclusions

ATP released from neuronal terminals and from astroglial cells mediates signaling within neuronal–glial networks. Effects of ATP are diverse and are determined by idiosyncrasies of ionotropic (P2X) and metabotropic (P2Y) receptors variably expressed in target cells. In pathological conditions, ATP may act as an important signal, triggering and controlling defensive reactions of microglia and reactive astrogliosis.

See also: Adenosine; Adenosine Triphosphate (ATP); Adenosine Receptor Mediated Functions; Calcium Waves: Purinergic Regulation; P2X Receptors; Pharmacology of Sleep: Adenosine; Presynaptic Receptor Signaling; Purinergic Receptors; Purines and Purinoceptors: Molecular Biology Overview; Synaptic Plasticity: Short-Term Mechanisms.

Further Reading

Abbracchio MP, Burnstock G, Boeynaems JM, et al. (2006) International Union of Pharmacology LVIII: Update on the P2Y G protein-coupled nucleotide receptors: From molecular mechanisms and pathophysiology to therapy. *Pharmacological Reviews* 58: 281–341.

Burnstock G (1972) Purinergic nerves. *Pharmacological Reviews* 24: 509–581.

Burnstock G (2003) Purinergic receptors in the nervous system. *Current Topics in Membranes* 54: 307–368.

Burnstock G, Campbell G, Satchell D, et al. (1970) Evidence that adenosine triphosphate or a related nucleotide is the transmitter substance released by non-adrenergic inhibitory nerves in the gut. *British Journal of Pharmacology* 40: 668–688.

Collo G, North RA, Kawashima E, et al. (1996) Cloning of P2X5 and P2X6 receptors and the distribution and properties of an extended family of ATP-gated ion channels. *Journal of Neuroscience* 16: 2495–2507.

Drury AN and Szent-Györgyi A (1929) The physiological activity of adenine compounds with special reference to their action upon mammalian heart. *Journal of Physiology (London)* 68: 213–237.

Edwards FA, Gibb AJ, and Colquhoun D (1992) ATP receptor-mediated synaptic currents in the central nervous system. *Nature* 359: 144–147.

Edwards FA, Robertson SJ, and Gibb AJ (1997) Properties of ATP receptor-mediated synaptic transmission in the rat medial habenula. *Neuropharmacology* 36: 1253–1268.

Fields RD and Burnstock G (2006) Purinergic signalling in neuron–glia interactions. *Nature Reviews Neuroscience* 7: 423–436.

Fiske CH and SubbaRow Y (1929) Phosphorous compounds of. muscle and liver. *Science* 70: 381–382.

Holton P (1959) The liberation of adenosine triphosphate on antidromic stimulation of sensory nerves. *Journal of Physiology* 145: 494–504.

Jahr CE and Jessell TM (1983) ATP excites a subpopulation of rat dorsal horn neurones. *Nature* 304: 730–733.

Khakh BS, Burnstock G, Kennedy C, et al. (2001) International union of pharmacology. XXIV. Current status of the nomenclature and properties of P2X receptors and their subunits. *Pharmacological Reviews* 53: 107–118.

Krishtal OA, Marchenko SM, and Pidoplichko VI (1983) Receptor for ATP in the membrane of mammalian sensory neurones. *Neuroscience Letters* 35: 41–45.

Lohmann K (1929) On the pyrophosphate fraction in muscle. *Naturwissenschaften* 17: 624–625.

North RA and Verkhratsky A (2006) Purinergic transmission in the central nervous system. *Pflugers Archives* 452: 479–485.

Pankratov Y, Lalo U, Krishtal O, et al. (2002) Ionotropic P2X purinoreceptors mediate synaptic transmission in rat pyramidal neurones of layer II/III of somato-sensory cortex. *Journal of Physiology* 542: 529–536.

Pankratov YV, Lalo UV, and Krishtal OA (2002) Role for P2X receptors in long-term potentiation. *Journal of Neuroscience* 22: 8363–8369.

Pankratov Y, Lalo U, Krishtal O, et al. (2003) P2X receptor-mediated excitatory synaptic currents in somatosensory cortex. *Molecular and Cellular Neuroscience* 24: 842–849.

Pankratov Y, Lalo U, Verkhratsky A, et al. (2006) Vesicular release of ATP at central synapses. *Pflugers Archives* 452: 589–597.

Pascual O, Casper KB, Kubera C, et al. (2005) Astrocytic purinergic signaling coordinates synaptic networks. *Science* 310: 113–116.

Silinsky EM and Gerzanich V (1993) On the excitatory effects of ATP and its role as a neurotransmitter in coeliac neurons of the guinea-pig. *Journal of Physiology* 464: 197–212.

Suadicani SO, Brosnan CF, and Scemes E (2006) $P2X_7$ receptors mediate ATP release and amplification of astrocytic intercellular Ca^{2+} signaling. *Journal of Neuroscience* 26: 1378–1385.

White TD (1978) Release of ATP from a synaptosomal preparation by elevated extracellular K^+ and by veratridine. *Journal of Neurochemistry* 30: 329–336.

Adolescent Brain Development and the Risk of Psychiatric Disorders

J N Giedd, National Institute of Mental Health, Bethesda, MD, USA

Published by Elsevier Ltd.

Introduction

Adolescence, the transition from childhood to adulthood, usually occurs in the second decade of life and is characterized by tumultuous changes in the body, brain, and behavior. The behavioral changes, famously characterized by GS Hall in 1904 as a time of *sturm und drang* (storm and stress), have been a prominent theme in art and literature. Across cultures and millennia, adolescence has been noted as a time of heightened passions, mood volatility, sensation seeking, risk taking, and conflict with parents.

Although most teens successfully navigate the transition from being dependent upon a caregiver to self-sufficiency, adolescence is a time of peak onset for the emergence of several classes of neuropsychiatric illnesses including anxiety, bipolar disorder, depression, eating disorder, psychosis, and substance abuse.

Advances in neuroimaging technology have created an unprecedented opportunity to explore the neurobiological substrates of these social, cognitive, behavioral, and pathological changes. In this article, we summarize adolescent brain changes, discuss the possible relationship to the emergence of psychopathology, and outline possible directions for future research.

Neurobiological Changes of Adolescence

Postmortem Data

Human In a series of postmortem studies conducted over the past 25 years, the lab of Peter Huttenlocher has examined over 50 autopsy brains spanning an age range from 28-week-old fetuses to 90-year-old adults. This work has provided seminal information about the characteristic of synapses across the life span, although only four of the specimens are from subjects in the adolescent age range. The investigations have targeted three different brain areas (frontal lobes, occipital, and auditory cortex) allowing assessment of regional variation. In the frontal cortex, synaptic density is similar to adult levels at birth, rises to 150–200% of adult levels during early childhood, and then declines from ages 7 to 16 years. The rate of decline for frontal synaptic density was noted to be protracted compared to auditory and occipital areas which stabilized at 11–12 years.

Primate Because of the paucity of human adolescent postmortem data, much of the speculation regarding adolescent brain changes has been inferred from studies of nonhuman primates. Studies of macaque monkeys indicate a striking decrease in cortical synapses during puberty with a loss of up to 30 000 synapses per second. This leads to a 50% reduction in average number of synaptic contacts per neuron compared to the prepubertal state.

Neurotransmitter Systems

Animal studies also demonstrate dynamic changes in neurotransmitter systems throughout childhood and adolescence. Dopamine receptors, like synapses, follow a developmental trajectory of low levels at birth, maximum density during prepuberty, followed by selective pruning during adolescence. The pattern is nuanced by sex and brain region. For instance, in rat striatum prepubertal dopamine receptor overproduction and adolescent pruning is much more pronounced in males than females and concurrent prepubertal D1 receptor overproduction in the nucleus accumbens is not followed by adolescent pruning. Consistent with the relatively late maturation of the prefrontal cortex in humans, D1 receptor pruning in rat prefrontal cortex does not begin until early adulthood, but then the pruning proceeds more robustly than in the striatum. The changing balance between cortical and subcortical dopamine activity may underlie the motivational, appetitive, and novelty seeking changes observed during adolescence.

Serotonin systems, which reciprocally interact with dopamine systems to modulate many behaviors, also display dynamic changes during adolescence with a study of male human postmortem brains demonstrating dramatically reduced 5HT1a binding during adolescence.

Receptor bindings for excitatory glutamate and inhibitory GABA neurotransmitters are higher during prepuberty than adulthood in rats and in nonhuman primates. In macaque monkeys, biochemical markers in a subtype of GABA neurons called chandelier cells, which connect to initial axon segments of multiple pyramidal cells to regulate pyramidal cell firing patterns within cortical columns, also follow the peak at puberty decline during adolescence pattern. Biochemical markers in another subtype of GABA neurons called wide arbor cells, which regulate firing of pyramidal cells in neighboring columns, increase steadily throughout from infancy to adulthood. Distribution of postsynaptic GABA receptor subtypes also undergoes substantial changes throughout pediatric development making this neurotransmitter system involved in working memory and other cognitive functions particularly volatile during adolescence.

Although developmental trajectories of these various neurotransmitters do not overlap precisely, the general pattern of increases during childhood followed by selective decreases during adolescence seems to hold. The relationship between adolescent cognitive and behavioral changes and these dynamically changing and interacting neurotransmitter systems is only beginning to be elucidated.

Structural Neuroimaging

Most of the information regarding brain anatomy in adolescents comes from magnetic resonance imaging (MRI) studies. MRI is particularly well suited for *in vivo* pediatric studies, because unlike conventional X-rays and computerized axial tomography it does not use ionizing radiation. This allows not only the initial scanning of children and adolescent but also of repeated scans over time.

The spatial resolution of MRI scans is constantly improving and greater resolution can be purchased with the currency of time in the scanner, but in current studies the usual size of the smallest volume with a single MRI signal value (i.e., voxel) is approximately 1 ml. This is worth noting with respect to interpretation of the MRI findings, because within a given 1 ml voxel there may be millions of neurons and trillions of synaptic connections. For example, one unit of the mouse cerebral cortex contains about 30% axons, 30% dendrites, 14% cell bodies, 12% dendritic spines, and 9% glia. However, given that a single MR voxel of cortical gray-matter in the human brain contains a mixture of cell types, it is impossible, at this point, to attribute the observed effects to any single cellular compartment.

White matter volumes increase throughout childhood and adolescence, although rates of increase vary by region and age. This increase in volume is thought to be due to ongoing myelination, the wrapping of an insulating material around axons by oligodendrocytes, which serves to increase greatly the speed of neuronal signal transmission and facilitates the integration of distributed neural networks.

An MR technology called diffusion tensor imaging (DTI), which enables assessment of directionality of fiber tracts, is being used to characterize further white matter development during childhood and adolescence. Pediatric DTI studies have begun to explore relationships between white matter development and cognitive capacities.

Gray matter volumes follow a distinctly different developmental path following an inverted U shaped trajectory. Gray matter volume trajectories vary by region but peak during adolescence for frontal, temporal, and parietal lobes. Particularly late to reach adult levels of cortical thickness are areas in the prefrontal cortex and posterior superior temporal region. Although specifics of the relationships between the anatomic changes and cognitive or emotional changes have not been well elucidated, discussions regarding the implications of the relatively late anatomical maturation of the prefrontal cortex (a key component of circuitry involved in judgment, decision making, and impulse control) have prominently entered social, educational, and judicial realms.

An important, but unresolved, question is the extent to which these gray matter volume changes reflect intracortical myelination versus dendritic and axonal arborization/pruning (the number and extent of neuronal branches). Although arborization/pruning and myelination processes both occur, understanding their relative contributions to the gray matter changes has implications for the conceptualization of brain development in health and illness, the design of future studies, and ultimately as a guide to interventions.

Functional Neuroimaging

Electroencephalography Several large sample electroencephalographic (EEG) studies have been conducted on children and adolescents and consistently demonstrate a cyclic reduction in EEG power during adolescence, particularly for the frontal lobes. The reduction in EEG power continues into the 20s although the rate of reduction is greater in adolescence. EEG studies also indicate profound sleep changes during adolescence with a 50% reduction of deep (stage 4) sleep and a 75% reduction in the peak amplitude of delta waves. EEG signals are thought to be generated by spatially coherent synaptic activity and to possibly reflect changes in synaptic density.

Positron emission tomography Positron emission tomography studies, because of the use of ionizing radiation, have been sparse for children and adolescents. The most cited data are from a study of 29 subjects with epilepsy ranging in age from 5 days to 15 years. Glucose utilization was 30% below adult levels at birth rising to 200% of adult values by the age of 4 years. Glucose utilization remained relatively stable from ages 4 to 10 years and then slowly declined to adult levels during adolescence.

Functional magnetic resonance imaging Functional magnetic resonance imaging (fMRI), which capitalizes on different magnetic properties of oxygenated versus deoxygeneated hemoglobin to assess indirectly blood flow (assumed to be related to neuronal activity), is increasingly being used in pediatric investigations of typically and atypically developing populations.

Although larger sample sizes and longitudinal studies will strengthen the certainty of the results, studies assessing language, social cognition, reward and motivational systems, and executive functions such as working memory and response inhibition have all demonstrated different activation patterns among children and adults. Consistent with the structural MRI findings, associative areas display greater developmental changes than sensorimotor during this developmental period.

A general pattern emerging from these studies is for neural activation during tasks to become less diffuse from childhood to adulthood. This pattern seen over several years in typical development seems to follow that seen for acute learning in adults as task relevant cerebral areas develop enhanced activation while other areas become less active.

Summary of Adolescent Brain Changes

Human postmortem data, animal studies, and neuroimaging all suggest that the tumultuous changes in behavior and cognition are paralleled by equally tumultuous changes in brain anatomy and physiology. The basic process of overproduction followed by selective/competitive elimination that shapes the developing nervous system *in utero* seems to continue to refine the central nervous system throughout adolescent development as prepubertal peaks in the number of synapses, glucose utilization, and neurotransmitter receptors are reduced during the teenage years. Concomitantly, aided by ongoing myelination, distributed brain modules become more and more integrated facilitating greater associative interactions and prefrontal cortex modulation of behavior.

Relationship to Psychopathology

Results of the National Comorbidity Survey Replication study, which entailed in-person household assessments of over 9000 people representative of the United States population (conducted from February 2001 to April 2003), indicate that the peak age of onset for having any mental health disorder is 14 years. Anxiety disorders, bipolar disorder, depression, eating disorder, psychosis, and substance abuse all most commonly emerge during adolescence. A hypothesis is that the emergence of certain psychopathology is related to anomalies or exaggerations of typical adolescent maturation processes.

Schizophrenia

The peak age of onset for schizophrenia is 18 years for males and 25 years for females. A relationship between this age of onset and changes in typical development was first proposed in 1982 by Irvin Feinberg. Based on his interpretation of the dramatic decreases in delta sleep of healthy adolescents as reflecting robust synaptic pruning, Feinberg postulated that schizophrenia may be a consequence of an exaggeration of typical synaptic elimination. Subsequently, several lines of evidence have lent support to this 'exaggeration of typical adolescent changes' hypothesis for schizophrenia. In addition to the exaggerated reductions in adolescent delta sleep, membrane phospholipids, prefrontal metabolism, and frontal cortical gray matter changes are all consistent with an exaggeration of changes seen in typical development. In a rare but phenomenologically similar to adult onset group of subjects with childhood onset schizophrenia (onset prior to age 12 years), the typical frontal gray matter reduction seen in subjects from ages 12 to 17 years was exaggerated fourfold in frontal areas. Direct evidence of a decrease in the number of synapses in schizophrenia, possibly impairing neuronal plasticity, comes from postmortem studies indicating decreased density of synaptic spines, reduction in neuropil, and decreased expression of the synaptic marker synaptophysin.

This notion of an exaggeration of typical adolescent changes is meant to be a further consideration in the neurodevelopmental hypothesis, not a comprehensive account of the etiology of schizophrenia. An increase in prenatal and perinatal adverse events and subtle cognitive, motor, and behavioral anomalies during childhood years before illness onset, all support earlier developmental disturbances underlying the abnormal maturational events during adolescence.

Although there is sparse data for other conditions of adolescent onset, conceptually the notion of anomalies in the dynamic processes of pruning or connectivity may extend to other disorders as well.

Substance Abuse

Across a wide array of mammalian species, adolescents exhibit increased risk taking, novelty seeking, and a greater valuation of social factors. These characteristics not only foster independence from the natal family unit but also increase the risk for harmful behaviors such as substance abuse. The most commonly abused substance is alcohol, and adolescent neurobiology may make them particularly prone to addiction. Evidence from rats indicates that adolescents are relatively insensitive to factors that may limit alcohol intake such as developing motor impairment, getting a 'hangover,' or becoming sedated (perhaps related to immaturity of the developing γ-aminobutyric acid (GABA)-A receptor systems).

Furthermore, the adolescent rat hippocampus is unusually susceptible to ethanol-induced inhibition of long-term potentiation making them more sensitive to the memory-impairing effect of alcohol. The mechanism for this effect appears to be largely mediated via alcohol's effect on *N*-methyl-D-aspartate (NMDA) receptors and juvenile-greater-than-adult impact occurs at the single cell level, is not confined to the hippocampus, and occurs at alcohol concentrations as low as from a single drink (5 mM).

Morphometric studies of humans suggest that some neural alterations predispose to risk, whereas others may be the result of the abuse. For instance, the right amygdala is smaller in youths with a family history of alcohol abuse even prior to the onset of problem drinking, whereas hippocampal volumes are reduced only after a history of alcohol use.

Increases in risk taking, novelty seeking, and priority on social factors along with a relative insensitivity to intake reducing factors and a greater sensitivity to ethanol-induced effects on hippocampal and neurotransmitter systems combine to make adolescence a particularly vulnerable time to develop substance abuse.

Affective and Anxiety Disorders

The most dramatic behavioral changes during adolescence are in the social realm with the emergence of sexuality and a shift from family to peer influences. This shift is reflected in a greatly heightened emotional responsiveness to social stimuli during adolescence. Nelson et al. have proposed a three-part model of social information processing involving an infancy/early childhood maturing node for detection of socially relevant environmental cues; an early adolescent maturing affective node which ascribes emotional significance to the cues; and a late adolescent/early adulthood maturing cognitive-regulatory node which inhibits prepotent responses and directs goal-directed behavior. The affective node (which includes limbic structures such as the amygdala, ventral striatum, septum, and hypothalamus) is preferentially sensitive to hormonal effects, and its maturation before the cognitive-regulatory node combined with the increasing importance of social factors may underlie the adolescent surge and gender specificity in the prevalence of anxiety and depressive disorders.

Structural MRI studies of depressed and anxious adolescents report structural anomalies in the superior temporal gyrus, ventral prefrontal cortex, and amygdala and an fMRI study of depressed and anxious adolescents reported anomalous amygdala responsive to social stimuli. Neuroimaging studies support this developmental lag between the affective and cognitive-regulatory nodes. In an fMRI face viewing study, adults, but not adolescents, were able to selectively engage the orbitofrontal cortex when asked to switch from an emotional assessment (i.e., how afraid does it make you feel?) to a nonemotional (i.e., how wide is the nose?).

Increasingly powerful emotional responses to social stimuli, abrupt changes in motivation and reward systems, and a developmental lag in the neural systems necessary to regulate and inhibit highly motivated behaviors in a goal directed manner may underlie the onset of anxiety and depressive disorders during adolescence.

Discussion

The relationship between typical adolescent brain changes and onset of psychopathology is not a unitary phenomenon, but an underlying theme may be conceptualized as 'moving parts get broken.' Many aspects of the complicated and reciprocally interconnected development trajectories may go awry predisposing people to the wide variety of disorders noted from epidemiological studies. Anxiety disorders, bipolar disorder, depression, eating disorder, psychosis, and substance abuse all most commonly emerge during adolescence. Other disorders such as autism or attention deficit hyperactivity disorder (ADHD) have earlier onsets, whereas Alzheimer's disease has a much later onset. The specificity of which disorders have an adolescent onset may offer clues as to the mechanisms involved. A greater understanding of the relationship between specific adolescent changes and the specific cognitive, behavioral, and emotional consequences may provide insight into preventive or treatment interventions.

The remaining section offers considerations for future research exploring the connection between typical adolescent brain changes and psychopathology.

Future Directions

Longitudinal Studies

An emerging theme from pediatric neuroimaging studies is that the path of brain development is often as important as the end point. For example, IQ is predicted by the developmental trajectory of cortical thickness, not by the adult size. Large individual variability in brain anatomy and function call for longitudinal study designs to capture the nuances of heterochronous developmental curves.

The first phases of longitudinal studies have mapped developmental trajectories for typical development, and less so for some psychiatric illnesses. The next phases should go beyond simply mapping brain growth and begin to discern the influences, for good or ill, on those trajectories.

Genetics

A common initial approach to assessing influences is to discern the relative effects of genetic versus nongenetic factors. This is best addressed through comparisons of monozygotic and dyzygotic twins. Results from an ongoing pediatric longitudinal neuroimaging project at the Child Psychiatry Branch of the National Institute of Mental Health indicate significant age by heritability interactions with gray matter heritability generally decreasing with age and white matter heritability generally increasing with age.

Heritability by age interactions may be related to timing of gene expression, which in turn may relate to timing of onset of illness. Postmortem human and animal studies indicate that developmental genes have diverse effects at various stages of brain development. Ongoing studies of specific gene effects on brain maturation may help to sharpen our understanding of brain development mechanisms and provide insight into the etiologies of various pathologies. Genetics may also provide biologically relevant subtypes of neuropsychiatric disorders which are obscured in current diagnostic schemes.

Sex Differences

The marked sex differences in age of onset, prevalence, and symptomatology for nearly every neuropsychiatric disorder may provide important clues as to pathophysiology. The most obvious outward physical manifestations of puberty are caused by changing levels of hormones. Perhaps this has contributed to the tendency to attribute all of the cognitive and behavioral changes of adolescence to 'raging hormones' as well. However, the relationship between hormones, brain, and behavior is complex, reciprocal, and poorly understood. Steroid hormones affect neuronal activity and morphology throughout development. Most neurons have receptors for adrenal and gonadal hormones receptors which when activated affect neurotransmitter function. Short-term effects are mediated by membrane receptors, whereas long-term effects alter gene expression via intraneuronal or nuclear receptors. Conversely, the dramatic hormonal changes of puberty are triggered by alterations in excitatory and inhibitory inputs to gonadotropin-releasing hormone neurons in the pituitary. Behaviorally, hormonal effects drive aggression and sexual interest but the impact on impulse control, logical problem solving, or other cognitive tasks has not been well established.

Social and cultural factors for boys and girls are profoundly different, and the relationship of these differences to manifest pathology should be explored. In the biological realm, sex differences likely stem directly from different genes on the X or Y chromosomes or indirectly through the effects of different hormone levels. Studies of subjects with sex chromosome variations (e.g., XO, XXY, XXYY, XXX, and XXXXY) or anomalous hormone levels (e.g., congenital adrenal hyperplasia, androgen insensitivity syndrome, and familial male precocious puberty) will be useful to sort out the relative contributions of gene and hormone effects. For instance, males with an extra X chromosome (XXY or Klinefelter's syndrome) have a high incidence of language disorders, ADHD, and social skills deficits that are reflected in cortical thickness changes consistent with reports in the literature for XY subjects with those disorders. Girls with 'congenital adrenal hyperplasia' (characterized by high levels of intrauterine testosterone) have an entirely different pattern of brain morphometric findings indicating differential effects of sex chromosomes and hormones.

Multimodal Imaging

Combining multiple imaging modalities on the same individuals, such as structural MRI, fMRI, DTI, magnetization transfer imaging, EEG, and magnetoencephalography (MEG), will synergistically enhance our ability to interpret the signals for each of the modalities. Being able to simultaneously examine interindividual variation from cellular to macroscopic levels will be instrumental in bridging gaps between genes, brain, and behavior.

Translational Research

Postmortem studies of animals that have been imaged will be needed to clarify the nature of changes driving the MRI findings. Of immediate relevance will be discerning the degree to which cortical gray matter changes as detected via MRI are related to arborization/pruning of neurons or to encroachment of white matter on the inner cortical border.

Integration with Social and Educational Science

Studies of adolescent behavior and decision making will need to be integrated better with social and educational science. Laboratory studies of teens using hypothetical situations in calm environments without peer influence may have little relevance for understanding real world decision making which is often in the context of intense physical or emotion arousal, conflicting priorities, and in the presence of peers.

Despite the shared goal of guiding people through the adolescent years safely and optimally prepared for the adult world little rigorous research has been conducted across the neuroscience and education science disciplines.

Summary

Adolescence is a time of tumultuous neurobiological and behavioral change. These changes are usually healthy and optimize the brain for the challenges ahead, but may also confer a vulnerability to certain types of psychopathology. The instruments and technologies to elucidate the relationship between specific neurobiological maturational processes and specific normative or pathologic changes are already in place. Applying these tools to understand when and how deviations from typical development occur may lead to improved treatments for disorders affecting a substantial number of people.

See also: Anxiety Disorders; Cognitive Dysfunction in Psychiatric Disorders; Diffusion Tensor Imaging (DTI); Dopamine – CNS Pathways and Neurophysiology; Electroencephalography (EEG); fMRI: BOLD Contrast; Gamma-Aminobutyric Acid (GABA); Genomics of Brain Aging: Twin Studies; Glutamate; Neuroimaging; Positron Emission Tomography (PET); Serotonin (5-Hydroxytryptamine; 5-HT): CNS Pathways and Neurophysiology; Stress Response: Sex Differences; Stress, Sex and Adolescent Nicotine Response; Substance Abuse and Dependence.

Further Reading

Casey BJ, Tottenham N, Liston C, and Durston S (2005) Imaging the developing brain: What have we learned about cognitive development? *Trends in Cognitive Sciences* 9: 104–110.

Feinberg I (1982) Schizophrenia: Caused by a fault in programmed synaptic elimination during adolescence? *Journal of Psychiatric Research* 17: 319–334.

Giedd JN, Blumenthal J, Jeffries NO, et al. (1999) Brain development during childhood and adolescence: A longitudinal MRI study. *Nature Neuroscience* 2: 861–863.

Giedd JN, Clasen LS, Lenroot R, et al. (2006) Puberty-related influences on brain development. *Molecular and Cellular Endocrinology* 254–255: 154–162.

Huttenlocher PR and Dabholkar AS (1997) Regional differences in synaptogenesis in human cerebral cortex. *The Journal of Comparative Neurology* 387: 167–178.

Keshavan MS, Anderson S, and Pettegrew JW (1994) Is schizophrenia due to excessive synaptic pruning in the prefrontal cortex? The Feinberg hypothesis revisited. *Journal of Psychiatric Research* 28: 239–265.

Kessler RC, Berglund P, Demler O, et al. (2005) Lifetime prevalence and age-of-onset distributions of DSM-IV disorders in the National Comorbidity Survey Replication. *Archives of General Psychiatry* 62: 593–602.

Lewis DA (1997) Development of the prefrontal cortex during adolescence: Insights into vulnerable neural circuits in schizophrenia. *Neuropsychopharmacology* 16: 385–398.

Luna B and Sweeney JA (2004) The emergence of collaborative brain function: FMRI studies of the development of response inhibition. *Annals of the New York Academy of Sciences* 1021: 296–309.

Paus T (2005) Mapping brain maturation and cognitive development during adolescence. *Trends in Cognitive Sciences* 9: 60–68.

Rakic P, Bourgeois JP, and Goldman-Rakic PS (1994) Synaptic development of the cerebral cortex: Implications for learning, memory, and mental illness. *Progress in Brain Research* 102: 227–243.

Shaw P, Greenstein D, Lerch J, et al. (2006) Intellectual ability and cortical development in children and adolescents. *Nature* 440: 676–679.

Spear LP (2000) The adolescent brain and age-related behavioral manifestations. *Neuroscience and Biobehavioral Reviews* 24: 417–463.

Steinberg L (2005) Cognitive and affective development in adolescence. *Trends in Cognitive Sciences* 9: 69–74.

Thatcher RW (1991) Maturation of the human frontal lobes: Physiological evidence for staging. *Developmental Neuropsychology* 7: 397–419.

Adrenal Steroids: Biphasic Effects on Neurons

M Joëls and H Karst, University of Amsterdam, Amsterdam, The Netherlands

Introduction

Stress activates the sympathoadrenomedullar system as well as the hypothalamic–pituitary–adrenal axis. As a consequence of the latter, large amounts of corticosteroid hormones are released from the adrenal glands. In addition to the stress-related release, there is also a circadian rhythmicity in the circulating hormone levels.

Corticosteroid hormones – cortisol in humans and corticosterone in most rodents – not only reach peripheral organs but also pass the blood–brain barrier, thus entering the brain. Within the brain, corticosteroid hormones bind to discretely localized intracellular receptors. Two receptors have been recognized: the high-affinity mineralocorticoid receptor (MR) and the lower affinity glucocorticoid receptor (GR). In view of the receptor affinities and circulating hormone levels, it is generally thought that in brain under rest, mainly MRs are activated. These are primarily localized in limbic regions, such as the hippocampus, lateral septum, and central amygdala, and in motor nuclei in the brain stem. During stress and at the circadian peak, GRs also become substantially activated, next to MRs. The GRs are much more ubiquitously distributed in brain.

Activated corticosteroid receptors translocate to the nucleus where they modulate the transcription of responsive genes. This means that corticosterone, through its gene-mediated actions, exerts a slow but persistent control over the cellular protein content, including molecules that are involved in cellular excitability and synaptic transmission. In addition to these slow actions, corticosteroid hormones can also affect neurotransmission through rapid, nongenomic pathways, but these are not considered here.

During the past decade, the effects of selective MR and GR activation on electrical properties of neurons have been mainly examined in the CA1 region of the hippocampus, where both receptors are co-localized. It was observed that MR and GR often induce opposite effects on a given parameter, resulting in biphasic effects of adrenal steroid hormones on neuronal properties.

Corticosteroid Hormones Affect Some but Not All Properties

Of the potential targets for corticosteroid hormone actions, some are extremely relevant to neurons. For instance, steroid modulation of ion channel function would largely affect the excitability and potentially the viability of neurons. Likewise, influences on neurotransmitter receptors and the signaling cascades linked to these receptors would have profound consequences for the information transfer in essential brain regions.

It was found that corticosterone affects some but not all of these properties in the CA1 region of the hippocampus. For instance, voltage-dependent calcium currents were found to be very sensitive to corticosteroid receptor occupation. By contrast, most of the tested potassium and sodium currents are far less sensitive to changes in the level of corticosterone. Of the G-protein-coupled receptors, particularly the serotonin$_{1A}$ (5-HT$_{1A}$) receptor is modulated by corticosteroid hormones, although to some extent other receptors (e.g., for acetylcholine or noradrenaline) are also changed in their function. The modulation of ionotropic receptors is more complex. Here, rapid nongenomic effects seem more prominent. However, a slow enhancement of glutamate receptor-mediated responses by corticosterone or stress has been described. It should be noted that corticosteroid effects on other endpoints (e.g., related to metabolism or cell structure (not discussed here)) will also indirectly affect the efficacy of synaptic transmission, including transmission mediated by glutamate.

Not only cell properties but also phenomena that depend on network function, such as long-term potentiation (LTP) and long-term depression (LTD), strongly depend on the circulating levels of corticosterone.

Biphasic Effects on Neural Properties

In the CA1 hippocampal area, conditions that result in predominant MR activation often induce effects opposite to those where both receptor types are activated. This was first demonstrated for calcium currents and cell properties that are regulated by the intracellular calcium level, such as adaptation of cell firing frequency. Under rest, with predominant MR occupation, influx through high-voltage activated calcium channels is limited. Hence, little adaptation in firing frequency is seen so that excitatory input is effectively transmitted. After stress (resulting in additional GR occupation), calcium influx slowly increases and adaptation of cell firing frequency becomes strong. In the absence of corticosterone, calcium influx through voltage activated channel is also large, as is the firing frequency adaptation. Overall, the dependence of calcium influx through voltage-dependent channels and related phenomena on the dose of corticosterone displays a U shape (**Figure 1**).

This dose–response relationship was subsequently also found for other properties of CA1 cells, most notably the responses mediated via 5-HT_{1A} receptors. These receptors are linked through a G-protein to inwardly rectifying potassium channels. Activation of the 5-HT_{1A} receptor therefore leads to opening of potassium channels and thus hyperpolarization of the membrane. Both in the absence of corticosterone and with high doses of the hormone, responses via 5-HT_{1A} receptors are large. By contrast, moderately low hormone levels lead to small 5-HT_{1A} receptor-mediated responses.

At the network level, U-shaped dose dependencies can also be observed for corticosteroid actions in the CA1 hippocampal area. Many studies have reported optimal induction of LTP in animals with low to moderate circulating levels of corticosterone. Both in the absence of the hormone and after considerable stress, it appears to be more difficult to induce LTP, while at the same time the occurrence of LTD is facilitated. It has been postulated that stress and corticosteroid hormones prime the circuits involved in LTP formation such that the threshold for subsequent potentiation is elevated. This may relate to behavioral observations that learning after a period of stress (unrelated to the learning situation) impairs memory

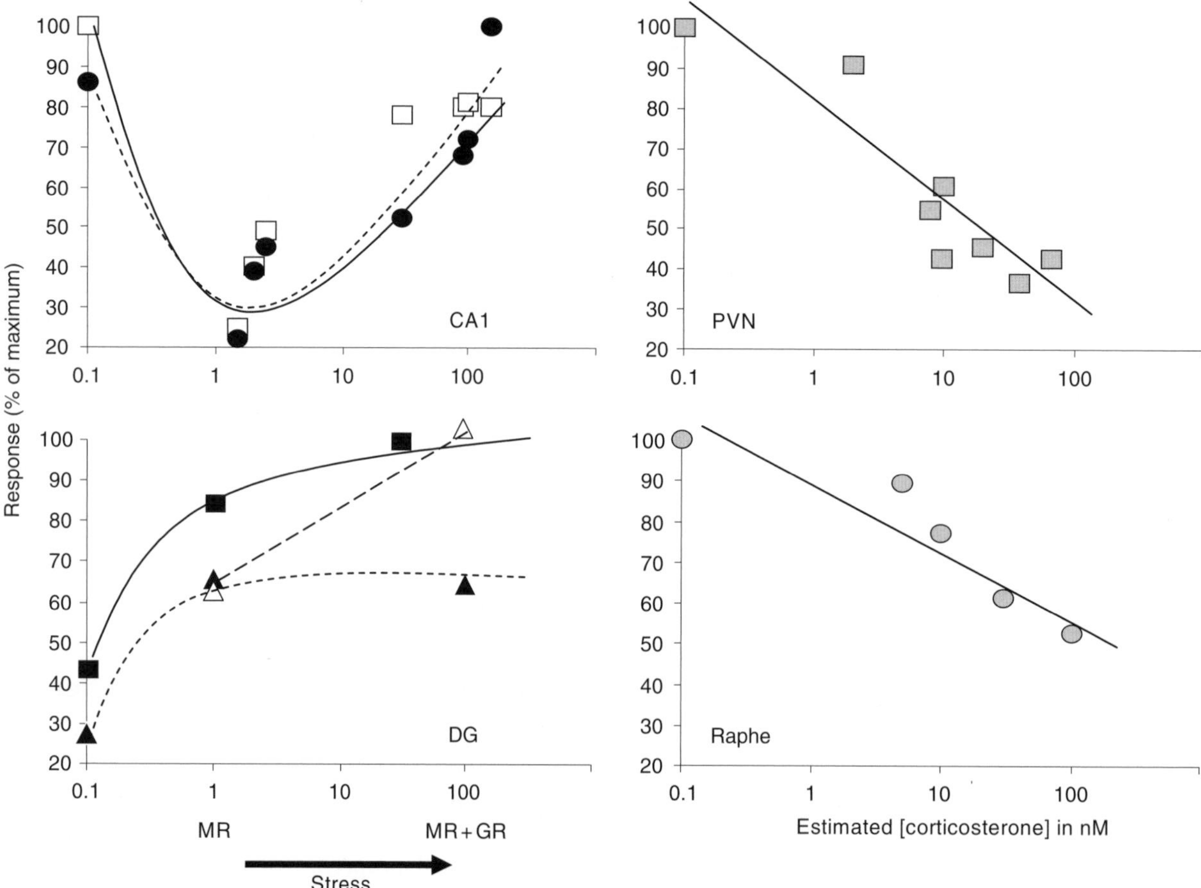

Figure 1 Adrenal corticosteroids cause different effects on neurons, depending on the concentration of the hormone that circulates. For instance, under rest (a situation that is associated with predominant mineralocorticoid receptor (MR) activation), calcium currents in CA1 neurons are small, whereas their amplitude increases several hours after stress (open squares). In the absence of the hormone, such as occurs after adrenalectomy, calcium currents are also large, resulting in a U-shaped dose dependency. A similar dose dependency has been described for responses of CA1 neurons to serotonin (solid circles). The differential responses are linked to the activation of two receptor types, with different affinity for corticosterone. However, U-shaped dose dependencies are not always observed. This can be due to the presence of only one receptor type, as illustrated for GABAergic responses in the paraventricular nucleus (PVN) or serotonergic responses in the dorsal raphe. In the dentate gyrus (DG), the dose–response relationship for glutamate-mediated field (squares) or single cell (solid triangles) responses also seems to be different from that of the CA1 region, despite the presence of MRs and glucocorticoid receptors (GRs) in both areas; after chronic stress (open triangles), GRs are responsive again. Adapted from Joëls M (2006) Corticosteroid effects in brain: U shape it. *Trends in Pharmacological Sciences* 27: 244–250, with permission from Elsevier.

formation. It should be noted, however, that stress which forms an intrinsic part of and occurs concurrently with the event to be remembered helps to consolidate this information.

Two Receptor Types Contribute to the Biphasic Responses in CA1 Neurons

With the use of selective corticosteroid receptor agonists and antagonists it was shown that the dose dependency is coupled to specific receptor occupation ratios. Collectively, the data show that a situation of predominant MR activation in the CA1 region, such as occurs under rest at the circadian trough, promotes ongoing activity and synaptic plasticity, yet at the same time permits only limited calcium influx through voltage-dependent channels. This is important because it allows cells to transfer excitatory information but protects them against calcium-dependent delayed damage. The reverse seems to be true for GR activation (always, of course, against a background of already activated MRs), as happens after stress. In the short term (i.e., over the course of several hours), GR activation will slowly suppress activity in the CA1 area. This is in line with the adaptational character of the hormone because this effect allows CA1 cells to normalize their activity after a temporary period of 'arousal.' However, GR activation also allows more calcium ions to flow into the cells. This may become detrimental when GR activation concurs with a period of heavy excitation (e.g., during an epileptic insult). In agreement, animal models have shown that hippocampal epileptogenesis and ischemic damage is exacerbated by exposure to high corticosteroid levels.

Do Corticosteroids Always Induce Biphasic Responses?

The U-shaped dose dependency is typical for the CA1 area but not for all regions in the brain. This is partly due to the differential distribution of corticosteroid receptors. Most neurons do express GRs, but the expression levels for MR are usually very low, with the exception of some limbic areas. Hence, cellular effects in most areas will be determined by one receptor type only, the GR. Although theoretically biphasic responses can be established through a single receptor system, this does not seem to be the case for neuronal effects of adrenal steroids. This is exemplified by responses measured in the paraventricular nucleus of the hypothalamus and the dorsal raphe, two areas in which GRs prevail. In these areas, a linear dose dependency is observed (**Figure 1**).

However, even in cells that do express both MRs and GRs, the dose dependency is not always U-shaped. For instance, glutamate receptor-mediated responses in the dentate gyrus are enhanced by MR activation (compared to the situation in which corticosterone is absent), but additional GR activation does not lead to noticeable changes.

Several factors could explain why, in this case, biphasic responses are not observed. First, the bioavailability of corticosterone may exhibit local differences. This could not only be due to regional differences in the expression of converting enzymes, such as 11-β-steroid dehydrogenase, but also to differences in transporters which determine the amount of corticosteroids transported over the plasma membrane.

Second, many variants of MR as well as GR have been described. Some of these variants have reduced transcriptional activity. Local differences in the expression of these variants could explain why high concentrations of corticosterone have little impact on cell function.

Finally, transcriptional activity of corticosteroid receptors to a considerable extent depends on protein–protein interactions with cofactors and other transcription factors. The cellular content of these additional molecules therefore also determines the efficacy with which GRs (and MRs) affect neuronal function. Because the cellular content of all of these factors can display large regional differences, it is not difficult to understand why biphasic responses do not always occur, even in the presence of the two receptor systems.

See also: Diabetes Type 2 and Stress: Impact on Memory and the Hippocampus; Gene Therapy and Protection from Stress-Induced Brain Damage; Stress and Neural Involvement in Metabolism; Stress and Cognition; Stress and Suicide; Stress and Vulnerability to Brain Damage; Stress, Dopamine, and Puberty; Stress, Cytokines and Depressive Illness; Stress, the HPA Axis and Depressive Illness.

Further Reading

De Kloet ER, Joëls M, and Holsboer F (2005) Stress and the brain: From adaptation to disease. *Nature Reviews Neuroscience* 6: 463–475.

Joëls M (1997) Steroid hormones and excitability in the mammalian brain. *Frontiers in Neuroendocrinology* 18: 2–48.

Joëls M (2006) Corticosteroid effects in brain: U shape it. *Trends in Pharmacological Sciences* 27: 244–250.

Joëls M, Pu Z, Wiegert O, Oitzl MS, and Krugers HJ (2006) Learning under stress: How does it work? *Trends in Cognitive Sciences* 10: 152–158.

Kim JJ and Diamond DM (2002) The stressed hippocampus, synaptic plasticity and lost memories. *Nature Reviews Neuroscience* 3: 453–462.

Kim JJ and Yoon KS (1998) Stress: Metaplastic effects in the hippocampus. *Trends in Neurosciences* 21: 505–509.

Lambert JJ, Belelli D, Peden DR, Vardy AW, and Peters JA (2003) Neurosteroid modulation of $GABA_A$ receptors. *Progress in Neurobiology* 71: 67–80.

Rose JD (2000) Corticosteroid actions from neuronal membrane to behavior: Neurophysiological mechanisms underlying rapid behavioral effects of corticosterone. *Biochemistry and Cell Biology* 78: 307–315.

Adrenergic Receptors

J P Hieble, Wayne, PA, USA

Introduction

The adrenoceptors (adrenergic receptors) mediate the diverse effects of the neurotransmitters of the sympathetic nervous system, norepinephrine and epinephrine, at virtually all sites throughout the body. During the past century, the adrenoceptors have been extensively studied by a variety of functional and molecular techniques, and they have been progressively subdivided. Currently, nine adrenoceptors have been cloned, and they have been divided into three major categories – the α_1-adrenoceptors, α_2-adrenoceptors, and β-adrenoceptors. Although there are pharmacological data that suggest the existence of additional adrenoceptor subtypes, it now seems likely that this results from multiple 'affinity states' of a particular adrenoceptor subtype, rather than from a new adrenoceptor protein yet to be cloned. Although the specific receptors involved have not all been identified, norepinephrine-releasing neurons within the central nervous system (CNS) play an important role in psychiatric disorders such as major depression and attention-deficit/hyperactivity disorder (ADHD).

The adrenoceptors belong to the large family of seven transmembrane-spanning, G-protein-coupled receptors, and there is 32–54% amino acid identity between the individual adrenoceptor proteins (**Figure 1**). The most likely sites for interaction of the catecholamines with conserved structural elements of the adrenoceptors have been identified.

α_1-Adrenoceptors

α_1-Adrenoceptors are widely distributed in both central and peripheral sites, and they are activated either by norepinephrine released from sympathetic nerve terminals or by epinephrine released from the adrenal medulla and some central adrenergic neurons. Peripheral α_1-adrenoceptors mediate a variety of functions, including contraction of smooth muscle, cardiac stimulation, cellular proliferation/apoptosis, and activation of hepatic gluconeogenesis and glycogenolysis. α_1-Adrenoceptors are widely distributed within the CNS, where their activation generally results in depolarization and increased neuronal firing. Most of the peripheral actions of α_1-adrenoceptors are mediated through the inositol phosphate pathway, although there is evidence for activation of adenylyl cyclase within the CNS.

Three distinct α_1-adrenoceptor proteins have been cloned; after some confusion in nomenclature, it has been established that these three recombinant α_1-adrenoceptors correspond to the pharmacologically defined α_{1A}, α_{1B}, and α_{1D} adrenoceptors in native tissues. Selective antagonists are available for each of the α_1-adrenoceptor subtypes. Multiple splice variants of the α_{1A}-adrenoceptor have been identified; however, they appear to have identical pharmacological characteristics. The contraction of several vascular and urogenital tissues has distinct pharmacology from other α_{1A}-mediated responses. This was thought to implicate an additional α_1-adrenoceptor, designated as the α_{1L}-adrenoceptor. This receptor was never cloned, and it now seems likely that the α_{1A}-adrenoceptor may have either α_{1A} or α_{1L} pharmacology, depending on the particular assay, tissue, or experimental conditions employed.

α_1-Adrenoceptor subtypes can be differentially distributed in certain tissues. This differential distribution can reflect either expression of receptor protein/message or pharmacological characteristics in functional assays. For example, the predominant α_1-adrenoceptor found in rat spleen is the α_{1B} subtype, and the functional pharmacological profile of α_1-adrenoceptor agonists (i.e., contraction) in this tissue has α_{1B}-adrenoceptor characteristics, as determined by the use of subtype selective antagonists. Although all of the α_1-adrenoceptor subtypes are present in most blood vessels, the pharmacological characteristics of agonist-induced contraction can vary, such that the responses may be α_{1A} (human microvessels and rat caudal artery), α_{1B} (rabbit or canine aorta), or α_{1D} (rat aorta).

Knockout and overexpression experiments have been reported only for the α_{1B}-adrenoceptor subtype. Knockout of the α_{1B}-adrenoceptor results in impaired vascular responsiveness to phenylephrine, demonstrated both in isolated blood vessels and through measurement of blood pressure responses in anesthetized mice. Cardiac-specific overexpression of the α_{1B}-adrenoceptor results in an impairment of cardiac function without cardiomyocyte hypertrophy.

The α_1-adrenoceptor subtypes are differentially distributed within the CNS. Studies in rat brain have shown α_{1B}-adrenoceptor mRNA to be concentrated in cerebral cortex, thalamus, raphe nuclei, cranial and spinal motor neurons, as well as the pineal gland. mRNA for the α_{1D}-adrenoceptor is localized to the olfactory bulb, cerebral cortex, hippocampus, dentate gyrus, reticular thalamic nucleus, motor neurons, and the inferior olivary complex. In the thalamus, the α_{1B}- and α_{1D}-adrenoceptors have a complimentary

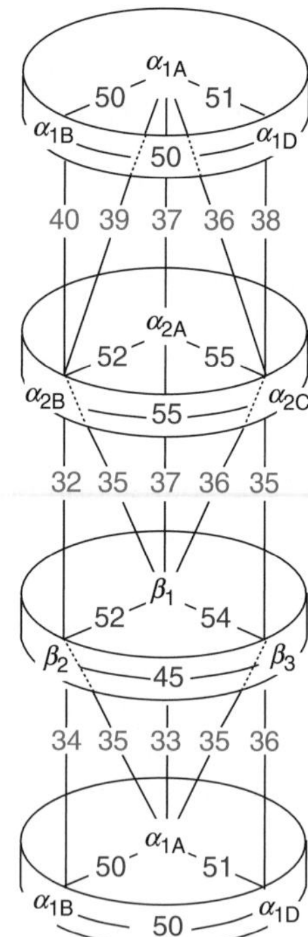

Figure 1 Structural similarity, based on overall amino acid identity, between the nine adrenoceptors. Based on this similarity, the adrenoceptors can be divided into three groups (α_1, α_2, and β), with each group further divided into three subtypes. Structural similarity is higher (45–55% identity) within either of the three groups than between individual members of different groups (33–40% identity).

distribution. Highest levels of α_{1A}-adrenoceptor mRNA are found in regions of the olfactory system, hypothalamic nuclei, and in regions of the brain stem and spinal cord related to motor function. Much less information is available on the distribution of α_1-adrenoceptors in the human brain. Although α_1-adrenoceptors are found in neocortex and dentate gyrus, their distribution differs substantially from that observed in the rat.

Centrally acting α_1-adrenoceptor agonists often potentiate motor stimulation or arousal produced by other pharmacological or behavioral challenges and will increase vigilance when administered alone. α_1-Adrenoceptor agonists can reverse sedation induced by norepinephrine depletion produced by a dopamine β-hydroxylase inhibitor or the cognitive impairment produced by a neurotoxin selective for noradrenergic neurons. Likewise, they can reverse cataplexy in a genetic canine model of narcolepsy. Based on studies in rats and primates, modafinil, currently marketed for the treatment of narcolepsy, may act through the activation of central α_1-adrenoceptors, although based on its structure and peripheral pharmacological effects, this drug would not be expected to be an α-adrenoceptor agonist.

Activation of central α_1-adrenoceptors, presumably in the hypothalamus, stimulates the secretion of several pituitary hormones in human volunteers. A centrally active α_1-adrenoceptor agonist, SDZ NVI-085 (naphtoxazine), improved the performance of patients with Parkinson's disease in some, but not all, tests for attentional deficits. SDZ NVI-085 has shown activity in many of the animal assays for central α_1-adrenoceptor activation.

The functional roles of each α_1-adrenoceptor subtype within the CNS are not completely defined. It has been proposed that both α_{1B}- and α_{2A}-adrenoceptors are present on cell bodies of presympathetic ganglionic neurons in the intermediolateral cell column of the thoracic spinal cord, mediating an excitatory and inhibitory action, respectively. This reciprocal action is consistent with the observation that central α_1-adrenoceptor stimulation will attenuate the sedative actions of an α_2-adrenoceptor agonist in the rat. Mice lacking the α_{1B}-adrenoceptor show different behavioral effects from wild-type mice, in which enhanced reactivity to new situations is observed. Learning behavior in a water maze was impaired, although the ability to escape onto a visible platform was unaffected.

Overexpression of either wild-type or constitutively active α_{1B}-adrenoceptors results in granulovacular neurodegeneration, beginning in the areas of the brain expressing highest levels of the α_{1b}-adrenoceptor (cortex, cerebellum, and hypothalamus) and progressing to other brain areas with age. These mice showed symptoms consistent with neuronal damage, including parkinsonian hindlimb symptoms associated with paralysis and tremor. Grand mal seizures were also observed by 12 months of age. The severity of the seizure disorder was proportional to the level of α_{1B}-adrenoceptor activation, with mice expressing constitutively active receptors being most affected. α_{1B}-Adrenoceptor blockade with terazosin partially reversed the behavioral pathology. However, knockout of the norepinephrine uptake transporter, which results in an increase in central noradrenergic activity, decreases the vulnerability of mice to seizures induced by several convulsant agents.

Correlation of affinity for recombinant α_1-adrenoceptors with their activity in the canine narcolepsy model suggests that the ability of α_1-adrenoceptor agonists to reverse cataplexy in this model results from α_{1B}-adrenoceptor activation. The presence of α_1-adrenoceptors in layers of human hippocampus having a concentration of glutaminergic synapses suggests the possibility for noradrenergic modulation of glutaminergic transmission. Activation of α_1-adrenoceptors in the nucleus tractus solitarius is involved in control of vagal activity to the stomach.

α_2-Adrenoceptors

α_2-Adrenoceptors are also widely distributed throughout the body. Perhaps the most extensively characterized action is the prejunctionally mediated inhibition of neurotransmitter release from many peripheral and central neurons. The demonstration that this effect had a different pharmacologic profile from actions mediated by postjunctional α-adrenoceptors, such as vascular contraction, originally led to the division of α-adrenoceptors into the α_1- and α_2-subtypes. Activation of prejunctional α_2-autoreceptors on sympathetic neurons results in a sympatholytic action. α_2-Adrenoceptors are also present at postjunctional sites, where they mediate actions such as smooth muscle contraction, platelet aggregation, and inhibition of insulin secretion. Many responses produced by α_2-adrenoceptor agonists are mediated through the inhibition of adenylyl cyclase as a consequence of interaction of the agonist–receptor complex with G_i, although other second messengers remain to be characterized.

Three α_2-adrenoceptor proteins have been cloned. These recombinant receptors result in four discrete pharmacological profiles since the α_{2A}-adrenoceptor appears to exist as species orthologs, with those of human, pig, and rabbit having a profile designated as α_{2A}, whereas those of rat, mouse, guinea pig, and cow exhibit pharmacologic profiles designated as α_{2D}. α_{2A}- and α_{2D}-adrenoceptor-mediated responses can be differentiated by the low sensitivity of the α_{2D}-adrenoceptor to blockade by the commonly used antagonists yohimbine and rauwolscine.

Studies in the rat demonstrate the existence of mRNA for all three α_2-adrenoceptor subtypes to be present in the CNS. α_{2A}-Adrenoceptor mRNA is most widely distributed, being found in the cerebral cortex, locus coeruleus, amygdala, hypothalamic paraventricular nucleus, nucleus tractus solitarii, ventrolateral reticular formation, spinal cord, and dorsal root ganglia. Message for the α_{2B}-adrenoceptor is found almost exclusively in the thalamus. The α_{2C}-adrenoceptor is found in olfactory bulb, islands of Calleja, cerebral and cerebellar cortex, hippocampal formation, and dorsal root ganglia. In human brain, radioligand binding assays demonstrate the presence of α_{2A}-adrenoceptor protein in frontal cortex, cerebellum, and hippocampal formation, with another subtype (α_{2B} or α_{2C}) predominant in neostriatum.

α_2-Adrenoceptors within the CNS have long been known to be involved in the antihypertensive action of clonidine and other α_2-adrenoceptor agonists. Although the involvement of a nonadrenergic imidazoline receptor in this action has been postulated, the failure of α_2-adrenoceptor agonists to produce a sympatholytic action in mice where the α_{2A}-adrenoceptor has been knocked out or mutated supports functional experiments in rats and rabbits suggesting an α_2-adrenoceptor-mediated action, at least for systemically administered agonists. In addition to inhibition of sympathetic outflow, the anesthetic and analgesic actions of α_2-adrenoceptor agonists appear to be mediated by the α_{2A}-adrenoceptor. The hypertensive action of α_2-adrenoceptor agonists likely results from activation of vascular α_{2B}-adrenoceptors. Consistent with the apparent role of the α_{2B}-adrenoceptor in the maintenance of vascular tone, mice lacking this receptor subtype failed to develop salt-induced hypertension. Hence, a selective α_{2A}-adrenoceptor agonist may be preferable as a centrally active drug, although sedative and antihypertensive actions can apparently not be dissociated by subtype selectivity. The α_{2A}- or α_{2D}-adrenoceptor subtype (depending on species) appears to be responsible for many other α_2-adrenoceptor-mediated responses, including the major component of prejunctional modulation of sympathetic neurotransmission.

Knockout of the α_{2C}-adrenoceptor has no apparent cardiovascular effect. However, whereas knockout or mutation of the α_{2A}-adrenoceptor subtype produces only partial attenuation of α_2-adrenoceptor-mediated inhibition of transmitter release, elimination of both α_{2A}- and α_{2C}-adrenoceptors results in complete loss of prejunctional modulation of adrenergic neurotransmission and induces pathologic effects related to excess adrenergic tone. Hence, the α_{2C}-adrenoceptor can also participate in prejunctional modulation of neurotransmission. The α_{2C}-adrenoceptor may also participate in cold-induced augmentation of α-adrenoceptor-mediated vasoconstriction.

α_{2C}-Adrenoceptors appear to be involved in the startle reflex and isolation-induced aggression since knockout of α_{2C}-adrenoceptors in mice resulted in an enhanced startle response and shortened attack latency, whereas overexpression of this receptor produced the opposite effect on both parameters. Similarly, the hypothermic response to an α_2-adrenoceptor agonist was attenuated in α_{2C}-adrenoceptor knockout mice, and this response was accentuated by α_{2c}-adrenoceptor overexpression. Behavioral despair in the forced swimming test was enhanced or inhibited by α_{2C}-adrenoceptor overexpression or knockout, respectively. Overexpression of the α_{2C}-adrenoceptor impairs the performance of mice in a water maze.

Despite clinical trials for several indications, no molecule specifically designed to block the α_2-adrenoceptor has been successfully developed as a drug, although α_2-adrenoceptor antagonist activity may play an important role in antidepressants such as mirtazepine.

In addition to their use as antihypertensives, selective centrally acting α_2-adrenoceptor agonists are used as adjuncts to general anesthesia in the treatment of opiate withdrawal, ADHD, and Tourette's syndrome. Their ability to alleviate many of the symptoms of opiate withdrawal most likely results from the ability of both opiate agonists and α_2-adrenoceptor agonists to inhibit locus coeruleus firing, although an action at spinal centers may also contribute. Based on data in a rat model, activation of α_2-adenoceptors may contribute to the activity of methylphenidate in ADHD.

β-Adrenoceptors

Important physiological consequences of β-adrenoceptor activation include stimulation of cardiac rate and force; relaxation of vascular, urogenital, and bronchial smooth muscle; stimulation of renin secretion from the juxtaglomerular apparatus; stimulation of insulin and glucagon secretion from the endocrine pancreas; stimulation of glycogenolysis in liver and skeletal muscle; and stimulation of lipolysis in the adipocyte. Prejunctional β-adrenoceptors are present on some central and peripheral nerve terminals, where their activation results in facilitation of stimulation-evoked neurotransmitter release. However, in contrast to the prejunctional α_2-adrenoceptors, these prejunctional receptors do not appear to have major physiologic significance. Most β-adrenoceptor-mediated actions involve stimulation of adenylyl cyclase through the interaction of the agonist–receptor complex with G_s.

Three β-adrenoceptor proteins have been cloned, and the characteristics of these recombinant receptors correspond with those of the three well-characterized β-adrenoceptors on native tissues, designated as β_1, β_2, and β_3. Many useful pharmacological tools are available for β-adrenoceptor characterization. These include agonists capable of selectively activating β_1-, β_2-, or β_3-adrenoceptors, as well as antagonists selective for each of the three subtypes. Although multiple β-adrenoceptor subtypes can participate in the cardiac stimulation produced by the catecholamines, gene knockout experiments demonstrate that the β_1-adrenoceptor subtype plays the major role. Renin release also appears to be mediated by the β_1-adrenoceptor. Bronchodilation is mediated primarily by the β_2-adrenoceptor. The β_3-adrenoceptor is responsible for lipolysis in white adipose tissue and thermogenesis in the brown adipose tissue found in rodents. Species differences appear to be important for the β_3-adrenoceptor since several selective β_3-adrenoceptor agonists can activate rodent, but not human, β_3-adrenoceptors.

In rat brain, mRNA for the β_1-adrenoceptor is widely distributed, whereas message for the β_2-adrenoceptor is concentrated in olfactory bulb, hippocampal formation, piriform cortex, and cerebellar cortex. The β_3-adrenoceptor does not appear to be present in the CNS. Radioligand assays using dissected rat brain have shown that β_1-adrenoceptors are typically associated with forebrain structures (cerebral cortex, striatum, and hippocampus), whereas β_2-adrenoceptors are most dense in the cerebellum. The β-adrenoceptors of rat cerebral cortex may be localized primarily on glial cells.

Central β-adrenoceptors appear to play a role in movement disorders. β-Adrenoceptor antagonists reduce tremor in MPTP-treated monkeys. Studies in human brain show dense β_1-adrenoceptor binding in striatum, which decreases in the late stages of Huntington's chorea. It is well established that chronic treatment with tricyclic antidepressants produces downregulation of central β-adrenoceptors, presumably as a consequence of increased synaptic levels of the catecholamine neurotransmitters as a result of blockade of neuronal reuptake. β-Adrenoceptors may be involved in the antinociceptive activity of the tricyclic antidepressants since β-adrenoceptor antagonists can reverse this activity. Other functions attributed to central β-adrenoceptors include adaptation to stress, memory and learning, control of respiration, and glial proliferation.

Knockouts of β_1-, β_2, and β_3-adrenoceptors have been reported. Mice lacking both β_1- and β_2-adrenoceptors have also been prepared. None of these mice have phenotypic changes reflecting deficits in CNS function, although detailed behavioral studies have not been reported. Knockout of the β_1-adrenoceptor has a marked effect on embryonic viability, although the few homozygous animals surviving appear normal. This effect on viability is not observed with knockout of the β_2- or β_3-adrenoceptor subtypes. Mice lacking the β_1 adrenoceptor fail to respond to the inotropic action of β-adrenoceptor agonists, confirming the importance of this subtype in the control of cardiac contractility. However, maximal exercise capacity is not reduced. Mice lacking the β_2-adrenoceptor have a normal response to exogenous β-adrenoceptor agonists and have even greater exercise capacity than wild-type mice. However, these animals become hypertensive during exercise and have a lower respiratory exchange ratio, suggesting influences of the β_2-adrenoceptor on energy metabolism. Mice lacking both β_1- and β_2-adrenoceptors have normal basal cardiovascular parameters and normal exercise capacity, although the ability of exercise or the administration of exogenous agonists to increase heart rate is blunted. Mice lacking the β_3-adrenoceptor show mild increases in body fat stores and do not show metabolic responses to a selective β_3-adrenoceptor agonist. In several cases, an increased responsiveness to one of the remaining β-adrenoceptors is observed when one of the subtypes is knocked out. Thus, there is increased β_1-adrenoceptor responsiveness in β_3-adrenoceptor knockout animals and enhanced β_3-adrenoceptor responsiveness in β_1/β_2 knockout animals. These observations support the concept that many physiological functions can be mediated by multiple β-adrenoceptor subtypes.

Selective overexpression of β_2-adrenoceptors in airway epithelium or smooth muscle results in decreased sensitivity to methacholine-induced bronchoconstriction and an increased sensitivity to a β-adrenoceptor agonists. Cardiac-directed overexpression of the human β_1-adrenoceptor results in the accumulation of fibrous tissue between cardiac myocytes, myocyte hypertrophy, and myofibrilar disarray. These changes result in cardiac dysfunction in older animals. Overexpression of human β_2-adrenoceptors in mouse heart enhances basal cardiac function due to constitutive activity of the expressed receptors. Fibrotic cardiomyopathy is observed in mice expressing high levels of the β_2-adrenoceptor, with the severity and rate of onset being dependent on the level of receptor expression. Overexpression of cardiac β_2-adrenoceptors exacerbates functional deterioration following pressure overload as a result of aortic stenosis.

Several β-adrenoceptor antagonists have also been evaluated for a variety of CNS applications; perhaps the most convincing results are from the use of these agents in treating anxiety of varying etiology, although it is not clear whether a central or peripheral site of action is involved.

Conclusions

Agonists or antagonists at central or peripheral adrenoceptors have several important therapeutic applications; most of these drugs interact with multiple adrenoceptor subtypes. Newly developed pharmacological tools will help to identify the functional role of specific subtypes. Although an action at multiple adrenoceptors may in some cases be beneficial, drugs interacting selectively with adrenoceptor subtypes could result in either an improvement in therapeutic index or the identification of new therapeutic opportunities.

See also: Brain Adrenergic Neurons; Norepinephrine: CNS Pathways and Neurophysiology; Norepinephrine: Adrenergic Receptors; Sympathetic Noradrenergic and Adrenomedullary Hormonal Systems in Stress and Distress.

Further Reading

Arnsten AFT and Dudley AG (2005) Methylphenidate improves prefrontal cortical cognitive function through α_2-adrenoceptor and dopamine D1 receptor actions: Relevance to therapeutic effects in attention deficit hyperactivity disorder. *Behavioral and Brain Functions* 1: 2.

Bjorklund M, Sirvio J, Riekkinen M, et al. (2000) Overexpression of alpha2C-adrenoceptors impairs water maze navigation. *Neuroscience* 95: 481–847.

Chotani MA, Flavahan SA, Mitra S, Daunt D, and Flavahan NA (2000) Silent α_{2C}-adrenergic receptors enable cold-induced vasoconstriction in cutaneous arteries. *American Journal of Physiology* 278: H1075–H1083.

Grijalba B, Callado LF, Javier-Meana J, Garcia-Sevilla JA, and Pazos A (1996) Alpha-2 adrenoceptor subtypes in human brain. A pharmacological delineation of [^{3}H] RX-821002 binding to membranes and tissue sections. *European Journal of Pharmacology* 310: 83–93.

Hein L, Altman JD, and Kobilka BK (1999) Two functionally distinct α-adrenergic receptors regulate sympathetic neurotransmission. *Nature* 402: 181–184.

Hermann GE, Nasse JS, and Rogers RC (2005) α-1 Adrenergic input to solitary nucleus neurons: Calcium oscillations, excitation and gastric reflex control. *Journal of Physiology* 562: 553–568.

Hieble JP, Bondinell WE, and Ruffolo RR, Jr. (1995) α- and β-Adrenoceptors: From the gene to the clinic. 1: Molecular biology and adrenoceptor subclassification. *Journal of Medicinal Chemistry* 38: 3415–3444.

Koshimizu TA, Tanoue A, Hirasawa A, Yamauchi J, and Tsujimoto G (2003) Recent advances in α_1-adrenoceptor pharmacology. *Pharmacology and Therapeutics* 98: 235–244.

Neumeister A, Charney DS, Belfer I, et al. (2005) Sympathoneural and adrenomedullary functional effects of α_{2C}-adrenoreceptor gene polymorphism in healthy humans. *Pharmacogenetics and Genomics* 15: 143–149.

Nicholas AP, Hökfelt T, and Pieribone VA (1996) The distribution and significance of CNS adrenoceptors examined with *in situ* hybridization. *Trends in Pharmacological Sciences* 17: 245–255.

Ruffolo RR, Jr., Bondinell WE, and Hieble JP (1995) α- and β-Adrenoceptors: From the gene to the clinic. 2: Structure–activity relationships and therapeutic application. *Journal of Medicinal Chemistry* 38: 3681–3716.

Sallinen J, Haapalinna A, MacDonald E, et al. (1999) Genetic alteration of the α_2-adrenoceptor subtype c in mice affects the development of behavioral despair and stress-induced changes in plasma corticosterone levels. *Molecular Psychiatry* 4: 443–452.

Spreng M, Cotecchia S, and Schenk F (2001) A behavioral study of alpha-1b adrenergic receptor knockout mice: Increased reaction to novelty and selectively reduced learning capacities. *Neurobiology of Learning & Memory* 75: 214–229.

Waeber C, Rigo M, Chinaglia G, Probst A, and Palacios JM (1991) Beta-adrenergic receptor subtypes in the basal ganglia of patients with Huntington's chorea and Parkinson's disease. *Synapse* 8: 270–280.

Zujic MJ, Sands S, Ros SA, et al. (2000) Overexpression of the α_{1B}-adrenergic receptor causes apoptotic neurodegeneration: Multiple system atrophy. *Nature Medicine* 6: 1388–1394.

Adult Cortical Plasticity

U T Eysel, Ruhr-University Bochum, Bochum, Germany

Introduction

Plasticity of the brain is a lifelong phenomenon. In general, cortical plasticity describes changes of neuronal wiring and function in the cerebral cortex in response to new challenges of external or internal origin. External stimuli are environmental conditions that impose new tasks on the system. Internal stimuli can be normal developmental demands or pathological lesions in the widest sense – structural, biochemical, or genetic – leading to severe functional impairment. Extrinsic as well as intrinsic neuronal activity can shape wiring and strength of connectivity. Developmental plasticity, as opposed to adult plasticity, has a stronger potential and can include axonal and dendritic growth at a larger scale. Many factors that guide normal development of the brain (e.g., growth factors, netrins) are downregulated after the critical periods of development. In response to lesions in adulthood, some factors can eventually be upregulated to again allow more extensive neuronal plasticity. In that case, neuronal plasticity is the response of the system in an attempt to repair or compensate loss of function. In development as well as in adult cortical plasticity, neuronal activity can change functional connections; however, the effectiveness of synaptic plasticity is usually reduced following postnatal development and critical periods. In the case of adult cortical plasticity, functional and structural adaptations remain, as a rule, spatially localized and restricted to the level of axonal terminals and synapses. New functional connections can be formed by increase of synaptic efficacy in a preexisting network. The principles of reorganization and remapping are common features of the adult auditory, somatosensory, and visual cortices. This article, however, focuses on plasticity in the visual system.

Transition from Visual Cortical Development into Adult Cortical Plasticity

Some principles of adult sensory cortex plasticity can be understood by comparing development in early postnatal life and reorganization after injury of sensory cortical maps. Sensory cortices are characterized by topographic maps that are arranged according to the neighborhood principle. Cells that are close in the sensory periphery are also close in the cortical representations, be it sound frequency in the auditory, body surface in the somatosensory, or retinal surface and visual space in the visual system. The representational maps remain stable throughout adulthood once they have formed through dependence on use in early postnatal life. Accordingly, the postnatal development of the visual cortex is characterized by use-dependent sculpturing of the necessary, correct connections. Characteristic cortical maps of retinal topography and stimulus specificity are formed. At the end of the critical periods, the functional connections are stabilized, and the programs that prevailed during this period, such as axon growth and guidance, removal of incorrect connections, strengthening of optimal, and weakening nonoptimal inputs (synaptic plasticity) are modified in a way to foster stability without sacrificing flexibility in adulthood. Finally, the ability to change synaptic weights by long-term potentiation (LTP) or long-term depression (LTD) persists throughout adult life into senescence. This is exemplified by the lifelong ability for perceptual learning. Use, disuse, and lesions are the strong stimuli for plasticity and reorganization in the adult visual cortex. For example, when challenged by central or peripheral lesions, the visual system displays characteristic patterns of remodeling. While the observed subcortical changes remain limited, more widespread reorganization is seen in the cortex. In the early phase, cortical reorganization appears more restricted and based on functional modifications of cortical connections governed by subtle changes in the balance of excitation and inhibition. In later stages, more extended reprogramming of connections, based on long-term changes in synaptic connectivity (LTP, LTD), is observed and can ultimately involve anatomical reorganization including local growth mechanisms.

There are many different types of adult cortical plasticity. One type is purely use-dependent and closely related to perceptual learning as observed in humans. This plasticity is operating in everyday life when cellular properties are modified in response to repetitive use in demanding tasks. This use-dependent plasticity is closely related to 'synaptic learning' based on mechanisms like LTP and LTD. Other types of visual cortical plasticity are observed after lesions; they are less specific and generally lead to changes in single cell properties and reorganization of cortical representations (maps). The following sections focus on adult use-dependent plasticity and two different types of lesion-induced plasticity in the visual cortex.

Use-Dependent Adult Visual Cortical Plasticity

Fast associative plasticity of visual cortical cells has been shown *in vitro* with electrical stimulation as well as *in vivo* by pairing of natural stimuli with artificial depolarization of single cells. The *in vitro* experiments were performed with intracellular recordings in brain slices of adult rats and cats and showed that adult visual cortical cells are capable of synaptic plasticity. High-frequency repetitive stimulation of specific inputs results in LTP, which is expressed in long-lasting increases of synaptic potentials.

To induce changes in cellular responses *in vivo* on a fast timescale, natural sensory stimuli have been paired with electrical or pharmacological stimulation of cells in the visual cortex of anesthetized animals. Properties such as orientation specificity (e.g., vertical vs. horizontal) or ocular dominance (left vs. right) were shifted toward the property reinforced by repetitive pairing with a facilitating stimulus during supervised learning. These experiments provide *in vivo* evidence that the modifiability of mature visual cortical connections follows the Hebbian rule, which requires both pre- and postsynaptic activity within a defined time window. While many experiments used electrical or pharmacological stimuli to facilitate synaptic plasticity, is was also shown in experiments with anesthetized animals that purely natural repetitive stimulation can induce long-lasting changes of visual cortex cell properties such as receptive field (RF) size and substructure. Visual costimulation of the central parts of an excitatory RF with an unresponsive region located just outside was applied to elicit 'associative' synaptic changes. With this paradigm visual cortical cells did specifically 'learn' to respond to an originally subthreshold region outside their excitatory RF or to change the subfield composition within their RF.

Recently, sophisticated *in vivo* studies have shed additional light on the possible properties and mechanisms of use-dependent adult visual cortical plasticity by examining short-term changes of orientation specificity induced by adaptation to certain orientations. Repulsive shifts of preferred orientation were observed that involved time-dependent processes of depression as well as enhancement. This kind of short-term plasticity was dependent on the location of cells within the layout of the orientation preference map of the visual cortex that is characterized by large coherent orientation domains and 'pinwheel centers' or singularities, where all orientations meet within a small circumscribed region. In the vicinity of singularities, the capacity for adaptive changes in response to oriented gratings is much stronger than in orientation domains. It has further turned out that certain forms of short-term plasticity in the adult visual cortex are stimulus timing-dependent. In respective experiments, plasticity of orientation selectivity *in vivo* depended on stimulus timing within a ± 40 ms window like the crucial parameters for spike timing-dependent plasticity. This suggests that underlying mechanisms are modifications of synaptic weights of intracortical connections.

Beyond the single cell level, use-dependent adult visual cortex plasticity can be shown in whole cortical networks. When optical imaging of intrinsic signals was combined with localized electrical intracortical microstimulation, profound changes in orientation preference maps were observed in the adult cat visual cortex. After a few hours of high-frequency focal electrical stimulation, the cortical representation of the orientation represented at the stimulation site increased because surrounding cells with previously different orientation preferences shifted toward the stimulated orientation at sites as distant as 4 mm. These widespread changes did not involve attention, special training, or reinforcement; they were purely dependent on high-frequency synchronized activation of the cortical network.

When monkeys are trained for several months to identify the orientation of a small grating, they show a marked increase in performance. Single cell recordings from primary visual cortex (V1) of these monkeys after successful training reveal significantly increased orientation sensitivity of only those cells and orientations that represented the trained orientation. This proves that use-dependent single cell plasticity in the adult visual cortex can be linked directly to behavior and perceptual learning.

Lesion-Induced Cortical Plasticity

Retinal lesions remove afferent input to thalamus and cortex. Cortical lesions destroy cortical cells that have a dual function, being targets for topographic retinal projections via the thalamus on one hand, and being the origin of intracortical connections on the other. These completely different types of lesion, retinal versus cortical, pose very different problems for reorganization and restitution of function to the visual system. In the case of retinal lesions, the option is to reinnervate intact cortical cells that have lost their sensory input; however, the topographically correct fibers are no longer available due to retinal destruction. In the case of cortical lesions, the retinal afferent information is still completely on hand, but the cortical target cells are lost. Accordingly, the options for reorganization are different, and so is the outcome of functional reorganization: after retinal lesions, a blind cortical region is filled in with retinal topography redundantly spreading across the cortical surface and resulting in many cortical cells newly representing the same retinal topography, whereas after cortical lesions, the surviving cells in the surround receive new suprathreshold information from geniculocortical afferents and acquire additional topographical representations from the retina. While the former fills in a cortical region of still existing but blind cells with new but topographically redundant activity, the latter enables true functional improvement by recovering previously lost afferent information for further cortical processing. The lesion-induced changes can be roughly subdivided into acute, subacute, and chronic effects. The pathology of cell death and functional suppression that is predominant in the acute phase (first day postlesion) is followed by events of neuronal plasticity in the subacute (second day to 1 week postlesion) and chronic phases (weeks to months after a lesion). It is interesting that the different types of lesions seem to trigger similar effects that lead to receptive field plasticity, topographical remapping, and a limited recovery of function.

Plasticity in the Primary Visual Cortex after Cortical Lesions

Focal lesions of the visual cortex lead to a local visual field loss (scotoma). The size of the scotoma mirrors cell death and functional depression of cortical neurons. However, the dimension of functional loss following cortical lesions can diminish in both animals and humans with time and especially when specific training is performed. A possible basis for restoration of function is a reactivation of 'silent' cells in the only partially damaged, functionally suppressed surround of the lesion (the penumbra) by strengthening of subthreshold synaptic inputs. Another possibility is the increase in size of receptive fields of intact cells at the border of the lesions, that is, an additional representation of lost parts of the visual field by cells that also maintain their original topographical representation of space and thereby enable the perception of previously lost parts of the visual field. It is interesting that in the first weeks after a lesion in the visual cortex, an increase of excitability is observed in its surround. This increased excitability seems to support the theory of increased plasticity of cells surrounding the lesion. Changes of receptive field size that do not occur spontaneously in the first days following a lesion develop months later close to the border of lesions in the cat visual cortex.

Faster changes in receptive field size can be experimentally induced by repetitive visual stimuli. This can be achieved even in neurons of the normal adult cat visual cortex. Repetitive stimulation associated with high response rates in the neurons under study can lead to a significant and stimulus-specific widening of receptive fields. The observed effects last from 20 min to several hours and are reminiscent of effects similar to short-term potentiation and LTP. Such LTP-like mechanisms might be enhanced because of locally increased excitability in the surround of visual cortical lesions and might be associated with the RF enlargements observed in the surround of cortical lesions. LTP is the best candidate to explain cellular learning and memory and can be studied in brain slices *in vitro* following high-frequency electrical stimulation of inputs. LTP is characterized by strengthened synaptic transmission over a span from more than 1 h to days and weeks. LTP was first described in the mammalian hippocampus and can also be observed in the visual cortex of mice, rats, and cats. During early postnatal development, when synaptic plasticity is highly expressed – as in the visual cortex during the critical period – very effective LTP is observed. In fact, lesion-induced reorganization after lesions in the visual cortex is associated with increased LTP, as was shown by increased synaptic plasticity in slices of the lesioned rat visual cortex *in vitro*.

Reorganization of the Primary Visual Cortex Following Retinal Lesions

Retinal lesions trigger chronic reorganization in the afferent visual pathway. This follows the interruption of visual signal flow from the destroyed retinal area to the visual cortex that locally deprives cells in the thalamus (lateral geniculate nucleus) and the primary visual cortex of their normal input. The excitability of cells in the visual thalamic relay nucleus (dorsal lateral geniculate nucleus, dLGN) that are primarily deafferented by a monocular photocoagulator lesion of the retina changes in a characteristic way: initially the cells show a complete loss of visual drive and a significant decrease in spontaneous activity; then spontaneous activity increases in deafferented cells, and some cells inside the deafferented region regain visual input originating from the retina directly adjacent to the lesion. Accordingly the RFs of these newly connected cells change their retinotopy and shift to a new position located at the border of the retinal lesion. This reorganization of the retinotopic map leads to a partial filling in of the scotoma on the subcortical level and is associated with a lateral spread of excitation in the dLGN of up to 300 μm. With the same type of lesion, a much more extended reorganization can be observed in the visual cortex. Apart from the longer distances involved, the reorganization was exactly of the same kind as observed in the dLGN: originally silenced cells regained visual input that originated from the border of the retinal lesion, a retinal region that did not excite these cells before. The same kind of plasticity was found in cat and monkey visual cortex. The reorganization gradually fills in cortical regions that were depleted of retinal inputs with new excitability over distances up to 5 mm. This distance conforms to the range of lateral horizontal connection in the visual cortex. This chronic retinotopic remapping is preceded by fast changes in visual RF properties that follow acute disuse induced by retinal lesions and are elicited as well when selective surround stimulation is applied to artificial retinal scotomata.

The cortical region affected by loss of input after retinal lesions constitutes an anatomically defined lesion projection zone (LPZ, **Figure 1(a)**). A characteristic imbalance of input occurs in and outside the LPZ as cells with normal afferent input are situated in direct proximity to visually silenced cells. In addition, excitation is upregulated and inhibition downregulated inside the LPZ. This situation leads to changes in cortical cell RF size and topography when the active cells gain influence on their inactive neighbors through the intracortical horizontal fiber systems. The filling in follows a characteristic spatiotemporal pattern in which hyperactive and visually hyperexcitable cells are first found close to the order of normal cortex in the LPZ, then progressively deeper inside the LPZ (**Figure 1(b)**). The long-range horizontal fibers are the most probable basis of a low-level perceptual filling in of retinal scotomata. In fact, artificial scotomata (produced in normal people by stabilized retinal images) take considerable time to fill in whereas scotomata are filled in immediately in patients with chronic

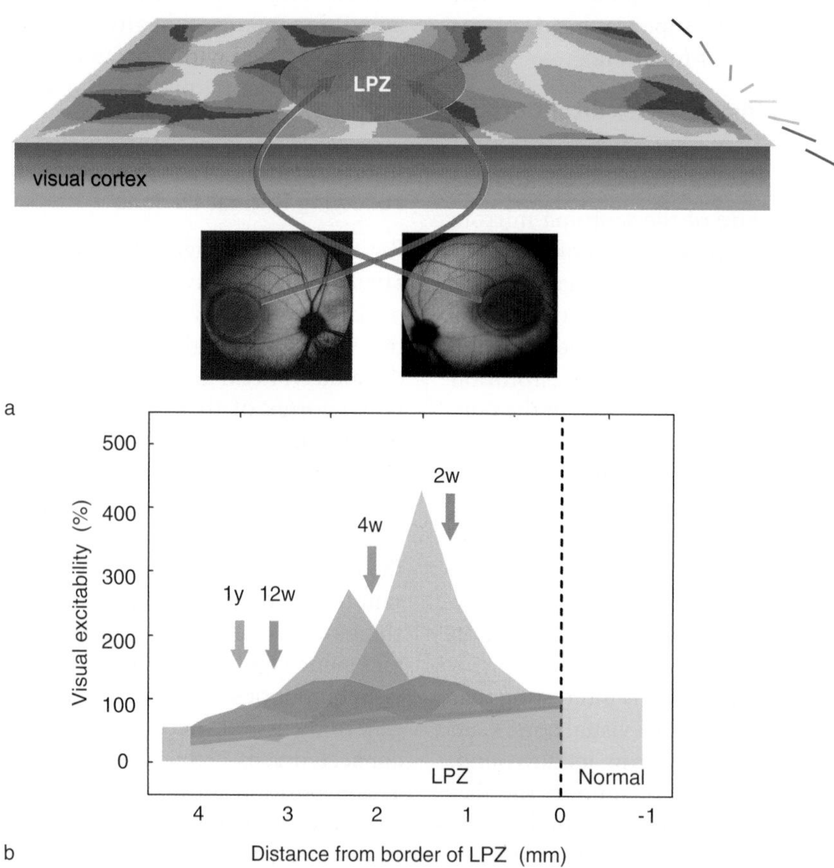

Figure 1 Spatiotemporal pattern of filling-in of the lesion projection zone (LPZ) in the cat visual cortex. (a) Schematic representation of the projection from the left and right eye retinae (with central 10° diameter photocoagulator lesions) to the primary visual cortex. The resulting binocular LPZ is surrounded by normal cortex as shown on top of a topographic map with color-coded orientation preferences. (b) Visual excitability inside the LPZ as a function of distance from the border between LPZ and normal cortex. Normal activity outside the LPZ is shown in yellow; the nonreorganized, and after 1 year visually nonresponsive center of the LPZ is shown in gray. Visual excitability in general moves gradually with time into the LPZ and is associated in the first weeks with a peak of significant hyperexcitability at the front of reorganization. Arrows indicate the first, well defined receptive fields plottable at the different time points. (b) Adapted from Giannikopoulos DV and Eysel UT (2006) Dynamics and specificity of cortical map reorganization after retinal lesions. *Proceedings of the National Academy of Sciences of the United States of America* 103: 10805–10810.

retinal lesions. This observation further supports the proposition that cortical connectivity is profoundly and permanently changed after chronic lesions. This proposition is corroborated by the above-mentioned experimental evidence that shows primary cortical reorganization taking place on the basis of functional changes in synaptic weights of preexisting connections but followed in a final step by anatomical stabilization associated with terminal sprouting, as observed more than 8 months after retinal lesions in the adult cat visual cortex.

It is interesting that when state-of-the-art high-resolution functional magnetic resonance imaging (fMRI) was applied in the visual cortex of adult monkeys, no changes of the visuotopic projection were observed. This seems to indicate that fMRI, a method prone to predominantly detect population responses and presynaptic activity, does not reflect the subtle changes found in suprathreshold performance of individual neurons. On the other hand, the lesion-induced remapping detected on the single-cell level shows close correlation with the psychophysical observations in humans with retinal lesions, indicating the functional relevance of cortical reorganization and filling in.

Mechanisms of Lesion-Induced Cortical Plasticity

Hyperactivity after Cortical and Retinal Lesions

Suppression and facilitation of activity was observed at different distances from the border of heat lesions in the visual cortex. Subnormal activity was seen in a region of <1 mm around the lesion whereas hyperactivity prevails at 1–2.5 mm from its border; at more distant positions (>2.5 mm) the activity was normal. This hyperactivity after heat lesions was already present 1 day after the lesion, became maximal after 3–7 days, and was still visible 30 days postlesion. One possible mechanism for the increase of activity in the surround of lesions is an imbalance between γ-aminobutyricacidergic (GABAergic) inhibition and glutamatergic excitation. This is supported by experimental evidence: in the immediate surround of neocortical lesions, inhibitory transmission is downregulated and a widespread reduction of GABA receptors is observed; then 1–5 days postlesion and up to 2 mm from the border of the lesions, the fast $GABA_A$- and late $GABA_B$-induced inhibitory postsynaptic potentials show reduction in amplitudes and peak currents.

At the same time, the glutamatergic excitation is increased, as substantiated by higher amplitudes and longer durations of *N*-methyl-D-aspartate (NMDA) receptor-mediated excitatory postsynaptic potentials. In addition, excitatory field potentials were significantly larger than normal. These enlarged potentials were NMDA-dependent as shown by blockade with d-amino-phosphonovaleric acid.

It is interesting that also after retinal lesions, a hyperactive zone was found within the LPZ in the visual cortex (**Figure 1(b)**), and an increase in glutamate immunoreactivity was observed at the same time. Hyperactivity and hyperexcitability appear to be a common feature associated with postlesion functional rewiring in the adult visual cortex following deafferentation.

The GABAergic and glutamatergic systems have been investigated with immunhistochemical methods 2 weeks after homonymous lesions in the retina (central 10°) and reveal a characteristic pattern of transmitter system immunoreactivity (IR) inside and in the surround of the deafferented region (LPZ). The glutamic acid decarboxylase (GAD) IR was downregulated inside the LPZ where the number of positive profiles was extremely reduced in the neuropil while the GAD IR in the cell somata remained rather unchanged. IR of the excitatory neurotransmitter glutamate revealed glutamate-positive cells in cortical layers II to VI of area 17. The retinal lesions caused a clear reduction (by 15–26%) in the number of glutamate-immunoreactive cells in the supra- and infragranular layers of the cortical LPZ, compared with normal cortex. Furthermore, close to the border of the LPZ inside the deafferented region, glutamate IR displayed a sharp increase (by 50–100%) throughout layers II to VI of area 17. This glutamate peak was largest in layer VI and had a width of 600–800 μm.

Both the changes in GAD and glutamate IR diminished with time. When the central and the peripheral portions of area 17 were compared in cats with postlesion survival times of more than 12 weeks, no significant differences in the number of glutamate IR-positive cells and in the GAD IR of the neuropil were observed. Obviously the initial imbalance of the excitatory and the inhibitory immunhistochemical reactions returns to normal once the functional reorganization is completed.

Enhanced LTP-Like Effects

Pharmacological studies in brain slices have shown that reduced inhibition and/or increased excitation facilitates the expression of LTP. This situation is mimicked in the visual cortex following lesions in which a decreased strength of inhibition and an increased excitation are observed. Accordingly, electrophysiological recordings in slices of rats with cortical lesions revealed a significantly elevated level of LTP at synapses from neurons located at a defined distance of 1–4 mm from the border of the lesion and up to 1 week following induction of the injury.

This enhanced synaptic plasticity is accompanied by changes in the intracellular calcium concentration $(Ca^{2+})_i$. Both the neuronal resting calcium concentration and the stimulus-evoked calcium influx are moderately increased at the border of the lesion. The origin of this increase of intracellular calcium was investigated by means of specific neuronal calcium permeation blockers for ionotropic NMDA and AMPA receptors, revealing an increase of Ca^{2+} influx mediated through both of these glutamate receptor types postlesion. While a previously present calcium permeability is just increased in the case of the NMDA receptors, there is a fundamental lesion-induced change in the functional properties of the ionotropic AMPA receptors. AMPA receptors are known to be calcium-impermeable under physiological conditions in rat neurons after postnatal day 15. However, after cortical lesions in animals more than 23 days old, a neuronal calcium influx was measured in the presence of pharmacological blockers of all known sources of intraneuronal calcium influx. This calcium influx was blocked by an antagonist for AMPA. This unusual calcium permeability mediated by AMPA receptors can be explained by a lesion-induced change in the specific composition of the AMPA receptor protein subunits: a reduction in the expression of the glutamate receptor 2 subunit, which is normally observed only in the young postnatal brain, is also observed after lesions in the adult.

A similar reversion to more juvenile patterns is observed for the NMDA receptors. Cortical lesions change the relative expression levels of the messenger RNA for the ionotropic NMDA receptor subunits NR2A and NR2B. Experimental data suggest an increased relative expression of the NR2B subunit of the NMDA receptor at the border of the lesion (due to downregulation of the amount of NR2A subunits). It is interesting that this relatively high amount of NMDA receptors containing the NR2B subunit has also been described in the early postnatal cortex in the phase when enhanced synaptic plasticity (LTP) is observed. Following cortical lesions the true relevance of NR2B subunits for the lesion-induced facilitation of LTP has been clearly shown: blockade of NR2B subunit-containing receptors with ifenprodil reduces the increased LTP in lesioned rat cortex to control values.

Calcium imaging with Fura2-AM (a fluorescent probe) combined with LTP induction can be utilized to record the stimulus-evoked and field potential-correlated calcium influx in the lesion-treated cortex. In fact, significantly increased calcium levels are observed during the induction of LTP by theta burst stimulation as well as 55 min thereafter. This increase of the stimulus-correlated calcium influx postlesion might explain the expression of facilitated LTP as elevated intracellular calcium levels directly correlate with the strength of synaptic LTP.

Growth Factors and Morphological Correlates of RF Plasticity

The neurotrophins brain-derived neurotrophic factor (BDNF), neurotrophin-3, nerve growth factor, and the insulin-like growth factor IGF-1 have been found elevated in the visual cortex as early as 3 days after binocular retinal lesions. The related neurotrophin receptors were elevated as well. Furthermore, increased transcription levels of calcium/calmodulin-dependent kinase II, microtubule-associated protein 2, and synapsins are observed in the area undergoing cortical reorganization. There is increased neuronal activity in regions with elevated BDNF expression. Thus, BDNF appears directly linked to the activity-dependent early unmasking of existing connections on one hand, and as a growth factor, it might also represent a link to the late morphological changes involving axonal sprouting.

See also: Activity in Visual Development; BDNF in Synaptic Plasticity and Memory; Developmental Synaptic Plasticity: LTP, LTD, and Synapse Formation and Elimination; Hebbian Plasticity; Long-Term Potentiation and Long-Term Depression in Experience-Dependent Plasticity; Metaplasticity; Neuronal Plasticity after Cortical Damage; Pain and Plasticity; Perceptual Learning and Sensory Plasticity; Plasticity, and Activity-Dependent Regulation of Gene Expression; Plasticity of Intrinsic Excitability; Somatosensory Plasticity; Spike-Timing Dependent Plasticity (STDP); Synapse Formation: Competition and the Role of Activity; Synaptic Plasticity: Neuronal Sprouting; Synaptic Plasticity: Short-Term Mechanisms; Visual Cortex: Mapping of Functional Architecture Using Optical Imaging.

Further Reading

Arckens L, Schweigart G, Qu Y, et al. (2000) Cooperative changes in GABA, glutamate and activity levels: The missing link in cortical plasticity. *European Journal of Neuroscience* 12: 4222–4232.

Chino YM, Smith EL III, Kaas JH, Sasaki Y, and Cheng H (1995) Receptive-field properties of deafferentated visual cortical neurons after topographic map reorganization in adult cats. *Journal of Neuroscience* 15: 2417–2433.

Darian-Smith C and Gilbert CD (1994) Axonal sprouting accompanies functional reorganization in adult cat striate cortex. *Nature* 368: 737–740.

Dreher B (2006) Reprogramming of striate and extrastriate visual cortices following retinal lesions. In: Lomber SG and Eggermont JJ (eds.) *Reprogramming the Cerebral Cortex*, pp. 3–46. Oxford, UK: Oxford University Press.

Eysel UT (2002) Plasticity of receptive fields on early stages of the adult visual system. In: Fahle M and Poggio T (eds.) *Perceptual Learning*, pp. 43–65. Cambridge, MA: MIT Press.

Eysel UT and Mittmann T (2006) Remodeling of cortical connections and enhanced long-term potentiation after lesions of the visual cortex. In: Lomber SG and Eggermont JJ (eds.) *Reprogramming the Cerebral Cortex*, pp. 61–71. Oxford, UK: Oxford University Press.

Giannikopoulos DV and Eysel UT (2006) Dynamics and specificity of cortical map reorganization after retinal lesions. *Proceedings of the National Academy of Sciences of the United States of America* 103: 10805–10810.

Gilbert CD (1998) Adult cortical dynamics. *Physiological Reviews* 78: 467–485.

Gilbert CD and Wiesel TN (1992) Receptive field dynamics in adult primary visual cortex. *Nature* 356: 150–152.

Kaas JH, Collins CE, and Chino YM (2003) The reactivation and reorganization of retinotopic maps in visual cortex of adult mammals after retinal and cortical lesions. In: Pessoa L and De Weerd P (eds.) *Filling-In: From Perceptual Completion to Cortical Reorganization*, pp. 187–206. Oxford, UK: Oxford University Press.

Kaas JH, Krubitzer LA, Chino YM, Langston AL, Polley EH, and Blair N (1990) Reorganization of retinotopic cortical maps in adult mammals after lesions of the retina. *Science* 248: 229–231.

Obata S, Obata J, Das A, and Gilbert CD (1999) Molecular correlates of topographic reorganization in primary visual cortex following retinal lesions. *Cerebral Cortex* 9: 238–248.

Smirnakis SM, Brewer AA, Schmid MC, et al. (2005) Lack of long-term cortical reorganization after macaque retinal lesions. *Nature* 435: 300–307.

Zur D and Ullman S (2003) Filling-in of retinal scotomas. *Vision Research* 43: 971–982.

Aggression: Hormonal Basis

N G Simon, Lehigh University, Bethlehem, PA, USA
C F Ferris, University of Massachusetts Medical School, Worcester, MA, USA

Characterization of the hormonal contribution to mammalian aggressive behavior and analogous emotional states in humans has followed a split path. Advances in the understanding of the molecular, cellular, and biochemical processes that mediate hormonal effects have enabled increasingly sophisticated models of behavioral regulation in animal models. Studies in humans, however, continue to face methodological limitations that limit progress in regard to cellular, systemic, and state-dependent processes, even as genome-based technologies identify polymorphisms that may impact on human aggression and aversive emotional states. Fortunately, trends in human research that recognize specific subtypes of aggression (e.g., hostility, impulsivity, and dominance) have provided a conceptual basis for a stronger link between animal models and human states. This more refined framework should facilitate defining the neuroendocrine contribution to pathological aggression, violent behavior, and aversive emotions.

Offensive aggression in males and females can serve as a useful model to elucidate hormonal influences on human behavior. In males, the central role of testosterone (T) is well-established. The hormone–aggression relationship in females, although more complex, involves dehydroepiandrosterone (DHEA), the most abundant hormone in humans and an androgenic neurosteroid. By defining regulatory pathways for normal, sex-typical aggression, a comparative basis for identifying changes in hormonal function that contribute to pathological aggression and aversive emotional states can be developed.

Hormonal Regulation: Shared Features in Males and Females

The neuromodulator hypothesis provides an overarching concept that bridges hormonal influences on aggressive behavior in both males and females. The model posits that hormones influence aggression by modulating neurochemical function. One strength of this hypothesis is that it bridges basic and clinical considerations related to hormone function, aggression, and emotional states. By considering hormonal processes that regulate aggression in adulthood and how these processes interact with representative neurochemical systems (e.g., serotonin (5HT), vasopressin (AVP), and γ-aminobutyric acid (GABA)) that are central to the expression of aggressive behavior, the potential utility of the neuromodulator hypothesis can be illustrated.

Metabolic processes critically influence hormonal effects in target cells and represent a second shared feature between neuroendocrine regulatory pathways in males and females. In males, aromatization and 5α-reduction of T profoundly influence both the development and expression of regulatory pathways. Defining the modulatory effect of DHEA on female-typical aggression parallels circumstances in the male in that multiple metabolites also may be involved. The biosynthesis and metabolism of DHEA has been described in the periphery, but it is not well characterized in the central nervous system (CNS) (**Figure 1**). The 3β-hydroxysteroid dehydrogenase (3β-HSD), hydroxysteroid sulfotransferase (HST), steroid sulfatase (SST), and CYP7B pathways potentially merit attention in relation to female typical aggression. The formation of androstenedione (AE) in response to 3β-HSD activity can increase the availability of more potent androgens, the relative activity of HST and SST determines the potential contributions of DHEA sulfate (DHEA-S) versus DHEA, and CYP7B family activity catalyzes the formation of 7α- and 7β-hydroxy DHEA in the brain. The contribution of these metabolites to female-typical aggression may involve both genomic and nongenomic effects. The demonstration of the direct androgenic effects of DHEA and the suggestion that more potent androgens are formed from DHEA establish a genomic component to the mechanism of action of DHEA. In regard to nongenomic effects of DHEA, the formation of the sulfate versus DHEA itself, and their differential potencies as modulators of $GABA_A$receptor function, further demonstrates the importance of metabolic processes.

Hormonal Regulation in the Adult CNS

Males

The ability of T to facilitate the display of intermale aggressive behavior is widely recognized. Research in adult males has shifted to a mechanistic emphasis to enable the characterization of cell/molecular processes underlying functional T-mediated pathways.

Hormonal substrates T can promote aggression in males through four distinct pathways (**Table 1**): androgen-sensitive, which responds to T itself or its 5α-reduced metabolite, dihydro testosterone (DHT);

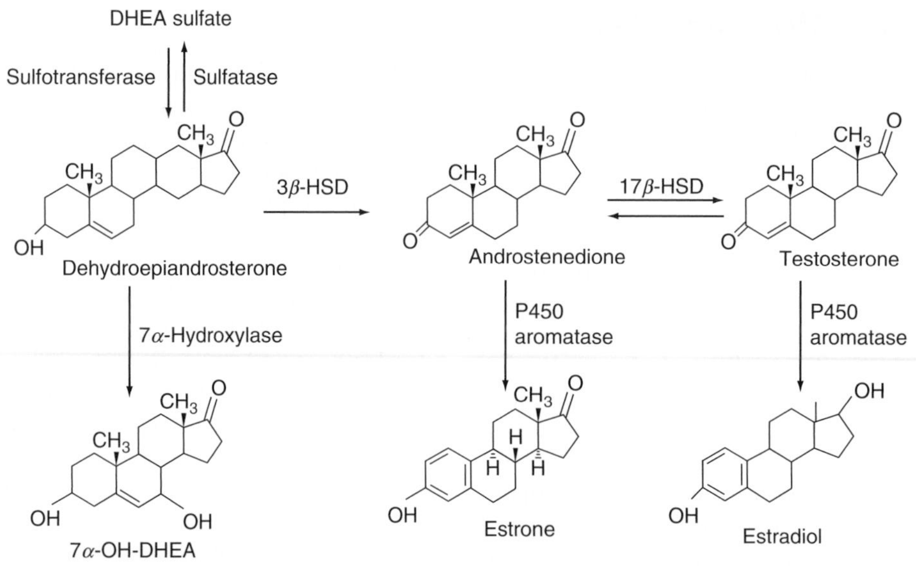

Figure 1 A summary of the various routes of DHEA metabolism in target tissues. Three pathways have been identified with DHEA as the initial substrate: (1) to DHEA sulfate, a reversible path involving hydroxysteroid sulfotransferase and steroid sulfatase, (2) to 7α- or 7β-hydroxy DHEA, which involves the CYP7B enzyme family, and (3) to androstenedione, which involves 3β-hydroxysteroid dehydrogenase (3β-HSD) and provides the possibility for the formation of more potent androgens and estrogens. DHEA, dehydroepiandrosterone. Reproduced from Simon NG, Mo Q, Hu S, Garippa C, and Lu S (2006) Hormonal pathways regulating intermale and interfemale aggression. *International Review of Neurobiology* 73: 99–123, with permission from Elsevier.

Table 1 Hormonal pathways in the adult male brain that facilitate the display of offensive intermale aggression[a]

Genotype	*Estrogen-sensitive*	*Androgen-sensitive*	*Synergistic (estrogen + androgen)*	*Direct T-mediated*
CF-1	+++	++	–	–
CFW	+++	–	–	–
CD-1	–	++	+++	–
C57BL/6J	–	–	–	+

[a]Plus (+) and minus (–) signs indicate relative sensitivity or insensitivity. Reprinted from Simon NG (2002) Hormonal processes in the development and expression of aggressive behavior. In: Pfaff DA, Etgen AM, Fahrbach SE, and Rubin RT (eds.) *Hormones, Brain, and Behavior*, pp. 339–392. San Diego: Academic Press.

estrogen-sensitive, which uses E_2 derived by aromatization; synergistic or combined, in which both the androgenic and estrogenic metabolites of T are used to promote behavioral expression; and direct T-mediated, which uses T itself. The full range of steroid-sensitive systems is not necessarily present in every male because genotype is the major determinant of the functional pathway(s). The most common system uses E_2, which supports a key role for aromatization and estrogen signaling in aggression. In males, all pathways share the feature of high sensitivity.

Neural steroid receptors

Androgen receptor. Major regions exhibiting abundant androgen receptors (ARs) in mammals include the bed nucleus of the stria terminalis (BNST), lateral septum (LS), medial preoptic area (MPOA), basal hypothalamus, and medial amygdala (MAMYG), all of which are part of the neuroanatomical substrate for conspecific aggression. These descriptive findings help define functional circuitry, but do not elucidate how AR regulation contributes to behavioral expression. The fact that AR protein level is controlled through a common, autoregulated mechanism in both male and female neural tissue exemplifies this point. This finding indicates that the rapid increase in AR protein level seen in response to androgen is not sufficient to produce parallel changes in behavioral responsiveness. An example of this point is that the activation of male-typical aggression in ovariectomized females requires 16–21 days of androgen treatment, whereas AR level increases dramatically within 24 h. Increased cellular AR content alters the transcription of other androgen-regulated genes, which over time leads to the expression of aggression. The extended time course required to induce male-like aggression in females thus may be a useful model for defining androgen-dependent circuitry and the role of trophic factors in the hormone-dependent elaboration and maintenance of regulatory pathways.

Estrogen receptor. There are two estrogen receptor (ER) subtypes, α and β. ERα appears to play a primary role based on knockout studies. Offensive attacks were rarely displayed by ER knockout (ERKO) males, whereas wild-type and heterozygous males showed significantly greater attack durations. ERβ seems to exert a more subtle, modulatory influence on estrogen signaling and male-typical aggression. Several potential ways that ERβ could effect aggressive behavior have been proposed based on behavioral changes in males fed diets containing high or low levels of soy phytoestrogens for 15 months. Specifically, males on a high-soy diet showed extreme agonistic behavior. Soy phytoestrogens preferentially bind to ERβ and are less active than E_2–ERα complexes in promoting gene expression. The behavioral and molecular results indicate that in normal males perturbations in estrogen signaling can lead to alterations in aggression. Potential mechanisms underlying the more extreme agonism potentially include decreased 5HT function in the dorsal raphe, where ERβ is the sole subtype thus far detected in primates, or enhanced ERα activity as a result of decreased negative modulation of ERα:ERβ heterodimers. These concepts are supported by *in vitro* assays, available immunochemical findings, and results with βERKO mice.

Neurochemical targets Aggression is a complex social behavior that involves virtually every known neurochemical system. In males, however, the roles of 5HT and AVP are widely recognized, as are hormonal influences on their function. The interactions between hormones and these two systems can be used to illustrate the neuromodulator hypothesis.

Hormonal modulation of serotonin function Lower serotonergic tone is associated with increased aggression, whereas enhanced serotonergic function reduces the expression of aggressive behavior. These relationships have been demonstrated in numerous mammalian species, including humans. Although there are different 5HT receptors, agonists with selective affinity for the 5HT1 subtype, particularly $5HT_{1A}$ and $5HT_{1B}$, specifically and selectively reduce offensive intermale aggression.

Testicular hormones most likely influence aggression by altering serotonin function at $5HT_{1A}$ and/or $5HT_{1B}$ sites in brain regions that either constitute, or project to, the neuroanatomical substrate for intermale aggression. To establish that 5HT function at these receptors is affected by gonadal hormones requires demonstrating that androgens or estrogens differentially affect the ability of $5HT_{1A}$, $5HT_{1B}$, or combined agonist treatments to alter the display of offensive intermale aggression. If this is shown, it then becomes critical to identify neuronal populations in which these effects are produced and the mechanisms through which androgens or estrogens influence $5HT_{1A}$ or $5HT_{1B}$ receptor function in these regions. Unfortunately, few studies have directly addressed this question in the specific context of intermale aggression, which represents a gap in the field. The effects of systemic treatment with serotonergic 1A and 1B agonists seem to be determined, at least in part, by the hormonal environment. If estrogens were present (as the aromatized metabolite of T), the ability of 1A or 1B agonists to inhibit offensive aggression was restricted. When aggression was promoted by a direct androgenic treatment like DHT, however, $5HT_{1A}$ and $5HT_{1B}$ agonists were highly effective in decreasing the expression of offensive behaviors.

Microinjection studies point to the underlying complexity associated with elucidating neuroendocrine modulation of 5HT function. For example, the introduction of selective $5HT_{1A}$ or $5HT_{1B}$ agonists into the LS in the presence of diethylstilbetrol (DES), a potent, specifically acting estrogen, had essentially no effect on behavior. When gonadectomized males were implanted with DHT, however, aggressive behavior was decreased with 1B-agonist microinjection alone or in combination with the 1A agonist 8-OH-DPAT. At the level of the LS, then, these observations suggest that an androgen-sensitive pathway can be attenuated by the action of serotonin at 1B receptor sites. In the MPOA, observed effects were robust. Significantly reduced aggression was seen in the presence of either androgen or estrogen with $5HT_{1A}$- or $5HT_{1B}$-agonist microinjections. The MPOA may thus be a major integrative site for T–5HT interactions in the regulation of T-dependent aggression.

The alteration by gonadal hormones of the ability of serotonergic 1A or 1B agents to effect T-dependent intermale aggression supports the neuromodulator hypothesis. Examples of comparable hormone–neurostransmitter interactions are more numerous in other systems, particularly in regard to reproductive behavior, anxiety, and mood disorders.

The findings of steroidal enhancement or repression of the ability of $5HT_{1A}$ and $5HT_{1B}$ agonists to attenuate offensive intermale aggression raise several potential mechanisms that require attention. Estrogens, for example, can alter $5HT_{1A}$ gene expression or influence ligand availability through effects on synthetic or degradative processes. Another possibility for establishing a direct effect on $5HT_{1A}$ gene function would be the identification of a functional estrogen responsive element (ERE) in the promoter region of the $5HT_{1A}$ receptor gene. Interestingly, both mouse and human $5HT_{1A}$ receptor genes contain an ERE. The spacer element represents a difference between the postulated motifs and the consensus sequence (i.e., 5 nucleotide (nt) rather than 3). However, nonconsensus EREs with different spacer lengths are responsive to estrogenic regulation. For example, the salmon gonadotropin-releasing hormone (GnRH) and BDNF genes have ERE motifs with either 8- or 9-nt spacers and can bind activated estrogen receptors *in vitro*.

Arginine vasopressin The structural and functional integrity of specific aspects of the AVP system in rats, hamsters, and other species is dependent on gonadal steroids. Direct evidence for a systematic relationship among AVP, aggression, and testosterone has been obtained from over 15 years of studies with male Syrian hamster (*Mesocricetus auratus*). Related investigations in rats, mice, voles, and other species, including humans, have broadened our understanding of this interrelationship by defining the ways that social structure interacts with the AVP system and by identifying the cellular and molecular processes involved in hormonal modulation of AVP synthesis and V1a receptor.

AVP can act on multiple brain areas to facilitate aggression through binding to V1a and, perhaps, V1b receptors. Microinjection of AVP into the anterior or ventrolateral hypothalamus of resident hamsters significantly increases biting attacks on intruders. Prairie voles show a dose-dependent increase in aggression toward intruders following intracerebroventricular (ICV) administration of AVP. Results from rats and humans show a positive correlation between aggression and AVP levels in cerebrospinal fluid (CSF). In hamsters, microinjections of Manning compound, a V1a antagonist, into the anterior hypothalamus causes a dose-dependent inhibition of aggression by a resident male toward an intruder and blocks aggression associated with the development of dominant-subordinate relationships. In male prairie voles, ICV injection of V1a receptor antagonist reduces aggression toward male intruders. SRX251, an orally active V1a antagonist with high affinity for the human receptor, causes a significant dose-dependent reduction in latency to bite and number of bites by a resident hamster toward an intruder. Although studies to date have focused on the blockade of the V1a receptor in the regulation of aggression, evidence from knockout mice suggests that the V1b receptor also may be involved. In support of a role for V1b receptor in aggression are results showing that treatment with a selective V1b antagonist, SSR149415, reduces offensive aggression of resident male hamsters.

T is essential for the modulation of aggression by AVP. Microinjections of AVP into the ventrolateral hypothalamus (VLH) did not activate offensive aggression in castrated male hamsters, whereas attack behavior was induced if animals were treated with T implants. The facilitative effect of T on the AVP system in hamsters appears to primarily involve maintenance of V1a receptor populations in the anterior and ventrolateral hypothalamus. Whether this is accomplished via a direct effect on V1a receptor gene expression is unknown because the promoter region for the hamster V1a gene has not yet been sequenced.

The androgenic and estrogenic metabolites of T differentially contribute to steroidal modulation of the AVP system. Estrogen treatment alone after castration partially restored AVP immunoreactivity, DHT alone was ineffective, and combined treatment produced the greatest effect. A direct effect on AVP gene expression appears to be an important mechanism because AVP mRNA also increases in response to circulating gonadal steroid levels. The recent identification of a putative androgen responsive element (ARE), ERE, and AP1 elements in the rat AVP gene promoter region strongly supports enhanced AVP gene transcription as a component of steroidal modulation of the AVP system in rats. Hormonal modulation of the AVP system in rats thus may be achieved through a mechanism different from that in hamsters, in which regulation of the V1a receptor appears to be the primary target for steroidal effects.

It seems reasonable to conclude that T modulates the AVP system. This can be achieved through different pathways, including regulatory effects on the AVP or V1a receptor genes. A specific role for estrogen, the product of aromatization, is indicated, but how the androgenic metabolites contribute has not yet been elucidated. Social structure and seasonality must be overlaid on these effects because they impose boundaries on the extent to which AVP contributes to offensive aggression. Recent studies suggest a potential molecular basis for these effects, at least in voles. Characterizing these interrelationships appears to represent one of the important steps needed for developing an integrative neurobiology of offensive aggression.

Females

The hormonal contribution to female-typical aggression appears to primarily involve DHEA, the most abundant steroid in humans and a neuroactive neurosteroid synthesized in the brain of humans and other mammals. DHEA, which is intrinsically androgenic and can be metabolized to more potent androgens (**Figure 1**), inhibits female-typical aggression when administered chronically. DHEA is a modulator of $GABA_A$ receptor function, and GABA is a principal neurochemical regulator of interfemale aggression. Collectively, these findings raise the possibility that the effects of DHEA on female-typical aggression are exerted through multiple mechanisms. Defining the relationship among DHEA, aggression, and aversive emotional states is a major issue confronting the field.

DHEA: Anti-aggressive mechanism of action The prevailing view concerning the modulation of female-typical aggression by DHEA is that it reduces bioavailable pregnenolone sulfate (Preg-S), a potent negative modulator of the $GABA_A$ receptor, which in turn enhances GABA function and the inhibition of offensive aggression. DHEA also can act at membrane sites rapidly to alter receptor conformation (a nongenomic effect) and effect longer-term processes directly or through neurosteroid metabolites (a genomic effect), which could then alter membrane receptor function.

The $GABA_A$ receptor has a pentameric structure with multiple binding sites, suggesting a number of possibilities for modulation by DHEA. Potential targets for DHEA include sites that bind GABA, the benzodiazepines, the Cl^- ionophore, barbiturates, and a still unidentified neurosteroid binding site.

The finding of direct androgenic effects of DHEA adds another level of complexity to the underlying mechanisms. DHEA exhibits characteristics of typical androgenic compounds, including upregulation of AR protein expression, conferring AR transcriptional activity and inducing dose-dependent increases in limbic system AR content. The androgenic activity of DHEA was confirmed by the observation that DHEA induced intracellular translocation of AR-green fluorescent protein (GFP) and the formation of nuclear clusters. When COS-7 cells transfected with an AR-GFP expression vector were treated with 10^{-7} M DHEA for 24 h, AR-GFP protein translocated from the cytoplasm into the nucleus and led to the formation of punctate fluorescent foci. Another potential mechanistic component tied to genomic effects of DHEA or its metabolites is an alteration of $GABA_A$ subunit structure, which potentially can influence the extent of modulation (**Figure 2**).

The combination of cell surface and direct androgenic effects of DHEA and its metabolites represents a cross-talk cellular signaling system potentially linked to its anti-aggressive effect. Defining the interrelationship among DHEA, its androgenic effects, the subunit structure of $GABA_A$ receptor, and attendant changes in function to elaborate the anti-aggressive mechanism of action of this neurosteroid and how it modulates the expression of female-typical aggression is a major challenge in the field.

Neuroanatomical Substrates

Characterizing the endocrine neuroanatomy of intermale and interfemale aggression requires elucidating critical sites where the modulatory effects of hormones are exerted on neurochemical function. The focus of the current discussion is on regions where these effects are produced directly in target cells, that is, where co-localization of steroid receptors and key neurotransmitters has been established. Clearly, there are indirect effects of steroids on neurochemical function that are probably of equal importance for developing a comprehensive description of neuroanatomical circuitry. However, the paucity of work on such effects precludes consideration at this time.

Males

Lesion and implant studies in rodents have identified subregions of the amygdala, LS, BNST, anterior hypothalamus, and MPOA as part of a presumptive steroid-sensitive circuit for this behavior. The interconnections among these regions are well documented, as is the presence of target neurons for estrogen and androgen. Conspecific offensive aggression in male rodents is dependent on an olfactory stimulus. There has been progress in defining the chemosensitive circuitry mediating intermale aggression in the hamster and rat. A functional pathway from the vomeronasal organ through the olfactory tracts to the MAMYG, BNST, LS, anterior hypothalamus and other limbic structures has been demonstrated in these model species. The integration of steroid receptor distribution with chemosensitive circuitry provides a basis for studies that can define modulatory effects of the hormonal environment.

Other considerations, however, constrain progress in defining neuroanatomical substrates. The identification of distinct androgen-sensitive and estrogen-sensitive pathways raises the possibility that these systems are at least partially independent. Next, aggression itself is a complex behavior with multiple components, many of which are nonreflexive. This aspect of conspecific aggression led to an approach in which investigators sought to identify regions that were either tied to distinct components of the behavioral sequence or appeared to serve as major integrative sites. Examples of the former include portions of the hypothalamus that constitute an attack area, as well as elegant studies of the neuroanatomy of defensive rage and predatory attack in the cat. However, findings in these studies are not linked to either androgenic or estrogenic effects on intermale aggression.

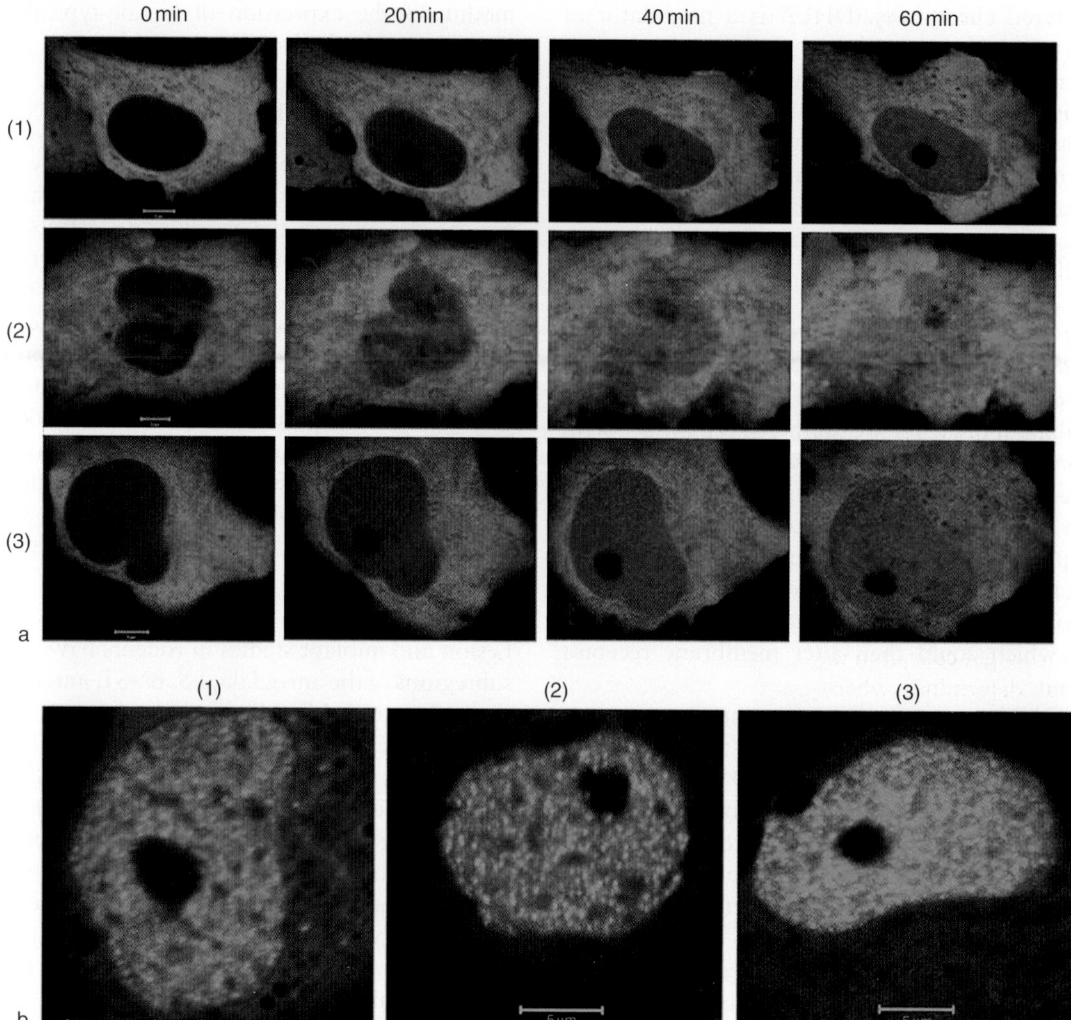

Figure 2 Androgen-dependent AR-GFP intracellular translocation in COS-7 cells: (a) typical intracellular trafficking of AR-GFP in COS-7 cells treated with three steroids, recorded at 0, 20, 40, and 60 min; (b) punctate nuclear distribution of AR-GFP proteins in COS-7 cells (scale = 5 μM). COS-7 cells were transfected with AR-GFP expression plasmid pEGFP-N1-AR. AR-GFP fusion proteins were detected in living cells by excitation with 488-nm line from an argon laser of a confocal microscope. In (a), cells were treated with. (1) 10^{-5} M DHEA, (2) 10^{-6} M Adiol, or (3) 10^{-6} M Adione. A very obvious accumulation of AR-GFP was observed in cells treated with the three compounds in 60 min. In (b), AR-GFP proteins formed fluorescent foci in the nucleus of some cells treated with the same three steroids for approximately 90 min. AR-GFP, androgen receptor-green fluorescent protein; DHEA, dehydroepiandrosterone. Reprinted from Mo Q, Lu SF, and Simon NG (2006) Dehydroepiandrosterone and its metabolites: Differential effects on androgen receptor trafficking and transcriptional activity. *Journal of Steroid Biochemistry and Molecular Biology* 99: 50–58.

Regarding integration, the MAMYG, MPOA, anterior hypothalamus, and LS seem to represent important sites based on findings in rodents. These investigations bridged findings on the chemical neuroanatomy of offensive aggression, in which serotonin and AVP have important regulatory roles, with manipulation of the steroidal environment. Interestingly, putative molecular linkages between androgens, estrogens, and subtypes of AVP and serotonin receptors have been identified. Determining whether these linkages are functional and their relationship to the physiology of conspecific aggression will represent a significant advance in defining neuroendocrine regulation.

An approach that may prove valuable for defining the androgen-sensitive system involves the adult female as a model. The 16–21 day treatment course required for behavioral activation may provide an opportunity to identify androgen-induced changes in neuronal structure and function that may be critical for the display of male-typical aggression.

Studies of neuromodulation by androgens and estrogens in male rodents may have implications for several aversive emotional states in humans. These include, for example, impulsive aggression and violent behavior, the extreme aggression that can accompany steroid or drug abuse, and the depression and anxiety that is sometimes associated with chronic stress.

Females

The neuroanatomical substrates regulating the display of interfemale aggression remain largely undefined. GABA function at the $GABA_A$ receptor, along with modulation by DHEA, appears to represent a critical regulatory component. Although this provides a basis for elucidating at least one major aspect of functional circuitry, the inherent complexity in an analysis of neuroanatomical substrates needs to be acknowledged. For example, the structural heterogeneity of $GABA_A$ across different brain regions, co-localization of $GABA_A$ with steroid receptors in subsets of target cells, and the cellular enzyme phenotype that determines whether neurosteroidal influences on $GABA_A$ are nongenomic, genomic, or both, suggest that multiple mechanisms and processes are at work even within a single neurotransmitter system. When these cell/molecular issues are viewed within the social structure needed to elicit this behavior and the duration of DHEA exposure required to influence interfemale aggression, the lack of progress in this area is perhaps more readily understood.

The analysis of neuroanatomical substrates for aggression in females potentially can advance understanding of aversive emotional states in humans. One potential area is clinically significant anxiety, in which the $GABA_A$ receptor is established as a major target for drug therapies. To the extent that interfemale aggression provides an opportunity to characterize regional differences in the function and modulation of $GABA_A$ receptor, there may be substantive implications for the development of new compounds that treat this disorder. Another area of interest is aging and associated mood changes, in which depression is more commonly seen. DHEA has attracted considerable interest as a replacement therapy. Advancing our understanding of neurosteroidal effects on the $GABA_A$ receptor can add to the understanding of cell/molecular processes that may be influenced by DHEA replacement.

Human Aggression and Aversive Emotional States

Defining the hormonal contribution to human aggressive behavior and aversive emotional states has proven to be a formidable task. Numerous factors that span methodological and conceptual considerations have limited substantive progress, although recent developments, particularly the use of more refined endpoints such as hostility, impulsivity, and dominance suggest that models with greater utility are likely to be forthcoming. Examples of methodological issues include, for example, the use of systemic T assays as a proxy for hormonal effects in the brain and the correlation of hormone levels with endpoints of aggression ranging from paper and pencil tests in college students to histories of violent acts. Conceptual issues include efforts to link hormone levels to a global, catch-all construct of aggression without discriminating between behavioral forms that may be influenced by hormones and those that are not; a view held for many years that female-typical aggression was necessarily also regulated by T; and, with the increasing application of genomic technologies, efforts to identify a single causal gene for aggression and violence. In addition, there is the inherent complexity that accompanies the interaction of physiological systems with social, cultural, cognitive, and experiential factors when attempting to discern neurobiological substrates of human behavior.

The considerations just described suggest that it may be premature to attempt a systematic test of the neuromodulator hypothesis in human males and females. It may be useful, however, to consider principles that have emerged from animal and human studies and use this information to develop more refined hypotheses that may help elucidate neuroendocrine contributions to human aggression, violent behavior, and aversive emotional states.

Characterizing hormonal contributions to sex-typical aggression can inform our understanding of pathological aggression, violence, and aversive states by identifying potential systems in which perturbations may engender atypical behavior. In previous sections, the roles of T, its major metabolites, and their interaction with the 5HT and AVP systems were considered in males, whereas in females the contributions of DHEA, its metabolites, and GABAergic function, particularly at the $GABA_A$ receptor, were presented to illustrate the neuromodulator hypothesis. The question, then, is whether analogous findings have been made in studies of human aggression and, if not, to identify research in those areas as potential priorities.

Males

Results regarding a correlation between T and aggression have been equivocal. This may be due to the breadth of models that have been assessed as correlates of T and a failure to recognize that systemic blood levels do not necessarily reflect intracellular processes, that is, the importance of E_2 and aromatization (except perhaps in extreme cases, such as large doses of anabolic steroids). It can be argued reasonably that dominance behavior, as well as assertiveness and novelty seeking, is influenced by T in human males. Important mechanistic considerations in determining these hormone-behavior relationships may well include aromatization rates and the dynamics of T secretion. The potential role of aromatization in human male aggression has received scant attention. Competitive interactions induce a consistent change in T levels – a rise in anticipation of competition and sustained elevation as a consequence of winning with lower levels in losers. These changes may produce effects on neurochemical function and subsequent behavior, leading to aversive states in the losing males.

Serotonergic function, with a few reported exceptions, is inversely related to impulsive aggression, hostility, and a life history of aggression. The repeated demonstration of this relationship through studies showing a blunted 5HT response to systemic challenge and lower 5HT metabolites in the CSF in impulsive/aggressive individuals provided the basis for the analysis of genes involved in 5HT function. Several of these studies have identified polymorphisms in the 5HT transporter and $5HT_2$ receptor that are associated with impulsive aggression and an increased risk of violent suicide, respectively. Intronic polymorphisms in the tryptophan hydroxylase 1 gene also have been suggested as a risk factor in suicidal behavior.

Elevated vasopressin function has been linked to impulsive aggression in human males. Higher levels of AVP in the CSF were positively correlated with higher scores aggression in borderline patients. An interesting feature of this study was that these same individuals also had reduced serotonergic function. Several investigations in animal models have demonstrated regulatory interactions among 5HT, AVP, and the expression of aggression. Because there is abundant evidence from studies in other mammals, including nonhuman primates, that establishes hormonal influence on 5HT and AVP function, an assessment of these interactions in humans may help develop an integrative neurobiology of inappropriate aggression and violent behavior.

An excellent example of the potential of the neuromodulator hypothesis and one that can be directly linked to aggression and aversive emotion is androgenic (as well as glucocorticoid) modulation of monoamine oxidase A (MAO-A), a key enzyme in the degradation of 5HT, dopamine, and norepinephrine. Specifically acting androgens regulate the MAO-A promoter via an ARE as well as via Sp1 sites. The association between MAO-A gene expression and impulsive aggression has been documented in humans.

Females

Far less is understood about the hormonal contribution to aggression in human females than in males. A significant factor in this state of affairs has been continuing efforts to identify a systematic relationship between T and various forms of aggression in women. Mixed to negative results generally have been obtained when T levels were correlated with competitive situations, delinquency, assertiveness, and hostility/irritability during the menstrual cycle. Although a focus on dominance-related behaviors might provide some insights, even this more restrictive perspective is uninformative. Women do not exhibit changes in T dynamics comparable to those seen in men in competitive settings, and changes in T levels during the menstrual cycle are not correlated with mood states associated with aggression. Interestingly, a relationship between T and postpartum depression recently was described, which demonstrates the need to carefully define forms of aggression and aversive states that are influenced by the hormonal environment.

Alterations in the serotonin system have been identified in women who exhibit impulsivity. This represents a potential area for the determination of hormonal modulation based on indirect evidence from several studies demonstrating estrogenic influences on serotonin function and mood. The potential role of AVP in human female aggression has not been determined, although results showing an influence of AVP on a broad range of social behaviors suggest that it is an area that merits attention. Overall, however, there is little clear evidence linking hormones, aggression, and aversive states in women.

The equivocal to negative results in human studies suggest that a different approach to the relationship between hormones and offensive aggression in women may be needed. Animal models suggest that sex-typical aggression in females is modulated by DHEA. If this pertains to humans, an emphasis on the GABA system and its modulation by neurosteroids might be useful.

Two examples illustrate the potential utility of this model. The $GABA_A$ receptor is a target for modulation by anabolic steroids. Decreased $GABA_A$ function through elevated androgen exposure may underlie the hostility/rage associated with steroid abuse. How DHEA, the most abundant androgen in women, may influence these responses in normal women remains to be determined. The second example is MAO-A gene regulation, in which DHEA or its metabolites can directly regulate expression and thus affect serotonergic (and other amine neurotransmitter) function. Together, these and related observations reinforce the potential modulatory role of DHEA in human female aggression and aversive emotions.

Conclusion

Knowledge of hormonal processes in humans linked to the expression of aggression and aversive emotional states during adulthood is limited. The importance of hormone metabolism in males has been demonstrated in animal models but remains an area ripe for study in humans. The intrinsic androgenicity of DHEA and the contribution of its metabolites also may represent important mechanistic steps in females, although this remains an open issue. Although several target neurochemical systems and candidate genes or gene polymorphisms have been identified, substantial work is needed to identify the full range of cellular processes that are affected as well as the genomic and nongenomic mechanisms that mediate these effects. Progress in this area is essential for defining the modulation of neurochemical function by gonadal hormones and neurosteroids. It is an essential step in developing a systems model that eventually should encompass gene regulation, functional circuitry, behavioral expression, and adaptation.

The neuromodulator hypothesis provides an integrative, systems-based framework for determining the hormonal contribution sex-typical aggression, pathological aggression, and aversive emotional states in humans. An interesting feature of the model is that hormonal systems seem to produce different effects in each sex. In males the net effect of T is neutral or facilitative, whereas in females DHEA may be inhibitory.

In males, critical studies defining the molecular interface among T, its metabolites, and components of 5HT, AVP, and no doubt other neurotransmitter systems are needed. In females, nongenomic and genomic modulation of $GABA_A$ receptor function requires characterization as well as the potential interface with MAO-A. This is only a partial list based on the representative systems covered here. Fully characterizing hormonal contributions will require defining multiple effects covering both genomic and membrane-level actions.

Finally, other important aspects of hormone function in aggression and aversive emotion should be noted. We chose to focus on gonadal steroids in males and an adrenal androgen/neurosteroid in females. Other areas of interest, for example, are the effects of corticosteroids and the dopamine system.

The complex nature of hormonal modulation and the need for refined models to assess endocrine contributions to human behavior has been demonstrated in animal models. Progress in defining forms of aggression (e.g., hostility, irritability, impulsivity, and dominance) is a significant improvement over grouping very different behaviors under aggression. An emphasis on a single gene or even a limited subset of physiological markers as the cause of aggression, although sometimes informative, is limiting. A systems perspective that recognizes when hormones may have a role; that physiological effects are modulatory; and that social structure, life events, and subsequent adaptations are reflected in alterations in cellular signaling pathways will be necessary to build an integrative neurobiology of human aggression and aversive emotional states.

See also: Aggression: Neurochemical and Molecular Mechanisms; Emotional Hormones and Memory Modulation; Emotional Control of the Autonomic Nervous System; Gene Expression Regulation: Steroid Hormone Effects; Hormones and Behavior.

Further Reading

Arango V, Huang YY, Underwood MD, and Mann JJ (2003) Genetics of the serotonergic system in suicidal behavior. *Journal of Psychiatric Research* 37: 375–386.

Archer J (2006) Testosterone and human aggression: An evaluation of the challenge hypothesis. *Neuroscience and Biobehavioral Reviews* 30: 319–345.

Baulieu EE (1997) Neurosteroids: Of the nervous system, by the nervous system, for the nervous system. *Recent Progress in Hormone Research* 52: 1–32.

Bethea CL, Lu NZ, Gundlah C, and Streicher JM (2002) Diverse actions of ovarian steroids in the serotonin neural system. *Frontiers in Neuroendocrinology* 23: 41–100.

de Almeida RM, Ferrari PF, Parmigiani S, and Miczek KA (2005) Escalated aggressive behavior: Dopamine, serotonin and GABA. *European Journal of Pharmacology* 526: 51–64.

de Vries G and Simerly RB (2001) Anatomy, development, and function of sexually dimorphic neural circuits in the mammalian brain. In: Pfaff DA, Etgen AM, Fahrbach SE, and Rubin RT (eds.) *Hormones, Brain, and Behavior*, pp. 137–191. San Diego: Academic Press.

Ferris CF (2005) Vasopressin/oxytocin and aggression. *Novartis Foundation Symposium* 268: 190–198; (discussion) 198–200, 242–153.

Gruber CJ, Gruber DM, Gruber IM, Wieser F, and Huber JC (2004) Anatomy of the estrogen response element. *Trends in Endocrinology and Metabolism* 15: 73–78.

Labrie F (2003) Extragonadal synthesis of sex steroids: Intracrinology. *Annals of Endocrinology* 64: 95–107.

Lee R and Coccaro E (2001) The neuropsychopharmacology of criminality and aggression. *Canadian Journal of Psychiatry* 46: 35–44.

Mehta AK and Ticku MK (2001) Unsulfated and sulfated neurosteroids differentially modulate the binding characteristics of various radioligands of GABA(A) receptors following chronic ethanol administration. *Neuropharmacology* 40: 668–675.

Meyer-Lindenberg A, Buckholtz JW, Kolachana B, et al. (2006) Neural mechanisms of genetic risk for impulsivity and violence in humans. *Proceedings of the National Academy of Sciences of the United States of America* 103: 6269–6274.

Mo Q, Lu SF, and Simon NG (2006) Dehydroepiandrosterone and its metabolites: Differential effects on androgen receptor trafficking and transcriptional activity. *Journal of Steroid Biochemistry and Molecular Biology* 99: 50–58.

Ogawa S, Choleris E, and Pfaff D (2004) Genetic influences on aggressive behaviors and arousability in animals. *Annals of the New York Academy of Sciences* 1036: 257–266.

Olivier B (2005) Serotonergic mechanisms in aggression. *Novartis Foundation Symposium* 268: 171–183; (discussion) 183–179, 242–253.

Ou XM, Chen K, and Shih JC (2006) Glucocorticoid and androgen activation of monoamine oxidase A is regulated differently by R1 and Sp1. *Journal of Biological Chemistry* 281: 21512–21525.

Roy-Byrne PP (2005) The GABA-benzodiazepine receptor complex: Structure, function, and role in anxiety. *Journal of Clinical Psychiatry* 66(supplement 2): 14–20.

Simon NG (2002) Hormonal processes in the development and expression of aggressive behavior. In: Pfaff DA, Etgen AM, Fahrbach SE, and Rubin RT (eds.) *Hormones, Brain, and Behavior*, pp. 339–392. San Diego: Academic Press.

Simon NG, Mo Q, Hu S, Garippa C, and Lu S (2006) Hormonal pathways regulating intermale and interfemale aggression. In: Bradley RJ, Harris RA, and Jenner P (eds.) *International Review of Neurobiology*, Vol. 73, pp. 99–124. London: Academic Press.

Trainor BC, Kyomen HH, and Marler CA (2006) Estrogenic encounters: How interactions between aromatase and the environment modulate aggression. *Frontiers in Neuroendocrinology* 27: 170–179.

Aggression: Neurochemical and Molecular Mechanisms

A Siegel, S Bhatt, and S Zalcman, University of Medicine and Dentistry of New Jersey – New Jersey Medical School, Newark, NJ, USA

Overview – Models of Aggressive Behavior

Aggressive behavior may be defined as behavior that threatens harm or leads to or causes harm, destruction, or damage to another organism. In this context, aggression is not a unitary phenomenon but, instead, reflects a variety of different behavioral processes that are contained under a single heading. A variety of animal research models of aggressive behavior have been used, which include fear-induced, maternal, intermale, irritable, sex-related, territorial, resident-intruder, and predatory aggression. With the exception of predatory aggression, these models of aggressive behavior share the following common features. They reflect a perceived or real threat, are aversive to the organism, are impulsive, display sympathetic signs, and are defensive in nature. Thus, these models can be reduced to the general category of affective (or defensive) aggression. In contrast, predatory aggression is quite different from the others, in particular because it requires planning, shows few autonomic signs, and is positively reinforcing to the organism. Consequently, the remainder of this article considers the neurobiological properties and functions associated with these two forms of aggressive behavior as determined from studies conducted mainly in the cat.

Affective defense (sometimes called defensive rage) occurs in nature in response to the presence of, or perceived presence of, a threatening stimulus such as another species within a cat's territory. It is characterized by arching of the back, retraction of the ears, piloerection, pupillary dilatation, marked hissing, and striking of the target species with the forepaw at the threatening object. This form of aggressive behavior can also be elicited by electrical or chemical stimulation of the midbrain periaqueductal gray (PAG) and medial hypothalamus.

In the animal kingdom, predatory (quite biting) attack behavior is manifested as hunting behavior. The attack is planned and is preceded by stalking of a specific prey object of another species, followed by biting of the back of the neck until the hunter kills the prey. In contrast to defensive rage, the cat displays few autonomic signs aside from mild pupillary dilatation. Predatory attack is induced following electrical stimulation of the lateral hypothalamus or ventrolateral aspect of the PAG and ceases immediately following termination of stimulation.

Regions and Pathways Mediating Defensive Rage and Predatory Attack

A principal goal in the study of the neurobiology of aggression and rage is to identify the underlying mechanisms of these behaviors. This objective requires knowledge of the anatomical substrates of these behaviors. The following discussion identifies the sites within the hypothalamus and midbrain from which each of these forms of aggression is elicited and the pathways mediating these behaviors to other regions of the brain stem and related areas of the central nervous system.

Defensive Rage Behavior

Defensive rage behavior can be elicited by electrical stimulation of wide regions along the rostrocaudal axis of the medial hypothalamus. Defensive rage is also elicited by electrical or chemical (i.e., glutamate analog) stimulation of the dorsolateral quadrant of mainly the rostral half of the PAG. Components or fragments of defensive rage can also be elicited from lower regions of the brain stem. These regions include the caudal PAG and pontine tegmentum and presumably lie along the descending pathways mediating this form of aggression.

The principal descending pathway from the medial hypothalamus subserving defensive rage behavior arises from the anterior medial hypothalamus, and its primary target is the dorsolateral aspect of the rostral half of the PAG. The functions of this pathway are mediated by glutamate acting on *N*-methyl-D-aspartate (NMDA) receptors in the PAG. Of particular interest is that other regions of the medial hypothalamus, such as the ventromedial nucleus, from which defensive rage can also be elicited, project rostrally to the region of the anterior medial hypothalamus from which the descending pathway to the PAG arises. Moreover, the anterior medial hypothalamus also receives significant inputs from components of the limbic system which modulate aggression and rage behavior. The converging inputs into the anterior medial hypothalamus thus enables this region to serve as a major site of integration for the expression of defensive rage behavior.

The second limb of the descending pathway for the expression of defensive rage behavior arises from the region of the dorsolateral PAG, which receives direct inputs from the anterior medial hypothalamus. The efferent projections of this region of the PAG are directed to structures that mediate autonomic and somatomotor components of defensive rage behavior. There are several routes by which autonomic functions are activated from the PAG. One pathway includes a projection to the locus ceruleus, which in turn projects to the intermediolateral cell column of the thoracic and lumbar spinal cord. Converging inputs to these sympathetic regions of spinal cord are also mediated through projections to the solitary nucleus, whose axons then project to the ventrolateral medulla and from there to the intermediolateral cell column of the thoracic and lumbar cord. There are several regions that mediate the somatomotor components of defensive rage behavior. One set of targets includes the motor nuclei of the trigeminal and facial cranial nerves, which are associated with jaw opening essential for the vocalization aspect of the defensive rage response. A second target includes the nuclei of the reticular formation, which comprise, in part, reticulospinal fibers directed toward alpha and gamma motor neurons. Those neurons directed to the cervical cord presumably affect movements of the upper limbs that make up the striking component of the rage response. It is the collective integration of these two components at the levels of the medial hypothalamus and PAG which make up the defensive rage response. A separate ascending projection of the dorsolateral PAG supplies the rostrocaudal extent of the medial hypothalamus, much of which relates to the expression of defensive rage. This projection probably serves as a substrate for a positive feedback mechanism, thus increasing the likelihood that this response can be prolonged under dangerous conditions, which is of survival value to the animal.

Predatory Attack Behavior

Predatory attack behavior can be elicited by electrical stimulation, most typically of the perifornical lateral hypothalamus, ventrolateral aspect of the PAG, and ventral tegmental area. The principal origin of the descending projections of the hypothalamus is the region of the perifornical lateral hypothalamus, from which predatory attack is elicited. This region supplies the ventrolateral aspect of the PAG, ventral tegmental area, central tegmental fields of the midbrain and pons, locus ceruleus, and motor and main sensory nuclei of the trigeminal complex. The projections to the trigeminal complex are significant in that they provide the anatomical substrate for the jaw-closing reflex critical for the culmination of biting attack. The projections to the brain stem tegmentum presumably provide the initial neuron in a series of descending projections to the lower brain stem and spinal cord which is essential for other motor aspects of the attack response such as stalking and striking at the prey object.

Anatomical and functional relationship between the medial and lateral hypothalamus Although defensive rage behavior and predatory attack clearly reflect distinctly different forms of aggression that use separate and nonoverlapping pathways, they also relate to one another in a unique manner. Within the medial hypothalamus and with respect to defensive rage behavior, there are two classes of neurons. One is a projection neuron (already described), whose target is the dorsolateral aspect of the PAG and that constitutes the descending pathway for this form of aggression. The second is a neuron with a short axon that supplies the lateral hypothalamus. It is γ-aminobutyric acid (GABA)ergic and inhibits neurons in the lateral hypothalamus associated with predatory attack. Likewise, there are at least two classes of neurons in the lateral hypothalamus with respect to predatory attack. The first is a neuron with a long axon that constitutes the descending pathway for the expression of predatory attack, and the second is a GABAergic neuron with a short axon which supplies the medial hypothalamus and inhibits neurons in the medial hypothalamus associated with defensive rage. The likely functional significance of the reciprocal inhibitory pathways linking the medial and lateral hypothalamus is as follows. Because these two responses are mutually exclusive, the effective expression of one requires the suppression of the other. It is intuitive that a successful act of predation can be accomplished only when a predator quietly approaches the prey object, which requires the suppression of hissing and related component responses. Similarly, when defensive rage is required following the presence of a threatening stimulus, elements of predation serve no function and therefore are suppressed in order for the affective components of the response to become manifest. Collectively, the neuroanatomical relationships between the medial and lateral hypothalamus thus provide the essential substrates which are of survival value to the animal.

Limbic Structure-Associated Pathways Modulating Aggression and Rage

The limbic system consists of the following structures: amygdala, hippocampal formation, septal area, prefrontal cortex, and cingulate gyrus. They possess several common anatomical and functional features which distinguishes them from other regions of the brain, including (1) they receive secondary or tertiary

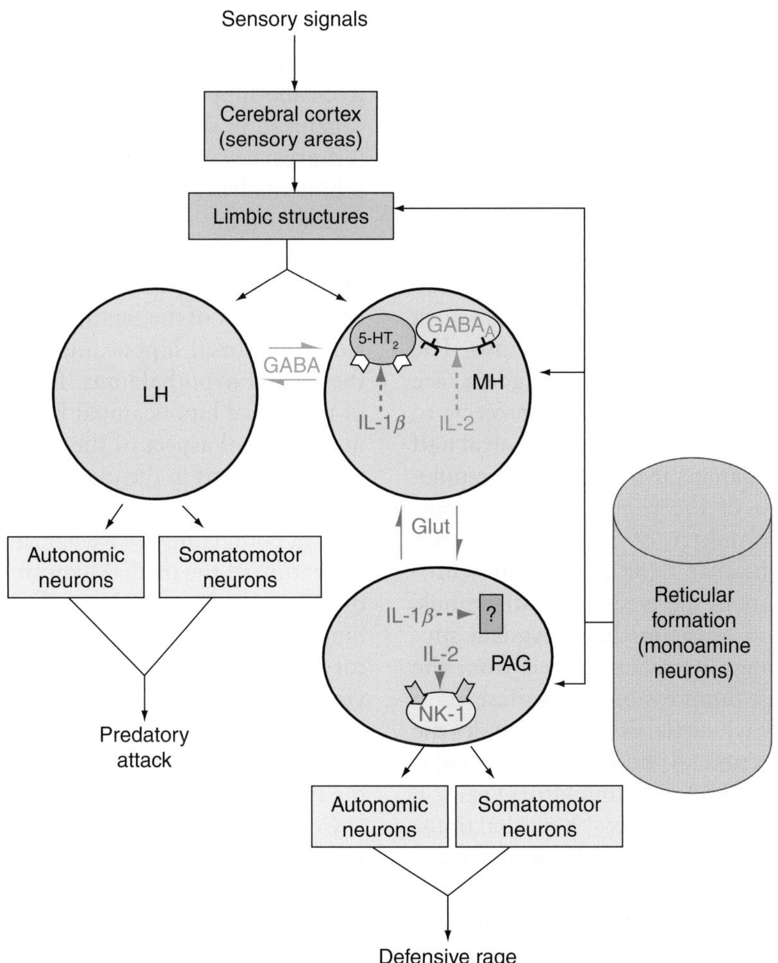

Figure 1 Schematic diagram depicting some of the critical neural mechanisms regulating aggression contained within limbic structures, such as the amygdala, hippocampus, and prefrontal cortex. These are activated by sensory signals that reach them through inputs from sensory regions of cerebral cortex, and these limbic neurons are further modulated by monoaminergic neurons situated within the reticular formation of the brain stem. Subsequent changes in levels of excitability within the limbic system are mediated through efferent pathways of limbic structures such as the fornix and stria terminalis to the hypothalamus, causing changes in excitability levels of hypothalamic neurons and, thus, directly affecting the neural mechanisms that control aggression and rage behavior. The expression of a predatory attack is generated in the lateral hypothalamus and the descending pathways, which engage both autonomic and somatomotor neurons of the lower brain stem. In a similar way, the expression of defensive rage is mediated through neurons in the medial hypothalamus and a glutamatergic descending pathway to the midbrain PAG, which, in turn, provides feedback through an ascending neuron to the medial hypothalamus. The medial and lateral hypothalami mutually inhibit one another through reciprocal GABAergic neurons. Also depicted in this diagram is the role of cytokines IL-1 and IL-2. IL-1 in the medial hypothalamus facilitates defensive rage, acting through 5-HT_2 receptors. IL-2 in the medial hypothalamus suppresses defensive rage, acting through $GABA_A$ receptors, but in the PAG this cytokine facilitates defensive rage, acting through NK_1 receptors. 5-HT, serotonin; GABA, γ-aminobutyric acid; IL, interleukin; LH, lateral hypothalamus; MH, medial hypothalamus; NK-1, neurokinin-1; PAG, periaqueductal gray. Reproduced from Zalcman S and Siegel A (2006) The Neurobiology of aggression and rage: Role of cytokines. *Brain, Behavior* and *Immunity* 20: 507–514, with permission from Elsevier.

sensory inputs which may vary among limbic structures, (2) they receive inputs from brain stem monoaminergic neurons, and (3) they project directly or indirectly to the hypothalamus and related structures of the brain stem. These combined sensory and monoaminergic inputs serve to activate the limbic structures to cause the powerful modulation of aggression and rage by virtue of their efferent projections to the efferent target structures (**Figure 1**). This section describes the modulating effects of limbic structures on aggression and rage and their associated pathways over which such modulation is mediated.

Amygdala

The amygdala, which consists of a complex of nuclei located in the rostral aspect of the temporal lobe, has received more attention than any other limbic structure with respect to its relationship to emotional behavior. These studies have revealed that the amygdala is not uniform in its effects on aggression and rage. Instead, the effects are dependent on both the form of aggression and region of amygdala considered.

Excitation of the region of amygdala, including the medial nucleus and medial aspect of the basal complex, in the cat potentiates defensive rage behavior elicited from the medial hypothalamus, whereas excitation of the lateral and central nuclei or lateral aspect of the basal complex suppresses this response. The potentiating effects of the medial amygdala are mediated over the stria terminalis, which projects to the bed nucleus of the stria terminalis and rostral half of the medial hypothalamus, including the dorsomedial region and shell of the ventromedial nucleus. A primary neurotransmitter of this pathway has been identified as substance P (SP), acting on neurokinin (NK)-1 receptors in the medial hypothalamus. In contrast, excitation of the medial amygdala suppresses predatory attack behavior elicited from the lateral hypothalamus. Suppression is manifest via a disynaptic pathway in which the first limb includes the stria terminalis projection to the medial hypothalamus and the second a GABAergic (inhibitory) neuron projecting from the medial to lateral hypothalamus. The inhibitory effects of the amygdala on defensive rage behavior are mediated through a descending projection to the midbrain PAG. The neurotransmitter has been shown to be enkephalin acting through μ-opioid receptors in the PAG. In a parallel manner, excitation of the lateral amygdala potentiates predatory attack. Although the pathway has not been experimentally identified, it is likely to include fibers of the ventral amygdalofugal pathway projecting to the lateral hypothalamus.

Hippocampal Formation and Septal Area

The hippocampal formation in rodents and felines is arranged in a manner that extends from its rostral tip, situated in proximity to the septal area caudally, beneath the corpus callusum, and parallel to the lateral ventricle, entering the temporal lobe where it passes ventrally and rostrally, and ending at a position just caudal to the amygdala. The dorsal region of hippocampus (near the septal pole) suppresses predatory attack, whereas the ventral region (near the temporal pole) facilitates this form of aggression. The modulating properties of the hippocampal formation on aggressive behavior is mediated through the septal area, a structure which may thus be viewed as a relay nucleus of the hippocampal formation.

The pathways over which hippocampal modulation of aggressive responses are mediated probably involve the precommissural fornix – the branch of the fornix that supplies the septal area. The projection from the hippocampal formation is topographically organized in that fibers arising from the dorsal (septal) aspect project to the medial aspect of the lateral septal nucleus, whereas progressively more caudal regions of the hippocampal formation (toward the temporal pole) project to more progressively lateral aspects of the lateral septal nucleus. In turn, the medial aspect of the septal area, which receives inputs from the dorsal hippocampal formation, projects to the medial hypothalamus. In this manner, activation of the dorsal hippocampal formation excites neurons in the medial aspect of the septal area, which in turn excites neurons in the medial hypothalamus. Because the medial hypothalamus communicates with the lateral hypothalamus via a short GABAergic neuron, activation of the medial hypothalamus either directly or indirectly through the medial septal area or dorsal hippocampal formation causes suppression of predatory attack. Activation of the medial hypothalamus via a projection from the medial aspect of the septal area provides the anatomical basis for septal area potentiation of defensive rage behavior elicited from the medial hypothalamus.

Concerning the anatomical basis by which the ventral hippocampal formation and lateral aspect of the lateral septal nucleus potentiate predatory attack behavior, the likely pathways include a direct projection to the lateral aspect of the lateral septal nucleus, which in turn projects to and (presumably) activates neurons in the lateral hypothalamus which mediate the expression of predatory attack behavior.

Prefrontal Cortex and Anterior Cingulate Cortex

The prefrontal cortex and adjoining regions of the anterior cingulate cortex exert the powerful suppression of predatory attack and rage behavior. There are a number of descending fiber systems from the prefrontal cortex that could provide the anatomical substrate for the modulation of aggression. These include a monosynaptic connection consisting of a small number of neurons that project directly to the hypothalamus from the prefrontal cortex and several multisynaptic pathways involving connections with either the amygdala or mediodorsal thalamic nucleus. Of these pathways, there is experimental evidence that the modulating effects from the prefrontal cortex and anterior cingulate gyrus on aggression and rage are mediated primarily through the multisynaptic pathway involving the mediodorsal thalamic nucleus. With respect to this pathway, the mediodorsal thalamic nucleus projects to the hypothalamus through a series of interneurons in the midline thalamus. The neurotransmitter for this system of neurons is not known but presumably includes a glutamate projection from the prefrontal cortex and anterior cingulate cortex.

Neurochemical and Molecular Mechanisms of Aggression

Neurotransmitters

An increasing body of knowledge has identified several different classes of neurotransmitters that play a role in regulating aggressive behavior. These include the following small molecule neurotransmitters: acetylcholine, GABA, and biogenic amines (dopamine, norepinephrine, and serotonin (5-HT)), and neuropeptides (opioid peptides, SP, cholecystokinin, and vasopressin). Some of these transmitters potentiate aggressive responses, whereas others have inhibitory properties. In addition, several of these neurotransmitters have been identified with specific neuroanatomical pathways, but others have not.

Excitatory neurotransmitters Studies conducted mainly in rodents and felines showed that cholinergic agents generally facilitate aggressive responses. These findings are based on the systemic application of agonists and antagonists that act through muscarinic receptors. Further studies supporting these findings have indicated that cholinergic agents produce their potentiating effects within the region of the medial hypothalamus and can, in fact, induce rage-like responses.

Both dopamine and norepinephrine have similar potentiating effects on both defensive rage and predatory attack. The mechanism presumably involves the activation of catecholaminergic neurons of the brain stem from such regions as the locus ceruleus for norepinephrine and the ventral tegmental area for dopamine, which project to widespread regions of the forebrain, including the hypothalamus and limbic system. Dopamine facilitation is mediated through dopamine D_2 and norepinephrine through α_2 receptors in the medial hypothalamus. Catecholaminergic facilitation of both defensive rage and predatory attack suggests that these neurotransmitters exert generalized potentiating effects on whatever ongoing responses are present during the epoch when these transmitters are activated. Serotonin distributed from brain stem raphe neurons to the PAG, hypothalamus and limbic system differentially affects defensive rage responses. Activation of 5-HT type 2 (5-HT_2) receptors in these regions facilitates defensive rage. The suppressing effects of 5-HT are indicated later in this article.

As already indicated, there are several pathways associated with the expression or modulation of aggressive or rage behavior whose primary neurotransmitters have been identified. These include glutamate neurons projecting from the medial hypothalamus to the PAG that act through NMDA receptors to mediate the expression of defensive rage, and SP neurons in the medial amygdala that project to the medial hypothalamus, which powerfully facilitates defensive rage and suppresses predatory attack (**Figure 1**). Glutamate neurons that project from the basal amygdala to the PAG, acting through NMDA receptors, also facilitate defensive rage behavior. Several peptides have also been identified that potentiate defensive rage behavior within the PAG. These include SP, acting through NK_1 receptors, and cholecystokinin (CCK), acting through CCK_B receptors. Within the anterior hypothalamus, it is well known that vasopressin (V) neurons play an important role in temperature regulation and drinking behavior. These neurons also potentiate aggressive responses in rodents, acting through V_{1A} receptors in the anterior hypothalamus. The pathway(s) associated with V functions in aggression have not been identified.

Inhibitory neurotransmitters Three neurotransmitters suppress defensive rage behavior. Activation of 5-HT_{1A} receptors in either the PAG or medial hypothalamus by 5-HT released from brain stem raphe neurons suppress this form of aggressive behavior. Powerful suppression of defensive rage behavior is mediated through μ-opioid receptors in the PAG. As already noted, these receptors are activated by enkephalinergic neurons arising in the central nucleus of amygdala. $GABA_A$ receptors in the medial and lateral hypothalamus suppress defensive rage and predatory attack behaviors, respectively. As previously indicated, $GABA_A$ receptors in the medial hypothalamus are activated by GABA neurons projecting from the lateral hypothalamus, and similarly, GABA neurons arising in the medial hypothalamus activate these receptors in the lateral hypothalamus (**Figure 1**). $GABA_A$ receptors in the PAG also suppress defensive rage behavior when activated by GABA neurons, the origin of which has not yet been identified.

Substances of abuse and psychotropic drugs It is well know that substances of abuse are associated with enhanced levels of aggressive behavior in both humans and animals. Such increases in aggression can be understood in terms of what is known about the link between neurotransmitters and aggression. Opioid peptides, when elevated, normally suppress aggressive behavior. Opioid withdrawal induces aggressive behavior, and this phenomenon can be understood in terms of the absence of critical levels of enkephalins in the central nervous system. Alcohol, another substance of abuse, has been associated with homicide, child abuse, marital assaults, rape, and verbal abuse. In the animal literature, parallel findings have been observed as dose-dependent increases in aggression have been reported in various species. Altered 5-HT levels have been associated with aggressive alcohol abusers. Other neurotransmitters may be involved, but the data are incomplete at this time. Cocaine, which results in heightened levels of aggression, has been shown to influence dopamine release and retard its removal from the brain. Cocaine's effects on aggression have also been associated with activation of 5-HT_3 receptors. Information concerning the pathways mediating the potentiating effects of substances of abuse on aggressive behavior is lacking in the literature.

Genes and Cytokines

Evidence from genetic studies Two methods have been applied to the study of the role of genetics in aggression and rage behavior. The first involves traditional breeding methods, and the second involves the use of genetic engineering to produce knockout mice in which specific receptors are absent.

Concerning the first approach, increased levels of aggression are present under the following conditions: (1) in animals selectively bred for heightened sensitivity to cholinergic agonists; (2) in animals bred in a manner producing higher levels of brain dopamine levels; and (3) in animals bred for selective loss of 5-HT axons. Thus, these findings have generally supported the findings obtained from pharmacological and neurochemical approaches summarized in the previous section. Concerning the second approach, mutant mice lacking the 5-HT_{1B} receptor or which display decreased 5-HT turnover in the brain have increased levels of aggressive behavior. In contrast, mice lacking the V1b receptor have reduced aggressive behavior.

Cytokines and aggression Cytokines are pleiotropic cell–cell signaling proteins or glycoproteins produced by various cells types in the periphery and in brain. They were initially described for their effects on immunity, including innate and acquired immunity. Cytokines also play fundamental roles in other functions, including development and feeding, among other processes. Cytokines exert their biological effects by binding their receptors, which in turn activates signaling pathways. Cytokine receptors are present on numerous cell types in periphery and the central nervous system, and they may act in autocrine, paracrine, and endocrine manners. Thus, a given cytokine can influence activity in the cell from which it is released, as well as local and distant cells. In the latter regard, some cytokines of peripheral origin may influence brain activity by transporting across the blood–brain barrier. Cytokines may also affect brain activity by entry through the circumventricular organs or by indirect routes, including the activation of sensory afferents and stimulation of cytokine release by brain endothelial cells. Because cytokines and their receptors are present in the brain, they can act as endogenous neuromodulators in regions associated with the expression of aggression and rage, including the hypothalamus and PAG.

A further possible linkage between cytokines and aggression is derived from observations obtained from human studies. Two types of studies have been conducted – the first concerning patient populations and the second involved experiments using healthy adult subjects. Studies involving patient populations, cytokine immunotherapy for the treatment of such disorders as acquired immunodeficiency syndrome (AIDS), cancer, and hepatitis C have shown that such therapy increased aggressive behavior as determined by measures of anger, hostility, and irritability. Similar conclusions were drawn from studies involving healthy adult subjects. In one study, it was demonstrated that experimentally induced enhancement of proinflammatory cytokines was correlated with increases in hostility scores. In another study, a positive correlation was drawn between the presence of hostile marital conflict and the increased production of plasma proinflammatory cytokines.

Animal studies have also indicated a relationship among the enhancement of cytokine production, immune cell activity, and aggression. To cite one example, animals displaying heightened levels of territorial behavior and aggression also revealed higher levels of interferon-γ (IFN-γ) and interleukin-2 (IL-2) production. Several questions can be raised from the results of these studies, including: (1) Which cytokines relate to the control of aggression and rage? (2) Where in the brain are these cytokines located? (3) What are the possible mechanisms underlying cytokine modulation of aggressive behavior? Recent studies have provided partial answers to these questions, indicating that IL-1 and IL-2 do potently affect defensive rage behavior and that the nature of these effects depends on the region of the brain where receptors for these cytokines are activated. In one study, it was demonstrated that IL-1β in the medial hypothalamus facilitates defensive rage behavior and that such facilitation is mediated through 5-HT_2 receptors. In contrast, activation of IL-2 receptors in the same region of the medial hypothalamus suppresses defensive rage behavior. The suppressive effects of IL-2 appear to be mediated through $GABA_A$ receptors. Thus, these two cytokines use different neurotransmitter mechanisms within the medial hypothalamus to differentially modulate defensive rage behavior. In addition, activation of IL-2 receptors in the PAG potentiates defensive rage behavior, indicating that the locus of activation of this receptor is critical to the nature of its effects on defensive rage behavior. The potentiating effects of IL-2 in the PAG on defensive rage are mediated through SP NK_1 receptors, a receptor that has known excitatory effects on defensive rage behavior (see **Figure 1**). These findings raise an additional question – namely, how can the same cytokine found in different regions of the brain alter defensive rage behavior in opposite ways? An answer to this and related questions concerning how cytokines modulate aggression and rage can only be answered by future studies directed at understanding the signaling pathways associated with these cytokines and their relationship to the neurochemical milieu of the various regions of the forebrain and brain stem associated with the expression and modulation of aggression and rage behavior.

Conclusion

Two principal forms of aggressive behavior – defensive rage and predatory attack – are induced following excitation of neurons in the hypothalamus and PAG. Defensive rage is elicited from the medial hypothalamus and dorsolateral PAG, whereas predatory attack is elicited from the lateral hypothalamus, ventral tegmental area, and ventrolateral PAG. The primary pathways over which defensive rage occurs include a projection from the medial hypothalamus to the PAG and from the PAG to autonomic and somatomotor neurons of the lower brain stem. Predatory attack is mediated over descending fibers from the perifornical lateral hypothalamus to the midbrain and pontine tegmentum as well as to motor nuclei of cranial nerves V and VII. The expression of these forms of aggression are modulated from limbic structures and their output pathways to the hypothalamus and PAG. These include projections from (1) the amygdala to the hypothalamus via the stria terminalis (containing SP) and ventral amygdalofugal pathways as through a descending pathway to the PAG (containing enkephalin); (2) the hippocampal formation to the hypothalamus via a relay in the septal area; and (3) the prefrontal and anterior cingulate cortex projections via the medial thalamus to the hypothalamus. Neurotransmitters (and neuromodulators) that are known to regulate aggression and rage include monoamines, excitatory and inhibitory amino acids, and various peptides, including cytokines.

See also: Aggression: Hormonal Basis; Emotional Hormones and Memory Modulation; Gene Expression Regulation: Steroid Hormone Effects; Hormones and Behavior.

Further Reading

Adamec RE (1990) Role of the amygdala and medial hypothalamus in spontaneous feline aggression and defense. *Aggressive Behavior* 16: 207–222.

Bandler R and Shipley M (1994) Columnar organization in the midbrain periaqueductal gray: Modules for emotional expression. *Trends in Neuroscience* 17: 379–389.

Hassanain M, Zalcman S, Bhatt S, and Siegel A (2005) Interleukin-1β (IL-1β) and IL-1β type 1 receptors in the medial hypothalamus in defensive rage behavior in the cat. *Brain Research* 1048: 1–11.

Meloy JR (1988) *The Psychopathic Mind: Origins, Dynamics and Treatment.* Northvale, NJ: Jason Aronson.

Meloy JR (1997) Predatory violence during mass murder. *Journal of Forensic Science* 42: 326–329.

Miczek KA and Fish EW (2005) Dopamine, glutamate and aggression. In: Schmidt WJ and Reith MEA (eds.) *Dopamine and Glutamate in Psychiatric Disorders*, pp. 237–266. Totawa, NJ: Human Press.

Miczek KA, Haney M, Tidey J, Vivian J, and Weerts E (1994) Neurochemistry and pharmacotherapeutic management of aggression and violence. In: Reiss A, Miczek KA, and Roth J (eds.) *Understanding and Preventing Violence*, vol. 2, pp. 245–514. Washington, DC: National Academy Press.

Monroe RR (1978) *Brain Dysfunction in Aggressive Criminals.* Lexington, MA: Lexington Books.

Nunn K (1986) The episodic dyscontrol syndrome in childhood. *Journal of Child Psychology and Psychiatry* 27: 439–446.

Petitto JM, Gariepy JL, Gendreau PL, Rodriguiz R, Lewis MH, and Lysle DT (1994) Differences in NK cell function in mice bred for high and low aggression: Genetic linkage between complex behavioral and immunological traits? *Brain Behavior and Immunity* 13: 175–186.

Siegel A (2005) *The Neurobiology of Aggression and Rage.* Boca Raton, FL: CRC Press.

Siegel A and Brutus M (1990) Neural substrates of aggression and rage in the cat. In: Epstein AE and Morrison AR (eds.) *Progress in Psychobiology and Physiological Psychology,* vol. 14, pp. 135–233. San Diego, CA: Academic Press.

Siegel A, Roeling TAP, Gregg TR, and Kruk MR (1999) Neuropharmacology of brain-stimulation-evoked aggression. *Neuroscience and Biobehavioral Reviews* 23: 359–389.

Vitiello B, Behar D, Hunt J, Stoff D, and Ricciuti A (1990) Subtyping aggression in children and adolescents. *Journal of Neuropsychiatry* 2: 189–192.

Zalcman S and Siegel A (2006) The neurobiology of aggression and rage. *Brain Behavior and Immunity* 20(6): 507–514.

Aging and Memory in Animals

P R Rapp, Mount Sinai School of Medicine, New York, NY, USA

Introduction

From a life-span perspective, aging comprises the late component of a genetically determined program of development, maturation, and senescence, interacting with a complex array of environmental factors. Commonly viewed as a process of deterioration, growing old is associated with sharply increased risk for many diseases and disabilities that compromise independent living, placing a heavy burden on families, caregivers, and society. Nonetheless, a majority of people successfully accommodate certain physical signs of aging, and in fortunate cases, old age can represent a rewarding period of new intellectual engagement, novel pursuits, and achievement. Such positive outcomes become increasingly unlikely in the face of failing cognitive function, and partly for this reason, disorders that lead to diminished mental capacity are among the most feared consequences of aging. The leading cause of dementia, Alzheimer's disease, ultimately results in a dense amnesia, gradually robbing patients of the lifetime of memories that define their personal history and identity. Even in the absence of disease, many people experience memory impairment that, although relatively mild, can cause considerable anxiety and compromise the quality of life. As the populations of industrialized countries rapidly age (**Figure 1**), we face a growing challenge of identifying ways to promote healthy cognitive aging and to maximize optimal functioning.

Research has illuminated many of the key features of age-related cognitive decline in humans; enabled by advances in *in vivo* brain imaging, it has begun to reveal how the neural systems organization of memory is altered. Studies of cognitive aging in humans are complicated by a variety of methodological factors, however, and they are limited by the range of applicable experimental approaches. A vexing issue is that individuals in preclinical stages of Alzheimer's disease and other disorders affecting cognition are difficult to identify with confidence. As a consequence, it is often unclear to what degree observed impairment is attributable to disease rather than normal nonpathological aging. Relating these deficits to underlying biological causes is also problematic, and although noninvasive imaging techniques continue to yield remarkable discoveries, defining the neurobiological mechanisms of cognitive aging requires experimental approaches not suitable for investigation in humans. Research in animal models has played a critical role in efforts to address these issues.

Cognitive Aging – A Neuropsychological Framework

There is a broad consensus on a number of central concepts concerning the neuropsychology of normal memory in young adults. Rather than comprising a unitary, monolithic capacity, multiple forms of memory can be distinguished according to the qualities of the information remembered, endurance, flexibility, and a variety of other characteristics. A major advance in this field is the recognition that the existence of multiple forms of memory follows from the organization of brain systems that mediate these capacities. Converging evidence from rats, monkeys, and amnesic humans, for example, indicates that normal memory for the facts and events of daily life critically requires a collection of anatomically related structures in the medial temporal lobe, including the hippocampus and adjacent parahippocampal cortical areas. Damage to the prefrontal cortex, by comparison, largely spares this declarative or episodic type of memory and, instead, leads to deficits in a variety of memory-related processes, including the strategic use of remembered information when confronted with novel circumstances and the ability to recall the source from which information was acquired.

Informed by this background, research detailing the specific profile of behavioral deficits that emerge during aging can provide a window on the functional status of the neural systems that mediate memory and other cognitive processes. The Morris water maze has been used widely in studies of this sort in rats, partly because memory assessed by this procedure critically requires the functional integrity of the hippocampus. Across trials in this task, animals learn the spatial location of an escape platform hidden in a pool of clouded water, and navigation is guided by the memory of the platform position in relation to cues surrounding the maze. Young adult rats acquire this task rapidly, learning to swim directly to the escape platform from any point around the perimeter of the apparatus. By roughly 24 months of age, acquisition rates are impaired in a substantial proportion of rats, and by comparison with young adults, spatial searching is less tightly focused on the escape location during probe testing, when the platform is removed from the apparatus. There are a variety of reasons to presume that this pattern reflects a genuine impairment in learning and memory rather than the effects of

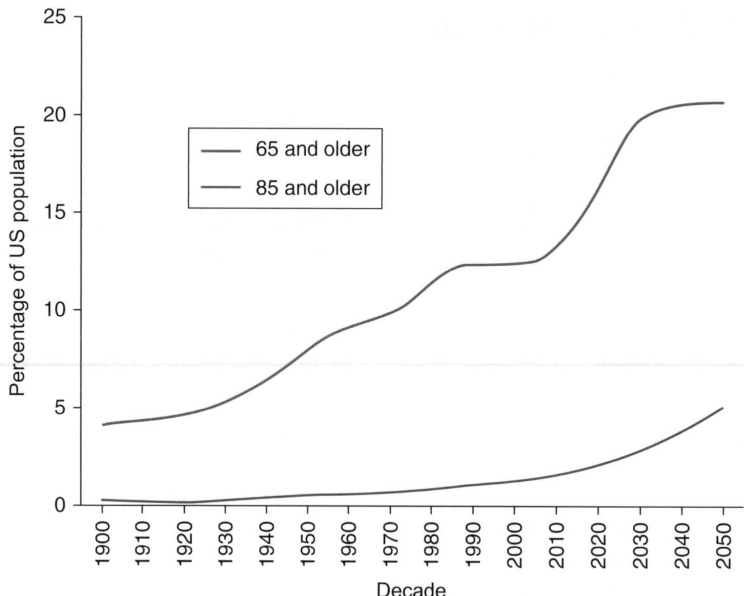

Figure 1 Society is aging rapidly; the historical and projected percentages of the US population that is over 65 and 85 years. Results compiled from US Census Bureau figures.

aging on sensorimotor function, motivation, and other performance factors. In a variant of testing that shares many of the same performance demands as the hidden goal task, rats swim to a visible escape platform that either protrudes slightly above the water's surface or is marked with a prominent visual cue. Performance in this cue-approach version of the water maze is relatively insensitive to aging, suggesting that, in the absence of a requirement to learn and remember information about the spatial location of the escape platform, age-related deficits in swimming ability, the motivation to escape, and visual acuity are not sufficient to prevent performance comparable to that of young adults. Notably, a qualitatively similar pattern of impaired spatial learning, together with preserved cue-approach performance, is seen in rats following the disruption of hippocampal function.

Age-related impairment in learning and memory for spatial information is not peculiar to rats, and parallel deficits have been documented in both humans and monkeys. The hippocampus is not restricted to processing spatial information, however, and in efforts to define how aging influences medial temporal-lobe memory, research in monkeys has relied predominantly on other sorts of assessments. In a particularly widely used procedure, that is delayed nonmatching-to-sample (DNMS), trials consist of two phases; the presentation of a sample item followed by a recognition test. A retention interval is imposed between the two, and monkeys demonstrate recognition by choosing the novel object (or nonmatch) when it is presented together with the previously viewed sample. Training typically proceeds by testing subjects with a short retention interval of 8 or 10 s until they master the nonmatching rule and subsequently introducing more challenging memory delays, up to 10 min or longer.

Visual object recognition in young adult macaque monkeys declines gracefully over increased retention intervals, confirming that the task effectively taxes memory. The average life span in rhesus macaques is less than 25 years, and monkeys at this age and older display modest but reliable deficits in visual recognition. Specifically, whereas older subjects score as well as controls at relatively short delays, recognition accuracy falters at longer intervals. This profile is significant because it demonstrates that sensorimotor function, motivation, and attention in the aged monkey are sufficiently intact to support normal performance and that age-related DNMS deficits emerge specifically in relation to increased demands on memory. In addition, this delay-dependent pattern of impairment is qualitatively similar to the deficits observed in young monkeys with experimental damage involving the hippocampal region. Thus, across both rats and monkeys, behavioral studies converge, concluding that memory mediated by the hippocampal system is vulnerable to normal aging.

Individual Variability in Cognitive Aging

An important insight from recent research is that the cognitive effects of aging vary considerably across individuals such that many experience little or no decline

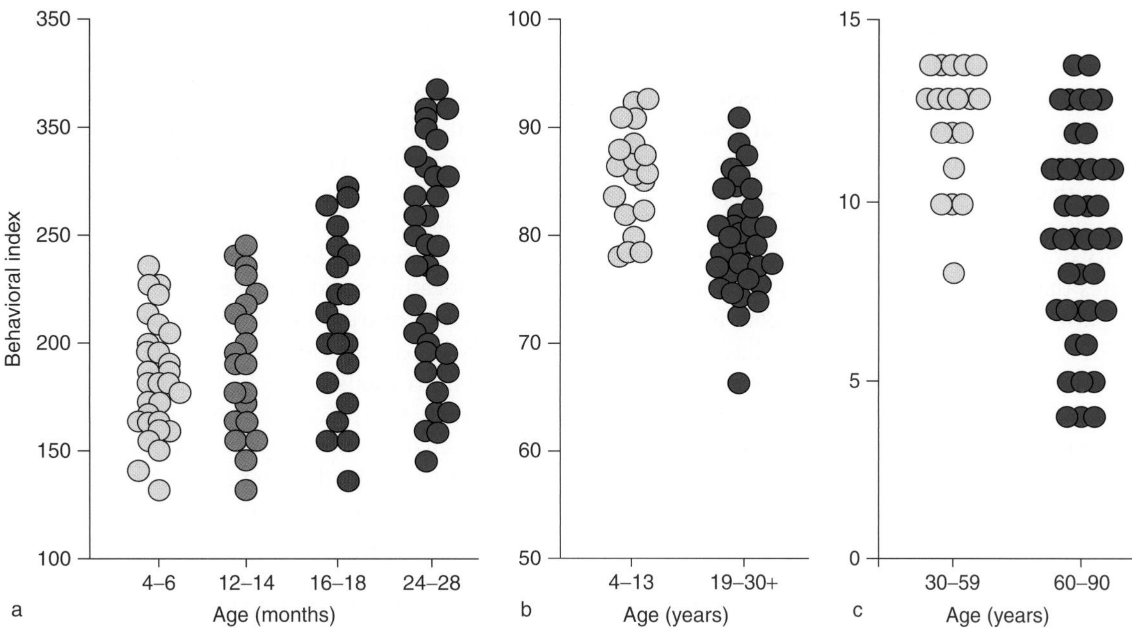

Figure 2 The status of memory varies considerably across aged individuals: (a) spatial memory in rats; (b) object recognition memory in monkeys; (c) delayed recall in humans. For rats (a), low scores represent better memory; in monkeys (b) and humans (c), high values reflect better performance. Symbols signify scores for individual subjects. Note that in all species the status of memory among the aged is distributed across a broad range, from individuals that score on a par with the best young adults to other individuals that exhibit substantial impairment. (a) Data courtesy of M Gallagher, Johns Hopkins University; (b) data courtesy of P Rapp, Mount Sinai School of Medicine; (c) data courtesy of M Albert, Johns Hopkins School of Medicine.

but others develop substantial impairment (**Figure 2**). Individual variability is widely recognized as a prominent feature of human aging, but it has often been considered a nuisance factor, making it difficult to draw firm conclusions about the likely course of decline. In addition, impaired function in some aged people is likely a prodromal harbinger of disease, obscuring individual variability associated with the process of normal aging itself. Here again, studies in animal models that are spared the spontaneous, dementia diseases of human aging have proved key.

The status of spatial learning and memory varies substantially among aged rats and is continuously distributed from subjects that perform on par with the best young adults to other aged individuals that display marked impairment, outside the range of adult control values (**Figure 2(a)**). Findings from monkeys are similar, indicating that roughly 40% of aged subjects exhibit marked recognition memory deficits, relative to young animals, as assessed by DNMS (**Figure 2(b)**). These findings mirror observations in humans, establishing the important concept that, in the absence of frank disease, marked cognitive decline is not an inevitable consequence of growing older (**Figure 2(c)**). In addition, the recognition of individual differences in the cognitive outcome of aging has enabled a powerful approach for exploring the neurobiological basis of dysfunction. Whereas aging research traditionally has comprised comparisons between groups distinguished on the basis of chronological age, an alternative approach involves the analysis of age-matched groups that differ according to neuropsychological indices of aging. As described in subsequent sections, studies adopting this approach have substantially advanced progress toward defining the neural alterations that specifically distinguish aged individuals with cognitive deficits from others of the same age with preserved function.

Neurophysiology of Cognitive Aging

Neuropsychological studies are an important source of evidence suggesting that memory mediated by the hippocampus is vulnerable to aging. Cognitive data alone, however, are a secondary proxy, and other sorts of approaches are needed to directly document the status of information processing in the aged hippocampus. An illuminating strategy used by a number of investigators involves recording the firing patterns of hippocampal neurons as freely moving rats navigate a maze apparatus. Under these conditions, neuronal activity in a large proportion of hippocampal pyramidal cells reliably and selectively increases when subjects move through a specific spatial location in the testing environment – a phenomenon referred to as place-field firing. Many features of place-field activity, including the incidence, selectivity, and specificity of location-related firing, appear normal in the aged hippocampus, even among rats with pronounced spatial learning and memory impairment. This sparing counts against the idea that deficits in memory result from generalized, nonspecific deterioration in the functional integrity of the hippocampus. In fact, recent evidence indicates that the influence of aging on hippocampal firing patterns is highly specific and coupled with the status of spatial memory measured behaviorally.

One means of defining the particular constellation of stimuli that controls place cell activity involves the use of probe conditions in which subsets of cues in the testing environment are manipulated independently. After mapping the place fields of neurons in a maze setting that is held constant, for example, the constellation of experimenter-defined tactile and odor cues on the surface of the apparatus can be rotated in the opposite direction of the extra-maze visual and auditory stimuli. Recorded under these conditions, a large percentage of neurons in the young adult hippocampus either stop firing or develop entirely new place fields, unrelated to the rotated configuration of either the intra- or extra-maze cues. Such remapping is thought to enable the flexibility of hippocampal information processing, providing the basis for encoding the relevant predictive relationships between familiar stimuli when they appear in novel circumstances. A similar incidence of flexible place-field remapping is observed in aged rats that perform normally on the hidden platform version of the Morris water maze. In contrast, aged rats with documented deficits in spatial learning and memory exhibit far less flexibility in place-field activity. Neuronal firing in this subpopulation appears abnormally rigid and disproportionately driven by the configuration of extra-maze cues in the testing environment relative to other available stimuli.

As a consequence of encoding a reduced scope of available information, it might be predicted that the aged hippocampus is prone to interference between environments that share a subset of cues. Consistent with this prediction, when aged rats repeatedly explore an environment containing cues that were also present in another familiar setting, place-field remapping occurs substantially more frequently than in young subjects (**Figure 3**). An orthogonal pattern of results is observed when unit activity is recorded in a highly familiar environment, followed by exploration of a novel setting, and subsequent return to the original environment. Place-field firing in memory-impaired aged rats initially persists across the familiar and novel conditions, and repeated exposure is needed to induce the remapping observed in young rats from the outset. Related findings also indicate that at least some features of age-related alteration in place-field firing are regionally selective across the principal cell layers of the hippocampus, appearing to be particularly pronounced in the CA3 field. Considered together, the picture emerges that prominent features of normal cognitive aging are coupled with reliable changes in key computational capacities of the hippocampus and that the specific nature of these alterations contributes to the spatial learning and memory impairment seen in a substantial proportion of aged individuals.

Although the precise cell biological basis of age-related change in hippocampal neurophysiology remains to be defined, recent research has considerably narrowed the range of plausible accounts. A particularly long-standing idea is that the complex constellation of neurobiological alterations that occurs during aging ultimately results in neuron death in vulnerable brain areas such as the hippocampus and that this loss of structural integrity is the proximal cause of cognitive aging. Enabled by the development and validation of efficient methods of morphometric quantification, however, studies taking advantage of these stereological tools have prompted a new consensus. Across all the animal models examined, including rats and monkeys, the findings consistently demonstrate that marked neuron loss in the hippocampus and associated components of the medial temporal-lobe memory system is not a necessary consequence of normal aging. Moreover, among investigations incorporating behavioral assessment prior to sacrifice and stereological analysis, age-related deficits in learning and memory frequently occur in the absence of detectable neuron loss in these structures (**Figure 4**). Alongside this preservation, synaptic connectivity that critically participates in hippocampal functional plasticity is compromised, and in some cases, these changes are correlated with individual variability in the cognitive outcome of aging.

Enduring changes in synaptic strength, mediated by long-term potentiation (LTP)-like mechanisms, are widely presumed to underlie the establishment of place-field firing and other electrophysiological correlates of behavior that are vulnerable to aging. Considerable attention has therefore focused on the possibility that deficits in LTP and other plasticity mechanisms might contribute to the cognitive outcome of aging. Many fundamental electrophysiological properties of neurons throughout the hippocampus are preserved during aging, including the resting membrane potential, the depolarization threshold for action potential firing, and, under maximal nonphysiological stimulation conditions, the induction of LTP. Studies using weaker stimulation, however, reveal that the threshold for LTP induction is increased in both the dentate gyrus

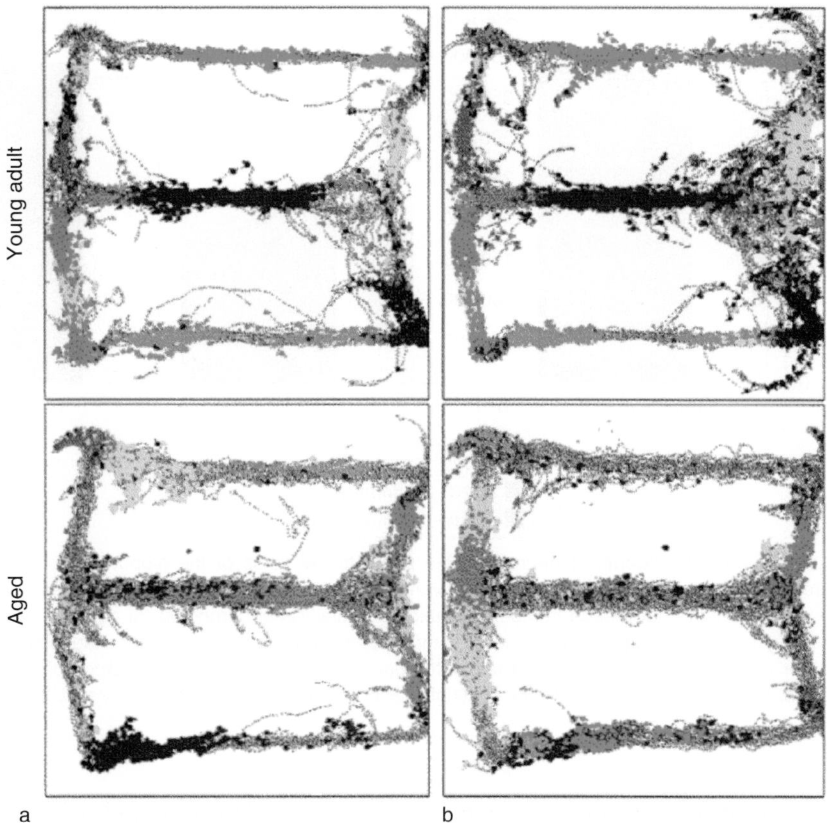

Figure 3 Neuronal firing of multiple individual hippocampal pyramidal neurons as young and older rats explored a figure-8 maze on two occasions: (a) maze exploration A; (b) maze exploration B. Light gray lines illustrate the path subjects took as they navigated the apparatus; the locations at which neurons fired above background levels are represented as color-coded dots. As shown, the coding of spatial location is prone to disruption in the hippocampus of aged rats with memory deficits. Note that in the young rat, the distribution of place-field firing is highly consistent across the two bouts of exploration. In contrast, on roughly 30% of such occasions, the aged hippocampus exhibited completely different place-field mappings, as though the familiar test setting was represented as multiple environments. Reprinted by permission from Macmillan publishers Ltd: *Nature* (Barnes CA, Suster MS, Shen J, and McNaughton BL (1997) Multistability of cognitive maps in the hippocampus of old rats. *Nature* 388: 272–275), copyright (1997).

and CA1 fields of the aged hippocampus. The long-term maintenance of synaptic enhancement, which requires gene expression, protein synthesis, and synaptic remodeling, is also compromised and declines toward the baseline more rapidly in aged subjects than young adults. A significant challenge for future research is to identify the relevant linkages between the effects of aging documented at different levels of analysis (determining whether synaptic alterations in critical circuitry account for changes in the computational capacities of hippocampal neurons) and, ultimately, age-related deficits in learning and memory.

Although it seems reasonable to speculate that age-related changes in synaptic plasticity contribute to observed alterations in hippocampal place-field firing and spatial learning and memory impairment, emerging evidence reveals considerable complexity in the underlying mechanism of these effects. LTP in the CA1 field of the hippocampus reflects the operation of two components: the *N*-methyl-D-aspartate receptor (NMDAR)-dependent pathway and a second, less well-studied form mediated by voltage-dependent Ca^{2+} channel (VDCC) activity. Using pharmacological manipulations to dissect the contribution of these components, current evidence indicates that, even under conditions in which the magnitude and endurance of compound LTP is similar, there is a shift toward a greater role of VDCC-mediated enhancement in the aged hippocampus compared to the young hippocampus. Related findings suggest that aging is also accompanied by a change in the bidirectional balance between synaptic enhancement and long-term depression (LTD). Prolonged low-frequency stimulation that is without effect in the young hippocampus produces an NMDAR-dependent LTD at the aged CA3 to CA1 synapses, and this vulnerability appears at least partly attributable to altered Ca^{2+} signaling. Studies in behaviorally characterized rats extend these observations, suggesting that, whereas age-related change in NMDAR-mediated LTD is unrelated to individual

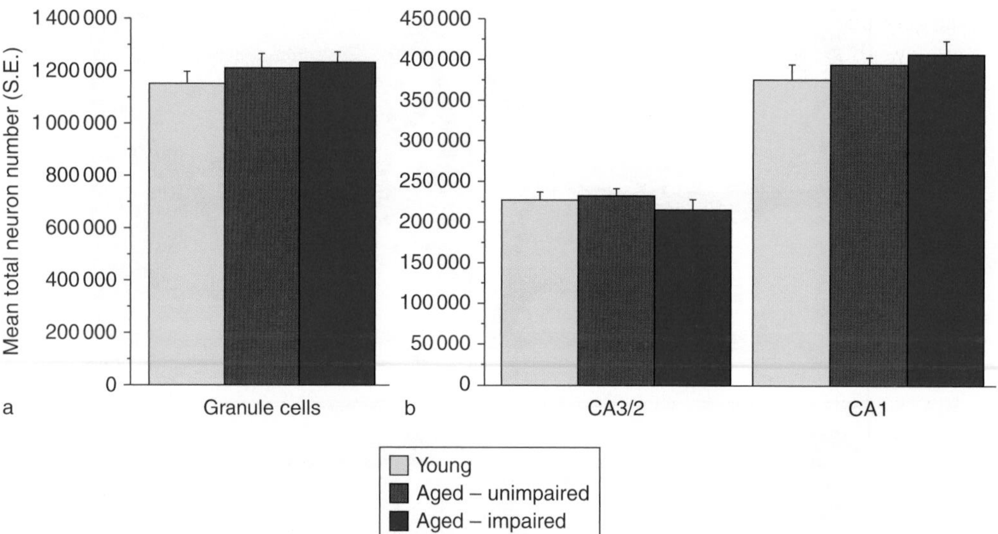

Figure 4 Estimated total number (unilateral) of neurons in the principal cell fields of the hippocampus for young rats and aged (unimpaired and impaired) rats based on their performance in a water maze test of spatial learning and memory: (a) granule cells; (b) CA3/2 and CA1. The results illustrate that neuron number is stable as a function of chronological age and cognitive status. Memory impairment during normal aging does not require neuron death in the hippocampus. Reproduced from Rapp PR and Gallagher M (1996) Preserved neuron number in the hippocampus of aged rats with spatial learning deficits. *Proceedings of the National Academy of Sciences of the United States of America* 93: 9926–9930, copyright 1996, National Academy of Sciences, USA.

differences in the cognitive outcome of aging, NMDAR-independent LTD is increased in magnitude and correlated with preserved spatial memory in aged rats. Together, these findings indicate that the basis of age-related decline in hippocampal memory is likely to involve dynamic interactions between multiple plasticity pathways and that active compensatory mechanisms are available in support of optimal hippocampal function during aging. Defining and harnessing these mechanisms through pharmacological or other means are important goals of current research efforts.

Horizons in Research on Cognitive Aging

Aging beyond the Hippocampus

This article has centered on medial temporal-lobe memory as a particularly well-studied system in animal models of age-related cognitive decline. This focus, however, should not be taken to suggest that this form of memory is affected in isolation. Indeed it is entirely clear that other systems are also vulnerable, prominently including cognitive capacities supported by the prefrontal cortex. Relevant research in rats has begun to emerge only recently, in part because the normative functional organization of the prefrontal cortex has proved challenging to characterize. Nonetheless, comparable to findings in humans, aged rats exhibit deficits in attentional set shifting, or the ability to update and modify behavior in response to altered task contingencies. This capacity is a member of a broad domain, termed executive function, that mediates the strategic use and manipulation of memory in support of adaptive behavior. Parallel findings in monkeys demonstrate age-related impairment on a nonhuman primate adaptation of the Wisconsin Card Sorting Test, a widely used neuropsychological assessment of executive function in humans. Another, particularly well-documented signature of cognitive aging is that older monkeys display robust deficits relative to controls on delayed-response tests that measure the spatiotemporal organization of memory. This procedure challenges animals to remember and select the location of a reward presented prior to a delay interval of several seconds or more. Because each location is rewarded frequently over the course of the many trials that make up daily test sessions, the opportunity for interference is substantial, placing considerable demands on the temporal organization of memory to distinguish information relevant to the current trial from target locations used earlier. Delayed-response performance, like other measures of executive function, critically requires the functional integrity of the prefrontal cortex. Thus, taken together, the available neuropsychological evidence prompts the conclusion that cognitive capacities supported by the prefrontal cortex are vulnerable to normal aging. In addition, because these signatures of aging are largely unrelated to the status of medial temporal memory, the emerging consensus is that cognitive decline reflects the independent vulnerabilities of multiple memory-related brain structures rather than a process of widespread generalized deterioration.

Intervention

Considerable attention in research on the biology of aging has been directed at efforts to extend the life span. Improving the quality of life is at least as important a goal, however, and in this regard, animal models can be valuable in informing research efforts to promote healthy cognitive aging. This perspective, emphasizing the importance of bringing into convergence evidence from multiple levels of analysis, from basic research to human clinical trials, is illustrated by a consideration of the results from the Women's Health Initiative (WHI). The WHI is among the largest double-blind, placebo-controlled clinical trials ever mounted in the United States and is aimed at defining the health and cognitive effects of hormone replacement therapy in postmenopausal women. A large body of basic research had provided ample reason to suspect that ovarian hormone replacement following menopause might offer significant benefit in terms of neuroprotection and cognitive function, and indeed, results from a number of relatively small observational studies in women supported this prediction. The surprising result of the WHI, in contrast, is that, in addition to significantly increasing the risk for a number negative health outcomes, including cancer and stroke, hormone replacement offered no benefit in terms of cognitive function. In fact, among the few who developed Alzheimer's disease, the incidence was nearly double in treated women compared with placebo controls.

Evidence from animal research suggests a number of factors that might contribute to these unexpected results. In rats, for example, the cognitive benefit of exogenous hormone replacement depends on initiating treatment soon after the onset of estrogen deficiency, and efficacy is compromised with delayed intervention. That there is a temporal window of opportunity for influencing cognitive function with ovarian hormone treatment is relevant to the WHI because, in that study, women averaged 65 years of age at the beginning of the trial, nearly 15 years older than the typical onset of menopause. The specific schedule of dosing may also be important, and adopting the standard clinical practice at the time, the regimen used in the WHI consisted of the chronic, daily administration of conjugated equine estrogen (CEE), with or without a synthetic progestin (medroxyprogesterone acetate (MPA)). Chronic, continuous treatment, however, may lead to the habituation or desensitization of the neurobiological responses that mediate the cognitive influence of estrogen. Supporting this speculation, acute dosing in rats is reportedly at least as, if not more, effective than continuous replacement. Findings in nonhuman primates are similar, demonstrating that a single injection of estradiol every 21 days in aged ovariectomized monkeys substantially reverses a key signature of cognitive aging relative to age-matched vehicle control values. The specific formulations used for replacement may also be critical, based on *in vitro* evidence that estradiol and progesterone produced by the ovaries can have very different, and sometimes opposite, cell biological effects compared to the CEE and MPA compounds typically prescribed in clinical practice. Although the precise influence of these and other variables remains to be defined, current research establishes the value of investigations in animal models for addressing these issues and of broader efforts aimed at the rational development of strategies to promote healthy cognitive aging.

See also: Aging of the Brain; Aging of the Brain and Alzheimer's Disease; Cognition in Aging and Age-Related Disease; Episodic Memory: Assessment in Animals; Executive Function and Higher-Order Cognition: Assessment in Animals; Functional Neuroimaging Studies of Aging; Gene Expression in Normal Aging Brain; Lipids and Membranes in Brain Aging; Long-Term Depression: Cerebellum; Long-Term Potentiation and Long-Term Depression in Experience-Dependent Plasticity; Long-Term Potentiation (LTP); Neuroendocrine Aging: Pituitary–Gonadal Axis in Males; Neuroendocrine Aging: Hypothalamic–Pituitary–Gonadal Axis in Women; Rodent Aging; Spatial Memory: Assessment in Animals; Synaptic Plasticity: Learning and Memory in Normal Aging.

Further Reading

Bachevalier J, Landis LS, Walker LC, et al. (1991) Aged monkeys exhibit behavioral deficits indicative of widespread cerebral dysfunction. *Neurobiology of Aging* 12: 99–111.

Barnes CA, Suster MS, Shen J, and McNaughton BL (1997) Multistability of cognitive maps in the hippocampus of old rats. *Nature* 388: 272–275.

Burke SN and Barnes CA (2006) Neural plasticity in the ageing brain. *Nature Reviews Neuroscience* 7: 30–40.

Gallagher M and Rapp PR (1997) The use of animal models to study the effects of aging on cognition. *Annual Review of Psychology* 48: 339–370.

Hedden T and Gabrieli JD (2004) Insights into the ageing mind: A view from cognitive neuroscience. *Nature Reviews Neuroscience* 5: 87–96.

Morrison JH and Hof PR (1997) Life and death of neurons in the aging brain. *Science* 278: 412–419.

Rapp PR and Gallagher M (1996) Preserved neuron number in the hippocampus of aged rats with spatial learning deficits. *Proceedings of the National Academy of Sciences of the United States of America* 93: 9926–9930.

Sherwin BB (2006) Estrogen and cognitive aging in women. *Neuroscience* 138: 1021–1026.

Tanila H, Shapiro M, Gallagher M, and Eichenbaum H (1997) Brain aging: Changes in the nature of information coding by the hippocampus. *Journal of Neuroscience* 17: 5155–5166.

Aging and Memory in Humans

A M Brickman and Y Stern, Columbia University, New York, NY, USA

Memory Systems

Memory is the explicit or implicit recall of information encoded in the recent or distant past. Current conceptualizations of memory, however, do not view the construct as a unitary system but rather divide it into hierarchical taxonomic modules based on duration of retention and the type of information that is being retrieved. Among the more fully elucidated conceptualizations of memory systems is one characterized by Larry Squire and colleagues, in which long-term memory is divided into declarative and nondeclarative subcomponents. Declarative, or explicit, memory refers to the ability to consciously recall facts (semantic memory), events (episodic memory), or perceptual information (perceptual memory). Nondeclarative memory requires the implicit recall of information and is usually divided into procedural, priming, or simple conditioning paradigms. Information that is retained on the order of seconds or minutes is usually referred to as short-term memory and is thought to represent a memory system distinct from long-term memory. Working memory, which comprises short-term memory, refers to the short-term store required to perform certain mental operations during retention. The following sections examine the impact of normal aging on different types of memory, as well as some of the potential moderators and mediators of cognitive aging. The information presented is organized hierarchically, following the memory systems just outlined.

Increased age puts an individual at risk for the development of neurodegenerative disorders, such as Alzheimer's disease (AD). Central to the dementia syndrome that characterizes AD is the gradual and progressive loss of long-term memory functions. Although the vast majority of older adults do not develop dementia, most experience some degree of cognitive change. Following the elucidation of memory systems and their component parts in the cognitive and cognitive neuroscience literature, there has been a recent interest in the impact of age on the different memory systems, independent of the devastating effects of dementia. Among well-screened individuals who do not meet diagnostic criteria for dementia, both cross-sectional and longitudinal studies demonstrate that the different memory component systems do not uniformly age; rather they show differential vulnerability to aging effects.

Long-Term Memory

Declarative Memory

Semantic memory As noted, semantic memory refers the recall of general or factual knowledge. Older adults commonly complain of subjective semantic memory problems when, for example, they report difficulty recalling the names of common objects or other well-learned information. Yet, despite these subjective complaints, semantic memory is among the more stable memory systems across the adult life span. The construct can be operationally defined by requiring subjects to define words or provide the answers to factual questions (e.g., the capital of a certain country), such as on the Vocabulary and Information subtests of the Wechsler Adult Intelligence Scales. Semantic memory is often included as part of the definition of 'crystallized intelligence,' which reflects an accumulation of information acquired over time and that is relatively impermeable to the effects of normal aging or mild brain disease.

It is well established that semantic memory shows very little decline in normal aging. In fact, semantic knowledge accumulation and memory increase into the sixth and seventh decades of life and may show only a gradual decline afterward. Much of our understanding of the impact of age on semantic memory has come from large-scale longitudinal and cross-sectional studies of normal aging. For example, longitudinal data from the Canberra Study , which followed a random sample of adults over the age of 70 years, demonstrated that crystallized abilities remained stable over approximately 8 years. This pattern of stability was apparent when age-associated differences (i.e., cross-sectional analysis) were considered as well. Similarly, Denise Park and her colleagues measured knowledge-based verbal ability, including three semantic memory tasks tapping word knowledge, in a large sample of healthy adults ranging in age from 20 to 92 years, and found a gradual increase in performance across the age groups. This finding is again consistent with the idea that semantic knowledge accumulates across the life span with little or no deleterious effects of normal aging.

Although there is little cross-sectional or longitudinal evidence to suggest that semantic memory changes significantly with normal aging, why are subjective complaints of semantic recall so common among older adults? One phenomenon, termed 'tip-of-the-tongue' (TOT), may explain this occurrence. TOT refers to the common experience in which individuals have the feeling that they know the correct information (e.g., a person's name, a relatively low-frequency word), yet they are unable to recall it explicitly. The frequency of the TOT phenomenon increases with advancing age and may underlie the perceived difficulty with semantic recall.

Episodic memory The contrast between episodic memory and semantic memory was introduced by Endel Tulving in the early 1970s. Episodic memory refers to the explicit recollection of events, the 'what,' 'where,' and 'when' of information storage, and though it is conceptually distinct from semantic memory, the two memory systems interact. Episodic memory binds together items in semantic memory to form conceptually related time-based events. For example, the explicit recall of a learned story about a cowboy requires episodic memory for the story and semantic memory, or prior knowledge, of the items contained within the story.

Unlike semantic memory, episodic memory declines considerably with age. Older adults, for example, when prompted, have more difficulty recalling what they had for breakfast than do younger adults. The formal observation that episodic memory is affected by aging has existed for decades and has been documented numerous times.

Episodic memory is typically tested by requiring persons to learn information explicitly (e.g., a list or story) and recall it after a delay period. The three aspects of episodic memory include the encoding phase, the storage phase, and the retrieval of the encoded and stored information. These three phases show differential aging effects. Older adults' overall difficulty on tasks of episodic memory may be partially accounted for by a more shallow depth of encoding, as compared to younger adults – that is, older adults recall less information because of more limited processing of the initial study stimuli. This idea is supported by findings of greater age-related decline in acquisition or early retrieval of new information than in the degree of forgetting (i.e., the amount of information lost relative to the amount of information encoded). However, on free-recall tasks of list-learning paradigms, older adults recall fewer absolute words than do their younger adult counterparts; when given the correct stimulus in a recognition paradigm, older adults tend to incorrectly endorse more distractor stimuli, or foils.

Several observations about episodic memory and aging have emerged from the recent literature. First, the pattern of age-related differences in episodic memory appears to be similar across several modalities and domains, such as story recall, paired associate learning, face and word recall, and recognition paradigms of verbal and nonverbal information. Second, cross-sectional data suggest that age-associated episodic memory decline begins as early as age 20 years and decline linearly until about age 60, at which time there is a more precipitous decline. Third, whether the amount of interindividual variability increases systematically as a function of age is still somewhat unclear. Greater variability with aging would suggest that episodic memory decline might be a marker for insipient brain pathology, rather than a primary aspect of normal aging.

There are several competing theories postulating the potential mechanisms for age-associated declines in episodic memory. They can generally be divided into four areas, including an age-related failure of memory monitoring, or metamemory; age-associated decreases in the depth of initial encoding; age-related impairment in processing of contextual information; and age-associated decline in a number of processing resources. While these theories have not been fully substantiated empirically, the latter has received the greatest amount of support in the cognition literature. Proponents of a 'resource reduction hypothesis' argue that central to age-associated changes in episodic memory is a reduction in primary cognitive resources such as attention or working memory, or a reduction in the ability to engage due to an age-associated diminution in attentional inhibitory control. Others argue that age-associated memory decline is not due to a reduction in available attentional resources *per se*, but rather to age-related declines in perceptual processing speed.

Source memory Episodic memory comprises the information that is being recalled as well as the context in which the information was learned. This latter aspect is referred to as source memory, and there is increasing evidence that even when older adults successfully recall information, they may have difficulty recalling the source in which the information was acquired. The phenomenon has been demonstrated in the identification of the temporal context in which an item was learned, as well as the spatial and perceptual context. Studies by Daniel Schacter and colleagues required older adults to listen to different speakers read different blocks of declarative statements, and found that memory for the source of the declarative sentences was disproportionately worse than was memory for the statements. This general finding has been well replicated and has been extended to demonstrate age-associated impairment in both specific-source memory and partial-source memory. For example, older adults have difficulty, relative to younger adults, remembering which of four people spoke a word (i.e., specific-source memory) as well as remembering partial information about the person who spoke the word, such as his or her gender (i.e., partial-source memory).

Nondeclarative Memory

Nondeclarative, or implicit, memory describes the memory system that allows learning outside of conscious awareness. It is generally divided into procedural memory and priming, each hypothesized to be mediated by a distinct neurobiological system. In general, nondeclarative systems are relatively spared across the adult life span, particularly compared to episodic memory, which shows the greatest aging effects.

Procedural memory Procedural or skill learning is one type of nondeclarative memory that refers to the nonconscious acquisition of motoric sequences. A common example of procedural memory is the process of learning how to drive an automobile. Initially, a novice driver needs to recollect consciously how to control each aspect of the automobile and the sequence in which to do so. With practice, the individual skill required to operate the vehicle enters into procedural memory, and control becomes relatively more automatic.

Little work has been conducted that explicitly examines the impact of chronological age on procedural memory, and results have been somewhat inconclusive; some studies show a decline with age and others show no apparent aging effect. A potential confounding factor in the examination of procedural memory among older adults is how much the experimental paradigm draws upon pure motoric speed and on other nonprocedural cognitive abilities, such as working memory, as well as whether the experimental paradigm addresses procedural learning versus long-term procedural memory. Studies that examine the effects of age on experts, such as pianists or typists, show a general trend of age-associated slowing, but a relative maintenance in measures of performance. Age-associated preservation of procedural memory may also be dependent on the amount of deliberate practice.

When considering age effects on procedural memory, it is important to dissociate performance time, learning, and memory for the task. For example, older adults tend to perform procedural memory tasks more slowly and learn procedural sequences at a slower rate, compared to younger adults. However, after initial acquisition, older adults will relearn a procedural memory task at rates similar to or more rapid than the rates at which younger adults relearn, even after a 2-year interval without practice. While procedural memory and aging remain somewhat understudied, there is some consensus that older adults have lasting preservation of procedural or motor memory. It is important, however, to distinguish between age-associated changes in motoric or perceptual processing speed and age-related changes in procedural memory; when accounting for aspects of the former, the latter appears to be relatively spared.

Priming Repetition priming is a special type of implicit memory that refers to the implicit impact that prior exposure to a stimulus has on later test performance. For example, an individual is more likely to complete the word stem STR__ with ONG (to form the word STRONG) than with EET (to form the word STREET) if he/she had previously studied the word STRONG. There is a long history of the examination of aging effects on priming in multiple modalities, with somewhat mixed results. In general, studies from the 1980s suggested that there are few differences between younger and older adults for priming across a number of tasks, such as word stem completion, picture naming, and word identification. There is evidence of a strong dissociation in the elderly between a preserved ability to perform implicit tasks and a deficit in performance on declarative tasks when compared to younger participants, although there are small, but reliable, age-associated decrements in priming for verbal abilities regardless of whether the dependent measure is accuracy or latency or whether the type of priming is item or associative.

Investigators such as John Rybash have distinguished among five types of priming, including (1) perceptual-item priming, (2) perceptual-associative priming, (3) conceptual-item priming, (4) conceptual-associative priming, and (5) perceptual-motor priming. Perceptual tasks are dependent upon implicit recall of properties of the stimuli presented during encoding, whereas conceptual tasks refer to priming for object meaning. There is some evidence that with normal aging there is relative preservation of performance on tasks that require perceptual priming and a relative decline in tasks that require conceptual priming. However, other investigators, such as Debra Fleischman and John Gabrieli, argue that studies that have shown no aging effect for priming may have had insufficient measurement reliability or power to detect age differences, and positive studies may have included a minority of individuals with insipient dementia. In contrast to Rybash, they argue that perceptual priming is vulnerable to aging effects, whereas conceptual priming remains relatively intact. Differences between these two conclusions may reflect differing positions on inclusion criteria for categories of priming.

For example, Fleischman and Gabrieli exclude all tasks that might include the processes that explicit memory retrieval tasks also engage. Further, they consider tasks such as category-exemplar generation to be conceptual, whereas Rybash considers these to be perceptual. In summary, although the literature is somewhat mixed, the effects of normal aging on repetition priming are of small magnitude or nonexistent, depending on modality.

Short-Term Memory

In contrast to long-term memory, in which information is stored for minutes to years, short-term memory refers to holding information in conscious awareness for the duration of seconds. While short-term memory is qualitatively a type of memory distinct from long-term memory, the two systems interact. For example, in order for information to be stored in long-term memory, the information must first exist within short-term memory. The exact mechanisms by which short-term memory is transferred to long-term storage have yet to be fully elucidated, but the 'modal model' of information transfer, proposed by Atkinson and Shiffrin in the 1960s, provided a theoretical framework that has dominated the cognitive literature for decades. Briefly, information flows from the environment to primary sensory (perceptual) stores and then into a short-term store, which can include rehearsal, coding, or decision. The capacity of the short-term store is limited and information can leave the system (i.e., forgetting), lead to a response output, or enter long-term storage, which is fairly permanent. When information is drawn from long-term memory, it exists in short-term memory while in conscious awareness. In terms of the capacity of short-term memory, a heuristic offered by Miller in the 1950s is that the average short-term memory span for a person is seven, plus or minus two items, or chunks. Although this heuristic has remained popular, more recent studies suggest that short-term memory span is closer to four items, or is contingent upon an individual's processing speed.

The term 'short-term memory' refers to the passive short-term store of information, but speaks little to the process by which information is retained in the store. It is difficult to discuss the effects of age on short-term memory without consideration of these processes. The term 'working memory' is typically applied to the process by which information is held in short-term memory and refers to the cognitive manipulation of information that is contained within short-term memory. An influential model that attempts to explain how information is stored within short-term memory was proposed by Baddeley and Hitch, in which a supervisory 'central executive' controls the information flow among three 'slave systems,' including the phonological or articulatory loop, the visuospatial sketchpad, and the episodic buffer. The first two systems rehearse phonological or visual information, respectively, and the third system integrates information from the other two systems and interfaces with long-term memory. The central executive acts to coordinate the flow of information among the slave systems, to direct attention to appropriate internal and external stimuli, and to suppress irrelevant stimuli.

Working memory paradigms require individuals to perform mental operations on items held in conscious awareness, such as reordering words or numbers. While there is little evidence that short-term memory *per se* significantly declines with normal aging, there is ample evidence that working memory abilities do. Within working memory, efficacy of inhibition and smaller span are particularly vulnerable to the effects of aging.

Working memory in aging studies is often operationally defined by tasks such as letter rotation, reading span, computation span, and line span. Performance on these tasks tends to show a linear age-associated decline that is similar to that seen for tasks of long-term episodic memory and speed of processing. Some research suggests that age-associated decline in working memory abilities mediates age-associated decline in other memory and cognitive domains. Indeed, statistical control of performance on tasks of working memory often attenuates the observed age-associated decline on other cognitive tasks.

Summary and Course

Normal age-associated memory decline is not uniform. Older adults evidence worse performance on long-term memory tasks compared to younger adults, but these differences are relatively greatest on tasks of declarative episodic memory. Semantic memory, on the other hand, remains relatively stable across the adult life span or may even increase as more semantic knowledge is accumulated with age. Similarly, working memory, or the manipulation of information that is held in conscious awareness, shows marked decline in with normal aging, and some theorists propose that working memory deficits mediate age-associated decline in other cognitive domains.

In terms of the course of memory changes across the adult life span, results from cross-sectional and longitudinal studies suggest that subtle memory changes can begin as early as the early or middle twenties and continue to decline linearly with age. Some authors distinguish between 'lifelong decline' and 'late-life decline.' Performance on tasks of episodic and working memory seems to begin to decline in the twenties and continues to decline linearly across the life span, which is supported by cross-sectional aging and cognition studies. Some longitudinal studies, however, suggest a curvilinear course of memory decline, with a more precipitous decline after about age 60 years, preceded by relatively little decline with age. Short-term memory store appears to remain relatively stable until about age 70, at which point it begins to drop, and, as noted, semantic abilities appear to remain relatively stable, at least until late life.

Moderators and Mediators of Cognitive Aging

A consistent observation in the aging and cognition literature is a greater amount of variability in memory performance with increasing age. This finding is evident both when age is considered as a cross-sectional variable and when individual age trajectories are examined over time; some adults experience great amounts of cognitive decline while others experience relatively little. Increased variability suggests that there are moderators and mediators of cognitive aging. Two factors in particular, including cerebrovascular or cardiovascular risk and cognitive reserve, have received considerable attention as modulators that may account for some of the increased variability, and these are potential targets for intervention or prevention of cognitive decline.

Cerebrovascular or cardiovascular risk factors may mediate the relationship between chronological age and the neurobiological changes that underlie cognitive aging. Increased blood pressure is associated with reduced psychomotor speed, visuoconstruction ability, learning, memory, and executive functioning. Similarly, both insulin resistance and diabetes are associated with diminished cognitive abilities in later life. Hyperlipidemia has also been identified as a potential cerebrovascular risk factor that negatively impacts cognition in older adults. While it is still somewhat unclear by what mechanism cerebrovascular or cardiovascular risk factors impact cognition, it is likely that they interact with age to have a cumulative effect on cognition in later life. Furthermore, intervention studies that target the vascular system, such as cardiovascular fitness regimens, are associated with increased cognitive functioning among older adults.

Cognitive reserve is another factor that may account for the increased variability observed in cognitive aging. The cognitive reserve hypothesis stems from the observation that there is a disconnection between degree of brain pathology and its cognitive manifestation and suggests that the brain actively attempts to cope with brain pathology by using existing processing approaches or by recruiting compensatory networks. Although cognitive reserve is often applied to clinical entities, such as AD or stroke, it may be operative among the normal aged. That is, cognitive reserve may moderate the relationship between normal age-associated neurobiological changes and their cognitive outcome. Indeed, proxies for reserve, such as measures of education or literacy and IQ, are related to the degree of age-associated memory decline. Declarative memory scores of older adults with lower levels of literacy decline at a greater rate than do those with higher levels of literacy. Like vascular risk factors, interventions that target cognitive reserve may improve the course of cognitive aging.

Other factors, such as genes and nutritional exposure, could potentially impact the course of memory decline with normal aging. It is also important to note that the possibility of inclusion of individuals with insipient dementia may have contaminated some studies that attempted to elucidate the pattern of age-associated memory decline, and future studies of normal aging should focus on disentangling normal cognitive decline from pathological aging at the earliest point possible. With the powerful capabilities of neuroimaging techniques, we can begin to understand the complex relationships among normal aging, neurobiology, and cognition while defining potentially modifiable moderators and mediators of cognitive aging.

See also: Aging of the Brain; Aging of the Brain and Alzheimer's Disease; Aging and Memory in Animals; Basal Forebrain and Memory; Cognition in Aging and Age-Related Disease; Episodic Memory; Functional Neuroimaging Studies of Aging; Humans; Lipids and Membranes in Brain Aging; Metal Accumulation During Aging; Short Term and Working Memory.

Further Reading

Atkinson RC and Shiffrin RM (1968) Human memory: A proposed system and its control processes. In: Spence KW and Spence JT (eds.) *The Psychology of Learning and Motivation: Advances in Research and Theory*, vol. 2, pp. 89–195. London: Academic Press.

Baddeley AD (2000) The episodic buffer: A new component of working memory? *Trends in Cognitive Science* 4: 417–423.

Christensen H (2001) What cognitive changes can be expected with normal ageing? *Australian and New Zealand Journal of Psychiatry* 35: 768–775.

Cowan N (2001) The magical number 4 in short-term memory: A reconsideration of mental storage capacity. *Behavioral and Brain Sciences* 24: 87–114; discussion 114–185.

Craik FIM and Salthouse TA (eds.) (2000) *The Handbook of Aging and Cognition,* 2nd edn. Mahwah, NJ: Lawrence Erlbaum.

Fleischman DA and Gabrieli JD (1998) Repetition priming in normal aging and Alzheimer's disease: A review of findings and theories. *Psychology and Aging* 13: 88–119.

Grady CL and Craik FI (2000) Changes in memory processing with age. *Current Opinion in Neurobiology* 10: 224–231.

Hedden T and Gabrieli JD (2004) Insights into the ageing mind: A view from cognitive neuroscience. *Nature Reviews Neuroscience* 5: 87–96.

Krampe RT and Ericsson KA (1996) Maintaining excellence: Deliberate practice and elite performance in young and older pianists. *Journal of Experimental Psychology: General* 125(4): 331–359.

LaVoie D and Light LL (1994) Adult age differences in repetition priming: A meta-analysis. *Psychology & Aging* 9: 539–553.

Light LL (1991) Memory and aging: Four hypotheses in search of data. *Annual Review of Psychology* 42: 333–376.

Park DC, Smith AD, Lautenschlager G, et al. (1996) Mediators of long-term memory performance across the life span. *Psychology & Aging* 11: 621–637.

Rybash JM (1996) Implicit memory and aging: A cognitive neuropsychological perspective. *Developmental Neuropsychology* 12: 127–179.

Salthouse TA (1993) Speed mediation of adult age differences in cognition. *Developmental Psychology* 29: 722–738.

Schacter DL, Kaszniak AW, Kihlstrom JF, et al. (1991) The relation between source memory and aging. *Psychology & Aging* 6: 559–568.

Small SA, Stern Y, Tang M, et al. (1999) Selective decline in memory function among healthy elderly. *Neurology* 52: 1392–1396.

Squire LR (2004) Memory systems of the brain: A brief history and current perspective. *Neurobiology of Learning and Memory* 82: 171–177.

Stern Y (2002) What is cognitive reserve? Theory and research application of the reserve concept. *Journal of the International Neuropsychological Society* 8: 448–460.

Stern Y (ed.) (2006) *Cognitive Reserve: Theory and Applications.* New York: Taylor & Francis.

Tulving E (2002) Episodic memory: From mind to brain. *Annual Review of Psychology* 53: 1–25.

Aging of the Brain

A B Scheibel, Brain Research Institute UCLA, Los Angeles, CA, USA

Possible Causal Factors

Experience indicates that all living things have finite lives, whether they be the ephemeral summer of the butterfly, a thousand days for the rat, the reported hundred-year-long lives of large tortoises, or the multi-thousand-year histories of giant redwoods and bristle-cone pines. The concept of genetic programming of the aging process seemed to receive support from Hayflick's studies of the late 1950s and early 1960s, in which young, actively dividing connective tissue cells raised *in vitro* appeared limited to a certain maximum number (50) of mitotic divisions, after which degenerative changes invariably set in. In retrospect, these classic experiments now seem less convincing, because of methodologic errors involved, and, although the concept of genetic control remains intuitively appealing, the case is considered far from proven. Nonetheless, the capacity for base excision DNA repair has been shown to be reduced in the brains of aging primates. Furthermore, the activity of polymerase beta in brain cells appears compromised with age. These findings may be bellwethers for future more widely ranging age-related changes in genomic capabilities.

A presently favored alternative theory operating either independently or, more likely, in tandem with the proposed genetic constraints envisages the possibility of progressive damage to the cell from external or internal factors, resulting in a cumulative pattern of dysfunctions. For example, toxic factors within the environment (i.e., chemical carcinogens and background radiation of terrestrial or galactic origin) may slowly affect the cell DNA content and protein-synthesizing machinery, leading to errors in synthesis with resultant progressive changes in cell structure and function. This summation of many small developing errors in the synthetic and enzymatic machinery of the cell (such as the progressive degradation of base excision DNA repair mentioned above) is conceived as mounting to a point beyond which conditions for cell life become impossible (error-catastrophe theory).

Obviously, this putative process might complement and potentiate a set of existing genetic instructions also directed toward the cell's eventual demise. However, normal oxidative metabolic activity of the cell may itself result in cumulative damage. Of particular interest is the development of a family of free radicals of oxygen (e.g., singlet oxygen, hydrogen peroxide, superoxide and hydroxyl radicals) that, through crosslinkage or cleavage reactions, may permanently alter the structure of the cell. In support of this are observations that mitochondrial free radical generation is lower in long-lived species than in short-lived ones. The varied but often overlapping phenomena of repeated stress/glucocorticoid-induced cell damage, especially in the hippocampus, excitotoxic activity of neurotransmitters like glutamate, and programmed cell death (apoptosis) may all contribute to the process. These ideas are still in early stages of development and exemplify the questions that surround the processes of aging in general, and of the nervous system in particular.

Structural Changes

A number of changes, both gross and microscopic, have been reported in the brains of the aged, although variation is the rule rather than the exception. The brain itself is often somewhat reduced in size and weight, especially beyond the eighth decade of life. Decrements of 10–15% are frequently quoted, although interindividual differences are large. Some aged brains show mild to moderate gyral atrophy and sulcal widening, but this is not the rule. The surrounding meninges are frequently more opaque and milky in appearance than those of the young brain, and may be adherent to underlying cortex. Isolated deposits of calcium are found in and around the pacchionian granulations near the vertex of the hemisphere. The ventricular system may show alterations in silhouette, and computed axial tomography and magnetic resonance imaging (MRI) scans often show enlarged ventricular shadows. A large and growing series of studies using T1- and T2-weighted MRIs reveal considerable variation in the aged brain. Although some apparently normal individuals in their seventh and eighth decades show signs of cortical atrophy, ventricular enlargement, and even textural changes in the cerebral white matter near the ventricles (leukoairosis), others show little or no discernible alteration. Recent MRI studies indicate that a slowly developing loss of gray matter, starting as early as the third decade, is accompanied by increases in white matter volume into the mid- or late 40s, followed then by degeneration. Similar studies at the regional level suggest decrement in size of the hippocampus bilaterally in men but not in women. Localized periventricular white matter changes, in particular, are increasingly thought to result from underlying processes such as hypertension or diabetes, and to have little or no relevance to normal brain aging.

Microscopically, a group of well-known stigmata is seen in routine Nissl or reduced silver-stained preparations. Their incidence varies rather widely among individuals, although there is a general tendency toward increased frequency with age. Neuronal cell bodies gradually accumulate masses of refractile granules with high lipid content, known as lipofuscins. The significance of these so-called aging pigments is not clear, and they have been found as early as the tenth year of life in cells of the inferior olive. The present consensus is that they represent the remains of lysosomal and mitochondrial membranes that have accumulated, due perhaps to gradual failure of mechanisms for turnover and reutilization. As such, they do not necessarily constitute a direct threat to the neuron except to the extent that they preempt increasing amounts of cytoplasmic space previously used for synthesis of glycoproteins, lipoproteins, and neurotransmitters. One interesting and opposing view maintains that the appreciable content of myoglobin and respiratory enzymes allows lipofuscin granules to serve a positive role in providing energy to neurons under conditions of low oxygen tension. Coarser vacuolization of the cytoplasm, usually concentrated in the area of development of the apical dendrite shaft in cortical pyramidal cells, is known as 'granulovacuolar degeneration of Simchowitz' and is of equally unknown etiology.

The neurofibrillary tangle is a structural alteration of neuronal cytoplasm that may involve soma, dendrites, and axon. Initially recognized by Alzheimer and Simchowitz as one of two defining microcriteria of individuals dying with dementia, it is also found in very small numbers in the normal aged. Electron microscope study reveals that the tangle consists of paired helical filaments, about 22 nm in width, which braid loosely around each other with a characteristic series of partial twists, each 80 nm from the next. Tangles are now believed to result from abnormal phosphorylation of tau protein, which is necessary to normal polymerization–depolymerization mechanisms of microtubule-associated proteins. Microtubules provide the infrastructure that facilitates virtually all movement of neurotransmitters, structural proteins, enzymes and endocellular organelles within the neuron and its extensive processes (axonal and dendritic transport). It is therefore easy to imagine how the buildup of neurofibrillary tangles can progressively starve and throttle the neuron. The appearance of tangles (tau aggregates) in the frontal lobe appears to correlate positively with cognitive loss. Experimental neurofibrillary tangle development is being extensively studied in primates, in small laboratory animals, and even in *Drosophila*.

Accompanying these characteristically intracellular alterations is the development of small foci of destruction in the surrounding neuropil, the senile plaque of Alzheimer. These are classically described as containing central cores of Congo red-positive, amyloid-like material, surrounded by radial auras of degenerating dendritic and axonal tissue and a halo of microglia. They constitute the second component of the two major histopathologic criteria of dementia and, like neurofibrillary tangles, are present in more restricted number in the brains of most aged individuals. The density of senile plaques has been said to correlate positively with the degree of cognitive impairment in dementia, although these data have recently been subject to question.

Intensive biochemical and immunocytologic analyses of these structures are presently under way. It is now believed that beta amyloid (a 40- to 42-amino-acid peptide) may represent an unusual by-product of a larger, normally present molecule, called 'amyloid precursor protein', and that production and deposition may result from aberrant enzymatic (secretase) activity. Some data suggest that accumulations of beta amyloid may drive molecular cascades that potentiate neurodegenerative changes, thereby representing a direct threat to the integrity of surrounding neuronal structures. Several experimental interventions are being explored in an effort to prevent its development or mitigate its effects.

All the histopathologic changes so far described appear most intensively in the limbic system, especially the entorhinal cortex and hippocampus. They are also found widely throughout the rest of central nervous system, however, including cerebral neocortex, diencephalon, brain stem, and spinal cord. Since interindividual variation is the rule in the aged brain, these descriptions represent a distillation and summary rather than the expected picture of the brain of any one aged individual.

The problem of neuronal loss has been hotly debated and is still not settled. The few quantitative studies dating from the late 1950s and 1960s suggested neuron loss of up to 30% in the normal aging cerebral neocortex. More detailed recent investigations, based on larger numbers of cases, indicate that such levels of neuronal loss, in normal aging at least, may have been excessive, based as they were on the study of a few areas in a limited number of brains. More broadly sampled data suggest a pattern of modest loss of neurons, primarily large cortical cells, with compensatory dendritic growth in adjacent neurons. The cholinergic cell masses of the basal forebrain and the noradrenergic cells of the locus coeruleus undergo undoubted change during the aging process, but even here, declining function may be more an expression of waning metabolic vigor and decreased axodendritic dimensions than of massive neuronal loss. This is in sharp contrast to dementing syndromes such as Alzheimer's disease, in which more than half of the complement of locus coeruleus cells may disappear.

Carefully controlled studies indicate that in the rat the only significant cerebral cell losses occur during the first 100 days of life in what might be considered a period of adaptation and fine-tuning to the environment. After this, there is little further discernable neuronal loss, even at 900 days of age, when rats are beginning to die of natural causes. Although the weight of evidence from both animal and human studies now increasingly points to negligible neuronal loss during the normal, uncomplicated aging process, a large number of cells appear to undergo changes in the dendritic (and axonal) extensions of the soma. Many neurons show progressive restriction and atrophy of their more peripheral dendrite branches and, especially in cortical pyramidal cells, among the basilar shafts. Accompanying this is irregular loss of dendrite spines and frequent beaded swellings along the remaining dendritic branches. These changes can be related in general terms to progressive loss of protein-synthesizing capabilities due, perhaps in part, to increasing incursions upon cytoplasmic space by lipofuscin deposits and neurofibrillary tangles. However, it has also been shown that the potential for neuronal growth is not lost during aging. Accompanying the progressive destruction of some dendritic systems, it appears that other neurons grow further dendritic extensions, thereby increasing their available synaptic areas (**Figure 1**). The concept of two types of neuronal response to aging, one involving dendritic retraction and one reactive dendritic expansion, brings with it a number of exciting implications for providing more effective and fulfilling lives for the elderly.

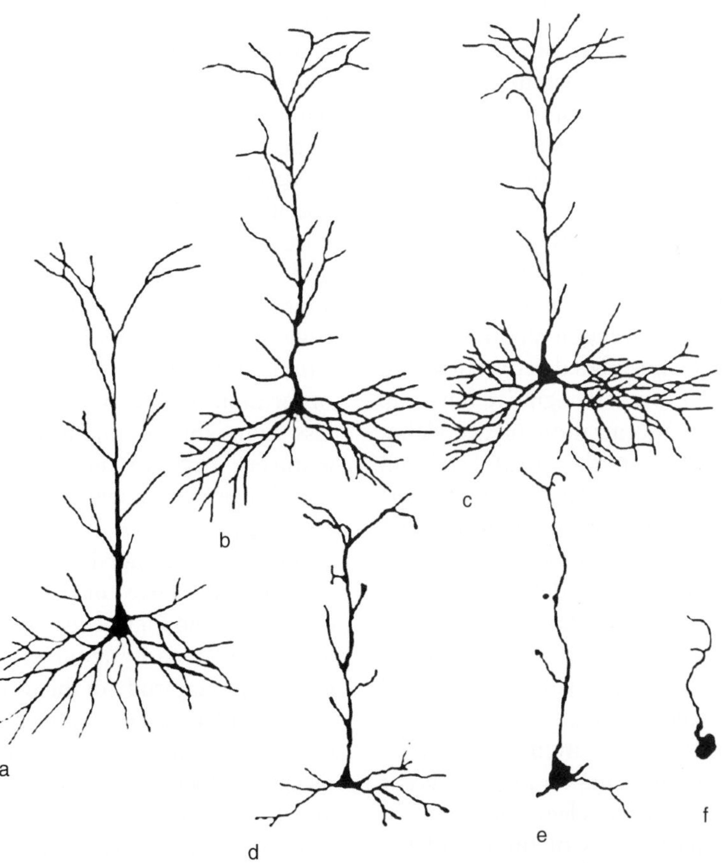

Figure 1 Two possible patterns of age-related alterations in cortical pyramidal cells. The normal mature neuron (a) may show regressive dendritic changes characterized by loss of basilar dendritic branches and eventual loss of the entire dendritic tree (d, e, f). Other neurons (b, c) may show progressive increase in dendritic branching. Drawing based on Golgi impregnations.

During the first half of the twentieth century, brain aging and dementia were usually associated with visible changes in the major arteries of the central nervous system. Cerebral arteriosclerosis was considered a necessary accompaniment to and, in fact, virtually synoymous with aging and senility. Careful studies during the 1950s showed the inaccuracy of such concepts, and today, large-vessel disease is seldom considered to contribute significantly to the picture of general brain aging. Recent investigations with positron emission tomography scanning methods and xenon clearance techniques have again called attention to the relative adequacy of blood flow and oxygen and glucose metabolism in the healthy aging brain. The scanning electron microscope and immunohistochemical studies have, coincidentally, focused interest on the microcirculation (capillary bed) of the brain and on the plexus of neural fibers that normally innervate their walls. Subtle changes in this delicate but all-pervasive system may prove to have significant impact on the maintained vigor or decline of brain structure and function.

Biochemical and Metabolic Changes

It is becoming clear that the aging process entails a broad range of biochemical and metabolic changes. Among these, alterations in neurotransmitter content and activity figure prominently. As many as 90% of neurons are involved, directly or indirectly, in cholinergic mechanisms, synthesizing, transporting, and releasing – or else being synaptically dependent upon – acetylcholine. Cholinergic systems are highly energy dependent, and age-related decreases in several important glycolytic enzymes have been reported. A link might thus conceivably be postulated between altered glycolytic energy mechanisms and cognitive function.

The most important concentration of acetylcholine-rich neurons is found in the basal forebrain area, in and around the nucleus basalis of Meynert and the ventral pallidum. Most studies indicate mild to moderate age-related cell loss in these areas. Associated with (although not entirely dependent upon) this loss are decreases in acetylcholine content in various portions of the brain, lower levels of the synthesizing enzyme, choline acetyltransferase, and decreased numbers of acetylcholine receptor binding sites. The relevance of these alterations to cognitive changes in the elderly, such as impairment of recent memory function, is not entirely clear. Of uncertain import, also, is the apparent loss of fluidity of the cytoplasmic membrane of the cell as the lipoprotein structure changes, a possible result of progressive diminution of choline content. Age-related increases in membrane microviscosity may bring with them a significant train of sequelae including, initially, a higher capacity for receptor binding followed by enhanced rates of receptor loss with eventual overall reduction in receptor binding affinity. Mechanisms responsible for these changes remain uncertain, but it has been postulated that increased microviscosity of the cell membrane leads first to increased exposure of those receptors out of the plane of the membrane, after which the uncovered receptors are progressively sloughed off into the surrounding medium. This suggests that membrane-bound proteins such as receptors and transporters are maintained in dynamic equilibrium by the extent of fluidity of the surrounding membrane lipids. As one example, immunohistochemical studies of a glucose transporter (GLUT 3) revealed significant loss (46%) in the hippocampal-dentate gyrus of the aged rat. Attempts to forestall or modify age-related losses in membrane fluidity through administration of active lipid fractions have been under investigation.

Catecholaminergic systems also show moderate to marked alteration in the brain. Levels of both dopamine and norepinephrine synthesis decrease, as do the numbers of adrenergic and dopaminergic receptors. Marked attenuation of norepinephrine synthesis in the hypothalamus may, in turn, be responsible, at least in part, for decrease in the synthesis of certain hypothalamic hormones.

Significant cell loss has been reported in the locus coeruleus, the single most important source of brain parenchymal norepinephrine. The substantia nigra and adjacent ventral tegmental area (VTA), major sources of dopamine, also undergo histologic alterations that include microvascular changes, iron deposition, and neuronal loss. The serotonin-rich systems of the raphe seem, on the other hand, to be somewhat less vulnerable. Resultant maintenance of minimally disturbed titers of serotonin in a setting of falling catecholaminergic levels may conceivably be related to a high incidence of disturbed sleep patterns and depression in the aged.

A broad range of age-related neuroendocrine changes is known to exist. Although such alterations are documented by innumerable anecdotal observations, regional study of the mechanisms involved is in its early phases. For instance, the importance of the ovary as a pacing factor in the aging female rat is receiving documentation in many laboratories. At an equally obvious level, the high likelihood of occurrence of benign prostatic hypertrophy and prostatic cancer in the aged male is being related in meaningful fashion to age-related alterations in testosterone metabolism that may lead to unbalanced growth stimulation. In a more global sense, the entire aging process can apparently be slowed by early restriction of nutritional input, which seems to delay pubescence and maturation. Finally, increasing imbalance among the various neurotransmitters, and in particular among those that are aminergic, appears to exert progressive impact upon the hypothalamus, pituitary, and pineal gland, leading to cascades of dysfunction that manifest themselves at every level of organization.

Although the picture presented appears to emphasize regressive elements, a much more positive picture of the normal aging brain can now be painted. Accumulated experience can help make up for age-related loss in the speed of cerebral processing or of memory recall.

Experimental evidence conclusively demonstrates the continued plasticity of even the very old brain. Continual environmental enrichment and challenge appear to enhance cognitive power in the aged. In fact, it has been shown that a subset of those genes whose expression is adversely affected by aging is oppositely affected by exposure to enriched environments. Under conditions of optimally maintained health, increasing numbers of individuals in their 70s and 80s and beyond continue to hold responsible positions, or otherwise distinguish themselves in commerce, the arts, and certain aspects of the sciences. In every case, the touchstone appears to be activity, involvement, and purpose.

See also: Aging of the Brain and Alzheimer's Disease; Alzheimer's Disease: An Overview; Apoptosis in Nervous System Injury; Cognition in Aging and Age-Related Disease; Dementia; Gene Expression in Normal Aging Brain; Metal Accumulation During Aging.

Further Reading

Barja G and Herrero A (2000) Oxidative damage to mitochondrial DNA is inversely related to maximum life span in the heart and brain of mammals. *FASEB Journal* 14(2): 312–318.

Buell S and Coleman P (1979) Dendritic growth in the aged human brain and failure of growth in senile dementia. *Science* 206: 854–856.

Coffey CE, Wilkinson WE, Parashos IA, et al. (1992) Quantitative cerebral anatomy of the aging human brain: A cross-sectional study using magnetic resonance imaging. *Neurology* 42(3 pt. 1): 527–536.

Flood D and Coleman C (1988) Neuron numbers and sizes in aging brains: Comparisons of human, monkey and rodent data. *Neurobiology of Aging* 9: 453–463.

Lee J, Duan W, Long JM, Ingram DK, and Mattson MP (2000) Dietary restriction increases the number of newly generated neural cells and induces BDNF expression in the dentate gyrus of rats. *Journal of Molecular Neuroscience* 15(2): 99–108.

Rowe J and Kahn R (1987) Human aging: Usual and successful. *Science* 237: 143–149.

Salmon E, Maquet P, Sadzot B, Degueldre C, Lemaire C, and Franck G (1991) Decrease of frontal metabolism demonstrated by positron emission tomography in a population of healthy elderly volunteers. *Acta Neurologica Belgica* 5: 288–295.

Scheibel A (1996) Structural and functional changes in the aging brain. In: Birren J and Schaie K (eds.) *Handbook of the Psychology of Aging*, 4th edn., pp. 105–128. San Diego: Academic Press.

Ylikowski R, Ylikowski A, Erkinjuntti T, Sulkava R, Raininko R, and Tilvis R (1993) White matter changes in healthy elderly persons correlate with attention and speed of mental processing. *Archives of Neurology* 50: 818–824.

Aging of the Brain and Alzheimer's Disease

D L Price, A V Savonenko, M Albert, J C Troncoso, and P C Wong, The Johns Hopkins University, Baltimore, MD, USA

Introduction

Over the past decade, major advances have been made in understanding the causes and mechanisms of age-associated alterations in memory and cognition, including mild cognitive impairment (MCI) and Alzheimer's disease (AD), which is the most common cause of dementia in the elderly. For many reasons (prevalence, lack of mechanism-based treatments, cost of care, and impact on individuals and caregivers), AD is one of the most challenging illnesses in medicine. However, extraordinary progress has been made in deliniating the pathology, biochemistry, and neurobiology of the disease; the clinical and pathological bases of MCI and AD; the utility of new diagnostic approaches and outcome measures; and the value of transgenic models of the genetic forms of familial AD and gene-targeted models which discloses mechanism-based targets for therapy of AD. These discoveries are leading to new treatments for this type of dementia.

Cognitive and Memory Impairments in the Elderly

Although many older individuals remain intellectually intact, a significant number of the elderly show declines in memory and cognitive abilities. These alterations are usually mild, occur relatively late in life, and involve speed of learning, complex problem solving, ability to retain large amounts of new information, and visuospatial skills. Vocabulary, information storing, and comprehension skills often remain relatively stable into old age. Some older individuals have mild cognitive impairments (MCIs) in which memory loss exceeds that expected for age, but they do not meet criteria for AD. However, among individuals with MCIs, there may be considerable heterogeneity (i.e., interindividual variability in the rate and severity of progression). The variability of alterations in memory with age makes it sometimes difficult to determine, after a single examination, whether an individual with mild age-associated memory impairments will remain relatively stable or will progress to severe dementia. The most accurate clinical approach to this problem is to assess cognitive abilities and memory performance repeatedly in the same patient over a period of time. Several neuropsychological tests are valuable in predicting AD: delayed recall tasks (recently learned information), as assessed by the California Verbal Test and Wechsler Memory Scale, and measures of executive functions (ability to organize and plan), as assessed by the Trail-Making Test and Self-Ordering Test. On the basis of clinical assessments correlated with postmortem examinations of the brain, MCI is now regarded as a transitional state between the normal functional state of older persons and early AD. Imaging studies have predictive value regarding the progression to AD.

Alzheimer's Disease: Clinical Features, Diagnostic Studies, Neuropathology and Biochemistry, and Current Treatments

AD is the most common cause of senile dementia, a term that refers to a syndrome occurring in older individuals that results in memory loss and cognitive impairments of sufficient severity to interfere with social, occupational, and personal functions. This type of dementia affects more than 4 million people in the United States. Because of increased life expectancy and the postwar baby boom, this population is the fastest growing segment of society. During the next 25 years, the number of people with AD in the United States will triple, as will the cost of care and treatment.

A majority of individuals with sporadic AD exhibit the first clinical signs during their seventh decade. However, some people develop disease in midlife; in these cases, a family history of the illness is more likely. In both the sporadic and familial forms of AD, affected individuals show difficulties with memory, problem solving, executive functions for language, calculation, visuospatial perceptions, judgment, and behavior; mental functions and activities of daily living are increasingly impaired. Some patients develop psychotic symptoms, such as hallucinations and delusions. In the late stages, individuals are mute, incontinent, bedridden, and usually die of intercurrent medical illnesses.

Other causes of dementia syndromes include cerebrovascular disease, alone or in combination with AD; Lewy body dementia; Parkinson's disease; alcoholism; drug intoxications; infections, such as with human immunodeficiency virus (HIV) (i.e., acquired immunodeficiency syndrome (AIDS)) and syphilis; brain tumors; vitamin deficiencies (e.g., B_{12}); thyroid disease; and a variety of other metabolic disorders. Because some of these disorders are treatable, it is important for physicians to exclude these entities before making a diagnosis of possible AD. At present, except for brain biopsy, it is not possible to definitively establish the diagnosis of early AD in living humans, and it is extremely important for the physician to exclude other causes of dementia syndromes, because some of these illnesses respond to specific treatments. Imaging studies are of great value in exclusions and are of increasing utility in supporting the diagnosis of AD. At present, the only available treatment for AD is for symptoms.

As mentioned previously, to establish a diagnosis of AD, clinicians rely on histories from patients and informants; physical, neurological, and psychiatric examinations; neuropsychologic testing; laboratory studies, including examinations of cerebrospinal fluid (CSF); and a variety of laboratory tests, including neuroimaging studies. The clinical profile, in concert with a variety of imaging and laboratory assessments (levels of CSF β-amyloid (Aβ) decrease and tau proteins increase), allows the clinician to make a diagnosis of possible or probable AD. Several imaging strategies and studies of biomarkers, particularly of CSF, are useful laboratory approaches. In early AD, magnetic resonance imaging (MRI) often discloses atrophy of specific regions of the brain, while positron emission tomography (PET) with [^{18}F]deoxyglucose, or single-photon emission computerized tomography (SPECT), demonstrates decreased glucose utilization and early reductions in regional blood flow in the parietal and temporal lobes. PET, following administration of brain penetrant ^{11}C-labeled Pittsburgh compound B (PIB), which binds to Aβ with high affinity, discloses patterns which are thought to reflect the Aβ burden in the brain. In concert, these various approaches as applied to patients should increase the accuracy of diagnosis in earlier stages of disease and allow assessments of the efficacies of new antiamyloid therapeutics. The presence of the apolipoprotein E (apoE) ε4 allele confers risk in late-onset disease (see later); apoE genotyping is a useful research tool, but it is not helpful for routine diagnostic purposes in individual patients.

AD is the result of abnormalities associated with dysfunction and death of specific populations of neurons, particularly those cells in neural systems participating in memory and cognitive functions. Abnormalities in the brain selectively involve neurons in the neocortex, entorhinal area, hippocampus, amygdala, nucleus basalis, anterior thalamus, and several brain stem monoaminergic nuclei (in particular the locus coeruleus and raphe complex). Dysfunction of neurons in these brain regions and circuits is reflected by the presence of cytoskeletal abnormalities (neurofibrillary tangles (NFTs)) in these cells, the presence of neuritic plaques in brain regions receiving inputs from these nerve cells, and reductions in transmitter markers of these neurons in their target fields. The cellular pathology is characterized by the presence of intracellular and extracellular protein or peptide aggregates: phosphorylated tau is assembled into the paired helical filaments (PHFs) within NFTs and in swollen neurites (around plaques); and Aβ peptides exist in extracellular β-pleated sheet conformations assembled into oligomers, which are the principal constituent of amyloid plaques. Ultimately, affected neurons die; often 'tombstone' tangles, amyloid deposits, and glial reactions remain to indicate sites of neurodegeneration.

The clinical manifestations of amnestic mild cognitive impairment (aMCI) and early Alzheimer's disease (eAD) result from abnormalities occurring among populations of neurons in neural systems/brain regions essential for memory, learning, and cognitive performance. Damaged circuits include those in the basal forebrain cholinergic system, the amygdala, the hippocampus, the entorhinal and limbic cortex, and the neocortex. In a recent study, the character, abundance and distribution of the lesions (i.e., diffuse plaques, neuritic plaques, and tangles) were correlated with clinical signs in cognitively characterized controls, individuals with aMCI, or cases of eAD. No differences were observed in the number of diffuse plaques among the groups. In persons with aMCI, tangles were significantly increased in regions of the ventral medial temporal lobe as compared to controls; individuals with eAD showed greater numbers of NFTs and neuritic plaques in both frontal and temporal regions. Individuals with aMCI exhibited increased numbers of neuritic plaques in neocortical regions as compared to controls, but not as compared to cases of eAD. Memory deficits correlated most closely with the abundance of NFTs in CA1 of the hippocampus and in the entorhinal cortex, a finding which leads to the interpretation that tangles were more important than amyloid deposition in the progression from normal function to aMCI to eAD, and that tangles in the medial temporal lobe play a key role in memory declines in aMCI. Data from multiple studies are interpreted to indicate that aMCI reflects a transitional state in the evolution of AD. Because the regional distributions of NFTs correlate most closely with the degree of clinical impairments from aged healthy controls to individuals with aMCI to cases of AD, the spread of NFTs beyond the medial temporal lobe is thought to be linked to the development of dementia.

As mentioned previously, clinical signs reflect cellular abnormalities within specific neural circuits responsible for memory and cognition. Affected neurons exhibit conformationally altered isoforms of tau comprising the PHFs in NFTs, neurites, and neuropil threads. Axonal pathologies include varicosities and terminal clubs, the latter representing disconnected synapses and transected axons (also observed in aged, memory-impaired rhesus monkeys with Aβ deposits, and in some transgenic monkey models of Aβ amyloidosis). The Aβ-containing neuritic plaques are sites of synaptic disconnection in regions receiving inputs from diseased populations of neurons; associated with these lesions are decrements in generic and transmitter-specific synaptic markers in the target fields of degenerating nerve cells and local astroglial and microglial responses (particularly associated with plaques). Ultimately, there is evidence of death of neurons, possibly by apoptosis. Thus, the clinical manifestations of aMCI and AD reflect a progressive disruption of synaptic communications involving subsets of neural circuits, first associated with degeneration of axon terminals and, later, death of neurons.

In one hypothetical model mechanistically linking Aβ and phosphorylated tau, Aβ42 species liberated at terminals oligomerize to form Aβ assemblies or Aβ-derived diffusible ligands (ADDLs), which are linked to synaptic damage. Subsequently, not yet identified retrograde signals, which originate at disconnected terminals, trigger the activation of kinases (or the inhibition of phosphatases) in cell bodies. The phosphorylation of tau at a variety of sites is associated with conformational changes in this protein and the formation of PHFs and, eventually, NFTs. Secondary disturbances in axonal transport can, in turn, compromise the functions and viability of neurons. Eventually, affected nerve cells die and extracellular tangles remain as markers of the nerve cells destroyed by disease.

These cellular abnormalities have profound clinical consequences. Abnormalities that damage the entorhinal cortex, hippocampus, and other circuits in the medial temporal cortex are presumed to be critical for memory impairments. Higher cognitive deficits, such as disturbances in executive functions for language, calculation, problem solving, and judgment, are believed to be related to pathology in the neocortex. Alterations in the basal forebrain cholinergic system may contribute to memory difficulties and attention deficits. The behavioral and emotional disturbances presumably may reflect involvement of the limbic cortex, amygdala, thalamus, and monoaminergic systems.

At present no cure exists for this devastating illness. Available treatments for AD target symptoms, and many present-day therapies focus on treating associated conditions such as depression, agitation, sleep disorders, hallucinations, and delusions. One therapeutic approach has been to try to support the functions of the basal forebrain cholinergic system, which is one of the circuits severely damaged in AD. Several strategies have been developed to influence this neurotransmitter system. Unfortunately, precursor loading (i.e., choline, lecithin) and muscarinic agonist approaches to improve cholinergic functions did not prove effective. Several acetylcholinesterase inhibitors, which prolong the half-life of the transmitter in the synaptic cleft, have been approved for the treatment of AD in the United States. These drugs are sometimes associated with side effects and they have, at best, a very modest effect on cognitive functions and the ability to perform activities of daily living. Clinical trials have tested the efficacies of anti-inflammatory compounds, estrogens, plant extracts, and antioxidants without great benefit. In an antiexcitotoxic approach, memantine, an *N*-methyl-D-aspartate (NMDA) receptor antagonist, has been suggested to reduce the rate of clinical deterioration of patients with moderate to severe AD.

Genetic Causes and Risk Factors for AD

Established genetic risk factors implicated in AD include: mutations in the *amyloid beta (A4) precursor protein* gene (*APP*; chromosome 21), mutations in *presenilin 1* (*PSEN1*; chromosome 14) and *presenilin 2* (*PSEN2*; chromosome 1), and the presence of a risk susceptibility allele (ε4) of *APOE* (chromosome 19), which predisposes to later onset AD and some cases of late-onset familial AD. Autosomal dominant mutations in *APP*, *PSEN1*, or *PSEN2* usually cause disease earlier than occurs in sporadic cases, with the majority of mutations in *APP*, *PSEN1*, and *PSEN2* influencing β-site APP cleaving enzyme 1 (BACE1) or γ-secretase cleavages of APP to increase the levels of all Aβ species or the relative amounts of toxic Aβ42.

A member of the *APP* gene family (including also *APLP1* and *APLP2*), *APP* encodes a type I transmembrane protein of unknown function which is abundant in the nervous system, rich in neurons, and transported rapidly anterograde in axons to terminals. In the central nervous system (CNS), BACE1 cleaves at the +1 and +11 sites; subsequently, the γ-secretase complex cleaves at a variety of sites (see later), which generate the N- and C-termini of Aβ peptides, respectively. A variety of *APP* mutations alter the processing of APP and influence increase in the production of Aβ peptides or the amounts of the more toxic Aβ42. The APP Swedish double mutation (*swe*) enhances manyfold the BACE1 cleavage at the N-terminus of Aβ (+1 site), resulting in substantial elevations in levels of all Aβ peptides. APP_{717} mutations influence γ-secretase cleavage to increase secretion of Aβ42, which is the most toxic peptide. In contrast, other mutations may promote local fibril formation and vascular amyloidosis. Moreover, individuals with increased *APP* gene dosage are prone to AD; familial duplications of *APP* or trisomy 21 (Down syndrome) are associated with extra copies of *APP* and develop AD pathology relatively early in life.

PSEN1 and *PSEN2* encode two highly homologous and conserved 43–50 kDa multipass transmembrane proteins that are involved in Notch1 signaling pathways critical for cell fate decisions. Presenilins are endoproteolytically cleaved by a 'presenilinase' to form an N-terminal ~28 kDa fragment and a C-terminal ~18 kDa fragment; both fragments are critical components of the γ-secretase complex. Nearly 50% of early-onset cases of familial AD are linked to >90 different mutations in *PSEN1*. A relatively small number of *PSEN2* mutations also cause autosomal dominant familial AD. The majority of abnormalities in *PSEN* genes are missense mutations that enhance γ-secretase activities and increase the levels of the Aβ42 peptides.

Biochemistry of Amyloidosis: APP and the Secretases

APP is cleaved by β- and γ-secretases, releasing the ectodomain of APP (APPs), liberating a cytosolic fragment termed APP intracellular domain (AICD), and generating several species of Aβ peptides. In the central (but not peripheral) nervous system, Aβ peptides are generated by sequential endoproteolytic cleavages by BACE1 (at the Aβ +1 and +11 sites) to generate APP-β C-terminal fragments (APP-βCTFs) and by the γ-secretase complex (at several sites, varying from Aβ 36 to 38, 40, 42, and 43) to form Aβ species peptides. The intramembranous cleavages of APP-βCTF by γ-secretase releases an AICD, which can form a complex with Fe65, a nuclear adaptor protein; Fe65 and Aβ or Fe65 alone (in a novel conformation) can gain access to the nucleus to influence gene transcription, a signaling mechanism analogous to that occurring in the Notch1 pathway (next to the Notch intracellular domain (NICD)). It has been speculated that the AICD signaling pathway may play a role in learning and memory. In other cells in other organs, APP is cleaved endoproteolytically within the Aβ sequence through alternative, nonamyloidogenic pathways: α-secretase (TNF-α converting enzyme, or TACE) cleaves between residues 16 and 17; BACE2 cleaves between residues 19 and 20 and 20 and 21. These cleavages, which occur in nonneural tissues, preclude the formation of Aβ peptides and serve to protect these cells/organs from Aβ amyloidosis.

BACE1, encoded by a gene on chromosome 11, is a transmembrane aspartyl protease that preferentially cleaves APP at the +1 and +11 sites of Aβ in APP and is critical for the generation of Aβ. BACE1 is present in the brain but is virtually undetectable in nonneural tissues; BACE1-specific immunoreactivities are readily localized in the hippocampus. Importantly, as compared to wild-type APP, BACE1 cleaves APP (swe) cleaved approximately 100-fold more efficiently at the +1 site, resulting in a greater increase in BACE1 cleavage products (elevating cell Aβ species) in the presence of this mutation. Significantly, *BACE1*-deficient mice show no overt developmental phenotype, and in cultures of $BACE1^{-/-}$ neurons, the secretion of A$\beta 1_{40/42}$ and A$\beta 11_{40/42}$ is abolished. Thus BACE1 is the principal neuronal β-secretase is responsible for the penultimate Aβ cleavage. In contrast, BACE2, which makes antiamyloidogenic cleavages at sites +19/+20 of Aβ, acts more like α-secretase; it does not appear to play a significant role in APP processing in neurons.

γ-Secretase, essential for the regulated intramembranous proteolysis of a variety of transmembrane proteins, is a multiprotein catalytic complex that includes PSEN1 and PSEN2 (described earlier); nicastrin (Nct), a type I transmembrane glycoprotein; and Aph-1 (anterior pharynx defective) and Pen-2 (presenilin enhancer), two multipass transmembrane proteins. Presenilin contains aspartyl residues that play roles in these intramembranous cleavages; substitutions at D257 in transmembrane (TM) segment 6 and at D385 in TM segment 7 reduce Aβ secretion and cleavage of Notch1 *in vitro*. The functions of the various γ-secretase proteins and their interactions in the complex are not yet fully defined. In one model, Aph-1 and Nct form a precomplex which interacts with presenilin; subsequently, Pen-2 enters the complex, where it is critical for the 'presenilinase' cleavage of presenilin into two fragments. In concert, this complex is responsible for γ-secretase cleavages of APP, Notch, and a variety of other transmembrane proteins. With regard to APP, this complex makes the final cleavage that generates a peptide (especially Aβ) and liberates AICD. Many familial AD-linked *PSEN* mutations promote the cleavages that lead to formation of Aβ42.

Animal Models of Aging and AD

Both spontaneously occurring and genetically engineered diseases in animals have been used to model features of AD.

For example, an age-related spontaneous disorder occurs in elderly rhesus monkeys (*Macaca mulatta*). With an estimated life span of more than 35 years (×2.8, for equivalent human age), these rhesus monkeys show cognitive and memory deficits that appear at the end of the second and at the beginning of the third decades of life. These animals develop virtually all of the brain abnormalities (amyloid deposits, neuritic plaques, scattered tangles, and modest reductions in transmitter markers) that are observed in older humans and, to a greater degree, in cases of AD. Genetically engineered models relevant to AD are discussed in the following section.

Genetic Models of Aβ Amyloidosis

In mice, expression of *APPswe* or APP_{717} (with or without mutant *PSEN1)* leads to an Aβ amyloidosis in the CNS. Mutant *APP/PSEN1* mice develop accelerated disease secondary to increased levels of Aβ (particularly Aβ42) associated with the presence of diffuse Aβ deposits and neuritic plaques in the hippocampus and cortex. Levels of Aβ peptides, particularly Aβ42, increase in brain with age, and oligomeric species (ADDLs, Aβ*56, etc.) appear in the CNS.

Over time, mice carrying mutant transgenes exhibit Aβ deposits, swollen neurites in proximity to these deposits, and neuritic plaques associated with glial responses. Some lines of mice show evidence of amyloid in vessels (congophilic angiopathy). In forebrain regions, the density of synaptic terminals and levels of several neurotransmitter markers are reduced. In some settings, there are deficiencies in synaptic transmission. Moreover, some lines of mice show evidence of degeneration of subsets of neurons.

Behavioral studies of lines of transgenic mice, including those generated by Dr. David Borchelt, disclose deficits in spatial reference memory (Morris water maze task) and episodic-like memory (repeated reversal and radial water maze tasks). At 6 months of age, *APPswe/PSEN1ΔE9* mice develop plaques, but all genotypes are indistinguishable from nontransgenic animals in all cognitive measures. However, in 18-month-old cohorts, *APPswe/PSEN1ΔE9* mice perform all cognitive tasks less well than do mice of all other genotypes. In these animals, amyloid burdens are high; decreases are detectable in levels of several neurotransmitter markers. The strongest relationships exist between deficits in episodic-like memory tasks and total Aβ loads in the brain. Collectively, these studies suggest that, in *APPswe/PSEN1ΔE9* mice, some form of Aβ (ultimately associated with amyloid deposition) can disrupt circuits critical for memory, particularly episodic-like memory. Some of these impairments have been linked to the presence of Aβ oligomers (see later) and can be reversed by antibody-mediated reductions of levels of brain Aβ. These mice do not recapitulate the complete phenotype of AD, but they are very useful subjects for research designed to examine behavioral consequences of Aβ amyloidosis in the CNS, to delineate disease mechanisms, and to test novel therapies.

Over the past decade, a variety of Aβ species, including monomers, oligomers, Aβ-derived diffusible ligands, structural assemblies, and amyloid deposits in neuritic plaques, have been suggested to play important roles in impairing synaptic communication. For example, in one study, when naturally secreted Aβ peptides were injected into the ventricular system of rats, long-term potentiation (LTP) was inhibited in the hippocampus; the activity of the peptide was completely blocked by the injection of a monoclonal Aβ antibody, a finding consistent with the concept that oligomers are the toxic moiety and that they are both necessary and sufficient to perturb learned behavior. Active immunization was less effective in rescuing function and correlated most closely with the levels of antibodies recognizing oligomers. More recently, studies of Tg2576 mice suggested that extracellular accumulations of a 56 kDa soluble amyloid assembly, termed Aβ*56, purified from the brains of memory-impaired mice, can interfere with memory when administered to young rats.

Models of Tau Abnormalities

The paucity of tau abnormalities in various lines of mutant *APP* mice may be related to differences in tau isoforms expressed in this species, as compared to humans. Early efforts to express mutant *tau* transgenes in mice did not lead to striking clinical phenotypes or pathology. More recently, mice overexpressing *tau* show clinical signs, attributed to degeneration of motor axons. When prion or Thy1 promoters are used to drive tau_{P301L} (a mutation linked to autosomal dominant frontotemporal dementia with parkinsonism), some brain and spinal cord neurons develop tangles. Mice expressing *APPswe/*tau_{P301L} exhibit enhanced tangle-like pathology in limbic system and olfactory cortex. Moreover, injection of Aβ42 fibrils into specific brain regions of tau_{P301L} mice increases the number of tangles in those neurons projecting to sites of Aβ injection. A triple transgenic mouse (3×Tg-AD), created by microinjecting *APPswe* and tau_{P301L} into single cells derived from monozygous $PSEN1_{M146V}$ knockin mice, develop age-related plaques and tangles as well as deficits in LTP which appear to antedate overt pathology. However, mice bearing both mutant *tau* and *APP* (or *APP/PSEN1*) or mutant *tau* mice injected with Aβ may not be ideal models of familial AD because the presence of the *tau* mutation alone is associated with the development of tangles and disease.

Targeting of Genes in the Amyloidogenic Pathway

To begin to understand the functions of some of the proteins thought to play roles in AD, investigators have targeted a variety of genes encoding BACE1, PSEN1, Nct, and Aph-1.

BACE1$^{-/-}$ Mice

These animals mate successfully and exhibit no overt pathology. *BACE1*$^{-/-}$ neurons do not cleave at the +1 and +11 sites of Aβ, and the production of Aβ peptides is abolished, establishing that BACE1 is the neuronal β-secretase required to generate the N-termini of Aβ. However, *BACE1*$^{-/-}$ mice show altered performance on some tests of cognition and emotion (see later); the former deficits can be rescued by overexpression of *APP* transgenes.

PSEN1$^{-/-}$ Mice

These embryos develop severe abnormalities of the axial skeleton, ribs, and spinal ganglia, a lethal outcome which resembles a partial *Notch1*$^{-/-}$ phenotype. *PSEN1*$^{-/-}$ cells show decreased levels of secretion of Aβ related to the fact that PSEN1 (along with Nct, Aph-1, and Pen-2) is a component of the γ-secretase complex that carries out the final (S3) intramembranous cleavage of Notch1. Without γ-secretase cleavage, the NICD is not released from the plasma membrane and cannot reach the nucleus to provide a signal to initiate transcriptional processes essential for cell fate decisions. Significantly, conditional *PSEN1/2* targeted mice show impairments in memory and synaptic plasticity in the hippocampus, raising the question (posed particularly effectively Dr. Jie Shen and colleagues) as to the roles of loss of presenilin function in neurodegeneration and AD.

Nct$^{-/-}$ Mice

These embryos die early and exhibit several patterning defects, including abnormal segmentation of somites; this phenotype closely resembles that seen in *Notch1*$^{-/-}$ and *PSEN1/2*$^{-/-}$ embryos. Importantly, *Nct*$^{-/-}$ cells do not secrete Aβ peptides, whereas *NctT*$^{+/-}$ cells show reduction of ~50%. The failure of *NctT*$^{-/-}$ cells to generate Aβ peptides is accompanied by accumulation of APP C-terminal fragments. Importantly, *Nct*$^{+/-}$ mice develop tumors of the skin, presumably related to reduced levels of signaling by Notch1, which appears to act as a tumor suppressor in the skin.

Aph-1a$^{-/-}$ Mice

Three murine *Aph-1* alleles (*Aph-1a*, *Aph-1b*, and *Aph-1c*) encode four distinct Aph-1 isoforms: Aph-1aL and Aph-1aS (derived from differential splicing of *Aph-1a*), Aph-1b, and Aph-1c. *Aph-1a*$^{-/-}$ embryos show patterning defects that resemble, but are not identical to, those of *Notch1*, *Nct*, or *PSEN* null embryos. Moreover, in *Aph-1a*$^{-/-}$-derived cells, the levels of Nct, presenilin fragments, and Pen-2 are decreased. There is an associated reduction in levels of the high-molecular-weight γ-secretase complex and a decrease in secretion of Aβ. In *Aph-1a*$^{-/-}$ cells, other mammalian Aph-1 isoforms can restore the levels of Nct, presenilin, and Pen-2.

Experimental Therapeutics

The availability of models of amyloidogenesis provides opportunities to ablate or knock down genes, to modulate cleavages, and to influence clearance; such experiments have set the stage for testing the influences of Aβ production, APP cleavage patterns, peptide neurotoxicity, and promotion of clearance and/or degradation of Aβ. Because it is not possible to discuss all experimental treatment in mouse models of Aβ amyloidosis, we focus on selected studies that illustrate the experimental strategies directed at specific therapeutic targets that we predict will provide mechanism-based therapeutic benefits to patients with AD.

Reductions in BACE1 Activity

Deletion of *BACE1* in *APPswe/PSEN1ΔE9* mice prevents both Aβ deposition and age-associated cognitive abnormalities that occur in this transgenic model. The *BACE1*$^{-/-}$ *APPswe/PSEN1ΔE9* mice do not develop the Aβ deposits or the age-associated abnormalities in working memory that occur in the *APPswe/PSEN1ΔE9* model of Aβ amyloidosis. Similarly, *BACE1*$^{-/-}$ Tg2576 mice show rescue from age-dependent memory deficits and physiological abnormalities. Moreover, Aβ deposits are sensitive to *BACE1* dosage and can be efficiently cleared from regions of the CNS when *BACE1* is silenced at these sites. Inhibitors of β-secretase, conjugated to carrier peptides, can inhibit *in vitro* and *in vivo* (following intraperitoneal injection into Tg2576 mice). Conditional expression systems or RNA interference (RNAi) silencing allow investigators to examine the pathogenesis of diseases and to assess the degrees of reversibility of the disease processes. Clearly, BACE1 is an attractive therapeutic target. However,

$BACE1^{-/-}$ mice manifest alterations in both hippocampal synaptic plasticity and in performance on tests of cognition and emotion; the memory deficits, but not emotional alterations, in $BACE1^{-/-}$ mice are prevented by co-expression of *APPswe/PSEN1ΔE9* transgenes, suggesting that APP processing influences cognition/memory and that the other potential substrates of BACE1 may play roles in neural circuits related to emotion. BACE1 and APP processing pathways appear to be critical for cognitive, emotional, and synaptic functions. Moreover, BACE1 cleaves reuregulin (NRC), an axonal signal for myelination; inhibition of β-secretase activity is an exciting therapeutic opportunity, but future studies should be alert to potential mechanism-based side effects that may occur with strong inhibition of the enzyme.

Inhibition of γ-Secretase Activity

Both genetic and pharmaceutical lowering of γ-secretase activity decreases production of Aβ in cell-free and cell-based systems and reduces levels of Aβ mutant mice with Aβ amyloidosis. Thus, γ-secretase activity is a significant target for therapy. Treated mice show reduced levels of Aβ and amyloid plaques. However, γ-secretase activity is also essential for processing of Notch, which is critical for lineage specification and cell growth during embryonic development. Significantly, inhibitors of γ-secretase activity reduce production of Aβ, but can also have profound effects on T and B cell development and on the appearance of intestinal mucosa (proliferation of goblet cells, increased mucin in gut lumen, and crypt necrosis). Significantly, $Nct^{+/-}$ *APPswe/PSEN1ΔE9* mice develop skin tumors, presumably, in part, because reduced γ-secretase acts, via Notch signaling, as a tumor suppressor in skin. Clinicians carrying out trials of this class of inhibitor will have to be alert to several potential adverse events associated with reduced activity of this enzyme complex.

γ-Secretase Modulation by Nonsteroidal Anti-inflammatory Compounds

Retrospective epidemiological studies suggested that significant exposure to nonsteroidal anti-inflammatory drugs (NSAIDs) reduces risk of AD, an outcome initially interpreted as related to suppression of the well-documented inflammatory process occurring in brains of cases of AD. However, more recent studies indicate that a subset of NSAID compounds in this class can modulate secretase cleavages (to shorter, less toxic Aβ species) without altering Notch or other APP processing outcomes. Short-term treatment of mutant mice appears to have some benefit in terms of lowering Aβ and plaque pathology. This strategy is now being evaluated in drug trials in humans.

Aβ Immunotherapy

Multiple lines of evidence, including studies of lesions of entorhinal cortex or perforant pathway (by lesioning cell bodies or axons/terminals transporting APP to terminals, respectively), indicate that removing the source of Aβ significantly reduces Aβ in target fields. Similarly, increasing local levels of degrading enzymes, including insulin-degrading enzymes (IDEs) and neprolysin (NEP), can cleave Aβ and can reduce levels of the Aβ peptide.

To date, the most exciting findings regarding clearance of Aβ are derived from studies using active and passive Aβ immunotherapy. In treatment trials in mutant mice and in rodents, both Aβ immunization (with Freund's adjuvant) and passive transfer of Aβ antibodies reduce levels of Aβ and the level of plaque burden. Although, the mechanisms of enhanced clearance are not certain, at least two not mutually exclusive hypotheses have been suggested: (1) a small amount of Aβ antibody reaches the brain, binds to Aβ peptides, promotes the disassembly of fibrils, and, via the Fc antibody domain, encourages activated microglia to enter the affected regions and remove Aβ; and (2) serum antibodies serve as a 'sink' to draw the amyloid peptides from the brain into the circulation, thus changing the equilibrium of Aβ in different compartments and promoting removal of Aβ from the brain. Whatever the mechanism, Aβ immunotherapy in mutant mice is successful in partially clearing Aβ, in attenuating learning and behavioral deficits in several cohorts of mutant *APP* mice, and in partially reducing tau abnormalities in the triple transgenic mice. However, several studies have documented that immunotherapy may be associated with brain hemorrhages related to congophilic angiopathy. In these settings, the presence of amyloid presumably can weaken vascular walls; potentially, removal of some intramural vascular amyloid could lead to rupture of damaged vessels and bleeding.

Although mutant mice that received immunotherapy were not reported to develop evidence of meningoencephalitis, a subset of patients in a clinical trial did manifest these problems. Individuals receiving vaccinations with preaggregated Aβ and an adjuvant (followed by a booster) develop antibodies that recognize Aβ in the brain and vessels. Unfortunately, although phase I trials with Aβ peptide and adjuvant vaccination were not associated with any adverse events, phase II trials detected complications (meningoencephalitis) in a subset of patients and were suspended. The pathology in the index case, consistent with T cell meningitis, was interpreted to show some clearance of Aβ deposits, but some regions contained a relatively high density of tangles, neuropil threads, and vascular amyloid. Aβ immunoreactivity was sometimes associated with microglia, and T cells were conspicuous in the subarachnoid space and around some vessels. In another case, there was significant reduction in amyloid deposits in the absence of clinical evidence of encephalitis. The trial was stopped. Assessment of cognitive functions in a small subset of patients (30) who received vaccination and booster immunizations disclosed that patients who generated Aβ antibodies (as measured by a new Aβ assay) had a slower decline in several functional measures. The events occurring in this subset of patients illustrate the challenges of extrapolating outcomes in mutant mice to human trials. Investigators continue to pursue the passive immunization approaches and are attempting to make new antigen/adjuvant formulations that do not stimulate T-cell-mediated immunologic attack.

Conclusions

Over many years, investigators have more accurately defined mild cognitive impairment and early AD, have developed diagnostic approaches, and have clarified the character and stages of pathology and related the findings to clinical signs. They have greatly enhanced our understanding of the mechanisms underlying the biochemistry of Aβ plaques and tau-related pathology. Following leads from human autopsy studies and from investigations of *in vitro* and *in vivo* models, investigators are now on the threshold of implementing novel treatments based on an understanding of the neurobiology, neuropathology, biochemistry, and genetics of AD. Moreover, a variety of tools, including amyloid imaging and measure of Aβ flux between various compartments, are now available to assess efficacy of treatment. It is anticipated that exciting discoveries over the next few years will lead to the design of new mechanism-based therapies that can be tested in models, and that these approaches will be introduced into the clinic for the benefit of patients with this devastating illness.

See also: Acetylcholinesterase Inhibitors and Alzheimer's Disease; Aging of the Brain; Alzheimer's Disease: Neurodegeneration; Alzheimer's Disease: An Overview; Alzheimer's Disease: Transgenic Mouse Models; Animal Models of Alzheimer's Disease; Axonal Transport and Alzheimer's Disease; Brain Glucose Metabolism: Age, Alzheimer's Disease and ApoE Allele Effects; Cholinergic System Imaging in the Healthy Aging Process and Alzheimer Disease; Dementia and Language; Dementia; Sleep Architecture.

Further Reading

Bard F, Cannon C, Barbour R, et al. (2000) Peripherally administered antibodies against amyloid β-peptide enter the central nervous system and reduce pathology in a mouse model of Alzheimer disease. *Nature Medicine* 6(8): 916–919.

Barten DM, Guss VL, Corsa JA, et al. (2005) Dynamics of β-amyloid reductions in brain, cerebrospinal fluid, and plasma of β-amyloid precursor protein transgenic mice treated with a γ-secretase inhibitor. *Journal of Pharmacology and Experimental Therapeutics* 312(2): 635–643.

Borchelt DR, Ratovitski T, VanLare J, et al. (1997) Accelerated amyloid deposition in the brains of transgenic mice coexpressing mutant presenilin 1 and amyloid precursor proteins. *Neuron* 19(4): 939–945.

Braak H and Braak E (1994) Pathology of Alzheimer's disease. In: Calne DB (ed.) *Neurodegenerative Diseases*, pp. 585–613. Philadelphia: W.B. Saunders.

Braak H and Braak E (1991) Neuropathological staging of Alzheimer-related changes. *Acta Neuropathologica* 82: 239–259.

Buxbaum JD, Thinakaran G, Koliatsos V, et al. (1998) Alzheimer amyloid protein precursor in the rat hippocampus: Transport and processing through the perforant path. *Journal of Neuroscience* 18(23): 9629–9637.

Cai H, Wang Y, McCarthy D, et al. (2001) BACE1 is the major β-secretase for generation of Aβ peptides by neurons. *Nature Neuroscience* 4(3): 233–234.

Cao X and Sudhof TC (2001) A transcriptionally [correction of transcriptively] active complex of APP with Fe65 and histone acetyltransferase Tip60. *Science* 293(5527): 115–120.

Cleary JP, Walsh DM, Hofmeister JJ, et al. (2005) Natural oligomers of the amyloid-β protein specifically disrupt cognitive function. *Nature Neuroscience* 8(1): 79–84.

Cummings JL (2004) Alzheimer's disease. *New England Journal of Medicine* 351(1): 56–67.

Davies P and Maloney AJF (1976) Selective loss of central cholinergic neurons in Alzheimer's disease. *Lancet* 2: 1403.

De Strooper B, Saftig P, and Craessaerts K (1998) Deficiency of presenilin-1 inhibits the normal cleavage of amyloid precursor protein. *Nature* 391(6665): 387–390.

DeMattos RB, Bales KR, Cummins DJ, et al. (2001) Peripheral anti-Aβ antibody alters CNS and plasma Aβ clearance and decreases brain Aβ burden in a mouse model of Alzheimer's disease. *Proceedings of the National Academy of Sciences of the United States of America* 98: 8850–8855.

DeMattos RB, Bales KR, Cummins DJ, et al. (2002) Brain to plasma amyloid-β efflux: A measure of brain amyloid burden in a mouse model of Alzheimer's disease. *Science* 295: 2264–2267.

Farzan M, Schnitzler CE, Vasilieva N, et al. (2000) BACE2, a β-secretase homolog, cleaves at the β site and within the amyloid-β region of the amyloid-β precursor protein. *Proceedings of the National Academy of Sciences of the United States of America* 97: 9712–9717.

Ghiso J and Wisniewski T (2004) An animal model of vascular amyloidosis. *Nature Neuroscience* 7(9): 902–904.

Goedert M and Spillantini M (2006) Neurodegenerative alpha-synucleinopathies and tauopathies. In: Siegel GJ, Albers RW, Brady S, and Price DL (eds.) *Basic Neurochemistry: Molecular, Cellular, and Medical Aspects,* 7th edn., pp. 745–759. Boston, MA: Elsevier Academic Press.

Hock C, Konietzko U, Streffer JR, et al. (2003) Antibodies against β-amyloid slow cognitive decline in Alzheimer's disease. *Neuron* 38(4): 547–554.

Jankowsky JL, Fadale DJ, Anderson J, et al. (2004) Mutant presenilins specifically elevate the levels of the 42 residue β-amyloid peptide *in vivo*: Evidence for augmentation of a 42-specific γ-secretase. *Human Molecular Genetics* 13(2): 159–170.

Jicha GA, Parisi JE, Dickson DW, et al. (2006) Neuropathologic outcome of mild cognitive impairment following progression to clinical dementia. *Archives of Neurology* 63(5): 674–681.

Killiany RJ, Hyman BT, Gomez-Isla T, et al. (2002) MRI measures of entorhinal cortex vs hippocampus in preclinical AD. *Neurology* 58(8): 1188–1196.

Klunk WE, Engler H, Nordberg A, et al. (2004) Imaging brain amyloid in Alzheimer's disease using the novel positron emission tomography tracer, Pittsburgh compound-B. *Annals of Neurology* 55: 1–14.

Klyubin I, Walsh DM, Lemere CA, et al. (2005) Amyloid β protein immunotherapy neutralizes Aβ oligomers that disrupt synaptic plasticity *in vivo*. *Nature Medicine* 11(5): 556–561.

Laird FM, Cai HS, Savonenko AV, et al. (2005) BACE1, a major determinant of selective vulnerability of the brain to Aβ amyloidogenesis is essential for cognitive, emotional and synaptic functions. *Journal of Neuroscience* 25(50): 11693–11709.

Lazarov O, Morfini GA, Lee EB, et al. (2005) Axonal transport, amyloid precursor protein, kinesin-1, and the processing apparatus: Revisited. *Journal of Neuroscience* 25(9): 2386–2395.

Lesne S, Koh MT, Kotilinek L, et al. (2006) A specific amyloid-β protein assembly in the brain impairs memory. *Nature* 440 (7082): 352–357.

Lewis J, Dickson DW, Lin W, et al. (2001) Enhanced neurofibrillary degeneration in transgenic mice expressing mutant tau and APP. *Science* 293: 1487–1491.

Li T, Ma G, Cai H, et al. (2003) Nicastrin is required for assembly of Presenilin/γ-secretase complexes to mediate notch signaling and for processing and trafficking of β-amyloid precursor protein in mammals. *Journal of Neuroscience* 23(8): 3272–3277.

Ma G, Li T, Price DL, et al. (2005) APH-1a is the principal mammalian APH-1 isoform present in γ-secretase complexes during embryonic development. *Journal of Neuroscience* 25(1): 192–198.

Markesbery WR, Schmitt FA, Kryscio RJ, et al. (2006) Neuropathologic substrate of mild cognitive impairment. *Archives of Neurology* 63(1): 38–46.

Martin LJ, Pardo CA, Cork LC, et al. (1994) Synaptic pathology and glial responses to neuronal injury precede the formation of senile plaques and amyloid deposits in the aging cerebral cortex. *American Journal of Pathology* 145(6): 1358–1381.

Morris JC, Storandt M, Miller JP, et al. (2001) Mild cognitive impairment represents early-stage Alzheimer disease. *Archives of Neurology* 58(3): 397–405.

Morris R and Mucke L (2006) Alzheimer's disease: A needle from the haystack. *Nature* 440(7082): 284–285.

Nestor PJ, Scheltens P, and Hodges JR (2004) Advances in the early detection of Alzheimer's disease. *Nature Medicine* 10(supplement): S34–S41.

Nicoll JA, Wilkinson D, Holmes C, et al. (2003) Neuropathology of human Alzheimer disease after immunization with amyloid-β peptide: A case report. *Nature Medicine* 9(4): 448–452.

Oddo S, Caccamo A, Shepherd JD, et al. (2003) Triple-transgenic model of Alzheimer's disease with plaques and tangles: Intracellular Aβ and synaptic dysfunction. *Neuron* 39(3): 409–421.

Petersen RC (2003) Mild cognitive impairment clinical trials. *Nature Reviews Drug Discovery* 2(8): 646–653.

Price DL, Tanzi RE, Borchelt DR, et al. (1998) Alzheimer's disease: Genetic studies and transgenic models. *Annual Review of Genetics* 32: 461–493.

Price JL and Morris JC (1999) Tangles and plaques in nondemented aging and "preclinical" Alzheimer's disease. *Annals of Neurology* 45(3): 358–368.

Rogaev EI, Sherrington R, Rogaeva EA, et al. (1995) Familial Alzheimer's disease in kindreds with missense mutations in a gene on chromosome 1 related to the Alzheimer's disease type 3 gene. *Nature* 376: 775–778.

Rovelet-Lecrux A, Hannequin D, Raux G, et al. (2006) APP locus duplication causes autosomal dominant early-onset Alzheimer disease with cerebral amyloid angiopathy. *Nature Genetics* 38(1): 24–26.

Savonenko AV, Laird FM, Troncoso JC, et al. (2005) Role of Alzheimer's disease models in designing and testing experimental therapeutics. *Drug Discovery Today* 2(4): 305–312.

Savonenko A, Xu GM, Melnikova T, et al. (2005) Episodic-like memory deficits in the APPswe/PS1dE9 mouse model of Alzheimer's disease: Relationships to β-amyloid deposition and neurotransmitter abnormalities. *Neurobiology of Disease* 18(3): 602–617.

Schenk D, Hagen M, and Seubert P (2004) Current progress in β-amyloid immunotherapy. *Current Opinion in Immunology* 16(5): 599–606.

Selkoe D and Kopan R (2003) Notch and presenilin: Regulated intramembrane proteolysis links development and degeneration. *Annual Review of Neuroscience* 26: 565–597.

Sheng JG, Price DL, and Koliatsos VE (2002) Disruption of corticocortical connections ameliorates amyloid burden in terminal fields in a transgenic model of Aβ amyloidosis. *Journal of Neuroscience* 22(22): 9794–9799.

Sisodia SS, Koo EH, Beyreuther KT, et al. (1990) Evidence that β-amyloid protein in Alzheimer's disease is not derived by normal processing. *Science* 248: 492–495.

Vassar R, Bennett BD, Babu-Khan S, et al. (1999) β-Secretase cleavage of Alzheimer's amyloid precursor protein by the transmembrane aspartic protease BACE. *Science* 286: 735–741.

Weggen S, Eriksen JL, Das P, et al. (2001) A subset of NSAIDs lower amyloidogenic Aβ42 independently of cyclooxygenase activity. *Nature* 414: 212–216.

Wolfe MS, Xia W, Ostaszewski BL, et al. (1999) Two transmembrane aspartates in presenilin-1 required for presenilin endoproteolysis and γ-secretase activity. *Nature* 398(6727): 513–517.

Wong PC, Li T, and Price DL (2005) Neurobiology of Alzheimer's disease. In: Siegel GJ, Albers RW, Brady S, and Price DL (eds.) *Basic Neurochemistry: Molecular, Cellular, and Medical Aspects*, 7th edn., pp. 781–790. Boston, MA: Elsevier Academic Press.

Wong GT, Manfra D, Poulet FM, et al. (2004) Chronic treatment with the γ-secretase inhibitor LY-411,575 inhibits β-amyloid peptide production and alters lymphopoiesis and intestinal cell differentiation. *Journal of Biological Chemistry* 279(13): 12876–12882.

Wong PC, Price DL, and Cai H (2001) The brain's susceptibility to amyloid plaques. *Science* 293: 1434–1435.

Aging: Brain Potential Measures and Reaction Time Studies

E Golob, Tulane University, New Orleans, LA, USA
H Pratt, Technion – Israel Institute of Technology, Technion City, Israel
A Starr, University of California at Irvine, Irvine, CA, USA

Brain and Cognitive Aging

Normal aging affects brain functions in a variety of ways. Some regions, such as prefrontal cortex, medial temporal lobe, and neuromodulatory systems (cholinergic, noradrenergic, and serotonergic), show marked age-related changes, whereas other regions, such as primary sensory and motor regions, show comparatively little change during normal aging. Age-related changes in memory function are also selective, which is related to regional differences in the impact of brain aging. In general, age-related declines are most evident in certain types of explicit memory tasks that require conscious retrieval of remembered information, but they are small or absent for many implicit memory tasks. For example, episodic memory is one type of explicit memory, and it is defined as memory for specific events and their context, such as remembering what one did last weekend. In contrast, implicit memory can be expressed even if a person does not consciously remember a specific learning episode or cannot articulate what was learned. An example of implicit memory is the gradual development of proficiency in various kinds of motor skills, such as riding a bicycle. Because the adverse effects of aging are most evident for explicit memory, this article focuses on event-related potential (ERP) and electroencephalogram (EEG) findings while subjects perform explicit memory tasks. For most studies of human aging, the comparison group of young subjects consists of young adult college students, typically between the ages of 18 and 25 years. Depending on the study, older subjects can range in age from the 60s, termed 'young-old,' to older than 90 years, termed the 'oldest-old.'

EEG and ERP methods

The EEG is a measure of continuous variations in voltage over time from electrodes placed on the scalp. It was first developed in the 1930s by Hans Berger, and it was one of the first methods for non-invasively measuring brain activity in humans. Event-related potentials can be used to define neural activity measured by the EEG that is time-locked to an event, such as the onset of a stimulus, the onset of a behavioral response, or their correlation with cognitive processing. An example of an ERP following presentation of an auditory stimulus is shown in **Figure 1(a)**. Depending on the component of interest, the EEG waveforms associated with tens to thousands of events (stimuli or behavioral responses) are averaged together to yield the ERP. The main ERP measures employed are peak amplitude and latency of individual components. Amplitude is conventionally defined as the voltage difference of a peak along the waveform relative to a baseline period, such as the mean amplitude during the 100 ms before stimulus presentation (**Figure 1(b)**). Latency is defined as the time between stimulus onset and the peak of a given component.

Processing Speed and Reaction Time

A basic observation of cognitive aging is that performance of a wide variety of reaction time tasks declines during the course of normal aging, even when subjects are in excellent health. This is often depicted using a Brinley plot, whereby reaction times on multiple tasks in young subjects are arranged in ascending order on the x-axis and plotted against the respective reaction times of older subjects performing the same tasks on the y-axis. Reaction times in young and older subjects show a linear relationship, with a slope >1 indicating larger absolute age effects on reaction time on tasks having longer overall reaction times.

Large-scale cross-sectional and longitudinal studies suggest that a general factor can account for most of the typical cognitive changes associated with normal aging ($>90\%$ of variance). Theorists have proposed that the general factor may indicate fluid intelligence or a 'resource' such as processing speed. Processing speed as a general factor is supported by observations that age differences in reaction time can predict age differences on more complicated tests, such as episodic memory or verbal fluency. Thus, slowing of reaction time appears to reflect a fundamental aspect of cognitive aging. As a practical matter, when comparing reaction times among young and older subjects performing two or more tasks, it is often worthwhile not only to express the absolute differences between tasks for each group but also to include a relative measure, such as percentage increase from task A to B in each group, to take into account age-related slowing of overall reaction time.

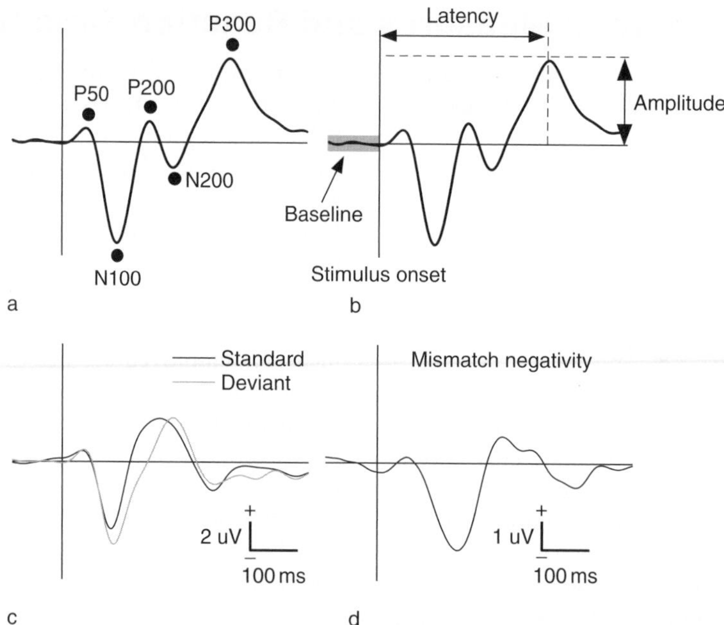

Figure 1 Examples of event-related potential components and measurement of amplitude and latency. (a) Example of auditory event-related potentials elicited by infrequent target stimuli during a target detection task in young subjects. The vertical line indicates stimulus onset. Voltage is plotted as a function of time, from a 100 ms prestimulus baseline to 500 ms after stimulus onset. Components are named by convention using polarity ('P' or 'N') followed by approximate latency. Thus, the 'P50' is a positive component having a latency of ~50 ms. The component labeled 'P300' refers to the P300b (or P3b) component. A different component called the P3a is elicited by distractor stimuli, often a novel sound other than the task-relevant target or nontarget stimuli. (b) Illustration of amplitude and latency measures of the P300 component. (c) Example of the mismatch negativity. Subjects passively listened to standard (90%) and deviant (10%) tone stimuli. Event-related potentials to standard and deviant stimuli are plotted together. Note that potentials to deviants are more negative than standards from ~100 to 240 ms after stimulus presentation, indicating the mismatch negativity. (d) Example of mismatch negativity difference waveform. To better visualize the mismatch negativity, the event-related potential waveform at each time point for standards can be subtracted from the deviant waveform. Note the differences in scale for (d) vs. (a)–(c). To be consistent with event-related potential examples in (a)–(c), negativity is plotted downward. However, in many publications the mismatch negativity is shown with negative polarity plotted upwards.

Sensory Memory

Sensory memory is a relatively automatic form of memory and has a duration of several seconds. In the auditory modality sensory memory is important for the perception of speech and various aspects of auditory scene perception. The mismatch negativity (MMN) is an ERP that has been extensively used to study sensory memory (**Figures 1(c)** and **1(d)**). The MMN is elicited by a stimulus that deviates from a previously established pattern of stimuli presented sequentially. The rationale for its use is that sensory memory can be probed by first establishing a standard sensory memory representation and then presenting a stimulus that is predicted to differ from the sensory memory representation in some respect. If the MMN is elicited, it can be inferred that the deviant aspect of the stimulus was not a property of the standard stimulus representation (mismatch). It is a way to infer memory representations by defining the 'limits' of the sensory memory representation, after which a stimulus is considered different.

Studies of the MMN in normal aging consistently report that when the deviant stimulus differs from the standard in terms of acoustic properties such as frequency or intensity, there are little or no age differences. However, if deviant stimuli differ in terms of timing, such as stimulus duration or interstimulus interval, then age differences are often evident, with smaller MMN amplitudes in older compared to young subjects.

Working Memory

Working memory is the process of maintaining and manipulating information that is relevant to performing a task for short periods of time (seconds to minutes). A typical example of maintaining information in working memory is remembering the digits of an unfamiliar phone number prior to placing the call. An example of manipulation is the addition of an area code before the phone number in working memory and reordering the digits to correctly dial the number.

Working memory is frequently studied using variations of a paradigm developed by Saul Sternberg, which provides the ability to document both behavioral and brain activity changes accompanying working memory retrieval. In the Sternberg task, subjects perform a series of working memory trials. In each trial a set of items, typically numbers or letters, are memorized in sequence. After a delay of a few seconds (typically 2–4 s), retrieval is assessed by presenting a probe item. Subjects then determine whether or not the probe was a member of the current memory set, and press one of two buttons to indicate their decision. When a range of memory loads are tested (from one to approximately six to eight items), reaction time shows a positive linear relationship with memory load, with longer reaction times to probes when more items are memorized. Thus, the Sternberg task is a useful tool to define systematically the impact of memory load on reaction time and ERP measures during encoding (memorized items) and memory retrieval (probe), which is important because limited capacity is a defining feature of working memory.

The slope of the memory load versus reaction time function is usually steeper in older subjects than in young subjects, indicating slower retrieval speed. Age effects are sometimes smaller when subjects are rigorously screened to exclude all but the healthiest older subjects or use young-older people (in their 50s), but retrieval speed is still slower in older groups relative to young adults in their 20s. Although not always examined, slower retrieval speed can persist even when there are controls for overall differences in reaction time, if present.

Most ERP studies using the Sternberg paradigm have focused on a prolonged late positive waveform present from approximately 200 to 800 ms that is elicited by probes. The most prominent component is often called the 'late positive wave' or 'P300'; it is similar to the P300 shown in **Figure 1** in response to targets, but it is longer in duration. Detailed analyses indicate that probes elicit a set of components comprising the late positive waveform and likely correlate with somewhat different cognitive processes during memory retrieval. Control tasks show that the late positive wave is not present when the same stimuli are presented as in the working memory task, but probes do not require comparison to memory. In young subjects, increased memory load is associated with increases in P300 latency and decreases in P300 amplitude. Older subjects show a similar pattern of results as a function of memory load, with overall increases in P300 latency and a weaker correlation between P300 latency and reaction time.

In auditory versions of the Sternberg task, a component called the 'sustained frontal negativity' is elicited by memorized items and probes when environmental sounds are used, especially when most probes match memorized items. It is unclear if the sustained frontal negativity is present for other modalities. The amplitude of the sustained frontal negativity is reduced in older subjects relative to young subjects. It has been suggested that a smaller sustained frontal negativity may be associated with age-related changes in prefrontal cortex function during working memory.

Studies have examined potentials that occur before the P300/late positive wave and sustained frontal negativity. The purpose of examining shorter latency components is to determine if modality-specific components associated with sensory cortical activity are also modulated by working memory demands. In the auditory modality, amplitude of the N100 component is affected by memory load during both encoding of items and retrieval, with linear reductions in amplitude as memory load is increased. Amplitude of the N100 to probes is also modulated by the details of encoding, such as the order of item presentation, and is not affected by memory load when memorized items are presented in the visual modality. Both results are consistent with psychological theories emphasizing a linkage between memory retrieval and specific processes during memory encoding, such that memory retrieval may reinstate aspects of perceptual processing.

Compared to young subjects, older subjects exhibit two main differences in auditory potentials in the Sternberg task. During encoding, young subjects have shorter N100 latencies with increasing memory load, whereas N100 latency in older subjects is invariant across memory loads. The second main age difference is that probe N100 amplitudes to memorized items show increases with greater memory load in older subjects, but the opposite is seen in the young as amplitudes decrease with memory load. In addition, N100 amplitudes to probes that were not in the memorized list also exhibit smaller changes in N100 amplitude as a function of memory load relative to young subjects. The significance of age-related changes in auditory potentials is unclear, but it has been proposed that changes in prefrontal cortex function may be relevant because prefrontal cortex can regulate the responsiveness of sensory cortex.

Declarative: Episodic Memory

ERP and EEG studies of episodic memory typically employ a recognition format instead of having subjects freely recall the memorized material. In recognition memory tests, the experimenter presents stimulus items to which the subject has to make a judgment, such as whether or not the item was previously memorized. A recognition format is preferred for ERP and EEG studies because the ability to precisely control stimulus timing is necessary to define neural responses, such as ERPs, that are time-locked to stimulus presentation. The Sternberg task described previously is also a recognition task because subjects judge whether the probe was a memorized item.

Target detection, or 'oddball,' tasks, first described by Samuel Sutton and colleagues in 1965, can be considered a simple recognition memory test. Subjects are presented a sequence of stimuli and are instructed to detect the infrequent occurrence of a predefined target stimulus. Subjects usually either press a button to each target or count the number of targets presented in the sequence. An example of ERPs recorded during a target detection task is shown in the top of **Figure 1**. Both target and nontarget stimuli elicit sensory potentials, such as P50, N100, and P200, in response to auditory stimuli. However, targets also elicit two additional components that are small or absent compared to nontargets: the N200 and P300 complex. N200 and P300 are associated with the active identification of targets because they are not present during passive listening tasks, such as those used to elicit the mismatch negativity. When the responses to nontargets are analyzed in detail as a function of the stimulus sequence history, a positive component at 250 ms can be seen when the likelihood for a target to appear is high. Thus, each stimulus in the train appears to be processed slightly differently depending on the expectations of the listener.

Many studies have shown that P300 changes substantially during normal development. P300 latency is shortest in early adulthood (teens) and has a gradual, progressive, increase in latency across age cohorts until at least the 10th decade of life (**Figure 2**). As a group, P300 latency to auditory stimuli for subjects in their 80s is approximately 100 ms longer than that of young adults in their early 20s (300 vs. 400 ms). The same pattern of age differences is observed for P300 to visual stimuli, which has somewhat longer latencies (by approximately 60 ms for all age groups) relative to auditory stimuli. When the distribution of voltage across the entire scalp at the time of the P300 peak is examined with an array of electrodes, the focus of maximum amplitude is more anterior in older relative to younger subjects. Age differences in P300 latency are probably not attributable to differences in sensory function because stimuli are easily distinguishable by all subjects, and age differences in P300 latency are not affected by manipulations of stimulus discriminability. Response preparation is also unlikely because, unlike most reaction time tasks, reaction times to targets show little change as a function of age in healthy subjects and P300 latency shows little or no change with response demands. Thus, although the precise cognitive correlates of P300 latency are uncertain, age differences are likely associated with memory-dependent stimulus classification processes.

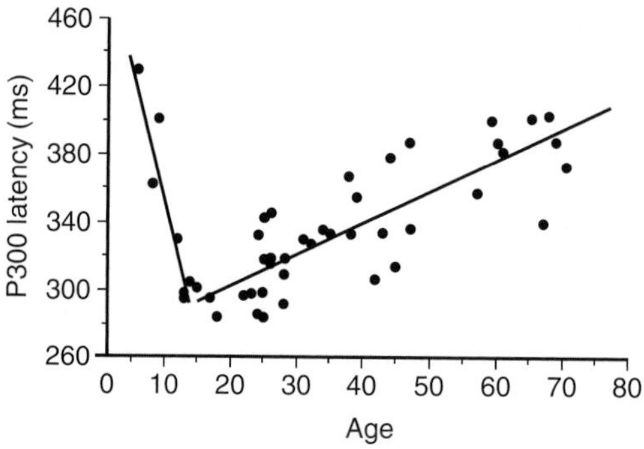

Figure 2 P300 latency as a function of age. Linear decreases in P300 latency are seen during childhood, with shortest latencies during late adolescence. Gradual increases in P300 latency are then observed throughout the adult years. An auditory target detection task was used to assess P300 latency to infrequent targets (20% stimuli) that differed from nontargets according to pitch. Circles indicate measures from individual subjects. Note that 'P300' refers to the P300b component. Data replotted from Goodin DS, Squires KC, Henderson BH, and Starr A (1978) Age-related variations in evoked potentials to auditory stimuli in normal human subjects. *Electroencephalography and Clinical Neurophysiology* 44(4): 447–458.

In most episodic memory studies subjects are requested to remember multiple items rather than just one target item as in target detection. Two main paradigms are used extensively in episodic memory studies using ERPs: study/test and continuous recognition. In the study/test paradigm, items are encoded during a study period and memory is later assessed during a test period. One advantage of the study/test paradigm is that it permits separate assessment of potentials during encoding and retrieval. In contrast, in continuous recognition tasks, subjects are presented a sequence of items and each item is typically judged as either old or new. Thus, encoding and retrieval processes are operative during each item presentation.

Presentation of items in both paradigms elicits positive slow wave potentials. The slow waves begin as early as ~250 ms and last for hundreds of milliseconds depending on the component and task. During retrieval in both study/test and continuous recognition previously studied items elicit slow waves that are more positive relative to new items – a phenomenon called the 'ERP old/new effect.' In the study phase of study/test tasks positive slow waves following item presentation are also present. Moreover, items that are later remembered, via either free recall or correct recognition, elicit more positive slow waves than items that are later not remembered. This finding is called the 'subsequent memory effect.' Here, we focus on episodic memory retrieval because only a few studies have examined subsequent memory effects as a function of aging, and the subsequent memory effect is sometimes not observed in young subjects.

There are two main positive slow waves associated with old/new effects in study/test paradigms: one that is largest over left parietal recording sites and another that is maximal over right frontal areas. The slow waves are also distinguished by onset time and duration, with the left parietal slow wave having an earlier onset and shorter duration compared to the right frontal slow wave. The two slow waves can be dissociated by manipulating task demands. The left parietal slow wave is associated with retrieval of item information, such as old versus new, whereas the right frontal slow wave is associated with item information and contextual information. Item information is specific knowledge presented during the study session, whereas contextual information reflects other aspects of the learning situation that are either not necessarily relevant to the memory test or are common to multiple items. For example, subjects may memorize words presented on two separate lists and be tested later using item recognition for the words (old/new words). In this example, the actual words on the lists would be considered item information, whereas remembering which of the two lists contained a given word would be considered contextual information.

The old/new effect at left parietal sites changes little with aging, suggesting that certain retrieval processes associated with item recognition may not change substantially during aging. In contrast, the right frontal slow wave does exhibit age differences. Although additional study is needed, the right frontal slow wave is usually attenuated in older subjects relative to young subjects when judgments of context are assessed. Age differences in right frontal slow wave amplitude are accompanied by behavioral differences, with less accurate judgments of item context in older subjects. Thus, behavioral and ERPs studies are consistent in showing that normal aging is associated with small differences in recognition memory for items, whereas substantial age differences are evident when memory for context is assessed.

Conclusions

This summary of findings on memory and aging using reaction time and electrophysiological measures underscores the benefit of combining behavioral and physiological tools. Behavioral measures reflect the overt product of cognitive and response selection. Different components of electrophysiological responses reflect specific aspects of covert brain processing. Thus, ERPs and EEG provide a tool to define covert memory processes and compare these memory processes as a function of normal aging.

See also: Aging of the Brain; Aging: Extracellular Space; Brain Volume: Age-Related Changes; Cognition in Aging and Age-Related Disease; Conditioned Reflex; Electroencephalography (EEG); Electrophysiology: EEG and ERP Analysis; Functional Neuroimaging Studies of Aging.

Further Reading

Allison T, Hume AL, Wood CC, and Goff WR (1984) Developmental and aging changes in somatosensory, auditory and visual evoked potentials. *Electroencephalography and Clinical Neurophysiology* 58(1): 14–24.

Friedman D (2000) Event-related brain potential investigations of memory and aging. *Biological Psychology* 54(1–3): 175–206.

Goodin DS, Squires KC, Henderson BH, and Starr A (1978) Age-related variations in evoked potentials to auditory stimuli in normal human subjects. *Electroencephalography and Clinical Neurophysiology* 44(4): 447–458.

Iragui VJ, Kutas M, Mitchiner MR, and Hillyard SA (1993) Effects of aging on event-related brain potentials and reaction times in an auditory oddball task. *Psychophysiology* 30(1): 10–22.

Naatanen R (1992) *Attention and Brain Function.* Hillsdale, NJ: Lawrence Erlbaum.

Sternberg S (1966) High-speed scanning in human memory. *Science* 153(736): 652–654.

Sutton S, Braren M, Zubin J, and John ER (1965) Evoked-potential correlates of stimulus uncertainty. *Science* 150(700): 1187–1188.

Aging: Extracellular Space

E Syková, Institute of Experimental Medicine, Academy of Sciences of the Czech Republic, and Charles University, Prague, Czech Republic

Gross Anatomic Aging Changes: Extracellular Space

Aging, Alzheimer's disease, and many degenerative diseases are accompanied by serious cognitive deficits, particularly impaired learning and memory loss. This decline in old age is a consequence of changes in brain anatomy, morphology, and volume, as well as functional deficits. Nervous tissue, particularly in the hippocampus and cortex, is subject to various degenerative processes, including a decreased number and efficacy of synapses, a decrease in transmitter release, neuronal loss, a decreased number of dendritic spines, astrogliosis, changes in astrocytic morphology, demyelination, deposits of β-amyloid, changes in extracellular matrix proteins, and, consequently, changes in the brain's diffusion parameters. These age-related changes in the morphology of neural elements have no universal pattern across the entire brain. Despite the fact that aging is associated with cognitive impairment, in most brain areas neuronal loss does not have a significant role in the observed deficits. It has been suggested that rather small, region-specific changes in dendritic branching and spine density are more characteristic changes in neuronal morphology.

Diffusion and Extrasynaptic Transmission

The gross anatomic changes and other subtle morphological changes not only affect the efficacy of signal transmission at synapses, but also the function of glia and extrasynaptic ('volume') transmission mediated by the diffusion of transmitters and other substances through the volume of the extracellular space (ECS). This mode of communication by diffusion (i.e., without synapses) provides a mechanism of long-range information processing in functions such as vigilance, sleep, chronic pain, hunger, depression, long-term potentiation (LTP), long-term depression (LTD), memory formation, and other plastic changes in the central nervous system (CNS). Neurons interact both by synapses and by the diffusion of ions and neurotransmitters in the extracellular space. Since glial cells do not have synapses, their communication with neurons is only mediated by the diffusion of ions and neuroactive substances in the ECS. Neurons and glia release ions, transmitters, and various other neuroactive substances into the ECS. Substances released nonsynaptically diffuse through the ECS and bind to extrasynaptic, usually high-affinity, binding sites located on neurons, axons, and glial cells.

Diffusion in the ECS is critically dependent on the structure and physicochemical properties of the ECS – the nerve cell microenvironment. These properties vary, however, around each cell and in different brain regions. Certain synapses ('private synapses') or even whole neurons are clearly tightly ensheathed by glial processes and by the extracellular matrix, forming so-called perineuronal nets, while others are left more 'naked.' The 'open' synapses are more easily reached by molecules diffusing in the ECS (**Figure 1**). On the other hand, many mediators, including glutamate and γ-aminobutyric acid (GABA), bind to high-affinity binding sites located on nonsynaptic parts of the membranes of neurons and glia. Mediators that escape from the synaptic clefts at an activated synapse, particularly following repetitive stimulation, diffuse in the ECS and can cross-react with receptors in nearby synapses. This phenomenon, called 'cross-talk' between synapses by the 'spillover' of a transmitter (e.g., glutamate, GABA, glycine), has been proposed to account for LTP, LTD, and memory formation in the rat hippocampus. The cross-talk between synapses, and the efficacy and directionality of volume transmission, are critically dependent on the diffusion properties of the ECS.

There is increasing evidence that changes in neuron–glia interaction (e.g., glial coverage and/or the retraction of glial processes from synapses) occur in many brain regions during physiological and pathological functional changes. The glial environment of neurons is likely to be a key factor in the regulation of intersynaptic communication mediated by glutamate. For example, most synaptically released glutamate is taken up by high-affinity transporters, such as GLT-1 and GLAST, that are located on surrounding astrocytes. Moreover, glial cells represent a diffusion barrier in the ECS, hindering the movement of neuroactive substances within the tissue. Long-term changes in the physical and chemical parameters of the ECS accompany many physiological (e.g., development, aging, lactation) and pathological (e.g., ischemia, epilepsy, hydrocephalus, demyelination, CNS trauma) states.

Measurements of the Diffusion Parameters of the ECS

The diffusion of substances in a free medium, such as water or diluted agar, is described by Fick's laws.

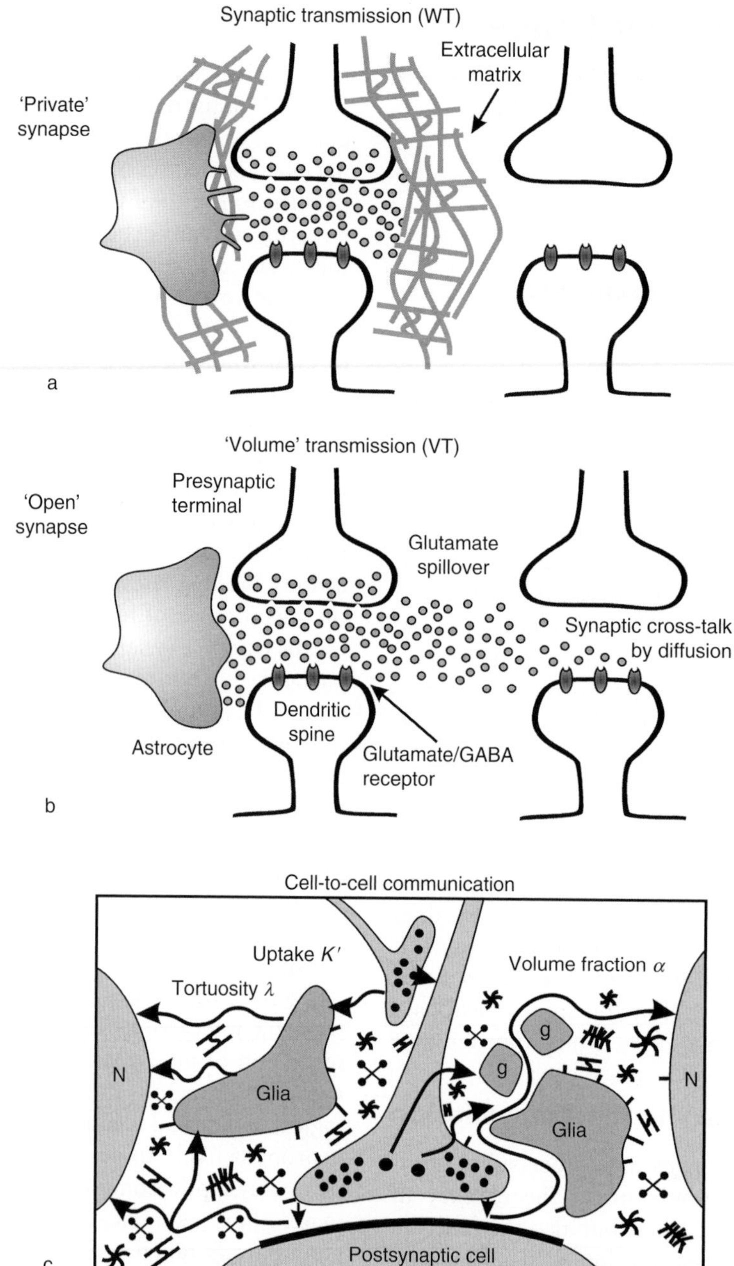

Figure 1 Concept of synaptic 'wired' transmission (WT) and extrasynaptic 'volume' transmission (VT). (a) Closed synapses are typical of synaptic transmission. The synapse is tightly ensheathed by glial processes and the extracellular matrix, forming perineuronal or perisynaptic nets. (b) An open synapse is typical of volume transmission. It allows the escape of transmitter – for example, glutamate and γ-aminobutyric acid (GABA) – from the synaptic cleft (spillover), diffusion in the extracellular space, and binding to receptors on nearby synapses. This phenomenon is known as 'cross-talk' between synapses. Spillover may also lead to plastic changes, inducing the formation of new synapses or eliciting the rearrangement of astrocytic processes around the synapse and the formation of diffusion barriers. (c) Schematic of central nervous system architecture, showing how it is composed of neurons (N), axons, glial cells (g), cellular processes, molecules of the extracellular matrix, and intercellular channels between the cells. The architecture affects the movement (diffusion) of substances in the brain, which is critically dependent on channel size, extracellular space tortuosity, and cellular uptake. Adapted from Syková E (2001) Glial diffusion barriers during aging and pathological states. *Progress in Brain Research* 132: 339–363.

In contrast to a free medium, the diffusion of any substance in the ECS is hindered by the ECS size, the presence of various obstacles, and also by cellular uptake. To take these factors into account, it was necessary to modify Fick's original diffusion equations. First, diffusion in the CNS is constrained by the restricted volume of the tissue available for the diffusing particles – that is, by the extracellular space volume fraction (α), which is a dimensionless quantity and is defined as the ratio between the volume of the ECS and the total volume of the tissue. It is now evident that the ECS in the adult brain amounts to about 20% of the total brain volume (i.e., $\alpha = 0.2$). Second, the free diffusion coefficient (D) in the brain is reduced by the tortuosity factor (λ). ECS tortuosity is defined as $\lambda = (D/\mathrm{ADC})^{0.5}$, where D is the free diffusion coefficient and ADC is the apparent diffusion coefficient in the brain. As a result of tortuosity, D is reduced to an apparent diffusion coefficient $\mathrm{ADC} = D/\lambda^2$. Thus, any substance diffusing in the ECS is hindered by membrane obstructions, glycoproteins, macromolecules of the extracellular matrix, charged molecules, and fine neuronal and glial cell processes. Third, substances released into the ECS are transported across membranes by nonspecific concentration-dependent uptake (k'). In many cases however, these substances are transported by energy-dependent uptake systems that obey nonlinear kinetics. When these three factors (α, λ, and k') are incorporated into Fick's law, diffusion in the CNS is described fairly satisfactorily.

At the present time, the real-time iontophoretic method is used to determine the absolute values of the ECS diffusion parameters and their dynamic changes in nervous tissue *in vitro* as well as *in vivo*. The second method that is also frequently used to study ECS volume and geometry is diffusion-weighted magnetic resonance imaging (DW-MRI). DW-MRI provides information only about the apparent diffusion coefficient of water. A relationship between an increase in the *ADC* of water and a decrease in ECS volume fraction has recently been found during pathological states as well as during aging.

The real-time iontophoretic method (**Figure 2(a)**) uses ion-selective microelectrodes to measure the diffusion of ions to which the cell membranes are relatively impermeable – for example, tetraethylammonium and tetramethylammonium ions (TEA^+, TMA^+) or choline. These substances are injected into the nervous tissue by pressure or by iontophoresis from an electrode aligned parallel to a double-barreled ion-selective microelectrode (ISM) at a fixed distance. Usually, such an electrode array is made by gluing together an iontophoretic pipette and a TMA^+-sensitive ISM with a tip separation of 130–200 μm. In the case of iontophoretic application, TMA^+ is released into the ECS by applying a current step of +100 nA with a duration of 40–80 s. The released TMA^+ is recorded with the TMA^+-sensitive ISM as a diffusion curve, which is then transferred to a computer to calculate α, λ, and k' from the modified diffusion equation. Values of the ECS volume, ADC, tortuosity, and nonspecific cellular uptake are extracted by a nonlinear curve-fitting simplex algorithm applied on the diffusion curves.

By introducing the tortuosity factor into diffusion measurements in nervous tissue, it soon became evident that diffusion is not uniform in all directions and is affected by the presence of diffusion barriers, including neuronal and glial processes, myelin sheaths, macromolecules, and molecules with fixed negative surface charges. This so-called anisotropic diffusion preferentially channels the movement of substances in the ECS in one direction, (e.g., along axons) and may, therefore, be responsible for a certain degree of specificity in volume transmission. Diffusion anisotropy was found in the CNS in the molecular and granular layers of the cerebellum, in the hippocampus, and in the auditory but not in the somatosensory cortex, and a number of studies have revealed that it is present in the myelinated white matter of the corpus callosum or spinal cord. It was shown that diffusion anisotropy in white matter increases during development. At first, diffusion in unmyelinated tissue is isotropic; it becomes more anisotropic as myelination progresses.

ECS Diffusion Parameters during Aging

Aging is not only accompanied by a decrease in the efficacy of signal transmission at synapses, but also by changes in the ECS diffusion parameters and consequently by a deficit in extrasynaptic transmission. Using the TMA^+ method, the ECS diffusion parameters were measured in the cortex, corpus callosum, and hippocampus (CA1, CA3, and dentate gyrus) of aged rats. The measurements revealed a significant decrease in ECS volume fraction. One of the explanations as to why α in the cortex, corpus callosum, and hippocampus of senescent rats and mice is significantly lower than in young adults could be astrogliosis in the aged brain. Increased glial fibrillary acidic protein (GFAP) staining and an increase in the size and fibrous character of astrocytes have been found in the cortex, corpus callosum, and hippocampus of senescent rats. Other changes could account for the decreases in λ and for the disruption of tissue anisotropy. In the hippocampus in the CA1 and CA3 regions, as well as in the dentate gyrus, changes in the arrangement of fine astrocytic processes were regularly found. These are normally organized in parallel in the x–y plane, but this organization totally disappears during aging. Moreover, decreased staining for chondroitin sulfate proteoglycans and for fibronectin suggests a loss of extracellular matrix macromolecules (**Figure 3**).

If diffusion in a particular brain region is anisotropic, then the correct value of the ECS volume

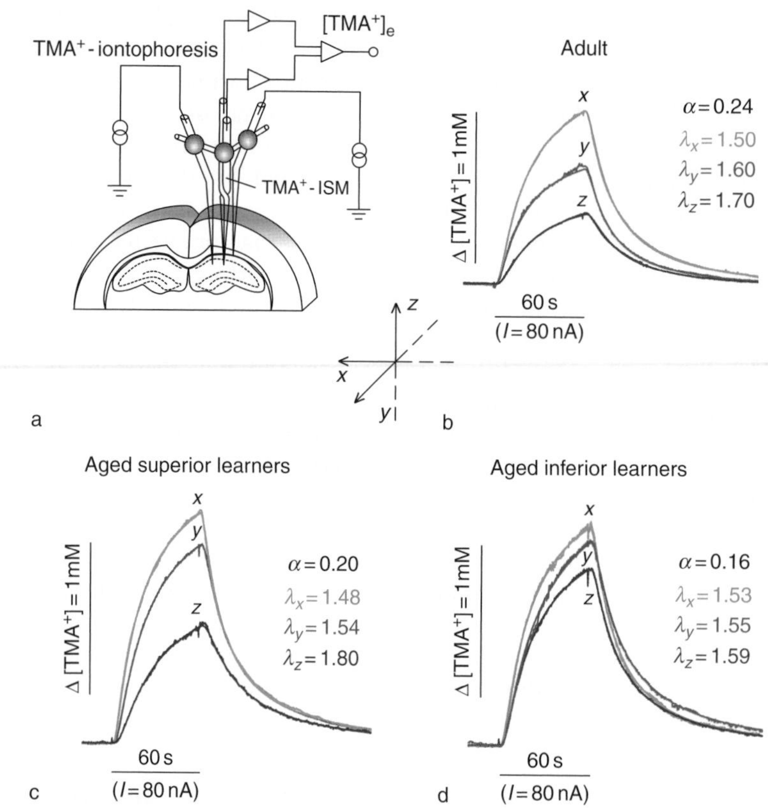

Figure 2 Experimental setup: tetramethylammonium ion (TMA^+) diffusion curves and typical extracellular space diffusion parameters α (volume fraction) and λ (tortuosity) in the hippocampus dentate gyrus of a young adult rat and an aged rat with memory impairment (rats were tested in a Morris water maze). (a) Schema of the experimental arrangement. A TMA^+-selective double-barreled ion-selective microelectrode (ISM) was glued to a bent iontophoresis microelectrode. The separation between electrode tips was 130–200 μm. (b) Anisotropic diffusion in the dentate gyrus of a young adult rat required measurements of TMA^+ diffusion curves (concentration–time profiles) along three orthogonal axes (*x*, mediolateral; *y*, rostrocaudal; *z*, dorsoventral). The slower rises in the *z* axis (compared to the *y* axis) and in the *y* axis (compared to the *x* axis) indicate a higher tortuosity and more restricted diffusion. The amplitudes of the curves show that the TMA^+ concentration, at approximately the same distance from the tip of the iontophoresis electrode, is much higher along the *x* axis than along the *y* axis and is even higher than along the *z* axis (λ_x, λ_y, λ_z). Note that the actual extracellular space volume fraction α is about 0.2 and can be calculated only when measurements are done along the *x*, *y*, and *z* axes. (c, d) Volume fraction and anisotropy decrease during aging. Note the greater decrease in volume fraction and that the anisotropy is almost lost in an aged rat with memory impairment (d). Note that the diffusion curves are higher, showing that α is smaller, and that their rise and decay times are longer when λ (diffusion barriers) increases. Adapted from Syková E, Mazel T, Hasenöhrl RU, et al. (2002) Learning deficits in aged rats related to decrease in extracellular volume and loss of diffusion anisotropy in hippocampus. *Hippocampus* 12: 469–479.

fraction cannot be calculated from measurements done only in one direction. For anisotropic diffusion, the diagonal components of the tortuosity tensor are not equal, and generally its nondiagonal components need not be zero. Nevertheless, if a suitable referential frame is chosen (i.e., if we measure in three privileged orthogonal directions), neglecting the nondiagonal components becomes possible, and the correct value of the ECS size can thus be determined. Therefore, TMA^+ diffusion was measured in the ECS independently along three orthogonal axes (*x*, transversal; *y*, sagittal; *z*, vertical) in the cortex, corpus callosum, and hippocampus of aged rats and mice. These studies revealed that the mean ECS volume fraction α is significantly lower in aged rats (26–32 months old), ranging from 0.17 to 0.19, than in young adults (3–4 months old), in which α ranges from 0.21 to 0.22. Nonspecific uptake k' is also significantly lower in aged rats. From **Figure 2** it is evident that the diffusion curves for the hippocampus are larger in the aged rat than in the young one – that is, the space available for TMA^+ diffusion is smaller. Although the mean α values along the *x* axis were not significantly different between young and aged rats, the values were significantly lower in aged rats along the *y* and *z* axes. This means that there is a loss of anisotropy in the aging hippocampus, particularly in the CA3 region and the dentate gyrus.

The three-dimensional pattern of diffusion away from a point source can be illustrated by

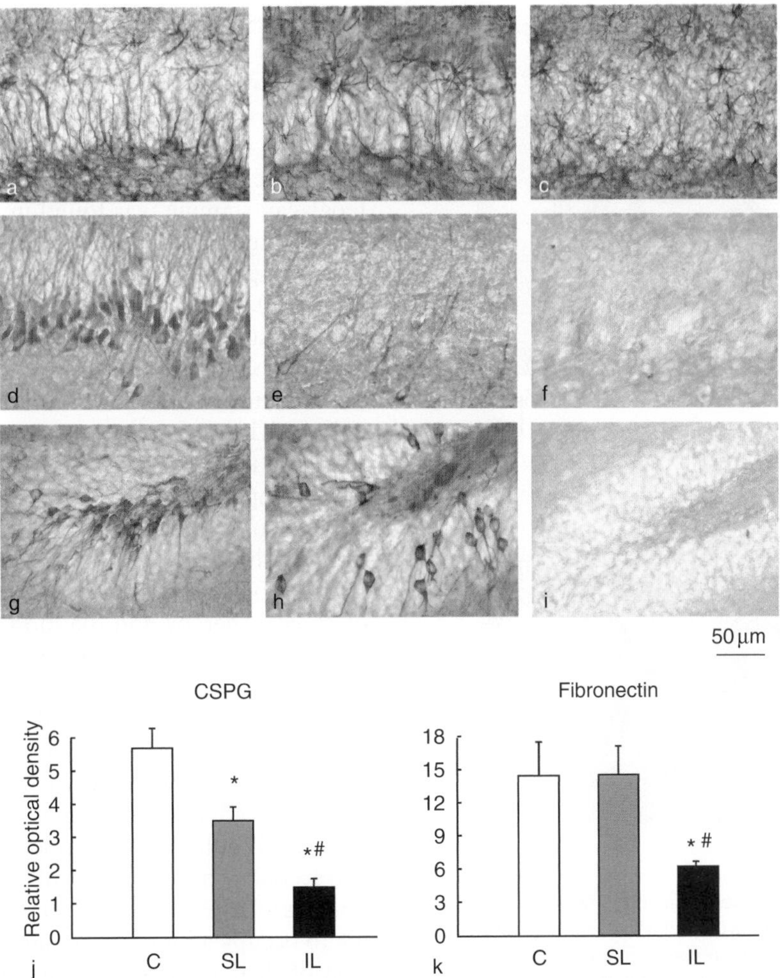

Figure 3 Structural changes in aged superior and inferior learners. (a–c) Astrocytes in the dentate gyrus (inner blade) of an adult rat, an aged superior learner (SL), and an aged inferior learner (IL). Note the loss of radial organization of the astrocytic processes in the aged IL. Staining for chondroitin sulfate proteoglycans (CSPG) in the hippocampal CA3 region (d–f), and for fibronectin in the dentate gyrus (g–i), shows a decrease in perineuronal staining in the aged SL and a loss in the aged IL. The graphs show the relative optical densities of CSPG (j) and fibronectin cell attachment fragment (k) immunoreactivity. Note the significant decrease in CSPG immunoreactivity between each group (C, control) and the decrease in fibronectin immunoreactivity in the aged IL only. *significant difference ($p < 0.05$) compared to control (adults). #significant difference ($p < 0.05$) compared to aged superior learners. Adapted from Syková E, Mazel T, Hasenöhrl RU, et al. (2002) Learning deficits in aged rats related to decrease in extracellular volume and loss of diffusion anisotropy in hippocampus. *Hippocampus* 12: 469–479.

constructing isoconcentration spheres (isotropic diffusion) or ellipsoids (anisotropic diffusion) for extracellular TMA^+ concentration. The surfaces in **Figure 4** represent the locations where the TMA^+ concentration first reached 1 mM, 60 s after its application in the center. The ellipsoid in the hippocampus of the young adult rat reflects the different abilities of substances to diffuse along the x, y, and z axes, while the sphere in the hippocampus of the aged rat shows isotropic diffusion. The smaller ECS volume fraction in aged rats is reflected in the sphere's being larger than the ellipsoid. Indeed, we found that there is a significant decrease in the ADC of many neuroactive substances, including the ADC of water (**Figure 5**), in the aging brain, which accompanies a decrease in ECS volume.

The hippocampus is well known for its role in memory formation, especially declarative memory. Indeed, it has been found that the degree of learning deficit during aging correlates with changes in α, λ, and k'. It is therefore reasonable to assume that diffusion anisotropy leads to a certain degree of specificity in extrasynaptic communication. There is a significant difference between mildly and severely behaviorally impaired rats (rats were tested in a Morris water maze), which is particularly apparent in the hippocampus. The ECS in the dentate gyrus of severely impaired rats (bad learners) is significantly smaller

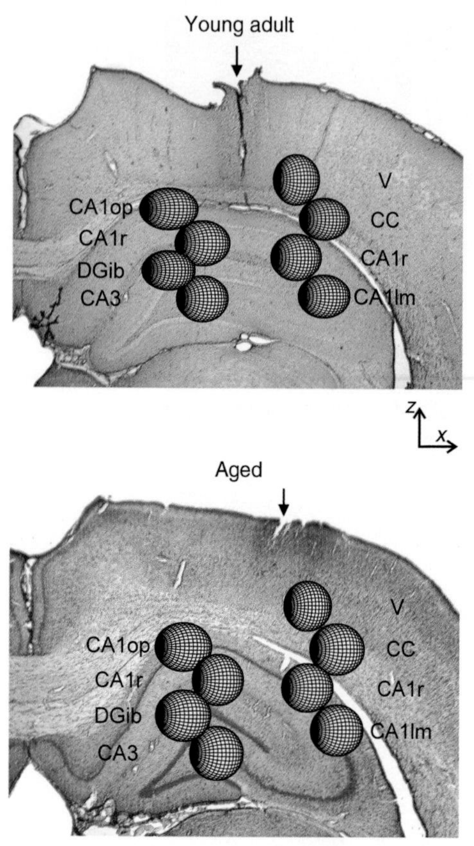

Figure 4 Diffusion parameters in a young adult (3 months old) rat and in an aged (28 months old) rat with a learning deficit in the Morris water maze. Data were recorded in anesthetized animals; microelectrode tracks were verified after the experiments (see arrows); isoconcentration surfaces for a 1 mM tetramethylammonium ion concentration contour 60 s after the onset of an 80 nA iontophoretic pulse. The surfaces were generated using the actual values of volume fraction and tortuosity. The ellipsoids represent anisotropic diffusion in a young adult rat. The larger sphere in an aged rat corresponds to isotropic diffusion and to a small extracellular space volume fraction. It demonstrates that diffusion from any given source will lead to a higher concentration of substances in the surrounding tissue and a larger action radius in aged rats than in young adults. Anisotropy is almost lost in the aged rat. Adapted from Syková E, Mazel T, Vargová L, et al. (2000) Extracellular space diffusion and pathological states. *Progress in Brain Research* 125: 155–178.

than in mildly impaired rats (good learners). Also, anisotropy in the hippocampus of bad learners, particularly in the dentate gyrus, is much reduced, while a substantial degree of anisotropy is still present in aged rats with good learning performance (**Figures 2** and **3**). Anisotropy might be important for extrasynaptic transmission by channeling the flux of substances in a preferential direction, and its loss may severely disrupt extrasynaptic communication in the CNS, which has been suggested to play an important role in memory formation.

The ensuing decrease in the ECS size could be explained by the disappearance of a significant part of the extracellular matrix, partial neuronal degeneration and loss, a decrease in the number of dendritic processes, demyelination, and changes in glia. Indeed, a decrease in fibronectin and chondroitin sulfate proteoglycan staining was found in the hippocampus of mildly impaired aged rats and almost a complete loss of staining was seen in severely impaired aged rats (**Figure 3**). Chondroitin sulfate proteoglycans participate in multiple cellular processes, such as axonal outgrowth, axonal branching, and synaptogenesis, which are important for the formation of memory traces. The observed loss of anisotropy in senescent rats could therefore lead to impaired cortical and, particularly, hippocampal function. Moreover, the decrease in ECS volume could be responsible for the greater susceptibility of the aged brain to pathological events, particularly to ischemia, and for the poorer outcome of clinical therapy and the more limited recovery of affected tissue after anesthesia or any severe insult.

The decrease in ECS volume fraction during normal aging might be attributed to the loss of extracellular matrix molecules such as chondroitin sulfate proteoglycans, fibronectin, tenascin-R, or highly sialylated cell adhesion molecules, such as polysialylated neural cell adhesion molecules (PSA-NCAMs). The extracellular matrix molecules act like a sponge, binding a large number of water molecules and tending to occupy a lot of space, due to the mutual repulsion of their numerous negatively charged residues. Thus, their loss during aging might lead to a decrease in ECS volume. ECS shrinkage might also be related to the general decrease in hydration during aging. Life can be viewed as a process during which the highly hydrated state of embryos and neonates is transformed into a gradually more and more dehydrated one. Among other things, this dehydration is associated with an increase in colloid density, resulting in a decrease in catalytic enzyme activity and possibly contributing to the accumulation and deposition of certain substances, both intra- and extracellularly. A reduction in ECS volume would thus contribute, together with neuronal shrinkage, to the decrease in cortical volume observed during aging. These changes might be partially compensated for by concomitant reactive gliosis and glial cell hypertrophy.

Significant differences in ECS volume fraction were also found in aged female mice compared to age-matched males; an even more pronounced difference was found for ADC_W. This was probably the first observation of a gender difference in these two parameters. However, other differences have also been described: female mice have higher numbers of astrocytes and microglia in the hippocampus than

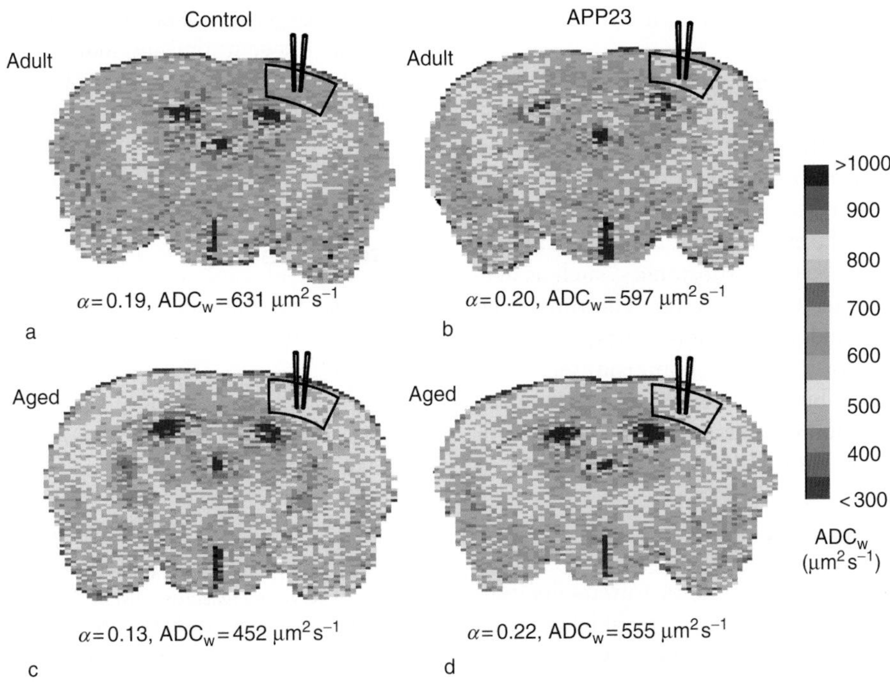

Figure 5 Apparent (water) diffusion coefficient (ADC$_w$) maps acquired in the brain of control and APP23 female mice. The mean value of ADC$_w$ was calculated in the areas indicated. In the corresponding region, tetramethylammonium ion measurements were performed on the same animals a few days later. The mean values of the ADC$_w$ and the extracellular space volume fraction (α) are given below each map. There were no differences between adult control (a) and adult transgenic (b) mice. (a, c) A decrease in ADC$_w$ and α was found during aging in control mice. In aged APP23 mice (d), there was an increase in both ADC$_w$ and α when compared with age-matched control mice (c). Adapted from Syková E, Voříšek I, Antonova T, et al. (2005) Changes in extracellular space size and geometry in APP23 transgenic mice – A model of Alzheimer's disease. *Proceedings of the National Academy of Sciences of the United States of America* 102: 479–484.

males do, and this difference increases with age. These changes might be related to the decrease in 17β-estradiol after estropause, though there is no direct evidence for this. Changes in diffusion parameters may serve as an important indicator of pathological processes, and therefore diffusion changes have been studied under many experimental pathological states using the TMA method as well as diffusion-weighted MRI, which is important for diagnostic purposes.

Changes in ECS Diffusion Parameters in an APP23 Mouse Model of Alzheimer's Disease

Neuronal loss and changes in synaptic transmission are generally considered the main reasons for memory impairment in Alzheimer's disease. However, changes in extrasynaptic transmission might also be important. Changes in ECS diffusion parameters were found in the cortex of transgenic APP23 mice, in which overproduction of mutated human amyloid precursor protein (APP) leads to amyloid plaque formation. In these animals, TMA measurements were compared with data obtained in the same animals using diffusion-weighted magnetic resonance imaging to measure the apparent diffusion coefficient of water (ADC$_W$) in the tissue (**Figure 5**). Measurements were performed *in vivo* in 6- to 8-month-old and 17- to 25-month-old hemizygous APP23 mice and in age-matched controls. In 6- to 8-month-old APP23 mice, the ECS volume fraction and ADC$_W$ were not significantly different from these factors in age-matched controls (mean α was 0.20, ADC$_W$ was 618 μm^2 s^{-1}). Aging in 17- to 25-month-old controls was accompanied by decreases in the ECS volume fraction (to 0.15) and ADC$_W$ (to 549 μm^2 s^{-1}) decreases significantly greater in females than in males, but there was no change in tortuosity. In aged, 17- to 25-month-old APP23 mice the ECS volume fraction increased to 0.22, and ECS tortuosity and ADC$_W$ increased compared to age-matched controls. These results confirm that in mice as well as in rats, aging leads to a decrease in ECS volume. In contrast, the deposition of β-amyloid is associated with an increase in ECS volume fraction, tortuosity, and ADC$_W$. The impaired navigation observed in transgenic APP23 mice in the Morris water maze correlated with their amyloid plaque load, which was twice as high in females (20%) as in males (10%). Obviously, the volume of the amyloid plaques, which are located extracellularly, is not included in the ECS volume fraction as measured by the TMA method. If the ECS volume fraction is about 20% and the amyloid plaque load in old female transgenic mice is also about 20%, the real ECS volume is the sum of both – about 40% of the total tissue volume.

Amyloid deposits, together with altered ECS diffusion properties, may therefore account for the impaired synaptic as well as extrasynaptic volume transmission and spatial cognition observed in old female transgenic mice. The enlargement of the ECS will contribute to a decrease in the extracellular concentrations of many important neuroactive substances, such as acetylcholine, dopamine, and serotonin. The functional radius of protein and peptide molecules, including nerve growth factors, will be compromised. It is therefore quite possible that there is a link between ECS volume, diffusion changes, and behavioral deficits in aged patients with Alzheimer's disease.

Concluding Remarks

The results obtained by diffusion measurements indicate that the degree of learning impairment in aged rats and mice correlates with changes in the ECS diffusion parameters. The changes in diffusion parameters during aging can therefore have an important functional significance. Anisotropy, which, particularly in the hippocampus and corpus callosum, may help to facilitate the diffusion of neurotransmitters and neuromodulators to regions occupied by their high-affinity extrasynaptic receptors, might have crucial importance for the specificity of signal transmission. The importance of anisotropy for the 'spill-over' of glutamate, for 'cross-talk' between synapses, and for LTP and LTD has been proposed. The observed loss of anisotropy in senescent rats could therefore lead to impaired cortical and, particularly, hippocampal function. The decrease in ECS size could be responsible for the greater susceptibility of the aged brain to pathological events, the poorer outcome of clinical therapy, and the more limited recovery of affected tissue after various insults.

See also: Aging of the Brain and Alzheimer's Disease; Alzheimer's Disease: An Overview; Alzheimer's Disease: MRI Studies; Amyloid: Vascular and Parenchymal; Animal Models of Alzheimer's Disease; Brain Volume: Age-Related Changes; Functional Neuroimaging Studies of Aging; Long-Term Depression (LTD): Metabotropic Glutamate Receptor (mGluR) and NMDAR-Dependent Forms; Long-Term Potentiation (LTP): Mossy Fiber cAMP-Dependent Presynaptic LTP; Long-Term Potentiation (LTP): NMDA Receptor Role; Synaptic Plasticity: Learning and Memory in Normal Aging.

Further Reading

Fuxe K, and Agnati LF (eds.) (1991) *Volume Transmission in the Brain. Novel Mechanisms for Neural Transmission.* New York: Raven Press.

Nicholson C and Phillips JM (1981) Ion diffusion modified by tortuosity and volume fraction in the extracellular microenvironment of the rat cerebellum. *Journal of Physiology (London)* 321: 225–257.

Nicholson C and Syková E (1998) Extracellular space structure revealed by diffusion analysis. *Trends in Neuroscience* 21: 207–215.

Syková E (1992) Ionic and volume changes in the microenvironment of nerve and receptor cells. In: Ottoson D (ed.) *Progress in Sensory Physiology*, pp. 1–167. Heidelberg: Springer.

Syková E (1997) The extracellular space in the CNS: Its regulation, volume and geometry in normal and pathological neuronal function. *Neuroscientist* 3: 28–41.

Syková E (2001) Glia and extracellular space diffusion parameters in the injured and aging brain. In: de Vellis J (ed.) *Neuroglia in the Aging Brain*, pp. 77–98. Totowa: Humana Press.

Syková E (2001) Glial diffusion barriers during aging and pathological states. *Progress in Brain Research* 132: 339–363.

Syková E (2003) Diffusion parameters of the extracellular space. *Israel Journal of Chemistry* 43: 55–69.

Syková E (2004) Extrasynaptic volume transmission and diffusion parameters of the extracellular space. *Neuroscience* 129: 861–876.

Syková E, Mazel T, Hasenöhrl RU, et al. (2002) Learning deficits in aged rats related to decrease in extracellular volume and loss of diffusion anisotropy in hippocampus. *Hippocampus* 12: 469–479.

Syková E, Mazel T, Vargová L, et al. (2000) Extracellular space diffusion and pathological states. *Progress in Brain Research* 125: 155–178.

Syková E, Voříšek I, Antonova T, et al. (2005) Changes in extracellular space, size, and geometry in APP23 transgenic mice – A model of Alzheimer's disease. *Proceedings of the National Academy of Sciences of the United States of America* 102: 479–484.

Zoli M, Jansson A, Syková E, et al. (1999) Intercellular communication in the central nervous system. The emergence of the volume transmission concept and its relevance for neuropsychopharmacology. *Trends in Pharmacological Sciences* 20: 142–150.

Aging: Invertebrate Models of Normal Brain Aging

M Artal-Sanz, K Troulinaki, and N Tavernarakis, Foundation for Research and Technology, Heraklion, Crete, Greece

Introduction

The free-living soil nematode *Caenorhabditis elegans* has been extensively used for studying the genetic regulation of aging, in part because of its short life span and genetic homogeneity. In the nematode, neuroendocrine signaling, nutritional sensing, and mitochondrial functions have been shown to play important roles in the determination of life span. However, aging in *C. elegans* is mainly controlled by a neuroendocrine system, the DAF-2/insulin signaling pathway, which also regulates the life span of flies and mammals, indicating that this pathway is a universal longevity regulator.

The insulin signaling pathway in *C. elegans* was first genetically identified for its effects on dauer larva formation (DAF). Dauer is an alternative developmental stage induced by harsh environmental conditions such as starvation, high population density, or high temperature. Under normal conditions *C. elegans* develops to reproductive adulthood through four larval stages (L1–L4), in 3 days. However, when conditions are adverse, larvae arrest development at the second molt, to enter the dauer stage. Dauers do not feed, are resistant to stress, and can survive up to several months. Dauer larvae are considered to be nonaging because postdauer life span is not affected by the duration of the dauer stage. In addition to insulin signaling, another neuronal pathway that regulates the choice between reproductive growth and dauer entry is the DAF-7 transforming growth factor-β (TGF-β) pathway.

The DAF-2 insulin/insulin-like growth factor-1 (IGF-1) receptor pathway is required for reproductive growth and metabolism, as well as for normal life span. *C. elegans* has a single transmembrane insulin receptor kinase, DAF-2. Upon ligand binding to DAF-2, the kinase domain of the receptor phosphorylates and activates AGE-1, which is a phosphatidylinositol 3-kinase (PI3K). Activated AGE-1 PI3K generates 3-phosphoinositides (PtdIns-3,4-P2 and PtdIns-3,4,5-P3), which are second messengers required for activation of downstream kinases. Downstream kinases include pyruvate dehydrogenase kinase and serine/threonine kinases (PDK-1, AKT-1, and AKT-2), which are protein kinase B (PKB) proteins. These protein kinases regulate the forkhead (FOXO) transcription factor DAF-16, which translocates to the nucleus depending on its phosphorylation level. Phosphorylated DAF-16 remains inactive in the cytoplasm, while upon dephosphorylation it enters the nucleus and exerts its effects on transcription. Thus, the insulin signaling pathway functions to block the nuclear localization of DAF-16. An antagonist of the DAF-2/AGE-1 signaling pathway is the DAF-18 phosphatase and tensin homolog (PTEN) lipid phosphatase (**Figure 1**).

Neuronal Insulin-Like Signaling

Mutations in *daf-2*, *age-1*, or in other genes positively regulated by *daf-2* result in constitutive developmental arrest at the dauer stage. Reducing the activity of genes that antagonize insulin signaling, such as *daf-18* PTEN or *daf-16* FOXO, suppresses the dauer-constitutive phenotype of insulin signaling mutants. While severe mutations in the insulin pathway induce dauer arrest, a milder reduction of insulin signaling results in longer life span. For example, when temperature-sensitive *daf-2* and *age-1* mutants are grown at the permissive temperature until past the dauer arrest decision point, and then shifted to higher temperatures, *daf-2* and *age-1* mutants show increased life span that is dependent on DAF-16. DAF-16 is the main downstream target and major effector of DAF-2/insulin-like signaling regulating *C. elegans* life span. Signaling via the DAF-2/insulin-like receptor antagonizes the FOXO transcription factor DAF-16 by promoting its phosphorylation. Similar modulation of insulin-like signaling pathways in the fruit fly and mouse also modify life span.

The insulin/IGF-1 signaling pathway was first linked to aging in *C. elegans* when mutations in *daf-2* were found to double the life span of the worm. Subsequent investigations, aimed at identifying in which specific cells insulin signaling controls animal aging, supported a prominent role for the nervous system. In these studies, the life span of mosaic animals that had lost *daf-2* activity in different cell lineages was examined. After fertilization, the first cell division produces the AB and the P_1 cells. The AB descendants produce most of the neurons, the hypodermis, the pharynx, the excretory glands, and the vulva, while the P_1 descendants give rise to the muscles, the germ line, the somatic gonad, the hypodermal cells, a few pharyngeal cells, and a few neurons. AB mosaics, which had lost *daf-2* activity in the AB lineage but were *daf-2*(+) in the P_1 lineage, lived twice as long

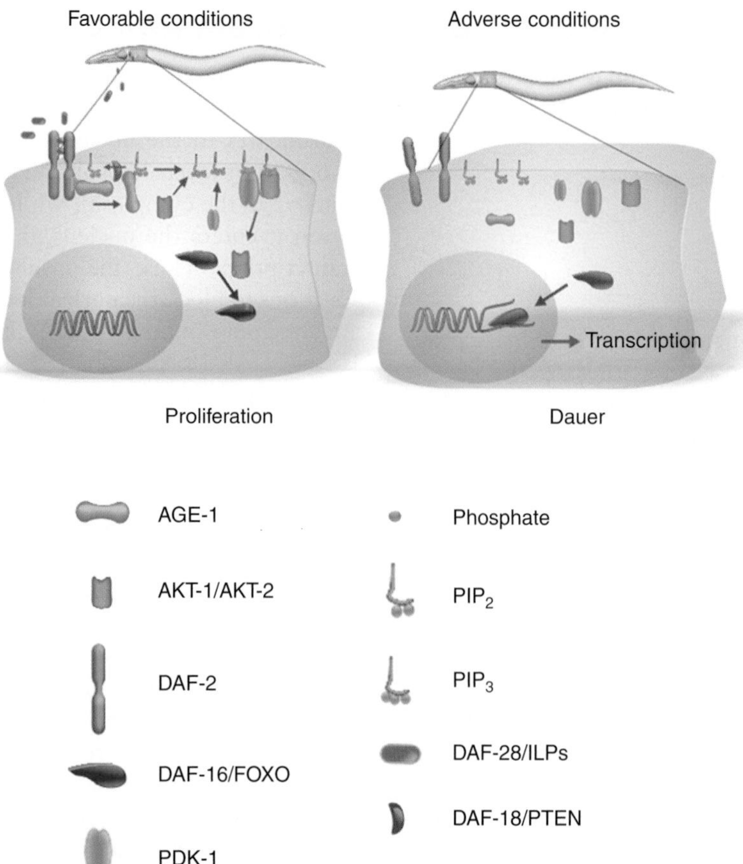

Figure 1 The *C. elegans* insulin-like signaling pathway genes regulate reproductive growth and aging. (Left): Under favorable growing conditions, insulin-like peptides (ILPs) are produced from sensory neurons, to promote the reproductive mode. Binding of ILPs to the DAF-2/insulin receptor results in the phosphorylation of AGE-1/PI3K. Activated AGE-1 generates the phosphoinositides required for the activation of downstream kinases (PDK-1 and AKT-1). This conserved signaling cascade phosphorylates the transcription factor DAF-16/FOXO, preventing its nuclear localization. Retention of DAF-16 in the cytoplasm leads to normal reproductive growth and aging. (Right): Under unfavorable conditions (e.g., crowding or starvation), insulin signaling is inhibited, resulting in the nuclear localization of DAF-16. In the nucleus, DAF-16 regulates the transcription of genes that induce dauer entry and extend life span. Protein nomenclature: AGE, advanced glycosylation end product; AKT, serine/threonine kinase; DAF, abnormal dauer formation; FOXO, forkhead transcription factors, group O; PDK, pyruvate dehydrogenase kinase; PIP, phosphoinositol phosphate; PTEN, phosphatase and tensin homolog. Illustration: Liesbeth de Jong.

as wild-type animals. Although genetic mosaic analyses of *daf-2* support the interpretation that DAF-2 signaling from the nervous system controls longevity, these experiments did not assign longevity control by *daf-2* to specific cell types. In a complementary approach, cell-type-restricted promoters were used to drive the expression of *daf-2* and *age-1* cDNAs in *daf-2* and *age-1* mutants, respectively. Transgenic expression of *daf-2* and *age-1* in neurons suppressed the life span extension phenotype of the corresponding *daf-2* and *age-1* mutants. Life span extension is not rescued when insulin signaling is restored in muscles or in intestinal cells. However, tissue-specific expression, genetic mosaic analysis, and RNA interference (RNAi) experiments indicate that *daf-16* FOXO activity in neurons accounts for not more than 20% of the longevity seen in *daf-16*(−); *daf-2*(−) double mutant animals. Instead, intestinal expression of *daf-16* is sufficient to extend the life span of these animals by 50–60%. These findings indicate that an intricate signaling network regulates aging in *C. elegans* and that neuronal insulin-like signaling controls life span by producing downstream signals that control aging of nonneuronal target tissues.

DAF-2 Insulin Receptor Function in the Nervous System

The *C. elegans* genome encodes more than 30 insulin-like ligands that might mediate input to the *daf-2* pathway through environmental cues, such as nutritional

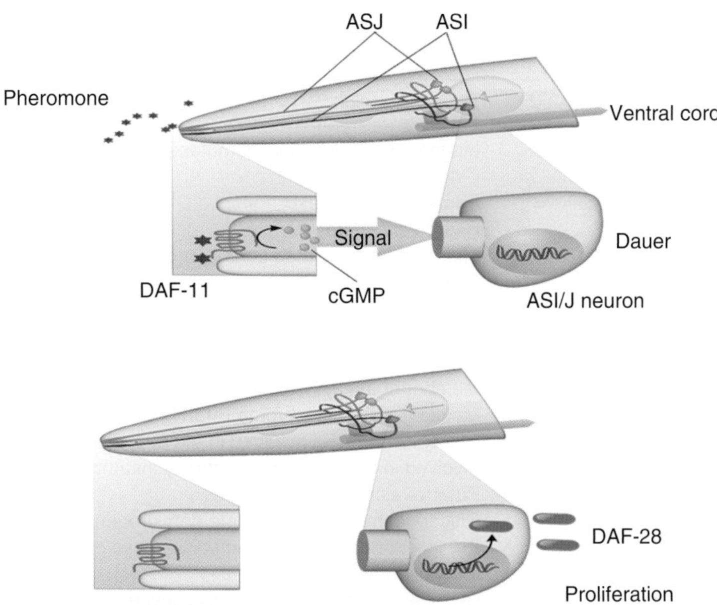

Figure 2 Sensory neurons couple environmental cues to the production of ILPs, such as DAF-28. In the ciliated sensory neurons ASJ and ASI, DAF-11/guanylyl cyclase transduces dauer pheromone signals to cyclic guanosine monophosphate (cGMP) gated ion channels to induce dauer arrest (upper panel). At low pheromone concentration, DAF-28 is produced to activate DAF-2/insulin signaling and reproductive growth (lower panel). Illustration: Liesbeth de Jong.

status or growth conditions. These insulin genes are mainly expressed in neurons, although they are also found in the intestine, epidermis, muscle, and gonad. Some of the insulin-like peptides (ILPs) have been shown to influence longevity. Thus, neuroendocrine control of aging may entail environmental cues that influence neuronal production of insulin-like ligands. For example, *daf-28* encodes an insulin-like protein, which when mutated causes dauer arrest and downregulation of DAF-2 signaling. *daf-28* is expressed in two sensory neurons (ASI and ASJ) that regulate dauer arrest. The presence of dauer pheromone is sensed by DAF-11/guanylyl cyclase, which in turn downregulates *daf-28*. Conversely, in the absence of dauer pheromone, DAF-28 is produced to induce reproductive growth (**Figure 2**). Although the dauer pheromone does not appear to influence aging, ILPs can act as either agonists or antagonists on DAF-2 to regulate metabolism, reproductive growth, and life span.

In support of this notion, mutations in two genes involved in Ca^{2+}-regulated secretion (*unc-64*, encoding syntaxin, and *unc-31*, encoding calcium-dependent activator protein for secretion, or CAPS), result in an increased life span that is dependent on *daf-16*. *unc-31* is expressed exclusively in neurons, and although *unc-64* is expressed in many secretory tissues, including the nervous system and the intestine, it is the function of *unc-64* in neurons that influences aging. In mammals, insulin secretion by β cells in the pancreas is a Ca^{2+}-regulated process. Therefore, a possible explanation is that the life span extension of *unc-64* and *unc-31* mutants is due to decreased secretion of a DAF-2 insulin-like ligand. Alternatively, *unc-64* and *unc-31* could regulate neurotransmitter input to other insulin-producing cells.

It has also been suggested that oxidative damage to neurons may be a primary determinant of life span. Loss of DAF-2 activity results in the activation of the FOXO transcription factor DAF-16, which controls the expression of antioxidant enzymes, such as superoxide dismutase (SOD) and catalase. Increased expression of these and possibly other free-radical-scavenging enzymes may protect neurons from oxidative damage. Thus, neuronal *daf-2* signaling might regulate animal life span by controlling the integrity of specific neurons that secrete neuroendocrine signals that might regulate the life span of target tissues. In support of this hypothesis, overexpression of Cu/Zn SOD exclusively in *Drosophila* motor neurons extends life span.

DAF-16 Targets

An important question that arises is how the insulin/IGF-1 pathway ultimately regulates aging. As noted earlier, the DAF-16/FOXO transcription factor is a downstream effector of insulin signaling and regulates a wide range of physiological responses by altering the expression of genes involved in metabolism and energy generation, as well as antimicrobial and cellular stress response genes (**Table 1**).

Table 1 Proteins implicated in neuroendocrine control of *C. elegans* life span, encoded by genes under control of the DAF-16/FOXO, transcription factor

Protein	*Brief description*
SOD-3	Manganese superoxide dismutase
CTL-1	Cytosolic catalase
CTL-2	Peroxisomal catalase
GST-4	Glutathione-*S*-transferase
MTL-1	Metallothionein-related cadmium-binding protein
OLD-1	Putative receptor tyrosine kinase
SCL-1	Secreted protein with sperm-coating protein (SCP) domain
LYS-7	Lysozyme
LYS-8	Lysozyme
DOD-1	Member of the cytochrome P450 family
DOD-13	Member of the cytochrome P450 family
DOD-16	Cytochrome P450, involved in the oxidation of arachidonic acid to eicosanoids
FAT-7	Involved in the biosynthesis of polyunsaturated fatty acid
DOD-11	Putative alcohol dehydrogenase
GPD-2	Glyceraldehydes-3 phosphate dehydrogenase
DAO-3	Putative tetrahydrofolate dehydrogenase/cyclohydrolase
DOD-14	Alcohol dehydrogenase
DOD-15	UDP-glucuronosyl and UDP-glucosyl transferases
GCY-6	Putative guanylyl cyclase
GCY-18	Guanylate cyclase catalytic domain
VIT-2	Vitellogenin
VIT-5	Vitellogenin
PES-2	Unknown function, putative role in ubiquitin-mediated protein degradation
PEP-2	Oligopeptide transporter
HSP16	Heat shock protein family
SIP-1	Heat shock protein
INS-7	Insulin/insulin-like growth factor-1 peptide
PNK-1	Pantothenate kinase
MRP-5	Adenosine triphosphate-binding protein, member of the subfamily C transporters

SOD is one of the most effective intracellular enzymatic antioxidants. This antioxidant enzyme catalyzes the dismutation of superoxide anions (O_2^-) to oxygen (O_2) and to the less reactive species, hydrogen peroxide (H_2O_2). *sod-3* encodes one of two manganese-containing SODs. It is localized in the mitochondrial matrix and is abundant in neural tissues. *sod-3* was one of the first recognized targets of DAF-16. *sod-3* mRNA is increased in *daf-2* mutants and undetectable in the absence of DAF-16; microarray experiments show that expression of this gene is at least tenfold higher in wild-type worms compared to DAF-16-deficient mutants. Moreover, SOD-3 is upregulated in long-lived strains, such as *age-1* and *daf-2* mutants, and in response to exogenously imposed oxidative stress in a DAF-16-dependent manner. Therefore, increased detoxification from damaging free radicals by this enzyme may contribute to life span extension.

DNA microarray analysis shows that *daf-16* affects the expression of stress response genes, the products of which directly influence aging. For example, expression of the catalase genes *ctl-1* and *ctl-2*, the glutathione-*S*-transferase gene *gst-4*, and the small heat shock protein genes increases when the activity of *daf-2* is reduced, whereas the expression of these genes decreases with the reduction of *daf-16* activity. All of these genes function to promote longevity, probably by preventing or repairing oxidative and other forms of macromolecular damage.

The gene *mtl-1*, which encodes the basally expressed form of metallothionein, is another known target of DAF-16. Metallothioneins are small cysteine-rich, metal-binding proteins protecting cells against heavy metal toxicity and reactive oxide species (ROS)-associated damage. They are induced under a variety of stress conditions, such as conditions involving metal ions, inflammation, glucocorticoids, or oxidative stress. The expression of metallothionein genes is increased in *daf-2* mutants, compared to wild-type animals. Elevated levels of these proteins as well as antioxidant enzymes are expected to decrease ROS-associated damage and hence extend life span.

An additional downstream signaling factor that is regulated by insulin/IGF-1 signaling and promotes longevity is OLD-1, a transmembrane tyrosine kinase. OLD-1 expression is increased in long-lived *daf-2* and *age-1* mutants. This increase is dependent on DAF-16. Moreover, OLD-1 is necessary for the increased longevity of *daf-2* and *age-1* mutants and its overexpression increases life span and stress resistance. *old-1* mutations render animals more sensitive to ultraviolet light, starvation, and heat stress. These results point to a positive regulatory role for OLD-1 in life span and stress resistance.

The gene *scl-1* (which encodes sperm-coating protein (SCP)-like extracellular protein) is another target of DAF-16 that is essential for life span extension. SCL-1 has an SCP domain and is homologous to the mammalian cysteine-rich secretory protein (CRISP) family. These proteins enter the secretory pathway and are either released or anchored extracellularly by a transmembrane domain or a glycophosphatidylinositol (GPI) anchor. Expression of *scl-1* is elevated in long-lived *daf-2* and *age-1* mutants and is required for their extension of life span, since downregulation of *scl-1* reduces both life span and stress resistance of these animals. However, *scl-1* is not expressed in *daf-16* mutants. *scl-1* expression correlates with dauer morphogenesis; *scl-1* expression increases (fivefold) as worms enter dauer morphogenesis and decreases (four- to sevenfold) during dauer exit. This implicates *scl-1* in both aging and dauer formation. CRISP family proteins are involved in host defense systems in various organisms, and several functions have been postulated for SCP domain proteins, including cell adhesion, ligand function, protease inhibition, and other enzymatic activities. SCP proteins function in a variety of biological processes that involve signaling, which can be reconciled with SLP-1 being a secreted protein. However, the biochemical function of SCL-1 remains unknown.

DNA microarray analysis has also revealed a number of other *daf-2*/*daf-16*-regulated genes with substantial effects on life span. These include antimicrobial genes encoding lysozymes (*lys-7* and *lys-8*). Lysozymes are upregulated in *daf-2* mutants, and RNAi with these genes suppresses longevity of *daf-2* mutants. Consistently, *daf-2* mutants are resistant to bacterial pathogens. Other genes induced in *daf-2* and repressed in *daf-16* mutants encode proteins potentially involved in the synthesis of steroid or lipid-soluble hormones (e.g., cytochrome P450s, estradiol dehydrogenases, esterases, alcohol/short-chain dehydrogenases, and UDP-glucuronosyltransferases), and genes involved in fatty acid desaturation.

Genes with expression decreased in *daf-2* mutants and increased in *daf-16* mutants include those (*gcy-6* and *gcy-18*) encoding two receptor guanylate cyclases that are expressed in neurons. Further, RNAi with these genes prolongs life span, indicating a role for insulin signaling in sensing the environment. Other genes in this group include vitellogenin genes (*vit-2* and *vit-5*; yolk protein/apolipoprotein-like) and some genes for proteases and metabolic enzymes, such as amino- and carboxypeptidases; also included are *pes-2* (encoding a protein associated with ubiquitin-mediated protein degradation) and an amino oxidase gene, an aminoacylase gene, and an oligopeptide transporter (*pep-2*) gene. Inhibition of several of these genes results in life span extension, leading to the suggestion that the life span extension of *daf-2* mutants may involve reduced turnover of specific proteins.

Studies utilizing computational tools have also identified putative DAF-16 transcriptional targets. The genomes of both *C. elegans* and *Drosophila* were surveyed for genes with DAF-16 binding sites. Orthologous genes identified were further examined in wild-type and *daf-2* mutants animals; *pnk-1* and *mrp-5* were among several putative targets. *pnk-1* encodes a pantothenate kinase that is involved in the biosynthesis of coenzyme A, which is key to fat metabolism. This gene is upregulated in *daf-2* mutants, and RNAi inactivation results in a dramatic decrease of life span of both wild-type and *daf-2* mutant worms. *daf-2* mutants have increased fat storage, which may reflect *pnk-1* upregulation. *mrp-5* encodes an adenosine triphosphate-binding cassette C transporter. Proteins of this family modulate the secretion of insulin and participate in the transport of glutathione and nucleoside analogs. Inactivation of this gene by RNAi extends life span. It has been proposed that MRP-5 could affect aging by regulating the secretion of insulin or the transport of glutathione, an enzyme required for the antioxidant defense of the organism.

It is known that both heat shock and reduced insulin signaling trigger the nuclear localization of DAF-16, where it promotes the expression of small heat shock protein (*shsp*) genes (e.g., *hsp-16.1*, *hsp-16.49*, *hsp-12.6*, and *sip-1*). The *C. elegans* transcription factor, heat-shock factor-1 (HSF-1), is also required for the life span extension, as observed in *daf-2* mutants. HSF-1 acts in response to heat stress and reduced insulin signaling to activate the expression of *shsp* genes, together with DAF-16. Interestingly, overexpression of the gene encoding HSP70F increases longevity, similar to overexpression of the *shsp* gene *hsp-16*. Heat shock proteins are involved in reparation of misfolded or damaged proteins and are essential for recovery of cells after heat treatment. This indicates that protein misfolding and aggregation are important factors in aging.

In addition to inducing the expression of genes involved in several processes, DAF-16 can also act as a transcriptional repressor. For example, DAF-16 inhibits the expression of *ins-7*. This gene encodes an insulin/IGF-1 peptide. Its expression is repressed in animals with reduced DAF-2 activity and is elevated in animals with reduced DAF-16 activity. RNAi for *ins-7* increases the life span of wild-type animals and the frequency of dauer formation. Thus, INS-7 behaves as a putative DAF-2 agonist. It is hypothesized that when DAF-2 is active, it inhibits the activity of DAF-16, which allows *ins-7* to be expressed. The production of INS-7 leads to further activation of DAF-2. However, when DAF-2 activity is reduced, DAF-16 is activated and inhibits the expression of *ins-7*.

Sensory Input and Neuroendocrine Signaling

C. elegans senses environmental cues through ciliated sensory neurons. Large numbers of genes required for the development and function of *C. elegans* sensory neurons have been identified. In addition to disrupting sensory neuron function, mutations in some of these genes increase life span.

Amphids are gustatory and olfactory neurons located at the *C. elegans* head. Gustatory neurons sense dauer pheromone, food, and amino acids in the environment. Olfactory neurons are responsible for sensing food-derived substances and volatile chemicals. To define which of these cells are involved in the regulation of life span, individual cells were ablated by a focused laser microbeam. Ablation of gustatory neurons revealed that only a specific subset (ASI and ASG, and not ADF, ASJ, and ASK) may influence life span. This was also the case for the olfactory neurons (AWA and AWC), since only the ablation of AWA extended life span, whereas, ablation of AWC had no effect. This was further confirmed by using mutants with specific defects in these neurons. Combined ablation of gustatory and olfactory neurons results in greater longevity, compared to ablation of either gustatory or olfactory neurons alone, suggesting that these neurons function in distinct pathways to control life span. Many mutations abrogating sensory neurons, including putative chemosensory receptors, extend the life span of *C. elegans*, largely in a *daf-16*-dependent manner. The life span extension caused by gustatory neurons depends on *daf-16*, indicating that these neurons might modulate the *daf-2* pathway. However, in the case of olfactory neurons, effects on life span are only partly dependent on *daf-16*. Many sensory neurons in *C. elegans* produce ILPs. Therefore, it is plausible that perception affects life span by influencing the activity of the insulin signaling pathway. However, these mutations show complicated interactions with the insulin/IGF-1 pathway. Double mutants did not live longer than the *daf-2* single mutant; instead their life span was shorter. This may indicate that sensory control of life span is only partially dependent on the *daf-2* pathway. Likewise, some gustatory and olfactory neurons enhance, whereas others reduce, longevity. Therefore, environmental signals that affect life span may be relayed by a *daf-2*-independent pathway.

A possible additional mechanism for influencing life span by sensory inputs is the regulation of lipid accumulation. *C. elegans* mutant strains with defects in neuroendocrine signaling show increased fat accumulation and extended life span. This is the case for *daf-2* insulin receptor mutants, the tryptophan hydroxylase *tph-1* serotonin defective mutant, and the *tub-1* (tubby ortholog) mutant. In *C. elegans*, *tub-1* is expressed in sensory neurons and *tph-1* is expressed in serotonergic neurons. In addition, a genome-wide RNAi screen identified several genes involved in food sensation and neuroendocrine signaling that, when knocked down, resulted in aberrant fat accumulation. These include genes for glutamate and dopamine receptors, as well as chemoreceptor and olfactory receptor genes. Consistently, *C. elegans* mutants with either structural or functional defects in nine specific ciliated neurons show increased fatty acid accumulation in the intestine. Interestingly, some mutations that increase lipid accumulation also lengthen life span. Thus, ciliated neurons may sense environmental cues and express neuropeptides and insulin ligands to regulate metabolism and life span. However, while the insulin signaling pathway is a major regulator of *C. elegans* fat storage (*daf-2* mutants show increased lipid accumulation), fat accumulation in these sensory mutants is independent of DAF-16/FOXO.

Other Neuroendocrine Mechanisms

In a chemical screen aimed at identifying drugs that delay aging, anticonvulsant medications were found to extend worm life span Anticonvulsants modulate neural activity in mammals and act presynaptically to modulate neuromuscular activity in the worm. Interestingly, anticonvulsants significantly increase the life span of *daf-2* loss-of-function mutants as well as that of *daf-16* mutants. This finding suggests that neural activity regulates aging by an additional mechanism independent of insulin-like signaling Similarly, these compounds further increased the life span of animals with mutations in genes important for the function of sensory neurons (*osm-3* and *tax-4*) and neurotransmission (*unc-31*, *unc-64*, and *aex-3*), highlighting the intricacy of the neuronal pathways that might be involved in the regulation of nematode aging.

The reproductive system also regulates aging in *C. elegans*. Ablation of the two germ line precursor cells, as well as mutations that reduce germ line proliferation, remarkably extend the *C. elegans* life span. This life span extension requires the presence of the somatic gonad, since ablation of both germ line and somatic gonad has no effect on life span. The life span extension of germ-line-ablated animals depends on DAF-16/FOXO and may be mediated hormonally, since it also depends on DAF-12, a nuclear hormone receptor, and DAF-9, a cytochrome P450 involved in the production of steroid hormones (3-keto sterols) that function as DAF-12 ligands. Another gene required for the increased life span associated with germ line loss is *kri-1*. The *kri-1* gene encodes a conserved protein with ankyrin repeats and is expressed in pharynx and intestine. Germ line ablation results in the nuclear localization of DAF-16 in the intestine (the worm's adipose tissue), and intestinal expression is sufficient to account for the observed longevity. KRI-1 and to a lesser extent DAF-12 and DAF-9 are required for the DAF-16 nuclear localization in the intestine of germ-line-defective animals. However, although loss of DAF-2 receptor activity promotes the nuclear localization of DAF-16 in many tissues, including the intestine, DAF-16 localization is not dependent on KRI-1, DAF-12, or DAF-9. Moreover, *kri-1* RNAi completely suppresses the life span extension of germ line mutants, while it has no significant effect on wild-type or *daf-2* mutants, indicating that modulation of life span by KRI-1 is specific to the reproductive-signaling pathway. Thus, the role of lipophilic hormones and KRI-1 on the nuclear localization of DAF-16 is germ line specific and independent of insulin signaling. Therefore, the germ line might possess a specific endocrine system to influence aging.

Nevertheless, neuronal and gonadal endocrine signaling mechanisms appear to interact in a complex manner. Somatic gonad ablation prevents the life span extension of germ-line-ablated wild-type animals, but it does not completely prevent life span extension in animals that lack olfactory neurons. Similarly, germ line ablation further extends the life span of *daf-2* mutants independently of whether or not the somatic gonad is present.

Concluding Remarks

The nervous system performs the task of sensing and integrating environmental cues into coordinated physiological responses that will ensure maximal survival and reproductive fitness. In *C. elegans*, food availability, temperature, and a secreted pheromone are some of the sensory inputs that regulate the decision of entering the metabolically active reproductive mode or shifting to the nonreproducing, nonfeeding dauer larva stage, with large amounts of stored fat.

Despite its apparent simplicity, *C. elegans* has a surprisingly sophisticated neuroendocrine system that regulates development, metabolism, and life span. Both, insulin-like and TGF-β signaling pathways act in parallel to regulate development and metabolism, with the insulin-like signaling pathway playing a major role in life span regulation. Importantly, the regulation of life span by insulin/IGF signaling is conserved across taxa, and reduction of insulin signaling has been shown to extend life span in worms, flies, and mammals. Similarly, the physiological processes involved in the aging process also appear to be conserved. For example, signals from the reproductive system also influence life span in mammals, and dietary restriction has been shown to extend life span in a wide variety of organisms. Likewise, sensory perception could also regulate life span in higher organisms, since blocking the sense of taste reduces insulin secretion in mammals, and the smell of food increases insulin levels in humans.

How physiological processes are coordinated by neuroendocrine signaling to meet the biological demands of an organism is still not completely understood. More than 30 ILPs are encoded in the *C. elegans* genome, some of which are agonists and others of which are antagonists. Some have been shown to influence aging, but many remain to be functionally characterized. It is of fundamental importance to understand which cells or tissues emit or receive signals to coordinate the aging process at the level of the whole organism. *C. elegans* has been instrumental for the discovery of conserved molecular pathways regulating aging. Its relatively short life span and its amenability for genetic and molecular analyses make it an ideal organism to pursue these studies further, aiming to ultimately understand why and how animals age.

See also: Aging of the Brain; Gene Expression in Normal Aging Brain; Insulin-Like Growth Factor Signaling and Actions in Brain; Neuroendocrine Aging: Pituitary Metabolism; Vesicular Neurotransmitter Transporters.

Further Reading

Aamodt E (2006) *The Neurobiology of Caenorhabditis elegans*. Amsterdam; Boston: Elsevier/Academic Press.

Ailion M, Inoue T, Weaver CI, et al. (1999) Neurosecretory control of aging in *Caenorhabditis elegans*. *Proceedings of the National Academy of Sciences of the United States of America* 96: 7394–7397.

Alcedo J and Kenyon C (2004) Regulation of *Caenorhabditis elegans* longevity by specific gustatory and olfactory neurons. *Neuron* 41: 45–55.

Antebi A (2004) Long life: A matter of taste (and smell). *Neuron* 41: 1–3.

Boulianne GL (2001) Neuronal regulation of lifespan: Clues from flies and worms. *Mechanisms of Ageing and Development* 122: 883–894.

Braeckman BP, Houthoofd K, and Vanfleteren JR (2001) Insulin-like signaling, metabolism, stress resistance and aging in *Caenorhabditis elegans*. *Mechanisms of Ageing and Development* 122: 673–693.

Evason K, Huang C, Yamben I, et al. (2005) Anticonvulsant medications extend worm life-span. *Science* 307: 258–262.

Finch CE and Ruvkun G (2001) The genetics of aging. *Annual Review of Genomics and Human Genetics* 2: 435–462.

Hekimi S and Guarente L (2003) Genetics and the specificity of the aging process. *Science* 299: 1351–1354.

Kenyon C (2005) The plasticity of aging: Insights from long-lived mutants. *Cell* 120: 449–460.

Libina N, Berman JR, and Kenyon C (2003) Tissue-specific activities of *Caenorhabditis elegans* DAF-16 in the regulation of lifespan. *Cell* 115: 489–502.

Mak HY, Nelson LS, Basson M, et al. (2006) Polygenic control of *Caenorhabditis elegans* fat storage. *Nature Genetics* 38: 363–368.

Murphy CT, McCarroll SA, Bargmann CI, et al. (2003) Genes that act downstream of DAF-16 to influence the lifespan of *Caenorhabditis elegans*. *Nature* 424: 277–283.

Schreibman MP and Scanes CG (1989) *Development, Maturation, and Senescence of Neuroendocrine Systems: A Comparative Approach*. San Diego: Academic Press.

Tatar M, Bartke A, and Antebi A (2003) The endocrine regulation of aging by insulin-like signals. *Science* 299: 1346–1351.

Thomas JH (1999) Lifespan. The effects of sensory deprivation. *Nature* 402: 740–741.

Walker DW, McColl G, Jenkins NL, et al. (2000) Evolution of lifespan in *Caenorhabditis elegans*. *Nature* 405: 296–297.

Wolkow CA (2002) Life span: Getting the signal from the nervous system. *Trends in Neurosciences* 25: 212–216.

Wood WB (1988) *The Nematode Caenorhabditis elegans*. Cold Spring Harbor, NY: Cold Spring Harbor Laboratory.

Yeoman MS and Faragher RG (2001) Aging and the nervous system: Insights from studies on invertebrates. *Biogerontology* 2: 85–97.

Relevant Websites

http://www.arclab.org – Aging Research Centre.

http://www.afar.org – American Federation for Aging Research.

http://www.ncoa.org – National Council on Aging.

http://www.nia.nih.gov – National Institute on Aging (U.S. National Institutes of Health).

http://sageke.sciencemag.org – Science of Aging Knowledge Environment, an interdisciplinary repository of issues related to aging (American Association for the Advancement of Science).

http://www.geron.org – The Gerontological Society of America.

Agnosia

D Tranel and N L Denburg, University of Iowa Hospitals and Clinics, Iowa City, IA, USA

'Agnosia,' a neurological term of Greek origin (a + Greek *gnosis*), signifies a lack of knowledge and is virtually synonymous with an impairment of recognition. In the traditional literature, two types of agnosia were commonly described. 'Associative' agnosia referred to as failure of recognition that results from defective activation of information pertinent to a given stimulus. 'Apperceptive' agnosia referred to a disturbance of the 'integration' of otherwise normally perceived components of a stimulus.

Teuber in 1968 gave a narrower definition, in which agnosia was synonymous with having "normal percepts stripped of their meaning." In this sense, agnosia is conceptualized as a disorder of memory, and only associative agnosia qualifies for this stricter definition. In practical terms, however, it has been useful to retain the concept of apperceptive agnosia and to maintain a distinction between apperceptive and associative agnosia. In both conditions, recognition is disturbed. In the apperceptive variety, the problem can be traced, at least in part, to faulty perception, usually in reference to aspects of higher-order perceptual capacities (it is not appropriate to use the term agnosia for conditions in which perceptual problems are severe and obviously preclude the patient's apprehension of meaningful information). In associative agnosia, perception is largely intact, and the recognition defect is strictly or primarily a disorder of memory.

The difficulties of trying to separate apperceptive and associative forms of agnosia underscore the fact that the processes of perception and memory are not discrete. Rather, those processes operate on a physiological and psychological continuum, and demarcation of a clear separation point at which perceptual processes end and memory processes begin is simply not possible. Many patients with recognition defects will have elements of both conditions, that is, high-level perceptual problems and disturbances in memory. Some, however, can be classified unequivocally into one type or the other. For these reasons, the following operational definitions are appropriate. Associative agnosia is a modality-specific impairment of the ability to recognize previously known stimuli (or new stimuli for which learning would normally have occurred) that occurs in the absence of disturbances of perception, intellect, or language, and is the result of acquired cerebral damage. The designation appreciative agnosia applies when the patient meets the preceding definition in all respects except that perception is altered.

The term 'agnosia' should be restricted to situations in which recognition impairments are confined to one sensory modality, for example, vision, or audition, or touch. When recognition defects extend across two or more modalities, the appropriate designation is 'amnesia'. As noted, the term agnosia should not be used for patients in whom recognition defects develop in connection with major disturbances of basic perception. Nor should the term be applied to patients with major impairments of intellect. Finally, the term agnosia-should be reserved for conditions that develop suddenly, following the onset of acquired cerebral dysfunction.

One other important distinction is between 'recognition' and 'naming.' The two capacities are often confused. It is true that recognition of an entity, under normal circumstances, is frequently indicated by naming (e.g., that is a 'groundhog' or that is 'Joe Montana'). Studies of brain-injured subjects, however, have shown clearly that recognition and naming are dissociable capacities, and the two terms should not be used interchangeably. Damage in the left inferotemporal region, for example, can render a patient incapable of naming a wide variety of stimuli, while leaving unaffected the patient's ability to recognize those stimuli. For the two preceding examples, for instance, the patient may produce the descriptions of 'that's a roly-poly animal that digs holes under barns and hibernates in the winter,' and 'that's the guy from Notre Dame who was a famous quarterback and won lots of football championships.' Both descriptions indicate unequivocal recognition of the specific entities, even if their names are never produced. In short, it is important to maintain a distinction between recognition, which can be indicated by any number of responses signifying that the patient understands the meaning of a particular stimulus, and naming, which may not, and need not, accompany accurate recognition. The patient with agnosia fails to experience familiarity with the stimulus, and is thus unable to evoke its meaning, use, or relevant relationships in both verbal and nonverbal terms.

In principle, agnosia can occur in any sensory modality, relative to any type of entity or event. In practice, however, some types of agnosia are far more frequent. 'Visual agnosia,' especially agnosia for faces ('prosopagnosia'), is the most commonly encountered form of recognition disturbance affecting a primary sensory modality. Visual agnosia is a disorder of recognition confined to the visual realm, in which a patient cannot arrive at the meaning of some or all categories of previously known nonverbal visual stimuli, despite normal or near-normal visual perception and intact alertness, attention, intelligence, and language. Most patients manifest a comparable defect in the anterograde compartment; that is, they cannot recognize new nonverbal, visual stimuli that would normally have been learned after adequate exposure.

The condition of 'auditory agnosia' is rarer, followed by the even less frequent 'tactile agnosia.' A frequently encountered condition which also conforms to the designation of agnosia is a disturbance in the 'recognition of illness,' or what has been termed 'anosognosia'.

It is important to distinguish anosognosia from several closely related conditions. One is 'anosodiaphoria', a term that refers to the condition in which a patient acknowledges, but fails to appreciate the significance of, acquired impairments in physical or psychological function. Although anosodiaphoria is not a true form of agnosia, in practice there is a certain degree of overlap between anosodiaphoria and anosognosia. In fact, it is common to observe that blatant forms of anosognosia, for example, denial of hemiplegia, tend to evolve over time, as the patient recovers, into various degrees of anosodiaphoria. Another condition refers to a disorder of body schema. Body schema disturbances are conditions in which patients become unable to localize various parts of their bodies. The most common manifestations are 'autotopagnosia,' 'finger agnosia,' and 'right–left disorientation' (the latter two being essentially partial forms of the first). Autotopagnosia refers to a condition in which the patient loses the ability to identify parts of the body, either to verbal command or by imitation. In its most severe form, the disorder affects virtually all body parts; however, this is quite rare, and it is far more common to observe partial forms of the condition, including deficits in finger localization (finger agnosia) and right–left discrimination.

Accurate detection and diagnosis of agnosia are important on several accounts. Both visual and auditory agnosia are strongly associated with the presence of bilateral cerebral disease, and the presence of one of these conditions can be a useful clue regarding the localization of brain dysfunction. This can be especially helpful in the early stages of acquired cerebral dysfunction, when even modern neuroimaging procedures may fail to detect a lesion. Such conditions furnish additional diagnostic clues because they are typically associated with cerebrovascular disease affecting the territories of the posterior or middle cerebral arteries. Furthermore, unilateral disease involving the dominant parietal lobe has recently been implicated in both tactile agnosia and apraxia. To avoid misdiagnosis, it is important to note that the complaints or behaviors of patients with agnosia can seem so bizarre as to raise questions about their veracity. There was, indeed, a time when it was doubted whether such conditions existed at all. That agnosic conditions do occur is no longer a contentious issue; nonetheless, clinicians may be skeptical of a patient who suddenly claims an inability to recognize familiar faces, despite normal vision, or of a patient who suddenly behaves as though all auditory information had lost its meaning. A particularly unusual case of agnosia was recently reported in which a child with sleep-induced electrophysiological abnormalities involving the occipito-temporal regions and episodic seizure disorder demonstrated stable defects in visual–spatial abilities, as well as visual agnosia.

Despite their relative rarity, agnosias have also proved to be important 'experiments of nature,' and they have assisted with the investigation of the neural bases of human perception, learning, and memory. Careful study of agnosic patients over the past couple of decades, facilitated by the advent of modern structural (computed tomography, magnetic resonance) and functional (positron emission tomography, functional magnetic resonance) neuroimaging techniques, and by the development of sophisticated experimental neuropsychological procedures, has yielded important new insights into the manner in which the human brain acquires, maintains, and retrieves various types of knowledge.

See also: Amnesia: Declarative and Nondeclarative Memory; Prosopagnosia; Recognition Memory; Shape Representation in Inferotemporal Cortex.

Further Reading

Bauer RM and Demery JA (2003) Agnosia. In: Heilman KM and Valenstein E (eds.) *Clinical Neuropsychology,* 4th edn., pp. 236–295. New York: Oxford University Press.

Caselli RJ (1991) Rediscovering tactile agnosia. *Mayo Clinicical Proceedings* 66: 129–142.

Crutch SJ, Warren JD, Harding L, and Warrington EK (2005) Computation of tactile object properties requires the integrity of praxic skills. *Neuorpsychologia* 43: 1792–1800.

Damasio AR and Damasio H (1994) Cortical systems for retrieval of concrete knowledge: The convergence zone framework. In: Koch C and Davis JL (eds.) *Large-Scale Neuronal Theories of the Brain*, pp. 61–74. Cambridge, MA: MIT Press.

Damasio AR, Damasio H, Tranel D, and Brandt JP (1990) Neural regionalization of knowledge access: Preliminary evidence. *Symposia on Quantitative Biology* 55: 1039–1047.

Damasio AR, Tranel D, and Damasio H (1990) Face agnosia and the neural substrates of memory. *Annual Review of Neuroscience* 13: 89–109.

Damasio H, Grabowski TJ, Tranel D, Hichwa RD, and Damasio AR (1996) A neural basis for lexical retrieval. *Nature* 380: 499–505.

Denburg NL and Tranel D (2003) Acalculia and disturbances of the body schema. In: Heilman KM and Valenstein E (eds.) *Clinical Neuropsychology,*, 4th edn., pp. 161–184. New York: Oxford University Press.

Eriksson K, Kylliainen A, Hirvonen K, Nieminen P, and Koivikko M (2003) Visual agnosia in a child with non-lesional occipito-temporal CSWS. *Brain & Development* 25: 262–267.

Teuber H-L (1968) Perception. In: Weiskrantz L (ed.) *Analysis of Behavioral Change*, pp. 274–328. New York: Harper & Row.

Tranel D and Damasio AR (1985) Knowledge without awareness: An autonomic index of facial recognition by prosopagnosics. *Science* 228: 1453–1454.

Tranel D and Damasio AR (1996) The agnosias and apraxias. In: Bradley WG, Daroff RB, Fenichel GM, and Marsden CD (eds.) *Neurology in Clinical Practice*, 2nd edn., pp. 119–129. Stoneham, MA: Butterworth Publishers.

Vignolo LA (1982) Auditory agnosia. *Philosophical Transactions of the Royal Society of London* 298: 49–57.

Warrington EK and McCarthy RA (1987) Categories of knowledge: Further fractionation and an attempted integration. *Brain* 110: 1273–1296.

Agonistic and Affiliative Signals: Resolutions of Conflict

R M Seyfarth and D L Cheney, University of Pennsylvania, Philadelphia, PA, USA

Vocal Displays

Most competitive interactions among animals take the form of vocal, visual, or gestural displays: loud calls, threatening postures, or other behaviors that allow rivals to assess each other without escalating the displays into potentially costly attacks. Regardless of the species involved, displays have one essential property: they involve signals of competitive ability that are difficult, if not impossible, to fake. Male red deer, for example, challenge and assess one another by the amplitude and pitch of their roars. These acoustic features are reliable indicators of size and endurance because only large males in excellent condition can produce loud, low-pitched roars. In much the same way, songbirds challenge and assess one another by the size of their song repertoires. Males with large and complex song repertoires are often older, more experienced, or in better condition than others. Repertoire size is therefore a reliable indicator of a male's age and condition, which in turn is correlated with his competitive ability.

Because natural selection favors the skeptical observer and acts against any individuals who allow themselves to be duped by traits that are unrelated to fighting ability, the only displays that persist over time are 'honest' indicators of a male's condition. Displays will always be more common than actual fighting because, regardless of the competitive ability he brings to the table, it invariably pays a male to display first, before the fight escalates and results in injury. Avoiding injury is of paramount importance because the cost of being injured almost always outweighs the benefits to be gained from any single dispute.

Male baboons' competitive displays take the form of violent chases and loud 'wahoo' calls. Wahoos are low-pitched calls that can be produced only by adult males in good condition. They are costly to produce, not just because of their loudness and low pitch, but also because males give them in long bouts, often as they run or jump through trees. A wahoo display is therefore an exhausting demonstration of a male's stamina and coordination. High-ranking males are more likely than low-ranking males to enter wahoo contests. They also give wahoos at the highest rate and produce wahoos with longer and louder *hoo* syllables than the wahoos of other males. As males age and fall in rank, they are less likely to enter wahoo contests, and the quality of their wahoos deteriorates substantially. The predictable relation between a male's rank and the quality of his wahoos allows competitors to assess each other without actual fighting. Males of very disparate ranks seldom engage in wahoo contests, presumably because subordinate males can assess, through their rival's wahoos and behavior, that they are outmatched. By contrast, wahoo contests involving males of similar rank – whose wahoos sound more alike – occur at high rates. Their contests also involve longer calling bouts, occur for unpredictable reasons, and are more likely to escalate to physical fights and wounding.

Vocalizations as Aggressive Signals

Wahoos allow male baboons to assess each other without escalating the contest to fighting. Other primate vocalizations also function to mediate and influence aggressive interactions. Female baboons, like females in most Old World monkey species, remain in their natal groups throughout their lives and maintain close bonds with their matrilineal kin. They assume dominance ranks similar to their mothers', with the result that the female dominance hierarchy is a stable hierarchy of matrilines. During aggressive disputes against lower-ranking opponents, female monkeys often form alliances both with their close relatives and with higher-ranking individuals. In baboons, most alliances do not involve physical intervention but occur in the form of vocal support: a female utters a series of threat grunts as she observes another female threaten another. In a series of playback experiments, subordinate female baboons were played the threat grunts of their opponent's relative within minutes after they had been threatened. As one control, they heard the same individual's threat grunt in the absence of a recent fight. After receiving aggression, subjects responded strongly to the threat grunts of their opponent's relative, and in the next hour they avoided both their opponent and the relative whose threat grunt they had heard. In contrast, if they had not recently been threatened, subjects ignored the threat grunts, and they did not try to avoid the signaler or her relative. In this case, subjects appeared to regard the threat grunts as directed as someone else.

Vocal 'alliances,' therefore, appear to serve the same function as alliances involving chases and threats. Rather than physically intervening, however, female baboons signal their support by communicating their willingness to do so.

The Resolution of Ambiguity

Communicative signals can also resolve the ambiguity surrounding less aggressive interactions. For example, when a high-ranking female baboon approaches a lower-ranking one, there is always some ambiguity about what will happen next. On the one hand, the high-ranking female might be attempting to groom the lower-ranking one. On the other hand, she might be attempting to supplant or threaten the lower-ranking female from a desirable resource. This uncertainty creates a dilemma, both for high-ranking females who want to behave in a friendly manner and for low-ranking females who are reluctant to give up a resource. In the face of such uncertainty, analyses based on game theory predict the evolution of low-cost, reliable signals that provide accurate information about the signaler's disposition and motivation. In baboons and other monkeys, grunts serve this function. They are individually distinctive and highly predictive of friendly behavior, and listeners respond accordingly. If a high-ranking female baboon grunts while approaching a lower-ranking one, the lower-ranking one is likely to remain seated. High-ranking females almost never threaten a lower-ranking female after grunting to her. Conversely, if the approaching female remains silent, the lower-ranking female usually moves away.

Finally, when aggression does occur, communicative signals can mollify its effects. Nonhuman primates are frequently aggressive toward one another, yet they live in relatively stable, cohesive social groups. A number of studies have shown that opponents mitigate the effects of aggressive competition by grooming or interacting in a friendly manner soon after they fight. In baboons, such interactions are usually preceded by a grunt from the dominant aggressor to her subordinate opponent. Dominant female baboons grunt to their subordinate victims following roughly 13% of all fights. These grunts appear to serve a reconciliatory function because subordinate victims are more likely both to approach their former opponent and to tolerate her approaches if her opponent has grunted to her than if she has remained silent.

Testing the Reconciliatory Hypothesis

The hypothesis that grunts alone, in the absence of other corroborating behavior, can serve a reconciliatory function, has been tested in a playback experiment that mimicked vocal reconciliation. Within minutes of being threatened by a higher-ranking female, subjects were played their opponent's grunt. As controls, subjects heard either no grunt or the grunt of an uninvolved, high-ranking female unrelated to their opponent.

After hearing their opponent's grunt, subjects approached their opponent and tolerated her approaches – by not moving away – at significantly higher rates than they did under baseline conditions, when no aggression had occurred. These results confirmed observational results, which indicated that rates of approaching and grunting increase substantially after conflict. Apparently, a reconciliatory grunt caused subordinate females not just to relax but to seek out their former opponent.

If, however, subjects heard either no grunt or the grunt of a different dominant female, they continued to avoid their opponent and retreated from her approaches. Subjects approached their former opponent after only 2% of the trials involving either no grunt or the control female's grunt. In contrast, they did so in 42% of the trials conducted after playback of their opponent's grunt.

Subjects did not, however, simply change their disposition toward *any* female whose grunt they heard, because hearing the control female's grunt caused no change in their behavior. After playback of the control female's grunt, subjects did not approach that female at higher rates or attempt to interact with her: the change in their behavior was specific to their opponent. They appeared to regard their opponent's grunt as directed at themselves, and they acted as if they regarded the control female's grunt as irrelevant and directed at someone else.

This is not to say, however, that aggression causes baboons to attend only to their opponent's vocalizations and to ignore all other individuals' calls. Although they appear to regard the grunts of unrelated females as irrelevant to the fight, this is not true of the grunts of their opponent's relatives. To test how kinship affects reconciliation, subjects were played the 'reconciliatory' grunt of their opponent's relative soon after being threatened. A grunt from a dominant female from a different matriline served as a control. Once again, subjects responded as if they assumed that the relative's vocalization was directed at themselves and related to the recent fight. Moreover, the calls seemed to serve a reconciliatory function. After hearing the grunt, subjects did not avoid their opponent or her relative; indeed, they often approached both the relative and the opponent. And when they did so, their first interaction with their opponent was less likely to be submissive, and more likely to be friendly, if they had heard a reconciliatory grunt from their opponent's relative than if they had heard a grunt from anyone else.

The Cognitive Mechanisms Underlying Reconciliation

What are the mechanisms that underlie vocal reconciliation in baboons? In its richest interpretation, 'reconciliation' implies that the dominant female recognizes that her victim is anxious and afraid – emotions that she herself is not experiencing – and grunts to alleviate the subordinate's anxiety. The subordinate, in turn, recognizes the dominant's intent to reconcile. In contrast, an alternative hypothesis based on learned contingencies makes no assumptions about animals' ability to recognize other individuals' mental states. It argues only that dominant females grunt to subordinates in order to influence their behavior. Through experience, dominant females learn that subordinate females are less likely to move away from them when they grunt than when they remain silent. Being sensitive to contingencies, subordinate females learn that grunts are associated with friendly behavior, so they do not move away when their former opponent grunts to them.

The playback experiments just described indicated that subordinate females were more likely to approach their former opponent and to tolerate her approaches if they heard her grunt than if they did not. In contrast, hearing the grunt of a different, unrelated dominant female produced no change in their behavior. They appeared to regard that call as directed at someone else. These results are consistent with the hypothesis that baboons have a rudimentary understanding of other individuals' intentions toward themselves, an ability that constitutes a crucial precursor to the attribution of other individuals' mental states. Alternatively, a recent interaction with a particular individual might simply prime baboons to attend preferentially to that individual's vocalizations. This more parsimonious explanation does not require that baboons make inferences about the directedness of calls.

A playback experiment to examine whether baboons' responses to another female's vocalizations are influenced by the nature of prior interactions with that individual again followed a matched-pair design. In the first condition, a subordinate female was played the threat grunts of a dominant female shortly after that female had threatened her. Because females sometimes threaten their victim again soon after their original fight, listeners might interpret these threat grunts as an indicator of renewed aggression directed at them. In the second condition, the same subordinate female was played the same dominant female's threat grunts shortly after the two had groomed. Because females almost never threaten a recent grooming partner, listeners might interpret the call as directed at someone else.

If baboons' responses to threat grunts are simply the result of priming by a recent interaction, subjects' responses after being threatened should have been the same as their responses after being groomed. If, however, listeners take into account the nature of recent interactions when making inferences about the intended recipient of a call, they should interpret the two threat grunts differently – directed at themselves after aggression, but directed at someone else after grooming. Specifically, subjects might respond more strongly to threat grunts after receiving aggression than after a grooming bout. We also predicted that a subject would be less likely to approach her former opponent and more likely to retreat from her approaches after being threatened by her than after having been threatened by a different dominant female. In contrast, when the subject heard the same female's threat grunts after grooming with her, she should have been just as likely to approach and to tolerate her approaches as she was after having been groomed by a different dominant female.

Subjects responded more strongly to a dominant female's threat grunts after being threatened by her than after grooming with her. After aggression, subjects were quicker both to look toward the speaker and to move away from the area. In the ensuing 15 min, they were also less likely to come near their former opponent than they were after they had been threatened by a different female. In contrast, when subjects heard the same female's threat grunt after a grooming bout, they were just as likely to tolerate her approaches as they were after a grooming bout with a different female. Finally, subjects were significantly more likely to approach and to tolerate the approaches of the dominant female if they heard her threat grunts after grooming than after a threat.

As in the experiments testing reconciliatory grunts, subjects' responses were specific to the calls of their former opponent. Hearing their opponent's threat grunt did not affect the likelihood that subordinate subjects would approach another, uninvolved dominant female or the likelihood that they would be supplanted when approached. Taken together, therefore, these experiments suggest that female baboons make inferences about the intended target of a vocalization, even in the absence of visual cues, and that the nature of prior interactions affects subsequent behavior. After a fight, the subordinate assumes that the dominant has aggressive intentions toward her. After grooming, or after hearing a reconciliatory grunt, she makes the opposite attribution.

It seems likely that baboons make inferences about the intended target of a call whenever they hear any vocalization. For example, when subjects heard the threat grunts of their opponent's relative soon after being threatened, they treated the grunts as a vocal 'alliance' directed at themselves. In contrast, when they heard the same threat grunts in the absence of aggression, they ignored the call and acted as if they assumed that the call was directed at someone else. Similarly, when subjects heard the 'reconciliatory' grunt of their opponent's relative after a fight, they often approached both their opponent and her relative. They did not do so, however, if they had heard the 'reconciliatory' grunt of another, unrelated dominant female. Here again, subjects behaved as if they assumed that a grunt from an opponent's relative must be directed at themselves, as a consequence of the fight. An unrelated female's grunt was irrelevant.

The ability to distinguish signals that are directed at oneself from those directed at someone else appears to be widespread in animals. For example, studies of 'eavesdropping' in birds indicate that listeners readily distinguish between songs that are directed at a third party and ones that are directed at themselves. To date, however, most of the evidence for this ability has come from studies in which individuals are interacting with only one or a few other animals and when factors such as the location of the signaler and the nature and pattern of the song provide information about the intended recipient. The challenge of inferring both the intended target of a signal and the signaler's probable behavior may be considerably more difficult in large social groups.

When deciding, Who, me? on hearing a vocalization, baboons must take into account the identity of the signaler (Who is it?), the type of call given (Was it friendly or aggressive?), the nature of their prior interactions with the signaler (Were they aggressive, friendly, or neutral?), and the correlation between past interactions and future ones (Does a recent grooming interaction lower or increase the likelihood of aggression?). Learned contingencies doubtless play a role in these assessments. Because listeners' responses depend on simultaneous consideration of all these factors, however, this learning is likely to be both complex and subtle.

Moreover, explanations based solely on learned contingencies seem unable to explain some aspects of baboons' behavior. For example, in the reconciliatory grunt experiments, subjects who heard their opponent's grunt following a fight were even more likely to approach their opponent than they were under baseline conditions, in the absence of a fight. If listeners' responses were guided solely by learned contingencies, they should have associated the call only with a low probability of aggression. Hearing the call should have returned their behavior to baseline tolerance levels, but it should not have induced them to approach their former opponent. Instead, females acted as if they interpreted their opponent's grunt as targeted specifically at them, as a directed signal of benign intent. They therefore specifically sought out their opponent.

When attending to vocal signals, female baboons appear to take into account, not only the signaler's identity and her probable subsequent behavior, but also the target of her attention. The ability to integrate these social cues simultaneously may represent a first critical step toward the recognition of other individuals' intentions and motives. In children, inferences about other individuals' intention constitute an early precursor to language learning and full mental state attribution. Do monkeys have a rudimentary understanding about other individuals' intentions toward themselves? If they do, this understanding would represent a crucial first step toward a communication system like language, in which speakers and listeners routinely assess each other's motives, beliefs, and knowledge.

See also: Animal Communication: Honesty and Deception; Communication Networks and Eavesdropping in Animals; Electrical Perception and Communication; Endocrinology of Animal Communication: Behavioral; Pheromones and other Chemical Communication in Animals; Primate Communication: Evolution; Seismic and Vibrational Signals in Animals; Visual Signaling in Animals.

Further Reading

Aureli F and de Waal F (eds.) (2000) *Natural Conflict Resolution.* Berkeley: University of California Press.

Cheney DL and Seyfarth RM (1997) Reconciliatory grunts by dominant female baboons influence victims' behaviour. *Animal Behaviour* 54: 409–418.

Engh AE, Hoffmeier RR, Cheney DL, and Seyfarth RM (2006) Who, me? Can baboons infer the target of vocalisations?. *Animal Behaviour* 71: 381–387.

Kitchen DM, Seyfarth RM, Fischer J, and Cheney DL (2003) Loud calls as an indicator of dominance in male baboons. *Papio cynocephalus ursinus. Behavioral Ecology & Sociobiology* 53: 374–384.

MaynardSmith J (1982) *Evolution and the Theory of Games.* Cambridge, UK: Cambridge University Press.

McGregor PK (ed.) (2005) *Animal Communication Networks.* Cambridge, UK: Cambridge University Press.

Searcy WA and Nowicki S (2005) *The Evolution of Animal Communication: Reliability and Deception in Signaling Systems.* Princeton, NJ: Princeton University Press.

Silk JB, Kaldor E, and Boyd R (2000) Cheap talk when interests conflict. *Animal Behaviour* 59: 423–432.

Agraphia

S W Anderson, D Tranel, and N L Denburg,
University of Iowa Health Care, Iowa City, IA, USA

Agraphia (a + Greek *graphein*) is a term used to denote an impairment of the ability to write caused by cerebral dysfunction. The term generally is reserved for instances of writing impairment due to a cerebral insult, acquired after the person has learned to write, that is, in someone who was previously literate. Writing impairments of developmental origin are usually considered part of dyslexia. Acquired impairments secondary to tremor, hemiparesis, or other basic motor defects are usually excluded from the definition of agraphia. Agraphia may involve impairment of one or multiple components of the psychomotor and linguistic aspects of writing, including grapheme formation, spelling, word selection, grammar, and spatial arrangement. Agraphia rarely occurs in isolation, and more often appears in the context of widespread language impairments, visuospatial defects, apraxia, dementia, or confusional states. The etiology may be virtually any cerebral disease, although stroke, traumatic brain injury, tumor, and Alzheimer's disease are the most frequent causes. Agraphia is often one of the early behavioral manifestations of cerebral dysfunction in acute confusional states of metabolic or toxic origin.

Much of the importance of agraphia in a clinical setting arises from the fact that collection of a writing sample is a simple, quick, and often highly informative procedure to include in a mental status examination, and generally can be obtained from even agitated or marginally cooperative patients. The task typically consists of having the patient copy a sentence, write a sentence to dictation, and spontaneously generate a sentence or two. Highly overlearned writing behaviors, such as writing one's name, may be preserved in agraphic patients and should not be used to infer normal writing ability.

The neural substrates of handwriting, a multifaceted and culturally transmitted skill which has risen to importance very late in the course of brain evolution, are far from 'hard wired.' Lesion studies have demonstrated variability across individuals in the nature and severity of agraphia stemming from damage to relevant brain areas, as well as considerable recovery of handwriting in the months following the onset of focal brain damage. The left dorsolateral frontal lobe, particularly the superior premotor region, is uniquely situated for the convergence of sequential motor activity of the dominant arm and hand together with activity of frontal and posterior language-related cortices, and damage to this region has been associated with severe and relatively circumscribed agraphia. However, agraphia also may arise from focal damage in left temporal and parietal regions, especially the perisylvian language-related cortices, as well as the left basal ganglia. **Figure 1** depicts a high left parietal lobe mass which resulted in agraphia.

Patients with severe agraphia may be unable to write even single letters. Milder versions of the condition may be manifest only as more subtle defects, such as an increased frequency of spelling errors. Written output, such as spelling, has been associated with damage to a network of brain regions in the left posterior frontal lobe involving Brodmann's area 44, 45, and 6. The ability to write numbers or other nonletter symbols is often defective in parallel with the impairment in writing letters and words, but in other cases these abilities may be dissociated. For example, number writing may be preserved even if letter writing is impaired. Spatial agraphia involves an inability to maintain appropriate placement of written material on a page; this most often occurs as a progressive upward slant from the left to right side of the page and is associated with superior parietal lesions of the dominant hemisphere. Micrographia involves a progressive diminution of letter size of the course of a sentence, and is associated with parkinsonism or, less commonly, Alzheimer's disease. **Figure 2** is an illustration of micrographia: the original stimulus to be copied is in the upper panel, and the patient's rendering of the stimulus is in the bottom panel. A dysexecutive agraphia, or decrease in an individual's ability to express ideas in writing, may also occur following nonfocal brain damage, such as in the case of dementia or traumatic brain injury affecting prefrontal cortex.

We offer the following brief case to illustrate the nature of the writing impairments observed in agraphia. A middle-aged man with a high school degree suffered multiple strokes resulting in damage to each of the left frontal, temporal, and parietal brain regions. Examples of his writing during the acute epoch are presented in **Figure 3**. Writing during the chronic epoch (i.e., 3 months following the neurological event) indicated improvement, although he continued to perform at a level well below expectations (**Figure 4**).

Most patients with aphasia also have agraphia which parallels their defect in spoken language, although in individual cases written or spoken production may be relatively preserved. The syndrome of alexia with agraphia involves letter distortion and

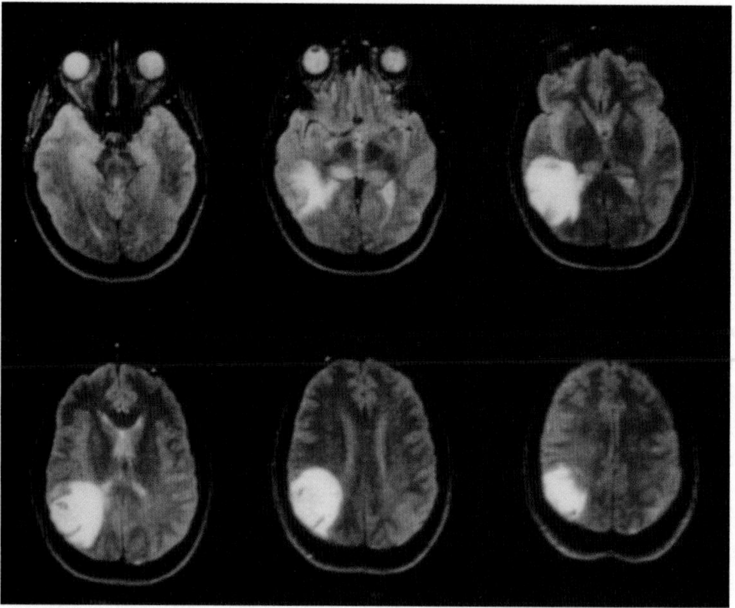

Figure 1

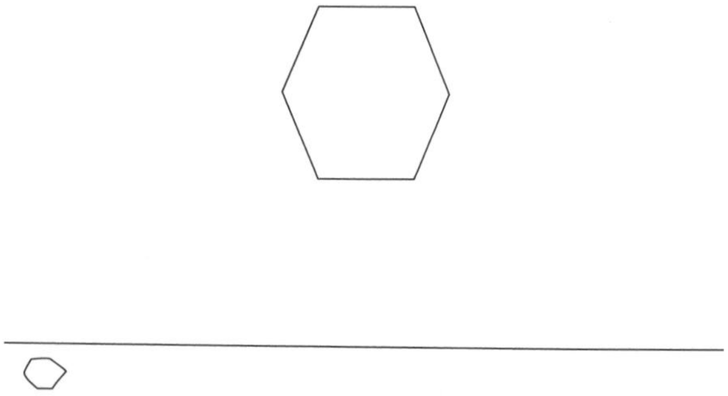

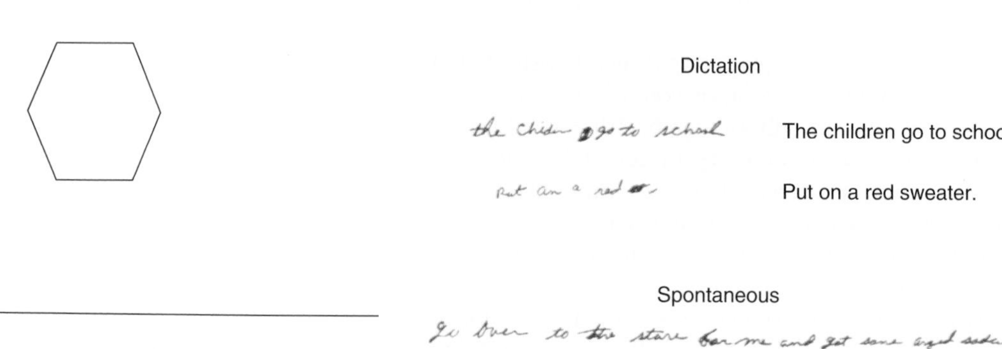

Figure 2

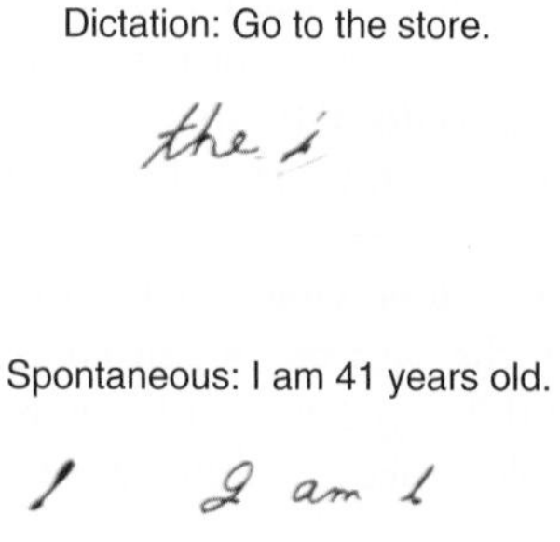

Dictation: Go to the store.
Spontaneous: I am 41 years old.

Figure 3

Dictation
The children go to school.
Put on a red sweater.
Spontaneous

Figure 4

impaired spelling, in addition to reading impairment, and is associated with lesions in the left angular or supramarginal gyrus. In the so-called Gerstmann syndrome, loosely associated with dominant parietal lobe damage, agraphia appears in a constellation with acalculia, finger agnosia, and left–right disorientation. Callosal agraphia affecting the left hand only can follow lesions in the anterior half of the corpus callosum. Phonological agraphia involves a specific defect in phoneme-to-grapheme translation, expressed as an isolated inability to write pronounceable pseudowords, and appears with lesions in various left perisylvian regions.

With regard to treatment, a recent study indicated that rehabilitation is generally beneficial to patients with acquired agraphia, regardless of lesion location.

See also: Aging of the Brain and Alzheimer's Disease; Alexia; Dyslexia: Neurodevelopmental Basis; Language: Cortical Processes; Reading; Stroke; Stroke: Neonate vs. Adult.

Further Reading

Anderson SW, Damasio H, and Damasio AR (1990) Troubled letters but not numbers: Domain specific cognitive impairments following focal damage in frontal cortex. *Brain* 113: 749–766.

Anderson SW, Saver J, Tranel D, and Damasio H (1993) Acquired agraphia caused by focal brain damage. *Acta Psychologica* 82: 193–210.

Ardila A and Surloff C (2006) Dysexecutive agraphia: A major executive dysfunction sign. *Internal Journal of Neuroscience* 116: 653–663.

Bub D and Chertkow H (1988) Agraphia. In: Boller T and Grafman J (eds.) *Handbook of Neuropsychology*, vol. 1. New York: Elsevier.

Chedru F and Geschwind N (1972) Writing disturbances in acute confusional states. *Neuropsychologia* 10: 343–353.

Hillis AE, Chang S, Breese E, and Heidler J (2004) The crucial role of posterior frontal regions in modality specific components of the spelling process. *Neurocase* 10: 175–187.

LaBarge E, Smith DS, Dick L, and Storandt M (1992) Agraphia in dementia of the Alzheimer type. *Archives of Neurology* 49: 1151–1156.

Leischner A (1969) The agraphias. In: Vinken PJ and Bruyn GW (eds.) *Handbook of Clinical Neurology*, vol. 4. Amsterdam: North Holland.

Rapp B (2005) The relationship between treatment outcomes and the underlying cognitive deficit: Evidence from the remediation of acquired dysgraphia. *Aphasiology* 19: 994–1008.

Roeltgen DP (2003) Agraphia. In: Heilman KM and Valenstein E (eds.) *Clinical Neuropsychology*, 4th edn., pp. 126–143. New York: Oxford University Press.

Alcoholism

C L Ehlers, The Scripps Research Institute, La Jolla, CA, USA
J A Chester, Purdue University, West Lafayette, IN, USA

Introduction

The majority of people who choose to drink alcohol restrict their intake to amounts that result in no serious health or social consequences. National guidelines for drinking suggest that two drinks a day for men and one drink for women are amounts that do not cause harm. Harmful or binge drinking is defined as drinking the equivalent of five drinks for men and four drinks for women over a 2 h period. Binge drinking typically results in intoxication and impairment as well as blood alcohol concentrations that exceed the legal limit for driving a motor vehicle ($0.08\,\mathrm{g\,dl^{-1}}$).

Diagnosis

The Diagnostic and Statistical Manual (DSM) of the American Psychiatric Association defines alcohol abuse as the repeated occurrence of problems associated with drinking, such as legal, social, or interpersonal problems, and the use of alcohol under hazardous conditions. 'Alcohol dependence' is a term that has been shown to be a more valid and reliable diagnosis. A diagnosis of alcohol dependence is made if a person has evidence of three of seven alcohol-related problems that occur over a 12-month or longer time period, including alcohol tolerance, alcohol withdrawal, alcohol use despite medical or psychological problems, preoccupation with obtaining, using, and recovering from drinking, a persistent desire to drink or unsuccessful efforts to cut down or control drinking, drinking larger amounts, or over a longer period than intended, and giving up or reducing important activities because of drinking. Since self-disclosure of symptoms and consequences of heavy drinking may not always be reliable, there has been a quest for valid biological markers of the disorder. Laboratory tests that can reliably index heavy drinking include blood alcohol, serum γ-glutamyltransferase (GGT), carbohydrate-deficient transferrin (CDT), and mean corpuscular volume (MCV) of erythrocytes. Possible new biomarkers include minor alcohol metabolites, such as phosphatidylethanolamine, 5-hydroxytryptophol, and possibly genetic markers, although such markers are as yet to be identified.

Epidemiology

Several national and worldwide surveys have found that the prevalence of alcohol use disorders differs based on gender and ethnicity. Alcohol abuse and dependence are twice as common in men as in women. Alcohol dependence appears to be higher in some European countries and Russia and lower in some countries in Asia. The prevalence of current alcohol abuse and dependence over a 1-year period, as estimated by US surveys, is approximately 5% and 4%, respectively. Alcohol abuse and dependence are also more common among younger people, and, in the US, among Anglo/Caucasians than in African, Asian, and Hispanic Americans. Native Americans/Alaska Natives, First Nation people of Canada, Australian Aborigines, New Zealand Maori, and some Pacific Islanders, as a group, have the highest prevalence of alcohol use disorders. It appears that both genetic and environmental factors contribute to the variance in alcohol dependence rates among men and women and between different ethnic groups.

Clinical Subtypes and Comorbidity

There has long been an effort to subtype alcoholics in order to better understand the etiology of the disorder and perhaps to design more effective treatments. A number of different subtypes have been proposed based on age of onset, severity of alcoholism, presence of severe withdrawal symptoms, comorbidity with other psychiatric disorders, and genetic versus environmental etiology. There is some general consensus that at least two general subtypes exist: (1) a type that is associated with antisocial traits, has a strong genetic component, and has an early age of onset of the disorder; and (2) a type that may occur later in life, is a milder form of the disorder, and may be more environmentally driven. Alcoholism is more common in patients with other psychiatric disorders (comorbidity). Schizophrenia, certain anxiety disorders, bipolar disorder, other drug dependencies (especially tobacco), and antisocial personality disorder have the highest co-occurrence with alcohol dependence. In some individuals alcohol can induce psychiatric illness, most commonly depression, which often resolves following abstinence. Other psychosocial consequences of drinking include increased risk for suicide, unplanned sex, accidents, interpersonal violence, and child abuse.

Clinical Course and Treatment of Alcoholism

Like many disorders, such as obesity or diabetes, the clinical course of alcoholism progresses in an orderly sequence, with the temporal emergence of several specific alcohol-related problems. These include problems with family/friends and work, physical fights, blackouts, morning drinking, tolerance, craving, psychological impairment, withdrawal, and health problems, all of which may appear over a 5- to 20-year course. Fluctuations between periods of heavy problematic drinking, controlled drinking, and abstinence are common occurrences in alcoholics. However, it has been demonstrated that a high degree of similarity exists in the clinical course of alcoholism between men and women and among different ethnic and treatment groups.

Severe alcoholism is associated with an alcohol withdrawal syndrome. Acute withdrawal from alcohol in a chronic heavy drinker can cause tremulousness, sleep disturbance, sweating, tachycardia, anxiety and depression, headache, nausea/vomiting, and, in severe cases, hallucinations, delirium tremens, and seizures. Signs and symptoms of acute withdrawal usually subside within 2 weeks; however, acute withdrawal may be followed by a protracted abstinence syndrome consisting of hyperarousal, symptoms of anxiety/depression, and insomnia that can persist for several months and may increase risk for relapse to drinking.

In the general population, perhaps 20% or more of alcoholics resolve their drinking problems without recourse to any treatment. Current treatment modalities that have demonstrated some degree of success include alcoholics anonymous, brief interventions, motivational therapy, cognitive behavioral therapy, and medications. Few medications have been shown, as yet, to be more effective than placebo or medical management in the treatment of alcoholism. Naltrexone has been demonstrated to show efficacy in reducing heavy drinking in some studies. Developing effective medications to treat alcoholism using specific neurochemical targets is a particularly strong focus of research.

Alcoholism-Associated Toxicity

Approximately 100 000 people in the United States die each year due to alcoholism. Alcohol poisoning is most likely to occur in young drinkers; rapid consumption of excessive amounts of alcohol can cause mortality through direct depression of the medullary respiratory center. Heavy drinking and alcohol dependence are associated with toxicity in a number of different organ systems. However, the organs with the highest alcohol-associated damage are the liver and brain. Alcoholism is associated with liver cirrhosis and end-stage liver disease, though only about 10% of alcoholics develop cirrhosis. It appears that while alcohol exposure is necessary to produce liver cirrhosis, it is not sufficient, and other genetic and environmental cofactors are necessary to induce the disorder. Alcoholic liver disease occurs in a continuum of severity that begins with fat infiltration that can progress to an inflammatory process involving cytokines (tumor necrosis factor-α) and activation of liver macrophagic-like cells (e.g., Kuppfer cells), and in its most severe form results in collagen synthesis by stellate cells during end-stage disease (cirrhosis). Alcohol can also damage the nervous system both directly and in association with vitamin deficiencies. Magnetic resonance image (MRI) scans of alcoholics in early recovery can demonstrate enlarged ventricles, enlarged sulci, and smaller than normal gyri. These findings can reverse to a certain extent over several months to years with continued abstinence. Neuropsychological studies have shown that alcoholic adults have poorer functioning on tests of memory, executive functioning, and visuospatial functioning. A severe form of alcohol neurotoxicity is Wernicke–Korsakoff syndrome (WK-S). WK-S involves thiamine deficiency and consists of ocular disturbances, ataxia, confusion, and short-term memory loss and is associated with necrotic lesions of the mammillary bodies, thalamus, and other brain stem areas. Other neurological and medical complications of chronic alcohol exposure include peripheral neuropathy associated with multiple vitamin B deficiencies, anterior lobe cerebellar degenerative disease, cardiomyopathy, esophageal varicosities, gastritis, pancreatitis, and myopathy.

Alcoholism and Neural Development

Alcohol may be more toxic during critical neurodevelopmental epochs. It also appears that adolescent brain may be particularly vulnerable to the toxic effects of alcohol caused by heavy bouts of binge drinking, although data in this area are just emerging. Exposure to alcohol *in utero* can result in a spectrum of deleterious effects that appear to exist along a continuum, ranging from subtle behavioral and cognitive dysfunctions to gross neuromorphological anomalies. The most severe result of *in utero* alcohol exposure is fetal alcohol syndrome (FAS). FAS is identified by characteristic facial anomalies, growth retardation, and central nervous system dysmorphology and behavioral abnormalities. Although heavy drinking is associated with FAS, not all women who drink excessively during pregnancy give birth to offspring with FAS. Both genetic and environmental factors, including nutritional status and the use of other drugs, contribute to the vulnerability to FAS.

Genetic and Environmental Influences on Alcoholism

Twin, family, and adoption studies have all provided evidence that perhaps half of the risk for alcoholism is genetically based. Studies of the heritability of alcohol drinking in twins suggest that the initiation of drinking may be more influenced by environmental factors, whereas development of dependence may have a larger genetic component. There are a number of environmental factors that can influence drinking and alcohol dependence. Efforts to prevent the development of alcohol-related morbidity and mortality often target environmental causes of the disorder. Reducing the availability of alcohol (e.g., by increasing cost or restricting sales) can reduce drinking. Social attitudes and legislation that stigmatize use of a substance can also influence consumption. Increasing the legal minimum age of drinking and lowering the legal blood alcohol concentration for driving also represent successful efforts that have reduced drinking. Restriction of underage drinking is particularly important; it has been demonstrated in several population samples that the earlier the age a person begins to drink, the greater the risk of developing alcohol dependence in their lifetime.

Alcoholism is not a single gene disorder but is rather polygenic, with vulnerability most likely arising from the simultaneous impact of functional variations at several genes. The genes that encode the structures of the enzymes that metabolize alcohol are the best characterized genetic factors that influence alcoholism risk. Alcohol is metabolized primarily in the liver, but also in the stomach and brain by several different enzyme systems. The two main enzymes involved in alcohol metabolism are alcohol dehydrogenase (ADH) and aldehyde dehydrogenase (ALDH). ADH breaks down alcohol to the toxic metabolite acetaldehyde, which is then broken down further by ALDH into the less toxic acetate. The genes that code for these enzymes are polymorphic and thus several different forms of the enzymes exist, each differing in their kinetic properties. This means that genetically determined differences in alcohol-metabolizing enzyme structures can cause alcohol to be broken down slower or faster, and acetaldehyde to either be quickly eliminated or to build up and thus increase risk for organ damage. One of the most common variants, a mutation in ALDH that is seen in approximately 40% of Asians, results in facial flushing, nausea, tachycardia, and headache following consumption of alcohol. Individuals with this mutation are at much less risk for developing alcohol dependence. Variants found in ADH, depending on ethnic origin (e.g., ADH1B in individuals of African ancestry), can modify response to alcohol and risk for alcohol dependence as well as alcohol-related organ damage.

A number of other genetic approaches have been taken in order to search for the etiology of alcohol use disorders and other alcohol risk phenotypes (the physical manifestation of a trait that is heritable). Several US research groups have identified alcohol-affected families and conducted genome scans, and evidence for linkage of alcoholism to a number of gene locations has been identified and replicated by several studies. The strongest evidence appears to point toward genes on chromosomes 1, 2, 4, 5, 6, 7, 9, 15, and 16. Another approach to studying the genetics of alcoholism has been to determine if there is an association between alcohol dependence and a polymorphism in a single identified gene that is suspected of contributing to the variance in the etiology of the disorder. These so-called candidate gene approaches have yielded preliminary evidence for a number of neuronal targets, including the γ-aminobutyric acid (GABA) system, dopamine, norepinephrine, cannabinoids, and opiates, as contributing to risk for alcohol dependence, although the studies have not always been robustly replicated. Future efforts in this area will employ the new technology of whole-genome association analyses, which will be a more powerful technique to identify multiple genes simultaneously over the entire genome.

The idea that some phenotypes, called endophenotypes, bear a close relationship to the biological processes that give rise to alcohol dependence has been given much attention in alcoholism research. A number of endophenotypes have been identified as being closely associated with alcoholism, including variants of the electroencephalogram, components of event-related potentials, alcoholism symptom clusters, personality traits, levels of alcohol craving, withdrawal, and sensitivity and tolerance to alcohol. Perhaps one of the most compelling endophenotypes is described as 'low level of response to alcohol.' Individuals with a low level of response to alcohol have a diminution of both objective and subjective responses to alcohol when they are given a standard number of drinks. This allows them to drink larger amounts of alcohol, on any one occasion, before feeling drunk. Prospective studies have demonstrated that individuals with the low level of response are significantly more likely to develop alcohol-related life problems than are matched individuals with higher levels of response to alcohol.

Alcohol Effects on the Central Nervous System

An initial hypothesis on how alcohol induces its intoxicating and addicting effects is that alcohol acts to fluidize membranes; however, these effects have been shown to be very small at pharmacologically relevant concentrations. Although no specific receptor for alcohol has been discovered, alcohol can bind to proteins in the cell membrane and alter receptor functions and intracellular signaling pathways that are coupled to membrane receptors. In the brain, alcohol decreases neuronal excitability by enhancing the inhibitory effects of the neurotransmitter GABA and by decreasing the excitatory effects of the neurotransmitter glutamate. Alcohol also has significant interactions with other primary brain neurotransmitters, including dopamine and serotonin, as well as with neuropeptide transmitters such as opioids, corticotropin-releasing factor/hormone (CRF/CRH), and neuropeptide Y (NPY) and endocannabinoids. The reinforcing effects of alcohol are thought to be due in large part to its ability to increase the activity of dopamine-containing neurons in the primary reward circuit in the brain, known as the mesolimbic dopamine pathway. Dopamine, in turn, can influence the function of serotonin to produce feelings of well-being and alterations in mood states. Alcohol also increases the release of endogenous opioid peptides, which reduce pain and produce euphoria. Long-term exposure to alcohol can result in relatively permanent changes in neuronal function resulting from alterations in gene expression and intracellular protein levels; this is termed neuroadaptation to alcohol. Neuroadaptation to alcohol is considered one of the primary brain mechanisms of addiction.

Preclinical Models of Alcoholism

A number of valid preclinical models have been developed to study particular characteristics of human alcoholism, including excessive alcohol intake, alcohol craving and relapse behavior, tolerance to alcohol, alcohol toxicity, and withdrawal. Many of these models are studied in combination with genetic strategies to characterize the role of genes, and their interaction with the environment, in shaping in animals those alcohol-related traits that may be relevant to alcoholism in humans.

Genetic Animal Models

Genetic animal models have provided important information about how certain alcohol-related traits, as well as other biological and behavioral traits (e.g., levels of dopamine in the brain, impulsivity, anxiety), may share a common genetic basis. For example, it has been clearly demonstrated that rodents with a genetic propensity toward low alcohol consumption display more pronounced signs of alcohol withdrawal. In addition, genetic propensity toward high alcohol consumption is associated with reduced sensitivity to alcohol's aversive effects. Many rodent lines have been genetically selected to show extreme differences on a particular alcohol-related trait, including voluntary alcohol consumption, the severity of alcohol withdrawal, the hypnotic effects of alcohol, the hypothermic effects of alcohol, and the locomotor-stimulant effects of alcohol. These lines show various neurobiological differences, including brain levels of serotonin, GABA, dopamine, opioids, CRF, and NPY that are associated with high or low alcohol-related behavioral responses for which they were selected. Many genetic animal models have been used to successfully identify single genes or groups of genes that are associated with individual differences in alcohol-related behaviors. One commonly used approach is a gene mapping strategy to determine if a particular chromosomal location containing a gene sequence or cluster of genes is associated with a particular alcohol-related behavior. These gene sequences can then be compared to homologous genes in the human genome. A number of candidate chromosomal locations and genes sequences have been identified, and some of the strongest data indicate that genes that regulate the GABA neurotransmitter system are associated with sensitivity to alcohol withdrawal responses. Converging evidence for a link between alcohol dependence and certain GABA gene variants has also been found in human genetic association and linkage studies.

To date, approximately 100 candidate genes have been studied for their role in regulating alcohol-related traits in preclinical models. For example, mice without functional dopamine D4 receptors (i.e., D4 knockout mice) show enhanced sensitivity to alcohol's locomotor stimulant effects, as well as to cocaine and methamphetamine. Mice with knockouts of the NPY receptor gene show decreased sensitivity to alcohol's effects and increased alcohol consumption. Mice that have a transgene that produces overexpression of CRF are more sensitive to the sedative effects of alcohol and drink less alcohol. Gene array technology has also been used to identify additional candidate genes, as well as to confirm some previously identified genes, that may be associated with risk for alcohol dependence.

Models of Alcohol Interactions with Stress and Hormones

Stress is one environmental factor that has long been thought to increase the risk for drinking in vulnerable individuals. Findings in animal models suggest that the effects of stress on alcohol drinking are complex and depend on many factors, such as the type and duration of stress exposure, the genetic makeup and sex of the animal, the age of first stress exposure, and whether the animal has previously been exposed to alcohol. Female rodents may be protected against the effects of stress under certain conditions. Male rodents with a genetic propensity toward high alcohol intake appear to be particularly vulnerable to stress-induced alcohol consumption compared to their alcohol-avoiding counterparts. These findings may be partly explained by genetic differences in the level of hypothalamic–pituitary–adrenal (HPA) axis hormones, such as CRH, NPY, and opioid peptides; such differences have been found in animals with innate preference or avoidance of alcohol. Primary HPA axis hormones released in response to stress also appear to increase the rewarding effects of alcohol in certain experimental paradigms. Mice with genetic disruption of the CRH1 receptor show stress-induced alcohol drinking, but only after repeated exposures to stress and prolonged alcohol consumption. This same CRH receptor subtype is associated with stress-induced relapse to alcohol-seeking behavior in rats genetically selected for high alcohol preference. Early exposure to stress may increase subsequent alcohol consumption in adult animals, and these effects appear to depend on interactions with gene variants. For example, the presence of certain genes that regulate serotonin function is associated with increased alcohol preference in female monkeys that were exposed to early-life stress. Animals exposed to alcohol during early periods of development show greater stress-induced increases in alcohol drinking than do animals not exposed to alcohol during development. These results parallel findings in humans that indicate greater stress-reactive drinking and a fivefold increased risk for adult alcoholism in people who began drinking alcohol before the age of 15 years.

Models of Alcohol-Induced Toxicity

Preclinical research using cell culture and animal models, in which variables can be studied under controlled conditions, has led to important knowledge of the mechanisms that contribute to alcohol-induced neuro- and hepatotoxicity. Alcohol damages the liver through a number of biological pathways that produce oxidative stress, disruptions in energy homeostasis, protein breakdown, inflammation, and collagen formation. Investigators have clearly shown the influence of nutritional factors in studies in which animals were fed alcohol directly into the stomach (intragastric feeding models). Animals fed alcohol-containing diets that were low in folate, protein, and carbohydrates, or high in polyunsaturated fats, had more pronounced liver damage as evidenced by greater activity of cytochrome P450 enzymes (CYP2E1 activity), scar tissue formation, and cell death. Conversely, supplementing the diets of alcohol-fed animals with folate or a phospholipid antioxidant has been shown to reduce CYP2E1 activity and associated signs of cell death and fatty cell accumulation. Cultured stellate cells from the livers of alcohol-exposed animals treated with a phospholipid antioxidant showed reduced collagen production, a process that leads to liver scarring. The role of various genes that may increase or protect against liver damage has been an important research area. For example, transgenic mice with overexpression of the CYP2E1 gene show greater alcohol-induced liver injury compared to wild-type controls. Several knockout mouse models in which particular genes have been inactivated (e.g., those encoding tumor necrosis factor and endotoxin receptors) are protected against alcohol-induced liver damage.

Preclinical studies have also provided models of alcohol-induced neurotoxicity. Oral self-administered alcohol over long periods has been shown to reduce synapses in cerebellum, although the effects are not robust. Early studies in cultured cortical neurons demonstrated that exposure to and withdrawal of high doses of alcohol could lead to glutamate-associated neuronal excitotoxicity. *In vivo* studies using a binge alcohol administration model have further shown that high-dose alcohol is selectively toxic and causes neurodegeneration in certain brain regions, such as entorhinal and perirhinal cortical areas in the rat. Binge alcohol treatment also increases neuronal death and reduces neurogenesis in the hippocampal dentate gyrus. The neurodegeneration seen in these studies does not appear to be directly related to glutamate neurotoxicity. Rather, it may involve mechanisms of oxidative stress, inflammatory cytokines, activation of nuclear factor-κB, and reductions in cAMP response element-binding protein (a transcription factor important for inducing key genes that improve resistance to insults, neuronal vitality, and growth). Another model of alcohol-induced brain damage using *in vivo* studies in the rat and organotypic brain slice cultures has indicated that neuronal degeneration is caused in part by traumalike neuroinflammatory processes involving brain edema and the release of proinflammatory signaling molecules such as tumor necrosis factor-α and arachidonic acid. Studies have also indicated that alcohol can inhibit new cell formation, although how neural degeneration relates to loss of regeneration is not as yet clear. Genetic factors may also influence the degrees of alcohol-induced toxicity, as rats from a selectively bred alcohol-preferring (P) line have a two- to threefold more brain damage in perirhinal and entorhinal cortices than do rats from a nonpreferring (NP) line, after exposure to binge alcohol treatment. Additionally, it has been demonstrated that alcohol-induced neurotoxicity is more extensive following adolescent and *in utero* exposure.

Models of Alcohol Therapeutics

Preclinical research using molecular and cellular techniques and whole-animal models has led to the identification of several biological targets in the brain for drugs to treat alcoholism. In 1995, naltrexone, a drug that blocks opioid receptors, was one of the first drugs approved in the US to treat alcoholism. Many animal studies have shown that naltrexone effectively reduces alcohol drinking and alcohol relapse-type behaviors. Acamprosate, a drug approved in 2004, has been shown to reduce alcohol drinking and signs of craving and withdrawal in animals. Acamprosate appears to work by interacting with the glutamate neurotransmitter system. A promising line of research suggests that targeting neuropeptide transmitter systems such as CRF and NPY may counteract many of the neuroadaptive responses seen after long-term alcohol exposure and may be particularly effective in reducing alcohol craving and relapse-drinking behavior. Many other drugs that target GABA, glutamate, serotonin, dopamine, and endocannabinoid neurotransmitter systems are being intensively studied in animal models of alcoholism to determine their potential for reducing the short- and long-term effects of alcohol and for treating alcohol dependence. Recent data suggest that one reason that only a certain proportion of people respond to medication therapy is in part due to the presence or absence of certain gene variants that regulate responses to the drug. Preclinical research in genetic animal models will advance the current research efforts aimed toward developing individualized drug therapy for alcoholism based on genetic background.

See also: Brain Damage: Functional Reorganization; Dementia; Fetal Alcohol Syndrome; $GABA_A$ Receptor Synaptic Functions; Gamma-Aminobutyric Acid (GABA); Hepatic Encephalopathy.

Further Reading

Adachi M and Brenner DA (2005) Clinical syndromes of alcoholic liver disease. *Digestive Diseases* 23: 255–263.

Chester JA and Cunningham CL (2002) GABA(A) receptor modulation of the rewarding and aversive effects of ethanol. *Alcohol* 26: 131–143.

Crabbe JC, Phillips TJ, Harris RA, et al. (2006) Alcohol-related genes: Contributions from studies with genetically engineered mice. *Addiction Biology* 11: 195–269.

Crews FT, Collins MA, Dlugos C, et al. (2004) Alcohol-induced neurodegeneration: When, where and why? *Alcoholism: Clinical and Experimental Research* 28: 350–364.

Ehlers CL, Gilder DA, Wall TL, et al. (2004) Genomic screen for loci associated with alcohol dependence in Mission Indians. *American Journal of Human Genetics* 129B: 110–115.

Goldman D, Oroszi G, and Ducci F (2005) The genetics of addictions: Uncovering the genes. *Nature Reviews Genetics* 6: 521–532.

Grant BF, Dawson DA, Stinson FS, et al. (2004) The 12-month prevalence and trends in DSM-IV alcohol abuse and dependence: United States, 1991–1992 and 2001–2002. *Drug and Alcohol Dependence* 74: 223–234.

Heilig M and Egli M (2006) Pharmacological treatment of alcohol dependence: Target symptoms and target mechanisms. *Pharmacology and Therapeutics* 111: 855–876.

Hesselbrock VM and Hesselbrock MN (2006) Are there empirically supported and clinically useful subtypes of alcohol dependence? *Addiction* 101(supplement 1): 97–103.

Lieber CS (2005) Metabolism of alcohol. *Clinics in Liver Disease* 9: 1–35.

McBride WJ and Li TK (1998) Animal models of alcoholism: Neurobiology of high alcohol-drinking behavior in rodents. *Critical Reviews in Neurobiology* 12: 339–369.

Niemela O (2007) Biomarkers in alcoholism. *Clinica Chimica Acta* 377: 39–49.

Room R, Babor T, and Rehm J (2005) Alcohol and public health. *Lancet* 365: 519–530.

Schuckit MA (2006) *Drug and Alcohol Abuse: A Clinical Guide to Diagnosis and Treatment*. New York: Springer.

Sullivan EV and Pfefferbaum A (2005) Neurocircuitry in alcoholism: A substrate of disruption and repair. *Psychopharmacology (Berlin)* 180: 583–594.

Relevant Website

http://www.niaaa.nih.gov– National Institute on Alcohol Abuse and Alcoholism (National Institutes of Health).

Alexia

D Tranel and N L Denburg, University of Iowa Hospitals and Clinics, Iowa City, IA, USA

Alexia is a neurological term of Greek origin (a + Greek *lex(is)*) that designates a partial or complete inability to read. There are a number of different subtypes of alexia, but all have in common the feature that the affected patient cannot read normally, so that reading is slow or impossible, and comprehension of read material is impaired. In the American literature, alexia is used nearly exclusively to designate acquired defects in reading, that is, reading impairments that occur as the result of a neurological condition in an individual who was previously literate. The British literature tends to use the term dyslexia to refer to this condition. Dyslexia in the American literature is typically applied to persons who have a developmental form of reading disorder – specifically, the not uncommon condition of being unable to learn to read normally, which becomes manifest during childhood and typically affects the acquisition of both reading and writing skills.

Stroke is the most common cause of acquired alexia, especially when the condition appears as an isolated or relatively isolated symptom. Other forms of neurological disease, including cerebral tumors, inflammatory processes, and head injury, can also cause alexia, and there are some cases of degenerative disease (e.g., Alzheimer's) in which a reading disturbance is a prominent early feature of the patient's cognitive dysfunction. Acquired alexia occurs in two main forms. One, termed alexia with agraphia, involves a reading impairment that is accompanied by writing defects; this form is common in patients with various aphasia syndromes. The other, termed alexia without agraphia (or 'pure alexia'), involves an isolated defect in reading that is not accompanied by impaired writing. Pure alexia is associated with lesions that disconnect both visual association cortices from the dominant, language-related temporoparietal cortices. Pure alexia can be caused by a single lesion strategically placed in the region behind, beneath, and under the occipital horn of the left lateral ventricle, by damaging pathways en route from the callosum and pathways en route from the left visual association cortex (**Figure 1**). Another setting is the combination of a lesion in the corpus callosum, which disconnects right-to-left visual information transfer, and a lesion in the left occipital lobe, which disconnects left visual association cortex from left language cortex (**Figure 2**). Such lesions are likely to produce a right hemianopia; this sign is a frequent, although not invariable, accompaniment of pure alexia. Another neuropsychological correlate of pure alexia is color anomia, that is, the inability to name colors.

The 'purity' of alexia without agraphia stems from the fact that patients with these lesions do not develop disturbances in writing or in other aspects of speech and linguistic functioning. This separates this type of alexia from the types of reading defects that are common in aphasic patients. (It is rather striking to observe patients who can write a sentence easily and then be unable to read what they have just written.) In this sense, pure alexia can be construed as a disturbance of visual pattern recognition. Patients with pure alexia are unable to read most words and sentences, and in severe cases cannot read even single letters. The problem is not one of visual acuity: the fact that the patients can see the sentences, words, and letters they cannot read can be readily demonstrated by having the patients copy those stimuli, a task that will be executed normally. Thus, most patients with pure alexia have normal visual acuity (although a quadrantanopia or hemianopia may be present), and most have normal recognition of nonverbal visual stimuli such as objects and faces.

In terms of treatment for pure alexia, there are two schools of thought. The first is to attempt to increase the speed and accuracy of letter-by-letter reading. The second, higher-level approach involves an abandoning of letter-by-letter reading for the adoption of whole word recognition, a more efficient approach.

As noted, the nonacquired, developmental varieties of reading disorder are subsumed under the term of developmental dyslexia. These reading impairments, almost invariably accompanied by writing weaknesses (especially spelling defects), constitute a learning disability which in fact is the most common form of learning disability in the general population. Some developmental dyslexias may be related to a disturbance of the cellular organization of language cortices, possibly originating *in utero*. A few older postmortem studies have suggested that abnormalities of cortical organization in the brains of dyslexic patients may have contributed to the reading disability. There is a strong association between developmental dyslexia and left-handedness. The older literature indicated that dyslexia was far more common in boys than girls, but recent and better-controlled studies have not supported this notion; the ratio may even be fairly close to 1:1.

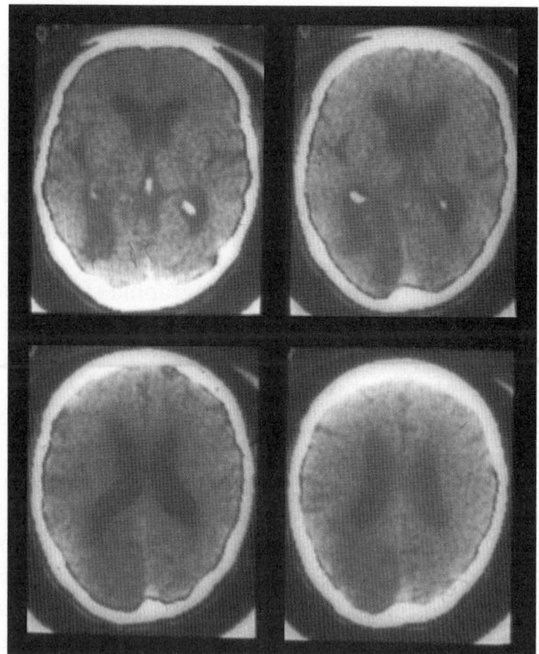

Figure 1 Illustration of the neuroanatomic findings of a patient with pure alexia, arising from a single lesion.

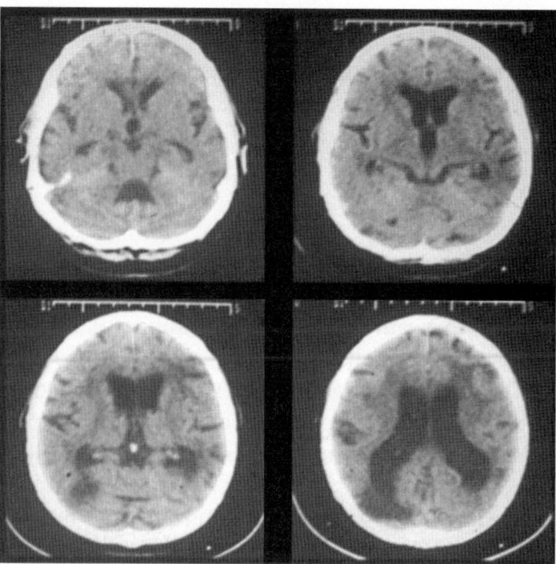

Figure 2 Illustration of the neuroanatomic findings of a patient with pure alexia, arising from a multiple lesion.

See also: Aging of the Brain and Alzheimer's Disease; Agraphia; Cerebrovascular Disease; Dementia; Dyslexia: Neurodevelopmental Basis; Reading; Stroke; Stroke: Neonate vs. Adult.

Further Reading

Benson DF, Brown J, and Tomlinson EB (1971) Varieties of alexia. *Neurology* 21: 951–957.

Coltheart M, Patterson K, and Marshall JC (1980) *Deep Dyslexia*. London: Routledge and Kegan Paul.

Geschwind N (1965) Disconnection syndromes in animals and man. *Brain* 88: 237–294, 585-644.

Greenblatt SH (1973) Alexia without agraphia or hemianopia: Anatomical analysis of an autopsied case. *Brain* 96: 307–316.

Petersen SE, Fox PT, Snyder AZ, and Raichle ME (1990) Activation of extrastriate and frontal cortical areas by visual words and word-like stimuli. *Science* 249: 1041–1044.

Tranel D (1994) Assessment of higher-order visual function. *Current Opinion in Ophthalmology* 5: 29–37.

Tranel D (1996) Disorders of color processing (perception, imagery, recognition, and naming). In: Feinberg T and Farah MJ (eds.) *Behavioral Neurology and Neuropsychology*, pp. 257–266. New York: McGraw-Hill.

Allometric Analysis of Brain Size

H J Jerison, University of California at Los Angeles, Los Angeles, CA, USA

Fossil 'brains' are fossils, but they are not brains. They are casts molded by the cranial cavity of fossils, but such endocasts can be made from any intact skull of a fossil or living species. Avian and mammalian endocasts look enough like brains to be treated as undissected brains. The fossil evidence is from the analysis of the external morphology of the brain as determined from endocasts. A few fossil endocasts are shown in **Figure 1**.

Analysis of brain evolution begins with measures of brain size. The size of the brain as a whole is a remarkably good estimator for the sizes of many of its parts, as indicated by **Figure 2**. These graphs show the relationship between the size of the whole brain and the sizes of the cerebellum and of the basal ganglia in living mammals. In a principal components factor analysis, gross brain size accounted for more than 85% of the variance in the size of major structures of the brain. In the sample of mammals for **Figure 2**, more than 99% of the variance in cerebellum and basal ganglia was accounted for by brain size. This encourages one to use gross brain size in the analysis of brain evolution.

In vertebrates other than birds and mammals, the brain does not fill the cranial cavity, and the evidence is more difficult to analyze. It is conventional to assume that in dinosaurs, as in living reptiles, the brain was approximately half the volume of the endocast. In fish and amphibians, it can be a much smaller fraction. The brain of *Latimeria*, the 'living fossil' fish from the coastal waters of southeast Africa and the Indian Ocean, is suspended in a sack of meninges and occupies only approximately 1% of the volume of the cranial cavity. In the megamouth shark, it is approximately 20% of the volume of the endocast. Precise quantitative evolutionary analysis of the fossil evidence of brain size is necessarily limited to birds and mammals, but it can be performed fairly reliably in reptiles and dinosaurs.

Quantitative Analysis: Convex Polygons

A quantitative analysis of the evolution of the brain is presented in **Figure 3**. Minimum convex polygons are drawn about much of the available data on living amniotes (mammals, birds, and reptiles). If one imagines the data on individuals in each of these classes of vertebrates as pins on a logarithmically scaled checkerboard, each polygon can be thought of as a string drawn to include all of the pins for a given class of vertebrates. The graph is based mainly on the sample described by Quiring in 1950, but other sources contributed. The polygons for living amniotes can be thought of as maps of brain–body space that indicate the present diversity of adaptations of brain size.

Quantitative analysis with convex polygons is equivalent to regression analysis in which lines are fitted to the clouds of points for each class. The polygons provide a realistic picture of the diversity of brain sizes. Given the logarithmic scaling for **Figure 3**, most of the within-species variability could be represented by the diameters of data points such as those representing archaic mammal species.

One often encounters a measure of encephalization, the 'encephalization quotient' (EQ), which is a quick way to characterize the relationship among species in relative brain size. To compute this measure, one must calculate the regression of the logarithms of brain size on body size, which is the allometric relationship: $\log Y = 2/3 \log X + \log 0.12$). EQ for a species is the residual from that regression as computed for that species. (Recall that a residual is the difference between a measured Y value and the Y value of the corresponding coordinate (X, Y) on the regression line.) Because the analysis is with logarithmic measures, the difference between the two logarithmic values is their quotient in linear measures. Therefore, EQ is the ratio of measured brain size to expected brain size, where expected brain size for an animal of given body size is determined by the allometric brain–body equation. Median EQ for these archaic mammals was 0.19 and ranged between 0.11 and 0.55.

The smallest mammal that contributed data for the convex polygon for living mammals in **Figure 3** was a 3.9 g pygmy shrew (*Sorex minitus*) with a 0.15 g brain. The heaviest mammal was a 59 059 kg finback whale (*Balaenoptera musculus*) with a 6800 g brain. The top vertex of the mammalian polygon is shared by a human (70 kg body and 1300 g brain) and a dolphin (142 kg body and 1735 g brain). Most of the data are from Quiring, whose species names are also used here.

I have seen other published measures that are slightly different from those in Quiring, but they do not significantly affect the picture of the present diversity in brain size. For example, blue whales have been reported as weighing more than 100 000 kg, the largest mammals that ever lived. Logarithmic scaling masks some these large differences in measurements, shifting the polygons only slightly if one uses sizes other than the ones

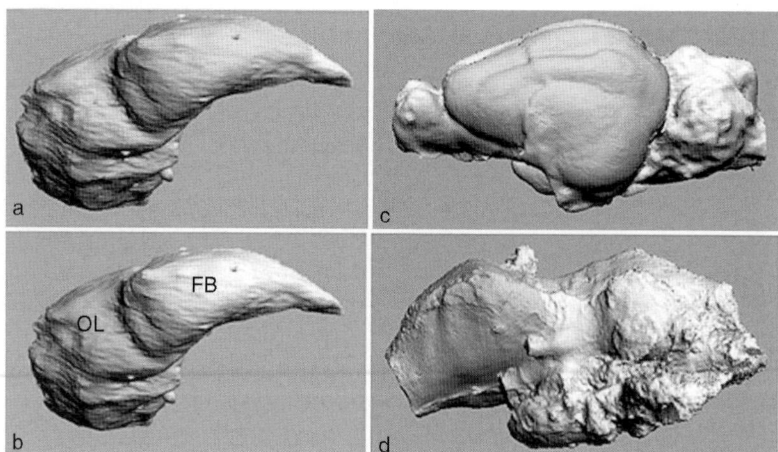

Figure 1 Fossil endocasts. (a) *Archaeopteryx lithographica*, the earliest bird, Late Jurassic, approximately 150 Ma. Facing right; length, 2.7 cm. Natural History Museum of London, Specimen BMNH 37001. (b) *Archaeopteryx*: FB, forebrain; OL, optic lobes (superior colliculi). (c) *Cynodictis*, Late Eocene carnivorous mammal, approximately 40 Ma; a fox-sized species ancestral to wolves and bears. Facing left; length, 5.3 cm. Field Museum of Natural History, Chicago, specimen FMNH PM 59013. (d) *Edmontosaurus*, a Late Cretaceous, approximately 70 Ma, duckbill dinosaur. Facing left; length, 16.4 cm. Museum of the Rockies Specimen, MOR Specimen 639.

in Quiring. If brain and body measurements for the largest blue whale had been available, the mammalian polygon would have been slightly longer than for dinosaurs, but it would have been approximately the same size and oriented in the same way. The relationships between the fossils and the living species are correctly represented in **Figure 3**.

The smallest bird for the living avian polygon was a 4.8 g hummingbird (*Amazilia tzacatl*) with a 0.2 g brain and the largest was a 123 kg ostrich (*Struthio camelus*) with a 42 g brain. Of living reptiles, the smallest was a 0.68 g lizard, *Phylodactylus gerrophygus*, with a 0.018 g brain. The largest was a 205 kg crocodile with a 14 g brain.

As one can see, when dinosaur data were added to the graph in **Figure 3**, it was natural to represent them by an extension of the reptilian polygon. The current consensus differentiates dinosaurs from other reptiles and has living birds as surviving small dinosaurs with feathers. The issue is not central to this article. It is appropriate to remember that organisms are mosaics of traits, and traits do not necessarily evolve at the same rate. The earliest bird evidently shared some brain features, such as brain size, with its living descendants; nonavian dinosaurs related to early birds may have been reptilian in brain size. There is one candidate, *Compsognathus*, a small dinosaur known from the same fossil quarry and approximately the same size as *Archaeopteryx*, but it has been impossible to prepare an endocast from this specimen. It would obviously be helpful to have the data. With respect to brain size, dinosaurs appear to have been more like reptiles than like birds.

The largest of the dinosaurs was the herbivorous *Brachiosaurus*, estimated to have weighed 87 000 kg with a brain size of approximately 155 g. The endocast of the famous carnivore, *Tyrannosaurus rex*, is also well-known – approximately 400 ml in volume, suggesting a 200 g brain. These are approximately the same measurements as those for *Edmontosaurus* (**Figures 1** and **3**; body weights estimated as 7700 kg for *T. rex* and 6000 kg for *Edmontosaurus*, and brain sizes estimated as approximately the same).

Measurements of gross brain size and gross body size in living species, appropriately analyzed, provide the background for understanding the evolution of the brain. Graphs such as those in **Figure 3** map present diversity. Adding fossil data then shows features of the evolution of the vertebrate nervous system. When the nervous system is analyzed in more detail for microscopic and neurochemical anatomy, one should keep in mind the evolutionary background provided by fossil endocasts, including the quantitative analysis of **Figure 3**.

Birds and Dinosaurs

The endocast of the fossil bird graphed at A on the bird polygon in **Figure 3** was excavated a few years after Darwin published his great book. It was immediately recognized as evidence of an important transition from reptiles to birds, supporting Darwin's views on natural selection. The present consensus separates dinosaurs from other reptiles and has *Archaeopteryx* as an avian dinosaur adapted for flight or gliding. Living birds are all viewed as surviving feathered

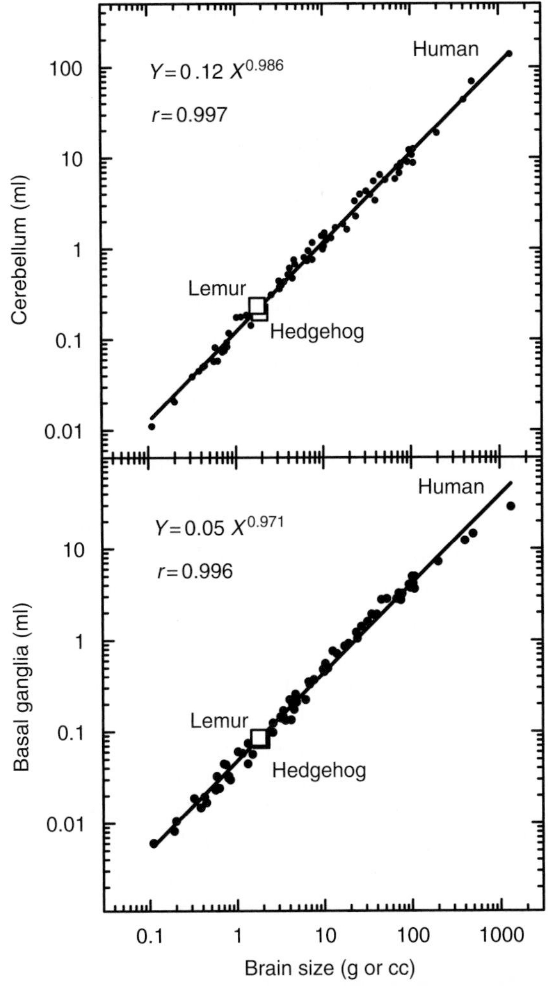

Figure 2 Brain size estimates the size of cerebellum and basal ganglia almost perfectly ($r > 0.99$). The marked points are from a desert hedgehog (*Hemiechinus*) with a 1.9 g brain and 250 g body and from a mouse lemur (*Microcebus*) with a 1.8 g brain and 54 g body, indicating that brain size, rather than body size, is the critical independent variable. Reprinted from Jerison HJ (1991) *Brain Size and the Evolution of Mind: 59th James Arthur Lecture on the Evolution of the Human Brain.* New York: American Museum of Natural History, with permission.

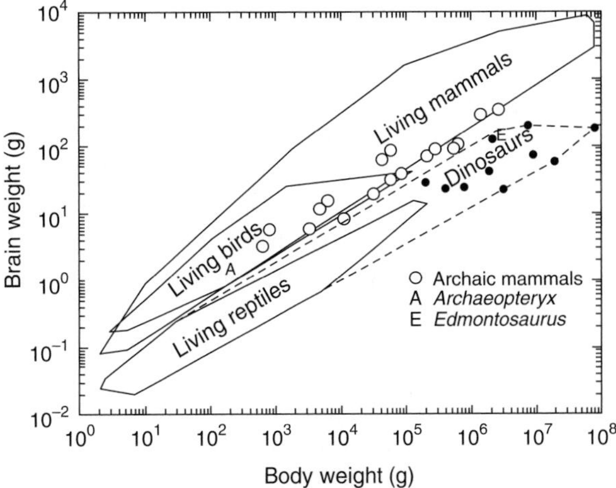

Figure 3 Brain–body relations in living and fossil amniotes. Data points for living species within each polygon are omitted; the convex polygons depict the present diversity of brain size in living amniotes. In drawing the polygons, each species was represented by a single point: 647 species of mammals, 219 species of birds, and 59 species of reptiles. Fossil data for 17 species of archaic mammals as published in Jerison in 1973 are added as open circles. Solid circles represent dinosaur data on 13 species. (For clarity, only 11 species are graphed because the omitted points masked the labels.) Data for the earliest bird, the late Jurassic *Archaeopteryx*, and the late Cretaceous duckbill dinosaur, *Edmontosaurus* (cf. **Figure 1**) are graphed as points A and E.

dinosaurs. The evidence of the brain, nevertheless, separates *Archaeopteryx* from dinosaurs, and the avian from the reptilian polygon of **Figure 3**.

External morphology also separates *Archaeopteryx* from dinosaurs and reptiles with respect to the brain. One superficial feature is the extent to which optic lobes (superior colliculi) are distinct from forebrain. Although on the brain these structures are as distinct on reptiles as on all living vertebrates, they are not distinct on endocasts. Comparing the endocasts of *Archaeopteryx* with that of *Edmontosaurus* illustrates the difference. It is not difficult to distinguish forebrain from optic lobes in the bird, but it is essentially impossible in the dinosaur.

A second superficial feature of endocasts of living birds is a large forebrain hump, the Wulst, which is a visual center in the brains of all living birds. It is probably homologous with mammalian visual cortex. *Archaeopteryx* had no Wulst. Besides *Archaeopteryx*, I know of only three fossil birds in which endocasts are good enough to show Wulst – an Eocene bird probably related to living curlews, a Miocene raptor, and the Pleistocene *Teratornis*. I see no evidence in Wulst in the Eocene curlew ('*Numenius gypsorum*'), but it is clearly evident in the later raptors, pretty much as in living birds. There should probably be a Wulst evident in at least some dinosaur brains. I have seen no convincing evidence of it in the dozen or so dinosaur endocasts that I have examined, but in personal communications, several colleagues have noted seeing signs of Wulst in at least one dinosaur, *Troodon*.

The distinction between forebrain and optic lobes in birds is quantifiable. Current three-dimensional imaging methods permit one to measure surface area, and it should be possible to measure visible surface area of forebrain and that of optic lobes. Qualitative observations are enough to show that forebrain covered less of the optic lobe in fossils than in living birds, but there are too few specimens to quantify the relationship at this time. A similar quantification on the size of neocortex was easily done for a sample of more than 100 mammalian endocasts, as discussed later.

The final point on birds concerns the relative size of the brain in *Archaeopteryx*. It lies within the convex polygon for living birds, but it is on the lower margin. Its brain was small compared to those of living birds of similar body size. Quiring reported a 282 g pigeon (*Columba livia*) with a 2.7 g brain. *Archaeopteryx* is usually estimated as slightly heavier, 300–400 g, but from its endocast its brain probably weighed approximately 1.5 g. This made it heavier than reptile brains at that body weight. It was just within the lower border of our polygon for living birds. Some encephalization beyond a dinosaurian grade is evident in the data on *Archaeopteryx*.

This summary shows how the fossil evidence contributes to our knowledge of the evolution of the brain even in very small samples – only four species – of fossil birds and their living descendants. Despite the small sample size, the evidence is convincing. It is somewhat inconsistent with the view of birds as surviving dinosaurs, which is based on a mosaic of traits. The brain is an exception. The best guess, widely shared on the basis of overall vertebrate morphology, is that the bird's brain evolved from a pattern shared with living reptiles and is probably fundamentally similar to that of dinosaurs. However, it departed from that pattern with respect to overall size and with respect to the extent to which its shape is reflected by endocasts from the cranial cavity.

Mammals

I reviewed the fossil evidence on mammalian brain evolution in 2006, emphasizing the evolution of neocortex. I summarize it here, and **Figure 3** also provides data on encephalization by comparing relative brain size in archaic and living species of mammals. In this section, as part of a review of the method of measurement (**Figure 4**), I also present the main evidence on neocorticalization (**Figure 5**).

The evolution of encephalization in mammals as shown in **Figure 3** is represented by the clustering of the data on archaic mammals about the lower edge of the convex polygon for living mammals. These brains of mammals that lived more than 40 Ma were between approximately 10% and 50% the size of those of living mammals of comparable body size. There has always been variation in encephalization. Many small-brained species are alive today and have enough gray matter to do quite well in the current world. The Virginia opossum (*Didelphis virginiana*) has approximately one-fifth as large a brain as a living animal of the same size, such as the cat (*Felis catus*).

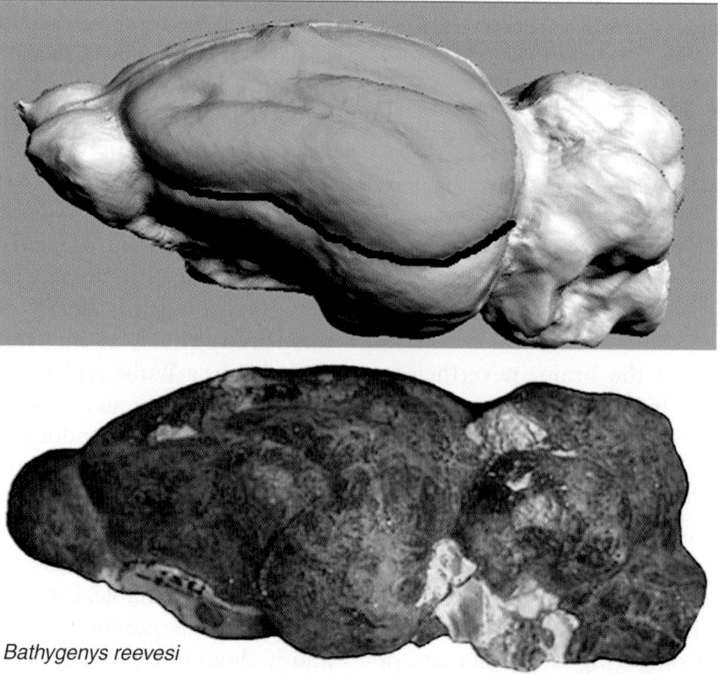

Figure 4 Endocast of a small Late Eocene artiodactyl, *Bathygenys reevesi*, 35 Ma; length, 5 cm. The animal, an oreodont, lived in what is now Big Bend National Park on the Rio Grande in Texas. (Top) Virtual image from a three-dimensional laser scan. Rhinal fissure is marked. (Bottom) Photograph of the same specimen (specimen UT 40209–431, University of Texas, Department of Paleontology).

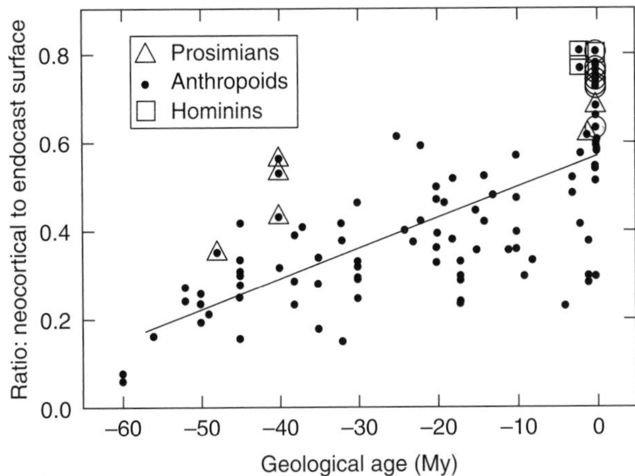

Figure 5 Relative size of neocortex in 106 species of mammals during the past 60 My. Prosimians are marked by triangles, hominins (including australopithecenes) by squares, and simians by circles. Regression of geological age on neocortical ratio: $Y = 0.007\,X + 0.57$; $r = 0.72$. In living mammals, 57% of the endocast surface area is devoted to neocortex; in living monkeys and humans, and in fossil hominins, the ratio averages approximately 75%. Reproduced from Jerison HJ (2006) What fossils tell us about the evolution of the neocortex. In: Kaas JH and Krubizer LA (eds.) *Evolution of Nervous System*, 4 vols. New York: Elsevier, with permission.

The European hedgehog (*Erinaceus europeaus*) has approximately one-third as large a brain. The prevalence of opossums and hedgehogs as road kill on US and European highways is evidence that these animals survive quite well if they stay out of traffic. Although the data in **Figure 3** indicate a selective advantage for an enlarged brain, the extent of the advantage differed in different lineages.

The endocast of only one species of mammal, the Eocene 'bear–dog' *Cynodictis*, was shown in the gallery of endocasts in **Figure 1**. It is an example of how brainlike an endocast can be. *Cynodictis* was not included among the 'archaic' mammals because it was a carnivore, a member of the same order Carnivora that survives today. The reader can verify that it was relatively small brained, however. Its coordinates are for a brain size of 12 g and an estimated 'foxlike' body size of 5000 g (EQ = 0.34). Living bears, wolves, cats, and dogs are 'average' mammals in encephalization (i.e., EQ = 1, approximately). If *Cynodictis* was an average carnivore of its time, there was a tripling in average brain size for carnivores during the 40 My of evolution covered here.

Mammalian endocasts often provide remarkably detailed pictures of the external topography of the brain. The pattern of gyri and sulci in *Cynodictis* is fairly typical for Eocene species, but its rhinal fissure may be difficult to identify. The endocast of a Late Eocene ungulate, *Bathygenys reevesi*, is presented in **Figure 4** to show the rhinal fissure less ambiguously.

Neocortex lies dorsal to the rhinal fissure and is distinguishable from paleocortex, which is ventral to the fissure. The top half of **Figure 4** is a snapshot of a rendered ditalized image of the endocast, and an area of the surface of this virtual endocast can be marked and measured. Such measurements were almost impossible before the development of digitizing methods with appropriate computer graphics software. The evolution of neocortex can then be determined on a series of mammals from different geological strata.

As I reported in 2006, a diverse sample of 84 fossil and 22 living species, a total of 106 mammalian endocasts, was scanned and measured. The sample included primates, carnivores, ungulates, rodents, insectivores, and marsupials as well as less familiar animals. Many of the species are from archaic orders that are now entirely extinct. To control for body size differences in brain size, neocorticalization was measured by the ratio of neocortex surface area to the surface area of the whole endocast. These ratios, represented by the points in the sample of 106 mammals, are graphed in **Figure 5**. Recognizing our special interest in human evolution, the primate points in the sample are identified by enclosing them in special symbols.

The most important result of this analysis is the relatively low correlation ($r = 0.72$) between the extent of neocorticalization in mammals and the passage of time. This is evidence of the diversity of specializations in mammals with respect to the brain. Many marsupials are only slightly more neocorticalized than the earliest mammals for which we have evidence. (The koala, despite its winning charms, holds the record in our sample as low beast on the living totem pole, with only 28% of its brain devoted to neocortex.) A variety of niches have been invaded by mammals, and they differ in the extent to which neocorticalization is important.

There are two results documented in **Figure 5**. First, there was a significant trend toward neocorticalization in mammals as a class. The measure of change in neocorticalizaion is the significant upward slope of the regression line. It averaged an approximately 7% increase in neocorticalization per 10 My. Variation about this trend reflects differences among species in selection for neocorticalization. Second, primates have always been more neocorticalized than other mammals, at least as suggested by this sample. It is worth noting that humans are average anthropoid primates on this dimension. The most neocorticalized of primate species in this sample was a langur, *Cercocebus albigena*, which was approximately the same as the human endocast in the sample, with 80% of its endocast surface area devoted to neocortex.

Marsupials were and are less neocorticalized than placental mammals. Furthermore, among species of carnivores in similar niches (Order Creodonta compared to Order Carnivora), the more neocorticalized species were probably at a selective advantage over less neocorticalized orders. At least they did not go extinct as less neocorticalized species did.

Conclusions

The results discussed here provide an overview of the evidence on brain evolution available from the fossil record. The evidence is primarily by inference from the data on brain size. The most complete evidence is on mammals, and it may be most useful for decisions about which species to study. It should be background for other analyses based on the anatomy, genetics, physiology, and neurochemistry of living brains.

See also: Bird Brain: Evolution; Brain Connectivity and Brain Size; Brain Development: The Generation of Large Brains; Brain Fossils: Endocasts; Brain Scaling Laws; Neocortex: Origins.

Further Reading

Butler AB and Hodos W (2005) *Comparative Vertebrate Neuroanatomy: Evolution and Adaptation,* 2nd edn. New York: Wiley–Interscience.

Dominguez PA, Milner A, Ketcham RA, Cookson MJ, and Rowe TB (2004) The avian nature of the brain and inner ear of Archaeopteryx. *Nature (London)* 430: 666–669.

Edinger T (1975) Paleoneurology, 1804–1966: An annotated bibliography. *Advances in Anatomy, Embryology and Cell Biology* 49: 12–258.

Hopson JA (1979) Paleoneurology. In: Gans C, Northcutt RG, and Ulinski P (eds.) *Biology of the Reptilia,* vol. 9, pp. 39–146. Academic Press: New York.

Jerison HJ (1973) *Evolution of the Brain and Intelligence.* New York: Academic Press.

Jerison HJ (1990) Fossil evidence on the evolution of the neocortex. *Cerebral Cortex* 8A: 285–309.

Jerison HJ (1991) *Brain Size and the Evolution of Mind: 59th James Arthur Lecture on the Evolution of the Human Brain.* New York: American Museum of Natural History.

Jerison HJ (2001) Epilogue: The study of primate brain evolution: Where do we go from here? In: Falk D and Gibson K (eds.) *Evolutionary Anatomy of the Primate Cerebral Cortex,* pp. 305–337. Cambridge, UK: Cambridge University Press.

Jerison HJ (2004) Dinosaur brains. In: Adelman G and Smith BH (eds.) *Encyclopedia of Neuroscience,* 3rd edn. Amsterdam: Elsevier [CD edition].

Jerison HJ (2006) What fossils tell us about the evolution of the neocortex. In: Kaas JH and Krubizer LA (eds.) *Evolution of Nervous System,* 4 vols. New York: Elsevier.

Karten HJ (1997) Evolutionary developmental biology meets the brain: The origins of mammalian cortex. *Proceedings of the National Academy of Sciences of the United States of America* 94: 2800–2804.

Nieuwenhuys R, ten Donkelaar HJ, and Nicholson C (1998) *The Central Nervous System of Vertebrates,* 4 vols. New York: Springer.

Quiring DP (1950) *Functional Anatomy of the Vertebrates.* New York: McGraw-Hill.

Stephan H, Frahm H, and Baron G (1981) New and revised data on volumes of brain structures in insectivores and primates. *Folia Primatologica* 35: 1–29.

Ulinski PS (1983) *Dorsal Ventricular Ridge: A Treatise on Forebrain Organization in Reptiles and Birds.* New York: Wiley.

Alternative Splicing in the Nervous System

J Y Wu, Northwestern University, Chicago, IL, USA
J A Potashkin, Rosalind Franklin University of Medicine and Science, North Chicago, IL, USA

Overview

Soon after the discovery of split genes in 1977, evidence began to accumulate to support the hypothesis that multiple mRNAs encoding different protein products could be produced from a single precursor mRNA (pre-mRNA). It was not until the completion of sequencing the human genome, however, that the extent of alternative splicing became apparent. The results from this major undertaking indicated that the number of protein-coding genes is far fewer than expected based on phenotypic diversity. Studies over the past 30 years have revealed the complex multi-dimensional networks of regulation in mammalian gene expression. Regulation can occur at the level of chromatin structure, transcription, pre-mRNA splicing/alternative splicing, alternative polyadenylation, RNA stability, RNA transport, RNA editing, and post-translational modifications. The following discussions focus on the role of alternative splicing in the nervous system and splicing mutations associated with neuro-degenerative diseases.

Pre-mRNA Splicing and Alternative Splicing

Pre-mRNA splicing, the process of removing introns from the nascent transcript and splicing together of exons to produce the functional messenger RNA (mRNA), occurs in a spliceosome, a macromolecular RNA–protein machine. In addition to the pre-mRNA and small nuclear ribonucleoprotein particles (snRNPs), the spliceosome contains over 200 protein factors. Essential to this process is the accurate recognition of the 5′ splice site (5′ss) and 3′ splice site (3′ss) by the spliceosome. In the assembled catalytically active spliceosome, the biochemical reactions of cleavages at splice sites and ligation of exons take place by a two-step transesterification mechanism.

In higher eukaryotes, multiple splicing isoforms can be produced from a single pre-mRNA by alternative usage of different splice sites. This process of alternative splicing can occur in a variety of patterns (**Figure 1**), including exon skipping/inclusion (cassette alternative exon), intron retention, alternative use of 5′ or 3′ splice sites, mutually exclusive exons, and more complex patterns that are coupled with alternative promoter selection or alternative polyadenylation. Alternative splicing may produce mRNA species with different stability or subcellular localization and/or transcripts that encode distinct protein products. As a result, alternative splicing is an extremely powerful and versatile mechanism for generating functional diversity from a limited number of genes.

In the nervous system, the process of alternative splicing is particularly remarkable because the brain has the highest frequency of alternative splicing of any tissue and it is becoming evident that regulation of alternative splicing in the brain has a significant impact on the development, function, and maintenance/repair of the nervous system. In addition, the largest group of genes that display tissue-specific splicing is expressed in the brain.

Molecular Mechanisms Regulating Alternative Splicing

In higher eukaryotes, especially mammals, pre-mRNA transcripts are usually long, containing multiple introns of variable sizes. Mammalian introns can be as large as 500 kb. In humans, the average size of exons is 150 nucleotides (nt), and that of introns is 3500 nt nucleotides. The basic splicing signals in mammalian pre-mRNAs are degenerate, with only two nucleotides (/GT at the 5′ss and AG/ at 3′ss) that are highly conserved. As a result, the nucleotide sequences surrounding the splice junctions usually contain only a limited amount of information, not sufficient for conferring the specificity required to achieve accurate splice site selection. Therefore, recognition not only of exon–intron junction sequences but also of the regulatory elements in intronic and exonic regions is important for defining splice junctions and maintaining splicing fidelity. This high degree of degeneracy in the splicing signals in mammalian pre-mRNAs provides the flexibility for alternative selection and pairing of different splice sites, a fundamental mechanism for regulating alternative splicing.

The intricate interactions between *cis*-elements in the pre-mRNA substrates and their *trans*-acting splicing factors determine the selective use of different splice sites. This network of interactions between pre-mRNA and *trans*-acting factors involves both snRNPs and non-snRNP splicing regulators. The *cis*-elements include splice sites and splicing regulatory sequences (splicing enhancers or silencers) inside either exons or introns. The sequences and sizes of the exons or introns, as well as secondary structures of the pre-mRNA, also influence splice site selection.

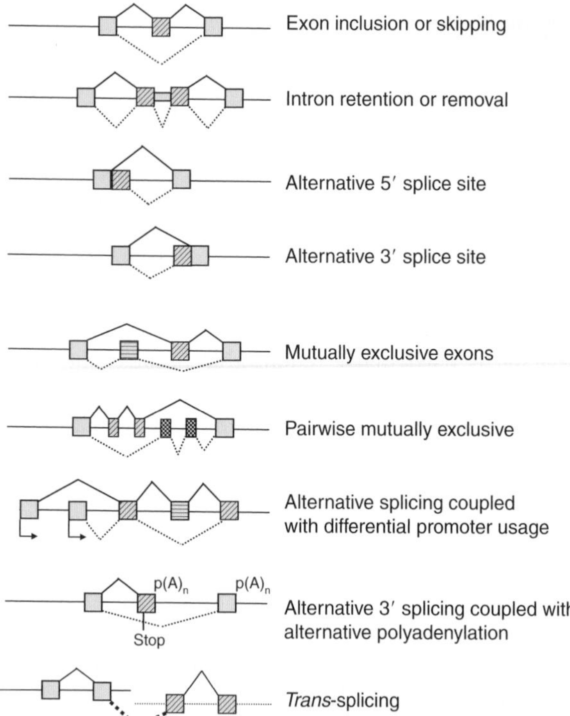

Figure 1 A range of alternative splicing patterns found in the nervous system. Most splicing events are *cis*-splicing, and trans-splicing may occur at a low efficiency in higher eukaryotes. Adapted from Wu JY, Yuan L, and Havlioglu N (2004). Alternatively spliced genes. In: Meyers RA (ed.) *Encyclopedia of Molecular Cell Biology and Molecular Medicine*, 2nd edn., vol. 1, pp. 125–177. Weinheim: Wiley-VCH.

The high level of degeneracy of mammalian splicing signals and the existence of large numbers of trans-acting factors allow versatile RNA–protein and RNA–RNA interactions during different stages of spliceosome assembly. The production and delicate balance of distinct isoforms generated by the alternative splicing of pre-mRNA transcripts are determined by combinatorial effects of multisite interactions among pre-mRNA, essential spliceosomal components, and regulatory factors.

A number of splicing regulators, either positive or negative, modulate alternative splicing. These splicing regulators contain several types of protein sequence motifs, including an RNA recognition motif (RRM) and other nucleic acid binding motifs (K-homology, or zinc finger) as well as a serine/arginine (SR)-rich domain or glycine-rich or RGG-rich sequences. Many SR domain-containing factors interact with exonic splicing enhancer elements to stimulate exon inclusion, whereas proteins of the heterogeneous nuclear ribonucleoprotein (hnRNP) family containing RRMs often act as splicing repressors by interacting with intronic splicing silencer sequences. A number of bifunctional splicing factors have been identified that can act either as activators or repressors for splicing, depending on their interactions with pre-mRNAs and other proteins. The concentration, distribution, composition, and state of modification of these regulatory factors determine whether they enhance or suppress the use of a particular splice site.

The *cis*-acting elements that regulate splicing are enhancer or silencer sequences present in either the intronic or exonic regions. Both types of elements may regulate splicing of a single pre-mRNA. Splicing enhancer and silencer elements may also overlap with each other. The splicing silencer sequences in several well-characterized genes, including those encoding c-src, caspase-2, and the γ-aminobutyric acid $(GABA)_A$ receptor $\gamma 2$ subunit, contain pyrimidine (U/C)-rich elements that interact with the polypyrimidine tract binding (PTB) protein (also referred to as hnRNP I). In these examples, PTB protein binds the repressor sequences in nonneuronal cells so that splice sites are skipped. The PTB gene also undergoes alternative splicing to generate three isoforms. In addition, a neuronal or brain-enriched PTB homolog (nPTB or brPTB) has been identified, and plays a role in relieving the suppression of inclusion of a neuron-specific N1 exon in c-src in neuronal cells. Splicing regulatory elements usually contain multiple binding sites for splicing regulators and function by recruiting other spliceosomal components to form RNP-like complexes. One of the common features of these *cis*-regulatory elements is that they are not simple sequence elements that act independently of each other. Splicing regulatory elements have been identified that can act to stimulate the splicing of one exon but repress another exon.

Neuron-Specific Splicing Regulators

A large number of splicing regulators are expressed in a wide range of tissues or cell types. Several RNA-binding proteins have been reported that are enriched in neurons or have neuron-specific isoforms. These neural splicing regulators include the Nova family (Nova-1 and Nova-2), Hu/ELAV family (Hud, Hel-N), Fox family (Fox-1 and Fox-2), STAR/GSG family, and CELF family (NAPOR/CUGBP2/ETR-3). Some of these factors were originally identified because of their association with neurological disorders.

Only a few of these neuron-specific splicing factors have been well studied. These include the K-homology (KH) domain RNA-binding proteins Nova-1 and Nova-2. Nova proteins were identified by their association with paraneoplastic opsoclonus myoclonus ataxia (POMA). In this syndrome, patients develop antibodies against Nova proteins, and this autoimmune response leads to a loss of inhibition of motor control in the spinal cord and brain stem. Nova-1 is expressed in the diencephalon, brain stem, and motor neurons of the spinal cord, whereas Nova-2 is present in the cerebral cortex, hippocampus, and dorsal spinal cord. Nova-1 binds to UCAUY elements, and Nova-2 interacts with GAGUCAU elements in their target RNAs (Y represents a pyrimidine). Nova-1 regulates the splicing of $GABA_A$ receptor $\gamma 2$ subunit and the $\alpha 2$ subunit of the glycine receptor (GlyR $\alpha 2$), and also autoregulates its own splicing. Nova has also been reported to co-regulate the splicing of several transcripts important for synaptic function, perhaps accounting for 7% of the brain-specific alternative splicing in the neocortex.

The Hu family is another group of splicing regulators that are autoimmune targets in POMA. Hu proteins bind to mRNAs, including many that are important for neuronal function, thus stabilizing the transcripts and increasing their translation. One member of the family of proteins, HuD, is a homolog of the fruit fly splicing regulator ELAV protein, which is essential for neurogenesis and plays a role in regulating splicing in neurons. Recently, Hu was reported to regulate splicing of calcitonin/calcitonin gene-related peptide (CGRP) pre-mRNA and facilitate exon 4 skipping by inhibiting the binding of TIA-1/TIAR, two ubiquitously expressed factors that promote the nonneuronal pathway.

The neuron-specific splicing of calcitonin/CGRP is also regulated by several other splicing factors, including Fox-1 and Fox-2. These proteins are expressed in muscle, heart, and brain tissues. In the brain, the Fox proteins are neuron specific. Fox-1 and Fox-2 promote exon inclusion in NMHC-B, c-src, FGFR2, and protein 4.1R transcripts. In contrast, within neurons Fox proteins repress exon 4 inclusion in calcitonin/CGRP pre-mRNA by binding to a UGCAUG element and blocking the constitutive splicing factor U2 auxiliary factor 65 ($U2AF^{65}$) from binding to the 3′ splice site upstream of exon 4. The splicing events, which result in the generation of calcitonin/CGRP, also involve several additional regulators, including Tra2β, SRp55, TIAR, SRp20, and PTB.

The STAR (signal transduction and activation of RNA)/GSG (GRP33, Sam68, GlD-1) protein family includes the mammalian Sam68-like protein SLM-1, which regulates splice site selection by binding to a purine-rich enhancer. Rat SLM-1 and SLM-2 are expressed primarily in neurons. The tissue-specific splicing factor SLM-1 is present in the dentate gyrus, whereas SLM-2 is found in the pyramidal cells of the CA1, CA3, and CA4 regions. It is not known if SLM proteins regulate neuron-specific splicing, but SLM-2 regulates the splicing of CD44 and may also regulate tau and tra2β1 pre-mRNA splicing.

The neuroblastoma apoptosis-related RNA-binding protein (NAPOR)/CUG-binding protein 2(CUGBP2)/embryonic lethal abnormal vision type RNA-binding protein 3(ETR-3) has three splice variants; NAPOR-3 is neuron specific while the other two forms are ubiquitously expressed. NAPOR-3 binds to UG-rich regions. The concentration of NAPOR in different brain regions is important for determining its function. Within the forebrain, where it is abundantly expressed, NAPOR promotes exon 5 skipping of E5 and exon 21 inclusion of in *N*-methyl-D-aspartate (NMDA) receptor R1 pre-mRNA.

In addition to these neuron-specific or brain-enriched splicing factors, ubiquitously expressed splicing factors may also play a role in the regulation of splicing in neurons. For example, several splicing factors recognize pyrimidine-rich sequences within the intron similar to the sequence recognized by PTB, including U2 auxiliary factor (U2AF), CUG-binding protein (CUG-BP), and hnRNP F. Partial replacement of general splicing factors by neuronal-enriched splicing factors may lead to neural specific splicing. Another example of a general splicing factor that regulates a neuronal splicing event is KSRP, a factor that binds to an intronic splicing enhancer element downstream of the neuron-specific c-src N1 exon. KSRP is present in both neuronal and nonneuronal cells, but it is more highly expressed in neuronal cells.

A good example of the complexity of neuronal gene alternative splicing regulation is that of exon 10 in the human tau gene. Mutations affecting this splicing event lead to frontotemporal lobe dementia (see later, neurological disorders). In this case, both exonic and intronic regulatory elements play important roles in controlling exon 10. In addition, both positive and negative *cis*-elements are involved and form a multidomain composite regulatory element. A number of *trans*-acting splicing regulators have been identified that interact with the splicing regulatory element.

It is not clear yet how the extremely complex alternative splicing events of different genes in the nervous system are coordinated. In some cases, the alternative splicing events of different genes appear to be co-regulated by the same protein. For example, the splicing regulator PTB can differentially recognize neural and nonneuronal substrates. The expression pattern of PTB in various regions of the brain at different developmental stages supports a role for PTB to act as one alternative splicing coordinator for different splicing target genes. Some regulatory elements have been identified that mediate splicing responses of neurons to extracellular stimuli. In general, very little is known about the molecular mechanisms controlling changes in alternative splicing in response to environmental signals.

Role of Alternative Splicing in the Nervous System

The regulatory role that alternative splicing plays in the nervous system is multifold. Alternative splicing has been reported for genes involved in almost every aspect of the nervous system, from neural development to the function and maintenance of the adult nervous system. The expression of a vast number of genes important for neuronal differentiation, function, and survival undergo alternative splicing. These genes encode trophic factors, neuronal guidance molecules, guidance cue receptors, transmitter receptors, ion channels, intracellular signal transduction molecules, and synaptic components. Similar to the role of splicing in nonneuronal cells, alternative splicing may affect transcript localization/stability or generate proteins of distinct functional activities.

The variety of receptors produced by alternative splicing in neurons is a classic example of how a few genes can produce many protein products. The genes encoding receptors for major neurotransmitters, including serotonin, dopamine, glutamate, and GABA, all undergo alternative splicing. This results in the production of receptors that differ in function and/or subcellular localization. For example, the NMDA R1 receptor has an alternatively spliced C1 exon. When C1 exon is included in the mRNA, the receptor localizes to the cell surface membrane; when C1 exon is skipped, the receptor is cytoplasmic. In addition, the regulation of C1 splicing modulates binding of NMDA R1 to neurofilaments and calmodulin. The splicing regulation of mGluR1 affects both localization and receptor activity. One variant, mGluR1a, has a long C-terminus that targets the protein to dendrites, whereas mGluR1b, containing a short C-terminus, localizes to axons.

Alternative splicing may affect how a receptor responds to drugs. For example, two splice variants of the $GABA_A$ receptor respond to benzodiazepine agonists in different manners. The splice variants of norepinephrine transporter have different rates of norepinephrine uptake and binding affinities to nisoxetine, a norepinephrine reuptake inhibitor. Additionally, the serotonin 5-HT_4 receptor has splice variants with different C-terminal domains that respond differently to agonists.

Alternative splicing is also important for the modulation of neuronal signaling. For example, alternative splicing of the gene encoding the α_{1N} N-type calcium channel produces a protein with a two-amino-acid insertion that results in a channel with altered kinetics. Another example is apolipoprotein E receptor 2 (Apoer2, a receptor for Reelin), which forms a functionally active complex with NMDA receptor in the postsynaptic densities of excitatory synapses. An alternatively spliced exon of Apoer2 is required for Reelin to induce phosphorylation of NMDA receptor and thus enhance long-term potentiation (LTP). Interestingly, the splicing of Apoer2 is regulated by synaptic activity.

In addition to modulating synaptic function, alternative splicing plays a role in synapse formation and plasticity by regulating the expression of proteins important for cell adhesion and cell–cell communication. Neural cell adhesion molecule NCAM1 is involved in synaptic plasticity, neurodevelopment, and neurogenesis. Single-nucleotide polymorphisms (SNPs) within NCAM1 splice sites that affect alternative splicing are associated with bipolar disorder and schizophrenia.

The role of alternative splicing in synaptic plasticity is only beginning to be understood. Recently, target genes for the splicing factor Nova were identified, including $GABA_B$ receptors and G-protein-activated inward rectifier potassium channel 2 (GIRK2) channels. These proteins mediate LTP of the slow inhibitory postsynaptic current (siPSC) in dendrites. In mice lacking Nova-2, the LTP of siPSCs, but not the excitatory postsynaptic currents, is absent. The current model is that Nova proteins control the splicing of a number of pre-mRNAs, resulting in the regulation of LTP of the siPSCs. The application of new techniques such as cross-linking immunoprecipitation (CLIP) and microarray assays may reveal additional examples of splicing factors regulating networks that control synaptic plasticity.

Neurological Disorders Associated with Dysregulation and Abnormalities of Pre-mRNA Splicing

Human diseases may result from defective or aberrant splicing caused by mutations in regulatory pre-mRNA sequences and splicing factors. Recent data suggest that approximately 31% of human diseases are caused by mistakes in pre-mRNA splicing, and 15% of disease-causing point mutations affect pre-mRNA splicing. These statistics most likely underestimate the extent of human diseases associated with abnormalities in splicing, because splicing enhancer or silencer sequences, very few of which have been clearly identified, may carry point mutations that disrupt the balance of splice variants and thus result in disease. In addition, environmental stimuli such as stress may change intracellular distribution or posttranslational modifications of splicing factors and thus affect pre-mRNA splicing. To demonstrate the role of splicing dysregulation in human diseases, a few well-studied examples are described in the following sections. In general, these diseases can be classified in two categories: those caused by *cis*-acting mutations in the affected genes and those associated with secondary splicing defects resulting from *trans*-acting mutations.

Neurological Diseases Caused by *cis*-Acting Splicing Mutations

Dementia Splicing mutations and aberrant pre-mRNA splicing have been associated with both sporadic and familial forms of dementia, including Alzheimer's disease. For example, splicing mutations have been identified in both presenilin-1 and presenilin-2 genes in Alzheimer's disease. Frontotemporal dementia and parkinsonism linked to chromosome 17 (FTDP-17) and related tauopathies are characterized by intraneuronal tau-containing deposits in affected brain regions. Microtubule-binding protein tau, a key player in microtubule stability, is critical for neuronal function. Mutations in human tau genes have been identified in FTDP-17 patients, including not only missense mutations affecting tau protein function but also splicing mutations causing imbalance of different tau splicing isoforms. Three exons in the human tau gene are alternatively spliced. In particular, exon 10 encodes one of four microtubule-binding repeat (R) domains, and alternative inclusion or exclusion of exon 10 leads to the formation of Tau4R or Tau3R, respectively. Exonic or intronic mutations that disrupt the balance of Tau4R versus Tau3R splicing isoforms lead to FTDP-17 and other forms of tauopathy with neurofibrillary tangles. In addition, splicing mutations altering exon 2 and exon 3 inclusion are associated with gliopathy and spinal cord degeneration. These observations demonstrate that an imbalance of different splicing isoforms can result in disease development or progression.

Muscular dystrophy Various mutations that affect alternative splicing of the dystrophin gene lead to Duchenne muscular dystrophy (DMD). Both exonic and intronic splicing mutations have been reported in DMD patients, including those inducing exon skipping and cryptic splice site activation. These mutations act either by disrupting splicing enhancers or by creating splicing silencers. In some cases, nonsense mutation(s) can also affect splicing by generating splicing silencer binding sites. For example, nonsense mutations in the dystrophin gene of some patients produced a binding site for hnRNP A1 (UAGACA) within the resultant pre-mRNA. These recent observations highlight the complexity of splicing errors and suggest that disease-causing splicing mutations may be underestimated.

Ataxia–telangiectasia and neurofibromatosis Ataxia–telangiectasia and neurofibromatosis type 1 are autosomal dominant neurological diseases with malignancy predisposition. Genetic studies have identified splicing mutations as common genetic defects in these patients, with approximately 50% patients carrying splicing mutations. These studies suggest the significant contribution of splicing defects to the pathogenesis of neurological disorders.

Spinal muscular atrophy Spinal muscular atrophy (SMA) is characterized by selective degeneration of motor neurons that leads to progressive paralysis. SMA is an autosomal recessive disease caused by loss of or mutations in the telomeric *survival of motor neuron 1* (*SMN1*) gene. The centromeric *SMN2* gene regulates the severity of the disease. Although *SMN1* and *SMN2* are almost identical, a single translationally silent nucleotide change in *SMN2* disrupts an exonic splicing enhancer and results in exon 7 skipping, with only about 20% of the *SMN2* transcripts producing a functional full-length SMN2 protein. This low level of full-length SMN2 protein is not sufficient to compensate for the loss of *SMN1*. A splice variant that retains intron 3 encodes an axonally localized truncated form of the protein called a-SMN. During development, a-SMN is expressed mainly in the axons of motor neurons and is only produced from the *SMN1* gene. It remains to be elucidated why the loss of *SMN1* primarily affects motor neuron function. Another interesting aspect of SMA is that the SMN1 protein plays a role in snRNP assembly and thus it may also be considered a *trans*-acting splicing disorder (see later).

Neurological Diseases Caused by *trans*-Acting Splicing Defects

A number of neurological syndromes or diseases are caused by *trans*-acting splicing defects. In these disorders, the mutations in genes encoding spliceosomal components or splicing regulatory factors lead to aberrant or defective splicing events.

Retinitis pigmentosa A common cause of blindness is retinitis pigmentosa, which is characterized by a loss of photoreceptor cells and progressive degeneration of the retina. A fraction of patients with autosomal dominant form of retinitis pigmentosa (adRP) have mutations in genes encoding spliceosomal proteins essential for spliceosomal assembly, including PRP3, PRP8, and PRP31. Similar to SMA, in these adRP patients, defects in the ubiquitously expressed RNA processing factors lead to specific neurological diseases. The mechanisms for the specificity of neurodegeneration in these diseases remain to be elucidated. Current models include haploinsufficiency and dominant-negative effects leading to cell-type-specific functional deficiencies.

Myotonic dystrophy Myotonic dystrophy ('dystrophia myotonica'; DM) is a multisystem syndrome with prominent muscular atrophy. Two types of DM have been revealed by genetic studies: DM1 is caused by the presence of a CUG expansion in the 3′ untranslated region (UTR) of the DM protein kinase pre-mRNA, and DM2 is caused by a CCUG expansion in intron 1 of the *CNBP* (*ZNF9*) gene. CUG/CCUG repeat expansion results in sequestration or upregulation of proteins interacting with CUG/CCUG, including muscle blind-like (MBNL) and CUG-binding protein 1 (CUG-BP1). This in turn causes a dysregulation in the splicing of a subset of pre-mRNA muscle and brain transcripts that require these regulatory factors. DM1 is also associated with changes in splicing of pre-mRNAs that encode amyloid precursor protein, NMDAR1, and tau, suggesting that aberrant splicing of these pre-mRNAs may be the basis of the central nervous system defects such as memory impairments.

Prader–Willi syndrome Patients with Prader–Willi syndrome (PWS) often exhibit behavioral problems such as obsessive–compulsive disorder and autism. A recent report has suggested an explanation for why these patients frequently respond to treatment with serotonin reuptake inhibitors. PWS is associated with deletions of the paternally inherited copy of chromosome 15q11–q13, a maternally imprinted locus encoding multiple copies of a small nucleolar RNA called HBII-52 snoRNA. A stretch of 18 nt in the HBII-52 snoRNA is complementary to the sequence within an alternatively spliced exon (Vb) of the 5-HT_{2C}R serotonin receptor pre-mRNA. HBII-52 snoRNA binds to and antagonizes a splicing silencer, facilitating the inclusion of exon Vb within the mRNA. Changes in splicing of 5-HT_{2C}R were detected in the hippocampal RNA in PWS patients. These data suggest that altered 5-HT_{2C}R pre-mRNA splicing caused by the loss of HBII-52 snoRNA contributes to the clinical manifestations in PWS patients.

Ataxias Fragile X-associated tremor/ataxia syndrome (FXTAS) is a newly recognized late-onset syndrome characterized by cognitive impairment along with tremor and gait problems. FXTAS is associated with moderate expansions of 55–200 CGG repeats in the region of the fragile X mental retardation (*FMR1*) gene encoding the 5′ UTR. Longer expansions (>200) of CGG repeats lead to the fragile X mental retardation syndrome with numerous neurological problems, including autism, learning disabilities, anxiety disorders, and mental retardation. The *FMR1* gene is regulated by alternative splicing and encodes RNA-binding proteins involved in regulation of protein synthesis at postsynaptic sites of dendrites and in the maturation of dendritic spines. The CGG trinucleotide repeat expansions in the 5′ UTR lead to transcription silencing and loss of function of *FMR1*. Intranuclear inclusions in neurons and astrocytes in the postmortem brains of FXTAS patients contain the RNA-binding proteins MBNL1 and hnRNPA2, suggesting that CGG expansions may affect the function of these splicing factors and that FXTAS may involve aberrant pre-mRNA splicing. Further research is needed to understand the role of MBNL1, hnRNPA2, and FMRP in disease development.

In another form of ataxia, spinocerebellar ataxia subtype 2 (SCA2), there is a CAG repeat expansion in the coding region of the gene encoding ataxin-2. Ataxin-2 protein contains RNA-binding domains and interacts with the splicing regulator *A2BP1* (*FOX1*). Disruption of the *A2BP1* locus by chromosomal translocation has been reported in patients with mental retardation and epilepsy. Triplet repeat expansions in the noncoding RNAs have been found in patients with other SCA subtypes, including SCA8, SCA10, and SCA12. These observations suggest the possible involvement of pre-mRNA splicing dysfunction in various forms of ataxia. Further research is necessary to understand if there are common molecular mechanisms among numerous repeat expansion diseases, why these diseases manifest differently in patients, and whether dysregulation of alternative pre-mRNA splicing contributes to the pathogenesis of these diseases.

Paraneoplastic neurologic disorders In paraneoplastic neurologic disorder (PND) patients, malignancies outside of the central nervous system (CNS) induce expression of neuron-specific splicing regulatory factors (SRFs). The autoimmune response generated against these neuron-specific SRFs ultimately results in the dysfunction of neurons that normally express these proteins. Two families of neuron-specific SRFs have been identified as PND target antigens, Hu and Nova. Nova-1 and Nova-2 regulate pre-mRNA splicing of genes encoding synaptic proteins (see earlier). Nova RNA targets correlate with the neurologic defects in PND patients, consistent with observations in Nova-1 knockout mice.

Conclusions and Future Perspective

Overwhelming evidence supports the concept that pre-mRNA splicing and alternative splicing play an important role in mammalian gene expression and that alternative splicing is a powerful mechanism for genetic diversity. This is particularly evident in the nervous system. Alternative splicing modulates activities of the majority of genes involved in the formation and function of the nervous system. A wide range of neurologic disease is caused by genetic mutations that affect pre-mRNA splicing. Some splicing mutations act *in cis* to cause splicing defects in genes containing the mutations. Other mutations affect spliceosomal components or splicing regulatory factors, which in turn lead to aberrant splicing or dysregulation of splicing of genes regulated by these splicing factors. Trinucleotide repeat expansion mutations can also affect pre-mRNA splicing by sequestering or altering splicing regulators that bind to the trinucleotide repeats and affect the splicing of the genes that are normally controlled by the splicing regulators. In addition, abnormal splicing has been reported in complex diseases such as psychiatric disorders. Although the important role of pre-mRNA splicing and alternative splicing regulation in the nervous system has been established, much work needs to be done to understand the molecular mechanisms underlying the splicing regulation and the molecular pathogenesis of splicing disorders.

Numerous environmental factors influence neuronal survival and function. Very little is known about how the nuclear splicing machinery responds to various extracellular signals, although many studies have documented altered pre-mRNA splicing in response to stresses such as pH change, osmotic or temperature shock, UV exposure, and oxidative stress. For example, brain ischemia induces the translocation of serine-arginine-rich splicing regulators from the nucleus to the cytoplasm and causes changes in alternative splicing. Further research is necessary to elucidate the molecular mechanisms by which environmental factors modulate pre-mRNA splicing and alternative splicing regulation.

The combined application of genetic, molecular, biochemical, and bioinformatic approaches will further our understanding of the role in pre-mRNA splicing and alternative splicing regulation in the nervous system. The ultimate goal is the development of strategies to correct splicing defects that cause human diseases.

See also: Gene Expression Regulation: Activity-Dependent; Gene Expression Regulation: Chromatin Modification in the CNS; Gene Expression Dysregulation in CNS Pathophysiology; Genetic Influence on CNS Gene Expression: Impact on Behavior; Genomic Disorder and Gene Expression in the Developing CNS.

Further Reading

Blencowe B (2006) Alternative splicing: New insights from global analyses. *Cell* 126: 27–46.

Dredge BK, Polydorides AD, and Darnell RB (2001) The splice of life: Alternative splicing and neurological disease. *Nature Reviews Neuroscience* 2: 43–50.

Faustino NA and Cooper TA (2003) Pre-mRNA splicing and human disease. *Genes & Development* 17: 419–437.

Grabowski PJ (1998) Splicing regulation in neurons: Tinkering with cell-specific control. *Cell* 92: 709–712.

Grabowski PJ and Black DL (2001) Alternative RNA splicing in the nervous system. *Progress in Neurobiology* 65: 289–308.

Jeanteur P (ed.) (2006) *Progress in Molecular and Subcellular Biology. Alternative Splicing and Disease,* vol. 44. Berlin, Heidelberg: Springer.

Kar A, Kuo D, He R, et al. (2005) Tau alternative splicing and frontotemporal dementia. *Alzheimer Disease, and Associated Disorders* 19(supplement 1): S29–S36.

Lee CJ and Irizarry K (2003) Alternative splicing in the nervous system: An emerging source of diversity and regulation. *Biological Psychiatry* 54: 771–776.

Lee JA, Xing Y, Nguyen D, et al. (2007) Depolarization and CaM kinase IV modulate NMDA receptor splicing through two essential RNA elements. *PLoS Biology* 5(2): e40 (doi:10.1371/journal.pbio.0050040).

Licatalosi DD and Darnell RB (2006) Splicing regulation in neurologic disease. *Neuron* 52: 93–101.

Potashkin J and Meredith G (2006) The role of oxidative stress in the dysregulation of gene expression and protein metabolism in neurodegenerative disease. *Antioxidant and Redox Signaling* 8: 144–151.

Ranum LP and Cooper TA (2006) RNA-mediated neuromuscular disorders. *Annual Review of Neuroscience* 29: 259–277.

Sharma S and Black DL (2006) Maps, codes, and sequence elements: Can we predict the protein output from an alternately spliced locus? *Neuron* 52(4): 574–576.

Stoilov P, Meshorer E, Gencheva M, et al. (2002) Defects in pre-mRNA processing as causes of and predisposition to diseases. *DNA and Cell Biology* 21: 803–818.

Ule J and Darnell RB (2006) RNA binding proteins and the regulation of neuronal synaptic plasticity. *Current Opinion in Neurobiology* 16: 102–110.

Wu JY, Yuan L, and Havlioglu N (2004) Alternatively spliced genes. In: Meyers RA (ed.) *Encyclopedia of Molecular Cell Biology and Molecular Medicine,* 2nd edn., vol. 1, pp. 125–177. Weinheim: Wiley-VCH.

Aluminum

S C Bondy, University of California at Irvine, Irvine, CA, USA

Introduction

Background and Chemistry

Aluminum (Al) is an abundant metal, composing 8.3% of the Earth's crust by weight. It is the thirteenth element in the periodic table and its electronic configuration is $[Ne]3s^2 3p^1$. It was not until 1825 that the metallic form of the element was first isolated. Aluminum is a simple trivalent cation incapable of undergoing valence changes. It is highly oxophilic, and aluminum within minerals is usually found surrounded by six oxygen atoms. Because of its charge and the fact that it is a Lewis acid, Al^{3+} has a high affinity for electrons and thus can be a good catalyst. However, this element is not known to have a biological function. Since aluminum has a high charge to radius ratio, thermodynamically the metal prefers electrostatic rather than covalent binding. The aqueous chemistry of aluminum is complicated because free Al^{3+} hydrolyzes to form a wide range of complexes with water. There are three main categories of aluminum species relevant to toxicological availability. At neutral pH, Al salts undergo extensive hydrolysis and $Al(OH)_3$ is produced. As the solutions age, $Al(OH)_4^-$ is also present, and this leads to precipitation of polymeric and colloidal aluminum.

Until recently it was generally believed that aluminum released into the environment is harmless. This is because, in solution, Al^{3+} salts form monomeric hydroxy compounds which become chemically inert polymeric and colloidal particles as the solution ages. Because of the formation of these insoluble aluminum species, it was assumed that absorption would be limited and thus the metal would be relatively innocuous. However, the metal has been shown to be toxic to both plants and animals, and there is a rising concern over the metal's potential adverse health effects.

Aluminum Prevalence in the Human Environment

In the Roman Empire, around 2000 years ago, aluminum salts were used for the purification of water, and in the Middle Ages, aluminum was combined with honey for the treatment of ulcers. Aluminum salts are still used today for water purification and medicinal purposes and are present in many antacids. Al is present in a variety of other medicines, including vaccines, buffered aspirin, and phosphate binders used in dialysis. Alum is used medically as a topical astringent and as a styptic to stop bleeding. Parenteral nutrition solutions are often high in aluminum salts.

The main source of inadvertent aluminum intake is food, and the element is found in many food additives, such as processed cheese, baked goods, and grain products. Alum is used as a crisping agent in pickle processing and may be present in baking powder. Aluminum-containing compounds are also used as preservatives, coloring agents, and leavening agents. Antiperspirants containing aluminum chlorohydrate are another source of exposure. Occupational exposure to aluminum can occur and is mainly found where the metal is being processed.

The presence of aluminum in drinking water is due both to natural sources and to widespread water purification procedures involving the use of Al as a coagulant. Water treatment generally increases the content of soluble, low molecular weight, chemically reactive, and possibly more readily absorbed aluminum species. The advent of modern industrial technology and the introduction of a variety of chemicals into the atmosphere have led to the formation of acid rains. Strong mineral acids such as sulfur and nitrogen oxyacids, found in acid rains, can solubilize aluminum from soil and mobilize the metal. Thus the increasing prevalence of acid rain can lead to the release of greater amounts of aluminum salts from insoluble minerals, leading to greater bioavailability.

Aluminum Entry into Nervous Tissues

The most common form of human exposure to Al^{3+} is absorption through the gastrointestinal tract. The rate of absorption is approximately 0.22%, and once in the blood, approximately 90% of the metal is bound to transferrin. Al^{3+} can pass the blood–brain barrier by receptor-mediated endocytosis of the Fe-carrier protein, and in rats, approximately 0.005% of the metal complex enters the brain.

Transdermal absorption of aluminum has been reported in mice after application of Al chlorohydrate, a substance commonly present in commercial underarm deodorants, and this led to an elevated aluminum content in the brain as well as in the serum. Airborne aluminum is capable of entering the brain through the olfactory neurons located in the roof of the nasal cavity. The axons of these sensory neurons form nerve bundles that pass through the cribriform plate of the ethmoid bone and synapse at the olfactory bulb. From here, there are projections to the olfactory cortex, and other cortical areas, including the hippocampus. Absorption of aluminum from the olfactory pathway has been shown in rats exposed to aluminum acetylacetonate to lead to accumulation of Al in the pons–medulla, olfactory bulb, and hippocampus, implying transneuronal migration.

Aluminum and Neurological Disease

Dialysis Encephalopathy

Acute exposure to aluminum can undoubtedly cause clinical neurotoxicity. Aluminum salts appear to be the cause of encephalopathy in uremic patients receiving chronic hemodialysis. In the past, such patients were routinely treated with aluminum-containing phosphate binders. In brain gray matter from a group of uremic patients on dialysis who died of neurologic syndrome, levels of aluminum were 25 ppm, compared to 6.5 ppm in a group of patients on dialysis who died of other causes, and 2.2 ppm in controls. Recovery from dialysis encephalopathy has been facilitated by application of an Al chelator. Demonstration of improved neurologic function following cessation of aluminum exposure further implicates Al as the cause of the dialysis encephalopathy syndrome. Other clinical procedures, such as bladder irrigation with 1% alum (a double salt of potassium and aluminum sulfate), can also cause symptoms of encephalopathy associated with elevated serum aluminum levels. A case study of a renal failure patient who developed Al-induced encephalopathy after using phosphate-binding Al-hydroxy gels for a prolonged period found increased proliferation of microglia and astrocytes upon postmortem analysis.

Evidence of Aluminum Neurotoxicity following Occupational Exposures

The development of an encephalopathy, characterized by cognitive deficits, incoordination, tremor, and spinocerebellar degeneration, among workers in the aluminum industry is further evidence that exposure to the metal can be profoundly deleterious. 'McIntyre Powder' (finely ground aluminum and aluminum oxide) was used as a prophylactic agent against silicotic lung disease between 1944 and 1979 in miners in northern Ontario. A morbidity prevalence study revealed that miners exposed to aluminum performed less well on cognitive examinations compared to unexposed workers. Furthermore, the likelihood of scores being in the impaired range increased with duration of exposure.

Other Studies Suggesting that Aluminum Salts Can Be Neurotoxic, and the Safety of Alum-Containing Vaccines

Prolonged feeding of infants with aluminum-containing intravenous solutions has led to impaired neurological function as judged by a reduction in the Mental Development Index. Treatment of aluminum-related bone disease with deferoxamine, which mobilizes bone Al and thus elevates serum Al^{3+}, has been reported to precipitate dementia.

Alum is present as an adjuvant in most vaccines, and abnormal neurological symptoms have been observed in several patients receiving intramuscular injections of Al-containing vaccines. The World Health Organization Vaccine Safety Advisory Committee has recognized that there may be a subset of predisposed individuals who are sensitive to Al-containing adjuvants. Macrophagic myofasciculitis secondary to intramuscular injection of aluminum hydroxide-containing vaccines shows both long-term persistence of aluminum hydroxide and an ongoing local immune reaction. These clinical findings are paralleled in several animal systems in which systemically administered aluminum can led to behavioral deficits, including incoordination and cognitive and morphological changes in the central nervous system (CNS) of treated animals. Since alum-containing vaccines have been in widespread use for a long time, the possibility of long-term hazard resulting from this is generally considered low.

Alzheimer's Disease

The presence of excess aluminum in the brains of patients with Alzheimer's disease (AD) has been described, but this issue remains controversial, since several conflicting reports exist. Thus the question as to whether the metal may play a role in AD and in other less common neurological disorders, such as the Guamanian parkinsonism–amyotrophic lateral sclerosis (ALS) constellation and Hallervorden–Spatz disease, is unresolved. However, chelation therapy, in order to reduce the aluminum burden in Alzheimer patients, has been reported as beneficial, and new Al-specific chelators for potential use in AD treatment have recently been developed.

There is a strong epidemiological relation between extended exposure to aluminum in drinking water and the incidence of AD. Increasing numbers of reports have related the aluminum content of drinking water with increased risk of developing AD. Residing in areas where aluminum concentrations in municipal drinking water are $100\,mg\,l^{-1}$ or greater appears to elevate AD incidence, and a dose–response correlation between increasing concentration of Al in drinking water and a higher risk of developing AD has been described in studies emanating from the USA, Canada, and France. A comprehensive literature survey has found 13 reports concerning a significant association between residing in areas where aluminum concentrations in the municipal drinking water are high and an increase in the incidence of AD, and a meta-analysis, integrating these results, reveals that this association is significant.

Correlative changes are never sufficient to imply causation. Proposals have been made that aluminum entry into the brain is a secondary epiphenomenon, consequent to damage to the blood–brain barrier. However, the finding that dialysis encephalopathy can be treated with deferoxamine with good results-suggests that Al is directly neurotoxic. Thus a causal relationship may exist between circulating Al^{3+} and dementia. Epidemiological findings are strengthened when confirmed by use of an animal model in which the number of potentially confounding variables can be minimized.

Mechanisms of Neurotoxicity

While Al exposure has repeatedly been shown to be correlated to neurodegenerative changes, the mechanisms that underlie this remain unresolved. There is considerable evidence that Al may accelerate the progression of two major and interrelated processes associated with neural senescence. These are oxidative stress and increased levels of basal unprovoked inflammatory activity within the brain.

Oxidative Stress

In both biological and nonbiological isolated systems, aluminum can potentiate production of reactive oxygen species (ROS) by the prooxidant metals iron (Fe) and copper (Cu). In view of the inability of aluminum ions to undergo the valence transitions, or to have a strong affinity for sulfhydryl groups, properties that characterize metals which have the potential to promote excess prooxidant activity within cells, this promotion of ROS generation by Al is puzzling. While the ability of aluminum to induce the generation of free radicals has been reported many times, this seems to depend on the concurrent presence of a transition metal such as copper or iron. These are essential elements and are present in most cell compartments. The mechanism by which this interaction occurs may be by the binding of these transition metals to the surface of colloidal aluminum.

Transition metals can be present at the solid–liquid interfaces of particles, due to their being absorbed onto the surface of insoluble particles, and under these conditions, their prooxidant nature can be enhanced. The ability of Fe and Cu to promote oxidative events based on their redox valence fluctuation is intensified by their incomplete sequestration. Particulates and colloidal compounds (including aggregated amyloid beta peptide) may partially complex prooxidant metals, and thus allow them to participate in Fenton reactions for a prolonged period. This has been reported for several mineral particulates. For example, silica that appears to be free of iron can promote ROS production, but this is prevented by acid washing. This illustrates that very low amounts of trace metals bound to particulate surfaces can initiate ROS generation. High molecular weight aluminum matrices are likely to have extensive solid–liquid interface surface area and could present an appropriate site for iron activation. The neurological hazard of aluminum may then be in part due to its ability to increase the rate of generation of ROS initiated by transition metals.

Other means by which Al may promote cellular ROS generation exist. Exposure of isolated cells of glial, but not neuronal, origin to aluminum sulfate increases formation of prooxidant species. It is possible that the glial-specific aluminum-induced increase in oxidative parameters is due to glial activation by extracellular aluminum complexes. The potentially prooxidant role of Al^{3+} is especially relevant in view of the reported presence of excess oxidative activity, both in the aged brain and, more specifically, in those regions of the Alzheimer's brain showing the most pronounced pathological changes. Iron and aluminum are both present at high concentrations in neurofibrillary tangles and may cooperate in the promotion of physiological aggregation of the amyloid beta peptide.

Inflammation

Elevated levels of intrinsic inflammation are associated with neural aging, and this is exacerbated in several neurodegenerative diseases. In age-related neurodegenerative disorders, such as AD and Parkinson's disease (PD), the further enhancement of age-related inflammatory processes is thought to significantly contribute to pathogenic events. The number of activated astrocytes is increased in AD and these are associated with senile plaques and with cerebral microvessels. In the hippocampus of AD patients, there is an upregulation of proinflammatory genes, and levels of cytokines are elevated in the brain as well as in cerebrospinal fluid and plasma of AD patients. AD is characterized by brain depositions of the toxic amyloid beta peptide (Aβ), which is generated from amyloid precursor protein (APP). In the brain of AD patients, reactive microglia, producing proinflammatory cytokines and acute-phase proteins, are associated with Aβ-containing neuritic plaques. Aluminum salts promote Aβ aggregation *in vitro*, and in the brains of transgenic mice overexpressing amyloid precursor protein, dietary exposure to Al can also exacerbate Aβ deposition and plaque formation.

It has recently been found that low levels of aluminum in drinking water of experimental animals

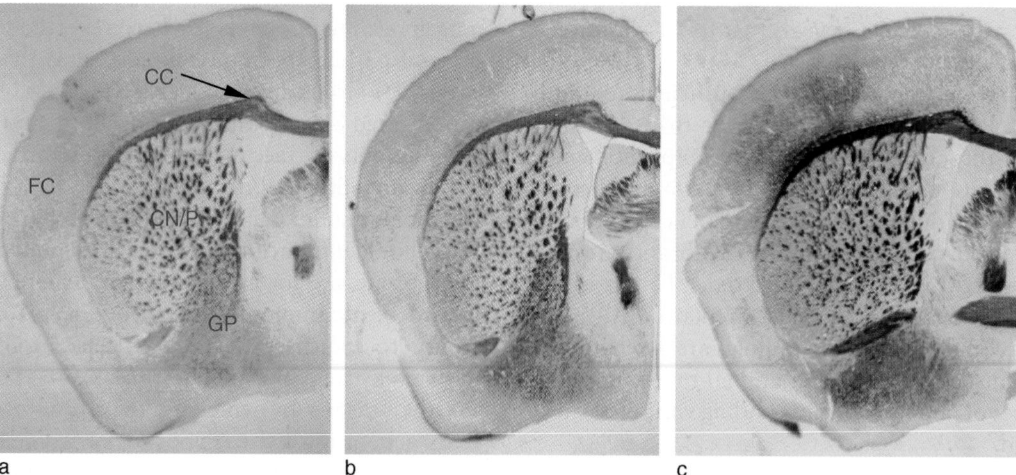

Figure 1 Microglial activation in the brain of aluminum lactate-exposed animals. (a) Control, (b) Al lactate (0.01 mM), (c) Al (0.1 mM); FC, frontal cortex; CC, corpus callosum; CN/P, caudate nucleus/putamen; GP, globus pallidus. 20× magnification.

elevates basal levels of inflammatory activity within the CNS. In this case, one of the initial events in the cascade leading to inflammatory responses is the activation of transcription factors. Levels of activated nuclear factor-kappa B (NF-κB) were increased in brains of mice treated with Al lactate (0.01, 0.1, and 1 mM) in drinking water for 10 weeks. Several inflammation-associated cytokines, such as tumor necrosis factor-α (TNF-α) and interleukin-1α (IL-1α), were increased in the brains of treated animals. The mRNA for TNF-α was also upregulated. Even the lowest concentration of Al tested increased inflammatory processes in the brain of mice. This level of Al is in the range present in some water supplies. The prevalence of AD is heightened in regions where similar concentrations of the metal are present in residential drinking water.

A pronounced increase in both astrocyte proliferation and microglial activation was observed in mice receiving extended treatment with low levels of Al in water, and these changes were marked in the striatum and globus pallidus as well as the cortex (**Figure 1**). Since no parallel changes in transcription factors or cytokine levels were observed in the serum or liver of treated animals, the proinflammatory effects of the metal may be relatively selective to nervous tissue. Although oxidative and inflammatory events are independent phenomena, they are related. In addition to the aforementioned indices of inflammation, rates of lipid peroxidation and levels of neuronal nitric oxide synthase (nNOS) were elevated after exposure to 100 μM Al.

In isolated systems, aluminum is able to promote aggregation of Aβ peptides, paired helical filaments, tubulin, and other proteins into insoluble complexes that resist proteolysis, but this has never been demonstrated in animals treated systemically with aluminum. However, direct intracerebral administration of aluminum salts can lead to accumulation of amyloid precursor protein. Aluminum-induced tau protein aggregates (neurofibrillary tangles) in the rabbit can be partially reversed by removal of aluminum with deferoxamine, and aluminum-induced proteinaceous tangles can be reversed by removal of aluminum with silicates.

The possibility that insoluble aluminum complexes may induce glial activation and macrophage activity is supported by the observation that, in rats intracerebroventricularly injected with Al^{3+}, complexes of the metal accumulate largely in the striatum, and this is accompanied by gliosis. Such findings obtained from study of Al-exposed humans or experimental animals can be supported by parallel experiments using isolated systems Treatment of cultured cells derived from glioblastoma, but not neuroblastoma, with aluminum lactate increases cell proliferation, cytokine secretion, and NF-κB activation. This suggests that Al effects on the brain are not systemically mediated but are consequent to activation of the pathway leading from receptors at the cell surface, to phosphorylation of transcription factor-activating kinases, and thence to augmentation of expression of proinflammatory genes (**Figure 2**).

Conclusions

As the median age of the population rises, protection of the performance of the brain during senescence becomes increasingly relevant. Thus it is important to identify those environmental factors which may accelerate undesirable changes in brain function. Since cortical Al^{3+} levels increase with age, a role for aluminum in the progression of brain aging is possible. In addition, the incidence of age-related

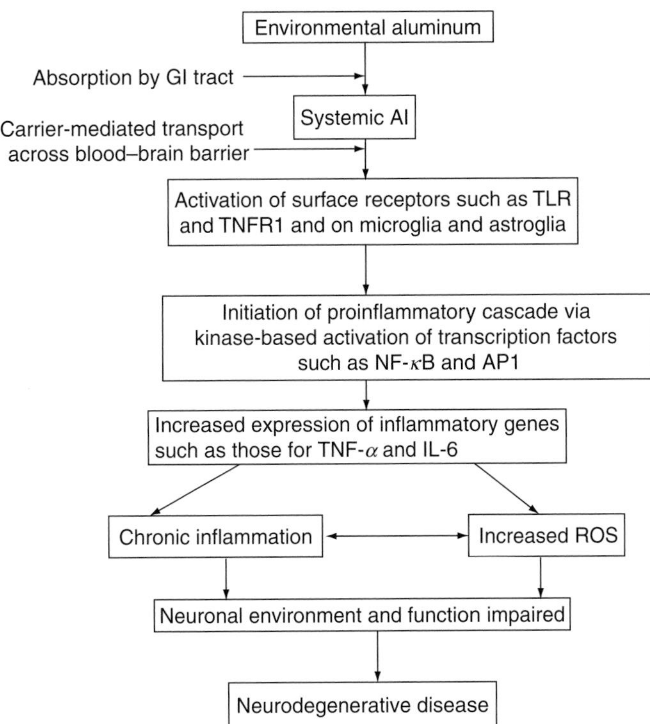

Figure 2 Proposed trajectory of aluminum (Al)-induced promotion of inflammatory processes within the central nervous system. GI, gastrointestinal; TLR, Toll-like receptor; TNFR1, tumor necrosis factor receptor 1; TNF-α, tumor necrosis factor-α; NF-κB, nuclear factor-kappa B; IL-6, interleukin-6; ROS, reactive oxygen species.

neurological diseases can be expected to rise with increasing longevity. The etiology of most neurodegenerative disorders is as yet unknown, but generally may not involve a primarily genetic component.

Aluminum compounds are undoubtedly a factor in the neurological sequelae of dialysis encephalopathy. The role of aluminum in brain aging and in chronic neurological disease is more uncertain. The proposal that excess levels of aluminum within the brain may contribute to the pathogenesis of AD has been extensively studied but remains controversial. This is also true of several other neurodegenerative diseases for which an intracerebral accumulation of aluminum has been reported. While many epidemiological correlations between Al exposure and neurological malfunction have been made, a clear mechanistic pathway that would account for this association is gradually emerging.

In the past, the problems encountered in making a definitive association between Al and CNS disorders have included the unavailability of a convincing aluminum-based animal model relevant to a specific CNS disease and absence of a clear theoretical and mechanistic framework for understanding how Al compounds could be deleterious in the CNS. The discussions herein were intended to illustrate that both of these limitations are likely to be overcome in the near future.

See also: Alzheimer's Disease: An Overview; Alzheimer's Disease: Molecular Genetics; Brain Composition: Age-Related Changes; Inflammation in Neurodegenerative Disease and Injury; Lipids and Membranes in Brain Aging; Metal Accumulation During Aging; Neurotoxins and their Neurotoxicology; Oxidative Damage in Neurodegeneration and Injury.

Further Reading

Becaria A, Campbell A, and Bondy SC (2004) Aluminum as a toxicant. *Toxicology and Industrial Health* 18: 309–320.

Campbell A, Becaria A, Lahiri DK, et al. (2004) Chronic exposure to aluminum in drinking water increases inflammatory parameters selectively in the brain. *Journal of Neuroscience Research* 75: 565–672.

Exley C (ed.) (2001) *Aluminum in Alzheimer's Disease*. New York, NY: Elsevier.

Flaten TP (2001) Aluminium as a risk factor in Alzheimer's disease, with emphasis on drinking water. *Brain Research Bulletin* 55: 187–196.

Miu AC and Benga O (eds.) (2006) Metal species and Alzheimer disease, *Journal of Alzheimer's Disease* 10: 133–341.

Perl DP and Moalem S (2006) Aluminum and Alzheimer's disease, a personal perspective after 25 years. *Journal of Alzheimer's Disease* 9(supplement 3): 291–300.

Yokel RA and McNamara PJ (2001) Aluminium toxicokinetics: An updated minireview. *Pharmacology and Toxicology* 88: 159–167.

Yang EY, Guo-Ross SX, and Bondy SC (1999) The stabilization of ferrous iron by a toxic β-amyloid fragment and by an aluminum salt. *Brain Research* 799: 91–96.

Alzheimer's Disease: An Overview

P I Moreira, University of Coimbra, Coimbra, Portugal
X Zhu and M A Smith, Case Western Reserve University, Cleveland, OH, USA
G Perry, University of Texas at San Antonio, San Antonio, TX, USA

Major Neuropathologic Hallmarks

The distinctive brain lesions, senile plaques (SPs) and neurofibrillary tangles (NFTs), used by Alois Alzheimer together with the clinical deficits to describe Alzheimer's disease (AD), are still used today as the defining features for diagnosis (**Figure 1**). In addition to these striking changes, there is variable cerebral cortical atrophy, particularly of the temporal and frontal lobes, and associated ventricular dilation, both of which are consequences of the neuron loss and astrocyte proliferation in affected regions.

Both NFTs and SPs are found in normal aged persons, but it is their quantitative increase that defines the pathologic diagnosis of AD. NFTs consist of abnormal 20 nm helical filaments with an 80 nm half-periodicity, termed paired helical filaments, and 12–15 nm straight filaments (**Figure 2**). NFTs usually occur in large numbers in the brain of a patient with AD, particularly in the entorhinal cortex; hippocampus; amygdala; association cortices of the frontal, temporal, and parietal lobes; and certain subcortical nuclei that project to these regions. The subunit protein of the paired helical filaments is the microtubule-associated protein tau τ. Paired helical filaments are not limited to the tangles found in the neuronal cell bodies but also occur in smaller bundles in many of the dystrophic neurites present around the amyloid plaques. Biochemical studies reveal that the tau protein present in paired helical filaments is a hyperphosphorylated, insoluble form of this normally highly soluble protein. The insoluble tau aggregates in the tangles are often complexed with ubiquitin, a feature they share with numerous other intraneuronal protein inclusions in etiologically diverse disorders, such as Parkinson's disease and diffuse Lewy body dementia. If this complexing with ubiquitin represents an attempt by the neuron to mark the altered tau proteins for degradation by the proteasome, it seems to be largely without benefit for the patient.

The amyloid-β (Aβ) peptides of SPs consist of 7–10 nm helical filaments (**Figure 2**) that share with paired helical filaments the ability to bind the dye Congo red and appear birefringent green when viewed under cross-polarized light, a property of β-pleated molecular sheets. SPs (or neuritc plaques) contain extracellular deposits of the Aβ42/40 (explained below) surrounded by dystrophic neurites (axons and dendrites), activated microglia, and reactive astrocytes. A large portion of the Aβ in these neuritic plaques is in the form of insoluble amyloid fibrils, but these are intermixed with a poorly defined array of nonfibrillar forms of the protein. Once protein sequencing established that Aβ was the subunit of fibrillar amyloid, Aβ immunohistochemistry revealed several deposits in brains of patients with AD that lacked the dystrophic neurites and altered glia that characterize the neuritic plaques. Such plain Aβ deposits, referred to as 'diffuse' plaques, exist mostly in a nonfibrillar (that is, 'preamyloid') form. The diffuse deposits are composed of Aβ42, which is far more prone to aggregation than the slightly shorter and less hydrophobic Aβ40. In healthy individuals, Aβ40 and Aβ42 make up 90% and about 10%, respectively, of the Aβ peptides that are normally produced by brain cells throughout life. Note that Aβ plaques do not occur simply in these two extreme forms (diffuse and neuritic) but rather as a continuum in which mixtures of nonfibrillar and fibrillar forms of the peptide can be associated with varying degrees of surrounding neuritic and glial alteration.

Neurotransmitter Deficits

Since the demonstration of a 90% drop in cortical choline acetyltransferase activity, there has been considerable interest in understanding whether neurotransmitter replacement therapy would be beneficial in the treatment of AD. The 'cholinergic hypothesis' states that decreased cholinergic transmission plays a major role in the expression of cognitive, functional, and possibly behavioral symptoms in AD. The cholinergic hypothesis rests on pathological, biochemical, and pharmacological observations. Cholinergic neurons in the ventral forebrain are depleted; many of those that remain contain NFTs. As a result of these pathological changes, there are decreases in biochemical indices of cholinergic function in neocortex and hippocampus that correlate with dementia severity. The hypothesis is further supported by an extensive literature from pharmacological studies using cholinergic agonists and antagonists, ablative lesions of the cholinergic pathway, and transgenic animal models that emphasize the close connection between cognition and the cholinergic neurotransmission. It has also been proposed that the cholinergic deficit plays a role not only in the cognitive symptoms but also in the behavioral changes observed in AD. A limitation of the cholinergic hypothesis is the lack of cholinergic deficit observed in early stages of AD or in patients with mild cognitive impairment.

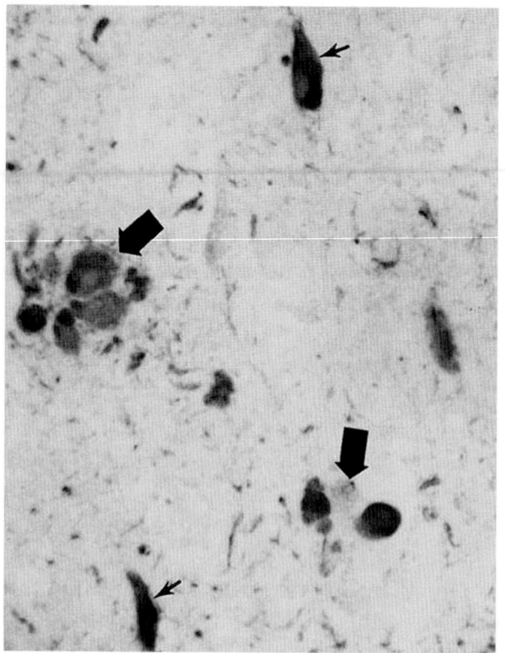

Figure 1 Characteristic pathologic lesions of Alzheimer's disease, neurofibrillary tangles (small arrows), and senile plaques (large arrows) are readily immunostained with antibodies to tau. (×760)

Other neurotransmitters may also be important in AD. For example, levels of three important neurotransmitters, serotonin, somatostatin, and noradrenaline, are reduced in the brains of some Alzheimer's patients. It has been suggested that these abnormalities are related to sensory disturbances and aggressive behavior. However, most neurotransmitter research in dementia continues to focus on acetylcholine because of its close ties to memory and reasoning.

Genetics

Autosomal dominant inheritance of mutations in the amyloid-β protein precursor (AβPP), PSEN1, and PSEN2 genes localized in chromosomes 21, 14, and 1, respectively, are responsible for familial early-onset AD (FAD). PSEN1 and PSEN2 encode for homologous polytopic membrane proteins, termed presenilin 1 and 2 (PS1 and PS2). To date, 15 missense mutations in AβPP, 199 in PS1, and 11 in PS2 have been reported to cause FAD. FAD mutations in PSEN1 cause most aggressive forms of AD, in some cases with onset younger than 30 years. Individuals with trisomy 21 (Down syndrome) have an extra copy of the AβPP gene and develop AD pathology as early as 20 years of age. Even though FAD-linked mutations in AβPP account for less than 5% of total AD cases, these autosomal dominant mutations are highly penetrant, and the clinical and pathologic symptoms of individuals with FAD mutations are nearly identical to those of patients

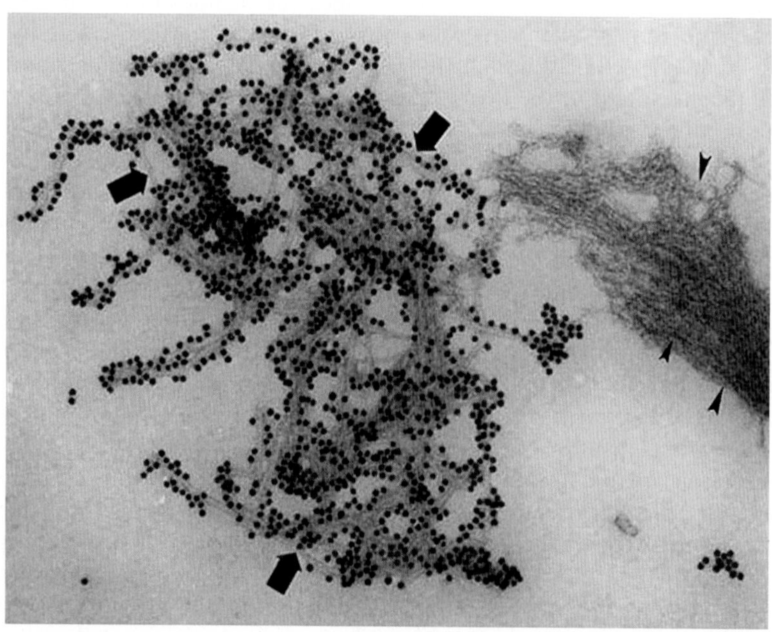

Figure 2 Structural and antigenic differences between paired helical filaments (large arrows) and amyloid-β (Aβ) filaments (arrowheads) are readily apparent in this negatively stained preparation. Heavy subunit of neurofilaments is localized by colloidal gold to paired helical filaments but not to Aβ filaments. (×105 000)

with late-onset sporadic AD. Biochemically, FAD-associated mutants in PSEN1 and PSEN2 lead to selective increase in the levels of Aβ42 species, which readily aggregate *in vitro* and are the initial Aβ species deposited in the brains of individuals with AD and Down syndrome. FAD-linked mutations in AβPP either increase Aβ42 or overall Aβ production or generate highly fibrillogenic Aβ variants. In addition to early-onset FAD mutations, it was also discovered that the presence of ε4 allele of the apolipoprotein E (APOE) gene is a risk factor for familial late-onset AD. APOE plays a vital role in the metabolism and clearance of Aβ along with α2-macroglobulin (α2M) and low-density lipoprotein receptor. The biochemical outcome of harboring the APOE ε4 allele includes increased Aβ aggregation and decreased Aβ clearance.

The number of genes involving distinct proteins suggests that the pathologic and clinical entity we call AD is not unique to abnormalities in a single gene locus or etiology. Significantly, the vast majority of AD is not linked to any established genetic abnormality.

Aβ and Its Protein Precursor

Understanding of the pathogenesis of Aβ deposition was greatly advanced by the sequencing of Aβ by G Glenner and C Wong in 1983. Subsequent cloning showed that Aβ is a 39–42-amino-acid fragment of a larger 695–770-amino-acid, membrane-spanning glycoprotein termed Aβ precursor protein (AβPP). The gene for AβPP resides as a single copy on the long arm of chromosome 21 and forms the locus for some cases of familial AD.

Aβ is proteolytically released from a large type 1 membrane glycoprotein of unknown function, the AβPP, via sequential cleavages by two aspartyl proteases, referred to as the β- and γ-secretases. The Aβ region of AβPP comprises the 28 residues just outside the single transmembrane domain, plus the first 12–14 residues of that buried domain. On this basis, Aβ was originally assumed to arise only under pathological circumstances, in that the second cleavage was thought to require some kind of prior membrane disruption to allow access of γ-secretase and a water molecule to the otherwise intramembranous region. This concept was disproved in 1992, when Aβ was shown to be constitutively released from AβPP and secreted by mammalian cells throughout life and thus occur normally in plasma and cerebrospinal fluid. This discovery enabled the dynamic study of Aβ production in cell culture and animal models, including examination of the effects of AD-causing genetic mutations. Moreover, high-throughput screening could now be conducted on cultured cells to identify Aβ-lowering compounds and determine their mechanism. Most AβPP molecules that undergo secretory processing are cleaved by α-secretase, rather than β-secretase, near the middle of the Aβ region. This releases the large, soluble ectodomain (APPs-α) into the medium and allows the resultant 83-residue, membrane-retained, C-terminal fragment (C83) to be cleaved by γ-secretase, generating the small p3 peptide. α-secretase acts on AβPP molecules at the cell surface although some processing also occurs in intracellular secretory compartments. The precise subcellular loci of the β- and γ-secretase cleavages are unclear but likely include early, recycling endosomes. The functional consequences of the proteolytic processing of AβPP remain ill-defined. The current leading hypothesis is that cleavage by α-secretase followed by γ-secretase enables the release of the AβPP intracellular domain (AICD) into the nucleus, where it may participate in transcriptional signaling. The APPs-α derivative secreted as a result of this processing appears to have distinct extracellular functions. For example, those APPs-α isoforms that contain an alternatively spliced Kunitz protease inhibitor domain function as serine protease inhibitors, including by inhibiting of Factor XIa in the coagulation cascade.

The importance of AβPP in the primary etiology of AD is fairly well established, with FAD associated with a number of mutations in the AβPP gene on chromosome 21. Some of the mutations in AβPP leading to AD have been related to AβPP processing, yielding more Aβ or the longer form of Aβ, which has a greater propensity to form Aβ fibrils. Additional support for the importance of AβPP in the pathogenesis of AD comes from Down's syndrome, in which an extra copy of chromosome 21 leads at midlife to a spectrum of pathologic changes similar to those found in AD.

Cytoskeletal Abnormalities

A fundamental process in the pathogenesis of AD is the breakdown of the cytoskeleton. The main component of NFTs is the paired helical filaments (PHF), which are mainly comprised of the protein tau in an abnormally phosphorylated status. Tau filaments accumulate in dystrophic neurites as fine neuropil threads or as bundles of PHF in neuronal bodies forming the NFTs, which become extracellular ghost tangles after the death of the neuron. The severity of dementia has been correlated with accumulation of NFTs in different brain regions, while with SP, such correlation has not been demonstrated. In AD, tau binds with lower affinity to microtubules, and it self-aggregates into aberrant structures, probably helped by other molecules. As a consequence of hyperphosphorylation, tau shows a loss of microtubule-binding capacity and is accumulated in neuronal bodies.

The finding that neurons are primarily responsible for AβPP production and also contain NFTs highlights a key issue in AD, the relationship of Aβ to NFTs. The application of monoclonal antibodies, immunoelectron microscopy, and antibody affinity techniques has shown that microtubule-associated protein tau and the heavy subunit of neurofilaments are major components of NFTs. Recent studies demonstrated a direct high-affinity interaction between tau and AβPP. More studies are required to understand the pathologic significance of tau-AβPP interaction in Aβ deposition.

Sparing solubility and heterogeneity of enriched fractions have hampered efforts to define NFTs quantitatively, and there is now a considerable effort to understand how the two identified posttranslational modifications of NFTs, phosphorylation and glycation, mediate the transformation of soluble tau into insoluble paired helical filaments. Indeed, although increased phosphorylation would lead to microtubule instability – a key hallmark of AD – phosphorylation does not mediate paired helical filament insolubility. Conversely, oxidation of tau confers the same solubility properties as paired helical filaments. Therefore, it is likely that oxidative cross-links play a role in NFT insolubility. Significantly, oxidation is one of the earliest changes in the disease that leads to neuronal dysfunction. It is not surprising, therefore, that therapeutic efforts to reduce oxidative stress slow disease progression and/or decrease the incidence of the disease.

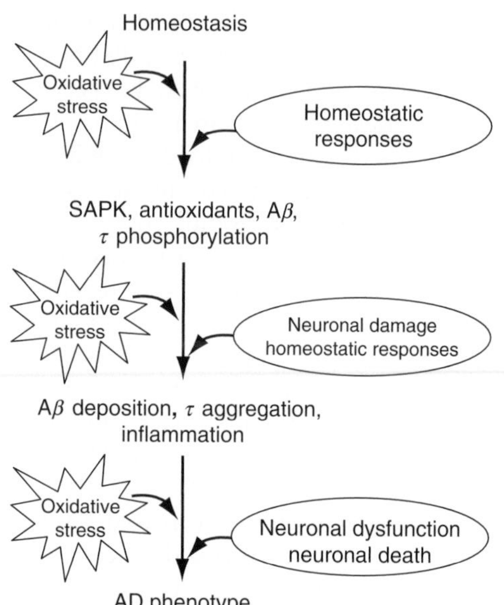

Figure 3 The sequence of events in Alzheimer's disease (AD). Oxidative stress, the earliest event occurring in AD, initially leads neurons to activate neuronal defenses, including stress-activated protein kinases (SAPK), tau (τ) phosphorylation, and amyloid-β (Aβ), to maintain homeostatic balance. However, given the chronic and insidious nature of oxidative stress, progression of the disease through Aβ deposition, tau aggregation, and inflammation overwhelms initial compensatory mechanisms and culminates in neuronal dysfunction and death, that is, the AD phenotype.

Oxidative Stress in AD

It has been shown that increased oxidative damage is a prominent and early feature of vulnerable neurons in AD. Over the past decade, an oxidative stress-related modification of macromolecules has been described in association with the susceptible neurons of AD. Such modifications include advanced glycation end products, nitration, lipid peroxidation adduction products, carbonyl-modified neurofilament protein, and free carbonyls, as well as glycation and glycoxidation products. Levels of these markers are initially elevated following some unknown triggering neuronal event, but these levels soon decrease as the disease progresses to advanced AD. Together these findings suggest that increased oxidative damage is not the terminal sequelae of the disease but instead plays an initial role. They also suggest that damage does not mark further destruction by reactive species and is instead marked by a broad array of increased cellular defenses. It can be argued that in AD, these defenses are responsible for the reduction of damage if we view AD in isolation. However, when seen in the context of other conditions in which reactive oxygen and nitrogen species are involved and damage is either limited or absent, such as Parkinson's disease, this result raises the question of whether oxidative damage noted in AD may be better thought of as homeostatic, that is, that oxidative damage could initiate signal transduction pathways to manipulate cellular responses to stress, which are characterized by increased levels of reactive oxygen and nitrogen species. Furthermore, there is evidence that in the first stage of AD development, Aβ deposition and hyperphosphorylated tau function as compensatory responses and downstream adaptations to ensure that neuronal cells do not succumb to oxidative damage (**Figure 3**).

Conclusions

The numerous genetic abnormalities as well as the high prevalence of sporadic disease suggest that AD has a defined pathogenesis stemming from numerous etiologies. Consequently, AD is a syndrome now defined by the most striking aspects of that pathogenesis, that is, NFTs and SP. Since oxidative damage is the earliest described cytopathological abnormality that occurs in vulnerable brain regions and selective neuronal populations, it is extremely important to decipher the causes and consequences of neuronal oxidative stress. Clarification of mechanistic viewpoints concerning AD pathophysiology may provide insights into efficacious therapeutics. In particular, investigation must uncover what the relationships are between increased oxidative stress and other facets of the disease such as regionally selective neuronal degeneration, Aβ deposition as SP, and tau phosphorylation and aggregation as NFTs.

See also: Acetylcholinesterase Inhibitors and Alzheimer's Disease; Aging of the Brain and Alzheimer's Disease; Alzheimer's Disease: Neurodegeneration; Alzheimer's Disease: Molecular Genetics; Alzheimer's Disease: Transgenic Mouse Models; Alzheimer's Disease: MRI Studies; Animal Models of Alzheimer's Disease; Axonal Transport and Alzheimer's Disease; Brain Glucose Metabolism: Age, Alzheimer's Disease and ApoE Allele Effects.

Further Reading

Glenner GG and Wong CW (1984) Alzheimer's disease: Initial report of the purification and characterization of a novel cerebrovascular amyloid protein. *Biochemical and Biophysical Research Communications* 120: 885–890.

Hendrie HC (1998) Epidemiology of dementia and Alzheimer's disease. *American Journal of Geriatric Psychiatry* 6: S3–S18.

Hernandez F, Engel T, Gomez-Ramos A, et al. (2005) Characterization of Alzheimer paired helical filaments by electron microscopy. *Microscopy Research and Technique* 67: 121–125.

Katzman R (1993) Clinical and epidemiological aspects of Alzheimer disease. *Clinical Neuroscience* 1: 165–170.

Khachaturian ZS (1985) Diagnosis of Alzheimer's disease. *Archives of Neurology* 42: 1097–1105.

Khachaturian ZS and Mesulam M-M (eds.) (2000) *Special Issue: Alzheimer's disease: A Compendium of Current Theories. Annals of the New York Academy of Sciences* 924.

Levy-Lahad E, Wasco W, Poorkaj P, et al. (1995) Candidate gene for the chromosome 1 familial Alzheimer's disease locus. *Science* 269: 973–977.

Moreira PI, Honda K, Zhu X, et al. (2006) Brain and brawn: Parallels in oxidative strength. *Neurology* 66: S97–S101.

Perry G (ed.) (1993) Alzheimer's Disease. *Clinical Neuroscience* 1: 163–224.

Perry G and Smith MA (1993) Senile plaques and neurofibrillary tangles: What role do they play in Alzheimer disease? *Clinical Neuroscience* 1: 199–203.

Post SG (2000) *The Moral Challenge of Alzheimer Disease: Ethical Issues from Diagnosis to Dying*, 2nd edn. Baltimore: Johns Hopkins University Press.

Roses AD (1994) Apolipoprotein E affects the rate of Alzheimer disease expression: β-Amyloid burden is a secondary consequence dependent on ApoE genotype and duration of disease. *Journal of Neuropathology and Experimental Neurology* 53: 429–437.

Selkoe DJ (1994) Normal and abnormal biology of the β-amyloid precursor protein. *Annual Review of Neuroscience* 17: 489–517.

Selkoe DJ (1994) Alzheimer's disease: A central role for amyloid. *Journal of Neuropathology and Experimental Neurology* 53: 438–447.

Selkoe DJ (2004) American College of Physicians, and American Physiological Society. Alzheimer disease: Mechanistic understanding predicts novel therapies. *Annals of Internal Medicine* 140: 627–638.

Sherrington R, Rogaev EI, Liang Y, et al. (1995) Cloning of a gene bearing missense mutations in early-onset familial Alzheimer's disease. *Nature* 375: 754–760.

Smith MA, Sayre LM, Monnier VM, et al. (1995) Radical ageing in Alzheimer's disease. *Trends in Neurosciences* 18: 172–176.

Smith MA, Siedlak SL, Richey PL, et al. (1995) Tau protein directly interacts with the amyloid β-protein precursor: Implications for Alzheimer's disease. *Nature Medicine* 1: 365–369.

Tanzi RE and Bertram L (2005) Twenty years of the Alzheimer's disease amyloid hypothesis: A genetic perspective. *Cell* 20: 545–555.

Relevant Websites

http://www.alz.org – Alzheimer Association.

http://www.alzforum.org – Alzheimer Research Forum.

http://www.molgen.ua.ac.be – Molecular Genetics Department, VIB, The Flanders Institute for Biotechnology.

http://www.nih.gov/nia – National Institute on Aging.

Alzheimer's Disease: Molecular Genetics

R Sherrington, Unité de Recherche en Neuroscience, Ste-Foy, QC, Canada
P H St. George-Hyslop, University of Toronto, Toronto, ON, Canada

Introduction

Alzheimer's disease (AD) is the most common form of degenerative dementia of the human central nervous system. It is defined clinically as a progressive loss of cognitive function with the onset of a slowly progressive impairment of memory during mid- to late adult life. The neuropathologic hallmarks of this disease include amyloid deposits, neurofibrillary tangles, astrocytic gliosis, and reductions in the number of neurons and synapses in many areas of the brain, but especially from the cerebral cortex and the hippocampus.

The etiology of AD is complex. Multiple epidemiologic studies have proposed a number of potential risk factors, including environmental (head trauma, smoking, and exposure to heavy metals such as aluminum), sociologic (depression and level of education), biologic (increasing age, hyperthyroidism, late maternal age), and family history (of Down syndrome or AD).

The repeated observation in multiple epidemiologic surveys that a positive family history is a strong risk factor for AD clearly suggest that genetic factors play a role in this disease. A simple autosomal dominant inheritance with age-dependent penetrance has been observed in many pedigrees, especially those with early age of onset. However, in many families multiply affected by AD, the pattern of inheritance is unclear and could fit one of several genetic models, including a high-frequency, low-penetrance single gene disorder or a multifactorial model in which several genes or nongenetic factors interact. Using the powerful tools of genetic linkage analysis, significant progress has been made in unraveling the genetic etiology. Four loci that play a role in the genetic susceptibility of AD have been identified, namely, genes on chromosomes 21 (β-amyloid precursor protein (βAPP)), 19 (apolipoprotein E (APOE)), 14 (presenilin 1 (PS1)), and 1 (presenilin 2 (PS2)).

β-Amyloid Precursor Protein

The βAPP gene was the first gene found to bear mutations capable of causing early-onset familial AD (FAD). The βAPP gene on chromosome 21 encodes an alternatively spliced single spanning transmembrane protein. The longer isoform, APP770 (770 amino acids), contains two exons, which encode a Kunitz protease inhibitor (KPI) domain and a 19-amino acid sequence homologous to the ox-2 antigen. The KPI-containing isoforms are expressed in most tissues. In contrast, the shorter isoform APP695 (695 amino acids), without the KPI domain, is predominantly expressed in the brain. The βAPP protein undergoes a series of endoproteolytic cleavages that give rise to Aβ peptide, a 4 kDa peptide resulting from proteolytic cleavages of the full-length protein from residues 672–711 that spans the last 28 residues of the extracellular domain and the first 14 residues of the transmembrane domain. One of these cleavages, which results from the action of a putative membrane-associated α-secretase, liberates the extracellular N-terminus of βAPP and is thus a nonamyloidogenic pathway, because this cleavage precludes the formation of Aβ peptide. The other cleavage pathway, which occurs in part in the endosomal-lysosomal compartment, involves the recently identified β-secretase (BACE) and the putative γ-secretases, which give rise to a series of peptides that contain the 40–42 amino acid Aβ peptide. Aβ peptides ending at residue 42 or 43 (long-tailed Aβ) are thought to be more fibrillogenic and more neurotoxic than Aβ ending at residue 40, which is the predominant isoform produced during normal metabolism of βAPP. The activity of these secretases, and especially g-secretase, giving rise to the more fibrillogenic and potentially neurotoxic long-tailed Ab1–42, appears to play an important role in the pathogenesis of AD. It has been suggested that processing of βAPP into Aβ and the subsequent accumulation of Aβ in the brain are central events in the pathogenesis of AD in both genetic and nongenetic forms. Although Aβ peptides are secreted by cells under normal physiologic conditions, how they induce neurodegeneration in AD is still unclear.

Five mutations in the βAPP gene (APP V7171, APP V717G, APP V717F, APP KM670/671NL, and APP A692G) can cause early-onset FAD. The mutations occur within or near the Ab peptide. Mutations in this gene account for only a small fraction of early-onset FAD cases (less than 0.1%). The function of the βAPP protein is unknown, although roles in cell adhesion, synaptic growth, and neural repair have been proposed. It has also been suggested that the βAPP may function as a G-coupled receptor linked through an interaction in the C-terminus to G_o protein.

The APOE Gene

The APOE gene is located on chromosome 19q13 and encodes the major apolipoprotein expressed in the central nervous system. APOE plays an important role in triglyceride-rich lipoprotein metabolism and regulation of cholesterol. It has also been suggested that APOE plays a role in repair, growth, and maintenance of myelin and axonal membranes. Studies implicating the APOE gene as an AD susceptibility locus derived from several intersecting lines of investigation (linkage analysis on late-onset AD families, APOE immunoreactivity in senile plaques and NFTs of patient with AD, and APOE binding to Aβ). Association studies revealed that inheritance of the APOE e4 allele (Cys112Arg) is associated with a copy number-dependent increase in risk of both sporadic and late-onset FAD and with a decrease in the age of onset in late-onset AD. Conversely, the e2 allele confers a protective effect against late-onset AD. However, it seems that the inheritance of the e4 allele (APOE4) may be neither necessary nor sufficient by itself to cause AD, and the predictive value of the APOE4 in presymptomatic people is unclear. The mechanism by which the inheritance of the APOE4 increases risk of AD and how this might be addressed therapeutically are still unclear. Moreover, it has been recently reported that APOE represents only 7–9% of the total variation in age at onset of AD, indicating that other loci may have greater importance than does APOE.

The PS Genes

Genetic linkage studies, using large numbers of pedigrees with early-onset autosomal dominant FAD, mapped a common locus (AD3) to chromosome 14q24.3. The AD3 locus is associated with a particularly aggressive form of early-onset (between 30 and 60 years of age) AD and accounts for up to 50% of early-onset FAD cases. Subsequent positional studies led to the isolation of a novel gene termed PS1. Mutational analysis of the PS1 gene have identified at least 70 different predominantly missense mutations in residues that are highly conserved in evolution. The PS1 is predicted to be an integral membrane protein with at least seven hydrophobic membrane-spanning domains (transmembrane domains (TM)) and two acidic hydrophilic domains located at the N-terminus and between TM6 and TM7. The amino acid sequence of the PS1 protein shows a weak homology to the *C. elegans* SPE-4 protein, which is thought to be involved in membrane budding and fusion events in a membrane-bound cytoplasmic organelle derived from the spermatocytes' Golgi network. Subsequently, a strong homology was found between PS1 and SEL-12, another *C. elegans* protein. SEL-12 facilitates signaling mechanisms in the LIN-12/Notch family of intercellular receptors, which are responsible for direct signal transmission from the cell surface to the nucleus during intercellular signal transduction specifying cell fate. These later data suggest that PS1 may have a role in the vertebrate Notch signaling pathway. This is further supported by the fact that PS1 knockout mice show developmental abnormalities similar to those of mice with targeted knockouts of the murine Notch1 gene.

Following the isolation of the PS1 gene causing the AD3 subtype of AD, a second gene, PS2, mapped to chromosome 1, was discovered on the basis of its strong nucleotide and amino acid sequence homology to PS1. The structural organization of the PS2 is very similar to that of PS1, including the transmembrane and acidic hydrophilic domains. These observations suggest that the PS1 and the PS2 genes are members of a gene family. Mutational analysis of the PS2 gene led to the discovery of six different mutations. The phenotype of these families shows a later onset than the phenotype associated with mutations in PS1 or βAPP genes (onset 50–70 years).

All the PS1 or PS2 mutations lead to significantly increased levels of secreted Aβ42 compared with wild-type controls. These observations suggest a dominant gain of function for the mutant PSs because even though the endogenous wild-type alleles are present, the level of Aβ42 is increased in cells expressing mutant cDNAs. How mutant presenilins alter APP processing is unknown. It has been proposed that PS1 (and probably PS2) is part of a complex that includes the recently discovered nicastrin, involved in the proteolytic cleavage of the C-terminal fragment of APP (mediated by γ-secretase). Other functions in intracellular signaling, suppression of apoptosis, and protein/membrane trafficking have been suggested for both PS1 and PS2.

Other AD Genes

Although four different genetic loci associated with inherited susceptibility to AD (βAPP, APOE, PS1, and PS2) have been identified, they account for only about half of the genetic forms of AD. Thus at least one and possibly several other genes remain to be identified. Several polymorphisms in different genes have been associated with AD. Recent findings suggest the presence of new AD genes on chromosomes 9, 10, and 12. However, the lack of replication of these associations in different AD pedigrees makes these results unclear at the moment.

Animal Models

Several transgenic murine model lines have been created using a variety of βAPP constructs and have met with varying degrees of success. The most successful model to date has been a transgenic line in which a mutant human βAPP minigene has been placed under the control of the platelet-derived growth factor receptor beta subunit promoter. This latter model demonstrates abundant Aβ deposition and synaptic pathology. Targeted null mutation (knockout) of the murine βAPP gene was created but is not very illuminating because it leads only to subtle phenotypes including minor weight loss, decreased locomotor activity, abnormal forelimb motor activity, and minor nonspecific degrees of reactive gliosis in the cortex.

Transgenic mice overexpressing wild-type or mutant PSs have been created. All mutants show an increase in Aβ42 production in the brain compared with wild-type. Despite the Aβ42 accumulation in the brain, these PS transgenic mice have not shown any major neuronal death. However, these mice showed reduced spontaneous alternation performance in a Y maze. PS1 knockout mice have also been generated. These knockout mice developed deformations of the caudal axial skeleton, hemorrhages in the central nervous system, and impairment in neurogenesis and usually did not survive beyond the first day after birth. These studies show that, at least, PS1 is required for proper formation of the axial skeleton and is involved in normal neurogenesis and survival of progenitor cells and neurons in specific brain subregions.

See also: Aging of the Brain; Aging of the Brain and Alzheimer's Disease; Alzheimer's Disease: An Overview; Alzheimer's Disease: Transgenic Mouse Models; Dementia; Dementia and Language; Gene Expression in Normal Aging Brain; Genomics of Brain Aging: Apolipoprotein E; Genomics of Brain Aging: Twin Studies.

Further Reading

Bertram L and Tanzi RE (2001) Dancing in the dark? The status of late-onset Alzheimer's disease genetics. *Journal of Molecular Neuroscience* 17(2): 127–136.

Farrer LA, Myers RH, Connor L, et al. (1991) Segregation analysis reveals evidence of a major gene for Alzheimer disease. *American Journal of Human Genetics* 48: 1026–1033.

Games D, Adams D, Alessandrini A, et al. (1995) Alzheimer-type neuropathology in transgenic mice overexpressing V717F beta-amyloid precursor protein. *Nature* 373: 523–527.

Goate AM, Chartier-Harlin MC, Mullan M, et al. (1991) Segregation of a missense mutation in the amyloid precursor protein gene with familial Alzheimer disease. *Nature* 349: 704–706.

Katzman R (1986) Alzheimer's disease. *New England Journal of Medicine* 314(15): 964–973.

Katzman R and Kawas C (1994) The epidemiology of dementia and Alzheimer disease. In: Terry RD, Katzman R, and Hick KL (eds.) *Alzheimer Disease*, pp. 105–122. New York: Raven Press.

Prince JA, Feuk L, Sawyer SL, et al. (2001) Lack of replication of association findings in complex disease: An analysis of 15 polymorphisms in prior candidate genes for sporadic Alzheimer's disease. *European Journal of Human Genetics* 9(6): 437–444.

Rogaev EI, Sherrington R, Rogaeva EA, et al. (1995) Familial Alzheimer's disease in kindreds with missense mutations in a novel gene on chromosome I related to the Alzheimer's disease type 3 gene. *Nature* 376: 775–778.

Saunders A, Strittmatter WJ, Schmechel S, et al. (1993) Association of apolipoprotein E allele e4 with late-onset familial and sporadic Alzheimer disease. *Neurology* 43: 1467–1472.

Selkoe DJ (1994) Normal and abnormal biology of β-amyloid precursor protein. *Annual Review of Neuroscience* 17: 489–517.

Selkoe DJ (2001) Alzheimer's disease: Genes, proteins, and therapy. *Physiological Reviews* 81(2): 741–766.

Sherrington R, Rogaev EI, Liang Y, et al. (1995) Cloning of a gene bearing missense mutations in early-onset familial Alzheimer's disease. *Nature* 375: 754–760.

Warwick DE, Payami H, Nemens EJ, et al. (2000) The number of trait loci in late-onset Alzheimer disease. *American Journal of Human Genetics* 66(1): 196–204.

Relevant Website

http://www.alzforum.org – Alzheimer Research Forum.

Alzheimer's Disease: MRI Studies

P M Thompson and A W Toga, UCLA School of Medicine, Los Angeles, CA, USA

Introduction

Alzheimer's disease (AD) is the leading cause of senile dementia, affecting 10% of those over age 65. The disease causes irreversible memory loss, behavioral and cognitive decline, personality changes, and a decreasing ability to cope with everyday life. Ironically, up to 30 years elapse between the onset of the cellular pathology that causes AD (amyloid plaques and neurofibrillary tangles in the brain) and the clinical changes that lead to diagnosis. To help understand how the disease emerges and progresses, imaging technology can be applied that is safe, repeatable, and widely available. Magnetic resonance imaging (MRI) scans of AD patients reveal profound anatomic changes: severe cortical and hippocampal atrophy, sulcal and ventricular enlargement, and reduced gray matter and white matter volume. These changes occur in a distinct spatial and temporal sequence, and correlate with cognitive and metabolic decline. If patients are scanned repeatedly with MRI as their disease progresses, dynamic maps can be reconstructed that reveal a shifting pattern of cortical changes. This spreading cortical atrophy mirrors the spread of the underlying pathology (as defined by tangle and amyloid plaque deposition). Repeat MRI scanning can monitor disease progression in individual patients and can evaluate how drugs oppose these changes. It can also clarify how anatomical deficits link with cognitive and behavioral deficits as they emerge in individuals and populations.

Impact

AD is a severe and growing public health crisis. The incidence of AD doubles every 5 years after age 60. It afflicts 1% of those aged 60 to 64 and 30% to 40% of those over 85. Without a cure, the number of AD victims will rise from 2.0 to 3.5 million now to an estimated 10 to 14 million by 2030. A number of promising AD treatments are now being developed. These range from acetylcholinesterase inhibitors, which ballast neurotransmitter function, to experimental vaccines, which directly attack the amyloid plaques that are a key element of AD pathology. Most therapeutic trials of new drugs in AD rely primarily on cognitive tests to determine efficacy. Neuroimaging, however, can be extremely beneficial in this research. It supplies a variety of biological markers that measure disease progression. With novel brain mapping techniques, the disease can be tracked as it spreads in the living brain. In Alzheimer's patients, MRI scans show prominent hippocampal atrophy. Diffuse tissue loss is also found in the medial temporal lobes. These deficits are progressive, and their magnitude correlates with cognitive decline. Because it is vital to detect the disease early, there is great interest in developing MRI measures that predict imminent transition to dementia, in healthy elderly subjects. For example, significant hippocampal volume deficits are found in subjects with mild memory impairments, who do not yet have dementia. These deficits shown on MRI can also help to predict how soon an elderly individual will develop AD. Reliable MRI predictors are especially valuable, as cholinergic drugs are most effective in the mildest phases, when widespread neuronal loss has not yet occurred.

Several neuroimaging measures can be used to characterize dementia. For instance, MRI scans can assess the integrity of medial temporal lobe structures involved in memory, such as the entorhinal cortex and hippocampus. The region and rate of atrophic brain changes can be measured as the disease progresses, as can the profile of cortical thinning and gray matter loss. The required three-dimensional MRI scans can be performed in approximately 10 min, on a conventional 1.5-Tesla scanner. Although MRI is not routinely used to diagnose AD, new techniques in brain image analysis can be applied to MRI scans to reveal how the disease emerges, and track how medications affect the disease process. MRI scans are often used in dementia research to (1) screen at-risk populations to find anatomical measures that might help predict each individual's likelihood of developing AD, (2) discriminate AD from normal aging and other dementias (such as frontotemporal and Lewy body dementias, which have distinct anatomic patterns), and (3) monitor disease progress and therapeutic response, gauging the effectiveness of drug treatments.

Brain Tissue Loss and Cognitive Decline

In the 1990s, MRI research in dementia focused on measuring medial temporal lobe structures. This was because AD pathology typically starts in the temporal cortex adjacent to the entorhinal cortex and quickly spreads to the entorhinal cortex before involving the hippocampus. This temporal lobe pathology persists for several years before spreading cortically to engulf the rest of the temporal, frontal, and parietal lobes. A more recent trend in dementia research has been to move from cross-sectional studies to dynamic measures. Serial MRI scans (acquired from the same patients repeatedly over time) can provide much greater power to detect pathological atrophy, because they provide a baseline reference point to calculate change. Fox et al. found that AD patients lose brain tissue at a faster 'rate' than age-matched controls. Evaluated with MRI for 5–8 years, AD patients lost brain tissue at a median rate of 2.20% per year (range, 0.82–4.19) versus 0.24% per year in controls (range, −0.35–0.64). These rates correlated with the rate of cognitive decline, reflected by worsening performance on the Mini Mental Status Examination (MMSE). In a recent 52 week clinical trial of milameline (a muscarinic receptor agonist), Jack et al. noted that hippocampal volume, measured with MRI, also tracked cognitive decline. Perhaps the most prominent sign of AD seen on an MRI scan is that the lateral ventricles are often greatly enlarged. Bradley et al. measured the ratio of the ventricular volume to the total brain volume (the ventricle-to-brain ratio (VBR) in 39 elderly subjects scanned with serial MRIs over 3 to 6 month intervals. The VBR rate of change was 15.6% ± 2.8% (mean ± SD) per year for AD patients compared with 4.3% ± 1.1% per year in controls ($P < 0.001$). VBR did not separate groups when measured at only a single time point, supporting the value of longitudinal assessments. Power calculations revealed that 135 subjects would be needed in each arm of a placebo-controlled clinical trial if this measure of AD progression were to detect a 20% reduction in the excess rate of atrophy over 6 months, with 90% power.

Gray Matter Deficits

Brain changes in AD can also be visualized using three-dimensional maps. **Figure 1** shows the spatial pattern of cortical gray matter loss in mild to moderate AD. This type of image is a composite map; it results from a sequence of image processing steps that compare scans of AD patients with matched healthy elderly subjects. With image analysis techniques, three-dimensional brain MRI scans can be split up

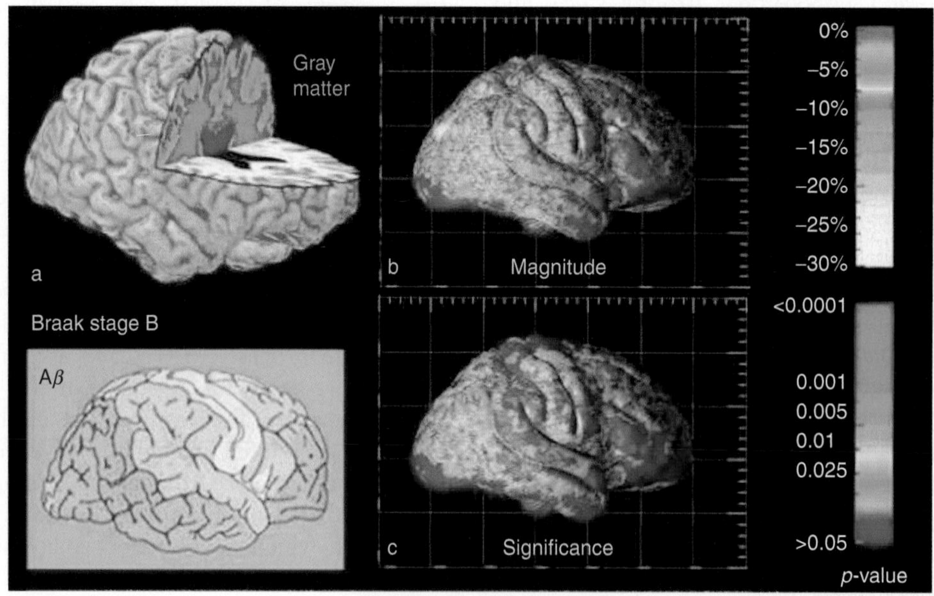

Figure 1 Gray matter deficits in early AD. Here the local amount of cortical gray matter (green colors, (a)) is compared across 26 patients with mild to moderate AD (age, 75.8 ± 1.7 years; MMSE score, 20.0 ± 0.9) and 20 matched elderly controls (age, 72.4 ± 1.3 years). At this stage of AD, 30% of the cortical gray matter has been lost in the temporoparietal regions (b). (c) Statistical significance of these deficits. The pattern of temporal lobe gray matter loss, seen on MRI, spatially matches the pattern of beta-amyloid (Aβ) deposition seen postmortem. The inset panel (Braak stage B) is adapted from data reported by Braak and Braak (1997). It shows regions with minimal (white), moderate (orange), and severe (red) Aβ deposition. Amyloid deposition and gray matter loss may not be synchronized, so these maps may represent different stages of AD; however, there is a clear spatial agreement in the severity of the deficits, between MRI and Aβ maps. Primary sensorimotor regions (white in the amyloid map), and the superior temporal gyri (blue colors in (c)) are spared relative to other temporal lobe gyri. These MRI patterns have been replicated in independent studies by Thompson PM, Mega MS, Woods RP, et al. (2001) Cortical change in Alzheimer's disease detected with a disease-specific population-based brain atlas. *Cerebral Cortex* 11: 1–16, Baron JC, Chetelat G, Desgranges B, et al. (2001) *In vivo* mapping of gray matter loss with voxel-based morphometry in mild Alzheimer's disease. *Neuroimage* 14: 298–309, and O'Brien JT, Paling S, Barber R, et al. (2001) Progressive brain atrophy on serial MRI in dementia with Lewy bodies, AD, and vascular dementia. *Neurology* 56: 1386–1388.

into regions representing gray matter, white matter, and cerebrospinal fluid. A measure is then computed that is related to the thickness of the cortical gray matter at each cortical location. Computer analyses then compare the amount of gray matter at each cortical location across subjects, while adjusting for potentially confounding factors, such as age and gender effects, and gyral patterning differences. Differences can be visualized locally in the form of color-coded statistical maps. These show how much gray matter volume is reduced in AD patients relative to healthy controls in each cortical region

Maps of Disease Progression

MRI scanning also reveals a dynamically spreading wave of gray matter loss in the brains of patients with AD. With novel brain mapping methods, the loss pattern can be visualized as it spreads over time from temporal and limbic cortices into frontal and occipital brain regions, sparing sensorimotor cortices. These shifting deficits are correlated with cognitive decline. As shown in **Figure 2**, cortical atrophy occurs in a well-defined sequence as the disease progresses, mirroring the temporal sequence of beta-amyloid (Aβ) and neurofibrillary tangle accumulation observed at autopsy. The trajectory of deficits also matches the sequence of metabolic decline typically observed with positron emission tomography.

To map AD progression, advancing deficits can be visualized as dynamic video maps that change over time, which distinguish different phases of AD and differentiate AD from normal aging. Frontal brain regions, spared early in the disease, show pervasive deficits later (>15% loss). Local gray matter loss rates (5.3 ± 2.3% per year in AD vs. 0.9 ± 0.9% per year in controls) are faster in the left hemisphere than the right, at least at this stage of AD. Transient

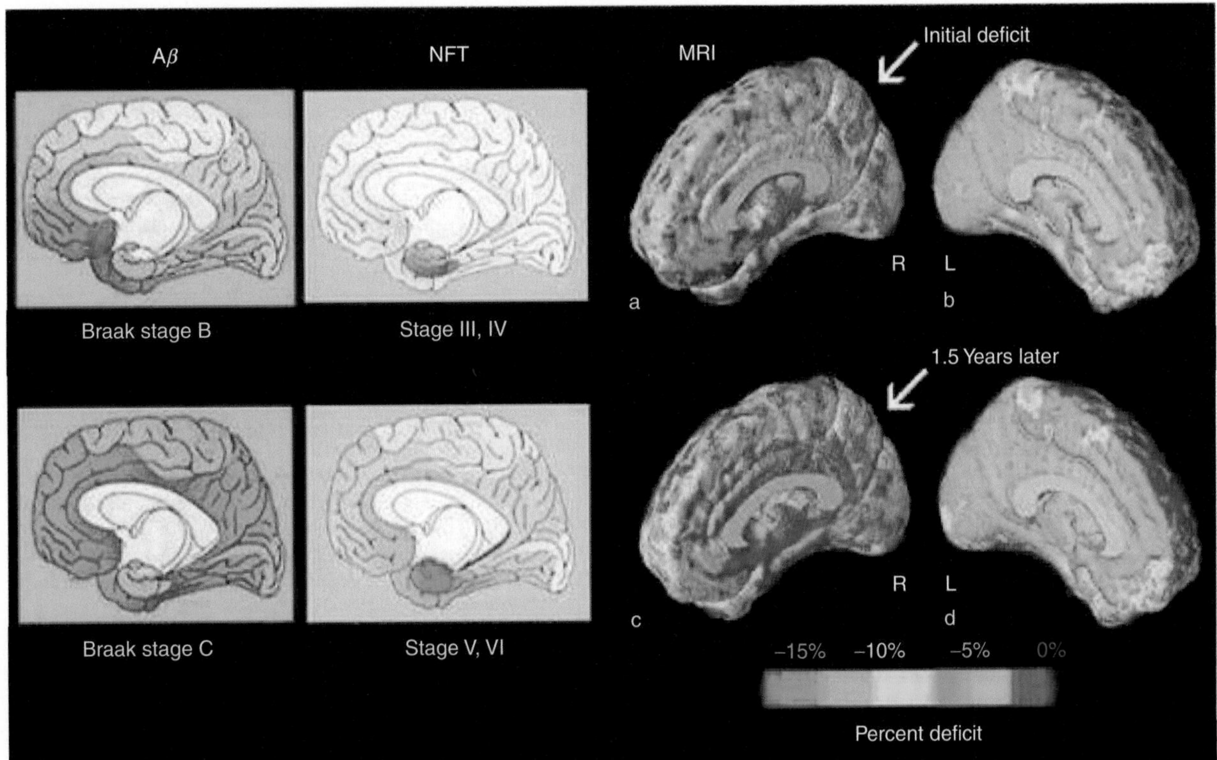

Figure 2 Gray matter deficits spread through the limbic system in moderate AD. Deficits during the progression of AD are detected by comparing average profiles of gray matter volumes between 12 AD patients (age, 68.4 ± 1.9 years) and 14 elderly matched controls (age, 71.4 ± 0.9 years). Colors show the average percent loss of gray matter relative to the control average. Profound loss engulfs the left medial wall (>15% (b), (d)). On the right, however, the deficits in temporoparietal and entorhinal territory (a) spread forward into the cingulate gyrus 1.5 years later (c), after a 5-point drop in average MMSE test scores. Limbic and frontal zones are prominently divided, with different degrees of impairment (c). MRI-based changes, observed in living patients, agree strongly with the spatial progression of Aβ and neurofibrillary tangle (NFT) pathology observed postmortem (Braak stages B and C and III to VI. NFT accumulation is minimal in sensory and motor cortices, but occurs preferentially in entorhinal pyramidal cells, the limbic periallocortex (layers II/IV), the hippocampus/amygdala and subiculum, the basal forebrain cholinergic systems and subsequently in temporoparietal and frontal association cortices (layers III/V). Left four panels adapted from Braak H and Braak E (1997) Staging of Alzheimer-related cortial destruction. *International Psychogeriatrics* 9(supplement 1): 257–271; discussion 269–272.

barriers to disease progression also appear. A frontal band (0–5% loss) is sharply delimited from the limbic and temporoparietal regions that show severest deficits in AD (>15% loss). This pattern is consistent with the hypothesis that AD pathology spreads centrifugally from limbic/paralimbic to higher-order association cortices. This degenerative sequence, observed as it develops in living patients, provides a quantitative, dynamic visualization of cortical atrophic rates in dementia, over a period of cognitive decline lasting 1.5 years.

What Is Gray Matter Atrophy?

Gray matter atrophy observed on MRI is linked with cognitive decline in AD and is attributable to several cellular processes. In healthy aging, age-related neuronal loss does not occur in most neocortical regions and appears specific to the frontal cortex and some hippocampal regions (e.g., CAI and the subiculum). In AD, however, there is substantial neuronal loss, with severe early losses in layer II of the entorhinal cortex.

Aβ and Neurofibrillary Tangle Maps

MRI-based maps of cortical atrophy agree strongly with postmortem maps of Aβ deposition (Aβ, **Figure 2**). Aβ is an insoluble protein that is a key feature of Alzheimer pathology. The spatial congruence of these two maps supports the hypothesis that Aβ deposition may participate in the cascade of events that leads to regional gray matter atrophy and neuronal cell loss. In both maps, primary sensorimotor cortices are relatively spared until late in the disease, and the superior temporal gyrus is less affected than other temporal lobe gyri. In early AD, intraneuronal filamentous deposits, or neurofibrillary tangles (NFTs), also accumulate within neurons. These deposits are composed of hyperphosphory lated tau protein. This cellular pathology disrupts axonal transport and induces widespread metabolic decline; it eventually leads to neuronal loss, observed as gross atrophy on MRI. Braak and Braak noted on autopsy that NFT distribution was initially restricted to entorhinal cortices, spreading to higher-order temporoparietal association cortices, then frontal, and ultimately primary sensory and visual areas. MRI scans suggest that a similar wave of cortical atrophy can be mapped in patients while they are alive. This provides a biological marker of disease progression that can monitor the effects of therapy.

It remains a mystery why brain changes in AD occur in this sequence. Braak and Braak suggested that the atrophic trajectory in AD is somewhat the reverse of the sequence in which cortical areas are myelinated during development. For example, primary sensory regions myelinate first and degenerate last, and temporal regions mature last but degenerate first in AD. This palindromic sequence is largely supported by the pattern of cortical changes observed on MRI. The selective vulnerability of specific cortical systems in AD may relate to differences in cellular maturational rates and/or plasticity. The most plastic systems may also be most vulnerable to AD.

Conclusion

MRI scans can measure brain change in AD, mapping the disease process in detail. MRI measures of disease progression, along with measures of genetic risk and abnormalities in specific neuropsychological tests, now provide key quantitative predictors to monitor brain degeneration and gauge how well it is decelerated or delayed in clinical trials.

See also: Acetylcholinesterase Inhibitors and Alzheimer's Disease; Aging of the Brain and Alzheimer's Disease; Alzheimer's Disease: Molecular Genetics; Axonal Transport and Alzheimer's Disease; Brain Glucose Metabolism: Age, Alzheimer's Disease and ApoE Allele Effects; Functional Neuroimaging Studies of Aging; Magnetic Resonance Spectroscopy; Neuroimaging; Numerical Intelligence: Neural Substrates; Perfusion MRI.

Further Reading

Arnold SE, Hyman BT, Flory J, et al. (1991) The topographical and neuroanatomical distribution of neurofibrillary tangles and neuritic plaques in the cerebral cortex of patients with Alzheimer's disease. *Cerebral Cortex* 1: 103–116.

Baron JC, Chetelat G, Desgranges B, et al. (2001) *In vivo* mapping of gray matter loss with voxel-based morphometry in mild Alzheimer's disease. *Neuroimage* 14: 298–309.

Braak H and Braak E (1997) Staging of Alzheimer-related cortical destruction. *International Psychogeriatrics* 9(supplement 1): 257–261; discussion 269–272.

Bradley KM, Bydder GM, Budge MM, et al. (2002) Serial brain MRI at 3–6 month intervals as a surrogate marker for Alzheimer's disease. *British Journal of Radiology* 75: 506–513.

Chetelat G and Baron JC (2003) Early diagnosis of Alzheimer's disease: Contribution of structural neuroimaging. *Neuroimage* 18: 525–541.

Convit A, de Asis J, de Leon MJ, et al. (2000) Atrophy of the medial occipitotemporal, inferior, and middle temporal gyri in non-demented elderly predict decline to Alzheimer's disease. *Neurobiology of Aging* 21: 19–26.

Fox NC, Crum WR, Scahill RI, et al. (2001) Imaging of onset and progression of Alzheimer's disease with voxel-compression mapping of serial magnetic resonance images. *Lancet* 358: 201–205.

Frisoni GB, Laakso MP, Beltramello A, et al. (1999) Hippocampal and entorhinal cortex atrophy in frontotemporal dementia and Alzheimer's disease. *Neurology* 52: 91–100.

Gomez-Isla T, Price JL, McKeel DW Jr., et al. (1996) Profound loss of layer II entorhinal cortex neurons occurs in very mild Alzheimer's disease. *Journal of Neuroscience* 16: 4491–4500.

Jack CR Jr., Slomkowski M, Gracon S, et al. (2003) MRI as a biomarker of disease progression in a therapeutic trial of milameline for AD. *Neurology* 60: 253–260.

Jobst KA, Smith AD, Szatmari M, et al. (1994) Rapidly progressing atrophy of medial temporal lobe in Alzheimer's disease. *Lancet* 343: 829–830.

Kaye J, Moore M, Kerr D, et al. (1999) The rate of brain volume loss accelerates as Alzheimer's disease progresses from a presymptomatic phase to frank dementia. *Neurology* 52(supplement): A569–A570.

Laakso MP, Lehtovirta M, Partanen K, et al. (2000) Hippocampus in AD: A 3-year follow-up MRI study. *Biological Psychiatry* 47: 557–561.

Malmgren R (2000) Epidemiology of aging. In: Coffey CE and Cummings JL (eds.) *Textbook of Geriatric Neuropsychiatry*, pp. 17–31. Washington, DC: American Psychiatric Press.

Mesulam MM (2000) A plasticity-based theory of the pathogenesis of Alzheimer's disease. *Annals of the New York Academy of Sciences* 924: 42–52.

Morrison JH and Hof PR (1997) Life and death of neurons in the aging brain. *Science* 278: 412–419.

O'Brien JT, Paling S, Barber R, et al. (2001) Progressive brain atrophy on serial MRI in dementia with Lewy bodies, AD, and vascular dementia. *Neurology* 56: 1386–1388.

Peters A, Morrison JH, Rosene DL, and Hyman BT (1998) Feature article: Are neurons lost from the primate cerebral cortex during normal aging? *Cerebral Cortex* 8: 295–300.

Scahill RI, Schott JM, Stevens JM, et al. (2002) Mapping the evolution of regional atrophy in Alzheimer's disease: Unbiased analysis of fluid-registered serial MRI. *Proceedings of the National Academy of Sciences of the United States of America* 99: 4703–4707.

Thal DR, Rub U, Orantes M, and Braak H (2002) Phases of A-beta deposition in the human brain and its relevance for the development of AD. *Neurology* 58: 1791–1800.

Thompson PM, Hayashi KM, de Zubicaray G, et al. (2003) Dynamics of gray matter loss in Alzheimer's disease. *Journal of Neuroscience* 23: 994–1005.

Thompson PM, Mega MS, Woods RP, et al. (2001) Cortical change in Alzheimer's disease detected with a disease-specific population-based brain atlas. *Cerebral Cortex* 11: 1–16.

Alzheimer's Disease: Neurodegeneration

N J Cairns, Washington University School of Medicine, St. Louis, MO, USA

Neuropathology

Alzheimer's disease (AD) is clinically, neuropathologically, and genetically heterogeneous. Subgroups include familial and sporadic forms; there are also phenotypic differences according to the gene and the defect within a single gene. Not infrequently, atypical cases may be seen, and AD may be found in combination with another neurodegenerative disease, most frequently dementia with Lewy bodies, which may also contribute to the cognitive deficits.

Patients with AD often die of bronchopneumonia and there is often little pathology outside the central nervous system (CNS). Depending on the stage of disease, the brain may appear unremarkable to the naked eye or may be grossly atrophic (**Figure 1**). Brain weight is typically reduced, often to less than 1000 g, from the average of 1250–1400 g. The atrophy is usually symmetrical and affects the frontal, temporal, and parietal lobes, with relative sparing of the sensorimotor cortices and occipital lobe, although all cortical areas may be affected in the most severe cases. Senile plaques and neurofibrillary tangles are frequently found with other lesions which are not specific to AD, as they are found in the aged brain, including granulovacuolar degeneration, Hirano bodies, gliosis, synaptic and neuronal loss, and white-matter and vascular changes.

The senile or neuritic plaque is one of the signature lesions of the AD brain (**Figure 2**). These complex extracellular structures range in size from 50 to 200 μm and are readily visualized by silver impregnation methods. In research settings and diagnostic laboratories, immunohistochemistry may be available, and these structures can be detected by antibodies raised against specific epitopes of the pathological protein (Aβ). Immunohistochemistry generally reveals much more extensive pathology than that seen by traditional silver and other staining methods and is now the method of choice for detecting the molecular pathology of most neurodegenerative diseases, not just of AD. The classical senile plaque consists of an amyloid core with a ring or crown, as seen in cross section with the light microscope, of argyrophilic axonal and dendritic processes, amyloid fibrils, astrocytic processes, and microglial cells. The neuritic processes of the senile plaque are often dystrophic and contain abnormal paired helical filaments (PHFs) made up largely of hyperphosphorylated tau protein. The amyloid is composed of 5–10 nm filaments of amyloid β-protein (Aβ), a 39- to 43-amino-acid (4 kDa) protein, a cleavage product of a transmembranous amyloid precursor protein (APP).

Aβ immunohistochemistry reveals a wider spectrum of plaque types than is seen with traditional staining methods. Aβ deposits include subpial, diffuse, ring-with-core, compact, vascular, and dyshoric deposits (**Figure 3**). Aβ plaques may be found throughout the brain, including the neocortex, amygdala, hippocampus, striatum, pallidum, nucleus basalis of Meynert, thalamus, midbrain, medulla oblongata, cerebellar cortex, and spinal cord. Immunohistochemistry using antibodies that recognize a full-length A$\beta_{1-42(43)}$ or a truncated Aβ_{1-40} detects most Aβ species in the AD brain. The predominant species in sporadic AD is A$\beta_{1-42(43)}$, which is present in most plaques types; A$\beta_{1-42(43)}$ is found only in diffuse plaques, indicating that this species may be more toxic. This widespread deposition of Aβ protein is evidence that AD is an Aβ-amyloidosis of the CNS. *In vitro* and *in vivo* models also provide evidence that soluble Aβ oligomers may also be injurious to neurons and disrupt synaptic function (**Figures 4–6**).

The second histological signature lesion of AD is the neurofibrillary tangle (NFT; **Figures 7** and **8**). The NFT is not specific to AD; NFTs occur in aging and other neurodegenerative diseases, including Down syndrome, dementia pugilistica, postencephalitic parkinsonism, amyotrophic lateral sclerosis–parkinsonism–dementia complex of Guam, subacute sclerosing panencephalitis, dementia with tangles with and without calcification, and myotonic dystrophy. NFTs also are characteristic of frontotemporal lobar degeneration (FTLD) with tauopathy, which includes corticobasal degeneration, progressive supranuclear palsy, and argyrophilic grain disease, but their ultrastructure and tau isoform patterns, as demonstrated by immunohistochemistry or biochemistry, differ from those seen in AD and the aging brain.

In AD, as in the aged brain of cognitively normal individuals, NFTs are typically numerous in the medial temporal lobe, but in AD their distribution is much more widespread and may involve neocortex and subcortical nuclei. NFTs are intraneuronal inclusions composed of cytoskeletal components, mainly tau, and their shape is determined by the type of neuron in which they develop – for example, in pyramidal neurons they are flame-shaped, while in subcortical neurons they may be globose in appearance. Mature NFTs are

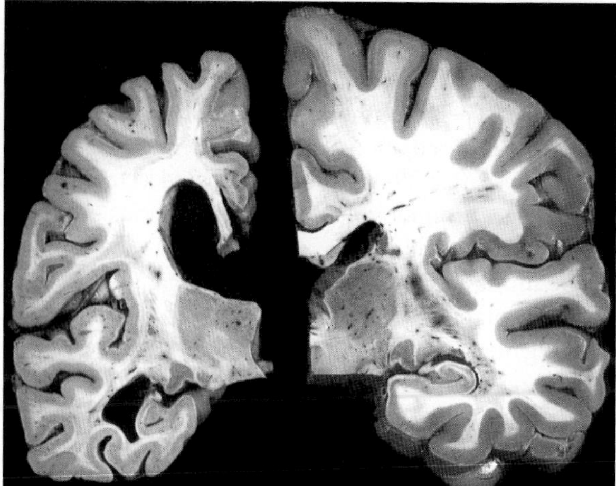

Figure 1 On the left, a coronal slice of the hemi-brain of a 70-year-old patient with severe Alzheimer's disease, showing atrophy: the lateral ventricle is enlarged with rounding of its angle, several gyri are narrowed, the hippocampus is small, and the lateral fissure is widened. On the right, a slice of the right hemi-brain of a normal age-matched individual.

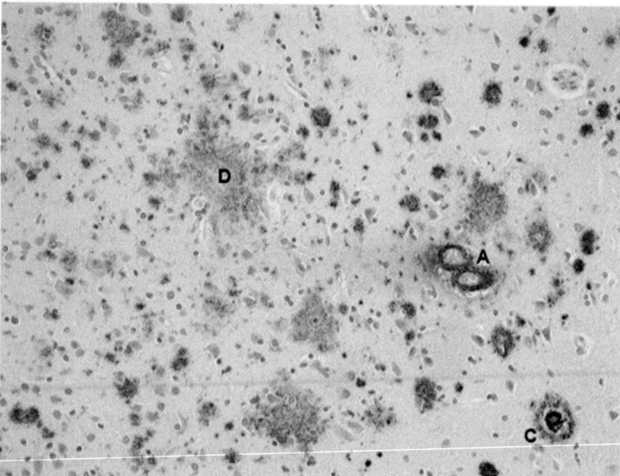

Figure 3 Extracellular amyloid β-protein deposits are present through all layers of the neocortex and have variable morphology: a diffuse plaque (D), a ring-with-core plaque (C), and cerebral amyloid angiopathy (A). Amyloid β-protein (10D5 antibody) immunohistochemistry.

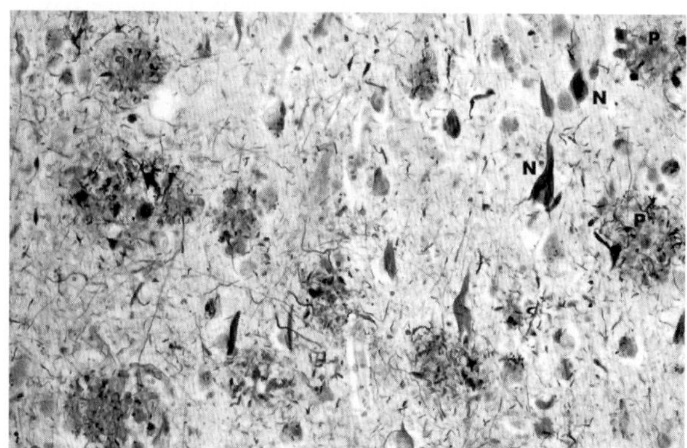

Figure 2 Neurofibrillary tangles (N) and neuritic plaques (P) in the hippocampus. Modified Bielschowsky silver impregnation.

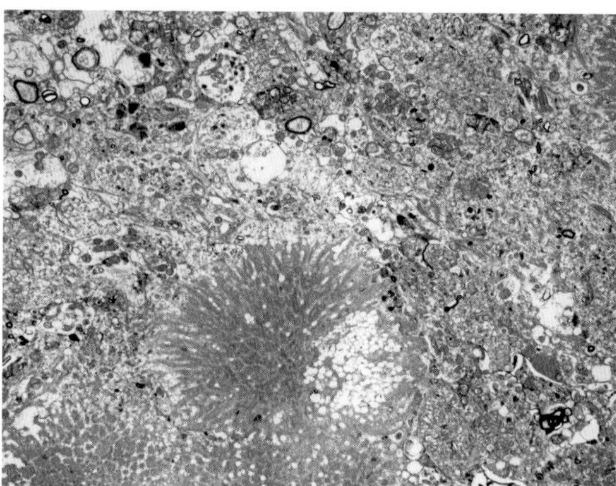

Figure 4 A low-power electron micrograph revealing a compact senile plaque.

argyrophilic, and immunohistochemistry reveals that NFTs are usually ubiquitinated, depending on the stage of evolution of the tangle; pretangles which are labeled by antiphosphorylated tau antibodies are not generally ubiquitinated, nor are they argyrophilic, indicating that the pretangle is an early stage in pathogenesis of this inclusion. Electron microscopy reveals that the NFT is composed largely of PHFs 10 nm in diameter, with cross-over points at every 80 nm, producing the periodicity of a double helix. NFTs also contain straight filaments with a mean diameter of 15 nm. The major component of the NFT is tau protein, a microtubule-associate protein that is expressed predominantly in axons, and at low levels in astrocytes and oligodendrocytes. Human tau proteins are encoded by a single-copy gene on chromosome 17q21. In adult human brain, alternative splicing of exons 2, 3, and 10 generates six tau isoforms, ranging from 352 to 441 amino acids in length, which differ by the presence of either three or four microtubule binding repeats (3R tau or 4R tau, respectively), of 31 or 32 amino acids each. Additionally, alternative splicing of exons 2 and 3 leads to the absence (0N) or presence of inserted sequences of 29 (1N) or 58 (2N) amino acids in the N-terminal third of the molecule. In the adult human brain, the ratio of 3R:4R tau isoforms is approximately 1:1, while in other FTLDs with tauopathy there is a preponderance of 3R tau (Pick's disease), 4R tau (corticobasal degeneration, progressive supranuclear palsy, and agyrophilic grain disease), or 3R, 4R, or 3R and 4R tau, as seen in cases of FTLD with

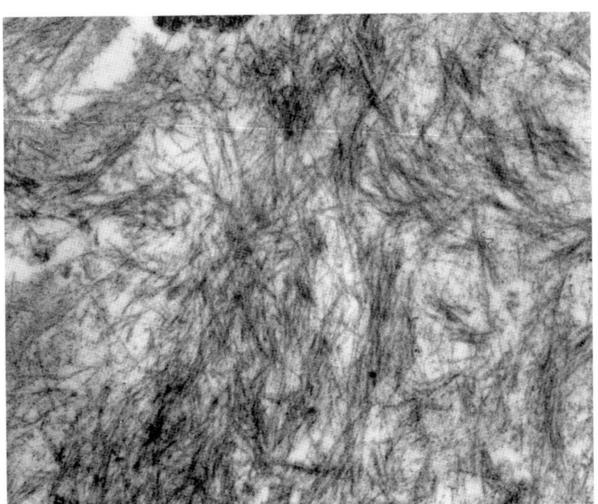

Figure 5 A high-power electron micrograph of loosely aggregated amyloid filaments of the corona of an amyloid β-protein plaque.

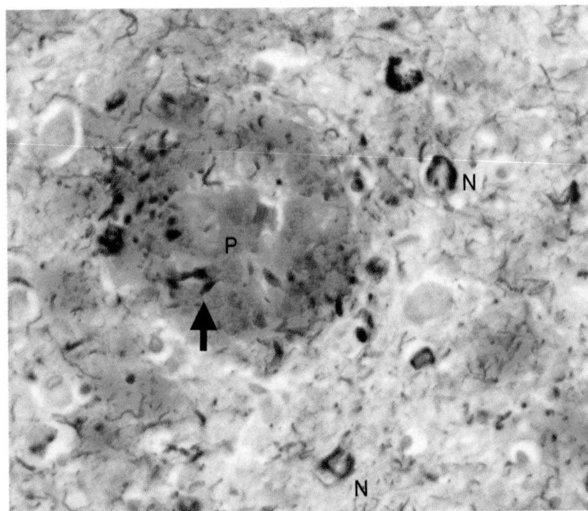

Figure 7 An amyloid β-protein plaque (P) and neurofibrillary tangles (N) in the superior temporal lobe of an Alzheimer's disease brain. The plaque contains tau-immunoreactive dystrophic neurites (arrow). Tau (black, PHF1 antibody) and amyloid β-protein (red, 10D5 antibody) immunohistochemistry.

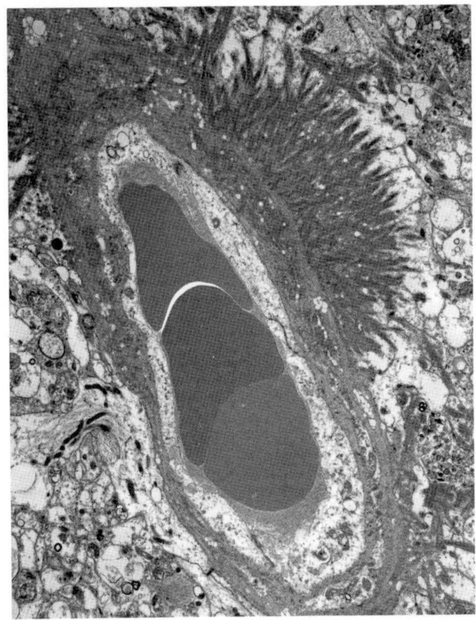

Figure 6 A low-power photomicrograph of dyshoric cerebral amyloid angiopathy with amyloid fibrils extending from the vessel wall and radiating into the surrounding brain parenchyma.

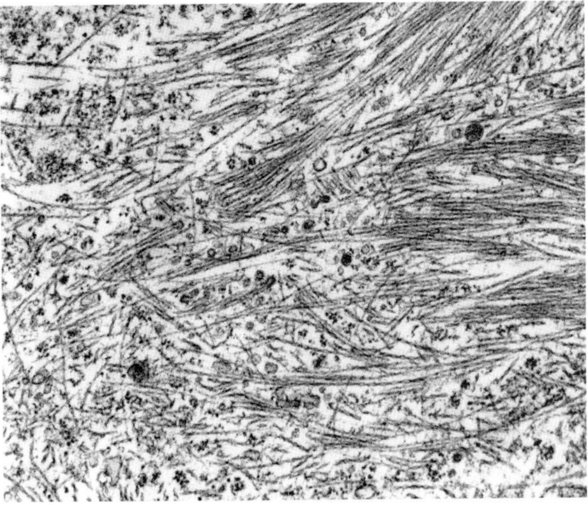

Figure 8 A high-power electron micrograph revealing paired helical filaments of a neurofibrillary tangle. Micrograph kindly provided by the late Professor LW Duchen, University of London, UK.

microtubule-associated protein tau (MAPT) gene mutation.

Tau binds to and stabilizes microtubules (MTs) and promotes MT polymerization. The function of tau as an MT binding protein is regulated by phosphorylation. Phosphorylation at approximately 30 of these sites has been identified in normal tau protein. Several protein kinases and protein phosphatases have been implicated in regulating the phosphorylation state and thus the function of tau. The phosphorylation sites are clustered in regions flanking the MT binding repeats, and increasing tau phosphorylation at multiple sites negatively regulates MT binding. All six isoforms of tau in AD are hyperphosphorylated, and phosphorylation-specific anti-tau antibodies may be used to identify three sites of tau pathology: the NFT, dystrophic neurites of senile plaques, and neuropil threads. In both sporadic and familial AD, tau is hyperphosphorylated, and it is this 'abnormal' tau that is the principal component of the filamentous aggregates in neurons and one of the pathological hallmarks of this disorder.

The Aβ Cascade Hypothesis

The initial observation that Aβ deposits are present in the blood vessels of young patients with Down syndrome (DS) was followed by the discovery that Aβ plaques and NFTs are present in adult DS and AD. These studies led to the hypothesis that a gene encoding Aβ was located on the long arm of chromosome 21. Subsequently, mutations in the APP gene were identified in a family with early-onset AD, and then additional mutations were reported. These neuropathological, biochemical, and genetic findings led to the hypothesis that increased Aβ production leads to a catastrophic cascade, including synaptic alterations, fibrilization, microglial and astrocytic activation, abnormal phosphorylation of tau proteins to form oligomers and the PHFs of NFTs, progressive synaptic and neuronal loss, loss of multiple neurotransmitters (especially acetylcholine), and ultimately to dementia (**Figure 9**).

The generation of Aβ peptides is an example of regulated intramembrane proteolysis. APP processing is regulated by either of two membrane-bound proteases: α-secretase (a type of metalloproteinase) and β-secretase, also called β-site APP-cleaving enzyme (BACE). The membrane-associated fragment caused by BACE cleavage may undergo further cleavage by the β-secretase complex, an aspartyl protease composed of presenilin-1 (PS1) or PS2, nicastrin, APH1 (anterior pharynx defective-1), and an anti-presenilin protein enhancer (PEN2). All four proteins are necessary and sufficient to form the γ-secretase complex. Further evidence for the importance of Aβ oligomerization in AD comes from the observation that mutations in PS1 and PS2, components of the γ-secretase complex, lead to enhanced production of Aβ_{1-42}. Neuropathological studies confirm that

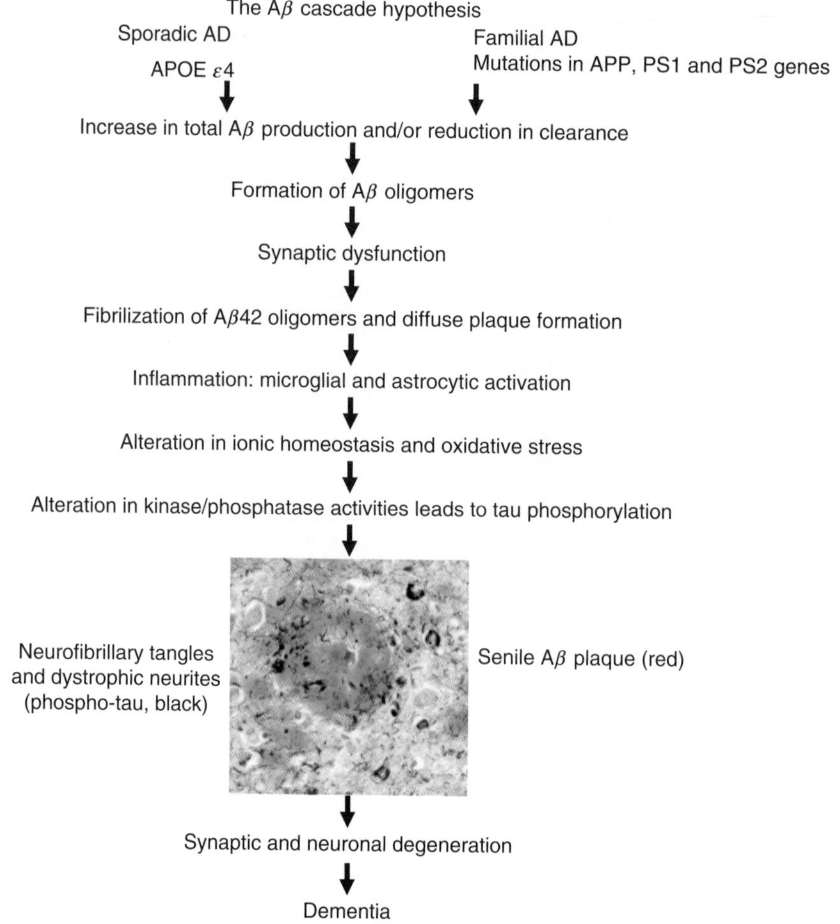

Figure 9 The amyloid β-protein (Aβ) cascade hypothesis of Alzheimer's disease (AD). APOE, apolipoprotein E; APP, amyloid precursor protein; PS1, PS2, presenilins.

$A\beta_{1-42}$ is present in diffuse plaques, an early stage in senile plaque pathogenesis. The molecular dissection of the $A\beta$ cascade has generated several potential targets for therapeutic intervention: inhibitors of β- and γ-secretases that generate $A\beta$, statins, $A\beta$ vaccines, and nonsteroidal anti-inflammatory drugs (NSAIDs) are some of the classes of compounds targeting $A\beta$-induced neurodegeneration.

The Tau Hypothesis

Several sporadic and familial neurodegenerative disorders, including AD, that are characterized clinically by dementia and/or motor dysfunction are characterized pathologically by abnormal intracellular accumulations of the microtubule-associated protein tau (MAPT); these disorders are collectively called tauopathies. The progressive accumulation of filamentous tau inclusions in the absence of other disease-specific neuropathological abnormalities provides evidence implicating tau dysfunction in disease onset and progression. However, the discovery of pathogenic *MAPT* gene mutations in the heterogeneous group of disorders known as frontotemporal lobar degeneration (FTLD) with *MAPT* mutation, also called frontotemporal dementia with parkinsonism linked to chromosome 17 (FTDP-17), provided confirmation of the central role of tau abnormalities in the etiology of neurodegenerative tauopathies. These findings have opened up novel areas for investigation into the mechanisms of tau dysfunction and the relationship of tau abnormalities to brain degeneration.

Several transgenic models of tau pathology have been generated by overexpressing human tau proteins in mice. However, these mice are either asymptomatic or develop pathology that is localized to the spinal cord and/or lacks many of the key features of tau-based disorders. In contrast, the introduction of the P301L mutation led to the development of transgenic mice that develop age- and gene dose-dependent accumulation of tau tangles in the brain and spinal cord, with associated nerve cell loss and gliosis as well as behavioral abnormalities. Similar to human disease, the tau aggregates are composed of only mutant human tau, further implicating the P301L change in promoting the selective aggregation of mutant tau. Other systems have also been developed to model various aspects of human tauopathies, including a transgenic mouse overexpressing the shortest human tau isoform; this mouse acquired age-dependent tau-pathology similar to that seen in FTLD with *MAPT* mutation. Overexpression of either wild-type or mutant tau in *Drosophila melanogaster* demonstrates features of tauopathy, including adult-onset progressive neurodegeneration with accumulation of abnormal tau. However, the neurodegeneration in this model occurred in the absence of NFT formation. More recent studies have demonstrated NFT-like pathology when tau is co-expressed with *shaggy*, a homolog of glycogen synthase 3-kinase (GSK3), an enzyme implicated in tau phosphorylation. Neurodegeneration and defective neurotransmission have also been demonstrated in *MAPT* transgenic *Caenorhabditis elegans*. In this model, panneuronal expression of normal and mutant tau resulted in altered behavior, accumulation of insoluble phosphorylated tau, age-dependent loss of axons and neurons, and structural damage to axonal tracts. These models recapitulate various features of the tauopathies and highlight targets for disease-modifying therapies, not only for AD but also for related tauopathies. Thus, microtubule-stabilizing drugs might ameliorate the sequestration of tau into NFTs, and drugs such as LiCl, in an animal model of tauopathy, that impede the abnormal phosphorylation of tau by inhibiting GSK3 indicate a fruitful area for drug discovery.

Timeline of Neurodegenerative Changes in AD

Many nondemented older individuals have few or no AD-type changes in the brain. However, about 30% of cognitively normal individuals over the age of 75 years have both diffuse and neuritic plaques throughout the neocortex, identical to those seen in individuals with symptomatic AD (**Figure 10**). The change in the brain that corresponds most closely to the clinical symptoms of those with very mild dementia (Clinical Dementia Rating = 0.5) is neuronal loss in the medial temporal lobe, including the entorhinal cortex and hippocampus; cognitively normal individuals, in contrast, have minimal or no loss of neurons from these areas. Thus, there is a 'preclinical' stage of AD in which there are sufficient plaques in neocortical areas for a neuropathological diagnosis of AD, but there is not sufficient synaptic and neuronal loss to produce cognitive symptoms. In familial cases, misfolding of $A\beta$ may start at a very young age, and only after two to three decades may result in the neuropathology of AD. Research efforts are now focusing on this preclinical stage of AD, which may extend years or even decades before the onset of clinical symptoms. The development of antecedent biomarkers that detect asymptomatic AD is a priority; efforts include clinical, cognitive, neuropsychological, structural and functional imaging, genetic, and proteomic studies. Tau and $A\beta$ protein levels in cerebrospinal fluid are potentially additional markers of preclinical

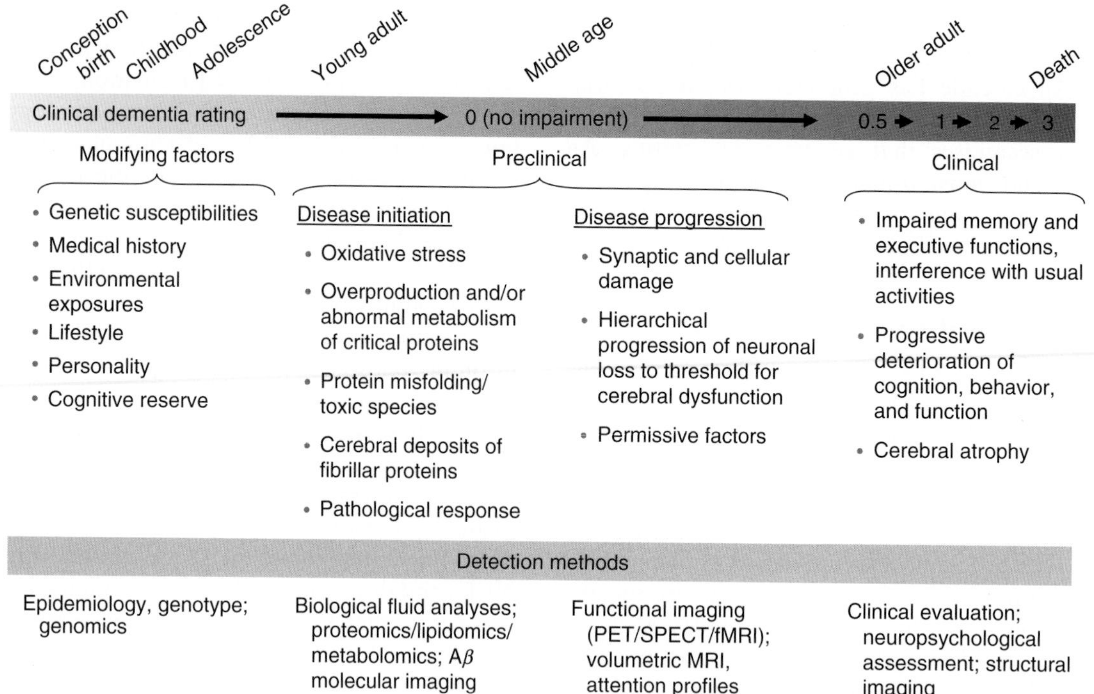

Figure 10 Hypothetical timeline for Alzheimer's disease. PET, positron emission tomography; SPECT, single-photon computed tomography; fMRI, functional magnetic resonance imaging.

AD, as is the Aβ imaging agent, Pittsburgh compound B (PIB). It is likely that a combination of biomarkers will provide greater diagnostic accuracy than is possible using any single analyte or measure.

Conclusions

The accumulations of filamentous Aβ species in the extracellular space, and of abnormally phosphorylated tau proteins within neurons, are the signature lesions of AD. Protein misfolding, fibrilization, and aggregate formation are not unique to AD, but are common features of a wide variety of sporadic and familial neurodegenerative disorders (synucleinopathies, tauopathies, trinucleotide repeat disease, and TDP-43 proteinopathies). These diseases are distinguished by the distinct topographic and cell type-specific distribution of inclusions. The biochemical and ultrastructural characteristics of the inclusions also reveal a significant phenotypic overlap. The discovery that multiple mutations in *APP*, *PS1*, and *PS2* genes lead to abnormal protein aggregation in AD demonstrates that neuronal Aβ mismetabolism is sufficient to produce neurodegenerative disease. Experimental evidence indicates that mutations lead to specific alterations in expression, function, and biochemistry of Aβ and tau proteins. The realization that AD pathology is present many years prior to the onset of clinical symptoms has led to dramatic progress in the search for antecedent biomarkers which herald disease onset. These new markers, once validated, will facilitate the identification of individuals at the preclinical or early stage of AD, and these individuals stand to benefit greatly from emerging therapeutic interventions.

See also: Acetylcholinesterase Inhibitors and Alzheimer's Disease; Aging of the Brain and Alzheimer's Disease; Alzheimer's Disease: An Overview; Alzheimer's Disease: Molecular Genetics; Alzheimer's Disease: Transgenic Mouse Models; Alzheimer's Disease: MRI Studies; Axonal Transport and Alzheimer's Disease; Axonal Transport and Neurodegenerative Diseases; Brain Glucose Metabolism: Age, Alzheimer's Disease and ApoE Allele Effects; Oxidative Damage in Neurodegeneration and Injury; Transgenic Models of Neurodegenerative Disease.

Further Reading

Braak H, Alafuzoff I, Arzberger T, et al. (2006) Staging of Alzheimer disease-associated neurofibrillary pathology using paraffin sections and immunocytochemistry. *Acta Neuropathologica* 112: 389–404.

Cairns NJ, Lee VM-Y, and Trojanowski JQ (2004) The cytoskeleton in neurodegenerative diseases. *Journal of Pathology* 204: 438–449.

Csernansky JG, Wang L, Swank J, et al. (2005) Preclinical detection of Alzheimer's disease: Hippocampal shape and volume predict dementia onset in the elderly. *NeuroImage* 25: 783–792.

Forman MS, Trojanowski JQ, and Lee VM-Y (2004) Neurodegenerative diseases: A decade of discoveries paves the way for therapeutic breakthroughs. *Nature Medicine* 10: 1055–1063.
Goedert M, Spillantini MG, Cairns NJ, et al. (1992) Tau proteins of Alzheimer paired helical filaments: Abnormal phosphorylation of all six brain isoforms. *Neuron* 8: 159–168.
Haass C and Selkoe DJ (2007) Soluble protein oligomers in neurodegeneration: Lessons from the Alzheimer's amyloid beta-peptide. *Nature Reviews – Molecular Cell Biology* 8: 101–112.
McKhann GM, Albert MS, Grossman M, et al. (2001) Clinical and pathological diagnosis of frontotemporal dementia: Report of the Work Group on Frontotemporal Dementia and Pick's Disease. *Archives of Neurology* 58: 1803–1809.
Mirra SS, Heyman A, McKeel D, et al. (1991) The Consortium to Establish a Registry for Alzheimer's Disease (CERAD). Part II. Standardization of the neuropathologic assessment of Alzheimer's disease. *Neurology* 41: 479–486.
Mirra SS and Hyman BT (2002) Ageing and dementia. In: Graham DI and Lantos PL (eds.) *Greenfield's Neuropathology,* 7th edn., pp. 195–271. London: Arnold.
Price JL, Davis PB, Morris JC, et al. (1991) The distribution of tangles, plaques and related immunohistochemical markers in healthy aging and Alzheimer's disease. *Neurobiology of Aging* 12: 295–312.
Price JL and Morris JC (1999) Tangles and plaques in nondemented aging and "preclinical" Alzheimer's disease. *Annals of Neurology* 45: 358–368.

Relevant Website

http://www.alzforum.org – Alzheimer Research Forum.

Alzheimer's Disease: Transgenic Mouse Models

K H Ashe, University of Minnesota, Minneapolis, MN, USA

Creation of Transgenic Mouse Models

Biological Foundation

Three scientific breakthroughs made the creation of the earliest transgenic mouse models of Alzheimer's disease (AD) possible. The first was the isolation and sequencing of the amyloid-β (Aβ) peptide in 1984. The second was the cloning of the amyloid precursor protein (APP) gene in 1987 and the elucidation of its role in generating the Aβ peptide. Third was the discovery of the first mutation in autosomal dominant familial AD in APP in 1991 and the subsequent realization that all autosomal dominant mutations causing AD enhance the ability of the Aβ protein to aggregate, either by increasing its overall production or by the generation of amino acid variants that potentiate its aggregation.

This information enabled investigators to create APP transgenic mice modeling AD. The earliest mouse models were developed in the first half of the previous decade.

Altogether about 20 such mice have been published, many of which show age-related amyloid plaque deposition and memory loss. However, APP transgenic mice are incomplete models of AD because they lack neurofibrillary tangles and develop few or no neurodegenerative changes, such as neuronal or synaptic loss. Transgenic mice that develop neurofibrillary tangles accompanied by significant neuronal loss emerged with the creation of tau transgenic mice.

Four important landmarks in tau biology made the generation of tau transgenic mice possible. First was the isolation and characterization in 1975 of tau, which is involved in promoting the aggregation and polymerization of tubulin to form microtubules. Second was the cloning in 1986 of the tau gene. Third was the recognition that tau is the principal protein forming the core of the paired helical filaments of neurofibrillary tangles. Fourth was the discovery in 1998 of mutations in tau linked to familial tauopathies, called frontotemporal dementia with parkinsonism (FDTP). These advances led to the development of tau transgenic mice in the first years of the current decade.

Technical Methodology

The laboratory mouse remains the genetic model organism that is evolutionarily closest to the human. Mice pose advantages over other species, such as rats, dogs, pigs, or primates, in several aspects: (1) it is relatively easy to house large numbers of them; (2) the mouse genome is better characterized than that of any other mammal except humans; (3) many genetically altered transgenic knockout and knockin lines and genetically characterized mouse strains are available for cross-breeding and further genetic analysis; (4) cognitive studies can be done in appropriate strains of mice, such as C57B6/SJL and 129S6, making them especially relevant to the study of AD; and (5) studies in invertebrates, while useful, will lack some relevance in pharmacological studies and pathogenic investigations. To create transgenic mice with salient features of AD, investigators must take care in several aspects of design and development. The most important factors are the selection of the gene or gene variant to be expressed, the promoter driving transgene expression, the background strain of the mice, and the levels and distribution of transgenic protein expression achieved in the brain. The mouse lines with the most robust neuropathological phenotypes have been developed using the method of pronuclear microinjection of transgenes into fertilized oocytes (**Figure 1**).

Both wild-type and variant human APP and tau genes have been used to generate mice that develop neuropathology related to AD. Transgenes containing mutations linked to autosomal dominant AD or FDTP generate more-robust neuropathology than transgenes encoding wild-type genes. Although mutations in presenilin-1 and presenilin-2 genes lead to early-onset AD in humans, mice expressing presenilin gene variants do not develop neuropathology *per se*. In the presence of human APP transgenes, however, the presenilin gene variants accelerate plaque deposition in transgenic mice.

Characteristics of Transgenic Mouse Models

APP Transgenic Mice

APP transgenic mice often develop age-related cognitive deficits and recapitulate many of the neuropathological features of AD, including amyloid plaques, oxidative stress, astrogliosis, microgliosis, cytokine production, and dystrophic neurites (**Figure 2**). However, many important neuropathological features of AD are conspicuously absent in APP transgenic mice, including neurofibrillary tangles and gross atrophy.

There are variable degrees of neuronal and synaptic loss among the various transgenic lines, but no line exhibits the severity of neurodegeneration found in

Alzheimer's patients, despite amyloid plaque loads that often exceed those in human brain specimens. For example, Tg2576 mice are virtually devoid of neuronal or synaptic loss, while synaptic loss is present in J20 and PDAPP mice, and there is some neuronal loss in APP23 mice with massive quantities of amyloid plaques. All four of these lines of mice eventually develop amyloid plaque loads that exceed the amount found in typical Alzheimer patient brains at autopsy. Although we do not understand the factors which account for the variations in neurodegeneration between the mouse lines, it is clear that the differences are not related to the quantity of amyloid plaque deposition.

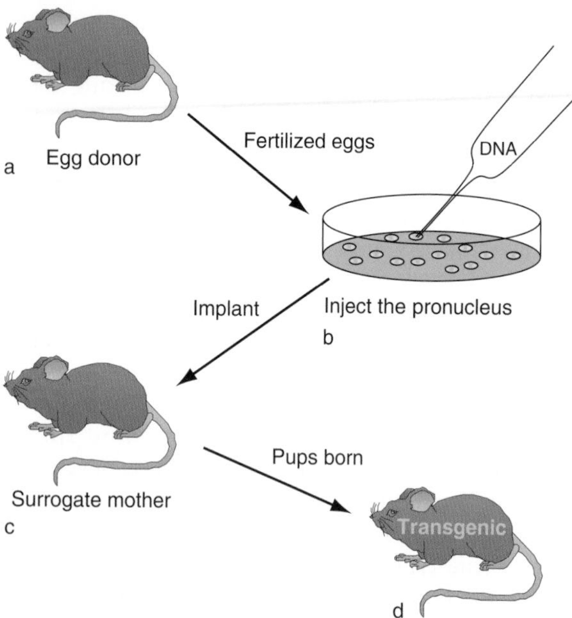

Figure 1 How transgenic mice are made. (a) Fertilized eggs (oocytes) are harvested from a donor mouse. (b) DNA encoding genes linked to Alzheimer's disease is injected into the pronucleus of the fertilized oocytes. (c) The oocytes are implanted into surrogate mothers, where they mature to full-term pups. (d) Of the pups that are born, a certain fraction, usually 10–20%, contain the transgene.

Tau Transgenic Mice

Tau transgenic mice generally develop age-related neurological abnormalities, neurofibrillary tangles, and neurodegeneration but invariably fail to form amyloid plaques. The earliest tau transgenic lines expressing tau ubiquitously in the neurons of the brain developed neurodegeneration mainly in the brain stem and spinal cord, leading to paralysis. Because the paralysis interfered with cognitive testing, tau transgenic mice in which variant tau expression was restricted to the forebrain were created and shown to develop age-related memory impairment.

Although the tau transgenic mice with the most robust neuropathological phenotypes express tau variants linked to FDTP (**Figure 3**), neurofibrillary tangles and neurodegeneration occur also in mice expressing wild-type human tau on a null tau background, suggesting a propensity for mouse tau to inhibit neurofibrillary tangle formation. Notably, these mice uniquely form the true paired-helical filaments characteristic of neurofibrillary tangles in AD. Straight filaments, rather than paired-helical filaments, are found in tau transgenic mice expressing FDTP tau variants. This distinction, once believed to be critical, has become less important with the discovery that neurofibrillary tangles contribute little or not at all

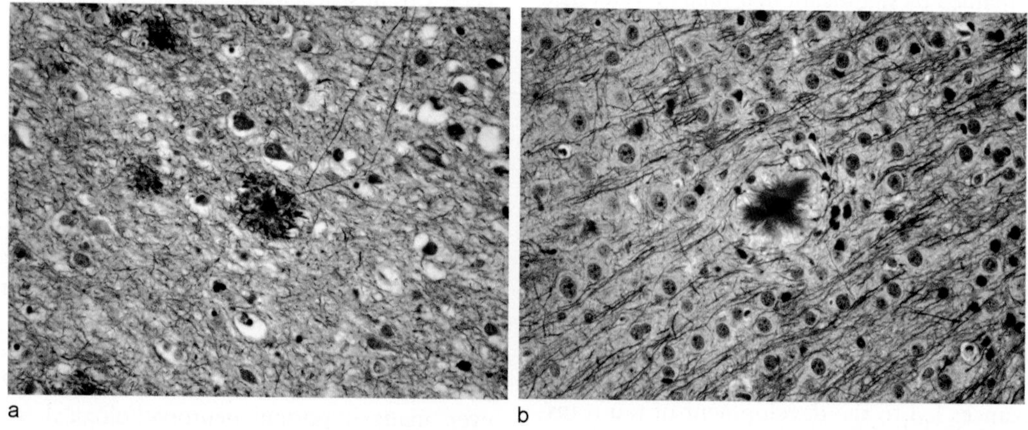

Figure 2 Amyloid plaques in human (a) and mouse (b). Amyloid plaques that are generated in amyloid precursor protein (APP) transgenic mice closely resemble the shape, composition, and size of those found in humans with Alzheimer's disease. The plaque on the left is from a human specimen, while the one on the right is from an aged APP transgenic mouse. Despite the similarities in plaques in the two species, neurodegenerative changes that are clearly present in the human specimen are largely absent in the mouse specimen. Both specimens were stained with the Bielschowsky silver stain. The photomicrographs were taken at the same magnification (40×). Photomicrographs were kindly provided by Dr. Martin Ramsden.

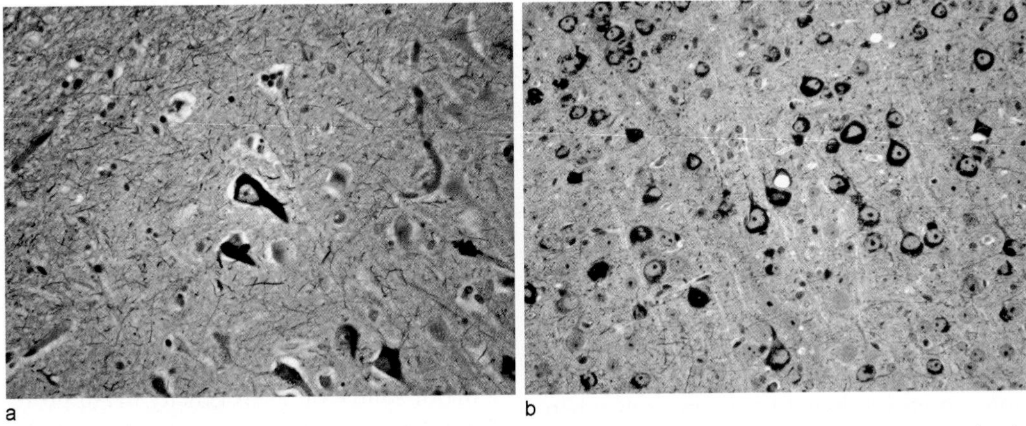

Figure 3 Neurofibrillary tangles in (a) human and (b) mouse. Neurofibrillary tangles that are generated in tau transgenic mice resemble those found in humans with Alzheimer's disease. The tangles on the left are from a human specimen, while those on the right are from an aged tau transgenic mouse. Both specimens were stained with the Bielschowsky silver stain. The photomicrographs were taken at the same magnification (40×). Photomicrographs were kindly provided by Dr. Martin Ramsden.

to neuronal death and memory impairment, as discussed more fully in the section titled 'Mechanism of memory loss in transgenic mouse models.'

Other Types of Transgenic Mice

Efforts to combine the cardinal neurological and neuropathological features of AD, namely amyloid plaques, neurofibrillary tangles, neurodegeneration, and cognitive impairment, within a single mouse resulted in the generation of transgenic mice expressing multiple tau and APP protein variants, sometimes in combination with presenilin 1 variants. Although some such mice develop amyloid plaques, neurofibrillary tangles, neuronal loss, and memory loss, they constitute essentially a hybrid of two separate disorders, AD and FDTP. A number of interesting studies have been carried out using these hybrid mice, but their relevance to AD is not entirely clear. A mouse exhibiting all the hallmarks of AD that does not depend on the use of genes not linked to familial AD remains an elusive creature.

Utility of Transgenic Mouse Models

Validity of Transgenic Mouse Models

The lack of several important features of AD in APP transgenic mice, such as neurofibrillary tangles and prominent neurodegeneration, along with the shortcomings of the tau transgenic mice and the APP-tau hybrid mice, have prompted some scientists, rightly, to challenge the validity of transgenic mouse models of AD. To address this challenge, criteria may be devised for validating Alzheimer's mouse models, three of which are discussed here. First, theoretical validity refers to whether the use of a given transgene is based on sound biological principles. Second, factual validity refers to how accurately various aspects of the human disease are represented. Third, predictive validity refers to whether studies using the model predict outcomes in human trials. Clearly, predictive validity is the main determinant of the value of a given mouse model. Thanks to rapid progress in bench-to-bedside research in AD, we know more about the predictive validity of APP transgenic mice than of transgenic mice modeling any other neurodegenerative illness.

Aβ immunotherapy illustrates this point. Young PDAPP mice vaccinated with Aβ fail to develop amyloid deposits and astrogliosis. This result, reported in 1999, quickly led to an international, multicenter, randomized trial of an experimental Aβ vaccine in Alzheimer patients. The vaccine cleared amyloid plaques away from large regions of brain in Alzheimer patients. However, neurofibrillary tangles were unaffected, and neuronal loss persisted. Furthermore, when the effects of the vaccine on cognitive function were examined, it became clear that memory decline may have been slowed, but it was not restored. The failure to restore memory function in Alzheimer patients is in marked contrast to the effects of either active or passive Aβ immunization on memory function in APP transgenic mice. Memory loss can be prevented, and preexisting memory loss can be reversed and memory function fully restored in several different lines of APP transgenic mice following Aβ immunization treatments.

These studies show that when the effects of the Aβ vaccine on amyloid plaques are studied, then the results in APP transgenic mice predict what occurs in humans. However, when memory function is examined, then there is a striking discrepancy between what occurs in the mice and what occurs in patients. Thus, the predictive validity of APP transgenic mice as regards amyloid plaques is good. In contrast, the predictive validity of APP transgenic mice in relation to memory function is poor.

Relevance of Transgenic Mouse Models

The value of transgenic models of AD is as good as our understanding of how closely the models mimic various stages of the illness. To appreciate the context in which these mice have helped us study AD, it is useful delineate the natural history of the illness. AD has a very insidious onset; we do not know precisely when neural dysfunction begins. Recent work suggests that the disease process may begin long before symptoms are present or neurodegeneration has occurred. By the time AD is clinically diagnosed, neurodegeneration is already under way.

APP transgenic mice and tau transgenic mice mimic different stages of AD. Each one is useful for exploring the progression of cognitive decline and the pathology of AD. The picture of minimal loss of neurons in many lines of memory-impaired APP transgenic mice suggests a closer resemblance to preclinical stages of AD than to the disease itself. In contrast, the substantial neuronal loss found in tau transgenic mice implies that they may be better models for studying the neurodegenerative aspects of AD.

Testing Experimental Therapies in Transgenic Mouse Models

A major effort in developing AD-modifying therapies has been directed at reducing the amyloid deposits and neurofibrillary tangles in the brain. Transgenic mice displaying plaques and tangles are useful models for testing the effects of therapies on these neuropathological abnormalities and have been used extensively for this purpose. However, as will be discussed in the next section, targeting molecules causing cognitive dysfunction and memory impairment may hold still better promise for treating the symptoms of AD.

Mechanism of Memory Loss in Transgenic Mouse Models

Ever since Alois Alzheimer described the amyloid plaques and neurofibrillary tangles that were evident when he examined the brain of a demented patient in 1906, the scientific research on what has come to be known as AD has focused on how the accumulation of tau and Aβ proteins form amyloid plaques and neurofibrillary tangles, which define the disease neuropathologically, rather than on the functional effects of these molecules on the brain. Why and how tau and Aβ molecules cause memory loss and dementia are only now becoming understood. Our knowledge of the manner in which these molecules disrupt brain function lags a century behind the neuropathological studies of tau and Aβ proteins because suitable tools for examining their effects on the workings of the living brain were lacking until recently. The creation of transgenic mouse models of AD helped fill this void and provided the vital reagents needed to begin to understand the adverse effects that the tau and Aβ proteins exert on memory and cognitive function.

In the Tg2576 APP transgenic mouse model of AD, a major source of heterogeneity in Aβ proteins arises from the aggregation of monomeric proteins to form higher-order structures. Monomeric and trimeric Aβ protein assemblies are located within neurons. In the extracellular space, monomers and trimers coexist with larger Aβ protein assemblies that have similar molecular weights as higher-order species, such as hexamers, nonamers, and dodecamers. Many of the monomeric and oligomeric forms of Aβ protein in the brain are soluble in aqueous buffers. In contrast, amyloid plaques are composed of Aβ fibrils that are insoluble in aqueous buffers and most detergents. In Tg2576 mice, amyloid plaques lag 3–6 months behind the initial loss of memory function, which occurs at 6 months of age, and they do not correlate with memory impairment, indicating that the plaques contribute little to brain dysfunction. Memory impairment is caused, instead, by a specific, soluble Aβ protein assembly, called Aβ*56 (read as A-beta star 56). Aβ*56 correlates strongly with memory dysfunction, and when purified from the brains of impaired Tg2576 mice and administered to young, healthy animals, it transiently disrupts memory function in the healthy subjects. Aβ*56 disrupts memory by a mechanism that does not involve synaptic or neuronal loss. Aβ*56 is present in brain tissue from Alzheimer patients and may contribute to memory loss associated with AD.

Neurofibrillary tangles contribute little, if at all, to neuronal loss and brain dysfunction in a line of transgenic mice modeling FDTP, called rTg4510. By dint of identical hyperphosphorylated tau species in the neurofibrillary tangles of rTg4510 mice and Alzheimer's brain tissue, it may be supposed that the effects of neurofibrillary tangles on dementia and neuronal loss in AD are similarly inconsequential. However, the molecular mechanisms by which tau induces memory impairment and kills neurons in rTg4510 mice or AD are unknown; it is likely that specific, as yet unidentified, tau* (tau star) molecules are responsible for disrupting brain function and interfering with neuronal viability. Finally, the relative contributions of Aβ*56, tau, and neurodegeneration to memory loss, cognitive impairment, and dementia remain unresolved.

See also: Acetylcholinesterase Inhibitors and Alzheimer's Disease; Aging of the Brain and Alzheimer's Disease; Alzheimer's Disease: Neurodegeneration; Alzheimer's Disease: An Overview; Alzheimer's Disease: Molecular Genetics; Animal Models of Alzheimer's Disease; Axonal Transport and Alzheimer's Disease; Brain Glucose Metabolism: Age, Alzheimer's Disease and ApoE Allele Effects; Lipids and Membranes in Brain Aging.

Further Reading

Andorfer C, Kress Y, Espinoza M, et al. (2003) Hyperphosphorylation and aggregation of tau in mice expressing normal human tau isoforms. *Journal of Neurochemistry* 86: 582–590.

Games D, Adams D, Alessandrini R, et al. (1995) Alzheimer-type neuropathology in transgenic mice overexpressing V717F beta-amyloid precursor protein. *Nature* 373: 523–527.

Hsiao K, Chapman P, Nilsen S, et al. (1996) Correlative memory deficits, Aβ elevation, and amyloid plaques in transgenic mice. *Science* 274: 99–102.

Ishihara T, Hong M, Zhang B, et al. (1999) Age-dependent emergence and progression of a tauopathy in transgenic mice overexpressing the shortest human tau isoform. *Neuron* 24: 751–762.

Iyadurai SJP and Ashe KH (2005) Creating APP transgenic lines in mice. In: Xia W and Xu H (eds.) *Amyloid Precursor Protein: A Practical Approach*, pp. 185–200. Boca Raton, FL: CRC Press.

Lesné S, Koh MT, Kotilinek L, et al. (2006) A specific amyloid-β assembly in the brain impairs memory. *Nature* 440: 352–357.

Lewis J, McGowan E, Rockwood J, et al. (2000) Neurofibrillary tangles, amyotrophy and progressive motor disturbance in mice expressing mutant (P301L) tau protein. *Nature Genetics* 25: 402–405.

Oddo S, Caccamo A, Shepherd JD, et al. (2003) Triple-transgenic model of Alzheimer's disease with plaques and tangles: Intracellular Abeta and synaptic dysfunction. *Neuron* 39: 409–421.

SantaCruz K, Lewis J, Spires T, et al. (2005) Tau suppression in a neurodegenerative mouse model improves memory function. *Science* 309: 476–481.

Amnesia: Declarative and Nondeclarative Memory

L R Squire, P J Bayley, and C N Smith, University of California at San Diego, San Diego, CA, USA

Introduction

Amnesia refers to difficulty in learning new information or in remembering the past. It is important to distinguish the amnesia that occurs following brain injury or disease (neurological amnesia) from the rarer functional (or psychogenic) amnesia that can occur as the result of an emotional trauma. Neurological amnesia has a variety of origins, including prolonged alcoholism, a temporary loss of blood supply or oxygen to the brain, and diseases such as herpes simplex encephalitis. All these conditions preferentially damage the medial temporal lobe or diencephalon. Neurological amnesia causes severe difficulty in learning new facts and events (anterograde amnesia). Amnesic patients also typically have some difficulty remembering facts and events that were acquired before the onset of amnesia (retrograde amnesia). Functional amnesia shows a different pattern of anterograde and retrograde memory impairment. Functional amnesia is characterized by a profound retrograde amnesia that is transient in some cases, and little or no anterograde amnesia is exhibited.

Functional Amnesia

Functional amnesia, also known as dissociative amnesia, is a dissociative psychiatric disorder that involves alterations in consciousness and identity. Although no particular brain structure or brain system is implicated in functional amnesia, the cause of the disorder must be due to abnormal brain function of some kind. Its presentation varies considerably from individual to individual, but in most cases, functional amnesia is preceded by physical or emotional trauma and occurs in association with some prior psychiatric history. Often, the patient is admitted to the hospital in a confused or frightened state. Memory for the past is lost, especially autobiographical memory and even personal identity. Semantic or factual information about the world is often preserved, though factual information about the patient's life may be unavailable. Despite profound impairment in the ability to recall information about the past, the ability to learn new information is usually intact. The disorder often clears, and the lost memories return. Occasionally, the disorder lasts longer, and sizable pieces of the past remain unavailable.

Etiology of Neurological Amnesia

Neurological amnesia results from a number of conditions, including Alzheimer's disease or other dementing illnesses, temporal lobe surgery, chronic alcohol abuse, encephalitis, head injury, anoxia, ischemia, infarction, and the rupture and repair of an anterior communicating artery aneurism. The common factor in all these conditions is that they disrupt normal function in one of two areas of the brain – the medial aspects of the temporal lobe and the diencephalic midline. Global amnesia results from bilateral damage, whereas material-specific amnesia results from unilateral damage. Typically, left-sided damage affects memory for verbal material, and right-sided damage affects memory for nonverbal material (e.g., the recall of faces and spatial layouts).

Anatomy

Well-studied cases of human amnesia and animal models of amnesia provide information about the neural connections and structures that are damaged in neurological amnesia. Damage limited to the hippocampus itself is sufficient to cause amnesia. For example, in one carefully studied case of amnesia (patient R.B.), the only significant damage was a bilateral lesion confined to the CA1 field of the hippocampus. The severity of memory impairment is exacerbated by additional damage outside the hippocampus. Thus, severe amnesia results when damage extends beyond the hippocampus to include adjacent structures in the medial temporal lobe, including the parahippocampal cortex, entorhinal cortex, and perirhinal cortex. Another well-studied case (H.M.) had surgery in 1953 to treat severe epilepsy. Most of the hippocampus and much of the surrounding medial temporal lobe cortices were removed bilaterally (the entorhinal cortex and most of the perirhinal cortex). Although the surgery was successful in reducing the frequency of H.M.'s seizures, it resulted in a severe and persistent amnesia.

Functional magnetic resonance imaging (fMRI) of healthy individuals who are engaged in learning and remembering reveals neural activity in the same structures that, when damaged, cause amnesia. It is also possible through structural imaging (MRI) to detect and quantify the neuropathology in amnesic patients. Many patients with restricted hippocampal damage have an average reduction in hippocampal volume of about 40%. Two such patients whose brains were available for detailed, postmortem neurohistological analysis (patients L.M. and W.H.) proved to have lost virtually all the neurons in the cornu ammonis (CA) fields of the hippocampus. These observations suggest that a reduction in hippocampal volume of approximately 40%, as estimated from MRI scans, likely indicates the near complete loss of hippocampal neurons. The amnesic condition is associated with neuronal death and tissue collapse, but the tissue does not disappear altogether because fibers and glial cells remain.

As questions about amnesia and the function of medial temporal lobe structures have become more sophisticated, it has become vital to obtain detailed, quantitative information about the damage in the patients being studied. In addition, single-case studies are not nearly so useful as group studies involving well-characterized patients. In the case of patients with restricted hippocampal damage, one can calculate the volume of the hippocampus itself as a proportion of total intracranial volume. One can also calculate the volumes of the adjacent medial temporal lobe structures (the perirhinal, entorhinal, and parahippocampal cortices), again in proportion to intracranial volume. Last, when there is extensive damage to the medial temporal lobe, it is important to calculate the volumes of lateral temporal cortex and other regions that might be affected. It is important to characterize patients in this way in order to address the kinds of questions now being pursued in memory research.

To understand the anatomy of human amnesia, and ultimately the anatomy of normal memory, animal models of human amnesia have been established in the monkey and in the rodent. In the monkey, following lesions of the bilateral medial temporal lobe or diencephalon, memory impairment is exhibited on the same kinds of tasks of new learning ability that human amnesic patients fail. Cumulative work with animal models suggests that the full medial temporal lobe memory system consists of the hippocampus and adjacent, anatomically related structures, including the entorhinal cortex, parahippocampal cortex, and perirhinal cortex (see **Figure 1**). When these adjacent structures are damaged, the severity of amnesia is greater than when only the hippocampus itself is damaged.

The important structures in the diencephalon are the mediodorsal thalamic nucleus, the anterior thalamic nucleus, the internal medullary lamina, the mammillary nuclei, and the mammillo-thalamic tract. Because diencephalic amnesia resembles medial temporal lobe amnesia in the pattern of sparing and loss, these two regions likely form an anatomically linked, functional system.

The Nature of Amnesia

It is important to appreciate that amnesic patients are not impaired at all kinds of memory. The major distinction is between declarative and nondeclarative memory. Only declarative memory is affected in amnesia. Declarative memory refers to the capacity to remember the facts and events of everyday life. It is the kind of memory that is meant when the term 'memory' is used in ordinary language. A declarative memory can be brought to mind as a conscious recol-

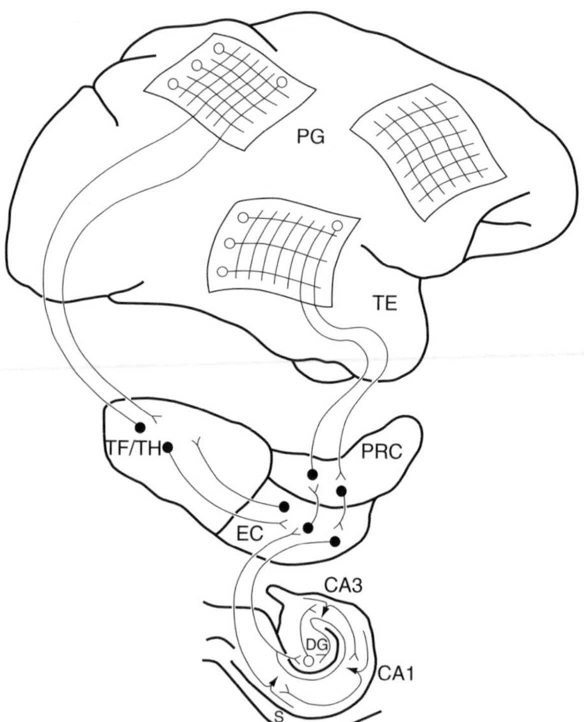

Figure 1 Schematic drawing of primate neocortex together with the structures and connections in the medial temporal region important for establishing long-term memory. The networks in the cortex show putative representations concerning visual object quality (in area TE) and object location (in area PG). If this disparate neural activity is to cohere into a stable long-term memory, convergent activity must occur along projections from these regions to the medial temporal lobe. Projections from neocortex arrive initially at the parahippocampal gyrus (TF/TH) and perirhinal cortex (PRC) and then at entorhinal cortex (EC), the gateway to the hippocampus. Further processing of information occurs in the several stages of the hippocampus, first in the dentate gyrus (DG) and then in the CA3 and CA1 regions. The fully processed input eventually exits this circuit via the subiculum (S) and the EC, where widespread efferent projections return to neocortex. The hippocampus and adjacent structures are thought to support the stabilization of representations in distributed regions of neocortex (e.g., TE and PG) and to support the strengthening of connections between these regions. Subsequently, memory for a whole event (for example, a memory that depends on representations in both TE and PG) can be revivified even when a partial cue is presented. Damage to the medial temporal lobe system causes anterograde and retrograde amnesia. The severity of the deficit increases as damage involves more components of the system. Once sufficient time has passed, the distributed representations in neocortex can operate independent of the medial temporal lobe. (This diagram is a simplification and does not show diencephalic structures involved in memory function.)

lection. Declarative memory provides a way to model the external world, and in this sense it is either true or false. The stored representations are flexible and can guide successful performance under a wide range of test conditions. Finally, declarative memory is especially suited for rapid learning and for forming and maintaining associations between arbitrarily different kinds of material (e.g., learning to associate two different words).

Anterograde Amnesia

Amnesia is characterized especially by profound difficulty in new learning. This impairment is referred to as anterograde amnesia. Amnesia can occur as part of a more global dementing disorder that includes other cognitive deficits, including impairments in language, attention, visuospatial abilities, and general intellectual capacity. However, amnesia can also occur in the absence of other cognitive deficits and without any change in personality or social skills. In this more circumscribed form of amnesia, patients have intact intellectual functions and intact perceptual functions, even on difficult tests that require the ability to discriminate between similar images containing overlapping features. Patients also have intact immediate memory (as measured, for example, by the ability to repeat a short string of digits). Their intact immediate memory explains why amnesic patients can carry on a conversation and appear quite normal to the casual observer. Indeed, if the amount of material to be remembered is not too large (e.g., a three-digit number), then patients can remember the material for minutes, or as long as they can hold it in mind by rehearsal. One would say in this case that the patients have carried the contents of immediate memory forward by engaging in explicit rehearsal. This rehearsal-based activity is referred to as working memory. The difficulty for amnesic patients arises when an amount of information must be recalled that exceeds immediate memory capacity (typically, when a list of eight or more items must be remembered) or when information must be recalled after a distraction-filled interval or after a long delay. In these situations, patients will remember fewer items than will their healthy counterparts.

Amnesic patients are impaired on tasks of new learning, regardless of whether memory is tested by free recall, recognition (e.g., presenting an item and asking whether it was previously encountered), or cued recall (e.g., asking for recall of an item when a hint is provided). In addition, the memory impairment involves not just difficulty in learning about specific episodes and events that occurred in a certain time and place (episodic memory), but also difficulty in learning factual information (semantic memory). Finally, the memory deficit is present regardless of the sensory modality in which information is presented (visual, auditory, olfactory, and so on).

Retrograde Amnesia

In addition to impaired new learning, amnesia also impairs memories that were acquired before the onset of amnesia. This type of memory loss is referred to as retrograde amnesia. Retrograde amnesia is usually temporally graded. That is, information acquired in the distant past (remote memory) is spared relative to more recent memory. The extent of retrograde amnesia can be relatively short and encompass only 1–2 years, or it can be more extensive and cover a much longer time. For example, an amnesic patient can have retrograde amnesia covering the previous one or two decades. In contrast, memories for the facts and events of childhood and adolescence can be intact. The severity and extent of retrograde amnesia is determined by the locus and extent of damage. Patients with restricted hippocampal damage have a limited retrograde amnesia covering a few years prior to the onset of amnesia. Patients with large medial temporal lobe damage have extensive retrograde amnesia covering decades.

The sparing of remote memory relative to more recent memory illustrates that the brain regions damaged in amnesia are not the permanent repositories of long-term memory. Instead, memories undergo a process of reorganization and consolidation after learning, during which time the neocortex becomes more important. During the process of consolidation, memories are vulnerable if there is damage to the medial temporal lobe or diencephalon. After sufficient time has passed, storage and retrieval of memory no longer require the participation of these brain structures. Memory is at that point supported by neocortex. The areas of neocortex important for long-term memory are thought to be the same regions that were initially involved in the processing and analysis of what was to be learned. Thus, the neocortex is always important, but the structures of the medial temporal lobe and diencephalon are also important during initial learning and during consolidation.

Spatial Memory

Discussions of amnesia have focused especially on the status of spatial memory because of the discovery of 'place cells' in the rodent hippocampus and the possible importance of the hippocampus in forming spatial maps. In human amnesia, spatial memory is impaired along with other forms of declarative memory. Patients have difficulty acquiring new spatial knowledge, and they are impaired in remembering recently acquired spatial knowledge. However, as is the case with other forms of declarative memory, remote spatial knowledge is intact. One well-studied patient with large medial temporal lobe lesions and severe amnesia (E.P.) was able to mentally navigate his childhood neighborhood, use alternate and novel routes to describe how to travel from one place to another, and point correctly to locations in the neighborhood while imagining himself oriented at some other location. These findings show that the medial temporal lobe is not needed for the long-term storage of spatial knowledge and does not maintain a spatial layout of learned environments that is necessary for successful navigation. Accordingly, the available data support the view that the hippocampus and related medial temporal lobe structures are involved in learning new facts and events, both spatial and nonspatial. Further, these structures are not repositories of long-term memory, either spatial or nonspatial.

Nondeclarative Memory

It is a striking feature of amnesia that many kinds of learning and memory are spared. Memory is not a unitary faculty of the mind but is composed of many parts that depend on different brain systems. Amnesia impairs only declarative memory and spares nondeclarative memory. Nondeclarative memory refers to a heterogeneous collection of abilities, all of which afford the capacity to acquire knowledge nonconsciously. Nondeclarative memory includes motor skills, perceptual and cognitive skills, priming, adaptation-level effects, simple classical conditioning, and habits, as well as phylogenetically early forms of experience-dependent behavior such as habituation and sensitization. In these cases, memory is expressed through performance rather than recollection, and performance does not require reflection on the past or even the knowledge that memory is being influenced by past events. For example, in the case of motor skills, one can learn how to ride a bicycle but be unable to describe what has been learned, at least not in the same sense that one might recall riding a bicycle on a particular day with a friend. Perceptual skills include such things as reading mirror-reversed print and searching a display quickly to find a hidden letter. In formal experiments, amnesic patients acquire perceptual skills at the same rate as individuals with intact memory, even though the patients may not remember performing the task.

Priming refers to an improved ability to identify a word or other item as a result of its prior presentation. For example, suppose that a line drawing of a dog, hammer, and airplane are presented in succession, with the instruction to name each item as quickly as possible. Typically, about 800 ms are needed to produce each name aloud. If in a later test these same pictures are presented intermixed with new drawings, the new drawings will still require about 800 ms to name, but now the dog, hammer, and airplane are named about 100 ms more quickly. The improved naming time occurs independent of whether one remembers having seen the items earlier. Furthermore, amnesic patients exhibit this effect at full strength, despite having a poor memory of seeing the items earlier. Priming effects of this kind can persist across intervals as long as several weeks. In formal experiments, severely amnesic patients had intact priming for recently presented words, even when the patients performed at guessing levels (50% correct) on tests that asked them to recognize which words were presented previously and which were not. This result shows that priming is fully independent of declarative memory.

Adaptation-level effects refer to changes in judgments about stimuli (e.g., their heaviness or size) that are caused by recent experience. For example, experience with light-weighted objects subsequently causes other objects to be judged heavier than they would be if the light-weighted objects had not been presented. Amnesic patients show this effect to the same degree as healthy individuals, though they have difficulty remembering what they have done.

Classical conditioning refers to the development of an association between a previously neutral stimulus and an unconditioned stimulus. One of the best-studied examples of classical conditioning in humans is eyeblink conditioning. In a typical conditioning procedure, a tone repeatedly precedes a mild air puff directed to the eye. After a number of pairings, the tone comes to elicit an eyeblink in anticipation of the air puff. Amnesic patients acquire the tone-air puff association at the same rate as healthy individuals do. In both groups, awareness of the temporal contingency between the tone and the air puff is unrelated to successful conditioning. Simple classical conditioning, where the tone overlaps with the air puff and terminates with it, is dependent on the cerebellum.

Habit learning refers to the gradual acquisition of associations between stimuli and responses, such as learning to make one choice rather than another. Habit learning depends on the neostriatum (basal ganglia). Many tasks can be acquired either declaratively, through memorization, or nondeclaratively, as a habit. For example, healthy individuals will solve many trial and error learning tasks quickly by simply engaging declarative memory and memorizing which responses are correct. In this circumstance, amnesic patients are disadvantaged. However, tasks can also be constructed that defeat memorization strategies, for example, by making the outcomes on each trial probabilistic. In such a case, amnesic patients and healthy individuals learn at the same

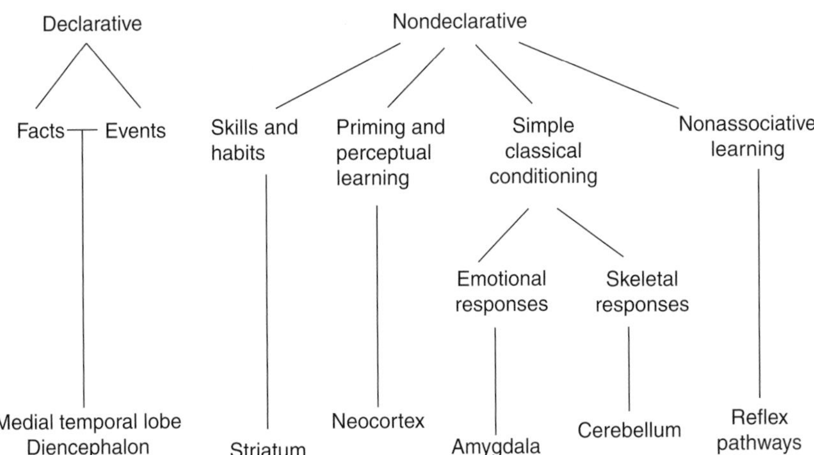

Figure 2 Classification of mammalian long-term memory systems. The taxonomy lists the brain structures thought to be especially important for each form of declarative and nondeclarative memory. In addition to the central role of the amygdala in emotional learning, it is able to modulate the strength of both declarative and nondeclarative learning.

gradual rate. It is also true that severely amnesic patients who have no capacity for declarative memory can gradually acquire trial-and-error tasks, even when the task can be learned declaratively by healthy individuals. In this case they succeed by engaging habit memory.

This situation is nicely illustrated by the eight-pair concurrent discrimination task, which requires individuals to learn the correct object for each of eight object pairs. Healthy individuals learn all eight pairs in a single test session. Severely amnesic patients acquire this same task over many weeks, even though at the start of each session they cannot describe the task, the instructions, or the objects. It is known that this task is acquired at a normal (slow) rate by monkeys with medial temporal lobe lesions and that monkeys with lesions of the neostriatum (basal ganglia) are impaired. Thus, humans appear to have a robust capacity for habit learning that operates outside awareness and independent of the medial temporal lobe structures that are damaged in amnesia.

These examples illustrate that nondeclarative memory is distinct from declarative memory. It is spared in amnesia, and it operates outside awareness. Nondeclarative forms of memory depend variously on the neostriatum, the amygdala, the cerebellum, and on processes intrinsic to neocortex (**Figure 2**).

Summary

The study of amnesia has illuminated the nature of memory disorders and has also led to a better understanding of the neurological foundations of memory. Experimental studies in patients, neuroimaging studies of healthy volunteers, and related studies in experimental animals continue to reveal insights about what memory is and how it is organized in the brain. As more is learned about the neuroscience of memory, about how memory works, more opportunities will arise for achieving better diagnosis, treatment, and prevention of diseases and disorders that affect memory.

See also: Animal Models of Amnesia; Episodic Memory; Functional Amnesia; Hippocampus: Computational Models; Memory Disorders; Recognition Memory; Semantic Memory.

Further Reading

Bayley PJ, Frascino JC, and Squire LR (2005) Robust habit learning in the absence of awareness and independent of the medial temporal lobe. *Nature* 436: 550–553.

Bayley PJ, Gold JJ, Hopkins RO, et al. (2005) The neuroanatomy of remote memory. *Neuron* 46: 799–810.

Eichenbaum H and Cohen NJ (2001) *From Conditioning to Conscious Recollection: Memory Systems of the Brain.* New York: Oxford University Press.

Gordon B (1995) *Memory: Remembering and Forgetting in Everyday Life.* New York: Mastermedia.

Milner B, Squire LR, and Kandel ER (1998) Cognitive neuroscience and the study of memory. *Neuron* 20(3): 445–468.

Schacter DL (1996) *Searching for Memory: The Brain, the Mind, the Past.* New York: BasicBooks.

Squire LR and Kandel ER (2000) *Memory: From Mind to Molecules.* New York: W. H. Freeman.

Squire LR and Schacter DL (eds.) (2002) *The Neuropsychology of Memory,* 3rd edn. New York: Guilford Press.

Squire LR, Stark CE, and Clark RE (2004) The medial temporal lobe. *Annual Review of Neuroscience* 27: 279–306.

Teng E and Squire LR (1999) Memory for places learned long ago is intact after hippocampal damage. *Nature* 400: 675–677.
Wais PE, Wixted JT, Hopkins RO, and Squire LR (2006) The hippocampus supports both the recollection and the familiarity components of recognition memory. *Neuron* 49: 459–468.

Relevant Websites

http://www.alzforum.org – Alzheimer's Research Forum.
http://whoville.ucsd.edu – Home page of Larry R. Squire.
http://www.nia.nih.gov – National Institute on Aging, Alzheimer's Disease Centers.

AMPA Receptor Cell Biology/Trafficking

P G R Hastie and J M Henley, University of Bristol, Bristol, UK

α-Amino-3-hydroxy-5-methyl-4-isoxazole propionic acid receptors (AMPARs) are ligand-gated ion channels that are crucial for normal brain function. They mediate nearly all fast excitatory synaptic transmission in the mammalian central nervous system (CNS). Changes in functional postsynaptic AMPARs mediate the two main forms of synaptic plasticity that are believed to underlie learning and memory. Long-term potentiation (LTP) involves the activity-dependent recruitment of AMPARs to the synapse and a concurrent increase in AMPA-mediated transmission. Conversely, long-term depression (LTD) is a decrease in synaptic AMPAR function. Given their importance, AMPARs are stringently regulated and the mechanisms of this regulation have been, and continue to be, the subject of intense study.

Subunit Composition of AMPARs

AMPARs are tetrameric assemblies of combinations of four individual subunits (glutamate receptor (GluR)1–4). Based on the length of the intracellular C-terminal domains of their predominant splice isoforms, GluR2 and GluR3 are classified as short and GluR1 and GluR4 are classified as long subunits. As discussed here, short and long subunits bind to different sets of intracellular proteins and are trafficked differently.

AMPAR Trafficking

There are three fundamental ways of targeting proteins in polarized cells:

1. Selective delivery in which cargo is sent directly to the target destination.
2. Selective fusion – exocytic machinery packaged with the cargo fuses with a particular membrane subregion.
3. Selective retention – cargo is removed from inappropriate membranes.

In general, dendritic sorting most likely relies on selective delivery because green fluorescent protein (GFP) attached to proteins containing dendritic sorting signals traffics directly to the somatodendritic compartment. Live imaging with GFP-labeled GluR1 and GluR2 suggests that AMPARs can be independently transported intracellularly rather than in clusters. This is consistent with a process of selective fusion acting as the regulated step in synaptic delivery.

Endoplasmic Reticulum to Synapse Trafficking Pathway

AMPARs are delivered to the synapse via the secretory pathway (**Figure 1**). A YxxLxxR motif targets the translated proteins to the endoplasmic reticulum (ER) for folding, assembly of dimers, and enzyme modification. GluR1 is able to rapidly exit the ER, but RNA editing in GluR2 delays the exit of this subunit. Therefore, a large pool of GluR2 is held available in the ER, ready to combine with other subunits, and thus most hippocampal AMPARs contain GluR2. An extreme N-terminal signal may further promote GluR1 exit from the ER because truncation mutants lacking these nine amino acids co-localize with ER markers despite forming heteromers with GluR2. It is believed that the assembly of dimers into tetramers (a dimer of dimers) leads to masking of the retention signals, allowing only correctly folded, desensitized AMPARs to exit the ER.

AMPARs next enter the *trans*-Golgi network (TGN) and become fully glycosylated. Passage through this pathway is dependent on the C-terminus of the receptor, a site for interaction with a number of proteins, particularly those containing postsynaptic density protein (PSD-)95/Dlg-A/ZO-1 (PDZ) domains. PDZ domains are protein interaction modules that, depending on the type of PDZ motif, can bind selectively to a range of peptide ligand sequences present on target proteins.

Anterograde trafficking of AMPARs occurs along microtubules. One candidate for mediating forward trafficking is mLIN-10/mint1/X11, the *Caenorhabditis elegans* ortholog of which, LIN-10, is required for the correct localization of GLR-1 receptors. The microtubule-associated kinesin motor protein KIF5 binds to the AMPAR binding protein (ABP) glutamate receptor interacting protein 1 (GRIP1) and is involved in selective dendritic delivery. Further evidence that AMPARs travel along microtubules comes from another GRIP1-interacting partner, liprin-α/SYD2, which links to the scaffold protein GIT1 and to microtubules via KIF1A. Interference with either of these interactions specifically reduces GluR2/3 dendritic clustering. In the case of GluR1, linkage to microtubules can occur through a synapse-associated protein (SAP-)97 interaction with KIF1Bα.

AMPAR synaptic delivery to and insertion at dendritic spines are two distinct processes, employing the

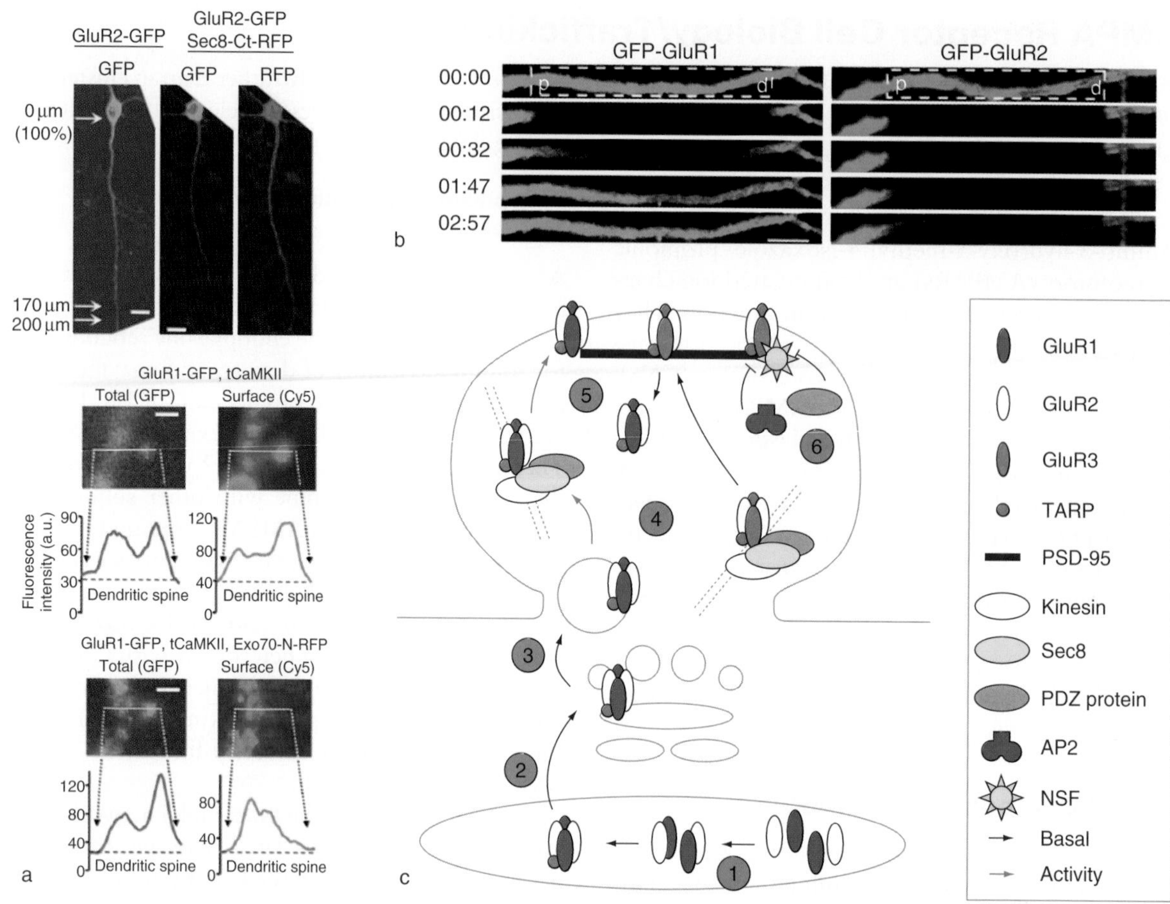

Figure 1 Delivery of AMPARs: (a) AMPAR synaptic delivery and insertion as two distinct processes; (b) differential transport rates of AMPAR subunits; (c) AMPAR synaptic delivery steps. In (a), a dominant negative C-terminal construct of the exocyst component Sec8 prevents the delivery of GluR2-GFP to the distal dendrites (upper panel). A dominant negative construct of another exocyst component, Exo70, prevents the surface insertion of CaMKII phosphorylated GluR1-GFP, causing the accumulation of correctly delivered AMPARs within the spine. In (b), 40 μm segments of cultured hippocampal dendrites expressing either GluR1-GFP or GluR2 were photobleached. Fluorescence recovery from both distal (d) and proximal (p) dendritic ends was greatest for GluR1-GFP, suggesting faster rates of transport and/or greater surface mobility for this subunit. (c) Shows the events in the AMPAR synaptic delivery: (1) individual AMPAR subunits enter the ER and form dimers and then a dimers-of-dimers; Q/R editing in GluR2 prevents GluR2 homomer assembly and thus most AMPARs contain GluR2; (2) ER exit is dependent on the correct assembly of the tetrameric complex, which must be able to enter a desensitized state to exit the ER; TARPs are also required for ER exit; (3) high-mannose sugars on the extracellular surface of the AMPAR are replaced by complex carbohydrates in the Golgi; AMPARs most likely pass through a sorting endosome before being delivered to the surface, with GluR1-containing AMPARs exiting this compartment more slowly than GluR1-lacking AMPARs; (4) anterograde trafficking of AMPARs occurs along microtubules as complexes containing either GluR1-K1F1b-SAP97 or GluR2-KIF5-GRIP; (5) synaptic activity is required for surface delivery of GluR1, which is accompanied by the delivery of slot proteins (e.g., PSD-95) to the synapse; GluR1-containing AMPARs are then replaced by constitutively cycling GluR2/3 heteromers; (6) delivery of AMPARs to the synapse is dependent on an interaction between GluR2 and NSF; this interaction prevents internalization due to AP2 and PICK1 binding to the receptor, thus stabilizing AMPARs at the synapse. AMPAR, α-amino-3-hydroxy-5-methyl-4-isoxazole propionic acid receptor; a.u., arbitrary units; CaMKII, calmodulin-dependent protein kinase II; ER, endoplasmic reticulum; GFP, green fluorescent protein; GluR, glutamate receptor; NSF, *N*-ethyl maleimide sensitive factor; PDZ, PSD-95/Dlg-A/ZO-1; PICK1, protein interacting with C kinase 1; PSD-95, postsynaptic density protein 95; Q/R, glutamine to arginine; RFP, red fluorescent protein; TARPS, transmembrane AMPAR interacting proteins. (a, middle and bottom) From Gerges NZ, Backos DS, Rupasinghe CN, Spaller MR, and Esteban JA (2006) Dual role of the exocyst in AMPA receptor targeting and insertion into the postsynaptic membrane. *EMBO Journal* 25: 1623–1634. (b) Adapted from Perestenko PV and Henley JM (2003) Characterisation of the intracellular transport of GluR1 and GluR2 α-amino-3-hydroxy-5-methyl-4-isoxazole propionic acid receptor subunits in hippocampal neurons. *Journal of Biological Chemistry* 278: 43525–43532.

exocyst, a multisubunit complex required for vectorial targeting of a subset of secretory vesicles. The exocyst complex acts to fuse vesicles to the plasma membrane. Proteins on one end recognize Rab GTPases on secretory vesicles, whereas proteins on the other end localize to the plasma membrane, probably by binding to Rho GTPases that mediate the activity of the cytoskeleton. The proteins in between may act as a structural core to transmit information between the two membranes. A dominant negative C-terminal construct of the exocyst component Sec8 prevents the delivery of GluR2-GFP to distal dendrites. A dominant negative construct of another exocyst component, Exo70, prevents the surface insertion of GluR2-GFP and calmodulin-dependent protein kinase (CaMK)II-phosphorylated GluR1-GFP, causing the accumulation of otherwise correctly delivered AMPARs within the spine.

Roles of Short and Long AMPAR Subunits

The long and short AMPAR subunits appear to serve different functions in AMPAR surface expression, and a subunit-specific AMPAR trafficking model has been proposed to explain how LTP is expressed and maintained. To summarize, the forward trafficking characteristics of the long-tail GluR1 subunit are dominant and AMPARs containing this subunit are inserted into the synapse during periods of activity. AMPARs lacking GluR1 default to the short-tail GluR2 subunit characteristics of constitutive cycling. GluR2-mediated cycling replaces synaptic AMPARs without changing synaptic efficacy. Therefore, GluR1-containing AMPARs are inserted during the initial phase of LTP, and these are then replaced by GluR2/3 heteromers. This hypothesis is based on the following experimental observations.

Constitutive Cycling of AMPARs Is *N*-Ethyl Maleimide-Sensitive Factor-Dependent

Plasma membrane insertion of $\text{GluR2}_{\text{short}}$ is much more rapid than for $\text{GluR1}_{\text{long}}$ under basal conditions, and the delivery of GluR2 continues when neuronal activity is blocked. However, surface expression of GluR2 is short lived; the blockade of exocytic events leads to a rapid run-down of AMPAR responses. GluR2-containing AMPAR insertion and/or stabilization at the synapse requires the hexameric adenosine triphosphatase (ATPase) *N*-ethyl maleimide-sensitive factor (NSF). NSF binds the GluR2 C-terminus upstream of the PDZ binding site between residues Lys844 and Gln853. Blocking this interaction with a peptide (pep2m) corresponding to the interaction site on GluR2 causes a decrease in surface GluR2, indicating the rapid internalization and recycling back to the membrane of GluR2-containing receptors. Pep2m also prevents *de novo* LTD. However, pep2m also blocks an overlapping binding site for the clathrin adaptor protein AP2, which interacts with GluR1–3 subunits and is required for AMPAR endocytosis. A more specific peptide, pep-R845A, which blocks only the GluR2–NSF interaction, leads to the loss of functional AMPARs but has no effect on LTD.

Activity-Dependent Insertion of AMPARs

Insertion of $\text{GluR1}_{\text{long}}$-containing AMPARs at the hippocampal synapses requires synaptic activity and occurs following *N*-methyl-D-aspartate receptor (NMDAR)-dependent LTP induction. Receptor phosphorylation is critical in the process, and in older neurons CaMKII activation is required, whereas protein kinase A (PKA) activation is more important in younger neurons. Although GluR1 is responsible for the majority of activity-dependent insertion in adults, GluR4 and the long C-terminal variant of GluR2 contribute significantly in younger animals when the expression of these subunits is prevalent.

In GluR1, the PDZ protein binding region comprises a –TGL motif. This is a type I PDZ ligand and therefore binds a different subset of proteins than the type II –SVKI motif present in short AMPAR subunits. SAP-97 directly interacts with the GluR1 PDZ ligand, and the overexpression of SAP-97 increases AMPAR surface delivery; but truncation mutants eliminating GluR1 binding do not affect AMPAR clustering and mice lacking the –TGL motif have normal levels of synaptic GluR1, suggesting that SAP-97 may be important in relieving intracellular retention interactions at this site.

There is close similarity between the signal transduction pathways for LTP and long-tailed AMPAR synaptic insertion. Activation of the Ras GTPase pathway during LTP leads to specific synaptic insertion of long-tailed AMPARs. Low-level activation leads to extracellular signal-regulated kinase kinase (MEK) and p42/44 mitogen-activated protein kinase (MAPK) signaling, leading to delivery of $\text{GluR2}_{\text{long}}$. Higher levels of activity stimulate the PI3K and protein kinase B (PKB) pathways leading to GluR1 delivery. This action of $\text{GluR2}_{\text{long}}$ may explain the persistent LTP in mice lacking the GluR1 PDZ ligand.

Maintenance of Increased Synaptic AMPARs Requires Slot Proteins

An important aspect of this model, which depends on an initial increase in surface expression due to long subunits but the maintenance of the increased expression via short subunits, is the need for marker or slot proteins to mark the potentiated synapses. Once trafficked to the PSD, these slot proteins remain there to maintain the increased expression of short subunit AMPARs unless a signal such as LTD induction triggers their removal. They therefore can act as placeholders, defining the number of AMPAR insertion sites at the synapse. If these were not added to the synapse, the internalized GluR1 could not be replaced by GluR2/3 because there would be no scaffolding structure for these new short AMPARs to attach to.

AMPARs Interact with a Complex Network of Scaffold Proteins

Cytoskeletal interactions at the synapse are both dynamic and complex, and there is likely to be redundancy between many of the binding partners. The most direct links with the actin cytoskeleton occur through long AMPAR subunits. Binding of protein 4.1N to the C-terminus of GluR1 and GluR4 upstream of the phosphorylation sites controls the surface expression of these subunits, possibly mediated through SAP-97 because the removal of the 4.1N binding site on this protein reduces AMPAR clustering. SAP-97 also links AMPARs to the actin cytoskeleton through myosin VI, although this minus-end-directed motor protein is involved in removal of AMPARs. A further interaction is provided by reversion-induced Lin11/rat Isl-1/Mec3 (LIM) domain-containing protein (RIL). Unusually, it is the LIM domain that specifically binds GluR1, whereas the PDZ domain links to the actin cytoskeleton via α-actinin, increasing the surface accumulation of AMPARs at the spines.

PSD-95 Regulates Synaptic AMPAR Content

Postsynaptic density protein of 95 kDa (PSD-95), also known as synaptic-associated protein 90 (SAP-90), represents 2.3% of the protein mass of the PSD (equating to $\sim$300 PSD-95 molecules per synapse). This protein is a member of the membrane-associated guanylate kinase (MAGUK) family due to the guanylate kinase (GK) homology domain in the C-terminus. The C-terminus also contains a src homology (SH3) domain. The N-terminus contains a consensus sequence for palmitoylation. Three PDZ domains make PSD-95 an effective scaffold protein.

PSD-95 is required for AMPAR synaptic localization, and there is a correlation between PSD-95 and AMPAR accumulation at the synapse. Expression of PSD-95-GFP increases the synaptic delivery of GluR1, thus occluding LTP, and increases the AMPAR-mediated currents in experience-deprived animals. Furthermore, PSD-95 gene transcription is upregulated by neuronal activity. An attractive model is that activity-dependent delivery of PSD-95 occurs prior to, and enables, the recruitment and insertion of GluR1-containing AMPARs and then acts as the slot protein for subsequent GluR2/3 AMPARs.

Transmembrane AMPAR Interacting Proteins Are Required to Anchor AMPARs to PSD-95

By definition, the five members of the transmembrane AMPAR interacting protein (TARP) family (γ2/stargazin, γ3, γ4, γ7, and γ8) bind AMPARs. This separates them from closely related proteins such as the γ1 calcium channel subunit. Mutations in γ2/stargazin lead to ataxia in waggler mice, which lack surface AMPARs on cerebellar granule cells. It has been suggested that TARPs constitute auxillary AMPAR subunits and that a type I PDZ domain at the C-terminus of TARPs enables interaction with PSD-95, SAP-97, and other MAGUKs. TARPs are involved in both AMPAR surface expression and lateral movement of AMPARs in the membrane to the synapse via an interaction with PSD-95. Whereas stargazin/γ2 is highly expressed in the cerebellum, γ8 is most highly expressed in the hippocampus and is not mutated in the waggler mice; this explains why they show locomotor but not learning deficiencies.

GRIP and ABP Anchor Short AMPAR Subunits

GRIP1 and ABP both contain a maximum of seven PDZ domains and interact with short AMPAR subunits through PDZ domains 4–5 and 3, 5, and 6, respectively. The number of PDZ domains and their ability to form homo- and heterodimers make GRIP1 and ABP ideal candidate scaffolds. Both proteins are enriched in dendritic spines and it is proposed that they anchor AMPARs at the synapse. Ser880Ala mutation in the GluR2 PDZ domain selectively prevents its interaction with GRIP/ABP, leading to a reduction in synaptic accumulation of GluR2 over time. Synaptic targeting was not affected, suggesting GRIP/ABP is specifically required for the synaptic retention of AMPARs. Thus, GRIP and ABP possibly work in tandem with PSD-95 to provide slot proteins for AMPAR insertion and retention at the synapse.

Neuronal Activity Regulated Pentraxin-Induced Clustering of AMPARs

The N-terminus of GluR1 is required for ER exit and synaptic targeting. Neuronal activity regulated pentraxin (NARP), an immediate-early gene released from the stimulated hippocampus, interacts at an extracellular site causing the clustering of GluR1–3 in heterologous cells and co-immunoprecipitates with these subunits, as well as clustering GluR1 in cultured spinal neurons. Endogenous NARP action varies between neurons. Hippocampal axons only secrete NARP at contacts with interneurons, and the exogenous application of NARP fails to cause clustering on pyramidal cells. Spinal neurons cluster AMPARs and NMDARs on cultured hippocampal interneurons where contacts are made on the dendritic shaft but not on pyramidal cells where contacts are made at dendritic spines.

AMPAR Endocytosis

The best understood model for internalizing proteins in neurons is clathrin-mediated endocytosis (CME). Hypertonic sucrose inhibits CME and prevents both agonist-induced GluR1 internalization and insulin-induced GluR2 internalization. Blocking dynamin activity prevents the internalization of GluR1 and GluR2 and blocks LTD. A peptide corresponding to the SH3 domain of amphiphysin also blocks LTD. Disruptions to specific components of the CME machinery (e.g., CPG2 and huntingtin interacting protein 1 (HIP1)) also caused reduced AMPAR internalization. Increased colocalization and interaction of AP2 with AMPARs following agonist treatment suggested AP2 recruits AMPARs to coated pits. This is supported by evidence that the proteins directly interact through the β-adaptin subunit of AP2. This interaction does not occur with the AP1 subunit responsible for cargo recruitment at intracellular sites. It is not clear if non-CME AMPAR endocytosis pathways exist, although lipid rafts, often associated with caveolar structures, are involved in stabilizing surface AMPARs.

Lateral Diffusion to Extrasynaptic Clathrin-Mediated Endocytosis Sites

The development of ecliptic GFP or pHlourins, which fluoresce only at pH > 7.0 allowed the measurement of dynamic changes to the surface expression of proteins. Because intracellular compartments are more acidic than the extracellular environment, super eclipitic phluorin (SEP) fluorescence is suppressed until the protein is surface expressed. This technology was used to show that, in response to NMDAR activation, endocytosis of synaptic GluR2 is preceded by the removal of extrasynaptic receptors (**Figure 2**). An obvious reason for this is that the endocytosis machinery is located at extrasynaptic sites. Quantum dot-tracking experiments show that the proportion of mobile GluR2 increases following neuronal activity, representative of increased lateral diffusion from synaptic anchors. This suggests that, rather than the regulation of endocytosis itself, the limiting step in generating LTD through internalization is the release of AMPARs from the PSD.

Protein Interacting with C Kinase 1 Releases AMPARs from the Synapse

Protein interacting with C kinase (PICK1) interacts with a number of neuronal receptors, channels, and transporters. A single PDZ domain enables an interaction with short AMPARs leading to LTD. This occurs because PICK1 competes for the GRIP/ABP binding site on GluR2, uncoupling the AMPARs from the synaptic scaffold. This is controlled by protein kinase C (PKC) phosphorylation of GluR2 at Ser880, which prevents GRIP/ABP but not PICK1 binding. Phosphorylation is facilitated by PICK1 itself, which binds PKCα to target the kinase to AMPARs. PKCα binding removes an intramolecular interaction, exposing the BAR domain of PICK1. A second phosphorylation switch with the same function has recently been found at Tyr876, and this residue must be Src phosphorylated in response to drug treatment for internalization to occur. NSF hydrolysis of adenosine triphosphate (ATP) disrupts the PICK1–GluR2 interaction in complexes containing α-SNAP. This stabilizes the AMPARs at the surface. β-SNAP, however, prevents this dissociation and causes receptor internalization. Thus, in addition to actively inserting AMPARs at the synapse, NSF activity prevents their internalization. The presence of a BAR domain in PICK1 suggests this protein can recruit AMPARs to clathrin-coated vesicles (CCVs) in constitutive endocytosis in which AP2 does not seem to be involved.

Ca^{2+} binding to PICK1 increases the affinity of the GluR2 interaction. This has obvious implications for LTD, in which Ca^{2+} concentration is elevated in spines. The increase in affinity may be another switch in the balance between constitutive cycling and long-lasting endocytosis of AMPARs. PICK1 is not the only Ca^{2+}-sensing molecule implicated in the endocytosis of AMPARs during LTD. The neuronal calcium-sensor hippocalcin binds AP2 in a Ca^{2+}-dependent manner, and the Ca^{2+}-sensing region of hippocalcin is required for the generation of LTD. It is thought that Ca^{2+} binding, which exposes a myristyl tail in hippocalcin, recruits AP2 to the membrane, enabling AMPAR sorting into CCVs. Hippocalcin also complexes with transferrin receptors but in a Ca^{2+}-independent manner, suggesting the function of this protein is modified for a specific role in AMPAR endocytosis.

Degradation of Synaptic Scaffolds Accompanies AMPAR Internalization

PSD-95 is ubiquitinated following synaptic NMDA activation. Truncated PSD-95 mutants lacking the ubiquitination motif prevent NMDA-induced GluR2 internalization. One function of ubiquitination is as a signal for proteasomal protein degradation, suggesting that the synaptic scaffold may have to be dismantled for AMPAR release. Although this has not yet been shown for the proteasome pathway, calpain cleavage of PSD-95, SAP-97, and GRIP1 has been shown. AMPARs are themselves cleaved at the C-terminus by calpain, allowing their release from the PSD.

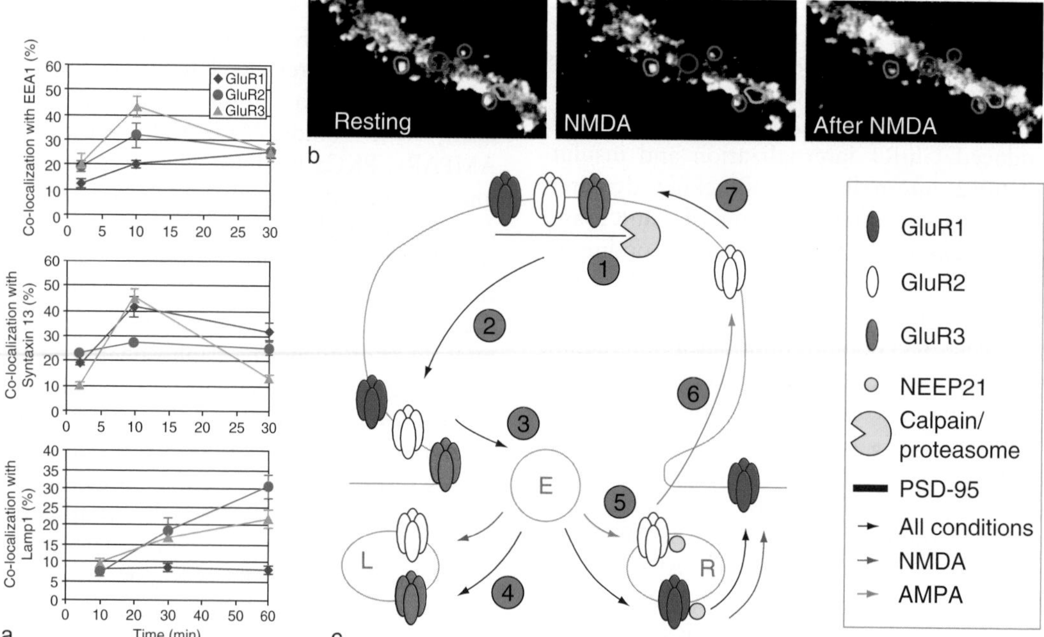

Figure 2 AMPAR internalization and recycling: (a) experiments using recombinant HA-tagged AMPAR subunits; (b) virally expressed pHluorin-GluR2 labeling resting, with NMDA, and after NMDA; (c) main steps in AMPAR recycling. In (a), experiments using recombinant HA-tagged AMPAR subunits demonstrate that they are differentially sorted on internalization. All HA-tagged AMPAR subunits are internalized following the application of 50 μM NMDA for 8 min. All subunits colocalize with the early endosome marker EEA1 after 10 min, but only GluR1 continues to accumulate in this compartment up to 30 min following NMDA treatment. Both GluR2 and GluR3 pass through the syntaxin 13-positive recycling endosome and accumulate in the Lamp1-positive late endosome prior to degradation, whereas GluR1 remains in the recycling endosome. In (b) under resting conditions virally expressed pHluorin-GluR2 labels both punctate spines (red, purple) and diffuse shaft regions (blue). Treatment with 50 μM NMDA causes an immediate loss of the diffuse staining, representative of GluR2 endocytosis. This is followed by the loss of punctate staining after approximately 10 min. (c) Shows a schematic of the main events in AMPAR recycling: (1) AMPARs are released from the PSD by switching binding partners accompanied by degradation of the PSD; (2) receptors exit the spine by lateral diffusion to sites of clathrin endocytic machinery; (3) clathrin-mediated exocytosis transports AMPARs to the early endosome; (4) under all conditions, GluR3 is sent to the late endosome prior to degradation, and GluR2 enters this pathway after NMDA treatment but is sent to the recycling endosome following AMPA treatment; (5) GluR1 constitutively enters the recycling endosome, but its rate of exit from this compartment is increased by NMDA; (6) GluR2 exit from the recycling endosome is promoted by an interaction with NEEP21; (7) GluR2 is rapidly recycled to the synapse following AMPA treatment or in the absence of activity. AMPA, α-amino-3-hydroxy-5-methyl-4-isoxazole propionic acid; E, early endosome; EEA1, early endosome marker; GluR, glutamate receptor; HA, peptide derived from human influenza haemagglutinin; L, late endosome; Lamp1, lysosome associated membrane protein 1; NEEP21, neuron-enriched endosomal protein of 21 kDa; NMDA, *N*-methyl-D-aspartate; PSD, postsynaptic density protein; R, recyclingendosome. (a) from Lee SH, Simonetta A, and Sheng M (2004) Subunit rules governing the sorting of internalized AMPA receptors in hippocampal neurons. *Neuron* 43: 221–236. (b) Adapted from Ashby MC, De La Rue SA, Ralph GS, Uney J, Collingridge GL, and Henley JM (2004) Removal of AMPA receptors (AMPARs) from synapses is preceded by transient endocytosis of extrasynaptic AMPARs. *Journal of Neuroscience* 24: 5172–5176.

Small GTPase Signaling Triggers AMPAR Endocytosis in LTD

NMDAR activation of the Rap1–p38MAPK pathway via the NR2B subunit causes internalization of GluR2/3 heteromers. Activation of the Rap2–c-Jun N-terminal kinase (JNK) pathway causes the internalization of GluR1/2 heteromers in depotentiation, which is the resetting of synapses that have undergone LTP back to the baseline. Rab5 overexpression specifically depresses AMPAR-mediated excitatory postsynaptic currents (EPSCs) through the removal of GluR1–3 from dendritic spine surfaces. A dominant negative mutant of Rab5 prevents LTD induction. Rab5 activation is likely to occur downstream of p38MAPK activation because this regulates its interaction with Rab-GDI. As would be expected from the localization of Rab5 to endosomes, these effects of overexpression occur downstream of Ser880 phosphorylation and release from the PSD.

Fate of Internalized AMPARs Is Subunit Specific

GluR1 internalizes rapidly and independently of activity, supporting the LTP model in which GluR1-containing subunits are inserted and then replaced by GluR2/3 heteromers. Overexpressed GluR2 and GluR3 are internalized on agonist treatment, leading to GluR2 being recycled, but NMDAR activation targets GluR2 to lysosomes. GluR1 is recycled and GluR3 is directed to endosomes independent of drug treatment (**Figure 2**). These subunit rules appear to be determined by the NSF binding site, with GluR2 being dominant over GluR1. NSF may promote the recycling of the receptors, and this further supports a model in which NSF activity or binding to GluR2 is reduced during LTD, allowing targeting to lysosomes. Overall, these data suggest that NMDAR activation during LTD leads to the degradation of GluR2-containing receptors, consistent with an activity-dependent depletion of AMPARs at the synapse. However, exposure to the sodium-channel blocker tetrodotoxin (TTX), which prevents action potentials and therefore evoked presynaptic glutamate release, reverses the effects of NMDA and AMPA. The physiological relevance of this is not clear, but it does highlight the importance of the prior experience of the synapse in determining the effect of plasticity protocols. It has been suggested that this might represent a form of homeostatic plasticity known as synaptic scaling, in which neuronal sensitivity is altered in line with the overall sensitivity of the network.

TTX-treated neuronal cultures have been used to show the role of proteins other than NSF involved in directing AMPARs through the endocytic pathway. Suppression of neuron-enriched endosomal protein of 21 kDa (NEEP21) reduces the rate of GluR1 recycling in response to NMDA. GluR2 internalizes to a NEEP21 positive compartment following NMDA treatment. The presence of Rab4 and syntaxin 13 in this compartment suggests that NEEP21 promotes trafficking to the recycling endosome. In neurons with suppressed NEEP21, AMPAR-mediated EPSCs are decreased, but this effect can be overridden with dynamin mutants to block endocytosis. Because LTP is blocked in the absence of NEEP21, this protein may also have a role in trafficking receptors from intracellular holding pools. The forward trafficking of GluR2 from the early endosome to the recycling endosome is mediated by the formation of a GluR2–GRIP1–NEEP21 complex. Although it is assumed that an as yet unidentified GluR1–NEEP21 complex interacting protein mediates the recycling of GluR1, the lack or low availability of such a protein may explain the retarded recycling of GluR1 to the cell surface.

Summary

The mechanisms of AMPAR trafficking, which in turn influence synaptic efficacy, are complex. However, overall patterns of AMPAR forward traffic, recruitment, and cycling have been established. The insertion of AMPARs at newly formed synapses and at synapses that have received a signal to undergo LTP depends on GluR1. Insertion of these GluR1-containing AMPARs is preceded and/or accompanied by an upregulation of slot proteins at the synapse that hold the AMPARs at the PSD until such time as they are replaced by constitutively cycling GluR2/3-containing AMPARs. During LTD, AMPARs are released from the synapse by alterations to their scaffold binding partners, which are then degraded. AMPARs are then endocytosed at sites separate from the PSD. The fate of the receptors endocytosed by LTD appears to differ. GluR2/3-containing receptors are degraded, whereas GluR1-containing receptors are slowly recycled to extrasynaptic sites. Although this model can explain many of the features of plasticity, clearly a great deal of work remains to be done to obtain a full mechanistic understanding of how AMPARs are regulated. Given the crucial role of these receptors in brain function and dysfunction, we believe that gaining this information is a goal of fundamental importance.

See also: AMPA Receptors: Molecular Biology and Pharmacology; AMPA Receptors: Disease; Long-Term Potentiation and Long-Term Depression in Experience-Dependent Plasticity; NMDA Receptors and Development; NMDA Receptors and Disease + C464; NMDA Receptors, Cell Biology and Trafficking; NMDA Receptor Function and Physiological Modulation; Receptor Trafficking; Transporter Proteins in Neurons and Glia.

Further Reading

Ashby MC, De La Rue SA, Ralph GS, Uney J, Collingridge GL, and Henley JM (2004) Removal of AMPA receptors (AMPARs) from synapses is preceded by transient endocytosis of extrasynaptic AMPARs. *Journal of Neuroscience* 24: 5172–5176.

Ashby MC, Ibaraki K, and Henley JM (2004) It's green outside: Tracking cell surface proteins with pH-sensitive GFP. *Trends in Neuroscience* 27: 257–261.

Bredt DS and Nicoll RA (2003) AMPA receptor trafficking at excitatory synapses. *Neuron* 40: 361–379.

Choquet D and Triller A (2003) The role of receptor diffusion in the organization of the postsynaptic membrane. *Nature Reviews Neuroscience* 4: 251–265.

Collingridge GL, Isaac JT, and Wang YT (2004) Receptor trafficking and synaptic plasticity. *Nature Reviews Neuroscience* 5: 952–962.

Dingledine R, Borges K, Bowie D, and Traynelis SF (1999) The glutamate receptor ion channels. *Pharmacological Reviews* 51: 7–61.

Gerges NZ, Backos DS, Rupasinghe CN, Spaller MR, and Esteban JA (2006) Dual role of the exocyst in AMPA receptor targeting and insertion into the postsynaptic membrane. *EMBO Journal* 25: 1623–1634.

Horton AC and Ehlers MD (2003) Neuronal polarity and trafficking. *Neuron* 40: 277–295.

Lee SH, Simonetta A, and Sheng M (2004) Subunit rules governing the sorting of internalized AMPA receptors in hippocampal neurons. *Neuron* 43: 221–236.

Malenka RC and Bear MF (2004) LTP and LTD: An embarrassment of riches. *Neuron* 44: 5–21.

Palmer CL, Cotton L, and Henley JM (2005) The molecular pharmacology and cell biology of alpha-amino-3-hydroxy-5-methyl-4-isoxazolepropionic acid receptors. *Pharmacological Reviews* 57: 253–277.

Perestenko PV and Henley JM (2003) Characterisation of the intracellular transport of GluR1 and GluR2 α-amino-3-hydroxy-5-methyl-4-isoxazole propionic acid receptor subunits in hippocampal neurons. *Journal of Biological Chemistry* 278: 43525–43532.

AMPA Receptors: Disease

R C Carroll and R S Zukin, Albert Einstein College of Medicine, Bronx, NY, USA

Introduction

Excessive glutamatergic signaling can result in neuronal death, or excitotoxicity, a mechanism thought to be involved in the pathology of a number of neurological diseases and disorders. Glutamate-induced cell death is often attributed to calcium-mediated toxicity. As *N*-methyl-D-aspartate (NMDA)-type glutamate receptors are highly permeable to calcium, much attention has been focused on their possible role in many forms of neurodegeneration. However, it has become increasingly evident that ion permeation through a subpopulation of non-NMDA-type ionotropic glutamate receptors, the α-amino-3-hydroxy-5-methyl-4-isoxazole propionic acid (AMPA)-type receptors (AMPARs), is likely to be the underlying cause of cell injury in several neurological disorders, including seizure, epilepsy, and amyotrophic lateral sclerosis (ALS).

AMPARs mediate fast excitatory synaptic transmission in the central nervous system (CNS). AMPARs are formed as tetramers from a combination of four subunits (GluR1 through 4 or GluRA through D). The messenger RNAs (mRNAs) encoding AMPARs are subject to RNA editing and alternative splicing, thereby increasing the molecular diversity of the receptors. The GluR2 subunit has profound impact on the biophysical characteristics of the heteromeric AMPARs. The presence of an arginine in place of glutamine in the pore-lining M2 region of the GluR2 subunit, resulting from mRNA editing, largely prevents the permeation of Ca^{2+} and Zn^{2+} ions (**Figure 1**). In contrast, AMPARs lacking the GluR2 subunit are highly permeable to Ca^{2+} and Zn^{2+}, have distinctly fast kinetics, and exhibit a characteristic inwardly rectifying current–voltage (I-V) relation as a result of a voltage-dependent block by intracellular polyamines (**Figure 1**). The presence of GluR2 also influences the kinetics, conductance, assembly, and trafficking of AMPARs . Therefore, regulation of GluR2 expression can have a substantial impact on synaptic function and neuronal survival.

Whereas the majority of synapses in the CNS express calcium-impermeable, GluR2-containing AMPARs, synapses onto aspiny neurons, as well as glia, throughout the CNS express Ca^{2+}-permeable AMPARs. In these cells, AMPAR-mediated Ca^{2+} signaling is often rapid and can be locally restricted by fast Ca^{2+} extrusion pumps. The function of calcium flux through these AMPARs may be similar to that of NMDA receptors (NMDARs) in other synapses, including contribution to synaptic plasticity. For example, at synapses of dorsal horn neurons in the spinal cord, calcium influx through AMPARs triggers long-term potentiation (LTP) of synaptic efficacy that is associated with nociceptive plasticity and chronic pain.

In most principal neurons of the neocortex, hippocampus, amygdala, and cerebellum, GluR2-lacking, Ca^{2+}-permeable AMPARs are expressed at very low density. However, the subunit composition and Ca^{2+} permeability of AMPARs can be dynamically regulated in both a cell- and a synapse-specific manner during development and in response to neuronal activity. Changes in AMPAR subunit composition can arise as a consequence of the regulated redistribution or trafficking of AMPAR subunits as well as through alterations in protein expression, including activity-dependent local protein synthesis of AMPARs in dendrites. Alterations in AMPAR subunit composition resulting in increased Ca^{2+} permeability are also known to be induced by a number of neuronal insults and diseases including seizures, ischemia, spinal cord injury, drugs of abuse, and ALS. Investigation of the molecular mechanisms underlying activity-dependent remodeling of synaptic AMPAR subunit composition and permeability have begun to highlight the importance of Ca^{2+}-permeable AMPARs in neuronal death associated with neurological disease.

GluR2-Lacking AMPARs in Neuronal Death

Ischemia

Transient global or forebrain ischemia is the complete disruption of blood flow to the brain such as occurs in patients during cardiac arrest, open heart surgery, or profuse bleeding. Transient global ischemia can also be induced experimentally in animal models, providing a clinically relevant experimental model that is well suited to investigation of both the molecular and cellular mechanisms of this disorder and its consequences. A surprising consequence of global ischemic insult is that while all brain regions are equally subject to oxygen deprivation, neuronal death is quite selectively localized. Hippocampal CA1 pyramidal neurons are particularly vulnerable to delayed cell death and are thought to likely contribute to the observed cognitive deficits. For many years it was thought that excessive calcium influx through NMDARs was a critical player in ischemia-induced neuronal death. However, studies involving administration of NMDAR and AMPAR-specific antagonists

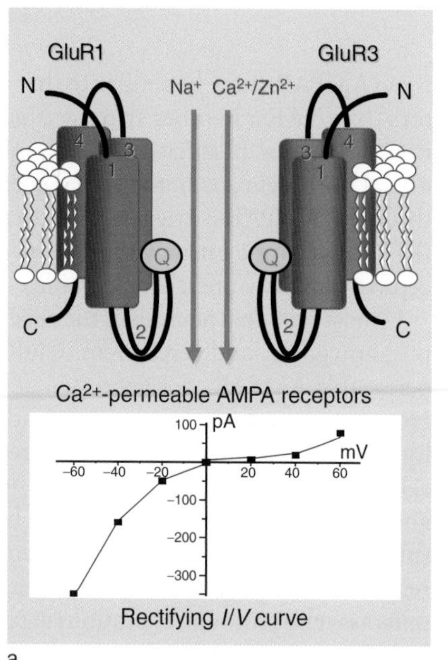

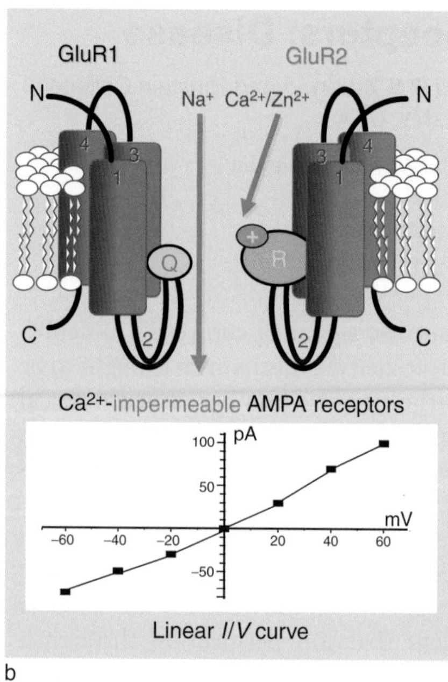

Figure 1 AMPA receptor subunit composition determines Ca^{2+} permeability. (a) GluR2-lacking AMPARs are highly permeable to Ca^{2+} and exhibit doubly rectifying current-voltage relationships. Rectification of GluR2-lacking AMPARs arises as a consequence of voltage-dependent channel block by intracellular polyamines. (b) GluR2-containing AMPARs are Ca^{2+}-impermeable and exhibit electrically linear current–voltage relationships. The presence of the GluR2 subunit in the heteromeric AMPAR channels limits Ca^{2+} and Zn^{2+} influx largely due to the presence of an arginine (R) in place of a glutamine (Q) residue at the Q/R RNA editing site. Reproduced from Liu SJ and Zukin RS (2007) Ca^{2+}-permeable AMPA receptors in synaptic plasticity and neuronal death. *Trends in Neurosciences* 30: 126–134, with permission from Elsevier.

suggest that AMPARs may play a greater role in the cell death associated with global ischemia. Consistent with this, global ischemia induces downregulation of AMPAR GluR2 expression in selectively vulnerable CA1 neurons and a switch in AMPAR phenotype from calcium impermeable to calcium permeable. It is now established that calcium influx through AMPARs is causally linked to the delayed death of hippocampal CA1 neurons.

While principal neurons of the hippocampus express GluR2-containing, Ca^{2+}-impermeable AMPARs under physiological conditions, ischemic insult reduces GluR2 mRNA and protein expression selectively in CA1 neurons. This change results in a long-lasting switch from GluR2-containing to GluR2-lacking AMPARs. By 42 h after ischemia, AMPAR excitatory postsynaptic currents (EPSCs) at CA1 synapses exhibit properties of GluR2-lacking AMPARs, including enhanced rectification, sensitivity to polyamines, and AMPAR-mediated Ca^{2+} influx. In addition to their role in mediating Ca^{2+} entry, GluR2-lacking AMPARs are thought to mediate a late rise in cytoplasmic Zn^{2+}, which reaches toxic levels by 48–72 h following transient global ischemia.

Regulated changes in synaptic AMPAR composition can occur through multiple mechanisms. Synaptic plasticity in cerebellar stellate cells results in rapid trafficking of AMPARs and selective incorporation of GluR2-containing AMPARs at synaptic sites, leading to a switch in the composition of synaptic AMPARs. In contrast, prolonged synaptic inactivity in hippocampal neurons alters AMPAR composition through increased local translation of GluR2-lacking AMPARs. Global ischemia induces a switch in AMPAR phenotype by regulation at the level of transcription. Ischemic insults induce expression of the transcriptional repressor (REST; **Figure 2**), which in turn silences expression of AMPAR GluR2 in selectively vulnerable CA1 neurons. REST binds to the promoter region of target genes, such as GluR2, and suppresses their expression through chromatin remodeling and epigenetic modifications (or marks). Acute knockdown of REST prevents the suppression of GluR2 and subsequent death of CA1 neurons following ischemia. In addition to changes in GluR2 expression, receptor trafficking and GluR2 RNA editing can be modified in response to neuronal insults. Ischemic insult enhances the clathrin-dependent internalization of GluR2-containing AMPARs and synaptic targeting of GluR2-lacking AMPARs to synapses of insulted hippocampal neurons via exocytosis, leading to a switch in the subunit composition of surface-expressed AMPARs (**Figure 3**). Global ischemia also inhibits activity of the RNA editing enzyme adenosine deaminase acting on RNA (ADAR)-2 and disrupts glutamine/arginine (Q/R) editing of GluR2,

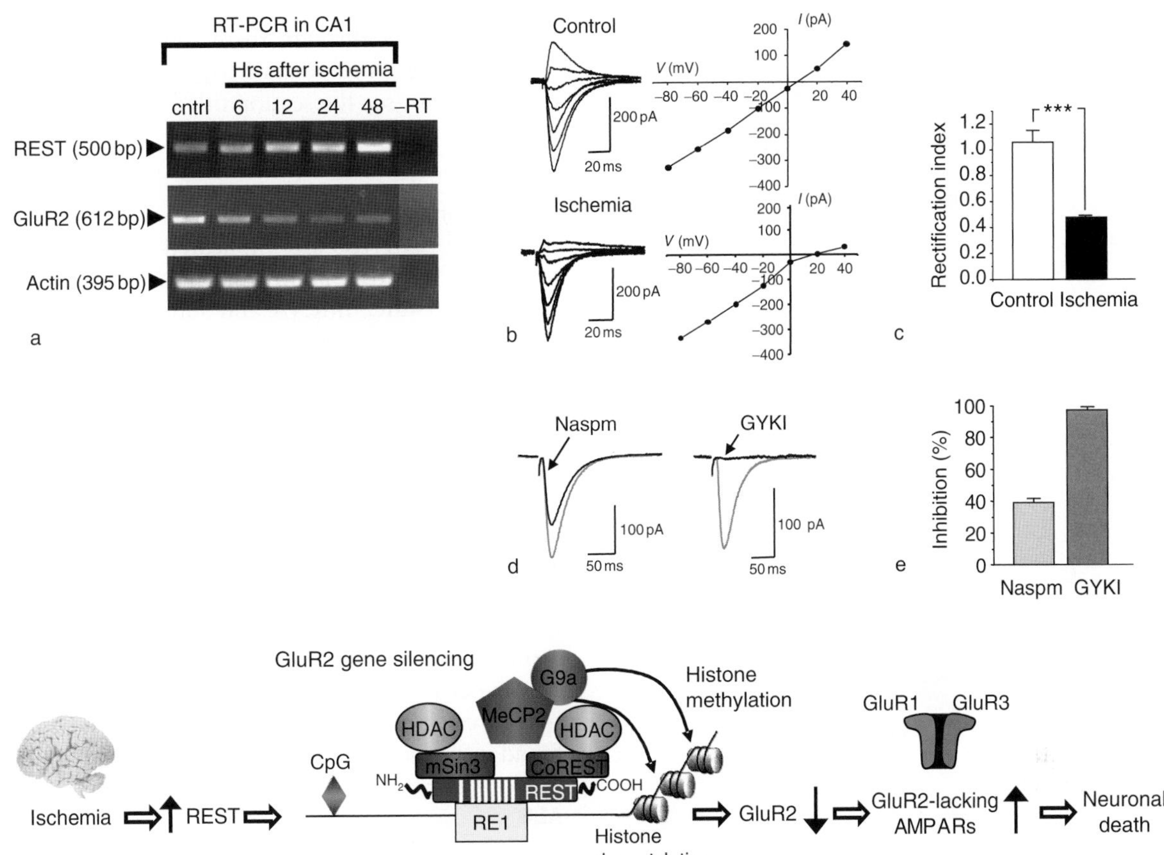

Figure 2 Ischemia silences glutamate receptor 2 (GluR2) expression through the activation of the transcriptional repressor (REST) in CA1. (a) Reverse transcription–polymerase chain reaction (RT-PCR) products amplified from REST, GluR2 and actin RNAs following global ischemia demonstrate the enhancement of REST and the suppression of GluR2 RNA levels. (b, c) Alpha-amino-3-hydroxy-5-methyl-4-isoxazole propionic acid receptors (AMPARs) at Schaffer collateral CA1 synapses in the postischemic hippocampus exhibit properties of Ca^{2+}/Zn^{2+}-permeable, GluR2-lacking AMPARs. (b) AMPA-excitatory postsynaptic currents (EPSCs) recorded from control slices exhibit a linear *I–V* relation whereas AMPA-EPSCs recorded from slices of animals at 42 h after ischemia exhibit inwardly rectifying currents. (c) The rectification index (the ratio of EPSC amplitude at +40 mV to that at 60 mV × 1.5), a measure of the extent of current/voltage relationship rectification, is reduced in ischemic versus control slices (***$p < 0.001$). (d) AMPA-EPSCs in ischemic slices are mediated in part by GluR2-lacking AMPARs as they are inhibited by Naspm, a specific antagonist of these receptors. The currents are completely blocked by the AMPA antagonist GYKI-53655 (right). (e) Summary of data in (d). (f) Scheme representing REST-dependent silencing of GluR2 gene expression in ischemic neurons. Ischemia triggers activation of the repressor REST. REST binds the RE1 element in the GluR2 promoter and silences GluR2 transcription. REST recruits co-repressors mSin3A and CoREST and histone deacetylases (HDACs), which silence gene expression via chromatin remodeling. CoREST recruits methyl-CpG binding protein 2 (MeCP2) and the histone methyltransferase G9a, which promotes DNA and histone methylation. Silencing of GluR2 expression reduces GluR2 subunit number, leading to assembly and insertion of functional, GluR2-lacking AMPARs at CA1 synapses of postischemic hippocampus. Error bars represent SEMs. Reproduced from Liu SJ and Zukin RS (2007) Ca^{2+}-permeable AMPA receptors in synaptic plasticity and neuronal death. *Trends in Neurosciences* 30: 126–134, with permission from Elsevier.

increasing calcium and zinc permeability. Direct delivery of ADAR2 restores Q/R editing and protects vulnerable neurons from cell death. Thus, a variety of processes, all of which enhance the expression of GluR2-lacking AMPARs, contribute to the susceptibility of select neurons to cell death in brain ischemia.

While it is well established that the subunit composition of AMPARs changes with transient global ischemia, understanding of the mechanisms leading to cell death has been more difficult to ascertain. Several lines of evidence link the increase in levels of Ca^{2+}-permeable, GluR2-lacking AMPARs following global ischemia directly to the death of CA1 hippocampal neurons. Reduction in the expression of GluR2 subunits results in enhanced death of pyramidal neurons even under physiological signaling conditions. Overexpression of Ca^{2+}-permeable GluR2(Q) channels or calcium-impermeable AMPARs exacerbates or protects against cell death of CA1 neurons following ischemic insult, respectively. Finally, application of two selective blockers of GluR2-lacking AMPARs, naphthylspermine and philanthotoxin, is

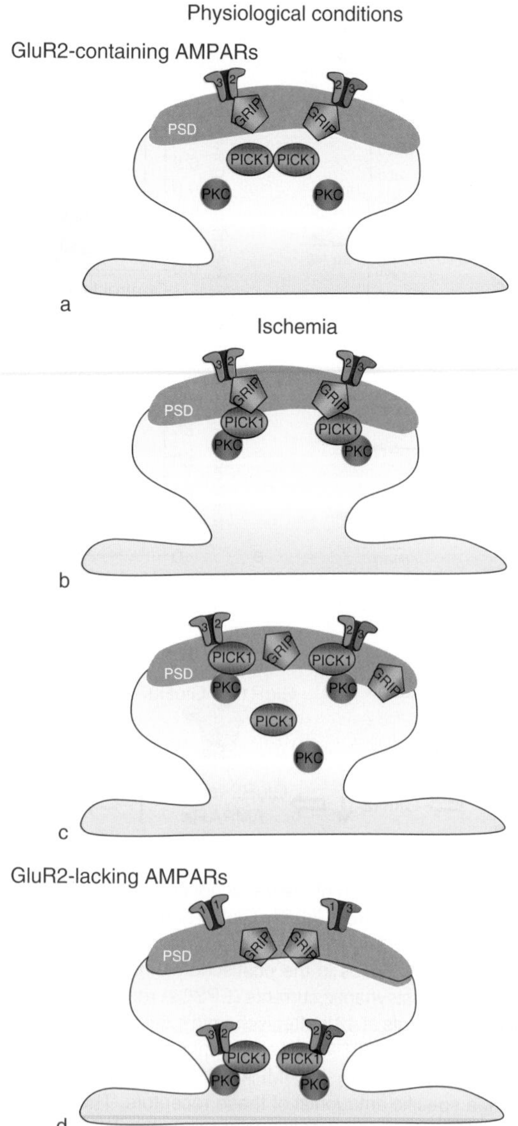

Figure 3 Ischemia-induced switch in AMPAR subtypes at CA1 synapses. In the basal state, GluR2-containing AMPARs are stabilized at the postsynaptic membrane via association with the GluR2/3-interacting AMPA receptor binding protein (ABP). (a) Ischemic insults activate protein kinase C (PKC) and promote binding of PKC to protein interacting with C kinase1 (PICK1), which in turn targets the PICK1-PKC complex to the ABP-GluR2 complex (b). (c) GluR2 binding to ABP is disrupted and its association with is enhanced. Interaction of GluR2-containing AMPARs with PICK1 promotes receptor internalization. (d) Ischemia promotes synaptic incorporation of GluR2-lacking AMPARs, which replace GluR2-containing AMPARs at insulted CA1 synapses. GRIP, GluR2-interacting protein. Reproduced from Liu SJ and Zukin RS (2007) Ca^{2+}-permeable AMPA receptors in synaptic plasticity and neuronal death. *Trends in Neurosciences* 30: 126–134, with permission from Elsevier.

neuroprotective in animal models of global ischemia. While it is likely that the increased calcium permeability into cells expressing GluR2-lacking AMPARs contributes to excitotoxic death, elevation of cytosolic Zn^{2+} via influx through these same receptors now appears to be an important trigger for cell death in these neurons (see the section titled 'Zinc').

Epilepsy

In addition to their role in global ischemia, Ca^{2+}-permeable AMPARs are implicated in the cell death following seizures. Following status epilepticus in adult rats, GluR2 mRNA and subunit expression is downregulated in vulnerable pyramidal neurons found in the CA3 and CA1 region of the hippocampus prior to the onset of their death. Status epilepticus induced in rat pups, however, which does not trigger neurodegeneration, does not cause downregulation of GluR2 expression, providing a further link between these events. Seizures also upregulate REST mRNA and suppress GluR2 promoter activity in hippocampal neurons. As has been observed in global ischemia, REST-dependent silencing of the AMPAR GluR2 gene is thought to contribute to the delayed neurodegeneration associated with seizures. As such, regulated expression of calcium-permeable AMPARs may represent a broad mechanism of insult-induced neuronal death.

ALS

Significant evidence also suggests a role for GluR2-lacking AMPARs in ALS, sometimes called Lou Gehrig's disease. ALS is a rapidly progressive, invariably fatal neurological disease that attacks the neurons responsible for controlling voluntary muscles. In ALS, both the upper motor neurons and the lower motor neurons degenerate or die, and muscles gradually weaken and waste away. Eventually, the ability of the brain to start and control voluntary movement is lost. Individuals with ALS lose their strength and the ability to move their arms, legs, and body and ultimately the ability to breathe. Approximately 5–10% of ALS cases are familial, and of these, one-fifth have been found to carry gain of function mutations in the superoxide dismutase-1 (SOD1) gene.

The selective loss of motor neurons in ALS has been linked to the elevated susceptibility of these neurons to excitotoxic damage. In particular, motor neurons *in vivo* and *in vitro* are subject to injury in response to exposure to glutamatergic agonists. Several factors may contribute to the vulnerability of these neurons. In particular, while these neurons express high levels of GluR2-lacking AMPARs, unlike many other neurons that express the Ca^{2+}-permeable receptors, they have low levels of Ca^{2+}-buffering proteins. The possibility that these receptors play an important role in excitotoxicity in motor neurons is supported by the fact that AMPA/kainate receptor agonists, but not NMDA receptor agonists, were highly effective at inducing cell death.

There are likely many factors that contribute to the pathology of ALS. GluR2-lacking AMPARs appear to play an important role. Some patients with sporadic ALS exhibit a reduced level of editing of the Q/R site of the GluR2 subunit, which controls AMPAR calcium permeability. This mutation is predicted to even further increase the level of calcium permeability through AMPARs in motor neurons of these patients. This deficiency seems to be restricted to the sporadic form of the disease, as GluR2 mRNA is fully edited in SOD1 transgenic rat models of familial ALS and in humans with spinal and bulbar muscular atrophy. However, evidence of a role for Ca^{2+}-permeable AMPARs in familial, SOD1-related forms of ALS has come from transgenic mouse models of the disease. Animals which do not express the GluR2 subunit do not show evidence of motor neuron disease, suggesting the absolute level of GluR2 is not important. However, if these animals are crossed with transgenic animals expressing the G93A gain of function SOD1 mutation, the motor neuron degeneration normally observed in the SOD1 animals is greatly accelerated. Similarly, crossing mice that have a mutation in GluR2 that prevents its editing with mice that have the SOD1 mutation aggravates the course of the motor neuron disease. Overexpression of GluR2, in contrast, delays the onset of motor neuron death in SOD1 mutant mice.

Increased levels of Ca^{2+}-permeable AMPARs may contribute to one apparently SOD1-independent form of juvenile onset ALS, termed ALS2. This autosomal recessive form of ALS is linked to a deficiency in the ALS2 gene that encodes the alsin protein. Alsin binds to a GluR2-interacting protein (GRIP1). In neurons from ALS2-deficient animals, GRIP1 localization at synapses is reduced. GRIP1 is believed to participate in the targeting and membrane stabilization of GluR2 subunits. Consistent with the mislocalization of GRIP1, GluR2-lacking AMPARs at synapses are increased in ALS2 mutant mice, resulting in an increased vulnerability of motor neurons to glutamate receptor-mediated excitotoxicity.

The exact mechanisms by which expression of Ca^{2+}-permeable AMPARs contributes to excitotoxic motor neuron injury are, as yet, unclear. However, a combination of factors likely makes the motor neurons susceptible to death in the course of the disease. Elevated glutamate levels, perhaps related to decreased clearance of glutamate by transporters, elevates intracellular calcium levels via influx through the high levels of calcium-permeable AMPARs in motor neurons. This may be exacerbated by mutations in the editing of GluR2. Because of the poor calcium buffering of motor neurons, calcium uptake into mitochondria can be elevated, resulting in the generation of reactive oxygen species. This again may be aggravated by mutations in SOD1 which can diminish mitochondrial function. The resulting oxidative stress may enhance the degeneration of the motor neurons characteristic of the progression of the disease (**Figure 4**).

Neuronal Injury

Tumor necrosis factor α (TNFα) is a proinflammatory cytokine reported to have both neuroprotective and neurotoxic functions. It is released following a number of neuronal insults, including trauma, ischemia, and multiple sclerosis. TNFα is released by astrocytes and microglia, and its receptors, TNF-R1 and 2, can be found on both neurons and glia in the CNS. TNFα, at high concentrations, has been found to facilitate glutamate-mediated neurotoxicity. Several mechanisms may account for this effect. TNFα has been found to inhibit glutamate uptake in astrocytes, leading to possible excessive glutamatergic activation. Additionally, TNFα directly enhances synaptic transmission mediated by glutamate receptors. For example, TNFα activation of TNF-R1 elevates the expression of synaptic AMPARs in hippocampal cultures and slices. Notably, this enhancement involves the preferential insertion of GluR2-lacking AMPARs, thereby enhancing Ca^{2+} permeability of synapses in hippocampal neurons (**Figure 5**). Such a rapid change in the calcium influx in neurons which normally express few calcium-permeable AMPARs would be predicted to contribute to enhanced glutamate pathogenicity and subsequent cell death, particularly when paired with a reduction in glutamate uptake. Future studies will be necessary to establish a direct role of these newly inserted AMPARs in the neurodegeneration associated with insults leading to the release of TNFα.

Mechanisms of Neuronal Death via GluR2-Lacking AMPARs

In all the cases of excitotoxic neurodegeneration discussed here, the exact mechanisms linking the expression or modification of Ca^{2+}-permeable, GluR2-lacking AMPARs to cell death are not completely understood. Elevated Ca^{2+} influx and increased intracellular Zn^{2+} levels are the most likely mediators of cell death. While not all the relevant downstream targets are known, a great deal is understood about the possible mechanisms by which these ions may contribute to toxicity in neurons.

Calcium

Neuronal homeostasis requires that the intracellular concentration of Ca^{2+} be maintained in the

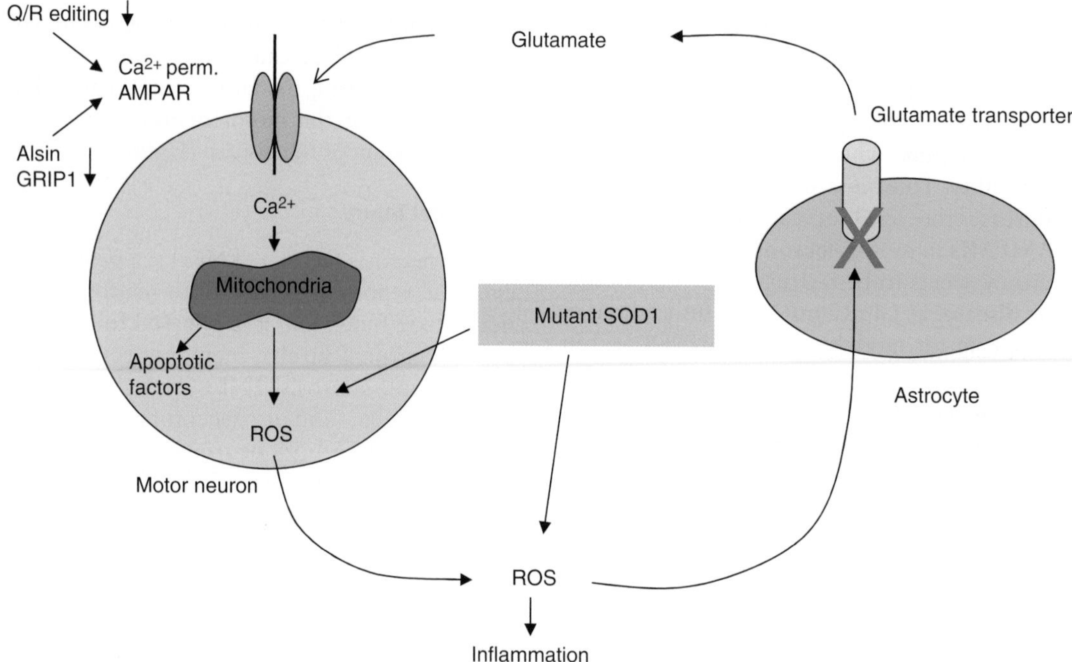

Figure 4 Potential mechanism of selective motor neuron death in amyotrophic lateral sclerosis (ALS). Elevated levels of extracellular glutamate due to reduced glutamate uptake or trauma result in the influx of nonphysiological levels of Ca^{2+} influx through Ca^{2+}-permeable (Ca^{2+} Perm.) α-amino-3-hydroxy-5-methyl-4-isoxazole propionic acid (AMPA) channels. The extent of influx may be exacerbated by decreased GluR2 editing in sporadic cases of ALS. Uptake of calcium by mitochondria results in reactive oxygen species (ROS) generation, and possibly apoptosis. The ROS can inhibit astrocytic glutamate transporters, causing additional increases in extracellular glutamate. In many familial cases of ALS, mutations in superoxide dismutase-1 (SOD1) are linked to cell death, perhaps in part through dysregulation of mitochondrial function. AMPAR, AMPA receptor; GRIP, GluR2-interacting protein; Q/R, glutamine/arginine. Reproduced from Kwak S and Weiss JH (2006) Calcium-permeable AMPA channels in neurodegenerative disease and ischemia. *Current Opinion in Neurobiology* 16: 281–287, with permission from Elsevier.

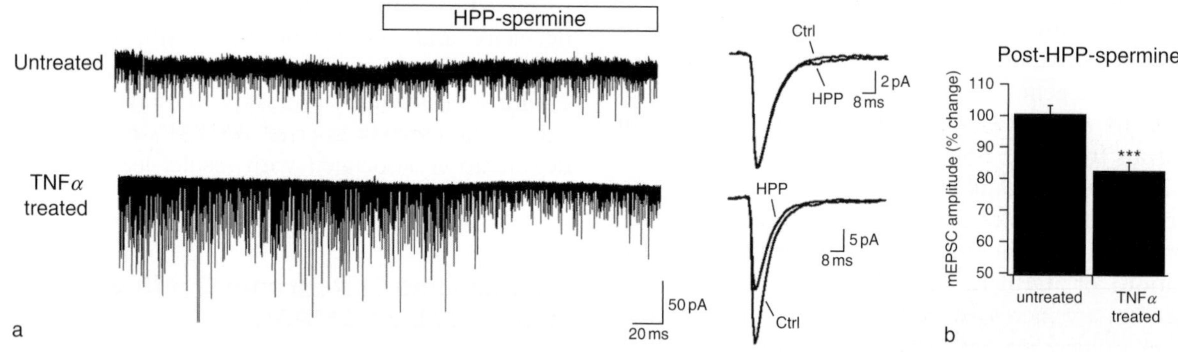

Figure 5 Tumor necrosis factor (TNF) α induces an increase in calcium-permeable α-amino-3-hydroxy-5-methyl-4-isoxazole propionic acid receptors (AMPARs) in hippocampal neurons. (a) Miniature excitatory postsynaptic currents (mEPSCs) from TNFα-treated cultured hippocampal neurons show enhanced blockade by the calcium-permeable AMPAR blocker *N*-(4-hydroxyphenylpropanoyl) (HPP)–spermine. (b) Averaged normalized mEPSC amplitudes after HPP-spermine for control and TNFα-treated cells demonstrate that the cytokine significantly enhances the expression of GluR2-lacking AMPARs. Error bar represents SEM. Reproduced from Stellwagen D, Beattie EC, Seo JY, and Malenka RC (2005) Differential regulation of AMPA receptor and GABA receptor trafficking by tumor necrosis factor-alpha. *Journal of Neuroscience* 25: 3219–3228, with permission from the Society for Neuroscience.

tens-of-nanomolar range. Unregulated or excessive influx of Ca^{2+} that can result from increases in Ca^{2+}-permeable AMPARs following acute insult or disease could initiate a series of damaging events in the cytoplasm and nucleus. Elevated cytosolic Ca^{2+} activates Ca^{2+}–adenosine triphosphatase, which depletes the energy stores of the cell. High Ca^{2+} levels also uncouple mitochondrial oxidative phosphorylation, causing acute swelling of dendrites and cell bodies and cell death. Ca^{2+} uptake by mitochondria can generate reactive oxygen species and can open the transition pore allowing the release of mediators of apoptosis, including cytochrome C. High cytosolic Ca^{2+} also causes excessive activation of nitric oxide synthase, which promotes generation of free radicals that destroy cell membranes, damage DNA, and can induce apoptosis. In addition to all these effects, Ca^{2+} activates a host of Ca^{2+}-sensitive transcription factors, phospholipases, endonucleases, and proteases, all of which may trigger specific events associated with necrosis or apoptosis.

Zinc

The transition metal Zn^{2+}, like Ca^{2+}, is an important neuronal signaling molecule under physiological conditions and a critical player in the cell death associated with ischemia and seizures. Zn^{2+} is co-localized with glutamate in presynaptic vesicles at a subset of excitatory synapses and is particularly high in mossy fiber tracts of the hippocampus. Zn^{2+} is co-released with glutamate spontaneously and in an activity-dependent manner. Under physiological conditions, Zn^{2+} release modulates the function of multiple neurotransmitter receptors, including NMDARs, AMPARs and γ-aminobutyric acid $(GABA)_A$ receptors. In the neuronal cytoplasm, Zn^{2+} acts as a functionally important component of metalloenzymes and zinc finger-containing transcription factors. Zn^{2+} can enter neurons via voltage-sensitive Ca^{2+} channels, NMDARs, GluR2-lacking AMPARs, and the Na^{+}/Zn^{2+} antiporter. Of these, GluR2-lacking AMPARs show the highest permeability to Zn^{2+}. Therefore, upregulation of the expression of these receptors in neurons can greatly enhance the levels of cytosolic Zn^{2+}.

Whereas relatively low concentrations of Zn^{2+} elicit neuronal death with the hallmarks of apoptosis, higher concentrations of Zn^{2+} elicit neuronal death characteristic of necrosis. Zn^{2+} induces neurotoxicity by a number of mechanisms similar to calcium, including production of free radicals, loss of mitochondrial membrane potential, formation of the mitochondrial transition pore, dysregulation of the electron transport chain, and disruption of glycolysis and energy production. It is interesting that Zn^{2+} has been found to be more potent than calcium at disrupting the function of mitochondria.

Studies indicate that Zn^{2+} at high concentrations is a critical mediator of the neuronal death associated with global ischemia, prolonged seizures, traumatic brain injury, and other brain disorders. In the first 1–2 h after global ischemia, Zn^{2+} acts at the level of mitochondria to promote the release of cytochrome c, apoptosis-inducing factor, and second mitochondria-derived activator of caspase/direct inhibitor of apoptosis-binding protein with low isoelectric point from the mitochondria into the cytosol. At delayed late time points following global ischemia or status epilepticus, Zn^{2+} accumulates in selectively vulnerable hippocampal neurons. Administration of Zn^{2+} chelators such as EDTA substantially reduces the rise in intracellular Zn^{2+} in vulnerable neurons and affords robust neuroprotection following both ischemia and epilepsy (**Figure 6**). These observations implicate Zn^{2+} as a critical mediator of the neuronal death associated with neural insults.

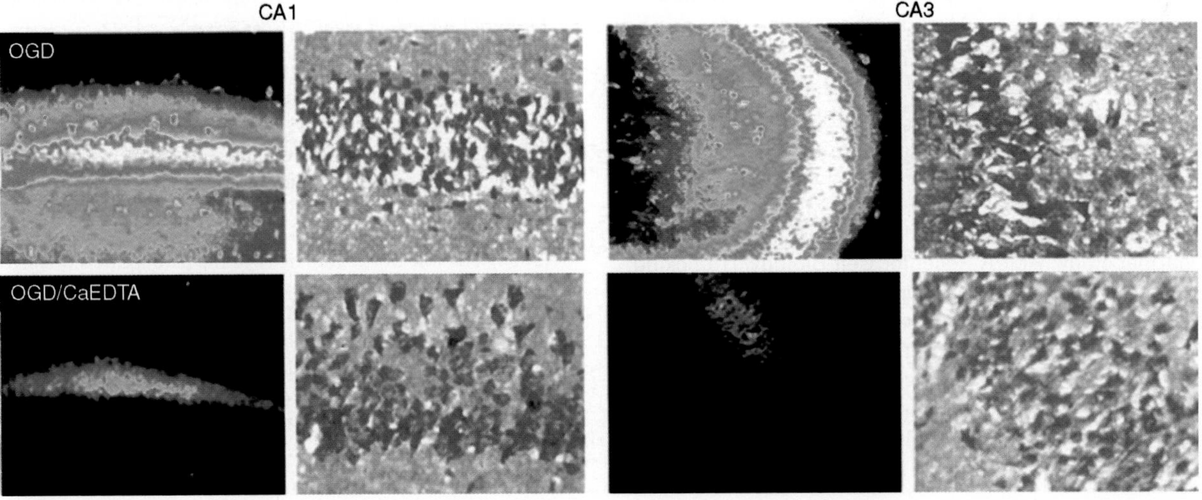

Figure 6 Chelation of zinc is neuroprotective in hippocampal pyramidal neurons following ischemic insult. Slices of adult mouse hippocampus were exposed for 5 min to oxygen glucose deprivation in the absence (OGD), or presence of the extracellular Zn^{2+} chelator Ca^{2+} EDTA (OGD + CaEDTA). Four hours after OGD, propidium iodide (PI) labeling was performed on of hippocampal CA1 (left four panels) and CA3 (right four panels) pyramidal cell layers to measure neuronal degeneration. Images in the left panel for each hippocampal region show pseudocolor images of PI labeling. Right panels show matched slices (25 mm) stained with toluidine blue. Zinc chelation significantly reduced the effects of ischemic injury on neurons. Reproduced from Yin HZ, Sensi SL, Ogoshi F, and Weiss JH (2002) Blockade of Ca^{2+}-permeable AMPA/kainate channels decreases oxygen-glucose deprivation-induced Zn^{2+} accumulation and neuronal loss in hippocampal pyramidal neurons. *Journal of Neuroscience* 22: 1273–1279, with permission from the Society for Neuroscience.

Summary

Rapidly increasing evidence supports the importance of Ca^{2+}-permeable AMPARs in neuronal death. Multiple mechanisms have been identified by which AMPAR subunit composition and consequently their Ca^{2+} permeability is modified in response to neuronal insults and disease. With ischemia and epilepsy, reductions in the protein expression of GluR2 subunits enhance the calcium and zinc permeability in select populations of neurons, leading eventually to cell death. Regulation of the transcription of the GluR2 gene by modulation of REST activity is emerging as an important means by which neural insults can affect the composition of synaptic AMPARs. In motor neuron disease such as ALS, defects in the editing of the GluR2 subunit or the normally elevated levels of Ca^{2+}-permeable AMPARs in motor neurons are linked to cell death. In each of these cases, the exact mechanisms underlying the increased levels of Ca^{2+}-permeable AMPARs are still unknown. However, the influx of calcium and zinc appear to mediate cell death through a variety of mechanisms, including disruption of mitochondrial function and integrity.

Ongoing studies will continue to define both how neuronal insult and disease couple to changes in AMPAR composition and how these changes in AMPARs lead to subsequent cell death. Because calcium-permeable AMPARs exhibit spatially restricted patterns of expression and are normally expressed at very low density by principal neurons of the hippocampus and neocortex, they may be a promising target for possible therapeutic approaches relevant to ischemia, epilepsy, and ALS. Already the ability of specific AMPAR antagonists and Zn^{2+} chelators to reduce neuronal death in models of neural insult provides an exciting glimpse into the possible development of neuroprotective treatments that could lie ahead.

See also: AMPA Receptors: Molecular Biology and Pharmacology; AMPA Receptor Cell Biology/Trafficking; Amyotrophic Lateral Sclerosis (ALS): Disease Mechanisms; Apoptosis in Nervous System Injury; Apoptosis in Neurodegenerative Disease; Glutamate Regulation of Dendritic Spine Form and Function; Intracellular Calcium and Neuronal Death; NMDA Receptors and Disease + C464.

Further Reading

Burnashev N, Monyer H, Seeburg PH, and Sakmann B (1992) Divalent ion permeability of AMPA receptor channels is dominated by the edited form of a single subunit. *Neuron* 8: 189–198.

Dingledine R, Borges K, Bowie D, and Traynelis SF (1999) The glutamate receptor ion channels. *Pharmacological Reviews* 51: 7–61.

Gorter JA, Petrozzino JJ, Aronica EM, et al. (1997) Global ischemia induces downregulation of GluR2 mRNA and increases AMPA receptor-mediated Ca^{2+} influx in hippocampal CA1 neurons of gerbil. *Journal of Neuroscience* 17: 6179–6188.

Kawahara Y, Ito K, Sun H, Aizawa H, Kanazawa I, and Kwak S (2004) Glutamate receptors: RNA editing and death of motor neurons. *Nature* 427: 801.

Kuner R, Groom AJ, Bresink I, et al. (2005) Late-onset motoneuron disease caused by a functionally modified AMPA receptor subunit. *Proceedings of the National Academy of Sciences of the United States of America* 102: 5826–5831.

Kwak S and Weiss JH (2006) Calcium-permeable AMPA channels in neurodegenerative disease and ischemia. *Current Opinion in Neurobiology* 16: 281–287.

Liu B, Liao M, Mielke JG, et al. (2006) Ischemic insults direct glutamate receptor subunit 2-lacking AMPA receptors to synaptic sites. *Journal of Neuroscience* 26: 5309–5319.

Liu SJ and Zukin RS (2007) Ca^{2+}-permeable AMPA receptors in synaptic plasticity and neuronal death. *Trends in Neurosciences* 30: 126–134.

Lo EH, Dalkara T, and Moskowitz MA (2003) Mechanisms, challenges and opportunities in stroke. *Nature Reviews Neuroscience* 4: 399–415.

Pellegrini-Giampietro DE, Zukin RS, Bennett MV, Cho S, and Pulsinelli WA (1992) Switch in glutamate receptor subunit gene expression in CA1 subfield of hippocampus following global ischemia in rats. *Proceedings of the National Academy of Sciences of the United States of America* 89: 10499–10503.

Porter BE, Cui XN, and Brooks-Kayal AR (2006) Status epilepticus differentially alters AMPA and kainate receptor subunit expression in mature and immature dentate granule neurons. *European Journal of Neuroscience* 23: 2857–2863.

Seeburg PH and Hartner J (2003) Regulation of ion channel/neurotransmitter receptor function by RNA editing. *Current Opinion in Neurobiology* 13: 279–283.

Stellwagen D, Beattie EC, Seo JY, and Malenka RC (2005) Differential regulation of AMPA receptor and GABA receptor trafficking by tumor necrosis factor-alpha. *Journal of Neuroscience* 25: 3219–3228.

Van Den Bosch L, Van Damme P, Bogaert E, and Robberecht W (2006) The role of excitotoxicity in the pathogenesis of amyotrophic lateral sclerosis. *Biochimica et Biophysica Acta* 1762: 1068–1082.

Yin HZ, Sensi SL, Ogoshi F, and Weiss JH (2002) Blockade of Ca^{2+}-permeable AMPA/kainate channels decreases oxygen-glucose deprivation-induced Zn^{2+} accumulation and neuronal loss in hippocampal pyramidal neurons. *Journal of Neuroscience* 22: 1273–1279.

AMPA Receptors: Molecular Biology and Pharmacology

S M Dravid, H Yuan, and S F Traynelis, Emory University, Atlanta, GA, USA

Published by Elsevier Ltd.

Introduction

Glutamate, one of the fundamental amino acid building blocks of proteins, is also a major excitatory neurotransmitter in the central nervous system (CNS). Neurons synthesize and package glutamate into presynaptic vesicles for release into the postsynaptic cleft. Synaptically released glutamate that diffuses across the 30 nm distance encounters a series of transmembrane postsynaptic proteins that comprise the glutamate receptor family. One class of glutamate receptors, metabotropic glutamate receptors, comprises transmembrane proteins that have extracellular clamshell-like domains that bind glutamate. When activated by glutamate binding, these G-protein-coupled receptors shift intracellular concentrations of signaling molecules to control a diverse set of cell properties. A second class of glutamate receptors, ionotropic glutamate receptors, comprises transmembrane proteins that contain an ion conduction path through the plasma membrane, as well as an array of clamshell-like extracellular ligand-binding domains, some of which bind to glutamate. Mammalian ionotropic glutamate receptors are ligand-gated ion channels encoded by 18 genes, and are subdivided into four major families on the basis of agonist pharmacology and sequence homology. These four receptor classes are known as amino-3-hydroxy-5-methyl-4-isoxazole propionic acid (AMPA), kainate, *N*-methyl-D-aspartate (NMDA), and δ receptors. This article focuses on the structure and function of AMPA receptors.

The AMPA receptor family is composed of four genes encoding the GluR1–GluR4 subunits (sometimes called GluRA–GluRD). In humans the chromosomal location of GluR1, GluR2, GluR3, and GluR4 encoding genes is 5q33, 4q32–33, Xq25–26, and 11q22–23, respectively. AMPA receptors were the first class of glutamate receptor cloned by screening a rat brain cDNA library for expression of kainate-activated ion channels in *Xenopus laevis* oocytes. After initial identification of GluR1, GluR2–GluR4 were rapidly identified by homology screening. There is about 70% sequence homology among different AMPA receptor subunits. A great deal of information now exists about AMPA receptor structure and function, and it could be argued that more is known about the structure of AMPA receptors than any other class of glutamate receptor.

Expression of AMPA Receptors

AMPA receptors are abundant and widely distributed in the central nervous system. Hippocampus, outer layer of cortex, basal ganglia, olfactory regions, lateral septum, and amygdala of the CNS are all enriched with GluR1, GluR2, and GluR3 subunits. In contrast, GluR4 expression is lower in many regions of the CNS except cerebellum, thalamus, and brain stem, where the expression is high. Immunoprecipitation studies have shown that the pyramidal cells of the hippocampus expressed AMPA receptors composed of GluR2 receptor in complex with either GluR1 or GluR3 subunits. GluR1 homomeric receptors, which have unique ion permeation properties, are thought to be expressed in select neuronal populations.

The expression of AMPA receptors is developmentally regulated. The GluR2 subunit appears as early as embryonic day 16 in rats whereas other receptors are upregulated later during the development. The GluR2 subunit can also be selectively altered during synaptic plasticity as well as during CNS injury, such as global ischemia. These changes in receptor subunit composition are known to change functional receptor properties.

AMPA receptors are present both postsynaptically and presynaptically. AMPA receptors are present on the synaptic membrane; however, 60–70% of AMPA receptors are present intracellularly. Glial cells also express AMPA receptors, which appear to be involved in glutamate-induced cell death. Activation of glial AMPA receptors also leads to release of ATP or nitric oxide.

Topology and Assembly of AMPA Receptor Subunits

All of the ionotropic glutamate receptors share a common topology, which consists of an extracellular N-terminal domain, a ligand-binding domain, three transmembrane domains (M1, M3, and M4), a cytoplasm-facing reentrant membrane loop (M2), and an intracellular C-terminal domain (**Figure 1(a)**). AMPA receptors are composed of approximately ~900 amino acids and have a molecular mass of ~105 kDa. The location of the N-terminus and C-terminus was first deduced by use of specific antibodies. Because the N- and C-terminal regions were located on the opposite ends of the polypeptide chain, it was proposed that AMPA receptors had an odd number of membrane-spanning domains. Further studies delineated the membrane topology, and showed that the M2 segment is a reentrant loop.

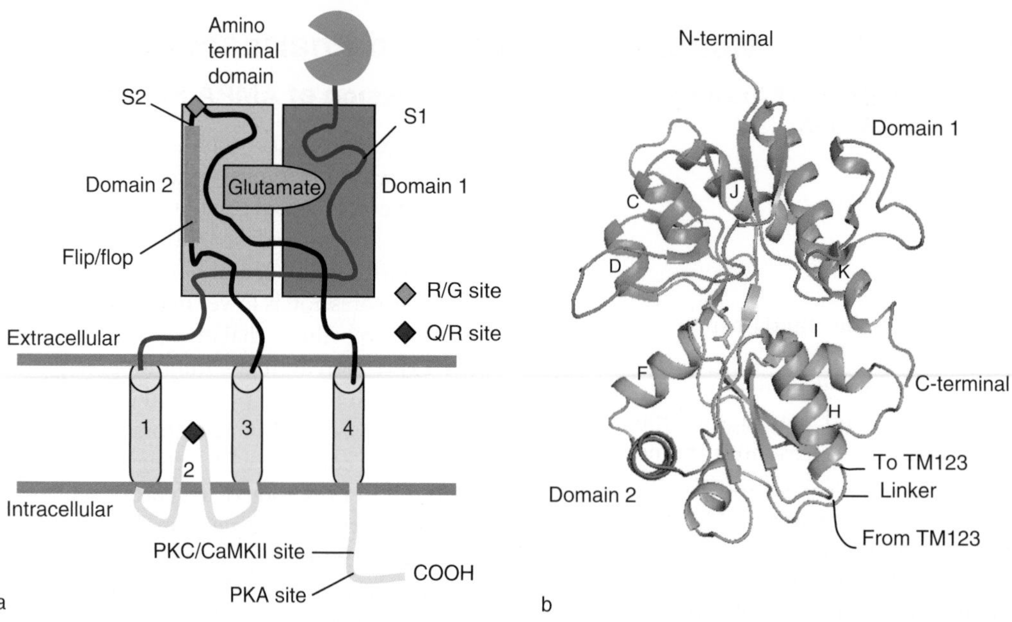

Figure 1 Structure of the AMPA receptor subunit. (a) Transmembrane topology of the AMPA receptor, indicating flip/flop site, Q/R site, R/G site, and phosphorylation sites (PKA, protein kinase A; PKC, protein kinase C; CaMKII, Ca^{2+}/calmodulin-dependent protein kinase II). (b) Crystal structure of the ligand-binding domain of the GluR2 subunit, with glutamate bound in the cleft formed by clamshells (Protein Data Bank code 1FTJ). The lime-colored structure is domain 1 and the violet-colored structure is domain 2 (TM, transmembrane domain).

This transmembrane topology shares a parallel organization to potassium channels, which are now considered a model for AMPA receptor membrane domain structure. The pore diameter of AMPA receptors is about 0.8 nm and, in contrast to potassium channels, permits the entry of Na^+ and, for some subunit combinations, Ca^{2+}.

The N-terminal domain in AMPA represents up to 45% of the mature polypeptide, but its function is poorly understood. Hypothesized functions of this domain include receptor assembly, allosteric modulation of the ion channel (similar to NMDA receptors), and binding of a second ligand. Contradictory to other subunits, the GluR4 subunit can form normally functioning homomeric channels even in the absence of the N-terminal domain.

The semiautonomous ligand-binding domain has been studied in detail using X-ray crystallography. The GluR2 agonist-binding domain is composed of two discontinuous peptide segments of approximately 150 amino acid residues. The first segment (S1) is adjacent and N-terminal to the M1 domain, whereas the second segment (S2) is located between M3 and M4 domains (**Figure 1(a)**). The agonist-binding domains of the AMPA receptors share similarities in sequence and structural arrangement with the ligand-binding site of several bacterial periplasmic amino acid-binding proteins.

Like potassium channels, AMPA receptors assemble as tetramers. Studies in recombinant and native receptors suggest that AMPA receptors assemble as dimer-of-dimers in a two-step manner. First, the monomers interact through the N-terminal domain to form dimers. Next, the dimers combine via the membrane domains to form the tetramer.

AMPA Receptor Function

A remarkable step forward in understanding the AMPA receptor structure and function occurred in 1995 when a water-soluble mini-receptor that included only the agonist-binding core was described. This was a fusion protein consisting of the S1 and S2 domains of GluR4 joined together via a short hydrophilic linker peptide that replaces the membrane-spanning regions. The engineered agonist-binding domain functionally reproduced the AMPA-binding properties of the GluR4 receptor. This concept paved the way for the subsequent production of soluble agonist-binding core and later generation of crystals for X-ray diffraction, which ultimately could be produced by careful refolding of the denatured ligand-binding core. The first crystal structure of the GluR2 ligand-binding domain complexed with kainate revealed a bilobed clamshell-like shape, with agonist bound deep in the cleft formed by the two lobes. Subsequent descriptions of crystal structures of the nonliganded form (apo state) as well as forms complexed to a variety of ligands were obtained. These studies revealed that the clamshell was geometrically opened widest in the apo state. In the glutamate-bound state, the clamshell was 21° more closed than in the apo state (**Figures 1** and **2**). Competitive antagonists of the receptor stabilized the open-cleft state.

Glutamate makes a number of hydrogen-bonded contacts within the binding pocket with both upper (D1) and lower (D2) domains. The upper domain of the clamshell-like structure is formed by segment S1 and the C-terminal portion of segment S2. The N-terminus of S2 forms the lower D2 domain. It has been proposed that when glutamate first encounters the ligand-binding pocket, it initially docks or interacts with the D1 domain, which then promotes the rotation of the D2 domain toward D1 and induces closure of the clamshell. The closed conformation is stabilized by glutamate, itself forming a cross-domain bridge, as well as a number of other hydrogen bonds that form between the domains during glutamate binding. Water appears to persist inside of the agonist binding pocket, particularly for some agonists that interact directly with it.

This closure of the cleft within the isolated ligand-binding domain is considered to be analogous to domain closure in the full-length native receptors. Domain closure has been hypothesized to pull the linker that connects the S1 and S2 domains. In native receptors, this pull or strain is considered to be the force responsible for rearrangement of the pore-forming membrane domains, leading to the opening of the transmembrane conduction path. Thus, each subunit can bind glutamate and undergo conformational changes that contribute to dilatation of the pore. In agreement with this view, single-channel analysis of intact receptor shows at least three conductance states that have been proposed to correspond to two, three, or four liganded subunits within a receptor complex. Pore conductance is therefore conceptually related to the fraction of subunit occupancy and activation within each receptor complex. This suggests that subunits can make incremental contributions to pore opening, gradually shifting the unitary conductance through the channel to higher levels as more subunits become activated. The observed concentration dependence of conductance levels supports this idea.

This view of fourfold rotational symmetry is also supported by the similarities of the pore-forming

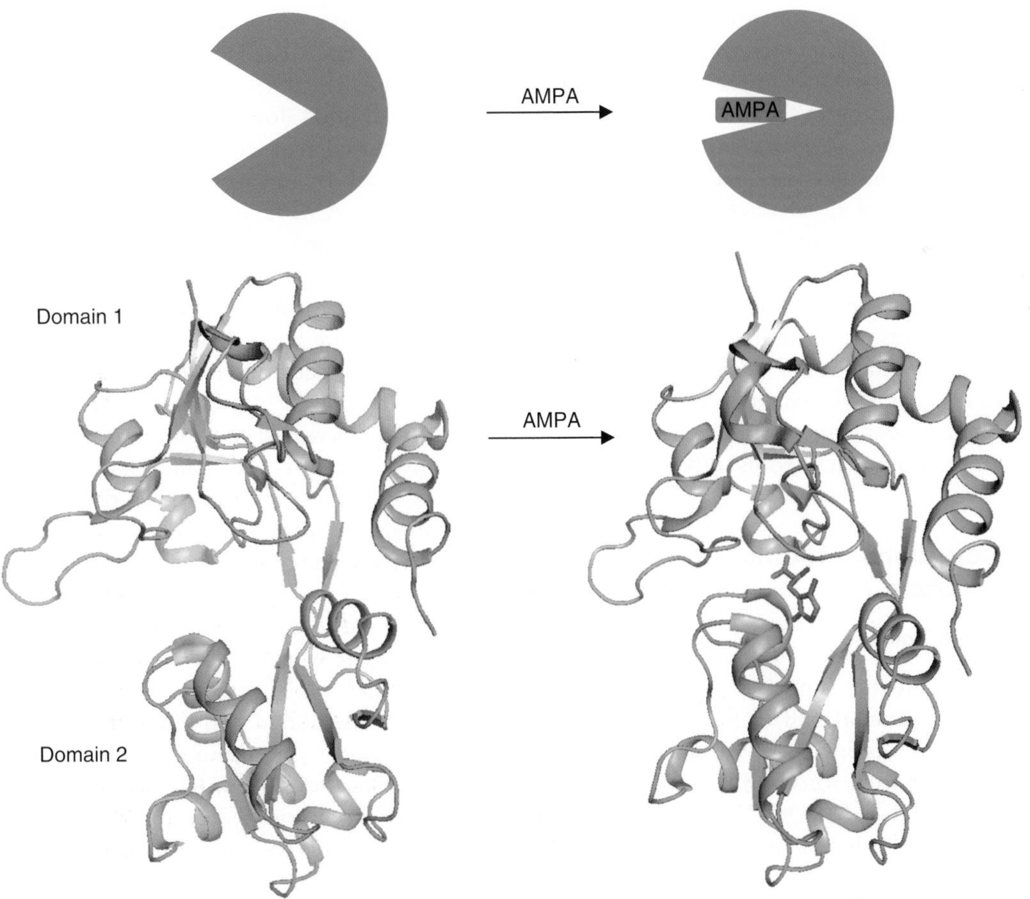

Figure 2 Conformational changes induced by agonist binding to the AMPA receptor. Crystal structure of the ligand-binding domain of the GluR2 subunit shown in nonliganded (apo –state; Protein Data Bank code 1FTO) and liganded (AMPA bound state; Protein Data Bank code 1FTM) states. Agonist binding in the cleft leads to closure of the clamshell (top).

AMPA receptor domain to potassium channels. However, several other features of AMPA receptors point to a twofold rotational symmetry for the tetrameric receptor complex. First, the glutamate-binding domains crystallize as dimers. Second, the D1–D1 dimer interface plays a role in channel activation by forming a structural scaffold that leads to the movement of D2. However, excessive tension can trigger rearrangement of the D1–D1 interface, leading to AMPA receptor desensitization. Mutations that affect desensitization lie on the interface. Additionally the AMPA receptor desensitization inhibitor cyclothiazide (see below) also acts on the dimer interface. Third, the reactivity of cysteine-modifying reagents on cysteine residues inserted into the M3 domain by site-directed mutagenesis fits well with a twofold symmetry rather than a fourfold symmetry. However, additional structural information about the full AMPA receptor complex would be needed to ascertain receptor symmetry.

Partial agonists are typically considered as ligands that induce a response at the maximally effective concentrations, which is lower than that of the endogenous ligand glutamate. The mechanism of action of partial agonists has been studied in AMPA receptors using a series of 5-substituted willardiines, which show lower efficacy compared to glutamate. The 5-substituted willardiines differ in only a single atom at the same position in the molecule. Structural and functional studies suggest that the degree of domain closure of the agonist binding cleft forms the basis for the partial agonist action at the AMPA receptor. Specifically, a combination of crystallographic and functional data show that the degree of domain closure of the clamshell is correlated with the efficacy of the agonist at an individual subunit such that agonists that induce less domain closure appear less effective in opening the channel. However, other conformational rearrangements of the agonist binding domain can also impact agonist efficacy.

Posttranscriptional Modification of AMPA Receptors

Alternative RNA Splicing

All four AMPA subunits undergo alternative RNA splicing in the C-terminal half of the M3–M4 loop, leading to so called flip/flop splice variants. The locations of flip/flop splicing for GluR1, GluR2, GluR3, and GluR4 are 742–793, 736–787, 740–791, and 737–788, respectively. In rats expression of the flip variant predominates up to postnatal day 8, but in adults both forms are expressed to a similar extent in many regions. The flip splice variant endows receptors with a diminished form of desensitization and a faster rate of recovery from desensitization as compared to the flop splice variant. GluR4-flop has the fastest desensitization (<1 ms at room temperature) among all the studied ionotropic glutamate receptors.

GluR2 and GluR4 subunits are also alternatively spliced at their C-terminal ends to generate short or long isoforms. A short isoform for GluR1 has not been reported, whereas GluR3 shows only the short isoform due to absence of splice sites. More than 90% of GluR2 is in the short form while the long form of GluR4 is predominant. The postsynaptic density/disc/zonula occludens-1 (PDZ) motif, which interacts with several proteins necessary for targeting of AMPA receptors to the synapse, is only present in the short form.

RNA Editing

One for the most interesting properties of AMPA receptor subunits is the consequence of differential RNA editing. The GluR2 subunit undergoes RNA editing at the so-called Q/R site (Q607) located at the tip of the reentrant loop of the second membrane-associated region. RNA editing leads to conversion of a glutamine codon to an arginine codon. This residue is a major determinant of the ionic selectivity of the pore. Replacement of the glutamine with an arginine leads to low calcium permeability of GluR2 subunit, low single-channel conductance, and a linear current–voltage relationship that reflects lack of block by intracellular polyamines (see later). Conversely, AMPA receptors lacking the GluR2 subunit have higher unitary conductances and show substantial Ca^{2+} permeability. The latter property may underlie some forms of plasticity at central synapses. Virtually all of the GluR2 subunit RNA is edited with high efficiency, and reduction in editing efficiency can lead to epilepsy and early death in mice. This may reflect the increased Ca^{2+} permeability of AMPA receptors, since it is well established that intracellular calcium overload can lead to cell death during glutamate-mediated neurotoxicity. Reduction in the efficiency with which GluR2 is edited has been proposed to play a role in schizophrenia, Huntington's disease, Alzheimer's disease, epilepsy, and malignant glioma, although more work will be required to evaluate these hypotheses.

Another site of RNA editing is the R/G site, where glycine replaces the arginine in GluR2–GluR4 subunits. The locations of the R/G site for GluR2, GluR3, and GluR4 are R764, R769, and R7665, respectively. This editing leads to reduced desensitization and rate of recovery from desensitization in GluR3 and GluR4 subunits. In adult mammals 50–90% of GluR2–GluR4 subunits are edited at the R/G site. GluR1 only has an arginine at the R/G site (R757).

Posttranslational Modification

Phosphorylation of AMPA Receptors

AMPA receptor function and trafficking is post-translationally regulated by protein kinases and phosphatases. AMPA receptors are closely localized to cyclic adenosine monophosphate (cAMP)-dependent protein kinase A (PKA) and calcineurin by the binding of the A-kinase anchoring protein (known as AKAP) to scaffolding proteins within the postsynaptic density. In hippocampal neurons PKA phosphorylation potentiates AMPA receptor function by increasing the open probability, increasing the mean open time of the channel, and facilitating receptor trafficking to the membrane surface. PKA phosphorylates Ser845 residue present at the intracellular C-terminal tail of the GluR1 receptor. Mutation of this residue to alanine (S845A) eliminates the ability of PKA to potentiate GluR1-mediated currents. Phosphorylation and dephosphorylation of the Ser845 site are one important component of the molecular events that underlie long-term potentiation (LTP) and long-term depression (LTD) of synaptic strength, respectively, at certain synapses.

AMPA receptors are also phosphorylated by Ca^{2+}/calmodulin-dependent protein kinase II (CaMKII), which is a major constituent of postsynaptic densities at the glutamatergic synapses and forms approximately 1% of the brain protein. CaMKII phosphorylates the Ser831 residue present exclusively in the GluR1 subunit among AMPA receptors. CaMKII enhances the current through native AMPA receptors in hippocampal neurons and increases the single-channel conductance of GluR1 receptors. This change in conductance reflects a shift in the proportion of subconductance to higher levels, rather than creation of new levels. This shift can be accounted for by changes in subunit-dependent gating, in which individual GluR1 subunits that are phosphorylated at Ser831 are more likely to become activated, increasing the probability that the receptor will be in the higher conductance states that require more activated subunits.

Protein kinase C (PKC) is a family of heterogeneous protein kinases activated by Ca^{2+} and/or phospholipids. PKC can be converted to a persistently active form by calpain-mediated proteolysis. GluR1 is phosphorylated by PKC at Ser831, which has similar effects on current response amplitude and conductance to CaMKII phosphorylation at this site.

The GluR2 subunit is phosphorylated by PKC at Ser880 and Ser863. PDZ domains are present in a number of proteins that are capable of meditating interaction with the AMPA receptors. The GluR2 Ser880 residue is present in the C-terminal sequence that is responsible for PDZ domain binding and its phosphorylation leads to decrease in binding to glutamate receptor-interacting protein 1 (GRIP 1), which can synaptically anchor GluR2. GluR2 Ser880 phosphorylation does not affect the binding to other targeting auxiliary proteins, such as PICK 1 (protein interacting with PKC 1). Phosphorylation by PKC at the Ser880 site facilitates the internalization of the GluR2 receptor.

Pharmacology of AMPA Receptors

AMPA Receptor Agonists

A number of natural products have been known for decades to have neurological or psychoactive effects, which were later attributed to actions on glutamate receptors. These include ibotenic acid, quisqualic acid, domoic acid, and willardiine (see **Figure 3** for structures). Ibotenic acid is derived from the mushroom *Amanita muscaria* (the fly mushroom, or toadstool). Quisqualic acid is obtained from the seeds and fruit of *Quisqualis chinensis*, domoic acid is a phycotoxin (algal toxin) associated with certain algal blooms and shellfish poisoning, and willardiine is from *Acacia willardiana* and *Mimosa asperata*. Although L-glutamate is the endogenous agonist of AMPA receptors, these receptors show higher selectivity for the synthetic ibotenic acid analog AMPA, thus deriving their name. Kainic acid can also activate AMPA receptors, although with 10- to 20-fold lower potency for GluR1 and GluR2 over subunits within the kainate receptor family, such as GluR5. AMPA receptors have lower affinity for glutamate compared to NMDA receptors.

Glutamate and AMPA act as full agonists and induce rapid desensitization, which is thought to involve rearrangement of the dimer formed by the ligand-binding domains. Kainic acid, which is the defining agonist for so-called kainate receptors composed of GluR5–GluR7 and KA1 and KA2, functions as a partial agonist and induces little desensitization on AMPA receptors. 5-Fluorowillardiine is not only more selective for AMPA receptors than AMPA itself (5-fluorowillardiine has a 70- to 150-fold higher affinity for GluR1 and GluR2 over GluR5), but also shows some selectivity for different AMPA receptor subunits.

AMPA Receptor Competitive Antagonists

Structurally different classes of competitive AMPA antagonists have been found which bind to the glutamate-binding site of the receptor. The first selective and useful AMPA receptor antagonists were 6-cyano-7-nitroquinoxaline-2,3-dione (CNQX) and 6,7-dinitroquinoxaline-2, 3-dione (DNQX). However,

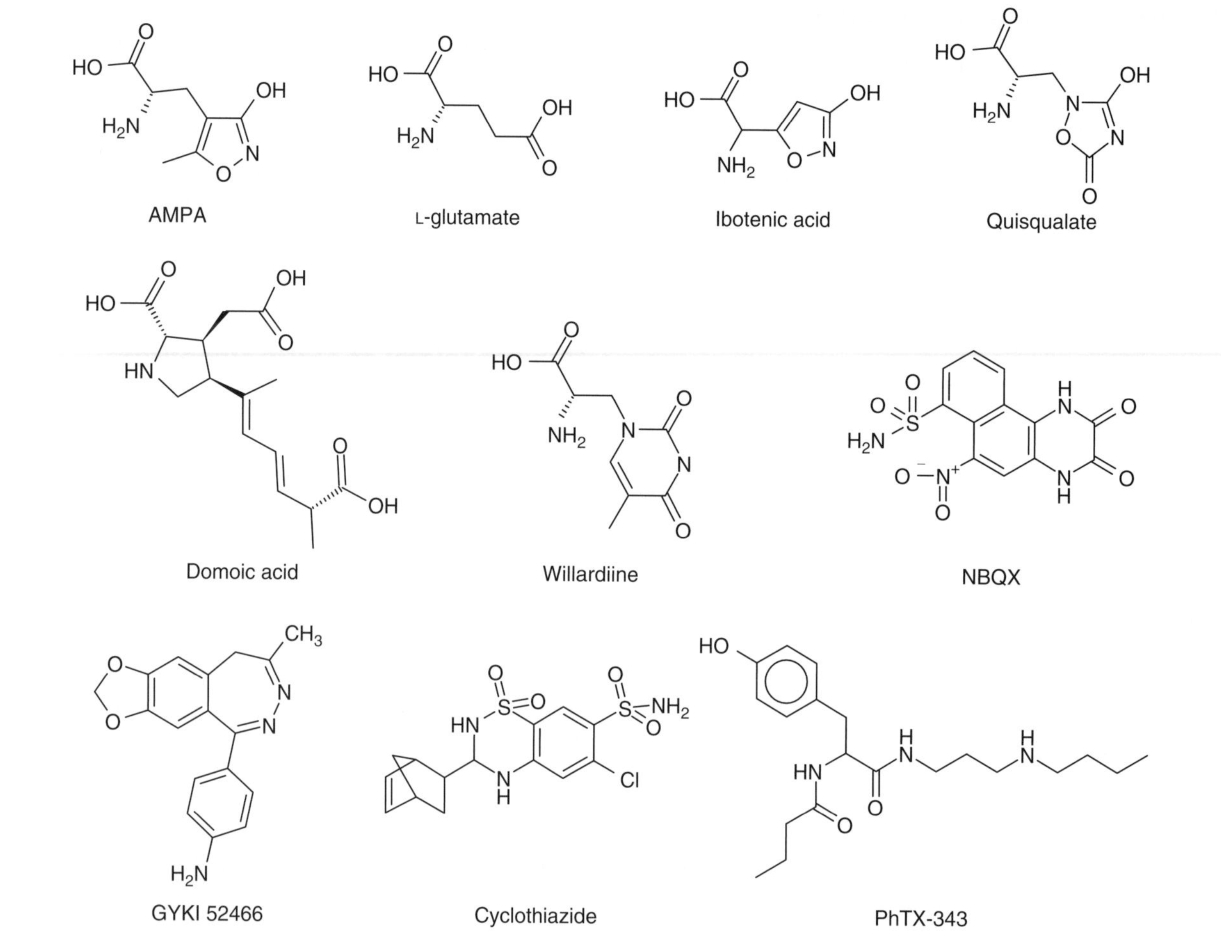

AMPA L-glutamate Ibotenic acid Quisqualate

Domoic acid Willardiine NBQX

GYKI 52466 Cyclothiazide PhTX-343

Figure 3 The structures of AMPA receptor agonists α-amino-3-hydroxy-5-methyl-4-isoxazole propionic acid (AMPA), L-glutamate, ibotenic acid, quisqualate, domoic acid, and willardiine; AMPA receptor antagonists 2,3-dihydroxy-6-nitro-7-sulfamoyl-benzo[*f*]quinoxaline (NBQX) and GYKI 52466; and AMPA receptor modulator cyclothiazide.

these quinoxaline derivates also have some affinity at the glycine-binding site of NMDA receptors. Newer antagonists have increased potency, higher AMPA receptor specificity, increased water solubility, and a longer duration of action *in vivo*. Examples of some of these competitive antagonists include 2,3-dihydroxy-6-nitro-7-sulfamoyl-benzo[*f*]quinoxaline (NBQX), 1,4,7,8,9,10-hexahydro-9-methyl-6-nitro-pyrido-[3,4-*f*]-quinoxaline-2,3-dione (PNQX), [2,3-dioxo-7-(1*H*-imidazol-1-yl)-6-nitro-1,2,3,4-tetrahydro-1-quinoxaliny]acetic acid (YM872), and [1,2,3,4-tetrahydro-7-morpholinyl-2,3-dioxo-6-(trifluoromethyl)quinoxalin-1-yl]methylphosphonate (ZK200775). Although some of these compounds have entered clinical trials for the treatment of ischemia, side effects and poor water solubility complicated clinical usage.

AMPA Receptor Noncompetitive Antagonists

The first noncompetitive AMPA receptor antagonist described was 1-(4-aminophenyl)-4-methyl-7,8-methylenedioxy-5*H*-2,3-benzodiazepine hydrochloride (GYKI 52466), which is structurally classified as a benzodiazepine. A related series of selective, noncompetitive antagonists were derived from three classes of compounds: 2, 3-benzodiazepine (GYKI 52466), phthalazine (GYKI 53784/LY303070, GYKI 53773/LY300164), and quinazolinone derivatives (CP-465,022 and CP-526,427). The noncompetitive AMPA antagonists are of interest as potential therapeutic agents because they function effectively even in the presence of high levels of glutamate. Some noncompetitive AMPA receptors antagonists have been evaluated as drug candidates in a number of clinical trials for various neurological disorders.

Positive Allosteric Modulators

When AMPA receptors are activated by glutamate they rapidly desensitize in 1 (or a few) ms (**Figure 4(a)**). This desensitization has been hypothesized to be triggered by domain closure around the agonist, which exerts strain on the ligand-binding dimer

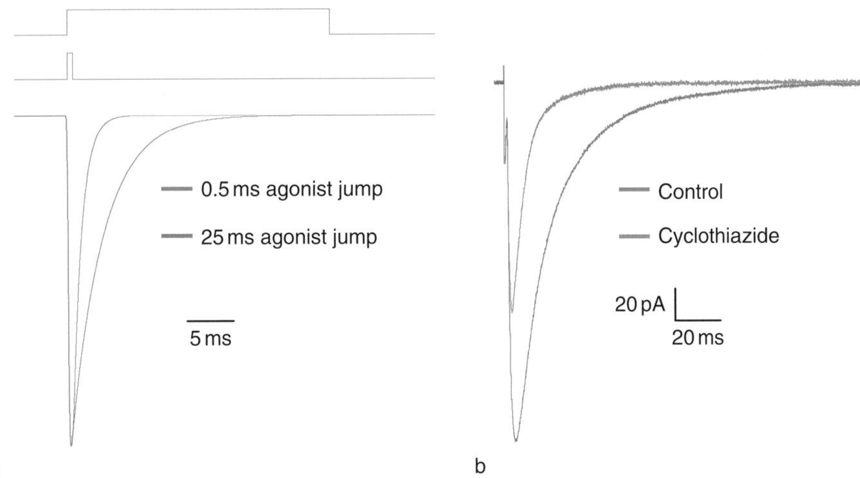

Figure 4 Electrophysiological properties of the AMPA receptor. (a) Simulated traces showing activation of GluR1 homomeric receptors by short (0.5 ms) and long (25 ms) pulses of 10 mM glutamate. The faster decay of current in the red trace activated by short pulse of glutamate represents deactivation. The relatively slower decay of current in green trace represents desensitization. (b) Evoked excitatory postsynaptic currents (eEPSC) from a pyramidal CA1 hippocampal neuron mediated by AMPA receptors in the presence of NMDA and GABA receptor antagonists. The eEPSC in green is in the presence of AMPA receptor desensitization inhibitor cyclothiazide. (a) Model used for simulation from Partin KM, Fleck MW, and Mayer ML (1996) AMPA receptor flip/flop mutants affecting deactivation, desensitization, and modulation by cyclothiazide, aniracetam, and thiocyanate. *Journal of Neuroscience* 16(21): 6634–6647.

interface. The strain can be relieved by opening of the channel or breakdown of the ligand-binding interface, which triggers desensitization. A number of AMPA receptor-positive allosteric modulators have been identified, sometimes referred to as AMPA receptor potentiators. These allosteric potentiators do not activate the AMPA receptors themselves, but enhance the agonist-evoked current by slowing the rate of desensitization (**Figure 4(b)**).

Drugs that positively regulate AMPA receptors fall into three classes: pyrrolidones (also known as ampakines, including aniracetam and piracetam), the benzothiadiazides (cyclothiazide, diazoxide), and biarylpropylsulfonamides (4-[2-(phenylsulfonylamino)ethylthio]-2,6-difluorophenoxyacetamide (PEPA), *N*-2-[4-(4-cyanophenyl)phenyl]propyl-2-propanesulfonamide (LY404187)). Both aniracetam and cyclothiazide almost completely eliminate desensitization of flip splice variants but only slow the entry into desensitized state for flop splice variant. PEPA potentiates AMPA receptor function by attenuating receptor desensitization without any effect on deactivation; Cyclothiazide and PEPA are approximately 100–500 times more potent compared to aniracetam in potentiating AMPA ion currents. PEPA has varying degrees of selectivity for the flop variants. LY404187 suppresses receptor desensitization with a distinct time dependence in the presence of agonist and shows high potency for flip splice variants. Accumulated evidence suggests that AMPA-positive modulators may act as cognitive enhancers, which could be useful as novel therapeutic agents for treating brain disorders such as Alzheimer's disease and schizophrenia.

The binding site of compounds that relieve desensitization appears to reside within the ligand-binding dimer interface. Crystallographic studies of aniracetam suggest that binding can stabilize the ligand-binding core dimerization. This supports the interpretation that desensitization involves breakdown of the dimer interface.

Pore-Blocking Molecules

The current–voltage relationship for AMPA receptors is controlled by subunit composition, with GluR2 endowing receptors with a linear current–voltage curve that reverses near 0 mV. By contrast, AMPA receptors lacking the GluR2 subunit show inward rectification, a term that means cations can flow through the channels into cells at negative potentials more easily than they can flow out of the channel at positive potentials. Inward rectification occurs due to the presence of intracellular polyamines such as spermine, which can block the flow of current out of the cell by entering and blocking the conduction path. A number of elongated and positively charged molecules such as spermine, spermidine, and polyamine toxins (ArgTX-636, PhTX-343, JSTX-3) are noncompetitive open-channel blockers of cation-conducting channels. Spermine is expressed in many CNS neurons and is implicated in regulation of cell division, protein synthesis, and perhaps specific functions in the nervous system. Polyamine toxins are nonoligomeric, low-molecular-weight compounds isolated from spiders and wasps.

The mechanism by which polyamines block AMPA receptors is thought to involve a deep binding site within the receptor pore, the binding to which appears voltage dependent. At positive potentials, polyamine binding is enhanced as the membrane potential favors entry of polyamines from the cell into the channel pore. At sufficiently positive potentials, polyamine block can be relieved as polyamines leave their block site to permeate the channel, joining the outward flow of cations. As described previously, GluR2 subunits that have been edited at the Q/R site at the apex of the reentrant loop have a positively charged arginine at the Q/R site in GluR2, as opposed to glutamine in GluR1, GluR3, and GluR4. The presence of an arginine at this site removes inhibition by intracellular polyamines and like molecules, rendering the current–voltage relationship linear.

See also: AMPA Receptor Cell Biology/Trafficking; AMPA Receptors: Disease; Glutamate Receptor Organization: Ultrastructural Insights; Kainate Receptors: Molecular and Cell Biology; Metabotropic Glutamate Receptors (mGluRs): Molecular Biology, Pharmacology and Cell Biology; NMDA Receptors, Cell Biology and Trafficking.

Further Reading

Armstrong N and Gouaux E (2000) Mechanisms for activation and antagonism of an AMPA-sensitive glutamate receptor: Crystal structures of the GluR2 ligand binding core. *Neuron* 28: 165–181.

Dingledine R, Borges K, Bowie D, et al. (1999) The glutamate receptor ion channels. *Pharmacological Reviews* 51: 7–61.

Erreger K, Chen PE, Wyllie DJA, et al. (2004) Glutamate receptor gating. *Critical Reviews in Neurobiology* 16: 187–225.

Gouaux E (2004) Structure and function of AMPA receptors. *Journal of Physiology* 554: 249–253.

Kew JN and Kemp JA (2005) Ionotropic and metabotropic glutamate receptor structure and pharmacology. *Psychopharmacology (Berlin)* 179(1): 4–29.

Marenco S and Weinberger DR (2006) Therapeutic potential of positive AMPA receptor modulators in the treatment of neuropsychiatric disorders. *CNS Drugs* 20: 173–185.

Mayer ML and Armstrong N (2004) Structure and function of glutamate receptor ion channels. *Annual Reviews in Physiology* 66: 161–181.

Nicoll RA, Tomita S, and Bredt DS (2006) Auxiliary subunits assist AMPA-type glutamate receptors. *Science* 311: 1253–1256.

Palmer CL, Cotton L, and Henley JM (2005) The molecular pharmacology and cell biology of alpha-amino-3-hydroxy-5-methyl-4-isoxazolepropionic acid receptors. *Pharmacological Reviews* 57(2): 253–277.

Partin KM, Fleck MW, and Mayer ML (1996) AMPA receptor flip/flop mutants affecting deactivation, desensitization, and modulation by cyclothiazide, aniracetam, and thiocyanate. *Journal of Neuroscience* 16(21): 6634–6647.

Relevant Website

http://www.iddb.com – Investigational Drugs Database.

Amphetamines

R Kuczenski, University of California at San Diego, La Jolla, CA, USA

Amphetamine is one of a group of powerful central nervous system psychostimulants which also includes, but is not limited to cocaine, methylphenidate, and methamphetamine. The psychostimulants are highly lipid soluble, and readily enter the brain to promote a wide range of behavioral effects. The prototypic drug, amphetamine (α-methylphenethylamine), was introduced in the 1930s, and during the next decade rapidly became popular both for its therapeutic potential and for its abuse liability. Within a few years of its introduction, behavioral changes ranging from its effects in improving the academic performance of 'problem children' and its antinarcoleptic actions to its ability to induce psychosis were recognized.

All of the amphetamine-like psychostimulants are capable of affecting the biogenic amine transmitters (dopamine, norepinephrine, serotonin), but in general, the most pronounced behavioral effects are mediated primarily by their interactions with dopamine neurons. However, it is important to recognize that relatively small structural modifications can significantly alter the pharmacology of the drug, for example, shifting the primary neurochemical effects to serotonin rather than dopamine (*p*-chloroamphetamine (PCA)) or producing powerful hallucinogens (e.g., 2,5-dimethoxy-4-methylamphetamine (DOM, 'STP')) (**Figure 1**).

Molecular Mechanisms

In vivo microdialysis studies in awake, behaving animals have revealed that all amphetamine-like stimulants promote a profound increase in extracellular dopamine in all brain regions that receive dopamine projections, and converging evidence has implicated these changes in the variety of behavioral effects induced by these drugs. Most data regarding the mechanisms of action of amphetamine are derived from studies of dopamine projections to striatal areas, including caudate and nucleus accumbens, because these regions receive dense dopamine innervation. In these regions at least, the quantitative features of the temporal profile for the amphetamine-(as well as cocaine-) induced increases in extracellular dopamine closely parallel the drug's pharmacokinetic profile. The dopamine responses in these regions have also provided the basis for structure–activity studies of potency among amphetamine derivatives. For example, amphetamine exists in two stereoisomeric forms, and in terms of its interaction with dopamine systems: the dextrorotatory S(+)– form is about threefold more potent than the levorotatory R(–)– form. The S(+)– form of the *N*-methylated amphetamine derivative, methamphetamine (*N*-methyl-α-methylphenethylamine), is about equipotent to S(+)– amphetamine, whereas the R(–)– form of methamphetamine is considerably weaker.

The specific mechanism underlying the stimulant-induced increase in extracellular dopamine depends on the type of psychostimulant (i.e., 'uptake blocker,' such as cocaine or methylphenidate, or 'releaser' such as amphetamine or methamphetamine). The normal physiological release of dopamine occurs through a calcium-dependent fusion of the synaptic vesicle with the plasma membrane, followed by the release of the transmitter content of the vesicle into the synaptic space. The released transmitter can interact with its target receptors, following which its active life in the synaptic/extracellular space is terminated by transport or uptake through the Na^+/Cl^--dependent dopamine transporter back into the nerve terminal. All amphetamine-like stimulants increase extracellular dopamine as a consequence of their direct binding to the dopamine transporter. First, by occupying the transporter, they compete for the uptake site with dopamine that has been released into the synaptic space, thereby interfering with the normal removal of transmitter from the synaptic space and back into the nerve terminal. As a consequence, the time during which released dopamine can interact with receptors is prolonged. This is the primary mechanism by which the so-called uptake blockers, including methylphenidate and cocaine, enhance dopamine neurotransmission. However, amphetamine (and methamphetamine) is also a substrate for the dopamine transporter. In addition to inhibiting dopamine uptake, the amphetamines can be transported into the nerve terminal. One consequence of the transport and accumulation of amphetamine within the nerve terminal is the subsequent release of cytoplasmic, nonvesicular dopamine from the nerve terminal and into the synaptic/extracellular space through a dopamine transporter-dependent process. The most common model for how amphetamine promotes this release of dopamine involves reversal of transporter function through a facilitated exchange diffusion process in which the transporter moves cytoplasmic dopamine into the synaptic space; that is, transport of amphetamine from the synaptic space and into the cytosol increases the availability of inward-facing

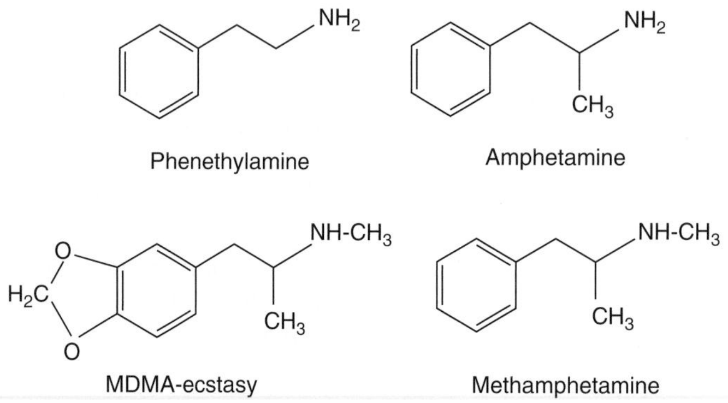

Phenethylamine Amphetamine

MDMA-ecstasy Methamphetamine

Figure 1 Phenylethylamine is the parent compound for many of the amphetamines.

transporter binding sites within the nerve ending to bind dopamine and then transport dopamine out. Most data support the participation of this process in amphetamine-induced dopamine release, but there are some characteristics of the release that cannot be fully explained only by facilitated exchange diffusion. In this regard, more recent evidence suggests that amphetamine can activate a channel-like mode of the dopamine transporter that supports efflux of dopamine out of the nerve ending. Dopamine efflux through the channel occurs in bursts that approximate a quantum of dopamine equivalent to vesicular release, and these bursts appeared to contribute 10–15% of the total amphetamine-induced release. Unlike amphetamine, dopamine itself does not stimulate this dopamine transporter channel-like activity. Thus, the potential physiological role of the channel-like mode of the transporter in the absence of amphetamine is unclear.

Both the facilitated exchange diffusion and channel mechanisms rely on the availability of dopamine in the cytosol. Although most of the nerve-ending dopamine is confined to vesicles, the enzymes for dopamine biosynthesis are outside the vesicle and in the cytoplasm. Thus, under physiological conditions, some transmitter would be available in the cytosol, in transit to the vesicle, to support the amphetamine-induced release process. However, because amphetamine is highly lipophilic, it can also readily enter the neuron by simple diffusion across the plasma membrane, in addition to its accumulation by the transporter, and once inside the nerve ending at sufficiently high concentrations, amphetamine exerts two actions that can augment the content of dopamine in the cytoplasmic pool. First, amphetamine promotes the redistribution of dopamine from the vesicle into the cytosol by collapsing the vesicular pH gradient. The interior of the vesicle is normally acidified and the electrochemical gradient contributes to the energy required by the vesicle to accumulate and retain transmitter. Amphetamine, a lipophilic weak base, accumulates in the vesicle and is protonated, thereby collapsing the pH gradient, disrupting transmitter storage, and enabling leakage of dopamine into the cytosol. Amphetamine also binds to the vesicular monoamine transporter to inhibit vesicular uptake of dopamine. Because of the relatively high concentration of transmitter within the vesicle, disruption of the storage process contributes substantially to the cytoplasmic pool of transmitter. Furthermore, at sufficiently high concentrations, amphetamine also inhibits monoamine oxidase, the primary catabolic enzyme for dopamine in the nerve terminal. Thus, through the combination of disrupted vesicular storage and inhibited catabolism, the catecholamine releasers enhance the availability of cytoplasmic transmitter to provide sufficient substrate for ongoing amphetamine-mediated, transporter-dependent dopamine release.

In addition to their interactions with the dopamine transporter, the amphetamine-like psychostimulants also exhibit some affinity for the norepinephrine and serotonin transporters, and increase extracellular concentrations of both of these transmitters to varying degrees, depending on the specific stimulant. For example, although methamphetamine and amphetamine have similar potency at the dopamine transporter, methamphetamine exhibits higher potency for affecting serotonin than does amphetamine. In contrast, methylphenidate, with little affinity for the serotonin transporter, has minimal effects on extracellular serotonin, even at relatively high doses, although it readily interacts with the norepinephrine transporter. Similar to their interactions with the dopamine transporter, cocaine and methylphenidate act at the norepinephrine and serotonin transporters strictly as uptake blockers, whereas amphetamine and methamphetamine can release both norepinephrine and serotonin, apparently through transporter-dependent mechanisms comparable to amphetamine-mediated, nonvesicular dopamine release. However, amphetamine-induced release of norepinephrine appears to be restricted to doses of the drug sufficiently high to disrupt vesicular storage of the transmitter. Because the final step in norepinephrine biosynthesis involves the intravesicular enzyme, dopamine β-hydroxylase, it has been argued that the cytoplasmic pool of norepinephrine is meager under normal conditions, and would not provide sufficient substrate to support amphetamine-induced release. Thus, at lower doses of amphetamine, drug-induced increases in extracellular norepinephrine likely occur through simple amphetamine blockade of uptake. At higher doses which result in sufficient intracellular concentrations of amphetamine to disrupt vesicular storage of norepinephrine, the cytoplasmic pool of this transmitter can be augmented to provide sufficient substrate for transporter-dependent release.

Behavioral Effects

Acute Administration

Amphetamine and the psychomotor stimulants exert a wide-range of dose-dependent behavioral changes ranging from therapeutic effects to toxic psychosis. These effects involve cognitive, affective, and motor processes. One fundamental behavioral effect of the psychomotor stimulants is that they induce arousal, or wakefulness. The term 'arousal' typically refers to the extent to which an organism is responsive to environmental stimuli. Most evidence suggests that both dopamine and norepinephrine each contribute to wakefulness, and in rodents psychostimulant-induced arousal appears to depend on the actions of the drug on both transmitters within a widely distributed network of subcortical regions, including the medial septal area, the medial preoptic area, and the lateral hypothalamus. These alerting effects of the amphetamines likely contribute to their ability to diminish fatigue and their efficacy in the treatment of narcolepsy, and may also play an indirect role in their ability to enhance cognitive function. In addition, however, at the lowest doses, amphetamine also facilitates a variety of cognitive effects, including enhanced information processing, sustained attention, and working memory, effects which may be important in the therapeutic efficacy of stimulants in the treatment of attention deficit/hyperactivity disorder. These influences on cognitive function appear to involve prefrontal cortical processes that depend on separate stimulant-induced contributions from both dopamine, acting at D1 receptors, and norepinephrine, acting at $\alpha 2$ receptors. However, catecholamine facilitation of cortical cognitive function exhibits an inverted U-shaped dose response, such that higher doses of stimulants disrupt these processes.

As the dose increases, amphetamine stimulates behavioral activation which, in rodents, initially presents as increased locomotion, including both horizontal and vertical movements. With further dose increases, the enhanced locomotion first becomes less varied and more perseverative, following repetitive paths and patterns, then ultimately, is replaced by highly repetitive stereotyped movements which become the predominant feature of the response. Depending on the dose, stereotyped movements can persist for prolonged periods (several hours) in the absence of locomotion, and, as the drug is metabolized, the stereotyped behaviors are replaced by a poststereotypy phase of enhanced locomotion. Comparable behavioral characteristics have been described for most mammals, including primates. The specific behaviors that occur during stimulant-induced stereotypy are highly species specific. For example, rats engage in repetitive sniffing, licking, and gnawing, focused on a particular object; pigeons engage in repetitive pecking, and dogs continuously pace. Human behaviors can appear as movement sequences without an apparent purpose, but are frequently more complex and manipulatory, such as repetitive sorting of objects, compulsive cleaning or bathing, or assembling and dissembling a mechanical object. Stimulant abusers have described stereotyped behaviors as highly pleasurable, funny, and relaxing. Individuals appear to perceive stereotypies as have a calming effect, and become irritated and/or anxious when the stereotypy is interrupted. Converging evidence, developed primarily in rodents, but extended to primates as well, implicates dopamine as the primary neurotransmitter in stimulant-induced motor effects. In particular, stimulant-induced increases in dopaminergic neurotransmission in the nucleus accumbens, a major projection area of the dopamine pathway originating in the ventral tegmental area, mediate the locomotor effects, whereas the caudate, receiving its dopamine projection from the substantia nigra, plays a critical role in the stereotypy effects of the drugs.

The amphetamines are particularly notorious for their ability to induce a state of pleasurable affect, elation and euphoria, and are therefore highly abused substances. Although slow-onset routes of drug delivery can promote an enhanced mood, the rapid increase in brain concentrations of the drug associated with intravenous administration, or inhaling the vapors of volatile methamphetamine hydrochloride through smoking, appears to result in a particularly intense pleasurable feeling, or flash, which is more often sought after by compulsive abusers. Subsequent to this stage of euphoria, the pleasurable effects of the drug can persist for prolonged periods because of the slow metabolism and elimination of the drug (amphetamine and methamphetamine have half-lives in humans of 6–12 h). Neuropharmacological studies have established an important role for dopamine in the reinforcing effects of amphetamine, and, in particular, the mesocorticolimbic dopamine system which projects from the ventral tegmental area to the nucleus accumbens and frontal cortex. For example, most drugs of abuse increase dopamine release in the nucleus accumbens, and laboratory animals will self-administer amphetamine directly into this brain region. In addition, selective destruction of this pathway eliminates amphetamine self-administration. As a consequence, studies directed at understanding psychostimulant addiction, drug craving, and relapse have focused on the adaptational changes in this dopamine pathway associated with chronic exposure to the drug.

Chronic Administration

Sensitization With the repeated, intermittent administration of amphetamine-like stimulants, the motor behaviors induced by the initial treatment with the drug exhibit a sensitized or augmented response to subsequent drug administration. Sensitization (sometimes called reverse tolerance) refers to a progressive enhancement of the locomotor and stereotypy responses with repeated exposure to any of this class of drugs, and the sensitized behavioral response profile persists for prolonged periods of abstinence. Psychomotor sensitization has been reported in all experimental mammalian species that have been examined, including primates, and some evidence suggests that sensitization can be observed in some stimulant response characteristics in humans, as well. In addition, cross-sensitization is evident among the amphetamine-like stimulants; that is, repeated treatment with amphetamine results in an enhanced response to subsequent administration of, for example, cocaine, but not in response to other stimulant-like drugs such as caffeine, whose mechanisms of action are distinct from the psychomotor stimulants. Sensitization does not appear to be a unitary phenomenon and does not simply reflect a shift to the left of the dose–response curve, because the augmented behavioral profile is not reproduced by a higher dose of the drug, and the duration of the response is not extended as would be expected with a higher dose. Rather, behavioral sensitization appears to arise from multiple neuroadaptations in multiple brain nuclei. The mechanisms underlying sensitization appear to include neural restructuring and neurochemical adaptations in the nucleus accumbens and prefrontal cortex. Prominent adaptations involve a long-lasting hyper-responsiveness of mesolimbic dopaminergic and cortical glutamatergic pathways. Because these regions are involved in reward and motivation, sensitization has been speculated to be a factor in the drug craving and compulsive drug-seeking behavior associated with amphetamine abuse, and the neuronal changes underlying the process of sensitization have generated interest as potential adaptations contributing to the process of addiction.

Tolerance Although stimulant-induced motor behaviors can exhibit augmentation with chronic stimulant exposure, depending on the pattern of drug administration, many effects of these drugs exhibit tolerance. For example, the potent anorexic effects of amphetamine, which appear to be mediated by its actions in the lateral hypothalamus, exhibit relatively rapid tolerance, which limits the drug's usefulness as an appetite suppressant. The sympathomimetic effects also exhibit tolerance with chronic exposure, as does the hyperthermia associated with high doses of the drug. Furthermore, the euphoric effects also appear to exhibit tolerance. The tolerance that develops to these effects does not appear to have a pharmacokinetic basis, based on the observations that chronic amphetamine and methamphetamine abusers exhibit drug metabolism profiles similar to those of naïve subjects. Tolerance to the euphoric effects likely contributes to the progressive dose escalation that almost inevitably occurs as those abusers who develop more severe abuse patterns attempt to maintain the intensity of their drug experience. That tolerance develops to the potentially toxic sympathomimetic effects enables compulsive users to self-administer the drug at exceedingly high levels, frequently as much as grams per day, without the lethality that would be experienced by naïve users.

Psychosis Prolonged exposure to high doses of the amphetamines can lead to the development of a paranoid, delusional psychotic state, which can be accompanied by auditory and tactile hallucinations, and which has occasionally been misdiagnosed as schizophrenia. It has been suggested that this drug-induced paranoid state, which frequently evokes considerable anxiety and intense fear, is responsible for the reckless, often aggressive antisocial behavior associated with high-dose stimulant intoxication. Although paranoid psychosis has been elicited with low dose and/or relatively short-term amphetamine exposure, most clinical evidence suggests that this pathological state emerges after chronic stimulant administration. Its full development appears to be related to the amount and duration of drug use, with clear psychosis and severe paranoia more commonly appearing during 'binge' or 'run' patterns of use, when individuals administer the drug throughout the day and night for several days. Some evidence suggests that paranoid psychosis may not occur during the first binge, but rather requires multiple binges before becoming clearly manifested. However, once elicited, this state appears progressively earlier in subsequent binge exposures. Most available data suggest that the paranoid symptoms dissipate within days after discontinuation of drug administration, and thus may include a component reflecting drug-induced pharmacodynamics.

Neurotoxicity Exposure of experimental animals to acute, high doses of amphetamine or methamphetamine results in relatively persistent alterations in dopamine neurons innervating the dorsal striatum. These changes, which have generally been referred to as neurotoxicity, appear to involve the loss of phenotypic markers of dopamine neurites in the striatum without apparent damage to the dopamine perikarya in the substantia nigra. The decrements include levels of dopamine, the dopamine biosynthetic enzymes, tyrosine hydroxylase and aromatic amino acid decarboxylase, and both the plasma membrane dopamine transporter, and the vesicular monoamine transporter. Loss of the vesicular monoamine transporter is frequently considered as an indicator of damage to the nerve terminals rather than reflecting adaptational changes to the high-dose stimulant treatment. The neurotoxicity is facilitated by the hyperthermia associated with acute high-dose stimulant exposure. Accumulating evidence indicates that drug-induced oxidative stress involving reactive oxygen/nitrogen species, and in particular, high levels of oxidized cytoplasmic dopamine consequential to the amphetamine-induced disruption of vesicular storage, plays an important role in the persistent alterations. In addition, microglial activation accompanies these alterations. Evidence identifying potential behavioral correlates of the drug-induced damage has not been forthcoming. Neuroimaging studies using positron emission tomography (PET) technology and radioligands for the dopamine transporter have also revealed relatively persistent decrements in this marker of striatal dopamine nerve terminals in long-term high-dose methamphetamine abusers. Likewise, limited postmortem data from methamphetamine abusers indicates decreases in the dopamine transporter as well as the dopamine biosynthetic enzymes, although no decrement in the vesicular monoamine transporter has been detected. Nevertheless, these results have been generally interpreted to suggest that methamphetamine abuse can lead to prolonged functional damage and/or injury to striatal dopamine terminals, and that these effects may contribute to the enduring impairments of cognitive and motor functioning that have been reported in abstinent methamphetamine abusers.

See also: Addiction: Neurobiological Mechanism; Attention Deficit Hyperactivity Disorder (ADHD): Methylphenidate (Ritalin) and Dopamine; Cognition: Neuropharmacology; Dopamine; Drug Addiction: Behavioral Neurophysiology; Drug Addiction: Cellular Mechanisms; 3,4-Methylenedioxymethamphetamine (MDMA, "Ecstasy"); Noradrenaline; Serotonin (5-Hydroxytryptamine; 5-HT): Neurotransmission and Neuromodulation; Stimulant and Wake-Promoting Substances.

Further Reading

Angrist B (1994) Amphetamine psychosis: Clinical variations of the syndrome. In: Cho AK and Segal DS (eds.) *Amphetamine and Its Analogs: Psychopharmacology, Toxicology, and Abuse*, pp. 387–414. New York: Academic Press.

Berridge CW (2006) Neural substrates of psychostimulant-induced arousal. *Neuropsychopharmacology.* 31: 2332–2340.

Davidson C, Gow AJ, Lee TH, and Ellinwood EH (2001) Methamphetamine neurotoxicity: Necrotic and apoptotic mechanisms and relevance to human abuse and treatment. *Brain Research. Brain Research Reviews* 36: 1–22.

Kuczenski R and Segal DS (1994) Neurochemistry of amphetamine. In: Cho AK and Segal DS (eds.) *Amphetamine and Its Analogs: Psychopharmacology, Toxicology, and Abuse*, pp. 81–113. New York: Academic Press.

Mehta MA, Sahakian BJ, and Robbins TA (2001) Comparative psychopharmacology of methylphenidate and related drugs in human volunteers, patients with ADHD, and experimental animals. In: Solanto MV, Arnsten AFT, and Castellanos FX (eds.) *Stimulant Drugs and ADHD Basic and Clinical Neuroscience*, pp. 303–331. New York: Oxford University Press.

Nichols DE (1994) Medicinal chemistry and structure–activity relationships. In: Cho AK and Segal DS (eds.) *Amphetamine and Its Analogs: Psychopharmacology, Toxicology, and Abuse*, pp. 3–41. New York: Academic Press.

Robinson TE and Berridge KC (1993) The neural basis of drug craving: An incentive-sensitization theory of addiction. *Brain Research. Brain Research Reviews* 18: 247–291.

Segal DS and Kuczenski R (1994) Behavioral pharmacology of amphetamine. In: Cho AK and Segal DS (eds.) *Amphetamine and Its Analogs: Psychopharmacology, Toxicology, and Abuse*, pp. 115–175. New York: Academic Press.

Sulzer D, Sonders MS, Poulsen NW, and Galli A (2005) Mechanisms of neurotransmitter release by amphetamines: A review. *Progress in Neurobiology* 75: 406–433.

Vanderschuren LJ and Kalivas PW (2000) Alterations in dopaminergic and glutamatergic transmission in the induction and expression of behavioral sensitization: A critical review of preclinical studies. *Psychopharmacology (Berlin)* 151: 99–120.

Amphibian Peptides

P Melchiorri and L Negri, University of Rome, Rome, Italy

The secretions of the olocryne granular glands of amphibian skin contain a wide variety of bioactive peptides, nearly all of which are ancestors of mammalian neuropeptides and hormones and are involved in basic cellular processes conserved throughout evolution.

A core of gene clusters within the granular gland cells of frog skin codes for these secreted peptides. A mild transdermal electrical stimulation of the frog releases granular gland contents on the skin surface as a result of glandular syncythia rupture produced by the contraction of myoepithelial cells surrounding the glands. These skin secretions contain the entire peptidome, transcriptome, and genome of the granular gland syncythia (olocryne secretion). The polyadenylated mRNAs constituting the secreted transcriptome and the peptides constituting the secreted proteome are protected from degradation by interactions with co-released amphipathic peptides and mucoproteins endowed with antimicrobial, RNAse-, and protease-inhibitory activities. Thus, amino acid sequencing of secreted peptides, nucleic acid sequencing of peptide-encoding mRNAs, and genomic information retrieving from the secreted DNA can be easily performed in secretion samples collected from few frogs on a regular basis and stored, lyophilized or frozen, for at least 6 years. This is a powerful method of determining evolutionary information on the ancestral sequences of biologically active peptides and proteins and understanding the sequence–function relationship of the human orthologs. Because difficult-to-synthesize bioactive peptides and proteins are currently obtained in large quantities from skin secretions of few amphibian specimens, extensive pharmacological studies have been performed with these amphibian molecules in order to elucidate the functional role of the mammalian orthologs. Whereas some of amphibian skin peptides represent analogs of already known mammalian peptide families (**Table 1**), others represent prototype peptides not encountered before in nature. In many instances, the discovery of new amphibian skin peptides led to the discovery of novel mammalian neuropeptides; notable examples are caerulein, bombesin, and sauvagin (**Table 2**). In the amphibian skin, opioid peptides are represented by two prototypes named dermorphin and deltorphin. They differ from mammalian opioid peptides by their amino acid sequences that selectively bind μ-opioid receptors (dermorphin) or δ-opioid receptors (deltorphin) (**Table 3**). Finally, a new amphibian peptide family, the Bv8-related peptides, has been recently identified (**Table 4**). Bv8-related peptides represent the ancestors of the novel chemokine-related mammalian peptides, the prokineticins.

Amphibian Skin Opioids

The first peptide family of amphibian opioids was discovered in 1981 and named dermorphins. Until the discovery of mammalian endomorphins by Zadina and colleagues, these peptides represented the most potent and selective μ-opioid receptor agonists identified in living organisms. Nine years later, deltorphins were discovered in the amphibian skin; these peptides are still the most potent and selective δ-opioid agonists available today.

A unique characteristic of amphibian opioid peptides (**Table 3**) is the presence in the second N-terminal position of a D-amino acid residue that confers on these compounds high resistance to enzyme degradation. Amphibian opioids have been found only in the skin of South American hylid frogs belonging to the subfamily Phyllomedusinae (*Phyllomedusa*, *Agalychnis*, and *Pachymedusa* species). Although pharmacologists discovered these opioids in Amazonian frogs comparatively recently, the Matses of the upper Amazonian basin knew of the pharmacological properties of amphibian skin opioids for a long time. For centuries they had habitually applied the dried skin secretions of *Phyllomedusa bicolor*, called *sapo* (the Spanish word for toad), to cuts in their skin during shamanistic hunting rituals. The abundance of deltorphins and dermorphins acting together with, or probably synergistically with, the other active peptides present in these secretions might have caused analgesia and behavioral excitation in the hunters.

Origin of the Amphibian Skin Opioids

The D-amino-acid-containing opioid peptides issue from precursors showing a common preproregion (a 22-residue signal peptide and a 18- to 25-residue acidic prosequence) with precursors of the peptide antibiotics dermaseptins (24- to 34-residue polycationic and α-helical amphipathic peptides). Of the three types of dermal glands (mucous, lipid, and granular) in the skin of *Phyllomedusa*, only the granular glands are specifically involved in the biosynthesis and secretion of dermaseptins and deltorphins. The granular glands are the largest glands in the *Phyllomedusa* skin; they lie deeper in the epidermis, are lined by epithelium that is more a syncytium, and are

Table 1 Amphibian and mammalian kinins

Amphibian peptides		*Mammalian peptides*
Aromatic tachykinins		
Physalaemus biligonigerus		Substance P (SP)
Physalaemin	pEADPNKFYGLM-NH_2	RPKPQQFFGLM-NH_2
Uperoleia rugosa		Endokinin A (EKA)
[Lys^5,Thr^6] physalaemin	pEADPKTFYGLM-NH_2	DGGEEQTLSTEAETWVIVALEEGAGPSIQLQLQEVKTGKASQFFGLM-NH_2
Uperoleia marmorata		Endokinin B (EKB)
Uperolein	pEPDPNAFYGLM-NH_2	DGGEEQTLSTEAETWEGAGPSIQLQLQEVKTGKASQFFGLM-NH_2
Uperoleia inundata		Endokinin-1 (EK-1)
Uperin	pEADPNAFYGLM-NH_2	GKASQFFGLM-NH_2
Kassina (Hylambates) maculata		Hemokinin-1 (HK-1)
Hylambatin	NPPDPNRFYGMM-NH_2	SRTRQFYGLM-NH_2
Rana margaratae		
Ranamargarin	DDASDRAKKFYGLM-NH_2	
Pseudophryne guntheri		
PG-SPI	pEPNPDEFYGLM-NH_2	
PG-SPII	pEPNPNEFYGLM-NH_2	
Agalychnis callidryas		
AC-AR1	GPPDPDRFYPGM-NH_2	
AC-AR2	GPPDPDKFYPGM-NH_2	
AC-AR3	pEPDPDKFYPGM-NH_2	
AC-AR4	GPPDPNKFYPVM-NH_2	
Aliphatic tachykinins		
Kasina senegalensis		Neurokinin A (NKA)
Kassinin	DVPKSDQFVGLM-NH_2	HKTDSFVGLM-NH_2
Kassina (Hylambates) maculata		Neurokinin B (NKB)
[G1u2,Pro5]kassinin	DEPKPDQFVGLM-NH_2	DMHDFFVGLM-NH_2
Phyllomedusa bicolor		Neuropeptide K (NPK)
Phyllomedusin	pENPNRFIGLM-NH_2	DADSSIEKQVALLKALYGHGQISHKRHKTDSFVGLM-NH_2
Pseudophryne guntheri		Neuropeptide γ(NPγ)
PG-KI	pEPHPDGFVGLM-NH_2	DAGHGQISHKRHKTDSFVGLM-NH_2
PG-KII	pEPNPDEFVGLM-NH_2	
PG-KIII	pEPHPNDFVGLM-NH_2	
Agalychnis callidryas		
AC-AL	GPPDPNKFIGLM-NH_2	
Bradykinins		
Rana temporaria		
Ranakinin-1	LLPIVGRPPGFSPFR	Bradykinin: RPPGFSPFR
Ranakini-2	RPPGFSPFRIA	Kallidin: KRPPGFSPFR
Ranakinin-3	RPPGFSPFRIAPAL	
Ranakinin-4	RPPGFSPFRIAPASI	
Phyllomedusa rhodei		
Phyllokinin	RPPGFSPFRIY(SO_3)	
Bombina orientalis		
Bombikinin	RPPGFSPFRGKFH	

surrounded by a layer of myoepithelial cells involved in the holocrine rapid discharge of secretory products collected in roundish granules. The granules do not bud off from the membranes on the Golgi apparatus but seem to be generated in the vacuoles of the vacuolated stage during gland development.

All the amphibian skin opioids contain the N-terminal sequence Tyr-D-Xaa-Phe, where the aromatic residues of Tyr^1 and Phe^3 are of L-configuration and the D-Xaa in the second position of the molecule is a D-amino acid (D-Ala or D-Met). The D-enantiomer is encoded, however, by the codon for the L-isomer in the precursor cDNA. Thus L-Xaa^2 must be converted to D-Xaa^2 by an unusual posttranslational reaction that presumably takes place in the precursor itself. Because [L-Xaa^2]-containing peptides have never been found in amphibian skin extracts, the epimerization mechanism probably involves a quantitative inversion of the chirality of the α-carbon of the amino acid residue rather than a racemization, which would yield an equimolar mixture

Table 2 Amphibian peptides that led to the discovery of mammalian neuropeptides

Amphibian peptides	*Amphibians*	*Mammalian neuropeptides*	*References*
Caerulein-related			
pEQDY(SO_3)TGWMDF-NH_2	*Litoria caerulea*	CCK8:	Erspamer (1994)
pEEY(SO_3)TGWMDF-NH_2	*Phyllomedusa sauvagei*	DY(SO_3)MGWMDF-NH_2	
pENDY(SO_3)LGWMDF-NH_2	*Kassina maculata*	Hexagastrin:	Bowie and Tyler (2006)
pEE(OMe)DY(SO_3) TGWMDF-NH_2	*Nyctimystes disrupta*	Y(SO_3)GWMDF-NH_2	
DY(SO_3)LGWMDF-NH_2	*Rana erythraea*	Eugenin: pEQDY(SO_3)VFMHPF-NH_2	
Bombesin-related			
pEQRLGNQWAVGHLM-NH_2	*Bombina bombina*	GRP:	Anastasi, Erspamer and Bucci (1971)
pEQRLGNQWAVGHFM-NH_2	*Bombina orientalis*	GNHWAVGHLM-NH_2	Spindel (2006)
pEQSLGNQWARGHFM-NH_2	*Bombina variegata*		
pEGRLGTQWAVGHLM-NH_2	*Alytes obstetrica*		
pEQWAVGHFM-NH_2	*Litoria aurea*	Neuromedin B:	
SNTALRRYNQWATGHFM-NH_2	*Rana pipiens*	GNLWATGHFM-NH_2	
Sauvagine-related		CRF:	
QGPPISIDLSLELLRKMIEIEKQE KEKQQAANNRLLLDTI	*Phyllomedusa sauvagei*	AEEPPISLDLTFHLLRE VLEMARAEQIAQQA HSNRKLMDIIGK	Erspamer (1994)

Table 3 Amphibian opioid peptides

Amphibian peptides	*Amphibians*	*Mammalian neuropeptides*	*References*
Opioids			
Dermorphins	*Phyllomedusa bicolor*		Negri and Melchiorri (2006)
μreceptor agonists	*Phyllomedusa sauvagei*		
YaFGYPS-NH_2, dermorphin	*Phyllomedusa rhodei*	Analogs not yet identified	
YaFGYPK-NH_2, [Lys^7]dermorphin			
YaFWYPN, [Trp^4,Asn^7]dermorphin			
Deltorphins			
δ-opioid antagonists	*Phyllomedusa burmeisteri*		
YaFDVVG-NH_2, [D-Ala^2]deltorphin I			
YaFEVVG-NH_2 [D-Ala^2]deltorphin II			
YmFHLMD-NH_2 [D-Met^2]deltorphin			

Lower case indicates D-amino acid residues.

of L- and D-isomers. Enzymes catalyzing the formation of D-amino acids are known so far only in yeast. From *Bombina* skin secretions, Kreil and colleagues recently purified a 52 kDa glycoprotein which catalyzes the reaction Ile-Ile-Gly to Ile-D-allo-Ile-Gly. The partial conversion of Ile to D-allo-Ile in peptide linkage proceeds without the addition of cofactors.

Amphibian Opioids and Opioid Receptors

Despite the common N-terminal tripeptide (Tyr-D-Xaa-Phe), the two families of amphibian opioids differ widely in receptor selectivity, although they bind to their own receptors with similar affinities. The N-terminal domain contains the minimum sequence essential for opioid receptor binding, whereas the C-terminal domain contains the address requisites for receptor selectivity. The presence of negatively charged amino acid residues (Asp 114 and Asp 147) that contribute to ligand binding within the putative transmembrane domains II and III of the μ-opioid receptor protein, may explain why positively charged dermorphins have high μ-opioid receptor selectivity and why amidation of their C-terminal group increases their affinity and potency. Similarly, the negatively charged

Table 4 Bv8-related peptides of frogs, fishes, lizards, and snakes

Peptides	*Primary structure*	*Species*
Amphibians		
	1-----------------------------------38	
Bv8	AVITGACDKDVQCGSGTCCAASAWSRNIRFCIPLGNSG	*Bombina variegata*
Bm8a	AVITGVCDRDAQCGSGTCCAASAFSRNIRFCVPLGNNG	*Bombina maxima*
Bm8b	AVITGVRDRDAQCGSGTCCAASAFSRNIRFCVPLGNNG	*Bombina maxima*
Bm8c	AVITGVCDRDAQCGSGTCCAASAFSRNVRFCVPLGNNG	*Bombina maxima*
Bm8d	AVITGVCDRDAQCGSGTCCAASAFSRNIRFCVPLGNNG	*Bombina maxima*
Bm8e	AVITGVCDRDAQCGSGTCCAASAFSRNIRFCVPLGNNG	*Bombina maxima*
Bm8f	AVITGVCDRDAQCGSGTCCAASAFSRNIRFCVPLGNNG	*Bombina maxima*
Bo8	AVITGACDRDVQCGSGTCCAASAWSRNIRFCVPLGNSG	*Bombina orientalis*
	39-----------------------------------77	
Bv8	EDCHPASHKVPYDGKRLSSLCPCKSGLTCSKSGEKFKCS	*Bmbina variegata*
Bm8a	EECHPASHKVPYNGKRLSSLCPCNTGLTCSKSGEKFQCS	*Bombina maxima*
Bm8b	EECHPASHKVPYNGKRLSSLCPCNTGLTCSKSGEKYQCS	*Bombina maxima*
Bm8c	EECHPASHKVPYNGKRLSSLCPCNTGLTCSKSGEKFQCS	*Bombina maxima*
Bm8d	EECHPASHKVPYNGKRLSSLCPCNTGLTCSKSGEKSQCS	*Bombina maxima*
Bm8e	EECHPASHKVPYNGKRLSSLCPCNTGLTCPKSGEKFQCS	*Bombina maxima*
Bm8f	EECHPASHKVPSDGKRLSSLCPCNTGLTCSKSGEKYQCS	*Bombina maxima*
Bo8	EECHPASHKVPYDGKRLSSLCPCKSGLTCSKSGAKFKCS	*Bombina orientalis*
Fishes		
	1-----------------------------------38	
Fugu-1	AVITGACERDVQCGLGLCCAVSLWLRGLRMCAPRGLEG	*Takifugu bimaculatus*
Fugu-2	AVITGACEKDSQCGGGMCCAVSLWIRSLRMCTPMGREG	*Takifugu chinensis*
	39-----------------------------------77	
Fugu-1	DECHPFSHKVPYPGKRQHHTCPCLPHLVCTRDRDSKYRC	*Takifugu bimaculatus*
Fugu-2	DDCHPMSHTVPFFGKRLHHTCPCLPNLSCIPMDEGRAKC	*Takifugu chinensis*
	78-------------94	
Fugu-1	TDDFKNVDLYEVGQTLR	*Takifugu bimaculatus*
Fugu-2	LSTYKYPDYYL	*Takifugu chinensis*
Snakes		
	1-----------------------------------38	
MIT-1	AVITGACERDLQCGKGTCCAVSLWIKSVRVCTPVGTSGE	*Dendroaspis polylepsis*
	39-----------------------------------77	
MIT-1	DCHPASHKIPFSGQRKMHHTCPCAPNLACVQTSPKKFKC	*Dendroaspis polylepsis*
	-80	
MIT-1	LSK	*Dendroaspis polylepsis*
Lizards		
	1-----------------------------------38	
VAR-1	AVITGACDKDLQCGEGMCCAVSLWIRSIRICTPLGSSGE	*Varanus varius*
VAR-2	AVITGACDKDLQCGEGMCCAVSLWIRSIRICTPLGSSGE	*Varanus varius*
	39-----------------------------------77	
VAR-1	DCHPLSHKVPFDGQRKHHTCPCLPNLVCGQTSPGKYKCL	*Varanus varius*
VAR-2	DCHPLSHKVPFDGQRKHHTCPCLPNLVCGQTSPGKHKCL	*Varanus varius*
	78---84	
VAR-1	PEFKNVF	*Varanus varius*
VAR-2	PEFKNVF	*Varanus varius*

C-terminal tetrapeptide of deltorphins enhances deltorphin selectivity for the δ-opioid receptor by electrostatic attraction to the positively charged binding site of the δ-opioid receptor (Arg 292) and electrostatic repulsion from the negatively charged μ-opioid receptor site. Electrostatic attraction within the peptide molecule may explain why deltorphins have a folded structure, whereas dermorphins have a more distended and flexible structure. Unlike the positively charged C-terminus of dermorphins, the negatively charged C-terminal tail of deltorphins comes into close contact with the positively charged N-terminal tripeptide of the molecule and folds the backbone, thus placing the Tyr^1 and Phe^3 aromatic rings in definite orientations best suited for δ-opioid receptor docking.

Dermorphin affinity for μ-opioid receptors and its opioid potency are 20 and 100 times higher respectively than those of morphine. Among other dermorphins, [Lys^7]dermorphin shows an affinity and selectivity for μ-opioid receptors 6 times higher than dermorphin and

100 times higher than morphine. Like μ-opioid agonists, dermorphins produce not only antinociception but also catalepsy, respiratory depression, constipation, tolerance, and dependence, albeit to a lower degree than morphine. Dermorphin-induced antinociception takes place at the supraspinal and spinal levels. Owing to its low central nervous system (CNS) permeability and bioavailability, the analgesic potency of dermorphin is about 250 times higher than that of morphine after intracerebroventricular (ICV) injection but comparable to that of morphine after subcutaneous (SC) injection. [Lys^7]dermorphin and some synthetic analogs bearing a hydrophilic group (a basic amino acid or a glycosyl residue) at the C-terminal end of the molecule enter the CNS in 7–10 times higher amounts than dermorphin, suggesting facilitated transport across the blood–brain barrier by a carrier or endocytosis.

Although D-Ala-deltorphins have δ-opioid binding affinity similar to D-Met-deltorphin, they consistently have the highest δ-opioid selectivity. The rank order of selectivity ($K_{i\delta}/K_{i\mu}$) is D-Ala-deltorphin-I = D-Ala-deltorphin-II $\gg$ D-Met-deltorphin. The high δ-opioid selectivity of D-Ala-deltorphins can be attributed to their C-terminal tetrapeptide sequence, in which the anionic residue plays an important role. Elimination of the charge at the fourth position normally results in opioids that have similar δ- and μ-opioid affinity and generally lack selectivity. Substitution of Gly in the fourth position permits D-Ala-deltorphins to assume a more extended conformation, dramatically increasing their μ-opioid affinity and potency. The hydrophobic qualities of the residues at the fifth and the sixth positions (Val5–Val6) are crucial in maintaining the affinity and selectivity of D-Ala-deltorphins for δ-opioid receptors, as evidenced in peptide analogs in which the aliphatic quality of the side chain was enhanced.

When D-Ala-deltorphins became available, pharmacological and biochemical studies provided evidence of distinct δ1- and δ2-opioid receptor subtypes in the rodent CNS. Studies comparing the binding properties of [^{3}H]deltorphin-I and D-Pen^2, D-Pen^{5}enkephalin (DPDPE) in rat brain synaptosomes, provided the first biochemical evidence for two δ-opioid receptor subtypes. In the mouse brain, homogenates and in rat brain slices, D-Ala-deltorphin-II binds preferentially to a population of δ-opioid receptors that develop after weaning and correspond to δ2-opioid receptor subtype. The accumulated evidence demonstrates that D-Ala-deltorphin-II exerts antinociceptive action at δ2-opioid receptor subtype, whereas D-Ala-deltorphin-I (and DPDPE) interact with greater specificity at δ1-opioid receptor subtype. Using selective δ-opioid antagonists, tolerance development, and δ-opioid receptor knockdown (antisense oligonucleotide-treated) mice and rats, numerous pharmacological studies also suggested that spinal and supraspinal antinociception produced by DPDPE/D-Ala-deltorphin-I and D-Ala-deltorphin-II in rats and mice are mediated by distinct δ-opioid receptor subtypes.

Despite pharmacological evidence of two distinct δ-opioid receptors, molecular biologists have not yet succeeded in cloning δ-opioid receptor subtypes. The available evidence implies that the relatively small change in the longer side chain on Glu, due to the methylene C–C bond, could influence the global conformation of D-Ala-deltorphin-II, exerting a role in selecting the postreceptor transduction pathway by differentially activating δ-opioid receptors.

D-Ala-deltorphin-II induces δ-opioid receptor-mediated analgesia in frogs and also in the invertebrate land snail (*Cepaea nemoralis*). When administered by intrathecal injection in rats, D-Ala-deltorphin-II produces a dose-related inhibition of nociceptive responses. Its inhibitory effect lasts 10–60 min, depending on the dose, and is naltrindole-reversible. Conversely, when injected ICV in rats, D-Ala-deltorphin-II was a weak partial agonist; only doses higher than 30 nmol produced some degree of antinociception and none of the doses elicited the maximum achievable response. Repeated injection of D-Ala-deltorphin-II induces tolerance to the antinociceptive effect. There is no cross-tolerance between antinociception induced by D-Ala-deltorphin-II and that generated either by μ- or by δ1-opioid receptor agonists. The finding that intrathecal injection of D-Ala-deltorphin-II has a higher analgesic effect than DPDPE is probably related more to the predominance of δ2-opioid receptors in the spinal cord than to a prevalence of supraspinal δ1-opioid receptors. Isobolographic analysis revealed that supraspinal/spinal antinociceptive interactions for both the δ1-opioid agonist DPDPE and the δ2-opioid agonist D-Ala-deltorphin-II were synergistic in many nociceptive tests, suggesting that compounds acting through δ-opioid receptors may have sufficient potency for eventual clinical applications.

Data suggest that the δ-opioid agonists play a predominantly modulatory role in antinociception rather than a primary role. In homozygote mice with a disrupted δ-opioid receptor gene, δ-opioid agonist-induced analgesia is reduced. Many studies showed that in mice and in rats the intensity of δ-opioid analgesia depends on coactivation of μ-opioid receptors by endogenous or exogenous opioids. Stress associated with the ICV injection may activate μ-opioid receptors through the release of endogenous opioids and thus potentiate the antinociceptive responses to δ-opioid agonists. Moreover, δ-opioid agonists can be regarded as potential drugs for the treatment of chronic pain: in rats; intrathecal administration of D-Ala-deltorphin-II dose-dependently antagonized the cold-water allodynia which developed after sciatic nerve injury, and ICV administration of D-Ala-deltorphin-II significantly reversed the hyperalgesic response associated with peripheral inflammation.

Injections of D-Ala-deltorphin-II into the rat brain ventricles, ventral tegmental area, and nucleus accumbens invariably increase locomotor activity and induce stereotyped behavior . The ambulatory activity is intermittent and usually intercalated by rearing events. The motor activity is antagonized by the δ-opioid-selective antagonist naltrindole, and by high doses of the μ-opioid antagonist naloxone, but is unaffected by the δ1-opioid-selective antagonist naloxonazine. ICV administration of 1.3 nmol/rat of D-Ala-deltorphin-II increases social contacts in rats. Local application of D-Ala-deltorphin-II to the nucleus accumbens, but not to the nucleus caudatus, increases extracellular dopamine concentrations (by up to 120%). It also stimulates locomotor activity and stereotypes. Repeated ICV injection of D-Ala-deltorphin-II in naive rats induces tolerance to the stimulant effects, whereas repeated daily injections or continuous infusion of morphine results in sensitization to the behavioral-activating effects of the δ-opioid agonist. Deltorphin improves memory consolidation in a passive avoidance apparatus in mice; this effect is abolished by naltrindole. D-Ala-deltorphin-II causes hypothermia in cold-adapted animals. In contrast to μ-opioid agonists, D-Ala-deltorphin-I, at low doses, stimulates respiratory activity in fetal lambs, and this effect is blocked by the simultaneous administration of naltrindole.

The discovery of the amphibian opioid peptides, apart from the intriguing problem of the occurrence of their analogs in the mammalian central and peripheral nervous systems, has provided new insights into the functional role of the μ- and δ-opioid systems.

Bv8-Prokineticins

A small protein, named Bv8 to indicate its origin from the skin secretion of *Bombina variegata* and its molecular weight (8 kDa), is the first amphibian member of the Bv8-prokineticin family. Homologs of Bv8 are present in the skin secretions of other amphibians such as *Bombina bombina*, *Bombina orientalis*, *Bombina maxima*, and of lizards and fishes of *Takifugu* species (**Table 4**). A Bv8 homolog mamba intestinal toxin (MIT-1 or VRPA) is a component of the venom of the black mamba, *Dendroaspis polylepsis*. The striking characteristics of these proteins are their identical N-terminal sequence, AVITG, and the presence of 10 cysteines with identical spacing, which define a five-disulfide-bridged motif called a colipase fold.

The high degree of identity between amphibian Bv8 peptides, fish peptides, and mamba MIT-1 (58%) suggested that similar peptides could also be present in other species, including mammals. In the mouse, rat, cattle, monkey, and human, cDNA cloning identified orthologs of Bv8 (**Table 5**). The two mammalian proteins similar to Bv8 were named prokineticin 1 (PK1 or EG-VEGF) and prokineticin 2 (PK2 or mBv8). A second form of PK2 has been identified and named PK2b (because of an insert of 21 basic amino acids in its sequence). The name prokineticin refers to the ability of these peptides to contract guinea pig ileum, a property shared with amphibian Bv8.

Bv8-Related Neuropeptides

Expression patterns of rodent and human mRNAs for PK1 and PK2 (PROK1 and PROK2) have been reported in peripheral tissues (dorsal root ganglia (DRG), gastrointestinal tract, endocrine glands, spleen, and human and murine leukocytes) and the CNS.

In the neonatal and adult rat and mouse brain, both PROK1 and PROK2 are clearly expressed in the olfactory bulb. Olfactory bulb neurogenesis may depend on PK2 signaling because PROK2-null mice display a marked reduction in the size of the olfactory bulb, a loss of normal olfactory bulb architecture, and an accumulation of neuronal progenitors in the rostral migratory stream. PROK1 and PROK2 are clearly expressed also in the Calleja islands and in suprachiasmatic nucleus (SCN). The PROK2 expression pattern in the SCN of mice and rats is rhythmic, following the circadian cycle (being lowest in the dark phase), and is severely blunted in mutant mice deficient in clock or cryptochrome genes. PK2, therefore, has been indicated as a candidate SCN clock output signal that regulates circadian locomotor rhythm. The reduction of locomotor rhythms in PROK2-null mice was apparent in both hybrid and inbred genetic backgrounds. PROK2-null mice also displayed significantly reduced rhythmicity for a variety of other physiological and behavioral parameters, including sleep–wake cycle, body temperature, feeding, circulating glucocorticoid and glucose levels, and the expression of peripheral clock genes. In addition, PROK2-null mice showed accelerated acquisition of food anticipatory activity during daytime food restriction. Thus PK2, acting as a SCN output factor, is important for the maintenance of robust circadian rhythms.

Prokineticins are also expressed in the neurons of the medial preoptic area, nucleus of solitary tract, trigeminal and facial nuclei, and DRG.

Table 5 Bv8-related prokineticins of mammals

Peptides	*Primary structure*	*Species*
		Mammals
	1--45	
mPK1	AVITGACERDIQCGAGTCCAISLWLRGLRLCTPLGREGEECHPGS	Mouse
mPK2	AVITGACDKDSQCGGGMCCAVSIWVKSIRICTPMGQVGDSCHPLT	Mouse
mPK2b	AVITGACDKDSQCGGGMCCAVSIWVKSIRICTPMGQVGDSCHPLT	Mouse
rPK1	AVITGACERDVQCGAGTCCAISLWLRGLRLCTPLGREGEECHPGS	Rat
rPK2	AVITGACDKDSQCGGGMCCAVSIWVKSIRICTPMGQVGDSCHPLT	Rat
rPK2b	AVITGACDKDSQCGGGMCCAVSIWVKSIRICTPMGQVGDSCHPLT	
bPK1	AVITGACERDVQCRAGTCCAVSLWLRGLRVCTPLGRAGEECHPGS	Cattle
bPK2	AVITGACDRDPQCGGGMCCAVSLWVKSIRICTPMGKVGDSCHPMT	Cattle
bPK2b	AVITGACDRDPQCGGGMCCAVSLWVKSIRICTPMGKVGDSCHPMT	Cattle
mkPK2	AVITGACDKDSQCGGGMCCAVSIWVKSIRICTPMGKLGDSCHPLT	Monkey
mkPK2b	AVITGACDKDSQCGGGMCCAVSIWVKSIRICTPMGKLGDSCHPLT	Monkey
hPK1	AVITGACERDVQCGAGTCCAISLWLRGLRMCTPLGREGEECHPGS	Man
hPK2	AVITGACDKDSQCGGGMCCAVSIWVKSIRICTPMGKLGDSCHPLT	Man
hPK2b	AVITGACDKDSQCGGGMCCAVSIWVKSIRICTPMGKLGDSCHPLT	Man
	46-----------------------------------86	
mPK1	HKIPFLRKRQHHTCPCSPSLLCSRFPDGRYRCFRDLKNANF	Mouse
mPK2	RKVPFWGRRMHHTCPCLPGLACLRTSFNRFICLARK	Mouse
mPK2b	RKSHVANGRQERRRAKRRKRKKEVPFWGRRMHHTCPCLPGLACLRTSFNRFICLARK	Mouse
rPK1	HKIPFFRKRQHHTCPCSPSLLCSRFPDGRYRCSQDLKNVNF	Rat
rPK2	RKVPFWGRRMHHTCPCLPGLACLRTSFNRFICLARK	Rat
rPK2b	RKSHVANGRQERRRAKRRKRKKEVPFWGRRMHHTCPCLPGLACLRTSFNRFICLARK	Rat
bPK1	HKVPFFRKRQHHACPCLPNLLCSRGLDGRYRCSTNLKNINF	Cattle
bPK2	RKVPFLGRRMHHTCPCLPGLACSRTSFNRYTCLAQK	Cattle
bPK2b	RKNHFGNGRQERRKRKRRRKKKVPFLGRRMHHTCPCLPGLACSRTSFNRYTCLAQK	Cattle
mkPK2	RKVPFLGRRMHHTCPCLPGLACLRTSFNRFICLARK	Monkey
mkPK2b	RKNNFGNGRQERRKRKRRKRKKEVPFGRRMHHTCPCLPGLACLRTSFNRFICLARK	Monkey
hPK1	HKIPFFRKRKHHTCPCLPNLLCSRFPDGRYRCSMDLKNINF	Man
hPK2	RKVPFFGRRMHHTCPCLPGLACLRTSFNRFICLAQK	Man
hPK2b	RKNNFGNGRQERRKRKRSKRKKEVPFFGRRMHHTCPCLPGLACLRTSFNRFICLAQK	Man

Bv8-Prokineticin Receptors

The two G-protein-coupled receptors for Bv8-PKs, prokineticin receptor 1 (PKR1) and prokineticin receptor 2 (PKR2) have an overall identity in their amino acid sequences of 85%, with most differences at the N-terminal, and are approximately 80% identical to the previously described mouse orphan receptor gpr73. In specific endothelial cells, neurons, and transfected cells expressing these receptors, Bv8-induced PKR activation stimulates Ca^{2+} mobilization and phosphoinositol turnover, indicating a $G_{q/o}$-protein coupling. Receptor binding studies showed that PKR2 is a MIT-preferring receptor; indeed, the affinity of MIT for PKR2 (in the picomolar range) is approximately 10 times higher than that of PK2 and 50 times higher than that of PK1. PKR1 is a MIT- and PK2-preferring receptor; the affinity of MIT for PKR1 (~5–10 times lower than for PKR2) is comparable to that of PK2 and 60 times higher than that of PK1. The affinity of Bv8 for the receptors is comparable to that of PK2 and is approximately 40 times higher than that of PK1.

In rat embryos from day 12, both receptors are highly expressed in the neuroepithelium lining ventricles, the olfactory bulb, the Gasser-ganglion, and DRG. One day after birth, receptors PKR2 are still expressed at high levels in the olfactory bulb, neuroepithelium lining the ventricles, striatum, hippocampus, thalamic and hypothalamic paraventricular nuclei, SCN, amygdala, and cortex, whereas receptors PKR1 are only found in the cortex. In adult rats, only PKR2 is abundantly or moderately expressed in several discrete brain regions. The presence of PKR2 mRNA and PKR2 protein in the nucleus arcuatus explains the anorexogenic effect of Bv8-prokineticins. The dipsogenic effect of Bv8 depends on its binding to PKR2 in the subfornical organ (SFO). Central and peripheral administration of Bv8 in rats induces growth hormone-releasing hormone (GRH), oxitocin, and vasopressin release and vasopressin-dependent antidiuresis, probably by acting on PKR2 in the paraventricular hypothalamic nucleus. Stimulation of PKR2 in the SCN by Bv8-prokineticins induces sleep in rodents.

Human Kallmann syndrome (KS), which combines anosmia, related to defective olfactory bulb morphogenesis, with hypogonadism due to gonadotropin-releasing hormone deficiency, appears to be related to mutations in the genes coding for PK2 and PKR2. In a cohort of 192 patients affected by KS, 10 different point mutations were identified in the genes encoding PKR2 and four were identified in the genes encoding PK2. The mutations in PK2 were detected in the heterozygous state, whereas PKR2 mutations were found in the heterozygous, homozygous, or compound heterozygous state. In addition, one of the patients heterozygous for a PKR2 mutation also carried a missense mutation in KAL1, thus indicating a possible digenic inheritance of the disease in this individual. These findings reveal that insufficient prokineticin-signaling through PKR2 leads to the abnormal development of the olfactory system and reproductive axis also in humans.

PKR1 and PKR2 mRNAs are expressed in the DRG of neonatal and adult rats. PKR1 is mainly expressed in small and medium-size neurons and PKR2 in large neurons. PKR proteins are present in the DRG, in the outer layers of the dorsal horns of the spinal cord, and in the peripheral terminals of nociceptor axons. Activation of nociceptor PKRs by Bv8 in rats and mice produces nociceptive sensitization to thermal and mechanical stimuli, without inducing any spontaneous, overt nocifensive behavior or local inflammation. A physiological role of Bv8-prokineticins as peripheral and central pain modulators is supported by the observation that mice lacking the PKRs or PK2 are less sensitive to noxious heat (hot-plate test) than are wild-type mice. PKR1-null mice also exhibited impaired development of hyperalgesia after tissue injury. The inflammatory agents CFA and mustard oil produced comparable paw edema in wild-type and PKR1-null mice, but induced inflammatory nociceptive sensitization to heat and pressure significantly lower in PKR1-knockout than in wild-type mice, demonstrating a role of prokineticins and their receptors in inflammatory pain.

The molecular mechanism of the prokineticin-induced hyperalgesia has been investigated in primary cultures of DRG neurons. Of the neurons responding to Bv8, 90% also responded to capsaicin, showing a very high degree of co-localization of functional PKRs with the heat and capsaicin-activated transient receptor potential cation channel subfamily (TRPV1). Half of the Bv8-responding neurons also express calcitonin gene-related peptide (CGRP) and substance P (SP) and release these neuropeptides on exposure to Bv8-prokineticins. Patch-clamp experiments showed that a brief exposure to Bv8 tremendously potentiated the inward current activated by capsaicin in DRG neurons, an effect that is lacking in neurons from PKR1-knockout mice. Bv8 causes translocation of protein kinase C (PKCε) to the neuronal membrane of nociceptors, and Bv8-activated PKCε sensitizes TRPV1 to activating agents by phosphorylating two serine residues of the channel. Indeed, the Bv8-induced capsaicin potentiation is partially inhibited by the PKC inhibitors staurosporine and RO 31–8220. These data indicate that Bv8-induced hyperalgesia results, at least in part, from increased sensitivity of TRPV1 to heat and acid and from the expression and release of excitatory transmitters (CGRP and SP) at the spinal dorsal horn.

In situ mRNA hybridization experiments demonstrated that PK2 is expressed by inflammatory cells, predominantly neutrophils, in the human inflamed tonsil and appendix and in the rat inflamed paw. PK2 released by inflammatory cells can bind and activate PKRs on the primary sensitive neurons contributing to inflammatory pain. Neutrophil extracts, fractionated using ionic exchange chromatography, gel filtration, and reverse phase (RP) chromatography, displayed Bv8-like activity, displaced ^{125}I-MIT binding from PKR1-transfected Chinese hamster ovary (CHO) cell membranes and produced the Bv8-characteristic hyperalgesia when injected intrathecally in rats.

Prokineticin receptors are potential targets for drugs which block the nociceptive information before it reaches the brain. Identifying of the structural determinants required for receptor binding and hyperalgesic activity of Bv8-prokineticins is thus mandatory for the design of PKR antagonists. The highly conserved N-terminal sequence AVITGA and the triptophan residue in position 24 in all members of the Bv8/PK family are required for biological activity; deletions and substitutions in these conserved residues produce antagonist molecules. Preclinical studies are currently being carried out on these mutated PK peptides as PKR antagonists endowed with analgesic properties.

See also: Invertebrate Neurohormone GPCRs; Mammalian Neuropeptide Families; Neuropeptide Signaling in Invertebrates; Neuropeptides: Pain; Opioid Peptides and Receptors.

Further Reading

Anastasi A, Erspamer V, and Bucci M (1971) Isolation and structure of bombesin and alytesin. *Experientia* 27: 166–167.

Bowie JH and Tyler MJ (2006) Bioactive peptides from Australian amphibians: caerulein and other neuropeptides. In: Kastin AJ (ed.) *Handbook of Biological Active Peptides*, pp. 283–289. USA: Academic Press Elsevier.

Erspamer V (1994) Bioactive secretions of the amphibian integument. In: Heatwole H and Barthalmus T (eds.) *Amphibian Biology, Vol. 1: The Integument*, pp. 178–350. Chipping Norton, NSW: Surrey Beatty & Sons.

Hu WP, Zhang C, Li JD, et al. (2006) Impaired pain sensation in mice lacking prokineticin 2. *Molecular Pain* 15: 35–40 [online].

Li JD, Hu WP, Boehmer L, et al. (2006) Attenuated circadian rhythms in mice lacking the prokineticin 2 gene. *Journal of Neuroscience* 26: 11615–11623.

Matsumoto S, Yamazaki C, Masumoto KH, et al. (2006) Abnormal development of the olfactory bulb and reproductive system in mice lacking prokineticin receptor PKR2. *Proceedings of the National Academy of Sciences of the United States of America* 103: 4140–4145.

Negri L, Lattanzi R, Giannini E, et al. (2002) Nociceptive sensitization by the secretory protein Bv8. *British Journal of Pharmacology* 137: 1147–1154.

Negri L, Lattanzi R, Giannini E, et al. (2006) Impaired nociception and inflamed pain sensation in mice lacking the prokineticin receptor PKR1: Focus on interaction between PKR1 and the capsaicin receptor TRPV1 in pain behavior. *Journal of Neuroscience* 26: 6716–6727.

Negri L, Lattanzi R, Giannini E, and Melchiorri P (2006) Modulators of pain: Bv8 and prokineticins. *Current Neuropharmacology* 4: 207–215.

Negri L and Melchiorri P (2006) Opioid peptides from frog skin and Bv8-related peptides. In: Kastin AJ (ed.) *Handbook of Biologically Active Peptides*, pp. 269–275. Boston: Academic Press.

Spindel ER (2006) Amphibian bombesin-like peptides. In: Kastin AJ (ed.) *Handbook of Biological Active Peptides*, pp. 277–281. USA: Academic Press Elsevier.

Vellani V, Colucci M, Lattanzi R, et al. (2006) Sensitization of transient receptor potential vanilloid 1 by the prokineticin receptor agonist Bv8. *Journal of Neuroscience* 26: 5109–5116.

Amygdala: Contributions to Fear

S Maren, University of Michigan, Ann Arbor, MI, USA

History

The physiological basis for emotions has intrigued mankind for centuries. Early Greek philosophers held emotion to be a product of visceral activity – a beating heart reflecting passion and love or a clenched gut fueling jealousy and sadness. Over the course of hundreds of years, however, the study of emotion has moved from the trunk to the head. There are many factors that encouraged this shift in thinking about the localization of emotion. Perhaps the most influential one was the discovery of profound emotional changes in a man named Phineas Gage, whose brain was penetrated by an iron spike in a railway construction accident in 1848. Amazingly, Gage survived the injury to his head, but he was not the same person after recovering from the wound. Relative to his gentle demeanor prior to the accident, he angered easily, used profanity, and could not maintain interpersonal relationships after his accident. Emotions, it would appear, had their roots in the brain.

Not long after Gage's condition was described, scientists were embarking on the first neurological studies of brain function in animals. Sanger Brown and E A Shafer at the University College, London, carried one of the first studies of this kind out in 1888. They systematically examined the behavioral consequences of surgical lesions to the monkey cerebral cortex, with a focus on damage to the temporal and occipital lobes of the brain. In one case, they produced a nearly complete and bilateral lesion of temporal cortex that also included underlying structures, including the amygdala (**Figure 1**). The animal in this case presented with profound behavioral changes. Although its sensory and motor function and general intellectual faculties appeared to be normal, the animal's emotional behavior changed considerably after the lesion. The monkey lost fear of other, dominant monkeys and was tame in the presence of people. It also did not remember an aversive event, readily returning to interact with a monkey that had defeated it in fight, for example.

This observation of emotional changes after temporal cortical damage languished for 50 years until it was rediscovered by Kluver and Bucy in 1937. They similarly observed altered emotional behavior, particularly a loss of fear, in monkeys with temporal cortical lesions. Lawrence Weiskrantz made a key advance in understanding the basis for fear reductions with these lesions in 1956. He showed that the damage to the amygdala, a brain structure underlying the temporal cortex, accounted for the fear loss, and that other changes in behavior were likely the consequence of the cortical damage. Weiskrantz was also the first to suggest that the amygdala has an important role in assigning emotional value (through learning) to stimuli that predict biologically important events (including noxious events). Hence, this seminal work pinpointed a specific brain area that was critical for the genesis of an emotion – in this case, fear. To this day, the amygdala remains a focus for studying the brain mechanisms of fear.

Anatomy of Fear

Fear is a universal emotion that not only has an obvious subjective component ('I feel afraid'), but also a unique behavioral and autonomic signature ('When I am afraid, I freeze in my tracks and my heart pounds in my chest'). The physiological and behavioral aspects of fear are readily studied in both humans and animals and these responses appear to share many similarities, among mammals at least.

Fear is either innate or learned. For instance, primates appear to have an innate predisposition to learn fear of snakes. Monkeys that are born and raised in captivity exhibit fear responses to live or rubber snakes, but not to belts or toy fire trucks (none of which they have seen before). This fear response depends on observing another monkey exhibit fear in the presence of snakes. Laboratory-born deer mice can differentiate predators and nonpredators from their typical habitat, despite never having contacted those animals before. This indicates that fear systems evolved as a defensive behavior system to ward off predators found in an animal's habitat, as well accommodating rapid learning about new threats in the environment. The utility of both types of fear is obvious. Animals that can respond to threats in the environment without requiring past experience with the threat are at an obvious advantage. And rapid learning about novel threats in the environment is similarly adaptive; an animal may only have one opportunity to learn about a threat before it is mortally injured (**Figure 2**).

Several lines of evidence indicate that the amygdala is involved in both innate and learned fear responses. The amygdala ('almond,' Greek) is an oval-shaped collection of neurons located deep within the temporal lobe of the vertebrate brain. It is prominent in

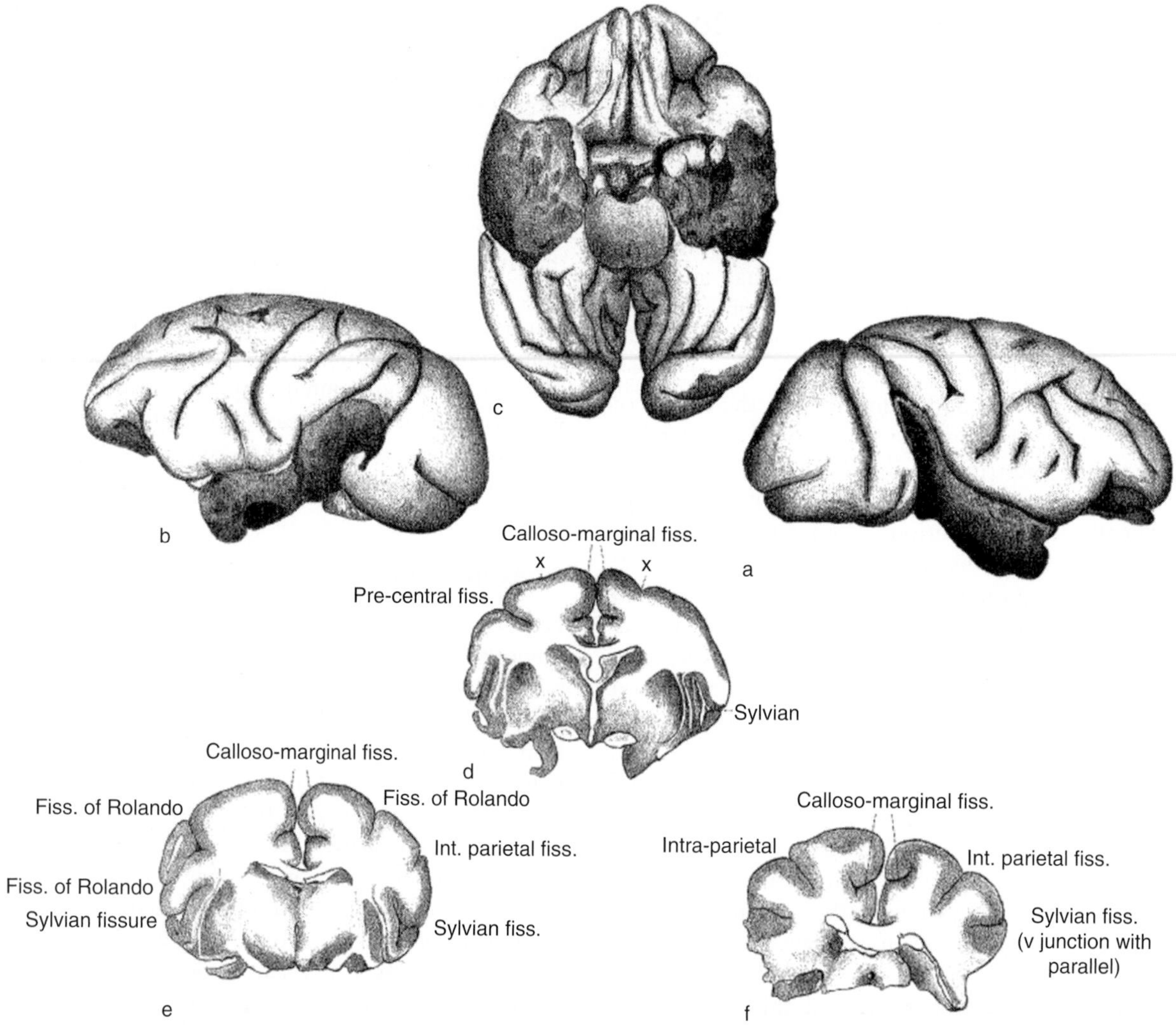

Figure 1 Temporal lobe lesions in a rhesus monkey. Reproduced from Brown S and Schäfer EA (1888) An investigation into the functions of the occipital and temporal lobes of the monkey's brain. *Philosophical Transactions of the Royal Society London, Series B* B179: 303–327, with permission.

Figure 2 A cat stalking and killing a rat. Reproduced from Caldecott R (1878) *The House that Jack Built*. London: Frederick Warne & Co. Ltd. Permission obtained under the Project Gutenberg License.

mammals, but there is also a homologous structure in amphibians and reptiles, among others. In all cases, the amygdala is not a monolithic structure in the brain; it is instead a collection of anatomically and functionally distinct nuclei with different developmental origins. It is therefore commonplace to describe the amygdala as the 'amygdaloid complex,' and to focus on particular nuclei with the amygdala.

In any case, the terms amygdala and amygdaloid complex are synonymous.

Of the ten or so amygdaloid nuclei, there are two major contributors to the fear response: the basolateral complex (including the lateral, basolateral, and basomedial nuclei) and the central nucleus (**Figure 3**). The basolateral complex receives considerable sensory information from subcortical and cortical areas. It in turn projects heavily to the central nucleus, which sends axons to other brain areas involved in generating specific fear responses, such as freezing (immobility), tachycardia, and stress hormone release. Damage to either the basolateral complex or the central nucleus can impair fear responses under a variety of conditions.

Innate Fear

Caroline and Robert Blanchard at the University of Hawaii first demonstrated that rats with amygdala lesions lose their innate fear of cats. Normal rats exhibited fear, particularly freezing behavior, in the presence of a cat. This response was markedly reduced by damage to the amygdala. However, the lesions, like those used in the early monkey studies, damaged both neurons in the amygdala, as well as axons passing through the amygdala from one brain area to another. More selective lesions of the amygdala, or reversible inactivation of amygdala neurons, suggest that the amygdala may not be as important in generating innate fear responses as once believed. For example, neurotoxic lesions (that spare axons) or reversible inactivation of neurons in the amygdala do not prevent fear (freezing behavior, in this case) to a predator odor (trimethylthiazoline, a component of fox feces). In primates, neurotoxic lesions of the amygdala do reduce fear responses to real or artificial snakes, but these effects are not nearly as pronounced as those produced by lesions that also damage axons. Similarly, humans with amygdala damage exhibit normal autonomic responses, such as a galvanic skin response, to aversive stimuli. Another brain area that may be involved in innate fear responses is the hypothalamus. Rosen and colleagues have shown that molecular markers for neuronal activity are increased in the rat hypothalamus, but not the amygdala, after exposure to a predator (a cat).

In sum, large lesions that encompass both neurons and axons passing through the amygdala produce profound disturbances in innate fear reactions, although more selective manipulations of amygdaloid neurons within discrete nuclei produce only mild impairments (if any) in innate fear responses. Indeed, rats with amygdala lesions exhibit normal fear reactions in a number of tests of fear and anxiety, such as the open field (which measures locomotor activity in a brightly lit novel environment) or the elevated-plus maze (in which animals avoid well-lit arms that are elevated above the floor in favor of dark and enclosed arms).

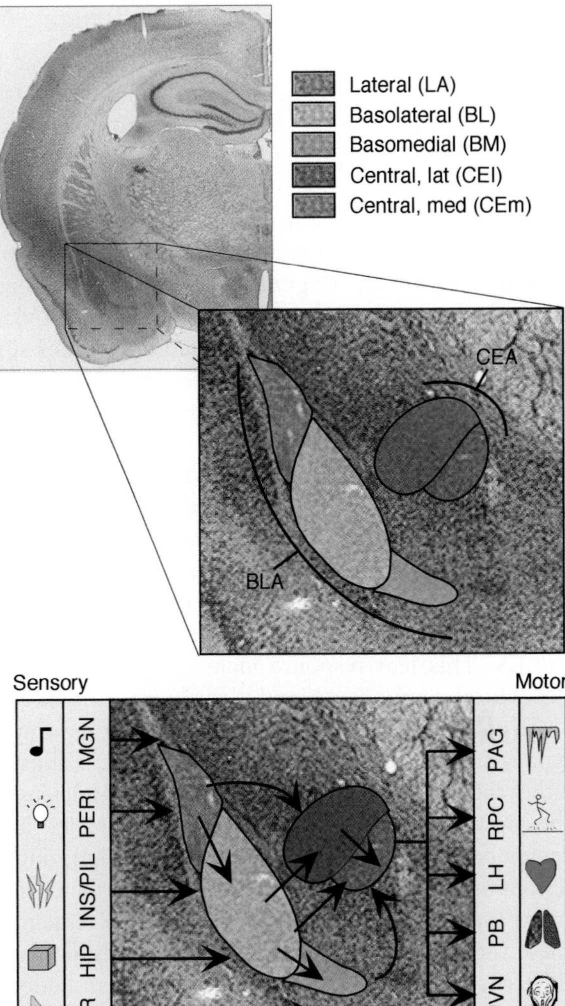

Figure 3 Anatomy of the amygdala. The amygdaloid nuclei can be roughly divided into two subsystems. These include the lateral, basolateral, and basomedial nuclei, which together form the basolateral complex (BLA) and the central nucleus (CEA). The BLA receives and integrates sensory information from a variety of sources. These include the medial and ventral divisions of the thalamic medial geniculate nucleus (MGN, auditory), the perirhinal cortex (PERI, visual), the insular cortex (INS, gustatory and somatosensory), the thalamic posterior intralaminar nucleus (PIL, somatosensory), the hippocampal formation (HIP, spatial and contextual), and the piriform cortex (PIR, olfactory). Intra-amygdaloid circuitry conveys the conditioned stimulus–unconditioned stimulus association to the CEA, where divergent projections to the hypothalamus and brain stem mediate fear responses such as freezing (periaqueductal gray, PAG), potentiated acoustic startle (nucleus reticularis pontis caudalis, RPC), increased heart rate and blood pressure (lateral hypothalamus, LH), increased respiration (parabrachial nucleus, PB), and glucocorticoid release (paraventricular nucleus of the hypothalamus, PVN). For simplicity, all projections are drawn as unidirectional connections, although in many cases these connections are reciprocal.

Learned Fear

As John Watson showed in the early 1920s, fear is readily learned. He conditioned the infant 'Little Albert' to fear white rats by pairing a noxious, loud noise each time he presented the boy with the rat. Albert, who had no fear of the rat before Watson's intervention, quickly came to be upset and scared by the animal, crying and reeling from it when Watson placed it in his lap (**Figure 4**). In this case, Albert learned to associate, through classical or Pavlovian conditioning, the white rat and loud noise. Not surprisingly, fear conditioning has become a standard model for understanding how fear is learned in animals and humans.

In the laboratory, fear conditioning proceeds by exposing animals to an innocuous stimulus (conditioned stimulus; CS) that is followed by a noxious event (unconditioned stimulus; US). After as little as a single conditioning trial, animals come to exhibit a learned fear response (conditional response; CR) to the CS. This fear response includes autonomic and behavioral components, and is easily quantified.

An enormous body of work indicates that the amygdala is critical for fear conditioning. In rats, lesions of either the basolateral complex or the central nucleus of the amygdala severely impair both learning and remembering conditional fear responses. Amazingly, lesions of the amygdala disrupt very old fear memories, up to 1-year-old memories in rats. The involvement of these amygdaloid nuclei in fear learning holds for many different stimuli (e.g., tones, lights, places, and odors) and responses (e.g., freezing, heart rate, analgesia, and acoustic startle). And like rats, humans with amygdala lesions (from either surgical interventions for epilepsy or from an amygdala-damaging genetic condition) show deficits in fear conditioning.

The inability of animals and humans with amygdala damage to learn new fears or remember old fears is not simply a problem with expressing fear. As mentioned previously, many fear responses survive amygdala damage. In fact, deficits in learning fear responses can be reproduced with temporary inactivation of the amygdala during the conditioning experience. And interfering with cellular events in the amygdala that are involved in synaptic plasticity shortly after a conditioning experience can impair memory formation even if the amygdala is functioning normally at the time the aversive event is experienced.

The critical region of the amygdala necessary for associating the CS and the US during fear conditioning is under debate. Considerable data suggest that CS and US convergence occurs in the basolateral complex (particularly the lateral nucleus), suggesting that it is where learning occurs in the amygdala (**Figure 5**). However, sensory information also reaches the central amygdala, and recent work suggests that it may be more important than previously recognized for learning new fears. Indeed, the two structures might work in parallel to mediate different aspects of what is learned during an aversive event.

Given the importance of the amygdala in learning and remembering fearful experiences, it is not surprising that the amygdala becomes active during fear conditioning. In humans, this has been revealed using functional magnetic resonance imaging. Blood flow in the amygdala increases when people view fearful facial expressions or when they are presented with conditioned stimuli that predict aversive unconditioned stimuli. In rats, amygdala activation is observed in the activity of single neurons. In both the basolateral complex and central nucleus, amygdala neurons increase their activity (fire more action potentials) to a newly learned fear stimulus. This activity is closely tied to what the animal has learned about the fear stimulus, and is not simply related to a state of fear itself.

The identification of the amygdala as a key brain area for emotional learning has allowed an unprecedented analysis of the molecular events involved in fear memory formation. Learning is believed to alter synaptic communication in the amygdala, possibly by inducing phenomena such as long-term potentiation, a long-lasting enhancement of synaptic transmission. In line with this possibility, drugs that prevent

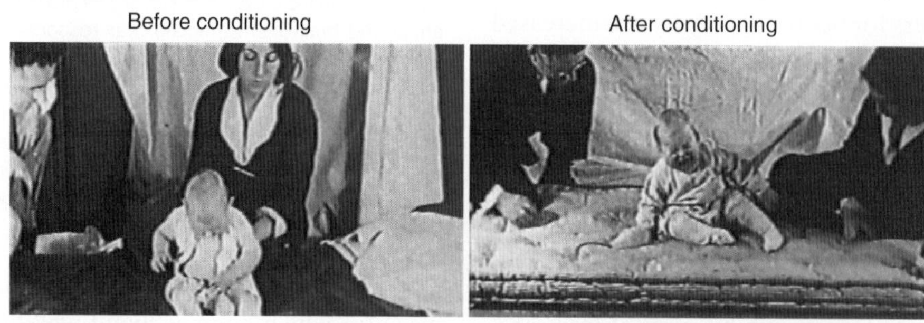

Figure 4 John B Watson conditions fear in 'Little Albert.' Reproduced from Watson JB (1920) *Experimental Investigation of Babies.* Chicago: Stoelting, with permission.

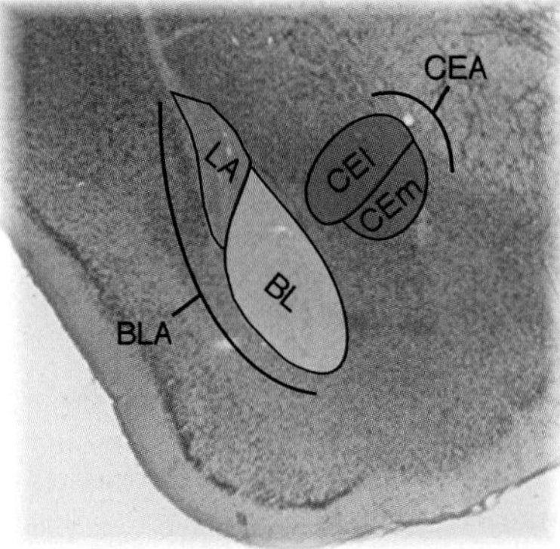

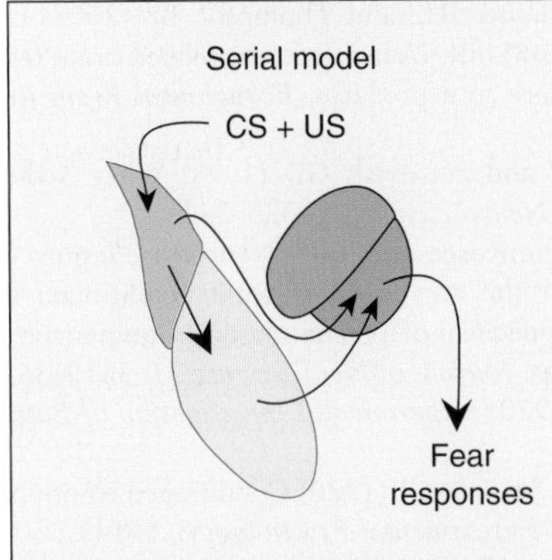

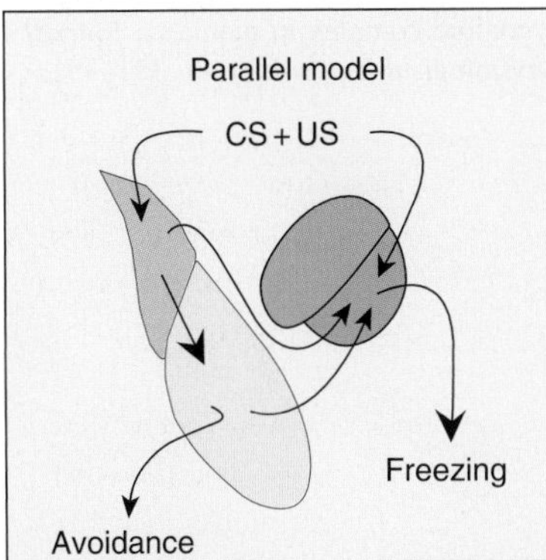

Figure 5 Different models for information processing in the amygdala during fear conditioning. BLA, basolateral complex; LA, lateral nucleus; BL, basolateral nucleus; CEA, central amygdala; CEl, central lateral nucleus; CEm, central medial nucleus; US, unconditioned stimulus; US, unconditioned stimulus. Reprinted from *Neuron*, Vol. 47, Maren S, Synaptic mechanisms of associative memory in the amygdala. 783–786, Copyright 2005, with permission from Elsevier.

long-term potentiation also impede fear learning when they are injected into the amygdala. Hence, fear memories appear to be represented as long-lasting synaptic changes among amygdala neurons.

Pathological Fear

The ability to engage the fear system and associated defensive behaviors after a threat is clearly adaptive. However, fear can also be pathological under some conditions, and excessive fear is at the root of many disorders of fear and anxiety, including panic disorder, posttraumatic stress disorder, and specific phobias, to name a few. In posttraumatic stress disorder, for example, a severe trauma yields a persistent and intrusive state of fear that has debilitating effects on the affected individual.

Pathological fear in posttraumatic stress disorder and other anxiety disorders is often associated with elevated activity in the amygdala. There is some evidence that elevated amygdala activity is the consequence of a loss of cortical inhibition. For example, prefrontal cortical activity is often inversely correlated with amygdala activity in patients with anxiety disorders.

Clinical interventions for anxiety disorders are varied, but many involve behavioral procedures, including exposure therapy to reduce fear in a controlled setting. In its essence, exposure therapy is an extinction procedure in which a fear stimulus is presented without its aversive consequence, in a safe environment. After fear conditioning, CRs can be extinguished by presenting the CS without the US for several trials. Importantly, there is considerable evidence that this procedure does not erase the fear memory, but rather results in new learning (e.g., CS is safe) that inhibits the fear memory. The persistence of the fear memory after extinction is demonstrated by the renewal of the fear when a CS is presented outside of the environment where extinction was performed, and by the spontaneous recovery of fear responses as time passes after the end of extinction training.

The new learning that accompanies extinction also requires the amygdala. Pharmacological manipulations in the amygdala that impair fear conditioning also impair extinction learning. Moreover, drugs that enhance the activity of amygdala neurons can facilitate extinction under some conditions. This may be an important complement to behavioral interventions for treating anxiety disorders.

In addition to the amygdala, there are other brain areas implicated in the suppression of pathological fear. Another medial temporal lobe structure, the hippocampus, is important for learning and remembering the places in which fear and extinction memories are established. The prefrontal cortex exhibits inhibitory control over the amygdala and may be involved in promoting extinction memory when there is conflict between safety and danger signals. It is clear that a wide network of neurons is engaged in both learning new fears and suppression of old ones.

See also: Amygdala: Structure and Circuitry in Primates; Amygdala: Structure and Circuitry in Rodents and Felines; Aversive Emotions: Molecular Basis of

Unconditioned Fear; Emotion: Neuroimaging; Extinction: Anatomy; Genetics of Human Anxiety and Its Disorders; Panic Disorder as an Emotional Disorder; Pharmacology of Fear Extinction; Phobia and Human Evolution; Prefrontal Cortex: Structure and Anatomy.

Further Reading

Adolphs R, Tranel D, Damasio H, et al. (1995) Fear and the human amygdala. *Journal of Neuroscience* 15: 5879–5891.

Blanchard DC and Blanchard RJ (1972) Innate and conditioned reactions to threat in rats with amygdaloid lesions. *Journal of Comparative and Physiological Psychology* 81: 281–290.

Brown S and Schäfer EA (1888) An investigation into the functions of the occipital and temporal lobes of the monkey's brain. *Philosophical Transactions of the Royal Society London, Series B* B179: 303–327.

Caldecott R (1878) *The House that Jack Built.* London: Frederick Warne & Co.

Cardinal RN, Parkinson JA, Hall J, et al. (2002) Emotion and motivation: the role of the amygdala, ventral striatum, and prefrontal cortex. *Neuroscience & Biobehavioral Reviews* 26: 321–352.

Choi JS and Brown TH (2003) Central amygdala lesions block ultrasonic vocalization and freezing as conditional but not unconditional responses. *Journal of Neuroscience* 23: 8713–8721.

Davis M (1992) The role of the amygdala in fear and anxiety. *Annual Review of Neuroscience* 15: 353–375.

Helmstetter FJ (1992) The amygdala is essential for the expression of conditional hypoalgesia. *Behavioral Neuroscience* 106: 518–528.

Kluver H and Bucy PC (1937) 'Psychic blindness' and other symptoms following bilateral temporal lobectomy in rhesus monkeys. *American Journal of Physiology* 119: 352–353.

LaBar KS and Cabeza R (2006) Cognitive neuroscience of emotional memory. *Nature Reviews Neuroscience* 7: 54–64.

LeDoux JE (2000) Emotion circuits in the brain. *Annual Review of Neuroscience* 23: 155–184.

Maren S (2001) Neurobiology of Pavlovian fear conditioning. *Annual Review of Neuroscience* 24: 897–931.

Maren S (2005) Synaptic mechanisms of associative memory in the amygdala. *Neuron* 47: 783–786.

Maren S and Quirk GJ (2004) Neuronal signalling of fear memory. *Nature Reviews Neuroscience* 5: 844–852.

Pare D and Collins DR (2000) Neuronal correlates of fear in the lateral amygdala: Multiple extracellular recordings in conscious cats. *Journal of Neuroscience* 20: 2701–2710.

Rosen JB, Adamec RE, and Thompson BL (2005) Expression of egr-1 (zif268) mRNA in select fear-related brain regions following exposure to a predator. *Behavioural Brain Research* 162: 279–288.

Swanson LW and Petrovich GD (1998) What is the amygdala? *Trends in Neurosciences* 21: 323–331.

Wallace KJ and Rosen JB (2001) Neurotoxic lesions of the lateral nucleus of the amygdala decrease conditioned fear but not unconditioned fear of a predator odor: Comparison with electrolytic lesions. *Journal of Neuroscience* 21: 3619–3627.

Watson JB (1920) *Experimental Investigation of Babies.* Chicago: Stoelting.

Watson JB and Rayner R (1920) Conditioned emotional reactions. *Journal of Experimental Psychology* 3: 1–14.

Weiskrantz L (1956) Behavioral changes associated with ablation of the amygdaloid complex in monkeys. *Journal of Comparative and Physiological Psychology* 49: 381–391.

Amygdala: Structure and Circuitry in Primates

R Kelly and L Stefanacci, Salk Institute for Biological Studies, La Jolla, CA, USA

Introduction

The primate amygdala comprises a group of cytoarchitecturally, histochemically, and connectionally heterogeneous nuclei in the medial temporal lobes. Clinically, impairments in amygdala function have been implicated in social, emotional, and perceptual disorders such as autism, depression, and schizophrenia. Numerous experimental studies support the hypothesis that the amygdala is critical for forming associations between environmental stimuli and their motivational and emotional significance. The anatomical and physiological substrates of the amygdala's role in learning have been most thoroughly described in fear conditioning paradigms, in which an animal develops a conditioned fear response to a previously neutral stimulus; studies in human and nonhuman primates have also demonstrated that the amygdala is important for forming positive associations. The amygdala's widespread interconnections with cortical and subcortical sites provide the substrate for its role in learning and behavior. Here we summarize these interconnections and suggest their functional significance.

Structure

The individual nuclei of the primate amygdala are distinguished by their histological profiles (**Figure 1**). The deep nuclei include the lateral, basal, accessory basal, and paralaminar nuclei. The superficial nuclei surround the deep nuclei and, for the most part, are situated on the medial aspect of the amygdala. They include the anterior and posterior cortical nuclei, the periamygdaloid cortex, the nucleus of the lateral olfactory tract, the medial nucleus, and the central nucleus. Other nuclei include the anterior amygdaloid area, which forms a transition between the amygdala and the substantia innominata; the amygdala–hippocampal area, which is situated between the amygdala and the hippocampal formation; and the intercalated nuclei, which are heterogeneous groups of cells that are embedded in white matter within the amygdala. Individual amygdala nuclei have been further partitioned based on immunohistochemical patterns, cell size, and cell packing density (**Figure 1**).

Intrinsic Connections

The lateral, basal, and accessory basal nuclei are the primary targets of cortical and subcortical projections to the amygdala. From these nuclei, information flows in a general medial and dorsal direction through the amygdala (**Figure 2**). A substantial portion of output from the lateral, basal, and accessory basal nuclei is directed to the central nucleus. The central nucleus, in turn, projects to regions of basal forebrain, hypothalamus, and brain stem that control autonomic and somatomotor responses. While the central nucleus is the primary source of amygdala output to subcortical sites, the basal and accessory basal nuclei are the main sources of amygdala output to cortical sites.

Cortical Connections

Sensory Cortex

All sensory systems project to the amygdala. The olfactory system provides the only direct primary sensory input to the amygdala. Olfactory inputs are directed to the anterior cortical nucleus, the periamygdaloid cortex, and the deep nuclei. Other sensory input originates from modality-specific association cortices and is directed primarily to the lateral nucleus. The bulk of unimodal sensory input to the amygdala arises from visual processing areas (Von Bonin and Bailey's areas TEO and TE) in the temporal lobe. Auditory information is relayed through auditory association cortex (area TA) in the superior temporal gyrus, and gustatory input and somatosensory information are relayed through anterior (agranular and dysgranular) and posterior (granular) insular cortices. Progressively higher level processing areas in visual and auditory association cortices originate more robust projections that terminate in a more widespread manner in the amygdala. This suggests that highly processed sensory information exerts a relatively greater influence in the amygdala. Sensory afferents terminate in a topographically organized manner in the lateral nucleus. Fibers relaying visual information terminate dorsally and rostrally, those relaying auditory information terminate ventrally and caudally, and fibers relaying somatosensory information terminate in a medial position.

The amygdala projects to all of the sensory cortical areas that innervate it. The projections arise primarily in the basal nucleus. Thus, sensory information accesses the amygdala by way of the lateral nucleus, and the return projections originate in the basal nucleus. This suggests that sensory information undergoes at least one stage of processing within the amygdala before returning to sensory cortex.

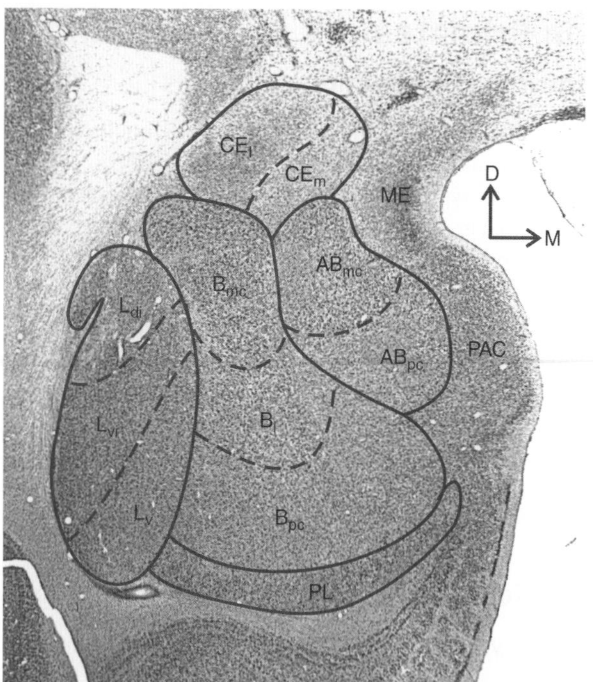

Figure 1 Nissl-stained coronal section through the midrostral–caudal level of the macaque amygdala. AB_{mc}, accessory basal nucleus, magnocellular division; AB_{pc}, accessory basal nucleus, parvicellular division; B_{mc}, basal nucleus, magnocellular division; B_i, basal nucleus, intermediate division; B_{pc}, basal nucleus, parvicellular division; CE_l, central nucleus, lateral division; CE_m, central nucleus, medial division; L_{di}, lateral nucleus, dorsal intermediate division; L_{vi}, lateral nucleus ventral intermediate division; L_v, lateral nucleus, ventral division; ME, medial nucleus; PAC, periamygdaloid cortex; PL, paralaminar nucleus; D, dorsal; M, medial.

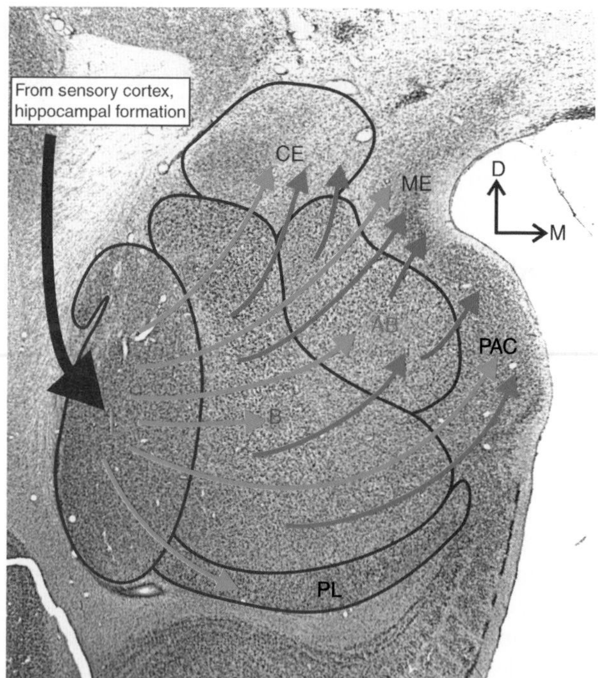

Figure 2 Nissl-stained coronal section through the midrostral–caudal level of the macaque amygdala. Arrows demonstrate the general lateral-to-medial flow of information from the lateral (red), basal (green), and accessory basal (blue) nuclei. The precise topographic organization and strength of the projections are not represented (see text for further details of extrinsic connectivity). AB, accessory basal nucleus; B, basal nucleus; CE, central nucleus; L, lateral nucleus; ME, medial nucleus; PAC, periamygdaloid cortex; PL, paralaminar nucleus; D, dorsal; M, medial.

Consistent with the pattern of sensory input to the lateral nucleus, output from the basal nucleus to sensory cortex is more robust to high-level processing areas in the temporal lobe.The amygdala also projects to areas of sensory cortex that do not project directly to the amygdala. For example, the basal nucleus projects weakly to visual areas V4, MT, and VI, to primary auditory cortex, and to ventral portions of somatosensory cortex, but these areas do not originate projections to the amygdala. Thus, the amygdala has direct access to early stages of sensory processing and may modulate early stages of perception.

Perirhinal and Parahippocampal Cortex

The perirhinal and parahippocampal cortices are major sources of polysensory input to the amygdala. The projection from the perirhinal cortex is directed primarily to the lateral and basal nuclei. The projection from anterior, polar regions of perirhinal cortex is more robust and terminates in a more widespread manner compared to posterior and ventral regions. The projection from the parahippocampal cortex is weaker and is directed primarily to the basal nucleus. The amygdala projection to perirhinal cortex arises from cells throughout the amygdala, particularly the lateral, basal, and accessory basal nuclei. The projection to the parahippocampal cortex arises almost exclusively from the magnocellular division of the basal nucleus.

Perirhinal and parahippocampal cortices are strongly interconnected to the hippocampal formation via the entorhinal cortex, and they receive input from a wide variety of cortical areas. Perirhinal cortex receives projections from visual association areas TE and TEO, dysgranular and granular insular cortex, area 13 of the orbitofrontal cortex, and parahippocampal cortex. The parahippocampal cortex receives projections from visual areas V4, TE, and TEO, and from polymodal association cortices, including the retrosplenial cortex and the dorsal bank of the superior temporal sulcus. The posterior parietal lobe, insular cortex, and prefrontal cortex also project to the parahippocampal cortex. Their connectional organization, together with behavioral studies in monkeys, suggests that the perirhinal cortex supplies the amygdala with information about object recognition, and the parahippocampal cortex relays information about visuospatial memory.

Entorhinal Cortex

The entorhinal cortex projects weakly to the basal nucleus. Efferent fibers from the entorhinal cortex pass through the lateral nucleus, but it is not clear if the fibers form synapses or terminal plexuses within the nucleus. The projection from the amygdala to the entorhinal cortex arises primarily from the lateral nucleus and is most robust passing to anterior portions of the cortex. Unlike its projections to other areas of cortex, the basal nucleus contributes only a weak projection to entorhinal cortex. Two-thirds of all cortical projections to the hippocampus are relayed through the entorhinal cortex. It is not known if the entorhinal cortex provides the same information to the amygdala as it does to the hippocampus.

Prefrontal Cortex

The prefrontal cortex, including anterior cingulate and orbital and medial prefrontal areas, projects primarily to the basal and accessory basal nuclei. The projection from caudal prefrontal cortex (areas 10 and 25) is more robust and terminates more widely in the amygdala relative to rostral prefrontal cortex.

The amygdala projects to the prefrontal cortical areas that innervate it. Cortically directed efferents arise primarily in the basal and accessory basal nuclei. The organization of the pathways connecting the amygdala and prefrontal cortex is partially overlapping. That is, there appear to be some point-to-point interconnections between the medial prefrontal cortex and the amygdala but not between the orbital prefrontal cortex and the amygdala.

The dorsolateral prefrontal cortex does not project to the amygdala. However, the basal nucleus projects weakly to Walker's areas 46, 9, 8, and 6 in the dorsolateral prefrontal cortex. These projections may allow the amygdala to influence working memory and visuomotor and somatomotor behaviors.

The amygdala is not directly connected to parietal cortex. The strength of amygdala–temporal connections, compared to the lack of amygdala–parietal connections, supports the notion that the identification of a stimulus (what?) is more relevant to amygdala processing than is its location (where?). However, visuospatial information from the parietal cortex may reach the amygdala through disynaptic connections with polysensory areas in temporal cortex, such as the parahippocampal cortex.

Subcortical Connections

Basal Forebrain

The amygdala makes a substantial projection to the basal forebrain, directed primarily to large cholinergic neurons in the nucleus basalis of Meynert and the horizontal limb of the diagonal band. The projection arises in the parvicellular portion of the basal nucleus, the accessory basal nucleus, and the central nucleus. The projection from the basal forebrain to the amygdala originates in cholinergic neurons of the nucleus basalis of Meynert and is directed primarily to the magnocellular portion of the basal nucleus. The magnocellular division of the basal nucleus receives the most substantial cholinergic innervation of any forebrain region.

Cholinergic neurons in the basal forebrain project to widespread regions of the cerebral cortex. They are sensitive to the motivational relevance of sensory cues and they have a modulatory effect on cortical activity and attentional processes. Therefore, the basal forebrain provides an additional route for the amygdala to modify cortical activity in response to sensory events.

Striatum

The amygdala makes a substantial projection to the striatum. The projection originates primarily in the basal nucleus and terminates in the ventral striatum, throughout the caudate nucleus and the rostral–caudal extent of ventral putamen. The amygdala–striatal projection is topographically organized, with the parvicellular division of the basal nucleus projecting to the medial portion of the ventral striatum and the magnocellular division of the basal nucleus projecting more extensively to the body and tail of the caudate nucleus and the ventral putamen. The amygdala innervates regions of the striatum that are also targeted by cortical areas that participate in somatosensory, motor, and cognitive operations. This suggests that motivational and emotional information from the amygdala interacts in the striatum to modulate habit and goal-directed behaviors. The striatum does not make a return projection to the amygdala.

Thalamus

The amygdala makes a substantial projection to the thalamus, originating primarily in the basal and accessory basal nuclei and terminating in the magnocellular division of the mediodorsal nucleus. The magnocellular division of the mediodorsal nucleus projects to the same regions of prefrontal cortex that receive direct input from the amygdala Thus there are direct and nondirect routes from the amygdala to the prefrontal cortex. However, axon terminals from the amygdala and thalamus have different laminar distributions in the prefrontal cortex.

The mediodorsal nucleus is part of a diencephalic memory system, and the strong projection from the amygdala to the mediodorsal nucleus suggests that the amygdala contributes to mnemonic operations in the thalamus. The mediodorsal nucleus does not make a return projection to the amygdala. The central and medial amygdala nuclei project to midline thalamic nuclei, and the central nucleus also makes a weak projection to medial pulvinar. These portions of thalamus project weakly back to the central nucleus.

Hypothalamus

The amygdala projects to the medial and lateral hypothalamic nuclei. The projection to the medial hypothalamus originates in the cortical nuclei and the basal and accessory basal nuclei. The projection to the lateral hypothalamus originates primarily in the central nucleus. Fear conditioning experiments in rats have shown that the projection from the central nucleus to the lateral hypothalamus is necessary for expression of the autonomic conditioned response. The return projection from the hypothalamus to the amygdala is much weaker and is directed primarily to the central, medial, basal, and accessory basal nuclei.

Brain Stem

Projections from the central nucleus extend through the midbrain (where they terminate in the substantia nigra and ventral tegmentum), the pons, and the medulla, and continue into the spinal cord. No other amygdala nuclei participate in this descending pathway. Along their route, fibers innervate a number of structures that control autonomic and somatomotor responses, including the periaqueductal gray, the dorsal vagal nuclei, the parabrachial nuclei, the nucleus of the solitary tract, and the nucleus ambiguous. Many of these nuclei project back to the amygdala.

Stimulation of the central nucleus or its targets results in autonomic and somatomotor responses, such as cessation of spontaneous activity, pupillary dilation, piloerection, posturing for attack, elevation of blood pressure, and changes in respiration and heart rate. This functional connectivity highlights the amygdala's role in driving responses to emotional events.

Chemoarchitecture

The amygdala is unique in the diversity of neuroactive substances that are distributed throughout it, including amino acids, catecholamines, acetylcholine, peptides, and hormones that exert excitatory, inhibitory, and modulatory effects on the amygdala and its efferent targets. The amygdala, and the central nucleus in particular, is a primary site of anxiolytic and anxiogenic action in the brain.

Function

The amygdala, by way of its extensive intrinsic and extrinsic connections, participates in multiple functional systems. Behavioral studies have demonstrated the importance of the amygdala, in conjunction with these systems, for (1) assigning emotional and motivational significance to a stimulus, (2) generating appropriate responses to environmental and social cues, and (3) modifying behavior based on expected outcomes.

The amygdala's circuitry suggests that it integrates incoming information from sensory, memory, and cognitive systems and provides feedback to each of these systems. This organization indicates that in addition to the aforementioned functions, the amygdala has the potential to modulate memory, learning, cognition, and perception.

See also: Amygdala: Structure and Circuitry in Rodents and Felines; Amygdala: Contributions to Fear; Aversive Emotions: Molecular Basis of Unconditioned Fear; Depression and the Brain; Emotional Learning in Humans; Emotional Hormones and Memory Modulation; Emotional Influences on Memory and Attention; Genetics of Human Anxiety and Its Disorders; Pharmacology of Fear Extinction.

Further Reading

Aggleton JP (ed.) (1992) *The Amygdala. Neurobiological Aspects of Emotion, Memory, and Mental Dysfunction.* New York: Wiley-Liss.

Bauman MD and Amaral DG (2005) The distribution of serotonergic fibers in the macaque monkey amygdala: An immunohistochemical study using antisera to 5-hydroxytryptamine. *Neuroscience* 136: 193–203.

Bauman MD, Lavenex P, Mason WA, et al. (2004) The development of social behavior following neonatal amygdala lesions in rhesus monkeys. *Journal of Cognitive Neuroscience* 16: 1388–1411.

Baxter MG and Murray EA (2002) The amygdala and reward. *Nature Reviews Neuroscience* 3: 563–573.

Emery NJ, Capitanio JP, Mason WA, et al. (2001) The effects of bilateral lesions of the amygdala on dyadic social interactions in rhesus monkeys (*Macaca mulatta*). *Behavioral Neuroscience* 115: 515–544.

Gallagher M and Chiba AA (1996) The amygdala and emotion. *Current Opinion in Neurobiology* 6: 221–227.

Ghashghaei HT and Barbas H (2002) Pathways for emotion: Interactions of prefrontal and anterior temporal pathways in the amygdala of the rhesus monkey. *Neuroscience* 115: 1261–1279.

LaBar KS and Cabeza R (2006) Cognitive neuroscience of emotional memory. *Nature Reviews Neuroscience* 7: 54–64.

Phelps EA (2006) Emotion and cognition: Insights from studies of the human amygdala. *Annual Review of Psychology* 57: 27–53.

Pitkanen A and Amaral DG (1993) Distribution of calbindin-d28k immunoreactivity in the monkey temporal lobe: The amygdaloid complex. *Journal of Comparative Neurology* 331: 199–224.

Pitkanen A and Amaral DG (1993) Distribution of parvalbumin-immunoreactive cells and fibers in the monkey temporal lobe: The amygdaloid complex. *Journal of Comparative Neurology* 331: 14–36.

Pitkanen A and Amaral DG (1994) The distribution of GABAergic cells, fibers, and terminals in the monkey amygdaloid complex: An immunohistochemical and *in situ* hybridization study. *Journal of Neuroscience* 14: 2200–2224.

Sadikot AF and Parent A (1990) The monoaminergic innervation of the amygdala in the squirrel monkey: An immunohistochemical study. *Neuroscience* 36: 431–447.

Stefanacci L and Amaral DG (2000) Topographic organization of cortical inputs to the lateral nucleus of the macaque monkey amygdala: A retrograde tracing study. *Journal of Comparative Neurology* 421: 52–79.

Stefanacci L and Amaral DG (2002) Some observations on cortical inputs to the macaque monkey amygdala: An anterograde tracing study. *Journal of Comparative Neurology* 451: 301–323.

Stefanacci L, Buffalo EA, Schmolck H, et al. (2000) Profound amnesia after damage to the medial temporal lobe: A neuroanatomical and neuropsychological profile of patient E.P. *Journal of Neuroscience* 20: 7024–7036.

Stefanacci L, Suzuki WA, and Amaral DG (1996) Organization of connections between the amygdaloid complex and the perirhinal and parahippocampal cortices in macaque monkeys. *Journal of Comparative Neurology* 375: 552–582.

Suzuki WA and Amaral DG (1994) Perirhinal and parahippocampal cortices of the macaque monkey: Cortical afferents. *Journal of Comparative Neurology* 350: 497–533.

Amygdala: Structure and Circuitry in Rodents and Felines

D Paré, Rutgers, The State University of New Jersey, Newark, NJ, USA

Introduction

The amygdala is a group of nuclei and cortex-like structures located in the rostromedial portion of the temporal lobe of the brain (**Figure 1**). As a group, amygdala nuclei receive sensory information of all modalities and project to all levels of the central nervous system, from the prefrontal cortex down to brain stem nuclei involved in cardiovascular control. The amygdala has become the focus of intense scrutiny in recent years, to a large extent because it participates in emotional expression and in the formation of emotional memories. Indeed, the amygdala plays a critical role in the acquisition of classically conditioned responses, both aversive and appetitive. Moreover, the amygdala mediates the facilitation of memory by emotions. It is thought that understanding the mechanisms that support the amygdala's involvement in these functions may shed light on the causes of human anxiety disorders and ways to alleviate them.

Main Components of the Amygdala: Structure, Cell Types, and Connectivity

As in primates, the feline and rodent amygdala can be subdivided into several components: superficial cortex-like fields (such as the periamygdaloid cortex, cortical nuclei, amygdalohippocampal area, and medial nucleus) and numerous deep nuclei that have been classified in various ways. Most accounts recognize a basolateral complex (BLA; including the lateral, basolateral, and basomedial or accessory basal nuclei), the central nucleus (itself typically subdivided in two or more subnuclei), and the intercalated cell masses. The nucleus of the lateral olfactory tract and anterior amygdaloid area will not be discussed here.

The heterogeneous composition of the amygdala has led some to suggest that the term 'amygdala' should be abandoned in favor of a functional classification that takes into consideration the distinct developmental and hodological profiles of its constituents. While this view has much merit, the term 'amygdala' will be used here, not only because most neuroscientists know what is meant when the term is used, but also because amygdala nuclei with different developmental origins are interconnected.

Two main structural differences become apparent when comparing the amygdala in primate and subprimate species (**Figure 1**). First, the relative size of the nuclei changes, with the primate BLA becoming much more voluminous than the central (CE) and medial nuclei. Second, because of the cortical expansion in monkeys, the amygdala rotates medially such that the CE nucleus shifts from a medial position in rodents to a dorsal position in primates.

Cortex-Like Nuclei of the Amygdala

Although the word 'nucleus' is often used when referring to these structures, the term is a misnomer because these nuclei display a laminar organization typical of cortex, including a cell-sparse molecular layer and a prevalent type of neurons with dendrites extending to the pia in a perpendicular manner. Many of these nuclei receive direct olfactory inputs from the main and/or accessory olfactory bulbs, are reciprocally connected with olfactory cortical areas, and contribute substantial projections to the several hypothalalamic nuclei, including the preoptic area, ventromedial nucleus, and lateral hypothalamus.

Basolateral Complex

Despite the random orientation of neurons in the BLA, its cellular composition is similar to that of the cerebral cortex. As in the cortex, two main cell types prevail in the BLA: multipolar spiny (often pyramidal-shaped) projection cells with highly collateralized axons, and a heterogeneous group of aspiny to sparsely spiny local-circuit neurons. Projection cells account for the majority of neurons, they use glutamate as a transmitter, and they contribute most, if not all, projections to other amygdala nuclei and the rest of the brain. There are at least four subtypes of BLA interneurons, all using the neurotransmitter γ-aminobutyric acid (GABA): parvalbumin-expressing neurons, somatostatin positive interneurons, large cholecystokinin interneurons, and small bitufted cells that display various degrees of co-localization of cholecystokinin, calretinin, and vasoactive intestinal polypeptide. Based on the evidence obtained in cortex, it is likely that these different types of interneurons structure the activity of principal cells in distinct ways because they target different postsynaptic domains and receive contrasting sets of inputs.

The cortical connections of the BLA are extensive. On the input side, the BLA receives cortical inputs about most sensory modalities, and these synapses act via α-amino-3-hydroxy-5-methyl-4-isoxazole propionic acid (AMPA) and *N*-methyl-D-aspartate (NMDA) glutamate receptors. Moreover, the BLA derives auditory and somatosensory information from the posterior thalamic complex. For these reasons, the

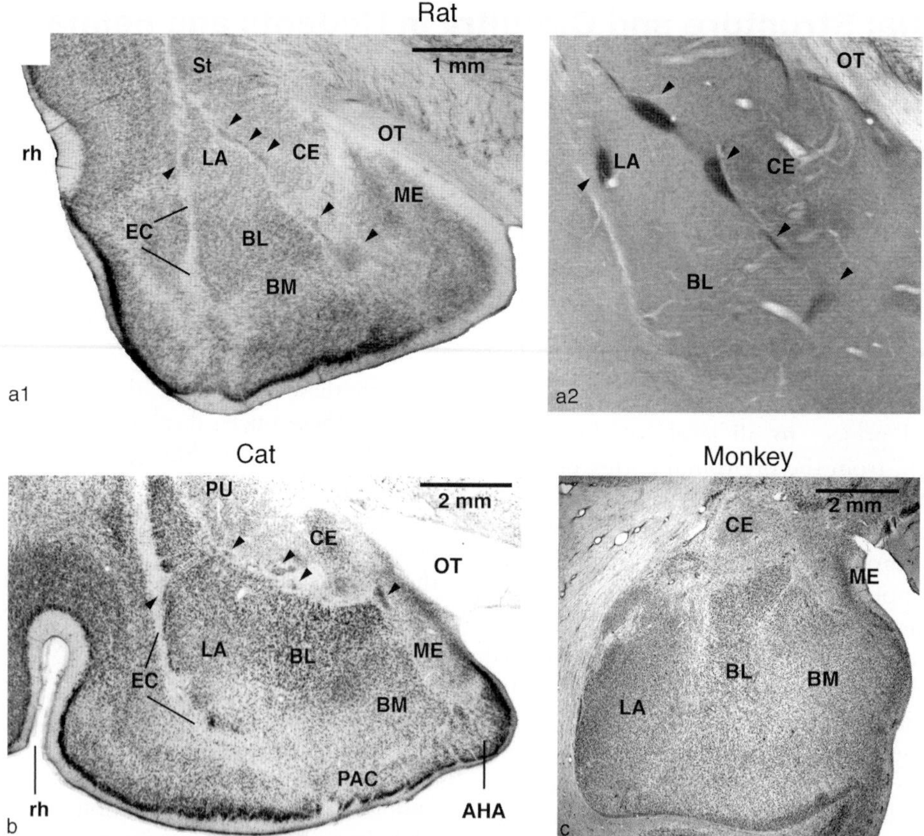

Figure 1 Comparison between the rat (a1 and a2), cat (b), and monkey (c) amygdala. All photomicrographs, with the exception of (a2), show thionin-stained coronal sections at the level of the amygdala. The distribution of μ opioid receptor immunoreactivity in the rat amygdala is shown in (a2). Arrowheads point to intercalated cell masses, where μ opioid immunoreactivity is most intense. AHA, amygdalohippocampal area; BL, basolateral nucleus; BM, basomedial nucleus; CE, central nucleus; EC, external capsule; LA, lateral nucleus; ME, medial nucleus; OT, optic tract; PAC, periamygdaloid cortex; PU, putamen; rh, rhinal sulcus; St, striatum.

BLA (and particularly its lateral nucleus) is typically considered as the sensory gateway of the amygdala. In addition, the BLA has major reciprocal connections with the rhinal cortices, medial prefrontal cortex, and subiculum.

One major species difference seen in the connectivity of the BLA resides in its cortical projections. Whereas the primate BLA projects to extensive neocortical territories, few BLA projections reach cortical fields lateral to the rhinal sulcus in felines and rodents. However, as in primates, the main thalamic projection site of the rodent BLA is the mediodorsal nucleus, but this projection does not exist in felines. Most other BLA connections are highly preserved in primate and subprimate species, including projections to the striatum, lateral hypothalamus, bed nucleus of the stria terminalis, and cholinergic basal forebrain. Last, it should be noted that in contrast to the CE, the BLA contributes few if any projections caudal to the mesencephalon.

Central Nucleus

Traditionally, the CE has been divided in two sections: lateral (CEl) and medial (CEm). However, in rats, CEl has been divided further and there is considerable disagreement regarding the exact borders of these subnuclei. To complicate matters further, some of the components included in the new classifications in fact correspond to cell groups, such as the intercalated cell masses, that were not traditionally included in the CE nucleus. Moreover, these various subnuclei do not appear to form distinct connections with the rest of the brain.

CEm and CEl each contain one prevalent cell type that is believed to use GABA as a neurotransmitter. In the CEm, neurons generally have a fusiform cell body and three to four primary dendrites that branch sparingly and bear a low to intermediate density of spines. In CEl, the main cell type is indistinguishable from the medium spiny neurons of the striatum. These cells generally have smaller somata than do principal CEm neurons and they usually have multiple primary dendrites that branch extensively and bear a high density of spines. Both CEm and CEl contain a low number of aspiny GABAergic neurons, thought to be local-circuit cells.

Even though the lateral nucleus of the amygdala (LA) is generally conceived as the main entry point of the amygdala for sensory inputs, the CE nucleus also has direct access to sensory information. Indeed, the sensory cortical areas that project to the BLA also typically project to CEl, albeit less intensely. This statement applies to the parietal (somatosensory inputs), piriform (olfactory inputs), and temporal (auditory inputs) and insular (gustatory and visceral inputs) cortices, as well as to several high-order cortical areas (polymodal inputs), such as the prefrontal cortex and rhinal cortices. In contrast, cortical projections to CEm are extremely sparse.

Besides cortical afferents, the CE draws sensory information from subcortical sources. These include afferents from the paraventricular thalamic nucleus (viscerosensory), and the posterior thalamic complex (nociceptive, auditory, polymodal). The posterior thalamic input ends in CEm. Importantly, the CE also receives nociceptive information from the trigeminal complex and spinal cord via the pontine parabrachial nucleus. Last, the BLA relays sensory information sensory from the cortex and thalamus to the CE.

In terms of outputs, CE contributes few if any cortical projections but is the principal amygdaloid source of brain stem projections. These brain stem efferents mainly originate in CEm, CEl projecting only to the parabrachial nucleus. The brain stem targets of CEm include structures controlling motor outputs (reticular formation of the mesencephalon and pons, periaqueductal gray), nuclei that participate in cardiovascular control (dorsal motor nucleus of the vagus, nucleus of the solitary tract), and modulatory cell groups that exert pervasive effects throughout the central nervous system (locus coeruleus, substantia nigra, pedunculopontine nucleus). In addition, CEl is the main amygdaloid source of projections to the cholinergic cell groups of the basal forebrain (substantia innominata, horizontal limb of the diagonal band). Last, the CE exerts indirect control over many of the same structures via the bed nucleus of the stria terminalis.

To summarize, the CE occupies a pivotal situation in the amygdala because it derives sensory inputs from a variety of sources and it targets modulatory systems as well somatomotor and visceromotor control structures. As a result, the CE is typically considered as the output station of the amygdala for fear responses.

Intercalated Cell Masses

In between the BLA (ventrolaterally) and the CE and medial nuclei (dorsomedially) is a fiber bundle that contains small GABAergic neurons in dense clusters known as the intercalated cell masses (**Figure 1(a)**). Laterally, this fiber bundle pierces the LA and merges with the external capsule, where intercalated cell masses can also be found. In addition, in most species, there is a larger intercalated cluster. This cluster caps the BLA rostrally in felines but is located medial to the BLA in rodents. The available evidence suggests that, irrespective of their position, intercalated clusters receive glutamatergic inputs from various cortical fields and from the BLA. However, their projection sites vary: the main intercalated cluster mainly projects to the substantia innominata and horizontal limb of the diagonal band of Broca, the lateral clusters project to the BLA, whereas the more medial clusters project to CE.

The principal types of intercalated neurons are small cells (8–19 μm in diameter) that use GABA as a neurotransmitter and have spiny dendrites, largely confined to the fiber bundle where their somata are located. Medially located intercalated neurons have two main axon collaterals: one is directed dorsally to the CE and one is directed medially to other intercalated cells. A second, rare type of intercalated cell is characterized by a very large soma (>40 μm in diameter) with extremely long aspiny dendrites that virtually span the entire mediolateral extent of the amygdala.

Electroresponsive Properties of Amygdala Neurons

Most of the data available on the intrinsic properties of amygdala neurons has been obtained in the BLA, CE, and intercalated cell masses. These findings are reviewed in this section.

Basolateral Complex

In terms of electroresponsive properties, three main types of projection cells have been distinguished in the BLA: (1) regular spiking cells that produce trains of action potentials with frequency accommodation, (2) bursting neurons that generate high-frequency spike bursts when depolarized just beyond spike threshold, and (3) late-firing cells that display a striking delay to firing following the onset of depolarizing current pulses.

Regular spiking cells are the prevailing type of projection neuron. Detailed investigations of their properties have revealed that regular spiking cells exhibit a continuum of spike frequency adaptation, from neurons that can fire in a sustained manner throughout depolarizing current pulses with only modest adaptation, to neurons that stop firing after a few spikes, or even that fire only once. These contrasting behaviors are not associated with different morphologies or membrane potentials, but result from variations in the density of calcium-activated potassium currents. In addition, regular spiking projection neurons generate intrinsic subthreshold membrane potential oscillations in the theta frequency range when depolarized just below firing threshold. These tetrodotoxin-sensitive oscillations allow BLA neurons to reverberate at the theta frequency during arousal.

Two types of aspiny neurons (i.e., presumed inhibitory interneurons) have been described so far. However, on the basis of findings obtained in the cerebral cortex, it is anticipated that future studies will reveal that many more physiological types of interneurons exist in the BLA. The most common type of interneuron is called a 'fast-spiking cell,' by analogy with the corresponding class of parvalbumin-immunoreactive cortical neurons. Like their cortical counterpart, fast-spiking neurons generate brief action potentials and can generate spikes at high rates without frequency accommodation. In addition, a few bursting aspiny neurons have been described. The peptide content of these two types of interneurons remains unknown.

Central Nucleus

Major species differences have been observed between the electroresponsive properties of CE neurons in rats, guinea pigs, and cats. In guinea pigs, the vast majority of neurons in the medial and lateral sectors of the CE exhibit a striking delay to firing when depolarized from hyperpolarized membrane potentials, hence the designation 'late-firing' neurons. This delayed firing onset is due to the activation of a slow voltage-dependent A-like potassium current. Late-firing neurons are believed to be projection neurons.

In contrast, late-firing neurons occur infrequently in rats and cats. In these species, the prevalent type of neuron shows a regular spiking firing pattern and expresses a hyperpolarization-activated mixed cationic current (I_H). Moreover, some rat and cat CE neurons are endowed with a T-current, giving rise to low-threshold spike bursts. Particularly in rats, the shape of spike afterhyperpolarizations (AHPs) has been used to classify CE neurons. Most rat CE neurons show medium-duration AHPs and modest spike frequency adaptation. A minority of CE neurons display long AHPs and pronounced adaptation. Both types appear to be present in CEm and CEl. Last, the CE also contains neurons that have relatively depolarized membrane potentials and a higher input resistance, and display fast-spiking or burst-firing patterns. These rare neurons are thought to be local-circuit cells.

Intercalated Cell Masses

Intercalated neurons display a regular spiking behavior. Moreover, as seen in the CE and BLA, they display a voltage- and time-dependent inward rectification in the hyperpolarizing direction, presumably because they express the hyperpolarization-activated mixed cationic current I_H. However, intercalated cells differ from principal cells of the BLA and CE in two important ways. First, intercalated neurons have an unusually high input resistance (around 800 MΩ), much higher than typically seen in principal BLA and CE cells (usually <200 MΩ). As a result, intercalated cells are more excitable and tend to have higher spontaneous firing rates.

In addition, intercalated cells express an unusual type of potassium current (slowly deinactivating; I_{SD}) that activates at subthreshold membrane potentials, shows inactivation in response to suprathreshold depolarizations, and recovers from inactivation with an extremely slow time course. As a result, the excitability of intercalated neurons is a function of their recent firing history. Indeed, following bouts of intense firing, the inactivation of I_{SD} causes an increase in input resistance and a membrane depolarization – in effect, increasing the likelihood that synaptic inputs will trigger action potentials. Moreover, because each spike renews the inactivation of I_{SD}, this state of increased excitability is self-sustaining.

Neuronal Interactions between the Basolateral Amygdala and Central Nucleus

A dominant theme of research on the amygdala has been its involvement in the acquisition of classically conditioned fear responses. In this model of emotional learning, a neutral sensory stimulus (conditioned stimulus; CS) acquires the ability to elicit fear responses after repeated pairings to a noxious stimulus (unconditioned stimulus; US). According to the prevailing model, convergence of sensory and nociceptive information in the LA would lead to a Hebbian potentiation of synapses conveying information about the CS. As a result, subsequent presentations of the CS would elicit larger responses in CE neurons. In turn, the CE would generate conditioned fear responses via its hypothalamic and brain stem projections. However, synaptic plasticity in the BLA is under tight inhibitory control. In fact, it was shown that inhibitory interneurons effectively gate the induction of long-term potentiation in principal cells. Thus, the following section considers the intrinsic inhibitory pressures that regulate synaptic excitability and associative plasticity within the BLA as well as the propagation of signals from the BLA to the CE.

Intrinsic Control of Neuronal Excitability and Plasticity within the Basolateral Amygdala

Although the cellular composition of the BLA is similar to that of the cerebral cortex, principal BLA neurons have extremely low firing rates (typically <1 Hz) compared to cortical neurons (average of 10 Hz). This dissimilarity is particularly puzzling because the BLA has an extremely divergent intrinsic connectivity. Indeed, principal BLA neurons have highly collateralized axons, each bearing numerous varicosities that form *en passant* glutamatergic AMPA–NMDA synapses, typically with the dendritic spines of other projection cells. In fact, based on the average length of the intervaricose segments (5–10 μm), it is estimated that for every millimeter of axon, each projection cell forms 100–200 excitatory synapses with other projection cells.

Considering this remarkable potential for divergence within the BLA, the low spontaneous firing rates of principal cells suggest that major inhibitory pressures regulate synaptic excitability within the BLA. In keeping with this, *in vivo* intracellular studies have revealed that the spontaneous activity of BLA neurons is dominated by large-amplitude hyperpolarizing potentials. These inhibitory potentials are generated by the combined influence of synaptically activated intrinsic membrane conductances and inhibitory synaptic conductances. First, principal cells exhibit a calcium-dependent potassium conductance that can be synaptically activated by calcium entry via NMDA receptors. Second, the activity of principal cells is tightly regulated by local-circuit neurons. In response to afferent stimulation, local-circuit cells generate biphasic inhibitory postsynaptic potentials (IPSPs) in principal BLA neurons via GABA-A and GABA-B receptors. In contrast, interneurons are subjected to far less inhibition than principal cells, because they receive fewer inhibitory synapses than principal cells and their GABA-A reversal potential is more depolarized. This results from the fact that the prevalent regulators of intracellular chloride are cation–chloride cotransporters that accumulate chloride in local-circuit cells and extrude chloride in principal neurons.

An intriguing property of the intrinsic inhibitory BLA network is that it is spatially heterogeneous. In the LA, for instance, feedback interneurons tend to inhibit projection cells located in the same coronal plane, their axons extending little rostrocaudally. As a group, the other types of interneurons also show spatial heterogeneity in the control of projection cells. They tend to generate more inhibition in neurons located laterally. Given that different types sensory inputs end in distinct compartments of the BLA, this implies that inscribed in the BLA network is a predisposition to preferentially attend to particular types of sensory information.

In addition to local-circuit neurons, principal LA neurons receive inhibitory inputs from intercalated clusters found in the external capsule. Interestingly, these two populations of GABAergic neurons display opposite responses to dopamine: local-circuit cells are excited whereas intercalated neurons are inhibited by dopamine. Yet, the net effect of dopamine is a disinhibition of principal LA neurons, resulting in a facilitation of long-term potentiation (LTP) induction at thalamic inputs.

Transfer of Sensory Inputs from the BLA to the CE

All BLA nuclei send glutamatergic projections to CE, exciting CE neurons through AMPA and NMDA glutamate receptors. As mentioned previously, it is believed that fear conditioning results in a potentiation of CS-evoked responses in the LA and that, as a result, subsequent presentations of the CS would elicit larger responses in the CE. However, tract-tracing studies indicate that there are no direct connections between the input (LA) and output (CEm) stations of the amygdala for conditioned fear responses. Indeed, the LA nucleus projects only to CEl, not to CEm, the subnucleus at the origin of most amygdala projections to brain stem structures mediating fear responses. Thus, it is likely that the LA transmits CS-related information to the CE indirectly. Consistent with this, the LA sends robust glutamatergic projections to the basal nuclei that, in turn, project to CEl and CEm. Moreover, post- but not pre-conditioning lesions of the basal nuclei block conditioned fear responses, suggesting that in an intact brain, the basal nuclei are critically involved, at least in transferring CS-related signals to the CE. However, the fact that preconditioning lesions of the basal nuclei do not prevent acquisition of conditioned fear responses suggests that CS-related information can reach the CE through other paths. Consistent with this, CEm is the target of auditory and somatosensory inputs from the posterior thalamus, and these inputs can undergo activity-dependent potentiation.

Gating of BLA Projections to the CE by Intercalated Neurons

En route to the CE, BLA axons contact GABAergic intercalated neurons that generate GABA-A IPSPs in CE neurons. Accumulating evidence suggests that despite their small size, intercalated cell clusters play a pivotal role in the synaptic economy of the amygdala. The available data indicate that there is a lateromedial correspondence between the position of intercalated cells, where they project in the CE, and the source of their BLA afferents. Moreover, intercalated clusters are interconnected, but these connections prevalently run in a lateromedial direction. Because of these connections, the amount of feedforward inhibition intercalated cells generate in the CE, and thus the amplitude of the excitatory postsynaptic potentials (EPSPs) generated by BL inputs, will depend on the spatiotemporal distribution of excitation in the BLA. Thus, inscribed in this network is the possibility of modulating behavior responsiveness as a function of current environmental contingencies. Moreover, because BLA inputs to intercalated cells exhibit activity-dependent plasticity, gating of BLA inputs to the CE by intercalated cells can be modified by experience.

Another factor adding flexibility to this gating function is the presence of glutamatergic inputs from various cortical areas to intercalated neurons. In particular, medial prefrontal inputs to intercalated cells have been implicated in the extinction of classically conditioned fear responses. Extinction refers to the gradual disappearance of conditioned fear responses when the CS is repeatedly presented without pairing to the US. Much behavioral evidence suggests that extinction does not result from unlearning of the original CS–US association, but from a new learning that competes with the first one. This view is based on the fact that extinguished fear memories can undergo spontaneous recovery over time, renewal (recovery when tested in a context different than the extinction context), and reinstatement (recovery when an unsignaled US is presented in the extinction context). Moreover, intra-amygdala injections of NMDA antagonists or of inhibitors of the mitogen-activated protein kinase (MAPK) pathway prevent extinction learning. At present, the processes underlying extinction learning remain obscure. However, mounting evidence suggests that an extrinsic input to the amygdala plays a critical role. Indeed, lesions of the medial prefrontal cortex impair recall of extinction. Moreover, in rats that have not been extinguished, electrical stimulation of the medial prefrontal cortex during the CS reduces conditioned responses. Finally, electrical stimulation of the medial prefrontal cortex produces an inhibition of brain stem-projecting CEm neurons. Because the medial prefrontal cortex sends little projections to CEm, it is believed that the inhibitory effects of medial prefrontal inputs on fear expression are mediated by intercalated neurons. One possibility is that prefrontal inputs, by depolarizing intercalated cells, facilitate NMDA-dependent plasticity of BLA inputs to intercalated neurons. As a result, subsequent CS presentations would evoke more inhibition in CE neurons.

The Amygdala Mediates the Facilitation of Memory by Emotions

Humans generally form more vivid memories of emotionally arousing events than of mundane experiences, and this trend is absent in individuals with amygdala lesions. In part, the impact of emotions on memory involves an enhancement of attention during encoding. Indeed, functional imaging studies have revealed that emotional stimuli activate the amygdala, that attention is enhanced by emotional material, and that this effect is absent in individuals with amygdala lesions. Consistent with this, animal studies have revealed that amygdala stimulation produces an electroencephalogram (EEG) activation associated with an orienting reaction.

However, animal work indicates that besides enhancing attention, the amygdala can also facilitate the consolidation of emotional memories. Overall, the available evidence suggests that stress hormones released in emotionally arousing conditions increase the activity of BLA neurons. In turn, the enhanced activity of BLA cells would facilitate memory consolidation in other brain areas via their widespread cortical and subcortical projections. For instance, posttraining intra-BLA injections of drugs that enhance or reduce BLA activity, respectively, facilitate or impair retention on a variety of emotionally charged learning tasks. Moreover, BLA lesions prevent the facilitation of memory by systemic injections of stress hormones. Importantly, although long-term memory can be affected by altering amygdala activity right after learning, manipulations of BLA excitability at later time points have no effects. These observations imply that the aforementioned memory-modulating effects of the amygdala manipulations do not result from alterations of memory storage in the amygdala, but rather in other structures that presumably constitute the storage site of particular forms of memories.

At present, the mechanisms underlying the facilitation of memory consolidation by amygdala activity remain unknown. One possibility is that the enhancement of BLA activity produced by emotional arousal leads to the recruitment of basal forebrain cholinergic neurons that project to the cerebral cortex. In turn, this amygdala-driven release of acetylcholine would facilitate plasticity in cortical networks. However, it is known that amygdala activity can also facilitate striatal-dependent memories. Since the striatum receives little if any inputs from basal forebrain cholinergic neurons, other mechanisms must be involved. A challenge for future studies will be to determine the distinguishing properties that allow amygdala synapses to facilitate synaptic plasticity in their targets.

See also: Amygdala: Structure and Circuitry in Primates; Amygdala: Contributions to Fear; Appetitive Systems: Amygdala and Striatum.

Further Reading

Dringenberg HC and Vanderwolf CH (1996) Cholinergic activation of the electrocorticogram: An amygdaloid activating system. *Experimental Brain Research* 108: 285–296.

Everitt BJ, Cardinal RN, Parkinson JA, et al. (2003) Appetitive behavior: Impact of amygdala-dependent mechanisms of emotional learning. *Annals of the New York Academy of Sciences* 985: 233–250.

LeDoux JE (2000) Emotion circuits in the brain. *Annual Review of Neuroscience* 23: 155–184.

Maren S and Quirk GJ (2004) Neuronal signaling of fear memory. *Nature Reviews Neuroscience* 5: 844–852.

McDonald AJ (1992) Cell types and intrinsic connections of the amygdala. In: Aggleton JP (ed.) *The Amygdala: Neurobiological Aspects of Emotion, Memory, and Mental Dysfunction*, pp. 67–96. New York: Wiley-Liss.

McDonald AJ (1998) Cortical pathways to the mammalian amygdala. *Progress in Neurobiology* 55: 257–332.

McGaugh JL (2004) The amygdala modulates the consolidation of memories of emotionally arousing experiences. *Annual Review of Neuroscience* 27: 1–28.

Pare D (2003) Role of the basolateral amygdala in memory consolidation. *Progress in Neurobiology* 70: 409–420.

Paré D, Royer S, Smith Y, et al. (2003) Contextual inhibitory gating of impulse traffic in the intra-amygdaloid network. *Annals of the New York Academy of Sciences* 985: 78–91.

Paré D, Quirk GJ, and LeDoux JE (2004) New vistas on amygdala networks in conditioned fear. *Journal of Neurophysiology* 92: 1–9.

Phelps EA (2004) Human emotion and memory: Interactions of the amygdala and hippocampal complex. *Current Opinion in Neurobiology* 14: 198–202.

Pitkänen A (2000) Connectivity of the rat amygdaloid complex. In: Aggleton JP (ed.) *The Amygdala: A Functional Analysis*, pp. 31–115. Oxford: Oxford University Press.

Quirk GJ, Garcia R, and Gonzalez-Lima F (2006) Prefrontal mechanisms in extinction of conditioned fear. *Biological Psychiatry* 60: 337–343.

Sah P, Faber ESL, De Armentia ML, et al. (2003) The amygdaloid complex: Anatomy and physiology. *Physiological Reviews* 83: 803–834.

Swanson LW and Petrovich GD (1998) What is the amygdala? *Trends in Neuroscience* 21: 323–331.

Weinberger NM (2004) Specific long-term memory traces in primary auditory cortex. *Nature Reviews Neuroscience* 5: 279–290.

Amyloid: Vascular and Parenchymal

R O Weller, R O Carare, and D Boche, University of Southampton School of Medicine, Southampton, UK

Introduction

Age brings many changes to the body and age is also a major risk factor for dementia. Alzheimer's disease is one of the major forms of dementia in the elderly and is characterized by the accumulation of amyloid beta (Aβ) peptides in the brain. Although there are no traditional lymphatics in the brain, interstitial fluid and solutes such as Aβ are eliminated from brain tissue along the basement membranes in the walls of capillaries and cerebral arteries. Here we review how aging of the brain and cerebral arteries may lead to the failure of elimination of Aβ from the brain and how this may contribute to the onset of dementia, particularly in Alzheimer's disease. Immunotherapy for the removal of Aβ from the brain is also briefly discussed.

Aging of the Brain and Cerebral Arteries

Changes occur progressively in the brain and cerebral arteries with advancing age, and age is a major risk factor for Alzheimer's disease and for cerebrovascular disease. Neurons in the brain are very long-lived cells, as they are established before birth and survive for the lifetime of the individual. With age, large neurons in the cerebral cortex and spinal cord accumulate waste products in the cytoplasm in the form of lipofuscin. Damaged mitochondria and membranes are broken down by lysosomal hydrolytic enzymes, and residual material accumulates in the cytoplasm of neurons as secondary lysosomes (lipofuscin). However, apart from a few familial disorders (lipofuscinoses), the lipofuscin in neurons does not seem to be a factor in dementia. Whereas cell organelles are broken down by lysosomes, cytoplasmic proteins are degraded by proteosomes. Damaged proteins become associated with ubiquitin and are delivered to proteosomes for degradation. As will be discussed later, there appears to be a failure of the ubiquitin–proteosomal system in dementias, particularly in Alzheimer's disease. In addition to the changes that occur within aging neurons, astrocytosis is seen in the brain, with advancing age reflecting the damage that occurs throughout life from head injuries, episodes of viral encephalitis, and ischemia.

Cerebral blood vessels and blood vessels elsewhere in the body change with age in a similar way. There is progressive stiffening of arterial walls, with an increase in collagen in the tunica intima and in the tunica media. Major cerebral arteries, such as the vertebral, basilar, and middle cerebral arteries, show progressive dilatation and tortuosity due to fibrosis of the intima and media. Generalized thickening and stiffening of cerebral arteries (arteriosclerosis) is also accompanied by patchy thickening and the accumulation of cholesterol within atherosclerotic plaques of the major cerebral arteries. Occlusion of large or small cerebral arteries may result from emboli arising in the heart or from thrombi associated with atherosclerosis of the carotid arteries. Large infarcts resulting from occlusion of cerebral arteries may result in strokes, with focal neurological signs such as hemiparesis, whereas multiple small infarcts may produce no overt neurological signs but can result in vascular dementia (see **Table 1**).

Changes in the Brain Associated with Dementia

Since the description in 1907 of the microscopic features in the brain in Alzheimer's disease, many pathological changes associated with different types of dementia have been described. As shown in **Table 1**, the pathology of dementias is classified broadly based upon presence of intracellular protein inclusions mainly within neurons (tau, synuclein, ubiquitin, and huntingtin), extracellular deposits of amyloid proteins in the brain and in blood vessel walls (Aβ, other amyloids, and prion protein), and small or large infarcts (multi-infarct or vascular dementia). (Lipofuscin has been included in **Table 1**, although excessive lipofuscin is a rare cause of dementia.)

Intracellular Protein Deposits

Two major proteins, tau and synuclein, accumulate within neurons in dementia patients. Both proteins are associated with ubiquitin, which suggests that disposal of these proteins through the ubiquitin–proteosomal system may fail with age and dementia.

Accumulation of Tau Protein

Tau protein (named after the Greek letter for 't') is a microtubule-associated protein that is concerned with axoplasmic transport in normal neurons. Hyperphosphorylated tau accumulates with ubiquitin in aging neurons as the neurofibrillary tangles that were identified by Alzheimer and others by the use of silver stains (**Figure 1**). Examination of postmortem

Table 1 Age changes in the brain associated with dementia[a]

Type of dementia	*Intracellular inclusions in neurons*	*Extracellular deposits*	*Cerebral amyloid angiopathy*	*Infarcts*
Aging brain	Lipofuscin +++ Tau +	Aβ +	Aβ +	+
Vascular dementia	Lipofuscin +++ Tau +	Aβ +	Aβ +	**++++**
Alzheimer's disease	Lipofuscin +++ **Tau** +++	**Aβ** ++++	**Aβ** ++++	+/++
Dementia with Lewy bodies	Lipofuscin +++ **Synuclein** +++ Tau ++	Aβ ++	Aβ ++	+/++
Creutzfeldt–Jakob disease	Lipofuscin +++ in elderly	**Prion protein** ++++	+/– (prion protein)	+
Huntington's disease	**Huntingtin** +++	Aβ +/–	+/– (Aβ)	+
Frontotemporal dementias	Lipofuscin +++ **Tau** ++++ **or ubiquitin only** +++	Aβ +/–	+/– (Aβ)	+

[a]Intracellular inclusions, extracellular protein deposits, and vascular damage in the aging brain and in different major types of dementia. The features used as the pathological diagnostic criteria in the various dementias are in boldface type.

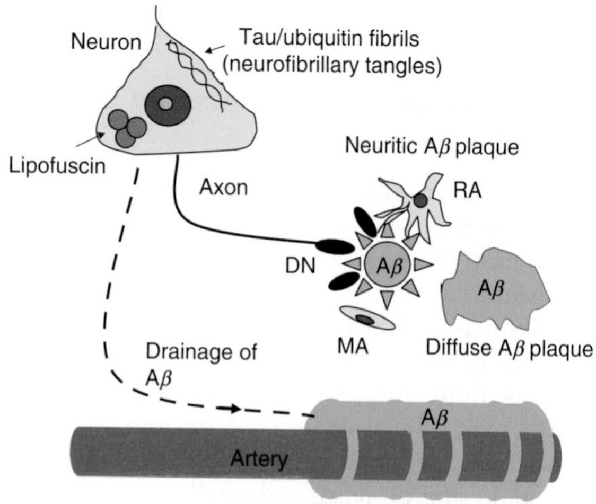

Figure 1 Histopathological features of Alzheimer's disease are characterized by the accumulation of neurofibrillary tangles composed of tau and ubiquitin within neurons, and by the deposition of amyloid beta (Aβ) in brain parenchyma as amyloid plaques and in blood vessel walls as cerebral amyloid angiopathy. Aβ in amyloid plaques is associated with microglial activation (MA) and reactive astrocytosis (RA). Swollen dystrophic neurons (DN) containing tau are associated with many of the amyloid plaques (neuritic plaques).

brains of aged patients with no impairment of cognition reveals some neurons with tau in their cytoplasm, particularly in the hippocampus. In Alzheimer's disease, neurofibrillary tangles composed of hyperphosphorylated tau and ubiquitin complexes are more widespread and more numerous, especially in the temporal cortex, hippocampus, and frontal lobes. Mutations in the tau gene on chromosome 17 are associated with a familial dementia characterized by widespread accumulation, in neurons, of neurofibrillary tangles containing tau protein.

Accumulation of Synuclein

Synuclein is a protein associated with synapses in the brain; it accumulates with ubiquitin as spherical structures (Lewy bodies) within neurons, particularly of the substantia nigra and locus coeruleus in Parkinson's disease. In dementia with Lewy bodies (see **Table 1**), synuclein and ubiquitin complexes (Lewy bodies) are observed in neurons of the cingulate gyrus, frontal and temporal cortices, and insula; synuclein also accumulates within neuronal processes, particularly in the hippocampus. There appears to be a failure in the ubiquitin–proteosomal system but the reasons for this are unclear in the majority of cases of dementia with Lewy bodies. Some familial cases of Parkinson's disease or dementia with Lewy bodies are associated with mutations in the α-synuclein or parkin genes.

Accumulation of Huntingtin

Inclusions of the protein huntingtin occur within nuclei of neurons in Huntington's disease (see **Table 1**) in which there is expansion of triplet repeats in the gene encoding huntingtin, on chromosome 4.

Deposition of Amyloid in the Extracellular Spaces of the Brain

A number of different proteins accumulate in the brains of elderly and demented patients as insoluble amyloid deposits in the extracellular spaces of gray matter, and, in many cases, in the walls of arteries as cerebral amyloid angiopathy. The most common of these proteins is amyloid beta (Aβ), which may accumulate in the brains of elderly people who show no evidence of cognitive decline; Aβ is deposited much more abundantly in the brain in Alzheimer's disease.

Other proteins, such as prion protein, accumulate in the gray matter in Creutzfeldt–Jakob disease, and a variety of other amyloid proteins, such as cystatin, transthyretin, and the British and Danish types of amyloid, accumulate in the extracellular spaces of the brain and in blood vessel walls in association with different types of familial dementia. In the following sections we explore the mechanisms by which amyloid proteins accumulate in the brain and blood vessel walls in elderly and demented patients, and in particular in Alzheimer's disease.

Accumulation of Amyloid Proteins in the Brain in Alzheimer's Disease

Alzheimer's disease is the commonest form of dementia in North America and Europe and has recently increased in incidence in many Asian countries. The disease is characterized pathologically by the accumulation of hyperphosphorylated tau within neurons and by the deposition of Aβ in the extracellular spaces of gray matter in the brain and in the walls of cerebral arteries (**Figure 1**). Overproduction of Aβ and failure of elimination of Aβ from brain tissue appear to be the major reasons why Aβ accumulates in the brain in old age and in Alzheimer's disease.

A number of familial dementias have been identified in which there is either an overproduction of Aβ, particularly of the less soluble form, Aβ_{1-42}, or there is a production of a chemically aberrant form of Aβ. Throughout life, Aβ is produced by neurons and virtually all other cells in the body by the cleavage of a transmembrane protein, amyloid precursor protein (APP), encoded on chromosome 21. Three major secretases (α, β, and γ) cleave APP to produce Aβ in a number of forms that differ in the number of amino acids in the Aβ peptide. The two major forms that accumulate in the brain and in the walls of arteries are Aβ_{1-40} (40 amino acids long) and Aβ_{1-42}, which contains 42 amino acids and is less soluble than Aβ_{1-40}. Aβ_{1-42} is mainly in the insoluble amyloid plaques within brain tissue and Aβ_{1-40} predominates in the blood vessel walls in cerebral amyloid angiopathy. In the familial forms of Alzheimer's disease, genetic mutations in the APP gene affect the cleavage sites of APP or the structure of Aβ. Mutations in the presenilin 1 and 2 genes appear to affect the γ-secretase cleavage of Aβ from APP. In these types of familial Alzheimer's disease, failure of elimination of the excessive amount of Aβ from the aging brain appears to be a major factor in the pathogenesis of the dementia.

Most cases of Alzheimer's disease are sporadic, with familial cases accounting for less than 10% of all cases of Alzheimer's disease. In the more common sporadic form of Alzheimer's disease, there is no firm evidence of overproduction of Aβ, so it is probable that failure of elimination of Aβ from the brain with advancing age is a major pathogenetic factor (**Figure 1**).

Elimination of Aβ from the Brain

Aβ is produced throughout life in the brain and it is present at an average concentration of 2.75 ng g^{-1} of cerebral cortex in clinically nondemented individuals; Aβ is greatly increased in the brains of patients with Alzheimer's disease. Studies in experimental mice and humans reveal a number of mechanisms by which Aβ is eliminated from the normal brain. Aβ is absorbed into the blood by several known pathways, such as that mediated by low-density lipoprotein receptor-related protein-1 (LRP-1), and it is also degraded by peptidases such as neprilysin, in the parenchyma of the brain. In addition, Aβ is eliminated from the brain with interstitial fluid along perivascular drainage pathways – effectively the lymphatics of the brain.

Perivascular Route for the Elimination of Interstitial Fluid and Aβ from the Brain

There is no traditional lymphatic system in the brain. In other tissues, such as lung and skin, there are well-defined thin-walled lymphatic vessels along which interstitial fluid, solutes, and inflammatory cells drain to regional lymph nodes. Although there are no true lymphatics in the brain, studies in experimental animals have shown that solutes injected into the brain do drain to lymph nodes in the neck. Thus, when a soluble fluorescent tracer of molecular mass equivalent to that of Aβ (4 kDa) is injected into gray matter of the brain, it initially spreads diffusely through the narrow extracellular spaces and rapidly enters basement membranes of capillary and artery walls, to drain out of the brain. It does appear that the perivascular route along basement membranes in the walls of capillaries and arteries is the pathway for lymphatic drainage of the brain.

Aβ is deposited in the walls of cerebral capillaries and arteries in aged humans and in Alzheimer's disease in a pattern that suggests Aβ also drains from the brain along basement membranes of capillary and artery walls. As Aβ progressively accumulates in the artery walls, smooth muscle cells in the tunica media die, so that in severe cerebral amyloid angiopathy the whole thickness of the artery is replaced by amyloid, thus increasing the likelihood of intracerebral hemorrhage.

Figure 2 summarizes the formation and elimination of interstitial fluid and metabolites from gray-matter

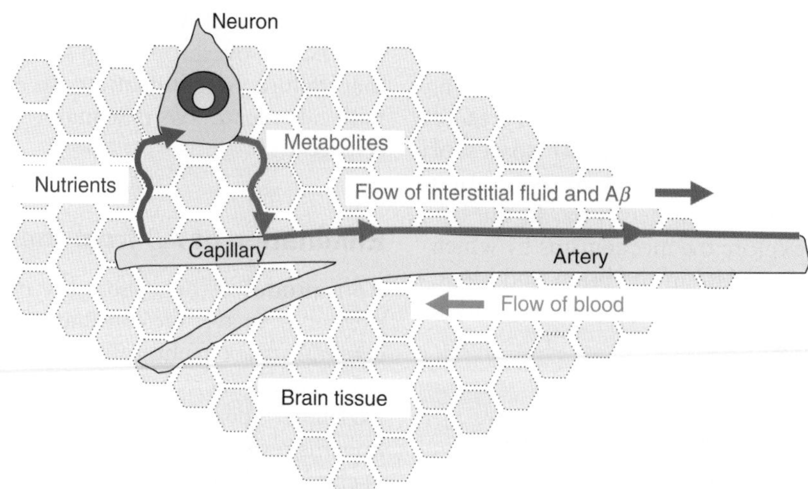

Figure 2 Perivascular drainage of interstitial fluid and metabolites from the brain. Experimental studies suggest that nutrients pass through the blood–brain barrier into capillary endothelial basement membranes and thence diffuse to neurons and other cells in the brain. Metabolites from these cells diffuse back to the capillary basement membranes through the narrow and restricted extracellular spaces of the cerebral cortex and other areas of gray matter in the brain. Once metabolites, including amyloid beta (Aβ), have entered capillary basement membranes, they pass by relatively unrestricted bulk flow into the basement membranes of arteries and out of the brain. The lymphatic flow along artery walls is probably driven by the reflection waves arising from arterial pulsations. Stiffening of artery walls with age may reduce the amplitude of arterial pulsations and reflection waves, thus interfering with the perivascular elimination of metabolites, including Aβ, from the brain.

areas of the brain. Nutrients pass through the blood–brain barrier into capillary basement membranes, which act as conduits or channels for the circulation of nutrients to neurons and other cells within the brain. Metabolites produced by cells in the brain pass through the very narrow extracellular spaces to drain out of the brain, with interstitial fluid, along basement membranes in the walls of capillaries and arteries.

Data from experimental and human studies suggest that interstitial fluid and solutes from the brain drain along artery walls to cervical lymph nodes just under the base of the skull. It is unlikely that lymphocytes and other inflammatory cells can migrate from the brain to lymph nodes along the narrow basement membrane pathways followed by interstitial fluid and solutes. Such failure of lymphocyte migration from the brain may be a factor in the well-known 'immunological privilege' of the brain. The perivascular pathway for drainage of interstitial fluid and solutes from the brain is largely separate from the cerebrospinal fluid (CSF).

Pathways for the Drainage of CSF

Drainage pathways for CSF have been extensively studied in both animals and humans. Tracers injected directly into the CSF of rodents drain largely through channels in the cribriform plate into lymphatics of the nasal mucosae and thence to cervical lymph nodes. In humans, however, CSF drains directly back into the blood through arachnoid granulations and villi in the walls of intracranial and spinal venous sinuses, and the route through the cribriform plate plays a very minor role.

Failure of Elimination of Aβ from the Brain with Advancing Age and Alzheimer's Disease

The various mechanisms and routes for the removal of Aβ from the brain in young animals and young humans appear to fail with advancing age (**Figure 3**). In mice, degradation of Aβ by neprilysin within the brain parenchyma, and absorption of Aβ into the blood by the LRP-1-mediated mechanism, are reduced with increasing age. Accumulations of insoluble amyloid Aβ in the interstitial fluid drainage pathways in artery walls in elderly individuals and in patients with Alzheimer's disease suggest that the perivascular route for elimination of Aβ also fails with age.

Detailed observations in tracer studies have shown that solutes drain along capillary basement membranes and then along basement membranes between the smooth muscle cells of artery walls. Using these data, theoretical mathematical models indicate that pulsation of artery walls is the motive force for the perivascular drainage of interstitial fluid and solutes from the brain. The model predicts that fluid and solutes would be driven along the vascular basement

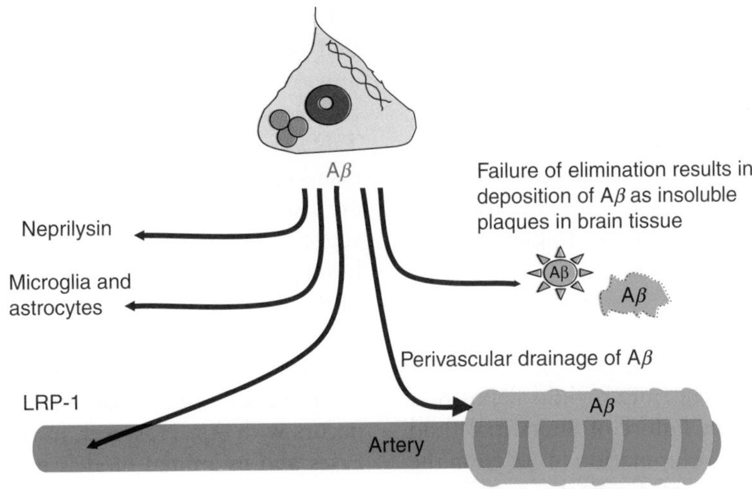

Figure 3 Pathways and mechanisms for the elimination of amyloid beta (Aβ) from the brain fail with advancing age, resulting in the accumulation of Aβ in brain parenchyma and in blood vessel walls. In the young, Aβ is absorbed into the blood by a mechanism mediated by low-density lipoprotein receptor-related protein-1 (LRP-1) and is degraded in brain parenchyma by neprilysin and other enzymes. Aβ also drains from the brain with interstitial fluid along capillary and artery walls. Failure of the LRP-1 mechanism and neprilysin with age and the failure of perivascular drainage of Aβ along stiff, aged arteries are associated with deposition of insoluble Aβ as plaques in brain parenchyma, and in blood vessel walls as cerebral amyloid angiopathy. Ultimately, levels of soluble Aβ, and possibly other metabolites in the brain, rise, and this is associated with decline of cognitive function and dementia in Alzheimer's disease.

membranes in the reverse direction to the blood flow during diastole by the reflection waves generated by the arterial pulse wave. The strength of the reflection waves driving the drainage of interstitial fluid and solutes from the brain would be proportional to the amplitude of the pulsations.

Why then does the perivascular drainage of fluid and solutes such as Aβ fail with age and in Alzheimer's disease? Age is not only a risk factor for Alzheimer's disease, it is also a risk factor for cerebrovascular disease in which there is fibrosis and stiffening of artery walls (arteriosclerosis), atherosclerotic plaque formation, and occlusion of arteries by thrombi or emboli. As cerebral arteries stiffen with age the amplitude of vessel pulsations would be reduced, and this would also reduce the strength of the reflection wave driving the interstitial fluid and solutes from the brain along the walls of arteries. Hypothetically, reduction in the motive force in stiffened arteries would lead to slowing of drainage of Aβ and induce the precipitation of insoluble fibrils of Aβ in the vessel walls. Such deposition would further impede the drainage of Aβ and eventually lead to the accumulation of soluble and insoluble Aβ and other metabolites in the brain.

Direct evidence that cerebrovascular disease and stiffening of artery walls impedes the elimination of Aβ along perivascular pathways is difficult to obtain. Nevertheless, thrombotic occlusion of small cortical arteries with complete abolition of vascular pulsations is associated with deposition of Aβ in the basement membrane of capillaries supplied by the occluded arteries. This suggests that once the pulsations in the vessel cease, Aβ can no longer be propelled along the drainage pathway from capillaries basement membranes into artery walls.

Consequences of Amyloid Deposition in the Brain and Vessel Walls

In cerebral amyloid angiopathy, there are two major complications of the deposition of Aβ in the walls of cerebral arteries. The acute complication is intracerebral hemorrhage and the more long-term complication is related to the onset of dementia, particularly of Alzheimer's disease. Soluble and insoluble forms of Aβ accumulate in the cerebral cortex, and the fluid content in subcortical white matter is increased.

Intracerebral Hemorrhage

There is a close association between intracerebral hemorrhage and cerebral amyloid angiopathy in the elderly. As increasing amounts of Aβ are deposited in the basement membranes of artery walls, smooth muscle cells in the tunica media are destroyed and the vessel wall becomes a brittle tube of insoluble Aβ, which makes it liable to rupture, causing an intracerebral hemorrhage. This complication occurs mainly in patients over the age of 75 years and can be fatal or may induce neurological deficits. Recent magnetic resonance imaging (MRI) studies have shown that the majority of intracerebral hemorrhages that have been attributed to cerebral amyloid angiopathy occur in the temporal and occipital lobes, supplied mainly by the posterior cerebral arteries.

Dementia

The other major consequence of amyloid deposition in blood vessel walls is the failure of interstitial fluid drainage and the failure of elimination of Aβ from the aging brain. The pathological diagnostic criteria for Alzheimer's disease are mainly based upon the presence of neurofibrillary tangles within neurons and the severity of deposition of insoluble Aβ as plaques in the cerebral cortex and hippocampus. As the elimination of Aβ fails, it is deposited as insoluble plaques in gray-matter areas of the brain. The more compact plaques induce microglial activation and are associated with axonal damage and dystrophic neurites; the term 'neuritic plaque' is used for these structures (**Figure 1**). Other plaques of Aβ are more diffuse and do not induce the same degree of brain tissue damage.

Deposits of insoluble Aβ in brain tissue impede the elimination of Aβ and other metabolites from the brain. Before entering the drainage pathways in capillary and artery walls, interstitial fluid and metabolites diffuse through the narrow extracellular spaces of gray matter; Aβ plaques block this diffusion. Eventually, drainage of solutes is severely restricted and levels of soluble Aβ in gray matter rise. Recent studies have emphasized that a high level of soluble Aβ in cerebral cortex and the severity of cerebral amyloid angiopathy correlate better with cognitive decline in patients with Alzheimer's disease than does the number of insoluble plaques of Aβ.

Why the failure of elimination of soluble Aβ results in dementia in Alzheimer's disease is still not clear. Many studies suggest that the toxicity of Aβ to neurons is not due to the accumulation of insoluble fibrils but rather is due to the presence of a soluble pool of Aβ, particularly in its oligomeric form. In transgenic mice, dodecameric Aβ assemblies are associated with an impairment of memory independently of Aβ plaques or neuronal loss. Thus increased levels of soluble oligomeric Aβ may be a factor inducing the onset of dementia in Alzheimer's disease. However, it is also possible that other metabolites do not drain adequately from gray matter, resulting in their accumulation in the extracellular spaces, a change in neuronal environment, neuronal malfunction, and dementia.

Creutzfeldt–Jakob disease is another dementia that is characterized by the deposition of insoluble amyloid protein in the extracellular spaces of brain parenchyma, but, in this case, it is prion protein (see **Table 1**). As in Alzheimer's disease and Aβ, the mechanisms underlying the neurotoxicity of the prion protein are also unclear. Recent studies in experimental models suggest that the insoluble prion protein deposits in brain parenchyma are not toxic but that the conversion of the normal prion protein to disease-associated isoforms within neurons results in the neurotoxic form of the protein. Accumulation of the prion protein within blood vessel walls of the brain is less common than occurs with Aβ. However, in some familial prion diseases and in animal models in which the protein has been truncated, prion protein is deposited within walls of the cerebrovasculature. Whether insoluble deposits of prion protein in brain parenchyma and in artery walls have the same effect of impairing interstitial fluid as do Aβ deposits is not yet certain.

Fluid in Subcortical White Matter

Cerebral amyloid angiopathy has an effect not only on gray matter areas in the brain in Alzheimer's disease, but also on the white matter. Accumulation of fluid in subcortical white matter is a frequent finding on MRI in patients with Alzheimer's disease, and this correlates with the severity of cerebral amyloid angiopathy. Blood vessels supplying the subcortical white matter arise from the leptomeningeal arteries on the surface of the brain, and it appears that deposition of Aβ in the walls of the leptomeningeal arteries impedes the drainage of fluid from subcortical white matter.

Genetic Factors in Alzheimer's Disease and Cerebral Amyloid Angiopathy

Familial Alzheimer's disease is associated with defined mutations in the APP and presenilin genes. Polymorphisms in apolipoprotein E (ApoE) are related to the development of cerebral amyloid angiopathy and sporadic Alzheimer's disease. ApoE is an amyloid-scavenging molecule regulating extracellular concentrations of Aβ through ApoE receptor internalization via the endosomal/lysosomal pathway. ApoE–Aβ complexes are transported from the interstitial fluid in the brain into blood through specific ApoE receptors present at the blood–brain barrier. However, co-localization of ApoE with Aβ in brain parenchyma and in artery walls suggests that ApoE–Aβ complexes are also transported with interstitial fluid along perivascular drainage pathways. In this way, ApoE may act as a chaperone molecule for Aβ. ApoE exists in three isoforms (ε2, ε3, and ε4); the ε4 isoform is the most important genetic risk factor for the development of sporadic Alzheimer's disease and has a strong link with cerebral amyloid angiopathy.

Immunotherapy for Alzheimer's Disease

Alzheimer-like transgenic mice have been generated to reproduce Aβ accumulation in brain parenchyma during their adult life in order to understand the mechanisms underlying Aβ deposition and to test treatments that may remove the amyloid deposits. Immunotherapy studies in Alzheimer-like transgenic mice have shown that active immunization against $A\beta_{1-42}$ in young mice prevents the accumulation of Aβ in brain parenchyma, and active immunization of older mice results in a decrease of Aβ in brain parenchyma. Following the success of Aβ immunotherapy in transgenic mice, a clinical trial was initiated in a cohort of patients affected with mild to moderate Alzheimer's disease. Patients were actively immunized against $A\beta_{1-42}$ protein; however, the clinical trial was halted when 6% of the immunized patients developed a complication of what appeared to be meningoencephalitis. Postmortem examination of the first immunized cases shows that insoluble Aβ plaques were removed from cortical areas by the antibody response generated following the immunization and by activated microglia. However, the tau pathology was not modified by Aβ immunotherapy, with the exception of the dystrophic neurites, which disappeared as Aβ plaques were removed.

One of the consequences of Aβ immunotherapy is an increase of the cerebral amyloid angiopathy. It appears, therefore, that removal of insoluble Aβ plaques from brain parenchyma allows access of Aβ to the perivascular drainage pathways. It is not clear whether this is the Aβ removed from the brain by the immunotherapy or whether it is the Aβ from the increased pool of soluble Aβ in gray matter in Alzheimer's disease. Whichever it is, it becomes entrapped in the perivascular drainage pathways. Such an increase in cerebral amyloid angiopathy also results in an increased frequency of microhemorrhages within the brains of the patients immunized against Aβ; this was also observed in experimental transgenic mouse studies. The first year of clinical follow-up of the immunized patients has shown a beneficial effect, with slowing of the decline of cognitive function, and further follow-up studies will allow clinical data and neuropathological features to be correlated. To date, $A\beta_{42}$ immunotherapy has not induced harmful effects in the large majority of patients, and when there have been complications they have not been lethal. A second approach involving passive immunization with specific antibody or peptide, to avoid the apparent meningoencephalitis, is currently in progress. However, it is still not known whether immunotherapy will be most useful as a means to prevent Aβ plaque formation in familial Alzheimer's disease and Down syndrome, or as a therapy in the established sporadic disease.

Conclusions

Aβ is produced by the brain throughout life by the selective cleavage from amyloid precursor protein. The mechanisms for removal of Aβ appear to fail in elderly humans, and if the failure is severe it is associated with Alzheimer's disease. Failure of elimination of Aβ along perivascular interstitial fluid drainage pathways may result from the stiffening of vessels that occurs with age and cerebrovascular disease in humans. If there is an increase in the production of Aβ and particularly the more insoluble form in familial Alzheimer's disease, the system of drainage may become overloaded and impeded at an early age. Understanding the mechanisms by which Aβ is eliminated from the brain, the possible chaperone molecules that are involved, and the exact reasons why elimination of Aβ fails is essential for the planning of future therapies for Alzheimer's disease. Future research may center upon the factors that induce Aβ to attach to vascular basement membranes. Prevention of such attachment may preserve the patency of the pathways for the drainage of interstitial fluid and solutes from the brain.

See also: Aging of the Brain and Alzheimer's Disease; Aging: Extracellular Space; Alzheimer's Disease: An Overview; Alzheimer's Disease: Molecular Genetics; Alzheimer's Disease: Transgenic Mouse Models; Alzheimer's Disease: MRI Studies; Animal Models of Alzheimer's Disease; Axonal Transport and Alzheimer's Disease; Axonal Transport and Huntington's Disease; Axonal Transport and Neurodegenerative Diseases; Brain Glucose Metabolism: Age, Alzheimer's Disease and ApoE Allele Effects; Dementia; Variant Creutzfeldt–Jakob Disease.

Further Reading

Abbott NJ (2004) Evidence for bulk flow of brain interstitial fluid: Significance for physiology and pathology. *Neurochemistry International* 45: 545–552.

Esiri MM, Lee VMY, and Trojanowski JQ (eds.) (2004) *The Neuropathology of Dementia,* 2nd edn. Cambridge: Cambridge University Press.

Gallagher PJ and van der Wal AC (2007) Blood vessels. In: Mills SE (ed.) *Histology for Pathologists,* 3rd edn. pp. 218–238. Philadelphia: Lippincott Williams & Wilkins.

Herzig MC, Van Nostrand WE, and Jucker M (2006) Mechanism of cerebral beta-amyloid angiopathy: Murine and cellular models. *Brain Pathology* 16: 40–54.

Lue LF, Kuo YM, Roher AE, et al. (1999) Soluble amyloid beta peptide concentration as a predictor of synaptic change in Alzheimer's disease. *American Journal of Pathology* 155: 853–862.

Nicoll JA, Wilkinson D, Holmes C, et al. (2003) Neuropathology of human Alzheimer disease after immunization with amyloid-beta peptide: A case report. *Nature Medicine* 9: 448–452.

Preston SD, Steart PV, Wilkinson A, et al. (2003) Capillary and arterial amyloid angiopathy in Alzheimer's disease: Defining the perivascular route for the elimination of amyloid beta from the human brain. *Neuropathology and Applied Neurobiology* 29: 106–117.

Roher AE, Kuo Y-M, Esh C, et al. (2003) Cortical and leptomeningeal cerebrovascular amyloid and white matter pathology in Alzheimer's disease. *Molecular Medicine* 9: 112–122.

Schley D, Carare-Nnadi R, Please CP, et al. (2006) Mechanisms to explain the reverse perivascular transport of solutes out of the brain. *Journal of Theoretical Biology* 238: 962–974.

Shibata M, Yamada S, Kumar SR, et al. (2000) Clearance of Alzheimer's amyloid-beta (1–40) peptide from brain by LDL receptor-related protein-1 at the blood–brain barrier. *Journal of Clinical Investigation* 106: 1489–1499.

Weller RO (2005) Drainage pathways of CSF and interstitial fluid. In: Kalimo H (ed.) *Pathology and Genetics. Cerebrovascular Diseases*, pp. 50–55. Basel: ISN Neuropath Press.

Weller RO and Nicoll JAR (2003) Cerebral amyloid angiopathy: Pathogenesis and effects on the ageing and Alzheimer brain. *Neurological Research* 25: 611–616.

Weller RO and Nicoll JA (2005) Cerebral amyloid angiopathy: Both viper and maggot in the brain. *Annals of Neurology* 58: 348–350.

Weller RO, Massey A, Newman TA, et al. (1998) Cerebral amyloid angiopathy: Amyloid beta accumulates in putative interstitial fluid drainage pathways in Alzheimer's disease. *American Journal of Pathology* 153: 725–733.

Wilcock DM, Rojiani A, Rosenthal A, et al. (2004) Passive immunotherapy against Abeta in aged APP-transgenic mice reverses cognitive deficits and depletes parenchymal amyloid deposits in spite of increased vascular amyloid and microhemorrhage. *Journal of Neuroinflammation* 1: 24.

Amyotrophic Lateral Sclerosis (ALS)

J M Bhatt and H Mitsumoto, Neurological Institute of New York, New York, NY, USA

Introduction

Amyotrophic lateral sclerosis (ALS) is a neurodegenerative disorder which affects motor neurons in the central nervous system. Its clinical and pathological entity was established by Jean Martin Charcot in 1874. In the United States, it is also known as Lou Gehrig's disease after the famous New York Yankee baseball player who was forced to retire and succumbed to this disease in 1941. The terms ALS and motor neuron disease (MND) are often interchangeably used, particularly in the United Kingdom and Europe. The classification and the clinical limits of the disease remain to be established. In general, motor neuron diseases represent a number of disorders which affect the motor neurons, and ALS is the prototype of all motor neuron diseases and is the most common motor neuron disease in adults. By definition, the first-order upper motor neurons in the premotor cortex and their corticospinal tracts and second-order lower motor neurons of the anterior horn in the spinal cord are affected, resulting in a fairly consistent clinical syndrome, course, and outcome. There are no established etiologies of the motor neuron degeneration as a cause of familial ALS except for the mutation in the super oxide dismutase 1 (SOD1) gene. Current research supports several, perhaps multifactorial, pathogenic mechanisms which may lead to a common path: a progressive motor neuron degeneration resulting in a clinical phenotype common to ALS. Classic ALS affects both upper and lower motor neurons, whereas progressive muscular atrophy affects only lower motor neurons and primary lateral sclerosis (PLS) affects upper motor neurons and the corticospinal tract. A disease variant which exclusively involves lower brain stem motor neurons is often called progressive bulbar palsy, but most cases are in fact bulbar-onset ALS, in which patients develop progressive weakness in muscles in other areas. Attention has focused on the overlap of ALS with other clinical syndromes, particularly frontotemporal dementia. The prevalence of cognitive and behavioral impairment in ALS patients varies from 30% to 50% depending on the study, but it is clearly more common than previously thought. Based on the current knowledge of disease involvement, it is clear that cortical neurons are widely affected, although ALS is still a distinctively motor neuron disease with regard to its clinical expression.

Epidemiology

Several regional studies throughout the world have shown a uniform incidence of ALS: 1–2.4 per 100 000 people with a prevalence of 5–7 per 100 000. The onset of disease occurs between 20 and 80 or more years of age, with a peak onset between 50 and 60 years. For unknown reasons, it is rare in the eldest segment of the population – those older than 85 years. ALS may occur sporadically in the general population or it may be inherited in families. Familial ALS is clinically indistinguishable from sporadic ALS and occurs in 5–10% of all ALS cases. Males are more affected by sporadic ALS than woman by a ratio of 3:2. Epidemiologic studies in industrialized nations show that mortality rates of ALS have increased during the past few decades, but this may be explained by improved case reporting or increased life expectancy.

There are several environmental and lifestyle factors which may predispose patients to ALS. Military service has been identified as a risk factor for ALS in veterans from the Persian Gulf War and among veterans in general across time periods. Studies have shown that physical activity may increase the risk of ALS; one investigation found increased mortality among a large cohort of professional Italian soccer players. Occupational exposures to electromagnetic fields, heavy metals (lead and mercury), and organic solvents have been associated with increased risk for developing ALS. Diets high in glutamate and fats have been shown to have a positive association with ALS, whereas diets rich in antioxidants (e.g., vitamin E) and fiber have demonstrated a possible protective effect. Investigations also demonstrate cigarette smoking as a risk factor for ALS, with mortality rates higher in smoking populations, which may be due to direct toxic injury of chemicals or the formation of free radicals. These environmental toxins and lifestyle habits likely contribute to ALS, along with other pathological mechanisms and genetic susceptibility.

ALS–Parkinsonism–Dementia Complex

In the 1950s, indigenous populations in Guam, the Kii peninsula of Japan, and other western Pacific islands had an astonishingly high incidence of ALS associated with parkinsonism and dementia – up to 100 times the worldwide rate. Since then, the incidence has declined to nearly the worldwide rate, suggesting a causative environmental factor. At one point, a high dietary intake of an excitotoxin called β-methylamino-L-alanine (BMAA), produced by cyanobacteria from seeds of a cycad tree, was suspected as a potential cause of Guamanian ALS. More recently, consumption of flying foxes, which feed on cycad seeds and thus increase the concentration of BMAA in a process of 'biomagnification' in this animal, is believed to be associated with this unusual edemic ALS–parkinsonism complex. Although this theory is attractive, solid epidemiological evidence is needed. Other proposed environmental agents of western Pacific ALS include low dietary and water consumption of calcium and magnesium in genetically susceptible individuals. The cause of Guamanian ALS–parkinsonism–dementia complex remains to be determined.

Familial or Hereditary ALS

Several novel mutations with different modes of inheritance have been discovered in families with ALS. The most commonly known mutation is the SOD1 mutation, which accounts for approximately 20% of familial ALS. Approximately 20% of familial cases contain a mutation of the Cu/Zn superoxide dismutase (SOD1/ALS1) gene located on chromosome 21q, which is transmitted in an autosomal dominant manner. More than 130 different mutations of SOD1 have been collectively identified in all five of its exons. Mutations occur at more than 65 of 153 possible amino acid positions and are mostly missense point mutations. Another autosomal dominant mutation in the senataxin (ALS4) gene, which transcribes a DNA/RNA helicase, has been associated with a juvenile-onset ALS on chromosome 9q34; average age of onset is 17 years. Autosomal recessive forms of juvenile-onset ALS have been found in a large Tunisian family with mutations of the alsin gene located on chromosome 2q33, which produces a GTPase exchange factor. Interestingly, other mutations in the alsin gene cause a milder PLS phenotype in families from Kuwait. A rare X-linked dominant form has also been identified on the Xp11–12 chromosome. Linkage studies in families with ALS and frontotemporal dementia (FTD) have identified a new locus at 9q21, and efforts are under way to identify a causative gene and its protein. After gene identification, knockout mouse models of ALS–FTD may be developed to investigate the pathogenic mechanisms of this exciting field.

Histopathology

Gross pathology of brains of patients with ALS exhibits atrophy of cerebral cortex in the precentral gyrus and surrounding cortices. The spinal cord typically shows atrophy of anterior roots. Histopathology shows loss of giant pyramidal cells and surrounding motor neurons of the motor cortex, motor neurons in cranial nerve nuclei, and spinal cord anterior horn cells. In addition, there is a loss of myelinated axons in the lateral and anterior corticospinal tracts. There is also diffuse extensive microglial and macrophage proliferation found by CD68 immunoreactive cells in spinal cord and cerebral cortex, suggesting a chronic microglial inflammatory reaction involved in neurodegeneration.

With regard to anterior horn cell pathology, there are a variety of inclusion bodies and evidence of cytoskeletal abnormalities in sporadic and familial ALS. Inclusion bodies may be found in the cytoplasm, such as eosinophilic hyaline bodies called Bunina bodies and ubiquinated skein-like inclusions, which, if found, seem highly characteristic of ALS pathology. Ubiquitin is a protein thought to mark cells for degradation via nonlysosomal proteolysis, and it is found in other neurodegenerative diseases. Phosphorylated neurofilament aggregates called spheroids are found as markedly swollen axons near the anterior horn cell body. In familial ALS, there are unique intracytoplasmic hyaline inclusions found in surviving motor neurons that contain SOD protein, neurofilaments, and peripherin. In general, the earlier and more aggressive the course of the disease, the greater the number of inclusions and spheroids. It is not known whether the various inclusion bodies are directly related to the pathogenesis of ALS or products of another process, yet these inclusions may potentially suggest a mechanistic importance of protein aggregate diseases. Notably, motor neurons controlling ocular motility and small motor neurons in the sacral spinal cord which innervate muscles controlling pelvic floor and bowel and bladder are largely spared in familial and sporadic cases. These neurons differ with regard to calcium metabolism, electrophysiological firing patterns, and androgen receptors.

Pathogenic Mechanisms

There have been several advances in the hypotheses that have been proposed to explain the cause of ALS. One important concept that has been rigorously scrutinized is the role of excessive neurotoxin, primarily involving glutamate and its receptors. ALS patients have an impaired ability to metabolize orally loaded glutamate; observed levels of serum and cerebrospinal fluid glutamate are increased, whereas glutamate levels in the tissue of brain and spinal cord are reduced in ALS patients. The abnormal levels have been linked to faulty glutamate transport proteins that have been observed in synaptosomal preparations from ALS patients, leading to a breakdown in glutamate absorption from the extracellular space into glial cells. The defective transport raises extracellular levels of glutamate at the synaptic cleft, allowing it to repeatedly stimulate glutamate receptors, specifically its cation channel receptor. The resulting calcium influx leads to a cascade of events including the activation of catabolic enzymes which produce reactive oxygen and nitrogen species that directly damage structural, enzymatic, and receptor proteins, lipid membranes, and DNA.

Significant discoveries have linked oxidative stress to motor neuron cell death via the production of free radicals. This hypothesis was strongly supported by the discovery of mutations in the SOD1 gene which is linked to 20% of diagnosed familial ALS cases. The initial discovery of SOD1 mutations in 1993 was a watershed moment for ALS research because it led to the swift development of transgenic ALS mice. The mutated SOD1 gene was transferred into a mouse genome, producing a readily available animal model. A number of different transgenic mice carrying mutant SOD1 genes are now available and they have undoubtedly broadened the scope and depth of fruitful ALS research. SOD1 is a dimeric protein that catalyzes superoxide radical O_2^- and H^+ to form H_2O_2. Most mutations in the SOD1 molecule do not involve the active site of the enzyme but affect the structure responsible for proper subunit folding or dimer contact. Hence, enzyme function is variable among SOD1 mutations, ranging from normal to absent. This observation suggests that the SOD1 protein is harmful not because of decreased catabolic activity but because of a toxic gain of function. Mutant SOD1 protein is often monomeric and highly unstable, leading to protein degradation and aggregation, which may interfere with axonal transport. Furthermore, mutant protein may cause copper molecules to detach, which may lead to harmful free radical formation. Free radicals are highly reactive species with unpaired electrons that oxidize cellular structures and make them unstable. Long-term accumulation of free radicals, particularly hydroxyl radicals, results in injury to DNA, cell membranes, and cellular protein. The cumulative DNA and cellular membrane damage may pass a threshold leading to programmed cell death.

Another hypothesis is that neuronal degeneration is caused by abnormal inflammatory responses. There is considerable activation and proliferation of central nervous system immune cells called microglia in regions of motor neuron loss and pyramidal tracts. In ALS, microglial cells markedly express proinflammatory cytokines in the spinal cord, such as tumor necrosis factor-α and cyclooxygenase-2. Activated cytokines and chemokines amplify the inflammatory response and recruit proteins that may produce a harmful cascade of events leading to motor neuron death. There is no known cause of the inflammation, and it is not known whether these inflammatory reactions are primary or secondary in nature.

Pathologic examination of motor neurons in ALS patients and in ALS mice models shows prominent protein aggregation in the form of cytoplasmic inclusions. Sometimes, inclusions are found in surrounding astrocytes as well. Such inclusions are a hallmark of several degenerative diseases, including Alzheimer's disease and Parkinson's disease. It is not clear if protein aggregates are directly toxic to motor neurons and their supporting cells, although several theories have been proposed. Most aggregates are intensely immunoreactive with antibodies to ubiquitin, which suggests dysfunction of the ubiquitin–proteasome system that is normally responsible for protein degradation. Proteasomes, which function to degrade damaged protein, may be overwhelmed with indigestible misfolded protein, or other important proteins may not function properly in the presence of the aggregates. Experiments have shows that aggregates affect the operation of folding chaperones, which help other vital proteins attain their appropriate three-dimensional conformations. There is evidence that inclusions may interfere with mitochondrial or peroxisomal function, leading to cellular demise. It is quite possible that protein aggregates are simply a feature of dying cells; this will require further investigation.

Neurofilament dysfunction has been a proposed mechanism for ALS ever since they were identified as pathologic landmarks in sporadic and familial forms. Neurofilaments are important structural proteins that confer motor neurons with their unique axonal diameters and length. Neurofilament mutations have been identified in a small percentage of sporadic ALS patients and in other neurodegenerative diseases, including infantile spinal muscular atrophy and hereditary sensorimotor neuropathy. However, genetic experiments in mice involving overexpression or deletion of neurofilament subunits have slowed the onset of motor neuron disease. In any case, neurofilament organization may be a significant risk factor, if not a direct cause, of motor neuron diseases.

The investigation of programmed cell death, called apoptosis, has shed light on motor neuron degeneration and potential targets for experimental therapy. Studies have shown activation of a family of enzymes called caspases months before neuron death and clinical manifestation of ALS in animal models. Caspases are proteasomes responsible for the breakdown of several cell components in programmed cell death. Inhibition of caspases has been shown to significantly slow disease onset and increase survival in mice genetically engineered to acquire motor neuron disease. Furthermore, increasing expression of factors that interfere with apoptosis has yielded similar promising results. The challenge for developing treatments that aim to retard the process of cell death is to selectively target corticospinal and motor neurons. Apoptotic proteins are widely expressed in cells outside the central nervous system, and identification of cell death mediators that specifically affect motor neurons will be important for future therapies.

All the aforementioned hypotheses of pathogenesis in ALS lead to a series of events resulting in neurodegeneration of motor neurons. It is still uncertain whether any one the hypothetical mechanisms described previously is the primary or secondary cause; however, it appears that these mechanisms are closely related to each other. Vigorous research efforts are under way.

Clinical Features and Clinical Course

There is strong evidence to support the hypothesis that the process of motor neuron degeneration precedes the appearance of identifiable weakness in patients with ALS. Electrodiagnostic testing in patients diagnosed early in the course of illness shows peripheral denervation and renervation in muscle. Thus, there is a preclinical stage of varying length before the onset of clinical symptoms. Patients usually present with complaints of focal muscle weakness in the arm or hand (arm onset, ~40%), weakness in the foot and difficulty walking (leg onset, ~35%), or problems with speech and swallowing (bulbar onset, ~25%). The localized weakness usually spreads to adjacent muscle groups before moving to other regions of the body, often on the same side. Most often, muscle weakness and atrophy begin in the small hand muscles in an asymmetric fashion, but weakness can be multifocal or proximal in distribution. Patients with familial ALS are approximately 10 years younger than those with sporadic ALS and present with prominent leg weakness, as opposed to arm weakness with sporadic ALS. Despite skeletal muscle paralysis, the ocular muscles, sensorium and bladder, bowel, and autonomic functions are, as a rule, well preserved. Cognitive dysfunction in ALS patients has been scrutinized in recent years because it may be much more common than previously thought. Indeed, as many as 30–50% of patients with ALS may exhibit cognitive symptoms ranging from mild cognitive impairment to executive and language dysfunction and personality and behavioral changes. Specifically, there is compelling clinical and pathologic overlap between motor neuron disease and frontotemporal dementia, which suggests ALS may be a multisystemic syndrome. It is unknown, however, if ALS with FTD is a spectrum of the same disease or if it represents a different disease.

The neurologic exam shows a combination of upper motor neuron signs, including abnormal hyperreflexia, pathological reflexes, and spasticity, and lower motor neuron signs, including muscle atrophy, muscle weakness, hyporeflexia, hypotonia, and fasciculations. Progressive impairment of motor function rapidly results in physical disability. Paralysis in certain muscles causes progressive impairment with speaking, swallowing, and respiration, resulting in severe chronic debilitation. Alveolar hypoventilation usually leads to terminal demise. Severe weight loss is common, sometimes out of proportion to muscle wasting. Fifty percent of patients die within 3 or 4 years of symptom onset, usually because of respiratory failure or aspiration pneumonia. On the other hand, more than 10% of patients may live more than 10 years. Some forms of ALS are slowly progressive over several decades. Hence, the clinical course is more variable than previously suspected.

In some instances, the diagnosis of ALS may be made by virtue of the history and neurologic examination alone. Often, the diagnosis is not readily apparent and further testing is required to rule out other diseases which may mimic ALS. There is no diagnostic test for ALS. Routine laboratory tests should be obtained at the onset of the workup to rule out other metabolic, inflammatory, hematological, immunological, or hormonal abnormalities. Anti-Gm1 antibodies may be sent to rule out a multifocal motor neuropathy, which is a treatable cause of muscle weakness. Nerve conduction studies and electromyography are also essential in distinguishing ALS from other neuromuscular diseases causing progressive weakness. Finally, neuroimaging is important to rule out any structural or infiltrating or neoplastic disorders of the brain and spinal cord that may resemble ALS, such as cervical spondylosis or a spinal cord tumor. If familial motor neuron disease is suspected, then SOD1 mutation analysis is warranted, but it should be remembered that SOD1 mutation is only positive in approximately 20% of patients with familial ALS. A small percentage of sporadic ALS patients are found to have SOD1 mutations. There are diagnostic guidelines known as the El Escorial criteria, named after the location of an important meeting of the World Federation in Neurology in 1990 and thereafter revised, which have set criteria for the clinical diagnosis of ALS. These criteria have not only facilitated making a uniform diagnosis of ALS in clinical trials and other clinical research, but they have also been used in clinical practice.

Management

Since the outcome of the disease is predictable in most cases, the diagnosis of ALS can be devastating to patients and their families. In making the diagnosis, it is important first to exclude all treatable diseases and, if not, all possible definable diseases before concluding that the disease is ALS. Sensitivity, care, and compassion by the physician are essential when discussing a diagnosis of ALS with patients and family caregivers, who should be available for such a discussion. Giving hope, a positive and compassionate attitude, and commitment to care for the patient are vital in such a desperate situation.

The first line of pharmacotherapy for ALS is an antiglutamate agent, riluzole. Two clinical trials that led to the approval of this medication by the Food and Drug Administration for the treatment of ALS showed statistically significant prolongation of survival by approximately 3 months in patients who took the medication. Patients' symptoms or motor dysfunction did not change. In the United States, less than two-thirds of patients take riluzole. Reasons for this limited use are high drug cost and modest benefits. Several retrospective studies following the drug's approval showed that patients who took riluzole lived longer than those who were in the pre-riluzole era or those who did not take riluzole. Prolongation of survival was as long as 12 months. However, the reliability of such retrospective studies is uncertain. Insurance coverage for riluzole is clearly an important factor for physicians who prescribe this drug.

Maximum symtomatic relief is also important in the treatment of ALS. For example, muscle relaxants can be given for spasticity and muscle cramps, anticholinergic or tricyclic antidepressants can control drooling, antidepressants can treat depression, and analgesics may ease pericapsular joint pains. Other management priorities include constipation, insomnia, and laryngeal spasm.

ALS poses a multitude of medical and psychosocial issues. For managing such a disease, patients are best cared for at a special multidisciplinary care center. At a typical ALS center, physical and occupational therapists evaluate the patient's physical impairment; discuss necessary exercises; and may recommend the use of splints, minor devices for assisting impaired manual dexterity, a neck brace, a wheelchair, or other home equipment to improve function in the activities of daily living. Nutritionists watch the patient's nutritional status by carefully monitoring weight, recommending changes in foods to accommodate dysphagia and maintaining necessary caloric needs. Eventually enteral tube feeding may be recommended. Speech therapists help to improve enunciation and swallowing during the early stages of the disease and later assist patients in using alternative communication methods, including communicators or computers. Social workers not only help to find resources for patients but also often provide psychologic support. Respiratory therapists or nurses assess patients' respiratory status to detect early respiratory difficulty. Physicians and neurologists are now paying serious attention to aggressively managing respiratory dysfunction and nutritional impairment. Nurses specializing in ALS play a key role in arranging all services at the clinic and provide psychologic and personal support to patients and family. A number of regionally and nationally known ALS centers provide such services throughout the United States. The Muscular Dystrophy Association and the ALS Association vigorously support such efforts.

At a certain point – ideally during the early stages of the disease – death and life-support issues (advance directives) should be discussed openly but sensitively. If a patient wishes to be sustained on a ventilator, a ventilator team should be consulted at some point. In patients with impending respiratory failure, caring physicians and pulmonologists recommend the use of a noninvasive positive pressure ventilator, which is a respiratory assist device that uses a mouth, nose, or mouth–nose interface. Positive pressure ventilation provides marked relief of respiratory distress and fatigue, helps with sleeping difficulty, and improves overall quality of life. To relieve terminal respiratory distress, pain, and overwhelming anxiety, palliative care is essential. Hospice services have been used at an increasing frequency. The use of morphine sulfate or other sedatives should be considered for those who do not wish to use life-support measures.

Clinical Trials and Emerging Therapies

Clinical trials in ALS can be an important part of patient management because they can provide hope to patients, their families, and even to the treating physicians. Well-designed clinical trials may shed new light on the pathogenesis of ALS, even if they do not confer a survival benefit. There has been substantial growth in the number of compounds suitable for current trials in ALS patients in the past decade. These medications include glutamate antagonists, antioxidants, anti-inflammatory agents, neurotrophic factors, and miscellaneous 'neuroprotective' compounds that have shown proven benefit in the mouse model. Unfortunately, many promising drugs that showed benefit in animal models of ALS have failed in human trials, including creatine, neurotrophic factors, gabapentin and vitamin E, cefecoxib, topiramate, and pentoxyfilline. As the number of studies has increased, the use of uniform diagnostic criteria, the availability of reliable and reproducible outcome evaluation techniques, and the study design have become critical issues and there appear to be continuous improvements. More potential pharmacological and nonpharmacological agents have become available in recent years. Further investigations are crucial for developing effective therapeutic agents.

There are quite a few exciting translational research advances on the frontier of ALS research that may deliver therapeutic options in the future. One such approach is called messenger RNA interference, which may potentially treat autosomal dominant forms of ALS. In short, double-stranded RNA molecules are designed to specifically target faulty RNA transcripts to prevent abnormal protein formation. This method allows normal alleles to express wild-type protein for the preservation of the cell and has been used successfully in cells expressing autosomal dominant SOD1 protein. At the same time, oligonucleotides that block the mutated SOD1 gene have been developed and used in SOD1 transgenic mice. This treatment does not address the majority of sporadic ALS cases, but it is a promising start. This treatment appears to delay the disease onset but does not prolong survival in animals. Nevertheless, such treatments may become available in patients with SOD1 mutations. There is also a burgeoning field of vaccine trials for neurodegenerative diseases. Experimental compounds that promote an anti-self cell-mediated response without risk for autoimmune disease have been tested as vaccines in mice. This heightened autoimmune response is protective rather than destructive. A compound known to treat multiple sclerosis called glatiramer acetate is being set up for human trials in ALS because it conferred neuroprotection to mice models of parkinsonism as a vaccine. The process of gene therapy, in which healthy genes are transferred into an individual's cells, has been considered for sporadic ALS. However, there are several practical and theoretical challenges that must be overcome. The correct gene must be identified and then delivered to the proper anatomic site in the motor neuron. Potential transfer candidates are the insulin-like growth factor-I gene and vascular endothelial growth factor gene, perhaps using adeno-associated virus. Furthermore, glial cells which support motor neurons nearby are also affected by mutations and are important in the cascade of events leading to degeneration. Another exciting development in ALS is potential therapy using stem cells. Embryonic or neurogenic stem cells may be able to replace lost motor neurons in animals or patients with ALS. Again, practical aspects of targeting and delivery in a hostile surrounding environment must be investigated in order for effective treatment to be developed.

The future of ALS research and therapy does have its challenges, but it holds great promise. One major difficulty is an overblown expectation created by the media publicizing experimental therapies as if they were readily available for human administration. New mutations are being discovered at a rapid rate and will undoubtedly further our understanding of ALS biology. In addition, the pathogenetic mechanisms of ALS are shared by other neurodegenerative diseases, and other ideas may emerge from those avenues. It will take time and patience before all these exciting developments become a reality.

See also: Amyotrophic Lateral Sclerosis (ALS): Disease Mechanisms; Animal Models of Motor and Sensory Neuron Disease; Apoptosis in Neurodegenerative Disease; Axonal Transport and ALS; Excitotoxicity in Neurodegenerative Disease; Infectious Agents in Neurodegenerative Disease; Inflammation in Neurodegenerative Disease and Injury.

Further Reading

Albert SM, Rabkin JG, Del Bene ML, et al. (2005) Wish to die in end stage ALS. *Neurology* 65: 68–74.

Anderson PM (2006) Amyotrophic lateral sclerosis associated with mutations in the CuZn superoxide dismutase gene. *Current Neurology and Neuroscience Reports* 6(1): 37–46.

Bruijn LI, Miller TM, and Cleveland DW (2004) Unraveling the mechanisms involved in motor neuron degeneration in ALS. *Annual Review of Neuroscience* 27: 723–749.

Mitsumoto H and Munsat T (2001) *Amyotrophic Lateral Sclerosis: A Guide for Patients and Families.* New York: Demos Medical.

Mitsumoto H, Przedborski S and Gordon PH (eds.) (2005) *Amyotrophic Lateral Sclerosis.* New York: Taylor & Francis.

Mitusmoto H, Bromberg M, Johnston W, et al. (2005) Promoting excellence in end-of-life care in ALS. *ALS and Other Motor Neuron Disorders* 6: 145–154.

Noonan CW, White MC, Thurman D, and Wong L (2005) Temporal and geographic variation in United States motor neuron disease mortality, 1969–1998. *Neurology* 64: 1215–1221.

Shaw PJ (2005) Molecular and cellular pathways of neurodegeneration in motor neurone disease. *Journal of Neurology, Neurosurgery & Psychiatry* 76: 1046–1057.

Amyotrophic Lateral Sclerosis (ALS): Disease Mechanisms

C Vande Velde and D W Cleveland, University of California at San Diego, La Jolla, CA, USA

Amyotrophic Lateral Sclerosis

Amyotrophic lateral sclerosis (ALS), known in the United States as Lou Gehrig's disease, was first described by the famous French scientist and physician Jean-Martin Charcot in 1869. ALS, the most common adult-onset motor neuron disease, refers to a heterogeneous group of neurodegenerative disorders characterized by the selective loss of upper and lower motor neurons (specialized cells that control movement) in the brain and spinal cord resulting in inevitable paralysis and death. The typical age of onset for most forms of ALS is between 50 and 60 years with an average survival of less than 3 years. This insidious disease is characterized by progressive muscle weakness, atrophy, and spasticity, and is traditionally viewed as lacking in cognitive impairment. However, recent literature documents a variety of cognitive deficiencies in a subset of ALS patients that occur subsequent to neuromuscular deficits. In general, the final fatal event is the loss of the motor neurons that innervate the respiratory muscles and diaphragm. The human impact of ALS is enormous as it significantly affects a patient's quality of life (the loss of speech and swallowing is inevitable). The life time risk for developing ALS is about 1 in 2000. At present, ALS patients suffer knowing there is no cure, and worse – no truly effective treatment exists to slow disease progression.

While 90–95% of all ALS cases lack an apparent genetic linkage, 5–10% are dominantly inherited disease. Of the familial cases, 15–20% are attributed to mutations in the ubiquitously expressed metalloenzyme copper/zinc superoxide dismutase (SOD1) whose endogenous function is to relieve the oxidative stress generated by normal cellular metabolism. At last count, more than 110 different mutations (scattered throughout the 153 possible positions!) have been identified in familial cases of ALS. Regardless of the nature of the mutation, all provoke the age-dependent and selective loss of motor neurons through acquisition of one or more as yet unidentified toxic properties, not through the loss of enzyme activity. The toxic property itself and the biological basis for the selectivity to motor neurons remain unknown. However, the discovery of disease-causing SOD1 mutations has paved the way for the creation of rodent models on which to model the disease, thus providing a tool to tease out the molecular events that trigger and/or modulate the disease and ultimately develop effective therapies. The extensive use of these models in ALS research has yielded six major themes in the study of motor neuron degeneration (**Figure 1**).

Mechanisms of Motor Neuron Degeneration

Excitotoxicity

Glutamate is an essential molecule which transmits an excitatory signal across a synapse to an awaiting motor neuron. Deregulated or excessive transmission of glutamate can result in a toxic increase in the intracellular calcium concentration within motor neurons. In general, extracellular glutamate is actively cleared from the synaptic space by closely juxtapositioned astrocytes via the activity of a glial-specific glutamate transporter, excitatory amino acid transporter 2 (EAAT2). In the case of ALS patients and rodent models, a focal loss of this particular transporter correlates well with the loss of motor neurons in the anterior horn of the spinal cord and is further corroborated by the expected increased level of glutamate in the fluid which bathes the brain and spinal cord (cerebrospinal fluid). Furthermore, at-risk motor neurons seem to express a reduced amount of calcium-binding proteins, and thus may be less well equipped to handle a wave of glutamate-mediated calcium influx. To date, excitotoxicity remains one of the few mechanistic links between sporadic and SOD1-mediated familial ALS. In fact, the biological basis for the only food and drug administration (FDA) approved ALS treatment, Riluzole (commercially known as Rilutek), is thought to be through the modulation of the glutamate-mediated excitotoxic response. Despite this, Riluzole offers a very limited extension in lifespan (of ~3 months) and a modest delay in the progression of the disease.

Mitochondrial Dysfunction

Mitochondria are the major energy-producing centers within cells and thus largely responsible for the maintenance of cellular metabolism and survival. The motor neuron is an energy-demanding cell – any interference with its mitochondrial energy production is usually detrimental to its survival. The visual observation of abnormal mitochondrial morphology in the motor neurons of ALS patients in early stages of degeneration was the first implication of

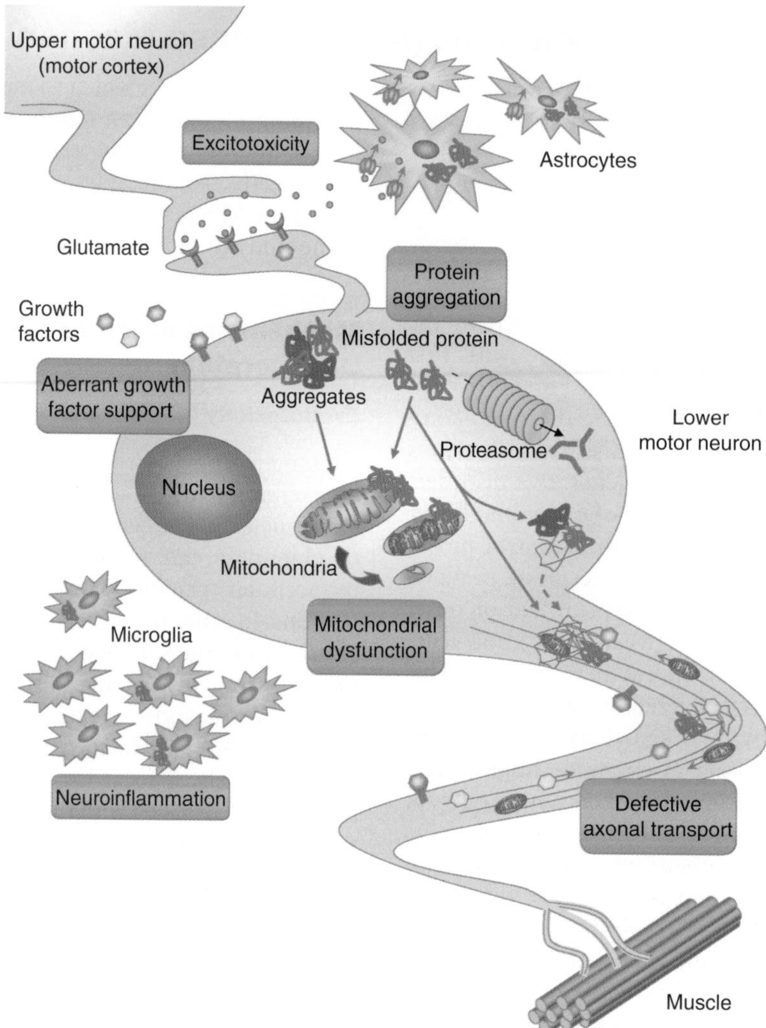

Figure 1 Mechanisms contributing to motor neuron degeneration in ALS.

mitochondria in ALS disease. Similar abnormalities, indicative of mitochondrial damage and dysfunction, were soon identified in rodent models of mutant SOD1-mediated ALS. Using both human (sporadic and familial) cases as well as transgenic models, there has been much focus on components involved in energy production, namely the electron transport chain which drives ATP (the primary energy currency in the cell) generation. While deficits in one or more components of this multicomplex system have been reported, mostly in tissues collected from disease end stage, there has been little agreement between groups on which complex is affected. Moreover, it is unclear whether these defects are causative or merely a consequence of the widespread degeneration of the motor neuron itself that is occurring at this late stage.

The proposal that mitochondrial damage is an initiating event in the failure of motor neurons has gained strong support through the analysis of disease in rodents that express ALS-linked SOD1 mutations. A common feature of all mutant SOD1 proteins (an abundant cytosolic protein) is their preferential association with spinal cord mitochondria purified from affected but not unaffected tissues. This was also observed in samples from familial (SOD1) patients. Moreover, in both cases, the normal endogenous SOD1 was largely excluded from the same spinal cord mitochondria. There is also an intriguing temporal correlation between mitochondrial association and disease progression for the mutant SOD1 proteins. This universal mitochondrial association of various mutant SOD1 proteins may be a critical determinant governing the tissue-specific degeneration of motor neurons (the obligate feature of all ALS cases).

Since optimal mitochondrial function (leading to efficient and sufficient energy production) is central to cellular stability/homeostasis, including the subsequent proposed mechanisms, understanding how mutant SOD1 proteins impact mitochondrial function is essential and currently of intense interest. Spinal cord mitochondria from presymptomatic mutant SOD1 animals may have a reduced capacity to buffer intracellular calcium fluxes, a concept that fits quite well with the earlier described mechanism of glutamate-mediated excitoxicity. An alternate proposal awaiting confirmation is a proposed direct interaction between mutant SOD1 proteins and BCL-2, an antiapoptotic protein residing on the mitochondrial surface. This proposed interaction may sequester BCL-2, thereby quenching its antiapoptotic activity by preventing BCL-2's interaction with proapoptotic proteins. Since mitochondrial abnormalities are widely reported in sporadic ALS, mitochondrial dysfunction or damage may represent a common pathway in both inherited and sporadic disease.

Protein Aggregation

ALS patient samples typically feature very large intracellular, cytoplasmic inclusions within motor neurons and occasionally in astrocytes. These inclusions are easily identifiable with classic immunohistochemical dyes, but their origin and composition is unresolved. In the mutant SOD1 models, these aggregates are intensely immunoreactive for SOD1 itself, as well as neurofilaments (a structural protein of the motor neuron) and ubiquitin (a component of the protein degradation machinery). Some of these aggregates are detergent resistant and contain misfolded SOD1 and protein folding chaperones (including HSP25 and $\alpha\beta$-crystallin) as well as other proteins including covalent adducts of SOD1 itself. These aggregates occur uniquely in the spinal cords of affected animals, correlate well with disease onset, and may actually be present at the mitochondrial surface.

The prevailing hypothesis is that these large aggregates/inclusions are local enrichments of some toxic species which damage motor neurons. The mechanisms which might contribute to this damage include aberrant chemistry by toxic isoforms of SOD1, co-sequestration of other cellular proteins into these large aggregates (leading to loss of their function), depletion of the protein folding machinery, saturation and eventual disruption of the protein degradation machinery (which has an affinity for misfolded proteins), and interference with mitochondrial function due to aggregation at the mitochondrial surface. The implication of the protein folding pathway is consistent with the co-association of a subset of chaperone proteins, including HSP25, HSP70, HSP40, and $\alpha\beta$-crystallin with SOD1 and SOD1-immunoreactive inclusions. Similar significance can be applied to the proposed defect in protein degradation. Specifically, decreased enzymatic activities of the proteasome, the main organelle involved in protein degradation, has been documented in spinal cord homogenates of mutant SOD1 animals. Finally, it remains controversial as to whether these inclusions are truly damaging. The alternate hypothesis is that these inclusions are in fact protective due to the sequestering of a soluble toxic species. This debate is not unique to ALS, recurring in all of the major examples of human neurodegenerative disease.

Neurofilaments and Axonal Transport

Motor neurons are among the largest and most highly asymmetric cells in nature, with meter long axons generating cell volumes that in humans are 5000 times that of a typical cell. These long processes are specified by an ordered array of neurofilaments, the most abundant structural proteins. Neurofilaments are assembled from neurofilament-L (NF-L), neurofilament-M (NF-M), and neurofilament-H (NF-H). Proper assembly of neurofilaments is essential to the establishment of correct axonal diameters. It is indeed only the largest caliber, neurofilament-rich motor neurons that are most at-risk in all ALS patients (both sporadic and familial) as well as mutant SOD1 rodent models. In the normal motor neuron, neurofilaments form a structural framework that determines axonal caliber (which in turn determines the speed of electrical signal conduction). Furthermore, as discussed earlier, accumulations of neurofilament subunits in motor neuron cell bodies is a frequent feature of ALS cases. In addition, there is some genetic evidence which indicates that mutations in the neurofilament genes may increase the risk of developing ALS. Indeed, mutations in NF-L have already been determined to be causative for one form of motor neuropathy, Charcot–Marie–Tooth disease (type II).

The active movement of cellular proteins and organelles along microtubule tracts in axons, from the cell body all the way to the far-established synapse (and back again), is mediated by ATP-consuming motor proteins. Components traveling anterogradely (toward the synapse) are moved by members of the kinesin family. Defects in the rate of anterograde transport of some components, including SOD1 and neurofilaments, have been documented in mutant SOD1 mice well before the onset of disease. Disturbances in the retrograde movement (from the synapse to the cell body) of cellular constituents, mediated by the single known retrograde motor dynein, have also been implicated in ALS models. In mice, the indirect disruption of dynein function uniquely in postmitotic neurons through disruption of an activation complex (named dynactin) results in impaired retrograde axonal transport and the eventual development of a late-onset, progressive motor neuron disease that is reminiscent of the ALS mutant SOD1 mouse models. Errors of axonal transport can be a primary cause of late onset motor neuron disease, as demonstrated by point mutations in dynein that provoke motor neuron degeneration in rodents. Interestingly, the introduction of these mutations into mutant SOD1 models leads to a delay in disease onset and extension in lifespan. The molecular basis for this observation is not understood, but perhaps may involve an interplay between the retrograde signaling of toxic and/or trophic factors.

Errors in axonal transport are likely contributors to human ALS. High concentrations of mitochondria and vesicles are found at nerve endings (the neuromuscular junctions where the motor neuron innervates the muscle) in ALS patient samples. Similar distal accumulations have also been reported in mutant SOD1 mouse models. However, what remains unclear is whether these distal accumulations of organelles are due to accelerated anterograde or impaired retrograde transport of these structures. Mitochondria are known to be highly dynamic organelles that are actively transported within cells and through both axonal and dendritic processes of neurons. While the delivery of these organelles is in itself an energy-consuming process, the inefficient delivery of these energy-generating organelles is likely to have significant effects on synaptic activity and neurotransmission, among other previously described cellular processes. The significance of accumulated synaptic vesicles at the nerve terminal remains to be determined but seems to suggest possible problems in synaptic transmission and/or vesicle recycling.

Growth Factor Signaling

Motor neurons require an abundance of both positive and negative signaling molecules to establish and maintain their connections. These signals include those produced by cells in close proximity to their cell bodies (astrocytes and microglia) and those intimately associated with their axons (Schwann cells and oligodendrocytes) and terminals (muscles). The contribution of neighboring cell types to motor neuron degeneration has come to the forefront in ALS research. It is now appreciated that even normal motor neurons (devoid of any disease-causing mutations) are damaged by other nonneuronal cells which do express disease-causing mutant SOD1. This apparent transfer of toxicity to healthy motor neurons may be mediated by a lack of positive trophic signaling from these damage-incurring nonneuronal cells. Candidate molecules that have been implicated include ciliary neurotrophic factor (CNTF), glial cell-line derived neurotrophic factor (GDNF), brain-derived neurotrophic factor (BDNF), and insulin-like growth factor-1 (IGF-1). While these neurotrophins have all been shown to protect motor neurons in culture and/or animal models, all have failed to significantly alter disease in patient trials. However, for IGF-1, CNTF, and GDNF, there are key concerns on whether delivery of appropriate levels of these agents to the nervous system was achieved.

An unexpected involvement of vascular endothelial growth factor (VEGF) has recently been discovered in ALS. VEGF is a growth factor classically thought to participate only in the development of new blood vessels (angiogenesis). Expression of the VEGF gene is regulated by a transcription factor that is itself able to sense low oxygen conditions (hypoxia) and thus upregulate VEGF expression, thereby restoring efficient vascular perfusion in the affected area. Most surprisingly, in a mouse model in which VEGF production was reduced (owing to a mutation in one of its promoter elements which provides normal basal levels of VEGF but does not permit its upregulation in response to hypoxia), a proportion of the mice develop a progressive, late-onset motor neuron disease that is neuropathologically similar to that observed in ALS mice. Furthermore, introducing this mutation of VEGF into mutant SOD1 animals exacerbates the disease. While a firm genetic link of VEGF to ALS pathogenesis remains to be established, an isoform of VEGF has been documented to be neurotrophic for motor neurons grown in culture. Furthermore, the delivery of VEGF whether by viral expression or purified protein into the central nervous system (CNS) of mutant SOD1 animals provides a significant delay in the onset of disease pathology and extends survival. While the exact mechanism involved in VEGF-mediated protection of motor neurons remains to be defined, a clinical trial involving the localized and direct delivery of purified VEGF into the CNS of ALS patients is currently underway.

Neuroinflammation/Glial Activation

It is now appreciated that cell types other than motor neurons contribute to disease initiation and progression. Indeed, the activation of surrounding glial cells, both the trophic-providing astrocytes and the injury-sensing microglia, has been widely reported in both rodent models and patient samples. Microglia share the same ancestry as peripheral tissue macrophages and thus are considered to be the resident immune cells of the CNS. Microglia sense injury in the CNS and rapidly become activated, initiating a neuroinflammatory pathway involving the release of cytotoxic molecules which affects all neighboring cells, including neurons and astrocytes. Microglial activation correlates well with disease and is observed in human ALS and in all rodent models of ALS and thus is strongly implicated in the disease process. The administration of the FDA-approved antibiotic minocycline to mutant SOD1 mice delays disease progression primarily by preventing microglial activation. Furthermore, the removal of toxic mutant SOD1 from microglial cells limits microglial activation and provides a substantial extension in lifespan in mice which are still expressing the mutant product in all other cell types. This substantial extension in lifespan is actually due to a significant slowing of disease progression after onset. Recognition that microglial activation affects the rate of disease progression offers a significant guidepost to the development of an effective therapy for ALS (as well as other neurodegenerative disease in which neuroinflammation is a component). A clinical trial in which minocycline is included in patient treatment plans is currently underway.

Summary

While many different mechanisms for disease initiation and progression in ALS have been proposed, these should not be seen as mutually exclusive. Many or all are likely to be important contributors that are intricately related to one another. While there may be more than one initiating event occurring within the motor neuron (mitochondrial dysfunction, excitoxicity, protein aggregation, and defective axonal transport), they are likely to converge on one (or at most a few) common final path(s) to motor neuron death. Trophic signaling and neuroinflammation are likely the primary elements mediating disease progression and thus are attractive targets for new therapeutic strategies.

See also: Amyotrophic Lateral Sclerosis (ALS); Axonal Transport and ALS; Axonal Transport and Neurodegenerative Diseases; Excitotoxicity in Neurodegenerative Disease; Inflammation in Neurodegenerative Disease and Injury; Neurofilaments: Organization and Function in Neurons; Oxidative Damage in Neurodegeneration and Injury; Transgenic Models of Neurodegenerative Disease.

Further Reading

Bachman SR, Bradley WG, and Moraes CT (2006) Mitochondrial involvement in amyotrophic lateral sclerosis: Trigger or target? *Molecular Neurobiology* 33: 113–131.

Boillée S, Yamanaka K, Lobsiger CS, et al. (2006) Onset and progression in inherited ALS is determined by motor neurons and microglia. *Science* 312: 1389–1392.

Bruijn LI, Miller TM, and Cleveland DW (2004) Unraveling the mechanisms involved in motor neuron degeneration in ALS. *Annual Review of Neuroscience* 27: 723–749.

Clement AM, Nguyen MD, Roberts EA, et al. (2003) Wild-type nonneuronal cells extend survival of SOD1 mutant motor neurons in ALS mice. *Science* 302: 113–117.

Cleveland DW and Rothstein JD (2001) From Charcot to Lou Gehrig: Deciphering selective motor neuron death in ALS. *Nature Neuroscience Reviews* 2: 806–819.

Strong MJ, Lomen-Hoerth C, Caselli RJ, Bigio EH, and Yang W (2003) Cognitive impairment, frontotemporal dementia, and the motor neuron diseases. *Annals of Neurology* 54: S20–S23.

Relevant Website

http://www.alsa.org – The ALS Association.

Angelman Syndrome

L M Bird, Rady Children's Hospital and University of California, San Diego, CA, USA

Introduction

Most autosomal genes are expressed from both parental copies (biallelic expression), regardless of the parental origin. A small number of genes are expressed from only one parent's allele, a phenomenon known as genomic imprinting. Angelman syndrome (AS) is one of the most well-understood syndromes of genomic imprinting. AS maps to the imprinted gene cluster on chromosome 15q11–q13 and is due to defective expression of the maternally inherited ubiquitin protein ligase 3A gene (*UBE3A/E6-AP*). There are at least four molecular classes of AS, demonstrating the genetic complexity of this disorder. Though AS is a monogenic disorder, it is a prototype for the so-called mixed epigenetic and genetic and mixed *de novo* and inherited (MEGDI) model of disease. Herein follows a summary of the progress made to date in elucidating the mechanisms of imprinting in AS.

Clinical Background

The clinical features of AS are microcephaly, functionally severe mental retardation, ataxic gait, seizures, and electroencephalographic abnormalities. The behavioral profile of AS includes exceptionally happy disposition with episodes of laughter; excessive mouthing of objects, drooling, and tongue thrusting; easy excitability and hypermotoric activity; reduced need for sleep, with sleep disturbance; and an affinity for water.

Seizures begin in infancy or early childhood in the majority of patients, and often show attenuation in adolescence. A wide variety of seizure types have been reported. Movement disturbances (tremors, jerkiness, poor coordination) contribute to the delayed acquisition of motor skills (sitting after 12 months; walking between 2 and 6 years). The incidence of nonambulation is said to be 10%, but this may no longer be true in the modern era of earlier diagnosis and prompt intervention. The early institution of physical therapy may change the natural history of scoliosis (said to occur in 50%) by improving truncal tone.

Virtually all AS patients exhibit a short attention span, and most are hypermotoric. Some aggressive, unwanted behaviors are displayed by the majority of patients, including biting, pinching, hair-pulling, and grabbing. Rarely are these behaviors intended to cause harm; they usually result from easy excitability, poor control over movements, reduced repertoire of need expression, and occasionally frustration over an inability to communicate effectively. Paroxysms of laughter are said to occur in AS, but laughter is rarely unprovoked, though often inappropriate to the triggering stimulus. A happy disposition characterizes most children, but irritability is common in infancy. Gastrointestinal difficulties such as gastroesophageal reflux may contribute to this irritability. AS patients have an apparent reduced need for sleep (5–6 h per night) and many have long or frequent periods of wakefulness during the night. With behavior modification and medication, these difficulties can be overcome in most patients, and sleep patterns improve with age.

Cognitive performance is usually in the range of severe functional impairment. The combination of deficits exhibited by individuals with AS make the commonly used developmental assessment tools difficult to apply, and these tests probably underestimate the abilities of AS children. Receptive language is far superior to expressive language, which is often completely absent. Nonverbal communication using a variety of systems (picture exchange cards; communication devices) is possible in a substantial proportion of AS patients. Poor fine motor coordination usually precludes the use of sign language.

AS adults are not able to live independently, but most are able to using feeding utensils and perform some household tasks. Dressing skills are dependent on the degree of fine motor difficulty. Daytime continence is possible with prompted voiding and habit training. Life expectancy appears to be normal.

Molecular Genetics of AS

Failure of maternally inherited *UBE3A* to be transcribed and translated into a functional protein causes AS. Four molecular mutation classes of AS have been defined: deletion, uniparental disomy (UPD), imprinting defect, and *UBE3A* mutation.

Deletion

Approximately 65% of AS cases exhibit a *de novo* (~4 Mb) microdeletion of chromosome 15q11–q13. Two major classes of deletions exist, one from breakpoint 1 (BP1) to breakpoint 3 (BP3) (class I) and the other from breakpoint 2 (BP2) to BP3 (class II) (see **Figure 1**). Rarely, patients may have a larger deletion that extends beyond the boundaries of BP1 and BP3 in one or both directions. Deletions are apparently mediated by homologous misalignment and meiotic recombination between low-copy-number repeats

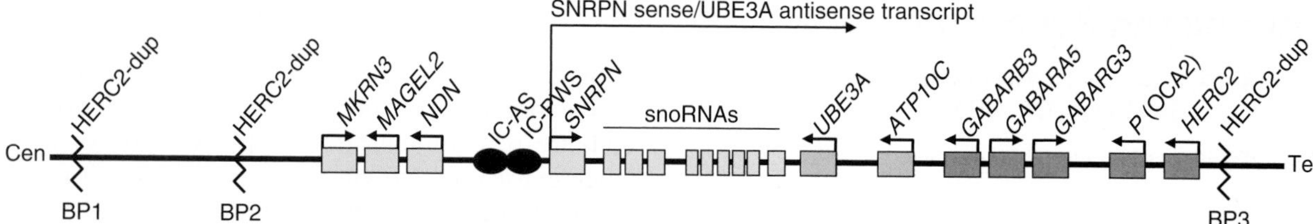

Figure 1 Genomic organization of human chromosome 15q11–q13. Blue loci are paternally transcribed, pink loci are maternally transcribed, and the gray loci are biparentally expressed. Arrows indicate direction of transcription. Black ovals denote bipartite imprinting center (IC-AS, IC-PWS). Cen, centromere; Tel, telomere; BP1, BP2, BP3, common break points. Adapted from Nicholls RD and Knepper JL (2001) Genome organization, function and imprinting in Prader–Willi and Angelman syndromes. *Annual Review of Genomils and Human Genetics* 2: 153–175.

(duplicons) that have been identified in proximal and distal 15q11–q13. These duplicons arose with the amplification of an ancestral gene, *HERC2*. Duplicons having 90–99% identity to the first 79 exons of *HERC2* are found in at least nine copies in 15q11–q13 as well as two copies on chromosome 16. There is one reported case of maternal germ line mosaicism producing familial recurrence of a deletion.

Uniparental Disomy

Paternal UPD accounts for approximately 5% of cases of AS, and in almost all cases, the UPD is isodisomic. The most likely origin of this event is maternal nondisjunction producing a monosomy 15 conception, with postzygotic rescue by duplication of the paternal chromosome 15. As in the deletion cases, this class of AS represents *de novo* mutational events and has a very low risk for recurrence.

Imprinting Defect

In approximately 10% of cases, an imprinting defect underlies AS. In this class of AS, a paternal imprint is erroneously assigned to the maternally inherited allele. Two types of imprinting defects are known: those due to a submicroscopic deletion of the imprinting center (IC) and those with no detectable mutation. Most submicroscopic IC deletions are familial, carrying a 50% risk for recurrence. Thus far all imprinting defects with undetectable mutations have been sporadic events. They are presumed to result from failure to establish or maintain the imprint during oogenesis, due to a stochastic event or perhaps (as yet unknown) environmental factors.

UBE3A Mutation

UBE3A was initially discarded as a potential candidate for the AS gene because it appeared not to be imprinted when studied in lymphocytes and fibroblasts, and its widespread expression ran counter to expectation since the AS phenotype is exclusively neurological. Establishing brain-only imprinting of *Ube3a* in mice resurrected *UBE3A*'s status as a candidate gene for AS. Analysis of *UBE3A* in AS patients with biparental contribution to 15q11–q13 and no imprinting abnormalities showed point mutations in several unrelated patients, identifying it as the gene responsible for the AS phenotype. Intragenic *UBE3A* mutations (insertion, deletion, nonsense, missense, and splice site mutations) cause approximately 10% of cases of AS. Substantial portions of *UBE3A* mutations are inherited from the mother's paternally acquired allele. In this circumstance, a 50% recurrence risk pertains.

Negative Molecular Studies

In addition to the four molecular classes described, approximately 10% of cases of AS have the clinical profile but no discernible molecular defect. Whole-gene or large intragenic deletions of *UBE3A* are not detected by the clinically available diagnostic tests (DNA methylation analysis of chromosome 15q11–q13 and *UBE3A* sequence analysis). Intragenic deletions have been reported in two families, but do not appear to be the explanation for the bulk of cases of AS with negative molecular studies. While some patients with the AS phenotype and negative molecular studies may represent misdiagnoses, it is likely that other molecular mechanisms affecting *UBE3A* expression will be found to explain this class of patients.

Molecular Biology of AS

UBE3A spans 120 kb of genomic DNA. Mutations have been detected throughout all regions of the gene, and there do not appear to be mutational hot spots. Sixteen exons have been identified; additional exons at the 5′ end of the gene are possible, as this region displays alternative splicing. Transcription of *UBE3A* occurs in a telomeric-to-centromeric direction. RNA transcripts of 5–6 kb include approximately 2 kb of 3′-untranslated region (UTR) sequence. There are at least seven mRNA species produced by alternative splicing at exons 5 and 6, at the 5′ end, and possibly at the 3′ end. The various transcripts produce three isoforms of UBE3A protein. Isoforms II and III have 20 and 23 additional amino acids, respectively, at the N-terminus, compared with isoform I. The function of the different isoforms of UBE3A remains unknown. There are tissue-specific variations in RNA splicing and isoform predominance, but the significance of these differences is unclear.

UBE3A encodes an E3 ubiquitin ligase, E6-AP (E6-associated protein). Though first characterized by its interaction with the E6 protein of papillomaviruses to promote degradation of p53, E6-AP does not maintain a stable association with p53. It participates in the ubiquitin pathway of protein degradation in proteosomes. In this pathway, ubiquitin is covalently attached to one or more lysine residues of proteins destined for destruction. Ubiquitination involves a three-step process (activation of ubiquitin by an E1 enzyme, transfer to an E2 conjugating enzyme, and covalent ligation of ubiquitin to the protein substrate by an E3 ligase), the third step of which is deficient in AS. The C-terminus of UBE3A is a functionally important and highly conserved domain that is shared by a family of proteins (hect domain, so named because it is homologous to the E6-AP carboxy terminus). The last six amino acids of the E6-AP C-terminus are essential for activity *in vitro*. It is unknown how substrate specificity is determined. E6-AP is found in all tissues that have been studied.

Clinical Difference in Classes of AS

Clinical differences between the molecular classes of AS have been recognized. As a group, patients with deletions tend to have a more severe phenotype (later onset of independent walking, earlier onset and increased severity of seizures, and complete absence of speech) compared to patients with UPD, imprinting defect, or *UBE3A* mutation. Though speech is usually absent in deletion AS patients, use of up to 20 words is reported in AS patients with other molecular classes of AS. The contribution of other genes in the deleted region to the increased phenotypic severity is not clear. Three genes for γ-aminobutyric acid (GABA) receptor subunits (*GABRB3*, *GABRA5*, and *GABRG3*) are located telomeric of *UBE3A* and are contained within the common class I and class II deletions. Though these genes show biallelic expression, a role for them in the genesis of epilepsy in AS has not been excluded.

Hypopigmentation occurs in some AS deletion patients. The (unimprinted) *P* gene located telomeric within the commonly deleted region is responsible for autosomal recessive oculocutaneous albinism type 2 (OCA2). Individuals with AS and OCA2 have been reported, the mechanism being a maternal chromosome deletion and paternal *P* gene mutation. Semidominant behavior of the *P* gene product (expression of a hypomorphic allele) has been offered as a potential explanation for the hypopigmentation seen in AS deletion patients. However, hypopigmentation is reported in patients with other classes of AS, including siblings with a maternally inherited intragenic deletion of exons 8–16 of *UBE3A*, so the contribution of the *P* gene to hypopigmentation remains speculative.

Within the deletion class of patients, those with larger class I deletions have lower cognitive scores and are more likely to merit a comorbid diagnosis of autism. Patients with atypical (larger) deletions are usually lower functioning.

Mechanisms of Genetic Imprinting

Genetic imprinting confers functional differences onto the genome such that expression of imprinted genes occurs only from the father or the mother. The mechanisms by which an imprint is established in the germ line, maintained during development and postnatal life, and reversed in the germ line during the subsequent generation are unknown. Additional complexities, such as tissue specificity of imprinting and changes in imprint with age, remain to be elucidated.

Mechanisms to control gene expression include insulators (DNA sequence elements that prevent interaction between nearby chromatin domains or prevent the spread of heterochromatin); transcriptional enhancer competition (promoters of linked imprinted genes competing for access to enhancers); modification of chromatin structure by histone acetylation, phosphorylation, and methylation; and DNA methylation. Each of these mechanisms provides a layer of information beyond what is contained within the nucleotide sequence, and thus contributes to the epigenotype. Imprinted genes cluster together in specific areas of the genome, an observation that suggests coordinate regulation by a regional element, known as the imprinting control region (ICR).

DNA methylation, the modification of DNA by the addition of a methyl group to the cytosine base of a CpG dinucleotide, is the most well studied of these epigenetic mechanisms. DNA methylation occurs in most, but not all, imprinted genes. Loss of DNA methylation occurs during a specific period of germ cell development, and is thought to represent erasure of the imprint from the previous generation. Reacquisition of DNA methylation occurs differentially during spermatogenesis and oogenesis, conferring the parent-of-origin specific imprint or epigenotype. It is not clear whether DNA methylation establishes the imprint or maintains it, or both.

Less is know about control of gene expression through covalent modification of histones and their association with DNA. In general, acetylation of histones H3 and H4 is associated with increased gene accessibility (and therefore increased expression). Methylation of Lys4 of H3 correlates with active gene expression; methylation of Lys9 of H3 usually accompanies gene silencing. These chemical modifications of histones attract effector proteins that apparently mediate the increased or decreased gene activity. It is not clear if DNA methylation directs the histone modifications, or if histone modification drives DNA methylation; it is possible that reciprocal interactions reinforce the effect of each.

Imprinting of the Prader–Willi Syndrome/AS Region

Both Prader–Willi syndrome (PWS) and AS are caused by defects of the imprinted gene cluster on chromosome 15q11–q13 (see **Figure 1**). Loss of expression of maternally derived *UBE3A* causes AS, with the clinical phenotype as described in the preceding sections. Loss of expression of paternally derived gene(s) from 15q11–q13 causes PWS. PWS is characterized by hypotonia and failure to thrive as a neonate, small hands and feet, short stature, hypogonadism, childhood-onset hyperphagia and obesity, and cognitive impairment. Whether a single gene or several genes are required to produce the PWS phenotype is still a matter of debate; that no PWS patient with a single gene defect has been reported suggests that PWS is truly a contiguous gene syndrome.

Differential methylation of chromosome 15q11–q13 has been thoroughly studied. Whether methylation directly or indirectly effectuates the imprint of genes within the region continues to be argued and investigated. Differential methylation provides the basis for the diagnostic testing used to confirm a suspected diagnosis of AS or PWS. The highly methylated maternal chromosome 15q11–q13 region can be distinguished from the (largely) unmethylated paternal contribution by Southern blot analysis or polymerase chain reaction (PCR) assay of the promoter region of *SNRPN*. Maternal-only contribution (due to paternal deletion, maternal UPD, or imprinting defect) is diagnostic for PWS. Exclusively paternal contribution is diagnostic for AS. A normal methylation profile does not rule out AS, however, since a substantial fraction of AS is due to maternally inherited *UBE3A* mutations.

The 15q11–q13 cluster is under the bidirectional control of the imprinting center (IC), the designation for the imprinting control region in this area. The IC was defined by the identification of small submicroscopic deletions in a subgroup of PWS and AS multiplex families with biparental inheritance yet uniparental imprint. The IC is thought to regulate in *cis* the establishment and maintenance of the imprint for the entire cluster. The IC has a bipartite structure, the PWS-IC and the AS-IC. All AS patients with IC deletions are missing the AS-IC (880 bp located 35 kb centromeric of PWS-IC). The function of the PWS-IC is to establish and maintain paternal gene expression. During oogenesis, the AS-IC functions to negatively regulate the PWS-IC (through CpG methylation) and prevent the paternal imprint from being established. Deletion of the AS-IC produces a paternal imprint of the entire 15q11–q13 region; when this is transmitted through the maternal germ line, AS results. A unique inversion case has shown that the PWS-IC and AS-IC must be closely and correctly related in order for the maternal imprint to be established.

MKRN3, *NDN*, and *SNURF-SNRPN* are expressed only from the paternal chromosome, where their promoter regions are unmethylated. *MAGEL2* is expressed only from the paternal chromosome, but its methylation status has not been clarified. Between *SNURF-SNRPN* and *UBE3A*, there are more than 70 snoRNA genes. These are noncoding genes that produce small nucleolar RNAs, the main function of which is to modify ribosomal RNA. Though not imprinted directly by methylation, these snoRNAs are imprinted indirectly because they are processed from the paternally expressed *SNURF-SNRPN* sense/*UBE3A* antisense transcript (discussed in the next section), which is under the control of methylation.

In sporadic imprinting defect cases of AS, the maternal grandfather's and maternal grandmother's chromosomes are equally likely to harbor the unmethylated paternal imprint, suggesting that erasure of the grandmaternal methylation imprint occurs normally, followed by failure to establish (in the case of the grandpaternal chromosome) or reestablish (in the case of the grandmaternal chromosome) the correct maternal imprint.

Much of what we know about methylation status and imprinting of the chromosome 15q11–q13 region comes from investigation of the orthologous region on mouse chromosome 7C. However, important differences exist. The *Frat3* gene is an evolutionarily younger gene which has joined the mouse PWS/AS region (by retrotransposition) and has acquired the differential methylation and expression of paternally expressed genes in the region. There is no homolog to *Frat3* in the human chromosome 15q11–q13 region. Transgenic studies in which human elements of the PWS/AS region are inserted in the mouse have shown that both the positive and negative regulatory elements of the imprinting machinery have diverged. The mouse equivalent of the human AS-IC has yet to be identified. Elucidating the mechanism of mouse PWS/AS region imprinting is likely to yield important insights, but this information may not be directly applicable to human *UBE3A* imprinting.

In addition to methylation, histone modifications play a role in establishing and maintaining the imprint. Hyperacetylation of histones H3 and H4, a modification usually signifying gene activation, is found at the PWS-IC on the paternal allele. Methyl-Lys4 histone H3, which is typically found at transcriptionally active loci, associates with PWS-IC and the promoters of *SNRPN* and *NDN* on the paternal allele. Dimethyl-Lys9 histone H3, which is generally found in inactive or heterochromatic regions, associates with the PWS-IC on the maternal chromosome.

Gene Repression of *UBE3A*

The IC is believed to regulate the imprint of 15q11–q13 through establishing or maintaining the methylation profile of the region. However, no site of differential methylation has been reported within the *UBE3A* gene or its vicinity. Though an as-yet-undiscovered site of allele-specific methylation in or around *UBE3A* remains a formal possibility, other mechanisms for imprint regulation are being proposed and evaluated. These include enhancer competition, chromatin state modification, and expression competition by antisense RNA. At present, there is no evidence for chromatin state modification by histone acetylation or methylation. The expression competition model proposes that the IC controls the methylation of an imprintor gene (distinct from the IC). The imprintor gene would be paternally expressed (in a tissue-specific manner) and would 'compete out' expression of *UBE3A*.

There is some evidence to support the expression competition model. A large (>500 kb) transcript oriented in an antisense manner and spanning the 3′ half of *UBE3A*, as well as the region downstream of *UBE3A*, has been identified. This antisense transcript (*UBE3A-AS*) is expressed only from the paternal allele in the brain, is not expressed in tissues with biallelic *UBE3A* expression, and is under the control of the PWS-IC. In this model, *UBE3A-AS* is the direct target of the IC and is imprinted primarily (in brain), and *UBE3A* is secondarily imprinted (in brain) through a repression mechanism. Based on other examples of imprinting regulation by antisense transcripts (e.g., *XIST*, *IGF2R*), *UBE3A-AS* is presumed to silence *UBE3A* by annealing in *cis* to the sense promoter. However, the general concept of *cis* antisense regulation of neighboring imprinted genes remains controversial because of conflicting evidence.

Recent evidence from the mouse indicates that *Ube3a-as* may be regulated by, rather than being a regulator of, *Ube3a*. Using a mouse model of AS generated by targeted disruption of exons 12 and 13 of *Ube3a*, Landers et al. observed that maternal transmission of the defective allele upregulated expression of *Ube3a-as* from the paternal allele. This evidence suggests that *Ube3a* expression, rather than being regulated by *Ube3a-as*, modulates the levels of the *Ube3a-as* and that this interaction occurs in *trans*. The mechanism of this modulation is unknown. Investigating the directionality of the interaction between *UBE3A* and *UBE3A-AS* in the human brain will be difficult given the scarcity of this resource.

Tissue-Specific Imprinting

UBE3A imprinting is tissue specific, in contrast to most other imprinted loci in the 15q11-q13 region. In the brain, there is preferential expression of *UBE3A* from the maternal allele, but a small amount of residual transcription (10%) from the paternal allele occurs. Another gene, *ATP10C*, is maternally expressed in a tissue-specific manner, but so far has not been shown to contribute to the AS phenotype.

The tissue-specific imprinting of *UBE3A* appears to be region specific, with the imprinted expression pattern being found in the hippocampus and cerebellum, and somewhat in the cerebral cortex. This regional specificity has been shown to be due to lineage specificity: *UBE3A* is expressed exclusively from the maternal allele only in neurons, and is biallelically expressed in glial cells and fibroblasts. Because the hippocampus and cerebellum are neuron-rich regions, these regions exhibit the imprint most strikingly.

The molecular basis of this neuron-specific imprinting is unknown. Two hypotheses have been put forward: (1) the epigenetic marks of neurons are the same as those in other tissues, but are interpreted differently by the neuronal transcription machinery; or (2) a specific additional epigenetic mark, a secondary imprint, is acquired during neuronal differentiation.

Summary

AS is a unique neurobehavioral disorder due to defective expression of maternally inherited *UBE3A*. The complex mechanisms of how the *UBE3A* imprint is established, maintained, and reversed remain to be elucidated, as does the mechanism of tissue specificity. *UBE3A* repression apparently does not involve differential methylation, but possibly is mediated by expression competition from an antisense transcript.

See also: Prader–Willi Syndrome; Ubiquitin–Proteasome System and Plasticity.

Further Reading

Boyes L, Wallace AJ, Krajewska-Walasek M, et al. (2006) Detection of a deletion of exons 8–16 of the UBE3A gene in familial Angelman syndrome using a semi-quantitative dosage PCR based assay. *European Journal of Medical Genetics* 49: 472–480.

Clayton-Smith J and Laan L (2003) Angelman syndrome: A review of the clinical and genetic aspects. *Journal of Medical Genetics* 40: 87–95.

Fridman C, Hosomi N, Varela MC, et al. (2003) Angelman syndrome associated with oculocutaneous albinism due to an intragenic deletion of the *P* gene. *American Journal of Medical Genetics* 119A: 180–183.

Guerrini R, Carrozzo R, Rinaldi R, et al. (2003) Angelman syndrome: Etiology, clinical features, diagnosis, and management of symptoms. *Pediatric Drugs* 5: 647–661.

Horsthemke B and Buiting K (2006) Imprinting defects on human chromosome 15. *Cytogenetic and Genome Research* 113: 292–299.

Jiang Y, Bressler J, and Beaudet AL (2004) Epigenetics and human disease. *Annual Review of Genomics and Human Genetics* 5: 479–510.

Jiang Y, Tsai T, Bressler J, et al. (1998) Imprinting in Angelman and Prader–Willi syndromes. *Current Opinion in Genetics & Development* 8: 334–342.

Johnstone KA, DuBose AJ, Futtner CR, et al. (2006) A human imprinting centre demonstrates conserved acquisition but diverged maintenance of imprinting in a mouse model for Angelman syndrome imprinting defects. *Human Molecular Genetics* 15: 393–404.

Kishino T (2006) Imprinting in neurons. *Cytogenetic and Genome Research* 113: 209–214.

Landers M, Calciano MA, Colosi D, et al. (2005) Maternal disruption of *Ube3a* leads to increased expression of *Ube3a-ATS* in *trans*. *Nucleic Acids Research* 33: 3976–3984.

Lossie AC, Whitney MM, Amidon D, et al. (2001) Distinct phenotypes distinguish the molecular classes of Angelman syndrome. *Journal of Medical Genetics* 38: 834–845.

Nicholls RD and Knepper JL (2001) Genome organization, function and imprinting in Prader–Willi and Angelman syndromes. *Annual Review of Genomics and Human Genetics* 2: 153–175.

Nicholls RD, Saitoh S, and Horsthemke B (1998) Imprinting in Prader–Willi and Angelman syndromes. *Trends in Genetics* 14: 194–200.

Rougeulle C and Lalande M (1998) Angelman syndrome: How many genes to remain silent? *Neurogenetics* 1: 229–237.

Sahoo T, Peters SU, Madduri NS, et al. (2006) Microarray based comparative genomic hybridization testing in deletion bearing patients with Angelman syndrome: Genotype–phenotype correlations. *Journal of Medical Genetics* 43: 512–516.

Soejima H and Wagstaff J (2005) Imprinting centers, chromatin structure, and disease. *Journal of Cellular Biochemistry* 95: 226–233.

Angiotensin Actions on and within Brain

M J McKinley and A M Allen, Howard Florey Institute, University of Melbourne, Melbourne, VIC, Australia
B J Oldfield, Monash University, Clayton, VIC, Australia

Overview

As well as having important actions on peripheral organs and the circulation, the renin-angiotensin system exerts influences on the central nervous system (CNS) via its effector peptides angiotensin II, III, and IV. Many of these actions, both central and peripheral, relate to maintaining the integrity of the cardiovascular system and body fluid and electrolyte homeostasis. They include peripheral vasoconstriction; stimulation of secretion of aldosterone, vasopressin, and adrenocorticotropic hormone (ACTH); inhibition of renin release; potentiation of sympathetic nerve activity; and stimulation of thirst and salt hunger. Intense study of these actions of angiotensin on both CNS and peripheral organs during the past 50 years has yielded much detailed knowledge of the renin-angiotensin system and its biological actions. In several aspects, however, understanding of this system is still incomplete.

Angiotensin II is generated both peripherally and within the CNS. While the peripherally generated peptide does not readily traverse the blood–brain barrier, it is still able to influence brain function by acting on specific angiotensin receptors in brain sites lacking this barrier: the circumventricular organs. Specific angiotensin receptors are also located in many other parts of the CNS that are inaccessible to circulating angiotensin because of its exclusion by the blood–brain barrier. Centrally generated angiotensin can act at these sites. Thus, in considering actions of angiotensin on the brain, it is necessary to recognize that the peptide products of separate angiotensin generating systems (one peripheral and the other central) can influence CNS function.

The receptors to which angiotensin II binds to exert its biological actions are classified into two subtypes, AT_1 and AT_2. Most of the known actions of angiotensin II on the brain are mediated by the AT_1 receptor, and it is widely distributed in discrete areas of many brain regions. High concentrations of AT_2 receptors occur in the fetal brain, suggesting that these receptors play a developmental role in the CNS. In the adult, the distribution of AT_2 receptors is more restricted; it is predominantly associated with regions involved in sensory processing.

Both AT_1 and AT_2 receptors have been cloned and belong to the 7-transmembrane-spanning G-protein-coupled receptor family. While it can associate with several G-proteins, the AT_1 receptor predominantly couples through $G_{q/11}$, leading to activation of inositol phosphate signaling pathways and elevated intracellular calcium. In addition, the AT_1 receptor is known to signal through G-protein-independent pathways such as production of reactive oxygen species via nicotinamide adenine dinucleotide phosphate oxidase. In neurons, angiotensin II-induced activation of AT_1 receptors leads to excitation, predominantly via inhibition of various outward cation conductances. The AT_2 receptor predominantly couples through G_i, leading to stimulation of phospholipase A_2 and production of arachidonic acid. In neurons this signaling also appears to modulate K^+ channel activity, but both increases and decreases in firing rates have been observed.

Circulating Angiotensin and Its Actions on the Brain

Generation of Circulating Angiotensin

Circulating angiotensin II is the product of the sequential actions of two enzymes, renin and angiotensin converting enzyme, also known as kininase II. Renin, an enzymatic hormone synthesized within the kidney, is released into the circulation from the renal juxtaglomerular apparatus in response to stimuli such as sodium depletion, reduced renal perfusion pressure, and increased renal sympathetic nerve activity. While circulating in the bloodstream, renin cleaves a decapeptide from the terminus of a large circulating plasma protein, angiotensinogen (synthesized by the liver). This decapeptide, known as angiotensin I, is the biologically inactive precursor of angiotensin II. Angiotensin I is then converted into the biologically active octapeptide angiotensin II by the action of angiotensin converting enzyme, located on cell membranes of vascular endothelial cells in several tissues, most abundantly in the lungs (see **Figure 1**). Once formed within the circulation, angiotensin II is able to reach regions of the brain that are devoid of a blood–brain barrier, the circumventricular organs, and act on its receptors there.

The three circumventricular organs at which angiotensin II from the circulation acts are the subfornical organ and organum vasculosum of the lamina terminalis (OVLT) in the forebrain and the area postrema in the hindbrain (see **Figure 2**). While both classes of angiotensin receptors (AT_1 and AT_2) are present within these three circumventricular organs, AT_1 receptor density is extremely high there. Studies using specific antagonists show that the responses (thirst, sodium

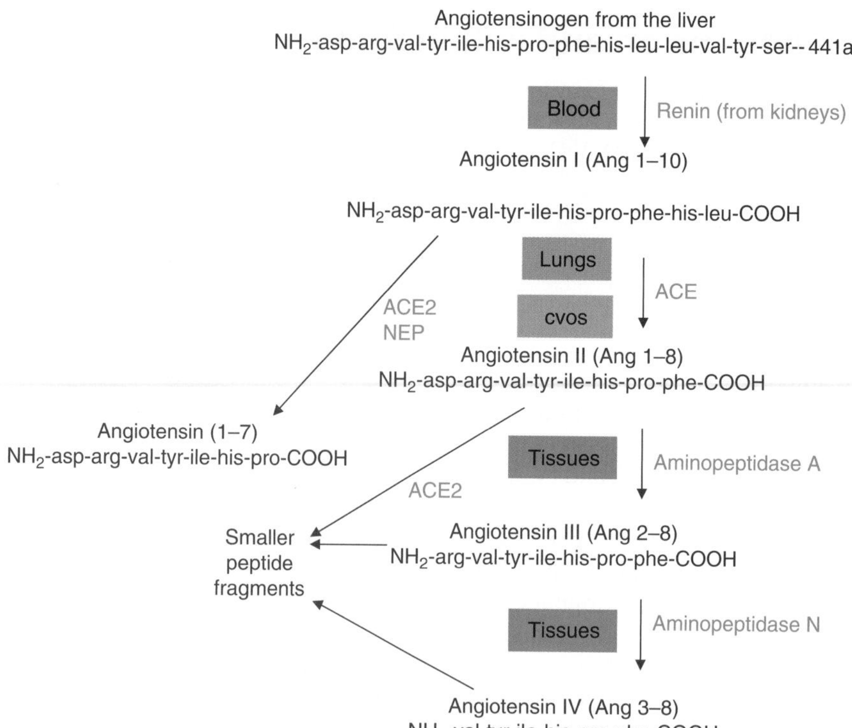

Figure 1 Pathways for the production of circulating angiotensin peptides. Enzymes are indicated in red lettering. ACE, angiotensin converting enzyme; cvos, circumventricular organs; NEP, neutral endopeptidase.

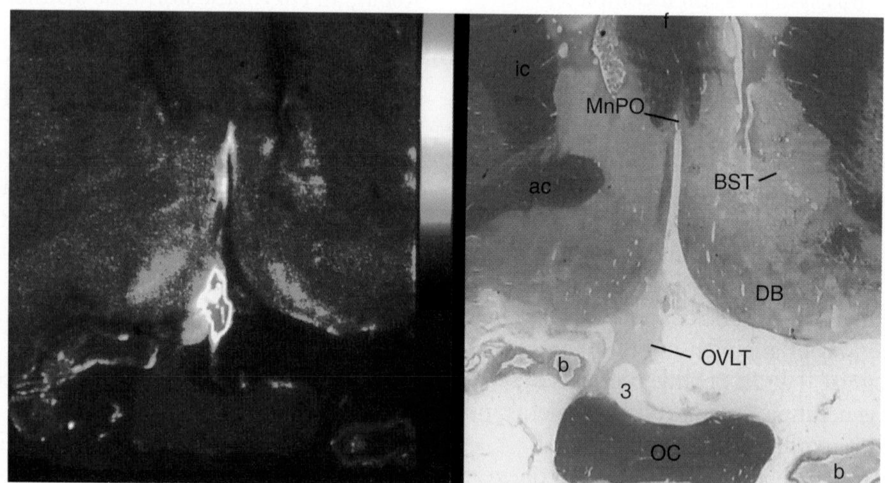

Figure 2 (Left) Angiotensin receptors in the lamina terminalis and other forebrain regions of the human brain. This computer-generated pseudocolor image of an autoradiograph shows high densities of radiolabeled angiotensin II to the vascular organ of the lamina terminalis (OVLT), dorsal part of the median preoptic nucleus (MnPO), bed nucleus of the stria terminalis (BST), and nucleus of the diagonal band. The color bar indicates lowest (blue) to highest (red) receptor-binding densities. (Right) Neuroanatomical features in the corresponding histological transverse section stained with cresyl-violet/luxol fast blue. ac, anterior commissure; b, blood vessel; f, fornix; ic, internal capsule; DB, nucleus of the diagnal band; oc, optic chiasma; 3, third ventricle optic recess.

appetite, vasopressin release, cardiovascular regulation) elicited by the action of blood-borne angiotensin II on these sites are exerted through AT_1 receptors. Extremely high concentrations of angiotensin converting enzyme are found within the circumventricular organs, and angiotensin II can be generated locally within these sites from angiotensin I delivered there from the bloodstream.

Centrally Mediated Cardiovascular Effects of Circulating Angiotensin

The first indication of an action of angiotensin on the brain was the discovery by RK Bickerton and JP Buckley in 1961 of a centrally mediated increase in blood pressure. These investigators showed that infusion of angiotensin II directly into the vertebral artery of dogs elicited a greater increase in arterial pressure than did an infusion of the same amount of angiotensin II into the general circulation. Because the vertebral artery supplied the hindbrain of the dog, they suggested that the pressor response caused by angiotensin infused into this blood vessel was due to an action on the medulla oblongata. Several investigators have shown subsequently that ablation of the area postrema, the only circumventricular organ in the hindbrain, prevents the pressor response to intravertebral arterial infusion of angiotensin II in dogs. Thus, it is likely that blood-borne angiotensin II acts on the area postrema to increase arterial pressure, and there is further evidence to show that it does so by increasing sympathetic vasoconstrictor nerve activity. This increased sympathetic vasoconstrictor activity may be mediated by neural connections between the area postrema, the adjacent nucleus of the solitary tract (NTS), and sympathetic premotor vasoconstrictor neurons in the rostral ventrolateral medulla (RVL). More recently, this idea has been refined to show that circulating angiotensin II acts on the area postrema to inhibit the baroreceptor reflex. Thus chronically administered angiotensin II resets the baroreceptor reflex so that sympathetic vasomotor activity remains elevated even when arterial pressure is high.

Circulating angiotensin may also exert influences on sympathetic vasomotor nerve activity and arterial pressure through AT_1 receptors in the forebrain circumventricular organs, the subfornical organ, and the OVLT. Polysynaptic neural pathways connect neurons in these two sites to both sympathetic premotor and preganglionic neurons in the RVL and the intermediolateral cell column of the spinal cord, respectively. There is evidence to show that the hypothalamic paraventricular nucleus is a relay point in these polysynaptic pathways connecting angiotensin-sensitive neurons in the subfornical organ and the OVLT to sympathetic vasomotor pathways. It should be noted that these forebrain actions of angiotensin II on sympathetic activity and arterial pressure have been implicated in the genesis of some forms of experimentally induced hypertension.

Thirst

Systemic infusion of angiotensin II initiates water drinking in experimental animals, and presumably this behavioral response reflects the stimulation of thirst by angiotensin II action on the brain. When the extracellular fluid is depleted, as for example with hemorrhage, excessive vomiting, or diarrhea, thirst is generated. Such thirst is independent of the osmoreceptors that signal dehydration of the intracellular space. In 1969, JT Fitzsimons showed that the renin-angiotensin system played a crucial role in water drinking resulting from loss of extracellular fluid. He showed that circulating angiotensin II induced water drinking in rats and that it probably acted synergistically with neural signals coming from low pressure volume receptors in the thorax, to initiate thirst. The dipsogenic response to infused angiotensin II was one of the first recognized behavioral effects elicited by a circulating peptide. Virtually all mammals that have been investigated increase water drinking if angiotensin II is administered systemically, although some species (including humans) need quite high blood concentrations of angiotensin II, probably above those encountered under physiological conditions, to stimulate thirst. Baroreceptor stimulation has an inhibitory influence on thirst, and as increased arterial pressure is another consequence of systemic infusion of angiotensin II, the blood levels of angiotensin II that cause drinking behavior may need to be high to overcome the antidipsogenic effect of increased blood pressure. An inhibitory effect of baroreceptor stimulation on thirst does not come into play when endogenous levels of angiotensin II increase under conditions of hypovolemia, however, because arterial pressure does not increase in these circumstances.

The dipsogenic action of angiotensin II may be additive with, potentiated by, or inhibited by factors that also act directly on the subfornical organ. The ovarian hormone relaxin (released during pregnancy) potentiates the dipsogenic action of angiotensin II, while atrial natriuretic peptide (released when extracellular volume is expanded) and estrogen have inhibitory influences on angiotensin-induced drinking. Circulating angiotensin and hypertonicity are additive dipsogenic stimuli. Thus, subjects in a dehydrated condition may experience thirst that results from a combination of both osmotic and angiotensin stimuli. As well as interactions with humoral factors, it is thought that afferent neural signals from the heart and possibly the kidney also interact with angiotensin's dipsogenic action to generate thirst in conditions of hypovolemia.

With the discovery of the dipsogenic action of blood-borne angiotensin II, the question arose as to how a circulating peptide hormone that did not pass across the blood–brain barrier could act directly on the brain to stimulate thirst and water drinking. JB Simpson and A Routtenberg were able to show in 1971 that the subfornical organ, a circumventricular organ lacking a blood–brain barrier, was the site at which angiotensin II in the circulation stimulated thirst. They demonstrated that the subfornical organ was exquisitely sensitive to locally injected angiotensin II for the induction of water drinking in animals and that ablation of this region abolished water drinking in response to systemic angiotensin II in the laboratory rat. Electrophysiological and *in vitro* autoradiographic binding investigations, as well as more-recent studies of immediate early gene *c-fos* expression, confirm the excitatory action of angiotensin II on neurons of the subfornical organ (see **Figure 3**).

The neural pathways subserving thirst that are activated subsequent to angiotensin stimulation of neurons within the subfornical organ remain largely unknown. The subfornical organ had efferent neural connections to many other brain sites that included regions such as the lateral hypothalamic area, hypothalamic paraventricular nuclei, OVLT, and median preoptic nucleus (MnPO), all of which have been implicated in thirst responses. The connection to the MnPO has been proposed to relay neural signals coming from angiotensin-sensitive neurons in the subfornical organ to cortical regions for the induction of thirst. Recently, imaging of brains of thirsty humans by positron emission tomography and functional magnetic resonance imaging techniques has drawn attention to the cingulate cortex and insula as cortical sites responsible for the generation of thirst.

Salt Appetite

A hunger for sodium salts develops slowly over a period of several hours when the extracellular fluid becomes depleted. Up to 24 h may elapse following salt and water depletion from the body before salt intake begins to increase. By comparison, thirst is a relatively rapid response to depletion of the extracellular fluid compartment. The most convincing evidence that activation of the systemic renin-angiotensin system is a humoral signal to the brain for salt hunger resulting from sodium depletion is the inhibition of this response by adequate doses of systemically administered angiotensin converting enzyme inhibitors (e.g., captopril) or angiotensin AT_1 receptor antagonists (e.g., losartan). Sodium intake that is blocked by angiotensin converting enzyme inhibitors can be restored by intravenous infusions of small amounts of angiotensin II, evidence that is consistent with a role for circulating angiotensin II in the genesis of sodium appetite. It is thought that the adrenal hormone aldosterone, secreted in response to sodium depletion, acts synergistically with angiotensin to stimulate salt appetite.

Blood-borne angiotensin II may act on AT_1 receptors in both the OVLT and the subfornical organ to influence a hunger for salt because ablation of either of these circumventricular organs reduces salt intake following sodium depletion. Evidence favors the OVLT as the site of action of angiotensin to stimulate salt appetite because microinjection of angiotensin II into the OVLT region (but not the subfornical organ) causes increased salt intake by rats. The subsequent efferent neural pathways distributing from the OVLT that mediate sodium appetite are still largely unknown. Angiotensin-sensitive neurons in the subfornical organ and the OVLT send efferent neural projections to a number of sites, including the bed nucleus of the stria terminalis. This region and the central nucleus of the amygdala have been shown in studies of lesions to have a role in sodium appetite. Thus, it is possible that neurons within the bed nucleus of the stria terminalis and the amygdala may relay signals from the OVLT to higher cortical regions to initiate salt appetite.

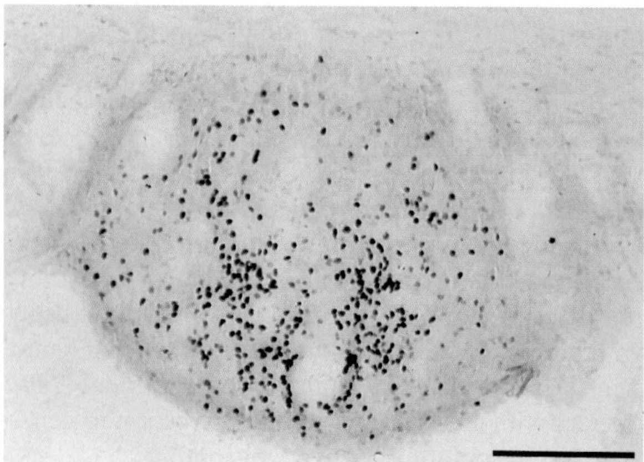

Figure 3 Angiotensin II stimulates neurons in the subfornical organ. Using immunohistochemistry, this transverse section of the rat subfornical organ shows expression of the proto-oncogene *c-fos* in the cell nucleus of neurons within the subfornical organ in response to an intravenous infusion of angiotensin II. Expression of *c-fos* indicates increased electrical activity of neurons. Such expression of *c-fos* is not observed in noninfused rats. Scale bar = 100 μm.

Vasopressin Secretion

Complementary to the dipsogenic action of circulating angiotensin II is its ability to stimulate the release of vasopressin from the posterior pituitary gland. Vasopressin, the antidiuretic hormone, acts on the collecting tubules of the kidney to concentrate urine, thereby reducing urine output and conserving body fluids. The vasopressin-secreting action of circulating angiotensin II is exerted via AT_1 receptors in both the subfornical organ and the OVLT. Some of the angiotensin-sensitive neurons in both these circumventricular organs project directly to the hypothalamic supraoptic and paraventricular nuclei (sites of vasopressin-synthesizing neurons), suggesting that this is the neural pathway that mediates angiotensin-induced vasopressin release. There is evidence also that neural signals from angiotensin-sensitive neurons are relayed from the subfornical organ and OVLT to the supraoptic and paraventricular nuclei via synaptic connections in the MnPO.

Renal Sympathetic Nerve Activity

Circulating angiotensin II is known to act directly on the sympathetic nerve endings in the kidney to potentiate the release of noradrenaline, but it also has an action on the brain to increase renal sympathetic nerve activity and thereby further promote renal fluid and sodium retention. Angiotensin-sensitive neurons within the subfornical organ and OVLT have polysynaptic neural connections to the renal sympathetic nerves. These neural connections are likely to mediate the effects of circulating angiotensin on renal sympathetic nerve activity, and neural signals from these regions are probably relayed to the kidney via synapses in the hypothalamic paraventricular nucleus and RVL.

Actions of Centrally Generated Angiotensin on the Brain

In addition to physiological responses stimulated by peripherally generated angiotensin II on the circumventricular organs of the brain, angiotensin derived from within the brain may exert a number of different actions on the CNS. These actions are mediated by receptors located at sites with an intact blood–brain barrier and therefore not accessible by circulating angiotensin peptides.

Generation of Angiotensin in the CNS

Several organs and tissues have intrinsic renin-angiotensin systems capable of generating angiotensin peptides, and the brain is no exception. The generation of angiotensin peptides within the CNS appears to be somewhat different from that in the periphery. The precursor protein angiotensinogen is abundant in the extracellular and cerebrospinal fluid of the brain, and astrocytes are the major cellular source of this brain angiotensinogen, as evidenced by the expression of messenger RNA encoding angiotensinogen in astrocytes. Angiotensinogen may also be synthesized within neurons, but the amount is small in comparison with that secreted constitutively by glia. Angiotensinogen-producing glial cells are found throughout the brain, although the hypothalamus and medulla have the highest concentrations. The details of the generation of angiotensin I, angiotensin II, the heptapeptides angiotensin III and angiotensin (1–7), and the hexapeptide angiotensin IV in the brain remain to be elucidated. Renin is present within the brain, but only in exceedingly small amounts. However, angiotensin-converting enzyme is abundant in many brain regions, and its location on external cell membranes facilitates the formation of angiotensin II in the extracellular domain of the brain. Immunohistochemical and electrophysiological evidence suggests that angiotensin II can be released from neurons as a neurotransmitter or neuromodulator. However, whether angiotensin II is generated within neurons or taken up from the brain extracellular space after synthesis in glia is not clear at present. It is also possible that angiotensin peptides diffuse within the brain extracellular fluid to receptor sites distant from where they were formed. Peptidases that generate angiotensin III, angiotensin (1–7), and angiotensin IV are present in the brain.

Angiotensin Receptors within the CNS

A clear indication that centrally generated angiotensin influences brain function is the multitude of different locations within the CNS that have a blood–brain barrier but still express angiotensin receptors, of either the AT_1 or the AT_2 subtype. Neuronal AT_1 receptors have been observed at both pre- and post-synaptic sites. In the hindbrain, AT_1 receptors are highly abundant in the NTS, dorsal motor nucleus of the vagus, rostral and caudal ventrolateral medulla, and medullary raphe nuclei (see **Figure 4**). In addition, the dorsal horn and intermediolateral cell column of the spinal cord express the AT_1 receptor. In the midbrain, high densities of AT_1 receptors are

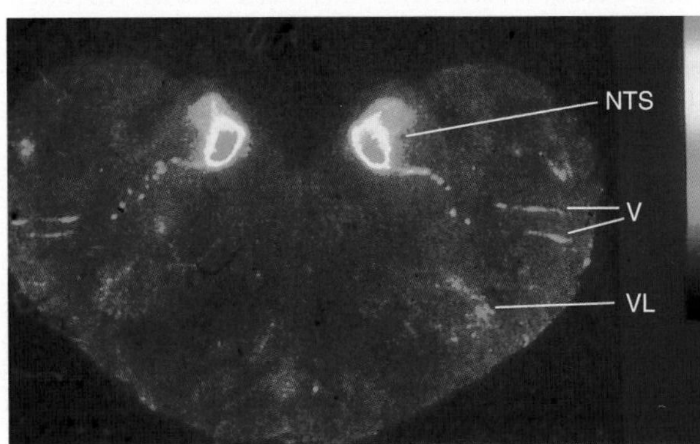

Figure 4 The location of angiotensin AT_1 receptors in the medulla oblongata of a sheep. This pseudocolor image of an autoradiograph of a transverse section shows the binding of radio-labeled angiotensin II to the nucleus of the solitary tract (NTS), ventrolateral medulla (VL) and fibers of the vagus nerve (v) coursing through medulla to the NTS. Highest density of binding is indicated by red and the diminishing gradation by the color bar. Magnification ×10.

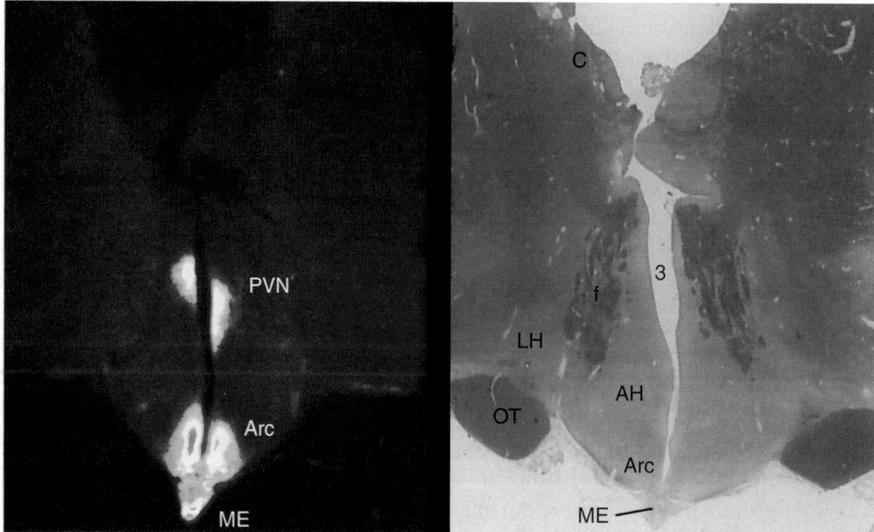

Figure 5 (Left) Location of angiotensin AT_1 receptors in the human hypothalamus. This computer-generated pseudocolor image of an autoradiograph shows high densities of binding of radiolabeled-angiotensin II to the hypothalamic paraventricular nucleus (PVN), arcuate nucleus (Arc), and median eminence (ME). (Right) Neuroanatomical features in the corresponding histological transverse section stained with cresyl violet/luxol fast blue. Magnification ×2.5. AH, anterior hypothalamic area; C, caudate nucleus; f, fornix; LH, lateral hypothalamic area; OT, optic tract.

found in the lateral parabrachial nucleus and substantia nigra pars compacta. In the diencephalon, high densities of angiotensin receptors occur in the hypothalamic paraventricular and supraoptic nuclei, the arcuate nucleus and median eminence, and the MnPO. In the forebrain, the amygdala, septum, caudate nucleus, and olfactory bulb are sites of AT_1 receptor expression (see **Figure 5**).

Effects of Centrally Administered Angiotensin II

Injection of angiotensin II or III into the lateral or third cerebral ventricles has been shown to elicit many physiological responses, often but not always similar to those obtained with peripherally administered angiotensin II. These responses include thirst and salt appetite, vasopressin secretion, and increased arterial pressure.

Thirst and Vasopressin Secretion

The initial evidence that brain angiotensin could play a role in thirst mechanisms came from AN Epstein and colleagues, who observed that administration of small amounts of angiotensin II directly into the brain could initiate water drinking within seconds. Central injection of angiotensin II also stimulates the secretion of vasopressin from the posterior pituitary gland, resulting in antidiuresis and reduced fluid loss in urine. These responses are similar to those obtained when angiotensin is infused systemically, and they are also blocked by AT_1 receptor antagonists. However, it is clear that the angiotensin-responsive neurons that mediate the responses to centrally injected angiotensin II are not the same as those in the subfornical organ that mediate the dipsogenic and vasopressin responses to circulating angiotensin, because responses to centrally administered angiotensin persist when the subfornical organ is ablated.

Although some neurons in the subfornical organ and OVLT are stimulated when angiotensin II is injected into the lateral cerebral ventricle, the MnPO is the major site of action of intracerebroventricularly infused angiotensin II to stimulate thirst and vasopressin secretion. The MnPO is situated in the anterior wall of the third ventricle (the lamina terminalis) between the subfornical organ and OVLT. Neurons situated in the MnPO have high concentrations of AT_1 receptors, and they receive afferent neural inputs from several other brain regions that are known to mediate fluid and electrolyte homeostasis, such as the subfornical organ, OVLT, lateral parabrachial nucleus, midbrain raphe, and ventrolateral medulla. Since angiotensin-responsive neurons in the MnPO are not directly affected by blood-borne angiotensin because of the blood–brain barrier, it is likely that the endogenous angiotensin that acts on neurons in the MnPO is derived from the brain and acts at the MnPO as a neurotransmitter to engage neural pathways subserving thirst and vasopressin secretion.

In regard to vasopressin release, direct projections from the MnPO to vasopressin-secreting cells of the supraoptic nucleus have been demonstrated, and ablation of the MnPO disrupts angiotensin-stimulated vasopressin secretion. These observations are consistent with an important role of angiotensin-sensitive neurons in the MnPO to relay signals to the supraoptic and paraventricular nucleus for vasopressin secretion. There are also data that indicate that angiotensin III may be the effector peptide responsible for central angiotensinergic influences on vasopressin release.

Sodium Appetite

Central administration of angiotensin II or renin is a potent stimulus to the ingestion of salt (sodium chloride) by mammals such as rodents, ruminants, and primates. The onset of this angiotensin-stimulated salt appetite is many hours later than the dipsogenic response to centrally administered angiotensin. Experiments in rodents and primates showing inhibition of salt intake by centrally injected angiotensin AT_1 receptor antagonist drugs are further evidence that brain angiotensin has a role in the generation of salt appetite. A powerful synergistic action between central angiotensin and mineralocorticoids in the stimulation of salt intake has been demonstrated. The locus of action of brain angiotensin to induce salt appetite is unclear at present.

Arterial Pressure

Angiotensin increases arterial pressure when microinjected into a number of different brain sites. These include the NTS, the hypothalamic paraventricular nucleus, and the RVL. These two latter regions have direct efferent connections to sympathetic preganglionic neurons in the intermediolateral cell column of the thoracic spinal cord, and stimulation of AT_1 receptors in the paraventricular nucleus or RVL increases arterial pressure by increasing sympathetic vasomotor nerve activity in the periphery.

Pharmacological blockade of brain angiotensin AT_1 receptors in sodium-depleted or hemorrhaged animals results in a failure to maintain arterial pressure. The same treatment drops blood pressure in a number of animal models of hypertension. These observations indicate that brain-derived angiotensin has a role in the regulation of arterial pressure in some physiological and pathophysiological conditions. An angiotensinergic synapse within the RVL may relay neural signals from the hypothalamic paraventricular nucleus to sympathetic preganglionic neurons in the spinal cord to influence arterial pressure.

Another angiotensin peptide, angiotensin (1–7), influences cardiovascular regulation when infused into the brain ventricles or RVL, probably via a non-AT_1 or non-AT_2 receptor. In the NTS, its action is to enhance the baroreceptor reflex, opposite the effect of angiotensin II, whereas in the RVL its action to increase sympathetic activity is similar to that of angiotensin II.

Natriuresis

Injection of angiotensin II into the lateral or third cerebral ventricles causes a rapid and large increase in sodium excretion by the kidneys. This natriuretic response is probably the result of a combination of the increased arterial pressure, reduced renal sympathetic nerve activity, and increased release of vasopressin that also result from the intracerebroventricular administration of angiotensin II. Intracerebroventricular injection of hypertonic saline also causes natriuresis. This effect is blocked by prior central administration of the angiotensin AT_1 receptor antagonist losartan, indicating a possible role of central angiotensinergic pathways in the osmoregulatory excretion of sodium by the kidneys. In combination with increased water intake and fluid retention caused by increased vasopressin levels, this natriuretic effect reduces the sodium concentration and osmolality of the extracellular fluid. Thus, it could be construed that central angiotensinergic action in the short term is attuned to reducing the tonicity of extracellular fluid and, as a consequence of osmotic forces, increasing intracellular volume.

ACTH Secretion

Following the intracerebroventricular injection of angiotensin II, corticotropin-releasing hormone (CRH), synthesized in the parvocellular part of the hypothalamic paraventricular nucleus, is secreted into the pituitary portal blood supply of the median eminence. Subsequently, this CRH stimulates ACTH secretion from the anterior pituitary gland into the systemic circulation. As a result of the action of circulating ACTH on the adrenal cortex, increased secretion of the glucocorticoid hormones cortisol and corticosterone from the adrenal gland occurs. CRH-containing neurons in the paraventricular nucleus that project to the median eminence express high levels of AT_1 receptors and are a likely site at which brain angiotensin influences the hypothalamic–pituitary–adrenal axis.

Body Temperature

Injection of angiotensin II into the cerebral ventricles reduces core body temperature in rats. Pharmacological blockade of brain angiotensin AT_1 receptors results in a higher core temperature of exercising rats or rats exposed to high ambient temperature. These data indicate that angiotensin endogenous to the brain probably has a role in thermoregulatory function. The thermoregulatory brain regions and mechanisms influenced by brain angiotensin have yet to be identified, although it is known that brain angiotensin may affect the redistribution of blood to the skin and therefore the ability to radiate heat from the body.

Memory

The angiotensin hexapeptide metabolite angiotensin IV has been implicated in memory because central infusions of angiotensin IV agonist drugs reverse experimentally induced memory deficits. The angiotensin IV binding site in the brain is the insulin-regulated amino peptidase (IRAP), and it has been proposed that angiotensin IV enhances memory and learning by inhibiting IRAP enzymatic activity. However, whether sufficient amounts of endogenous angiotensin IV can be generated in the brain to influence memory is undetermined.

Cerebral Blood Flow

In addition to its neurally mediated responses, angiotensin II causes vasoconstriction of cerebral blood vessels and may alter cerebrovascular autoregulation. Administration of AT_1 receptor antagonists improves recovery of brain function following focal cerebral ischemia in animal stroke models by improving vasodilation and restoring cerebrovascular autoregulation.

See also: Angiotensin II; Autonomic Nervous System: Central Cardiovascular Control; Blood Pressure: Baroreceptors; Circumventricular Organs; Circumventricular Organs in Neuroendocrine Control; Neuronal Angiotensin; Osmoregulation; Salt Appetite; Vasopressin/Oxytocin and Receptors.

Further Reading

Allen AM, Oldfield BJ, Giles ME, Paxinos G, McKinley MJ, and Mendelsohn FAO (2000) Localization of angiotensin receptors in the nervous system. In: Quirion R, Bjöorklund A, and Hökfelt T (eds.) *Handbook of Chemical Neuroanatomy, Vol. 16: Peptide Receptors,* Part I, pp. 79–124. Amsterdam: Elsevier.

Andersson B, Eriksson L, Fernandez O, Kolmodin C-G, and Oltner R (1972) Centrally mediated effects of angiotensin II on arterial pressure and fluid balance. *Acta Psychiatrica Scandinavica* 85: 398–407.

Davisson RL (2003) Physiological genomic analysis of the brain renin-angiotensin system. *American Journal of Physiology* 285: R498–R511.

DiBona GF (2001) Peripheral and central interactions between the renin-angiotensin system and the renal sympathetic nerves in control of renal function. *Annals of the New York Academy of Sciences* 940: 395–406.

Fitzsimons JT (1998) Angiotensin, thirst and sodium appetite. *Physiological Reviews* 78: 583–686.

Johnson AK and Thunhorst R (1997) The neuroendocrinology of thirst and sodium appetite: Visceral sensory signals and mechanisms of central integration. *Frontiers in Neuroendocrinology* 18: 292–353.

Llorens-Cortes C and Mendelsohn FA (2002) Organisation and functional role of the brain angiotensin system. *Journal of the Renin-Angiotensin-Aldosterone System* 3(supplement 1): S39–S48.

McKinley MJ, Albiston AL, Allen AM, et al. (2003) The brain renin-angiotensin system: Location and physiological roles. *International Journal of Biochemistry & Cell Biology* 35: 901–918.

McKinley MJ, McAllen RM, Davern P, et al. (2003) The sensory circumventricular organs. *Advances i n Anatomy, Embryology, and Cell Biology* 172: 1–127.

Saavedra JM, Ando H, Armando I, Bregonzio G, Jezova M, and Zhou J (2004) Brain angiotensin II, an important stress hormone: Regulatory sites and therapeutic opportunities. *Annals of the New York Academy of Sciences* 1018: 76–84.

Unger T, and Scholkens BA (eds.) (2004) *Handbook of Experimental Pharmacology 163. Angiotensin,* 2 vols. Berlin: Springer.

Von Bohlen und, Halbach O, and Albrecht D (2006) The CNS renin-angiotensin system. *Cell and Tissue Research* 326: 599–616.

Washburn DL and Ferguson AL (2001) Membrane properties of subfornical organ neurons. *Clinical and Experimental Pharmacology & Physiology* 28: 575–580.

Angiotensin II

M J McKinley, University of Melbourne, Melbourne, VIC, Australia
B J Oldfield, Monash University, Clayton, VIC, Australia

Introduction

The renin–angiotensin system, through the actions of its major effector peptides angiotensin II and angiotensin III (angiotensin peptides), has major influences on the regulation of bodily water and salt balance. Consideration of the role in the brain of angiotensin's action on fluid and electrolyte homeostasis is incomplete unless viewed in the context of its peripheral actions and overall involvement in fluid and electrolyte homeostasis. In this regard, the influences of the renin–angiotensin system range from the hormonal actions of circulating angiotensin II on the kidney, vasculature, adrenal gland, and brain to likely roles of angiotensin peptides as neurotransmitters within the brain. In addition to regulating fluid and electrolyte balance, there are powerful influences of angiotensin on the cardiovascular system and diverse biological actions of locally generated angiotensin peptides in various organs and tissues throughout the body.

Circulating angiotensin II, generated under conditions of sodium depletion, hypovolemia, or hypotension, acts on specific receptors in the adrenal gland, kidney, and brain to regulate fluid and electrolyte homeostasis. Within the adrenal gland, angiotensin peptides act on cells in the zona glomerulosa to stimulate secretion of aldosterone. This steroid hormone acts on the distal tubule of the kidney to cause sodium retention. As well, circulating angiotensin II has direct effects on the kidney, to influence renal sodium and water handling and to provide an inhibitory feedback regulation of renin secretion. Circulating angiotensin II acts on the brain to initiate water drinking (presumably as a response to stimulation of thirst), salt hunger, vasopressin secretion, and increased sympathetic nerve activity. These responses complement the peripheral actions of angiotensin II (vasoconstriction, cardiac stimulation, sodium retention resulting from aldosterone action, and renal Na reabsorption) in restoring the capacity of the cardiovascular system to function optimally. The responses are consistent with the notion that the renin–angiotensin system is the main humoral signal coordinating fluid and electrolyte homeostasis.

In examining the roles of angiotensin peptides in the regulation of fluid and electrolyte homeostasis by the brain, consideration of the actions of angiotensin II derived from within the brain, as well as that circulating in the systemic bloodstream, is also necessary. It is relevant to appreciate the physiological conditions and mechanisms by which angiotensin peptides are generated both in the bloodstream and locally within the brain. Circulating angiotensin peptides do not readily pass across the blood–brain barrier, and therefore the actions of blood-borne angiotensin II on the brain are limited to brain sites lacking this barrier – these are the circumventricular organs (CVOs), such as the subfornical organ and organum vasculosum of the lamina terminalis (OVLT) in the forebrain, and area postrema in the medulla oblongata. In the rest of the brain (separated from blood-borne angiotensin II by the blood–brain barrier), angiotensin peptides may be generated independently of renin secretion from the kidneys. These centrally generated angiotensin peptides may interact with angiotensin receptors that are situated in many parts of the central nervous system (CNS).

Generation of Systemic Angiotensin II

Generation of angiotensin peptides in the bloodstream is dependent on the action of renin, an enzyme that is synthesized in the kidney as preprorenin and processed to its circulating active form, renin. The existence of renin within the kidney had been known since its discovery in 1898. However, it was not until many years later that angiotensin was identified as the effector peptide responsible for the pressor action of circulating renin. Subsequently, angiotensin I was purified and its amino acid sequence determined.

Circulating angiotensin peptides are formed when the aspartyl protease renin is released by the juxtaglomerular apparatus of the kidney into the circulation to act on a circulating protein, angiotensinogen, to enzymatically cleave the decapeptide molecule angiotensin I (**Figure 1**). (Angiotensinogen has a 452-amino-acid sequence, is synthesized in the liver, and is also known as renin substrate.) Angiotensin I has no biological actions. However, the octapeptide angiotensin II is generated for distribution in the bloodstream when another protease, angiotensin converting enzyme (ACE; also known as dipeptidyl peptidase or kininase II), located on the luminal surface of endothelial cell membranes in many tissues, especially in the lungs, cleaves two more amino acids from the decapeptide. Angiotensin II is the major, but not the exclusive effector peptide of the renin–angiotensin system. Another enzyme, aminopeptidase A, removes a further amino acid from angiotensin II to cause the

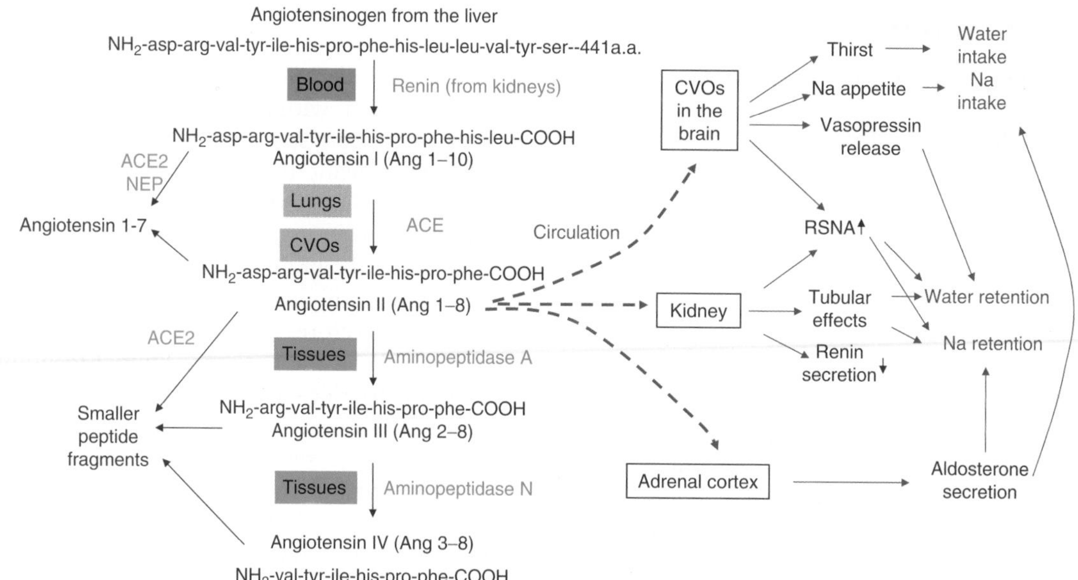

Figure 1 Diagram of the enzymatic reactions that lead to the generation of angiotensin peptides in the peripheral circulation and the circumventricular organs (CVOs) of the brain. Enzymes are indicated in red lettering. The main actions of circulating angiotensin II on salt and water balance are shown on the right. ACE, angiotensin converting enzyme; Ang, angiotensin; NEP, neutral endopeptidase; RSNA, renal sympathetic nerve activity.

formation of the heptapeptide angiotensin III. This molecule is also biologically active and can interact with angiotensin AT_1 receptors in the kidney, adrenal gland, and brain to stimulate sodium reabsorption, aldosterone secretion, and thirst. Further degradation of angiotensin II and III occurs in tissues, with the subsequent formation of the hexapeptide angiotensin IV and smaller peptide fragments that do not act on AT_1 receptors and do not appear to have any actions on salt and water balance.

Stimuli for Renin Secretion and Angiotensin Generation

As previously mentioned, the renin–angiotensin system brings mechanisms into play that both restore depleted fluid and electrolytes (thirst and sodium appetite) and prevent further losses of body water and sodium levels. As sodium chloride is the major ionic component of extracellular fluid, it can be seen that the renin–angiotensin system is a major factor in preventing extracellular fluid depletion. Renin secretion and the subsequent generation of angiotensin II occur when the cardiovascular system is compromised. Reduction of extracellular fluid volume (hypovolemia) is a potent stimulus to renin secretion from the kidney. Hypovolemia can result from sodium depletion caused by loss of fluid and electrolytes from the gastrointestinal tract, excessive sweating, or prolonged exposure to a low-salt diet. Renin secretion also results when there is a decrease in arterial blood pressure. The kidney responds with renin release to three main signals that signify depletion of the extracellular fluid or cardiovascular dysfunction. These signals are a reduction in the renal perfusion pressure, a reduction in the sodium load within the macula densa of the renal distal tubule, and neural signals from the brain in the form of increased activity in the sympathetic nerve supply to the kidney.

Angiotensin Receptors

Subtypes and Structure

The chemical structures of the specific receptors on which angiotensin II and III act in the brain and the periphery to regulate fluid and electrolyte homeostasis were discovered in 1991. Two main classes of angiotensin receptors have been identified and designated as the AT_1 and AT_2 receptors; they have 30% peptide sequence homology. In rodents (but not humans), the AT_1 receptor has two subtypes, AT_{1A} and AT_{1B}, that differ by only a few amino acids. The AT_1 and AT_2 receptors are both membrane-bound G-protein-coupled receptors, with seven membrane-spanning domains, three intracellular and extracellular loops, an intracellular C-terminus, and extracellular N-terminus. Although an AT_4 receptor was proposed as a site of action of the hexapeptide angiotensin IV, in the brain the site of the binding of this peptide has been shown to be insulin-regulated aminopeptidase (IRAP).

Location of AT_1 Receptors in the CNS

The location of angiotensin binding sites within the brain (of the rat) was mapped in 1984, and these sites were later identified as regions containing either AT_1 or AT_2 receptors. The distribution of these receptors in the human brain is similar, although not identical, to those in rat brain. One of the more remarkable features of the distribution of the AT_1 receptor in the brain is the presence of such receptors in virtually all brain regions known to participate in the neural regulation of body fluid and electrolyte homeostasis and cardiovascular control. AT_1 receptors can be located on either pre- or postsynaptic neuronal sites in these regions. In the human brain, high concentrations of AT_1 receptors are found in many regions. These regions include the subfornical organ and OVLT, sensory CVOs that lack a blood–brain barrier and are therefore susceptible to blood-borne angiotensin II. A number of other regions – such as the median preoptic nucleus (MnPO), septum, and bed nucleus of the stria terminalis in the forebrain; supraoptic, paraventricular, and arcuate nuclei in the hypothalamus; lateral parabrachial nucleus, substantia nigra, and locus coeruleus in the midbrain; and nucleus of the solitary tract and ventrolateral medulla in the hindbrain – are rich in AT_1 receptors (**Figure 2**). Angiotensin AT_1 receptors are also present in the intermediolateral cell column and dorsal horn of the spinal cord and in dorsal root ganglia.

Intracellular Signaling

Angiotensin II acts on neuronal AT_1 receptors to stimulate firing of action potentials. While the intracellular signaling events to which AT_1 receptors are coupled have been studied in only a few brain sites, AT_1 receptor signaling occurs via G-protein (G_q) coupled to the third intracellular loop. Engagement of G_q activates phospholipase Cβ, with subsequent formation of diacylglycerol and inositol triphosphate leading to increased intracellular calcium ions, and activation of protein kinase C and calcium/calmodulin kinase II (CaMKII). These kinases open Ca and close K channels, resulting in neuronal excitation. There is recent evidence that superoxide formation has a role in transducing angiotensin influences on neurons.

The density of AT_1 receptors in the CNS is not static, and they can be up- or downregulated in some regions in response to alterations in sodium and water balance, as well as to other metabolic changes. Specific pharmacological antagonists to both AT_1 (e.g., losartan, candesartan, irbesartan) and AT_2 (e.g., PD123319, CGP42112A) receptors have been developed, and AT_1 receptor antagonists are

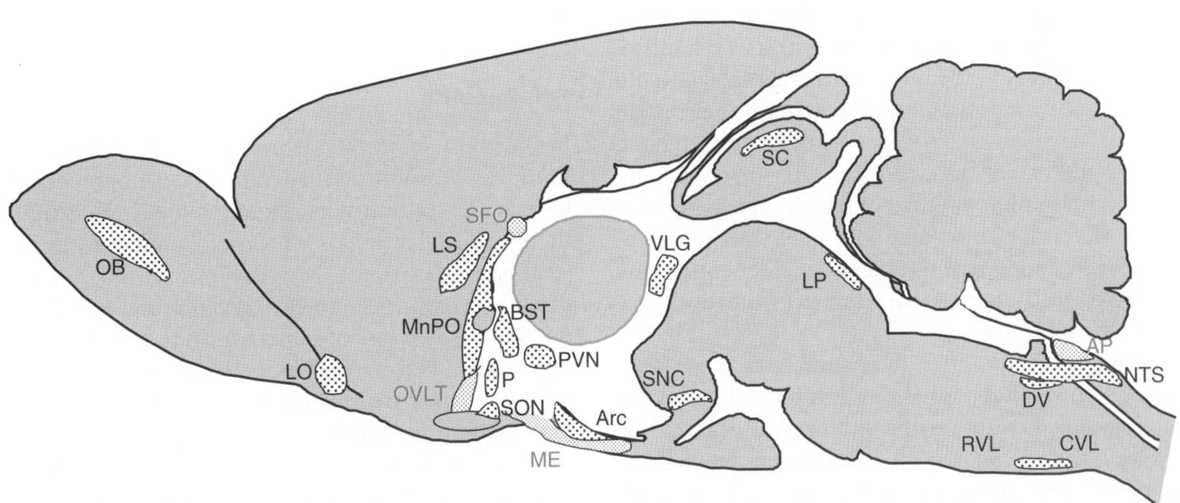

Figure 2 Diagram of the locations of angiotensin AT_1 receptors in the mammalian brain. Circumventricular organs that lack a blood–brain barrier, and are therefore exposed to circulating angiotensin II, are shown in red. Results from several species, including humans, have been projected onto a midsagittal diagram of the rat brain. AP, area postrema; Arc, arcuate nucleus; BST, bed nucleus of the stria terminalis; CVL, caudal ventrolateral medulla; DV, dorsal motor nucleus of vagus; LO, lateral olfactory tract; LP, lateral parabrachial nucleus; LS, lateral septum; ME, median eminence; MnPO, median preoptic nucleus; NTS, nucleus of the solitary tract; OB, olfactory bulb; OVLT, organum vasculosum of the lamina terminalis; P, periventricular nucleus; PVN, hypothalamic paraventricular nucleus; RVL, rostral ventrolateral medulla; SC, superior colliculus; SFO, subfornical organ; SON, supraoptic nucleus; SNC, substantia nigra pars compactus; VLG, ventral lateral geniculate nucleus.

now in wide clinical use. Earlier peptide analog antagonists (e.g., saralasin, sarile, sarthran) blocked both types of angiotensin receptors.

Central Actions of Circulating Angiotensin

In 1961, an increase in arterial pressure in dogs was observed in response to infusion of angiotensin II into the blood supply of their brain. This pressor response was the first recognized action on the brain of angiotensin. Subsequently, other brain-mediated responses of angiotensin have been discovered, including thirst, vasopressin release, sodium appetite, and sympathetic activation.

Thirst

Depletion of the extracellular compartment of the body stimulates thirst. Such thirst is independent of osmoreceptors, the sensors that signal dehydration of the intracellular compartment to initiate thirst and vasopressin secretion. In 1969, JT Fitzsimons showed that circulating angiotensin II was a potent dipsogen, and that it probably acted synergistically with neural signals coming from low-pressure volume receptors in the thorax to initiate thirst.The dipsogenic action of angiotensin II was one of the earliest recognized behavioral responses elicited by a circulating peptide. Virtually all mammals tested increased water drinking when angiotensin II was administered systemically, although some species, including humans, need supraphysiological levels of angiotensin II in blood to stimulate thirst and drinking behavior. Baroreceptor stimulation has an inhibitory influence on thirst, and, as increased arterial pressure is another consequence of systemic infusion of angiotensin II, the blood levels of angiotensin II needed to cause drinking behavior may be high to overcome the antidipsogenic effect of increased blood pressure in normovolemic subjects. An inhibitory effect of baroreceptor stimulation on thirst does not come into play when endogenous levels of angiotensin II increase under conditions of hypovolemia, because arterial pressure does not rise in this condition.

The dipsogenic action of angiotensin II may be additive with, potentiated, or inhibited by factors that also act directly on the subfornical organ. The ovarian hormone relaxin (released during pregnancy) potentiates the dipsogenic action of angiotensin II, while atrial natriuretic peptide (released when extracellular volume is expanded) and estrogen have inhibitory influences on angiotensin-induced drinking. Angiotensin and hypertonicity are additive dipsogenic stimuli. Thus, individuals in a dehydrated condition, occurring as a consequence of water deprivation, may experience thirst that results from a combination of both osmotic and angiotensin stimuli. As well as interactions with humoral factors, it is thought that afferent neural signals from the heart and possibly the kidney also interact with angiotensin's dipsogenic action to generate thirst in conditions of hypovolemia.

With the discovery of the dipsogenic action of blood-borne angiotensin II, the question arose as to how a peptide hormone that did not pass across the blood–brain barrier could stimulate drinking behavior. The answer is that angiotensin II from the circulation stimulates thirst by acting at the subfornical organ, a CVO without a blood–brain barrier. The subfornical organ is exquisitely sensitive to locally injected angiotensin II for the induction of water drinking in animals, and ablation of this region abolished water drinking in response to systemic angiotensin II. Subsequent electrophysiological and *in vitro* autoradiographic binding investigations, as well as studies of immediate-early gene *c-fos* expression, confirm the excitatory action of angiotensin II on subfornical neurons.

The neural pathways that are activated to initiate thirst subsequent to angiotensin stimulation of the subfornical organ remain largely unknown to this point. The subfornical organ has efferent neural connections to many other brain sites including the lateral hypothalamic area, hypothalamic paraventricular nuclei, OVLT, and median preoptic nucleus, all of which have been implicated in thirst responses. The connection to the median preoptic nucleus has been proposed to relay dipsogenic signals coming from angiotensin stimulation of the subfornical organ to cortical regions for the induction of thirst.

Salt Appetite

In addition to thirst, depletion of the extracellular fluid also results in the gradual development of a hunger for sodium salts. Experimentally, depletion of the extracellular fluid is achieved in animals by diuretic treatment, peritoneal dialysis, subcutaneous injection of colloid (e.g., polyethylene glycol), low-salt diet, or uncompensated loss of saliva from a fistula. Up to 24 h may elapse before the onset of increased salt intake following such treatments. In comparison, thirst and increased water intake are relatively rapid responses (minutes to hours) to extracellular fluid depletion. The most convincing evidence that activation of the systemic renin–angiotensin system is the humoral signal to the brain for salt hunger subsequent to sodium depletion is the inhibition of this response by systemically administered ACE inhibitors (see later, for discussion of doses) such as captopril or angiotensin AT_1 receptor antagonists. Salt intake that is blocked by ACE inhibitors can be restored by intravenous infusions of small amounts of angiotensin II, evidence that is consistent with a role for circulating angiotensin II in the genesis of sodium appetite. It is thought that aldosterone acts synergistically with angiotensin to stimulate salt appetite.

Circulating angiotensin II probably acts on AT_1 receptors in both the OVLT and the subfornical organ to induce a hunger for salt, because ablation of either of these CVOs can reduce salt intake following sodium depletion. Interestingly, microinjection of angiotensin II into the OVLT region, but not the subfornical organ, stimulates salt intake. Since there are reciprocal neural connections between these two CVOs, it is possible that these connections relay signals relevant to salt appetite, and that is why ablation of either of these regions reduces salt hunger. The subsequent efferent neural pathways from these two CVOs that mediate sodium appetite are still largely unknown.

Vasopressin Secretion

Shortly after the discovery of the dipsogenic action of angiotensin, it was shown that systemic infusion of angiotensin II increased the secretion of the antidiuretic hormone vasopressin from the pituitary gland. Again, angiotensin II from the circulation acts on AT_1 receptors mainly in the subfornical organ, but also in the OVLT, to initiate vasopressin release. Some of the angiotensin-sensitive neurons in both of these CVOs project directly to the hypothalamic supraoptic and paraventricular nuclei (sites of vasopressin-synthesizing neurons), implying that this neural pathway mediates angiotensin-induced vasopressin release. Angiotensin-stimulated signals from these CVOs may also be relayed to the supraoptic and paraventricular nuclei via the MnPO.

Renal Sympathetic Nerve Activity

Sodium depletion and hypovolemia result in increased activity of the sympathetic nerves supplying the kidney. Increased renal sympathetic nerve activity (RSNA) leads to reduced excretion of sodium and water in urine. Circulating angiotensin II is known to act directly on the sympathetic nerve endings in the kidney to potentiate the release of noradrenaline. It also has an action on the brain to increase RSNA, and thereby further promote renal fluid and sodium retention. There are angiotensin-sensitive neurons within the subfornical organ and OVLT that have polysynaptic neural connections to the renal sympathetic nerves. These connections are likely to mediate the effects of circulating angiotensin on RSNA, and there is evidence that neural signals from these CVOs are relayed to the kidney via synapses in the hypothalamic paraventricular nucleus and rostral ventrolateral medulla. There is evidence that angiotensin action on the area postrema may also influence renal sympathetic outflow.

Brain-Derived Angiotensin

Angiotensin Generation in the Brain

The generation of angiotensin peptides within the brain contrasts with their formation in the periphery. While details are less clear regarding the central formation of angiotensin II, it is clear that the precursor protein angiotensinogen is abundant within the extracellular and cerebrospinal fluids of the brain. Astrocytes are the major cellular source of this brain angiotensinogen, and the expression of mRNA encoding angiotensinogen in astrocytes confirms the central origin of astrocytic angiotensinogen. There is also evidence that some angiotensinogen may also be synthesized within neurons, but the amount is small in comparison to that secreted constitutively by glia. Angiotensinogen-producing glial cells are found within all divisions of the CNS, but there appears to be a preponderance of these cells in the hypothalamus and medulla, divisions that are of major significance for fluid and electrolyte homeostasis and cardiovascular regulation. A major unresolved question is the manner in which the centrally synthesized angiotensinogen is processed to angiotensins I, II, and III within the brain. The amount of renin that can be detected in the brain is quite small, casting doubt on whether renin is the main processing enzyme for angiotensin I formation in the CNS. On the other hand, ACE is plentiful in several parts of the brain, and its location on external cell membranes facilitates the formation of angiotensin II in the extracellular domain of the brain. Immunohistochemical and electrophysiological evidence suggests that angiotensin II can be released from neurons as a neurotransmitter/neuromodulator. The question of how angiotensin II arises within such angiotensinergic neurons is unresolved, as are the enzymes and cellular domains responsible.

Generation of angiotensin peptides in the subfornical organ and OVLT deserves separate analysis from that described in the previous paragraph, because these sites have extremely high concentrations of ACE and they lack a blood–brain barrier. As a consequence, in rat brain, blood-borne angiotensin I gains access to the ACE in these CVOs, with subsequent local generation of angiotensin II. It is possible that this locally generated angiotensin II in the subfornical organ and OVLT is in close proximity to the neurons that engage neural pathways that mediate thirst, vasopressin secretion, and salt appetite. Such local generation of angiotensin II in the subfornical organ is the probable explanation of a paradoxical stimulation of water and salt intake that is observed in rats when they are treated with amounts of ACE inhibitors such as captopril or enalopril sufficient to block peripheral formation of angiotensin I, but not the high concentrations of ACE in the CVOs. As a result of peripheral blockade of ACE, circulating angiotensin I levels increase (due to disruption of feedback inhibition of angiotensin II on the kidney), and when this angiotensin I reaches the subfornical organ and OVLT, it is converted to angiotensin II locally, with subsequent stimulation of thirst, vasopressin secretion, and sodium hunger.

Actions of Brain-Derived Angiotensin Peptides on Fluid and Electrolyte Balance

Injections of angiotensin II into the cerebral ventricles or discrete brain sites stimulate copious drinking of water within seconds. Centrally injected angiotensin also initiates a salt appetite, but with a latency of many hours, and can have a powerful synergistic action with mineralocorticoid steroids in the stimulation of salt intake. Central injection of angiotensin II also is a strong stimulus for the secretion of vasopressin from the posterior pituitary gland, resulting in antidiuresis and reduced fluid loss in urine. These actions are similar to those obtained when angiotensin is infused systemically, and they are also blocked by AT_1 receptor antagonists. However, it is clear that the angiotensin-responsive neurons that are mediating the responses to centrally injected angiotensin II are not the same as those in the subfornical organ that mediate the responses to circulating angiotensin, because responses to centrally administered angiotensin persist when the subfornical organ is ablated. As well, not all responses to centrally administered angiotensin II are consistent with maintaining a positive sodium balance. In fact, the initial response of the kidney to centrally administered angiotensin is to increase the excretion of sodium in urine (natriuresis), thus depleting the extracellular fluid of sodium and secondarily stimulating salt appetite. In combination with increased water intake and fluid retention caused by increased vasopressin levels, this natriuretic effect reduces the sodium concentration and osmolality of the extracellular fluid. Thus, it could be construed that central angiotensinergic action in the short term is attuned to reducing the tonicity of extracellular fluid, and as a consequence of osmotic forces, increasing intracellular volume.

Although some neurons in the subfornical organ and OVLT are stimulated when angiotensin II is injected into the lateral cerebral ventricle, the MnPO is the major site of action of intracerebroventricularly infused angiotensin II to stimulate thirst, vasopressin secretion, and natriuresis. The MnPO is situated in the anterior wall of the third ventricle (the lamina terminalis) between the subfornical organ and OVLT. Neurons situated in the MnPO have high concentrations of AT_1 receptors, and they receive afferent neural inputs from several other brain regions that are known to mediate fluid and electrolyte homeostasis, such as the subfornical organ, OVLT, lateral parabrachial nucleus, midbrain raphe, and ventrolateral medulla. Since angiotensin-responsive neurons in the MnPO are not directly affected by blood-borne angiotensin because of the blood–brain barrier, it is likely that the endogenous angiotensin that acts on neurons in the MnPO is derived from the brain, and acts at the MnPO as a neurotransmitter to engage neural pathways subserving thirst, vasopressin secretion, and natriuresis.

Direct projections from the MnPO to vasopressin-secreting cells of the supraoptic nucleus have been demonstrated, and ablation of the MnPO disrupts angiotensin-stimulated vasopressin secretion. These observations are consistent with an important role of angiotensin-sensitive neurons in the MnPO to relay signals to the supraoptic and paraventricular nuclei for vasopressin secretion. There are also data that indicate that angiotensin III may be the effector peptide responsible for central angiotensinergic influences on vasopressin release.

While this discussion has emphasized the importance of AT_1 receptors in fluid and sodium homeostasis, it would be imprudent to totally discount AT_2 receptors in this regard. There is evidence that central administration of AT_2 receptor antagonists can cause a small increase in dipsogenic and pressor responses to centrally administered angiotensin II, suggesting an antidipsogenic and depressor effect of AT_2 receptor stimulation, opposite to the effects of AT_1 receptor stimulation.

See also: Angiotensin Actions on and Within Brain; Atrial Natriuretic Peptide: Fluid/Mineral Balance; Blood Pressure: Baroreceptors; Circumventricular Organs; Fluid and Electrolyte Homeostasis: Clinical Disease; Neurohypophyseal System; Neuronal Angiotensin; Osmoregulation; Salt Appetite; Thirst; Vasopressin/Oxytocin and Receptors.

Further Reading

Allen AM, Oldfield BJ, Giles ME, et al. (2000) Localization of angiotensin receptors in the nervous system. In: Quirion R, Björklund A, and Hökfelt T (eds.) *Handbook of Chemical Neuroanatomy, Volume 16: Peptide Receptors. Part I*, pp. 79–124. Amsterdam: Elsevier Science B.V.

Andersson B, Eriksson L, Fernandez O, et al. (1972) Centrally mediated effects of angiotensin II on arterial pressure and fluid balance. *Acta Physiologica Scandinavica* 85: 398–407.

Davisson RL (2003) Physiological genomic analysis of the brain renin–angiotensin system. *American Journal of Physiology* 285: R498–R511.

Denton DA (1982) *The Hunger for Salt.* Berlin: Springer.

DiBona GF (2001) Peripheral and central interactions between the renin–angiotensin system and the renal sympathetic nerves in control of renal function. *Annals of the New York Academy of Sciences* 940: 395–406.

Fitzsimons JT (1969) The role of a renal thirst factor in drinking induced by extracellular stimuli. *Journal of Physiology* 201: 349–368.

Fitzsimons JT (1998) Angiotensin, thirst and sodium appetite. *Physiological Reviews* 78: 583–686.

Johnson AK and Thunhorst R (1997) The neuroendocrinology of thirst and sodium appetite: Visceral sensory signals and mechanisms of central integration. *Frontiers in Neuroendocrinology* 18: 292–353.

Llorens-Cortes C and Mendelsohn FA (2002) Organisation and functional role of the brain angiotensin system. *Journal of the Renin–Angiotensin–Aldosterone System* 3(supplement 1): S39–S48.

McKinley MJ, Albiston AL, Allen AM, et al. (2003) The brain renin–angiotensin system: Location and physiological roles. *International Journal of Biochemistry & Cell Biology* 35: 901–918.

McKinley MJ, McAllen RM, Davern P, et al. (2003) The sensory circumventricular organs. *Advances in Anatomy, Embryology, and Cell Biology* 172: 1–127.

Sakai RR, Nicolaidis S, and Epstein AN (1986) Salt appetite is suppressed by interference with angiotensin II and aldosterone. *American Journal of Physiology* 251: R762–R768.

Stricker EM and Sved AF (2002) Control of vasopressin secretion and thirst: Similarities and dissimilarities in signals. *Physiology & Behavior* 77: 731–736.

Unger T and Scholkens BA (eds.) (2004) *Handbook of Experimental Pharmacology, Volume 163. Angiotensin*, 2 vols. Berlin: Springer.

Washburn DL and Ferguson AL (2001) Membrane properties of subfornical organ neurons. *Clinical and Experimental Pharmacology and Physiology* 28: 575–580.

Animal Communication: Honesty and Deception

S L Vehrencamp, Cornell University, Ithaca, NY, USA

Definitions

Communication is defined as the transmission of information from a sender to a receiver via signals. True communication occurs when the information encoded in the signal enables the receiver to make decisions about behavioral responses and actions that subsequently benefit both the receiver and the sender. There are two ways in which this mutually beneficial exchange can break down in such a way that one party exploits the other: deception and eavesdropping.

Deception is the provision of inaccurate information by the sender such that the sender benefits from the interaction but the receiver pays the cost of a wrong decision. Types of deceit include 'exaggeration' or 'bluff' (using a signal whose rank among ordered alternatives is different from that for the corresponding condition values), 'lies' (use of the wrong signal among an unordered set of alternatives), and 'withholding information' (not giving a signal when appropriate). The evolution of such deceptive signaling is the focus of this discussion.

Eavesdropping typically occurs when a third-party receiver detects a signal directed at another receiver and uses the information to make decisions. The effect of eavesdropping on the sender could be positive or negative. For example, a countersinging interaction between two territorial male birds could provide information for eavesdropping females to make mate-choice decisions and for eavesdropping males to learn about the presence, relative dominance, or fighting ability of future rivals.

Both deception and eavesdropping can also take place in interspecific interactions. There are numerous examples of predatory species that mimic the mate attraction signal of their prey, but in such deceptive interactions between two different species, there is no selection pressure on the predator to be honest, and there is little the prey can do to avoid being exploited. The prey, as receiver, can try to improve its discrimination between true mates and impostors, but this process will simultaneously select for better mimicry by the predator. Similarly, predators can eavesdrop on the signals of their prey to locate their next meal. Prey can attempt to reduce the conspicuousness of these signals to predators while maximizing conspicuousness to conspecifics, but predators will counter with increased sensitivity. Such costs of signaling do affect signal design and receiver discrimination but are outside of this discussion of signal honesty. However, as shown below, some signals do appear to transmit honest information to heterospecifics for the benefit of both parties and provide a strong test of honest signaling models.

Honest signals, whether visual, vocal, olfactory, or tactile, are those in which some characteristic of the display (e.g., presence/absence, alternative forms, or a continuously varying parameter) is reliably associated with some attribute of the sender or its environment about which receivers want to know. Identifying this attribute specifies the kind of information transmitted by the signal. To fully demonstrate an honest signaling system, we would need to show that receivers not only attend to the signal and its variants but also benefit from knowing this information to make decisions. No signal is likely to be perfectly accurate. Senders make errors, but if errors lead to both positive and negative payoffs for the sender, we would not call it deception. Receivers will tolerate a certain level of sender error and deception as long as the signal is honest on average (see the section titled 'Do animals cheat?').

Context of Deception

When the sender and receiver in a signaling exchange both rank the payoffs of alternative receiver responses in the same order, selection will favor the accurate exchange of information within the limits of encoding, transmitting, and receiving error. However, sender and receiver often have conflicting interests because they rank the payoffs of alternative receiver responses differently. Under these conditions, animal senders will be tempted to provide misleading information so that the receiver performs the act that most benefits the sender. The strength of the selective pressure to deceive depends on the signaling context and the degree of conflict between the parties. Conflict of interest is greatest when two more or less equal competitors both desire the same nonsharable resource. Each would like the other to back down without a fight, and each would benefit from persuading the other that it is the better fighter by any means possible, including bluff. In the mate attraction context, both male and female benefit from mating with the correct species and therefore agree about the accurate transmission of species information. But females may want to mate only with a high-quality male, putting pressure on low-quality males to hide or exaggerate their quality. Similarly, an offspring in a brood of siblings may exaggerate its need for food to the parent in order to garner a larger share of the food for itself. Even in cooperative groups, where all members benefit from coordinated flock cohesion, two individuals may disagree about the direction the flock should move and lie to ensure that their directive is heeded.

A Brief History

In the early days of ethology, signals were shown to evolve through the ritualization of behaviors that are or were functionally appropriate to the contexts in which the signals are now given. Signals were believed to be honest indicators of underlying motivations because they were derived from physiologically or anatomically linked sources. With the rise of evolutionary game theory in the 1970s, this notion of signal honesty was questioned. Why should a sender give an honest signal? Similarly, if senders wear their emotions on their shirtsleeves, what prevents a clever receiver from using that information to exploit the sender? Richard Dawkins and John Krebs suggested that senders were best characterized as deceitful manipulators trying to mask their true intentions and trick receivers into actions benefiting senders. Receivers in turn were best viewed as mind readers trying to discount false signals, anticipate the true intent of the sender, and thus identify their own best countermove. This scenario leads to a never-ending arms race, with increasing deceit and concealment of true intentions by senders parried by increased discrimination and exploitation by receivers. Except when sender and receiver have common interests, the resulting signals would be largely deceitful and uninformative.

Amotz Zahavi challenged this pessimistic view of signal honesty. He asserted that receivers have the upper hand and ought not respond to signals unless they carry some guarantee of honesty. One guarantee is to require that the signal impose a cost such that deceitful senders cannot afford to produce an exaggerated signal or that they produce it only in an ineffective way. Signals characterized by such costs are called handicap signals. Although Zahavi's idea was viewed skeptically at first, subsequent game theory models demonstrated the evolutionary feasibility of handicap signaling, and the handicap principle is now widely accepted.

Since the mid-1980s, several dozen game theory models of biological communication have been developed, each depicting different signaling contexts and comprising different game structures and sets of assumptions. The common feature in all the models that found at least some conditions for stable signaling was the assumption of some type of cost imposed on dishonest senders. Without such costs, senders become dishonest, receivers ignore the signals, and no evolutionarily stable state with informative communication signals can be attained. These costs ranged from signal production costs to receiver retaliation costs, reputation costs, and various types of physical and physiological constraints. The realization that the type of cost affects both the form of a signal and the specific kind of information it can encode then led to the useful classification of signals based on the type of cost. Independent attempts to classify signals in this way have largely converged and provide a very powerful framework for understanding the evolution and diversity of animal communication signals.

Classifying Signals Based on the Type of Cost That Guarantees Honesty

Table 1 summarizes the simplest scheme for the relationship between the type of cost that maintains signal honesty, the form or design of the signal, and the kinds of information the signal can encode. Three signal categories have been distinguished here: handicap, index, and conventional signals.

Table 1 Signal classification based on the type of cost that maintains signal honesty

Signal class	*Cost*	*Signal design*	*Information*
Handicap	Signal production, time lost, risk of predation	Graded display, intensity correlated with sender quality	Fighting ability, stamina, condition, territory quality, need, motivation
Index	Physical or physiological constraints	Form linked to sender attributes	Body size, age, strength, pointing at intended receiver
Conventional	Retaliation by receivers	Antithetical discrete or graded display, arbitrary form	Motivation to escalate versus retreat, aggressive intentions, fighting ability, condition

Based on Maynard Smith J and Harper D (2003) *Animal Signals*. Oxford, UK: Oxford University Press; Vehrencamp SL (2000) Handicap, index, and conventional signal elements of bird song. In: Espmark Y, Amundsen T, and Rosenqvist G (eds.) *Animal Signals: Signalling and Signal Design in Animal Communication*, pp. 277–300. Trondheim: Tapir Publishers; Hurd PL and Enquist M (2005) A strategic taxonomy of biological communication. *Animal Behaviour* 70: 1155–1170.

Handicap Signals

The key concept of the handicap principle is that the cost imposed on the sender should be one that 'uses up' the particular sender attribute about which the receiver wants information. Receivers are selected to pay attention only to those types of signals that impose costs linked to the type of information receivers need. Thus the form of the signal is linked to its information content, typically a continuously varying signal parameter correlated with some sender attribute. Directional selection pressure from receivers favoring the most costly signal variants often results in extreme elaboration and exaggeration of the display character. Handicap signals are strategic signals, in the sense that all senders can produce all signal variants in principle, but senders of poor quality or condition tend not to produce the more intense variants because of the high signaling cost. **Figure 1** illustrates the classical handicap model, in which poor quality senders pay a higher cost for a given intensity of display. Their best option is to display at a lower intensity than a high-quality individual does, so signal intensity is reliably correlated with sender quality. Such signals are also called quality indicators and condition-dependent signals. Handicapping costs can be subdivided into production costs paid at the time of display and developmental costs paid earlier to grow the structures and organs needed for displaying. Several examples below show how such costs can be linked to useful information for receivers.

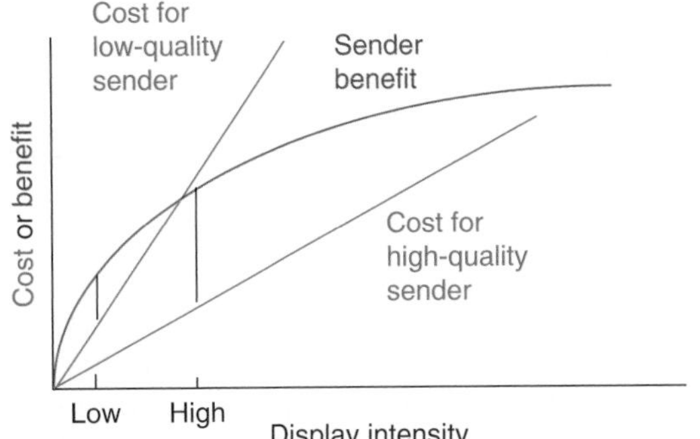

Figure 1 A graphical model for a handicap signal of sender quality. A key assumption is that displaying is costly, but high-quality senders pay a lower cost for a given display intensity than low-quality senders do. High- and low-quality senders receive the same benefit for a given display intensity. The optimal display level occurs at the point where the difference between benefit and cost is greatest. At equilibrium, high-quality senders display at a higher intensity than low-quality senders do, so display intensity is correlated with sender quality or condition. From Johnstone RA (1997) Evolution of animal signals. In: Krebs JR and Davies NB (eds.) *Behavioural Ecology: An Evolutionary Approach*, 4th edn., pp. 155–178. Oxford, UK: Blackwell Science.

Signals with energetically expensive production costs can inform receivers about the health, vigor, or foraging abilities of senders. For example, vocal and visual mate attraction signals must be repeated again and again. Female preference for males that not only produce high-quality displays but also repeat them at a high rate will increase the selection pressure on males to perform at the highest possible energetic level they can sustain. Unhealthy or poor-quality males cannot maintain such an expensive display rate, and this fact will be detectable to females. Choosy females will therefore benefit by selecting vigorous mates who are good genetic fathers or better parental providers or who possess food-rich territories (**Figures 2** and **3**). Repetitive countercalling contests between rival males can facilitate the assessment of relative condition, fighting ability, or motivation without their having to engage in a physical fight (**Figure 4(a)**). As a final example, energetic jumping displays are given by gazelles to approaching predators. Only individuals in good condition can afford to perform these actions well, and this provides honest information to the predators, which discourages them from chasing the more able displayers (**Figure 4(b)**).

Developmental costs are borne before the onset of display and can reveal either nutritional condition during development or more intrinsic aspects of genetic quality. The development of exaggerated display organs can also result in high maintenance costs for the bearer. A well-documented example is the elongated tail feathers of the barn swallow (*Hirundo rustica*) studied by Anders Møller. Females prefer males with longer tails, pairing much more quickly with males having artificially enlarged tails than with males with shortened tails. Long tails are a handicap for males; barn swallows are aerial foragers that capture flying insects on the wing. Artificial tail elongation increases the drag on the tail and reduces agility and foraging efficiency. Males with naturally long tails are stronger, healthier, and more parasite-resistant individuals who can not only grow long tails and cope with the foraging handicap but also pass on their parasite resistance to their genetic offspring. Females thus obtain better-quality offspring by selecting long-tailed males (**Figure 5**). Another good example of a handicap signal with a development cost is red coloration in birds and fish. Bearers of such color patches may also sustain the maintenance cost of increased conspicuousness to predators. Males in many songbird species possess repertoires of song types that they learn in the first months of life, and females have been shown to prefer larger repertoires and more complex songs. One promising explanation for this preference is the developmental stress

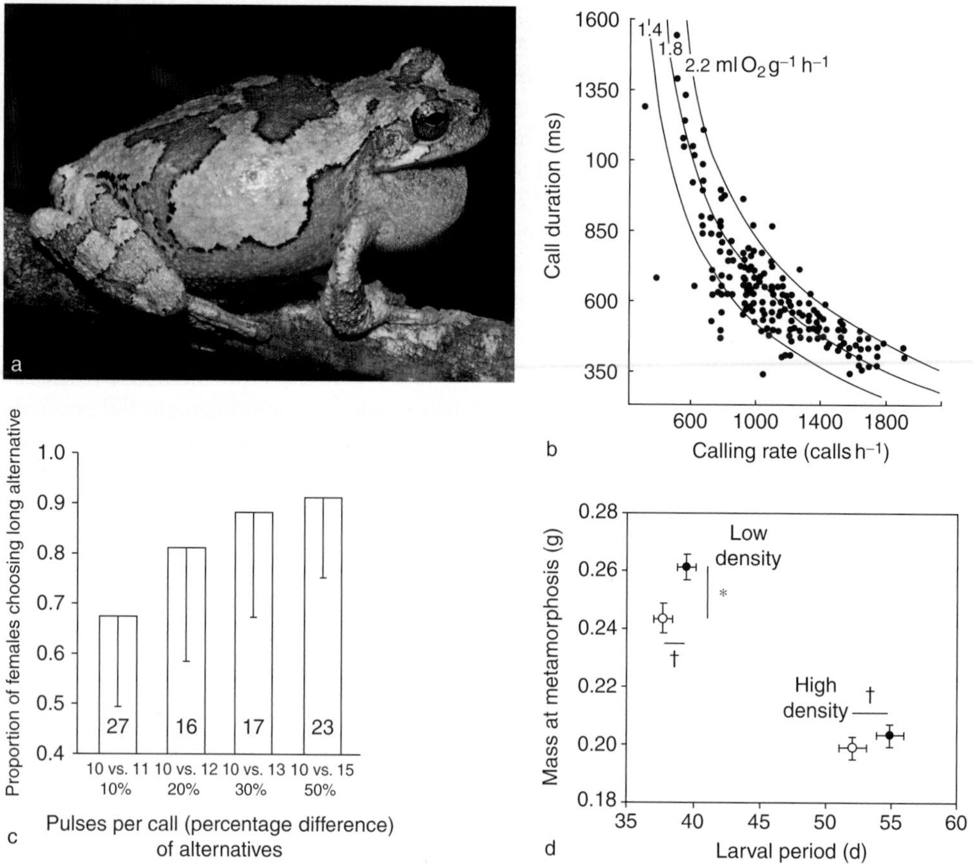

Figure 2 Call duration in the gray tree frog (*Hyla versicolor*) (a) is a costly handicap signal of male quality. (b) Calling is energetically expensive, with call duration and calling rate showing a trade-off in the field. An increase in both parameters is associated with an increase in oxygen consumption. Males increase their call duration when singing in dense choruses and when interacting vocally with other males. (c) Females strongly prefer males and experimental stimuli with longer call duration. (d) The offspring of females mated to long-calling males exhibit greater mass at metamorphosis than offspring of females mated to short-calling males, especially in contexts of low food density. In (c), error bars show the lower 95% confidence limits, and numbers in each bar indicate number of females in the sample. (d) The offspring of females mated to long-calling males (filled symbols) exhibit greater mass at metamorphosis than offspring of females mated to short-calling males (open circles), especially in contexts of low food density; error bars represent ± 1 standard error, asterisk indicates significance at $P < 0.05$, and daggers indicate $P < 0.10$. (a) Photo courtesy of H Carl Gerhardt. (b) From Wells KD and Taigen TL (1986) The effect of social interactions on calling energetics in the gray treefrog (*Hyla* versicolor). *Behavioral Ecology and Sociobiology* 19: 9–18. (c) From Gerhardt C, Tanner SD, Corrigan CM, and Walton HC (2000) Female preference functions based on call duration in the gray tree frog (*Hyla versicolor*). *Behavioral Ecology* 11: 663–669. (d) From Welch AM (2003) Genetic benefits of a female mating preference in gray tree frogs are context-dependent. *Evolution* 57: 883–893.

hypothesis, which argues that young birds require good nutrition to develop the brain nuclei for song learning and production.

Two other types of handicap models involve signaling costs but provide different kinds of information. Models of sender need, developed with the context of begging in mind, require a signal whose intensity is correlated with increasing immediate costs (energetic or predation risk). The key features that stabilize the correlation between signal intensity and need are the greater benefit of food for a needier sender and the fact that the beggar sender and the donor receiver are genetic relatives. The other handicap model variant was developed in the context of aggressive signals, in which the performance of a signal makes the sender vulnerable to attack by the receiver. This has been called a vulnerability handicap or an interaction handicap because it represents a hybrid between the handicap model and the conventional signal model (see the section titled 'Conventional signals'). Signals in this category typically vary in their tactical impact such that a strong signal places the sender at greater risk of injury but is more effective in threatening the opponent.

Index Signals

Handicapping is not the only mechanism for generating honest signals. Recent models suggest that if coevolving senders and receivers can hit on a cost-free signal that reliably correlates with important sender

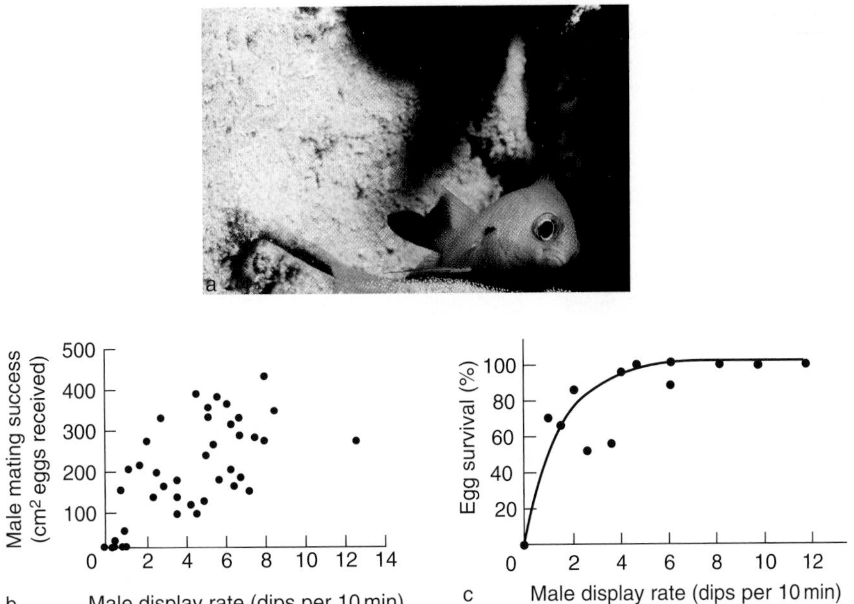

Figure 3 Example of display intensity as an indicator of male parental ability in the bicolor damselfish *Stegastes partitus* (a). This courting male is in the process of fertilizing recently laid eggs visible at the bottom. (b) Females prefer males that display at a high rate. Displaying is energetically costly and uses up calories, but males that are in good condition can afford high display rates and are more successful at guarding eggs (c). (a) Courtesy of Ken Clifton, with permission. (b, c) From Knapp RA and Kovach JT (1991) Courtship as an honest indicator of male parental quality in the bicolor damselfish, *Stegastes partitus*. *Behavioral Ecology* 2: 295–300.

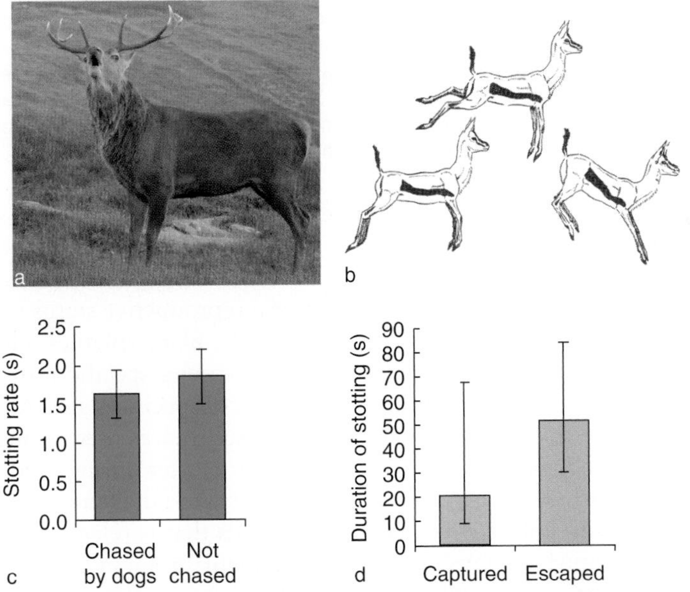

Figure 4 Examples of costly handicap signals in contexts other than mate attraction. (a) Male red deer (*Cervus elaphus*) engage in extended roaring contests with rivals. Roaring is an energetically expensive behavior that uses the actions and muscles employed during fighting. It is thought that only those males in good enough condition to fight can afford to roar at a winning level (Clutton-Brock TH and Albom SD (1979) The roaring of red deer and the evolution of honest advertisement. *Behaviour* 69: 145–170). (b) Stotting in Thomson's gazelles (*Gazella thomsoni*) involves leaping off the ground with all four legs held stiff and straight while running from predators (Walther FR (1969) Flight behaviour and avoidance of predators in Thomson's gazelle (*Gazella thomsoni* Guenther 1884). *Behaviour* 34: 184–221). They are far more likely to stot when chased by coursing wild dogs (78%) than when hunted by stalking cheetahs (9%). Wild dogs concentrated their chases on individuals that stotted at slower rates (c). Captured gazelles also stotted for shorter durations than gazelles that escaped (d). Gazelles were less likely to stot, and stotted at slower rates, during the dry season when food was lower and animals were in poorer condition. Stotting is therefore believed to be an honest signal to coursing predators of the prey's ability to escape capture. The error bars in (c) are ±1 standard deviation and in (d) are interquartile ranges. (a) Photo courtesy of Alison Donald. (b) Reproduced from Walther FR (1969) Flight behaviour and avoidance of predators in Thomson's gazelle (*Gazella thomsoni* Guenther 1884). *Behaviour* 34: 184–221, with permission from Koninklijke Brill NV. (c, d) From FitzGibbon CD and Fanshawe JH (1988) Stotting in Thomson's gazelles: An honest signal of condition. *Behavioral Ecology and Sociobiology* 23: 69–74.

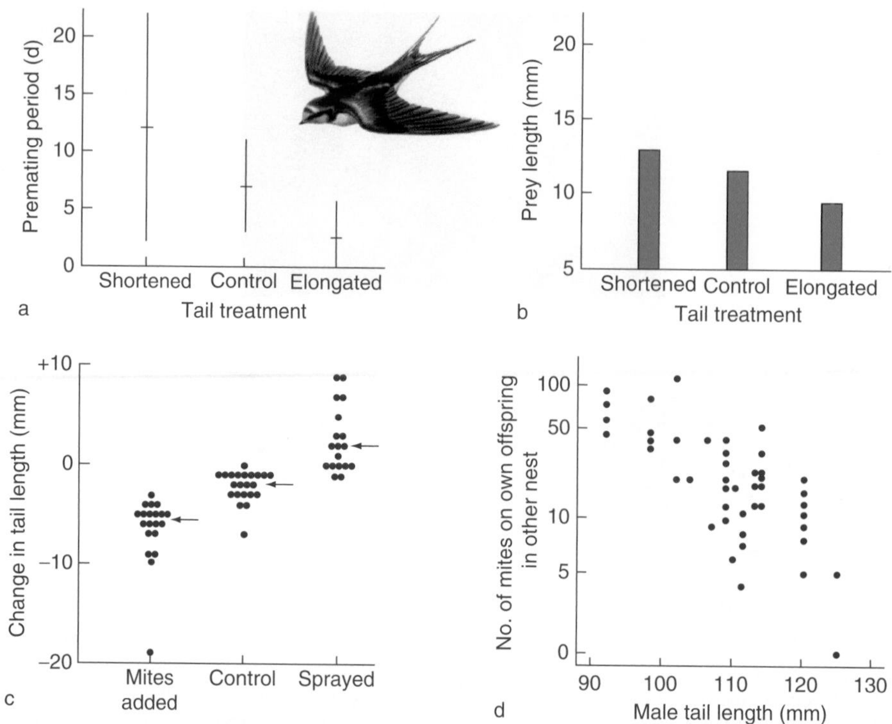

Figure 5 Tail length as a male quality indicator in the barn swallow (*Hirundo rustica*). (a) Females prefer (pair earlier with) males with experimentally lengthened tails. Error bars are ±1 standard deviation. (b) The cost of a lengthened tail in this aerial forager is increased drag, which reduces foraging efficiency and survivorship; males with naturally long tails suffer less of an aerodynamic cost than males with naturally short tails. (c) Long-tailed males possess fewer blood-sucking mites than do short-tailed males; removing the mites results in better tail growth. (d) Nestlings of males with longer tails and fewer parasites pass on their resistance to their offspring, even when they are raised in foster nests. (a) From Møller AP (1988) Female choice selects for male sexual tail ornaments in the monogamous swallow. *Nature* 332: 640–642. (b) from Møller AP, Lope F, and Lopez Caballero JM (1995) Foraging costs of a tail ornament. *Behavioral Ecology and Sociobiology* 37: 289–296. (c, d) from Møller AP (1990) Effects of hematophagous mite on the barn swallow *Hirundo rustica*: A test of the Hamilton and Zuk hypothesis. *Evolution* 44: 771–784, with permission from Blackwell publishing.

attributes, it will be favored over the costly signal. Some signals are constrained by physiology, anatomy, or physical principles to be honest indicators of certain types of sender attributes. Such signals are called unbluffable or index signals. Like handicaps, the form of the signal is strongly linked to the sender attribute of interest to the receiver. Unlike handicaps, they are not strategic signals, because some senders simply cannot produce some signal variants.

Examples of index signals include low-frequency vocal threats that are obligatorily linked with body size in many vertebrates, tailbeating displays in fish, and side-by-side head-high postures in ungulates. Similarly, push-pull and mouth-wrestling forms of ritualized fighting are reliable index signals of weight and strength. Signals that are linked with aging processes or health, such as silver backs in gorillas and song repertoire size in birds, indicate age, experience, and ability to survive (**Figure 6**). In song bird species with age-restricted (predispersal) song learning, song-type sharing between newly settled males and their neighbors is inversely correlated with dispersal distance and could serve as an index of fighting ability if good fighters are more likely to gain territories close to their tutors. Olfactory signals that are directly derived from reproductive hormones and by-products should also be classified as index signals. They directly reveal reproductive status in both males and females. Similarly, plant volatiles released by herbivores feeding on species-specific host plants are often used to attract conspecific mates. Although still controversial, alarm substances used by many schooling fish to disperse or cluster during predator attacks appear to be an antibacterial agent sequestered in the vacuoles of skin cells that is released on injury and would therefore qualify as an index signal.

Another cost-free constraint that leads to a type of index signal is an informational constraint. In this case, use of the signal is constrained by having access to some information. An example is stalked prey staring at a hidden predator, thereby signaling to the predator both its alerted state and the futility of continuing the hunt. The signal can be performed only by a sender who knows the location of the hidden predator.

Both index and handicap signals may have associated with them traits that make direct assessment of the relevant sender attributes easier or more accurate. Typically these are color patch markings or body

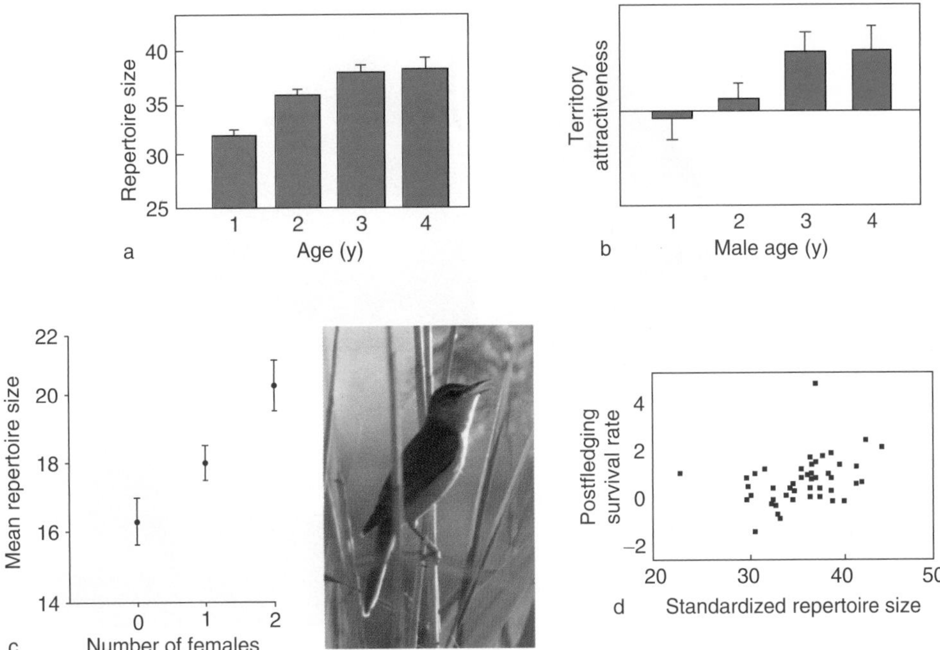

Figure 6 Repertoire size in an open-ended learner, the great reed warbler *Acrocephalus arundinaceus*, as an index of age. (a) Repertoire size increases with age for the first few years of life. (b) Older males possess better territories, as measured by occupancy and vegetation cover. (c) Males with larger repertoires (and better territories) attract more females in this polygynous species. Nine of ten females who sought extrapair fertilizations selected a neighboring male with a larger repertoire than their social mate. (d) The offspring of males with larger repertoires had higher survival, including return rate after migration, suggesting that larger-repertoire males pass on their inherently higher quality to their offspring. Error bars in (a), (b), and (c) are ± 1 standard error. (a, b) From Hasselquist D (1994) Male attractiveness, mating tactics and realized fitness in the polygynous great reed warbler. PhD Thesis, Lund University, Sweden. (c) From Catchpole CK, Leisler B, and Dittami J (1986) Sexual differences in the responses of captive great reed warblers (*Acrocephalus arundinaceus*) to variation in song structure and size. *Ethology* 73: 69–77. (d) From Hasselquist D, Bensch S, and von Schantz T (1996) Correlation between male song repertoire, extra-pair paternity and offspring survival in the great reed warbler. *Nature* 381: 229–232. Photo courtesy of Bengt Hansson.

structures, called 'amplifiers,' that enhance visual perception of the physical attribute or movement display. Many fish have contrasting lines demarcating their body margins, which make assessment of body size easier. Similarly, a contrasting triangular marking on the abdomens of spiders makes assessment of nutritional condition, and thus fighting ability, easier. Contrasting stripes on the head in many birds, or outlining of the ears in many mammals, make it easier for a visual receiver to assess where a sender is looking.

Conventional Signals

Some communication signals are neither costly to produce nor obligatorily linked with physical properties of senders. The code by which these signals are associated with contexts is an arbitrary convention, and therefore these are called conventional signals. If there is no conflict of interest between sender and receiver, senders will not be tempted to cheat and conventional signals can be honest and stable without further guarantees. However, conventional signals are seen even during conflicts of interest, and in this case there must be a stabilizing cost that maintains honesty. The cost is receiver retaliation, which must come in the form of receiver skepticism or a retaliation rule in which receivers approach and check the sincerity of senders. In the most compelling model of conventional signaling with conflict of interest, both contenders must signal, and retaliate when both give a similarly strong signal, but retreat when the opponent gives a stronger one. Conventional signals are often discrete, with alternate antithetical signals for the strong versus weak message. This type of signal can convey information about aggressive motivation as well as fighting ability.

Prominent color patches in some birds and lizards are the classic example of this type of signal (**Figure 7**). The size or hue of the patch is correlated with the dominance rank of the individual, hence their designation as badges of status. Large badge size deters aggressive challenges by small-badged individuals. The cost of possessing a large badge is aggressive retaliation from other large-badged individuals. The evolution of such signals must be accompanied by frequent testing of the honesty of other individuals with a badge size similar to one's own and avoiding or ignoring individuals with larger or smaller badges.

Figure 7 Badge of status in house sparrows, *Passer domesticus*. (a) The black bib on the chest of males varies greatly in size. (b) Size of the patch is strongly correlated with dominance rank. (c) Males with experimentally enlarged patches were involved in more aggressive encounters than were controls, and their attackers had larger than average patches. In Harris sparrows with a similar black chest patch, males with normally large bibs whose patch size was experimentally reduced had to fight very hard, but eventually won. Error bars in (c) are ±1 standard error. (a) Photo courtesy of Kevin McGowan. (b) From Møller AP (1987) Variation in badge size in male house sparrows. *Passer domesticus*: Evidence for status signalling. *Animal Behaviour* 35: 1637–1644. (c) From Møller AP (1987) Social control of deception among status signalling house sparrows *Passer domesticus*. *Behavioral Ecology and Sociobiology* 20: 307–311.

Such a rule makes it very dangerous and costly for a low-status individual to cheat by sporting a large badge. Certain vocal signals, such as song type matching and song type switching rate, are also believed to be conventional signals of aggressive intention.

Receiver skepticism is a type of receiver retaliation cost that can potentially stabilize honesty of conventional signals among group-living animals that recognize each other and interact repeatedly. Thus individuals who are observed to signal deceptively are remembered and tagged with a poor reputation or lack of trust, which may result in their failing to obtain certain benefits in the future. Reputation models with repeated interactions can even maintain honesty of cheap conventional signals when receivers have conflicting interests.

Do Animals Cheat?

Most signals are believed to be honest most of the time because conflicts of interest are minimal or appropriate costs are imposed on cheaters. But dishonest signaling does occur. Receivers will tolerate a certain amount of deception as long as they obtain a net gain by attending to signals (i.e., the probability that the signal is accurate times the payoff of making the right decision, plus the probability that the signal is inaccurate times the payoff of making the wrong decision, is positive). In theory, there are several contexts in which senders may be able to get away with some dishonesty. One context is a high level of perceptual error on the part of receivers, which favors less accurate mapping of sender quality on signal characteristics. A second context arises from a nonequilibrium outcome of dynamic coevolution between sender and receiver. A third model results when there are several classes of sender, each with its own costs and benefits of signaling, and receivers cannot distinguish between sender types; honest sender types must be sufficiently common relative to dishonest types to maintain the signaling system. A fourth situation occurs when receivers pay a very high cost for failing to respond to a signal, which leaves receivers vulnerable to deceit if senders figure out a way to benefit from the response. Documented examples of dishonesty fit several of these models.

Deceitful bluffing and exaggeration of body size or condition have been documented for competitive signals in several species. For example, mantis shrimp (*Neogonodactylus bredini*) defend their burrow from intruders with a claw-spreading threat display, which is effectively given even when the resident has recently molted and has a soft exoskeleton. In snapping shrimp (*Alpheus heterochaelis*), chela (claw) size is positively correlated with body size and ability to win fights, but individuals with chelae larger than expected for their body size give more open-chela displays than individuals with smaller-than-expected chelae. Males of the mealworm beetle (*Tenebrio molitor*) release a pheromone that attracts female mates, but males whose immune system has been experimentally challenged with a foreign material implant increase their pheromone production and are even more attractive to females than normal males are. These examples may fit the multiple-sender model above, in which receivers cannot accurately distinguish the intrinsic quality of senders and honest senders are more common than dishonest ones.

A good example of an outright lie has been described in birds foraging in flocks; one individual may give a false alarm call to scare competitors away from a rich food find. Initially it was believed that senders would not 'cry wolf' too often, because receivers would learn to ignore the signal and it would cease to be effective in true alarm contexts. However, several studies have shown that the incidence of alarm calls in the absence of a predator is quite high, 55–63%. Although a certain amount of alarm unreliability is caused by errors (e.g., by young individuals who haven't learned which heterospecifics are truly dangerous), senders have been shown to give false alarms in which they clearly benefit, and the likelihood of false alarming varies with the costs and benefits to senders. The reason such a high level of dishonesty can be sustained is clearly a result of the high cost of a miss to receivers (i.e., death) relative to the cost of falsely fleeing and losing a bite of food, supporting the fourth model above. False alarm calls are also given by some male birds and squirrels during the fertile period of female mates to interrupt extra-pair copulations with competitors.

The withholding of information is a much more difficult form of deceit to document than is the provision of false information. Not only must one demonstrate a reliable association between a specific context, a signal, and a specific response, but one must also show that there is some behavioral flexibility on the part of senders to either give or fail to give the signal in different circumstances in a way that benefits the sender. This last point has been termed the 'audience effect' of signaling. Food calls may be given by males in some species (falsely at times) to attract mates, but food calls are always withheld in the presence of rival males. Food calls may also be withheld in group-living species that expect group members to share rich food finds. In this case, there needs to be a significant cost of withholding information to sustain mostly honest notification of food. Such a cost has been described only in primates who fail to advertise a rich food find. If other group members catch an individual feeding on such a find without having called, the individual is aggressively punished. But receivers may have a difficult time detecting a lie in this case, supporting the first model above.

In conclusion, low levels of dishonesty may persist in many signaling systems, but signals must be reliable enough on average to justify receiver response. Otherwise, receivers will ignore signals, senders will no longer benefit from giving dishonest information, and the signals will disappear from the species' repertoire. A variety of costs and constraints maintain signal honesty, and signals must be costly in a way that explains why they provide reliable information.

See also: Communication Networks and Eavesdropping in Animals; Game Theory and the Economics of Animal Communication; Pheromones and other Chemical Communication in Animals; Sexual Selection and the Evolution of Animal Signals; Signal Transmission in Natural Environments; Signal Design Rules in Animal Communication; Visual Signaling in Animals; Vocal Communication in Birds.

Further Reading

Catchpole CK, Leisler B, and Dittami J (1986) Sexual differences in the responses of captive great reed warblers (*Acrocephalus arundinaceus*) to variation in song structure and size. *Ethology* 73: 69–77.

Clutton-Brock TH and Albon SD (1979) The roaring of red deer and the evolution of honest advertisement. *Behaviour* 69: 145–170.

Fitch WT and Hauser MD (2003) Unpacking "honesty": Vertebrate vocal production and the evolution of acoustic signals. In: Simmons AM, Fay RR, and Popper AN (eds.) *Acoustic Communication*, pp. 65–137. New York: Springer.

FitzGibbon CD and Fanshawe JH (1988) Stotting in Thomson's gazelles: An honest signal of condition. *Behavioral Ecology and Sociobiology* 23: 69–74.

Gerhardt C, Tanner SD, Corrigan CM, and Walton HC (2000) Female preference functions based on call duration in the gray tree frog (*Hyla versicolor*). *Behavioral Ecology* 11: 663–669.

Hasselquist D (1994) Male attractiveness, mating tactics and realized fitness in the polygynous great reed warbler. PhD Thesis, Lund University, Sweden.

Hasselquist D, Bensch S, and von Schantz T (1996) Correlation between male song repertoire, extra-pair paternity and offspring survival in the great reed warbler. *Nature* 381: 229–232.

Hasson O (1994) Cheating signals. *Journal of Theoretical Biology* 167: 223–238.

Hurd PL and Enquist M (2005) A strategic taxonomy of biological communication. *Animal Behaviour* 70: 1155–1170.

Johnstone RA (1997) Evolution of animal signals. In: Krebs JR and Davies NB (eds.) *Behavioural Ecology: An Evolutionary Approach*, 4th edn., pp. 155–178. Oxford, UK: Blackwell Science.

Knapp RA and Kovach JT (1991) Courtship as an honest indicator of male parental quality in the bicolor damselfish, *Stegastes partitus*. *Behavioral Ecology* 2: 295–300.

Maynard Smith J and Harper D (2003) *Animal Signals*. Oxford, UK: Oxford University Press.

Møller AP (1987) Social control of deception among status signalling house sparrows. *Passer domesticus*. *Behavioral Ecology and Sociobiology* 20: 307–311.

Møller AP (1987) Variation in badge size in male house sparrows. *Passer domesticus*: Evidence for status signalling. *Animal Behaviour* 35: 1637–1644.

Moller AP (1988) Female choice selects for male sexual tail ornaments in the monogamous swallow. *Nature* 332: 640–642.

Møller AP (1990) Effects of hematophagous mite on the barn swallow. *Hirundo rustica*: A test of the Hamilton and Zuk hypothesis. *Evolution* 44: 771–784.

Møller AP, Lope F, and Lopez Caballero JM (1995) Foraging costs of a tail ornament. *Behavioral Ecology and Sociobiology* 37: 289–296.

Vehrencamp SL (2000) Handicap, index, and conventional signal elements of bird song. In: Espmark Y, Amundsen T, and Rosenqvist G (eds.) *Animal Signals: Signalling and Signal Design in Animal Communication*, pp. 277–300. Trondheim: Tapir Publishers.

Walther FR (1969) Flight behaviour and avoidance of predators in Thomson's gazelle (*Gazella thomsoni* Guenther 1884). *Behaviour* 34: 184–221.

Welch AM (2003) Genetic benefits of a female mating preference in gray tree frogs are context-dependent. *Evolution* 57: 883–893.

Wells KD and Taigen TL (1986) The effect of social interactions on calling energetics in the gray treefrog (*Hyla versicolor*). *Behavioral Ecology and Sociobiology* 19: 9–18.

Animal Intelligence: The Search for Animal Intelligence

A A Wright, University of Texas Health Science Center, Medical School at Houston, Houston, TX, USA

Introduction

Discoveries of obvious intelligent behavior by animals are appearing at an accelerated rate. Although examples of ape and monkey intelligence have been known for some time, even avians are casting off their 'bird brain' label. Pigeons, for example, can navigate by landmarks, sun compass, magnetic compass, infrasound sources, and possibly by cues from polarized and ultraviolet light. They can compute a navigational fix to determine their location (if unknown) to find 'home' which would be impossible with a compass alone. Galapagos finches and New Caledonian crows (**Figure 1**) select and fashion tools (e.g., cactus spines, twigs, and leaves) to 'fish' out insects and grubs in tree-limb holes/cracks or under leaf detritus and carry their favorite tools around with them. Clark's nutcracker birds store thousands of pine seeds in hundreds of cache sites (**Figure 2**), with sites varying yearly and often covered with snow during retrieved (**Figure 3**). Alex, a parrot, judges pairs of objects in the laboratory that differ in color, material, and/or shape and answers (in English) to the verbal query 'what's different' or 'what's same.'

Despite such astonishing demonstrations of animal intelligence, there is no answer to the question, What is animal intelligence? Theorists have proposed hierarchies of animal intelligence based on learning rates, learning phenomena (e.g., latent inhibition and tool use), discrimination learning (e.g., two-choice 'learning set' and serial discrimination reversals), abstract concept learning, and self-awareness, but none have withstood the test of time. Consequently, most researchers have turned to the study of specific cognitive phenomena (in laboratory settings, natural settings, social settings, etc.). Nevertheless, comparative cognition invariably gravitates to 'who can do what better.' Inevitably, the focus is on species differences and speculation returns to hierarchies, if but limited hierarchies (e.g., a branch of the evolutionary tree such as corvid or paridae birds). Even single-species studies are often (implicitly) compared to humans (e.g., "... the first demonstration in a nonhuman animal"). By contrast, if the behavioral and neurological mechanisms/processes were known, then similarities and differences should be apparent without resorting to a hierarchy ('better than'). In terms of mechanisms, similarities are as important as differences. Similar mechanisms can have somewhat different outcomes or vice versa. For example, a single test of some cognitive ability (e.g., abstract concept learning) might lead to a premature conclusion that some species (e.g., pigeon) does not possess this cognitive ability (or cognitive module). By contrast, if the study had manipulated critical parameters of concept learning, then it might have been revealed that this difference was only a quantitative difference, not the qualitative absence as concluded. These are among the issues that have guided our selection and discussion of animal cognitive phenomena. Some areas of research that have been considered relevant to animal intelligence are mentioned only in passing (encephalization quotient, consciousness, insight, associative learning, self-recognition in mirrors, animal communication, and animal language) because neither theory nor experiments have proven decisive and the field has moved on, and in other cases it is still too early to assess the reliability/validity and relevance to the field (e.g., theory of mind: seeing–begging, knower–guesser, and deception, social theory of intellect: dominance and imitation).

Stimulus-Bound Concepts

The movie '2001: A Space Odyssey' opens with the allegorical scene of primates discovering tool use and the 'touch stone' of intelligence. Tool use has long been considered a behavior whereby humans exploited their (greater) intelligence, passed down tool use over generations, eventually resulting in tools such as the wheel. Some animals in the wild, such as chimpanzees, use twigs to fish for termites and anvil and hammers to break open hard-shelled fruits/nuts. In laboratories, chimpanzees move stools and swing sticks to get hanging fruit or use rakes to pull food within reach, but this behavior does not seem to be learned by imitation. Interpretation of the tube-trap and stick problem (trap hole on one side of the food) is even less clear. Apes (chimpanzees, bonobos, and orangutans) and capuchin monkeys show successful use of the stick solving this problem, but capuchins, at least, apparently learn by trial and error without understanding which direction of stick poke will avoid the trap. With humans and apes atop the hierarchy of tool use, this seemed to be a promising index of intelligence until birds were also shown to use tools. Tool use by Galapagos finches and New Caledonian crows occurs in the laboratory and New Caledonian crows even construct tools, a feat

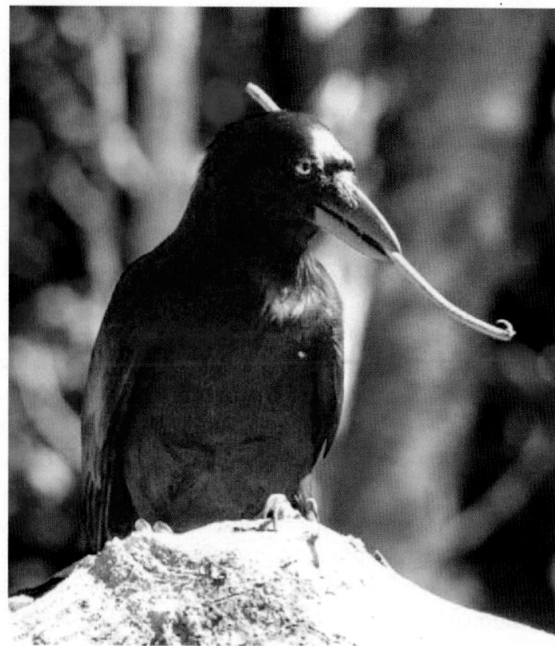

Figure 1 A New Caledonian crow with hooked tool for coaxing food items out of holes and cracks in tree limbs.

Figure 2 A Clark's nutcracker preparing to cache pinyon nuts. Courtesy of A Kamil.

Figure 3 A Clark's nutcracker in winter. Courtesy of A Kamil.

not shown for apes. But here too, captive finches do not seem to learn this tool use by imitation. If these species do not learn tool use by imitation, then the issue becomes how they learn it (i.e., learning mechanisms/processes). The same can be said for other remarkable animal behaviors that, contrary to early speculation, are apparently not learned by imitation (e.g., pilfering cream from milk bottles by English blue tits, pilfering food from backpacks by New Zealand keas, and washing potatoes by Japanese snow monkeys).

Navigation is typically based on relationships among landmarks and/or vector sources (magnetic, sun, and infrasound compasses). The nutcracker's cache retrieval is thought to be dependent on differential weightings of landmarks by a specialized cognitive module. Similar arguments have been made for rats and their spatial abilities. Some of the most remarkable feats of navigation, however, are by insects – animal species that would not stand very high on anybody's intellectual hierarchy. Honey bees, for example, receiving 'dance' communications about a food source strategically evaluate this information according to their knowledge. If the communicated location is an impossible location (e.g., middle of a lake), then they do not fly. If it is adjacent to an island in the middle of the lake, then they will alter the 'flight plan' and fly to the island. At issue is, do honey bees, or any animals, have cognitive maps (i.e., a mental representation of spatial relationships for flexible use)? Navigation in most instances (ants, wasps, bees, and rats) can often be explained by landmarks joined together by movements and local cues resulting in path integration and vector sums – a none too shabby intellectual feat in itself.

Animals (apes, monkeys, and parrots) can learn about number relationships. Monkeys learn to match similar numbers of objects in pictures independent of object size or cumulative area of the objects. They generalize this relational learning to new objects, new sizes, and new numbers and, in some instances, outside the range of numbers used to train the accurate behavior. A gray parrot vocally responds (in English) to the query "how many" and can accurately identify the number of one object type (e.g., keys) mixed with other objects (e.g., rocks). Apes accurately point to Arabic numerals (4, 5, 6, etc.) corresponding to the number of objects even if the objects (e.g., oranges) are hidden in several places. Monkeys too have been shown to identify correct addition (e.g., lemons) by looking longer at incorrect additions. These latter looking-time procedures have the advantage that they require little training, but it is not always clear what cognitive process accounts for such looking-time differences.

Transitive inference has been considered unique to some of the most intelligent species (e.g., humans, apes, and dolphins). Animals discriminate between successive pairs of stimuli taken from a hierarchical sequence of arbitrary stimuli designated by the following letters: A, B, C, D, E. For each successive pair, the choice of the stimulus highest on the hierarchy is rewarded (+) (e.g., A+ vs. B−, B+ vs. C−, and C+ vs. D−). At issue is whether performance will be accurate on a test of B versus D following training. That is, have they learned the hierarchy and can infer that B should be chosen over D without ever been trained on this pair? Although extensive training and a single test example can make for sometimes tricky interpretations, the evidence does indicate that monkeys, rats, jays, and pigeons in addition to humans and nonhuman primates seem to have this ability. Such information adds to the growing list of animal cognitive skills but it hardly lends itself to ranking animals on anything remotely approaching intelligence, and it is not clear whether this task is particularly well suited to investigating the underlying processes.

Animals can learn stimulus relationships and classify/categorize stimuli bound by a category, prototype, or equivalence class. Even pigeons can learn categories (go/no-go tasks) such as person, water, trees, fish, oak leaves, alphabetical letters, and impressionist paintings and as many as four categories (e.g., person, flower, car, and chair) simultaneously. They can accurately transfer to novel category examples. They learn the go/no-go discriminations more rapidly and more accurately if the pictures are grouped by category than when the same pictures are arbitrarily assigned, but they can learn (i.e., memorize) as many as 830 arbitrarily assigned pictures. Pigeons also learn artificial categories (e.g., artificial 'seeds' and geometrical colored shapes and backgrounds). Artificial categories are more amenable to feature analysis (i.e., controlling mechanisms/processes), but these results have often been at odds with feature theory (from human categorization).

Abstract Concept Learning

Abstract concepts are rules about stimulus relationships (e.g., identity). Animals are tested in the laboratory for abstract concept learning in matching-to-sample (MTS) tasks (e.g., sample plus two comparison/choice stimuli) and same–different (S/D) tasks (pairs of stimuli). Rhesus monkeys and apes tested with either junk objects or slide-picture stimuli have for decades shown complete/full concept learning (equivalent training and transfer performance) when tested with novel stimuli. Chimpanzees allowed to play with six different tin cups and six different metal locks showed immediately excellent performance with other stimuli in an MTS task, interpreted as spontaneous abstract concept learning. Although there is little doubt about primates' (apes, baboons, rhesus monkeys, and capuchin monkeys) abstract concept learning ability, other species have been considerably more problematical. Theories of animal cognitive ability and intelligence have been based on which species can (e.g., apes), cannot (e.g., avians), or partially (e.g., dolphins have this ability with auditory stimuli and monkeys with visual stimuli, but not vice versa) learn abstract concepts. Recently, however, this has changed. Dolphins and monkeys have been shown to learn abstract concepts with both visual and auditory stimuli. Even more damning to the hierarchical theories of abstract concept learning is that parrots, crows, rooks, nutcrackers, jays, and pigeons have all showed good (sometimes full) abstract concept learning. Indeed, the amazing feats of intelligent behavior by avians discussed here and elsewhere in this article are even more significant because of the different neural architecture for avians. For example, hemispheric lateralization of function in avians is at least as strong as it is in humans, yet avians have neither a neocortex nor a corpus callosum (and apparently little interhemispheric communication) and these two structures have been in the past thought to be highly correlated with intelligence. Since intelligence is thought to have evolved just like other bodily traits, if intelligence were related to brain size or certain neural structures, then the size of these brain structures should not be modifiable by experience. However, there is abundant evidence (mostly from rats) that size of the hippocampus, corpus callosum, and overall brain size can be altered by experience.

Regarding abstract concept learning, the pigeon is by far the species of greatest focus as well as the most controversial species. In MTS tasks, pigeons have shown full concept learning with cartoon stimuli; they can learn MTS in any of three different (strategic) ways: configural learning, if–then rule learning, or relational learning. Only relational learning leads to abstract concept learning, whereas the other two strategies produce stimulus-specific learning. Moreover, rats learning MTS with different odors (e.g., cinnamon) apparently learn the task relationally (the strategic process necessary for abstract concept learning), exhibit some characteristics of declarative/episodic memory, and require an intact hippocampus to do so.

Pigeons have shown evidence of S/D concept learning with computer-icon arrays, colored geometrical-shape arrays, and travel-slide pictures. In the first case, the degree of concept learning varied with array entropy, but when the array elements were reduced, concept learning diminished and all but disappeared with the minimum of two elements. Work in my laboratory has shown that pigeons tested with travel-slide stimuli are capable of learning the S/D abstract concept with pairs of pictures. No concept learning was found with an 8-item training set. However, successive doublings of the training set improved concept learning to the point where concept learning (i.e., novel stimulus transfer) became equivalent to baseline performance with 256 or more training pictures (**Figure 4**). Similar experiments with capuchin and rhesus monkeys showed more rapid concept learning, and they achieved full concept learning with 128 pictures. These three species showed qualitative similarity in their ability to achieve full S/D abstract concept learning, but they also showed quantitative differences in their rate of concept learning. Pigeons required more exemplars to learn the rule (i.e., larger training set) than monkeys. Further analyses from successive cycles of training set expansion, learning, and transfer testing ruled out generalization (from the training stimulus pairs to the novel transfer pairs) as the mechanism responsible for transfer and concept learning, adding further evidence that these species have learned the higher order concept of S/D.

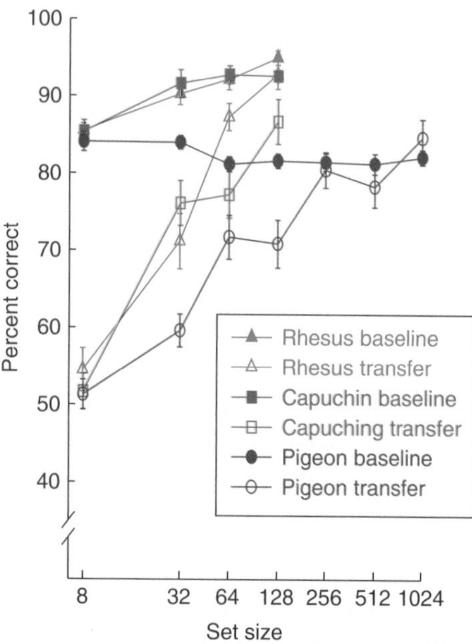

Figure 4 Mean percentage correct performance and standard errors of the mean for baseline performance (solid symbols) and transfer performance (open symbols) at each training set size for rhesus monkeys (triangles), capuchin monkeys (squares), and pigeons (circles). On each trial, a pair of vertically aligned pictures was presented simultaneously along with a white rectangle to the right of the lower picture. If the pair was the same picture, then a response to the lower picture was correct and rewarded. If the pair was two different pictures, then a response to the white rectangle was correct and rewarded. Incorrect choices were unrewarded. All subjects were trained with the same set of eight pictures (100 trials daily) until they were accurate at better than 80% correct and then tested with novel picture pairs (10 novel trials plus 90 training trials) for six sessions. The training set size was then doubled and transfer was tested with novel stimulus pairs after 85% correct performance. This sequence of training-set doubling, retraining, and retesting was repeated several times until transfer was equivalent to baseline (training) performance for each species.

'Higher order' is the watchword for species hierarchies of learning and concept learning in particular. Despite little agreement on most issues, most would probably agree that relations among relations (analogical reasoning) is the highest (most abstract, most difficult) form of abstract concept learning. Supporting the claim of exclusivity would be early evidence that the only animal species capable of this relational feat are chimpanzees because they alone have the ability to recode (tokens and artificial language) abstract relations. However, like previous claims of exclusivity of abstract concept learning, here too failures of other species in analogical reasoning may be the result of training and task failures not cognitive capacity failures. Indeed, baboons have shown similar results using icon arrays, and rhesus monkeys were able to accurately match musical tunes transposed one or two octaves with the first relation being the tunes (a relation among notes) and the second relation being that between the original and transposed tune.

Memory

There is no doubt that many animals have good memory, even by human standards. In addition to the amazing memory ability of nutcrackers to retrieve cached seeds in the natural alpine setting, laboratory studies of monkey memory have shown >70% correct accuracy for sample pictures in a delayed MTS task for 24 h. This good monkey memory occurred with trial-unique pictures so that the animals would not be confused (i.e., proactive interference) from seeing these pictures on previous trials. Although demonstrations of excellent animal memory capture our imagination, research on animal memory has shifted to studies of the memory processes (e.g., rehearsal, decay, consolidation, and interference), different types of memory (episodic, recollection, and familiarity), and brain mechanisms responsible for these different aspects and/or types of memory.

Related to how monkeys in the 24-h memory test might be confused by repeated pictures is what/how animals remember. Do they remember the actual presentation (episode) of the item at the time of test (i.e., episodic memory), or are they just capable of (a vague sense of) familiarity? Familiarity is considered to be a more primitive memory process (hierarchically simpler), associated with less intelligent species, than is episodic memory. Episodic memory and familiarity may be different kinds of memory possibly mediated by different cognitive modules involving different specialized brain areas and neural circuits. Most research on different memory brain areas has been conducted with monkeys and rats and has focused on the medial temporal lobe structures (hippocampus and perirhinal and parahippocampal cortices) for processing of information, the amygdala for emotional significance, and the prefrontal cortex for awareness and planning based on one's own memory (metacognition) and how the memory information will be used. Despite the vast amount of work, only tentative glimpses of a coordinated picture are beginning to emerge. Perhaps the perirhinal cortex might be most critical for object memory, whereas the parahippocampal cortex might be most critical for spatial memory. The hippocampus and prefrontal cortex may be critical for episodic memory through their role in integrating and abstracting information from many other brain areas, including the perirhinal and parahippocampal cortices, thereby associating context information (e.g., 'when and where' information) with the primary memory events ('what' information) and producing evidence of recollection/ episodic memory as opposed to item familiarity.

Although human participants can report "I remember" versus "I just know" to indicate episodic versus familiarity memory, respectively, more objective and clever procedures are required for testing animals for these different types of memory. Although it would seem logical for animal research on episodic/familiarity memory to have been done with (nonhuman) primates due to their relative evolutionary proximity to humans, the best and most clever work is actually being done with avians. An additional advantage of working with avians on this issue is that success will have more profound implications because most hierarchical theorists would have apes and/or monkeys being capable of episodic memory – not the proverbial bird brain. Scrub jays have been shown to cache and recover perishable wax worms and nonperishable peanuts in distinctive halves of sand-filled ice cube trays (**Figure 5**). After 4-h delays, the jays first recovered the more desirable wax worms, but after 124-h delays they first recovered peanuts because they had learned that wax worms would deteriorate in 124 h. Thus, they remembered what (peanuts vs. wax worms), when (4 vs. 124 h), and where (which tray side) the foods were stored – requirements of episodic memory. Many

Figure 5 A scrub jay retrieving previously cached wax worms, a food preferred over peanuts at short cache-retrieval times before they deteriorate. Courtesy of NS Clayton.

supporting experiments have replicated these remarkable findings and further explored the processes responsible for this episodic-like memory. Nevertheless, some experts are skeptical that these amazing memory feats by scrub jays may be merely the result of stimulus–response associative learning. However, vast numbers of associative learning studies have shown that associative learning is much more than a collection of stimulus–response reflexes or habits. Rats and other animals in classical conditioning experiments anticipate the future and can even add and subtract anticipated time through complex combinations of forward and backward primary and secondary conditioning. They learn expectations of certain outcomes (what), at certain times (when), and in certain places (where). In these situations, expectations develop over multiple trials (i.e., conditioning) and become reference or generic memory. By contrast, the jays cache a particular item, in a particular place, at a particular time – hence a single episode. Part of the single-episode success may be the act of caching. Caching would be expected to enhance memory for a particular episode, much like the generation effect in human memory. However, the same can be said for taste-aversion memory, which can be a single associative event, a particular stimulus (food taste), and a particular time (as much as several hours or more) before getting sick. Moreover, it has been shown that taste aversion is cognitively flexible (not an automatic stimulus–response association) by manipulations of which of several foodstuffs become associated with sickness. Few would question whether taste aversion is episodic memory because of numerous reports of humans, sometimes decades later, recalling exactly when and under what conditions some previously liked food was inadvertently associated with illness. These so-called 'autonoetic' (self-knowing) reports are considered by some to be the hallmark of episodic memory which, by design or theory, would seem to secure humans as the only animal that definitively possesses episodic memory.

Nevertheless, a delayed-MTS experiment showed that rhesus monkeys, at least, can be consciously aware (metacognition) of their own memories. During retention delays monkeys opted for a memory test and a chance for highly desired reward (peanuts) or else a less desired but certain pellet. Surprise tests showed that they were more accurate when they opted for the test, with the interpretation that they knew when their own memory was good. The key to this interpretation was having the monkeys monitor their memory before the test was presented and thereby avoid confounds of recognition familiarity.

The issues of whether animals have episodic memory, metacognition, or autonoetic experiences aside, animals most certainly have working memory. Indeed, how could any animal learn anything without working memory? Working memory is what is remembered on a particular trial; reference memory is the rules of the task. Good memory observed in natural settings has been tested in laboratory analogues. Nutcrackers have been tested in large arenas where they made as many as 25 caches and accurately recovered these cached seeds 6 months later from 180 possible sites (sand-filled cups). Other caching birds (e.g., tits and chickadees) have been tested in arenas filled with artificial tree limbs where the birds cached seeds in several cloth-covered tree limb holes and later remembered these cache sites from as many as 97 available sites. Many caching birds (scrub jays and nutcrackers) will re-cache if they are observed by another bird or alter their retrieval order (e.g., ravens) to beat the would-be pilferer to the cache site. In at least one case (scrub jays), they re-cache only if they themselves have a history of pilfering from others. Thus, they impute to the observing bird (typically a conspecific) their own pilfering intentions – a prototypical example of so-called 'theory of mind.'

Laboratory tests (e.g., radial arm maze analogues and MTS video monitor images) of single-item memory of some of these caching birds have not always found such remarkably good memory compared to other related species that do not cache. As previously noted, perhaps the act of caching is critical to the good memory. Moreover, the size of the hippocampus does not seem to be highly correlated with the degree of caching or performance on laboratory memory tasks. Current work seems to be more focused on how memory works for these bird species (i.e., processes/mechanisms of landmark navigation) rather than which one has the best memory.

Memory is seldom limited to single items. Certainly in the real world there are collections or streams of things to remember. The study of (human) memory began with memory lists. The hallmark of list memory studies is serial position function (SPF) and its primacy and recency effects (good memory for beginning and ending items, respectively), often revealing a U-shaped SPF. Although animals are notoriously difficult to train in list memory studies, U-shaped SPFs with visual stimuli have been shown for apes, rhesus monkeys, squirrel monkeys, capuchin monkeys, rats, and pigeons. Research on some of the mechanisms/processes responsible for the SPFs has shown that with short four-item lists there is a strong recency effect (last items) at very short retention intervals and little or no primacy effect. As the retention interval is lengthened, the primacy effect develops and the recency effect wanes. Research in my laboratory has shown that these same changes occur for rhesus monkeys, capuchin monkeys, pigeons, and humans (**Figure 6**). The qualitative similarity in these SPFs is complemented by a quantitative difference where the time course of changes takes place in approximately 10 s for pigeons, 30 s for monkeys, and 100 s for humans. The developing primacy effect is counterintuitive because it is an (absolute) increase in memory performance with increasing retention delay. SPFs for auditory stimuli have been shown for dolphins and rhesus monkeys. Rhesus monkeys were tested with short natural/environmental sounds at the same retention delays and list lengths as other rhesus monkeys with visual memory. The auditory SPFs were opposite to the visual SPFs, showing initial strong primacy effects and no recency effects. The recency effects grew with retention delay and the primacy effects waned. Other tests revealed that interference among the list items produced retrieval inhibition and that this inhibition dynamically changed with retention interval. These animal SPFs show some of the mechanisms/processes involved in visual and auditory list memory of these species. They also show that memory is not a unitary system, memory does not decay with time, and rehearsal does not produce the primacy effect – all long-standing tenets of the so-called modal model of memory.

The two examples of work from my laboratory have shown qualitative similarity in abstract concept learning and visual memory processing across such diverse species as humans, New and Old World monkeys, and pigeons. Single tests of these cognitive abilities would likely have shown differences that might have been taken as evidence for a species' differences in

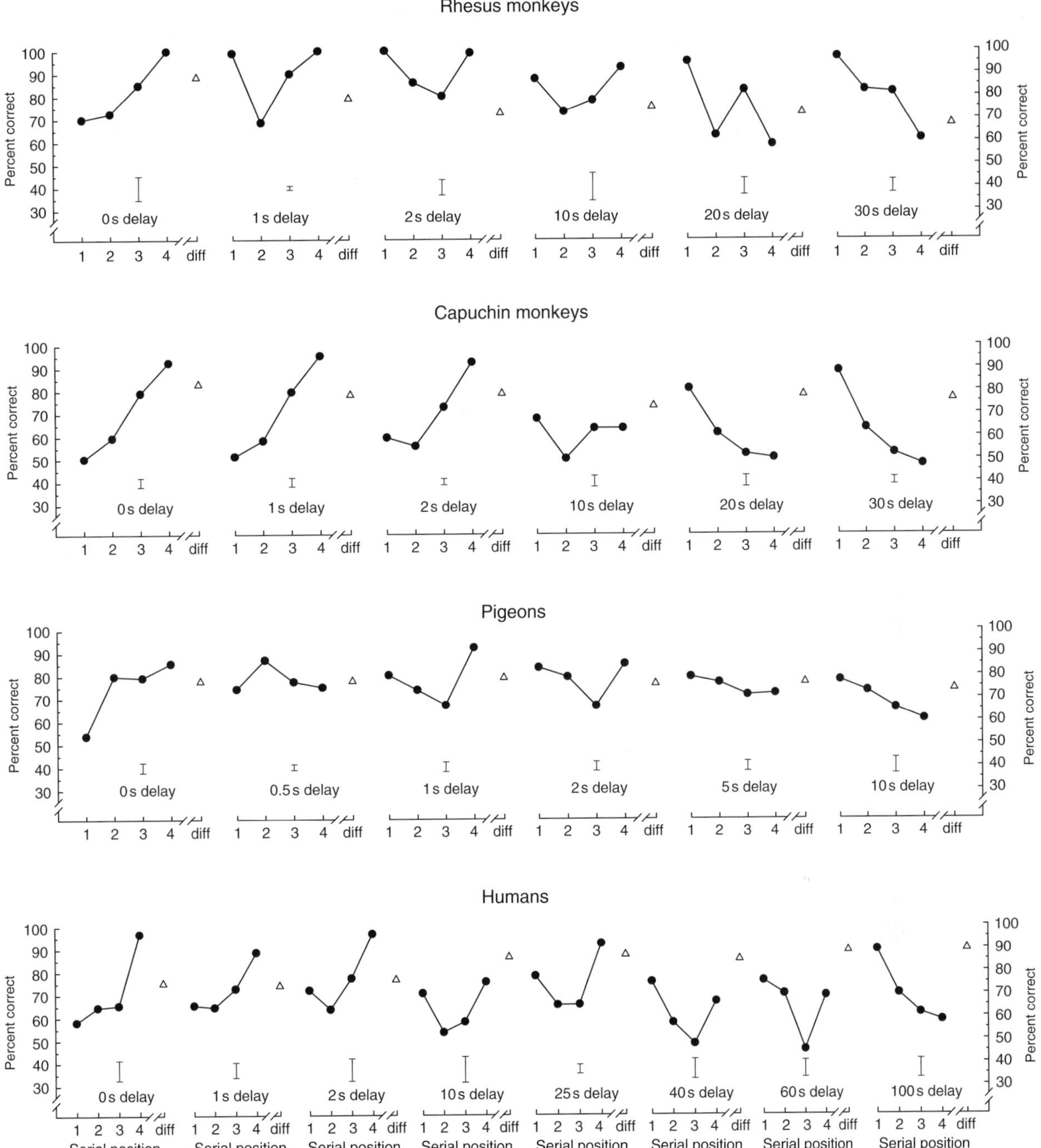

Figure 6 Serial-position memory functions for capuchin monkeys, rhesus monkeys, pigeons, and humans. On each trial, four pictures were briefly presented followed by a retention delay which is indicated in the lower portion of each panel. Subjects responded to indicate their memory of whether a single test picture was or was not in the list. The serial position functions show the results for trials in which the test matched one of the list pictures. (Serial position 4 was the last list item presented.) On half the trials, the test matched no list picture and that performance is shown as 'diff' (different). Test items for monkeys and pigeons were travel slides selected from more than 3000 items; test items for humans were kaleidoscope pictures selected from more than 550 items. The items for all subjects were unique on each daily test session. The error bars are the average standard errors for the four serial positions.

cognitive capability, evolved cognitive modules, and perhaps intelligence. There are differences among species to be sure, but in these cases the differences were quantitative differences. Some of the species (e.g., pigeons) required more exemplars of the rule to learn the abstract concept than monkeys, and in terms of memory, some species had differences in the time course changes of their serial position functions.

See also: Episodic Memory: Assessment in Animals; Executive Function and Higher-Order Cognition: Assessment in Animals; Numerical Intelligence: Neural Substrates; Procedural Learning in Animals; Reasoning and Problem Solving: Models; Referentiality and Concepts in Animal Cognition; Spatial Memory: Assessment in Animals.

Further Reading

Clayton NS, Bussey TJ, and Dickinson A (2003) Can animals recall the past and plan for the future? *Nature Reviews Neuroscience* 4: 685–691.

Emery NJ (2006) Cognitive ornithology: The evolution of avian intelligence. *Philosophical Transactions of the Royal Society B* 361: 23–43.

Olson DJ, Kamil AC, Balda RP, and Nims PJ (1995) Performance of four seed-caching corvid species in operant tests of nonspatial and spatial memory. *Journal of Comparative Psychology* 109: 173–181.

Rogers LJ and Kaplan G (eds.) (2004) *Comparative Vertebrate Cognition.* New York: Kluwer Academic/Plenum.

Shettleworth SJ (1998) *Cognition, Evolution, and Behavior.* New York: Oxford University Press.

Wasserman EA and Zentall TR (eds.) (2006) *Comparative Cognition.* New York: Oxford University Press.

Wright AA (1997) Concept learning and learning strategies. *Psychological Science* 8: 119–123.

Wright AA (1998) Auditory and visual serial position functions obey different laws. *Psychonomic Bulletin and Review* 5: 564–584.

Wright AA and Katz JS (2006) Mechanisms of *same/different* concept learning in primates and avians. *Behavioural Processes* 72: 234–254.

Animal Models of Alzheimer's Disease

J Koenigsknecht-Talboo and D M Holtzman,
Washington University, St. Louis, MO, USA

Introduction

Alzheimer's disease (AD) is the most common cause of dementia. The pathological hallmarks of this neurodegenerative disease include extracellular structures called diffuse and neuritic plaques as well as amyloid angiopathy. In addition, there are intracellular structures known as neurofibrillary tangles and neuropil threads. The plaques are comprised primarily of aggregated, nonfibrillar, and fibrillar forms of amyloid-β (Aβ) peptide, a 38–43-amino-acid peptide that is a normal cleavage product of the amyloid precursor protein (APP). The neurofibrillary tangles and neuropil threads are composed of hyperphosphorylated, aggregated forms of the microtubule-associated protein tau. In addition to neuropil threads, there is also neuritic dystrophy consisting of large axonal and dendritic varicosities, surrounding fibrillar amyloid deposits in the brain parenchyma. Neuronal cell death and synaptic degeneration also occur in AD. Furthermore, gliosis invariably occurs whereby microglial cells and astrocytes accumulate and cluster around plaques of deposited Aβ. Genetic mutations in the APP and presenilin (PS) genes have been identified and are associated with rare forms of early onset, autosomal-dominant familial AD as well as cerebral amyloid angiopathy (CAA). The normal processing of APP or the sequence of Aβ is altered in the presence of these mutations, thus leading to abnormal production or an increased aggregation propensity of Aβ and its subsequent accumulation. While mutations in tau do not result in AD, they do cause certain forms of fronto-temporal dementia and tangle accumulation. These mutations appear to alter the likelihood that tau will aggregate leading to a tauopathy. Advances in research and further understanding of AD have increased in part due to the generation of animal models that mimic certain aspects of AD, CAA, and fronto-temporal dementia that take place in humans. Introduction of mutations in APP, PS, and tau using transgenic, knockout, and knockin techniques in mouse models has resulted in the production of animals that develop aspects of AD-like pathology and other changes (**Figure 1**). These transgenic animals directly test whether expression of mutant genes leads to AD-like pathology and behavioral phenotypes, provide insight into molecular mechanisms, and are useful for developing better diagnostic methods as well as testing the efficacy of new treatments.

The presence of a few important criteria is vital to model aspects of AD in animal models. An AD animal model should exhibit at least one if not more than one pathological hallmark of AD. Another criterion is observation of cognitive and behavioral deficits. In models that contain mutations that cause familial AD or that result in expression of genes that lead to AD-like changes, changes in phenotype should be more severe in genetically altered than in age-matched wild-type animals and occur in an age-dependent fashion. The findings seen in a good animal model should be reproducible and be confirmed by several different labs. The generation of a comprehensive animal model of AD has proved to be a challenge for various reasons. AD is a complex disease that thus far involves at least three proven causative genes and one gene that is a known risk factor. In addition, there are also other genes that are required for various aspects of AD pathology. Another obstacle for constructing an AD model is mimicking a human disease that normally occurs in the seventh decade of life or later in mice, which live only 2–3 years. It is also extremely difficult to reproduce all the complexities of human behaviors in rodents. Various model organisms have been used to try and replicate aspects of human AD. Each organism has its strengths and weaknesses in regard to this endeavor. Although the anatomy of a mouse brain is similar to other mammals including humans in many regions, the brain of a rodent is also very different than a primate brain. Mice can be induced to develop plaques and tangles in region-specific areas; however, mice do not normally develop plaques and tangles. Aged rhesus monkeys have been shown to contain similar brain pathology as AD including the presence of aggregated amyloid, reactive glia, and enlarged distal axons. However, monkeys are expensive to maintain and experiments can take a long time to complete. Both worms and flies provide the benefits of easy and fast breeding in addition to the ability to employ powerful genetic tools. Yet, the brain anatomy of worms and flies is vastly different than in humans and AD-like behavior is difficult to address in these organisms. Therefore, much of the advancement in AD research in animal models has come from studies utilizing genetically manipulated mice. These models are reviewed here.

Genetically Modified Animal Models

APP Mouse Models and Aβ-Related Pathology

While multiple different APP transgenic mice have been produced, several examples are described below

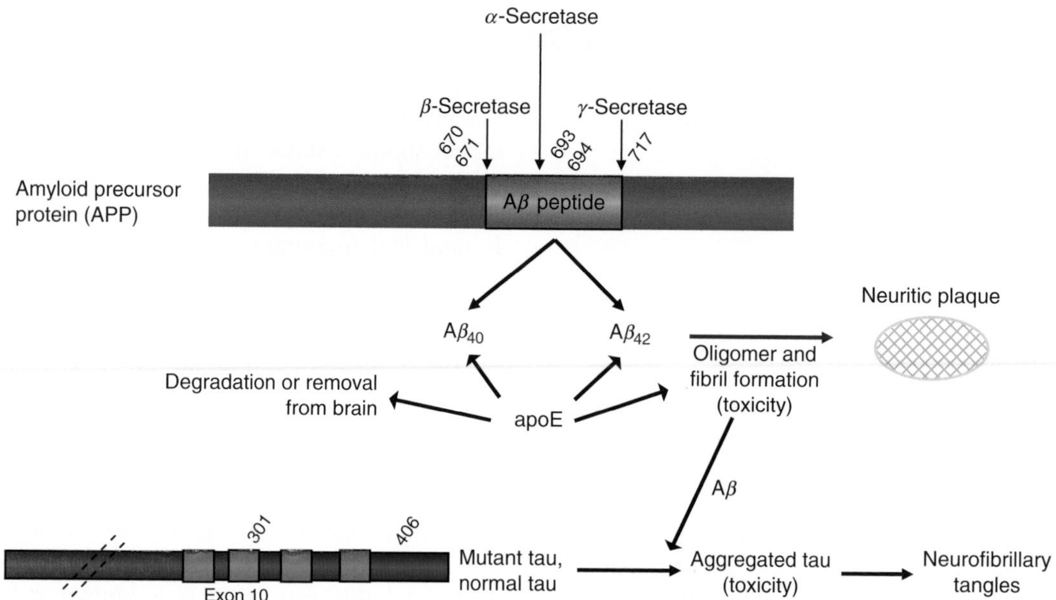

Figure 1 Model for Aβ and tau leading to cell toxicity. Aβ is generated by cleavage of APP by β-secretase and γ-secretase. Oligomers and fibrils of Aβ form and likely lead to neuronal and synaptic damage. Aβ also is hypothesized to induce tau aggregation and toxicity. The numbered sites in APP and tau correspond to amino acids where mutations are present that cause familial autosomal dominant AD or fronto-temporal dementia, respectively, that have been introduced into the mouse models described herein.

to illustrate some key features of the different types of models. One of the first APP transgenic mouse models to develop AD-like pathology described was the PDAPP mouse model of AD. The PDAPP mouse was generated by employing a platelet-derived growth factor (PDGF)-β promoter to induce overexpression of a transgene construct of human APP that contains the V717F mutation predominantly in neurons in the central nervous system (CNS). This point mutation in APP results in a mutation first found in families in Indiana which causes a form of autosomal dominant familial AD. Human APP levels were reported to be 10 times higher than endogenous mouse APP levels in this animal model. PDAPP mice develop age-dependent Aβ accumulation and deposition in diffuse as well as fibrillar neuritic plaques in the hippocampus, cerebral cortex and corpus callosum, as well as to a lesser extent in cerebral arterioles. This pattern of deposition corresponds with the pattern of Aβ plaques revealed in the brains of AD patients. Aβ deposition generally develops between 6 and 9 months and continues to accumulate with age. Different forms of Aβ deposition occur in this animal model. Diffuse plaques with nonfibrillar as well as compacted plaques containing fibrillar Aβ are both observed. In this model, only about 10% of the plaques are fibrillar, as defined with dyes such as Thioflavin-S or Congo red. In addition to Aβ deposition, these animals also exhibit robust gliosis in which plaques are associated with reactive astrocytes and microglial cells. Neuritic dystrophy has been reported in the PDAPP mice; however, studies have not reported a loss of neurons in this model. A decrease in synaptic and dendritic density in the hippocampus has also been observed in this model. It is not clear whether this is a developmental or neurodegenerative phenotype in this model. While behavioral studies have revealed some cognitive deficits in aspects of learning and memory in these mice prior to AD pathology, there are also age-dependent and pathology-dependent behavior deficits in learning and memory, making this likely to be a useful model for certain aspects of cognitive deficits in AD as well as a model of the Aβ-related pathological changes.

The Tg2576 transgenic mouse model was also one of the first APP transgenic mouse models developed. Also known as APPsw, the Tg2576 mouse model was generated by overexpression of the human APP695 construct that contains a double Swedish mutation. This mutation was identified in a large Swedish family that presented with early onset AD. The Swedish mutations (K670N/M671L) are driven by the hamster prion protein in this model. This mouse model expresses 5–6 times the amount of human APP as endogenous mouse APP. Senile plaques with dense cores of fibrillar amyloid in addition to diffuse amyloid deposits have been reported in the Tg2576 animal model. Deposition of amyloid generally occurs between 9 and 12 months of age in Tg2576 mice. Clusters of astrocytes and microglia surround many of the amyloid deposits. These cells induce an inflammatory response surrounding the areas of amyloid plaques. Despite the presence of amyloid deposits, neurofibrillary tangles are not present in any of the APP transgenic models including the Tg2576 mouse model. Neuritic dystrophy is present in this and other APP transgenic mice but there is no marked neuronal loss. Neuronal loss in the immediate vicinity of fibrillar amyloid has been reported. CAA has been shown to be more prevalent in the Tg2576 model than in PDAPP mice. Notable CAA has also been observed in the APP23 transgenic mouse model along with micro- and sometimes macrohemorrhages. Cognitive decline has been demonstrated to increase with age in this mouse model. For example, spatial learning in a water maze was shown to be impaired in Tg2576 mice. Also, results of a spatial alternation task in this mouse model were consistent with hippocampal damage. These behavior deficits have been reported to occur just prior to the onset of plaque deposition, beginning at ~6 months of age. Learning and memory was normal at 3 months of age in the water maze. Deficits in synaptic plasticity such as long-term potentiation (LTP) have been reported in the hippocampus and the dentate gyrus. Since development of the PDAPP and Tg2576 mouse models, numerous other APP transgenic mouse models that develop Aβ deposition and associated changes have been developed.

The PSAPP mouse model incorporates a PS mutation and was generated by crossing an APP mutated mouse (Tg2576) with a PS1 mutated mouse (M146L). This mouse model exhibits accelerated plaque pathology as well as robust gliosis most likely due to enhanced Aβ42 generation or an altered 42/40 ratio due to the overexpression of a PS1 mutation responsible for familial AD. Abundant diffuse and fibrillar Aβ deposits form in the cortex and hippocampus of these double transgenic mice much earlier than their Tg2576 singly transgenic littermates. By 3 months of age, these mice contain fibrillar plaques within the cortex. Amyloid deposition accumulates and plaque size increases and extends to the hippocampus by 6–8 months of age. This double transgenic mouse model preferentially expresses Aβ42, and brains of these mice have been reported to have a 41% increase in Aβ42 as compared to singly transgenic Tg2576 mice. Neurofibrillary tangles have not been reported in the PSAPP mouse model. Despite the accelerated amyloid pathology in PSAPP mice, the clearest cognitive deficits appear at 15–17 months of age as reported by some. Other models combining expression of a PS mutation together with APP have been developed.

Mouse Models with Mutations in the Aβ Region

A few mouse models have been generated that contain mutations within the Aβ sequence of APP. For example, the APP Dutch mouse model was created by inducing a point mutation, E693Q, in human APP. This mutation, in humans, results in an autosomal dominant form of CAA. Expression of APP Dutch results in a higher level of Aβ40 accumulation than Aβ42. Similar to the human cases, amyloid is deposited predominantly in cerebrovascular plaques in APP Dutch animals in contrast to the parenchymal Aβ deposits that are characteristic of other APP transgenic animal models. The APP Dutch animal has been crossed with animals containing the G384A PS1 mutation, which is known to increase Aβ42 production. The expression of the G348A PS1 mutation in these mice results in a shift of amyloid deposition from the vessels into the brain parenchyma in the cortex and hippocampus. It is hypothesized that a high Aβ42/40 ratio favors parenchymal deposition, whereas a low Aβ42/40 ratio favors vascular Aβ deposition. Similar to the Dutch mutation, the Iowa mutation (D694N) also leads to Aβ accumulation in the brain vasculature. These two mutations have been reported to augment fibrillogenic properties of Aβ *in vitro* and may also result in abnormal clearance of Aβ from the brain leading to its accumulation.

The Arctic mutation, E693G, is another mutation that is within the Aβ sequence and causes an AD phenotype in humans. Transgenic mice that contain the Arctic mutation in addition to the Swedish mutation have been reported to develop intraneuronal Aβ aggregates prior to the deposition of extracellular Aβ. The intraneuronal accumulation of Aβ is not grossly fibrillar. The Arctic lines promote accumulation of shorter species of Aβ. This was demonstrated by a higher Aβ1-38/Aβ1-40 ratio and a lower Aβ1-42/Aβ1-40 ratio. The Arctic mutation is extremely amyloidgenic; one of the Arctic lines generated was reported to contain extracellular neuritic plaque deposition of Aβ as early as 2 months of age. A significant difference between the Arctic and the Dutch mutations is the location of Aβ deposition. The Dutch mutation results in amyloid deposition primarily in vessels; however, the Arctic mutation generally results in parenchymal Aβ plaques within the brain.

All of the APP models discussed result in production of different ratios and amounts of Aβ species, making it difficult to determine the specific effects of Aβ40 versus Aβ42 on pathology and behavior. In addition, in some models the overexpression of APP may be responsible for some of the changes in the animals. An important advancement in AD animal models was achieved by creating mice that specifically produce Aβ40 or Aβ42, known as the BRI-Aβ models. This model employs fusion constructs that contain 243 amino acids of BRI protein followed by a sequence that encodes either Aβ40 or Aβ42. Aβ is produced when the BRI protein is cleaved by furin in a vesicular compartment in cells. Aβ40 is selectively produced, expressed, and secreted by BRI-Aβ40 mice, whereas Aβ42 is selectively generated, expressed, and secreted by BRI-Aβ42 mice. Only BRI-Aβ42 mice develop plaques demonstrating the critical importance of Aβ42 in the seeding and propagation of Aβ deposition. These models will be useful in sorting out specific effects of Aβ40 or Aβ42 without the confound of APP overexpression.

ApoE Mouse Models

The only proven genetic risk factor for AD is apoE genotype. The ε4 allele of apoE has been demonstrated to be a risk factor for late onset AD and CAA, whereas the ε2 allele of apoE has been shown to be protective against AD. apoE avidly binds Aβ and co-localizes with neuritic plaques and CAA. Evidence suggests that one of the main reasons for the linkage between Aβ and apoE results from the ability of apoE to bind Aβ, acting as a chaperone to influence Aβ metabolism and structure.

In regard to a potential role for apoE in normal brain structure and function, apoE knockout mice have been shown in some studies but not others to contain less synaptophysin and staining with dendritic markers. Interestingly, when either PDAPP or Tg2576 mouse models are bred with apoE knockout mice, the resultant APP transgenic mice lacking apoE exhibit a striking decrease in Aβ deposition compared to animals that express apoE, thereby demonstrating the importance of apoE in amyloid deposition. In addition, the Aβ deposits in mice lacking apoE are not fibrillar until very old ages and contain little to no neuritic dystrophy, microglial inflammation, CAA, or CAA-associated hemorrhages. This suggests that apoE plays an important role in induction of fibrillar Aβ and its consequences.

To examine the effects of individual human apoE isoforms on the brain and on AD mouse models, mice have been generated that contain a specific apoE isoform knocked into the endogenous mouse *apoe* locus. These mice display apoE expression predominantly in glial cells, the normal pattern of endogenous apoE expression. The apoE2, apoE3, and apoE4 mice all express similar levels of apoE. Other mice have been generated that express specific apoE isoforms in astrocytes with a GFAP promoter (GFAP-apoE) and in neurons with a neuron-specific endolase promoter (NSE-apoE). Each of these mice were bred with mouse apoE knockout mice to eliminate the confound of endogenous mouse apoE. In the NSE-apoE4 mice, C-terminal truncated fragments of apoE accumulate in an age-dependent manner. Fragment accumulation occurs to a lesser extent in NSE-apoE3 mice and is not seen in the GFAP-apoE mice. Tau phosphorylation also accumulates in an age-dependent manner in NSE-apoE4 mice. As with apoE fragment accumulation, tau phosphorylation occurs to a lesser degree in NSE-apoE3 mice and is not observed in the GFAP-apoE mice. Co-cultures of hippocampal neurons and astrocytes derived from GFAP-apoE mice displayed greater neurite outgrowth in GFAP-apoE3 mice than in GFAP-apoE4 mice. Importantly, in regard to effects on Aβ, using human apoE isoform-specific transgenic or knockin mice bred to different APP transgenic mice, strong isoform-specific effects of apoE are seen. The isoform-specific effects on Aβ are such that earlier and more Aβ-related pathology occurs in the pattern E4 > E3 > E2. In addition to the isoform-specific pattern, all human apoE isoforms delay the development of Aβ deposition relative to the presence of murine apoE or no apoE. This suggests that human apoE plays a role in Aβ clearance or transport in addition to its role in fibrillogenesis. While the effects of apoE on Aβ are profound, experiments with animals that develop both Aβ deposition and tauopathy will be required to determine if apoE also influences tangle formation.

Tau Mouse Models

Several AD animal models have focused on APP/Aβ; however, other models have been developed that examine other aspects of AD pathology. The first tau transgenic model was generated prior to the identification of pathogenic tau mutations. This model expresses the longest human brain tau isoform under the control of the human Thy-1 promoter. Although this model mimicked some aspects of AD including hyperphosphorylation of tau, neurofibrillary tangles did not develop in these animals. This AD animal model may represent an early AD-like phenotype that occurs prior to tangle formation.

Pathogenic mutations were later identified in the tau protein that cause fronto-temporal dementia and parkinsonism linked to chromosome 17 (FTDP-17). One tau mouse model employs the most common FTDP-17 mutation, P301L, driven by the prion protein promoter. The JNPL3 mouse develops neurofibrillary tangles in addition to astrogliosis, most prominently in the brain stem and spinal cord. Although tangles do not develop in the cortex or hippocampus of JNPL3 mice, there is pretangle pathology in these brain regions. Contrary to the APP mouse models of AD, this mouse model does not develop amyloid plaques. The majority of these mice developed motor and behavioral deficiencies by 10 months of age. The behavioral deficits described in JNPL3 mice include delayed righting reflex, decreased locomotion, and muscular weakness. Another tau mouse model contained a P301L mutation; however, this model employed the Thy1.2 promoter. These mice expressed P301L tau in neurons in the cortex and hippocampus. Hyperphosphorylation of tau occurs in these mice and neurofibrillary tangles are present. In addition, this mouse model also develops numerous TUNEL-positive neurons in the cortex suggesting neuronal damage.

Another tau mutation, R406W, was utilized to make a tau transgenic mouse model. This transgenic mouse expressed the R406W mutation driven by the α-calcium-calmodulin-dependent kinase-II promoter. Hyperphosphorylated tau inclusions are present in neurons in the forebrain of these animals by18 months of age. These tau inclusions were closely associated with disruptions in microtubules. Associative memory was reported to be impaired in this mouse model through contextual and cued fear conditioning tests. This tau mouse model was not reported to have obvious motor deficits, as has been the case with other tau mutations. A variety of mice with tau mutations causing fronto-temporal dementia have now been generated that have pathological features similar to those described above, although the location of the pathology varies depending on the promoter utilized as well as the mutation.

Htau mice were generated by crossing mice that express a human tau transgene and tau knockout mice. These mice express the normal sequence of human tau. The htau mice therefore express six isoforms of human tau, however, does not express mouse tau. These mice develop pathology despite the fact that they contain nonmutant tau. Hyperphosphorylated tau accumulates within the cell bodies and dendrites of neurons in the hippocampus. Tau aggregated and accumulated inside cells within the cortex and the hippocampus of this mouse model.

Recently, a novel tau model was described that expresses a mutant tau which can be suppressed with doxycycline. These mice overexpress the P301L tau mutation in the forebrain. The expression of mutant tau resulted in the accumulation of hyperphosphorylated tau and a loss of neurons that increased as the animals aged. Spatial memory was also impaired in association with age. Blocking tau expression of tau caused memory function to improve and neuron numbers to stabilize. Importantly, neurofibrillary tangles continued to accumulate, after the mutant tau was suppressed despite improved cognitive performance. Results from this mouse model therefore suggest that a soluble form of tau may be contributing strongly to toxicity.

Other Mouse Models

An advance in AD mouse models was made with the introduction of a novel triple transgenic model (3x Tg-AD) which progressively developed both amyloid plaques and neurofibrillary tangles in brain regions that are affected in AD. The 3x Tg-AD model contained a PS1 mutation (M146V), an APP mutation (APPswe), and a tau mutation (P301L). These mice were generated by microinjecting two transgenes into single cell embryos derived from homozygous M146V knockin mice. One benefit of this method is that the APP and tau transgenes are co-integrated and transmit together to the offspring. Plaque formation was shown to develop in these animals prior to the production of tangles. This pathology is consistent with the amyloid cascade hypothesis. The neocortex of 3x Tg-AD mice has been reported to contain intracellular Aβ immunoreactivity as early as 3–4 months of age which is followed by Aβ aggregation in the hippocampus at 6 months of age. In a similar region-dependent manner, extracellular Aβ deposits formed in the cortex prior to the hippocampus as early as 6 months of age and were easily detectable at the age of 12 months. In contrast to plaque formation, tau pathology of tangle formation appears first in the hippocampus and later is observed in the cortical area of the brains of 3x Tg-AD mice. Synaptic dysfunction was also observed in an age-dependent manner in this mouse model. Interestingly, removal of Aβ by immunotherapy in these mice reverses early changes in tau such as hyperphosphorylation but not later changes such as tangle formation.

Another mouse model that results in both tau and amyloid pathology has been generated by crossing the Tg2576 mouse with the P301L tau mouse model. The resulting mutant tau and APP mice, TAPP mice, develop both amyloid plaques and neurofibrillary tangles. The plaques generated in the TAPP mice were similar in number and distribution as was reported in Tg2576 mice. Aβ plaques were detected as early as 6 months of age and increased as the animals aged. Neurofibrillary tangles were seen in the TAPP mice as early as 3 months of age; however, they were more numerous as the animals aged. There was an acceleration of tau pathology in the TAPP mice as compared to singly transgenic mice only expressing tau with the P301L mutation suggesting that Aβ may somehow exacerbate tangle pathology. Although plaques and tangles were shown to be located in the same brain regions, the neurofibrillary tangles did not form in the immediate environment of the amyloid plaques. Tangle pathology also occurs in brain areas not affected by Aβ deposition in TAPP mice. Dystrophic neurites were observed around the amyloid deposits. Similar to the P301L mutated mice, the TAPP mice demonstrated some disturbances in motor skills. It was also observed in TAPP mice that neurofibrillary tangle pathology developed earlier and more robustly in female mice than in male mice.

Conclusions

Various animal models have been developed which recapitulate one or more aspects of AD. While many of these models are useful, none of the models encompass all of the pathology and cognitive deficits observed in human AD. A key feature of AD that is not readily demonstrated in the mouse models is the presence of plaques and tangles combined with progressive and severe neuronal loss that is seen in certain brain regions in human AD. More models with AD-like pathology can also be bred with mice in which other genes are altered to assess their effects on aspects of AD-like pathology or behavior. Alternatively, expression of genes with viral vectors can be another way to both assess effects of potential gene therapies as well as better model aspects of AD. Important information has been gained from crossing AD models with apoE models and with mice lacking or overexpressing β-secretase, α-secretase, ABCA1, neprilysin, and insulin degrading enzyme. There are many other proteins that potentially play a role in AD that can be assessed in a similar manner. Although there is not yet a perfect AD animal model, valuable information on AD pathogenesis and potential therapies such as secretase inhibitors and immunotherapy have been gleaned from the current available models. It is likely that future improvements in current models as well as a better understanding of AD pathogenesis will continue to be made as the full complement of genes involved in AD is unraveled. Rodent models will likely play a major role in this endeavor.

See also: Acetylcholinesterase Inhibitors and Alzheimer's Disease; Alzheimer's Disease: Neurodegeneration; Alzheimer's Disease: An Overview; Alzheimer's Disease: Transgenic Mouse Models; Axonal Transport and Alzheimer's Disease; Dementia.

Further Reading

Allen B, Ingram E, Takao M, et al. (2002) Abundant tau filaments and nonapoptotic neurodegeneration in transgenic mice expressing human P301S tau protein. *The Journal of Neuroscience* 22(21): 9340–9351.

Bales KR, Verina T, Dodel RC, et al. (1997) Lack of apolipoprotein E dramatically reduces amyloid beta-peptide deposition. *Nature Genetics* 17(3): 263–264.

Brendza RP, O'Brien C, Simmons K, et al. (2003) PDAPP; YFP double transgenic mice: A tool to study amyloid-beta associated changes in axonal, dendritic, and synaptic structures. *The Journal of Comparative Neurology* 456(4): 375–383.

Cheng IH, Palop JJ, Esposito LA, et al. (2004) Aggressive amyloidosis in mice expressing human amyloid peptides with the Arctic mutation. *Nature Medicine* 10(11): 1190–1192.

Davis J, Xu F, Deanes R, et al. (2004) Early-onset and robust cerebral microvascular accumulation of amyloid beta-protein in transgenic mice expressing low levels of a vasculotropic Dutch/Iowa mutant form of amyloid beta-protein precursor. *The Journal of Biological Chemistry* 279(19): 20296–20306.

Fagan AM, Murphy BA, Patel SN, et al. (1998) Evidence for normal aging of the septo-hippocampal cholinergic system in apoE (–/–) mice but impaired clearance of axonal degeneration products following injury. *Experimental Neurology* 151(2): 314–325.

Fryer JD, Taylor JW, DeMattos RB, et al. (2003) Apolipoprotein E markedly facilitates age-dependent cerebral amyloid angiopathy and spontaneous hemorrhage in amyloid precursor protein transgenic mice. *The Journal of Neuroscience* 23(2): 7889–7896.

Games D, Adams D, Alessandrini R, et al. (1995) Alzheimer-type neuropathology in transgenic mice overexpressing V717F beta-amyloid precursor protein. *Nature* 373(6514): 523–527.

Gotz J, Chen F, Barmettler R, and Nitsch RM (2001) Tau filament formation in transgenic mice expressing P301L tau. *The Journal of Biological Chemistry* 276(1): 529–534.

Herzig MC, Winkler DT, Burgermeister P, et al. (2004) Abeta is targeted to the vasculature in a mouse model of hereditary cerebral hemorrhage with amyloidosis. *Nature Neuroscience* 7(9): 954–960.

Holcomb L, Gordon MN, Mcgowan E, et al. (1998) Accelerated Alzheimer-type phenotype in transgenic mice carrying both mutant amyloid precursor protein and presenilin 1 transgenes. *Nature Medicine* 4(1): 97–100.

Holtzman DM, Bales KR, Tenkova T, et al. (2000) Apolipoprotein E isoform-dependent amyloid deposition and neuritic degeneration in a mouse model of Alzheimer's disease. *Proceedings of the National Academy of Sciences of the United States of America* 97(6): 2892–2897.

Hsiao K, Chapman P, Nilsen S, et al. (1996) Correlative memory deficits, Abeta elevation, and amyloid plaques in transgenic mice. *Science* 274(5284): 99–102.

Lewis J, Dickson DW, Lin W-L, et al. (2001) Enhanced neurofibrillary degeneration in transgenic mice expressing mutant tau and APP. *Science* 293(5534): 1487–1491.

Lewis J, McGowan E, Rockwood J, et al. (2000) Neurofibrillary tangles, amyotrophy and progressive motor disturbance in mice expressing mutant (P301L) tau protein. *Nature Genetics* 25(4): 402–405.

Lord A, Kalimo H, Eckman C, Zhang X-Q, Lannfelt L, and Ng Nilsson L (2006) The Arctic Alzheimer mutation facilitates early intraneuronal Abeta aggregation and senile plaque formation in transgenic mice. *Neurobiology of Aging* 27(1): 67–77.

Oddo S, Caccamo A, Shepherd JD, et al. (2003) Triple-transgenic model of Alzheimer's disease with plaques and tangles: Intracellular Abeta and synaptic dysfunction. *Neuron* 39(3): 409–421.

Santacruz K, Lewis J, Spires T, et al. (2005) Tau suppression in a neurodegenerative mouse model improves memory function. *Science* 309(5733): 476–481.

Sturchler-Pierrat C, Abramowski D, Duke M, et al. (1997) Two amyloid precursor protein transgenic mouse models with Alzheimer disease-like pathology. *Proceedings of the National*

Academy of Sciences of the United States of America 94(24): 13287–13292.

Sun Y, Wu S, Bu G, et al. (1998) Glial fibrillary acidic protein-apolipoprotein E (apoE) transgenic mice: Astrocyte-specific expression and differing biological effects of astrocyte-secreted apoE3 and apoE4 lipoproteins. *The Journal of Neuroscience* 18(9): 3261–3272.

Tatebayashi Y, Miyasaka T, Chui D-H, et al. (2002) Tau filament formation and associative memory deficit in aged mice expressing mutant (R406W) human tau. *Proceedings of the National Academy of Sciences of the United States of America* 99(21): 13896–13901.

Relevant Website

http://www.alzforum.org – Alzheimer Research Forum.

Animal Models of Amnesia

M Moss, Boston University School of Medicine, Boston, MA, USA

Early Contributions to the Study of Memory: The Pioneers

Perhaps the first major contribution to the formal literature on human memory came from the French psychologist Theodule Ribot in his treatise on the *Diseases of Memory* that produced his 'law' which states that older memories are better remembered than more recent memories. But it is also difficult to introduce this subject without mention of the seminal contribution by William James in his *Principles of Psychology*. James made the case that the processing of memory had a neural substrate and that memory could be fractionated into primary (short-term) and secondary (long-term) memory, concepts we still struggle with today. He also introduced what we now consider the two major systems of memory processing: declarative memory and habit (procedural memory). Nor can any introduction to the subject of memory ignore the contribution of Hermann Ebbinghaus, who made the first major attempt at the quantification and measurement of memory, which culminated in his work *On Memory*. Sergei Korsakoff described a disorder of memory associated with the effects of alcohol and that now bears his name, and Vladimir Bekhterev was the first to point to the possible relationship of memory impairment with damage to the temporal lobe.

While these individuals were laying the foundation for the study of human memory, others began to apply this body of work to animal models. Most notable among these were EL Thorndike, a student of James, and Ivan Pavlov, with his work on classical conditioning. It was also at this time that Shepard Franz began using formal training paradigms in his work on the effects of brain lesions in animals. His student, Karl Lashley, extended this work to a level that many now consider to have constituted the seminal phase of the utility and value of animal models in neuropsychology.

Amnesia: The Paradox and Animal Models

Perhaps the most informative and seminal case study in the field of amnesia and one that initiated major activity using animal models of amnesia was that of patient H.M. In 1953, H.M. underwent surgery for treatment of intractable seizures. The procedure entailed bilateral resection of the medial temporal lobes, including the anterior two-thirds of the hippocampus, the amygdala, and hippocampal gyrus (**Figure 1**). Although the surgery was successful in ameliorating the frequency and intensity of his seizures, H.M. was left with a devastating impairment in his anterograde short-term memory and to a limited extent his retrograde memory, extending to at least several months preceding the procedure.

Following the surgery, H.M. was able to remember events over very brief intervals. However, it was the fate of this immediate memory that was imperiled. After a period as short as 30 s, or with brief distraction, recall of immediate events was, in virtually all cases, permanently erased. Despite this striking condition, it soon became evident through formal experimentation that H.M. retained some learning capacity. H.M. was able to perform certain motor skill and perceptual learning tasks and could retain much of this learning for as long as a year. It was this case that cemented the theoretical notions that the ability to remember is subserved by at least two memory systems. It was Cohen and Squire who dubbed the impaired and spared memory in H.M. (and other amnesiac patients) as declarative and procedural systems, respectively. Accordingly, declarative memory has been ascribed to medial temporal lobe structures and their anatomic connections to polymodal association areas of the cerebral mantle, whereas procedural memory has been ascribed to nonmedial temporal regions, with the neostriatum and cerebellum as the leading proposed neural substrates.

To date, although studies in animals have explored the neural basis of nondeclarative memory, most have focused primarily on localizing and fractionating the anatomic and functional neuronal substrate of declarative memory. Hence, the thrust of research has focused on the hippocampal formation and the adjacent parahippocampal cortices (including the entorhinal, perirhinal, and parahippocampal cortices; see **Figure 2**).

Rodent Models of Memory Function

In light of the anatomic and connectional similarities of the medial temporal lobe across species, both rodents and nonhuman primates (as well as canines and felines) have been utilized to assess various aspects of memory function. To a great extent, behavioral paradigms that are equivalent in design have been adapted to each of these species. One such behavioral task is delayed nonmatching to sample.

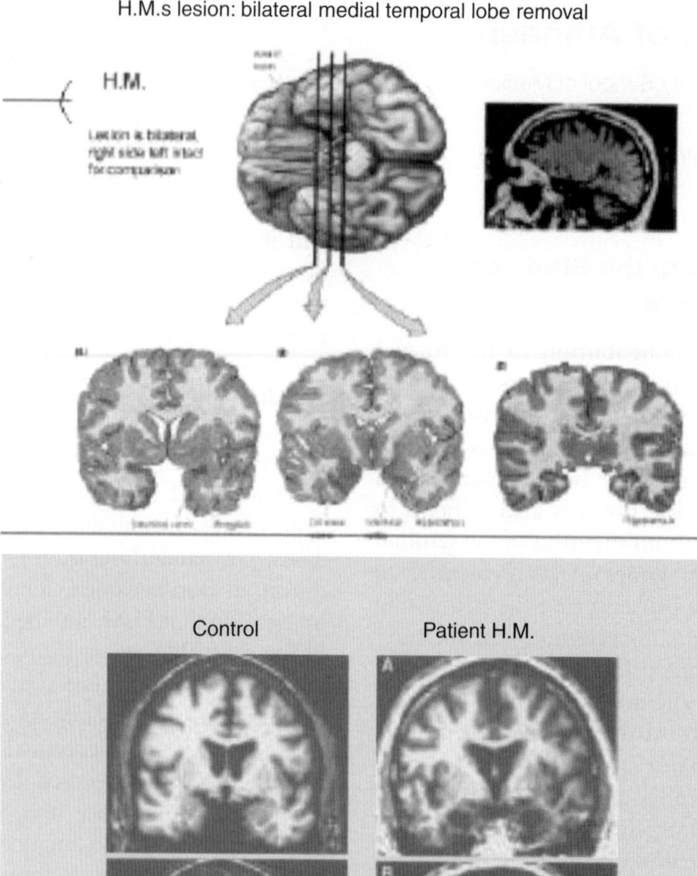

Figure 1 (Above) Representation of area of resection of medial temporal lobe areas in patient H.M. (Below) (A) Magnetic resonance image of coronal section of H.M.'s brain at rostral level showing resection of the amygdala (right) relative to normal control. (B) More-caudal level of H.M.'s brain showing removal of hippocampal formation (right) relative to normal control brain.

In this task, animals must respond to a sample stimulus and then, after a delay, must choose between a novel stimulus and the previously presented familiar one (**Figure 3**). Rats with damage to the hippocampus can learn the delayed nonmatching to sample task as efficiently as controls with delays of a few seconds but evidence a marked impairment when delays are extended to longer than a minute. These findings support the view that the hippocampus is essential in the ability to form associations between given stimuli and the context in which the stimuli appear.

Perhaps the most widely used method of assessing memory in rodents relies on the use of the Morris Water Maze. In the typical paradigm, the rat (or mouse) is placed into a pool of opaque water that contains an escape platform hidden a few millimeters below the water's surface. Visual cues, such as colored shapes, are placed around the pool in plain sight of the animal. When released, the rat swims around the pool in search of an exit. Several variables can be measured, including the time spent in each quadrant of the pool, the time taken to reach the platform (latency), and the total distance traveled. Rats with damage to the hippocampus are markedly impaired in this task. In contrast, rats with the same lesion can perform successfully on the task when the platform is visible above the water's surface. The data have accumulated to show that removal of the hippocampus in rats results in a marked impairment in tasks requiring spatial navigation and in those that specifically require the animal to remember recent locations it has visited. This effect appears more robust with

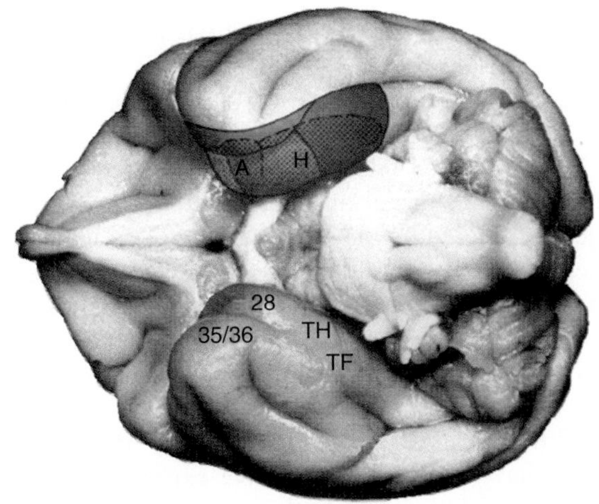

Figure 2 Ventral view of the rhesus monkey brain. Left hemisphere (top of photograph) depicts entorhinal cortex (purple), perirhinal cortex (blue), and parahippocampal cortices (green). Positions of subcortically positioned amygdala (A) and hippocampus (H) are depicted by dotted line. Corresponding Brodmann and Von Bonin and Bailey nomenclature is shown on the right hemisphere (bottom of photograph).

Figure 3 Photograph showing rhesus monkey in the Wisconsin General Testing Apparatus. The monkey is displacing the 'novel' object to obtain a reward (raisin) during the recognition portion of a trial on the delayed nonmatching to sample task.

damage to the hippocampus itself rather than to the surrounding rhinal cortices.

Taking advantage of rodents' exquisitely sensitive olfactory system, Eichenbaum and colleagues have used a novel paired associate odor task to demonstrate that the hippocampus is essential for memory function. In this task, rats first were rewarded for identifying an odor (A) that was paired with another given odor. Several such pairs were readily learned by normal rats as well as those with damage to the hippocampus. This basic stimulus–response learning task could easily be seen as supported by the nondeclarative memory system. On another phase of the task, rats had to infer an association between two odors that were not directly paired but that shared a common associate odor. Whereas control rats learned this task readily, those with hippocampal lesions evidenced marked impairment. Together with the results of related 'transitive' tasks, the data strongly support the view that the hippocampus is key to mediating the connection and expression of memory in a relational context. Eichenbaum has also shown that the hippocampus is essential in mediating sequential events, a working component for episodic memory. Rats with lesions of the hippocampus retain their capacity to recognize individual odors but are severely impaired in their ability to remember the sequential order of odors presented to them.

Nonhuman Primate Models of Memory Function: Medial Temporal Lobe

While evidence from human cases strongly implicated the hippocampal formation in subserving normal memory function, it was unclear whether specific components of this structure had selective roles in memory function. Indeed, the hippocampus represents the endpoint of the transition from a six-layer cortex of the adjacent parahippocampal gyrus to a three-layer cortex in the hippocampus itself. Two of these regions, the perirhinal and parahippocampal cortices, are the recipients of input from secondary sensory association areas, including visual, somatosensory, and auditory regions, that, in turn, project to the entorhinal cortex, which is the major cortical afferent to the hippocampus. The perirhinal, entorhinal, and parahippocampal regions were damaged in patient H.M. (**Figure 1**), and these regions had also been removed in the surgical approach to the hippocampal formation in the early stages of nonhuman primate studies. As a result, it was difficult to distinguish between deficits traditionally attributed to damage to the hippocampus proper and those that arose from damage to the rhinal and adjacent cortices.

Several studies over the past 10 years have been aimed at fractionating the specific contributions of the hippocampus and rhinal cortices. Findings regarding the effect of damage to the hippocampus in nonhuman primates have been equivocal, particularly regarding the most commonly used task in assessing memory, delayed nonmatching to sample (DNMS). Results from studies of the effects on DNMS performance of damage to the hippocampus of nonhuman primates range from virtually no effect to marked impairment, particularly when delays greater than 30 s are used. However, when certain experimental design variables are taken into account, the contribution of the hippocampus to this task is better revealed.

One striking methodological difference in these studies of memory in nonhuman primates is the point in the experimental paradigm at which the tasks are initially learned. In studies in which monkeys with hippocampal lesions were both trained on the DNMS task and tested on the delay conditions following surgery, a marked impairment obtained. In contrast, in monkeys that were trained on the DNMS task preoperatively and tested postoperatively for retention of performance, little or no impairment was found. This difference in procedure may play an important role in the results because it very likely affects how and where the initial memory is mediated.

Another task used to assess memory, and one that has been widely used with human participants, is the delayed recognition span test (DRST). The DRST is a short-term memory test that requires the participant to identify, trial by trial, the new stimulus within an increasing array of serially presented stimuli. The task is administered using different classes of stimulus material in order to help characterize recognition memory deficits across several stimulus domains. For the spatial condition, identical discs are used as stimuli. A disc is placed over one of 18 wells, which is baited. A screen is raised, and the monkey is allowed to displace the disc to obtain the reward. The screen is then lowered, and a second disc is placed on the board over a baited well while the first disc is returned to its original position over the now unbaited well. After 10 s, the screen is raised, and the monkey is required to displace the new disc in order to obtain the reward. Each successive correct response is followed by the addition of a new disc until the monkey makes an error (i.e., chooses one of the previously chosen discs). With the occurrence of the first error, the trial is terminated, and the number of discs on the test tray minus 1 (i.e., the number of correct responses) constitutes the recognition span score for that trial. The object condition is administered in a similar fashion. Within each trial, the position of the previously correct stimuli is changed in a pseudorandom fashion so that the animal has to identify the new stimulus on the basis of visual cues without the use of spatial cues. Ten trials are administered each day for 10 days (see **Figure 4**).

Monkeys with bilateral damage restricted to the hippocampal formation made with ibotenic acid injection guided by magnetic resonance imaging are, like patients with Alzheimer's disease, markedly impaired in both the spatial and the object conditions of the DRST. They achieve lower spans than controls, whether the spans are unique or have been presented repeatedly over several sessions.

With regard to the rhinal cortices, Murray and others have performed extensive experiments using neurotoxic lesion methods to produce selective lesions to the perirhinal cortex in the monkey. Damage to this region produces impairment in stimulus recognition and in the maintenance of stimulus–stimulus associations for individual objects that, taken together, suggests that the perirhinal cortices are important for the accurate representation of visual stimuli and have profound effects on both perceptual and memory functions.

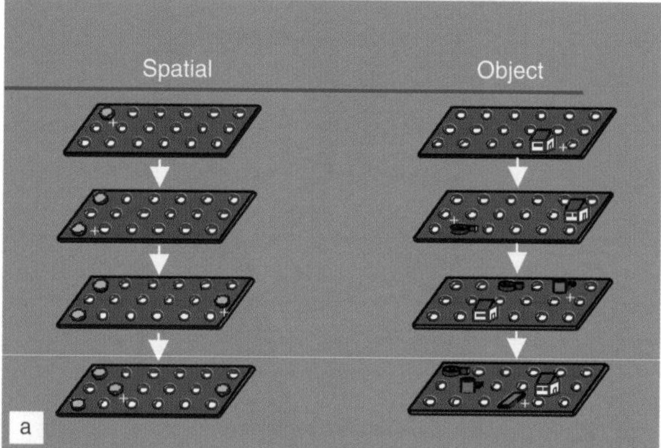

Figure 4 Schematic of the delayed recognition span test (DRST). Left: spatial condition; right: object condition.

Nonhuman Primate Models of Memory Function: Prefrontal Cortices

In the 1930s, researchers began to realize that the frontal lobe was involved in short-term memory processes. Jacobsen, taking the lead from findings of Brown and Schafer, advanced the notion that the frontal lobes may have a role in cognitive function. Jacobsen found that monkeys with damage to the prefrontal cortex were markedly impaired on a delayed response task. The findings were received as the first convincing evidence that the frontal lobes played a role in short-term memory function. Over the past 70 years, evidence for a role of the prefrontal cortices in memory function has been limited. It has become evident form the work of Goldman-Rakic that a small region of the prefrontal cortex plays a key role in working memory and that this area may be modality specific to spatial organization. It has been unclear whether damage to a specific region of the prefrontal cortex (e.g., dorsolateral region (area 9, area 46), ventrolateral region, or ventromedial region) has a specific role in memory or is related more to alteration of response control or some other aspect of so-called executive function.

More-recent evidence has shown that combined damage to areas 9 and 46 of the dorsolateral prefrontal cortex in the monkey results in an impairment on postoperative acquisition and delay conditions of the DNMS task. This deficit is quite severe, as monkeys with dorsolateral prefrontal cortex lesions take more trials to learn the task than do those with lesions of the hippocampus by a factor of almost 2 and perform as badly as monkeys of advanced age.

As is the case with findings from the DNMS task in nonhuman primates following lesions to the hippocampus, the findings following damage to the dorsolateral prefrontal cortex are not consistent. While it may be the case that combined, but not separate, lesions of dorsal prefrontal cortex that include both areas 9 and 46 are necessary to produce a measurable impairment on this task, it is equally plausible that preoperative versus postoperative acquisition of the task may represent the critical difference in findings.

Medical Temporal and Prefrontal Interactions?

While the exact temporal and anatomical sequence of memory processing remains to be elucidated, it is plausible that brain regions associated with visual recognition memory function (such as the hippocampal complex) and other medial temporal lobe structures, along with various parts of the prefrontal cortex, are differentially active in the course of acquisition and retention of qualitatively different cognitive tasks. Further, it is possible that declarative memory function may represent a functionally linked sequential network of mnemonic processing involving both medial temporal and prefrontal areas. According to this notion, one might speculate that during the initial acquisition of the nonmatching principle, both the hippocampus and the dorsal prefrontal cortex are active in processing and storing rules of the nonmatching paradigm. Hence, one would predict that damage to either structure alone would produce impairment in a postoperative acquisition paradigm, and both this report and previous work are consistent with this notion. Accordingly, one might predict that damage sustained by either structure alone following initial acquisition of tasks which employ the nonmatching rule would result in unimpaired or mildly impaired performance on the tasks, and indeed published reports of studies of such a postoperative retention paradigm are consistent with this. Moreover, one would predict that only combined damage to both the medial temporal lobe and the prefrontal cortex would produce significant impairment in such a postoperative retention paradigm, but this has yet to be tested. While these questions have yet to be tested directly, the notion of a complex interaction between prefrontal and medial temporal lobe systems in the acquisition and retention of recognition memory tasks can be tested empirically by making lesions of both the dorsal prefrontal cortex and medial temporal lobe concurrently in a postoperative retention paradigm.

Conclusion

While much progress has been made, greater effort must be made to bring tasks of memory function in animals into parallel with those in humans. It is also here that the great value of animal models can be brought to bear. One can more easily parcellate the anatomic and functional components of memory in animal models with carefully constructed experimental designs and then turn to mapping these onto studies of human memory.

See also: Aging and Memory in Animals; Amnesia: Declarative and Nondeclarative Memory; Episodic Memory: Assessment in Animals; Executive Function and Higher-Order Cognition: Assessment in Animals; Functional Amnesia; Spatial Memory: Assessment in Animals.

Further Reading

Beason-Held LL, Rosene DL, Killiany RJ, and Moss MB (1999) Hippocampal formation lesions produce memory impairment in the rhesus monkey. *Hippocampus* 9(5): 562–574.

Clark RE, West AN, Zola SM, and Squire LR (2001) Rats with lesions of the hippocampus are impaired on the delayed nonmatching-to-sample task. *Hippocampus* 11: 176–186.

Cohen NJ and Squire LR (1980) Preserved learning and retention of pattern analyzing skills in amnesics: Dissociation of knowing how and knowing that. *Science* 210: 207–210.

Eichenbaum H (2004) Hippocampus: Cognitive processes and neural representations that underlie declarative memory. *Neuron* 44: 109–120.

Eichenbaum H, Otto T, and Cohen NJ (1994) Two functional components of the hippocampal memory system. *Behavioral and Brain Sciences* 17: 449–472.

Goldman PS and Rosvold HE (1970) Localization of function within the dorsolateral prefrontal cortex of the rhesus monkey. *Experimental Neurology* 27: 291–304.

Mahut H, Zola-Morgan S, and Moss MB (1982) Hippocampal resections impair associative learning and recognition memory in the monkey. *Journal of Neuroscience* 2: 1214–1229.

Mishkin M (1978) Memory in monkeys severely impaired by combined but not separate removal of the amygdala and hippocampus. *Nature* 273: 297–298.

Murray EA (1990) Representational memory in nonhuman primates. In: Kesner RP and Olton DS (eds.) *Neurobiology of Comparative Cognition*, pp. 127–155. Hillsdale, NJ: Erlbaum.

Squire LR (1992) Memory and the hippocampus: A synthesis from findings with rats, monkeys, and humans. *Psychological Review* 99: 195–231.

Zola-Morgan S and Squire LR (1986) Memory impairment in monkeys following lesions limited to the hippocampus. *Behavioral Neuroscience* 100(2): 155–160.

Animal Models of Huntington's Disease

J Alberch, E Pérez-Navarro, and J M Canals,
University of Barcelona, Barcelona, Spain

The generation of animal models is necessary to understand the pathophysiology of neurodegenerative disorders, which will allow the development of new therapeutic strategies. Therefore, it is important to know when and what triggers the signals that induce the neurodegenerative process. The ideal animal model would have reproducible and well-defined behavioral abnormalities and neuropathological features, such as selective loss or dysfunction of specific neuronal populations, all of which must be the closest match to the human pathology. However, neurodegenerative disorders are chronic with different severity and onset, suggesting that there are different factors that modulate to activate the pathogenic mechanisms leading to selective neuronal dysfunction or death. Therefore, different animal models are generated to reproduce different stages of the disease. The generation of animals with a rapid onset and progression will help to determine the earliest molecular and cellular changes associated with the disease, although they can be too aggressive to test new treatments. However, animals with a late onset and milder disease can provide greater specificity and are potentially more useful to test neuroprotective approaches.

Huntington's disease (HD) is a devastating neurodegenerative disorder characterized by irrepressible abnormal movements, cognitive deterioration, and psychiatric disturbances. There is also striking specificity of neuronal loss localized in the striatum (caudate nucleus and putamen), which is the main coordinator of motor activity. The most sensitive cell populations are the striatal medium-sized spiny neurons that project to the globus pallidus and substantia nigra. Moreover, in more advanced stages of the disease, there is also a loss of cortical cells, affecting predominantly the large pyramidal neurons in layers III, V, and VI. Together with the anatomical and neurochemical substrate of HD, the etiology of the neurodegenerative process is well known. The disease is caused by an expansion of a CAG repeat region in the gene encoding huntingtin (IT15) and it is transmitted in an autosomal dominant manner. This mutation results in an increased stretch of glutamines in the N-terminal portion of the protein, which is widely expressed in the brain and peripheral tissues. Asymptomatic individuals have 35 or fewer CAG repeats, with longer expansions causing the illness. There is an inverse relationship between CAG repeat number and the age of onset of symptoms. Longer expansions (>55 repeats) usually give rise to a juvenile form of the disease. However, for medium-length expansions, the age of onset is unpredictable for any individual, which strongly suggests the existence of modifier genes or others factors that influence the age of onset of HD. Among these factors, excitotoxicity, mitochondrial dysfunction, neurotrophic factors, or transcriptional dysregulation have been proposed to modulate the neurodegenerative process.

The mutation has been known for over a decade, but there is still no effective treatment for HD because the pathogenic mechanism is not well understood. Although mutant huntingtin is expressed in all body tissues, the mutation preferentially induces the degeneration of specific neuronal populations in the striatum and cortex. Thus, the development of animal models for HD is required to study the pathogenic mechanisms and to test new treatments.

The first animal models for HD were achieved in the 1980s by either excitotoxic lesioning or metabolic impairment. The intrastriatal injection of glutamate receptor agonists, such as quinolinic acid, induces in rats a selective degeneration of striatal projection neurons, whereas interneurons are relatively spared. Similar results are observed after systemic administration of the mitochondrial toxin 3-nitropropionic acid. These models have been very useful because they produce a similar neuropathological profile as that observed in HD. However, they do not provide any data about the participation of mutant huntingtin in the degenerative process.

Other models have been developed in *Drosophila* and *Caenorhabditis elegans*. It has been shown that flies and worms expressing mutant huntingtin or polyQ peptides alone present a progressive degeneration. These models can be a powerful approach for genetic screens to identify genes that alleviate or modify the disease. However, the mutant protein is not expressed in neuronal types that degenerate in human HD. Hence, cell-specific effects are missing in these models.

However, breakthrough in HD research has been the development of transgenic mice expressing expanded polyglutamine repeats in the huntingtin gene. The first transgenic mice were developed in 1996. These models have been followed by many new HD transgenic and knockin mice that differ with regard to the type of mutation expressed, portion of the protein included in the transgene, promoter employed, and expression levels of mutant protein and background strain, making each of them unique and related to different degrees of the human pathology. These mouse models can fit in three broad categories: (1) mice that express exon-1 fragments of the human huntingtin gene containing polyglutamine expansions; (2) mice that express the full-length human HD gene; and (3) mice with pathogenic CAG repeats inserted into the endogenous CAG expansion (knockin mice) (**Figure 1**).

N-Terminal Exon-1 Transgenic Mouse Models

R6/1 and R6/2 transgenic mice were the first mouse models developed to study HD. They both express exon 1 of the human HD gene with 115 and 150 CAG repeats, respectively. The transgene expression in these mice is driven by the human huntingtin promoter. The resulting levels of transgene expression are around 31% and 75% of the endogenous huntingtin

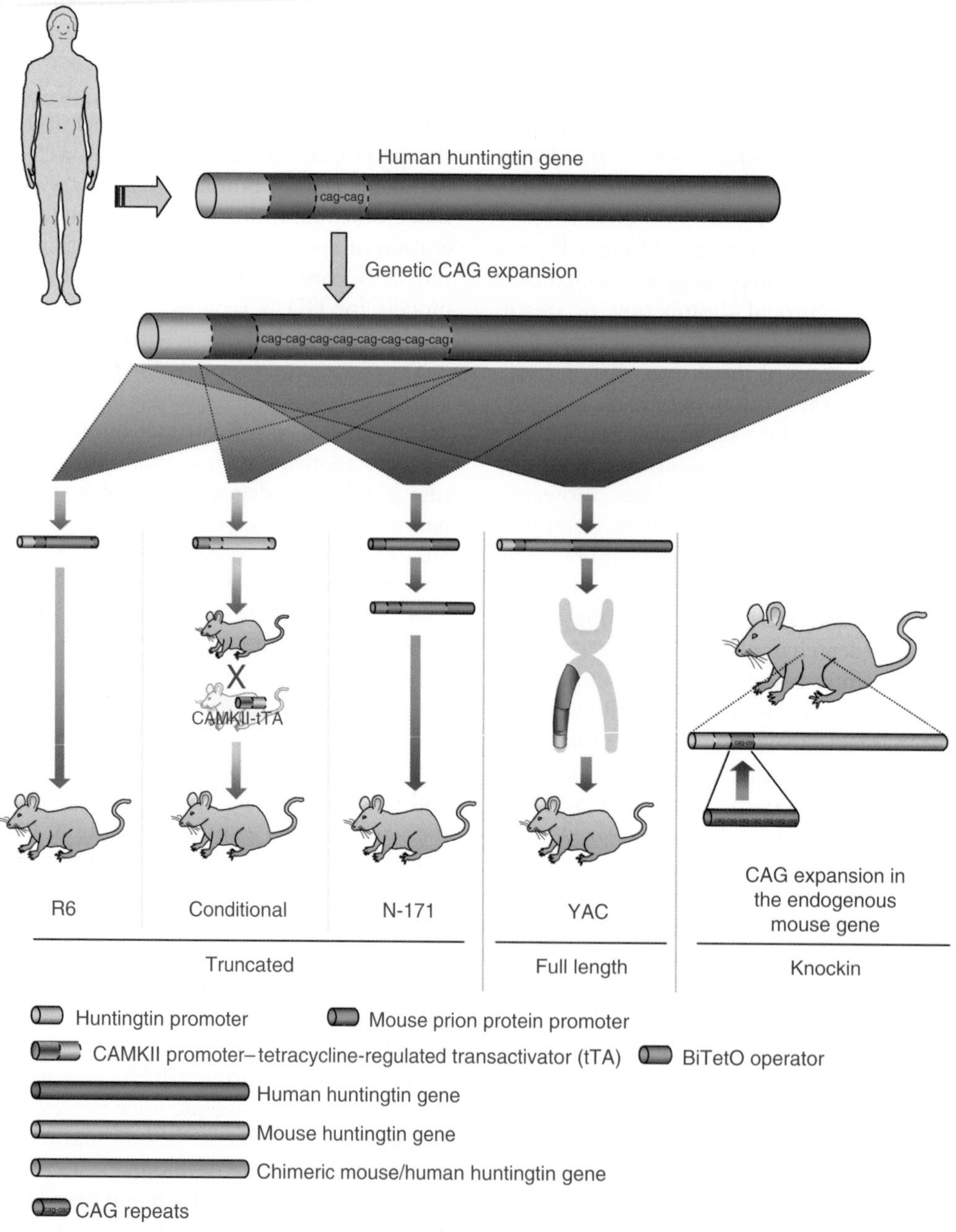

Figure 1 Schematic representation of the human huntingtin gene and constructs used for the generation of mouse models of Huntington's disease. The CAG repeats are localized in the exon 1 of the huntingtin gene, and the normal range is between 19 and 35. In transgenic mouse models the huntingtin gene (A fragment: R6, Conditional, N-171 or full length: yeast artificial chromosome (YAC)) with an expanded CAG repeat is randomly inserted into the genome. Knockin mouse models are obtained by gene targeting of the endogenous mouse huntingtin gene.

in the R6/1 and R6/2, respectively. In accordance with human pathology, R6 mice with the longest CAG repeats (R6/2) have an earlier onset and more severe symptoms like a juvenile form of HD. R6/2 mice are severely impaired by 8–12 weeks of age, whereas R6/1 mice are around 15–20 weeks. A typical sign in these models is the abnormal paw-clasping response. When suspended by the tail, normal mice spread their four limbs, whereas R6 mice clasp their hind- and forelimbs tightly against their thorax and abdomen (**Figure 2**). Although the pathophysiology of this behavior is not understood, the paw-clasping test is often used in studies examining novel treatments. Other changes in motor function are stereotypical hindlimb grooming, changes in gait patterns, and the emergence of some involuntary movements. As a result, their motor coordination progressively deteriorates, which can be detected as a reduction in the time they can stay on a rotating rod (rotarod).

Neuropathological analysis shows that brain and striatal volume are markedly reduced in R6 mice. However, the significant reduction in brain volume is the result of atrophy of individual neurons with a massive decreased neuropil, because neuronal death is minimal in these mice. Comparison of the time courses of cell death and behavioral anomalies shows that changes in motor activity precede any evidence of neuronal death in these mice, by a matter of several weeks. However, cell death can be observed if the neurotrophin brain-derived neurotrophic factor (BDNF) is reduced in R6/1 mice. The generation and characterization of a double-mutant mouse obtained by crossing R6/1 mice and BDNF heterozygote mice demonstrate that the levels of endogenous BDNF together with mutant huntingtin modulate the onset and severity of motor dysfunction and the degeneration of striatopallidal neurons. Thus, the different levels of endogenous BDNF in various mouse models of HD may also regulate the severity of the phenotype.

The most prominent morphological feature of HD, in addition to the reduction in brain volume, is the development of neuronal intranuclear and intracytoplasmatic inclusions. These are aggregations formed by polyglutamine repeats in the N-terminal huntingtin fragment. They also contain other components such as ubiquitin, and are located within neuronal cell bodies, the nucleus, and their arbors. These inclusions were first observed in animal models with mutant huntingtin and later confirmed in postmortem brains of patients with HD. In contrast to the lack of cell death in early stages in R6 mice, the presence of huntingtin- and ubiquitin-positive aggregates in R6/2 mice is observed as early as postnatal day 1 in the striatum and other brain areas, with the greatest number at 90 days. Thus, there is a close correlation between the degree of motor impairment and the number of striatal neurons exhibiting intranuclear inclusions of mutant huntingtin in this model. However, the function of these aggregates is under discussion, because there is evidence suggesting that they can be either toxic or neuroprotective.

As observed in HD patients, R6 mice also show cognitive deficits in advance of classical motor symptoms. R6/2 mice between 3.5 and 8 weeks of age

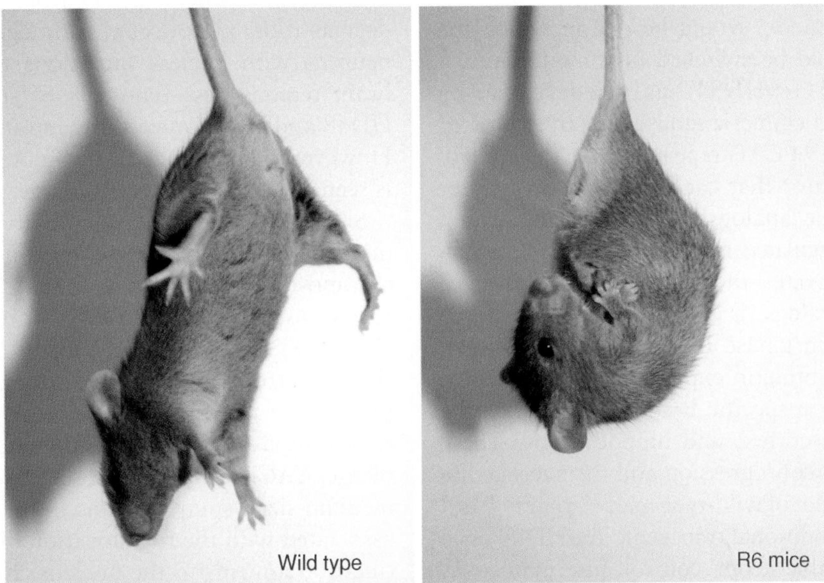

Figure 2 Clasping phenotype of R6 mice. Photomicrographs showing 30-week-old wild-type and R6 mice during the tail suspension test. R6 mice display a characteristic full body clasp.

display progressive learning impairment on cognitive tasks sensitive to frontostriatal and hippocampal function, as shown by the Morris water maze, two-choice swim tank, and T-maze tests.

Although the R6 mice are the most extensively studied, other transgenic mouse lines expressing the truncated huntingtin gene with expanded CAG repeats have been generated. N-171 mice express a cDNA encoding an N-terminal fragment (171 amino acids) of human huntingtin with 82, 44, or 18 glutamines. The expression of the transgene is directed by the mouse prion protein promoter, which drives the expression of foreign genes in virtually every neuron of the central nervous system. Mice expressing relatively low steady-state levels of N-171 huntingtin with 82 glutamine repeats (N-171-82Q) develop behavioral abnormalities, including loss of coordination, tremors, hypokinesis, hind limb clasping, and abnormal gait by 3 or 4 months of age, before dying prematurely (4–6 months). In mice exhibiting these abnormalities, diffuse nuclear labeling to the N-terminus of huntingtin, intranuclear inclusions, and neuritic aggregates were found in cortex, striatum, hippocampus, cerebellum, and amygdala as early as 4 weeks of age.

Other lines have also been generated expressing the N-terminal one-third of huntingtin with expanded (HD46, HD100) glutamine repeats under the regulation of the rat neuron-specific enolase promoter. These HD mice exhibit motor deficits at 3 months of age with an exacerbation throughout their lifespan. Neuropathological features, such as accumulation of huntingtin, dysmorphic dendrites, and atrophy in the striatum and cortex, precede the onset of behavioral deficits.

An interesting model would be one in which the mutated gene could be switched off or on. Thus, a conditional model (Tet/HD94) has been developed by the expression of a chimeric mouse/human exon 1 of the HD gene with 94 CAG repeats, under the control of a BiTetO operator that can be switched off after adding tetracycline analogs to the drinking water. The tetracycline-regulated transactivator (tTA), which constitutively activates the BiTetO in the absence of tetracycline analogs, is under the control of the calcium/calmodulin kinase IIα (CAMKII) promoter. Hence, mutant huntingtin expression is restricted to neurons localized in specific brain areas such as the striatum, septum, cortex, and hippocampus. These animals have a slow progression and the average life span is similar to that of wild-type mice (2 years). Until 4 weeks of age, conditional transgenic Tet/HD94 mice are indistinguishable from control littermates. At 4 weeks, some mice begin to clasp, and by 8 weeks all mice are clasping. In addition to this abnormal behavior, at 20 weeks Tet/HD94 mice begin to show a mild tremor, and at 36 weeks they are hypoactive. Neuropathological analysis shows that the brain and striatal volume at 18 weeks of age is smaller than in wild-type mice. This is accompanied by a gross enlargement of the ventricles, although cell death is not observed. The long survival of these animals has allowed the study of aged animals, revealing that at 17 months of age neuronal death is present in the striatum. Interestingly, shutting off expression of the mutant transgene produces significant changes in the neuropathological phenotype of these Tet/HD94 mice. When the gene is turned off, mutant huntingtin is rapidly cleared from the mouse brain and the motor symptoms are reversed. In aged animals, gene silencing can still be beneficial and full motor recovery is possible despite neuronal loss. This model provides evidence that treatments blocking mutant huntingtin expression in the early or even advanced stages of HD might enable the reversal of the clinical symptoms (**Table 1**).

Full-Length Transgenic Mouse Models

Several mouse models have been generated using a full-length huntingtin gene as the transgene. Mice that express a full-length huntingtin cDNA clone with either 48 (HD48) or 89 (HD89) repeats driven by the cytomegalovirus (CMV) promoter show a progressive motor phenotype. Feet clasping following suspension by their tails is observed in all HD48 and HD89 mice at 8 weeks of age. At 24 weeks of age, decreased motor activity is observed which progresses to locomotor deterioration and akinesia. Importantly, in the hypokinetic stage, neuronal loss, gliosis, and degeneration also occur. Although the number of neurons with nuclear inclusions varies in different brain regions, less than 1% of striatal neurons of HD48 and HD89 mice show intranuclear inclusions. However, the presence of polyglutamine aggregation is seen as early as 12 weeks of age.

Similar features have been observed in other mouse models obtained by the expression of yeast artificial chromosome (YAC) that spans the entire human HD gene with 46 (YAC46), 72 (YAC72), or 128 (YAC128) CAG repeats. YAC46 and YAC72 mice show electrophysiological abnormalities prior to the presence of nuclear inclusions or neurodegeneration indicating a cytoplasmic dysfunction. By 12 months of age, YAC72 mice have a selective degeneration of medium-sized spiny neurons in the striatum. This is associated with the translocation of N-terminal huntingtin fragments to the nucleus. Neurodegeneration can be present in the absence of macro- or micro-aggregates, showing that aggregates are not essential

Table 1 Mouse models of Huntington's disease

	Model design			*Abnormal motor behavior (onset)*	*Neuropathology*		*Life span*
	Promoter	*Gene size*	*CAG repeats*		*Aggregates/inclusions*	*Cell loss*	
Transgenic models							
R6 Mangiarini et al. (1996)	IT15	Exon 1	115 (R6/1)	Clasping (15–21 w) Decline in rotarod performance (13–20 w)	Observed in cortex and striatum from 8 w	Dark cell degeneration in cortex and striatum Reduction of medium-sized spiny neurons in the striatum	32–40 w
			144 (R6/2)	Clasping (8 w) Locomotor hyperactivity (3 w) Hypoactivity (8 w) Decline in rotarod performance (8–12 w)	Present in cortex and striatum from 3–4 w	Dark cell degeneration in cortex and striatum Reduction of medium-sized spiny neurons in the striatum	13–16 w
HD Laforet et al. (2001)	Rat neuron specific enolase	3-kb N-terminal fragment	48	Clasping (3–6 m) Impaired performance in rotarod (3–6 m)	Few intranuclear inclusions in cortex and striatum	Reduced number of neurons in the striatum in some animals	>12 m
			100	Clasping (3–6 m) Impaired performance in rotarod (3–6 m)	Intranuclear inclusions in cortex and striatum	Reduced number of neurons in the striatum in some animals	
N-171 Schilling et al. (1999)	Mouse prion protein	First 171 amino acids from exon 1	44	Normal up to 2 years of age	No	No	Normal
			82	Clasping (not reported) Impaired performance in rotarod (3 m) Abnormal gait (3–4 m) Hypoactivity (5 m)	Cytoplasmatic aggregates at 6 m Nuclear inclusions in the cortex, striatum, hippocampus, cerebellar granular cells, and amygdala at the end stage	Not severe loss of neurons in cortex, striatum, hippocampus, and cerebellum	10–50 w
YAC46 YAC72 Hodgson et al. (1999)	IT15	Full length	46 72	Normal up to 20 m Clasping (nondescribed) Hyperactivity (7 m) Circling (9 m)	No Cytoplasmic microaggregates in striatal neurons Nuclear inclusions: high number of neurons in the striatum, olfactory tubercle, and nucleus accumbens; small number of neurons in the septum and granule cell layer of the cerebellum	No Apoptotic cells in the striatum (12 m)	>12 m >12 m
YAC128 Slow et al. (2003)			128	Hyperkinesia (3 m) Motor deficit on the rotarod (6 m) Hypokinesia (12 m)	Inclusions in striatal and cortical neurons	Reduced number of striatal neurons	>12 m
HD94-tet off Yamamoto et al. (2000) Diaz-Hernandez et al. (2005)	CAMKIIα-tTA	Exon 1	94	Clasping (4 w) Mild tremor (20 w) Hypoactivity (36 w) Progressive motor decline (10 m)	Nuclear inclusions in cortex and striatum Extranuclear aggregates in cortex and with higher occurrence in the striatum	Dark cell degenerating neurons in the striatum (14 m) Decreased number of striatal neurons (17 m)	Normal

Continued

Table 1 Continued

	Model design			Abnormal motor behavior (onset)	Neuropathology		Life span
	Promoter	Gene size	CAG repeats		Aggregates/inclusions	Cell loss	
Knockin models							
HdhQ92 Wheeler et al. (2000)	Hdh		90	No clasping (up to 17 m)	Nuclear inclusions in striatum, olfactory tubercle, and cortex In older mice, nuclear aggregates are also observed in septum, olfactory bulb, nucleus accumbens, cerebellar granule cell layer, and hippocampus	No	Normal
Hdh94 Menalled et al. (2002a)	Hdh		94	No clasping (up to 24 m) Increased rearing (2 m) Decreased locomotion (4 m)	Nuclear microaggregates in the striatum Nuclear inclusions in the striatum (old mice) Other brain areas nonexamined	No	Normal
HdhQ111 Wheeler et al. (2000)	Hdh		109	Subtle gait deficits (24 m)	Nuclear inclusions in striatum, olfactory tubercle, and cortex In older mice nuclear aggregates are also observed in septum, olfactory bulb, nucleus accumbens, cerebellar granule cell layer, and hippocampus	No	Normal
HdhQ140 Menalled et al. (2003)	Hdh		140	Hyperactivity (1 m) Decreased locomotor activity (4 m) Gait anomalies (12 m)	Nuclear microaggregates in striatum, olfactory tubercle, cortex, and olfactory bulb Nuclear inclusions in striatum, olfactory tubercle, cerebellum, and layer II of piriform cortex Neuropil aggregates in striatum, globus pallidus, substantia nigra pars reticulata, cortex, cerebellum, olfactory tubercle, and olfactory bulb	Not reported	Normal
Hdh150 Lin et al. (2001)	Hdh		150	Gait and rotarod deficits (15–40 w) Clasping (15–40 w) Hypokinesia (40 w)	Nuclear inclusions in the striatum and nucleus accumbens; less frequent in cortex, hippocampus, and cerebellum; sparse in the thalamus, hypothalamus, and brain stem	No loss of striatal neurons	Normal

w, weeks; m, months.

to the initiation of neuronal death. YAC128 mice reveal a hyperkinetic phenotype first manifested at 3 months of age, followed by a progressive motor deficit on the rotarod at 6 months with eventual progression to hypokinesis by 12 months of age. These behavioral changes are followed by striatal atrophy clearly evident by 9 months, cortical atrophy at 12 months, and a progressive loss of striatal neurons accompanied by a decrease in their soma area. In contrast to N-terminal exon-1 transgenic models, the motor deficit in the YAC128 mice is highly correlated with neuronal loss. These mice demonstrate that initial neuronal cytoplasmic toxicity is followed by cleavage of huntingtin, nuclear translocation of huntingtin N-terminal fragments, and selective neurodegeneration.

Mild cognitive deficits have also been described in YAC128 mice that precede motor onset and progressively worsen with age. Rotarod testing reveals a motor learning deficit at 2 months, which progresses until 12 months of age, when YAC128 mice are unable to learn the rotarod task. These mice also show deficits in procedural learning, memory, sensorimotor gating, and strategy shifting.

Knockin Models

Knockin mouse models have been generated by homologous recombination of CAG repeats in the endogenous mouse huntingtin gene. Therefore, knockin mice carry the mutation in its appropriate genomic and protein context, making them more faithful genetic models of the human condition than other transgenic lines, which have the transgene inserted randomly. Several knockin mice have been developed with different CAG expansions and they exhibit similar neuropathological phenotype, although at different ages. All these mice have a late onset and slow progression. However, there are differences in the magnitude of motor abnormalities in knockin and transgenic mice, but both lines show a shift from hyper- to hypoactivity. Thus, for example, knockin mice with 94 repeats display a biphasic motor behavior characterized by increased motor activity at 2 months, followed by hypoactivity at 4 months of age. Gait abnormalities are observed at 24 months. HdhQ111 mice do not differ from wild-type animals in paw clasping and rotarod at 15 months of age, but at 24 months motor function assessed by tunnel walks reveals a subtle gait deficit. In animals with longer CAG repeats, similar patterns of motor abnormalities are observed although at early ages. Gait abnormalities are already observed at 12 months of age. None of the knockin mice develop neuronal loss, although reactive gliosis is observed in the striatum. Nuclear staining and aggregates of the mutant huntingtin are only observed in various knockin models when the mice are old. In mice with 94 repeats, intranuclear inclusions are restricted to the striatum, whereas with longer expansions (140 and 150 CAG repeats) the distribution is more widespread.

Conclusion

In the previous years, several models have been developed showing different phenotypes and timing depending on the genetic construct, length of CAG repeats, and background strain. They all produce aspects of the pathology observed in HD patients, such as movement abnormalities, specific atrophy of the striatum and cortex, and the presence of intranuclear and cytoplasmatic aggregates of mutant huntingtin. Hence, the availability of these mouse models has been an important step in research aimed at dissecting the pathogenesis of HD. Each model can be useful for different types of studies. Knockin models accurately replicate the underlying genetic defect of HD, but they have a slow progression with no robust motor deficits, and brain atrophy is only observed in aged animals. The transgenic mice expressing full-length mutant huntingtin have a less aggressive course of the disease than mice that express a truncated form. The varying degrees of striatal atrophy, brain weight loss, and motor deficit in these animal models allow the study of different stages and intensities (juvenile or adult) of the pathology. Therefore, the different animal models of HD are powerful tools to understand the pathogenic mechanisms and for the assessment of neuroprotective and other therapeutic interventions.

See also: Axonal Transport and Huntington's Disease; Cell Replacement Therapy for Huntington's Disease; Huntington's Disease; Huntington's Disease: Neurodegeneration; Transgenic Models of Neurodegenerative Disease; Triplicate Repeats: Huntington's disease.

Further Reading

Beal MF and Ferrante RJ (2004) Experimental therapeutics in transgenic mouse models of Huntington's disease. *Nature Reviews Neuroscience* 5: 373–384.

Canals JM, Pineda JR, Torres-Peraza JF, et al. (2004) Brain-derived neurotrophic factor regulates the onset and severity of motor dysfunction associated with enkephalinergic neuronal degeneration in Huntington's disease. *Journal of Neuroscience* 24: 7727–7739.

Diaz-Hernandez M, Torres-Peraza J, Salvatori-Abarca A, et al. (2005) Full motor recovery despite striatal neuron loss and formation of irreversible amyloid-like inclusions in a conditional mouse model of Huntington's disease. *Journal of Neuroscience* 25: 9773–9781.

Hodgson JG, Agopyan N, Gutekunst CA, et al. (1999) A YAC mouse model for Huntington's disease with full-length mutant huntingtin, cytoplasmic toxicity, and selective striatal neurodegeneration. *Neuron* 23: 181–192.

Laforet GA, Sapp E, Chase K, et al. (2001) Changes in cortical and striatal neurons predict behavioral and electrophysiological abnormalities in a transgenic murine model of Huntington's disease. *Journal of Neuroscience* 21: 9112–9123.

Lin CH, Tallaksen-Greene S, Chien WM, et al. (2001) Neurological abnormalities in a knockin mouse model of Huntington's disease. *Human Molecular Genetics* 10: 137–144.

Mangiarini L, Sathasivam K, Seller M, et al. (1996) Exon 1 of the HD gene with an expanded CAG repeat is sufficient to cause a progressive neurological phenotype in transgenic mice. *Cell* 87: 493–506.

Menalled LB and Chesselet MF (2002) Mouse models of Huntington's disease. *Trends in Pharmacological Science* 23: 32–39.

Menalled LB, Sison JD, Dragatsis I, Zeitlin S, and Chesselet MF (2003) Time course of early motor and neuropathological anomalies in a knockin mouse model of Huntington's disease with 140 CAG repeats. *Journal of Comparative Neurology* 465: 11–26.

Menalled LB, Sison JD, Wu Y, et al. (2002) Early motor dysfunction and striosomal distribution of huntingtin microaggregates in Huntington's disease knockin mice. *Journal of Neuroscience* 22: 8266–8276.

Rubinsztein DC (2002) Lessons from animal models of Huntington's disease. *Trends in Genetics* 18: 202–209.

Schilling G, Becher MW, Sharp AH, et al. (1999) Intranuclear inclusions and neuritic aggregates in transgenic mice expressing a mutant N-terminal fragment of huntingtin. *Human Molecular Genetics* 8: 397–407.

Slow EJ, van Raamsdonk J, Rogers D, et al. (2003) Selective striatal neuronal loss in a YAC128 mouse model of Huntington disease. *Human Molecular Genetics* 12: 1555–1567.

Wheeler VC, White JK, Gutekunst CA, et al. (2000) Long glutamine tracts cause nuclear localization of a novel form of huntingtin in medium spiny striatal neurons in HdhQ92 and HdhQ111 knockin mice. *Human Molecular Genetics* 9: 503–513.

Yamamoto A, Lucas JJ, and Hen R (2000) Reversal of neuropathology and motor dysfunction in a conditional model of Huntington's disease. *Cell* 101: 57–66.

Relevant Website

http://www.hdfoundation.org – The Hereditary Disease Foundation.

Animal Models of Inherited Retinal Degenerations

J Stone, Australian National University, Canberra, ACT, Australia
J R Heckenlively, University of Michigan, Ann Arbor, MI, USA

Introduction

The retinal dystrophies are genetic diseases whose mutations target retinal neurons, particularly photoreceptors. The retina has a highly interdependent cellular structure, and diseases that initially affect a particular specialized cell, such as rod or cone photoreceptors, will subsequently damage adjacent cells. The effect of the mutation is readily apparent in higher mammals as they are visually disabled. For mice, however, blindness has not been a selective disadvantage. Wild mice blinded by the *rd1* mutation, for example, have been found all over the world, and the viral insert in rd1 mice is estimated to be more than a million years old.

In many panretinal degenerations, the photoreceptors are the prime target of the mutation, and remodeling of synapses may occur in adjacent cells after the photoreceptors die. Disorders of the retinal pigment epithelium (RPE), the cellular layer adjacent to the outer tips of photoreceptors, which provides essential metabolic support to rods and cones, can also lead to photoreceptor death. Inherited retinal dystrophies (IRDs) have been reported in many mammals, commonly mice, rats, dogs, and cats, and are a clinically important cause of blindness in humans. Genetic mutations have been engineered extensively in mice and, less frequently, in the rat, pig, frog, and fish. The number of mutations identified as causing IRDs is large; the numbers have been documented best in humans (**Figure 1**), in which thousands of mutations in more than a hundred genes have been identified. The functions of these IRD genes are diverse, and their identification or engineering in animal models has made possible the systematic study of their mutational effects on the physiology of the retina and viability of retinal neurons.

Identification of mutations targeting other retinal cell types is still expanding and has been documented best in humans and mice. Abnormalities of bipolar cells, Müller cells, and retinal astrocytes are all known to cause minor to serious disease. As examples, mutations affecting primarily bipolar cells lead to forms of congenital stationary night blindness, while mutations affecting retinal astrocytes are associated with retinal vasculopathies. Because the large majority of IRDs affect photoreceptors either primarily or at a very early stage, much work in animal models has been dedicated to understanding the function and fragility of photoreceptors, with the aim of identifying key mechanisms that will provide a rationale for therapy.

Organization of the Retina

The mammalian retina has two classes of photoreceptors, rods and cones. The rods function at low-light levels, providing highly sensitive, low-resolution, black-and-white vision. In brighter conditions, such as daylight, the photodetection mechanisms of the rods are saturated, and rods contribute only minimally to visual function. In these brighter conditions, cones are the functional photoreceptor, providing high-resolution color vision. In many primates, including humans, the retina has evolved a specialized macula, or foveal region, where cones are the dominant photoreceptor class and retinal circuitry is specialized to provide very high resolution of detail and color. Together, rods and cones allow vision over a five-log-unit range of light intensities, from near-darkness to bright daylight. Rods and cones have highly organized transmission pathways to the output cells of the retina, the ganglion cells. However, rod and cone pathways interact, beginning with direct electrical synapses between rods and cones.

The photoreceptors of mammalian retina form, counterintuitively, the outer layers of the retina. Further, they are oriented with their photosensitive elements, the outer segments, oriented most externally. This requires light to pass through the inner layers of the retina, which contain neurons processing the electrical signals generated by rods and cones. It has evolved to allow the tips of photoreceptors to be embedded in a bounding epithelium, the RPE. This relationship of rod and cone outer segments to the RPE is critical for the viability and function of photoreceptors. The RPE plays a role in the recycling of the breakdown products of phototransduction and a distinct but equally critical role in phagocytosing membrane segments discarded daily by the outer segments.

The retina forms a delicate, thin (300 μm) neural membrane lining the inner surface of the back of the eyeball. It is protected by the robust outer wall of the eyeball (the sclera) and by specialized ocular media, particularly the vitreous body, which fills much of the cavity of the eyeball. The metabolic needs of the photoreceptors are high and are met by the flow of nutrients from the choroidal circulation, a vascular bed that forms between the RPE and the

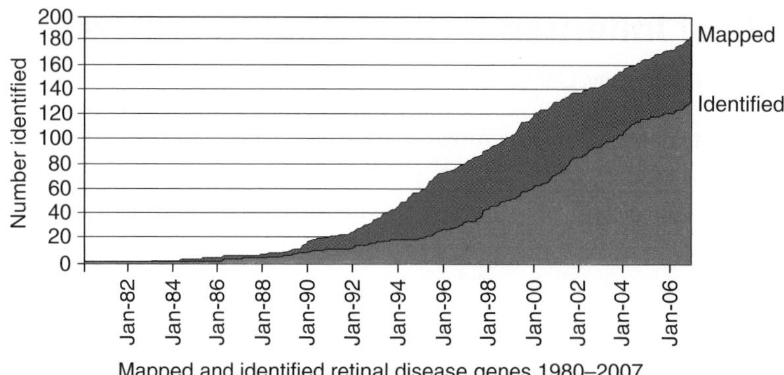

Figure 1 Cumulative graph of the discovery of genes whose mutations cause 'inherited retinal dystrophies'. 'Mapped' means that the chromosomal location of the gene has been defined. 'Identified' means that the specific gene sequence is known. From Daiger SP, Sullivan LS, and Browne SJ (2007) *RetNet: Retinal Information Network*, http://www.sph.uth.tmc.edu/RetNet/ (accessed Jul. 2008).

sclera. The metabolic needs of the inner layers of retina are provided in many species, including the human and the animal species used to model the IRDs, by a retinal circulation spreading through the inner retina from the head of the optic nerve.

Embryologically, the retina and the optic nerve are extensions of the diencephalon of the forebrain and are the only part of the brain to lie outside the cranium. Thus the retina forms a thin layer of central nervous tissue, bounded physiologically by the tight junction-linked endothelium of the retina vasculature and RPE. Within this compartment, the environment of retinal neurons is maintained by radially oriented glial cells, the Müller cells, which play a structure role in forming the limiting membranes of the retina and also participate in the recycling of transmitters and maintaining the ionic environment essential for electrical signaling by retinal neurons. At the inner surface of the retina, an additional layer of glial cells, the astrocytes, is found. These are immigrants to the retina from the optic nerve; they spread over the inner retinal surface, forming a template for the formation of the retinal vasculature.

Mutations causing IRDs have now been described affecting all these major elements of retinal structure: the photoreceptors, the other retinal neurons, the RPE, Müller cells, and astrocytes.

Interaction of Genetically Induced Photoreceptor Death with Normal Photoreceptor Death

Study of rodent models has demonstrated how mutation-induced stresses on photoreceptors interact with normal developmental stress to produce the adult form of disease. In neonatal life, for example, a phase of photoreceptor death has been identified in a developmental 'critical period' soon after the eyes open and photoreceptors commence function. During this period, photoreceptors are subject to hypoxia-induced stress, which culls the population to a sustainable level. In several forms of inherited dystrophy (e.g., the Royal College of Surgeons (RCS) rat, the *rd* mouse, the transgenic P23H rat lines), the high rates of photoreceptor death that characterize the degeneration begin during the normal critical period and persist at varying rates into adulthood. That is, the degeneration induced by a range of mutations (disturbance of recycling in the RCS rat, of a phosphodiesterase (PDE) function in the *rd* mouse, of rhodopsin in the P23H rat) exacerbates a normal process of photoreceptor culling.

In the adult, similarly, there is extensive evidence that the normal (wild-type) retina suffers photoreceptor attrition throughout life, in all mammals examined (rodents and primates). This normal attrition has been documented best in humans, in which the death of rods and cones occurs throughout life, at startlingly high rates – as many as 500 photoreceptors a day, or 150 000 per annum. The loss does not normally become important functionally until the later decades of life, when age-related degeneration can be severe. Again, the inherited dystrophies can be understood as an acceleration of a basal, naturally occurring photoreceptor loss. Photoreceptors operate in a stressful environment of high metabolism, powerful membrane currents, active ion pumps, and rapid turnover of the outer segment membrane. Retinal dystrophies occur when a mutation decreases the evolved stability of adult photoreceptors.

Age-related changes are not unique to the retina; age-related deterioration of many aspects of brain function is well documented. In various ways, the mutations that cause the IRDs destabilize the interactions between retinal neurons and accelerate a naturally occurring loss of photoreceptors.

Insights into the Diversity of Genetically Induced Photoreceptor Death

The diverse forms of the IRDs, listed now on several databases, can be classified in several ways: according to their mode of inheritance (autosomal vs. sex-linked, dominant vs. recessive), by age of onset, by which neuron (rod, cone, or both) is the primary target of the mutation, or by whether the mutation affects more than one sense organ, as in Usher's syndrome in humans, which involves deafness as well as vision loss. As the diversity of mutations that accelerate photoreceptor death has become better known, the function of the gene product has become an alternative way of describing the diversity of the IRDs.

One recent classification, for example, identifies four broad classes of gene products affected by IRD mutations: (1) phototransduction proteins (e.g., rhodopsin, cyclic guanosine monophosphate (cGMP) PDE, arrestin), (2) structural proteins (e.g., peripherin, the adhesion molecule RS1, genes interacting with the photoreceptor cilium), (3) proteins involved in metabolism of photoreceptors and the associated RPE (e.g., phosphorylases such as MertK and RPE65), and (4) transcription factors. This list of the functional classes of IRD genes is likely to grow to include molecular chaperones (proteins which facilitate the correct folding of other proteins); genes that function in every cell but have evolved into retina-specific isoforms, a recent example being inosine monophosphate dehydrogenase 1 (IMPDH1); and ubiquitously expressed, non-retina-specific genes, an example being pre-messenger RNA (mRNA) processing genes, whose products are part of the spliceosome (see **Figure 2**).

Why Are Photoreceptors Selectively Vulnerable to Mutations in Ubiquitously Expressed Genes?

The selective vulnerability of photoreceptors to mutations in ubiquitously expressed genes is one of the scientific challenges of IRDs, and answers to it are being worked out disease by disease. For example, *IMPDH1* is a highly conserved, ubiquitously expressed gene. The gene product is the enzyme IMPDH1, which regulates the rate-limiting step of guanine synthesis. Although the gene is highly conserved and ubiquitously expressed, the clinical consequences of mutations are

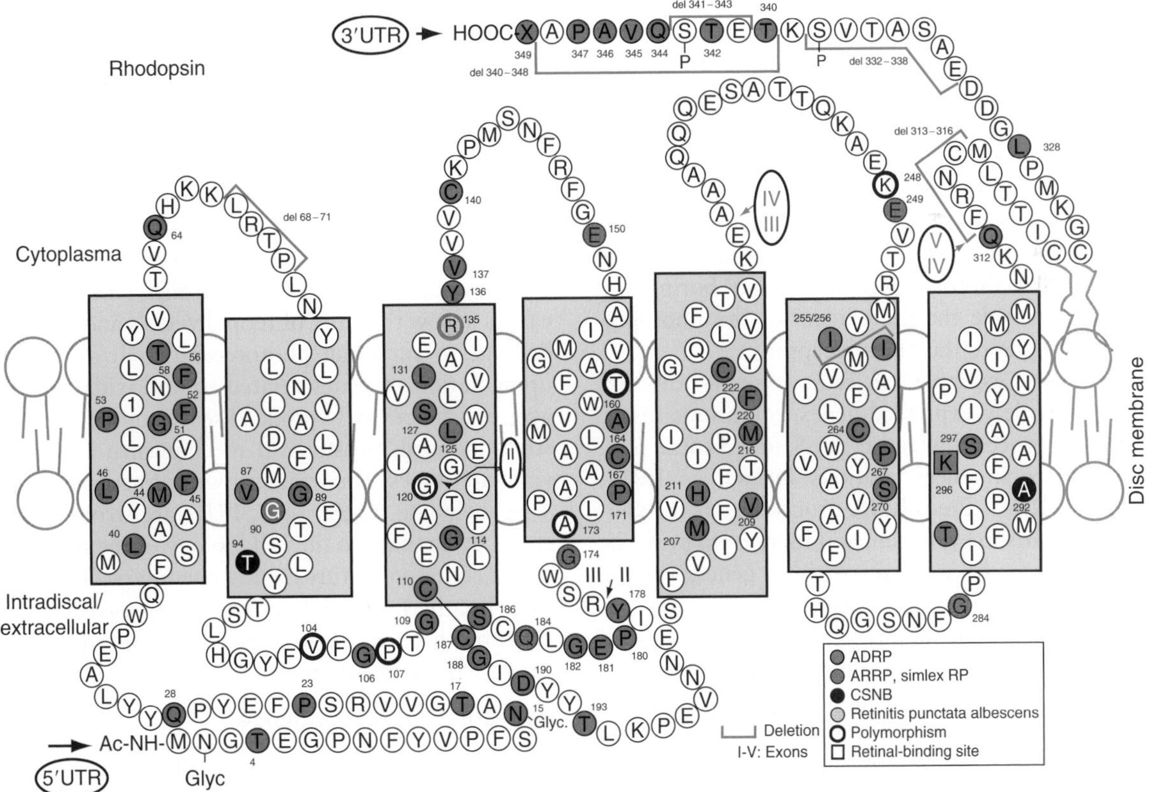

Figure 2 A representation of the rhodopsin protein, embedded in the membrane of the rod outer segment. Sites are shown at which mutations (single-amino-acid substitutions, deletions) have been identified which cause autosomal dominant and recessive forms of retinitis pigmentosa (ADRP, ARRP), congenital stationary night blindness (CSNB), or other conditions, such as retinitis punctata albescens. The diagram emphasizes the large number of mutations that persist into the adult phenotype of a single protein and cause photoreceptor degeneration. del, deletion; UTR, untranslated region. From Preising M (2007) *Retina International Scientific Newsletter*.

retina-specific – the RP10 form of retinitis pigmentosa (RP) and, less commonly, Leber's congenital amaurosis or congenital RP. The selective vulnerability of photoreceptors has been traced to the circumstance that the retina has evolved specific isoforms of the gene, and mutations can be specific to retinal isoforms. The retina also expresses IMPDH1 at high levels, increasing its vulnerability. Diseases related to mutations in nonretinal IMPDH1 have not been reported.

One of the earliest-discovered retinal degeneration mutations in mouse, *rd1*, has a murine viral insert and a second nonsense mutation in exon 7 of the *PDE6B* (β-subunit of cGMP PDE) gene. Again, this enzyme is expressed ubiquitously, but rods and cones have evolved a specific isoform (PDE6), so mutations can be retina specific. The enzyme catalyzes the hydrolysis of cGMP to GMP; because cGMP (but not GMP) gates the ion channels of the outer segment, which conduct the dark current of the photoreceptor, the enzyme's normal action, triggered as part of the phototransduction cascade, is to reduce dark current, creating an electrical signal. The mutation inactivates the enzyme and raises cGMP levels, leaving the channels gated open. This results in elevated intracellular Ca^{2+} levels, which induce apoptotic death, causing the rapid degeneration of rods. *Mertk* is a receptor tyrosine kinase gene expressed in the RPE. Mutations cause a photoreceptor-specific degeneration, first recognized in the naturally occurring RCS rat strain but subsequently identified as a human IRD. Photoreceptors are specifically vulnerable because of their dependence on the RPE for a number of functions, including the phagocytosis of outer segment membrane. This active discarding of membrane and the dependence of photoreceptors on a neighboring epithelium to recycle the ingredients of the photopigment are unique to the photoreceptor.

In another example of retina-specific diseases caused by widely expressed genes, mutations in the pre-mRNA processing factors 31, 8, and 3 cause photoreceptor-specific degenerations known as RP11, RP13, and RP18. Pre-mRNA processing in the spliceosome occurs in every cell, and there is no evidence of a retina-specific isoform of these genes, yet the degenerations are photoreceptor specific.

Thus, the high vulnerability of photoreceptors results in some cases from the evolution of photoreceptor-specific isoforms of the gene and in other cases seems related to photoreceptor-specific parameters, such as the dependence of photoreceptors on the RPE or the high dark current of rods. One common factor is likely the high degree of specialization of photoreceptors, which underlies their high metabolism, the evolution of retina-specific isoforms of some genes, and a multiple dependence on surrounding cells. Contributing to the large number of retinal dystrophies observed is the fact that photoreceptors are not essential for fetal or postnatal development, allowing a large number of retinal dystrophies to be expressed in adults.

Interactions with Retinal Cell Biology

Despite the diverse causes of photoreceptor degeneration, the IRDs show sufficient common symptoms to give some meaning to the clinical name given to many human forms of IRD, RP. The thinning and degeneration of the RP retina result in pigmentary invasion of the RP retina, which becomes apparent on ophthalmic inspection. The vessels of the retina thin, and the retina is intriguingly resistant to hypoxia-driven diseases, such as diabetic retinopathy. Work in rodent models has shown that the retinal vessel thinning, and presumably the resistance to hypoxia, is caused by a chronic rise in choriocapillaris-derived oxygen levels as the photoreceptor population is depleted.

The photoreceptors are the only part of the central nervous system to lack intrinsic vessels, for optical reasons. They are supplied by oxygen and glucose diffusing from the capillaries of the choroidal circulation, the choriocapillaris. The choriocapillaris lies close to the outer retina, separated from the photoreceptors by the RPE and its basement membrane (Bruch's membrane). Critically, however, the choriocapillaris vessels do not lie in the retina, cannot sense the levels of oxygen and other metabolites in the tissue they supply, and therefore cannot adjust their flow to the needs of the photoreceptors. As a consequence, when the photoreceptor population is depleted and its consumption of oxygen falls, oxygen levels in outer retina rise chronically. This hyperoxia causes a slow thinning of retinal vessels and oxidative damage to surviving photoreceptors. It also causes upregulation of genes related to antioxidative damage; resistance to hypoxic disease; upregulation of stress-induced, prosurvival mechanisms; and at the very late stages, abnormal growth of neurons.

In their late stages, then, IRDs are strongly influenced by changes to the cell biology of the retina that create common features to the phenotype. The hyperoxia becomes cumulative, the progressive death of photoreceptors adding to the hyperoxia and reinforcing the oxidative damage.

Rod–Cone Interdependence

The fragility of cones in many IRDs in which the mutation affects a protein (such as rhodopsin) expressed specifically in rods is clinically important and mechanistically puzzling. Good cone vision provides much of the value and pleasure of vision: the ability to read, see faces, and see what others can see in daylight. Cone vision is typically unstable after rods are depleted, however. Clinically, cone vision can persist for years, even decades, after rod vision fades, but in most cases it is eventually lost. Animal models such as the *rd1* mouse or the rhodopsin-mutant P23H-3 rat confirm that cones are destabilized by rod loss.

The study of IRDs in animals has given insight into the mechanisms of this interdependence, providing evidence that rods and cones rely on diffusible trophic factors for their viability. In mice, evidence has been developed that rods release a 'rod-derived cone viability factory,' identified as a polypeptide consisting of 109 amino acids and essential for the survival of cones. As the rod population is depleted or damaged, cones are starved of this factor and die progressively. A second mechanism, known as the oxygen toxicity hypothesis, came from the study of the effects of photoreceptor depletion in the RCS rat and P23H-3 rhodopsin-mutant transgenic rat and evidence that direct hyperoxia (inhaling oxygen-enriched air) is directly toxic to photoreceptors. Because the supply of oxygen to photoreceptors from the choroidal circulation is unregulated, oxygen accumulates in the photoreceptor layers of retina as a retinal dystrophy progresses. This rise is chronic and results in oxidative damage to the surviving photoreceptors. A third idea is that as the rods disappear, the supporting matrix they provide to the photoreceptor layer is lost, and the cones are distorted or crushed.

Evidence has emerged from mouse IRDs of a converse dependence. In many mouse 'cone' dystrophies, in which cones are affected first, the rods typically show a panretinal degeneration 6–12 months after the cones are lost. There appears to be a co-dependence of rods and cones, neither population being stable without the other.

Pathways to Therapy

The growth of knowledge about the molecular genetics of IRDs has been rapid in the past 15 years, and work on animal IRDs has expanded understanding of the cell biology of the degenerating retina, the effects of oxidative stress on the retina, and the retina's ability to protect its cells against oxidative stress. These bodies of knowledge are being brought together to test numerous avenues to therapy. Some of the major approaches are described in the next sections.

Gene replacement therapy Gene replacement therapy involves delivering a functioning gene to supplement or replace the mutant gene in cells surviving in the degenerating retina. Research in animal models indicates that it is likely to be most useful in IRDs in which mutations such as a deletion or stop codon leave the targeted cell without the needed gene product. Placing small genes in vectors such as lentiviruses or adeno-associated viruses has become routine, and delivery to the retina is usually achieved by subretinal or occasional vitreal injections. A number of successful treatments have been pioneered in animal models of IRDs, including dogs as well as rats and mice. The degeneration of the retina has been slowed or stopped, the function of the retina has been partially restored, and the improvements have been shown to be long-lasting. The use of gene replacement therapy is theoretically more difficult in missense mutations in which the gene product is abnormal, since the abnormal product may continue to be produced and to disrupt the cell even though a correct product is present. In this circumstance, a two-step procedure has been proposed, the first step being to knock out the mutant gene, and the second, to add the normal gene. Because of the diversity of mutations in the IRDs, therapeutic approaches will have to be customized to the mutation and its pathologic effects. Several of the achromatopsia mutations appear to be amenable to possible gene therapy intervention since the cones survive, even though the photopic electroretinogram (ERG) becomes nonrecordable (see **Figure 3**).

Stem cell or progenitor cell therapies The term 'stem cell therapy' refers to interventions in which embryonic stem cells, or more-differentiated progenitor cells, are introduced to the retina with the hope that they will differentiate into functional neurons, replacing lost cells. Most such interventions reported to date have not been successful. The introduced cells have failed to differentiate into the needed retinal neurons or to integrate into the circuitry of the retina. Often the introduced cells do not survive, and there is no sign of repair of the photoreceptor degeneration. Recently, rod embryonic precursor cells harvested from neonatal mouse retina were shown to differentiate into rods and to integrate morphologically and functionally as rods into the retinas of the peripherin-deficient *rds* mouse and of a rhodopsin knockout strain. Provided the precursor cells were harvested in the first few days of postnatal life, during the period of rod genesis, appropriate synaptic connections were formed within the retina, and the electroretinogram showed restoration of physiologic responses. Other studies using bone marrow-derived progenitor cells have shown reparative properties to inner retinal layers of *rd*10 mice but no significant restoration of photoreceptors. These studies suggest that successful stem or progenitor cell interventions can be attained when the stage of development of the introduced cell is carefully controlled.

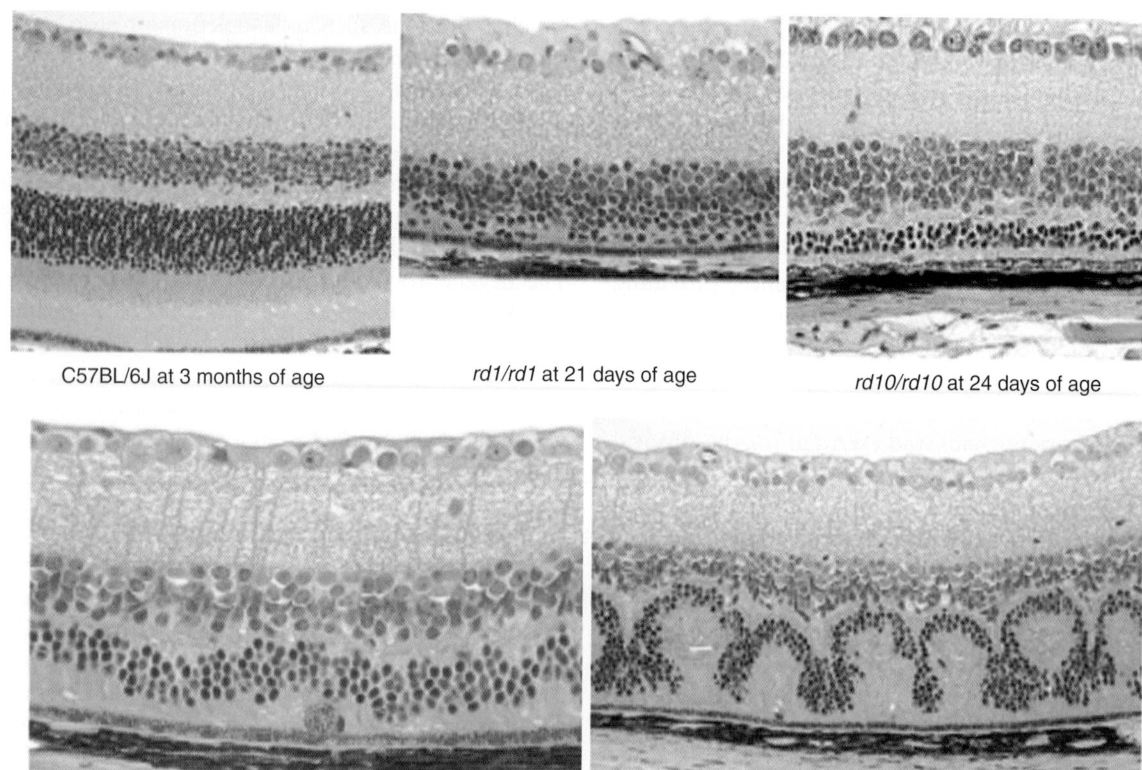

Figure 3 Retinal histologic phenotypes in the mouse, from a set of known mutations. The C57BL/6J is a nondegenerative, wild-type retina, used as a background for many engineered mutations. The retinal layer is normal. The *rd1/rd1* mouse is homozygous for a widely occurring viral insert in exon 7 of the β component for a retina-specific isoform of cGMP phosphodiesterase (PDE6). The outer nuclear layer (the layer of photoreceptor somas) is severely and quickly depleted by 1 month of age. The *rd10/rd10* mouse is homozygous for a missense point mutation in exon 13 of the same gene. The outer nuclear layer depleted, and distorted in the *rd6/rd6* and *rd7/rd7* mouse strains; these IRDs are associated respectively with a 4 bp deletion in the splice donor sequence of intron 4 in the *Mfrp* gene, and a 380-nt deletion from the coding region of a photoreceptor-specific nuclear receptor gene, *Nr2e3*. From Chang B, Dacey MS, Hawes NL, et al. (2006) Cone photoreceptor function loss-3, a novel mouse model of achromatopsia due to mutation in Gnat2. *Investigative Opthalmology & Visual Science* 47: 5017–5021.

Retinal transplantation Many studies using animal models have reported the transplantation of sheets of retinal cells under or on degenerating retina, with the transplanted retinal segments sometimes surviving as the host retina degenerates. No studies have shown that the transplanted cells have integrated into the circuitry of the retina sufficiently to transmit visual information to the brain.

Pharmaceutical interventions Large numbers of antiapoptotic drugs are known, and the various arms of the apoptosis pathway appear to be receptive to pharmacologic interventions. This therapeutic opportunity is attracting attempts to deliver drugs that regulate or inhibit the pathways that trigger apoptotic mechanisms. These drugs have been well tested in other tissues, but few have been tested in the retina.

More progress has been made using endogenous protective factors, proteins, and cytokines produced by the retina itself. Studies of the effectiveness of fibroblast growth factor-2 (FGF-2), ciliary neurotrophic factor (CNTF), and brain-derived neurotrophic factor (BDNF) were pioneered in animal models of retinal dystrophies, both inherited and stress-induced. These factors are endogenous, in the sense that the cells of the retina itself express them. When expressed, they protect the retina, especially photoreceptors, against stress, such as the stress of bright light. Further, their expression is stress inducible. They can be viewed, therefore, as inbuilt self-protective mechanisms, triggered by stress.

Their use as therapy has been inhibited by concerns about the possibility that the factors will have actions beyond photoreceptor protection, exerting trophic effects on blood vessels, on the tunica fibrosa of the eyeball, and on the differentiation of retinal neurons. Further, the factors can reduce the responsiveness of photoreceptors to light. This has been shown most clearly for FGF-2 and for CNTF, which binds to the outer segment, reduces rhodopsin expression, and by altering the arrestin–rhodopsin ratio, downregulates phototransduction. Whether this downregulation is intrinsic to the protective action of CNTF is not known.

Nevertheless, work to test the value of these factors continues. BDNF is proving effective in protecting photoreceptors in an animal model of the collateral damage caused during photodynamic therapy of human eyes, and trials of CNTF delivered by an encapsulated intraocular implant of cells engineered to produce the factor have passed the initial safety stage. The discovery of these stress-inducible protective mechanisms shows how powerfully the retina can intervene to prevent photoreceptor death caused by both genetic mutations and environmental factors such as bright light. Further understanding of the mechanisms of action may allow their protective actions to be isolated and applied therapeutically.

Light restriction One of the earliest ideas of treatment for IRDs was to reduce the light reaching the retina, for example by wearing sunglasses. Early trials of light restriction in humans gave partly encouraging, partly disappointing outcomes, probably because the effects vary with the mutation. Some forms of rhodopsin mutation, the P23H mutation, for example, make photoreceptors particularly vulnerable to light. A recent survey of rodent IRD models showed that in the majority (15 of 20 strains summarized), dark rearing slows the degeneration and light exposure accelerates it. In other strains, the degeneration appeared to be unaffected by light exposure. In no case did light restriction make the degeneration worse.

In one rodent strain, the rhodopsin-mutant P23H-3 transgenic rat, light restriction has been shown, not just to slow the degeneration, but to allow photoreceptors to recover functional responsiveness to light; that is, the visual loss was partially reversed. The recovery results from the ability of damaged but surviving photoreceptors to rebuild their severely shortened outer segments and resume function. This self-repair ability may be a feature of wild-type photoreceptors; it has been described in the context of photostasis, the ability of photoreceptors to shorten and lengthen their outer segments as ambient light rises and falls, to keep the daily photon catch constant.

For many of the light-responsive strains, the underlying genetic mutation has been shown to occur in humans and cause a form of IRD. The animal data indicate, however, that light restriction is likely to be effective only in some forms of the disease.

Antioxidant therapy There is growing interest in the use of antioxidants, added to the diet or given by injection, in preventing retinal degenerations. Supplements such as β-carotene, vitamins C and D, and zinc have shown promise in reducing complications and may slow the onset and progress of age-related macular degenerations. Prefeeding with dietary β-carotene and saffron has been shown to be effective in protecting rat photoreceptors from light-induced damage, with saffron providing better functional preservation. In mouse IRDs, antioxidants given by injection have been shown to slow the death of cones after rods have died in the rd1 mouse and to slow the death of rods in rd1, rd10, and Q334ter mouse strains. These are distinct mutations, *rd*1 and *rd*10 in cGMP PDE and Q334ter in rhodopsin. The effectiveness of antioxidants in these several cases suggests that oxidative stress is a factor in most degenerations, as predicted by the oxygen toxicity hypothesis.

Near-infrared radiation The irradiation of tissue by light in the far- to near-infrared (NIR) range (670–880 nm) has been shown to improve the recovery of heart muscle from ischemic damage, to promote healing of opportunistic infections in immune deficiency states, and to attenuate degeneration in the injured optic nerve. Mechanistic studies have shown that NIR interacts with cytochrome oxidase, the mitochondrial enzyme which sequesters oxygen from extracellular fluid into oxidative phosphorylation pathways, triggering signaling pathways which result in improved adenosine triphosphate production; upregulation of the expression of protective, antioxidant factors; and improved cell survival. In the retina, NIR has been shown to protect photoreceptors against toxin-induced mitochondrial dysfunction, and preliminary studies have shown a marked slowing of the developmental stage of photoreceptor death in the P23H-3 rat.

The dosages required for these effects are low (fluence of $4\,\mathrm{J\,cm^{-2}}$ for periods of minutes) and can be delivered noninvasively. Further, the mitochondrion is the organelle in which oxygen is sequestered into oxidative metabolism, with the production of damaging oxygen radicals. It is also the source of many cell death signals triggered by stress. This suggests that NIR is acting a critical site, where oxidative stress can lead to cell death.

The Range of Animal Models of Human Diseases

This article has touched on information from only a fraction of the scores of animal models in which mutations cause IRDs. The great majority of these models are murine. Mouse strains with spontaneous or engineered mutations now provide mutation-specific models of many human IRDs for which the gene responsible has been identified (see **Figure 4**). Indeed, novel IRDs recognized first in the mouse have provided candidates for gene discovery when a new human family presents with an IRD of unknown cause. More generally, the value of animal models, and particularly the mouse forms, seems likely to grow, for several reasons. The entire genomes of human and mouse are now known,

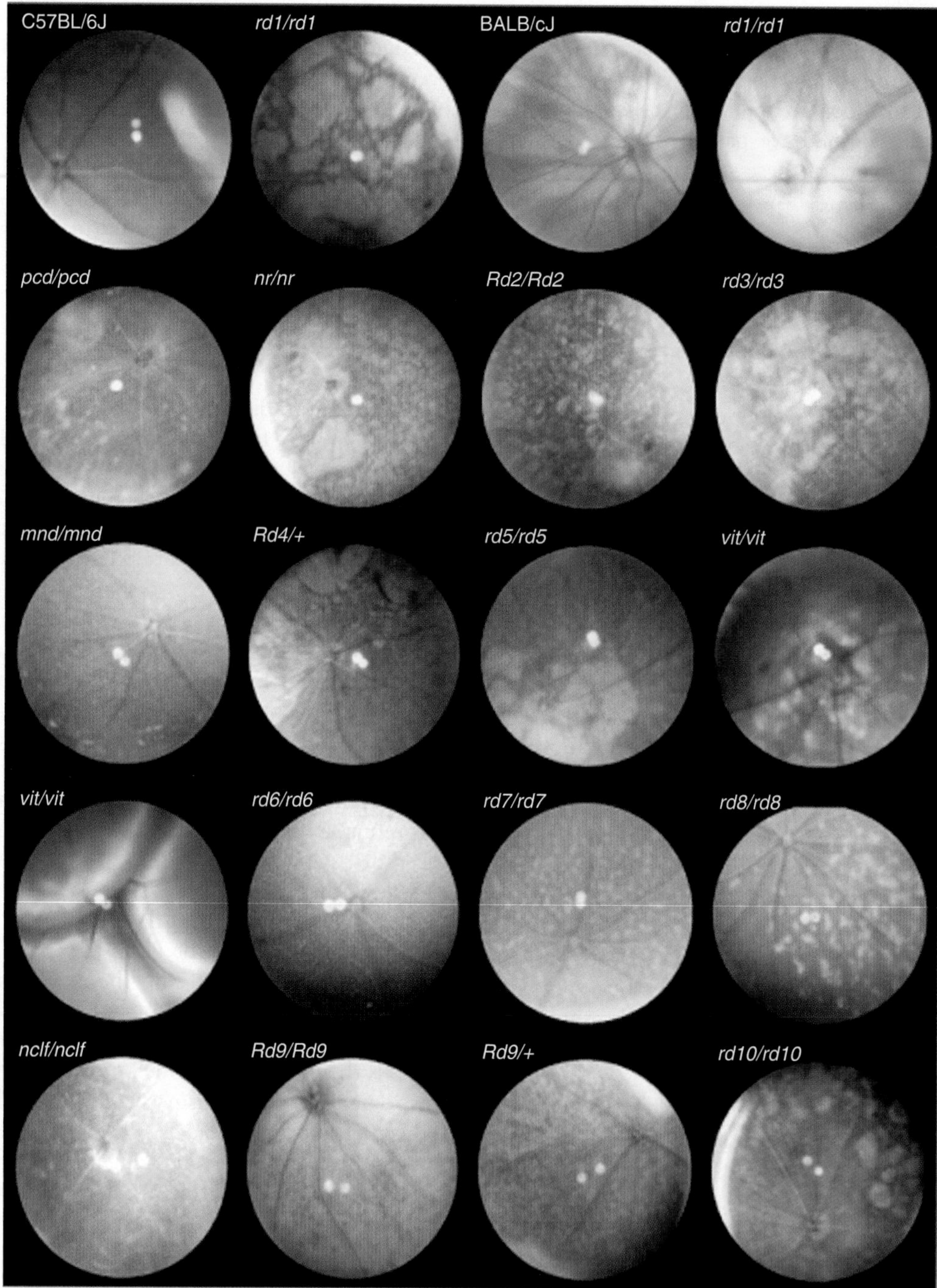

Figure 4 The appearance of the fundus of the mouse eye as seen through an indirect ophthalmoscope. The appearance of the fundus varies characteristically with the form of the inherited retinal dystrophy (IRD). This approach is noninvasive and a valuable screening tool for disease. The C57BL/6J and Balb/CJ retinas are nondegenerative wild types, the former pigmented, the latter albino. The lack of pigment gives the Balb/C fundus its characteristically pink (albino) appearance. The retinal blood vessels radiating from the optic disc are a prominent feature of the fundus. The fundi of IRD retinas occasionally show abnormal spots, thinning of the retinal vessels, and abnormalities of pigmentation. Most have their counterparts in the human IRDs. From Chang B, Hawes NL, Hurd RE, Davisson MT, Nusinowitz S, and Heckenlively JR (2002) Retinal degeneration mutants in the mouse. *Vision Research* 42: 517–525.

and the homology between mouse and human genomes is very high, of the order of 95%. The physiology, structure, and organization of the retina are very similar in the two species, so that the mouse retina is a strong model for the primate retina, except that it lacks a central specialization comparable to the fovea centralis or macula lutea found in primates such as the human. The techniques of engineering transgenic and knockout strains are most highly developed for the mouse. The life cycle of the mouse is naturally short, allowing the relatively rapid observation of developmental and aging processes. Finally, because the mouse genome has been extensively mapped, known markers of gene location are abundant, reducing the time needed to identify mutations (see **Figure 5**).

The use of other animal models has complemented this reliance on the mouse. The success of gene therapy in two dog strains has been very influential in establishing the potential of this approach, and the larger eyes of larger species provide better models for testing the delivery of drugs and engineered gene therapy constructs to the human eye. The study of

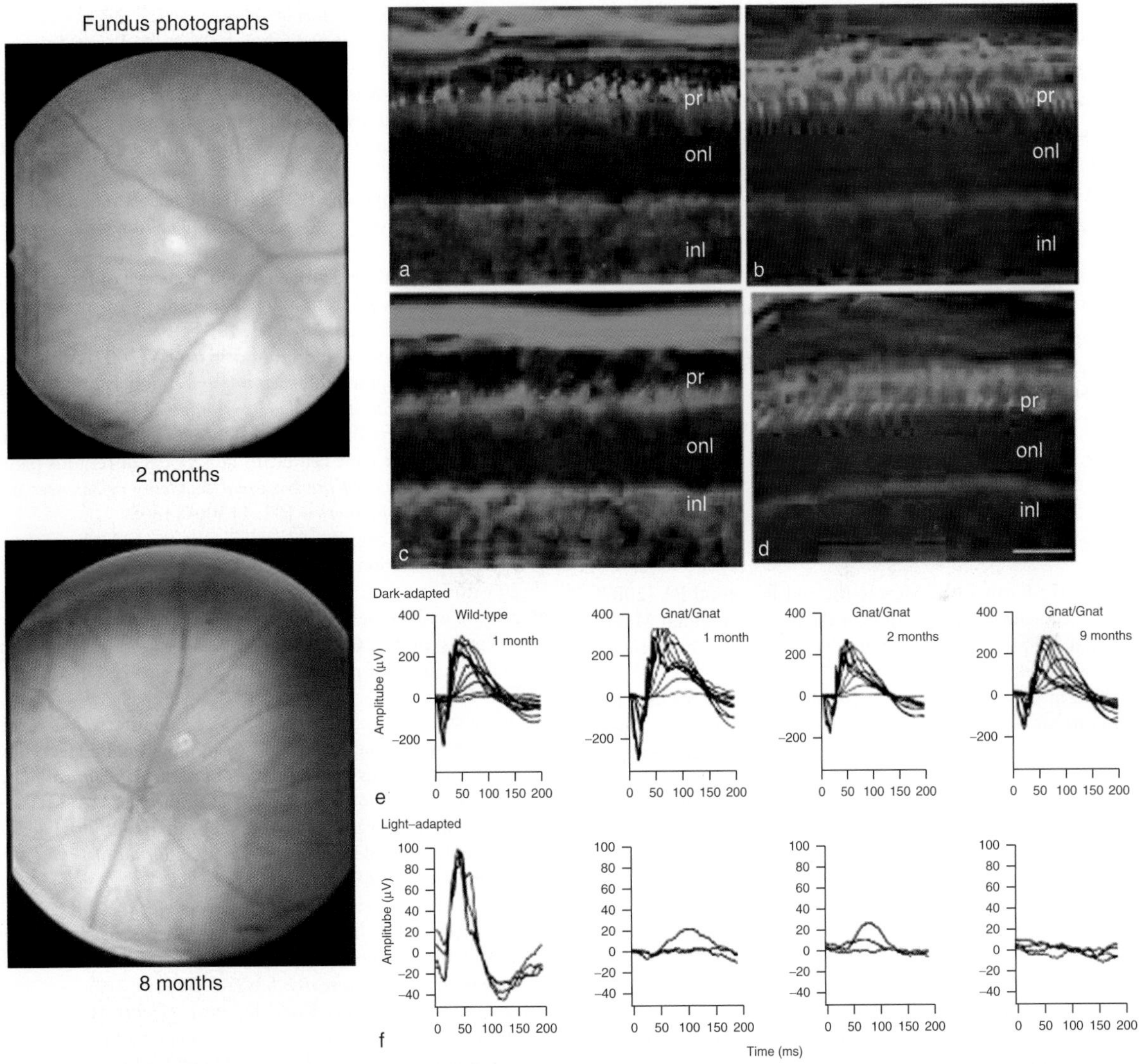

Figure 5 Mutation in mice strain with *GNAT2*, guanine nucleotide binding protein alpha-subunit of transducin, which is necessary for hyperpolarization of the cone photoreceptor. Fundus photograhs of GNAT2 mutant at 2 months and 8 months show progressive vascular attenuation in albino mouse with mutation. Histology Figures 5(a)–5(d), shows staining of α-transducin receptors in wild type (a) and mutant mouse (c) at 4 weeks, and cone staining with rhodamine-conjugated PCNA in wild type (b) and mutant (d). The GNAT-2 mutant shows the presence of large numbers of cones even though the α-transducin expression is greatly reduced and the photopic ERG is very abnormal. The retention of cones in the face of poor cone function suggests patients with this disease may respond to early gene therapy, before degeneration occurs. Scale bar = 50 μm. From Chang B, Dacey MS, Hawes, NL, et al. (2006) Cone photoreceptor function loss-3, a novel mouse model of achromatopsia due to a mutation in Gnat2. *Investigative Ophthalmology & Visual Science* 47: 5017–5021.

fish retina has demonstrated mechanisms of lifelong retinal neurogenesis, which underlie the lifelong growth of the retina in teleosts. In particular, Müller cells act as stem cells, able to generate the full spectrum of retinal cells in order to maintain retinal growth or repair lesions. This knowledge has led to a search for comparable mechanisms in the adult mammalian retina, which is normally unable to regenerate. Evidence is growing that mammalian Müller cells can be induced experimentally to act as stem cells, raising the possibility that regeneration can be induced in human retina to replace photoreceptors lost in a dystrophy.

Because until recently their pathophysiology has been unknown, IRDs have proved intractable to therapy over the century since they were recognized in humans. A range of therapeutic approaches, and many animal models, will be needed as the knowledge that has accumulated about the pathogenesis of IRDs is translated into long-awaited therapies.

See also: Photoreceptor Adaptation; Photoreceptors: Physiology; Phototransduction; Retinal Ganglion Cells: Anatomy; Retinal Ganglion Cells: Receptive Fields; Retinal Horizontal Cells; Retinal Pharmacology: Inner Retinal Layers; Retinal Development: An Overview; Retinitis Pigmentosa.

Further Reading

Bernardos RL, Barthel LK, Meyers JR, and Raymond PA (2007) Late-stage neuronal progenitors in the retina are radial Muller glia that function as retinal stem cells. *Journal of Neuroscience* 27: 7028–7040.

Bowne SJ, Liu Q, Sullivan LS, et al. (2006) Why do mutations in the ubiquitously expressed housekeeping gene IMPDH1 cause retina-specific photoreceptor degeneration? *Investigative Ophthalmology & Visual Science* 47: 3754–3765.

Chang B, Dacey MS, Hawes NL, et al. (2006) Cone photoreceptor function loss-3, a novel mouse model of achromatopsia due to a mutation in Gnat2. *Investigative Ophthalmology & Visual Science* 47: 5017–5021.

Chang B, Hawes N, Davisson M, and Heckenlively J (2007) Mouse models of RP. In: Tombran-Tink J and Barnstable CJ (eds.) *Retinal Degenerations: Biology, Diagnostics, and Therapeutics*, pp.149–160. Totowa, NJ: Humana Press.

Chang B, Hawes NL, Hurd RE, Davisson MT, Nusinowitz S, and Heckenlively JR (2002) Retinal degeneration mutants in the mouse. *Vision Research* 42: 517–525.

Cideciyan AV, Jacobson SG, Aleman TS, et al. (2005) *In vivo* dynamics of retinal injury and repair in the rhodopsin mutant dog model of human retinitis pigmentosa. *Proceedings of the National Academy of Sciences of the United States of America* 102: 5233–5238.

Clarke G, Collins RA, Leavitt BR, et al. (2000) A one-hit model of cell death in inherited neuronal degenerations. *Nature* 406: 195–199.

Clarke G, Heon E, and McInnes RR (2000) Recent advances in the molecular basis of inherited photoreceptor degeneration. *Clinical Genetics* 57: 313–329.

Coleman H and Chew E (2007) Nutritional supplementation in age-related macular degeneration. *Current Opinion in Ophthalmology* 18: 220–223.

Daiger SP, Sullivan LS, and Browne SJ (2007) *RetNet: Retinal Information Network*, http://www.sph.uth.tmc.edu/RetNet/ (accessed Jul. 2008).

Daiger SP, Sullivan LS, Browne SJ, and Rossiter BJF (2008) RetNet: Cloned and/or mapped genes causing retinal diseases. Available at http://www.sph.uth.tmc.edu/RetNet.

Doonan F, Donovan M, and Cotter TG (2005) Activation of multiple pathways during photoreceptor apoptosis in the *rd* mouse. *Investigative Ophthalmology & Visual Science* 46: 3530–3538.

Duncan JL, Paskowitz DM, Nune GC, et al. (2006) Retinal damage caused by photodynamic therapy can be reduced using BDNF. *Advances in Experimental Medicine and Biology* 572: 297–302.

Hitchock P and Raymond PA (2004) The teleost retina as a model for developmental and regeneration biology. *Zebrafish* 1: 257–272.

Jozwick C, Valter K, and Stone J (2006) Reversal of functional loss in the P23H-3 rat retina by management of ambient light. *Experimental Eye Research* 83: 1074–1080.

Komeima K, Rogers BS, and Campochiaro PA (2007) Antioxidants slow photoreceptor cell death in mouse models of retinitis pigmentosa. *Journal of Cellular Physiology* 213(3): 809–815.

Komeima K, Rogers BS, Lu L, and Campochiaro PA (2006) Antioxidants reduce cone cell death in a model of retinitis pigmentosa. *Proceedings of the National Academy of Sciences of the United States of America* 103: 11300–11305.

MacLaren RE, Pearson RA, MacNeil A, et al. (2006) Retinal repair by transplantation of photoreceptor precursors. *Nature* 444: 203–207.

Mervin K and Stone J (2002) Developmental death of photoreceptors in the C57BL/6J mouse: Association with retinal function and self-protection. *Experimental Eye Research* 75: 703–713.

Nusinowitz S and Heckenlively J (2006) Evaluating retinal function in the mouse retina with the electroretinogram. In: Heckenlively J and Arden G (eds.) *Principles and Practice of Clinical Electrophysiology of Vision*, pp. 899–910. Cambridge, MA: MIT Press.

Paskowitz DM, LaVail MM, and Duncan JL (2006) Light and inherited retinal degeneration. *British Journal of Ophthalmology* 90: 1060–1066.

Preising M (2007) *Retina International Scientific Newsletter.*

Rakoczy P, Yu MJ, Nusinowitz S, Chang B, and Heckenlively JR (2006) Mouse models of age-related macular degeneration. *Experimental Eye Research* 82: 741–752.

Stone J, Maslim J, Valter-Kocsi K, et al. (1999) Mechanisms of photoreceptor death and survival in mammalian retina. *Progress in Retinal and Eye Research* 18: 689–735.

Stone J, Mervin K, Walsh N, Valter K, Provis J, and Penfold P (2005) Photoreceptor stability and degeneration in mammalian retina: Lessons from the edge. In: Penfold P and Provis J (eds.) *Macular Degeneration: Science and Medicine in Practice*, pp. 149–165. Berlin: Springer.

Sullivan LS, Heckenlively JR, Bowne SJ, et al. (1999) Mutations in a novel retina-specific gene cause autosomal dominant retinitis pigmentosa. *Nature Genetics* 22: 255–259.

Valter K, Bisti S, and Stone J (2003) Location of CNTFR on outer segments: Evidence of the site of action of CNTF in rat retina. *Brain Research* 985: 169–175.

Wu WW, Wong JP, Kast J, and Molday RS (2005) RS1, a discoidin domain-containing retinal cell adhesion protein associated with X-linked retinoschisis exists as a novel disulfide-linked octamer. *Journal of Biological Chemistry* 280: 10721–10730.

Relevant Websites

http://www.retina-international.org – Retina International Animal Model Database.

http://www.retina-international.org – Retina International Mutation Databases.

http://www.sph.uth.tmc.edu – RetNet: Retinal Information Database.

Animal Models of Motor and Sensory Neuron Disease

P Bomont and D W Cleveland, University of California at San Diego, La Jolla, CA, USA

Motor and Sensory Tracts

Movements of many parts of the body, from the lips and the eyelids to the hands and toes, have their origin in the brain in a specialized region called the motor cortex (**Figure 1**). Upper motor neurons receive information there that they transmit to spinal motor neurons, often through an intermediate interneuron. Firing of the lower motor neuron triggers muscle contraction. Upper motor neuron degeneration induces brisk tendon reflexes, spasticity, and hyperreflexia, whereas denervation of the muscle upon lower motor neuron degeneration or loss leads to muscle weakness and loss of tone.

The sensory system works in the opposite direction with information flowing to the brain from the periphery. Receptors, primarily located at the surface of the skin, provide initial sensory input (touch, pressure, vibration perception, pain, and temperature) to the axons of the sensory neurons which carry these signals into the spinal cord. The information is then transferred to the neurons located in the sensory cortex, where the information is decoded by the brain.

Gene Deletion ('Knockout') and Transgenic Mouse Models

Two main principles are used to reproduce inherited human pathologies in the mouse (**Figure 2**). When the human disease is recessive, a situation usually caused by the absence of one particular protein, a genetic mimic in mice can be constructed by disrupting the corresponding gene in the mouse genome (**Figure 2(a)**). These models are frequently called 'knockout models,' referring to the disruption or removal of the targeted gene and its encoded protein. A variation of this method, called 'conditional knockout,' permits the removal of the gene of interest only in chosen tissues (**Figure 2**, box 1). This alternative is preferred in cases in which the deletion of the gene from the entire animal induces lethality in early development.

For dominant neurodegenerative disorders, disease is usually caused by the gain of a toxic property resulting from a modification in the protein (an example of this is amyotrophic lateral sclerosis (ALS) caused by mutation in the gene encoding superoxide dismutase). Two approaches are possible for generating a mouse model. The mouse gene can be replaced by the mutated one (**Figure 2(a)**), thereby creating a gene 'knock in' model. Alternatively, random insertion of the abnormal gene in the mouse genome can be achieved to produce a transgenic mouse (**Figure 2(b)**). The latter is largely preferred because it can be constructed more quickly. Importantly, transgene insertion can be readily achieved in mice or rats. An advantage to rat models in neuroscience is that the larger size permits more complex surgical approaches and tissue dissections. Additionally, for some classes of neurons, those derived from rats are more easily grown in culture. Finally, rats are amenable to some behavioral analyses that are more difficult in the mouse.

Neurodegeneration in Human

Motor Disease

Amyotrophic lateral sclerosis ALS, more familiarly known in the United States as Lou Gehrig's disease, is the most prominent adult motor neuron disease, with a typical age of disease onset of 50–60 years. Most (90%) incidences of ALS are referred to as sporadic because there is no family history of disease and hence no evidence for a major genetic component. Disease involves the selective alteration and death of both the upper and the lower motor neurons, resulting in spasticity, hyperreflexia, and progressive weakness of skeletal muscles, atrophy, and death due to respiratory muscle paralysis within 1–5 years after onset.

For the incidences (~10%) of ALS with a genetic origin, determination of the gene(s) responsible has been significantly challenged by the wide heterogeneity of the disease, genetically and clinically. Many genetic loci have been associated with different forms of motor neuron degeneration. Most prominently, mutations in the gene encoding for Cu/Zn superoxide dismutase (SOD1), an enzyme used to detoxify an aberrant oxygen species, have been shown to account for 20% of the familial cases of ALS. Dominantly inherited, analysis of enzymatic activity of mutated SOD1 in ALS patients together with genetic manipulations in mice have clearly revealed that SOD1-mediated toxicity is not due to a reduction in activity of this enzyme but, rather, due to a gain of a new toxic property or properties. Accordingly, transgenic rodent models have been generated by inserting into their genome a human SOD1 gene carrying different ALS-causing mutations. Many models have been created to dissect the disease-associated toxic mechanism(s), the most extensively used being transgenic for the

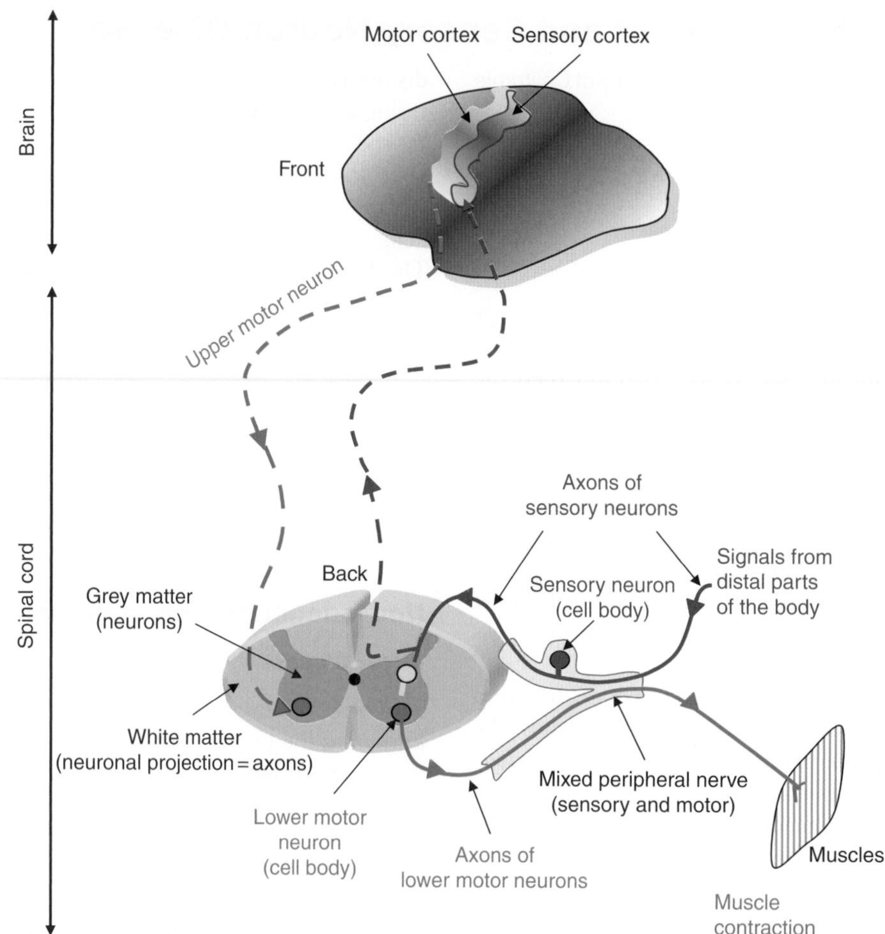

Figure 1 Motor and sensory pathways. Sensory neurons receive information from different parts of the body and transmit it to the spinal cord and then to the brain through the ascending axons of sensory neurons (dark blue). The motor pathway originates in the motor cortex, where upper motor neurons (light blue) connect to transmit information to the lower motor neurons in the spinal cord, which in turn send and deliver signals to the muscles, inducing their contraction.

amino acid substitution mutations G93A (glycine substituted to alanine at position 93), G37R (glycine to arginine at position 37), and G85R (glycine to arginine at position 85). Although none of these completely recapitulate all features of the human disease, they all develop progressive motor neuron degeneration with slightly different symptoms and disease progressions that are characteristic for each mutation, with onset and survival determined by the level of mutant expression. All of these mice develop motor neuron degeneration, limb weakness associated with neurofilament misaccumulation, impaired axonal transport, and axonal swelling.

Mouse and rat models for ALS have been extensively used to test proposals for a wide range of pathological mechanisms. These include impairment of the chaperone/degradation machinery, dysfunction of mitochondria, oxidative stress, alteration of cytoskeletal architecture and axonal transport, glutamate-mediated excitotoxicity, aberrant growth factor signaling, microglial cell involvement, and inflammation. Another aspect particularly important to the identification of therapeutic targets is whether other cell types contribute to motor neuron degeneration. Motor neurons require support provided by several cell types in the central nervous system which also express the mutant SOD1. Use of a transgene that can be eliminated selectively from motor neurons has shown that mutant damage in motor neurons initiates disease onset and an early phase of disease progression. A similar approach has proven that mutant damage within microglial cells, the immune cells of the central nervous system, has no effect on disease onset but accelerates disease progression.

Additional forms of human ALS-like motor neuron disease, all associated with motor neuron degeneration and death, have been designated ALS2–ALS8. Significant divergence in the age of onset and clinical presentation has led to disagreement as to whether these should be classified as forms of ALS. This is

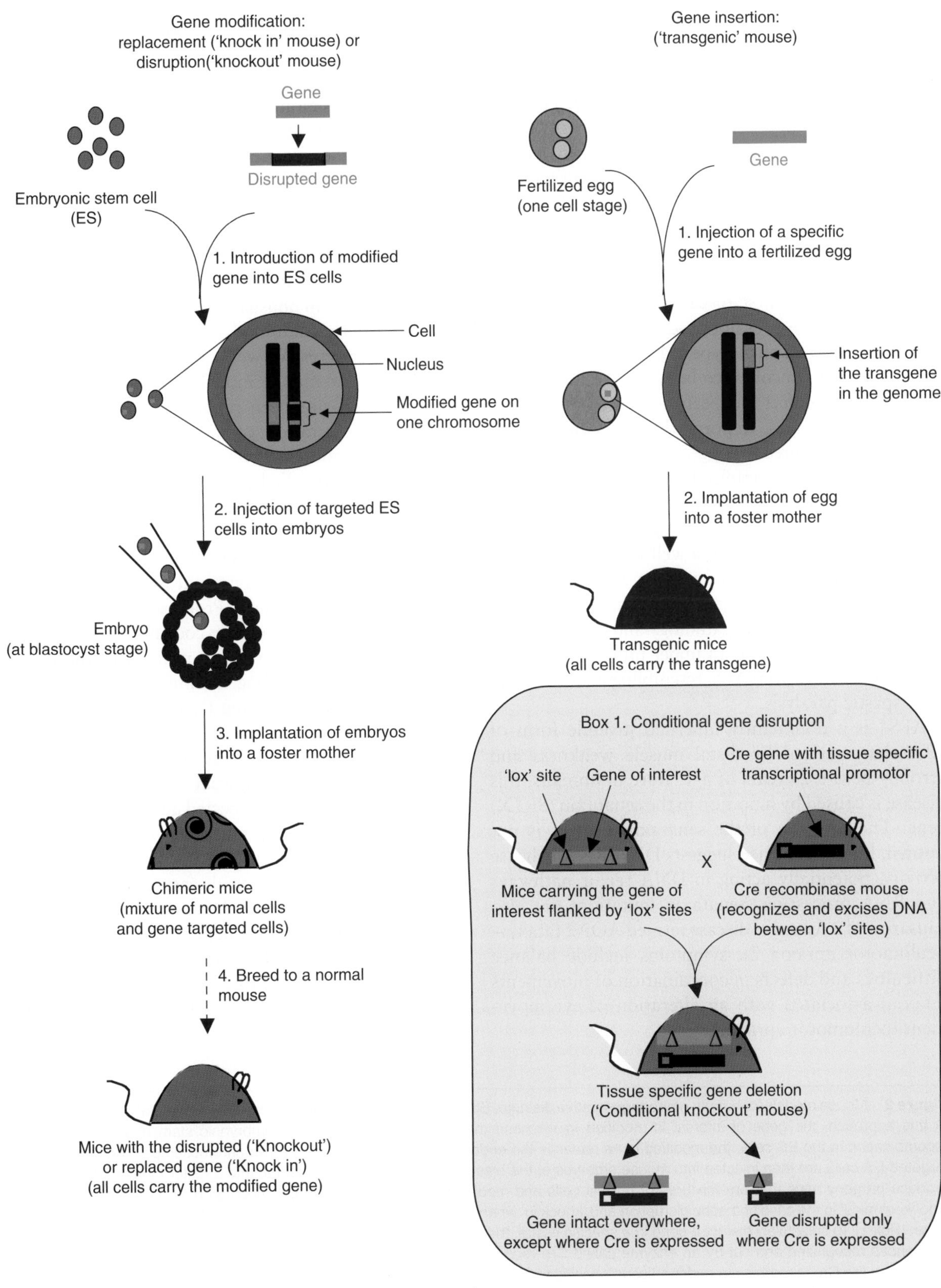

Figure 2 Continued

especially so for ALS2, an infantile and juvenile-onset form of motor neuron disease which is caused by loss of an apparent guanine exchange factor enzyme, termed alsin, that can activate one or more enzymes from the family of small GTPases known as G-proteins. Proposed G-protein partners of alsin include Rab5 and Rac1. Alsin has been shown to modulate processes such as cytoskeleton organization and transport of membrane cargoes.

Recessive AL2 motor neuron disease has been reproduced in mice by deleting the corresponding gene (**Figure 2(a)**). This first revealed that the absence of alsin in the mouse does not produce a juvenile disease that has as severe a phenotype as found in human. The lack of perfect concordance between the human and mouse genetic defect has been proven to be true for many other diseases. Importantly, lower motor neurons (those that directly trigger contraction of skeletal muscles) are almost completely spared in mice deleted of the ALS2 gene, which do not develop muscle weakness. Analysis of voluntary running, however, revealed that these mice move slowly, a well-recognized clinical sign of an upper motor neuron defect. Additionally, the upper motor neurons degenerate. Thus, the animal model reinforces that the pathology and the phenotype of loss of ALS2 provoke an upper motor neuron disease quite distinct from ALS, resembling instead hereditary spastic paralysis.

ALS4 is a dominantly inherited juvenile form of ALS characterized by distal muscle weakness and atrophy before 25 years of age. Rare in humans, this disease is caused by mutation in the senataxin (SETX) gene. The function of the senataxin protein is not known, but its sequence suggests DNA/RNA helicase activity, potentially acting in DNA repair pathways and RNA processing. Senataxin mutations are also causative of a recessive disease termed AOA2 (ataxia–oculomotor apraxia 2); symptoms include balance difficulties and defects in coordination of movements (ataxia) associated with an alteration of eye movement (oculomotor apraxia).

Axonal transport has been shown to play an essential role in neuronal survival. Two families of motor proteins, the kinesins and dynein, which power transport of cargoes along the microtubules away from the nerve cell body (anterograde) or returning components from the nerve ending to the cell body (retrograde), respectively, have been shown to cause or contribute to motor neuron disease in humans. This has been replicated in mice. In humans, mutation in the subunit p150 of the dynactin complex (which interacts with the dynein complex) has been identified to be causative of an unusual form of lower motor neuron disease in early adulthood, characterized by vocal fold paralysis, progressive weakness, and muscle atrophy.

Although a mouse model of the p150 dynactin mutation has not been reported, it is widely established that alteration of the dynein–dynactin complex causes motor deficits. Indeed, overexpression of dynamitin (also called p50 of the dynactin complex) has been shown in mice to disrupt dynein–dynactin interaction leading to late-onset slowly progressive motor neuron degeneration, characterized by muscle weakness, trembling, abnormal gait, and deficits in strength and endurance. Progressive alterations of muscle function and motor coordination have been linked in mice to distinct mutations in the dynein gene. The corresponding mutants, *Loa* (Legs at odd angles) and *Cra1* (cramping 1), develop motor deficits, confirming that alteration in the dynein motor complex and its presumed effects on moving components through axons are of unusual importance for motor axons. A completely unexpected and unexplained finding is that both dynein mutations delay disease progression and increase the life span of mice that develop early onset motor neuron disease caused by expression of an ALS-causing SOD1 mutant. Understanding what is causing the motor deficits in the *Loa* and *Cra1* mice will certainly help us to understand what leads to this amelioration of disease in SOD1 mutant mice and might help to identify new therapeutic approaches for ALS.

Figure 2 Mouse models for human neurodegenerative disease. Strategies for gene modification or replacement. (a) Gene modification. In this approach, the gene of interest is modified *in vitro* and then introduced into (brown) embryonic stem (ES) cells. Following recombination in the ES cells, the modified gene replaces the endogenous gene on one of the two chromosomes. The corresponding targeted ES cells are then injected into mouse embryos at the blastocyst stage and are implanted into a foster mother. Those females produce progeny mice that are mixtures of normal cells and modified cells. Mice carrying the modified gene in all cells (also called 'knockout mice' in the case of a gene disruption and 'knock in' when the gene is replaced) are obtained by standard breeding with normal mice. (Box 1) To selectively disrupt a gene in a specific tissue, the gene region to be deleted is flanked by two lox sites, which are small sequences recognized and cut by an enzyme called Cre recombinase. Breeding mice carrying the modified lox–gene–lox with mice expressing the Cre recombinase under a tissue-specific transcriptional promoter generates the gene disruption only in the tissues that express the Cre recombinase. (b) Gene insertion: 'transgenic' mouse. An exogenous gene is inserted randomly into the mouse genome after direct injection of the gene into the one cell stage fertilized egg. After implantation into a foster mother, the oocyte matures into mice whose cells all carry the transgene.

Spinal muscular atrophy After cystic fibrosis, spinal muscular atrophy (SMA) is the most common autosomal recessive disorder in humans and represents the most common genetic cause of infant mortality. In this disorder, specific neuronal loss of lower motor neurons in the spinal cord results in atrophy of proximal muscles of the trunk and the limbs. SMA cases are classified into three groups: Type I SMA is the most severe phenotype, with an age of onset before 6 months and death occurring before 2 years of age; type II SMA is intermediate, with an age of onset between 6 and 18 months; and type III SMA, which initiates after 18 months, is the mildest form, slowly progressing with a normal life span.

Identifying the genetic cause of SMA was challenging because of the complex and unstable nature of the region of human chromosome 5 where the defective gene resides. For all disease types, symptoms are caused by inactivating mutations in the *SMN1* (survival of motor neuron 1) gene, but the variability in the disease severity depends on the number of copies (ranging from one to five) of an almost identical copy gene, the *SMN2*, located just next to *SMN1*. Each copy of the *SMN2* gene produces only a small amount of functional product so that only a single copy leads to fatal, severe type I disease, and five copies lead to the more benign type III disease. Although the role of SMN protein in lower motor neuron survival is not fully understood, it has been implicated in processing within the nucleus of initial RNA copies of each expressed gene into their mature forms capable of translation into the corresponding proteins. A second likely function is as an RNA-binding protein for RNA transport into axons.

Because the mouse genome contains only the *SMN1* gene and no *SMN2* gene, mouse models for SMA have been attempted by deleting the *SMN1* locus. These animals are not viable, demonstrating an essential role for SMN1 in cell survival. Selective deletion of the *SMN1* gene solely from neurons (**Figure 2**, box 1) produced mice with motor abnormalities and skeletal muscle denervation secondary to motor axon loss, culminating in death at a mean age of 25 days. This is not a very satisfying model, but it does demonstrate an essential role for SMN1 in neurons. SMN1 expression has also been modified only in muscle cells. This has revealed that SMN expression in muscle prevents the SMA phenotype and increases the life span of the animals. This supports targeting muscle as a therapeutic strategy in SMA. Because in human the severity of disease is modulated by the *SMN2* gene, the *SMN2* gene has been introduced (by a transgene) to the mice with the 'neuronal' deletion of *SMN1*. This showed that *SMN2* gene number ameliorates SMN1 absence, closely mimicking what occurs in human disease.

Motor and Sensory Disease

Charcot–Marie–Tooth Charcot–Marie–Tooth (CMT) diseases refer to a heterogeneous class of neuropathies that affect not only motor but also sensory nerves in the peripheral nervous system. In addition to the weakness and atrophy of the distal limb muscles, patients experience impaired sensation and absence of deep tendon reflexes. Representing the most common inherited disorder of the peripheral nervous system, this group of diseases, which are sometimes transmitted through dominant, recessive, or X-linked modes of inheritance, has been classified into two subgroups. The demyelinating form (also called CMT1) results in an impairment of the myelin sheath produced by Schwann cells. CMT2 forms are characterized by degeneration of the axon. The genetic heterogeneity of the disease has led to the identification of at least 17 genes involved in both demyelinating and axonal forms of CMT. The CMT1-causing genes (*PMP22*, *MPZ*, *LITAF/SIMPLE*, and *EGR2*) have revealed an essential role of myelin compaction, the regulation of myelin protein degradation, and the transcriptional control of myelination-specific genes in demyelination.

Axonal survival has been shown to depend on many different pathways. Among these, mutation in the gene encoding the mitochondrial GTPase mitosfusin 2 as a cause of CMT has proven an essential role of mitochondrial fusion/transport in axon survival. Maintenance of axonal architecture and transport (for motor neuron survival) is particularly important for sustaining neuron function. Forms of CMT2 are also caused by mutation in the neurofilament subunit NF-L or the motor kinesin KIF1bβ. Impaired intracellular trafficking is also a common theme in axonal degeneration, as revealed by mutation in the regulator of vesicle trafficking Rab7 and the phosphatase MTMR2, which modulate membrane trafficking.

CMT1A is the most frequent form of CMT. Instead of a typical disease-causing mutation that alters or eliminates the encoded protein, CMT1A disease is caused by a duplication of the genomic region surrounding an otherwise normal *PMP22* gene so that affected individuals have an extra copy of the normal gene. Transgenic rat and mouse models containing multiple copies of the *PMP22* gene have proven that overexpression of only this gene is sufficient to cause peripheral demyelination and the symptoms associated with the human disease. This also demonstrated an increased disease severity with increased expression of PMP22. These animal models have been shown to be very useful to test the first rational experimental therapies of CMT1A. Indeed, a synthetic antagonist of the nuclear progesterone receptor was shown to reduce PMP22 expression and to ameliorate the clinical severity in those animals. Moreover, administration of ascorbic acid, an essential factor of *in vitro* myelination (already approved by the US Food and Drug Administration for other clinical indications), has prolonged survival and restoration of myelination in those models, suggesting its use in human clinical trials for CMT1A.

Giant axonal neuropathy Factors that establish the cytoarchitecture of the axon and its associated axonal transport components are critical for neuronal survival. Disorganized cytoskeletal intermediate filaments constitute the hallmark of another neurodegenerative disease termed giant axonal neuropathy (GAN). This is a rare, recessively inherited condition characterized by a structural deficit in both the central nervous system and the peripheral motor and sensory axons. With disease onset in infancy and marked by gait instability and frequent falls, patients develop diminution of deep tendon reflexes, muscle atrophy, and muscle weakness that progressively evolves to sensory loss and loss of ambulation. Other symptoms, including ataxia, language deficits, and mental retardation, reveal a later impairment of the central nervous system.

GAN is a progressive, fatal disease, with life expectancy of less than 30 years. Identification of the *GAN* gene led to recognition that its encoded protein, gigaxonin, plays a critical role in the ubiquitination machinery whose function, among others, is to tag proteins for degradation. Other errors in other components of this degradation apparatus are implicated in several neurodegenerative disorders, the most prominent of which is Parkinson's disease. Gigaxonin seems to link degradation and impairment of the axonal cytoskeleton. Loss of gigaxonin prevents normal degradation of proteins involved in the assembly and dynamics of microtubules, the protein polymers that serve as the tracks for transport of components up and down the axon. Whether altered microtubules properties are the disease causing damage and, if so, how this provokes the specific aggregation of intermediate filaments seen in patients are not known.

Developing animal models for rare disorders such as GAN is particularly crucial since insights from direct inspection of patient materials are very limited. The study of the mechanisms underlying neurodegeneration through construction and analysis of genetic models mimicking the loss of gigaxonin is at its earliest stages. Mice deleted of the *GAN* gene develop a progressive deterioration of motor function and ataxia, a sign of central nervous system impairment. As in human disease, the absence of gigaxonin generates axonal loss and neurofilament aggregation. Alteration of cytoskeleton architecture has been further confirmed by the decreased density of axonal microtubules and the increased abundance of proteins that affect microtubule assembly.

Conclusion

Since essentially all forms of motor or sensory neuron degeneration are incurable, mouse and rat models represent an indispensable resource to address the origin of neuronal dysfunction, the mechanisms of degeneration, and the determinants of the selectivity of death of subpopulations of neurons. In turn, these key issues will power the design of strategies to ameliorate these life-long, frequently fatal conditions. Thus, for example, for ALS, this fundamental research has led to the development of many experimental therapies in these animals, including antiglutamate drugs, neurotrophic growth factor delivery, SOD1 silencing, and stem cell therapy, which will be tested in human clinical trials. Identifying the pathways of neuron survival will help in understanding the basis of motor/sensory neuron diseases in humans for which the majority of genetic causes remain unknown.

See also: Amyotrophic Lateral Sclerosis (ALS): Disease Mechanisms; Amyotrophic Lateral Sclerosis (ALS); Axonal Transport and ALS; Neurofilaments: Organization and Function in Neurons.

Further Reading

Boillee S, Vande Velde C, and Cleveland DW (2006) ALS: A disease of motor neurons and their nonneuronal neighbors. *Neuron* 52: 39–59.

Boillee S, Yamanaka K, Copeland NG, et al. (2006) Motor neurons and microglia as determinants of disease onset and progression in inherited ALS. *Science* 312: 1389–1392.

Bomont P, Cavalier L, Blondeau F, et al. (2000) The gene encoding gigaxonin, a new member of the cytoskeletal BTB/kelch repeat family, is mutated in giant axonal neuropathy. *Nature Genetics* 26(3): 370–374.

Chen YZ, Bennett CL, Huynh HM, et al. (2004) DNA/RNA helicase gene mutations in a form of juvenile amyotrophic lateral sclerosis (ALS4). *American Journal of Human Genetics* 74(6): 1128–1135.

Ding J, Allen E, Wang W, et al. (2006) Gene targeting of GAN in mouse causes a toxic accumulation of microtubule-associated protein 8 and impaired retrograde axonal transport. *Human Molecular Genetics* 15(9): 1451–1463.

Monani UR (2005) Spinal muscular atrophy: A deficiency in a ubiquitous protein; A motor neuron-specific disease. *Neuron* 48(6): 885–896.

Rosen DR, Siddique T, Patterson D, et al. (1993) Mutations in Cu/Zn superoxide dismutase gene are associated with familial amyotrophic lateral sclerosis. *Nature* 362(6415): 59–62.

Shy ME (2004) Charcot–Marie–Tooth disease: An update. *Current Opinion in Neurology* 17(5): 579–585.

Animal Models of Parkinson's Disease

K E Soderstrom, G Baum, and J H Kordower,
Rush University Medical Center, Chicago, IL, USA

Introduction

Parkinson's disease (PD) is the second-most-common neurodegenerative disorder (after Alzheimer's disease), affecting more than 1 000 000 people in the United States. While there is widespread degeneration in the central and peripheral nervous systems in PD, the hallmark pathology remains the dopaminergic striatal insufficiency secondary to degeneration of dopaminergic neurons in the substantia nigra (SN). While a small percentage of PD is of a genetic origin, the vast majority of cases are sporadic and of unknown origin. It is particularly challenging to model a disease whose etiology is, for the most part, unclear. Therefore, PD researchers have attempted to model specific aspects of PD pathology independent of etiology. These attempts have included pharmacological models aiming to replicate striatal dopamine (DA) depletion and neurotoxic models that replicate nigral cell loss. While excellent for evaluating the role of the nigrostriatal system in PD and for testing therapeutic compounds, most of these models are limited in their ability to replicate such important aspects of the human disease as Lewy body formation, nondopaminergic systems affected in PD, and the progressive nature of the disease. Novel models emerging in the field have attempted to model these important facets of the human disease more accurately. Genetic models, the lipopolysaccharide model, and the use of aged animals as a PD model are all emerging alternatives to traditional DA-centric models of PD. The validity, strengths, and weaknesses of these traditional and novel animal models of PD will be addressed in this article, as well as the importance of animal models in general in our understanding and treatment of PD.

PD

PD is a progressive neurodegenerative disorder clinically characterized by the cardinal symptoms of cogwheel rigidity, resting tremor, bradykinesia, stooped posture, and shuffling gait. As stated above, there is a loss of dopaminergic cells in the SN pars compacta (SNpc) that results in insufficient DA innervation of the basal ganglia and subsequent increased inhibition of excitatory thalamo-cortical connections. Lewy bodies, intracellular inclusions principally containing α-synuclein, are also found in the remaining nigral neurons of PD patients.

While motor symptoms are often the most obvious deficits exhibited in PD, it is clear that numerous nonmotor and peripheral nervous system symptoms can significantly affect patient quality of life. Braak and co-workers recently used α-synuclein immunohistochemistry to hypothesize that PD pathology begins in the lower brain stem and olfactory bulb and then progresses in a caudal to rostral fashion. This progressive nature of PD and the involvement of nondopaminergic pathways in its development have traditionally been unrepresented in animal models of PD, despite the growing realization of their role in the human disease.

DA Depletion

The ultimate result of cell loss and cell dysfunction in the SN is the depletion of the neurotransmitter DA in the basal ganglia. This insufficient DA innervation is principally localized to the postcommissural putamen and results in the overdrive of globus pallidus and subthalamic nuclear outputs. The resulting inhibition of thalamocortical function results in the characteristic bradykinesia experienced by PD patients. Researchers first attempted to model PD by replicating this DA depletion in animals. Early studies using pharmacological interventions aimed to selectively deplete monoamine neurotransmitters lost in PD patients. Of these pharmacological models, one of the earliest and most utilized has been the reserpine model, which prevents the storage of DA in presynaptic terminals.

Reserpine

In the late 1950s, Arvid Carlsson demonstrated that the drug reserpine could cause reversible parkinsonism in animals, in a manner similar to that seen in humans. Reserpine blocks the storage of monoamines in catecholaminergic neurons, resulting in DA depletion in the striatum. Treating rabbits with reserpine causes a behavioral syndrome that modeled idiopathic PD and helped elucidate the critical role of DA in the pathogenesis of this disease. This new concept radically changed clinical neurology and lead to the discovery that the DA precursor, levodopa, as well as other prodopaminergic drugs, could profoundly benefit the lives of PD patients. Indeed, decades later, administration of levodopa remains the gold-standard treatment for PD patients.

While clearly groundbreaking, the reserpine model has significant limitations. Most obviously, reserpine produces only biochemical lesions, and therefore reserpine-treated animals do not model the neuronal dysfunction and degeneration seen in the human disease. In addition, the behavioral effects of reserpine are transient. While this can be advantageous in some circumstances in understanding nigrostriatal function, PD is a progressive chronic disorder, and if possible, progressive functional deficits are most often required for study. Last, reserpine depletes all monoamines, not exclusively DA. While norepinephrine depletion from the locus coeruleus is a facet of PD degeneration, depletion of all monoamines does not accurately model PD.

Nigral Cell Loss

While pharmacological models helped elucidate the role of DA in PD, their inability to model the histopathology of the disease made them poor models for evaluating many novel therapies. More beneficial have been models that aim to replicate the loss of dopaminergic nigral neurons observed in Parkinson's patients. Recognized as a primary neuropathology responsible for PD, SNpc cellular loss has been the prime focus of neurotoxic models that aim to replicate this cellular pathology.

6-Hydroxydopamine

The DA analog 6-hydroxydopamine (6-OHDA), because of its similarity in molecular structure, can be taken up into dopaminergic terminals via the DA transporter. Once inside the cell, it is metabolized, resulting in the production of hydrogen peroxide and free radicals. Ultimately these toxic molecules induce neuronal death via mitochondrial dysfunction.

Like DA itself, 6-OHDA is not able to cross the blood–brain barrier and therefore must be delivered directly to the brain of experimental animals via stereotaxic surgery. In 1968, Ungerstedt and colleagues demonstrated the utility of 6-OHDA lesions as animal models of PD. In their study, 6-OHDA was unilaterally injected into the medial forebrain bundle, extensively depleting the nigrostriatal pathway on one side. Ungerstedt noted that lesioned animals rotated toward the side of their lesions spontaneously as well as after administration of the dopaminergic drug d-amphetamine. Conversely, apomorphine, a drug that acts upon upregulated DA receptors on the side ipsilateral to the lesion, induces rotations contralateral to the lesion.

The number of rotations performed by a lesioned animal can be quantified to serve as an index of the integrity of nigrostriatal function. Experimental therapeutic strategies, such as neural or stem cell transplantation and gene therapy, can use the number of rotations an animal performs as an index of the intervention's efficacy.

Since these initial observations, there have been modifications and improvements on the 6-OHDA model although it is still frequently used in its original form. By varying the position and extent of the lesion, different stages of human PD can be modeled. The late stages of PD have been modeled using 6-OHDA lesions to the medial forebrain bundle and SN (**Figure 1**). Using this model, neuronal loss is detected as soon as 12 h postinjection and peaks at 48 h. In addition, striatal fibers are found to degenerate between 1 and 7 days after 6-OHDA delivery, ultimately resulting in more than 90% striatal DA depletion. This provides an ideal environment to evaluate cellular replacement strategies. Alternatively, 6-OHDA delivery to the striatum can result in levels of DA depletion more representative of early-stage PD. Kirik and colleagues demonstrated the location of striatal injections, either 'terminal' (within the caudate–putamen) or 'preterminal' (at the caudate–putamen boundary), greatly affected the resulting lesion, with preterminal injections creating greater levels of DA depletion. In addition, they found variable reductions in tyrosine hydroxylase-positive (TH+) fiber densities and TH+ SN neurons after either single or multiple 6-OHDA intrastriatal injections. As injection site number

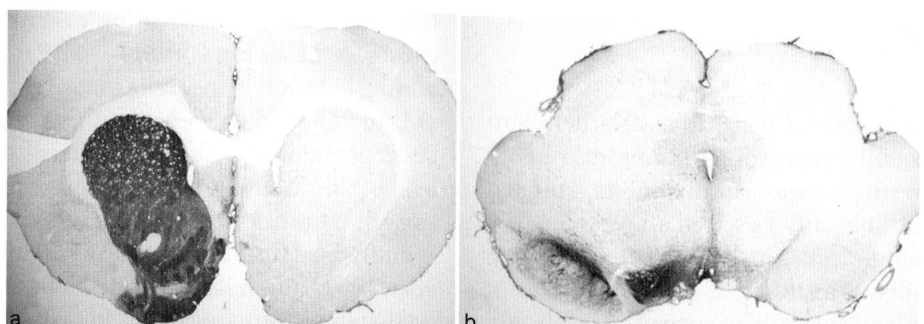

Figure 1 Tyrosine hydroxylase (TH)-stained coronal sections of a 6-OHDA-treated rat striatum (a) and substantia nigra (b) in which the dopaminergic neurotoxin was injected into the right medial forebrain bundle. Note the comprehensive loss of TH-immunoreactive fibers staining in the striatum (a) and the loss of TH-immunoreactive neurons in the nigra ipsilateral to the lesion.

increased, so too did DA depletion. A single preterminal injection site elicited a 60–70% decrease in TH+ fiber density in the rostral striatum and a 25% decrease in TH+ fiber density in the caudal striatum. In contrast, four preterminal site injections resulted in decreases as high as 80% and 90% in TH+ fiber density in the rostral and caudal striatum, respectively. These intrastriatal 6-OHDA models that provide varying degrees of DA depletion have been particularly effective in the evaluation of neuroprotective strategies for human PD.

The 6-OHDA lesion produces a range of functional deficits that have been assessed by means of a number of behavioral tests. Rotational analyses in response to prodopaminergic drugs such as amphetamine or agonists such as apomorphine have been used extensively to assess the extent of an animal's lesion and of its subsequent recovery after therapeutic behavioral analyzes. Nonpharmacological interventions such as open field activity in which a rat is observed for a time within an activity-monitoring chamber are also altered after 6-OHDA. Analysis of the animal's movements can provide a researcher with information about the animal's overall activity, rearing behavior, and any stereotypic behavior. The cylinder test (or limb use asymmetry test) consists of rating the use (and disuse) of each forelimb while an animal navigates itself around a plexiglass cylinder. While an unlesioned rat will typically use each limb equally, a rat with a 6-OHDA lesion will preferentially use the limb ipsilateral to its lesion. The adjusting steps test and its variations involve the restriction of an animal's hindpaws and one forepaw. The unbound forepaw is then moved slowly sideways across a flat surface. In an unlesioned rat, both forepaws will take steps to adjust to the sideways motion. However, in the lesioned animal, the forepaw contralateral to the lesioned striatum will drag along without adjustment. These and many other behavioral tests have been instrumental in the assessment of 6-OHDA lesion-induced deficits and in determining the functional efficacy of therapeutic interventions.

To date, no model of PD has been used as often or, arguably, as effectively as the 6-OHDA model. However, its limitations must also be acknowledged and considered when one is interpreting data obtained from its use. As with any transmitter-specific model, the 6-OHDA lesion does not replicate the constellation of neurotransmitter-based pathologies seen in human PD. The depletion of norepinephrine, due to degeneration of the locus coeruleus seen prior to SN degeneration in human PD patients, is not replicated in the 6-OHDA model. In addition, dysfunction of serotonergic and peptidergic systems seen in PD patients is not mimicked in 6-OHDA-treated animals. Also problematic is the timing of the degenerative process. Whereas Parkinson's is now thought to be a progressive degenerative disorder beginning in the olfactory bulbs and caudal brain stem and progressing over many years rostrally to the midbrain, 6-OHDA lesions are regionally selective and exhibit cell loss in a matter of days to weeks. Finally, while lesioned animals show some of the cellular and behavioral deficits seen in the human disease, 6-OHDA-treated animals fail to develop Lewy bodies or α-synuclein aggregates, features that are pathoneumonic for PD. While 6-OHDA remains a convenient and effective model for the analysis of the nigrostriatal dysfunction and evaluation of many novel therapies, these important shortcomings must be taken into account when translating information obtained from the 6-OHDA-treated animal from the laboratory to the clinic.

1-Methyl-4-Phenyl-1,2,3,6-Tetrahydropyridine

One of the best models of PD results from the administration of 1-methyl-4-phenyl-1,2,3,6-tetrahydropyridine (MPTP), which selectively targets dopaminergic neurons in the brain. Its validity as a model is supported by the fact that MPTP is a definitive, albeit rare, environmental cause of PD in humans. Its neurotoxic effects were first reported in the early 1980s by Bill Langston and co-workers after drug users exposed to MPTP began to develop Parkinsonian symptoms. Taken up primarily by astrocytes, MPTP is converted to its metabolite MPP+. MPP+ can then be taken up by DA neurons, where it exerts its toxic effect through interactions with cytosolic proteins and through the interruption of the complex I component of the electron transport chain.

The MPTP mouse model Due to the relative economy and small size of mice, the cost and convenience of using them have made the MPTP mouse model a widely used model of PD. In addition, the ability to genetically modify the mouse genotype has allowed for the analysis of possible genetic factors that may contribute to MPTP sensitivity, a procedure presently not possible in primate models of PD. In this regard, transgenic α-synuclein knockout mice show an increased resistance to MPTP toxicity, while transgenic mice overexpressing α-synuclein show increased susceptibility.

As with 6-OHDA, varying regimens of MPTP delivery can produce varying degrees of nigrostriatal dysfunction. Acute dosing, consisting of a single injection or a short series of high dose injections ($\sim$10–40 mg kg^{-1}), can produce lesions that result in mild to moderate cell loss and adequately represent early stages in human PD. An early study of acute

MPTP toxicity found a 33% reduction of TH+ SN neurons and a 90% depletion in striatal DA in the C57/bl mouse after a single acute injection of MPTP (40 mg kg^{-1}). Alternatively, chronic dosing, consisting of long series of low-dose injections (~5–20 mg kg^{-1}), results in more-robust lesions representative of late-stage human PD. Jackson-Lewis and colleagues have demonstrated as much as 70% nigral cell loss and 90% striatal DA depletion in C57/BL6 mice after a chronic MPTP dosing regimen (80 mg kg^{-1} total). Chronic administration is also associated with lower levels of spontaneous recovery. Research has suggested that neuronal loss and DA depletion incurred after MTPT administration occur through different mechanisms, with acute dosing resulting in necrotic cell loss and chronic dosing leading to apoptotic cell loss. It has also been shown that MPTP toxicity in the mouse can be enhanced through the delivery of MPTP-enhancing agents such as probenecid. In one study, the chronic delivery of MPTP (25 mg kg^{-1}; 10 doses over 5 weeks) with probenecid (250 mg kg^{-1}) to C57/bl mice resulted in the accumulation of α-synuclein immunoreactive aggregations, although these were not phenotypically similar to the Lewy bodies observed in human PD.

A number of behavioral tests have been developed to assess deficits and recovery of function in MPTP-treated mice. As with the 6-OHDA-treated rat, the open-field test has been used extensively in the mouse to assess overall behavioral activity. The rotarod test has also been used to assess the coordination and balance of MPTP-treated and control mice. In this test, mice are taught, prior to MPTP administration, to run on a rod rotating at varying speeds. The time that lesioned mice are able to remain on the rod then decreases with MPTP-induced nigrostriatal dysfunction. MPTP also impairs mouse performance on the pole test, a measure of bradykinesia that involves placing a mouse, head facing upwards, at the top of a textured pole. The amounts of time it takes the mouse to rotate its body toward the ground and to descend the pole have been taken as indices of the mouse's functional impairment.

While the MPTP mouse model has been invaluable in an understanding of the mechanisms of MPTP toxicity, there are some limitations to its uses. Due to the small size of mice, MPTP is typically administered to them systemically. While systemic delivery has the advantage of producing a bilateral lesion more representative of human PD, it is also associated with higher levels of animal morbidity and increases the animal maintenance required of caregivers. In addition, mice may show robust and spontaneous improvement after the initial lesions, making determinations of therapeutic efficacy difficult. These factors, combined with the mouse's insensitivity to MPTP relative to primates, result in the need for high doses to create stable lesions of clinical relevance in the mouse. Also problematic is the variability of MPTP sensitivity that exists between mouse strains. Muthane and colleagues have found that C57/bl mice innately possess fewer TH+ SN cells and furthermore show an increased sensitivity to the toxic effects of MPTP when compared with the CD-1 strain of mouse. Finally, while MPTP-treated mice show some occurrence of motor dysfunction, they have not been particularly useful in the analysis of parkinsonian behavior and recovery after therapeutic intervention.

The MPTP primate model Nonhuman primates are much more sensitive to MPTP toxicity and require smaller doses to create adequate stable nigrostriatal lesions. MPTP has been used experimentally in monkeys to create both bilateral and unilateral lesions. Systemic administration delivered via intraperitoneal, subcutaneous, intravenous, or intramuscular injections creates bilateral lesions that successfully model the cardinal symptoms associated with human PD. However, because these delivery methods are associated with increased animal morbidity and maintenance, researchers have also used primates that receive unilateral MPTP lesions delivered via intracarotid artery (ICA) injection (**Figure 2**). Despite the necessity of a more invasive surgical procedure, ICA lesions can be more attractive to researchers. Also, despite a small degree of damage to the contralateral hemisphere, it remains relatively intact in unilaterally injected animals and may serve as a useful control and allow for the occurrence of rotational behavior.

In an attempt to combine the benefits of ICA and systemic MPTP treatment and to avoid the weaknesses of each delivery method, recent studies have created an 'overlesioned' model in which an initial ICA unilateral injection is followed by subsequent intravenous injections. This approach elicits an asymmetrical bilateral lesion. In one study looking at the effects of overlesioning in rhesus monkeys, a comprehensive cell loss was seen in the SN ipsilateral to the ICA injection, with a subtotal cell loss in the contralateral SN. Researchers noted a positive correlation between the animals' behavioral deficits and the extent of their lesion contralateral to the ICA injection. The preferential degeneration of one hemisphere over another produced by the overlesioning method is much like the hemispheric biases seen in human PD patients. It is important to note that the deficits in these animals are bilateral and remain stable and that the animals are able to maintain themselves and do not require heroic veterinary intervention.

Again, the method of delivery and dosage of the MPTP toxin may be manipulated to create lesions representative of varying stages of human PD. Acute

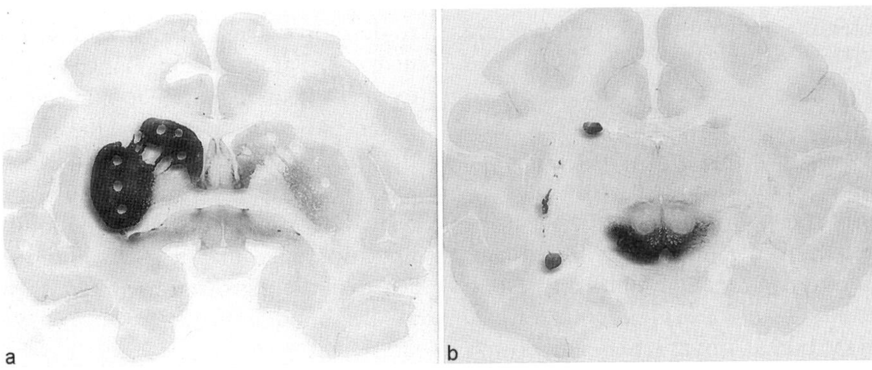

Figure 2 Tyrosine hydroxylase (TH)-stained coronal sections through the striatum (a) and substantia nigra (b) from a rhesus monkey that received an injection of MPTP in the right internal carotid artery. Note the comprehensive loss of TH-immunoreactive fibers staining in the striatum (a) and the loss of TH-immunoreactive neurons in the nigra ipsilateral to the lesion.

MPTP administration results in a significant nigral cell loss, modeling late-stage human PD. In contrast, chronic MPTP administration results in more-moderate cell loss, resembling the middle stages of human PD. In this regard, squirrel monkeys display a 70% nigral cell loss and 95% striatal DA depletion after acute MPTP treatment compared with 40% nigral cell loss and 60–70% striatal DA depletion after chronic systemic administration. Chronic MPTP administration may also be used to model the more progressive nature of cell loss and behavioral deficit observed in the human disease.

As with idiopathic PD, MPTP-induced deficits can be evaluated using clinical rating scales. These assess the severity of symptoms such as tremor, posture, gait, bradykinesia, balance, defense reactions, freezing, and gross motor skills. Fine motor skills testing can also be analyzed through a number of behavioral tests. The forelimb reaching test involves training the monkey to reach toward a recurring target object on which a reward is given, usually food. The time it takes for the monkey to reach toward its reward is an index of its nigrostriatal function. A similar test is the bar-pressing test, in which monkeys are made to press a lever bar to receive a food reward. Tests have also been developed to assess the cognitive functioning of MPTP-treated monkeys. In the barrier–detour retrieval test, monkeys are made to retrieve a food reward from a transparent box with an opening on one side. The monkey must determine which side is the open side of the box before receiving the reward. Both the speed at which the monkey can attain the reward and the number of errors that occur for it to do so can be taken as indices of cognitive functioning.

The effectiveness of MPTP to create lesions of the nigrostriatal pathway can be remarkably variable. One cause for this variability is the age of the animal exposed to the toxin, with older animals showing a much greater sensitivity to MPTP toxicity than their younger counterparts do. This finding may be explained by older individuals' increased levels of MAO-B, the enzyme required for the conversion of MPTP to its toxic metabolite, MPP+. In addition, studies have shown a higher sensitivity to MPTP toxicity in males than in females, suggesting a possible protective effect of estrogen on dopaminergic neurons. Animal species is also a critical predictor of MPTP susceptibility, with some species, such as the nonhuman primate, showing a high sensitivity to MPTP toxicity and others, like the rat, showing little to no effect.

MPTP has done a remarkably better job than 6-OHDA in reproducing many of the nondopaminergic deficits observed in human PD. MPTP-treated monkeys show depletions in norepinephrine and serotonin metabolites and after chronic exposure, cell loss in the locus coeruleus similar to findings in human patients. While true Lewy body formations are not seen in MPTP models of PD, accumulations of α-synuclein protein have been noted in the brains of baboons and squirrel monkeys after MPTP administration. MPTP treatment has been shown to elicit a resting tremor akin to the cardinal symptom manifested in human patients. Resting tremor has been observed in the MPTP-treated vervet monkey although it has not been reported in any other primate models. While the above cases demonstrate that protein aggregation and tremor are possible after MPTP exposure, it should be stressed that these important Parkinson's pathologies are not consistently found to occur within this model.

Environmental Toxin Models

The success of MPTP in replicating the neuropathology of human PD has led researchers to examine other neurotoxins that may similarly affect the central nervous system. Two promising models to emerge have been the rotenone and combined paraquat and maneb models.

Rotenone Rotenone, an organic pesticide that interrupts complex I of the electron transport chain, also elicits mitochondrial dysfunction and ultimately cell loss in the nigrostriatal pathway. Rotenone may be delivered systemically to create a bilateral lesion or via intranigral infusion to create a unilateral lesion. Like other neurotoxic models, bilateral lesions, though similar to the human disease, frequently result in high animal morbidity, leading researchers to occasionally favor unilateral surgeries. After rotenone exposure, rats can show progressive nigral degeneration, ubiquitin and α-synuclein inclusions in remaining nigral cells, and behavioral deficits similar to those observed in human patients. Because rotenone exerts its effects in a DA transporter (DAT)-independent fashion, it elicits mitochondrial dysfunction in nondopaminergic systems, although its neurotoxic effects are DA neuron selective.

While rotenone can accurately model human nigrostriatal degeneration in PD, the great variability in individual sensitivity to its effects has hindered its use in research. In one study, only 46% of animals treated with rotenone (2 mg kg^{-1} daily for 21 days) showed decreased TH reactivity, with only one animal showing any nigral cell loss. Recently another study has found that while C57/bl mice displayed some functional dysfunction after chronic rotenone treatment, they failed to develop the nigral degeneration associated with the rotenone rat model. This substantial variability among individuals and species is a major obstacle for the use of the rotenone model for the evaluation of therapies for PD.

Paraquat and maneb Paraquat (PQ; 1,1′-dimethyl-4-4′-bipyridinium) is another pesticide that has been examined as a possible neurotoxic animal model for PD. Paraquat is initially converted to a PQ cation, which is then reoxidized to from a parent compound as well as superoxide radicals. This allows for further redox cycling and further oxidative stress. Paraquat is often administered in combination with the fungicide manganese ethylene-bis-dithiocarbamate (maneb). Exposure to maneb has been linked to the development of parkinsonian symptoms in humans, and its combination with paraquat as an animal model has been useful in the investigation of the role of environmental toxins in the etiology of human PD.

Rodent studies have found significant nigral cell loss after combination paraquat and maneb exposure. Mice receiving biweekly PQ (10 mg kg^{-1}) and maneb (30 mg kg^{-1}) injections had 25–47% TH+ SN cell loss. PQ-exposed mice have also been shown to develop α-synuclein inclusions in remaining nigral neurons. Paraquat and maneb have also been used extensively in research looking at effects of gestational exposure to toxins. Researchers have found that mice exposed to maneb during gestation show increased sensitivity to PQ toxicity in adulthood, providing support for the 'multihit' hypothesis of PD and suggesting that an initial early insult can sensitize the nigrostriatal system to subsequent 'hits' in adulthood.

Protein Aggregation

The presence of Lewy bodies, α-synuclein-rich intracellular inclusions, in the brains of parkinsonian patients has suggested a role for protein aggregation in the pathogenesis of PD. This has been further supported by the discovery that genetic mutations in the α-synuclein gene can cause genetic forms of PD. In addition, increased levels of α-synuclein in SN neurons have been observed in response to toxin exposure and oxidative stress, suggesting that protein aggregation may have a pivotal importance in the development of idiopathic PD as well. Despite evidence to suggest the central role of protein aggregation in PD development, animal models have traditionally been very poor at exhibiting this important pathology. However, with the development of transgenic animals and the advancement of viral vector delivery systems have come new models for this previously overlooked pathology.

Transgenic Animals

Transgenic models of PD have been emerging that attempt to recreate the known genetic causes of PD. Particularly interesting have been models that knock out, overexpress, and mutate the α-synuclein gene to examine the role of protein aggregation in PD neurodegeneration. In addition, researchers have looked to mutate proteins involved in the ubiquitin–proteosome pathway known to be affected in familial cases to examine the role of proteosome dysfunction in PD.

Alpha-synuclein transgenic models Discovery of the A53T and A30P mutations of the α-synuclein gene in familial PD have led researchers to attempt to manipulate its expression in animals in order to evaluate its role in the human disease. Knockout mice are viable and show decreased striatal DA levels and reduced rearing in the open field, despite DA metabolism's being unaffected. These findings may be explained by the hypothesis that α-synuclein plays a role in synaptic vesicle function. Conversely, mice overexpressing α-synuclein develop intraneuronal inclusions and show decreases in striatal DA levels, despite showing no nigral cell loss. These effects are generally directly correlated to the amount of α-synuclein overexpressed. Mice expressing mutated forms of α-synuclein have shown varying degrees of pathology, dependent on the forms of construct utilized. Although transgenic animals expressing human α-synuclein with the platelet-derived growth factor *β* promoter show behavioral deficits and increased SN inclusions, other constructs have shown very little pathology or behavioral deficit.

A major strength of the transgenic α-synuclein models for PD is that expression alterations are not limited to dopaminergic neurons but are expressed throughout the nervous system. In addition, the slow accumulation of protein that occurs in transgenic animals provides a better model of the progression seen in human disease. However, despite these models; ability to replicate the Lewy body inclusions seen in human PD, a major failing is their inability to model SNpc cell loss, making them suboptimal for the evaluation of many novel therapies for PD. In addition, technical restrictions limit the use of transgenic models to the phylogenetically distant mouse, preventing these models' immediate application to the clinic.

Parkin and ubiquitin C-terminal hydrolase L1 transgenic models The discovery of mutations to parkin and ubiquitin C-terminal hydrolase L1 genes, both components of the ubiquitin–proteosome system, in familial cases of PD has led researchers to examine the role of proteosome dysfunction in PD and to develop models that target these genes.

Viral Vector-Delivered *α*-Synuclein

While transgenic models have been limited to mice, adeno-associated (AAV) and lentiviral delivery systems have enabled the selective overexpression of α-synuclein within the nervous system of higher-ordered species. In rats, delivery of wild-type or mutant AAV-human α-synuclein to the SNpc resulted in cytoplasmic inclusions, behavioral dysfunction, and nigral cell loss. Similar cellular pathology was found in rats receiving lentiviral-delivered wild-type and mutant forms of human α-synuclein. The overexpression of α-synuclein via viral vectors has also been evaluated in the nonhuman primate. Marmosets receiving mutant AAV-human α-synuclein showed a 32–41% reduction of DA neurons in the SN. Behavioral deficits observed included head positional bias and rotation asymmetry, both of which are commonly seen in PD toxicity models.

The ability to overexpress α-synuclein has given researchers the opportunity to evaluate the role of protein aggregation in PD. In addition, it offers a relatively slowly progressing model with which to analyze the gradually progressive nature of the human disease, an advantage not offered by most toxicity models. While it is a seemingly powerful model that can encompass the majority of PD pathologies, the selective overexpression of α-synuclein to one specific brain region limits the observation of nonmotor effects such as olfactory impairments, gastrointestinal dysfunction, depression, sleep disturbances, and cognitive impairments that result from α-synuclein aggregation in other brain regions.

Other PD Models

Lipopolysaccharide

The immune response has been suggested as a major contributor to the progressive nature of nigral cell loss in PD. Researchers have tried to replicate this increased immune activation through exposure to the bacteriotoxin lipopolysaccharide (LPS). Stereotaxic injections of LPS have been used to induce nigral cell loss in the adult animals, and prenatal LPS exposure has been shown to elicit decreased DA neuron number in offspring. In addition, these animals show locus coeruleus cell loss, suggesting nondopaminergic effects, as well as intracellular inclusions. Researchers have also found that prenatal LPS treatment further prolongs the immune response to secondary LPS exposure in adulthood, suggesting a role for neuroinflammation in the 'multihit' hypothesis of PD.

Aged Animals

Aging is the primary risk factor for the development of PD. In normal aging, striatal DA is lost, and there is a loss of TH immunoreactive and DA transporter immunoreactive nigral neurons. These losses are associated with age-related increases in α-synuclein. While controversial, emerging evidence suggests that PD represents the extreme end on a continuum of aging. Some experimenters therefore have utilized aged animals themselves as models for PD. Taken together, these findings suggest that aged animals may represent a useful model for evaluation of therapies aimed at the earliest stages of PD.

Animal Models of Dyskinesia

In addition to mimicking the pathology of PD, models have been used to study treatment-related side effects of PD. In this regard, the laboratory of Angela Cenci has led the field in demonstrating the use of the 6-OHDA rat model of PD for the study of levodopa-induced dyskinesias. Dyskinesias are disabling, involuntary movements that commonly arise in PD patients after prolonged levodopa therapy. While dyskinesias can take several years to develop in patients, rats may be primed with levodopa to exhibit abnormal involuntary movements in a matter of weeks. The affordability and convenience of the rat model make it ideal for the evaluation of the molecular and cellular causes of this serious side effect.

While demanding much more cost and care, MPTP-treated monkeys have also been one of the best and most widely used animal models in which to study levodopa-induced dyskinesias because of their phylogenetic proximity to humans. Unlike the 6-OHDA-treated rat, the MPTP-treated primate model shows similarities to human anatomy and physiology.

Conclusions

The modeling of PD in animals has been invaluable to our understanding of the human disease. By selectively targeting certain aspects of PD, researchers have been able to evaluate potential therapies aimed at specific pathologies. Pharmacological models, such as reserpine-treated animals, have provided a better understanding of the role of DA in PD. Through its experimental use, the reserpine model has served to evaluate DA replacement strategies such as levodopa treatment, still routinely used effectively to alleviate parkinsonian symptoms. Neurotoxin models, such as the 6-OHDA and MPTP models, replicate the nigral cell loss seen in PD patients and have been used successfully to evaluate neuroprotective strategies and novel therapies such as cell transplantation. A new wave of neurotoxic models, such as the rotenone and paraquat and maneb models, have been invaluable in the investigation of the role of environmental toxins in the etiology of PD.

While the experimental value of the models cannot be understated, they have so far failed to provide a comprehensive model of the human disease. Aspects of human PD that have been particularly poorly modeled by traditional models are Lewy body formation, pathologies of nondopaminergic systems, and the progressive nature of the disease. Novel genetic models mimic the previously unexamined pathology of protein aggregation in PD, allowing for the investigation of its role in PD development. LPS exposure provides an excellent model to observe the role of neuroinflammation in PD and the progressive loss of dopaminergic cells. Finally, aged animals may represent a truly global model for analyzing the very early stages of PD.

See also: Aging of the Brain; Cell Replacement Therapy: Parkinson's Disease; Deep Brain Stimulation and Parkinson's Disease; Dopamine; Dopamine in Perspective; Dopaminergic Agonists and L-DOPA; Parkinsonian Syndromes; Parkinson's Disease: Alpha-Synuclein and Neurodegeneration; Transgenic Models of Neurodegenerative Disease.

Further Reading

Braak H, Del Tredici K, Rub U, de Vos RA, Jansen Steur EN, and Braak E (2003) Staging of brain pathology related to sporadic Parkinson's disease. *Neurobiology of Aging* 24(2): 197–211.

Carlsson A (1959) The occurrence, distribution and physiological role of catecholamines in the nervous system. *Pharmacology Review* 11: 490–493.

Chu Y and Kordower JH (2007) Age-associated increases of alpha-synuclein in monkeys and humans are associated with nigrostriatal dopamine depletion: Is this the target for Parkinson's disease? *Neurobiology of Disease* 25(1): 134–149.

Fleming SM, Zhu C, Fernagut PO, et al. (2004) Behavioral and immunohistochemical effects of chronic intravenous and subcutaneous infusions of varying doses of rotenone. *Experimental Neurology* 187(2): 418–429.

Herrera AJ, Castano A, Venero JL, Cano J, and Machado A (2000) A single intranigral injection of LPS as a new model for studying the selective effects of inflammatory reactions on the dopaminergic system. *Neurobiology of Disease* 7(4): 429–447.

Jackson-Lewis V, Jakowec M, Burke RE, and Przedborski S (1995) Time course and morphology of dopaminergic neuronal death caused by the neurotoxin 1-methyl-4-phenyl-1,2,3,6-tetrahydorpyridine. *Neurodegeneration* 4(3): 257–269.

Jellinger KA (1991) Pathology of Parkinson's disease: Changes other than the nigrostriatal pathway. *Molecular Chemical Neuropathology* 14: 153–197.

Jeon BS, Jackson-Lewis V, and Burke RE (1995) 6-Hydroxydopamine lesion of the rat substantia nigra: Time course and morphology of cell death. *Neurodegeneration* 4: 131–137.

Kirik D, Annett L, Burger C, Muzycka N, Mandel R, and Bjorklund A (2003) Nigrostriatal α-synucleinopathy induced by viral vector-mediated overexpression of human α-synuclein: A new primate model of Parkinson's disease. *Proceedings of the National Academy of Sciences of the United States of America* 100(5): 2884–2889.

Kirik D, Rosenblad C, and Bjorklund A (1998) Characterization of behavioral and neurodegenerative changes following partial lesions of the nigrostriatal dopamine system induced by intrastriatal 6-hydroxydopamine in the rat. *Experimental Neurology* 152: 259–277.

Langston JW, Ballard P, Tetrud JW, and Irwin I (1983) Chronic parkinsonism in humans due to a product of meperidine-analog synthesis. *Science* 219: 979–980.

Oiwa Y, Eberling JL, Nagy D, Pivirotto P, Emborg ME, and Bankiewicz KS (2003) Overlesioned hemiparkinsonian non human primate model: Correlation between clinical, neurochemical and histochemical changes. *Frontiers in Bioscience* 8: a155–a166.

Olanow CW and Tatton WG (1999) Etiology and pathogenesis of Parkinson's disease. *Annual Review of Neuroscience* 22: 123–144.

Przedborski S, Jackson-Lewis V, Naini AB, et al. (2001) The parkinsonian toxin 1-methyl-4-phenyl-1,2,3,6-tetrahydropyridine (MPTP): A technical review of its utility and safety. *Journal of Neurochemistry* 76: 1265–1274.

Przedborski S, Levivier M, Jiang H, et al. (1995) Dose-dependent lesions of the dopaminergic nigrostriatal pathway induced by intrastriatal injection of 6-hydroxydopamine. *Neuroscience* 67: 631–647.

Przedborski S and Vila M (2003) The 1-methyl-4-phenyl-1,2,3,6-tetrahydropyridine mouse model: A tool to explore the pathogenesis of Parkinson's disease. *Annals of New York Academy of Science* 991: 189–198.

Schultz W (1988) MPTP-induced parkinsonism in monkeys: Mechanism of action, selectivity and pathophysiology. *General Pharmacology* 19: 153–161.

Thiruchelvam M, Brockel BJ, Richfield EK, Baggs RB, and Cory-Slechta DA (2000) Potentiated and preferential effects of combined paraquat and maneb on nigrostriatal dopamine systems: Environmental risk factors for Parkinson's disease. *Brain Research* 873(2): 225–234.

Ungerstedt U (1968) 6-Hydroxydopamine induced degeneration of central monoamine neurons. *European Journal of Pharmacology* 5: 107–110.

van der Putten H, Wiederhold KH, Probst A, et al. (2000) Neuropathology in mice expressing human alpha-synuclein. *Journal of Neuroscience* 20: 6021–6029.

Animal Models of Stroke

L C Hoyte and A M Buchan, University of Oxford, Oxford, UK

Introduction

Among western countries, such as the United States, Canada, and the United Kingdom, stroke is the third leading cause of death and a major source of disability. The risk of having a stroke or a transient ischemic attack increases dramatically with age. These facts, combined with the current shift in age demographics toward a more elderly population in western countries, ensure that stroke will continue to be a health care burden. For example, in the United States, the estimated cost of stroke (from direct and indirect sources) in 2006 was predicted to be $57.9 billion (American Heart Association). It is vital to reduce brain damage and improve quality of life for stroke patients.

Human stroke occurs in two forms: ischemia and hemorrhage. The ischemic stroke results from an occlusive clot, this being from either a thrombus or an embolus that blocks blood flow within an artery of the brain. During a focal ischemic insult, there is a difference in cerebral blood flow (CBF) to the brain depending on the proximity to the clot and the level of collateral flow. The area that is within the main occluded territory is generally termed the core of the infarct, while the surrounding tissue is termed the penumbra. Within the core infarct there is extensive tissue damage, with pannecrosis, gliosis, inflammatory infiltration, breakdown of the blood–brain barrier, cytotoxic and vasogenic edema, and excitotoxic cell death. The penumbra is the area of the brain that has reduced CBF and metabolism, but damage is fundamentally reversible with reperfusion. Hemorrhagic stroke is due to a cerebral vessel bursting and the resultant bleeding into the brain. Hemorrhagic stroke can be further subdivided into two main categories: intracerebral hemorrhage (bleeding from cranial blood vessels into the brain tissue) and subarachnoid hemorrhage (bleeding from cranial blood vessels into the subarachnoid space).

A vast majority of human strokes are ischemic in nature; 85% of all strokes in western countries are ischemic, while 15% are hemorrhagic, however, there is a shift to increased hemorrhage rate among Asian populations. Due to the fact that most strokes are ischemic, experimental stroke research primarily attempts to discover ways to reduce morbidity and mortality among the aging population.

Can Animal Models Mimic Human Stroke?

Most experimental stroke modeling is conducted using rodent models, with surgical procedures used to induce stroke. The rodent brain has a smooth cortex (lissencephalic), and different components of white and gray matter as compared to the human brain. As well, the rodent and human immune systems have marked differences. For example, a prevalence of infiltrating neutrophils is seen in rats following stroke, while in humans generally there is predominantly a macrophage infiltration. However, many rodent species, with the gerbil being a notable exception, possess a complete circle of Willis, similar to the human circulation.

This similarity in circulation allows for experimenters to examine stroke and CBF in the rodent brain and relate it to the human condition. Fundamental brain regions are similar from animal to human, allowing for insight into stroke recovery and plastic adaptation of the brain. However, there has been a plethora of failed human clinical trials; the animal work frequently did not translate to human stroke treatment in the past. Particularly, it has been seen that many experimental results were confounded by alterations in animal physiology; such that the improved outcome was due to an innate physiological response, and was not due to the specific pharmacological action of the drug on its original target. A classic example of such a phenomenon is the use of the *N*-methyl-D-aspartate (NMDA) noncompetitive receptor blocker MK-801. Treatment with this compound was found to be highly efficacious in reducing infarct volumes in the rat cerebral ischemia, but it was later proved that this drug caused a persistent hypothermia in the global model and that an improved collateral flow to the brain in the middle cerebral artery (MCA) occlusion models was responsible for improved outcome. This compound was moved to human clinical trial, without success, and there have been many more, similar examples since. This has led to some despair among researchers that animal modeling will not advance the treatment of human stroke.

However, there has been evidence of successful translation of animal work to the clinic. The only US Food and Drug Administration (FDA)-approved drug for the treatment of acute cerebral ischemia in the human, tissue plasminogen activator (tPA), was first shown to be efficacious in a rabbit model of small clot emboli. Currently there is clinical testing on another compound (NXY-059, a free radical spin trap agent) that shows positive experimental results and has also shown some promise in the clinic.

Cerebral Ischemia Models: Global versus Focal Ischemia

Cerebral ischemia accounts for approximately 85% of the strokes seen in western society. As such, these are extensively studied experimentally. The two forms of cerebral ischemia are global (either complete or incomplete) and focal (either multifocal or focal) (**Figure 1**).

Global Cerebral Ischemia

Global cerebral ischemia is used to mimic the clinical situation of cardiac arrest, and is also widely used to examine the selective vulnerability of neurons to cerebral ischemia. It is thought that any treatment that prevents the loss of these highly susceptible neurons might be well suited to act as a neuroprotective agent. Complete global ischemia, which can be produced by decapitation, cardiac arrest, neck cuff, or aortic occlusion, results in a loss of blood flow to the entire brain. Incomplete global ischemia; also called transient, severe forebrain ischemia, results in oligemia throughout the forebrain, including the cortex, striatum, and hippocampus. The blood flow to the brain stem is preserved in this model, allowing for the maintenance of heart rate and breathing by the animal, and possibly sparing the systemic effects of such complete global ischemia models as cardiac arrest.

Complete and incomplete global ischemia models are extremely severe, with very exceptional drops in cerebral blood flow. Consequently, the length of the ischemia in these models is generally very short in comparison with focal models. Global ischemia results in a characteristic cell death, without infarct, but with selective neuronal death. Susceptible neurons, such as the CA1 pyramidal cells of the hippocampus and select neurons of the striatum, and cortical layer III, V, and VI, die when exposed to the insult. The complexity arises from this model when one considers that adjacent to the CA1 pyramidal cells in the hippocampus are the CA3 cells – neurons that seem to have similar characteristics in location and cell type, but are resistant to the insult. It has been postulated that the CA3 are more robust than the CA1 due to a difference in the incoming afferents to these areas of the hippocampus, and to differences in the regulation of α-amino-3-hydroxy-5-methyl-4-isoxazole propionic acid (AMPA) receptor subunit GluR2 following global ischemia.

Complete global ischemia The most commonly studied form of complete global ischemia is the cardiac arrest model. This model is used in both gyroencephalic (convoluted cortex) and lissencephalic (smooth cortex) animals. Frequently, in canine models, it is achieved with exsanguinations, followed by mechanical

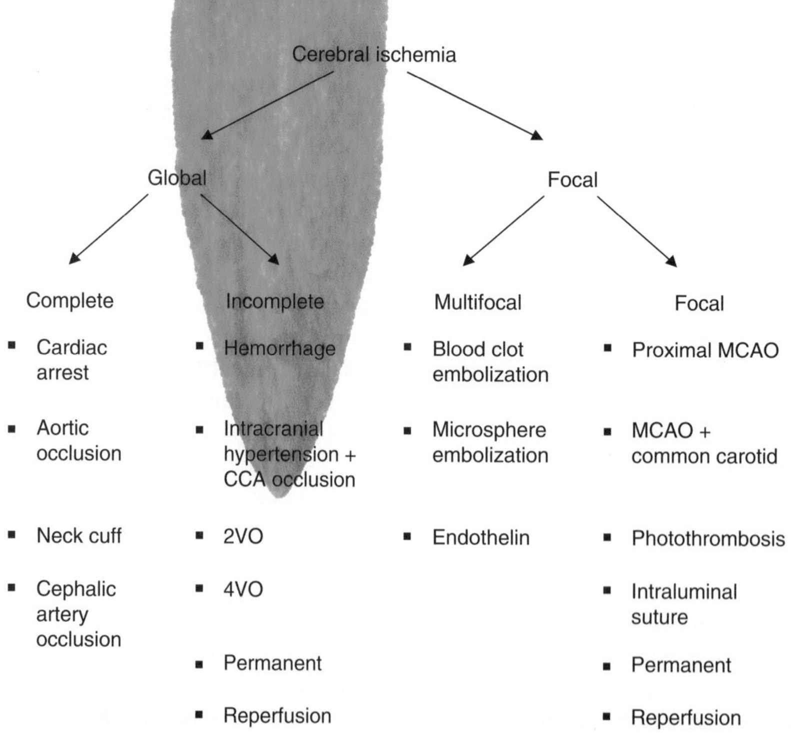

Figure 1 Methods used to generate ischemia in animal models. These include both focal and global models and permanent and reperfusion models. MCAO, middle cerebral artery occlusion; 2VO, two-vessel occlusion; 4VO, four-vessel occlusion. Adapted from Traystman RJ (2003) Animal models of focal and global cerebral ischemia. *ILAR Journal* 44: 85–95 (p. 87). Reprinted with permission from ILAR Journal, Institute for Laboratory Animal Research, National Academy of Sciences, 500 Fifth Street NW, Washington, DC, 2001.

or chemical resuscitation after the specified arrest time. Arrest of the heart can also be achieved by electrical fibrillation, in the cat, with consequent resuscitation. In rodents, the cardiac arrest model is frequently performed using KCl to arrest the heart, with resuscitation. Cardiac arrest models can be used to examine the efficacy of rapid induction of cerebral hypothermia, by aortic or jugular flushing with ice-cold saline. This can rapidly reduce brain temperature and improve outcome in dogs, even following very severe ischemia. The main drawback of the cardiac arrest models is that they generate systemic ischemia, which may, in turn, make interpretation of the cerebral ischemia more challenging due to damage occurring to other body organs.

Incomplete global ischemia Incomplete cerebral ischemia results in transient, but severe, forebrain ischemia, followed by reperfusion. This avoids the systemic effects of cardiac arrest, while still achieving very low cerebral blood flows of less than 10% of baseline. The use of four-vessel occlusion (4VO) is widespread in rat experiments, and can be achieved by occlusion of both vertebral arteries as well as both common carotid arteries (**Figure 2**). This ischemic insult results in a selective neuronal death that is dependent on ischemia duration. Pyramidal cells of the CA1 region of the hippocampus are especially sensitive to global ischemia, and duration of ischemia has a direct effect on progression of CA1 cell death; shorter durations of ischemia result in slower progression of CA1 damage. The model of global forebrain ischemia creates a lack of blood flow to the hippocampus, striatum, and cortex, while sparing flow to the brain stem. This oligemia can be reversed after 5, 10, 15, or 30 min and a recovery of blood flow and ATP levels is observed immediately. The degree of oligemia is not influenced by infusion of drug during the insult; however, hypothermia, even 6 or 24 h postischemia, is successful in increasing the long-term survival of the sensitive CA1 neurons. One major caveat to the 4VO model is the necessity of a 2-day procedure to generate the ischemia. This model relies on a preparatory day, which constitutes a surgery to isolate and prepare the common carotid arteries (CCAs) for the following day's occlusion along with a cauterization of the vertebral arteries and placement of sutures to limit collateral flow. The animals are fasted overnight, and the following day are reanesthetized to allow for occlusion of the CCAs. Some groups wish to avoid this preparatory day, and to achieve ischemia on the first day. This has led to the development of two-vessel occlusion (2VO) with hypotension.

The process of 2VO with hypotension involves a bilateral occlusion of the common carotid arteries, along with a systemic hypotension by exsanguinations. This model must achieve a level of hypotension sufficient to generate the global ischemia – reduction of blood pressure to 50 mmHg is sufficient in the rat. This generates a similar pattern of selective neuronal death as is seen in the 4VO and the cardiac arrest models, and is highly sensitive to hypothermia. 2VO is also used in the mouse, frequently without systemic hypotension to generate global ischemia, but the duration of ischemia tends to be longer to

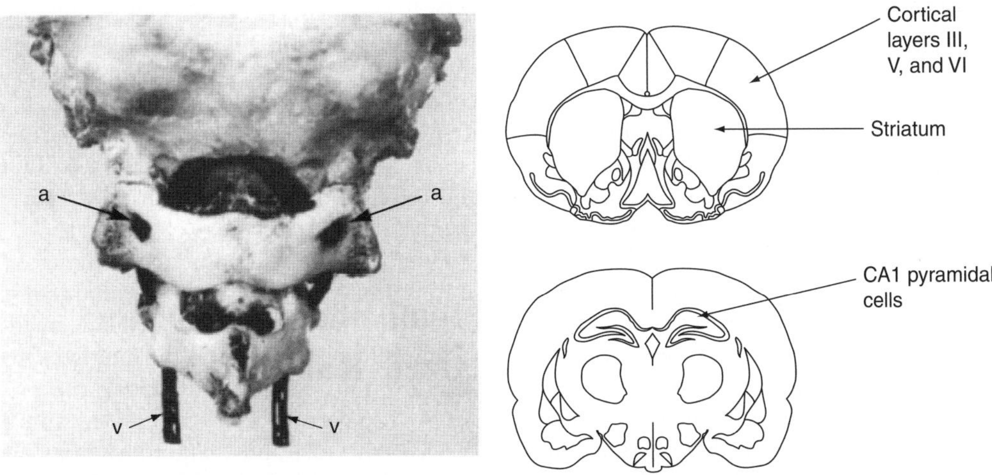

Figure 2 The four-vessel occlusion for incomplete cerebral ischemia in rats. Vertebral arteries (v) are cauterized at the alar foramina (a) in the preparatory surgery, the animals are fasted overnight, and the following day bilateral common carotid artery occlusion generates the transient, but severe, forebrain ischemia. CA1 pyramidal neurons are very susceptible to the ischemia, along with select neurons of the striatum and cortical layers III, V, and VI; this model shows selective neuronal death. Adapted from Pulsinelli WA and Brierley JB (1979) A new model of bilateral hemispheric ischemia in the unanesthetized rat. *Stroke* 10: 267–272.

compensate for the increased flow due to the posterior circulation.

Gerbils can be used to examine the effects of global ischemia, with the 2VO model, but without systemic hypotension. This is because, in general, gerbils lack a posterior communicating artery and do not have a complete circle of Willis. However, new evidence suggests the presence of a posterior communicating artery in some gerbils from specific suppliers, so experimenters must be careful in the assumption that bilateral CCA occlusion will generate global ischemia. The use of the gerbil model has verified the selective vulnerability of CA1 pyramidal cells, and has confirmed the protection of the CA1 by hypothermia. This protection has been shown to markedly improve survival of the CA1 for 18–20 months. One major advantage of this model is the relative ease of preparation and execution of the ischemia.

Currently, a new model for global cerebral ischemia has been developed in the mouse; this model mimics the effects of 4VO in the rat. The model involves the bilateral occlusion of the common carotid arteries as well as an occlusion of the basilar artery. It preserves blood flow to the brain stem while a transient oligemia is induced in the striatum, hippocampus, and cortex. The difficulties with induction of global ischemia in the mouse include an increased risk of respiratory failure due to manipulation of the trachea, which is necessary to access the basilar artery, and the possible breaking of the basilar artery following removal of the microaneurysm clip. The main advantage of generating this global model in mice is to study different transgenic strains following ischemia. This should allow for an in-depth examination of specific genes that might be implicated in selective neuronal death.

Focal Cerebral Ischemia

Focal cerebral ischemia is used to study the tissue damage that results from stroke. Generally, focal cerebral ischemia models examine the blockage of one artery of the brain (e.g., the MCA). Focal cerebral ischemia can also be generated with multiple focal insults, such as occur with embolization. These models result in classic infarcts with core and penumbra areas. The core of the infarct is the area that is closest to the occlusion; this area has the lowest cerebral perfusion and the tissue contained within this area is presumed to be unsalvageable. Seminal work from the late 1970s demonstrated that there was a difference in cerebral perfusion depending on the distance from the occlusion, and the area surrounding the core infarct was termed the ischemic penumbra. The neurons within the penumbral area are structurally-intact, but metabolically paralyzed. There is a direct relationship to the cerebral blood flow and the electrical activity of the brain, such that during cerebral ischemia, the electrical activity of the brain will disappear; however, this loss of electrical activity occurs at a less severe ischemia than is necessary to generate failure of the cell's baseline functions. Thus, the core infarct contains cells that have become structurally and metabolically compromised, while neurons in the penumbra retain structural integrity. The infarct loses the ability for autoregulation, and over time, develops into a glial scar.

Focal ischemia models are considered to have either a single focus (focal) or to have multiple foci (multifocal). The focal insult is generally created by surgical occlusion of an artery of the brain; however, there are cases whereby an artery was occluded by direct placement of clot into the artery. Multifocal insults are produced by injection of emboli (either clot or microspheres) into the brain. This form of insult results in multiple infarcts and is far more variable than focal ischemia.

Middle cerebral artery occlusion By far the most common form of focal ischemia is achieved by occlusion of the MCA. This was first attempted by Harvey and Rasmussen in 1951, when they surgically occluded the MCA with a clip to produce a large infarct and marked hemiparesis in dogs. The same two experimenters showed that the duration of ischemia was important for the development of infarct in the monkey – a 50 min ischemia resulted in an infarct similar to that of permanent ischemia, while ischemia durations of less than 30 min did not produce gross cortical damage. Currently, MCA occlusion is used to induce stroke in many animal species, including rats, mice, pigs, cats, dogs, and nonhuman primates. Rodents are by far the most commonly used experimental animal in MCA occlusion studies due to ethical and financial constraints of using higher order mammals. The occlusion of the MCA can be induced either proximally, such as in the case of intraluminal suture techniques, or more distally by a clipping of the distal MCA.

One of the more commonly used methods for generating MCA ischemia is the intraluminal suture technique. This method achieves ischemia by introduction of an occlusive suture into the internal carotid artery, whereby it is advanced to block the origin of the MCA. The carotid artery is accessed from the ventral cervical surface, and craniotomy is avoided in this model. The occlusion is generated by accessing and isolating the common carotid artery, external carotid artery (ECA), and internal carotid artery (ICA). Originally, the suture was passed directly into the ICA through a small puncture, while the arteries (CCA,

ICA, and ECA) were tied off to prevent bleeding. Further refinements were established to allow for introduction of the suture into the ECA and temporary placement of the suture in the CCA, followed by a movement of the suture into the ICA. This refinement necessitated the permanent ligation and cutting of the ECA, so that the artery could be moved, allowing for advancement of the suture into the ICA. Upon introduction of the suture into the ICA, it is advanced until it blocks the proximal MCA (**Figure 3**). Various small alterations in this technique exist. They include using different sizes of sutures to block the MCA with a heat-blunted tip, silicone coating of the suture tip, or the use of polylysine-coated sutures. The intraluminal suture model induces severe ischemia in the striatum with milder cortical ischemia. Commonly, the intraluminal suture model results in ischemia within the hypothalamus, and, in the rat, this results in hyperthermia. This hyperthermia can markedly enhance infarct progression and increases the variability of this model. Proximal MCA occlusion in the mouse is conducted in a manner similar to that in the rat, which again results in a severe ischemia of the striatum with milder ischemia in the cortex. The intraluminal mouse model can be variable due to the differences in patency of the posterior communicating artery among strains, and even among animals of the same strain. The possible use of transgenic animals is a great advantage of using the mouse model. However, differences in patency of posterior communicating arteries make the use of transgenic mice more difficult if the background strain used in not the same as the control strain.

Mice are more delicate than rats and frequently succumb to the stress of surgical procedures, and the survival rate of experimental mice is less than that of rats. Physiology of rodents must be carefully controlled, and following focal ischemia by intraluminal suture, mice tend to become hypothermic. This results in smaller infarct volumes and better histological and behavioral scores as compared to temperature-regulated mice.

In rats, direct placement of a clot into the proximal MCA has been shown to generate consistent ischemia, while introducing an occlusive thrombus. In this model, the clot is placed into the MCA by introducing tubing containing the clot into the ECA and advancing it through the ICA, in a manner similar to the placement of the suture in the intraluminal technique. However, the clot is ejected from the tubing, effectively obstructing the MCA. The main advantage of this model is the placement of the clot within the MCA. This should allow for testing of the efficacy of thrombolytic drugs on clot lysis in this model, which is not possible in the intraluminal suture models.

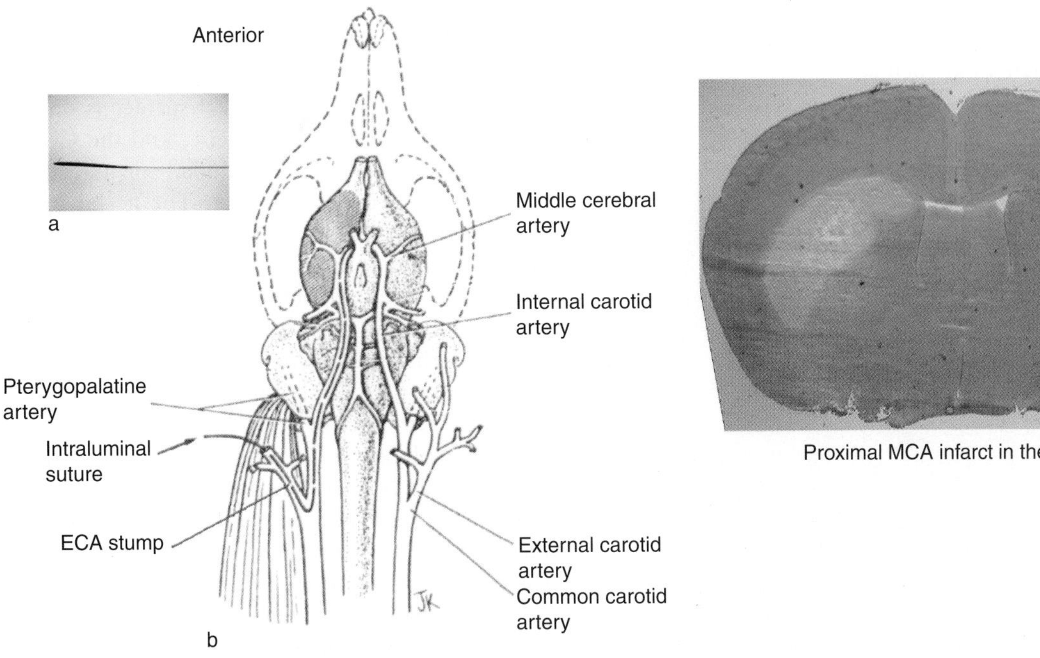

Figure 3 The intraluminal suture middle cerebral artery (MCA) occlusion. The obstructive filament, often coated to reach a specified diameter (a), is introduced in the external carotid artery (ECA) and advanced briefly into the common carotid artery. The ECA is cut to allow for manipulation of the suture. The suture is withdrawn slightly, is maneuvered into the internal carotid artery, and then advanced until it blocks the proximal middle cerebral artery (b). This model generally creates a large striatal infarct, with some cortical involvement. Adapted from Longa EZ, Weinstein PR, Carlson S, et al. (1989) Reversible middle cerebral artery occlusion without craniectomy in rats. *Stroke* 20: 84–91.

Distal MCA occlusion is achieved by accessing the distal MCA, either by craniotomy over the parietal/frontal cortex, in rodents; or by orbital rim approach following enucleation, in gyroencephalic mammals. In the rat model, the skin between the eye and the ear is cut, and the underlying muscles are dissected to expose the skull. Two small burr holes are drilled into the skull, one to access the artery for monitoring of CBF and the other for occlusion of the artery. The distal MCA is then directly visualized through the operating microscope and the more distal burr hole is used for occlusion of the MCA. The dura over the MCA is opened and the MCA is occluded along with bilateral or ipsilateral occlusion of the CCA, depending on the strain. This model was first described in 1986 by Chen and colleagues, who occluded the MCA by ligation with consequent occlusion of one or both CCAs, generating large cortical insults. A newer model of distal MCA occlusion was created that also generated a large cortical insult, but which relied on the placement of a microsurgical clip on the distal MCA, increasing the ease of producing ischemia with reperfusion (**Figure 4**). One major advantage of this clip model is that it generates the ischemia with only brief periods of anesthesia during the clip placement and removal. Occlusion of one or both CCAs is dependent on rat strain: Wistars require both CCAs to be occluded for successful ischemia, while spontaneously hypertensive rats require only ipsilateral occlusion. The cortical infarction induced by these models correlates with a breakdown of the blood–brain barrier and edema.

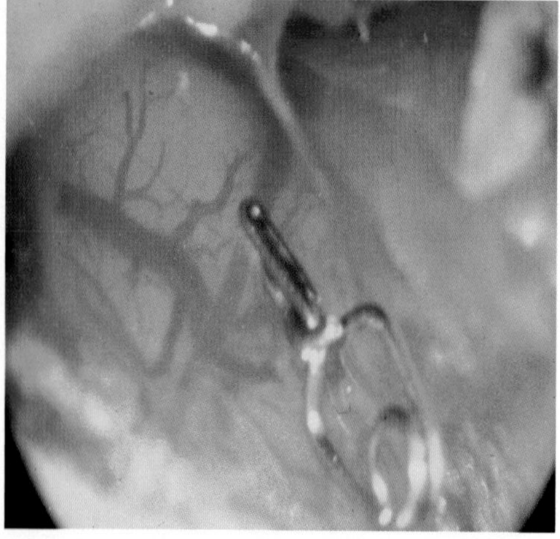

Figure 4 Distal middle cerebral artery occlusion; clip placement on the middle cerebral artery. This model requires a craniotomy and incision through the dura to access the distal middle cerebral artery. The rat is anesthetized during the clip placement and occlusion of one or both common carotid arteries, depending on the strain. The animal is awake during the ischemia, but is reanesthetized for clip removal.

Gyroencephalic models generally attempt to address questions regarding ischemia in an animal with brain structure more similar to that of humans. A new model to assess MCA occlusion in higher mammals is achieved in the miniature pig; the approach to the MCA is over the frontotemporal cortex, accessing the MCA via an orbital rim osteotomy. The cat has also been used to study focal ischemia and the effects of hypothermia on the brain. Generally, the nonhuman primate model, such as in the baboon, generates occlusion of the MCA by transorbital approach. The MCA is located following enucleation and an inflatable silastic balloon cuff is placed around the MCA. This allows for inflation of the balloon and induction of ischemia in an awake animal. Nonhuman primates are important to assess complex behaviors and recovery following stroke and allow for an assessment of injury in a brain that is very similar to a human brain. Nonhuman primate models should only be used to evaluate neuroprotective agents when positive results in rodent models generate enthusiasm before transferal into clinical trial.

Common carotid artery occlusion Occlusion of the common carotid artery can be used to generate focal ischemia in the gerbil, the adult rat with systemic hypotension, or the young rat with hypoxia. By far the simplest method for generating a focal insult is by clipping one of the CCAs in the gerbil, due to their lack of the posterior communicating artery, but investigations must be carefully conducted due to the possibility of gerbils having a PCA, which might confound results. The ventral cervical surface is opened surgically, under general anesthetic, and the CCA is isolated. The artery is then occluded for a specified period of time, with reperfusion. The gerbil model has been used to evaluate motor deficits and examine magnetic resonance image changes following focal ischemia. The main advantage of the gerbil model is the ease of inducing focal ischemia; however, to limit variability in this model, one must be careful to ensure that the posterior communicating arteries are not present in the supplies of gerbils.

Occlusion of one CCA in the rat with systemic hypotension can be used to generate focal insults with minimal surgical invasiveness. The ventral cervical surface of the rat is opened, and the CCA located and isolated. This artery is then occluded and systemic hypotension is achieved by exsanguinations to a blood pressure of 40 mmHg. This model approaches the ease of the gerbil model, but introduces more difficulty due to the removal of blood to generate hypotension. In neonatal rats, it is possible to generate a focal insult by a combination of hypoxia and ischemia. Again, the ventral cervical surface is opened and the CCA located and dissected. The artery is occluded and hypoxia is created by placing the rat in a low-oxygen environment. This model is very important for modeling neonatal or fetal ischemia in humans; among children, ischemia *in utero* or during parturition is a common mechanism for brain damage.

Pial strip The pial strip method for creating a localized, specific cortical lesion is achieved by removal of the meninges over a particular area of the cortex. This method is generally used in the rat, and is able to generate lesions consistent in size and location. A midline incision in the scalp in created and a small craniotomy is used to access the meninges. The dura and pia mater are removed, thus removing the collateral vessels from that particular portion of cortex, which will go on to die. The main advantage of the pial strip is the reproducibility and the ability to selectively place the lesion on a specific brain region (i.e., motor cortex), creating a specific deficit. This allows for greater power in behavioral studies, and assessment of different treatments in very specific and precise motor behavioral tests. A disadvantage of this model is that it is not a classical focal ischemia/reperfusion model, with a core and penumbral areas of CBF reduction and infarct, so it may be difficult to assess some aspects of ischemic cell death.

Photothrombosis Photothrombosis is a method to induce stroke in rodents; it relies on the injection of light-sensitive dye followed by a laser pulse to create clot within the artery. To generate ischemia in this model, a photosensitizing dye is first injected intravenously into the animal. Following this, the skin and muscle over the skull are retracted, and the area of the brain to undergo ischemia is irradiated through the intact skull, using green light at 560 nm. This model has been shown to create thrombotic plugs and stasis of the red blood cells in vessels, with platelet aggregates. The associated infarction is due to the formation of a thrombus within the vessels and generates a reproducible ischemia, making this model very attractive for investigations requiring a thrombus. As well, this process is much less invasive than the other surgical models and the application of the laser can be directed to ensure ischemia in one particular brain area, such as the motor cortex, making this an attractive model for consequent behavioral testing. A major disadvantage of this model is that it is a permanent ischemia. There is no reperfusion aspect to the injury, limiting the evaluation of reperfusion injury that is possible with other models. This model has also been adapted for use in the mouse, using rose bengal and a krypton laser beam (568 nm). The region of the MCA is irradiated along with a ligation of the CCA to generate a consistent, reproducible ischemia.

Embolic The embolic model is created by introducing a cannula containing occluding substances, such as a clot or microspheres, into the CCA, and injecting these into the artery; the occluding substances generate ischemia in different locations dependent on their size. These embolic models are frequently used in rabbit ischemia. Generally, the rabbit model of focal ischemia induces an occlusion through the injection of a clot into the common carotid artery. This includes both small microclot and larger clot models. The rabbit small clot embolus model (RSCEM) introduces obstructive emboli into random small arteries of the brain, resulting in a heterogeneous stroke, while the rabbit large clot embolus model (RLCEM) causes an obstructive embolus in the MCA. Following embolus injection into the rabbit, behavioral changes can be used to assess damage. The RSCEM can also be used to determine clot load needed to affect 50% of rabbits, which can be used to assess neuroprotection. Because this model involves thrombus formation, it is very useful for determining the effects of proposed thrombolytics. In fact, tissue plasminogen activator was first evaluated as useful therapy for ischemic stroke in the rabbit model.

Rabbit physiology is very stable following the preparatory surgery to induce RSCEM or RLCEM. This is in contrast to the rat and mouse models, which require constant monitoring to maintain physiology within normal levels.

Endothelin The endothelin (ET-1) model of cerebral ischemia is achieved by stereotaxic injection of endothelin, a potent vasoconstrictor that occludes arteries in the area of injection. The application of ET-1 onto the cerebral cortex has been shown to create an ischemic injury that is reproducible, with consistent lesions following injection, and which shows a dose-dependent reduction of flow. ET-1 can be used to induce ischemia in several different brain locations, depending on the delivery. The drug can be applied topically to the cerebral cortex, directly over the MCA, or injected into various brain regions, including the white matter, by advancing a needle into the brain to the correct region using stereotaxic coordinates. Upon application of the ET-1 there is a rapid, but not immediate, reduction in blood flow, which shows gradual reperfusion over time. This mimics the aspect of human stroke without thrombolysis. Increased variability can be introduced into the MCA occlusion model of ET-1 ischemia due to the difficulty in accurate application over the MCA.

Another difficulty with the MCA occlusion model is the necessity for craniotomy, which might cause some trauma to the cortex. The intracerebral injections limit this trauma because craniotomy is not required for the injections. A final consideration must be made on the appropriate application site for experimental study, because certain application sites (i.e., MCA) do not allow for a sufficient motor cortex infarction, necessary for assessment of behavior.

Hemorrhagic Stroke Models

Hemorrhagic stroke models usually attempt to address the effects of cerebral bleeding into either the intracerebral tissue or in the surrounding meninges. These models are usually achieved by either an injection of autologous blood into the brain or injection of collagenase, a chemical that breaks down the vessel walls.

The injection of collagenase into the cerebral tissue will break down type 4 collagen, a major component of the vasculature, and create intracerebral hemorrhage, with a core infarct and a surrounding penumbral area. The core of the hemorrhage has high levels of fluid, red blood cells, and fibrin.

Injection of autologous blood can be performed in various brain areas, including directly into the cerebral tissue or into spaces between the meninges. Following creation of subarachnoid hemorrhage in the monkey, there is a consequent chronic vasospasm. Immediately following hemorrhaging into the brain, there is a consequent reduction of perfusion pressure to the brain and also a reduced CBF. The newest model to generate hemorrhage is by ultrashort laser pulses, which will allow for creation of small, discrete hemorrhage.

Conclusion

Stroke modeling, both ischemic and hemorrhagic, is constantly evolving in an effort to create animal models that most closely mimic human stroke. While these models approximate the human condition, they are not absolute. Thus, it is important to show the efficacy of treatment in more than one model, and for extended durations of animal survival, with extensive behavioral testing. All animal physiology must be carefully monitored and kept within the physiologic levels to ensure that the outcome seen is due to the treatment and not to physiologic perturbations. A clear understanding of the models themselves and what they are capable of testing is also required, and the suitability of the chosen animal model for testing a hypothesis must be carefully considered.

See also: Map Plasticity and Recovery from Stroke; Stroke; Stroke: Injury Mechanisms; Stroke: Neonate vs. Adult.

Further Reading

Astrup J, Siesjo BK, and Symon L (1981) Thresholds in cerebral ischemia – the ischemic penumbra. *Stroke* 12: 723–725.

Barber PA, Demchuk AM, Hirt L, et al. (2003) Biochemistry of ischemic stroke. *Advances in Neurology* 92: 151–164.

Carmichael ST (2005) Rodent models of focal stroke: Size, mechanism, and purpose. *NeuroRx* 2: 396–409.

del Zoppo GJ (1998) Clinical trials in acute stroke: Why have they not been successful? *Neurology* 51: S59–S61.

Dirnagl U, Iadecola C, and Moskowitz MA (1999) Pathobiology of ischaemic stroke: An integrated view. *Trends in Neurosciences* 22: 391–397.

Fisher M (2004) The ischemic penumbra: Identification, evolution and treatment concepts. *Cerebrovascular Disease* 17(supplement 1): 1–6.

Fukuda S and del Zoppo GJ (2003) Models of focal cerebral ischemia in the nonhuman primate. *ILAR Journal* 44: 96–104.

Hoyte L, Kaur J, and Buchan AM (2004) Lost in translation: Taking neuroprotection from animal models to clinical trials. *Experimental Neurology* 188: 200–204.

Longa EZ, Weinstein PR, Carlson S, et al. (1989) Reversible middle cerebral artery occlusion without craniectomy in rats. *Stroke* 20: 84–91.

Pulsinelli WA and Brierley JB (1979) A new model of bilateral hemispheric ischemia in the unanesthetized rat. *Stroke* 10: 267–272.

Richard GA, Odergren T, and Ashwood T (2003) Animal models of stroke: Do they have value for discovering neuroprotective agents? *Trends in Pharmacological Sciences* 24: 402–408.

Traystman RJ (2003) Animal models of focal and global cerebral ischemia. *ILAR Journal* 44: 85–95.

Xi G, Keep RF, and Hoff JT (2006) Mechanisms of brain injury after intracerebral haemorrhage. *Lancet Neurology* 5: 53–63.

Zivin JA, Fisher M, DeGirolami U, et al. (1985) Tissue plasminogen activator reduces neurological damage after cerebral embolism. *Science* 230: 1289–1292.

Animals and the Biology of Music

W T Fitch, University of St. Andrews, St. Andrews, UK

Introduction

In recent years, there has been a resurgence of interest in the biology and evolution of human cognitive specializations, including language, speech, and music. One of the dominant issues in all of these fields is the extent to which unusual human abilities depend on general characteristics of vertebrates, mammals, or primates that predated human evolution and to what extent something 'uniquely human' is based on mechanisms which themselves are unique to our species. Logically, an answer to this question requires comparative data: we cannot know what is unique to our own species without first examining others. Whether there are uniquely human mechanisms is a question to be answered by first identifying relevant mechanisms for the domain (e.g., pitch and rhythm perception, vocal learning, and instrumental abilities in the case of music) and then carefully examining multiple animal species to determine whether the trait is present in any of them. Unique mechanisms can only be identified by a failure to discover any corresponding mechanism in a nonhuman species. Currently, few of the mechanisms that underlie human cognitive specializations appear to be unique to humans, and it seems clear that the bulk of human cognitive functioning is based on mechanisms shared in some form with other species.

When mechanisms are found which are shared, we can further address issues of when and why a particular mechanism appeared in evolutionary time. Addressing 'when' requires the examination of homologies among related species (traits shared by descent from a common ancestor who possessed some version of the trait). For example, the hair cells of the inner ear are shared by all vertebrates, and thus we can infer that the common ancestor of all living vertebrates possessed hair cells. A different trait, a complex of three middle ear bones (malleus, incus, and stapes) connecting the tympanum to the cochlea, is found in all living mammals and thus represents a mammalian homology, present in the ancestral mammal. A second type of similarity, often called analogy, arises quite differently by a process of convergent evolution. Such traits as bipedalism in man and birds, wings on birds and bats, or vocal imitation in marine mammals and humans are analogous traits because they were not present in the last common ancestor of these species. Although such traits are not, by themselves, helpful for determining when the trait arose, they can offer valuable clues to help answer the 'why' question: convergent evolution often results from two species independently 'discovering' the same solution to a similar problem (e.g., flight). Thus, both homology and analogy have different complementary roles to play in understanding the evolutionary history of a trait. Understanding cognitive evolution requires that we take a broad comparative approach, and it will be incomplete if focused only on primates.

This article considers the biology and evolution of music from a broad comparative perspective. The evolution of music has been seen as a biological puzzle since Darwin, who wrote in 1871, "As neither the enjoyment nor the capacity of producing musical notes are faculties of the least use to man in reference to his daily habits of life, they must be ranked amongst the most mysterious with which he is endowed." As Darwin was well aware, some important clues to resolving this mystery come from animals related to humans only distantly, such as birds. This is because some core characteristics of human music production are shared with many bird species but not with nonhuman primates. Bird 'song' is thus an example of convergent evolution, and it may provide a means of testing hypotheses about evolutionary function of complex, learned acoustic communication. Music, like spoken language, has two sides that are intertwined: production and perception. This article focuses on the former because without music production there can be no music perception. This discussion is also restricted to vertebrates and acoustic communication due to space limitations. In humans, music is often accompanied by dance, and in some languages there is not even a separate word for these two activities; unfortunately, little is known about the neural mechanisms underlying dance even in our own species, and next to nothing is known about similar conventionalized movement patterns in nonhuman animals. As discussed later, even this somewhat restricted focus on 'musical' vertebrate acoustic communication provides a rich source of comparative data.

Terms and Definitions

The term 'music' is notoriously difficult to define because of the huge range and diversity of musical practices in our own species. For comparative purposes, such suggestions as Blacking's "humanly organized sound" or McDermott and Hauser's "structured sounds produced directly or indirectly by humans" are clearly inappropriate because they render the concept of animal music impossible, by definition. We can avoid this problem by defining two core subdivisions of human music – song and drumming – which can be unambiguously defined but which have clear parallels in the animal world. Although vocal and instrumental music are often combined, the distinction between them is quite an obvious one among human musical cultures, with virtually all cultures possessing both but with musicians often specializing in producing one or the other form of music. As discussed later, the search for animal parallels of these two subdivisions of music leads in quite different directions.

'Song' in nonhumans can be defined simply as complex, learned vocalizations (for humans, we must add 'nonspeech' to this definition because human speech also constitutes a complex, learned vocalization, but one used for linguistic purposes of expressing propositional meanings). Complexity can be quantified in various ways, such as Kolmogorov or encoding complexity, but more subjective definitions may be appropriate for many purposes than any absolute threshold because what is 'complex' among birds or primates may be different for frogs or fish. Learning can most clearly be demonstrated in laboratory experiments, in which individuals are raised in the absence of auditory stimulation (and do not develop normal song) or given particular songs as input which they then acquire, perhaps with some modification or permutation. By this metric, song exists in humans, birds, and several groups of marine mammals.

'Drumming' can be defined as nonvocal sounds produced by striking an object or some part of the body against another object (the ground, a tree, or, in a few cases, as in human hand clapping or gorilla chest beating, against a different part of the drummer's body). By this definition, drumming has evolved independently in birds, the African great apes, and various rodents and other small mammal species. To my knowledge, there are no other instances of nonvocal communicative sounds in the animal kingdom that could be considered 'instrumental,' although there are many instances of mechanically produced sounds that rely on specialized anatomical mechanisms, such as cricket wing scraping or feather snapping or bill clapping in some bird species. Such uses of specialized body parts for sound-making is in some sense analogous to vocalization rather than to instrumental music; more important, there is no evidence that such mechanical sounds are learned and thus directly relevant to human instrumental music. Therefore, the discussion of nonhuman instrumental music will be restricted to drumming as just defined.

Animal Song

By the definition of song as complex, learned vocalizations, song has evolved convergently in many different vertebrate clades. Birds are the most prominent case, and the ability of birds to learn novel sounds from their environment (e.g., a different species or human speech) has been noted since Aristotle. At least three clades of birds appear to have independently evolved song: parrots, hummingbirds, and the oscine passerines, termed 'songbirds.' It has become clear that several groups of marine mammals can also learn complex novel signals. Among the cetaceans, these include the odontocetes (toothed whales such as dolphins and killer whales) and the mysticetes (baleen whales such as humpback and right whales). Among the pinnipeds, there is good evidence for complex vocal learning among the phocid or 'true' seals and weaker evidence for the otariid or 'eared' sea lions and fur seals, and walruses. Finally, of course, humans have an ability to learn complex vocal signals, which can be divided into linguistic speech signals and musical song signals. Most mammals tested, from guinea pigs to primates, can gate or inhibit vocal signals in an operant context and thus have some voluntary control over their vocal output, and there is evidence for a modicum of control over particular acoustic components of the call (e.g., fundamental frequency or amplitude). Beyond such basic modifications of innate calls, however, no nonhuman primate has been shown to learn novel signals of any complexity. Similarly, some bat species can match aspects of other bats' signals, but these too are rather simple and thus do not constitute song by the definition adopted here. Finally, a number of other species (e.g., elephants) may have vocal learning of complex signals, but data remain sparse.

Birds are the group best known for their vocal learning, and people have been fascinated by 'talking' birds that mimic human speech for centuries. Careful isolation experiments have demonstrated beyond reasonable doubt that many different songbirds must be exposed to normal song of their species if they are to sing normally themselves. In many species, this input must occur within a certain 'sensitive period' early in life, or normal song cannot be attained. For some species, song 'crystallizes' after this period ends, and the bird will sing in the same way for the rest of its life. In other 'open-ended learners,' the bird retains the ability to learn new song as an adult and often produces new songs each breeding season. Prototypically, birdsong is a male activity, concentrated during the breeding season and under strong hormonal control. In these cases, song often has a dual function: it serves both to repel other males (a territorial function) and to attract females (a courtship/mating function). However, it should not be supposed that these are the only functions of song, nor that only male birds sing. In many species, females also sing (either alone or in duets with their mate). Such well-known species as North American cardinals (*Cardinalis cardinalis*) or European robins (*Erithacus rubecula*) have female song, and duetting may even be typical of sedentary birds living in the tropics. Song can also have various other functions besides mating and territoriality. For example, in many parrot species, both male and female birds sing throughout the year; although the function of parrot song remains poorly understood, it seems to have more to do with building and cementing social relationships than with territoriality or mating. In black-headed Grosbeaks (*Pheucticus melanocephalus*), both parents sing, apparently to maintain contact with their fledgling offspring. The great diversity of bird species with learned song (more than 4000 species) renders sweeping generalizations invalid, and this great diversity allows us to test adaptive and phylogenetic hypotheses about the evolution of birdsong.

Another vertebrate group well-known for its complex songs is the cetaceans, particularly the baleen whales. Although it has been suggested that sailors hearing whale songs through the hulls of their boats gave rise to the Greek myth of sirens' song, whale song was not discovered by science until the advent of underwater hydrophone-based listening systems during the 1940s. It is now clear that many cetaceans are highly vocal. Whereas odontocetes use their vocalizations as the basis for echolocation, and use learned 'whistles' in a variety of communicative situations, the complex songs of baleen whales show the clearest parallel to birdsong and to human music. The songs of the humpback whale (*Megaptera novaeangliae*) are by far the best studied, but some other whales, such as the right whale, also have quite complex, geographically varied vocalizations. The larger baleen whales, such as fin or blue whales, also produce impressive low-frequency vocalizations (20 Hz or below), but these are quite simple and thus do not constitute song by our definition. In all of these Mysticete species, it appears that only males produce vocalizations, and these are concentrated in the time preceding and during the mating season. Because it is difficult to observe whale behavior, much less is known about the function of whale song than about birdsong, but most scientists believe that these vocalizations serve roles similar to that of song in most songbirds and have a dual mate-attraction and competitor-repulsion function. An interesting difference is that, at any given time and place, all the males sing the same song, with the same complex pattern of shared motifs. However, this song changes throughout the mating period, and over a period of decades it changes beyond recognition. This pattern shows that humpback whales, at least, are open-ended learners, capable of adjusting or modifying their song throughout their life span.

The vertebrate group most recently discovered to be capable of learning complex vocalizations is the pinnipeds or seals. Of the main groups of pinnipeds (otariids or eared seals, phocid or earless seals, and the odobenid walruses), the evidence for vocal learning is strong only for phocids. Many phocids produce complex underwater vocalizations that are traditionally termed 'song.' However, the other two pinniped groups have considerable vocal control and are also likely to be vocal learners. The most striking demonstration of vocal learning comes from a male harbor seal (*Phoca vitulina*) named Hoover who lived for many years at the New England Aquarium in Boston. Hoover was orphaned at a very young age and spent the beginning of his life living with some Maine fisherman before being donated to the aquarium, where, upon attaining sexual maturity, he began producing strikingly accurate imitations of human speech, including a convincing New England dialect. Hoover's repertoire was quite limited, and he showed no indication of having any sense of the meaning of the phrases he uttered (which often seemed to be directed at other seals). Beyond this isolated but striking case, the evidence for song learning in phocids is typically more circumstantial, but many phocids show clear geographic differences in song that do not appear to be genetically based and thus are consistent with song learning. Phocid seal 'songs' are almost entirely produced by males during the mating season and thus appear to have a function similar to that of birdsong. Although phocids can and do vocalize in air (e.g., this is an important component of mother–infant communication), the complex vocalizations termed song are typically produced underwater, making behavioral observations and experiments challenging. This, combined with the relatively recent discovery of vocal learning in this group, means that we know frustratingly little about vocal control and vocal learning in this group. However, seals are easily trained and adapt readily to captivity (they are found in zoos and aquariums throughout the world), suggesting that the study of vocal learning in this group has a bright future.

Mechanisms Underlying Complex Learned Vocalizations

Humans possess direct connections between neurons in motor cortex and the primary motor neurons that control the larynx; because such connections are lacking in other primates, this may represent one key aspect of our increased control over vocalization. Similarly, songbirds have direct monosynaptic connections between neurons in the forebrain and the hypoglossal motor neurons that control their vocal organ, consistent with the hypothesis that monosynaptic connections are an important component of the increased vocal control seen in this group. This hypothesis obviously predicts similar connections in seals or cetaceans, but I know of no anatomical work that tests this prediction. An interesting fact about the vocal production mechanism may provide some clues to the evolution of such motor control mechanisms: In most singing vertebrates, vocalizations are produced by an evolutionarily novel production mechanism rather than the larynx. This is true of all birds, which use the syrinx (a complex organ at the base of the trachea) to produce sounds, and of odontocete cetaceans, which appear to use a novel nasal sac system to produce sounds rather than the larynx. Sound production in the other major cetacean group, the large baleen whales, remains poorly understood. Only two 'singing' groups have evolved adequate control over the primitive tetrapod laryngeal vocal source: humans and seals. In both of these groups, the sound production mechanism is the standard one seen in most other vertebrates: vibrating vocal cords housed within the larynx produce sounds which are then filtered by the vocal tract. This unusual distribution suggests that it may be easier, in an evolutionary sense, to bring a novel production system under complex telencephalic control than to 'emancipate' the primitive laryngeal system from the unlearned and largely innate vocalizations it normally produces. Of course, this puts a special focus on the vocal learning mechanisms of seals, which are more likely to provide insights into the function of the human system than are those of birds or odontocetes. Unfortunately, very little is known about pinniped neuroanatomy and neurophysiology, so this potential remains unexplored.

Animal Drumming

Although animal song has received far more attention, the production of sounds by striking a body part against the substrate or an object has evolved in parallel in an equally diverse group of vertebrates. The most common object struck is the ground: several groups of mammals have convergently evolved a propensity to strike the earth with their limbs to signal alarm or threat (e.g., deer, elephant shrews, agoutis, wallabies, and various desert rodents), and a few use their head or teeth in the same way (e.g., mole rats *Spalax* and mole voles *Microtus*). Although these displays are mostly limited to simple 'thumps,' substrate drumming with elaborate, modifiable rhythms has evolved in some of the kangaroo rats in the genus *Dipodomys* (order Rodentia, family Heteromyidae). Of these, only the banner-tailed kangaroo rat *Dipodomys spectabilis* has been studied in detail by Jan Randall and colleagues. Kangaroo rats of both sexes use footdrumming with their hind legs to drum out complex rhythms that have multiple functions. The most common form appears to serve as a territorial ownership advertisement, although footdrumming patterns can also be used as mating displays and as a snake alarm call. Playback studies demonstrate that kangaroo rats listen to these sounds and respond to them by footdrumming, typically waiting for a pause in the other rat's drumming to begin their own display. Footdrumming patterns consist of a series of individual 'drums' arranged into short bursts termed 'footrolls' which themselves form larger sequences. These patterns are individually distinctive, and rats recognize their neighbors, responding more strongly to novel interlopers. Different kangaroo rat species have different footdrumming patterns. However, when changing territories, rats often change their footdrum signatures to be more different from those of their new neighbors, indicating that these patterns are ontogenetically flexible, suggesting voluntary control. Although the rats have no special adaptations for sensing vibration, they have ears modified for low-frequency hearing. Randall and Lewis have shown convincingly that these signals can be effectively transmitted seismically through ground-borne vibrations which are then converted into airborne sound within the rat's burrows, where environmental acoustic noise (e.g., wind noise) is silenced. This phenomenon may explain why drumming is so common among fossorial or ground-dwelling mammals. In summary, footdrumming in *D. spectabilis* represents a relatively elaborate communication system, with at least an element of learning involved, that provides an interesting parallel to many functional aspects of birdsong.

Another group of species which drum communicatively are the woodpeckers (Order Piciformes, Family Picidae). Although their normal foraging method (rapid, powerful pounding of wood with their chisel-like beaks) generates sound as an unavoidable by-product, many woodpeckers also engage in a communicative drumming in which they seek out particularly resonant hollow trees and drum out species-specific patterns that serve both territorial and mating advertisement functions (these are particularly common during the spring mating season). The different species-specific patterns of these drummed signals are readily audible to human listeners and can be used in species identification; playback studies with several species have demonstrated that listening birds respond most strongly to the pattern of their own species. Although individual birds must certainly learn which trees to drum, there is no evidence that the patterns are learned. In an interesting example of partial convergence, palm cockatoos *Probosciger atterimus* also drum on hollow logs as a mating and territorial signal; however, they do so with a stick held in the beak. This is the only example of animal drumming known in which a 'drum stick' is used, and which thus it involves a two-part 'instrument.'

Ape Drumming

Another fascinating example of drumming is found in our nearest living cousins and thus provides the only example that is potentially homologous to drumming in our own species: the drumming behavior of the African great apes (chimpanzees, bonobos, and gorillas). I know of no reports of drumming, or even auditory gestures such as repeated clapping, in orangutans, suggesting that this behavior is a homology in the African great ape clade (including humans). Surprisingly, this interesting and conspicuous behavior has received little detailed scientific attention. Perhaps the most prominent and well-known variant of ape drumming is the chest-beating display of adult male silverback gorillas, which is a typical component of aggressive displays that is often preceded by vocalizations and may be followed by branch shaking and charging at the intruder. Drumming in chimpanzees typically seems to have a similar function: it occurs at the culmination of a pant–hoot display. Although chimpanzees do sometimes make percussive sounds by striking their own bodies (e.g., hand clapping), their drumming typically involves repeatedly striking the hands, feet, or both against particularly resonant trees. This sound can carry quite far through the forest and be audible when the chimpanzee's voice is not. Similar drumming behavior has been noted in the bonobo, or pygmy chimpanzee. Thus, all three of the African great apes feature drumming as one component of a multimodal display. However, drumming is not associated exclusively with agonistic or territorial contexts: it is also a component of playful interactions. This type of drumming is particularly prominent in juvenile gorillas, which appear to use drumming as an affiliative signal, often accompanied by the species-typical 'play face.' Drumming frequently initiates a bout of chasing, wrestling, or other rough-and-tumble play in young gorillas, and it is sometimes exchanged in bouts by the two playing individuals. Drumming in these contexts is far more flexible than that seen in silverback displays. Although chest beating is still common, playing gorillas will also drum on hollow objects or the ground, on their own feet, bellies, or heads, or even on the body of their playmate. Young chimpanzees also engage in flexible, relaxed drumming, although this is more often seen in solitary play than in an interactive, dyadic context. This type of playful, flexible drumming, observed in both sexes, seems a more plausible parallel to the affiliative, bonding aspects typical of human music making than the relatively stereotyped aggressive displays seen in adult males. Unfortunately, little is known about the details of ape drumming (e.g., concerning ontogeny and learning, timing and isochronicity, manual and hemispheric lateralization, or other key factors); thus, the degree to which the mechanisms underlying it are true homologs remains unknown. Even the degree to which apes can synchronize or 'entrain' to an externally-generated pulse remains unstudied. However, the drumming behavior of African apes shows enough parallels to human drumming, and thus to human instrumental music, to be clearly worthy of more detailed investigation.

Conclusions

There appear to be two possible parallels to human music in the animal world: song and drumming. Song has evolved in parallel in birds and marine mammals, and it appears to serve similar functions in many of these species. However, the diversity of behavior of singing birds renders any simple generalizations impossible, and this diversity provides an ample source of comparative data for testing phylogenetic and functional theories. In addition, the ease with which birds can be kept in the laboratory means that our understanding of the mechanistic basis of bird song learning, at both neural and genetic levels, is rapidly increasing. By comparison, almost nothing is known about the neural mechanisms underlying vocal learning in whales or seals, and this seems likely to remain the case for the large baleen whale, which produces the most complex songs. However, seals are an example of a poorly studied group that could potentially provide rich dividends for our understanding of the mechanistic basis for vocal learning in mammals. Song learning in seals is particularly relevant to human song and speech because (unlike birds or toothed whales) seals use essentially the same vocal production mechanism (larynx and vocal tract) as humans. Seal vocalization thus provides a neglected but potentially extremely promising field for future research.

Regarding drumming, many vertebrates produce sound by striking some body part against a substrate, but the use of this technique to produce complex, learned signals is less common. In the best studied example, kangaroo rats of both sexes drum out individually distinctive patterns that can be adapted to the drums of new neighbors and thus appear to be learned. Similar patterns of behavior are likely to be present in several other desert rodents as well. Whereas all of the behaviors just described have evolved by a process of convergent evolution, and thus are (at best) analogs of human musical behaviors, there is one case of animal 'music' that can be plausibly considered a potential homolog of human instrumental music: the drumming behavior of the African great apes. Given the vast amount of information on other aspects of ape behavior, this topic has been oddly neglected, and so the detailed similarities and differences between human drumming and that of gorillas, chimpanzees, and bonobos remain to be empirically determined. However, applications of noninvasive imaging techniques such as positron emission tomography and functional magnetic resonance imaging to nonhuman primates raise the exciting possibility of examining the neural basis of rhythm perception (and perhaps production) in apes, and thus of exploring the function of a neural system homologous to but simpler than that used by humans in musical rhythm processing.

See also: Audiovocal Communication in Bats; Bird Song Systems: Evolution; Birdsong Learning: Evolutionary, Behavioral, and Hormonal Issues; Birdsong Learning; Birdsong: The Neurobiology of Avian Vocal Learning; Game Theory and the Economics of Animal Communication; Referentiality and Concepts in Animal Cognition; Signal Design Rules in Animal Communication; Vocal Communication in Birds.

Further Reading

Arcadi C, Robert D, and Boesch C (1998) Buttress drumming by wild chimpanzees: Temporal patterning, phrase integration into loud calls, and preliminary evidence for individual distinctiveness. *Primates* 39: 505–518.

Dodenhoff DJ, Stark RD, and Johnson EV (2001) Do woodpecker drums encode information for species recognition? *Condor* 103: 143–150.

Fitch WT (2006) The biology and evolution of music: A comparative perspective. *Cognition* 100: 173–215.

Janik VM and Slater PB (1997) Vocal learning in mammals. *Advances in the Study of Behavior* 26: 59–99.

Jurgens U (2002) Neural pathways underlying vocal control. *Neuroscience and Biobehavioral Reviews* 26: 235–258.

Justus TC and Hutsler JJ (2005) Fundamental issues in the evolutionary psychology of music: Assessing innateness and domain specificity. *Music Perception* 23(1): 1–27.

Marler P and Slabbekoorn H (2004) *Nature's Music: The Science of Birdsong*. New York: Academic Press.

McDermott J and Hauser MD (2005) The origins of music: Innateness, uniqueness, and evolution. *Music Perception* 23(1): 29–59.

Ralls K, Fiorelli P, and Gish S (1985) Vocalizations and vocal mimicry in captive harbor seals, *Phoca vitulina*. *Canadian Journal of Zoology* 63: 1050–1056.

Randall JA (2001) Evolution and function of drumming as communication in mammals. *American Zoologist* 41: 1143–1156.

Rehding A (2000) The quest for the origins of music in Germany circa 1900. *Journal of the American Musicological Society* 53(2): 345–385.

Riebel K (2003) The 'mute' sex revisited: Vocal production and perception learning in female songbirds. *Advances in the Study of Behavior* 33: 49–86.

Schaller GB (1963) *The Mountain Gorilla*. Chicago: University of Chicago Press.

Schusterman RJ and Van Parijs SM (2003) Pinniped vocal communication: An introduction. *Aquatic Mammals* 29: 177–180.

Wallin NL, Merker B, and Brown S (2000) *The Origins of Music*. Cambridge, MA: MIT Press.

Anterior–Posterior Spinal Cord Patterning of the Motor Pool

J S Dasen, New York University School of Medicine, New York, NY, USA

Introduction

Locomotor behaviors in vertebrates depend on the formation of selective connections between motor neurons and their synaptic targets in the spinal cord and periphery. The precision in which spinal circuits are assembled is critical in controlling the precise temporal activation of muscle groups in the limbs. Many of the spinal circuits that control simple locomotor behaviors, such as the alternation of left and right limbs or the coordination extensor and flexor muscles during locomotion, are established during the early stages of embryonic development. Locomotor circuits appear to be shaped initially independent of sensory experience, suggesting a high degree of genetic determinism in their formation. A critical feature of all spinal locomotor circuits is the establishment of precise connections between motor axons and muscle targets in the limb.

Motor neurons share certain basic features that distinguish them from other classes of neurons in the spinal cord, and they also acquire specialized properties that allow them to make selective connections with target cells. For example, all spinal motor neurons possess axons that project outside the spinal cord and release acetylcholine as the primary form of neurotransmission. In many if not all other regards, motor neurons are a highly diverse class of neuron. This diversity is most apparent in the motor neuron subtypes that innervate skeletal muscles in the limb, where each muscle is innervated by dedicated groups of motor neurons called motor pools. The typical vertebrate limb contains more than 50 muscle groups, and each of these targets is innervated by a unique pool of motor neurons.

Anatomical Organization of Motor Neuron Cell Bodies and Their Axonal Projections

During development, motor neurons acquire subtype identities that define their position within the spinal cord and determine their ability to selectively innervate peripheral targets. Motor pool identities emerge over a series of sequential stages, and each step restricts the potential of motor axons to innervate alternate targets. An early step in the differentiation of motor neurons involves the segregation of their cell bodies into longitudinally arrayed columns, each column containing motor neurons that project their axons to a common peripheral target. Five major columnar groups of motor neurons are generated within the spinal cord, four of which localize to specific positions along the rostrocaudal axis (**Figure 1**). Of particular importance to the specification of motor pool fates is the generation of the lateral motor column (LMC) because the acquisition of an LMC identity directs motor axons toward the limb. Motor neurons within the LMC are generated selectively at brachial (forelimb) and lumbar (hindlimb) levels of the spinal cord and contain the motor pools that innervate specific limb muscles.

Motor neurons within the LMC can be further delineated on the basis of how their axons initially project into the developing limb bud. The cell bodies of motor neurons within the LMC that project dorsally or ventrally into the limb are segregated from one another and define two divisional identities within the LMC. Medially positioned LMC motor neurons (LMC_m) project ventrally within the limb bud mesenchyme while laterally positioned LMC neurons (LMC_l) project dorsally. These divisional subtypes define two coherent subgroups of motor neurons within the LMC and define an initial choice point for motor axons projecting into the limb.

Motor pools are organized within the columnar and divisional identities of motor neurons. A motor pool is defined as the group of motor neurons that project to a single muscle target in the limb. In many cases the cell bodies of motor neurons are clustered in discrete nuclei, although the physiological relevance to motor pool clustering is still uncertain. Anatomical studies of the position of motor pools in the spinal cord have revealed that each motor pool occupies a stereotypical position in the spinal cord. A motor pool within the LMC typically spans two to three segments of the spinal cord, and the number of motor neurons within a given pool is proportional to size of the muscle it innervates.

Classical Studies on Motor Neuron Development in the Chick

Many of the insights into the mechanisms controlling the synaptic specificity of motor neurons emerged from classical manipulations of the neural tube in chick embryos. The idea that motor neurons have intrinsic properties that allow them to selectively innervate specific muscle targets in the limb was supported by studies in which motor neurons were displaced from their normal position within the spinal cord. In one set of experiments, lumbar-level neural tube was rotated

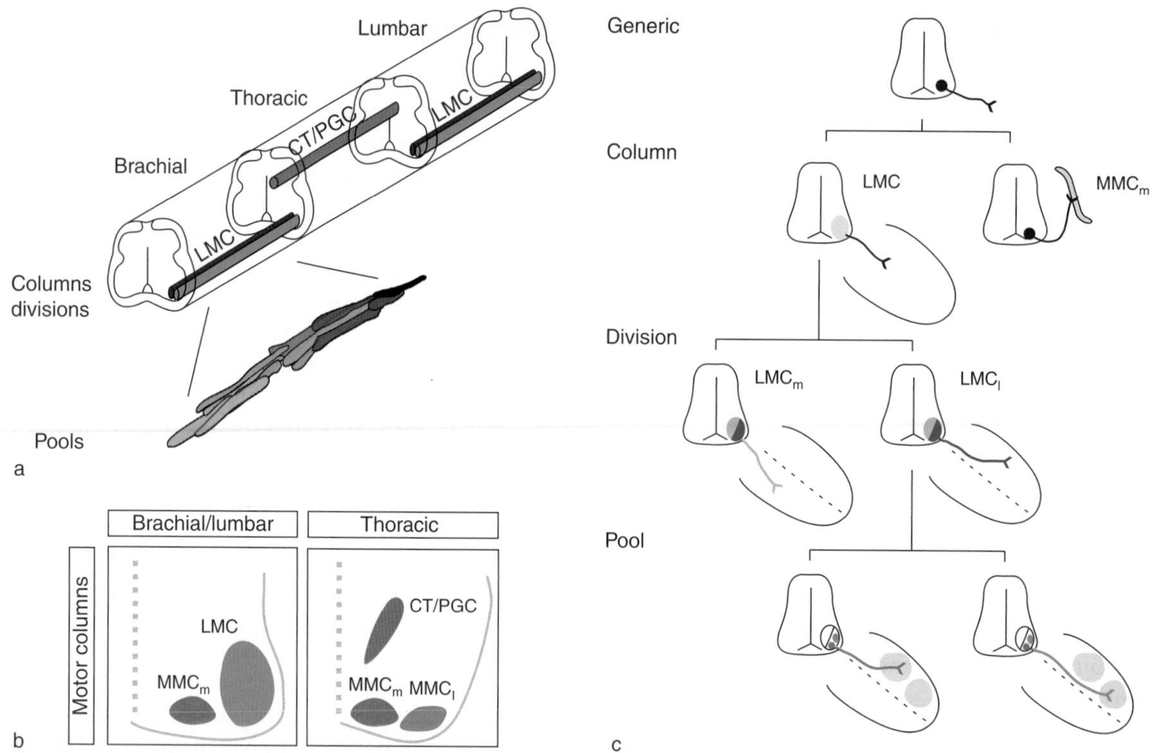

Figure 1 Motor neuron organization within the spinal cord and projection patterns of motor axon. (a) Motor columns and motor pools are generated at specific locations along the rostrocaudal axis. The cell bodies of motor neurons that send axons to the limb are contained within the lateral motor column (LMC) at brachial and lumbar levels of the spinal cord. Preganglionic motor neurons called the column of Terni (CT) in chick or preganglionic columns (PGC) in mouse are found at thoracic levels. The medial and lateral divisions of the LMC are depicted in different colors. Motor pools are generated at specific rostrocaudal positions within the LMC. (b) Schematic of a cross section of brachial/lumbar and thoracic regions of a chick spinal cord, showing the position of motor columns. LMC neurons are found at brachial and lumbar levels of the spinal cord. Motor neurons within the medial division of the medial motor column (MMC_m) are generated at all rostrocaudal levels of the spinal cord and extend axons to axial muscles. At thoracic levels, two segmentally restricted columns are generated, a set of lateral MMC (MMC_l) neurons that project their axons to body wall muscles and CT/PGC neurons that project axons to sympathetic neuronal targets. (c) Stages of motor neuron specification. Generic motor neurons share features common to all motor neurons, such as the projection of axons away from the spinal cord. Columns are sets of motor neurons that project to distinct regions in the periphery. Divisions are binary subdivisions of columns. Pools are subsets of motor neurons within the LMC that innervate a single muscle target in the limb. (a) Derived from Dasen JS, Tice BC, Brenner-Morton S, and Jessell TM (2005) A Hox regulatory network establishes motor neuron pool identity and target muscle connectivity. *Cell* 123: 477–491.

along the rostrocaudal axis, and the projection patterns of the displaced motor neurons were assessed using retrograde labeling assays. After inversion of the neuron tube along the rostrocaudal axis, motor neurons were still capable of finding their appropriate muscle targets even though they entered the limb from inappropriate positions. Thus, the position in which a motor axon enters the limb is not the primary determinant of its projection pattern or the selection of its synaptic target. In addition, these observations are consistent with the view that aspects of motor pool identity are specified prior to limb innervation.

The intrinsic properties of neurons within a pool presumably allow motor axons to differentially respond to guidance cues in the limb. In experiments in which mirror-image duplications of the limb musculature have been generated, motor axons within a single pool innervate both the normal muscle target and the duplicated muscle. These observations reinforce the idea that motor neurons have intrinsic properties that respond to positional cues and also underscore the importance of limb guidance cues in defining motor axon trajectories. A major challenge over the past 20 years has been to identify the molecules expressed within motor neurons and the limb mesenchyme that control motor axon projection patterns and synaptic specificity.

Positional Information and Cell Type Specification in the Spinal Cord

Motor neurons, like other cell types in the spinal cord, acquire their identities in response to positional cues acting along the dorsoventral and rostrocaudal axis of the neural tube. The pathways controlling motor neuron subtype identity involve both extrinsic signals, typically in the form of secreted morphogens, and intrinsic signals, in the form of cell type-specific transcription factors. In the mechanisms that control motor pool specification, the intrinsic signals are particularly relevant because transcription factors are differentially expressed by motor neuron subtypes and presumably regulate the downstream genes involved in motor axon guidance decisions.

A major function of extrinsic signals is to establish unique patterns of transcription factor expression in naïve neuronal cell types. Depending on the relative position of neural progenitors from the source of a secreted signal, cells within the neural tube are exposed to different levels of morphogens. In the ventral neural tube, distinct classes of progenitors are specified in response to secreted signals originating from surrounding mesoderm, including sonic hedgehog from the notochord and floor plate and retinoic acid from the paraxial mesoderm. Graded sonic hedgehog and retinoic acid signaling induces the patterned expression of transcription factors in progenitor cells along the dorsoventral axis (**Figure 2**). These initial patterns of transcription factor expression are further refined through the selective cross-repressive interactions between pairs of transcription factors that act to sharpen the boundaries between progenitor domains and ensure that each progenitor expresses a unique transcription factor profile.

Progenitor cells expressing a specific pattern of transcription factors give rise to distinct classes of postmitotic neurons, including motor neurons. After leaving the cell cycle, all spinal motor neurons express transcription factors that are critical for generic features of their identity. Transcription factors expressed by early postmitotic motor neurons include the homeodomain proteins Hb9, Lhx3, Isl1, and Isl2. Genetic analysis of mice lacking these transcription factors has revealed they are required in each of the subsequent steps in motor neuron differentiation as mice lacking these transcription factors show defects in motor neuron columnar and pool specification. Some of the transcription factors required for early aspects of motor neuron specification are subsequently used in the further diversification of motor neuron subtypes. For example, the LIM homeodomain protein Lhx3 is initially required in all motor neurons and also has later function in the specification of the nonsegmentally restricted motor column that projects to axial muscle.

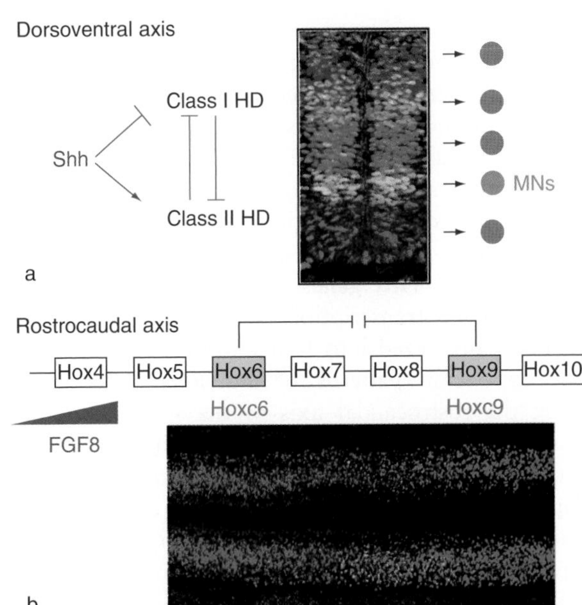

Figure 2 Specification of motor neuron (MN) subtypes along the dorsoventral and rostrocaudal axes. (a) Along the dorsoventral axis of the neural tube, motor neurons are generated in response to the graded activities of sonic hedgehog (Shh). Shh regulates the expression of transcription factors in progenitor cells. Class I transcription factors are induced by Shh, whereas class II transcription factors are repressed. Selective cross-repressive interactions between class I and class II transcription factors sharpens the boundaries between progenitor domains. Each of these progenitor domains gives rise to a distinct class of neuron, including motor neurons. Retinoic acid from the paraxial mesoderm also influences the pattern of transcription factors in neural tube progenitors (not shown). (b) Along the rostrocaudal axis, graded fibroblast growth factor (FGF) signaling induces the expression of chromosomally linked *Hox* genes in the neural tube. Genes located at one end of the cluster are expressed more rostrally, and genes at the opposite end are expressed caudally, in response to higher levels of FGF. The initial domains of Hox expression are further refined through selective cross-repressive interactions. Mutual exclusion of two Hox proteins is shown in top view of a longitudinal section through the spinal cord.

Establishing Patterns of *Hox* Gene Expression along the Rostrocaudal Axis

Although the transcriptional programs mediated by signaling along the dorsoventral axis of the neural tube define how motor neurons as a generic class are specified, additional signaling pathways are necessary for their further diversification into columnar and pool subtypes. Both the columnar and the pool identity of motor neurons requires that certain motor neuron subtypes are generated at specific rostrocaudal positions within the spinal cord. This allows motor neurons to be generated in proximity to their peripheral targets. For example, LMC neurons are generated selectively at limb levels of the spinal cord, while preganglionic column of Terni (CT) motor neurons are generated at thoracic levels in proximity to their synaptic targets in the autonomic nervous system. The columnar and pool identities of motor neurons have been typically defined by their cell body position within the spinal cord, their axonal trajectories, and late profiles of LIM homeodomain transcription factor expression. However, the early signaling events that control motor neuron columnar and pool specification have only recently been explored.

How are motor neuron subtypes generated at distinct segmental levels of the spinal cord? One class of transcription factors known to be critical for establishing differences in cell identity along the rostrocaudal axis consists of members of the *Hox* gene family. In vertebrates, *Hox* genes encode a large family of homeodomain transcription factors consisting of 39 members organized into four chromosomal clusters. The expression patterns of individual *Hox* genes along the rostrocaudal axis of the spinal cord are closely related to the chromosomal position of a gene within a cluster – a principle called spatial colinearity (**Figure 2**). *Hox* genes at the 3′-end of a chromosomal cluster are expressed more rostrally in the embryo than genes at the 5′-end, which are expressed more caudally.

Similar to patterning events along the dorsoventral axis, *Hox* gene expression along the rostrocaudal axis is controlled by the actions of secreted signaling molecules which act in graded manner. Genes within a *Hox* cluster are sequentially activated in response to the activities of several signaling molecules, including fibroblast growth factors (FGFs), retinoids, and members of the transforming growth factor-β superfamily. Graded FGF signaling in particular appears to be important for the initial induction of *Hox* gene expression at brachial, thoracic, and lumbar levels of the spinal cord. An organizing region at the posterior end of the embryo, called Hensen's node in chick, is a source of FGF signaling, and as the node regresses caudally, more posterior regions of the spinal cord are exposed to FGF in higher concentration and over longer periods of time. Although FGF signaling appears to be essential in establishing the initial pattern of *Hox* gene expression in the neural tube, the final pattern observed in motor neurons is dictated largely by regulatory interactions between *Hox* genes.

Hox Proteins Function in the Specification of Segmentally Restricted Motor Columns

Several early studies have provided suggestive evidence that *Hox* genes are involved in specifying motor neuron subtype identities along the rostrocaudal axis. Experimental manipulation of mesodermally derived signals is known to affect the specification of segmentally restricted motor columns, such as LMC and CT motor neurons, and these changes in columnar fate are accompanied by alterations in the pattern of Hox protein expression. In addition, *Hox* genes are known to control cell type specification along the rostrocaudal axis of the hindbrain, and certain *Hox* mutants show defects in the projection of motor axons in the limb. However, one of the difficulties in relating the expression patterns of Hox proteins to motor neuron subtype identities was the lack of molecular markers that are specific for columnar subtypes.

The identification of genes expressed by segmentally restricted motor columns has permitted the analysis of the early steps in motor neuron diversification. LMC and CT motor neurons are generated at specific rostrocaudal levels and express unique molecular markers. LMC neurons can be defined by expression of retinaldehyde dehydrogenase-2 (RALDH2), an enzyme involved in retinoid synthesis, while CT motor neurons in chick selectively express bone morphogenetic protein-5, a member of the transforming growth factor-β superfamily. The expression of specific Hox proteins coincides with the position in which these molecularly defined columnar subtypes

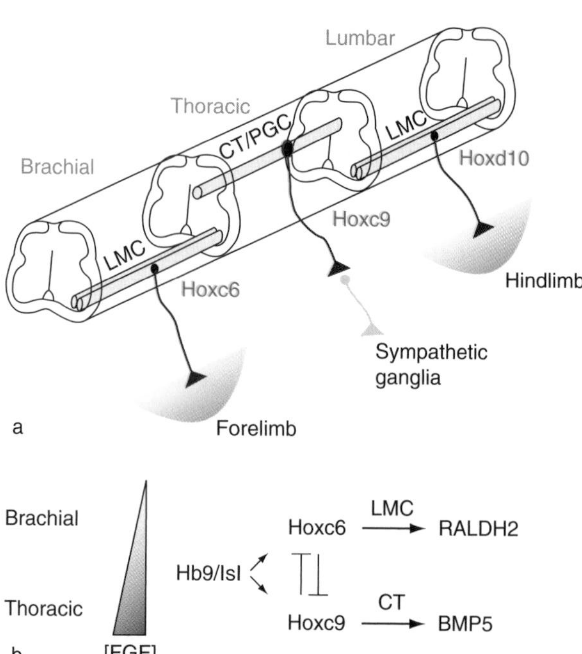

Figure 3 The role of Hox proteins in generating segmentally restricted motor neuron (MN) columnar subtypes. (a) Hoxc6, Hoxc9, and Hoxd10 are expressed in MNs (defined by Hb9 and ISL expression) at distinct rostrocaudal levels of the spinal cord. Each motor column has a distinct peripheral target. (b) Model indicating the roles of fibroblast growth factor (FGF) signaling and Hox expression in the specification of MN columnar identity. FGF establishes an initial pattern of Hox expression in the spinal cord. Cross-repressive interactions between Hoxc6 and Hoxc9 proteins refine the distinct Hox profiles of lateral motor column (LMC) and column of Terni (CT) neurons. Hoxc6 activity in brachial MNs directs expression of retinaldehyde dehydrogenase-2 (RALDH2; a marker for LMC neurons) and late features of LMC identity, whereas Hoxc9 activity in thoracic MNs directs bone morphogenetic protein-5 (BMP5) expression (a marker for CT neurons). Derived from Dasen Js, Liu J-P, and Jessel TM (2003) Motor neuron columnar fate imposed by sequential phases of Hox-c activity. *Nature* 425: 926–933.

are generated. Hoxc6 is expressed by brachial LMC neurons, Hoxc9 by thoracic CT neurons, and Hoxd10 by lumbar LMC neurons (**Figure 3**). The correlation in the pattern of Hox expression with LMC and CT columnar subtypes suggests that the same signals that control Hox expression may also specify columnar fates and that Hox proteins may function in the specification of columnar identities.

Consistent with a model in which *Hox* gene expression is controlled by graded FGF signaling, elevation of FGF levels at brachial levels of the spinal cord *in vivo* induces a pattern of Hox expression characteristic of thoracic levels. This switch in Hox expression patterns is accompanied by a conversion of brachial LMC neurons to a CT cell fate, as the CT marker bone morphogenetic protein-5 is ectopically expressed by brachial-level motor neurons. These effects of FGF on the columnar identity of motor neurons appear to be directly mediated by changes in *Hox* expression. Misexpression of Hoxc9 at brachial levels is sufficient to convert LMC neurons to CT neurons, while expression of Hoxc6 or Hoxd10 at thoracic levels can convert CT neurons to LMC neurons. In addition to these changes in motor neuron identity based on the expression of molecular markers, switching the pattern of Hox expression leads to alterations in the peripheral pattern of motor axon connectivity. For example, conversion of LMC to CT neurons forces limb-level motor neurons to project to sympathetic chain ganglia.

Hox proteins can transcriptionally cross-repress each other's expression, ensuring that columnar subtypes are generated only at specific segmental levels (**Figure 3**). Thus some of the mechanistic principles that govern dorsoventral patterning of the neural tube are shared in patterning along the rostrocaudal axis. One important difference is that while transcriptional cross-repression along the dorsoventral axis occurs in neural progenitors, transcriptional cross-repressive interactions between different Hox proteins occur predominantly in postmitotic cells. Nevertheless, these findings reinforce the view that cross-repressive interactions have critical roles in generating cellular diversity in the central nervous system.

Hox Transcription Factors and Motor Pool Identity

Studies on the columnar identity of motor neurons have provided evidence that Hox proteins contribute to neuronal specification along the rostrocaudal axis of the spinal cord. Many additional Hox proteins are expressed by motor neurons in patterns that do not coincide with the rostrocaudal positional boundaries of motor columns, suggesting additional roles in motor neuron diversification. The requirement for more than 50 motor pools to innervate each of the muscles in a vertebrate limb suggests a significant number of Hox proteins would need to be expressed by LMC neurons. Accordingly, of the 39 *Hox* genes, 21 are expressed in discrete subpopulations of motor neurons at brachial, thoracic, and lumbar levels of the spinal cord in a manner consistent with a role in motor pool specification.

In trying to understand the developmental programs that control motor pool specification, two organizational features of motor pools are particularly relevant (**Figure 4**). First, each motor pool occupies a stereotypic rostrocaudal position within the spinal cord. Thus some aspects of motor pool organization parallel the pattern of motor neuron columnar organization. Second, within a single segmental level, multiple motor pools can be present, and therefore some aspects of motor pool differentiation appear to emerge independent of the early signals that confer rostrocaudal positional information.

Identification of Motor Pools at Early Stages by Transcription Factor Expression

One of the difficulties in trying to define the pathways that control motor pool identity has been relating their anatomical organization with the patterns of Hox protein expression at the time that pools are specified. Motor pools have been defined classically by means of retrograde labeling assays performed after motor axons have reached their muscle targets, yet pool identities appear to be specified several days earlier in development. In addition, the pattern of Hox expression by motor neurons is complex; of the 11 Hox proteins expressed by brachial LMC neurons, a single motor neuron may express up to four different ones. To explore a role for Hox proteins in motor pool identity, what was needed was a set of molecular markers that are expressed by individual pools early in development.

Analysis of transcription factor expression has revealed that brachial and lumbar LMC neurons express an assortment of pool-specific transcription factors. Within the brachial LMC, anatomically defined motor pools can be molecularly defined by expression of the ETS transcription factor Pea3, the runt-related protein Runx1, and the POU-domain factor Scip (Pou3f1). Expression of these transcription factors defines motor pools that occupy stereotypic positions along the rostrocaudal axis and within a given segment. The expression of these pool-specific transcription factors has been used to ascertain the potential contribution of Hox proteins to the anatomical positioning and peripheral connectivity of motor pools.

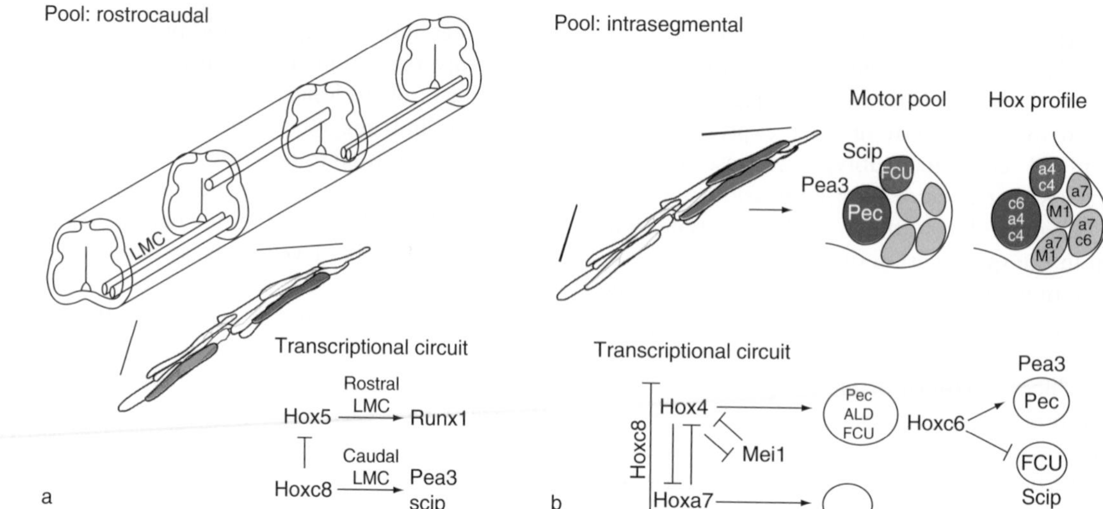

Figure 4 Regulatory networks of Hox proteins control motor pool identity and connectivity. (a) Hox proteins determine the rostrocaudal position of motor pools within the lateral motor column (LMC). At brachial levels of the spinal cord, cross-repressive interactions between Hox5 proteins and Hoxc8 establish the boundary between molecularly defined motor pools. Hox5 proteins are required to generate the motor pool that expresses the transcription factor Runx1 in rostral LMC neurons. Hoxc8 is required in caudal LMC neurons to generate the motor pools that express the transcription factors Pea3 and Scip. (b) Intrasegmental specification of motor pool identity. At a single axial level of the spinal cord, approximately six to ten pools are generated. Motor pools projecting to the pectoralis (Pec) and flexor carpi ulnaris (FCU) can be molecularly defined by expression of the transcription factors Pea3 and Scip, respectively. Both Pec and FCU pools express unique profiles of Hox expression. For simplicity, Hoxa4 and Hoxc4 are shown as Hox4, Hoxa5, and Hoxc5 are shown as Hox5; Mei1 is abbreviated as 'M1', Hoxc6 as 'c6', etc. The patterns of Hox expression in the Pec and FCU pools are established through a transcriptional circuit driven by Hox repressive interactions. Derived from Dasen JS, Tice BC, Brenner-Morton S, and Jessell TM (2005) A Hox regulatory network establishes motor neuron pool identity and target muscle connectivity. *Cell* 123: 477–491.

Hox Proteins and Motor Pool Rostrocaudal Positional Specification

Like motor columns, each motor pool occupies a stereotypic rostrocaudal position within the spinal cord. The position at which motor pools are specified appears to be determined by the pattern of Hox protein expression along the rostrocaudal axis. Brachial LMC neurons, defined by Hoxc6 and RALDH2 expression, can be further subdivided along the rostrocaudal axis by differential expression of Hox3, Hox4, Hox5, Hox7, and Hox8 proteins. For example, motor neurons in the rostral half of the brachial LMC express Hox5 proteins (Hoxa5 and Hoxc5) while motor neurons in the caudal half express Hoxc8. At the boundary between these two regions, expression of Hox5 and Hoxc8 are mutually exclusive.

The exclusive domains of Hox5 and Hoxc8 protein expression within the brachial LMC define a positional boundary between certain motor pools (**Figure 4**). The motor pool defined by expression of Runx1 is generated within the domain of Hox5 expression, whereas the pools expressing Pea3 and Scip are generated within the domain of Hoxc8 expression. Altering the pattern of Hox expression in these territories leads to changes in motor pool identity, defined by a switch in the molecular profile of pool-specific transcription factors. In addition, changing the pattern of Hox expression alters the peripheral connectivity of motor axons. For example, misexpression of Hoxc8 in the domain of Hox5 expression induces expression of Pea3 in rostral LMC neurons, and these ectopically generated Pea3 motor neurons project to the normal muscle target of this motor pool.

The mechanisms by which the rostrocaudal boundaries of motor pools are established by Hox proteins largely follow the mechanisms of motor neuron columnar identity; motor pool boundaries are established through selective cross-repressive interactions between Hox proteins. Rostral misexpression of Hoxc8 extinguishes expression of Hox5 proteins in a cell-autonomous manner; removing Hoxc8 expression from caudal LMC neurons leads to an expansion in the domain of Hox5 expression. In addition to Hox5 and Hoxc8, several other Hox proteins have rostrocaudal positional boundaries within the brachial LMC, and these boundaries likely correspond to the rostral and caudal limits of other motor pools.

Hox Proteins and the Intrasegmental Diversification of Motor Pool Identities

Multiple motor pools are generated at a given segmental level of the spinal cord, and this intrasegmental diversification appears to be established independent of the early patterning signals acting along the rostrocaudal axis of the neural tube. Lineage analysis of motor neurons generated from a single rostrocaudal position has revealed that the fate of progenitors is not fixed because progenitors can give rise to motor neurons that occupy several different motor neuron subtypes. In addition, motor neurons that occupy a pool are initially dispersed within the spinal cord, and only relatively late in development do they cluster into discrete nuclei. Together, these observations suggest that there is no prepattern to motor pool specification at intrasegmental levels but that each motor neuron acquires a specific pool identity on a cell-by-cell basis.

Hox proteins appear to contribute to the intrasegmental diversification of motor pool identities. At late stages of development, the pattern of Hox protein expression becomes progressively restricted to distinct motor pools that occupy a given segment. The expression of the transcription factors Pea3 and Scip in brachial LMC neurons has been used to explore the function of Hox proteins in the intrasegmental diversification of motor pools because pools expressing these transcription factors are generated in overlapping segmental positions. The late patterns of Hox protein expression in pools defined by Pea and Scip expression suggest a model in which the intrasegmental diversification is driven by cross-repressive interactions between Hox4, Hox6, and Hox7 proteins (**Figure 4**). Experimental manipulation of the intrasegmental pattern of Hox expression alters the pattern of Pea3 and Scip expression by motor pools and the connectivity of their motor axons to their peripheral muscle targets.

The mechanisms that define the pattern of Hox expression at intrasegmental levels of the spinal cord appear to be distinct from those that control the pattern of Hox expression along the rostrocaudal axis. Along the rostrocaudal axis, cross-repressive interactions between Hox proteins are apparent shortly after motor neurons leave the cell cycle. In contrast, at intrasegmental levels, individual LMC neurons appear to express an initial cohort of Hox proteins based on their position along the rostrocaudal axis. Through continuous cross-repressive interactions that may be biased for the expression of one Hox protein over another, motor pools eventually express specific combinations of Hox proteins.

One corollary of this model is that biases in the efficacies of Hox cross-repression may explain differences in motor pool size. The number of motor neurons allocated to a specific pool is proportional to the size of the muscle it innervates. These differences in pool size emerge independent of trophic signals from the limb and thus appear to be intrinsically determined during an early phase in motor pool specification. Understanding the mechanisms controlling the differences in motor pool size may be relevant in understanding the allocation of cell numbers to specific neuronal fates in other regions of the nervous system.

Conclusions

Together, the studies described above provide evidence that members of the *Hox* gene family have important roles in the specification of motor pool subtypes. Yet many questions remain as to how Hox proteins contribute to the intrinsic programs that determine the specificity of motor neuron connectivity. Hox proteins function in many contexts throughout the embryo, and within the nervous system the same Hox factor can be expressed by multiple classes of neurons. These observations raise the question of how Hox proteins control gene expression in individual neuronal subtypes. Studies in *Drosophila* and other model systems have provided evidence that the specificity of Hox function is determined through interactions with other DNA-binding proteins. It remains to be determined whether Hox target specificity in motor neurons is controlled through interactions with additional motor neuron-restricted transcription factors.

Although these studies of motor pool specification have helped define some of the transcriptional networks that determine motor neuron identity, the pathways downstream of Hox proteins are less clear. The studies described earlier suggest that the combinatorial expression of Hox proteins in motor pools controls the expression of pool-specific transcription factors. Many of these pool-specific transcription factors may in turn control the expression of receptors which guide motor axons to specific muscle targets. Alternatively, Hox proteins themselves may directly regulate the expression of guidance molecules that are involved in intermediate choice points for motor axons projecting along the major axes of the limb.

Hox factors are expressed by several classes of neurons in addition to motor neurons, including sensory neurons and interneurons. One of simplest circuits in the spinal cord is the monosynaptic stretch reflex circuit, which in its most basic form consists of a motor neuron, a sensory neuron, and a muscle target. One possibility is that the precision of connections in this circuit arises from matching profiles of gene expression in motor neurons and sensory neurons. The expression of Hox factors in muscle sensory neurons parallels the segment-specific Hox patterns in motor neurons, raising the intriguing possibility that Hox factors are involved in the formation of sensorimotor circuits. The analysis of Hox function in sensorimotor connectivity provides a starting point for exploring more-complex levels of spinal circuitry, such as the local networks of interneurons and motor neurons that give rise to rhythmic patterns of activity in the spinal cord and form the central pattern generators required for coordinate locomotor behaviors.

See also: Hox Gene Expression; Motor Neuron Specification in Vertebrates; Segmentation: Segmental Boundaries, Establishment and Morphogenesis (EPH); Segmentation: Spinal Cord Segmentation and A–P Somite Patterning; Transcriptional Networks and the Spinal Cord.

Further Reading

Bel-Vialar S, Itasaki N, and Krumlauf R (2002) Initiating Hox gene expression: In the early chick neural tube differential sensitivity to FGF and RA signaling subdivides the HoxB genes in two distinct groups. *Development* 129: 5103–5115.

Briscoe J, Pierani A, Jessell TM, and Ericson J (2000) A homeodomain protein code specifies progenitor cell identity and neuronal fate in the ventral neural tube. *Cell* 101: 435–445.

Dasen JS, Liu J-P, and Jessell TM (2003) Motor neuron columnar fate imposed by sequential phases of Hox-c activity. *Nature* 425: 926–933.

Dasen JS, Tice BC, Brenner-Morton S, and Jessell TM (2005) A Hox regulatory network establishes motor neuron pool identity and target muscle connectivity. *Cell* 123: 477–491.

Ensini M, Tsuchida TN, Belting HG, and Jessell TM (1998) The control of rostrocaudal pattern in the developing spinal cord: Specification of MN subtype identity is initiated by signals from paraxial mesoderm. *Development* 125: 969–982.

Jessell TM (2000) Neuronal specification in the spinal cord: Inductive signals and transcriptional codes. *Nature Reviews Genetics* 1: 20–29.

Kmita M and Duboule D (2003) Organizing axes in time and space: 25 years of colinear tinkering. *Science* 301: 331–333.

Landmesser LT (2001) The acquisition of motoneuron subtype identity and motor circuit formation. *International Journal of Developmental Neuroscience* 19: 175–182.

Leber SM, Breedlove SM, and Sanes JR (1990) Lineage, arrangement, and death of clonally related motoneurons in chick spinal cord. *Journal of Neuroscience* 10: 2451–2462.

Lin JH, Saito T, Anderson DJ, Lance-Jones C, Jessell TM, and Arber S (1998) Functionally related MN pool and muscle sensory afferent subtypes defined by coordinate ETS gene expression. *Cell* 95: 393–407.

Liu JP, Laufer E, and Jessell TM (2001) Assigning the positional identity of spinal MNs: Rostrocaudal patterning of Hox-c expression by FGFs, Gdf11, and retinoids. *Neuron* 32: 997–1012.

Stirling RV and Summerbell D (1988) Specific guidance of motor axons to duplicated muscles in the developing amniote limb. *Development* 103: 97–110.

Tsuchida T, Ensini M, Morton SB, et al. (1994) Topographic organization of embryonic MNs defined by expression of LIM homeobox genes. *Cell* 79: 957–970.

Antipsychotic Drugs

J P McEvoy, Duke University, Durham, NC, USA

Introduction

Chlorpromazine, the first of the conventional neuroleptic drugs, was synthesized in the early 1950s as a potential antihistamine. When Henri Laborit gave it to patients being prepared for surgery, their spontaneous movements became rare and slow. Initiative, interest, and affect diminished. They tended to rest, but could readily be aroused, and their intellectual functioning remained intact. Because of these effects, Laborit suggested that chlorpromazine be tried in psychotic patients, targeting agitation, aggression, and excitement; Jean Delay and Pierre Deniker did such trials, and these target features decreased. Unexpectedly, in these patients, the power and intrusiveness of hallucinatory perceptions and delusional beliefs also declined. Contact between the patient and the environment improved. Thoughts and behaviors became more organized. However, chlorpromazine and the other conventional neuroleptic drugs also produced dose-dependent neurological side effects, including dystonic contractions of facial, neck, and trunk muscles, a state of Parkinson-like muscular rigidity and tremors, and distressing motor restlessness. After months or years with these initial, reversible neurological side effects, some patients developed (tardive) dyskinetic or choreiform movements that persisted even if the conventional neuroleptic drug was discontinued.

The fundamental pharmacological action underlying the antipsychotic and neurological effects of the conventional neuroleptic drugs is the antagonism of dopamine neurotransmission. Stereospecific binding sites for conventional neuroleptic drugs in the brain proved to be dopamine (D2) receptors. Antagonism at D2 receptors leads to an initial increase in the firing of presynaptic dopamine neurons, followed by the induction of depolarization block in these neurons. The neurons fire and fail to repolarize; their capacity for phasic ('burst') firing is markedly diminished.

Animal models for psychosis are not available. The initial animal models used to screen for antipsychotic drugs were based on the neurological side effects of chlorpromazine, and all of the conventional neuroleptic drugs share dopamine D2 receptor antagonism and the same neurological side effects. Although the conventional neuroleptic drugs produce clinically meaningful reductions in psychopathology in most treated patients, few patients become symptom free or able to return to full functioning, and approximately 25% have substantial residual psychopathology. Most patients maintained on conventional neuroleptic drugs have at least mild neurological side effects, often with accompanying subjective distress.

Clozapine was the first 'atypical' antipsychotic. Clozapine was atypical because it did not produce neurological side effects in animal models. Initial clinical experience with clozapine suggested that it had a wider scope and greater degree of therapeutic efficacy than did the conventional neuroleptic drugs, and no propensity to cause neurological side effects. However, cases of agranulocytosis led to the withdrawal of clozapine from the market in 1975. In 1988, a definitive study demonstrated clozapine's significantly greater therapeutic effects in patients who had garnered little benefit from conventional neuroleptic drugs. Clozapine was then made available for use in patients unresponsive to other antipsychotic drugs, with the requirement for close monitoring of patients' white blood cell counts throughout the duration of treatment.

Clozapine has a complex pharmacology, including actions at numerous neurotransmitter receptors. Investigations are ongoing to determine if and how each of these actions contributes to clozapine's distinct clinical effects. Clozapine is an antagonist at type 2 serotonin (5-hydroxytryptamine; 5-HT_2) receptors, a property that ameliorates neurological side effects and improves sleep. A series of compounds (the newer atypical antipsychotic drugs) that combine antagonism (or partial agonism in the case of aripiprazole) at dopamine D2 receptors with antagonism at serotonin 5-HT_2 receptors proved to have equivalent or slightly superior therapeutic efficacy and a lower propensity for neurological side effects than did the conventional neuroleptic drugs. However, treatment with clozapine and several of the newer atypical antipsychotic drugs is associated with weight gain, lipid elevations, and an increased incidence of diabetes mellitus, resulting in increased risk for cardiovascular disease.

Conventional Neuroleptics

It is necessary to understand the dose–response relationships of conventional neuroleptic drugs to use them well.

Treatment of Acute Psychotic Exacerbations

Fixed-dose studies suggest that the dose–therapeutic response relationship for neuroleptic treatment is to be found between the daily equivalent of 100 and 700 mg of chlorpromazine. Higher doses are not only unlikely to be more effective but may also yield inferior outcomes because of increased risk of neurological side effects. A simplistic argument could be made to treat all acutely psychotic patients with the equivalent of 700 mg chlorpromazine (~15 mg haloperidol) daily, since this dose will cover the minimum therapeutic dose range needed for essentially all potentially responsive patients. However, the group dose–response relationship for neurological side effects closely overlaps that for antipsychotic benefit, with a median effective dose in the range of 200–400 mg chlorpromazine equivalents daily. At a chlorpromazine dose of 700 mg daily, most patients experience coarse neurological side effects.

In 1961, Haase proposed that the lowest neuroleptic doses at which individual patients develop a slight increase in rigidity, detectable only by careful examination ('crossing the neuroleptic threshold'), are also the lowest doses at which patients attain maximum therapeutic benefit, and that no greater therapeutic benefit accrues to patients receiving higher doses – only increments in neurological side effects. Simply stated, those few patients who develop slight stiffness at 100 mg chlorpromazine daily are the same few patients who will respond therapeutically to that dose, and so on for the patients initially developing rigidity with each progressive dose increment. There is not a single minimum effective dose for all patients at which sufficient dopamine D2 blockade is achieved to induce depolarization block. The improvements in group mean therapeutic benefit seen across the dose range of 100–700 mg chlorpromazine daily reflect the accrual of more patients responding (i.e., more 'lights being turned on'), rather than all patients showing greater therapeutic benefit (rheostats increasing the brightness of all lights).

Positron emission tomography permits measurement of the occupancy of dopamine D2 receptors by conventional neuroleptic drugs. In keeping with the neuroleptic threshold hypothesis, antipsychotic efficacy becomes available when ~65% of D2 receptors are occupied, whereas coarse neurological side effects appear when >75% are occupied. Simply stated, coarse neurological side effects imply excessive dosing of these drugs.

Preventing Psychotic Relapse

The haloperidol dose that 'just' blocks a rise in plasma growth hormone levels in response to the oral administration of bromocriptine ($0.5\,\mathrm{mg\,kg^{-1}}$) approximates the minimum dose with the maximum therapeutic effect in preventing relapse in chronic schizophrenia. When the haloperidol doses of 16 patients with chronic schizophrenia (originally maintained on doses >20 mg daily) were lowered to 10 mg daily, 3 patients escaped from blockade. Two more escaped from blockade when the haloperidol dose was reduced to 5 mg daily, and 6 more escaped from blockade at 2.5 mg daily. The remaining 5 escaped when haloperidol was discontinued. No patient lost therapeutic antipsychotic benefit until there was an escape from blockade. After escaping from blockade, 9 of the 16 patients exhibited exacerbated psychosis. Lowest effective maintenance doses also appear to be between 100 and 700 mg of chlorpromazine equivalents daily.

Long-Acting Injectable Preparations

Because of the limited acknowledgment of illness and need for treatment that characterizes chronic psychotic disorders, long-acting injectable preparations of conventional neuroleptic drugs were developed. These preparations, which provide real-time information as to whether patients are taking prescribed antipsychotic treatment, have been used in maintenance treatment studies. If active treatment with one of these preparations is replaced by placebo injections, the 1-year relapse rates are >80%.

When 126 patients with chronic schizophrenia were assigned to blinded maintenance treatment with either standard-dose (12.5–50 mg every 2 weeks) or very-low-dose (1.25–5 mg every 2 weeks) fluphenazine decanoate (FD), cumulative 1-year relapse rates were significantly higher for the very-low-dose patients (56%) than for the standard-dose patients (7%). However, the majority of the patients who relapsed in the very-low-dose group responded quickly to additional antipsychotic medication and did not require rehospitalization or experience severe disruption of their social/occupational performance.

There were no differences in cumulative relapse rates at the end of 1 year among 66 outpatients with schizophrenia randomly assigned to low (5 mg every 2 weeks) or standard (25 mg every 2 weeks) dose maintenance treatment with FD, but at 2 years there was significantly better survival in the 25 mg dose group (64%) than in the 5 mg dose group (31%). If clinicians were permitted to make a dose adjustment up to 10 mg every 2 weeks in the low-dose group and up to 50 mg in the standard-dose group, there were no significant advantages for the standard dose in survival. Patients assigned to the standard dose treatment, compared to those given low doses, had more neurological side effects and more subjective distress during the early months of the study.

Among 70 outpatients with schizophrenia randomly assigned to double-blind maintenance treatment, either with the standard dose at which they had been stabilized (on average, 25 mg every 2 weeks) or with 20% of that dose (on average, 4 mg every

2 weeks), patients assigned to reduced doses and who were living in stressful situations (high expressed emotion families) experienced more minor but aborted episodes in the second year, with no increase in rehospitalization rate. Fewer neurological side effects occurred in the reduced-dose group, and, over time, reduced-dose patients were significantly more improved in their social/occupational role performance than were standard-dose patients.

The 1-year relapse rates (>80%) with placebo are progressively reduced as FD doses are increased through the very low (1.25–5 mg) and low (5 mg) dose ranges; relapse rates begin to match those associated with standard doses in the range of 10–12.5 mg every 2 weeks. For the vast majority of patients, no additional relapse prevention accrues at higher 'standard doses,' but more objective neurological side effects, subjective distress, and dampening of social and occupational performance develop.

Tardive Dyskinesia

Tardive dyskinesia (TD) is a syndrome of choreiform or athetoid abnormal involuntary movements that increase with emotional arousal, decrease with relaxation, and disappear during sleep. Limb dyskinesias are more common in younger patients, whereas orofacial dyskinesias are more common in older patients. Although abnormal involuntary choreiform movements and orofacial dyskinesias were described in patients with schizophrenia before the introduction of the conventional neuroleptics, use of the conventional neuroleptics has increased the prevalence by approximately 20–30%, also increasing the severity of the movements in some patients. During continuous treatment with conventional neuroleptic drugs, the incidence rate is 4–6% of unaffected patients per year. Older age, longer duration of treatment, and taking a conventional neuroleptic drug at doses that produce neurological side effects are factors associated with development of TD. Patients whose conventional neuroleptic dose cannot be reduced to a level at which neurological side effects are minimal without losing therapeutic benefit should be switched to a newer atypical antipsychotic, in particular, quetiapine or olanzapine.

In most cases, TD is not progressive, and even with continued treatment with an antipsychotic drug there may be improvement over time, especially if the lowest effective doses can be employed. However, a subgroup of patients will develop severe and disabling forms of TD; they can be identified early by their extreme sensitivity to neurological side effects. When prescribing a conventional neuroleptic drug, clinicians must document: (1) that patients have been informed of the risk for TD and have consented to treatment, (2) that examinations for neurological side effects and early signs of TD, using standard scales and procedures, are completed at regular intervals during treatment, and (3) that reasonable steps are taken if evidence of neurological side effects or TD appear.

Clozapine

Clozapine offers therapeutic efficacy to approximately 50% of patients who get little benefit from conventional neuroleptic drugs, but has a substantial burden of side effects. Among 268 patients with treatment-resistant schizophrenia randomly assigned to 6 weeks of blinded treatment with clozapine up to 900 mg daily or chlorpromazine up to 1800 mg daily, 30% of clozapine-treated patients, but only 4% of chlorpromazine-treated patients, met predetermined 'responder' criteria. Clozapine-treated patients had significantly greater reductions in Brief Psychiatric Rating Scale total scores than did chlorpromazine-treated patients.

Clozapine produced significantly greater reductions of positive psychopathology than did haloperidol among 39 outpatients who had shown partial therapeutic responses to conventional neuroleptics. Eight of 19 clozapine-treated patients, but only 1 of 20 haloperidol-treated patients, met responder criteria. Smaller advantages for clozapine were seen on negative psychopathology, with little benefit to deficit-state patients.

In a study of 227 chronically hospitalized patients randomly assigned to clozapine or continuation of standard care for up to 2 years, those patients assigned to clozapine had significantly greater reductions in neurological side effects, disruptive behaviors, and hospitalization than did patients continued on standard care. There were no differences in psychopathology or quality of life. Both groups were equally likely to be discharged, but, once discharged, patients assigned to clozapine were less likely to be readmitted. Among 423 treatment-resistant patients with schizophrenia randomly assigned to up to 1 year of treatment with either clozapine or haloperidol, patients assigned to clozapine were significantly more likely to complete the year of treatment (57%) than were patients assigned to haloperidol (28%), had lower psychopathology scores and had fewer inpatient days.

In addition to producing superior antipsychotic efficacy, clozapine reduces hostility and violent behaviors, suicidal behaviors, substance use, and the likelihood of incarceration, relative to treatment with other antipsychotics. It is the only antipsychotic that can reduce excessive water drinking and hyponatremia that endanger some patients with chronic schizophrenia. However, clozapine can have severe, life-threatening side effects. Approximately 1 in 100 patients treated with clozapine will develop agranulocytosis, which, if not detected, can result in overwhelming infection and death. Monitoring of white blood cell counts is required by the Food and Drug Administration as a condition of treatment with clozapine. Approximately 1 in 400 patients treated with clozapine develops myocarditis, usually within the first 2–4 weeks of treatment. Monitoring of inflammatory markers (e.g., eosinophil counts, erythrocyte sedimentation rate), indicators of cardiac damage (e.g., creatine phosphokinase level, troponin level), and the electrocardiogram, weekly for the first 4 weeks, may permit early detection, leading to discontinued treatment. Clozapine can precipitate epileptic seizures in high-risk individuals. Clozapine can produce substantial weight gain, lipid elevations, and diabetes mellitus; these effects are associated with increased risk for cardiovascular disease. Less dangerous side effects include bed-wetting that can be corrected by ephedrine (25 mg) at bedtime, drooling that reflects a decreased swallowing rate rather than excessive saliva production, and marked sedation.

Newer Atypical Antipsychotic Drugs

In 8-week registration trials, risperidone (4–8 mg daily) produced greater reductions in affective and negative psychopathology than did haloperidol (10–20 mg daily), with less neurological side effects. In a maintenance trial in patients with chronic schizophrenia, risperidone-treated patients (mean dose, xx mg daily) had better survival (1-year relapse risks, 34% vs. 60%), greater reductions in psychopathology, and better performance on neurocognitive tests than did haloperidol-treated patients (mean dose yy mg daily). Risperidone showed little advantage over low-dose haloperidol in the treatment of first-episode patients, and was not significantly better than haloperidol for treatment-resistant patients. Risperidone elevates prolactin levels more than other antipsychotics do; sensitive individuals may develop breast engorgement, milk production, loss of menstrual cycle, and sexual dysfunction.

In short-term registration trials, olanzapine (10–15 mg daily) was superior to haloperidol in reducing affective and negative psychopathology. In an extension trial involving patients who had responded to olanzapine or haloperidol acutely, the 1-year relapse rate was estimated to be 20% for olanzapine and 28% for haloperidol. However, in a large collaborative study comparing olanzapine versus haloperidol plus prophylactic benztropine (an anticholinergic anti-Parkinson's drug), little advantage could be demonstrated for olanzapine. First-episode patients treated with olanzapine had less affective and negative psychopathology than did haloperidol-treated patients, and better preservation of brain volume over 2 years of treatment. Olanzapine produced greater reductions in psychopathology compared to haloperidol in treatment-resistant patients. Olanzapine has been associated with more weight gain, more lipid elevation, and more insulin resistance compared to the other antipsychotic drugs, except clozapine.

In two large registration trials, quetiapine was not superior to either haloperidol or chlorpromazine in reducing psychopathology, but had substantially less risk for producing neurological side effects. Ziprasidone was not superior to conventional neuroleptic comparators in registration trials. Ziprasidone produced less neurological side effects. In a 28-week maintenance trial, ziprasidone was equivalent to haloperidol on overall psychopathology, but was associated with less negative psychopathology.

In short-term registration trials, aripiprazole was not superior to haloperidol in therapeutic efficacy. In a 52-week maintenance treatment trial, aripiprazole-treated patients had better survival, and less affective and negative psychopathology, than did haloperidol-treated patients. Aripiprazole was not more efficacious than perphenazine, a conventional neuroleptic drug, for treatment-resistant patients.

The newer atypical antipsychotic drugs are not equivalent in efficacy; only olanzapine and risperidone have evidence of greater therapeutic efficacy than is noted with the conventional neuroleptic drugs. Any advantages for the newer atypical antipsychotic drugs in therapeutic efficacy or survival disappear if the conventional neuroleptic dose is less than 12 mg daily of haloperidol (~600 mg daily of chlorpromazine), although an advantage of less neurological side effects persists.

Large Comparative Trials

To delineate the relative effectiveness and tolerability of the conventional neuroleptic drugs, the newer atypical antipsychotic drugs, and clozapine, the National Institute of Mental Health in the United States and the National Health Service in Great Britain supported large comparative trials. In the Clinical Antipsychotic Trials of Intervention Effectiveness (CATIE) Phase 1 Schizophrenia trial, 1460 patients with schizophrenia were randomly assigned to blinded treatment with olanzapine (up to 30 mg daily), quetiapine (up to 800 mg daily), risperidone (up to 6 mg daily), ziprasidone (up to 160 mg daily), or perphenazine (up to 32 mg daily). Time to treatment discontinuation for any cause, the primary outcome measure for the trial, was longer for olanzapine (median 9.2 months) than for quetiapine (4.6 months), risperidone (4.8 months), ziprasidone (3.5 months), or perphenazine (5.6 months), and olanzapine produced the greatest initial reductions in psychopathology. Olanzapine appeared to be more effective than the other drugs studied, and there were no significant differences in effectiveness between the conventional neuroleptic, perphenazine, and the other newer atypical antipsychotic drugs. However, olanzapine was associated with the greatest weight gain and increases in cholesterol, triglycerides, and glycosylated hemoglobin levels. Rates of discontinuation for specific side effects differed; olanzapine was associated with more discontinuation for weight gain or metabolic side effects, and perphenazine was associated with more discontinuation for neurological side effects.

In the CATIE Phase 1b trial, 114 patients who had discontinued treatment with perphenazine in the CATIE Phase 1 trial were randomly assigned to continued blinded treatment with either olanzapine, quetiapine, or risperidone. Median time to discontinuation for any cause was significantly longer with quetiapine (9.9 months) and olanzapine (7.1 months) than with risperidone (3.6 months). Individual sensitivity to neurological side effects likely contributed to discontinuation from perphenazine in the Phase 1 trial; survival in Phase 1b was in keeping with the newer atypical antipsychotic drugs' relative propensities to produce neurological side effects (i.e., quetiapine < olanzapine < risperidone).

In the CATIE Phase 2 ziprasidone trial, patients who had discontinued treatment with a newer atypical antipsychotic drug were randomly assigned to treatment with ziprasidone (50%; this was the newest atypical antipsychotic drug at the time, and had the promise to produce less weight gain and metabolic disturbances than produced by the other newer atypical antipsychotic drugs) or a newer atypical antipsychotic (50%; olanzapine, quetiapine, or risperidone), different from the one they had received in Phase 1. Median times to treatment discontinuation were significantly longer for risperidone (7.0 months) and olanzapine (6.3 months) than for quetiapine (4.0 months) and for ziprasidone (2.8 months). Patients receiving ziprasidone did have the greatest weight loss and reductions in lipid levels.

In the Cost Utility of the Latest Antipsychotic Drugs in Schizophrenia Study (CUtLASS 1), 227 patients with chronic schizophrenia were randomly assigned to 1 year of open-label treatment with either a conventional neuroleptic drug or newer atypical antipsychotic drug (clinicians could select the specific drug they preferred from the assigned class). There was no significant difference between the two drug classes on the primary outcome measure, quality of life scale scores. However, only 31 of 58 (53%) of the patients who received sulpiride (the most commonly prescribed conventional neuroleptic drug; not available in the United States), but 37 of 50 (74%) of the patients who received olanzapine (the most commonly prescribed newer atypical antipsychotic drug), completed the full year of treatment on their initially assigned antipsychotic.

In the CATIE Phase 2 Clozapine trial, 99 patients who prospectively failed to improve with an assigned newer atypical antipsychotic drug in Phase 1 were randomly assigned to clozapine (50%) or to another newer atypical antipsychotic (50%; olanzapine, quetiapine, or risperidone), different from what they had received in Phase 1. Clozapine produced longer median time to treatment discontinuation (10.5 months) than was seen with the newer atypical antipsychotic drugs (range 2.7–3.3 months) and greater reductions in psychopathology.

In the CUtLASS 2 trial, 136 patients with treatment-resistant schizophrenia were randomly assigned to clozapine or one of the newer atypical antipsychotic drugs for up to 1 year. Clozapine-treated patients had significantly greater reductions in psychopathology than did patients treated with a newer atypical antipsychotic.

Summary

The fundamental pathophysiology of schizophrenia remains poorly understood. One manifestation of that pathophysiology appears to be abnormal functioning of dopamine tracts that extend from the midbrain to ramify through limbic structures. These dopamine tracts, like those that extend from the substantia nigra to ramify through movement-related structures, appear to be tone setting. If the latter (motor) tracts have decreased functioning, a Parkinson-like decreased readiness to move ensues; if they function excessively, abnormal, involuntary, disorganized movements intrude into normal activity. Psychotic exacerbations are believed to reflect excessive functioning of the limbic dopamine tracts, resulting in the abnormal, involuntary, disorganized application of limbic weighting (emotional valence, veracity, apparent connectiveness) that intrudes into normal thought, resulting in psychotic psychopathology.

All antipsychotic drugs decrease dopamine neurotransmission. The conventional neuroleptics do this solely by blocking D2 dopamine receptors; the brain concentrations that affect limbic dopamine tracts are very close to the brain concentrations that affect the motor dopamine tracts. If doses of these drugs are restricted to levels at which the first evidence of

Parkinson-like rigidity is detected by careful examination, many patients reap therapeutic benefit without coarse neurological side effects. The newer atypical antipsychotics add antagonism at serotonin 5-HT_2 receptors to D2 antagonism, resulting in fewer propensities for neurological side effects, emotional dulling, and subjective distress. Olanzapine, and to a lesser degree quetiapine and risperidone, produce weight gain and increases in lipids and insulin resistance.

Clozapine and, to a lesser degree, olanzapine offer greater control of psychotic psychopathlogy than do the other antipsychotics. It remains unclear what pharmacological mechanisms contribute to their advantage. It is intriguing that the therapeutic response and weight gain produced by these drugs are correlated, and the possibility that some of their metabolic effects contribute to therapeutic benefit should be investigated. When prescribing clozapine or one of the newer atypical antipsychotic drugs that produce weight gain and metabolic abnormalities, clinicians should document that (a) they have informed patients and their family members of these risks, (b) they monitor weight and fasting glucose and lipids, and, if abnormalities occur, (c) they respond to them reasonably.

See also: Cognitive Deficits in Schizophrenia; Cognitive Dysfunction in Psychiatric Disorders; Dopamine; Dopamine in Perspective; Dopamine – CNS Pathways and Neurophysiology; Dopamine Receptors and Antipsychotic Drugs in Health and Disease; Dopaminergic Agonists and L-DOPA; Neuroleptics; Psychotherapeutic Approaches to Psychiatric Disorders; Schizophrenia: Epidemiology, Clinical Features, Course and Outcome; Schizophrenia: Genetics.

Further Reading

A Task Force, Report of the American Psychiatric Association (1992). *Tardive Dyskinesia.* Washington, DC: American Psychiatric Association.

Csernansky JG, Mahmoud R, and Brenner R (2002) A comparison of risperidone and haloperidol for the prevention of relapse in patients with schizophrenia. *New England Journal of Medicine* 346: 16–22.

Davis JM, Chen N, and Glick ID (2003) A meta-analysis of the efficacy of second-generation antipsychotics. *Archives of General Psychiatry* 60: 553–564.

Kapur S, Zipursky R, Jones C, et al. (2000) Relationship between dopamine D2 occupancy, clinical response, and side effects: A double-blind PET study of first-episode schizophrenia. *American Journal of Psychiatry* 157: 514–520.

Geddes J, Freemantle N, Harrison P, et al. (2000) Atypical antipsychotics in the treatment of schizophrenia: Systematic overview and meta-regression analysis. *British Medical Journal* 321: 1371–1376.

Grace AA, Bunney BS, Moore H, et al. (1997) Dopamine-cell depolarization block as a model for the therapeutic actions of antipsychotic drugs. *Trends in Neuroscience* 20: 31–37.

Haase HJ and Janssen PAJ (1985) *The Action of Neuroleptic Drugs.* New York: Elsevier Science Publishing Company.

Healy D and Savage M (1998) Reserpine exhumed. *British Journal of Psychiatry* 172: 376–378.

Hirsch SR and Weinberger DR (eds.) (2003) *Schizophrenia, Part Three: Physical Treatments.* Oxford: Blackwell Publishing.

Jones PB, Barnes TR, Davies L, et al. (2006) Randomized controlled trial of the effect an quality of life of second-vs first-generation antipsychotic drugs in schizophrenia (CUtLASS 1). *Archives of General Psychiatry* 63: 1079–1087.

Kane J, Honigfeld G, Singer J, et al. (1988) Clozapine for the treatment-resistant schizophrenic; a double-blind comparison with chlorpromazine. *Archives of General Psychiatry* 45: 789–796.

Leucht S, Barnes TR, Kissling W, et al. (2003) Relapse prevention in schizophrenia with new-generation antipsychotics: A systematic review and exploratory meta-analysis of randomized, controlled trials. *American Journal of Psychiatry* 160: 1209–1222.

Leucht S, Wahlbeck K, Hamann J, et al. (2003) New generation antipsychotics versus low-potency conventional antipsychotics: A systematic review and meta-analysis. *Lancet* 361: 1581–1589.

Lieberman JA, Stroup TS, McEvoy JP, et al. (2005) Clinical Antipsychotic Trials of Intervention Effectiveness (CATIE): Effectiveness of antipsychotic drugs in patients with schizophrenia. *New England Journal of Medicine* 353: 1209–1223.

McEvoy JP, Hogarty GE, and Steingard S (1991) Optimal dose of neuroleptic in acute schizophrenia: A controlled study of the neuroleptic threshold and higher haloperidol dose. *Archives of General Psychiatry* 48: 739–745.

Volavka J, Czobor P, Sheitman B, et al. (2002) Clozapine, olanzapine, risperidone, and haloperidol in the treatment of patients with chronic schizophrenia and schizoaffective disorder. *American Journal of Psychiatry* 159: 255–262.

Anxiety Disorders

J J Benson-Martin, University of Cape Town, Cape Town, South Africa
D J Stein, University of Cape Town, Cape Town, South Africa; and Mount Sinai School of Medicine, New York, NY, USA
E Hollander, Mount Sinai School of Medicine, New York, NY, USA

Introduction

Almost a century ago, Freud coined the term 'anxiety neurosis,' explained this condition in terms of unconscious drives, and introduced psychoanalytic treatment. Since then there have been important advances in classifying the anxiety disorders, in delineating their underlying psychobiology, and in providing effective intervention. Advances in neuroscience, including rigorous animal models of anxiety, increased understanding of genetic variation, and functional brain imaging have all contributed to knowledge of the anxiety disorders. Advances in pharmacotherapy and psychotherapy, and increased public awareness of the anxiety disorders, have contributed to better outcomes. In this article we briefly review some of these advances.

Phenomenology

Current diagnostic systems attempt to make a distinction between anxiety as a normal emotion and pathological anxiety. Normal anxiety is universally experienced and is characterized by a sense of apprehension (often diffuse and unpleasant), and by accompanying autonomic phenomena. These phenomena vary between individuals but may take the form of abdominal discomfort, restlessness, perspiration, palpitations, etc. Pathological anxiety, on the other hand, consists of excessive fear, worry, physical arousal, and avoidance. The symptoms of pathological anxiety cause clinically significant distress or impairment in important areas of functioning such as social and occupational functioning.

Modern classification systems list several different anxiety disorders, including generalized anxiety disorder, obsessive–compulsive disorder, panic disorder, social anxiety disorder (or social phobia), and specific phobia (see below). The fourth edition of the *Diagnostic and Statistical Manual of Mental Disorders* (DSM) provides specific diagnostic criteria for each of these conditions and is the most widely used nosological system in anxiety disorders research. The *International Classification of Diseases* (ICD-10) similarly has a section on anxiety disorders, although there are some differences (e.g., obsessive–compulsive disorder is not considered as one of the anxiety disorders).

Epidemiology

Nationally representative epidemiological surveys have indicated that anxiety disorders are the most prevalent of the common psychiatric disorders. They are also among the most disabling and costly to society. For example, the Epidemiological Catchment Area (ECA) study in the United States found that specific phobia was the most common psychiatric disorder and obsessive–compulsive disorder (OCD) was the fourth most common. Subsequently, the National Comorbidity Survey (NCS) and the National Comorbidity Survey-Revised (NCS-R) similarly reported that anxiety disorders are the most prevalent of the psychiatric disorders.

Similar findings have been reported in Europe and in the developing world. For example, a review of 27 epidemiological studies carried out in the EU from 1990 to 2004 found that anxiety disorders are the most frequent psychiatric disorders in Europe with a median 12 month prevalence of 12%. The World Mental Health Survey, a set of epidemiological surveys recently conducted around the globe, has again emphasized the prevalence of the anxiety disorders, as well as their early age of onset, high levels of subsequent comorbidity, and considerable underdiagnosis and undertreatment.

Studies in primary care and clinical settings confirm many of the epidemiological findings. Thus, the anxiety disorders have an early onset and chronic duration, they are often associated with subsequent comorbid conditions (particularly depression, substance use disorders, and other anxiety disorders), and they have a significant negative impact on social life, on occupational function, and on family relations. Given that these disorders are characterized by an early age of onset, they can also impact adversely on normal developmental processes. The high costs of the anxiety disorders are due mostly to indirect costs (e.g., loss of occupational productivity, nonpsychiatric medical treatment) rather than to direct treatment costs. There is a clear need for higher awareness and earlier recognition and intervention.

Diagnosis and Evaluation

DSM-IV and ICD-10

The DSM-IV-TR lists 12 anxiety disorders: (1) panic disorder (PD) with agoraphobia, (2) panic disorder without agoraphobia, (3) agoraphobia without history of panic disorder, (4) specific phobia, (5) social phobia (SP) (social anxiety disorder), (6) OCD, (7) posttraumatic stress disorder (PTSD), (8) acute stress disorder, (9) generalized anxiety disorder (GAD), (10) anxiety disorder due to a general medical condition, (11) substance-induced anxiety disorder, and (12) anxiety disorder not otherwise specified. A thirteenth category, mixed anxiety-depressive disorder (MAD), is currently in the appendix of *Disorders in Need of Further Study.*

There are specific diagnostic criteria for each of the major anxiety disorders (PD, SP, OCD, PTSD, and GAD). In each disorder, there are similar exclusionary diagnostic criteria which emphasize that the diagnosis should not be found to be caused by the physiological effect of a substance or of a general medical condition. The symptoms should also not be better accounted for by another mental disorder. Lastly, the symptoms must cause clinically significant distress or impairment in important areas of functioning such as social and occupational functioning.

In the ICD-10, anxiety disorders are categorized under the class of neurotic, stress-related, and somatoform disorders. This includes the following: (1) phobic anxiety disorders (agoraphobia, social phobias, and specific phobias); (2) other anxiety disorders (including panic disorder, generalized anxiety disorder, and MAD; (3) OCD; (4) dissociative (conversion) disorders; and (5) somatoform disorders. A further category called reactions to severe stress includes (1) acute stress reaction and (2) PTSD. The ICD-10 is less rigorous in its formulation of operational criteria than the DSM.

Categorical and Dimensional Approaches

Although the DSM system focuses on categories, dimensional approaches to measuring anxiety symptoms may also be useful. Arguably, categorical and dimensional approaches to psychiatric diagnosis and evaluation have a complementary role. For many clinical purposes, it may be useful to know whether a diagnosis is present or absent. However, dimensional approaches may be valuable for describing the way in which symptoms fall to a variable extent on different spectrums, the way in which different symptom spectrums intersect or overlap, and the relationship between symptom spectrums and underlying psychobiological mechanisms.

The tripartite model of anxiety symptoms (Clark and Watson) is one influential dimensional approach to the symptoms of anxiety and depression. The relevant dimensions are (1) negative affect – general distress such as worry, tension, and irritability; (2) positive affect – the level of pleasurable agreement with the environment; and (3) autonomic hyperarousal. The Tripartite model proposes that negative affect is shared by both anxiety and mood disorders, while autonomic hyperarousal is peculiar to anxiety. An absence of positive affect (mainly, anhedonia) differentiates mood disorder from anxiety disorders. Spectrum concepts have also been applied to each of the anxiety disorders.

Rating Scales

A number of self-report and interviewer-administered rating scales have been developed to diagnose and evaluate anxiety disorders. Structured diagnostic instruments allow reliable diagnosis of the anxiety disorders and include the Anxiety Disorders Interview Schedule for DSM-IV-TR. Symptom severity measures allow reliable assessment of the characteristic symptoms of a disorder and include the Yale-Brown Obsessive–Compulsive Scale (for OCD), the Liebowitz Social Anxiety Scale (for SAD), the Clinician Administered PTSD Scale (for PTSD), the Hamilton Anxiety Rating Scale (for GAD), and the Panic Disorder Severity Scale (PDSS). The development of reliable symptom measures has contributed to the ability of randomized controlled trials to differentiate between effective medications and placebo.

Neurobiology

Significant advances have been made in understanding the psychobiology of anxiety disorders. Lesion studies have long allowed the study of the neurocircuitry of anxiety disorders, and modern structural and functional brain imaging modalities have provided new insights. Anxiety and its disorders are regulated by a range of neurotransmitters and neurochemical systems, and advances in neurogenetics and neuroproteomics should increasingly allow the specific molecular mechanisms in the neurocircuitry underpinning anxiety disorders to be delineated.

Neurocircuitry

Basic studies of fear have demonstrated that the amygdala and related neurocircuitry play a key role. By virtue of afferent and efferent connections to a range of structures, the amygdala is able to play a key role in mediating the fear response. Sensory input reaches the lateral nucleus of the amygdala, while efferents from the central nucleus include the dorsal motor nucleus of the vagus (parasympathetic activation), the locus coeruleus (release of norepinephrine and autonomic arousal), the hypothalamic–pituitary–adrenal (HPA) axis (neuroendocrine activation), the parabrachial nucleus (hyperventilation), the trigeminal and facial nerve nuclei (fearful facial expression), and the para-aqueductal gray matter (escape or freezing behavior).

A range of other structures are also important in mediating fear responses. The hippocampus plays a crucial role in memory of the context of fear conditioning. The medial prefrontal cortex may be able to inhibit limbically mediated fear responses, and so likely plays a key role in fear extinction. Thus, it can be hypothesized that structures such as the amygdala, hippocampus, and anterior cingulate play a key role in mediating the various anxiety disorders that present in the clinical setting.

Indeed, early lesion studies noted that damage to the amygdala was associated with a loss of fear responses (Kluver–Bucy syndrome), while temporal epilepsy could be associated with increased anxiety. More recently, functional and structural imaging studies have confirmed amygdala hyperactivation in a number of different anxiety disorders. Hippocampal volume appears to be decreased in a number of clinical conditions, including PTSD, and there is some evidence that reduced volume correlates with memory dysfunction. Successful treatment of the anxiety disorders may be accompanied by changes in frontal circuits involved in the suppression of limbic circuitry.

Neurochemistry

A range of neurochemical systems are found in the neurocircuitry thought to mediate anxiety disorders. Noradrenaline, serotonin, glutamate, and γ-aminobutyric acid (GABA) may play a particularly important role in mediating anxiety, and are certainly targets for a number of anxiolytic medications. Nevertheless, a range of other systems, including dopamine, and various neuropeptides are also likely to play an important role in anxiety symptoms. Here we provide only a brief overview of some of the key neurotransmitter systems that are relevant to the anxiety disorders.

The majority of norepinephrine cell bodies are located in the locus coeruleus in the pons. There is evidence for the overactivity of the norepinephrine system in anxiety disorders such as panic disorder and PTSD. For example, there are increased behavioral, physiologic, and neurochemical responses to intravenous challenge with yohimbine, an α 2-adrenoreceptor antagonist, in these disorders. Effective treatment of these conditions with selective serotonin reuptake inhibitors (SSRIs) may decrease yohimbine sensitivity and enable appropriate regulation of α 2-adrenoreceptor function.

Activation of the 5HT system can produce both anxiogenic and anxiolytic effects. Given the complexity of the serotonergic system, it is difficult to make simple generalizations about the relationship between specific receptors and anxiety levels. Nevertheless, stimulation of $5HT_1$ receptors in the hippocampus appears to be anxiolytic, while stimulation of $5HT_{2A}$ receptors in the forebrain can be anxiogenic. Certainly, a range of clinical research with serotonergic agonists, and with serotonin depletors, has confirmed the importance of this system in the mediation of anxiety disorders. SSRIs may work in part by desensitization of auto-receptors, so allowing increased neurotransmission at the level of the synapse.

Glutamate is the main excitatory neurotransmitter and GABA the main inhibitory neurotransmitter in the mammalian brain. The *N*-methyl-D-aspartate (NMDA)/glutamate receptor may play a role in fear conditioning, while the GABA receptor is a key target of benzodiazepines, which are among the most widely prescribed anxiolytics. Molecular imaging technologies will increasingly allow the assessment of a range of neuroreceptors in patients with anxiety disorders, and preliminary work has already indicated that GABA receptors are decreased in anxiety disorders such as PTSD.

Corticotrophin-releasing hormone (CRH) and cortisol receptors are found not only in the paraventricular nucleus of the hypothalamus but also in multiple areas in the cortex. Both play a central role in the mediation of fear and anxiety. In animal studies, early adversity results in long-lasting changes in HPA axis function. In humans with PTSD, there appears to be overproduction of central CRH, with low levels of plasma cortisol, suggesting enhanced negative feedback. This is particularly interesting in view of prior findings that depression is characterized by increased cortisol, and a blunted central response. There is growing interest in the use of CRH antagonists in the potential treatment of anxiety disorders, and trials of these agents are under way.

Neurogenetics

In animals, particular strains may demonstrate increased levels of anxious and avoidant behavior. In humans, family studies have indicated that the anxiety disorders demonstrate significant heritability. Thus, there is growing interest in delineating the neurogenetics of specific behavioral endophenotypes that underpin susceptibility to particular anxiety disorders. For example, behavioral inhibition is a temperament that may predispose to a number of subsequent anxiety disorders. Genetic studies have suggested that variants in the promoter region of the serotonin transporter gene and HPA axis gene variants may play a role in mediating behavioral inhibition.

There is growing interest in assessing the complex relationships between genes, environments, neurocircuitry, and behavior. For example, subjects with the short allele of the serotonin transporter gene promoter region are more likely to have an increased amygdala response to negative facial expressions on functional magnetic resonance imaging. Similarly, while subjects with this particular gene variant are slightly more likely to have anxiety symptoms, in the presence of significant life stressors, they are much more likely to develop a depressive disorder. Future work would likely focus on assessing a range of different gene polymorphisms and life events in attempting to predict psychopathology.

Treatment

There have been significant advances in both the pharmacotherapy and psychotherapy of the anxiety disorders in recent decades. There is now a substantial database of randomized controlled trials of both psychotropics and specific psychotherapies. Although there remain important gaps in our knowledge, the field now has effective first-line interventions for the major anxiety disorders. There is also increased understanding of how current therapies are able to normalize the underlying neurocircuitry of the anxiety disorders. In the future, additional work is needed on how best to combine different treatment modalities, long-term treatments, and intervention for special populations such as children and adolescents, and the elderly. There is also growing interest in preventive measures, which may ultimately prove effective in forestalling onset of anxiety disorders.

Pharmacotherapy

Benzodiazepines have long been a mainstay of the pharmacotherapy of anxiety disorders. However, these agents have important limitations, including safety problems, and current guidelines emphasize antidepressants, and the SSRIs in particular, as first-line agents for the treatment of major anxiety disorders. There are multiple studies of SSRIs in each of these disorders, and meta-analysis of this work confirms both efficacy and safety. There are subtle differences in the appropriate dose and duration of SSRI treatment across anxiety disorders. Functional imaging studies have provided information on how these agents normalize the underlying neurocircuitry of anxiety disorders.

Although the introduction of the SSRIs represents an important advance, not all patients respond to these agents. There is only a relatively small database of studies on the pharmacotherapy of the treatment-resistant anxiety disorder patient. Options include switching to a different SSRI, to a different class of antidepressant, or augmenting an antidepressant with an additional agent. In the case of OCD, there is substantial evidence that augmentation of a SSRI with low doses of a dopamine receptor blocker can be effective. There are also relatively few studies of SSRIs in children and adolescents with anxiety disorders, although these are encouraging in terms of indicating efficacy and safety.

Psychotherapy

Although early therapists adopted a psychodynamic approach to treatment of the anxiety disorders, there has been relatively little empirical work to support the efficacy of these interventions. In contrast, later cognitive-behavioral perspectives, have been rigorously studied. A large series of controlled studies supports the recommendation of current treatment guidelines to use cognitive-behavioral therapy (CBT) as a first-line psychotherapy in the management of major anxiety disorders. These interventions emphasize the value of behavioral exposure to anxiety-provoking stimuli, and the importance of restructuring cognitive processes and structures. There are subtle differences in techniques used across the anxiety disorders. Functional imaging studies have provided information on how CBT techniques are able to normalize the underlying neurocircuitry of the anxiety disorders.

The introduction of CBT represents an important advance. Although less studied in younger patients, CBT may also be useful in the treatment of child and adolescent anxiety disorders. Nevertheless, not all patients are willing to subject themselves to this kind of treatment, and not all patients respond to this form of therapy. There is only a relatively small database of studies on the psychotherapy of the treatment-resistant anxiety disorder patient. Options include combining psychotherapy with pharmacotherapy. A fascinating series of recent studies, based on the basic neuroscience of fear conditioning, have combined D-cycloserine, a partial agonist at the NMDA receptor, with CBT, in an attempt to increase the efficacy of CBT techniques.

Novel Treatments

Neurosurgical treatments such as anterior cingulotomy and anterior capsulotomy have been used in patients with severe treatment-refractory OCD. The efficacy of these treatments in some cases is consistent with our growing understanding of the neurocircuitry of OCD (where frontostriatal circuitry appears to play a particularly important role). Transcranial magnetic stimulation (TMS) and deep brain stimulation (DBS) are novel procedures under development for treatment-refractory anxiety disorders, including OCD. Further work is needed before these can be recommended for use in routine clinical practice.

Conclusion

In this article, we have briefly discussed the phenomenology, psychobiology, pharmacotherapy, and psychotherapy of the anxiety disorders. There have been significant advances in each of these areas of investigation in recent decades; we now understand some of the basic mechanisms underlying the pathogenesis of the anxiety disorders, and we have a range of effective treatments. Nevertheless, much additional work is needed to delineate fully the neurocircuitry and molecular basis of the anxiety disorders. Hopefully, such advances will lead to further improvements in our therapeutic armamentarium.

See also: Amygdala: Structure and Circuitry in Primates; Amygdala: Contributions to Fear; Anxiety: Drug Therapy; Genetics of Human Anxiety and Its Disorders; Obsessive–Compulsive Disorder; Panic Disorder; Panic Disorder as an Emotional Disorder; Posttraumatic Stress Disorder: Neurobiology; Prefrontal Cortex: Structure and Anatomy; Psychotherapeutic Approaches to Psychiatric Disorders; Stress and Cognition.

Further Reading

Baldwin DS, Anderson IM, Nutt DJ, et al. (2005) Evidence-based guidelines for the pharmacological treatment of anxiety disorders: Recommendations from the British Association for Psychopharmacology. *Journal of Psychopharmacology* 19(6): 567–596.

Bartz JA and Hollander E (2006) Is obsessive-compulsive disorder an anxiety disorder? *Progress in Neuro-Psychopharmacology & Biological Psychiatry* 30(3): 338–352.

Carey PD, Warwick J, Neihaus DJ, et al. (2004) Single photon emission computed tomography (SPECT) of anxiety disorders before and after treatment with citalopram. *BMC Psychiatry* 4: 30.

Caspi A, Sugden K, Moffett TE, et al. (2003) Influence of life stress on depression: Moderation by a polymorphism in the 5-HTT gene. *Science* 301: 386–389.

Charney DS (2003) Neuroanatomical circuits modulating fear and anxiety disorders. *Acta Psychiatrica Scandinavica* 108(Supplement 417): 38–50.

Charney DS (2004) Psychobiological mechanisms of resilience and vulnerability: Implications for successful adaptation to extreme stress. *American Journal of Psychiatry* 161: 195–216.

Clark LA and Watson D (1991) Tripartite model of anxiety and depression: Psychometric evidence and taxomic implications. *Journal of Abnormal Psychology* 100(3): 313–336.

Furmark T, Tilfors M, Garpenstrand H, et al. (2004) Serotonin transporter polymorphism related to amygdala excitability and symptom severity in patients with social phobia. *Neuroscience Letters* 362: 189–192.

Furmark T, Tillfors M, Marteindottir I, et al. (2002) Common changes in cerebral blood flow in patients with social phobia treated with citalopram or cognitive-behavioral therapy. *Archives of General Psychiatry* 59: 425–433.

Gorman JM, Kent JM, Sullivan GM, and Coplan JD (2000) Neuroanatomical hypothesis of panic disorder, revised. *American Journal of Psychiatry* 157: 493–505.

Greenberg PE, Sisitsky T, Kessler RC, et al. (1999) The economic burden of the anxiety disorders in the 1990s. *Journal of Clinical Psychiatry* 60: 427–435.

Kessler RC (2001) The epidemiology of pure and comorbid generalized anxiety disorder. A review and evaluation of recent research. *Acta Psychiatrica Scandinavica* 406: 7–13.

Krystal JH, D'Souza DC, Sanacora G, Goddard AW, and Charney DS (2001) Advances in the pathophysiology and treatment of psychiatric disorders: Implications for internal medicine. *Medical Clinics of North America* 85(3): 559–577.

Maier W, Gansicke A, Freyberger HJ, et al. (2000) Generalized anxiety disorder (ICD-10) in primary care from a cross-cultural perspective: A valid diagnostic entity? *Acta Psychiatrica Scandinavica* 101: 29–36.

Mathew SJ, Coplan JD, and Gorman JM (2001) Neurobiological mechanisms of social anxiety disorder. *American Journal of Psychiatry* 158: 1558–1567.

Miller LA, Taber K, Gabbard GO, and Hurley RA (2005) Neural underpinnings of fear and its modulation: Implications for anxiety disorders. *The Journal of Neuropsychiatry and Clinical Neurosciences* 17: 1–6.

Mogotsi M, Kaminer D, and Stein DJ (2000) Quality of life in the anxiety disorders. *Harvard Review of Psychiatry* 8: 273–282.

Starcevic V (2006) Anxiety states: A review of conceptual and treatment issues. *Current Opinion in Psychiatry* 19: 79–83.

Stein DJ (2004) *Clinical Manual of Anxiety Disorders*. Washington, DC: American Psychiatric Pub., Inc.

Yehuda R (2002) Post-traumatic stress disorder. *New England Journal of Medicine* 346: 108–114.

Anxiety: Drug Therapy

C K Haas, A Shekhar, and A W Goddard, Indiana University School of Medicine, Indianapolis, IN, USA

Introduction

Anxiety disorders are a significant mental health concern. The National Institute of Mental Health reports that anxiety is the most common mental health problem in the United States today, affecting at least 19 million people. Approximately one in four individuals in the United States reports a lifetime history of at least one anxiety disorder, according to the National Comorbidity Survery, and anxiety disorders may be more chronic than either substance abuse or affective disorders. Greenberg and colleagues found that anxiety disorders afflict 15.7 million people in the United States each year and that 75% of these sufferers (11.7 million) also have at least one comorbid psychiatric condition. In addition, international community surveys, such as the Zurich Cohort Study, the World Health Organization, the World Mental Health 2000 Initiative, and the Netherlands Mental Health Survey and Incidence Study, have yielded comparable lifetime prevalence rates of anxiety disorders. There is a disparity in the distribution of anxiety disorders between men and women: women (30.5% lifetime prevalence) are more likely to have an anxiety disorder than are men (19.2% lifetime prevalence). In addition, anxiety disorders are not limited simply to the adult population; the prevalence of anxiety disorders in youth appears to be quite similar to that reported in adults. The economic costs of anxiety disorders are also quite staggering. In 1998 the United States spent an estimated $\$63.1 \times 10^9$ for costs associated with anxiety disorders. Costs include not only the cost of treatment but also the loss of productivity. All anxiety disorders except specific phobia are associated with lost productivity or absenteeism from work. Because anxiety disorders are common and quite costly, the need for effective treatments is apparent.

This article begins with a discussion of the neural fear circuitry and the known and hypothesized sites of action of the major classes of agents used for the treatment of panic disorder, generalized anxiety disorder (GAD), social anxiety disorder (SAD), posttraumatic stress disorder (PTSD), and obsessive–compulsive disorder (OCD). For each anxiety disorder presented, treatment selection, efficacy, and implementation are discussed.

Mechanisms of Drug Action

Because the fear circuit is so complex, it is not surprising that many neurochemical systems modulate the stress response. Included in this list are the serotonergic (5-hydroxytryptamine; 5-HT), noradrenergic (NE), γ-aminobutyricacidergic (GABAergic), glutamatergic, and peptidergic systems. The neurobiology and neuroanatomy of anxiety and fear are complex and beyond the scope of this article. However, it is important to note that, within the fear circuit, the amygdala serves an important role in coordinating the cognitive, affective, neuroendocrine, cardiovascular, respiratory, and musculoskeletal components of fear and anxiety responses. Not only is the amygdala involved in modulating the fear response; in regard to the cognitive components of fear and anxiety, it is also involved in fear conditioning and fear extinction. The amygdala balances incoming stimuli from the environment presented by the thalamus and sensory cortex with memories of past experience presented by the frontal cortex and hippocampus. The amygdala balances inputs from these areas and then orchestrates the anxiety and panic response by stimulating effector brain areas (locus coeruleus, hypothalamus, periaqueductal gray, and parabracheal nucleus) involved in the expression of fear symptoms. The medications used to treat anxiety disorders are thought to modify dysfunction within this circuit, thereby producing anxiolysis.

Serotonergic Mechanisms

There is evidence that 5-HT mechanisms may have an important role in mediating the anxiolytic effects of several classes of therapeutic agents. Important among these options are the selective serotonin reuptake inhibitors (SSRIs), which are thought to exert their long-term therapeutic effects by enhancement of 5-HT neurotransmission. Successful SSRI treatment of panic disorder and PTSD, disorders in which NE system overactivity has been implicated, could be mediated via selective 5-HT-NE interactions. For example, chronic SSRI administration may enhance function of inhibitory 5-HT neurons projecting from the dorsal raphe nuclei to locus coeruleus-NE neurons, thus reducing cell firing of these NE neurons. In addition, effects of SSRIs in the fear circuit may be mediated through inhibitory effects on glutamatergic neurons that regulate locus coeruleus (LC)–NE neuron excitability. It has also been reported that 5-HT, by activating GABA inhibitory interneurons, can modulate glutamatergic inputs to the amygdala, thus inhibiting glutamatergic activation of the amygdala.

Thus, SSRIs may have anxiolytic effects at the level of the amygdala itself. Serotonin release in the prefrontal cortex is implicated in 'coping' with stress and thus could provide additional pathways to reduce anxiety. Other potential SSRI therapeutic effects within the fear circuit may be via 5-HT/NE and 5-HT/cholecystokinin (CCK) interactions in frontal cortical brain regions. Other serotonergic medications, such as the monoamine oxidase inhibitors (MAOIs), also increase net 5-HT neurotransmission with chronic administration and therefore may have anxiolytic mechanisms similar to those of the SSRIs. In other anxiety disorders, such as OCD, the therapeutic mechanism of the SSRIs and the SRI clomimpramine may involve different pathways. For example, pathophysiologic theories of OCD hypothesize that there are dysfunctional interactions between 5-HT and dopaminergic systems that result in neuronal hyperactivity in the orbitofrontal cortex. The anti-OCD action of SSRIs, therefore, may be partly caused by enhancement of 5-HT-related inhibition of orbitofrontal cortex neurons.

At this time, multiple subclasses of 5-HT receptors have been identified. The different subclasses of 5-HT receptors likely result in the various therapeutic effects and side effect profiles of the serotonergic anxiolytics. Anxiolytic and antidepressant effects seems to be associated with 5-HT_1 receptor stimulation. 5-HT_2 receptor stimulation produces sleep disruption, sexual dysfunction, and an increase in anxiety and agitation. Acute stimulation of 5-HT_{2C} receptors by the projections from the raphe to limbic cortex may cause the acute mental agitation, anxiety, or induction of panic attacks that can be observed during initiation of SSRI therapy. 5-HT_3 receptor stimulation produces nausea, diarrhea, and headache but is also implicated in amygdala synaptic plasticity. All three of these 5-HT receptors are found on the postsynaptic neurons. 5-HT_{1A} receptors are found on the cell body of the presynaptic serotonergic neurons and function as somatodendritic autoreceptors that, when stimulated by 5-HT, exert an inhibitory effect that reduces cell firing and thus reduces the release of 5-HT into the synapse. The 5-HT_{1A} receptors are also found postsynaptically. The 5-HT_{1A} partial agonist buspirone is thought to exert therapeutic effect via action at the presynaptic 5-HT_{1A} receptor. As already mentioned, 5-HT_2 receptor stimulation is thought to produce an increase in anxiety and agitation. Trazodone is a robust 5-HT_{2A} receptor antagonist, and its anxiolytic effect may be due to this antagonism. In addition, stimulation of 5-HT_{2A} receptors by 5-HT may reduce the potency of 5-HT at 5-HT_{1A} receptors. Trazodone may be used to boost SSRI efficacy by blocking 5-HT_{2A} receptors so that they are less efficient in inhibiting cell firing through 5-HT_{1A} action. The atypical antipsychotics may produce anxiolytic effects via a similar mechanism in which 5-HT_{2A} receptor antagonism increases 5-HT transmission by desensitizing 5-HT_{1A} autoreceptors, thus increasing 5-HT release. Trazodone belongs to the chemical class of antidepressants known as phenylpiperazines, which also includes nefazodone. Nefazodone's anxiolytic properties, like trazadone's, may be due to antagonism of the postsynaptic 5-HT_{2A} receptor. Nefazodone differs from trazodone in that nefazodone also weakly blocks the reuptake of 5-HT. Another serotonergic agent with anxiolytic properties is mirtazapine. Mirtazapine's anxiolytic properties are thought to be elicited via blocking postsynaptic 5-HT_{2A1} and 5-HT_{2C} receptors. Mirtazapine also blocks postsynaptic 5-HT_3 receptors, which helps prevent the typical serotonergic side effect of nausea.

Noradrenergic Mechanisms

The LC–NE system coordinates fear-induced sympathetic nervous system activation that results in the anxiety symptoms of tachycardia, increased blood pressure, diaphoresis, piloerection, and pupil dilation. NE dysregulation is a key component of Gray's model of pathological anxiety. This model is based on the concept of 'behavioral inhibition,' a constellation of behaviors seen in animals facing an unpredictable stimulus. When exposed to unexpected stimuli, the septohippocampal system assesses the level of threat the stimuli poses. If appropriate, the behavioral inhibition circuit is activated. Within this circuit, when a stimulus is not deemed significant, the septum suppresses hippocampal activation, acting as an 'all-clear' signal. Increased NE output to the septum inhibits signals from the septum to the hippocampus, resulting in prolonged hippocampal activation. Noradrenergic hyperactivity results in chronic activation of this system and may produce the chronic state of fear equivalent to GAD in humans. Another important site of action for norepinephrine is the amygdala. Norepinephrines, acting via β-adrenergic receptors, have been demonstrated to be critical for formation of fear or emotional memory formation. Thus, hyperactive noradrenergic neurotransmission in the amygdala is more likely to enhance fearful memory formation and may contribute significantly to the development of severe anxiety disorders such as PTSD and agoraphobia. Preclinical evidence has strongly suggested that blocking β-receptors at the time of or immediately after a traumatic stimulus interferes with subsequent fear memory formation. This has recently been supported clinically: Treatment with propranalol within a short period after an acute trauma resulted in significantly reduced likelihood of developing PTSD. Thus, NE system dysregulation may apply to other chronic anxiety syndromes, such as PTSD, panic disorder with severe anticipatory anxiety, and agoraphobia. Antianxiety agents that have the capacity to regulate NE function in these systems, such as the tricyclic antidepressants, MAOIs, and serotonin norepinephrine reuptake inhibitors (SNRIs), have documented clinical efficacy in the disorders mentioned above. In addition, the novel selective NE reuptake inhibitor reboxetine has been found to be anxiolytic in patients with panic disorder. These agents can inhibit NE function by downregulating β-receptors, reducing LC firing rates, and reducing NE turnover. The decrease in firing rates of NE neurons in the LC by selective NE reuptake inhibitors (NRIs) such as the tricyclic imipramine and reboxetine results from the inhibition of NE transporters on the cell body of NE neurons, leading to an accumulation of synaptic NE in the vicinity of α_2-adrenergic autoreceptors, which exert a negative feedback action on NE neuronal firing. With prolonged treatment and in the presence of NE reuptake inhibition, the firing rate of NE neurons does not recover, because the cell body α_2-adrenergic autoreceptors do not desensitize. This is also the case with prolonged administration of MAOI inhibitors when the α_2-adrenergic autoreceptors fail to desensitize and so there is a decrease in NE neuron firing. Chronic administration of dual reuptake inhibitors of 5-HT and NE produces adaptive changes to 5-HT neuronal firing similar to those of SSRIs and attenuation of NE neuron firing, as seen with NRIs.

GABAergic and Glutamatergic Mechanisms

GABA

As discussed earlier, within the fear circuit, the amygdala is situated as a central structure in the coordination of fear and anxiety responses. GABA, a major inhibitory neurotransmitter found in most parts of the brain, reduces fear in animals after its local infusion into the amygdala. In addition, a variety of pharmacologic agents that increase GABA neurotransmission have similar anxiolytic effects when administered directly into the amygdala. These include the full benzodiazepine (BZD)/$GABA_A$ agonists, the benzodiazepines, the GABAergic anticonvulsants such as sodium valproate, gabapentin, and pregabalin, and partial BZD/$GABA_A$ agonists such as bretazenil and abecarnil. With respect to full BZD/$GABA_A$ receptor agonists, the $gamma_2$ subunit of the $GABA_A$ receptor is necessary for benzodiazepine-mediated anxiolysis, and the other subunit of the $GABA_A$ receptor has profound influence on the efficacy of benzodiazepine modulation. Since preclinical and clinical data implicate decreased benzodiazepine receptor binding in panic disorder and GAD, benzodiazepines may restore homeostasis in the GABA system by correcting these deficits. The GABA system may also influence anxiety levels by mediating the release of other neurotransmitters within the fear circuit (e.g. NE, CCK, 5-HT). In addition, there is evidence that GABA/NE interactions may be an important therapeutic mechanism in panic disorder, as benzodiazepine treatment blocks panicogenesis induced by yohimbine. It is conceivable, then, that within the fear circuit, enhancement of GABA function is a common therapeutic pathway for many classes of effective anxiolytics.

Glutamate

It is well known that glutamate is the primary excitatory neurotransmitter within the brain. Regulation of the amygdala neuronal excitability is achieved by a balance between glutamate-induced excitation and GABA-mediated inhibition. Increased amygdala activation has been associated with various anxiety states. In addition, studies have demonstrated that basal levels of anxiety in rats can be decreased by the suppression of excitatory input to the basolateral amygdala by either glutamate antagonists or lesioning of the basolateral amygdala. Thus modulation of glutamatergic function in the amygdala, and other structures within the fear circuit, is likely to promote anxiolysis. Recent animal studies investigating glutamate receptor antagonists showed that both metabotropic glutamate receptor 5 antagonists MPEP and MTEP and the mGlurR1 antagonist LY456236 produced anxiolytic-like effects similar to those of the benzodiazepine chlordiazepoxide. In rodent models, chronic stress leads to repeated activation of the basolateral amygdala, disrupting the balance of GABAergic inhibition and glutamatergic excitation. This changes the sensitivity of the basolateral amygdala neurons and results in the development of a chronic anxiety-like state. The increased sensitivity of the amygdala may cause nonsalient stimuli to elicit anxiety, fear, or neuroendocrine responses leading to syndromes such as panic and PTSD. It is hypothesized that disruption of the inhibitory tone in the basolateral amygdala and a heightened responsivity of the basolateral amygdala could also result in assigning exaggerated salience to novel social stimuli, resulting in a social anxiety-like condition. Modulation of this system by glutamatergic and GABAergic medications may help restore the normal balance between glutamate excitation and GABA inhibition within the amygdala.

Decreasing glutamatergic excitation within the fear circuit, however, may not be the only glutamatergic mechanism that results in anxiolysis. Glutamate is critically involved in the neuronal plasticity underlying learning and memory. Recent work on the neural basis of fear extinction suggests that extinction is an active learning process in which glutamate plays a significant role. Evidence suggests fear extinction may be mediated by activation of a hippocampal-prefrontal circuit that inhibits amygdala-generated fear responses. Acute enhancement of glutamate function during extinction learning can facilitate this type of learning. For example, in animal and human paradigms, acute administration of the partial agonist at the glycine site of the NMDA receptor, D-cycloserine, facilitates extinction of fears. Moreover, glutamate mediates neurotoxicity and cell death and is thought to be related to injury of hippocampal neurons in patients with PTSD. The drug lamotrigine, which has been shown to decrease presynaptic release of glutamate, may also help modulate the release of glutamate in anxious patients and thus decrease glutamatergic damage to hippocampal neurons, thereby attenuating symptoms of PTSD.

Corticotropin Releasing Hormone and Peptidergic Systems

Corticotropin-releasing hormone (CRH) is significantly involved in effecting the stress response. During stress, CRH levels are increased in the hypothalamus, resulting in increased release of cortisol and dehydroepiansdrosterone. Equally important are the extrahypothalamic effects of CRH throughout the fear circuit. Neurons containing CRH receptors are located throughout the brain, including the prefrontal and cingulate cortices, central nucleus of the amygdala, bed nucleus of stria terminalis, nucleus accumbens, periaqueductal gray, brain stem nuclei such as the major NE-containing nucleus, the LC, and the serotonin nuclei in the dorsal and median raphe. Recent research suggests that corticotrophin-releasing factor (CRF)-mediated glutamate excitation within the amygdala induces long-term synaptic plasticity and increases the excitability of basolateral amygdala neurons, contributing to the etiology of several anxiety disorders such as panic disorder and PTSD. This line of inquiry has led to CRF antagonist drug development programs with the goal of producing a novel class of anxiolytic. This is not an exhaustive list of peptides; neuropeptide Y, CCK, and orexin are currently being investigated to determine their possible role in anxiogenesis.

Pharmacotherapeutic Strategies

GAD

GAD is characterized by excessive anxiety and worry about several events or activities for most days during at least a 6-month period. Patients with GAD have difficulty controlling the worry, and the worry is associated with somatic symptoms such as muscle tension, irritability, difficulty sleeping, and restlessness. When practitioners select agents to treat GAD, it is important that they consider the risk–benefit ratio for the patient. Patients with low levels of anxiety should be encouraged to reserve medications for when they are clearly needed. Comorbidity tends to be the rule rather than the exception in instances of GAD, and therefore comorbid diagnoses may be crucial in determining treating agents. Major depression (>40–60%) or dysthymia (40%) are the most common disorders comorbid with GAD. Medications with anxiolytic and antidepressant actions would therefore be most appropriate first-line therapies.

The newer 5-HT and 5-HT/NE reuptake inhibitors have anxiolytic and antidepressant efficacy and have become first-line agents for the pharmacologic management of GAD (see **Table 1**). Paroxetine and escitalopram are equally efficacious in treating GAD; however, one study has found that significantly fewer patients withdrew from escitalopram than from paroxetine treatment because of adverse events. SSRIs are safe, nonaddictive, and generally very well tolerated, with only mild adverse side effects, such as sleep disturbance, headache, nausea, and sexual dysfunction. The delay in onset of therapeutic effect with SSRIs as anxiolytics is usually 2–4 weeks, similar to that seen in SSRI treatment of depression; however, once SSRIs begin to work, the therapeutic effect can be dramatic. Venlafaxine, an SNRI antidepressant, has been shown to be efficacious in both short- and long-term treatment of GAD. Side effects associated with venlafaxine are similar to those of SSRIs.

Because many GAD patients have comorbidity with mood and other anxiety disorders, selection of 5-HT/NE agents as first-line treatment is a good initial strategy over other treatment options such as benzodiazepines and buspirone. Benzodiazepines have a rapid onset of efficacy; however, questions have been raised about their long-term utility. One study has shown that the effect of the benzodiazepine diazepam is not significantly different statistically from that of placebo after the initial 4–6 weeks of treatment. Other concerns about benzodiazepine treatment include difficulty in tapering patients; failure to ameliorate worry (the principal cognitive symptom of GAD); possible exacerbation of depressive symptoms in comorbid patients; dependence issues; and side effects, including sedation, hypnotic effects, and cognitive impairment. Also, a higher rate of recurrence has been suspected following benzodiazepine treatment. Buspirone, a partial 5-HT_{1A} receptor agonist, has been shown to be more efficacious than benzodiazepines in

Table 1 Selective serotonin reuptake inhibitors (SSRIs) and serotonin norepinephrine reuptake inhibitors (SNRIs)

Agent	*Normal dosage range (mg day^{-1})*	*$T_{1/2}$ (h)*	*FDA anxiety indications*	*GAD[a,b]*	*PD[a]*	*SAD[a]*	*OCD[a]*	*PTSD[a]*
SSRIs								
Fluoxetine	20–80	24–72	OCD, PD	C	A	A	A	A
Paroxetine	20–60	21	OCD, PD, GAD, SAD, PTSD	A	A	A	A	A
Sertraline	50–200	24	OCD, PD, SAD, PTSD	A	A	A	A	A
Fluvoxamine	50–300	15	OCD	C	A	A	A	C
Citalopram	20–60	35		C	A	C	A	C
Escitalopram	10–20	27–32	OCD, PD, GAD, SAD, PTSD	A	A	A	A	C
SNRIs								
Venlafaxine	75–225	5–11	Extended release – PD, GAD, SAD	A	A	A	B	B
Duloxetine	40–60	12	GAD	A	C	D	D	C

[a]Level of evidence supporting use in anxiety disorders: A, established treatment (replicated, large-scale randomized clinical trials (RCTs)); B, promising treatment approach (e.g. one or more RCTs of moderate size; single site); C, limited evidence (open trials; case series literature); D, very limited evidence (only preclinical evidence, theoretical prediction, or anecdotal reports).
[b]Limited evidence available evaluating fluoxetine, fluvoxamine, and citalopram in GAD; however, efficacy is likely similar to that of other SSRIs.
FDA, US Food and Drug Administration; GAD, generalized anxiety disorder; PD, panic disorder; SAD, social anxiety disorder; OCD, obsessive–compulsive disorder; PTSD, posttraumatic stress disorder; $T_{½}$, half-life.
Data from Allgulander C, Hackett D, and Salinas E (2001) Venlafaxine extended release (ER) in the treatment of generalized anxiety disorder: Twenty-four-week placebo controlled dose-ranging study. *British Journal of Psychiatry* 179: 15–22; Asnis GM and Kohn SR, Henderson M, and Brown NL (2004) SSRIs versus non-SSRIs in post-traumatic stress disorder: An update with recommendations. *Drugs* 64(4): 383–404; Bielski RJ, Bose A, and Chang CC (2005) A doubleblind comparison of escitalopram and paroxetine in the long-term treatment of generalized anxiety disorder. *Annals of Clinical Psychiatry* 17(2): 65–69; Blomhoff S, Haug TT, Hellstrom K, et al. (2001) Randomised controlled general practice trial of sertraline, exposure therapy and combined treatment in generalized social phobia. *British Journal of Psychiatry* 179: 23–30; Bradwejn J, Ahokas A, Stein DJ, Salinas E, Emilien G, and Whitaker T (2005) Venlafaxine extended-release capsules in panic disorder: Flexible-dose, double-blind, placebo-controlled study. *British Journal of Psychiatry* 187: 352–359; Brady K, Pearlstein T, Asnis GM, et al. (2000) Efficacy and safety of sertraline treatment of posttraumatic stress disorder: A randomized controlled trial. *Journal of the American Medical Association* 283(14): 1837–1844; Brady K, Pearlstein T, Asnis GM, et al. (2000) Efficacy and safety of sertraline treatment of posttraumatic stress disorder: A randomized controlled trial. *Journal of the American Medical Association* 283(14): 1837–1844; Davidson J, Baldwin D, Stein DJ, et al. (2006) Treatment of posttraumatic stress disorder with venlafaxine extended release: A 6-month randomized controlled trial. *Archives of General Psychiatry* 63(10): 1158–1165; Davidson JR, Bose A, and Wang Q (2005) Safety and efficacy of escitalopram in the long-term treatment of generalized anxiety disorder. *Journal of Clinical Psychiatry* 66(11): 1441–1446; Davidson JR, Foa EB, Huppert JD, et al. (2004) Fluoxetine, comprehensive cognitive behavioral therapy, and placebo in general social phobia. *Archives of General Psychiatry* 61: 1005–1013; Davidson JR, Rothbaum BO, van der Kolk BA, et al. (2001) Multicenter, double-blind comparison of sertraline and placebo in the treatment of posttraumatic stress disorder. *Archives of General Psychiatry* 58(5): 485–492; Davidson JRT (2006) Pharmacotherapy of social anxiety disorder: What does the evidence tell us? *Journal of Clinical Psychiatry* 67(supplement 12): 20–26; Gelenberg AJ, Lydiard RB, Rudolph RL, et al. (2000) Efficacy of venlafaxine extended-release capsules in nondepressed outpatients with generalized anxiety disorder: A 6-month controlled trial. *Journal of the American Medical Association* 283: 3082–3088; Goddard AW, Brouette T, Almai A, et al. (2001) Earlly coadministration of clonazepam with sertraline for panic disorder. *Archives of General Psychiatry* 58: 681–686; Goodman WK (2004) Selecting pharmacotherapy for generalized anxiety disorder. *Journal of Clinical Psychiatry* 65(supplement 13): 8–13; Gorman JM (2003) Treating generalized anxiety disorder. *Journal of Clinical Psychiatry* 64(supplement 2): 24–29; Hollander E, Allen A, Steiner M, et al., for the paroxetine OCD Study Group (2003) Acute and long-term treatment and revention of relapse of obsessive-compulsive disorder with paroxetine. *Journal of Clinical Psychiatry* 64: 1113–1121; Hollander E, Koran LM, Goodman WK, et al. (2003) A double-blind, placebo-controlled study of the efficacy and safety of controlled-release fluvoxamine in patients with obsessive-compulsive disorder. *Journal of Clinical Psychiatry* 64: 604–647; Liebowitz MR, DeMartinis NA, Weihs K, et al. (2003) Efficacy of sertraline in severe generalized social anxiety disorder: Resultes of a double-blind, placebo-controlled study. *Journal of Clinical Psychiatry* 64: 785–792; Liebowitz MR, Stein MB, Tancer M et al. (2002) A randomized, double-blind, fixed-dose comparison of paroxetine and placebo in the treatment of generalized social anxiety disorder. *Journal of Clinical Psychiatry* 63(1): 66–74; Lydiard RB, and Bobes J (2000) Therapeutic advances: Paroxetine for the treatment of social anxiety disorder. *Depression and Anxiety* 11: 99–104; Marshall RD, Beebe KL, Oldam M, et al. (2001) Efficacy and safety of paroxetine treatment for chronic PTSD: A fixed-dose, placebo-controlled study. *American Journal of Psychiatry* 158(12): 1982–1988; Marshall RD, Beebe KL, Oldham M, and Zaninelli R (2001) Efficacy and safety of paroxetine treatment for chronic PTSD: A fixed dose, placebo-controlled study. *American Journal of Psychiatry* 158: 1982–1988; Martenyi F and Soldatenkova V (2006) Fluoxetine in the acute treatment and relapse prevention of combat-related post-traumatic stress disorder: Analysis of the veteran group of a placebo-controlled, randomized clinical trial. *European Neuropsychopharmacology* 16(5): 340–349; Martenyi F, Brown EB, Zhang H, Koke SC, and Prakash A (2002) Fluoxetine v. placebo in prevention of relapse in post-traumatic stress disorder. *British Journal of Psychiatry* 181: 315–320; Montgomery SA, Sheehan DV, Meoni P, Haudiquet V, and Hackett D (2002) Characterization of the longitudinal course of improvement in generalized anxiety disorder during long-term treatment with venlafaxine XR. *Journal of Psychiatric Research* 36: 209–217; Pallanti S, Hollander E, and Goodman W (2004) A qualitative analysis of nonresponse: Management of treatment-refractory obsessive-compulsive disorder. *Journal of Clinical Psychiatry* 65(supplement 14): 6–10; Pallanti S, Hollander E, Biestock C, et al. (2002) Treatment non-response in OCD: Methodological issues and operational definitions. *International Journal of Neuropsychopharmacology* 5: 181–191;

ameliorating the cognitive aspects of GAD; however, more-recent research has suggested that, like benzodiazepines, buspirone may not be effective in the long-term treatment of anxiety. Other medications that are efficacious in GAD are imipramine and trazodone, which have 5-HT_{2A} agonist properties. A comparative randomized clinical trial (RCT) in 1993 demonstrated that imipramine and trazodone were as effective as diazepam in treatment of GAD.

Panic Disorder

Panic disorder with or without agoraphobia is characterized by the spontaneous, unexpected occurrence of panic attacks and has a lifetime prevalence of 3.5%. Optimal treatment of panic disorder has the aim of reducing or eliminating panic attacks and improving overall function and quality of life for patients. Of all the anxiety disorders, panic disorder is the most responsive to treatment and has a good prognosis.

When practitioners evaluate pharmacotherapy options for the treatment of panic disorder, it is important that they consider the predicted efficacy and tolerability of the agents as well as drug–drug interactions and safety in overdose. A 1995 meta-analysis showed that the effect size of the three major antipanic drug classes, tricyclic antidepressants, SSRIs, and benzodiazepines, are essentially equal and fall in the medium range of effect size. In addition, 5-HT/NE reuptake inhibitors have demonstrated short-term efficacy in panic disorder in many controlled clinical trials. The tricyclic antidepressants, specifically imipramine, were the gold standard in drug therapy of panic disorder for many years until the development of SSRIs and SNRIs. Approximately 60–70% of patients respond to imipramine; however, tolerability of imipramine is less than that of SSRIs. Unpleasant side effects of tricyclic antidepressants include orthostasis, activation, anticholinergic effects, and weight gain. This heavy side effect burden, as well as the potential for toxicity in overdose, has resulted in the relegation of tricyclic antidepressants to second-line agents for panic disorder.

Currently the SSRIs are considered first-line medication treatments for panic disorder because of their demonstrated efficacy and safety. SNRIs, such as venlafaxine, have demonstrated efficacy in the short-term treatment of panic disorder. In addition, SSRIs and SNRIs treat many of the comorbidities associated with panic disorder, including depression, social phobia, GAD, and PTSD. One drawback to SSRIs and SNRIs is the excessive activation, with potential exacerbation of panic symptoms, during the initiation of these medications – an occurrence presumably related to sudden increase in synaptic level of 5-HT and NE. Compared with tricyclic antidepressants, however, SSRIs have less weight gain and fewer anticholinergic side effects, as well as a comparatively benign cardiovascular profile.

Benzodiazepines are considered third-line agents in the treatment of panic. Despite this fact, they are still a widely used treatment option, with one study reporting that the most common type of treatment for panic disorder remains benzodiazepine monotherapy. In general, the high-potency, short half-life benzodiazepines are preferred now in clinical practice

Phelps NJ and Cates ME (2005) The role of venlafaxine in the treatment of obsessive-compulsive disorder. *Annals of Pharmacotherapy* 39(1): 136–140; Physicians' Desk Reference (2007) 61st edn. Montvale, NJ: Thomson PDR; Pollack MH (2005) The pharmacotherapy of panic disorder. *Journal of Clinical Psychiatry* 66(supplement 4): 23–27; Rickels K, Mangano R, and Khan A (2004) A double-blind, placebo-controlled study of a flexible dose of venlafaxine ER in adult outptatients with generalized social anxiety disorder. *Journal of Clinical Psychopharmacology* 24: 488–496; Rocca P, Fonzo V, Scotta M, et al. (1997) Paroxetine efficacy in the treatment of generalized anxiety disorder. *Acta Psychiatrica Scandinavica* 95: 444–450; Rosenbaum JF, Pollock RA, Jordan SK, Pollack MH (1996) The pharmacotherapy of panic disorder. *Bulletin of the Menninger Clinic* 60(2 supplement A): A54–A75; Sheehan DV (1999) Venlafzine extended release (XR) in the treatment of generalized anxiety disorder. *Journal of Clinical Psychiatry* 60(supplement 22): 23–28; Sheehan DV (2001) Attaining remission in generalized anxiety disorder; venlafaxine extended release comparative data. *Journal of Clinical Psychiatry* 62(supplement 9): 26–31; Sheehan DV (2002) The management of panic disorder. *Journal of Clinical Psychiatry* 63(supplement 14): 17–21; Sinajkic A, Weine S, Djuric-Bijedic Z, et al. (2001) Sertraline, paroxetine, and venlafaxine in refugee posttraumatic disorder with depression symptoms. *Journal of Traumatic Stress* 14(3): 445–452; Stein MB, Fryer AJ, Davidson JRT, et al. (1999) Fluvoxamine treatment of social phobia (social anxiety disorder): A double blind, placebocontrolled study. *American Journal of Psychiatry* 156: 756–760; Stein MB, Liebowitz MR, Lydiard RB, et al. (1998) Paroxetine treatment of generalized social phobia (social anxiety disorder): A randomized controlled trial. *Journal of the American Medical Association* 280: 708–713; Stocchi F, Nordera G, Jokinen RH, et al., for the Paroxetine Generalized Anxiety Disorder Study Team (2003) Efficacy and tolerability of paroxetine for the long-term treatment of generalized anxiety disorder. *Journal of Clinical Psychiatry* 64: 240–258; Tucker P, Zaninelli R, Yehuda R, Ruggiero L, Dillingham K, and Pitts CD (2001) Paroxetine in the treatment of chronic posttraumatic stress disorder: Results of a placebo-controlled, flexible-dosage trial. *Journal of Clinical Psychiatry* 62(11): 860–868; Van Ameringen MA, Lane RM, Walker JR, et al. (2001) Sertraline treatment of generalized social phobia: A 20-week, double-blind, placebo-controlled study. *American Journal of Psychiatry* 158: 275–281; Yaryura-Tobias JA, and Neziroglu FA (1996) Venlafaxine in obsessive–compulsive disorder (letter). *Archives of General Psychiatry* 53: 653–654; and Zohar J, Amital D, Miodownik C, et al. (2002) Double-blind placebo-controlled pilot study of sertraline in military veterans with posttraumatic stress disorder. *Journal of Clinical Psychopharmacology* 22(2): 190–195.

due to simpler metabolism and fewer active metabolites. Multiple studies have documented the efficacy of alprazolam for the treatment of panic disorder. Benzodiazepines have the benefits of a favorable side effect profile and a rapid onset of action, producing improvement within 1 week in one study. On the other hand, drawbacks to benzodiazepines include the development of physiologic dependence with sustained use and the risk of abuse in persons predisposed to substance abuse. In contrast, SSRIs and SNRIs have very low abuse or dependence potential. Recent work has shown that coadministration of a benzodiazepine such as clonazepam with an SSRI such as sertraline accelerates the response of patients with moderate to severe panic disorder and may reduce early antidepressant-related stimulation. Thus benzodiazepines may be used effectively for short-term stabilization and SSRIs for longer-term management of symptoms.

In particularly resistant cases of panic disorder, a tertiary option is the use of MAOIs, which have efficacy that is similar to or exceeds that of tricyclic antidepressants. Anticonvulsants such as valproate, gabapentin, and lamotrigine are considered fourth-line agents for treatment of panic disorder. Small studies have shown efficacy of valproic acid, pregabalin, and gabapentin in panic disorder. If a panic disorder patient has comorbid bipolar disorder, one of the anticonvulsants may be preferable because of the risk of inducing mania with an antidepressant (see **Table 2**).

SAD

SAD is an anxiety syndrome characterized by excessive fears of humiliation or embarrassment in various social situations. It is generally agreed that two subtypes of SAD exist: the generalized subtype, in which the patient fears a number of performance and social interaction situations, and the specific subtype, in which the fear is confined to performance situations

Table 2 Anticonvulsants

Agent	*Usual daily dose range (mg day^{-1})*[b]	*$T_{1/2}$ (h)*	*GAD*[a]	*PD*[a]	*SAD*[a]	*OCD*[a]	*PTSD*[a]
Divalproex	Trough plasma concentration between 50 and 125 μg ml^{-1}	9–16	D	C	C	D	C
Carbamazepine	Further research needed	25–65 initially, then 12–17[c]	D	C	C	D	C
Topiramate	Further research needed	21	D	E	C	E	C
Gabapentin	Further research needed; 900–1800 mg day^{-1} and given in divided doses (3 times a day)	5–7	D	C	B	C	C
Lamotrigine	Further research needed	25–32	D	C	D	D	B

[a]Level of evidence supporting use in anxiety disorders: A, established treatment (replicated, large-scale randomized clinical trials (RCTs)); B, promising treatment approach (e.g., one or more RCTs of moderate size; single site); C, limited evidence (open trials; case series literature); D, very limited evidence (only preclinical evidence, theoretical prediction, or anecdotal reports); E, evidence of negative therapeutic effects or no evidence of efficacy.

[b]No established doses specifically for anxiety treatment.

[c]Because Tegretol induces its own metabolism, half-life is variable; autoinduction is completed after 3–5 weeks of a fixed dosing regimen. Initial half-life values range from 25–65 h, decreasing to 12–17 h on repeated doses.

GAD, generalized anxiety disorder; PD, panic disorder; SAD, social anxiety disorder; OCD, obsessive–compulsive disorder; PTSD, posttraumatic stress disorder; $T_{1/2}$, half-life.

Data from Asnis GM, Kohn SR, Henderson M, and Brown NL (2004) SSRIs versus non-SSRIs in post-traumatic stress disorder: An update with recommendations. *Drugs* 64(4): 383–404; Bartzokis G, Lu PH, Turner J, Mintz J, and Saunders CS (2005) Adjunctive risperidone in the treatment of chronic combat-related posttraumatic stress disorder. *Biological Psychiatry* 57(5): 474–479; Butterfield MI, Becker ME, Connor KM, et al. (2001) Olanzapine in the treatment of post-traumatic stress disorder: A pilot study. *International Clinical Psychopharmacology* 16: 197–203; Hamner MB, Faldowski RA, Ulmer HG, Frueh BC, Huber MG, and Arana GW (2003) Adjunctive risperidone treatment in post-traumatic stress disorder: A preliminary controlled trial of effects on comorbid psychotic symptoms. *International Clinical Psychopharmacology* 18(1): 1–8; McDougle CJ, Goodman WK, Leckman JF, et al. (2000) A double-blind, placebo-controlled trial of risperidone addition in serotonin reuptake inhibitor-refractory obsessive-compulsive disorder. *Archives of General Psychiatry* 57: 794–801; Pallanti S, Hollander E, and Goodman W (2004) A qualitative analysis of nonresponse: management of treatment-refractory obsessive-compulsive disorder. *Journal of Clinical Psychiatry* 65 (supplement 14): 6–10; Petty F, Brannan S, Casada J, et al. (2001) Olanzapine treatment for posttraumatic stress disorder: An open-label study. *International Clinical Psychopharmacology* 16: 331–337; Physicians' Desk Reference (2007) 61st edn. Montvale, NJ: Thomson PDR; Schmidt AW, Lebel LA, Howard HR Jr., and Zorn SH (2001) Ziprasidone: A novel antipsychotic agent with a unique human receptor biding profile. *European Journal of Pharmacology* 425: 197–201; Shapira NA, Ward HE, Mandoki M, et al. (2004) A double-blind, placebo-controlled trial of olanzapine addition in fuoxetine-refractory obsessive-compulsive disorder. *Biological Psychiatry* 550: 553–555; and Stein MB, Kline NA, and Matloff JL (2002) Adjunctive olanzapine for SSRI-resistant combat-related PTSD: A double-blind, placebo-controlled study. *American Journal of Psychiatry* 159: 1777–1779.

(e.g., public speaking anxiety, writing in front of others). Individuals with generalized SAD report greater anxiety and avoidance than do individuals with specific SAD, and individuals with generalized SAD usually require pharmacotherapy.

SSRIs and the SNRI venlafaxine are considered first-line treatments for SAD. Multiple RCTs have shown paroxetine to produce a significant improvement over placebo in patients with SAD and to reduce phobic avoidance, anticipatory fear, and disability. In addition, numerous studies have demonstrated the efficacy of venlafaxine in the treatment of SAD. At this time there is no evidence indicating that either SSRIs or an SNRI is more efficacious in SAD. Irreversible MAOI agents, such as phenelzine, are effective for the treatment of SAD. However, these drugs are no longer considered first-line therapy because of the necessity of patients following a low-tyramine diet to prevent hypertensive crisis. Evidence also exists supporting the efficacy of the benzodiazepines clonazepam and bromazepam. There is also some evidence suggesting the efficacy of the atypical antipsychotic olanzapine (see **Table 3**) and $\alpha_2\delta$ calcium-channel blockers such as gabapentin (see **Table 2**) and pregabalin in treating SAD, but further investigation is needed.

OCD

OCD is a chronic psychiatric disorder characterized by recurrent persistent thoughts (obsession) and/or repetitive compulsory behaviors (compulsions).

Table 3 Atypical antipsychotics

Agent	*Quetiapine*	*Risperidone*	*Olanzapine*	*Ziprasidone*	*Aripiprazole*
Receptor binding profile					
D_2	+	++++	++	++++	++++ (partial agonist)
5-HT_{2A}	+	+++++	++++	+++++	++++
5-HT_{2C}	−	++++	++++	+++++	++
5-HT_{1A}	+	+	−	++++	++++
5-HT_{1D}	−	+	+	++++	N/A
α_1-adrenergic	++	++++	++	++	++
M_1-muscarinic	++	−	++++	−	−
H1-histaminergic	++++	++	++++	++	++
NE reuptake	+	−	−	++	N/A
Usual dosage range (mg day^{-1})[b]	100–400	2–6	5–15	20–80	10–30
$T_{1/2}$(h)	6	20	21–54	7	75–94
GAD[a]	D	D	D	D	D
PD[a]	D	D	C	D	D
SAD[a]	C	D	C	D	D
OCD[a]	B	A	B	D	C
PTSD[a]	C	B	C	D	D

[a]Level of evidence supporting use in anxiety disorders: A, established treatment (replicated, large-scale randomized clinical trials (RCTs)); B, promising treatment approach (e.g., one or more RCTs of moderate size; single site); C, limited evidence (open trials; case series literature); D, very limited evidence (only preclinical evidence, theoretical prediction, or anecdotal reports).
[b]No established doses specifically for anxiety treatment.
GAD, generalized anxiety disorder; PD, panic disorder; SAD, social anxiety disorder; OCD, obsessive–compulsive disorder; PTSD, posttraumatic stress disorder; N/A, not applicable; D_2, dopamine D_2 receptor; (+++++), very high affinity; (++++), high affinity; (++), moderate affinity; (+), low affinity; (−), negligible affinity.
Data from Asnis GM and Kohn SR Henderson M, and Brown NL (2004) SSRIs versus non-SSRIs in post-traumatic stress disorder: An update with recommendations. *Drugs* 64(4): 383–404; Berlant J and van Kammen DP (2002) Open-label topiramate as primary or adjunctive therapy in chronic civilian posttraumatic stress disorder: A preliminary report. *Journal of Clinical Psychiatry* 63(1): 15–20; Davidson JRT (2006) Pharmacotherapy of social anxiety disorder: What does the evidence tell us? *Journal of Clinical Psychiatry* 67 (supplement 12): 20–26; Goodman WK (2004) Selecting pharmacotherapy for generalized anxiety disorder. *Journal of Clinical Psychiatry* 65(supplement 13); 8–13; Hertzberg MA, Butterfield MI, Feldman ME, et al. (1999) A preliminary study of lamotrigine for the treatment of posttraumatic stress disorder. *Biological Psychiatry* 45(9): 1226–1229; McGrath MJ, Campbell KM, Parks CR, and Burton FH (2000) Glutamatergic drugs exacerbate symptomatic behavior in a transgenic model of comorbid Tourette's syndrome and obsessive-compulsive disorder. *Brain Research* 877(1): 23–30; Pallanti S, Hollander E, Biestock C, et al. (2002) Treatment non-response in OCD: Methodological issues and operational definitions. *International Journal of Neuropsychopharmacology* 5: 181–191; Pallanti S, Hollander E, and Goodman W (2004) A qualiative analysis of nonresponse: Management of treatment-refractory obsessive-compulsive disorder. *Journal of Clinical Psychiatry* 65: (supplement 14): 6–10; Pande AC, Pollack MH, Crockatt J, et al. (2000) Placebo-controlled study of gabapentin treatment of panic disorder. *Journal of Clinical Psychopharmacology* 20: 467–471; Physicians' Desk Reference (2007) 61st edn. Montvale, NJ: Thomson PDR; Woodman CL and Noyes R Jr. (1994) Panic disorder: Treatment with valproate. *Journal of Clinical Psychiatry* 55: 134–136; Rosenbaum JF, Pollock RA, Jordan SK, and Pollack MH (1996) The pharmacotherapy of panic disorder. *Bulletin of the Menninger Clinic* 60(2 supplement A): A54–A75.

SRIs have been the primary agents of choice in the pharmacotherapy of OCD, suggesting that the pathophysiology of OCD might be related to disturbance in the functioning of brain serotonin systems (see **Table 1**). This hypothesis was formulated in the late 1960s, when clinicians observed that the serotonergic tricyclic antidepressant clomipramine relieved OCD symptoms but other tricyclics did not. Since then, clomipramine's efficacy compared with placebo has been confirmed by numerous double-blind controlled trials. Side effects such as dry mouth and constipation are a result of the anticholinergic blockade associated with clomipramine. Orthostatic hypotension is the result of adrenoreceptor blockage. Serotonin-related side effects seen with SSRIs, such as nausea, tremor, impotence, and anorgasmia, are also seen with clomipramine. Safety concerns include risk of seizures at higher doses and prolongation of QT interval.

A large percentage of patients with OCD (40–60%) will be nonresponsive to treatment with SRIs. Various strategies have been studied to treat these patients. Pharmacotherapeutic strategies include switching from a serotonergic agent to one with a different mechanism of action or augmenting with another medication. Addition of conventional and atypical antipsychotics to an SRI in treatment-resistant OCD has also been studied (see **Table 3**). A 1994 RCT demonstrated efficacy of augmenting fluvoxamine with haldol in patients with OCD. In this study, most of the benefit of haloperidol addition to fluvoxamine occurred in the OCD patients with a coexisting chronic tic disorder. The atypical antipsychotic risperidone has also been studied as an augmenting agent to SRIs and found to be efficacious; patients with and without comorbid tics responded at an equal rate (see **Table 3**).

It is hypothesized that OCD may be partly due to excessive forebrain glutamate output. This hypothesis has led to the study of glutamatergic agents such as riluzole (initially approved as a neuroprotective agent for patients with amyotrophic lateral sclerosis with an inhibitory effect on glutamate release) and an mGluR2/3 agonist, LY354740, that acts on autoreceptors to inhibit release of glutamate. Preliminary data from an open trial of riluzole augmentation in the treatment of patients with refractory OCD suggest promising efficacy of this agent.

PTSD

The goal of pharmacological therapy in treating PTSD is to reduce symptom distress, strengthen resilience, and restore function. The core symptoms of PTSD are intrusive reexperiencing of the original trauma (e.g., through nightmares and flashbacks), avoidance of stimuli associated with the trauma, numbing, estrangement, anhedonia, and hyperarousal. Secondary symptoms associated with PTSD are impaired functioning, poor resilience stress, and comorbid conditions.

Unfortunately, there are no therapies that are highly effective in chronic PTSD. Currently, SSRIs are the most prescribed drugs for PTSD, based on data from RCTs demonstrating modest efficacy (see **Table 1**). The percentage of patients who respond to SSRIs is between 40% and 60%. The SNRI venlafaxine has been demonstrated to be effective and well tolerated in the treatment of PTSD in a large ($N=329$) RCT. The antidepressant nefazodone is unique in that its mechanism is presynaptic serotonin and norepinephrine reuptake inhibition as well as 5-HT_{2A} and 5-HT_{2C} postsynaptic antagonism. A small ($N=15$), randomized, double-blind, placebo-controlled trial has demonstrated the efficacy of nefazodone in treating patients with PTSD. Nefazodone has demonstrated potent anxiolytic properties in patients with depression, and preliminary studies suggest nefazodone is effective in treating insomnia, making this agent attractive to clinicians treating patients with PTSD. Mirtazapine is an antagonist of postsynaptic 5-HT_2 and 5-HT_3 receptors and a potent antagonist of α_2-autoreceptors, and it has antihistaminergic effects. As a result, mirtazapine has antidepressant, anxiolytic, and sedative properties that have led investigators to study its efficacy in PTSD. At this time, small, open-label studies suggest mirtazapine may be efficacious in PTSD, but confirmation via double-blind, placebo-controlled studies is needed.

The anticonvulsants (carbamazepine, valproic acid, topiramate, lamotrigine, and gabapentin) have therapeutic potential for PTSD. The rationale for their use is that patients with PTSD may have affective instability due to a kindling process at the level of the amygdala and medial temporal lobe. The anticonvulsants have been shown to be efficacious in treating affective instability in epilepsy and in nonepileptic disorders, and many anticonvulsants have antikindling properties. There has been only one double-blind, placebo-controlled study of the efficacy of an anticonvulsant in patients with PTSD (see **Table 2**). Large-scale controlled studies are needed to confirm these promising reports.

Recent work in the neurobiology of anxiety has led to the hypothesis that the central-acting β-adrenergic receptor antagonist propranolol may be beneficial in preventing PTSD if administered early after exposure to trauma. Propranolol blocks recognition and recall of emotionally significant memories and has been shown to block reconsolidation (the conversion of labile, short-term memory into long-term memory). One promising smaller RCT has been done evaluating its efficacy in the secondary prevention of PTSD.

See also: Anxiety Disorders; Aversive Emotions: Genetic Mechanisms of Serotonin; Corticotropin-Releasing Hormone: Integration of Adaptive Responses to Stress; Gamma-Aminobutyric Acid (GABA); Mood Stabilizers; Norepinephrine: CNS Pathways and Neurophysiology; Obsessive–Compulsive Disorder; Panic Disorder; Panic Disorder as an Emotional Disorder; Posttraumatic Stress Disorder: Neurobiology; Posttraumatic Stress Disorder as an Emotional Disorder; Stress, Sex and Adolescent Nicotine Response.

Further Reading

Anand A and Shekhar A (2003) Brain imaging studies in mood and anxiety disorders: Special emphasis on the amygdala. *Annals of the New York Academy of Sciences* 985: 370–388.

Asnis GM, Kohn SR, Henderson M, and Brown NL (2004) SSRIs versus non-SSRIs in post-traumatic stress disorder: An update with recommendations. *Drugs* 64(4): 383–404.

Blier P and Abbot FV (2001) Putative mechanisms of action of antidepressant drugs in affective and anxiety disorders and pain. *Journal of Psychiatry & Neuroscience* 26(1): 37–43.

Blier P and Szabo ST (2005) Potential mechanisms of action of atypical antipsychotic medications in treatment-resistant depression and anxiety. *Journal of Clinical Psychiatry* 66(supplement 8): 30–40.

Boone ML, McNeil W, Masia CL, et al. (1999) Multimodal comparisons of social phobia subtypes and avoidant personality disorder. *Journal of Anxiety Disorders* 13: 271–292.

Brawman-Mintzer O (2001) Pharmacologic treatment of generalized anxiety disorder. *Psychiatry Clinics of North America* 24(1): 119–137.

Bruce SE, Vasile RG, Goisman RM, et al. (2003) Are benzodiazepines still the medication of choice for patients with panic disorder with or without agoraphobia? *American Journal of Psychiatry* 160: 1432–1438.

Charney D and Nestler E (2004) *Neurobiology of Mental Illness*, 2nd edn. Oxford, UK: Oxford University Press.

Davidson JRT (2004) Long-term treatment and prevention of post-traumatic stress disorder. *Journal of Clinical Psychiatry* 65(supplement 1): 44–48.

Davidson JRT (2004) Use of benzodiazepines in social anxiety disorder, generalized anxiety disorder, and posttraumatic stress disorder. *Journal of Clinical Psychiatry* 65(supplement 5): 29–33.

Davidson JRT (2006) Pharmacotherapy of social anxiety disorder: What does the evidence tell us? *Journal of Clinical Psychiatry* 67 (supp 12): 20–26.

Davis M and Myers KM (2002) The role of glutamate and gamma-aminobutyric acid in fear extinction: Clinical implication for exposure therapy. *Biological Psychiatry* 52: 998–1007.

Garakani A, Mathew SJ, and Charney DS (2006) Neurobiology of anxiety disorders and implications for treatment. *Mount Sinai Journal of Medicine* 73(7): 941–949.

Gelpin E, Bonne O, Peri T, et al. (1996) Treatment of recent trauma survivors with benzodiazepines: A prospective study. *Journal of Clinical Psychiatry* 57(9): 390–394.

Goddard AW, Brouette T, Almai A, et al. (2001) Early coadministration of clonazepam with sertraline for panic disorder. *Archives of General Psychiatry* 58: 681–686.

Goddard AW and Charney DS (1997) Toward an integrated neurobiology of panic disorder. *Journal of Clinical Psychiatry* 58(supplement 2): 4–11.

Goodman WK (2004) Selecting pharmacotherapy for generalized anxiety disorder. *Journal of Clinical Psychiatry* 65(supplement 13): 8–13.

Gorman JM (2003) Treating generalized anxiety disorder. *Journal of Clinical Psychiatry* 64(supplement 2): 24–29.

Keck PE, Strawn JR, and McElroy SL (2006) Pharmacologic treatment considerations in co-occurring bipolar and anxiety disorders. *Journal of Clinical Psychiatry* 67(supplement 1): 8–15.

McDougle CJ, Goodman WK, Leckman JF, et al. (2000) A double-blind, placebo-controlled trial of risperidone addition in serotonin reuptake inhibitor-refractory obsessive-compulsive disorder. *Archives of General Psychiatry* 57: 794–801.

Mellman TA, Bustamante V, David D, et al. (2002) Hypnotic medication in the aftermath of trauma. *Journal of Clinical Psychiatry* 63(12): 1183–1184.

Morgan CA III, Krystal JH, and Southwick SM (2003) Toward early pharmacologic posttraumatic stress intervention. *Biological Psychiatry* 53: 834–843.

Nutt D, Bell C, Masterson C, and Short C (2001) *Mood and Anxiety Disorders in Children and Adolescents: A Psychopharmacological Approach*. London: Martin Dunitz.

Pallanti S, Hollander E, Biestock C, et al. (2002) Treatment non-response in OCD: Methodological issues and operational definitions. *International Journal of Neuropsychopharmacology* 5: 181–191.

Pallanti S, Hollander E, and Goodman W (2004) A qualitative analysis of nonresponse: Management of treatment-refractory obsessive-compulsive disorder. *Journal of Clinical Psychiatry* 65(supplement 14): 6–10.

Pitman RK, Sanders KM, Zusman RM, et al. (2002) Pilot study of secondary prevention of posttraumatic stress disorder with propranolol. *Biological Psychiatry* 51: 183–188.

Pollack MH (2005) The pharmacotherapy of panic disorder. *Journal of Clinical Psychiatry* 66(supplement 4): 23–27.

Rainnie DG, Bergeron R, Sajdyk TJ, Patil M, Gehlert DR, and Shekhar A (2004) Corticotrophin releasing factor-induced synaptic plasticity in the amygdala translates stress into emotional disorders. *Journal of Neuroscience* 24(14): 3471–3479.

Roy-Byrne PP, Craske MG, and Stein MB (2002) Panic disorder. *Lancet* 369: 1023–1032.

Sheehan DV (2002) The management of panic disorder. *Journal of Clinical Psychiatry* 63(supplement 14): 17–21.

Shekhar A, Truitt W, Rainnie D, and Sajdyk T (2005) Role of stress, corticotrophin releasing factor (CRF) and amygdala plasticity in chronic anxiety. *Stress* 8(4): 209–219.

Stutzmann GE and LeDoux JE (1999) GABAergic antagonists block the inhibitory effects of serotonin in the lateral amygdala: A mechanism for modulation of sensory inputs related to fear conditioning. *Journal of Neuroscience* 19: RC8.

Westenberg HGM, Boer JAD, and Murphy DL (1996) *Advances in the Neurobiology of Anxiety Disorders*. New York: Wiley.

Apelin

A Reaux-Le Goazigo, X Iturrioz, and C Llorens-Cortes, INSERM U691 Collège de France, Paris, France

Discovery

The apelin story began in 1993 with the cloning of the complementary DNA for the apelin (APJ) receptor (putative receptor protein related to the type 1 angiotensin receptor) from a human genomic library, which was subsequently cloned in rodents. The human receptor is 380 amino acids long and was identified as a member of the family of orphan seven-transmembrane domain G-protein-coupled receptors (GPCRs). Its amino acid sequence was found to be 31% identical to that of the human type 1 angiotensin (AT1) receptor, but it did not bind radiolabeled angiotensin II (AngII), and stimulation of the rat APJ receptor AngII did not modify cyclic adenosine monophosphate (cAMP) production, demonstrating that this receptor was not an angiotensin receptor subtype. It therefore remained an orphan GPCR for which the endogenous ligand had to be isolated. The APJ receptor was deorphanized in 1998 when apelin, its endogenous ligand, was isolated from bovine stomach tissue extracts.

Structure and Processing of the Apelin Precursor

Apelin is a 36-amino-acid peptide (apelin 36) generated from a 77-amino-acid precursor, proapelin (**Figure 1**), which has been isolated from various species. The human proapelin gene is located on the X chromosome at locus Xq25–q26.1 and contains three exons, with the coding region spanning exons 1 and 2. The 3′-untranslated region also spans two exons (2 and 3). This may account for the presence of transcripts of two different sizes (~3 and ~3.6 kb) in various tissues. The alignment of proapelin amino acid sequences from cattle, humans, rats, and mice has demonstrated strict conservation of the C-terminal 17 amino acids, known as apelin 17 or K17F. *In vivo*, proapelin gives rise to various molecular forms of apelin, probably through the action of prohormone convertases, due to the presence of pairs of basic residues in proapelin. In rat brain and plasma, the predominant forms of apelin are the pyroglutamyl form of apelin 13 (pE13F) and, to a lesser extent, K17F. In rat lung, testis, and uterus and in bovine colostrum, apelin 36 predominates, whereas in the rat mammary gland, both apelin 36 and pE13F have been detected.

Apelin Receptor Internalization and Signaling Cascades

The rat and human apelin receptors are negatively coupled to adenylate cyclase activity. In Chinese hamster ovary (CHO) cells that stably expressed the human or rat apelin receptor, the most potent inhibitors of forskolin-induced cAMP production were found to be apelin 36, K17F, apelin 13 (Q13F), and pE13F, whereas apelin fragments R10F and G5F were inactive. In an Ala scan (individual amino acids of the primary sequence are replaced with an alanine residue) of pE13F, or in N- or C-terminal deletions of K17F, it was shown that the arginine residues in positions 2 and 4 or the leucine in position 5 in pE13F played a critical role in binding affinity or in the inhibition of cAMP production. Interestingly, apelin 36, K17F, and pE13F are also potent inducers of rat and human apelin receptor internalization and suppress the hypotensive effect of K17F, suggesting that apelin receptor endocytosis is required for initiation of a second wave of signal transduction different from adenylate cyclase coupling and is responsible for the biological effect. Deletion of the C-terminal phenylalanine of K17F abolishes internalization without affecting the adenylate cyclase coupling of the apelin receptor. Apelin 36, K17F, and pE13F also increase intracellular calcium mobilization in both NTera-2 human teratocarcinoma (NT2N) cells, which differentiate into postmitotic neurons following retinoic acid stimulation, and RBL-2H3 cells derived from rat basophils stably expressing the human apelin receptor. In CHO cells expressing the mouse apelin receptor, apelin activates extracellular signal-regulated kinases (ERKs) via a pertussis toxin-sensitive G-protein and Ras-independent pathway. In addition, ERK activation by apelin is mediated by an unidentified isoform of protein kinase C (PKC). Apelin was also shown to activate p70S6 kinase in human umbilical vein endothelial cells (HUVECs) or in CHO cells expressing the mouse apelin receptor via two intracellular cascades, one ERK dependent and the other phosphatidylinositol 3-kinase (PI3K) dependent.

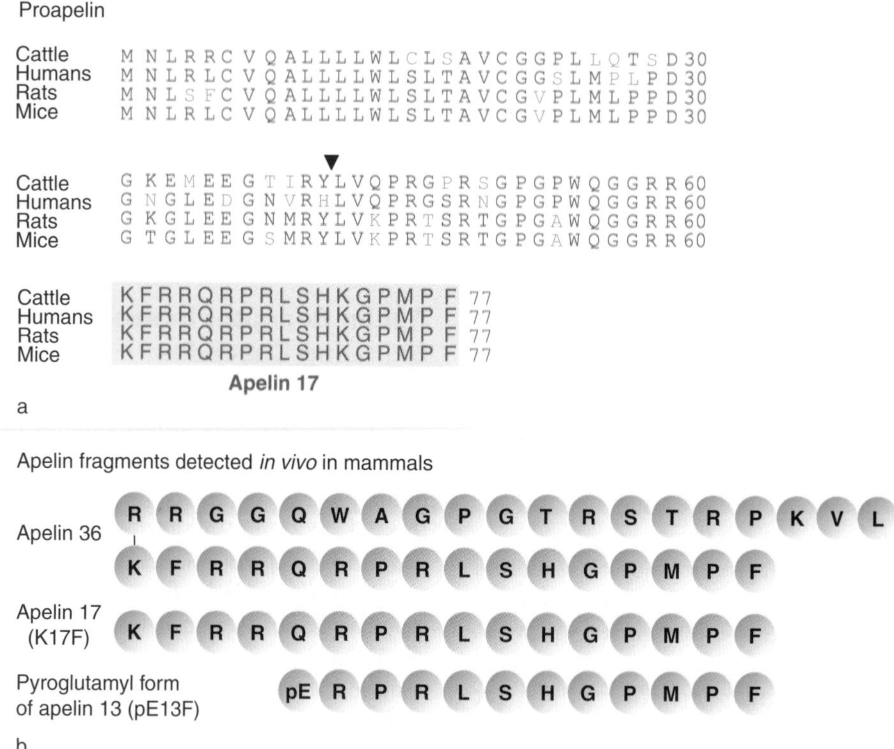

Figure 1 (a) Amino acid sequence of the apelin precursor, proapelin, in cattle, humans, rats, and mice. The first amino acid of apelin 36 is indicated by an arrowhead and the apelin 17 sequence is indicated by pink shading. (b) Apelin fragments detected *in vivo* in mammals: apelin 36, apelin 17 (K17F), and the pyroglutamyl form of apelin 13 (pE13F).

Distribution of Apelin and Its Receptor in Adult Rat Brain

Topographical Distribution of Apelin Immunoreactivity

The precise central topographical distribution of apelin immunoreactivity shows that apelin-immunoreactive (IR) neuronal cell bodies are particularly abundant in the hypothalamus and medulla oblongata, involved in neuroendocrine control, food intake, body fluid homeostasis, and the regulation of cardiovascular functions. These cell bodies are predominantly detected in the hypothalamic supraoptic nucleus (SON) and the magnocellular part of the paraventricular nucleus (PVN), the arcuate nucleus, and the lateral reticular and ambiguus nuclei. Conversely, apelin-IR nerve fibers are much more widely distributed in many brain regions than are neuronal apelin cell bodies. Apelin-IR nerve fibers innervate the mesencephalon, the pons, the medulla oblongata, and several circumventricular organs, such as the vascular organ of the lamina terminalis (OVLT), the subfornical organ (SFO), the subcommissural organ, and the area postrema (**Figure 2**). The density of IR nerve fibers and apelinergic nerve endings is highest in the inner layer of the median eminence and in the posterior pituitary, suggesting that the apelin neurons of the SON and PVN, like the magnocellular AVP and ocytocin neurons, project into the posterior pituitary via the internal layer of the median eminence. Double immunofluorescence staining confirms this finding, showing that apelin co-localizes with AVP and ocytocin in magnocellular hypothalamic neurons.

Expression of Apelin Receptor mRNA

Apelin receptor mRNA expression, like its ligand, is widely distributed throughout the rat central nervous system. Apelin receptor mRNA is present in the piriform and entorhinal cortices, the nucleus of the lateral olfactory tract, the septum, the hippocampus, and structures containing monoaminergic neuronal cell bodies (pars compacta of the substantia nigra, dorsal raphe nucleus, and locus coeruleus). The apelin receptor is particularly abundant in the apelin-rich hypothalamic nuclei (including the SON, PVN, and arcuate nucleus). The anterior and intermediate lobes of the pituitary are also highly labeled, as well as the pineal gland. Furthermore, like its ligand, apelin receptors (and type 1A and type 1B AVP receptors) are expressed by magnocellular AVP neurons, suggesting an interaction between AVP and apelin (**Figure 3**).

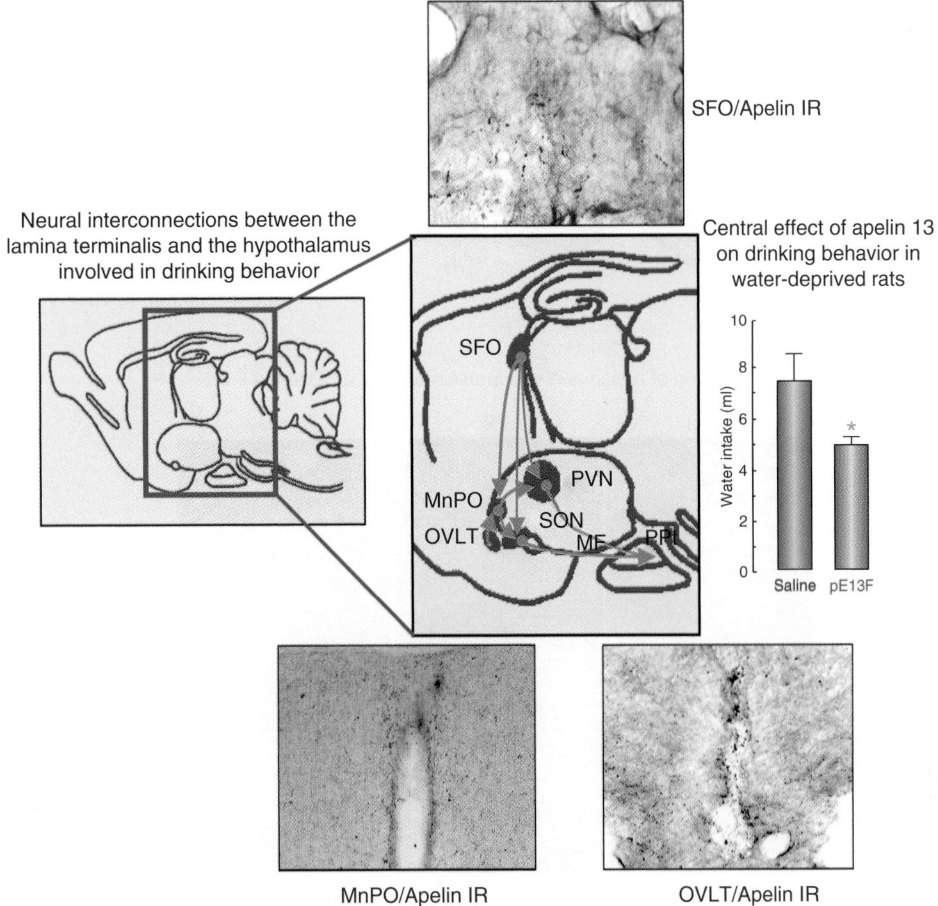

Figure 2 Apelin immunoreactivity in the lamina terminalis: involvement in drinking behavior. Numerous apelin-immunoreactive nerve fibers have been detected in the lamina terminalis, including the subfornical organ (SFO), vascular organ of the lamina terminalis (OVLT), and median preoptic nucleus (MnPO), nuclei involved in the regulation of drinking behavior. Apelin administered into the lateral ventricle in 24 h-water-deprived rats significantly decreased water intake. ME, median eminence; PPlt, posterior pituitary; PVN, hypothalamic paraventricular nucleus; SON, supraoptic nucleus. Adapted from Johnson AK, Cunningham JT, and Thunhorst RL (1996) Integrative role of the lamina terminalis in the regulation of cardiovascular and body fluid homeostasis. *Clinical and Experimental Pharmacological Physiology* 23 (2): 183–191; from Reaux A, De Mota N, Skultetyova I, et al. (2001) Physiological role of a novel neuropeptide, apelin, and its receptor in the rat brain. *Journal of Neurochemistry* 77(4): 1085–1096, Blackwell Publishing and reproduced from Reaux A, Gallatz K, Palkovits M, et al. (2002) Distribution of apelin-synthesizing neurons in the adult rat brain. *Neuroscience* 113(3): 653–662, with permission from Elsevier.

Apelin: Physiological Actions within the Brain and Anterior Pituitary Gland

Involvement of Central Apelin in the Regulation of Food Intake

Apelin-IR cell bodies and nerve fibers have been detected in several hypothalamic nuclei involved in the control of food intake, including the PVN and the suprachiasmatic, arcuate, ventromedial, and dorsomedial nuclei (**Figure 4**). Whether apelin acts in the brain to modulate feeding behavior is controversial. Central administration of apelin decreases food intake in both fed and starved rats. Similar effects have been reported when apelin is given nocturnally to rats, but the reverse occurs during daytime administration, suggesting a circadian-dependent mechanism of action. In contrast to these findings, another report has shown that intracerebroventricular administration of apelin has either little or no effect on food intake.

Role of Peripheral Apelin in Feeding and Digestion

Apelin precursor mRNA and apelin immunoreactivity have also been detected in the periphery in rats, in the gastrointestinal tract (in neuroendocrine chromogranin A-sensitive cells and in mucosal epithelium). In a murine intestinal enteroendocrine cell line (STC-1 cells) producing and secreting cholecystokinin, apelin stimulates both cholecystokinin secretion and the proliferation of gastric cells. These data suggest that apelin in the gastrointestinal tract may play a role in both exocrine and endocrine functions. Apelin has also been shown to be a novel adipokine released

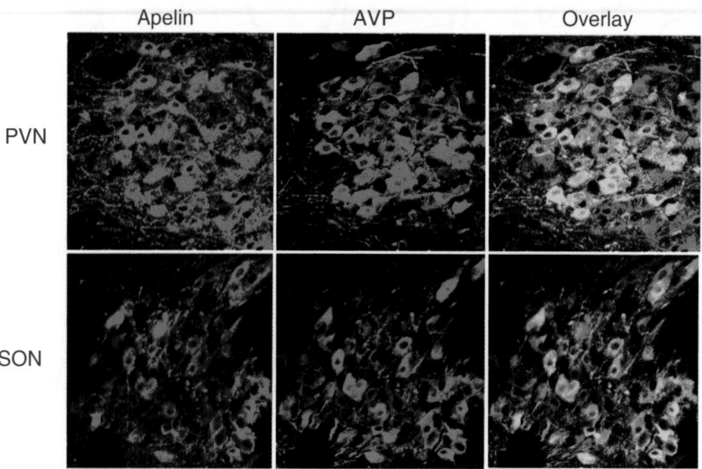

Figure 3 Detection of apelin immunoreactivity and apelin receptor mRNA expression in the rat hypothalamus: co-localization with vasopressin. Distribution of apelin receptor mRNA expression in the rat supraoptic nucleus (SON) and paraventricular nucleus (PVN) by *in situ* hybridization. Apelin receptor mRNA is highly expressed in both nuclei and is synthesized by magnocellular arginine vasopressin-immunoreactive (AVP IR) neurons (arrowhead). Confocal images illustrating the distribution of apelin-immunoreactive cell bodies within the rat SON and PVN. A high co-localization of apelin (red) and AVP (green) has been detected in both SON and PVN magnocellular neurons. PIR, piriform cortex; NLOT, nucleus of the lateral olfactory tract; OC, optic chiasm. Adapted from De Mota N, Lenkei Z, and Llorens-Cortes C (2000) Cloning, pharmacological characterization and brain distribution of the rat apelin receptor. *Neuroendocrinology* 72(6): 400–407 (S. Kargel AG, Basel); from Reaux A, De Mota N, Skultetyova I, et al. (2001) Physiological role of a novel neuropeptide, apelin, and its receptor in the rat brain. *Journal of Neurochemistry* 77(4): 1085–1096, Blackwell Publishing; and from Reaux-Le Goazigo A, Morinville A, Burlet A, et al. (2004) Dehydration-induced cross-regulation of apelin and vasopressin immunoreactivity levels in magnocellular hypothalamic neurons. *Endocrinology* 145(9): 4392–4400, Copyright 2004, The Endocrine Society.

from fat cells and upregulated by insulin and obesity. In experimental animal models of obesity, plasma apelin levels are significantly higher than normal, but only in states associated with hyperinsulinemia.

Apelin appears to play a role in the peripheral and central regulation of food intake. However, given the conflicting nature of the data obtained to date on the central effects, further investigations are required to highlight the mode of action of apelin in the control of this function.

Involvement of Apelin in Regulation of the Hypothalamic–Adrenal–Pituitary Axis

The existence of an apelinergic system within the adult male rat anterior pituitary gland has been recently reported. Apelin is highly co-expressed in corticotrophs and to a much lower extent in somatotrophs, and a high expression of apelin receptor mRNA is also found in corticotrophs. These morphological data suggest a local interaction between apelin and adrenocorticotropic hormone (ACTH). In an *ex vivo* perifusion system of anterior pituitaries, apelin significantly increased basal ACTH release and induced a dose-dependent increase in K^+-evoked ACTH release, outlining the potential role of apelin as an autocrine/paracrine-acting peptide on ACTH release.

Moreover, the detection of apelin-immunoreactive nerve fibers together with apelin receptor mRNA expression in the parvocellular part of the PVN, and the stimulatory action of apelin on corticotropin-releasing

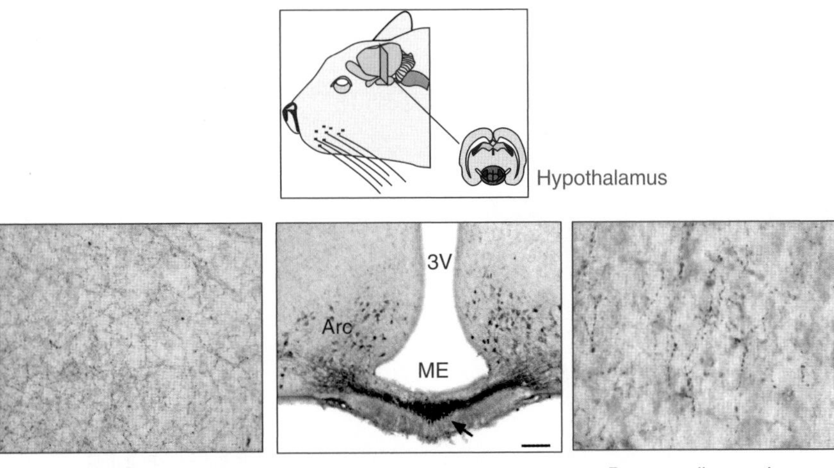

Figure 4 Apelin immunoreactivity in hypothalamic structures involved in the regulation of food intake. The arcuate, ventromedian, and dorsomedian nuclei are brain structures involved in the control of food intake. The presence of numerous apelin-immunoreactive cell bodies in the arcuate nucleus and the presence of numerous apelin nerve fibers in both ventromedian and dorsomedian nuclei suggest that apelin could be involved in the feeding behavior. Adapted from Reaux A, Gallatz K, Palkovits M, et al. (2002) Distribution of apelin-synthesizing neurons in the adult rat brain. *Neuroscience* 113(3): 653–662, with permission from Elsevier.

hormone (CRH) release, indicate that apelin could modulate ACTH release via an indirect action at the level of the hypothalamus. Central injection of apelin is shown to significantly increase plasma ACTH and corticosterone release, at least in part via a stimulatory action on CRH release. Thus, the stimulation of ACTH secretion by apelin occurs by a direct action in the anterior pituitary and by an indirect action via stimulation of CRH release in the hypothalamus. In agreement with the involvement of apelin in the regulation of corticotrophs, it has been reported that in adult rats submitted to acute stress (restraint stress), known to increase the activity of the hypothalamic–adrenal–pituitary (HPA) axis, apelin receptor mRNA expression is increased in the parvocellular division of the PVN. Moreover, dexamethasone, a glucocorticoid agonist, drastically decreases apelin mRNA levels in 3T3-L1 mouse adipocytes, whereas in adrenalectomized rats, apelin receptor mRNA expression is increased, suggesting that glucocorticoids downregulate the expression of apelin and its receptor.

Involvement of Apelin in Maintenance of Body Fluid Homeostasis

The magnocellular neurons release AVP, an antidiuretic vasoconstrictor peptide, into the fenestrated capillaries of the posterior pituitary in response to changes in plasma osmolality and volemia. The recent reports of co-localization of AVP and apelin in the magnocellular neurons of the hypothalamus and the presence of receptors for AVP and apelin on these same neurons suggest a potential apelinergic response to these stimuli.

Independently of the feedback control exerted by AVP on its own release, apelin may also regulate AVP release. This hypothesis has been tested in lactating rats exhibiting a reinforced phasic pattern of AVP neurons during lactation, thereby facilitating systemic AVP release to maintain body water content for optimal milk production. In this model, the intracerebroventricular injection of apelin inhibits the phasic firing activity of AVP neurons, thereby decreasing AVP release into the bloodstream, leading to an increase in aqueous diuresis (**Figure 5**). Similarly, a marked decrease in systemic AVP release is observed following the intracerebroventricular injection of apelin in mice deprived of water for 24 h, a condition known to increase AVP neuron activity. These data suggest that apelin is probably released from the SON and PVN AVP cell bodies and inhibits AVP neuron activity and release by means of a direct action on the apelin autoreceptors expressed by AVP/apelin-containing neurons. This mechanism probably involves apelin acting as a natural inhibitor of the antidiuretic effect of AVP. The co-localization and opposite biological actions of these two peptides raise questions concerning how these two peptides are regulated to maintain body fluid homeostasis. For the purpose of addressing these questions, the effect of water deprivation on the neuronal content and release of both apelin and AVP were studied.

Water deprivation increases apelin receptor mRNA expression in the magnocellular part of the PVN and SON. Water deprivation largely increases apelin immunoreactivity within both the SON and the PVN and also results in the *de novo* appearance of apelin-IR

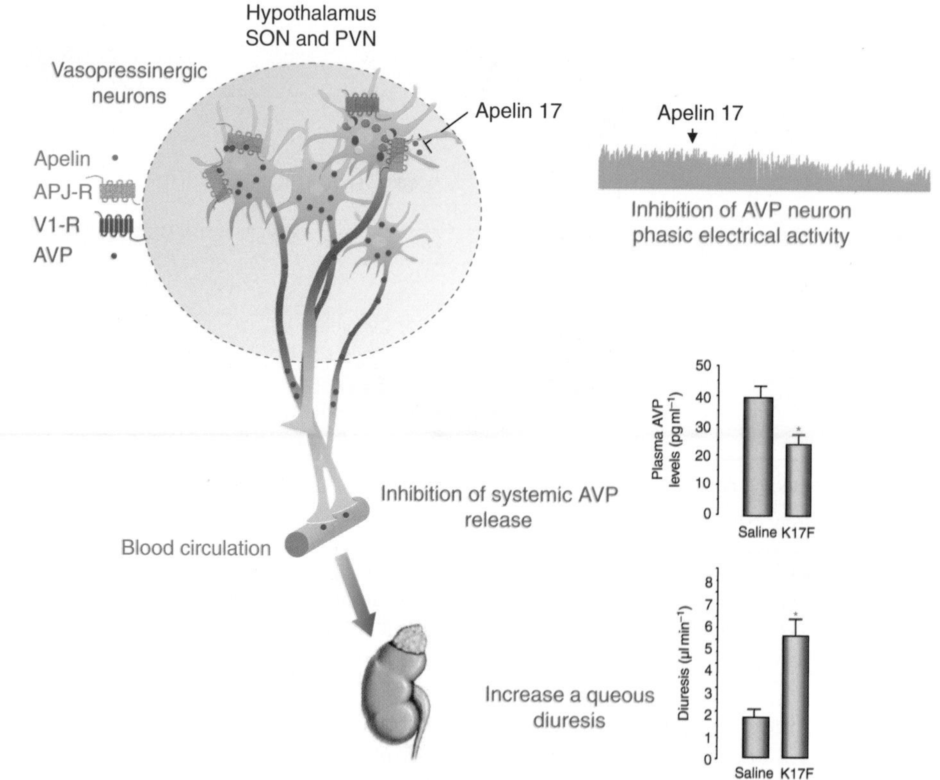

Figure 5 Apelin, a potent diuretic neuropeptide counteracting the effects of arginine vasopressin (AVP) through inhibition of AVP neuron activity and AVP release. In rodents, apelin and its receptor (APJ-R) co-localize with AVP and AVP receptors (V1-R) in the supraoptic nucleus (SON) and paraventricular nucleus (PVN) magnocellular neurons. In lactating animals, the central injection of apelin 17 induces a gradual and sustained inhibition of the phasic electrical activity of AVP neurons, thereby decreasing systemic AVP secretion and increasing aqueous diuresis. Adapted from De Mota N, Reaux-Le Goazigo A, El Messari S, et al. (2004) Apelin, a potent diuretic neuropeptide counteracting vasopressin actions through inhibition of vasopressin neuron activity and vasopressin release. *Proceedings of the National Academy of Sciences of the United States of America* 101: 10464–10469, Copyright (2004), National Academy of Sciences, USA.

neurons in accessory magnocellular nuclei, which contain approximately one-third of the oxytocin and AVP neurons projecting to the posterior pituitary. Among these is the nucleus circularis, which, like the SON and PVN, receives inputs from the SFO, a structure involved in the control of water balance. The increase in apelin immunoreactivity could be due in part to AVP endogenously released in response to dehydration, since chronic treatment with a V1 receptor antagonist significantly reduces the dehydration-induced increase in apelin immunoreactivity in magnocellular neurons.

In rats deprived of water for 24 h, the large increase in hypothalamic apelin content is mirrored by a decrease in plasma apelin levels, suggesting that, under these conditions, apelin accumulates within AVP neurons, rather than being released. The apelin response to dehydration is therefore the opposite of that of AVP, which is released faster than it is synthesized. This interpretation implies that apelin and AVP are released differentially by the magnocellular AVP neurons in which they are produced. Consistent with this hypothesis, high-resolution confocal microscopic images of magnocellular hypothalamic neurons show a marked segregation of apelin and AVP immunoreactivity within SON and PVN neurons. This interpretation is further supported by the presence of apelin-positive/AVP-negative and AVP-positive/apelin-negative varicosities along the same hypothalamo-hypophyseal axons (**Figure 6**).

These opposite regulatory patterns of apelin and AVP suggest that these molecules act in concert to maintain body fluid homeostasis. During dehydration, increases in the somatodendritic release of AVP optimize the electrical phasic activity of AVP neurons, facilitating the release of AVP into the bloodstream, whereas apelin accumulates in these neurons, rather than being released into the bloodstream and, probably, into the nuclei. Thus, decreases in the local supply of apelin to magnocellular AVP cell bodies may facilitate the expression by AVP neurons of an optimized

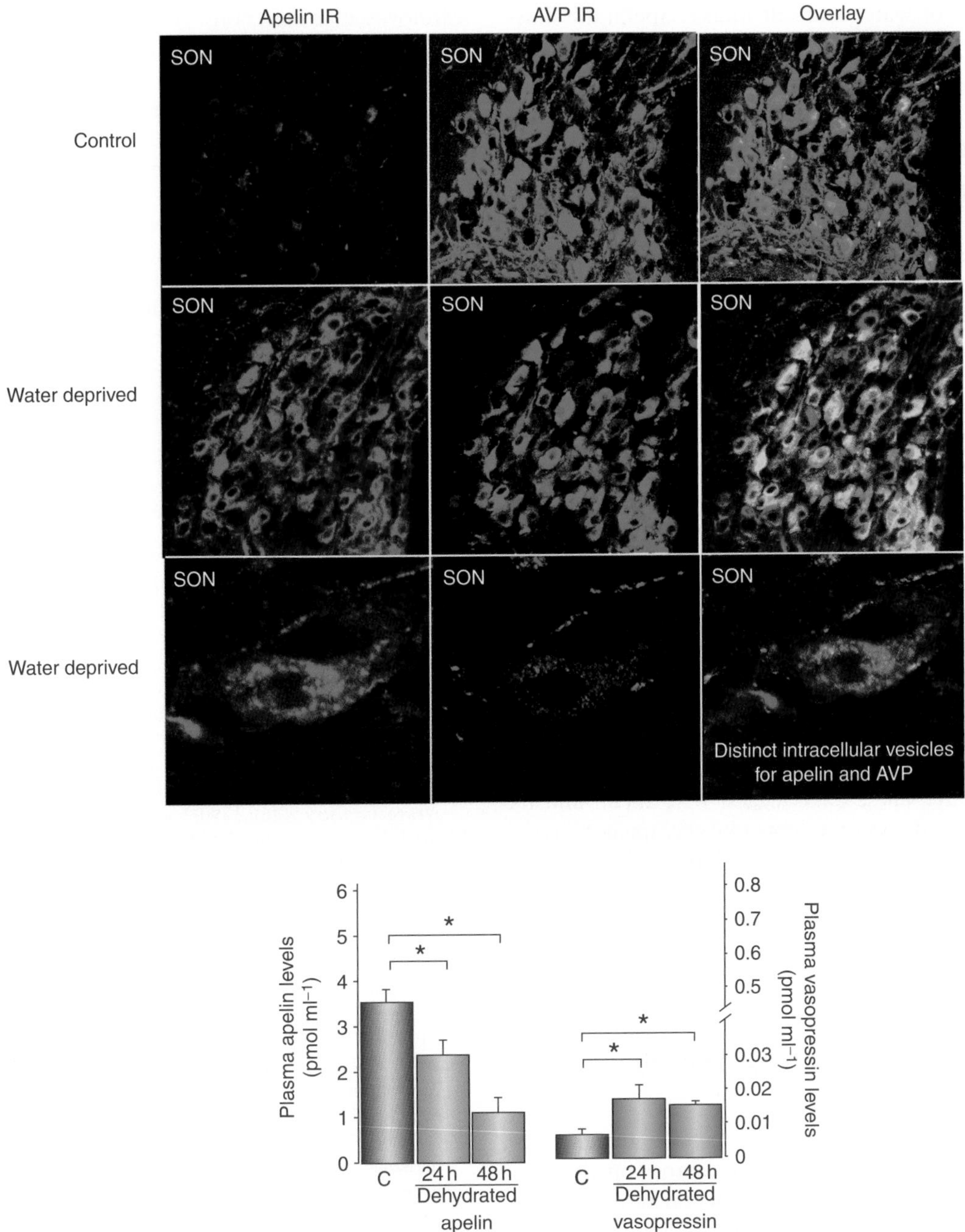

Figure 6 Apelin and vasopressin are conversely regulated during water deprivation. Water deprivation induces a large increase of the number of apelin-immunoreactive cells and in the density of apelin immunofluorescence in the supraoptic nucleus (SON), as compared to control rats with free access to water. In contrast, the number of arginine vasopressin (AVP)-immunoreactive cell bodies is decreased in water-deprived animals. In parallel, water deprivation, which increases systemic AVP release, decreases plasma apelin concentration. The reverse regulation of apelin and AVP during dehydration can occur, since both peptides are detected within a distinct pool of cytoplasmic vesicles. Adapted from De Mota N, Reaux-Le Goazigo A, El Messari S, et al. (2004) Apelin, a potent diuretic neuropeptide counteracting vasopressin actions through inhibition of vasopressin neuron activity and vasopressin release. *Proceedings of the National Academy of Sciences of the United States of America* 101: 10464–10469, Copyright (2004) National Academy of Sciences, USA; and Adapted from Reaux-Le Goazigo A, Morinville A, Burlet A, et al. (2004) Dehydration-induced cross-regulation of apelin and vasopressin immunoreactivity levels in magnocellular hypothalamic neurons. *Endocrinology* 145(9): 4392–4400, Copyright 2004, The Endocrine Society.

phasic activity, by decreasing the inhibitory actions of apelin on these neurons. Antagonistic regulation of apelin and AVP has a biological purpose, making it possible to maintain the water balance of the organism by preventing additional water loss via the kidneys. Consistent with a role for apelin in the control of water balance, which depends not only on AVP secretion regulating renal water fluid loss, but also on the regulation of water and salt intake, apelin administered centrally clearly and significantly decreases water intake in rats deprived of water for 24 h.

Peripheral Cardiovascular Actions

Apelin also has cardiovascular effects. The mRNA encoding apelin receptors has been detected in the endothelial cells of large conduit arteries, coronary vessels, and endocardium of the right atrium. The injection of apelin into the bloodstream decreases arterial blood pressure (BP), via a mechanism dependent on nitric oxide (NO) production. In normotensive or hypertensive rats, apelin increases the contractile force of the myocardium via a positive inotropic effect. Moreover, apelin receptor knockout mice display an enhanced vasopressor response to systemic AngII, suggesting a counterregulatory action of apelin on AngII. Acute administration of apelin *in vivo* reduces left ventricular preload and afterload while increasing the compliance and contractile reserve of the heart. Chronic treatment with apelin *in vivo* increases cardiac output without causing ventricular hypertrophy. Apelin immunoreactivity has been found to increase in the plasma of patients in the early stages of heart failure and then to decrease during later, more severe stages of heart failure. These data suggest that apelin and its receptor could constitute potential therapeutic targets in the treatment of heart failure. Indeed, the administration of a nonpeptide agonist of the apelin receptor might improve the contractile performance of the myocardium while reducing cardiac loading and increasing aqueous diuresis in patients with heart failure.

Conclusions and Pathophysiological Implications

The identification/discovery of apelin as the endogenous ligand of the orphan APJ receptor constitutes a major advance, both for fundamental research and, potentially, for clinical practice. It demonstrates the validity of the 'deorphanization' approach to orphan receptors for the identification of new bioactive peptides and new potential therapeutic targets. The experimental data obtained to date demonstrate that apelin, by inhibiting the phasic electrical activity of AVP neurons and the systemic secretion of AVP, increases aqueous diuresis. In the periphery, apelin decreases arterial BP and increases the contractile force of the myocardium. Overall, these data show that this new circulating vasoactive neuropeptide may play a key role in the maintenance of water balance and cardiovascular functions. The development of nonpeptide agonists of the apelin receptor, based on the knowledge of the structures of apelin and its receptor, could lead to new therapeutic tools for the treatment of the syndrome of inappropriate secretion of AVP, thirst disorders, and heart and kidney failure.

See also: Angiotensin II; Atrial Natriuretic Peptide: Fluid/Mineral Balance; Blood Pressure: Baroreceptors; Circumventricular Organs; Cyclic AMP (cAMP) Role in Learning and Memory; Invertebrate Neurohormone GPCRs; Osmoregulation; Thirst; Vasopressin/Oxytocin and Receptors.

Further Reading

Ashley EA, Powers J, Chen M, et al. (2005) The endogenous peptide apelin potently improves cardiac contractility and reduces cardiac loading *in vivo*. *Cardiovascular Research* 65 (1): 73–82.

Boucher J, Masri B, Daviaud D, et al. (2005) Apelin, a newly identified adipokine up-regulated by insulin and obesity. *Endocrinology* 146(4): 1764–1771.

De Mota N, Lenkei Z, and Llorens-Cortes C (2000) Cloning, pharmacological characterization and brain distribution of the rat apelin receptor. *Neuroendocrinology* 72: 400–407.

El Messari S, Iturrioz X, Fassot C, et al. (2004) Functional dissociation of apelin receptor signaling and endocytosis: Implications for the effects of apelin on arterial blood pressure. *Journal of Neurochemistry* 90(6): 1290–1301.

Kleinz MJ and Davenport AP (2005) Emerging roles of apelin in biology and medicine. *Pharmacological Therapeutics* 107(2): 198–211.

Lee DK, Cheng R, Nguyen T, et al. (2000) Characterization of apelin, the ligand for the APJ receptor. *Journal of Neurochemistry* 74: 34–41.

Llorens-Cortes C and Beaudet A (2005) Apelin, a neuropeptide that counteracts vasopressin secretion. *Medecine et Sciences (Paris)* 21(8–9): 741–746.

Masri B, Knibiehler B, and Audigier Y (2005) Apelin signalling: A promising pathway from cloning to pharmacology. *Cell Signaling* 17(4): 415–426.

O'Carroll AM and Lolait SJ (2003) Regulation of rat APJ receptor messenger ribonucleic acid expression in magnocellular neurones of the paraventricular and supraoptic nuclei by osmotic stimuli. *Journal of Neuroendocrinology* 15(7): 661–666.

O'Dowd BF, Heiber M, Chan A, et al. (1993) A human gene that shows identity with the gene encoding the angiotensin receptor is located on chromosome 11. *Gene* 136: 355–360.

Szokodi I, Tavi P, Foldes G, et al. (2002) Apelin, the novel endogenous ligand of the orphan receptor APJ, regulates cardiac contractility. *Circulation Research* 91(5): 434–440.

Taheri S, Murphy K, Cohen M, et al. (2002) The effects of centrally administered apelin-13 on food intake, water intake and pituitary hormone release in rats. *Biochemical and Biophysical Research Communications* 291: 1208–1212.

Tatemoto K, Hosoya M, Habata Y, et al. (1998) Isolation and characterization of a novel endogenous peptide ligand for the human APJ receptor. *Biochemical and Biophysical Research Communications* 251: 471–476.

Aphasia: Sudden and Progressive

M-M Mesulam, Northwestern University Medical School, Chicago, IL, USA

Introduction

Aphasia is an acquired disorder of language caused by brain damage. It is diagnosed when deficits are detected in naming, word choice, word comprehension, spelling, or syntax. Dysarthria and mutism do not, by themselves, lead to a diagnosis of aphasia. In approximately 90% of right handers and 60% of left handers, aphasia occurs only after lesions of the left hemisphere. For the majority of the population, the left hemisphere is therefore said to be dominant for language function. In some individuals no hemispheric dominance for language can be discerned, and in others (including a small minority of right handers) there is a right hemisphere dominance for language. A language disturbance occurring after a right hemisphere lesion in a right hander is called crossed aphasia.

Language is controlled by a large-scale distributed network, usually located within the left hemisphere, revolving around two perisylvian nodes. One extends into the temporoparietal junction and is known as Wernicke's area; the other extends into the inferior frontal gyrus and is known as Broca's area. These two areas are connected with each other as well as with multiple regions of the temporal, parietal, and frontal lobes. Wernicke's area can be considered the semantic-lexical pole of the language network, whereas Broca's area can be considered its syntactic-phonological pole. These specializations are relative rather than absolute. The network as a whole links sensory patterns corresponding to the words we hear and read into the distributed associations that encode their meaning. Damage to the language network can therefore have two major consequences. In some patients, the meanings of words cannot be decoded. In others, thoughts and experiences cannot be translated into statements that have the appropriate syntactic structure or semantic content. The location of the damage within the language network determines the type of aphasia experienced by the patient.

There are two major groups of acquired aphasias: those caused by cerebrovascular accidents and those caused by degenerative diseases. Aphasias caused by cerebrovascular accidents start suddenly and display maximal deficits at the onset. The underlying lesion is relatively circumscribed and associated with a total loss of neural function at the lesion site. These are the 'classic' aphasias in which relatively reproducible relationships between lesion site and aphasia pattern can be discerned. Aphasias caused by neurodegenerative diseases have an insidious onset and relentless progression so that the symptomatology changes over time. Since the neuronal loss within the areas encompassed by the neurodegeneration is partial and since it tends to include multiple components of the language network, distinctive clinical patterns and clinico-anatomical correlations are less obvious.

Clinical Examination

The clinical examination of language should include the assessment of naming, spontaneous speech, comprehension, repetition, reading, and writing. A deficit of naming (anomia) is the single most common finding in aphasic patients. When asked to name common objects (e.g., a pencil or wristwatch), the patient may fail to come up with the appropriate word, may provide a circumlocutious description of the object ('the thing for writing'), or may come up with the wrong word (paraphasia). If the patient offers an incorrect but legitimate word ('pen' for 'pencil'), the naming error is known as a semantic paraphasia; if the word approximates the correct answer but is phonetically inaccurate ('plentil' for 'pencil'), the error is known as a phonemic paraphasia. Asking the patient to name body parts, geometric shapes, and component parts of objects (e.g., lapel of coat, cap of pen) can elicit mild forms of anomia in patients who can otherwise name common objects. In most anomias, the patient cannot retrieve the appropriate name when shown an object but can point to the appropriate object when the name is provided by the examiner. This is known as a one-way (or retrieval-based) naming deficit. A two-way naming deficit exists if the patient can neither provide nor recognize the correct name, indicating the presence of a word comprehension impairment.

Spontaneous speech is described as fluent if it maintains appropriate output volume, phrase length, and melody or as nonfluent if it is sparse and halting and if average utterance length is less than four words. The examiner should also note whether the speech is paraphasic or circumlocutious; whether it shows a relative paucity of substantive nouns and action verbs compared to function words (prepositions, conjunctions); and whether word order, tenses, suffixes, prefixes, plurals, and possessives are used appropriately. Comprehension can be tested by assessing the patient's ability to follow the conversation, by asking yes/no questions ('Can a dog fly?,' 'Does it snow in summer?'), or by asking the patient to point to appropriate objects ('Where is the source of illumination in this room?'). Statements with embedded clauses or passive voice construction ('If a tiger is eaten by a lion, which animal stays alive?') help to assess the patient's ability to comprehend complex syntactic structure. Commands to close or open the eyes, stand up, sit down, or roll over should not be used to assess overall comprehension since appropriate responses aimed at such axial movements can be preserved in patients who otherwise have profound comprehension deficits.

Repetition is assessed by asking the patient to repeat single words, short sentences, or strings of words such as 'No ifs, ands, or buts.' The testing of repetition with tongue-twisters such as 'hippopotamus' or 'Irish constabulary' provides a better assessment of dysarthria than aphasia. Aphasic patients may have little difficulty with tongue-twisters but have a particularly hard time repeating a string of function words. It is important to make sure that the number of words does not exceed the patient's attention span. Otherwise, the failure of repetition becomes a reflection of the narrowed attention span (verbal working memory) rather than an indication of an aphasic deficit. Reading should be assessed for deficits in reading aloud as well as comprehension. Writing is assessed for spelling errors, word order, and grammar. Alexia describes an inability to read or comprehend written words; agraphia (or dysgraphia) is used to describe an acquired deficit in the spelling or grammar of written language.

Aphasias of Cerebrovascular Origin

Aphasias of cerebrovascular origin can be divided into central syndromes, which result from damage to the epicenters of the language network (Broca's and Wernicke's areas), and disconnection syndromes, which arise from lesions that interrupt the functional connectivity of these centers with each other and with the other components of the language network. The syndromes outlined in the following sections are idealizations; pure syndromes occur rarely.

Wernicke's Aphasia

In Wernicke's aphasia, comprehension is impaired for spoken and written language. Language output is fluent but is highly paraphasic and circumlocutious. The tendency for paraphasic errors may be so pronounced that it leads to strings of neologisms, which form the basis of what is known as jargon aphasia. Speech contains large numbers of function words (e.g., prepositions, conjunctions) but few substantive nouns or verbs that refer to specific actions. The output is therefore voluminous but uninformative. The patient does not seem to realize that his or her language is incomprehensible and may appear angry and impatient when the examiner fails to decipher the meaning of a severely paraphasic statement. Patients with Wernicke's aphasia cannot express their thoughts in meaning-appropriate words and cannot decode the meaning of words in any modality of input. This aphasia therefore has expressive as well as receptive components. Repetition, naming, reading, and writing are also impaired. The lesion site most commonly associated with Wernicke's aphasia is in the posterior portion of the language network and tends to involve at least parts of Wernicke's area.

Broca's Aphasia

In Broca's aphasia, speech is nonfluent, labored, interrupted by many word-finding pauses, and usually dysarthric. It is impoverished in function words. Abnormal word order and the inappropriate deployment of bound morphemes (word endings used to denote tenses, possessives, or plurals) lead to a characteristic agrammatism. Speech is telegraphic and pithy but quite informative. Output may be reduced to a grunt or single word ('yes' or 'no'), which is emitted with different intonations in an attempt to express approval or disapproval. In addition to fluency, naming and repetition are also impaired. Comprehension of spoken language is intact, except for syntactically difficult sentences with passive voice structure or embedded clauses.

Global Aphasia

In global aphasia, speech output is nonfluent, and comprehension of spoken language is severely impaired. Naming, repetition, reading, and writing are also impaired. This syndrome represents the combined dysfunction of Broca's and Wernicke's areas and usually results from strokes that involve the entire middle cerebral artery distribution in the left hemisphere. Most patients are initially mute or say a few words, such as 'hi' or 'yes.'

Conduction Aphasia

In conduction aphasia, speech output is fluent but paraphasic, comprehension of spoken language is intact, and repetition is severely impaired. Naming and writing are also impaired. Reading aloud is impaired, but reading comprehension is preserved. The lesion sites spare Broca's and Wernicke's areas but may induce a functional disconnection between the two such that neural word representations formed in Wernicke's area and adjacent regions cannot be transmitted to

Broca's area for assembly into corresponding articulatory patterns. Occasionally, a Wernicke's area lesion gives rise to a transient Wernicke's aphasia that rapidly resolves into a conduction aphasia. The paraphasic output in conduction aphasia interferes with the ability to express meaning, but this deficit is not nearly as severe as the one displayed by patients with Wernicke's aphasia.

Nonfluent Transcortical Aphasia (Transcortical Motor Aphasia)

The features of nonfluent transcortical aphasia are similar to those of Broca's aphasia, but repetition is intact and agrammatism may be less pronounced. The lesion site disconnects the intact language network from prefrontal areas of the brain and usually involves the anterior watershed zone between anterior and middle cerebral artery territories or the supplementary motor cortex in the territory of the anterior cerebral artery.

Fluent Transcortical Aphasia (Transcortical Sensory Aphasia)

The clinical features of fluent transcortical aphasia are similar to those of Wernicke's aphasia, but repetition is intact. The lesion site disconnects the intact core of the language network from other temporoparietal association areas. Infarctions in the posterior watershed zone are common causes.

Isolation Aphasia

Isolation aphasia, a rare syndrome, represents a combination of the two transcortical aphasias. Comprehension is severely impaired, and there is no purposeful speech output. The patient may parrot fragments of heard conversations (echolalia), indicating that the neural mechanisms for repetition are at least partially intact. This condition represents the pathologic function of the language network when it is isolated from other regions of the brain. Broca's and Wernicke's areas tend to be spared, but there is damage in surrounding frontal, parietal, and temporal cortex. Lesions are patchy and can be associated with anoxia, carbon monoxide poisoning, or complete watershed zone infarctions.

Anomic Aphasia

Anomic aphasia may be considered the 'minimal dysfunction' syndrome of the language network. Articulation, comprehension, and repetition are intact, but confrontation naming, word finding, and spelling are impaired. Speech is enriched in function words but impoverished in substantive nouns and verbs denoting specific actions. Language output is fluent but paraphasic, circumlocutious, and uninformative. The lesion sites can be anywhere within the left hemisphere language network, including the middle and inferior temporal gyri.

Pure Word Deafness

In pure word deafness, the most common lesions are either bilateral or left-sided in the superior temporal gyrus. The net effect of the underlying lesion is to interrupt the flow of information from the unimodal auditory association cortex to the language network. Patients have no difficulty understanding written language and can express themselves well in spoken or written language. They have no difficulty interpreting and reacting to environmental sounds since primary auditory cortex and subcortical auditory relays are intact. Since auditory information cannot be conveyed to the language network, however, the patient reacts to speech as if it were in an alien tongue that cannot be deciphered. Patients cannot repeat spoken language but have no difficulty naming objects.

Pure Alexia without Agraphia

Pure alexia without agraphia is the visual equivalent of pure word deafness. The lesions (usually a combination of damage to the left occipital cortex and to a posterior sector of the corpus callosum known as the splenium) interrupt the flow of visual input into the language network. There is usually a right hemianopia, but the core language network remains unaffected. The patient can understand and produce spoken language, name objects in the left visual hemifield, repeat, and write. However, the patient acts as if illiterate when asked to read even the simplest sentence because the visual information from the written words (presented to the intact left visual hemifield) cannot reach the language network. Objects in the left hemifield may be named accurately because they activate nonvisual associations in the right hemisphere, which, in turn, can access the language network through transcallosal pathways anterior to the splenium. Patients with this syndrome may also lose the ability to name colors, although they can match colors. This is known as a color anomia. The most common etiology of pure alexia is a vascular lesion in the territory of the posterior cerebral artery.

Neurodegenerative Aphasia

The language network can also become the target of dementia-causing neurodegenerative diseases. Dementia is a generic term used to designate a neurodegenerative disease that impairs intellect and behavior to the point at which customary activities of daily living become compromised. Alzheimer's disease is the single most common cause of dementia. The neuropathology of Alzheimer's disease causes the earliest and most profound neuronal loss in memory-related parts of the brain such as the entorhinal cortex and the hippocampus. This is why progressive forgetfulness for recent events and experiences is the cardinal feature of Alzheimer's disease. In time, the neuronal pathology in Alzheimer's disease spreads to the language network, and a progressive aphasia becomes added to the progressive amnesia. There are other patterns of dementia, however, in which neurodegeneration initially targets the language rather than the memory network of the brain, leading to the emergence of a progressive aphasia that becomes the most prominent aspect of the clinical picture during the initial phases of the disease. Primary progressive aphasia (PPA) is the most widely recognized syndrome with this pattern of selective language impairment.

Clinical Presentation and Diagnosis

The patient with PPA comes to medical attention because of word-finding difficulties, abnormal speech patterns, and spelling errors of recent onset. PPA is diagnosed when other mental faculties such as memory for daily events, visuospatial skills (assessed by tests of drawing and face recognition), and comportment (assessed by history obtained from a third party) remain relatively intact, when language is the only major area of dysfunction for the first few years of the disease, and when structural brain imaging does not reveal a specific lesion, other than atrophy, to account for the language deficit. Impairments in other cognitive functions may also emerge, but the language dysfunction remains the most salient feature and deteriorates most rapidly throughout the illness.

Language in PPA

The language impairment in PPA varies from patient to patient. Some patients cannot find the right words to express thoughts; others cannot understand the meaning of heard or seen words; still others cannot name objects in the environment. The language impairment can be fluent (i.e., with normal articulation, flow, and number of words per utterance) or nonfluent. The single most common sign of PPA is anomia, manifested by an inability to come up with the right word during conversation and/or an inability to name objects shown by the examiner. Asking the patient to name geometric shapes, body parts, or components of common objects reveals early stages of anomia. Many patients remain in an anomic phase throughout most of the disease and experience a gradual intensification of word-finding deficits to the point of near mutism. Others, however, proceed to develop distinct forms of agrammatism and/or word comprehension deficits. The agrammatism consists of inappropriate word order and misuse of small grammatical words. One patient, for example, sent the following e-mail to her daughter: 'I will come my house in your car and drive my car into chicago . . . You will back get your car and my car park in my driveway. Love, Mom.' Comprehension deficits, if present, start with an occasional inability to understand single low-frequency words and gradually progress to encompass the comprehension of conversational speech.

The impairments of syntax, comprehension, naming, or writing in PPA are no different than those seen in aphasias of cerebrovascular causes. However, they form slightly different patterns. According to a classification proposed by Gorno-Tempini and colleagues, three variants of PPA can be recognized: an agrammatic variant characterized by poor fluency and impaired syntax, a semantic variant characterized by preserved fluency and syntax but poor single-word comprehension, and a logopenic variant characterized by preserved syntax and comprehension but frequent word-finding pauses during spontaneous speech. The agrammatic variant is also known as progressive nonfluent aphasia, and the semantic variant as semantic dementia. The three variants display overlapping distributions of neuronal loss, but the agrammatic variant is most closely associated with atrophy in the anterior parts of the language network (where Broca's area is located), the semantic variant with atrophy in the temporal components of the language network, and the logopenic variant with atrophy in the temporoparietal component of the language network.

Pathophysiology

Patients with PPA display progressive atrophy (indicative of neuronal loss), electroencephalographic slowing, decreased blood flow (measured by single photon emission computed tomography) and decreased glucose utilization (measured by positron emission tomography) that are most pronounced within the language network of the brain (**Figure 1**). The abnormalities may remain confined to left hemisphere perisylvian and anterior temporal cortices for many years. The clinical focality of PPA is thus matched by the anatomical selectivity of the underlying pathological process.

Neuropathology

Approximately 30% of patients have shown the microscopic pathology of Alzheimer's disease, presumably with an atypical distribution of lesions. In the majority of cases, the neuropathology falls within

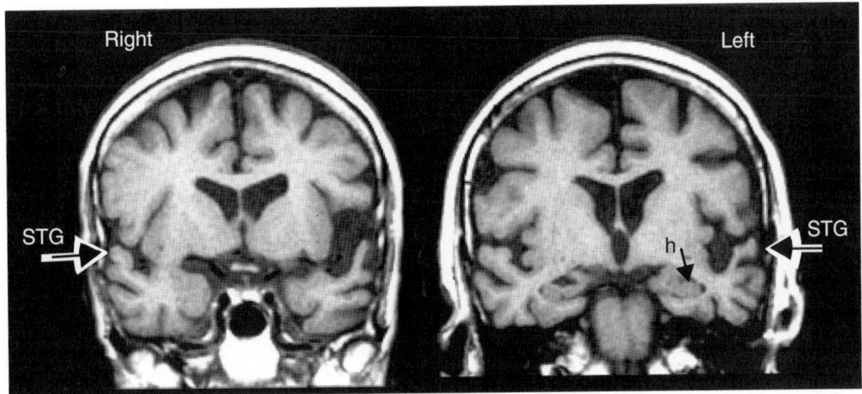

Figure 1 Coronal magnetic resonance image (MRI) of a primary progressive aphasia patient showing neuronal loss in the language network as indicated by atrophy of the left superior temporal gyrus (STG). The hippocampus (h) is intact.

the family of frontotemporal lobar degenerations and displays focal neuronal loss, gliosis, tau-positive inclusions, Pick bodies, and tau-negative ubiquitin inclusions. Apolipoprotein E and prion protein genotyping have shown significant differences between patients with typical clinical patterns of Alzheimer's disease and those with a diagnosis of PPA. The intriguing possibility has been raised that a personal or family history of dyslexia may be a risk factor for PPA, at least in some patients, suggesting that this disease may arise from a background of genetic or developmental vulnerability affecting language-related areas of the brain.

See also: Agraphia; Alexia; Alzheimer's Disease: An Overview; Brain Damage: Functional Reorganization; Brain Injury: Functional Recovery After; Brain Trauma; Cerebrovascular Disease; Dementia and Language; Language: Auditory Processes; Language: Aphasia; Lexical Impairments Following Brain Injury; Sentence Production; Sentence Comprehension; Speech Perception: Adult; Stroke; Word Production; Word Recognition; Word Learning.

Further Reading

Chawluk JB, Mesulam MM, Hurtig H, et al. (1986) Slowly progressive aphasia without generalized dementia: Studies with positron emission tomography. *Annals of Neurology* 19: 68–74.

Gorno-Tempini ML, Dronkers NF, Rankin KP, et al. (2004) Cognition and anatomy in three variants of primary progressive aphasia. *Annals of Neurology* 55: 335–346.

Kertesz A, Hudson L, Mackenzie IRA, and Munoz DG (1994) The pathology and nosology of primary progressive aphasia. *Neurology* 44: 2065–2072.

Knibb JA, Xuereb JH, Patterson K, and Hodges JR (2006) Clinical and pathological characterization of progressive aphasia. *Annals of Neurology* 59: 156–165.

Li X, Rowland LP, Mitsumoto H, et al. (2005) Prion protein codon 129 genotype is altered in primary progressive aphasia. *Annals of Neurology* 58: 858–864.

Mesulam M-M (2001) Primary progressive aphasia. *Annals of Neurology* 49: 425–432.

Mesulam M-M, Johnson N, Grujic Z, and Weintraub S (1997) Apolipoprotein E genotypes in primary progressive aphasia. *Neurology* 49: 51–55.

Mesulam M-M, Johnson N, Krefft TA, et al. (2007) Progranulin mutations in primary progressive aphasia – The PPA1 and PPA3 families. *Archives of Neurology* 64: 43–47.

Rogalski E and Mesulam M-M (2007) An update on primary progressive aphasia. *Current Neurology and Neuroscience Reports* 7: 388–392.

Sonty SP, Mesulam M-M, Thompson CK, et al. (2003) Primary progressive aphasia: PPA and the language network. *Annals of Neurology* 53: 35–49.

Weintraub S, Rubin NP, and Mesulam MM (1990) Primary progressive aphasia. Longitudinal course, neuropsychological profile, and language features. *Archives of Neurology* 47: 1329–1335.

Apoptosis in Nervous System Injury

B A Miller, J C Bresnahan, and M S Beattie,
University of California at San Francisco, San Francisco, CA, USA

CNS Injury – Overview

The central nervous system (CNS) is generally considered to be refractory to repair and regeneration; although recently it has been realized that neurogenesis does occur in the adult CNS. Nevertheless, the loss of neurons and glial cells after CNS injury results in lifelong deficits. The most recognizable forms of CNS injury include traumatic brain injury (TBI) and spinal cord injury (SCI; **Figure 1(a)**), but CNS injury also results from ischemic stroke, hemorrhage, infection, and other conditions that cause neural cell death. Since there is little replacement of neurons after damage to the CNS, protection is critical, as evidenced by the evolution of the human skull and vertebral column. It is important to note that in addition to neurons, glial cells (astrocytes, oligodendrocytes, and microglia), along with vascular cells and progenitor cells, are all lost in CNS injury. The final neurological outcome of an individual with CNS injury will depend upon the summation of injury-induced degeneration and repair processes. Therefore, it is important to understand how cells die after CNS injury, which is by a combination of the two principal forms of cell death, necrosis and apoptosis.

Apoptosis – Introduction

Apoptosis, or programmed cell death, occurs naturally during development, and has recently gained attention as an important factor in central nervous system disease and injury. Apoptotic cell death is differentiated from passive, necrotic cell death by both morphological and biochemical features. Apoptotic cells shrink and undergo cellular blebbing, forming small membrane-bound debris which is phagocytosed by immune cells. In contrast, necrotic cells swell and lyse, spilling their contents into the extracellular space. A classic feature of apoptotic cell death is chromatin condensation and fragmentation, as DNA is broken up into smaller fragments by enzymes that are activated as part of the apoptotic program. This DNA fragmentation produces a characteristic pattern that can be visualized by gel electrophoresis, referred to as DNA laddering. A commonly used assay to detect apoptotic DNA fragmentation is terminal deoxynucleotidyl transferase dUTP nick-end labeling, commonly referred to as TUNEL. This assay labels 3′-OH DNA ends that are produced during apoptotic DNA cleavage. Nuclear features of apoptosis can also be seen histologically as fragmented or shrunken nuclei when cells are labeled with DNA-binding dyes such as Hoechst.

Apoptosis occurs normally during embryogenesis. It is interesting to note that this programmed cell death occurs coincidently with cell proliferation and maturation during development, and apoptosis and regeneration also occur simultaneously after CNS injury. Many essential developmental events outside the nervous system take place via apoptosis, such as the elimination of self-recognizing B and T cells and the loss of tissue to create independent digits. Apoptosis also takes place throughout the developing CNS where many different cell types undergo apoptosis, such as Purkinje cells of the cerebellum, retinal ganglion cells, and oligodendrocytes in the optic nerve, cortex, and spinal cord. Apoptosis during CNS development may take place in order to control cell number and regulate the size of CNS tissues.

The biochemical definition of apoptosis is based on the induction of specific intracellular signaling cascades that culminate in enzymatic activation leading to cell death. The enzymes most widely linked to apoptotic cell death are caspase enzymes. Caspase enzymes are cysteine proteases that are present in cells as zymogens until activated. Caspase enzymes are homologs of *ced* (*Caenorhabditis elegans* death) family genes that were originally described in the nematode *C. elegans*, in which they mediate programmed cell death. Once activated, caspases cleave both nuclear and cytoskeletal proteins. Other caspase substrates include antiapoptotic proteins and other regulatory proteins that influence cell survival. Many different events can trigger caspase activation, and different caspase enzyme subtypes can activate each other or undergo autocatalysis. Caspase activation can be detected with antibodies specific to the activated form of caspase enzymes in Western blots and isolated cell or tissue samples. Specific caspase inhibitors can also be used to block apoptotic pathways in experimental preparations. Calpains comprise another set of proteolytic enzymes that play a role in apoptosis. Calpains are calcium-dependent enzymes that can be activated in CNS disease and trauma and are also thought to contribute to apoptotic cell death.

Despite widely accepted morphological and molecular markers of apoptosis, such as DNA fragmentation and caspase activation, the distinction between apoptosis and necrosis is not always clear. It is likely that apoptosis and necrosis exist along a continuum where

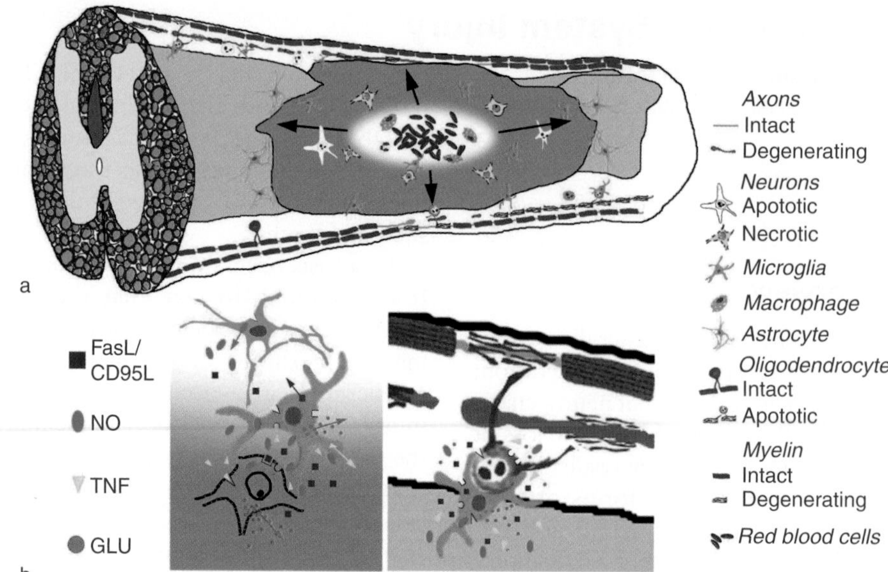

Figure 1 Apoptosis plays a role in the pathogenesis of spinal cord injury, which provides an example of the complex dynamics of central nervous system injuries. (a) Immediately after cord injury, cells are mechanically disrupted and die at the center of the injury site (white area), where petechial hemorrhages form, causing vascular elements to be released into the tissue. Over the next few hours to days, the lesion center expands (dark gray area) and additional tissue is lost via both necrosis and apoptosis. Inflammatory cells, including both microglia and macrophages, are recruited to remove cellular debris. Subsequently, the lesion cavity becomes walled off by reactive astrocytes, which contribute to the chronic scar that forms after spinal cord injury. Distant from the lesion, microglial activation is observed in the axonal tracts undergoing Wallerian degeneration. Both oligodendrocytes and microglia are observed to undergo apoptosis in these regions, as well as in the thin rim of spared white matter around the lesion center. The loss of oligodendrocytes in these areas possibly causes demyelination of surviving axons and contributes to functional loss after injury. (b, c) Examples of secreted molecules – FasL/CD95L, nitric oxide (NO), tumor necrosis factor (TNF), glutamate (GLU) – and their target cells that are thought to play a role in cell death after injury. Activated microglia are observed to be in close apposition to dying neurons (b) and apoptotic oligodendrocytes (c), suggesting an active role for microglia in inducing cell death.

the same initial insult could cause either apoptosis or necrosis, depending on the strength of the stimulus. It is likely that more severe injury results in necrosis, as cells are unable to execute coordinated apoptotic signaling cascades. It is also possible that necrotic cells can initiate apoptotic signaling cascades even as they die by necrosis; thus biochemical events of apoptosis could occur within a cell that ultimately dies via necrosis.

Molecular Mechanisms of Apoptosis in CNS Injury

Developmental apoptosis is likely a result of competition between individual cells for prosurvival signals that are in limited supply, whereas apoptosis after CNS injury is thought to be predominantly due to proapoptotic signals released after injury. Since apoptosis occurs over time, and is meditated by specific signaling cascades, apoptotic cell death could be a viable target for therapies to reduce pathological neuronal and glial loss in CNS injury and disease. Several common mediators of cell death are present in both CNS injury and disease. Among the best studied of these are excitotoxicity, inflammation, and free radicals. Excitotoxicity classically refers to the overactivation of excitatory ionotropic glutamate receptors on neurons, leading to excess intracellular calcium accumulation culminating in cell death. Though oligodendrocytes do not conduct action potentials, oligodendrocytes do express glutamate receptors and are also vulnerable to excitotoxicity. Both neurons and oligodendrocytes have been shown to undergo apoptotic cell death as a result of excitotoxicity, but excitotoxic cell death can be necrotic as well. A key event in excitotoxic cell death, either apoptotic or necrotic, is an increase in intracellular Ca^{2+}. Ca^{2+} that enters cells through ionotropic glutamate receptors can activate nitric oxide synthase enzymes that lead to intracellular free radical production. Additionally, excess intracellular Ca^{2+} can accumulate in mitochondria and cause increased production of the free radical superoxide, which can then rapidly combine with nitric oxide to form the highly reactive free radical peroxynitrite. Intracellular free radicals have been shown to activate apoptotic signaling cascades and can also directly damage cellular contents. While intracellular free radical formation has been linked to apoptosis, free radical-mediated cell death can be necrotic as well. The net concentration or specific species of intracellular free radicals produced may determine whether a cell dies via apoptosis or necrosis.

One way intracellular free radicals can induce apoptosis is through their effects on mitochondria. Free radicals have been linked to the release of cytochrome *c* from the mitochondrial intermembrane space, which initiates what is referred to as the intrinsic pathway of apoptosis. Cytochrome *c* released into the cytosol can combine with apoptosis-inducing factor-1, which, in the presence of adenosine triphosphate or deoxyadenosine triphosphate, can lead to the activation of caspase-9. Caspase-9 can subsequently activate caspase-3, which is considered to be one of the final effectors of apoptosis.

Inflammation has also been linked to apoptosis in neurological diseases and injury. CNS inflammation can be induced by microglia within the CNS or by cells from the peripheral immune system. Blood–brain barrier (BBB) breakdown occurs after stroke and CNS trauma and could provide a route for peripheral immune cells to access the CNS. Both microglia and peripheral immune cells can produce tumor necrosis factor-α (TNF-α) and other proinflammatory cytokines which can induce apoptosis in both neurons and oligodendrocytes. The pathway by which TNF-α induces apoptosis is known as the extrinsic apoptotic pathway. This pathway is initiated by TNF-α binding to receptors on the cell surface, which causes the receptors to trimerize. Once TNF-α receptors have trimerized they activate caspase-8 via adapter molecules that bind to the intracellular domain of the TNF-α receptors. Caspase-8 can subsequently activate caspase-3, leading to a point of convergence for the extrinsic and intrinsic apoptotic pathways. TNF-α can also induce free radical production, which can lead to apoptotic cell death. Since excitotoxicity and inflammation are present simultaneously in CNS disease and trauma, free radical injury could become a central mediator of cell death in numerous disease states. Although TNF-α has been implicated in apoptosis of both neurons and oligodendrocytes, some studies have shown TNF-α to have a beneficial role in CNS injury and repair. This is just one example of how factors that may induce apoptosis after CNS injury may also have beneficial effects. Further research into the complicated signaling cascades that are activated by TNF-α and other factors present after CNS injury will help to better separate the beneficial and detrimental roles of these molecules.

Another pathway that may induce apoptosis in multiple diseases and trauma is the p75 receptor signaling pathway. p75 is a low-affinity neurotrophin receptor that is part of the TNF receptor superfamily. p75 can bind to several different neurotrophins and increases the response of the neuronal TrkA receptor to nerve growth factor (NGF), which promotes neuronal growth. However, in the absence of the TrkA receptor, p75 activation via NGF or pro-NGF leads to apoptosis. Injuries to the CNS increase p75 levels, and cells undergoing apoptosis after facial nerve axotomy and SCI upregulate expression of p75. Activation of the p75 receptor by neurotrophins that are produced after injury may be another mechanism which induces apoptosis after CNS trauma.

When considering the multitude of factors that may induce apoptosis after CNS injury, it is important to realize that these factors may be activated concurrently or sequentially following one particular injury. There are many points at which these signaling pathways may converge, such as one signaling pathway amplifying the expression of receptors for another, or two or more separate pathways inducing activation of the same caspase pathways. This mechanistic overlap makes it likely that numerous antiapoptotic treatments may be needed to reduce cell death after CNS injury. Furthermore, the mechanisms of apoptosis described here are by no means the only inducers of apoptosis after CNS injury, as molecules such as interferon-γ and Fas/CD95 and its ligand have also been linked to apoptosis after CNS injury. In the future, a more thorough understanding of the extracellular and intracellular signaling events that lead to apoptosis may lead to better treatments for CNS injury.

Neuronal Apoptosis in CNS Injury

Neuronal apoptosis has been described in several CNS disorders and trauma. Apoptotic neuronal death is seen in both human amyotrophic lateral sclerosis (ALS) and animal models of ALS. Apoptotic neurons have also been detected in regions of cell death in Huntington's disease, Alzheimer's disease, Parkinson's disease, and human immunodeficiency virus (HIV)-induced encephalitis. Since these diseases are chronic and may be due to genetic mutations, it would be expected that programmed apoptotic cell death plays a role in neuronal loss in these conditions. However, apoptosis has also been shown to play a role in acute neurological injuries such as stroke, TBI, and SCI. After ischemic stroke, cells are quickly lost in the core of the lesion, due to energy deprivation. It is thought that this cell loss is primarily necrotic. Surrounding the necrotic core of a stroke lesion, there is an area referred to as the penumbra, in which cells are lost over time. Neuronal death in the penumbra is thought to be apoptotic, and may be more amenable to therapeutic intervention than is acute neuronal loss in the ischemic core.

One of the factors that may lead to neuronal apoptosis in stroke is reperfusion injury. The return of blood flow to an area affected by stroke can paradoxically lead to increased damage. One mechanism behind reperfusion injury is thought to be intracellular free radical generation due to the sudden availability of oxygen to tissue previously deprived of oxygen and glucose. Neurons in a previously hypoxic area may not be able to deal with a sudden increase in oxygen availability and therefore be paradoxically injured when blood flow returns to an ischemic area. The return of blood flow to a previously ischemic area allows for the execution of energy-dependent signaling cascades that mediate apoptosis. Additionally, reperfusion into a damaged area can bring peripheral immune cells into the area, particularly when the BBB has been damaged. This combination of increased free radical production, initiation of apoptotic signaling pathways, and immune infiltration can lead to apoptosis.

Apoptosis is present in animal models of TBI and SCI and is thought to contribute to neuronal loss in human traumatic injury. Apoptotic neurons have been detected in the cortex and hippocampus following experimental TBI. The loss of even a small percentage of neurons in the hippocampus could have a significant effect on quality of life after TBI, when memory impairment is often a result of injury. After SCI, even a small amount of motor neuron loss in critical areas of the cord can lead to functional deficits such as breathing difficulties or loss of hand function. In experimental models of CNS injury, some treatments that reduce apoptosis have been shown to result in better outcomes.

Oligodendrocyte Apoptosis in CNS Injury

In addition to neurons, oligodendrocyte progenitor cells (OPCs) and mature oligodendrocytes undergo apoptosis in CNS trauma and disease. In periventricular leukomalacia (PVL), OPCs are lost as a consequence of hypoxia, ischemia, or intrauterine infection. This loss of OPCs, with the resulting deficiency of mature oligodendrocytes, is thought to be one of the causes of spastic cerebral palsy. Examination of brains from preterm infants has shown apoptosis associated with white-matter lesions and OPC loss. Experimental models of PVL also show OPCs undergoing apoptosis. Mature oligodendrocytes die in multiple sclerosis (MS), and examination of brains of MS patients has revealed apoptotic oligodendrocytes. Experimental models of MS also support oligodendrocyte apoptosis as a feature of this disease. Numerous *in vitro* studies designed to model PVL and MS have shown TNF-α, free radicals, and glutamate receptor activation to induce OPC and oligodendrocyte apoptosis.

Oligodendrocytes also die after CNS trauma such as TBI and SCI. In SCI, it has been shown that oligodendrocytes undergo apoptosis for several weeks after the initial injury. This delayed death of oligodendrocytes is thought to reduce the effectiveness of neural conduction in the spared rim of white matter that often exists in injured spinal cords, and to impair optimal function of neural pathways that may have survived the initial injury. Though many studies have demonstrated delayed oligodendrocyte apoptosis after SCI, the events that cause this remain unknown. It has been shown that axonal transection can lead to oligodendrocyte cell death in the optic nerve, and oligodendrocyte survival *in vitro* can be enhanced by the presence of specific growth factors and hormones. Therefore, it is possible that oligodendrocyte apoptosis in SCI is due to a loss of trophic support from axons as axons degenerate after injury, similar to developmental apoptosis that occurs due to limited availability of growth factors.

An alternate hypothesis is that oligodendrocyte apoptosis after SCI is a result of the presence of substances that are released after injury, rather than a decrease in the amount of survival factors present. After SCI, elevated levels of TNF-α and glutamate are present in the spinal cord. Both of these molecules have been shown to induce oligodendrocyte cell death *in vitro* (**Figure 2**). *In vivo,* glutamate antagonists have been shown to protect oligodendrocytes after SCI. Though TNF-α is likely to contribute to oligodendrocyte apoptosis after SCI, some studies have shown that knocking out TNF-α or TNF-α receptors results in worse functional outcome and less regeneration after SCI. These findings emphasize that signals that promote apoptosis in some settings may have beneficial roles in other settings or at other times postinjury. One possible source for both glutamate and TNF-α after SCI could be microglia, the resident immune cells of the CNS.

Role of Microglia in Neuronal and Oligodendrocyte Apoptosis

Microglia are known to react quickly in response to CNS injury or disease, proliferating and migrating into an injury site. Microglia secrete a wide array of molecules that have been shown to be toxic to neurons and oligodendrocytes, including glutamate, free radicals, and TNF-α. Microglial activation has been

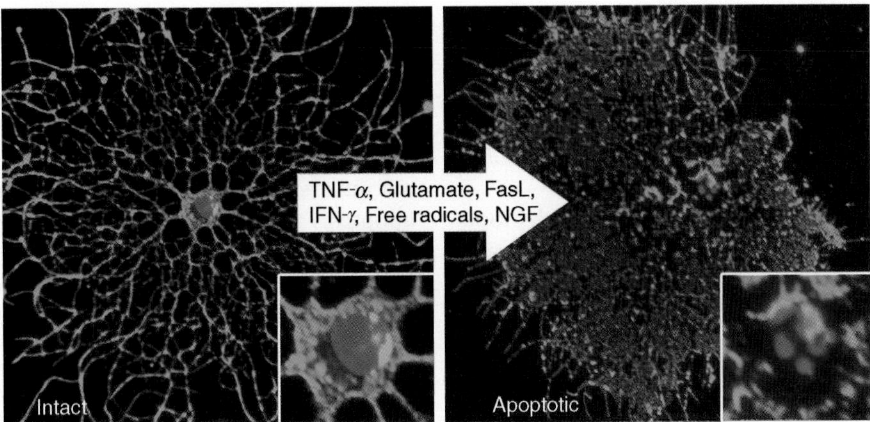

Figure 2 These oligodendrocytes, grown *in vitro*, have been labeled with antibodies to the oligodendrocyte marker galactocerebroside (green) and with Hoechst, which labels DNA and cell nuclei (blue). The nuclei of these cells show the difference between a cell with an intact nucleus (left panel, inset) and a cell with a fragmented apoptotic nucleus (right panel, inset). Some of the factors that have been shown to induce oligodendrocyte apoptosis are identified within the arrow. TNF-α, tumor necrosis factor-α; IFN-γ, interferon-γ; NGF, nerve growth factor.

linked to neuronal apoptosis in several diseases and TBI. Minocycline, a drug that decreases microglial activation, has been shown to reduce neuronal apoptosis in animal models of brain injury. Additionally, treatments antagonizing TNF-α and free radicals, both of which can be produced by microglia, have been shown to improve outcomes in experimental TBI.

It has been shown that activated microglia contribute to oligodendrocyte death in models of MS and PVL. Microglia-induced oligodendrocyte damage may be further exacerbated by the fact that many factors that lead to oligodendrocyte death also lead to microglial activation, such as glutamate and proinflammatory cytokines. Microglia may exert their effects on oligodendrocytes and OPCs both at a distance and via direct cell–cell interaction. *In vitro*, microglia have been shown to be capable of inducing OPC death without being in direct contact with OPCs. However, *in vivo*, microglia have been observed in close proximity to dying oligodendrocytes after SCI (**Figure 1(c)**). This proximity of microglia to oligodendrocytes after CNS injury may increase the influence of microglia on oligodendrocyte and OPC survival, as it has been shown *in vitro* that physical contact between microglia and oligodendrocytes is a key factor in mediating oligodendrocyte death via microglial TNF-α. While it is possible that microglia induce oligodendrocyte apoptosis by TNF-α, glutamate, or free radical generation, it is also possible that activated microglia near apoptotic oligodendrocytes are only removing the debris left from oligodendrocyte death, similar to apoptosis during development. The relationship between apoptotic oligodendrocytes and activated microglia remains under investigation.

Microglia are also observed to undergo apoptosis in neurological injury. Microglial apoptosis occurs after SCI, peripheral nerve injury, and *in vitro* as a result of activation. This may be caused by excessive free radical production by activated microglia, which could lead to self-induced apoptosis. Microglial apoptosis after CNS injury may be a beneficial self-regulatory process that reduces the number of microglia and the amount of damage microglia do to other CNS cells. On the other hand, microglia have also been suggested to play a beneficial role after CNS injury, and microglial apoptosis may have a detrimental effect on recovery.

Role of Astrocytes in Apoptosis after CNS Injury

Unlike neurons, oligodendrocytes, and microglia, astrocytes are rarely observed to undergo apoptosis after CNS injury. However, astrocytes are thought to play an important role in determining if other cells undergo apoptosis after injury. Astrocytes are connected by gap junctions to form a syncitium that can allow large numbers of astrocytes to react quickly and in unison to signals within the CNS. Under normal conditions, astrocytes protect other cells in the CNS by scavenging excess glutamate, free radicals, and potassium from the extracellular environment. Though not classically considered immune cells, astrocytes can exhibit many of the functions of immune cells after CNS injury and may contribute to the apoptosis of other cell populations. Astrocytes react to CNS injury by becoming activated, much like microglia – altering their cytoskeletal structure and secreting cytokines and free radicals (**Figure 1(b)**). Astroglial cytokines and free radicals can activate apoptotic pathways in neurons and oligodendrocytes and also induce microglial activation. Astrocytes also contribute to the formation of the glial scar, which persists chronically after CNS injury and is thought to impair regeneration.

Astrocytes are found in close association with blood vessels in the CNS, and along with endothelial cells constitute the BBB that restricts access of peripheral blood cells and cytokines to the CNS. BBB breakdown occurs in multiple sclerosis, stroke, TBI, and SCI. A compromised BBB allows peripheral immune cells into the injured CNS, where inflammation is thought to induce apoptosis of neuronal and glial cells but may also contribute to repair after injury. Astrocytes are considered more resilient than are neurons and oligodendrocytes in the setting of CNS trauma, and astrocyte apoptosis has not been studied to the same extent as has neuronal or oligodendrocyte apoptosis, although astrocytes do die after CNS injury.

Summary

Apoptosis, or programmed cell death, is morphologically and biochemically distinct from necrosis and is carried out by several intracellular signaling cascades. Neuronal and glial apoptosis processes are thought to play a role in many different neurological diseases and injury. Several molecules may mediate apoptosis in the CNS, including cytokines, neurotrophins, glutamate, and free radicals. The inflammatory response in CNS injury and disease may contribute to apoptosis but may also be important for tissue repair. Since apoptosis occurs over time, and is mediated by distinct intracellular signaling events, apoptotic cell death may be a viable target for therapies designed to improve outcomes after CNS injury. Antiapoptotic therapies have shown promise for the treatment for CNS injury. Treatments such as minocycline, mitochondrial permeability inhibitors, and caspase inhibitors have shown beneficial effects on cell survival and functional outcome in models of stroke, SCI, and excitotoxicity. The only currently approved clinical treatment for spinal cord injury in the United States is methylprednisolone, which has shown to have anti-apoptotic effects, although there is controversy surrounding its use. There are currently no clinically approved antiapoptotic treatments for TBI or stroke, and much research continues to focus on developing clinically efficacious treatments for these CNS injuries.

Although much effort has been directed at reducing apoptosis after CNS injury, apoptosis after neurotrauma may have a beneficial role as well, by removing dysfunctional cells and making way for the repair and regeneration that can occur after injury. Apoptosis may also play a role in regulating the immune response to injury and preventing excess inflammation. Although apoptosis has only recently been identified as an important event after CNS injury, our understanding of the mechanisms and mediators of apoptosis is expanding rapidly. The development of therapies that reduce neuronal and glial apoptosis may lead to more effective treatment and recovery after CNS injury.

See also: AMPA Receptors: Disease; Apoptosis in Neurodegenerative Disease; Astrocyte: Response to Injury; Axonal Injury in Demyelinating Disease and CNS Injury; Blood-Brain Barrier and Neurovascular Mechanisms of Neurodegeneration and Injury; Brain Injury: Functional Recovery After; Brain Trauma; Excitotoxicity in Neurodegenerative Disease; Glial Responses to Injury; Inflammation in Neurodegenerative Disease and Injury; Microglial Response to Injury; Mitochondrial Dysfunction in Nervous System Injury; Neurogenesis and Neural Precursors, Progenitors, and Stem Cells in the Adult Brain; Sensorimotor Plasticity and Control of Movement Following Spinal Cord Injury +D18; Spinal Cord Regeneration and Functional Recovery: Strategies; Spinal Cord Injuries; Stroke: Injury Mechanisms.

Further Reading

Baptiste DC and Fehlings MG (2006) Pharmacological approaches to repair the injured spinal cord. *Journal of Neurotrauma* 23: 318–334.

Barres BA, Jacobson MD, Schmid R, et al. (1993) Does oligodendrocyte survival depend on axons? *Current Biology* 3: 489–497.

Beattie MS, Hermann GE, Rogers RC, et al. (2002) Cell death in models of spinal cord injury. *Progress in Brain Research* 137: 37–47.

Beattie MS (2004) Inflammation and apoptosis: Linked therapeutic targets in spinal cord injury. *Trends in Molecular Medicine* 10: 580–583.

Casaccia-Bonnefil P (2000) Cell death in the oligodendrocyte lineage: A molecular perspective of life/death decisions in development and disease. *Glia* 29: 124–135.

Casha S, Yu WR, and Fehlings MG (2001) Oligodendroglial apoptosis occurs along degenerating axons and is associated with FAS and p75 expression following spinal cord injury in the rat. *Neuroscience* 103: 203–218.

Crowe MJ, Bresnahan JC, Shuman SL, et al. (1997) Apoptosis and delayed degeneration after spinal cord injury in rats and monkeys. *Nature Medicine* 3: 73–76.

Friedlander RM (2003) Apoptosis and caspases in neurodegenerative diseases. *New England Journal of Medicine* 348: 1365–1375.

Park E, Velumian AA, and Fehlings MG (2004) The role of excitotoxicity in secondary mechanisms of spinal cord injury: A review with an emphasis on the implications for white matter degeneration. *Journal of Neurotrauma* 21: 754–774.

Selmaj KW and Raine CS (1988) Tumor necrosis factor mediates myelin and oligodendrocyte damage *in vitro*. *Annals of Neurology* 23: 339–346.

Shuman SL, Bresnahan JC, and Beattie MS (1997) Apoptosis of microglia and oligodendrocytes after spinal cord contusion in rats. *Journal of Neuroscience Research* 50: 798–808.

Snider BJ, Gottron FJ, and Choi DW (1999) Apoptosis and necrosis in cerebrovascular disease. *Annals of the New York Academy of Sciences* 893: 243–253.

Sugawara T, Fujimura M, Noshita N, et al. (2004) Neuronal death/survival signaling pathways in cerebral ischemia. *NeuroRx* 1: 17–25.

Zajicek JP, Wing M, Scolding NJ, et al. (1992) Interactions between oligodendrocytes and microglia. A major role for complement and tumour necrosis factor in oligodendrocyte adherence and killing. *Brain* 115: 1611–1631.

Relevant Websites

http://www.nih.gov – National Institutes of Health Special Interest Groups (Apoptosis Interest Group).

http://www.edc.pitt.edu/neurotrauma – National Neurotrauma Society (Gainesville, FL, USA).

http://www.sfn.org – Society for Neuroscience, briefing on cell suicide.

Apoptosis in Neurodegenerative Disease

K A Roth and J J Shacka, University of Alabama at Birmingham, Birmingham, AL, USA

Apoptosis

In 1972, Kerr and colleagues coined the term apoptosis, which is Greek for 'falling off,' as in petals from a flower or leaves from a tree. This term was used to define a type of cell death characterized morphologically by the discrete budding of self-contained fragments from the cell membrane following a death stimulus and was distinct from necrotic (passive or accidental) cell death. Apoptosis, known also as type I cell death, is now defined by distinct morphological criteria, including pyknosis (condensation or reduction in size) of the cell and/or cell nucleus along with nuclear fragmentation and hyperchromatosis. The DNA of apoptotic nuclei is typically digested into 200–300 bp fragments which can be observed by enzymatic labeling of nicked DNA or by electrophoretic detection of DNA laddering. The plasma membrane of apoptotic cells remains relatively intact, with the dying cell budding off discrete packages of itself into 'apoptotic bodies' that are ultimately phagocytosed. This efficient and discrete phagocytosis in the absence of a ruptured plasma membrane generally limits the inflammatory response around an apoptotic cell and is unlike necrosis, which is characterized by a rupture of the plasma membrane and subsequent leakage of cell contents into the surrounding extracellular space with a resultant inflammatory response. In addition, unlike necrosis, apoptosis is an energy-dependent process and often requires new gene transcription and protein translation for cellular execution.

Molecular Regulation of Apoptosis

Many seminal discoveries have been made in the past quarter century that have enhanced our understanding of cell death. Using the nematode *Caenorhabditis elegans* as a model, Ellis and Horvitz showed for the first time in the 1980s that apoptotic cell death was regulated by distinct genes. Subsequent cloning studies demonstrated that mammals also express homologs of these genes with similar function, although it has been found that the regulation of mammalian cell death is more complex than in the nematode. In 1984, a novel oncogene, B-cell lymphoma-2 (Bcl-2), was isolated and was found to have prosurvival or antiapoptotic functions. Mammalian cell apoptosis can be classified into distinct pathways, extrinsic (death-receptor mediated) or intrinsic (mitochondrial), and involves the participation of an extensive family of Bcl-2 family proteins (described below) and the activation of caspases (see **Figure 1**).

Caspases

Caspases are cysteine-containing proteases that cleave substrates after aspartic acid residues and critically regulate both the transduction of apoptotic stimuli and the final execution of apoptotic death. There are 14 mammalian and 12 human caspases that can be divided into two basic groups, the interleukin-1β converting enzyme (ICE)-like and the CED-3-like (named after the *C. elegans* homolog) families. The ICE-like caspases, which include caspase-1, caspase-4, caspase-5, and caspase-12, are involved in the proteolytic processing of cytokines and as such are related indirectly to neuron cell death as a function of modulating inflammatory responses. The CED-3-like caspases are directly involved in the apoptotic signaling cascade and can be further divided into two subgroups, the initiator caspases (caspase-2, caspase-8, caspase-9, and caspase-10) and the effector caspases (caspase-3, caspase-6, and caspase-7). Both initiator and effector caspases exist as inactive zymogens within the cell prior to receipt of a death stimulus. Initiator caspases are activated early in the apoptosis cascade by conformational change resulting from interaction with other apoptosis-associated proteins. Effector caspases are activated by the catalyzed cleavage from 'activated' initiator caspases. Both initiator and effector caspases have been shown to be proteolytically cleaved, but initiator caspases do not typically require cleavage for their activation. It is the selective cleavage of substrates by caspases which induces in large part the appearance of the morphological features of apoptosis.

Bcl-2 Family

There are three major subgroups of proteins in the Bcl-2 family that are classified by composition of their Bcl-2 homology (BH) domains and by function (anti- vs. proapoptotic). It is widely believed that the relative 'ratio' of antiapoptotic versus proapoptotic Bcl-2 family members influences the response of neurons to apoptotic stress. Members of the antiapoptotic subgroup include Bcl-2, Bcl-X_L, Bcl-w, and Mcl-1 and possess four BH domains, except for Mcl-1, which lacks the BH4 domain. The importance of the BH1 and BH4 domains in the maintenance of antiapoptotic activity of Bcl-2 and Bcl-X_L has been emphasized by molecular studies, where the targeted

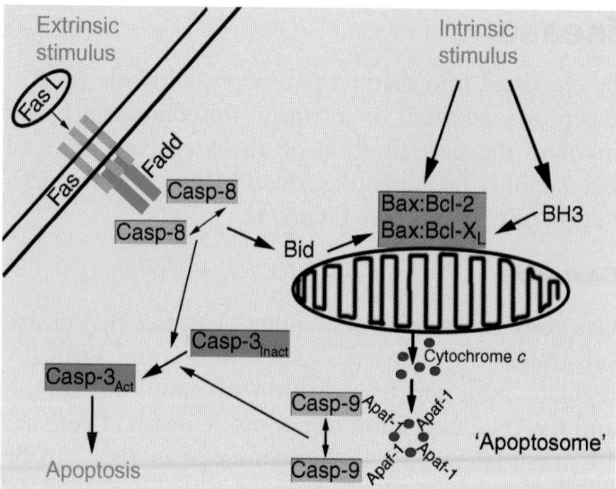

Figure 1 Molecular regulation of cell death in mammals by the extrinsic (death receptor-mediated) versus intrinsic (mitochondrial) apoptotic pathways. Extracellular death receptor ligands (e.g., Fas ligand) bind to and activate their receptors (e.g., Fas), which in turn induces recruitment of adaptor proteins (Fadd) to the complex with subsequent recruitment, dimerization, and activation of caspase-8, an initiator caspase. Activation of caspase-8 in turn can directly cleave and activate caspase-3 resulting in apoptotic death. In addition, caspase-8 can cleave Bid, a proapoptotic, BH3-domain-only member of the Bcl-2 family, to truncated Bid which interacts with Bax at mitochondria and causes the subsequent activation of the intrinsic apoptotic pathway. Intrinsic death stimuli (e.g., DNA damage) can cause changes in expression and/or function of anti- and/or proapoptotic members of the Bcl-2 family. The ratio of anti- to proapoptotic Bcl-2 family members is thought to regulate the release of cytochrome *c* from mitochondria, which in turn binds with Apaf-1 and caspase-9 in a protein complex termed the 'apoptosome.' The binding of caspase-9 in the apoptosome induces its forced dimerization and subsequent activation, which in turn induces activation of caspase-3. The activation of caspase-3 and other effector caspases is responsible for the cleavage of downstream substrates that promotes in part the appearance of apoptotic morphology and ultimate cell death.

deletion of these regions has been shown to abrogate their antiapoptotic activity. Bcl-2 is expressed at high levels during nervous system development and remains high in the adult spinal cord and in sympathetic neurons, but decreases in relative expression in the adult brain. Upregulation of Bcl-2 has been documented in many neurodegenerative diseases, occurring possibly as an antiapoptotic neuron survival response or in reactive glial cell elements. Bcl-X$_L$ and Bcl-w are expressed at high levels in adult central nervous system (CNS) and have also been shown to be altered in neurodegenerative disease; in contrast, Mcl-1 is thought to play a more focused role in CNS development.

The proapoptotic, multidomain subgroup consists of Bax, Bcl-2-homologous antagonist/killer (Bak), Bok, and Bcl-X$_S$. These proteins share three homologous BH domains (BH1–3) except for Bcl-X$_S$, which lacks BH1 and BH2 domains but rather possesses a BH4 domain. Both Bax and Bak regulate the release of cytochrome *c* from mitochondria, a critical event in activation of the intrinsic apoptotic pathway. Upregulation of Bax has been documented in many neurodegenerative diseases. In contrast, while present in the CNS and similar in function to Bax, studies have indicated a limited role for Bak in promoting apoptotic neuron death. Bcl-X$_S$ is expressed at low levels in the mammalian nervous system and a regulatory role for Bok in CNS apoptotic death is yet to be described. The proapoptotic BH3 domain-only subgroup includes Bcl-2-associated death protein (Bad), Bid, Bcl-2 interacting mediator of cell death (Bim), Bcl-2 19 kDa interacting protein (BNip), death protein 5/harakiri (DP5/Hrk), Noxa, and p53-upregulated modulator of apoptosis (Puma). Although members of this subgroup are comparably the most diverse, their regulation of cell death depends critically upon the BH3 domain and the presence of Bax and/or Bak. The response of BH3 domain-only proteins to an apoptotic stress is regulated in part by distinct protein modifications, such as phosphorylation (Bim), dephosphorylation (Bad), induction of expression (Bim, DP5, Noxa, Puma), or cleavage (Bid). Altered regulation of BH3 domain-only proteins has been implicated in many neurodegenerative diseases, which are discussed below.

Extrinsic versus Intrinsic Death Pathway

There are two distinct pathways of caspase-dependent cell death, the extrinsic or death receptor-mediated pathway and the intrinsic or mitochondrial-mediated pathway. The extrinsic pathway involves binding of a death receptor ligand (e.g., Fas ligand or TNF-α) to a cell surface death receptor, which in turn stimulates receptor dimerization and the association of adaptor proteins to a newly formed complex that facilitates the association-induced activation of caspase-8. In the intrinsic pathway, a death stimulus regulates in part the levels and/or function of pro- versus antiapoptotic Bcl-2 family members, which induces the mitochondrial release of cytochrome *c* into the cytosol and in turn facilitates the energy-dependent activation of caspase-9 through formation of the 'apoptosome' complex with the protein Apaf-1. Both the intrinsic and extrinsic pathways converge at the level of effector caspase activation. Interestingly, cross-talk between these two pathways occurs via the cleavage of Bid by activated caspase-8 and the translocation of truncated Bid to the mitochondria, where it associates with the proapoptotic protein Bax to induce the release of cytochrome *c* and resultant activation of the intrinsic pathway.

Apoptosis and Neurodegeneration

A variety of evidence for the occurrence of apoptotic death in human neurodegenerative disease has been cited to support the hypothesis that apoptosis-associated molecules such as Bcl-2 and caspase family members play a role in disease pathogenesis. Although the extent and quality of data vary for specific neurodegenerative diseases (see below), much of the data is correlational and fraught with technical and methodological obstacles inherent in the study of human postmortem brain tissue. Poorly validated antibodies to Bcl-2 and caspase family members and nonspecific 'markers' of apoptosis such as terminal deoxytransferase-mediated deoxyuridine triphosphate nick end labeling (TUNEL) have further complicated the analysis of the biological significance of apoptosis in neurodegenerative disease onset and progression. To date, specific mutations in apoptosis-associated genes have not been linked to any human neurodegenerative disease, raising concern about the primacy of altered apoptosis regulation in disease pathogenesis. Despite these caveats, it is clear that alterations in apoptosis-associated molecules occur in a variety of neurodegenerative diseases, that apoptotic death is one of several morphological forms of neuron death detected in human neurodegenerative disease, and that pharmacological inhibition of apoptotic death offers some hope for at least temporary preservation of neuron survival.

Alzheimer's Disease

Alzheimer's disease (AD) is the most common human neurodegenerative disease and is the most prevalent form of dementia among elderly patients. Neuropathological hallmarks of AD include mature amyloid plaques and neurofibrillary tangles, which are considered the classical features of AD neuropathology and were used to identify the first case of AD more than 100 years ago. In addition to plaques and tangles, AD brains show decreased synapse density and profound neuron loss. To date, the majority of research efforts in AD has focused on mechanisms of plaque and tangle formation and their contribution(s) to neuron dysfunction and ultimate neuron death. Amyloid plaques are caused by the aberrant cleavage of the plasma membrane-bound amyloid precursor protein (APP), a soluble form of amyloid beta (Aβ), which ultimately deposits in the extracellular space in its fibrillar form. However, the formation of amyloid plaques correlates weakly at best with the symptoms and progression of AD. Neurofibrillary tangles are caused in part by the hyperphosphorylation of the cytoskeletal protein tau and more closely follow AD progression. Many studies have indicated that a loss in synapse density and neuron number correlate highly with dementia in AD, although the definitive causes of synapse and neuron loss in AD remain speculative. Whether neuron loss precipitates a decrease in synaptic integrity or if a decrease in neurotrophic support at the level of the synapse precipitates neuron death is not entirely clear in AD. Significant decreases in immunoreactivity for the synaptic protein synaptophysin have been detected in early AD prior to pronounced neuron loss, suggesting that neuronal dysfunction precedes neuron death. However, some brain regions, including the nucleus basalis of Meynert, which contains cell bodies of neurons which project to the cortex, show little neuron loss yet still exhibit AD neuropathology, suggesting that synapse function and cell death may not be directly related. Nevertheless, a better understanding of cell death pathways in AD combined with determining the events that trigger cell death in AD will hopefully generate more effective treatment strategies that preserve both neuron numbers and function.

A role for apoptosis in AD brain has been suggested previously by numerous studies. Techniques such as TUNEL staining have been developed that detect *in situ* the fragmentation of DNA, and studies of AD brain show a marked increase in numbers of such 'apoptotic' nuclei relative to control brains. However, subsequent studies have indicated that these detection methods are not specific for apoptotic nuclei, that is, necrotic nuclei have also been shown to exhibit fragmented DNA and are highly sensitive to fixation artifact. In addition, the large number of apoptotic neurons determined by these methods does not correlate well with the chronic neurodegenerative process of AD, which would predict low numbers of apoptotic neurons at any one time that are efficiently phagocytosed and degraded. Thus, despite findings of numerous studies, the morphological assessment of apoptotic neurons in AD remains inconclusive.

Many studies have indicated an altered regulation of Bcl-2 family members in AD brain. Both proapoptotic (Bak, Bad) and antiapoptotic (Bcl-2 and Bcl-X_L) Bcl-2 family members have been shown to be upregulated in AD brain. The increased expression of Bcl-2 and Bcl-X_L may represent a chronic, antiapoptotic stress response of compromised neurons to compensate for increases in Bax. Based on these findings, a proposed therapeutic strategy in AD is the pharmacological delivery of agents that effectively increase the ratio of anti- to proapoptotic Bcl-2 family members in neurons. However, Bcl-2 family members are also expressed in glial cells, and increases in Bcl-2 have been shown to occur in reactive astrocytes in AD brain. Thus, expression data alone may provide an incomplete picture of the role of Bcl-2 family members and apoptosis in AD pathogenesis. Although they are clearly altered in AD brain, careful studies of Bcl-2 family members in the early stages of AD are required to differentiate possible etiologic from compensatory alterations in Bcl-2 family members specific for neurons.

In vitro models of AD have also implicated altered regulation of Bcl-2 family members. Treatment with Aβ, which causes the death of cultured neurons, induces the expression of proapoptotic Bcl-2 family members including Bax and DP5 while decreasing the expression of Bcl-2 and Bcl-w. In addition, the overexpression of Bcl-2, Bcl-X_L, or Bcl-w and the targeted deletion of Bax prevent Aβ-induced neuron death *in vitro*, which further implicates potential neuroprotective strategies *in vivo*. Future studies are necessary to validate the relative contribution of Bcl-2 family members for Aβ-induced neurotoxicity *in vivo*.

Caspase activation is a critical component of the apoptotic cascade and is largely responsible for the cytological changes that define apoptotic morphology. As such, many *in vivo* studies of human AD brain and *in vitro* models of AD have focused on levels and/or activation of caspases. An increased expression of caspase-1 has been reported in AD brain extracts, which may indicate activation of proinflammatory cytokines such as interleukin-1β that is regulated by active caspase-1 and may lead to the indirect death of neurons in AD. Caspase-12 and caspase-4 are endoplasmic reticulum (ER)-associated proteins that mediate in part ER stress-induced apoptosis. While human isoforms of caspase-4 exist and have been shown along with caspase 12 to induce Aβ-induced neuron apoptosis, caspase-12 was shown recently not to be associated with ER stress-induced apoptosis in humans. In addition, caspase-4 has been shown to cleave APP and may thus be involved in the disease-associated formation of Aβ. Although little, if any, evidence currently exists for its activation in AD, future studies of apoptosis in AD brain warrant further investigation of caspase-4.

A clear role for caspase-2, caspase-8, or caspase-9 in AD has not yet emerged. Of the effector caspases, caspase-7 exists in very low levels and thus is unlikely to play a role in AD pathogenesis. Both caspase-6 and caspase-3 are abundant in human brain and several studies have suggested that caspase-6 may be involved in regulating both APP processing and apoptosis of human neurons. The majority of evidence for caspase activation in apoptosis has focused on caspase-3. Many studies have reported an increased expression and activation of caspase-3 in AD brain, but many conflicting reports have also reported little difference in levels or activity of caspase-3 in AD versus control brains. Caspase-3 has been shown to be activated in cultured neurons treated with Aβ; however, neurons deficient in caspase-3, while lacking apoptotic morphology, are not protected from the death-promoting effects of Aβ, which suggests that caspase-independent death pathways may also exist in AD (see below). While it remains entirely possible that caspase-dependent neuron death is important in AD, a comprehensive analysis of previous studies suggests a degree of caution in defining its relative contribution to AD pathogenesis. A potential disease-promoting role for nonlethal caspase-3 activation in AD neurons has also been proposed. Although cleaved 'activated' caspase-3 is typically associated with apoptotic cell death, cleaved caspase-3 immunoreactivity is consistently observed in AD neurons undergoing granulovacuolar degeneration, a nonapoptotic neurodegenerative process (**Figure 2**). The significance of this observation is currently unclear since granulovacuolar degeneration occurs in only a small fraction of degenerating neurons in the AD brain. If caspases are indeed involved in both sublethal and terminal events in AD neurons, caspase inhibitors may yet prove to be an effective therapy for preserving neuron function in AD.

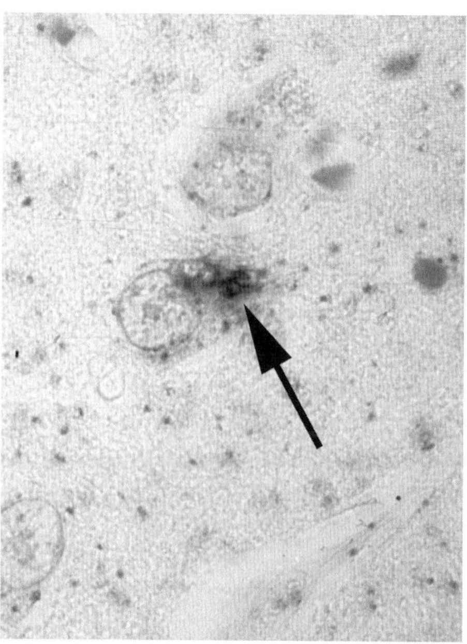

Figure 2 Immunohistochemical detection of activated caspase-3 in AD brain shows strong labeling (dark deposit, indicated by an arrow) in hippocampal pyramidal neurons bearing features of granulovacuolar degeneration. Note that despite the presence of activated caspase-3 immunoreactivity, the nucleus appears intact and the neuron lacks apoptotic histological features.

Parkinson's Disease

Parkinson's disease (PD), the second most common neurodegenerative disease after AD, is characterized by progressive motor impairment and is defined neuropathologically by the substantial loss of dopamine-containing neurons of the substantia nigra that project to the striatum along with the development of neuronal inclusions called Lewy bodies. Lewy bodies are cytoplasmic aggregates containing many proteins including α-synuclein, ubiquitin, and neurofilaments and occur in all affected brain regions of PD. The majority of PD cases are idiopathic, but approximately 5–10% of PD patients show familial patterns of inheritance, and to date several PD-linked gene mutations have been identified, such as those that encode for proteins including α-synuclein, parkin, and DJ-1.

The contribution of apoptosis in PD has been controversial both due to a limited number of reports suggesting alterations in apoptotic markers and morphology and to reports indicating little, if any, contribution of apoptosis to the neurodegenerative process in PD brain. This may be due to the chronic nature of the disease whereby only a few select neurons undergo apoptosis at any one time. Studies have indicated an upregulation in proapoptotic Bax and also antiapoptotic Bcl-2 and Bcl-X_L in PD brain in comparison to age-matched controls. Increases in Fas ligand and activity of caspase-8 have been reported in PD brain, which may indicate the contribution of the extrinsic apoptotic death pathway. In addition, some studies have indicated an increase in the activity of caspase-3 in the PD substantia nigra along with apoptotic nuclear morphology, but it is not clear whether this is due to activation of the extrinsic and/or intrinsic apoptotic pathways or relates to disease pathogenesis.

Analysis of mutant PD genes, either in genetically modified mice or in cultured dopaminergic neurons, has determined that mutant α-synuclein or the targeted deletion of parkin induce apoptotic neuron death whereas overexpression of the wild-type genes prevent apoptosis. In addition, treatment with the dopaminergic neurotoxins such as 6-hydroxydopamine and 1-methyl, 4-phenyl, 1,2,3,6-tetrahydropyridine (MPTP) induce alterations in Bcl-2 family members, including increases in the expression of proapoptotic Bax and PUMA and decreases in expression of Bcl-2. MPTP-induced neurodegeneration can also be spared by the targeted deletion of Bax. Furthermore, the activation of both initiator (caspase-8 and caspase-9) and effector (caspase-3) caspases has been demonstrated in cultured dopaminergic neurons and in neurons of mice treated with MPTP. Thus, data from both *in vivo* and *in vitro* models of PD suggest a link between PD and apoptotic death and raise the possibility that inhibition of apoptotic death may prove beneficial in PD treatment strategies.

Huntington's Disease

Huntington's disease (HD) induces chronic, progressive motor dysfunction and is caused by a mutation generating expanded CAG repeats in the *huntingtin* gene. In general, HD onset and prevalence is proportional to the number of CAG repeats. In the normal population, the number of CAG repeats varies from 6 to 35, but an increased risk is associated with between 36 and 41 CAG repeats and those with 42 or greater CAG repeats almost always manifest HD. The striatum is the principal area affected in HD with reports from postmortem studies demonstrating as much as 95% loss of striatal GABAergic, medium spiny neurons. In more severe cases of HD, other brain regions including the cortex, thalamus, globus pallidus, and hippocampus may also be affected.

Similar to AD and PD, there are conflicting reports of apoptosis in HD brain. Early reports of apoptosis in HD brain indicated increased TUNEL reactivity, but electrophoretic analysis of DNA fragments in HD brain did not reveal a characteristic apoptotic pattern, which possibly suggests a role for additional death processes, such as necrosis in HD brain. Regardless, there are scattered reports of caspase-8, caspase-9, and caspase-3 activation in HD brain, in addition to an increase in cytosolic cytochrome *c*, suggesting possible activation of both the extrinsic and intrinsic apoptotic pathways in HD. Increased levels of Bax and Bcl-2 have also been reported in HD brain specific to the caudate nucleus, which may be interpreted as an antiapoptotic survival response to an increase of Bax in stressed neurons. More studies are needed to determine the role of extrinsic versus intrinsic apoptotic death pathways in HD brain and the potential altered regulation of Bcl-2 family members in this devastating disease.

Animal models of HD, however, show more convincing evidence of a proapoptotic tone of Bcl-2 family members in affected neuronal populations. One of the most well-studied mouse models of HD is the R6/2 transgenic mouse, which expresses only exon 1 of the *huntingtin* gene with 145 CAG repeats, shows a dramatic HD phenotype, and dies by 4 months of age. R6/2 mice show increases in mitochondrial Bim and Bax, a decrease in phosphorylated Bad, and activation of caspase-1 and caspase-3. Levels of Bcl-2 or Bcl-X_L are unchanged in these mice, but the overexpression of Bcl-2 has been shown to slow disease progression. Treatment with 3-nitroprussic acid (3-NP), a chemical inhibitor of mitochondrial succinate dehydrogenase used to model HD, induces a selective striatal lesion when administered *in vivo* that increases the ratio of Bax/Bcl-2 and Bax/Bcl-X_L, thus shifting their balance to a proapoptotic state.

Amyotrophic Lateral Sclerosis

Amyotrophic lateral sclerosis (ALS) is a fatal disease that affects one to two persons per 100 000 and is characterized by the progressive death of upper and lower motor neurons and eventual loss of motor function. Approximately 10% of ALS cases are familial, and of these 20% are linked to mutations in the gene encoding superoxide dismutase 1 (SOD1). It is well accepted that a percentage of dying motor neurons in ALS undergoes apoptosis, and apoptotic morphology has been identified in motor neurons of the human spinal cord and motor cortex. However, like most chronic neurodegenerative diseases, a major challenge in the study of apoptosis in ALS is the relative rapidity of the apoptotic process in the face of a disease with a slow time course, such that at any one time very few motor neurons are actively undergoing apoptosis. As such, there are reports indicating little, if any, evidence for increased apoptotic morphology in ALS cases when compared to control brains.

Studies of postmortem ALS spinal cord suggest a shift toward proapoptotic members of the Bcl-2 family, such as increased expression of Bax and DP5/Hrk and decreased expression of Bcl-2. Co-immunoprecipitation studies have also indicated an increase in Bax–Bax interactions and a decrease in Bax–Bcl-2 interactions in ALS tissue, which also suggests a shift to a proapoptotic state. A decrease in Bcl-2 may be of greater significance in ALS than in other neurodegenerative diseases since Bcl-2 remains high after neurodevelopment in the spinal cord relative to the brain, which suggests an important regulatory function of Bcl-2 in postmitotic spinal cord motor neurons. Analysis of symptomatic, transgenic mice with mutant SOD1 also shows decreased expression of Bcl-2 and Bcl-X_L and increased expression of Bax and Bad. Truncated Bid is also evidenced in mutant SOD1 mice, which may suggest participation of the extrinsic apoptotic pathway in ALS.

In vitro studies of cultured motor neurons from mutant SOD1 mice indicate a role for Fas-mediated activation of caspase-8 and the extrinsic apoptotic pathway, but these same studies also report release of cytochrome *c* from mitochondria, suggesting activation of the intrinsic apoptotic pathway. Reports of altered regulation of Bax and Bcl-2 in human ALS would clearly support a role for the intrinsic apoptotic pathway in ALS. While there is no current evidence supporting the direct activation or altered expression of caspase-8 in human ALS, recent analysis of sera obtained from human ALS patients indicates an increase in antibodies to Fas that induced death of cultured motor neurons, which further suggests the contribution of the extrinsic apoptotic pathway. In addition to caspase-8, *in vitro* and animal models of ALS have shown activation of caspase-1, caspase-12, caspase-9, and caspase-3, while reports of caspase activation in human ALS tissue are limited to activation of caspase-1 and caspase-3. While good evidence exists supporting a role for apoptosis in ALS, more studies are needed to determine its relative contribution to the overall disease pathogenesis.

Nonapoptotic Neurodegeneration

The reported inconsistencies for the role of apoptosis in neurodegenerative disease have led to an increasing number of studies investigating potential mechanisms of caspase-independent neuron death. Apoptosis-inducing factor (AIF) is a small protein that resides normally within the intermembrane space of mitochondria, and upon certain death stimuli translocates into the cytosol and ultimately the nucleus where it contributes to DNA fragmentation and chromatin condensation. The death-promoting effects of AIF have been shown to be independent of caspase activation, although the mitochondrial release of AIF has been shown to be Bax dependent. A role for AIF in cell death has been implicated in many neurodegenerative diseases including AD, PD, and ALS. Calpains are cytosolic proteases that are activated upon ER stress-induced intracellular calcium influx. While calpains have been shown to act downstream of caspases, many studies indicate caspase-independent, calpain-induced cell death. Calpain activation has been implicated in AD, PD, HD, and ALS. For instance, in an *in vitro* model of AD, the activation of calpains has been implicated in the Aβ-induced, caspase-independent degeneration of axons. α-Synuclein has also been shown to be a substrate of activated calpain, which may lead to its aberrant deposition in PD.

The increased aggregation and deposition of specific cytoplasmic proteins in neurodegenerative diseases has implicated a role for disturbances in lysosomal function and/or type II autophagic cell death. Autophagy is a normal cellular process whereby vesicles shuttle cytoplasmic and organellar macronutrients through a series of fusion events to the lysosome for ultimate proteolytic degradation and recycling. Autophagic cell death is characterized by an aberrant accumulation of vacuoles that result either as a response to metabolic stress or to a disruption in lysosomal function, concomitant with morphological features of both apoptosis and necrosis. An accumulation of autophagic vacuoles has been clearly documented in AD, PD, and HD, which suggests that aberrant autophagic stress plays a significant role in the pathogenesis of these diseases.

Age-related lysosome disturbances may contribute to a compromise in protein degradation that allows for the accumulation of specific proteins in neurodegenerative disease. In addition, a compromise in the integrity of lysosome membranes may also induce leakage of lysosomal proteases, such as cathepsins, into the cytosol. Cathepsins have been shown to induce Bid- and Bax-dependent activation of the intrinsic apoptotic pathway and Bax-dependent release of AIF into the cytosol. In addition, cathepsins can also act as efficient effector proteases similar to that of effector caspases, thus creating another potential mechanism of caspase-independent death in neurodegenerative disease.

Summary

Apoptotic death is a well-characterized and tightly orchestrated cell biological process whose involvement in normal brain development is unquestionable. Apoptotic neurons have been detected to varying degrees in virtually all human neurodegenerative diseases, and alterations in the expression of numerous apoptosis-associated genes and proteins have been widely reported. However, the significance of these findings is unclear, and separating correlation from causation has proven difficult. It remains possible that apoptosis-associated molecules and sublethal intracellular events contribute to neuron dysfunction early in disease pathogenesis, prior to significant loss of neurons, but definitive data on this conjecture are currently lacking for most neurodegenerative diseases. Neuroprotective agents that can inhibit both apoptotic and nonapoptotic neuron death may ultimately be used in conjunction with drugs designed to decrease disease-specific death stimuli, resulting in effective combination therapy to preserve neuron function in AD, PD, HD, ALS, and other neurodegenerative diseases.

See also: Alzheimer's Disease: Neurodegeneration; Amyotrophic Lateral Sclerosis (ALS): Disease Mechanisms; Animal Models of Motor and Sensory Neuron Disease; Animal Models of Alzheimer's Disease; Animal Models of Huntington's Disease; Apoptosis in Nervous System Injury; Autonomic and Enteric Nervous System: Apoptosis and Trophic Support During Development; Autophagy and Neuronal Death; Huntington's Disease: Neurodegeneration; Parkinson's Disease: Alpha-Synuclein and Neurodegeneration.

Further Reading

Boatright KM and Salvesen GS (2004) Mechanisms of caspase activation. *Current Opinion in Cell Biology* 15: 725–731.

Clarke PG (1990) Developmental cell death. *Anatomy and Embryology* 181: 195–213.

Ellis HM and Horvitz HR (1986) Genetic control of programmed cell death in the nematode *C. elegans*. *Cell* 44: 817–829.

Hickey MA and Chesselet MF (2003) Apoptosis in Huntington's disease. *Progress in Neuro-Psychopharmacology and Biological Psychiatry* 27: 255–265.

Kerr JF, Wyllie AH, and Currie AR (1972) Apoptosis: A basic biological phenomenon with wide-ranging implications in tissue kinetics. *British Journal of Cancer* 26: 239–257.

Putcha GV, Harris CA, Moulder KL, et al. (2002) Intrinsic and extrinsic pathway signaling during neuronal apoptosis: Lessons from the analysis of mutant mice. *Journal of Cell Biology* 157: 441–453.

Roth KA (2001) Caspases, apoptosis and Alzheimer disease: Causation, correlation, and confusion. *Journal of Neuropathology and Experimental Neurology* 60: 829–838.

Sathasivam S, Ince PG, and Shaw PJ (2001) Apoptosis in amyotrophic lateral sclerosis: A review of the evidence. *Neuropathology and Applied Neurobiology* 27: 257–274.

Shacka JJ and Roth KA (2004) Regulation of neuronal cell death and neurodegeneration by members of the Bcl-2 family: Therapeutic implications. *Current Drug Targets – CNS and Neurological Disorders* 4: 25–39.

Tatton WG, Chalmers-Redman R, Brown D, and Tatton N (2003) Apoptosis in Parkinson's disease: Signals for neuronal degradation. *Annals of Neurology* 53: S61–S70.

Appetitive Systems: Amygdala and Striatum

J A Parkinson, Bangor University, Bangor, UK

Introduction

Imagine going to a grocery store for some milk. While strolling the aisles, a carton of Madagascan Vanilla Custard catches your eye; it sounds delicious and would be great to have lashings of this custard with some seasonal fruit for dessert this evening. Into the basket it goes, and before you know it you will be sitting at home indulging your extravagant desires. Even though there was no intention to buy the custard prior to entering the shop, the mere exposure to the salient packaging has captured attention, elicited appetitive arousal, established the Madagascan Vanilla Custard as an important goal for prospective behavior, and even generated an expectancy for the consequences of such a course of action. This simple example captures the essence of what interactions between the amygdala and striatum are believed to underlie.

Major advances in our understanding of emotion have come about through studying model systems in animals, such as fear conditioning in rats. Similarly, our understanding of positive emotions and motivation has also been furthered through the study of appetitive processes in animals and in particular anticipatory, or preparatory, responses to food, sex, and drugs. Thus, whereas fear can be conceived as the anticipation of an unpleasant, or aversive, event, appetitive behavior can be conceived as the anticipation of a pleasant, or desirable, event. The amygdala appears to be a critical mechanism for these anticipatory processes. As far back as the middle of the last century it was suggested that the amygdala complex processed stimulus–reward associations, in that it allowed sensory representations of the world to be linked with affective values. More recently, in the 1980s, researchers suggested that the ventral striatum (VS), a structure that receives signals from the amygdala, acted as a limbic–motor interface allowing motivational states and emotional information to influence action selection. Both of these seminal propositions are broadly accepted; this article describes contemporary neuroscientific research which has refined these ideas and provided a more precise indication of how these two structures interact. Neither the amygdala nor striatum is a unitary structure, and their function can be better understood through an understanding of their anatomical structure and relationships. Indeed, it is likely that subregions of the amygdala and striatum play dissociable yet complementary roles in affective processing.

Anatomical Routes for Amygdalar–Striatal Interactions

Although the amygdala was traditionally viewed as a single structure, detailed cytoarchitectural, neurochemical, and connectional analyses have now supported a multicompartmental view with at least three major subdivisions: (1) a phylogenetically newer cortex-like region (basolateral amygdala regions (BLA)), (2) part of the olfactory system, and (3) an extension of the striatum (central amygdala region (CEA)). The first two regions use glutamate as their primary extrinsic neurotransmitter and can be considered parts of the cortex. The last region shares many features with the neighboring striatum, such as having a dense preponderance of γ-aminobutyric acid (GABA) and a similar pattern of neuropeptide expression. Neuroscientific evidence supports interactions primarily between the BLA and the CEA with the ventromedial regions of the striatum within the context of two organizing models of anatomical connectivity and structure: corticostriatal circuitry and extended amygdala.

Corticostriatal Circuitry

The striatum is primarily viewed as a major input station of the basal ganglia, a system that is critically modulated by dopamine and which supports sensory-motor initiation and control. The basal ganglia consists of a number of independent and parallel circuits (see **Figure 1**) which are topographically organized. These cortico-striatal-pallidal-thalamic loops (or corticostriatal loops, for the purpose of this article) are capable of selection and amplification, mapping the patterns of sensory input on to response output within a particular organizing context. In effect, these loops can maintain and remap the process of response selection.

The number of functional corticostriatal circuits is not fully resolved and is a matter of considerable debate. Current thinking acknowledges (1) at least one sensorimotor circuit running through lateral striatum (putamen); (2) at least one dorsomedial (caudate) circuit underlying cognitive, or goal-directed, functions; and (3) a ventral striatal circuit that underlies stimulus-induced motivation and is implicated in excessive appetite and addiction. The BLA provides

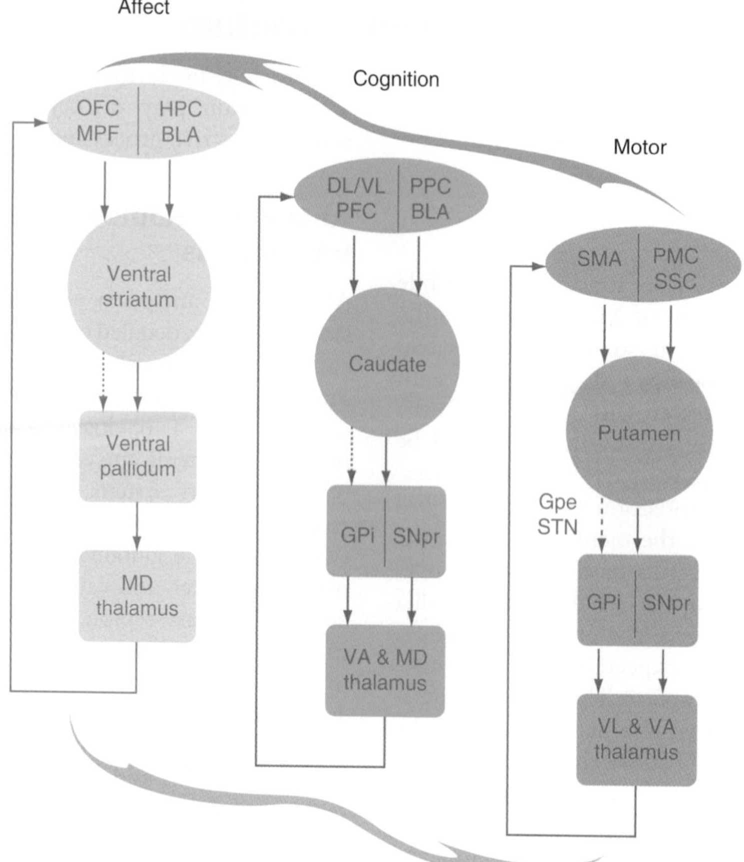

Figure 1 Three segregated, parallel corticostriatal loops are shown with their putative function. Each basal ganglia circuit includes cortical input, striatum, pallidum, and thalamus. The precise manner in which parallel loops interact is not known, although hierarchical control over behavior from medial (affective) to lateral (sensorimotor) circuits is achieved by descending influence of each circuit over the midbrain dopamine (DA) innervation of the next circuit laterally (indicated by red arrow) enabling integration of value and goal information, knowledge-based strategy, and sensorimotor coordination in the pursuit of goal-directed behavior. Further, each circuit may provide feedback (lateral to medial) at the level of cortex (indicated by blue arrow). Finally, local within-loop ensembles may show lateral inhibition across loops (not shown). BLA, basolateral amygdala; DL/VL PFC, dorsolateral and ventrolateral prefrontal cortex; GPe, external segment of the globus pallidus; GPi, internal segment of the globus pallidus; HPC, hippocampal formation; MD, mediodorsal; MPF, medial prefrontal cortex; OFC, orbitofrontal cortex; PMC, premotor cortex; PPC, posterior parietal cortex; SMA, supplementary motor area; SNpr, substantia nigra, pars reticulata; SSC, somatosensory cortex; STN, subthalamic nucleus; VA ventral anterior; VL, ventrolateral.

direct input across the medial striatum, including both the VS and caudate circuits. There are also indirect connections from the CEA to ascending arousal systems and thence to the striatum (see **Figure 2**). Efferent projections from the VS show a marked divergence across the medial–lateral axis. The lateral core subregion, which merges into overlying caudate, is the more typically striatal, projecting to the ventral pallidum. The medial shell subregion projects not only to the ventral pallidum but also to the lateral hypothalamus and bed nucleus of the stria terminalis and is considered part of the extended amygdala.

Extended Amygdala

The concept of the extended amygdala reflects the developmental similarity between cell groups of the amygdala, specifically the striatal-like centromedial nuclei and surrounding tissue such as the bed nucleus of the stria terminalis. A topographic continuum of cells courses through the forebrain and forms a loop around the internal capsule. In contrast to traditional output systems of the corticostriatal circuits, the extended amygdala is typified by its downstream connections to autonomic and endocrine centers, allowing the coordination of species-typical adaptive responses (labeled brain stem effectors in **Figure 2**).

Function of Amygdala–Striata Interactions

Overview

Along with the orbital prefrontal cortex (OFC), communication between the amygdala and striatum underlies the anticipation of emotional events and expectations of the consequences of choices and

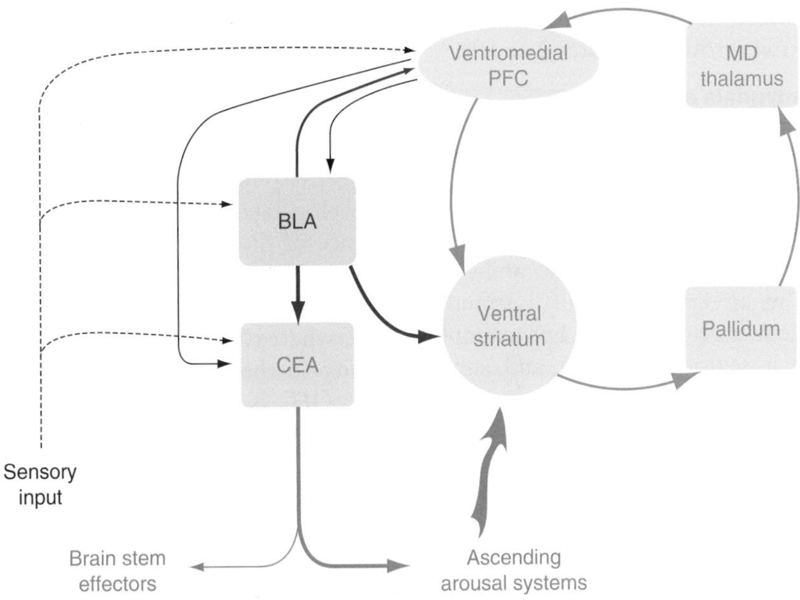

Figure 2 Affective corticostriatal circuit emphasizing two anatomical routes through which the amygdala interacts with the striatum. Orbital and medial prefrontal (ventromedial PFC) afferents converge with amygdala inputs (BLA; blue arrow) at the level of the VS. Modulation of corticostriatal circuitry by dopamine (and other ascending arousal systems) also depends on signals from the amygdala (CEA; red arrow). This interaction between CEA and VS has been implicated in two related processes: (1) providing a reinforcement signal for associative learning and (2) modulating the vigor of a motivated instrumental response (also see the section titled 'Mesolimbic dopamine contributes to both motivational learning and performance' and **Figure 3**). BLA, basolateral amygdala; CEA, central amygdala; MD, mediodorsal; PFC, prefrontal cortex; VS, ventral striatum. Reproduced from Parkinson JA, Cardinal RN, and Everitt BJ (2000) Limbic cortico-ventral striatal systems underlying appetitive conditioning. In: Uylings HBM, Van Eden CG, De Bruin JPC, et al. (eds.) *Progress in Brain Research*, vol. 126, pp. 263–285. Amsterdam: Elsevier Science, with permission from Elsevier.

behavior. Initial insights into amygdala function came from observations of the effects of gross damage in nonhuman primates. Such animals seemed incapable of learning the association between an object and its emotional, or motivational, value (such as the sight of a banana and its value as a tasty food). This impairment was described as being an inability to form stimulus–reward associations; that is, it is not an impairment in stimulus processing *per se* nor in the ability to experience the hedonic impact of the food outcomes once tasted, but in the relationship between the two: the ability to predict, indeed expect, a particular outcome given its associated stimulus.

Anatomical distinctions within the amygdala have been supported through functional studies. The BLA shares connectivity and structure with the prefrontal and temporal cortex and has been shown to support mnemonic representations of value (e.g., evidence for long-term potentiation underlying associations between a sensory stimulus and its affective value). The BLA may therefore contribute to a system that provides a signal of the affective value of specific objects and events in the world (blue arrows in **Figures 2** and **3**). The CEA, which sits downstream of the BLA appears to play a greater role in the acquisition and performance of anticipatory conditioned

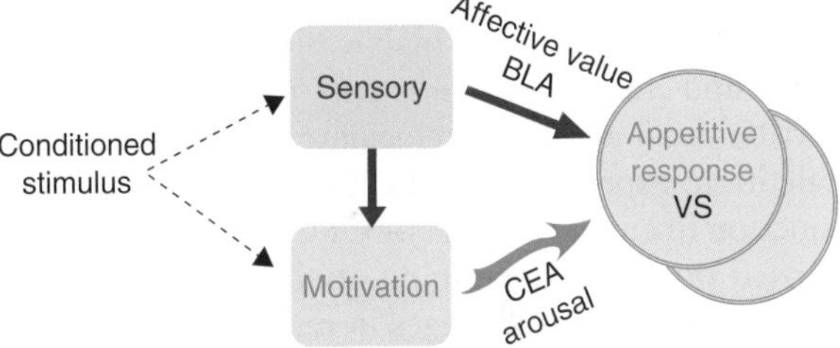

Figure 3 Psychological processes by which affect-laden stimuli can influence cognition and appetitive instrumental behavior. Associations between a conditioned stimulus and an affective outcome can be formed directly between the specific sensory features or with the more general motivational features. The CEA provides the general motivational arousal elicited by conditioned stimuli, whereas the BLA provides the direction for motivational responses based on the specific sensory and motivational properties of a stimulus. BLA, basolateral amygdala; CEA, central amygdala; VS, ventral striatum. Reproduced from Balleine BW (2005) Neural bases of food-seeking: Affect, arousal and reward in corticostriatolimbic circuits. *Physiology and Behavior* 86: 717–730, with permission from Elsevier.

responses, that is, in the way that reflexive, species-typical responses can be triggered by environmental stimuli in appropriate circumstances (freezing to threat or orienting to a stimulus-predicting reward). The CEA also has access to ascending arousal systems and may influence affective learning and performance through this route (red arrows in **Figures 2** and **3**).

The Basolateral Amygdala and Stimulus–Affect Learning

Sophisticated behavioral tests can be employed to dissect distinct psychological processes that an individual may employ to achieve a goal. One such paradigm, outcome devaluation, involves the ability to react to a change in affective value of a stimulus after its initial value has been learned. For example, after having eaten a seafood bouillon and subsequently becoming ill, most people would avoid choosing the same, or similar, dish in future forays to the restaurant, even without tasting the food. This demonstrates an ability to represent the value of food and subsequently modulate or change that value property as a result of new experiences. Witness a more dynamic version at mealtimes when, after experiencing a filling main course, an individual may readily desire a sugary dessert, that is, a transient reduction in value of the main course but a maintenance of value for sugary dessert, sensory-specific satiety. Our choices reflect the changing incentive values of the competing foods. Animal studies of food valuation and revaluation (including outcome devaluation) have been employed to explore the neural mechanisms of goal-directed behavior.

Within this broad paradigm, researchers have employed various neuroscientific techniques including regional lesions, regional inactivations, electrophysiology, and functional magnetic resonance imaging (fMRI) in an attempt to elucidate the neural basis of valuation, expectation, and choice. Disruption in the function of either the BLA or OFC in rats or primates carried out prior to training abolishes the ability of animals to adapt to changes in reinforcer value (e.g., devaluation) on such tasks. In contrast, these lesions do not appear to disrupt motoric ability, general activity, or hedonic responses to rewards themselves. Electrophysiological recording of neurons in awake behaving rodents and primates has further refined our understanding: neurons in the BLA develop selective firing both to the delivery of a particular affective outcome and to the stimulus that predicts it. Further, some neurons that fire to the reinforcer also develop responding during the delay between a behavioral response and the subsequent outcome. As such, the former neurons appear to code the relationship between stimulus and outcome (a form of passive expectancy), whereas others code the expected nature of the outcome following a response (a more active response-based expectation). If the affective nature of the outcomes is reversed, so that a stimulus that used to predict a pleasant outcome now predicts an unpleasant one, the responding of the anticipatory neurons also changes, critically demonstrating that the neurons are tracking the valence of the outcome, not just its sensory properties. Similar activity has also been observed in OFC neurons. However, the selectivity in neuronal firing in the OFC appears to develop later than in the BLA and tracks the emergence of discriminated behavior. Further, OFC neurons tend not to reverse with changing outcome contingencies and, instead, simply lose their selectivity, whereas previously silent neurons acquire selectivity to the new contingencies. This indicates that the OFC is more tightly linked to action selection and the BLA is more tightly linked to stimulus value.

Human neuroimaging studies have arrived at similar conclusions when adopting paradigmatic approaches employed in animal work. Such studies have further shown that activity in the amygdala correlates with the magnitude of arousal elicited by affective stimuli (including pictures, words, tastes and smell of foods, pictures of sexual stimuli, and positively valenced pictures of social interactions). In other words, the level of activity in the amygdala reflects the subjective perception of affective value. OFC activity also responds to the predicted stimulus value, but appears to code for different response options and their expected consequences.

The Central Amygdala and Conditioned Anticipatory Arousal

Much of the foregoing discussion has centered on the BLA and its role in stimulus–affect learning. The CEA appears to play a different role in the learning and performance of appetitive behavior. Damage to the CEA does not impair the ability of animals to perform the sort of tasks that we have described, such as incentive learning and devaluation. As such, many appetitive motivational processes are dissociable between these two amygdala subregions. Lesions of the CEA do abolish learned orienting responses to affective stimuli and also abolish conditioned approach behavior to cues that predict appetitive events – in general terms, the acquisition of attentional and motivational anticipatory responses. The CEA also provides direct input to specific neurotransmitter systems of the ascending arousal systems such as dopamine (DA). Disruption of the CEA–DA–dorsal striatum system impairs attentional learning, whereas manipulations of the CEA–DA–VS system affect the acquisition of motivated approach.

The CEA and BLA can also be dissociated in their influence on the performance of instrumental behavior (see **Figure 3**). One of the most powerful forms of motivation is the process by which everyday stimuli can induce motivational arousal, even in the absence of a biological need, and influence the vigor and direction of behavior. Indeed, theories of excessive appetite, including overeating and relapse to drug use, posit a major role for stimulus-induced arousal and desire. An important animal model of such processes examines the way affective stimuli influence ongoing behavior (termed Pavlovian to instrumental transfer (PIT)). PIT can have its effects specifically; that is, the perceived value of a stimulus matches a particular motivational need (known as sensory-specific PIT). Alternatively, it can elicit conditioned motivation, producing a general arousal of behavior irrespective of motivational needs (known as general PIT). Amygdalar–striatal circuitry is critical for both forms of PIT; whereas the CEA is important for the general arousal elicited by a stimulus, the BLA provides specific direction for stimulus-induced motivation (as shown in **Figure 3**). Further, the VS provides the route for expression of PIT potentially through a combination of direct projections from the BLA and also from downstream projections of the CEA to ascending arousal systems, including DA; intra-VS infusions of DA agonists enhance the PIT response, and systemic blockade of DA abolishes the motivational response.

The Striatum as a Site of Integration and Selection

The VS was famously described as a limbic–motor interface using information about the value of stimuli in the world to guide behavior. Another avenue of research into VS function relates to its involvement, and in particular its dopaminergic innervation, in drug reward. The VS was once thought to underlie the hedonic component of reward; damage to this structure reduced drug-related responses in rodents on tasks modeling drug reward. However, these tasks sometimes confounded dissociable psychological processes such as drug-seeking, anticipatory arousal, and actual drug ingestion, making it difficult to be certain what the effects of VS or dopaminergic manipulations were. Indeed, ingestion of drugs (or food or copulation) is rarely abolished by striatal damage or by manipulation of the striatal DA system. Instead, rises in extracellular VS DA are predominantly seen in anticipation of appetitive events, and manipulations reducing the function of the VS most often reduce anticipatory motivation for reinforcers (again shown with recreational drugs, food, and sex). Other neurochemical systems, notably opioids, may contribute to consummatory/hedonic processes.

Electrophysiological studies in rats and primates have shown that different classes of neuron in the medial striatum code for reward anticipation and for stimulus–response mapping. Both the VS and overlying caudate receive input from the OFC and amygdala, and it is perhaps no surprise that each class of these neurons also appears to exist in one or other of these afferent structures; in this case, the amygdala preferentially shares stimulus–outcome coding, whereas the OFC provides information on responses and their consequences. The VS and overlying caudate therefore serve as integration sites for this information in order to allow appropriate behavior to be initiated and maintained. The distinction between these two corticostriatal circuits is that, whereas the VS circuit underlies stimulus-bound motivational arousal, the caudate circuit contributes to cognitive aspects of goal-directed behavior such as attention and flexibility. Further, the medial and orbital prefrontal cortex (PFC), afferents to the striatum, have been implicated in extinction and habit control (medial) and in inhibition and reversal (orbital). As such, input from the PFC to the striatum appears to provide top-down control of response selection, particularly in allowing flexible behavior as a result of changing environmental contingencies or changing internal motivational and emotional states.

Mesolimbic Dopamine Contributes to Both Motivational Learning and Performance

Seminal electrophysiological work demonstrated that DA neurons in the midbrain, as well as the VS neurons to which they project, show phasic firing patterns to unexpected affective outcomes. If a stimulus predicts the outcome, then neuronal firing begins to track the predictive stimulus rather than the outcome. The omission of an expected stimulus or outcome also elicits phasic firing. This pattern appears to code for a prediction error – a reinforcement signal – which supports learning by predicting the likelihood of an outcome given that a behavioral response has been made or a conditioned stimulus has been presented. Many learning models have incorporated this signal, although at its simplest, DA may act as a reinforcement teaching signal to stamp in the association between a stimulus and the appropriate response, that is, to help create stimulus–response mappings through identifying predictive relationships in the world.

In other work, gross manipulation of DA levels in the VS modulate the vigor, or effort, with which animals respond for reinforcement. Indeed, human subjective experience of psychostimulant drugs (such as cocaine and amphetamine, which enhance DA function in the VS) often includes increased feelings of vigor and behavioral output. Rats with reduced VS DA select less effortful courses of action to gain reinforcement and choose less preferred foods if greater effort is required to get the preferred food. Infusions of DA agonists into the VS specifically potentiate responding for reward-related stimuli and outcomes, rather than potentiating all behavior, and enhance the desire (or wanting) for valued outcomes.

In summary, phasic DA signals appear to underlie simple affective learning, whereas tonic levels of DA in the extracellular matrix of the VS provide the energy or vigor for responding. It has also been suggested that tonic levels of DA play a permissive role in response selection, and so response vigor may simply be a computational consequence of the response selection process. Interestingly, damage to the CEA abolishes the enhancing effect of intra-VS infusions of DA and impairs the acquisition of appetitive approach behavior (which is also dependent on the VS). As a result, several authors have speculated that the CEA may play a role in stimulus–response learning via influence on the phasic firing of DA neurons and also a role in the selection and arousal of performance by modulating a more tonic level of extracellular DA in the VS (red arrows in **Figures 2** and 3). Finally, the extracellular levels of DA appear to be critically modulated by incoming afferents such as the BLA and hippocampal formation and may therefore relate to calculations of value and of context. Again, the VS and its dopaminergic innervation can be seen as a mechanism of integration between sensory input and appropriate behavioral output.

Amygdalar–Striatal Interactions and Drug Addiction

It is likely that drugs exert their effects on emotion and motivation through the same neural circuits that underlie natural rewards, although perhaps in a more potent manner. It is not surprising then that certain aspects of the addictive process relate to amygdalar–striatal function. Drugs and associated drug stimuli (including paraphernalia, incidental stimuli, and contexts) acquire incentive value through association with drug taking. Such stimuli themselves generate affective arousal and lead to increased anticipation and desire for drugs. This process of stimulus-induced arousal and desire can contribute to two important aspects of drug addiction: (1) the abnormal control over motivation and behavior that drugs produce, such as focusing on drug-seeking behavior to the detriment of other natural motivational needs; and (2) the high relapse rates in individuals attempting to abstain from drug use. Stimulus-induced craving is thought to be a major contributor to relapse and the maintenance of abuse. Drug-associated stimuli can therefore be seen as possessing an abnormal level of control over goal-directed processes, can suppress (healthy) inhibition, and can lead to maladaptive motivated behavior.

The motivational impact of drug-associated cues depends on the VS and its dopaminergic innervation. Disruption of VS function, or of incoming glutamatergic signals, reduces drug seeking, probably by reducing motivational arousal normally elicited by affective cues. Indeed, neuroadaptations in amygdalar–striatal circuitry have been observed following chronic drug use, such as sensitization of the DA system; such adaptations are now being documented with food stimuli. Indeed, the release of endogenous opioids in the VS is seen following consumption of energy-dense foods, just as similar changes in opioid levels are observed following heroin self-administration in rats. This growing link between drug and natural rewards highlights the commonality in neural substrates of appetitive motivational control.

Summary

Through experience, individuals learn the affective structure of the world, enabling stimuli endowed with value to provide an important impetus for behavior. The amygdala appears to encode the expected value of a predicted event, and the VS contributes to the formation and expression of motivational responses. These processes are maintained by DA signals that support value-based learning and also modulate the direction and energy underlying behavior. The OFC most likely contributes by providing a top-down control over goal-directed response selection, including functions such as inhibition, flexibility, and choice.

One of the major functions of the brain is to predict and time events. A specific appetitive function of amygdalar–striatal interactions is to predict events of adaptive (emotional and motivational) significance. The position of the amygdala and VS at the anatomical head of corticostriatal circuitry allows them a preeminent role in enabling affect-laden stimuli to influence an individual's choice of goals and course of action.

See also: Amygdala: Structure and Circuitry in Primates; Amygdala: Structure and Circuitry in Rodents and Felines; Amygdala: Contributions to Fear; Conditioning: Theories; Dopamine Control of Arousal; Dopamine; Drug Addiction: Behavioral Neurophysiology; Drug Addiction: Behavioral Pharmacology of Drug Addiction in Rats; Emotion Systems and the Brain; Goal-Directed Behavior Theories; Prefrontal Contributions to Reward Encoding; Procedural Learning: Striatum; Psychopharmacology of Reward and Appetite in Rats; Reward Systems: Human; Striatum: Internal Physiology.

Further Reading

Alexander GE, Delong MR, and Strick PL (1986) Parallel organization of functionally segregated circuits linking basal ganglia and cortex. *Annual Review of Neuroscience* 9: 357–381.

Balleine BW (2005) Neural bases of food-seeking: Affect, arousal and reward in corticostriatolimbic circuits. *Physiology and Behavior* 86: 717–730.

Baxter MG and Murray EA (2002) The amygdala and reward. *Nature Reviews Neuroscience* 3: 563–573.

Berridge KC (2004) Motivation concepts in behavioral neuroscience. *Physiology and Behavior* 81: 179–209.

Cardinal RN and Everitt BJ (2004) Neural and psychological mechanisms underlying appetitive learning: Links to drug addiction. *Current Opinion in Neurobiology* 14: 156–162.

Daw ND, Niv Y, and Dayan P (2005) Uncertainty-based competition between prefrontal and dorsolateral striatal systems for behavioral control. *Nature Neuroscience* 8: 1704–1711.

Haber SN, Fudge JL, and McFarland NR (2000) Striatonigrostriatal pathways in primates form an ascending spiral from the shell to the dorsolateral striatum. *Journal of Neuroscience* 20: 2369–2382.

Holland PC and Gallagher M (2004) Amygdala–frontal interactions and reward expectancy. *Current Opinion Neurobiology* 14: 148–155.

Kelley AE, Baldo BA, Pratt WE, and Will MJ (2005) Corticostriatal-hypothalamic circuitry and food motivation: Integration of energy, action and reward. *Physiology and Behavior* 86: 773–795.

Parkinson JA, Cardinal RN, and Everitt BJ (2000) Limbic cortico-ventral striatal systems underlying appetitive conditioning. In: Uylings HBM, Van Eden CG, De Bruin JPC, et al. (eds.) *Progress in Brain Research,* vol. 126, pp. 263–285. Amsterdam: Elsevier Science.

Redgrave P, Prescott TJ, and Gurney K (1999) The basal ganglia: A vertebrate solution to the selection problem? *Neuroscience* 89: 1009–1023.

Salamone JD, Correa M, Mingote SM, and Weber SM (2005) Beyond the reward hypothesis: Alternative functions of nucleus accumbens dopamine. *Current Opinion in Pharmacology* 5: 34–41.

Schultz W and Dickinson A (2000) Neuronal coding of prediction errors. *Annual Review of Neuroscience* 23: 473–500.

Swanson LW and Petrovich GD (1998) What is the amygdala? *Trends in Neurosciences* 21: 323–331.

Yin HH and Knowlton BJ (2006) The role of the basal ganglia in habit formation. *Nature Reviews Neuroscience* 7: 464–476.

Apraxia: Disease

G Goldenberg, Krankenhaus Munchen Bogenhausen, Munich, Germany

Introduction

The term apraxia denotes 'higher-order' disturbances of motor control. Apraxia differs from more mundane disorders of motor skill like hemiparesis or ataxia by its bilaterality: unilateral hemisphere lesions cause apraxia of both limbs. Consequently, apraxia can be demonstrated on the limb ipsilateral to the lesioned hemisphere and contrasts with normal motor skill of the same limb in other situations. This definition embraces a wide variety of heterogeneous disturbances, but traditionally diagnosis and research on apraxia concentrate on three domains of human motor actions: imitation of gestures, performance of communicative gestures on command, and skillful use of tools and objects.

A seminal model of apraxia was elaborated by the German psychiatrist Hugo Liepmann about a hundred years ago. He distinguished between two sequential stages of high-level motor control. In the first stage, a concept or mental image of the intended action is elaborated, and in the second stage, this mental image is translated into appropriate motor commands. Liepmann thought that generation of mental images of the intended actions is accomplished by posterior regions of the brain and that their translation into motor commands depends on integrity of fiber paths linking posterior brain regions to the motor cortex. Parietal lesions can sever these fiber paths and are hence a locus of predilection for causing apraxia. Liepmann's most important empirical contribution to the study of apraxia was the observation that it is predominantly a symptom of left brain damage (LBD). Most afflicted patients are aphasic, but the severities of aphasia and apraxia are not tightly correlated, and there are LBD patients who display only one of them. This led Liepmann to speculate that there is left hemisphere dominance for motor control in addition to its well-established dominance for speech and language.

It may appear somewhat untimely to start this article with a theory proposed at the start of the twentieth century, but indeed Liepmann's ideas have remained influential and have shaped the study of apraxia until now. Three basic assumptions have received different interpretation but have remained essentially unchallenged until recently (see **Figure 1**):

- There is a sequentially ordered 'praxis system,' interruption of which underlies all manifestations of apraxia.
- This praxis system is located in the left hemisphere.
- The parietal lobe has a predominant role within this system.

This article discusses each of the manifestations of apraxia separately, with an emphasis on the role of LBD and particularly of left parietal lesions. Then it questions the existence of a unique praxis system and discusses the usefulness of classifying apraxia as 'ideational' or 'ideomotor.'

Imitation of Gestures

Defective imitation is an easily demonstrable and impressive symptom of unilateral brain damage. Even if patients are severely aphasic, they usually understand the instruction to imitate and obviously strain to copy the gesture made by the examiner, but although they use the otherwise normally skillful hand ipsilateral to their brain lesion, they produce insecure, hesitant, and searching movements which end up in positions quite grossly different from those demonstrated.

The presence and severity of impairment may depend on the kind of gesture that is being tested. A first major divide is between meaningful and meaningless gestures. If the examiner demonstrates a meaningful gesture, patients may understand the meaning and reproduce the gesture out of a repertoire of meaningful gestures stored in long-term memory rather than copying the shape of the gesture. By contrast, meaningless gestures usually have no representation in long-term memory, at least not when tested for the first time. Their imitation must translate the shape of the gesture from visual perception of another person's body to motor commands configuring their own body without support from the repertoire of habitual gestures. The mechanisms employed for meaningless gestures can also support imitation of meaningful gestures, as their shape can be copied without taking account of their meaning. Intermingled presentation of meaningful and meaningless gestures can blur their distinction because participants then tend to make a strategic choice of using the same route for both kinds of gestures and treat meaningful gestures as if they were meaningless. Studies which tested imitation of meaningful and meaningless gestures separately brought forward robust evidence that the two routes

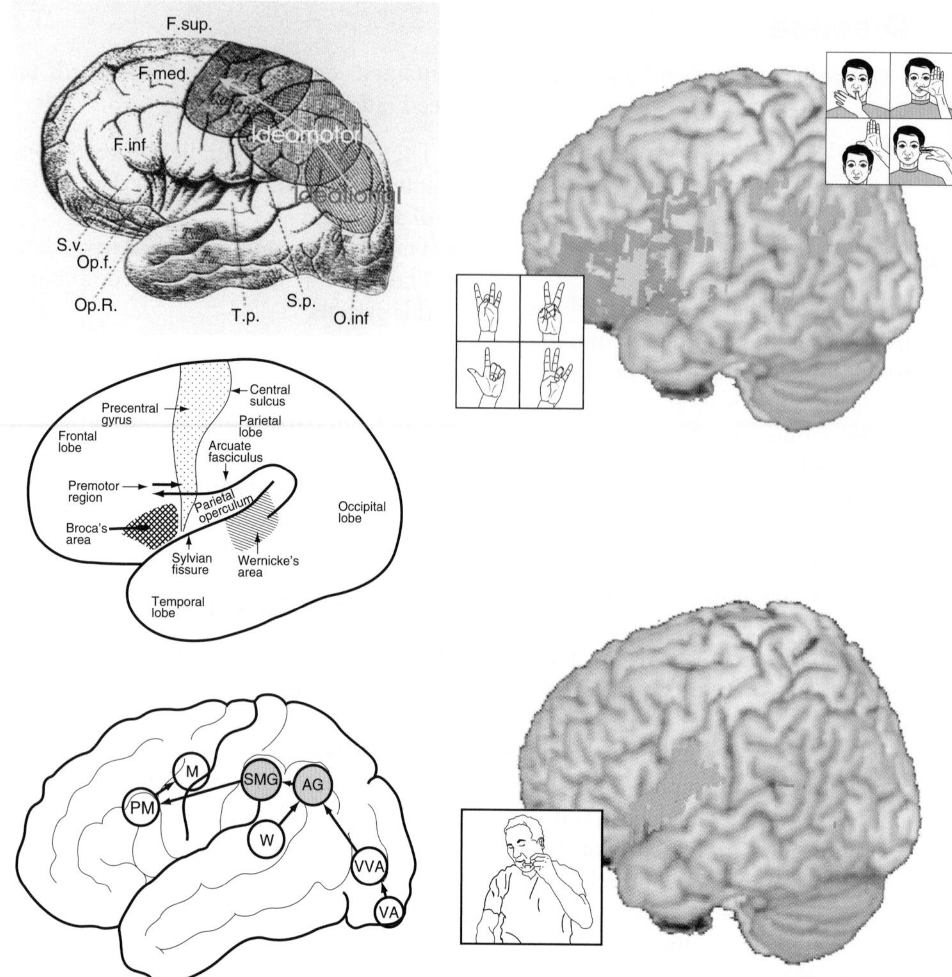

Figure 1 (Left side) Three versions of the 'praxis system.' They concur in the assumption of a posterior to anterior stream of action control with a central role for the parietal lobe. (Right side:) MRI-lesion subtraction identifying left hemisphere regions, damage to which is crucial for different manifestations of apraxia. (Above) Imitation of hand postures is susceptible to lesions in inferior parietal lobe and the parietotemporal junction, while imitation of finger postures is mainly affected by lesions in the inferior frontal lobe. (Below) Disturbed pantomime of object use is predominantly a symptom of left inferior frontal lesions. These lesion locations are hardly compatible with any version of the 'praxis system.' Reproduced from Goldenberg G and Karnath HO (2006) The neural basis of imitation is body-part specific. *Journal of Neuroscience* 26: 6282–6287. Copyright 2006 by the Society for Neuroscience.

of imitation have different neural underpinning. In patients with left parietal lesions, impressive dissociations have been demonstrated between severe impairment of imitating meaningless gestures and perfect imitation of meaningful gestures. Due to the possibility of compensating defective imitation of meaningful gestures by treating them like meaningless ones, the reverse dissociation is less easily demonstrated, but lesion analysis in a group study of imitation has provided evidence that imitation of meaningful gestures depends primarily on integrity of left medial temporal rather than parietal regions.

Meaningless gestures are not a uniform class either. One important distinction is between single gestures and sequences of several gestures. Imitation of single gestures and sequences is strongly correlated in patients with LBD, whereas defective imitation of gesture sequences contrasting with normal imitation of single gestures has been demonstrated in patients with right hemisphere lesions, Parkinson's disease, or supranuclear palsy. It thus appears that LBD affects imitation of single gestures and sequences equally, but that sequences are also vulnerable to lesions elsewhere.

Another feature of the imitated gesture which may be important for the laterality of interfering lesions is the body part involved. In the 1980s, a lively debate arose over whether imitation of whole body and axial movements such as bending the head down or raising the shoulder is generally spared in patients with LBD who commit errors on imitation of limb gestures. The debate has been settled by the conclusion that imitation of axial movements is indeed less error prone than imitation of limb movements but that their preservation is due to the lower complexity of the axial gestures. There are far fewer degrees of freedom for axial postures than there are for postures of the limbs.

A dissociation which can less easily be referred to different task difficulty has been demonstrated between gestures of the fingers, the whole hand, and the foot. LBD patients have difficulties with all three kinds. By contrast, patients with right brain damage (RBD) have severe difficulties with finger postures and some difficulties with foot postures but imitate hand postures nearly as perfectly as healthy controls. Within the left hemisphere, different locations are important for imitation of hand and finger postures: whereas imitation of hand postures depends on integrity of inferior parietal lobe and parietotemporal junction, imitation of finger postures is predominantly compromised by lesions affecting the inferior frontal lobe and subjacent white matter (**Figure 1**).

Production of Communicative Gestures on Command

Gestures are a pervasive component of human communication. Their dependence on simultaneous verbal communication varies from gestures which emphasize or modulate the meaning of simultaneous oral speech to sign languages which can completely replace oral language. Diagnosis and research on apraxia traditionally concentrate on a rather small sector of this wide array, namely, gestures which can transmit meaning independently of speech but which are not normally combined into syntactic structures replacing speech, as is the case for sign languages. Such gestures can be 'emblems' which convey a conventionally defined message, like thumb up for approval or the nose thumb for mockery, or pantomimes of object use in which an object and its use are indicated by performing the motor action of use with the empty hand. The majority of research concerns pantomime rather than emblematic gestures. One of the reasons for this preference may be simple pragmatism: comprehension of the verbal instruction to demonstrate a meaningful gesture out of its natural communicative context is notoriously problematic in patients with aphasia. For pantomime of tool use, comprehension can be facilitated by first demonstrating correct pantomimes and then showing pictures of the single objects whose use should be demonstrated. By contrast, there is hardly any way to bypass defects of verbal comprehension when instructing patients which emblematic gesture they should demonstrate. Comprehension of the instruction to pantomime remains nonetheless dubious in severely aphasic patients, who in response to the command to pantomime the use of an object try to name or describe it, to grasp it for actual use, or to outline with the finger a more or less recognizable shape on the table. Independence of apraxic errors from language comprehension becomes plausible when patients make searching movements for the correct grip or movement or when their pantomime displays some but not all distinctive features of the intended pantomime (e.g., pantomiming drinking from a glass with a narrow grip not accommodated to the width of the pretended glass). In severe cases, patients may produce stereotyped circling or swaying movements of the hand, which might be taken to indicate but not specify movement of the object in peripersonal space.

Defective production of communicative gestures to command is invariably bound to left hemisphere lesions and virtually always associated with aphasia. This tight association could lead to the belief that defective production of communicative gestures is a sequela of language impairment or that one common basic disorder underlies both aphasia and defective gesturing. The latter idea was formulated before Liepmann by another German psychiatrist, Finkelnburg, who conceived of such a basic disorder as 'asymbolia,' that is, a general inability to comprehend and produce communicative signs. Such proposals are, however, contradicted by observation of single cases of left-handed patients in whom RBD impaired production of meaningful gestures to command but did not cause aphasia. In these patients, the left hemisphere is dominant for language and the right for communicative gestures. The possibility of such dissociation argues against any hypothesis claiming identity of the neural mechanisms subserving production of speech and communicative gestures.

A recent study subtracted lesions of LBD patients with intact pantomime from those of LBD patients with defective pantomime. Contrary to traditional beliefs, the proportion of parietal lobe lesions was the same in both groups, whereas inferior frontal lobe lesions were distinctly more frequent in patients with defective pantomime. The independence of communicative gestures from integrity of the parietal lobes is supported by single case reports of 'visuoimitative' apraxia. In these patients, defective imitation of meaningless gestures stands in contrast to preserved production (and in some of them, also imitation) of communicative gestures. These patients invariably had left parietal lobe damage. Their preserved production of pantomimes endorses the conclusion that integrity of the left parietal lobe is not obligatory for correct pantomime of tool use.

Use of Single Tools

Defective use of single tools or objects is less frequent than either defective imitation or defective production of communicative gestures. Affected patients may try to cut paper with closed scissors, write with the wrong end of the pencil, press the knife into the loaf without moving it to and fro, press the hammer on the nail without hitting, or close the paper punch on top of the sheet without inserting the sheet. The responsible lesions are always left-sided and patients are aphasic. Lesions are usually large. They involve the parietal lobes but are not confined to them.

More-detailed insights into the location of responsible lesions come from studies which distinguish between different components of tool use. Conventional tools such as hammers, screwdrivers, knives, or scissors have in common that they are familiar to most persons and that their function is based on transparent mechanical relationships between tools and object. It has been proposed that knowledge about their correct use can be based on either of these properties. As the tools are familiar, knowledge about their prototypical use is stored in semantic memory. Such knowledge presumably specifies the typical purpose of the tool, the object it is associated with, and the motor action of its use. As their function is based on transparent mechanical relationships, possible functions can also be deduced directly from structural properties. Other than retrieval of knowledge from semantic memory, direct inference of function from structure permits detection of nonprototypical uses of familiar tools and discovery of possible uses of novel tools. For example, when there is no hammer for driving in a nail, one may use pliers because they have a flat and rigid surface which can transmit the power of the beat onto the nail.

The observation that only patients with extensive LBD fail use of single conventional tools has led to the conclusion that both sources of knowledge about their correct use are based on left hemisphere function. This conclusion is corroborated by experimental studies which tested integrity of each of the alternative sources. Knowledge about prototypical tool use is a prerequisite for pantomime of tool use (see above). The observation that pantomime is intact in patients with RBD thus indicates that the right hemisphere does not contribute to its retrieval. Likewise, only left brain-damaged patients commit errors when shown the picture of a tool and asked to select in multiple choice the object that is typically associated with the tool or another tool which serves the same purpose. Evidence that the capacity to infer possible function from structure is also bound to left hemisphere integrity comes from studies which required patients to detect the function of novel tools or to select an alternative tool when the tool typically used for a task is absent, such as selecting a coin for turning a screw when there is no screwdriver. Again, patients with left, but not with right, hemisphere damage have difficulties.

Dissociations between both sources of correct use of single familiar objects have been documented between patients with two different degenerative diseases, corticobasal degeneration, and semantic dementia. Whereas the bulk of cortical pathology affects the parietal and superior frontal lobes in corticobasal degeneration, it is concentrated on the temporal and inferior frontal lobes in semantic dementia. Patients with corticobasal degeneration fail completely when asked to find out the use of novel tools but can associate familiar tools with their corresponding object or another tool serving the same purpose. By contrast, patients with semantic dementia who have lost any knowledge about the purpose and prototypical use of common tools can be astonishingly good in inferring possible functions from examination of the structure of the tool. For example, they may find out know that a nail-clipper can be used for cutting but do not know that it is used for cutting nails.

In conclusion, left parietal lobe integrity seems to be necessary for direct inference of function from structure, which is one source of correct tool use. Retrieval of knowledge about prototypical use from semantic memory does not depend on the parietal lobes but may be contingent on integrity of temporal and possibly also inferior frontal regions. Defective use of single tools becomes manifest only when left hemisphere lesions are large enough to destroy both sources.

Multistep Actions with Multiple Tools and Objects

In daily life one is rarely handed a tool and asked to perform its prototypical action on an adequately prepared recipient. Usually the use of the single tool is embedded in a chain of actions involving several tools and objects, frequently including technical devices, and aiming at a superordinate goal transgressing and modulating the purposes of each single action step. Furthermore, daily life may require the parallel completion of two or more multistep actions, such as when preparing the components of a meal.

Such multistep actions tax mental capacities beyond retrieval of knowledge about the use of single tools. They require maintenance of the ultimate goal of the whole action as well as of a record of already completed and outstanding action steps. The completion of steps must be monitored to secure that the next step is not initiated before the previous one is completed. The order in which outstanding steps are initiated may necessitate preplanning for consideration of sequential constraints.

In contrast to the exclusive association of defective use of single familiar tools with LBD, multistep actions are affected also by lesions of the right hemisphere, by diffuse brain damage, and by frontal lobe damage. This lack of localizing specificity seems to fit well with the involvement of rather difficult-to-localize functions such as attention, executive function, or working memory capacity. Alternatively it might indicate that the cognitive complexity of multistep actions renders them vulnerable to multiple cognitive deficiencies and that failure on multistep actions has different causes in different patient groups.

The Praxis System

The left side of **Figure 1** shows Liepmann's original version of a left hemisphere praxis system together with two influential modern accounts, by Geschwind and Heilman. They concur in postulating a posterior-to-anterior processing stream leading to motor execution of skillful motor actions. In Liepmann's version this stream had its origin in a visual image of the action generated in occipital regions. Geschwind emphasized the production of gestures to verbal command and put the start of the processing stream into Wernicke's area in the posterior temporal lobe as this area plays a pivotal role in language comprehension. Heilman's model embraced both possibilities but shifted the occipital source nearer to primary visual cortex to account for imitation of visually presented gestures. All these versions postulate a crucial role of parietal lesions. Liepmann's and Geschwind's models are 'hodological.' They ascribe the importance of parietal lesions to the interruption of fiber tracts passing through the parietal lobe on their way from the origin to the target of the praxis system. By contrast, Heilman contended that parietal cortex stores 'temporo-spatial engrams' needed for skillful execution of motor actions.

Of course, all authors were aware that apraxia is not a unitary syndrome but can affect different domains of actions to different degrees. They postulated, however, that such dissociations can be accounted for by different distributions of lesions within the praxis system. The most influential version of such a division had already been proposed by Liepmann. He distinguished between ideational apraxia, afflicting the concept of the action, and ideomotor (in Liepmann's original terminology, 'ideo-kinetic') apraxia, resulting from an interruption of the stream from posterior to motor regions of the brain. He thought that a hallmark of ideomotor apraxia is its restriction to movements made without external objects. "The guidance of the left hand by mental images of the shape of the movement is disturbed. In the majority the disturbance becomes manifest only if patients are required to imitate or to make expressive movements or manipulations without object." Later authors have suggested amendments to this classification and have introduced additional terms, but Liepmann's original division is still the most popular. It classifies defective imitation, together with defective production, of communicative gestures as ideomotor apraxia, and defective use of tools and objects as ideational apraxia. The empirical findings reviewed above do not support such classification: pantomime of tool use has another neural substrate than imitation, and the neural substrate of imitation varies according to which kind of gesture is imitated. The neural substrate of tool use varies according to whether single tools or multistep actions involving several tools and objects are probed.

The central role of left parietal lobe lesions in apraxia does not fare better when the empirical findings are considered. Although some manifestations of apraxia are indeed bound to left parietal lesions, others are predominantly caused by inferior frontal regions. Even the association of apraxia with LBD is limited: for imitation of finger positions and for multistep actions, RBD is as deleterious as LBD. Obviously, there is no such entity as a left hemisphere praxis system, damage to which explains all manifestations of apraxia. It rather seems as if the disorders subsumed under the heading of apraxia are manifestations of different, more or less related, cognitive deficits. Such 'deconstruction' of apraxia does not diminish interest in its exploration. On the contrary, liberating imitation of gestures, production of communicative gestures, and use of tools and objects from their forced unification into a unique praxis system has advantages both for clinical practice and scientific research. For clinical practice, equivocal classifications will be replaced by usable descriptions of what patients can and cannot do. For research, abandoning the traditional classification into ideomotor and ideational apraxia clears the stage for discussing the differences and commonalities between the cognitive and neuronal basis of the affected domains of action.

See also: Apraxia: Sensory System; Basal Ganglia: Motor Functions; Cerebellar Lesions and Effects on Posture, Locomotion and Limb Movement; Finger Movements: Control; Map Plasticity and Recovery from Stroke; Posterior Parietal Cortex and Arm Movement; Posterior Parietal Cortex and Tool Usage and Hand Shape.

Further Reading

De Renzi E (1990) Apraxia. In: Boller F and Grafman J (eds.) *Handbook of Clinical Neuropsychology*, vol. 2, pp. 245–263. Amsterdam: Elsevier.

Geschwind N (1975) The apraxias: Neural mechanisms of disorders of learned movements. *American Scientist* 63: 188–195.

Goldenberg G (2003) Apraxia and beyond: Life and works of Hugo Karl Liepmann. *Cortex* 39: 509–525.

Goldenberg G (2008) Apraxia. In: Goldenberg G and Miller B (eds.) *Handbook of Clinical Neurology, 3rd Series, Vol 88: Neuropsychology and Behavioral Neurology*, pp. 323–338. Edinburgh: Elsevier.

Goldenberg G (in press) How the mind moves the body: Lessons from apraxia. In: Morsella E, Bargh JA, and Gollwitzer PM (eds.) *The Psychology of Action*, vol. 2. New York: Oxford University Press.

Goldenberg G, Hermsdörfer J, Glindemann R, Rorden C, and Karnath HO (2007) Pantomime of tool use depends on integrity of left inferior frontal cortex. *Cerebral Cortex* 17: 2769–2776.

Goldenberg G and Karnath HO (2006) The neural basis of imitation is body-part specific. *Journal of Neuroscience* 26: 6282–6287.

Hartmann K, Goldenberg G, Daumüller M, and Hermsdörfer J (2005) It takes the whole brain to make a cup of coffee: The neuropsychology of naturalistic actions involving technical devices. *Neuropsychologia* 43: 625–637.

Heilman KM and Rothi LJG (1993) Apraxia. In: Heilman KM and Valenstein E (eds.) *Clinical Neuropsychology*, pp. 141–164. New York: Oxford University Press.

Rothi LJG, Ochipa C, and Heilman KM (1997) A cognitive neuropsychological model of limb praxis and apraxia. In: Rothi LJG and Heilman KM (eds.) *Apraxia: The Neuropsychology of Action*, pp. 29–50. Hove, UK: Psychology Press.

Schwartz MF, Buxbaum LJ, Montgomery MW, et al. (1999) Naturalistic action production following right hemisphere stroke. *Neuropsychologia* 37: 51–66.

Spatt J, Bak T, Bozeat S, Patterson K, and Hodges JR (2002) Apraxia, mechanical problem solving and semantic knowledge: Contributions to object usage in corticobasal degeneration. *Journal of Neurology* 249: 601–608.

Tessari A, Canessa N, Ukmar M, and Rumiati RI (2007) Neuropsychological evidence for a strategic control of multiple routes in imitation. *Brain* 130(4): 1111–1126.

Apraxia: Sensory System

L J Buxbaum, Moss Rehabilitation Research Institute and Thomas Jefferson University, Philadelphia, PA, USA
H B Coslett, University of Pennsylvania School of Medicine and Center for Cognitive Neuroscience, Philadelphia, PA, USA

Historical Background

Limb praxis is subserved by a complex multicomponent system that provides a processing advantage to previously experienced, purposive movements. The term 'apraxia' was introduced by Steinthal in 1871. While this word, derived from Greek, means 'without action', the term is used to describe a decrease or disorder in the ability to perform skilled movements. Scientific and clinical interest in the disorder dates to the early twentieth century, when Liepmann reported patients with cerebral lesions who were unable to gesture to command or, in some instances, to imitation. Subsequently, Liepmann and Maas (1907) described a patient with a lesion of the corpus callosum who was unable to produce gestures with the left hand to verbal command. On the basis of these findings, Liepmann proposed that the left hemisphere was "dominant" for gesture in that it supported the learned "movement formulae" or "time-space-form picture of the movement" which specified the timing, trajectory, and content of learned movements.

Liepmann's ideas were extended by Geschwind (1965), who proposed a specific left hemisphere-based neural circuitry for movement representations. On his account, failure to produce a movement to command was attributable either to a disruption of Wernicke's area, with resultant failure to understand the command, or to a disconnection of the posterior language areas from motor cortex. A failure to imitate movements was attributed to a lesion involving the arcuate fasciculus, which was assumed to connect the visual association cortex to motor cortices.

Limb Apraxia Subtypes

Hugo Liepmann's description of three major forms of apraxia brought about a 'paradigmatic shift' in our understanding of motor control. These three types were 'limb kinetic apraxia' (LKA), 'ideational apraxia' (IA), and 'ideomotor apraxia' (IMA). To this triad, Gonzalez-Rothi, Heilman, and colleagues added another type, termed 'conceptual apraxia.' These types of apraxia are described briefly below.

LKA

Patients with LKA perform actions with slow, stiff, clumsy movements and exhibit impairment on tasks requiring rapid independent finger movements, such as rotating a coin between the thumb, index, and middle finger. Errors are more apparent in distal (finger) movements than in proximal movements. LKA is associated with lesions that include the primary motor cortex, premotor cortex, or descending corticospinal tract. It frequently occurs in patients with stroke and in degenerative disorders such as progressive supranuclear palsy and corticobasal degeneration.

IA

IA is defined as an impaired ability to carry out a sequence of acts that leads to a goal and that incorporates multiple objects, such as making a sandwich or lighting a candle. For example, a patient with IA might attempt to seal an envelope prior to inserting the letter. Another type of error exhibited in this disorder is illustrated in **Figure 1**.

IA is most frequently induced by bilateral damage and degenerative dementia. Injury to the frontal lobes is often also associated with temporal order processing deficits as well as impaired working memory, and thus one of the critical foci of dysfunction in IA may be in frontal-subcortical systems. The strongest predictor of errors in multistep, naturalistic action is overall severity of cognitive impairment and not lesion location.

Conceptual Apraxia

Patients with conceptual apraxia make content errors in complex action – that is, they substitute incorrect objects or movements in their actions. For example, patients with conceptual apraxia may eat with a toothbrush. In some cases, underlying deficits in knowledge of specific tools or objects or the association of tools and objects has been demonstrated; these patients may misuse objects because they have lost knowledge regarding the function of the object. Conceptual apraxia also frequently co-occurs with IMA, and it has not been established whether these disorders can be reliably disambiguated.

Deficits in conceptual action knowledge have been associated with the dominant posterior parietal lobe and/or temporal parietal junction. On the other hand, errors apparently attributable to conceptual deficits frequently occur in patients whose lesions entirely spare brain regions typically associated with conceptual action knowledge (e.g., right parietal cortex).

IMA

IMA is a common disorder of complex skilled action not attributable to weakness, incoordination, or other elemental sensory or motor impairments. It is typically observed in individuals who have suffered left hemisphere strokes; IMA is observed in the actions of the 'unimpaired' left hand of approximately 50% of patients with left hemisphere stroke and commonly persists for at least 1 year after stroke. IMA is also common in Alzheimer's disease and in corticobasal degeneration. In stroke, it is usually a consequence of damage to the left inferior parietal lobe (and, on occasion, adjacent intraparietal sulcus and superior temporal gyrus) but has also been observed following left dorsolateral prefrontal, callosal, and subcortical damage (see **Figure 2**).

IMA is usually diagnosed on the basis of spatiotemporal errors in the production of gesture pantomime both to sight of objects and on imitation of others. That is, IMA is typically seen when a patient is asked to show how an object (e.g., scissors) would be used or when the patient is asked to copy a gesture produced by the examiner. Kinematic analyses have revealed that IMA patients pantomime skilled

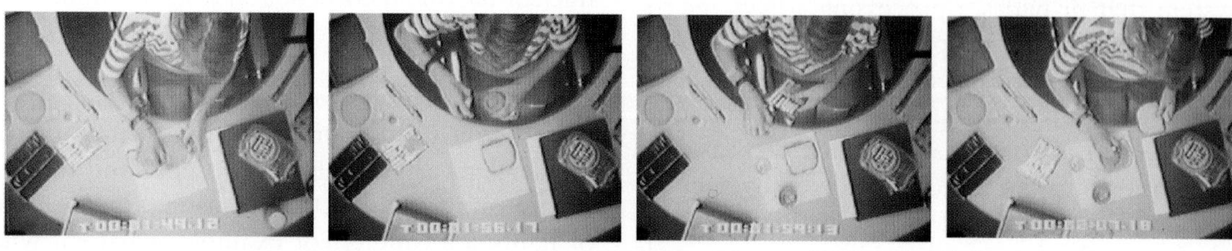

Figure 1 Photographs of a patient with conceptual apraxia making a sandwich with meat and mustard. She correctly places meat on a slice of bread, closes the sandwich, and opens a mustard jar. She replaces the mustard jar, reaches into a package of marking pens, retrieves a yellow marker, and proceeds to color the meat yellow.

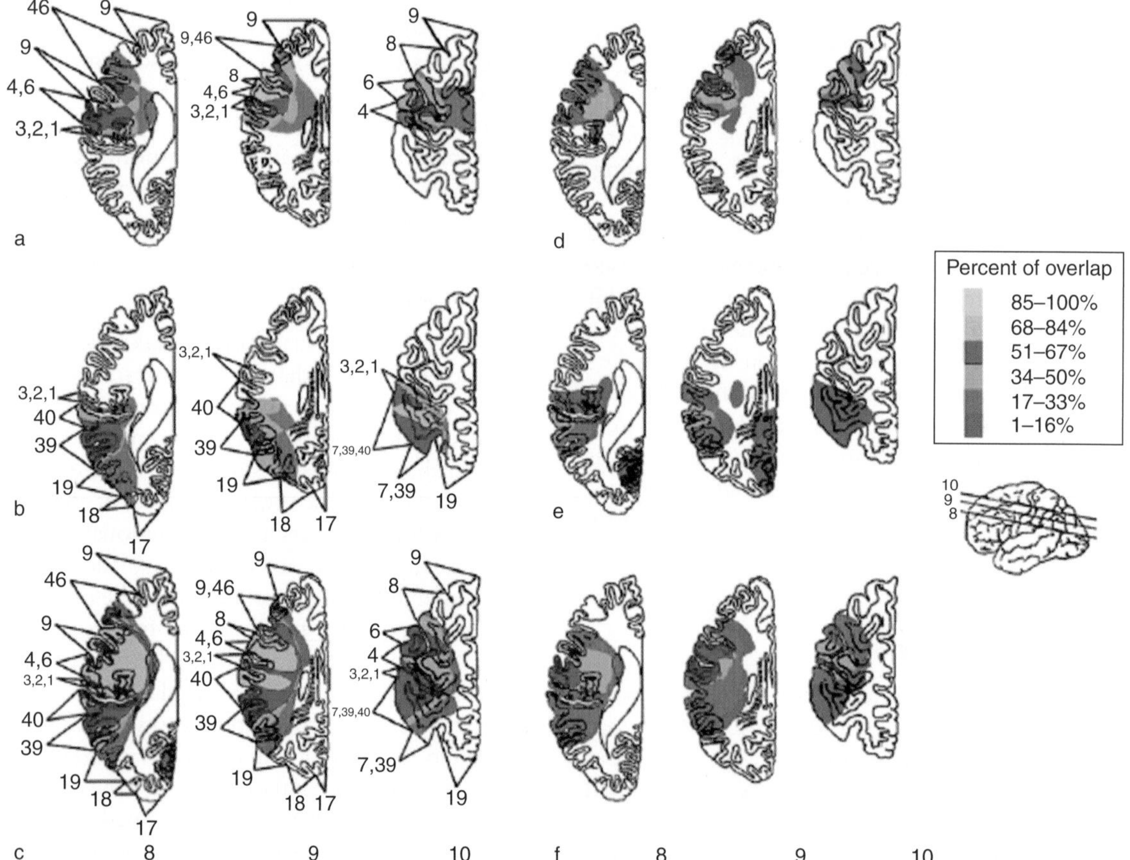

Figure 2 Maximal lesion overlap from 17 apraxic patients is shown in dorsolateral and inferior parietal regions. From Haaland KY, Harrington DL, and Knight RT (2000) Neural representations of skilled movement. *Brain* 123: 2306–2313.

Figure 3 Typical errors in ideomotor apraxia. Top: Three still photographs from a videotape showing a sequence of postures produced by an apraxic patient in imitating a sawing movement. Note the typical hand posture error comprised of repeated hand opening with arm extension, despite the fact that the model maintained a closed grip throughout. Bottom left: Typical 'body-part-as-object' error in a toothbrushing pantomime. Bottom center: Typical arm posture error in imitating a scissoring movement. The model's movement was produced perpendicular to the body wall (from near to far) whereas the patient's movement proceeded left to right. Bottom right: Typical amplitude error in imitation of a hammering gesture. The model to be imitated demonstrated a large swing with peak amplitude at shoulder height, whereas the patient's maximal amplitude was at elbow height.

tool-use movements with abnormal joint angles and limb trajectories and uncoupling of the spatial and temporal aspects of movement (see **Figure 3**). Spatiotemporal errors persist to a lesser degree with actual tool use. The deficit is not restricted to meaningful movements and has also been observed in meaningless postures and sequences. IMA is also associated with cognitive deficits in declarative knowledge of the manipulation actions appropriate to objects, impairments in mechanical problem solving, deficits in motor planning, and difficulty learning new gestures.

The disorder may be attributed either to damage to stored spatiotemporal gesture representations in the left parietal lobe, sometimes called 'visuokinesthetic engrams,' or to disconnection of intact movement representations from motor output. The integrity of gesture representations is thought by many investigators to bear on the integrity of gesture recognition. In the case of damage to the representations, patients have impaired knowledge of the appropriate motor action to perform, as evidenced by deficits in gesture recognition (representational IMA). In the case of disconnection of intact engrams, patients have unimpaired knowledge of appropriate gestures, as evidenced by intact gesture recognition and ability to discriminate correct from incorrect gestures, but nevertheless perform with spatiotemporal errors (dynamic IMA).

In representational IMA, inability to discriminate correctly from incorrectly performed meaningful object-related hand movements correlates strongly with ability to produce the same movements, suggesting that the same representations may underlie both. In addition, representational IMA patients are significantly more impaired when producing object-related than symbolic, nonobject-related movements. This in turn suggests that the damaged system underlying representational IMA is specialized for movements related to skilled object use.

Disconnection and Dissociation Apraxias

Several apraxia patterns indicate that aspects of input to and output from the skilled action system are dissociable. Verbal–motor dissociation apraxia refers to a pattern of impairment in which patients are unable to gesture in response to command despite adequate comprehension and unimpaired ability to gesture to imitation. Heilman and colleagues posited that the lesion responsible for this apraxia subtype was in the angular gyrus, but they were unable to obtain neuroimaging data. Another reported pattern is seen in the tactile–motor and visuomotor dissociation apraxias, in which patients fail to gesture appropriately when holding tools or viewing tools, respectively, despite unimpaired object recognition and better gesture performance in the unaffected modality. In response to these and other patterns of dissociation, Gonzalez-Rothi, Heilman, and colleagues proposed a detailed diagrammatic model of IMA. Theoretical 'lesions' at various loci in the model appear to explain many of the observed dissociations.

Outstanding Issues in Diagnosis of Apraxia Subtype

Relevance of Recognition and Imitation Deficits for Diagnosis of IMA versus IA or Conceptual Apraxia

Historically, gesture recognition and imitation have both been used to distinguish between IMA and IA/ conceptual apraxia. In Liepmann's account, patients with IA fail to reliably activate gesture engrams. Consequently, they perform normally when provided with the 'idea' of the movement; that is, when they are asked to imitate the movement of another person. Liepmann believed that, by contrast, patients with IMA suffered a disconnection of an intact idea (time–space–form picture of the movement) from motor innervatory patterns. Thus, on Liepmann's account, providing IMA patients with the 'idea' in the form of a gesture to imitate would not be of benefit.

On many contemporary accounts, it is representational IMA patients who fail to reliably activate gesture representations and who therefore may be able to imitate gestures. Recent evidence indicates that imitation may be accomplished either via gesture engrams (the so-called indirect or semantic route), or by way of a 'direct route' to action that enables imitation without access to meaning. (The direct route may bear a relationship to the putative 'mirror neuron' system discussed below.) Therefore, the ability to imitate may depend on the integrity of each of these routes. On the other hand, there is evidence that the direct route is not used for meaningful gestures even when it is intact, suggesting that there may be obligatory activation of the semantic route whenever familiar gestures are viewed.

Disagreement persists on whether gesture recognition problems signify IMA or IA. However, recent evidence from monkeys and humans indicates that the same representations are likely used for action recognition and production. In the macaque, cells in the inferior parietal lobule and in a sector of premotor cortex corresponding to Brodmann's areas 44 and 45 in humans respond both when the monkeys produce actions and when they observe the same specific actions performed by others ('mirror neurons'). In humans, there are strong correlations between action production and recognition for the same items. This suggests that gesture recognition problems may reflect degraded or inaccessible sensorimotor representations, a characteristic of IMA.

Relationship of Object Knowledge to Gesture Representations

Continued work is required to clarify the relationship of knowledge of appropriate object-oriented actions to the gesture engram system. In an influential paper on IA, De Renzi and Lucchelli proposed that the problem underlying deficient object use was a loss of knowledge of the manner in which objects are to be used, which they characterized as a semantic deficit. This emphasis on 'manner' of manipulation raises questions about the role gesture engrams might play in object-use knowledge. In contrast, other investigators view deficient recognition of the gestures associated with objects (that is, the manner of use) to be a symptom of the representational type of IM, and not IA.

Recent evidence suggests that different types of object knowledge may bear different relationships to apraxia. For example, knowledge of object function and knowledge of manner of object manipulation are dissociable. One might have knowledge of the function of an object (e.g., a knife is for cutting things) without knowledge that knives are often used with a back-and-forth, sawlike gesture. Patients with IMA tend to have the latter type, but not necessarily the former. Some reports indicate that patients with IMA may also be impaired in mechanical problem solving or the ability to infer function from structure.

Evidence for a relationship between function knowledge and performance on tasks involving multiple objects is equivocal. One potential source of confusion is that the relationship between functional knowledge and object use is sometimes assessed with single objects and sometimes with tasks involving multiple objects. In several studies using single object tests in patients with semantic dementia, a disorder with a predilection for the temporal lobes, a strong relationship has been reported. In other investigations, no relationship between single object use and functional knowledge has been found. There is stronger evidence that function knowledge is not well correlated with performance on tasks involving multiple objects. There are patients who make 'conceptual' errors on these tasks but who perform normally on semantic tests of functional and associative object knowledge, and others who perform nearly normally on tasks involving multiple objects or in real-life action, despite considerable semantic deficits.

Functional Implications of Limb Apraxia

Historically, most clinicians and researchers regarded IMA as a clinical oddity that had little significance in the real world. It appears that this view was derived from the notion that IMA was present when gestures to command and imitation were tested but improved when actions with actual objects were examined. A number of recent studies, however, have suggested that IMA is associated with deficits in activities of daily living. At least in some studies, participants with IMA are more likely to be impaired in object use, particularly in complex tasks, than nonapraxic participants who have suffered a stroke.

IMA in View of Recent Developments in the Motor Control Literature

Imitation

With the discovery of mirror neurons in the macaque premotor and parietal cortex that respond to both observed and performed actions, action imitation has emerged as an area of considerable interest in the neuroscience community. One important question concerns the degree to which imitation failure in IMA reflects damage to the mirror neuron system. Indeed, the neuroanatomic loci of lesions leading to IMA overlap considerably with the localization of mirror neurons (see **Figure 4**). In addition, as noted, imitation and recognition impairments show a strong correspondence in IMA. On the other hand, IMA due to left parietal lesions frequently disrupts object-related (transitive) imitation far more than nonobject related, symbolic (intransitive) imitation. In addition, there is evidence of body part specificity in IMA imitation disruption that is not easily accommodated by putative damage to a mirror neuron system. Left hemisphere IMA patients tend to be significantly more impaired in imitation of hand postures than of finger positions, and in general, IMA appears to affect arm more than leg imitation. These dissociations could be accommodated by positing that effector-specific populations of mirror neurons might reside in different cortical regions in each hemisphere, but to this point there is little evidence for this possibility. Future investigations addressing these issues are required.

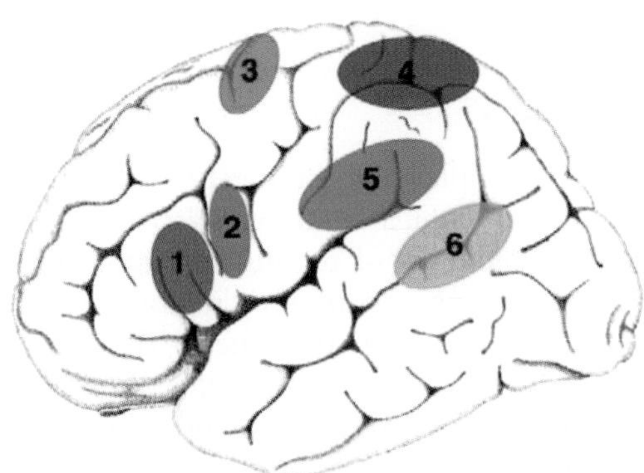

Figure 4 Mirror neuron cortical regions. From Brass M and Heyes C (2005) Imitation: Is cognitive neuroscience solving the correspondence problem? *Trends in Cognitive Sciences* 9(10): 489–495.

Object-Related Action

Recent evidence from single cell recordings in monkeys indicates that populations of neurons in the inferior premotor cortex (in an area with probable homolog of areas 44 and 45 in humans) as well as in the anterior intraparietal sulcus are active in response to objects that are graspable by the monkey observer. These have been termed 'canonical' neurons. Complementary studies using functional magnetic resonance imaging and transcranial magnetic stimulation in humans are consistent in suggesting that similar regions in the human brain are responsive to the structural properties (i.e., shape and size) of graspable objects. These populations appear to encode hand movement parameters (e.g., finger thumb aperture) for object grasping. In this context, there is considerable recent evidence that IMA patients, while intact in their ability to position the hand in response to object structure, are disproportionately impaired in hand shaping for functional object manipulation. The relationship of this pattern of performance to the 'canonical' neuron system is an additional area of interest for future investigation.

Spatiomotor Frames of Reference for Action

At least two different frames or reference or coordinate systems have been proposed for action. Many investigators have proposed that action may be planned in workspace-specified extrinsic coordinates. On this account, movements are planned with respect to a target that is coded in external space. Reaching to grasp a target would entail the creation of a spatial vector describing a desired movement's direction and amplitude. An alternative hypothesis proposes that movement control may occur in body-specified intrinsic coordinates; on this account, a movement plan would specify the positions of the shoulder, elbow, and wrist that would be needed to get the hand to the target. Extensive evidence for both types of control has led to a third group of accounts proposing that control is an interactive process that uses both extrinsic and intrinsic coordinate frames, depending in part on the demands of the task.

Recent evidence from IMA patients indicates that movements that may putatively rely strongly on extrinsic control (i.e., object-directed movements) are accurate, whereas movements not having external referents (i.e., body-directed movements) are characterized by spatial errors in hand configuration, wrist angle, hand orientation, and hand location. The possibility that IMA may in part reflect deficient coding of action in a body-centered framework is an area of active investigation in several laboratories.

Feedforward and Feedback-Driven Processes in IMA

The process of motor control is commonly subdivided into planning and online correction components. Planning is the preparation of a movement before movement initiation, whereas online correction refers to the adjustment of the movement plan during movement execution. There is evidence that IMA may be attributable in part to deficits in planning actions with relatively intact online correction. IMA patients are impaired in motor imagery, thought by several investigators to be a proxy for motor planning stages of action. They are also abnormally disrupted when visual feedback of movement is unavailable. This suggests that such patients may rely abnormally on visual feedback in the performance of skilled action.

Treatment of Limb Apraxia

The current literature on apraxia treatment is sparse. Approximately ten treatment efforts have been reported; in many cases, there is but a single study devoted to each treatment approach. The studies uniformly fall into the category of Phase I studies in which feasibility is assessed in small numbers of patients. Thus, it is difficult at this stage to draw conclusions about treatments that may hold particular promise.

In general, the few reported treatment approaches can be grouped into three categories: (1) studies that attempt to directly ameliorate deficient object-related gesture production with a variety of visual and tactile cues and feedback, (2) studies focusing on providing corrective feedback for errors in naturalistic multistep action, and (3) studies that attempt to prevent error from occurring (errorless learning approach). All the studies report at least some treatment benefit, but several difficulties obscure the interpretation of results. For example, apraxia type is frequently poorly characterized. Although gesture recognition is clearly an important index of the integrity of gesture representations (which in turn may have important implications for rehabilitation strategies), recognition testing is usually not performed. Only a few studies report generalization to untreated stimuli (or behaviors), maintenance of treatment effects, or impact on daily activities.

There is preliminary evidence based on these few studies that limb apraxia is amenable to treatment. The purpose of Phase I research, however, is to develop hypotheses, protocols, and methods; establish safety and activity; determine the best-outcome measures; identify responders versus nonresponders; determine optimal intensity and duration; and determine why the treatment is producing an effect. This suggests that further systematic inquiry is required to satisfy the objectives of Phase I research.

Testing for Limb Apraxia

Apraxia cannot be assessed in patients whose comprehension or cognitive deficits prevent them from understanding the task or whose visual deficits preclude identification of an object or gesture; before one tests for apraxia, these disorders must, therefore, be excluded. In order to identify apraxic patients and distinguish between the different types of apraxia described above, a testing battery should include at least the following components:

1. Assessment of manual dexterity (e.g., rotation of coin between fingertips).
2. Testing of gesture to command and to sight of object.
3. Imitation of meaningful and meaningless gestures.
4. Assessment of intrinsic egocentric spatial coding – that is, the ability to imitate meaningless static positions of the body such as holding the dorsum of the left hand against the right cheek.
5. Assessment of extrinsic egocentric spatial coding by reaching to touch or grasp objects.
6. Tests of functional semantic knowledge (e.g., which two of three pictured objects – paper clip, rubber band, and door lock – are used for the same purpose).
7. Tests of manipulation knowledge (e.g., which two of three pictured objects – saw, clothes iron, and watering can – is used with the same or similar gesture).
8. Perform a familiar multistep task such as preparing a cup of instant coffee.
9. Recognize gestures by naming a gesture or selecting which of two gestures is correctly performed.

Conclusions

Apraxia is a complex and heterogeneous disorder that has important clinical and scientific implications. Recent investigations of the disorder that are motivated by emerging accounts of motor control and planning are beginning to explicate the processing impairments underlying the apraxic disorders. The accumulating knowledge offers promise not only for the development of treatments for apraxia but also for greater understanding of the procedures by which actions are generated and the underlying neural basis.

See also: Apraxia: Disease; Basal Ganglia: Motor Functions; Computational Approaches to Motor Control; Finger Movements: Control; Kinematics and Dynamics; Motor Skill Learning; Motor Sequences; Motor Timing; Motor Psychophysics; Neural Coding of Spatial Representations; Sensorimotor Integration: Attention and the Premotor Theory; Sensorimotor Integration: Models; Sensorimotor Control of Manipulation.

Further Reading

Buxbaum LJ, Johnson-Frey SH, and Bartlett-Williams M (2005) Deficient internal models for planning hand–object interactions in apraxia. *Neuropsychologia* 43: 917–929.

Buxbaum LJ and Saffran EM (2002) Knowledge of object manipulation and object function: Dissociations in apraxic and nonapraxic subjects. *Brain and Language* 82: 179–199.

Derenzi E and Lucchelli F (1988) Ideational apraxia. *Brain* 111: 1173–1185.

Goldenberg G (2003) Apraxia and beyond: Life and work of Hugo Liepmann. *Cortex* 39: 509–524.

Goldenberg G and Hagmann S (1997) The meaning of meaningless gestures: A study of visuo-imitative apraxia. *Neuropsychologia* 35: 333–341.

Gonzalez Rothi LJ, Ochipa C, and Heilman KM (1997) *A Cognitive Neuropsychological Model of Limb Praxis and Apraxia.* Hove: Psychology Press.

Haaland KY, Harrington DL, and Knight RT (1999) Spatial deficits in ideomotor limb apraxia: A kinematic analysis of aiming movements. *Brain* 122: 1169–1182.

Haaland KY, Harrington DL, and Knight RT (2000) Neural representations of skilled movement. *Brain* 123: 2306–2313.

Haaland KY, Prestopnik JL, Knight RT, and Lee RR (2004) Hemispheric asymmetries for kinematic and positional aspects of reaching. *Brain* 127: 1145–1158.

Heilman KM, Rothi LJ, and Valenstein E (1982) Two forms of ideomotor apraxia. *Neurology* 32: 342–346.

Jeannerod M (2003) Simulation of action as a unifying concept for motor cognition. In: Johnson-Frey SH (ed.) *Taking Action: Cognitive Neuroscience Perspectives on Intentional Acts*, pp. 139–163. Cambridge, MA: MIT Press.

Johnson-Frey S, Newman-Norland R, and Grafton S (2005) A distributed network in the left cerebral hemisphere for planning everyday tool use skills. *Cerebral Cortex* 15: 681–695.

Klatzky R, Pellegrino J, Mccloskey B, and Lederman S (1993) Cognitive representations of functional interactions with objects. *Memory & Cognition* 21: 294–303.

Leiguarda RC and Marsden DC (2000) Limb apraxias: Higher-order disorders of sensorimotor integration. *Brain* 123(5): 860–879.

Peigneux P, Van Der Linden M, Garraux G, et al. (2004) Imagining a cognitive model of apraxia: The neural substrate of gesture-specific cognitive processes. *Human Brain Mapping* 21: 119–142.

Rizzolatti G and Matelli M (2003) Two different streams form the dorsal visual system: Anatomy and functions. *Experimental Brain Research* 153: 146–157.

Roy EA and Square PA (1985) Common considerations in the study of limb, verbal and oral apraxia. In: Roy EA (ed.) *Neuropsychological Studies of Apraxia and Related Disorders*, pp. 111–164. Amsterdam: North-Holland.

Rumiati RI and Humphreys GW (1998) Recognition by action: Dissociating visual and semantic routes to action in normal observers. *Journal of Experimental Psychology: Human Perception and Performance* 24: 631–647.

Schwartz MF and Buxbaum LJ (1997) Naturalistic action. In: Rothi LJG and Heilman KM (eds.) *Apraxia: The Neuropsychology of Action.* East Sussex: Psychology Press.

Schwoebel J, Buxbaum L, and Coslett HB (2004) Representations of the human body in the production and imitation of complex movements. *Cognitive Neuropsychology* 21: 285–299.

Sirigu A, Duhamel J-R, Cohen L, Pillon B, Dubois B, and Agid Y (1996) The mental representation of hand movements after parietal cortex damage. *Science* 273: 1564–1568.

Artificial Intelligence

J Feldman

This article is reproduced from 'Artificial Intelligence in Cognitive Science' in the *International Encyclopedia of the Social & Behavioral Sciences*, Vol 2, pp. 792–796, © 2001, Elsevier Science Ltd.

Overview

Artificial intelligence (AI) and cognitive science are two distinct disciplines, with overlapping methodologies but with rather different goals. AI is a branch of computer science and is concerned with construction and deployment of intelligent agents as computer programs, and also with understanding the behavior of these artifacts. The core scientific goal of AI is to understand the basic principles of intelligent behavior that apply equally to animal and artificial systems. Almost all of the work is mathematical or computational in character and much of the literature is technique oriented.

Cognitive science is an explicitly interdisciplinary field that has participation not only from AI, but also from linguistics, philosophy, psychology, and subfields of other social and biological sciences. The unifying goal of cognitive science is to understand and model human intelligence, using the full range of findings and methodologies of the complementary disciplines. As one would expect, a wide range of techniques from the mathematical, behavioral, social, and biological sciences are employed. Cognitive science, in contrast with AI, is defined more by phenomena than by methodology. There are research groups that are active in both AI and cognitive science, but they tend to produce different types of reports for journals and conferences in the two areas.

Shared Origins in the Postwar Cognitive Revolution

Both AI and cognitive science evolved after 1950, and in their early development were more tightly integrated than at present. For much of the first half of the twentieth century, the Anglo-American study of cognition was dominated by the behaviorist paradigm, which rejected any investigation of internal mechanisms of mind. The emergence of both AI and cognitive science was part of a general postwar movement beyond behaviorist theories, which also included new approaches in linguistics and the social sciences. The idea of computational models of mind was a central theme of what is sometimes called the 'postwar cognitive revolution.' One of the leading early AI groups, under the leadership of Allen Newell and Herbert Simon at Carnegie Mellon, was explicitly concerned with cognitive modeling using the symbolic processes of AI. The current version of this continuing effort is the symbolic cognitive architecture known as Soar (the term derived originally from the concept 'state, operator, and result,' but those currently involved in Soar development prefer not to represent the term as an acronym). Another traditional symbolic approach to modeling intelligence is known as ACT (an acronym for 'adaptive control of thought'). However, there is very little work in contemporary AI that is explicitly focused on modeling human behavior as opposed to intelligent systems in general. There is some continuing work on human and machine game playing, but it is not integrated into the fields of AI and cognitive science.

The central idea of AI is computational modeling of intelligent behavior – its main contribution to cognitive science. The basic notion of a computational model is now commonplace in all scientific fields and many other aspects of contemporary life. One builds a detailed software model of some phenomenon and studies the behavior of the model, hoping to gain understanding of the original system. Much of the work in AI has the engineering goal of producing practical systems, and there is no sharp boundary between AI and other applied fields of computer science and engineering. AI techniques are now commonplace in the full range of business, scientific, and public applications. While all fields use computational models, researchers, in computer science in general and AI in particular, invent and study computational techniques for constructing models, presenting the results of simulations, and understanding the limitations of the simulation. AI has traditionally studied the modeling of the most complex phenomena – those relating to intelligence. Because of the technical challenges arising in the construction of these complex simulations, many innovations in computing have arisen in AI and then have been more widely applied.

Domain-Focused Research That Cuts Across AI and Cognitive Science

The relationship between AI and cognitive science is further complicated by the fact that there are currently several distinct research fields that cut across both disciplines, but have separate journals, meetings, etc. The most prominent of these research areas are speech, language, vision, and neural networks. Each of these fields has thousands of practitioners, many of whom are interested in AI, cognitive science, or both. Appropriately, each of these areas is represented by a vast body of literature. As AI and cognitive science have grown, specialized areas such as language and vision modeling have become largely independent, but they do continue to share the development of underlying methodologies. The main areas that have remained as core AI include knowledge representation and reasoning, planning, and problem solving. The study of learning has evolved somewhat differently (see later). There are some common scientific paradigms that cut across all of these fields, and these are discussed in the following sections as they relate to the social and behavioral sciences. The common thread linking AI to cognitive science is reliance on computational models of different kinds.

Role of Formal Logic in AI and Cognitive Science

To a great extent, the early development of AI was based on symbolic, as opposed to numerical, modeling. This led to the introduction of some novel representations such as those of Soar, but the main effect of this was to align AI with formal logic for much of its early history. In fact, much of the driving force for the creation of the field called cognitive science came from people who saw mathematical logic as its unifying theme. This remains a fruitful approach to AI and constitutes one major area of overlap with cognitive science. Mathematical logic is elegant and well developed and can be shown to be, in some sense, general enough to represent anything that can be described formally. There remains a significant community of linguists, philosophers, and computer scientists for whom logic is the only scientific way to study intelligence.

However, the twentieth century was not kind to categorical, deterministic theories in any field, and cognitive science was no exception. A central issue in the study of cognition has been how to describe the meaning of words and concepts. In formal logic, a concept is defined by a set of necessary and sufficient conditions. For example, a bachelor might be defined as a male who never married. The limitations of classical, all-or-none, categories were already recognized by Wittgenstein, who famously showed that concepts such as 'game' could not be characterized by necessary and sufficient conditions, and were better described as family resemblances. Even the definition of 'bachelor' becomes graded when we consider cohabitation, or how old a male needs to be before being considered a bachelor.

Starting in the 1960s, a wide range of cognitive science studies by Rosch and others showed the depth and complexity of human conceptual systems and their relation to language. This helped give rise to the subfield called cognitive linguistics, which overlaps with cognitive science, but has its own paradigms, journals and conferences. The graded and relational nature of human categories undermined the attempt to create a unified science of mind based on formal logic. Another attack on the formalist program arose from the growing understanding of the neural basis of intelligence. A crucial insight was that basic human concepts are grounded in direct experience and that more abstract concepts are mapped metaphorically to more embodied ones. This undercuts the formalist program, and also leads to a separation of AI, which studies intelligence in the abstract, from cognitive science, which is explicitly concerned with human minds. While much excellent work continues to be done using logic, it is now generally recognized that there are a wide range of phenomena that are better handled by biologically based models and/or some form of numerical, often probabilistic, modeling. For a variety of reasons, the movement to quantitative numerical models followed somewhat different paths in AI and cognitive science, but there are some recent signs of reconvergence.

Learning and the Connectionist Approach to Cognitive Science

As it happens, while the formalists were trying to establish a cognitive science based on formal logic, an antithetical neural network movement was also developing, and this approach has become a major force in cognitive science. The two contrasting approaches to cognitive modeling, neural modeling and logic, were mirrored in the two different methods by which early computer scientists sought to achieve AI. From the time of the first electronic computers around 1950, people dreamed of making them 'intelligent' by two quite distinct routes. The first, 'conventional' AI, is to build standard computer programs as models of intelligence. This remains the dominant paradigm in AI and has had considerable success. The other approach is to try to build hardware that is as brainlike as possible and have it learn the required behavior. The history of this 'neural modeling' approach has been well described. After some promising early mathematical results on learning in simple networks, the neural learning approach to modeling intelligence fared much less well for three decades and had little scientific or applied success. Around 1980, a variety of ideas from biology, physics, psychology, and computer science and engineering coalesced to yield a 'new connectionist' approach to modeling intelligence that has become a core field of cognitive science, and also the basis for a wide range of practical applications. Among the key advances was a mathematical technique (back-propagation) that extended the early results on learning to a much richer set of network structures.

Connectionist computational models are almost always computer programs, but programs of a different kind than those used in, for example, word processing or symbolic AI. Connectionist models are specified as a network of simple computing units, which are abstract models of neurons. Typically, a model unit calculates the weighted sum of its inputs from upstream units and sends to its downstream neighbors an output signal that is a nonlinear function of its inputs. Learning in such systems is modeled by experience-based changes in the weights of the connections between units. The basic connectionist style of modeling is now being used in three quite different ways – in neurobiology, in applications, and in cognitive science. Neurobiologists who study networks of neurons employ a wide range of computational models, from very detailed descriptions of the internal chemistry of the neuron to the abstract units just described. The use of connectionist neural models in practical applications is part of the reconvergence with AI and is discussed in the final section of this article.

In cognitive science, connectionist techniques have been used for modeling all aspects of language, perception, motor control, memory, and reasoning. This universal coverage represents a potential breakthrough; previously, the computational models of, for example, early vision and problem solving used entirely different mathematical and computational techniques. Since the brain is known to use the same neural computation throughout, it is not surprising that neurally inspired models can be applied to all behavior. Unfortunately, the existing models are neither broad nor deep enough to ensure that the current set of mechanisms will suffice to bridge the gap between structure and behavior, but the work remains productive.

Connectionist models in cognitive science fall into two general categories, often called structured and layered networks (also called parallel distributed processor, or PDP, networks). Most modelers are primarily interested in learning, which is modeled as experience-driven change in connection weights. There is a great deal of research studying different models of learning with and without supervision, different rules for changing weights, etc. Because of the focus on what the network can learn, any prewired structure will weaken the results of the experiment. The standard approach is to use networks with unidirectional connections arranged in completely connected layers, sometimes with a very restricted additional set of feedback links. This kind of network contains a minimum of presupposed structure and is also amenable to efficient learning techniques, such as the aforementioned back-propagation method. Most researchers using totally connected layered models do not believe that the brain shares this architecture, but there is an ongoing controversy about the implications of PDP learning models for theories of mind (see the section titled 'Nature and nurture: rules versus connections').

Structured connectionist models are usually less focused on learning than on the representation and processing of information. Essentially all the modeling done by neurobiologists involves specific architectures, which are known from experiment. For structured connectionist models of cognitive phenomena, the underlying brain architecture is rarely known in detail and sometimes not at all at the level of neurons and connections. The methodology employed is to experiment with computational models of the behavior under study that are consistent with the known biological and psychological data and are also plausible in the resources (neurons, computing time, etc.) required. This methodology is very similar to what are called 'spreading activation' models, widely used in psycholinguistics. Some studies combine structured and layered networks, or investigate learning in networks with an initial structure that is tuned to the problem area or the known neural architecture.

Nature and Nurture: Rules versus Connections

Perhaps the most visible contribution to date of connectionist computational models in cognitive science has been to provide a new jousting ground for contesting some age-old issues on the nature of intelligence. Much of the debate has been published in *Science* magazine, which suggests that it is considered to be of major importance by the US scientific establishment. The nature versus nurture question concerns how much of some trait, usually intelligence, can be accounted for by genetic factors, and how much depends on postnatal environment and training. Some PDP connectionists have taken very strong positions, suggesting that learning can account for everything interesting. In the particular case of grammar, an important group of linguists and other cognitive scientists take an equally extreme nativist position, suggesting that humans only need to choose a few parameters to learn grammar. A related issue is whether human grammatical knowledge is represented as general rules or just appears as the rulelike consequences of PDP learning in the neural network of the brain. There is ample evidence against both extreme positions, but the debate continues to motivate a great deal of thought and experiment.

Current and Future Trends

Although the fundamental split between the AI focus on general methods and the cognitive science emphasis on human intelligence remains, there are a growing number of areas of overlapping interest. As already discussed, quantitative neural models are playing a major role in cognitive science. It turns out that the mathematical and computational ideas underlying learning in neural networks have found application in a wide range of practical problems, from speech recognition to financial prediction. The basic idea is that, given current computing power, back-propagation and similar techniques allow large systems of nonlinear units to learn quite complex probabilistic relationships using labeled data. This general methodology overlaps not only with AI but also with mathematical statistics, and is part of a unifying area called computational learning theory. There is also a large community of scientists and engineers who identify themselves as working on neural networks and related statistical techniques for various scientific and applied tasks, along with conferences and journals to support this effort.

While probability was entering cognitive science from the bottom-up through neural models, AI experienced the introduction of probabilistic methods from general theoretical considerations, which only later led to practical application. As discussed earlier, the limitations of formal logic became well recognized in the 1960s. Over the subsequent decades, AI researchers, led by Judea Pearl, at the University of California (Los Angeles), developed methods for specifying and solving large systems of conditional probabilities. These belief networks are now widely used in applications ranging from medical diagnosis to business planning. A growing field that involves both symbolic and statistical techniques is 'data mining,' processing large historical databases to search for relationships of commercial or social importance. Recent efforts to learn or refine belief networks from labeled data are another area of convergence of AI and cognitive science in computational learning theory. Of course, the explosion of Internet activity is affecting AI along with the rest of the computing field. Two AI application areas that seem particularly important to cognitive science are intelligent Web agents and spoken-language interaction.

As the range of users and activities on the Internet continues to expand, there is increasing demand for systems that are both more powerful and easier to use. This is leading to increasing efforts on the human–computer interface, including the modeling of user plans and intentions – clearly overlapping with traditional concerns of cognitive science. One particularly active area is interaction with systems using ordinary language. Whereas machine recognition of individual words is relatively successful, dealing with the full richness of language is one of the most exciting challenges at the interface between AI and cognitive science, and a problem of great commercial and social importance.

Looking ahead, we can be confident that the increasing emphasis on intelligent systems will continue. From the scientific perspective, it is very likely that most of the interdisciplinary research in cognitive science will remain focused on specialized domains such as language, speech, and vision. General issues including representation, inference, and learning will continue to be of interest and will constitute the core of the direct interaction between AI and cognitive science. With the rapid advances in neurobiology, both fields will increasingly articulate with the life sciences, with great mutual benefits.

See also: Animal Intelligence: The Search for Animal Intelligence; Cognitive Neuroscience: An Overview; Cognitive Control and Development; Connectionist Models; Connectionist Models of Language Processing; Executive Function and Higher-Order Cognition: Computational Models; Hippocampus: Computational Models; Memory: Computational Models; Numerical Intelligence: Neural Substrates.

Further Reading

Elman J, Bates E, and Johnson M (1996) *Rethinking Innateness: A Connectionist Perspective on Development (Neural Network Modeling and Connectionism).* Cambridge, MA: MIT Press.

Lakoff G (1987) *Women, Fire, and Dangerous Things: What Categories Reveal About the Mind.* Chicago, IL: University of Chicago Press.

McClelland JL and Rumelhart DE (1986) *Parallel Distributed Processing.* Cambridge, MA: MIT Press.

Newell A (1990) *Unified Theories of Cognition.* Cambridge, MA: Harvard University Press.

Pearl J (1988) *Probabilistic Reasoning in Intelligent Systems.* San Mateo, CA: Morgan Kaufmann.

Posner MI (ed.) (1985) *Foundations of Cognitive Science.* Cambridge, MA: MIT Press.

Regier T (1996) *The Human Semantic Potential.* Cambridge, MA: MIT Press.

Russell SJ and Norvig P (2000) *Artificial Intelligence.* Upper Saddle River, NJ: Prentice-Hall.

Astrocyte: Calcium Signaling

J W Deitmer, FB Biologie, Kaiserslautern, Germany
A Araque, Cajal Institute, Madrid, Spain

Introduction

Ca^{2+} signaling has been recognized as one of the major second messenger steps in most cell types, including astrocytes, the major macroglial cell type in vertebrate nervous systems. Astrocytes are endowed with a large number of metabotropic receptors in their cell membrane, most of which are coupled to G-protein-mediated activation of phospholipase C (PLC), and hence the formation of inositol trisphosphate (IP_3) and diacylglycerol. The second messenger IP_3 diffuses within the cytosol, binds to IP_3 receptors in the membrane of the endoplasmic reticulum (ER), and gates Ca^{2+} channels, resulting in a cytosolic Ca^{2+} rise due to release of Ca^{2+} from the ER. This can be a single Ca^{2+} transient with or without a shoulder or plateau phase, repetitive Ca^{2+} transients, so-called Ca^{2+} oscillations, or irregular Ca^{2+} rises, depending on the species of primary messenger (neurotransmitter, hormone, or growth factor) and its concentration. These Ca^{2+} signals may spread along the cell and across cell boundaries to neighboring astrocytes in the form of Ca^{2+} waves, and they can be evoked or modulated by neuronal activity. The signaling pathway leading to a cytosolic Ca^{2+} rise, common to many cell types, may be regarded as a type of 'excitation' in electrically inexcitable cells such as astrocytes. The cytosolic Ca^{2+} signaling may initiate a variety of glial functions and interactions with other cells, for example, by Ca^{2+}-dependent release of transmitters (gliotransmitters) or by affecting vasoconstriction of blood vessels in the brain. This article reviews the types of cytosolic Ca^{2+} signaling in astrocytes, their different modes of initiation, and their functional significance for astrocytes and glia–neuron communication.

Identification of Astrocytes by Ca^{2+} Signals

Identification of astrocytes, in culture as in the tissue, has been established by using staining of glial-specific marker proteins by immunocytochemical techniques. Most astrocytes express an intermediate filament, the glial fibrillar acidic protein (GFAP), particularly in their cellular processes, and a Ca^{2+}-binding protein S100β, which do not occur in neurons. Hence, these two proteins are often detected as glial-specific markers in astrocytes, although they may not be exclusively restricted to this cell type. Brain slices incubated with the Ca^{2+}-sensitive dye (e.g., Fluo-4 AM) can be fixed with EDAC (1-ethyl-3-(3-dimethyl-aminopropyl)carbodiimide) and then stained for GFAP after experiments in which intracellular Ca^{2+} was measured to identify the cells of interest (**Figures 1(a)** and **1(b)**).

The use of fluorescent tags to highly expressed genes of glial cells by creating transgenic mice has produced fluorescent astrocytes for identification. In these mice, astrocytes were labeled by the green fluorescent protein (GFP) or the enhanced green fluorescent protein (EGFP) under the control of the human GFAP promoter (**Figures 1(c)** and **1(d)**). The level of EGFP expression, however, was variable in different brain areas. High expression was found in the cerebellum, but low expression was found in retina and hypothalamus, where some astrocytes were not labeled at all. Oligodendrocytes and neurons were not stained by these fluorescent tags. The red fluorescent dye sulforhodamine 101 was specifically taken up by a type of astrocytes ('protoplasmic'), which also expressed the EGFP, and was used to quantify morphological characteristics and the close association of astrocytes with cortical microvasculature.

When recording cytosolic Ca^{2+}, astrocytes can be identified by an erratic Ca^{2+} rise at low external K^+ concentrations (<1 mM), observed in rat and mouse astrocytes, but not in neurons. When superfusing a cell culture or brain slices with a low-K^+ solution for 3–5 min, more than 80% of cells that show an immunoreaction to GFAP and S100β antibodies respond with a monophasic or repetitive Ca^{2+} rise (**Figure 1(e)**). This Ca^{2+} response persists after depleting the intracellular Ca^{2+} stores, but it is suppressed in the absence of external Ca^{2+}, indicating that it results from Ca^{2+} influx. Since this low-K^+-induced Ca^{2+} rise can be inhibited by micromolar amounts of Ba^{2+}, it has been suggested that an inward rectifier K^+ channel becomes permeable to divalent cations. This idea is supported by using $K_{ir}4.1$ knockout mice, in which the low-K^+-induced Ca^{2+} response is significantly reduced. The $K_{ir}4.1$ appears to be highly expressed in oligodendrocytes and some astrocytes, in which this channel is believed to be involved in the clearance of extracellular K^+.

In summary, various methods of identifying glial cells may not be completely reliable on their own, but the use of a combination of two or three of these methods can identify astrocytes with a high degree of certainty.

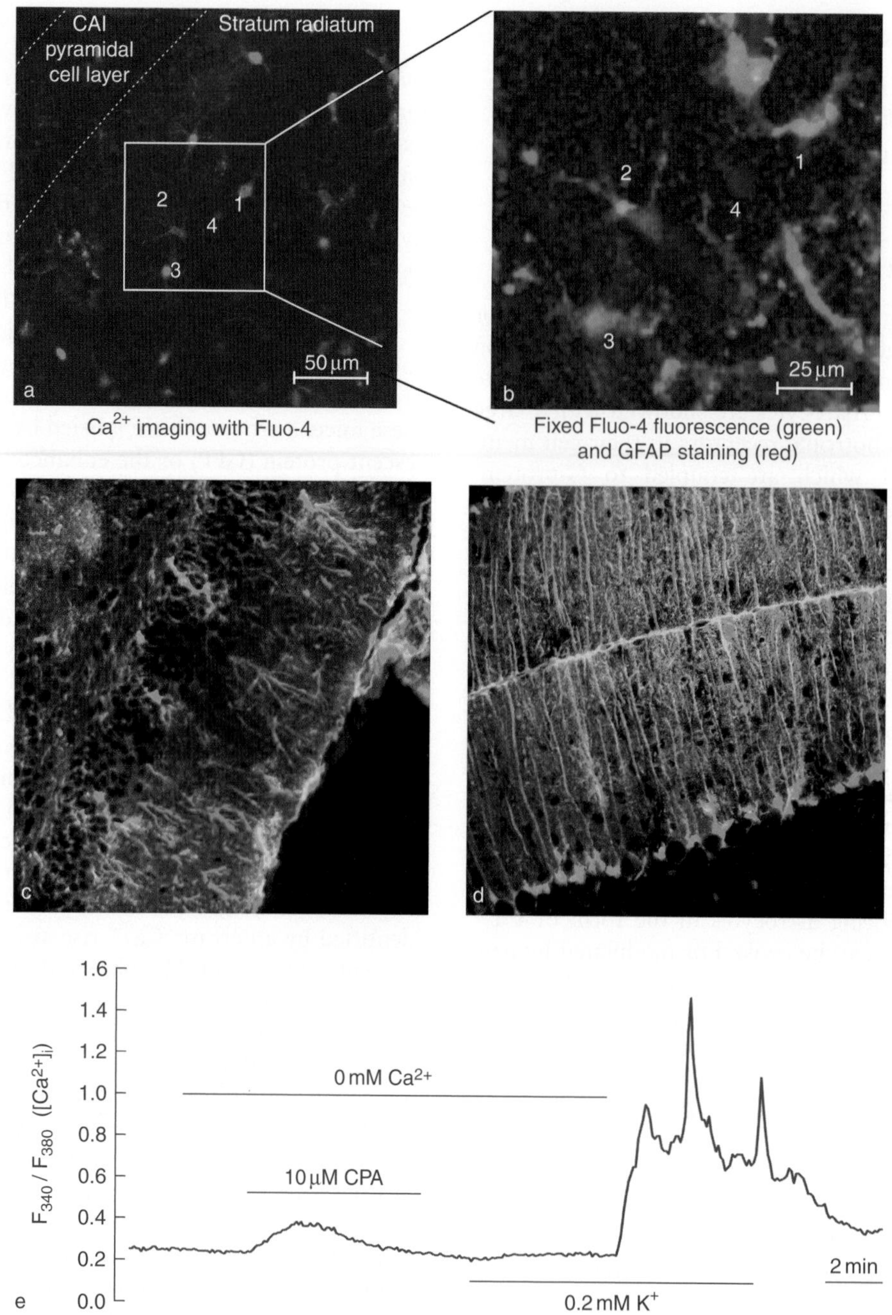

Figure 1 Identification of astrocytes. (a, b) Confocal images of a brain slice after loading cells with Fluo-4 (a) and after fixation of the fluorescence with EDAC (1-ethyl-3-(3-dimethyl-aminopropyl)carbodiimide) and staining with antibodies against GFAP (b). (c, d) Confocal images of GFAP-labeled cryosections from hippocampus with dentate gyrus (c) and from cerebellum showing molecular layer with Bergmann glial cell bodies and radial processes with endfeet (d) of transgenic mice. Scale value for (c) and (d) is same as in (a). (e) Low-K^+-induced Ca^{2+} responses cultured rat astrocytes after depleting intracellular Ca^{2+} stores with CPA (cyclopiazonic acid) are due to Ca^{2+} influx. (a, b) From Beck A, Nieden RZ, Schneider HP, and Deitmer JW (2004) Calcium release from intracellular stores in rodent astrocytes and neurons *in situ*. *Cell Calcium* 35: 47–58. (c, d) From Nolte C, Matyash M, Pivneva T, et al. (2001) GFAP promoter-controlled EGFP-expressing transgenic mice: A tool to visualize astrocytes and astrogliosis in living brain tissue. *Glia* 33: 72–86. (e) From Dallwig R, Vitten H, and Deitmar JW (2000) A novel barium-sensitive calcium influx into rat astrocytes at low external pottassium. *Cell Calcium* 28: 247–259.

Spontaneous Ca^{2+} Transients and Oscillations

Spontaneous Ca^{2+} transients and oscillations have been reported in both neurons and glial cells, in culture, and *in situ* and *in vivo*. Ca^{2+} oscillations are defined as repetitive rises of cytosolic Ca^{2+} and may occur intermittently with single Ca^{2+} transients and with intervals of up to several minutes (**Figure 2(a)**). These Ca^{2+} signals can propagate as waves along the

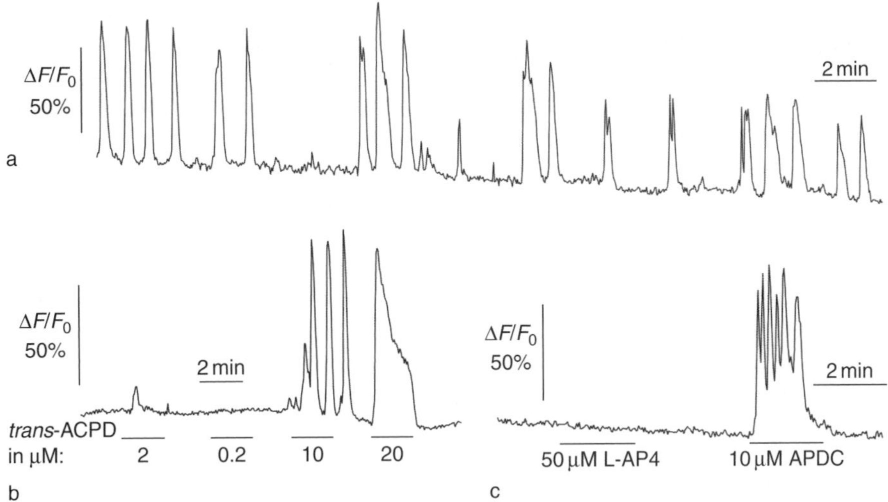

Figure 2 Ca^{2+} oscillations in acute rat hippocampal slices. (a) Continuous recording of spontaneous Ca^{2+} signals in an astrocyte. (b, c) Different concentrations of t-ACPD (b), an agonist of metabotropic glutamate receptors of group I, evoking single Ca^{2+} transients at low concentrations (1 μM), larger Ca^{2+} transients with shoulder at higher concentrations (20 μM), and Ca^{2+} oscillations at intermediate concentrations (5–10 μM). The metabotropic glutamate receptor agonist of group II APDC, but not the agonist of group III L-AP4 (c), can also elicit Ca^{2+} oscillations. From Zur Nieden R and Deitmer JW (2006) The role of metabotropic glutamate receptors for the generation of calcium oscillations in rat hippocampal astrocytes *in situ*. *Cerebral Cortex* 16: 676–687.

cell processes and even beyond cell boundaries to neighboring glial cells. Ca^{2+} transients and Ca^{2+} oscillations are usually evoked by release of Ca^{2+} from intracellular stores (ER) and hence are blocked by inhibiting the Ca^{2+}-ATPase of the ER and/or by IP_3 receptor antagonists. These spontaneous Ca^{2+} transients/oscillations occurred in 35% of astrocytes in the hippocampus of juvenile rats and in 82% of hippocampal astrocytes in juvenile mice with a rate of approximately 1.3 transients/min.

These Ca^{2+} signals require operating mechanisms for refilling the intracellular Ca^{2+} stores, which include store-operated calcium entry (SOCE) through channels in the plasma membrane. Spontaneous Ca^{2+} signals in astrocytes are not immediately affected by neuronal activity, but their synchronous occurrence may be affected after blocking neuronal activity by tetrodotoxin. As discussed later, various neurotransmitters, such as glutamate, noradrenaline, histamine, and ATP, as well as mechanical stimulation can evoke transients, oscillations, and waves of cytosolic Ca^{2+} in astrocytes (**Figures 2(b)** and **2(c)**).

Spontaneous Ca^{2+} transients and oscillations have been associated with nonsynaptic release of glutamate (e.g., via the cystine–glutamate exchanger) and hence with the glutamate background level in the extracellular space, acting on metabotropic glutamate receptors of groups I and II in the astrocyte membrane. Thus, the frequency of these spontaneous Ca^{2+} signals may help to sense the level of extracellular glutamate in nervous tissue. The elaboration of the exact mechanism and of other functions of spontaneous Ca^{2+} transients and oscillations in astrocytes, and their mode of initiation and maintenance, require more experimental analysis.

Molecular Mechanisms of Ca^{2+} Signaling

Calcium signaling in both excitable and nonexcitable cells is based on the existence of a relatively low concentration of cytosolic Ca^{2+} (usually 50–100 nM), and a Ca^{2+} concentration gradient between the cytosol and the extracellular space or the lumen of intracellular organelles (e.g., the ER, where Ca^{2+} concentration may be as high as hundreds of micromolar or even a few millimolar). Cellular Ca^{2+} signaling, manifested as rapid intracellular Ca^{2+} rises, may result from several Ca^{2+} sources. Cytosolic Ca^{2+} levels can increase via Ca^{2+} influx from the extracellular spaces across the plasma membrane and by Ca^{2+} release from intracellular stores. Calcium influx may occur through specific Ca^{2+}-permeable ligand-gated channels, voltage-gated Ca^{2+} channels that are activated by membrane depolarization, and Ca^{2+}-permeable, store-operated channels that are activated by depletion of intracellular Ca^{2+} stores and that help to refill these stores.

Whereas electrically excitable cells use the electrochemical gradient across the plasma membrane to effectively increase the intracellular Ca^{2+} levels due to the high expression of specific ligand- and voltage-gated ion channels that allow the rapid and massive influx of Ca^{2+} ions, nonelectrically excitable cells such as astrocytes mainly use the Ca^{2+} stored in the ER as the main source for cytoplasmic Ca^{2+} signaling. Ionotropic glutamate/α-amino-3-hydroxy-5-methyl-4-isoxazole propionic acic (AMPA) receptors, known to be expressed by astrocytes from different brain regions, including cortex, hippocampus, cerebellum, and retina, lack the GluR2 subunit, being therefore permeable to Ca^{2+}. However, aside from this special case, Ca^{2+} influx through AMPA receptors does not seem to play a major role in astrocyte Ca^{2+} signaling because in most brain regions astrocytes express AMPA receptors that contain the GluR2 subunit, such as Bergmann glial cells in the cerebellar cortex, and hence display a low Ca^{2+} permeability. Activation of other ligand-gated channels permeable to Ca^{2+}, such as ionotropic purinergic receptors (P2X receptors) or nicotinic acetylcholine receptors containing the α7 subunit, has been shown to lead to influx of Ca^{2+} in cell culture preparations. However, studies *in situ* have failed to observe Ca^{2+} responses in astrocytes under these conditions, which further suggests that Ca^{2+} influx through ligand-gated channels is not a major pathway for astrocyte cytosolic Ca^{2+} signaling.

Calcium influx through voltage-gated Ca^{2+} channels in the plasma membrane constitutes another source for intracellular Ca^{2+} elevation in different cell types. Astrocytes in culture express a variety of voltage-gated channels, including Ca^{2+} channels; however, studies performed *in situ* indicate that functional voltage-gated Ca^{2+} channels are mainly expressed by immature astrocytes, and this expression gradually disappears during development. Hence, the low, if any, level of expression of these channels in mature astrocytes does not serve to increase the intracellular Ca^{2+} levels to any greater extent. Therefore, unlike neurons and other electrically excitable cells that express a high density of ligand- and voltage-gated channels that are permeable to Ca^{2+} and allow an efficient rise of cytosolic Ca^{2+}, astrocytes employ different cellular mechanisms to increase their cytosolic Ca^{2+} concentration.

Astrocytes express a wide variety of functional receptors for many neurotransmitters (including glutamate, norepinephrine, γ-aminobutyric acid (GABA), histamine, ATP, adenosine, and acetylcholine), most of which belong to the metabotropic receptor family. These receptors are associated with G-proteins that upon activation stimulate phospholipase C, which leads to the formation of the intracellular second messenger IP_3. Activation of specific receptors in the ER by IP_3 evokes Ca^{2+} release from the IP_3-sensitive Ca^{2+} stores, thus increasing the cytosolic Ca^{2+} concentration. This cellular mechanism can represent one important mode of neuron-to-astrocyte communication, in which the astrocyte Ca^{2+} signal is controlled by neurotransmitter(s) released during synaptic activity.

In addition to IP_3 receptors, the membrane of the ER of most cells also contains ryanodine receptors, a different type of Ca^{2+}-permeable receptor that is activated by cytosolic Ca^{2+}. An increase in the cytosolic Ca^{2+} leads to the opening of the ryanodine receptors, which can be directly stimulated by caffeine and hence induce the release of Ca^{2+} from the internal stores, a process known as Ca^{2+}-induced Ca^{2+} release. The presence of these receptors in astrocytes, however, is controversial; moreover, the lack of caffeine-sensitive Ca^{2+} release, as found in astrocytes in culture and *in situ*, and in contrast to neurons, suggests a minor, if any, role of these receptors in the astrocyte Ca^{2+} activity.

The maintenance of the high intraluminal Ca^{2+} concentration in the ER depends on Ca^{2+} entry using specific Ca^{2+}-permeable plasma membrane channels, called store-operated channels, that are regulated by the filling state of the ER and that serve to replenish the stores with Ca^{2+}. This SOCE, also called capacitative Ca^{2+} entry, has been observed in many different cell types and recorded electrophysiologically as a persistent current, called I_{CRAC} for 'Ca^{2+} release-activated current.' Although I_{CRAC} has not been identified in astrocytes, SOCE has been clearly demonstrated in these cells by Ca^{2+} imaging studies. SOCE can be detected as a sustained plateau of the intracellular Ca^{2+} levels that depends on the previous Ca^{2+} depletion of intracellular Ca^{2+} stores and that requires the presence of extracellular Ca^{2+}. A diffusible messenger termed Ca^{2+} influx factor (CIF) has been suggested to act through activation of inducible, Ca^{2+}-independent phospholipase A_2 activity. It has been proposed that the EF-hand protein stromal interaction molecule 1 (STIM1), which migrates from the ER to the plasma membrane and activates calcium release-activated channel (CRAC) channels upon Ca^{2+} store depletion, functions as this CIF. An exciting question is whether this mechanism of Ca^{2+} influx serves as a further Ca^{2+} signal beyond the simple role of refilling internal stores.

It is the interplay of the different cellular mechanisms (Ca^{2+} influx and release, Ca^{2+} extrusion and uptake, and Ca^{2+} buffering) that control the intracellular Ca^{2+} levels that is crucial for the generation and shape of astrocyte Ca^{2+} signaling. Mechanisms that are involved in regulating cytosolic Ca^{2+} contribute to shaping the kinetics and amplitude of Ca^{2+} transients. Ca^{2+}-binding proteins, such as parvalbumin, calcineurin, and calmodulin, that are present in the cytoplasm may serve to buffer cytosolic Ca^{2+} concentration. However, this is a low-capacity system that is more effective as a mediator of Ca^{2+}-dependent cellular processes. Energy-consuming mechanisms are involved in the effective Ca^{2+} clearance. Sarco/endoplasmic reticulum Ca^{2+}-ATPases pump Ca^{2+} ions into the ER using ATP as an energy source. Ca^{2+}-ATPases and Na^{+}/Ca^{2+} exchangers present in the plasma membrane use ATP or the electrochemical gradient of Na^{+} across the plasma membrane, respectively, to extrude Ca^{2+} against a steep concentration gradient across the plasma membrane. Finally, mitochondria are dynamic Ca^{2+} stores that can also sequester Ca^{2+} through the activity of the Ca^{2+} uniporter, with a relatively low affinity but a large capacity to store Ca^{2+} transiently, and protect the cell from cytosolic Ca^{2+} overload, which may lead to cell death.

Intra- and Intercellular Ca^{2+} Signaling

The astrocyte Ca^{2+} signal can be manifested as Ca^{2+} elevations with different temporal patterns (**Figure 2**). In some cases, astrocytes display a single, fast Ca^{2+} transient that eventually can be followed by a more sustained response. In other cases, astrocytes exhibit irregular Ca^{2+} oscillations that can be modulated in amplitude, duration, and frequency by different stimuli, including the synaptic activity level. The astrocyte Ca^{2+} rise may act as an intra- and intercellular signaling mode that can propagate within and between astrocytes, signaling to different regions of the cell and to different cells (**Figure 3**).

The astrocyte Ca^{2+} signal that occurs spontaneously or is evoked by synaptic activity can be initiated in spatially restricted areas, called 'microdomains,' of the astrocytic processes. From these discrete regions, the Ca^{2+} signal can propagate along the process to other regions of the astrocyte, generating a wave of intracellular Ca^{2+}. A thorough characterization of the intracellular waves is yet to be accomplished; however, the spatial spread of the intracellular wave seems to be

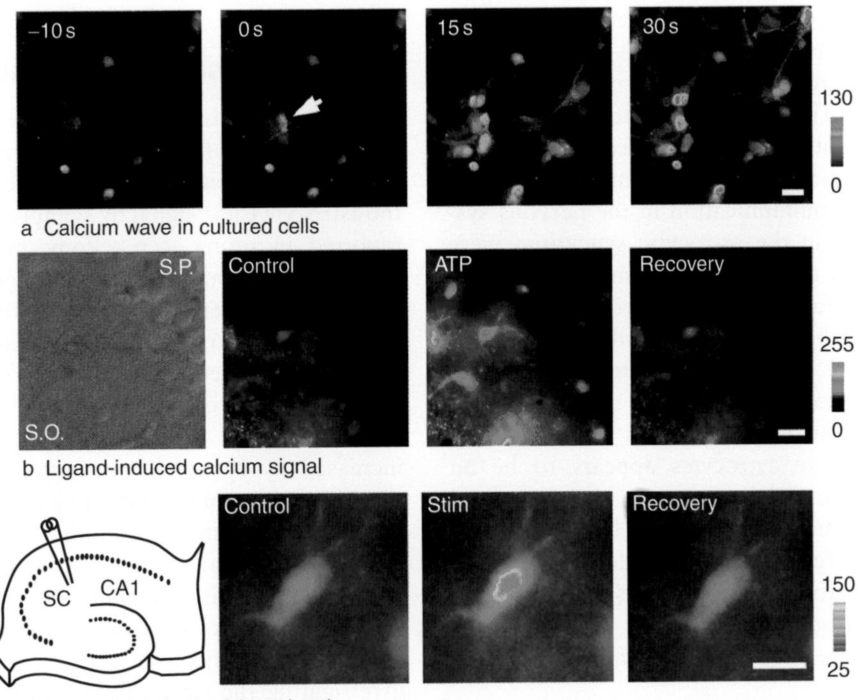

Figure 3 Ca^{2+} signaling in astrocytes evoked by different stimuli. (a) Intercellular Ca^{2+} wave in cultured rat hippocampal astrocytes. Astrocytes were bulk loaded with the fluorescent Ca^{2+} indicator Ca^{2+}-Green 1. Images in pseudocolor mode represent fluorescence intensity of Ca^{2+}-Green 1 emission taken during and after mechanical stimulation of a cell (arrow; time zero) at the times indicated. Mechanical stimulation increased intracellular Ca^{2+} in the directly stimulated cell as well as in neighboring, unstimulated astrocytes by a propagating Ca^{2+} wave. (b) Infrared differential interference contrast image and pseudocolor images representing fluorescence intensities of a Fluo-3-filled rat hippocampal slice in control, after ionophoretical application of ATP for 5 s, and after recovery. S.O., stratum oriens; S.P., stratum pyramidale. (c) Responses evoked by Schaffer collateral (SC) stimulation in rat hippocampal astrocytes. Schematic drawing of the hippocampal slice preparation and pseudocolor images representing fluorescence intensities of a Fluo-3-filled astrocyte in control, 10 s after stimulation of the SC (at 30 Hz for 5 s), and after recovery. Scale bar = 20 μm (a, b), 10 μm (c). (a) Adapted from Perea G and Araque A (2005) Glial calcium signalling and neuron–glia communication. *Cell Calcium* 38: 375–382. (b, c) Adapted from Perea G and Araque A (2005) Properties of synaptically evoked astrocyte calcium signal reveal synaptic information processing by astrocytes. *Journal of Neuroscience* 25: 2192–2203.

under the control of synaptic activity. The intracellular compartmentalization and the control of the propagation of the Ca^{2+} signal are highly relevant for cellular functions and the astrocyte-to-neuron communication. Since 'gliotransmitters' released from astrocytes through Ca^{2+}-dependent mechanisms may act as modulators of synaptic transmission, the mechanisms controlling the intracellular Ca^{2+} wave may be of great significance because they will determine the degree of extension of the signal that triggers the neuromodulatory effects of astrocytes. The fact that the Ca^{2+} signal evoked by synaptic activity is initiated in restricted regions of the astrocyte reveals that the neuron-to-astrocyte communication is a localized event that does not result from uncontrolled spillover of neurotransmitter but more likely from a point-to-point signaling between synaptic terminals and astrocytic processes.

Astrocytes can also communicate with adjacent cells by generating intercellular Ca^{2+} waves, which are propagated along cell processes onto neighboring astrocytes (**Figures** 3(**a**) and 3(**b**)). Different stimuli, such as mechanical stimuli or exogenous, focal, application of neurotransmitters, may induce intracellular Ca^{2+} increases that can propagate as Ca^{2+} waves between astrocytes in cultured cells as well as organotypic and acute brain slices. These waves can extend for long distances ($<500\,\mu m$) at relatively low speed ($\sim 14\,\mu m\,s^{-1}$), and they may constitute a novel form of slow, long-distance cellular communication in the nervous system. Ca^{2+} waves in the astrocytic syncytium were originally thought to spread as a result of gap junction-mediated diffusion of IP_3 between cells. Later studies proposed that, in addition to diffusion of IP_3, other mechanisms that involve extracellular messengers, such as extracellular ATP, may contribute to the propagation of astrocyte Ca^{2+} waves. ATP released from active astrocytes appears to be an important mediator of long-range Ca^{2+} signaling, whereas shorter range signaling may be mediated by gap junctions. More experimental evidence is required to determine whether synaptic processes in different areas that lack immediate neuronal connection are modulated by these astrocytic Ca^{2+} waves.

Ca^{2+} Responses to Transmitters and Other Signaling Molecules

Elevations of the cytosolic Ca^{2+} concentration in astrocytes may be triggered by a large number of signaling molecules, including hormones, prostaglandins, and many neurotransmitters. As described previously, astrocytes express a wide variety of functional receptors for most neurotransmitters, and many of them belong to the metabotropic family; that is, they are coupled to second messenger pathways that lead to the IP_3-mediated Ca^{2+} release from intracellular stores. Activation of these receptors by exogenous application of messengers such as glutamate, norepinephrine, 5-hydroxytryptamine, histamine, acetylcholine, dopamine, nitric oxide (NO), ATP, and GABA can increase the intracellular Ca^{2+} level in astrocytes (**Figure 3c**). These Ca^{2+} signals can be linked to initiating cellular activity, such as transmitter release, a rise in the K^+ conductance of the cell membrane, or mitochondrial energy production.

Astrocytes may also respond with intracellular Ca^{2+} elevations to other signaling molecules such as chemokines, a class of small proteins that bind to G-protein-coupled receptors that were originally identified as inflammatory mediators of leukocyte chemotaxis but later shown to possess additional roles beyond neuroinflammation. Various types of chemokine receptors are expressed in cells of the nervous system, including astrocytes, in which the chemokine stromal cell-derived factor-1 has been shown to increase astrocytic Ca^{2+} through activation of the receptor CXCR4.

Ca^{2+} Responses to Neuronal Activity

Astrocyte receptors can be activated by neurotransmitters released from synaptic terminals to initiate astrocyte Ca^{2+} signaling (**Figure 3c**). The control of the astrocyte Ca^{2+} signal by synaptic activity has been reported in retina, cerebellum, hippocampus, and cortex and has been shown to be mediated by glutamate, GABA, acetylcholine, noradrenaline, and NO. Further studies are required to elucidate the involvement of other neurotransmitter systems in the neuron-to-astrocyte communication.

The fact that the astrocyte Ca^{2+} level may be increased by synaptic transmitters does not imply that the Ca^{2+} signal simply reflects synaptic activity; rather, it represents a complex signal that can be tuned by neuronal activity. The amplitude and duration of Ca^{2+} transients as well as the frequency of Ca^{2+} oscillations in astrocytes may vary depending on the level of synaptic activity. Moreover, studies investigating the synaptic control of astrocyte Ca^{2+} signaling have demonstrated that astrocytes display integrative properties for synaptic information processing because astrocytes can discriminate between the activity of synaptic terminals belonging to different axon pathways. Moreover, the synaptically evoked astrocyte Ca^{2+} signal can be bidirectionally modulated by the interaction of different synaptic inputs, being potentiated or depressed depending on the level of synaptic activity. This modulation may also control the intracellular spread of the Ca^{2+} signal in the form of waves, which may have important consequences on brain function by regulating the spatial extension of the influence of a single astrocyte on different synapses.

Functional Significance of Ca^{2+} Signaling

There are numerous processes in the nervous system that may be initiated and/or modulated by cytosolic Ca^{2+} rises in astrocytes, as has been reported for many other cell types. On the other hand, astrocytic Ca^{2+} signaling is subject to modulation by neuronal activity, particularly at synapses, and therefore is one of the prime mechanisms by which reciprocal neuron–glia signaling is established. One of the prime and most consequential processes initiated by Ca^{2+} rises in astrocytes is the Ca^{2+}-dependent release of transmitters, so-called 'gliotransmitters.' Some of these gliotransmitters and glia-derived messengers, such as glutamate, ATP, tumor necrosis factor-α, or D-serine, have been shown to modulate neuronal excitability, synaptic transmission, and cerebrovascular microcirculation. Some of these transmitters can be released by a Ca^{2+}-dependent mechanism, although Ca^{2+}-independent modes of release have also been reported, such as through large membrane anion channels, purinergic $P2X_7$ channels, or gap junction hemichannels. Therefore, these gliotransmitters may serve as feedback signals that modulate neuronal activity. Astrocyte-induced modulation of synaptic transmission has been observed in cultured cells as well as in tissue slices of several brain areas. Since astrocytes respond to neurotransmitters with calcium elevations that can extend relatively large distances and that induce release of gliotransmitters that modulate neurotrasnmission, astrocyte Ca^{2+} signaling may represent a form of brain information pathway that establishes a functional link between distant synapses.

Neuronal excitability has also been shown to be modulated by glutamate released from astrocytes both *in vitro* and *in situ*. Calcium-dependent glutamate release from astrocytes may have a strong impact on brain pathophysiology because it can lead to the synchronized activity of clusters of neurons, and it may be responsible for the generation of epileptiform activity in neurons.

Other messenger molecules that have been reported to be released from astrocytes in a Ca^{2+}-dependent fashion include NO (e.g., mediated by activation of metabotropic purinergic P2Y receptors). This short-lived volatile messenger, which acts both in an ortho- and a retrograde way, may be a link to other second messenger cascades in neighboring neurons and glial cells, such as the cyclic guanosine monophosphate (cGMP)-mediated pathway. Moreover, NO may enhance SOCE and thereby feeds back on the cellular competence to generate Ca^{2+} signals due to repeated Ca^{2+} release from intracellular stores.

Ca^{2+} signaling in astrocytes has also been associated with the control of cerebral blood flow. Ca^{2+} transients and Ca^{2+} oscillations are propagated to the astrocytic endfeet, which ensheath part of the endothelial layer of blood capillaries, and can elicit vasodilation and vasoconstriction by the release of lipid metabolites. Inhibition of cyclooxygenase-1 activity blocked astrocyte-triggered vasodilation, supporting the notion that Ca^{2+}-dependent release of prostaglandins from astrocytes mediates the control of local blood flow. This modulation may be initiated by neuronal activity in restricted brain areas by evoking astrocyte Ca^{2+} signals that in turn mediate the control of microcirculation and hence the supply of oxygen and glucose.

Furthermore, cytosolic Ca^{2+} rises in astrocytes may gate or modulate ion channels in the cell membrane, particularly K^+ channels, which may be involved in K^+ clearance from the extracellular space following neuronal activity. The list of cellular functions mediated by Ca^{2+} signals in astrocytes will certainly grow in the future. It is the regional control of cellular functions that may be regulated or modulated by Ca^{2+} signals in a unique way. By invading different cellular processes of astrocytes, by being regenerative, and by being a versatile mode of signaling, this form of 'glial excitation' has many assets for allowing a complex dialogue between glial cells and neurons, and it may contribute critically to information processing and metabolism in the brain.

Summary

Cytosolic Ca^{2+} signals in astrocytes can be spontaneous or evoked, and they are often initiated by activation of metabotropic receptors, resulting in Ca^{2+} release from intracellular Ca^{2+} stores. Whereas voltage-dependent Ca^{2+} influx is rare in astrocytes, the refilling of the intracellular Ca^{2+} stores requires store-operated Ca^{2+} entry, a major influx pathway of Ca^{2+} into astrocytes. Astrocyte Ca^{2+} signals are generated in the form of transients or oscillations, which can be evoked by neurotransmitters, hormones, cytokines, and growth factors, often in response to neuronal activity. These Ca^{2+} signals are propagated along cellular processes and can travel from the point of origin across the tissue as Ca^{2+} waves. When propagated into astrocytic endfeet, they may release cyclooxygenase products to control cerebral blood flow. Astrocyte Ca^{2+} signaling may lead to the release of transmitters and may be one of the key mechanisms allowing a complex dialogue between neurons and glial cells.

See also: Astrocyte: Response to Injury; Astrocyte: Neurotransmitter and Hormone Receptors; Astrocyte: Identification Methods; Calcium and Signal Transduction; Neurotransmitter Release from Astrocytes.

Further Reading

Araque A and Perea G (2004) Glial modulation of synaptic transmission in culture. *Glia* 47: 241–248.

Beck A, Nieden RZ, Schneider HP, and Deitmer JW (2004) Calcium release from intracellular stores in rodent astrocytes and neurons *in situ*. *Cell Calcium* 35: 47–58.

Berridge MJ (2005) Unlocking the secrets of cell signalling. *Annual Review of Physiology* 67: 1–21.

Castonguay A, Levesque S, and Robitaille R (2001) Glial cells as active partners in synaptic functions. *Progress in Brain Research* 132: 227–240.

Dallwig R, Vitten H, and Deitmer JW (2000) A novel barium-sensitive calcium influx into rat astrocytes at low external potassium. *Cell Calcium* 28: 247–259.

Deitmer JW, Verkhratsky A, and Lohr C (1998) Calcium signalling in glial cells. *Cell Calcium* 24: 405–416.

Fiacco TA and McCarthy KD (2006) Astrocyte Calcium elevation: Properties, propagation and effects on brain signalling. *Glia* 54: 676–690.

Haydon PG (2001) Glia: Listening and talking to the synapse. *Nature Reviews Neuroscience* 2: 185–193.

Kettenmann H and Ransom BR (2005) *Neuroglia*, 2nd edn. New York: Oxford University Press.

Nedergaard M, Ransom BR, and Goldman SA (2003) New roles for astrocytes: Redefining the functional architecture of the brain. *Trends in Neurosciences* 26: 523–530.

Newman EA (2004) Glial modulation of synaptic transmission in the retina. *Glia* 47: 268–274.

Nolte C, Matyash M, Pivneva T, et al. (2001) GFAP promoter-controlled EGFP-expressing transgenic mice: A tool to visualize astrocytes and astrogliosis in living brain tissue. *Glia* 33: 72–86.

Perea G and Araque A (2005) Glial calcium signalling and neuron–glia communication. *Cell Calcium* 38: 375–382.

Perea G and Araque A (2005) Properties of synaptically evoked astrocyte calcium signal reveal synaptic information processing by astrocytes. *Journal of Neuroscience* 25: 2192–2203.

Porter JT and McCarthy KD (1997) Astrocytic neurotransmitter receptors *in situ* and *in vivo*. *Progress in Neurobiology* 51: 439–455.

Singaravelu K, Lohr C, and Deitmer JW (2006) Regulation of store-operated calcium entry by calcium-independent phospholipase A_2 in rat cerebellar astrocytes. *Journal of Neuroscience* 26: 9579–9592.

Verkhratsky A, Orkand RK, and Kettenmann H (1998) Glial calcium: Homeostasis and signalling function. *Physiological Reviews* 78: 99–141.

Volterra A, Magistretti PJ, and Haydon PG (eds.) (2002) *The Tripartite Synapse: Glia in Synaptic Trasnmission*. New York: Oxford University Press.

Volterra A and Meldolesi J (2005) Astrocytes, from brain glue to communication elements: The revolution continues. *Nature Reviews Neuroscience* 6: 626–640.

Volterra A and Steinhauser C (2004) Glial modulation of synaptic transmission in the hippocampus. *Glia* 47: 249–257.

Zur Nieden R and Deitmer JW (2006) The role of metabotropic glutamate receptors for the generation of calcium oscillations in rat hippocampal astrocytes *in situ*. *Cerebral Cortex* 16: 676–687.

Astrocyte: Identification Methods

A Scheller and F Kirchhoff, Max Planck Institute of Experimental Medicine, Göttingen, Germany

Introduction

Astrocytes are the major glial cell type of the central nervous system (CNS). Originally, they were described as nerve glue (*Nervenkitt*) by the German neuropathologist Rudolf Virchow in 1858. Work during the past few decades, however, demonstrated that astrocytes play numerous important roles in brain function. Astrocytes are responsible for nourishing neurons, regulation of extracellular ion and transmitter homeostasis, and maintenance of the blood–brain barrier. In addition, they act as guiding structures for neuronal migration during development, contribute to neurotransmitter metabolism, and participate in repair and regeneration processes. Furthermore, it has been shown that astrocytes are in close contact to active synaptic terminals, modulate synaptic transmission, or possess even neurogenic potential. The current view regards astrocytes as equal partners of neurons.

Methods to identify astrocytes are based on three of their basic properties: unique cellular structure, selective gene expression, and characteristic physiological membrane properties (**Figure 1**).

Astroglial Structure

In the late nineteenth century, the neuroanatomists Karl Weigert, Albert von Kölliker, and Michael von Lenhossek were the pioneers who described astrocytes as distinct cellular elements in the CNS by taking advantage of simultaneous improvements of the light microscope and staining techniques. In the beginning, differential binding properties of dyes such as hematoxylin, carmine, nigrosin, and a series of aniline derivatives were used to distinguish cell types and subcellular structures of the brain. Some of these dyes are still in use today. Later, the silver and gold impregnation technique developed by Camillo Golgi and ingeniously applied by Santiago Ramón y Cajal in the late nineteenth and early twentieth century fully revealed the polarized three-dimensional structure of astrocytes as we know it today. From a small soma less than 10 μm in diameter, major processes extend which give rise to a highly arborized network of finer processes. From these endings, numerous membrane protrusions emanate and enwrap synaptic terminals. In addition, astroglial endfeet of distinct major processes contact brain capillaries by interaction with endothelial cells and pericytes.

In the 1950s and 1960s, the complete fine structure of astrocytes was uncovered by Sanford Louis Palay and colleagues using electron microscopy (EM). This technique highlighted the differences of major proximal processes radiating from the small astroglial soma and distal irregular side branches. Whereas the first are characterized by the presence of intracellular fiber bundles, the distal endings of the other processes are, beside some glycogen granules, almost devoid of electron-dense material and can be as thin as 20 nm. Together with the synapses, these lamellar extensions form functional units throughout the nervous system, which led to the concept of tripartite synapses in which the pre- and postsynaptic terminals are enwrapped by thin astroglial compartments. Reconstruction of hundreds of serial EM sections revealed the extreme complexity of the astroglial process network.

Cell-Type-Specific Gene Expression

Glial Fibrillary Acidic Protein: The Main Constituent of Astroglial Intermediate Filaments

In the 1970s, Lawrence Eng and colleagues purified the intracellular astroglial fibers from gliotic autopsy material. The 50 kDa glial fibrillary acidic protein (GFAP) was identified as its main constituent. This discovery represented an important milestone in astroglia research. Since GFAP turned out to be highly immunogenic, a series of mono- and polyclonal antibodies could be generated. These reagents significantly facilitated subsequent immunohistochemical studies analyzing astrocytes *in situ*. Today, GFAP is the most common marker for astrocytes. Due to the availability and reliability of antibodies against GFAP, it was simple to selectively label astrocytes in normal, as well as in pathological brain tissue. In subsequent years, morphologically distinct astroglial subtypes such as fibrous astrocytes in the white matter or protoplasmic astrocytes in gray matter were identified. Now, at the light microscopic level, primarily using fluorescence microscopy, the direct contacts of astrocytes to blood vessels, the parallel orientation of astroglial fibers along myelinated axons, or the radial interdigitation of cerebellar Bergmann glia fibers through the parallel fiber–Purkinje cell network can readily be visualized. GFAP immunostaining is highly relevant to understanding brain development and pathology. In early work using GFAP antibodies, Amico Bignamy and Doris Dahl demonstrated that in the three mouse mutants *reeler*, *staggerer*, and *weaver*,

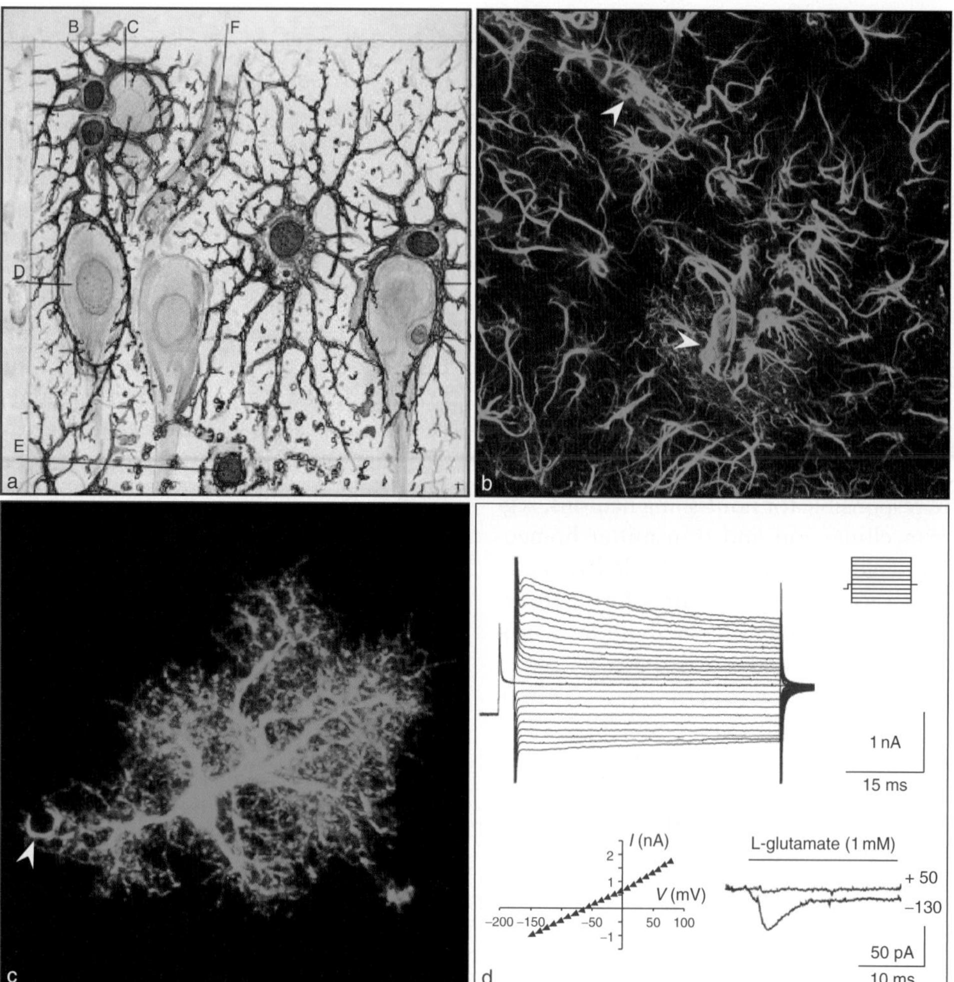

Figure 1 Ways to identify astrocytes. (a) Reproduction of an original drawing of Ramon y Cajal showing gray matter astrocytes visualized by the silver impregnation technique developed by Camillo Golgi (original deposited in the Cajal Institute, CSIC, Madrid). The major astroglial processes intimately enwrap neuronal cell bodies, whereas others contact brain capillaries. (b) Immunostaining of the astroglia-specific intermediate filament protein GFAP recorded by confocal laser scanning microscopy. The arrowheads indicate blood vessels on which GFAP-rich astroglial endfeet impinge. In contrast to the Golgi stain, only GFAP-rich cytoskeletal structures can be revealed. (c) Two-photon laser scanning micrograph of a single cortical astrocyte expressing the enhanced green fluorescent protein in a TgN(hGFAP-EGFP) transgenic mouse. Due to the cytosolic expression of EGFP, very fine details of the complete three-dimensional structure can be recognized. It is evident that astrocytes are highly polarized. Whereas some processes contact the brain capillaries (arrowhead), others are highly branched, forming an extended network of fine membrane protrusions which enwrap neuronal synapses. (d) Electrophysiological characterization of astroglial membrane properties by whole-cell patch clamp recordings. Astrocytes are characterized by a high potassium conductance with a linear current-to-voltage relationship seen when a series of de- and hyperpolarizing voltage steps are applied. In addition, transporter currents can be identified. EGFP, enhanced green fluorescent protein; GFAP, glial fibrillary acidic protein; hGFAP, human glial fibrillary acidic protein. (d) Adapted from Matthias K, Kirchhoff F, Seifert G, et al. (2003) Segregated expression of AMPA-type glutamate receptors and glutamate transporters defines distinct astrocyte populations in the mouse hippocampus. *Journal of Neuroscience* 23: 1750–1758.

Bergmann glia change their morphology with aging, although its formation was not delayed. The authors not only recognized the value of GFAP antibodies but also concluded that mouse strains with neurological mutations are valuable models for neuronal interactions and for the analysis of astroglial responses in neuronal degeneration. Currently, GFAP antibodies are also used for the diagnosis of astroglial tumors and the analysis of astroglia development and differentiation. Since its expression is strongly upregulated upon CNS insults such as traumatic brain injury or neuronal degeneration as in Alzheimer's disease, GFAP is not only the most common marker for gliosis in all types of brain pathology but also a bona fide marker of astrocytes in all vertebrate species from zebra fish to humans. However, *in situ*, particularly in some brain regions such as the cortex, the level of GFAP is significantly downregulated. Numerous astrocytes appear GFAP negative, although other labeling techniques clearly reveal their abundant presence. Interestingly, some astrocytes, those that contact brain capillaries or touch the pia, display a significant lesser reduction in GFAP expression.

S100β and Glutamine Synthetase as Cytosolic Markers of Astrocytes

Another important marker for astrocytes is S100β. This 11 kDa protein belongs to the superfamily of S100 proteins from which 20 human genes are known. S100 proteins can form calcium-binding, homo- and heterodimer complexes. In addition, they display a remarkable heterogeneous tissue distribution. In mammals, S100β is found predominantly in the brain. It has been implicated in stimulation of neurite outgrowth, neuronal survival during development, and neuroinflammation. S100β is a soluble protein which is expressed throughout the astroglial cytoplasm, but it can also be released into the extracellular space to act as a neurotrophic factor. Due to its cellular localization, antibodies directed against S100β reveal the complete three-dimensional structure of astrocytes, whereas GFAP immunostainings primarily visualize only the major processes. Unfortunately, the expression of S100β is less selective than GFAP. In some areas of the brain, neurons or oligodendrocytes can also be labeled by S100β antibodies.

Similar to S100β, glutamine synthetase (GS) of the brain is located mainly in astroglia. Its primary role is the detoxification of ammonia by generation of glutamine from glutamate. In addition, it serves the metabolism of amino acid neurotransmitters, γ-aminobutyric acid (GABA) and glutamate. The distribution of glial glutamate receptors and glial glutamate transporters parallel to expression of glutamine synthetase suggests a functional coupling between glutamate uptake and the metabolic regulation. Glutamine is released into the extracellular space and is used by neurons as glutamate precursor. GS is considered to be a marker enzyme for astroglia in general. In the adult brain, glutamine synthetase immunoreactivity is observed on astroglial processes surrounding excitatory synapses of most CNS regions. It is also expressed in Müller cells of the retina and embryonic radial glia. However, GS expression has been documented for oligodendroglial subpopulations as well.

Glutamate Transporters GLAST and GLT-1: The Most Abundant Membrane Proteins of Astrocytes

L-Glutamate is the most important excitatory neurotransmitter in the mammalian nervous system. High-affinity glutamate transporters in the plasma membrane terminate the neuronal transmission by removing glutamate from the extracellular space. This mechanism is essential to keep the glutamate concentration in the synaptic cleft below neurotoxic levels. By controlling extracellular glutamate during development, glutamate transporters participate in the regulation of cell differentiation, migration, and synaptogenesis. GLAST (EAAT1) and GLT-1 (EAAT2) are the two astroglial transporters, whereas EAAT3–EAAT5 are primarily expressed on neurons. Although both GLAST and GLT-1 are expressed in almost all brain regions and even in the same astrocyte, GLAST is primarily expressed in Bergmann glia of the cerebellum and GLT-1 predominates in cortex and hippocampus. High-affinity antibodies directed against GLAST and GLT-1 have been instrumental not only to visualize the ultrastructure of astroglial membranes but also to correlate the expression density with uptake currents observed in electrophysiological studies.

Transgenic Labeling of Astrocytes

Advances in molecular biology made provided the possibility to generate transgenic mice in which astrocytes could be labeled not only selectively but also permanently in subsequent generations. Artificial minigenes can be composed of astroglia-selective promoters and reporter genes such as β-galactosidase or the green fluorescent protein. These expression cassettes are then used for transgenesis by injection into fertilized mouse oocytes.

After isolation of the mouse astroglia-specific GFAP gene, in 1991 a 15 kb-long transgene was constructed which included all introns and 2 kb of 5′-flanking and 1.4 kb of 3′-flanking DNA. The *Escherichia coli* β-galacosidase gene (lacZ) was inserted into the first exon as ectopic reporter of the promoter activity. With this construct, transgenic mice were generated by nonhomologous gene recombination. Incubation of brain slices with the β-galacosidase substrate 5-bromo-4-chloro-3-indolyl-β-D-galactopyranoside (X-gal) revealed abundant reporter gene expression, primarily in astrocytes. However, variable levels of reporter activity were also observed in neurons. Subsequent *in vitro* studies identified regulatory elements which are essential for astroglial expression. Interestingly, using a 2.2 kb 5′ sequence of the human GFAP promoter for transgenesis revealed a more reliable labeling of astrocytes at high expression levels. Reporter activity of transgenic lines with this promoter mimicked to a large degree the endogenous GFAP levels. The transgenes were expressed beginning on embryonic days 12.5–13.5 and continued to be expressed until adulthood. Also, acute brain lesions resulted in increased reporter expression in parallel to normal gliotic scar reactions. However, not all astrocytes are labeled by transgenic reporter gene expression. Similar to the normal mouse GFAP levels, transgene expression is most abundant in young mice – in cerebellar Bergmann glia and in astrocytes which contact either pia, the capillaries, or the ventricular walls. The downregulation in cortical brain areas is most dramatic and not understood.

Approximately 10 years ago, transgenic mice were generated in which the green fluorescent proteins S65T-GFP or later EGFP were used to label astrocytes. The unique advantage of fluorescent proteins is buried in their structure, in which the fluorophor is spontaneously generated from three residues of the polypeptide chain (by cyclization and oxidation). No exogenous cofactors are needed as labeling dyes. GFAP–GFP transgenic mice revolutionized the identification of astrocytes in brain tissue. For the first time, it was possible to unequivocally identify astrocytes in all brain regions without prior immunohistochemical labeling, which normally requires tissue fixation. Live astrocytes could be analyzed in acutely isolated brain slices or even *in vivo*. Via fluorescence-activated cell sorting, fluorescent astrocytes could be isolated und subsequently analyzed in cell culture. Using this approach, the neurogenic potential of embryonic radial glia cells was revealed. In addition, acutely isolated tissue slices isolated from brain regions such as the brain stem or hippocampus served for the analysis of the dynamic interaction of astroglial processes with neighboring active synaptic terminals. Time-lapse recordings demonstrated highly motile lamellipodia and filopodia emanating from major process branches. This work strengthened the view of astrocytes as regulators of neuronal transmission. However, astrocytes do not just modulate neuronal function. *In vivo*, the pivotal role of astroglial endfeet associated with neighboring capillaries could be demonstrated.

Transgenic labeling with the human GFAP promoter has been very helpful in identifying cells with astroglial properties in other species such as the zebra fish. Another important approach to label astrocytes uses viral infections. Baculo-, lenti-, adeno-, and adeno-associated viruses carrying GFAP promoter elements have been used to deliver genes of interest into astrocytes. Depending on brain region and tropism of the viral substrain, even preferential infection of astrocytes can be achieved.

Promoter elements of other astroglial genes such as S100β or the gap junction protein connexin 43 have been used successfully to generate transgenic mice. In addition, a series of mice were derived from nonhomologous recombination using bacterial artificial chromosomes carrying large genomic regions of astroglial genes such as the glutamate transporters GLAST and GLT-1.

Interestingly, Fritjof Helmchen and colleagues discovered that the common red fluorescent dye sulforhodamine 101 (SR101; also known as Texas Red) can be used to label astrocytes nontransgenically. Injection of SR101 into the cortical parenchyma *in vivo* or incubation of acutely isolated brain slices in SR101 lead to widespread labeling of all astrocytes in a given area. After uptake, the dye is distributed via gap junctions within the astroglial syncytium. Similarly, green fluorescent calcium indicator dyes such as Oregon Green or Fluo4 have been used to unspecifically label all brain cells simultaneously. Currently, costaining with SR101 allows the analysis of neuronal and astroglial network activities simultaneously. Whereas neurons are stained only by the fluorescent calcium indicator, the astrocytes are highlighted by SR101. An additional advantage of SR101 labeling is that it is species -independent. SR101 has successfully been used to label astrocytes in CNS tissues of rodents, as well as primates.

Inducible genetic labeling of astrocytes has been developed. Astroglial expression of the fusion protein of the Cre DNA recombinase and a modified ligand-binding domain of the human estrogen receptor (CreERT2) was induced either by the GFAP promoter (TgN(GFAP-CreERT2)) or by homologous recombination into the GLAST gene (TgH(GLAST-CreERT2)). Such mice must then be cross-bred to mice which carry a reporter cassette consisting of an ubiquitous promoter, a loxP-flanked stop sequence, and a reporter gene such as lacZ or GFP. Temporal control of astroglial labeling can be achieved by intraperitoneal injection of the estrogen receptor antagonist tamoxifen into these double-transgenic mice. After tamoxifen-induced gene recombination, all CreERT2-positive cells and their progeny are permanently labeled. Such experiments have been performed to investigate the role of astroglial cells as putative neural stem cells during development or disease progression.

Electrophysiological Membrane Properties

Another method to identify astrocytes is based on their electrophysiological properties since they express specific membrane currents composed of ion channels, transporters, and ionotropic receptors. Identification of astrocytes according to their membrane properties was first achieved by Kuffler and colleagues in the 1960s in the amphibian optic nerve and later verified by many other groups in the mammalian

CNS. Mature astrocytes express high levels of symmetrical in- and outwardly rectifying potassium currents with a linear current-to-voltage relationship. The astroglial resting membrane potential is usually very negative (−80 mV or more).

Beside imaging, the most common approach to study the physiological properties of astrocytes is the patch clamp technique. In the early 1990s, patch clamp recordings of astrocytes relied either on a unique localization of astroglial cell populations in defined brain regions (e.g., Bergmann glia in the cerebellum) or on postrecording labeling techniques. Both approaches have major drawbacks. Phase or differential interference contrast does not always allow the distinction of astrocytes from neighboring neurons in acutely isolated brain slices, particularly from interneurons, which very often have similar cell somata. Subsequent cell identification is often hampered by distortion of the cell integrity after patch pipette retraction. The availability of transgenic mice with astroglial GFP expression not only facilitated the physiological analysis but also contributed significantly to define a set of unique membrane properties for astrocytes. In this way, astroglial subpopulations could be distinguished in various brain regions. Patch clamp recordings in GFAP–GFP transgenic mice confirmed the findings of Kuffler and others: a dominance of the membrane conductance by high potassium currents with a linear current-to-voltage relationship and negative membrane potentials of more than −80 mV. In addition, they abundantly express glutamate transporters and are coupled within a syncytial network. Other astroglial cells are characterized by expression of voltage-gated sodium and outwardly rectifying potassium currents. Studies performed on almost all brain regions provide a complex picture of the astrocyte. Although in the past astrocytes were regarded as passive elements of the nervous system acting plainly as sink for potassium or the excitatory transmitter glutamate, AMPA- and NMDA-type glutamate receptors have been found on astroglial cells, as well as GABA receptors and transporters. However, not all receptors, transporters, and ion channels are found on all astrocytes. The current picture reveals astrocytes as a class of cells with some common characteristics and many heterogeneous properties which can be developmentally regulated and brain region dependent.

During patch clamp recordings, astrocytes can also be visualized by simultaneous diffusion of fluorescent dyes from the pipette solution into the cell. Since the majority of astrocytes are coupled via gap junctions consisting primarily of the connexin proteins cx43 and cx30, the complete astroglial syncytium can be revealed.

Conclusion

Since their first description in the nineteenth century, the functional characterization of astrocytes has depended significantly on the development of novel labeling techniques. The current techniques allow the unequivocal identification of astrocytes under almost all experimental conditions: in tissue culture, in postmortem brain sections, or even *in vivo*. We can now identify living astrocytes using transgenic mice with fluorescent protein expression or using astroglia-selective vital dye staining. In addition, we can study the plasticity of astrocytes *in vivo* using advanced imaging techniques such as two-photon laser scanning microscopy. These studies are complemented by the analysis of their signaling molecule repertoire of receptors and transporters at the single-cell level using the whole-cell patch clamp technique. Finally, we can now selectively control gene activity in astrocytes using the inducible Cre/loxP system. This latter approach will help us to understand the impact of a given gene expressed in astrocytes for the living animal.

See also: Astrocyte: Neurotransmitter and Hormone Receptors; Calcium Homeostasis in Glia; Glial Cells: Astrocytes and Oligodendrocytes During Normal Brain Aging; Intermediate Filaments; Neurotransmitter Release from Astrocytes; Potassium Homeostasis in Glia.

Further Reading

Brenner M, Kisseberth WC, Su Y, Besnard F, and Messing A (1994) GFAP promoter directs astrocyte-specific expression in transgenic mice. *Journal of Neuroscience* 14: 1030–1037.

Halassa MM, Fellin T, and Haydon PG (2007) The tripartite synapse: Roles for gliotransmission in health and disease. *Trends in Molecular Medicine* 13: 54–63.

Hirrlinger PG, Scheller A, Braun C, Hirrlinger J, and Kirchhoff F (2006) Temporal control of gene recombination in astrocytes by transgenic expression of the tamoxifen-inducible DNA recombinase variant CreERT2. *Glia* 54: 11–20.

Kuffler SW, Nicholls JG, and Orkand RK (1966) Physiological properties of glial cells in the central nervous system of amphibia. *Journal of Neurophysiology* 29: 768–787.

Lee Y, Su M, Messing A, and Brenner M (2006) Astrocyte heterogeneity revealed by expression of a GFAP-LacZ transgene. *Glia* 53: 677–687.

Matthias K, Kirchhoff F, Seifert G, et al. (2003) Segregated expression of AMPA-type glutamate receptors and glutamate transporters defines distinct astrocyte populations in the mouse hippocampus. *Journal of Neuroscience* 23: 1750–1758.

Misgeld T and Kerschensteiner M (2006) *In vivo* imaging of the diseased nervous system. *Nature Reviews Neuroscience* 7: 449–463.

Nimmerjahn A, Kirchhoff F, Kerr JN, and Helmchen F (2004) Sulforhodamine 101 as a specific marker of astroglia in the neocortex *in vivo*. *Nature Methods* 1: 31–37.

Palay SL, Peters A, and Webster H (1991) *The Fine Structure of the Nervous System: Neurons and Their Supporting Cells.* New York: Oxford University Press.

Somjen GG (1988) Nervenkitt: Notes on the history of the concept of neuroglia. *Glia* 1: 2–9.

Astrocyte: Neurotransmitter and Hormone Receptors

C Steinhäuser, University of Bonn, Bonn, Germany
H Kettenmann, Max Delbrück Center for Molecular Medicine, Berlin, Germany

Background

Until about 20 years ago, the expression of neurotransmitter receptors in the nervous system was considered an exclusive property of neurons. Astrocytes were viewed as electrically silent elements which would not participate in the information processing of the central nervous system (CNS). This view was challenged by a series of studies in cell culture demonstrating that astrocytes had the potential to express transmitter receptors. In the culture dish, astrocytes seemed to express almost all receptors for important transmitter systems. Two exceptions remained: *N*-methyl-D-aspartate (NMDA) and glycine receptors. In the early 1990s, the studies were extended to acute brain slices, and it became evident that astrocytes expressed functional receptors in this more intact preparation. It is interesting to note that the two receptor types lacking in culture, NMDA and glycine receptors, were now found to be functionally expressed by some glial cells from defined regions of the CNS. This article describes the receptor repertoire expressed by astroglial cells. However, a basic limitation has to be mentioned here. It is becoming clear that different types of cells with astroglial properties coexist within a given brain region and that the properties of these cells vary in different areas. So far, we have only rudimentary understanding of this glial diversity, and most of the previous studies describing astroglial properties did not identify the specific cell type. This article refers to different types of cells with astroglial properties as astrocytes.

Ionotropic Glutamate Receptors

Functional α-amino-3-hydroxy-5-methyl-4-isoxazole propionic acid (AMPA) receptors have been described in astrocytes *in situ* in various brain regions. In the hippocampus, two types of cells with astroglial properties were shown to coexist, with one type expressing and the other lacking AMPA receptors. While in most brain regions astrocyte AMPA receptors display a low Ca^{2+} permeability, Bergmann glial cells in the cerebellum lack the glutamate receptor 2 (GluR2) subunit and allow passage of divalent cations. The physiological relevance of this subunit combination has been demonstrated by experimentally inducing expression of GluR2 in Bergmann glial cells, which results in neuronal rearrangement. The pharmacological properties and single channel conductance of astroglial AMPA receptors mimic those of their neuronal counterparts. In the hippocampus, astroglial receptors possess an intermediate Ca^{2+} permeability and are primarily assembled from the subunits GluR1, GluR2, and GluR4. Receptor activation leads to reversible blockage of K^+ channels, through intracellular accumulation of Na^+. Early in postnatal development, astrocytes co-express AMPA receptors with variable Ca^{2+} permeability while later on, only receptors with low Ca^{2+} permeability, carrying GluR2, are formed. In addition, changes in receptor splicing occur within the first 3 postnatal weeks: in contrast to hippocampal neurons, the portion of flip splice variants increases in astrocytes, resulting in prolonged receptor opening. Joint stimulation of AMPA and metabotropic glutamate receptors (mGluRs) in hippocampal and neocortical astrocytes triggers Ca^{2+} oscillations and mediates glutamate release. Non-NMDA receptor activation in astrocytes was reported to trigger ATP release from these cells, which in turn caused homo- and heterosynaptic suppression in CA1 neurons of the hippocampus. Recently, cells with astroglial properties in the hippocampus were shown to receive direct synaptic input from glutamatergic neurons. In these glial cells, nerve fiber stimulation led to the activation of AMPA receptor-mediated postsynaptic currents. Fiber stimulation in the cerebellum leads to ectopic presynaptic release of glutamate, which rapidly activates AMPA receptors in Bergmann glial cells. While the physiological impact of this rapid form of neuron-to-glia signaling is poorly understood, increasing evidence suggests that dysfunctional astroglial AMPA receptors are involved in the pathogenesis of neurological disorders.

Clear evidence for a functional expression of kainate receptors in astrocytes is lacking. Transcripts of all five kainate receptor subunits, GluR5-7, KA1, and KA2, have been found in adult bovine white matter astrocytes *in situ*. Antibodies against GluR5-7, GluR6,7, and KA2 confirmed the expression of kainate receptor protein, but functional receptors have yet to be identified.

Some astrocytes may express functional NMDA receptors. Bergmann glial cells express messenger RNAs (mRNAs) for NR2B, and a physiological study in acute brain slices reported NMDA-induced membrane currents in Bergmann glial cells. However, these responses displayed unusual properties (e.g., no Mg^{2+} and glycine sensitivity) and indirect effects cannot be excluded. Evidence for the presence of functional

NMDA receptors *in situ* comes from cortical astrocytes. In these cells, astroglial NMDA responses were observed that resembled NMDA response in neurons, including sensitivity to extracellular Mg^{2+} and blockade by MK-801. In contrast, another study described Mg^{2+}-insensitive NMDA receptors in cortical astrocytes. Moreover, activation of NMDA receptor currents were observed in hippocampal astrocytes after ischemia *in vivo*, due to upregulation of the NR2B subunit.

mGluRs

The mGluR family consists of eight members which couple to G-proteins and can be classified into three groups: group I comprises mGluR1 and mGluR5, and their activation leads to stimulation of phospholipase C (PLC), increase in inositol triphosphate (IP_3), and release of Ca^{2+} from internal stores, and group II (mGluR2, mGluR3) and group III (mGluR4, mGluR6–8) receptors couple to adenylate cyclase. mGluR3 and mGluR5 are the predominant subtypes expressed by astrocytes. mGluR1 has been reported in hippocampal astrocytes and in the spinal cord. Accordingly, activation of these receptors led to an increase in intracellular Ca^{2+} and inhibition of cyclic adenosine monophosphate (cAMP) accumulation, although Gs-coupled cAMP stimulation has also been reported. Stimulation of astroglial mGluRs leads to intracellular Ca^{2+} oscillations and Ca^{2+} wave propagation within the astrocyte network, activates Ca^{2+}-dependent K^+ channels, and induces prostaglandin-mediated glutamate release from astrocytes that activates neuronal receptors. These responses are likely to occur under physiological conditions because astroglial mGluR activation and subsequent Ca^{2+} responses could be evoked by electrical stimulation of fiber tracts in acute brain slices and through sensory stimulation *in vivo*. Activation of mGluRs induced other astrocyte responses, including swelling, activation of phospholipase D and glutamine synthetase, release of arachidonic acid, cAMP-dependent block of K^+ currents, modulation of proliferation, and regulation of the expression of the glutamate aspartate transporter (GLAST) (**Figure 1**). Moreover, mGluR-induced intracellular Ca^{2+} elevations in astrocytes are involved in the regulation of cerebral blood flow. Together, these findings suggest that mGluRs play a pervasive role in modulating astrocyte function and intercellular communication in the CNS. Dramatic changes in astrocyte mGluR expression occur after spinal cord injury and in neurological disorders.

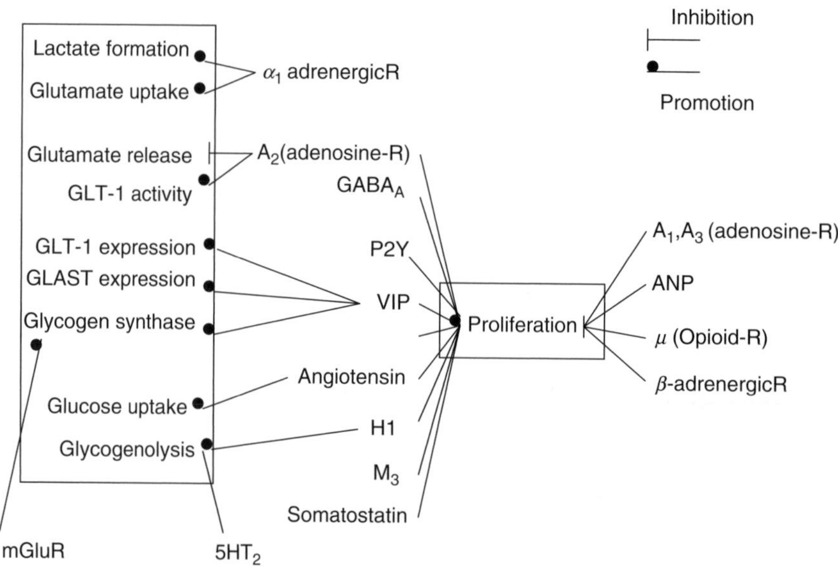

Figure 1 Astrocyte receptors linked to proliferation and glutamate/glucose metabolism. This figure summarizes the influence of different astrocytic receptors on glutamate and glucose metabolism and transport and combines this information with the impact of receptors on proliferation. It becomes evident that only proliferation-promoting receptors also control glutamate or glucose metabolism and transport, in contrast to receptors which inhibit astrocyte proliferation. It is evident that only proliferation-promoting receptors also control glutamate or glucose metabolism and transport, in contrast to receptors that inhibit astrocyte proliferation. A_1, A_2, and A_3, adenosine receptors; α_1 adrenergicR, α_1 adrenergic receptor; ANP, atrial natriuretic peptide; β-adrenergicR, β adrenergic receptor; GLAST, glutamate aspartate transporter; $GABA_A$, γ-aminobutyric $acid_A$; GLT-1, glutamate transporter 1; H_1, histamine receptor; $5\text{-}HT_2$, 5-hydroxytryptamine-2; M_3, muscarinic acetylcholine receptor; mGluR, metabotropic glutamate receptor; P2Y, a type of purinergic receptor; R, receptor; VIP, vasoactive intestinal polypeptide.

γ-Aminobutyric Acid Receptors

Ionotropic γ-aminobutyric acid ($GABA_A$) receptors are abundantly expressed by astrocytes. *In situ*, functional $GABA_A$ receptors have been demonstrated in Bergmann glial cells, astrocytes of the spinal cord, optic nerve, retina, hippocampus, and pituitary gland. Astrocytic $GABA_A$ receptors form Cl^- channels with a conductance of about 30 pS. In contrast to mature neurons, $GABA_A$ receptor activation leads to depolarization in astrocytes. Indeed, local stimulation of interneurons in the hippocampus, which form functional synapses with astroglial cells, led to $GABA_A$ receptor-mediated glial depolarization. In astrocytes of the pituitary gland, postsynaptic potentials were activated by neuronally released GABA. Besides opening Cl^- channels, $GABA_A$ receptor activation in astrocytes triggers long-lasting blockade of K^+ conductances, thereby augmenting depolarization and activating voltage-gated Ca^{2+} channels with subsequent cytosolic Ca^{2+} increase.

Bergmann glial cells in cerebellar slices showed GABA responses which were benzodiazepine-insensitive. Immunocytochemistry revealed that these cells lack the γ subunit; they express instead the δ subunit. In contrast, hippocampal astrocytes are benzodiazepine-sensitive, indicating functional heterogeneity among astrocyte subtypes. In Bergmann glial cells, $GABA_A$ receptors are downregulated during postnatal maturation. In these cells, immunolabeling preferentially identified receptors along the processes. There is evidence that $GABA_A$ receptor activation promotes differentiation of astrocytes. In culture, GABA triggers formation of processes and leads to a more complex morphological shape of the astrocytes. A similar effect has been observed in the hypothalamus *in situ*: GABA released from neurons activated astroglial $GABA_A$ receptors and induced differentiation.

$GABA_B$ Receptors

The most convincing demonstration of functional expression of astroglial $GABA_B$ receptors was obtained in acute hippocampal slices. Here, these receptors mediated intracellular Ca^{2+} elevations that were accomplished by GABA release from inhibitory interneurons. This indicates that astrocytes *in situ* can sense inhibitory neuronal activity via $GABA_B$ receptors. It has also been demonstrated that $GABA_B$ receptor activation in astrocytes has an impact on neuronal function by potentiating inhibitory postsynaptic currents in pyramidal neurons. Other studies reported that astroglial $GABA_B$ receptor activation mediates heterosynaptic neuronal depression. Thus, in the hippocampus, astrocytic $GABA_B$ receptors are part of a functional circuit involving inhibitory neurons, astrocytes, and pyramidal neurons.

Purinergic Receptors

Astrocytes express a large number of purinergic receptor subtypes linked to different effector mechanisms. The presence of adenosine receptors *in situ* has been established in the hippocampus and in acutely isolated cortical astrocytes. In both preparations, activation of adenosine receptors led to the accumulation of intracellular Ca^{2+} via release from cytoplasmic stores. All three types of adenosine receptors (A_1, A_2, A_3) can be expressed by astrocytes. A_2 receptor activation increased internal cAMP concentration and stimulated astroglial proliferation, whereas stimulation of A_1 and A_3 receptors had opposite effects. A_1 receptor activation stimulated the production of neurotrophic factors (nerve growth factor (NGF), S100β, transforming growth factor-β (TGF-β)) and enhanced mGluR5-induced intracellular Ca^{2+} responses. Receptor subtypes can have opposing actions: while A_1 receptor stimulation inhibited purine release, A_2 receptor activation stimulated release *in situ*. A_3 receptor stimulation not only inhibits proliferation but also leads to apoptosis and induces reorganization of the astroglial cytoskeleton. Thus, adenosine triggers a complex response in astrocytes which can lead either to proliferation or to apoptosis, depending on the expression levels of A_1 and A_3 receptors, respectively. A_{2a} receptor signaling is linked to the control of glutamate levels in the extracellular space: Receptor activation inhibited glutamate transporter 1 (GLT-1)-mediated glutamate uptake and facilitated glutamate release, leading to increased extracellular levels of the transmitter.

Both types of ATP receptors, P2X and P2Y, have been identified in astrocytes. In cell culture, activation of P2X receptors leads to the opening of a cationic conductance. So far, however, P2X receptor-mediated cationic conductances have not been observed in astrocytes *in situ* despite positive immunolabeling for various P2X subunits. Only in Müller cells of the human retina have $P2X_7$-induced currents been recorded, a property that seems to be species specific. Here, $P2X_7$ receptor activation triggered Ca^{2+}-activated K^+ (BK) channels which might be involved in $P2X_7$-dependent regulation of proliferation.

P2Y-mediated signaling has been identified in Bergmann glial cells in acute slices of cerebellum, in freshly isolated astrocytes from cortex and hippocampus, and in Müller cells and astrocytes from the retina. P2Y receptor activation leads to a G_q-PLC mediated increase in IP_3 and Ca^{2+} release from intracellular stores. Moreover, these receptors are coupled to pertussis toxin (PTX)-sensitive G_i/G_o proteins linked to the synthesis of phospholipase A2, arachidonic acid, and protein kinase C (PKC), extracellular signal-regulated kinase, cyclooxygenase 2, release of arachidonic acid and prostaglandin E2, stimulation of mitogen-activated protein kinases (MAPKs), and induction of immediate early genes. Astrocytes of the postnatal hippocampus express the $P2Y_1$ and $P2Y_2$ receptor subtypes, with $P2Y_2$ increasing during development. P2Y receptor activation is also involved in the induction of astrogliosis *in vivo* since lesion-induced astrogliosis could be blocked by P2Y receptor antagonists.

Glycine Receptors

Functional glycine receptors have been detected in astrocytes from spinal cord slices. A combined patch clamp reverse transcription polymerase chain reaction (RT-PCR) approach revealed the expression of $\alpha1$ and β subunits. As in neurons, glycine activated a Cl^- conductance. It is interesting to note that glycine receptor expression has not been reported for cultured astrocytes and is also not a common property of astrocytes *in situ*: Bergmann glial cells, for instance, do not express functional glycine receptors.

Acetylcholine Receptors

α-Bungarotoxin-sensitive nicotinic acetylcholine (ACh) receptors containing the $\alpha7$ subunit were described for cultured astrocytes. Receptor activation led to Ca^{2+} influx through the receptor channels and Ca^{2+}-induced Ca^{2+}-release from caffeine-sensitive stores. This subunit was also localized to astrocytes of human cerebellum *in situ*. This receptor was upregulated in astrocytes from the brains of Alzheimer's patients, suggesting a role in the pathogenesis of this disease.

The expression of muscarinic ACh (mACh) receptors by astrocytes is well established. There is evidence for the expression of transcripts for the mACh receptor subtypes M_1–M_5 in cultured astrocytes. Acetylcholine induces IP_3 formation, release of Ca^{2+} from cytoplasmic stores and inhibits adenylate cyclase. As tested in culture, acetylcholine induced an increase in astroglial proliferation due to activation of M_3 receptors and phosphorylation of MAPK. Neuronal activity can stimulate astrocytic mACh receptors: stimulating cholinergic fibers projecting from the septum to the hippocampus triggered Ca^{2+} responses in hippocampal astrocytes. Excessive neuronal activity leads to upregulation of mACh receptors *in vivo*, as has been shown in amygdala kindling, indicating that neuronal activity can regulate astrocyte receptor expression. During development, mACh receptor responses and receptor protein expression is upregulated.

Adrenergic Receptors

Both α- and β-adrenergic receptors have been described in astrocytes in culture and *in situ*. The α_1 receptors trigger the release of Ca^{2+} from internal stores. They affect astrocyte function by inhibiting gap junction coupling via PLC. This is in line with the finding that mechanically induced Ca^{2+} waves in hippocampal astroglial cells are inhibited after stimulation of α_1 receptors. Moreover, α_{1b} receptor activation stimulated glutamate uptake in cultured astrocytes. There is also a hint for the control of metabolic activity by α_1 receptors because receptor activation induced lactate formation. The α_1 receptors regulate expression of α_2 receptors and thus form a complex feedback circuit. Both α_1 and α_2 receptors have been identified *in situ*; α_1 receptors are expressed postnatally in the hippocampus (after postnatal day 8 in rat). In the hippocampus, α_2 receptors are located on astroglial processes, near terminals forming asymmetric excitatory synapses. Neuronal stimulation in acute cerebellar slices led to the release of noradrenaline from afferent fibers and activation of α_1 receptors in Bergmann glial cells.

Both subtypes of β receptors, β_1 and β_2, are expressed by astrocytes *in situ*, with β_2 being more prominently expressed under pathological conditions. In astrocytes of rat visual cortex, β receptors appear after the second postnatal week. In the injured brain, adrenergic receptors are differentially regulated. While α_1 receptors decrease in areas of neuronal degeneration and gliosis, β receptors are upregulated, as evidenced by binding assays. The change in β receptor density and astrogliosis seems to be functionally linked. Blockade of β receptors suppressed glial scar formation, indicating that adrenergic receptors are part of the cascade leading to astrocyte activation. This suggestion is substantiated by the finding that β_2 receptors are activated or upregulated in the transected optic nerve *in vivo*, confirming that β-adrenergic signaling is an important feature of the astrocyte response to injury. Further support comes from the observation that stimulation of β receptors leads to astrogliosis and cell proliferation in the optic nerve *in vivo*. This effect might be mediated via the control of growth factor expression (e.g., fibroblast growth factors 1and 2, BDNF, and ciliary neurotrophic factor (CNTF)). In contrast, an inhibition of proliferation on β-adrenergic stimulation was observed in cultured astrocytes, which, however, does not contradict gliosis. The involvement of β receptors in pathologic processes is further substantiated by the finding that their stimulation leads to production of amyloid precursor protein, changes in morphology, and increase in glial fibrillary acidic protein (GFAP).

Stimulation of β_1 receptors is also accompanied by enhanced glycogen levels which could be part of the gliotic response. The observation that β adrenergic receptor stimulation leads to a cAMP-mediated inhibition of astroglial inwardly rectifying K^+ channels points to a role in the regulation of the extracellular K^+ homeostasis.

Dopamine Receptors

D1 receptor activation leads to a PTX-sensitive cAMP formation via protein kinase A stimulation. Prolonged receptor activation led to a reduction of astroglial dopamine sensitivity. D2 receptors in astrocytes are linked to the actin cytoskeleton via interaction with filamin A. A recent study provided first evidence for the presence of dopamine receptors on astrocytes *in vivo*. Ligand binding and ultrastructural analysis found strong expression of D2 receptors in GFAP-positive cortical astrocyte processes surrounding interneurons. These receptors were indeed functional since their activation induced intracellular Ca^{2+} elevations in the astrocytes.

Histamine Receptors

In cultured astrocytes, binding sites for H_1 and H_2 subtypes of histamine receptors have been identified which couple to G_q and G_s types of G-proteins, respectively. Accordingly, H_1 receptors couple to PLC, leading to IP_3 production and Ca^{2+} release while H_2 receptor activation in astrocytes is linked to adenylate cyclase and intracellular cAMP accumulation. H_1 receptors were mainly found on the processes of astrocytes, and stimulation led to locally restricted Ca^{2+} responses. Activation of histamine H_1-receptor enhances neurotrophic factor secretion from cultured astrocytes. Histamine also acts as a mitogen in cortical astrocytes. Activation of H_1 receptors has been shown to stimulate glycogen breakdown. Although *in situ* hybridization has not yet unequivocally associated histamine receptors with astrocytes, pharmacological studies indicate their expression in cerebellum and hippocampus. H_1 receptor activation triggered an increase in $Ca^{2+}{}_i$ due to release from thapsigargin-sensitive intracellular pools. Based on ultrastructural studies suggesting glial apposition to histaminergic neurons, astrocytes were supposed to respond to histamine released from hippocampal neurons.

Serotonin Receptors

RT-PCR in human and rat brain cultures revealed that astrocytes can express various serotonin receptors. In cell culture, stimulation of 5-hydroxytryptamine $(5HT)_2$ led to phosphatidylinositol (PI) hydrolysis, cAMP accumulation, and upregulation of glycogenolysis. A distinct expression of 5HT subtypes also occurs *in situ* since the subtypes $5HT_{1A}$, $5HT_{2A}$, and $5HT_{5A}$ have been identified in different brain areas. 5HT receptors have been speculated to be involved in pathogenesis. The $5HT_{5A}$ subtype is upregulated in gliosis, and the $5HT_{2A}$ subtype is enhanced in schizophrenia. Stimulation of 5HT receptors leads to the release of S100β from astrocytes, and this release has been hypothesized to have an impact on the ongoing pathologic event. $5HT_{5A}$ transcripts are developmentally regulated: Receptor mRNA is detected prior to birth, and expression peaks at postnatal day 20 in the rat.

Angiotensin Receptors

Expression of angiotensin (AT) receptors in astrocytes *in situ* seems to be restricted to white matter. Antibody staining identified AT_1 and AT_2 receptors in astrocytes of white matter tracts in cerebellum and subcortical regions, in the optic nerve, and in the corpus callosum, while no immunoreactivity was found in grey matter. The diversity in expression is also reflected when astrocytes are harvested from different brain areas: AT_1 receptors were found in astrocytes from medulla oblongata and cerebellum, whereas astrocyte cultures from hypothalamus and cortex did not express functional receptors. Activation of astroglial AT receptors stimulates PI hydrolysis and PKC and triggers Ca^{2+} release from internal stores. Receptor activation as studied in culture stimulated proliferation, increased glucose uptake, and induced prostaglandin release.

Atrial Natriuretic Peptide Receptors

All three types of atrial natriuretic peptide (ANP) receptors have been identified in cultured astrocytes: ANP_A and ANP_B (biological receptor), as well as ANP_C (clearance receptor). So far, only the ANP_A receptor has been clearly identified *in situ* (in the brain stem). ANP receptors, without subtype specification, have been found in several other brain structures, including olfactory bulb, hippocampus, and amygdala. In astrocytes, the clearance receptor has a possible biological function: ANP inhibits MAPK via ANP_C. MAPK is stimulated by growth factors such as endothelin-3 (ET-3) or platelet derived growth factor, and thus ANP_C counteracts their activation. This antimitogenic action of ANP in astrocytes is mediated via inhibition of ET-3 induced G-protein activation. In consequence, since these growth factors stimulate proliferation, ANP acts as an antiproliferative substance on astrocytes. Accordingly, if astrocytes are stimulated with a proliferative agent such as AT_2 or other growth factors, ANP acts as an antagonist to restrict cell number. The specific role of ANP_A and ANP_B receptors is less evident. ANP has been shown to stimulate cyclic guanosine monophosphate production and activates guanylate cyclase in astrocytes.

Bradykinin Receptors

So far, functional bradykinin receptors (B_1 and B_2 receptors) have been identified only in cultured astrocytes. Receptor activation leads to an increase in intracellular Ca^{2+}, induced by release from intracellular stores, and to a blockade of a constitutive K^+ conductance. Pharmacological data indicate that this block is mediated via B_2 receptors. B_2 receptor-induced increase in intracellular Ca^{2+} leads to the release of glutamate and aspartate from astrocytes, which stimulates neuronal glutamate receptors, substantiating that astrocytes can directly modulate neuronal activity.

Endothelin Receptors

Astrocytes in culture and in tissue express endothelin B receptors. Stimulation of these receptors is linked to control of proliferation, in particular in the context of brain injury. There is even evidence that pathologic conditions such as a hypoxic insult increase endothelin B receptor expression. Endothelin stimulates astrocytic neurotrophin-3 and BDNF production. Since astrocytes are also capable of releasing endothelins, an autocrine pathway is possible which might be relevant in pathology.

Opioid Receptors

Early work indicated the presence of opioid receptors on tanycytes and pituicytes. There is evidence for the expression of all three subtypes of the opioid receptors, μ, δ, and κ, both in cell culture and *in situ*. Activation of μ and κ receptors activation DNA synthesis, proliferation, and astroglial growth. It is interesting to note that the inhibitory action of opioids on astrocyte proliferation was restricted to GFAP-positive astrocytes while S100β-positive cells were not affected. Receptors κ and δ are linked to Ca^{2+} signaling, and their stimulation inhibited forskolin-induced cAMP stimulation. Circumstantial evidence indicates that the effect of opioids on astroglial proliferation is Ca^{2+}-dependent. Indeed, application of opioid receptor agonists induced an increase in intracellular Ca^{2+}, but the underlying mechanisms are not yet fully resolved. Receptors μ and δ were shown to be preferentially expressed on astrocytic processes *in situ*. While μ receptors are more frequently present on immature astrocytes, the expression of δ and κ receptors is increased in adult tissue.

Oxytocin and Vasopressin Receptors

Both vasopressin and oxytocin can trigger Ca^{2+} responses in astrocytes, indicating the presence of their receptors. Neuronal production and release of TGF-β led to an increase in astroglial oxytocin receptor mRNA in astrocytes, indicating that neurons were able to upregulate the astrocyte receptor. V_{1A} vasopressin receptors were identified by binding and biochemical studies. There is evidence that vasopressin is important for Cl^- homeostasis and volume regulation of astrocytes. An increased Cl^- uptake was observed on vasopressin application, presumably mediated by V_{1A} receptors.

Somatostatin Receptors

Prominent expression of somatostatin (sst) receptors in astrocytes of hippocampus, amygdala, and hypothalamus *in situ* has been identified in binding studies on brain slices. The subtypes have been more stringently identified in culture, and it is evident that astrocytes express the sst_{2A} splice variant. Receptor activation is linked to inhibition of cAMP accumulation and leads to enhanced proliferation rates. Moreover, it leads to a reduction in forskolin-induced interleukin-6 (IL-6) release.

Tachykinin Receptors

Astrocytes express all subtypes of tachykinin receptors, NK_1 to NK_3. In cell culture, activation of NK_1 receptors leads to membrane depolarization. This depolarization is due to blockade of a constitutive K^+ conductance and opening of Cl^- channels. Patch clamp recordings demonstrate that channel activity is linked to the tachykinin receptor via intracellular pathways. Receptor localization on the light and electron microscopic level in different species, including humans, has identified NK_2 and NK_3 receptors in astrocytes of various areas of the CNS, such as spinal cord, cortex, and hippocampus. NK_2 receptors have been found to cluster close to axon terminals in spinal cord. Substance P, as a prominent ligand, enhances secretion of IL-6 and prostaglandin E_2 on IL-1β stimulation in cultured spinal cord astrocytes. Moreover, lipopolysaccharide-induced secretion of tumor necrosis factor-α and IL-1 from cultured astrocytes was augmented by the ligand known as substance P. A proinflammatory function of tachykinin receptors in reactive astrocytes has been suggested since a lesion-induced upregulation of the receptors was observed in astrocytes of transected optic nerve *in vivo*.

Thyrotropin-Releasing Hormone Receptors

Thyrotropin-releasing hormone (TRH) receptors have been identified in culture and *in situ* by immunocytochemistry and Western Blot. Astrocytes express the TRH_1 subtype. In the adult rat spinal cord, expression *in situ* is stronger in white than in grey matter. The receptors are present in TRH-synthesizing astrocytes, indicating that this hormone acts in an autocrine and paracrine fashion.

Vasoactive Intestinal Polypeptide Receptors

Vasoactive intestinal polypeptide (VIP) receptors are expressed by astrocytes, and the activation leads to Ca^{2+} signaling. All three subtypes of VIP receptor have been identified in cultured astrocytes. VIP receptor activation induces release of IL-6 and neurotrophic factors and stimulates proliferation. There is a link between receptor activation and glutamate uptake: Stimulation of VIP receptors leads to the promotion of GLT-1 and GLAST expression. This finding is in line with another study showing that VIP enhances glutamate uptake in astrocytes. Together, these data indicate that VIP can increase the strength of glutamate-mediated neurotransmission. In addition, VIP receptors may play an important role in regulating energy metabolism. VIP depletes astrocyte glycogen initially, followed by delayed reaccumulation to a level beyond baseline. These effects are mediated by regulating a number of related genes, such as glycogen synthase, via the transcription factor family CCAAT enhancer binding protein. However, there is not yet convincing evidence of the presence of VIP receptors in astrocytes *in situ*.

See also: AMPA Receptors: Molecular Biology and Pharmacology; Astrocyte: Identification Methods; Dopamine Receptors and Antipsychotic Drugs in Health and Disease; $GABA_A$ Receptors: Molecular Biology, Cell Biology and Pharmacology; $GABA_B$ Receptors: Molecular Biology and Pharmacology; Glycine Receptors: Molecular and Cell Biology; Metabotropic Glutamate Receptors (mGluRs): Molecular Biology, Pharmacology and Cell Biology; Nicotinic Acetylcholine Receptors; Purinergic Receptors; Serotonin (5-Hydroxytryptamine; 5-HT): Receptors; Somatostatin and Receptors; Substance P/Tachykinins and its/their Receptors; Vasoactive Intestinal Peptide and Pituitary Adenylate Cyclase Activating Peptide Receptors; Vasopressin/Oxytocin and Receptors.

Further Reading

Hatton GI and Parpura V (2004) *Glial Neuronal Signaling*. Boston: Kluwer.

Kettenmann H and Ransom BR (2005) *Neuroglia*. New York: Oxford University Press.

Matthias K, Kirchhoff F, Seifert G, et al. (2003) Segregated expression of AMPA-type glutamate receptors and glutamate transporters defines distinct astrocyte populations in the mouse hippocampus. *Journal of Neuroscience* 23: 1750–1758.

Schipke CG and Kettenmann H (2004) Astrocyte responses to neuronal activity. *Glia* 47: 226–232.

Seifert G, Schilling K, and Steinhäuser C (2006) Astrocyte dysfunction in neurological disorders: A molecular perspective. *Nature Reviews Neuroscience* 7: 194–206.

Seifert G and Steinhäuser C (2001) Ionotropic glutamate receptors in astrocytes. *Progress in Brain Research* 132: 287–299.

Verkhratsky A, Orkand RK, and Kettenmann H (1998) Glial calcium: Homeostasis and signaling function. *Physiological Reviews* 78: 99–141.

Volterra A, Magistretti PJ, and Haydon PG (2002) *The Tripartite Synapse*. Oxford, UK: Oxford University Press.

Volterra A and Meldolesi J (2005) Astrocytes, from brain glue to communication elements: The revolution continues. *Nature Review. Neuroscience* 6: 626–640.

Wang X, Lou N, Xu Q, et al. (2006) Astrocytic Ca(2+) signaling evoked by sensory stimulation in vivo. *Nature Neuroscience* 9: 816–823.

Astrocyte: Response to Injury

C J Feeney and P K Stys, Hotchkiss Brain Institute, University of Calgary, Calgary, AB, Canada

Introduction

Astrocytes are the most numerous cell type in the central nervous system (CNS) and make up about 50% of human brain volume. A central role of glial cells (in particular astrocytes) in regulation of the extracellular ionic environment has been well established, with key functions including K^+ buffering, homeostasis of H^+ and Ca^{2+}, detoxification of ammonia, and free radical scavenging. In addition, astrocytes are important in the production and release of growth factors, in supportive roles such as supply of energetic metabolites (lactate, pyruvate, and tricarboxylic acid cycle metabolites), neurotransmitter uptake, and precursor provision. Very recent data have even identified a role for astrocytes in promoting myelination of central axons in response to electrical activity. Thus it is clear that healthy astrocyte functioning is of paramount importance to proper CNS operation, and indeed survival, after insults such as ischemia and traumatic injuries.

In general, a perception predominates that astrocytes are relatively resistant to insults such as those caused by ischemia (oxygen–glucose deprivation; OGD) and oxidative stress. Recent work, however, points to significant astrocyte dysfunction and also cell death in response to these injuries. This article focuses on the various pathophysiological changes that can occur in astrocytes in response to hypoxia and ischemic-like insults.

How Do Astrocytes Fare in the Face of Ischemic-Like Insults?

Hypoxia

Cultured astrocytes have a well-documented prolonged resistance to exposure to lowered O_2 tension. In general, these studies utilize cultured cortical astrocytes derived from fetal or early neonatal-stage rodents. The elimination of oxygen alone (often termed *in vitro* hypoxia) has been shown to be relatively innocuous to these cultures, illustrating the astrocytes' ability to maintain viability for many hours of O_2 deprivation. These cells are able to survive such extended periods of hypoxia through the utilization of glycolytic adenosine triphosphate (ATP) production. Prolonged oxygen deprivation (12–24 h) is not, however, without effect on the cultured astrocyte, as subsequent reoxygenation of cultures leads to pronounced reactive oxygen species (ROS) production, mitochondrial damage, and cell death, at least in part, via apoptosis.

OGD (*In Vitro* Ischemia)

As with hypoxia, cultured astrocytes can also survive many hours of OGD, often termed *in vitro* ischemic conditions. Astrocyte cultures show an age-dependent resistance to OGD, with no significant cell death occurring in astrocyte cultures exposed to OGD for 6–12 h. Younger cultures (1 week *in vitro*) were found to be much more resistant to such OGD than older astrocyte cultures (3 weeks old). As well, the elimination of exogenous glucose also leads to relatively slow astrocyte demise, as cultured astrocytes can survive tens of hours in the absence of glucose when oxygen is present. Again, immature (younger) astrocyte cultures were found to be more robust than mature cultures. These findings suggest that maturational increases in metabolic rates in the developing brain *in vivo* and increases in oxidative metabolism in astrocytes *in vitro* with increasing age of culture confer a greater susceptibility of these cells to hypoxic/ischemic damage.

In Vivo Studies ('True' Ischemia)

Numerous published reports point to the susceptibility of neuronal cells to ischemia-induced injury *in vivo*, while far fewer have focused on astrocytic susceptibility. Those that have reported on the nature of astrocytic vulnerability to ischemia most generally document a greater resistance of astrocytes to ischemia compared with their neuronal counterparts. For example, a number of studies have shown that a brief transient cerebral ischemia sufficient to produce substantial neuronal cell death in CA1 of the hippocampus produces little to no astrocytic cell death. A more substantial insult, such as prolonged (2 h) or permanent middle cerebral artery occlusion (MCAO), has been shown to lead to rapid and significant astrocyte injury within 30–60 min, in some cases preceding neuronal demise. It should be noted that it is not clear whether these findings reflect a differential vulnerability of certain astrocyte populations or a different (i.e., faster) rate of cellular demise (necrosis vs. apoptosis). Finally, some have shown that particular subsets of astrocytes may be more susceptible to ischemic insults than others. For example, protoplasmic astrocytes appear to be more rapidly compromised than fibrous astrocytes in response to ischemia.

Astrocytic Responses to Ischemic-Like Insults

Mitochondrial Dysfunction

Ischemia is well known to induce rapid and sustained loss of mitochondrial function in numerous cell types. Of particular relevance in this article is the rapid and seemingly complete loss of neuronal mitochondrial potential ($\Delta\Psi_M$) in response to OGD. Thus, in cultured neuronal cell models, hippocampal slice cultures, and acute brain slices, neurons lose their mitochondrial potential within minutes of an ischemic-like insult. While prolonged exposure to oxygen and substrate deprivation must inevitably lead to mitochondrial dysfunction, conflicting evidence exists as to the early responses (if any) of astrocytic mitochondria to OGD. For example, some have shown that cortical astrocyte cultures exposed to 1 h of OGD do not lose their $\Delta\Psi_M$ or experience ATP depletion, nor does the exposure induce free radical production or decrease nicotinamide adenine dinucleotide phosphate (critical in the maintenance of reduced glutathione levels in the cell), whereas the same insult does so in neuronal cultures. Contrary to this finding, others have clearly shown that both cultured neurons and astrocytes lose their mitochondrial potential within 45–60 min when exposed to OGD. Of importance is that astrocytes that showed 'rapid' loss of $\Delta\Psi_M$ were cocultured with neurons, and the astrocytes cultured alone showed massive $\Delta\Psi_M$ loss only at 90 min of OGD (see **Figure 1**). The rapid mitochondrial dysfunction induced in these astrocytes was found to be, at least in part, related to the induction of permeability transition, as cyclosporin A partially protected astrocytes against $\Delta\Psi_M$ loss. It is interesting that astrocytes that find themselves amid neurons (e.g., cocultures or slice cultures) respond to OGD with a more profound mitochondrial dysfunction than those in isolation. In addition, the reestablishment of astrocyte $\Delta\Psi_M$ after ischemia-induced collapse appears to be gradual in nature. On 'reperfusion' with O_2 and glucose, astrocytic $\Delta\Psi_M$ is restored to 'normal' after several hours, possibly owing to some long-lasting structural changes in the astrocytic mitochondria themselves.

Whether these differences in astrocytic response are a result of differential culturing environments or a synergistic interaction between astrocytes and neurons leading to a heightened astrocytic sensitivity remains to be firmly established. What is clear, though, is that the loss of astrocytic $\Delta\Psi_M$ would likely have detrimental effects on CNS recovery after an ischemic-like insult. As astrocyte mitochondrial dysfunction proceeds, the homeostatic functions astrocytes perform would be compromised, thereby leading to dysregulation of K^+ and H^+ ion concentrations, neurotransmitter uptake, and metabolite transfer to neighboring neurons. While short-term loss of astrocyte mitochondrial function may be well tolerated by the astrocytes themselves (by utilization of glycogen stores, alternate energetic metabolites), the relatively modest capacity of energy stores would soon lead to cellular compromise in both neuronal and nonneuronal compartments.

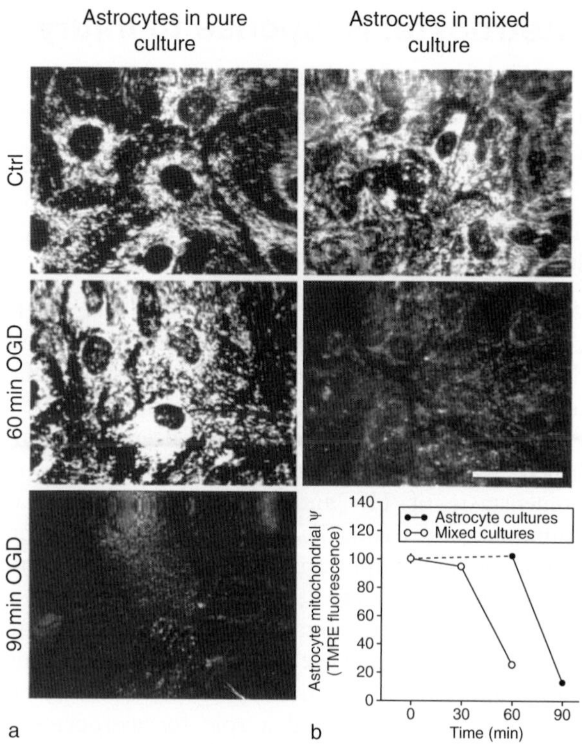

Figure 1 Astrocytes in pure and cocultures with neurons (mixed culture) lose mitochondrial potential in response to oxygen–glucose deprivation (OGD). Decreases in the mitochondrial potential dye tetramethylrhodamine ethylester (TMRE) reflect loss of mitochondrial potential ($\Delta\Psi_M$), which occurs faster in the cocultured astrocytes. Scale bar = 20 μm. From Reichert SA, Kim-Han JS, and Dugan LL (2001) The mitochondrial permeability transition pore and nitric oxide synthase mediate early mitochondrial depolarization in astrocytes during oxygen–glucose deprivation. *Journal of Neuroscience* 21(17): 6608–6616, copyright 2001 by the Society for Neuroscience.

Reactive Oxygen Species

The production of injurious free radical compounds (e.g., ROS) has been linked to several neuropathological states, such as ischemic insults, Parkinson's disease, and Alzheimer's disease. Regardless of the exact mechanism and species of ROS produced, the oxidative stress these compounds initiate leads to the compromise of metabolic pathways, eventually leading to structural breakdown of cellular components.

ROS-initiated oxidative stress has been shown to be an essential process in excitotoxicity-induced degeneration. In ischemia, the neuronal ROS species produced in response to glutamate-dependent increases in cytosolic Ca^{2+} have been shown to cause elevated mitochondrial free Ca^{2+} concentrations, and this is perhaps an initiator of primary mitochondrial dysfunction in neurons after reperfusion. In astrocytes, this type of sustained elevation of intracellular Ca^{2+} concentration has been documented. Thus, activation of voltage-gated Ca^{2+} channels and receptor-mediated release of endoplasmic reticulum (ER) stores has been reported after ischemic-like insults (see **Figure 2**), sufficient to load astrocytic mitochondria. In addition, it has been demonstrated that astrocytes isolated from neuronal elements experience a less significant elevation in intracellular Ca^{2+}, suggesting the possibility that neuronally derived factor(s) may promote astrocytic dysfunction (or stress) after ischemic-like insults.

Most of the available data suggest that astrocytes are much more resistant to oxidative stress than are neuronal cell types. This may stem from the fact that they have been reported to contain higher concentrations of antioxidants such as α-tocopherol and ascorbate and higher levels of glutathione and the enzymes involved in glutathione metabolism. This, coupled with a relatively low level of membrane polyunsaturated lipids, suggests that these cells would be quite resistant to ROS-induced lipid peroxidation and oxidative stress. In light of these findings, many have suggested that astrocytes, being more competent in scavenging ROS, play a key role in the detoxification of oxidative processes in the CNS and the promotion of neuronal survival. In fact, astrocytes have been shown to be resistant to and 'neuro-protective' in the face of oxidative stress induced by the free radical generator H_2O_2.

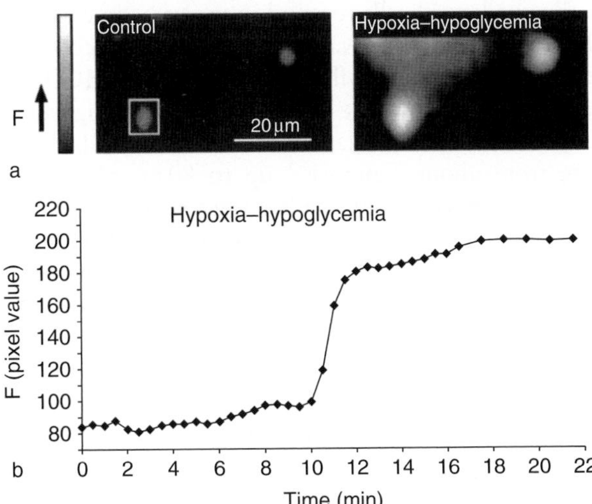

Figure 2 (a) Astrocytes from hippocampal slices were loaded iontophoretically with fluorescent Ca indicators. (b) Brief oxygen–glucose deprivation (termed Hypoxia–Hypoglycemia in the figure) leads to an abrupt rise of $[Ca^{2+}]_i$ in these cells. F, calcium orange fluorescence (arbitrary units). From Duffy S and MacVicar BA (1996) *In vitro* ischemia promotes calcium influx and intracellular calcium release in hippocampal astrocytes. *Journal of Neuroscience* 16(1): 71–81.

Astrocytes are, however, not impervious to the effects of ROS or sustained oxidative stress. These cells will respond to an oxidative insult with an increase in their intracellular Ca^{2+} concentrations and loss of $\Delta\Psi_M$ and thereby oxidative phosphorylation. While some have shown that elevated intracellular Ca^{2+} is not a requisite for oxidative stress-induced astrocytic mitochondrial dysfunction, others have reported that increases in mitochondrial ROS production can lead to increased Ca^{2+} release from internal stores (ER), which can in turn lead to mitochondrial Ca^{2+} loading and subsequent dysfunction (via mitochondrial permeability transition induction) under oxidative stress (see **Figure 3**).

Thus, it may be that very local increases in Ca^{2+} concentration, at sites of ER/mitochondrial contact shown in other cell types, could lead to this destructive cycle of Ca^{2+} release, mitochondrial Ca^{2+} loading, and further ROS production, leading to progressive mitochondrial potential collapse and necrotic cell death.

Glutamate Uptake

Astrocytes are well known for their ability to sequester extracellular neurotransmitters such as glutamate and are thus vital in maintaining excitatory–inhibitory network balances in brain activity. Glutamate is the primary excitatory neurotransmitter in the mammalian CNS, and excess release and unmanaged uptake have been linked to a number of neuropathological states, such as epilepsy, trauma, and ischemia. The existence of astrocytic high-affinity glutamate transporters (EAAT1 and EAAT2) is well established and is required for the rapid termination of transmitter action at presynaptic receptors. The uptake of glutamate into astrocytes is not chemically favored, given a ~1000-fold higher intracellular glutamate concentration in astrocytes. Thus, the uptake of glutamate is coupled to the electrogenic inward transport of Na^+ and H^+ in exchange for outward K^+ flow, which is dependent on a large Na^+ gradient generated by the action of the membrane Na^+/K^+-ATPase. Thus, under conditions of low ATP concentrations (e.g., mitochondrial dysfunction, ischemia), ionic gradients collapse, and these glutamate transporters become conduits for glutamate release (see **Figure 4**).

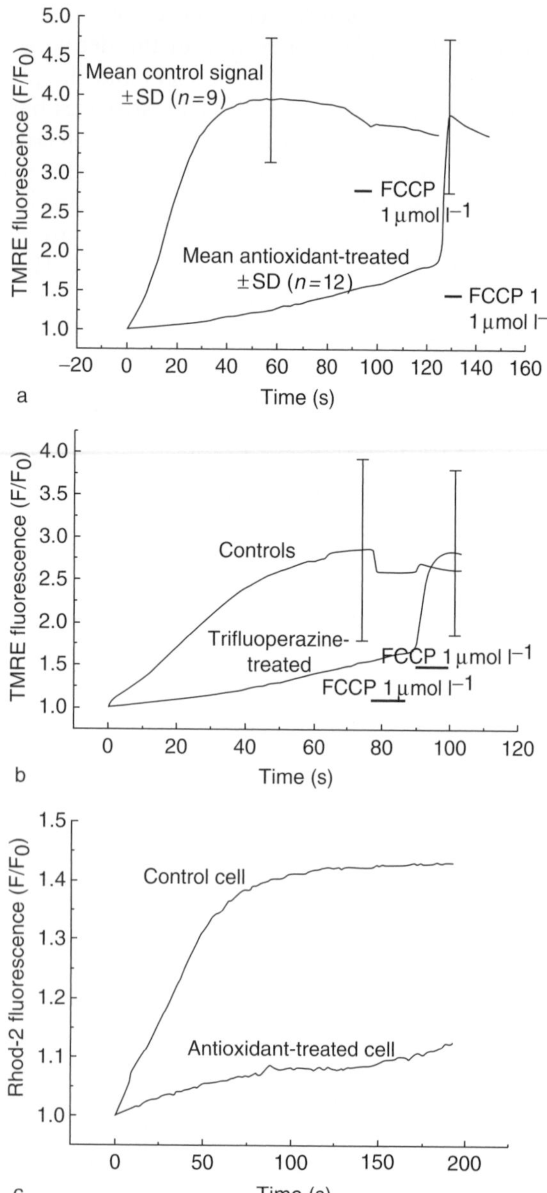

Figure 3 Reactive oxygen species (ROS)-induced astrocyte mitochondrial dysfunction. The antioxidants ascorbic acid, catalase, and Trolox, the spin trap TEMPO (a), and the mitochondrial permeability transition inhibitor trifluoperazine (b) were effective in reducing the rate of mitochondrial depolarization as measured by tetramethylrhodamine ethylester (TMRE; increasing fluorescent signal indicates loss of mitochondrial membrane potential). Large fluorescence increases with the addition of the protonophore FCCP indicate prior preservation of mitochondrial polarization, as with the antioxidants, or mitochondrial permeability transition inhibition. (c) ROS-induced astrocytic mitochondrial Ca^{2+} loading was significantly reduced with antioxidant treatment. In all experiments, ROS production was initiated ($t = 0$) by brief exposure to excitation light. F/F_0, fluorescence ratio compared to time zero. Modified from Jacobson J and Duchen MR (2002) Mitochondrial oxidative stress and cell death in astrocytes: Requirement for stored Ca^{2+} and sustained opening of the permeability transition pore. *Journal of Cell Science* 115(Pt. 6): 1175–1188.

The functional abilities of astrocyte glutamate transporters under less than complete ischemic conditions is unclear. While some have shown that astrocyte glutamate transporters are inhibited by incomplete ischemia, others have reported an initial stimulation or a maintained activity for substantial periods (hours) of ischemia. These differences may be attributable to degrees of acidosis experienced by the astrocytes, with greater decreases in pH leading to a more profound reduction in glutamate uptake. Thus, the extent of the ischemia experienced by the astrocytes, and the precise environment the cells are exposed to, will likely influence the activity of their glutamate transporter activity.

The effects of increased glutamate concentrations in astrocytes under pathologic conditions may not be without consequences to the astrocytes themselves. Both glutamate-induced swelling and cell death have been illustrated in cultures of astrocytes and are likely as a result of glutamate-stimulated K^+ uptake, and glutathione depletion and oxidative stress, respectively.

Cell Swelling

Ischemic-type insults are also accompanied by significant astrocytic swelling, occurring initially at astrocytic end-feet processes, then involving cell body regions. Astrocytes are the primary regulators of brain K^+ homeostasis. The removal of 'excess' extracellular K^+ is accomplished by two types of processes: K^+ spatial buffering and K^+ uptake, which work to rapidly redistribute high K^+ loads from active regions. Under ischemic-like environments, metabolically challenged cells respond to increases in extracellular K^+ with a rapid water accumulation, preceded by energy-independent influx of K^+, Cl^-, and HCO_3^- in response to membrane depolarization (see **Figure 5**). The load of extracellular K^+ can be substantial, climbing from about $3\,mmol\,l^{-1}$ up to $80\,mmol\,l^{-1}$ with ischemia. Recent evidence has suggested a significant role for the Na–K–Cl cotransporter (i.e., NKCC1) in ischemic-induced astrocytic swelling. Also, the combined action of the inwardly rectifying K channels (i.e., $K_{ir}4.1$) and the water channel protein aquaporin-4 (AQP4) has been shown to be particularly important in astrocyte K^+ siphoning at perivascular astrocytic end-feet. Thus, under the conditions of escalating extracellular $[K^+]$ observed in ischemia, homeostatic water influx is increased along with K^+ uptake.

Astrocytic swelling *per se* may not be lethal for the cells involved, as the reestablishment of normal extracellular $[K^+]$ results in repolarization of cell membranes. One of the consequences of continued K^+-induced cell swelling is the release of compounds such as glutamate and aspartate, via volume-regulated

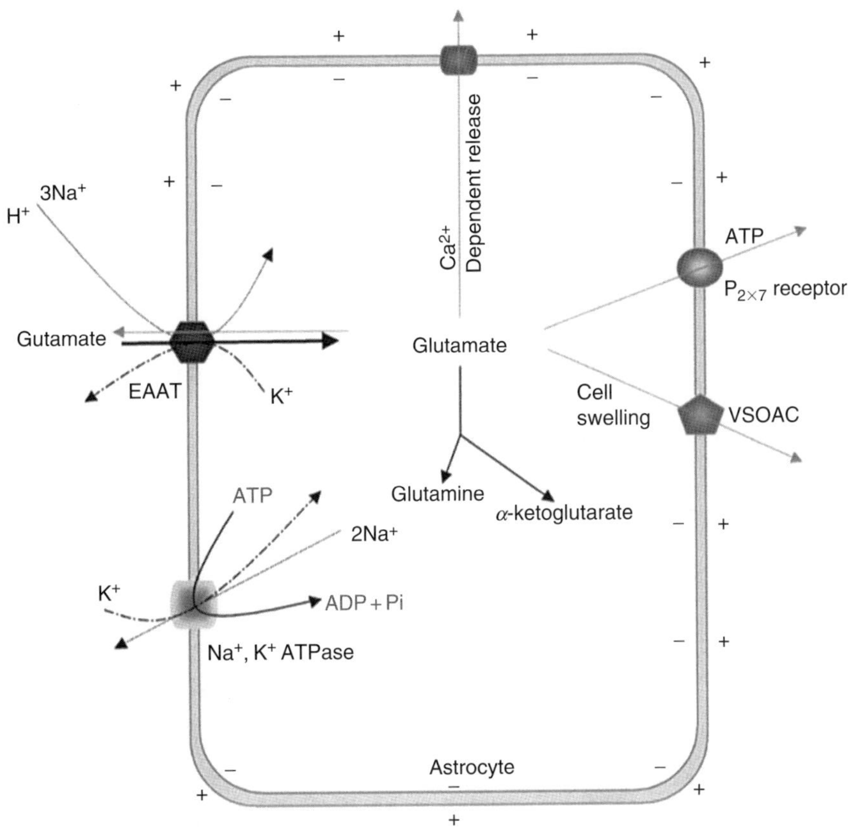

Figure 4 Glutamate handling in the ischemic astrocyte. In response to an ischemic-like insult, astrocyte mitochondria depolarize (with perhaps various time courses, depending on cell type, microenvironment, etc.), and glycogen stores are depleted. As energy metabolism is compromised, adenosine triphosphate (ATP) levels decline, and ATP-supported ionic membrane gradients collapse, leading to a halt in uptake and the release of glutamate through uptake reversal. EAAT, excitatory amino acid transporter; VSOCA, volume-sensitive organic anion channels. From Chen Y and Swanson RA (2003) Astrocytes and brain injury. *Journal of Cerebral Blood Flow and Metabolism* 23(2): 137–149.

anion channels, in an effort to reduce intracellular water loads. In fact, this adaptive homeostatic pathway may be a dominant mechanism of excitatory amino acid release in ischemic insults and may, ironically, lead to astrocyte-mediated neuronal excitotoxicity (see **Table 1**).

Apoptosis

Cell death, both *in vitro* and *in vivo*, may progress by two mutually nonexclusive ways. Thus, necrotic cell death is characterized by cell and organelle (mitochondria) swelling, leading to a rapid depletion of energy levels, breakdown of homeostatic regulation, and cell membrane rupture and release of the intracellular contents. In contrast, apoptotic ('programmed') cell death can exhibit varying degrees of chromatin condensation and DNA fragmentation, general membrane preservation (with possible blebbing), cysteine aspartate specific protease (caspase) activation, and cell shrinkage.

Astrocyte apoptosis has been demonstrated more clearly in focal ischemia, rather than global ischemic models. Thus, a MCAO for a few hours has been shown to cause extensive DNA laddering and activation of procaspase-12 in glial fibrillary acidic protein-positive cells (presumably astrocytes). In addition, a permanent occlusion of the same artery has been shown to upregulate numerous proapoptotic caspases in both neurons and astrocytes. It should be noted that others have shown that the morphological/ultrastructural changes that occur in astrocytes susceptible to ischemia resemble necrosis rather than apoptosis. From younger rat models, very strong evidence exists of an apoptotic astrocyte cell death in response to focal ischemic insults. Focal ischemia has been shown to upregulate Bax, to release cytochrome *c*, and to activate caspase-3 and -9 in astrocytes. However, it has also been shown that these types of insults do not lead to apoptotic astrocyte cell death. For instance, MCAO in adult mice leads to the induction of active caspase-3 and DNA laddering in neurons but not in astrocytes.

Cultured astrocytes can be induced to undergo apoptosis. The list of apoptotic inducers in astrocytes is long. In general, these factors and treatments lead

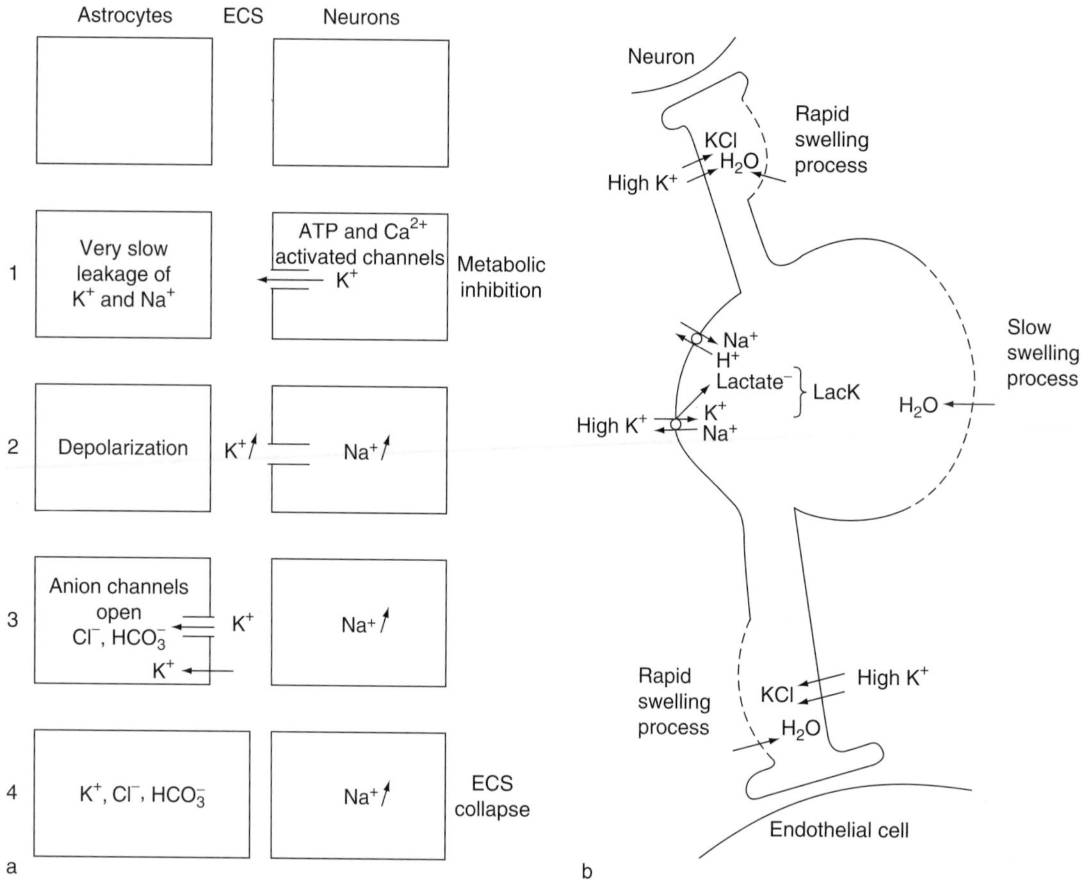

Figure 5 Ischemia-induced inhibition of Na^+/K^+ adenosine triphosphatase leads to progressive increases in extracellular $[K^+]$ (1) and membrane depolarization (2); continued elevation in $[K^+]_O$ leads to opening of anion channels and passive fluxes of K^+, Cl^-, and HCO_3^- (3), and water accumulates in astrocytes, causing them to swell (4). As ischemia continues, the short-term benefits of K^+ clearance become outweighed by the detrimental effects of reversal of glutamate uptake and/or cell swelling-induced excitatory amino acid release and possible effects on neuronal excitability (see **Table 1**). (b) A rise in extracellular K^+ depolarizes astrocytes, which in turn activate astrocytic anion channels. Passive, Donnan-mediated astrocytic K^+, Cl^-, and HCO_3^- accumulation causes significant cellular swelling. ATP, adenosine triphosphate; ECS, extracellular space; LacK, lactate–pottasium. (b) From Walz W, Klimas-Zewski A, and Paterson IA (1993) Glial swelling to ischemia: A hypothesis. *Developmental Neuroscience* 15(3–5): 216–225.

Table 1 Astrocytic pottasium homeostasis

Coping mechanism	*Primary beneficial effects*	*Secondary detrimental effects*
Increased spatial buffering (gap junction coupling)	↑ K^+ clearance	↑ Spreading depression and secondary injury
Alkalinization (Na^+/HCO) and sustained Na^+/K^+ ATPase[a] (glycogen)	↑ K^+ clearance	↑ Na^+ entry and glutamate release (reversal of uptake)
Increased passive K^+ uptake and cell swelling	↑ K^+ clearance	↑ Release of excitatory amino acids (regulatory volume decrease; RVD)

[a]ATPase, adenosine triphosphatase.
From Leis JA, Bekar LK, and Walz W (2005) Potassium homeostasis in the ischemic brain. *Glia* 50(4): 407–416.

to increases in intracellular calcium concentrations, oxidative stress, mitochondrial dysfunction, and protease activation. Of particular interest for this article, a number of studies have pointed to the involvement of oxidative stress in astrocytic apoptosis. The ROS producer H_2O_2 causes a loss of mitochondrial potential, cytochrome *c* release, and caspase-3 activation, a pathway likely to involve activation of transcription factor nuclear factor κB as a death-promoting factor in astrocyte apoptosis. The importance of elevated cytosolic calcium levels to this process has been established. Strategies to limit these increases of intracellular calcium, such as inhibition of astrocyte Na^+–Ca^{2+} exchanger and intracellular Ca^{2+} chelation, lead to reduction in astrocytic apoptosis *in vitro*.

Conclusions

Research over the past two decades has illustrated the importance of astrocyte functioning in management of the CNS microenvironment. Alterations in astrocytic physiology during pathological states such as ischemia and traumatic brain injury could have profound implications for the progression of these insults. Understanding the precise mechanism(s) of astroctyic demise or dysfunction should help in directing future clinical interventions aimed at reducing the extent of neuropathological damage.

See also: Apoptosis in Nervous System Injury; Astrocyte: Calcium Signaling; Astrocyte: Neurotransmitter and Hormone Receptors; Astrocyte: Identification Methods; Glial Responses to Injury; Neurotransmitter Release from Astrocytes.

Further Reading

Alberdi E, Sanchez-Gomez MV, and Matute C (2005) Calcium and glial cell death. *Cell Calcium* 8(3–4): 417–425.

Almeida A, Delgado-Esteban M, Bolanos JP, and Medina JM (2002) Oxygen and glucose deprivation induces mitochondrial dysfunction and oxidative stress in neurones but not in astrocytes in primary culture. *Journal of Neurochemistry* 81(2): 207–217.

Bender AS, Schousboe A, Reichelt W, and Norenberg MD (1998) Ionic mechanisms in glutamate-induced astrocyte swelling: Role of K^+ influx. *Journal of Neuroscience Research* 52(3): 307–321.

Benjelloun N, Joly LM, Palmier B, Plotkine M, and Charriaut-Marlangue C (2003) Apoptotic mitochondrial pathway in neurones and astrocytes after neonatal hypoxia-ischaemia in the rat brain. *Neuropathology and Applied Neurobiology* 29(4): 350–360.

Chen CJ, Liao SL, and Kuo JS (2000) Gliotoxic action of glutamate on cultured astrocytes. *Journal of Neurochemistry* 75(4): 1557–1565.

Chen Y and Swanson RA (2003) Astrocytes and brain injury. *Journal of Cerebral Blood Flow and Metabolism* 23(2): 137–149.

Duffy S and MacVicar BA (1996) *In vitro* ischemia promotes calcium influx and intracellular calcium release in hippocampal astrocytes. *Journal of Neuroscience* 16(1): 71–81.

Erecinska M and Silver IA (1990) Metabolism and role of glutamate in mammalian brain. *Progress in Neurobiology* 35(4): 245–296.

Fern R (2001) Ischemia: Astrocytes show their sensitive side. *Progress in Brain Research* 132: 405–411.

Ishibashi T, Dakin KA, and Stevens B, et al. (2006) Astrocytes promote myelination in response to electrical impulses. *Neuron* 49(6): 823–832.

Jacobson J and Duchen MR (2002) Mitochondrial oxidative stress and cell death in astrocytes: Requirement for stored Ca^{2+} and sustained opening of the permeability transition pore. *Journal of Cell Science* 115(Pt. 6): 1175–1188.

Juurlink BH, Hertz L, and Yager JY (1992) Astrocyte maturation and susceptibility to ischaemia or substrate deprivation. *Neuroreport* 3(12): 1135–1137.

Kimelberg HK (2005) Astrocytic swelling in cerebral ischemia as a possible cause of injury and target for therapy. *Glia* 50(4): 389–397.

Leis JA, Bekar LK, and Walz W (2005) Potassium homeostasis in the ischemic brain. *Glia* 50(4): 407–416.

Somjen GG (1979) Extracellular potassium in the mammalian central nervous system. *Annual Review of Physiology* 41: 159–177.

Swanson RA, Farrell K, and Stein BA (1997) Astrocyte energetics, function, and death under conditions of incomplete ischemia: A mechanism of glial death in the penumbra. *Glia* 21(1): 142–153.

Takuma K, Baba A, and Matsuda T (2004) Astrocyte apoptosis: Implications for neuroprotection. *Progress in Neurobiology* 72(2): 111–127.

Walz W (1989) Role of glial cells in the regulation of the brain ion microenvironment. *Progress in Neurobiology* 33(4): 309–333.

Atomic Force Microscopy Methodologies

R M Twyman, University of York, York, UK

Introduction

Atomic force microscopy (AFM) is a technique used to produce quantitative topographic images of surfaces at a magnification of up to 10^7 (subnanometer scale, some 10^3 times better than possible using an optical microscope). The underlying principle is surface scanning using a sharp tip or stylus and the measurement of atomic forces between the stylus and the specimen. The stylus on an atomic force microscope is sensitive to both attractive and repulsive surface forces in the 0.1 μN range, but in practice the repulsive forces are the most stable, and these are the most widely used to obtain images. Scanning can be carried out in a vacuum, but a major advantage of AFM over scanning electron microscopy (SEM) is that AFM can be performed in air or with the sample immersed in liquid. Also, in contrast to SEM, no pretreatment of the sample, such as gold coating, is required. The resulting data also provide a true three-dimensional profile, not a two-dimensional image of a surface, as obtained by SEM. The data produced by AFM are a collection of points characterized by a constant force between the tip and the specimen, analogous to isobars on a weather map.

AFM is based on a related technique called scanning tunneling microscopy (STM), and some instruments can be fitted with swappable detectors to facilitate both types of analysis. The main difference between AFM and STM is that the latter requires a conducting surface that can sustain a stable tunneling current, making it unsuitable for most applications in neuroscience. Since AFM works on nonconducting surfaces, it is suitable for the analysis of cells, membranes, surface structures, protein complexes, individual proteins, and nucleic acids. Applications in neurobiology include the measurement of forces corresponding to receptor–ligand interactions, structural analysis of neuronal membranes, cell adhesion, and protein folding.

The information collected by surface scanning is quantitative in three dimensions, with vertical resolution on the order of 1 nm and horizontal resolution on the order of 0.1 nm. Samples that provide the best data have a smooth, flat surface, whereas rough or tilted samples generate poor images. Data are presented as topographical maps, often shaded or rendered in false color to indicate contours. Software can be used to rotate the image, generate virtual sections through the sample, and measure distances between features.

Atomic Forces and Their Measurement

The force between the stylus and the surface of the specimen is almost zero when the two are separated by a distance equivalent to more than a few hundred atoms. As the stylus gets closer to the surface, an attractive force builds up based primarily on van der Waals interactions. This attractive force is maximal when the tip of the stylus and the surface of the specimen are a few atoms apart, and then it is rapidly dispelled by a repulsive force as the two are pushed closer together. Contact equilibrium is established at the point where the net force between the stylus and the specimen surface is zero. At this point, any further movement toward the surface must involve an external force which causes elastic deformation and possible damage to the tip and the specimen.

If the stylus is maintained at a constant height relative to the surface at its starting position, then any peaks and troughs in the specimen's surface landscape encountered during scanning could result in loss of contact or collision, neither of which is desirable. Therefore, in most applications, and certainly in those where a variable surface landscape is anticipated, the stylus is mounted on a cantilever which allows its absolute height to be varied during scanning. A feedback mechanism is used to maintain a constant distance between the tip and the sample, and it is these changes in height that are used to map the specimen surface (**Figure 1**). The most common approach to measuring the small forces produced by surface scanning is to fabricate by microlithography a silicon nitride cantilever beam with an integral silicon nitride pyramidal tip and measure displacement by laser reflection, capacitance, or the detection of tunneling currents. Mechanical scanning is accomplished by using a piezoelectric transducer; if it is fabricated in the form of a hollow cylinder, with one electrode on the interior wall and four separate electrodes on each quadrant of the exterior wall, motion in three dimensions can be achieved. One end of the cylinder is fixed to the microscope base, and the specimen is mounted on the opposite end. A voltage applied between the interior electrode and all the outer electrodes will cause a dimensional change along the axis of the cylinder, or z-direction. Potentials applied between opposite quadrants of the outer electrodes will cause bending of the tube and subsequent deflection of the free end of the tube in the x- or y-direction. Typical ranges are 10 μm along the z-axis and 100 μm

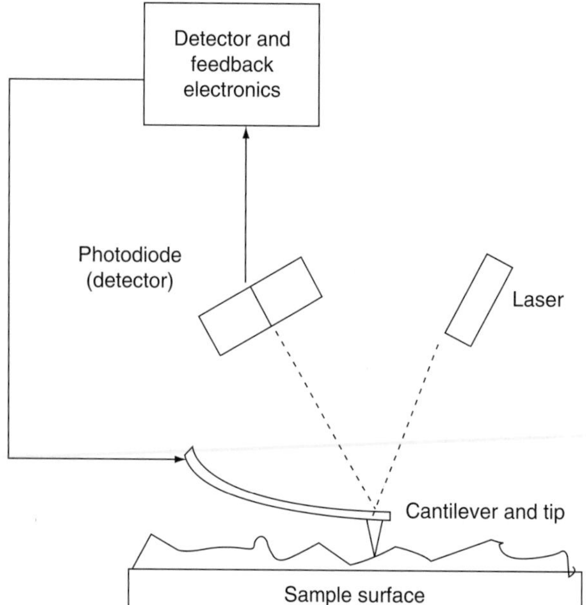

Figure 1 The principle of atomic force microscopy. Reproduced from Morris WG (2001) Atomic force microscopy In: *Encyclopedia of Materials: Science and Technology*, pp. 365–371. London: Elsevier, with permission from Elsevier.

along the x- and y-axis. For atomic resolution, actuators with smaller ranges in x- and y-direction are preferred because the scan may cover only a few nanometers. Alternatively, a tripod configuration can be used with three separate piezoelectric crystals, each responsible for scanning along one of the three axes. This can address image distortion issues that are caused by cross-talk when a traditional tube scanner is used.

Scanning Modes

The description above refers to atomic force microscopes operating in what is known as contact mode, in which the stylus remains in constant contact with the surface at the contact equilibrium point and differences in repulsive forces can be measured as the stylus tracks across the specimen. One disadvantage of this mode is that the measurements are somewhat prone to noise, although cantilevers that are more flexible than usual are employed to optimize the deflection signal. Another problem with contact mode is that under ambient conditions, a meniscus layer tends to form on most surfaces, and this strongly influences the interaction between tip and specimen. The tip can be held in contact with the specimen by the surface tension, and at the point of contact where the meniscus covers the specimen and stylus, attractive forces of up to 1 μN can be experienced. The repulsive force between the tip and specimen balances the surface tension, so the tip is almost always in the repulsive force mode when operating in air. If the liquid molecules remain between the tip and specimen, they can alter the force strength, which causes image distortion. This can be addressed through the use of liquid cells such that both the tip and the specimen are totally immersed in a liquid to eliminate forces generated by liquid–air surface tension.

An alternative approach which addresses the problems of static measurements is carrying out AFM in dynamic mode. Instead of remaining in static contact with the sample beneath, the stylus in dynamic mode oscillates at or near its resonance frequency. Feedback is provided, not by the z-displacement of the cantilever beam, but by differences in oscillation amplitude, phase, and resonance frequency relative to the external reference oscillation caused by tip–sample interaction forces. There are several varieties of dynamic mode, depending on which characteristics of the oscillating probe are measured. In frequency modulation, the changing oscillation frequency provides sensitive data about the forces between the stylus and sample, and very stiff cantilevers can be used that provide stability close to the sample and allow high-resolution scanning. Amplitude modulation is the most widely used mode; it uses changes in the oscillation amplitude or phase to provide the feedback for image creation. Amplitude modulation is particularly useful for addressing the problems caused by surface tension when scanning is done in air: The intermittent contact allows the stylus to be kept close enough to the sample for short-range repulsive forces to become detectable while the restoring force provided by the cantilever spring is strong enough to break contact with the surface meniscus. This mode is also known as tapping AFM because the stylus taps against the surface rather than dragging along it.

Image Collection

Feedback from sample scanning is fed into a computer which converts the forces into image data. However, before the sample is scanned, the same feedback system must be used to determine the correct operational position of the tip. The operating distance is established in two phases, a coarse stage, controlled by stepping motors, and a fine stage, in which the sample is moved toward the tip with a piezoelectric actuator. Data collection at this point is used to measure the atomic forces between tip and sample solely to establish the contact equilibrium point prior to the initiation of the scan.

Once scanning begins, the most straightforward way to present the data is height versus distance along a given line such that repeated scans along parallel courses with a given vertical or horizontal offset can generate a crude topographical map. This results in a top–down topographical view showing the x–y area with different contours along the z-axis shown in gray scale (or rendered in false color) to indicate peaks and troughs. The maximum value of the gray scale can be changed, which in turn will change the contrast of the image, but it does not actually change the quantitative height information. The gray scale can be set to optimize the contrast in each individual image, making most of the details visible while minimizing regions of saturated black or white. Alternatively, it is easier to compare images if they are displayed with the same maximum gray-scale values. On a given specimen, wide-area scans will frequently require a larger maximum gray-scale value than small-area scans.

More-sophisticated computer programs can convert the simple topographical map into a three-dimensional perspective that can be viewed from any angle. The relative scale of the z-axis can also be increased or decreased with respect to the surface area so that surface features can be stretched and magnified. However, while enabling surface features to be studied in greater detail, this technique would also lead to the distortion and overemphasis of inconsequential features and artifacts.

Besides being an alternative to gray scale, the addition of color to an AFM image can help enhance certain image features, such as edges and clefts, which may be difficult to identify in monotone. At higher resolutions, coloring can be used to identify functionally important features of molecules, such as particular amino acid residues in proteins or points of contact between a receptor and its ligand. False color can also be used for trivial reasons, for instance, to add shadows to three-dimensional landscapes as if the surface were illuminated by a directional light source. While color can be used for clarification in many circumstances, the overuse of color has the counterproductive effect of making images more difficult to interpret.

Quantitative Data

One disadvantage of AFM compared with SEM is the relatively long time it takes to collect image data, in part reflecting the necessity to carry out several scans along the horizontal and vertical axes to establish topography. This can also lead to some loss of data integrity through a phenomenon known as thermal drift, which makes it difficult to collect accurate quantitative data, such as to measure distances between features on the surface landscape. Another potential source of inaccuracy is hysteresis of the piezoelectric material. Several AFM designs with faster scan times have been developed to overcome the data problems caused by thermal drift, including video-AFM and the use of feature-oriented scanning and closed-loop scanners which collect and process images in near real time. Software has also been developed to correct distortions caused by thermal drift and hysteresis.

The accuracy of calibration can also limit the accuracy of any quantitative information extracted from the image, so the instrument must be calibrated regularly using a specially constructed mesh with 1-μm periodicity along both the x- and y-axis as well as a peak-to-valley distance of 0.2 μm. Corrections for other more complex types of distortion, described by terms like pinch, punch, pincushion, shear, and swirl that are familiar to Adobe Photoshop users, may also be necessary.

Further software fixes can be used to process images and enhance the information content. For example, the Fourier transform can be applied to a height profile to determine the frequency of regularly spaced surface features, and it can help clean up and smooth out the features of rough, noisy images, thus preserving the relevant features while removing artifacts.

Limitations and Drawbacks

Although AFM has many advantages over SEM in terms of the convenience of sample preparation and scanning, there are several limitations which restrict the applications of this method and which are the subject of current research. One significant issue is the limited image area and depth. The SEM can image an area more than 100 times the size of that possible using an AFM because it has a displacement of several millimeters along the x- and y-axis, whereas the maximum useful horizontal and vertical displacement possible with an AFM is about 150 μm. Newer piezodrivers extend this to approximately 250 μm, but such wide scans often result in greater image distortion, which must be corrected with software. Mechanical stages permit the cantilever–detector assembly to be positioned within a 5-mm^2 area so that larger areas can be covered with adjoining multiple images when needed.

The depth of field for an SEM is also in the millimeter range, whereas for AFM it is a few micrometers. This means that although the AFM has great sensitivity for small changes in elevation, it has a very limited vertical range, restricting the AFM to the analysis of surfaces which are relatively smooth and flat when observed using other methods and making the analysis of convex surfaces challenging.

The resolution and characteristics of an AFM image are also dependent on the integrity of the stylus. The radius of curvature at the tip must be on the order of 30 nm to achieve the best spatial resolutions, but extended use can cause wearing and damage which will generate artifacts in image data. For example, a worn tip with a flat region on it may generate artifactual square features on the image, and a new tip or a damaged tip with transferred material on it may contain rough protrusions (asperities) that generate shadows.

See also: Cognition: An Overview of Neuroimaging Techniques; Neuroimaging.

Further Reading

Eibl RH and Moy VT (2005) Atomic force microscopy measurements of protein–ligand interactions on living cells. *Methods in Molecular Biology* 305: 439–450.

Gadegaard N (2006) Atomic force microscopy in biology: Technology and techniques. *Biotechnic & Histochemistry* 81: 87–97.

Hansma HG (2001) Surface biology of DNA by atomic force microscopy. *Annual Reviews of Physical Chemistry* 52: 71–92.

Li G, Xi N, and Wang DH (2006) Probing membrane proteins using atomic force microscopy. *Journal of Cellular Biochemistry* 97: 1191–1197.

Meyer E, Hug H-J, and Bennewitz R (2004) *Scanning Probe Microscopy: The Lab on a Tip*. Berlin: Springer.

Morita S, Wiesendanger R, and Meyer E (2002) *Noncontact Atomic Force Microscopy, Nanoscience and Technology*. Berlin: Springer.

Morris VJ, Kirby AR, and Gunning AP (1999) *Atomic Force Microscopy for Biologists*. London: Imperial College Press.

Shao Z, Mou J, Czajkowsky DM, Yang J, and Yuan J-Y (1996) Biological atomic force microscopy: What is achieved and what is needed. *Advances in Physics* 45: 1–86.

Zlatanova J and van Holde K (2006) Single-molecule biology: What is it and how does it work? *Molecular Cell* 24: 317–329.

Atrial Natriuretic Peptide: Fluid/Mineral Balance

W K Samson and M M White, Saint Louis University School of Medicine, St. Louis, MO, USA

Introduction

Originally isolated from extracts of the cardiac atrium, atrial natriuretic factor (ANF) was the first peptide identified in what is now known to be a family of homologous peptides, products of unique genes, which formed the basis of our initial understanding of the heart as an endocrine organ and the science of cardiovascular endocrinology. Released primarily in response to stretch (increased venous return or increased afterload), ANF and its homolog B-type natriuretic peptide (BNP) exert potent adrenal, vascular, and renal actions all seemingly coordinated to unload the vascular tree (e.g., inhibition of aldosterone release, increased glomerular filtration, natriuresis, diuresis, and inhibition of sympathetic tone). A third member of the family of natriuretic peptides (NPs), C-type natriuretic peptide (CNP), is predominantly produced in the vascular endothelium and shares some of the actions of ANF and BNP. All three peptides are present in the central nervous system (CNS) and exert potent pharmacologic actions there. Evidence for a coordination of the CNS and peripheral actions of these peptides is quite convincing. Thus, complementary actions in several organ systems indicate that these peptides are important contributors to the physiologic regulation of fluid and electrolyte homeostasis.

Discovery of the Natriuretic Peptides

It was recognized more than 40 years ago that cardiac myocytes were more than simply contractile elements, based on the histological identification of secretory granules in the cytoplasm. In 1981, pioneering studies by AJ deBold and colleagues in Kingston, Ontario, provided evidence that those granules contained a vasoactive substance that not only caused hypotension when injected into bioassay animals but also stimulated increased urine volume (diuresis) and urinary sodium excretion (natriuresis). Two years later, TG Flynn and AJ deBold provided the presumptive amino acid sequence for what is now known as ANF, the factor extracted from those myocytes which is the renotropic agent identified in their bioassay. Soon thereafter, K Kangawa and colleagues in Osaka identified the second member of the family, BNP, and in 1990 that same group reported the sequence of CNP. All three peptides are products of unique genes and, largely because of their structural homology, they share common receptors with varying degrees of affinity (**Figure 1**). The type A NP (NPR-A) and NPR-B receptors are membrane-bound guanylyl cyclases (GCs) that signal through the formation of cGMP. They are also referred to as GC-A and GC-B, respectively. A third receptor, NPR-C, shares some structural homology with the NPR-A and NPR-B in the extracellular domains, but it is devoid of the intracellular guanylyl cyclase domain. This receptor was originally thought to function as primarily a biologically inactive, clearance receptor; however, it is now recognized that ligand binding to this receptor results in inhibition of cAMP formation, probably via a G_i protein, which in isolated hypothalamic neurons results in inhibition of L-type calcium channels. The three receptors display selective ligand binding (**Table 1**).

Control of Natriuretic Peptide Production and Release

The major stimulus for the release of ANF and BNP from the heart is volume expansion (i.e., stretch). However, numerous neurohumoral factors stimulate release, including endothelin, angiotensin, adrenal steroids, inflammatory cytokines, growth factors, prostaglandins, thyroid hormones, and alpha-adrenergic agents. Although constitutive release of the peptides occurs under normal conditions, production and secretion are upregulated in volume overload states such as congestive heart failure or after ischemic events (e.g., acute myocardial infarction). In fact, cardioprotective effects of the peptides imply that the upregulation of ANF and BNP production and secretion represents a physiologically relevant compensatory mechanism that protects cardiac function (anti-mitogenic actions and increased coronary artery perfusion) and facilitates restoration of normal vascular volume (renotropic actions and effects on capillary permeability). It is not surprising, then, that both ANF and BNP have been examined as potential therapeutic agents for use immediately following acute myocardial infarction and in congestive heart failure. In addition, the increased secretion of the peptides into the circulation in these pathologies has led to the use of rapid assays for the peptides as diagnostic tools.

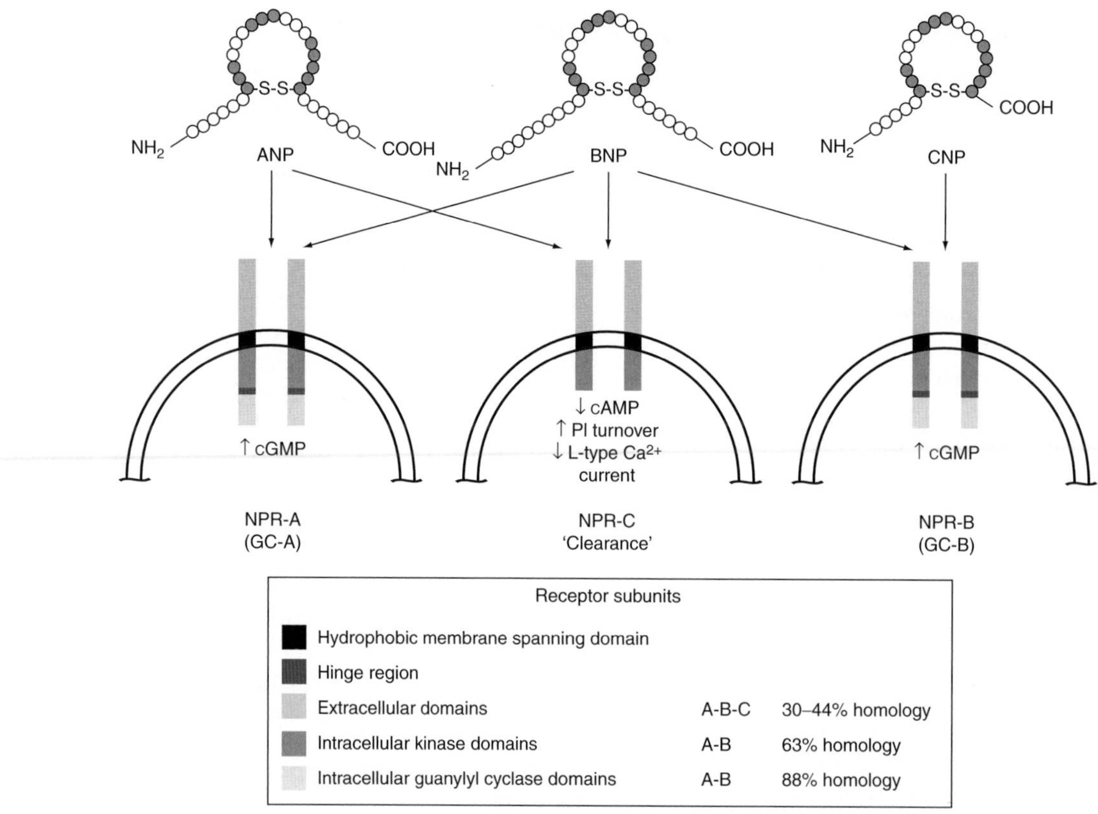

Figure 1 Structure of the natriuretic peptides and their receptors. Signaling mechanisms recruited by the three receptors are indicated.

Table 1 Natriuretic peptide receptor: order of binding potency

NPR-A (GC-A)	ANF $\geq$ BNP > CNP
NPR-B (GC-B)	CNP $\gg$ ANF $\geq$ BNP
NPR-C ('clearance')	ANF $\geq$ CNP $\geq$ BNP

CNS Localization of NPs and NP Receptors

Soon after the initial description of ANF in heart, it became clear that neurons in the CNS also produce the peptide and that it exerts actions within brain that complemented its peripheral actions. Indeed, the first example of a CNS action of ANF was the demonstration that the peptide inhibits arginine vasopressin (AVP) secretion. ANF also acts in brain to inhibit thirst and sodium appetite. The anatomic framework for these pharmacologic actions was established with the demonstration by autoradiography of ANF binding in CNS sites known to be important in the control of fluid and electrolyte homeostasis, cardiovascular/renal function, and the hypothalamic–pituitary–adrenal axis.

Immunoreactive perikarya were identified in lateral septal nucleus, bed nucleus of the stria terminalis, and medial amygdaloid nucleus in the telencephalon. In diencephalon, ANF-positive cell bodies were localized in numerous sites, including organum vasculosum lamina terminalis (OVLT), periventricular and medial preoptic nuclei, preoptic suprachiasmatic nucleus, parvocellular paraventricular nucleus, arcuate nucleus, ventro- and dorsomedial hypothalamic nuclei, habenula, zona incerta, mammillary nuclei, and the dorsal aspect of the lateral hypothalamic area. Dense innervation of these same diencephalic sites as well as the subfornical organ (SFO), external lamina of the median eminence, periventricular thalamus, and the medial lemniscus has been described. Peptide localization to amygdala and hypothalamus implied a role for ANF in the hypothalamic control of thirst, salt appetite, and AVP secretion and even hinted at possible actions of the peptide in the anterior pituitary gland.

In midbrain, ANF-positive cell bodies were identified in interpeduncular nucleus, periaqueductal gray, ventral tegmental area, dorsal raphe, and substantia nigra pars lateralis. ANF-immunoreactive perikarya in the interpeduncular nucleus were surrounded by a dense network of ANF-positive fibers that extended into the neighboring medial longitudinal fasciculus. In hindbrain, ANF-positive cell bodies were detected in dorsal parabrachial nucleus, nucleus of the spinal tract of V, the dorsal tegmental nucleus, pedunculopontine tegmental nucleus, lateral tegmental nucleus, Barrington's nucleus, and locus coeruleus. Caudally, cells were stained positively in nucleus prepositus hypoglossi and nucleus tractus solitarius (NTS). Immunoreactive fibers were abundantly present in NTS and dorsal motor nucleus of the vagus (DMV). A more diffuse pattern of fiber staining was observed in spinal tract of V, nucleus ambiguus, central gray, and locus coeruleus. The presence of ANF-positive parikarya and nerve fibers in these brain stem sites suggested a potential role for the peptide in autonomic function, both the afferent (e.g., baroreceptor input) and the efferent (both sympathetic and parasympathetic) components.

As with the initial studies identifying CNS sites of ANF immunoreactivity, early work from several groups identified receptors for the peptide based on binding studies. It is impossible with this approach to identify the specific subtype of receptor since ANF binds to all three cloned NP receptors. However, these localization studies further suggested a role for endogenous ANF (or BNP or CNP) in the CNS control of fluid and electrolyte homeostasis and autonomic function. Dense autoradiographic binding was reported in olfactory structures and at least two circumventricular organs, the SFO and area postrema (AP). In hypothalamus, abundant binding was described particularly in the paraventricular and arcuate nuclei. These findings, as well as those of peptide localization studies, predicted subsequent findings that revealed potent actions of ANF on AVP and prolactin secretion, respectively. The medial preoptic area also displayed autoradiographic binding sites, perhaps providing an anatomic framework for the subsequent demonstration of the effects of the NPs on the hypothalamo–pituitary–gonadal axis. In hindbrain, binding was detected in NTS, AP, DMV, locus coeruleus, and raphe nuclei and broadly dispersed throughout the reticular formation.

The development of receptor subtype selective probes led to the hypothesis that the major NP receptor within the brain was the NPR-B, in agreement with data implying that the most abundant NP produced in CNS was CNP. In fact, the majority of the NP binding sites behind the blood–brain barrier are NPR-B and NPR-C receptors, whereas those present in areas without the barrier (e.g., the SFO) are predominantly the NPR-A receptor. This finding implies that circulating ANF or BNP of cardiac origin affects neural function at those barrier-free sites, whereas the majority of the pharmacologic actions described to date for the various NPs when administered behind the blood–brain barrier (i.e., into the cerebroventricles or directly into the brain interstitium) reflect an action on the NPR-B receptor. Indeed, the opposing actions of ANF and CNP on water drinking support this hypothesis.

Pharmacologic Effects of the NPs in Brain

The major focus of the initial studies on the actions of the NPs in brain was anatomic localization. Thus, early reports detailed the effects of intracerebroventricular (i.c.v.) administration of ANF on thirst, salt appetite, AVP secretion, and blood pressure regulation. With the eventual realization that most of the NP produced in brain was in fact CNP, those initial studies were repeated using CNP as the pharmacologic agent. In general, these studies can be summarized as follows: NPs act, with one notable exception, either at circumventricular organs or in CNS sites behind the blood–brain barrier to complement the peripheral actions of the peptides on intravascular volume, plasma osmolality, and blood pressure. As in the periphery, many of these CNS actions rely on the ability of the peptides to abrogate the actions of angiotensin II (Ang II).

NPs and Vasopressin Secretion

In 1985, it was reported that the diuretic action of ANF in kidney mirrored an action of the peptide in brain to inhibit AVP secretion. Dehydration- and hemorrhage-induced AVP release, as well as pharmacologically driven (i.e., Ang II-stimulated) AVP secretion, was inhibited by ANF. The site of action was the hypothalamus since ANF inhibited AVP release from isolated hypothalamo-neurohypophyeal explants but not from isolated neural lobes. In addition, ANF inhibited neuronal activity in the paraventricular nucleus. Whether these data reflect a physiologic action of endogenously produced ANF or CNP is not known; however, dehydration, a clear stimulus for AVP production and secretion, also increased ANF-like immunoreactivity in the rat hypothalamus.

There is electrophysiologic evidence for direct effects of ANF on synaptic transmission in hypothalamic explants. Osmotic stimulation of the OVLT increased frequency of spontaneous, glutamatergic excitatory postsynaptic potentials (EPSPs) in magnocellular neurons, and the amplitude of the glutamatergic EPSPs evoked by electrical stimulation of the OVLT was reduced by ANF. The effect was mimicked by 3′,5′-dibutyryl cGMP, implying mediation by either the NPR-A or NPR-B receptors. These effects were presynaptic since ANF did not alter depolarization of the magnocellular cells in the presence of glutamate receptor agonists. Direct, postsynaptic effects of the NPs on magnocellular neurons of the paraventricular and supraoptic nuclei have been demonstrated. CNP and the NPR-C receptor selective agonist cANF-4–23 produced significant inhibitions of L-type calcium currents that were mimicked by application of a peptide that activates G_i proteins and prevented by nicarpidine pretreatment. In current clamp mode, CNP decreased the number of action potentials resulting from depolarizing stimuli and decreased repolarization time. Although the morphologic phenotype of the cells recorded was magnocellular elements dispersed from or located within these nuclei, the chemical identity was not established and thus these results may have been generated in both vasopressin- and oxytocin-expressing cells. Importantly, these experiments demonstrate the contribution of the G_i signaling pathway in the inhibitory effect of NPs acting on the 'clearance' receptor (NPR-C) in hypothalamus.

ANF and Thirst

The first descriptions of a CNS action of ANF were reports that the peptide inhibits water drinking in response to dehydration or Ang II administration. These pharmacologic actions of ANF have a physiologic correlate since pretreatment of rats i.c.v. with an ANF antiserum exaggerated water drinking in response to Ang II administration or water deprivation. These studies imply the SFO as the site of the antidipsogenic action, which is consistent with the presence of the NPR-A receptor there. The location of the SFO at the confluence of the lateral and third cerebroventricles predicts exposure of the receptors to both circulating and i.c.v.-administered ANF. Earlier localization studies described ANF-positive nerve fibers innervating the SFO; thus, ANF of central origin may act there as well, explaining the results of the passive immunoneutralization study.

However, not all NP effects on thirst are inhibitory. CNP, when administered i.c.v., increased water drinking in response to Ang II and dehydration. This dipsogenic action of CNP was in all likelihood expressed on the NPR-B receptor since both ANP and the NPR-C receptor selective ligand, cANF 4–23, inhibited water intake in these models. It should be remembered that CNP is the main NP produced in brain and that the NPR-B and NPR-C receptors are abundant in blood–brain barrier-protected sites. Thus, although circulating ANF, or BNP, may exert antidipsogenic actions in SFO (intravenous infusion of ANF did inhibit vasopressin secretion), the hypothalamic effect of the NPs on thirst, particularly those of endogenous CNP, may be exerted via the NPR-B and therefore be stimulatory in nature. Thus, the CNS actions of brain-derived NPs on thirst may not complement the diuretic action of ANF in kidney.

ANF and Salt Appetite

The natriuretic action of ANF in kidney is complemented by an action in brain to inhibit salt appetite. The physiologic relevance of this CNS action has been established. Acute and chronic i.c.v. administration of ANF inhibits salt appetite induced by NaCl deprivation in normotensive animals and in spontaneously hypertensive rats (SHRs). The results in the SHR model were striking since these were *ad libitum*, two-bottle preference protocols and the effect on salt appetite was not present in the Wistar Kyoto controls. In addition, the effect to inhibit salt appetite was not related to any change in blood pressure in these hypertensive animals. Both the SHR and Wistar Kyoto animals consumed less water in response to ANF infusion, demonstrating selectivity of action of ANF on thirst versus salt appetite during *ad libitum* fluid ingestion.

ANF is not the only neuropeptide reported to act within brain to inhibit salt appetite. A physiologically relevant role for oxytocin and for adrenomedullin in the control of salt intake has been established. In fact, oxytocin neurons may mediate the inhibitory effect of adrenomedullin, implying the importance of both peptides in the neural network responsible for the control of salt appetite.

A series of experiments examined the potential interactions of ANF and oxytocin on salt appetite. Rats were pretreated with cytotoxin (ricin A chain) conjugates designed to compromise cells expressing either oxytocin or ANF receptors or they were treated with unconjugated ricin A chain (nontoxic control). Salt appetite in these animals was studied following a combination of hypovolemia and hyperosmolar challenges using either 2.0 M mannitol (which raises osmolality but lowers plasma sodium levels) or 1.0 M NaCl (which raises both osmolality and plasma sodium level). The induction of hypovolemia stimulates both water and NaCl ingestion in control animals and the superimposition of the hyperosmolar load inhibits salt appetite, even in the face of the hypovolemic challenge. In contrast, exaggerated saline drinking in response to hypertonic mannitol administration was observed in animals in which either the oxytocin or the ANF receptor expressing cells were compromised by cytotoxin targeting. Thus, in the absence of either oxytocin or ANF signaling in brain, these animals could no longer appropriately terminate their appetite for salt in the hypovolemic but hyperosmotic state induced by mannitol.

On the other hand, in animals stimulated to drink water and saline by the colloid pretreatment, hyperosmotic challenge with sodium chloride resulted in appropriate inhibition of saline drinking in controls and in rats pretreated with the oxytocin–ricin A chain cytotoxin conjugate. However, exaggerated saline intake was observed in animals unable to respond to the signal of endogenous ANF (ANF–ricin A chain

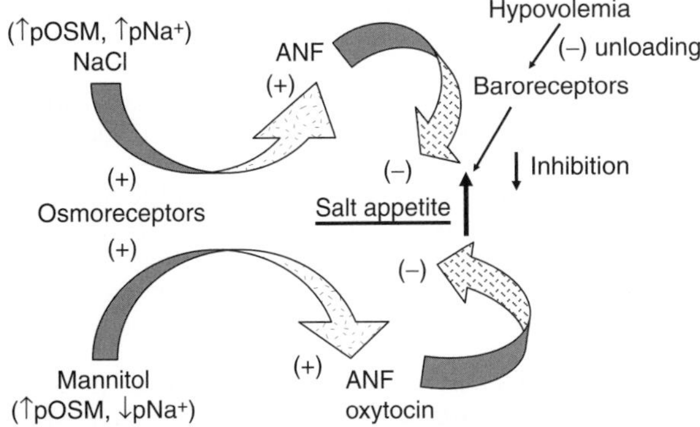

Figure 2 In rats with thirst and salt appetite induced by hypovolemia, hyperosmotic stimulation of the osmoreceptors with mannitol inhibits salt appetite by recruitment of central oxytocin and/or ANF. Hyperosmotic stimulation of the osmoreceptors with NaCl, in contrast, selectively activates central ANF pathways that inhibit salt appetite. pNa^+, plasma sodium concentration; pOSM, plasma osmolality.

pretreated rats). Thus, a divergence appeared between the inhibitory actions of oxytocin and ANF on saline drinking depending entirely on the nature of the hyperosmotic stimulus (**Figure 2**).

In a final experiment, pharmacologically driven salt appetite was similarly affected in both oxytocin and ANF receptor compromised rats when administered a dose of Ang II. This finding implies that Ang II acts not only to stimulate water and saline drinking but also to stimulate the central release of both oxytocin and ANF, which then buffer the effect of Ang II on salt appetite. Interestingly, the dipsogenic effect of Ang II was not enhanced in the ANF–ricin A chain conjugate pretreated animals, which might have been expected since ANF inhibits water intake stimulated by Ang II in control animals. This finding implies that the sensitivity of the inhibitory effect of ANF on salt appetite is much greater than that for thirst.

ANF and Central Autonomic Control

The initial localization of ANF binding in CNS sites known to be important in central cardiovascular control predicted the actions of ANF on blood pressure regulation in conscious rats. Indeed, the inhibitory effect of ANF on Ang II-stimulated mean arterial blood pressure in conscious rats has been demonstrated and extended to include an inhibitory effect of ANF on the blood pressure elevation observed following central hypertonic saline injection. Potential sites of action of ANF to lower blood pressure and reduce heart rate were identified to be the NTS and rostral ventrolateral medulla (RVLM). These two areas receive afferent ANF-positive input from neurons in the PVN and lateral hypothalamus and the lateral parabrachial nucleus. In addition to inhibitory effects of ANF on putative sympathoexcitatory neurons of the NTS and RVLM, the peptide has been demonstrated to stimulate preganglionic parasympathetic neurons in the dorsal motor nucleus of the vagus.

In addition to inhibiting the central action of Ang II on blood pressure, thirst, and vasopressin secretion, ANF may act in brain stem sites to reduce the action of Ang II on renal sympathetic nerve activity. In this way, the peripheral natriuretic actions of ANF, exerted directly in the tubule or via blockade of the peripheral renin–angiotensin system, may be complemented by CNS actions as well. Indeed, when ANF is administered centrally, diuresis and natriuresis occur.

Summary

The natriuretic peptides exert CNS actions that complement those expressed in the periphery, particularly in heart, kidney, and adrenal gland. These actions seem well coordinated to protect the cardiovascular system against volume or pressure overload. The physiologic relevance of the CNS actions has been established in animal models; however, it is not clear that these effects are expressed under physiologic conditions in humans. Although there is evidence for coordinated production and release of ANF in brain and heart, the conditions under which CNS production of ANF or CNP may vary have not been exhaustively catalogued. On the other hand, circulating levels of ANF and BNP are elevated in several disease states, and thus physiologically or pathologically relevant actions of the peptides may be exerted in the circumventricular organs under those conditions. Therefore, therapeutic use of NP analogs or enzyme inhibitors that prolong the plasma half-life of the endogenous peptides may result in CNS actions on thirst, vasopressin secretion, salt appetite, or autonomic function.

See also: Angiotensin II; Blood Pressure: Baroreceptors; Circumventricular Organs; Fluid and Electrolyte Homeostasis: Clinical Disease; Natriuretic Peptides; Neurohypophyseal System; Osmoregulation; Salt Appetite; Thirst.

Further Reading

Ackermann U and Azizi N (2000) Increased central AT1 receptor activation, not systemic vasopressin, sustains hypertension in ANP knockout mice. *American Journal of Physiology* 278: R1441–R1445.

Blackburn RE, Samson WK, Fulton RJ, Stricker EM, and Verbalis JG (1995) Central oxytocin and ANP receptors mediate osmotic inhibition of salt appetite in rats. *American Journal of Physiology* 269: R245–R251.

John SW, Veress AT, Honrath U, et al. (1996) Blood pressure and fluid–electrolyte balance in mice with reduced or absent ANP. *American Journal of Physiology* 271: R109–R114.

Levin ER, Gardner DG, and Samson WK (1998) Natiuretic peptides. *New England Journal of Medicine* 339: 321–328.

Lopez MJ, Wong SK, Kishimoto I, et al. (1995) Salt-resistant hypertension in mice lacking the guanylyl-cyclase-A receptor for atrial natriuretic peptide. *Nature* 378: 65–68.

McGrath MF, Kuroski de Bold M, and de Bold AJ (2005) The endocrine function of the heart. *Trends in Endocrinology and Metabolism* 16: 469–477.

Melo LG, Steinhelper ME, Pang SC, Tse Y, and Ackermann U (2000) ANP in regulation of arterial pressure and fluid–electrolyte balance: Lessons from genetic mouse models. *Physiological Genomics* 3: 45–58.

Munagala VK, Burnett JC, and Redfield MM (2004) The natriuretic peptides in cardiovascular medicine. *Current Problems in Cardiology* 29: 707–769.

Okada J, Takayama K, Xiong Y, and Miura M (1994) Influence of humoral control peptides on medullary vasomotor control neurons: Microstimulation and double-labeling studies using SHR and WKY rats. *Journal of the Autonomic Nervous System* 49: 171–182.

Potter LR and Hunter T (2001) Guanylyl cyclase-linked natriuretic peptide receptors: Structure and regulation. *Journal of Biological Chemistry* 276: 6057–6060.

Rademaker MT and Richards AM (2005) Cardiac natriuretic peptides for cardiac health. *Clinical Science* 108: 23–36.

Richard D and Bourque CW (1996) Atrial natriuretic peptide modulates synaptic transmission from osmoreceptor afferents to the supraoptic nucleus. *Journal of Neuroscience* 16: 7526–7532.

Rose RA, Anand-Srivastava AB, Giles WR, and Bains JS (2005) C-type natriuretic peptide inhibits L-type Ca^{2+} current in rat magnocellular neurosecretory cells by activating the NPR-C receptor. *Journal of Neurophysiology* 94: 612–621.

Samson WK and Levin ER (eds.) (1997) *Natriuretic Peptides in Health and Disease*. Totowa, NJ: Humana Press.

Trachte G (2005) Neuronal regulation and function of natriuretic peptide receptor C. *Peptides* 26: 1060–1067.

Attention and Eye Movements

E Kowler, Rutgers University, Piscataway, NJ, USA

Eye movements are inextricably linked to visual attention because both are the principal tools available for selecting interesting portions of visual scenes for enhanced perceptual and cognitive processing. Selection is crucial for animals that rely on foveal images for making decisions about the visual environment or for guiding action. As a result, most of what is known about eye movements and attention derives from studies of human and nonhuman primates.

Eye movements are often taken to mark the path of attention through a scene, where attention refers to the internal (or covert) distribution of processing resources. Eye movements and attention are assumed to serve useful purposes connected to the visual task, an assumption that has fueled decades of efforts to use eye movements to study how people search, read, study pictures of scenes, or carry out all manner of visually guided actions involving reaching, pointing, manipulating objects, walking, or driving. All these tasks require serial processing of different parts of a scene, and all routinely call upon eye movements.

For eye movements to be useful, they must be able to bring the line of sight to relevant locations quickly and accurately, and to keep it there, with no deliberate effort or laborious decision stages beyond those already incorporated into the task itself. In order to meet these requirements – speed, accuracy, and low cognitive load – eye movements must be sensitive to both the physical structure of the visual array and to the momentary needs of the task. To see why this is the case, consider the two extreme possibilities. On the one hand, if eye movements were completely under the control of the stimulus, with gaze attracted to the most vivid or salient portions in a scene, cognitive load would be minimal, but the eye would continually be dragged off to useless and irrelevant locations. It would be impossible to choose targets according to the needs of the task. (Nevertheless, there are many models based on just this assumption.) On the other hand, too much emphasis on volition, choice, or effort to control gaze would create undue cognitive demands and interfere with the ongoing flow of the task. The effort devoted to eye movement control would be a continual distraction from the job of recognizing objects and making task-related decisions and plans. Over the past decades, oculomotor scientists have struggled to find ways of integrating these two extreme modes of eye movement control. The focus of these attempts, in behavioral, psychophysical and neurophysiological studies, has been the attempt to find and understand the relationship between the control of eye movements and the mechanisms of visual attention.

What Is Attention?

Before considering the relationship between eye movements and attention, it is useful to briefly summarize some aspects of attention as they apply more generally to perception.

In his classical treatment of attention, William James acknowledged that, although everyone knows what attention is ("experience is what I agree to attend to", p. 402) the question of what attention does is difficult to answer. Attention is credited with being able to enhance certain experiences at the expense of others, but at the same time it must do so without distorting the nature of these experiences. Attention, James said, might make it possible to notice a faint sound or light, but paying attention to faint sounds or lights does not make them suddenly appear unusually loud or bright.

Modern researchers grappling with James's paradox have devised a number of clever experimental paradigms to explore the effects of attention on perception in many different contexts. Attention can be summoned to locations or to objects by various sorts of cues or signals, or it can be drawn away by imposition of a competing task. Studies using these techniques have led to broad agreement about several characteristics of attention:

1. Attention can lower detection thresholds and increased perceived contrast, producing improvements resembling those produced by modest increases in the contrast of the stimulus itself. These effects may correlate with attention-induced changes in cell firing patterns (e.g., increased contrast gain) at neural levels as early as V1 or the lateral geniculate nucleus (LGN).

2. Improvements in perceptual thresholds can be produced by paying attention either to a selected location (the 'attentional spotlight'), to a selected feature, or to a selected object. In the case of feature-based attention, the attentional benefits can spread across broad regions of space; in the case of object-based attention, the benefits of attention can extend to the different features (color, size, orientation) making up the attended object.

3. Attention to an object serves to overcome potential perceptual interference from visual objects

nearby. The importance of attention in the presence of competing stimuli has contributed to the view that attention should be viewed as an internal processing resource that can be distributed among features, objects or regions, as needed, to accomplish task goals. The importance of competition for limited resources has also been reflected at the neural level (areas V4 or the middle temporal area (MT), for example) with observations showing that, in the presence of multiple stimuli falling in a cell's receptive field, attention determines the relative contribution of each to the response of the cell.

4. Attention may also control access to limited capacity visual short-term or visual working memory, including the visual memory that preserves information from one fixation to the next.

All four of these characteristics – demonstrated in experiments that took pains to eliminate a role for eye movements, often by restricting stimulus presentations to very brief intervals of time – have influenced thinking and research about the role of attention in oculomotor control and the role played by eye movements and attention in visual processing.

Attention and Gaze Stabilization

One of the most important functions of eye movements is to maintain stable gaze. Even modest rotations of the head, if not compensated by counterrotations of the eye in the orbit, can create sufficient motion of the retinal image to impair visual resolution, carry important visual details away from the fovea, and create the illusion that the environment itself is moving. The oculomotor system is equipped with sophisticated systems that use both visual and vestibular signals to compensate for head movements and maintain stable gaze. These traditional full-field gaze stabilization reflexes, however, are not ideally suited to environments containing multiple objects at various distances and locations, viewed by observers who continually pick up and handle some of these objects, and who continually shift their own position in myriad ways. These situations require the maintenance of a stable gaze on only one thing at a time, and this means overriding the global visual or vestibular gaze stabilization reflexes, or at least restricting the portion of the scene on which they are allowed to operate.

In the early twentieth century, physicist and philosopher Ernst Mach described the situation well. The problem, said Mach, is that the retinal motion generated by head movements and body movements as we move through the environment may "exert a peculiar motor stimulus upon the eye, and draw our attention and our gaze after them" (p. 146). But this attraction can be overridden:

> No special apparatus is necessary for observing the foregoing phenomena. They are to be met with on all hands. I walk forward by a simple act of the will. My legs swing to and fro without my having to attend to them particularly. My eyes fixed steadfastly upon my goal without suffering themselves to be drawn aside by the motion of the retinal images consequent upon progression. All this is brought about by a single act of the will. . . . The same process must also be set up if the eyes are to resist for any length of time the stimulus of a mass of moving objects. (Mach, 1906/1959: 146)

Mach's impression that we can override flows of motion on the retina and maintain stable gaze on chosen targets has been confirmed and extended many times in the modern oculomotor laboratory. It is possible, for example, to maintain gaze on even very small stationary targets superimposed on large and vivid moving backgrounds. It is also possible to smoothly track a target moving across a stationary background (although this has proven to be the more difficult task) and to track a target moving in the presence of a nearby moving nontarget. In the case of the target and nontarget, it may require about 100 ms or so for complete selectivity to be achieved. Observations such as these have been made in many studies since 1930, with different kinds of backgrounds and with fixation targets of different sizes and retinal locations, each time showing that it is possible to maintain fixation on selected targets and override the effects of the more plentiful motion signals originating from the background (**Figure 1**).

The ability to select the target for smooth eye movements has also been demonstrated with superimposed transparent sheets of dots, each moving at a different velocity (**Figure 1(b)**). With such patterns, attending to a discrete location would not be sufficient to select one sheet as the target because dots are superimposed everywhere. Thus, effective selection with transparent motions requires attending to a selected object or surface rather than to a discrete location within the pattern. Interestingly, the selection of one of the transparent surfaces does not disrupt the perceptual impression of transparency, nor does it disrupt perceptual interactions between the superimposed fields. For example, if one of the fields is moving and the other is stationary, the stationary field seems to be moving opposite in direction to the other, ignored moving field. The percept of induced motion shows that the selection of the target for the eye movements leaves intact basic operations of sensory and perceptual motion processing. This raises the possibility that the selective mechanism that determines the target for

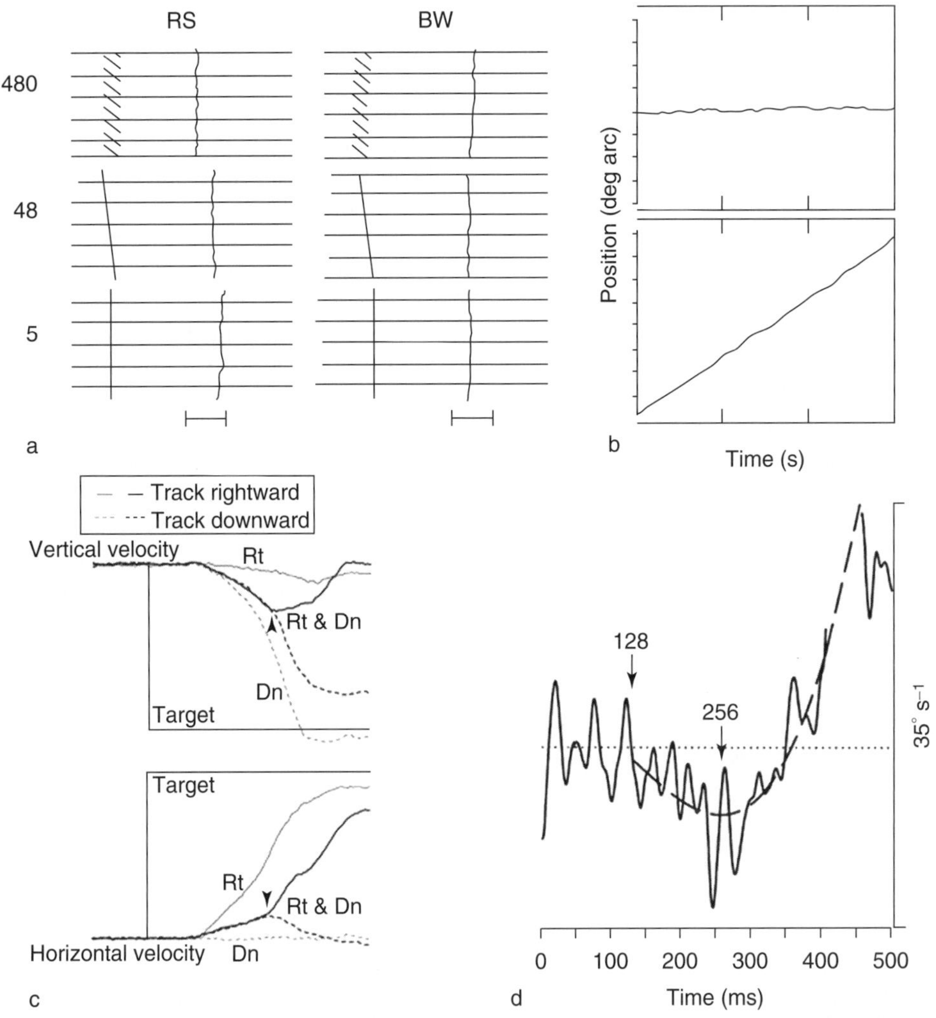

Figure 1 Smooth eye movements and attention: (a) examples of horizontal smooth eye movements maintaining a stable line of sight on a stationary fixation target superimposed on a background moving grating; (b) example of the independence of smooth eye movements from background motion with stationary and moving dots; (c) average eye velocity in response to a single target moving either downward (top) or rightward (bottom) or to a pair of targets, one moving rightward and the other downward; (d) horizontal eye velocity in the presence of two visual arrays of noise elements moving in opposite directions, one located above and the other below the line of sight. In (a), the data are from two observers, RS and BW. The left trace in each panel shows the stimulus motion (5, 48, or 480 min arc s^{-1}); the right trace shows horizontal eye movements. Horizontal lines are 1 s time markers, with time beginning at the bottom. The horizontal bar below the panels for each observer represents a 1° eye rotation. In (b), each trace shows horizontal smooth eye movements with two full-field superimposed arrays of random dots, one array stationary and the other moving at 1.2° s^{-1}. In the upper panel, the observer was told to fixate the stationary array and, in the lower panel, to pursue the moving array. There was virtually no effect of the unattended field on performance. In (c) average eye velocity is in response to a single target moving either downward (top) or rightward (bottom) (thin eye traces, labeled Rt or Dn) or to a pair of targets, one moving rightward and the other downward (bold traces labeled Rt and Dn). Traces labeled Target shows the stimulus which traveled over 20° for 400 ms. After 150 ms, one of the target motions of the pair disappeared, leaving only a single stimulus moving downward (top) or rightward (bottom). Eye velocity when both target motions were present was the average of the two target velocities until about 80 ms after one of the motions was removed (marked by the arrows) when the traces diverged. In (d), the eye trace shows the change in pursuit direction when first one, and then the other, moving field is selected as the pursuit target. (a) Reproduced from Murphy BJ, Kowler E, and Steinman RM (1975) Slow oculomotor control in the presence of moving backgrounds. *Vision Research* 15: 1263–1268, with permission from Elsevier. (b) Reproduced from Kowler E, Van der Steen J, Tamminga EP, and Collewijn H (1984) Voluntary selection of the target for smooth eye movements in the presence of superimposed, full-field stationary and moving stimuli. *Vision Research* 24: 1789–1798, with permission from Elsevier. (c) From Lisberger SG and Ferrera VP (1997) Vector averaging for smooth pursuit eye movements initiated by two moving targets in monkeys. *Journal of Neuroscience* 17: 7490–7502. (d) From Liston D and Krauzlis RJ (2003) Shared response preparation for pursuit and saccadic eye movements. *Journal of Neuroscience* 23: 11304–11314.

smooth eye movements may use its own dedicated attentional filter whose operation has minimal consequences for perception.

The Role of Attention in Perception versus the Role of Attention in Smooth Eye Movements

One way to find out whether smooth eye movements and perception share a common attentional mechanism is to expand the eye movement experiment to include a concurrent perceptual task. In one representative situation, an observer is presented with two sets of moving targets and asked to smoothly track one set and ignore the other (**Figure 2**). In this situation, perceptual judgments made about the tracked target are more accurate than those made about the untracked background (even after taking into account any differences in retinal position and retinal velocity), implying that a single filter and a single attentional decision determine the strength of signals reaching both perceptual and oculomotor systems. But there are important differences between eye movements and perception. Whereas the smooth tracking eye movements are nearly perfect, with virtually no effect of the unselected background motion, perceptual judgments about the unselected background are above chance, showing that some perceptual registration of the background remains. Thus, there is an asymmetry between the effects of attention on perception and eye movements. Analogous results have been obtained from neurons in the extrastriate cortical areas MT and medial superior temporal region (MST), the presumed sources of the motion signals that guide smooth eye movements and serve perception. Neurons in these areas respond more vigorously to attended and tracked motion than to unattended untracked motion, but the differences in firing patterns are small and do not explain the high degree of selectivity that can be demonstrated in smooth pursuit eye movements (although they are in general agreement with the magnitude of selectivity observed

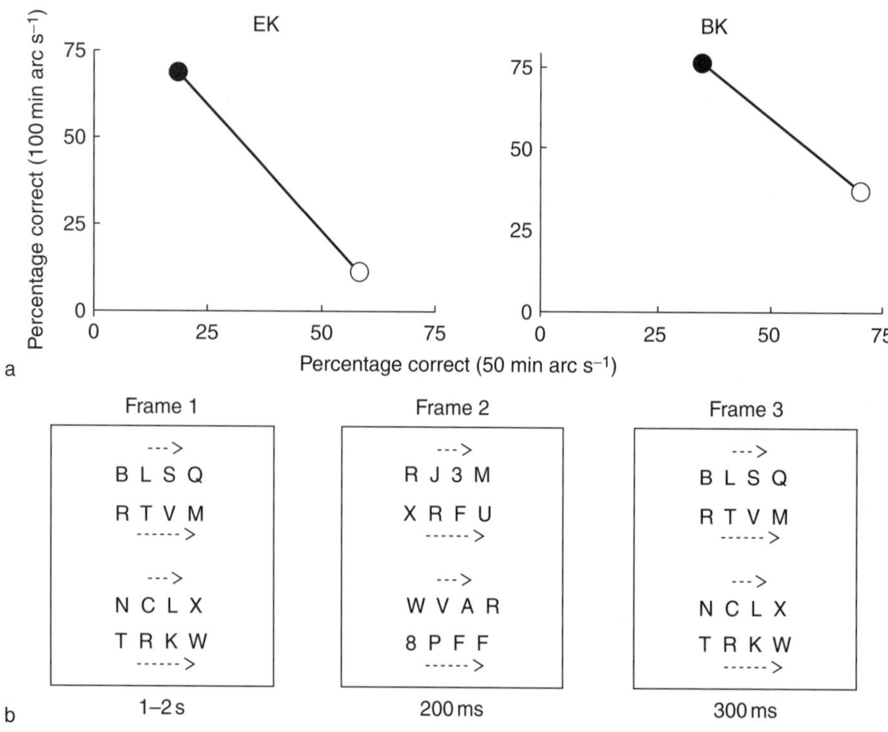

Figure 2 Smooth eye movements with a concurrent perceptual task: (a) perceptual performance for two observers; (b) representative sequence of display frames for a trial. The three display frames in (b) show four rows of letters moving rightward with the top and third rows moving at half the velocity of the second and bottom rows. Letters remained within the confines of the frame so that as portions reached an invisible boundary on the right they re-appeared immediately on the left side of the display. Vertical eye position remained in the gap between rows 2 and 3 while subjects attended and tracked either the faster or slower pair of rows. Frames 1 and 3 are masks. Frame 2 is the critical display, containing two numerals that had to be identified at the end of the trial. Tracking gains were >0.8. The perceptual performance shown in (a) for two observers depended on instructions. When observers attended and tracked the slower pair of rows (open symbols), identification accuracy was better for the slower rows than the faster rows. When they attended and tracked the faster pair of rows (closed symbols), identification improved for the faster pair at the expense of the slower pairs. Separate analyses showed that these results were not due to differences in retinal speed of tracked and untracked rows and were due solely to attention. Reproduced from Khurana B and Kowler E (1987) Shared attentional control of smooth eye movements and perception. *Vision Research* 27: 1603–1618, with permission from Elsevier.

at the perceptual level). The observed differences between the effects of attention on eye movements and perception, and the small effects observed in the motion areas MT and MST during the selective tracking tasks, imply that a given distribution of attention across target and background has different consequences for perception and for eye movements. This rules out early selection models, in which attention operates primarily at an early sensory level of visual processing shared by perception and eye movements.

The ability to eliminate almost all influence of the unselected motion signals from the eye movements suggests the involvement of a winner-take-all (WTA) network operating on the set of attentionally weighted motion signals emerging from areas MT and MST. WTA networks are often used in models of attention or perception and are effective ways to remove weak, and potentially interfering, signals. Although such a network operating downstream from sensory motion areas could account for the perfect or near-perfect selectivity observed for eye-tracking performance, an unconstrained WTA network places a strict limit on the options. It would not, for example, allow people to divide attention between a target and background and to track at a velocity intermediate between the two. An alternative to the WTA network is an executive controller that determines the relative weights assigned to different sets of sensory motion signals. A single executive controller could be used by both perceptual and oculomotor systems with different effects on each. For example, assigning maximum weight to a target could completely attenuate the representation of nontargets in neural areas devoted to smooth tracking and only partially attenuate the nontarget representation in perceptual areas. Different assignments of weight would produce different performance outcomes. Given that the neural circuitry that controls smooth eye movements includes several high-level cortical areas (frontal eye field (FEF), supplementary eye field (SEF), and lateral intraparietal area (LIP)) that also play roles in perceptual attention, there is considerable opportunity for executive intervention in the attentional weighting of motion signals from target and background.

Summary: Attention and Gaze Stabilization

Selective attention plays a crucial role in the control of smooth eye movements by ensuring that stable gaze can be maintained on either stationary or moving targets, regardless of the signals present in the background. Attention ensures that eye movements are controlled only by signals originating from the selected object. Performance of concurrent oculomotor and perceptual tasks have shown that a single attentional decision affects both – it is not possible to maintain gaze on one object while fully attending another. Unselected objects nevertheless remain perceptible and continue to generate neural signals in areas sensitive to motion. This outcome rules out early selection models of attentional control and raises still unanswered questions of how a common attentional decision can have different consequences for smooth eye movements and for perception.

Attention and Saccadic Shifts of Gaze

Saccadic eye movements are used to inspect the visual environment, with saccadic shifts of gaze occurring anywhere from once every several seconds to several times a second. Visual processing occurs during the pauses between saccades. Unlike smooth eye movements, which cannot be generated without some sort of representation of smooth motion, saccadic eye movements can be made at will to look at arbitrary locations.

Saccades are typically directed to useful or informative locations and are often taken to disclose the path of attention during performance of natural tasks. Attention, however, need not be locked to the fovea even as the eye jumps from place to place. Attention to locations remote from the line of sight is valuable for broad surveys of the scene and is indispensable for ensuring that saccades land on the selected object, regardless of the visual structure of the immediate surround.

Eye Position as an Overt Marker of the Locus of Attention

Figure 3 shows examples of saccadic patterns made during various visual tasks: reading, counting, geometry problem-solving, and visual scene recall. **Figure 4** shows eye movements during visual search. **Figure 5** replicates and extends the well-known experiment described by Yarbus in his 1967 book *Eye Movements and Vision*, in which people were asked to inspect the same painting in order to answer different questions. The scan patterns differed depending on the question, showing that the physical features of the pictures do not by themselves determine the landing locations of the eye but that there is considerable influence of purpose and intent. **Figure 6** looks at this issue from another angle, showing scan patterns obtained when looking at two versions of the same photograph, where in one version the image has been filtered and the fine detail removed. Although the task is the same in each, the scatter of the fixated locations is different.

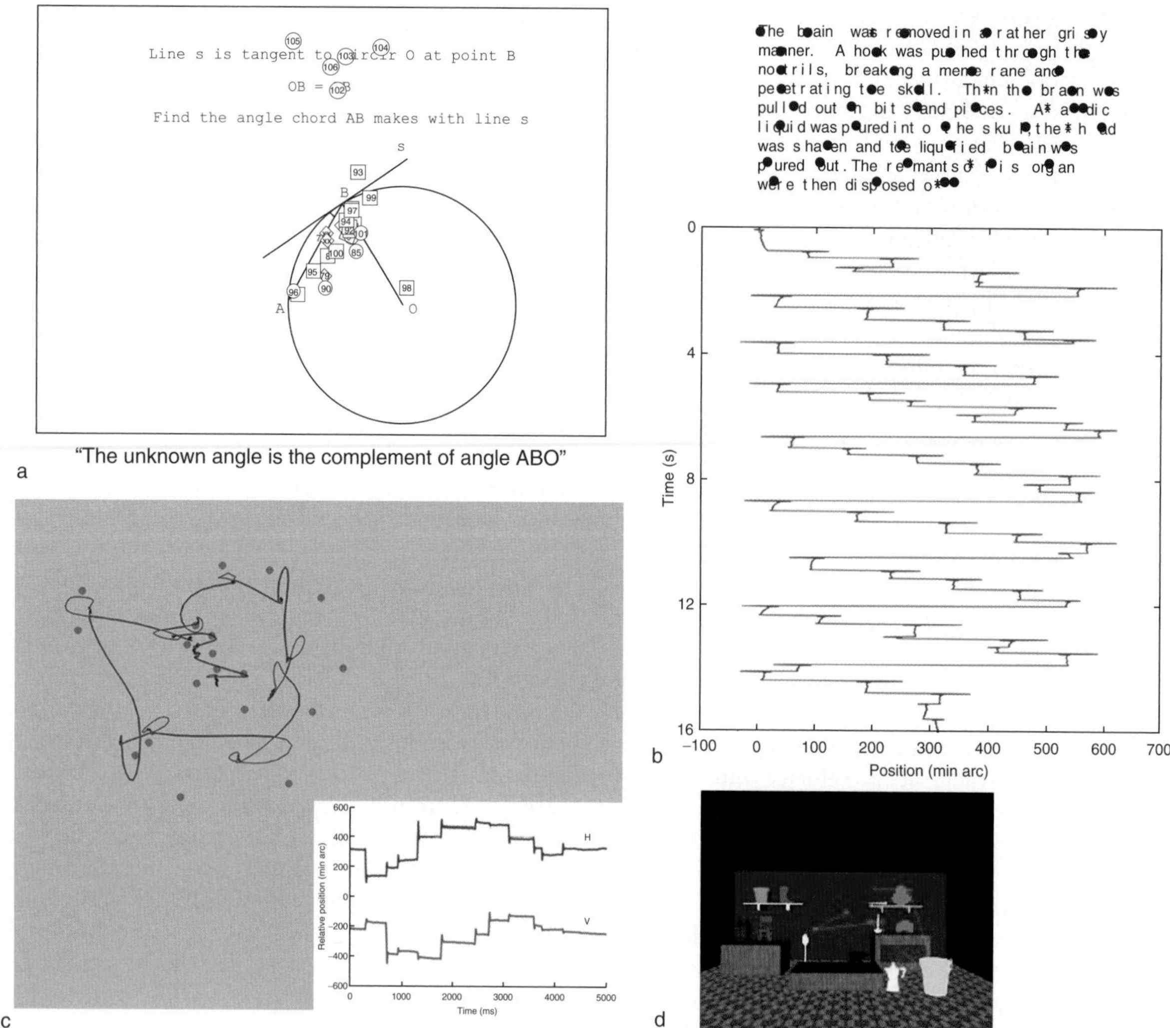

Figure 3 Examples of eye movements during various visual and cognitive tasks: (a) fixation positions of one observer during one stage of solving a problem in geometry; (b) eye movements during reading; (c) eye movements while counting an array of dots; (d) eye movements during a visual recall task. In (a), the problem is presented above the diagram and the excerpt from the protocol is below it. Important features of the diagram were refixated at regular intervals. Analyses of the timing and locations of fixations showed frequent revisits, consistent with a working memory capacity of five visual features of the problem. In (b), the same pattern of eye movements is shown as individual fixation locations superimposed on the text (top) and as a trace of eye positions over time (bottom). Eye position is on the abscissa, and time moves from top to bottom. These eye movements are typical of reading in that saccades were about seven characters long and intersaccadic pauses were about 275 ms. More than 40 readings of this same text showed little change in the pattern of eye movements and a high level of consistency in the distribution of landing locations. In (c), eye movements are shown both superimposed on the dot array and also as horizontal and vertical traces over time. In (d), subjects had to report the contents of the picture after periods of scanning ranging up to 4 s. Analyses showed that memory for the content of the pictures was preserved across successive viewings separated by several minutes and several intervening viewed scenes. (a) Reproduced from Epelboim J and Suppes P (2001) Eye movements during problem solving in geometry. *Vision Research* 41: 1561–1574, with permission from Elsevier. (b) Reproduced from Schnitzer BS and Kowler E (2006) Eye movements during multiple readings of the same text. *Vision Research* 46: 1611–1632, with permission from Elsevier. (d) Reproduced from Melcher D and Kowler E (2001) Visual scene memory and the guidance of saccadic eye movements. Vision Research, 41: 3597–3611, with permission from Elsevier.

Similar efforts to document sequences of fixations in different tasks have been made in more complex and dynamic situations, in which observers move about the environment, handling various objects for some specific purpose. The sequences of fixation positions of the two eyes are required when observers perform three different tasks: tapping a set of rods in prescribed order, grooming a fellow primate, and assembling a doll.

Data such as these have always been intriguing. They prove that ordinary visual experience, which seems smooth, continuous, and flowing, is really made up of sequences of brief, still frames and snapshots, whose detailed contents escape awareness. We do not notice

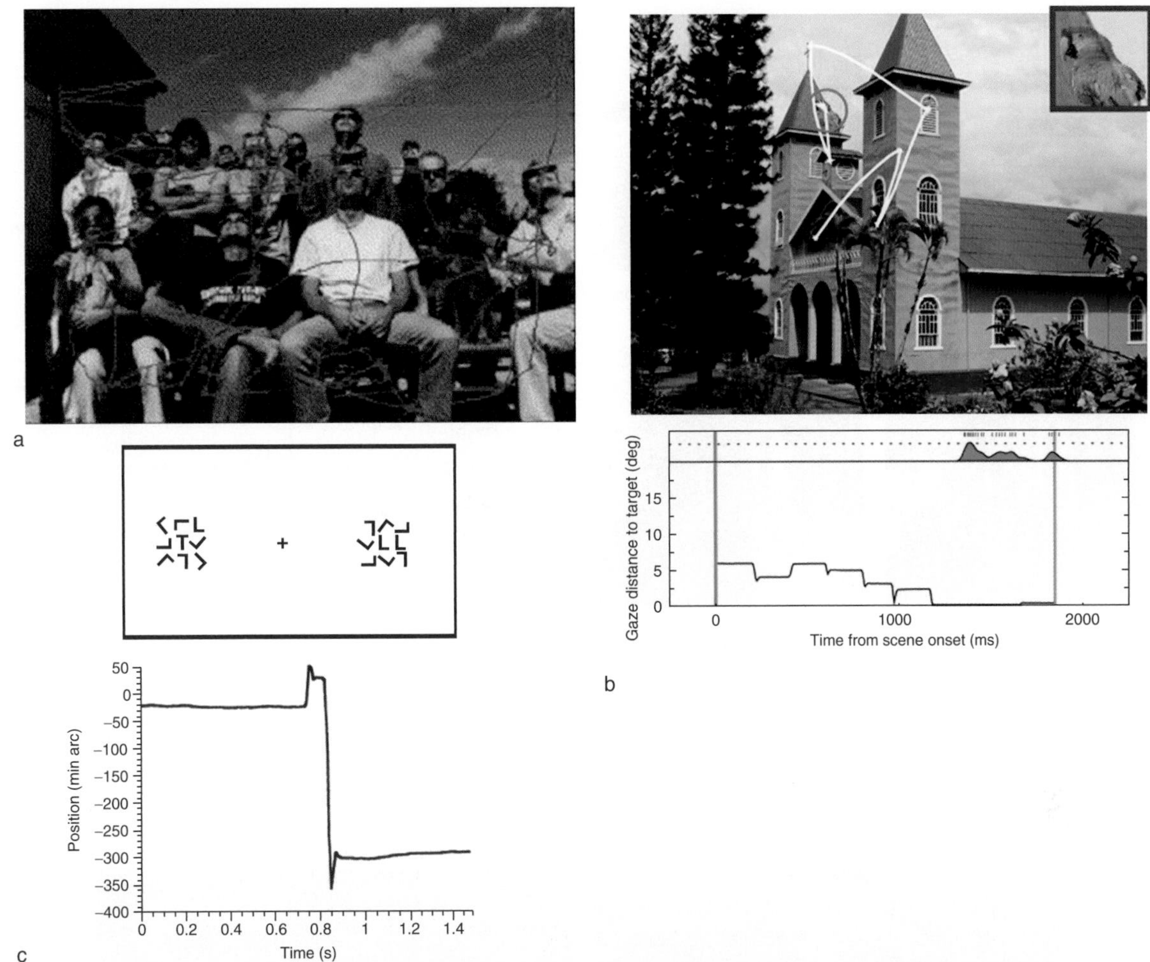

Figure 4 Eye movements during visual searche: (a) search for small gray crosses superimposed on the scene; (b) search in monkey for a small embedded target (reproduced in inset); (c) search for a target letter T embedded in one of two clusters of L's. In (a), analyses showed little or no bias to avoid returning to previously fixated location. In (b), eye traces show the distance to the target over time. Spike density functions show little response until just before the animal fixates the hidden target. Selection of the target as a saccadic goal was shown to be a necessary condition for the firing bursts. In (c), the probability that one of the clusters contained the T (0.8/0.2) was disclosed by a brightness cue. The eye traces show that often subjects looked first at the cluster with the lower probability, particularly when the cluster was closer to the line of sight, and then quickly with an unusually short latency interval (<100 ms) made a second saccade to the high probability cluster containing the target letter. (a) Reproduced from Hooge ITHC, Over EAB, van Wezel RJA, and Frens MA (2005) Inhibition of return is not a foraging facilitator in saccadic search and free viewing. *Vision Research* 45: 1901–1908, with permission from Elsevier. (b) Reproduced from Sheinberg DL and Logothetis NK (2001) Noticing familiar objects in real world scenes: The role of temporal cortical neurons in natural vision. *Journal of Neuroscience* 21: 1340–1350, with permission from the Society for Neuroscience (c) Reproduced from Araujo C, Kowler E, and Pavel M (2001) Eye movements during visual search: The costs of choosing the optimal path. *Vision Research* 41: 3613–3625, with permission from Elsevier.

each individual glance because we remember only what is most essential – the cumulative number of dots counted or the meaning of a word or phrase – or we use what we see to guide an immediate action. Observing sequences of eye movements reveals the sequential nature of vision and provides a view of underlying goals and strategies that would be hard to recover through other means.

Decisions, Attention, and Saliency Maps

Understanding the rules that govern saccadic planning in everyday life tasks is an enormous undertaking that must be based on models of the task, combined with knowledge of the relevant visual, cognitive, and motor capacities that limit performance and affect the strategies adopted. Although we still lack comprehensive and general models, progress has been made toward the goal. One of the influential constructs developed to account for the eye movements made during scanning of a scene is the salience map. As originally proposed by Koch and Ullman, the salience map is derived from the output of banks of visual filters, modeled after those present in early levels of the

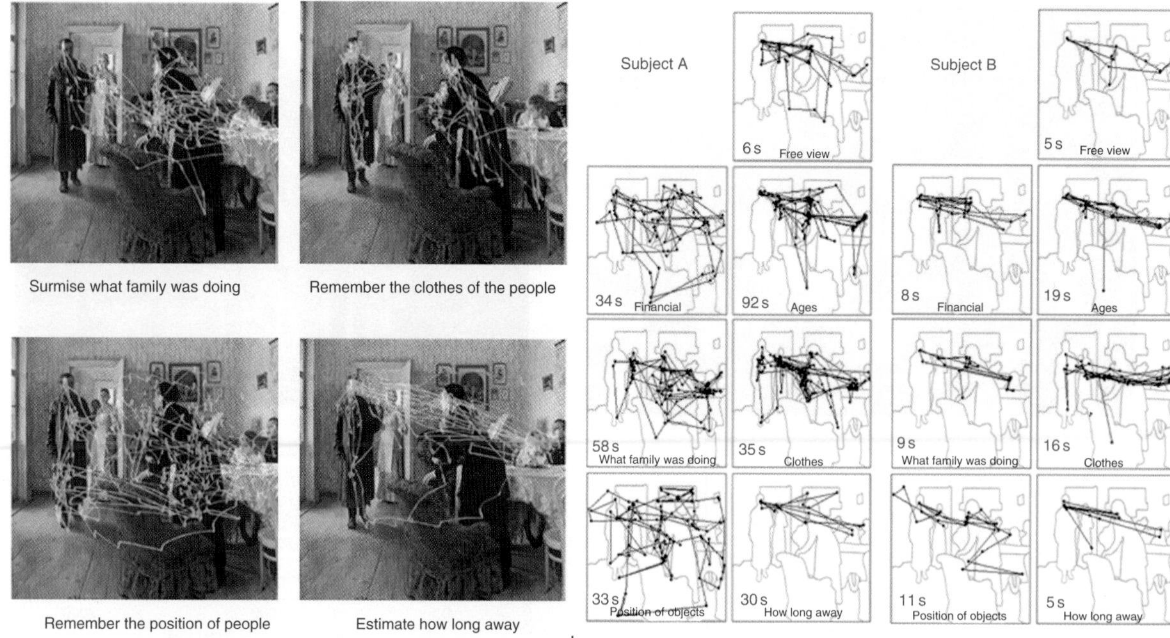

Figure 5 Effect of task motivation on eye movements: (a) people looking at Repin's painting *The Unexpected Visitor* altered the choice of fixation positions depending on the question they had to answer; (b) a replication of Yarbus's experiment, in which subject A shows the same sensitivity to task as Yarbus demonstrated but subject B tends to look in the same places regardless of the task. (a) Adapted from Yarbus A (1967) *Eye Movements and Vision.* New York: Plenum. (b) Figure provided by J Pelz and reproduced with permission. For discussion see Lipps M and Pelz JB (2004) Yarbus revisited: Task-dependent oculomotor behavior [abstract]. *Journal of Vision* 4: 115a, with permission from ARVO.

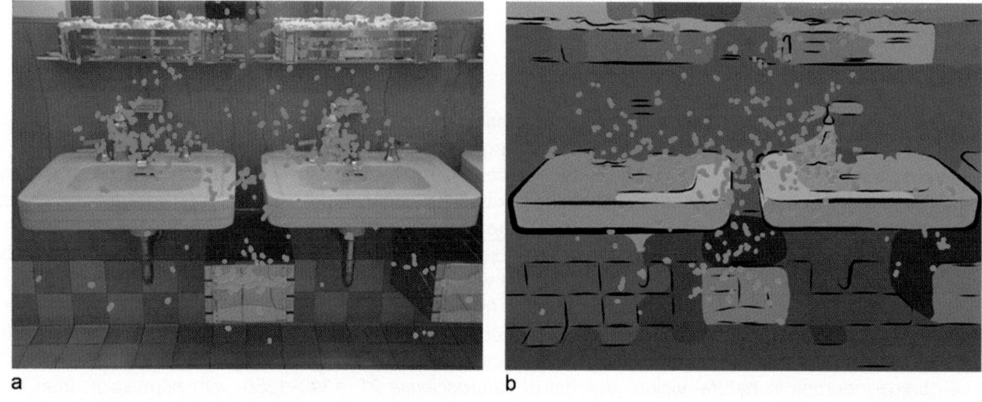

Figure 6 Fixations made by 10 viewers while looking at two variants of the same photograph: (a) original photo; (b) version in which computer graphics techniques were used to create a drawing-like version of the scene and remove detail throughout the image. Subjects were instructed to remember the picture for a subsequent recognition task. In (a), fixations are tightly clustered in a few key areas. In (b), fixations are more uniformly distributed. Figures provided by A Santella; original photo courtesy of Philip Green spun. For discussion see DeCarlo D and Santella A (2002) Stylization and abstraction of photographs. *ACM Transactions on Graphics* 21: 769–776; Santella A and DeCarlo D (2004) Visual interest and NPR: An evaluation and manifesto. In: *Proceedings of the Third International Symposium on Nonphotorealistic Animation and Rendering*, pp. 71–78. Annecy, France, 7–9 June. New York: ACM Press.

visual system (V1). These filters respond on the basis of local feature contrast (luminance, color, or orientation contrast) present in the scene. Models for the generation of salience maps integrate all three types of feature contrasts, taking the effects of retinal eccentricity into account. Saliency maps have been used to predict where people direct saccades and how they distribute visual attention (i.e., processing capacity) across a scene even when eye movements are not made.

In order to use a global salience map to predict a sequence of fixations, two additional assumptions have to be made. The first is that a WTA network (similar to that already discussed in the context of smooth eye movements) establishes the ordering of the fixations, with the region with the highest salience level designated as the next one to be fixated. The second assumption is that the salience map is subjected to inhibition of return; that is, once a location is examined its strength in the salience map will abruptly diminish, thus allowing the line of sight to proceed to new places and avoiding the need to store a list of already-scanned locations. Armed with a salience map and these two supporting assumptions, it has been possible to generate precise predictions about where people look during unconstrained free-viewing. The predictions have proven to be reasonably accurate, at least when computed salience is compared to the global aggregate of fixated locations over some period of inspection. Nevertheless, the approach has not been without controversy.

The main criticism of the salience map approach is the obvious one, recognized and discussed even by proponents of the approach. Strict control of eye movements or attention based solely on physical salience becomes counterproductive as soon as people are faced with real-world tasks (**Figures 4–6**). In real-world tasks, people look at the objects and details that are important, regardless of the physical salience levels. Gaze and attention may be drawn to novel or unexpected objects or to highly familiar objects, depending on the situation. None of these characteristics is represented by physical salience. Physical salience also does not capture the long-term changes in gaze preferences that develop as the contents of the scene are learned, nor does it capture effects of motivation. Do you look at something because it attracts your interest, because it contains some new information you need to perform a task, or because you need to confirm a detail that you already know is present? All of these so-called top-down variables that influence gaze and attention can be viewed as displacing the concept of a saliency map or, alternatively, as imposing an additional set of weights on the physical salience levels.

Neurophysiological studies using a variety of active saccadic tasks involving visual search or visual discriminations have addressed a portion of the problem of incorporating top-down variables by showing that neurons in areas related to saccadic planning (LIP, FEF, and superior colliculus (SC)) may be able to encode the top-down salience level at a given location. The evidence for this assertion is that neurons in these areas show enhanced activity when objects in their receptive fields contain information relevant to the guidance of an upcoming saccade, even when the location of enhanced activity does not correspond to the location of the saccadic target. Visual areas (the V4 or inferotemporal visual area (IT), for example), also show sensitivity to top-down salience. In these areas, neurons tuned to specific features show effects of attention, as well as patterns of presaccadic enhancement. Thus, there has been considerable progress in identifying brain areas that modulate responses to objects depending on their immediate behavioral significance, and these areas may constitute the core of a network that directs attention and saccades to objects or locations relevant to a task.

Incorporating top-down factors and physical salience levels into a single map may be a physiologically plausible means of deriving a single message to guide saccades and attention. Alternatively, there may be multiple maps that compete for control. Studies of eye movements during visual search have provided some support for the notion of competing maps. In these studies, saccades are drawn to physically salient locations that are not likely to contain the search target, followed quickly by corrective saccades to more appropriate, but less salient, places (**Figure 4(c)**). These rapid saccadic sequences could represent the results of the competition between different maps for momentary control of the neural centers that produce saccades. Although such competition seems inefficient, the amount of time lost by an occasional errant saccade may be small enough that an investment of time and resources in more prudent saccadic planning may not be warranted.

Saccades and Attention

Saliency – either stimulus-driven or top-down – is said to predict the distribution of both attention (internal processing resources) and saccades over time and space. Is there any important or necessary distinction between the distribution of attention and saccades? According to premotor theories of attention, saccadic eye movements and shifts of attention are essentially the same process, originating in the same neural areas, with an attention shift occurring only when the saccade is suppressed or inhibited. The premotor theory is supported by finding several neural areas (SC and FEF) whose activity can evoke or is correlated with both attention shifts and saccades. Nevertheless, it is troubling for premotor theories that attention can be distributed in parallel across the visual field or aligned symmetrically about the line of sight; spatial patterns do not readily map on to trajectories of saccades.

If we view attention not as a subthreshold saccade but rather as an internal processing resource that influences both perception and motor control, we can then ask how attention and saccades interact. Attention plays an important role in saccadic guidance in that, although it is possible to shift attention independently of the saccade, it does not appear possible to plan and carry out an accurate saccade without shifting some attention to the saccadic goal.

Dual-task studies, similar to those described earlier for smooth eye movements, have examined the distribution of visual attention prior to saccades, using various psychophysical tasks as indicators of local attentional strength. These dual-task studies require observers to do two things at once (in as brief a time as possible), namely, prepare to look at a specified object and identify a perceptual target presented while the preparation of the saccade is in progress. The perceptual target appears either at the saccadic goal or elsewhere. These experiments can succeed only if care is taken that the two tasks are done in the same brief interval of time because otherwise observers can decide to delay saccades in order to improve perceptual performance.

Dual-task experiments have shown that perceptual performance is better at the saccadic goal than elsewhere (see illustrative study in **Figure 7**), showing that saccades and attention do not functional independently. Yet, as was the case with smooth eye movements, perceptual performance at nongoal locations remained well above chance even when efforts were made to produce the saccades as quickly and accurately as possible (i.e., with the same latency and accuracy as observed with no concurrent perceptual task). Moreover, it only required a modest sacrifice in saccadic performance (10–20% latency increases) to achieve pronounced improvement in perceptual performance at nongoal locations.

Similar connections between attention and saccades are found during the execution of sequences of saccades (**Figure 8(a)**). Best perceptual performance is found at the saccadic goal, with elevations in perceptual thresholds at other locations, the amount of elevation depending on when and where the perceptual target appears. The pattern of activity observed during saccadic sequences suggests that the reduction in the attentional levels at nongoal locations is important for providing sufficient attentional contrast to guide an accurate saccade and prevent the line of sight from landing at salient, but incorrect, locations during scanning tasks. Thus, the main role of top-down attention in saccadic guidance may be to suppress the attraction of physically salient, but irrelevant, locations.

When the scan path to be followed is marked by a feature difference, rather than being followed from memory, the distribution of attention changes significantly. Attention spreads to locations along the feature-cued path, including targets of subsequent, as well as previous saccades (feature-based attention) (**Figure 8 (b)**). An interesting question raised by these results is how the saccades remain accurate when so many locations are attended at once. As was the case with smooth eye movements, there are two main alternatives. The first is a WTA network that directs the eye to the location where attentional strength is greatest. The second alternative requires the intervention of a central executive controller that imposes its own attentional weights. The controller could, for example, focus all active effortful attention on the upcoming target. The perceptual enhancement observed elsewhere along the featured-cued path could result from a passive process that facilitates the spread of attention across space to the selected features.

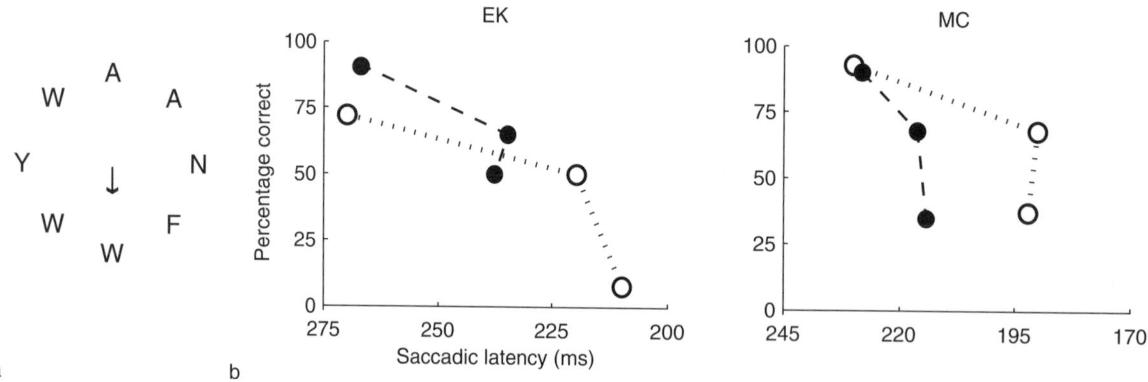

Figure 7 Saccades and attention: (a) the task display; (b) attentional-operating characteristics for two subjects showing the trade-off of saccadic and perceptual performance under two different conditions: saccade to a randomly chosen location (filled circles) and saccade to the same location on each trial (open circles). Shown in (a), the display contained eight randomly chosen letters. The task was to make a saccade (downward, in this case) while simultaneously identifying the letter in the right-hand position. Display timing was adjusted so that the critical letters appeared only during the saccadic preparation interval. In (b), The three data points in each function show performance under three instructions: (1) minimize saccadic latency (lowermost symbols), (2) sacrifice latency for improved perceptual accuracy (uppermost symbols), and (3) perform at a level intermediate between these two extremes. Achieving best perceptual identification accuracy required a sacrifice in saccadic latency (analysis of saccadic precision showed analogous results). Perceptual identification reached near-perfect levels with latency increases of only 10–20%. Reproduced from Kowler E, Anderson E, Dosher B, and Blaser E (1995) The role of attention in the programming of saccades. *Vision Research* 35: 1897–1916, with permission from Elsevier.

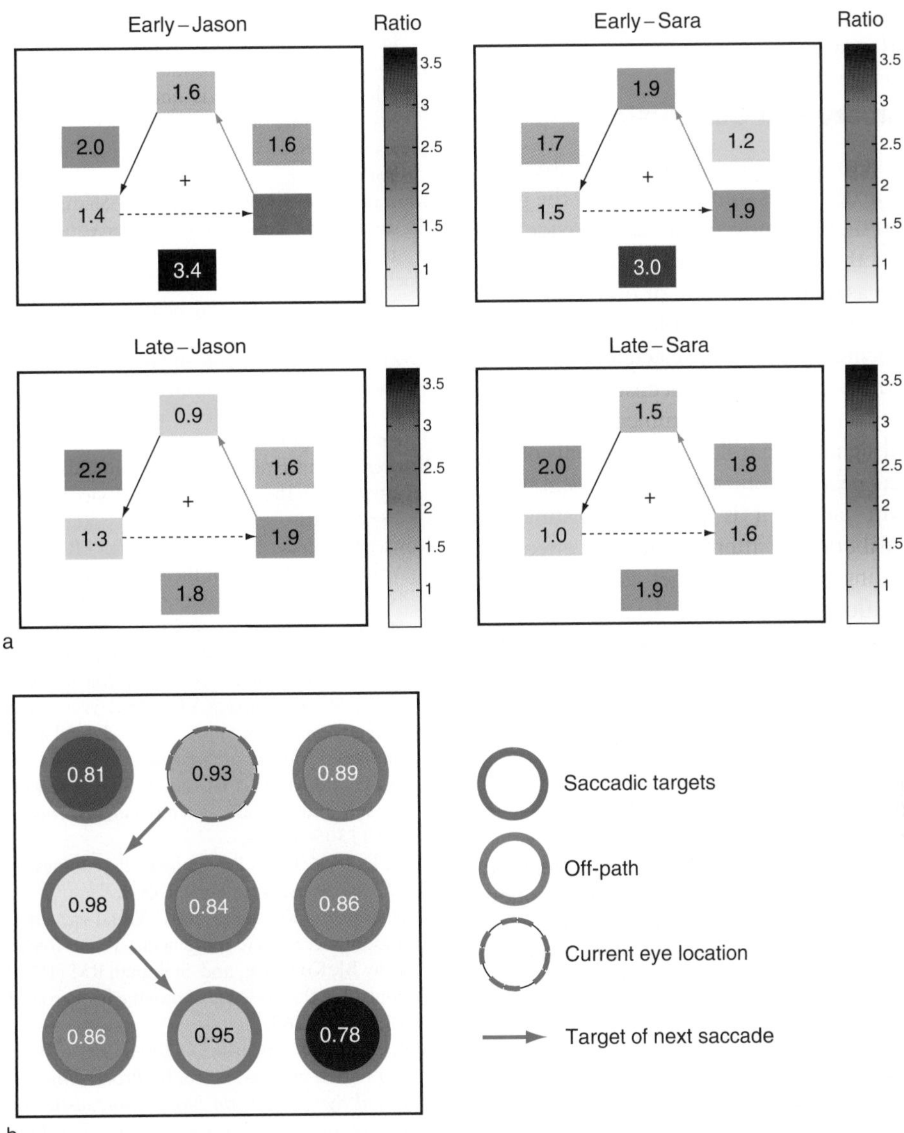

Figure 8 Attention during performance of sequences of saccades: (a) repetitive sequences of saccades; (b) nonrepetitive sequences of saccades. In (a), two subjects made saccades to rectangular boxes following a triangular pattern (shown by arrows) in which saccades were made to every other box. During a randomly selected fixation pause, an oriented Gabor patch was shown in one of the boxes either during the early (top row) or later (bottom row) portion of the pause. Current fixation position is in the top box. The colors of the boxes and the numbers in the boxes show the ratios of contrast threshold for correct orientation identification relative to thresholds during steady maintained fixation. Thresholds were significantly elevated at all locations except the currently fixated position (top of each graph) and target of the next saccade (lower left box in each graph). In (b), arrows represent a segment of a saccadic path. Current eye position is the top circle (dotted outline). Shading and numbers in the circles show percentage of correct orientation identification (two alternative forced-choice) for a Gabor patch shown briefly in one of the circles during a randomly selected intersaccadic pause. Perceptual performance is best for locations on the saccadic path, including the targets of the next two saccades. Performance for on-path locations is better than for off-path locations, including off-path locations of equal or smaller retinal eccentricity. (a) Reproduced from Gersch TM, Kowler E, and Dosher B (2004) Dynamic allocation of visual attention during the execution of sequences of saccades. *Vision Research* 44: 1469–1483, with permission from Elsevier (b) From Gersch TM (2007) *Attentional Filter Used during Scanning Influent Both Perception and Memory*. Doctoral thesis, Rutgers University.

Summary: Attention and Saccades

The picture of saccades and attention that emerges from these psychophysical and behavioral studies, which is in agreement with neurophysiological work, is that saccades and attention normally operate in a connected fashion, with saccades landing at the object that is the main focus of attention. Nevertheless, attending to the goal of saccades still allows significant perceptual processing at different locations, perhaps with the aid of other mechanisms that passively distribute attention to relevant locations in parallel across the visual array.

Overall Summary and Conclusion

The connection between eye movements and attention described in this article gives some support for the prevalent assumption that eye movements can provide reliable indicators of the main focus of attention in active visual tasks, such as reading, search, or object manipulation. Attention plays a crucial role in oculomotor control by enhancing signals from selected targets relative to those in the background, thus ensuring that eye movements are planned and programmed on the basis of selected visual signals, regardless of the visual configuration of objects nearby. Following the classical treatment of William James, it is useful to distinguish between the distribution of attention and the effects of attention. Attending to a given target can completely eliminate the influence of visual backgrounds on eye movements (smooth or saccadic) while continuing to allow perceptual processing, albeit at reduced levels. This asymmetry between perception and eye movements has the desirable consequence of allowing eye movements to be accurate without disrupting global perceptual analysis across the visual field.

See also: Attentional Functions in Learning and Memory; Contextual Interactions in Visual Perception; Perception and Eye Movements; Saccades and Visual Search; Saccadic Eye Movements; Sensorimotor Integration: Attention and the Premotor Theory; Vision for Action and Perception; Visual Attention.

Further Reading

Araujo C, Kowler E, and Pavel M (2001) Eye movements during visual search: The costs of choosing the optimal path. *Vision Research* 41: 3613–3625.

Corbetta M, Akbudak E, Conturo TE, et al. (1998) A common network of functioning areas for attention and eye movements. *Neuron* 21: 761–773.

DeCarlo D and Santella A (2002) Stylization and abstraction of photographs. *ACM Transactions on Graphics* 21: 769–776.

Deubel H and Schneider WX (1996) Saccade target selection and object recognition: Evidence for a common attentional mechanism. *Vision Research* 36: 1827–1837.

Epelboim J and Suppes P (2001) Eye movements during problem solving in geometry. *Vision Research* 41: 1561–1574.

Gardner JL and Lisberger SG (2001) Linked target selection for saccadic and smooth pursuit eye movements. *Journal of Neuroscience* 21: 2075–2084.

Gersch TM (2007) *The Attentional Filter Used during Scanning Influences Both Perception and Memory.* Doctoral Thesis, Rutgers University.

Gersch T, Kowler E, and Dosher B (2004) Dynamic allocation of attention during sequences of saccades. *Vision Research* 44: 1469–1483.

Hayhoe M and Ballard D (2005) Eye movements in natural behavior. *Trends in Cognitive Science* 9: 188–194.

Hooge ITHC, Over EAB, van Wezel RJA, and Frens MA (2005) Inhibition of return is not a foraging facilitator in saccadic search and free viewing. *Vision Research* 45: 1901–1908.

James W (1890/1950) *Principles of Psychology*, Vol. 1, p. 402. New York: Dover.

Khurana B and Kowler E (1987) Shared attentional control of smooth eye movements and perception. *Vision Research* 27: 1603–1618.

Koch C and Ullman S (1985) Shifts in selective visual attention: Towards the underlying neural circuitry. *Human Neurobiology* 4: 219–227.

Kowler E (1990) The role of visual and cognitive processes in the control of eye movement. In: Kowler E (ed.) *Eye Movements and Their Role in Visual and Cognitive Processes*, pp. 1–70. Amsterdam: Elsevier.

Kowler E, Anderson E, Dosher B, and Blaser E (1995) The role of attention in the programming of saccades. *Vision Research* 35: 1897–1916.

Kowler E, Van der Steen J, Tamminga EP, and Collewijn H (1984) Voluntary selection of the target for smooth eye movements in the presence of superimposed, full-field stationary and moving stimuli. *Vision Research* 24: 1789–1798.

Lipps M and Pelz JB (2004) Yarbus revisited: Task-dependent oculomotor behavior [abstract]. *Journal of Vision* 4: 115a.

Lisberger SG and Ferrera VP (1997) Vector averaging for smooth pursuit eye movements initiated by two moving targets in monkeys. *Journal of Neuroscience* 17: 7490–7502.

Liston D and Krauzlis RJ (2003) Shared response preparation for pursuit and saccadic eye movements. *Journal of Neuroscience* 23: 11305–11314.

Mach E (1906/1950) *Principles of Psychology, Vol. 1, The Analysis of Sensations.* New York: Dover.

Melcher D and Kowler E (2001) Visual scene memory and the guidance of saccadic eye movements. *Vision Reearch* 41: 3597–3611.

Murphy BJ, Kowler E, and Steinman RM (1975) Slow oculomotor control in the presence of moving backgrounds. *Vision Research* 15: 1263–1268.

Santella A and DeCarlo D (2004) Visual interest and NPR: An evaluation and manifesto. In: *Proceedings of the Third International Symposium on Nonphotorealistic Animation and Rendering*, pp. 71–78. Annecy, France, 7–9. June New York: ACM Press.

Schnitzer BS and Kowler E (2006) Eye movements during multiple readings of the same text. *Vision Research* 46: 1611–1632.

Sheinberg DL and Logothetis NK (2001) Noticing familiar objects in real world scenes: The role of temporal cortical neurons in natural vision. *Journal of Neuroscience* 21: 1340–1350.

Sperling G and Dosher BA (1986) Strategy and optimization in human information processing. In: Boff KR, Kaufman L, and Thomas JP (eds.) *Handbook of Perception and Human Performance Vol.1: Sensory Processes and Perception.* New York: Wiley.

Steinman RM, Menezes W, and Herst AN (2005) Handling real forms in real life. In: Jenkin MRM and Harris LR (eds.) *Seeing Spatial Form*, pp. 187–212. New York: Oxford University Press.

Yarbus A (1967) *Eye Movements and Vision.* New York: Plenum Press.

Relevant Website

http://www.philip.greenspun.com – Philip Greenspun.

Attention Deficit Hyperactivity Disorder

L Tamm, University of Texas Southwestern Medical Center, Dallas, TX, USA

Introduction

Attention-deficit/hyperactivity disorder (ADHD) is the most prevalent disorder in childhood. Children with ADHD demonstrate age-inappropriate difficulties maintaining attention and concentration, as well as hyperactivity and impulsivity. These deficits occur in multiple settings and result in impairments in academic performance and social functioning. In approximately 60% of children with ADHD, symptoms persist into adolescence and may even continue into adulthood. Negative sequelae of ADHD include academic underachievement, substance abuse, mood disorders, antisocial behavior, and higher rates of accidents and hospitalizations. This article provides an overview of ADHD, covering diagnosis, prevalence, neurobiology and etiology, and current empirically supported treatments.

Diagnosis

All children are sometimes restless, sometimes act without thinking, and sometimes daydream. Because everyone shows these behaviors at one time or another, the guidelines for determining whether a person has ADHD are very specific. When hyperactivity, distractibility, poor concentration, or impulsivity begin to affect performance in school, social relationships with other children, or behavior at home (i.e., cause functional impairment), ADHD may be suspected.

While there are no biological, physiological, or genetic markers or independently valid tests that can reliably identify the disorder, the scientific consensus is that a diagnosis of ADHD can be made reliably using well-tested diagnostic interview methods. Currently, the *Diagnostic and Statistical Manual of Mental Disorders*, Fourth Edition (DSM-IV), is utilized as the basis for diagnostic process in ADHD in the United States. The DSM-IV includes categorizations for Predominantly Inattentive Type, Predominantly Hyperactive – Impulsive Type, and Combined Type within the ADHD diagnosis. **Table 1** lists the DSM-IV symptoms for ADHD.

Determining if a child has ADHD is a multifaceted process. Many biological and psychological problems can contribute to symptoms similar to those exhibited by children with ADHD. For example, anxiety, depression, and certain types of learning disabilities may cause similar symptoms. Thus, a comprehensive evaluation is necessary to establish a diagnosis, rule out other causes, and determine the presence or absence of coexisting conditions. Such an evaluation should include a clinical assessment of the individual's academic, social, and emotional functioning and developmental level. A careful history should be taken from the parents, the teachers, and, when appropriate, the child. Checklists for rating ADHD symptoms and ruling out other disabilities are often used by clinicians.

Information about each of the DSM-IV ADHD symptoms is collected to determine presence or absence of the disorder. In addition, the behaviors must create significant difficulty in at least two areas of life, such as home, social settings, school, or work. Symptoms must be present for at least 6 months. Different symptoms may appear in different settings, depending on the demands the situation may pose for the child's self-control. Symptoms of ADHD will appear over the course of many months, often with the symptoms of impulsiveness and hyperactivity preceding those of inattention, which may not emerge for a year or more, or until the child begins school, where demands on attention are increased.

Neuropsychological or psychological assessment of executive functioning is also useful, as children with ADHD typically perform poorly on tests of executive function, including inhibition and working memory tasks. A medical exam by a physician is important and should include a thorough physical examination, including hearing and vision tests, to rule out other medical problems that may be causing symptoms similar to ADHD (e.g., thyroid problems).

In children and teenagers, the symptoms must be more frequent or severe than in other children the same age. In very young children, more stringent criteria may be applied (e.g., increasing the duration of symptoms from 6 to 9 months, considering moderate to severe symptomatology by multiple raters, and using higher ratings of impairment). Diagnosing ADHD in an adult requires an examination of childhood academic and behavioral history as well as reviewing current symptoms. In adults, the symptoms must affect the ability to function in daily life and persist from childhood.

Prevalence

The prevalence rates for ADHD vary depending on diagnostic practices (i.e., structured interview, inclusion

Table 1 DSM-IV criteria for ADHD

I. Either A or B:

A. Six or more of the following symptoms of inattention have been present for at least 6 months, to a point that is disruptive and inappropriate for developmental level:

- Inattention

1. Often does not give close attention to details or makes careless mistakes in schoolwork, work, or other activities.
2. Often has trouble keeping attention on tasks or play activities.
3. Often does not seem to listen when spoken to directly.
4. Often does not follow instructions and fails to finish schoolwork, chores, or duties in the workplace (not due to oppositional behavior or failure to understand instructions).
5. Often has trouble organizing activities.
6. Often avoids, dislikes, or doesn't want to do things that take a lot of mental effort for a long period of time (such as schoolwork or homework).
7. Often loses things needed for tasks and activities (e.g., toys, school assignments, pencils, books, or tools).
8. Is often easily distracted.
9. Is often forgetful in daily activities.

B. Six or more of the following symptoms of hyperactivity-impulsivity have been present for at least 6 months to an extent that is disruptive and inappropriate for developmental level:

- Hyperactivity

1. Often fidgets with hands or feet or squirms in seat.
2.Often gets up from seat when remaining in seat is expected.
3.Often runs about or climbs when and where it is not appropriate (adolescents or adults may feel very restless).
4. Often has trouble playing or enjoying leisure activities quietly.
5. Is often 'on the go' or often acts as if 'driven by a motor.'
6. Often talks excessively.

- Impulsivity

1. Often blurts out answers before questions have been finished.
2. Often has trouble waiting one's turn.
3. Often interrupts or intrudes on others (e.g., butts into conversations or games).
4. Some symptoms that cause impairment were present before age 7 years.
5. Some impairment from the symptoms is present in two or more settings (e.g., at school/work and at home).
6. There must be clear evidence of significant impairment in social, school, or work functioning.
7. The symptoms do not happen only during the course of a pervasive developmental disorder, schizophrenia, or other psychotic disorder. The symptoms are not better accounted for by another mental disorder (e.g., mood disorder, anxiety disorder, dissociative disorder, or a personality disorder).

II. Based on these criteria, three types of ADHD are identified:

ADHD, Combined Type: if both criteria 1A and 1B are met for the past 6 months.

ADHD, Predominantly Inattentive Type: if criterion 1A is met but criterion 1B is not met for the past 6 months.

ADHD, Predominantly Hyperactive–Impulsive Type: if criterion 1B is met but criterion 1A is not met for the past 6 months.

of data from both parents and teachers), sampling techniques, and sampling populations (school, clinic, community). A behavioral definition of ADHD, based on symptoms shown at one point in time (which does not indicate actual psychiatric disorder), is found in 10–20% of the general population in several countries. A psychiatric definition in the DSM tradition, with specific inclusion criteria for symptom onset, duration, pervasiveness, and impairment, is found in 5–10% of the general population, whereas this frequency is 1–2% with the International Classification of Diseases (ICD) tradition, which restricts diagnosis to the full syndrome with limited comorbidity. The prevalence rate for adults is estimated at 4%, for preschoolers, 2–5%; and for the school-aged population, 3–5% using the DSM.

ADHD has been extensively studied in the US, and the predominance of American research in this field and apparent differences in the prevalence of ADHD or hyperkinesis, as defined by ICD, has also led to the impression that ADHD is largely an American disorder. However, a review of the prevalence literature indicates the rate of ADHD is at least as high in many non-US children as in US children, with the highest prevalence rates being seen when using DSM-IV diagnoses, compared with previous versions (DSM-III and DSM-III-R).

With regard to gender, boys are nearly three times as likely as girls to receive a diagnosis of ADHD. With regard to race and ethnicity, prevalence rates seem to be somewhat lower in Hispanic and African-American families compared to White non-Hispanic families, who were almost twice as likely to receive a diagnosis of ADHD. These prevalence differences do not seem to be accounted for by birth weight, income, and insurance coverage.

Neurobiology

The neurobiology of ADHD is not completely understood, however, imbalances in dopaminergic and noradrenergic systems have been implicated in the core symptoms that characterize this disorder. Although there are inconsistencies among studies, there is converging evidence to suggest that deficits in frontal lobe function and the connections between the frontal lobe and key subcortical regions underlie this disorder. Because of the complexity of prefrontal circuitry, it is not yet clear whether the prefrontal abnormalities in ADHD are secondary to abnormalities of the prefrontal cortex or to brain areas with prefrontal projections.

Structural imaging studies, with computerized tomography or magnetic resonance imaging (MRI), suggest structural brain abnormalities among ADHD patients, with the most common findings being smaller whole brain volumes, and smaller gray-matter volumes in frontal cortex, cerebellum, and subcortical structures (e.g., caudate, globus pallidus). These differences have been shown to remain stable with age for ADHD patients independent of medication status.

Functional magnetic resonance imaging (fMRI) studies are consistent with structural studies in implicating the frontosubcortical systems. The caudate, putamen, and globus pallidus subcortical structures are part of the neural circuitry underlying motor control, executive functions, inhibition of behavior, and the modulation of reward pathways. These frontal–striatal–pallidal–thalamic circuits provide feedback to the cortex for the regulation of behavior.

Functional imaging studies also implicate the cerebellum and corpus callosum in the pathophysiology of ADHD. The cerebellum contributes significantly to cognitive functioning, presumably through cerebellar-cortical pathways involving the pons and thalamus. The corpus callosum connects regions of the two cerebral hemispheres. Size variations in the callosum and volume differences in the number of cortical neurons might degrade communication between these two hemispheres, which might account for some of the cognitive and behavioral symptoms of ADHD.

Etiology

No single etiology for ADHD has been determined to date. A vast majority of studies indicate a strong genetic contribution to ADHD, with heritability rates estimated at 0.6 to 0.8 and an approximately fivefold elevated risk for ADHD in first-degree relatives. However, it is clear that other factors contribute. In a diathesis–stress model, a genetic vulnerability or predisposition (diathesis) interacts with the environment and life events (stressors) to trigger behaviors or psychological disorders. A diathesis–stress model probably best explains the etiology of ADHD.

Environmental Agents

Alcohol and cigarette use during pregnancy has been shown to increase risk for ADHD in the offspring of that pregnancy. High levels of lead in the body also seem to be associated with risk for ADHD, particularly in preschoolers. However, lead does not account for the majority of ADHD cases, and many children with high lead exposure do not develop ADHD. Furthermore, recent regulations prohibiting lead in paint have resulted in significantly fewer cases of lead exposure (except for those who live in older buildings).

Food and Food Additives

Popular lore suggests that there may be a relationship between various foods, food additives, and sugar and ADHD symptoms. However, no controlled trial has demonstrated a causal relationship between ADHD and food intake. For example, 12 double-blind, placebo-controlled studies of sugar challenges failed to provide any evidence that sugar ingestion leads to untoward behavior in children with ADHF or in normal children. Likewise, no studies have found any negative effect of candy or chocolate on behavior. Furthermore, diet restrictions were only shown to help approximately 5% of children with ADHD (most of these with food allergies), and sugar restrictions were not shown to have any impact on core ADHD symptoms.

Genetics

Data from family, twin, and adoption studies show that genes play a significant role in the etiology of ADHD. The familial risk of ADHD is two- to eightfold greater in parents and siblings of ADHD probands. Adoption studies have demonstrated that biological relatives of nonadopted children have higher rates of symptoms than do adopted relatives. Furthermore, adopted relatives show a risk similar to that of controls.

Twin studies show perhaps the most compelling data for understanding heritability. Monozygotic twins share 100% of their genes, whereas fraternal twins and other siblings share 50% of their genes. Therefore, heritability can be computed by determining the extent to which monozygotic twins are more concordant for ADHD compared with fraternal twins. A pooled analysis of 20 international twin studies has revealed the heritability of ADHD to be 0.76, thus demonstrating that ADHD is one of the most heritable psychiatric disorders. Although the mode of inheritance of ADHD is unknown, it is likely to be polygenic based on its modest relative risks and high population prevalence. The individual risk contribution per gene, therefore, may be quite small.

Two approaches are used to evaluate the genetic etiology of ADHD: (1) the genome scan, which examines all chromosomal locations without *a priori* guessing as to which genes underlie ADHD, and (2) the candidate gene approach, which examines one or more genes based on theoretical considerations and empirical evidence from neurobiological studies. Few genome-wide linkage scan studies have been conducted, in part due to the large number of study participants required for sufficient power. The convergence of data from the studies which have been conducted implicate chromosomal regions 5p13, 16p13, and 17p11, although other linkage studies have implicated 7p, 10q26, 12q23, 15q, 15q15, 5p12, and 11q22–25. There are inherent challenges to drawing inferences from few studies with relatively small samples, in which it is likely that the genomic regions suggestive of linkage will differ appreciably across studies for statistical reasons alone. That is, although there may be other reasons for the discrepant findings, such as differences in the populations sampled or in the assessment or diagnostic methods used, the stochastic fluctuations associated with few studies of small sample size are sufficient to cause such discrepancies. Additional research is required to substantiate and replicate these linkages, as well as their contribution to ADHD.

Candidate gene studies have particularly focused on neurotransmitters such as dopamine and norepinephrine. These neurotransmitters are found in abundance in the prefrontal cortex and are thought to mediate executive function and hyperactive/impulsive behaviors. Furthermore, these neurotransmitters are implicated in the mechanism of action in pharmacological agents used to treat ADHD. Thus, genes that produce these neurotransmitters and arbitrate their function are logical candidates for investigation in ADHD probands.

Candidate genes for neurotransmitter systems may include (1) precursor genes that affect the rate at which neurotransmitters are produced from precursor amino acids (e.g., tyrosine hydroxylase for dopamine), (2) receptor genes that are involved in receiving neurotransmitter signals (e.g., genes corresponding to the five dopamine receptors, DRD1, DRD2, DRD3, DRD4, and DRD5), (3) transporter genes that are involved in the reuptake of neurotransmitters back into the presynaptic terminal (e.g., DAT1), (4) metabolite genes that are involved in the metabolism or degradation of these neurotransmitters (e.g., the gene for catechol-O-methyl-transferase, COMT), and (5) genes that are responsible for the conversion of one neurotransmitter into another (e.g., dopamine β-hydroxylase, which converts dopamine into norepinephrine). The majority of candidate gene studies in ADHD have implicated dopamine transport (DAT), dopamine β-hydroxylase (DBH), dopamine receptor D4 (DRD4), and dopamine receptor D5 (DRD5) in the pathogenesis of ADHD. By far, the gene most strongly implicated in ADHD is the seven-repeat allele of the DRD4 gene, confirming a strong dopamine component in the pathogenesis of ADHD. In addition, emerging evidence for other candidate genes has implicated serotonin transporter 5-hydroxytryptamine (5-HT), serotonin receptor 1B (HTR1B), and synaptosomal-associated protein 25 (SNAP-25).

Although genetic studies have contributed significantly to the understanding of ADHD, none of the studies accounted for more than 5% of ADHD. It is likely that several gene polymorphisms with moderate to small effect sizes contribute to the phenotype of ADHD; different combinations of such predisposing variants presumably underlie ADHD in different individuals. It is also possible that some of the discrepancies in the genetic studies could be due to differences in geographic, clinical characteristics and sample size used in the experiments. Therefore, large samples for molecular genetic studies are mandatory to detect these polymorphisms. Accordingly, several of today's findings have to be regarded as preliminary. The understanding of ADHD's neurobiology may be advanced by new technologies, such as single nucleotide polymorphism (SNP)-based genome scans performed with gene chips comprising 10 000–1 000 000 SNPs, as well as using more sophisticated animal model designs.

Assessments of comorbid psychiatric disorders in family studies further support a polygenic theory for ADHD. Results of analyses from independent samples of children with ADHD suggest that (1) ADHD and major depression share common familial vulnerabilities (e.g., children with ADHD often have mothers with depression) and (2) ADHD children with comorbid conduct disorders and bipolar disorders might be a distinct familial subtype of ADHD. Thus, dividing ADHD samples into those with conduct and bipolar disorders and those without might help define more familially homogeneous subgroups.

Treatment

Stimulant Medication

The most common treatment for ADHD is medication, with the vast majority of children receiving some form of stimulant medication (e.g., methylphenidate or Ritalin). Stimulant medication is effective in 70–80% of children receiving medication, and safety and efficacy have been well documented. Of those children who do not respond to one stimulant medication, approximately 20–25% respond to an alternative stimulant medication. Recent trends in stimulant medication treatments include long-acting formulations (to reduce stigma and need to administer medication in the school setting) and patches (to reduce the need for oral administration and decrease the dosing schedule).

The frontosubcortical systems pathways associated with ADHD are rich in catecholamines, which are involved in the mechanism of action of stimulant medications used to treat this disorder. Stimulants, such as methylphenidate, seem to reduce core ADHD symptoms (i.e., inattention, hyperactivity, impulsivity) by inhibiting the dopamine transporter and blocking dopamine and norepinephrine reuptake into the presynaptic neuron, thereby increasing the release of these monoamines into the extraneuronal space. Changes in dopaminergic and noradrenergic function seem to be necessary for the clinical efficacy of pharmacologic treatments of ADHD. A plausible model for the effects of medications in ADHD suggests that, through dopaminergic and/or noradrenergic pathways, these agents increase the inhibitory influences of frontal cortical activity on subcortical structures. In contrast, effects on serotonin metabolism seem to be marginally related to the clinical efficacy of ADHD treatments, and serotonergic drugs have few effects in mitigating core symptoms of the disorder.

Side effects Most side effects of the stimulant medications are minor and are usually related to the dosage of the medication being taken. Higher doses produce more side effects. The most common side effects are decreased appetite, insomnia, increased anxiety, and/or irritability. Some children report mild stomach aches or headaches. A few children develop tics during treatment. Recent studies show consistent use of stimulant medication is associated with mild growth suppression (particularly height). Rare but potentially severe adverse effects, including sudden cardiac death and cancer, following long-term treatment have been reported; however, these effects have not been adequately demonstrated to be of significant concern at this time.

Nonstimulant Medication

In some individuals it may be necessary to try nonstimulant medication treatments. Reasons include (1) stimulants do not relieve symptoms, (2) stimulants cause intolerable side effects, (3) comorbid medical problems contraindicate treatment with stimulants, or (4) comorbid psychiatric conditions suggest treatment with nonstimulants.

A common nonstimulant alternative is atomoxetine (Strattera), which is a selective norepinephrine reuptake inhibitor. Its mechanism of action in ADHD seems to be the highly specific presynaptic inhibition of norepinephrines. Atomoxetine does not appear to affect the dopamine systems as directly as do the stimulants. Given its pharmacokinetic half-life of 5 h, it is generally dosed twice a day. Atomoxetine may also reduce tics and may be effective in children with ADHD who have comorbid anxiety. The Food and Drug Administration has issued warnings regarding rare side effects of hepatotoxicity and suicidal ideation.

Antihypertensives (e.g., Tenex), designed to treat high blood pressure, can also help control aggressive and impulsive behaviors in some people and are a second line of treatment for ADHD. Alternatively, antidepressants (e.g., Wellbutrin) may be needed if psychostimulants do not improve symptoms.

Nonmedication Interventions

Stimulant medications are particularly effective in remediating ADHD symptomatology. However, medications do not cure ADHD, nor do they increase knowledge or improve academic skills. The medications help the child to use those skills he or she already possesses, and allow the child sufficient self-regulation to learn new skills. Moreover, there is not compelling evidence to suggest that the stimulants improve the rather guarded long-term prognosis of the disorder. Thus, nonmedication treatments are also warranted in the treatment of ADHD. The most common and successful treatments for ADHD include behavior therapy (reinforcement and environmental controls), parent training, and social skills training. Due to the large evidence base consisting of rigorous experimental investigations, behavioral parent training and behavioral classroom interventions have been designated 'empirically supported' psychosocial treatments for ADHD.

Treatment sequencing Nonmedication interventions should be started early in a child's life and, ideally, before treatment with stimulant medication. Parent training should be considered a first-line treatment for ADHD, particularly with younger children, as there is evidence that such intervention may prevent later school problems, in addition to improving overall family functioning and parental ability to manage behavioral problems. At school entry, behavioral classroom interventions should be implemented as soon as possible. If treatment with stimulant medication is initiated, it should be administered in conjunction with these effective nonmedication treatments for ADHD. A large multisite randomized control trial showed that maintenance of gains was better in children who received combination behavioral and medication treatment compared to those receiving medication only. Furthermore, combined treatments are effective in targeting other dysfunctions associated with ADHD, including oppositional behavior and anxiety.

Behavioral parent training A large evidence base exists for the use of parent behavioral interventions to reduce ADHD symptoms, to improve parenting skills and parent sense of competence, and to diminish family distress. Behavioral parent training interventions are based on a foundation of social learning principles that teach the child socially acceptable behavior. Parents and primary caregivers are trained in contingency management strategies, emphasizing behavior modification, cues, and consequences, reward systems, and discipline. Parents learn how to identify and manipulate the antecedents and consequences of a child's behavior (functional behavioral analysis), target and monitor problematic behaviors, reward positive and prosocial behavior through praise, positive attention, and tangible rewards, and decrease unwanted behavior through planned ignoring, time out, response cost, and other nonphysical discipline techniques. These approaches focus on reducing any positive reinforcement (e.g., parental attention) provided to the child for engaging in disruptive/defiant behavior, while simultaneously increasing the reinforcement parents provide for appropriate and compliant behavior. Parents are trained to provide punishment contingent on the display of disruptive or unacceptable behavior in a predictable, contingent, and immediate (to the behavior which precedes it) manner. Courses typically include training in specific evidence-based techniques for giving commands, reinforcing adaptive and positive social behaviors of the child while ignoring minor inappropriate behaviors to reduce or eliminate them, training in techniques for establishing and enforcing rules and establishing time-out procedures, training on initiating a point system with reward and response cost, and training in how to enforce contingencies across settings, problem-solving techniques, and strategies for maintenance and relapse prevention.

Behavioral classroom interventions As with parent training, behavioral classroom interventions generally involve regular consultation with the child's teacher regarding the use of behavior modification strategies. Consultation usually begins with psychoeducation about ADHD and identification of specific target behaviors, based upon a functional assessment of behavior (i.e., examination of antecedents, behaviors, and consequences). Teachers are then instructed regarding the use of specific behavioral techniques, including praise, planned ignoring, effective commands, and time out, as well as the daily report card (DRC) and/or more extensive individualized or classroom-wide contingency management programs. The consultant and teacher collaboratively develop specific behavioral interventions. Research suggests that teachers widely implement behavioral classroom interventions that target ADHD symptoms and associated functional difficulties, such as complying with classroom rules, engaging in appropriate interactions with classmates, displaying disruptive behavior, and complying with teacher commands. Furthermore, direct contingency management strategies employed in the classroom setting have been shown to be more effective than is traditional outpatient treatment for ADHD-related behaviors. A meta-analysis found that behavioral classroom interventions showed a very large effect size (ES = 1.44) on measures of treatment outcome, with a larger effect on child behavior than on academic or clinic performance.

Social skills training Peer rejection of children with ADHD is quite common and may lead to serious long-term consequences. Further, children with ADHD who overcome their social problems have better outcomes in the long term than do those children who continue to experience problems with peers. Thus, investigators have focused on designing psychosocial interventions that specifically target peer relationships. Social skills training interventions include instruction in social skills, social problem solving, and behavioral competencies. In addition, the interventions attempt to enhance social competence by encouraging close friendships, and decreasing undesirable and antisocial behaviors. Interestingly, although children with ADHD are less socially effective than are their peers, they typically do not perceive themselves as such. Thus, the primary goal of social skills training is to promote prosocial behaviors that include cooperation, communication, participation, and validation. Social skills training represents the most common approach to treating social problems in children, with groups typically being conducted at a clinic, a summer treatment program, or in school-based settings, and often including parent and teacher participation. The literature shows that combining social skills training with other interventions (e.g., behavior management, parent training) is more effective than is providing it as a stand-alone treatment.

Alternative therapies The following promising interventions must be tested in controlled, randomized trials in order to become established as empirically supported or evidence-based treatments.

• *Academic interventions*. Academic interventions for ADHD focus primarily on manipulating antecedent conditions such as academic instruction or materials in order to improve both behavioral and academic outcomes. Task and instructional modifications, peer tutoring, computer-assisted instruction, and strategy training have all been shown to improve outcomes in ADHD in preliminary studies. A recent meta-analysis of school-based interventions for children with ADHD found that both behavior management and academic interventions had similar positive effects on ADHD-related behaviors, although the impact of academic interventions on academic performance was more difficult to discern due to the relatively few studies employing academic outcome measures. Additional research is warranted to document the efficacy of academic interventions.

• *Attention training (ATT)*. ATT is an innovative intervention based on the concept that adaptive activities can be directed to enhance skills related to sustained and executive attention functions. ATT originated from the field of cognitive rehabilitation. ATT activities emphasize many different aspects of attention, including sustained, selective, divided, alternating, working memory, self-regulation, overall body awareness, and metacognitive strategies. Most ATT approaches are adaptive, continually increasing in difficulty to challenge a child and provide practice in the domain being trained. Improvements in executive function, untrained academic skills, sustained attention, cognitive function, inhibition, and intelligence have been reported in various ATT studies with children and adolescents. Furthermore, transfer of improvements (or generalization) to untrained areas in individuals with ADHD has also been reported. These studies provide support for the notion that ATT may positively impact the developing attention skills of children with ADHD, and such increases may generalize to ecologically valid assays of real-world effortful task performance and expression of ADHD symptoms. However, more controlled clinical trials are needed before ATT can be endorsed as an effective, reliable treatment.

• *Biofeedback/neurofeedback*. Among the alternative treatments suggested for ADHD are electroencephalographic (EEG) biofeedback (neurofeedback or neurotherapy) and metacognitive therapy. These approaches are a form of behavioral training aimed at developing skills for self-regulation of brain activity. EEG biofeedback attempts to treat ADHD by decreasing g their slow wave activity and/or increase their fast-wave EEG activity, often using behavioral principles such as operant condition (i.e., positive reinforcement). Essentially, electrodes are positioned on the patient's head; a computer detects EEG activity and provides auditory or visual feedback in the targeted frequency bands. When the person is producing the desired EEG pattern, the computer provides a positive response or reward. After 20–50 training sessions, the person is hypothesized to be able to produce the desired brain activity independently via his or her own awareness of physiological processes. Despite some promising results, treatment effects may be due to nonspecific or placebo effects. Despite anecdotal and preliminary evidence for these alternative treatments, reviews of the applicability of neurofeedback for ADHD generally conclude that more controlled clinical trials are needed before it can be endorsed as an effective, reliable treatment. Studies are necessary to address flaws such as small sample size, diagnostic status of participants, lack of control or placebo procedures, failure to use blinding techniques as to subject status, practice effects, and failure to use random assignment. Additionally, few studies have been published in journals that are subject to rigorous peer review. Additional research is required to document the efficacy of this approach, as well as the effective agents of change in ADHD. Thus a National Institutes of Health consensus report on ADHD treatment noted that the empirical evidence for biofeedback and neurofeedback treatments was uneven and recommended more controlled studies before this treatment could be endorsed.

See also: Attention Deficit Hyperactivity Disorder (ADHD): Methylphenidate (Ritalin) and Dopamine; Attention: Models; Attentional Mechanisms in Ventral Pathway; Attentional Networks; Attentional Networks in the Parietal Cortex; Attentional Functions in Learning and Memory; Sensorimotor Integration: Attention and the Premotor Theory; Visual Attention.

Further Reading

Bush G, Valera EM, and Seidman LJ (2005) Functional neuroimaging of attention-deficit/hyperactivity disorder: A review and suggested future directions. *Biological Psychiatry* 57: 1273–1284.

Chronis AM, Jones HA, and Raggi VL (2006) Evidence based psychosocial treatments for children and adolescents with attention deficit/hyperactivity disorder. *Clinical Psychology Review* 26: 486–502.

Conner DF (2002) Preschool attention deficit hyperactivity disorder: A review of prevalence, diagnosis, neurobiology, and stimulant treatment. *Developmental and Behavioral Pediatrics* 23: S1–S9.

Himelstein J, Schulz KP, Newcorn JH, et al. (2000) The neurobiology of attention-deficit hyperactivity disorder. *Frontiers in Bioscience* 5: 461–478.

Hinshaw SP (2007) Moderators and mediators of treatment outcome for youth with ADHD: Understanding for whom and how interventions work. *Ambulatory Pediatrics* 7: 91–100.

Khan SA and Faraone SV (2006) The genetics of ADHD: A literature review of 2005. *Current Psychiatry Reports* 8: 393–397.

Krain AL and Castellanos FX (2006) Brain development and ADHD. *Clinical Psychology Review* 26: 433–444.

Spencer TJ, Biederman J, and Mick E (2007) Attention-deficit/hyperactivity disorder: Diagnosis, lifespan, comorbidities, and neurobiology. *Ambulatory Pediatrics* 7: 73–81.

Waldman ID and Gizer IR (2006) The genetics of attention deficit hyperactivity disorder. *Clinical Psychology Review* 26: 396–432.

Relevant Websites

http://www.chadd.org – Children and Adults with Attention Deficit/Hyperactivity Disorder.

http://www.nimh.nih.gov – National Institute of Mental Health, National Institutes of Health, US Department of Health and Human Services: Attention Deficit/Hyperactivity Disorder booklet.

Attention Deficit Hyperactivity Disorder (ADHD): Methylphenidate (Ritalin) and Dopamine

C J Vaidya, Georgetown University and Children's National Medical Center, Washington, DC, USA
P S Lee, Georgetown University, Washington, DC, USA

Attention-deficit hyperactivity disorder (ADHD) is observed in 3–9% of school-aged children and 4% of adults worldwide at higher rates in males than females (2.5:1) and in children older than 9 years relative to younger children. Although family, twin, and adoption studies indicate high heritability (0.76), the mode of transmission is unknown but suspected to be polygenic. Molecular genetic studies suggest that susceptibility to ADHD involves multiple small-effect genes coding for proteins involved in catecholaminergic transmission. Catecholaminergic etiology is consistent with the treatment of choice for ADHD – psychostimulants such as amphetamines and methylphenidate hydrochloride (MPH) that increase synaptic levels of dopamine and norepinephrine by somewhat different mechanisms. Psychostimulants are highly effective for temporary alleviation of symptoms, starting at 30 min and peaking 60–90 min following oral administration of immediate-release formulations. Attempts to elucidate the therapeutic efficacy of MPH for ADHD have shaped current hypotheses about the central role of catecholamines, particularly dopamine (DA), in the neuropathophysiology of ADHD.

Behavioral Profile of ADHD

Clinical Characteristics

Based on the *Diagnostic and Statistical Manual of Mental Disorders–Fourth Edition* (*DSM-IV*), diagnostic criteria for ADHD include parent and teacher reports of symptoms of inattention (e.g., making careless mistakes and difficulty concentrating), hyperactivity (e.g., fidgeting and excessive running or climbing), and/or impulsivity (e.g., difficulty waiting one's turn and talking excessively). These symptoms must appear prior to age 7 years, persist for at least 6 months in two settings (e.g., school and home), and cannot be explained by other psychiatric or neurological conditions. Symptom presentation varies widely among children and although the definition of subtypes in *DSM-IV* (e.g., predominantly inattentive, predominantly hyperactive/impulsive, and combined type with both inattention and hyperactivity/impulsivity) recognizes the heterogeneity, it does not fully account for it.

Phenotypic heterogeneity in ADHD stems from a variety of sources. First, assessment of symptoms varies across clinical settings such that primary care physicians rely on parent/teacher reports and direct observation of behaviors but specialty care settings (e.g., neuropsychology services) supplement this information with performance-based measures of cognitive functioning, academic achievement, and symptoms (e.g., Test of Variables of Attention (TOVA)) (**Table 1**). Performance-based measures alone, however, have limited discriminant validity for diagnostic use. Second, the severity or nature of symptom presentation may depend on gender (e.g., greater motor restlessness in boys but excessive talking in girls), age (e.g., reduced motor restlessness in adolescence relative to childhood), and context (e.g., reduced symptoms in structured environments). Third, ADHD symptoms co-occur with other psychiatric conditions such as mood disorders (e.g., depression and anxiety), conduct disorder, oppositional defiant disorder, obsessive–compulsive disorder, Tourette's, developmental dyslexia, autism spectrum disorders, as well as nonpsychiatric conditions such as sleep disorders. This complex pattern of comorbidity is difficult to resolve because ADHD symptoms could be primary or secondary or define a distinct phenotype that includes the comorbid disorder.

Cognitive Characteristics

It is commonly believed that the primary cognitive domain impaired in ADHD is executive function, an umbrella term that subsumes multiple dissociable 'top-down' operations that control attention and actions in the service of goal-directed behavior. Operations that have been the focus of ADHD research include inhibitory control (e.g., suppressing prepotent or irrelevant responses), working memory (e.g., temporary maintenance and/or manipulation of information), sustained attention or vigilance, and switching of task set (e.g., adapting to changing task demands). Studies using neuropsychological tasks of executive function that involve multiple operations (e.g., Tower of Hanoi, a problem-solving task requiring set maintenance and switching, working memory, and inhibition of inappropriate responses) indicate reduced performance in ADHD relative to control subjects. A meta-analysis of 83 studies involving 6700 subjects reported reductions in all executive operations sampled with effect sizes ranging from 0.4 to 0.7. Furthermore, effect sizes were larger for studies of spatial working memory

Table 1 Neuropsychological and laboratory measures of clinical symptoms exhibited by children with ADHD

Symptom	*Measure*	
	Neuropsychological	*Laboratory*
Inattention	Test of Variables of Attention (TOVA) Continuous Performance Test (CPT) Test of Everyday Attention for Children (TEA-Ch)–Score! TEA-Ch–Sky Search/Map Mission Finger-Windows test–spatial span Letter/number cancellation Symbol Search subtest of WISC-IV	N-back working memory Sternberg memory search
Impulsivity	TEA-Ch–Walk, Don't Walk TEA-Ch–Opposite Worlds Stroop task Eriksen flanker task Antisaccade task	Stop Signal task Go/No-go task
Hyperactivity	None	None

WISC-IV, Wechsler Intelligence Scale for Children–Fourth Edition.

($d = 0.85-1.14$). Drawing conclusions from these studies about which executive operations may be core deficits in ADHD is challenging because most neuropsychological tasks do not measure executive operations in a process-pure manner.

An influential theory of ADHD has proposed that a core deficit of response inhibition underlies executive dysfunction in ADHD. Response inhibition can be measured in a relatively process-pure manner using tasks that require subjects to withhold a prepotent response to a cue that occurs unpredictably following initiation of a response on the Stop Signal task or infrequently on the Go/No-go task. A meta-analysis of 17 studies involving 1200 children reported longer Stop Signal reaction times, an index of poor response inhibition, in ADHD relative to control children (effect size, $d = 0.58$). Use of response inhibition tasks in frontal lobe-damaged patients and in functional neuroimaging studies of healthy adults and children indicates involvement of dopamine-rich frontal–striatal circuitry. In ADHD children, those frontal–striatal regions are recruited to a lesser extent relative to controls. Thus, converging behavioral and neuroanatomical findings provide evidence for involvement of dopamine-rich regions underlying response inhibition deficits in ADHD.

Three lines of evidence have called into question the view that executive dysfunction is the cardinal cognitive deficit in ADHD. First, effect sizes of group differences in executive task performance are modest ($d = 0.50–1.0$) and suggest considerable overlap between score distributions of ADHD and control groups. In a study of five executive function measures, only 10% of ADHD children were impaired on all operations, whereas 21% of ADHD and 53% of control children were unimpaired on all operations. Such results have led researchers to conclude that executive dysfunction is neither necessary nor sufficient to account for all cases of ADHD.

Second, a variety of nonexecutive processes show impairments in ADHD subjects. On Go/No-go tasks, responses on Go trials that do not require inhibitory control are slower in ADHD than control children ($d = 0.58$). Reaction time performance of ADHD children deteriorates over the course of the task, is often slow on initial trials suggesting underarousal, and is more variable from trial to trial. Atypical response characteristics may relate to problems with temporal processing because ADHD children perform poorly on tasks of time estimation, time duration, and motor timing.

Third, motivational deficits have gained attention in ADHD with the observation that ADHD subjects are delay averse – that is, they tend to prefer small immediate rewards than to wait for larger delayed rewards. Delay aversion is posited to reflect atypical functioning of dopaminergic reward systems with motor hyperactivity reflecting compensatory behavior in response to unavoidable delays. Indeed, results of functional imaging study of reward function in ADHD showed that relative to controls, adolescents with ADHD showed reduced ventral striatal activation during anticipation of monetary rewards; furthermore, it was negatively associated with severity of impulsive/hyperactive symptoms. Impaired executive function and delay aversion appear to relate independently to ADHD because when tested in the same group of subjects, the two deficits were uncorrelated but together accounted for 90% of the cases. Thus, executive and motivational dysfunction may represent distinct phenotypes of ADHD.

Summary

The behavioral profile of ADHD suggests considerable phenotypic heterogeneity that is not fully captured by current diagnostic taxonomy and cognitive models of ADHD. At the clinical end, reliance on subjective reports of the child's behavior is unlikely to provide fine-grained symptom profiles unless supplemented by performance-based measures derived from cognitive models of ADHD. At the cognitive end, executive and motivational function appears to be independently affected in ADHD. A unifying factor, however, is that dopaminergic pathways figure centrally in the functional anatomy of both executive and motivational functions.

Evidence for Dopaminergic Dysfunction in ADHD

Whereas convergent lines of evidence suggest abnormal dopaminergic transmission in ADHD, its specific nature is debated. Furthermore, it is believed that the pathophysiology of ADHD is complex, with some interaction between dopaminergic and noradrenergic systems. This view comes from findings showing that pure DA agonists such as levadopa do not attenuate ADHD symptoms. Although the nature of its interaction with the noradrenergic system remains to be specified, strides made in understanding dopaminergic dysfunction in ADHD have been useful for elucidating the neurobiology of ADHD. Sources of evidence suggesting dopaminergic dysfunction in ADHD are discussed next.

Therapeutic Action of MPH

The primary source of evidence for hypothesizing dopaminergic dysfunction in ADHD comes from the mechanism of action of MPH. MPH blocks the dopamine transporter (DAT) that reuptakes DA following release, thereby increasing synaptic levels of DA. Indeed, ligand-based imaging studies in primates and humans showed that MPH bound to DAT in concentrations that were highest in the striatum and lower in thalamus, cortex, and cerebellum, leading to increases in magnitude of extracellular DA. At clinically relevant doses of $0.25–0.5\,\mathrm{mg\,kg^{-1}}$, MPH blocked 50–60% of DAT in the striatum. MPH-induced (dose, $0.3\,\mathrm{mg\,kg^{-1}}$) increases in extracellular DA in the striatum were positively correlated with severity of impulsivity and inattention on a performance-based measure (e.g., TOVA) in ADHD adolescents. Thus, *in vivo* imaging in humans provides evidence for increased synaptic DA following inhibition of DAT by MPH and its association with ADHD symptom expression.

Hyper- and hypodopaminergic status in ADHD In light of the pharmacodynamics of MPH, two lines of evidence have been important in shaping current views about dopaminergic dysfunction in ADHD. First, measurements of homovanillic acid (HVA), a metabolite of DA in cerebrospinal fluid (CSF), correlated positively with hyperactivity ratings from parents and teachers, and CSF levels of HVA decreased following MPH (and other stimulants) treatment. This finding suggests that ADHD reflects increased dopaminergic activity. Second, ligand-based imaging studies of DAT, the primary target of MPH, showed increased binding in the striatum in ADHD relative to control subjects in several studies. This finding suggests that ADHD reflects reduced dopaminergic activity due to increased DA reuptake resulting from higher DAT availability.

Both trait- and state-related factors may underlie increased DAT availability in ADHD. Trait-related factors resulting in higher DAT availability include 'hypertrophy' of dopaminergic neurons due to insufficient pruning during development and/or genetic differences. Alternately, increased DAT availability in ADHD may be state related, resulting from adaptive processes compensating for greater DA release, reduced vesicular storage, or other differences in DA transmission. In some studies, DAT availability in the striatum did not differ between ADHD and control subjects but, rather, was reduced in the midbrain of ADHD subjects. Differences in sample characteristics (e.g., nicotine use) and properties of ligands (e.g., detection of internalized or external DAT) may have contributed to the inconsistent findings across studies. It is also likely that heterogeneity among individuals may emerge from some combination of trait- and state-related factors.

Working hypothesis of dopaminergic dysfunction in ADHD Increased DA release and increased DAT availability in ADHD have been reconciled in a model by Seeman and Madras that considers both tonic and phasic dopaminergic activity. Phasic activity reflects acute DA release stimulated by a nerve impulse, whereas tonic activity reflects chronic DA accumulation in synaptic space. The two types of DA activity are reciprocal such that tonic DA activity stimulates autoreceptors on the presynaptic neuron that serve to attenuate phasic DA release. Seeman and Madras posit that tonic DA activity is reduced in ADHD. Findings of elevated DAT in ADHD provide some support for this view because greater reuptake of synaptic DA is likely to result in reduced tonic DA activity. Furthermore, it is posited that low tonic DA activity is unlikely to stimulate presynaptic autoreceptors, resulting in higher phasic release of DA. Greater phasic DA release is consistent with findings of greater HVA levels in CSF of hyperactive children. The therapeutic action of MPH in Seeman and Madras's model is mediated through increased tonic DA and reduced phasic DA. Specifically, they posit that the blockade of DAT reuptake by MPH increases tonic DA activity that leads to increased autoreceptor stimulation, which in turn attenuates phasic DA release. Seeman and Madras's model, therefore, posits that ADHD is characterized by both hypo- and hyperdopaminergic status that is normalized by MPH.

Variability in MPH response Two properties of MPH effects, context dependency and individual variability, should be considered by any model of ADHD pathophysiology. First, anecdotal and clinical reports have emphasized that efficacy of MPH varies across situations. Indeed, in a placebo-controlled study, symptom improvements in ADHD children were greater in a classroom than playground setting. Ligand-based imaging of D2 receptors in healthy adults suggests that context dependency in MPH response is a property of dopaminergic function. Specifically, oral administration of MPH at clinically relevant doses (20 mg) increased extracellular DA in the striatum selectively for a mathematical task with monetary incentives. DA levels were associated with subjects' reports of greater interest and engagement in the mathematical task following MPH administration. These effects were not observed following MPH and placebo administration for a neutral task (viewing nature scenes without monetary incentives) and were not observed on placebo for the mathematical task. Thus, these results suggest stimulus selectivity in MPH-induced enhancement of attentional performance and interest and associated DA modulation.

Second, a majority (60–70%) but not all ADHD individuals respond favorably to MPH. Whereas responders and nonresponders did not differ in symptom characteristics, nonresponders had lower than normal DAT availability. Furthermore, at similar levels of DAT blockade, the magnitude of extracellular DA increase induced by MPH varied across subjects due to differences in rates of DA cell firing. Thus, it is important to consider etiological factors underlying individual differences in DAT availability and DA cell firing.

Candidate Dopaminergic Genes

In light of the molecular targets of stimulant medications and results of transporter imaging studies in ADHD, molecular genetic studies have searched for susceptibility genes relating to catecholamine function. Studies of the dopaminergic system have focused on polymorphisms of D2, D5 receptors – DAT, dopamine beta-hydroxylase (DBH), tyrosine hydroxylase, catechol-O-methyltransferase (COMT), and monoamine oxidase A (MAO-A).

Although findings are not conclusive, the most consistent evidence for association with ADHD comes from polymorphisms of the dopamine transporter gene (DAT1) and D4 receptor gene (DRD4). DAT1, located on chromosome 5p15.3, contains a 40-base pair variable number of tandem repeats ranging from 3 to 13, with 9 and 10 being most common, in the 3′-untranslated region. Greater prevalence of the 10-repeat allele in subjects with ADHD has been reported by many studies but not by all studies. DRD4 contains a 48-base pair variable number of tandem repeats in exon 3, and greater prevalence of the 7-repeat allele in ADHD subjects has been reported by some studies but not by all. Furthermore, some studies have found association of ADHD with other DRD4 alleles (e.g., 2–5 repeat). Mixed findings across studies may result from differences in allelic variability associated with ethnicity of samples, study designs for ascertaining linkage, methods of ADHD diagnosis, and from limited statistical power due to small sample sizes. Examination of odds ratios (a statistic assessing magnitude of association) pooled across studies, a method that circumvents statistical limitations of small samples, showed that polymorphisms of several dopaminergic genes (DRD4, DRD5, DAT, and DBH) confer a statistically significant, albeit small, risk for ADHD.

Genetic polymorphisms have the potential to elucidate dopaminergic pathophysiology of ADHD because allelic variation in DAT1 and DRD4 influences DA transmission. DAT expression was greater for the 10-repeat allele in *in vitro* studies but findings of *in vivo* studies are mixed, reporting higher, lower, and similar striatal DAT expression in homozygous carriers of the 10-repeat allele relative to carriers of the 9-repeat allele. *In vitro* studies of DRD4 alleles indicate that response to DA was reduced for the 7-repeat allele. Selective ligands for D4 receptors are currently unavailable and, therefore, it is unknown whether D4 receptor density differs between carriers of the 7-repeat and other DRD4 alleles.

Variability in ADHD symptoms and MPH response Allelic variation in DAT1 and DRD4 relates to ADHD symptoms selectively. Homozygosity of the 10-repeat DAT1 allele was associated with higher hyperactivity symptoms, whereas that of the 7-repeat DRD4 allele was associated with higher inattention symptoms, in ADHD children and their parents. Furthermore, children homozygous for the 10-repeat

DAT1 allele performed worse on performance measures of impulsivity, the Test of Everyday Attention for Children (TEA-Ch) Opposite Worlds task and errors of commission on the Continuous Performance Task, relative to heterozygous ADHD children. Also, homozygous 10-repeat DAT1 children exhibited greater intraindividual response time variability than heterozygous children. It is noteworthy that DAT is primarily expressed in the striatum, whereas D4 is abundant primarily in prefrontal cortex. Thus, selective effects of DAT1 and DRD4 polymorphisms on inattention and impulsivity in ADHD may reflect functional characteristics of the striatum and prefrontal cortex, respectively.

Pharmacogenetic studies suggest that the efficacy of MPH depends on DAT1 and DRD4. Based on *in vitro* studies indicating that DAT expression was greater for the 10-repeat allele, carriers of that allele should show a better response to MPH. Indeed, homozygosity of the 10-repeat DAT1 allele was associated with a better response to MPH in some studies. Other studies, however, did not replicate this finding. Discrepant findings among studies probably relate to differences in sample selection and characteristics of treatment regimens, such as dosage, duration, and assessment of MPH response. Based on the *in vitro* observation that DA response is blunted for the 7-repeat allele, carriers of that allele should show a reduced MPH response. Indeed, a study examining variability in dose–response to MPH found that carriers of the 7-repeat allele required higher doses to attain normalization of symptoms relative to noncarriers. Thus, examination of allelic variability in MPH effects is a worthwhile approach to elucidate etiological heterogeneity in DA function in ADHD, but larger well-controlled studies are needed to draw definitive conclusions.

Functional Brain Imaging of MPH Effects

Functional magnetic resonance imaging (fMRI) performed with methylphenidate challenge is a noninvasive way to visualize dopaminergic function *in vivo*. fMRI capitalizes on changes in blood oxygenation secondary to neural activity evoked by a cognitive challenge. Results of several studies indicate that frontal and striatal activation during performance of attentional and inhibitory tasks differs between ADHD and control subjects. Comparison of fMRI measures in the same subjects during performance of a cognitive challenge, once with and once without administration of MPH, provides an index of changes in neural activity induced by alterations in catecholamine levels (generally referred to as pharmacological fMRI). Evidence from animal studies has validated the ability of fMRI to index metabolic changes induced by dopaminergic stimulation. Specifically, fMRI signal in response to amphetamine or its analog, CFT (2B-carbomethoxy-3B-[4-flurophenyl] tropane, WIN 36 528), was correlated with changes in extracellular DA concentrations in the striatum (confirmed by microdialysis) and was lost following denervation of the striatum by 6-OHDA lesioning. Furthermore, the fMRI signal changes were restricted to regions with high expression of DA receptors, such as striatum and cingulate and frontal cortex (confirmed by receptor-labeled positron emission tomography). Thus, fMRI with MPH challenge provides a noninvasive assay for dopaminergic function relevant to cognitive deficits observed in ADHD.

Findings of pharmacological fMRI studies indicate that MPH has selective effects on brain regions involved in executive function in subjects with ADHD. Results of these studies suggest that MPH normalizes activation in striatal structures but not in other brain regions. Striatal activation was reduced relative to control children and increased following administration of MPH during performance of response inhibition and divided attention tasks. However, other regions with reduced activation in ADHD subjects, such as middle temporal and anterior cingulate gyri, did not change following administration of MPH. Furthermore, effects of MPH on prefrontal activation in ADHD subjects differed across studies such that activation was increased during a difficult response inhibition condition, did not change during an easy response inhibition condition, and was reduced during working memory performance. Thus, effects of MPH on metabolism in circuitry involved in executive function are not uniform across regions.

The influence of MPH on brain metabolism appears to vary by symptom expression. In contrast to increased striatal activation following MPH administration, fMRI imaging during response inhibition in control children showed that MPH reduced striatal activation. These opposite effects of MPH in ADHD and control children were not observed elsewhere in the brain. Furthermore, these opposite effects did not relate to performance differences because MPH improved response inhibition to the same extent in both ADHD and control groups. Results of other studies indicate that MPH increased blood volume in children who were more hyperactive, whereas it decreased blood volume in those who were less hyperactive, in the cerebellar vermis and putamen. In light of evidence relating symptom expression to functional properties of DA transmission, MPH effects on brain metabolism probably reflect differences in baseline dopaminergic function.

Summary

Convergent evidence from studies of MPH effects and *in vivo* imaging suggests that DA transmission is dysfunctional in ADHD and may relate to polymorphisms of dopaminergic genes. The most direct evidence for presynaptic DA dysfunction in ADHD comes from ligand-based imaging studies showing elevated DAT density in subjects with ADHD. Whether this is a trait related to abnormal neuronal pruning or a state reflecting adaptive response to abnormalities in other aspects of DA transmission remains to be resolved. Resolution may come from molecular genetic studies because inheritance of the 10-repeat allele of the DAT gene has been linked to individual differences in DAT expression and susceptibility to ADHD. Polymorphism of D4 receptor genotype also confers some risk for ADHD, suggesting that postsynaptic DA function is also part of the ADHD pathophysiology. The leading model of DA dysfunction in ADHD posits that the disorder involves both lower and higher dopaminergic activity. Specifically, Seeman and Madras hypothesize that ADHD reflects reduced tonic but increased phasic DA activity. Efficacy of MPH in this model is mediated by reduced phasic DA activity via autoreceptor stimulation due to increased tonic DA activity that results from increased synaptic DA following DAT blockade by MPH. However, MPH response varies depending on contextual and subject-related factors, a finding that is not accounted for by current models of ADHD. Use of pharmacological fMRI provides a noninvasive assay of DA function in ADHD. Its use has revealed that changes in neural activity induced by MPH during executive functions serve to 'normalize' striatal function in ADHD subjects but depend on symptom severity.

Future Direction: Endophenotypes for ADHD

Researchers agree that phenotypic heterogeneity of ADHD is the primary limiting factor in elucidating the pathophysiology of ADHD. Currently, most researchers endorse an approach that characterizes individual differences in terms of endophenotypes rather than symptom profiles. Endophenotypes refer to cognitive functions that can be specified on performance-based measures that are heritable and draw upon neurophysiology relevant to the disorder. Current research has identified response inhibition measured by Stop Signal or Go/No-go tasks as a possible endophenotype for ADHD because functional variability relates to dopaminergic modulation in prefrontal–striatal regions and to heritability via dopaminergic genes. Other candidate endophenotypes for ADHD are intraindividual response variability on reaction time tasks and a shortened delay gradient on tasks involving rewards. Pharmacological fMRI is a promising tool in the search for endophenotypes for ADHD because it can characterize individual variation in the functional anatomy of a cognitive function with respect to MPH response. Furthermore, candidate genes can be incorporated into this approach by examining whether MPH effects on prefrontal–striatal function vary for polymorphisms of dopaminergic genes. A characterization of ADHD phenotypes with an approach that unifies genetic, neurophysiological, and cognitive factors is also likely to be promising for the development of objective diagnostic criteria and for treatment planning.

See also: Amphetamines; Attention Deficit Hyperactivity Disorder; Cognition: An Overview of Neuroimaging Techniques; Dopamine – CNS Pathways and Neurophysiology; Dopamine in Perspective; Neuroimaging.

Further Reading

Castellanos FX, Sonuga-Barke EJ, Milham MP, and Tannock R (2006) Characterizing cognition in ADHD: Beyond executive dysfunction. *Trends in Cognitive Science* 10(3): 117–123.

Faraone SV, Perlis RH, Doyle AE, et al. (2005) Molecular genetics of attention-deficit/hyperactivity disorder. *Biological Psychiatry* 57(11): 1313–1323.

Faraone SV, Sergeant J, Gillberg C, and Biederman J (2003) The worldwide prevalence of ADHD: Is it an American condition? *World Psychiatry* 2(2): 104–113.

Grace AA (2001) Psychostimulant actions on dopamine and limbic system function: Relevance-related behavior and impulsivity. In: Solanto MV, Arnsten AFT, and Castellanos FX (eds.) *Stimulant Drugs and ADHD: Basic and Clinical Neuroscience*, pp. 134–157. New York: Oxford University Press.

Krause KH, Dresel SH, Krause J, la Fougere C, and Ackenheil M (2003) The dopamine transporter and neuroimaging in attention deficit hyperactivity disorder. *Neuroscience & Biobehavioral Reviews* 27(7): 605–613.

Madras BK, Miller GM, and Fischman AJ (2005) The dopamine transporter and attention-deficit/hyperactivity disorder. *Biological Psychiatry* 57(11): 1397–1409.

Nigg JT, Willcutt EG, Doyle AE, and Sonuga-Barke EJ (2005) Causal heterogeneity in attention-deficit/hyperactivity disorder: Do we need neuropsychologically impaired subtypes? *Biological Psychiatry* 57(11): 1224–1230.

Sonuga-Barke EJ (2002) Psychological heterogeneity in AD/HD—A dual pathway model of behaviour and cognition. *Behavioural Brain Research* 130(1–2): 29–36.

Spencer TJ, Biederman J, Madras BK, et al. (2005) *In vivo* neuroreceptor imaging in attention-deficit/hyperactivity disorder: A focus on the dopamine transporter. *Biological Psychiatry* 57(11): 1293–1300.

Swanson JM and Volkow ND (2001) Pharmacokinetic and pharmacodynamic properties of methylphenidate in humans. In: Solanto MV, Arnsten AFT, and Castellanos FX (eds.) *Stimulant Drugs and ADHD*. New York: Oxford University Press.

Vaidya CJ (2002) Application of pharamacological fMRI to developmental psychiatric disorders. *Developmental Science* 5(3): 310–317.

Vaidya CJ, Austin G, Kirkorian G, et al. (1998) Selective effects of methylphenidate in attention deficit hyperactivity disorder: A functional magnetic resonance study. *Proceedings of the National Academy of Sciences of the United States of America* 95(24): 14494–14499.

Volkow ND, Wang GJ, Fowler JS, et al. (2004) Evidence that methylphenidate enhances the saliency of a mathematical task by increasing dopamine in the human brain. *American Journal of Psychiatry* 161(7): 1173–1180.

Willcutt EG, Doyle AE, Nigg JT, Faraone SV, and Pennington BF (2005) Validity of the executive function theory of attention-deficit/hyperactivity disorder: A meta-analytic review. *Biological Psychiatry* 57(11): 1336–1346.

Attention: Models

P H E Tiesinga, University of North Carolina at Chapel Hill, Chapel Hill, NC, USA
T J Sejnowski, Salk Institute and University of California at San Diego, La Jolla, CA, USA

Introduction

Our sensory environment is represented in the cortex as the electrical activity of billions of neurons. The neural activity in the visual pathway reflects not only the current visual environment but also the goals and expectations of the organism as well. Suppose that one is at the airport picking up a friend who is wearing a red coat. In general, red objects will activate specific neurons in the visual cortex, but, because they are now relevant to the current goal, their responses will be enhanced and preferentially processed. Selective attention is a general strategy for selecting the currently relevant information out of the overwhelming amount of information that enters the brain via the senses. This article reviews models for selective attention in the visual system at the level of single neurons and the local circuit they are embedded in.

Parametric Models for Stimulus Response Properties of Early Visual Neurons

The response of a visual neuron depends both on the stimulus that is driving it (bottom-up) and attention (top-down). The purely sensory responses of cortical neurons were originally characterized in anesthetized cats and primates and with simple stimuli such as oriented bars and gratings. In the primary visual cortex (V1), many neurons respond best to a stimulus of a specific orientation placed at a specific location in the visual field. The receptive fields (RF) of middle temporal (MT), V2, and V4 cortical areas downstream to V1 are much larger than those of V1 neurons and involve more complex combinations of features.

When an oriented bar is placed at the center of the RF of a V1 simple cell, the mean firing rate varies approximately as a bell-shaped curve with the stimulus orientation (**Figure 1(a)**) that can be approximated mathematically as

$$R(\theta) = R_0 + A\,\exp\left(\frac{(\theta - \theta_{\text{pref}})^2}{2\sigma^2}\right) \qquad [1]$$

where R is the firing rate in response to a stimulus of orientation θ, θ_{pref} is the neuron's preferred orientation, σ is the tuning width (related to the half width at half height), R_0 is the neuron's firing rate in the absence of the stimulus, and A represents the strength of firing rate modulation. Studies using anesthetized cats and macaques as well as alert macaques to investigate the orientation selectivity of V1 neurons have revealed spontaneous activity R_0 ranging from zero to about 20 Hz and A varying from a few hertz to 100 Hz. The tuning width σ could be as low as 10° with a median value of 30° across all layers in V1 in alert macaque monkeys.

Luminance contrast is the absolute difference in luminance values across the stimulus expressed relative to the background. It is a measure of stimulus strength, and it takes values between zero and 100%. The firing rate increases monotonically with stimulus contrast (**Figure 1(b)**). Mathematically, this relationship is summarized in terms of the contrast response function (CRF), a commonly used expression for which is

$$R(c) = R_0 + R_m \frac{c^n}{c_0^n + c^n} \qquad [2]$$

where c is the contrast, R_m is the maximum firing rate, c_0 is the contrast for which the firing rate above baseline is half R_m, and n is the power of the nonlinearity. Typical values for cat and primate V1 are $c_0 \approx 10\%$, $n \approx 2$. The CRF has three regimes. For low contrast ($c \ll c_0$), the firing rate increases supralinearly as a power of the contrast c. The largest rate of change in firing rate is at medium contrast, around c_0, and the firing rate saturates at high contrasts ($c \gg c_0$).

Orientation tuning is approximately contrast invariant. For a range of contrast values, the orientation-tuning curve has approximately the same shape but a different maximum firing rate (the parameter A). Mathematically, this implies that the rate above baseline can be written as a product of the orientation-tuning curve and the contrast response curve (**Figure 1(c)**):

$$R(\theta, c) = R_0 + R_m \frac{c^n}{c_0^n + c^n} \exp\left(\frac{(\theta - \theta_{\text{pref}})^2}{2\sigma^2}\right) \qquad [3]$$

This functional form reflects a normalization model. A given stimulus activates not only the neuron that is recorded but also a pool of other neurons. The recorded neuron receives input proportional to the orientation-tuned output of a filter L and the

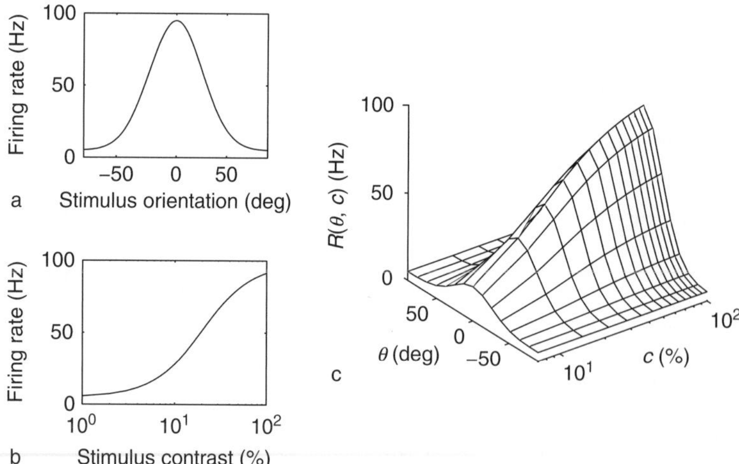

Figure 1 (a) Orientation-tuning curve. (b) Contrast response function. (c) The firing rate, according to eqn [3], as a function of orientation θ and contrast c.

contrast c. The combined activity of neurons in the pool is independent of stimulus orientation, but it increases with contrast as c^n. This activity divisively normalizes the response of the recorded neuron. The normalization model can account for a large number of experimentally measured responses. Model studies and *in vitro* experiments show that a number of biophysical mechanisms yield divisive normalization. An alternative means of achieving normalization involves the short-term plasticity of the synapses onto the neuron.

Parametric Models for Attentional Modulation of Early Visual Responses

The modulation by selective attention of neural responses has been studied in awake nonhuman primates. The strongest modulation of individual neurons was recorded in the MT, V4, and inferotemporal cortical areas, which all are downstream of V1. In one experimental paradigm, two equivalent arrays of stimuli are presented, typically on opposite sides of the vertical meridian. The monkey is rewarded for detecting a subtle change in the stimulus at a cued location. It is not rewarded for reporting a change in a stimulus at the noncued location. Since this is a difficult task, the monkey needs to pay attention to the cued location in order to observe and respond to the stimulus change. The neuron that is recorded has its receptive field at one of the two locations. Hence, the response of that neuron to the same stimulus with attention can be compared with when attention is directed at the location outside the neuron's RF.

Attention modulates the firing rate. For spatial attention, the firing rate typically increases when the neuron's receptive field overlaps with the area where attention is directed. The neuron receives feed-forward synaptic inputs and transforms them into a firing rate. Attention could act in several ways: it could make the feed-forward input stronger, or it could change the gain of the input-to-output transformation. The former is referred to as contrast gain and the latter as response gain. In **Figure 2(a)**, illustrating contrast gain, the CRF with attention (left curve) is plotted together with the curve for attention directed away from the RF (right curve). Attention shifted the curve to the left, which implies that a neuron responds to a stimulus as if it had a contrast higher than its actual (veridical) contrast. Contrast gain has three distinguishing properties. First, with attention the neuron will respond to low-contrast values that it did not respond to without attention. Second, the largest change in firing rate will occur for moderate firing rates corresponding to contrasts around c_0. Third, the saturation firing rate remains the same, but it is reached for a lower value of the contrast than without attention.

Response gain is illustrated in **Figure 2(b)**. The CRF with attention is obtained by multiplying the CRF without attention by a gain factor larger than unity. Response gain can be distinguished from contrast gain. First, if the neuron does not respond to a low-contrast stimulus without attention, neither will it do so with attention. Second, the largest change in firing rate will occur at the highest value of the contrast where the neuron fires at its highest rate. Third, the saturation rate will be higher for attention compared with no attention.

Experiments in a spatial attention paradigm in cortical areas MT and V4 of macaque show that attentional modulation of the CRF is best described as contrast gain. This matches the results of psychophysical experiments: Human observers report that the perceived contrast of a stimulus is higher if it is in the focus of attention. Other experiments on the

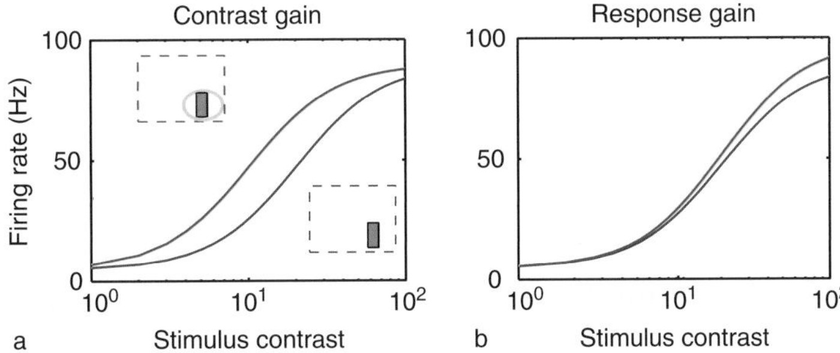

Figure 2 Attentional modulation of the contrast response function when attention is directed away from (blue) or directed toward (green) the receptive field of the recorded neurons. Illustration of (a) contrast gain and (b) response gain.

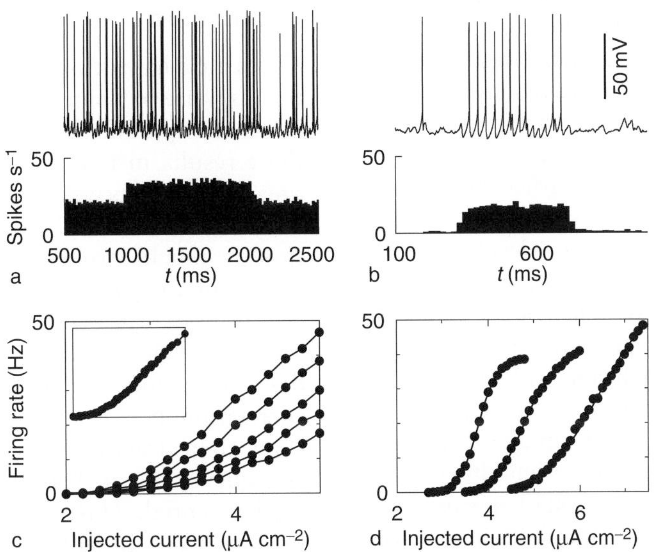

Figure 3 Response gain and contrast gain can be mediated by modulation of inhibitory synaptic inputs. (a, b): (top) Example voltage trace and (bottom) spike time histogram based on 500 trials. (a) The model neuron was driven by synchronous inhibitory and asynchronous excitatory inputs. During the interval between 1000 and 2000 ms, the jitter of the inhibitory volleys was decreased from 4 to 2 ms. During the period of increased inhibitory synchrony (low jitter), the firing rate was increased from 22 to 35 Hz. (b) The model neuron was driven by synchronous inhibition and a constant depolarizing current. During the interval between 300 and 700 ms, the jitter of the inhibitory volleys was decreased from 8 to 2 ms. The inhibitory inputs act as a gate. During periods of low inhibitory synchrony (high jitter), the neuron does not spike, whereas during periods of high inhibitory synchrony, it does spike. (c) and (d) The firing rate vs. current curves. Response gain: in (c) the synchronous inhibitory volleys consisted on average of ten presynaptic spikes, the jitter was, from top to bottom: 1, 2, 3, 4, and 5 ms. Inset: the five curves were made to overlap by scaling the firing rate and by small shifts in the current. Contrast gain: in (d) the synchronous inhibitory volleys consisted on average of 50 presynaptic spikes, and jitter was, from left to right, 1, 3, and 5 ms. Reproduced from Tiesinga PH, Fellous JM, Salinas E, Jose JV, and Sejnowski TJ (2004) Inhibitory synchrony as a mechanism for attentional gain modulation. *Journal of Physiology-Paris* 98: 296–314, with permission from Elsevier.

modulation of orientation tuning curves were best described as a response gain with attention. Overall, there is evidence in support of response gain, as well as contrast gain, depending on the specific details of the behavioral task by which attentional modulation was studied. In experiments designed to study feature-based attention in area MT, effects consistent with response gain were found. The neuron's response was multiplied by a gain factor that monotonically increases with the difference between the attended feature (direction, color, or shape) and the feature value preferred by the neuron. A similar effect of feature-based attention was obtained in fMRI experiments with human participants. The response of MT to a moving random dot pattern was increased when it was task-relevant. In macaque visual cortex, attention led in some, but not all, cases to an increase in baseline rate (R_0).

Thus, the response to a stimulus of contrast c and orientation can be described by eqn [3]. In this expression, attention 'a' enters via the baseline rate $R_0(a)$ and the contrast threshold $c_0(a)$. The effects of attention are consistent with either response gain or contrast gain depending on which stimulus parameter is varied. This form makes a prediction that has recently been confirmed experimentally: the response to a stimulus of nonoptimal orientation should saturate at the same contrast as a stimulus with optimal orientation.

Biophysical Models for Attentional Response Modulation

A biophysical model must account for the inputs a neuron receives from three sources: feed-forward input that represents stimulus aspects (bottom-up), modulatory feedback input (top down), and recurrent connections within the same cortical area (but potentially different layers). The recurrent activity depends both on bottom-up and top-down inputs. Furthermore, the bottom-up input includes the influence of top-down projections on the earlier cortical area. Response mechanisms can be studied theoretically and *in vitro* at the single-neuron level under the assumption that the three sources of synaptic inputs can be split up in two sets of independent inputs, one of which depends only on the stimulus identity and contrast, the other only on the locus of attention. The rate of stimulus-related synaptic inputs depends, as a bell-shaped tuning curve, on the stimulus orientation, with excitatory inputs dominating inhibitory inputs. The attention-related inputs consist of approximately equally strong excitatory and inhibitory inputs. Thus the main contribution of attention consists of modulating fluctuating inputs. Two mechanisms have been proposed that are based on the idea that attention modulates the amount of fluctuation. First, in gain modulation by balanced synaptic inputs, attention causes a proportional change in rates of excitatory and inhibitory inputs such that the variance of the membrane potential changes but its mean remains the same. Second, in gain modulation by correlation, there is a change in correlation between inputs such as the synchrony of inhibitory inputs or the delay between excitatory and inhibitory inputs. Both of these proposed mechanisms are supported by *in vitro* experiments. The basic paradigm is as follows: in the experiments, feed-forward inputs are presented by a constant depolarizing input current I injected at the soma, whereas the modulatory inputs are injected at the soma using dynamic clamp. The firing rate versus current curve (f–I) is then constructed by calculating the firing rate for a set of different values of I in the presence of the modulatory inputs. Chance and co-workers have shown that an increase in the level of balanced synaptic inputs yields a decrease in the gain of the f–I curve. Tiesinga et al. have shown that an increase in inhibitory synchrony also leads to an increase in gain of the f–I or a shift in the f–I along the current ordinate (**Figure 3**).

The single-neuron studies show how the response could be modulated, but they do not provide a mechanism at the local circuit level by which the necessary changes in the modulatory inputs come about. Furthermore, in the model as well as *in vitro*, the inputs were somatic, but the modulatory and feed-forward inputs may impinge at different locations on the neuron's extensive dendritic tree, which could introduce more complex dynamical interactions.

Discussion

This article has focused on the effects of attention on the response of a single neuron to a simple stimulus. Current experimental investigations are extending these results in two ways. First, in higher cortical areas, the receptive fields are so large that they accommodate multiple stimuli. Attentional modulation is much stronger when there are multiple stimuli in a receptive field. Phenomenological models are presently being developed to quantitatively account for these data. Second, even when attention does not change the firing rate, it may change the correlation between neurons. Experiments in V4 show that attention increases the coherence in the γ-frequency range between neurons and the local field potential. This suggests an important role for the synchrony in networks mediating the effects of attention. Future theoretical research is directed at exploring the synchrony hypothesis at the level of the cortical circuit.

See also: Attention and Eye Movements; Attentional Networks; Attentional Functions in Learning and Memory; Neglect Syndrome and the Spatial Attention Network; Sensorimotor Integration: Attention and the Premotor Theory; Visual Attention; Visual Motion Models.

Further Reading

Albrecht DG, Geisler WS, Frazor RA, and Crane AM (2002) Visual cortex neurons of monkeys and cats: Temporal dynamics of the contrast response function. *Journal of Neurophysiology* 88: 888–913.

Bichot NP, Rossi AF, and Desimone R (2005) Parallel and serial neural mechanisms for visual search in macaque area V4. *Science* 308: 529–534.

Chance FS, Abbott LF, and Reyes AD (2002) Gain modulation from background synaptic input. *Neuron* 35: 773–782.

Ferster D and Miller KD (2000) Neural mechanisms of orientation selectivity in the visual cortex. *Annual Review of Neuroscience* 23: 441–471.

Fries P, Reynolds JH, Rorie AE, and Desimone R (2001) Modulation of oscillatory neuronal synchronization by selective visual attention. *Science* 291: 1560–1563.

Maunsell JH and Treue S (2006) Feature-based attention in visual cortex. *Trends in Neurosciences* 29: 317–322.

Reynolds JH and Chelazzi L (2004) Attentional modulation of visual processing. *Annual Review of Neuroscience* 27: 611–647.

Salinas E and Sejnowski TJ (2001) Correlated neuronal activity and the flow of neural information. *Nature Reviews Neuroscience* 2: 539–550.

Tiesinga PH, Fellous JM, Salinas E, Jose JV, and Sejnowski TJ (2004) Inhibitory synchrony as a mechanism for attentional gain modulation. *Journal of Physiology* 98: 296–314.

Williford T and Maunsell JH (2006) Effects of spatial attention on contrast response functions in macaque area V4. *Journal of Neurophysiology* 96: 40–54.

Attentional Functions in Learning and Memory

M Sarter and C Lustig, University of Michigan, Ann Arbor, MI, USA

Attention and Memory: Conceptual Issues

Attention describes a set of cognitive processes which act to optimize stimulus detection, discrimination, and processing. Attention operates in part by top-down tuning of sensory systems to facilitate the detection of selected stimulus characteristics, such as location and modality, by switching the cortical processing from associational to input modes, and by allocating attentional resources to these operations. Different forms of attention have been categorized, such as sustained, selective, and divided attention. Sustained attention describes the subject's state of readiness to detect rarely and unpredictably occurring changes in the stimulus situation over extended periods of time. Selective attention brings targeted information into the focus of consciousness, while suppressing the detection and processing of nontarget signals. Divided attention emphasizes the allocation and the management of limited attentional resources in situations that require attention to multiple stimuli or tasks.

The assumption that attended stimuli are encoded more effectively into memory than less-attended ones is straightforward and supported by substantial evidence. However, theoretical frameworks that more fully describe the relationships between attention and learning and memory (L&M) have remained unexpectedly rare. Furthermore, empirical analyses of interactions between attention and L&M focused on selected aspects of attention and a rather small number of experimental paradigms. For example, increased demands on the division of attention impair stimulus encoding, but generally do not impede the retrieval of previously learned information. Conversely, high demands on memory operations impair attentional capacities, particularly the ability to filter nontarget stimuli from being processed. A second major paradigm used in research on the interactions between attention and L&M concerns the impaired learning of extradimensional shifts, as such shifts require the processing of a previously unattended stimulus dimension (e.g., shape, when one had been attending to color).

Attentional processes and capacities represent a cluster of variables contributing to the efficacy of L&M. At the extreme, it is difficult to envision meaningful acquisition of declarative information in the absence of attention. Furthermore, levels of attentional performance vary considerably, and this variation affects the rate of learning and thus the efficacy of memory. Therefore, it is not unexpected that brain mechanisms mediating attentional functions and capacities are also part of the neuronal circuitry mediating L&M.

Functional neuroimaging studies provide strong evidence for the interplay between attention, learning, and memory. Meta-analytic and experimental studies reveal large overlaps between prefrontal and parietal regions activated in attention and memory tasks that are not shared with other (e.g., language processing) domains. Dividing attention at learning leads to reduced activity in prefrontal brain regions associated with subsequent memory, and has long been known to reduce memory performance. Conversely, recent evidence suggests that higher levels of tonic or baseline activity in regions involved in stimulus perception (e.g., parahippocampal regions for perception of scenes) are associated with higher levels of attention to those stimuli, and with better subsequent memory than those preceded by lower tonic activity. Memory also serves to guide attention: memory-guided visual search activates similar brain regions as does cue-guided visual search, and may even provide greater performance benefits than a visual cue.

In summary, attention, learning, and memory are highly integrated, dynamically interactive processes. This interactivity is reflected both in heavy overlap between the brain systems involved in attention and those involved in L&M, and in the mutual influences of attention on memory processing. The following sections describe the neurobiological mechanisms of attention–memory interactions in more detail.

Neuronal Macrosystems Mediating Attentional Functions and Capacities: Major Research Themes

Both lesion studies and functional neuroimaging provide evidence for a distributed cortical network of prefrontal and posterior parietal regions involved in attention. The specific regions involved vary according to the nature of the attentional task. Demands on sustained attention, irrespective of the modality of targets, activate frontal and parietal regions primarily in the right hemisphere. Selective attention and demands for nontarget filtering recruit cingulate and other prefrontal regions, again primarily in the right hemisphere. Bilateral frontoparietal regions typically are activated by tasks requiring the division of attention between multiple targets.

Ongoing research, particularly experiments employing functional magnetic resonance imaging (fMRI), continues to refine the attribution of aspects of attentional performance to specific cortical networks. Results from this research confirm the conceptualization of anterior and posterior attention systems. Posner and Peterson proposed that the anterior attention system, consisting mainly of cingulate and other prefrontal regions, mediates target detection or the processes involved in the subject's consciousness of a signal's presence through producing a response documenting its detection. The posterior attention system involves the posterior parietal cortex and collicular and thalamic regions and controls management of the visual-attentional space.

Essential insights into the neuronal mechanism mediating attentional processes have been gained from research focusing on bottom-up versus top-down selection of stimuli. Bottom-up selection is a function of the intrinsic properties of the stimulus; stimuli compete and cooperate for detection and processing depending on their salience. Bottom-up, stimulus-driven attention is associated with increases in activity in sensory and sensory-associational regions representing the actual stimuli and their location. In addition, increased activity in frontoparietal networks indicates that stimulus-driven attentional control may also influence the executive management of attentional priorities and resources. In contrast, top-down selection is a function of experience, expectations, or instructions. It involves the tuning of receptive field properties, (sustained) anticipatory activity, and suppression of activity in irrelevant regions or modalities, all in order to optimize the detection and processing of expected targets. Top-down attentional control is generally thought to be executed by prefrontal modulation of parietal networks.

The actual mechanisms allowing prefrontal regions to initiate top-down effects, and the cognitive mechanisms and neuronal circuits mediating such top-down effects have remained largely unclear. Likewise, the neuronal circuitries mediating the bottom-up enhancement of stimulus processing, usually in a highly topographic fashion, are not known. As discussed further below, evidence on the role of neurotransmitter-specific cortical input systems provides the basis for hypotheses about the neuronal mechanisms modulating the detection of stimuli as a function of top-down, voluntary attention versus the bottom-up influences based on stimulus properties (see also **Figure 1**).

Evidence from recent fMRI experiments questioned the widely held view that the suppression of irrelevant information or modalities represents an important aspect of prefrontally controlled top-down regulation of attentional mechanisms. These studies demonstrated that prefrontal mechanisms act to amplify detection of task-relevant information and did not find consistent evidence for top-down suppression of irrelevant stimuli. However, other investigators have found evidence that instructions to ignore one stimulus set in order to more successfully memorize another caused brain activity associated with the ignored stimulus set to be suppressed below a perceptual baseline. As will be further discussed below, the view that selective detection is primarily mediated via enhancement of target stimuli, with little contribution of suppression of nontargets, corresponds with the attentional functions of the cortical cholinergic input system.

Intuitively, motivation is an important factor for engaging top-down mechanisms in order to combat fatigue- or distractor-related decline in attentional performance, to stabilize impaired performance, or to recover from performance impairments. The effects of motivational processes on top-down attention and the neuronal mechanisms via which motivation accesses the anterior attention system to facilitate top-down effects in challenging situations are emerging as an important theme. Increasing the incentives for attentional performance is associated with enhanced activity in frontoparietal attentional networks, and with activation in additional limbic regions involved in processing errors and response outcomes. Hypotheses concerning the interactions between motivational and attentional processes focus on the regulation of frontal–parietal attention systems via interactions with ventral striatal circuitry known to process reward, reward expectation, and prediction errors.

While motivation is a critical factor for top-down attention, stimuli associated with affective information exhibit superior potency for the bottom-up capture of attentional resources. For example, people in a crowded bar manage to ignore the TV until an emotionally charged symbol or action shifts their attention effectively from ongoing activities to the screen. The neuronal mechanisms allowing such stimuli to capture attentional resources and then drive the subsequent engagement of top-down mechanisms are not well known. However, it is safe to hypothesize, as illustrated in **Figure 1**, that such stimuli access the brain's main attention systems via reciprocal connections with prefrontal regions (not shown) and with ascending neuronal systems crucially involved in the mediation of attentional functions.

Neurotransmitter-Specific Projection Systems in Attention and Learning

Substantial evidence supports the hypothesis that the cortical cholinergic input system contributes

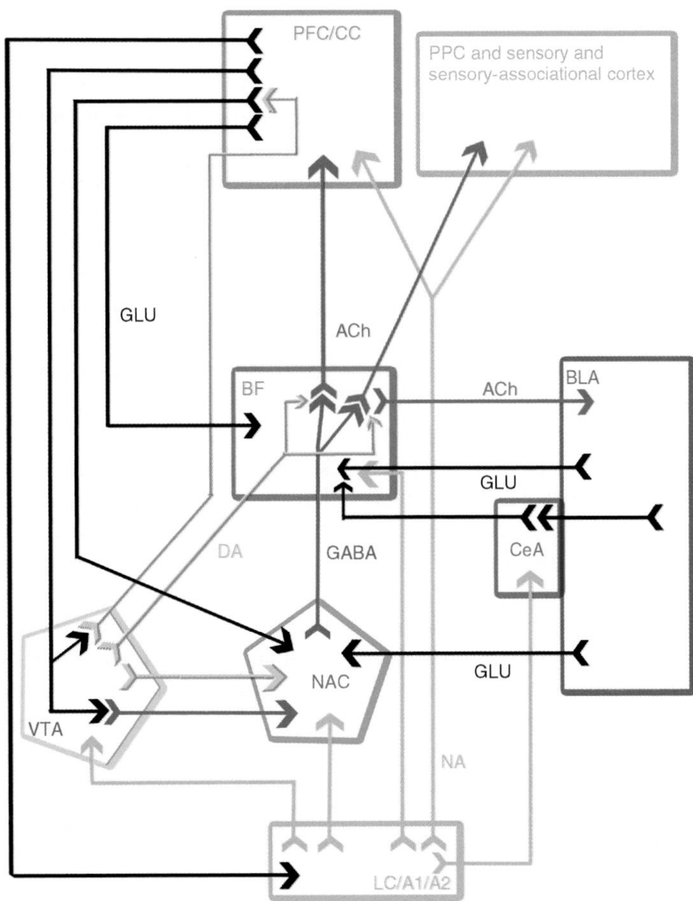

Figure 1 Schematic illustration of the main components of a neuronal network mediating attentional performance and the motivated activation of the cortical cholinergic input system in order to counteract the effects of challenges on attentional performance. The illustration depicts and emphasizes certain direct neuronal projections on the basis of anatomical evidence and importance in the present context. As discussed in the text, neurons using acetylcholine (ACh) as transmitter originate from basal forebrain (BF) regions and innervate all cortical regions, including the prefrontal cortex (PFC), cingulate cortex (CC), and somatosensory and posterior associational regions, including the posterior parietal cortex (PPC). Increases in cortical cholinergic activity contribute to the recruitment of anterior and posterior attention systems and the cholinergic amplification of input processing in sensory and sensory-associational regions. Cholinergic modulation of the PFC and CC is also involved in the implementation of top-down mechanisms, and the cholinergic inputs to posterior cortical regions are a component of the prefrontal efferent circuitry mediating such top-down effects. Increases in top-down effects that are a result of challenges on attentional performance, as well as the subjects' motivation to stabilize residual performance or regain normal levels, depend in part on the regulation of BF neurons by direct projections from the dopaminergic (DA) ventral tegmental region (VTA) and indirectly, via the nucleus accumbens (NAC) and its GABAergic (GABA; γ-aminobutyric acid) projections to BF cholinergic neurons. These midbrain and ventral-striatal circuits mediate incentive information and, via feedback from the CC and PFC, information about errors and unpredicted reward contingencies. Therefore, cholinergic modulation by outputs from these regions forms the basis for the integration between motivational and attentional processes. Corresponding with this hypothesis is evidence indicating that control over cortical cholinergic activity by ventral-striatal regions is confined to prefrontal regions. In addition to signals predicting reward or reward loss, stimuli associated with salient affective qualities exhibit potent attention-capturing qualities. The cholinergic and glutamatergic (GLU) reciprocal connections between the basolateral amygdala (BLA), as well as additional projections from the amygdala to the BF via the central amygdaloid nucleus (CeA), are thought to be involved in the mediation of such processes. Finally, ascending noradrenergic (NA) projections originating from the locus coeruleus (LC) and the catecholaminergic cell groups in the medulla (A1/A2) innervate major components of this circuits and also receive afferent feedback from the PFC. Thus, ascending noradrenergic projections are hypothesized to contribute to specific aspects of attentional processing in parallel with those attributed to the BF cholinergic system.

essentially to the mediation of attentional performance. Cholinergic inputs contribute to both top-down and bottom-up modulation of detection processes, and represent a link between motivational systems and cortical attention systems. Originating from basal forebrain regions, cholinergic neurons receive main inputs from midbrain and telencephalic regions, including the prefrontal cortex, and innervate all cortical areas and layers.

Removal of cholinergic inputs results in persistent impairments in sustained, selective, and divided attentional abilities. Furthermore, attentional tasks selectively increase the release of acetylcholine (ACh) from cortical terminals. Evidence suggests that the right hemispheric cortical cholinergic input system is dominant for sustained attention performance; this finding corresponds with evidence from human fMRI studies and from patients with lateralized cortical damage (as described above).

Recent data indicate that in the prefrontal cortex, a transient cholinergic signal is elicited specifically by successfully detected targets, but not by targets that were missed. The temporal characteristics of these cholinergic signals confirm that essential aspects of the detection process are mediated via increases in cholinergic neurotransmission in the prefrontal cortex. These aspects include disengagement from noncontingent ongoing behaviors or from internal, associational processing, as well as the initiation of target-associated behavioral responses. These findings correspond with Posner's original hypothesis that the anterior attention system mediates the detection of attention-demanding targets (as described above).

The cortical cholinergic input system is involved both in the activation of the top-down attention system itself and in the implementation of its downstream effects. Cholinergic inputs to prefrontal regions, including cingulate cortex, specifically contribute to the activation of the anterior attention system, thereby enhancing target detection and activating top-down mechanisms. From there, cholinergic inputs to more posterior cortical regions, including parietal cortex, serve as a branch of the prefrontal efferent circuitry that implements top-down influences on input processing in sensory and sensory-associational regions.

The idea that cholinergic function can be separated into top-down and bottom-up components is supported by the dissociable interactions of these components with other brain systems. For example, ventral-striatal modulation of cortical cholinergic activity is limited to the prefrontal cortex. As described above, ventral-striatal modulation, particularly by the nucleus accumbens, is considered critical in situations characterized by the motivated, top-down increase in attentional effort. The modulation of cholinergic inputs to prefrontal regions corresponds with the role of those prefrontal cholinergic inputs in activating top-down effects.

In contrast, bottom-up, stimulus-driven recruitment of the cortical cholinergic input system is believed to involve recruitment of the entire cholinergic projection system, thereby directly influencing sensory processes while also allowing the stimulus to influence the executive management of attentional priorities. Such broad recruitment of the basal forebrain cholinergic system by salient stimuli is mediated in part by ascending noradrenergic projections targeting the cholinergic neurons in the basal forebrain. Noradrenergic neurons in the brain stem are wired to receive information about visceral correlates of salient stimuli and thus 'import' information about salience to forebrain regions. Support for the hypothesis that noradrenergic–cholinergic interactions are involved in bottom-up capturing of attentional resources by salient stimuli was generated by experiments demonstrating, for example, that the cortical processing of salient stimuli is attenuated by loss of cholinergic inputs to the cortex or blockade of receptors for noradrenaline in the basal forebrain.

However, the ascending noradrenergic system may exert additional and more specific contributions to attentional performance. There are direct prefrontal projections to noradrenergic neurons in the brain stem, and noradrenergic neurons of the locus coeruleus are activated by attended stimuli. Both of these findings indicate that this ascending system is involved in attentional functions that parallel, at least in part, the attentional functions attributed to the cortical cholinergic input system. The overlaps, interactions, and dissociations between the attentional functions of basal forebrain cholinergic and brain stem noradrenergic systems represent an important research topic.

Likewise, the interactions between ascending dopaminergic, cholinergic, and noradrenergic systems likely are essential mechanisms in the mediation of attentional processing. The processing of stimuli capable of predicting reward, including the processing of prediction errors, involves mesolimbic dopaminergic systems. Reward and error signals are capable of capturing significant attentional resources, particularly in response to prediction errors. Thus, interactions between dopaminergic systems and the two other ascending modulatory systems may be necessary to optimize arousal and attentional processing. The dopaminergic recruitment of basal forebrain cholinergic neurons is based on tegmental projections to the basal forebrain, and indirectly on dopaminergic projections to the prefrontal cortex and nucleus accumbens and the innervation of the basal forebrain by these regions (see **Figure 1**). The dopaminergic regulation of the cortical cholinergic input system supports the integration of motivation- and attention-related processing. The prefrontal regulation of brain stem noradrenergic neurons and noradrenergic–cholinergic links represents a second major branch of the neuronal systems which, in concert, regulate attentional processes and resources (see **Figure 1**).

Ascending Modulatory Systems Mediating Attention: Involvement in Learning

Failure to attend and thus to detect a stimulus involves a failure to produce a representation of this stimulus for encoding. Thus, it is not unexpected that the available evidence on brain regions involved in attention matches, at least partly, the prefrontal–parietal neuronal systems critical for L&M. Indeed, given the critical role of attention for effective encoding, efforts designed to dissociate brain regions involved in attention versus learning pose conceptual challenges. It appears that a system of cortical regions processes stimuli at levels giving rise to awareness about these stimuli; however, the source of such processing of stimuli, whether it is a result of attentional or mnemonic operations, is not indexed in these regions.

The overlapping roles of ascending modulatory systems in attention and L&M are less well documented. In fact, several animal studies on the effects of loss of cortical cholinergic inputs have suggested that L&M do not depend on the integrity of these neurons, and that therefore the neuronal circuitries mediating attention and L&M could be dissociated. However, the lack of effects of, for example, lesions of the basal forebrain cholinergic system on L&M in animals may reflect, at least in part, the limited degree to which conventional animal tasks for L&M assess attention-requiring encoding and retrieval of declarative information. By contrast, more recent studies using tasks that require involving attention to stimuli or attention shifts find that learning is readily impaired by manipulations of levels of cortical cholinergic neurotransmission. Recent studies in humans have begun to simultaneously measure the cognitive and hemodynamic effects of drugs modulating cholinergic neurotransmission ('pharmaco-fMRI'). These studies indicated that drug-induced increases in cholinergic transmission enhance attentional mechanisms and the selectivity of stimulus processing during encoding. These results confirm the overlapping role of the cholinergic system in attention and learning, at least in situations requiring the encoding of attended stimuli. Similar conclusions have been drawn based on data from experiments in animals and humans on the role of the noradrenergic system in attention and learning.

Attention and Learning: Relevance for Aging and Neurodegenerative and Neuropsychiatric Disorders

Normal and pathological aging, including mild cognitive impairments and the dementias, are characterized by coinciding impairments in attentional functions and learning.

In nondemented aging, reduced attentional function is typically thought to be the major source of age differences in forgetting. There is some debate as to the nature of age deficits in attention function, with some theories emphasizing age differences in the amount of attentional resources, whereas others emphasize age differences in the allocation of attention, particularly deficits in inhibition. Across these perspectives, there is broad agreement that the controlled, top-down aspects of attention are those that show the greatest declines, whereas more automatic, bottom-up influences are relatively spared. Decreases in top-down attention are often ascribed to atrophy and reduced function in prefrontal cortex and basal ganglia structures, the structures that typically show the largest volume reductions in both longitudinal and cross-sectional studies of normal aging. These changes are thought to be largely responsible for age differences in most areas of cognition, although recent longitudinal evidence suggests that changes in select medial temporal lobe structures may be greater than suggested by earlier cross-sectional studies. Moreover, age-related attentional impairments may be a result of dysregulation and eventual degeneration of basal forebrain cholinergic projections to cortical regions.

The idea that the reduced availability of attentional resources is a major contributor to age differences in L&M receives substantial support from studies showing that asking young adults to divide their attention between multiple tasks (thus reducing the attentional resources available to any single task) often leads to results similar to those of older adults under single-task conditions. This is especially the case when the performance measure is subsequent memory for material studied under divided attention. Like older adults under single-task conditions, young adults studying under divided attention conditions show less activity in left frontal brain areas associated with later successful memory than do young adults studying under single-task conditions. On behavioral tests, the memory costs of divided attention are often similar for young and older adults, but older adults show larger impairments on the secondary task.

On the positive side, reducing the demand for top-down control often improves both memory and brain function for older adults. For example, intentional memory instructions ('memorize the words') require participants to engage top-down control to choose and implement a strategy for processing words that will support their later memory. Older adults typically perform much worse than do young adults under these conditions, and show less activation in left prefrontal cortex areas involved in subsequent memory. By contrast, when given instructions that guide attention to the memory-supporting semantic aspects of a stimulus ("Does the word mean something abstract, or concrete?"), memory is improved, and older adults often show prefrontal activations that are at least as great as those shown by young adults.

While initial findings focused on age-related underactivations of frontal brain regions involved in controlled attention and memory, more recent data show that in many cases, older adults show more activation, or activate additional brain regions that young adults do not. In many cases, this additional activation is linked to better performance, especially on memory tests. Patterns of under- and overactivation of frontal brain regions by older adults may be related to failures of top-down, 'proactive' control, and later 'reactive' attempts to compensate.

A recent meta-analytic review suggested that reduced or disrupted activations may be more common in right anterior frontal regions (often thought to be involved in high-level controlled strategic processing), perhaps reflecting a failure to adequately engage top-down processing to organize efficient task execution. By contrast, left frontal regions were more often associated with greater activation by older adults, perhaps reflecting greater engagement of lower-level processes in an attempt to compensate. Further supporting the idea of an age-related shift to reactive control, memory-related frontal brain activations have an extended time course for older adults under conditions with high control demands (remembering words studied once), but equivalent time courses to young adults under conditions with low control demands (remembering words studied 20 times). This shift is specific to frontal regions; parietal brain regions involved in successful recognition do not show a similar sensitivity to control demand.

Whereas attention and dopamine function are the focus of most research on normal aging, deficits in memory and cholinergic function are the characteristic features of Alzheimer's disease (AD). However, AD is also associated with marked deficits in attention. For example, the ability to divide attention between two ongoing tasks is substantially impaired in AD, even when performance on the individual tasks has been matched with that of healthy controls.

Attention deficits are not easily detected in the earliest stages of the disease, which is primarily marked by memory deficits. However, difficulties with attention precede the breakdowns in language and visuospatial function that occur at the moderate and severe stages of the disease. It is not clear which aspect of attention (selective, divided, or sustained) is the first to deteriorate in AD. Several reports indicate that very demanding selective or divided attention tasks show declines early in the disease process, whereas simpler sustained attention tasks only show deficits at later stages or under conditions (e.g., stimulus degradation) that increase difficulty. However, there have also been reports of preserved performance on selective attention tasks in patients with reduced sustained attention performance.

Within a domain, the detection of AD-related deficits in attention may depend not only on disease severity but also on the specific processes being tested. In selective attention, patients are often able to shift and engage attention appropriately in response to a valid cue. However, they appear to be impaired at disengaging attention from an inappropriate location following an invalid cue, especially if the task involves discrimination of a target stimulus from nontargets rather than simple detection.

For both normal aging and AD, a heuristic description is that attention performance deficits become more obvious with increased complexity or control demand. Functions that require little or no top-down control generally show little or no deficit. Examples include the engagement of spatial attention by the sudden onset of a peripheral cue, or the inhibition of attention's return to a previously attended location.

Deficits can be especially large on so-called 'executive function' tests, which require top-down control in order to override automatic responses driven by bottom-up stimulus characteristics (e.g., moving one's attention away from a peripheral cue in the antisaccade task, responding to ink color rather than word identity in the Stroop task) or to maintain and rapidly switch between multiple response rules (e.g., dual-task procedures). The memory problems of AD may contribute to poor performance on tasks designed to measure attention, as patients may have difficulty both with remembering the response rule and with its execution.

The specific abnormalities in the regulation of integrity of neuronal systems responsible for the age- or dementia-related impairments in attention and encoding are still not well understood. However, structural and regulatory age differences in basal forebrain cholinergic projection have been extensively documented in brains of humans and animals. Furthermore, the development and maintenance of forebrain cholinergic neurons depend on nerve growth factor (NGF) signaling via several neurotrophic receptors expressed selectively by cholinergic neurons. The availability of such receptors is dramatically reduced in the brains of subjects with mild cognitive impairments. Given the central role of the cholinergic system in attention (as described above), these reductions support the hypothesis that the attentional and related cognitive impairments observed in these patients are due, at least to a significant extent, to dysregulation of forebrain cholinergic neurons. Subsequent loss of forebrain cholinergic neurons during the onset and progression of dementia contributes essentially to the severity of impairments in attention and memory.

Impairments in both the activation of relevant stimuli and the inhibition or filtering of irrelevant stimuli are among the fundamental cognitive dysfunctions of schizophrenia. These disruptions of attentional abilities are reflected in impaired L&M. For example, the exhaustion of attentional resources for encoding of relevant stimuli contributes to impaired maintenance and updating of patients' memory, thereby perhaps contributing to the development of aspects of positive symptoms. Abnormal metabolic responses to cognitive challenges have been observed in frontoparietal networks, although the specific neuronal mechanisms underlying the attentional dysfunctions of schizophrenia are not fully understood. However, available evidence strongly suggests an abnormally reactive dopamine system in schizophrenia. Given the role of dopaminergic–cholinergic interactions in the motivational regulation of attentional performance (as described above), both cortical cholinergic recruitment of the anterior attention system and the cholinergic mediation of input processing in sensory and sensory-associational regions are expected to be highly abnormal in schizophrenia. Evidence from animal models of this disease corresponds with this hypothesis, as does the demonstration of down-regulated muscarinic receptors in the cortex of schizophrenic patients. It is widely assumed that abnormalities in the development of cortical, particularly prefrontal, circuits represent a primary neuropathological foundation for schizophrenia. These abnormalities may in turn lead to the dysregulation of the ascending, bottom-up modulatory systems (as described above). Moreover, the dysregulated release of these neuromodulators interacts with defective cortical target circuits, collectively mediating the attentional and encoding dysfunctions characteristic for schizophrenia.

See also: Aging and Memory in Humans; Attentional Networks; Attentional Networks in the Parietal Cortex; Basal Forebrain and Memory; Cognition: An Overview of Neuroimaging Techniques; Cognitive Deficits in Schizophrenia; Executive Function and Higher-Order Cognition: Neuroimaging; Humans; Memory Representation; Prefrontal Cortex: Structure and Anatomy; Prefrontal Contributions to Reward Encoding; Psychophysics of Attention; Reward Systems: Human.

Further Reading

Aston-Jones G and Cohen JD (2005) An integrative theory of locus coeruleus-norepinephrine function: Adaptive gain and optimal performance. *Annual Review of Neurosciences* 28: 403–450.

Buckner RL (2004) Memory and executive function in aging and AD: Multiple factors that cause decline and reserve factors that compensate. *Neuron* 44: 195–208.

Cabeza R and Nyberg L (2000) Imaging cognition II: An empirical review of 275 PET and fMRI studies. *Journal of Cognitive Neuroscience* 12: 1–47.

Corbetta M and Shulman GL (2002) Control of goal-directed and stimulus-driven attention in the brain. *Nature Reviews Neuroscience* 3: 201–215.

Craik FIM and Byrd M (1982) Aging and cognitive deficits: The role of attentional resources. In: Craik FIM and Trehub S (eds.) *Aging and Cognitive Processes*, pp. 191–211. New York: Plenum.

Duclukovic NM and Wagner AD (2006) Attending to remember and remembering to attend. *Neuron* 49: 784–787.

Fernandes MA and Moscovitch M (2000) Divided attention and memory: Evidence of substantial interference effects at retrieval and encoding. *Journal of Experimental Psychology: General* 129: 155–176.

Kruschke JK (2003) Attention in learning. *Current Directions in Psychological Sciences* 12: 171–175.

Mesulam M (2004) The cholinergic lesion of Alzheimer's disease: Pivotal factor or side show? *Learning and Memory* 11: 43–49.

Mufson EJ, Kroin JS, Sendera TJ, and Sobreviela T (1999) Distribution and retrograde transport of trophic factors in the central nervous system: Functional implications for the treatment of neurodegenerative diseases. *Progress in Neurobiology* 57: 451–484.

Perry RJ, Watson P, and Hodges JR (2000) The nature and staging of attention dysfunction in early (minimal and mild) Alzheimer's disease: Relationship to episodic and semantic memory impairment. *Neuropsychologia* 38: 252–271.

Reuter-Lorenz PA and Lustig C (2005) Brain aging: Reorganizing discoveries about the aging mind. *Current Opinion in Neurobiology* 15: 245–251.

Sarter M, Bruno JP, and Givens B (2003) Attentional functions of cortical cholinergic inputs: What does it mean for memory? *Neurobiology of Learning and Memory* 80: 245–256.

Sarter M, Gehring WJ, and Kozak R (2006) More attention must be paid: The neurobiology of attentional effort. *Brain Research Reviews*. 51: 145–160.

Sarter M, Hasselmo ME, Bruno JP, and Givens B (2005) Unraveling the attentional functions of cortical cholinergic inputs: Interactions between signal-driven and top-down cholinergic modulation of signal detection. *Brain Research Reviews* 48: 98–111.

Sarter M, Nelson CL, and Bruno JP (2005) Cortical cholinergic transmission and cortical information processing following psychostimulant-sensitization: Implications for models of schizophrenia. *Schizophrenia Bulletin* 31: 117–138.

Wager TD, Jonides J, and Reading S (2004) Neuroimaging studies of shifting attention: A meta-analysis. *Neuroimage* 22: 1679–1693.

Attentional Mechanisms in Ventral Pathway

L Chelazzi, C Della Libera, and E Santandrea,
University of Verona, Verona, Italy

Introduction

Perhaps the most obvious form of visual selective attention is when individuals turn their gaze toward a salient or otherwise interesting object in their surroundings to align it with the high-resolution fovea of the retina. This allows more detailed processing of the fixated object at the expense of competing objects falling on peripheral regions of the retina. However, it is well established that selective attention can be aimed at extrafoveal locations and objects, thus effectively decoupling the high-resolution power of the fovea from enhanced central processing due to selective attention.

There appear to be several computational reasons why the brain implements selective attention mechanisms. First and foremost, selective attention can be viewed as the mechanism that mediates selection of the next target for preferential, foveal analysis. In this vein, selective attention primarily assists the oculomotor system to optimize sensory sampling of the visual environment, given the current goal. Second, motor systems in general, including those for reaching and grasping movements, are physically constrained and can act only on one (or a few) objects at any given moment. Therefore, selective attention is needed to focus processing onto a single object in order to plan coherent behavioral responses targeted at the selected object. Third, it is probably impossible, or perhaps simply disadvantageous, for the memory systems of the brain to store each and every single object and event occurring within a crowded environment. Therefore, selective attention is necessary to gate access of perceptual representations to memory systems. Finally, perceptual awareness is inherently limited in nature, and it unfolds serially, with a single perceptual representation gaining dominance at any instant in time. Therefore, selective attention is needed to allow entrance of the selected representation into working memory and conscious perception. On top of all the reasons above, students of vision and selective attention raise two further reasons why selective attention may be indispensable, and they are probably related. One reason is that processing of incoming retinal input must be focused on a single object at a time simply because, otherwise, processing and recognition of all objects simultaneously would overcome the limited processing capacity of the system. In addition, and more specifically, directing attention toward a single object at a time might serve the important function of aiding the correct conjoining of all its elemental features, therefore preventing the erroneous binding of features belonging to separate objects in a cluttered scene. In brief, selective attention appears to be a key mechanism aiding efficient object recognition, perceptual awareness, goal-directed behavior, and selective memory storage.

Given the key role of selective attention in visual processing, in particular its role in building and gating object representations, it comes as no surprise that much of the relevant experimental work over the past two decades has been devoted to the investigation of the neuronal correlates of selective attention along the ventral pathway of cortical visual processing. The ventral pathway originates at the level of primary visual cortex, or V1, and nearby secondary visual cortex, or V2, and further extends through extrastriate area V4 and posterior inferotemporal cortex, or area TEO, to culminate in a relatively vast cortical territory occupying the middle and anterior segments of the inferotemporal (IT) cortex. The ventral pathway represents a network of interconnected areas, largely organized according to a hierarchical scheme, whereby object representations are created with an ascending level of complexity and representational invariance. Ultimately, patterns of activity within IT cortex are now known to represent with remarkable speed and efficiency the various objects we are able to recognize. Key nodes along this pathway are represented by area V4 and the various sectors of IT cortex, and therefore this article will focus on these nodes of the pathway.

Studying the Manifestations versus the Control

In speaking of selective attention, one should distinguish between the manifestations and the causal control mechanisms. Specifically, one may use the term 'selective attention' to refer to the modulation of sensory processing along the ventral pathway in relation to concurrent changes in behavioral performance. In this case, the term would index the manifestations of visual selective attention at the neuronal, as well as at the behavioral, level. In contrast, one may use the term selective attention to refer to the signals that, within a given behavioral context, bring about the manifestations of attention considered above. The latter signals may or may not originate within the visual system, and the available evidence suggests that in most cases they do not. This article mainly concentrates on the manifestations of selective attention within the ventral pathway but also briefly discusses available evidence concerning the signals impinging on the ventral pathway to exert attentional control. Several forms of selective attention are distinguished, including enhanced processing of individual attended items, selective processing among competing items (or biased competition), feature-based guidance of target selection in visual search, and finally, feature-selective attention.

Enhanced Processing of Attended Objects: The Beneficial Effects of Spatially Directed Attention

Psychophysical studies of human observers have documented robust effects of attention on visual sensitivity at selected regions of space. Sensitivity has been shown to increase at attended versus unattended locations in the visual field, with relatively shorter reaction times to detect an item at the attended location, as well as greater accuracy. In particular, attentional facilitation entails better detection of faint, low-contrast stimuli and improved discrimination of their features, as if attention to the stimulus led to enhancement of signal strength. In turn, these effects are reminiscent of those produced by an increase in stimulus contrast, and it has been reported recently that indeed attention increases perceived stimulus contrast. Consistent with these behavioral results, single-unit recording studies in the behaving macaque have found enhanced neuronal responses (e.g., in area V4) to a single stimulus presented inside the receptive field (RF) of the recorded neuron when the animal's attention is aligned with the stimulus location, relative to when attention is directed elsewhere in the visual field. As a result, stimuli at an attended location engender stronger central representations than unattended stimuli do. These neural effects likely represent part of the mechanism underlying enhanced behavioral performance, as previously described. Overall, however, enhancement of neuronal responses to individual stimuli presented inside the RF is not very strong, typically on the order of 20%. Furthermore, the effect has not been found in all reported studies, and a possible account of this variability is offered below.

Recent studies of neuronal responses in area V4 have shed further light on the above modulation of responses to single RF stimuli as a function of attention. If the effects of attention are akin to those brought about by increased stimulus contrast, then one might predict that directed attention changes the contrast response function of neurons. Neurons at many stages of the visual system produce increasing responses as a function of stimulus contrast, up to a plateau, and the function takes the form of a sigmoid. If attention acts by increasing the effective contrast of the RF stimulus, then one predicts a leftward shift in the contrast response function of the neurons. In line with this prediction, it was found that attention directed to an RF stimulus causes a leftward shift of the sigmoid relative to when the stimulus is unattended. As a consequence, responses to an attended stimulus will not differ reliably from those to an unattended stimulus at or beyond the point of saturation in the contrast response function. Instead, effects of attention will be greatest within – or just below – the dynamic range of the contrast response function of the neuron. These findings may explain, at least in part, why not all single-cell recording studies have found enhanced responses to attended compared with unattended single RF stimuli along the ventral pathway, including area V4, since attentional effects may be minimal, if any, when stimuli of high contrast are employed. In summary, prevailing evidence indicates that stimuli presented at attended locations will elicit greater responses compared with stimuli at ignored locations. However, the effect is relatively large with stimuli of low contrast, whereas it tends to decrease with contrast of the stimulus.

Notice that there is an important difference between the effects of directed attention and contrast on responses of neurons in visual cortex, including area V4; namely, while attention mimics the effect of contrast in terms of response magnitude, it does not do so in terms of response onset latencies. Response latency has been shown to increase considerably for low-contrast stimuli, whereas no detectable change in response latency is associated with manipulations of attention. This imposes some caution in likening effects of attention to changes in effective stimulus contrast.

An important question is whether directed attention, in addition to changing the strength of neuronal responses, will modify neuronal tuning for the stimulus features, for example, stimulus orientation. This has been addressed in a number of studies, and the prevailing view is that tuning properties of neurons are relatively immune to the influence of directed spatial attention, although they may be modified as a result of extensive discrimination training with perceptual learning protocols. Instead, spatially selective attention has been shown to cause a multiplicative scaling of tuning curves. Responses throughout the tuning curve will be multiplied by a constant factor, with no appreciable changes in the filter properties of neurons. Again, this is similar to the known effect on tuning curves of varying stimulus contrast. Nonetheless, it is conceivable that at the population level, a gain modulation of tuning curves allows finer encoding of features at an attended location than at an unattended one, for instance by reducing the signal-to-noise ratio at the attended location.

Top-Down Control: Biases and Baseline Shifts

Given the distinction between manifestations of attention in sensory processing areas and control signals, researchers have sought evidence for control signals that may cause the manifestations of attention summarized previously. A key feature of these signals is that they ought to precede onset of task-relevant stimuli, that is, they should be present while the animal is attending to a given visual field location in preparation for performing a task on some relevant item. In practice, people have compared baseline activity of the neurons during the waiting period of the task between conditions in which the animal's attention was directed toward the RF of the studied neuron versus when attention was directed toward some location outside the RF. Single-unit recording studies have shown that neurons in area V4 (and V2) display elevated baseline firing during periods when the animal is attending to a location inside the RF of the neuron in anticipation of RF stimulus onset. It is interesting that analogous changes in baseline activity depending on the direction of spatial attention can be observed even when one compares attention to different locations inside the single RF, provided that the locations to be compared are not equally sensitive. Specifically, baseline activity has been found to covary with the strength of the visually evoked responses at any given location within the RF. Increases in baseline activity are typically small in absolute terms, on the order of a few spikes per second. However, in fractional terms, they can amount to a 50% increase in firing rate in the absence of visual stimulation. Therefore they represent a substantial percentage increment in neural activity over a relatively large population of neurons, those neurons with RFs encompassing the attended location.

The accepted account of elevated baseline activity due to spatial attention is that it reflects the influence of an incoming signal, originating in areas of the brain responsible for exerting control over spatially directed attention. In the case of area V4, these likely include areas in the posterior parietal and prefrontal cortices, although subcortical sources (e.g., the superior colliculus) have also been implicated. Recent evidence obtained with low-current electrical microstimulation has directly demonstrated that signals of this sort may originate at the level of the frontal eye fields, and they are capable of enhancing visual responses within area V4 at selected visual field representations. The same type of microstimulation was also shown to improve the animal's performance in a demanding stimulus detection task. Analogous effects on behavior have been obtained with electrical microstimulation of the superior colliculus, although it remains to be established whether microstimulation of the superior colliculus would also enhance neuronal responses in area V4 or other areas along the ventral pathway. In summary, the available data suggest that a number of cortical and subcortical regions are involved in delivering control signals for spatially directed attention, and they largely overlap with critical nodes of the circuit controlling saccadic eye movements. These signals are likely responsible for elevated baseline firing in areas along the ventral pathway when attention is directed to a location inside the RF of the recorded neuron. The elevated baseline firing, in turn, may be part of the mechanism that confers to the neurons increased sensitivity to visual stimulation.

Coherent Firing at the Population Level

Recent work has shown that attention to an RF location may entail, not only elevated baseline firing and enhanced responses to an RF stimulus, but also increased synchronization of firing among the relevant neurons. Increased synchronization may or may not take the form of oscillatory activity, but typically it does. Therefore evidence is rapidly accruing to indicate that when attention is directed to a given location in the visual field, neurons with RFs encompassing that location will entertain an enhanced coherent firing, usually in the gamma-band frequency range, around 50 Hz. In turn, increased synchronization of firing at the attended location may enhance synaptic transmission downstream of the considered neural population, effectively amplifying transmission of information in a spatially selective manner. It is interesting that this effect has also been observed under task conditions in which there was no consistent change in the magnitude of visual responses as a function of spatial attention. Therefore, enhanced processing and spike transmission at the attended location may take the form of increased firing, increased synchronization, or both. It remains to be established whether there are specific task parameters that lead preferentially to one or the other manifestation of attentional modulation.

Competitive Interactions among Multiple Visual Stimuli

A special problem for perceptual and attentional mechanisms to solve is one in which multiple stimuli are presented together (crowding) and an individual must select the relevant stimulus while at the same time discarding any potential distracter, in particular nearby distracters. In neurophysiological terms, this translates to conditions in which multiple stimuli impinge simultaneously onto the RF of an individual neuron and they compete for controlling the neuron's firing pattern. It has been demonstrated that neurons along the ventral pathway, including area V4 and the IT cortex, produce responses to two or more stimuli falling inside their RF that approximate the average of the responses elicited by the component stimuli presented in isolation. In other words, neurons in areas of the ventral pathway seem to be incapable of clutter invariance, a property that, if present, would allow neurons to encode the single most preferred stimulus inside the RF while automatically discarding other nearby stimuli, effectively implementing a MAX operation. In contrast, it appears that multiple stimuli falling inside a single RF, or in its immediate surroundings, compete for the encoding capacity of the neuron, and the neuron's firing is ambiguous as to which stimulus is encoded. It seems that competitive interactions among multiple RF stimuli are only weakly affected, if at all, by the specific nature of the stimuli involved, including the degree of their similarity (but see the discussion of luminance contrast in the section titled 'Top-down versus bottom-up in selective attention'), except that the greatest competitive interactions occur with stimuli far apart in their ability to drive a neuron's visual response, as when a highly preferred or a null stimulus is involved. Under these circumstances, the presence of the null or ineffective stimulus can drive the response to the preferred stimulus well below the level that it would have elicited if presented alone. An important notion that has emerged from these studies is that a stimulus, which causes only modest changes in firing rate when presented in isolation, can nevertheless exert a profound (suppressive) influence on the neuronal firing when presented in combination with an effective stimulus, thus demonstrating clear-cut decoupling between effectiveness of stimuli in driving a response from the given neuron and their effectiveness in determining the firing rate of the same neuron. In terms of the latter property, an ineffective stimulus can be no less effective than a highly preferred, or optimal, stimulus in exerting control over the neuron's activity. In area V4, competitive interactions of this sort have been shown to span a limited extent of visual space, covering the RF size and extending only a small distance beyond the boundary of a neuron's RF. In contrast, competitive interactions sometimes span a much larger extent of the visual field in IT cortex, including portions of the visual hemifield ipsilateral to the recorded hemisphere. However, it has been reported that competitive interactions in IT cortex are much weaker, or nearly absent, when competing stimuli are placed across the vertical meridian, as if competitive interactions could not come about at full strength when they involve the midline commissures (e.g., the corpus callosum). Selective attention mechanisms are needed to resolve these competitive interactions – the core notion of the biased competition model of attention.

Resolving the Competition: Selection and Filtering

The biased competition model of attention has been highly influential over the past decade, as it can account for a great deal of experimental observations obtained with a variety of approaches and techniques, in both human and animal studies of perception and attention. Mathematical and neural network implementations of the model have been developed. The model rests on two tenets. First, as discussed in the previous section, multiple stimuli falling within the RF of a given neuron (or in its immediate surround) compete for controlling the neuron's firing rate. The most compelling evidence of this takes the form of suppressed responses to an effective stimulus falling inside the RF of a cell when it is paired with a second, ineffective stimulus for the cell, with responses to the pair approaching an average of the responses elicited by each of the two stimuli in isolation. Second, competition among stimuli can be resolved when a signal biases the competitive interaction in favor of either stimulus in the pair, thus causing the cell's firing to be primarily determined by the favored stimulus. When this occurs, selective attention is enacted: one of the competing stimuli is selected; the other is filtered out of the RF, or ignored. In cell physiological terms, a neuron's firing to multiple stimuli impinging on its RF as if only one of them were present – the favored one – corresponds to selecting a salient or otherwise relevant stimulus while discarding distracters.

As already considered, the biasing signal for spatially directed attention may take the form of elevated baseline activity of the relevant neural population, but the proposal has been made that increased synchronization of firing across the population of neurons with RFs encompassing the attended location may as well bias competition in favor of the relevant stimulus location. It remains to be established to what extent increased baseline firing and enhanced synchronization are related phenomena in functional terms. Regardless of this, we have already mentioned that likely sources of signals biasing competition in favor of the attended location include cortical areas such as the frontal eye field and lateral intraparietal area, in the frontal and posterior parietal cortex, respectively, as well as subcortical structures, such as the superior colliculus. Future work might well reveal that other parts of the brain, at the cortical and subcortical level, play as important a role in controlling spatial attention.

Selection of a relevant object (or target) among competing stimuli can be achieved not only on the basis of its location in space, but also on the basis of its feature composition. For example, in visual search tasks, an observer is asked to find a target object among irrelevant distracters. Under some conditions, the target may be found easily, at no increasing cost as a function of the number of distracters, such as when it is characterized by some unique property (known as 'pop-out'). In contrast, under less efficient conditions, locating the target may take some effort and increasing time as a function of the number of distracters. By means of search tasks of the latter kind, it has been shown that neurons in areas V4 and IT may contribute significantly to the search process. In particular, as the search process unfolds, neurons in both areas come to encode the target but much less, or not at all, the distracters. Specifically, while neural activity shortly after search array onset to some extent represents all items in the array, later on, in anticipation of the behavioral response, only the target item activates the neural population which is selective for its constituent features, while neural populations activated by the features of the distracters are strongly suppressed. This form of selective attention has been shown to engage underlying mechanisms similar to those engaged by spatially directed attention, except that here, selection is guided by feature information. It has been further suggested that control signals for feature-based selection of a target object likely originate in at least partly different brain regions from those involved in delivering control signals for spatially selective attention. The proposal has been made that feature information specifying the target item and guiding its ultimate selection is represented within brain networks responsible for holding object feature information online during the execution of the task (working memory).

Behavioral evidence obtained following lesion or deactivation of area V4 (and/or TEO) in the monkey is in full agreement with the biased competition model of attention. This work has elegantly shown that, when selective attention mechanisms are knocked out, the animal is at the mercy of stimulus salience. In other words, when multiple stimuli are presented and the animal must select a high-salience target among low-salience distracters, behavior is largely unimpaired. Conversely, when the animal is required to select a low-salience target among high-salience distracters, performance shows a dramatic drop. Consistent with a key role of area V4 in the implementation of attention mechanisms, this deficit has been observed following lesion of area V4 in the macaque, as well as following damage to the homolog of area V4 in the human brain. These findings suggest that, when the mechanisms for cognitively mediated selection are compromised, such as can be obtained through damage to area V4 (and TEO), competitive mechanisms and selection are primarily controlled by the intrinsic salience of objects, the topic of the next section.

Top-Down versus Bottom-Up in Selective Attention

There is now evidence at the single-cell level that competitive interactions among multiple stimuli falling inside the RF of an individual V4 neuron are directly modulated by stimulus salience, such as can be obtained by varying the luminance contrast of the stimuli. As already noted, with attention directed well outside the RF of the recorded neuron (e.g., to the opposite visual hemifield), adding an ineffective stimulus reduces responses of V4 (and IT) neurons to a concurrently presented effective stimulus for the neuron. It has also been recently shown that the suppressive effect is progressively stronger as the luminance contrast of the ineffective stimulus is increased, with the contrast of the effective stimulus held constant at an intermediate level (40%). Although the suppressive effect increases with contrast of the ineffective stimulus, notice that at the same time, the ineffective stimulus presented alone elicits a progressively larger, albeit weak, visual response when its contrast is increased. This again indicates a remarkable dissociation between the efficacy of a stimulus to drive a visual response from a neuron and efficacy of the same stimulus to control the neuron's firing. A stimulus that, for its feature composition, may be largely ineffective in driving a visual response from a given neuron, can nonetheless be highly effective in determining the neuron's response, due to its salience, or strength, such as its high luminance contrast. Moreover, within the same experimental context, attention directed to the ineffective stimulus in the pair has been shown to further enhance the suppressive effect exerted by this stimulus to the point that attention to a high-contrast ineffective stimulus almost completely dominates the cell's firing, namely, it almost completely silences the cell. These findings indicate that competitive interactions are entertained automatically within visual cortex and that competition can be resolved in favor of a high-salience (e.g., high-contrast) stimulus in bottom-up, in the absence of top-down signals reflecting the current volitional control on selective attention.

Feature-Based Attention

As noted previously, selective attention can be directed toward a specific spatial location, or it can be guided by feature information specifying the target-defining properties. Furthermore, behavioral evidence in humans indicates that feature-based attention can affect processing throughout the entire visual field, in a parallel fashion. Consistent with this, single-unit recordings from area V4 of the macaque have revealed the correlates of this form of nonspatial selection. It has been discovered that neuronal responses to any potential target in the visual field – that is, any element that shares one or more of the target-defining features, including the target itself – are enhanced as the search process progresses, long before the animal actually locates the designated target. In other words, this form of feature-based attention is able to 'highlight' all the objects in the visual array that are potentially relevant for the task at hand. Essentially, the mechanism allows privileged processing of these objects, while other objects are effectively filtered out in parallel across the visual array. Although findings of this kind have come in slightly different flavors in the literature, perhaps related to specific characteristics of the experimental protocols, all converge to indicate that among the entire population of neurons in area V4 activated by the array elements, the neurons firing at the highest rate will be those directly stimulated by a feature in the RF that matches the feature preference of the neurons (e.g., red) while the animal is searching for a target item defined by the same feature (e.g., red). Evidently, depending on the currently relevant features, a specific control signal can target the neuronal populations with RFs anywhere in the visual field that are selective for the corresponding features.

Gating-Feature Information, or Feature-Selective Attention

Unlike the form of feature-based attention discussed in the previous section, feature-selective attention is engaged under task conditions in which an individual is asked to identify, or otherwise respond to, a specific object feature while at the same time ignoring other features of the same object. This form of feature-selective attention, therefore, entails that the unity of perceptual objects be broken down in order to cope with the current task. Feature-selective attention plays an important role in many real-life situations, for instance when an individual wishes to sort, or classify, objects on the basis of one elemental feature (e.g., color) and other features (e.g., shape and texture) must be ignored. In addition, this type of feature-selective processing is tapped by a number of classical neuropsychological tests, such as the Stroop test and the Wisconsin card sort test. In both cases, performance must be guided by selective feature information, and interference from the irrelevant feature or features must be blocked. The neuronal underpinnings of the latter form of feature-selective attention have been systematically explored in a recent single-unit recording study in which the activity of V4 neurons was recorded while an animal was attending to either one or the other feature of differently colored, oriented bars. It was found that, under these task conditions, responses of V4 neurons to otherwise identical stimuli are modulated depending on the component feature of the stimulus being currently attended. Most important, it turns out that a large fraction of the recorded neurons are able to cluster the attended features of the stimuli into one or the other of two behaviorally relevant response categories, indicating that area V4 may be important in the process of converting selected feature information into a categorical code available to guide the animal's behavioral responses.

Conclusions

Research over the past 25 years has allowed impressive progress in the understanding of the brain mechanisms underlying the ability to concentrate mental resources on a single location or object at any given time – an essential component of the ability to implement goal-directed behavior. Fundamental pieces of evidence have come from neurophysiological investigations in the awake, behaving macaque monkey. Science is very close to a full understanding of what it means at the single-neuron level to pay selective attention to a specific location or object, or object feature, including the fine details of the circuitry that brings about attentional modulation of firing in visual cortical areas, as well as the source and nature of the signals that control the same circuitry, thus initiating attention-related phenomena at the neuronal and behavioral level. The investigation of the neuronal correlates of visual selective attention along the ventral pathway of cortical visual processing has been particularly successful at identifying specific ways in which mechanisms for selective attention are intertwined with perceptual mechanisms for feature analysis and object recognition.

See also: Attention and Eye Movements; Attention: Models; Attentional Networks; Attentional Networks in the Parietal Cortex; Attentional Functions in Learning and Memory; Decision-Making and Vision; Neglect Syndrome and the Spatial Attention Network; Psychophysics of Attention; Vision for Action and Perception; Visual Attention.

Further Reading

Bichot NP, Rossi AF, and Desimone R (2005) Parallel and serial neural mechanisms for visual search in macaque area V4. *Science* 308: 529–534.

Chelazzi L, Duncan J, Miller EK, and Desimone R (1998) Responses of neurons in inferior temporal cortex during memory-guided visual search. *Journal of Neurophysiology* 80: 2918–2940.

Desimone R and Duncan J (1995) Neural mechanisms of selective visual attention. *Annual Review of Neuroscience* 18: 193–222.

De Weerd P, Peralta MR III, Desimone R, and Ungerleider LG (1999) Loss of attentional stimulus selection after extrastriate cortical lesions in macaques. *Nature Neuroscience* 2: 753–758.

Fries P, Reynolds JH, Rorie AE, and Desimone R (2001) Modulation of oscillatory neuronal synchronization by selective visual attention. *Science* 291: 1560–1563.

Luck SJ, Chelazzi L, Hylliard SA, and Desimone R (1997) Neural mechanisms of spatial selective attention in areas V1, V2, and V4 of macaque visual cortex. *Journal of Neurophysiology* 77: 24–42.

Maunsell JH and Cook EP (2002) The role of attention in visual processing. *Philosophical Transactions of the Royal Society of London, Series B: Biological Sciences* 357: 1063–1072.

Maunsell JH and Treue S (2006) Feature-based attention in visual cortex. *Trends in Neuroscience* 29: 317–322.

McAdams CJ and Maunsell JH (1999) Effects of attention on orientation-tuning functions of single neurons in macaque cortical area V4. *Journal of Neuroscience* 19: 431–441.

Mirabella G, Bertini G, Samengo I, et al. (2007) Neurons in area V4 of the macaque translate attended visual features into behaviorally relevant categories. *Neuron* 54: 303–318.

Moore T and Armstrong KM (2003) Selective gating of visual signals by microstimulation of frontal cortex. *Nature* 421: 370–373.

Motter BC (1994) Neural correlates of attentive selection for color or luminance in extrastriate area V4. *Journal of Neuroscience* 14: 2178–2189.

Reynolds JH and Chelazzi L (2004) Attentional modulation of visual processing. *Annual Review of Neuroscience* 27: 611–647.

Reynolds JH, Chelazzi L, and Desimone R (1999) Competitive mechanisms subserve attention in macaque areas V2 and V4. *Journal of Neuroscience* 19: 1736–1753.

Zoccolan D, Cox DD, and DiCarlo JJ (2005) Multiple object response normalization in monkey inferotemporal cortex. *Journal of Neuroscience* 25: 8150–8164.

Attentional Networks

N U F Dosenbach and S E Petersen, Washington University in St. Louis School of Medicine, St. Louis, MO, USA

Attention is the brain's ability to selectively allocate cognitive resources to those stimuli, responses, memories, and trains of thought that are behaviorally most relevant, at the expense of less relevant ones. As William James pointed out in 1890, "Each of us literally chooses, by his ways of attending to things, what sort of universe he shall appear to himself to inhabit."

Attention can have widespread effects on behavior. It has been known since the time of William James that attention can improve our ability to perceive, conceive, distinguish, remember, and respond. For example, attending to the spatial location at which a visual target will occur decreases the time it takes to respond to it, even when subjects only covertly shift their attention to the cued target location in the absence of overt eye or head movements.

Attention has been shown to alter neural activity. Single-unit recording studies in macaque monkeys placed stimuli in the receptive fields of single cells and compared two conditions: (1) when the monkey was attending to the stimulus and (2) when it was attending somewhere else in the visual field. The classical finding has been that selectively attending to a stimulus increases the neuronal response to it. Similar effects have been seen in neuroimaging studies of selective attention in humans. Selectively attending to the color, shape, and motion of an object, for example, increases activity in extrastriate visual regions specialized for processing these features.

The attention-driven modulation of neural activity is not believed to be inherent to early sensorimotor regions of the brain. Instead, it is thought that so-called biasing signals from higher order source regions influence moment-to-moment processing in sensorimotor regions such as extrastriate visual cortex.

The notion that anatomically separate brain regions may control the selective allocation of attention was initially driven by studies of neglect patients. Unilateral lesions, particularly in temporoparietal and frontal cortex, often cause patients to neglect the contralateral half of extrapersonal space, a condition known as spatial neglect. Such patients may only dress the ipsilesional half of their body, only eat the food on the ipsilesional half of their plate, and only attend to stimuli in the ipsilesional half of their visual field. Studies showed neglect to be a primary deficit of attention, not of sensorimotor processing.

Principles of the Human Attention System

The combination of connectional anatomy, electrophysiology, lesion research, cognitive psychology, and positron emission tomography (PET) imaging allowed Posner and Petersen to build a theoretical cognitive neuroscience model of human attention. This model proposed attention to be the emergent property of a network of functional areas. Posner and Petersen formulated three principles of attentional networks that still seem relevant: (1) The brain's attention system is anatomically separate from those downstream systems that process specific inputs independent of whether these inputs are being attended to or not; (2) attention is the emergent property of networks of distinct anatomical areas, not a single area; and (3) these areas and networks carry out separable attentional functions.

Posner and Petersen focused on three putative classes of attentional processes: alerting, orienting, and detecting targets for conscious processing. Much research since then has expanded and refined this initial set of proposed classes.

Alerting

Alerting is thought to constitute the most basic attentional function. It appears to consist of distinct subfunctions or processes. Intrinsic alertness describes the ability to maintain certain levels of arousal in the absence of cues. It is thought to reflect general increases in excitability, mediated by top-down control signals. Changes in intrinsic alertness can be assessed over longer periods of time (minutes to hours) by measuring simple reaction times to perceptual stimuli that occur without warning.

Alerting subjects to an upcoming target decreases reaction time and error rate. Phasic alerting effects can be documented by comparing reaction time and neural activity on uncued trials and trials for which a nonspecific warning cue provided temporal information about the upcoming trial (**Figure 1**(a)). Phasic alerting may ready task-specific processing pathways for the next stimulus.

Alerting is thought to be supported by the widespread cortical distribution of the brain's norepinephrine system arising in the locus coeruleus (LC-NE) of the midbrain. The reticular thalamus may relay the effect of LC-NE activity to the cerebral cortex. Imaging studies of phasic alerting indicate that the thalamus strongly responds to alerting cues. LC-NE neurons are most active during wakefulness and become silent during rapid eye movement sleep.

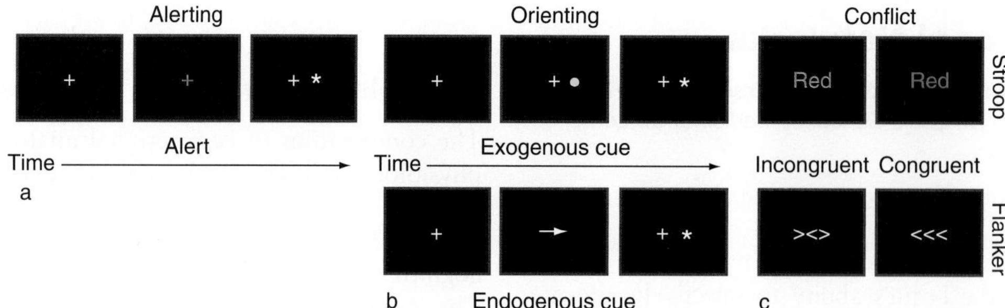

Figure 1 Examples of tasks commonly used to study attentional phenomena. (a) When asked to respond to a target (asterisk) by pressing a button, simple warning cues (e.g., color change of the fixation cross) that phasically alert subjects to an upcoming target can decrease reaction times and error rates. Such warning cues do not carry any spatial information about the target, only temporal information. (b) An exogenous orienting cue, such as a peripheral flash of light, will automatically capture selective attention and facilitate responses at the cued location, starting 50 ms after the cue. An endogenous orienting cue, such as a centrally placed arrow pointing to the right, will also facilitate responses at the cued location, but only after approximately 150 ms. Endogenous cueing effects are mediated by top-down mechanisms, whereas the effects of exogenous cues are thought to be entirely bottom-up or stimulus driven. (c) When the stimulus properties are incongruent, such that the task demands come in conflict with well-trained stimulus-driven responses, performance worsens. In the classic Stroop task, subjects are asked to report the ink color of words. When the word meaning (red) and ink color (green) are incongruent, the overtrained response of reading the word interferes with naming the ink color. In the Eriksen–Flanker task, subjects are asked to report the direction of a central arrow. Responses are faster when flanking arrows point in the same direction as the central arrow (congruent) than when they point in the opposite direction (incongruent). Since executive control processes are needed to overcome stimulus-driven responses, the Stroop, Eriksen–Flanker, and other conflict tasks have been used to study the brain's executive control networks. However, overcoming conflict from incongruent stimuli is only one of the many aspects of executive control.

Salient stimuli cause phasic LC-NE activations and norepinephrine release. LC-NE neurons project to parietal cortex, primary motor cortex, the pulvinar nucleus of the thalamus and the superior colliculus. Anterior cingulate cortex (ACC) and orbitofrontal cortex send strong projections to the LC. Therefore, it has been argued that LC activity might, in part, be controlled by the ACC.

Shifting Selective Attention: Cue Interpretation and Orienting

Attention can be selectively directed toward locations in space; intervals of time; features such as frequency, volume, color, and motion; semantic categories; abstract concepts; memories; and different output modalities. Although spatial attention has been the focus of many experimental studies, it may have its own specific mechanisms and represent a special case. It is important to keep in mind selective attention in its broader sense, which also includes selecting specific internal representations and response configurations.

In studies of spatial attention, the most commonly used analogy for focused selective attention is that of a spotlight. Although simple, this analogy captures the dynamics of selecting specific types of information initially proposed by Posner: disengaging attention from its current focus, moving attention, and engaging it at a new focus. Early PET and functional magnetic resonance imaging (fMRI) studies implicated regions in dorsal parietal and dorsal frontal cortex (DFC) under conditions of selective attention. Since then, lateral and medial parts of posterior parietal cortex (PPC), especially the intraparietal sulcus (IPS) and, more inferiorly, the temporoparietal junction (TPJ), have consistently been associated with shifts of selective attention. Regions in DFC, potentially constituting the human homologue of the frontal eye field, as well as parts of more ventral frontal cortex (VFC) are also widely believed to be important for selective attention.

A distinction can be made depending on whether the focus of attention is shifted involuntarily by a salient stimulus or voluntarily through top-down mechanisms. Stimulus-driven selective attention effects can be measured by flashing a cue on a screen (exogenous cue) and comparing how quickly subjects respond to targets at the cued location compared to a second location (**Figure 1(b)**). As early as 50 ms after the cue, reaction times are facilitated at the cued location. This is in contrast to situations in which cognitive or endogenous cues are used, such as a centrally placed arrow pointing toward one side of the visual field (**Figure 1(b)**) or the word 'right.' If the endogenous cue is valid and correctly predicts the location of the upcoming target, responses are also facilitated. However, these effects develop more slowly and are not measured until 150 ms after the cue.

Based largely on single-unit recording and event-related fMRI studies that can dissociate cue from target-related activity, Corbetta and Shulman proposed that stimulus-driven and goal-oriented shifts of selective attention might be mediated by partially distinct brain networks.

In this model, the right TPJ and parts of the right VFC appear to mediate stimulus-driven shifts of selective attention. The TPJ is relatively unresponsive to endogenous cues that provide information about an upcoming target, such as its likely location or direction of motion. Yet, the right TPJ and VFC are strongly activated by unexpected targets that are thought to automatically capture selective attention. This response to unexpected targets appears to be independent of the task's input and output modalities. Corbetta and Shulman have conceptualized the stimulus-driven attention control system as a circuit breaker that can disrupt the current attentional focus and redirect selective attention in the absence of voluntary control. This apparent right lateralization of stimulus-driven attention control is consistent with the clinical finding that neglect is most commonly due to right hemisphere lesions.

In contrast, bilateral regions in the IPS and DFC appear to be important for the voluntary engagement of selective attention. The IPS and DFC consistently show greater activity when subjects have been endogenously cued toward a specific target property or location than when they have been cued to passively view a display. Furthermore, the bilateral IPS and DFC show anticipatory pretarget activity that is time-locked to the presentation of the endogenous cue. This cue-related activity is more extended in duration than cue-related activity in occipital cortex, which likely reflects purely visual processing of the cue. It has been demonstrated that the IPS and DFC are also active when visuospatial attention is shifted based on long-term memory in the absence of cues.

Goal-directed selective attention signals in dorsal parietal and frontal cortex are not limited to visual tasks. Studies have demonstrated dorsal frontoparietal activity during endogenous shifts of selective attention between different locations in auditory space and different auditory features. Regions in PPC are also activated by voluntary shifts of selective attention between audition and vision. Furthermore, dorsal parietal and frontal regions are thought to covertly orient goal-directed selective attention toward specific movements. Such selective motor attention leads to reaction time facilitation analogous to selective visual attention.

In addition, DFC is thought to select specific movements in anticipation of a response. Single-unit recordings in adjacent parts of macaque PPC, as well as frontal cortex, have documented effector specific anticipatory activity triggered by endogenous cues. Dorsal frontoparietal regions may even be important for selectively attending to specific sensorimotor transformations or stimulus-response mappings, especially when they are simple and well practiced. This is consistent with task-switching experiments that show PPC and dorsal frontal regions to be active when subjects switch between different tasks while the attentional focus remains constant for both input and output.

Executive Control

Executive control of selective attention is believed to be the third major function of attentional networks. Executive control is composed of distinct processing classes related to the instantiation, maintenance, monitoring, and adjustment of attentional sets. The clearest evidence that humans have some voluntary control over the selective allocation of cognitive resources comes from the fact that we can choose to perform many different operations on a given stimulus. Top-down control is critically important for the flexibility of behavior because it allows us to implement arbitrary criteria for input and output selection, as well as stimulus categorization. Without executive control, humans would be automatons limited to a finite number of preset stimulus-response mappings.

The term executive implies that this system is informed about the current task goals. The brain's executive control system likely transmits top-down biasing signals to downstream moment-to-moment information processors, such as visual and sensorimotor cortex. The executive system should have some knowledge of the organizational states of downstream processors. Thus, executive control systems also include a bottom-up component. Executive control regions receive ongoing performance feedback signals that can be used to adjust top-down signals for improved task performance. Behavioral observations have provided insight into top-down and bottom-up signals related to executive control and their interplay. Task-switching paradigms, for example, have shown that under certain conditions reaction times increase on postswitch trials. It is thought that this response slowing reflects the time it takes executive control to clear the previous attentional set and implement a different one.

Despite the importance of executive control systems for behavioral flexibility, their control over the selection of processing pathways is incomplete. When voluntary task goals come in conflict with strong prepotent stimulus-response mappings, behavioral performance worsens. Several well-studied paradigms, such as the Stroop and Eriksen–Flanker tasks, demonstrate the behavioral decrements caused by such conflict (**Figure 1(c)**). For these tasks, reaction times increase when the stimulus properties are such that the task demands are incongruent with a well-learned prepotent response. In the Stroop task, for example, subjects are instructed to name the color of ink in which a word is written. Reaction times are slowest when the ink color and the meaning of the word are incongruent – for example, the word 'green' written in red ink. This response slowing is likely caused by competition between the voluntarily adopted task goals and the prepotent trained behavior of automatically reading the word.

Because top-down control over information processing is incomplete, executive systems are thought to receive feedback information about action outcomes so that top-down biasing signals can be adjusted as needed. The executive control system is thought to adjust the attentional set when performance, as indexed by slow reaction times or errors, is poor. Reaction time changes that are dependent on the nature of the previous trial are thought to reflect such trial-by-trial adjustments of control parameters. Often, reaction times will be systematically slower on the trial following an error, suggesting top-down adjustments of the attentional set. In the Stroop task, the average response time to an incongruent stimulus is faster if the preceding stimulus was incongruent than if it was congruent (**Figure 1(c)**). It has been proposed that adjustments of top-down biasing signals triggered by incongruent stimuli facilitate performance on subsequent trials.

Regions in dorsal anterior cingulate cortex/medial superior frontal cortex (dACC/msFC), dorsolateral prefrontal cortex (dlPFC), and anterior prefrontal cortex (aPFC) have been consistently associated with different executive control processes by a wide range of experimental approaches. The anterior insula/frontal operculum (aI/fO) has also been shown to play an important role in executive control. Although the brain's executive control network comprises all of these regions, each one of them likely carries out a slightly different executive control function.

The dACC/msFC is thought to play a central role in the exertion of executive control. It is believed to be essential for relating goal-oriented behavior to its outcomes. Evidence for the notion that the dACC/msFC carries out several important executive control processes comes from a variety of sources. Lesions of the dACC/msFC can lead to difficulties with the voluntary initiation and maintenance of complex behaviors, whereas more automatic stimulus-driven behavior is spared. In addition, dACC/msFC lesions can affect one's ability to correct task performance following an error.

Since executive control maintains the task goals, sustained neural activity is thought to be its hallmark feature. Maintenance activity has been measured in the dACC/msFC between a cue and the subsequent trial, as well as sustained across a whole block of trials. The dACC/msFC is activated by instructional cues that provide information about the task demands indicating that it is important for the instantiation of attentional sets. It also shows activity related to switching between tasks. Meta-analyses of neuroimaging studies have shown the dACC/msFC to be active for a wide variety of cognitive operations, which is consist with the idea that the dACC/msFC exerts executive control independent of the specific processing domain.

Besides signals thought to be related to the exertion of top-down control, a range of apparent feedback signals have also been measured in the dACC/msFC. Human neuroimaging studies have shown activation differences in the dACC/msFC related to errors, error likelihood, conflict, monetary loss and gain, pain, dread, social rejection, and expectancy violations. In addition, the dACC/msFC also carries domain-independent target detection signals.

Several experimental approaches besides human fMRI have shown that the dACC/msFC may help monitor behavior for errors. Event-related potentials (ERPs), for example, show a negative difference wave localized to the dACC/msFC when correct and error trials are compared. This effect, termed error-related negativity (ERN), is time-locked to the occurrence of the error. It has also been suggested that the dACC/msFC monitors conflict on a trial-by-trial basis. ERP studies have shown a difference wave when comparing congruent and incongruent stimuli localized to the dACC/msFC, labeled conflict-related-negativity (CRN; N450). Event-related fMRI studies have shown greater activity for high-conflict (incongruent) than low-conflict (congruent) trials in the dACC/msFC independent of input and processing domain.

It has been suggested that closely adjacent frontal midline structures may carry out different executive control functions. More posterior and dorsal msFC may implement attentional sets, whereas a slightly more anterior and ventral region in the dACC helps guide behavior by integrating actions and their outcomes. Imaging studies that compared self-guided actions to experimenter-guided ones showed greater activity for self-guided actions in the msFC. In contrast, the dACC did not show a preference for self-initiated movements. Instead, activity related to contingency-based learning and decision making has been documented in the dACC.

A region on the border of the anterior insula (aI) and frontal operculum (fO) has shown executive control properties very similar to the dACC/msFC. Imaging studies have shown the aI/fO to be coactivated with the dACC/msFC across a wide range of tasks.

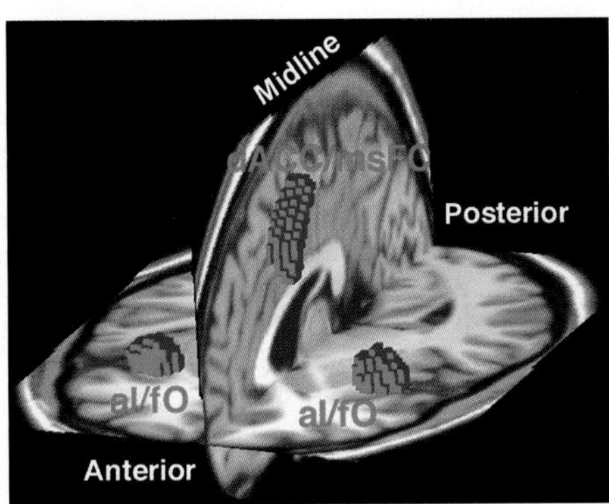

Figure 2 The dorsal anterior cingulate cortex/medial superior frontal cortex (dACC/msFC) has consistently shown activity related to attentional control. A cross-studies analysis of fMRI experiments documented attention-related signals in the dACC/msFC and bilateral anterior insula/frontal operculum (aI/fO). Across tasks, the dACC/msFC and aI/fO showed cue-related alerting/orienting as well as sustained task set-maintenance activity and error-related feedback signals important for executive control. Reproduced from Dosenbach NU, Visscher KM, Palmer ED, et al. (2006) A core system for the implementation of task sets. *Neuron* 50: 799–812, with permission from Elsevier.

Activation maps reveal that the bilateral aI/fO carries goal-maintenance and instantiation activity, as well as many of the same performance monitoring signals as the dACC/msFC (**Figure 2**). Similar to dACC/msFC lesions, strokes of the aI can lead to a reduction of self-initiated goal-directed activity.

It stands to reason that some brain regions may integrate executive control processes related to the instantiation, maintenance, monitoring, and adjustment of top-down biasing signals. Consistent with this notion, a cross-studies analysis showed that the dACC/msFC and aI/fO carried activity related to cue processing and the maintenance and monitoring/adjustment of attentional sets across a wide range of tasks. Therefore, it has been proposed that the bilateral aI/fO and dACC/msFC could form a domain-independent core of the human executive control network (**Figure 2**).

Regions in aPFC and dlPFC appear to support somewhat different executive control functions from the dACC/msFC and aI/fO. It is thought that the aPFC is especially important for the generation and maintenance of strategic plans, especially when they involve subgoaling and the integration of different types of information. It has also been proposed that aPFC neurons may adaptively code for different stimulus categorizations depending on the task goals. Humans with aPFC lesions often have difficulties solving complex tasks that require strategic planning. The maintenance of task contingencies, rules, and plans seems to be especially impaired by aPFC lesions. A series of fMRI experiments have documented sustained signals in the aPFC likely related to the maintenance of attentional sets and complex rules. However, this attentional set-maintenance activity was not common across all tasks. aPFC maintenance signals seem to be selectively recruited by more complex categorizations. Consistent with these findings, a meta-analysis of neuroimaging studies showed that the contrasts highlighting lateral aPFC had compared activity on trials with slow reaction times to activity on faster trials. In addition, neuroimaging studies have also shown activity related to switching between different task sets and stimulus categorizations in aPFC.

dlPFC may be specifically important for trial-by-trial adjustments of attentional control parameters. Consistent with this idea, an ERP study has shown that right dlPFC lesions increase errors and alter the ERN. Some dlPFC lesion patients fail to correct their errors and show no improvements during extended performance of complex cognitive tasks. Human fMRI studies have also shown greater dlPFC activity on error than correct trials. In contrast to the dACC/msFC, the dlPFC does not appear to carry conflict-related signals as measured by fMRI. According to the conflict-monitoring hypothesis, dlPFC functions as 'active memory in the service of control.' This hypothesis postulates that the dACC/msFC sends signals about needed adjustments in control parameters to the dlPFC, which helps to adjust attentional control parameters accordingly. In contrast, based on ERP data it has been argued that the dlPFC may be sending signals to the dACC/msFC, which in turn adjusts top-down biasing signals. Perhaps dlPFC maintains information about planned control adjustments from one trial to the next, whereas the dACC/msFC and aI/fO maintain the basic task parameters for the entire time period during which subjects are performing a specific task. The dlPFC may send adjustment-related signals directly to downstream processors, or it may transmit them to the dACC/msFC for downstream implementation.

Integration of Attentional Processes

The conceptual separation of attentional processes related to alerting, selective attention shifting (orienting), and executive control was initially based on differences in the behavioral effects of alerting, visuospatial cueing, and conflict. Although brain regions preferentially carry signals related to alerting, selective attention shifting, and executive control, these functions likely do not occur completely independently from each other. The brain's putative alerting, attention shifting, and executive control networks must communicate.

The executive control network needs to receive information about the meaning of sensory cues from both the alerting and the attention shifting networks so that it can implement the appropriate attentional settings. Conversely, the executive control network is thought to maintain the task goals according to which the attention-shifting network directs endogenous selective attention. It has even been shown that some stimulus-driven shifts in selective attention are contingent on the underlying attentional set. A stimulus that would not normally capture selective attention may do so if the subject is already searching for a feature shared by the stimulus. Voluntary changes in intrinsic alertness indicate that the executive control network can, at least indirectly, affect LC-NE activity.

Some data suggest that dACC/msFC and aI/fO may integrate several different executive control functions. It seems possible that dACC/msFC and aI/fO may also receive warning signals from the brain's alerting network, as well as information about the meaning of cues and attentional shifts from regions in PPC and frontal cortex. Perhaps specific nodes in each of the attentional networks form bridges between them that allow for an integration of attentional functions. Although distinct networks of brain regions support alerting, attention-shifting, and executive control functions, attention is likely a property that emerges from the interactions between these networks. One speculative idea is that the integration of attentional functions may be helped by the implementation of a basic goal-oriented task mode. This focused attention mode may stand in a mutually exclusive push–pull relationship with the brain's default mode.

See also: Attention and Eye Movements; Attention: Models; Attentional Functions in Learning and Memory; Attentional Mechanisms in Ventral Pathway; Attentional Networks in the Parietal Cortex; Decision-Making and Vision; Executive Function and Higher-Order Cognition: Definition and Neural Substrates; Frontal Lobe Syndrome; Neglect Syndrome and the Spatial Attention Network; Prefrontal Cortex: Structure and Anatomy; Psychophysics of Attention; Spatial Cognition and Executive Function; Visual Attention.

Further Reading

Aston-Jones G, Rajkowski J, and Cohen J (1999) Role of locus coeruleus in attention and behavioral flexibility. *Biological Psychiatry* 46: 1309–1320.

Braver TS, Reynolds JR, and Donaldson DI (2003) Neural mechanisms of transient and sustained cognitive control during task switching. *Neuron* 39: 713–726.

Bunge SA, Wallis JD, Parker A, et al. (2005) Neural circuitry underlying rule use in humans and nonhuman primates. *Journal of Neuroscience* 25: 10347–10350.

Corbetta M, Kincade MJ, Lewis C, et al. (2005) Neural basis and recovery of spatial attention deficits in spatial neglect. *Nature Neuroscience* 8: 1603–1610.

Corbetta M and Shulman GL (2002) Control of goal-directed and stimulus-driven attention in the brain. *Nature Reviews Neuroscience* 3: 201–215.

Corbetta M, Shulman GL, Miezin FM, and Petersen SE (1995) Superior parietal cortex activation during spatial attention shifts and visual feature conjunction. *Science* 270: 802–805.

Dosenbach NU, Visscher KM, Palmer ED, et al. (2006) A core system for the implementation of task sets. *Neuron* 50: 799–812.

Duncan J and Owen AM (2000) Common regions of the human frontal lobe recruited by diverse cognitive demands. *Trends in Neurosciences* 23: 475–483.

Eriksen CW and Hoffman JE (1973) The extent of processing of noise elements during selective encoding from visual displays. *Perception and Psychophysics* 14: 155–160.

Gehring WJ and Knight RT (2000) Prefrontal–cingulate interactions in action monitoring. *Nature Neuroscience* 3: 516–520.

James W (1890) *The Principles of Psychology.* New York: Holt.

Lau HC, Rogers RD, Haggard P, and Passingham RE (2004) Attention to intention. *Science* 303: 1208–1210.

Miller EK and Cohen JD (2001) An integrative theory of prefrontal cortex function. *Annual Review of Neuroscience* 24: 167–202.

Posner MI and Petersen SE (1990) The attention system of the human brain. *Annual Review of Neuroscience* 13: 25–42.

Posner MI, Snyder CRR, and Davidson BJ (1980) Attention and the detection of signals. *Journal of Experimental Psychology: General* 109: 160–174.

Raz A and Buhle J (2006) Typologies of attentional networks. *Nature Reviews Neuroscience* 7: 367–379.

Rushworth MF, Walton ME, Kennerley SW, and Bannerman DM (2004) Action sets and decisions in the medial frontal cortex. *Trends in Cognitive Science* 8: 410–417.

Sakai K and Passingham RE (2003) Prefrontal interactions reflect future task operations. *Nature Neuroscience* 6: 75–81.

Stroop JR (1935) Studies of interference in serial verbal reactions. *Journal of Experimental Psychology* 18: 643–662.

Weissman DH, Gopalakrishnan A, Hazlett CJ, and Woldorff MG (2005) Dorsal anterior cingulate cortex resolves conflict from distracting stimuli by boosting attention toward relevant events. *Cerebral Cortex* 15: 229–237.

Attentional Networks in the Parietal Cortex

G H Patel, B J He, and M Corbetta, Washington University School of Medicine, St. Louis, MO, USA

Introduction

The brain is continuously flooded with information from different senses: vision, audition, touch, smell, and taste. As the inflow of sensory information is much greater than what the brain can process at any given time, one of the fundamental interests of neuroscience is to understand which brain mechanisms are responsible for the selection of those few bits of information that are relevant to the ongoing goals of an individual, and how irrelevant information is filtered out. Attended objects tend to be perceived and remembered much better than unattended objects are, and only objects that are attended become the target of motor plans, such as when we look at and reach for an apple. Accordingly, attention is defined as the ensemble of psychological and neural operations that mediate selection of sensory stimuli and link them to response and memory systems.

The parietal lobe, from the Latin *paries*, or 'walls of a house,' is the part of the brain that sits between the occipital, temporal, and frontal lobes, and that includes the lateral superior part of each hemisphere. The parietal lobe is situated between sensory (visual, tactile, and auditory) areas and contains cells that respond predominantly to behaviorally relevant stimuli. The parietal lobe is also active when individuals prepare to look or move toward a stimulus, and is heavily connected with the frontal lobe, in which the actions are planned. The parietal lobe of the brain is therefore well suited to perform operations that are neither strictly sensory nor motor, but rather operations that integrate sensory and motor information. Because of the integrative nature of the parietal lobe, the functions of the cortical areas contained within it are complex and diverse, and damage to the parietal lobe often results in multifaceted deficits.

In this article we first describe the anatomical organization of the parietal lobe; next, we turn to what we know about the functional organization of parietal cortex, and finally, we consider some of the behavioral deficits that arise after it is damaged.

Organization and Connectivity

The parietal lobe is bordered on the posterior and ventral sides by visual and auditory cortex, respectively, and the anterior portion of the parietal lobe is occupied by somatosensory cortex. Anterior to the parietal lobe is the frontal lobe, much of which is devoted to the planning and execution of movements. The parietal lobe is divided into smaller parts based on gross anatomy. The most prominent anatomical feature of the parietal lobe is the intraparietal sulcus (IPS), which runs anterior–posteriorly along the lateral aspect of the parietal lobe. The IPS divides the parietal lobe into two lobules: the superior parietal lobule (SPL), which encompasses the lateral aspect of the parietal lobe dorsal to the IPS as well as the medial wall, and the inferior parietal lobule (IPL), which encompasses all of the parietal lobe ventral to the IPS. The part of the parietal lobe on the medial wall is also often called the precuneus.

Using histological techniques, investigators have subdivided the parietal lobe in several ways. Perhaps one of the best known schemes for subdividing parietal cortex was proposed by Brodmann in 1909. In this scheme, the human parietal cortex is subdivided into seven areas: areas 1–3, which comprise the anterior edge of the parietal lobe and cover somatosensory cortex; area 5, which is immediately posterior to area 2 and covers the anterior part of the SPL; area 7, which is posterior to area 5 and covers much of the lateral and medial SPL and some of the IPL; and areas 39 and 40, which cover much of the IPL (see **Figure 1(a)**). Beyond these coarse divisions, however, not much was known for many years about the anatomical organization of parietal cortex, in part because invasive techniques necessary to trace connections and study the function of this part of the brain in humans were not available. As a result, most of the detailed information that we have on the parietal lobe's organization and on other parts of the human brain comes from anatomical and physiological studies of nonhuman primates, especially macaques, the brains of which share many of the same sensory and motor functions of the human brain.

The gross and histological organization of macaque parietal lobe is similar to that of the human parietal lobe in many ways: an IPS separates the parietal lobe into an SPL and IPL, somatosensory cortex makes up the anterior edge of the parietal lobe, and visual areas lie along the posterior edge. Areas 1, 2, 3, 5, and 7 are also present in macaque; however, the relative position of areas 5 and 7 is more ventral, given that areas 39 and 40, which occupy the ventral part of the human IPL, are not present in the macaque (see **Figure 1(b)**). Despite these potential differences, the macaque continues to serve as a useful model of how human parietal areas may be involved in attention.

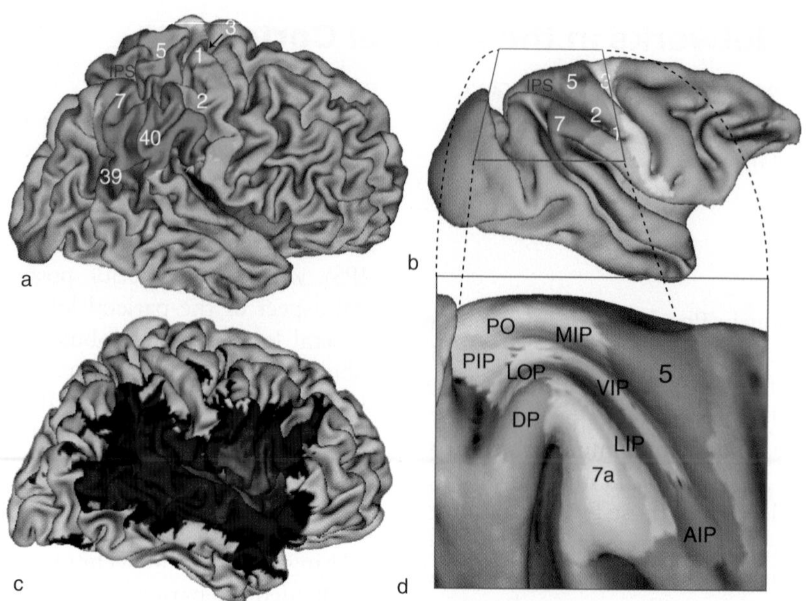

Figure 1 (a) Human Brodmann areas 1–3, 5, 7, 39, and 40, and the intraparietal sulcus (IPS). (b) Macaque Brodmann areas. (c) Cortical damage underlying neglect. (d) Areas in and around the macaque IPS (PO, parietal-occipital; MIP, medial intraparietal; PIP, posterior intraparietal; LOP, lateral occipitoparietal zone; VIP, ventral intraparietal; DP, dorsal prelunate; AIP, anterior intraparietal; 5, area 5; 7a, area 7a). Adapted from Lewis JW and Van Essen DC (2000) Mapping of architectonic subdivisions in the macaque monkey, with emphasis on parieto-occipital cortex. *Journal of Comparative Neurology* 428: 79–111.

In the 1970s, the advent of anatomical methods to trace connections between areas allowed for the partition of posterior parietal cortex into multiple areas based on their profile of feed-forward and feedback connections. Feed-forward connections refer to connections from lower to higher levels of a sensory hierarchy of cortical areas, whereas feedback connections refer to connections from higher to lower levels. There are several proposed schemes to divide the parietal cortex. Along the IPS, one common scheme divides the cortex into ten areas: dorsal prelunate (DP), posterior intraparietal (PIP), parietal-occipital (PO), the lateral occipitoparietal zone (LOP), medial intraparietal (MIP), ventral intraparietal (VIP; divided into medial and lateral), lateral intraparietal (LIP; divided into dorsal and ventral), anterior intraparietal (AIP), area 5, and area 7a (see **Figure 1(d)**). Other similar schemes have also been proposed. Tracer studies have shown that these areas receive input from other brain structures involved in the processing of vision, audition, and sensation, and may send output to the same sensory areas and/or to various motor and motor-planning areas. The IPS areas residing more posteriorly and laterally receive most of their input from visual areas and send output to oculomotor structures, whereas the more medial and anterior areas also receive input from somatosensory areas and send output to hand and arm motor areas.

For an area to be involved in the selective operations of visual spatial attention, it is likely to be connected to the visual areas responsible for the processing of incoming stimuli. It is also likely to be connected to areas involved in the planning and execution of saccades, since after the selection of a stimulus of interest, a saccade is often made to bring the stimulus into the fovea for further scrutiny. Area LIP in the macaque fits this description. It occupies the caudal half of the lateral bank of the IPS, and is demarcated histologically from surrounding parietal areas by increased myelination of layers 3–5. This area receives feed-forward connections from and sends feedback connections to many extrastriate visual areas, including areas V4, V3a, and middle temporal complex (MT), an area known to be involved in processing visual motion. LIP is also heavily interconnected with oculomotor structures, such as the superior colliculus in the midbrain, the pulvinar in the thalamus, and the frontal eye fields (FEFs) in prefrontal cortex.

Area 7a has a profile of connections with cortical areas similar to that of LIP, except rather than FEF, it is connected with area 46 in prefrontal cortex. Area 46 is an area known to be involved in spatial working memory, implying that 7a may also play a role in working memory and attention. Other posterior parietal areas, such as DP and PO, connect mainly with extrastriate areas and more anterior IPS areas, such as LIP, indicating that they play a role in relaying information from visual cortex to the higher level planning areas in IPS.

In general, the many anatomical studies of parietal cortex over the past several decades have shown that it appears to be divided into many areas, each of which appears to play a role in translating sensory information into motor plans. But what exactly this role entailed remained a mystery until the technology to study these areas *in vivo* was developed, first in macaques and then later in humans.

Function

In 1975 Mountcastle and colleagues first reported on electrophysiological recordings of macaque parietal cortex neurons while the monkey performed various tasks. They found that neurons in the IPS "appear[ed] to direct visual attention to objects of interest and motivational power, and to issue commands for maintaining directed fixation of the object when it [was] stationary, and to track it if it mov[ed]." In the decades since these early experiments, electrophysiological recordings of parietal cortex neural activity have determined that the SPL and IPL are subdivided into many smaller functional areas that appear to roughly correspond with the subdivisions determined by histological techniques.

Each of these areas appears to extract the spatial information from one or more sensory inputs, and then play some role in transforming that information into a general movement plan in the same spatial coordinates. Because of its involvement in spatial processing, the parietal cortex is said to be part of the 'where' pathway (as opposed to the ventral 'what' pathway, which appears to be involved in processing the identity of a stimulus regardless of spatial location). For instance, the area LIP is involved in the planning of upcoming saccadic eye movements. The neurons in this area respond transiently to visual stimuli presented within a specific part of the visual field, which is known as a receptive field. Moreover, if a saccade is to be performed to that stimulus at some point in the future, the neurons continue to fire until the saccade has been executed. In this way, they are said to represent the planned target of the saccade until the action is completed, even if the visual stimulus is no longer in the receptive field. This characteristic is often used to define the boundaries of LIP in electrophysiological recording experiments. LIP neurons have large visual receptive fields that cover on average a quarter of the visual field, and these receptive fields move with the position of gaze (gaze centered). They are also more likely to respond to stimuli in the contralateral visual hemifield than in the ipsilateral, and may contain a coarse but continuous map of the contralateral visual field (otherwise known as being retinotopically organized). LIP neurons, then, appear to represent the plan of an upcoming saccadic eye movement in a rough spatial map of the visual environment.

LIP neurons will also respond to a visual stimulus that is not the target of a planned saccade, but is otherwise task relevant. For instance, if the monkey is instructed to maintain fixation on a central point, and to indicate with his hand when a visual stimulus in the periphery dims just slightly, the LIP neurons representing the location of this stimulus continue to fire until the task is completed, and this increased level of firing is directly correlated to increased performance of whatever perceptual decision needs to be made at that location. Moreover, chemically deactivating LIP will result in the reduction of the monkey's ability to discriminate an oddball stimulus from other stimuli (such as a red circle among a field of green circles). It appears, then, that LIP neurons also represent the current locus of visuospatial attention.

Are these two purported functions of LIP neurons in conflict, or do they represent two ways of saying the same thing? The answer to this has been the source of a contentious debate. Thus far we have seen that LIP is involved both in covert shifts of attention (marking a location of interest so that processing of objects in that location can be enhanced) and in overt shifts of attention (making a saccade to a location so that it can be studied in more detail). One side of the debate holds that all covert shifts of attention represent potential saccades, and that the enhanced processing at that location is merely a side effect of planning a movement to that location. As evidence for this view, they point to other parietal areas, which appear to play the same planning role for other effectors, such as arm and hand movements. The 'intentional map,' then, might be a general principle of what the parietal cortex contributes to a sensorimotor transformation. The other side, however, contends that the LIP is a 'salience map' of the visual world, marking locations of interest for any number of systems to use, including the visual attention and oculomotor system. Part of the evidence for this view is that LIP neurons track the location of salient objects independently of stored oculomotor plans. While the differences between these two sides may seem minute, the resolution of this debate will give us insight into both what specific calculations parietal cortex neurons are performing on incoming information and in what terms the brain represents the external world – as a distorted version of reality, skewed toward the most interesting and relevant stimuli, or always in terms of a motor plan for interacting with the world?

Other parietal areas in the macaque may also play roles in selective attention, though these areas have not been as thoroughly investigated as LIP. Areas caudal to LIP, such as DP, appear to play a role in visuospatial processing, though probably at an earlier stage than LIP (again fitting with their profiles of connections). Area 7a, which covers much of the IPL lateral to LIP, also appears to be involved in visuospatial processing. Like LIP, its neurons have large receptive fields, though it appears to evenly represent both the contralateral and ipsilateral hemifields. The neurons in this area appear to respond to the novel appearance of behaviorally relevant visual stimuli, but not as many of the 7a neurons have presaccadic activity as compared to LIP cells. This may indicate that rather than representing the current focus of attention, 7a neurons may be involved in detecting novel but potentially relevant stimuli, a counterpart of sorts to LIP. How other macaque parietal areas are involved in spatial attention is less clear, as they have generally been studied under the rubric of motor planning. Another important organizing principle for thinking about parietal cortex is to consider how space is coded in this region of the brain. There is evidence that parietal cortex subregions may be specialized for coding spatial location away from the body (or extrapersonal space) or near the body (or peripersonal space).

This wealth of information about monkey's parietal cortex until recently did not have a counterpart in the human work, because studies were limited to clinical observations (see later). This state of affairs changed in the late 1980s with the advent of positron emission tomography (PET), and then again in the 1990s with the advent of functional magnetic resonance imaging (fMRI). These neuroimaging technologies allowed, for the first time, *in vivo* studies of neural activity with sufficient spatial and temporal resolution to discern areas of the brain that were active in different tasks. Early PET studies confirmed that in humans the parietal lobe is also part of the dorsal 'where' pathway and is involved in the control of both spatial attention and eye movements. Several subsequent fMRI studies have also shown that, like in the macaque, the parietal region activated during covert shifts of attention largely overlaps with the region activated during saccades. This region includes much of the cortex in and around the IPS. Parts of this region are also activated during nonspatial shifts of attention, such as changing the focus of attention from the direction of moving dots to the color of the moving dots. Because of this profile of activity, the region along the human IPS has been thought of as generally homologous to the macaque IPS.

However, due to the combination of relatively poor spatial resolution of fMRI and the high degree of variability in the location of sulci and gyri among human individuals, it has been difficult to subdivide this region into functional areas as has been done in the macaque. Two foci of neural activity are consistently activated in different tasks requiring shifting or maintenance of attention. The first is in the SPL, on the dorsal–medial bank of the IPS. Because it is thought to be homologous to the macaque LIP, it is sometimes called human LIP (hLIP). Like macaque LIP, hLIP appears to be gaze centered in its spatial reference frame, responds more strongly to stimuli in the contralateral visual hemifield than in the ipsilateral, and appears to be loosely retinotopically organized. This area is activated if there is an attention shift to a peripheral location, and activity is sustained in this area if either attention remains focused on, or a saccade is being planned to, the peripheral location. Activity in this area seems to track both the locus of visuospatial attention and the target of an upcoming eye movement, extending the debate about parietal function to the human parietal cortex. While much of the functional profile of hLIP appears to be broadly similar to that of macaque LIP, the lack of direct comparisons of histological data or neural activity during tasks requiring shifts of attention prevents a more thorough assessment of the homology of these areas.

Areas posterior to hLIP are also often activated in tasks requiring shifts of spatial attention. One of the most prominent is in the fundus of the caudal section of the IPS, and is variously termed ventral IPS (vIPS) or V7. This area is preferentially activated by attending to the contralateral hemifield, but appears to be more visual in nature than hLIP. V7 most likely represents an earlier stage of the 'where' pathway that probably relays visual information to hLIP, and is a potential homolog to one of the posterior parietal areas in the macaque.

While the aforementioned areas in the human SPL are activated during voluntary and involuntary shifts of attention, the IPL appears to play a different role in attention. Regions in the IPL are only activated in response to the appearance of a salient, behaviorally relevant object, and even more so if attention has to be reoriented from another location (known as stimulus-driven reorienting of attention). Moreover, while activity in the SPL areas is increased during sustained covert shifts of attention, activity in the IPL is suppressed. It has been proposed that these areas play a role in stimulus-driven attention, especially when the focus of attention is captured by a salient and relevant novel stimulus. In macaques, area 7a (on the macaque IPL) is one possible candidate for a homolog to the human IPL.

Another important feature of human parietal cortex is that it is anatomically and functionally asymmetrical. Several studies have now shown that functional regions of the supramarginal and angular gyrus at the temporoparietal junction (TPJ) in humans show hemispheric asymmetries of attention (right-dominant) or verbal memory (left-dominant) tasks. This asymmetry is even more dramatic when one considers the behavioral effects of stroke in the middle cerebral artery (MCA) distribution that feeds the parietal cortex. Whereas left MCA strokes commonly result in aphasia (language deficits), strokes of the right MCA commonly result in a syndrome termed 'neglect,' which is a collection of spatial attention–perception–premotor deficits (see **Figure 1(c)**).

An important goal for the future will be to reconcile monkey and human data in a common evolutionary framework that takes into account both commonalities and differences in functional organization. An important technological advance that will help this line of research is the development of fMRI studies in awake-behaving monkeys.

Lesions

Damage to the right IPL often results in spatial neglect, a common syndrome following stroke in which patients are biased toward attending to the right side of visual space more than to the left side. Neglect can leave patients unable to perform simple tasks requiring spatial attention, and occurs in about 30% of all stroke-affected individuals.

Patients with neglect often act as if part of the left side of their world does not exist. This inattention to the left side can occur in extrapersonal space or about the patient's own body, and can occur in different reference frames: gaze-centered frame (left with respect to the center of gaze), body-centered frame (left with respect to the midline of the body), or world-centered frame (left with respect to the environment). Less commonly, patients can manifest so-called object-centered neglect, where the left side of an object is ignored, no matter if it is presented in the left or right side of the visual space. For instance, if a page of text with three columns is presented, a patient with body-centered or world-centered neglect will miss the words on the left side of the page, whereas a patient with object-centered neglect will miss the left side of each column or the left side of each word.

Functionally, neglect patients may 'forget' to shave, groom, or dress the left side of body. They also have major problems with driving a car. Severely affected patients might also display a tonic rotation of the body or eyes toward the right, due to a coexisting motor imbalance. Some of these patients may still be able to detect a stimulus presented alone in the left hemifield, but when presented with stimuli simultaneously to both visual fields, patients tend to see only those in the right hemispace. This deficit is referred to as 'extinction.' Extinction is often tested by snapping fingers in one or both of the patient's visual fields and asking the patient to point to which hand they saw or heard snapping. When fingers are snapped simultaneously in both visual fields, neglect patients will inexorably point to the hand in their right visual field. This test is one of the many routine exams in clinical settings to assess the severity of neglect. Other commonly used neuropsychological tests for diagnosis of neglect include line bisection, in which patients with neglect bisect a horizontal line to the right of the true center; a series of cancellation tasks, in which neglect patients commonly miss marking targets on the left side of a paper; clock drawing, in which neglect patients may draw only the right half of a clock; and a baking-tray task, in which neglect patients usually cluster cookies on the right side of a baking tray.

In addition to the lack of awareness of the left side of space, neglect patients can also manifest deficits in planning hand or eye movements toward the left side of space. In this type of neglect, patients may not demonstrate a bias in awareness to either visual field, but will be less inclined to reach or look for objects located in the left visual field. This form of neglect is called motor or premotor neglect, or directional hypokinesia, to emphasize its relationship to action. In addition to lateralized (i.e., left worse than right space) deficits, a number of so-called nonlateralized deficits (i.e., similar across left and right visual fields) have also been described, including impairment of spatial working memory, sustained attention, and an overall lower level of alertness.

An important focus of current research is to assess the functional effect of different deficits, their importance for final outcome, their recovery over time, and their localization in the brain. Shall we think of neglect as a homogeneous or heterogeneous syndrome? What are the regions more responsible for one or another deficit? We know that spatial neglect can occur as a clinical syndrome for lesions in many parts of the brain, including parietal, temporal, frontal cortex, basal ganglia, thalamus, and in various white-matter tracts. It is still unknown how these different lesions may or may not cause different behavioral profiles of neglect, and how the effect of lesions in the brain relates to the functional organization observed in healthy individuals.

An alternative view considers neglect not to be the effect of damage to specific cortical areas, but rather the result of a distributed and combined anatomical–functional dysfunction of large parts of parietal and frontal cortices and their connections. In general, the areas that are damaged in neglect are located ventrally in the brain, including the IPL regions specialized for stimulus-driven attention. Conversely, damage to more dorsal IPS and SPL regions does not produce strong neglect; rather, these lesions produce primarily eye- or arm-movement planning problems, which is also consistent with the results of functional imaging studies in healthy volunteers. However, only a few patients with damage restricted to the SPL have been studied.

One model of neglect postulates that damage by stroke or trauma to the IPL in the region of the TPJ could give rise to both spatial and nonspatial deficits by interrupting input to the SPL attentional areas from the TPJ. This disruption would then reduce the brain's ability to detect important sensory events in both visual fields. In addition, this lack of input into ipsilesional dorsal areas may induce a relative interhemispheric imbalance in the dorsal attentional areas that code for spatial locations and eye and arm movements. The imbalance due to the competitive nature of interhemispheric processing will lead to a relative hyperorienting toward right space, with consequent problems of attention and detection in left space. This model of spatial bias based on interhemispheric competition is also supported by a number of animal studies in which neglect has been cured by inactivating homologous parietal regions of the intact hemisphere.

Conclusion

Parietal cortex is subdivided into different areas, based on both anatomical and functional criteria; these areas have connections with both sensory processing and motor planning regions of the brain. Some of these areas appear to be involved in the selection and representation of locations and objects of interest in the immediate environment. The spatial information in parietal cortex can then be 'read out' by other brain areas, both for the planning of movements and for the enhancement of sensory processing. Attention can be directed to a location or object by a premeditated plan, or by the appearance of a novel stimulus that may require additional processing. Area LIP in the macaque and its putative homolog in the SPL of humans appear to be involved in representing the locations of interest, whereas IPL areas appear to play the role of reorienting attention whenever a novel stimulus enters awareness. Damage to these areas results in the syndrome of neglect, which fundamentally involves a lack of awareness for spatial information, as well as problems with vigilance and motor planning.

See also: Attention and Eye Movements; Attention: Models; Attentional Functions in Learning and Memory; Attentional Mechanisms in Ventral Pathway; Attentional Networks; Decision-Making and Vision; Neglect Syndrome and the Spatial Attention Network; Parietal Cortex and Spatial Attention; Psychophysics of Attention; Visual Attention.

Further Reading

Andersen RA and Buneo CA (2002) Intentional maps in posterior parietal cortex. *Annual Review of Neuroscience* 25: 189–220.

Colby CL and Goldberg ME (1999) Space and attention in parietal cortex. *Annual Review of Neuroscience* 22: 319–349.

Corbetta M and Shulman GL (2002) Control of goal-directed and stimulus-driven attention in the brain. *Nature Reviews Neuroscience* 3: 201–215.

Desimone R and Duncan J (1995) Neural mechanisms of selective visual attention. *Annual Review of Neuroscience* 18: 193–222.

Egeth HE and Yantis S (1997) Visual attention: Control, representation, and time course. *Annual Review of Psychology* 48: 269–297.

Goldman-Rakic PS (1988) Topography of cognition: Parallel distributed networks in primate association cortex. *Annual Review of Neuroscience* 11: 137–156.

Hillis AE (2006) Neurobiology of unilateral spatial neglect. *Neuroscientist* 12: 153–163.

Husain M and Rorden C (2003) Non-spatially lateralized mechanisms in hemispatial neglect. *Nature Reviews Neuroscience* 4: 26–36.

Kastner S and Ungerleider LG (2000) Mechanisms of visual attention in the human cortex. *Annual Review of Neuroscience* 23: 315–341.

Lewis JW and Van Essen DC (2000) Mapping of architectonic subdivisions in the macaque monkey, with emphasis on parieto-occipetal complex. *Journal of Comparative Neurology* 428: 79–111. Avaliable at URL: http://sumsdb.wustl.edu:8081/sums/directory.do?id=679531.

Mesulam MM (1999) Spatial attention and neglect: Parietal, frontal and cingulate contributions to the mental representation and attentional targeting of salient extrapersonal events. *Philosophical Transactions of the Royal Society of London, Series B: Biological Sciences* 354: 1325–1346.

Mountcastle VB, Lynch JC, Georgopoulos A, et al. (1975) Posterior parietal association cortex of the monkey: Command function for operations within extrapersonal space. *Journal of Neurophysiology* 38: 871–908.

Orban GA, VanEssen D, and Vanduffel W (2004) Comparative mapping of higher visual areas in monkeys and humans. *Trends in Cognitive Science* 8: 315–324.

Pashler HE (1998) *The Psychology of Attention*. Cambridge, MA: MIT Press.

Posner MI and Petersen SE (1990) The attention system of the human brain. *Annual Review of Neuroscience* 13: 25–42.

Attractor Network Models

X-J Wang, Yale University School of Medicine, New Haven, CT, USA

Introduction

The term attractor is being increasingly used by neurophysiologists to characterize stable, stereotyped spatiotemporal neural circuit dynamics. Examples include readily identifiable rhythmic activity in a central pattern generator; well-organized propagation patterns of neuronal spike firing in cortical circuits *in vivo* or *in vitro*; self-sustained persistent activity during working memory; and neuronal ensemble representation of associative long-term memory. In these examples, the rich and complex neural activity patterns are generated largely through regenerative mechanism(s), and emerge as collective phenomena in recurrent networks. In part, interest in attractor networks arises from our growing appreciation that neural circuits are typically endowed with an abundance of feedback loops and that the attractor theory may provide a conceptual framework and technical tools for understanding such strongly recurrent networks.

The concept of attractors originates from the mathematics of dynamical systems. Given a fixed input, a system consisting of interacting units (e.g., neurons) typically evolves over time toward a stable state. Such a state is called an attractor because a small transient perturbation alters the system only momentarily; afterward, the system converges back to the same state. An example is illustrated in **Figure 1**, in which a neural network is described by a computational energy function in the space of neural activity patterns and the time evolution of the system corresponds to a movement down hill, in the direction of decreasing the computational energy. Each of the minima of the energy function is thus a stable (attractor) state of the system; a maximum at the top of a valley is an unstable state. Such a depiction is not merely schematic, but can be rendered quantitative for certain neural models.

The characterization stable and stereotyped is sometimes taken to imply that an attractor network is not sensitive to external stimuli and is difficult to reconfigure. On the contrary, as was shown by recent studies, attractor networks are not only responsive to inputs, but may in fact to instrumental to the slow time integration of sensory information in the brain. Moreover, attractors can be created or destroyed by (sustained) inputs; hence, the same network can serve different functions (such as working memory and decision making), depending on the inputs and cognitive control signals. The attractor landscape of a neural circuit is readily modifiable by changes in cellular and synaptic properties, which form the basis of the attractor model for associative learning.

In this article, we first introduce the basic concepts of dynamical systems, attractors, and bistability using simple single-neuron models. Then, we discuss attractor network models for associative long-term memory, working memory, and decision making. Modeling work and experimental evidence are reviewed and open questions are outlined. We show that attractor networks are capable of time integration and memory storage over timescales much longer than the biophysical time constants of fast electrical signals in neurons and synapses. Therefore, strongly recurrent attractor networks are especially relevant to memory and higher cognitive functions.

The Neuron Is a Dynamical System

A passive nerve membrane is a dynamical system described by a simple resistance–capacitance (RC) circuit equation:

$$C_{\mathrm{m}}(\mathrm{d}V_{\mathrm{m}}/\mathrm{d}t) = -g_{\mathrm{L}}(V_{\mathrm{m}} - E_{\mathrm{L}}) + I_{\mathrm{app}} \qquad [1]$$

where V_{m} is the transmembrane voltage, C_{m} the capacitance, g_{L} the leak conductance (the inverse of the input resistance), E_{L} the leak reversal potential, and I_{app} the injected current. In the absence of an input, the membrane is at the resting state, $V_{\mathrm{ss}} = E_{\mathrm{L}}$, say -70 mV. This steady state is stable; if V_{m} is transiently depolarized or hyperpolarized by a current pulse, after the input offset it will evolve back to V_{ss} exponentially with a time constant $\tau_{\mathrm{m}} = C_{\mathrm{m}}/g_L$ (typically 10–20 ms). Thus, V_{ss} is the simplest example of an attractor. More generally, for any sustained input drive I_{app}, the membrane always has a steady-state $V_{\mathrm{ss}} = V_{\mathrm{L}} + I_{\mathrm{app}}/g_{\mathrm{L}}$, given by $\mathrm{d}V_{\mathrm{m}}/\mathrm{d}t = 0$ (i.e., V_{ss} does not change over time). The behavior of this passive membrane changes quantitatively with input current, but remains qualitatively the same; the dynamics is always an exponential time course (determined by τ_{m}) toward the steady-state V_{ss}, regardless of how high is the current intensity I_{app} or how large is the capacitance C_{m} or the leak conductance g_{L}. Moreover, the response to a combination of two stimuli I_1 and I_2 is predicted by a linear sum of the individual responses to I_1 or I_2 presented alone. These characteristics are generally true for a linear dynamical system, such as a differential equation with only linear dependence on V_{m}.

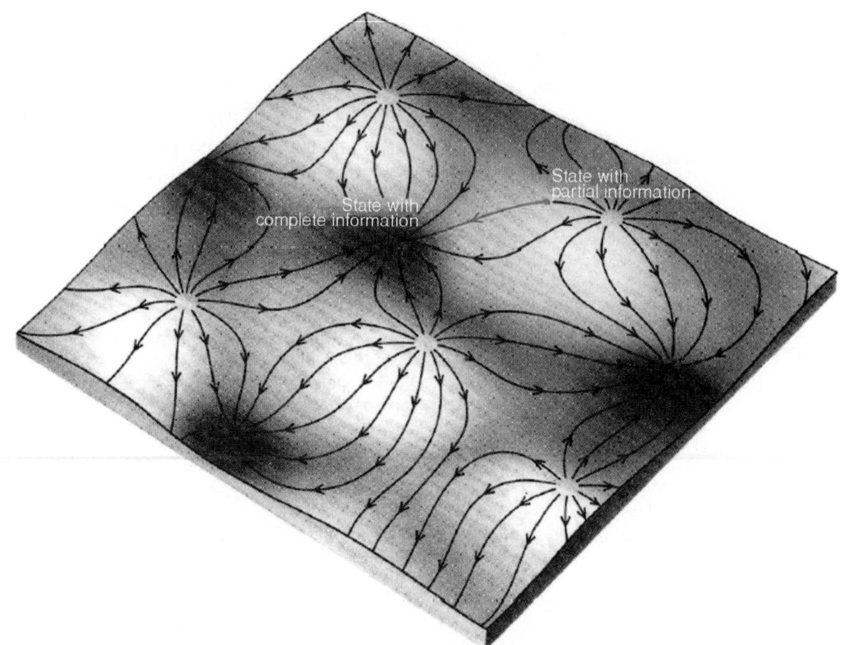

Figure 1 Schematic of an attractor model of neural networks. Computational energy function is depicted as a landscape of hills and valleys plotted against the neural activity states (on the *XY* plane). The synaptic connections and other properties of the circuit, as well as external inputs, determine its contours. The circuit computes by following a path that decreases the computational energy until the path reaches the bottom of a valley, which represents a stable state of the system (an attractor). In an associative memory circuit, the valleys correspond to memories that are stored as associated sets of information (the neural activities). If the circuit is cued to start out with approximate or incomplete information, it follows a path downhill to the nearest valley (red), which contains the complete information. From Tank DW and Hopfield JJ (1987) Collective computation in neuronlike circuits. *Scientific American* 257: 104–114.

More interesting behaviors become possible when nonlinearity is introduced by the inclusion of voltage-gated ionic currents. For instance, if we add in the RC circuit a noninactivating sodium current $I_{\mathrm{NaP}} = g_{\mathrm{NaP}} m_{\mathrm{NaP}}(V_{\mathrm{m}})(V_{\mathrm{m}} - E_{\mathrm{Na}})$, where the conductance exhibits a nonlinear (sigmoid) dependence on V_{m}, the membrane dynamics becomes

$$C_{\mathrm{m}}(\mathrm{d}V_{\mathrm{m}}/\mathrm{d}t) = -g_{\mathrm{L}}(V_{\mathrm{m}} - E_{\mathrm{L}}) - g_{\mathrm{NaP}} m_{\mathrm{NaP}}(V_{\mathrm{m}})(V_{\mathrm{m}} - E_{\mathrm{Na}}) + I_{\mathrm{app}} \quad [2]$$

This system is endowed with a self-excitatory mechanism: a higher V_{m} leads to more I_{NaP}, which in turn produces a larger depolarization. If g_{NaP} is small, the weak positive feedback affects the membrane dynamics only slightly (**Figure 2(b)**, red lines). With a sufficiently large g_{NaP}, the steady state at $V_{\mathrm{Down}} \simeq -70\,\mathrm{mV}$ is still stable because at this voltage I_{NaP} is not activated. However, the strong positive feedback gives rise to a second, depolarized plateau potential (at $V_{\mathrm{Up}} \simeq -20\,\mathrm{mV}$) (**Figure 2(b)**, blue lines). Therefore, the membrane is bistable; a brief input can switch the system from one attractor state to another (**Figure 2(a)**). As a result, a transient stimulus can now be remembered for a long time, in spite of the fact that the system has only a short biophysical time constant (20 ms). Unlike linear systems, in a nonlinear dynamical system gradual changes in a parameter (g_{NaP}) can give rise to an entirely new behavior (bistability).

Attractor states are stable under small perturbations, and switching between the two can be induced only with sufficiently strong inputs. How strong is strong enough? The answer can be found by plotting the total ion current $I_{\mathrm{tot}} = I_{\mathrm{L}} + I_{\mathrm{NaP}} - I_{\mathrm{app}}$ against V_{m}, called the *I–V* curve (**Figure 2(b)**, top, with $I_{\mathrm{app}} = 0$). Obviously, a V_{m} is a steady state if $I_{\mathrm{tot}}(V_{\mathrm{m}}) = 0$ (thus $\mathrm{d}V_{\mathrm{m}}/\mathrm{d}t = 0$). As seen in **Figure 2(b)** (blue lines), the two attractors (filled circles) are separated by a third steady state (open circle). The third state is unstable – if V_{m} deviates slightly from it, the system does not return but converges to one of the two attractors. Indeed, if V_{m} is slightly smaller, I_{tot} is positive (hyperpolarizing), so V_{m} decreases toward V_{Down}. Conversely, if V_{m} is slightly larger, I_{tot} is negative (depolarizing), so V_{m} increases toward V_{Up}. Therefore, an external input must be strong enough to bring the membrane potential beyond the unstable steady state to switch the system from one attractor state to the other.

It is worth noting that the attractor landscape depends not only on the strength of the feedback mechanism (the value of g_{NaP}) but also sustained inputs. As shown in **Figure 2(c)**, with a fixed g_{NaP} a constant applied current I_{app} shifts the *I–V* curve up

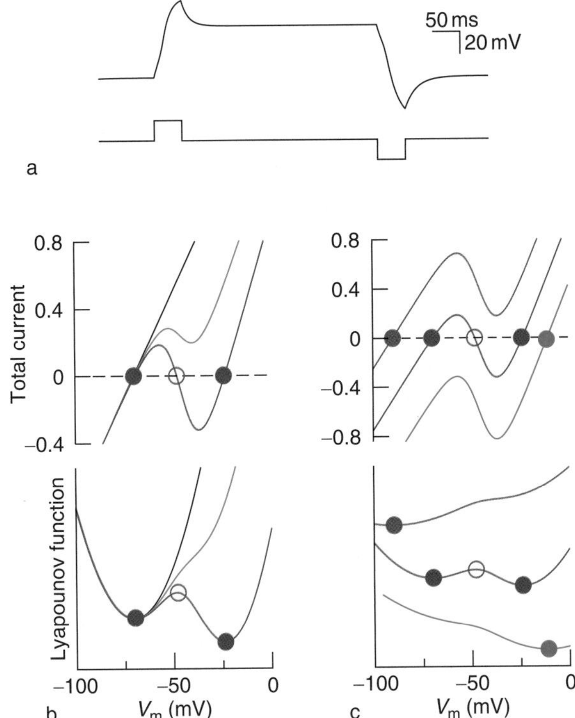

Figure 2 Positive feedback and attractor dynamics in a simple neural membrane model: (a) bistability with $g_{NaP} = 0.015\,\mu S$; (b) I–V curve (top) and computational energy function (Lyapounov function) (bottom); (c) bistability modulated by inputs, with $g_{NaP} = 0.015$ fixed, I–V curve (top) and computational energy function (Lyapounov function) (bottom). The membrane voltage is described by an RC circuit with the addition of a fast noninactivating sodium current $I_{NaP} = g_{NaP} m_{NaP}(V_m)(V_m - E_{Na})$, where $m_{Nap}(V_m) = 1/(1 + \exp[-(V_m + 45)/5])$ is a sigmoid function of V_m. The interplay between I_{NaP} and membrane depolarization produces an excitatory regenerative process. In (a), the system is initially at rest ($V_{Down} \simeq -70\,mV$). A depolarizing current pulse switches the membrane to a plateau potential ($V_{Up} \simeq -20\,mV$), which persists after the input offset. A second, hyperpolarizing current pulse switches the membrane back to the resting state. In (b), I–V curve (upper) is the total current $I_{tot} = I_L + I_{NaP} - I_{app}$ as a function of V_m, with $I_{app} = 0$); a steady state is given by an intersection with $I_{tot} = 0$. In the computational energy function (lower) $U(V_m)$, a steady state corresponds to a maximum (unstable) or a minimum (stable). For $g_{NaP} = 0$ (black, passive membrane) or $g_{NaP} = 0.08$ (red), there is only one steady state ($\simeq -70\,mV$). For $g_{NaP} = 0.015$ (blue), there are three steady states; two are stable (filled circles) and the third is unstable (open circle). In (c), the injected current intensity is varied (blue, violet, and green for $I_{app} = 0$, -0.5, and 0.5 nA, respectively). Other parameter values are $C_m = 0.5\,nF$, $g_L = 0.025\,\mu S$, $V_L = -70\,mV$, and $V_{Na} = 55\,mV$. The energy function $U(V_m)$ is defined by rewriting the circuit Eqn. (2) as $dV_m/dt = F(V_m) = -dU/dV_m$; hence, the energy function $U(V_m)$ is the integral of $-F(V_m)$. For instance, with $g_{NaP} = 0$, Eqn. (2) is reduced to Eqn. (1) and can be rewritten as $dV_m/dt = (V_{ss} - V_m)/\tau_m = F(V_m)$. By integrating $-F(V_m)$, we have $U(V_m) = (V_{ss} - V_m)^2/(2\tau_m) + U_0$, with an arbitrary constant U_0. Therefore, $U(V_m)$ is a parabola with V_{ss} at the bottom of the valley of the energy function (**Figure 1(b)**, black). For $g_{NaP} = 0.015$ (**Figure 1(b)**, blue), the energy function $U(V_m)$ displays two valleys, at V_{Down} and V_{Up}, separated by a peak at the unstable steady state. RC, resistance-capacitance; V_m, membrane voltage.

or down. Either a hyperpolarization or depolarization can destroy the bistability phenomenon. This simple example demonstrates that neuronal bistable dynamics can be readily reconfigured by external inputs. This is generally true for neural networks as well and has important computational implications. Attractors need not be steady states. In neurons, a plateau potential is typically not stable as a steady state. Instead, on depolarization the Hodgkin–Huxley-type sodium and potassium currents produce repetitive action potentials, which represent another (oscillatory) type of attractor behavior. The attractor nature of periodic spiking is shown in **Figure 3** using the classic Hodgkin–Huxley model; regardless of the initial states, the membrane system always converges to the same periodic attractor state. If the system is perturbed by a transient stimulus, it resumes the same firing pattern after the stimulus offset, except for a shift of spiking time or the phase of the periodic attractor state. Thus, a periodic attractor is robust (the amplitude and periodicity are not sensitive to transient perturbations). At the same time it is sensitive to phase shift; hence, the clock can be readily reset. This general property of nonlinear oscillators is a key to understanding synchronization among neurons and coherent brain rhythms. It is also a cornerstone of the neurobiology of biological clocks, such as the circadian rhythm, sleep cycle, and central pattern generators for locomotion such as walking, swimming, and breathing. In each of these systems, the rhythm is amenable to being reset by a transient input which leads to a phase shift in time but otherwise does not alter the stereotypical activity pattern in the network.

The oscillatory attractor illustrates that the behavior of a dynamical system (the Hodgkin–Huxley model) is determined by direct observables (the membrane voltage) as well as internal dynamical variables (ion channel-gating variables). The space of all dynamical variables form a phase space of the system, which is typically multidimensional. A neural circuit in the mammalian brain consists of many thousands of cells; its phase space is enormous. The dynamics of such systems can be very complex, exhibiting a wide gamut of spatiotemporal activity patterns. It is generally not possible to define an energy function for such systems. Nevertheless, the concept of computational energy landscape is still helpful for developing intuitions about attractor networks.

Synaptic Plasticity and Associative Memory

Bistable switches or, more generally, multiple attractors can be realized on all scales, from the molecular

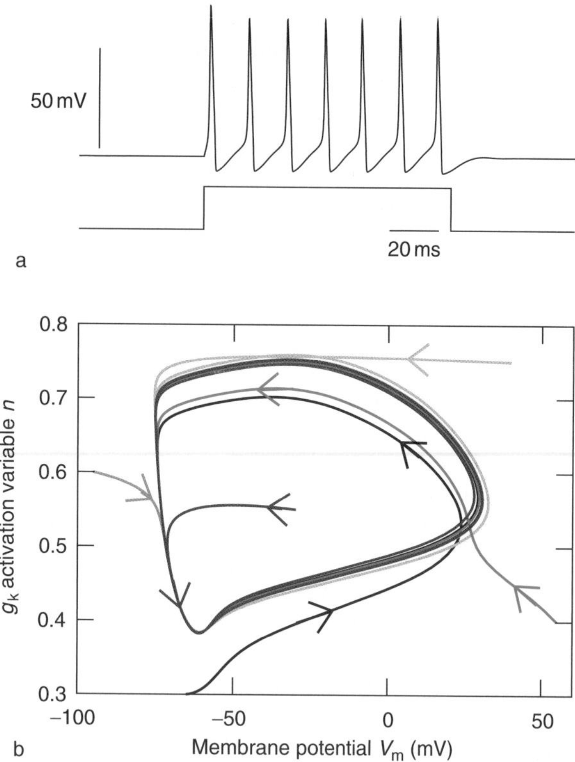

Figure 3 Oscillatory attractor in the original Hodgkin–Huxley model of action potentials: (a) repetitive firing of action potentials with a current pulse $I_{app} = 9\,\mu A/cm^2$; (b) potassium activation variable n plotted against V_m, showing that the oscillation forms a closed trajectory in this projected phase space. The model has a leak current I_L, a fast sodium current I_{Na}, and a noninactivating potassium current I_K. Its dynamics is described by four coupled differential equations (the membrane voltage V_m, the activation and inactivation gating variables m and h for I_{Na}, and the activation variable n for I_K). In (b), the trajectory roughly consists of three portions: the upstroke of an action potential (when both V_m and n increase), the downstroke (when V_m decreases while n keeps increasing and then starts to decrease), and the refractory period (when n continues to decrease while V_m starts to increase). Different colors correspond to five different initial conditions (with different V_m and n values at $t = 0$), in all cases the system dynamically evolves into the oscillatory attractor state. The Hodgkin–Huxley model exhibits bistability between a steady resting state and an oscillatory state, which is not shown for the sake of clarity.

machinery of individual synapses and the electrical activity of single neurons to large neural circuits. It is well known that synapses that form connections between neurons are highly plastic, and experience-dependent synaptic modifications are believed to be a physiological substrate of learning and memory. A single synapse comprises hundreds of proteins that interact with one another in a highly connected signal transduction network; therefore, a synapse is a dynamical system. The biochemical time constants in such a network, and the typical protein lifetimes, range from seconds to hours. So how can synapses store memories that may be retained for many years? One possibility is that the expression of memory maintenance involves changes in the molecular composition of the synapse that are mechanically stable over a long time. Alternatively, memories could be stored in molecular switches, with two or more states that are stable over durations beyond the intrinsic molecular time constants. Recent physiological experiments have yielded evidence for switchlike behavior during synaptic modifications. Molecular studies and modeling have revealed several candidate protein kinases that may exhibit switchlike behavior at single synapses, such as calcium/calmodulin-dependent protein kinase II (CaMKII), protein kinase C (PKC), and mitogen-activated protein kinase (MAPK). Memory switches do not necessarily have an infinite lifetime because an active state may be turned off by subsequent synaptic changes during ongoing neural activity by protein turnover or by molecular fluctuations due to the small synaptic volume (approximately 0.1 fl). Nevertheless, the key point here is that, if individual synapses exhibit stable self-sustained active states (attractors), the lifetime of a memory trace is not directly limited by the biochemical time constants of synaptic signaling pathways.

Whereas synaptic plasticity provides a structural basis for memory formation, the stored information is encoded in a distributed manner by the synaptic connection patterns in a neural circuit. Theoretically, it has been proposed that associative memories are learned through the creation of stable neural activity patterns (attractors). In this view, a memory network has many attractor states, each representing a particular memory item, and has its own basin of attraction within which other states evolve dynamically into the attractor state (**Figure 1**). The stability of attractor states ensures that memory storage is robust against small perturbations. These memory states are imprinted in the network by long-lasting changes of synaptic connections through Hebbian learning, and much theoretical work has been devoted to the analysis of storage capacity (the number of memory items that can be stored and retrieved reliably as a function of the network size). Memories thus established are associative because a partial cue brings the network into the basin of attraction of an attractor with the information content close to that of the sensory cue. Thus, memory retrieval can be done by association between a cue and the corresponding memory item, and the recall process is error-correcting (incomplete information still leads to the correct memory retrieval). This capability for pattern completion is a hallmark of associative memory. At the same time, an attractor model is also capable of pattern separation, in the sense that two slightly different input patterns near the boundary of two basins of attraction may drive the network to two distinct attractor patterns, leading to the retrieval of two separate memories. To experimentally test the attractor model, many neurons must be simultaneously monitored so that distributed activity patterns in memory circuits can be assessed. One of such circuits is the hippocampus, which is known to be critical to the formation of episodic memory and spatial memory. In rodents, during exploration in a familiar environment, pyramidal place cells in the hippocampus are activated when the animal passes through a specific location, called a place field. This spatial selectivity is characterized by a bell-shaped tuning curve (neuronal firing rate as a function of the animal's location). Place fields of hippocampal cells cover the entire surface of the environment with about an equal density, so that the neural ensemble firing can be decoded to read out the animal's location in space. The hippocampus thus implements a neuronal representation of a spatial map. Moreover, when the animal is exposed to another environment, the place fields of cells undergo great changes (remapping) until a representation of the new environment is established in the hippocampus. Therefore, the hippocampus stores distinct maps, with each map being activated as the rat enters a different environment. It has been proposed that these spatial representations reflect distinct attractor states, based on the observation that activity of place cells is preserved in the dark (without external visual cues). The attractor model has recently been directly tested by combining a clever navigation task design with simultaneous recording from many single cells in the hippocampus. In one experiment, rats were trained to forage in two distinct environments, a square and a circle that differed in color, texture, and shape. This led to the remapping of place fields in the majority of cells recorded in the hippocampal area CA1. Then, in probe trials, the rat commuted between a series of environments of the same color and texture, but the shape morphed gradually between the square and the circle. These environments were chosen randomly from trial to trial. Remarkably, it was found that the place fields of the recorded cells abruptly and coherently changed from squarelike to circlelike (**Figure 4**). This observation provides strong support for the attractor model, which predicts that the hippocampal coding of space should exhibit both pattern completion (minor changes from a familiar environment – a square or a circle – do not alter place fields) and pattern separation (similar inputs intermediate between squarelike and circlelike result in drastically different place fields, due to a switch between the two learned maps).

Persistent neural firing observed in a brain area may not necessarily be generated locally but is a mere reflection of mnemonic activity elsewhere. The area CA1 is commonly viewed as an readout circuit, whereas memories are believed to be stored upstream. It has long been hypothesized that autoassociative memories are formed in the area CA3, which projects to CA1 and is endowed with strong recurrent excitatory connections among pyramidal cells, a prerequisite for the generation of attractor states. Furthermore, recent discoveries have drawn attention to the entorhinal cortex, the major input area for the hippocampus. Cells in the dorsolateral entorhinal cortex fire whenever the rat is on any vertex of a triangular lattice spanning the whole surface of the environment. This grid-cell activity is preserved in the dark, when the visual cues are absent. These findings have led to the proposal that the entorhinal cortex embeds an attractor network with a gridlike representation of the environment and that the hippocampus transforms this periodic firing pattern in the input into a nonperiodic firing pattern that encodes the animal's current spatial position. If so, it remains to be seen whether the hippocampus still needs to operate as an attractor network and whether spatial navigation is subserved by interconnected attractor circuits.

In order for an animal to navigate without reference to external cues, the neural instantiation of a spatial map must be constantly updated by the integration of linear and angular self-motion. Whether an attractor model is capable of carrying out such computations robustly is not well understood. Finally, little is known about precisely how the attractor landscape in the space of neuronal firing patterns is shaped by the details of the animal's training process. Another study, using a somewhat different learning and probe procedure but also with morphed environmental shapes, found that place fields of CA3 and CA1 cells switched more gradually and less coherently. This finding does not necessarily mean that the attractor paradigm is incorrect; we expect from computational work that a different learning history gives rise to different sets of attractor states. Elucidation of the general principles and cellular mechanisms of this learning process in future experiments and modeling will help determine whether the attractor model represents a sound theoretical framework for the neurobiology of learning and memory.

Persistent Activity and Working Memory

Attractor models have also been applied to working memory, our brain's ability to actively hold information online for a brief period of time (seconds). Neurons that maintain working memory must be manifestly active in a sustained manner. How can such persistent activity be generated in the absence

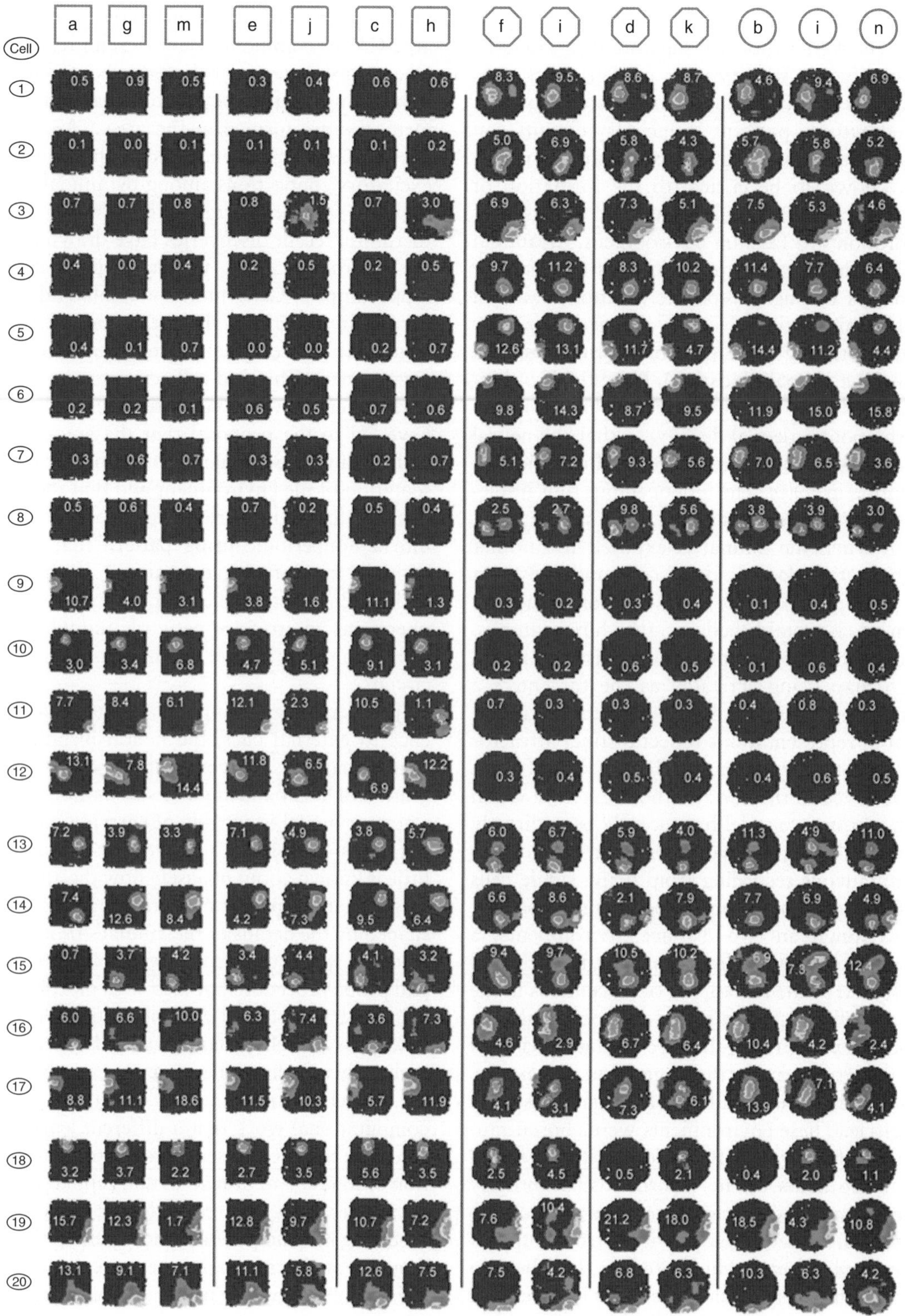

Figure 4 Abrupt and coherent expression of spatial representation of hippocampal neurons. A rat explores a series of environments with morphed shapes (top icons) between a square (the three left-most columns) and a circle (the three right-most columns). Twenty single cells were simultaneously recorded, shown in rows from top to bottom. Warm colors indicate locations (place field) of the rat when the cell's spiking activity is high. Each field is scaled to peak firing rate shown in red. The 17 of 20 simultaneously recorded place cells with different (remapped) firing patterns in the square and the circle, almost all switch from the squarelike to the circlelike pattern between the h and f octagons. Eight cells had fields in the circle but not the square (cells 1–8); four had fields in the square but not the circle (9–12); five fired in both but in different places (13–17); and three did not reach the criterion for remapping (18–20). From Wills TJ, Lever C, Cacucci F, Burgess N, and O'Keefe J (2005) Attractor dynamics in the hippocampal representation of the local environment. *Science* 308: 873–876.

of direct external input? R Lorente de Nó in the 1930s and DO Hebb in the late 1940s proposed that the answer lies in the feedback loop connections. Thus, in a working memory network, every cell receives excitatory drive from both afferent inputs and intrinsic synaptic connections. Inputs activate neurons in a selective cell assembly; the triggered spike activity reverberates through excitatory synaptic circuit, which is enough to sustain an elevated firing when the inputs are withdrawn. This general idea has been made rigorous in attractor models, according to which a working memory circuit exhibits multiple attractor states (each coding a particular memory item) that coexist with a background (resting) state. All the attractor states are self-maintained and relatively stable in the face of small perturbations or noise. Yet memory states can be turned on or switched off by brief external stimuli.

Stimulus-selective neural persistent activity has been observed in awake animals performing delayed-response tasks that depend on working memory. For example, in a delayed match-to-sample task, two visual objects are shown consecutively separated by a time gap of a few seconds, and the subject is asked to judge whether the two stimuli are the same. Or after the presentation of the first object and the delay, an array of visual objects are shown and the subject must indicate by a motor response which of them is identical to the first visual cue. In both cases, the subject's performance relies on the working memory about the first object across the delay. The stored information involves a collection of discrete items. Other delayed-response tasks engage working memory of an analog quantity, such as spatial location or stimulus amplitude. While a monkey was performing such a task, the neurons in the prefrontal, posterior parietal, inferotemporal, and premotor cortices were found to exhibit elevated persistent activity that was selective to stimuli. Three types of mnemonic coding has been observed: (1) object working memory cells are tuned to one or a few of discrete memory items, (2) spatial working memory cells typically exhibit a bell-shaped (Gaussian) tuning function of the spatial location or directional angle, and (3) cells that store parametric working memory of magnitudes (e.g., vibration stimulus frequency) are characterized by a monotonic tuning function of the encoded feature.

These experimental observations lend support to the attractor model inasmuch as stimulus-selective persistent firing patterns are sustained internally in the absence of direct sensory input, are dynamically stable, and are approximately tonic in time (e.g., across a delay). However, experiments show that delay neural activity is often not tonic but exhibits time variations such as ramping up or ramping down. How to account for the heterogeneity and time courses of mnemonic persistent activity represents a challenge to the attractor network model. Moreover, it remains an open question as to what cellular or circuit mechanisms are responsible for the generation of persistent activity. This question is now being addressed using biologically constrained models of persistent activity. **Figure** 5 shows such a recurrent network model for spatial working memory in which spiking neurons and synapses are calibrated by the known cortical electrophysiology. The key feature is an abundance of recurrent connections (loops) between neurons, according to a Mexican-hat-type architecture – localized recurrent excitation between pyramidal cells with similar preference to spatial cues and broader inhibition mediated by interneurons (**Figure** 5(**a**)). In a simulation of a delayed oculomotor task (**Figure** 5(**b**)), the network is initially in a resting state in which all cells fire spontaneously at low rates. A transient input drives a subpopulation of cells to fire at high rates. As a result, they send recruited excitation to one another via horizontal connections. This internal excitation is large enough to sustain elevated activity, so the firing pattern persists after the stimulus is withdrawn. Synaptic inhibition ensures that the activity does not spread to the rest of the network, and persistent activity has a bell shape (bump attractor). At the end of a mnemonic delay period, the cue information can be retrieved by reading out the peak location of the persistent activity pattern; and the network is reset to the resting state. In different trials, a cue can be presented at different locations. Each cue triggers a persistent firing pattern of the same bell shape but with the peak at a different location (**Figure** 5(**c**)). A spatial working memory network thus displays a continuous family of bump attractors.

Biophysically realistic models have specific predictions about the circuit properties required for the generation of stimulus-selective persistent activity. In particular, it was found that a network with strong recurrent loops is prone to instability if excitation (positive feedback) is fast compared to negative feedback, as is expected for a nonlinear dynamical system in general. This is the case when excitation is mediated by the α-amino-3-hydroxy-5-methyl-4-isoxazole propionic acid receptors (AMPARs), which are approximately two to three times faster than inhibition mediated by γ-aminobutyric acid A ($GABA_A$) receptors (time constant, 5–10 ms). The interplay between AMPARs and $GABA_A$ receptors in a excitatory-inhibitory loop naturally produces fast network oscillations. In a working memory model, the large amount of recurrent connections, needed for the generation of persistent activity, often leads to excessive oscillations that are detrimental to network stability.

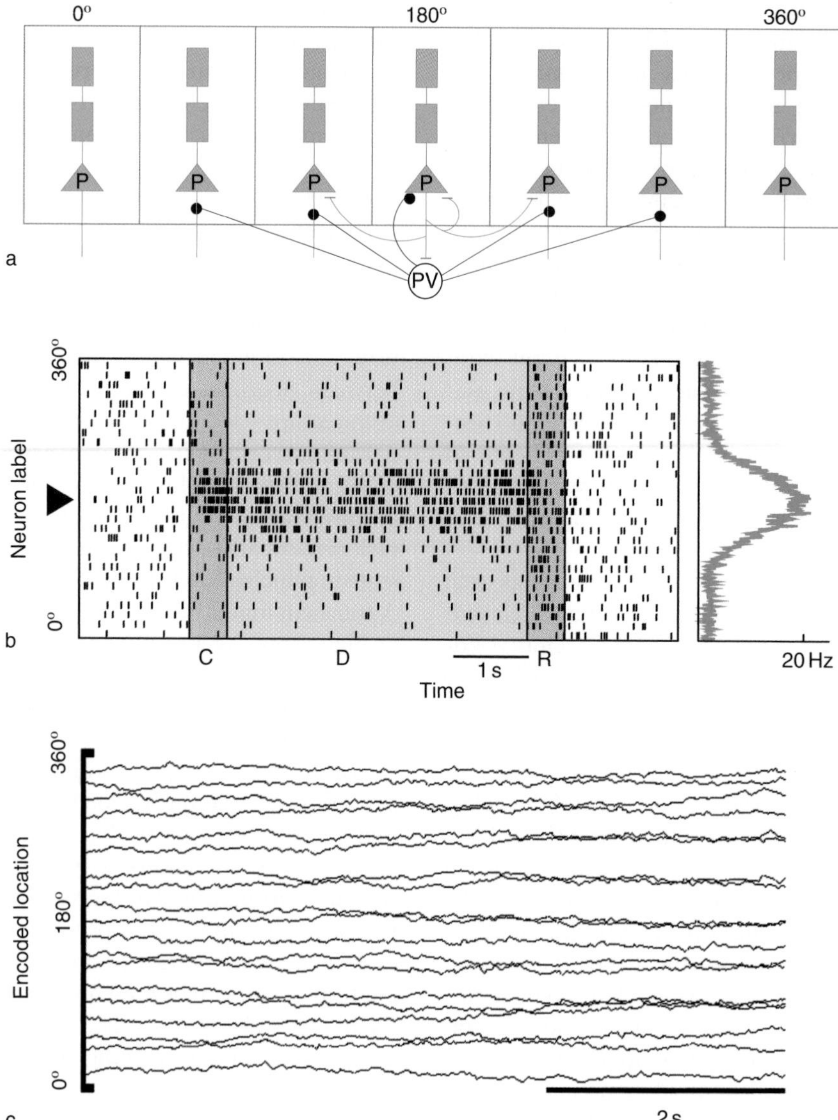

Figure 5 Bump attractor model for spatial working memory: (a) Mexican-hat-type connectivity of the model circuit; (b) spatiotemporal activity pattern of the pyramidal cell population in a simulation of delayed oculomotor response task; (c) temporal evolution of the peak location of mnemonic persistent activity pattern in 20 trials with transient stimuli at different locations. In (a), pyramidal (P) neurons are arranged according to their preferred cues (0–360°). Recurrent excitatory connections are strong between cells with similar cue preference and decrease with the difference in their preferred cues. Local excitation is counteracted by broad synaptic inhibition mediated by paravalbumin (PV) expressing inhibitory interneurons. In (b), each dot is an action potential; the elevated and localized neural activity is triggered by the cue stimulus at 180° and persists during the delay period. On the right is shown the spatial pattern, in which the average firing rate during the delay period is plotted vs. the neuron's preferred cue. The memory of the spatial cue is stored by the peak location of this bell-shaped persistent activity pattern (bump attractor). In (c), the remembered cue (defined by the peak location of the network activity pattern) as a function of time during a 6 s delay period, for a number of trials, each with a different initial spatial cue. R, response period. Panel (b) From Compte A, Brunel N, Goldman-Rakic PS, and Wang X-J (2000) Synaptic mechanisms and network dynamics underlying visuospatial working memory in a cortical network model. *Cerebral Cortex* 10: 910–923; panel (c) from Renart A, Song P, and Wang X-J (2003) Robust spatial working memory in a heterogeneous network model with homeostatic synaptic scaling. *Neuron* 38: 473–485.

Working memory function can be rendered stable if excitatory reverberation is slow, that is, contributed by the *N*-methyl-D-aspartate (NMDA) receptors (time constant 50–100 ms) at recurrent synapses. Thus, the model predicts a critical contribution of NMDA receptors to working memory. Other processes with time constants of hundreds of milliseconds, such as short-term synaptic facilitation or intrinsic ion channels in single cells, may also contribute to reverberatory dynamics underlying working memory.

On the other hand, the feedback mechanism cannot be too slow. An alternative to the attractor model is the scenario in which persistent activity actually is not stable but represents slowly decaying afterdischarges mediated by some intracellular mechanisms such as second-messenger pathways or kinetics of protein kinases. However, this scenario predicts that triggering inputs must be long lasting (lasting for seconds), which is incompatible with physiological experiments in which working memory states have been shown to be switchable quickly by brief external inputs (a few hundreds of a millisecond). The recurrent (attractor) network mechanism achieves the stability and long persistence time of memory storage, as well as rapid flexible memory encoding and erasure, that are behaviorally desirable.

Winner-Take-All and Vector Averaging

Attractor dynamics naturally instantiate winner-take-all (WTA); when several neural pools, each selective for a different sensory stimulus, are simultaneously activated by the presence of several stimuli, competition through recurrent attractor dynamics may ultimately lead to an output pattern with a high firing activity in one neural pool while all the other neural pools are suppressed. WTA could subserve a number of neural processes, such as categorical decision making. In a continuous neural network, such as the ring model shown in **Figure 6**, WTA operation is not always categorical but depends on the similarity between simultaneously presented stimuli. For instance, if two inputs are very close to one another, instead of WTA the network's response is expected to be a vector average of the two individual responses to the stimuli presented alone. Moreover, in a working memory network, WTA often involves different stimuli that occur at different times; this interaction across temporal gaps is possible because an earlier input can trigger persistent activity that interacts with responses to later inputs.

These points are illustrated in **Figure 6**, which shows the interaction of a remembered stimulus and a distractor in the continuous (bump) attractor network model of spatial working memory (same as in **Figure 5**). Significantly, the to-be-remembered stimulus (located at θ_S) and the distractor (θ_D) have the same amplitude but presented at different times, one at the beginning of a trial and the other during the delay. As we intuitively expect, the impact of a distractor depends on its strength (saliency). If the stimulation amplitude is sufficiently large, the distractor is powerful enough to overcome the intrinsic dynamics of the recurrent circuit, and the network is always perturbed to a location close to the intervening stimulus (**Figure 6(a)**, top; red curve in **Figure 6(b)**). In this case, the network can be reset by every new transient stimulus and keeps a memory of the last stimulus in the form of a refreshed selective persistent activity state. On the other hand, if external input is moderately strong relative to recurrent synaptic drive, a distractor triggers only a transient response and the memory of the initial cue is preserved (**Figure 6(b)**, bottom). The resistance to distractors can be understood by the fact that, in a memory delay period, active neurons recruit inhibitions which project to the rest of the network. Consequently, those cells not encoding the initial cue are less excitable than when they are in the resting state and hence are less responsive to distracting stimuli presented during the delay. Moreover, WTA takes place only when the spatial locations of the initial cue and the later distractor are sufficiently distant from one another (more than 90°) (blue curve in **Figure 6(b)**). When they are closer, the activated neural pools (selective for θ_S and θ_D) overlap one another and the peak location of the resulting population activity is roughly the vector average of θ_S and θ_D (blue curve in **Figure 6(b)**). Hence, the same network can perform both WTA and vector-averaging computations, depending on the similarity of stimuli.

Time Integration and Categorical Decision Making

Cortical areas that are engaged in working memory – such as the prefrontal cortex – are also involved in other cognitive functions such as decision making, selective attention, and behavioral control. This suggests that microcircuit organization in these areas is equipped with the necessary properties to subserve both the internal representation of information and dynamical computations of cognitive types. As it turns out, models originally developed for working memory can account for decision-making processes as well. An example is shown in **Figure 7** from model simulations of a visual motion-discrimination experiment. In this two-alternative forced-choice task, monkeys are trained to make a judgment about the direction of motion (say, left or right) in a stochastic random dot display and to report the perceived direction with a saccadic eye movement. A percentage of dots (called motion strength) move coherently in the same direction, so the task can be made easy or difficult by varying the motion strength (from close to 100 to 0%) from trial to trial. While a monkey is performing the task, single-unit recordings revealed that neurons in the posterior parietal cortex and prefrontal cortex exhibit firing activity correlated with the animal's perceptual choice. For example, in a trial

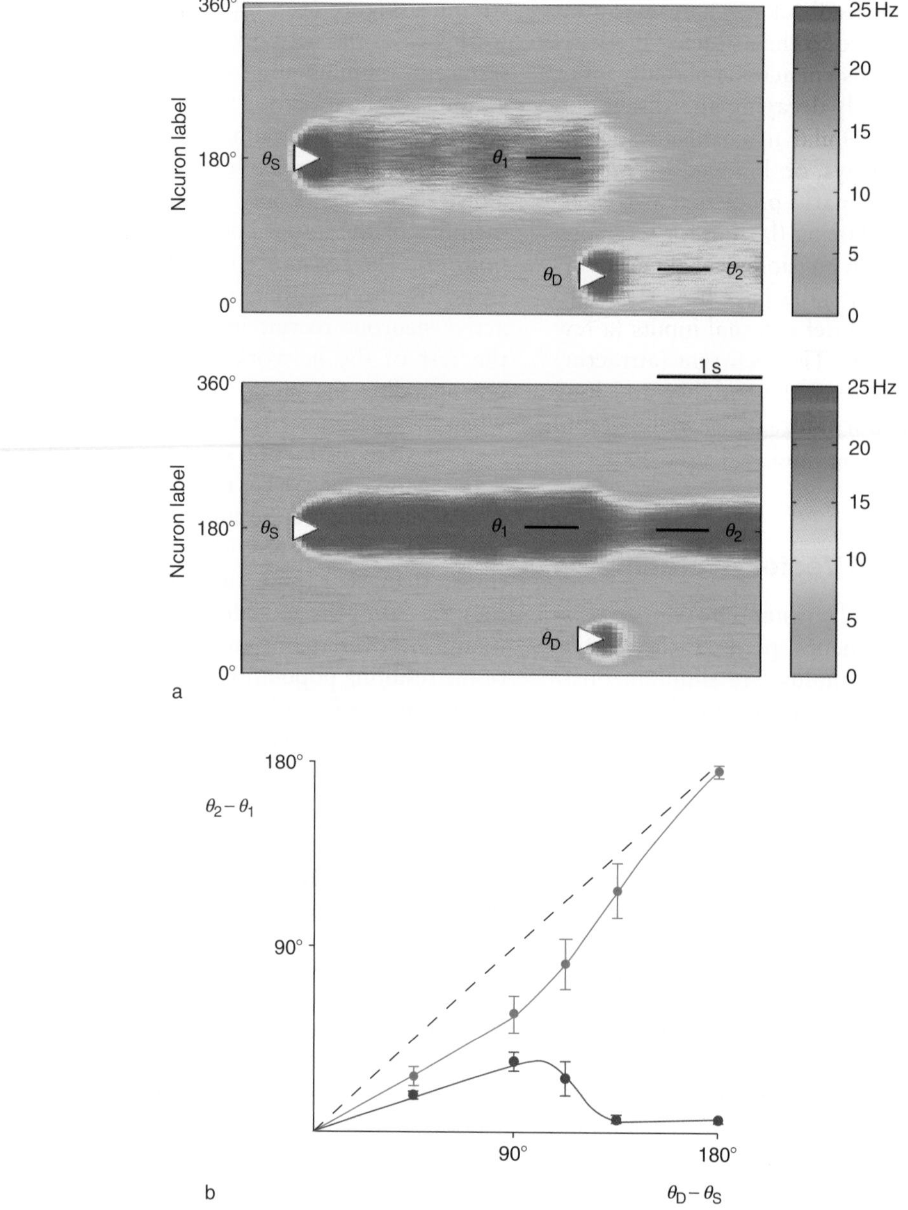

Figure 6 Winner-take-all and vector averaging in the continuous attractor network model of spatial working memory (**Figure 5**): (a) two sample simulations; (b) dependence of network distraction on the distance between the cue and distractor and on the stimulation intensity. In (a), the network spatiotemporal firing pattern is plotted in a color-coded map. A cue is presented initially at θ_S, triggering a tuned persistent activity. After a 2.5 s delay, a distractor stimulus is presented at θ_D, with the same intensity and duration as the cue stimulus. The peak location of the population activity pattern is computed just before the distractor (θ_1) and 500 ms after the distractor (θ_2). The network is completely distracted by a strong stimulus (top), but is resistant to a distractor of moderate intensity (bottom). In (b), the distracted angle $\theta_2-\theta_1$ is plotted vs. the distraction angle $\theta_D-\theta_S$ for several distractor cues. The dashed line indicates perfect distraction (as in (a), top), whereas points on the *x*-axis imply the complete absence of distraction (as in (a), bottom). The red curve indicates large input intensity; the blue curve indicates moderate input intensity. In the latter case, the network exhibits a switch from winner-take-all (when $\theta_D-\theta_S$ is large) to vector averaging (when $\theta_D-\theta_S$ is small). Adapted from Compte A, Brunel N, Goldman-Rakic PS, and Wang X-J (2000) Synaptic mechanisms and network dynamics underlying visuospatial working memory in a cortical network model. *Cerebral Cortex* 10: 910–923.

in which the motion strength is low (say, 6.4%), if the stimulus direction is left but the monkey's choice is right, the response is incorrect. In that case, cells selective for right display a higher activity than those selective for left. Hence, the neural activity signals the animal's perceptual decision rather than the actual sensory stimulus. This experiment can be simulated using the same model designed for working

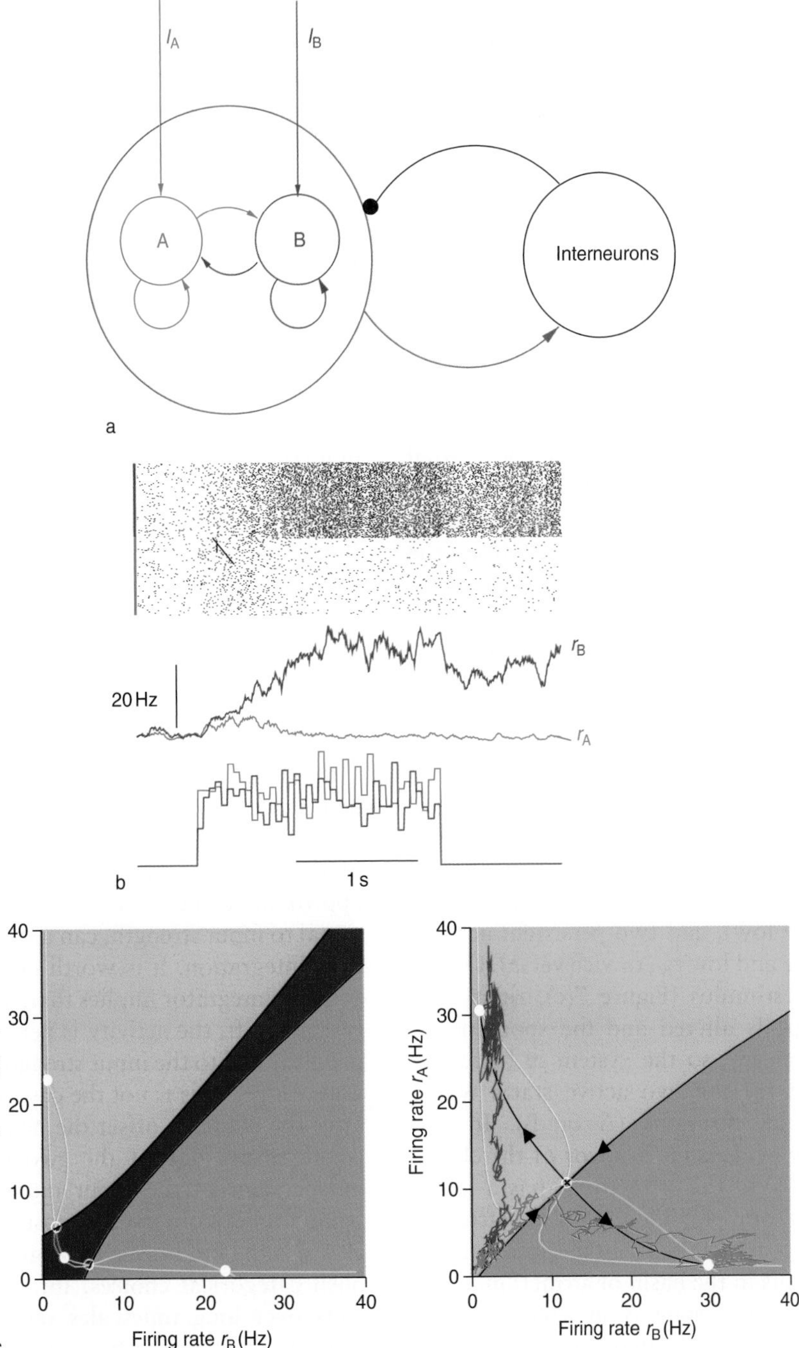

Figure 7 A spiking neuron circuit model for two-alternative forced-choice tasks: (a) model scheme; (b) network simulation with $c' = (I_A - I_B)/(I_A + I_B) = 6.4\%$; (c) decision dynamics shown in the two-dimensional plane where firing rates r_A and r_B are plotted against one another. In (a), there are two groups of spiking pyramidal cells, each of which is selective to one of the two directions (A = left, B = right) of random moving dots in a visual motion-discrimination experiment. Within each pyramidal neural group, there are strong recurrent excitatory connections which can sustain persistent activity triggered by a transient preferred stimulus. The two neural groups compete through feedback inhibition from interneurons. In (b), the population firing rates r_A and r_B exhibit an initial slow ramping (time integration) followed by eventual divergence (categorical choice). In this sample trial, although the input A is larger, the decision is B (an erroneous choice). As shown in (c), in the absence of stimulation (left), there are three attractor states (filled circles) and two unstable steady states (open circles); colored regions are the basins of attractions (maroon for the resting state; orange and brown for the persistent activity states). With a stimulus of $c' = 6.4\%$ in favor of choice A (right), the decision trajectory is shown for two trials (correct trial in blue; error trial in red). Panels (a,b) from Wang X-J (2002) Probabilistic decision making by slow reverberation in neocortical circuits. *Neuron* 36: 955–968; panel (c) from a simulation by KF Wong using the model published in Wong KF and Wang X-J (2006) A recurrent network mechanism for time integration in perceptual decisions. *Journal of Neuroscience* 26: 1314–1328.

memory. The only difference between a working memory simulation and a decision simulation is that in a delayed-response task only one stimulus is presented but for a perceptual discrimination task conflicting sensory inputs are fed into competing neural subpopulations in the circuit. This is schematically depicted in **Figure 7(a)**, in which the relative difference in the inputs $c' = (I_A - I_B)/(I_A + I_B)$ mimics the motion strength in the visual motion discrimination experiment. **Figure 7(b)** shows a simulation with $c' = 6.4\%$. At the stimulus onset, the firing rates of the two competing neural populations, r_A and r_B, initially ramp up together for hundreds of milliseconds before diverging from one another when one increases while the other declines. The perceptual choice is decided based on which of the two neural populations wins the competition. Therefore, consistent with the physiological observations from the monkey experiment, the decision process proceeds in two steps. Sensory data are first integrated over time in a graded fashion, which in the model is instantiated by the NMDA receptor-dependent slow reverberation. This is followed by WTA competition produced by synaptic inhibition, leading to a categorical (binary) choice.

Figure 7(c) shows attractor dynamics in the decision space, where r_A is plotted against r_B. In the absence of stimulation (**Figure 7(c)**, left), three attractors coexist (filled circles): a spontaneous state (when both r_A and r_B are low), and two persistent activity states (with high r_A and low r_B, or vice versa). On the presentation of a stimulus (**Figure 7(c)**, right), the attractor landscape is altered and the spontaneous steady state disappears, so the system is forced to evolve toward one of the two active states which represent perceptual decisions (A or B). In this graph, the sensory evidence is in favor of the choice A (with $c' = 6.4\,\%$), so the attractor A has a larger basin of attraction (orange) than that of the attractor B (brown). The system is initially in the spontaneous state which now falls in the basin of attraction A and evolves toward decision state A in a correct trial (blue). However, at low c', the bias is not strong, and noise can induce the system's trajectory to travel across the boundary of the two attraction basins, in which case the system eventually evolves to decision state B in an error trial (red). The crossing of a boundary between attraction basins is slow, which explains why the reaction times are longer in error trials than in correct trials, as was observed in the monkey experiment. After the offset of the stimulus, the system's configuration reverts to that in **Figure 7(c)** (left). Because a persistently active state is self-sustained, the perceptual choice (A or B) can be stored in working memory for later use, to guide behavior. In this way, the attractor dynamics model offers an unified account for working memory and decision-making computations.

Concluding Remarks

In summary, the language of attractors is natural for describing the electrical activity of neurons and neural circuits. It provides a plausible theoretical framework for both short-term working memory and long-term associative memory. Significantly, nonlinearity due to feedback loops makes it possible that graded changes of a cellular or synaptic parameter lead to the emergence of qualitatively different behaviors (e.g., with or without persistent activity). The functional implications of this insight are potentially far-reaching because it suggests that cortical areas dedicated to distinct functions (e.g., sensory processing vs. working memory) may share a same canonical cortical circuit layout but with subtle differences in the cellular/molecular makeup, connectivity properties, and neuromodulatory influence.

Qualitatively speaking, working memory requires neurons to convert a transient input pulse into a sustained persistent activity, such as a time integral of the stimulus. Similarly, in perceptual decisions, approximate linear ramping activity, at a rate proportional to input strength, can also be conceptualized as time integration. It is worth noting, however, that a genuine integrator implies that, after a transient input is turned off, the activity is persistent at a firing rate proportional to the input strength, spanning a continuous range. This is not the case in **Figure 7**, in which after the stimulus offset the neural activity is binary (representing one of the two categorical choices), independent of the input motion strength. This is what has been observed in posterior parietal neurons, and it is the kind of neural signals needed to accomplish categorical choices. Integration by neural circuits over long timescales represents an important topic of active research.

An open question concerning working memory is whether persistent activity is primarily generated by local circuit dynamics, a single-cell property, or a large-scale network composed of several brain regions. For both working memory and long-term memory, a major challenge is to elucidate the biological substrates that underlie the robustness of continuous attractors and integrators. Furthermore, there is a dichotomy between rapidly switchable attractors, on the one hand, and an intracellular signaling network with multiple time constants, on the other hand. The interplay between cellular processes and collective network dynamics will turn out to be an exciting topic in future research.

Acknowledgments

I thank K-F Wong for making **Figure** 7(c). This work is supported by the NIH grant MH62349.

See also: Hodgkin–Huxley Models; Neural Integrator Models; Prefrontal Cortex: Structure and Anatomy; Prefrontal Cortex; Short Term and Working Memory; Visual Associative Memory; Working Memory: Capacity Limitations.

Further Reading

Amari S (1977) Dynamics of pattern formation in lateral-inhibition type neural fields. *Biological Cybernetics* 27: 77–87.

Amit DJ (1992) *Modelling Brain Function: The World of Attractor Neural Networks*. Cambridge, UK: Cambridge University. Press.

Ben-Yishai R, Bar-Or RL, and Sompolinsky H (1995) Theory of orientation tuning in visual cortex. *Proceedings of the National Academy of Sciences of the United States of America* 92: 3844–3848.

Bhalla US and Iyengar R (1999) Emergent properties of networks of biological signaling pathways. *Science* 283: 381–387.

Brunel N (2005) Network models of memory. In: Chow CC, Gutkin B, Hansel D, Meunier C, and Dalibard J (eds.) *Methods and Models in Neurophysics*, pp. 407–476. Amsterdam: Elsevier.

Camperi M and Wang XJ (1998) A model of visuospatial working memory in prefrontal cortex: Recurrent network and cellular bistability. *Journal of Computational Neuroscience* 5: 383–405.

Hopfield JJ (1982) Neural networks and physical systems with emergent collective computational abilities. *Proceedings of the National Academy of Sciences of the United States of America* 79: 2554–2558.

Machens CK, Romo R, and Brody CD (2005) Flexible control of mutual inhibition: A neural model of two-interval discrimination. *Science* 307: 1121–1124.

McNaughton BL, Battaglia FP, Jensen O, Moser EI, and Moser MB (2006) Path integration and the neural basis of the "cognitive map." *Nature Reviews Neuroscience* 7: 663–678.

Miller P, Zhabotinsky A, Lisman J, and Wang X-J (2005) The stability of a stochastic CaMKII switch: Dependence on the number of molecules and protein turn over. *PLoS* 3: 705–717.

O'Connor DH, Wittenberg GM, and Wang SS (2005) Graded bidirectional synaptic plasticity is composed of switch-like unitary events. *Proceedings of the National Academy of Sciences of the United States of America* 102: 9679–9684.

Seung HS (1996) How the brain keeps the eyes still? *Proceedings of the National Academy of Sciences of the United States of America* 93: 13339–13344.

Strogatz SH (1994) *Nonlinear Dynamics and Chaos: With Applications to Physics, Biology, Chemistry and Engineering*. Reading, MA: Addision-Wesley.

Tsodyks M (1999) Attractor neural network models of spatial maps in hippocampus. *Hippocampus* 9: 481–489.

Wang X-J (2001) Synaptic reverberation underlying mnemonic persistent activity. *Trends in Neuroscience* 24: 455–463.

Wang X-J (2002) Probabilistic decision making by slow reverberation in neocortical circuits. *Neuron* 36: 955–968.

Zhang K (1996) Representation of spatial orientation by the intrinsic dynamics of the head-direction cell ensemble: A theory. *Journal of Neuroscience* 16: 2112–2126.

Audiovocal Communication in Bats

J S Kanwal, Georgetown University, Washington, DC, USA

Introduction

The advanced status of audiovocal communication in bats, in general, rivals the sophisticated communication strategies reported in mammalian species with dramatically larger brains, such as primates and cetaceans. Audiovocal communication fulfills an important survival function for insectivorous bats. These bats forage under low light conditions, and in some species social interactions and breeding may occur in complete darkness, where coordination via communication sounds is critical. With technical advances enabling a rapid analysis of complex sounds, there has been a proliferation of information on audiovocal communication in several species of bats. The social behavior of bats, however, is relatively difficult to study in their natural environment and much of our knowledge of audiovocal social communication in bats has been obtained from observations of interactions between conspecifics in a colony of captive, free-flying bats. In some species, such as *Carollia*, field studies and observations made on captive bats indicate that under seminatural conditions, their social organization and interactions remain relatively unchanged. The vocal apparatus and vocal control circuits in the brain of insectivorous bats are especially well developed and underlie their ability to emit complex vocalizations. Their auditory apparatus and the accompanying neural machinery are equally advanced at all levels, from the elaborate structure of the ear pinnae to the specializations of the cochlea and hair cells in the periphery and of neurons in the central nervous system.

Audiovocal communication can be studied at three distinct levels. First is the acoustic level, which requires a clear understanding of the acoustic structure of communication sounds. Second is the behavioral level, which is intricately connected to the social interactions between conspecifics, and the third is the sensorimotor (audiovocal) neural level. Audiovocal communication in bats, as in other animals, evolves through communication behavior as an adaptive interface of the organism with its biological and physical environment. This is especially well developed in many insectivorous, microchiropteran bat species since sound continues to be a viable means of signaling when vision and vision-mediated tactual interactions are eliminated or minimized in the darkness of caves and the night sky.

The ability to produce a rich variety of sounds for communication, especially via vocal learning, appears to be independent of brain size among mammals and birds. For example, songbirds with relatively small brains are more advanced in their vocal learning skills than other species with larger brains, such as ostriches and penguins. Whereas echolocation has been studied extensively in many species of bats, auditory communication is only now being studied in an increasing number of species. This article summarizes data on a few species of bats that have been extensively studied and/or exhibit characteristics that provide new representative examples of advancements in audiovocal communication among the taxonomic group of Chiroptera.

Echolocation versus Communication

Frequently, in the published literature, both echolocation and communication sounds emitted by bats have been referred to as 'calls,' blurring the important and clear differences in the function of these two types of signals. Whereas echolocation sounds are produced nearly continuously during flight and occasionally during roosting, communication sounds are largely produced in a roost and during flight-mediated social interactions. Spectrally, echolocation sounds may not be very different from some types of communication sounds, but as the name implies, echolocation sounds are meaningless to the emitter until paired with their respective echoes. They are also relatively stereotypic, with systematic differences observed in their structure during different phases of food search and insect capture in the case of insectivorous bats. Communication sounds, as a group, cover a wide range of acoustic forms, varying from pure tone-like to various levels of frequency modulations and noise bursts as well as mixtures of these three basic acoustic elements. Using 'discriminant' or 'principal components' analysis, most calls can be grouped into discrete clusters, with each cluster representing a wide range of variations in multiple parameters (**Figure 1**). Clearly, the factors shaping the acoustic structure of communication sounds in bats are different from those that shape their echolocation sounds. For example, the latter are greatly influenced by the foraging strategy of a species and its ecological niche. Variation in communication sounds may follow motivational–structural rules as first proposed by Morton in 1977.

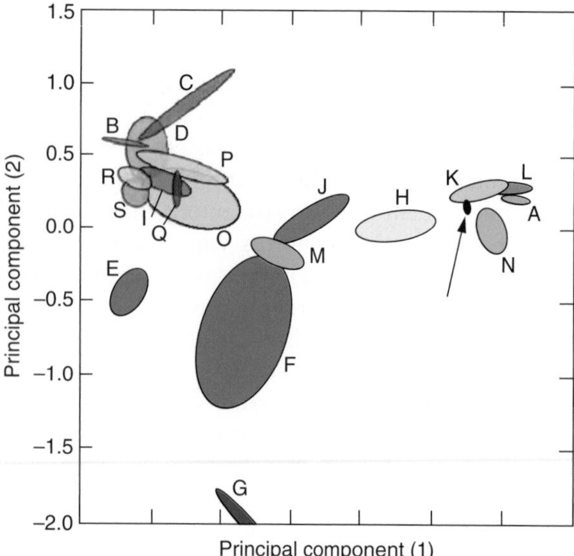

Figure 1 Scatter plot obtained from a principal-components analysis for mustached bat social calls. Bivariate sample ellipsoids drawn at the 75% confidence interval based on scatter plots of the first two principal components of a three-dimensional solution for the multivariate parameters of syllables (plotted with letters). Ellipsoids that overlap in the first two components are separated by the third component. Echolocation pulses (arrow) are the most tightly grouped of all vocalizations. Letters correspond to the following syllable abbreviations (see **Table 1**): A, TCFs; B, QCFs; C, QCFl; D, tQCF; E, NNBs; F, NNBl; G, rBNB; H, bUFM; I, bDFM; J, cDFM; K, sAFM; L, dAFM; M, sHFM; N, dRFM; O, fRFM; P, fSFM; Q, qSFM; R, sWFM; S, WFMl. Adapted from Kanwal JS, Matsumura S, Ohlemiller K, and Suga N (1994) Analysis of acoustic elements and syntax in communication sounds emitted by mustached bats. *Journal of the Acoustical Society of America* 96: 1229–1254, with permission.

To emphasize the functional distinction between the two types of sounds, echolocation sounds are best referred to as 'echolocation pulses' rather than calls, and communication sounds are best referred to as social calls or simply calls. This distinction is consistent with the neurophysiological literature on processing of echolocation signals. Per this terminology, neurons in the auditory system are tuned to combinations of various parameters of a pulse–echo pair, with the pulse being the relatively brief, stereotypic sound produced by a bat for echolocation. Suga and co-workers have shown that this tuning extracts important information relevant for echolocation (e.g., target range during insect pursuit).

Eavesdropping

Eavesdropping or 'secretly listening in' may be considered as a passive form of communication or gathering information from the vocal signals of conspecifics. An echolocating animal gives away a lot of information about itself, its position, and flight path by the timing and structure of its echolocation pulses. High-intensity echolocation pulses may also reveal the identity of the emitter and possibly the presence of a food source to an attentive conspecific. Therefore, echolocation signals can provide useful cues to eavesdroppers that may use this information for various purposes. For example, it has been shown that instead of focusing on one's own echolocation pulse and echo information, bats may use the echolocation pulses of conspecifics to obtain positional information about preferred roost mates. Möhres made this observation in the horseshoe bat, *Rhinolophus ferrumquinuum*, in 1967. Similarly, *Myotis lucifugus* has been experimentally shown to use the presence of a high density of echolocation pulses emitted by conspecifics in the field as an indicator of various resources such as a food source at a particular location. This is intriguing because several neural mechanisms have been enunciated that show how CF–FM bats (species with constant frequency (CF) and frequency modulated (FM) elements in their echolocation pulse) may avoid hearing and therefore jamming of their echolocation signal by conspecifics when thousands of them located in a roost or in close proximity emit high-intensity (~120 dB SPL) echolocation pulses at the same time. Accordingly, cortical neurons of at least some bat species, such as mustached bats, are sharply tuned and highly specialized in that they respond well only to specific parameters in the echo following the pulse. Interestingly, the same neurons are not as sharply tuned to the frequency of the fundamental in the echolocation pulse. Therefore, it is plausible that individuals can eavesdrop on the fundamental frequency in echolocation pulses emitted by conspecifics to locate a food source or be alerted to other foraging and social activities, such as mating. Young bats have been shown to be more responsive than adults to playback presentations of echolocation pulses. This could result from a general increased level of curiosity and playfulness observed in the young of many mammalian species. In some larger bat species, adults use echolocation pulses produced by foraging conspecifics to achieve spacing of at least 50 m from conspecifics on feeding grounds. This type of transfer or gathering of information from echolocation sounds is best labeled as an indirect one-way form of communication.

Mother–Infant Interactions

In bats, echolocation pulses, sometimes considered as a form of autocommunication or communication with the self, may also inform conspecifics and modify their behavior in some contexts. In these contexts, they function as calls. A *bona fide* use of echolocation pulselike sounds for the purpose of communication is during mother–infant interactions in which the mother may use them to identify its pup and vice versa. Among all communication behaviors that involve the production of sounds, mother–infant communication is one of the best studied. Calls produced during these interactions have been studied in at least 24 different species. Since orientation or echolocation signals are among the first to be produced by infants in many species, these play a major role in communication via antiphonal calling between infants and their respective mothers in a bat colony. Some of the studies documenting this behavior were conducted in the little brown bat, *M. lucifugus*, and in Japanese horseshoe bats, *Rhinolophus ferrumequinum Nippon*. In addition, Habersetzer has shown that *Rhinopoma hardwicki*, found in India, can decrease the slope of their normal echolocation pulses when emitting them in the roost compared to the pulse emitted in flight. In noctilionids (*Noctilo leporinus*) and a few other species, two bats on a collision course drop the frequency sweep of the echolocation pulse by one octave to warn an approaching bat and avoid a collision. This drop in frequency may activate neurons that are tuned broadly to low frequencies and are not engaged in processing echolocation information. For purposes of echolocation, the second harmonic of the echolocation pulse in mustached bats has the most energy so it can provide information from the echo returning from a target. Sometimes, however, mustached bats produce bursts of echolocation pulselike sounds in which the predominant energy is shifted to the fundamental instead of the second harmonic. These variants of the echolocation pulse are produced only in the presence of conspecifics.

Echolocation pulses likely coevolved by modification of a communication call, perhaps associated with foraging, since the latter are also present in nonecholocating bats and other mammalian species. A scaled multidimensional representation of several acoustic parameters present in calls and echolocation pulses of mustached bats shows that echolocation pulses share parameters with several different calls (**Figure 2**). The observation that echolocation pulses and their variants can serve a communicative role in various contexts even in modern-day bats, as shown by Fenton and Gould, is consistent with this idea. The evolutionary relationship between echolocation and communication sounds, however, needs to be investigated in a number of different species.

Acoustic Structure of Social Calls

From an acoustics standpoint, all natural sounds, including communication calls, consist of three basic types of elementary sounds, namely CF sounds, FM sounds, and noise the burst (NB)-type sounds. Pure tones are rare in nature; accordingly, very few calls are strictly tonal, although many have a clear harmonic structure (**Figure 2**). Most complex calls consist of combinations of tones and/or noise bursts; the latter can be either broad- or narrowband, whereas the former contain multiple (harmonic) bands of predominant frequencies. Additional complexities may be present in both the spectral and the temporal domains. Both animal and human sounds may contain varying levels of noisiness, making it difficult to identify distinct classes of calls. Also, sounds with two different fundamental frequencies may be emitted at the same time and independently modulated. The simple syllables in mustached bats are defined on the basis of the statistically significant differences in the overall spectrographic pattern of calls so that any one pattern is present in only one call type. These basic patterns of acoustic structure are also present in the calls of the other species, such as monkeys and cats, as well in the phonemes in speech sounds, although the vocal frequency range and audiogram in these species is shifted down by nearly an order of magnitude and the meaning of each sound is obviously species specific.

Social calls in bats are grouped into categories or call types using different criteria depending on an investigator's method of analysis. Traditionally, social calls have been grouped according to the social behavior with which they are associated – for example, aggressive or warning calls, mating and song flight calls, distress calls, or isolation and direction calls. In a few cases, they may be grouped according to the subjective description of the sound (e.g., a twitter, scream, or whistle), but this anthropocentric categorization is not very meaningful because, unlike bird calls, many of the acoustic elements in bat calls may lie in the ultrasonic range and are therefore not heard. A more accurate, although behaviorally less-meaningful way is to categorize calls on the basis of the acoustic elements present within a call and the geometric pattern these may take. For example, calls may be considered as being largely CF, FM, or NB call types (**Table 1**). This relatively objective classification scheme has been used to define the calls of at least two bat species in detail and allows one to visualize roughly the spectrographic structure of calls simply by knowing its name. This also allows a more objective way to compare call structures in related species of bats from an acoustics standpoint, which can be later used to compare their behavioral correlates as well. Prefixes such as long and short give additional information about the call. Dissociating call names and behaviors also makes it possible to assign a call to two or more different behaviors with which it may be associated. Alternatively, variants of the same

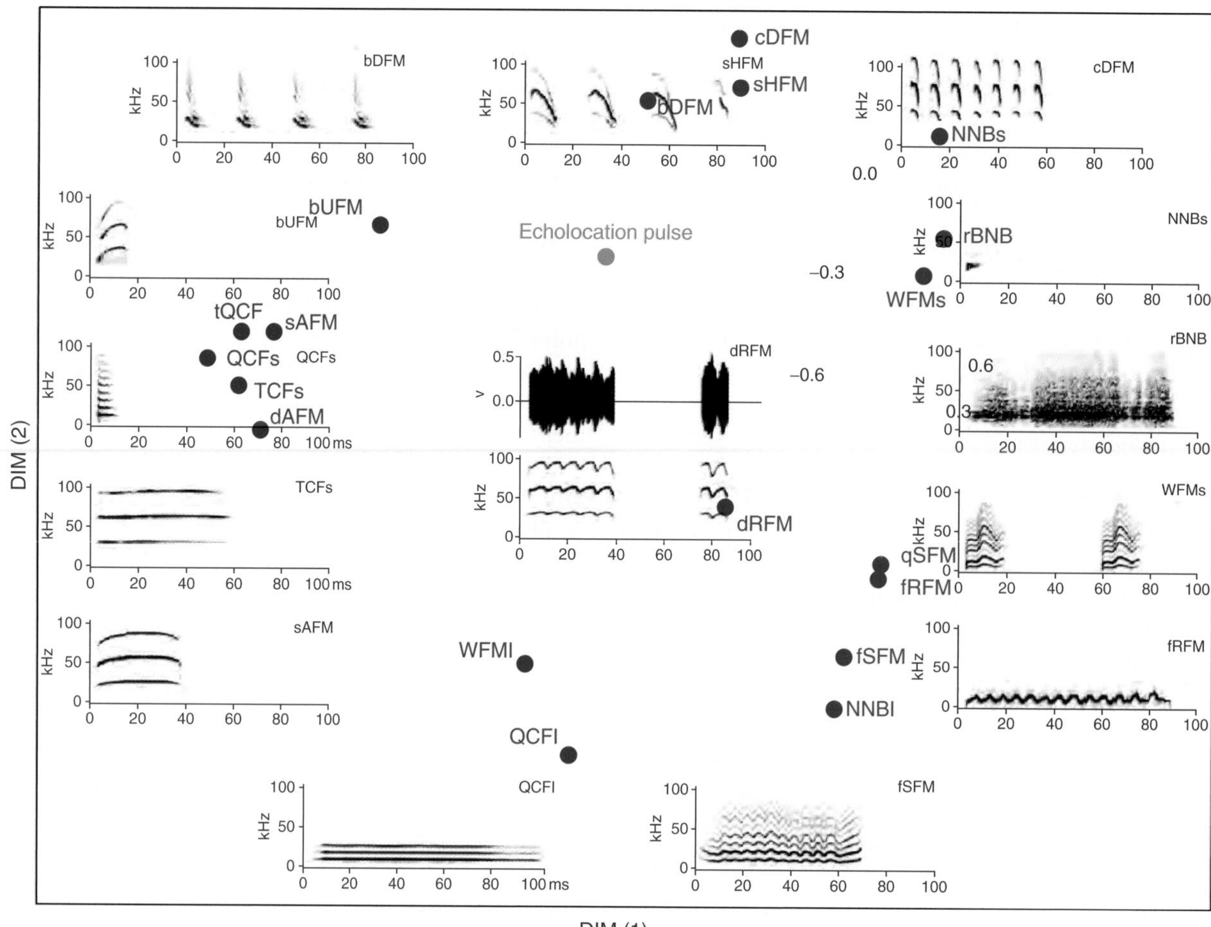

Figure 2 Multidimensionally scaled (MDS) configuration of syllable types (blue dots) and echolocation pulses (red dot) obtained from mean values of nine parameters of mustached bat calls. The structural similarities captured by the plot are illustrated by superimposed spectrograms of several syllable call types roughly corresponding to its position in the two-dimensional MDS plot. More than 90% of the variation was captured for the first three dimensions. All spectrograms are drawn to the same scale. Spectrogram for the descending, rippled FM (dRFM) is shown at the center of the contour plot close to the location of the echolocation pulse. Note the separation of CF and CF-like syllables from FM syllables. Adapted from Kanwal JS, Matsumura S, Ohlemiller K, and Suga N (1994) Analysis of acoustic elements and syntax in communication sounds emitted by mustached bats. *Journal of the Acoustical Society of America* 96: 1229–1254, with permission.

call type may be used in two different contexts. This usage of one call type in more than one behavior has been shown in primates and a few other species. Relatively complete acoustical and statistical analysis of vocalization repertoires has been performed for the mustached bat, *Pteronotus parnellii*, and the horseshoe bat, *R. ferrumequinum*. Earlier studies also provided information on different call types produced by *Carollia perspicallata*, *Pteropus poliocephalus*, *Myotis lucifugus*, and the Mexican free-tailed bat, *Tadarida brasilensis*. Also, the social calls of 16 species of European bats have been recorded to compare their structure and usage.

Syllables and Syntax

For social communication, two or more different categories of sounds may be combined in the time domain to generate 'composites' that do not belong to any one single class of elemental sounds. This ability has been demonstrated in at least two different species of bats. Furthermore, simple syllables are combined within composites according to some rules, which for primate vocalizations has been referred to as a 'phonetic-like' syntax and in birdsong as syllabic syntax, although in the latter case short silent intervals are included as part of a syllable. From recordings of mustached bat calls, only 11 of the 19 simple syllables reported to date in a single subspecies are used to construct composites. For example, the bent upward FM (bUFM) component in a composite is never followed by an attached syllable whose fundamental frequency is lower than the mean terminal frequency of the bUFM. Similarly, a fixed, sinusoidal FM (fSFM) component in a composite typically

Table 1 Abbreviations of syllable types

Abbreviation	*Syllable name*
TCFs	short, True CF
QCFs	short, Quasi CF
QCFl	long, Quasi CF
tQCF	trapezoidal, Quasi CF
NNBs	short, Narrowband NB
NNBl	long, Narrowband NB
rBNB	rectangular, Broadband NB
bUFM	bent, Upward FM
bDFM	bent, Downward FM
cDFM	checked, Downward FM
sAFM	single, Arched FM
dAFM	double, Arched FM
sHFM	single, Humped FM
dRFM	downward, Rippled FM
fRFM	fixed, Rippled FM
fSFM	fixed, Sinusoidal FM
qSFM	quasi, Sinusoidal FM
WFMs	short, Wrinkled FM
WFMl	long, Wrinkled FM

precedes a short, quasi CF (QCFs) or a QCFs-like component and is not attached to syllables with very high (>20 kHz) fundamental frequency. In mustached bat calls, only the duration parameter of a simple syllable appears to be significantly altered when that syllable is combined within a composite (e.g., bUFM). This total duration of a syllable is probably constrained by the lung volume and respiratory rate of the animal; therefore, the duration of one of the components in a composite may be significantly modified.

Simple syllabic calls emitted by mustached bats may also be combined in different forms and contexts to produce a train or 'stream' of sounds. Repetition of similar syllables during such vocalization bouts frequently indicates the urgency of a desired response. In mouse pups, wriggling (isolation) calls emitted at high repetition rates and sometimes with increasing amplitudes trigger a retrieval response on the part of maternal females. This type of sequencing of sounds is less common in animals compared to speech in humans, except in the context of song as in songbirds, bats, dolphins, and whales or chorus calls in some species of frogs and primates. It has been postulated both for bats and for some species of primates that these types of syllable combinations in the time domain follow elementary syntactical rules not unlike those in the production of a stream of speech sounds.

Call Variability

Categorization of call types strictly on the basis of the acoustic structure of calls is accompanied by quantification of the variability in the acoustic structure of each call type. Data indicate that for most calls there is considerable variability among different examples of the same call type, as uttered by different individuals, in terms of both tempo and spectral position (**Figure 3**). The source of this variability may be the sex, mood, and/or identity of the emitter as well as the social and environmental context in which a call type is emitted. The receivers may use variations in calls to gauge motivational levels of others and to modulate their own vocal behavior so as to interact with conspecifics. Alarm calls may be used to signal an environmental threat, allowing others to behave in the most adaptive manner. Variations in the tonality, pitch, and frequency modulation may also be used to identify individuals and their mood in the dark, especially when olfactory cues are unavailable from a short distance away. Males, females, and infants differ in their fundamental frequencies or pitch. Yet, what is remarkable about auditory communication is that pitch-shifted variants of the same call may reveal the identity/mood of the emitter without altering its contextual identity or meaning. The receiving individual must therefore be capable of ignoring information that is irrelevant identifying a call and derive a caller-invariant representation of each call type. At the same time, complementary information within a call type can be used to define caller identity and other factors that produce the variation. Neuroscientific studies are needed to understand how the brains of bats and other animals extract the different information-bearing elements from the acoustic structure of a call.

Geographic Dialects and Vocal Learning

Geographic variation has been studied most extensively in bird species; for example, trills of the blue tit, *Parus caeruleus*, vary geographically depending on environmental factors, competition, and food availability. Among mammals, much of the research on geographic variation of communication calls has focused on primate species; for example, differences in the acoustical repertoires of free-ranging langurs, *Presbytis entellus*, have been documented and attributed to phylogenetic adaptations to their respective habitats or to the different social organizations of the langur populations. Although geographic variation has been shown to exist in the echolocation calls of certain bat species such as the hoary bat *Lasiurus cinereus* and mustached bat, no such comparative study has been done on bat communication calls even though they represent the vast majority of a bats' vocal repertoire. There is only preliminary evidence of the existence of dialects for calls in bats. These data derive from an analysis of calls

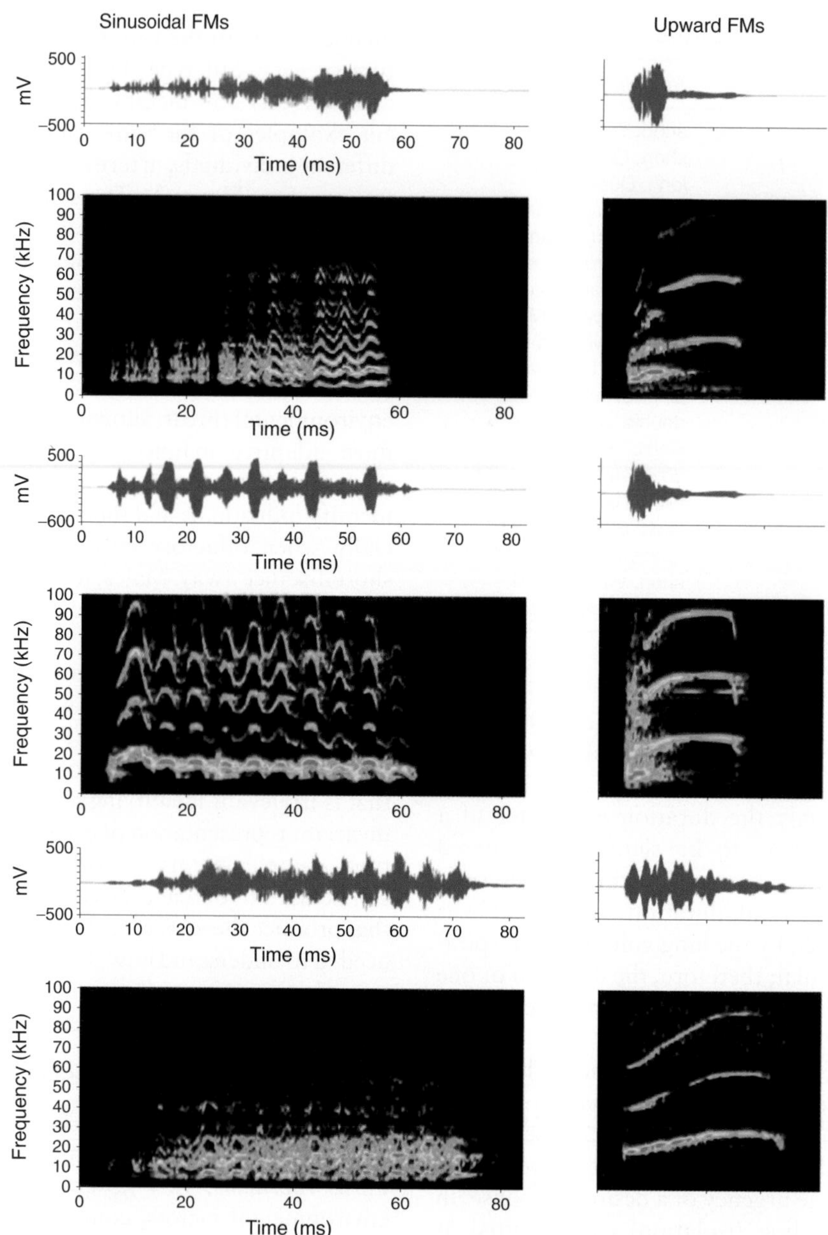

Figure 3 Amplitude envelops and spectrograms of variants of two different call types (same timescale), fixed, sinusoidal FM (fSFM) and bent, upward FM (bUFM), emitted by mustached bats. Multiple parameters are modified between different examples of a call type.

in mustached bats collected from different islands in the Caribbean (Jamaica, Puerto Rico, and Trinidad). The vocal repertoire of Jamaican mustached bats includes four types of CF calls: short, true CF calls (TCFs), long and short quasi-CF calls (QCFl and QCFs, respectively), and a trapezoidal quasi-CF (tQCF). Of these, Trinidad bats have been observed to vocalize only the QCFl and QCFs types of CF calls. Mustached bats obtained from Puerto Rico, *Pteronotus pamellii puertoricensis*, demonstrate a surprising paucity of CF calls and instead emit a CF subsyllable as part of a frequently occurring composite syllable. This call, an upward FM sweep preceding the CF segment, resembles a reversed echolocation pulse, with its harmonics at 30, 60, and 90 kHz and maximal intensity in the second harmonic. Additional CF calls may also be emitted by this species, but they were not readily recorded under seminatural laboratory conditions. Males of *Saccopteryx bilineata* exhibit variation at both the geographic and the individual level in the acoustic profile of their vocalizations.

Dialects usually develop within an ontogenetic time frame and their presence within subspecies of a taxonomic group may be taken as indirect evidence of vocal learning. Several species in four separate vertebrate groups have been shown to exhibit vocal learning.

These include songbirds, whales and dolphins, elephants, and bats. Of these, songbirds and cetaceans have been shown to have dialects and exhibit vocal learning. Vocal learning in bats has been demonstrated both in infants and in adults. Interestingly, in the primate species studied so far, vocal learning of species-specific calls has not been demonstrated.

In bats, studies of mother–pup interactions in *Carollia* show that the pup learns to match the sinusoidal FM call of the mother. Other forms of vocal learning in pups include adaptation or tuning of the resting frequency in echolocation pulses. Studies in horseshoe bats, *R. ferrumequinum*, showed that vocal development hinges on the perfection of the laryngonasal junction and consists of a shift from oral to nasal calls, low to higher frequencies, and noisy to pure tone type of calling. One of the major hurdles to parental care in bats that are highly social is that the pups grow up in crèches that consist of dozens to millions of young ones, many of which are born at approximately the same time. These pups mingle with each other while the parents are out foraging. To find and feed its young is therefore a difficult task for the mother. In evening bats, *Nycticeius humeralis*, it has been shown that the infants produce an isolation call whose distinctive pattern of frequency modulation can be recognized across the life span of the individual even though several other parameters may change with age. Moreover, these patterns are variable between individuals and repeatable within individuals. Thus, heritable signatures provide potential information about genetic relatedness in these species. This method of identification of young ones by mothers on the basis of vocal signatures has been demonstrated in birds, such as penguins, as well as dolphins.

Spear-nosed bats, *Phyllostomus hastatus*, produce different types of screech calls during foraging and for coordination of hunting movements, and these are highly group distinctive, clustering around a group average. When bats from different groups are housed together, they modify their screech calls to converge on a new unique screech call that is indicative of the new group identity. Thus, the structure of screech calls shifts following changes in the group composition. Since these signature calls result from socially mediated learning rather than from shared genes, they may serve to facilitate recognition of unrelated social group members. These group-distinctive calls can also vary significantly within different roosts of the same species even within a relatively confined geographic space. For example, distinct variations in the 'screech calls' of *P. hastatus* were obtained from three separate caves in Trinidad. Because the emergence of regional song dialects in birds is considered to be a result of vocal song learning, these data from *P. hastatus* suggest that differences on the group level may represent a microcosm of larger-scale geographic dialect patterns. Thus, signature calls are used not only to indicate identity of infants for promoting mother–infant interactions but also as group-distinctive calls to signify the identity of the social group.

Social Communication Behavior

From a behavioral perspective, acoustic social signals in bats are grouped into aggressive or warning calls, mating and song flight calls, distress or isolation calls, and directive or attraction calls. Individuals of the leaf-nosed bat, *Carollia perspicillata*, rely heavily on acoustic signals for social communication. Many acoustically distinct call types have been recorded from observations in a free-flying, captive colony of *Carollia*. FM signals that are produced by both females and infants have a communicative function during mother–infant interactions. The FM signals do not change much in this species as the infant matures. Whines, warbles, and trills are emitted in the context of male-to-female interactions within a harem. Whines, which are similar to the fixed rippled FM (fRFM) in mustached bats, are emitted as a male approaches a female during mating.

False vampire bats, *Megaderma lyra*, interact socially in various contexts and emit different types of communication sounds. Unlike some of the other species studied, many of the auditory communication behaviors observed in *Megaderma* occur in flight. One of these behaviors is labeled 'grumbling flight' and involves several bats hovering for a few seconds in a head-to-head formation emitting a series of short downward FM sounds that may terminate in a shallow downward FM. The spectral structure of this sound sequence is remarkably similar to the checked downward FM (cDFM) sounds observed in mustached bats. This may be preceded by an initial side-to-side flight by two or more bats in an excited state. Hovering behavior accompanied by calling is also exhibited during mating by males of *S. bilineata* and in *Carollia*. Reports on social interactions during flight are usually based on chance observations apart from honk calls, which are emitted when flying bats are about to collide.

Many of the Megachiropteran species that do not echolocate are nevertheless highly vocal and emit loud calls when disturbed as well as in other social contexts. Detailed studies of their social calls have been made in only a few species. For example, the Australian flying foxes (Pteropodidae) emit at least

22 different types of calls, rivaling the variety of calls emitted by primates. The distance calls emitted by *Pteropus* species are threat, alarm, or warning calls.

Singing and Babbling

The occurrence of singing and babbling in some species also attests to the specialized audiovocal abilities of bats. Like songbirds and some species of whales and dolphins, a few species of bats have been shown to sing and engage in courtship displays. Singing in bats was first reported by Vaughan in 1976 in *Cardioderma cor*, an African megaderma bat species. Singing is strongly seasonal in nature and probably establishes territoriality of the foraging bat. False vampire bats, *M. lyra*, engage in song flight behavior. This behavior is displayed only by the dominant males and can occur at any time of the year. It consists of a stereotyped flight pattern that is continuously accompanied by vocalization. The behavior is clearly one aspect of courtship and is directed only at nonlactating females within the group. In the only study describing this behavior, it was also elicited by the introduction of new females to the colony. The song flight in *Megaderma* consists of three stages that have been named introductory, advancing, and final flight. Spectrally distinct segments or strophes accompany each of these stages.

Singing is used to establish roosting territories in the sac-winged bat, *S. bilineata*, which employs an unusually large vocal repertoire. In this species, males emit tonal calls while interacting with females and other types of calls consisting of composites when actively defending their territories. Their songs consist of short repeated tones that do not appear to have any obvious context other than advertising the quality of singing by a male. The phenomenon of babbling has also been demonstrated in male and female pups of this species. As with infant babbling in humans, these pups produce renditions of all known adult vocalization types during bouts of vocalizations. These appear to be independent of a distinct social context and provide evidence of training to communicate effectively (**Figure 4**).

Motivational Basis of Call Structure

A study of social calls produced by 16 different European species suggests that calls commonly emitted by each of these species can be grouped into four major sonographic structures or types. Type A (long, narrowband NB) calls, also labeled as a 'squawk,' are generally agonistic or aggressive calls and may also function as a threat display. Type B calls (train of steep downward FM), such as a 'repeated trill' and a 'curved cheep,' express different levels of irritation. Type C (downward gliding FM) calls are characteristically used for interactions between females and pups. Pups continuously emit these isolation calls in the absence of their mothers, and the mothers use these calls to recognize their pups upon their return from foraging. Type D (humped, wrinkled, and sinusoidal FM) calls are involved in mate attraction behaviors and may also function during agonistic interactions between conspecifics. Both sexes of

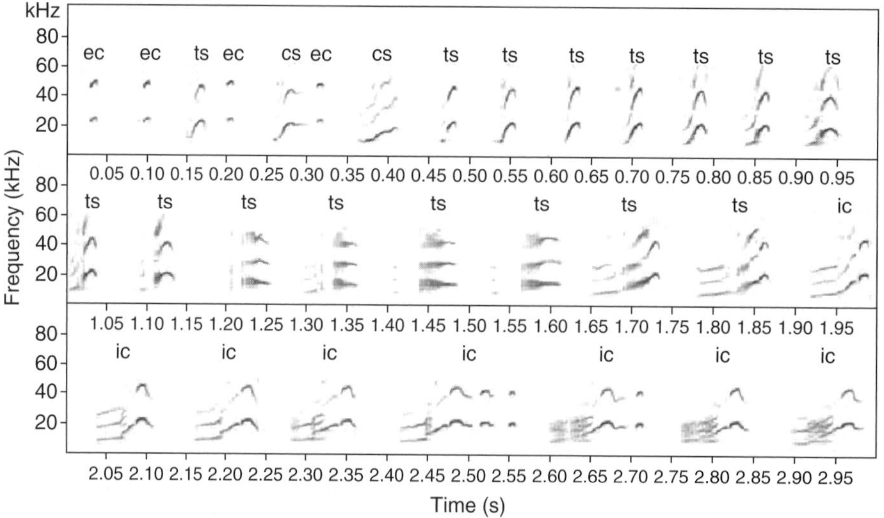

Figure 4 Sequence of spectrograms of babbling bouts from an individual of the sac-winged bat, *Saccopteryx bilineata*, depicting spectral frequency as a function of time: echolocation pulses (ec) and different elements of territorial song (ts), courtship song (cs), and isolation calls (ic) in a single babbling bout. Adapted from Knörnschild M, Behr O, and von Helversen O (2006) Babbling behavior in the sac-winged bat (*Saccopteryx bilineata*). *Naturwissenschaften* 93: 451–454, with permission.

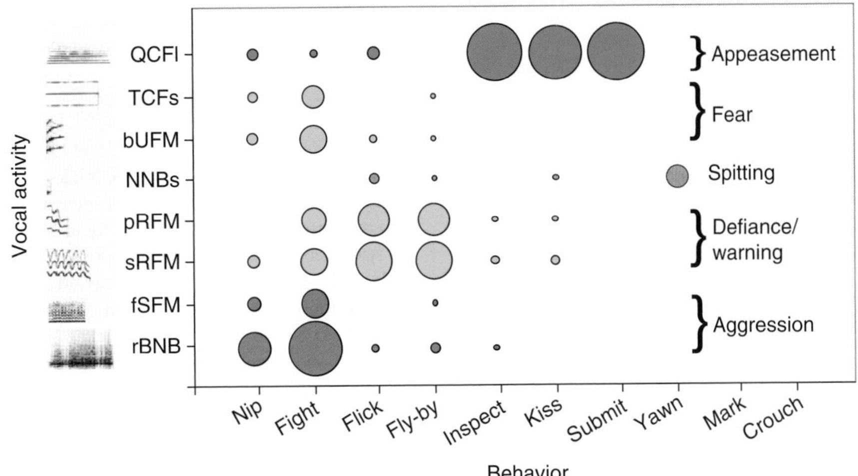

Figure 5 A bubble plot indicating the association of call types with behavioral states or emotions that influence group cohesion. Color coding is from red for aggression to green for appeasement. An act or state of appeasement impacts positively on the affiliation between two individuals and may be triggered either spontaneously or because of an impending uncertainty. A state of fear implies increased awareness to external stimuli in preparation for fight or flight and preempts a sensation of pain or stress. A state of defiance or to warn implies readiness to combat when confronted by another individual and may be triggered by either fear or insecurity; a state of aggression is one possible outcome of a state of defiance and warning. The size of the bubble is proportional to the normalized (percentage of total events) frequency of occurrence of a call type. The light gray circle indicates an 'organic' (spitting) sound that is sometimes associated with yawning but is not a vocalization. No calls were emitted during yawning, marking, and crouching. Fighting, which can be a complex and long-lasting behavior, produced the largest variety of call types. Adapted from Clement MJ, Dietz N, Gupta P, and Kanwal JS. (2006) Audiovocal communication and social behavior in mustached bats. In: Kanwal JS and Ehret G (eds.) *Behavior and Neurodynamics for Auditory Communication*, pp. 57–84. Cambridge, UK: Cambridge University Press, with permission.

M. lyra, an Old World microchiropteran species, use a low, multiharmonic 'grumble' as an aggressive call, and males use a mix of tonal CF and FM calls in a display for females. Male *S. bilineata* use harsh, broadband calls to threaten other males and direct tonal calls toward females.

Studies of the European species as well as mustached bats indicate that the acoustic structure of calls emitted by bats appears to follow the rules of the motivation–structure hypothesis for signal design as proposed by Morton in 1977 and August and Anderson in 1987 for various avian and mammalian species. Accordingly, harsh broadband calls in bats are widely used during aggression, although buzzes and trills also fill this role (**Figure 5**). Tonal calls in bats are typically used between mothers and pups, whereas more complex calls are used during mate attraction behavior.

Conclusion

Audiovocal communication in bat species is as advanced as that of any nonhuman mammal. In fact, speech in humans may owe its origin to selection pressures that shaped behavioral and neural adaptations for audiovocal communication in ancestors common to modern-day bats and humans.

See also: Communication Networks and Eavesdropping in Animals; Echolocation I: Behavior; Echolocation II: Neurophysiology; Signal Design Rules in Animal Communication.

Further Reading

August PV and Anderson JGT (1987) Mammal sounds and motivation–structural rules: A test to the hypothesis. *Journal of Mammalogy* 68: 1–9.

Balcombe JP (1990) Vocal recognition of pups by mother Mexican free-tailed bats: Do pups recognize their mothers? *Animal Behaviour* 39: 980–986.

Barclay RM, Fenton MB, and Thomas DW (1979) Social behavior of the little brown bat, *Myotis lucifugus*: II. Vocal communication. *Behavioral Ecology and Sociobiology* 6: 137–146.

Boughman JW (1998) Vocal learning by greater spear-nosed bats. *Proceedings of the Royal Society of London, Series B* 265: 227–233.

Clement MJ, Dietz N, Gupta P, and Kanwal JS (2006) Audiovocal communication and social behavior in mustached bats. In: Kanwal JS, and Ehret G (eds.) *Behavior and Neurodynamics for Auditory Communication*, pp. 57–84. Cambridge, UK: Cambridge University Press.

Davidson SM and Wilkinson GS (2002) Geographic and individual variation in vocalizations by male *Saccopteryx bilineata* (Chiroptera: Emballonuridae). *Journal of Mammalogy* 83: 526–535.

Eser KH and Schmidt U (1989) Mother–infant communication in the lesser spear-nosed bat, *Phyllostomus discolor* – Evidence for acoustic learning. *Ethology* 82: 156–168.

Fenton MB (1984) Echolocation: Implications for ecology and evolution of bats. *Quarterly Review of Biology* 59: 33–53.

Fenton MB (1985) *Communication in Chiroptera*. Bloomington: Indiana University Press.

French B and Lollar A (1998) Observations on the reproductive behavior of captive *Tadarida brasiliensis mexicana* (Chiroptera: Molossidae). *Southwest Nature* 43: 484–490.

Gould E (1971) Studies of maternal–infant communication and development of vocalizations in the bats *Myotis* and *Eptesicus*. *Communications in Behavioral Biology* 5: 263–313.

Habersetzer J (1981) Adaptive echolocation sounds in the bat *Rhinopoma-hardwickei* – a field-study. *Journal of Comparative Physiology* 144: 559–566.

Kanwal JS, Matsumura S, Ohlemiller K, and Suga N (1994) Analysis of acoustic elements and syntax in communication sounds emitted by mustached bats. *Journal of the Acoustical Society of America* 96: 1229–1254.

Knörnschild M, Behr O, and von Helversen O (2006) Babbling behavior in the sac-winged bat (*Saccopteryx bilineata*). *Naturwissenschaften* 93: 451–454.

Ma J, Kobayasi K, and Metzner M (2007) Common themes in communication call design in five species of echolocating bats. Eighth International Congress of Neuroethology, Vancouver, Canada, Abstract no. PO328, p. 208.

Matsumura S (1981) Mother–infant communication in a horseshoe bat (*Rhinolophus ferrumequinum nippon*). Vocal communication in three-week-old infants. *Journal of Mammalogy* 62: 20–28.

Möhres FP (1967) Communicative characters of sonar signals in bats. In: Busnel RG (ed.) *Animal Sonar Systems Biology and Bionies*, vol. 2, pp. 939–945. NATO Advanced Study Institute, Joy-en-Josas: INRA-CNRS.

Morton ES (1977) On the occurrence and significance of motivation–structural rules in some bird and mammal sounds. *American Naturalist* 111: 855–869.

Nelson JE (1964) Vocal communication in Australian flying foxes (Pteropodidae; Megachiroptera). *Zeitschrift fur Tierpsychologie* 121: 857–870.

Pfalzer G and Kusch J (2003) Structure and variability of bat social calls: Implications for specificity and individual recognition. *Journal of Zoology* 261: 21–33.

Porter FL (1979) Social behavior in the leaf-nosed bat, *Carollia perspicillata*: II. Social communication. *Zeitschrift fur Tierpsychologie* 50: 1–8.

Thomas DW, Fenton MB, and Barclay RMR (1979) Social behavior of the little brown bat, *Myotis lucifugus*: I. Mating behavior. *Behavioral Ecology and Sociobiology* 6: 129–136.

Wilkinson GS and Boughman JW (1998) Social calls coordinate foraging in greater spear-nosed bats. *Animal Behaviour* 55: 337–350.

Auditory/Somatosensory Interactions

S E Shore, University of Michigan, Ann Arbor, MI, USA

Introduction

When stimulated with an auditory (sound) or somatosensory (touch) stimulus, distinct sensations are perceived that are modality specific (e.g., pitch and itch). However, sensory modalities are also able to influence each other through anatomical convergence of more than one modality onto single neurons. Such 'multisensory' neurons are evident throughout the nervous system, as first-order neurons in the brain stem, second-order neurons in the midbrain and higher-order neurons in the cerebral cortex (**Figure 1**). Multisensory neurons are usually excited by stimuli from more than one modality and can often 'integrate' this bimodal information by responding with a greater (or lesser) number of action potentials to combined stimulation than to either modality alone. The corresponding behavioral consequence of these cross-modal interactions is an increase in the probability of a correct response to an event accompanied by a decrease in reaction time for bimodal stimulation. In some cases, a neuron may respond only to one modality but its response may be suppressed or enhanced by the simultaneous addition of a stimulus from another modality. This form of integration has been demonstrated in the auditory cortex, inferior colliculus (IC), and cochlear nucleus (CN).

Somatosensory Influences on Auditory Cortex

Figure 1 is a schematic of the distribution of multisensory neurons in the rat sensory neocortex. It is evident from this figure that the majority of multisensory interactions occur at the borders between primary sensory regions, a feature that is attaining increasing support across several species. Many of the multisensory neurons in these border regions are capable of integrating information from each modality. The auditory cortex of primates and cats contains a core region of three primary areas surrounded by surrounding 'belt' and 'para-belt' regions of secondary areas (**Figure 2**). Neurons in the belt regions respond to both auditory and somatosensory stimuli.

Connections between Auditory and Somatosensory Cortex

Auditory cortex receives input from somatosensory cortex The anatomical basis for these responses is provided by tract tracing studies demonstrating that the belt regions receive inputs from somatosensory cortex as well as from auditory cortex core regions. The caudomedial belt region (CM) especially is densely connected with the retroinsular (RI) area of the somatosensory cortex, the superior temporal sulcus, and the posterior parietal and entorhinal cortex. Bimodal auditory and somatosensory responses have been reported in both CM and RI in response to electrical stimulation of the median nerve or mechanical stimulation of the upper body. Magnetic resonance imaging reveals an analogous belt region in humans, the planum temporale, which also shows evidence of auditory somatosensory convergence. Interestingly, there is sparse or no input to these belt regions from the medial geniculate complex, supporting the concept that the major functions of these belt areas are integrative.

Responses of bimodal neurons in auditory cortex Bimodal (or 'multisensory') neurons in auditory cortex typically are excited by stimulation of either modality and often integrate this information by producing responses that are substantially greater than the sum of the responses to each modality alone (**Figure 3**). However, in some cases, responses of auditory cortical neurons to sound can be suppressed by stimulation of somatosensory inputs to this area, even when the somatosensory stimulation by itself does not elicit a response (**Figure 3**). These subthreshold cross-modal interactions may arise through activation of inhibitory interneurons in the auditory cortex that receive somatosensory input, which, in itself, designates a neuron as 'multisensory.'

Neurons in the belt region of marmoset monkeys show a particular tendency to respond preferentially to the vocalizations of other monkeys, and do not respond to self-vocalizations. This preferential responsiveness to external vocalizations coupled with the suppression of self-vocalizations may be initiated at earlier stages in the auditory system, since neurons in the IC and even CN show similar tendencies (see below). The anatomical substrates for preservation of these patterns across levels is indeed present in the projection patterns that originate in the dorsal cochlear nucleus (DCN), terminate in the ICX, which in turn projects to the dorsal/posterior geniculate nucleus (MG). Whereas the primary (lemniscal) auditory pathway projects mainly to the core regions of auditory cortex (A1) via the ventral MG, projections to the belt and parabelt regions arise largely from the dorsal/posterior MG, and also other thalamic regions that demonstrate bimodal responses to auditory and

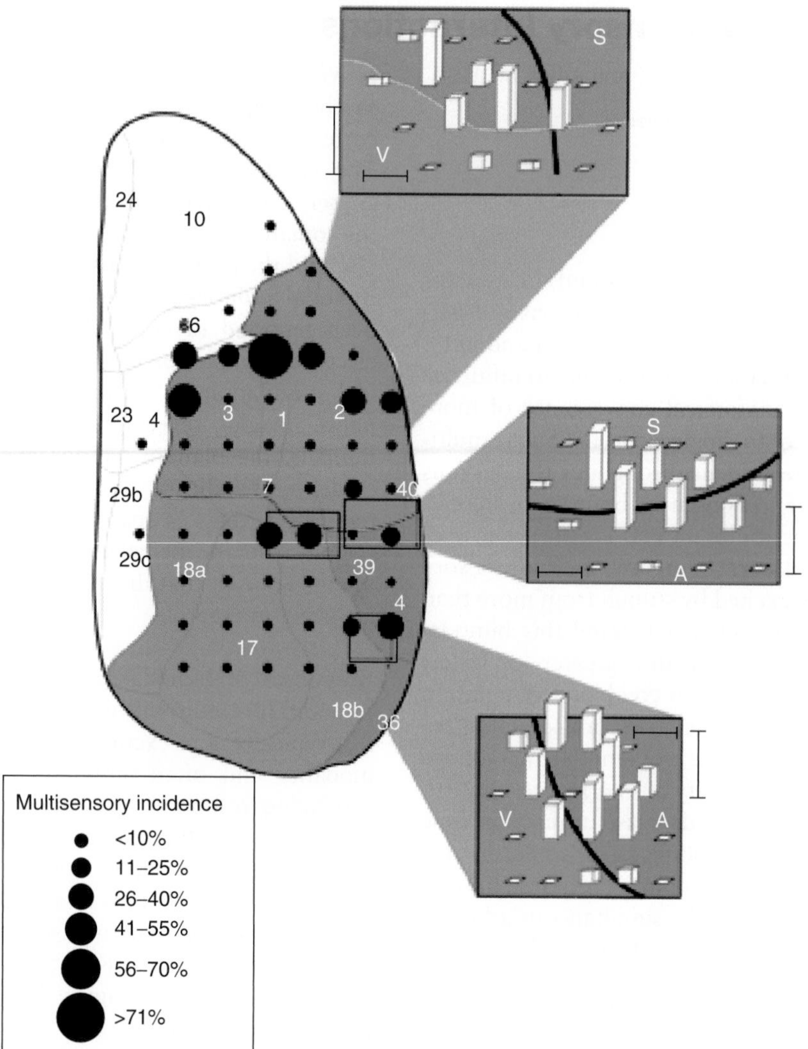

Figure 1 The distribution of multisensory neurons in rat sensory neocortex. The line drawing depicts the dorsal surface of cortex. Numbers and solid lines designate major subdivisions (17) (parietal, red shading; temporal, green shading; and occipital, blue shading). Filled circles show locations of electrode penetrations in a coarse-grain analysis that was conducted in 22 animals, and circle size indicates the relative incidence of multisensory neurons at each site. Insets show the results of higher-resolution sampling through each of the transitional regions that was conducted in a total of nine animals. Bar height indicates the relative incidence of multisensory neurons. Horizontal scale bar = 250 μm, and vertical scale bar = 50% multisensory incidence. V, visual cortex; A, auditory cortex; S, somatosensory cortex. Reproduced from Wallace MT, Ramachandran R, and Stein BE (2004) A revised view of sensory cortical parcellation. *Proceedings of the National Academy of Sciences of the United States of America* 101: 2167–2172. Copyright (2007) National Academy of Sciences, USA.

somatosensory stimuli, such as the Po, Sg, Lim, and PM. It is paradoxical that these latter areas also project to A1, in which there is no evidence for bimodal integration. This suggests that the bimodal responses observed in CM are either solely the result of inputs from the RI somatosensory area, or co-incident inputs from several of these regions. CM sends projections to the IC, terminating primarily in ventrolateral regions including the external cortex, which, itself, demonstrates bimodal properties (see below).

Activity-dependent plasticity in cortical auditory–somatosensory interactions Determination of the size of a cortical sensory field may be genetically determined depending on the environmental demands. For example, the duck billed platypus has a large portion of its cortex assigned to somatic inputs from the bill, while squirrels devote much of the cortex to visual processing. However, recent studies suggest that activity from sensory receptors can also play a role in the determination of cortical fields and their connections. In congenitally deaf mice, the auditory

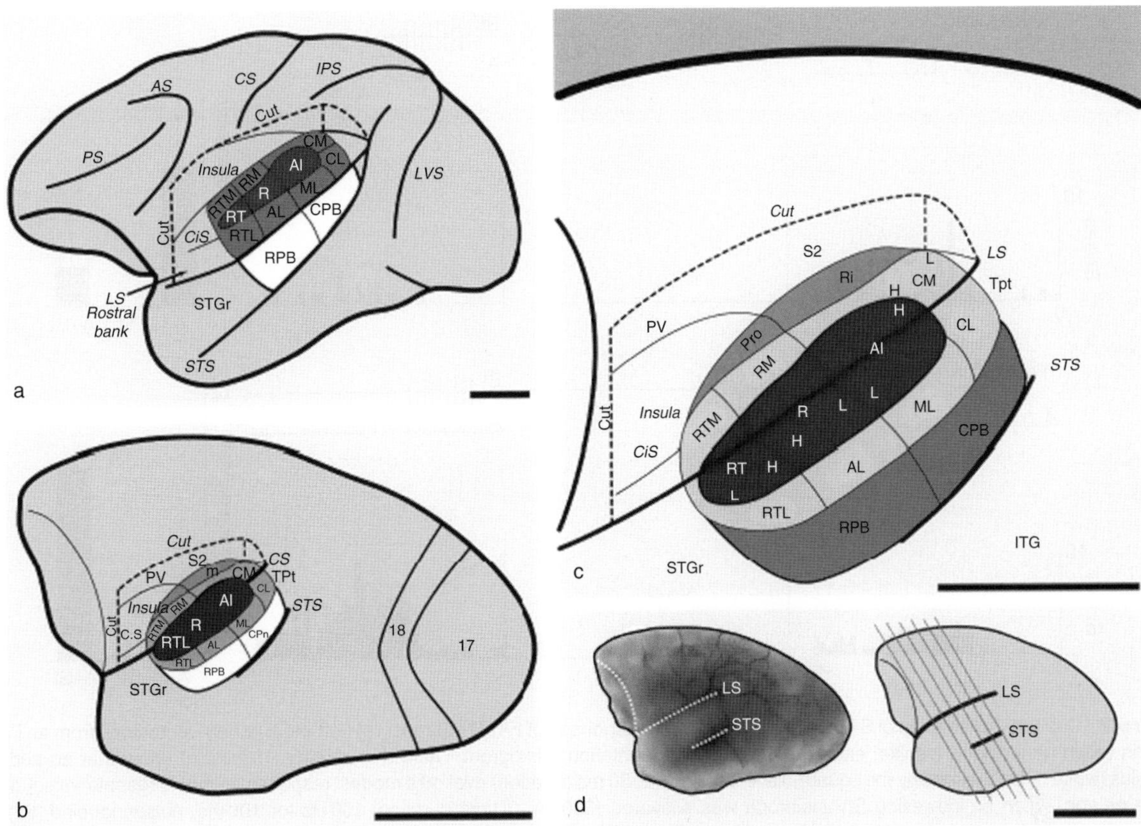

Figure 2 Schematic models of macaque (a) and marmoset (b, c) monkey auditory cortex. The lateral sulcus (LS) of the left hemisphere was graphically opened (cut) to reveal the locations of auditory cortical areas on its lower bank. The circular sulcus (CiS) was flattened to show the position of the rostromedial (RM) and rostrotemporal medial (RTM) areas that occupy its lateral wall. The upper bank of the LS was partly opened (cut) to show the locations of the retroinsular area (Ri) in the fundus, second (S2) and parietoventral (PV) somatosensory areas on the upper bank, and insula. The three areas that comprise the core region of the auditory cortex (dark shading) are located on the lower bank (A1, auditory area 1; R, rostral; RT, rostrotemporal). The core is surrounded by seven or eight areas that belong to the belt region (light shading) (CM, caudomedial; CL, caudolateral; ML, middle lateral; RM, rostromedial; AL, anterolateral; RTM, rostrotemporal medial; RTL, rostrotemporal lateral). The proisocortex area (Pro) is a putative addition to the medial belt. The core and lateral belt regions are mostly contained within the lateral sulcus in macaques but extend onto the superior temporal gyrus (STG) in the marmoset. On the surface of the STG are two areas that make up the parabelt region (medium shading; RPB and CPB, rostral and caudal parabelt). The rostral part of the STG (STGr) extends to the temporal pole. The temporal parietotemporal area (Tpt) occupies the caudal end of the STG and extends onto the supratemporal plane within the LS. Tonotopic gradients within areas are indicated by H (high frequency) and L (low frequency). Other sulci shown include the arcuate sulcus (AS), central sulcus (CS), intraparietal sulcus (IPS) inferior temporal gyrus (ITG), principal sulcus and superior temporal sulcus (STS). (d) Photographic image of the marmoset left hemisphere and schematic showing the plane of section (diagonal lines) used in the present study for histological processing. Scale bar = 10 mm (a, b, d); 5 mm (c). From de la Mothe LA, Blumell S, Kajikawa Y, and Hackett TA (2006) Cortical connections of the auditory cortex in marmoset monkeys: Core and medial belt regions. *Journal of Comparative Neurology* 496: 27–71.

cortical area that normally contains unimodal auditory neurons is replaced largely by neurons responsive to somatosensory stimuli and to a lesser extent by neurons responsive to visual or combined visual–somatosensory stimulation. Furthermore, most of the receptive fields of these neurons represent the face as opposed to the limbs, trunk, or tail. This extensive multisensory plasticity, which affects the deprived cortical area, as well as the other sensory cortices, is evident also in humans through the use of functional magnetic resonance imaging (FMRI) studies. In congenitally deaf mice, cortical areas that would normally be responsive to auditory stimulation are responsive to both somatosensory and visual stimulation. Mechanisms underlying these cortical alterations may be partly determined by alterations in subcortical connections observed in deaf animals.

Somatosensory Influences on IC

The IC is a major synaptic station of the auditory midbrain. Almost all ascending and descending

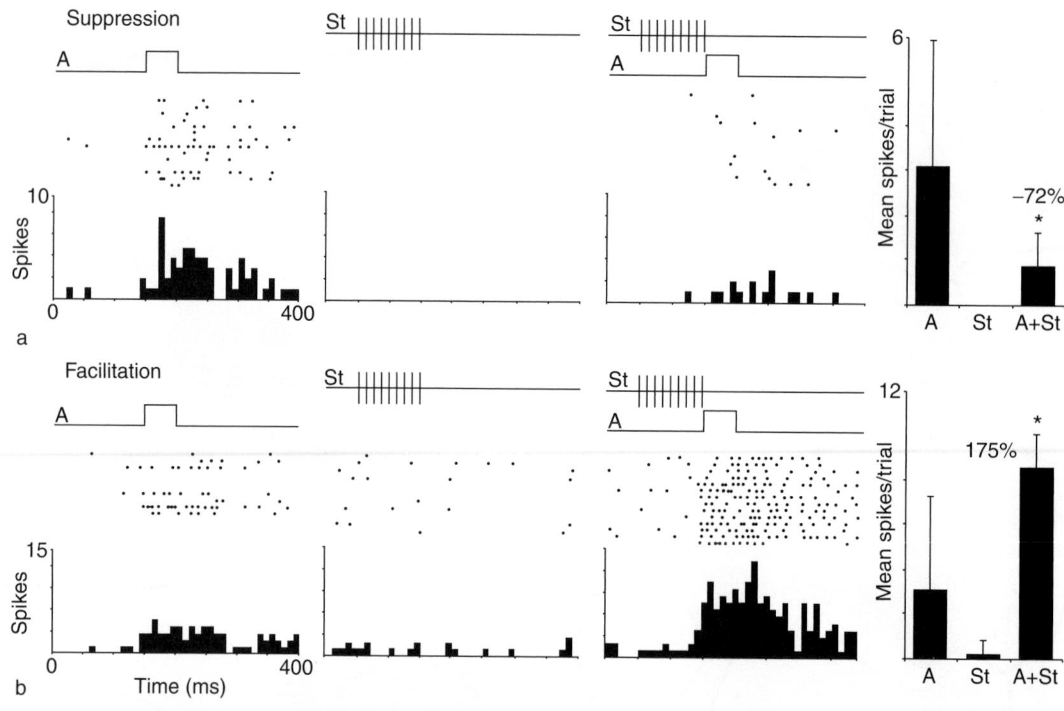

Figure 3 Subthreshold effects of SIV activation on auditory responses of FAES neurons. (a) (left-most panel), recordings from an FAES neuron (each raster dot = 1 spike; each raster row = 1 presentation; histogram = activity summary, 10 ms bin) show that an auditory stimulus (white noise denoted by the square-wave, 55 dB SPL, 50 ms duration) evoked a modest response on most presentations. For the same neuron, when an indwelling SIV electrode was activated (600 μA, 0.1 ms duration, 100 Hz for 100 ms, pulses labeled 'St'), no neuronal activity was recorded. However, when the two stimulation elements were combined ('St + A'), the response to the auditory stimulus was substantially diminished. The bar histogram to the far right summarizes these results, showing that combined SIV and auditory stimulation significantly (*$P < 0.05$, paired *t*-test) reduced the response to the auditory stimulus alone by 72%. These same conventions are used in (b), where the auditory stimulus alone (far left) elicited a modest response, but the SIV stimulus did not affect the neuron. However, when the SIV and auditory stimuli were combined, there was a vigorous and reliable response. The 'bar histogram' summarizes these results and indicates that combined SIV-auditory stimulation produced a large (175%) and significant facilitation of neuronal activity. From Meredith MA, Keniston LR, Dehner LR, and Clemo HR (2006) Crossmodal projections from somatosensory area SIV to the auditory field of the anterior ectosylvian sulcus (FAES) in cat: Further evidence for subthreshold forms of multisensory processing. *Experimental Brain Research* 172(4): 472–484.

auditory pathways synapse here. The IC is comprised of three subdivisions: the central nucleus (CIC), surrounded by a dorsal cortex (DCIC) and a laterally located external cortex (ICX). The CIC forms part of the 'primary' or 'lemniscal' auditory pathway that originates in the auditory nerve and terminates in primary auditory cortex. It receives input from the majority of auditory ascending projections and displays sharply tuned, low-threshold neuronal properties. The DCIC and ICX form part of the 'nonclassical' auditory pathway that receives not only auditory inputs, but also nonauditory projections, including somatosensory, and descending inputs from the auditory cortex and CIC. In contrast to the sharply tuned neuronal responses in CIC, neurons in the ICX are broadly tuned and respond preferentially to spectrally complex stimuli such as vocalizations. ICX neurons are further distinguished by their responsiveness to multimodal stimulation.

Connections between IC and Somatosensory Nuclei

The anatomical bases for responses to multimodal stimulation are projections that originate in the spinal trigeminal nuclei (Sp5), the cuneate nuclei, and the medullary reticular formation, and terminate primarily in the ICX.

Inputs from the trigeminal nuclei The dorsomedial and marginal areas of Sp5I (interpolaris), and the deep and marginal layers of Sp5C (caudalis), project directly to the ventral ICX (ICXV). In addition to their involvement in the modulation of pain, these subdivisions of the Sp5 receive low-threshold nonnociceptive afferents from head and face, such as those sensitive to gentle pressure and vibrissa deflection. These regions also receive inputs from vocal tract/intra oral structures including the temporomandibular joint and tongue muscles. Furthermore,

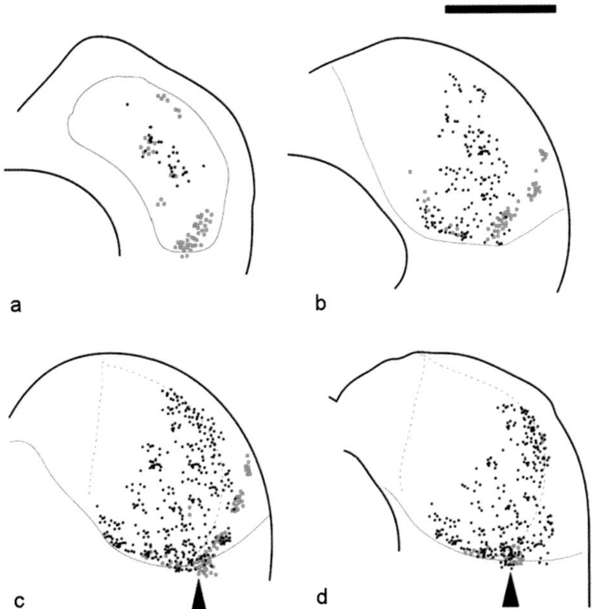

Figure 4 Sp5 terminations in ICX overlap with those from the CN. (a–d) Rostral to caudal sequence. Red squares represent the Sp5 terminations and black dots represent the CN terminations. Most of the convergence was observed in the ICXV (triangles in c, d) and some convergence in the rostral ICX. Each dot/square represents 1–3 terminal endings. Scale bar = 2 mm (b) (applies to (a–d)). From Zhou J and Shore S (2006) Convergence of spinal trigeminal and cochlear nucleus projections in the inferior colliculus of the guinea pig. *Journal of Comparative Neurology* 495: 100–112.

the terminals of these projections overlap with direct inputs to the ICX from the DCN (**Figure 4**). As a region receiving auditory inputs from ascending auditory and somatosensory pathways, it is ideally placed as the primary IC region for processing multisensory information.

Bimodal Responses of Neurons in ICX

Neurons in the ICX of guinea pigs respond to both acoustic and tactile stimuli. In some ICX neurons, electrical stimulation of the Sp5 can modulate (usually suppress) sound-evoked responses, even when Sp5 stimulation alone does not evoke responses. This subthreshold suppression of sound-evoked activity by somatosensory stimulation is similar to that described above in the ectosylvian cortex and the DCN (see below), the latter projecting to the ICX. Furthermore, recordings from the ICX in awake monkeys demonstrate that these neurons distinguish between the vocalizations of other monkeys and self-vocalizations: external vocalizations produce vigorous neuronal responses, while self-produced vocalizations suppress neuronal responses to spontaneous rate or below (**Figure 5**). These studies suggest that bimodal neurons in the ICX may play a role in the suppression of self-generated sounds. Anatomical support for this hypothesis can be obtained from findings that the regions of Sp5I and Sp5C that project to ICX receive somatosensory inputs from vocal structures. Thus, sounds associated with chewing, vocalization, and respiration may be suppressed by the concurrent activity from the somatosensory system, which likely originates in the proprioceptive afferents or broad dynamic range mechanoceptors in deep tissues during these activities.

Somatosensory Influences on the CN

The CN is the obligatory nucleus in which all ascending information from the cochlea forms synaptic connections with the auditory brain. Once considered a simple 'relay nucleus,' the CN performs complex computations to transform the incoming signals from the auditory nerve. The well-established somatosensory influences on higher auditory structures have only recently been extended to include the CN where it is now evident that somatosensory and auditory inputs converge. Neurons both in the ventral and dorsal CN can be modified by activating somatosensory neurons.

Connections between CN and Somatosensory Nuclei

Inputs to the CN from somatosensory structures originate in the trigeminal nerve, trigeminal nucleus, dorsal column, dorsal column nuclei, and indirectly from the medullary reticular formation that receives input from the trigeminal nerve (**Figure 6**).

Inputs from the trigeminal system The trigeminal nerve conveys sensory information from the head and face to the central trigeminal sensory nuclear complex via the ophthalmic, maxillary, and mandibular branches of the trigeminal nerve. The ophthalmic nerve innervates the forehead, upper eyelid, and extra ocular muscles; the maxillary nerve supplies the upper lip, the lower eyelid, and the upper jaw and roof of the mouth; and the mandibular nerve innervates the lower lip, the mucous membranes of the lower jaw, the floor of the mouth, and the anterior two-thirds of the tongue. Trigeminal neurons that project directly to the CN are located in the medial portion of the trigeminal ganglion (TG) at the origin of the ophthalmic nerve, as well as in the mandibular division of the ganglion. The axons of TG projection neurons are thin (~1 μm) and typically form en passant boutons that are concentrated in the ipsilateral marginal cell areas that contain small cells and granule cells. Some terminals are located around cell bodies in the magnocellular regions of ventral

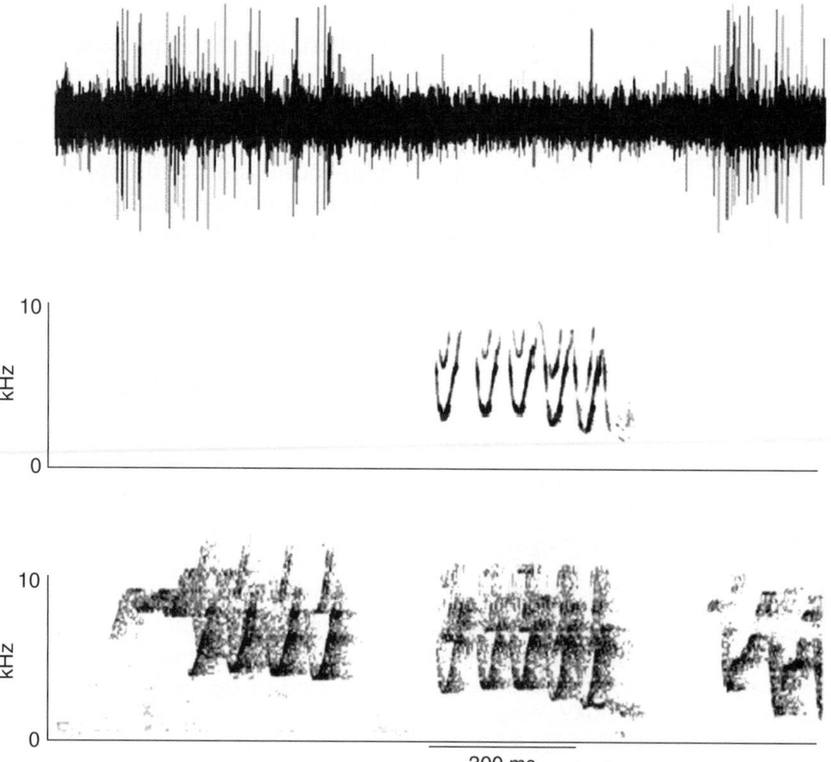

Figure 5 Neuronal activity in the external nucleus of the IC during self-produced vocalization and perception of vocalizations produced by other animals. Upper trace: extracellular recording. Middle trace: vocalization of the experimental animal in the form of a sonagram. The signal was picked up by the piezo-electric skull vibration sensor and transmitted telemetrically. Lower trace: sonagrams of the vocalizations recorded with the room microphone. Vocalization 2 stems from the experimental animal, vocalizations 1 and 3 from group mates. Reproduced from Tammer R, Ehrenreich L, and Jurgens U (2004) Telemetrically recorded neuronal activity in the inferior colliculus and bordering tegmentum during vocal communication in squirrel monkeys (*Saimiri sciureus*). *Behavioural Brain Research* 151: 331–336, with permission from Elsevier.

cochlear nucleus (VCN). Electron microscopic evaluation of these endings reveals round synaptic vesicles and asymmetric synaptic specializations indicative of excitatory synapses.

The spinal trigeminal nucleus (Sp5) is comprised of three nuclei: pars oralis (Sp5O), pars interpolaris (Sp5I), and pars caudalis (Sp5C). Each subdivision receives nociceptive and nonnociceptive afferents from the head and face that are sensitive to pressure and vibrissa deflection. The Sp5 also receives proprioceptive inputs from vocal tract/intra oral structures including the temporo-mandibular joint and tongue muscles. Projections from Sp5 to the CN originate primarily in Sp5I and Sp5C (see **Figure 6**). The paucity of projection cells in the subnucleus gelatinosus, which receives nociceptive afferents, suggests that the neurons in Sp5 that project to the auditory system convey primarily vocal structure mechanosensory information, and not pain information. Fibers and terminal endings of the Sp5 projection are small or medium in size with *en passant* or large, irregular terminal swellings. Postsynaptic targets include the dendrites of granule cells in the marginal regions of VCN and principal cells in the molecular and deep layers of DCN and magnocellular regions of VCN. As observed for TG projections to the CN, electron microscopic evaluation reveals round synaptic vesicles and asymmetric synaptic specializations indicative of excitatory synapses.

Projections from other somatosensory structures to the CN Somatosensory innervation to the CN also originates in the dorsal root ganglion of the spinal cord and dorsal column nuclei (cuneate and gracile) of the brain stem, which receive proprioceptive afferents from head, trunk, and limbs. These projection cells project also to the granular cell domain of the CN, terminating on the dendrites of granule and other small cells. Electron microscope examination indicates that these terminals are excitatory, containing round

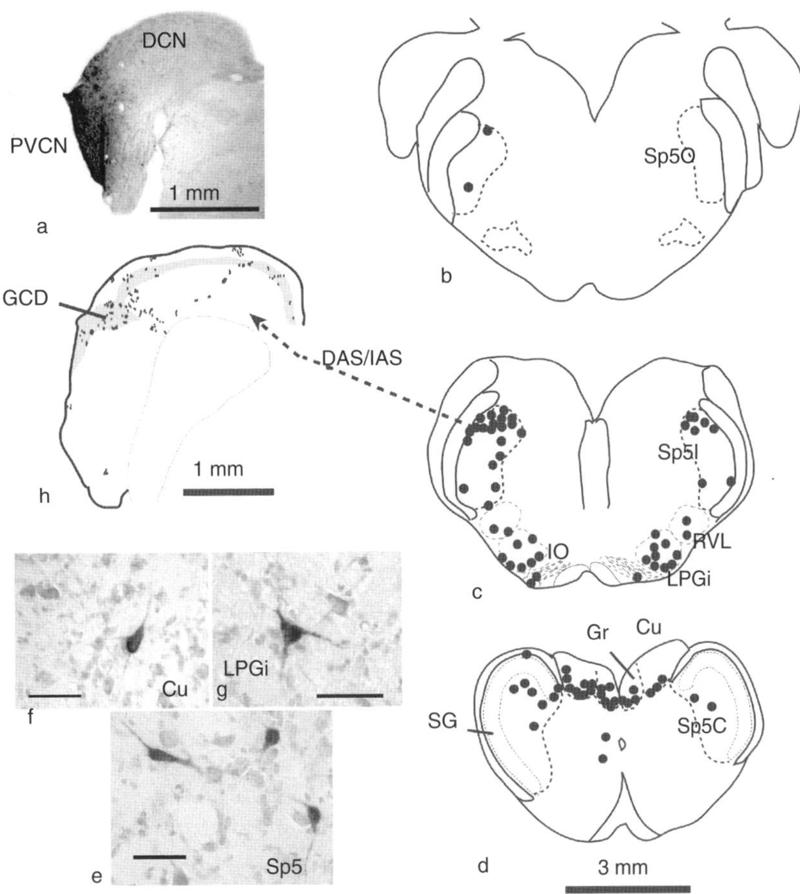

Figure 6 Projections from brain stem somatosensory regions to the CN in the guinea pig. (a)–(g) Retrograde labeling in the brain stem after an injection of biotinylated dextran amine (BDA) into the CN. (a) Photomicrograph of the injection site. The injection site is restricted to the granule cell domain of the PVCN. (b)–(d) Drawings of 1 mm transverse sections across the medulla. Each dot represents one labeled cell. The labeled neurons are located primarily in the Sp5I and Sp5C. Few labeled cells are located in the SG (d). Labeled neurons are also seen in the medullary reticular formation (RVL and LPGi, (c)), inferior olive (IO, (c)), and dorsal column nuclei (Gr and Cu, (d)). Projection neurons in Sp5 have polygonal or elongated somata (e). Projection neurons in dorsal column nuclei and reticular formation are multipolar (f and g). (h) Terminal labeling in the CN after placement of an anterograde tracer into Sp5I. Most Sp5 fibers enter the CN via DAS/IAS and terminate primarily in the granule cell domain (gray area), but also in deep DCN. Each dot represents one to three labeled terminal endings. Scale bars = 25 μm (e–g). CN, cochlear nucleus; Cu, cuneate nucleus; DAS, dorsal acoustic striae; DCN, dorsal cochlear nucleus; GCD, granule cell domain; Gr, gracile nucleus; IAS, intermediate acoustic striae; IO, inferior olive; LPGi, lateral paragigantocellular reticular nucleus; PVCN, posteroventral cochlear nucleus; RVL, rostral ventrolateral reticular formation; SG, subnucleus gelatinosus; Sp5, spinal trigeminal nucleus; Sp5C, pars caudalis of Sp5; Sp5I, pars interpolaris of Sp5; Sp5O, pars oralis of Sp5. Reprinted from Shore SE and Zhou J (2006) Somatosensory influence on the cochlear nucleus and beyond. *Hearing Research* 216–217: 90–99, with permission from Elsevier.

synaptic vesicles and forming asymmetric synapses. Some of the proprioceptive information conveyed by these systems, such as pinna and head position, might provide important spectral cues for sound localization. In conjunction, neurons that innervate the muscle spindles of the extra ocular muscles are located in the ophthalmic division of TG, which also projects to the CN. It is thus likely that trigeminal inputs to the CN provide proprioceptive information related to eye-to-head orientation while dorsal column inputs provide head-to-space and eye-to-head orientation, for the purpose of locating sound sources important for prey tracking or recognizing danger.

Responses of CN Neurons to Stimulation of Somatosensory Pathways

Some of the terminations from the TG that have characteristics of excitatory synapses end on somata in the magnocellular regions of VCN. Thus, it is not unexpected that TG stimulation primarily excites cells in the VCN. However, granule cells in the marginal layers of the CN are the major recipients of somatosensory terminations. The glutamatergic parallel fiber axons of granule cells throughout the CN terminate on the apical dendrites of the principal output neurons, fusiform cells as well as inhibitory interneurons, the cartwheel and superficial stellate cells. Thus, activation of granule cells by somatosensory stimulation could excite principal cells directly through their apical dendrites or inhibit them via cartwheel or stellate cells. In addition, direct activation of cells in the deep DCN may occur. Likely candidate cells for this activation are vertical cells or giant cells.

Responses of CN neurons to somatosensory stimulation in the absence of sound Stimulation of the dorsal column nuclei and manual manipulation of the pinna produce a complex pattern of inhibition and excitation of DCN neurons. In the cat, stimulation of pinna regions evokes stronger responses in DCN than stimulation of other areas of the face, leading to the suggestion that dorsal column nuclear input to the DCN may be involved in sound localization, which in the cat is aided by the mobility of the pinna. DCN neurons can be affected in a similar manner by stimulation of somatosensory nuclei innervating the vibrissae and peripheral nerves innervating the neck, perhaps providing feedback regarding head movements or position as an aid to sound localization.

Stimulation of the TG, Sp5 or reticular formation (RF) produces similar complex patterns of inhibition and excitation in DCN neurons. Units responding to trigeminal stimulation show primarily pauser-buildup, buildup or chopper responses to best frequency (BF) toneburst stimulation, responses that are associated with fusiform cells and giant cells in the deeper layers of DCN. The latencies of inhibitory responses are consistently longer (by >2 ms) than excitatory responses, consistent with the inhibition being mediated through an additional interneuron such as the cartwheel or superficial stellate cell.

Multisensory integration in the CN: responses to combined somatosensory and acoustic stimulation When exposed to bimodal stimulation of the auditory and the trigeminal nerves, DCN neurons are capable of multisensory integration as described above for cortical and IC neurons. More than half of DCN output neurons demonstrate multisensory integration, of which about two-thirds are suppressive and about one-third enhancing. **Figure 7** shows PST histograms from one DCN unit to broadband noise (BBN) stimulation (1) and combined trigeminal-BBN stimulation (2). A strong suppression of the firing rate to the noise burst occurs when it is preceded by trigeminal stimulation (2).

Significance of bimodal integration in the CN and beyond The suppression of responses to sound by activating the trigeminal system provides evidence that the DCN may be involved, not only in sound localization, but also may act as an adaptive filter to reduce neuronal responses to body-generated sounds. The stimulus paradigm illustrated in **Figure 7** simulates a 'natural' condition in which the CN would receive overlapping auditory and somatosensory information, as would occur in chewing, respiration, and self-vocalization. The findings that trigeminal stimulation can strongly suppress sound-evoked

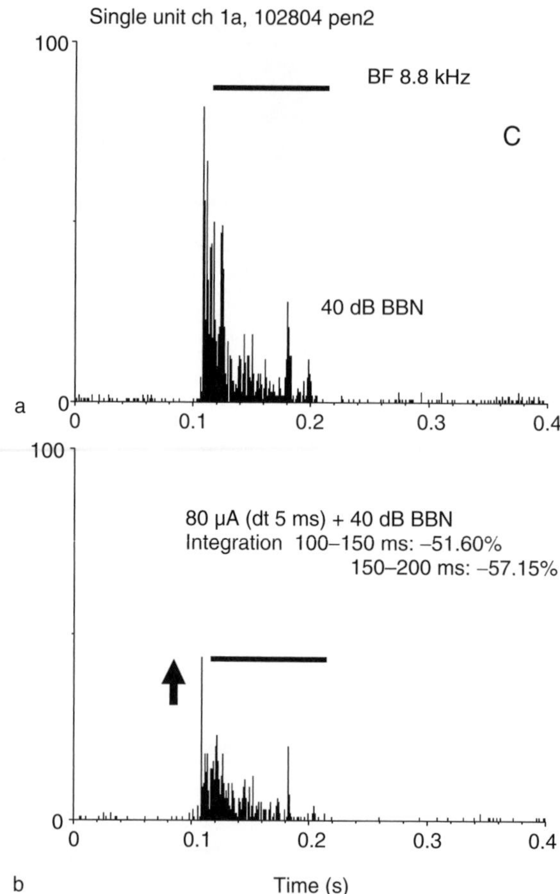

Figure 7 TG stimulation can suppress responses to sound in single units. (a) Unimodal stimulation. Poststimulus time histogram of responses of an isolated single unit to a broadband noise (BBN) stimulus (40 dB SPL, 100 ms). (b) Bimodal stimulation. Poststimulus time histogram of responses of the same single unit to the BBN noise stimulus preceded by electrical stimulation of the TG (onset 5 ms preceding BBN, 80 μA, 100 μs/phase). Arrow indicates onset of electrical stimulation at 95 ms; solid bar indicates 100 ms duration of BBN, 200 presentations. Bin width 0.5 ms. Multisensory integration to the bimodal stimulus is calculated for times 100–150 ms or 150–200 ms. Suppression of more than 50% occurs for both measures in this unit indicating maximal trigeminal suppression of activity to the BBN at the beginning of the response and continuing throughout its duration. Bin width 1 ms, 100 repetitions. From Shore S (2005) Multisensory integration in the dorsal cochlear nucleus: Unit responses to acoustic and trigeminal ganglion stimulation. *European Journal of Neuroscience* 21: 3334–3348.

activity, and even modify the temporal firing pattern evoked by a noise stimulus after the addition of a trigeminal stimulus, are consistent with the behavior of other cerebellar-like systems with granule cell – parallel fiber circuits that subtract predictable stimulation produced by the animal's own movements. For example, in the electrosensory lateral line of weakly electric fish, electric fields generated by the animal's own respiration are 'subtracted out' in principal cells by motor and proprioceptive information carried via the parallel pathways analogous to those seen in the mammalian DCN. Thus, an additional function of the parallel fiber inputs to the DCN, arising in the trigeminal system, may be to suppress internally generated sounds produced by chewing, respiration, and also self-vocalization. These combined acoustic-somatosensory activations would occur on an ongoing basis and therefore represent expected signals, as contrasted to environmentally generated or unexpected signals. The bimodal integration, in which trigeminal input enhances the responses of both VCN and DCN units to noise stimuli may be important in improving signal-to-noise ratios when attention is directed to a particular location. Thus, the CN could act as an 'adaptive filter' to suppress self-generated sounds, and also enhance perception of behaviorally relevant sounds, such as the vocalizations of other animals, generated externally, and therefore not combined with internal somatosensory stimulation.

These findings add to the emerging view that the 'merging of the senses' begins at early stages of sensory processing and not, as previously thought, at later stages of cortical hierarchical schemes. Early studies of multisensory integration in the superior colliculus add to the view that multisensory integration serves the purpose of enhancing the detection of behaviorally relevant stimuli and speeding up reaction time, orientation, and accuracy of responses to these stimuli.

See also: Auditory Cortex Structure and Circuitry; Auditory System: Central Pathways; Canal–Otolith Interactions; Cochlear Development; Somatosensory Perception; Somatosensory Cortex; Visual–Vestibular Interactions.

Further Reading

de la Mothe LA, Blumell S, Kajikawa Y, and Hackett TA (2006) Cortical connections of the auditory cortex in marmoset monkeys: Core and medial belt regions. *Journal of Comparative Neurology* 496: 27–71.

Jain R and Shore S (2006) External inferior colliculus integrates trigeminal and acoustic information: Unit responses to trigeminal nucleus and acoustic stimulation in the guinea pig. *Neuroscience Letters* 395: 71–75.

Meredith MA, Keniston LR, Dehner LR, and Clemo HR (2006) Crossmodal projections from somatosensory area SIV to the auditory field of the anterior ectosylvian sulcus (FAES) in cat: Further evidence for subthreshold forms of multisensory processing. *Experimental Brain Research.* 172(4): 472–484.

Shore S (2005) Multisensory integration in the dorsal cochlear nucleus: Unit responses to acoustic and trigeminal ganglion stimulation. *European Journal of Neuroscience* 21: 3334–3348.

Shore SE and Zhou J (2006) Somatosensory influence on the cochlear nucleus and beyond. *Hearing Research* 216–217: 90–99.

Stein B and Meredith M (1993) *The Merging of the Senses.* Cambridge, MA: MIT Press.

Tammer R, Ehrenreich L, and Jurgens U (2004) Telemetrically recorded neuronal activity in the inferior colliculus and bordering tegmentum during vocal communication in squirrel monkeys (*Saimiri sciureus*). *Behavioural Brain Research* 151: 331–336.

Wallace MT, Ramachandran R, and Stein BE (2004) A revised view of sensory cortical parcellation. *Proceedings of the National Academy of Sciences of the United States of America* 101: 2167–2172.

Zhou J and Shore S (2004) Projections from the trigeminal nuclear complex to the cochlear nuclei: A retrograde and anterograde tracing study in the guinea pig. *Journal of Neuroscience Research* 78: 901–907.

Zhou J and Shore S (2006) Convergence of spinal trigeminal and cochlear nucleus projections in the inferior colliculus of the guinea pig. *Journal of Comparative Neurology* 495: 100–112.

Auditory Cortex Structure and Circuitry

D Barbour, Washington University in St. Louis, St. Louis, MO, USA

Introduction

The auditory system comprises neuronal circuits responsible for evaluating distant objects through mechanical (acoustic) vibrations transmitted to the observer via air or water. Acoustic vibrations are converted into neuronal signals by the ear, and these signals undergo multiple processing stages so that the organism can infer relevant information about the vibrations' source. Auditory cortex (ACx) of mammals represents the highest level of the conventionally described auditory system and one of the final stages of auditory processing before acoustic information has been sufficiently decoded to achieve cognitive awareness or to be acted upon in complex fashion. Many subcortical circuits extract various features from acoustic signals prior to ACx processing. These circuits have received the bulk of auditory research attention and consequently are better understood than are cortical circuits.

ACx represents the region of mammalian neocortex dedicated predominantly to the processing of auditory information and is located on the lateral surface of the brain, typically lateral or anterolateral to primary somatosensory cortex and anterior to primary visual cortex. Generally, ACx occupies a prominent position on the lateral surface of the telencephalon, and in primates (including humans) lies on the temporal lobe just inferior to and within the lateral sulcus (see **Figure 1**). In heavily convoluted primate brains, such as that of humans, much of ACx lies within the operculum of the lateral sulcus and is thus not visible from the external surface of the brain. Neuroanatomical studies indicate that ACx comprises several distinct areas that vary by species and that have individual functions that are still rather poorly understood. Organization of ACx demonstrates considerably more variation among species than does that of subcortical auditory nuclei.

Neuronal projections to the auditory cortex arise largely from the thalamus and other cortical areas, although inputs from other structures have been demonstrated. The various divisions of the medial geniculate body (MGB) of the thalamus represent ACx input arising directly from the ascending auditory pathway. These forward projections and (generally) reciprocal connections among distinct cortical areas define the region collectively referred to as ACx. Many of these distinct areas have unclear functions and await further physiological investigation to determine their contributions to auditory processing. Projections from ACx target other cortical areas, such as the frontal lobe, as well as each of the subcortical auditory nuclei.

Organization of Auditory Cortex

Thalamic Input

Distinct cortical areas are generally identified by architectonic differences and confirmed by neuroanatomical tracing studies and neurophysiological properties. Nonspecialized mammals have 10–15 distinct areas identifiable as ACx by their connections. These areas are typically subdivided into groups based upon their proximity and the types of subcortical projections they receive. One major source of thalamic input to ACx arises from the ventral division of the medial geniculate body (MGv), often referred to as the lemniscal pathway. This pathway appears to transmit strictly auditory information at short latencies using projection neurons having small receptive field sizes and arranged topographically in the thalamus. These neurons receive the bulk of their ascending input from topographically arranged neurons in the central nucleus of the inferior colliculus (ICc). Of the distinct thalamocortical pathways, the ventral pathway projects to fewer cortical areas and connects to these areas with relatively patchy distributions predominantly in cortical laminae 3b and 4.

Another major source of thalamic input to ACx arises from the dorsal division of the medial geniculate body (MGd), often referred to as one component of the paralemniscal or extralemniscal pathway. In some species, MGd has been shown to comprise multiple nuclei, each with potentially distinct cortical targets. The dorsal pathway contrasts with the ventral pathway in most if not all anatomical and physiological features. Compared with the ventral pathway, the dorsal pathway receives different, nontonotopic inputs from the inferior colliculus, projects largely to distinct areas of cortex, and sends less dense axonal collaterals more widely throughout most cortical laminae, although the largest projections are also to layers 3b and 4. The neurons composing this pathway exhibit less topographic organization, broader tuning, longer and more variable response latencies, and tendencies to habituate to ongoing stimuli. The projection patterns of the individual nuclei of the dorsal division appear to be more diverse than those of the ventral division.

The medial division of the medial geniculate body (MGm) comprises diverse cell types that represent

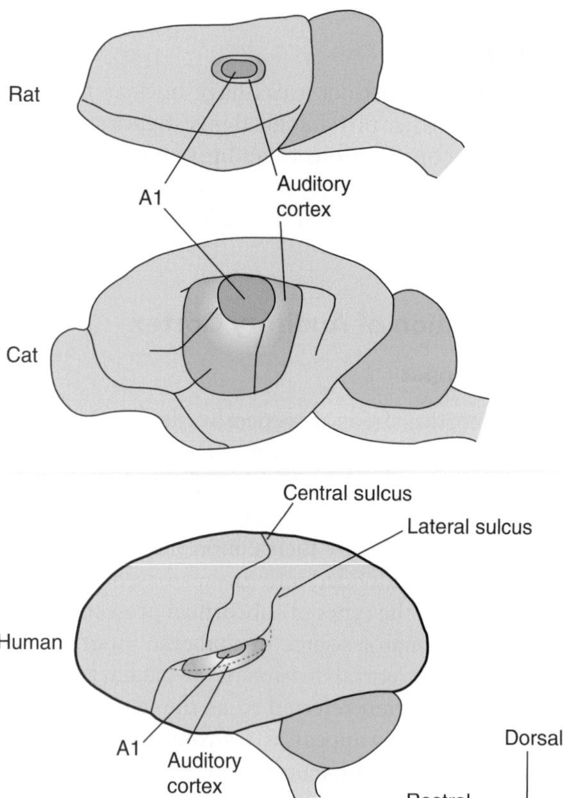

Figure 1 Approximate locations of auditory cortex on the surface of rat, cat, and human cortex. Some areas in the human brain are visible only by cutting away overlying brain tissue. A1 is the primary auditory cortex. Drawings not to scale.

another paralemniscal pathway. The MGm receives largely nontonotopic ascending auditory input and projects rather diffusely to multiple auditory cortical areas as well as nonauditory cortical areas and the amygdala. Its neurons typically respond at relatively short latencies (although with potential cell-to-cell variation) and some project with large axons, possibly indicating a fast connection from thalamus to cortex. Medial pathway projections to ACx are densest in cortical lamina 1a and possibly 6 and otherwise tend to be very sparse and fairly evenly distributed over the remaining laminae.

Other subcortical structures project to ACx in at least some species, although relatively little is known about the nature and effects of most of these projections. These projections likely underlie associative, learning, and sensory integration functions.

Area Parcellation

Distinct areas of auditory cortex are typically discerned by cytoarchitectural differences with adjacent areas and the pattern of thalamic input. Additional criteria for distinctiveness typically include variation in functional responses, although most of the anatomically determined areas of auditory cortex have yet to be explored physiologically in any detail. Primary auditory cortex (A1) receives its largest subcortical input from MGv, as typically do several other areas, the designations of which vary by species. Determination of A1 follows historically from its local architectonic distinctiveness, MGv input, and similarity in appearance to other koniocellular primary sensory areas. In addition, certain developmental stages in some species yield differential staining properties of A1, with stains differentiating a variety of molecules, such as cytochrome oxidase, acetylcholinesterase, parvalbumin, and neurofilament protein. Homology of areas labeled A1 among multiple species may be open to reinterpretation because of the plurality of cortical areas receiving substantial organized MGv input, particularly in primates.

Nomenclature of both area names and groupings varies by species. In primates, three groups of areas have been delineated based largely upon architecture, thalamocortical connections, interconnections with one another, and physiology. The core group stretches from anteriolateral to posteromedial along the lateral sulcus, contains tonotopically organized areas (including A1), and receives most of its thalamocortical input from MGv (see **Figure 2**). Two additional area groups situated just medial and lateral to the core are collectively referred to in modern parcellation schemes as ‘the belt.’ Belt areas receive little MGv input and substantially greater MGd and MGm input than do the core areas. These areas appear to be tonotopically organized but are difficult to study with classical, simple sound stimuli. Lateral to the lateral belt areas lie the parabelt areas, which receive no MGv projections and scattered MGd and MGm input. The internal organization of parabelt areas remains largely unknown.

Connections of Auditory Cortex

Local Structure of Primary Auditory Cortex

A1 represents both the most studied auditory cortical area and is the focus of the most research. For many years even such a basic organizational structure as tonotopy – the topographic representation of the cochlea – of A1 was debated because of the great variety of functional responses encountered during physiological experiments. Many modern lines of evidence, however, demonstrate tonotopy in A1, but other potential topographic structures remain unclear. In tonotopic areas, each sound frequency is represented in a narrow strip along the surface of the cortex

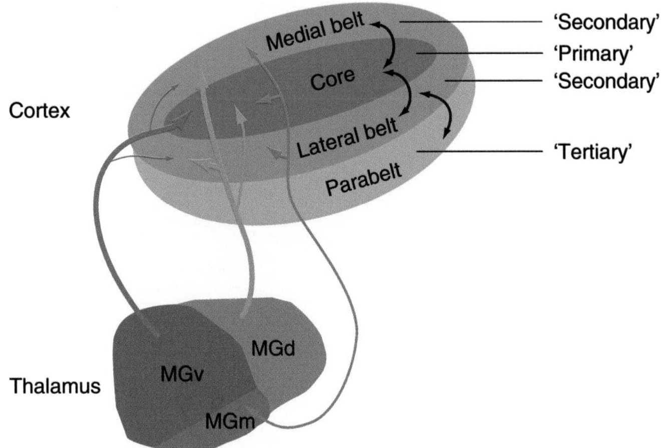

Figure 2 Thalamic input schematic to primate auditory cortex, including corticocortical connections. Arrow size approximates projection size; MGv, ventral division of medial geniculate; MGd, dorsal division of medial geniculate; MGm, medial division of medial geniculate.

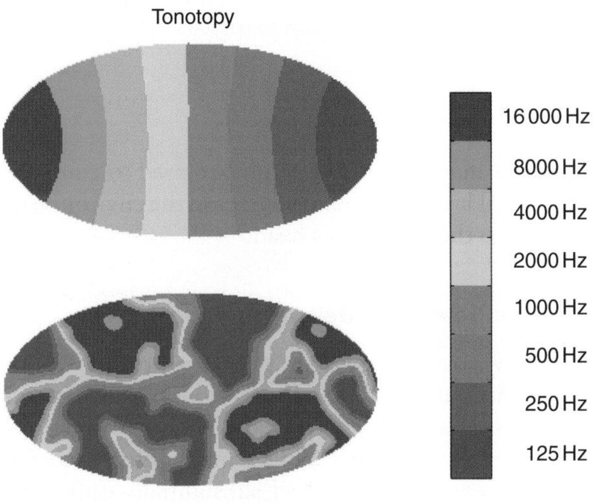

Figure 3 Model tonotopic and nontonotopic representations of sound frequency in a cortical area.

(see **Figure 3**), and many attempts have been made to identify other sound properties that may be mapped along each of these strips. Many sound properties are known to be mapped nonrandomly onto the surface of A1, and computational studies imply that this relatively large number (>5) makes it unlikely that any one of these parameters is mapped orthogonally to the tonotopic axis.

Local horizontal connections in A1 tend to be patchy along the lengths of the isofrequency strips; when coupled with the patchiness of the MGv inputs, this finding implies that nonfrequency processing subunits may be spatially segregated in A1. Spatial segregation of processing appears to be true for frequency, as well, and evidence exists for substantial local frequency interactions in A1, currently of unknown function but potentially for the sensitive detection of frequency modulation sweeps and extraction of harmonicity underlying the perception of pitch – both important components of species-specific vocalizations. The patchiness of other stimulus parameters, such as neuronal receptive field bandwidth, is predicted by computational topographic models.

Local connections within cortical columns of A1 remain to be further elucidated, but, as with other primary sensory areas, it is likely that specificity of interlaminar connections in A1 will become more apparent with further experimentation. Layer 4 cells tend to receive most of their excitatory input from other layer 4 cells. Layer 2/3 pyramidal cells appear to receive most of their excitatory input from either other layer 2/3 cells or layer 4 cells, reminiscent of the lemniscal/paralemniscal projection dichotomy apparent in rodent barrel cortex. How these local intracolumnar connections interface with thalamocortical and corticocortical connections remains unknown, but potentially they represent multiple modes of information transfer through A1 to other cortical areas.

Interconnections between Areas of Auditory Cortex

Auditory cortical interconnections follow the parcellation created by thalamocortical projections. The core areas in primates, which receive their major thalamocortical input from MGv, tend to interconnect densely. Core areas also interconnect with nearby belt areas, but not with parabelt areas. Belt areas appear to interconnect with most or all adjacent auditory areas (**Figure 4**). In terms of hierarchical auditory

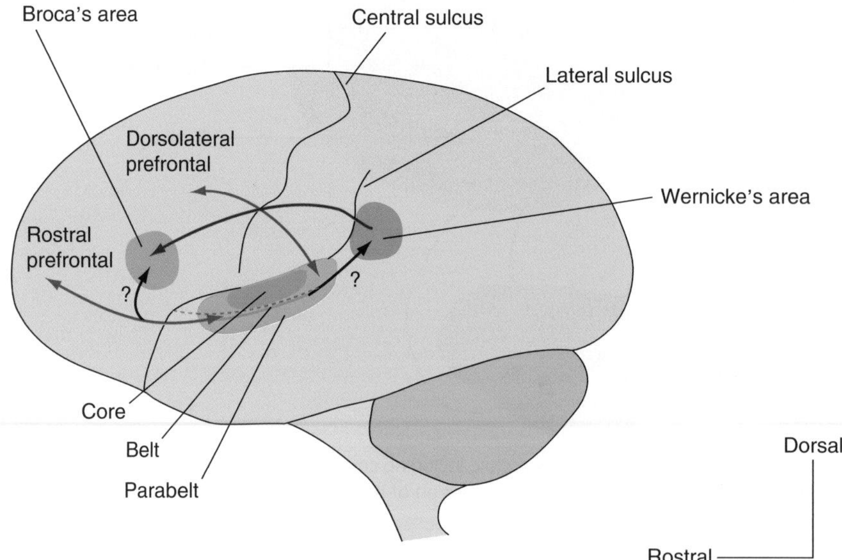

Figure 4 Projections from lateral belt to frontal and parietal lobes in primates. Frontal lobe projections are from experiments in monkeys. Wernicke's and Broca's areas have been demonstrated only in humans, but auditory cortex projections to Wernicke's and Broca's areas have not been conclusively demonstrated in humans.

cortical processing, therefore, the belt areas appear to represent an obligatory intermediate step. This arrangement of cortical area interconnection may be unique to the auditory system.

Projections among auditory cortical areas tend to be topographic in nature, connecting subregions of each tonotopic area that correspond to similar frequencies, as assessed by physiological recordings. A similar topography appears to exist even for areas that show little or no clear frequency tuning, implying that the topographic nature of the projections reflects cortical circuits implementing more than simple frequency extraction. The persistence of topographical projections in absence of clear frequency tuning may ultimately provide some insight into the sensory transformations occurring between interconnected areas of auditory cortex.

Topography of projections across the midline to the contralateral auditory cortical areas follows the same general guidelines for ipsilateral connections, including area specificity and topography. Layer 2/3 pyramidal neurons project to other auditory cortical areas and to the contralateral hemisphere, as do some layer 5a pyramidal cells and possibly some cell types in layer 4. The proportion of auditory interhemispherical projections far outweighs the proportions of similar connections in other sensory modalities, indicating that ACx of both hemispheres represents fairly integrated processing circuits. It should be noted that unlike the visual system, many subcortical auditory nuclei receive input from both sensory organs (ears), as many of these circuits are used to compute the spatial layout of sound sources in the environment relative to the organism's head.

Cortical Projections Beyond Auditory Cortex

Auditory cortical areas in primates are known to project to various nonauditory frontal and parietal lobe areas shown to be involved in cognitive functions rather than audition. While core auditory cortex areas appear to interconnect substantially only with one another and nearby belt fields, the lateral belt and parabelt areas have been shown in monkeys to project to multiple targets in the frontal lobe, in addition to interconnections with one another. At least two projection pathways from lateral belt and parabelt areas to the frontal lobe have been elucidated in macaque monkeys. The anterior lateral belt and parabelt areas appear to send projections predominantly to the rostral prefrontal areas, while the posterior belt and parabelt areas appear to target mainly the dorsolateral prefrontal cortex and adjacent areas. The parabelt areas appear to send more projections to the frontal lobe than do the belt areas. All of the projections appear to be topographic in nature, with nearby auditory cortex neurons projecting to nearby frontal lobe neurons.

It has been proposed, based upon primate functional imaging studies, visual cortex projections, and limited electrophysiology, that the posterior belt/parabelt projections to the dorsolateral prefrontal cortex constitute a 'dorsal stream' representing predominantly spatial information. Conversely, the anterior belt/parabelt projections to the rostral prefrontal areas have been proposed to constitute a 'ventral stream' representing predominantly identifying information about the physical objections producing the sounds. Alternate interpretations for the information transmitted by these projections also exist, including representations of speaker identity. It is likely that future experimentation will reveal a greater complexity of information processing and transfer through auditory cortex to higher cognitive cortical areas than is implied by the 'two-stream' model.

Projections to Subcortical Nuclei

The ascending auditory pathway appears to have multiple projection streams that all converge initially in the inferior colliculus (IC) and again in the MGB. In other words, the IC and MGB are obligatory synapses for ascending auditory projections. In contrast, descending corticofugal projections, which represent a numerically large proportion of total auditory projections, project monosynaptically all the way to the earliest auditory brain stem nuclei. It appears, however, that few individual cortical neurons project to multiple subcortical auditory stations.

Most of the descending corticothalamic projections tend to mirror the ascending projections. In other words, thalamic nuclei that tend to send large projections to a particular auditory cortical area also tend to receive large projections from that cortical area. Like the ascending projections, these descending projections tend to be topographic, even for areas currently without clear functional organization. Not surprisingly, core areas such as A1 tend to project largely to MGv, although substantial divergence in corticothalamic projections exist such that A1 sends some projections to multiple nuclei of the MGB.

Descending corticocollicular projections appear to be organized topographically, much as are the corticothalamic pathways, but are smaller in total number and tend to target paralemniscal IC nuclei. Corticofugal projections to the brain stem represent a numerically small subset of the overall descending projections, terminating largely in the superior olivary complex (SOC) and the cochlear nuclei (CN). Most of these descending projections to all subcortical targets appear to be ipsilateral.

The cortical laminar locations of the corticofugal cells are infragranular (laminae 5 and 6) and tend to be located in specific sublaminar distributions. Corticothalamic cells tend to reside in layer 5b and throughout layer 6. Corticocollicular cells tend to reside in laminae 5 in regions that contain relatively few corticothalamic projections, although some IC-projecting cells are found in deep layer 6b. Few cortical cells appear to project to both the MGB and the IC. Cortical projections to SOC and CN are typically located in deep layer 5b.

The lack of extensive populations of cortical neurons sending projections to multiple subcortical stations argues that these corticofugal projections represent parallel descending pathways, each performing a distinct function. Details of the nature of these functions await further experimentation, but apparently dynamic and plastic modification of spectral tuning and neuronal dynamic range represents a major role of at least some of these pathways.

Other Circuitry

Other interconnections with ACx not described in detail here exist to various degrees throughout development and in the adult. Most of these connections are presumed to have modulatory roles on the encoding of acoustic stimuli, but most remain to be studied. Cholinergic projections from the basal forebrain play an active role in cortical plasticity, particularly in the developing animal but to some extent also in the adult. Dopaminergic, serotonergic, and norepinephrinergic projections to auditory cortex also exist and have yet to be studied extensively. The thalamic reticular nucleus (TRN) receives both ascending and descending auditory collateral input and appears to negatively modulate MGB responses. Interconnections with the limbic system, especially prevalent in nonprimary auditory cortex, presumably play a critical role in adding proper behavioral context to acoustic perception and memories, although little is known about how this phenomenon actually occurs. The auditory system at several levels, including cortex, interfaces with other sensory systems, most likely to integrate perception of space and object identity.

Organization of Specialized Auditory Cortex: Echolocating Bats

Extensive neuroanatomical and neurophysiological research on the acoustic processing and behavior of echolocating bats over the past half-century has revealed a tightly integrated neuronal system exquisitely adapted for one critical behavior. When seeking prey, the echolocating bat emits a series of vocalizations with stereotypical energy signatures. These energy signatures are modified by the size, shape, and velocity of the target and are reflected back to the bat's ears. Bats process this information extremely rapidly and use it to alter their own flight velocity to intercept desirable targets.

The subcortical auditory nuclei and auditory cortical areas of several species of echolocating bats have been studied successfully, thanks largely to the presence of their prey-seeking ethology and their stereotyped vocalizations. Use of species-specific vocalizations to probe physiological responses has resulted in marked advances in understanding the hierarchical processing of echolocating calls throughout the auditory system. In fact, more is known about the encoding of this specialized behavior in ACx than about any other auditory cortical phenomenon.

Knowledge of bat auditory processing of more general acoustic processing tasks, such as passive listening or species-specific vocalizations, generally does not surpass the equivalent knowledge in less specialized species. In fact, the auditory areas that seem to respond in such specific ways in echolocation tasks often respond in rather complex ways to more general tasks – much as is the case for other mammals. Nevertheless, it is hoped by many that adopting similar experimental techniques of studying behaving animals with relevant tasks and acoustic stimuli, such as species-specific vocalizations, will eventually lead to an improved knowledge of acoustic coding in these less specialized species.

Physiology of Unspecialized Auditory Cortex

Early studies of A1 using cortical surface electrodes revealed an apparent cochleotopic (tonotopic) topography. Later studies using penetrating microelectrodes recoding single neurons revealed much greater complexity than the surface electrodes hinted at and fueled claims that no such tonotopy existed. Tonotopy of A1 was eventually demonstrated unequivocally, and today many lines of evidence demonstrate this feature. The difficulty of determining such a basic organizational structure as tonotopy foreshadowed the difficulties in studying physiological responses from a cortical region where complexity, variability, and lack of clear organizing principles appear to dominate. Many attempts to assign unique processing features to A1 and other auditory cortical areas have been made, although a viable comprehensive theory regarding even the major role of A1 remains elusive.

The extraction of sound pitch or the perception of a single fundamental frequency for complex sounds represents a well-known psychophysical phenomenon. Imaging studies that show a functional signal in or near A1 in response to acoustic stimuli eliciting a sensation of pitch fail to reveal a similar source of subcortical activity. It is possible that one role of auditory cortex is to reassemble acoustic information decomposed in previous processing stages into relevant percepts, including the dimension of pitch. It has been speculated that pitch may represent only one dimension of an inherent musical processing ability implemented by circuits in ACx designed to process naturally occurring stimuli.

Experimentally, individuals exposed to many examples of repeated sounds with an occasional low-probability sound typically reveal a particular electroencephalogram (EEG) signal localized to auditory cortex in a phenomenon called mismatched negativity (MMN). Research into this phenomenon has led to theories that auditory cortex may carry recent acoustic history in its firing pattern in order to reassemble multiple streams of information over time, such as melodies, vocalization patterns, or sentences. Auditory stream analysis is a popular theory for the role of auditory cortex, although many of the properties of phenomena such as MMN can be found in subcortical structures or even simpler neural networks, such as the retina.

Different auditory cortex areas exhibit different responses to classes of species-specific vocalizations, leading to the theory that cortex is responsible for extracting information relevant to vocalizations. No models have been proposed for how this type of hierarchical processing may take place, and data upon which to build such models are currently sparse.

Nonauditory cortical areas that are known to be involved with the processing of human speech and language include Wernicke's area at the temporoparietal junction and Broca's area in the lateral frontal cortex (see **Figure 4**). These areas are located in the same cerebral hemisphere as one another and define the dominant hemisphere for humans, who have high degrees of brain lateralization. (Most individuals are left-hemisphere dominant.) Wernicke's area is involved with language comprehension, and impairment of function in this area induces a language deficit or aphasia that results in fluent speech lacking syntactical meaning. Broca's area is involved with speech production, and impairment of function in this area induces an aphasia that results in nonfluent, halting speech. These areas are connected by a projection pathway called the arcuate fasciculus, and impairment of function of this pathway preserves language comprehension and speech production but induces a so-called conduction aphasia whereby affected individuals have difficulty repeating complex utterances. (Conduction aphasias are likely to be caused by damage to other structures, as well.) Nonhuman primate analogs of Wernicke's and Broca's areas have yet to be convincingly demonstrated. It is assumed that auditory cortex connects to Wernicke's and Broca's areas in humans, but these projections have not been fully characterized.

Outstanding Questions

The auditory system differs from other sensory systems in the number and complexity of its subcortical nuclei and circuits. Bilateral information is represented at a very early stage in the auditory pathway, largely to calculate sound direction by circuits accurately measuring timing differences to microsecond resolution. Many other stimulus features appear to be extracted subcortically as well, including those that would be useful in acoustic recognition tasks. An obvious question arising from this observation is the appropriate analogy between auditory cortical areas and the processing regions of other sensory systems, notably the somatosensory and visual systems. All three of these systems project through the thalamus, but A1 is situated at least four or five spiking synapses from the periphery, unlike other primary sensory areas. The cytoarchitectural appearance of koniocellular primary auditory cortex, its cellular arrangement, and its subcortical interconnections appear to match well with those of the other two sensory systems, so the anatomical analogy seems appropriate.

On the other hand, statistical evaluations of the physiological filter elements most efficient in encoding natural stimuli indicate that the auditory nerve may actually generate an output analogous to that of primary visual cortex. If the assumption is correct – that the role of the early visual system is to neuronally construct efficient filters to encode visual scenes – then this comparison has some merit. Much of the subcortical auditory circuitry has no relevance for visual processing, especially given that binocular interactions are first seen in primary visual cortex.

Describing the responses of A1 neurons has been challenging and at times frustrating for decades – even more so for nonprimary auditory cortical areas. Is it appropriate to think of A1 in nonspecialized mammals as equivalent to a visual association area that requires appropriate stimuli and behavioral contexts to successfully study? Or perhaps a reductionist view is more appropriate: A1 may be responsible for creating particular 'filters' used to deconstruct stimuli and pass on to other areas for further processing, as V1 and bat A1 appear to do.

Questions and proposals such as these reflect at least part of the future research on high-level sensory coding in the auditory system, and begin to be resolved with continued efforts to discern what unique processing is performed by A1.

See also: Auditory Scene Analysis; Auditory System: Efferent Systems to the Auditory Periphery; Auditory System: Central Pathways; Auditory/Somatosensory Interactions; Cochlear Development; Language: Auditory Processes; Musical Illusions; Sensory Aging: Hearing; Sound Localization: Neural Mechanisms.

Further Reading

Doucet JR, Molavi DL, and Ryugo DK (2003) The source of corticocollicular and corticobulbar projections in area Te1 of the rat. *Experimental Brain Research* 153(4): 461–466.

Doucet JR, Rose L, and Ryugo DK (2002) The cellular origin of corticofugal projections to the superior olivary complex in the rat. *Brain Research* 925(1): 28–41.

Griffiths TD (2003) Functional imaging of pitch analysis. *Annals of the New York Academy of Sciences* 999: 40–49.

Hu B (2003) Functional organization of lemniscal and nonlemniscal auditory thalamus. *Experimental Brain Research* 153(4): 543–549.

Huang CL and Winer JA (2000) Auditory thalamocortical projections in the cat: Laminar and areal patterns of input. *Journal of Comparative Neurology* 427(2): 302–331.

Kaas JH and Hackett TA (2000) Subdivisions of auditory cortex and processing streams in primates. *Proceedings of the National Academy of Sciences of the United States of America* 97(22): 11793–11799.

Kaur S, Lazar R, and Metherate R (2004) Intracortical pathways determine breadth of subthreshold frequency receptive fields in primary auditory cortex. *Journal of Neurophysiology* 91(6): 2551–2567.

Kaur S, Rose HJ, Lazar R, et al. (2005) Spectral integration in primary auditory cortex: Laminar processing of afferent input. *in vivo and in vitro*. *Neuroscience* 134(3): 1033–1045.

Lee CC and Winer JA (2005) Principles governing auditory cortex connections. *Cerebral Cortex* 15(11): 1804–1814.

Lewicki MS (2002) Efficient coding of natural sounds. *Nature Neuroscience* 5(4): 356–363.

Penagos H, Melcher JR, and Oxenham AJ (2004) A neural representation of pitch salience in nonprimary human auditory cortex revealed with functional magnetic resonance imaging. *Journal of Neuroscience* 24(30): 6810–6815.

Peretz I and Zatorre RJ (2005) Brain organization for music processing. *Annual Review of Psychology* 56: 89–114.

Romanski LM, Bates JF, and Goldman-Rakic PS (1999) Auditory belt and parabelt projections to the prefrontal cortex in the rhesus monkey. *Journal of Comparative Neurology* 403(2): 141–157.

Schreiner CE, Read HL, and Sutter ML (2000) Modular organization of frequency integration in primary auditory cortex. *Annual Review of Neuroscience* 23: 501–529.

Schwartz O and Simoncelli EP (2001) Natural signal statistics and sensory gain control. *Nature Neuroscience* 4(8): 819–825.

Smith PH and Populin LC (2001) Fundamental differences between the thalamocortical recipient layers of the cat auditory and visual cortices. *Journal of Comparative Neurology* 436(4): 508–519.

Suga N (1990) Biosonar and neural computation in bats. *Scientific American* 262(6): 60–68.

Suga N (1994) Processing of auditory information carried by species-specific complex sounds. In: Gazzaniga MS (ed.) *The Cognitive Neurosciences*, pp. 295–313. Cambridge, MA: MIT Press.

Suga N, Xiao Z, Ma X, et al. (2002) Plasticity and corticofugal modulation for hearing in adult animals. *Neuron* 36(1): 9–18.

Winer JA (2005) Decoding the auditory corticofugal systems. *Hearing Research* 207(1–2): 1–9.

Winer JA, Miller LM, Lee CC, et al. (2005) Auditory thalamocortical transformation: Structure and function. *Trends in Neuroscience* 28(5): 255–263.

Auditory Cortex: Models

S A Shamma and J B Fritz, University of Maryland, College Park, MD, USA

Introduction

The auditory cortex plays a critical role in the perception and localization of complex sounds. Incoming acoustic stimuli arrive in the ears, are transformed into a neural code in the cochlea, and ascend in an interwoven and bidirectional network of processing centers, passing through the cochlear nuclei and the superior olivary complex, the lateral lemniscus, the inferior colliculus (IC), and the medial geniculate body (MGB) en route to auditory cortex. Despite the rapidly expanding knowledge of the neuroanatomy and connectivity of the auditory cortex, relatively little is known about its functional organization, especially compared with the visual and the motor systems. One exception has been the auditory cortex of the mustache bat, in which functional maps have been reported, serving the stereotypical echolocating behavior of this species. In other mammals, it is more difficult to isolate a natural auditory behavior and its associated relevant stimulus features with comparable specificity. Nevertheless, a few tasks have been broadly accepted as being vital for all species, such as sound localization, timbre recognition, pitch perception, and of course survival and reproduction! For each, evidence of various functional and stimulus feature maps has been found or postulated, a significant number of them in the past few years. This article elaborates on a few examples of such organization. In each example, the goal is to determine how and whether models of the underlying neural networks can deepen our understanding of the auditory cortex.

Auditory Cortical Fields

The layout and neural structure of the auditory cortex is in many respects similar to that of other sensory cortices. For instance, based on physiological and cytoarchitectonic criteria, and patterns of connectivity, it is subdivided in the cat into a primary auditory field (A1) and several other surrounding fields, such as the anterior auditory field and the secondary auditory cortex. The number and specific arrangement of surrounding fields varies among different species, reflecting presumably the complexity of the animal's acoustic behavior (and the number of anatomical and physiological mapping studies performed). For ballpark figures, both the macaque and the cat have at least 14 auditory cortical fields, the ferret has at least seven, the mustache bat has at least 12, the Mongolian gerbil has at least eight, and the rodent has at least five. The most obvious physiological organizational feature of the core and other tonotopic fields is the presence of a systematic frequency sensitivity map inherited originally from the cochlea. In A1, responses are spatially ordered based on the tone frequency to which they respond best (or their best frequency, BF). The A1 area devoted to biologically significant frequencies may be enlarged, as exemplified by the 'acoustic fovea' in the Doppler-shifted constant frequency (DSCF) area in the mustache bat. But there may also be an increase in frequency representation in A1 to encode the learned importance of sound. Within the isofrequency bands, there is an interleaved modular organization that includes tuning curve bandwidth, binaural properties, and intensity threshold and tuning, which suggests significant differences in local network architecture. In the awake animal, A1 neurons respond vigorously to the onset of a tone and show sustained responses to preferred stimuli. In some cortical belt areas, although there is a poor response to pure tones, cochleotopic frequency organization can be demonstrated by the use of bandpassed noise stimuli. In other fields, responses are even less frequency-selective, more adaptive, or totally absent to single tones, preferring instead more spectrally or temporally complex stimuli. A sudden change in these response patterns or in the gradual spatial order of the tonotopic map is usually taken to signify a border between different fields. Finally, A1 and possibly other fields are further subdivided into smaller regions, serving perhaps different functional roles. One promising approach to the study of cortical functionality is illustrated by a set of recent reversible lesion studies.

General Cortical Mechanisms

The auditory cortex shares with other cortical structures basic neuronal mechanisms and architectures that are thought to be actively involved in a wide range of perceptual processes. For example, the input layers in A1 contain spectrotemporal receptive fields that share strong analogies to those of the visual cortex. This suggests that common organizational and functional principles underlie the excitatory and inhibitory thalamocortical projections and the representation of sensory cues in the primary areas. Anatomical and physiological support of this hypothesis was provided by experiments in which optic nerve projections to the auditory thalamus (MGB) of newborn ferrets resulted in the eventual development of classical visual sensitivity and receptive fields in the target A1 that supported functional visual behavior. However, one curious finding from intracellular recordings in the auditory cortex is the apparent complete overlap of the inhibitory and excitatory inputs (which may be as important for increasing temporal precision as for spectral receptive field shape). Although some of the rich spectral structure of the receptive field is already established in the thalamus, it appears that the thalamocortical input from the lemniscal pathway preferentially confers responses to BF of cortical neurons. However, it is also clear that there are highly influential horizontal fiber corticocortical inputs that also shape spectral receptive field properties, especially for non-BF responses, thus creating a much more complex spectral receptive field structure for A1 neurons.

Adaptation and synaptic depression are two important physiological properties of cortical responses in general that have found strong resonance in psychoacoustics. In particular, they have been hypothesized to be the neural correlates of perceptual phenomena such as 'forward masking,' 'buildup of perceptual streams' in auditory scene analysis (see below), and the 'multiple looks hypothesis' for the integration of cues leading to detection. Interesting possible physiological correlates of these perceptual phenomena have been recently described. One intriguing discovery is that stimulus-specific adaptation takes place on multiple timescales ranging from hundreds of milliseconds to tens of seconds, which may play a role in encoding auditory memory.

Another important property of auditory cortex is redundancy reduction in representation of complex spectrotemporal stimuli compared with stimulus-induced redundancy observed in IC and MGB. Neural correlation in spiking patterns between adjacent neurons is more likely when there is overlap of spectrotemporal receptive fields or when the difference in characteristic frequency is small.

Finally, rhythmic gamma oscillations (20–120 Hz) abound in the auditory cortex, much like those found in other areas. The functional role of oscillatory synchronization of neural activity remains mysterious but may enhance coincidence detection or noise tolerance or play a role in plasticity and attention.

Encoding of Timbre

Timbre is the perceptual attribute of sound responsible for recognizing and classifying complex sounds, such as the distinctions among speech phonemes and different instruments playing at the same pitch. While a multitude of acoustic cues underlie this ability, the most important is the shape of the spectral profile of sound and its evolution in time. The spectral profile emerges early in the auditory system as the sound is analyzed into different frequency bands, in effect distributing its energy along the one-dimensional tonotopic axis. However, in the central auditory system, the spectral profile is expanded into a two-dimensional response pattern, with each frequency represented by an entire sheet of neurons – the so-called isofrequency axis. Many hypotheses have been proposed as to the nature of this added dimension and how the response patterns in the primary auditory cortex might be representing the spectral profile and its dynamics. One early hypothesis is the best-intensity model, motivated primarily by the strongly nonmonotonic responses as a function of stimulus intensity observed in many cortical cells. One can view such a cell's response as being selective to (or encoding) a particular intensity. Consequently, a population of such cells, tuned to different frequencies and intensities, can provide an explicit representation of the spectral profile by their spatial pattern of activity. The most compelling example of such a representation is in the DSCF area of A1 in the mustache bat. However, an extension of this hypothesis to multicomponent stimuli has not been demonstrated in any species.

The second model is inspired by extensive data and ideas gained from physiological and psychoacoustical experiments over the past decade. Specifically, insight has been gained from measurements of the so-called spectrotemporal response fields (STRFs) of A1 cells. An STRF summarizes the dynamics and sensitivity of a cell or, more precisely, the impulse response of the cell at each frequency. Thus, an STRF displays the excitatory and inhibitory interactions that give the cell its selectivity to spectrotemporal patterns. Some STRFs are responsive (excited or suppressed) over a broad range of frequencies, exceeding an octave, while others are quite narrowly tuned. Dynamically, some STRFs' responses decay rapidly after an impulse, while others last twice as long. Finally, this combined time–frequency sensitivity can take more complex forms that are 'inseparable,' as in oriented STRFs that are sensitive to frequency modulations (FM).

STRFs have been measured in many ways, including reverse correlation with random tone chords or spectrotemporally modulated noise. Another method is the 'ripple analysis method' that employs broadband noise with sinusoidally modulated spectrotemporal envelopes with different parameters. Ripples serve the same function as regular sinusoids in measuring the transfer function of linear filters, except that they are two dimensional (spectral and temporal). A1 cells respond well to ripples and are usually selective to a narrow range of ripple parameters that reflect details of their receptive fields. By compiling a complete description of the responses of a cell to all ripple densities and velocities, it is possible by an inverse Fourier transform to compute the corresponding STRF.

From a functional perspective, the rich variety of STRFs found in A1 implies that each STRF acts as a modulation-selective filter of its input spectrogram, specifically tuned to a particular range of spectral resolutions (also called scales) and a limited range of temporal modulations (or rates). The collection of all such STRFs then would constitute a filter bank spanning the broad range of psychoacoustically observed scale and rate sensitivity in humans and animals. Evidence of the importance of spectrotemporal modulations in the perception of complex sounds has come from experiments in which systematic degradations of the speech signal were correlated with the gradual loss of intelligibility. In fact, the relationship between the temporal modulations and speech intelligibility has long been codified in the formulation of the widely used speech transmission index and the spectrotemporal modulation index, which assesses the integrity of both the spectral and temporal modulations in a signal as a measure of intelligibility.

Pitch Representation in the Cortex

A sound complex consisting of several harmonics is heard with a strong pitch at the fundamental frequency of the harmonic series, even if there is no energy at all at that frequency. This percept has been variously called the missing fundamental, virtual pitch, or residue pitch. A large number of psychoacoustical experiments have been carried out to elucidate the nature of this percept and its relationship to the physical parameters of the stimulus. Basically, all models fall into one of two camps. The first believes that the pitch is extracted explicitly from the harmonic spectral pattern. This can be accomplished in a variety of ways, for instance by finding the best match between the input pattern and various harmonic templates assumed to be stored in the brain. The second group claims that the pitch is extracted from the periodicities in the time waveform of responses in the auditory pathway that can be estimated, for example, by computing their autocorrelation functions. In these latter models, some form of organized delay lines is assumed to exist in order to do the computations, much like those that seem to exist in the FM–FM area of the mustache bat.

In all pitch models, however, it is assumed that the extracted pitch is finally represented as a spatial map in higher auditory centers. This is because many studies have confirmed that neural synchrony to the repetitive features of a stimulus, whether it is the waveform of a tone or its AM modulations, becomes progressively worse toward the cortex. It is a remarkable aspect of pitch that, despite its fundamental and ubiquitous role in auditory perception, only a few reports exist of physiological evidence of spatial pitch maps, and none have been independently confirmed. One source is functional magnetic resonance imaging (fMRI) and magnetoencephalography (MEG) scans of the human auditory cortex. Another is from single-unit and multiunit responses in various precortical auditory structures and recently in the primate auditory cortex. One key difficulty in all experiments seeking to demonstrate physiological correlates of pitch is the cochlear nonlinearity that produces distortion components at the fundamental of the upper harmonics and that unintentionally excites low BF cells. This 'artifact' has cast a shadow of doubt over all discoveries of physiological pitch maps because of the experimental difficulties in avoiding or masking it.

Of course, the difficulty in finding a spatial pitch map in the auditory cortex may be due to the fact that it does not exist! This possibility is counterintuitive, given the results of ablation studies that show that bilateral cortical lesions in the auditory cortex severely impair the perception of pitch of complex sounds but do not affect the fine discrimination of frequency and intensity of simple tones. Another possibility is that the maps sought are not at all as straightforward as imagined. For example, harmonic complexes may evoke stereotypical patterns that are distributed over large areas in the auditory cortex and not localized, as the simple notion of a pitch map implies. Finally, it is also possible that A1 simply functions as one stage that projects sufficient temporal or spectral cues for later cortical stages to extract the pitch explicitly.

Models of Sound Localization

It has been recognized for many years that the auditory cortex (and especially the A1) is involved in sound localization. Detailed physiological studies further confirmed that A1 cells are sensitive to manipulations of the binaural stimulus, such as the interaural level difference (ILD) and interaural time difference, a sensitivity that has its origins early in the auditory pathway, where the first convergence of binaural inputs occurs. In fact, a combined auditory–visual map of auditory space has been found in the superior colliculus of several mammals and in the IC of the barn owl. However, the exact role of the auditory cortex in the perception of auditory space remains unclear since no obvious topographic organization of auditory space has been found. Instead, responses from the two ears have been found to alternate between excitatory–excitatory (EE) and excitatory–inhibitory (EI) interactions in bands along the isofrequency planes of A1. One possible functional model that utilizes such maps assumes that EI cells are tuned to particular ILDs and hence encode the location of a sound source based on this cue. EE cells would in contrast encode the absolute level of the sound. However, there is little evidence to support this hypothesis in the sense that neither EE nor EI cells are particularly stable encoders of specific ILD or absolute sound levels. An alternate hypothesis is that these cells encode the absolute levels of the stimulus at each ear rather than the difference and average binaural levels, as previously postulated. Location coding has recently been modeled by an opponent process theory in which sound source locations are represented by population differences in activity of broadly tuned contralateral and ipsilateral space neurons. Another hypothesis uses a likelihood approach to spatial coding.

Finally, it has also been proposed that A1 units encode the spatial location of a stimulus by unique patterns of temporal firing, ones that can be discerned by more elaborate pattern recognition neural networks. Recent findings reveal a role for auditory cortex in long-term plasticity of auditory localization. Attentional mechanisms may operate in a much shorter time frame; for example, a recent study in the barn owl suggests that focused spatial attention may tighten auditory spatial receptive fields in the IC.

Neural Correlates Auditory Scene Analysis

Humans and many animals can attend to one sound source and stream (or segregate) it from a background of many sources rapidly and with no learning or prior exposure to the specific sounds. For humans, this is the essence of the well-known 'cocktail party' phenomenon, in which a person can effortlessly conduct a conversation with a new acquaintance in a crowded and noisy room. For frogs, songbirds, or penguins, this ability is vital for locating a mate or an offspring in the midst of a loud chorus.

The question of how the acoustic scene is parsed by the auditory system into auditory objects and streams is one of the most fundamental in perceptual science. Despite its importance, the study of the underlying neural mechanisms and their computational models remains in its infancy. Several studies have looked for neural correlates of auditory streaming in humans, using electroencephalography (EEG), MEG, and fMRI. The earlier studies focused on the mismatch negativity (MMN) as a potential electro- or magnetoencephalographic index of streaming and found indications that certain features of the MMN (its presence, amplitude, or latency) are related to how the evoking sequence was perceptually organized. However, because the experimental manipulations in these studies always involved changes in the physical stimulus characteristics, differences in the MMN across experimental conditions could reflect factors other than streaming *per se*. Other EEG investigations of streaming have overcome some of these criticisms by exploring the properties of streaming and the systematic differences that emerge when listeners were in different auditory perceptual states. A recent study examines the way in which attending to one sound stream modifies the extent to which the unattended sounds are processed. Some investigators have also recently begun to use fMRI in order to identify what regions of the brain are activated during the perception of one or two auditory streams. Findings thus far point to overall activation in nonprimary auditory cortex and parieto-occipital regions.

While fMRI (and to some extent MEG and EEG) studies may reveal what regions of the brain are involved during streaming, they provide only limited insight into the actual neuronal mechanisms underlying auditory streaming and are hence only indirectly helpful in the formulation of computational and functional models of auditory scene analysis. A more fine-grained view is achieved through single- and multiunit recording studies in primates, bats, and birds. The results have generally pointed to a model in which enhanced segregation of neural activity plays a critical role in mediating the streaming percept and to mechanisms that are fairly basic to cortical physiology, including 'forward suppression' or 'adaptation,' synaptic depression, and spiking threshold. Despite these encouraging results, two criticisms can be made of all neural studies of streaming so far. First, none has provided behavioral tests of the percepts that the animals were experiencing with the stimuli used in the experiments. Second, all the above studies involved awake but nonbehaving animals. This may be particularly important for streaming-related issues as it is possible that attention is critically involved in at least some aspects of auditory stream formation. Combined physiological and behavioral experiments, although generally difficult to perform, are not impossible. Such experiments have been successfully performed and have led to important insights into the neural mechanisms of perception.

There are numerous computational models of auditory scene analysis that can partially disentangle mixtures of sources. These models vary substantially in flavor, complexity, and details, depending on the specific phenomena they address. For instance, computational models of streaming with relatively simple periodic tonal sequences, noise, and other abstract stimuli have emphasized data-driven low-level processing and the notion that peripheral cochlear channeling is crucial in stream formation. Auditory scene segmentation and edge detection have been modeled based on a study of amplitude transients, and sparse overcomplete linear representations in A1 have been proposed to be the basis for solving the cocktail party problem. By contrast, models for segregating mixtures of speech, music, and other complex nonperiodic sources have sought to include more-complex perceptual approaches, including a role for context and expectations, integration of grouping cues over relatively long times, and even explicit training with information beyond that normally available to a listener, such as prior exposure to clean exemplars of the voices in the mixture.

An alternate computational view of streaming is a purely data-driven process that, at the same time, incorporates top-down perceptual processes such as expectations. In this view, the auditory representation of sound undergoes an iterative clustering process that integrates cortical dynamics and Kalman-filtering prediction to form the final streams. Integrating bottom-up feedforward transformations with top-down expectations provides a computational account of (1) how low-level grouping cues such as harmonicity, onsets, and binaural cues can influence the formation of perceptual streams; (2) how cortical dynamics can give rise to the observed dynamics of streaming in simple stimuli as well as musical and speech rhythms; and (3) why the 'old-plus-new' heuristic has proven exceptionally influential in explaining the perceptual organization of complex sounds.

Emerging Functional Views of the Cortex

The classical conceptual view of the auditory system (and other sensory systems) as a sequence of feedforward transformations and feature extractions leading to the ultimate 'cognitive' recognizers is rapidly being replaced by a more integrated view that takes into account (1) the adaptive nature of its receptive fields and their sensitivity to contextual cues and stimulus 'meaning'; (2) the vast networks of feedback from higher centers down to the earliest processing stages; (3) the substantial multimodal interactions among different sensorimotor pathways; (4) interactions with reward, limbic, and cognitive pathways; and (5) the global rhythmic brain activity that may reflect and (in turn) enhance the attentional state of the animal. Consideration of all these important factors is essential for a complete understanding of active auditory processing. One hypothesis is that A1 is not a feature analyzer or detector but rather that the role of auditory cortex is to organize these disparate feature properties into discrete auditory objects in the context of a coherent auditory scene.

The Chameleon Brain

Long-term auditory experience or learning has been shown to cause profound global effects, such as reshaping of tonotopic maps, and significant local effects by transforming receptive field properties of neurons in A1. A computational model has been developed to describe the experience-dependent reorganization of tonotopic maps. Convergent studies of plasticity in the auditory, visual, and motor systems have recently also demonstrated the capacity for dynamic modulation of representational maps and shown that cortical cells in these systems can undergo rapid, task-dependent, and context-specific changes of their receptive field properties during attentive behavior. The key elements of this form of adaptive plasticity appear to be (1) directed attention to salient task-related cues, which leads to (2) selective functional reconfiguration of the underlying cortical circuitry that occurs simultaneously with task performance and causes (3) changes in receptive field properties of individual neurons and the cortical ensemble, which may enhance behavioral performance in the current task. These findings suggest that cortical receptive fields are not fixed but may be constantly adapting and reorganizing dynamically to meet the challenges of an ever changing environment and new behavioral demands and may play an important role in information processing and storage. In this functional model, each primary sensory cortical neuron participates in multiple behavioral contexts, and it is likely that its receptive field properties are differentially modified by top-down influences in each case; its network connectivity may also be reconfigured in an immediate and reversible manner as the animal switches between behavioral states. In this way, the same neuronal ensemble can mediate entirely different perceptual functions. The basic adaptive mechanisms that underlie this plasticity may be similar in perceptual and motor learning and also during optimal performance of a previously learned task. It is interesting that it appears that flexibility in task-dependent processing of similar acoustic stimuli is a fundamental principle, not only at the level of single cells and local networks, but also at the level of hemispheric activation.

Multimodal Influences

Recent neuroanatomical and neurophysiological studies have shown a convergence of multisensory (visual and somatosensory) inputs to auditory cortex. Task-related motor and reward-related responses have also been reported in auditory cortex. It is interesting that nonauditory inputs appear to be gated by task relevance in the auditory cortex and IC. These results suggest that rather than a purely unisensory processing stream, responses in the auditory cortex must be understood as an interwoven tapestry of relevant multimodal contextual inputs.

Feedback and Top-Down and Cognitive Influences

It is well established that cognition is simultaneously bottom-up and top-down, involving interactions between bottom-up, sensory-driven information and top-down, attentional memory and executive processes that modulate the bottom-up processing. This bottom-up versus top-down distinction is roughly consistent with the anatomical and physiological evidence of a cortical architecture that abounds with forward and backward axonal projections in the neocortex and associated structures (such as the thalamus, the amygdala, the striatum, and the hippocampus). In fact, there is increasing evidence that most higher brain functions, including the brain's ability to learn from experience, depend on the integration of such forward and feedback signals. Consequently, a complete understanding of how auditory cortical responses encode the acoustic environment must take into account the behavior of the animal within it.

For instance, one simply cannot obtain a true understanding of auditory cortical responses to a threatening sound by merely playing it to an anesthetized animal. Instead, when an animal recognizes and escapes threatening sounds, it enters a highly aroused and attentive state in which it categorizes its predators' calls as salient foreground targets to distinguish them from harmless background sounds. Simultaneously, it likely integrates other sensory cues (visual, olfactory) as well as its stored acoustic memories into its auditory judgment of the nature of the calls. In addition, neuronal correlates of category formations in the prefrontal cortex would likely feed back and adapt the receptive fields of the auditory cortex so as to enhance the perception of the target sounds (against the background) and subsequently to generate an appropriate multimodal representation of the scene and plan motor actions to respond to the threat. All these interactions significantly alter auditory responses in the cortex, and a massive descending corticofugal feedback system dynamically reshapes cortical inputs, perhaps influencing precortical structures all the way down to the cochlea, and hence must be taken into account when dissecting the nature of auditory cognition.

See also: Auditory Cortex Structure and Circuitry; Auditory Scene Analysis; Auditory System: Efferent Systems to the Auditory Periphery; Auditory System: Central Pathway Plasticity; Auditory Evoked Potentials; Auditory System: Central Pathways; Auditory Localization; Sound Localization: Neural Mechanisms; Speech Perception: Neural Encoding; Temporal Processing in the Auditory Pathway.

Further Reading

Bakin JS and Weinberger NM (1996) Induction of a physiological memory in the cerebral cortex by stimulation of the nucleus basalis. *Proceedings of the National Academy of Sciences of the United States of America* 93: 11219–11224.

Bregman AS (1990) *Auditory Scene Analysis: The Perceptual Organisation of Sound.* Cambridge, MA: MIT Press.

Carlyon RP (2004) How the brain separates sounds. *Trends in Cognitive Neurosciences* 8: 465–471.

Clarey J (1992) Physiology of thalamus and cortex. In: Popper AN and Fay RR (eds.) *The Mammalian Auditory Pathway: Neurophysiology,* pp. 232–334. New York: Springer.

Edeline JM (2003) The thalamo-cortical auditory receptive fields: Regulation by the states of vigilance, learning and the neuromodulatory system. *Experimental Brain Research* 153: 554–572.

Elhilali M, Chi T, and Shamma S (2003) Intelligibility and the spectrotemporal representation of speech in the auditory cortex. *Speech Communication* 41: 331–348.

Fritz JS, Shamma S, Elhilali M, and Klein D (2003) Rapid task-dependent plasticity of spectrotemporal receptive fields in primary auditory cortex. *Nature Neuroscience* 6: 1216–1223.

Ghazanfar A and Schroeder C (2006) Is neocortex essentially multisensory? *Trends in Cognitive Neurosciences* 10: 278–285.

Goldstein J (1973) An optimum processor theory for the central formation of pitch of complex tones. *Journal of the Acoustical Society of America* 54: 1496–1516.

Kaas JH and Hackett TA (2005) Subdivisions and connections of the auditory cortex in primates: A working model. In: König R, Heil P, Budinger E, and Scheich H (eds.) *The Auditory Cortex: A Synthesis of Human and Animal Research,* pp. 7–26. Mahwah, NJ: Lawrence Erlbaum.

Knudsen EI and Konishi M (1978) A neural map of auditory space in the owl. *Science* 200(4343): 795–797.

Li C-SR, Padoa-Schioppa C, and Bizzi E (2001) Neuronal correlates of motor performance and motor learning in the primary motor cortex of monkeys adapting to an external force field. *Neuron* 30: 592–607.

Naatanen R, Tervaniemi M, Sussman E, Paavilainen P, and Winkler I (2001) "Primitive intelligence" in the auditory cortex. *Trends in Neurosciences* 24: 283–288.

Sur M, Garraghty PE, and Roe AW (1988) Experimentally induced visual projections into auditory thalamus and cortex. *Science* 242: 1437–1441.

Ulanovsky N, Las L, Farkas D, and Nelken I (2004) Multiple time scales of adaptation in auditory cortex neurons. *Journal of Neuroscience* 24: 10440–10453.

Auditory Evoked Potentials

J R Melcher, Harvard Medical School, Boston, MA, USA

Introduction

When sounds reach our ears, they are transduced by peripheral hearing structures into electrical signals in the auditory nerve. The nerve transmits these signals to the central auditory system, where they are processed by neurons in the brain stem, thalamus, and cerebral cortex. The result of this processing is the perception of sound. Many thousands of neurons are engaged in the transformation from sound to perception. Neuronal activity produces in the surrounding brain tissue a voltage that can be recorded from electrodes on the surface of the head. A measurement of the voltage produced collectively by the active neurons is an auditory evoked potential (AEP).

AEPs Produced by Brief Stimuli

Many sounds produce a measurable AEP, but AEPs produced by brief sounds, such as acoustic pulses (called clicks), brief tones, and short bursts of broadband noise, are especially robust. **Figure 1** shows a stylized AEP in response to a click. The response consists of a series of voltage fluctuations continuing for hundreds of milliseconds following the sound stimulus. In response to a single stimulus presentation, most, if not all, of these fluctuations are swamped by the voltage produced by ongoing activity in the brain (i.e., the electroencephalogram). An AEP is extracted from this ongoing signal by averaging the voltage following many presentations of the same stimulus.

The AEP illustrated in **Figure 1** is based on recordings in people. However, AEPs have also been recorded in many other species, including mice, guinea pigs, rats, cats, and dolphins.The AEP waveform differs across species, reflecting, for instance, interspecies differences in head size, axonal path lengths, physical location and orientation of cells in the head, and the specific auditory cell types present in the auditory pathway. In all cases, however, there is a major commonality: the AEP comprises a time-varying voltage following the stimulus.

The AEP produced by a brief stimulus is generally divided into three parts:

1. The auditory brain stem response (ABR) consists of the shortest latency fluctuations occurring within approximately 10 ms of the stimulus (green in **Figure 1**). The voltage fluctuations of the ABR are each ~1 ms in width. As the name of this response indicates, these fluctuations are generated by auditory neurons in the brain stem.
2. The middle latency response (MLR) consists of a series of voltage fluctuations immediately following the ABR and preceding the long latency response (red in **Figure 1**). The voltage fluctuations comprising the MLR are much broader than are the fast fluctuations of the ABR. The MLR partly reflects cellular activity in the thalamus.
3. The late latency response (LLR) begins 50–100 ms after the stimulus (blue in **Figure 1**). It includes one of the most widely studied components of the AEP, the N1 or N100. The complete response consists of a complex of positive and negative fluctuations generated by the cortex.

Dependencies on Stimulus Parameters

The ABR, MLR, and LLR all depend on physical aspects of the sound stimulus. For instance, increases in the rate of stimulus presentation generally decrease response amplitude. Increases in stimulus intensity generally increase the amplitude of responses and decrease the latency (i.e., shorten the time between stimulus and response). These rate and intensity dependencies are partly attributable to the behavior of the auditory nerve, which generates the first wave of the ABR and provides input, directly or indirectly, to neurons contributing to the remainder of the AEP. For example, individual auditory nerve fibers are more likely to discharge in response to high-intensity stimuli and will do so with a shorter latency. Wave I of the ABR, which reflects the summed response from all auditory nerve fibers, is correspondingly greater in amplitude and shorter in latency, the typical intensity dependence for AEPs.

Dependencies on Arousal and Attention

The MLR and LLR differ from the ABR in their sensitivity to whether individuals exposed to the stimulus are asleep or awake, whether they are paying attention to the stimuli, and whether they are anesthetized. For instance, the LLR can change substantially across different stages of sleep. It also shows distinct changes in amplitude when individuals perform a task that requires concentrated listening (i.e., attention) to the stimuli used to evoke the LLR. The ABR is highly resistant to changes in state, persisting, for instance,

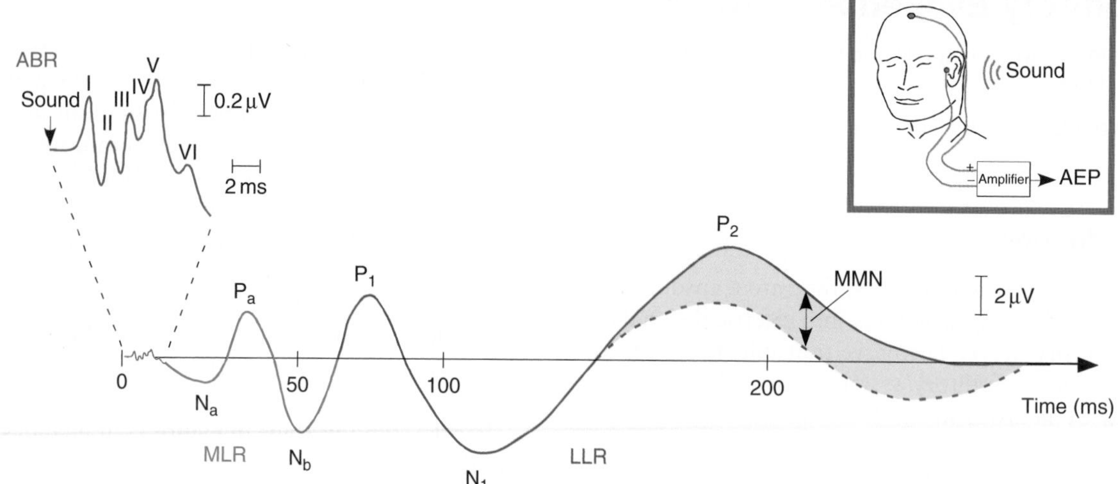

Figure 1 Stylized auditory evoked potential (AEP) in humans. The illustration is modeled after responses to brief stimuli recorded between two electrodes: at the vertex or top of the head and near the stimulated ear (inset at top right). Stimulus presentation is at 0 ms (at arrow in top waveform). The entire AEP includes the auditory brain stem response (ABR; green), the middle latency response (MLR; red), and the late latency response (LLR; blue). The ABR has been expanded (top left) so that its individual wave components (labeled I–V) can be resolved. The dashed trace indicates how the response to the same stimulus would differ if presented as the deviant stimulus in an oddball paradigm. The dashed waveform, minus the solid, is the mismatch negativity (MMN).

even when individuals are anesthetized deeply enough for surgery. The resilience of the ABR has made it an effective tool for monitoring the health of the auditory system during neurosurgery (e.g., to remove a tumor from the vestibulocochlear nerve) and for infant hearing tests (ideally performed during sleep).

Neural Generators of AEPs

While the ABR, MLR, and LLR differ in many ways (amplitude, latency, duration of individual voltage fluctuations) and are generated by different neurons, their mechanism of generation is fundamentally the same, since all represent the net potential produced by large populations of neurons. Several factors determine whether any given neuronal population will contribute substantially to the overall AEP:

1. *The total number of neurons in the population.*
2. *The amplitude of the potential produced by individual neurons at the AEP recording electrodes.* This amplitude can be dependent on neuronal morphology. For example, when current flows across the membrane of a neuronal cell body and returns through the dendrites, a so-called closed field results if the dendrites uniformly surround the cell body (i.e., zero potential at the AEP electrodes), whereas an open field results if the dendrites are distributed nonuniformly (i.e., nonzero potential).
3. *Similarity of morphology and orientation across neurons.* For example, two oppositely oriented neurons next to one another will produce equal but opposite (i.e., canceling) potentials.
4. *The degree to which activity is temporally synchronized across neurons.* The potentials produced by the individual neurons of a population will only add effectively if the neuronal responses to the AEP stimulus are similar in latency (i.e., temporally aligned with one another).

A variety of approaches have been used to identify neural generators of the AEP. One widely used approach involves making focal brain lesions in animals and examining the effect on the AEP. Lesioning techniques used for this purpose include making surgical cuts, passing electric current, or injecting a neurotoxin. The effects of neurotoxic lesions, combined with modeling, have led to the conclusion that most of the click-evoked ABR in cats is generated by neurons in two of the many pathways of the auditory brain stem. These pathways originate with the bushy cells of the cochlear nucleus, the most peripheral nucleus of the central auditory system and recipient of direct inputs from the auditory nerve. They extend centrally to the superior olive, nuclei of the lateral lemniscus, and inferior colliculus. In cats, there are two types of bushy cells, spherical and globular, which mark the beginning of the two pathways involved in ABR generation. The pathways emanating from bushy cells have specialized features for maintaining the timing of neuronal activity (e.g., large-diameter axons and unusually large terminal endings called endbulbs), and hence maintain the high degree of temporal synchronization across neurons present in the auditory nerve. This temporal synchronization is one of several features that make bushy cells (spherical and globular), as well as the direct and indirect targets of these cells, robust contributors to the cat ABR. In contrast to cats, the globular bushy cell population in humans is proportionately quite small, and according to some is nonexistent. Therefore, the human ABR (after wave II) may largely reflect activity in just one pathway consisting of spherical bushy cells and their primary and secondary targets (e.g., medial superior olive, inferior colliculus).

The potential contributed to the AEP by brain stem neurons likely reflects discharge activity on the cell body and axon. (This is consistent with the brevity of the fluctuations comprising the ABR, ~1 ms.) In contrast, the traditional view of cortically generated potentials (e.g., P1, N1, P2, of the LLR) is that they arise from synaptic currents flowing between the cell body and dendrites of pyramidal cells (a synaptic origin is consistent with the longer duration of the fluctuations comprising the LLR). The cell bodies of pyramidal cells, lying in the deeper layers of cortical gray matter, give rise to an apical dendrite extending into the superficial layers and oriented perpendicular to the cortical surface. The morphology of pyramidal cells (resulting in an open, or dipole, field) and the consistent dendritic orientation across pyramidal cells mean that large groups of synchronously activated pyramidal cells are capable of generating a robust evoked potential.

The model for pyramidal cells just described is generally assumed by one of the most extensively used approaches for identifying cortical generators of AEPs. The approach involves (1) measuring the AEP at many finely spaced locations over the head (e.g., 100 or more), (2) modeling the underlying cortical activity using the dipole model suggested by the morphology and dendritic orientation of pyramidal cells, (3) assuming an electric model for the head, and (4) determining the strength, location, and orientation of dipole(s) that yield a best fit between the measured AEP and AEP calculated based on the head and dipole models. The location of AEP generators estimated by this approach can be further refined using information from other imaging modalities, including structural and functional magnetic resonance imaging (MRI). The results are vivid spatial mappings of cortical activity unfolding over time.

Magnetic Analog of the AEP

When neurons responding to sound produce an AEP, they also produce a magnetic field recordable outside the head. While small (about 100 000 times less than the magnetic fields in our everyday surroundings), these auditory evoked magnetic fields (AEMFs) can be measured routinely when specialized, low-noise magnetic detectors are used. Both AEPs and AEMFs are produced directly by neuronally generated currents in the brain. This distinguishes AEP and AEMF recordings from other modalities for measuring brain function, such as functional MRI (fMRI) and positron emission tomography (PET), which detect hemodynamic changes accompanying changes in neuronal activity, rather than neuronal activity itself.

While both AEPs and AEMFs are a direct reflection of neuronal activity, the two types of recordings differ and are, in many respects, complementary. For instance, AEMFs, but not AEPs, are preferentially sensitive to cortical activity in pyramidal cells having a particular dendritic orientation relative to the surface of the head (dendrites parallel to the head surface, creating what is called a tangential dipole source). Much of the gray matter on the human superior temporal lobe contains pyramidal cells with this orientation, including the crown of Heschl's gyrus, where primary auditory cortex is typically located, and planum temporale, which contains nonprimary areas. Because the orientation of auditory cortex is well suited for magnetic recordings, AEMFs have been used extensively to investigate human auditory cortical function.

Other, Specialized AEPs

In order to obtain a specific indicator of activity in limited neuronal populations, responses are sometimes derived from multiple AEP measurements. The binaural difference potential (BDP) illustrates the selectivity provided by a derived response; it specifically reflects activity involved in comparing sounds between the two ears. A BDP is derived by summing the ABRs produced by left and right monaural stimulation and subtracting the ABR produced by stimulating both ears simultaneously. For click stimuli, the summed monaural and binaural ABRs show no difference during waves I–III, but diverge during waves IV/V. This difference is the BDP. A BDP would not be detected if sounds at the two ears were processed independently; the BDP reflects an interaction between the two ears. The coincidence of the BDP with the ABR indicates that it is generated by brain stem neurons.

Animal lesion data and the prominence of a particular binaural nucleus in the brain stem of humans, the medial superior olive (MSO), have led to suggestions that MSO neurons contribute substantially to the human BDP.

While AEPs can be elicited by discrete stimuli presented in silence, they can also be elicited by brief changes in ongoing sound. One example is the lateralization shift response (LSR), a cortically generated potential produced by changing the interaural time difference of binaurally presented continuous noise. A paradigm for eliciting this response involves initially presenting identical noise to the two ears (the noise is heard diffusely, but centered in the head), then, over a brief interval, temporally shifting the noise at one ear relative to other (by ~1 ms), and then quickly shifting it back (the diffuse, centered noise is still heard without interruption, but a brief sound lateralized to the ear with the leading noise is also heard). The result is an evoked response 100–300 ms after the shift, the LSR. Since the noise presented to either ear alone continues without disruption, it does not produce an AEP. The LSR only occurs when the noise is presented to both ears. Thus, like the BDP, the LSR specifically reflects binaural processing within the auditory central nervous system, although at a cortical rather than brain stem stage.

In addition to depending on the physical features of the stimuli used to evoke them, evoked potentials can also depend on the context in which the stimuli are presented. An example of a context-dependent AEP is the mismatch negativity (MMN). An MMN is elicited in a paradigm that involves presenting a standard stimulus (e.g., a tone burst) at regular intervals but, every so often, replacing the standard with a deviant sound (e.g., a tone burst of slightly different frequency; also called a rare stimulus). Responses to the deviant sound presented in the context of this paradigm (called an oddball paradigm) are different from the responses to the same stimulus presented over and over again on its own. The difference is the MMN (difference between the solid trace, which is the stimulus presented on its own, and the dashed trace, which is the stimulus presented as deviant, in **Figure 1**). The MMN, generated largely by the auditory cortex, can be thought of as a form of sensory memory, 'sensory' because it originates within cortex that specializes in encoding one of our senses, and 'memory' because a history of the stimuli appears to be retained by cortex; each incoming sound is compared to its predecessors and an MMN is produced when they differ. An MMN can be recorded even when individuals are distracted from the stimuli (e.g., by watching a movie), indicating that the MMN reflects, at least in part, neuronal processing prior to conscious perception.

AEP Correlates of Auditory Perception

Since AEPs can be recorded in awake people while they listen to sounds, they offer a way to understand the neural processing that determines how sounds are perceived. Two examples of this important role for AEPs follow.

The first example concerns the binaural difference potential and its correlation with certain perceived aspects of binaurally presented clicks. When clicks of equal intensity are presented simultaneously to the two ears, they are perceived as a single, fused object that is centered with respect to the head. As the clicks presented to one ear are made greater in intensity or shorter in latency, compared to the clicks presented to the other ear, the perceived object moves to the side of the more intense or leading clicks. As the interaural click delay is increased beyond about 1 ms, the object stops moving, and eventually splits in two such that two separate clicks are perceived. Studies have compared these perceptual properties with measurements of the BDP in human listeners. They showed that (1) the BDP was only detectable when the clicks at the two ears were fused into a single object and (2) the latency of the BDP covaried with the perceived location of the object. These correlations between percept and BDP indicate that the neuronal activity underlying the BDP contains a representation of both the fusion and the location of binaurally presented sounds. This implies that binaural fusion and perceived sound location may be at least partially encoded in the brain stem, where the BDP is generated, and, interestingly, well before cortical processing stages, where the activity dictating our perceptions is usually assumed to reside.

The second example concerns the N1 of the LLR and its correlation with the perceived rate of successive sounds and the segregation of sounds into distinct perceptual streams. A nice illustration of stream segregation, and the changes in perceived rate that sometimes accompany it, is provided by a sequence of tone bursts alternating in frequency (ABAB..., where A and B denote different frequencies). When the frequency difference between tones is small, a single, high-rate sequence is heard. However, when the frequency difference is large, the tones perceptually segregate into two lower rate sequences, comprising the high- and the low-frequency tones, respectively. For this and similar tone sequences, N1 evoked by the tones changes with increasing frequency separation in a manner predictable from the perceived rate and typical AEP rate dependencies described earlier (see the section titled 'Dependencies on stimulus parameters'). In particular, N1 is greater for large frequency separations (perceived rate is low), and less for smaller ones (perceived rate is high). Especially interesting are the changes in N1 that occur when there is a change in percept without a corresponding change in physical stimulus. Such changes can be seen using tone sequences with an intermediate frequency separation, chosen such that the percept spontaneously fluctuates between 'one high-rate stream' and 'two low-rate streams.' In experiments selectively averaging N1 during each of these two perceptual conditions, N1 (or more specifically, its AEMF analog) was found to be greater during the perception of the low-rate streams. In other words, N1 was again greater during the perception of low-, as compared to high-rate sounds, but this time the difference in N1 occurred in the absence of physical alterations to the stimulus. The implication is that some of the neural activity underlying N1 specifically encodes the percept elicited by sound, rather than physical attributes of sound.

These examples are two of many in which auditory evoked responses have been used to identify, and study, the neural activity leading to human sound perception.

See also: Auditory Cortex Structure and Circuitry; Auditory System: Central Pathways; Evoked Potentials: Recording Methods; Sound Localization: Neural Mechanisms.

Further Reading

Burkard RF, Don M, and Eggermont JJ (2006) *Auditory Evoked Potentials: Basic Principles and Clinical Application.* Philadelphia: Lippincott Williams & Wilkins.

Butler RA (1968) Effect of changes in stimulus frequency and intensity on habituation of the human vertex potential. *Journal of the Acoustical Society of America* 44: 945–950.

Cohen D and Cuffin BN (1983) Demonstration of useful differences between magnetoencephalogram and electroencephalogram. *Electroencephalography and Clinical Neurophysiology* 56: 38–51.

Ridgway SH, Bullock TH, Carder DA, et al. (1981) Auditory brainstem response in dolphins. *Proceedings of the National Academy of Sciences of the United States of America* 78: 1943–1947.

Dale A and Halgren E (2001) Spatiotemporal mapping of brain activity by integration of multiple imaging modalities. *Current Opinion in Neurobiology* 11: 202–208.

Furst M, Levine RA, and McGaffigan PM (1985) Click lateralization is related to the B component of the dichotic brainstem auditory evoked potentials of human subjects. *Journal of the Acoustical Society of America* 78: 1644–1651.

Gutschalk A, Micheyl C, Melcher JR, et al. (2005) Neuromagnetic correlates of streaming in human auditory cortex. *Journal of Neuroscience* 25: 5382–5388.

Hall JW III (1992) *Handbook of Auditory Evoked Responses.* Boston: Allyn and Bacon.

Hari R (1990) The neuromagnetic method in the study of the human auditory cortex. In: Grandori P, Hoke M, and Romani GL (eds.) *Auditory Evoked Magnetic Fields and Electric Potentials*, pp. 222–282. Basel: Karger.

Llinás R and Nicholson C (1974) Section IV. Analysis of field potentials in the central nervous system. In: Rémond A (ed.) *Handbook of Electroencephalograpy and Clinical Neurophysiology*, vol. 2B, pp. 61–83. Amsterdam: Elsevier.

Melcher JR and Kiang NYS (1996) Generators of the brainstem auditory evoked potential in cat III: Identified cell populations. *Healing Research* 93: 52–71.

Melcher JR (1996) Cellular generators of the binaural difference potential in cat. *Hearing Research* 95: 144–160.

Näätänen R (1995) The mismatch negativity: A powerful tool for cognitive neuroscience. *Ear & Hearing* 16: 6–18.

Näätänen R and Picton T (1987) The N1 wave of the human electric and magnetic response to sound: A review and an analysis of the component structure. *Psychophysiology* 24: 375–425.

Nunez P and Srinivasan R (2005) *Electric Fields of the Brain*, 2nd edn. New York: Oxford University Press.

Picton TW, McEvoy LK, and Champagne SC (1991) Human evoked potentials and the lateralization of a sound. *Acta Otolaryngology (Stockholm)* 491(supplement): 139–144.

Auditory Localization

M Konishi, California Institute of Technology,
Pasadena, CA, USA

Basic Concepts and Terms

Complex sounds consist of simple 'spectral' components, which are represented by sine or cosine functions. Different spectral components are distinguished by frequency, amplitude, and phase (**Figure 1**). Amplitude is also referred to as intensity or level in psychoacoustic literature. Phase is a circular variable ranging from 0° to 360°. The time to go from 0° to 360° is called the period (T). The number of periods in a second is called frequency (F), so that $F = T^{-1}$. The mammalian and avian inner ear breaks down complex sounds into their frequency components, and primary auditory neurons carry information about the amplitude and phase of each frequency band. The narrow range of frequency to which a neuron responds is not coded by the number of nerve impulses *per se* but by its site of innervation on the inner ear basilar membrane. The sound intensity of each frequency band is coded by the rate of nerve impulses emitted by each neuron tuned to that frequency range. The phase of each frequency is coded by 'phase locking,' in which each primary auditory neuron fires impulses at or near a specific phase angle, such as 90° (**Figure 2**).

Many animals can determine the direction from which sound signals arrive. The two main cues for sound localization are the interaural time difference (ITD) and the interaural level difference (ILD). An ITD results when a particular point or phase angle of a sound wave reaches one (near) ear earlier than the other (far) ear (**Figure 3**). When a source is located directly in front, sound reaches both ears simultaneously. The ITD is zero in this case. As the source moves to one side, sound arrives earlier at the ear on that side and later at the other ear. Such a difference can occur in different aspects of sound waves, including sound onset times, phase angles, and points on the pattern of amplitude modulation (cf. **Figure 2**). Of these potential cues, the use of phase differences by humans and animals has been most extensively studied. Note that the same ITD corresponds to different binaural phase disparities in different frequency channels. An ILD results from a greater loss of sound energy by distance, absorption, and reflection en route to the far ear than to the near ear.

Behavioral Studies of Auditory Localization

The simplest way to test one's ability to localize sounds is to use earphones. When the ITD is zero in signals delivered by earphones, humans perceive not two but a single phantom source at the midpoint between the ears or sometimes straight ahead outside the head. The brain fuses the two separate signals into one internal representation. As the ITD is varied, the source appears to shift toward the ear at which the signal or waveform arrives earlier. Humans are also very sensitive to differences in the quality of sound signals between the two ears. The clarity and size of a perceived phantom source depend on the degree of similarity between the sounds. Similarity is measured by the degree of correlation between the sounds in the two ears. A sound signal on a television screen looks like a complex curve across the screen. If the same curve is drawn twice, two curves obviously match point by point, that is, they are perfectly 'correlated.' Binaural correlation is 1 when the signal waveform is identical in the two ears. Addition of random noise to correlated signals reduces the degree of binaural correlation. The amount of difference between correlated signals and random noise determines the degree of correlation. Partially decorrelated signals produce a blurred source, and the blur increases as the signals are further decorrelated. In one series of experiments, human participants used an electronic delay device to bring a sound source to the midpoint between the ears from an arbitrary initial locus. As the degree of correlation became smaller than about 0.3, the participants' choice of delays became more variable; that is, the standard deviations of the mean delays chosen by the participants increased.

Finding an animal model for human sound localization has not been easy, because most laboratory animals are difficult to train to perform perceptual tests such as those mentioned above. The barn owl is an ideal animal for the study of sound localization, however, because both its brain and its behavior are organized to perform this task. The owl uses its ears to pinpoint prey in total darkness. Owls listening through specially made earphones rapidly turn their heads in the direction that is predictable from the ITD contained in the signals. Thus, owls also perceive a single phantom source from separate sounds delivered to the two ears. Tests with partially decorrelated noise signals in owls also showed that the standard deviations of mean localization scores remained almost

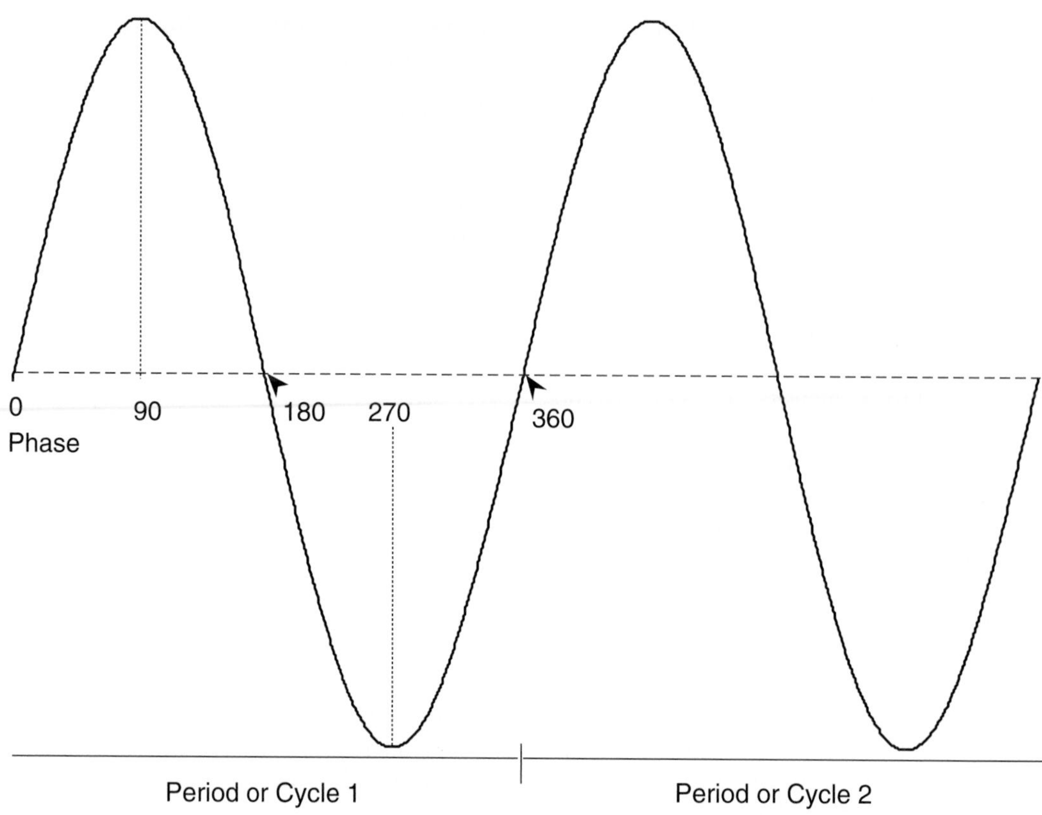

Figure 1 Simple acoustic terms. Complex sounds such as human voice consist of simple components, which can be represented by sine and cosine waves. The amplitude of a sine wave such as shown in this figure varies periodically as a function of time, which is expressed in terms of phase, because the amplitude changes cyclically. The period is the time to go from 0° to 360°. The number of periods or cycles per second is called frequency.

constant until the degree of correlation decreased to about 0.3. The curves for the standard deviations closely resembled those obtained for humans. The necessity for binaural correlation suggests that the brain needs it to compute the ITD from waveforms (phase) instead of signal onset times (cf. **Figure 2**). The theory of binaural cross-correlation assumes that the human brain performs a process similar to mathematical cross-correlation to compute the ITD. The results obtained in owls are also consistent with this theory.

A simple relationship between ITD and location holds when signal bandwidth is broad. With narrow-band signals such as tones, both humans and owls may perceive sounds coming from directions other than those of real sources. This phenomenon is called 'phase ambiguity.' If the sound incidence angles encoded by ITDs are known, phase-ambiguous ITDs are predictable from the following relationship: the true ITD and phase-ambiguous ones ITD $\pm nT$, where T is the period of the stimulus tone and n is an integer. Thus, if the tonal frequency is 8 kHz and $n = 1$, T is $1/8000 = 125$ μs. Because the range of ITDs that barn owls experience is from 0 to about 150 μs, they have to distinguish only 0 from 125 μs because its multiples ($n = 2$, etc.) on both sides of the midline are outside the 0–150 μs range. Neither humans nor owls perceive illusory sound sources when signal bandwidth is broader than a certain value, which is 300 Hz in humans and 3 kHz in owls. This difference is correlated with the frequency range in which each species can use the ITD for localization. Owls can use the ITD for sound localization at frequencies as high as 8.5 kHz, whereas the highest frequency for humans is about 1200 Hz.

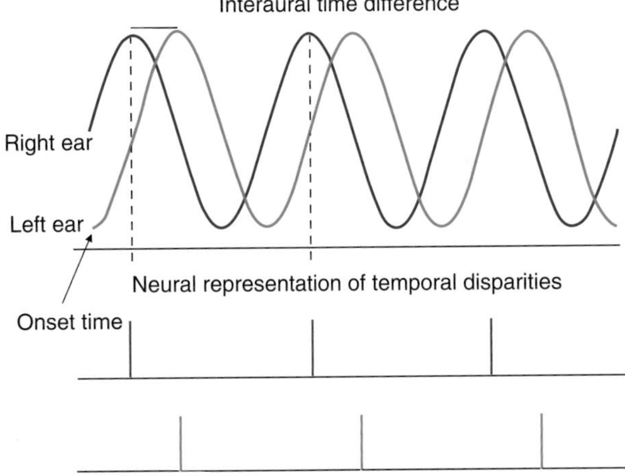

Figure 2 Interaural time difference and phase locking. Phase or time differences between the two ears are important cues for sound localization. Primary auditory neurons of many animals produce impulses at or near a particular phase angle. This phenomenon is called phase locking. This figure shows a highly simplified example in which each neuron fires exactly at 90°. The horizontal lines indicate time, and the vertical bars on them represent nerve impulses. Since a single neuron cannot fire more than about 800 impulses per second, the phase of a high-frequency tone is thought to be encoded by multiple neurons. The height of waves is called amplitude. Differences in amplitude between the two ears, as in this figure, are also available for sound localization.

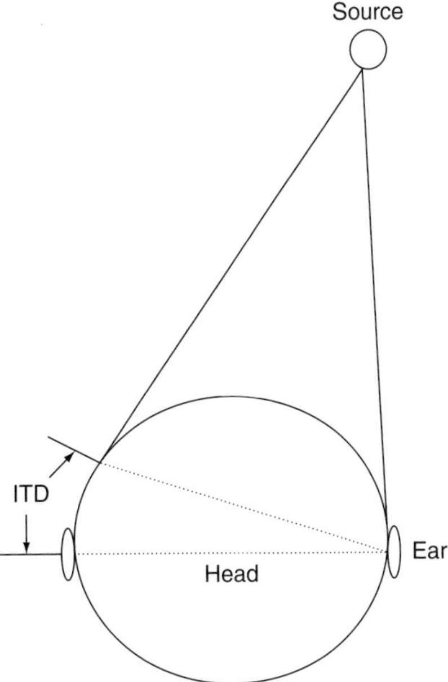

Figure 3 Interaural time differences. An interaural time difference results when a particular point of a sound wave reaches one (near) ear earlier than the other (far) ear. The reference point can be a sound onset time, a phase angle, or a point on the pattern of amplitude modulation.

Humans can use both the ITD and the ILD for localization in the horizontal plane, but barn owls use the ILD for the vertical plane. Owls do not have any external ear, but the feather ruff that surrounds their face serves as a sound-collecting device. The facial ruff consists of left and right halves that are separated by a tall ridge of feathers along the midline of the face. In barn owls, the left ear opening is located higher in the ruff than the right one. This asymmetry and other differences between the left and right halves of the ruff make the left and right ears more sensitive to sound coming from below and above eye level, respectively. This difference in the directional sensitivity of the ear causes the ILD to vary as a function of source elevation. In owl species with symmetrical ears, the ILD does not change along the vertical plane. Given a constant size of the head and sound collectors, shorter wavelengths and thus higher frequencies produce greater intensity differences than do lower frequencies. The barn owl needs relatively high frequencies (5–8 kHz) for accurate localization in elevation. Thus, to localize sound in two dimensions, the owl must be able to derive both interaural cues from the same high-frequency signal. Mammals including humans, which use both cues for azimuth, detect phase-based ITD only in low-frequency ranges and ILD in high-frequency ranges. Humans can also use monaural spectral cues for localization in the vertical plane.

Parallel Brain Pathways for Processing ITD and ILD

The owl's auditory system processes the two cues in separate pathways, and a similar design has been suggested for humans (**Figure 4**). Birds have two anatomically separate hindbrain areas on each side called nucleus angularis and nucleus magnocellularis. Each primary auditory fiber divides into two branches, one projecting to nucleus angularis and the other to nucleus magnocellularis. Neurons of nucleus magnocellularis 'phase lock' to each spectral component of complex sounds, whereas those of nucleus angularis generally do not. Thus, each cochlear nucleus is more specialized to process one cue than the other cue. The two pathways that start from the cochlear nuclei are anatomically separate until they converge in the lateral shell of the inferior colliculus (cf. **Figure 4**).

Detection and Coding of the ITD

Temporal disparities between left and right trains of phase-locked impulses provide the data for extracting

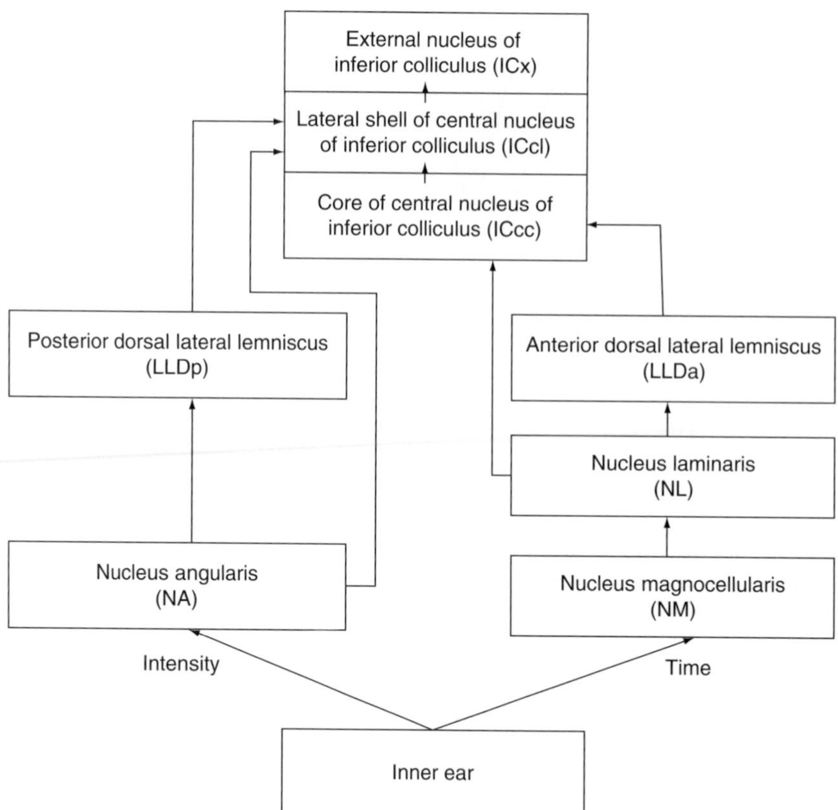

Figure 4 Parallel pathways for time and intensity. This figure shows only the brain areas whose role in sound localization is known and the connections that indicate ipsilateral or contralateral projections. The owl's auditory system processes interaural time differences (ITDs) and interaural level differences (ILDs) in separate parallel pathways. Each primary auditory fiber (nVIII) divides into two branches; one innervates cochlear nucleus angularis (NA) and the other, cochlear nucleus magnocellularis (NM). These two nuclei are the starting stations for the two pathways. Nucleus laminaris (NL), which receives inputs from the left and right magnocellular nuclei, is the first site for processing ITD. NL projects contralaterally to one of the lemniscal nuclei (LLDa) and the core of the central nucleus of the inferior colliculus. The core projects to the lateral shell (LS) on the contralateral side, where the two pathways meet. Different frequency bands converge on single neurons in the external nucleus of the inferior colliculus (ICx). This area contains a map of auditory space which is composed of neurons selective for combinations of ITD and ILD.

ITDs (cf. **Figure 2**). A model for the detection of ITDs uses delay lines and an array of coincidence detectors to measure the degree of match between the two trains (**Figure 5**). Loyd Jeffress was the first to propose this idea, in 1949, when little was known about the neurophysiology of the auditory system. He reasoned that the place of coincidence detectors in the array encodes the direction of sound sources. This model involves a transformation of codes from time to place, meaning that neither the rate nor the pattern of firing by single neurons encodes ITDs, although somewhere there must be mechanisms to compare the rates of discharge along the array. In the barn owl's auditory system, the axons from nucleus magnocellularis serve as delay lines, and the somata of nucleus laminaris serve as coincidence detectors (cf. **Figure 5**). The magnocellular axons convey phase-locked impulses to an array of laminaris cells with delays. These delays increase in the opposite direction for the ipsilateral and contrateral inputs because the length of axonal paths for the two sides increases toward the opposite ends of the array. When impulses travel to a laminaris cell by a longer path on one side than on the other side, they reach the cell at different times. In general, a left–right difference in impulse arrival times (Ta) consists of an ITD in the stimulus and a difference in impulse conduction delays (Tc). Hence, $\mathrm{Ta} = \mathrm{ITD} + \mathrm{Tc}$. Coincidence means $\mathrm{Ta} = 0$, which is obtained when $-\mathrm{ITD} = \mathrm{Tc}$; that is, ITD equals Tc in magnitude but is opposite in sign. Because the length of axonal paths varies systematically in both left and right inputs, Tc should also vary regularly. This argument is consistent with the finding that neurons sensitive to different ITDs are systematically arranged in the owl's nucleus laminaris. These neurons respond mostly to sounds coming from the contralateral side, as indicated in **Figure 5**.

Laminaris neurons are, however, not all-or-nothing coincidence detectors, for their responses decrease gradually from maximum to minimum as stimulus

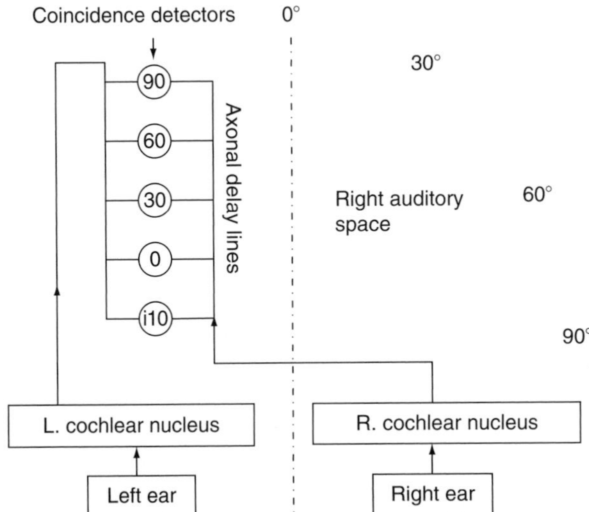

Figure 5 Neural encoding of interaural time differences in owls. The neural network that detects and encodes interaural time differences consists of cochlear nucleus (magnocelluaris) and nucleus laminaris on each side of the brain. The network consists of delay lines, which are axons of magnocellular cells, and coincidence detectors, which are laminaris neurons, as shown by 0, 30, and so forth, in circles. The coincidence detectors fire when impulses from the two sides arrive simultaneously. The length of axons to the coincidence detectors varies systematically from one end of the neuron array to the other. The longer the axon, the longer the time for nerve impulses to travel. The right and left sets of axons provide every possible pair of conduction delays, which range from 0 μs to about 150 μs in barn owls. An ITD of 90 μs, right side leading, must be delayed by 90 μs so that the neural impulses reach coincidence detectors simultaneously. In the owl, the left set of coincidence detectors encodes mostly sound locations on the right side. Thus, neuron 60 in the figure responds to sounds coming from 60° (corresponding to ~120 μs) on the right side. Jeffress used the term 'place coding' to emphasize the point that neither the number of nerve impulses nor the temporal pattern of their discharge encodes ITDs but the place of the coincidence detectors in the array. Therefore, if the neuron at 0 fires, the individual perceives a sound source straight ahead.

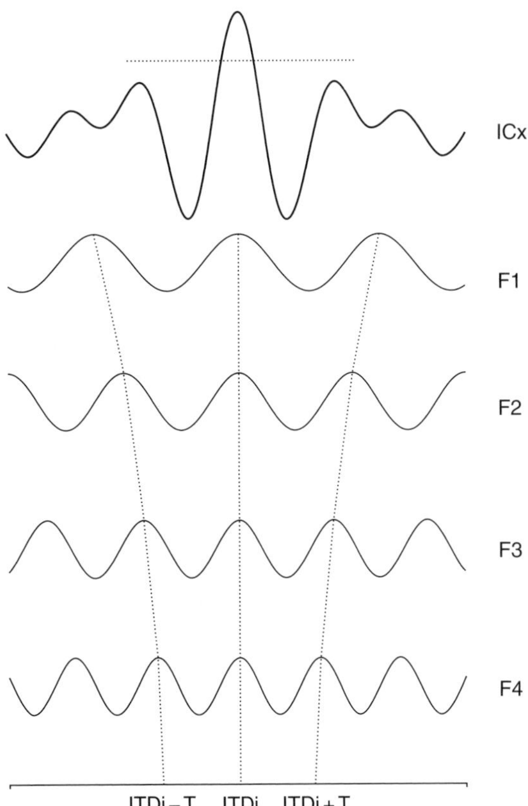

Figure 6 Detection of frequency-independent interaural time difference (ITDi). Single neurons in all stations below the external nucleus of the inferior colliculus (ICx) show phase ambiguity (responding to ITDi ± *nT*, where *T* is the period of the stimulus tone and n is an integer). Although ICx neurons also show phase-ambiguous responses to ITD carried by tones or narrow-band signals, they respond to a single ITD carried by broadband signals such as noises. An ICx neuron receives inputs from contralateral lateral shell of the central nucleus of the inferior colliculus (ICcl) neurons tuned to the same ITD from different frequency channels. The peak representing this ITD is the highest when ITD curves for different frequencies (F1, F2, etc.) are aligned and summed. This process can be seen by recording postsynaptic potential from an ITD neuron. The firing threshold (horizontal dotted line) of an ICx neuron appears to be set such that it is crossed only by the highest peak representing the frequency-independent ITD.

timing changes from perfect coincidence to out of phase by 180°. Laminaris neurons respond best to a single interaural time difference (ITDi) which is independent of frequency and its phase equivalents (ITDp). The relationship between these two variables is ITDp = ITDi ± nT, where T is the period of the stimulus tone and n is an integer. This phenomenon is the neural equivalent of phase ambiguity. Laminaris neurons fire maximally when they receive binaurally synchronous trains of phase-locked impulses. However, the coincidence of the two trains recurs every time ITD is changed by an integer multiple of the period of the stimulus tone, causing the laminaris neurons to discharge maximally. Nucleus laminaris projects both to the core of the central nucleus of the inferior colliculus (ICc) and to one of the lemniscal nuclei (the anterior dorsal lateral lemniscus (LLDa); cf. **Figure 4**). The core projects to the contralateral lateral shell of the central nucleus of the inferior colliculus (ICcl).

Resolution of Phase Ambiguity

Human psychoacousticians proposed a scheme that would solve the problem of phase ambiguity. ITD is independent of frequency, whereas ITD ± T varies with frequency because T changes with frequency. Therefore, comparison of ITD responses of neurons across different frequency bands should discriminate between frequency-dependent and -independent ITDs. How this comparison is carried out in the human brain has been a subject of debate. In the owl, one can observe the process at the level of single neurons in the inferior colliculus (**Figure 6**). In the external nucleus (ICx), different frequency bands carrying ITD information converge on single neurons. We can explain how the convergence of different frequency bands results in the identification of the frequency-independent ITD. When ICx neurons receive signals from lateral shell neurons tuned to different frequencies, they gather input from those tuned to the same ITD. However, because the afferent neurons respond to ITD $\pm T$, they confer this property on the ICx neurons. When the ICx neuron adds the inputs (postsynaptic potentials) across all frequency bands, the peaks at the frequency-independent ITD add, whereas the peaks and troughs of F1–Fn partially cancel each other out. The ICx neuron fires impulses only when the sum of the inputs crosses the threshold, as shown by the horizontal dotted line in **Figure 6**. This linear summation across frequencies is not the only method that the owl's auditory system uses. Evidence indicates that nonlinear integration such as inhibition is also involved.

Processing of ILDs

The nucleus angularis, the first station in the intensity-processing pathway, projects to a lemniscal nucleus called the posterior dorsal lateral lemniscus (LLDp), in which all neurons are excited by sound in the contralateral ear and inhibited by sound in the ipsilateral ear (**Figure 4**). The excitatory input comes from the contralateral nucleus angularis, and the inhibitory input from the other LLDp. The louder the sound in the contralateral ear is, the greater the response, whereas the louder the sound in the ipsilateral ear is, the greater is the degree of inhibition. Thus, the response of an LLDp neuron varies as a function of the difference in sound level between the two sides, even though the neuron still responds to monaural stimulation of the contralateral ear alone. Furthermore, the threshold and degree of inhibition change systematically from the dorsal to ventral direction of LLDp. This gradient suggests a map of ILDs in this nucleus.

Convergence of Time and Intensity Pathways

The lateral shell of the inferior colliculus (LS) is the site of convergence between the time and intensity pathways (**Figure 4**). The LS receives direct inputs from nucleus angularis and LLDp and indirect inputs from nucleus laminaris and LLDa via the core of the inferior colliculus. LS neurons respond to combinations of ITD and ILD within a narrow range of frequencies, indicating that the convergence of the two pathways occurs initially in each frequency band. The LS projects to the adjoining external nucleus of the inferior colliculus (ICx), where neurons respond to broad frequency ranges (3–9 kHz). ICx neurons do not respond to ITD or ILD alone but require specific combinations of the two cues, and unfavorable combinations may induce inhibition.

A Map of Auditory Space

The tuning of ICx neurons to specific pairs of ITD and ILD makes these cells selective for the direction of sound propagation because such combinations are obtained only when sound comes from particular areas in an owl's auditory space. These areas are called receptive fields, and cells with such areas are called space-specific neurons. ICx neurons are systematically arranged according to the location of their receptive fields, thus forming a neural map of auditory space (**Figure 7**). The topographical representation of auditory space is due, not to anatomical projections of the inner ear, as with the retinotopic map of the visual field, but to central synthetic processes. Nevertheless, the auditory space map projects onto the retinotopic map of the optic tectum. The tectal neurons making up the bimodal map respond to both auditory and visual stimuli coming from the same area in space; that is, their auditory and visual receptive fields are mutually aligned. Similar maps are found in the superior colliculus of the ferret, although the mechanisms underlying these maps appear to differ from those in owls.

Owl species with symmetrical ears also have space-specific neurons in their ICx, but these cells map only azimuth, because their receptive fields do not have elevational boundaries, although the cells are tuned to combinations of ITD and ILD. In these owls, the ILD does not change as a function of sound elevation, because the highest frequency audible to them is too low for this change to occur.

A Motor Map for Sound Localization

The overt behavior that characterizes sound localization in owls is rapid head rotation. The bimodal map of space in the optic tectum projects to a motor map for gaze control, which is found in the optic lobe. Neurons of the motor map innervate the medullary nucleus involved in the control of head rotation. The tectal motor map encodes the direction, speed, and amplitude of head rotations. Experiments reveal that the map contains separate circuits for head rotations in azimuth and elevation.

Another new development in the study of sound localization in owls concerns the role of the forebrain

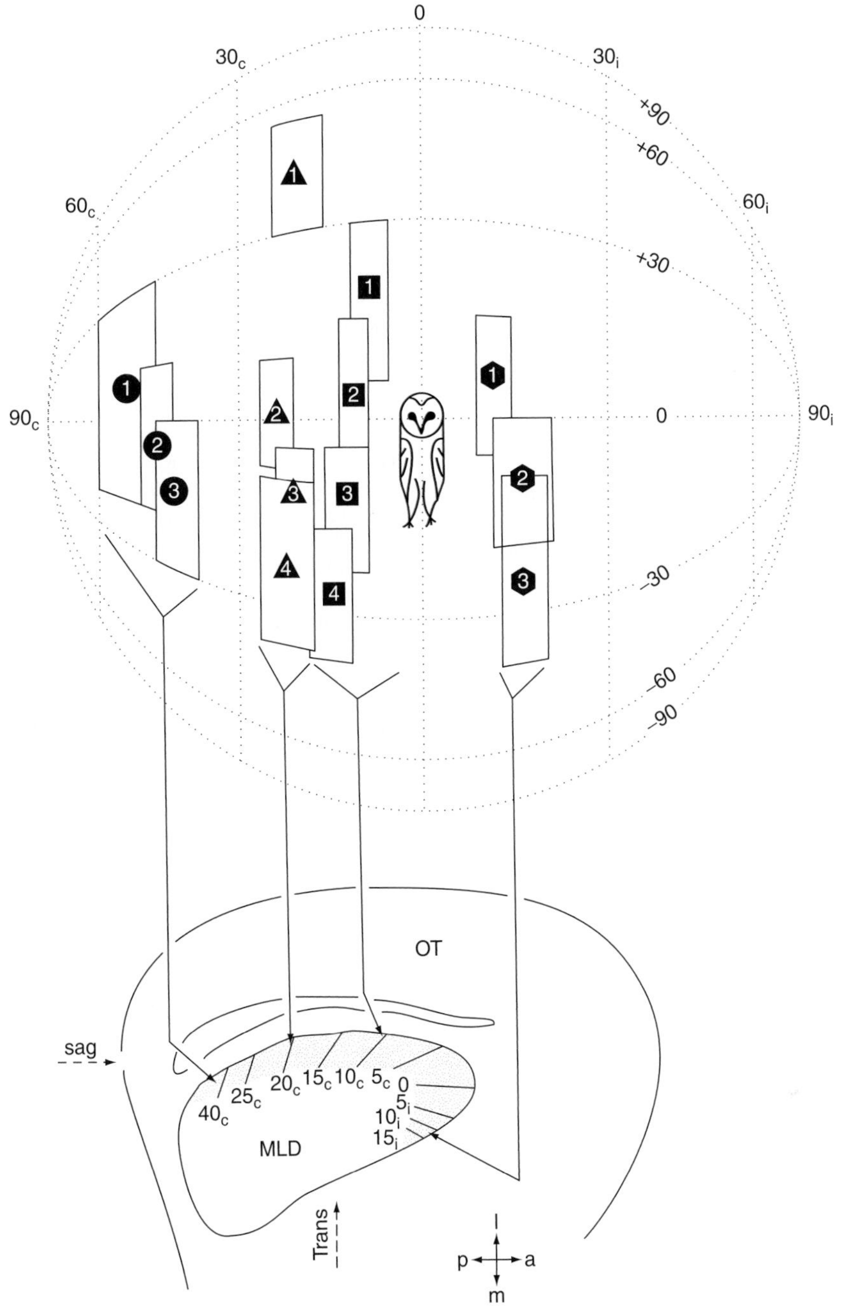

Figure 7 A neural map of auditory space. Space-specific neurons have spatial receptive fields the centers of which are shown by rectangular boxes with numbers indicating the sequences in which the neurons were recorded during a single electrode penetration. The shaded area in the bottom panel is the external nucleus of the inferior colliculus (ICx). The map represents mostly the contralateral (c) side of the auditory space, although a small extension to the ipsilateral (i) side is shown. The auditory space map is not due to topographic projections of sensory surfaces such as the body surface, the retina, and the cochlea but is created by means of neural computations. Nevertheless, the owl's auditory map projects to the optic tectum (OT) to join the topographically made map of the retina to form a bimodal map of space in which each neuron responds to both light and sound coming from the same direction.

in this behavior. Lesions of the auditory space map affect sound localization for a while, but the owl's ability to localize sound persists. Ablations of both the map and the nucleus ovoidalis, the avian homolog of the medial geniculate body, completely abolish the ability of owls to localize sound. The nucleus ovoidalis receives inputs from all parts of the inferior colliculus except ICx. Many ovoidalis neurons are tuned to combinations of ITD and ILD. Nucleus ovoidalis projects to the forebrain auditory area field L, which in turn projects to the archistriatum (new term: acropallium). This structure contains many neurons tuned to combinations of ITD and ILD. But neither field L nor archistriatum neurons form an auditory space map. Focal electrical stimulation of the archistriatum induces head rotations even when the tectal motor map is disabled. It is interesting that lesions of the archistriatum cause owls to turn their head toward sound sources like an automaton. Barn owls determine and memorize sound locations before orienting toward them. When the archistriatum of barn owls is inactivated with a drug, the owls can no longer remember the computed sound location, even though they can still localize the sound source if it continues to emit the signal. Thus, one of the functions of the owl's archistriatum may be storage of working memory.

See also: Auditory Scene Analysis; Auditory System: Central Pathway Plasticity; Auditory System: Giant Synaptic Terminals, Endbulbs and Calyces; Auditory Cortex: Models; Localizing Signal Sources; Sound Localization: Neural Mechanisms.

Further Reading

Blauert J and Lindemann W (1986) Spatial mapping of intercranial auditory events for various degrees of interaural coherence. *Journal of the Acoustical Society of America* 79: 806–813.

Carr CE and Konishi M (1990) A circuit for detection of interarual time differences in the brain stem of the barn owl. *Journal of Neuroscience* 10: 3227–3246.

Du Lac S and Knudsen IE (1990) Neural maps of head movement vector and speed in the optic tectum of the barn owl. *Journal of Neurophysiology* 63: 131–149.

Jeffress LA (1949) A place theory of sound localization. *Journal of Comparative and Physiological Psychology* 41: 35–39.

Jeffress LA, Blodgett HC, and Deatherage BH (1962) Effect of interaural correlation on the precision of centering a noise. *Journal of the Acoustical Society of America* 32: 1122–1123.

Knudsen IE (2002) Instructional learning in the auditory localization pathway of the barn owl. *Nature* 417: 322–328.

Konishi M (2000) Study of sound localization by owls and its relevance to humans. *Comparative Biochemistry and Physiology. A, Comparative Physiology* 126: 459–469.

Konishi M (2003) Coding of auditory space. *Annual Review of Neuroscience* 26: 31–55.

Peña JL and Konishi M (2001) Auditory spatial receptive fields created by multiplication. *Science* 292: 249–252.

Saberi K, Takahashi Y, Farahbod H, and Konishi M (1999) Neural bases of an auditory illusion and its elimination in owls. *Nature Neuroscience* 2: 656–659.

Saberi K, Takahashi Y, Konishi M, Albeck Y, Arthur BJ, and Farahbod H (1998) Effects of interaural decorrelation on neural and behavioral detection of spatial cues. *Neuron* 21: 789–798.

Stern RM and Trahiotis T (1997) Models of binaural perception. In: Gilkey RH and Anderson TR (eds.) *Binaural and Spatial Hearing*, pp. 499–531. Mahwah, NJ: Erlbaum.

Takahashi T and Konishi M (1986) Selectivity for interaural time difference in the owl's midbrain. *Journal of Neuroscience* 6: 3413–3422.

Takahashi TT and Konishi M (1988) Projections of the cochlear nuclei and nucleus laminaris to the inferior colliculus of the barn owl. *Journal of Comparative Neurology* 274: 190–211.

Takahashi TT and Konishi M (1988) Projections of nucleus angularis and nucleus laminaris to the lateral lemniscal nuclear complex of the barn owl. *Journal of Comparative Neurology* 274: 212–238.

Volman SF and Konishi M (1990) Comparative physiology of sound localization in four species of owls. *Brain, Behavior and Evolution* 36: 196–215.

Auditory Scene Analysis

A S Bregman, McGill University, Montreal, QC, Canada

The Scene Analysis Problem

In the natural world that surrounds us, and in which we evolved, it is rarely the case that only one sound is present at a time. What reaches our ears is a mixture of all the sounds present at a given moment: a voice speaking, the recorded music in the background, a car passing by, a bird singing, etc. If you ask people how they can hear an individual sound in this mixture, they say that they just focus their attention on whatever they want to listen to. This answer, however, ignores a fundamental difficulty. The mixture that reaches each of our ears is a single pressure waveform that is the moment-by-moment arithmetic sum of the pressure patterns that arise from individual events. This summed wave does not have written on it how many sounds contributed to it or how each sound is buried in it. Yet our auditory systems, and those of other animals as well, have the capacity to find the individual sounds in the mixture – a capacity called auditory scene analysis (ASA).

Every sound of finite duration can be thought of as the sum of a set of frequency components of different amplitudes and phases (a spectrum). An example of a mixture is shown in **Figure 1** in the form of a spectrogram (showing time on the x-axis, frequency on the y-axis, and the intensity at any time-by-frequency position as darkness). This representation is relevant to ASA because there is evidence that the first stage in the neural processing of an incoming acoustic signal involves analyzing its frequency composition. The ASA problem is equivalent to finding a set of spectrograms, which when superimposed and summed, gives us the observed spectrogram. This decomposition would be made easier if natural sounds were compact in frequency or in time, but generally, they are not. Think of two voices heard at the same time. Far from being compact in frequency or time, the frequency components of the two are intermingled on both dimensions.

It should be evident that there are a virtually infinite number of ways in which this spectrogram can be decomposed into two or more component spectrograms, if the decomposition is not guided by principles. But what sort of principles? Engineers have used mathematical procedures, such as principal components analysis, that solve the ASA problem decisively, but only in very restricted circumstances. Since no single computation will always solve the problem in a broad range of environments, animals are forced to work with sets of principles that are called 'heuristics' because they are very helpful but are not guaranteed to always work. These heuristics exploit regularities in the world such as the following: since it is unlikely that two unrelated sounds in the world will start at exactly the same time, if the auditory system registers a set of frequency components whose onsets are approximately simultaneous, it is highly probable that they are parts of the same sound and therefore should be grouped together.

One cannot be sure that every type of animal solves the problem in the same way. Some undoubtedly have specialized mechanisms to extract critical sounds (for detecting prey, predators, or mates in the mixture). The specialized mechanisms may well be neural circuits that act as detectors for the required information, responsive to certain time and frequency relations, and unresponsive to sounds that lack these relations. However, larger-brained animals deal with sound in much more complicated ways, and can learn about the dangers and affordances of new sounds, but only if they can extract them from mixtures. For these animals, including ourselves, it is of great value to have general methods for decomposing a mixture into its component sounds, regardless of whether the latter are familiar or not.

Primitive and Schema-Based Processes

There seem to be at least two types of brain mechanisms involved in the grouping of information. The first is a set of primitive (unlearned) bottom-up processes that is probably shared with nonhuman animals. As described by Albert Bregman, in his book *Auditory Scene Analysis*, these processes group the information by using similarities and discontinuities in the signal, similar to those described by Gestalt psychology in their study of vision.

The second type of mechanism consists of a set of brain processes (called schemas) for dealing with frequent and important patterns in the environment. They may be completely, or in part, innate, but large-brained animals such as humans can modify them and develop new ones through learning. These are the mechanisms that permit recognition of conspecific animal sounds, familiar words, melodies, etc., and probably assist in segregating these patterns from their backgrounds. Schemas are extremely numerous in humans, numbering at least in the hundreds of thousands for a particular adult (consider just the schemas for the tens of thousands of

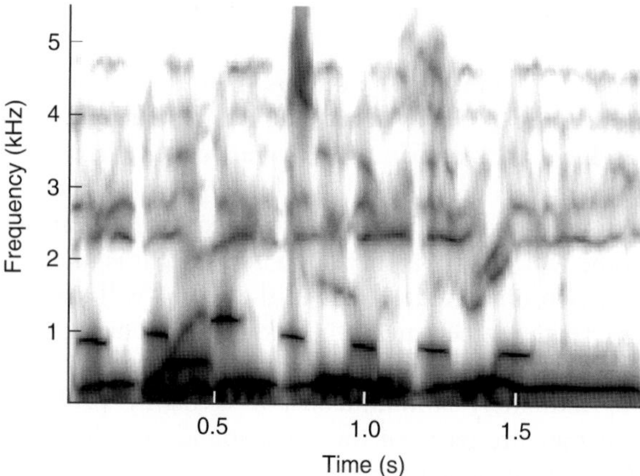

Figure 1 Spectrogram of a mixture of four sound sources: the words, "one, two, three"; a voice singing "da-da-da"; a person whistling; and a computer fan. Reproduced from Bregman AS and Woszczyk W (2003) Controlling the perceptual organization of sound. In: Greenebaum K and Barzel R (eds.) *Audio Anecdotes: Tools, Tips, and Techniques for Digital Audio*, pp. 35–66, with permission from A K Peters, Ltd.

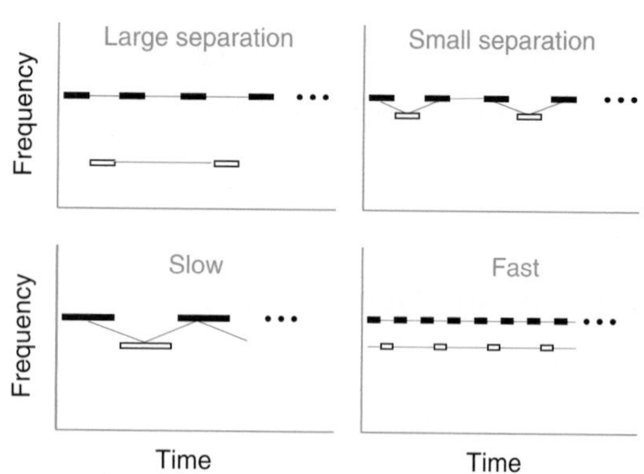

Figure 2 Diagram of four conditions in the galloping pattern of a higher tone (H) and a lower one (L) played repetitively in a galloping pattern: HLH–HLH–HLH–···. The connecting lines indicate the perceived streams. The top two panels show the effects of frequency separation; the bottom panels show the effects of speed. Reproduced from Bregman AS and Woszczyk W (2003) Controlling the perceptual organization of sound. In: Greenebaum K and Barzel R (eds.) *Audio Anecdotes: Tools, Tips, and Techniques for Digital Audio*, pp. 35–66. with permission from A K Peters, Ltd.

words that an adult can recognize). They can operate in conjunction with attention, as when we are trying to listen to a familiar voice in a crowded room, or prior to attention, as when, in that same room, our own name pops out of the background sound, attracting our attention. Despite the probable importance of schema-based perceptual processes in isolating familiar patterns from their contexts, the present discussion focuses on the primitive processes of ASA.

ASA as Grouping

The decomposition of the information in the mixed spectrogram of **Figure 1** can be viewed as a problem of grouping. This grouping has two aspects, simultaneous and sequential. Simultaneous grouping determines which parts of the complex information, presented simultaneously to the senses, should be allocated to the same description of an environmental event. One can think of this as sharing out the energy of the spectrum that is present at a given moment. In the spectrogram, this represents the grouping of energy on the *y*-dimension, and even on the *z*-dimension (intensity), since all the energy at a single frequency-by-time location may not have come from a single environmental sound.

Sequential grouping is responsible for answering the question as to which spectral components should be connected up over time (the horizontal dimension of the spectrogram) to yield a representation of a distinct train of events in the environment. The two types of groupings interact, but it is convenient, for exposition, to describe them one at a time.

Sequential Integration and Segregation

A laboratory phenomenon called stream segregation, or streaming, illustrates sequential integration. The reader can view an illustration of a pattern of sounds in **Figure 2**. The stimulus is a rapid cycle composed of two tones, a higher-frequency one (H) and a lower-frequency one (L), formed into triplets, separated by short silences: HLH–HLH–HLH–(repeated). The H and L tones are far apart in frequency. This cycle gradually speeds up. At slow rates, the HLH triplets are heard as repeating units with a galloping rhythm; but as the sequence gets faster, the high tones seem to segregate from the low ones and we hear two parallel sequences, one consisting of a sequence of high tones and the other, a slower sequence of low tones. These two streams of sound seem to be going on at the same time, yet are perceived independently. Attention seems to be enable to focus on one stream or the other, but not on both at the same time. When the high and low tones are close to one another in frequency, the perception of the gallop persists even at higher speeds, and the sequence remains as a single coherent stream. This shows the importance of frequency separation, as well as speed, in causing the H and L tones to form separate streams. These effects are illustrated in **Figure 2**. This stream formation is a form of ASA. When the streams are segregated, the system is betting that there are two sources of sound in the environment, not one. While sequential grouping of the sensory data in a complex environment may not operate in as simple a way as in the streaming phenomenon of the laboratory, it is thought that this phenomenon is a glimpse of the sequential process of ASA in a pure form. Consequently, this artificial version of ASA has been used extensively to study sequential grouping.

The streams need not be as simple as those of **Figure 2**. One can interleave the notes of a melody with distractor tones, destroying our ability to recognize it. However, if the melody and the distractors are separated in pitch range, one can hear each sequence in a separate auditory stream. Incidentally, the number of concurrent streams is not restricted to two: the upper limit is unknown. If one uses only musical notes, varying only their pitch, the limit seems to be three or four. However, if one were to add quite different sounds to the mixture, such as the clicking of a clock, the ringing of a telephone, spoken digits, and so on, the number would undoubtedly be higher.

Factors Contributing to Sequential Segregation

We can think of the tones in each panel of **Figure 2** as laid out on a two-dimensional surface (time by frequency) and can imagine that grouping occurs as a result of relative proximity on this surface. When the frequency separations are small and the time separations large (at low speeds), the time dimension – having the greater range of values – will dominate the grouping, and tones will group with their nearest temporal neighbors, regardless of the small frequency differences. However, when the temporal separations are small (high speeds) and the frequency separations large, then frequency differences will be dominant and tones will group with tones that are close in frequency. It is as if the auditory system forms clusters in the frequency-by-time space that minimize the distances within clusters and maximize the distances between them.

However, two dimensions, for example, the frequency and time separations of pure tones, are not enough to represent all possible differences between sounds. Other differences also contribute. The segregation of complex (as opposed to pure) tones can be based on (1) differences in fundamental frequency F_0 (even when the harmonics are restricted to the same frequency range); (2) differences in spectral shape, with F_0 held constant, perceived as differences in timbre; (3) differences in spatial location; (4) differences in intensity; and (5) differences in amplitude envelope (e.g., rise and fall times).

The effects of these differences combine. For example, if, in a rapid sequence of alternating A and B tones, A and B are different in two ways, say in timbre and in spatial location, they will segregate more readily than if different in only one of these ways. It is as if the best clusters are formed in a multidimensional similarity space, including time as one of its dimensions.

A separate factor that contributes to grouping is continuity. If changes from A to B in the repeating sequence ABAB... are abrupt, the A's and B's are more likely to form separate streams than if the properties of A smoothly transform into those of B. For example, if A and B differ in frequency, then frequency glides joining A and B tend to hold the tones together as a single stream of sounds.

Cumulative Effects

There is a gradual increase in the tendency of A and B to form their own streams with increased numbers of repetition of the A–B alternation. However, the alternation of two tones in a regular pattern is not a requirement for the buildup of such streams. Two sets of tones all different, but separated into two distinct frequency bands, can also segregate into separate streams. The segregation tendency dies away gradually during a silent gap and starts building up again after the silence.

It has been proposed by Stuart Anstis and Shinya Saida that the effects of repetition can be explained by the existence of frequency-transition detectors, whose function is to integrate successive tones into a single stream. Repetition of ABAB... transitions lead to the habituation of these detectors, so that they can no longer perform this integrative function. A problem with this theory is that it appears that any perceptible difference at all between two tones may promote their segregation into two streams; so the habituation theory would require the auditory system to have a very large number of types of transition detectors.

An alternative theory, functional, rather than physiological, argues that the default condition of grouping is to assign all incoming sounds to a single stream, but the repeated occurrence of tones in different frequency regions builds up evidence that the sounds are coming from two different sources and should be assigned to separate perceptual streams.

Van Noorden's Two Boundaries

Stream segregation seems to take two forms, shown by the two curves in **Figure 3**, based on the data of Leon van Noorden who studied segregation of the HLH–HLH–... pattern. The x-axis represents time between adjacent tones, while the y-axis represents the frequency separation of the H and L tones. Each of the two curves represents the time-by-frequency threshold between hearing one versus two streams. At frequency–time values above the curve one hears two streams, and below the curve one stream. However, there are two different curves. The upper

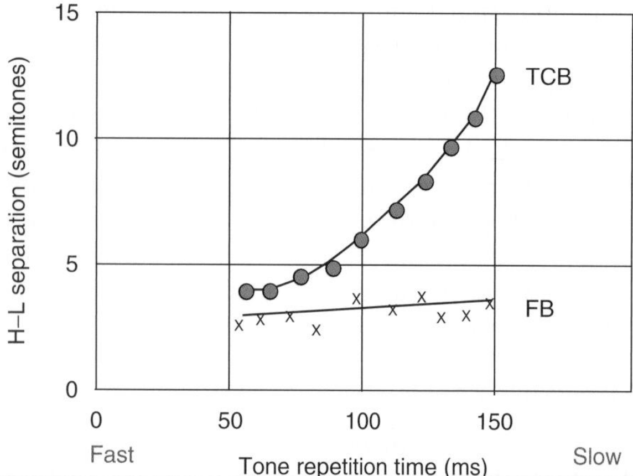

Figure 3 Curves by Van Noorden showing the thresholds for stream segregation. For each curve, the area above it shows the region (in frequency-by-time) in which two streams are heard; in the area below it, only one stream is heard. The temporal coherence boundary (TCB) is obtained when the listener is trying to hear all the tones as a single stream. The fission boundary (FB) is found when the listener is trying to hear the tones in two separate streams. The tonal pattern used was a galloping sequence of the form LHL–LHL–···, where L and H are lower- and higher-frequency tones, respectively. Adapted with the permission of LPAS Van Noorden.

one, the temporal coherence boundary (TCB), was obtained when the listeners were trying to hold onto a single-stream percept, and the lower one, labeled fission boundary (FB), was obtained when they were trying to hear two streams. Two facts are evident: the first is that when trying to hold onto a single stream they could do so at higher frequency separations and speeds than when they were not. This is not surprising. More remarkably, when they were trying to segregate the streams, the rate of presentation had a very small effect.

The different shapes of the two curves reveal the activity of two different mechanisms of segregation. When trying to hold onto all the tones as a single stream, the process that interferes with this goal is probably the primitive, bottom-up process of grouping. When trying to hear separate high and low streams, the process employed is one of selective attention, trying to hear either the high or the low tones – a top-down process, which is limited only by the listener's ability to discriminate the higher and lower tones at high speeds.

Effects of Sequential Grouping

When streams are strongly segregated from one another, the effects are numerous: (1) melodies and rhythms seem to include only within-stream sounds and (2) the temporal relations between concurrent streams become uncertain. Indeed, it appears that for any pattern of sound to be clearly perceptible, it must involve only the elements of a single stream.

Simultaneous Integration

Cues Favoring the Grouping of Simultaneous Components

One of the most important effects on the allocation of spectral energy to simultaneous sounds is 'harmonicity.' Many natural sounds including the vowel-like segments of human speech, and the pitch-possessing portions of the calls of other animals, have repetitive waveforms. So also do certain manufactured sounds, such as the sound of a violin. Repetitive waves have a pitch, and are composed of harmonics – frequencies that are integer multiples of the lowest frequency (the fundamental). The auditory system has mechanisms for detecting and grouping a subset of frequency components that are multiples of the same fundamental (a harmonic series). Furthermore, it can find more than one harmonic series at a time, allowing the listener to hear concurrent sounds that have different pitches. Both humans and computers depend strongly on harmonicity to segregate simultaneous components. For example, it is easier for us to segregate a man's voice from a woman's than from another man's, and computer models of the segregation of speech from other interfering sounds typically use the harmonic structure of the vocalic sounds as the main grounds for finding a stream of speech in the mixture.

Harmonicity is not the only basis for segregating concurrent sounds. Another important one is synchrony or asynchrony of onset. It is very likely that all the frequency components from a single environmental sound will start together, and it is very unlikely that the components of unrelated sounds will start at the same moment. So synchrony and asynchrony are used by the ASA system to determine whether or not frequency components should be allocated to the same sound. A third factor is the frequency separation of concurrent components. Components that are further apart are less likely to be treated as part of the same sound.

Another difference between components that is used by ASA is their difference in spatial location. This difference, by itself, does not powerfully group and segregate sets of concurrent components (although it is very powerful in sequential stream segregation). However, when other principles, such as asynchrony of onset, act to segregate concurrent components, spatial differences seem to multiply the strength of this segregation. It is surprising that spatial differences have such a limited effect in human ASA. Engineers use a technique called blind separation, in which the separation of voices is primarily or even exclusively based on spatial separation. However, spatial information is not always reliable in natural environments (e.g., when the sounds are coming around a corner). Perhaps this is why spatial cues are not used by humans as a primary basis for grouping simultaneous components.

Achievement of Stability in the Face of Unreliable Acoustic Evidence

It is not only spatial cues that are unreliable. While harmonicity is a good cue, many sounds are not harmonic. There is no pitch involved in footsteps, accelerating cars, scratches, or bumps; yet these sounds are informative and we need to know how to group their components across the spectrum and over time. Because even the best cues are not always reliable, the ASA system assesses many cues and allows them to reinforce, or compete with, one another in controlling the decisions about grouping (as if the various cues could vote for their own preferred organizations).

Another method that the ASA system uses because cues are less than perfectly reliable is the conservative strategy of maintaining an existing interpretation until evidence piles up that it is wrong. The system seems to start off with the hypothesis that all acoustic input is part of a single sound. As evidence builds up, an organization in terms of a number of distinct sounds emerges. Maintaining a stable percept, despite the fact that cues can suffer interference, means not altering the grouping of the sounds upon encountering a brief drop in intensity, or a momentary change in interaural spatial cues, as the listener passes behind an obstruction. If each brief glitch caused a reorganization, our perceptions would be highly unstable. Hanging on to an existing organization is a valuable strategy in a world that is more stable than the cues about it are.

What, then, are the perceptual effects of simultaneous grouping – sometimes called fusion? The most obvious one is on the number of perceived sounds and the distinctness of their qualities. When segregated, we hear more distinct sounds, each with its own qualities of pitch, timbre, loudness, location, etc., whereas when fused, the full set of frequency components create the qualities of a single global unit, which, at any given moment, has only one pitch, one loudness, one spatial location, and one timbre.

Competition between Sequential and Simultaneous Grouping

Although we have been looking separately at the two major dimensions of grouping – sequential and simultaneous (the horizontal and vertical dimensions, respectively, of a spectrogram) – it is important to recognize that each of these affects the other. Often they compete. This competition is illustrated in **Figure 4**, a simplified spectrogram whose horizontal and vertical dimensions show time and frequency, respectively. A, B, and C are three pure-tone components. First A occurs alone, followed by B and C together, and this pattern is repeated cyclically. The BC spectrum can be interpreted as a complex tone with two frequency components or else as two simple tones, B and C, that happen to occur at about the same time. One can hear the total cycle as a two-tone stream formed of A and B, repeating over and over, accompanied by a one-tone stream, consisting of repetitions of C. Alternatively, one can hear a pure tone, A, alternating with a rich tone, BC.

This is a good stimulus for demonstrating the competition of sequential and simultaneous organizations. First, we can increase the sequential grouping of A with B by moving A closer to B in frequency. This grouping not only increases the AB sequential grouping, but also weakens the BC simultaneous grouping, so that the BC spectrum is heard as less rich. Conversely, if the simultaneous grouping of B with C is manipulated by altering the BC asynchrony, not only is the fusion of B and C affected, but also the sequential grouping of A with B. As the BC fusion becomes weaker due to greater BC asynchrony, B becomes more available to form a

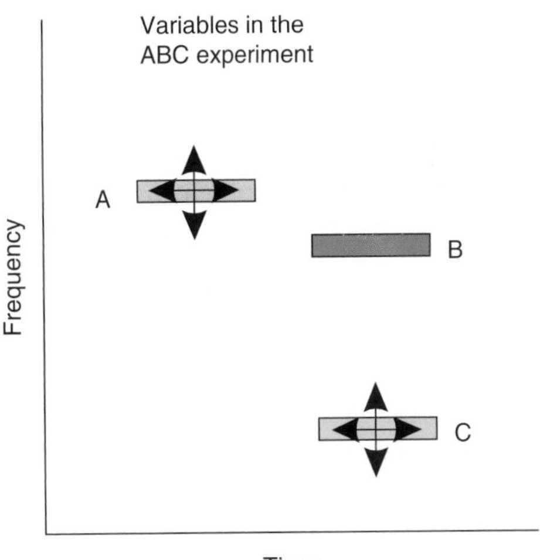

Figure 4 Diagram of a stimulus pattern in which a pure tone, A, alternates repeatedly with a pair of pure tones, B and C. The arrows show that the temporal positions and frequencies of both A and C may be varied. As A comes closer to B in frequency or time, it captures the latter more strongly into a sequential stream. As C comes closer to B in frequency or becomes more synchronous with it, it captures B more strongly into a fused perceptual unit.

sequential pure-tone stream with A. It is as if A and C were competing to make a connection with B.

The Old-Plus-New Heuristic

Another example of an interaction between simultaneous and successive organization is the old-plus-new heuristic. In natural environments, sounds do not strictly follow one another in time, nor are they exactly superimposed on one another. Instead they are typically overlapped. The old-plus-new heuristic uses the moment when a new sound enters a mixture to derive a very good description of the newly entering sound. It works as follows: when a spectrum suddenly becomes more intense or more complex, the auditory system carries out an analysis of this spectrum to determine whether the spectrum just before the change – or one very similar to it – is still present after the change. If so, it treats the changed spectrum as the sum of a continuing old spectrum plus a new sound (hence the name old-plus-new). Since it can compute the difference between the earlier and later spectra, it can derive, with some accuracy, the properties of the new sound. Note that it is not necessary that the earlier spectrum contains only a single sound, as long as no sound within it drops out at exactly the instant that the new sound begins. So the moment of onset of a new sound plays a critical role in identifying it and it would be surprising if the nervous system did not have mechanisms that were specialized for the analysis of sudden increases in spectral intensity or complexity.

Explanations

Explanations of stream segregation and other manifestations of ASA have fallen into two categories: functional and neurophysiological. The discussion up to this point has been about the job of the ASA system and which of its properties allows it to carry out its job. What cues does it use? How does it put them together? How does evidence accumulate? How do top-down and bottom-up processes interact? The concept of an ASA system is not actually a theory but a set of concepts and observations that can act as constraints on the form of such a theory.

An example of a physiological theory is one offered by Leo van Noorden and a closely related one proposed by William Hartmann and Douglas Johnson. The basic argument is that for stream segregation to occur, the members of the two streams must activate nonoverlapping populations of hair cells in the cochlea of the listener (Van Noorden), or, different frequency channels in the cochlea (Hartmann and Johnson). It is argued that the reason that streams can form on the basis of frequency differences is because each frequency has its strongest effect on a different population of hair cells. The reason that streams can form on the basis of differences in ear-of-arrival is that there are different populations of hair cells (or frequency channels) in the cochleas of the two ears.

This theory is attractive because it is simple and makes the peripheral sensing apparatus responsible for stream segregation. If true, it would open streaming to investigation by well-understood physiological techniques. Unfortunately, a number of findings fail to support this theory. The most clear-cut evidence comes from an experiment in which complex tones containing exactly the same spectral components were made to sound different in timbre and pitch by manipulating the phases of their components. When two tones that had different phase relations among their components were rapidly alternated in the galloping pattern described earlier (ABA–ABA– ···), each tone formed its own stream, despite the fact that it activated the same frequency channels within the cochlea as the other tone did.

In some trivial sense, the external sense organ is indeed responsible for stream segregation. After all, unless sounds differ in some way at the early receptive levels, the rest of the brain will never hear about it. However, this is far from saying that the neural computation that establishes the separate streams is close to the sensory periphery. One reason to believe that stream segregation, and in general, ASA, is computed higher up in the brain is that the instructions given to listeners, which activate top-down processes, can affect grouping in ambiguous cases. Another is that various auditory differences among the set of incoming sounds (e.g., pitch, spatial location, onset asynchronies) have a combined effect on the segregation of streams. Yet these cues may be provided by different sorts of feature analyzers. Bringing together the 'votes' of the different feature analyses has to be done at a level beyond the one at which the features are first detected, and at a level at which the effects of prior learning can play a role (probably in auditory cortex).

While adequate physiological explanations of ASA have not yet been forthcoming, physiological methods have been able to attack questions that are not easily answered using behavioral methods. One such question is whether the creation of auditory streams (the brain's representation of distinct environmental sounds) out of a sequence of sounds can precede the involvement of attention or requires the participation of attention. It is very hard to solve this problem using behavioral methods, because whenever human listeners have to carry out a task, such as reporting whether a particular set of sounds forms a separate stream, this task focuses their attention on the sounds. Elyse Sussman and her colleagues have studied the involvement of attention in stream formation by recording event-related potentials (ERPs) from scalp-mounted electrodes when people are exposed to sequences of sounds while their attention is distracted by a visual task. She has used a component of the ERP called mismatch negativity (MMN) to detect whether the sounds are forming separate streams. The results suggest that at least three separate streams can be formed while the listener is not paying attention to the sounds, but as soon as attention is focused on one stream, the MMN evidence for the existence of the other streams disappears. So at least in some circumstances, stream segregation may be pre-attentive.

The ERP method is particularly well suited for studying auditory organization in persons who are not capable of reporting on their experiences, such as young infants. Do newborns organize sound the way adults do? If they do not, it suggests that the ASA methods may have to be learned. István Winkler and his colleagues played sequences of sounds to sleeping newborn infants 2–5 days of age. There was clear evidence for auditory stream segregation. This supports the idea that the mechanisms of ASA are primitive, in the sense of being inborn, and can give an initial boost to the infants' early learning about the important sounds in their environments, preventing them from memorizing the accidental combinations of properties exhibited by fortuitous combinations of sounds.

ASA in Other Animals

Nonhuman just like human animals live in a world where sounds come mixed with others. Yet they must respond only to certain sounds in the mixture. An important motivation for studying ASA in nonhuman animals is that its physiological basis could be investigated. Accordingly, ASA has been studied in a number of species, including birds, bats, frogs, fish, pinnapeds, and insects. However, a word of caution is in order. ASA is an accomplishment, not a mechanism. Even if different animals succeed in partitioning the sound mixture to the extent needed to detect their particular predators, prey, or potential mates, this does not mean that each of them does it by the same mechanism that the others do or that humans do. It is probable that many species have mechanisms to extract specific features and patterns that are important to the particular type of animal involved, and probably prepare certain actions that are appropriate responses to the detected patterns. Even humans may have them: some theorists have claimed that we possess specific mechanisms for extracting speech patterns from mixtures. Another example is that bats use specific bands of acoustic energy in the echolocation of their prey and have neural mechanisms specialized in the extraction of specific patterns from the echo data.

Macaque monkeys are large-brained animals and might be expected to share the general ASA mechanisms with humans. In one study with macaques, a rapid alternation of tones of two different frequencies, A and B, was delivered to them while recordings were made from their auditory cortexes. In humans, as the rate of alternation of A and B is increased, stream segregation increases. In the macaque auditory cortex, cells that respond best to A also respond a little to B. But as the rate of A–B alternation increases, these A-sensitive cells stop responding to B. It is as if the rate increase had caused stream segregation to occur, so that the cells that responded to A no longer 'saw' B. The same finding has also been reported in bats and starlings. Of course, even if this effect really is part of the stream-segregation mechanism, it does not automatically imply that the A-sensitive cell is, by itself, responsible for stream segregation. It may merely be a point at which one of the effects of stream segregation can be detected by an outside observer.

An interesting physiologically motivated neural network model, the ARTSTREAM model of Stephen Grossberg and his colleagues, has attempted to give a foundation for ASA in terms of neural computation. The process is much more complex than the activity of feature-sensitive cells.

It is possible to conceptualize ASA as the binding together of acoustic information of various types, in order to create acoustic objects, be they sequences or single sounds. Remember that more than one acoustic object (or stream) is being formed at the same time. So the brain has to register, for example, that it is the loud sound that has the rich timbre, and the soft sound that has the purer timbre, and not the reverse. It is possible that this binding of the right features to individual sounds could be carried out by temporarily recruiting some cells, activated by all the to-be-bound features, to represent the object as a whole (sort of a Hebbian phase sequence).

However, the existence of binding need not imply the convergence of all the information about the acoustic object to a common pool of neurons. This may not be the way 'objectness' is encoded in the brain. It has been proposed by DeLiang Wang, following the approach of Christoph von der Malsberg, that individual auditory features activate oscillatory processes in the brain and that their binding occurs when the oscillations representing particular features are driven into synchrony.

Computational Auditory Scene Analysis

Wang's theory, implemented in the form of a computational model, is one of many such models, inspired by the findings about human ASA. A new field of research, known as computational auditory scene analysis (CASA), attempts to create computational models of sound segregation. Most such models focus on the strongest cues for the grouping of sensory input: spatial location and harmonicity. However, a beginning has been made by Darryl Godsmark and Guy J Brown, on a more open system that allows a multiplicity of features to influence the final organization, using a blackboard architecture.

There are important practical benefits in creating effective computational systems for ASA: computer systems have great difficulty in recognizing speech mixed with other sounds, so a system that segregated sounds in the course of recognizing them would have a greater chance of successful recognition.

Conclusions

ASA affects all perceptible features of sound. The perceived loudness, position in space, pitch, rhythm, and timbre of sounds all depend on how the sensory input is organized. As we have become increasingly aware of this fact, the topic of ASA has stimulated research in psychophysics, cognitive science, biology, neuroscience, mathematical and computational modeling, speech, hearing science, audio engineering, and the psychology of music.

See also: Auditory Cortex Structure and Circuitry; Auditory System: Efferent Systems to the Auditory Periphery; Auditory System: Central Pathways; Auditory/Somatosensory Interactions; Auditory Cortex: Models; Cochlear Mechanics; Dichotic Listening Studies of Brain Asymmetry; Language: Auditory Processes; Otoacoustic Emissions; Sensory Aging: Hearing.

Further Reading

Alain C, Reinke K, He Y, Wang C, and Lobaugh N (2005) Hearing two things at once: Neurophysiological indices of speech segregation and identification. *Journal of Cognitive Neuroscience* 17(5): 811–818.

Bee MA and Klump GM (2004) Primitive auditory stream segregation: A neurophysiological study in the songbird forebrain. *Journal of Neurophysiology* 92: 1088–1104.

Bregman AS (1990) *Auditory Scene Analysis: The Perceptual Organization of Sound.* Cambridge, MA: MIT Press (paperback 1994).

Bregman AS and Ahad P (1996) *Demonstrations of Auditory Scene Analysis: The Perceptual Organization of Sound.* Audio compact disk. Montreal: Authors (distributed by MIT Press).

Cooke MP and Brown GJ (1993) Computational auditory scene analysis: Exploiting principles of perceived continuity. *Speech Communication* 13(3–4): 391–399.

Darwin CJ and Carlyon RP (1995) Auditory grouping. In: Moore BCJ (ed.) *Hearing*, pp. 387–424. San Diego, CA: Academic Press.

Ellis DPW (1999) Using knowledge to organize sound: The prediction-driven approach to computational auditory scene analysis and its application to speech/nonspeech mixtures. *Speech Communication* 27(3–4): 281–298.

Fishman YI, Reser DH, Arezzo JC, and Steinschneider M (2001) Neural correlates of auditory stream segregation in primary auditory cortex of the awake monkey. *Hearing Research* 151: 167–187.

Grossberg S, Govindarajan KK, Wyse LL, and Cohen MA (2004) ARTSTREAM: A neural network model of auditory scene analysis and source segregation. *Neural Networks* 17(4): 511–536.

Hartmann WM and Johnson D (1991) Stream segregation and peripheral channeling. *Music Perception* 9(2): 155–184.

Hulse S (2002) Auditory scene analysis in animal communication. In: Slater PJB, Rosenblatt JS, Snowdon CT, and Roper TJ (eds.) *Advances in the Study of Behavior,* vol. 31, pp. 163–200. San Diego, CA: Academic Press.

Klump G (2005) How does the hearing system perform auditory scene analysis? In: Van Hemmen JL and Sejnowski T Jr. (eds.) *23 Problems in Systems Neuroscience,* ch. 15. Oxford: Oxford University Press.

Moss CF and Surlykke A (2001) Auditory scene analysis by echolocation in bats. *Journal of the Acoustical Society of America* 110(4): 2207–2226.

Rosenthal DF and Okuno HG (eds.) (1998) *Computational Auditory Scene Analysis.* Mahwah, NJ: Erlbaum.

Sussman E (2005) Integration and segregation in auditory scene analysis. *Journal of the Acoustical Society of America* 117(3, Pt 1): 1285–1298.

Van Noorden LPAS (1975) Temporal coherence in the perception of tone sequences. Doctoral disseration, Eindhoven University of Technology, Eindhoven, The Netherlands (available online from http://aboxandia.tue.nl/extra1/PRF2A/7707058.pdf).

Wang D (1996) Primitive auditory segregation based on oscillatory correlation. *Cognitive Science* 20: 409–456.

Wang D and Brown GJ (2006) *Computational Auditory Scene Analysis: Principles, Algorithms and Applications.* Piscataway, NJ: Wiley–IEEE Press.

Winkler I, Kushnerenko E, Horváth J, et al. (2003) Newborn infants can organize the auditory world. *Proceedings of the National Academy of Sciences of the United States of America* 100(20): 11812–11815.

Yost WA (2004) Determining an auditory scene. In: Gazzaniga MS (ed.) *The Cognitive Neurosciences,* 3rd edn., pp. 385–396. Cambridge, MA: MIT Press.

Auditory System: Central Pathway Plasticity

D R F Irvine, Monash University and Bionic Ear Institute, Melbourne, VIC, Australia

Introduction

One of the most striking discoveries in sensory neuroscience in the past 40 years has been the fact that the functional response characteristics of single neurons in sensory central nervous system (CNS) structures, and the functional organization of those structures, can be modified by changes in the organism's sensory experience. This capacity for functional change is commonly referred to as 'plasticity.' Plasticity is demonstrated experimentally by changing either the organism's sensory environment or the significance for the organism of particular stimuli in that environment, or by modifying either the ancillary structures that deliver energy to the receptors, the receptor function itself, or the transmission of information from the receptors to the CNS. The first demonstrations of such plasticity were of changes that occurred only during restricted 'critical' periods during development, and it was initially assumed that the capacity for plastic change was restricted to these periods. More recently, however, it has been established that considerable plasticity is retained in adult sensory systems. The study of this plasticity has been complemented by major changes in the way in which the receptive fields (RFs) of sensory neurons are conceptualized. The evidence for both developmental and adult plasticity in central auditory system (CAS) structures, and on the nature of the mechanisms involved, is reviewed in the following sections.

It should be emphasized that the definition of 'plasticity' as changes in the structural and functional characteristics of neurons that occur in response to altered patterns of input is intended to distinguish it from changes that occur as passive consequences of the altered input or as a direct consequence of changes in the organism's state (e.g., emotional state, age). In the auditory system, for example, changes in the frequency tuning of auditory nerve fibers and central neurons occur as a direct (or passive) consequence of destruction of the outer hair cells (and thus of the cochlear amplifier), but these changes are not manifestations of plasticity. Plasticity involves some form of dynamic change in neural properties as a consequence of the altered input. Although the distinction between passive and plastic changes as a consequence of altered input is usually easily made, there are borderline cases where the distinction is difficult.

Developmental Plasticity

Effects of Neonatal Cochlear Ablation

Neonatal ablation of one cochlea in mammals, which eliminates afferent input from that ear to the CAS, results in structural and functional changes in the projections from the intact ear to binaural neurons in the CAS. Axons from the ventral division of the cochlear nucleus (CN) normally terminate in restricted regions of the major nuclei of the superior olivary complex (SOC). Following neonatal cochlear ablation, CN neurons on the side of the intact ear project to, and make synapses on, SOC regions normally innervated by the CN on the side of the ablated cochlea. Similarly, the terminal fields in the central nucleus of the ipsilateral inferior colliculus (IC) of axons from the CN on the side of the intact ear are much larger than those in normal animals. Both of these results indicate that, following unilateral cochlear ablation, axons from the CN on the intact side sprout to innervate additional territory. These structural changes are associated with increased excitatory responses to stimulation of the intact ear in the IC and primary auditory cortex (AI) ipsilateral to that ear. Although the effects of cochlear ablation are apparently restricted to a critical period preceding the onset of hearing, there is recent evidence indicating that CN lesions after the onset of hearing result in similar structural changes in the SOC.

Changes in Maps of Auditory Space

In barn owls, and in many mammals, the deep layers of the superior colliculus (SC) (or optic tectum) contain maps of auditory and visual space. These maps are in register: bimodal neurons, or auditory and visual neurons at the same locus, have corresponding auditory and visual spatial RFs (**Figure 1(a)**). The auditory space map is derived via orderly projections from the external nucleus of the IC (ICX). It is based on neural sensitivity to interaural time and level differences (ITDs and ILDs, respectively) and, in mammals, on sensitivity to spectral cues produced by the effects of the head and pinnae on the sound field. If a sound-attenuating plug is placed in one external meatus, the values of the interaural disparities associated with any given spatial location are altered. Such ear plugging in adult animals results in the auditory and visual space maps being misaligned, and the owls fail to recover normal localization, for as long as the plug is in position. When the ear plug is removed, the maps are aligned and the owls localize with normal accuracy. However, when barn owls or ferrets are

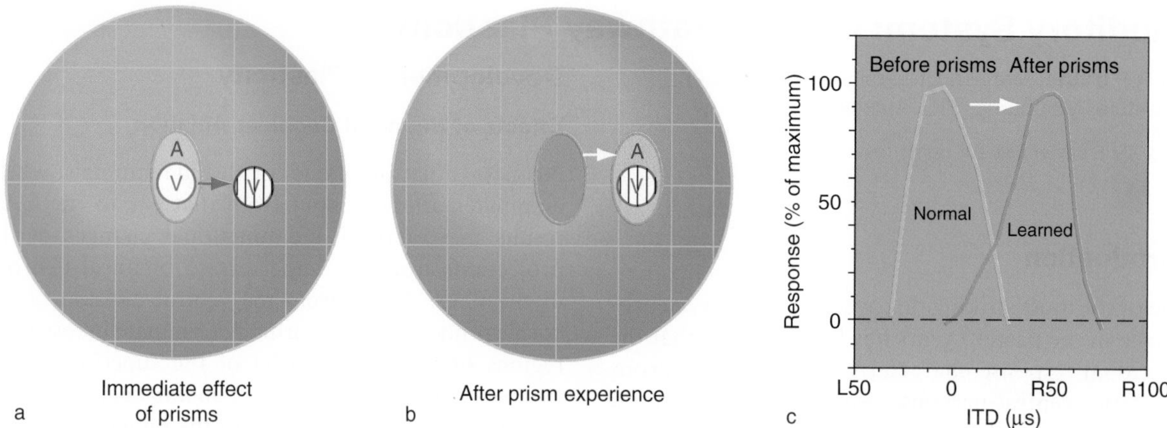

Figure 1 Plasticity of auditory tuning in the optic tectum of a juvenile barn owl, resulting from prism experience. Most neurons in the optic tectum respond to both auditory and visual stimuli and have auditory and visual receptive fields that are mutually aligned in space. (a) An example of the effect of 23° prisms on the location of a neuron's visual receptive field (encircled V; prismatically displaced field is striped). The globe represents space relative to the owl's line of sight. The auditory receptive field (A) is orange. (b) After the owl has experienced prisms for 8 weeks, the auditory receptive field has shifted to align with the prismatically displaced visual receptive field. (c) Plasticity of interaural time difference (ITD) tuning. ITD tuning of 2 units, both with visual receptive fields located at 0° azimuth, with the prisms removed (as shown in a), measured before (normal) and after (learned) 8 weeks of prism experience. On the ITD scale, L values represent left ear leading; R values represent right ear leading. Adapted from Knudsen EI (2002) Instructed learning in the auditory localization pathway of the barn owl. *Nature* 417: 322–328, with permission.

reared from infancy with one ear plugged, the visual and auditory maps in the adult are in register. This registration between the maps has been shown in barn owls to reflect the fact that the tuning to ITDs and ILDs of neurons in ICX and SC has changed to maintain alignment of the auditory and visual RFs. This change in neural tuning is associated with the appearance of novel projections from ICC to ICX: axons sprout into, and form synapses in, regions outside their normal projection zone. Immediately after ear plugging, young owls mislocalize sounds, but in parallel with the map changes they recover accurate localization ability. When the plug is removed, they exhibit systematic localization errors, and normal localization is recovered only if plug removal occurs within a critical period. Analogous developmental plasticity occurs when owls are raised wearing prismatic spectacles that displace the visual field in the horizontal plane: the tuning of SC neurons to ITDs (which serve as the cue for azimuthal location in barn owls) shifts to maintain the correspondence between auditory and prismatically displaced visual RFs (**Figure 1**). The plasticity demonstrated in these studies using experimental manipulations of sensory input is analogous to plastic changes that must occur in all animals during development, as the size of the head and pinnae increases and the nature of the ITDs, ILDs, and spectral transformations associated with particular spatial locations consequently changes.

Adult Plasticity

Plasticity Induced by Restricted Cochlear Damage

Lesions of a restricted region of the cochlea (which result in a partial hearing loss) in adult animals result in a reorganization of the frequency map in the AI. The nature of this reorganization is that the region deprived of its normal input by the cochlear lesion is occupied by an enlarged area in which neurons have characteristic frequency (CF) at the frequency represented at the edge(s) of the cochlear lesion (**Figure 2**). The thresholds, latencies, and sharpness of frequency tuning of neurons in this expanded representation at their new CF indicate that the changed frequency organization is not a passive consequence of the peripheral lesion but instead reflects a dynamic process of reorganization. Such reorganization has been described in a number of species, including nonhuman primates, and as a consequence of cochlear lesions produced by different procedures (mechanical, noise trauma, ototoxic drugs). After mechanical cochlear lesions in adults, similar reorganization is seen in the major auditory thalamic nucleus (the ventral division of the medial geniculate nucleus), but reorganization either does not occur or is weak in the ICC, suggesting that this form of CAS plasticity is a characteristic of thalamocorticothalamic circuitry. However, reorganization is seen in both the AI and the ICC of animals partially deafened neonatally (by ototoxic injections), suggesting that the

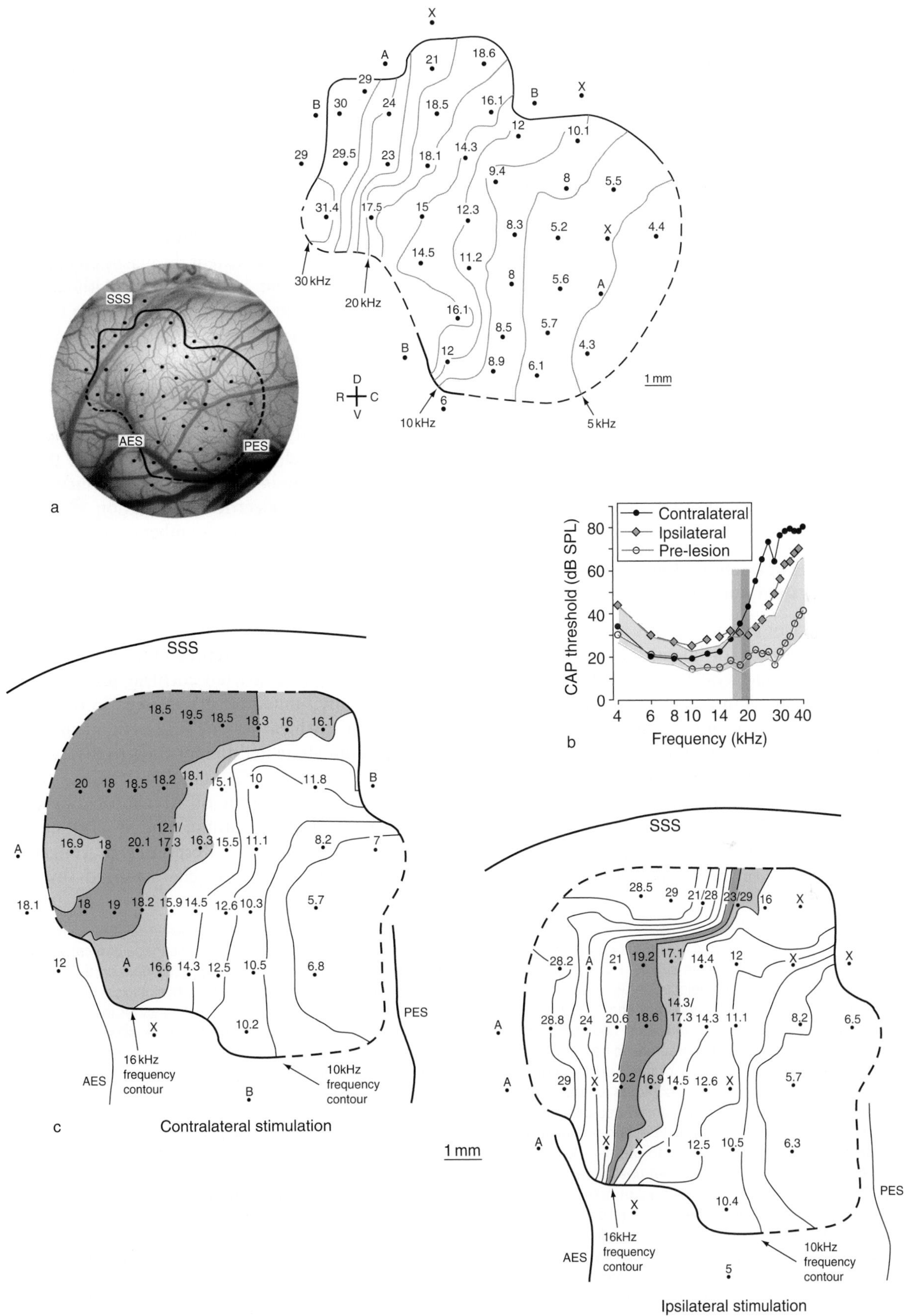

Figure 2 Continued

capacity for such reorganization is greater during development than in adults.

In all mammals, auditory cortex comprises a number of fields, commonly classified as 'core' (AI and one or more other tonotopically organized fields), and as surrounding 'belt' and 'parabelt' fields. Studies of injury-induced cortical reorganization have almost exclusively focused on the AI, with incidental observations on a second core field in some cases. The reason for this is mainly technical, in that the precise tonotopy in the core fields makes it easier to both identify and quantify reorganization. It is likely, however, that the capacity for plastic change is at least as great, and probably greater, in the less precisely organized belt and parabelt fields. Finally, reorganization analogous to that seen in the AI after restricted cochlear lesions is seen in visual cortex after restricted retinal lesions and in somatosensory cortex after digit amputation or after a variety of other procedures for eliminating input from a restricted portion of the skin. It is clear that this capacity for map reorganization in adults is a general characteristic of sensory systems.

Plastic Effects of Cochlear Electrical Stimulation in Profoundly Deaf Animals

The occurrence of plastic changes in the CAS of profoundly deaf animals (or humans) can obviously not be investigated using acoustic stimuli. However, electrical stimulation via intracochlear electrodes, which directly activates surviving spiral ganglion neurons, can be used to assess the responsiveness of the CAS in such cases. This stimulation is directly comparable to that provided by cochlear implants, which have been used to restore functional hearing to many thousands of profoundly deaf humans. Plasticity in the CAS of congenitally deaf or neonatally deafened animals has been demonstrated by comparing responses to cochlear electrical stimulation in adult animals that have received chronic electrical stimulation, for some specified period, with those in unstimulated controls. In some cases the stimulation is presented over weeks or months prior to the final recording, in which case differences undoubtedly reflect adult plasticity. In other cases, the stimulation is initiated at an early stage of development (as soon as it is possible to carry out implant surgery), but continues during adulthood, so it is unclear to what extent the plastic changes occur during development or adulthood. Although there appear to be no studies directly comparing the effects of cochlear electrical stimulation restricted to an early developmental period with that restricted to adulthood, it is likely that the capacity for plastic change is in fact greater in the former.

In congenitally deaf or neonatally deafened adult animals, the cochleotopic organization of the IC is normal, but that in the AI is rudimentary at best. Chronic cochlear electrical stimulation via a multichannel electrode (i.e., at a number of cochlear places) can restore (or maintain) basic cochleotopy. Stimulation of a restricted region of the cochlea results in an expansion of the representation of that cochlear region in the IC and in the AI. This effect is analogous to the expansion of the representation of lesion-edge frequencies after restricted cochlear lesions, but the fact that only limited effects are seen in the IC after mechanical cochlear lesions suggests that subcortical changes contribute in different ways to the two forms of plasticity.

The temporal pattern of AI neuronal responses to cochlear electrical stimulation is also modified by chronic stimulation. Responses in congenitally deaf adult animals are characterized by small

Figure 2 (a) Frequency map of the primary auditory cortex and digital photograph of the exposed cortical surface of a cat with normal hearing (AES, anterior ectosylvian sulcus; PES, posterior ectosylvian sulcus; SSS, suprasylvian sulcus). The dots on the photograph indicate the sites at which microelectrode penetrations were made, and the solid black line indicates the physiological boundary of the auditory cortex as defined from the data shown in the map. In the frequency map, the characteristic frequency of the neuron cluster recorded in each penetration is indicated above the dot other penetrations are labeled X' (no response to acoustic stimulation), A (acoustically responsive, but characteristic frequency could not be determined), B (broadly tuned), or I (inhibitory response). The line defining the physiological boundary of the auditory cortex is broken where this boundary was not determined unequivocally. Thin lines indicate iso-characteristic frequency contours (characteristic frequency identified by figures at lower boundary of the auditory cortex) fitted to the data at 2.5 kHz intervals using an inverse-distance smoothing function; R, C, D, and V indicate rostral, caudal, dorsal, and ventral directions, respectively. (b) Compound action potential (CAP) audiograms for a chronically lesioned cat. CAP thresholds (in decibels sound pressure level; dB SPL) prior to cochlear lesioning (prelesion) in the lesioned ear (open circles), and for both ears, lesioned (contralateral; filled circles) and normal (ipsilateral; diamonds), at the time of cortical mapping. The shaded area gives mean normal ± 1 SD CAP thresholds from a large pool of cats ($n > 50$). The lesion produced a steeply sloping hearing loss with edge frequency in the region of 18 kHz. The normal ear has a mild high-frequency loss at the time of cortical recording, but all thresholds are close to 1 SD above the mean. The vertical colored bars indicate the frequency bands at the edge of the lesion for which the representations (in c) are enlarged. (c) Frequency maps of the auditory cortex in the hemisphere contralateral to the (unilateral) cochlear lesion for stimulation of the contralateral (lesioned) ear and the ipsilateral (normal) ear in the cat for which the CAP audiogram is shown in b; conventions as in a. Blue and purple bands indicate the area of cortex containing neurons with characteristic frequencies in the range 16–18 and 18–20 kHz, respectively. Modified from Kamke MR, Brown M, and Irvine DRF (2004) Basal forebrain cholinergic input is not essential for lesion-induced plasticity in mature auditory cortex. *Neuron* 48: 675–686, with permission.

middle-latency responses and the complete absence of long-latency responses. Chronic cochlear electrical stimulation (initiated early in development) results in increased amplitude and a broader spatial distribution of middle-latency responses, and the appearance of long-latency responses. The temporal resolution of IC responses, as manifested in response latency and in the ability to follow the frequency of pulse trains, is also degraded by prolonged deafness, but chronic intracochlear electrical stimulation can reverse this degradation, and for some stimulation parameters, produce temporal resolution better than that seen in control animals.

Plasticity Associated with Various Forms of Learning

Changes in the stimulus selectivity of neurons in the higher levels of the auditory pathway as a consequence of behavioral conditioning procedures which alter the significance of particular stimuli for the organism were the first demonstrations of CAS plasticity in adult animals, and this remains the most active area of research on this topic. The most frequently reported finding in studies in which a tonal stimulus serves as the conditioned or discriminative (henceforth 'training') stimulus is that the response of cortical neurons at the training frequency increases, while the response at other frequencies decreases, such that the training frequency becomes the neurons' 'best' frequency (i.e., the frequency evoking the largest response) (**Figure 3**). Such effects have been described in the AI and in belt (secondary) auditory cortical fields, and in the auditory thalamus. These studies have commonly used fear conditioning procedures, but appropriate controls, and the specificity of the effects to the training frequency, establish that the observed effects are manifestations of plasticity rather than consequences of changes in state variables. Similar changes in the spectrotemporal receptive fields of auditory cortical neurons have been described in animals trained to detect a target tone of a particular frequency embedded in a sequence of broadband noiselike stimuli.

Another form of auditory learning that might involve CAS plasticity is perceptual learning, the improvement in discriminative capacity with training that is a common observation in all sensory modalities, both in psychophysical experiments and in everyday life. The specificity of many forms of visual perceptual learning to particular features of the training stimuli, or to the region of the receptor to which the stimuli are presented, led to the proposal that the learning might involve changes in neural tuning in primary sensory cortex. In the auditory system, the evidence from animal studies for changes in the AI associated with perceptual learning on frequency discrimination tasks is equivocal, and it is possible that the learning involves changes in higher-order decision-making processes. However, evidence for auditory cortical changes is provided by reports that auditory evoked responses to musical tones (measured by either electro- or magnetoencephalography) are of greater amplitude in trained musicians than in nonmusicians. Although this correlation could simply reflect the fact that people with this characteristic are more likely to undertake musical training, the fact that in some studies the effect is specific to the trained instrument (i.e., it is timbre specific), and that similar enhancement is seen when nonmusicians undergo training, suggests that the correlational evidence reflects plasticity.

The cortical responses exhibiting plasticity in these human studies are mainly localized to non-core (viz., belt and parabelt) areas of auditory cortex, in contrast to most of the animal evidence which has been derived from the AI. As noted previously, it is likely that the capacity for plastic change is generally greater in non-core than in core sensory cortical fields.

Two forms of auditory perceptual learning that are of great practical significance relate to speech perception. The first is the effect of language experience on the perception of speech sounds, and thus on language acquisition, during a critical period of development. The second is the improvement in speech discrimination shown by people with cochlear implants over the months and years following implantation. It is not clear, however, whether these forms of learning reflect plasticity in the CAS or in regions involved in higher-order cognitive processing of auditory information.

Plasticity Associated with Environmental Enrichment

One of the most intensively studied aspects of structural brain plasticity over more than 50 years has been the effects of so-called environmental 'enrichment': the brains of rats raised in enriched environments are structurally different in many respects from those of rats raised in (admittedly often impoverished) standard conditions. There has been only a limited investigation of the effects of environmental enrichment on the response characteristics of auditory cortical neurons, but the available evidence indicates that a combination of generalized and specifically acoustic enrichment results in changes in the response strength, RF characteristics, and temporal response properties of AI neurons.

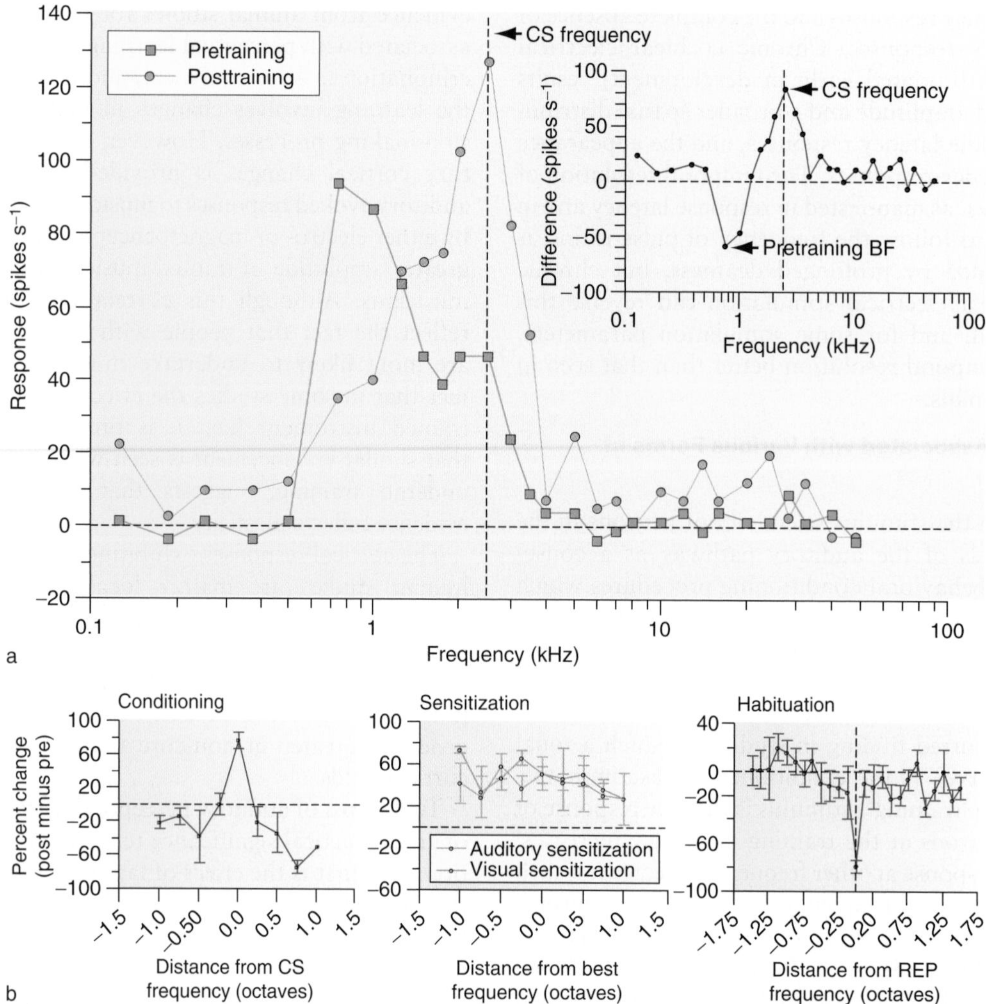

Figure 3 The effects of classical conditioning on the frequency selectivity of neurons in the primary auditory cortex of the guinea pig. (a) Receptive field plasticity of a single auditory cortex neuron is manifested in isolevel functions obtained (at 70 dB sound pressure level) before and after tone–shock conditioning (pretraining and posttraining, respectively). The best frequency (that evoking the maximum response) shifts from below 1 kHz to the frequency of the conditioned stimulus (CS). The inset shows the difference in frequency selectivity (posttraining minus pretraining), with the maximum increase in response at the frequency of the CS. (b) Normalized group difference functions (posttraining minus pretraining) showing change in response as a function of the distance of the probe frequency from the frequency used in behavioral training in conditioning, sensitization, and habituation conditions. Conditioning (left) produces a selective increase in response at the CS frequency, with reduced responses at most other frequencies. Sensitization training produces a nonselective increase in response at all frequencies, both for auditory sensitization (tone–shock unpaired) and visual sensitization (light–shock unpaired), showing that this nonassociative effect is transmodal. Habituation (repeated presentation of the same tone alone) produces a selective decrease in response at that frequency. Adapted from Weinberger NM (2004) Specific long-term memory traces in primary auditory cortex. *Nature Reviews Neuroscience* 5: 279–290, with permission.

Plasticity Induced by Microstimulation

Studies in a range of species indicate that focal electrical stimulation of a region of the AI can change the frequency selectivity of neurons around the stimulation site. In some experiments, the frequency selectivity of neurons in the tonotopically corresponding area of the central nucleus of the IC has also been shown to change as a consequence of cortical microstimulation. Centrifugal fibers from the AI to IC have been shown to play a role in the latter form of plasticity.

Plasticity Induced by Direct Activation of Neuromodulatory Systems

The neocortex receives diffuse extrathalamic projections from five different subcortical cell groups which use different neurotransmitters. These systems act to modulate the sensitivity of cortical neurons and have been implicated in cortical plasticity. The most extensively studied is the system of cholinergic fibers originating in the basal forebrain (BF). Pairing electrical stimulation of the BF with tonal stimulation at a particular frequency has been reported to shift the frequency selectivity of cortical neurons toward the paired frequency and to result in an enlarged representation of that frequency in AI. These effects of experimental activation of the BF cholinergic system suggest that it is almost certainly involved in the use-related plasticity in the auditory cortex, as described in previous sections. This hypothesis is supported by a substantial body of evidence that the cholinergic basal forebrain plays an important modulatory role in various forms of auditory learning in both animals and humans. Lesions of the cholinergic BF do not, however, prevent injury-induced reorganization of AI tonotopy.

Cellular Mechanisms of CAS Plasticity

The evidence reviewed so far for the plasticity of the RFs of neurons in the CAS has been complemented by changing conceptions of RFs in all sensory modalities over the past two decades. In the early period of sensory neuroscience, RFs were described simply in terms of the range of the given stimulus parameters that evoked an increase in action potential frequency above the background resting activity. In the absence of spontaneous activity, these studies could not identify inhibitory components of RFs. Subsequent studies using forward-masking paradigms or the iontophoretic application of inhibitory transmitter antagonists revealed that many neurons at all levels of the CAS receive suprathreshold excitatory input, which is masked by inhibition, outside the 'classically' defined RF. More recently, intracellular recordings *in vivo* have revealed that many cortical neurons receive subthreshold excitatory and inhibitory inputs over an even broader range of stimulus values beyond their 'classical' RFs, much of which is derived from intrinsic corticocortical connections.

The cellular mechanisms underlying plasticity in adult sensory systems undoubtedly comprise a hierarchy of changes, differing in time course and complexity, and different forms of plasticity might involve different subsets of these mechanisms. The supra- and subthreshold inputs outside the aforementioned classic RF provide a broad matrix of inputs from which short-term RF changes can be sculpted. The first stage of injury-induced plasticity in adult sensory cortices appears to be an immediate expansion of RFs, the rapidity of which indicates that it reflects an unmasking of normally inhibited excitatory inputs. A similar unmasking is involved in space map plasticity in the barn owl. The subsequent establishment of smaller RFs in injury-induced plasticity presumably involves changes in the efficacy of both excitatory and inhibitory synapses (i.e., in synaptic weights). These changes are likely to involve an increase in the efficacy of previously subthreshold inputs, although changes of this sort have not yet been experimentally demonstrated. In both developmental plasticity in the barn owl and adult plasticity in the mammalian visual and somatosensory systems, there is evidence for the strengthening of excitatory synapses via *N*-methyl-D-aspartate (NMDA) receptor-mediated long-term potentiation (LTP). LTP itself, and this stage of plasticity, are also thought to involve the activation of previously 'silent' synapses. As discussed previously, there is evidence that neuromodulatory systems contribute to plastic changes in the cortex, and that centrifugal fibers are involved in some forms of subcortical plasticity.

Finally, on a longer timescale, structural changes in axon collaterals and in dendritic orientation have been shown to contribute to some forms of cortical plasticity. Long-term changes in the barn owl involve axonal sprouting and synaptogenesis, and axonal sprouting is also involved in the changes observed after unilateral cochlear ablation in neonatal animals. Axonal sprouting of horizontal fibers in the superficial layers of the cortex has been shown to be associated with visual cortical plasticity after retinal lesions in adults. In the barrel cortex of adult rats, vibrissal deafferentation, which results in an expansion of the influence of intact vibrissae into the area in which the deafferented vibrissae were represented, has been shown to result in changes in the dendritic orientation of neurons in this area toward the sources of this new input. Structural changes of this sort have not yet been shown in injury-induced auditory plasticity in adults, but the similarity of many forms of adult plasticity in different sensory systems suggests that such changes are likely to be involved.

Functional Significance of Central Auditory System Plasticity

Auditory developmental plasticity and learning-induced plasticity are undoubtedly adaptive, in that they enhance the organism's ability to adjust to altered patterns of input and, in the case of human language acquisition and adaptation to a cochlear implant, to make important auditory discriminations. It is less clear that adult injury-induced plasticity is adaptive, as the organism remains deaf in the frequency range affected by the cochlear lesion. Although there is evidence for enhanced frequency discrimination ability at lesion-edge frequencies in humans with cochlear damage (and associated hearing losses) of the sort that results in cortical reorganization in animal studies, it is unlikely that this slight change constitutes any form of compensation for the deafness. It is likely that this form of plasticity is an extreme manifestation of the processes that underlie other forms of plasticity in response to altered input, but it is also possible that similar processes are involved in recovery of function after central damage such as that produced by stroke.

The possibility that some forms of learning impairment might involve deficiencies in aspects of auditory processing that exhibit plasticity, such that these deficiencies can be modified by training, has resulted in the recent development of a number of auditory training software packages. As discussed previously, the success of cochlear implants undoubtedly rests in part on the plasticity of CAS structures, and plasticity therefore contributes to the therapeutic effects of these devices.

It is also possible that some forms of CAS plasticity might have pathological consequences. In the somatosensory system, there is evidence that cortical reorganization might be responsible for some features of phantom limb experiences in amputees. It has similarly been suggested that tinnitus might be a consequence of cortical reorganization consequent on a cochlear lesion.

See also: Adult Cortical Plasticity; Auditory Cortex Structure and Circuitry; Auditory System: Central Pathways; Auditory Cortex: Models; Auditory Localization; Cortical Plasticity and Learning: Mechanisms and Models; Perceptual Learning: Neural Mechanisms; Perceptual Learning and Sensory Plasticity; Somatosensory Plasticity.

Further Reading

Atwood HL and Wojtowicz JM (1999) Silent synapses in neural plasticity: Current evidence. *Learning and Memory* 6: 542–571.

Buonomano DV and Merzenich MM (1998) Cortical plasticity: From synapses to maps. *Annual Review of Neuroscience* 21: 149–186.

Calford MB (2002) Dynamic representational plasticity in sensory cortex. *Neuroscience* 111: 709–738.

Engineer ND, Percaccio CR, Pandya PK, et al. (2004) Environmental enrichment improves response strength, threshold, selectivity, and latency of auditory cortex neurons. *Journal of Neurophysiology* 92: 73–82.

Gilbert CD (1998) Adult cortical dynamics. *Physiology Review* 78: 467–485.

Gu Q (2002) Neuromodulatory transmitter systems in the cortex and their role in cortical plasticity. *Neuroscience* 111: 815–835.

Hayes EA, Warrier CM, Nicol TG, et al. (2003) Neural plasticity following auditory training in children with learning problems. *Clinical Neurophysiology* 114: 673–684.

Irvine D, Brown M, Martin R, et al. (2005) Auditory perceptual learning and cortical plasticity. In: Koenig R, Heil P, Budinger E, et al. (eds.) *The Auditory Cortex: A Synthesis of Human and Animal Research*, pp. 409–428. Mahwah, NJ: Lawrence Erlbaum.

Irvine DRF and Wright BA (2005) Plasticity in spectral processing. In: Malmierca M and Irvine DRF (eds.) *Auditory Spectral Processing*, pp. 435–472. San Diego: Elsevier.

Kaas JH and Florence SL (2001) Reorganization of sensory and motor systems in adult mammals after injury. In: Kaas JH (ed.) *The Mutable Brain*, pp. 165–242. Amsterdam: Harwood.

Kamke MR, Brown M, and Irvine DRF (2004) Basal forebrain cholinergic input is not essential for lesion-induced plasticity in mature auditory cortex. *Neuron* 48: 675–686.

Knudsen EI (2002) Instructed learning in the auditory localization pathway of the barn owl. *Nature* 417: 322–328.

Kuhl PK (2000) A new view of language acquisition. *Proceedings of the National Academy of Sciences of the United States of America* 97: 11850–11857.

Parks TN, Rubel EW, Popper AN, et al. (eds.) (2004) *Plasticity of the Auditory System*. New York: Springer.

Suga N and Ma X (2003) Multiparametric corticofugal modulation and plasticity in the auditory system. *Nature Reviews Neuroscience* 4: 783–794.

Syka J (2002) Plastic changes in the central auditory system after hearing loss, restoration of function, and during learning. *Physiological Reviews* 82: 601–636.

Weinberger NM (2004) Specific long-term memory traces in primary auditory cortex. *Nature Reviews Neuroscience* 5: 279–290.

Auditory System: Central Pathways

J C Middlebrooks, University of Michigan, Ann Arbor, MI, USA

Introduction

The central auditory system consists of interconnected processing stations at brain stem, midbrain, thalamic, and cortical levels (**Figure 1**). Input to the central auditory system originates in the cochlea of the inner ear. Active mechanical processes in the cochlea perform a frequency analysis of sound such that an inner hair cell at any point along the cochlear spiral is tuned to a particular frequency, with hair cells at the base of the cochlear spiral tuned to high frequencies and hair cells at the apex tuned to low frequencies. The systematic map of high-to-low frequency onto basal-to-apical cochlear place is known as 'tonotopic organization' ('tono' referring to frequency and 'topic' referring to cochlear place). Tonotopic organization is preserved in the projection to the brain and forms a functional organizing principle throughout all levels of the central auditory pathway.

The frequency-tuned input from each cochlea is carried to the brain by the left and right auditory nerves, each nerve containing ~30 000 fibers (in humans). Approximately 95% of auditory nerve fibers are 'type I' afferents that contact only one inner hair cell, thereby inheriting the frequency tuning of that hair cell. Action potentials in auditory nerve fibers tend to fire in phase with low-frequency sounds, up to about 4 kHz, and fire in a non-phase-locked manner to higher-frequency sounds. Based on this frequency-tuned, partially phase-locked auditory-nerve input, the central auditory system performs a multitude of tasks, including the following.

Spectral analysis. The auditory system preserves and enhances the frequency analysis that begins in the cochlea. It distinguishes sound spectra that characterize communication calls (including speech sounds) and environmental sounds. Whereas single auditory nerve fibers and some central neurons are tuned to single frequencies, many neurons in the central pathway are selective for complex spectra that contain multiple frequency components.

Intensity analysis. The auditory system discriminates sound pressure levels over nearly six orders of magnitude of pressure (i.e., 120 dB). Single auditory nerve fibers have dynamic ranges of only ~20–50 dB between the threshold sound level that barely elicits a reliable response and the saturated level at which further increases in level produce no further increase in response. Because of the limited dynamic ranges of single fibers, the full dynamic range of hearing must involve multiple nerve fibers that differ in threshold and in frequency sensitivity.

Temporal analysis. The auditory system analyzes temporal features of sounds on multiple timescales, from tens of microseconds for sound localization, to tens of milliseconds for discrimination of certain speech sounds (e.g., /da/ vs. /ta/), to hundreds of milliseconds for segmentation of speech syllables. It detects periodicity on the scale of tens to hundreds of cycles per second to distinguish pitch or to recognize voices.

Efferent control of the cochlea. The central auditory system enhances processing of signals in noise and protects the ear against intense sounds by modulating cochlear mechanics and the sensitivity of afferent fibers. This task involves pathways that originate in the auditory cortex and brain stem, ending with olivocochlear efferent fibers that terminate on outer hair cells and afferent fibers in the cochlea.

Sound localization. The auditory system computes the location of a sound source by integrating acoustical cues provided by the interaction of the incident sound wave with the head and external ears. The principal cues for localization are the relative timing and level of sounds at the two ears (for left/right localization) and direction-dependent spectral shapes imposed by the directional filter properties of the external ears (for up/down and front/back localization). Location is a fundamental attribute of a sound source, and sound localization is a topic that is amenable to both physiological and psychophysical research. For those reasons, the mechanisms for sound localization are among the most thoroughly studied of central auditory mechanisms.

Central auditory processing is accomplished by parallel processes involving anatomically discrete, physiologically specialized pathways. In some cases, one can point to a particular pathway that performs a specific task. In other cases, functions are carried out in multiple parallel pathways, and the result of a specific auditory function is distributed widely across multiple areas of the auditory and multimodal cortex.

Auditory Brain Stem

Cochlear Nucleus

The auditory nerve terminates in the cochlear nucleus complex at the junction of the medulla and pons (**Figure 2**). The cochlear nucleus comprises two

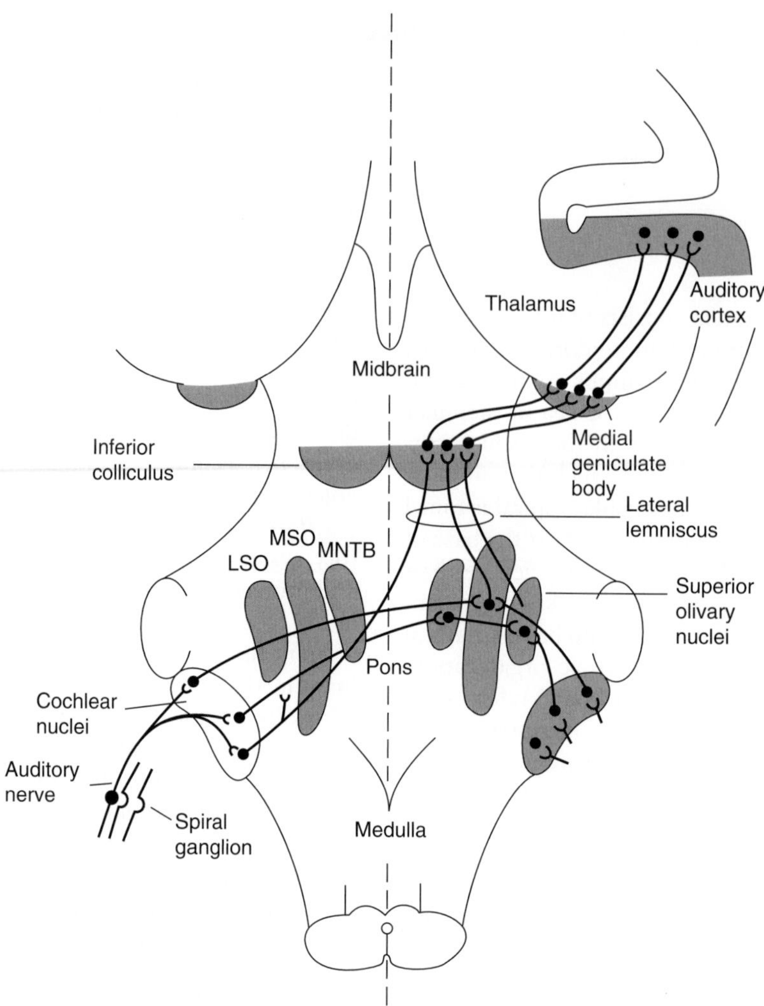

Figure 1 Central auditory pathway. All structures are bilaterally symmetrical, although the illustrated connections show only the pathways to the right auditory cortex. LSO, lateral superior olive; MSO, medial superior olive; MNTB, medial nucleus of the trapezoid body. Reproduced from Imig TJ, Irons WA, and Samson FR (1990) single-unit selectivity to azimuthal direction and sound pressure level of noise bursts in cat high-frequency primary auditory cortex. *Journal of Neurophysiology* 63: 1448–1466, with permission from Elsevier.

physically distinct divisions: the dorsal cochlear nucleus (DCN) and the ventral cochlear nucleus. The ventral cochlear nucleus is further divided, on the basis of the pattern of termination of auditory nerve fibers, into the posteroventral cochlear nucleus (PVCN) and anteroventral cochlear nucleus (AVCN). Each auditory nerve fiber divides into two major branches as it enters the cochlear nucleus. The ascending branch terminates in the AVCN. The descending branch gives off terminals in the PVCN, then continues to the DCN where it terminates. In that way, each auditory nerve fiber forms three terminal regions. The fibers terminate in a tonotopic pattern within each region, resulting in three maps of sound frequency within the cochlear nucleus on each side of the brain.

The cochlear nucleus complex traditionally has been regarded as a strictly monaural structure, responding only to stimulation of the ipsilateral ear. Recent studies, however, demonstrate a projection from the contralateral ear by way of the contralateral cochlear nucleus. The contralateral input is predominantly inhibitory, although a period of conductive hearing loss can result in an increase in excitatory inputs. These observations indicate that the cochlear nucleus plays at least a minor part in binaural hearing.

The cochlear nucleus contains many distinct classes of neurons that are distinguished by the morphology of cell body and dendrites, by patterns of auditory nerve synapses, and by intrinsic physiological properties. In the AVCN and PVCN, some of the prominent projection neurons (i.e., cell types that send projections out of the cochlear nucleus) are the spherical bushy cells, the globular bushy cells, the octopus cells, and the multipolar (or stellate) cells. Spherical bushy cells, found in the AVCN, have large spherical cell

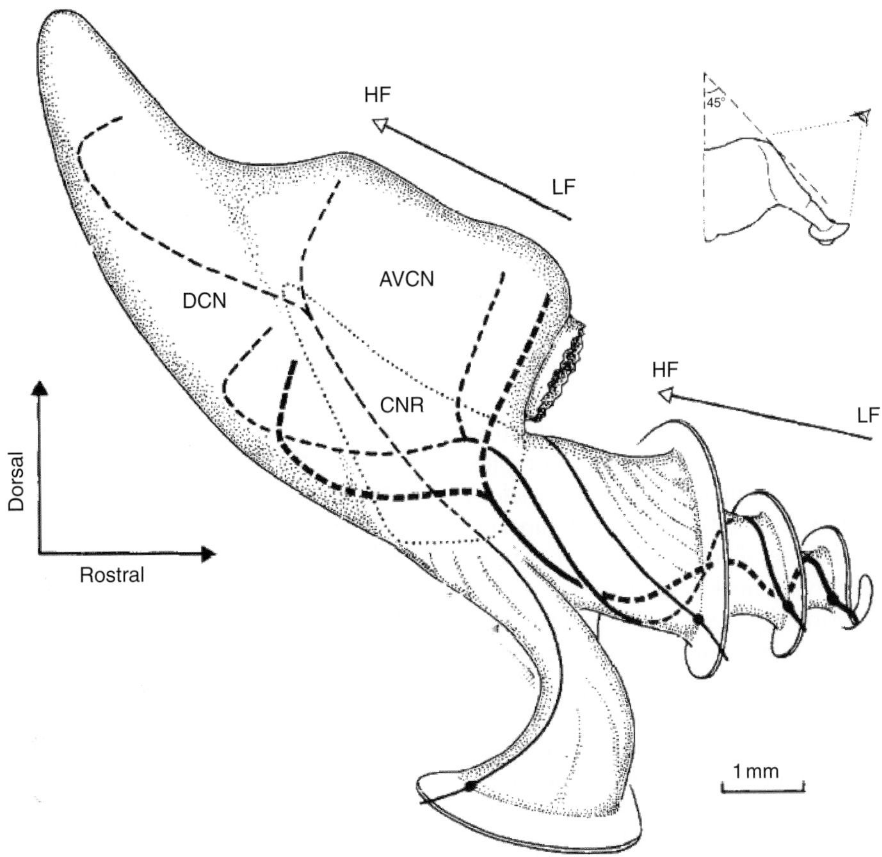

Figure 2 Cochlear nucleus complex and auditory nerve. This diagram from the right side of the cat shows auditory nerve fibers originating at four levels of the spiral ganglion. Each fiber forms major ascending (rostral) and descending (caudal) branches in the cochlear nucleus complex, resulting in maps of characteristic frequency from low frequency (LF) to high frequency (HF). The posterior ventral cochlear nucleus (PVCN), not shown here, lies deep to the cochlear nerve root (CNR). DCN, dorsal cochlear nucleus; AVCN, anterior ventral cochlear nucleus. Arnesen AR and Osen KK (1978) The cochlear nerve in the cat: Topography, cochleotopy, and fiber spectrum. *Journal of Comparative Neurology* 178: 661–678.

bodies that receive input from auditory nerve fibers by way of one or two large axosomatic endings, the endbulbs of Held. These synapses transmit temporal information with great fidelity, so the frequency selectivity and temporal firing patterns of spherical bushy cells closely resemble the responses of auditory nerve fibers – these are 'primary-like' firing patterns. Globular bushy cells receive axosomatic endings from a few auditory nerve fibers and respond with a slightly modified primary-like temporal pattern ('primary with notch'). Octopus cells, found in a restricted portion of the PVCN, receive input onto their dendrites from a large number of auditory nerve fibers spanning a wide frequency range, resulting in wide frequency response areas. The responses of octopus cells are tightly locked to the onsets of sounds. Multipolar cells are widely distributed throughout the AVCN and PVCN. They have elaborately branched dendritic trees that receive numerous synaptic boutons from auditory nerve fibers. They respond to sound with a periodic 'chopping' spike pattern; the chopping rate is determined by the electrical properties of the cell membrane rather than by the period of the sound.

The major projection cells of the DCN are the fusiform cells (also known as pyramidal cells) and some of the giant cells. They receive input directly from auditory nerve fibers as well as projections from granule cells and multipolar cells in the ipsilateral PVCN and AVCN. Fusiform cells and giant cells have complex frequency response areas consisting of multiple excitatory and inhibitory frequency regions. One likely role of neurons with complex frequency response areas is in detecting spectral shape cues for the up/down or front/back location of a sound source. That is, the external ear filters sounds differentially depending on the angle by which the sound enters the ear. Recognition of particular direction-dependent sound spectra, perhaps performed by the DCN, provides cues to vertical sound location. Indeed, experimental lesions of the DCN result in deficits in sound localization behavior in the vertical dimension.

The ascending projections from the cochlear nucleus form three major fiber bundles. The 'trapezoid body' is formed by fibers from the AVCN, destined primarily for the superior olivary complex or the inferior colliculus (IC). The 'intermediate acoustic stria' is formed primarily by fibers from the PVCN that will project to the contralateral IC and nuclei of the lateral lemniscus. The 'dorsal acoustic stria' contains the projection fibers of the DCN, which terminate primarily in the contralateral IC and nuclei of the lateral lemniscus.

Superior Olivary Complex

The superior olivary complex in the pons is the first major site of binaural convergence. The complex comprises three prominent nuclei: the medial nucleus of the trapezoid body (MNTB), the medial superior olive (MSO), and the lateral superior olive (LSO), as well as several periolivary nuclei. Neurons within the periolivary nuclei primarily make connections within the superior olivary complex or send projections to the cochlea and cochlear nucleus.

The MSO and LSO each have distinct roles in binaural hearing, which is important for sound localization in the horizontal dimension. Horizontal sound localization has long been understood in the context of a 'duplex theory' in which low-frequency sounds are localized on the basis of differences in the time of arrival of sounds at the two ears ('interaural time differences'; ITDs) and high-frequency sounds are localized on the basis of differences in sound level at the two ears ('interaural level differences'; ILDs). Based on the distinctive physiology of constituent neurons, one can associate ITD and ILD sensitivity with the MSO and the LSO, respectively. ITDs result from differences in the length of the path from a lateral sound source to the near and far ear. Humans utilize ITDs cues in the range of $\pm 700\,\mu s$ with resolution of $\sim 10\,\mu s$; ITD cues are most salient at frequencies below ~ 1 kHz. ILDs result from the tendency of the head to shadow the ear on the side away from the sound source. Humans process ILD cues for localization in the range of ± 30 dB with resolution of ~ 1 dB; ILDs are most informative at frequencies >3 kHz.

The MSO has a tonotopic organization that primarily represents low frequencies. Individual neurons receive direct inputs from spherical bushy cells on both sides of the brain. The MSO cells are sensitive to the relative timing of inputs originating at the two ears, responding strongly to action potentials that arrive simultaneously at the MSO. A recent study in gerbils indicates that an inhibitory input from the MNTB acts to shift the ITD tuning of MSO neurons so that maximum spike rates are produced by binaural sounds leading at the contralateral ear and that maximum 'changes' in spike rates are associated with near-zero ITDs. That arrangement would result in maximum spatial acuity for sounds near the midline of auditory space, which is what is observed psychophysically.

The tonotopic organization of the LSO primarily represents high frequencies. LSO neurons receive excitatory inputs directly from spherical bushy cells of the ipsilateral AVCN. The contralateral pathway to the LSO is indirect, involving an excitatory projection from globular bushy cells in the contralateral AVCN to the ipsilateral MNTB, then an inhibitory projection from the ipsilateral MNTB to the LSO. The net effect is that LSO neurons are inhibited by sounds at the contralateral ear and excited by sounds at the ipsilateral ear. The binaural sensitivity of LSO neurons results in sensitivity to differences in sound levels at the two ears. Analogous to the situation in the MSO, the greatest change in LSO spike rates occurs for ILDs passing through zero, corresponding to sound-source locations near the midline of auditory space.

The 'olivocochlear' system consists of diffusely organized neurons in the superior olivary complex and their efferent fibers that terminate in the cochlea. There are lateral and medial components. Generally, the neurons of the lateral olivocochlear system are located in or around the LSO. The majority of these neurons have fine unmyelinated fibers that travel in the uncrossed olivocochlear bundle toward the ipsilateral cochlea. They eventually enter the cochlea by way of the vestibular nerve. Fibers of the lateral system terminate on the dendrites of primary afferent fibers near their synapses with inner hair cells. The function of the lateral efferent system is not well understood, but is thought to serve a protective role. The neurons of the medial olivocochlear system are located in the medial and rostral periolivary nuclei. Their projection fibers are myelinated, and the majority travel in the crossed olivocochlear bundle toward the contralateral cochlea. The fibers terminate with cholinergic synapses on outer hair cells. Activity in the medial olivocochlear system modulates the active mechanical feedback that outer hair cells contribute to the process of frequency analysis in the cochlea.

Nuclei of the Lateral Leminiscus

The ascending projections of most brain stem auditory nuclei coalesce on each side of the brain in a major fiber bundle known as the lateral lemniscus. Amid those fibers are nuclear groups known as the 'nuclei of the lateral lemniscus.' The nuclear groups vary widely among species, but generally one can distinguish ventral (VNLL), intermediate (INLL), and dorsal

(DNLL) nuclei of the lateral lemniscus. The VNLL receives its major input from the contralateral cochlear nucleus. Neurons in the VNLL typically are excited by stimulation of the contralateral ear and show no response to ipsilateral stimulation. Tonotopic organization in the VNLL is rather indistinct. The INLL also receives predominately contralateral inputs and is excited by stimulation of the contralateral ear. The INLL is particularly well developed in echolocating bats, in which it is thought to contribute to processing of temporal characteristics of sounds. The DNLL is a strongly binaural nucleus. It receives bilateral input from the AVCN and LSO, ipsilateral input from the MSO, and contralateral input from the opposite DNLL. High-frequency neurons in the DNLL display sensitivity to ILDs, similar to the responses of LSO neurons. Many DNLL neurons that respond to lower frequencies also show ITD sensitivity similar to that of neurons in the MSO.

The fibers of the lateral lemniscus, originating in bilateral cochlear nuclear complexes, superior olivary complexes, and nuclei of the lateral lemniscus, terminate in a key structure of the midbrain, the IC.

Midbrain

The major midbrain structures involved in hearing are the IC and superior colliculus (SC). The major divisions of the IC are the central nucleus (ICC), the lateral nucleus (roughly equivalent to the 'external nucleus' in other nomenclatures), the dorsal cortex, and the dorsomedial nucleus (**Figure 3**). Nearly all ascending auditory pathways synapse in the ICC

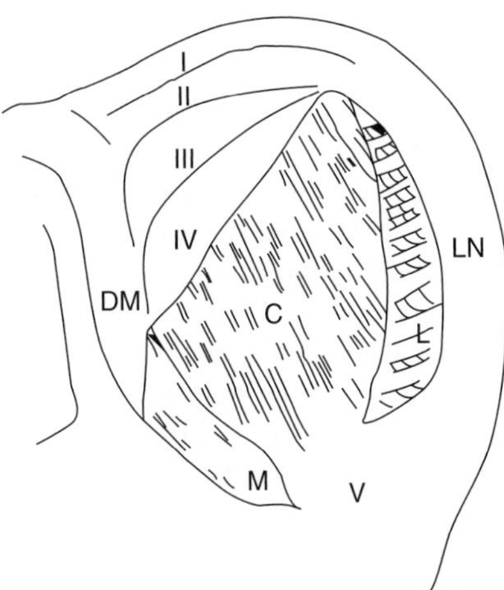

Figure 3 Subdivisions of the inferior colliculus, shown in transverse section. DM, dorsomedial nucleus; C, M, and L, central, medial, and lateral parts of the central nucleus; LN, lateral nucleus; I, II, III, IV layers of the dorsal cortex. Oliver DL and Morest DR (1984) The central nucleus of the inferior colliculus in the cat. *Journal of Comparative Neurology* 222: 237–264.

en route from the brain stem to the forebrain. The ICC shows a prominent laminar organization formed by sheets of disk-shaped neurons and by the distribution of dendrites and axons. These sheets form functional 'isofrequency laminae' in that neurons within a particular sheet have similar characteristic frequencies. Perpendicular to the plane of the isofrequency lamina, there is a steady tonotopic progression in characteristic frequency, with characteristic frequencies increasing from low to high frequencies along a roughly dorsolateral to ventromedial axis.

Each isofrequency lamina in the ICC receives projections from multiple brain stem nuclei. Some areas of isofrequency laminae receive projections that converge from multiple sources, whereas other areas are dominated by inputs from a particular brain stem source. The partial segregation of inputs within particular regions of isofrequency laminae permits parallel processing of specific attributes of sounds. The segregation of inputs is manifest as a diversity of response properties. The frequency response areas of neurons, for instance, show three major classes. Type V neurons generally have low characteristic frequencies and have V-shaped tuning curves similar to those of primary afferents. Type I neurons have narrow excitatory frequency bands that are flanked by inhibition. The type O neurons show sharply tuned excitatory responses to low-level tones and inhibition to high-level tones – this suggests a direct input from the DCN. Binaural response properties also vary, showing strictly contralateral excitatory responses (suggestive of a direct input from the contralateral cochlear nucleus), contralateral excitation and ipsilateral inhibition (suggestive of input from the contralateral LSO), sensitivity to interaural phase differences in low-frequency tones (suggestive of input from the MSO), or other more complex characteristics. The major output from the ICC to the forebrain travels through a fiber bundle known as the brachium of the IC.

Subdivisions of the IC outside of the central nucleus integrate input from the central nucleus of the IC, from descending cortical projections, and/or from the somatosensory system. Convergence of information across frequencies in those subdivisions partially obscures tonotopic organization. Neurons from the lateral nucleus of the IC and/or the nucleus of the brachium of the IC provide auditory input to a sensorimotor integrative structure of the midbrain, the SC.

The SC has the task of integrating auditory localization information with information from other sensory modalities to initiate orienting movements of the eyes and head. It receives visual, somatosensory, and auditory input. The ascending auditory input from the IC produces relatively short-latency responses of

SC neurons to the onsets of sounds. There is also descending input from auditory cortical areas that is more likely to have a modulatory role. The SC is organized functionally to form a dynamic map of motor error. That is, the receptive fields of neurons reflect the disparity between the angle of gaze (determined by head and eye position) and an auditory, visual, or somatic target. Studied under anesthetized conditions, this dynamic motor map can be seen as a complex of mutually aligned auditory, visual, and somatic sensory maps. In response to acoustic noise bursts presented from a moveable loudspeaker, single neurons show spatial receptive fields that are restricted in the horizontal and vertical dimensions. The centers of auditory spatial receptive fields vary according to the locations of neurons in the SC such that the horizontal dimension of auditory space maps onto the rostrocaudal dimension of the SC and the vertical dimension maps onto the mediolateral dimension. It is important to appreciate that the auditory spatial topography in the SC is not a simple representation of the auditory sensory epithelium – the cochlea maps frequency, not space. Instead, the auditory spatial map in the SC must reflect the integrated activity of multiple brain stem nuclei that are sensitive to specific auditory spatial cues, such as interaural phase difference, ILD, and direction-dependent spectral shape cues. Auditory spatial topography is particularly evident in the barn owl in a homologue of the SC, the optic tectum.

Auditory Thalamus and Cortex

The auditory forebrain consists of the medial geniculate body (MGB) of the thalamus and multiple auditory cortical areas in the temporal lobe. The MGB and cortical areas are linked by reciprocal thalamocortical and corticothalamic projections. The MGB has three major divisions: ventral, dorsal, and medial. The ventral division receives direct projections from the central nucleus of the IC. As in the central nucleus of the IC, neurons of the ventral division are sharply tuned for frequency and there is a prominent laminar tonotonic organization. Indeed, the ascending pathway from the ICC to the ventral division of the MGB to the primary auditory cortex constitutes the highest-fidelity representation of sound frequency in the form of tonotopically organized neurons with narrow frequency tuning curves. The inputs to the dorsal division of the MGB are primarily from IC subdivisions outside of the central nucleus. Dorsal division neurons tend to be broadly tuned for frequency. Dorsal division neurons project to the 'belt' areas of the auditory cortex (defined below), outside of the tonotopically organized 'core' regions. The medial division of the MGB receives convergent inputs from many auditory sources as well as some somatosensory input. The medial division projects rather diffusely to multiple auditory cortical areas. A rather unique population of very-large-diameter myelinated axons originates in the medial division and terminates in the superficial lamina of the auditory cortex.

The auditory cortex has a hierarchical organization consisting of core, belt, and parabelt levels (**Figure 4**). The core area receives the principal tonotopic projection from the ventral division of the MGB. The core area contains the primary auditory cortex (area A1) and, in some species, one or more additional tonotopically organized fields. The core area in the macaque monkey, for instance, contains two or three maps of frequency representations: in area A1, in a rostral area, and possibly in a rostrotemporal area. The core areas exhibit a pronounced columnar organization in which radial cell columns extending through the cortical layers share common characteristic frequencies and binaural response properties.

The core area of the auditory cortex is surrounded by multiple belt areas. The belt areas receive feed-forward projections from the core as well as direct projections from the MGB, primarily from the dorsal division. Neurons in belt areas tend to respond more strongly to noise bands than to pure tones and often show complex frequency response areas with multiple excitatory and inhibitory regions. At least in primates, the parabelt areas form a third level of the hierarchy, with parabelt areas receiving inputs from belt areas but not from core areas. The parabelt areas project widely to temporal, parietal, and frontal cortical areas for association of auditory processes with other sensory and motor processes.

Auditory responses in the cortex reflect a chain of processing that is a minimum of five synaptic levels beyond the hair cells of the cochlea. One might expect that these many levels of processing would result in cortical neurons that are highly selective for specific sounds or specific features of sounds. Contrary to that expectation, most cortical neurons tend to show graded responses to multiple stimulus parameters. In regard to sound localization, for instance, there is no evidence in the cortex of a point-to-point map like that found in the SC. Instead, the responses of many cortical neurons vary systematically according to sound-source location through as much as 360° of location so that individual neurons represent auditory space panoramically. As a result, the cortical representation of any point in auditory space is distributed widely through the cortex. Our current understanding points to the representation of any sound by the coordinated activity of widespread populations of cortical neurons. The way in which

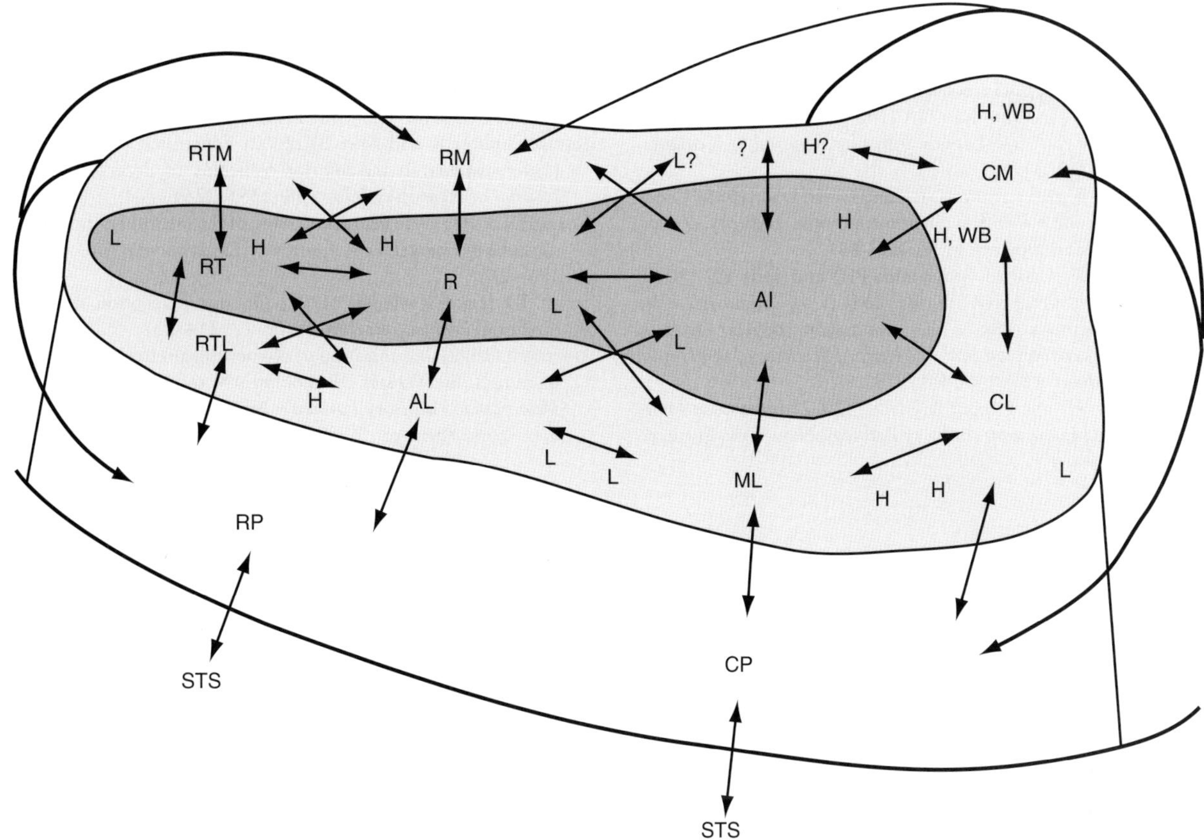

Figure 4 Auditory cortical areas in the macaque monkey. The auditory area lies on the dorsal surface of the temporal lobe; this is a view from the left hemisphere. Dark, medium, and light shading indicate core, belt, and parabelt areas, respectively. Broad and thin arrows indicate heavy and light connections, respectively. H, L, and WB indicate areas responsive to high, low, and wideband stimuli, respectively. Core areas: AI, primary auditory cortex; R, rostral; RT, rostrotemporal. Belt areas: RTL, lateral rostrotemporal belt; RTM, medial rostrotemporal belt; AL, anterior lateral belt; RM, rostromedial region; ML, middle lateral belt; CL, caudal lateral belt; CM, caudal medial belt. Parabelt areas: RP, rostral parabelt; CP, caudal parabelt. STS, superior temporal sulcus. Reproduced from Kass JH, Hackett TA, and Tramo MJ (1999) Auditory processing in primate cerebral cortex. *Current Opinion in Neurobiology* 9: 164–170, with permission from Elsevier.

patterns of activity in neural populations are translated to auditory perception and behavior remains a key topic in ongoing research.

See also: Auditory Cortex Structure and Circuitry; Auditory System: Giant Synaptic Terminals, Endbulbs and Calyces; Auditory/Somatosensory Interactions; Cerebral Cortex; Cochlear Development; Echolocation I: Behavior; Reticular Formation and the Brain Stem; Sound Localization: Neural Mechanisms; Temporal Processing in the Auditory Pathway.

Further Reading

Arnesen AR and Osen KK (1978) The cochlear nerve in the cat: Topography, cochleotopy, and fiber spectrum. *Journal of Comparative Neurology* 178: 661–678.

Brand A, Behrend O, Marquardt T, McAlpine D, and Grothe B (2002) Precise inhibition is essential for microsecond interaural time difference coding. *Nature* 417: 543–547.

Davis K (2005) Contralateral effects and binaural interactions in dorsal cochlear nucleus. *Journal of the Association for Research in Otolaryngology* 6: 280–296.

Ehret G and Romand R (eds.) (1997) *The Central Auditory System*. New York: Oxford University Press.

Hackett TA, Stepniewska I, and Kaas JH (1998) Subdivisions of auditory cortex and ipsilateral cortical connections of the parabelt auditory cortex in macaque monkeys. *Journal of Comparative Neurology* 394: 475–495.

Huang CL and Winer JA (2000) Auditory thalamocortical projections in the cat: Laminar and areal patterns of input. *Journal of Comparative Neurology* 427: 302–331.

Kaas JH, Hackett TA, and Tramo MJ (1999) Auditory processing in primate cerebral cortex. *Current Opinion in Neurobiology* 9: 164–170.

Loftus WC and Sutter ML (2001) Spectrotemporal organization of excitatory and inhibitory receptive fields of cat posterior auditory field neurons. *Journal of Neurophysiology* 86: 475–491.

May BJ (2000) Role of the dorsal cochlear nucleus in the sound localization behavior of cats. *Hearing Research* 148: 74–87.

Middlebrooks JC and Green DM (1991) Sound localization by human listeners. *Annual Review of Psychology* 42: 135–159.

Middlebrooks JC and Knudsen EIA (1984) Neural code for auditory space in the cat's superior colliculus. *Journal of Neuroscience* 4: 2621–2634.

Middlebrooks JC, Xu L, Eddins AC, and Green DM (1998) Codes for sound-source location in non-tonotopic auditory cortex. *Journal of Neurophysiology* 80: 863–881.

Oertel D, Bal R, Gardner SM, Smith PH, and Joris PX (2000) Detection of synchrony in the activity of auditory nerve fibers by octopus cells of the mammalian cochlear nucleus. *Proceedings of the National Academy of Sciences of the United States of America* 97: 11773–11779.

Oertel D, Fay RR and Popper AN (eds.) (2002) *Integrative Functions in the Mammalian Auditory Pathway.* New York: Springer.

Oliver DL and Morest DK (1984) The central nucleus of the inferior colliculus in the cat. *Journal of Comparative Neurology* 222: 237–264.

Pickles JO (1988) *An Introduction to the Physiology of Hearing*, 2nd edn. London: Academic Press.

Ramachandran R and May BJ (2002) Functional segregation of ITD sensitivity in the inferior colliculus of decerebrate cats. *Journal of Neurophysiology* 88: 2251–2261.

Sparks DL (1989) The neural encoding of the location of targets for saccadic eye movements. *Journal of Experimental Biology* 146: 195–207.

Strutt JD (Lord Rayleigh) (1907) On our perception of sound direction. *Philosophical Magazine* 13: 214–232.

Warr WB (1992) Organization of olivocochlear efferent systems in mammals. In: Webster DB, Popper AN and Ray RR (eds.) *The Mammalian Auditory Pathway: Neuroanatomy*, pp. 410–448. New York: Springer.

Auditory System: Efferent Systems to the Auditory Periphery

M C Brown, Massachusetts Eye & Ear Infirmary, Boston, MA, USA

The auditory central pathway contains a number of descending systems that allow higher centers to modulate the processing performed at lower centers. The actions and functions of these descending systems are generally not known. Progress toward an understanding of descending systems has progressed most rapidly on the descending systems to the auditory periphery, since these are most accessible to investigation. This article reviews the two main efferent systems to the auditory periphery: (1) the olivocochlear neurons that innervate the organ of Corti in the inner ear and (2) the middle ear muscles that are linked to the ossicles of the middle ear.

Olivocochlear Neurons

Hair cell receptors within the auditory, vestibular, and lateral line systems receive an abundant efferent innervation from the central nervous system. This characteristic sets them apart from receptors in other sensory modalities. The auditory receptors receive innervation from the olivocochlear (OC) system, named because of its origin in the brain stem's superior olivary complex and termination in the cochlea (inner ear) (**Figure 1**). Two separate systems of OC neurons exist. Lateral olivocochlear (LOC) neurons originate in and near the lateral superior olive and project to auditory-nerve dendrites beneath the inner hair cells. Medial olivocochlear (MOC) neurons originate in the more medial parts of the complex and project to outer hair cells (**Figure 1**). LOC neurons are located predominantly on the same side of the brain as the cochlea that they innervate, whereas MOC neurons are distributed bilaterally.

Little is known about the functional role of the LOC neurons, although they may excite auditory nerve fibers. LOC neurons may be divided into subgroups that release acetylcholine, gamma-amino butyric acid (GABA), calcitonin gene-realated peptide, and/or dopamine. Much more is known about MOC neurons. They release acetylcholine, which binds to an alpha 9/alpha 10 nicotinic cholinergic receptor located in the outer hair cell membrane. Calcium inflow activates a calcium-activated potassium channel to open, which hyperpolarizes the hair cell. This effect decreases the outer hair cell electromotility, so they provide less amplification of the movement of the organ of Corti. The sensory inner hair cells thus receive less stimulation and respond

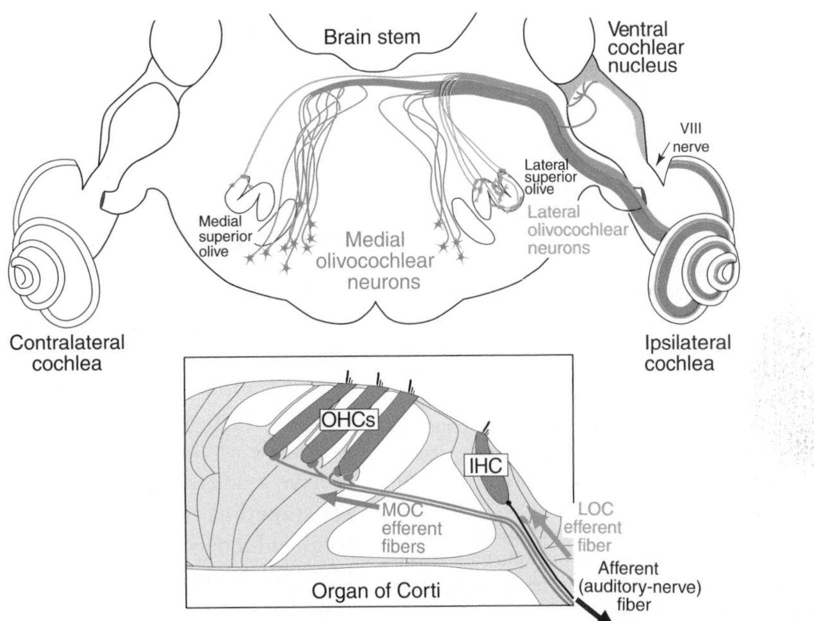

Figure 1 Neural pathways of the medial and lateral olivocochlear neurons in the brain stem and cochlea (inset). Only the efferent pathways to the right (ipsilateral) cochlea are illustrated. Large arrows indicate the direction of spike transmission. An afferent auditory-nerve fiber (type I) from an inner hair cell is illustrated in black; auditory nerve fibers (type II) from outer hair cells are not illustrated. Adapted from Liberman MC (1990) Effects of chronic cochlear de-efferentation on auditory-nerve response. *Hearing Research* 49: 209–224, with permission from Elsevier.

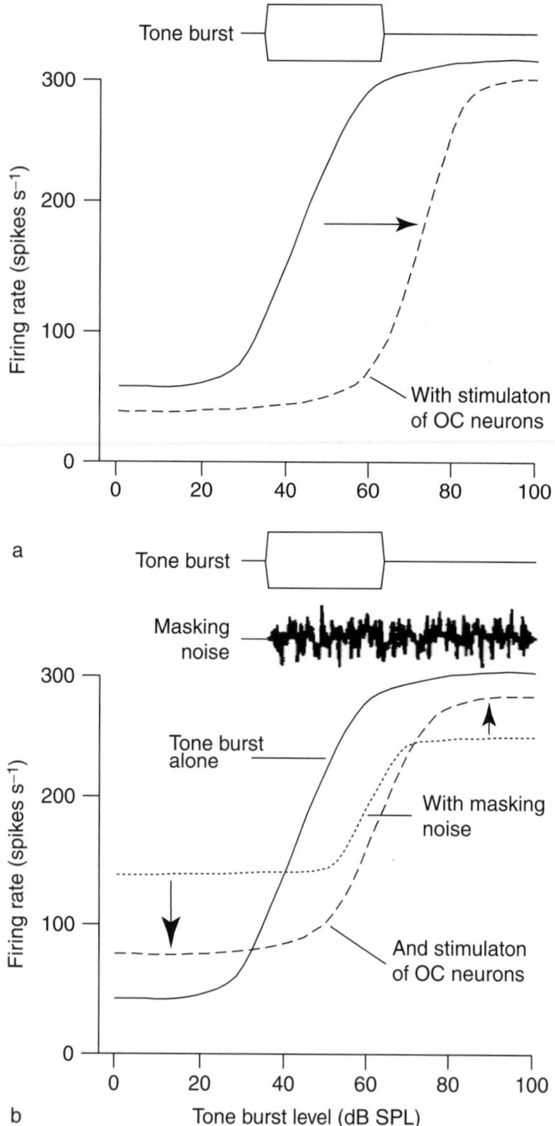

Figure 2 Rate vs. level functions for an auditory-nerve fiber. Insets above the graphs show the stimuli. (a) Response to tone bursts alone (solid curve) increases up to saturated rate, where further increases in level cannot be signaled by the fiber. Response to tone bursts plus stimulation of the MOC neurons (dashed curve). The effect of olivocochlear stimulation is to shift the function to the right (arrow), shifting the dynamic range of the fiber to higher sound levels. (b) Response to tone bursts in a continuous noisy background is 'compressed' (dotted function, solid curve reproduced from panel 'a'). Here, the fiber responds to the ongoing noise, boosting its rate even at low tone burst levels, and causing adaptation and less response at high tone burst levels. In this condition, stimulation of the MOC neurons (dashed curve) decreases the response to the ongoing noise (left arrow), decreasing adaptation and allowing a higher response to the transient tone burst (right arrow). The effect is to 'uncompress' the function, allowing the fiber to better signal changes in level of the tone.

less. Less excitatory neurotransmitter is released and there is less response of the auditory nerve fibers. The effect is a shift in the rising portion of the rate/level function (**Figure 2(a)**), with little effect on spontaneous rate and usually not much change on saturated rate. Overall, the most sensitive range of hearing is shifted to higher levels. Thus, controlling the dynamic range of hearing is an important function of the MOC neurons.

MOC neurons are likely to perform an additional function: to reduce the masking effects of noise on transient signals. To understand the role of MOC neurons, the effects of noise on rate versus level functions of auditory-nerve fibers will be briefly reviewed. The rate-level function to a transient tone burst alone is shown in **Figure 2** (see solid lines). The effect of noise is a 'compression' of this function (compare dotted vs. solid functions in **Figure 2(b)**). The compression is caused at low tone burst levels by the fiber's response to the ongoing noise. Also, at high tone burst levels, the fiber is in an adapted state because of its response to the noise and thus responds less to the tone burst. MOC neurons 'unmask' this compressed level function by decreasing the response to the ongoing noise. This decreases the noise response at low tone burst levels (left arrow in **Figure 2(b)**). Because there is less response to the noise, the fiber is less adapted and can better signal the tone burst when it comes along at moderate and high levels (right arrow on **Figure 2(b)**). Overall, the function is less compressed and better able to signal changes in the tone even when masked by noise.

One additional function of both MOC and LOC neurons is to protect the inner ear from damage due to high-level sounds. However, these systems have a relatively slow time course of action. Thus, their protective effects are probably mostly seen for sounds of long duration rather than for brief, impulsive sounds. Finally, OC neurons may mediate selective attention, for example, by decreasing cochlear sensitivity during a visual task.

Middle Ear Muscles

Two muscles attach to the ossicles of the middle ear: the tensor tympani that attaches to the malleus and the stapedius that attaches to the stapes (**Figure 3**). The tensor tympani is innervated by a branch of the fifth cranial nerve, whereas the stapedius is innervated by a branch of the seventh cranial nerve. The muscles receive a dense innervation from motor neurons. The cat stapedius receives almost 1200 motor neurons in total, which is about one motor neuron per muscle fiber. Such high innervation ratios are only found in muscles requiring a high degree of control such as the muscles controlling movements of the eye. This innervation ratio implies a high degree of central control on the action of these muscles.

The effect of contraction of the middle ear muscles is generally to reduce transmission of sound through

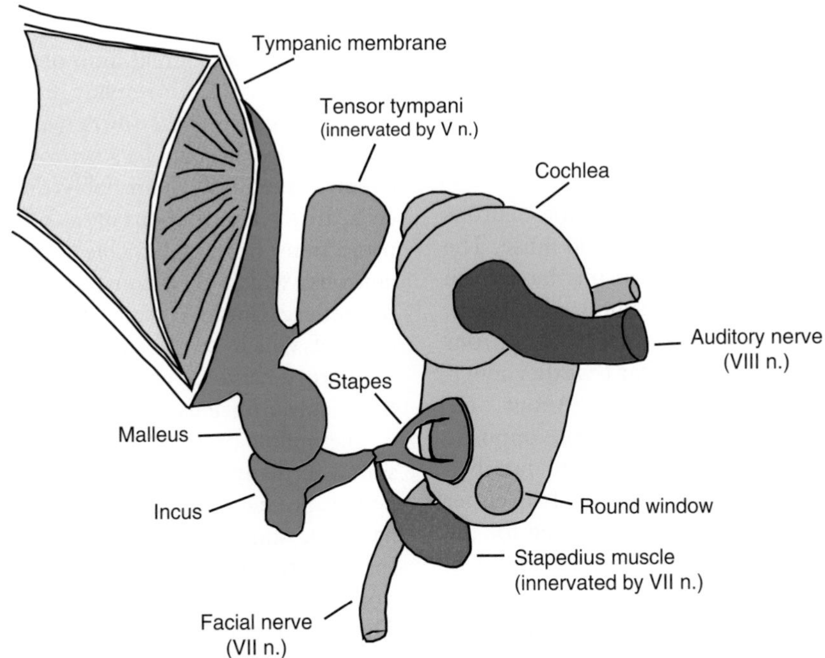

Figure 3 Diagram of the two middle ear muscles and their attachments to the ossicles. The stapedius muscle is attached to the stapes; it is innervated by a branch of the VII nerve. The tensor tympani is attached to the malleus; it is innervated by a branch of the Vth cranial nerve. The action of both muscles is to decrease sound transmission through the middle ear.

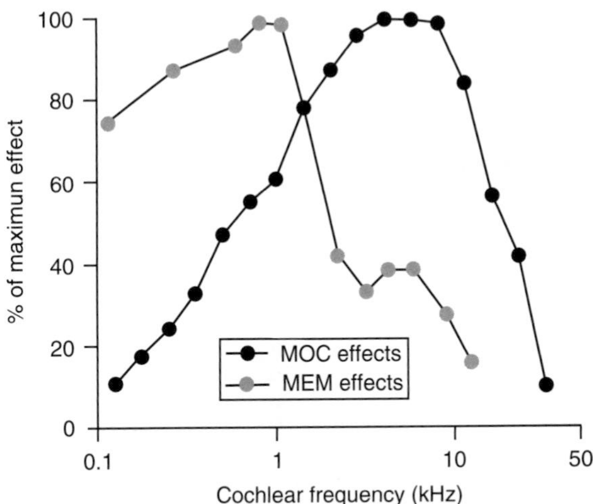

Figure 4 Relative strengths of the MOC and MEM effects. MOC data are the effects on auditory-nerve fibers during electrical activation of the OC bundle; MEM data are the changes in sound transmission during electrical stimulation of the stapedius muscle. All data are from anesthetized cats. Reproduced from Liberman MC and Guinan JJ, Jr. (1998) Feedback control of the auditory periphery: Anti-masking effects of middle ear muscles vs. olivocochlear efferents. *Journal of Communication Disorders* 31: 471–483, with permission from Elsevier.

the middle ear. The maximal effect is about 20 dB. While this is generally the same effect as stimulation of the OC neurons, the frequency ranges of the two systems are different and complimentary. Whereas the middle ear muscle contractions cause loss of sensitivity mainly for low frequencies, the action of the MOC neurons is mainly at middle and high frequencies (**Figure 4**). The middle ear muscles probably have generally similar functions to the OC systems. The middle ear muscle (MEM) reflex is particularly effective at decreasing the masking of auditory-nerve fibers tuned to high frequencies in the presence of low-frequency noise. This is an important situation because much environmental noise contains low frequencies. In addition, activity in the middle ear muscles of some animals precedes their own vocalizations, evidence that this system prevents self-stimulation.

The action of both MOC neurons and MEM motor neurons causes changes in otoacoustic emissions (OAEs). These emissions are sounds that emanate from the ear either spontaneously or evoked by sound stimulation. OAEs are thought to be generated by outer hair cells, which are electromotile. OAEs can be recorded with a microphone placed in the ear canal in animal studies as well as in human subjects. They can hence be used as a 'window' to indicate changes taking place in the outer hair cells. In one type of experiment, an OAE is recorded in one ear in response to a low-level sound stimulus and a reflex elicitor is presented to the opposite ear to evoke a reflex via a crossed central pathway (see below). As both MOC neurons (via their direct endings) and MEM muscles (via changing the sound transmission through the middle ear) can influence the outer hair cells, both systems must be considered when changes in OAEs are observed.

Olivocochlear and Middle Ear Muscle Reflexes

Both MOC neurons and stapedius motor neurons respond to sounds such as pure tones and noise. The response characteristics of these neurons have been determined in recordings from anesthetized animals. The majority (about two-thirds) of MOC neurons respond to sound presented in the ear that they innervate. These units are called Ipsi units. About one-third of the units respond to sound in the opposite ear and are called Contra units. A small number, called Either-Ear units, respond to sound in either ear. Regional separation has been established for the Ipsi and Contra groups in labeling experiments. Ipsi units are located in the side of the brain stem opposite the cochlea they innervate, whereas Contra units are located on the same side as the cochlea they innervate (**Figure 5**). MOC neurons are sharply tuned to sound frequency. Their projections onto the cochlea are tonotopic and generally follow the cochlear frequency mapping set by the cochlear tuning. Thus, an MOC neuron that responds to high sound frequencies feeds back onto a high-frequency location in the cochlea.

The 'wiring diagram' of the MOC reflex is fairly well established. The afferent limb of this reflex is the type I auditory-nerve fibers sending messages centrally from inner hair cells. These fibers synapse in the cochlear nucleus. Cochlear nucleus neurons within the posteroventral subdivision, probably the stellate/multipolar cells, are the interneurons of the reflex. There are direct projections from the cochlear nucleus to the MOC neurons, which then constitute the efferent limb of the reflex. Thus, the reflex consists of three neurons each of which has been fairly well identified: type I auditory-nerve fiber, stellate/multipolar neuron of the cochlear nucleus, and MOC neuron.

The middle ear muscle with the biggest response to sound in many species (including humans) is the stapedius. Studies in animals demonstrate that like MOC neurons, stapedius motor neurons are divided into Ipsi, Contra, and Either-Ear groups. There is an additional group that responds only to sound in both ears. These different subgroups show regional clustering around and rostral to the facial motor nucleus. All motor neurons are located on the same side as the innervated muscle. For the middle ear muscle reflex, less is known about the interneurons. The reflex pathway probably consists of three or four neurons. The reflex interneurons within the cochlear nucleus have not been specified. The stapedius motor neurons

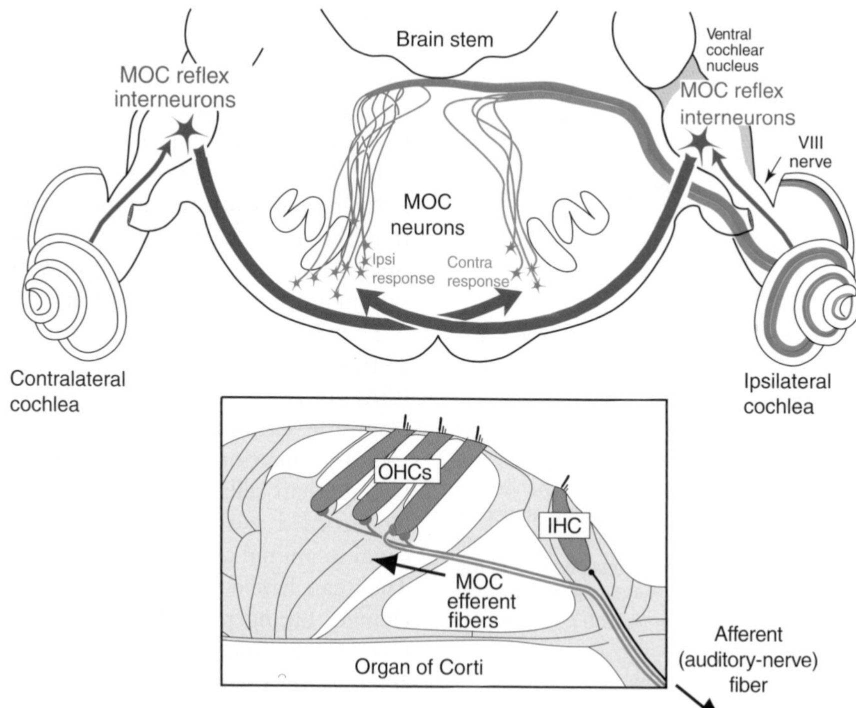

Figure 5 Neural pathways for the MOC reflex. The reflexes begin with activation of the auditory-nerve fibers. This information is transmitted to the cochlear nucleus, where interneurons are activated. MOC Ipsi neurons (blue color) receiving crossing information from the cochlear nucleus of the opposite side and in turn project back to the cochlea on that side. This ipsilateral reflex loop is thus double-crossed. MOC Contra neurons (red color) receive crossing information from the opposite cochlear nucleus but their axons in turn project back to the cochlea on the same side. This contralateral reflex loop is thus single crossed.

differ from MOC neurons in that they are fairly broadly tuned to sound frequency. They also have fairly high thresholds to sound (75–90 dB SPL), whereas MOC neurons can respond at much lower sound levels.

The simple reflex pathways shown in **Figure 5** are under the control of higher centers. Evidence in support of this idea is that OC neurons receive inputs from higher centers such as the inferior colliculus and auditory cortex. In fact, work by Suga in the bat has shown that electrical stimulation of the bat cortex results in changes in cochlear physiology, a demonstration of the effectiveness of the higher centers on this brain stem pathway. The descending projections may enable the brain stem reflexes to be actively controlled – for instance, turned off when listening to low-level sounds near threshold and left on in noisy situations or perhaps when vocalizing. The details of this higher control remain to be worked out.

See also: Auditory Cortex Structure and Circuitry; Auditory System: Central Pathways; Auditory Cortex: Models; Hair Cells: Sensory Transduction; Hair Cell Differentiation.

Further Reading

Delano P, Elgueda D, Hamame C, and Roblest L (2007) Selective attention to visual stimuli reduces cochlear sensitivity in chinchillas. *The Journal of Neuroscience* 27(15): 4146–4153.

de Venecia RK, Liberman MC, Guinan JJ, Jr., and Brown MC (2005) Medial olivocochlear reflex interneurons are located in the posteroventral cochlear nucleus. *Journal of Comparative Neurology* 487: 345–360.

Guinan JJ, Jr. (1996) The physiology of olivocochlear efferents. In: Dallos P, Popper AN, and Fay RR (eds.) *The Cochlea*, pp. 435–502. New York: Springer.

Kobler JB, Guinan JJ, Jr., Vacher SR, and Norris BE (1992) Acoustic reflex frequency selectivity in single stapedius motoneurons of the cat. *Journal of Neurophysiology* 68: 807–817.

Liberman MC (1990) Effects of chronic cochlear de-efferentation on auditory-nerve response. *Hearing Research* 49: 209–224.

Liberman MC and Guinan JJ, Jr. (1998) Feedback control of the auditory periphery: Anti-masking effects of middle ear muscles vs. olivocochlear efferents. *Journal of Communication Disorders* 31: 471–483.

Suga N, Xiao Z, Ma X, and Ji W (2002) Plasticity and corticofugal modulation for hearing in adult animals. *Neuron* 36: 9–18.

Warr WB (1992) Organization of olivocochlear efferent systems in mammals. In: Webster DB, Popper AN, and Fay RR (eds.) *The Mammalian Auditory Pathway: Neuroanatomy*, pp. 410–448. New York: Springer.

Auditory System: Giant Synaptic Terminals, Endbulbs, and Calyces

D K Ryugo, Johns Hopkins University School of Medicine, Baltimore, MD, USA
G A Spirou, West Virginia University, Morgantown, WV, USA

Introduction

The inescapable relationship between form and function in biological systems has focused attention on the giant synaptic endings in the central auditory system. Two of these giant endings are the subject of this article: endbulbs of Held that arise from myelinated auditory nerve fibers and calyces of Held that arise from globular bushy cells of the cochlear nucleus. These endings form hundreds of synapses with their targets and are implicated in fail-safe synaptic transmission that faithfully couples neural activity to environmental auditory events. The importance of this linkage is the preservation of neural timing that codes for all aspects of sound. Sounds only have significance to us when they occur over time. This preservation of timing within auditory signals enables the translation of prosodic utterances into perceptible speech and the processing of two cues used to localize sounds in space, interaural time differences (ITDs) and interaural level (intensity) differences (ILDs).

The ITD and the ILD pathways contain some of the largest, fastest, and most powerful synaptic endings in the central nervous system. The ITD pathway is initiated by the myelinated auditory nerve fibers that give rise to large and highly branched axosomatic endings called the endbulbs of Held. One or two of these endbulbs converge onto the cell body of a spherical bushy cell in the anteroventral cochlear nucleus (AVCN) that, in turn, transmits high-fidelity temporal information bilaterally to structures in the superior olivary complex. The ILD pathway is initiated by myelinated auditory nerve fibers that give rise to many smaller endbulbs that converge onto globular bushy cells in the AVCN. Globular bushy cells project to the contralateral medial nucleus of the trapezoid body (MNTB) and form a giant axosomatic synaptic ending called the calyx of Held. These two giant endings represent the key components for preserving timing in the auditory system.

Endbulbs of Held: Morphology

Large endings of auditory nerve fibers were originally described in Golgi-stained material by Hans Held in 1893. They were characterized in kittens as axosomatic, spoon-shaped endings with many filopodia that marked the end of the anterior branch of the fiber (**Figure 1**, left). When endings of the anterior branch were labeled in the same location of the adult cat, however, they appeared vastly different (**Figure 1**, right). The question was whether the differences were due to staining techniques or age of the animal studied. These endings were analyzed in an age-graded series of cats using the Golgi stain and horseradish peroxidase (HRP), demonstrating that the structural differences were due to the age of the animal examined (**Figure 2**). Auditory nerve fibers from every vertebrate examined have exhibited large axosomatic endbulbs of Held. These include the red-eared turtle, alligator lizard, chickens, owls, guinea pigs, rats, cats, monkeys, and humans. Endbulbs are found in the anteroventral cochlear nucleus and make synaptic contact with neurons known as spherical bushy cells.

The phylogenetically conserved ending implies functional significance for the processing of acoustic information. The size of the ending suggests powerful synaptic drive on the postsynaptic cell. The faithful transmission of presynaptic activity to the postsynaptic cell ensures that neural activity in the brain is yoked in time to acoustic events. The precision of the timing forms the substrate for sound localization as well as for translating sequences of sounds into perceptible speech.

Synapses: Presynaptic Endings

The reliability of transmission at this synapse was inferred from the observation that a small positive potential, called the prepotential, occurred 0.5 ms prior to the action potential. The prepotential was interpreted to represent the depolarization of the endbulb, and the coupling between prepotential and spike resulted in the notion of a fail-safe synapse that may nonetheless be modulated by inhibition. This ending has between 400 and 1500 release sites as calculated through serial section electron microscopy. The release sites of endbulbs have the morphology typical of normal synapses in the central nervous system (**Figure 3**). They are characterized by the presynaptic assemblage of clear, round synaptic vesicles, approximately 50–55 nm in diameter, that accumulate across from convex, dome-shaped postsynaptic densities (PSDs) of the postsynaptic membrane.

The postsynaptic density is visible as a small (on the order of 0.5 μm in diameter or less) aggregation of electron-dense material on the cytoplasmic side of the

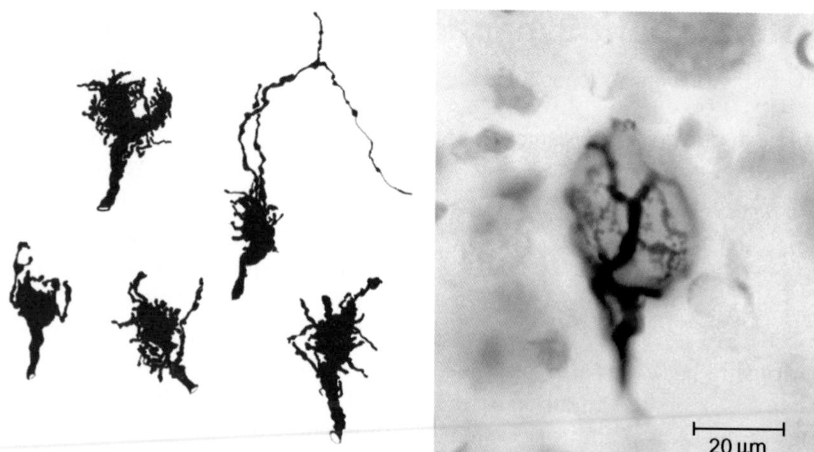

Figure 1 Left panel: Golgi-stained endbulbs of Held from neonatal kittens. They appear as spoon-shaped swellings with filopodia, identical to the original description by Hans Held in 1893. Right panel: HRP-labeled endbulb from an adult cat. The contrast in morphology of neonatal and adult cat endbulbs raised the question as to whether the differences were due to staining methods or age of the animals.

target membrane. The density is composed of neurotransmitter receptors, ion channels, signal transduction proteins, cytoskeletal proteins that anchor receptors to the postsynaptic site, and adhesion molecules for proper alignment of pre- and postsynaptic membranes. All of this machinery is essential for chemical synaptic transmission. The main demand on this transmission is speed. The auditory system processes sound for spatial localization, identification, and communication. Inherent to these functions is the ability to faithfully transmit rapid changes in the acoustic signals. The amino acid glutamate is the candidate for facilitating rapid transmission.

The unmistakable identification of glutamate as the neurotransmitter of the auditory nerve has been impeded by the widespread distribution of glutamate in nervous tissue and the inability to measure its release during synaptic activity. As a result, research strategies have been to analyze the effects of pharmacologic agonists and antagonists of glutamate, to study the properties of glutamate receptor subunits, and/or to characterize glutamate transporters. Quantitative immunohistochemistry demonstrated greater labeling over primary auditory nerve endings compared to nonprimary endings (containing flat or pleomorphic synaptic vesicles) or glia. Moreover, potassium-induced depolarization that depleted glutamate significantly lowered such staining in endbulb terminals. The presence of glutamate and one of its metabolic precursors in the endbulb and the complementary distribution of glutamate receptors (GluRs) in cochlear nucleus neurons represent strong inferential evidence that glutamate is the neurotransmitter. The involvement of glutamate in auditory nerve synapses is greatly strengthened by observations that large postsynaptic currents in cochlear nucleus slice recordings are blocked by glutamate receptor antagonists.

Synapses: Postsynaptic Targets

Receptors that reside in the PSD of the target cell seem to be primarily of the ionotropic type that are formed by several subunits surrounding a central ion pore. Immunogold and immunoperoxidase methods demonstrate a strong reaction for GluR2/3 and GluR4. Since GluR1 shows little or no immunolabeling at endbulb synapses, it is inferred that most synaptic α-amino-3-hydroxy-5-methyl-4-isoxazole propionic acid (AMPA) receptors are composed of GluR3 and GluR4 subunits. These subunits facilitate fast transmission because of rapid desensitization. Moreover, calcium permeability due to the lack of GluR2 could account for the rapidly decaying responses of the postsynaptic spherical bushy cells. Immunolabeling of metabotropic glutamate receptors (mGluR1α) is modest at the endbulb synapse.

N-Methyl-D-aspartate (NMDA) receptors feature a voltage-dependent calcium channel. Opening of the NMDA channel occurs as a result of depolarization via the AMPA receptors. There are a number of different NMDA receptor subunits, including NR1, NR2A–D, and NR3. Receptors are composed of NR1 plus one or more variants of NR2, the combination of which determines physiological properties. The role of these receptors at the endbulb synapse is not known; they are present in the developing auditory system but they diminish by weaning.

The spiking pattern of auditory neurons is determined in part by the type of voltage-sensitive potassium channels expressed. The activation and deactivation

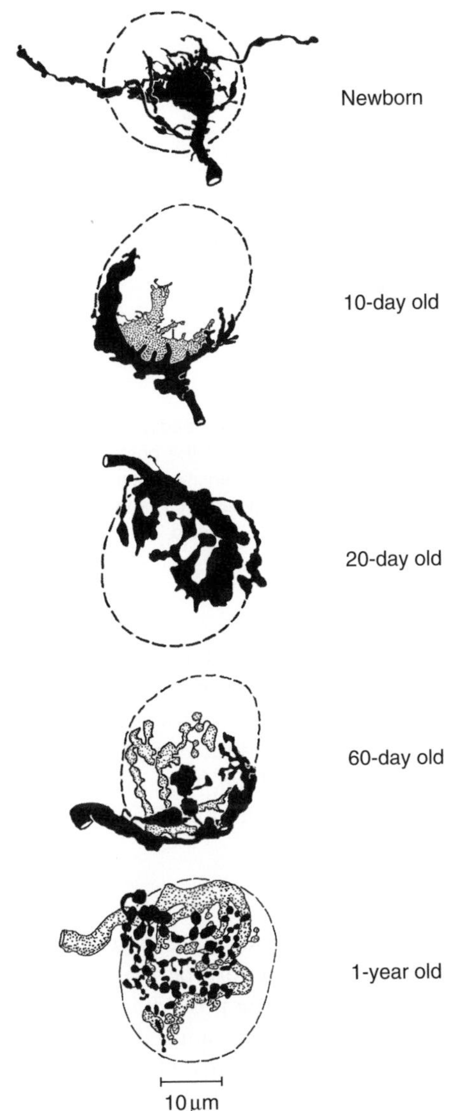

Figure 2 Developmental sequence of endbulbs of Held. Endbulbs were stained using either the Golgi or HRP method. Note the progression of endbulb structure, from a spoon-shaped ending to a highly branched arborization. This result emphasizes the importance of age on structure and presumably function.

kinetics of potassium channels provide for rapid repolarization of action potentials. These potassium channels are abundant in spherical bushy cells. The fast response is necessary because auditory neurons maintain timing while responding to high rates of synaptic input. The likely candidates for the low voltage-activated and high voltage-activated potassium currents in auditory neurons are Kv1.1 and Kv3.1, respectively. The combination of AMPA receptors and these potassium channels accounts for the brevity and rapidity of auditory neuronal signaling.

Physiology

The time-varying nature of natural sounds requires that accurate identification depends on the ability of auditory neurons to encode timing. Vertebrates have the capacity to discriminate between pitches that differ by as little as 2 μs in the period of the sound waves, and to resolve differences on the order of 10 μs in the time of arrival of a sound at the two ears. The duration of single action potentials is significantly longer than these timing differences, so it is remarkable that there are biological mechanisms that reliably preserve and transmit this information. There are, however, a number of specializations in auditory neurons that facilitate the processing demands.

One to two endbulbs of Held converge upon single spherical bushy cells in the anteroventral cochlear nucleus. It is estimated that a single auditory nerve fiber will activate between 2000 and 10 000 AMPA receptors. Excitatory postsynaptic potentials (EPSPs) in bushy cells are brief and have very rapid rise times. The consequences are (1) threshold is reached quickly, (2) action potentials occur reliably and with little temporal jitter, (3) endbulb depolarization produces only one postsynaptic spike, and (4) refractory period is short. As a result, spherical bushy cells exhibit the ability to 'follow' repeated stimulation up to 300 Hz, which is near the maximum firing rate of auditory nerve fibers, but there is also 'enhancement' – these cells perform better than the fibers. These factors contribute to the coupling in time of neural discharges to acoustic events by the auditory system.

Plasticity

The efficiency of synaptic transmission at the endbulb synapse is well known and is determined by the amplitude and time course of evoked excitatory postsynaptic currents (EPSCs). Both spontaneous and evoked EPSCs, however, exhibit large fluctuations in amplitude and time course. Because receptor desensitization did not play a role in this variability, it was proposed that variability in response characteristics was due to intrinsic fluctuations in release probability and numbers of available receptors. Modulation of release probability is achieved by varying calcium and protein kinase C levels, which in turn can elicit short-term plasticity of responses.

Evidence for long-term plasticity arises in different forms. Auditory nerve fibers exhibit separate groupings on the basis of spontaneous discharge rates and thresholds to activation. Low-spontaneous-rate (SR) fibers have high thresholds, represent approximately 40% of the population of myelinated fibers, and give

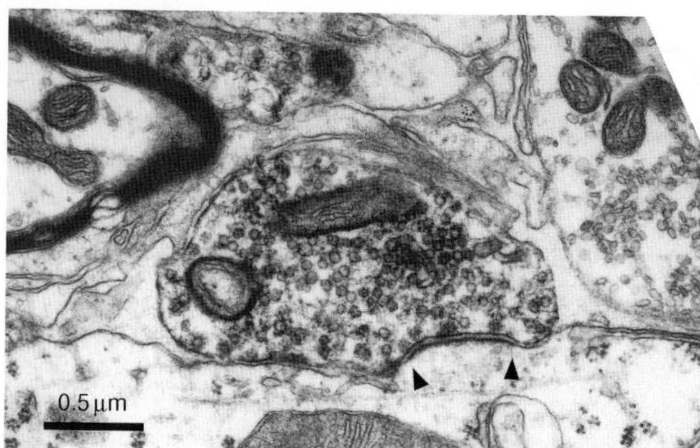

Figure 3 Electron micrograph through a section of endbulb containing a synapse (flanked by arrowheads). The ending contains mitochondria and many clear, round synaptic vesicles. At the synapse, there is the signature arch of the postsynaptic membrane accompanied by a fuzzy membrane thickening known as the postsynaptic density. The postsynaptic density contains neurotransmitter receptors, ion channels, and structural proteins. Reproduced from Ryugo DK, Huchhton DM, Pongstapron T, and Niparko JK (1997) Ultrastructural analysis of primary endings in deaf white cats: Morphologic alterations in endbulbs of Held. *Journal of Comparative Neurology* 385: 230–244, with permission. Copyright (1997) by Wiley-Liss. Inc.

rise to endbulbs having very complicated arborizations. In contrast, high-SR fibers have low thresholds, represent roughly 60% of the population of myelinated fibers, and give rise to endbulbs have less complicated shapes (**Figure 4**). Under normal conditions, these two populations of endbulbs exhibit different morphology.

In the event of congenital deafness, however, the auditory nerve fibers have little spike activity, and the resulting endbulbs appear withered (**Figure 5**). Moreover, the endbulb synapses of congenitally deaf cats have a pathologic appearance when compared to those of normal hearing cats (**Figure 6(a)**, asterisks). Those of congenitally deaf cats lack synaptic vesicles and show hypertrophied, flattened postsynaptic densities (**Figure 6(b)**, arrowheads). These synapses could be rescued if miniaturized human cochlear implants were surgically placed into congenitally deaf cats (**Figure 6(c)**, asterisks). The implanted cats were stimulated 7 h per day, 5 days per week for 10–12 weeks using the same processing strategy and programming software used in children. Environmental sounds had biological significance because the implanted cats could routinely be 'called' for a food reward. Since the cochlear nucleus gives rise to all ascending auditory pathways, the preservation of the endbulb–bushy cell circuit would support the faithful transmission of temporal cues contained within auditory signals. It is proposed that the changes in endbulb synapses by cochlear implants represent one key to the development of integrative and cognitive brain functions reflected in aural and oral communication in deaf children.

Endbulbs of Held serve as a near one-to-one pipeline from a single inner hair cell to a single spherical bushy cell. They are found in a wide variety of vertebrates, emphasizing evolutionary pressure for their preservation. They are known for their reliability but also exhibit structural and functional plasticity under different conditions. Much remains to be learned about these important synaptic endings.

Calyces of Held

The calyx of Held is perhaps the largest nerve terminal in the mammalian central nervous system (CNS) and covers 25–50% of the postsynaptic somatic surface (**Figure 7**). It was recognized as the terminal portion of an axon by Hans Held (1893), whose name became associated with this structure via reference to the 'calyces of Held' by other neuroanatomists of his time. This structure figured in the great debate over the neuron doctrine that blazed at the end of the nineteenth century. Held and Camillo Golgi, two leaders in the nascent field of neuroanatomy, adhered to the theory that the nervous system was a continuous reticulum, resembling a nerve net. Held interpreted rodlike structures seen in the postsynaptic neuron to be the injection of calyx fibrils into the cytoplasm of its target. Ramón y Cajal claimed that the rodlike structures belonged strictly to the postsynaptic neuron. His analysis of the calyx as supporting the integrity of individual nerve cells was eventually proved correct.

The advent of the electron microscope and single-neuron recording techniques in the mid-twentieth century opened a second phase of study of this large terminal and its functional role in hearing. Round synaptic vesicles, structural evidence for an excitatory connection, fill the terminal and associate with

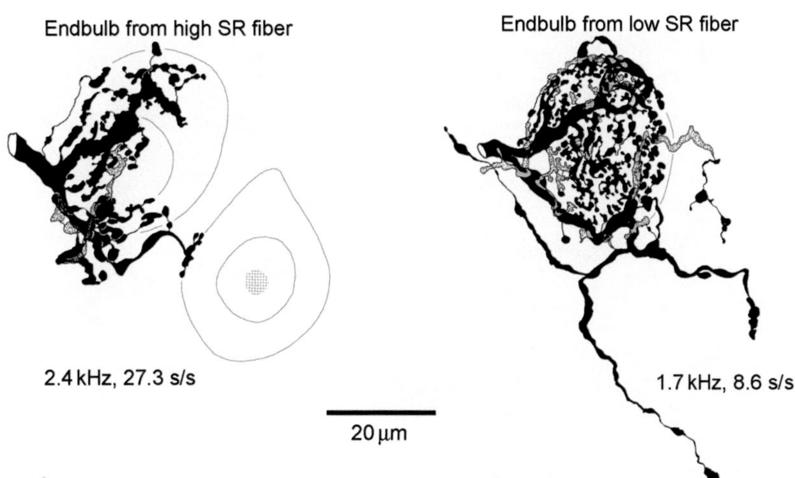

Figure 4 Camera lucida drawings of HRP-labeled endbulbs stained by intracellular injections. Two groups of auditory nerve fibers may be distinguished on the basis of spontaneous discharge rate (SR) and sensitivity. Because they are found at all frequency ranges, they are hypothesized to underlie separate aspects of stimulus processing. Note that the endbulb from the low-SR fiber exhibits more complex branching compared to that of the high-SR fiber. Reproduced from Sento S and Ryugo DK (1989) Endbulbs of Held and spherical bushy cells in cats: Morphological correlates with physiological properties. *Journal of Comparative Neurology* 280: 553–562, with permission. Copyright 1989 by Alan R. Liss.

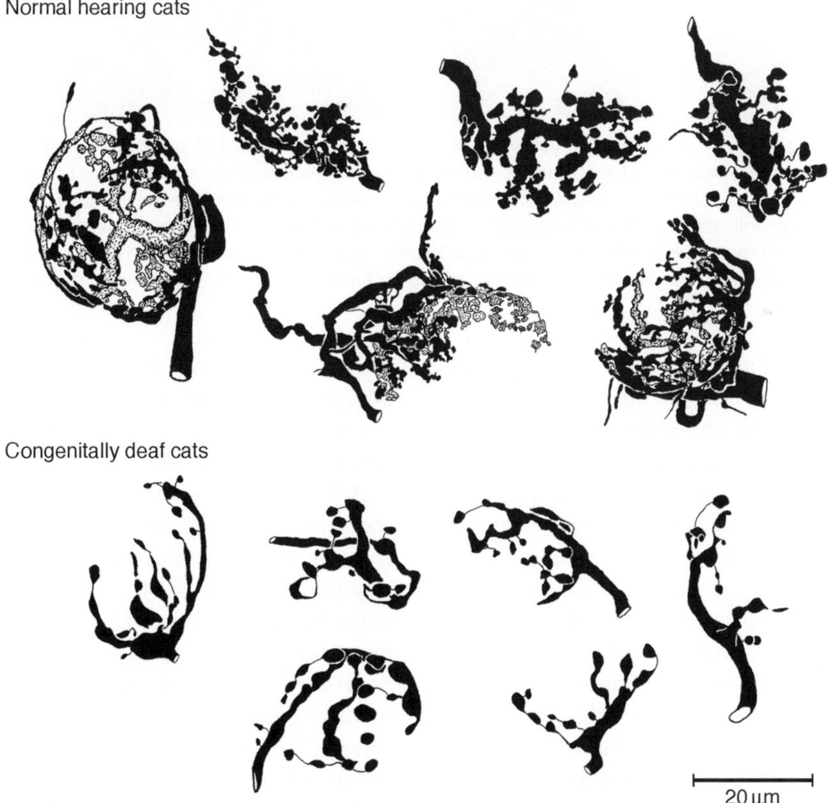

Figure 5 Camera lucida drawings of HRP-labeled endbulbs from normal hearing cats and congenitally deaf cats. There is no spontaneous or driven spike activity in auditory nerve fibers of congenitally deaf cats. Note that the endbulbs of deaf cats appear withered and atrophic. Spike activity has an influence on endbulb shape. Reproduced from Ryugo DK, Rosenbaum BT, Kim PJ, Niparko JK, and Saada AA (1998) Single unit recordings in the auditory nerve of congenitally deaf white cats: Morphological correlates in the cochlea and cochlear nucleus. *Journal of Comparative Neurology* 397: 532–548, with permission. Copyright 1998 by Wiley-Liss, Inc.

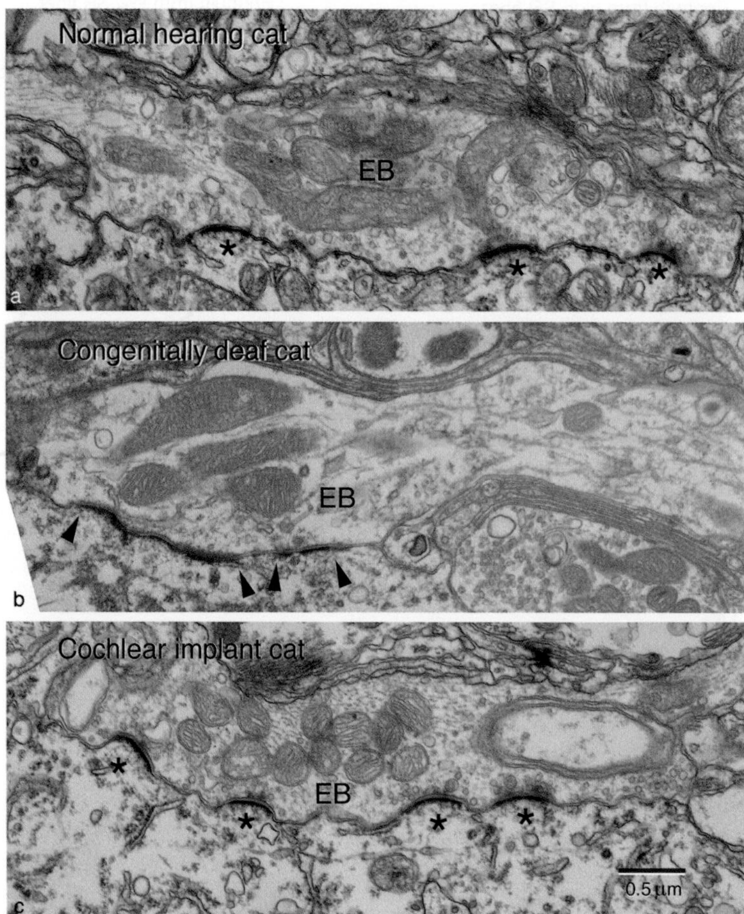

Figure 6 Electron micrographs of endbulbs (EB; yellow) illustrate the restorative effect of activity on synapses. (a) The endbulb of a normal hearing cat exhibits the typical arched postsynaptic density with accumulations of synaptic vesicles (asterisks). (b) The endbulb of congenitally deaf cats, by contrast, contains few synaptic vesicles and the postsynaptic densities are flat and hypertrophied (arrowheads). (c) The endbulb of a congenitally deaf cat that received stimulation from a cochlear implant exhibits synapses with normal morphology (asterisks). Reproduced from Ryugo DK, Kretzmer EA, and Niparko JK (2005) Restoration of auditory nerve synapses in cats by cochlear implants. *Science* 310: 1490–1492, with permission. Copyright 2005 by American Association for the Advancement of Science.

multiple synaptic sites, later shown to range in number from 400 in mice to 600 in rat to 2500 in cat. Resembling the endbulb, the calyx was shown to exhibit a physiological signature in extracellular recordings, whereby a short prepotential preceded an action potential in the postsynaptic neuron, located in the medial nucleus of the trapezoid body within the superior olivary complex, by about 0.5 ms. Each postsynaptic action potential is typically associated with a prepotential. Intracellular recording and labeling of single neurons unambiguously associated the calyx nerve terminal with its origin from the globular bushy cell located in the ventral cochlear nucleus, and revealed that globular bushy cells innervate other cell groups in the superior olivary complex via small bouton endings (**Figure** 8). As described in the previous section of this article, globular bushy cells are themselves contacted by multiple large endings of auditory nerve fibers. The ability of bushy cells to encode temporal fine structure in the incident acoustic wave and their involvement in brain stem circuits that mediate sound localization implicated the calyx in functional roles for localizing sound in space.

A third phase of investigation began in 1994 with the first intracellular recording directly from the calyx in brain slices using patch electrode techniques (**Figures** 9(**a**) and 9(**b**)). Over the ensuing years the calyx has been studied as a model for glutamatergic neurotransmission throughout the brain. The ability to record pre- and postsynaptic signals simultaneously (**Figure** 9(**c**)) has made this a powerful preparation. To date, bulk labeling techniques using tract tracing molecules, viral transfection, and immunocytochemistry have not revealed a single instance of more than one calyx terminating onto a single postsynaptic neuron. Contrary to the pervasiveness of endbulbs of Held, calyces are typically mammalian structures. Their presence in humans, however, can be

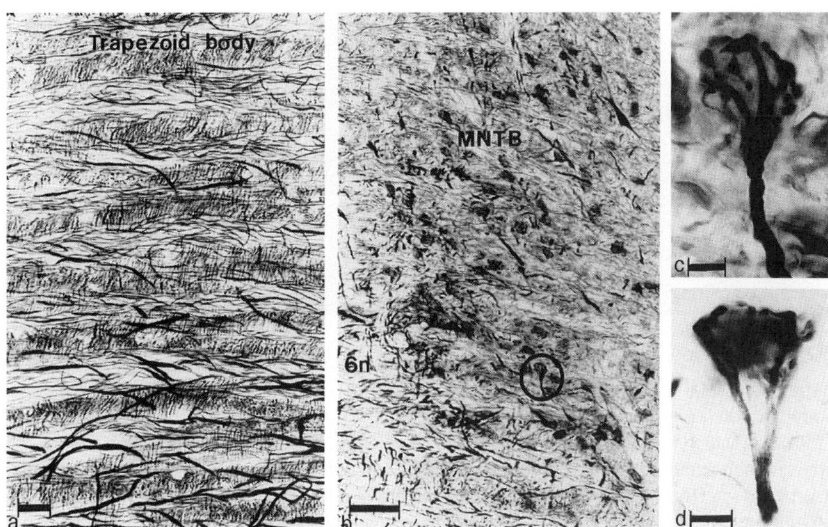

Figure 7 Globular bushy cells in cats are immunolabeled by antisera to the putative calcium-binding protein PEP-19. (a) Thick, immunolabeled fibers in the ventral trapezoid body belong to globular bushy cells. (b) Calyces of Held are immunolabeled throughout the medial nucleus of the trapezoid body (MNTB), and are easily seen because the MNTB neurons are immunonegative. (c) The large-diameter calyceal axon branches near its postsynaptic target. (d) In some cases, large branches occur further from the soma. Scale bar = 50 μm (a), 100 μm (b), 10 μm (c, d). Reproduced from Berrebi AS and Spirou GA (1997) PEP-19 immunoreactivity in the cochlear nucleus and superior olive of the cat. *Neuroscience* 83(2): 535–554, with permission from Elsevier Science Ltd.

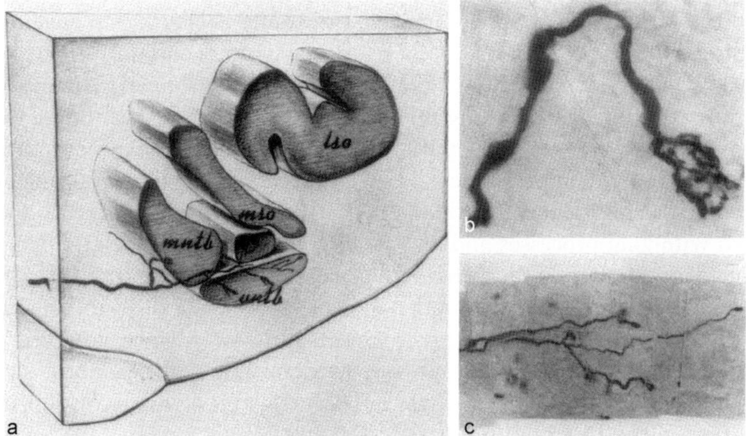

Figure 8 Branching pattern and terminal profiles of a single globular bushy cell input to the superior olivary complex of the cat. (a) Artist's rendering of the branching and terminal profile for the labeled terminals shown in (b) and (c). A thick section through the superior olive in the transverse plane is pictured, viewed from a caudal perspective (mso, lso: medial and lateral superior olive). The large-diameter axon branches and delivers a calyx into the medial nucleus of the trapezoid body (mntb). The calyx is pictured in panel (b). The parent axon turns rostrally, but also branches further within the superior olive. These branches terminate as boutons; those within the ventral nucleus of the trapezoid body are pictured in panel (c). Reproduced from Spirou GA, Brownell WA, and Zidanic M (1990) Recordings from cat trapezoid body and HRP labeling of globular bushy cell axons. *Journal of Neurophsyiology* 63(5): 1169–1190, with permission. Copyright 1990 by The American Physiological Society.

questioned, since their postsynaptic cell group, the MNTB, has not been unambiguously identified in the human brain stem.

Internal Structure

Mature calyces resemble clawlike structures of complex geometry. Each calyx is formed from a large-diameter axon that branches repeatedly, sometimes via thin necks, into single swellings or series of swellings. Early in development the calyx is less branched (see later), and hence derived its name in reference to a floral calyx, the collection of green sepals that surround the petals of the flower. All parts of the calyx except the necks contain active zones and are filled with round synaptic vesicles. The large size of the calyx and number of active zones pose challenges for neurotransmission at high rates, where ionic

Figure 9 Intracellular recordings from the calyx and its postsynaptic target can be performed simultaneously. (a) Differential interference contrast (DIC) image of a medial nucleus of the trapezoid body (MNTB) cell (asterisk) and calyx (arrow points to calyx covering the right side of the cell body). A patch pipette is positioned to record from the calyx. (b) The calyx pictured in panel (a) is filled through the patch pipette with fura dye, which fluoresces in the presence of calcium. This technique confirms the selective recording from the calyx. (c) DIC image showing simultaneous recording from the MNTB cell (left pipette electrode) and the calyx (right pipette electrode). Scale bar = 10 μm (a, b), 15 μm (c). Reproduced from Billups G and Forsythe ID (2002) Presynaptic mitochondrial calcium sequestration influences transmission at mammalian synapses. *Journal of Neuroscience* 22(14): 5840–5847, with permission. Copyright 2002 by the Society for Neuroscience.

gradients must be maintained and neurotransmitter rapidly cleared from the synaptic cleft. Early electron microscopic studies revealed regions of the calyx that are apposed to the postsynaptic membrane and that are studded with synaptic sites and puncta adherentia. These contacts form directly onto the somatic membrane or onto somatic appendages, which are hairlike protrusions of the postsynaptic somatic membrane. Interdigitated with these contact regions are separations of the pre- and postsynaptic membranes, called extended extracellular spaces (EESs; **Figure 10**). These spaces can be infiltrated with glial processes and may clamp extracellular ionic concentration in the cleft and facilitate diffusion of neurotransmitter away from the cleft. These factors minimize desensitization of postsynaptic AMPA receptors because glutamate transporters, which recover synaptically released molecules, are found on glial membranes in the EES and in greater number on glial membranes surrounding the calyx. The fenestrated geometry of mature calyces also supports rapid diffusion from the synaptic cleft. A specialized organelle complex, called the mitochondrion-associated adherens complex (MAC), tethers mitochondria to a punctum adherens within 200 nm of the presynaptic membrane. This complex can be ringed by synaptic sites, and also is hypothesized to support high-fidelity synaptic transmission.

Development

The 1:1 innervation ratio between calyces and their postsynaptic targets provides a developmental model for studying the precise formation of topographic connections between cell groups. The calyx proceeds through four stages of growth: (1) appearance of the migratory growth cone in the vicinity of the MNTB, (2) expansion of growth cone processes into a single large ending called a protocalyx, (3) further expansion of the protocalyx into a spoon-shaped structure called the young calyx, which envelopes about 50% of the postsynaptic somatic surface, and (4) the refinement of the spoon-shaped structure into a more clawlike calyx (**Figure 11**). The timing for these stages of calyx formation has been most studied using rodent animal models. Postsynaptic neurons of the MNTB are born between embryonic days 12 and 14 and within a day or two, axons of ventral cochlear nucleus cells have crossed the midline into the region where the MNTB will arise. Since the MNTB and other cell groups of the superior olivary complex do not coalesce into recognizable entities until P17, innervating axons may appear in this vicinity prior to the postsynaptic targets.

Young calyces appear in relative abundance by P5, so physiological studies of the calyx typically commence at this age. Synapses onto MNTB cells, evident in the electron microscope as early as the day of birth, generate postsynaptic currents which grow in amplitude from 50 pA at P0 to as large as 10 nA by P4. The growth in postsynaptic current accelerates between P2 and P4. Serial section electron microscopy reveals that protocalyces appear at P2, and can expand into young calyces within 48 h. This rapid growth period prompted the question of whether competition among inputs was occurring at the earliest age of young calyx formation at P4. Minimal stimulation techniques suggest multiple inputs onto a minority of MNTB neurons (15%). Three-dimensional reconstructions of MNTB cells and their inputs from electron micrographs reveal multiple inputs onto a

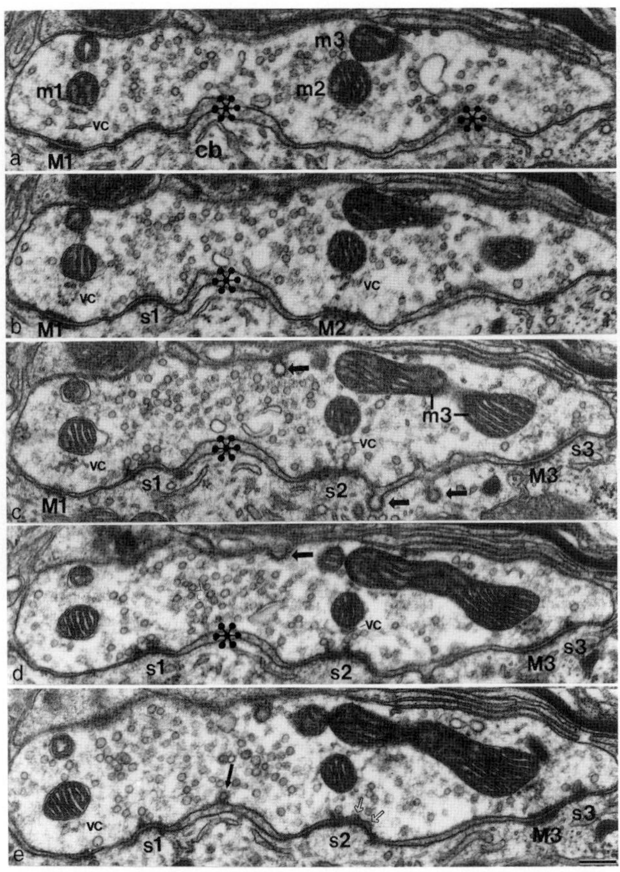

Figure 10 Serial electron micrographs from a calyceal swelling contacting a cell body (cb) in a cat. Synapses (s) and mitochondrion-associated adherens complexes (MACs; labeled M) lie adjacent to one another. These functional contacts can be separated by extended extracellular spaces (asterisk), where pre- and postsynaptic membranes are separated. MACs are comprised of a mitochondrion (m), vesicular chain of membrane between the mitochondrion and cell membrane (vc), and punctum adherens (adjacent to M label). Individual synapses and MACs are numbered. Arrows indicate coated, and presumably endocytosing, vesicles. Scale bar (shown in panel E for all panels) = 0.25 μm. Reproduced from Rowland KC, Irby NK, and Spirou GA (2000) Specialized synapse-associated structures within the calyx of Held. *Journal of Neuroscience* 20: 9135–9144.

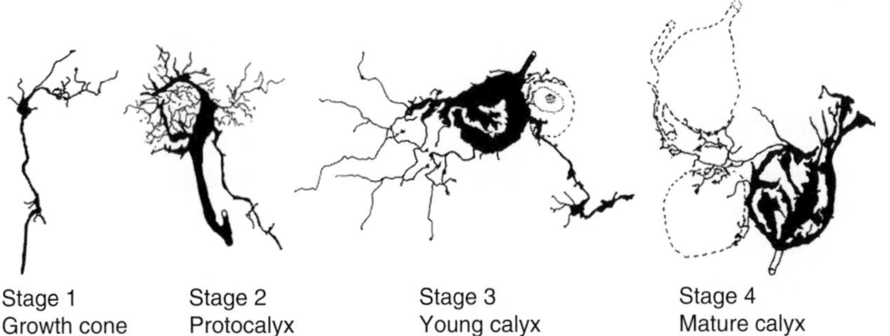

Figure 11 Stages of calyx growth, defined using Golgi stain of pouch-young opossum. Stage 1: Migratory growth cones enter the region of the medial nucleus of the trapezoid body. Stage 2: Growth cones expand into a protocalyx. Stage 3: The protocalyx further expands into a cup-shaped structure called a young calyx, which envelops much of the postsynaptic cell body. Stage 4: The young calyx is pruned into a highly branched, mature calyx. Reproduced from Morest DK (1968) The growth of synaptic endings in the mammalian brain: A study of the calyces of the trapezoid body. *Zetschrift fur Anatomie und Entwicklungsgeschichte* 127: 201–220, with kind permission of Springer Science and Business Media.

similarly small percentage of cells (12%; **Figure 12**). Therefore, specificity in matching of axonal inputs to their targets occurs at early stages of nerve terminal formation. In other neural systems, axons typically compete for innervation territories, and these are refined via experience-driven processes. In the calyx system, the cochlea is only beginning to transduce sound when young calyces are formed, so calyx

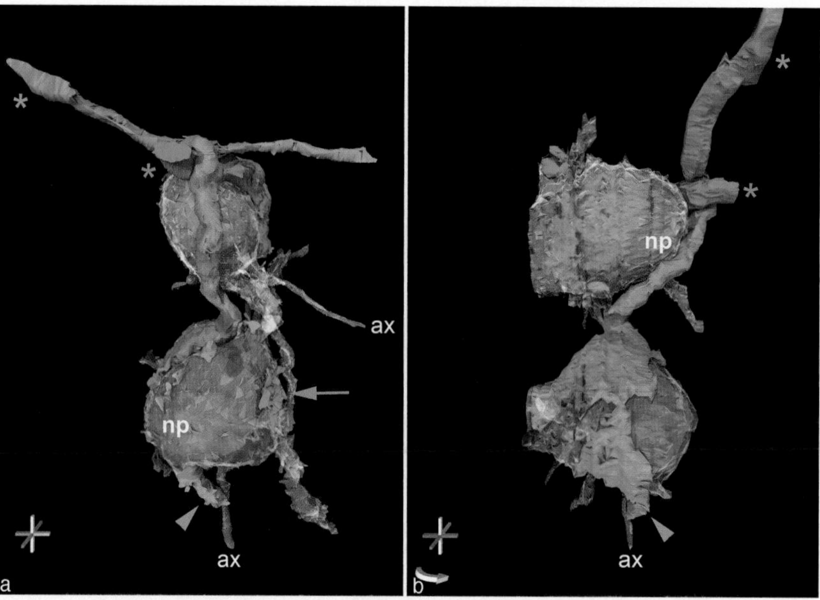

Figure 12 A 3-D reconstruction of two neighboring MNTB cells at P4. (a, b) Two views of a pair of representative MNTB cells that were contacted by single large terminals. Panel (b) is a 90° clockwise rotation of panel (a) about the vertical axis. Rotated scale axes and arrows are provided for reference. MNTB cells are transparent and colored tan (top cell) or brown (bottom cell) for distinction of overlapping dendrites and axons. Cell bodies are ovoid in shape and have two to three dendrites (only proximal dendrites, an axon (ax), and an eccentric nucleus (red) are included in the reconstruction). The blue calyx terminal consists of two large pieces linked by a narrow bridge (panel (b)) and has a solitary process extending away from the cell body (blue arrowhead). The axon (ax) of this terminal was traced to its branch point from a trapezoid body fiber. The green terminal on the top cell extends a collateral that contacts the bottom cell (green arrow in panel (a)). Innervation territories avoid the nuclear pole (np) of the MNTB cells (top cell, panel (b); bottom cell, panel (a)). The nuclear pole is defined by the close apposition of the nuclear membrane and cell membrane. Orange terminals could not be linked to other inputs onto the cell or to an axon. Asterisks indicate the axons for each terminal of the corresponding color. Scale axes are 2 μm from origin to end, 4 μm in total length. Note that scale axes in the 3-D perspective views apply most accurately to the middle depth of field. Reproduced from Hoffpauir BK, Grimes JL, Mathers PH, and Spirou GA (2006) Synaptogenesis of the calyx of Held: Rapid onset of function and one-to-one morphological innervation. *Journal of Neuroscience* 26(14): 5511–5523, with permission. Copyright 2006 by the Society for Neuroscience.

growth is an experience-independent process. Competition among calyx-forming inputs may occur earlier among the growth cones or protocalyces.

Short-Term Synaptic Plasticity

In mammals, the mechanisms that underlie synaptic plasticity have proved difficult to study, especially in the CNS. However, the calyx of Held has offered a solution to this quandary. The power of simultaneous pre- and postsynaptic recording is increased when these techniques are combined with other approaches, such as pharmacologic manipulations and fluorescence imaging. Important information has been gleaned from the calyceal connection with the MNTB cell, tempered by the reminder that studies are conducted during the maturation process of the terminal and its postsynaptic partner, and often not at physiological temperatures. Most studies employ the rat as the animal model for physiological studies, and recordings are made from animals between P5 and P15. This age range brackets the onset of hearing in rodents between P8 and P12 and includes the early stages of myelination.

The calyx-to-MNTB neuron connection has been considered an efficient converter of excitation into inhibition, although recent *in vivo* demonstrations of prepotentials that do not elicit postsynaptic spikes challenge that view. Cellular studies *in vitro* reveal modulatory events, such as direct glycinergic and GABAergic effects on the calyx, that also provide a different picture of synaptic processing and integration by MNTB cells. Within a week after its formation, the calyx exhibits a nearly adultlike ability to follow high rates of stimulation with good temporal fidelity. Seemingly contrary to this observation, postsynaptic EPSCs depress rapidly during repetitive stimulation, although depression is lessened by 2 weeks of age. The calyx connection offers an experimental paradigm to relate presynaptic Ca^{2+} levels, which are strongly linked to short-term plasticity, to postsynaptic currents in a mammalian nerve terminal. Experimental manipulations of Ca^{2+} concentrations using fluorophores, Ca^{2+} chelators, and caged Ca^{2+} have led to quantitative relationships between Ca^{2+} concentration within the calyx and vesicle release. Resting Ca^{2+} levels are estimated to increase at the active zone to approximately 10–20 μM following a single action potential. In adult animals, Ca^{2+} enters the terminal primarily via P/Q-type Ca^{2+} channels. Facilitation of postsynaptic currents can be elicited by prepulse depolarization of the calyx terminal or by posttetanic potentiation. During potentiation, postsynaptic current amplitudes follow presynaptic Ca^{2+} levels, which increase by an order of magnitude during the tetanic stimulus. Ca^{2+} levels are thought to couple with highly nonlinear Ca^{2+} sensors to facilitate vesicle release. These studies provide strong evidence for a causal relationship between Ca^{2+} levels and potentiation. Although the vesicle release probability is likely to vary among active zones, the average release probability across the multiple active zones is estimated to be on the order of 0.25, with a readily releasable pool of vesicles numbering between 1500 and 4000. Given the sensitivity of the vesicle release mechanism to small changes in Ca^{2+}, action potential amplitude and duration must be tightly controlled, especially during high rates of activity. Specializations that mediate these properties include the exclusion of sodium channels from the terminal and the positioning of these channels along a lengthy heminode of the axon leading into the calyx. In sum, many mechanisms operate in concert to affect Ca^{2+} levels at microdomains surrounding individual active zones and regulate vesicle release across hundreds of synaptic sites.

Conclusion

The large axosomatic endings, the endbulbs and calyces of Held, are more than independent, synaptic transmission machines. Recent evidence suggests that they are subject to modulatory influences. Glycine acts presynaptically to depolarize the terminal and facilitate neurotransmitter release. Retrograde signaling via endocannabinoids occurs at these endings, perhaps enhancing the sensory experience. Certain second messenger systems can influence vesicle mobilization, and activation of metabotropic glutamate receptors can affect vesicle release probability and the size of the readily releasable pool of vesicles during trains of activity. The next few years promise advancement of our knowledge of the mechanistic elements of synaptic transmission at these large synaptic terminals. Studies at the cellular level will lead to progress at the molecular level to complement *in vivo* exploration of the role of these remarkable structures in the neural encoding of natural sounds.

See also: Auditory System: Central Pathway Plasticity; Auditory System: Central Pathways; Auditory Localization; Cochlear Mechanics; Cochlear Development; Deafness; Sound Localization: Neural Mechanisms; Temporal Processing in the Auditory Pathway.

Further Reading

Berrebi AS and Spirou GA (1997) PEP-19 immunoreactivity in the cochlear nucleus and superior olive of the cat. *Neuroscience* 83(2): 535–554.

Billups G and Forsythe ID (2002) Presynaptic mitochondrial calcium sequestration influences transmission at mammalian synapses. *Journal of Neuroscience* 22(14): 5840–5847.

Choe S (2002) Potassium channel structures. *Nature Reviews Neuroscience* 3: 115–121.

Ehret G and Romand R (eds.) (1997) *The Central Auditory System*. Oxford: Oxford University Press.

Forsythe ID and Barnes-Davies M (1997) Synaptic transmission: Well-placed modulators. *Current Biology* 7: R362–R365.

Geisler CD (1998) *From Sound to Synapse*. New York: Oxford University Press.

Hoffpauir BK, Grimes JL, Mathers PH, and Spirou GA (2006) Synaptogenesis of the calyx of Held: Rapid onset of function and one-to-one morphological innervation. *Journal of Neuroscience* 26(20): 5511–5523.

Kopp-Schneinpflug C, Dehmel S, Dörrscheidt GJ, et al. (2002) Interaction of excitation and inhibition in anteroventral cochlear nucleus neurons that receive large endbulb synaptic endings. *Journal of Neuroscience* 22: 11004–11018.

Lu Y, Monsivais P, Tempel BL, et al. (2004) Activity-dependent regulation of the potassium channel subunits Kv1.1 and Kv3.1. *Journal of Comparative Neurology* 470: 93–106.

Morest DK (1968) The growth of synaptic endings in the mammalian brain: A study of the calyces of the trapezoid body. *Zetschrift fur Anatomie und Entwicklungsgeschichte* 127: 201–220.

Niparko JK, Kirk KI, Mellon NK, et al. (eds.) (2000) *Cochlear Implants: Principles & Practices*. Philadelphia: Lippincott Williams & Wilkins.

Oleskevich S, Youssoufian M, and Walmsley B (2004) Presynaptic plasticity at two giant auditory synapses in normal and deaf mice. *Journal of Physiology* 560: 709–719.

Petralia RS, Wang Y-X, Zhao H-M, et al. (1996) Ionotropic and metabotropic glutamate receptors show unique postsynaptic, presynaptic, and glial localizations in the dorsal cochlear nucleus. *Journal of Comparative Neurology* 372: 356–383.

Rowland KC, Irby NK, and Spirou GA (2000) Specialized synapse-associated structures within the calyx of Held. *Journal of Neuroscience* 20: 9135–9144.

Ryugo DK and Parks TN (2003) Primary innervation of the avian and mammalian cochlear nucleus. *Brain Research Bulletin* 60: 435–456.

Ryugo DK, Rosenbaum BT, Kim PJ, Niparko JK, and Saada AA (1998) Single unit recordings in the auditory nerve of congenitally deaf white cats: Morphological correlates in the cochlea and cochlear nucleus. *Journal of Comparative Neurology* 397: 532–548.

Ryugo DK, Huchhton DM, Pongstapron T, and Niparko JK (1997) Ultrastructural analysis of primary endings in deaf white cats: Morphologic alterations in endbulls of Held. *Journal of Comparative Neurology* 385: 230–244.

Ryugo DK, Kretzmer EA, and Niparko JK (2005) Restoration of auditory nerve synapses in cats by cochlear implants. *Science* 310: 1490–1492.

Sento S and Ryugo DK (1989) Endbulbs of Held and spherical bushy cells in cats: Morphological correlates with physiological properties. *Journal of Comparative Neurology* 280: 553–562.

Spirou GA, Brownell WA, and Zidanic M (1990) Recordings from cat trapezoid body and HRP labeling of globular bushy cell axons. *Journal of Neurophysiology* 63(5): 1169–1190.

Trussell LO (1997) Cellular mechanisms for preservation of timing in central auditory pathways. *Current Opinions in Neurobiology* 7: 487–492.

VonGersdorff H and Borst JG (2002) Short-term plasticity at the calyx of Held. *Nature Reviews Neuroscience* 3: 53–64.

Auditory Systems in Insects

H C Hughes and S S Wang, Dartmouth College, Hanover, NH, USA

Many people might not give much thought about whether insects can hear. After all, none of them has anything that look like ears. On the other hand, anyone who has heard the cascade of cricket songs that begin promptly at dusk might suppose that crickets can hear – it is unlikely all that singing is for our benefit alone. Careful research has demonstrated all of the insects that produce various forms of calls or songs hear those songs very well. These include crickets, grasshoppers, and bushcrickets. However, even more surprising is the fact that many insects that appear silent to us also hear. Among these silent listeners are various species of moths and flies.

Insects are the most diverse and most numerous animals on Earth. Entomologists have identified approximately 900 000 different species of insects, but they estimate that 80% of the species remain undiscovered. The actual number of insect species probably lies somewhere between 3 million and 30 million! There are probably 10 quintillion (10 000 000 000 000 000 000) individual insects alive at this moment – that comes to an astonishing 200 million insects for every living human on earth. Many of those insects can hear.

This article describes the variety of insects that can hear, the function of insect hearing, the anatomical organization of insect auditory systems, and the ways in which the insect nervous system processes acoustic information.

The Physical Properties of Sound Waves

A sound is the propagation of a mechanical disturbance through an elastic medium. For terrestrial animals, that medium is air. A vibrating object displaces air molecules, and those displacements are transmitted through the air as variations in air pressure. Very close to the vibrating object, the vibrations produce high-velocity movements of air molecules that are called near-field sounds. These near-field sounds are detected in some insects by very fine mechanoreceptors that are actually displaced in a manner analogous to the way shafts of wheat bend with the wind. These molecular movements become negligible once the sound has moved more than several wavelengths of distance from the source however. As a result, far-field sounds exist only as variations in air pressure. Far-field sounds are detected using a different kind of mechanoreceptor that is sensitive to variations in air pressure.

The velocity of sounds varies with atmospheric conditions, but under typical conditions sound waves travel at approximately 340 m s^{-1} (1225 km h^{-1} or 761 mph). Sounds have three basic properties that any hearing animal needs to be able to identify. The first property is the frequency, which is rate with which the pressure changes as the sound moves past the listener. The unit of frequency, called hertz (Hz), is the number of pressure cycles that occur in 1 s. The second important property of sound waves relates to the amplitude of the pressure variations. This is the intensity of the sound, and is related to the perceptual quality we call loudness. The third important feature of a sound is the location relative to the listener – whether the sound is to the left or the right, in front of or behind the listener. The small size of insects complicates the process of sound localization, which has led to the evolution of some very clever adaptations.

Diversity of Auditory Mechanisms Within the Class Insecta

There are 29 different orders of insects and a sense of hearing has been identified in seven: Neuroptera (green lacewings), Lepidoptera (moths and butterflies), Coleoptera (tiger beetles), Dictyoptera (praying mantis and American cockroach), Orthoptera (migratory locusts, field crickets, and katydids), Hemiptera (bees, wasps, ants, sawflies), and Diptera (flies).

Diversity of ear location A remarkable aspect of insect hearing is the fact that the ears can be found on virtually any part of the body. Like vertebrates, the insect body is based on a segmental plan, and each of these segments has stretch receptors and other kinds of mechanoreceptors that sense movements of various organs within each segment. It appears that insect ears evolved from a mutation in genes that control the development of these so-called chordotonal mechanoreceptors, and those mutations could occur in any of the body segments. Thus, insect ears evolved independently in many different species, and these independent mutations produced ears at many different locations in different species. Insect ears can be found on the wing base, abdomen, metathorax, base of forehead, hindwing, forewing base, cervical membranes, metathoracic leg, first abdominal segment, prothoracic leg, mesothorax, and ventral prosternum (see **Figure 1**). Crickets and katydids have eardrums

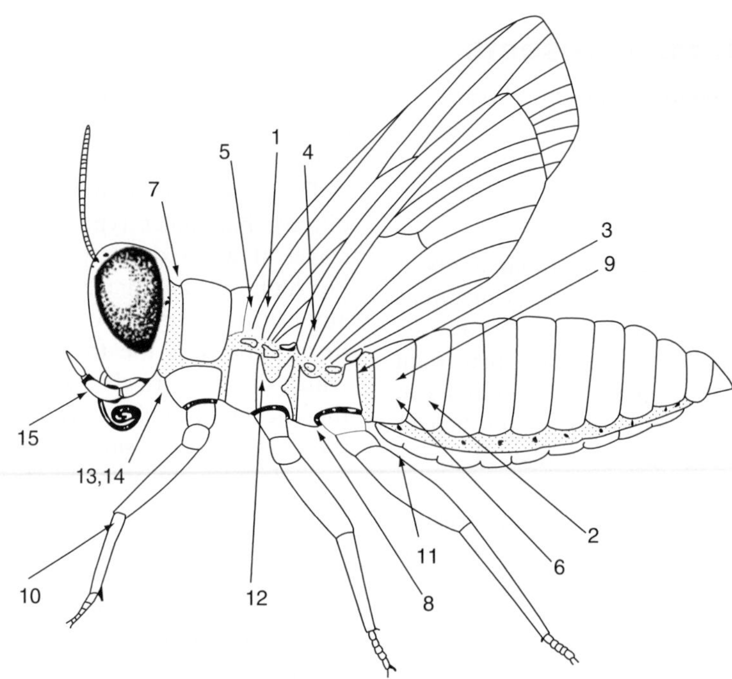

Figure 1 The variety of locations of tympanal organs as illustrated in a 'generic insect'. The locations indicated are: 1, wing base; 2, abdomen; 3, metathorax; 4, base of hind-wing; 5, fore-wing base; 6, abdomen; 7, cervical membranes; 8, ventral metathorax/metathoracic leg; 9, first abdominal segment; 10, prothoracic leg; 11, abdomen; 12, mesothorax; 13 and 14, ventral pro-sternum; 15, pilifer organ. Reprinted from Hoy RR and Robert D (1996) Tympanal hearing in insects. *Annual Review of Entomology* 41: 443–450 (figure 1, p. 438).

on their forelegs, grasshoppers and locusts have eardrums on their first abdominal segments, and cicadas have exposed eardrums on the abdomen.

Diversity of ears Although all insect auditory receptors are a variety of chordotonal receptor, there are two basic types of insect ears: tympanal and flagellar. Tympanal auditory organs are pressure detectors and are designed to detect far-field sounds. The basic principle of their operation is similar to the tympanic membrane (ear drum) of vertebrate ears. The variations in air pressure provide a force that is applied to the surface of the tympanum, which makes it move. Those movements are detected by the auditory receptors. Three morphological features characterize most tympanal organs. First, the tympanal membrane appears as a localized thinning of the external cuticle at the location of the hearing organ (see **Figures 2** and **3**). Second, an air-filled sac of tracheal origin is situated close to the internal surface of the tympanum. Third, the presence of a chordotonal organ composed of a sensory neuron with a dendrite embedded within a specialized support structure called the scolopidia.

Flagellar auditory organs function as particle velocity detectors for near-field sounds. They are known as Johnston's organs, and consist of a fine cilia that protrudes from the insect's body and bends in response to the motion of air molecules. In mosquitoes, the antennas are covered with hairs that serve as near-field sound detectors. Johnston's organs are generally used to detect the wing-beats of conspecifics or nearby predators.

The function of auditory receptor cells is to convert the acoustic mechanical energy into an electrochemical response in the nervous system. In tympanal organs, the sensory cell dendrite and its associated scolopidia are attached to the tympanal membrane, and undergo a mechanical disturbance when sound waves displace the tympanal membrane. The detailed biophysical processes that underlie transduction tympanal ears is not yet understood.

Purposes of Audition in Insects

Insects use audition to communicate with members of their own species, detect predators, and, in some cases, detect prey. Many insects, including grasshoppers, crickets, bushcrickets, katydids, fruit flies, bees, and cicadas, produce acoustic signals that serve to attract mates, proclaim territories, and exchange information such as the location of a food source. Some moths, lacewings, beetles, mantids, and flies can hear the ultrasonic sonar signals emitted by predatory bats. Detection of ultrasonic signals usually initiates evasive maneuvers in these species.

It is usually found that the auditory systems are specialized for processing species-specific communications. Thus, the auditory receptors often show their

Figure 2 Tympanal organs on the forelegs of two species of insects. (a) Southern mole cricket *Scaptericsus borellii.* (b) Oblong-winged katydid *Amblycorypha oblongifolia.* (a) Photo reproduced with permission of TJ Walker, University of Florida. http://buzz.ifas.ufl.edu/341ppl.htm. (b) Photo reproduced with permission of TG Forest, University of North Carolina – Asheville. http://buzz.ifas.ufl.edu/007pmy2.htm.

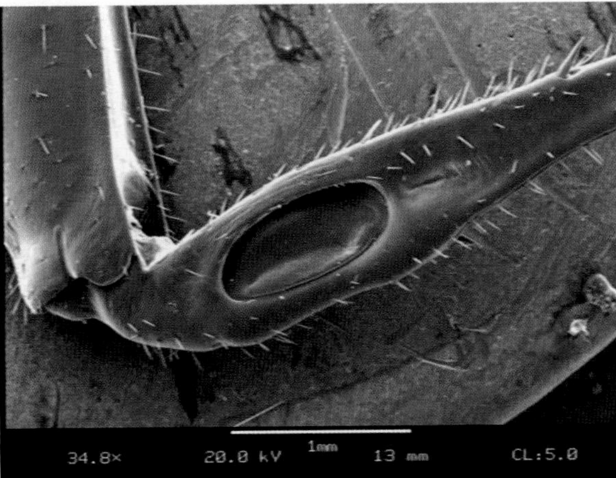

Figure 3 A scanning electron micrograph of tympanal organ of a katydid. Photo reproduced with permission of WE Conner, Lake Forest University. http://www.wfu.edu/biology/batsandbugs/bugs_ears.htm.

greatest sensitivity for the frequencies used in intraspecies auditory signaling. Other important aspects of insect auditory systems appear to maximize the range over which auditory communications will be effective. Cicadas and crickets produce mating calls that appear to be detected by females at distances between 10 and 20 m. Amazingly, one form of African grasshopper produces a mating call that can be heard by prospective mates at distances up to 1.9 km. An equivalent range in human auditory communication would allow us to converse with people that were over 40 km away.

In most species, the males produce the mating calls, usually by rubbing their wings or legs against each other – a technique called stridulation. Potentially receptive mates detect these calls with their tympanal organs. Most insect calls are comprised of rapid variations in the amplitude of what is termed a carrier frequency. Both the carrier frequency and the temporal patterning or rhythm of the bursts varies with different species. Thus, the receiving insect must encode both the carrier frequency and the rhythm, so that communication is confined to members of the same species. Examples of the mating calls of several cricket species are illustrated in **Figure 4**. The females respond to these calls by either approaching (positive phonotaxis) or turning away from them (negative phonotaxis). The calls of other species are either ignored or result in negative phonotaxis. Scientists use these phonotaxic responses to determine which aspects of a mating call the females find most attractive. The ability to orient toward the source of a mating call demonstrates that the insects can both recognize conspecific calls and localize their source. This requires specialized neural processes that are described below.

Insects that broadcast their presence by emitting these calls and songs take a certain risk: potential predators might hear them. In some instances, conspecific communication attracts not only mates but also deadly parasites. The female tachinid fly (*Ormia ochracea*) has sternal tympanal ears for acute directional hearing, which allows them to locate singing male crickets. Once a cricket is located, the female fly climbs onto the host and deposits small and undetectable larvae that enter the cricket's body and develop into large maggots that devour the cricket from the inside out (see **Figure 5**).

Audition in insects is also important for predator detection. Moths use their metathoracic tympanal organs to detect the sonar signals of bats that prey upon them. These organs are remarkably sensitive to ultrasonic sound waves, as they can detect a

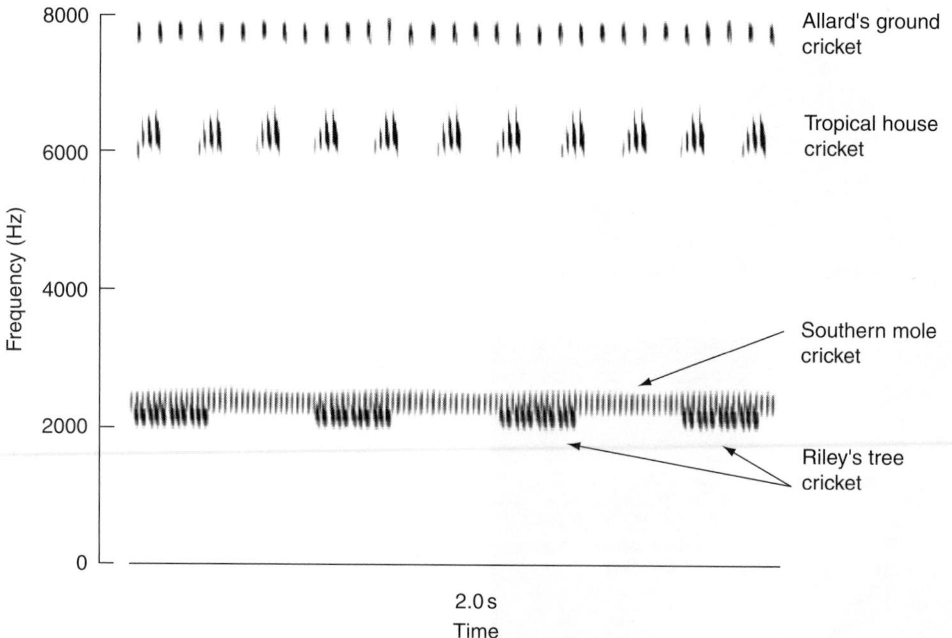

Figure 4 Sound spectrograms of four different species of crickets. Each song consists of brief pulses centered on a carrier frequency. The carrier frequencies and the temporal patterns differ for each of the four species although the carrier frequencies of Riley's tree cricket and Southern mole cricket are very similar. The ranges of Riley's tree cricket and the southern mole cricket do not overlap. The range of the tropical house cricket overlaps with both Riley's and Southern mole cricket. The spectrograms were obtained from http://buzz.ifas.ufl.edu and are reprinted here with permission of TJ Walker.

Figure 5 *Ormia ochracea* (on top) laying a larva on a host crcket. See text for further details. Obtained from http://hoylab.cornell.edu/.

movement of the tympanal membrane of as little as 0.000 000 000 001 m, which is substantially smaller than the diameter of a hydrogen atom. Detection of bat sonar signals initiates evasive countermaneuvers. Other nocturnal insects, including green lacewings, field crickets, praying mantises, locusts, katydids, and beetles display startle responses or fold up their wings and fall from the sky when exposed to simulated bat sonar signals.

Drosophilas (fruit flies) use flagellar organs to discriminate species-specific courtship songs within a close range. Similarly, bees use their Johnston's organs in the flagellar antennas to decode the acoustic signals of the waggle dance and song within the hive.

Central Processing of Auditory Information in Insects

Insects have a much more primitive central nervous system (CNS) than most vertebrate species. However, like the vertebrate nervous system, the insect CNS is an elaboration upon a basic segmental design. Each body segment has a corresponding ganglia (collection of nerve cells), and the ganglia are interconnected via nerves. Sometimes the ganglia fuse into larger groups. The head ganglia is the insect's brain.

As in all nervous systems, we can identify three main classes of neurons: sensory neurons (afferent neurons), interneurons, and motor neurons (efferent neurons). The sensory neurons convey information from the various sensory receptors to the CNS. The motor neurons have cell bodies in the ganglia, with axons that project to and innervate muscles. In some cases, sensory neurons make direct synaptic contact with motor neurons. These direct sensory–motor connections mediate the simplest forms of reflexes. Frequently, however, the relationship between sensory inputs and motor responses are more complicated and requires more elaborate neural circuitry. In those cases, circuits of interneurons lie between the sensory afferents and motor efferents. In general, the auditory system of insects is composed of a relatively small number of sensory neurons and a larger set of interneurons. The neurons that actually innervate the auditory receptor are called primary auditory afferents: they are the neurons that actually generate neural impulses in response to a mechanical disturbance in the auditory receptor. In some species, each ear may be innervated by a single sensory neuron. Typical insect ears have 50–100 primary auditory afferents. In general, more primary afferents allow for more elaborate processing of the auditory signal. By way of comparison, the human auditory system contains about 30 000 primary afferents from each ear. The primary afferents make synaptic contact with interneurons. Interneurons that receive an input from primary afferents are often called second-order afferents. The patterns of synaptic connections between neurons are what enable the computations that are the hallmark of all nervous systems. In general, each higher-order set of interneurons continues a trend of increasing response specificity.

The human auditory system can be considered a general purpose processor of auditory information. In contrast, the insect auditory system appears specialized for extracting very specific information about the location of possible mates, predators, or prey. Yet the insect nervous system still has to extract information about the frequencies, loudness, temporal patterns, and locations of biologically relevant sounds.

Frequency Coding

Each species of animal hears a limited range of frequencies, although the range differs in different species. The range of human hearing goes from 20 to 20 000 Hz. In general, insect auditory systems respond to a higher and broader band of frequencies, but generally with less resolution than humans. Many insects can hear sounds from around 1000 Hz to frequencies as high as 40 000 Hz. In general, those that can hear very high frequencies (30 000 Hz or higher) evolved this ability to detect when they are in the sonar beam of predatory bats.

Studies of the phonotaxic responses of crickets and other insects have shown that they will orient toward a simulated mating call with a carrier frequency of say 5000 Hz, but turn away from an identical call if the carrier frequency is changed to 15 000 Hz. Thus, the cricket can discriminate the calls on the basis of a difference in frequency. The way auditory systems achieve frequency discrimination is by having different auditory fibers respond somewhat selectively to different frequencies. This is known as frequency tuning. Because different neurons are tuned to different frequencies, a change in frequency produces a corresponding change in the ensemble of active neurons – we can say that identity of responding neurons is the neural code for frequency.

Because insects have relatively few primary auditory afferents, their ability to discriminate frequencies is much more limited than humans. Insects that detect the ultrasonic emissions of bats usually possess only two primary afferents that are tuned to those high frequencies – one for each tympanal organ. These insects cannot perform any detailed analysis of the bat sonar signal; they simply function to detect those signals and make a rough determination of the bat's location. This is an example of a special purpose auditory system in which a minimal amount of information processing needs to be performed in order to produce a very specific motor response (avoidance).

Other insects have more than one primary afferent, so in principle they could discriminate different frequencies. As early as 1966, Michelsen discovered that locusts' auditory receptors are tuned to different sound frequencies. This gives these insects their ability to discriminate between sound frequencies. Since Michelsen's original discovery, frequency-sensitive auditory receptors have been found in other insects such as bushcrickets and crickets. It is not known how frequency selectivity is achieved in insect primary auditory afferents, but it appears that the mechanism is somewhat different that the one employed in vertebrate auditory systems. The mechanism in insects might be related to anatomical specializations of the support structures surrounding the receptor dendrites, or it may related to the biophysical properties of the receptor cell membrane or of the tympanal membrane itself. Most insects can discriminate among different frequency bands within their own hearing range, and the hearing range is generally determined by the evolutionary needs of the particular species.

For many insects, the neurons that are tuned to different frequencies are organized in a specific spatial arrangement – adjacent cells prefer nearby frequencies. This is called tonotopic organization. Tonotopic organization implies that the nervous system contains something analogous to a frequency map, so that the location of the activity within the map is part of a neural code for frequency. Take, for example, the locust ear: it has four receptor groups, each group is activated by distinctive sound frequencies and the groups are arranged in an orderly tonotopic map. Auditory receptors in the bushcricket ears and cricket ears are also tonotopic. In crickets, a large clustering of neurons tuned to the carrier frequency of the calling-song frequency results in an overrepresentation of the carrier frequency within the array of primary afferents. This overrepresentation of particularly important frequencies is commonly found in many different sensory systems in many different species. Tonotopic organization aids in the formation of appropriate connections between neurons because neural pathways are often topographically organized (adjacent afferents make synaptic contact with adjacent neurons at the next synaptic station).

The principle of increasing complexity in higher-order sensory neurons is well illustrated in the insect auditory system. In crickets, two identified interneurons, designated AN1 and AN2, receive inputs from primary afferents and project to the brain. AN1 neurons are most sensitive to low frequencies, whereas AN2 neurons are sensitive to high frequencies and the species-typical calling song frequency. In field crickets, AN1 neurons are tuned to the frequency of the calling and aggression song, while AN2 neurons are tuned to the higher frequencies of the courtship song. In female katydids, interneurons TN-1 respond to both the frequencies of male songs and the frequencies of bat calls. However, TN-1 neurons also function as temporal pattern analyzers: they respond to the bat call pulse rate but not to male katydid pulse rate. This suggests that the TN-1 neurons function as bat detectors. The patterns of connectivity between primary afferents and interneurons appear to perform a primitive and automatic form of auditory stream segregation. This allows the insects to quickly identify and differentiate behaviorally important sound signals, even when frequency content alone is insufficient to make the distinction.

Loudness Coding

Studies of positive phonotaxis have shown that female crickets prefer louder mating calls to softer ones. Thus, they can discriminate the calls based on loudness. Loudness is usually represented by the frequency with which the responding neurons generate action potentials. As the intensity of the sound increases, the frequency of action potentials increases as well. This is known as a rate code for intensity.

Encoding Temporal Patterns

Most insect songs do not have patterned changes in frequency. Unlike vertebrate communication signals, insect songs are monotonous and lack melodies. The main information that distinguishes the individual song type from the others is in the rhythm and temporal structure of amplitude modulations of the carrier frequency. Thus, although an analysis of carrier frequency is performed, the insect nervous system must also be capable of recognizing species-specific temporal patterns of the song. The structure is defined by (1) duration and shapes of sound pulses, (2) the spacing between the sound pulses, and (3) the organization of sound pulses into higher-order groupings. A typical song consists of repetitions of a serious of brief sounds called pulses or syllables (see **Figure 4**). By subtly varying the temporal structures of synthetic calling songs and noting the phonotaxic responses they produced, previous studies were able to establish the aspects of temporal patterns most important for song recognition.

For many crickets, song models with the species-typical pulse period induce the strongest phonotaxic responses. In some cases, different parameters of structural features trade off against one another, suggesting a parallel process mechanism where the output response is the summation of all the feature information in the input. Similarly, grasshoppers use temporal structures to decode the meaning of songs.

Using synthetic song models, behavioral experiments established the properties of the temporal-pattern filters used to identify conspecific acoustic signals. This filtering is for the most part performed in the CNS. In crickets, neither the primary afferents nor the second-order interneurons demonstrate selectivity for temporal pattern. In the cricket, neurons that are selective for the temporal pattern of conspecific songs have so far only been found in the brain. The general trend is that higher-order interneurons tend to respond selectively to temporal patterns that produce positive phonotaxis in behavioral experiments. In contrast, the responses of lower-order neurons faithfully reproduce the temporal pattern of both effective and ineffective temporal patterns. It is not yet known how this specificity for temporal pattern in achieved. One interesting possibility that these higher-order interneurons receive two copies of the temporal pattern, but one is delayed relative to the other. If the delays are carefully calibrated, the two copies of temporal pattern will arrive at the interneuron in the same temporal phase. In this case, they can interact synergistically, but only for the correct species-specific calling pattern.

Sound Localization

Detecting the presence of a potential mate or a potential predator is of little value if the location of the mate or the predator cannot be determined. Thus, sound localization is an essential function for any auditory system. Consider the process of mate finding in short-horned grasshoppers. The male sings, and, if he is lucky, a receptive female sings a response song. Depending on the direction of the response song, the male jumps either straight ahead, or to the left or right. The male continues walking for a short distance and sings again, listens for a reply, and jumps again. By repeating this sequence, the male eventually finds the female. Female field crickets use a similar strategy when they walk toward the source of a mating call while repeatedly determining its direction. Various measurements of the direction of orienting provide scientists with detailed information concerning the accuracy of sound localization in different insects. Many of them can distinguish sound sources that are about 20° apart. By contrast, the minimum sound separation that humans can discriminate is about 1°. However, the localization accuracy of certain insects is equivalent to the limits of human sensitivity.

Cues for Sound Localization

It has been known since the latter part of the nineteenth century that there are two cues available for detecting the direction of sound waves. The first is diffraction and the second is the time of arrival. Both cues require comparisons between two detectors (ears) in different locations. No matter where an animal's ears are located, they are almost always on opposite sides of the body. Diffraction refers to the bending of waves around an occluding object. Diffraction is heavily dependent on the size of the occluding object relative to the wavelength of sounds, and the small size of insect bodies complicates the problem of sound localization. Significant diffraction occurs when the distance between the ears is greater than one-tenth of the wavelength of the sound. In this case, the sound bends around the body, which produces changes in both the amplitude and the phase of the sound wave arriving at each ear. When the wavelengths of the sound are very small relative to the size of the head, less diffraction occurs, which means that the sounds do not bend around the body as readily and this creates a sound shadow: the ear that is farther from the source receives a less intense signal that the ear that is nearer to the source. The time of arrival cue simply results from the fact that the sound must travel farther to arrive at the more distant of the two ears, so it arrives at the distant ear later in time. Because their bodies are so small, the time of arrival cue is not useful to the insects – the small distances between their two ears means that the difference in arrival times would be less than a few microseconds, which is too small a difference for the nervous system to detect. The insect's small bodies mean that, unless the frequencies are exceedingly high (and have correspondingly short wavelengths), their bodies do not produce much of a sound shadow. In such cases, the sound waves diffract around the insect's body and arrive with very little attenuation of sound pressure on the other side. The extra distance does introduce a small change in phase however.

The small size of insects makes sound localization especially difficult for them to achieve. Insects (and some small vertebrates, including frogs and small birds) have evolved an ingenious solution to the problem. Instead of using a simple pressure-sensitive receiver like the human ear, the insects have evolved a pressure difference receiver. The key element of a pressure difference receiver is that the sound is applied to both sides of the tympanal membrane, rather than just the outside surface, as occurs in larger vertebrates. Of course, this means that there must be an air-filled cavity that extends from the outside of the animal's body all the way through to the inside of the tympanum. This is easier to achieve when the ear is on a foreleg, but when the ear is located on one of the body segments, the canal must extend completely through the segment (e.g., from the left side of the body to the inside of the right tympanum). This allows the sound to apply a force on both the outside and the inside of the membrane. When the sound waves reach both the outer- and inner-surface tympanal membrane, the difference in phase that is produced by the sound arrival time difference between the two surfaces provides information about directionality using the pressure difference receiver. Careful measurements of sound pressure at the external (front) and internal (back) surfaces of a grasshopper's eardrum demonstrated that the movement of the tympanum was well approximated by the sum of two vectors. One vector represents the pressure on the outside of the membrane and the other represents the pressure on the inside, after it passed through the air-filled cavity that runs from the opposite side of the body to inside of the tympanal membrane. Using this two-input model, and combining the pressure information obtained by the ipsilateral and the contralateral tympanum, the direction of the sound source can be determined.

The pressure difference receiver thus provides a difference in the response of the two tympanal membranes that varies with the location of the sound source. It is therefore a cue to location of the source. To extract the information inherent in this cue, a comparison between the two ears is required. That comparison is performed by interneurons that receive inputs from the primary auditory afferents arising from each ear.

Intracellular recordings confirm that some interneurons in the auditory pathway integrate excitatory input from the ipsilateral ear with the inhibitory input from the contralateral side. In some cases, there are mechanical linkages between the two ears that serve to amplify the interaural signals even before any nervous processing takes place. These specializations endow small insects with enough sound localization ability to do what needs to be done: find mates and avoid predators.

The auditory system of insects appears in many ways rudimentary when compared to many vertebrate species. However, it demonstrates a number of ingenious adaptations that have permitted the insects to condense a great deal of auditory processing capacity into a very small and efficient package. It is a marvel of biological engineering, and has helped insects become the most numerous animals on the planet.

See also: Auditory Cortex Structure and Circuitry; Auditory System: Central Pathways; Auditory Cortex: Models; Auditory Localization; Sound Localization: Neural Mechanisms.

Further Reading

Eberl DF (1999) Feeling the vibes: Chordotonal mechanisms in insect hearing. *Current Opinion in Neurobiology* 9(4): 389–393.

Green DM (1976) *Introduction to Hearing.* Mahwah, NJ: Lawrence Erlbaum Associates.

Helversen D and Helversen O (1983) Species recognition and acoustic localization in acridid grasshoppers: A behavioral approach. In: Huber F and Markl H (eds.) *Neuroethology and Behavioral Physiology*, pp. 95–107. Berlin: Springer.

Hoy RR and Robert D (1996) Tympanal hearing in insects. *Annual Review of Entomology* 41: 443–450 (figure 1, p. 438).

Hoy RR (1998) Acute as a bug's ear: An informal discussion of hearing in insects. In: Hoy RR, Popper AN, and Fay RR (eds.) *Comparative Hearing: Insects*, pp. 1–17. New York: Springer.

Hughes HC (2001) *Sensory Exotica: A World Beyond Human Experience.* Cambridge, MA: MIT Press.

Jack JE (2004) The structure and function of chordotonal organs in insects. *Microscopy Research and Technique* 63(6): 315–337.

Michelsen A (1998) Biophysics of sound localization in insects. In: Hoy RR, Popper AN, and Fay RR (eds.) *Comparative Hearing: Insects*, pp. 18–62. New York: Springer.

Pollack GS (1998) Neural processing of acoustic signals. In: Hoy RR, Popper AN, and Fay RR (eds.) *Comparative Hearing: Insects*, pp. 139–196. New York: Springer.

Pollack GS (2000) Who, what, where? Recognition and localization of acoustic signals by insects. *Current Opinion in Neurobiology* 10(6): 763–767.

Rayleigh JWS (1877) *The Theory of Sound,* 2nd edn. London: Macmillan.

Robert D, Amoroso J, and Hoy RR (1992) The evolutionary convergence of hearing in a parasitoid fly and its cricket host. *Science* 258(5085): 1135–1137.

Romer H (1998) The sensory ecology of acoustic communication in insects. In: Hoy RR, Popper AN, and Fay RR (eds.) *Comparative Hearing: Insects*, pp. 63–97. New York: Springer.

Schul J and Sheridan RA (2006) Auditory stream segregation in an insect. *Neuroscience* 138: 1–4.

Young D and Hill KG (1997) Structure and function of the auditory system of the cicada, *Cystosoma Saundersil. Journal of Comparative Physiology A* 117: 23–45.

Autism

G Dawson and M Murias, University of Washington, Seattle, WA, USA

Autism Spectrum Disorder

It is now recognized that autism is part of a broad and heterogeneous group of disorders, referred to as autism spectrum disorders (ASD), that involve impairments in reciprocal social interaction and communication, and the presence of restricted, stereotyped, and repetitive interests and behaviors. Impairments in the domain of social relationships are considered a core feature of ASD and can be manifest as a lack of social and emotional reciprocity; atypical eye–eye-gaze, facial expression, and gestures; lack of interest in and/or difficulty relating to others, particularly peers; and a failure to share enjoyment and interests with others. Asperger syndrome, a subtype of ASD, is characterized by intact use of formal language skills, such as vocabulary and syntax, along with impairments in social use of language (e.g., conversational skills), social relationships, and idiosyncratic and consuming interests. Diagnosis of most forms of ASD is based on behavioral observation and developmental history. Although the diagnosis of ASD is quite stable from early childhood to adulthood, outcome is extremely varied. Whereas approximately 25% of individuals do not develop communicative speech and the majority function in the mentally retarded range, a substantial proportion of individuals have average to above average intelligence and are able to live independent and productive lives. IQ and language ability in childhood are the strongest predictors of long-term outcome.

A strong focus on current research is early recognition of autism. Screening questionnaires for toddlers exist that can be used by health professionals. Studies suggest that infants at risk for autism show very few behavioral symptoms at 6 months but that by 12 months, many core autism symptoms are apparent for the majority of infants. Evidence for this comes from both prospective and retrospective home videotape studies, which have shown that by 12 months, infants later diagnosed with autism exhibit poor eye contact and reduced joint attention, communicative babbling, and social reciprocity. Behavioral screening methods for detection of infants at risk for autism appropriate for use with 12-month-olds are currently being developed. Scientists are searching for biomarkers that can identify infants at risk for autism, but no reliable biomarkers have yet to be discovered. A subgroup, estimated to be approximately 25%, shows autistic regression, characterized by apparently normal development until the second or third year of life, followed by a loss of language and social skills and onset of autism symptoms, such as repetitive behaviors. Recent genetic linkage studies suggest that the susceptibility genes associated with autistic regression differ from those associated with early-onset autism.

Prevalence and Etiology

The prevalence of ASD has increased significantly in the past few decades. A large epidemiological study conducted by the Centers for Disease Control (CDC) found a prevalence rate of 34 per 10 000 among 3- to 10-year-old children in metropolitan Atlanta, Georgia (USA). However, this rate of 34 per 10 000 is likely to be an underestimate. Higher functioning individuals may have been missed and younger children may not have been identified. The rate reported for 5- to 8-year-olds in the CDC study – that of 41–45 per 10 000 – may be more accurate, and is similar to other surveys that report a prevalence rate of 60 per 10 000. Autism affects males at rates three to four times higher than females.

Genetic Influences

Evidence for genetic influences on autism is strong: 5–10% of autism cases are due to an identifiable medical disorder with a known inheritance pattern, such as fragile X syndrome, untreated phenylketonuria (PKU), tuberous sclerosis, and neurofibromatosis. Fragile X syndrome in particular accounts for about 8% of cases of autism.

Heritability estimates for idiopathic autism range from 91 to 93%, with concordance rates for autism of 60–95% for monozygotic (MZ) twins versus 3–7% concordance in siblings and dizygotic (DZ) twins. In contrast, rates in the general population are ~0.1%. When a broader phenotype is used, concordance increases for both MZ (88–91%) and DZ (9–30%) twins. The rapid decrease in concordance rates from identical twins to siblings, together with the differential risk rates for male versus female siblings, suggests interactions (epistasis) among several genes, estimated to be between 5 and 10 or more. Over 100 genes have been tested as candidates for autism susceptibility loci. Promising candidate genes include *Engrailed 2* (*EN2*) where a positive association was reported in two independent samples. *SLC6A4* has been tested as a candidate gene in multiple studies that have yielded positive but not entirely consistent results. Several mutations have been described in the neuroligin 3 and neuroligin 4 genes, though these appear to be an extremely rare cause of autism. A recent study implicates reduced *MET* gene expression in autism susceptibility. A number of genome scans for linkage have been reported using autism families. While no single region has been unambiguously associated with autism, coincident signals for specific chromosomal regions have been identified in multiple studies. The 2q, 7q, and 17q regions are associated with the strongest signals. The challenges posed by mapping the risk alleles in autism have led some scientists to consider epigenetic mechanisms, including genomic imprinting and epimutations. Involvement of such mechanisms in ASD is supported by the central role of epigenetic regulatory mechanisms in the pathogenesis of Rett syndrome and fragile X syndrome, both single-gene disorders associated with ASD.

The complex behavioral expression of ASD also poses challenges to genetic research. ASD presentation is heterogeneous, involving at least three symptom domains (communication, social, and restrictive behaviors/flexibility), with evidence that the autism phenotype extends beyond classic autism to 'lesser variant' phenotypes. Parents and siblings of individuals with autism exhibit higher than normal rates of autism-related impairments, as do more extended relatives. Older unaffected siblings exhibit impairments in receptive and expressive language skills, as well as specific learning disabilities. A family history study of 195 parents and 137 siblings of autistic probands found that 12–20% of siblings demonstrated broader phenotype impairments, including language and communication deficits, social impairments, and/or learning disabilities, as compared to only 2–3% of siblings of individuals with Down syndrome. Relatives of autistic probands from multiplex families (with $\geq$2 affected children with autism) are at greater risk of having broader phenotype impairments than are relatives of probands from families with only one child with autism. A study of 149 families of autistic probands reported excess rates of broader phenotype deficits in the 3095 relatives studied, including second- and third-degree relatives. Like autism, broader phenotype characteristics are more common in males than in females.

The use of quantitative measures of autism in genetic studies may be more informative than discrete diagnoses. Most linkage analyses of autism have used qualitative diagnoses. There have been few genetic studies that have attempted to measure autism-related traits along a continuum. In one study of 201 autism multiplex families (694 individuals), two traits, social motivation and restricted activities/flexibility, showed the highest heritability as well as strong genetic correlation, suggesting that these traits are promising for gene mapping and share a genetic basis. Both clinical progress and basic science likely will profit from the identification of endophenotypes. Diagnostic categories of autism were created for clinical classification and are based on behavioral symptoms that show considerable heterogeneity across individuals. Such symptoms might not be the best phenotype for genetic studies, because of the complexity of the cascade of events leading from genes to these behavioral symptoms. Intermediary measures that assess neural pathways and function might be more useful. The use of electrophysiological endophenotypes, in addition to clinical diagnoses, has led to successful identification of genes in other complex psychiatric disorders, such as alcoholism. Furthermore, similar endophenotypic approaches are in widespread use in other areas, such as in genetic studies of cardiovascular disease and migraine. Previous studies have identified elevated rates of specific neurocognitive impairments and atypical electrophysiological responses to social stimuli (e.g., faces) in family members of autistic probands. These include impairments on tests of executive function, reading ability, pragmatic language ability, and face processing, and atypical event-related brain potentials to facial stimuli. Future linkage studies will increasingly focus on these more refined, quantitative endophenotypes.

Brain Development

Cerebral Enlargement

In the first two years of life, head circumference (HC) measurements accurately reflect brain volumes. At birth, HC measurements in children with autism are not atypically large; however, by adulthood, approximately 20% of individuals with autism exhibit larger than average HC. A period of exceptionally rapid head growth occurs during the first year of life in children who are later diagnosed with autism, as well as in many younger siblings of children with autism who are at risk for the disorder. Patterns of HC trajectory appear to be an early risk marker for the development of ASD symptoms. Increased rate of growth may index aberrant processes during early development, and may precede the onset of autism-specific symptoms. HC is easily measured in the course of routine check-ups and is potentially useful as an early biological risk marker for ASD.

Imaging studies have consistently found increases in overall brain volumes in children and adults with ASD. Results from a longitudinal imaging study of brain growth demonstrate that the cerebral enlargement found in ASD children at 3–4 years old were no longer significant at 6–7 years of age, compared to age-matched developmentally delayed and typically developing children. However, within individuals, this growth curve can be quite variable. Alterations in early cellular developmental processes, such as failure of apoptosis or synaptic pruning, have been hypothesized to account for early cerebral enlargement, but recent longitudinal magnetic resonance spectroscopic imaging (MRI) findings of brain chemical changes in ASD children have failed to support such mechanisms.

Structural Findings

Abnormalities of the medial temporal lobe (MTL) structures, in particular the amygdala, have been strongly implicated in ASD and autism-related symptom expression. Recent reviews examining structural and functional imaging findings have highlighted the potential role of early amygdalar developmental deficits impacting social perception and social behavior, which are hallmark symptoms of ASD. Imaging studies have identified volumetric amygdalar abnormalities, which, similar to cerebral volume, vary as a function of age. An imaging study of 3- to 4-year-old males with ASD exhibited disproportionate right amygdalar enlargement, which was more accentuated in children with more severe symptoms. A subsequent cross-sectional study that assessed amygdalar volume across a broad age range also observed amygdalar enlargement among young, but not older, individuals with ASD.

A fairly consistent finding in quantitative and diffusion tensor MRI studies is reduced callosal volumes in ASD. In particular, several studies have found the genu, the anterior portion of the callosum containing axons that interconnect portions of the frontal lobes. This suggests changes in cortical connectivity may be associated with ASD.

Volumetric MRI evidence for hippocampal differences in ASD has been inconsistent; however, postmortem studies have identified abnormalities of neuron size and density within the hippocampus. Although there is not compelling evidence of direct hippocampal involvement in core symptoms of ASD, the hippocampus has been implicated in the degree of mental retardation associated with ASD, and strong links between mental retardation and increased seizure risk are present in ASD. A 3-D hippocampus surface topography study in 3- to 4-year-olds with ASD found that a characteristic pattern of hippocampal shape alterations was strongly associated with lower intellectual ability and poorer performance on neuropsychological tests specifically tapping the MTL, but not to tests of prefrontal function. The shape alteration found in these ASD children, who did not have seizures at the time they were imaged, was similar to a pattern found to characterize adults with MTL epilepsy.

The basal ganglia system is associated with a variety of functions that are altered in ASD, including motor control, cognition, and emotions. In particular, the presence of repetitive and stereotyped behavior in ASD suggests abnormalities in basal ganglia circuitry. Several MRI studies have demonstrated increased mean basal ganglia volume in ASD.

Cerebral water content can be characterized using quantitative MRI T2 relaxation times, and the technique has recently been used to investigate possible pathologic processes in ASD that are associated with cerebral white matter and cortical gray matter. One study compared T2 relaxation in 3- to 4-year-old children with ASD with age-matched children with typical development and developmental delay. Whole-brain cortical gray matter T2 was prolonged in the children with ASD, whereas whole-brain white matter T2 did not differ. These findings may reflect brain mechanisms involving neuroinflammation, which is typically accompanied by edema and would result in increased T2 relaxation. The presence of autoimmune markers, such as microglial activity, elevation in cytokines, and increases in glial fibrillary acidic protein, could provide a plausible mechanism for the prolongation of gray matter T2 observed in the ASD sample. These results support altered gray matter cytoarchitecture or pathology in ASD.

Both gray and white matter deviations are present in ASD, and volumetric MRI data suggest that the patterns of alteration are strongly age related. In 2- to 3-year-old children with ASD, MRI measurements indicate that cerebral gray and white matter volumes appear abnormally large. Enlargements of gray and white matter volumes in frontal and temporal lobes is apparent by 2 to 4 years of age, with white matter enlargements maximal in the frontal and parietal lobes. Gray and white matter differences are apparently less pronounced in older children. In ASD children aged 6 to 11, total white matter volumes have appeared enlarged, with fewer differences found in gray matter. Recent studies have observed white matter increases among 6- to 12-year-olds, with short-range white matter underlying frontal cortex showing the greatest enlargements, while deeper long-range fibers (callosal and anterior–posterior projections) did not differ between ASD and controls. Cerebellar developmental abnormalities appear to be more complex. In a study of 2- to 4-year-old ASD children, cerebellar gray matter volume did not appear enlarged. In the same children, however, cerebellar white matter volume was dramatically greater than in controls. Among older ASD patients, reduction in cerebellar vermis has been a fairly consistent finding.

Neurochemical Imaging

Magnetic resonance spectroscopy (MRS) allows noninvasive, *in vivo* study of cellular neurochemistry. *N*-Acetyl aspartate (NAA), found only in the nervous system, is a sensitive marker for neuronal integrity as well as for homeostatic interactions between neurons and glia. Increases in NAA are thought to parallel myelination and to reflect neuronal maturation. In ASD, decreases in NAA have been reported, but the specific brain regions contributing to these reductions have been inconsistent. One MRS study of 3- to 4-year-olds with ASD revealed regional and global decreases in NAA, and lower levels of other chemicals, and these difference appeared largely specific to gray matter. Recent longitudinal findings suggest that this pattern of chemical alterations in ASD generally persists between 3–4 and 6–7 years of age.

Autopsy Findings

Sparse tissue availability has made postmortem studies difficult to perform; however, several studies have described anatomic abnormalities in ASD brains. Consistent findings have been observed in the limbic system, cerebellum, and related inferior olive. Decreased cell sizes and increased cell packing density have been shown in hippocampal, amygdala, and entorhinal cortex. Cerebellum findings include significantly reduced numbers of Purkinje cells, primarily in the posterior inferior regions of the hemispheres. In ASD, frontal white matter may be compromised by neuroinflammatory responses, with evidence for inflammation noted especially at the junction of the cortex and white matter. Pathologies among cortical minicolumns have been reported in ASD, in which minicolumns appear narrower and more numerous. Minicolumns appear to be packed closer together in ASD brains, but the total number of cells does not differ from that in controls.

Postmortem studies in ASD have also documented histopathological features reflecting reduced numbers of neurons in the amygdala and especially the lateral nucleus of the amygdala, particularly in adults. Taken together with volumetric MRI studies, the apparent lower number of neurons in the amygdala suggests that the amygdala appears to undergo an abnormal pattern of development that includes early enlargement and a decreased number of neurons in adulthood. This suggests that either fewer amygdalar neurons are generated during early development, or a normal or even excessive number of neurons form in early development, with subsequent degeneration.

Functional Brain Imaging Studies

Functional brain imaging studies in autism have focused on understanding the neurofunctional bases of the core social impairments and the well-documented impairments in complex, higher order, abstract thinking found in this disorder.

Social Brain Circuitry

Specific areas of the brain, including the occipital and temporal cortices (e.g., superior temporal sulcus, fusiform face area), the amygdala, the orbito-frontal cortex, and the anterior cingulated cortex, are critically involved in social processing, forming the 'social brain circuitry.' Each of these brain regions subserves different aspects of social processing, and each has been implicated in autism. Many of the early social impairments in autism, such as eye contact, joint attention, responses to emotional displays, and face recognition, involve the ability to attend to and process information from faces. Face processing may play a fundamental role in the dysfunction of the brain systems underlying the other impairments in social cognition in autism. The neural systems that mediate face processing come online very early in life. Behavioral studies have demonstrated face processing impairments in children, adolescents, and adults with autism, including deficits in face discrimination and face recognition. Functional magnetic resonance imaging (fMRI) studies have demonstrated abnormal activation of the fusiform face area during face processing in individuals with autism. Electrophysiological studies also document atypical responses to faces in children and adults with autism. Specifically, autism appears to be associated with an atypical N170 response to faces and to emotional facial expressions. Thus, face processing impairments may be one of the earliest indicators of abnormal brain development in autism. Several fMRI studies also have documented abnormalities in amygdala activation during facial expression processing tasks. Current models of abnormalities in social brain circuitry in autism posit impairments in the amygdala-fusiform system.

A recent area of investigation in autism involves brain regions related to motor imitation. Imitative deficits in individuals with autism are consistently observed and imitation deficits are one of the core impairments of autism. In typically developing children and adults, passive observation of actions performed by others activates some of the same cortical regions that are activated during execution of action, likely contributing to action recognition. Electroencephalogram (EEG) studies with typical adults and children show that the mu (8–13 Hz) rhythm is attenuated both when individuals execute and simply observe actions.

Recent studies with children and adults with autism found that attenuation of mu was present during the execution, but not during the observation, of movement. Mu wave attenuation has been interpreted as reflecting a neural system for action observation–execution matching, also referred to as a 'mirror neuron' system. fMRI studies have shown that observation, execution, and imitation of actions in humans generate activity in the premotor cortex and Broca's area (homolog for area F5 in monkeys). It has been hypothesized that the action observation–execution matching system serves as the neural substrate for imitation, understanding of others' intentions, development of empathy, theory of mind, and even the evolution of language, and thus may be important for understanding the neurofunctional basis of ASD.

Functional Connectivity

Functional MRI and EEG studies indicate that autism is associated with abnormal cortical connectivity, particularly underconnectivity of the frontal regions with other higher cortical regions (e.g., the parietal region during an executive function task, the left latero-superior temporal area during a sentence comprehension task). Underconnectivity, as measured by fMRI, has been found during a wide range of tasks, including executive function tasks, sentence comprehension, visual motor performance, and resting states. Reduced corpus callosum volume has been found to be correlated with lower connectivity in individuals with ASD. A recent EEG study utilized measures of EEG coherence to assess functional connectivity in adults with ASD during a resting state. Functional connections can be estimated in EEG signals by assessing the consistency of phase relationships between them via coherence measurements. High coherence between two EEG signals reflects synchronized neuronal oscillations and suggests functional integration, while low coherence suggests independently active neural populations. This study documented reduced long-range EEG coherence in the alpha frequency range, and increased short-range EEG coherence in the theta range in adults with ASD. These alpha range findings suggest that the frontal lobe has selectively weak functional connections with the rest of the cortex, and may be consistent with metabolic studies showing reduced correlated blood oxygenation between frontal and other regions.

Summary and Future Directions

In autism, the overall cerebral volume appears to increase at an unusually rapid rate during early childhood, suggesting that the disorder must be caused by prenatal or early postnatal events. Clear evidence of functional abnormalities in several regions involved in the social brain has been documented. Such evidence includes abnormal activation of the fusiform gyrus, superior temporal sulcus, amygdala, and orbit-frontal regions. Current studies are utilizing event-related potential (ERP) paradigms to study the development of social and language brain circuitry in infants at risk for autism. Of particular interest is whether the abnormalities previously documented in older children with autism are related to a lack of critical input to early developing neural systems during the first few years of life. Ongoing studies of young children with ASD are examining the effects of early behavioral interventions, designed to increase social engagement and communication, on functional cortical organization and neural responses (e.g., ERPs to social and linguistic stimuli). Furthermore, recent pathology studies have provided new insights into underlying mechanisms that might help explain anatomical abnormalities, such as enlarged cerebral volume, and functional impairments, such as reduced functional connectivity. Evidence for postnatal processes in abnormal brain development in autism, such as postnatal abnormalities in head growth, suggest the possibility that autism could be amenable to treatment, both medical and behavioral, during early life, and that such very early treatment might prevent or significantly ameliorate the onset of autism symptoms. As autism susceptibility genes are identified, the hope is that these genes, in combination with other behavioral, electrophysiological, and MRI indices, might allow very early identification of autism, thus offering the potential of prevention or reduction of the full-blown syndrome. Meanwhile, behavioral interventions that are appropriate for very young children with autism are becoming increasingly sophisticated and effective, at least for a substantial subgroup of children with this complex and challenging disorder.

See also: Cognition: An Overview of Neuroimaging Techniques; Development of Behavior; Executive Function and Higher-Order Cognition: Neuroimaging; Fragile X Syndrome.

Further Reading

Baron-Cohen S and Belmonte MK (2005) Autism: A window onto the development of the social and the analytic brain. *Annual Review of Neuroscience* 28(1): 109.

Belmonte MK, Cook EH Jr., Anderson GM, et al. (2004) Autism as a disorder of neural information processing: Directions for research and targets for therapy. *Molecular Psychiatry* 9(7): 646–663.

Casanova MF (2006) Neuropathological and genetic findings in autism: The significance of a putative minicolumnopathy. *The Neuroscientist* 12(5): 435.

Courchesne E, Redcay E, and Kennedy DP (2004) The autistic brain: Birth through adulthood. *Current Opinion in Neurology* 17(4): 489–496.

Dawson G, Webb S, and McPartland J (2005) Understanding the nature of face processing impairment in autism: Insights from behavioral and electrophysiological studies. *Developmental Neuropsychology* 27(3): 403–424.

Dawson G, Webb S, Schellenberg G, et al. (2002) Defining the phenotype of autism: Genetic, brain, and behavioral perspectives. *Development and Psychopathology* 14: 581–611. (Special Issue: Multiple Levels of Analysis.)

Dawson G, Webb SJ, Wijsman E, et al. (2005) Neurocognitive and electrophysiological evidence of altered face processing in parents of children with autism: Implications for a model of abnormal development of social brain circuitry in autism. *Development and Psychopathology* 17: 679–697.

DiCicco-Bloom E, Lord C, Zwaigenbaum L, et al. (2006) The developmental neurobiology of autism spectrum disorder. *Journal of Neuroscience* 26(26): 6897–6906.

Friedman SD, Shaw DW, Artru AA, et al. (2003) Regional brain chemical alterations in young children with autism spectrum disorder. *Neurology* 60(1): 100.

Herbert MR (2005) Large brains in autism: The challenge of pervasive abnormality. *The Neuroscientist* 11: 417–440.

Just MA, Cherkassky VL, Keller TA, et al. (2004) Cortical activation and synchronization during sentence comprehension in high-functioning autism: Evidence of underconnectivity. *Brain* 127(8): 1811.

Pardo CA, Vargas DL, and Zimmerman AW (2005) Immunity, neuroglia, and neuroinflammation in autism. *International Review of Psychiatry* 17(6): 485–495.

Schultz RT (2005) Developmental deficits in social perception in autism: The role of the amygdala and fusiform face area. *International Journal of Developmental Neuroscience* 23(2–3): 125–141.

Schumann CM and Amaral DG (2006) Stereological analysis of the amygdala neuron number in autism. *The Journal of Neuroscience* 26: 7674–7679.

Sparks BF, Friedman SD, Shaw DW, et al. (2002) Brain structural abnormalities in young children with autism spectrum disorder. *Neurology* 59: 184–192.

Wassick TH, Brzustowics LM, Barlett CW, et al. (2004) The search for autism disease genes. *Mental Retardation Developmental Disabilities Review* 10(4): 272–283.

Autoimmune Autonomic Neuropathy

S Vernino, University of Texas Southwestern Medical Center, Dallas, TX, USA
P A Low, Mayo Clinic College of Medicine, Rochester, MN, USA

Introduction

Acquired autonomic neuropathy can involve dysfunction of peripheral autonomic nerves or ganglia and can take many forms. Chronic progressive autonomic neuropathy may occur in the context of a more diffuse peripheral neuropathy associated with medical conditions such as diabetes or amyloidosis. A chronic, slowly progressive, idiopathic presentation of peripheral autonomic failure without sensorimotor neuropathy has customarily been called pure autonomic failure (a degenerative neurological disorder first described by Bradbury and Eggleston), but this arbitrary classification probably encompasses disorders with several different pathophysiologies, including chronic autoimmune autonomic neuropathy (AAN). Many cases of subacute autonomic neuropathy, where the maximal autonomic deficit occurs within 3 months, appear to be caused by neurological autoimmunity. Subacute autonomic neuropathies can be further divided into dysautonomia associated with sensory and motor neuropathy (acute inflammatory neuropathies such as Guillain–Barré syndrome; GBS), autonomic neuropathy associated with malignancy (paraneoplastic AAN), and AAN.

Autoimmune Autonomic Neuropathy

Clinical Description

The presentation of AAN (formerly known as acute pandysautonomia, idiopathic subacute autonomic neuropathy, or pure autonomic variant of GBS) is often highly characteristic. The typical syndrome occurs in a previously healthy individual in whom autonomic failure develops over the course of a few days or weeks. The most common pattern is severe generalized autonomic failure. Sympathetic failure is manifested as severe orthostatic hypotension and anhidrosis, and parasympathetic failure as dry mouth, dry eyes, sexual dysfunction, constipation, urinary retention, impaired pupillary light response, and fixed heart rate. Gastrointestinal dysmotility is common (70% of patients) and manifests as anorexia, early satiety, postprandial abdominal pain and vomiting, constipation, or diarrhea. Orthostatic hypotension, anhidrosis, and gastrointestinal dysfunction each occur in more than 50% of patients.

We now recognize that some cases of AAN have a more insidious onset. In those cases, the autonomic syndrome may be difficult to distinguish from an autonomic-predominant form of multiple-system atrophy or from pure autonomic failure. Clinical features which can help differentiate these disorders are presented in **Table 1**. Even if the time course is unclear, the presence of prominent gastrointestinal dysmotility as well as impaired pupillary light reflexes suggests AAN.

Due to loss of postural autonomic reflexes in AAN, a common chief complaint is light-headedness and fainting when upright. Some patients are unable to stand for more than a few minutes, and the most severely affected may even faint when sitting up. In addition to gravitational pooling of blood in the dependent limbs, impaired autonomic innervation of the splanchnic-mesenteric vascular bed contributes significantly to failure to maintain blood pressure when standing. When these patients stand up (or are brought to a head-up posture on a tilt table), there is a progressive decline in blood pressure without a compensatory rise in heart rate. This pattern is indicative of diffuse autonomic failure. Due to failure of the cardiovagal control of heart rate, there may be a slight resting tachycardia and minimal change of heart rate during deep breathing, Valsalva maneuver, or standing.

Gastrointestinal symptoms are another common and disabling symptom in AAN. Patients may have severe gastroparesis, which leads to early satiety and nausea after eating. Some patients regurgitate undigested food many hours after eating. A feeding jejunostomy tube may be required to provide nutrition distal to the stomach. Lower bowel hypomotility is also frequent, typically causing severe constipation. Abdominal X-ray may suggest intestinal obstruction with dilated loops of small bowel or distended colon. In some cases, laparotomy and bowel resection have been performed. The gastrointestinal symptoms lead to abdominal pain and weight loss due to inability to maintain nutrition.

Heat intolerance due to loss of sweating can be potentially life-threatening for patients in warm, dry climates. Patients are at risk for hyperthermia, and so must avoid extreme heat and cool their skin with water. Urinary retention (neurogenic bladder) is common in patients with AAN. Many other autonomic symptoms occur but are rarely life-threatening. These should be recognized as features of autonomic failure and not ignored. Dry eyes and dry mouth due

Table 1 Autoimmune autonomic neuropathy (AAN), pure autonomic failure (PAF), and multiple system atrophy (MSA)

Parameter	*AAN*	*PAF*	*MSA*
Onset	Subacute or insidious	Insidious	Insidious
First symptom	Multiple	Orthostatism	Neurogenic bladder
Gastrointestinal symptoms	Common	Absent	Uncommon
Pupillary involvement	Common	Absent	Uncommon
Central nervous system involvement	Absent	Absent	Present
Somatic neuropathy	Mild/minimal[a]	Absent	Present in 15–20%
Pain	Often present	Absent	Absent
Autonomic findings	Widespread	Limited	Relatively widespread
Progression	Often monophasic	Slow	Inexorably progressive
Prognosis	Relatively good	Relatively good	Poor
Lesion	Postganglionic	Postganglionic	Preganglionic; central
Supine plasma Norepinephrine	Reduced	Markedly reduced	Normal
Electromyogram	Usually normal	Normal	Usually normal
Ganglionic AChR antibody	Positive (50%)	Negative[b]	Negative

[a]More common in paraneoplastic cases.
[b]Chronic AAN may be indistinguishable from PAF.

to loss of parasympathetic secretomotor function are occasionally misdiagnosed as sicca or Sjögren's syndrome. Patients with impaired pupillary light reflexes may complain of blurry vision and light sensitivity. Impotence is a common early autonomic symptom in men.

In the laboratory, objective autonomic abnormalities are usually obvious and include orthostatic hypotension, tonic pupils, and lack of heart rate variability. The spectrum and severity of autonomic failure, however, varies from patient to patient. Less common presentations are a selective cholinergic (parasympathetic) failure, selective adrenergic neuropathy, or isolated gastrointestinal dysmotility.

Some case reports have demonstrated a dramatic and complete spontaneous autonomic recovery. In larger case series, the typical course is monophasic worsening followed by stabilization or partial remission without recurrences. Many patients report spontaneous improvement, but recovery is typically incomplete. Only a third of patients experience major functional improvement of autonomic deficits.

It is apparent that a subset of patients present with a more insidious and progressive form of AAN rather than with the typical subacute monophasic presentation. Some of these patients are readily identified as AAN by their pattern of autonomic symptoms or by the presence of specific serum autoantibodies (especially ganglionic acetylcholine receptor (AChR) antibodies; see later). Seronegative patients with an insidious presentation may be difficult to distinguish from degenerative forms of dysautonomia. Features which help differentiate AAN from pure autonomic failure and multiple-system atrophy are shown in **Table 1**.

Associated Features

When subacute autonomic neuropathy occurs in isolation (AAN), the syndrome follows a viral prodrome in about 60% of cases, with a flulike illness or upper respiratory infection being the most frequent association. Specific preceding viral infections have been reported, with Epstein–Barr virus being the most common association. Other recognized antecedent events include minor surgical procedures or routine immunizations. The spinal fluid protein is often elevated. Motor and sensory nerve abnormalities are minimal or absent. Neuropathic symptoms, such as tingling in distal extremities, are reported by about 25% of patients, but these symptoms are not accompanied by objective signs or electrophysiologic evidence of sensory neuropathy.

Ganglionic AChR Antibodies

About 50% of patients with the typical clinical features of AAN have high titers of autoantibodies directed against the ganglionic AChR. This receptor is a pentameric transmembrane complex consisting of two AChR $\alpha3$ subunits in combination with AChR β subunits. The $\alpha3$-type ganglionic AChR mediates fast synaptic transmission in all peripheral autonomic ganglia and is homologous but genetically and immunologically distinct from the AChR at the neuromuscular junction (**Figure 1**). Serum ganglionic AChR antibody levels in AAN cases correlate with the severity of autonomic neuropathy clinically and with the severity on laboratory testing. Although the finding of high levels of ganglionic AChR antibody is specific for the diagnosis of AAN, a negative antibody test does not rule out the diagnosis.

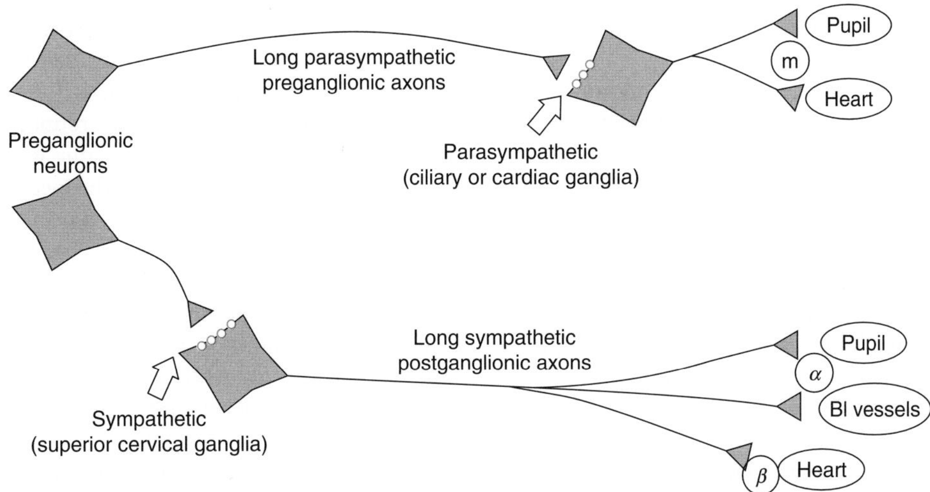

Figure 1 Autonomic system. The two major limbs of the peripheral autonomic nervous system both have a synapse in peripheral autonomic ganglia. Ganglionic synaptic transmission is mediated by acetylcholine, which binds to neuronal receptors on the postganglionic neuron (ganglionic AChR). Postganglionic autonomic neurons innervate multiple targets. Antibodies against ganglionic AChR could interfere with synaptic transmission in autonomic ganglia and cause diffuse autonomic failure (α, $\alpha3$ subunits; β, β subunit). m, muscarinic; α, alpha-adrenergic receptors; β, beta-adrenergic receptors.

Clinically, patients with high levels of ganglionic AChR antibodies more often have a subacute onset and generally show more prominent cholinergic dysautonomia (sicca complex, pupillary abnormalities, gastrointestinal dysmotility, and bladder symptoms) compared to seronegative autonomic neuropathy patients. Recognition of ganglionic AChR antibodies has allowed for the serological detection of AAN and led to a better appreciation of the spectrum of this disorder, including the observation that some cases are characterized by insidious symptom onset and gradual progression without antecedent event. Such chronic cases may be initially indistinguishable from degenerative forms of autonomic failure. Some features that help distinguish AAN from degenerative forms of autonomic failure are shown in **Table 1**. A high serum level of ganglionic AChR antibody is most characteristic of idiopathic AAN. Low levels of antibodies are found in a minority of patients with limited forms of acquired dysautonomia. Ganglionic AChR antibodies are found in 5–10% of patients with postural tachycardia syndrome, idiopathic gastrointestinal dysmotility, or Lambert–Eaton myasthenic syndrome.

Treatment

The mainstay of treatment for AAN is symptomatic, including volume expansion for blood pressure support, bowel management, and supplemental moisture for dry eyes and mouth. Acetylcholinesterase inhibitors may be used to alleviate neurogenic orthostatic hypotension, and this class of drugs might be particularly appropriate to treat AAN where the pathophysiology is presumed to be impaired ganglionic cholinergic synaptic transmission.

Based on evidence of an autoimmune pathogenesis and the presence of a specific autoantibody, it is reasonable to consider plasma exchange or intravenous immune globulin. Single case reports of a successful intravenous immune globulin therapy have all involved early therapeutic intervention. It remains to be determined whether this treatment is effective for patients with a chronic progressive or chronic stable course. Plasma exchange has more recently been used to successfully treat AAN and was effective even in one patient with severe autonomic failure for over 10 years.

Animal Models of Experimental Autoimmune Autonomic Neuropathy

Several animal models of autoimmune or antibody-mediated autonomic neuropathy have been developed, and these have allowed insights into autonomic pathophysiology in humans. The most convincing way to prove an antibody-mediated disease is to transfer the disease to a healthy animal by administering antibody (passive transfer). AAN can be induced in mice by passive transfer of immunoglobulin G (IgG) from affected rabbits or humans. An animal model of experimental autoimmune autonomic neuropathy (EAAN) can also be induced in rabbits by immunization with the ganglionic AChR. Rabbits with EAAN manifest symptoms of chronic autonomic failure similar to those seen in AAN patients, including impaired blood pressure and heart rate variability, impaired pupil constriction, decreased lacrimation, and gastrointestinal hypomotility. As in patients with AAN, more severe experimental autonomic dysfunction occurs with higher antibody levels.

Closer evaluation of the EAAN animal models provides some insight into the pathophysiology of AAN. Histologic studies of EAAN indicate that this autoimmune form of autonomic neuropathy is associated with a loss of ganglionic AChR from peripheral (ganglionic) autonomic neurons. Most likely, the binding of specific ganglionic AChR antibodies causes internalization and accelerated degradation of the synaptic receptors. This leads to failure of ganglionic synaptic transmission without significant loss of autonomic neurons. Since there is no permanent neuronal loss, EAAN and AAN may be reversible. Indeed, mice treated with ganglionic AChR antibodies demonstrate signs of transient autonomic failure and then recover fully.

Paraneoplastic AAN

Paraneoplastic neurological disorders are remote immunological effects of systemic malignancy that can affect any part of the nervous system. Sensory peripheral neuropathy, cerebellar ataxia, and limbic encephalitis are some well-recognized syndromes. A paraneoplastic autonomic disorder can take several different forms. As with other paraneoplastic disorders, the symptoms usually precede the diagnosis of cancer, and the tumors, when found, are limited in stage or only locally metastatic (regional lymph nodes). The autonomic symptoms, therefore, cannot be attributed to direct effects of the tumor, to nonspecific consequences of chronic illness, or to chemotherapy-induced neuropathy. In many cases, paraneoplastic dysautonomia is accompanied by symptoms of limbic encephalitis or elements of other paraneoplastic syndromes.

Paraneoplastic autonomic neuropathy can be a subacute diffuse autonomic failure indistinguishable from idiopathic AAN. More commonly, the autonomic syndrome specifically targets the enteric autonomic nerves and ganglia causing severe gastrointestinal dysmotility (gastroparesis or intestinal pseudoobstruction). Esophageal dysmotility (including achalasia) has also been reported. Patients present with nausea, early satiety, bloating, abdominal pain, constipation, and weight loss. Patients have difficulty getting adequate calorie intake and may even have difficulty maintaining fluid intake. This is complicated by the loss of appetite and taste that often accompanies cancer and chemotherapy. Radiographs show dilated loops of bowel, and motility studies reveal delayed gastric emptying, diffuse intestinal hypomotility, and absent or incoordinated motor complexes. Endoscopy and exploratory laparotomy fail to identify an obstruction. Pathologically, paraneoplastic dysmotility has been associated with inflammation and destruction in the myenteric plexus throughout the gut, with reduction in neurons and axons, and lymphocytoplasmic infiltration.

Several different autoantibody specificities may be encountered in patients with paraneoplastic autonomic neuropathy. The best recognized serological finding is the antineuronal nuclear antibody type 1 (ANNA-1, also known as anti-Hu). Small-cell lung carcinoma is found in at least 80% of patients who are seropositive for ANNA-1. Autoantibodies against collapsing-response mediator protein (CRMP-5) can be found with many neurological syndromes, including autonomic neuropathies. CRMP-5 antibodies are associated with small-cell lung carcinoma as well as thymoma. Antibodies specific for neuronal cation channels (ganglionic or muscle AChR, voltage-gated N-type or P/Q-type calcium channels, or neuronal potassium channels) are also commonly found. A minority of patients will lack any currently recognized autoantibody marker. Small-cell lung carcinoma is the tumor most commonly associated with paraneoplastic autonomic neuropathy. Dysautonomia, especially gastrointestinal dysmotility, may also be encountered with thymoma with or without associated symptoms of myasthenia gravis.

Treatment of paraneoplastic autonomic neuropathy generally consists of supportive symptomatic treatments to alleviate the most problematic symptoms, orthostatic hypotension and gastrointestinal dysmotility. Paraneoplastic dysmotility is usually refractory to treatment with motility-enhancing agents or surgical decompression. Many patients require jejunostomy feeding tubes or intravenous nutrition. Gastrostomy feeding is usually not tolerated because of gastroparesis and vomiting. Every effort should be made to locate and treat the underlying malignancy. In some cases, autonomic function improves once the malignancy is effectively treated. Plasma exchange or intravenous immunoglobulin has also been reported to be effective in individual cases.

Other Autoimmune Disorders Affecting Autonomic Function

Autonomic symptoms are commonly encountered in other recognized autoimmune disorders. Dry mouth, impotence, and constipation are common in patients with Lambert–Eaton syndrome, a presynaptic disorder of neuromuscular junction transmission caused by antibodies against the P/Q-type voltage-gated calcium channel. Autonomic instability (manifested as fluctuations in heart rate and blood pressure) and gastrointestinal dysmotility are often encountered in patients with acute inflammatory demyelinating polyradiculoneuropathy (GBS syndrome). Autonomic overactivity is a typical feature of autoimmune neuromyotonia and Morvan syndrome. Both disorders are associated with antibodies against voltage-gated potassium channels and present with muscle stiffness and spontaneous muscle twitching. Future studies are likely to uncover other ion channels and other proteins in the autonomic nervous system that are targets for autoimmunity.

See also: Autonomic and Enteric Nervous System: Apoptosis and Trophic Support During Development; Autonomic Disorders; Autonomic Failure; Autonomic Nervous System: Central Control of the Gastrointestinal Tract; Autonomic Nervous System; Blood Pressure: Baroreceptors; Parasympathetic Nervous System; Sympathetic Nervous System.

Further Reading

Buckley C and Vincent A (2005) Autoimmune channelopathies. *Nature Clinical Practice Neurology* 1: 22–23.

Etienne M and Weimer LH (2006) Immune-mediated autonomic neuropathies. *Current Neurology & Neuroscience Reports* 6: 57–64.

Klein CM, Vernino S, Lennon VA, et al. (2003) The spectrum of autoimmune autonomic neuropathies. *Annals of Neurology* 53: 752–758.

Lee HR, Lennon VA, Camilleri M, et al. (2001) Paraneoplastic gastrointestinal motor dysfunction: Clinical and laboratory characteristics. *American Journal of Gastroenterology* 96: 373–379.

Lennon VA, Ermilov LG, Szurszewski JH, et al. (2003) Immunization with neuronal nicotinic acetylcholine receptor induces neurological autoimmune disease. *Journal of Clinical Investigation* 111: 907–913.

Mathias CJ and Bannister R (eds.) (2006) *Autonomic Failure.* Oxford: Oxford University Press.

Robertson D and Low PA (eds.) (2004) *Primer of the Autonomic Nervous System.* Amsterdam: Elsevier Academic Press.

Schroeder C, Vernino S, Birkenfeld AL, et al. (2005) Plasma exchange for primary autoimmune autonomic failure. *New England Journal of Medicine* 353: 1585–1590.

Vernino S, Ermilov LG, Sha L, et al. (2004) Passive transfer of autoimmune autonomic neuropathy to mice. *Journal of Neuroscience* 24: 7037–7042.

Vernino S, Low PA, Fealey RD, et al. (2000) Autoantibodies to ganglionic acetylcholine receptors in autoimmune autonomic neuropathies. *New England Journal of Medicine* 343: 847–855.

Autonomic and Enteric Nervous System: Apoptosis and Trophic Support During Development

R O Heuckeroth, Washington University School of Medicine, St. Louis, MO, USA

Introduction

Programmed cell death (PCD) is an evolutionarily conserved process in multicellular organisms that is important for morphogenesis during development and for the maintenance of tissue homeostasis in organs with ongoing cell proliferation. PCD also occurs in many tissues in response to injury. Apoptotic cell death is one form of PCD that provides a mechanism to remove individual cells from tissues without damage to adjacent cells. Defects in molecular mechanisms that regulate cell death cause both cancer and abnormal development. Because many aspects of apoptosis have been reviewed over the past decade, this article focuses on the developmental role of apoptosis in the sympathetic, parasympathetic, and enteric nervous system and provides only a brief overview of apoptotic mechanisms.

Apoptotic Cells and Molecular Mechanisms

Apoptosis is a form of cell death characterized by cell shrinkage, plasma membrane blebs, nuclear chromatin condensation, and DNA fragmentation. Phosphatidylserine, usually restricted to the inner face of the plasma membrane in live cells, also partially redistributes to the extracellular face of the plasma membrane. These changes largely result from activation of a set of cysteine-aspartyl-specific proteases called caspases. Caspases can be activated via either an extrinsic pathway or an intrinsic pathway. The extrinsic pathway is activated by binding of extracellular ligands to cell surface death receptors like Fas (Apo-1 or CD95), tumor necrosis factor receptor-1 (TNFR-1/p55/CD120a), interferon receptor, and TRAIL (TNF-related apoptosis-inducing ligand or Apo2-L) receptor. This binding leads to the recruitment of a variety of proteins that activate procaspase-8 or procaspase-10. In contrast, the intrinsic pathway is characterized by cytochrome c release from mitochondria. Cytochrome c in the cytoplasm induces heptamerization of Apaf-1, that then binds to and activates procaspase-9. These initiator caspases (-8, -9, and -10) activate the effector caspases (caspase-3 and caspase-7). The effector caspases have many substrates (estimated to include 0.5–5% of cellular proteins), induce DNA cleavage and mitochondrial permeabilization, and activate additional proteolytic cascades that eventually result in cell death.

Apoptosis in the Nervous System and the Discovery of Neurotrophic Factors

PCD is common in most parts of the developing nervous system, where 20–80% of all neurons produced during embryogenesis die before adulthood. Typically, these cells are eliminated by PCD at the time that they would normally innervate their targets. In this setting, PCD is thought to occur to allow matching of the neuronal population to the size of the innervation target via target-derived, trophic factor-dependent cell survival. This mechanism also efficiently eliminates cells with abnormal migration or abnormal axon targeting because cells that fail to innervate appropriate targets do not receive adequate trophic factor. The observation that developing sympathetic and sensory neurons were dependent on exogenous trophic factor for survival led to the discovery of nerve growth factor (NGF) by Rita Levi-Montalcini and Stanley Cohen. This discovery prompted work that in turn resulted in the discovery of the related proteins brain-derived neurotrophic factor (BDNF), neurotrophin-3 (NT-3), and NT-4/5. An *in vitro* sympathetic neuron survival assay was also critical for discovery of neurturin (NRTN) and for the identification and characterization of the related proteins glial cell line-derived neurotrophic factor (GDNF), artemin (ARTN), and persephin (PSPN). Furthermore, molecular mechanisms of neonatal sympathetic neuron survival *in vitro* have been studied extensively and form the basis of much of what we know about neuronal apoptosis. In the rat, for example, superior cervical ganglion (SCG) sympathetic neurons are NGF dependent from embryonic day (E)16 to postnatal day (P) 7. NGF deprivation during this time results in PCD within 48 h. With this assay, investigators demonstrated that apoptosis is triggered after NGF withdrawal by increased c-Jun protein levels and c-Jun phosphorylation, increased translocation of the proapoptotic protein Bax from the cytosol to mitochondria, and Bax-induced cytochrome c release that causes apoptosis via the intrinsic pathway. Release of Smac/DIABLO and HtrA2/Omi from mitochondria is also important to relieve caspase inhibition by inhibitor of apoptosis proteins so that caspases can become activated.

Lessons from Mutant Mice

The neurotrophic factors NGF, BDNF, NT-3, GDNF, NRTN, and ARTN are critical for the normal development and maintenance of the peripheral nervous system. Depending on the specific cell type and developmental time point, these factors determine nervous system structure and function via a number of distinct effects including preventing or in some cases promoting cell apoptosis. These proteins have other important roles as well:

- Promotion of cell survival
- Promotion of cell proliferation
- Support for neuronal differentiation
- Promotion of directed cell migration
- Promotion of neurite extension
- Axon guidance

Each of these actions is mediated by binding of trophic factors to specific cell surface receptors. There is also some ability for specific trophic factors to activate multiple receptors at least *in vitro*. Preferred receptor–ligand interactions, alternate receptor–ligand interactions, and a summary of interactions that are important *in vivo* are presented in **Figure 1**. Briefly, NGF, BDNF, and NT-3 bind to and activate the transmembrane tropomyosin-related kinase (Trk) receptors TrkA, TrkB, and TrkC, respectively. In addition, NT-4 activates TrkB, whereas NT-3 can directly activate TrkA and TrkB, but with lower efficiency than it activates TrkC. All these NTs also bind to the low-affinity receptor p75NTR. The interaction with p75NTR inhibits the activation of Trk receptors by their nonpreferred ligands and in some cases improves intracellular signaling. However, in the absence of specific Trk receptor activation, binding of proneurotrophins to p75NTR promotes cell death. For example, pro-BDNF binding to p75NTR on sympathetic neurons, which express TrkA but not TrkB, promotes cell death, but NGF promotes cell survival by activating both TrkA and p75NTR. This dual role for p75NTR in sympathetic neurons is thought to improve the specificity of axon targeting by eliminating cells whose axons innervate the wrong target and arrive at a source of BDNF but not of NGF. The proteins GDNF, NRTN, ARTN, and PSPN form a separate family of trophic factors called the GDNF-related ligands (GFLs). GFLs all activate the transmembrane tyrosine kinase Ret. Instead of binding directly to Ret, however, GFLs activate Ret by binding to a glycosylphosphatidylinositol-linked cell surface receptor (GFRα1–4). Each GFL preferentially binds to a specific GFRα protein, with GDNF, NRTN, ARTN, and PSPN interacting best with GFRα1, GFRα2, GFRα3, and GFRα4, respectively. There is some *in vitro* cross-talk between GFLs and nonpreferred GFRα proteins, as indicated in **Figure 1**, but these interactions generally require higher trophic factor concentrations and do not appear to be physiologically important *in vivo*. However, receptor

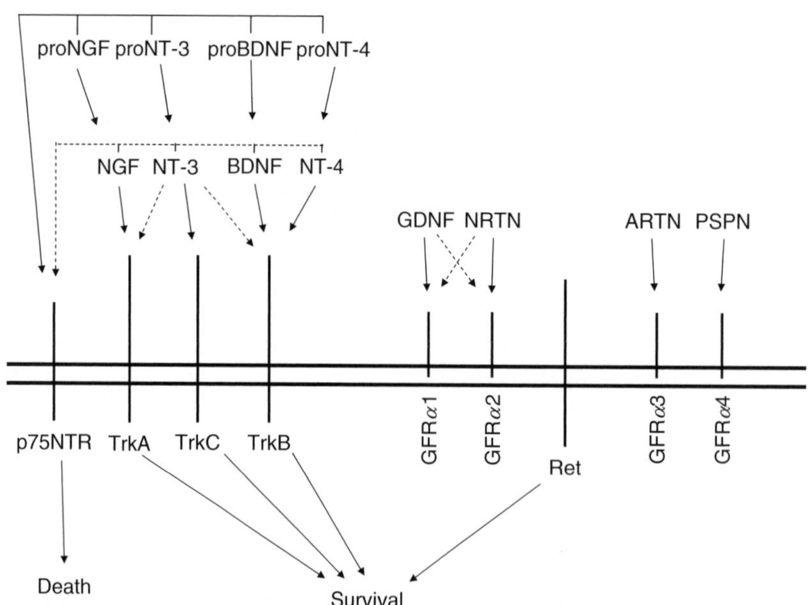

Figure 1 Summary of neurotrophic factors and their receptors. Solid arrows show preferred receptor–ligand interactions. Dashed arrows show nonpreferred interactions between ligands and receptors. For neurturin (NRTN) and glial cell line-derived neurotrophic factor (GDNF), the nonpreferred interactions do not appear to be important *in vivo*. For neurturin (NT)-3, nonpreferred interactions are important *in vivo*. Interactions of neurotrophins with p75NTR are also likely to be physiologically relevant. ARTN, artemin; BDNF, brain-derived neurotrophic factor; GFR, growth factor receptor; NGF, nerve growth factor; PSPN, persephin.

cross-talk appears important in the pharmacological response to exogenously administered factors, presumably acting at supraphysiological levels.

Given the complexity of these systems, determining the relevance of specific trophic factors and their receptors for neuronal development and function has required a combination of *in vitro* studies and the analysis of mutant animals. While many conclusions can be drawn from this work, several important themes have emerged:

1. Patterns of trophic factor receptor and ligand expression correlate well with trophic factor dependence *in vivo* but incompletely predict function.
2. For many developing neurons, trophic factor dependence changes during development such that cells initially dependent on one factor become dependent on a different trophic factor at a later developmental stage.
3. Trophic factors may promote both survival and proliferation early in development or may direct precursor migration. At later stages, these factors may support survival (e.g., at the time of target innervation) or provide trophic support without being essential for survival.
4. Trophic factor dependence *in vitro* does not imply developmental stage-specific programmed cell death *in vivo*. Cell death *in vivo* requires both trophic factor dependence and trophic factor deficiency.
5. Trophic factor receptor expression also does not imply trophic factor-dependent survival since specific cell populations may respond to multiple trophic factors simultaneously.

A few examples will clarify the strategies employed to determine cell number in the autonomic and enteric nervous system (ENS). Effects of specific receptor and ligand mutations on sympathetic, parasympathetic, and enteric neurons are summarized in **Table 1**.

Trk Receptors and Ligands Regulate Cell Death *In Vivo* in the Sympathetic Nervous System

The sympathetic chain in the mouse embryo arises at E11.5, and at that time, TrkC is expressed, but TrkA is not detectable. By E13.5, when the SCG forms, both TrkC and TrkA are detected in the SCG. TrkA expression becomes more robust by E15.5 and continues to be expressed at high levels throughout postnatal development. TrkC levels in the SCG fall significantly by P0, when only a few cells have detectable TrkC expression. In contrast, TrkB is not detected in sympathetic neurons at any developmental stage.

The pattern of TrkA receptor gene expression correlates well with the normal role of TrkA and NGF in development of the mouse sympathetic nervous system. In TrkA–/– mice, for example, neuron number is normal in the sympathetic chain at E11.5 and in the SCG and sympathetic chain at E13.5. By E15.5, there are 15% fewer SCG neurons in TrkA-deficient mice than wild-type (WT). By E17.5, TrkA–/– mice had 35% fewer SCG neurons than WT mice did, and almost all SCG neurons are lost in TrkA–/– mice by P9. At each of the periods investigated, the reduction in SCG neuron number in TrkA–/– mice was attributable to increased cell death. Similarly, in NGF-deficient mice, all SCG neurons are absent by P14, and increased cell death was observed at P3. These changes in neuron number demonstrate the critical role for NGF/TrkA signaling in sympathetic neuron survival and are consistent with results obtained in the 1960s with exposure to neutralizing antibodies to NGF. They also define a period of trophic factor-dependent cell death that correlates well with the timing of trophic factor dependence of SCG neurons in primary culture. Moreover, the striking similarity of the SCG phenotype for NGF–/– and TrkA–/– mice highlights the importance of this preferred ligand–receptor interaction.

In contrast to these results with TrkA–/– and NGF–/–mice, however, the effect of TrkC or NT-3 deficiency is less easily predicted based on gene expression patterns. For example, although TrkC is expressed in the sympathetic nervous system as early as E11.5, the SCG is normal at all developmental stages in TrkC–/– mice. This contrasts with the 50% reduction in SCG neurons found in NT-3-deficient mice. The time during development when NT-3 is required for sympathetic neuron survival, however, has been controversial, with an early report suggesting increased SCG precursor cell death between E11 and E17 and a later study using more animals failing to reproduce this observation and demonstrating increased programmed cell death after birth. Furthermore, injection of anti-NT-3 antibodies into neonatal rats caused a 60–80% loss of SCG neurons, suggesting a role for NT-3 in postnatal SCG survival. Together these observations demonstrate the complexity of determining how specific trophic factors regulate neuronal apoptosis, especially in cells with multiple trophic factor receptors. They further suggest that although NT-3 critically regulates SCG apoptosis, it may perform this function primarily via TrkA receptor activation.

The role TrkB and BDNF in SCG neuron survival is even more remarkable. Since TrkB is not expressed in the sympathetic nervous system, it might initially have been predicted that the SCG would be normal in

Table 1 Neuron number in autonomic and enteric ganglia of mutant mice

	Sympathetic		*Parasympathetic*				*Enteric*
	SCG	*Chain ganglia*	*Ciliary*	*Submandibutar*	*Sphenpalatine*	*Otic*	
TrkA KO	100% loss by P9						
NGF KO	100% loss by P7						
TrkB KO	Normal						
BDNF KO	36% increase						
TrkC KO	Normal						Selective neuron loss
NT-3 KO	50% loss						Selective neuron loss
p75NTR KO	Increased						
Ret KO	Abnormal, but variable	Abnormal	48% loss	30–47% loss	99–100% loss	99–100% loss	99% loss
GFRα1 KO	Normal			33% loss	99–100% loss	99–100% loss	99% loss
GDNF KO	30% loss		40% loss	36% loss	99% loss	86% loss	99% loss
GFRα2 KO	Normal			42% loss	Normal	40% loss	
NRTN KO	Normal		50% loss	45% loss	Normal	Normal	Normal
GFRα3 KO	Abnormal, but variable	Abnormal					
ARTN KO	Abnormal, but variable	Abnormal					Normal

BDNF−/− mice. Remarkably, SCG neuron number is increased by 36% in BDNF−/−compared to WT animals at P15. This was hypothesized to occur because pro-BDNF promotes sympathetic neuron cell death by binding to p75NTR. Indeed, WT SCG neuron number decreases by 42% because of PCD between P0 and P23, but SCG neuron number increases between P0 and P23 in P75NTR−/− mice. These results suggest that both BDNF and p75NTR are required for naturally occurring cell death in the SCG of WT mice. Further support for this hypothesis is the observation that mice deficient in both p75NTR and TrkA have markedly reduced sympathetic neuron apoptosis compared to TrkA−/− animals. This mechanism allows for efficient elimination of cells with misguided axons that encounter pro-BDNF and reinforces the importance of TrkA activation for SCG survival.

Ret Signaling Promotes Sympathetic Neuron Precursor Migration and Supports Axon Extension Required for Sympathetic Neurons to Innervate Targets and Encounter Trk Ligands

Ret activation via GFRα3 and ARTN is essential for normal sympathetic nervous system development. In this case, however, Ret signaling appears to be important for neuronal precursor migration and axon extension, but not directly for cell survival. This conclusion is based on the following observations. Ret−/− mice have smaller-than-normal SCG ganglia that appear in a more caudal position than in WT animals. Furthermore, sympathetic fiber innervation of targets in the nasal mucosa, eye, and skin of Ret−/− mice is almost completely absent, while sympathetic innervation density of the submandibular salivary gland is significantly reduced. In addition, there are defects all along the sympathetic chain in Ret−/− mice, including smaller-than-normal ganglia, abnormally located ganglia, and reduced target innervation. Detailed evaluation of the mechanism of these defects demonstrated that Ret−/− sympathetic neuron precursor lineage commitment appears normal, but differentiation of these cells is delayed, and neuronal cell death increases between E16.5 and P0 compared with WT mice. This increased cell death was particularly interesting since Ret is expressed in all sympathetic precursors at E11.5, but most of these cells lose Ret expression by E15.5. Furthermore, although increased cell death occurred in Ret−/− SCG, it did not occur preferentially in Ret-expressing cells.

Additional analysis demonstrated that the Ret ligand ARTN is expressed in blood vessels along the normal route of sympathetic axons where ARTN potently stimulates SCG axon growth and is a chemoattractant that directs axon pathfinding. It is interesting that ARTN- and GFRα3-deficient mice have variable defects in the SCG. Specifically, SCG size was near normal in ARTN−/− and GFRα3−/− mice if the ganglia were normal and in position and peripheral targets were innervated, but SCG size was markedly diminished and apoptosis was increased in mice with abnormally located SCG and reduced peripheral target innervation. Defects in ARTN−/− and GFRα3−/− mice were detected as early as E10.5, when sympathetic neuron precursors emerge from the neural crest, demonstrating an important role for Ret/ARTN/GFRα3 signaling in precursor migration and neurite extension. Overall, these data suggest that increased cell death in the sympathetic nervous system of Ret−/−, ARTN−/−, and GFRα3−/− mice, compared with WT mice, occurs because of failure to properly innervate targets and subsequent deficiency in target-derived trophic factor (i.e., NGF deficiency). The Ret ligand GDNF also appears to be important for SCG development; since GDNF−/− mice have been reported to have a 30% loss of SCG neurons. Although GDNF supports the survival of cultured SCG neurons, the precise role of GDNF for SCG neuron survival *in vivo* has not been defined. Furthermore, mice missing the preferred GDNF receptor GFRα1 have a normal SCG at P0, which suggests that GDNF effects on the SCG are mediated via an alternate signaling pathway. These analyses demonstrate the complexity of interpreting mouse phenotypes and primary culture data and the importance of careful mechanistic time-course studies to define the role of specific trophic factors and their receptors in neuron survival. They also demonstrate robustly that programmed cell death in response to trophic factor deficiency at the time of target innervation provides a powerful mechanism to ensure survival of only the sympathetic neurons that correctly innervate their targets.

Programmed Cell Death in the Parasympathetic Nervous System

Cell death in parasympathetic neurons is much less well studied than in sympathetic neurons, but these cells do not appear to rely on neurotrophins (NGF, BDNF, or NT-3) for survival. Furthermore, although ciliary ganglion cell survival is supported by ciliary neurotrophic factor (CNTF) *in vitro*, CNTF does not appear to perform this function *in vivo* during normal development. Instead, Ret activation by GDNF and NRTN appears to be the most important determinant of parasympathetic neuron number. Unfortunately, a detailed analysis of cell apoptosis and proliferation within the parasympathetic nervous system of mutant mice has not been performed for most cell populations. Nonetheless, several important conclusions can be drawn about the role of GDNF and NRTN in parasympathetic neurons:

1. Different parasympathetic neuron populations respond differently to the loss of Ret signaling. For example, otic and sphenopalatine ganglia are essentially completely absent from Ret–/– and GDNF–/– mice, but the reduced cell number in the ciliary and submandibular ganglia of these mice is much less dramatic (**Table 1**).
2. Reduced number of ganglion cells does not imply increased apoptosis. This is demonstrated by analysis of the sphenopalatine ganglion in GDNF and GFRα1–/– mice. In these animals, the sphenopalatine ganglia are absent as early as E12.5. Furthermore, bromodeoxyuridine labeling in GDNF–/– mice demonstrated reduced proliferation of sphenopalatine ganglion cell precursors, but terminal transferase deoxyuridine triphosphate nick end labeling (TUNEL) analysis failed to demonstrate any increase in cell death. Thus, at least in the sphenopalatine ganglion, GDNF/Ret signaling is required for precursor proliferation and not for survival of mature neurons.
3. Reduced parasympathetic ganglion cell survival may result from abnormal ganglion cell precursor migration. This is illustrated by the otic ganglion in GDNF–/– mice. Like the sphenopalatine ganglion, most otic ganglion cells are absent in GDNF–/– animals. In this case, however, otic ganglion cell precursors migrate abnormally, and TUNEL staining demonstrated increased cell death in the abnormally migrating precursors.

Thus, while programmed cell death may be important for parasympathetic nervous system development, it is much less well understood than in the sympathetic nervous system. In particular, mechanisms for eliminating cells with abnormal axon targeting have not been well documented in the parasympathetic nervous system. Indeed, target innervation by parasympathetic sphenopalatine neurons is dramatically reduced in NRTN–/– and GFRα2–/– mice, but sphenopalatine ganglion numbers are normal, which suggests that these cells are not dependent on target-derived trophic factors for survival.

Mechanisms Governing Neuron Number in the ENS

The ENS is a complex network of neurons and glia within the bowel wall that controls intestinal motility, responds to sensory stimuli, and regulates intestinal secretion and blood flow. To perform these functions, there are roughly as many neurons in the ENS as in the spinal cord, and the ENS comprises many distinct neuron subtypes that differ in function, transmitter phenotype, and pathways of axon pathfinding. Mechanisms of axon targeting in the ENS, however, and mechanisms to ensure that enteric neurons have correctly innervated their targets, are not yet understood. Furthermore, the ENS presents challenges that do not occur in other regions of the nervous system. This is especially true within the myenteric plexus since specific subtypes of myenteric neurons must extend their axons either orally (toward the mouth) or aborally (toward the end of the bowel) for the gut to function normally. Although there must be axon guidance cues present during development to direct these axons, adjacent regions of the bowel wall appear remarkably similar in the mature organism. Thus, unlike the sympathetic nervous system, where targets of innervation are far from the neuronal cell bodies and target-derived trophic factor dependence is an excellent mechanism for ensuring that only correctly targeted neurons survive, it is difficult to imagine how apoptosis could be used in the ENS to eliminate neurons whose axons project in the wrong direction. It is easier to imagine that apoptotic pathways could be important for enteric neurons projecting outside the muscular gut wall (e.g., to villi), but this has not yet been investigated.

With these ideas in mind, it is perhaps not surprising that apoptosis does not appear to occur within the developing or adult ENS of WT mice. This is not to suggest that enteric neurons are trophic factor-independent. Indeed, TrkA, TrkB, TrkC, p75NTR, Ret, GFRα1, and GFRα2 are all expressed within the ENS. Furthermore, enteric neurons undergo apoptotic cell death in primary culture when they are deprived of trophic factors and clearly respond to a variety of neuronal survival factors, including GDNF, NRTN, and NT-3, *in vitro*. In addition, Ret–/–, GFRα1–/–, and GDNF–/– mice miss essentially all enteric neurons from small bowel and colon. In fact, defective Ret signaling is the most commonly identified etiology of distal colon aganglionosis in humans (i.e., Hirschsprung disease). NT-3- and TrkC-deficient animals also have abnormal ENS development, with striking reduction in some subpopulations of enteric neurons. In contrast to these results, NRTN–/– and GFRα2–/– mice have a normal density of enteric neurons, but as in the parasympathetic nervous system, enteric neurons of NRTN–/– and GFRα2–/– mice are smaller than normal, with reduced neuronal projections, at least in some subtypes of neurons. Finally, apoptosis commonly occurs in ENS precursors of mice with a variety of mutations that cause intestinal aganglionosis, including in Ret–/–,

$\mathrm{Sox10}^{\mathrm{Dom}}/\mathrm{Sox10}^{\mathrm{Dom}}$, and Phox2b−/− animals. Thus, since ENS precursors depend on trophic factors for survival, but apoptosis does not appear to occur during normal ENS development, these factors must be produced in an adequate supply in WT animals to prevent programmed cell death. This further implies that ENS precursor proliferation must be carefully regulated to avoid producing more neurons than can be supported by available trophic factor. One way this occurs is that the availability of GDNF directly determines the rate of ENS precursor proliferation. Both increases and decreases in GDNF availability alter enteric neuron number via changes in precursor mitotic rates.

Apoptosis in Neuronal Injury

In addition to the physiologic role of apoptosis during normal development, cellular injury may also cause cell death via apoptotic pathways. In the ENS, for example, apoptosis has been reported in age-related myenteric neuron loss, anti-HuD-associated paraneoplastic syndrome, colitis-induced neuronal injury, and diabetes-associated ENS injury. Furthermore, at least in mouse models of diabetes, apoptosis can be reduced by providing additional trophic factor (GDNF) *in vivo*. It is interesting that in contrast to the effect of diabetes on the ENS, sympathetic neuron cell death does not appear to occur in diabetic rats. Thus, once again, the importance of apoptotic pathways in the autonomic nervous system and the ENS is specific to the age of the animal, the apoptotic trigger, and the neuronal subtypes evaluated. Presumably, these differences in the extent of apoptosis in different regions of the nervous system and at different times during life reflect both the availability of trophic factors to support survival and the abundance of intracellular pro- and anti-apoptotic proteins.

Summary

Programmed cell death is important for tissue morphogenesis during development and for the maintenance of tissue homeostasis during adult life. In part, apoptosis is valuable because it provides a way for the organism to specifically eliminate single cells that are no longer needed. This is important in the nervous system, where correct axon targeting and matching of the target size to the number of innervating nerve fibers is critical for function. Because both too much and too little cell death could be detrimental, carefully regulated intracellular and extracellular control mechanisms have been established to control PCD. Many of these mechanisms have been studied in detail in the sympathetic nervous system, where target-derived trophic factors are required for neuron survival during a defined developmental period when neurons are innervating their targets. Because trophic factors are active at the axon tip and are produced in the axon target this strategy effectively eliminates neurons that fail to innervate an appropriate target. Similar strategies are employed in most regions of the nervous system.

Mechanisms of PCD in the autonomic nervous system and the ENS highlight several important themes. First, for PCD to be useful for tissue morphogenesis, the axon tip and neuron cell body must be in different environments. For this reason, PCD tends to occur as neurons are innervating their final targets instead of when they first begin to extend axons. Because many parasympathetic neuron cell bodies are embedded in their targets and many enteric neurons project axons to targets in an environment similar to that around the cell body, it is more difficult to see how PCD could be used in the parasympathetic nervous systen and the ENS to ensure proper axon targeting. Indeed, it is difficult to find evidence that PCD occurs as a part of normal development in either the ENS or the parasympathetic nervous system. Second, for PCD to be effective, different targets need to produce different trophic factors, and subsets of neurons must respond to only a limited array of trophic factors. Even more effective would be a strategy to actively eliminate cells exposed to the 'wrong' trophic factor. This explains the variety of trophic factors and receptors present in the nervous system and the role of p75NTR to induce cell death in the absence of Trk receptor activation. Indeed, given the wide array of neuron subtypes and targets, it is remarkable that the nervous system can be established with so few trophic factors and receptors. It is probably for this reason that developing neurons may respond to a combination of trophic factors or change trophic factor dependence during development. Finally, most postmitotic neurons are needed throughout the life of the organism. For this reason, resistance to apoptosis in neurons that have correctly innervated their targets is important for longevity. In the nervous system, this is accomplished by limiting trophic factor-dependent cell survival to a small developmental period. It is important to note that this requires changes in the intracellular machinery for apoptosis once targets are innervated.

Tremendous advances have been made over the past decade in understanding the role of apoptosis in development and disease, but many challenges remain. In particular, it would be valuable to develop additional strategies to prevent neuronal cell death after injury. Ideally, these strategies should target specific cell populations since global inhibition of apoptosis is likely to be carcinogenic. Continued detailed analysis of the molecular mechanisms of apoptosis and survival in different defined cell populations is therefore critical to allow targeted therapy and reduce disease-related morbidity and mortality.

See also: Apoptosis in Nervous System Injury; Apoptosis in Neurodegenerative Disease; Enteric Nervous System: Neurotrophic Factors; Neurotrophic Factor Therapy: GDNF and CNTF; Neurotrophic Factor Therapy: NGF, BDNF and NT3; Parasympathetic Nervous System; Programmed Cell Death.

Further Reading

Airaksinen MS and Saarma M (2002) The GDNF family: Signalling, biological functions, and therapeutic value. *Nature Reviews Neuroscience* 3(5): 383–394.

Baloh RH, Enomoto H, Johnson EMJ, and Milbrandt J (2000) The GDNF family ligands and receptors: Implications for neural development. *Current Opinion in Neurobiology* 10: 103–110.

Enomoto H, Crawford PA, Gorodinsky A, Heuckeroth RO, Johnson EM Jr, and Milbrandt J (2001) RET signaling is essential for migration, axonal growth, and axon guidance of developing sympathetic neurons. *Development* 128(20): 3963–3974.

Enomoto H, Heuckeroth RO, Golden JP, Johnson EM Jr, and Milbrandt J (2000) Development of cranial parasympathetic ganglia requires sequential actions of GDNF and neurturin. *Development* 127: 4877–4889.

Fagan AM, Zhang H, Landis S, Smeyne RJ, Silos-Santiago I, and Barbacid M (1996) TrkA, but not TrkC, receptors are essential for survival of sympathetic neurons *in vivo*. *Journal of Neuroscience* 16(19): 6208–18.

Francis N, Farinas I, Brennan C, et al. (1999) NT-3, like NGF, is required for survival of sympathetic neurons, but not their precursors. *Developmental Biology* 210(2): 411–427.

Gewies A (2003) ApoReview: Introduction to apoptosis. http://www.celldeath.de/encyclo/aporev/aporev.htm (accessed 3 August 2007).

Gianino S, Grider JR, Cresswell J, Enomoto H, and Heuckeroth RO (2003) GDNF availability determines enteric neuron number by controlling precursor proliferation. *Development* 130(10): 2187–2198.

Honma Y, Araki T, Gianino S, et al. (2002) Artemin is a vascular-derived neurotropic factor for developing sympathetic neurons. *Neuron* 35(2): 267–282.

Majdan M, Walsh GS, Aloyz R, and Miller FD (2001) TrkA mediates developmental sympathetic neuron survival in vivo by silencing an ongoing p75NTR-mediated death signal. *Journal of Cell Biology* 155(7): 1275–1285.

Newgreen D and Young HM (2002) Enteric nervous system: Development and developmental disturbances: Part 2. *Pediatric and Developmental Pathology* 5(4): 329–349.

Oppenheim RW (1991) Cell death during development of the nervous system. *Annual Review of Neuroscience* 14: 453–501.

Putcha GV, Harris CA, Moulder KL, Easton RM, Thompson CB, and Johnson EM Jr. (2002) Intrinsic and extrinsic pathway signaling during neuronal apoptosis: Lessons from the analysis of mutant mice. *Journal of Cell Biology* 157(3): 441–453.

Putcha GV and Johnson EM Jr. (2004) Men are but worms: Neuronal cell death in *C. elegans* and vertebrates. *Cell Death and Differentiation* 11(1): 38–48.

Reichardt LF (2006) Neurotrophin-regulated signalling pathways. *Philosophical Transactions of the Royal Society of London, Series B: Biological Sciences* 361(1473): 1545–1564.

Sastry PS and Rao KS (2000) Apoptosis and the nervous system. *Journal of Neurochemistry* 74(1): 1–20.

Wikipedia (2007) Apoptosis. http://en.wikipedia.org/wiki/Apoptosis (accessed 3 August 2007).

Yuan J and Yankner BA (2000) Apoptosis in the nervous system. *Nature* 407(6805): 802–809.

Autonomic Disorders

E L Phillips and P D Donofrio, Vanderbilt University Medical Center, Nashville, TN, USA

Introduction

The autonomic nervous system regulates cardiovascular, gastrointestinal, genitourinary, thermoregulatory, exocrine, and pupillary function to maintain internal homeostasis. The autonomic system has two major components, the sympathetic and parasympathetic divisions. The sympathetic system sends efferents from T1–L2 to end organs, producing a 'fight-or-flight' response. This causes an increase in heart rate, cardiac contractility, and blood pressure; dilated coronary arteries and pupils; bronchodilation; piloerection; sweating; constriction of intestinal sphincters; and ejaculation. The parasympathetic system's efferents emanate from cranial nuclei and S2–S4 nerve roots to end organs and regulate 'rest and digest' functions. Increased parasympathetic activity results in heightened gastric secretions and peristalsis, decreased heart rate and cardiac contractility, bronchoconstriction, constricted pupils, bladder contraction, and maintenance of erection.

Both of these systems are controlled by higher central nervous system (CNS) centers. The nucleus tractus solitarius (NTS) is considered one of the main central processors. It receives input from many central structures, including the forebrain, hypothalamus, hippocampus, amygdala, periaqueductal gray, raphe nucleus, and pontomedullary reticular formation. The NTS also receives sensory afferent information from the periphery, including signals from internal nuclei (i.e., chemoreceptors, osmoreceptors, thermoreceptors, and baroreceptors). The efferents from the NTS project to the brain stem and spinal cord.

The sympathetic system has short preganglionic and long postganglionic neurons (**Figure 1**). Preganglionic sympathetic fibers originate in the interomediolateral cell column in lamina VII of the spinal cord from levels T1 to L2. From there, they project to paravertebral and prevertebral ganglia, also called the sympathetic chain or trunk. This chain runs from cervical to sacral levels on each side of the cord. The prevertebral ganglia are located in the celiac plexus and include the celiac ganglia, superior mesenteric ganglion, and inferior mesenteric ganglion.

The sympathetic preganglionic nerves secrete acetylcholine, and most of the sympathetic postganglionic fibers secrete norepinephrine. The exception is the sympathetic postganglionic fibers that innervate sweat glands and adrenal medulla, which secrete acetylcholine.

The parasympathetic preganglionic fibers originate from cranial nuclei 3, 7, 9, and 10 and sacral parasympathetic nuclei in the lateral gray matter from S2–S4 (**Figure 2**). The preganglionic fibers project to terminal ganglion in or near the effector organs. In contradistinction to the sympathetic system, the parasympathetic system has long preganglionic fibers and short postganglionic fibers. The parasympathetic preganglionic and postganglionic fibers secrete acetylcholine.

Symptoms and Signs of Autonomic Dysfunction

Autonomic dysfunction may result in many different symptoms. One common symptom is orthostatic hypotension, which patients may describe as a feeling of dizziness, lightheadedness, weakness, confusion, blurred vision, or syncope after a postural change. To test for orthostasis, the blood pressure must be measured while lying, sitting, and standing. If there is a decrease in systolic pressure of at least 20 mmHg or a decrease in diastolic pressure of at least 10 mmHg within 3 min of standing accompanied by symptoms of cerebral hypoperfusion, then the patient meets criteria for orthostatic hypotension proposed by the American Autonomic Society and the American Academy of Neurology. Orthostasis associated with an appropriate rise in heart rate is consistent with hypovolemia, which is seen with dehydration, medication side effects, or diuretics. In sympathetic adrenergic failure, the characteristic feature is a drop in blood pressure without an appropriate rise in heart rate.

Intolerance to position changes can be seen without a drop in blood pressure. In positional tachycardia syndrome, there is an increase in heart rate of more than 30 beats per minute when changing from a reclined to upright position. This is most commonly observed in women younger than age 50 years. Patients complain of dizziness, usually without syncope, during postural change or modest exertion.

Gastrointestinal dysfunction is a common complication of autonomic neuropathy. Patients may have symptoms arising from the proximal gastrointestinal tract, including dysphagia and gastroparesis, or from the distal gastrointestinal tract, including chronic intestinal pseudoobstruction, constipation, diarrhea, and fecal incontinence. Patients may complain of

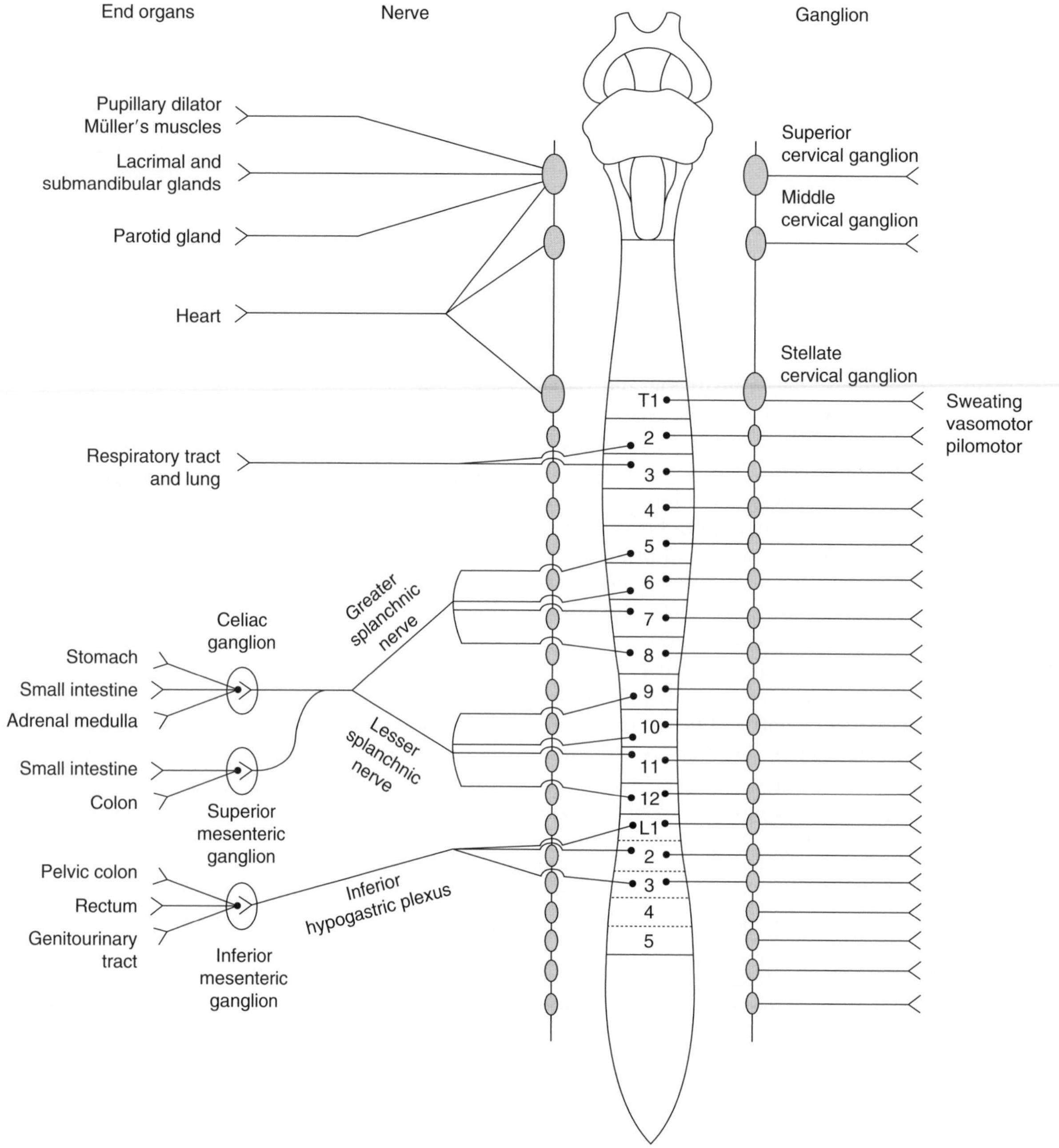

Figure 1 Anatomic framework of the sympathetic system. Reprinted from Donofrio PD (2001) Autonomic disorders. *The Neurologist* 7: 220–233, with permission.

heartburn, nausea, early satiety, loss of appetite, postprandial emesis, bloating, belching, constipation, or diarrhea.

Autonomic dysfunction can be associated with a variety of urologic complications, including bladder atony, nocturia, frequency, urgency, incontinence, and retention. The differences in symptoms depend on the location of the lesion in the sympathetic or parasympathetic system. Lesions above the sacral cord can cause detrusor overactivity. In this situation, the parasympathetic system dominates and causes the detrusor muscle to contract, resulting in increased intravesicular pressure, which the patient senses as imminent micturition and can lead to urge incontinence. When there is a lesion between the pons and the sacral cord, detrusor–sphincter dyssynergia occurs. This is characterized by a nonrelaxing sphincter during detrusor contraction. Dyssynergia can lead

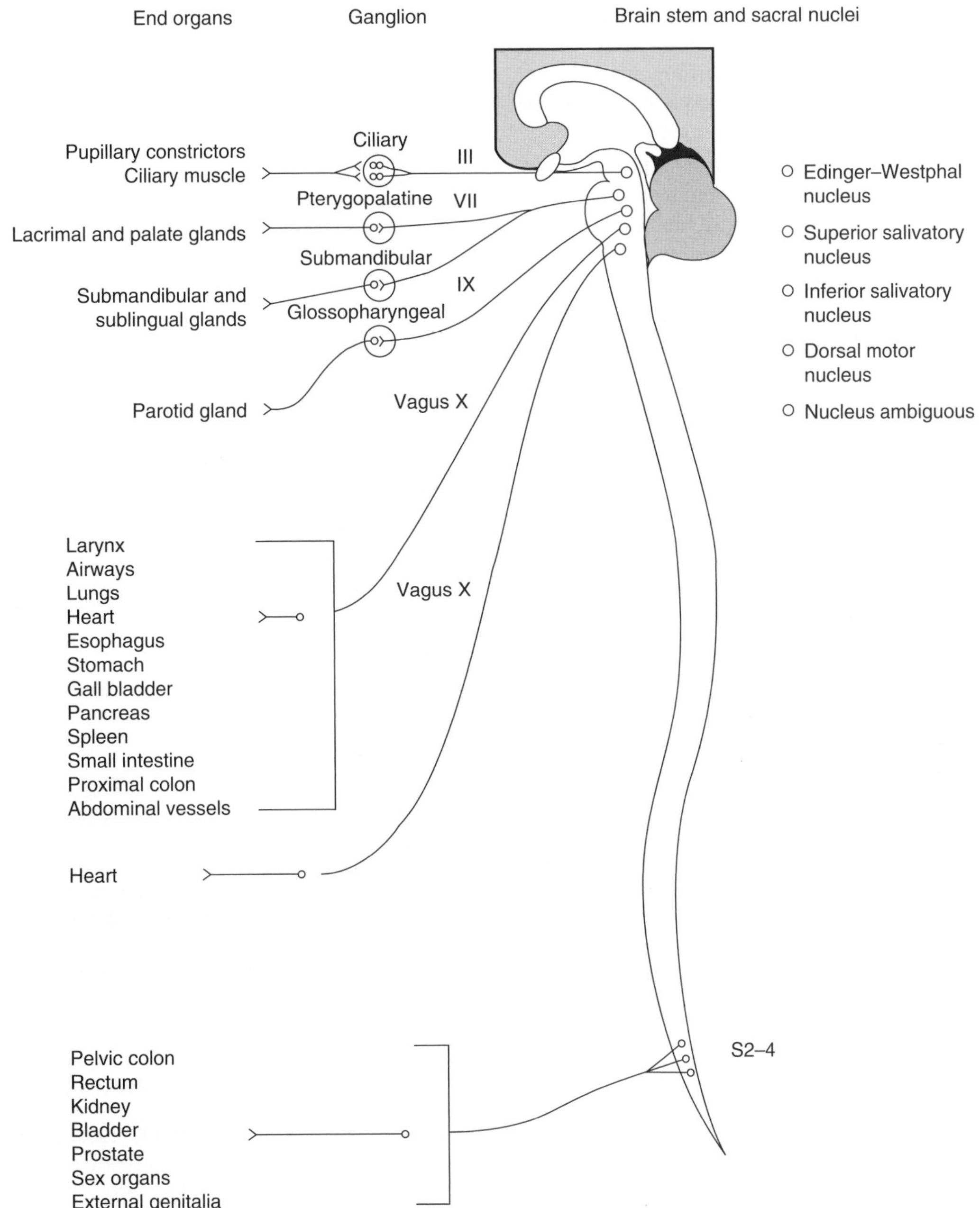

Figure 2 Anatomic framework of the parasympathetic system. Reprinted from Donofrio PD (2001) Autonomic disorders. *The Neurologist* 7: 220–233, with permission.

to incomplete bladder emptying, frequent urinary tract infections, and chronic renal failure from hydronephrosis. Bladder atony and overflow incontinence occur with more peripheral lesions of autonomic system. In this situation, the parasympathetic pelvic nerves are affected and are unable to cause contraction of the detrusor muscle.

Often, erectile dysfunction is one of the earliest symptoms of autonomic disease in men and is usually followed by ejaculatory failure. Women usually describe decreased libido, decreased ability for orgasm, and vaginal dryness.

When pupillomotor function is affected, patients complain of blurry vision, poor night vision, or a glare after bright lights. Patients with parasympathetic dysfunction have dry eyes and dry mouth, secondary to secretomotor paralysis of the salivary and lacrimal glands.

Another complication seen with autonomic dysfunction is an abnormality of sweating, either hypohidrosis (extreme dryness) or hyperhidrosis (excessive perspiration). Patients may not notice a change in sweating; therefore, probing questioning regarding sweating on hot days, after a hot bath, or during a fever may be necessary.

When vasomotor instability is present, patients may complain of extremely cold hands and feet. On examination of the distal extremities, patients may manifest changes in the appearance of the skin, including mottling, redness, or pallor, or trophic changes in the skin, nails, and hair, including thinning of the skin, brittle nails, and thickening or thinning of the hair.

Diseases Affecting the Autonomic System

Many diseases affect the autonomic nervous system; therefore, a systematic approach to classifying them is needed. The simplest approach is to divide them into two groups – those that affect the CNS and those that affect the peripheral nervous system. This section briefly discusses a few of these diseases.

Progressive autonomic failure, Parkinson's disease, and Shy–Drager syndrome are all neurodegenerative disorders that cause autonomic dysfunction. Progressive autonomic failure usually presents in men between the ages of 40 and 60 years. Onset is usually insidious, with slow progression leading to death within 10–15 years. Clinical findings are typically limited to the autonomic system. Patients with Parkinson's disease usually present with resting tremor, rigidity, bradykinesia, and postural instability and later develop autonomic symptoms. Shy–Drager syndrome typically presents with significant autonomic symptoms along with parkinsonian features; however, tremor is usually minimal. Other diseases of the CNS associated with autonomic dysfunction include multiple sclerosis, strokes, brain stem tumors, spinal cord lesions, tabes dorsalis, and Adie syndrome (**Table 1**).

Diseases of the peripheral autonomic system can further be categorized depending on the coexistence of a somatic peripheral neuropathy. Some of the peripheral neuropathies commonly associated with autonomic symptoms include diabetes, Guillain–Barré syndrome, Riley–Day syndrome, infectious neuropathies, amyloidosis, and acute intermittent porphyria (**Table 2**).

The most common cause of autonomic neuropathy in developed countries is diabetes mellitus. The neuropathy in diabetes typically presents as a distal sensorimotor polyneuropathy, and the autonomic neuropathy presents late in the course of the illness.

Guillain–Barré syndrome (acute inflammatory demyelinating polyradiculoneuropathy) is an acquired monophasic illness that begins as an acute ascending demyelinating sensorimotor polyneuropathy. Autonomic dysfunction occurs in approximately 60% of patients with Guillain–Barré syndrome as cardiac arrhythmias, postural hypotension, transient hypertension, urinary retention, and ileus.

Table 1 Central nervous system disorders affecting autonomic function

Progressive autonomic failure (idiopathic orthostatic hypotension)
Multiple-system atrophy (Shy–Drager syndrome)
Parkinson's disease
Spinal cord lesions
Wernicke's encephalopathy
Cerebrovascular disease
Brain stem tumors
Multiple sclerosis
Adie's syndrome
Tabes dorsalis

Adapted from McLeod JG and Tuck RR (1987) Disorders of the autonomic nervous system: Part I. Pathophysiology and clinical features. *Annals of Neurology* 21: 419–430.

Table 2 Peripheral autonomic neuropathy with somatic neuropathy

Autonomic dysfunction clinically important
Diabetes mellitus
Amyloidosis
Guillain–Barré syndrome
Acute intermittent porphyria
Familial dysautonomia (Riley–Day syndrome, HMSN III)
Chronic sensory and autonomic neuropathy
Infections: HIV, leprosy, Chagas disease, diphtheria, and Lyme disease
Autonomic dysfunction usually clinically unimportant
Hereditary sensory and autonomic neuropathy type I, II, IV, and V
Fabry's disease
Toxic neuropathies (vincristine sulfate, acrylamide, heavy metals, perhexiline maleate, organic solvents, marine toxins, vacor, cisplatin, paclitaxel, amiodarone, pentamidine)
Metabolic (B_{12} deficiency, alcohol-induced neuropathy, uremia)
Connective tissue diseases (systemic lupus erythematosus, rheumatoid arthritis, Sjogren's disease)
Chronic inflammatory demyelinating polyneuropathy

HIV, human immunodeficiency virus; HMSN, hereditary motor and sensory neuropathy.
Adapted from McLeod JG and Tuck RR (1987) Disorders of the autonomic nervous system: Part I. Pathophysiology and clinical features. *Annals of Neurology* 21: 419–430.

Familial dysautonomia (Riley–Day syndrome or HMSN III) is an autosomal recessive neuropathy with prominent autonomic symptoms, most commonly observed in Ashkenazi Jewish children. It presents in infancy with loss of pain and temperature sensation, hypotonia, and hyporeflexia. Patients have absence of tear production and lingual fungiform papillae. Autonomic symptoms can present at any time during the disease course.

Infectious neuropathies have been described in patients with human immunodeficiency virus (HIV), Chagas disease, leprosy, and diphtheria. The autonomic dysfunction associated with HIV infection can arise from direct viral effects, malnutrition, or drug side effects. Chagas disease is caused by the protozoan *Trypanosoma cruzi*. Severe gastrointestinal and cardiovascular autonomic dysfunction is typically observed during the chronic phase of the illness. Leprosy is contracted from the infectious agent *Mycobacterium leprae*. In the cooler parts of the body, there is loss of pain and temperature sensation along with focal anhidrosis. Several weeks after a cutaneous or pharyngeal diphtheria infection, a toxin-mediated neuropathy develops. This can present with multiple cranial mononeuropathies, hyporeflexia, sensory ataxia, and accommodation paralysis in the setting of a preserved light response.

In amyloidosis there is a deposition of insoluble protein fibers in β-pleated sheets within many tissues and organs. The neuropathy found in primary and hereditary amyloidosis is frequently associated with an autonomic neuropathy. However, secondary amyloidosis is not usually associated with autonomic dysfunction. Amyloidosis can be diagnosed by gingival biopsy, aspiration of subcutaneous fat pad, or a biopsy of rectal mucosa.

Acute intermittent porphyria is an autosomal dominant disease due to a defect in the enzyme porphobilinogen deaminase, which leads to a buildup of porphobilinogen and aminolevulinic acid. Most patients present between 18 and 40 years of age with intermittent attacks of abdominal pain, neuropathy, and constipation lasting several days. The neuropathy is predominantly motor and autonomic. Diagnosis is confirmed by an elevated urine porphobilinogen. Treatment of an acute attack consists of hematin $4\,\mathrm{mg\,kg^{-1}\,day^{-1}}$ for 4 days.

The peripheral autonomic neuropathies not typically associated with a peripheral neuropathy include botulism, pure adrenergic neuropathy, acute pandysautonomia, cholinergic dysautonomia, acute paraneoplastic autonomic neuropathy, and chronic idiopathic anhidrosis (**Table 3**).

Botulism is a neuromuscular junction disorder caused by the binding of *Clostridium botulinum* toxin to the presynaptic terminal. This toxin prevents the release of acetylcholine, leading to autonomic symptoms and a severe quadriplegia from failure of the neuromuscular junction. Dilated pupils with poor response to direct light and accommodation are classic features, in contrast to diphtheria, in which the light reflex is not affected.

Acute pandysautonomia is a disorder of both sympathetic and parasympathetic systems. The etiology is likely an autoimmune process or postviral reaction since the onset is usually after a viral illness. Approximately 50% of patients will have an autonomic ganglionic acetylcholine receptor antibody.

Table 3 Peripheral autonomic neuropathy without significant sensorimotor neuropathy

Acute and subacute autonomic neuropathy
Acute pandysautonomia
Cholinergic dysautonomia
Acute paraneoplastic autonomic neuropathy
Chronic idiopathic anhidrosis
Pure adrenergic neuropathy
Botulism

Adapted from McLeod JG and Tuck RR (1987) Disorders of the autonomic nervous system: Part I. Pathophysiology and clinical features. *Annals of Neurology* 21: 419–430.

Acute cholinergic neuropathy is similar to acute pandysautonomia, but it only affects the postganglionic cholinergic neurons. Pure adrenergic neuropathy is very rare, affecting postganglionic sympathetic neurons.

Acute paraneoplastic autonomic neuropathy is frequently associated with anti-Hu antibodies. Small cell lung cancer is the most common malignancy associated with these antibodies. Other antibodies detected with paraneoplastic autonomic neuropathies include Purkinje cell cytoplasmic antibodies type 2 and antibodies to collapsing response-mediator protein 5.

Chronic idiopathic anhidrosis is an acquired neuropathy producing diffuse anhidrosis without other signs of autonomic failure. When exposed to heat or after exercise, patients become hot, lightheaded, flushed, and weak because of their inability to release heat through sweating.

Autonomic Testing

Patients with autonomic signs or symptoms are candidates for autonomic testing if verification is necessary. Depending on the clinical presentation of the patient, evaluation should begin with hematologic, chemistry, autoantibody, and pathologic studies. **Table 4** lists some of the more commonly used tests for evaluating autonomic dysfunction.

Further testing can be pursued to define the physiology of the sympathetic and parasympathetic systems. All patients who are scheduled to undergo autonomic testing need to take certain measures in preparation for the testing. They should not consume any caffeine within the previous 24 h. They should avoid nicotine and food for 3 h prior to testing. Medications that affect autonomic function should be avoided for at least 48 h prior to testing, and preferably 5 half-lives before testing. Medications to be avoided include antihypertensives, volume contractors (e.g., diuretics), volume expanders (e.g., fludrocortisones), anticholinergics (e.g., antihistamines, antidepressants, and decongestants), and analgesics. Also discontinued should be anything that mechanically changes blood flow (e.g., compressive stockings and corsets) 24 h prior to testing.

One means to assess the function of the sympathetic and parasympathetic systems is by evaluating cardiovascular physiology. As mentioned previously, activation of the sympathetic system leads to tachycardia, elevation of blood pressure, increased cardiac contractility, and dilated coronary arteries. Conversely, stimulation of the parasympathetic system reduces heart rate and cardiac contractility. The baroreflex arc modulates these two systems. The baroreflex arc consists of afferent input from arterial baroreceptors in the aortic arch, carotid sinus, thoracic arteries, cardiac mechanoreceptors, and pulmonary stretch receptors. Activation of the baroreceptors leads to increased afferent input to central processors. This results in a decrease in sympathetic and increase in parasympathetic activity. Conversely, when the afferent activity decreases, there is an increase in sympathetic and decrease in parasympathetic activity.

Table 4 Laboratory assessment of autonomic neuropathy

Chemistry, hematology, and pathology
- Complete blood count and differential
- Fasting blood glucose, glucose tolerance test if diabetes suspected
- HIV testing
- Immunoelectrophoresis of blood and urine
- Plasma norepinephrine (supine and standing)
- Porphyria investigations
  - Urinary porphyrin concentration (24 h urine collection for aminolevulinic acid, porphobilinogen, and porphyrins)
  - Erythrocyte porphobilinogen deaminase activity (patients with type I and type III disease have approximately 50% enzyme activity)
- Genetic testing for inherited neuropathies
- Fat aspirate, rectal biopsy, or gingival biopsy for amyloid

Autoantibody assessment
- Anti-nuclear antibody
- Rheumatoid factor
- Anti-Ro/SS-A
- Anti-La/SS-B
- Antibodies to neuronal nicotinic acetylcholine receptor
- Antibodies to P/Q-type calcium channel
- Antibodies to acetylcholine receptor
- Paraneoplastic antibodies
  - Anti-Hu (type 1 anti-neuronal nuclear antibody, ANNA-1)
  - Purkinje cell cytoplasmic antibodies type 2 (PCA-2)
  - Collapsin response-mediator protein 5 (CRMP-5)

Adapted from Freeman R (2005) Autonomic peripheral neuropathy. *Lancet* 365: 1259–1270.

Valsalva Maneuver

The Valsalva maneuver evaluates the sympathetic and parasympathetic responses of the baroreflex arc. The patient exhales into a manometer and maintains a pressure of 40 mmHg for 15–20 s. Heart rate and blood pressure are monitored 30 s before and 2 min after testing. There are four phases of the Valsalva maneuver (**Table 5**). During phase 1, there is an increase in intrathoracic pressure that mechanically causes a brief increase in blood pressure and decrease in heart rate. In early phase 2, there is a reduction of venous return and a subsequent decrease in stroke volume, causing a decrease in blood pressure. In late phase 2, the decreased blood pressure activates the baroreflex that causes a sympathetically mediated increase in heart rate and blood pressure back toward baseline levels. When the patient terminates the Valsalva maneuver, blood refills the pulmonary vasculature. This causes the change seen in phase 3 – a temporary further decline in blood pressure. During phase 4, there is an increase in venous return, which leads to a compensatory decrease in heart rate and increase in blood pressure that may overshoot baseline blood pressure. By monitoring the heart rate during the Valsalva maneuver, the Valsalva ratio can be measured. This is calculated by dividing the highest heart rate during Valsalva by the lowest heart rate usually occurring during phase 4. In autonomic failure, the reflex bradycardia and blood pressure overshoot is typically absent. Valsalva ratio varies with age, but a ratio less than 1.1 is abnormal at any age.

Sinus Arrhythmia (Deep Metronomic Breathing; Heart Rate Response to Respiration)

Respiratory sinus arrhythmia is a normal phenomenon; the term relates to the increase in heart rate with

Table 5 Phases of Valsalva

Phase	*Timing*	*Physiology*	*Heart rate*	*Blood pressure*
1	Early valsalva	Increase in intrathoracic pressure	Decrease	Increase
2 – early		Decreased venous return	Decrease	Decrease
2 – late		Baroreflex	Increase	Increase
3	End of valsalva	Refills pulmonary vasculature	Increase	Decrease
4		Increase in venous return	Decrease	Increase

inspiration and decrease during expiration. Measuring this variability in heart rate is a reliable way to assess the parasympathetic innervation of the heart mediated by the vagus nerve. The heart rate is monitored while the patient is asked to breathe six respirations per minute (each breath lasting 10 s) for seven or eight cycles. After resting for 5 min, this sequence is repeated again. Age, pulmonary disease, hypocapnia, hyperventilation, cardiac failure, tachycardia, and CNS depression can all lead to a decrease in sinus arrhythmia.

There are several ways to assess sinus arrhythmia. One way is to measure heart rate variation, which is determined by taking the difference between the average heart rate during inspiration from the average heart rate during expiration. Heart rate variation greater than 15 beats per minute (bpm) is considered normal. Less than 10 bpm is abnormal in younger patients (<40 years), and less than 5 bpm is abnormal at any age. Another measure includes calculating the expiratory/inspiratory ratio (E/I ratio). This is the longest R-R interval during expiration divided by the shortest R-R interval during inspiration. The E/I ratio varies depending on age. However, an E/I ratio less than 1.1 is abnormal at any age.

Sympathetic Skin Test

The sympathetic skin response (SSR) can be used to evaluate the integrity of the entire sympathetic sudomotor response, including the central and peripheral components. The preoptic anterior hypothalamus, spinal cord, viscera, and skin have thermoreceptors that input to the posterior hypothalamus. The posterior hypothalamus then determines a set point. Efferent output is directed through the brain stem down to the intermediolateral cell column. The preganglionic sympathetic cholinergic fibers synapse on the paravertebral sympathetic ganglia. Postganglionic sympathetic cholinergic neurons terminate on eccrine sweat glands. The resulting sweat response produces the SSR.

SSRs occur spontaneously or can be provoked by a stimulus. The stimulus can be direct, such as electrical activation of the peripheral C-fibers, or indirect (auditory, visual, startle, cough, or inspiratory gasp), which activates the centrally mediated sympathetic sudomotor response leading to a change in skin potential. The recording electrodes are usually placed on the palms and soles, with references on the dorsal surfaces. The skin potential is the difference in voltage between two skin surfaces. This potential is largely determined by the sweat glands and the epidermis. The initial change in potential is likely due to sweating and the following changes are likely secondary to changes in skin potential. Normally, the SSR amplitude is 0.5–3 mV and the latency is 1.5 s in the hand and 2 s in the foot. The SSR is easy to perform using conventional electromyography equipment; however, interpreting the results is challenging due to a large variability in responses. Most electromyographers agree that an absent SSR is abnormal. However, there are no well-accepted criteria for interpreting responses that have a reduction in amplitude or latency. Soliven et al. published reference values of 0.791 ± 0.346 mV for the palm and 0.388 ± 0.227 mV for the sole of the foot. The SSR decreases with age, likely due to the reduction in the number of sweat glands during aging. Other factors that interfere with the amplitude include skin creams and cold, dry, or oily skin.

Quantitative Sudomotor Axon Reflex Test

The quantitative sudomotor axon reflex test (QSART) assesses the postganglionic cholinergic sympathetic innervation of eccrine sweat glands. This test is performed by applying acetylcholine to the skin through a multicompartmental sweat cell. Through iontophoresis, acetylcholine binds to the muscarinic receptors on the sweat gland, causing an immediate sweat response. Acetylcholine also binds to the postganglionic nicotinic receptors, which causes an antidromic activation. When the antidromic activation reaches a branch point, it travels orthodromically to release acetylcholine in other sweat glands near the initial iontophoretic stimulation. This is the indirect axon reflex sweat gland response. One of the compartments of the sweat cell measures the sweat production from this indirect axon reflex using a sudorometer. This recording is usually done prior to, during, and after iontophoresis. The indirect axon–reflex sweat response usually occurs 1 or 2 min after iontophoresis. The usual recording sites for the QSART include the proximal dorsal foot (sural nerve), the lateral proximal calf (saphenous nerve), the distal medial leg (peroneal nerve), and the middle forearm (ulnar nerve). A normal response in men is 2 or 3 $\mu l^{-2} cm^{-2}$ and that in women is 0.25–1.2 $\mu l^{-2} cm^{-2}$, with some variation depending on the site tested.

QSART evaluates postganglionic sympathetic sudomotor function in different regions of the body. The most distal sites may have a small or absent response in the early stages of a small fiber neuropathy, whereas proximal sites have a normal QSART. However, as the disease progresses, the more proximal sites will begin to show abnormal responses. When the lesion is preganglionic or central, such as in spinal cord lesions or in multiple system atrophy, the QSART response should be normal at all sites. However, if the preganglionic lesion has been present for an extended period of time, the QSART response may be abnormal. Theories for this phenomenon include a transsynaptic defect or sweat gland atrophy.

Other abnormal findings in central lesions include resting sweat activity (RSA) and persistent sweat activity (PSA). RSA is defined as sweat response at a skin temperature that is below the normal temperature for a sweat response. PSA is the continuation of the sweat response even after the stimulus is removed. Both RSA and PSA are thought to arise from damaged sympathetic sudomotor axons that spontaneously fire. This is commonly seen in early or painful neuropathies, reflex sympathetic dystrophy, hyperhidrosis, or anxious patients.

Treatment of Autonomic Dysfunction

If possible, treatment of autonomic dysfunction should be directed at the underlying disease. For example, Guillain–Barré syndrome can be treated with plasma exchange or intravenous immunoglobulin (IVIG). In Lambert–Eaton myasthenic syndrome (LEMS), the underlying malignancy should be treated. Immunosuppressants, plasma exchange, IVIG, and 3,4-diaminopyridine can improve symptoms in LEMS. A trivalent antitoxin exists for the treatment of botulinum; however, it is only beneficial if given within the first 30 min of exposure. In porphyria, hematin, phenothiazines, and meperidine may provide some relief. Tight glycemic control can prevent and stabilize diabetic autonomic neuropathy. Unfortunately, many autonomic diseases are not treatable, and therefore management must be symptomatic.

Orthostatic Hypotension

Treating orthostatic hypotension can be as simple as increasing the intake of salt (10 g of sodium per day) and fluids (2–2.5 l per day in adults), wearing elastic stockings, rising slowly, and avoiding hot environments, large meals, and alcohol. Sleeping with the head of the bed elevated 10–20 degrees attenuates nocturnal diuresis, which decreases the morning symptoms of orthostatic hypotension.

If simple lifestyle changes are not effective, then multiple medications can be prescribed (**Table 6**). Fludrocortisone (0.1–1.0 mg qday) is a mineralocorticoid that increases blood volume. Side effects include peripheral edema, hypokalemia, and headache. It should not be used in patients with chronic heart failure.

Midodrine, a direct peripherally acting α_1-adrenoreceptor agonist, can be added to fludrocortisone. It causes arterial and venous constriction and has a duration of action of 4 h. Midodrine does not cross the blood–brain barrier, so central sympathomimetic side effects are absent. Initial dosing is 2.5 mg tid titrating up to 10 mg tid. Ephedrine and pseudoephedrine are direct and indirect α_1-adrenoreceptor agonists. They not only act directly on the α_1-adrenoreceptor but also cause norepinephrine release from postganglionic sympathetic neurons. Because of the indirect action, these medications are best used in patients with partial or incomplete lesions. They may produce central sympathomimetic side effects, including anxiety, tachycardia, and tremulousness.

Table 6 Pharmacotherapy of orthostatic hypotension

Mineralocorticoids	
Fludrocortisone	0.1–1.0 mg po daily
Sympathomimetic agents	
Midodrine	2.5–10 mg po tid-qid
Clonidine	0.1–0.3 mg po bid
Ephedrine	15–45 mg po tid
Yohimbine	2.5–5.4 mg po tid
V2 receptor agonists	
Desmopressin (DDAVP)	100–400 mg po at bedtime or 5–40 μg intranasally at bedtime
Erythropoietin	25–75 U kg^{-1} sc three times a week
Dopamine blocking agents	
Metoclopramide	10–15 mg po up to qid
Prostaglandin synthetase inhibitors	
Indomethacin	25–50 mg po tid
Ibuprofen	300–400 mg po tid
Somatostatin analog	
Octreotide	25–50 μg sc tid

Erythropoietin has been shown to improve symptoms in patients with orthostatic hypotension. The exact mechanism for this is unknown, but it is likely explained by changes in blood viscosity, an increase in hematocrit and central blood volume, and neurohumoral effects on blood vessels. Desmopressin acts on V2 receptors in the kidney to inhibit water excretion and to increase volume expansion. Clonidine is an α_2 agonist which reduces sympathetic tone centrally; thus, in healthy patients it decreases blood pressure. However, in severe central autonomic dysfunction, central sympathetic receptors are depleted and postsynaptic α_2 receptors in the periphery may be activated by clonidine, leading to increased blood pressure. Caution must be used when treating orthostatic hypotension with clonidine because of its propensity to cause hypotension.

When treating orthostatic hypotension, one must be cognizant of coexisting supine hypertension. Treatment of supine hypertension in patients who have orthostatic hypotension includes elevating the head of the bed or using short-acting antihypertensives at night. In this situation, captopril or nitrates could be good choices. If supine hypertension becomes a problem, ambulatory monitoring may be needed to design a treatment plan.

Gastrointestinal Atonomic Dysfunction

If gastroparesis is related to diabetes, controlling hyperglycemia can improve gastric motility. Patients should be advised to eat frequent small meals with decreased fat content. Agents that promote motility can be used, including metoclopramide, domperidone, and erythromycin. Rarely, a jejunostomy tube is needed.

Bowel hypomotility initially should be treated with increased fluid intake, increased fiber intake (5–25 g per day), and bulk agents (bran and psyllium). Pharmacological treatment usually begins with stool softeners (docusate) and laxatives (polyethylene glycol or lactulose) (**Table 7**). Lactulose is a disaccharide and should be avoided in diabetics since a small amount may be absorbed and cause elevations in blood glucose. Osmotic laxatives (sodium and magnesium salts) should not be used for extended periods due to a concern for producing fluid and electrolyte imbalance. Stimulant laxatives (senna, cascara, bisacodyl, and phenolphthalein) should also be prescribed only on a short-term basis since these agents can cause smooth muscle atrophy and neuronal damage.

Neurogenic diarrhea may be treated with clonidine. Psyllium hydrophilic mucilloid may be used as a bulk-forming agent. Loperamide may also be helpful to treat diarrhea. In diabetic diarrhea, bowel stasis may lead to bacterial overgrowth, in which case treatment with tetracycline is useful. For sphincter incontinence, phenylephrine ointment 1% may be applied topically to improve sphincter tone for approximately 4 h.

Sexual Dysfunction

Erectile dysfunction requires a thorough medical history to exclude other causes, especially medications. One treatment option is a phosphodiesterase type 5 inhibitor, which inhibits the breakdown of cyclic GMP. This results in relaxation of smooth muscle and increased blood flow in the corpus cavernosum. Sildenafil 50 mg, tadalafil 20 mg, and vardenafil 20 mg are the three drugs currently marketed in the United States for erectile dysfunction. These medications are contraindicated in patients with unstable heart disease, orthostatic hypotension, significant hypotension, or in patients on nitrates. Other treatments for erectile dysfunction include intracorporeal injections of phentolamine, papaverine, or prostaglandin E. Vasoactive drugs can be delivered transurethrally. Mechanical devices such as constricting rings or a vacuum erection device are other options. If these therapies do not provide benefit, then penile prosthetic implants may be considered. Women with autonomic dysfunction may benefit from vaginal lubricants and estrogen creams. **Table 8** lists treatments for erectile dysfunction.

Table 7 Pharmacotherapy of bowel hypomotility

Bulk agents	
Bran	1–6 tablespoons po daily
Psyllium	2.2–45 g po daily
Methylcellulose	1 tablespoon po daily to tid
Stool softeners and lubricants	
Mineral oil	15–45 ml po prn
Docusates	50–200 mg po daily
Osmotic laxatives and cathartics	
Lactulose	10–40 g po daily
Sorbitol	30–150 ml po prn
Magnesium citrate	150–300 ml po daily
Contact cathartics	
Bisacodyl	10–15 mg po prn
Senna	2–8 tabs po daily
Prokinetic agents	
Metoclopramide	10–15 mg po up to qid
Domperidone	10–20 mg po qid
Erythromycin	250 mg tid

Urinary Dysfunction

Treatment of incontinence should begin with scheduled voiding every 3 or 4 h. Combining this plan with Valsalva or suprapubic pressure can increase bladder contractions and bladder emptying. In many cases, pharmacologic treatment is needed (**Table 9**). Anticholinergics and tricyclic antidepressants may be used to help decrease detrusor muscle activity. If oral medications do not provide relief, Botox injections into the detrusor muscle can be used to decrease detrusor overactivity by inhibiting acetylcholine release. Alpha-blockers may be used for sphincter dysfunction. Occasionally, desmopressin can provide relief by decreasing urine production for 6–8 h. Often, clean intermittent catheterization or indwelling catheterization is required. Many surgical procedures may be performed to treat incontinence, including bladder neck suspension, augmentation cystoplasty, and artificial sphincter and urinary diversion with stoma collection bag.

Table 8 Pharmacotherapy of erectile dysfunction

Phosphodiesterase type 5 inhibitors	
Sildenafil	50 mg po qday prn
Tadalafil	5–20 mg po qday prn
Vardenafil	5–20 mg po qday prn
Intracorporeal injections	
Phentolamine	
Papaverine	
Prostaglandin E	
Transurethral suppository	
Alprostadil	
Vacuum erection device	
Penile prosthesis	

Table 9 Pharmacotherapy of urinary dysfunction

Adrenergic blocking agents (sphincter dysfunction)	
Prazosin	1–5 mg po bid to tid
Terazosin	1–5 mg po daily
Anticholinergic agonists (detrusor overactivity)	
Oxybutynin chloride	5 mg po tid
Dicyclomine hydrochloride	20 mg po tid
Tricyclic antidepressants (detrusor overactivity)	
Imipramine	25 mg po qday to tid
Detrusor Botox injections	

Anhidrosis

Anhidrosis usually leads to dry skin, which can be treated with emollients. It may produce hyperthermia, which can be prevented by avoiding hot environments. If hyperpyrexia occurs, fans, cool drinks, and cool baths can be helpful.

Hyperhidrosis

Treatments for hyperhidrosis include anticholinergics, such as trihexyphenidyl and propantheline. However, their use is limited by anticholinergic side effects. Glycopyrrolate does not cross the blood–brain barrier and might have less side effects. Beta-blockers and benzodiazepines can be of some benefit. Localized hyperhidrosis may be treated with topical anticholinergic creams (hyoscine hydrobromide or glycopyrrolate), tap water iontophoresis, or Botox injections. Surgical treatment includes percutaneous endoscopic transthoracic sympathectomy, with ablation of prevertebral sympathetic ganglia from T2 to T4.

See also: Autoimmune Autonomic Neuropathy; Autonomic Failure; Autonomic Nervous System: Neuroanatomy; Autonomic Nervous System: Central Control of the Gastrointestinal Tract; Autonomic Nervous System; Erectile Dysfunction; Parasympathetic Nervous System; Sympathetic Nervous System.

Further Reading

Apostolidis AN and Fowler CJ (2003) Evaluation and treatment of autonomic disorders of the urogenital system. *Seminars in Neurology* 23: 443–452.

Chelimsky G and Chelimsky TC (2003) Evaluation and treatment of autonomic disorders of the gastrointestinal tract. *Seminars in Neurology* 23: 453–458.

Donofrio PD and Caress JB (2001) Autonomic disorders. *The Neurologist* 7: 220–233.

Freeman R (2003) Treatment of orthostatic hypotension. *Seminars in Neurology* 23: 435–442.

Freeman R (2005) Autonomic peripheral neuropathy. *Lancet* 365: 1259–1270.

Gibbons C and Freeman R (2004) The evaluation of small fiber function – Autonomic and quantitative sensory testing. *Neurology Clinics* 22: 683–702.

Goldstein DS, Robertson D, Esler M, Straus SE, and Eisenhofer G (2002) Dysautonomias: Clinical disorders of the autonomic nervous system. *Annals of Internal Medicine* 137: 753–763.

Hilz MJ and Dutsch M (2006) Quantitative studies of autonomic function. *Muscle and Nerve* 33: 6–20.

Low AL (1993) *Clinical Autonomic Disorders: Evaluation and Management*. Boston: Little, Brown.

Low PA (2003) Testing the autonomic system. *Seminars in Neurology* 23: 407–421.

Mansoor GA (2006) Orthostatic hypotension due to autonomic disorders in the hypertension clinic. *American Journal of Hypertension* 19: 319–326.

Mathias CJ (2003) Autonomic diseases: Clinical features and laboratory evaluation. *Journal of Neurology, Neurosurgery and Psychiatry* 74: 31–41.

Mathias CJ (2003) Autonomic diseases: Management. *Journal of Neurology, Neurosurgery and Psychiatry* 74: 42–47.

Mathias CJ (2004) Role of autonomic evaluation in the diagnosis and management of syncope. *Clinical Autonomic Research* 14: 45–54.

McLeod JG and Tuck RR (1987) Disorders of the autonomic nervous system: Part I. Pathophysiology and clinical features. *Annals of Neurology* 21: 419–430.

Autonomic Dysfunction: Drug-Induced

A L Tonkin, University of Adelaide, Adelaide, SA, Australia

Introduction

Many drugs and other compounds are capable of causing autonomic dysfunction by a variety of mechanisms. Acute exposure may cause autonomic overactivity in some cases, but most drug-induced autonomic dysfunction results in reduced autonomic activity. This article focuses primarily on drugs and other compounds that reduce autonomic activity, either directly by transiently inhibiting autonomic nerve function or indirectly by causing autonomic neuropathy. It concludes with a discussion of compounds that can cause an abnormal transient increase in autonomic activity.

Direct Inhibition of Autonomic Activity

Drugs that reduce autonomic activity are very widely used in clinical medicine, particularly in management of cardiovascular disorders but also in areas such as gastroenterology and urology. Many different sites of action have been targeted, including central autonomic neurons, autonomic ganglia, postganglionic neurons (neurotransmitter synthesis, storage, and release), and receptors on effector cells. Both the sympathetic and parasympathetic systems have provided fertile ground for the development of inhibitory drugs for a variety of therapeutic purposes. These drugs are described in detail in pharmacological textbooks and only a brief summary is provided here.

Drugs That Reduce Both Sympathetic and Parasympathetic Activity

Drugs acting on preganglionic nerve fibers and autonomic ganglia Since both sympathetic and parasympathetic preganglionic cells utilize acetylcholine as their neurotransmitter, drugs that act on preganglionic cholinergic neurons or nicotinic receptors in autonomic ganglia have very wide-ranging autonomic effects and are usually poorly tolerated. Some of these, such as hexamethonium, have been used in the past as antihypertensives, but with very poor tolerability records. Because of their lack of specificity, none are used as first-line therapeutic agents currently.

Drugs That Reduce Sympathetic Activity

Antagonists of the sympathetic nervous system may act at central autonomic pathways, and/or peripheral α- or β-adrenoceptors. Many of these have been clinically important at various times in the management of hypertension. In general, their effects on autonomic function are well known and underpin their therapeutic efficacy.

Centrally acting drugs Inhibitors of central sympathetic outflow have been used in the management of hypertension for several decades. These drugs include α-methyldopa, clonidine, and, more recently, moxonidine. They act as α_2-receptor agonists, reducing sympathetic outflow from the central nervous system (CNS) by stimulating presynaptic α_2-receptors that control the release of noradrenaline (NA). Clonidine and moxonidine also bind to imidazoline receptors, which may mediate some of the effects of these drugs, although this is a controversial area.

Many other centrally acting drugs have autonomic effects, the pathogenesis of which remains poorly understood. Levodopa and dopamine D_2 receptor agonists, such as bromocriptine and pergolide, can cause severe orthostatic hypotension, particularly in the early phase of therapy of Parkinson's disease, which itself is associated with abnormal cardiovascular autonomic control. Combination anti-parkinsonian therapy, such as L-dopa with selegiline or bromocriptine, can cause more severe autonomic dysfunction than single drug treatment. The effects of L-dopa and the D_2 agonists are presumed to be related directly to an agonist effect at presynaptic D_2 receptors, resulting in reduced sympathetic outflow from the vasomotor centers of the brain stem. Other centrally acting drugs with similar clinical effects include the centrally acting anorectic drugs such as dexfenfluramine and the nonspecific monoamine oxidase inhibitors. Selegiline, an inhibitor of monoamine oxidase B (MAO-B), is metabolized to amphetamine compounds which deplete NA from sympathetic terminals, and selegiline can sometimes cause orthostatic hypotension acutely.

Peripherally acting drugs The vast majority of clinically useful sympathetic antagonists act on postsynaptic receptors. Three different major types of receptors are involved in mediating sympathetic activity: α- and β-adrenoceptors that respond to NA and adrenaline, and muscarinic receptors present in sweat glands which, uniquely, are innervated by cholinergic postganglionic neurons that are part of the sympathetic nervous system. Muscarinic receptors will be considered further in the section on the parasympathetic division of the autonomic nervous system.

Drugs can reduce sympathetic activity relatively selectively by antagonizing the effects of NA at

Table 1 Drugs antagonizing peripheral adrenoceptors

Receptor type	*Example antagonist*	*Common indications*
α_1-Receptors and α_2-receptors	Phenoxybenzamine (nonselective, irreversible)	Pheochromocytoma
	Phentolamine (nonselective, reversible)	
α_1-Receptors	Prazosin (selective, reversible)	Hypertension, bladder outflow obstruction
	Chlorpromazine	Psychosis
	Prochlorperazine	Nausea and vomiting
	Amitriptyline	Depression
α_2-Receptors	Yohimbine	Not used
β_1-Receptors and β_2-receptors	Propranolol (nonselective)	Migraine prophylaxis, essential tremor
β_1-Receptors	Metoprolol (selective)	Angina, hypertension
β_2-Receptors	No selective agents available	

peripheral receptors. **Table 1** shows examples of such drugs, with their common indications for use. Many of these inhibitors of sympathetic transmission are used with the specific intention of reducing sympathetic activity to achieve therapeutic outcomes, such as reduction in blood pressure or heart rate, or relief of bladder obstruction. However, in practice, the drugs that most commonly lead to clinical presentations suggesting autonomic dysfunction are those that used for other effects, and they inhibit sympathetic activity as an adverse effect. Symptoms are particularly likely to occur when such drugs are used in the elderly. Some important examples are discussed here.

The clinical effects of sympathetic inhibition depend upon the balance of α- and β-adrenoceptors affected by the particular drug. Drugs with peripheral α-antagonist properties (such as the phenothiazines and tricyclic antidepressants) commonly cause orthostatic hypotension, particularly in elderly individuals. These drugs can also cause exacerbation of urinary incontinence, particularly in elderly women. This effect may have significant consequences, including institutionalization, if the causative agent is not identified and removed.

β-Blocker drugs are well known to have negative inotropic effects, and may sometimes lead to significant exacerbations of cardiac failure. In addition, they have negative chronotropic effects, a particular risk when administered in conjunction with other drugs that slow atrioventricular nodal conduction, such as digoxin or verapamil. Such a combination may cause clinically significant atrioventricular conduction delay, even in individuals with normal baseline electrocardiograms (ECGs). On the other hand, metoprolol and other β-blockers improve long-term outcome in chronic cardiac failure, and many patients with heart failure can tolerate β-blockade if the drug is introduced at a low dose and up-titrated slowly. Exacerbation of symptoms of obstructive airways disease is also well recognized as a complication of the use of β-blockers in susceptible people, although cardioselective β-blockers can be used safely in some patients with mild to moderate reactive airways disease.

Local administration of drugs with autonomic effects may also cause systemic symptoms. A well-recognized example of this phenomenon is the local instillation of timolol to the eye for the treatment of glaucoma, with subsequent exacerbation of bronchospasm or cardiac failure due to its systemic β-adrenoceptor antagonist effect.

Drugs That Reduce Parasympathetic Activity

Reduced parasympathetic activity occurs frequently in response to drug therapy. In some cases the drug is specifically intended to inhibit the parasympathetic system, but many drugs commonly used for other purposes also have peripheral antimuscarinic effects. Common examples are listed in **Table 2**.

Anticholinergic drugs are sometimes given with the aim of producing muscarinic antagonism. A common example is the use of benztropine in the management of parkinsonian symptoms, either due to Parkinson's disease or the movement disorders associated with antipsychotic use. Recently there has been an increase in the use of muscarinic receptor antagonists, such as oxybutynin and tolterodine, in the management of urinary incontinence due to detrusor instability. This condition is found most commonly in elderly people, who are particularly sensitive to antimuscarinic adverse effects such as dry mouth, sleep disruption, confusion, memory impairment, constipation, nausea, dizziness, and headache. The advent of more selective antagonists with less capacity to cross the blood–brain barrier may improve tolerability.

Protriptyline and amitriptyline are the most potent antimuscarinic receptor antagonists among the tricyclic antidepressants, and should be considered as possible contributors to presentations such as urinary retention or constipation. The newer antidepressant drugs, such as the selective reuptake inhibitors

Table 2 Drugs antagonizing peripheral muscarinic receptors

Example antagonist drug/drug group	*Usual indication*
Antimuscarinics	Prevention of extrapyramidal motor symptoms Bladder instability; nocturnal enuresis or urge incontinence
Phenothiazines	Psychosis; sedation in acute delirium
Tricyclic antidepressants	Depression
Antihistamines	Atopy, allergic reactions
Quinidine, disopyramide	Ventricular arrhythmias
Antispasmodics	Gastrointestinal motility disorders

fluoxetine and venlafaxine, are much less potent antagonists at muscarinic receptors. Class I antiarrhythmic agents such as quinidine and disopyramide also have significant antimuscarinic effects.

In all of these cases the drug is administered systemically and can be expected to have widespread autonomic effects. These include blurred vision, which is more symptomatic in young people with preserved lens accommodation, as well as dry mouth and eyes, severe constipation, and urinary retention, particularly in elderly men with prostatic hypertrophy. At very high doses, gastric emptying and gastric secretion may also be inhibited, leading to epigastric discomfort. The concurrent use of more than one anticholinergic drug is most common in the elderly and any change in CNS functioning in particular, which may range from a subtle impairment of memory to an acute anticholinergic-induced delirium, should be a stimulus to a careful review of drug therapy. At higher doses, anticholinergic drugs can precipitate frankly psychotic reactions with hallucinations.

Topical or local use of drugs with inhibitory effects on the parasympathetic nervous system can also lead to significant symptoms. The local use of botulinum toxin for conditions such as cervical dystonia or hyperhidrosis has been associated with a significant risk of dose-dependent anticholinergic adverse effects such as dryness of the mouth, accommodation difficulties, reduced sweating, swallowing difficulties, constipation, and urinary retention.

Other Compounds That Directly Reduce Autonomic Activity

A wide variety of naturally occurring and synthetic compounds also influence autonomic function.

Nicotine Nicotine has a variety of pharmacological actions and may be a source of toxicity in tobacco smokers. Nicotine binds to nicotinic receptors in autonomic ganglia, as well as at neuromuscular junctions and in the CNS. At usual doses associated with smoking there is autonomic stimulation, with effects including increased heart rate, blood pressure, cutaneous vasoconstriction, and reduced baroreflex sensitivity. Exposure to toxic doses (e.g., in tobacco workers or those exposed to nicotine-containing pesticides) can result in ganglionic paralysis, leading to reduction in sympathetic, parasympathetic, and neuromuscular transmission, with bradycardia, hypotension, coma, and eventual respiratory muscle paralysis. The role of nicotine in the pathogenesis of the known toxic effects of long-term tobacco smoking is unclear, but it is possible that autonomic dysregulation may contribute to the well-documented susceptibility of smokers to a variety of cardiovascular disorders.

Anticholinesterase compounds Compounds with anticholinesterase activity, such as organophosphorus esters and carbamate compounds, are commonly used in agriculture as insecticides. Acute ingestion by humans causes inhibition of the enzyme that removes acetylcholine from synaptic clefts, thus increasing neurotransmission at autonomic ganglia and parasympathetic terminals (discussed later). In some circumstances, however, these compounds may reduce autonomic function. Such a situation is observed after acute exposure to sarin gas, a highly toxic organophosphorus compound originally developed during World War II and more recently used in terrorist attacks. The autonomic effects of sarin are long-lasting (up to several weeks) in some individuals, and observations on workers involved in handling similar compounds during World War II have indicated that autonomic dysfunction may persist for years after acute exposure.

Herbal remedies Many freely available proprietary herbal remedies contain alkaloids with anticholinergic activity, such as atropine and scopolamine, and there have been case reports of anticholinergic poisoning resulting in confusion, dry mouth, tachycardia, and dilated pupils.

Illicit drugs Acute suppression of vasomotor outflow may occur following ingestion of amphetamines, with effects that include pronounced drowsiness and severe orthostatic hypotension, possibly due to NA depletion, with preservation of normal vagal function. The chronic use of amphetamine derivatives has been shown to cause significant impairment of parasympathetic cardiovascular control, still detectable about a week after last ingestion of the drug.

Contamination of illicit drugs may also lead to acute autonomic dysfunction. A number of cases of acute anticholinergic poisoning have been reported in individuals who had self-administered material sold as street 'heroin,' but which had been mixed with scopolamine. The clinical picture of hallucinations, tachycardia, pupillary dilatation, dry skin and mucous membranes, and urinary retention responds rapidly to treatment with the anticholinesterase compound, physostigmine.

Venoms Much progress has been made over recent years in understanding the molecular structure and mechanisms of action of the toxic components of venoms of various types. Many have acute inhibitory effects on autonomic activity. For example, some snake toxins bind to muscarinic receptors, leading to parasympathetic inhibition; others bind to ganglionic nicotinic receptors, leading to reduced sympathetic and parasympathetic transmission, and others bind to neuromuscular nicotinic receptors, causing flaccid paralysis.

Components of venoms from snakes and other venomous creatures, such as scorpions, are also known to interact with monoaminergic neurotransmission at multiple sites, including synthesis, release, or reuptake of monoamines, or agonist or antagonist activity at postsynaptic receptors. Venoms of cone snails and green mamba snakes, for example, contain small peptides that are antagonists at α-adrenoceptors. Many other neurotoxins have stimulatory effects and will be discussed later.

Autonomic Neuropathy Leading to Impaired Autonomic Function

Drug- or toxin-induced autonomic neuropathy is an important differential diagnosis in the assessment of patients with evidence of subacute or chronic autonomic dysfunction. Many neurotoxic compounds have selective effects on particular components of the nervous system, and not all compounds that cause peripheral neuropathy affect the autonomic system. Examples are arsenic, which causes a primarily sensory neuropathy without autonomic dysfunction, and lead, which commonly causes motor neuropathy and CNS effects (encephalopathy, behavioral changes) with little, if any, effect on autonomic function.

In specific clinical situations, damage to the autonomic nervous system is induced for therapeutic purposes by introduction of neurolytic agents into specific regions. Agents such as phenol are used for this purpose, and the resulting autonomic dysfunction includes the intended effect (e.g., in sympathectomy for peripheral vascular disease or hyperhidrosis) in addition to any unintended sequelae. More commonly, autonomic neuropathy is an unintended complication of exposure to drugs or chemicals in a therapeutic or occupational context.

Cytotoxic Agents Used in Cancer Chemotherapy

Vinca alkaloids Vincristine and vinblastine are used in the chemotherapy of hematological malignancies such as lymphomas and leukemias. Vincristine, and to a lesser extent, vinblastine, frequently cause a dose-dependent peripheral sensory neuropathy due to axonal degeneration, which is common at cumulative doses over 40 mg. Motor symptoms are rare but autonomic involvement occurs in more than a third of patients, usually manifest as orthostatic hypotension, constipation, paralytic ileus, erectile dysfunction, and urinary retention. Most of the neurological symptoms are reversible after months or years but in some cases are irreversible.

Platinum compounds The platinum compounds, including cisplatin, carboplatin, and oxaliplatin, react with DNA and cause apoptotic cell death in rapidly dividing cells, such as malignant cells. Platinum compounds frequently cause a predominantly sensory peripheral neuropathy associated with axonal degeneration and secondary atrophy of the dorsal root, but autonomic involvement is uncommon, probably affecting less than 10% of recipients of cisplatin.

Taxanes Paclitaxel and docetaxel, chemotherapeutic agents with activity against breast and ovarian cancer, commonly cause a peripheral sensorimotor neuropathy, probably by preventing the normal dissociation of axonal microtubles involved in fast axonal transport, although the precise mechanism of neurotoxicity remains unknown. Autonomic involvement is relatively uncommon, but paclitaxel has been reported to cause acute severe orthostatic hypotension associated with both parasympathetic and sympathetic dysfunction. Paralytic ileus and orthostatic hypotension are the most common manifestations of paclitaxel-induced autonomic neuropathy, and patients with diabetes mellitus may be more susceptible to this complication.

Other Therapeutic Drugs

Perhexiline maleate Perhexiline, used as a last-line agent in the management of refractory angina, is known to cause a peripheral sensorimotor neuropathy. In a few of these cases the neurotoxicity extends to the autonomic nervous system, causing postural hypotension and abnormal heart rate control.

Amiodarone Chronic treatment of cardiac arrhythmias with amiodarone causes peripheral neuropathy (predominantly sensory) in a small proportion of patients. In some of these there may also be autonomic dysfunction, manifest as orthostatic hypotension.

Pentamidine Pentamidine, commonly used in the management of *Pneumocystis carinii* pneumonia secondary to AIDS, has been associated with acute autonomic insufficiency manifested as severe orthostatic hypotension with a fixed heart rate and loss of baroreflex function.

Occupational Exposure

Organic solvents Parasympathetic function appears to be disturbed in some workers exposed to a variety of organic solvents, including hydrocarbons, alcohols, ketones, esters, and ethers, although long-term exposure to toluene has not been associated with peripheral or autonomic neuropathy. Long-term exposure (over years to decades) to low levels of carbon disulfide, for example in the viscose/rayon industry, may cause mild parasympathetic dysfunction, but not sufficient to cause clinical effects.

Acrylamide Acrylamide, which is widely used in paper manufacture, water treatment, building construction, and laboratory research, is known to cause a neuropathy affecting somatic and autonomic nerves. Clinically detectable autonomic disturbances are uncommon in humans but have been studied extensively in experimental animals, in which damage to sympathetic vasomotor nerves as well as parasympathetic fibers has been demonstrated.

Heavy metals Chronic exposure to lead is known to cause a peripheral neuropathy with axonal degeneration, particularly affecting motor fibers. Very long-term exposure may also result in mild sensory and autonomic polyneuropathy, and similar effects, mainly involving the parasympathetic system, have also been reported following mixed exposure to lead, zinc, and copper.

Other Forms of Exposure

Alcohol Alcohol ingestion has both acute and chronic effects on autonomic function in humans. Acute exposure to alcohol reduces parasympathetic function, possibly due to alterations in central cardiovascular control centers rather than a direct effect on vagal function. Chronic alcohol abuse causes an axonal neuropathy, affecting both somatic (sensory and motor) and autonomic function. About 30% of chronic alcohol abusers have detectable abnormalities in the parasympathetic control of heart rate and, in others, sympathetic function may also become affected, resulting in orthostatic hypotension and anhidrosis. Long-term follow-up of chronic alcohol abusers suggests that the presence of autonomic abnormalities on clinical testing is associated with an increased mortality, possibly related to cardiac dysregulation.

Pesticides Most insecticides in common use in agricultural, industrial, or domestic settings are neurotoxic to the target organisms. In general they are not species selective, and can also affect mammalian nervous systems, the outcome depending on the level of exposure in relation to body size. Organochlorine insecticides (such as DDT) affect sensory and motor nerves with few, if any, autonomic effects after acute or chronic exposure.

Organophosphorus compounds (discussed earlier in relation to their acute anticholinesterase activity) may also cause various subacute or chronic neuropathies following a single high-intensity exposure, as in the case of the use of organophosphorus-based chemical weapons in warfare, or during chronic exposure – for example, in manufacturing processes. Long-term chronic low-level exposure to organophosphorus ester insecticides may produce persistent autonomic dysfunction. Organophosphorus compounds bind to acetylcholinesterase for a much longer period than do the more recently developed carbamate compounds, and there is little evidence of prolonged toxicity with the latter group.

Botulinum toxin Although rare, botulism due to ingestion of contaminated food is an important differential diagnosis for the presentation of acute autonomic neuropathy, particularly when there is predominant parasympathetic dysfunction. The toxin acts on cholinergic synapses to prevent the calcium-mediated release of acetylcholine from the presynaptic terminal. Its effects are seen primarily at neuromuscular junctions, causing paralysis, and, to a lesser extent, at autonomic ganglia, where the effect is to block autonomic transmission acutely.

Compounds That Increase Autonomic Activity

Several therapeutic drugs are used to produce increased autonomic activity in situations such as asthma, heart conduction disorders, and hypotension. Excessive autonomic activity is frequently seen in individuals with denervation hypersensitivity (e.g., due to preexisting autonomic neuropathy) during treatment with autonomic agonists. Examples include the use of sympathetic α-adrenoceptor agonists in orthostatic hypotension, and the parasympathomimetic agent, bethanechol, in the management of bladder atony. In other situations, a toxic compound may cause abnormally increased autonomic activity in an individual with previously normal autonomic function.

Sympathetic Stimulation

Drugs used in psychiatry Acute hyperactivity of the sympathetic division of the autonomic nervous system is seen occasionally as a consequence of exposure to psychoactive drugs, both therapeutic and illicit. Neuroleptic malignant syndrome is thought to be mediated by changes in central dopaminergic transmission, causing increased sympathetic outflow manifested as hyperthermia, muscle rigidity, and instability of blood pressure and heart rate. It occurs in fewer than 1% of patients receiving antipsychotic drugs, but can be fatal if untreated. A similar clinical picture, with occasional fatalities, occurs in some individuals who ingest the so-called designer drug Ecstasy (3,4-methylenedioxymethamphetamine (MDMA)) and other amphetamine derivatives that stimulate the sympathetic system.

An interaction between antidepressant drugs, which inhibit NA and/or serotonin reuptake from the synapse, and tyramine, which releases catecholamines from storage vesicles, can result in a marked increase in neurotransmitter concentrations within sympathetic synapses. Clinically there is marked sympathetic hyperactivity with sweating, tachycardia, and severe hypertension, which may lead to intracranial hemorrhage. A similar picture can be seen following the use of the combination of a monoamine oxidase inhibitor and a serotonin uptake inhibitor (such as fluoxetine). Known as the 'CNS serotonergic syndrome,' it comprises general CNS overactivity, muscle spasms, hyperthermia, and autonomic instability, resulting in hyper- or hypotension, tachycardia, and profuse sweating. Fatalities have been reported. The mechanism is believed to be an increase in serotonergic activity, particularly involving serotonin (5-HT_{1A}) receptors in the brain stem and spinal cord.

Ketamine The neurolept anesthetic/analgesic agent, ketamine, primarily an *N*-methyl-D-aspartate (NMDA) receptor antagonist, has an interesting combination of stimulatory and inhibitory cardiovascular and respiratory effects. In isolated organ experiments, in the absence of an intact autonomic nervous system, it has a direct negative inotropic effect on the myocardium. However, in the intact animal or human, this is counterbalanced by a central stimulatory effect, probably mediated by increased sympathetic outflow, which results in an elevation in blood pressure, heart rate and cardiac output, and bronchodilatation.

Venoms and toxins Although snake venom may have a variety of autonomic effects, it is very likely that much of the autonomic activation following snake bite is fear related, resulting in nausea, vomiting, diarrhea, syncope, tachycardia, and cold, clammy skin, even in the absence of actual envenomation. Some venoms, such as those of the stonefish, South American viper snake, and various spiders and scorpions, contain agonists for α- and β-adrenoceptors, which are likely to contribute to venom-induced hypertension and tachycardia. Some venomous creatures, including snakes and scorpions, cause prominent sympathetic hyperactivity, resulting in the potentially fatal clinical syndrome sometimes known as 'autonomic storm.' Components of venoms of scorpions, Australian funnel-web spiders, and the box jellyfish contain neurotoxins that induce massive release of NA, resulting in sympathetic hyperstimulation, which may cause severe hypertension, hypertensive encephalopathy, tachycardia, and cardiac arrhythmias.

Parasympathetic Stimulation

Therapeutic drugs Denervation hypersensitivity has been observed in association with autonomic neuropathy in patients given bethanechol, which produced a markedly exaggerated effect. Anticholinesterase treatment for Alzheimer's disease frequently leads to adverse effects attributable to increased cholinergic neurotransmission in the periphery, such as nausea, vomiting, and diarrhea. More rarely, more serious cholinergic effects, such as heart block, hallucinations, and confusion, may occur. The available drugs appear to be similar in their spectrum of adverse effects.

Other compounds Similar muscarinic stimulatory effects are seen after poisoning with wild mushroom species that contain neurotoxins with muscarinic agonist activity. The effects occur rapidly and, predictably, include lacrimation, salivation, nausea, vomiting, abdominal pain, bronchospasm, headache, miosis, blurred vision, bradycardia, and hypotension. A similar clinical picture is seen acutely following anticholinesterase poisoning with agricultural chemicals such as organophosphorus compounds, which inhibit acetylcholinesterase. The inhibition results in the typical picture of increased secretions, bronchoconstriction, miosis, abdominal cramps, and bradycardia (all related to muscarinic receptor stimulation), and hypertension, muscle fasciculations, tremor, and eventual muscle paralysis (due to nicotinic receptor stimulation at autonomic ganglia and at neuromuscular junctions).

Less common causes of this constellation of features include acute exposure to snake venom toxins known as fasciculins, which also have anticholinesterase activity, both at neuromuscular junctions, where they produce a prolonged muscular contraction, and at autonomic ganglia, where they increase autonomic transmission. Some scorpion toxins have a similar effect, most commonly by binding to neuronal voltage-sensitive sodium and potassium channels and increasing the spontaneous release of acetylcholine, although some venom components have direct muscarinic agonist effects.

Conclusion

While the mechanisms of action of drugs which interact reversibly with autonomic nerve fibers or receptors as part of their pharmacodynamic activity are generally well known, more remains to be established about the effects of drugs and other compounds that cause autonomic neuropathy. Drugs, chemicals, and toxins should always be considered as a possible cause of autonomic dysfunction or autonomic neuropathy in patients presenting with the clinical manifestations of sympathetic and/or parasympathetic dysfunction.

See also: Autoimmune Autonomic Neuropathy; Autonomic Disorders; Autonomic Failure; Autonomic Nervous System; Parasympathetic Nervous System.

Further Reading

Bouhaddi M, Vuillier F, Fortrat JO, et al. (2004) Impaired cardiovascular autonomic control in newly and long-term-treated patients with Parkinson's disease: Involvement of L-dopa therapy. *Autonomic Neuroscience: Basic and Clinical* 116: 30–38.

Dressler D and Benecke R (2003) Autonomic side effects of botulinum toxin type B treatment of cervical dystonia and hyperhidrosis. *European Neurology* 49: 34–38.

Gerhardt U, Hans U, and Hohage H (1999) Influence of smoking on baroreceptor function: 24 hour measurements. *Journal of Hypertension* 17: 941–946.

Gold BS, Dart RC, and Barish RA (2002) Bites of venomous snakes. *New England Journal of Medicine* 347: 347–356.

Gouzoulis-Mayfrank E, Thelen B, Habermeyer E, et al. (1999) Psychopathological, neuroendocrine and autonomic effects of 3,4-methylene dioxyethylamphetamine (MDA), psilocybin and *d*-methamphetamine in healthy volunteers. *Psychopharmacology* 142: 41–50.

Gwee MCE, Nirthanan S, Khoo H-E, et al. (2002) Autonomic effects of some scorpion venoms and toxins. *Clinical and Experimental Pharmacology and Physiology* 29: 795–801.

Jamal GA, Hansen S, and Julu POO (2002) Low level exposures to organophosphorus esters may cause neurotoxicity. *Toxicology* 181–182: 23–33.

Jerian SM, Sarosy GA, Link CJ, et al. (1993) Incapacitating autonomic neuropathy precipitated by taxol. *Gynecologic Oncology* 51: 277–280.

Kay GG and Granville LJ (2005) Antimuscarinic agents: Implications and concerns in the management of overactive bladder in the elderly. *Clinical Therapeutics* 27: 127–138.

Ketch T, Biaggioni I, Robertson RM, et al. (2002) Four faces of baroreflex failure: Hypertensive crisis, volatile hypertension, orthostatic tachycardia, and malignant vagotonia. *Circulation* 105: 2518–2523.

Nicolosi C, DiLeo R, Girlanda P, et al. (2005) Is there a relationship between somatic and autonomic neuropathies in chronic alcoholics? *Journal of Neurological Sciences* 228: 15–19.

Quasthoff S and Hartung HP (2002) Chemotherapy-induced peripheral neuropathy. *Journal of Neurology* 249: 9–17.

Smit AAJ, Wieling W, Voogel AJ, et al. (1996) Orthostatic hypotension due to suppression of vasomotor outflow after amphetamine intoxication. *Mayo Clinic Proceedings* 71: 1067–1070.

Russell JW and Reading PH *Acute Autonomic Neuropathy* and *Chronic Autonomic Neuropathies* Available at: http://www.medlink.com/medlinkcontent.asp.

Tjeerdsma G, Szabo BM, van Wijk LM, et al. (2001) Autonomic dysfunction in patients with mild heart failure and coronary artery disease and the effects of add-on β-blockade. *European Journal of Heart Failure* 3: 33–39.

Tonkin AL and Frewin DB (1999) Drugs, chemicals and toxins that alter autonomic function. In: Mathias CJ and Bannister R (eds.) *Autonomic Failure: A Textbook of Clinical Disorders of the Autonomic Nervous System,* 4th edn. Oxford: Oxford University Press.

Autonomic Dysregulation During REM Sleep

S M Caples and V K Somers, Mayo Clinic, Rochester, MN, USA

Sleep Stage Transitions and the Autonomic Nervous System in Normal Individuals

In humans, wakefulness normally transitions to sleep with the onset of non-rapid eye movement (NREM) sleep, which comprises the majority of total sleep time. Compared with wakefulness, NREM sleep is characterized by cardiovascular stability marked by a reduction in blood pressure (BP) and heart rate (HR). Slowing of the HR is primarily under the influence of vagal activation that is characteristic of NREM sleep. Because of the limited parasympathetic influences on the peripheral vasculature, the decrease in BP is probably a result of concurrent sympathetic withdrawal, as supported by animal models of sympatholysis. During NREM sleep, there is a prominent sinus arrhythmia as cardiac output couples with the respiratory rhythm. With inspiration, as venous return to the heart increases, there is an acceleration of HR followed by a deceleration with exhalation. BP is tightly controlled on account of high baroreceptor gain, which is particularly responsive to BP increments during NREM sleep, serving to ensure the maintenance of stable low BP, referred to as the 'dipping' phenomenon. The absence of this decrement in BP ('nondipping') has, in some studies, been associated with cardiovascular morbidity and mortality – outcomes perhaps mediated by autonomic dysregulation.

Progression of NREM sleep from stage 1 through 4 is characterized by increasingly synchronous cortical and brain stem electrical discharges as exhibited by low-frequency (LF), high-amplitude output on electroencephalography (EEG). This natural sequence is accompanied by further reduction in sympathetic neural traffic such that by stages 3 and 4 (slow wave), the output may be half that encountered in wakefulness. Stage 2 sleep is punctuated by characteristic 'K-complexes' – high-amplitude EEG discharges associated with transient increases in peripheral sympathetic neural activity and an attendant rise in BP (**Figure 1**). Bursts of vagal activity sometimes accompany the transition out of NREM sleep, resulting in HR pauses.

Sharply distinct from NREM sleep, rapid eye movement (REM) sleep occurs cyclically (four to six times) throughout sleep, is more concentrated during the second half of a sleep cycle, and accounts for approximately 25% of total sleep time. With the exception of the respiratory diaphragm and extraocular muscles, REM sleep is heralded by skeletal muscle atonia along with high-frequency (HF), low-amplitude waves on EEG. Because the EEG may otherwise resemble that of wakefulness, REM sleep has been referred to as 'paradoxical sleep.' REM sleep may be subdivided into tonic and phasic stages, the latter of which features characteristic darting eye movements in association with pontogeniculo-occipital EEG spikes, and muscle twitches, predominantly seen in the extremities and face, which break through skeletal muscle atonia.

As discussed later, REM sleep, particularly the phasic component, is considered a state of autonomic instability, marked by surges in both sympathetic and vagal activity with attendant variations in HR, BP, and peripheral vascular resistance. Animal and human studies have demonstrated more concentrated autonomic changes associated with phasic REM sleep, during which intense electrical discharges have been recorded from the brain stem. As a testament to the complexity of the origins of REM sleep and its associated autonomic findings, measurable physiologic parameters representative of autonomic fluctuations have been shown to precede, by several seconds, characteristic eye movements and an active, desynchronized EEG.

Dreaming

Dreaming is thought to occur primarily during REM sleep. Human experiments show that the majority of subjects awoken during REM sleep report a dream experience, compared with a minority reporting the same when awoken from NREM sleep. It has been postulated that the paradoxically active cerebral cortex associated with REM sleep relates to dreams, during which powerful emotions such as anger and fear are frequently generated. Since psychological stress has been associated with cardiac ischemia and rhythm disturbances during wakefulness, it may follow that similar vulnerability could occur during dreaming, as suggested by Mac William in 1923 and also by case reports of patients awakening from dreaming with angina pectoris. There is evidence that brain electrical discharges may directly impact cardiac activity and are perhaps mediated by autonomic tone. Early animal experiments, confirmed by Verrier, have shown that electrical stimulation of certain areas of the brain could induce ventricular arrhythmias and that β-adrenergic blockade, but not vagotomy, reduced vulnerability.

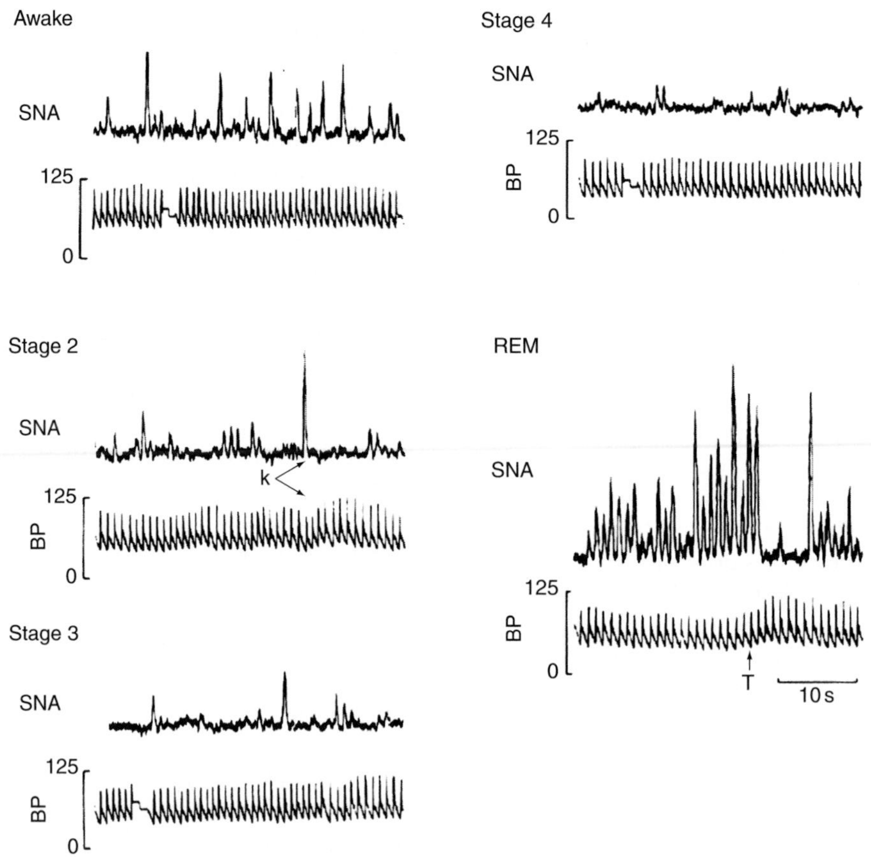

Figure 1 Recordings of sympathetic nerve activity (SNA) and mean blood pressure (BP) in a single subject while awake and while in stages 2, 3, 4, and REM sleep. Note the progressive reduction in SNA and BP as sleep progresses through the stages of NREM sleep. The 'K' during stage 2 corresponds to simultaneous K-complexes noted on the electroencephalogram. In contrast, REM sleep is marked by higher frequency and amplitude of SNA as well as variability in BP. The 'T' during REM sleep refers to a muscle twitch characteristic of phasic REM sleep, associated with abrupt inhibition of SNA and an increase in BP. Reproduced from Somers VK (1993) Sympathetic-nerve activity during sleep in normal subjects. *New England Journal of Medicine* 328(5): 303–307.

Other Physiologic Responses to REM Sleep

In addition to directly observed cardiovascular irregularities, REM sleep is associated with other physiologic perturbations which may indirectly contribute to autonomic dysfunction and, therefore, cardiovascular risk. The respiratory diaphragm usually escapes skeletal muscle atonia, but loss of tone in accessory muscles of respiration and those of the upper airway may pose an increased workload. The ventilatory pattern fluctuates during REM sleep. As a consequence, even though respiratory rate may increase during REM sleep, tidal volumes and minute ventilation usually decrease compared with NREM sleep. There are blunted ventilatory responses to hypoxia and hypercapnia and a heightened arousal threshold, all contributing to prolonged reductions in blood oxygen content, particular in those with impaired respiratory pump function (neuromuscular weakness) or gas exchange capabilities (chronic obstructive lung disease). Hypoxemia stimulates the chemoreflex, resulting in further sympathetic neural outflow. Increased upper airway resistance related to muscle atonia is operative in the pathogenesis of obstructive sleep apnea (OSA), as detailed later. Partial loss of thermoregulatory control during REM sleep further disrupts homeostasis. Finally, cortical influences related to dream activity during REM sleep may contribute to the irregular breathing pattern, along with increased cortical and brain stem electrical discharges seen in association with dreaming.

Cardiovascular Reflexes Mediating Autonomic Activity

A number of reflexes feed back to the autonomic nervous system to influence cardiac and vascular function during periods of stress or physical exertion.

From the standpoint of sleep, these mechanisms may be particularly important during changes in stage of sleep (NREM to REM), position, or with the occurrence of arousals, all of which may take place repeatedly throughout a sleep cycle and are increasingly common with advancing age, a population vulnerable to cardiovascular disease. Dysregulation of these reflexes may be operative in the pathophysiology of certain disease states, such as sleep disordered breathing.

The Chemoreflexes

The peripheral arterial chemoreceptors, the most important of which are located in the carotid bodies, respond primarily to changes in the blood partial pressure of oxygen. Hypoxemic stimulation elicits a brain stem-mediated increase in respiratory muscle output. By virtue of connections between the carotid bodies and sympathetic ganglia, there is an increase in sympathetic outflow to peripheral blood vessels. Homeostasis is preserved by within-breath activation of pulmonary stretch receptors, which are mediated through the vagus nerve, restraining the overall adrenergic response. This sequence is often part of a normal physiologic response under conditions of hypoxia, such as at high altitude. However, an exaggerated ventilatory response, with attendant further increases in sympathetic tone, has been found in certain disease states, such as heart failure and OSA.

The Arterial Baroreflex

The arterial baroreflex system, composed of sensory receptors in the aortic arch and carotid sinuses that relay signals to the brain stem, provides powerful beat-by-beat negative feedback regulation of arterial BP. Increases in BP stretch the receptors and dampen efferent sympathetic output to cardiac and vascular smooth muscle, thereby relatively increasing cardiac parasympathetic tone. The acute increases in BP associated with peripheral vasoconstriction are partially tempered by the baroreflex. This may occur simultaneously with activation of cardiac vagal drive resulting in bradycardia, collectively referred to as the 'diving reflex,' so-called because of its detailed description in diving mammals and also described in humans. Baroreceptor sensitivity appears to be heightened during NREM sleep, but findings in very small studies that measure REM sleep are inconsistent, with some showing a blunting of baroreceptor sensitivity, no change, or a time-dependent increase in sensitivity. Influences of comorbid illnesses, such as hypertension or heart failure, on baroreceptor function are well documented.

Measures of Autonomic Cardiovascular Regulation

Although there are a number of methods by which to measure autonomic neural output, there have been, with some exceptions, concordant findings of parasympathetic predominance during NREM sleep and sympathetic surges with variable states of parasympathetic tone during REM sleep. A commonly utilized method, heart rate variability (HRV), is defined as the oscillation in the interval between consecutive heartbeats (the R-R interval) as well as the oscillations between consecutive instantaneous heart rates. During NREM sleep, a near sinusoidal modulation of HR variation occurs due to the normally occurring respiratory sinus arrhythmia. Spectral analysis allows quantification of the short-term oscillatory components of cardiovascular variability, which are organized in two frequency bands – the LF (approximately 0.1 Hz) and the HF (>0.15 Hz) respiratory bands. The HF components of R-R variability primarily reflect vagal modulation of cardiac rhythm, whereas the LF rhythm has been attributed, albeit not without controversy, to cardiac sympathetic modulation.

Utilizing HRV analysis, a number of studies have demonstrated increased parasympathetic tone and sympathetic withdrawal with the transition from wakefulness to NREM sleep. These studies also demonstrate evidence for increased sympathetic output with the onset of REM sleep, although not all studies have shown reductions in parasympathetic neural output during REM sleep.

Invasive and technically demanding, microneurographic recording of sympathetic nerve activity (MSNA) involves the direct recording of postganglionic sympathetic nerve traffic which may originate from muscle and skin. MSNA has yielded fairly concordant sleep stage-dependent findings. Peripheral sympathetic neural activity, HR, and BP all decrease in concert from sleep onset and progression of NREM sleep to deep, synchronized slow-wave sleep, such that by stage 4, sympathetic measured output was half that encountered in wakefulness. REM sleep, on the other hand, has been associated with SNA exceeding that encountered during wakefulness, particularly during clusters of eye movements (**Figure 1**). Muscle twitches that break through REM atonia have been associated with abrupt attenuation of sympathetic output, probably due to baroreceptor-mediated inhibition during transient rises in BP. That REM sleep is associated with minimal changes in HR and BP may relate to the dissociation of peripheral muscle and cardiac sympathetic activity and underscores the complexities of cardiac sympathovagal balance

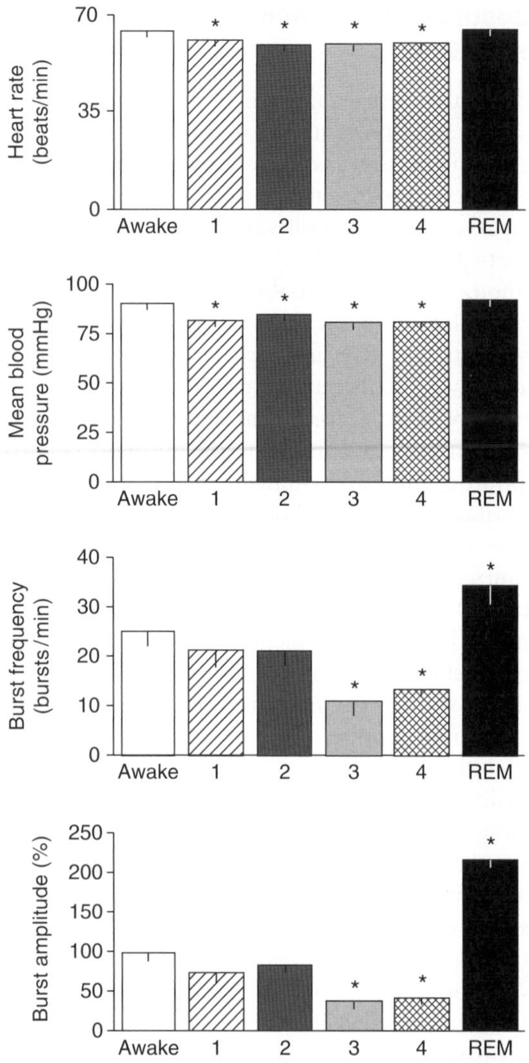

Figure 2 Physiologic indices in a group of normal subjects during wakefulness and NREM and REM sleep. Burst frequency and amplitude refer to sympathetic nerve activity (SNA) measured by intraneural microneurography. Heart rate (HR) and blood pressure (BP) were significantly lower during all stages of NREM sleep than during wakefulness, and SNA was significantly lower during slow-wave sleep (asterisk denotes $p < 0.001$). During REM sleep, sympathetic activity increased significantly, but BP and HR were similar to those recorded during wakefulness. Reproduced from Somers VK (1993) Sympathetic-nerve activity during sleep in normal subjects. *New England Journal of Medicine* 328(5): 303–307.

(**Figure 2**). Thus, this technique provides insight into regional sleep stage-dependent autonomic mechanisms, although it may not necessarily reflect global sympathetic tone.

Other methods for measuring autonomic neural activity, all with their own limitations, include arterial baroreflex function, plasma and urine catecholamine levels, peripheral arterial tonometry, and pulse transit time.

REM Sleep and Heart Rate/Rhythm

Animal experiments have demonstrated marked HR instability during REM sleep, where increases of more than 30% are seen in association with phasic REM sleep, accompanied by an elevation in BP. The frequency of HR surges is higher during phasic REM than tonic REM sleep, with particular concentration during sleep epochs in which an associated muscle twitch was recorded. In animals with coronary artery stenosis, REM sleep causes further decrements in coronary artery blood flow. That these findings were prevented with sympathectomy in cats and dogs suggests an effect attributable to sympathetic overdrive rather than parasympathetic withdrawal. In rats, REM, particularly during phasic episodes, is associated with a higher incidence of bradycardia.

In humans, the average HR is higher during REM than NREM sleep, as is sympathetic neural output, although there is evidence that sympathetic activation is not a global phenomenon during REM sleep. Case reports and series suggest that healthy young adults, sometimes athletes, have periods of exaggerated vagal tone during REM sleep, manifesting as first- or second-degree atrioventricular (AV) block or sinus pauses that may exceed 9 s in duration. Further insight comes from feline experiments demonstrating the highest occurrence of these pauses during the transition from slow-wave sleep to REM sleep as well as during phasic REM sleep. Reversal of the pauses has been reported with atropine or vagotomy. REM sleep therefore appears to be marked by fluctuations in autonomic activity, both to the cardiac conduction system and to peripheral blood vessels.

Patients with established cardiac conduction disease may have distinct pathophysiologic mechanisms during REM sleep, suggested by a small group of subjects with preexisting AV block demonstrating shortening of AV delay during periods of REM sleep. Ambulatory cardiac monitoring (non-polysomnographic) data in healthy elderly populations demonstrate less common occurrences of sinus pauses and AV blocks during sleep episodes, with relatively common supraventricular and ventricular ectopic beats. The circadian variation in paroxysmal atrial fibrillation noted by some suggests a potential influence of the autonomic nervous system. Analysis of HRV in patients with atrial fibrillation demonstrated abrupt shifts in sympathovagal balance preceding the onset of the arrhythmia. Finally, gender-selective effects of REM sleep on cardiac electrical activity were demonstrated by Lanfranchi et al., who measured prolongation of the QT interval exclusive to women.

Arousals/Morning Awakening

Arousal from sleep, either spontaneous or induced by exogenous stimuli, is associated with sympathetic neural surges, leading to increases in HR and BP. Sympathetic tone, however, remains elevated long after return of HR and BP to prearousal levels. Nonspecific arousals and sleep fragmentation increase with advancing age, where the presence of comorbid cardiovascular disease is prevalent and provides a milieu potentially more vulnerable to the effects of sympathoexcitation. Morning awakening induces a stepwise activation of the sympathoadrenal system, with increased HR, BP, and blood catecholamines, and further increases occur with position change and physical activity. Population-based studies have shown that the risk of sudden death from cardiac causes in the general population is significantly greater during the morning hours after waking (i.e., from 6 a.m. to noon) than during other times of the day or night. Because REM is concentrated in the second half of a night's sleep, it is conceivable that autonomic dysregulation associated with REM sleep, among other mechanisms, may be an important contributor, particularly in those with a vulnerable predisposition, such as in the setting of ischemic heart disease. OSA, perhaps due to autonomic dysregulation, has been associated with disruption of this day–night pattern of sudden death.

Obstructive Sleep Apnea

The hallmark of OSA is repetitive upper airway collapse resulting in oxyhemoglobin desaturation. Central nervous system arousals intervene with termination of each breathing event, followed by compensatory hyperventilation. As a result of upper airway/pharyngeal muscle atony as well as reduced lung functional residual capacity, REM sleep usually manifests more upper airway instability and profound deoxygenation than that which occurs during NREM sleep. Further contributing to prolonged hypoxia is the high arousal threshold characteristic of REM sleep. These multiple physiologic stressors all contribute to acute and repetitive surges in sympathetic neural output, the detection of which by measurement of peripheral arterial tone has been used as a diagnostic tool. Some studies report augmented sympathetic responses when central nervous system arousals are accompanied by hypoxia, hypercapnia, and upper airway occlusion. Further evidence for autonomic imbalance in patients with OSA is the finding that the usual decrements in sympathetic activity seen during NREM sleep are lost.

It is possible that there are cumulative effects of long-standing repetitive upper airway events that lead to disruption of autonomic homeostatic mechanisms. In fact, there is abundant evidence for autonomic dysregulation in OSA, in which high levels of MSNA during daytime normoxic wakefulness have been measured (**Figure 3**). HRV is altered in OSA, independent of comorbid illnesses including obesity. Sleep apneics have abnormal baroreceptor function and exaggerated peripheral and central chemoreflex responses, resulting in additive sympathetic stimulation. Deficits in compensatory neural responses to upper airway collapse have been detected in humans with OSA.

Treatment of OSA, typically with the application of continuous positive airway pressure (CPAP) which prevents upper airway collapse, has been shown to acutely attenuate MSNA. Longer term treatment with CPAP reduces daytime sympathetic neural traffic and may also restore, to some degree, autonomic regulatory mechanisms, including baroreceptor function and HRV. Such effects may mediate the salutary effects of CPAP on cardiovascular diseases commonly associated with OSA, such as hypertension and heart failure. Sympathetic neural output may be particularly important in the pathophysiology of heart failure, with OSA having an additive effect and CPAP treatment resulting in acute reductions in sympathetic activity that remain measurable after 3 months of therapy. The long-term implications of OSA treatment on HF outcomes are unknown.

Sleep Deprivation/Short Sleep Duration

National surveys have reported increasingly shorter sleep time per night; as such, there has been intense interest in the systemic effects of short sleep duration. In tightly controlled human experiments, acute restriction of sleep time has been shown to result in metabolic dysregulation, a finding which, in part, could have a physiologic basis in autonomic control. Large population-based studies suggest a higher risk of hypertension in those who self-report shorter sleep duration. Evidence for mediation through autonomic tone has come from experiments demonstrating REM rebound following acute sleep deprivation in young adults and associated increased sympathetic tone by HRV analysis. The blunting of the reduced ventilatory response to hypoxia and hypercapnia associated with sleep deprivation could further contribute to autonomic imbalance.

REM Sleep and Autonomics in the Setting of Cardiovascular Disease

Whereas the acute surges in sympathetic activity associated with REM sleep may be well tolerated in

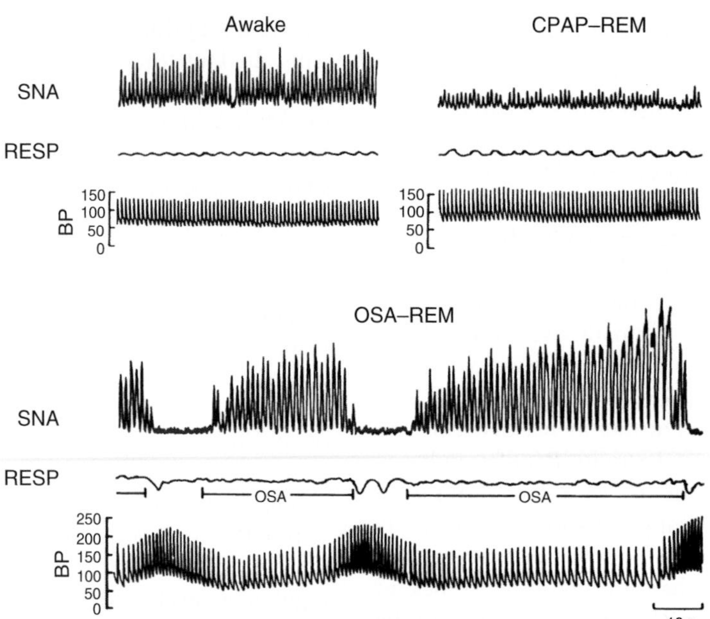

Figure 3 Recording of sympathetic nerve activity (SNA) by microneurography, respiratory excursion (RESP), and intra-arterial blood pressure (BP) in a subject with obstructive sleep apnea (OSA) when awake, during obstructive apneas while in REM sleep, and with elimination of obstructive apnea by continuous positive airway pressure (CPAP) therapy during REM sleep. SNA is high even during normoxic wakefulness, with further increases during REM sleep with obstructive apnea. Note increases in BP as apnea progresses and instantaneous inhibition of SNA with ventilation. Elimination of apneas by CPAP is associated with decreased SNA. Reproduced from Somers VK (1995) Sympathetic neural mechanisms in obstructive sleep apnea. *Journal of Clinical Investigation* 96(4): 1897–1904.

healthy individuals, those with preexisting cardiovascular disease may be vulnerable to adrenergic-mediated vasoconstriction and increments in HR and BP. Polysomnographic recordings of patients with nocturnal angina show ST segment changes on electrocardiography occurring with HR surges associated with REM sleep, as well as with changes in sleeping position that may accompany the transition between NREM and REM sleep. Some patients with diabetes and a recent history of myocardial infarction (MI) have abnormal HRV profiles consistent with higher nocturnal sympathetic tone or reduced vagal tone, which may help explain a higher rate of unfavorable cardiovascular outcomes in these patients. It should also be noted that NREM sleep, which may be associated with hypotension and reduced coronary perfusion pressure, may not be risk-free in susceptible individuals with coronary artery stenosis or after MI. The normal dipping of BP associated with sleep is thought to result from the dominance of parasympathetic tone during NREM sleep over the course of the night. The blunting or absence of dipping suggests a possible role of autonomic dysregulation and has been associated with a higher risk of stroke and congestive heart failure.

Conclusion

There are undoubtedly complex interactions between the autonomic nervous system and REM sleep. The study of cardiovascular function in health and of disease states such as OSA has furthered our understanding of such mechanisms. Additional research will be needed to better characterize these complexities to determine if targeted sleep interventions have any impact on clinical outcomes.

See also: Autonomic Nervous System; Endocrine Function During Sleep and Sleep Deprivation; Immune Function During Sleep and Sleep Deprivation; Metabolic Syndrome and Sleep; Sleep Mentation in REM and NREM: A Neurocognitive Perspective; Sleep Apnea; Sleep Deprivation: Neurobehavioral Changes; Sleep Deprivation and Brain Function; The AIM Model of Dreaming, Sleeping, and Waking Consciousness; Thermoregulation During Sleep and Sleep Deprivation.

Further Reading

Adlakha A (1998) Cardiac arrhythmias during normal sleep and in obstructive sleep apnea syndrome. *Sleep Medicine Reviews* 2(1): 45–60.

Bonnet MH (1997) Heart rate variability: Sleep stage, time of night, and arousal influences. *Electroencephalography and Clinical Neurophysiology* 102(5): 390–396.

Elsenbruch S (1999) Heart rate variability during waking and sleep in healthy males and females. *Sleep* 22(8): 1067–1071.

Guilleminault C (1984) Sinus arrest during REM sleep in young adults. *New England Journal of Medicine* 311(16): 1006–1010.

Narkiewicz K (1999) Enhanced sympathetic and ventilatory responses to central chemoreflex activation in heart failure. *Circulation* 100(3): 262–267.

Narkiewicz K (1999) Selective potentiation of peripheral chemoreflex sensitivity in obstructive sleep apnea. *Circulation* 99(9): 1183–1189.

O'Driscoll DM (2004) Cardiovascular response to arousal from sleep under controlled conditions of central and peripheral chemoreceptor stimulation in humans. *Journal of Applied Physiology* 96(3): 865–870.

Somers VK (1989) Influence of ventilation and hypocapnia on sympathetic nerve responses to hypoxia in normal humans. *Journal of Applied Physiology* 67(5): 2095–2100.

Somers VK (1993) Sympathetic-nerve activity during sleep in normal subjects. *New England Journal of Medicine* 328(5): 303–307.

Somers VK (1995) Sympathetic neural mechanisms in obstructive sleep apnea. *Journal of Clinical Investigation* 96(4): 1897–1904.

Tank J (2003) Relationship between blood pressure, sleep K-complexes, and muscle sympathetic nerve activity in humans. *American Journal of Physiology – Regulatory, Integrative and Comparative Physiology* 285(1): R208–R214.

Van de Borne P (1994) Effects of wake and sleep stages on the 24-h autonomic control of blood pressure and heart rate in recumbent men. *American Journal of Physiology – Heart and Circulatory Physiology* 266(2): H548–H554.

Verrier RL (1996) Sleep, dreams, and sudden death: The case for sleep as an autonomic stress test for the heart. *Cardiovascular Research* 31(2): 181–211.

Autonomic Failure

E M Garland and D Robertson, Vanderbilt University, Nashville, TN, USA

Autonomic Dysfunction

The clinical features of autonomic dysfunction are pervasive, involving the cardiovascular, gastrointestinal, urogenital, thermoregulatory, sudomotor, and pupillomotor systems. Much of the testing evaluates the integrity of the cardiovascular reflexes. Under normal conditions, cardiovascular reflexes act through the autonomic nervous system to maintain blood pressure and cerebral perfusion at appropriate levels. For example, when stretch or baroreceptors detect a decrease in blood pressure, they relay this information to the central nervous system, which decreases parasympathetic tone and increases sympathetic outflow. Increases in norepinephrine (NE) secretion and peripheral resistance restore blood pressure and provide adequate cerebral perfusion. Disturbances in one or several parts of this reflex arc may result in syncope, a sudden, transient loss of consciousness with spontaneous recovery that may be associated with hypotension, bradycardia (reduced heart rate), and loss of postural tone. Conservative estimates suggest that 30% of the general population has experienced at least one syncopal spell and that syncope is responsible for over 1% of hospital admissions. Infrequently, syncope may be a sign of impaired cardiovascular reflexes resulting from autonomic failure.

Disorders of autonomic failure include central neurodegenerative diseases such as multiple-system atrophy (MSA), primary peripheral autonomic nervous system degeneration (as in autonomic neuropathy or pure autonomic failure (PAF)), and congenital diseases such as familial dysautonomia (FD) and dopamine β-hydroxylase (DBH) deficiency. Patients with severe autonomic failure experience profound hypotension either after assuming the upright position or after food consumption (orthostatic and postprandial hypotension, respectively). In baroreflex failure, exaggerated blood pressure and heart rate fluctuations are exacerbated by emotional or physical stress, whereas patients with postural tachycardia syndrome (POTS) endure orthostatic symptoms and enhanced increases in heart rate with standing. Although Parkinson's disease (PD) is primarily a disorder of motor control, cardiac sympathetic dysregulation has been demonstrated and symptoms of autonomic failure occur in 20–50% of patients. Other systemic illnesses, such as diabetes and amyloidosis, may disturb autonomic function.

Pure Autonomic Failure

Clinical Manifestations

PAF, or Bradbury–Eggleston syndrome, is a persistent, degenerative disorder of the peripheral autonomic nervous system presenting in middle to late life, and affecting men more often than women. The disorder appears to be confined to the sympathetic and parasympathetic nervous systems; the adrenal medulla is relatively spared. Failed attempts to identify the cause of a patient's autonomic failure trigger a diagnosis of PAF.

Although the initial feature of PAF in men is often erectile dysfunction, patients predominantly present with orthostatic hypotension. The orthostatic hypotension, which may be described as unsteadiness, dizziness, or faintness upon standing, is worse in the morning, after meals or exercise, or in hot weather. Patients also complain of pain in the neck or back of the head, relieved by lying down. The orthostatic hypotension is often accompanied by supine hypertension. Nocturia may cause the patient to get up as many as 5 times per night. Urinary hesitancy, urgency, dribbling, and occasional incontinence might also occur. Some patients develop signs of neurogenic urinary retention associated with repeated urinary tract infections. A sudden decline in daily functions in a patient with PAF suggests an infection, usually of the urinary tract. Hypohidrosis, or at least an asymmetrical distribution of sweating, is typical in PAF. Basal metabolic rate may be reduced.

The pathology of PAF has not been completely elucidated, but lesions in the sympathetic ganglia and deficient catecholamine uptake and catecholamine fluorescence in sympathetic postganglionic neurons have been observed. Cell loss has also been reported in the intermediolateral column of the spinal cord, and Lewy bodies, which stain for ubiquitin and α-synuclein, have been found in ganglia, as well as in peripheral autonomic nerves and the central nervous system. Neuroimaging studies confirm the absence of functional sympathetic nerve terminals in the myocardium of patients with PAF.

PAF patients have greatly reduced levels of catecholamines (**Figure 1**) while lying down and have little increase upon standing. Plasma and urinary NE levels are usually considerably below normal, as is plasma dihydroxyphenylglycol (DHPG; an intraneuronal metabolite of NE). Plasma levels of epinephrine are

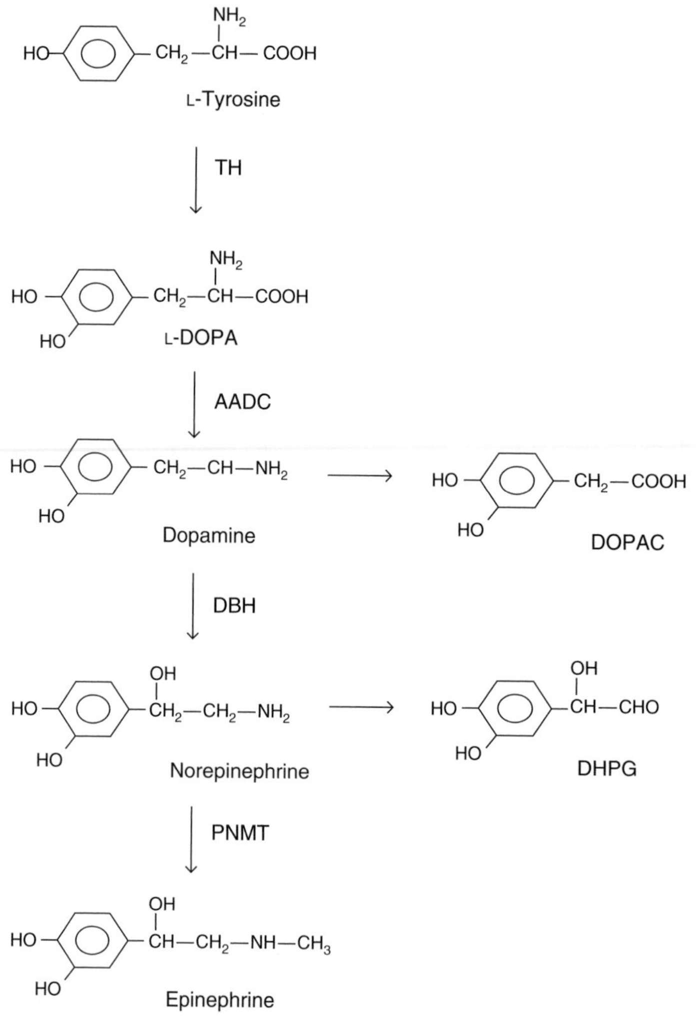

Figure 1 Catecholamine biosynthetic pathway. TH, tyrosine hydroxylase; L-DOPA, levodopa; AADC, aromatic amino acid decarboxylase; DBH, dopamine β-hydroxylase; PNMT, phenylethanolamine-*N*-methyltransferase; DOPAC, 3,4-dihydroxyphenylacetic acid; DHPG, dihydroxyphenylglycol.

also reduced but usually to a lesser extent than NE. Reductions in plasma 3,4-dihydroxyphenylalanine (DOPA) and dopamine (DA) levels are consistent with diminished catecholamine synthesis. In PAF, drugs that release NE from sympathetic terminals have little effect on plasma NE or blood pressure, whereas drugs that directly act on adrenoreceptors produce an exaggerated response.

Diagnosis

The definitive diagnosis of orthostatic hypotension is usually based on a drop in systolic blood pressure of $\geq$20 mmHg and diastolic blood pressure of $\geq$10 mmHg after at least 1 min of standing. A decrease in systolic blood pressure of 50 mmHg or greater is not uncommon in patients with PAF. However, a diagnosis of PAF cannot be excluded based on a single normal upright blood pressure measurement. Several measurements of orthostatic blood pressure should be made.

PAF should be distinguished from two other disorders of autonomic failure: MSA and idiopathic PD. PAF is less progressive and generally less disabling than these other syndromes. For a diagnosis of PAF, there should be no indication from the history or physical examination of cerebellar, striatal, pyramidal, or extrapyramidal dysfunction. Symptoms of hoarseness and sleep apnea are also highly suggestive of MSA. PAF has not been definitively associated with any genetic or environmental causes, nor is it infectious.

Management

Patients with PAF have a generally good outlook; many live for 20 years or more after the onset of their disease. The most common cause of death in these patients is pulmonary embolus or recurrent infection. The incidence of both myocardial infarction and stroke appears to be significantly reduced.

Some patients with PAF find that leg crossing helps forestall presyncopal symptoms during upright posture. Other nonpharmacological treatment options include high-salt diet, consumption of small meals, elevating the head of the bed, squatting, abdominal compression, bending forward, and compression stockings. Pharmacological treatment is directed mainly at the orthostatic hypotension. Medication options include fludrocortisone, somatostatin analogs, yohimbine, midodrine, erythropoietin, and other vasopressor agents. Patients should be encouraged to avoid factors that can exacerbate orthostatic hypotension, such as large meals, alcohol, drugs, straining during micturition and defecation, and exposure to a warm environment. Finally, a recently discovered treatment option for hypotension is water. Sixteen ounces of water can raise blood pressure by as much as 40 mmHg, with a peak at 30 min after ingestion. Consistent with residual sympathetic tone contributing to supine hypertension, clonidine can be effective at decreasing nighttime blood pressure and natriuresis.

Multiple-System Atrophy

Clinical Manifestations

MSA is a sporadic, progressive, neurodegenerative disease characterized by extrapyramidal, pyramidal, cerebellar, and autonomic dysfunction in any combination. MSA is classified according to predominant features: Shy–Drager syndrome, when autonomic failure predominates; striatonigral degeneration (MSA-P), when extrapyramidal features prevail; and olivopontocerebellar atrophy (MSA-C), for predominantly cerebellar features. The autonomic dysfunction in MSA can be viewed as a primarily central defect with a generally intact peripheral autonomic nervous system.

The average age of onset of MSA is in the sixth decade. Men are affected more frequently than are women, with a male:female ratio of 3–9:1. Up to 75% of MSA patients present with complaints related to autonomic dysfunction, with motor symptoms generally developing within 2 years. Patients who initially have principally movement symptoms develop autonomic dysfunction within 5 years of diagnosis. The most common major autonomic problem of MSA patients is orthostatic hypotension. Other symptoms and signs of autonomic failure include syncope, erectile dysfunction, urinary incontinence or retention, frequent urinary tract infections, constipation, and decreased thermoregulatory sweating. Orthostatic hypotension can be worsened by nighttime pressure natriuresis induced by associated supine hypertension, which occurs in approximately 60% of patients with MSA.

Parkinsonian signs of MSA-P include progressive akinesia and rigidity, with facial stare, and tremor. MSA-C is characterized by gait ataxia with a wide-based gait, cerebellar oculomotor disturbances, intention tremor, and poor performance of rapid, alternating movements. Other symptoms and signs associated with neurodegeneration in MSA include hoarse voice, headache, neck pain, change in writing style, slurred speech, sleep apnea, and recurrent aspiration.

Diagnosis

MSA is classified as possible, probable, or definite on the basis of features and criteria in three clinical domains: autonomic and/or urinary dysfunction, parkinsonism, and cerebellar dysfunction (**Table 1**). Possible MSA is diagnosed when one criterion and two features from separate clinical domains are found. The diagnosis of probable MSA requires autonomic and/or urinary dysfunction and poorly

Table 1 Clinical diagnosis of MSA

Clinical domain	*Incidence (%)*	*Criterion*	*Features*
Autonomic or urinary dysfunction		Orthostatic hypotension (decrease by ≥30/15 mmHg) or persistent urinary incontinence with erectile dysfunction or both	Orthostatic hypotension ≥20/10 mmHg, urinary incontinency, incomplete bladder emptying
Parkinsonism	87	Bradykinesia and rigidity, postural instability, or tremor	Bradykinesia, rigidity, postural instability, tremor; minimal or reduced response to L-DOPA
Cerebellar dysfunction	54	Gait ataxia and limb ataxia, ataxic dysarthria, or sustained gaze-evoked nystagmus	Gait ataxia, ataxia dysarthria, limb ataxia, sustained gaze-evoked nystagmus
Corticospinal tract dysfunction	49	Not a criterion in diagnosing this disease	Extensor plantar responses with hyperreflexia

levodopa (L-DOPA)-responsive parkinsonism or cerebellar dysfunction. MSA can be definitively diagnosed only at autopsy, by the loss of neuronal and oligodendroglial cells in numerous regions of the brain and spinal cord and by the presence of ubiquitin-containing glial cytoplasmic inclusions (GCIs) in both glial cells and neurons. GCIs stain for α-synuclein, and MSA is classified as a 'synucleinopathy' along with PD and dementia with Lewy bodies.

In patients whose initial symptoms resemble those of PD, MSA should be suspected if there is a poor response to L-DOPA, if there is loud snoring, if disease progresses rapidly to wheelchair confinement, and if prominent urinary tract symptoms are present. Neuroimaging studies have enabled distinction between MSA and PD. All patients with PD and orthostatic hypotension have markedly decreased uptake and retention of sympathoneuronal imaging agents in the heart, whereas all patients with MSA have intact cardiac sympathetic innervation, as measured by sympathetic neuroimaging. Patients with MSA who present with only autonomic and/or urinary dysfunction may be incorrectly diagnosed with PAF. The distinctive features of MSA compared to PAF are presented in **Table 2**.

The etiology of MSA is unknown, but a genetic component has been considered unlikely because no familial clustering has been reported for the disease. The results of genetic association studies, however, indicate that genetic polymorphisms associated with interleukin-1A, interleukin-8, intercellular adhesion molecule-1, tumor necrosis factor, and α1-antichymotrypsin increase the risk for MSA. An autoimmune process has been suggested to cause the neurodegeneration, although neurotoxicity related to oxidative stress, environmental neurotoxins, and infection has also been proposed.

Management

Patients rarely survive longer than 10 years after a diagnosis of MSA. The most common causes of death are pulmonary embolus, apnea, and intercurrent infection. No current therapy can reverse or halt progression of MSA. Management involves treatment of the depression, tremor and gait disturbances, supine hypertension, orthostatic hypotension, and urinary tract problems. Orthostatic and postprandial hypotension and supine hypertension respond to the interventions mentioned for PAF. Treatment of the movement-disorder component of MSA is less effective than when the same drugs used in PD are administered.

DBH Deficiency

Clinical Manifestations

DBH deficiency, also known as NE deficiency, is a rare disorder due to the congenital absence of the enzyme that converts DA to NE. It is characterized by normal parasympathetic and sympathetic cholinergic function in the absence of sympathetic noradrenergic function. The prevalence of DBH deficiency is unknown. Fewer than 20 cases, all of Western European descent, have appeared in the literature.

Patients with DBH deficiency have experienced vomiting, dehydration, hypotension, hypothermia, and hypoglycemia perinatally. Orthostatic symptoms tend to worsen during adolescence, leading to evaluation and eventual diagnosis. Physical examination of individuals with DBH deficiency reveals poor cardiovascular regulation with a low normal supine blood pressure, a low supine heart rate, and profound orthostatic hypotension with an inadequate compensatory orthostatic tachycardia. The orthostatic hypotension may be accompanied by presyncopal

Table 2 Comparison of MSA with PAF

Characteristic	*Multiple-system atrophy*	*Pure autonomic failure*
Site of autonomic impairment	Mainly preganglionic, central	Mainly postganglionic; peripheral
Prognosis	Poor; median survival 6.5–9.5 years	Good; slow disease progression with survival >10–15 years
Parkinsonian and cerebellar symptoms	Common	Not present
Neuropathological marker	Glial cytoplasmic inclusions in glia and neurons	Lewy bodies in neurons
Orthostatic hypotension	Common	Common
Respiratory disturbances	Common	Uncommon
Urinary tract disturbances	Early in disease	Late in disease
Supine plasma norepinephrine	Normal	Very low
Orthostatic rise in norepinephrine	Subnormal	Subnormal
Response to norepinephrine-releasing drugs	May be increased	Low
Postsynaptic adrenoreceptor sensitivity	Mildly increased	Increased

Table 3 Common clinical features of DBH deficiency

Severe orthostatic hypotension
Postprandial hypotension
High palate
Anemia
Impaired ejaculation
Ptosis of eyelids
Hyperflexible or hypermobile joints
Nasal stuffiness

symptoms, including dizziness, blurred vision, dyspnea, nuchal discomfort, and occasionally chest pain. Although patients are profoundly affected by the defects in autonomic regulation of the cardiovascular system, there is little evidence of central nervous system dysfunction. Other clinical features that have been reported in DBH deficiency are included in **Table 3**.

Patients with DBH deficiency are unique in that they have minimal or undetectable plasma NE and epinephrine and a five- to tenfold elevation of plasma DA. In addition, metabolites of NE are low or absent in plasma, urine, and cerebrospinal fluid, whereas the DA precursor, DOPA, and metabolites of DA are elevated. The plasma DA concentration responds to various physiological and pharmacological stimuli, similarly to NE in normal individuals. For example, a change from supine to upright posture normally doubles plasma NE levels, but in patients with DBH deficiency, plasma NE remains undetectable while the plasma DA concentration increases two- to threefold. These observations are consistent with release of DA into the synapse in place of NE. Responses to α_1-adrenoceptor agonists and β-adrenoceptor agonists are exaggerated in this disorder.

DBH deficiency is inherited in an autosomal recessive manner. A number of variants in both the coding and noncoding regions of the *DBH* gene have been identified in patients and in unaffected family members but not in unrelated control individuals or in patients with other autonomic disorders.

Diagnosis

Although DBH deficiency appears to be present from birth, the diagnosis is not generally recognized until adolescence or later. Measurement of plasma catecholamines (NE, DA, and their metabolites) is required to verify the diagnosis. Results of autonomic function testing are consistent with sympathetic noradrenergic failure and adrenomedullary failure but intact vagal and sympathetic cholinergic function. DBH enzyme activity varies over a wide range in healthy individuals, and most individuals with low plasma DBH enzyme activity do not have DBH deficiency. Not only is DBH enzyme activity undetectable in the blood of individuals with DBH deficiency, but immunoassay and immunocytochemistry also indicate an absence of DBH protein.

Management

The life expectancy of patients with DBH deficiency is unknown. At least two patients are now in their fifties, and no reported patient has died as yet. For the most part, treatment of DBH deficiency is supportive and directed at relieving orthostatic symptoms. When it was determined that these patients lacked NE, an effective treatment was developed in the form of droxidopa (dihydroxyphenylserine; DOPS). Droxidopa is metabolized directly to NE by L-aromatic amino acid decarboxylase, bypassing DBH. In patients with DBH deficiency, it increases blood pressure and restores plasma NE to the normal range. Plasma DOPA is also lowered to normal levels by droxidopa, but DA and its metabolites remain somewhat elevated and plasma epinephrine continues to be at or below the limits of detection.

Individuals with DBH deficiency do not respond as well to standard medications used for autonomic failure. Fludrocortisone and indomethacin have been mildly effective at raising blood pressure. Phenylpropanolamine elicits some pressor response, presumably due to the denervation hypersensitivity of vascular α-adrenoreceptors.

Familial Dysautonomia

Clinical Manifestations

FD, or Riley–Day syndrome, is a disease of inadequate development and progressive neuronal degeneration that affects central and peripheral components of the autonomic nervous system as well as sensory neurons. It is an autosomal recessive disorder seen almost exclusively in individuals of Ashkenazi heritage. FD is one of a group of rare neurodevelopmental disorders termed hereditary sensory and autonomic neuropathies. More than 500 patients had been diagnosed with FD by 2002.

Patients with FD have gastrointestinal dysfunction, vomiting crises, recurrent pneumonia, decreased pain and temperature perception with intact sensitivity to visceral pain, and cardiovascular instability (**Table 4**). Patients can exhibit both extreme hypertension and profound and rapid orthostatic hypotension without appropriate compensatory changes in heart rate. Cardiac autonomic dysfunction may be related to hypoplastic cervical and thoracic sympathetic ganglia. Dysautonomic crises with vomiting, hypertension,

Table 4 Clinical features of FD

System	*Common features*
Sensory	Decreased pain and temperature sensation
	Impaired vibration sense
	Intact visceral sensation
Autonomic nervous	Dysphagia
	Esophageal and gastric dysmotility
	Gastroesophageal reflux
	Vomiting crises
	Aspiration
	Insensitivity to apoxia
	Orthostatic hypotension
	Hypertensive crises
	Blotching
	Reduced or absent tears
	Corneal analgesia
Central nervous	Emotional lability
	Normal intellect
Orthopedic	Short stature
	Spinal curvature
Other	Ashkenazi Jewish extraction
	Lack of lingual fungiform papillae
	Decreased patellar reflexes
	Ataxia
	Pneumonia

tachycardia, sweating, blotching of skin, puffy hands, and behavior abnormalities may be triggered by emotional or physical stress.

FD adversely affects growth and development, and the incidence of scoliosis is as high as 85%. Intelligence is normal. An age-related decline in renal function is frequently observed. In individuals with FD, while plasma NE might be normal when patients are supine, the orthostatic increase in NE is reduced. Plasma levels of the NE metabolite DHPG are also low. Plasma DOPA is elevated by about 25%, consistent with stimulation of tyrosine hydroxylase activity, and DA may exhibit dramatic surges during 'crises.' The high DOPA:DHPG ratio in FD is not seen in other autonomic disorders. An exaggerated hypertensive response to infused NE is consistent with supersensitivity resulting from a decrease in the number of postganglionic terminals on peripheral blood vessels.

Diagnosis

The clinical diagnosis of FD is supported by the presence of five 'cardinal' criteria: lack of overflow tears, absent lingual fungiform papillae, decreased patellar reflexes, no axon flare following intradermal histamine, and documentation of Ashkenazi Jewish extraction. However, a definitive diagnosis can now be made using genetic testing of the *IKBKAP* gene.

Management

With greater understanding of the disorder and more effective treatment, quality of life and survival are improving for patients with FD. However, as these patients age, they frequently complain of poor balance, unsteady gait, and difficulty with concentration. Psychiatric disorders may develop, and renal function may decline. Cardiovascular autonomic function may worsen. Due to their increased risk of aspiration, FD patients can develop chronic lung disease.

Treatment of FD is presently supportive and directed toward the various symptoms. Diazepam is used during crises to treat vomiting, hypertension, and agitation. Clonidine can also be administered to help control blood pressure. The orthostatic hypotension that occurs between crises may respond to standard interventions, such as increased salt and fluid intake as well as use of fludrocortisone and midodrine. Decreased sensation requires vigilance, and orthopedic therapy may be needed for scoliosis.

Baroreflex Failure

Clinical Manifestations

Baroreflex failure, a rare disorder, is a result of damage along the baroreflex arc at the baroreceptors, along the glossopharyngeal or vagal nerves, or in the brain stem nuclei. It is generally a consequence of neck surgery, head or neck irradiation for carcinoma, brain stem lesions, or bilateral carotid body tumors.

Acute interruption of the baroreflex arc, with damage isolated to the afferent limb, may be associated with severe, sustained hypertension, tachycardia, and headache. Diaphoresis and apnea may occur, especially in the first 48 h postoperatively. Labile, episodic hypertension (volatile hypertension), on the other hand, is often seen in patients who develop baroreflex failure gradually or during a more chronic phase of the disorder. Surges in blood pressure are elicited by mental or physical stress, during which sympathetic outflow is increased. These patients experience sensations of warmth or flushing, dizziness or light-headedness, palpitations, headache, and diaphoresis. Tremulousness, anxiety, and irritability may also be present. In patients with baroreflex failure, periods of volatile hypertension can be interrupted by hypotensive episodes, especially during periods of quiet, sedation, or sleep, when sympathetic outflow is diminished.

In some cases of baroreflex failure, with interruption of efferent vagal function, tachycardia is provoked by mild sympathetic activation. In other patients, with selective baroreflex failure (Jordan syndrome), the lesion is in the afferent input from the carotid sinus to the nucleus tractus solitarii, and efferent sympathetic and parasympathetic output remains intact. Malignant vagotonia with hypotension, bradycardia, and asystole characterize selective baroreflex failure. Accompanying symptoms include fatigue and dizziness, with possible syncope. Plasma NE levels in baroreflex failure tend to parallel the blood pressure changes. During periods of relatively low sympathetic activity, NE may be normal (111–360 pg ml^{-1} (0.66–2.13 nM)). During hypertensive-tachycardic episodes, plasma NE may rise to levels as high as 2260 pg ml^{-1} (13.36 nM).

Diagnosis

The key diagnostic feature in baroreflex failure is a parallel increase in blood pressure and heart rate with stress, and a parallel decrease with sedation or rest. In addition, pressor agents do not cause a reflex decrease in heart rate nor does reflex tachycardia occur after vasodilators. A history of neck surgery, irradiation, or some other traumatic event would support this diagnosis.

Management

The primary goal of therapy of patients with baroreflex failure is to decrease the frequency and magnitude of surges in blood pressure and heart rate. Clonidine is the treatment of choice for blood pressure surges. Once symptoms have been well controlled, diazepam can be instituted to control stress and decrease central sympathetic input. Medication, such as fludrocortisone, may also be required to prevent hypotensive episodes. In patients with selective baroreflex failure, a pacemaker may be needed. Finally, because of the excessive levels of plasma NE in baroreflex failure, agents that prevent release of NE may also be helpful, and agents that can increase synaptic NE should be avoided.

Postural Tachycardia Syndrome

Manifestations

POTS differs from other forms of autonomic failure in that upright posture is associated with little or no decrease in blood pressure but marked orthostatic tachycardia. Also known as orthostatic intolerance (OI), POTS is relatively common (approximately 500 000 Americans are affected) and tends to occur in individuals between 15 and 50 years old, with a female:male ratio of 5:1. POTS is the most frequently encountered dysautonomia among patients referred to centers specializing in autonomic disorders.

Posture-related symptoms in POTS are consistent with impaired cerebral perfusion and include lightheadedness, clouding of thought, blurred vision, anxiety, substernal chest pain, fatigue, palpitations, and occasionally syncope. Warm dry feet, excessive orthostatic blood pooling and dusky skin in the lower leg, reduced galvanic skin response, abnormal sweating on the extremities, and hyperresponsiveness of leg veins to NE are other clinical signs of POTS. The supine heart rate is usually normal or slightly elevated, whereas patients typically have a greater than 30 beats per minute increase in heart rate on standing. Blood pressure in patients is similar to that in healthy individuals in both supine and upright postures. Supine plasma catecholamines are high, and their responses to standing are exaggerated.

Pathophysiology and Diagnosis

POTS is a heterogeneous disorder, and a number of different pathophysiologies underlie symptoms, perhaps with elevated sympathetic drive to the cardiovascular system as a final common pathophysiological mechanism in the majority of patients. Some cases of POTS may result from a patchy impairment of autonomic neuronal function that allows the heart to remain normally innervated in the face of impaired innervation of the distal extremities. High plasma NE levels support the view that increased sympathetic nervous system activation can underlie POTS in some patients. These patients may have reduced plasma renin activity and aldosterone, as well as supine and dynamic orthostatic hypovolemia. A low absolute or orthostatic blood volume would be expected to elicit tachycardia on standing. Other potential pathophysiological mechanisms include excessive venous pooling, ganglionic-receptor-binding antibodies, exaggerated gravity-dependent fluid shift, cardiac β-adrenoceptor hypersensitivity, and diminished cardiovagal baroreflex sensitivity with impaired central autonomic regulation. The onset of POTS is often preceded by a recent viral infection. Patients can undergo extensive clinical evaluation by a variety of specialists, but the basis of POTS remains obscure in most afflicted individuals and it remains undiagnosed in many patients.

A small percentage of patients diagnosed with POTS have norepinephrine transporter (NET) deficiency, a condition that was first described in 2000. The substantial orthostatic tachycardia in NET deficiency is accompanied by an increase in the plasma NE concentration to almost 4 times its value in the supine position. The relatively low concentrations of plasma DHPG, relative to NE, as well as decreased NE clearance, increased NE spillover and tyramine resistance, are consistent with impaired NET function. A missense mutation has been identified in the NET gene that renders the transporter nonfunctional. In case-control studies, genes encoding the β1- and β2-adrenoreceptors, nitric oxide synthase, and endothelin have been associated with the development of POTS and the magnitude of orthostatic responses. Contributions of these genes have been relatively minor, however, and a number of other genes are likely to contribute to POTS.

Management

A majority of patients with POTS have a relatively mild disorder, which improves over time. Those patients with potentially precipitating events, such as a viral infection, appear to do better overall than others. Volume loading and increases in sodium intake blunt the orthostatic tachycardia, normalize the plasma NE response to standing, and improve orthostatic symptoms in POTS. Effective therapeutic regimens include orthostatic 'exercise,' water, fludrocortisone, low-dose propranolol, clonidine, midodrine, and pyridostigmine.

See also: Autoimmune Autonomic Neuropathy; Autonomic Disorders; Autonomic Nervous System: Cardiovascular Control; Autonomic Nervous System: Metabolic Function; Autonomic Nervous System: Neuroanatomy; Autonomic Nervous System: Central Cardiovascular Control; Autonomic Nervous System; Blood Pressure: Baroreceptors; Dopamine; Dopamine in Perspective; Dysautonomia: Familial; Sympathetic Nervous System.

Further Reading

Axelrod FB (2004) Familial dysautonomia. *Muscle and Nerve* 29: 352–363.

Garland EM, Hahn MK, Ketch TP, et al. (2002) Genetic basis of clinical catecholamine disorders. *Annals of the New York Academy of Sciences* 971: 506–514.

Gilman S, Low PA, Quinn N, et al. (1999) Consensus statement on the diagnosis of multiple system atrophy. *Journal of the Neurological Sciences* 163: 94–98.

Goldstein DS (2001) *The Autonomic Nervous System in Health and Disease*. New York: Marcel Dekker, Inc.

Goldstein DS, Robertson D, Esler M, et al. (2002) Dysautonomias: Clinical disorders of the autonomic nervous system. *Annals of Internal Medicine* 137: 753–763.

Grubb BP (2005) Neurocardiogenic syncope and related disorders of orthostatic intolerance. *Circulation* 111: 2997–3006.

Hilz MJ and Dutsch M (2006) Quantitative studies of autonomic function. *Muscle and Nerve* 33: 6–20.

Ketch T, Biaggioni I, Robertson R, et al. (2002) Four faces of baroreflex failure: Hypertensive crisis, volatile hypertension, orthostatic tachycardia, and malignant vagotonia. *Circulation* 105: 2518–2523.

Man in't Veld AJ, Boomsma F, Lenders J, et al. (1988) Patients with congenital dopamine beta-hydroxylase deficiency. A lesson in catecholamine physiology. *American Journal of Hypertension* 1: 231–238.

Mathias CJ (2003) Autonomic diseases: Clinical features and laboratory evaluation. *Journal of Neurology, Neurosurgery & Psychiatry* 74: iii31–iii41.

Mitsky V and Robertson D (1995) Failure of the autonomic nervous system. *Comprehensive Therapy* 21: 529–534.

Parikh SM, Diedrich A, Biaggioni I, et al. (2002) The nature of the autonomic dysfunction in multiple system atrophy. *Journal of the Neurological Sciences* 200: 1–10.

Robertson D and Biaggioni I (1995) *Disorders of the Autonomic Nervous System*. Luxembourg: Harwood Academic Publishers.

Robertson D, Flattem N, Tellioglu T, et al. (2001) Familial orthostatic tachycardia due to norepinephrine transporter deficiency. *Annals of the New York Academy of Sciences* 940: 527–543.

Robertson D, Haile V, Perry SE, et al. (1991) Dopamine beta-hydroxylase deficiency. A genetic disorder of cardiovascular regulation. *Hypertension* 18: 1–8.

Autonomic Nervous System

J B Furness, University of Melbourne, Parkville, VIC, Australia

The autonomic nervous system (ANS) is the system of neurons that control peripheral organs, other than striated muscle that is under voluntary control (**Figure 1**). Thus the ANS controls the visceral organs of the thoracic, abdominal, and pelvic cavities, which includes the lungs, heart, digestive organs, kidneys, urinary bladder, and internal generative organs. It also controls endocrine and exocrine glands, the blood vessels that supply all organs, and, within the eye, the diameter of the iris (**Table 1**).

The purpose of the ANS is to adjust the activities of these organs so that they function at levels that are most favorable to the state of the body and to its environment. The ANS is thus one of two systems, the other being the endocrine system, that control the functions of the internal and surface organs, including the skin. The two control systems would be better thought of as one, because they act in synergy to control organs.

Autonomic control of organs is through reflexes and through cortical control centers. To elicit a reflex, the relevant states of organs must be detected. This detection is through visceral afferent neurons, that are properly regarded as part of the ANS. Many visceral afferent neurons also communicate other information, for example, pain from the viscera, satiety from the digestive tract, or temperature. Thus, autonomic visceral afferent neurons, while part of the ANS, may carry signals to the central nervous system that serve other functions. Some autonomic visceral afferent neurons (e.g., baroreceptor neurons and intrinsic primary afferent neurons of the intestine) seem to have roles exclusively in autonomic reflexes.

Since the late nineteenth century and early twentieth century, it has been common to divide the ANS into three divisions – the sympathetic, parasympathetic, and enteric divisions. There were pragmatic reasons for this separation of parts of what is essentially one control system. The efferent (motor) autonomic outflows from the central nervous system have gaps, that is, there are nerves emerging from the brain and spinal cord that do not carry autonomic motor pathways (**Figure 1**). Autonomic fibers are absent from the first two cranial nerves, olfactory and optic; they are then present in cranial nerves III, VII, IX, and X. Then a gap in outflow occurs, with no autonomic contribution to cranial nerves XI and XII or the cervical nerves (there can rarely be a contribution to C7). The next group of nerve roots, T1 to L2 or -3 (the thoracolumbar outflows), all have autonomic components, and then there is a small gap where there are few autonomic fibers, which become prominent again in sacral roots 2–4. The outputs are thus considered as cranial, thoracolumbar, and sacral. The number of segments covered by the divisions varies slightly between mammalian species; the levels quoted are for human.

Some of the pathways that emerge from the cranial autonomics slow the heart and dilate blood vessels, and there are also dilator nerves in the sacral outflows. Thus, these were grouped for classification purposes as the craniosacral autonomic. Conversely, there are cardio-accelerator and vasoconstrictor pathways in the thoracolumbar outflows. The nerves of the thoracolumbar outflows were termed sympathetic because they were proposed to maintain a sympathy between the organs. The autonomic nerves of the craniosacral division of the ANS were later called parasympathetic, because of their opposite effects on the cardiovascular system, being associated with slowing or lowering cardiovascular function.

The division of the ANS into sympathetic and parasympathetic has led to enormous misconceptions, the most serious being the concept that the two divisions are somehow in opposition to each other. This is quite a wrong idea. Autonomic nerves, whatever their anatomical origin, act in concert to control visceral organs and the vasculature.

The third division of the ANS is the enteric division, which is the system of autonomic ganglia and nerve fibers that is contained within the walls of the digestive organs. This is given a status as a separate division because it contains complete reflex circuits which can operate in the absence of connections with the central nervous system. In terms of numbers of neurons, the enteric is the largest autonomic division. In humans, it contains 200–600 million neurons.

The motor pathways of the ANS that arise in the central nervous system pass through autonomic ganglia and make synapses on the way to the organs that they innervate, with the exception of the pathways to the adrenal glands. Synaptic transmission in autonomic ganglia is mediated primarily by acetylcholine (ACh) that acts on nicotinic receptors. The use of nicotine and other drugs that block these receptors has been important in analyzing the nerve pathways. Neurons with cell bodies in the central nervous system that make synapses in peripheral ganglia are known as autonomic (sympathetic or parasympathetic)

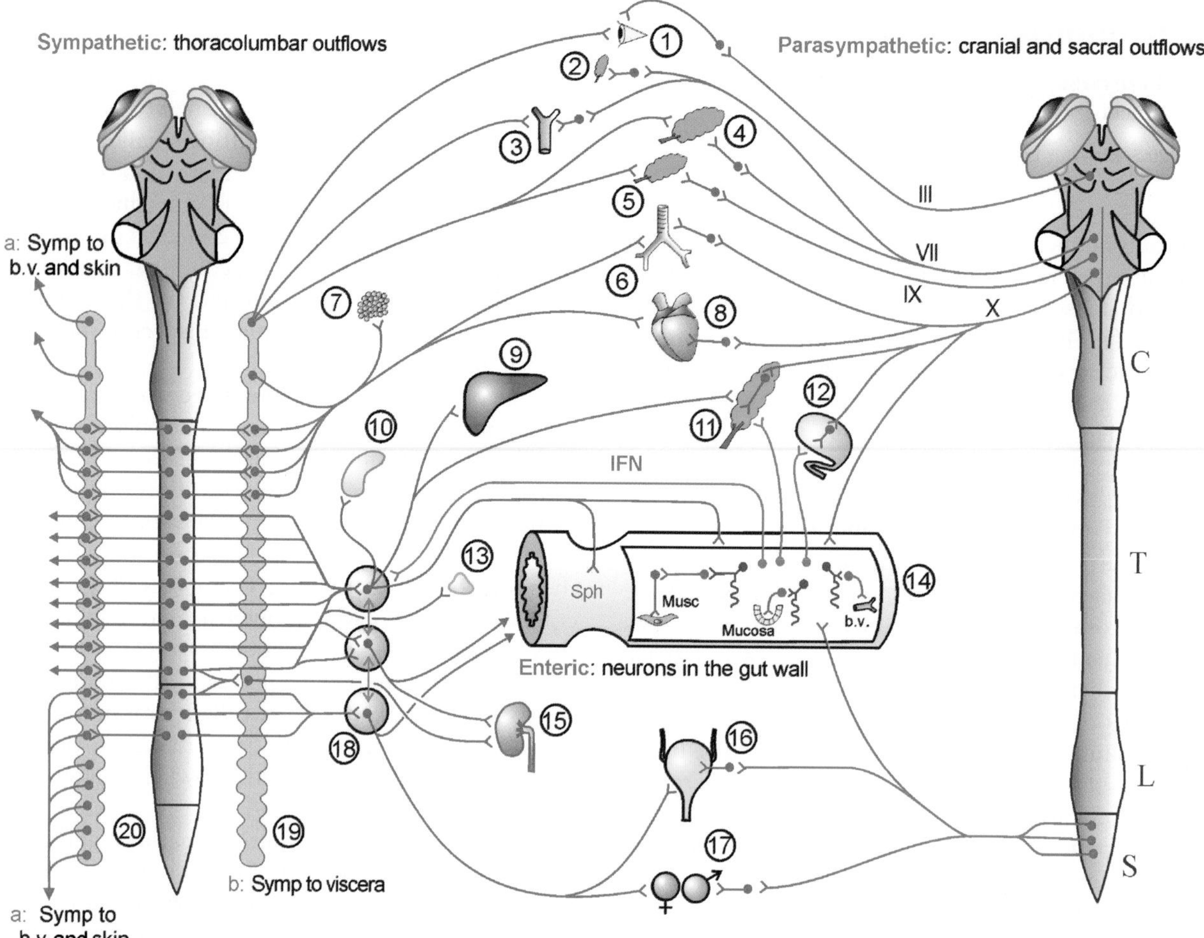

Figure 1 Schematic representation of the efferent (motor) pathways of the ANS, including illustration of some peripherally confined primary afferent neurons (autonomic visceral afferent neurons relaying signals to the central nervous system (CNS) have been omitted to simplify the diagram). The brain stem and spinal cord are represented twice, on the left to show sympathetic connections and on the right to show parasympathetic connections. The enteric division is within the gut wall (center of diagram (14)). *Sympathetic outflows.* (a) To the left side of the spinal cord, sympathetic chain (paravertebral) ganglia are represented; some neurons of these ganglia supply blood vessels (b.v.) throughout the body and effectors in the skin (sweat glands pilomotor muscles). These pathways have synapses in the paravertebral ganglia. For simplicity of illustration, pathways that run rostrally and caudally within the sympathetic chains are not illustrated. (b) On the right side of the cord are the connections that pass first through the sympathetic chains and then through prevertebral ganglia and plexuses (18) to supply visceral organs, as well as pathways that supply structures in the head and neck and intracranial arteries. Synapses occur in either prevertebral and paravertebral ganglia. *Parasympathetic outflows.* These emerge from cranial and sacral levels and innervate structures in the head, neck, abdomen and pelvis, but not in the limbs. *Enteric neurons.* These are represented within the outline of the intestine (14). The enteric reflex circuits contain intrinsic primary afferent neurons (IPANs, purple), interneurons and motor neurons (red). As illustrated, these control muscle (musc), the secretory epithelium of the mucosa and blood vessels (b.v.). Sph: this indicates the sphincter regions of the intestine; these are controlled by enteric and extrinsic neurons. *Peripherally confined neural connections between organs.* These are marked in blue. One of the neurons that contribute to these circuits is the intestinofugal neuron (IFN) that projects from the intestine to prevertebral ganglia. *Target tissues and organs.* (1) Eye; (2) lacrimal glands; (3) intracranial arteries; (4) and (5) salivary glands; (6) airways; (7) brown fat; (8) heart; (9) liver; (10) spleen; (11) pancreas; (12) gallbladder; (13) adrenal gland; (14) tubular gastrointestinal tract; (15) kidney; (16) urinary bladder; (17) genital organs; (18) prevertebral ganglia and plexuses; and (19, 20) sympathetic chains (paravertebral ganglia and their interconnections). *Spinal cord levels.* C, cervical; T, thoracic; L, lumbar; S, sacral.

preganglionic neurons. The neurons with which they connect are called postganglionic neurons. Enteric neurons are innervated by parasympathetic preganglionic neurons and sympathetic postganglionic neurons.

The ANS is associated with specific collections of neurons within the central nervous system. A series of nuclei in the brain stem (including the Edinger–Westphal nucleus, the salivatory nuclei, and the dorsal motor nucleus of the vagus) contain the cell bodies of preganglionic neurons. Within the spinal cord, autonomic cell groups are in the intermediolateral column nuclei at levels corresponding to autonomic outflows,

Table 1 A summary of functions controlled by the ANS

Heart rate, force, and conduction
Arterial diameter (all vascular beds)
Mesenteric venous capacity
Pupillary diameter, accommodation of lens
Exocrine gland secretion: lacrimal gland, salivary glands, gastric glands, exocrine pancreas, sweat glands, glands of genital organs
Endocrine secretion: adrenal medulla, endocrine pancreas
Secretion into organs: intestinal water and electrolyte secretion, pulmonary secretion, nasal secretion
Gastrointestinal wall movement
Gall bladder contraction and biliary tract motility
Regulation of the urinary bladder and control of micturition
Tracheal and bronchial diameter
Contraction of vas deferens, vagina, other internal genitalia, dartos muscle
Penile erection, clitoral and labial engorgement
Fat mobilization
Immune system[a]
Piloerection

[a]See especially Jänig (2006).

from the upper thoracic to mid-lumbar levels and in the sacral spinal cord. The central autonomic motor nuclei receive inputs from autonomic integrative cell groups, including the nucleus of the tractus solitarius, autonomic cell groups of the ventrolateral medulla, the paraventricular nucleus, and other cell groups within the hypothalamus. At higher levels, important autonomic control is exerted through the parabrachial nuclei, amygdala, insular cortex, and cingulate cortex. At these higher levels, it is no longer possible to distinguish the centers as simply autonomic. They are also involved in endocrine control and affective behavior, both of which are closely related to autonomic function.

See also: Autonomic and Enteric Nervous System: Apoptosis and Trophic Support During Development; Autonomic Disorders; Autonomic Failure; Autonomic Nervous System: Neuroanatomy; Autonomic Neuroplasticity: Development; Parasympathetic Nervous System; Sympathetic Nervous System.

Further Reading

Blessing WW (1997) *The Lower Brainstem and Bodily Homeostasis.* Oxford: Oxford University Press.

Furness JB (2006) *The Enteric Nervous System.* Oxford: Blackwell Publishing.

Furness JB (2006) The organisation of the autonomic nervous system: Peripheral connections. *Autonomic Neuroscience* 130: 1–5.

Jänig W (2006) *The Integrative Action of the Autonomic Nervous System.* Cambridge, UK: Cambridge University Press.

Loewy AD and Spyer KM (1990) *Central Regulation of Autonomic Functions.* 390pp. Oxford: Oxford University Press.

Morrison SF (2001) Differential control of sympathetic outflow. *American Journal of Physiolology* 281: R683–R698.

Saper CB (2002) The central autonomic nervous system: Conscious visceral perception and autonomic pattern generation. *Annual Review of Neuroscience* 25: 433–469.

Autonomic Nervous System Development

D F Newgreen, Murdoch Childrens Research Institute, Parkville, VIC, Australia
M J Howard, Medical University of Ohio, Toledo, OH, USA
R Nishi, University of Vermont College of Medicine, Burlington, VT, USA

Origin of the Autonomic Nervous System from the Neural Crest

All neurons and glia of the autonomic nervous system – the sympathetic, parasympathetic, and enteric nervous systems – are derived from the neural crest (NC). The NC, a migratory cell population unique to vertebrates, arises during early embryogenesis, becoming microscopically distinguishable in a rostral-to-caudal wave in human embryos approximately from embryonic (E) days E18 to E40, in mice from E8.5 to E11, and in chicks from E1.3 to E3.5. The induction of the NC has been elucidated by, among others, Marianne Bronner-Fraser's group (bird embryos) and Robert Mayor's group (frog embryos). This cell population is induced by earlier signals from neighboring tissues, and it consists of epithelial cells at the border between the medial neural ectoderm, which gives rise to the central nervous system (CNS), and the more lateral epidermal ectoderm, which contributes to the skin. This involves signaling via secreted growth factors: fibroblast growth factors (FGFs), Wnts, and transforming growth factor (TGF)-β family members, especially the bone morphogenetic proteins (BMPs). This sets up a pattern of gene activity that specifies the NC cells as different from the ectodermal cells on each side. This involves expression in the nascent NC cells of genes such as *Pax7* and *Zic3* (in mice) for transcription factors (which control the expression of other genes). In a positive feedback, the gene for the inducer BMP-4 is also activated in the NC. This is rapidly followed by the transient expression of transcription factor genes such as *Snail1* (in mice; *SNAI2/SLUG* in birds and reptiles), the forkhead gene *Foxd3* (mice), and the Sox group E genes (in mice, *Sox8*, *Sox9*, and *Sox10*).

The transcription factors encoded by these genes promote the onset of cell migration, a classic example of the epithelial–mesenchymal transition. NC cells lose the homophilic cell–cell adhesion molecule N-cadherin first at translational then at transcriptional levels. This is accompanied by reorganization of the cytoskeleton into a motile form and by the ability to adhere to and also digest extracellular matrix involving, among others, expression of genes for RhoA (actin modulator), integrin (matrix adhesion molecule), and matrix metalloproteases (MMPs). This allows the formerly coherent epithelial cells to assume a quasi-individualistic migratory behavior.

The fate of the NC cells after migration has been demonstrated in most detail by Nicole Le Douarin and colleagues using chick–quail orthotopic transplantation as a device for labeling and tracking cells. These form not only the peripheral autonomic nervous system, but also many other neural and related endocrine cells, and even connective tissues. These fates can be mapped to position of origin along the neuraxis (**Figure 1**). The differentiation competence of NC cells has been tested by heterotopic transplantation. These experiments indicate that competence is broader than the fate, and the particular lines of differentiation are controlled by the tissues with which the NC cells interact. However, superimposed on this are also some early-imposed restrictions on differentiation. More rostrally originating cells have a wider range of options; this is epitomized by the restriction of connective tissue competence to cranial NC cells. This early spatial restriction also extends to some autonomic competences such as enteric neurogenesis capacity. These cells can be elicited when NC cells are combined with intestinal tissues, but only cranial NC cells are competent to produce large numbers of neurons. In contrast, sympathetic neurons can differentiate from all of the NC, including cranial levels that are not fated to do so.

The nascent NC cells, like the forming CNS, express multiple homeobox genes in a nested rostro-to-caudal pattern. There are four clusters of these genes (A, B, C, and D), each cluster being variously numbered (from 1 to 13). The equivalently numbered genes of each cluster are similar and tend to be co-expressed. The most rostral and earliest formed neural tissues including NC, at forebrain and midbrain levels, express no homeobox genes. Progressively higher number homeobox genes are expressed in the NC at progressively more caudal levels. Thus, *HOXB1* is expressed maximally at hindbrain levels rostral to the ear, *HOXB4* in the hindbrain caudal to the ear, and *HOXB9* at the level of the neck. A combinatorial code of homeobox gene expression is important in specifying an early positional memory that influences later differentiation processes. It is possible that the rostral-to-caudal patterned restrictions on later differentiation in early NC cells, noted earlier, are regulated by the homeobox code.

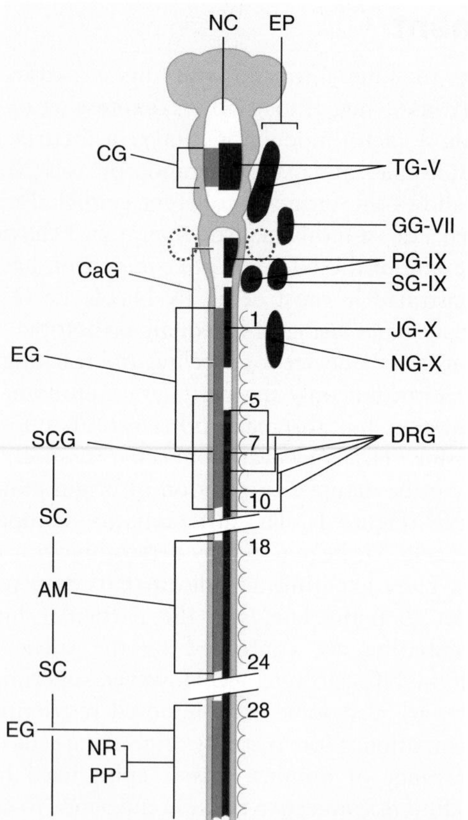

Figure 1 Scheme of neural crest (NC) derivatives mapped onto the neuraxis (light blue). Autonomic nervous system derivatives (red) include ciliary ganglion (CG), cardiac ganglion (CaG), enteric ganglia (EG), superior cervical ganglia (SCG), sympathetic ganglia (SC), nerve of Remak (NR; avian only), pelvic ganglia (PP), and adrenomedullary (AM) gland ganglia. Related sensory neural structures (dark blue), dorsal root sensory ganglia (DRG), and epibranchial placodes (EP) give rise to cranial sensory ganglia. The somites are numbered, and the otic (ear) vesicle position is circled. Adapted from Newgreen DF and Erickson CA (1986) The migration of neural crest cells. *International Review of Cytology* 103: 89–145.

Migratory Morphogenesis of Neural Crest Cells

The outstanding characteristic of NC cells is their migratory morphogenesis. The sympathetic and parasympathetic cells require a migration of only several 100 μms from their NC origin, and in mouse and chick embryos this takes about half a day. Trunk NC cells that form the sympathetic system migrate in lateroventral chains through the mesenchymal cell mass of the rostral half of the adjacent sclerotome (a subregion of the somitic body segments) to reach the dorsal aorta (**Figure 2**). Recent live cell imaging *in vivo* has shown that these cells then turn rostrally and caudally parallel to the aorta, so that cells from one segmental level become distributed over four or more segments.

The enteric nervous system is by far largest division of the autonomic nervous system (ANS) in cell numbers and is much more widely distributed. Yet it arises from more restricted NC regions, mostly stemming from the vagal neuraxis (in the caudal hindbrain and slightly overlapping the trunk) with a numerically minor component from lumbosacral levels (**Figure 1**). The vagal cells migrate over and through the vagal-level somites to the nearby foregut (esophageal, gastric, and duodenal primordia), then migrate through the dense gut mesenchyme of the midgut (future small intestine and cecum) and hindgut (future colon). This colonization takes about 4 weeks in humans and 4–5 days in mouse and chick embryos. Live cell imaging from Heather Young and Miles Epstein and their colleagues has shown that these cells also migrate as dynamic chains (**Figure 3**). Biomathematical modeling and experimental studies by Kerry Landman and co-workers have indicated that proliferation in the NC wavefront is a major driver of colonization of the gastrointestinal tract. However, completion of colonization to the distal colon is made more difficult, since the gut is simultaneously elongating. Failure to complete colonization leads to Hirschsprung's disease, in which the distal gut lacks neural ganglia, cannot perform peristalsis, and, after birth, becomes massively distended with fecal contents proximal to the aganglionic region. Mutations in many genes predispose to neural dysplasias of the intestine, and many of these defects are Hirschsprung-like (**Table 1**). Many of these genes control NC cell numbers by affecting cell proliferation or survival; this may lead to an apparent migration defect.

Molecular Control of Migration

Migration routes of NC cells are controlled by receptor-mediated interactions with their surroundings, including extracellular matrix molecules, growth factors, and molecules on the surface of non-NC cells. Initial migration is into extracellular matrix that forms a migratory substrate. Important matrix molecules include fibronectin, laminins, and collagens, for which NC cells possess multiple integrin class adhesion and signaling receptors. Treatment with peptides and antibodies that block matrix interactions can cause NC cell migration to stall *in vitro* and *in vivo*. Chemoattraction of NC cells may also occur: the glial cell line-derived neurotrophic factor (GDNF) appears to attract enteric NC cells in *in vitro* tests, and is produced by intestinal mesenchyme cells (**Figure 4**). However, analysis is complicated, because growth factors often have several functions; in this

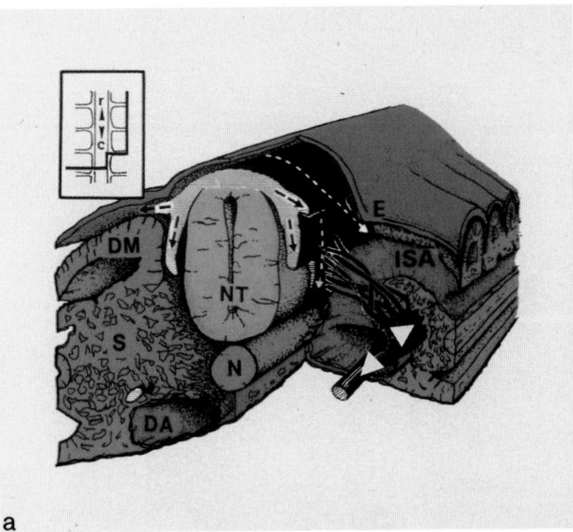

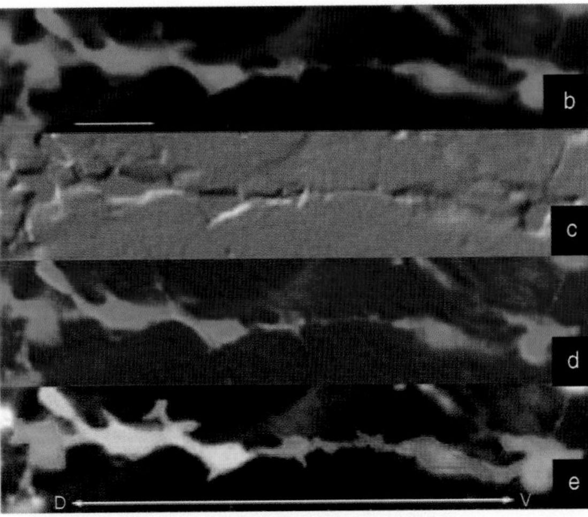

Figure 2 (a) Cutaway scheme of neural crest cells migrating as chains through the rostral sclerotome (S) to the dorsal aorta (DA), where they turn rostrally and caudally (arrows; see inset: r, rostral; c, caudal). Other areas depicted are the demomyotome (DM), epidermal ectoderm (E), intersegmental artery (ISA), notochord (N), and neural tube (NT). (b–e) Single neural crest cell chain, extending from neural tube to dorsal aorta. (b) Green fluorescent protein-labeled neural crest cells. Scale bar = 20 μm. (c) Embossed image of chain. (d) Fluorescence and embossed overlay. (e) Individual cells within chain colored separately; D↔V, dorsal to ventral. (b–e) From Kasemeier-Kulesa JC, Kulesa PM, and Lefcort F (2005) Imaging neural crest cell dynamics during formation of dorsal root ganglia and sympathetic ganglia. *Development* 132: 235–245.

example, as well as a chemotactic role, GDNF is a required survival factor, a potent mitogen, and a differentiation driver for NC cells possessing its receptor, Ret (the product of *ret proto-oncogene*).

Negative regulators of migration are of equal importance to positive regulators. These molecules repel NC cells from regions such as the perinotochordal zone, the caudal half of the somites, parts of the gut mesenchyme, and elsewhere. These repulsive molecules include ephrinB1 (receptor on NC cells: EphB3), semaphorins 3A and 3F (receptors: neuropilins 1/2), and slit 2 (receptor: Robo), the large chondroitin sulfate proteoglycan aggrecan, the glycoprotein tenascin-C, and the growth factor sonic hedgehog (Shh).

Migrating NC cells contact each other, and these contacts are important for migration and directional choice. Isolated NC cells show little translocation. NC cells at the forefront of migration, and at certain 'decision points,' extend filopodia simultaneously in many directions, and move erratically, but NC cells in contact in chains move more consistently. Major cell–cell adhesion molecules N-cadherin and N-CAM are reduced on NC cells in the migratory phase, but lower affinity cadherins may maintain the transient adhesions observed. Interference with the cell adhesion molecule L1CAM on mouse NC cells disrupts the chainlike connections of these cells in the intestine and delays migration. Interestingly, *L1CAM* is a modifier gene for Hirschsprung's disease genes (see **Table 1**), and since it is located on the X chromosome, it may contribute to the 4:1 male:female ratio of Hirschsprung's disease.

Gangliogenesis

Less is known about gangliogenesis in the ANS than about the process of migration, but upregulation of the homophilic cell adhesion molecules N-cadherin and N-CAM often occurs, and genetic disruption of these systems leads to less compact aggregates. In the forming of sympathetic ganglia, initially hemisegmental NC cells migrate longitudinally across segmental boundaries to form a relatively uniform chain (**Figure 2(a)**). This chain then, via N-cadherin upregulation, forms a string of aggregates. These aggregates assemble at the anterior half of each hemisegment, and this is based on avoidance of repulsive interactions with ephrinB1 that is increasingly expressed in the alternate posterior hemisegment.

In the enteric nervous system the ganglia are uniformly small and regularly spaced in two dimensions, in two layers in the gut mesenchyme. These ganglionated plexuses form at a distance from the gut endoderm. When Shh expression is reduced, neurons are found closer to the endoderm, suggesting that Shh is involved in establishing a domain that excludes NC-derived cells. The cell behavior in ganglion formation, visualized by time-lapse microscopy, suggests that the uniformity of size and spacing could result from a balance between contact-mediated NC cell cohesion and repulsion-at-a-distance between

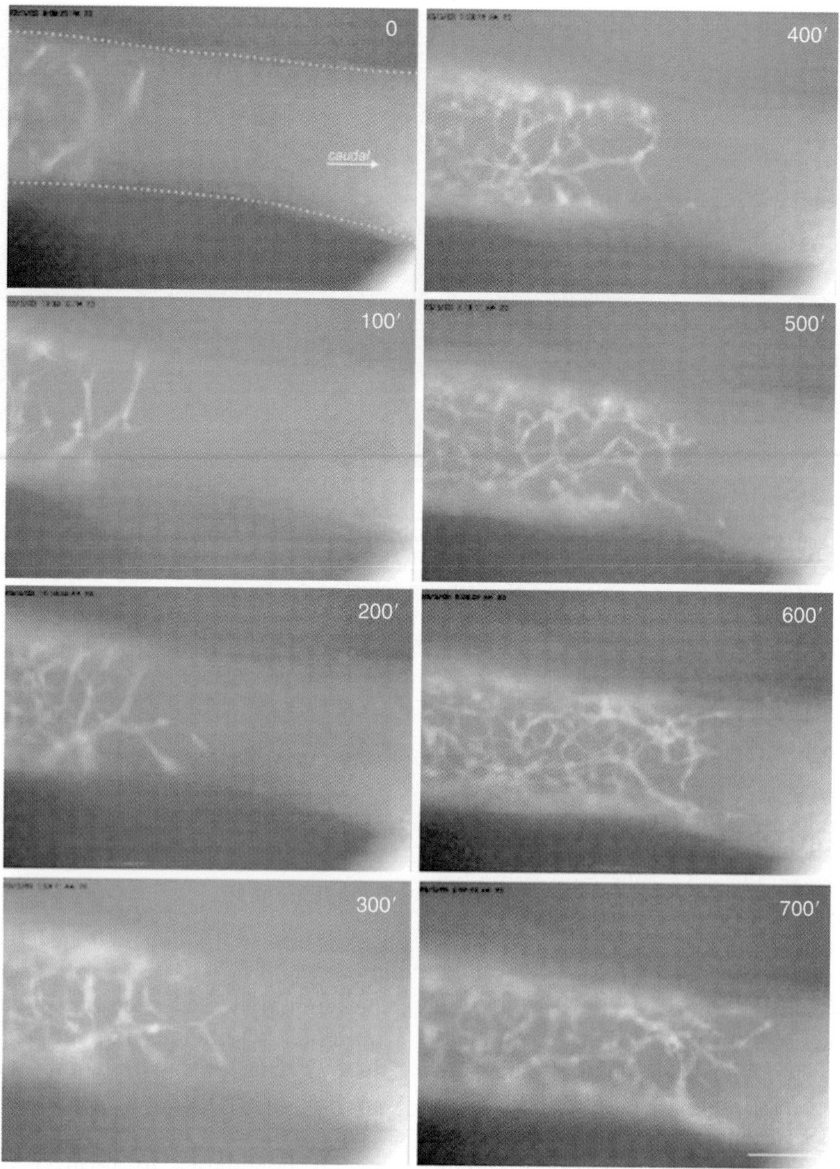

Figure 3 Selected time-lapse frames of the hindgut of an E12.5 mouse showing the caudal progression of green fluorescent protein-expressing neural crest cells. Time is noted in minutes. Most of the cells are present in intersecting chains that follow a variety of trajectories. Scale bar = 100 μm. From Young HM, Bergner AJ, Anderson RB, et al. (2004) Dynamics of neural crest-derived cell migration in the embryonic mouse gut. *Developmental Biology* 270: 455–473.

like cells. However, the molecules responsible for this reorganization are not yet known.

Sympathetic Ganglia: Neuronal Differentiation and Connections

The sympathetic chain ganglia, derived from trunk NC (**Figure 1**), innervate all of the organs, smooth muscle, skeletal muscle, and glands, and serve to modulate their function. Innervation proceeds in a rostral-to-caudal pattern, with more rostral neurons innervating rostral structures and caudal neurons innervating caudal organs.

The preganglionic motor neurons that innervate sympathetic chain ganglia arise at thoracic levels of the spinal cord. Their axons leave the spinal cord via the ventral roots and project both rostrally and caudally. The rostral-to-caudal projection pattern is determined by the location of the cell bodies along the neural axis. Several factors contribute to motor column-specific motor neuron identity as well as generation of a segment-specific projection pattern. The generation of neurons in the Column of Terni (designation for preganglionic motor neurons in the chick) depends upon loss of expression of the homeodomain

Table 1 Genes implicated in enteric nervous system formation and dysplasias

Gene	*Human chromosomal location*	*Phenotype of ENS in mice in which the gene is homozygously inactivated*
RET	10q11.2	Absence of neurons from small and large intestines
GDNF	5p12-p13.1	Absence of neurons from small and large intestines
GFRAI	10q25	Absence of neurons from small and large intestines
EDNRB	13q22	Absence of neurons from distalmost large intestines
EDN3	20q13.2-q13.3	Absence of neurons from distalmost large intestines
ECE1	1p36.1	Absence of neurons from distalmost large intestines
PHOX2B	4p12	Absence of neurons from entire gastrointestinal tract
SOX10	22q13	Absence of neurons from entire gastrointestinal tract
PAX3	2q37	Absence of neurons from small and large intestines
ASCL1	12q22-q23	Absence of neurons from esophagus
IHH	2q33-q35	Absence of neurons from parts of the small intestine and colon
SHH	7q36	Ectopic neurons within mucosa
ZEB2 (*SIP1*)	2q22	Absence of neurons from distalmost large intestine
TLX2	2p13.1	ENS hyperplasia in colon and hypoplasia in small intestine

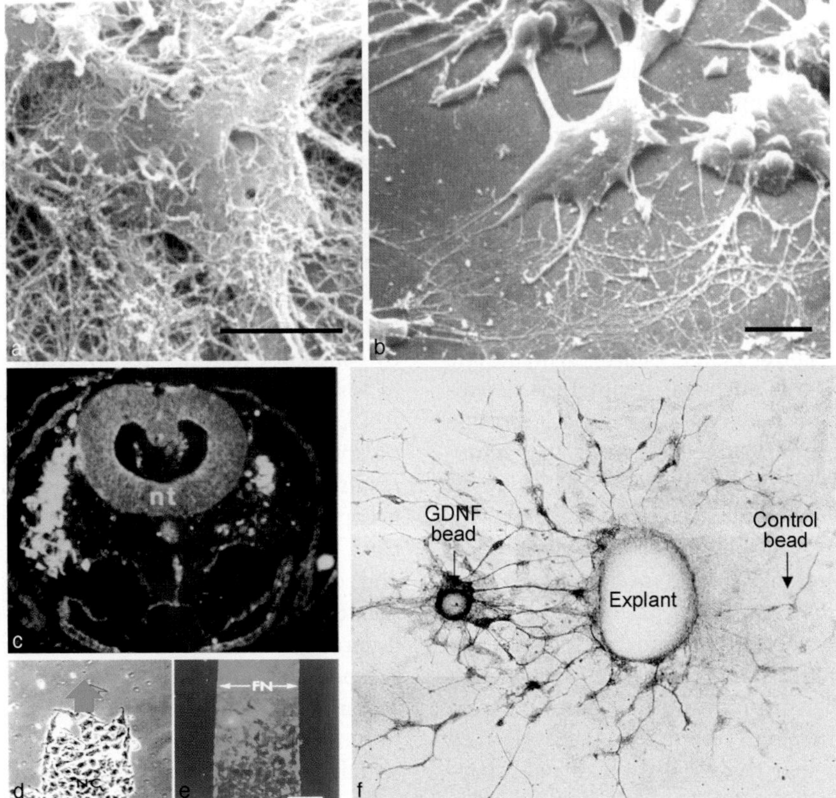

Figure 4 The microenvironment influences neural crest cell migration. (a) Scanning electron micrograph of neural crest cells *in vivo*, migrating in a fibrillar fibronectin extracellular matrix. (b) Fibrillar fibronectin strongly promotes neural crest cell migration in *in vitro* assays. (c) Transverse section of the head of a chick embryo injected on the right side (*) with a function-blocking antibody to the fibronectin receptor. The neural tube (nt) is labeled. Neural crest outgrowth (white area) labeled with the HNK-1 antibody to avian neural crest cells is reduced on the injected side. (d, e) Neural crest (NC) cells migrating *in vitro* (d, arrow) accurately follow stripes of fibronectin (FN); (e) shows same field shown in (d), but is immunolabeled (red) for fibronectin. (f) Enteric neural crest cells labeled with PGP9.5 antibody migrate *in vitro* from a gut segment explanted onto a collagen gel. This migration is strongly biased toward a bead loaded with glial-derived neuronal factor (GDNF), compared to migration toward a control bead. Scale bars = 20 μm (a), 20 μm (b), 50 μm (c). (a) From Newgreen DF (1985) Control of the timing of commencement of migration of embryonic neural crest cells. *Experimental Biology and Medicine* 10: 209–221; (c) from Bronner-Fraser M (1985) Alterations in neural crest migration by a monoclonal antibody that affects cell adhesion. *Journal of Cell Biology* 101: 610–617; (f) adapted from Young HM, Hearn CJ, Farlie PG, et al. (2001) GDNF is a chemoattractant for enteric neural cells. *Developmental Biology* 229: 503–516.

DNA-binding proteins MNR2, Lim3, and HB9. Few molecular markers that distinguish visceral from somatic motor neurons have been identified, but expression of BMP-5 appears to be specific for neurons in the Column of Terni. Although mechanisms underlying the segment-specific projection pattern observed in sympathetic preganglionic fibers are not completely understood, it appears that soluble signals present in the somitic mesoderm contribute to segment-specific identity. Correct patterning is necessary in order to ensure that preganglionic input is received by the appropriate postganglionic neurons. Following early phases of development, preganglionic inputs are necessary for regulation of neurotransmitter biosynthesis in postganglionic neurons.

Postganglionic sympathetic neurons are located in three anatomically distinct sets of ganglia. The majority of the principal sympathetic neurons are located in the paravertebral ganglia that are bilaterally distributed along each side of the spinal cord. The prevertebral ganglia are located at the midline; these cells lie anterior (ventral) to the dorsal aorta. The previsceral (terminal) ganglia, in a pattern more similar to parasympathetic ganglia, are situated in close proximity to several organs in the pelvis, mainly the bladder and rectum. Interestingly, although noradrenaline (norepinephrine) is the major neurotransmitter utilized by principal sympathetic ganglion neurons, similar to that found in the enteric nervous system, there is chemical coding of the neurotransmitters and neuropeptides expressed by sympathetic ganglion neurons. This chemical template is dependent upon the ganglia in which the cell bodies reside and the identity of their target. Neurons that innervate sweat glands (sudomotor) and the periosteum of bone are cholinergic; these represent 10–15% of sympathetic ganglion neurons. There is now general agreement that for some rather specialized neurons that neurotransmitter identity is plastic and changes with development once target innervation is complete. In addition to the now classic case of the rodent footpad, where the neurons are initially noradrenergic and become cholinergic in response to target-derived factors, it appears that the expression of neuropeptides in the prevertebral and previsceral ganglia also depends upon target for their expression. Neuropeptides, including neuropeptide Y (NPY) and vasoactive intestinal polypeptide (VIP), have been co-localized primarily with noradrenaline (NPY) and acetylcholine (VIP), and additional molecules such as substance P, dopamine, and serotonin have been found co-localized with these neurotransmitters in neurons of the previsceral and prevertebral chains.

Choice of neurotransmitter is an attribute that is acquired early in development and is dependent upon instructive cues encountered by the NC-derived precursor cells during their migration from the neural tube, as well as at sites along the dorsal aorta where these cells segregate into ganglia (**Figure 5**). There are two sources of instructive soluble signals required for appropriate migration and differentiation of NC-derived precursor cells as sympathetic ganglion neurons. The neural tube synthesizes and secretes an as yet unidentified member of the TGF-β family that is necessary for migration and differentiation as sympathetic neurons. BMP is a second required instructive factor synthesized and secreted by the dorsal aorta. BMP is an essential determinant of the noradrenergic phenotype. BMP is a proximal signal that induces expression of a network of DNA-binding proteins required for both neurogenesis and cell type-specific expression of noradrenergic marker genes. The core network of DNA-binding proteins that support differentiation of NC-derived precursor cells as sympathetic ganglion neurons includes the homeodomain (HD) proteins Phox2b and Phox2a, the basic helix–loop–helix (bHLH) DNA-binding proteins achaete-scute homolog 1 (MASH1 in mouse, CASH1 in chick), and HAND2, and the zinc finger protein GATA3 (GATA2 in chick). These proteins function together for both cell determination and differentiation; it is common to find that bHLH and HD proteins function together in networks of cross-regulated DNA-binding proteins. An essential function for both Phox2b and HAND2 has been demonstrated by gene knockout in mice. Deletion of Phox2b results in loss of autonomic neurons in each branch of the ANS; this DNA-binding protein is a master regulator of specification for the ANS. Deletion of HAND2 results in loss of sympathetic ganglion neurons with no apparent effect on parasympathetic ganglion neurons. Interestingly, loss of HAND2 affects migration and differentiation of NC-derived precursors that will contribute to the enteric nervous system. The differential effects on HAND2 in the generation of sympathetic, compared to enteric, neurons suggests that additional instructive cues necessary for specification of neurotransmitter phenotypic characteristics remain to be elucidated. The roles of various molecules in sympathetic, parasympathetic, enteric, and sensory neuron differentiation are illustrated in **Figure 6**.

Although neurotransmitter specification and expression are early developmental events, this aspect of phenotypic choice and expression is remarkably plastic. The adult neurotransmitter phenotype for some neurons is not achieved prenatally. Sympathetic ganglion neurons innervating the eccrine sweat glands and periosteum are noradrenergic prior to birth. Expression of noradrenergic characteristics does not depend upon target innervation; this sets

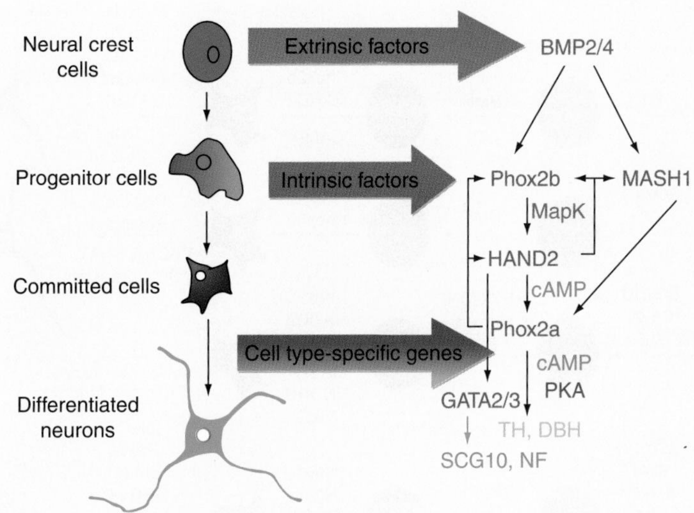

Figure 5 Specification and differentiation of peripheral autonomic neurons are dependent upon the interplay between cell extrinsic and cell intrinsic factors. Initial instructive cues from the neural tube influence neural crest cells that then respond to bone morphogenetic proteins (BMP2/4) derived from the dorsa aorta or retro-orbital mesenchyme. Induction of paired-like homeobox 2b (Phox2b) and mammalian achaete-scute homolog 1 (MASH1) proteins is followed by the induction of heart and neural crest derivatives-expressed protein 2 (HAND2) and Phox2a, resulting in expression of genes encoding pan-neuronal proteins (SCG10, superior cervical ganglion-10 protein (also called stathmin-like 2 protein); NF, neurofilament protein) and cell type-specific proteins (TH, tyrosine hydroxylase; DBH, dopamine β-hydroxylase; choline acetyltransferase and vesicular acetylcholine transporter, not shown). The recognized transcription factors required for the differentiation of noradrenergic sympathetic and cholinergic parasympathetic neurons are the same, with the notable exception of HAND2. Consequently, it is possible that this diversity could be in part the result of exclusive expression of HAND2 in precursors of noradrenergic sympathetic ganglion neurons. Cross-regulation of transcription factor expression suggests patterns of regulation based on generation of local gradients. MapK, mitogen-activated protein kinase; cAMP, cyclic adenosine monophosphate; PKA, protein kinase A; GATA2/3, GATA transcription factors 2 and 3. From Howard MJ (2005) Mechanisms and perspectives on differentiation of autonomic neurons. *Developmental Biology* 277: 271–286.

these neurons apart from other cholinergic neurons whose neurotransmitter phenotype is determined early in development and does not depend upon instructive target-derived signals. Interestingly, in the adult, the neurotransmitter phenotype of these neurons is altered following target innervation. This switch in neurotransmitter expression from noradrenergic to cholinergic depends upon retrograde transport of a target-derived factor that affects expression of a number of genes; the cholinergic differentiation factor has not been definitively identified, but is likely cardiotrophin 1. In addition to expression of choline acetyltransferase (ChAT), these neurons also express the vesicular acetylcholine transporter (VAChT), the choline transporter (ChT), as well as VIP. The expression of each of these molecules can be increased (induced) in sympathetic ganglion neurons by a variety of cytokines signaling through gp130. Expression of ChAT and VAChT is coordinately regulated and these two gene products comprise the cholinergic gene locus. Expression of both the cholinergic gene locus and VIP is regulated, in part, by the zinc finger DNA-binding protein REST. Although expression of ChAT, VAChT, ChT, and VIP appear to increase in parallel, relatively little is known about the underlying transcriptional mediators or how expression of these genes is coordinately regulated.

As development proceeds, sympathetic ganglion neurons become dependent upon nerve growth factor (NGF) for their maturation and survival. Precursor cells and young neurons express the neurotrophin receptor TrkC and have some dependence upon neurotrophin-3 (NT-3) for their survival. Mature postmitotic sympathetic ganglion neurons express TrkA and later acquire dependence upon NGF. These trophic factors are target derived and are retrogradely transported to the neuron somata, where they affect many cellular processes involved in cell death, cell survival, and maturation. Adult sympathetic ganglion neurons depend upon both NGF and NT-3. Together these trophic factors affect neuron survival. In addition, NGF influences axon targeting, the size and extent of dendritic arbors, and expression of the biosynthetic enzymes required for synthesis of noradrenaline. A little appreciated aspect of aging, dependent in part upon NGF and NT-3, is a decrease in levels of noradrenaline as well as a decrease in the innervation of many sympathetic ganglion target tissues, including heart, spleen, and cerebral blood vessels. Although not well understood, it appears that there are

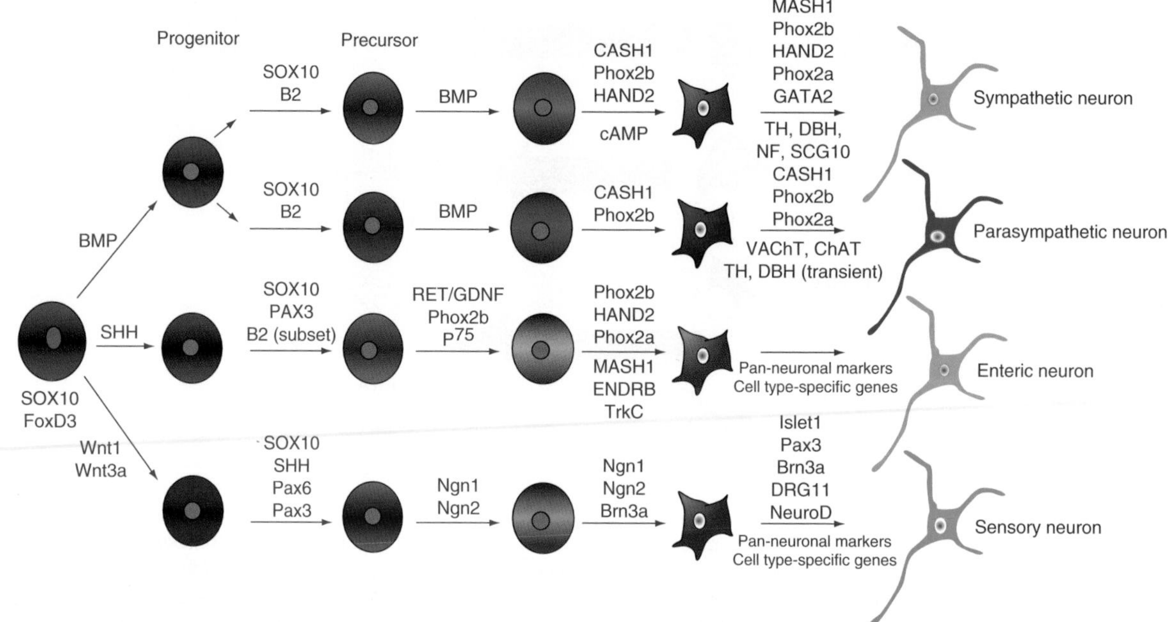

Figure 6 Analyses *in vitro* and *in vivo* have identified a series of growth factors and transcriptional regulators (markers) affecting different stages of neurogenesis of neural crest-derived progenitor cells. This schematic diagram summarizes work from many laboratories and includes some information not explicitly described in the body of the text. The compiled data provide a roadmap for following important hallmark events in the development of autonomic, enteric, and sensory neurons. Neural crest cells segregated from the neuroepithelium can be identified by the expression of FoxD3 and Sox10 (among other markers); cells expressing FoxD3 give rise to neurons and not melanocytes. Sox10 maintains multipotency in neural crest-derived cells as well as neurogenic potential. In the enteric nervous system, Sox10 and Pax3 together regulate Ret, which is required for normal development of these neurons. Progenitor cells differentiate into sympathetic, parasympathetic, enteric, or sensory neurons, in part dependent upon instructive signals encountered early at or near the time of egress from the neural tube. Additionally, extrinsic cues encountered during migration or at sites where neural crest-derived cells differentiate influence patterns of gene expression. In autonomic ganglia, expression of HAND2 appears to select cells as noradrenergic sympathetic ganglion neurons, as well as functioning in cell type-specific gene expression. In the sensory neuron lineage, the POU domain transcription factor Brn3a, expressed downstream of neurogenins 1 and 2 (Ngn1, Ngn2), regulates a large array of genes influencing cell death, neurotransmitter expression, and axon guidance. The signaling molecule sonic hedgehog is necessary for the expression of neurogenin. In the enteric nervous system, HAND2 is expressed downstream of Phox2b in all segments of the developing gut; the function of HAND2 in development of enteric neurons is unknown. The neurotrophin receptor TrkC is expressed early by neural crest-derived cells, the potential of which is restricted to neuronal or glial lineages as well as a subset of enteric neurons. From Howard MJ (2005) Mechanisms and perspectives on differentiation of autonomic neurons. *Developmental Biology* 277: 271–286.

alterations in the transcriptional and translational regulation of expression of both NGF and NT-3 coincident with the changes in function and survival of sympathetic ganglion neurons with increased age.

Parasympathetic Ganglia: Neuronal Differentiation and Connections

Of the components of the ANS, the least is known about parasympathetic ganglia because they are small, diffusely structured ganglia embedded within the tissues that they innervate (**Figure 7**). Parasympathetic ganglia are derived from the NC; however, the detailed axial origins of many of the ganglia are not known. The cranial parasympathetic ganglia, including the ciliary, lacrimal, otic, and sphenopalatine, arise from the cranial NC rostral to the otic (ear) placode. Caudal to this, the cardiac ganglion originates from the NC near the otic placode to the third somite, overlapping the region giving rise to the enteric nervous system (see later). Most of the trunk NC contributes the sympathetic ganglia (as previously mentioned), but the lumbosacral NC also gives rise to the parasympathetic pelvic ganglia and, in birds, the ganglion of Remak near the hindgut that is thought to be an extension of the pelvic ganglia (these origins are diagrammed in **Figure 1**).

The specification and differentiation of parasympathetic ganglia have many pathways in common with sympathetic ganglia (see **Figure 6**): both rely on signaling by BMP and require the expression of the transcription factors MASH1, Phox2a, and Phox2b. Specification of the cholinergic phenotype typical of parasympathetic neurons may occur prior to migration and is reinforced by suppression of the

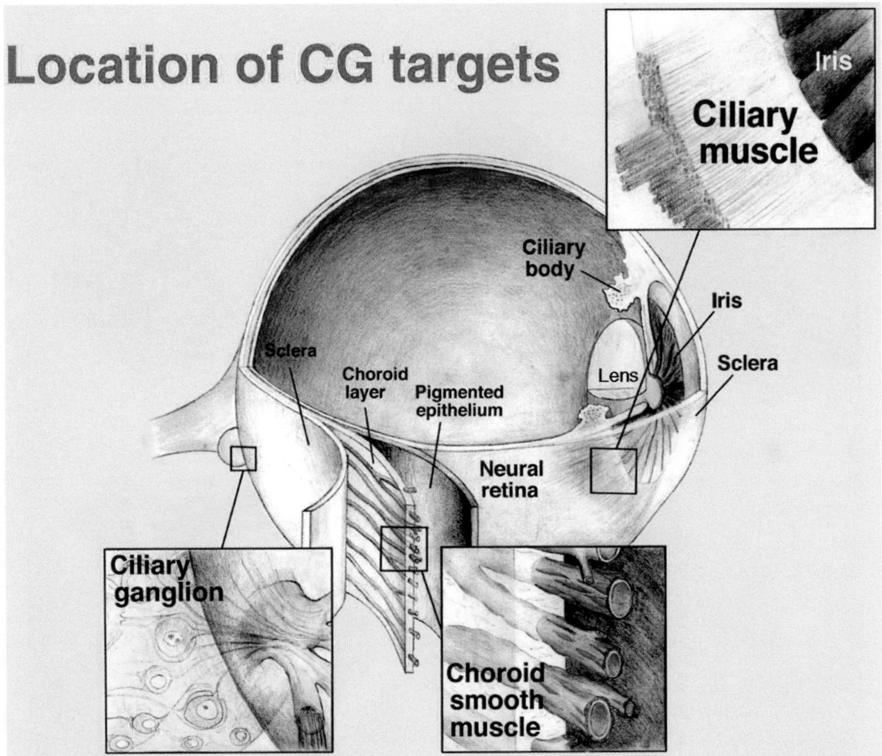

Figure 7 The ciliary ganglion (CG) and its targets of innervation, illustrating the relatively close relationship between parasympathetic ganglia and their innervation targets. The schematic shows the location of the ciliary ganglion and its target tissues, the vascular smooth muscle of the choroid layer, and the striated muscle of the iris and ciliary body. The enlargements show the details of the histological structures of the tissues. The ciliary ganglion contains small, unmyelinated neurons that innervate the vascular smooth muscle (choroid neurons) and large, myelinated neurons that innervate the iris and ciliary muscle (ciliary neurons).

transcription factor HAND2 (dHAND), which is necessary for the expression of catecholamines. In the examples that have been best studied (ciliary ganglion, cardiac ganglion, and submandibular ganglion), the precursors of parasympathetic ganglia migrate to their targets of innervation prior to the onset of organogenesis of the particular organ. As morphogenesis commences, neuronal differentiation occurs. For example, as the salivary gland epithelium branches and forms ducts, many of the neurons associated at the base of the organ rudiment are already postmitotic and extend axons as epithelial buds form (**Figure 8**). Likewise, cardiac ganglion neurons undergo neurogenesis within the developing heart, and ciliary ganglion neurons elaborate axons as the optic vesicle forms the eye and its associated structures.

This intimate differentiation of parasympathetic neurons along with their targets of innervation suggests that trophic interactions reciprocally guide development. In fact, the expression of GDNF and the closely related molecule neurturin in target tissues is essential for the differentiation and development of a number of parasympathetic ganglia. These include ciliary, otic, sphenopalatine, submandibular, lacrimal, penile, and pancreatic islet neurons. In many instances, differentiation of parasympathetic ganglia appears to require sequential action of GDNF, followed by neurturin.

By far the best-studied parasympathetic ganglion is the avian ciliary ganglion (see **Figure 7**). This ganglion contains two populations of principal neurons: ciliary neurons, which innervate the iris and ciliary muscle, and choroid neurons, which innervate the arterial smooth muscle in the choroid layer. Both populations differentiate, innervate their respective target tissues, and undergo programmed cell death in midgestation (between E6 and E14 of chick development). Target tissues control the degree of programmed cell death that occurs as well as other aspects of neuronal differentiation, such as the expression of the neuromodulatory neuropeptide somatostatin, and the expression of ion channels, such as the calcium-activated potassium channel and the nicotinic acetylcholine receptor (**Table 2**). Afferents from the accessory oculomotor nucleus of the midbrain also regulate differentiation

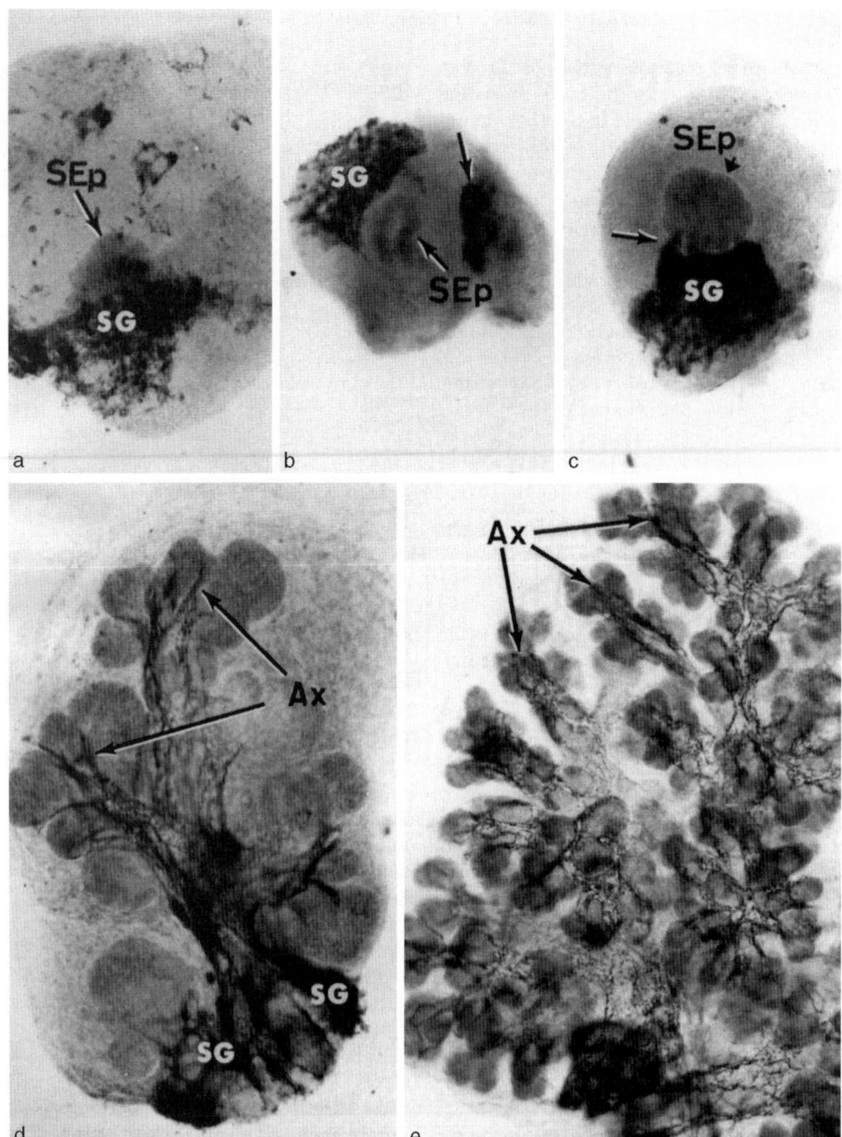

Figure 8 Early development of parasympathetic nerves matches the morphogenesis of the target, shown in whole mounts of the submandibular salivary gland stained for acetylcholinesterase activity. (a) At E12, the salivary gland epithelial bud (SEp) is surrounded by darkly staining salivary ganglion cells (SG); (b) E12, looking down on the preparation shown in a; (c) early E13, when the lobule of the SEp has just begun to form clefts (arrowhead) and axons begin to extend into the epithelium (arrow); (d) E14 gland; the epithelium has grown and has begun to branch extensively, and axons (Ax) course over the epithelium and travel in the clefts between the buds; (e) a portion of an E15 gland; axon outgrowth continues to parallel the growth of the epithelium. From Coughlin MD (1975) Early development of parasympathetic nerves in the mouse submandibular gland. *Developmental Biology* 43: 123–139.

of the ciliary ganglion neurons. These inputs influence cell death and differentiation through nicotinic cholinergic activation of ciliary neurons. Recently, nicotinic activation has been tied to the expression of a chloride transporter, which changes the magnitude of the chloride gradient across the plasma membrane, thereby influencing γ-aminobutyric acid (GABA)ergic signaling in neurons as well as axonal morphology. Whether the principles of development uncovered by studies of the ciliary ganglion apply to other parasympathetic ganglia is a matter of ongoing investigation.

Enteric Ganglia: Neuronal Differentiation and Connections

The enteric nervous system (ENS) is by far the largest division of the ANS, and the most complex. The enteric nervous system consists of numerous small ganglia placed as nodes in a lattice of interconnections. Two

Table 2 Anterograde and retrograde influences of development in the avian ciliary ganglion

Effector molecule	*Mode*	*Neuronal property*	*Effect*
ACh (all nAChRs)	Anterograde	Chloride transporter	Increase
Neuregulin	Anterograde	Ca^{2+}-activated K^{+} channels	Increase
Neuregulin	Anterograde	nAChRs	Increase
ACh (α7 nAChRs)	Anterograde	Neuronal survival	Decrease
ACh (all nAChRs)	Anterograde	Neuronal survival	Increase
CNTF	Retrograde	Neuronal survival	Increase
GDNF	Retrograde	Neuronal survival	Increase
TGF-β4	Retrograde	Ca^{2+}-activated K^{+} channels	Increase
TGF-β3	Retrograde	Ca^{2+}-activated K^{+} channels	Decrease
Activin	Retrograde	Somatostatin	Increase
Unknown	Retrograde	nAChRs	Increase

such layers occur, the myenteric plexus and the submucosal plexus, with extensive radial connections. The ENS has a full reflex circuitry with sensory, motor, and interneurons, and to a degree can function without CNS input, which derives from visceral motor neurons in vagal and sacral levels of the CNS (the same levels from which the enteric NC precursors arise). At least 15 neurotransmitters (but not noradrenaline) and neuromodulatory peptides occur in mature enteric ganglia, and each small ganglion has several neuron types differentiated by neurochemical code, morphology, and projection pattern. How so many different neuronal types are specified in each ganglion is not well understood. As explored by Michael Gershon and co-workers, differences in timing of differentiation of different neuron classes could be involved, and neuron type-related differences in median birth days does occur, as determined by ascertaining neuronal birth days using pulse-delivered tritiated thymidine or bromodeoxyuridine (BrdU). However, neurons of each class are born over a time range, and the ranges for different classes overlap strongly. Moreover, some neuron types, such as 5-hydroxytryptamine neurons, are specified prior to or early in NC migration. As with other parts of the ANS, Phox2b is a decisive requirement, and deletion of HAND2 has revealed its essential role in the specification and differentiation of enteric nervous system neurons that express VIP.

Differentiation of enteric neurons has similarities to that of the neurons of the rest of the ANS (**Figure 6**). For example, it is marked by the loss of the transcription factor Sox10 and the appearance of the neuron-specific transcription factor family member HuC/D. Sox10 remains present in enteric precursors and enteric glia, and the latter also express glial fibrillary acidic protein (GFAP) like CNS glia, but unlike glia of the other ANS divisions. One of the first neuron classes to appear is nitric oxide synthase-expressing neurons, and in mice these cells are initially transiently catecholaminergic, showing dopamine (DOPA) metabolism such as DOPA decarboxylase and dopamine-β-hydroxylase. These early differentiating neurons cease to migrate but extend axons distally, in parallel with the migration of vagal enteric NC cells.

Unlike the rest of the ANS and CNS, the enteric neuron population size does not normally overshoot and then undergo apoptotic pruning. Instead, the population proliferates up to a density set by the gut tissue. Instrumental in this so-called logistical growth is GDNF. This factor, via its receptor Ret on NC cells and its co-receptor GDNF family receptor-α (GFR-α) has a complex role, since it is not only a mitogen for enteric NC cells, but also induces differentiation into postmitotic neurons. *In vitro* experiments suggest that the small peptide endothelin-3, via its receptor EdnrB on NC cells, retards the differentiation effect of GDNF while preserving or even increasing its mitogenic function. Mutation of all the genes for these can cause Hirshsprung's disease (**Table 1**) either by directly decreasing GDNF mitogenic signaling (for mutations in *RET*, *GFRA1*, and *GDNF*) or by removal of the brake on GDNF-stimulated mitotic withdrawal (for mutations in *EDN3*, *EDNRB*, and *ECE1*).

See also: Autonomic Nervous System: Neuroanatomy; Autonomic Nervous System; Enteric Nervous System Development; Enteric Nervous System: Neurotrophic Factors; Nerve Growth Factor; Neural Crest.

Further Reading

Anderson RB, Newgreen DF, and Young HM (2006) Neural crest and the development of the enteric nervous system. *Advances in Experimental Medicine and Biology* 589: 181–196.

Bertrand N, Castro DS, and Guillemot F (2002) Proneural genes and the specification of neural cell fates. *Nature Reviews Neuroscience* 3: 517–530.

Bronner-Fraser M (1985) Alteration in neural crest migration by monoclonal antibody that affects cell adhesion. *Journal of all Biology* 101: 610–617.

Caughlin MD (1975) Early development of parasympathetic nerves in the mouse submandibalar gland. *Developmental Biology* 43: 123–139.

Enomoto H, Heuckeroth RO, Golden JP, et al. (2000) Development of cranial parasympathetic ganglia requires sequential actions of GDNF and neurturin. *Development* 127: 4877–4889.

Guillemot F, Lo L-C, Johnson JE, et al. (1993) Mammalian achaete-scute homolog-1 is required for the early development of olfactory and autonomic neurons. *Cell* 75: 463–476.

Howard MJ (2005) Mechanisms and perspectives on differentiation of autonomic neurons. *Developmental Biology* 277: 271–286.

Kasemeier-Kulesa JC, Kulesa PM, and Lefcort F (2005) Imaging neural crest cell dynamics during formation of dorsal root ganglia and sympathetic ganglia. *Development* 132: 235–245.

LeDouarin NM, Creuzet S, Couly G, et al. (2004) Neural crest cell plasticity and its limits. *Development* 131: 4637–4650.

Muller F and Rohrer H (2002) Molecular control of ciliary neuron development: BMPs and downstream transcriptional control in the parasympathetic lineage. *Development* 129: 5707–5717.

Newgreen DF (1985) Control of the timing of Commencement of migration of embryonic neural crest cells. *Experimental Biology and Medicine* 10: 209–221.

Newgreen DF and Erickson CA (1986) The migration of neural crest cells. *International Review of Cytology* 103: 89–145.

Nishi R (2003) Target-mediated control of neural differentiation. *Progress in Neurobiology* 69: 213–227.

Simpson MJ, Landman KA, Hughes BD, et al. (2006) Looking inside an invasion wave of cells using continuum models: Proliferation is the key. *Journal of Theoretical Biology* 243: 343–360.

Stanke M, Duong CV, Pape M, et al. (2006) Target-dependent specification of the neurotransmitter phenotype: Cholinergic differentiation of sympathetic neurons is mediated *in vivo* by gp130 signaling. *Development* 133: 141–150.

Young HM, Bergner AJ, Anderson RB, et al. (2004) Dynamics of neural crest-derived migration in the embryonic mouse gut. *Development Biology* 270: 455–473.

Young HM, Hearn CJ, Farlie PG, et al. (2001) GDNF is a Chemo attractant for enteric neural cells. *Developmental Biology* 229: 503–516.

Autonomic Nervous System: Cardiovascular Control

J A Armour, University of Montreal, Montreal, QC, Canada

Introduction

The function of the neuronal hierarchy involved in cardiovascular control is ultimately to match cardiac output to whole body blood flow demands. The peripheral and central components that make up this hierarchy interact on an ongoing basis to coordinate regional cardiac indices such that regional demands for blood flow are met in an efficient manner. Dysfunction of any population within this hierarchy can result in cardiac or vascular malfunction, such as the induction of atrial or ventricular arrhythmias or the genesis of essential hypertension.

The cardiovascular neuronal hierarchy is a massively parallel and, for the most part, stochastic control system such that stable control usually occurs in the absence of obvious cause and effect. Apparently that is because the individual components that make up this hierarchy display emergent properties, while interacting in a highly optimized fashion to tolerate normal cardiovascular perturbations. On the other hand, the functional interconnectivity of the hierarchy is so organized that the whole can be catastrophically disabled by cascading failures initiated by sometimes relatively minor abnormal inputs. In this article, an overview of the anatomy and the function of its afferent and efferent components are presented first, followed by how these neurons putatively interact to coordinate autonomic motor neuron outputs to the heart and peripheral vasculature. It is proposed that defining the relevance of each of its populations may be essential to understanding the role of the whole during the evolution of cardiovascular disease.

Peripheral Autonomic Neurons Regulating the Heart and Major Blood Vessels

Cardiovascular Afferent Neurons

Pain associated with myocardial ischemia is frequently referred to a patient's left upper limb and/or anterior thoracic wall. Thus, it has been assumed that most cardiac afferent neurons are located in left-sided, cranial thoracic dorsal root ganglia. Anatomical evidence indicates that cardiac afferent neurons are distributed relatively evenly throughout both nodose ganglia and the right and left C7 to T4 dorsal root ganglia. Cardiac afferent neurons are also located in intrathoracic ganglia, including those intrinsic to the heart. The sensory neurites of such neurons are distributed in the walls of the major veins adjacent to the heart, all four chambers of the heart, and along the arch of the aorta. Select ones also lie concentrated in the region of the carotid bulbs.

Cardiac Motor Neurons

Cardiac parasympathetic efferent neurons The somata of parasympathetic efferent preganglionic neurons that synapse with cholinergic efferent postganglionic neurons on the heart are located primarily ventral lateral to the nucleus ambiguous of the medulla; lesser numbers are located in the dorsal motor nucleus and the intermediate zone in between these two medullary nuclei. Cardiac efferent preganglionic neurons in individual medullary loci project axons to parasympathetic efferent postganglionic neurons distributed throughout all major atrial and ventricular ganglionated plexuses.

Cardiac sympathetic efferent neurons Sympathetic efferent preganglionic neurons in the spinal cord that are involved in cardiac regulation project axons via the T1 to T5 rami to synapse with cardiac sympathetic efferent postganglionic neurons located in the cranial poles of the stellate ganglia, throughout the right and left middle and superior cervical ganglia, as well as in mediastinal ganglia adjacent to the heart. They also project to sympathetic efferent postganglionic neurons in intrinsic cardiac ganglia.

Intrathoracic Local Circuit Neurons

Interneurons are interposed between cardiac afferent and efferent neurons in the various intrathoracic ganglia. As these project axons to neurons not only within their own ganglia, but also those in other intrathoracic ganglia, they have been called local circuit neurons. Some of these larger-diameter neurons (i.e., up to ~30 μm) form rosettes within intrathoracic ganglia. The somata of many of them are frequently found in the periphery of an intrinsic cardiac ganglion, their centrally projecting dendrites lying adjacent to one another to form synaptic connections with other neurons of that rosette. Such anatomy may represent the substrate for local information processing within one ganglion. Many intrathoracic neurons, including those on the heart, display immunoreactivity to multiple peptides. These data indicate that intrathoracic cardiac neurons can be involved in a multiplicity of interactions that collectively coordinate regional cardiac indices (cf. section on cardiac neuroaxis).

Peripheral vascular sympathetic efferent neurons Sympathetic efferent preganglionic neurons in the spinal cord synapse with sympathetic efferent postganglionic neurons located not only in the superior cervical, middle cervical, and stellate ganglia, but abdominal sympathetic chain and midline ganglia. Many of their sympathetic efferent postganglionic neurons project axons to the major vessels (arteries and veins) in the limbs, for instance, to control regional blood flow throughout the body.

Physiology of Cardiovascular Neurons

Cardiovascular Afferent Neurons

Much of the activity generated by neurons throughout the cardiovascular neuroaxis is dependent on tonic inputs arising from the various cardiovascular sensory neurons depicted in the earlier sections. Thus, the activity generated by many central and peripheral cardiac neurons is ultimately reflective of the transduction capabilities of cardiac and major vascular mechano- and/or chemosensory neurons. For instance, most (~75%) nodose ganglion cardiac afferent neurons transduce chemical stimuli. About 35% respond to mechanical stimuli while fewer (~10%) are multimodal in nature, responding to mechanical and chemical stimuli. In contrast, the vast majority of dorsal root ganglion cardiac afferent neurons (~95%) are multimodal in nature. That capacity permits ventricular sensory neurites associated with individual dorsal root ganglion afferent neurons to transduce not only regional ventricular dynamics (**Figure 1**), but altered release of local chemicals such as occurs during ventricular ischemia as well. The varied transduction capabilities displayed by individual dorsal root ganglion afferent neurons presumably permit such a relatively limited population of afferent neurons to simultaneously transduce multiple cardiac signals to spinal cord neurons.

Cardiac sensory neurites are also associated with afferent neuronal somata in intrathoracic extracardiac (**Figure 2**) and intrinsic cardiac ganglia. Most of these are multimodal in nature, responding thereby to regional ventricular ischemia. Limited populations of intrathoracic cardiac afferent neurons also transduce the postischemia (reperfusion) state, becoming activated early on during reestablishment of regional coronary arterial blood flow.

With respect to cardiac chemotransduction, adenosine is known to be released in increasing quantities from the ischemic myocardium. This chemical activates the sensory neurites associated with many ischemia-sensitive cardiac afferent neurons. In fact, it has been proposed that purinergic dorsal root ganglion cardiac afferent neurons are involved in the genesis of symptoms in patients subjected to myocardial ischemia.

Numerous mechanosensory neurites associated with other intrathoracic afferent neuronal somata are located on the major intrathoracic vessels or coronary arteries. These transduce local vascular wall deformation subsequent to underlying blood pressure changes

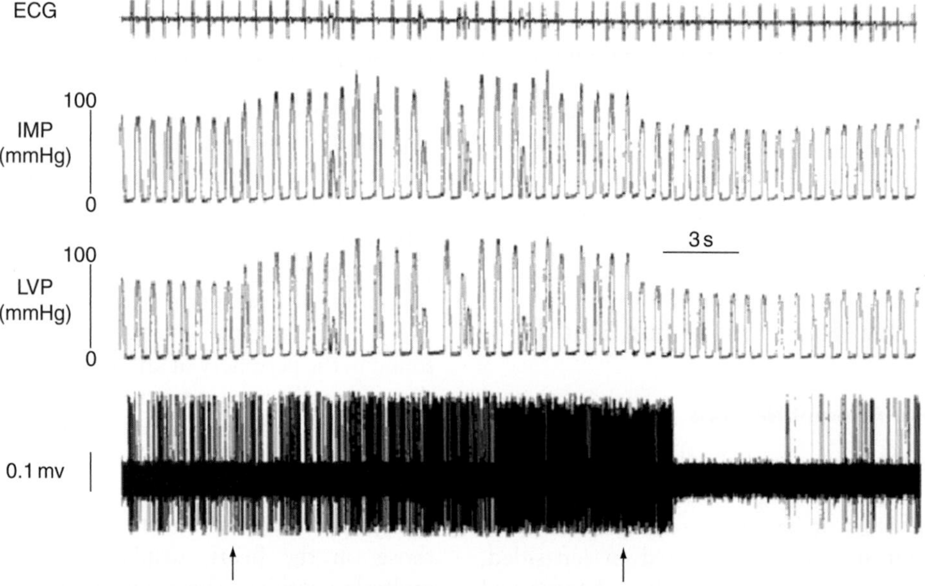

Figure 1 Behavior of a T3 left-sided dorsal root ganglion afferent neuron associated with left ventricular mechanosensory neurites in response to altered left ventricular dynamics. In this anaesthetized canine preparation, afferent neuronal activity was increased by partial occlusion of descending aorta (between arrows, below), as left ventricular intramyocardial (IMP) and chamber (LVP) systolic pressures rose. The lowest trace represents the afferent neuronal activity. ECG, electrocardiogram; LVP, left ventricular cavity pressure.

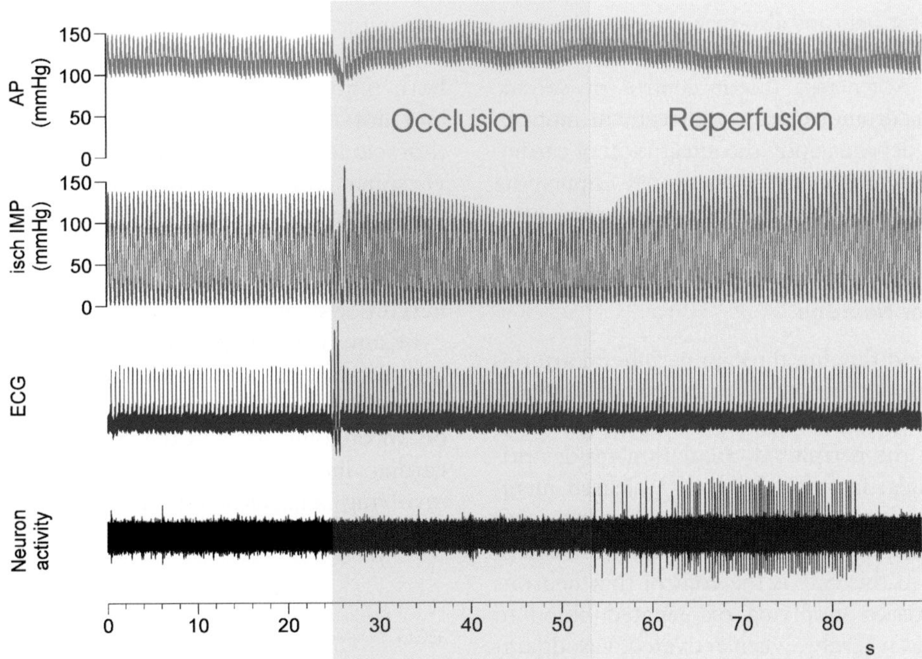

Figure 2 Left-sided middle cervical ganglion afferent neurons associated with left ventricular sensory neurites were activated in an anesthetized canine immediately following (early reperfusion; green section), but not during occlusion of its left anterior descending coronary artery. Occlusion (pink section lasting for 28 s) reduced left ventricular wall regional dynamics (ischemic LV intramyocardial systolic pressure), while aortic pressure (AP) remained stable. The lowest trace represents raw middle cervical ganglion neuronal activity. (These data are presented in Acrobat Distiller format for video viewing with sound.)

with the degree of fidelity that many carotid artery baroreceptor neurons projecting to the nucleus solitarius exhibit. In sum, cardiovascular afferent neuronal somata located in various ganglia continuously transduce not only arterial dynamics (i.e., carotid artery and aortic baroreceptors), but also the heart's regional mechanical and chemical environment.

Intrathoracic Local Circuit Neurons

The cardiac and major vascular milieu transduced by intrathoracic afferent neurons influence some cardiac motor neurons on a beat-to-beat basis, primarily via intrathoracic local circuit neurons. Intrathoracic, extracardiac, and intrinsic cardiac local circuit neurons process information arising not only from cardiac and major intrathoracic vascular mechanosensory and/or chemosensory afferent neurons, but also descending information from sympathetic (spinal cord) and parasympathetic (medullary) efferent preganglionic neurons. Due to the fact that some receive direct inputs from cardiac and aortic mechanosensory neurons, they may display activity that is phase-related to the cardiac cycle. Their capacity to processing cardiac sensory inputs persists even after all of their central neural connections have been chronically eliminated. That is one reason why it has been proposed that much of the daily transduction of cardiovascular milieu to cardiac motor neurons can occur independent of the central nervous system, relying to a considerable degree on intrathoracic local circuit neurons.

Cardiac Motor Neurons

Parasympathetic efferent neurons Medullary efferent preganglionic neurons innervate parasympathetic postganglionic neurons distributed throughout the atrial and ventricular ganglionated plexuses. When activated, these cholinergic motor neurons suppress not only atrial rate and force and atrioventricular nodal conduction, but also ventricular contractile force. Cardiac cholinergic efferent neurons are under the particular, but not exclusive, reflex control of arterial mechanosensory inputs (particularly those arising from the carotid artery mechanosensory neurites). Thus, some of these neurons generate activity that reflects changing arterial pressure. Indeed, their effectiveness in modulating cardiac rate frequently relates to arterial dynamics as a result of the short-latency reflexes that they initiate.

Sympathetic efferent neurons Sympathetic efferent postganglionic neurons in each intrathoracic ganglion receive inputs from sympathetic efferent preganglionic neurons in the caudal-cervical and cranial-thoracic spinal cord. These postganglionic neurons, in turn, innervate widely divergent regions of the heart. Cardiac adrenergic motor neurons also receive inputs from intrathoracic sensory neurons via intrathoracic local circuit neurons. Adrenergic motor control of cardiac chronotropism, dromotropism, and regional inotropism ultimately depends upon the integration of cardiovascular sensory and central neuronal inputs via complex information processing within the intrathoracic neuroaxis.

Vascular Motor Neurons

Control of blood flowing through peripheral arteries and veins resides in intrinsic (local metabolic effects) and extrinsic (neurohumoral) mechanisms. Classically, the control of the peripheral circulation resides primarily with local factors (cf. locally released metabolites). Sympathetic efferent neurons that innervate arteries and veins act to constrict most major vessels. An exception to this rule is the control that neurons exert over arteries supplying oxygenated blood to skeletal muscles whereby, when activated, vasodilatation occurs that increases local perfusion.

In the coronary circulation, intrinsic mechanisms predominate. In the skin extrinsic control predominates, while in skeletal muscles both are important. Arterial baroreceptors concentrated on the carotid arteries and along the arch of the aorta, acting via medullary, spinal cord, and intrathroacic reflexes, play a key role in short-term adjustments of arterial pressure. Central chemosensory neurons, when activated, for instance, by local hypoxemia, exert profound influences on arterial resistance and venous capacitance, in addition to cardiac work. In order to maintain normal cardiovascular status, a balance of these cardiac and vascular reflex forces must be achieved via ongoing neurohumoral balance.

The Cardiovascular Neuroaxis

The target organ nervous system constantly interacts with neurons in intrathoracic and cervical ganglia, as well as central neurons, to process cardiovascular sensory information via multi-tiered reflexes. Excitatory and inhibitory reflexes within the cardiac neuroaxis coordinate cardiac motor neuronal outputs that regulate regional cardiac indices on a beat-to-beat basis. Neurons in the intrathoracic neuronal hierarchy interact via a host of chemicals that include acetylcholine, butyrylcholine, alpha- and beta-adrenoceptor agonists, histamine, nitric oxide donors, peptides such as angiotensin II (**Figure 3**), purinergic agents (adenosine and adenosine 5′-triphosphate (ATP)), excitatory and inhibitory amino acids, and serotonin.

Some intrinsic cardiac local circuit neurons receive inputs from both sympathetic and parasympathetic efferent preganglionic neurons. In other words, populations of neurons on the heart process inputs from both 'limbs' of the autonomic efferent nervous system and not necessarily in a reciprocal fashion. Intrathoracic local circuit neurons receive indirect inputs via spinal cord neurons derived from sensory neurites in extrathoracic tissues (i.e., in tissues of the neck or each of the four limbs) such that alterations in the extrathoracic milieu can influence the intrinsic cardiac nervous system.

In sum, it has been proposed that the mammalian intrinsic cardiac nervous system functions collectively as a 'little brain,' processes these multiple inputs on an ongoing basis in the coordination of regional cardiac indices. Short-latency reflexes (40–100 ms) involving neurons within and near the target organ influence cardiac motor neurons throughout each

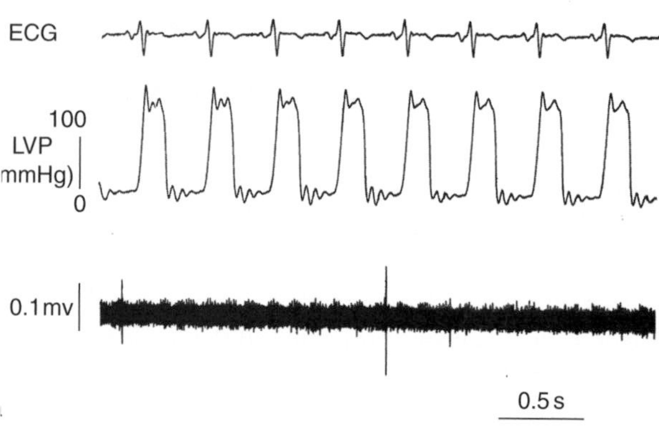

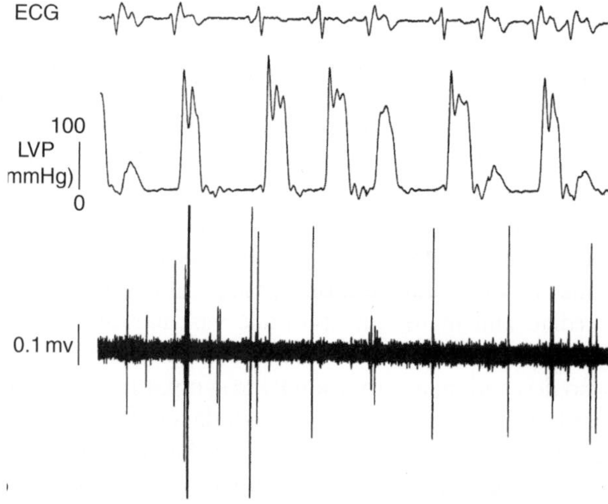

Figure 3 Angiotensin II-induced changes in intrinsic cardiac neuronal activity in an anesthetized, canine preparation. Spontaneous activity generated by canine right atrial neurons *in situ* was enhanced following local application of angiotensin II (10 μl of a 10 μm solution over a 5 s period between panels). Note that previously inactive neurons became active and that ventricular ectopy was induced. ECG, Lead II electrocardiogram; LVP, left ventricular cavity pressure. The lowest trace represents neuronal activity.

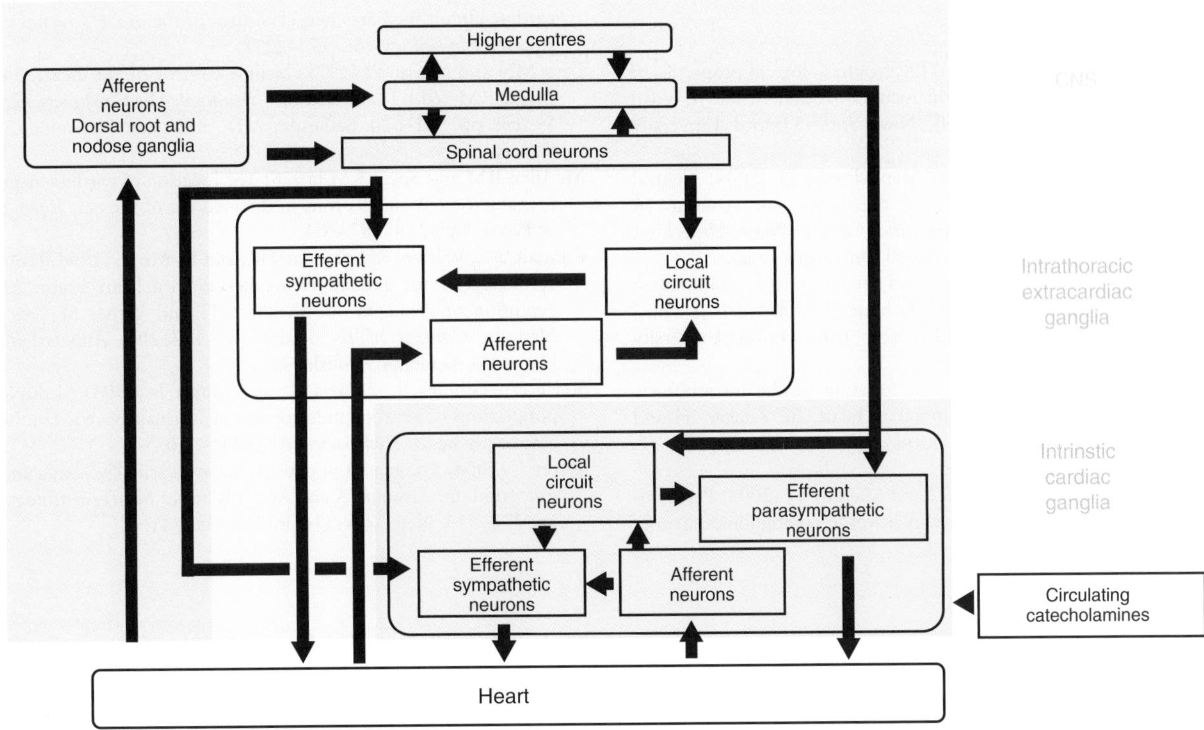

Figure 4 Schematic representation of the locations of afferent, local circuit, and efferent (parasympathetic and sympathetic) neurons in peripheral autonomic ganglia involved in regulating cardiac function, along with some of their putative interconnections. This peripheral neuronal hierarchy is under the tonic influence of medullary and spinal cord neurons, as well as circulating catecholamines.

phase of the cardiac cycle. On the other hand, longer latency cardiocardiac reflexes (300 ms to 2 s) involving central neurons influence cardiodynamics over longer timescales, that is, during subsequent cardiac cycles.

Clinical Implications

All of the above indicate that the cardiac neuroaxis can be represented as a massively parallel and, for the most part, stochastic control system that is involved in transducing the cardiac mechanical and chemical milieu to cardiac motor neurons on a beat-to-beat basis (**Figure 4**). The varied intrathoracic neuronal interactions so engendered are under the tonic influence of central cardiovascular reflexes that rely to a considerable extent on arterial baroreceptor initiated reflexes. The unforeseen relevance of such intrathoracic reflexes lies in their stability such that relatively little demand may be placed on central neurons in the daily maintenance of adequate cardiac output. In fact, the interactions that occur among their varied neural elements appear to be optimized to tolerate normal cardiac perturbations. As individual units of the whole collectively display emergent properties, the whole cannot be represented as a simple positive and negative reflex control system.

On the other hand, such an organizational arrangement means that it can be catastrophically disabled by cascading failures initiated by relatively minor abnormal inputs as that is not a design feature of the system. Reductions in regional coronary arterial blood supply affect the function of various intrinsic cardiac neuronal populations in either a direct (ischemic damage) or indirect (altered sensory inputs) manner. The latter occurs when regional cardiac sensory inputs arising from an ischemic myocardium are transduced throughout this hierarchy. Excessive activation of some populations may have devastating effects on the capacity of cardiac motor neurons to coordinate regional cardiac indices. For instance, it has been proposed that excessive activation of inputs to the intrinsic cardiac nervous system may be a predisposing factor for cardiac arrhythmia formation (cf. **Figure 3**). Because neurons within this hierarchy communicate in part via angiotensin II or β-adrenergic receptors, therapy targeting these receptors may indeed modify cardiomyocyte function not only directly but also indirectly by influencing the cardiac neuroaxis.

See also: Autonomic Nervous System: Central Cardiovascular Control; Cardiovascular Function: Central Nervous System Control; Cerebrovascular Disease; Cotransmission; Noradrenaline; Parasympathetic Nervous System; Sympathetic Nervous System.

Further Reading

Adams DL and Cuevas J (1994) Electrophysiological properties of intrinsic cardiac neurons. In: Armour JA and Ardell JL (eds.) *Neurocardiology*, pp. 1–60. New York: Oxford University Press.

Andresen MC, Kuntz DJ, and Mendelowitz D (2004) Central nervous system regulation of the heart. In: Armour JA and Ardell JL (eds.) *Basic and Clinical Neurocardiology*, pp. 187–219. New York: Oxford University Press.

Ardell JL (1994) Anatomy and function of mammalian intrinsic cardiac neurons. In: Armour JA and Ardell JL (eds.) *Neurocardiology*, pp. 95–114. New York: Oxford University Press.

Armour JA (1991) Anatomy and function of the intrathoracic neurons regulating the mammalian heart. In: Zucker IH and Gilmore JP (eds.) *Reflex Control of the Circulation*, pp. 1–37. Boca Raton, FL: CRC Press.

Gray AL, Johnson CI, Ardell JL, and Massari VJ (2004) Parasympathetic control of the heart. A novel interganglionic intrinsic cardiac circuit mediates neural control of the heart. *Journal of Applied Physiology* 96: 2273–2278.

Levy MN and Martin PJ (1979) Neural control of the heart. In: Berne RM (ed.) *Handbook of Physiology: The Cardiovascular System*, pp. 581–620. Bethesda, MD: American Physiological Society.

McAllen RM and Spyer KM (1976) The location of cardiac vagal preganglionic motorneurons in the medulla of the cat. *Journal of Physiology* 258: 187–204.

Randall WC, Wurster RD, Randall DC, and Xi-Moy S (1996) From cardioaccelerator and inhibitory nerves to a heart brain: An evolution of concepts. In: Shepherd JT and Vatner SF (eds.) *Nervous Control of the Heart*, pp. 173–200. Amsterdam: Harwood Academic Publishers.

Steele PA, Gibbins IL, Morris JL, and Mayer B (1994) Multiple populations of neuropeptide containing intrinsic neurons in the guinea-pig heart. *Neuroscience* 62: 241–250.

Sylvén C (2004) The genesis of pain during myocardial ischemia and infarction. In: Armour JA and Ardell JL (eds.) *Neurocardiology*, pp. 298–314. New York: Oxford University Press.

Autonomic Nervous System: Carotid Body and Chemoception

C A Nurse, McMaster University, Hamilton, ON, Canada
V A Campanucci, McGill University, Montreal, QC, Canada

Introduction

The mammalian carotid bodies are small, oval-shaped sensory organs that sense chemical components in arterial blood and initiate corrective reflex adjustments so as to maintain homeostasis. For example, they can detect changes in partial pressure of the blood gases O_2 and CO_2, that is, PO_2 and PCO_2, acidity (pH), osmolarity, and glucose concentrations. They can also sense temperature, further emphasizing their role as polymodal sensors of blood-borne stimuli. The bilateral location of these organs at the bifurcation of the common carotid artery is strategic, since they are ideally positioned to sample blood chemicals just before they reach the brain, a critical organ that is very sensitive to both O_2 and glucose deprivation. Consistent with its principal role as an arterial chemosensor, the carotid body is richly vascularized and receives an enormous blood flow, reputed to be the highest per unit weight of all body tissues and organs. It receives sensory (afferent) innervation from neurons located in the petrosal ganglion (PG) via the carotid sinus nerve (CSN), a branch of the glossopharyngeal nerve (GPN). Additionally, it receives 'efferent' innervation from autonomic neurons embedded within the GPN and CSN. The functional and structural properties of the carotid body, and the roles of these nervous elements in carotid body physiology, are discussed in more detail in subsequent sections.

Carotid Body as an Arterial Chemosensor

Since the pioneering studies of Heymans and collaborators in the 1930s, it has been well established that the carotid body is activated by a fall in oxygen tension or PO_2 (i.e., hypoxia) in arterial blood. Additionally, increases in blood PCO_2 (i.e., hypercapnia) and decreased pH (i.e., acidosis) have long been recognized as potent carotid body stimuli. These three stimuli can separately excite the carotid body, and initiate corrective changes in ventilation such that blood levels of PO_2, PCO_2, and pH are restored to normal. Interactions between PO_2 and PCO_2 are known to occur at the level of the carotid body and are synergistic, resulting in a more than additive chemosensory response when the two stimuli are combined. Though central chemoreceptors located in the brain stem are the main CO_2/pH sensors and regulators of minute-to-minute control of ventilation in the whole animal, carotid bodies appear to be the main PO_2 sensors. Moreover, they contribute to respiratory adjustments due to increased blood PCO_2, and are at least responsible for the initial changes in ventilation during metabolic acidosis, caused by changes in blood acidity at constant PCO_2 levels. These reflex responses originate at carotid body receptor cells, which are directly innervated by sensory fibers of the CSN (**Figure 1**). These receptor cells, commonly known as glomus or type I cells, release neurotransmitters that excite the afferent endings of the CSN, causing an increase in sensory discharge. This increase in discharge is relayed centrally, where it influences the 'central pattern generator,' a local neuronal network in a region of the brain stem (i.e., nucleus tractus solatarius) responsible for controlling breathing rhythm and ultimately contraction of the diaphragm muscles. Thus, conditions that result in low blood oxygen (hypoxemia), for example, during cardiovascular disease or ascent to high altitude, may cause an increase in CSN sensory discharge, especially when arterial PO_2 falls from its normal level of $\sim$95 mmHg to values below 60 mmHg. The hypoxia-induced increase in CSN sensory discharge can be readily demonstrated *in vitro*, using an isolated but intact rat carotid body-attached nerve (CSN) preparation (**Figure 2**). *In vivo*, increased CSN activity ultimately leads to an increase in frequency and depth of breathing, such that blood PO_2 is restored toward normal levels, thereby correcting the initial disturbance.

More recent studies indicate the carotid body is also a glucosensor, capable of detecting low blood glucose (hypoglycemia) and activating a counter-regulatory neuroendocrine response. Together with the pancreas, liver, and portal vein, it contributes to a class of physiologically important glucosensors located outside the brain. Though central hypothalamic neurons are also known to sense hypoglycemia, their physiological significance has been questioned. An important role of the carotid body as a systemic glucosensor was initially demonstrated by the observation that in order to obtain an appropriate counter-regulatory neuroendocrine response to insulin-induced mild hypoglycemia, the innervation of the carotid body by the CSN needed to be intact. Additionally, insulin-induced mild hypoglycemia in adult rats results in a significant increase in spontaneous ventilation that is abolished if the CSNs are transected. Though recent studies have demonstrated that low glucose can stimulate carotid body receptors *in vitro*, the underlying mechanisms are not completely understood.

Anatomical Organization of the Carotid Body

The principal cell components of the carotid body are clusters of endocrine-like glomus or type I cells that are intimately associated with glial-like, sustentacular type II cells. Type I cells are polymodal receptor cells that derive from the sympathoadrenal sublineage of the embryonic neural crest. Like sympathetic neurons and adrenal chromaffin cells, they synthesize catecholamines, of which dopamine is the one that predominates in type I cells. Immunolocalization of tyrosine hydroxylase (TH), the rate-limiting enzyme in catecholamine biosynthesis, has frequently been used to identify and characterize type I cells in carotid bodies of several species *in situ* (**Figure 1**). At the ultrastructural level, the occurrence of large cytoplasmic dense-cored vesicles is the most recognizable feature of these cells. Another morphological characteristic of clustered type I cells is that they form chemical and electrical synapses (i.e., gap junctions) with each other, and chemical synapses with sensory terminals of the petrosal afferent fibers. Additionally, electrical synapses have also been reported between type I and type II cells. The function of the type II cells is presently unclear, though it is possible they help coordinate activity within the cluster. Blood vessels, with fenestrated capillaries, are intimately associated with type I clusters, and this presumably facilitates delivery of the sensory stimuli to the receptor sites. Few autonomic ganglion-like cells have also been found inside the carotid body structure but their physiology is not well understood.

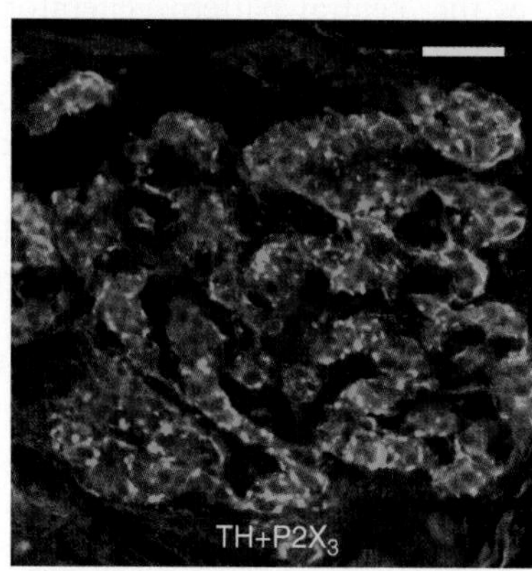

Figure 1 Innervation of chemoreceptor type I cells by petrosal sensory nerve terminals. In this tissue section from a 2-week-old rat carotid body, type I cell clusters are immunopositive for tyrosine hydroxylase (TH; red fluorescence), the rate-limiting enzyme in catecholamine biosynthesis. Nerve terminals surrounding type I cells are immunopositive for the purinergic receptor $P2X_3$ subunit (green fluorescence). Calibration bar represents 50 μm.

Neural Inputs to the Carotid Body

The main sensory innervation to the carotid body originates from neurons located in the PG (**Figure 3**). Their peripheral branches travel along the glossopharyngeal and CSNs to the carotid body (**Figure 3**), whereas their central projections terminate in the nucleus tractus solatarius of the brain stem. In rats, the chemoafferent neurons that directly innervate type I cells are located in the more distal region of the ganglion (**Figures 3** and **4(a)**). The majority of these chemoafferent neurons express TH (**Figure 4(a)**), and synthesize dopamine similar to carotid body type I cells. Neurons in the more proximal region of the ganglion send fibers that terminate near blood vessels and express other peptidergic markers such as substance P and calcitonin

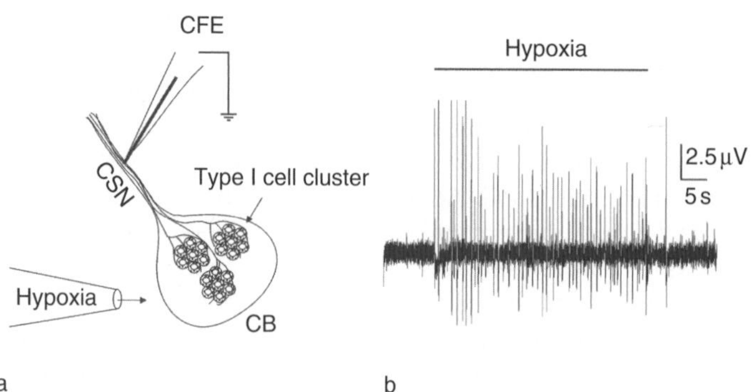

Figure 2 Hypoxia-induced increase in CSN activity in the isolated rat carotid body-attached nerve preparation *in vitro*. A schematic diagram showing an extracellular carbon fiber electrode (CFE) applied to CSN while a hypoxic solution is superfused over the preparation (a). A recording trace of the hypoxia-induced increase in CSN sensory discharge is shown in (b).

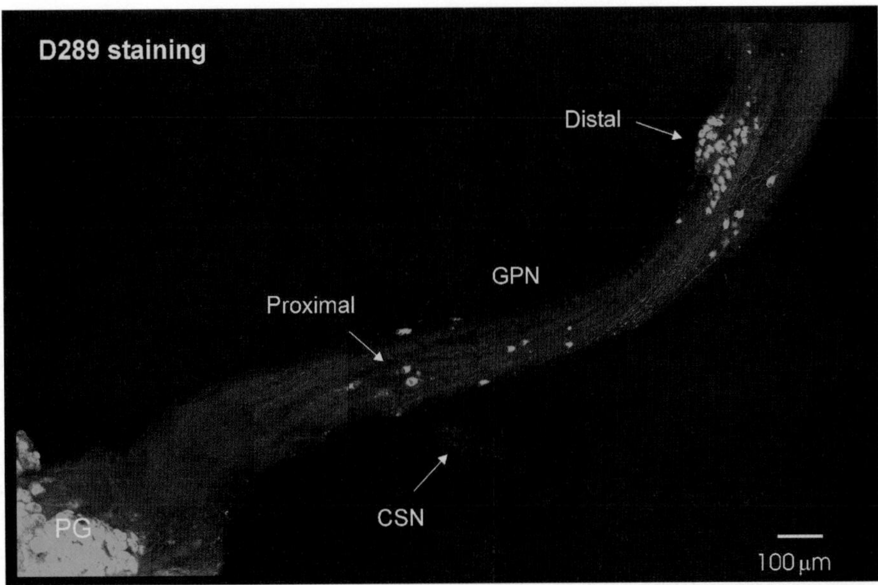

Figure 3 Vital staining of neuronal elements in rat PG and GPN. This whole-mount preparation was stained by the vital styrylpyridinium dye D289, following intraperitoneal injection. Sensory neurons in distal region of the PG, and two groups of autonomic neurons in the GPN are stained. The proximal GPN neuronal population is located near the branch point with the CSN; the distal population is located further along the GPN. The carotid body, normally attached to the CSN, was removed in this preparation. Reproduced from Campanucci VA and Nurse CA (2005) Biophysical characterization of whole-cell currents in O2-sensitive neurons from the rat glossopharyngeal nerve. *Neuroscience* 132(2): 437–451, with permission from Elsevier.

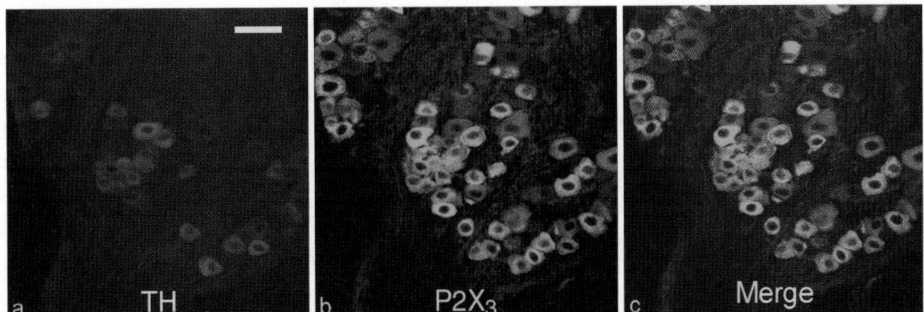

Figure 4 Immunostaining of carotid body chemoafferent neurons in the distal region of the PG. These chemoafferent neurons comprise a subpopulation that are mostly immunopositive for TH (red fluorescence) as illustrated in (a). Many petrosal neurons in the same section are immunopositive for the $P2X_3$ purinergic receptor subunit (green fluorescence) (b), including the TH-positive neurons. Co-localization of the two markers in chemoafferent neurons are shown in the merged images (c). Calibration bar represents 50 μm.

gene-related peptide. Additionally, autonomic neurons that express neuronal nitric oxide synthase (nNOS) are found in discrete populations within the glossopharyngeal and CSNs, and supply efferent innervation to the carotid body. These neuronal groups can be conveniently labeled and visualized in whole-mount preparations following injection of the vital styryl pyridinium dye D289 (**Figure 3**). Their role in chemoreceptor inhibition is discussed in a later section.

Transduction of Chemical Signals in Arterial Blood

While significant progress has been made toward our understanding of the transduction mechanisms in type I receptor cells during hypoxia, gaps in knowledge still remain. The prevailing view is that the hypoxia-induced receptor potential in type I cells is due to closing of K^+ channels that are open at rest. The particular type of PO_2-regulated K^+ channel may differ among species, and it is likely that more than one K^+ channel subtype is involved in the hypoxic response even in the same cell. However, the identity of the O_2-sensor and the signaling pathways leading to K^+ channel closure are not completely resolved, and it is plausible that a single cell may use more than one pathway depending on the PO_2 level. Candidates for the O_2 sensor include heme proteins of the mitochondrial electron transport chain, heme oxygenase II, and AMP kinase. These are thought to generate second messenger signals (e.g., changes in ADP/ATP ratio, CO) during hypoxia, leading to K^+ channel closure and membrane depolarization (**Figure 5**). The latter in turn causes opening of voltage-gated Ca^{2+} channels, Ca^{2+} entry, and neurotransmitter release.

Transduction of CO_2 stimuli (hypercapnia) and acidosis also appear to involve K^+ channel inhibition, and the second messenger signal is thought to be an acidic intracellular pH (pH_i). In the case of hypercapnia, acidification of pH_i is catalyzed by intracellular carbonic anhydrase present in type I cells. The signaling pathways involved in low glucose sensing by type I cells are not completely understood, though inhibition of K^+ channels may also contribute.

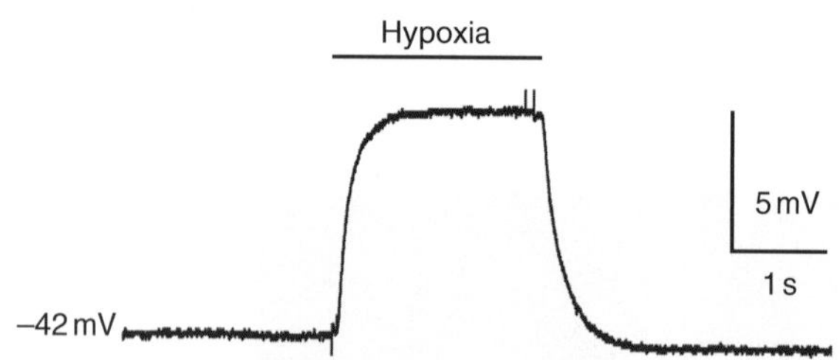

Figure 5 Hypoxia-evoked receptor potential in a type I cell. Current clamp recording showing depolarizing receptor potential evoked in a type I cell following exposure to acute hypoxia ($PO_2 \sim 5$ mmHg). The cell was a member of a cluster and the hypoxic stimulus was applied by rapid perfusion during the period indicated by the horizontal bar.

Role of Neurotransmitters in Sensory Processing

The neurotransmitter mechanisms involved in processing of chemosensory stimuli in the carotid body have attracted much attention, and in recent years substantive progress has been made. Traditional approaches to this problem relied on attempts to pharmacologically block the chemosensory discharge in the isolated, perfused, and/or superfused sinus nerve-carotid body preparation *in vitro* (see **Figure 2**). One disadvantage of this preparation is that synaptic events at the nerve terminal are not directly monitored, and block of spike activity could occur while subthreshold transmission still remains intact. A cell-culture model based on reconstruction of the chemosensory synapse *in vitro* has proved especially powerful in elucidating these neurotransmitter events. In co-cultures containing dissociated rat petrosal neurons and type I cell clusters, neurofilament-positive neuronal processes ramify over the culture substrate and contact TH-positive type I cells (**Figure 6(a)**). Importantly, in some cases, the petrosal soma is fortuitously situated adjacent to a type I cluster (**Figure 6(b)**). This allows the monitoring of electrical events, including subthreshold potentials, close to the synaptic sites. In such coculture models, it is possible to show that the same functional chemosensory neuron can be robustly activated by different stimuli including hypoxia, hypercapnia, and low glucose (**Figure 7(a)**). These models led to the first suggestion that corelease of the neurotransmitters ATP and

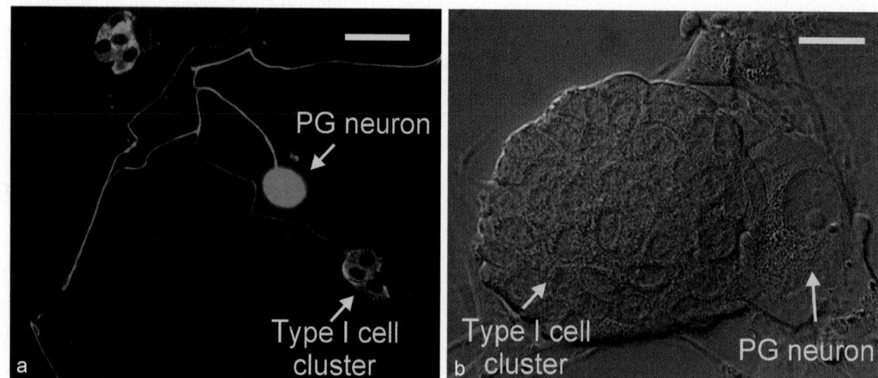

Figure 6 Coculture model of rat type I cell clusters and dissociated petrosal neurons. In (a), immunostaining of the coculture revealed positive neurofilament (NF) immunoreactivity in the petrosal neuron (PN) soma and its processes (red fluorescence), and positive TH immunoreactivity in two type I cell clusters (green fluorescence). Note that neuronal process appears to contact lower type I cell cluster. In (b), the differential interference contrast image shows a living type I cell cluster juxtaposed to a petrosal neuron, as typically used in electrophysiological recordings of synaptic interactions (see **Figure 7**). Calibration bar represents 20 μm in (a) and 10 μm in (b).

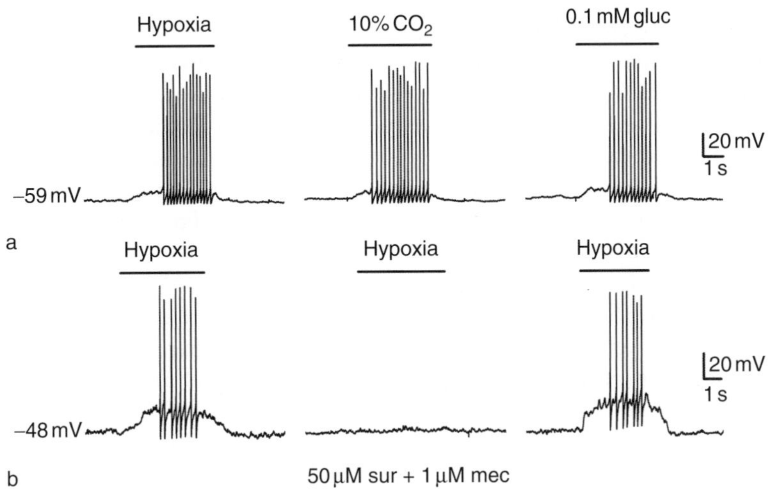

Figure 7 Chemosensory signaling in coculture and role of ATP and ACh as cotransmitters. In all traces, current clamp recordings of membrane potential were obtained from petrosal neurons juxtaposed to type I clusters as in **Figure 6(b)**. In (a), the same chemosensory unit was excited by three stimuli, that is, hypoxia (left), isohydric hypercapnia (10% CO_2; pH = 7.4) (middle), and low glucose (0.1 mM) (right). Stimuli were applied during the period indicated by upper horizontal bar. In (b), hypoxic chemotransmission was reversibly abolished by a combination of purinergic (50 μM suramin) and nicotinic (1 μM mecamylamine) receptor blockers.

acetylcholine (ACh) was the main mechanism mediating hypoxic chemotransmission in the rat carotid body. Thus, in many cases, the hypoxic response recorded in cocultured neurons could be partially inhibited by purinergic (e.g., suramin) and nicotinic (e.g., mecamylamine) blockers applied separately, and completely inhibited when both blockers were present together (**Figure 7(b)**). Similarly, the hypoxia-induced sensory discharge in the isolated rat sinus nerve-carotid body preparation can be blocked by a combination of purinergic and nicotinic blockers, suggesting that the corelease mechanism is not an artifact of culture conditions. The neurotransmitter ATP, released from type I cells during hypoxia, acts on ligand-gated, purinergic P_2X receptors located on petrosal afferent nerve endings. Consistent with this idea, confocal immunofluorescence studies indicate that purinergic P_2X_2 and P_2X_3 subunits are localized on sensory nerve terminals apposed to type I cell clusters *in situ* (e.g., **Figure 1**). Moreover, co-localization of P_2X_2 and P_2X_3 receptor immunofluorescence in TH-positive chemoafferent neurons can be demonstrated in the distal region of the PG (**Figures 4(a)–4(c)**). A pivotal role of P_2X_2 subunits in hypoxia chemotransmission has been elegantly demonstrated with the use of P_2X_2 receptor knockout ($P_2X_2^{-/-}$) mice. These animals showed a markedly attenuated ventilatory response to hypoxia, and there was a dramatic reduction in hypoxia-induced nerve activity in the *in vitro* sinus nerve-carotid body preparation from $P_2X_2^{-/-}$ mice. Surprisingly, P_2X_3 receptor-knockout ($P_2X_3^{-/-}$) mice behaved similar to wild-type mice in their ventilatory response to hypoxia, though the resting chemoafferent activity in the sinus nerve was significantly reduced in these animals. Further, the hypoxia-induced sinus nerve discharge was further attenuated in $P_2X_{2/3}^{Dbl-/-}$ mice relative to $P_2X_2^{-/-}$ mice. The combined data from these knockout mice indicate an essential (and sufficient) role for P_2X_2 subunit in the ventilatory response to hypoxia, with smaller contributions from the P_2X_3 subunit. Comparable knockout experiments have not been reported to validate the more controversial role of ACh in chemotransmission, though such experiments are complicated by the presence of several and similar nicotinic ACh receptor subunits in both petrosal neurons and type I cells. The possibility that other neurotransmitters (e.g., 5-hydroxytryptamine (5-HT)) may contribute to the fast chemoexcitatory response in some cases has not been ruled out since ligand-gated 5-HT_3 receptors are expressed by many petrosal neurons.

In addition to the fast-acting neurotransmitters discussed, there is evidence for a complex regulation of carotid body sensory discharge by slower-acting neuromodulators that signal through G-protein-coupled pathways. These include positive regulators which facilitate chemoreceptor output such as 5-HT acting via presynaptic 5-HT_{2A} receptors, and adenosine acting via presynaptic A_{2A} receptors on type I cells. On the other hand, negative regulators appear to inhibit chemoreceptor output including dopamine acting on presynaptic D2 receptors and γ-aminobutyric acid (GABA) acting via presynaptic $GABA_B$ receptors. Histamine is also released during carotid body excitation and though various histamine receptors (H1, H2, and H3) have been identified in the organ, its physiological role is presently unclear. To add to this complexity, there is evidence that released ATP can also act presynaptically on P2Y receptors located on both type I and type II cells, and ACh can produce opposing excitatory and inhibitory actions via nicotinic and muscarinic type I cell autoreceptors respectively. Neuropeptide modulators including substance P and enkephalins appear to have stimulatory and inhibitor actions respectively on carotid body discharge. The reasons for such a plethora of neuroactive chemicals in such a tiny organ remain obscure, but may be the basis for chemoreceptor plasticity that results in short- and long-term changes in organ sensitivity, following exposure to chronic or intermittent stimuli, for example, hypoxia.

Modulation of Sensory Discharge by Nitric Oxide Produced via Autonomic Neurons

In addition to the chemical neuromodulators of carotid body sensory discharge discussed in the previous section, gaseous modulators including nitric oxide (NO) are also involved. Sources of NO include endothelial cells, which express endothelial nitric oxide synthase, and efferent autonomic neurons within the GPN/CSN nerves that express nNOS (**Figures 3** and **8(a)**). The nNOS-containing fibers from autonomic GPN neurons or 'paraganglia' travel along the CSN and are a key component in the efferent inhibition of carotid body chemoreceptors via NO release. This idea had been suggested for many years, and was supported by the observations that electrical stimulation of the CSN, or carotid body exposure to hypoxia, caused an elevation in NO production that was prevented by specific NOS antagonists. Also, nNOS mutant mice displayed augmented ventilatory responses to hypoxia, and NO donors were reported to inhibit the carotid body chemosensory responses to hypoxia, as well as L-type Ca^{2+} currents in type I cells. Despite these compelling data for an inhibitory role of NO in chemoreception, the mechanisms that underlie activation of GPN neurons during hypoxia were not completely understood. Recent data have shed new light on this issue with the demonstration that isolated GPN neurons are O_2 sensitive and depolarize during hypoxia due to inhibition of background K^+ channels. This provides a potential pathway for Ca^{2+} entry, nNOS activation, and NO synthesis and release during hypoxia. Additionally, GPN neurons express at least four different types of P_2X receptor subunits (P_2X_2, P_2X_3, P_2X_4, and P_2X_7), allowing them to respond to a wide range of ATP concentrations (**Figure 8(b)**). This is particularly important since, as discussed earlier, ATP is a key excitatory neurotransmitter involved in carotid body chemoreceptor function. Therefore, ATP released from type I cells during hypoxia could excite GPN nerve terminals, leading to nNOS activation. The latter could occur following the rise in intracellular Ca^{2+} due to Ca^{2+} entry through Ca^{2+}-permeable P_2X receptors and/or voltage-dependent Ca^{2+} channels activated by ATP-induced membrane depolarization (**Figure 8(b)**). Indeed, the direct involvement of ATP on efferent inhibition of carotid chemoreceptors was recently demonstrated in a novel coculture model of type I cells and GPN neurons. In these conditions, application of ATP (or hypoxia) to GPN neurons caused a robust type I cell hyperpolarization that was prevented by preincubation with the NO scavenger carboxy 2-(4-carboxyphenyl)-4,4,5,5-tetramethyl-imidazoline-1-oxyl-3-oxide potassium (PTIO) or the NOS inhibitor L-nitro arginine methyl ester. Significantly, the NO donor sodium nitroprusside alone, but not ATP, could strongly hyperpolarize type I cells grown without GPN neurons in otherwise similar cultures. Therefore ATP, released from type I cells or erythrocytes during hypoxia, could lead to activation of GPN neurons and/or their

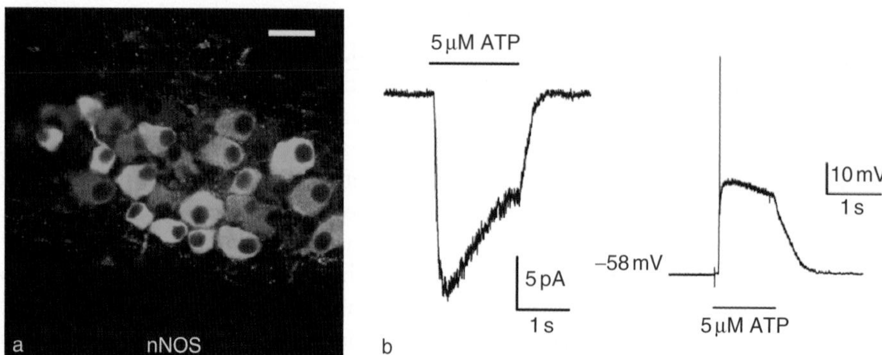

Figure 8 ATP sensitivity in nNOS-positive autonomic neurons of the distal population of GPN neurons. In (a), GPN neurons in a whole-mount of the GPN are immunopositive for nNOS (green fluorescence). In isolated GPN neurons in culture (b), ATP (5 μM) induced an inward current during voltage clamp at −60 mV (left), and membrane depolarization with superimposed spike activity during current clamp recording (right). Calibration bar represents 30 μm.

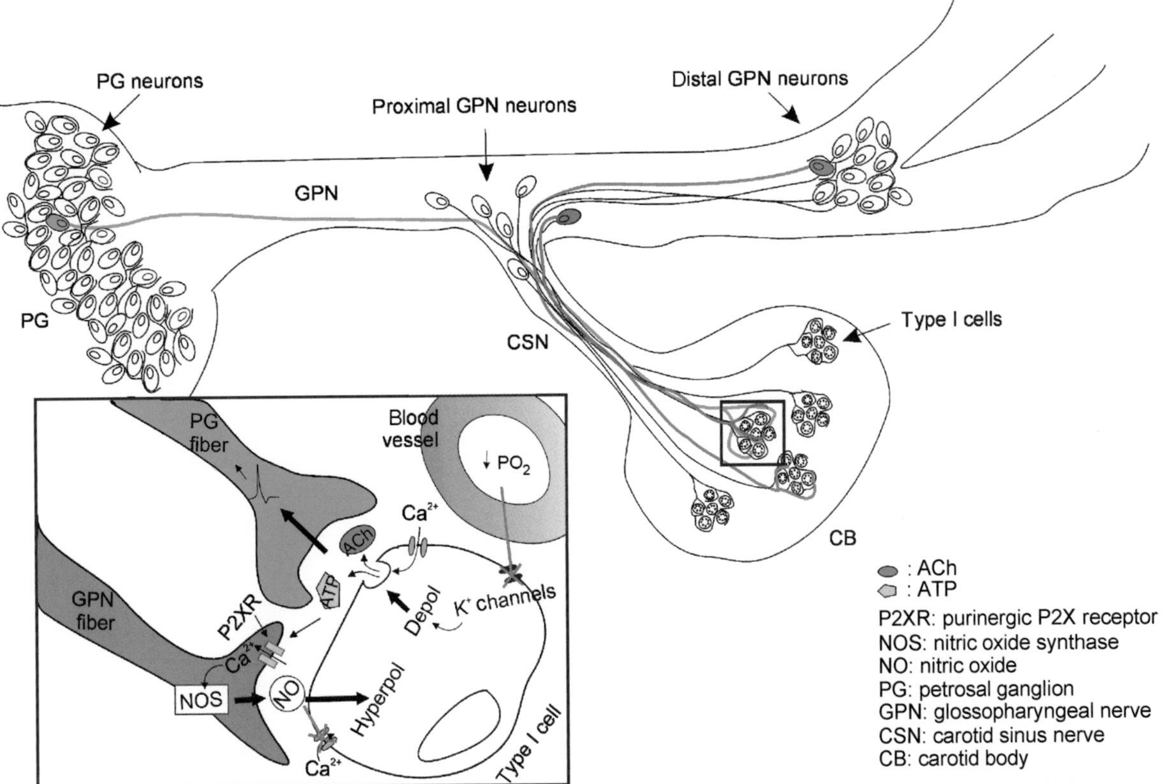

Figure 9 Summary diagram of afferent and efferent innervation of rat carotid body type I cells. Note that PG neurons (green) send afferent fibers that synapse with type I cell clusters. Also, proximal (red) and distal (blue) GPN autonomic neurons supply efferent fibers that terminate near type I clusters. During a decrease in PO_2 (hypoxia), ATP and ACh are coreleased from type I cells and interact with $P2X_2$–$P2X_3$ and nicotinic ACh (nAChR) receptors respectively (not shown) on petrosal terminals (green), causing CSN excitation (lower inset). This ATP (or ATP released from red blood cells during hypoxia) can also act on P2X receptors on GPN 'efferent' terminals (red), causing a rise in intracellular calcium, activation of nNOS, followed by synthesis and release of NO. Finally, NO causes type I cell hyperpolarization, suppression of neurotransmitter release, and inhibition of CSN sensory discharge.

terminals via multiple P_2X receptors, leading to nNOS activation, NO production, and chemoreceptor inhibition. The diagram in **Figure 9** summarizes a proposed mechanism for carotid body efferent inhibition.

Carotid Body Plasticity

A remarkable feature of the carotid body is its ability to adapt to conditions of chronic or intermittent hypoxia. Chronic hypoxia can lead to time-dependent changes in the respiratory system including alterations in carotid body sensitivity to hypoxia. Ventilatory acclimatization to hypoxia involves both morphological and functional changes in the carotid body. For example, there is organ enlargement accompanied by both hypertrophy and hyperplasia of type I cells, as well as neovascularization. Functionally, there is an alteration in ion channel expression and neurotransmitter or neuromodulator functions in type I cells. These result in an increased sensitivity of the organ to hypoxia. Similarly, intermittent hypoxia, as occurs during recurrent apnea in humans, results in a selective enhancement of carotid body sensitivity to hypoxia, without obvious morphological changes. These adaptations also appear to involve changes in neurotransmitter functions, though these are not fully understood. The importance of understanding these mechanisms is underscored by the fact that intermittent hypoxia during recurrent apneas can lead to serious cardiovascular abnormalities in humans.

See also: Adenosine Triphosphate (ATP); Autonomic and Enteric Nervous System: Apoptosis and Trophic Support During Development; Autonomic Nervous System: Metabolic Function; Autonomic Nervous System; Gamma-Aminobutyric Acid (GABA); Nitric Oxide; Role of NO in Neurodegeneration.

Further Reading

Campanucci VA, Zhang M, Vollmer C, and Nurse CA (2006) Expression of multiple P2X receptors by glossopharyngeal neurons projecting to rat carotid body O_2 chemoreceptors: Role in nitric oxide-mediated efferent inhibition. *Journal of Neuroscience* 26: 9482–9493.

Iturriaga R and Alcayaga J (2004) Neurotransmission in the carotid body: Transmitters and modulators between glomus cells and petrosal ganglion nerve terminals. *Brain Research Brain Research Reviews* 47: 46–53.

Lopez-Barneo J (2003) Oxygen and glucose sensing by carotid body glomus cells. *Current Opinion in Neurobiology* 13: 493–499.

Nurse CA (2005) Neurotransmission and neuromodulation in the chemosensory carotid body. *Autonomic Neuroscience: Basic & Clinical* 120: 1–9.

Prabhakar NR (2006) O_2 sensing at the mammalian carotid body: Why multiple O_2 sensors and multiple transmitters. *Experimental Physiology* 91: 17–23.

Prabhakar NR and Peng YJ (2004) Peripheral chemoreceptors in health and disease. *Journal of Applied Physiology* 96: 359–366.

Rong W, Gourine AV, Cockayne DA, et al. (2003) Pivotal role of nucleotide $P2X_2$ receptor subunit of the ATP-gated ion channel mediating ventilatory response to hypoxia. *Journal of Neuroscience* 23: 11315–11321.

Autonomic Nervous System: Central Cardiovascular Control

M C Andresen, Oregon Health & Science University, Portland, OR, USA
D Mendelowitz, George Washington University, Washington, DC, USA

Introduction

Vital nourishment of all tissues is assured by a properly regulated cardiovascular system. Consistent and coordinated contractions of the heart provide pressure for the circulation system and vessel conduits distribute the blood to flow throughout the body. Thus, the goal of cardiovascular control is to maintain and allocate blood flow that meets tissue demands over a diverse range of circumstances. Cardiac contractions are normally initiated by electrical pacemaker activity within the sinus node of the right atrium, and without external stimuli, intrinsic pacemaking automatically maintains rates that are fairly constant. However, the autonomic nervous system is responsible for adjustments that meet changing circulatory demands such as exercise or postural challenges. Thus, the nervous system orchestrates optimal adjustments in the pump function of the heart. Both rate and the timing of events between atria and ventricles are altered by released neurotransmitters to facilitate efficient pumping. Neurotransmitter activation of smooth muscle in blood vessels adjusts vessel caliber and thus flow patterns. These cardiovascular adjustments are accomplished by neurotransmitters released by autonomic nerves. Overall, the level of arterial blood pressure at all times is a dominant factor in driving the circulation of blood. Multiple neural reflexes rapidly control blood pressure to accommodate a wide range of systemic and regional demands and priorities. Acute postural change, for example, can require changes in pump rate at the very next heart beat, and such rapid adjustments use the autonomic nervous system to quickly fine-tune the heart and the vascular distribution of blood.

The organization of the central nervous system (CNS), directed toward control of the cardiovascular system, follows a general, prototypic pattern for these reflex pathways. Afferent information arises from a network of afferent neurons distributed within key portions of the cardiovascular system. The peripheral endings of cardiovascular afferent neurons are unequally distributed and are particularly concentrated in the centralmost portions of the vascular tree (aorta, common carotid artery, vena cava, and jugular vein) and within the heart. Mechanosensitive afferent neurons (e.g., arterial baroreceptors) transduce local vessel and chamber stretch as an index of filling or local pressure. This 'filling' information plays a key role in rapid cardiovascular adjustments on a beat-to-beat basis. Afferent neurons arising from the heart as with many other visceral organs have cell bodies in cervical ganglia associated with the IX and X cranial nerves (nodose, jugular, and petrosal ganglia). Additional cardiovascular afferents arise from neurons of the dorsal root ganglia and often these spinal afferent neurons send axons that travel via the sympathetic nerve trunks and have been termed sympathetic afferents in the literature. Visceral afferents dominate reflexes that adjust the autonomic outflow back to the organ of origin, but often additionally contribute to the control of entirely different organs (e.g., respiratory afferents that alter cardiac control). Both cardiac and arterial mechanosensory neurons initiate rapidly responding, relatively short pathways for parasympathetic efferent adjustments as well as longer latency reflexes via the spinal cord that modulate sympathetic efferent neurons. Cardiovascular reflex pathways predominantly involve networks of neurons in the brain stem and spinal cord. Control of heart rate and arterial pressure is the primary focus of this overview.

The arterial baroreceptor reflex (baroreflex) control of heart rate is perhaps one of the most basic autonomic regulatory reflexes and serves as a central pathway prototype for reflexes arising from many other cranial visceral afferents. The parasympathetic reflex control of the heart is fundamentally better understood than control of arterial pressure or venous capacitance. The consensus view of the core circuit for the cardiac parasympathetic control suggests that this reflex arc relies on structures confined to the medulla. Baroreceptor afferents project to brain stem neurons in the nucleus of the solitary tract (NTS). NTS neurons, in turn, contact output neurons within the nucleus ambiguus (NA) that exit the brain stem to contact final parasympathetic motor neurons within the heart itself. While this constitutes the simplest pathway of cardiovascular control, additional, more convoluted, brain stem pathways exist as well as indirect contributions arising from ascending spinal viscerosensory pathways.

Medulla Oblongata

The arterial baroreflex is organized as a classic negative feedback system in which increases in arterial blood pressure activate mechanosensitive arterial baroreceptors (**Figure 1**) that trigger reflex responses

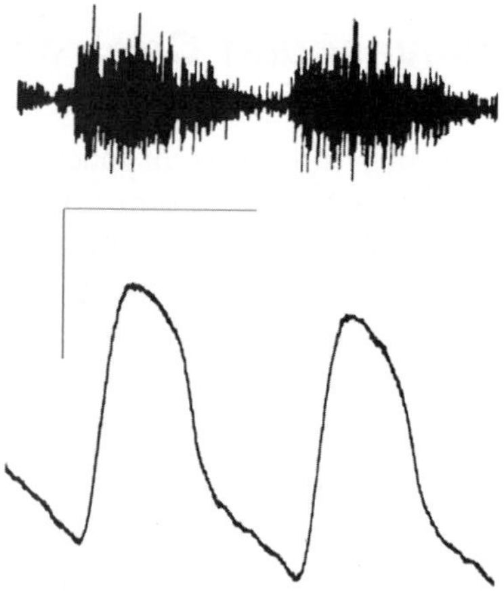

Figure 1 Pulse synchronous arterial baroreceptor discharge. Aortic depressor nerve activity recorded from whole nerve in a conscious rabbit (upper trace) rises abruptly with the systolic phase of femoral arterial blood pressure (lower trace). Scales are 20 mmHg (vertical) and 500 ms (horizontal).

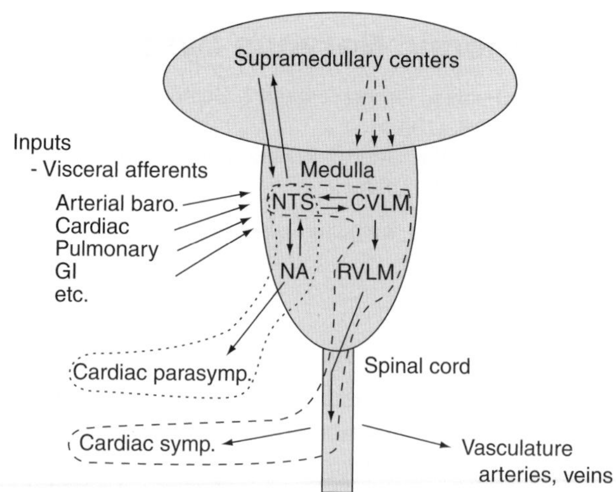

Figure 2 Schematic of core baroreflex pathways for cardiovascular control. Reflex responses are initiated from visceral primary afferent neurons such as cardiac afferent and arterial baroreceptor neurons (left) that converge on initial central neurons (second-order neurons) within the nucleus of the solitary tract (NTS). From there, information diverges to two key autonomic centers along the parasympathetic pathway to nucleus ambiguus (NA) and along the sympathetic pathway to the caudal ventrolateral medulla (CVLM). The parasympathetic pathway (dotted outline) leaves NA to target the intrinsic cardiac network including postganglionic neurons at the heart. The sympathetic regulatory pathway (dashed outline) goes on to neurons in the rostral ventrolateral medulla (RVLM) before descending to the spinal cord at the intermediolateral cell column and out to the sympathetic postganglionic neurons. NTS is reciprocally connected to a wide range of CNS regions (bidirectional arrows) including many supramedullary regions that modulate reflex function but are not part of the essential core circuit. Forebrain regions, for example, often act to modify activity within the core brain stem circuits. A wide range of pathways (excitatory and inhibitory) contacts neurons of the core baroreflex pathway. Note that the negative feedback and slowing of the heart rate occurs peripherally in the parasympathetic pathway by actions of the postganglionic neurotransmitter acetylcholine, but centrally in the sympathetic pathway as CVLM-released GABA inhibits RVLM neurons (and sympathetic activity).

that ultimately reduce arterial blood pressure. The discharge of arterial baroreceptors is conducted along cranial nerve axons (IX and X) projecting directly into the brain stem to neurons in the NTS where they make the first synapse in the baroreflex pathway (**Figure 2**). Processed afferent signals exit NTS to activate vagal preganglionic neurons located primarily in NA that project to cardioinhibitory parasympathetic postganglionic neurons in the heart. NTS neurons also activate sympathoinhibitory neurons located in the caudal ventrolateral medulla (CVLM). Thus, activity along dual, cardiac directed autonomic pathways is oppositely regulated to produce negative feedback control of cardiac pumping by (1) increased release of acetylcholine by parasympathetic postganglionic neurons and (2) decreased release of norepinephrine by sympathetic postganglionic neurons.

In the resting, conscious state, midcollicular decerebration does not alter the gain (sensitivity) of this heart rate baroreflex, suggesting that brain stem mechanisms are sufficient for reflex function. Although brain structures above the medulla are not required, information from the forebrain and other higher CNS sites modulate the baroreflex during various behaviors including stressful emotional responses. Thus, the core baroreflex circuit appears to reside below the pons and contains both short and longer latency pathways. The shortest anatomical pathway activates cardiac parasympathetic efferent neurons by engaging, at a minimum, (1) primary afferent neurons associated with cardiovascular mechanoreceptors, (2) second-order NTS neurons, (3) cardiac parasympathetic preganglionic motor neurons located primarily in the NA, and (4) parasympathetic postganglionic neurons intrinsic to the heart. Such a parasympathetic pathway is responsible for relatively short-latency (<100 ms) cardiovascular-cardiac reflexes that adjust heart rate within one cardiac cycle (**Figure 2**).

Primary Afferent Neurons of the Arterial Baroreflex

Cardiovascular regulation depends on the constant monitoring of arterial pressure, central venous pressure, and cardiac rate and force. Populations of cardiovascular-selective sensory neurons supply a constant stream of information and these viscerosensory neurons belong to cellular/molecular phenotypic classes based on axon myelination that are broadly similar to sensory neurons of the skin (somatic afferents). Baroreceptors are concentrated near the bifurcation of the common carotid artery (carotid sinus) and just outside the left ventricle in the arch of the thoracic aorta. Neurites from arterial baroreceptors lie within the adventitia, the outermost layer of the artery wall.

This position within the artery wall means that they are stretched with each expansion of the artery wall that accompanies each expulsion of blood from the heart. A-type baroreceptors rapidly conduct responses via myelinated fibers (2.5–25 m s^{-1}) in brief bursts of activity during each pressure pulse (**Figure 1**). A-type baroreceptors discharge within the normal range of blood pressures even at rest and their discharge activity is quite regular and encodes mean arterial pressure, as well as the rate of rise and amplitude of the pressure pulse. A much larger number of arterial baroreceptors are C-type and conduct more irregular discharge patterns along slowly conducting (<1 m s^{-1}), unmyelinated axons. C-type baroreceptors are often silent at resting blood pressures and become active with sparse, intermittent discharge only at higher pressures than the A-type baroreceptors. A minority of C-type baroreceptors is activated at normal arterial pressures. The relationship between the amplitude of sensory terminal stretch and the frequency of evoked discharge defines the 'baroreceptor sensitivity.' This is not a fixed relationship, but displays 'plastic' changes both acutely and chronically in the face of sustained blood pressure changes. Under physiological and pathological conditions, the roles played by specific afferent neuronal populations can become 'reset' and change along with changes in cardiovascular regulation.

Solitary Tract Nucleus – Arterial Baroreflex Second-Order Neurons

Neurons within dorsal medial portions of the caudal NTS mark the central first step of the medullary baroreceptor reflex. Sensory inputs from baroreceptor afferent neurons release glutamate as their primary neurotransmitter to activate non-*N*-methyl-D-aspartate (NMDA) receptors on second-order NTS neurons. Inhibitory transmission via γ-aminobutyric acid (GABA) is prominent at second-order neurons in the medial NTS with inhibitory postsynaptic potentials arriving shortly after visceral afferent glutamatergic excitatory synaptic potentials. Since a majority of these GABAergic neurons are themselves directly coupled to viscerosensory afferent inputs, this organization provides effective short-latency feedback inhibition within this cardiovascular-cardiac reflex. These two amino acid transmitters activate ionotropic (ion channel coupled) receptors to convey fast transmission but both glutamate and GABA, in addition, activate metabotropic receptors that are coupled to intracellular second messengers. Thus, heterosynaptic cross talk between neurotransmitters is common and provides afferent-linked mechanisms that are very sensitive to the frequency content of afferent patterns of discharge. A wide range of accessory transmitters such as peptides like ATP, substance P, and calcitonin gene-related peptide are also involved in such transmission, although their precise roles are less certain.

Cardiac Parasympathetic Neurons

Cardiac parasympathetic motor neurons exert powerful control on cardiodynamics. Heart rate is dominated by the activity of the cardioinhibitory parasympathetic nervous system (**Figure 2**). Cardiac parasympathetic neurons are tonically active at rest but little, if any, cardiac sympathetic activity is commonly present under normal conditions. Preganglionic cardiac vagal neurons have cell bodies located mostly in the NA and, to a lesser extent, in the dorsal motor nucleus of the vagus. These parasympathetic preganglionic neurons send axons that exit the brain stem via the vagus nerves to contact postganglionic neurons located in cardiac ganglia. These cardiac ganglia are found within the connective and fatty tissue that surround the right atrium and vena cava. Parasympathetic postganglionic fibers emerge from these ganglia to make extensive synaptic contacts with the nearby sinoatrial and atrioventricular nodes of the heart where they release the neurotransmitter acetylcholine.

Activity within the arterial baroreflex is a major determinant of the tonic activity of the parasympathetic efferent preganglionic neurons that target the heart. During normal states, parasympathetic efferent preganglionic neuronal activity occurs during specific phases of the cardiac cycle initiated by this fast-responding reflex. Heart rate is dominated by the activity of the cardioinhibitory parasympathetic nervous system. Measures of these ongoing, tonic, and rapid adjustments in parasympathetic activity (e.g., heart rate variability) are indicators of cardiac health and, when absent, can predict an important risk of cardiac sudden death. During increases in arterial pressure, the initial reflex-induced slowing of the heart (bradycardia) is caused primarily, if not exclusively, by increases in cardiac vagal nerve activity. During decreases in arterial pressure, the baroreflex-induced cardiac acceleration (tachycardia) is caused mostly by decreases in parasympathetic along with somewhat later-arriving increases in sympathetic nerve activity. When both parasympathetic and sympathetic activities are present, parasympathetic activity generally dominates the control of heart rate. Increases in parasympathetic activity evoke a cardiac slowing that is more pronounced when there is a high level of sympathetic firing. When there is a moderate or high level of parasympathetic activity, changes in sympathetic firing elicit negligible changes in heart rate.

The CNS system normally controls the parasympathetic activity reaching the heart. Preganglionic cardiac vagal fibers tonically fire with a cardiac pulse synchronous pattern that is reduced during inspiration. Interruption of links to the CNS by cutting the cardiac vagal preganglionic fibers leads to an increase in heart rate demonstrating that the remaining, intact cardiac parasympathetic postganglionic neurons by themselves cannot sustain normal cardiac rates. Similarly, synaptic blockade silences parasympathetic neurons within the brain stem indicating an absence of pacemaker-like activity such as repetitive or phasic depolarizations or action potential. However, even small depolarizing currents (100 pA) evoke sustained, repetitive firing in cardiac vagal neurons with minimal spike frequency adaptation. The voltage-gated currents and firing characteristics of cardiac vagal neurons enable them to closely follow fast synaptic drive as well as to integrate long-lasting modulatory influences. Together, the characteristics of both these preganglionic and postganglionic cardiac parasympathetic neurons suggest that heart rate at rest is determined by activity within the parasympathetic pathways that arises from the integration of afferent and other drives to produce normal levels of cardiac vagal tone.

The activity of cardiac vagal preganglionic neurons is controlled by three major classes of synaptic inputs: glutamatergic, GABAergic, and glycinergic. NTS neurons provide a major glutamatergic input pathway that activates both NMDA and non-NMDA postsynaptic currents in cardiac vagal neurons. Owing to baroreceptor activation of NTS neurons, this glutamatergic pathway to cardiac vagal neurons likely constitutes the essential link between increases in blood pressure and the parasympathetic component of the arterial baroreceptor reflex. Additionally, NTS activation triggers GABAergic inputs to cardiac vagal neurons. Electrical stimulation of afferents in the vagus nerve evokes both GABAergic and glutamatergic responses in cardiac vagal neurons. In addition, the pungent active ingredient in hot peppers, capsaicin, can chemically activate and then, with prolonged exposure, inactivate visceral sensory C-type afferents including baroreceptors. Blockade of C-type afferents by capsaicin increases the latency of the GABAergic response without changing the latency of the glutamatergic responses. Such findings suggest that an inhibitory GABAergic pathway is involved in patterning of cardiac vagal activity to produce bursting synchronous with the cardiac cycle.

Cardiac vagal nerve activity also has pronounced respiratory modulation that mediates respiratory sinus arrhythmia, a pattern of slowing and accelerating heart rate in phase with lung inflation and deflation. Cardiac vagal fibers fire most rapidly in the postinspiration period, and often fall silent during inspiration. A form of this firing activity of cardiac vagal neurons during the phases of respiration persists in slices of the medulla that contain cardiac vagal neurons, pre-Bötzinger complex, and local circuits for motor output generation, and inspiratory motor discharge patterns can be recorded in hypoglossal cranial nerves (**Figure 3**). The firing of cardiac vagal neurons is inhibited during inspiratory activity.

Nicotinic cholinergic mechanisms have important actions on cardiac vagal neurons. During inspiration,

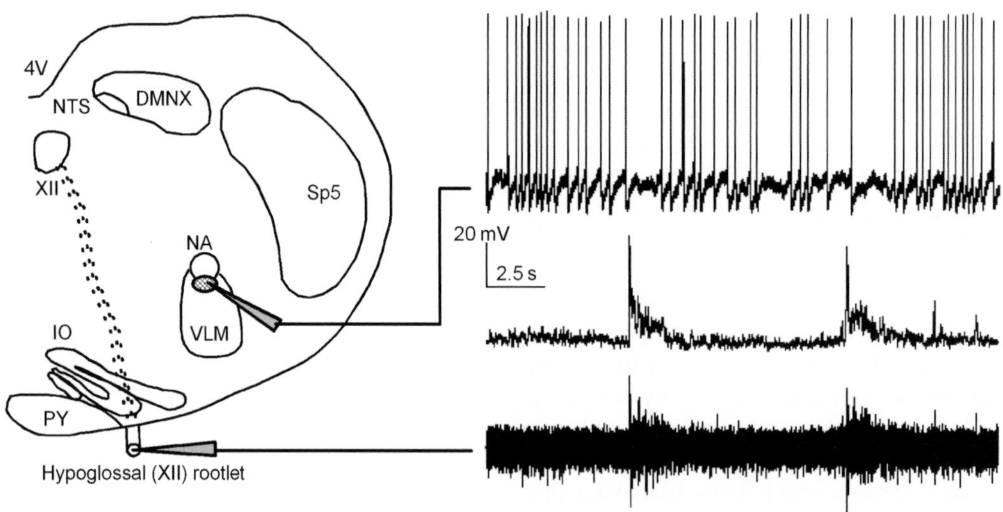

Figure 3 The medullary slice preparation that contains the cardiac vagal neurons, pre-Bötzinger complex, and respiratory hypoglossal motor neurons with intact hypoglossal cranial nerves is illustrated. Inspiratory activity recorded from the hypoglossal rootlet is shown bottom trace, right, and integrated, middle trace, right. The firing activity of a cardiac vagal neuron, induced to fire repetitively with a constant injection of inward current (~50 pA), top right, is inhibited during each inspiratory burst.

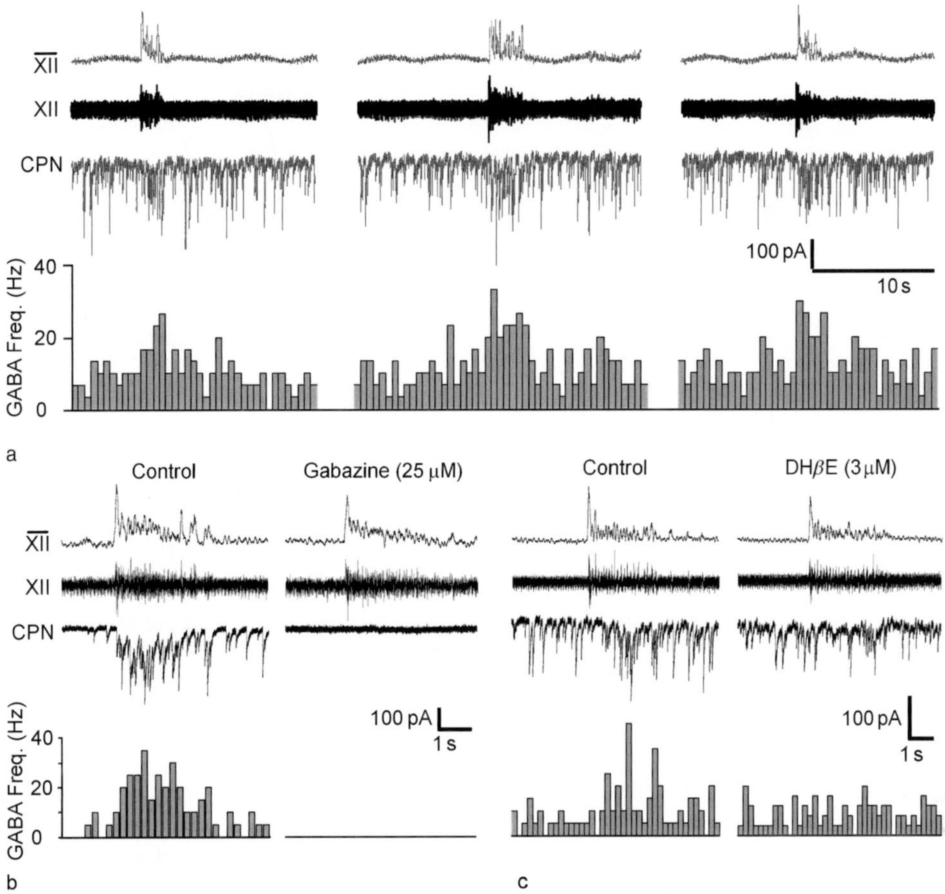

Figure 4 Inspiratory-related bursting activity was recorded from the hypoglossal rootlet (XII) and rectified and adjacent averaged ($\overline{XII}$). GABAergic inhibitory postsynaptic currents (IPSCs) were isolated by focal application of the NMDA, non-NMDA, and glycine receptor antagonists AP-5 (50 μM), CNQX (50 μM), and strychnine (1 mM), respectively. (a) During inspiratory activity, the frequency of GABAergic IPSCs in cardiac vagal neurons was significantly increased ($p < 0.05$). (b) The $GABA_A$ antagonist gabazine blocked all IPSCs. (c) This inspiratory-related increase in GABA synaptic frequency was abolished by focal application of the nicotinic receptor antagonist dihydro-β-erithroidine (DhβE) at a concentration selective for the α4β2 receptor subtype DHβE did not significantly change GABA synaptic frequency between inspiratory bursts.

the frequency of both GABAergic and glycinergic synaptic events increases substantially (**Figure 4**). A nicotinic antagonist, selective for a specific subset of nicotinic receptors (α4β2), abolishes this respiratory-evoked increase in GABAergic events supporting the critical role of cholinergic inputs in shaping cardiac parasympathetic control. This is a highly specific interaction since this intervention has no effect on the increase in glycinergic synaptic events during inspiration. Together, such findings provide a basic outline of interactions between glutamate, glycine, and acetylcholine in the control of heart rate in this core NTS–NA pathway (**Figure 5**).

Spinal Cord

Sympathetic preganglionic neurons related to the cardiovascular system are located within the intermediolateral cell column of the spinal cord and constitute a second major autonomic effector pathway. These sympathetic preganglionic neurons send processes from the spinal cord to activate postganglionic neurons within peripheral ganglia. The postganglionic sympathetic neurons themselves send axons to contact visceral organs and blood vessels where they release the neurotransmitter norepinephrine. Activity in these sympathetic neurons influences all aspects of cardiac function including heart rate, cardiac conduction, the strength of cardiac contraction, and ventricular stroke volume. Many sympathetic motor neurons are directed to blood vessels resulting in activation of smooth muscle and changes in vessel diameter and local arterial resistance or venous capacitance.

Although reflexes within the medulla exert substantial influence on cardiovascular sympathetic efferent neurons, at present these reflexes are incompletely defined. Sympathetic premotor neurons within the

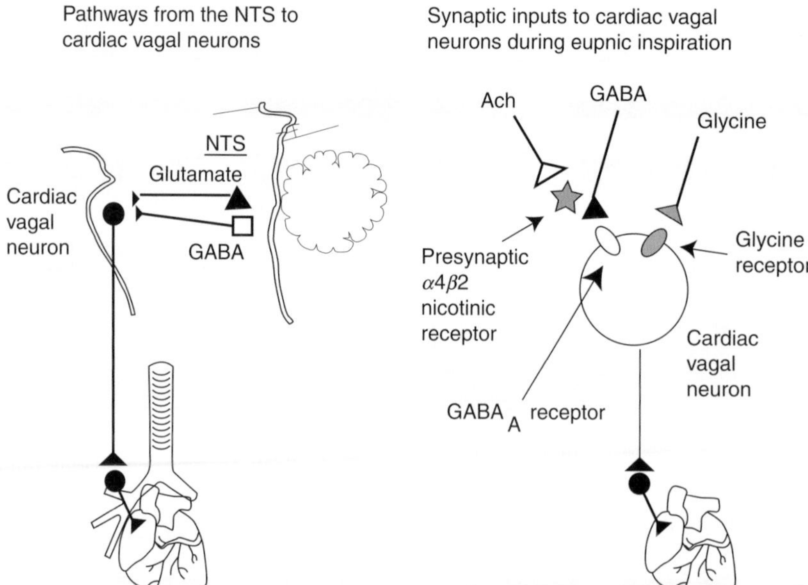

Figure 5 Neurotransmitter actions in the core reflex pathway in the control of cardiac parasympathetic neurons. NTS sends glutamatergic and GABAergic but not glycinergic pathways to cardiac vagal neurons (left). Inspiratory activity increases both GABAergic and glycinergic inputs to cardiac vagal neurons (right). Nicotinic $\alpha 4\beta 2$ receptors presynaptically facilitate GABAergic, but not glycinergic, neurotransmission.

rostral ventrolateral medulla (**Figure 2**) send axons to modulate spinal cord cardiovascular sympathetic preganglionic neurons. An important determinant of cardiovascular sympathetic regulation arises from sympathoinhibitory neurons lying within the CVLM that secrete GABA onto sympathetic premotor neurons within the rostral ventrolateral medulla. The full range of mechanisms of the control of sympathetic premotor neurons remains to be established. However, baroreflex pathways provide substantial sympathoinhibitory input and thus remain part of a functionally negative feedback system that, for example, reduces heart rate by inhibiting (i.e., withdrawing) ongoing norepinephrine release. Thus, NTS neurons that receive tonic inputs from cardiovascular afferent neurons activate sympathoinhibitory neurons that, in turn, phasically depress sympathetic premotor neuronal activity. These premotor neurons, in turn, project to sympathetic efferent preganglionic neurons in the intermediolateral cell column of the spinal cord. The cranial thoracic spinal cord preganglionic neurons project to cardiac sympathetic efferent postganglionic neurons distributed throughout the intrathoracic and superior cervical ganglia. Interestingly, such spinal cord sympathetic motor neurons also receive inputs from cardiac afferent neurons in dorsal root ganglia. Unlike the medullary cardiovascular reflexes depicted above, activation of spinal cardiac afferent neurons results in the activation of spinal cord sympathetic premotor neurons involved in enhancing cardiac rate and contractile force – a positive feedback mechanism. Spinal cord sympathetic preganglionic motor neurons also innervate peripheral blood vessels and influence arterial vascular resistance and venous capacitance. The interrelationship between positive and negative feedback pathways controlling cardiovascular sympathetic motor neurons is poorly understood, although often suspected to be involved during pathophysiological circumstances.

Higher Centers

Under many normal and pathophysiological conditions, the gain or sensitivity of the heart rate baroreflex can be altered by influences arising from supramedullary brain structures such as the hypothalamus or amygdala. Under stress and even during physiological states, these supramedullary inputs transform the operation of the baroreflex to accommodate integrative responses for exercise or strong emotional behaviors. Thus, cardiovascular regulation including the major baroreflex pathways is often an integral component of major homeostatic and highly complex patterned responses. These descending inputs from higher center neurons modulate both the fast-responding parasympathetic reflexes that provide beat-to-beat adjustments in cardiac performance as well as the longer latency cardio-cardiac reflexes involving spinal cord (sympathetic) neurons.

Clinical Implications

As a powerful, classic negative feedback system, the baroreflex provides rapid cardiovascular adjustments to the changing demands on the heart and great vessels. Cardiovascular feedback is constantly responding and remodeling to altered physiological status such as occurs during prolonged bed rest or zero gravity conditions. Substantial remodeling and reshaping of these reflex pathways also occurs in pathological states such as essential hypertension or heart failure. Prominent alterations in neuronal structures and neurotransmitter density within the NTS are associated with cases of sudden infant death syndrome, for example. The quantitative performance characteristics of the cardiac baroreflex have been linked to postmyocardial infarction survival as well as risk of sudden death from lethal cardiac arrhythmias. The sensitivity of the multiple reflexes involved in blood pressure regulation is reshaped at all stages of the cardiovascular neuroaxis during development of hypertension. The mechanisms for integration of all of the central neuronal reflexes involved in the maintenance of normal arterial pressure are not fully understood but presumably neural system remodels during the evolution of diseases hold keys to understanding essential hypertension as well as the development of novel and optimally effectively targeted therapies to treat such syndromes.

See also: Autonomic Nervous System: Cardiovascular Control; Autonomic Nervous System; Blood Pressure: Baroreceptors; Cardiovascular Function: Central Nervous System Control; Cerebrovascular Disease; Parasympathetic Nervous System.

Further Reading

Andresen MC and Kunze DL (1994) Nucleus tractus solitarius: Gateway to neural circulatory control. *Annual Review of Physiology* 56: 93–116.

Andresen MC, Kunze DL, and Mendelowitz D (2004) Central nervous system regulation of the heart. In: Armour JA and Ardell JL (eds.) *Basic and Clinical Neurocardiology*, 2nd edn., pp. 187–219. New York: Oxford University Press.

Armour JA (1976) Instant to instant reflex cardiac regulation. *Cardiology* 61: 309–328.

Cole CR, Blackstone EH, Pashkow FJ, Snader CE, and Lauer MS (1999) Heart-rate recovery immediately after exercise as a predictor of mortality. *New England Journal of Medicine* 341: 1351–1357.

Kinney HC, Filiano JJ, and Harper RM (1992) The neuropathology of the sudden infant death syndrome. A review. *Journal of Neuropathology and Experimental Neurology* 51: 115–126.

Kunze DL and Andresen MC (1991) Arterial baroreceptors. In: Zucker IH and Gilmore JP (eds.) *Excitation and Modulation, Reflex Control of the Circulation*, pp. 141–166. Boca Raton: CRC Press.

LaRovere MT, Pinna GD, Hohnloser SH, et al. (2001) ATRAMI investigators: Baroreflex sensitivity and heart rate variability in the identification of patients at risk for life-threatening arrhythmias: Implications for clinical trials. *Circulation* 103: 2072–2077.

Loewy AD (1990) Central autonomic pathways. In: Loewy AD and Spyer KM (eds.) *Central Regulation of Autonomic Functions*, pp. 88–103. New York: Oxford University Press.

Mayr U and Keele S (2000) Changing internal constraints on action: The role of backward inhibition. *Journal of Experimental Psychology: General* 129: 4–26.

Mendelowitz D (1999) Advances in parasympathetic control of heart rate and cardiac function. *News in Physiological Sciences* 14: 155–161.

Spooner PM, Albert C, Benjamin EJ, et al. (2001) Sudden cardiac death, genes, and arrhythmogenesis: Consideration of new population and mechanistic approaches from a national heart, lung, and blood institute workshop. Part I & II. *Circulation* 103: 2447–2452.

Autonomic Nervous System: Central Control of the Gastrointestinal Tract

W Jänig, Christian-Albrechts-Universität zu Kiel, Kiel, Germany

Introduction

The functions of the gastrointestinal tract (GIT) are ingestion, transport, and enzymatic breakdown of its content, reabsorption, and secretion of fluid (regulation of fluid balance), defense of the body against poisonous substances, and evacuation of waste. These functions are exerted by groups of effector cells of the GIT, such as smooth nonvascular muscle cells, secretory epithelia, interstitial cells of Cajal, endocrine cells, and cells related to the immune system (gut-associated lymphoid tissue). The enteric nervous system (ENS) is a specialized autonomic nervous system, consisting of motor neurons, intrinsic primary afferent neurons, and interneurons; the ENS regulates the activity of these effector cells, leading to spatial and temporal coordination of the functions of the GIT independent of the brain.

The brain adapts the intestinal functions at any moment to the behavior of the organism. This control is particularly powerful at the oral site (esophagus, stomach) and at the anal site (hindgut). It is exerted by various parasympathetic pathways, most of them originating in the dorsal motor nucleus of the vagus (DMNX) and the sacral spinal cord, and various sympathetic nonvasoconstrictor pathways originating in the thoracolumbar spinal cord. The vasculature of the GIT is, in addition to being locally controlled by enteric neurons, under direct control of the brain by visceral vasoconstrictor neurons. This control is particularly important in the regulation of arterial blood pressure (**Figure 1**).

Information from the GIT to the brain is transmitted via visceral afferent neurons and hormones secreted by gastrointestinal (GI) endocrine cells. Afferents projecting through the vagus nerves convey detailed information on various functions of the GIT. Visceral afferent neurons projecting from the GIT to the spinal cord are involved in evacuation and continence of the hindgut, in nociception and pain (including protective reflexes), and in some other functions. The endocrine signals derive from endocrine cells in the mucosa of the GIT (most of them in the anterior part) and pancreas and act on the brain mainly via the area postrema (AP) in the medulla oblongata and/or on the arcuate nucleus (ARN) in the hypothalamus. These GI hormonal signals are particularly important in the long- and short-term regulation of metabolism as well as of food intake, and therefore of hunger, satiety, and body weight (**Figure 1**).

The biological global functions of the control of the GIT by the brain are (1) adaptation of the functions of the GIT to the behavior of the organism and (2) regulation of energy balance and food intake (this includes regulation of body weight, hunger, and satiety). Pathobiologically, these global neural regulations by the brain may fail and result in (1) dysregulation of transport and fluid balance (secretion, reabsorption), being obvious at the oral and anal section of the GIT, where the control by the brain is particularly powerful, (2) dysregulation of energy balance, food intake, body weight, hunger, and satiety, and (3) psychosomatic diseases related to the GIT.

In the following sections, the neuronal pathways by which the central nervous system (CNS) controls the GIT are described, concentrating mainly on the medulla oblongata.

Afferent and Autonomic Pathways Connecting GIT and Brain

The parasympathetic systems have their dominant effects and overall access to target cells at the oral and anal sites of the GIT. Otherwise they do not interfere with single groups of effector cells in the GIT but influence enteric sensorimotor programs. The sympathetic nonvasoconstrictor systems act throughout the GIT, do not interfere directly with single groups of effector cells, and also modulate the enteric sensorimotor programs. The numbers of preganglionic parasympathetic and sympathetic neurons involved in GI functions is low (in the range of $\leq 1\%$!) in comparison to the numbers of enteric neurons.

Visceral Afferent Neural Pathways

The GIT gives detailed information about mechanical and intraluminal chemical events by the activity in visceral afferents to the brain. These visceral afferent neurons consist of three sets with different although overlapping functions:

1. About 80% of all axons in the vagal branches innervating the GIT are afferent. They are involved in regulation of motility, digestion, secretion, food intake, and possibly protection of the GIT. Their activation also triggers general sensations associated with the GIT, such as satiety, hunger, and fullness, and they are involved in the generation of emotional feelings. They are not directly involved in the generation

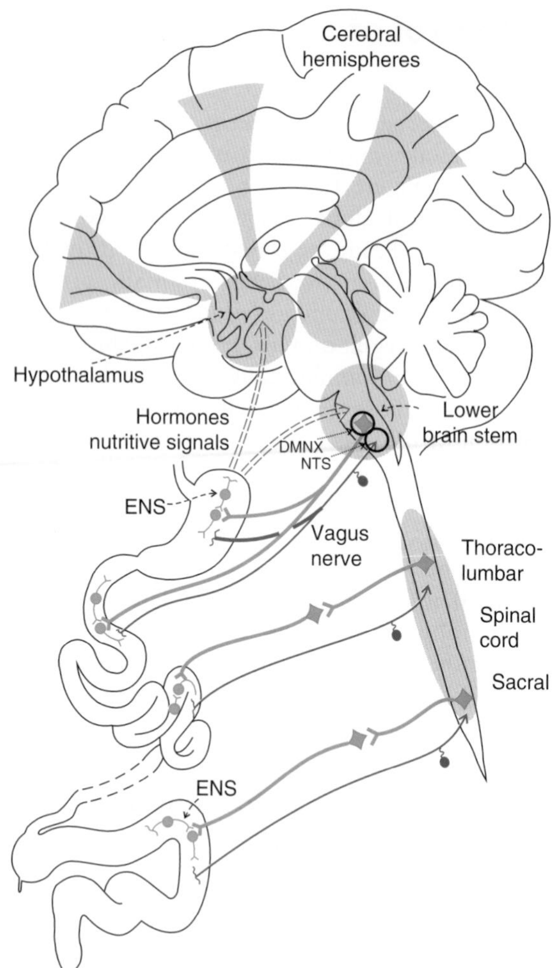

Figure 1 Afferent and efferent pathways connecting GIT (ENS, enteric nervous system) and brain. Afferent connections: vagal afferents to the lower brain stem (NTS, nucleus tractus solitarii); spinal afferents to the thoracolumbar or sacral spinal cord; and GI hormones acting on the neurons in the dorsal vagal complex via the AP in the lower brain stem and/or on the hypothalamus (mainly on the neurons of the ARN); nutritive signals (glucose and lipids in the blood) acting on neurons in the ARN and possibly the dorsal vagal complex. Efferent connections: preganglionic vagal (parasympathetic) neurons in the dorsal motor nucleus of the vagus (DMNX); sympathetic pathways originating in the intermediate zone of the thoracolumbar spinal cord; parasympathetic (spinal) pathways originating in the intermediate zone of the sacral spinal cord.

of visceral pain, although they participate in the control of transmission of nociceptive impulses in the spinal cord and possibly elsewhere. Most of them exhibit relative specificity in their responses to physiological stimulation. They respond to mechanical (distension, contraction, shearing stimuli acting on the mucosa) or/and intraluminal chemical stimuli. The chemical stimuli are mediated by enterochromaffin cells releasing serotonin, or by enteroendocrine cells releasing cholecystokinin (CCK), for example. They project viscerotopically to the nucleus tractus solitarii (NTS), the oral part of the anterior GIT being represented rostrocaudally and the caudal part (e.g., the cecum in the rat), caudomedially.

2. Thoracolumbar visceral afferent neurons innervating the GIT are involved in local and centrally organized protective reflexes, in visceral pain and discomfort, and in shaping emotions.

3. Sacral visceral afferent neurons innervating the GIT are involved in the same functions as are thoracolumbar visceral afferent neurons, but in addition they are important in the regulation of pelvic organs and in nonpainful sensation associated with the hindgut.

Afferent Hormonal and Nutritive Signals Related to the GIT

The brain receives hormonal and nutritional signals from the GIT or from cell groups (e.g., adipose tissue) related functionally to the GIT. These signals are important in the regulation of metabolism and food intake of the organism and therefore also of hunger, satiety, and body weight. Hormonal signals either act on neurons of the dorsal vagus complex (DVC) via the AP as satiety signals – CCK, glucagon-like peptide-1 (GLP-1), pancreatic polypeptide (PP), and peptide YY (PYY) (see **Table 1**) – or on neurons in the ARN of the hypothalamus as adipositas-related signals (leptin from the fat tissue, insulin from the endocrine pancreas, PYY from the ileum–hindgut) to inhibit food intake and metabolism or to initiate food intake (ghrelin from the stomach). Nutritional afferent signals are glucose and lipids (free fatty acids) in the blood. They act on neurons of the ARN of the hypothalamus as adipositas-related signals and possibly also on the lower brain stem via the AP.

Parasympathetic Pathways

Rather little is known about parasympathetic preganglionic DMNX neurons that project to the GIT. The DMNX is practically a motor nucleus of the GIT, and the neurons are specialized with respect to various motility, secretomotor, and enteric endocrine functions of the GIT. They may also be involved in protection of the GIT, innervating enteric circuits that regulate GI defense. They innervate motor neurons and interneurons of the ENS. **Figure 2** summarizes the way in which excitation of parasympathetic preganglionic neurons in the DMNX may generate or modulate different motility patterns of the GIT. This somewhat hypothetical scheme shows that the brain does interfere, via the preganglionic parasympathetic neurons, with many different functions in which the ENS is involved, and aids in understanding the complex peripheral mechanisms that are activated by the different functionally defined classes of preganglionic

Table 1 Integration of endocrine and neural signals in the DVC in the control of GI functions

Peptide	*Main origin*	*Released by/ during*	*Effect*	*Activation/inhibition of neurons*	*Global function(s)*
CCK*+	I cells in duodenum and small intestine	Protein, fat digests in duodenum	Gastric emptying↓, exocrine pancreas↑, food intake↓, satiety↓	Activation of neurons in the cNTS	Enhancement of satiety and digestion
CRH	Circulating, CRH neurons of the paraventricular nucleus of the hypothalamus and Barrington's nucleus	Stress	Gastric motility↓, acid secretion↓, transit time small intestine↓, transit time large bowel↓	Activation of inhibitory (NANC) pathways	Response of the GIT during stress
GLP-1*#	L cells of distal small intestine	Food ingestion (proportional to meal caloric intake)	Insulin secretion↑ (glucose dependent), glucagon↓, gastric emptying↓, caloric intake↓, satiety↑	Activation of neurons of DMNX (probably NANC pathway)	Enhancement of satiety, facilitation of disposal of ingested nutrients
PP*	Endocrine cells in pancreas	Fasting in anticipation of feeding (in cephalic phase)	Gastric motility↑, gastric secretion↑, gastric transit↑	Activation of secretomotor and motility-regulating neurons in DMNX (excitatory pathways)	Anticipatory preparation of feeding
PYY*#	Endocrine cells in the ileum, colon–rectum	Fatty acids in the ileum	Acid secretion↓, gastric motility↓, pancreatic secretion↓, food intake↓	Inhibition of secretomotor and motility-regulating neurons in DMNX (excitatory pathways)	Antagonizes effect of TRH/5-HT neurons, inhibits activity of stomach
TNF-α+	Macrophages, T lymphocytes, glia cells	Inflammation of the GIT	Gastric functions↓, nausea, emesis, anorexia	Activation of neurons in NTS and DMNX (NANC pathway)	Defense of the GIT
TRH	Circulating (thyroid gland), TRH/5-HT neurons in VMM	Cold stress	Gastric secretion↑, feeding behavior↑, metabolism↑	Activation of neurons in DMNX (excitatory pathways)	Coordination of thermoregulation and regulation of metabolism

References: general (Stanley et al., 2005); CCK (Moran and Kinzig 2004); CRH (Lewis et al., 2002; Taché et al., 2001); GLP-1 (Yamamoto et al., 2003); PP (Browning et al., (2005); Rogers et al., 1995; PYY (Yang et al., 2000); TNF-α (Emch et al., 2000; Emch et al., 2002; Hermann and Rogers 1995; Hermann et al., 2002; Travagli and Rogers 2001); TRH (Rogers et al., 1980; Rogers et al., 1995; Travagli and Rogers 2001; Yang et al., 2000).

Key to notation and abbreviations: *, also present in central neurons and/or vagal afferents (CCK); #, also acting on ARN; +, effects also, or mainly mediated, by vagal afferents; CCK, cholecystokinin; cNTS, central nucleus tractus solitarii; CRH, corticotropin-releasing hormone; DMNX, dorsal motor nucleus of the vagus; GIT, gastrointestinal tract; GLP-1, glucagon-like peptide-1; 5-HT, 5-hydroxytryptamine; NANC, nonadrenergic noncholinergic; PP, pancreatic polypeptide; TNF-α, tumor necrosis factor-α; TRH, thyrotropin-releasing hormone; VMM, ventromedial medulla; PYY, peptide YY.

parasympathetic neurons. Parasympathetic preganglionic neurons in the DMNX may influence the motility of the GIT (above all, stomach, duodenum, and ileum) via enteric neurons that excite the smooth muscle directly, enteric neurons that excite the smooth muscle cells via interstitial cells of Cajal, enteric neurons that inhibit smooth muscle cells, and enteric neurons that activate endocrine cells to release a hormone (e.g., gastrin), which in turn influences smooth muscle cells and enteric neurons involved in regulation of motility.

In addition to vagal preganglionic neurons integrated in regulation of motility of the GIT, there exist other types of parasympathetic preganglionic neurons that innervate enteric secretomotor neurons in the submucosal plexus or enteric neurons supplying endocrine cells of the GIT (e.g., G cells secreting gastrin and S cells secreting secretin; endocrine cells in the pancreas secreting insulin, glucagon, or PP). The cell bodies of these preganglionic neurons are possibly also situated in the DMNX. The organization of these peripheral pathways is poorly understood. Finally, preganglionic parasympathetic neurons that innervate the liver are involved in regulation of hepatic glucose production. Activation of these neurons inhibits hepatic glucose production.

The sacral (parasympathetic) spinal pathways to the colon–rectum are important to regulate continence

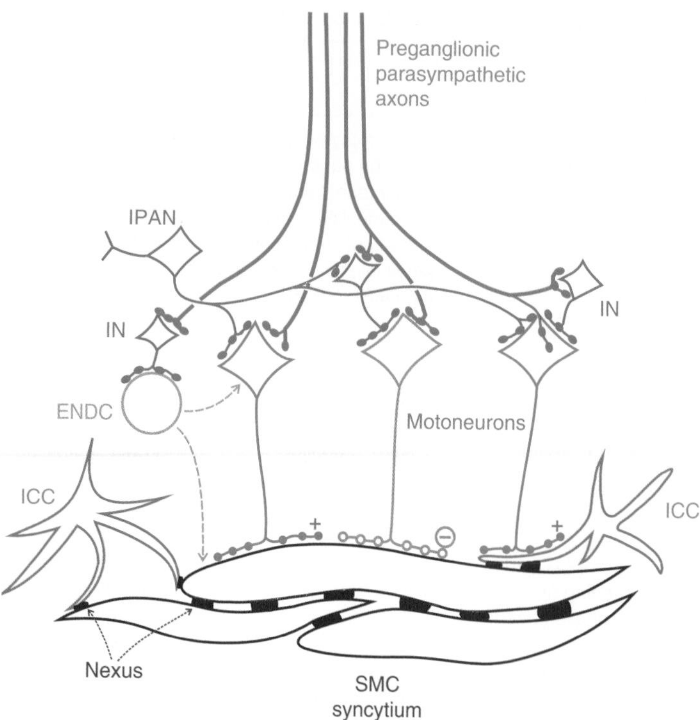

Figure 2 Diagram depicting the relation between vagal parasympathetic preganglionic neurons (cell bodies in the DMNX) and neurons of the ENS in regulation of motility. Various parasympathetic (vagal) pathways are involved in regulation of the motility of the smooth muscle syncytia of the GIT. ENDC, endocrine cell; ICC, interstitial cell of Cajal; IN, interneuron; IPAN, intrinsic primary afferent neuron; SMC, smooth muscle cell. +, excitation; −, inhibition. Adapted from Jänig W (2006) *The Integrative Action of the Autonomic Nervous System. Neurobiology of Homeostasis.* Cambridge, NY: Cambridge University Press.

and evacuation of the hindgut. Most postganglionic parasympathetic neurons do not innervate the target cells directly, but instead innervate enteric neurons in the myenteric plexus (but not in the submucosal plexus) or in the serosal plexus.

Sympathetic Pathways

The postganglionic sympathetic neurons are located in the prevertebral ganglia, and some are also in the paravertebral ganglia. Most postganglionic visceral vasoconstrictor neurons innervating the vasculature (arteries, arterioles, veins) of the GIT seem to be located in the paravertebral ganglia. They receive synapses from preganglionic neurons in the intermediate zone of the spinal cord but not from peripheral intestinofugal neurons. Furthermore, the postganglionic visceral vasoconstrictor neurons do not receive synaptic input from collaterals of spinal peptidergic primary afferent fibers (**Figure 3**).

The population of sympathetic nonvasoconstrictor neurons consists of motility-regulating neurons and secretomotor neurons (**Figure 3**). These two groups of neurons are most likely further subdifferentiated with respect to various functions of the GIT. Most of the motility-regulating neurons and secretomotor neurons innervate neurons of the myenteric or submucosal plexus, respectively. Activation of these sympathetic neurons inhibits enteric neurons; this inhibition occurs mainly presynaptically by decrease of release of excitatory transmitter, but also postsynaptically. A few postganglionic sympathetic secretomotor neurons innervate the mucosa directly and a few postganglionic sympathetic motility-regulating neurons innervate the nonsphincteric smooth muscles, leading in both cases to inhibition (decrease of contraction or secretion) when activated. Sphincter muscles are directly innervated by sympathetic postganglionic fibers and contract when these fibers are excited.

Postganglionic nonvasoconstrictor neurons in the prevertebral ganglia receive synaptic input from enteric intestinofugal neurons and form extraspinal reflex circuits. These intestinofugal neurons are cholinergic and may have vasoactive intestinal peptide co-localized. In the guinea pig, collaterals of spinal peptidergic afferents containing substance P (SP) appear to influence only postganglionic neurons in prevertebral ganglia that express somatostatin (in addition to noradrenaline). Only these postganglionic neurons, which have secretomotor function, express the tachykinin neurokinin-receptor 1 for SP (**Figure 3**).

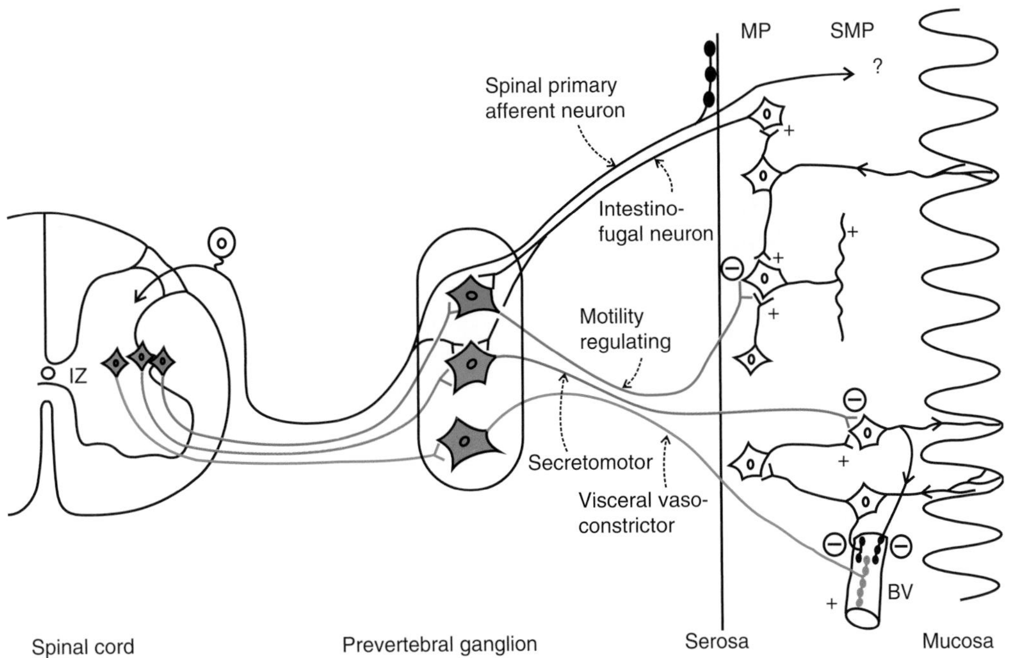

Figure 3 Sympathetic outflow to the small intestine. Most sympathetic postganglionic neurons projecting to the intestine are situated in the prevertebral ganglia. These neurons consist of three classes, each receiving separate synaptic inputs from functionally distinct classes of preganglionic neurons located in the intermediate zone (IZ) of the thoracolumbar spinal cord: vasoconstrictor neurons (red) innervating blood vessels, secretomotor neurons (blue) innervating neurons of the submucosal plexus (SMP), and motility-regulating neurons (green) innervating the myenteric plexus (MP). The vasoconstrictor neurons do not get synaptic input from collaterals of spinal primary afferent neurons (peptidergic) or from enteric intestinofugal neurons (cholinergic). The cell bodies of many visceral vasoconstrictor neurons are also located in paravertebral ganglia. The two latter classes of neurons, secretomotor and motility regulating, exert (mainly pre- but also postsynaptically) inhibitory effects on enteric neurons. They are probably divided into functional subtypes. The postganglionic nonvasoconstrictor neurons integrate preganglionic and peripheral synaptic inputs. They receive synaptic input from intestinofugal neurons of the ENS (cholinergic nicotinic). Postganglionic secretomotor neurons additionally receive peptidergic synaptic input from collaterals of spinal primary visceral afferent neurons. BV, blood vessels; +, excitation; −, inhibition. Adapted from Furness JB (2005) *The Enteric Nervous System*. Oxford: Blackwell Science Ltd.

Activity of sympathetic secretomotor neurons is related to the balance of fluid volume and electrolytes in the body. Decrease of both leads to activation of these neurons and consequently to inhibition of chloride and bicarbonate secretion by the intestinal mucosa. This is followed by a decreased transport of sodium and water into the intestinal lumen and conservation of body water. The central circuits linked to sympathetic secretomotor and motility-regulating neurons are unknown. They are unlikely to be identical to those linked to the visceral vasoconstrictor pathway.

Organization of the Dorsal Vagus Complex

The topographical organization of vagal afferents projecting to the NTS and of the efferent neurons projecting to the anterior part of the GIT is the anatomical basis of the neural control of the GIT (stomach, duodenum, small intestine) by the CNS. The NTS subnuclei related to the GIT and the ventrally located DMNX, including the AP, are also called the dorsal vagal complex (DVC) (see **Figure 4(a)**).

Functional Anatomy

Nucleus tractus solitarii Afferents from the subdiaphragmatic GIT project preferentially to the gelatinous, medial and commissural nuclei, but not to the central, interstitial, intermedial and ventrolateral nuclei of the NTS. Afferents from the esophagus project to the central nucleus of the NTS. The projection of afferent neurons to the NTS in the rat from the soft palate, pharynx, esophagus, stomach, and cecum is organized topographically, with limited overlap. This organization has three distinct characteristics: (1) afferents from different sections of the alimentary canal project to different subnuclei of the NTS (**Figure 4(a)**); (2) the afferent projections have a mediolateral and caudorostral segregation (**Figure 4(b)**); and (3) the organization of these afferent projections

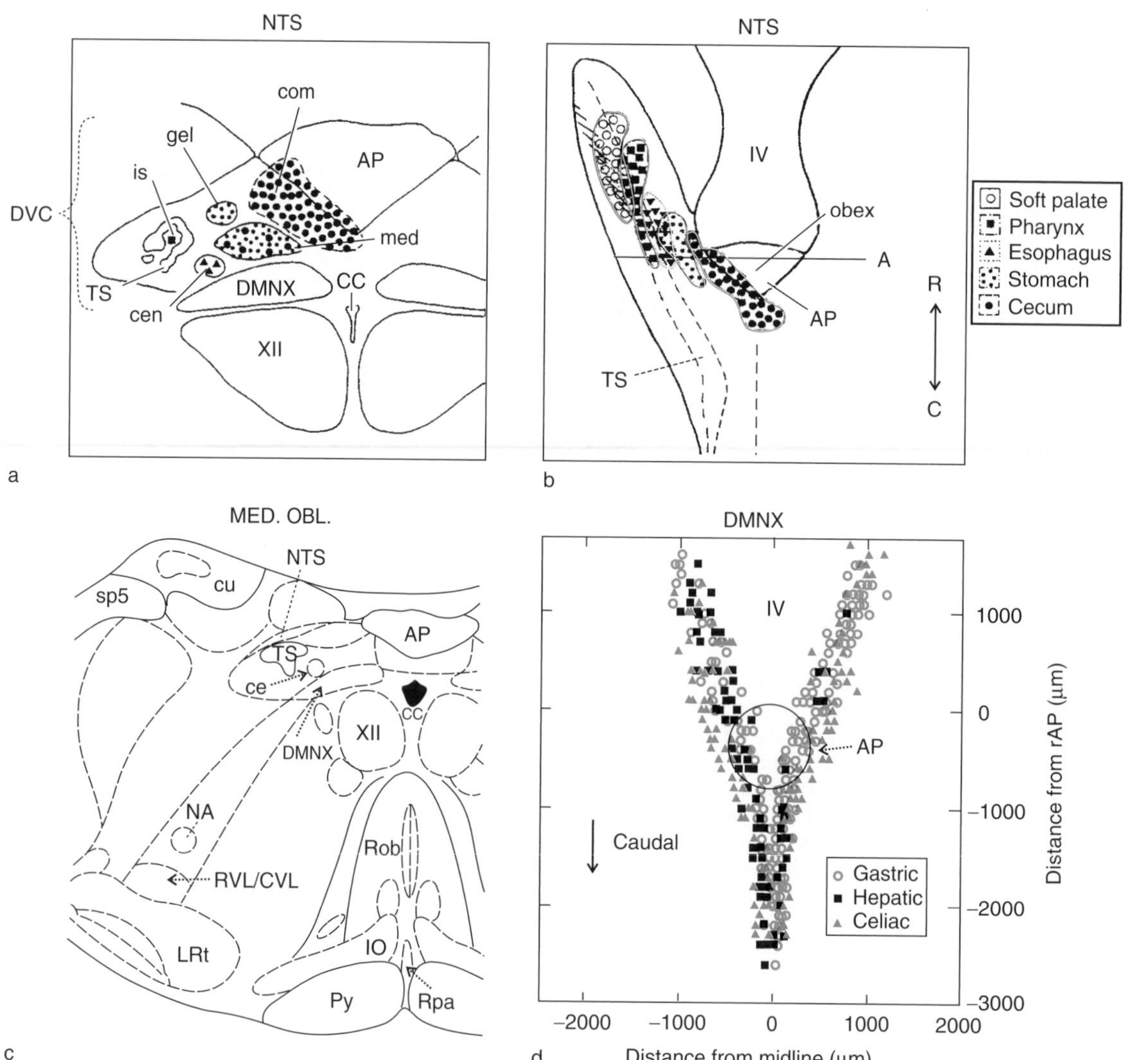

Figure 4 Topographic projection of vagal afferents from different sections of the GIT to the nucleus tractus solitarii (NTS) and topographic organization of preganglionic neurons supplying the GIT in the dorsal motor nucleus of the vagus (DMNX) of the rat. (a) Schematic transverse section through the dorsomedial medulla oblongata at the level of the area postrema (AP) containing the NTS and DMNX (location: see line A in (b)); nuclei of the NTS are central (cen), commissural (com), gelatinous (gel), interstitial (is), and medial (med). (b) Horizontal (longitudinal) section through the NTS in the rat, showing the rostrocaudal and mediolateral topography of the projections of afferents from the rostral part of the GIT up to the cecum. (c) Transverse section through the medulla oblongata at the level of the AP, showing the location of the DMNX in relation to other nuclei, in particular the NTS, and the AP. (d) The distribution of preganglionic neurons projecting through the gastric (circles), celiac (triangles), and hepatic (squares) branches of the subdiaphragmatic vagus nerves, as viewed in the horizontal plane. The neurons were labeled with Fast Blue injected intraperitoneally in rats 4–6 days before perfusing the animals. Various combinations of abdominal vagal branches (two gastric, two celiac, one hepatic) were transected, leaving one or two vagal branches intact. Fast Blue is taken up by the terminals of the intact vagal preganglionic axons and transported to the cell bodies. Preganglionic cell bodies were visualized in horizontal 100-μm-thick sections of the DMNX under ultraviolet epifluorescence, and were intracellularly filled via a glass micropipette with the dye Lucifer yellow. In this way, cell bodies, dendrites, and axons of the neurons were visualized and reconstructed. CC, central canal; cu, cuneate nucleus; DVC, dorsal vagal complex; IO, inferior olive; LRt, lateral reticular nucleus; NA, nucleus ambiguus; Py, pyramid; Rob, raphe obscurus; Rpa, raphe pallidus; RVL/CVL, rostral ventrolateral medulla/caudal ventrolateral medulla; sp5, spinal trigeminal tract; TS, tractus solitarius; IV, fourth ventricle; XII, hypoglossus. (a, b) Adapted from Altschuler SM, Ferenci DA, Lynn RB, et al. (1991) Representation of the cecum in the lateral dorsal motor nucleus of the vagus nerve and commissural subnucleus of the nucleus tractus solitarii in rat. *Journal of Comparative Neurology* 304: 261–274. (c) Adapted from Paxinos G and Watson C (1998) *The Rat Brain.* San Diego: Academic Press. (d) Modified from Fox EA and Powley TL (1992) Morphology of identified preganglionic neurons in the dorsal motor nucleus of the vagus. *Journal of Comparative Neurology* 322: 79–98.

is related to the mediolateral organization of the more ventrally located DMNX preganglionic neurons that project to the GIT.

The projection pattern of the GI vagal afferents to the subnuclei of the NTS is distinct from the projection pattern of cardiovascular afferent neurons and afferent neurons innervating the respiratory tract, although there is no simple relation between the subnuclei in the NTS and the projecting afferents from the three systems (cardiovascular, GI, respiratory). Furthermore, it is unknown whether functionally distinct types of vagal afferents from the GIT (mechanosensitive, chemosensitive, etc.) exhibit distinct projection patterns.

Synaptic transmission from vagal afferents to the NTS neurons is excitatory and the transmitter is glutamate. In the rostrocaudal area of the NTS, to which vagal afferents project, some 40–50% of the NTS neurons are synaptically activated by electrical stimulation of the subdiaphragmatic vagal nerve. Inhibitory responses are practically absent. Neurons activated from the GIT do not seem to receive additionally convergent synaptic input from arterial baro- or chemoreceptor afferents.

Dorsal motor nucleus of the vagus Preganglionic parasympathetic neurons innervating the proximal part of the GIT (stomach, duodenum, and small intestine, including pancreas and liver) are located in the DMNX. This nucleus is the motor nucleus of the foregut, in which the stomach has the highest representation. In the rat, it contains about 10 000 neurons (5000 neurons on each side), 7500 neurons being preganglionic and projecting to the GIT; the remaining neurons are interneurons or other types of preganglionic neurons projecting to visceral organs in the thoracic cavity.

The preganglionic DMNX neurons projecting through the subdiaphragmatic vagus nerves to the GIT are topographically organized into rostrocaudal cell columns. The neurons projecting in the gastric branches are located medially and the neurons projecting in the celiac branches are located laterally. Neurons projecting in the hepatic branch are located in a cell column between the medial (gastric) and the lateral (cecal) cell column of the left DMNX. The dendrites of these preganglionic neurons are almost confined to the cell columns projecting rostrocaudally, but also extend dorsally to the subnuclei of the NTS (**Figures 4(c)** and **4(d)**). Most preganglionic parasympathetic neurons innervating the lower esophageal sphincter probably are also located in a medial rostrocaudal cell column of the DMNX.

Intestino-Intestinal Reflex Circuits in the DVC

The intimate anatomical relationship between the NTS and the DMNX, both forming, together with the AP (see later), the DVC, implies that afferents from the GIT form disynaptic contacts with the preganglionic neurons projecting to the GIT. Systematic neurophysiological analysis of preganglionic DMNX neurons that project to the GIT have shown that several functional types of preganglionic neuron can be identified by their reflex responses to physiological stimuli applied to stomach, duodenum, or small intestine – for example, distension of organ sections (esophagus, stomach, duodenum) and intraluminal chemical stimulation. The patterns of reflex responses are correlated with other properties of the neurons, such as the rate of spontaneous activity, morphology of dendrites, and peptide content. At least four functional types of preganglionic neurons have been identified in the DMNX, indicating the functional differentiation of the parasympathetic pathways that are represented in the DMNX and project to the GIT.

The predominant synaptic transmission between the NTS neurons and the preganglionic DMNX neurons projecting to the GIT is inhibitory, the transmitter being mainly γ-aminobutyric acid. Other synaptic connections between NTS neurons and DMNX neurons are excitatory, the transmitter probably being glutamate. Gastric relaxation induced by esophageal distension is mediated by a reflex pathway involving the central nucleus of the NTS and the DMNX. NTS neurons involved in this reflex are noradrenergic, and this receptive relaxation reflex is significantly reduced after blockade of α-adrenoceptors (α_1, α_2) in the DMNX. Of the DMNX neurons projecting to the stomach, 75% are either excited or inhibited by noradrenaline. Thus noradrenaline is believed to be the transmitter mediating this reflex in the DMNX.

For most of these parasympathetic pathways, it is at present uncertain or a matter of speculation as to which type of target cell or functionally distinct enteric circuit they innervate. However, it is reasonable to assume that circuits of the ENS, which are related to nonvascular smooth muscles, exocrine glands, endocrine cells, or even the vasculature, receive differential signals from parasympathetic preganglionic neurons in the DMNX (see **Figure 2**). Thus it is also not far-fetched to assume that there are many more than four functional types of preganglionic parasympathetic neurons in the DMNX. Based on cyto- and chemoarchitectonic criteria, the DMNX in the human contains nine types of neuron.

Integration of Endocrine Signals by the AP in Neural Control of GI Functions

The reflex neural circuits in the DVC are also modulated by bloodborne signals that reach the neurons of the DVC by way of the AP and neighboring parts of the DVC. These signals are, for example, hormones from the GIT, such as GLP-1, PP, or PYY, or other hormones, such as corticotropin-releasing hormone (CRH) or thyrotropin-releasing hormone (TRH), or cytokines (e.g., tumor necrosis factor-α(TNF-α)). The AP and neighboring medial parts of the DVC do not have a vascular diffusion barrier, thus the blood–brain barrier is absent (**Figure** 5), allowing the hormonal signals (peptides) to have free access to the neural circuits of the DVC. The dendrites of both NTS neurons and DMNX neurons project into the area that lacks a blood–brain barrier. The circuits in the DVC are furthermore influenced by neuronal circuits that use neuroendocrine hormones as transmitters or neuromodulators (such as TRH or CRH).

Table 1 shows examples that demonstrate the close integration, in the lower brain stem, of autonomic (parasympathetic) systems and neuroendocrine systems in the regulation of GI functions. Each of the peptides listed has multiple functions, related to the GIT, that are mediated by peripheral neural and hormonal pathways and central pathways involving the DVC. With the exception of TNF-α, these peptides are also localized in neurons or act additionally or mainly via vagal afferents on the DVC (e.g., CCK). They may also act on the hypothalamus via circumventricular organs. Some circulating peptides acting mainly on the hypothalamus in the context of regulation of metabolism and satiety, such as leptin, ghrelin or insulin, are not listed in **Table 1**. However, these peptides may also act on the neural circuits of the DVC via the AP. It is important to note that each peptide listed in **Table 1** (see last column in the table) represents an organizing principle for global functions in which the brain–gut axis is involved.

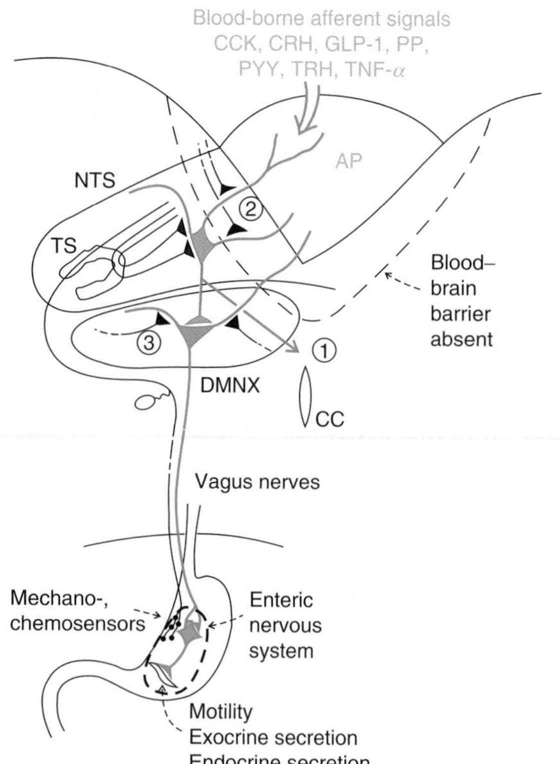

Figure 5 Dorsal vagal complex basic vago-vagal reflex circuits involved in regulation of functions of the GIT, and their modulation by hormones via the area postrema (AP). Neurons in the nucleus tractus solitarii (NTS) form inhibitory or excitatory synapses with preganglionic neurons in the dorsal motor nucleus of the vagus (DMNX) and are synaptically activated from the periphery by mechano- and/or chemosensitive vagal afferent neurons. The preganglionic neurons project through the vagus nerves to the ENS and have various functions. NTS neurons project additionally to various regions of the brain stem, hypothalamus, and telencephalon (1). Both neurons in the NTS and neurons in the DMNX are under multiple synaptic controls from brain stem, hypothalamus, and telencephalon (2 and 3). Signals in the blood (e.g., hormones) have access to the vago-vagal circuits via the AP and neighboring areas in which the blood–brain barrier is absent. CC, central canal; CCK, cholecystokinin; CRH, corticotropin-releasing hormone; GLP-1, glucagon-like peptide-1; PP, pancreatic polypeptide; PYY, peptide YY; TNF-α, tumor necrosis factor-α; TS, tractus solitarius; TRH, thyrotropin-releasing hormone. Adapted from Rogers RC, McTigue DM, and Hermann GE (1995) Vagovagal reflex control of digestion: Afferent modulation by neural and endoneurocrine factors. *American Journal of Physiology* 268: G1–G10, and Travagli RA and Rogers RC (2001) Receptors and transmission in the brain–gut axis: Potential for novel therapies. V: Fast and slow extrinsic modulation of dorsal vagal complex circuits. *American Journal of Physiology* 281: G595–G601.

Control of the DVC by Upper Brain Stem and Forebrain: A Concept

In order to account for the many types of functional reflexes mediated by the preganglionic neurons in the DMNX – as defined by the excitatory or inhibitory effect on the different target cells (different types of smooth muscle cells, exocrine glands, endocrine glands etc.) – and generated by the different types of vagal afferent neurons from the GIT, Powley and co-workers have created an interesting hypothesis on the spatial organization of the DVC. This hypothesis is based on the mediolateral representation of the GIT in the DMNX and on the rostrocaudal organization of the projections of vagal afferents to the NTS. It is proposed that these two layers of the DVC form a sensorimotor lattice that would allow for many specific reflexes elicited in the parasympathetic preganglionic neurons, by stimulation of distinct types of afferents from the GIT, and for various combinations (patterns) of reflexes.

The functionally distinct reflex circuits formed in the DVC between the afferents from and the preganglionic neurons to the GIT are the basic building blocks for the brain to control GI functions (**Figure 6**). The idea of the sensorimotor lattice of the DVC as proposed by Powley is a heuristic model and an approximation and does not apply to all situations. For example, esophageal distension elicits proximal gastric relaxation, the basis being the 'receptive relaxation reflex' arc consisting of vagal afferents from the esophagus, the second-order neurons in the pars centralis of the NTS (NTS_{cen}) that receive synaptic input from these afferents, and the vagal preganglionic neurons projecting to the stomach (probably two pathways). It has been shown that the neurons in the NTS_{cen} project extensively throughout the rostrocaudal DMNX. Thus the interesting hypothesis of the sensorimotor lattice in the DVC has to be tested rigorously before it can be accepted universally.

Anatomical studies have shown that several nuclei in the brain stem, hypothalamus, and telencephalon have reciprocal connections with the circuits of the DVC. The functions of most of these neural connections are poorly understood. This situation is conceptually quite similar to the role of spinal autonomic circuits in the control of spinal autonomic final pathways by supraspinal centers. Thus, these reflex pathways of the DVC are under modulatory control of neurons in the medulla oblongata and supramedullary brain centers (so-called 'executive' neurons; for example, neurons in the ventrolateral and ventromedial medulla, in the parabrachial and Kölliker–Fuse nuclei and A5 area, in the paraventricular nucleus of the hypothalamus and lateral hypothalamus, in the central nucleus of the amygdala and bed nucleus of the stria terminalis, and in the insula and medial prefrontal cortex), which also receive detailed afferent information from the GIT (via the NTS), from other visceral organs, and from somatic body domains. Executive neurons and basic autonomic circuits in the DVC associated with the GIT represent the internal state of the organism as far as the GIT is concerned (**Figure 6**).

An essential component of this internal state of the organism is the feedback by hormonal and nutritive signals from the GIT to the DVC via the AP and to the hypothalamus (mainly via the ARN). This hormonal feedback consists of several components (see **Table 1** for the DVC) and is integrated by the brain into the regulation of the metabolic (nutritional) state, the thermoregulatory state, the fluid matrix, the reproductive state, and the protective state of the organism.

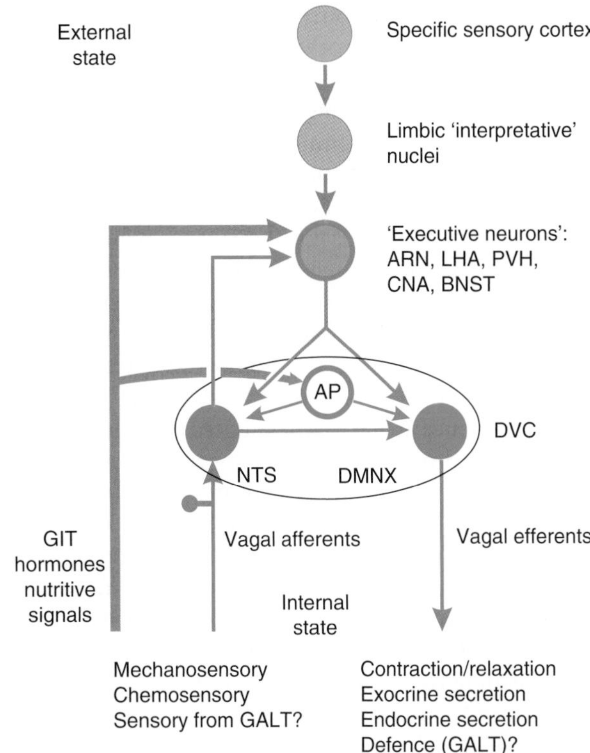

Figure 6 Proposed relationship between gastrointestinal vago-vagal reflex pathways, autonomic 'executive' neuronal circuits, limbic 'interpretative' neuronal circuits, and exterosensory systems (special cortical sensory systems). Several functionally specific vago-vagal reflex pathways organized in the dorsal vagal complex (DVC) of the medulla oblongata are the basic neuronal building blocks of the control of the gastrointestinal tract (GIT) by the brain. Vagal afferents measure mechanosensory, chemosensory, and other sensory events – for example, in the gut-associated lymphoid tissue (GALT) – and project to the nucleus tractus solitarii (NTS); vagal preganglionic neurons are located in the dorsal motor nucleus of the vagus (DMNX) and are involved in regulation of motility, exocrine secretion, endocrine secretion, and probably other events related to protection of the GIT (e.g., those associated with the GALT). Multiple hormonal afferent inputs from the GIT occur via GI hormones to the DVC, via the area postrema (AP), and to the hypothalamus, mostly via the arcuate nucleus (ARN) of the hypothalamus. Nutritive signals in the blood (glucose, lipids) act via the AP and ARN on the neural circuits in the DVC and the hypothalamus, respectively. Neurons of the 'executive' centers – for example, ARN, lateral hypothalamic and parafornical area (LHA), paraventricular nucleus of the hypothalamus (PVH), central nucleus of the amygdala (CNA), and bed nucleus of the stria terminalis (BNST) – evaluate the state of the internal milieu (by way of inputs from visceral afferents and hormonal inputs) as well as the current or anticipated behavioral state (via input from limbic nuclei that evaluate the significance of exteroceptive signals). These executive centers adapt the internal state (e.g., GI functions) to the behavioral state of the organism. Adapted from Rogers RC and Hermann GE (1992) Central regulation of brainstem gastric vago-vagal control circuits. In: Ritter S, Ritter RC, and Barnes CD (eds.) *Neuroanatomy and Physiology of Abdominal Vagal Afferents*, pp. 99–134. Boca Raton: CRC Press.

These regulations are homeostatic, involving the lower brain stem and the hypothalamus. However, they are adapted to the external state of the organism by the telencephalon, as occurring, for example, during heavy exercise, extreme environmental temperature changes, food and fluid deprivation, and exposure to toxic compounds or invasion by bacteria (sepsis, toxemia). The temporary adaptation of the homeostatic regulation to potentially life-threatening situations is called allostasis. The extra costs to meet these allostatic adaptations are called the allostatic load. Disease may result if this allostatic adaptation of the homeostatic regulations either does not occur or cannot be switched off and is too strong once it has been initiated and is not any longer needed or is switched on by trivial events that are not threatening for the organism.

The internal state is adapted to the behavior of the organism by cortical and limbic system structures that monitor and represent the external state of the organism (**Figure 6**). However, the internal state also modulates the central representation of the external state, leading to changes of sensory perception, body feelings, and emotional responses. This concept shows that there is a close integration between the homeostatic regulation of GIT functions and the higher nervous system functions related to body perception, emotions, and adaptation of behavior. The highly specific autonomic reflex pathways in the brain stem, hypothalamus, and cerebral hemispheres are the basis of this integration. This general concept of the control of GI functions by the brain has been propagated and worked out by Rogers and co-workers, and is the biological basis for the changes of GI function during various behaviors, including stress.

See also: Autonomic Nervous System: Gastrointestinal Control; Autonomic Nervous System: Neuroanatomy; Autonomic Nervous System; Gastrointestinal Tract role in Neural Control of Metabolism, Food Intake and Body Weight: A Summary; Parasympathetic Nervous System; Sympathetic Nervous System.

Further Reading

Altschuler SM, Ferenci DA, Lynn RB, et al. (1991) Representation of the cecum in the lateral dorsal motor nucleus of the vagus nerve and commissural subnucleus of the nucleus tractus solitarii in rat. *Journal of Comparative Neurology* 304: 261–274.

Berthoud HR and Neuhuber WL (2000) Functional and chemical anatomy of the afferent vagal system. *Autonomic Neuroscience* 85: 1–17.

Browning KN, Coleman FH, and Travagli RA (2005) Effects of pancreatic polypeptide on pancreas-projecting rat dorsal motor nucleus of the vagus neurons. *American Journal of Physiology* 289(2): G209–G219.

Emch GS, Hermann GE, and Rogers RC (2002) Tumor necrosis factor-alpha inhibits physiologically identified dorsal motor nucleus neurons *in vivo*. *Brain Research* 951: 311–315.

Emch GS, Hermann GE, and Rogers RC (2005) TNF-alpha activates solitary nucleus neurons responsive to gastric distension. *American Journal of Physiology* 279: G582–G586.

Fox EA and Powley TL (1992) Morphology of identified preganglionic neurons in the dorsal motor nucleus of the vagus. *Journal of Comparative Neurology* 322: 79–98.

Furness JB (2005) *The Enteric Nervous System*. Oxford: Blackwell Science Ltd.

Hermann G and Rogers RC (1995) Tumor necrosis factor-alpha in the dorsal vagal complex suppresses gastric motility. *Neuroimmunomodulation* 2: 74.

Hermann RC, et al. (2002) *American Journal of Physiology* 283: G634.

Jänig W (2005) Vagal afferents and visceral pain. In: Undem B and Weinreich D (eds.) *Advances in Vagal Afferent Neurobiology*, pp. 461–489. Boca Raton: CRC Press.

Jänig W (2006) *The Integrative Action of the Autonomic Nervous System. Neurobiology of Homeostasis*. Cambridge, NY: Cambridge University Press.

Lewis MW, Hermann GE, Rogers RC, and Travagli RA (2002) *In vitro* and *in vivo* analysis of the effects of corticotropin releasing factor on rat dorsal vagal complex. *Journal of Physiology* 543: 13.

McEwen BS and Wingfield JC (2003) The concept of allostasis in biology and biomedicine. *Hormones and Behavior* 43: 2–15.

Moran TH and Kinzig KP (2004) Gastrointestinal satiety signals. II: Cholecystokinin. *American Journal of Physiology* 286: G183.

Paton JF and Kasparov S (2000) Sensory channel specific modulation in the nucleus of the solitary tract. *Journal of the Autonomic Nervous System* 80: 117–129.

Paxinos G and Watson C (1998) *The Rat Brain*. San Diego: Academic Press.

Powley TL, Berthoud HR, Fox AP, et al. (1992) The dorsal vagal complex forms a sensory-motor lattice: The circuitry of gastrointestinal reflexes. In: Ritter S, Ritter RC, and Barnes CD (eds.) *Neuroanatomy and Physiology of Abdominal Vagal Afferents*, pp. 55–79. Boca Raton: CRC Press.

Rogers RC and Hermann GE (1992) Central regulation of brainstem gastric vago-vagal control circuits. In: Ritter S, Ritter RC, and Barnes CD (eds.) *Neuroanatomy and Physiology of Abdominal Vagal Afferents*, pp. 99–134. Boca Raton: CRC Press.

Rogers RC, Hitoshi K, Larry B, and Donald N (1980) Afferent projections to the dorsal motor nucleus of the vagus. *Brain Research Bulletin* 5: 365–373.

Rogers RC, McTigue DM, and Hermann GE (1995) Vagovagal reflex control of digestion: Afferent modulation by neural and endoneurocrine factors. *American Journal of Physiology* 268: G1–G10.

Stanley S, Wynne K, McGowan B, et al. (2005) Hormonal regulation of food intake. *Physiological Reviews* 85: 1131–1158.

Taché Y, Martinez V, Mulugeta M, and Wang L (2001) Stress and the gastrointestinal tract. III: Stress-related alterations of gut motor function: Role of brain corticotropin-releasing factor receptors. *American Journal of Physiology* 280: G173–G177.

Travagli RA, Hermann GE, Browning KN, et al. (2006) Brainstem circuits regulating gastric function. *Annual Review of Physiology* 68: 279–305.

Travagli RA and Rogers RC (2001) Receptors and transmission in the brain-gut axis: Potential for novel therapies. V: Fast and slow extrinsic modulation of dorsal vagal complex circuits. *American Journal of physiology* 281: G595–G601.

Udem B and Weinreich D (2005) *Advances in Vagal Afferent Neurobiology.* Boca Raton: CRC Press.

Yamamoto H, Kishi T, Lee CE, et al. (2003) Glucagon-like peptide-1-responsive catecholamine neurons in the area postrema link peripheral glucagon-like peptide-1 with central autonomic control sites. *Journal of Neuroscience* 23: 2939.

Yang H, Kawakubo K, Wong H, Ohning G, and Tache Y (2000) Peripheral PYY inhibits intracisternal TRH-induced gastric acid secretion by acting in the brain. *American Journal of Physiology* 279: G575–G581.

Yang H, Yuan PQ, Wang L, and Taché Y (2000) Activation of the parapyramidal region in the ventral medulla stimulates gastric acid secretion through vagal pathways in rats. *Neuroscience* 95: 773–779.

Autonomic Nervous System: Central Respiratory Control

A V Gourine and K M Spyer, University College London, London, UK

Introduction

Breathing is the vital automatic rhythmic activity that ensures appropriate rates of oxygen uptake and carbon dioxide elimination. Neurophysiological data indicate that the respiratory rhythm is a coordinated alternation of inspiratory, postinspiratory, and expiratory activities. Their coordination controls lung ventilation and is also important for vocalization, swallowing, and motor control. Respiratory activity in mammals is generated by a network of neurons located within the lower brain stem – in the medulla and pons. The respiratory neuronal network has to be fully operational before birth and must ensure adaptation and survival of the organism in variable environmental and physiological conditions. In this article we discuss the central nervous mechanisms controlling breathing (**Figure 1**). We describe the anatomy of the respiratory network, discuss mechanisms of respiratory rhythm generation and cardiorespiratory interactions, and focus on the control of breathing exerted by chemo- and mechanoafferent inputs. We also give a brief overview on how central nervous control of respiratory activity is modified in some common physiological states.

Central Nervous Control of Respiratory Activity

Anatomical Organization

The core neuronal network responsible for the generation and shaping of the respiratory rhythm, as well as transmitting this rhythm to the spinal motor neurons controlling respiratory muscles, is located in the lower brain stem – within a bilaterally organized dorsal respiratory group and ventral respiratory column (VRC) of neurons in the pons and the medulla oblongata (**Figure 2**).

The VRC is essential and sufficient to generate the respiratory rhythm. It extends rostrocaudally in the ventrolateral medulla oblongata in close proximity to the nucleus ambiguus. The VRC consists of 'distributed' neuronal circuits, involving assemblies of adjacent classes of respiratory neurons. The VRC can be divided into several functional compartments, including (from rostral to caudal) the overlapping retrotrapezoid nucleus/parafacial respiratory group (RTN/pFRG), the Bötzinger complex, the pre-Bötzinger complex (PBC), and rostral and caudal ventral respiratory groups (**Figure 2**). These functional divisions of the VRC have been identified using specific neurochemical markers and physiologically by monitoring changes in respiratory activity following activation and/or blockade of neurons in these regions of the brain stem. Lesioning of the PBC has identified it as the principal kernel for generating respiratory activity. It contains neurons that are essential and sufficient for generation of the basic inspiratory rhythm. The PBC can be visualized immunohistochemically using antibodies directed against neurokinin-1, μ-opioid receptors, somatostatin, or connexins. Recent evidence suggests that RTN/pFRG may contribute to the generation of the respiratory rhythm – it appears to produce an expiratory rhythm, in addition to inspiratory activity that is generated by the PBC. It has been suggested that the activity of the latter is dominant in mammals at rest, when there is little or no active expiration (discussed in more detail later). The other functional compartments of the VRC are involved in shaping the basic rhythm (produced by PBC), generation of the respiratory pattern and relaying this pattern to cranial and spinal resistance and respiratory motor neurons. Neuronal interactions within the dorsolateral pontine parabrachial and Kölliker–Fuse nuclei (pontine respiratory group; referred to as 'pneumotaxic center' in the older literature) (**Figure 2**) provide input to the VRC and play an important role in reflex regulation and shaping of respiratory activity patterns.

Activity of the brain stem respiratory network is powerfully influenced by peripheral afferent inputs originating from the arterial chemoreceptors, which are sensitive to changes in the arterial pH and levels of oxygen and carbon dioxide ($[H^+]$, PaO_2, $PaCO_2$), and from the mechanoreceptors that provide information about the mechanical status of the respiratory tract. Both arterial chemoreceptor afferents and afferents originating from the pulmonary mechanoreceptors terminate in the dorsal medullary nucleus tractus solitarii (NTS). NTS neuronal circuits which involve neurons within the dorsal respiratory group (located in the ventrolateral NTS) integrate these afferent inputs, modulate respiratory activity patterns in relation to these inputs, and contribute in particular to promoting the transition from inspiration to expiration.

The primary central CO_2 chemosensitive areas are located on the ventral surface of the medulla just beneath the VRC and in the close proximity to the brain stem respiratory rhythm and pattern generator (**Figure 2**). There is evidence suggesting that the PBC, pontile regions, and NTS may also contain functional respiratory CO_2 chemoreceptors.

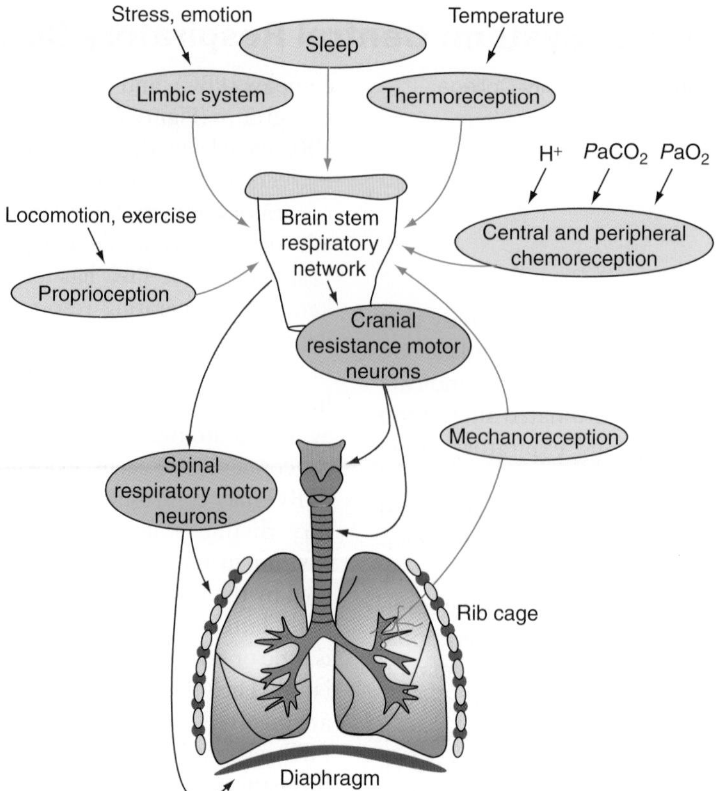

Figure 1 Regulation of breathing by the central nervous system: main afferent inputs and motor output. Refer to the text for details.

Generation of the Respiratory Rhythm

Neurophysiological data indicate that the medullary respiratory neuronal network oscillates in a three-phased respiratory pattern (**Figure 3**). These three phases are defined as inspiration, postinspiration (passive expiration), and expiration (active expiration). From a motor control perspective these phases represent successive activation of different respiratory muscles. During inspiration, the diaphragm and other inspiratory muscles contract. During postinspiration, contraction of the inspiratory muscles ceases and laryngeal muscles contract to control the outflow of the air from the lungs. Expiratory intercostal and abdominal muscles contract if there is an active expiration. Breathing in mammals is dominated by inspiration, and at least two antagonistic phasic activities are necessary – for example, inspiratory and postinspiratory activities – during rapid shallow breathing. These phases are usually supplemented by a third, expiratory activity, which is either weak or absent during quiet breathing. The respiratory network is dominated by synaptic inhibition, which determines the phases of the respiratory cycle. Most of the medullary respiratory neurons show evidence of synaptic inhibition at some time of the respiratory cycle. When synaptic inhibition is blocked experimentally, phase relationships between inspiratory, postinspiratory, and expiratory activities are markedly disturbed; however, as discussed later, generation of the respiratory rhythm *per se* is not completely abolished.

This brain stem respiratory neuronal network has the ability to generate different respiratory patterns (**Figure 3**). Normal breathing, or eupnea, is characterized by a slow augmenting sequential activation of the respiratory muscles innervated by cranial and spinal nerves during inspiration and expiration. Gasping is an inspiratory pattern characterized by a rapid decrementing synchronous activation of cranial and spinal respiratory muscles with minimal expiratory activity. Gasping is expressed in pathological conditions – for example, during severe hypoxia – and serves as a powerful mechanism of autoresuscitation. Emotions, partial lung collapse, and some other conditions increase the incidence of large-amplitude inspirations or sighs (augmented breaths), which occur on top of the 'normal' eupneic inspirations. There is evidence that the PBC neuronal network isolated within a brain stem slice is capable of generating all these qualitatively different breathing patterns, suggesting that normal breathing, gasps, and sighs may result from reconfiguration of activity of

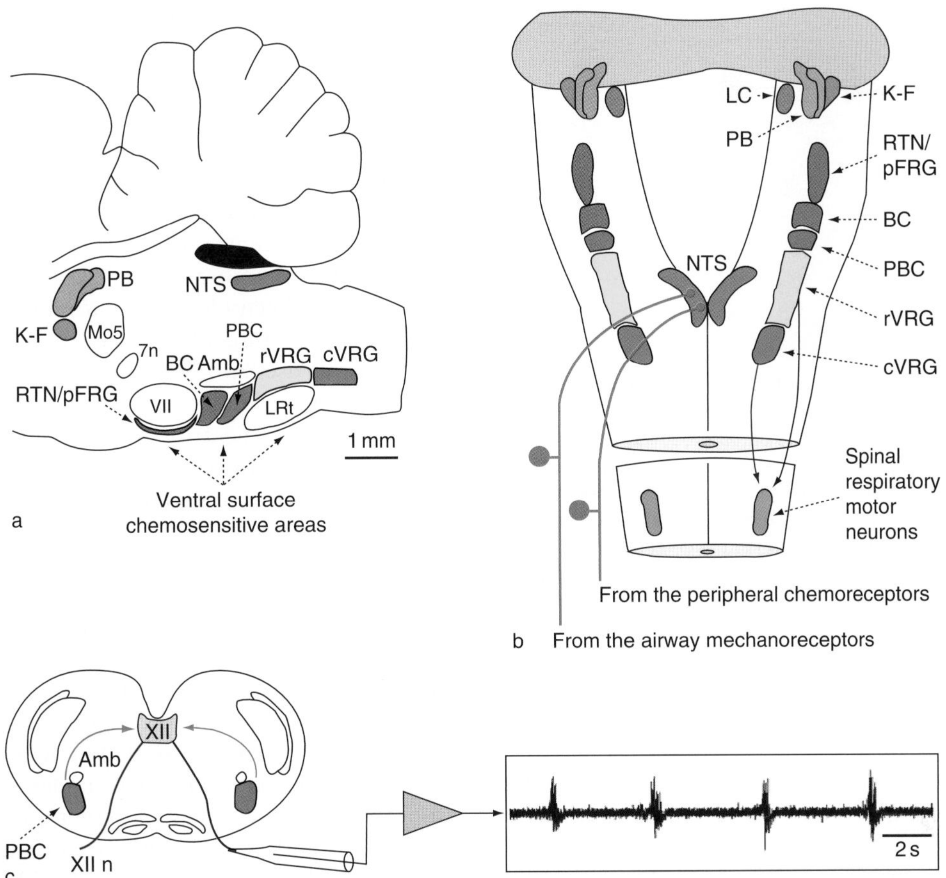

Figure 2 Neuroanatomy of the brain stem respiratory network. Sagittal (a) and horizontal (b) views of the rat brain stem, showing locations of the main groups of respiratory neurons in the mammalian central nervous system. (c) Diagram of a 'rhythmic' 300- to 700-μm-thick transverse slice prepared from the neonatal rat medulla oblongata and containing PBC, which retains rhythmic activity after *in vitro* isolation. Rhythm generated by the PBC is transmitted to the XIIth nucleus allowing recordings of inspiratory motor discharges with a suction electrode from the rootlet of the hypoglossal nerve (XII n). VII, facial nucleus; XII, hypoglossal nucleus; Amb, nucleus ambiguus; BC, Bötzinger complex; cVRG, caudal ventral respiratory group; K-F, Kölliker–Fuse nucleus; LC, locus coeruleus; LRt, lateral reticular nucleus; Mo5, motor trigeminal nucleus; NTS, nucleus tractus solitarii; PB, parabrachial nucleus; PBC, pre-Bötzinger complex; RTN/pFRG, retrotrapezoid nucleus/parafacial respiratory group; rVRG, rostral ventral respiratory group.

the same rhythmogenic neuronal circuits. Underlying synaptic and cellular (membrane) mechanisms responsible for generation of different respiratory patterns appear to be quite distinct. During severe hypoxia, gasping is induced when synaptic inhibitory mechanisms (essential for eupnea) are suppressed. It has also been shown that gasping relies on persistent sodium current (which is augmented by hypoxia), whereas the generation of eupnea may not require this cellular mechanism.

As mentioned previously, the PBC is believed to contain the principal kernel, or 'noeud vital,' for producing basic inspiratory rhythm (**Figure 2**). The central role of the PBC in neurogenesis of breathing was demonstrated in experimental animals from evidence showing that the rhythmic respiratory output from the brain stem disappears when the PBC is destroyed. All PBC neurons which discharge with preinspiratory firing pattern (i.e., those neurons which fire during inspiration and for a short period before that) express neurokinin-1 receptors. In the adult rat there are approximately 300 such neurons per side, and selective experimental bilateral elimination of >80% of them results in an abnormal (ataxic) breathing pattern. Less extensive loss of neurokinin-1-expressing PBC neurons leads to a significant disruption of breathing during sleep.

The PBC neuronal circuitry continues to generate rhythmic inspiratory activity when isolated *in vitro* within coronal or sagittal brain stem slices prepared from the brains of neonatal rodents (**Figure 2**). Although the respiratory network is known to be dominated by synaptic inhibition, *in vitro* respiratory rhythm continues after blockade of postsynaptic inhibition, suggesting that pacemaker neurons may generate this rhythm. Indeed, respiratory pacemaker-like neurons have been identified within the PBC, supporting the pacemaker theory for central respiratory

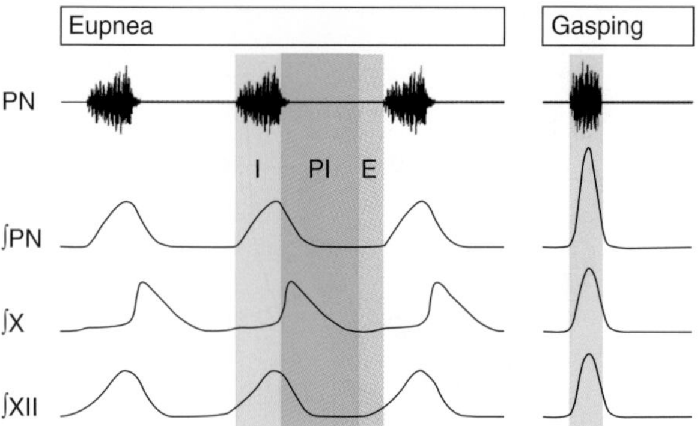

Figure 3 Three-phased respiratory activity. Schematic showing the activity patterns of spinal (phrenic) and selected cranial (vagus and hypoglossal) nerves during normal breathing (eupnea) and hypoxic gasping. Medullary respiratory neurons which discharge during either inspiration, postinspiration, or expiration can be recorded in various functional divisions of the VRS. Note how activity patterns of cranial nerves change during hypoxia-induced gasping (refer to the text for details). PN, activity of the phrenic nerve; ∫PN, integrated activity of the phrenic nerve; ∫X, integrated activity of the vagus nerve; ∫XII, integrated activity of the hypoglossal nerve; I, inspiratory phase; PI, postinspiratory phase; E, expiratory phase.

rhythm generation. This theory postulates that neurons with voltage-dependent pacemaker properties (they show spontaneous rhythmic oscillations in membrane potential) are responsible for generation of the respiratory rhythm. Voltage-independent pacemaker neurons have also been identified, suggesting that the PBC pacemaker populations are heterogeneous, with various burst-generating conductance mechanisms. However, the proportions of both populations of pacemaker-like neurons within the region are rather low. Furthermore, a stable rhythm still remains in slices in which the individual key ion channels for pacemaker activity are blocked, suggesting that respiratory rhythmic activity could be generated by neuronal networks exhibiting oscillatory behavior through chemical and electrotonic synaptic interactions.

At present many investigators favor a unifying 'hybrid pacemaker-network model' for respiratory rhythm generation; this model suggests that under normal conditions regular respiratory activity results from excitatory and inhibitory synaptic interactions between respiratory neurons within a network which shape the drive of the pacemaker units embedded within this network. Finally, when developmental changes in the operational conditions of the respiratory network are also taken into account, a 'maturational network-bursting theory' has been developed, which proposes that the principles for respiratory rhythm generation change significantly as the network matures. This concept emerged when data obtained using various neonatal and adult preparations were analyzed systematically. Understanding the connectivity within the PBC respiratory network has also been aided significantly by mathematical modeling. However, the detailed neuronal organization of the PBC remains to be very difficult to dissect experimentally.

There is evidence indicating that there might be more than one central respiratory rhythm generator in the mammalian brain stem. A second respiratory oscillator has been identified within the VRC, in the RTN/pFRG (**Figure 2**). The functional role of this more rostral oscillator has been a subject of debate. However, the available data suggest that it may be responsible for generation of the expiratory rhythm, in addition to the inspiratory rhythm generated by the PBC. Brain stem transection caudal to the RTN/pFRG, but rostral to the PBC (i.e., removal of the RTN/pFRG input; see **Figure 2**), does not affect inspiratory rhythm, but completely and irreversibly eliminates active expiration. Conversely, partial blockade of the PBC inspiratory oscillator activity by opiates disturbs inspiratory rhythm but does not affect expiratory activity, revealing the input and importance of the expiratory oscillator located in the RTN/pFRG. It was proposed that in mammals under normal conditions the activities of these different respiratory oscillators are tightly coupled but are dominated by the inspiratory rhythm generated by the PBC. It was also suggested that in other vertebrate species in which breathing is dominated by expiration (amphibians and reptiles have a breathing pattern of active expiration and passive inspiration), the RTN/pFRG oscillator may play a dominant role in neurogenesis of the respiratory activity. However, this hypothesis has not been tested experimentally.

The role of the pons in generation of eupnea has been a subject of debate for many years. A role of the pontile mechanisms in rhythmogenesis of normal respiratory activity is supported by the evidence obtained in anesthetized experimental animals showing that interruption of the vagal afferent input (vagotomy) and complete removal of the pons acutely eliminate eupnea and evoke gasping. This was observed in many mammalian species examined (but not in rodents) and suggested that pontile mechanisms may play an important role in the neurogenesis of eupnea. However, in unanesthetized animals, eupneic respiratory activity continues following transections at midpontine level or caudally through the facial (VIIth) nucleus. Similarly, studies in vagotomized and anesthetized juvenile rats have demonstrated that complete transection of the brain stem just rostral to the VIIth nucleus (i.e., complete removal of the pons; see **Figure 2**) does not significantly affect respiratory activity. Furthermore, as discussed previously, in the absence of pontile input, medullary neuronal circuitry appears to be capable of generating both gasping and eupneic inspiratory activity.

The basic respiratory rhythm generated by the respiratory oscillators is subsequently shaped, modified, and transmitted to bulbospinal inspiratory (located within the rostral ventral respiratory group) and expiratory (within the caudal ventral respiratory group) premotor neurons, which in turn relay the resultant respiratory pattern to spinal motor neurons innervating inspiratory and expiratory muscles, respectively.

Motor Output

The diaphragm (major inspiratory muscle in mammals) and intercostal muscles are innervated by the motor neurons located in the midcervical and thoracolumbar segments of the spinal cord. Inspiration occurs when phrenic motor neurons innervating the diaphragm and thoracic motor neurons innervating the external intercostal muscles are activated by descending command from the VRC. Expiratory motor neurons, which innervate internal intercostal muscles, are responsible for active expiration. There are mono- and polysynaptic pathways, which transmit respiratory pattern from the brain stem respiratory network to the spinal respiratory motor neurons.

Adequate ventilation also depends on airway patency, which is regulated by the 'resistance muscles' controlled by several cranial nerves. Striated muscles of the pharynx and larynx as well as smooth muscles of the trachea and bronchi all modulate airway resistance. The glottis works as a valve and plays an important role in regulating upper airway patency in eupnea. Glottal abductor and adductor muscles, which contract during inspiration and postinspiration, respectively, are innervated by the recurrent and superior laryngeal nerves. Different laryngeal motor neurones which discharge during inspiratory and postinspiratory phases of the respiratory cycle have been identified in the medulla and have been found to receive synaptic input from the respiratory network. Contraction of the glottal adductor muscles in postinspiration slows outflow of the air from the lungs, thus increasing the time of gas exchange and preventing lung collapse by maintaining its functional residual capacity. Laryngeal activity is also important for other related activities such as vocalization and swallowing.

The smooth muscles of the airways are controlled by parasympathetic (vagal and glossopharyngeal) and nonadrenergic, noncholinergic innervations. Vagal preganglionic neurons are located in the dorsal vagal nucleus and nucleus ambiguus of the medulla oblongata. The latter neurons discharge predominantly during the inspiratory phase of the respiratory cycle, contributing to an increase in airway smooth muscle tension during inspiration. Most reflex changes in the airway resistance, as well as maintenance of resting airway tone, are mediated via the excitatory cholinergic parasympathetic smooth muscle innervation.

Respiratory Afferent Inputs

The medullary respiratory network receives modulatory inputs continuously from the other parts of the brain as well as afferent inputs from the periphery (**Figure 1**) and ensures appropriate ventilation of the lungs in variable environmental and physiological conditions. Peripheral arterial chemoreceptors (sensitive to changes in $P\mathrm{aO_2}$ and $P\mathrm{aCO_2}$/pH), brain stem CO_2/pH chemoreceptors, and mechanoreceptors located in the airways control the activity of the brain stem respiratory network. Respiratory activity is also influenced by the activity of cardiovascular and other afferents; however, the inputs arising from chemo- and mechanosensors are of primary importance.

Chemosensory inputs Experiments conducted in animals in which the carotid and aortic bodies have been denervated have shown different functional roles for the peripheral and central chemoreceptors. Hypoxia fails to stimulate ventilation centrally, suggesting that carotid and aortic bodies are the primary oxygen-sensitive sites. In contrast, although peripheral chemoreceptors are sensitive to changes in $P\mathrm{CO_2}$ and pH, the ventilatory response to CO_2 is largely preserved in animals after denervation of the peripheral chemoreceptors.

In adult mammals the specialized chemosensitive neurosecretory glomus cells of the carotid body (located strategically at the place where the common carotid artery bifurcates) are the primary O_2 sensors. Glomus cells monitor changes in $P\mathrm{O_2}$ and, when stimulated by hypoxia, release excitatory mediator(s) to activate afferent fibers of the carotid sinus nerve. Information about arterial $P\mathrm{O_2}$ relays to the brain via the glossopharyngeal nerve and is processed by second-order neurons in the NTS. Under basal conditions ($P\mathrm{aO_2}$ $\sim$75–100 mmHg) the drive from the peripheral chemoreceptors is low. When $P\mathrm{aO_2}$ decreases below 60 mmHg, carotid chemoreceptor activity and ventilation increase exponentially. This ensures adequate supply of O_2 to the brain, which is highly sensitive to hypoxia.

CO_2 provides the major tonic drive for ventilation. Even a small increase in $P\mathrm{aCO_2}$ results in a profound increase in respiratory activity to restore the normal level of CO_2 in the arterial blood (~40 mmHg). Up to 80% of the CO_2-evoked respiratory response is mediated by the action of CO_2 at the chemoreceptors located within the lower brain stem. The primary central CO_2 chemosensors are believed to be located on the ventral surface of the medulla oblongata (**Figure 2**), although other regions of the brain stem, such as the PBC, the medullary raphe nuclei, the NTS, and the locus coeruleus, have been reported to contain functional respiratory chemoreceptors. Ventral surface CO_2 chemosensors are located just beneath and in a close proximity to the VRC. There is evidence suggesting that upon stimulation these chemosensors release the purine nucleotide adenosine triphosphate (ATP), which may act on the VRC respiratory neurons to evoke appropriate changes in breathing. There is also evidence indicating that respiratory neurons within the VRC could be intrinsically sensitive to CO_2.

Mechanosensory inputs Mechanoreceptors are found throughout the entire respiratory tract. They supply the brain stem respiratory network with information about the mechanical status of the lungs and chest, contribute to the control of respiratory activity, and initiate protective reflexes such as cough. Mechanoreceptors in the lungs and airways have both myelinated and unmyelinated nerve fibers. These fibers travel to the brain mainly within glossopharyngeal and vagus nerves and terminate in the NTS. Slowly adapting pulmonary stretch receptors are activated when the lungs inflate and play a critical role in termination of inspiration and prolongation of expiration (Breuer–Hering reflex). These receptors are located in the airway smooth muscles and have large myelinated fibers. The information is transmitted to the pontine nuclei (pneumotaxic center) and VRC via the NTS second-order relay neurons located within the dorsal respiratory group. Other mechanoreceptors – rapidly adapting pulmonary stretch receptors and bronchopulmonary C-fiber receptors – initiate defensive respiratory reflexes in response to inhaled irritants or rapid lung inflations or deflations. Lung congestion and edema are considered to be the main stimuli for pulmonary C-fiber afferents.

Other inputs Inputs arising from the limbic system influence the activity of the brain stem respiratory network and evoke modifications of breathing during stress, affective behavior, etc. Activation of proprioceptors in muscles, tendons, and joints stimulates breathing and contributes to the increases in ventilation during locomotion and exercise. There are also pathways originating from the neocortex (some of them going directly to the spinal motor neurons innervating the respiratory muscles) which allow voluntary control of breathing.

Plasticity

Breathing is continually adjusted to ensure appropriate ventilation of the lungs in changeable environmental and physiological conditions. Respiratory motor output exhibits significant plasticity, both short-term, when changes in the activity outlast the duration of the stimulus (e.g., hypoxic episode) by seconds and minutes, and long-term, when these changes persist for several hours. Respiratory plasticity can be elicited by hypoxia, hypercapnia, exercise, deafferentation, neural injury, and some other stimuli. Several neurotransmitters, such as serotonin, dopamine, and nitric oxide, have been proposed to be responsible for the respiratory neuroplasticity. For example, serotonin acting within the brain stem as well as at the spinal level (phrenic nucleus) appears to play a central role in the long-term facilitation of breathing triggered by recurrent hypoxic episodes.

Interactions between the Respiratory and Cardiovascular Systems

Activities of the cardiovascular and respiratory systems are intimately linked (**Figure 4**). Respiratory and cardiovascular rhythms are controlled in relation to one another, ensuring adequate ventilation–perfusion matching within the lungs, thus maintaining optimal respiratory gas exchange. Respiratory sinus arrhythmia – increases in heart rate during inspiration due to a rhythmic partial withdrawal of the inhibitory vagal tone – represents a classical example of cardiorespiratory integration. In variable physiological conditions appropriate changes always occur in both systems. Stimulation of the 'respiratory afferents' has a profound effect on the activity of the cardiovascular system. Similarly, stimulation of the 'cardiovascular afferents' modifies central respiratory drive (for example, increases in blood pressure reduce respiratory activity by stimulation of the arterial baroreceptors). Rhythmic respiratory modulation of vagal parasympathetic and sympathetic activities can be recorded from the appropriate efferent nerves. It is probably not a coincidence that anatomically both cardiac vagal preganglionic neurons and presympathetic neurons are located in the ventrolateral medulla at sites in a close proximity to the brain stem respiratory network. The evidence of central cardiorespiratory integration can be seen at the level of individual neurons. For example, cardiac vagal preganglionic neurons of the nucleus

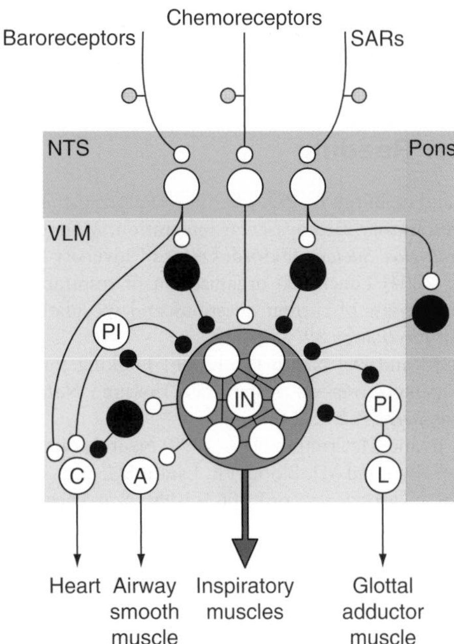

Figure 4 Diagramatic representation of central nervous mechanisms of cardiorespiratory interactions, showing some hypothesized central pathways of the pontomedullary cardiorespiratory network controlling inspiratory activity and motor output in selected cranial nerves. Connections shown may involve mono- or polysynaptic pathways. Open circles (o) depict excitatory connections. Filled circles (•) show inhibitory connections. A, airway vagal preganglionic neurons; C, cardiac vagal preganglionic neurons; IN, neuronal circuitry generating inspiratory activity; L, laryngeal motoneurones ('postinspiratory'); NTS, nucleus tractus solitarii; PI, postinspiratory neurons; SARs, slowly adapting pulmonary stretch receptors; VLM, ventrolateral medulla.

ambiguus receive powerful inhibitory inputs during inspiration as well as excitatory inputs during postinspiration, and these rhythmic changes in their activity underlie respiratory sinus arrhythmia.

Central Nervous Control of Breathing in Some Common Physiological States

Sleep

Sleep–wake cycles have profound influence on the activity of the brain stem respiratory network. Respiratory sensitivities to arterial levels of PO_2 and PCO_2 are reduced during sleep, and the apneic threshold is raised. There is a mild decrease in ventilation, resulting in an increase in arterial level of PCO_2 by around 2–5 mmHg. Withdrawal of a wakefulness-associated excitatory drive to the VRC is believed to be responsible for this decrease in respiratory activity during sleep. Therefore, breathing during sleep is often disordered and frequently interrupted by periods of apnea even in healthy individuals. The incidence of central sleep apneas increases significantly with age. Sleep is also associated with increases in upper airway resistance, which in severe cases can lead to complete upper airway obstruction and profound hypoxia (see later). Because sleep is an oscillatory state, the resultant changes in respiratory activity may be recurrent and/or amplified with repeated episodes.

Exercise and Locomotion

An adequate increase in lung ventilation is crucial to support the metabolic demands of the exercise. Respiratory activity increases even before the onset of the exercise, independently of inputs from the respiratory chemoreceptors or muscle proprioceptors. Hypothalamic structures are believed to be the source of this 'central anticipatory command' to the brain stem respiratory center. The major inputs into the hypothalamus originate from the cortical motor areas associated with locomotion. During exercise (and locomotion in general), respiratory and locomotor patterns are coupled. Respiratory activity is entrained by the somatic afferent input which ascends to the VRC via the lateral parabrachial nucleus of the pons and causes resetting of the respiratory rhythm.

Temperature Regulation

Low and high ambient temperatures are known to have a profound effect on ventilation. During cold exposure, as well as during fever, heat production increases and this increase in metabolism is accompanied by appropriate increases in ventilation. During a febrile response, mediators of inflammation (e.g., cytokines) may also increase respiratory activity by their action on the peripheral chemoreceptor sites and/or within the brain stem respiratory network. At high ambient temperatures many animal species increase the frequency of the respiratory movements to facilitate evaporative heat loss from the respiratory tract (polypnea and panting). These adaptive reflex modifications in breathing can be evoked from the hypothalamic thermoregulatory center following activation of the peripheral (skin) thermoreceptors, even in the absence of changes in core body temperature.

Emotions and Stress

Emotions and stress result in a significant alterations in breathing patterns. Inputs from the limbic system to the brain stem regions concerned with central respiratory control evoke changes in breathing activity during stress, affective behavior, etc.

Hypoxia

The initial increase in respiratory activity in response to hypoxia is evoked by activation of the peripheral chemoreceptors. If hypoxia is sustained, this increase in breathing is followed by a decline in central respiratory output. Structures located in the hypothalamus, mesencephalon, and the brain stem have been shown to be involved in this secondary respiratory decline. In neonatal experimental animals, brain stem transection or lesions in the lateral upper pons help to maintain respiratory activity that is normally abolished during sustained hypoxia. Several brain sites within the thalamus, hypothalamus, pons, and the medulla have been proposed to function as central O_2 sensors. Their physiological role, however, remains largely unclear.

Summary

Breathing is controlled by a network of neurons located within the lower brain stem. In adult mammals the basic rhythm of respiratory activity is generated by a network of neurons within the so-called PBC complex of the ventral medulla oblongata and is then subsequently shaped, modified, and transmitted to bulbospinal premotor neurons which relay the resultant respiratory pattern to spinal motor neurons innervating respiratory muscles. The activity of the respiratory network is influenced by continuous modulatory inputs from the other parts of the brain and afferent inputs from the periphery. Peripheral and central chemoreceptors (which supply information about arterial pH and levels of O_2 and CO_2), mechanoreceptors located in the respiratory tract (which provide information about the mechanical status of the lungs and chest), proprioceptors in muscles, tendons, and joints (which are responsible for coupling of respiratory and locomotor patterns), as well as from the higher centers of the central nervous system, such as the limbic system, are all essential for adaptive changes (both short- and long-term) in the respiratory motor output to ensure appropriate ventilation of the lungs in variable environmental and physiological conditions.

See also: Autonomic Nervous System: Cardiovascular Control; Autonomic Nervous System: Respiratory Control; Autonomic Nervous System: Neuroanatomy; Autonomic Nervous System: Central Cardiovascular Control; Brainstem Respiratory Circuits; Cardiovascular Function: Central Nervous System Control; Respiration.

Further Reading

Daly M and de Burgh B (1997) Peripheral arterial chemoreception and respiratory-cardiovascular integration. *Monograph for the Physiological Society.* Oxford: Oxford University Press.

Duffin J (2004) Functional organization of respiratory neurons: A brief review of current questions and speculations. *Experimental Physiology* 89: 517–529.

Feldman JL and Del Negro CA (2006) Looking for inspiration: New perspectives on respiratory rhythm. *Nature Reviews Neuroscience* 7: 232–242.

Feldman JL and McCrimmon DR (1999) Neural control of breathing. In: Zigmond MJ, Bloom FE, Landis SC, et al. (eds.) *Fundamental Neuroscience*, pp. 1063–1090. San Diego: Academic Press.

Feldman JL, Mitchell GS, and Nattie EE (2003) Breathing: Rhythmicity, plasticity, chemosensitivity. *Annual Review of Neuroscience* 26: 239–266.

Jordan D (2001) Central nervous pathways and control of the airways. *Respiration Physiology* 125: 67–81.

Lieske SP, Thoby-Brisson M, Telgkamp P, et al. (2000) Reconfiguration of the neural network controlling multiple breathing patterns: Eupnea, sighs and gasps. *Nature Neuroscience* 3: 600–607.

Putnam RW, Filosa JA, and Ritucci NA (2004) Cellular mechanisms involved in CO_2 and acid signaling in chemosensitive neurons. *American Journal of Physiology – Cell Physiology* 287: C1493–C1526.

Ramirez JM and Viemari JC (2005) Determinants of inspiratory activity. *Respiratory Physiology & Neurobiology* 147: 145–157.

Richter DW and Spyer KM (2001) Studying rhythmogenesis of breathing: Comparison of *in vivo* and *in vitro* models. *Trends in Neurosciences* 24: 464–472.

Rybak IA, Shevtsova NA, Paton JF, et al. (2004) Modeling the ponto-medullary respiratory network. *Respiratory Physiology & Neurobiology* 143: 307–319.

Smith JC, Butera RJ, Koshiya N, et al. (2000) Respiratory rhythm generation in neonatal and adult mammals: The hybrid pacemaker-network model. *Respiration Physiology* 122: 131–147.

Speck DF, Dekin MS, Revellette WR, et al. (eds.) (1993) *Respiratory Control – Central and Peripheral Mechanisms.* Lexington: The University Press of Kentucky.

St-John WM and Paton JF (2004) Role of pontile mechanisms in the neurogenesis of eupnea. *Respiratory Physiology & Neurobiology* 143: 321–332.

Taylor EW, Jordan D, and Coote JH (1999) Central control of the cardiovascular and respiratory systems and their interactions in vertebrates. *Physiological Reviews* 79: 855–916.

Autonomic Nervous System: Central Thermoregulatory Control

C J Gordon, U.S. Environmental Protection Agency, Research Triangle Park, NC, USA

The Thermoregulatory System

Temperature regulation in mammals and birds has evolved with autonomic and behavioral motor responses, termed thermoeffectors, to defend the core body temperature against changes – heat gain from and heat loss to the environment, as well as heat production from exercise and fever. The body's core temperature (i.e., in brain, heart, lungs, viscera, etc.) is usually maintained within narrow limits and is distinct from the thermal shell, which represents the skin and mucosal surfaces of the body that engage in heat exchange with the environment. The thermal shell also includes those tissues under the surfaces whose temperature may deviate from the core due to heat exchange with the environment.

Behavioral versus Autonomic Mechanisms

Thermoregulation is indeed a unique homeostatic system because it relies on higher level central nervous system (CNS) processes for the conscious sensation and elicitation of corrective motor responses. There is little conscious awareness of most other homeostatic processes, such as those involved in the regulation of blood pressure, respiration, blood pH, and other systems, whereas we are almost always aware and keenly sensitive to changes in our external thermal environment. We continuously strive to maintain a comfortable thermal environment, primarily by using behavioral thermoregulation. Behavioral thermoregulatory mechanisms in humans, such as adding or removing clothing, adjusting a thermostat, or setting a forced convective environment with a fan, all involve high-level, cerebral processing. Other species of mammals and birds will also preferentially use behavioral rather than autonomic processes to thermoregulate. The regulation of body temperature via behavioral mechanisms involves little metabolic energy and is thus preferred over that of autonomic mechanisms. Most species of the reptile, amphibian, and fish phyla depend entirely on behavioral mechanisms to thermoregulate.

Behavior alone is insufficient to maintain a regulated core temperature under the wide range of thermal environments that may be inhabited by birds and mammals. Behavioral thermoregulation can be suppressed during sleep, sickness, injury, and other conditions where behavior cannot be used to provide an optimum thermal environment. With the evolution of a CNS-mediated control of autonomic thermoeffectors, birds and mammals have developed a tightly controlled core temperature that is maintained day and night from soon after birth until the impending point of death. The thermal milieu is largely responsible for the expansion of mammal and bird fauna into environments that are uninhabitable for most other species.

The Autonomic Thermoregulatory System

The autonomic aspect of the temperature regulatory system is generally defined as the involuntary responses to heat and cold stress which modify the rates of heat production and heat loss. The motor responses, termed thermoeffectors, include sweating, panting (thermal tachypnea), or saliva spreading to increase evaporative heat loss; shivering, and nonshivering thermogenesis to increase heat production; adjustments in circulation of blood to the skin and other sites for heat exchange; and involuntary bristling of the fur (piloerection) to curtail heat loss. From this definition, one should note that autonomic thermoeffectors are not limited to responses controlled solely by sympathetic and parasympathetic efferents. Autonomic systems to defend the body core against heat and cold stress are recruited when behavioral options are unavailable or ineffective for maintaining thermal homeostasis. The autonomic systems for thermoregulation use organ systems that have evolved for other purposes: the cardiovascular system for controlling skin blood flow and heat loss, the respiratory system for evaporative heat loss in panting species (dog, cattle, sheep), and salivary glands of the digestive system in rodents that spread saliva on the fur for evaporative water loss. Brown adipose tissue (BAT) appears to be the only organ that has evolved with the main purpose of thermoregulation.

Basic Neural Pathways

Our understanding of the neural networks involved in the autonomic control of thermoeffectors has seen remarkable progress in the past decade. The cutaneous vasculature of the rat tail and rabbit ear and the sympathetic activation of BAT are key experimental models to elucidate the final motor pathways for the control of heat loss and heat production. Decades of research using microelectrodes to record the activity of thermoregulatory neurons *in vivo* and *in vitro* have led to a thorough understanding of the sensory and integrative aspects of the thermoregulatory system in the peripheral and CNSs. Microinjection of neurotransmitters, modulators, and hormones into specific locations of the CNS has provided data that tie together the neurophysiological and neurochemical mechanisms of thermoregulatory control.

A general model for thermal sensation, integration, and elicitation of thermoeffector responses can be applied to most mammals (**Figure 1**). Sensory inputs from the temperature of the skin, spinal cord, and other locations have been shown to impinge on thermoregulatory neurons in the preoptic area and anterior hypothalamus (POAH) that serve as sensors of core temperature and show exaggerated changes in firing rate with either an increase or decrease in POAH temperature. A relatively large percentage of neurons in the POAH are warm-sensitive. It is thought that cold-sensitive neurons increase their firing rate with POAH cooling because they are inhibited by synaptic connections from warm-sensitive neurons. Many neurons in the POAH and other select areas of the CNS (spinal cord, and/or other locations in the core) are sensitive to changes in skin temperature. There are also populations of neurons that are insensitive to changes in local temperature but respond vigorously to elevations or reductions in skin temperature. All together, the integration of the afferent thermal sensory processes with core temperature in a network of warm-sensitive, cold-sensitive, and thermal-insensitive neurons leads to the generation of effector signals that control the autonomic effectors (see later). These same integrated signals within the POAH are also thought to drive behavioral thermoregulatory processes through activation of higher level cortical pathways. Thermoregulatory neurons in the POAH are also influenced by nonthermal factors that are associated with or induce changes in the core temperature. For example, circulating reproductive hormones that control menstrual- and estrous-related

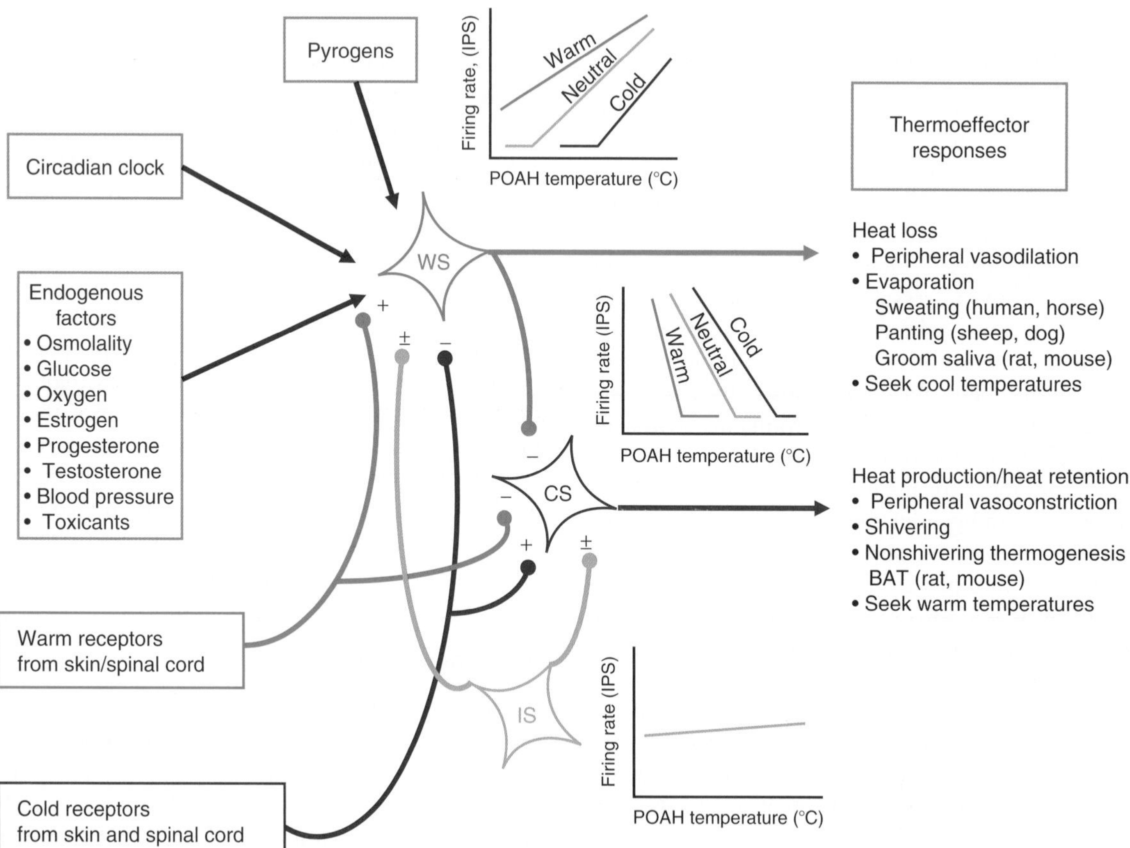

Figure 1 A neural model showing synaptic connections of warm-sensitive (WS), cold-sensitive (CS), and thermally insensitive neurons in the preoptic area and anterior hypothalamus (POAH). Activity of WS and CS neurons is modulated by ambient temperature, resulting in effector signals to modulate the activity of autonomic and behavioral motor outputs of the thermoregulatory system. For example, a WS neuron has higher activity when exposed to warm ambient temperature combined with elevated POAH temperature. IPS, Impulses/sec; + indicates stimulation; − indicates inhibition; ± indicates stimulation or inhibition.

rhythms of the core temperature have been shown to selectively alter activity of warm- and cold-sensitive neurons in the POAH. Glucose, oxygen, blood pressure, and tissue osmolarity have also been found to alter activity of some thermoregulatory neurons. The circadian pacemaker located in the suprachiasmatic nucleus also modulates activity of warm- and cold-sensitive neurons, leading to a 24 h regulated rhythm in core temperature. Environmental toxicants and other insults can affect the activity of the thermoregulatory centers.

The integration of the peripheral and central thermal sensory capacity of the thermoregulatory system was elucidated years ago with the application of stereotaxically implanted thermodes to heat or cool the POAH or other areas of the CNS. For example, heating of the POAH stimulates a heat loss response characterized by reduction in shivering (if the animal is exposed to cold), peripheral vasodilation, and sweating or panting. The animal behaves as if it was hot and seeks cooler temperatures. The thermoregulatory response to POAH heating can reach a magnitude such that heat loss exceeds heat production and the animal becomes hypothermic. Cooling the POAH area elicits a thermogenic response, cutaneous vasoconstriction to restrict heat loss, and the animal seeks warmer ambient temperatures. Studies employing POAH thermodes and manipulations of ambient temperature contributed to the concept of reciprocal inhibition of warm and cold neural pathways in the POAH.

Integrative neurons in the CNS are generally 5–10 times more sensitive to a change in local temperature than to changes in skin temperature. That is, a 1 °C reduction in brain temperature has about the same effect on a thermoeffector response as would a 5–10 °C reduction in skin temperature. When maintained within ambient temperatures where core temperature is controlled (i.e., zone of normothermia), autonomic and behavioral thermoeffectors are controlled primarily by changes in the shell or skin temperature. However, thermal sensors in the CNS and other core sites are called upon in the response to the hyperthermia from exercise and fever as well as when thermoregulation begins to fail during excessive cold or heat stress. The thermal sensory properties of neurons in the CNS of mammals and birds may be considered a vestige of evolution whereby the 'lower vertebrates' rely primarily on CNS thermal sensors to elicit thermoeffector responses.

Integration in the CNS

Neurons in the POAH integrate thermal information as well as nonthermal stimuli that can have a bearing on thermoregulation. Blood pH, blood pressure, oxygen level, osmotic tonicity, and glucose levels can all influence the activity of POAH neurons in a manner that would be expected based on their thermoregulatory effects. For example, dehydration has been shown to increase the threshold core temperature for evaporative water loss in mammals exposed to heat stress. This is an adaptive mechanism that conserves precious body water under conditions of prolonged heat exposure but also results in a higher regulated core temperature. Warm-sensitive neurons in the POAH respond to osmotic stress *in vitro* with a reduction in thermal sensitivity, a response that essentially raises the set point or threshold temperature to activate sweating. Overall, one can visualize a network of warm-sensitive, cold-sensitive, and thermally insensitive neurons in the POAH as well as in other locations in the CNS; all interact with nonthermal stimuli to generate effector signals that drive a series of thermoeffectors to control autonomic thermoregulation.

Thermal-sensitive neurons in the POAH are integrated with the control of blood pressure. Baroreceptor neurons that monitor blood pressure and blood volume have projections to thermal-sensitive neurons. The capacity of thermoregulatory neurons to integrate pressure/volume of the circulatory system is adaptive in times of cardiovascular crisis. For example, a reduction in blood pressure has been shown to preferentially stimulate warm-sensitive neurons and suppress cold-sensitive neurons in the POAH. This should result in a heat dissipatory response and subsequent reduction in core temperature. Rats subjected to hemorrhagic shock (an acute loss in blood volume) become hypothermic and select cooler ambient temperatures, an indication of a reduction in the set point. The hypothermic response improves the ability to survive the hemorrhagic shock, suggesting that the integration of thermal and baroreceptor inputs in the POAH is an adaptive response.

Set Point

The concept of a thermostat with a set-point temperature as depicted in **Figure 2** is an extremely useful analogy for explaining how autonomic and behavioral thermoeffectors are controlled by the CNS. Thermal physiologists define set point as follows: "The value of the regulated variable which a healthy organism tends to stabilize by the processes of regulation. When external or internal interferences tend to alter the regulated variable (i.e., body temperature), the resulting thermoeffector activities counter the alterations." In other words, if an organism uses its thermoeffectors to maintain core temperature at

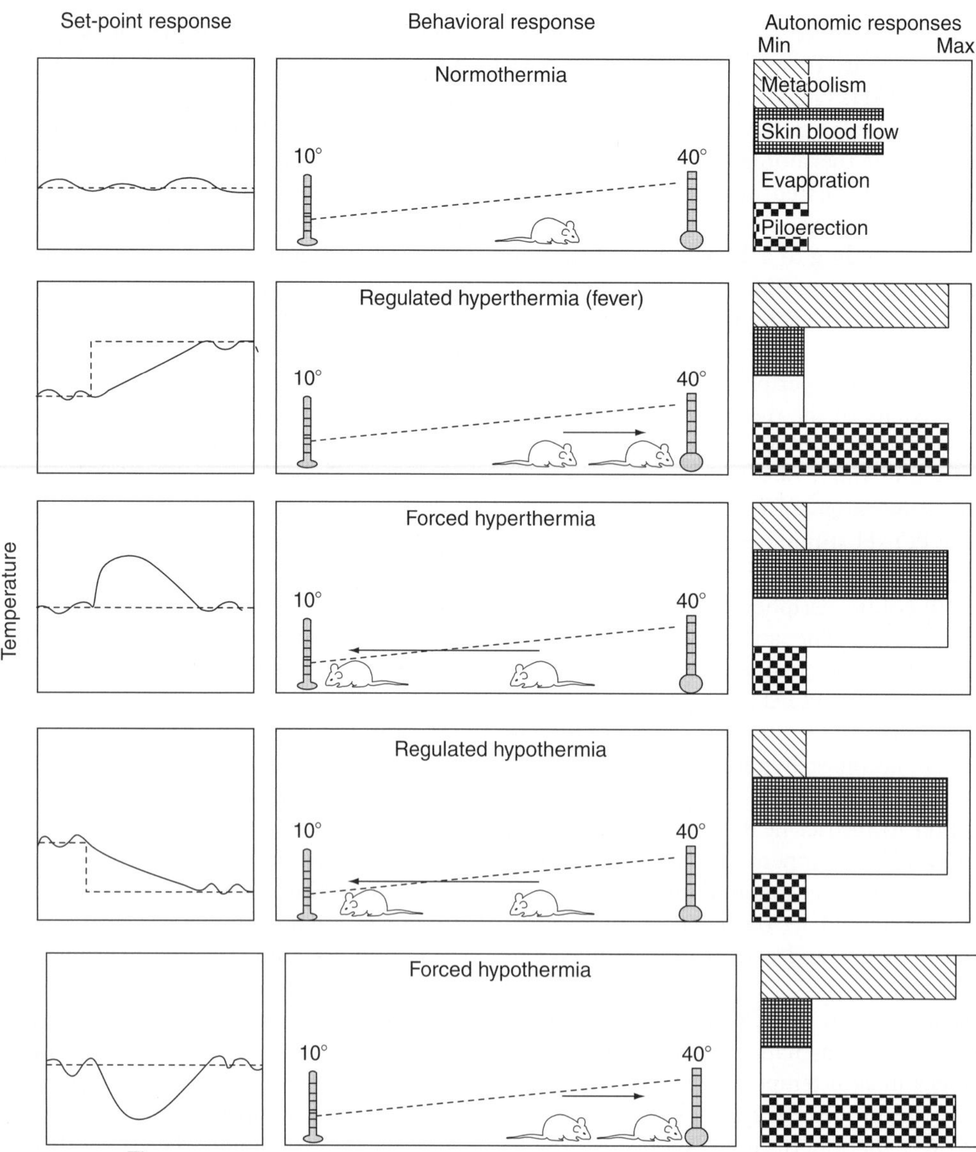

Figure 2 Summary of behavioral and autonomic responses of a homeotherm when subjected to manipulation of body and set-point temperature. Graphs on the left represent the relationship between the set point (dashed line) and core temperature (solid line).

37.5 °C, then it is assumed that its set point or reference temperature is set at 37.5 °C. The set point for temperature regulation may change with certain endogenous and environmental stimuli, such as fever, starvation, and dehydration. The circadian variation in core temperature is considered to be a result of a 24-hr oscillation in the set-point temperature.

The set-point theory allows one to distinguish between stimuli that elicit regulated changes in thermoregulatory control from ones that simply impart deficits in thermoeffector function (**Figure 2**). The thermoregulatory system of homeotherms and poikilotherms attempts to maintain a core temperature equal to the set-point temperature. This process is continuous in homeotherms but is intermittent in poikilotherms when options are available to behaviorally thermoregulate (e.g., shuttling between sun and shade). In a thermoneutral environment for a healthy homeothermic species, set-point temperature is equal to core temperature and thermoeffectors for heat gain and heat loss are balanced and maintained at a minimal level of activity. The animal has a normothermic body temperature and selects an ambient temperature that is comfortable and associated with minimal energy expenditure. Normothermy means core temperature is controlled within ±1 standard deviation of the range associated with normal, resting, thermoneutral conditions.

Autonomic Thermoeffectors

Brown Adipose Tissue and Nonshivering Thermogenesis

Nonshivering thermogenesis, defined as the heat produced from metabolic processes not involving contracting muscles, has been one of the most intensively studied topics in thermal physiology. A key strategy for mammals subjected to chronic cold exposure is to either reduce heat loss by adding insulation or accelerate oxygen consumption and produce heat, but without muscular contraction. The former mechanism is not a viable option in relatively small mammals that are limited in the amount of insulative fur that can be added from summer to winter. Smaller species of mammals must either develop better thermogenic capacity in the winter, which also requires more foraging for food, or hibernate, to avoid the metabolic costs of thermoregulation in the winter. The heat produced from shivering is effective at counteracting the heat loss from cold exposure, but shivering is metabolically inefficient and uncomfortable. Ideally, nonshivering thermogenesis is the most effective way to adapt to a cold environment. Many tissues, including heart and liver, from cold-acclimated mammals, compared to warm-climate species, have increased aerobic capacity and higher concentrations of enzymes involved in cellular respiration.

BAT is a key thermoeffector for the autonomic control of nonshivering thermogenesis in rodents. There is a renewed interest in understanding the control of BAT thermogenesis because of possible links between the function of this metabolically active tissue and human obesity. BAT function in rodents is intimately related to total energy balance. The neural and hormonal contributions to obesity (e.g., leptin) are now approached in rodent models with consideration to the neural control of BAT thermogenesis. The activity of thermoregulatory neurons in the POAH modulates the activity of BAT thermogenesis through descending excitatory and inhibitor pathways utilizing the neurotransmitters glutamine and γ-aminobutyric acid (GABA), respectively (**Figure 3**). For example, microinjection of bicuculine, a $GABA_A$ receptor antagonist, into the rostral raphe pallidus evokes a marked elevation in sympathetic nerve activity and a prolonged rise in BAT temperature, reflecting an explosive rise in its metabolism and heat production. Currently, the focus is on warm-sensitive POAH neurons that evoke tonic inhibition in the dorsomedial hypothalamus. Releasing this inhibition by exposure to a cold environment, selective cooling of the POAH, or induction of a fever blocks warm-sensitive neuron activity in the POAH, leading to BAT thermogenesis.

BAT is found in other small mammals and also occurs in appreciable amounts in the newborn of large mammals, including cows, goats, and humans. Because the ability to shiver is not developed until about 10 days of age, BAT is the main source for heat production in newborn rats and mice. BAT is also critical to stress-induced hyperthermia and energy balance. In addition to the possible role in obesity discussed previously, the neural control of BAT is becoming a crucial model to study how dysregulation of thermogenesis contributes to hyperthermia and heat stroke. Rodents and other species respond to stressors such as noxious stimuli, placement in a novel environment, and restraint with a marked elevation in body temperature. Such an autonomic response prepares the animal adequately to respond to potential danger and to escape.

Peripheral Vasomotor Tone

The rate of dry heat loss (i.e., nonevaporative) from the skin to the surrounding environment is proportional to the difference between skin and air temperature. The neural control of the distribution of warm blood in the core to the thermal shell represents an effective means of regulating heat loss under a wide range of ambient temperature. The regulation of heat loss through the variation in skin blood flow is most effective in locations that are concentrated with arterial–venous anastomoses (AVAs). AVAs are under sympathetic control and are capable of shunting large amounts of blood to the skin for heat dissipation. For example, the hands and feet of humans are replete with AVAs. Indeed, blood flow to the extremities of humans can vary by 100-fold and mean skin blood flow varies by about tenfold, depending on ambient conditions and body temperature.

The rat and mouse tail and rabbit ear are also key sites for the regulation of dry heat loss. The rat tail and rabbit ear abound in AVAs, have little or no fur, and a relatively high surface area:volume ratio, making them ideal sites for regulating heat loss. Studies of the sympathetic control of peripheral vasomotor tone in rabbit and rat have led to the development of neural models involving neurons of the raphe magnus/pallidus and parapyramidal area (**Figure 4**). Neurons from these regions innervate preganglionic, sympathetic cutaneous vasomotor neurons. Their activation is critical to evoke vasoconstriction when exposed to cold temperatures. These pathways are also essential for eliciting peripheral vasoconstriction when subjected to nonthermal, stressful stimuli such as a startle, noxious fumes, or pain. Interestingly, the thermoregulatory aspects of vasomotor control involve subtle differences in activation from the CNS as compared to other

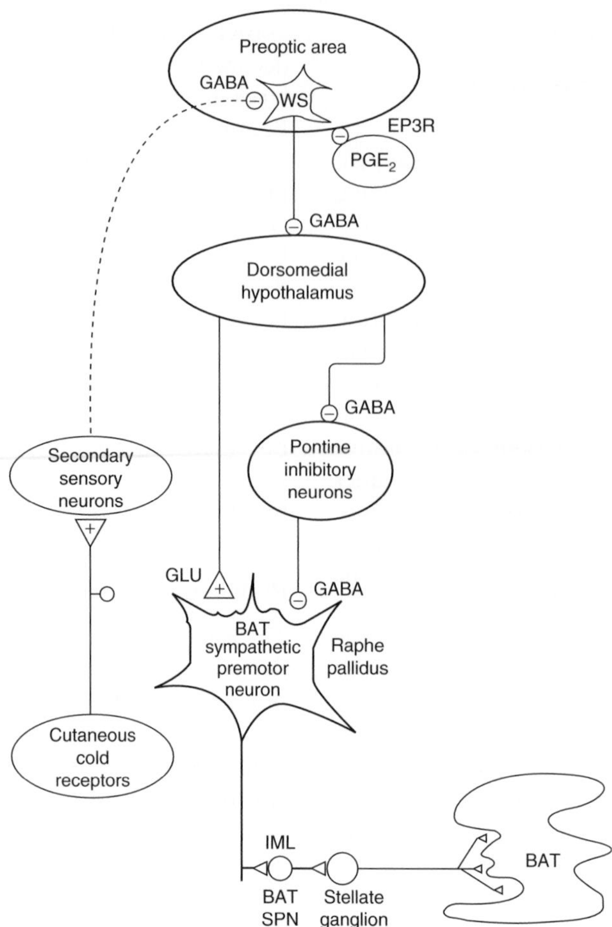

Figure 3 Neural model for the organization of the central pathways that control brown adipose tissue (BAT) thermogenesis. Warm-sensitive (WS) neurons in the POAH are proposed to impart tonic inhibition of the dorsomedial hypothalamus that normally activates BAT thermogenesis through glutamate (GLU) excitatory pathways, which drive BAT sympathetic premotor neurons in the raphe pallidus. γ-Aminobutyric acid (GABA) pathways inhibit BAT thermogenesis via pontine inhibitory pathways (IML, intermediolateral column; SPN, sacral parasympathetic nucleus). Prostaglandin E_2 (PGE_2), a final modulator of fever, inhibits activity of WS neurons, resulting in a marked rise in BAT thermogenesis to raise core temperature (EP3R, prostaglandin E receptor subtype).

nonthermal stimuli. Overall, thermal stimuli, from the core and periphery, are thought to be integrated in the POAH area. The resulting effector signals are further processed in the aforementioned areas of the brain stem and spinal cord to elicit the proper thermoeffector signals to regulate heat loss by modulating skin blood flow.

Evaporative Heat Loss

The effectiveness of the transfer of heat from the body to the surrounding environment by radiation, convection, and conduction decreases with increasing ambient temperature. Thus, evaporative heat loss becomes the key thermoeffector for heat loss if thermal homeostasis is to be maintained in warm environments. Insensible or passive evaporation of water accounts for approximately 20% of the total heat loss in mammals maintained in a comfortable thermal environment. However, when exposed to heat stress, passive water loss is insufficient to maintain thermal homeostasis. Active mechanisms of evaporative water loss include sweating, panting, or saliva spreading, depending on the species. Each thermoeffector achieves a rise in evaporative heat loss, to thermoregulate in the heat, but with unique neural and anatomical mechanisms.

Rodents exposed to heat stress will actively groom saliva on their tail and fur to increase evaporative heat loss, allowing for effective regulation of core temperature. Grooming saliva is in fact a complex thermoeffector involving the autonomic activation of salivary glands combined with higher level participation by cerebral processes to apply the saliva to the skin. The submaxillary salivary gland of the rat is selectively stimulated by heat stress via activation of parasympathetic cholinergic neurons. Mammals that sweat, including humans, secrete copious amounts of sweat on the skin from sweat glands that are controlled by the sympathetic nervous system. Sympathetic nerve

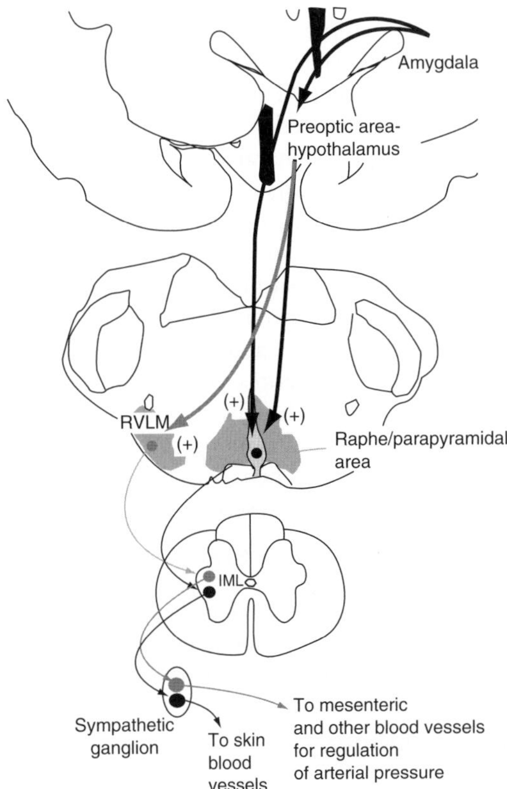

Figure 4 Proposed CNS pathways that mediate control over peripheral vasomotor tone. Thermally sensitive neurons in the POAH modulate activity of the raphe/parapyramidal area and rostral ventrolateral medulla (RVLM) neurons. IML, Intermediolateral column.

fibers innervating human sweat glands are uniquely cholinergic and are under control of thermoregulatory pathways in the POAH (see **Figure 1**). Panting species such as dog and sheep achieve an increase in evaporative water loss by moving air rapidly across mucosal surfaces to maintain a high rate of evaporative heat loss. The neural control of respiration in the medulla in panting species can be modulated by the thermoregulatory centers in the POAH and possibly other sites of the CNS.

Neuroendocrine Pathways

The hypothalamic–pituitary axis is critical in the autonomic control of thermoregulation. The combined stimulation of cold-sensitive neurons and suppression of warm-sensitive neurons in the POAH elicits effector signals that trigger a cascade of neuroendocrine events to raise heat production. Skin cooling or direct cooling of the POAH triggers the release of thyrotropin-releasing hormone from the hypothalamus, leading to the release of thyroid-stimulating hormone (TSH) from the anterior pituitary gland. TSH stimulates the release of thyroxine (T_4) from the thyroid gland, which is deiodinated to triiodothyronine (T_3) that is bound to cellular and nuclear receptors, stimulating heat production. Corticotropin-releasing factor is also released from the hypothalamus during POAH heating and cooling and stimulates the release of adrenocorticotropic hormone from the anterior pituitary, leading to release of corticosteroid hormones from the adrenal cortex. Corticosteroids are critical for the maintenance of thermal homeostasis during cold stress, heat stress, and fever.

Pathology

Menopausal Hot Flashes

Millions of women worldwide undergoing menopause experience the thermal discomfort of hot flashes. The hot flash is a feeling of thermal discomfort that is accompanied by a widespread thermal dissipation, including sweating and cutaneous vasodilation. Recent application of functional magnetic resonance imaging (fMRI) in symptomatic women suggests that the insular cortex and anterior cingulate play a critical role in the response to hot flashes. These areas of the CNS show a remarkable rise in activity in women as they exhibit hot flashes. Interestingly, the hypothalamus does not emerge in the fMRI as a key region of activation during a hot flash. Hence, the hot flash is an autonomic motor response with obvious conscious sensation elicited through the cortex.

Aging

Aging is associated with generalized dysfunction of the autonomic nervous system characterized by impaired baroreflex function, reduced parasympathetic reflexes, increase in blood pressure, and reduced parasympathetic reflexes. With these deficits, it is not surprising to find impaired autonomic control of thermoregulatory processes with aging. Aging seems to have no effect on baseline body temperature in humans and is associated with a minor attenuation in the amplitude of the circadian temperature rhythm. Aged humans do have reduced capacity to generate heat during cold exposure and diminished ability to reduce skin blood flow, which together imply greater susceptibility to hypothermia during cold exposure. Older adults subjected to heat stress are unable to produce the increase in sweating and rise in skin blood flow as seen in younger adults, which could mean a greater susceptibility to hyperthermia during heat stress. There is increased mortality in the aged from hyperthermia and hypothermia, but this incidence appears to be attributed to nutritional, medication, and disease effects and not to a direct malfunction of thermoregulation.

Fever

The development of and recovery from a fever are mediated through circulating inflammatory cytokines that modulate the CNS control of autonomic and behavioral thermoeffectors, and through immune systems. The fever pathway is initiated by exposure to an exogenous pyrogen, which is defined as any substance that causes a fever when it enters the body. The most potent pyrogens are infectious agents or their components (e.g., bacterial endotoxins such as lipopolysaccharide). These pyrogens activate circulating lymphocytes and/or fixed macrophages (e.g., Kupffer cells) to produce anti-inflammatory and pro-inflammatory cytokines that enter the circulation. Interleukins (IL-1 and IL-6) are key pro-inflammatory cytokines that play a major role in the development of an elevated body temperature. Anti-inflammatory cytokines, such as IL-10 and tumor necrosis factor-α, act as antipyretics (or cryogens) and serve, among many functions, to limit the magnitude of the fever. Circulating cytokines are thought to enter the hypothalamus and preoptic area via fenestrated (i.e., leaky) areas of the organovasculosum lamina terminalis or by active transport mechanisms. Constitutive and inducible cyclooxygenases are activated by IL-1 and IL-6, leading to the conversion of arachidonic acid to prostaglandins and other products. Prostaglandin E_2 (PGE_2) is a final mediator of fever. When released into the POAH, PGE_2 selectively stimulates the activity of cold-sensitive neurons and suppresses the activity of warm-sensitive neurons, leading to an explosive elevation in BAT thermogenesis and marked reduction in skin blood flow. The fever, in conjunction with an increase in white blood cell motility and lymphocyte proliferation, enhances the host's defense to the infectious agent. The high body temperature of the host also suppresses growth of some pathogens. The fever pathway can also be activated by other traumatic insults that cause cellular injury, such as a severe sunburn, physical injury to the CNS, exposure to chemical toxicants, and stroke.

Toxicology

Exposure to pesticides, heavy metals, air pollutants, and many other xenobiotic chemicals can alter thermoregulatory function. Organophosphate-based insecticides, which are similar in structure to nerve gases, inhibit acetylcholinesterase (AChE) activity, leading to a marked reduction in body temperature. Inhibition in AChE is thought to stimulate the warm-sensitive neurons and/or suppress cold-sensitive neurons in the POAH (see **Figure 1**), leading to a regulated hypothermic response in rodents. The relatively large body mass of adult humans impedes the development of the hypothermic response characteristic of small rodents. Humans and other large species may nonetheless exhibit a brief period of mild hypothermia following intoxication. Both humans and rodents exposed to organophosphates can sustain a prolonged fever following exposure that persists for several days. Air pollutants such as ozone and carbon monoxide also induce marked hypothermia in rats and mice. The hypothermic response has been shown to alleviate the toxic symptoms and prolong survival, suggesting that the autonomic and behavioral thermoregulatory response is an adaptive response.

See also: Autonomic Nervous System: Neuroanatomy; Energy Homeostasis: Adiposity Signals; Energy Homeostasis: Thermoregulation; Sympathetic Nervous System; Thermoregulation During Sleep and Sleep Deprivation; Thermoregulation: Autonomic, Age-Related Changes.

Further Reading

Blessing WW (2003) Lower brainstem pathways regulating sympathetically mediated changes in cutaneous blood flow. *Cellular and Molecular Neurobiology* 23: 527–538.

Brown JW, Whitehurst ME, and Carroll RG (2005) Thermoregulatory set point decreases following hemorrhage in rats. *Shock* 23: 239–242.

Freedman RR, Benton MD, Genik RJ II, et al. (2006) Cortical activation during menopausal hot flashes. *Fertility and Sterility* 85: 674–678.

Gordon CJ (1993) *Temperature Regulation in Laboratory Rodents.* New York: Cambridge University Press.

Gordon CJ (2005) *Temperature and Toxicology: An Integrative, Comparative, and Environmental Approach.* Boca Raton, FL: CRC Press.

Gordon CJ, Mohler FS, Watkinson WP, et al. (1988) Temperature regulation in laboratory mammals following acute toxic insult. *Toxicology* 53: 161–178.

Hori T, Nakashima T, Koga H, et al. (1988) Convergence of thermal, osmotic and cardiovascular signals on preoptic and anterior hypothalamic neurons in the rat. *Brain Research Bulletin* 20: 879–885.

Kenney WL and Munce TA (2003) Aging and human temperature regulation. *Journal of Applied Physiology* 95: 2598–2603.

Morrison SF (2004) Central pathways controlling brown adipose tissue thermogenesis. *News in Physiological Science* 19: 67–74.

Nalivaiko E and Blessing WW (2001) Raphe region mediates changes in cutaneous vascular tone elicited by stimulation of amygdala and hypothalamus in rabbits. *Brain Research* 891: 130–137.

The Commission for Thermal Physiology of the International Union Of Physiological Sciences (2001) Glossary of terms for thermal physiology. *Japanese Journal of Physiology* 51: 245–280.

Autonomic Nervous System: Central Urogenital Control

W C de Groat, University of Pittsburgh Medical School, Pittsburgh, PA, USA

Introduction

Many functions of the urogenital system are controlled by complex neural pathways in the brain and spinal cord. These central pathways in turn regulate the activity of peripheral autonomic (sympathetic and parasympathetic) and somatic nerves that innervate the smooth muscle, striated muscle, epithelial cells, and exocrine glands in the urogenital organs. Some urogenital functions (penile erection) are purely involuntary and are mediated by reflex pathways in the spinal cord or brain stem; whereas others (micturition) are more complex, involving voluntary control by the cerebral cortex. This article provides a brief review of the central neural circuitry and neurotransmitters involved in (1) the control of urine storage and release by the lower urinary tract and (2) the control of male sexual organs.

Innervation of the Urogenital Organs

The lower urinary tract and sex organs are innervated by three sets of peripheral nerves that arise at the level of the lumbosacral spinal cord (**Figure 1**). Visceral structures (urinary bladder, urethra, cavernous tissue of the penis and clitoris, vas deferens, and exocrine glands) receive an innervation from both divisions of the autonomic nervous system. Parasympathetic efferent axons that originate from preganglionic neurons in the sacral spinal cord pass through the pelvic nerves to synapse on peripheral ganglion cells that, in turn, innervate the organs. Activation of parasympathetic pathways induces a bladder contraction, urethral relaxation, and penile and clitoral erection. Sympathetic efferent axons that originate from preganglionic neurons in the rostral lumbar segments of the spinal cord project to ganglion cells in the prevertebral and paravertebral ganglia, which then send axons through various nerves to the target organs. Activation of sympathetic pathways relaxes the bladder, contracts the bladder neck and urethra, produces seminal emission, and can elicit penile erection or detumescence. Somatic efferent axons originate in motor neurons in the sacral spinal cord and pass through the pudendal nerves to striated muscles of the urethral sphincter and periurethral striated muscles (bulbocavernosus and ischiocavernosus) involved in ejaculation.

Afferent axons innervating the urogenital organs arise in the lumbosacral dorsal root ganglia and are contained in the three sets of peripheral nerves. The most important afferents for initiating micturition are those which travel in the pelvic nerve to the sacral spinal cord. These afferents consist of small myelinated (A-δ) and unmyelinated (C) axons that convey impulses from tension receptors and nociceptors in the bladder wall. Afferent activity arising from mechanoreceptors in the penis is carried in the pudendal nerves and triggers penile erection and emission–ejaculation.

Anatomy of Central Nervous Pathways Controlling the Urogenital System

The reflex circuitry controlling the urogenital organs consists of four basic components: primary afferent neurons, spinal efferent neurons, spinal interneurons, and neurons in the brain that modulate spinal reflex pathways.

Pathways in the Spinal Cord

Afferent projections in the spinal cord Afferent pathways from the bladder project through the pelvic nerves into Lissauer's tract at the apex of the dorsal horn in the caudal lumbosacral spinal cord and then send collaterals laterally and medially around the dorsal horn into laminae V–VII and X at the base of the dorsal horn (**Figure 2**). The lateral pathway terminates in the region of the sacral parasympathetic nucleus and also sends some axons to the dorsal commissure (**Figure 2**). Pudendal afferent pathways from the penis and clitoris project into the deeper layers of the medial lumbosacral dorsal horn.

Efferent neurons Parasympathetic preganglionic neurons innervating the urogenital organs are located in the intermediolateral gray matter (laminae V–VII) in the caudal lumbosacral segments of the spinal cord (**Figures 3** and **4**), whereas sympathetic preganglionic neurons are located in medial (lamina X) and lateral sites (laminae V–VII) in the rostral lumbar spinal cord. Motor neurons innervating the periurethral striated muscles are located in lamina IX in the ventral horn (**Figure 4**).

Spinal interneurons Interneurons involved in spinal reflex circuits controlling urogenital functions have been identified by retrograde transneuronal transport following injection of pseudorabies virus into the urinary lower urinary tract or genital organs of the rat. Large populations of interneurons are located just dorsal and medial to the preganglionic neurons as well as in the dorsal commissure and lamina I (**Figure 2(c)**). The spinal neurons involved in processing afferent

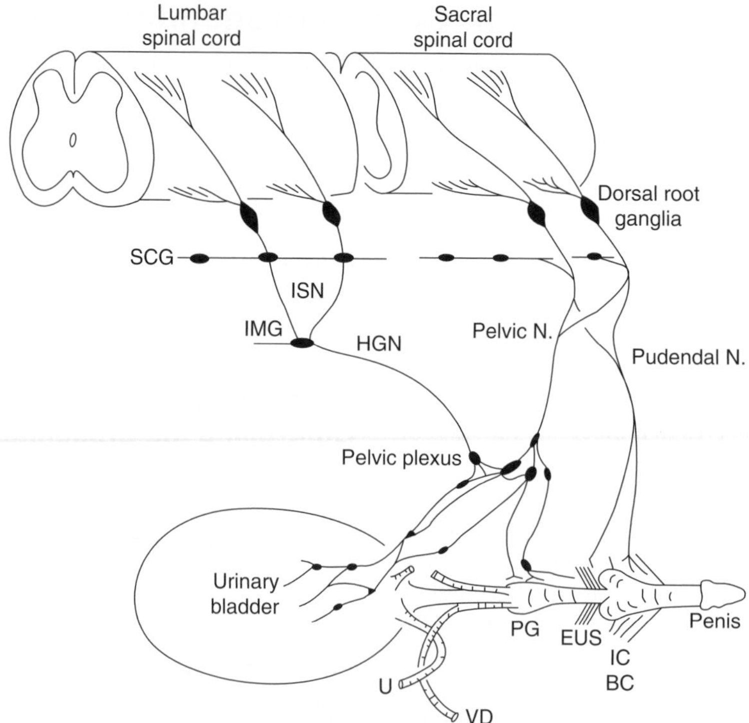

Figure 1 Diagram showing the sympathetic, parasympathetic, and somatic innervation of the urogenital tract of the male cat. Sympathetic preganglionic pathways emerge from the lumbar spinal cord and pass to the sympathetic chain ganglia (SCG) and then via the inferior splanchnic nerves (ISN) to the inferior mesenteric ganglia (IMG). Preganglionic and postganglionic sympathetic axons then travel in the hypogastric nerve (HGN) to the pelvic plexus and the urogenital organs. Parasympathetic preganglionic axons, which originate in the sacral spinal cord, pass in the pelvic nerve to ganglion cells in the pelvic plexus and to distal ganglia in the organs. Sacral somatic pathways are contained in the pudendal nerve, which provides an innervation to the penis, the ischiocavernosus (IC), bulbocavernosus (BC), and external urethral sphincter (EUS) muscles. The pudendal and pelvic nerves also receive postganglionic axons from the caudal sympathetic chain ganglia. These three sets of nerves contain afferent axons from the lumbosacral dorsal root ganglia. U, ureter; PG, prostate gland; VD, vas deferens.

input from the lower urinary tract have been identified by the expression of the immediate-early gene, c-*fos* (**Figure 2(b)**). In the rat, chemical or mechanical stimulation of the bladder and urethra increases the levels of Fos protein primarily in the dorsal commissure, in the superficial dorsal horn, and in the area of the sacral parasympathetic nucleus (**Figure 2(b)**). Some of these interneurons make local connections in the spinal cord and participate in segmental spinal reflexes, whereas others send ascending projections to regions in the brain that are involved in the supraspinal control of urogenital function.

Pathways in the Brain

In the rat, transneuronal virus tracing methods have identified many populations of neurons in the brain that are involved in the control of bladder, urethra, and urethral sphincter. Labeled areas include Barrington's nucleus (the pontine micturition center (PMC)); medullary raphe nuclei, which contain serotonergic neurons; the locus coeruleus, which contains noradrenergic neurons; periaqueductal gray (PAG); and the A5 noradrenergic cell group (**Figure 3**). Several regions in the hypothalamus (including the paraventricular nucleus and medial preoptic nucleus) and the cerebral cortex also exhibit virus-infected cells. Neurons in the cortex are located primarily in the medial frontal cortex. Similar areas in the brain have been identified after injection of pseudorabies virus into the penis or prostate gland.

Other anatomical studies in which anterograde tracer substances were injected into brain areas and then identified in terminals in the spinal cord are consistent with the virus tracing data (**Figure 4**). Tracer injected into the paraventricular nucleus of the hypothalamus labeled terminals in the sacral parasympathetic nucleus as well as the sphincter motor nucleus. On the other hand, neurons in the anterior hypothalamus project to the PMC. Neurons in the PMC in turn project primarily to the sacral parasympathetic nucleus, the lateral edge of the dorsal horn, and the dorsal commissure. Conversely, projections from neurons in the lateral pons, an area implicated in the control of the urethral sphincter, terminate rather selectively in

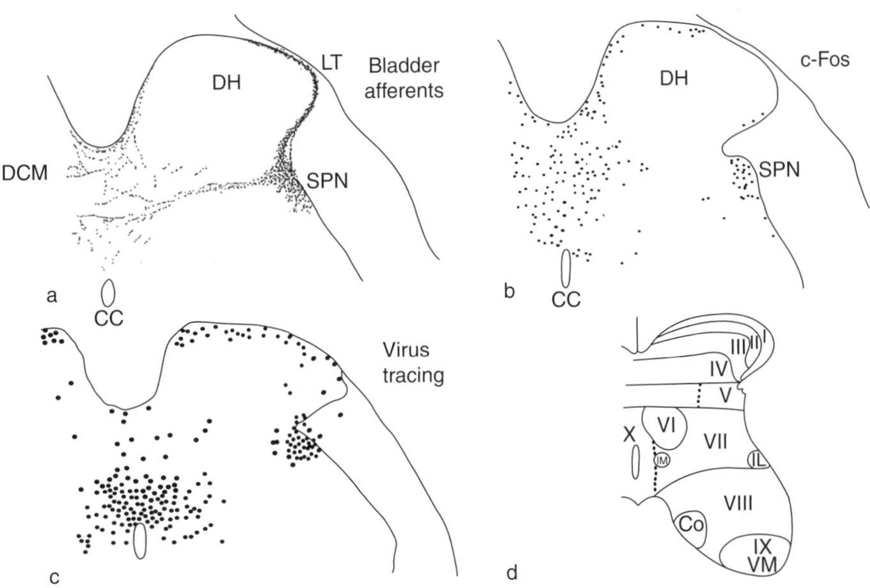

Figure 2 Comparison of the distribution of rat bladder afferent projections to the L6 spinal cord (a) with the distribution of rat c-Fos-positive cells in the L6 spinal segment following chemical irritation of the lower urinary tract (b) and the distribution of interneurons in the L6 spinal cord labeled by transneuronal transport of pseudorabies virus injected into the urinary bladder (c). Afferents are labeled by wheat germ agglutinin-conjugated horseradish peroxidase injected into the urinary bladder. c-Fos immunoreactivity is present in the nuclei of cells. (d) The laminar organization of the cat spinal cord. DCM, dorsal commissure; DH, dorsal horn; LT, Lissauer's tract; SPN, sacral parasympathetic nucleus; CC, central canal.

the sphincter motor nucleus, also known as Onuf's nucleus (**Figure 4**).

Central Neural Control of the Lower Urinary Tract

The central pathways controlling lower urinary tract function are organized as simple on–off switching circuits that maintain a reciprocal relationship between the urinary bladder and urethral outlet (**Figure 5**). The principal reflex components of these switching circuits are listed in **Table 1** and illustrated in **Figure 6**. Intravesical pressure measurements during bladder filling in both humans and animals reveal low and relatively constant bladder pressures when bladder volume is below the threshold for inducing voiding (**Figure 5**). The accommodation of the bladder to increasing volumes of urine is primarily a passive phenomenon dependent upon the intrinsic properties of the vesical smooth muscle and quiescence of the parasympathetic efferent pathway. In addition, in some species urine storage is also facilitated by sympathetic reflexes that mediate an inhibition of bladder activity as well as closure of the bladder neck and proximal urethra (**Table 1**, **Figure 6**). During bladder filling the activity of the sphincter electromyogram also increases (**Figure 5**), reflecting an increase in efferent firing in the pudendal nerve and an increase in outlet resistance which contributes to the maintenance of urinary continence. Both sympathetic and pudendal nerve reflex activity are activated by low-level bladder afferent activity induced by bladder distension.

The storage phase of the urinary bladder can be switched to the voiding phase either involuntarily (**Figure 5(a)**) or voluntarily (**Figure 5(b)**). The former is readily demonstrated in the human infant (**Figure 5(a)**), when, at a certain level of bladder filling, increased afferent firing from bladder tension receptors produces firing in the sacral parasympathetic pathways and inhibition of sympathetic and somatic pathways. The expulsion phase consists of an initial relaxation of the urethral sphincter (**Figure 5(a)**) followed in a few seconds by a contraction of the bladder, an increase in bladder pressure, and flow of urine. Relaxation of the urethral outlet is mediated by activation of a parasympathetic reflex pathway to the urethra (**Table 1**) that triggers the release of nitric oxide, an inhibitory transmitter, as well as by removal of adrenergic and somatic excitatory inputs to the urethra.

Storage and voiding reflexes require the integrative action of neuronal populations at various levels of the neuraxis (**Figure 6**). Reflexes mediating the excitatory outflow to the sphincters and the sympathetic inhibitory outflow to the bladder are organized at the spinal level (**Figure 6(a)**), whereas the parasympathetic outflow to the bladder has a more complicated central organization involving spinal and spinobulbospinal pathways (**Figures 6(b)** and **7**).

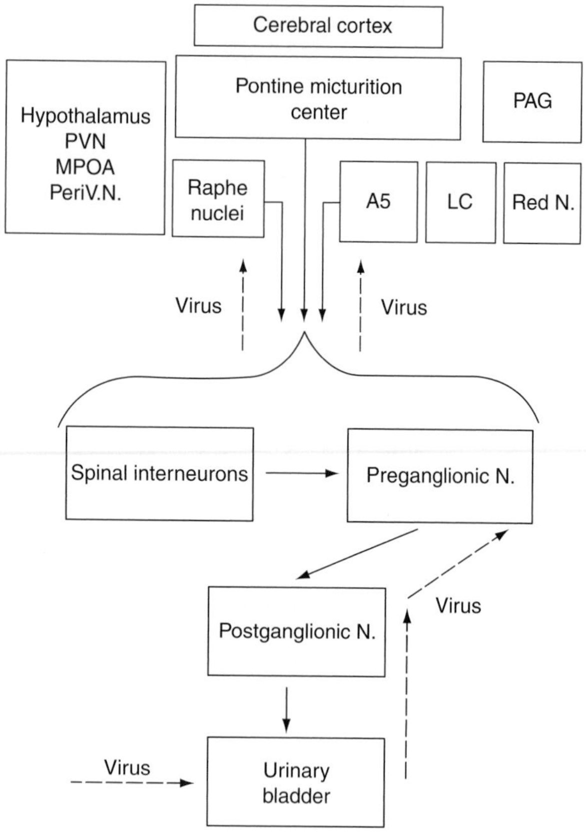

Figure 3 Structures in the brain and spinal cord of the adult and neonatal rat, labeled after injection of pseudorabies virus into the urinary bladder or the urethra. Virus is transported transneuronally in a retrograde direction (dashed arrows). Normal synaptic connections are indicated by solid arrows. At long survival times the virus can be detected in neurons at specific sites in the spinal cord and brain, extending to the PMC in the pons (i.e., Barrington's nucleus) and to the cerebral cortex. Other sites in the brain labeled by the virus are the paraventricular nucleus (PVN), medial preoptic area (MPOA), and periventricular nucleus (PeriV.N.) of the hypothalamus; the periaqueductal gray (PAG); the locus coeruleus (LC) and subcoeruleus; the red nucleus (N); the medullary raphe nuclei; and (A5) the noradrenergic cell group.

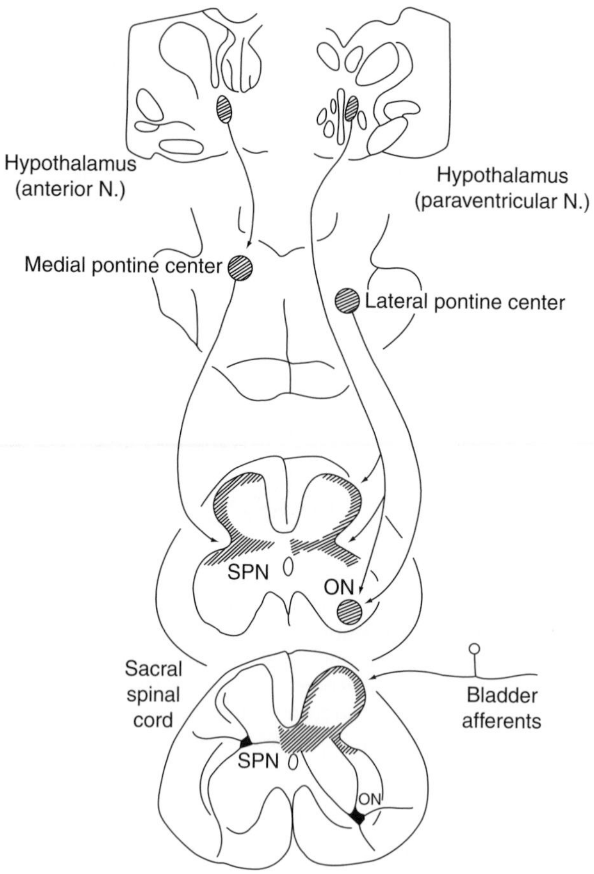

Figure 4 Neural connections between the brain and the sacral spinal cord that may be involved in the regulation of the lower urinary tract in the cat. The lower section of spinal cord shows the location and morphology of a preganglionic neuron in the sacral parasympathetic nucleus (SPN), a sphincter motor neuron in Onuf's nucleus (ON), and the sites of central termination of afferent projections from the urinary bladder. The upper spinal cord section shows the sites of termination of descending pathways arising in the medial PMC, the lateral pontine storage center, and the paraventricular nuclei of the hypothalamus. Section through the pons shows the projection from the anterior hypothalamic nuclei to the PMC.

Organization of Urine Storage Reflexes

Sympathetic storage reflex Although the integrity of the sympathetic input to the lower urinary tract is not essential for the performance of micturition, it does contribute to the storage function of the bladder. Surgical interruption or pharmacological blockade of the sympathetic innervation can reduce urethral outflow resistance and reduce bladder capacity. Sympathetic reflex activity is elicited by a sacrolumbar intersegmental spinal reflex pathway that is triggered by vesical afferent activity in the pelvic nerves (**Figure 6(a)**). This reflex pathway is inhibited when bladder pressure is raised to the threshold for producing micturition. The inhibitory response is abolished by transection of the spinal cord at the lower thoracic level, indicating that it originates at a supraspinal site. Thus, the vesicosympathetic reflex represents a negative feedback mechanism that allows the bladder to accommodate larger volumes during bladder filling, but is turned off during voiding to allow the bladder to empty completely (**Figure 6(b)**).

Urethral sphincter storage reflex Motor neurons innervating the striated muscles of the urethral sphincter exhibit a tonic discharge that increases during bladder filling (**Figure 5**). This activity is mediated in part by low-level afferent input from the bladder. During micturition the firing of sphincter motor neurons is inhibited. This inhibition is dependent in part on supraspinal mechanisms, since it is less prominent after spinal cord injury.

Organization of Voiding Reflexes

Spinobulbospinal micturition reflex pathway Micturition is mediated by activation of the sacral parasympathetic efferent pathway to the bladder and the urethra, as well as by reciprocal inhibition of the somatic pathway to the urethral sphincter (**Table 1, Figure 6(b)**). Studies in animals using brain-lesioning and electrophysiological techniques revealed that the micturition reflex is mediated by a spinobulbospinal pathway that passes through the PMC in the rostral brain stem (**Figures 6(b)** and 7). Electrical stimulation in the PMC evokes bladder contractions and voiding, whereas application of inhibitory agents directly to the PMC suppresses micturition. Single-unit recording in the PMC has revealed increased neuronal activity during bladder distension. It has been proposed that the micturition reflex pathway functions as an on–off switch that is activated by a critical level of afferent activity arising from tension receptors in the bladder and is, in turn, modulated by inhibitory

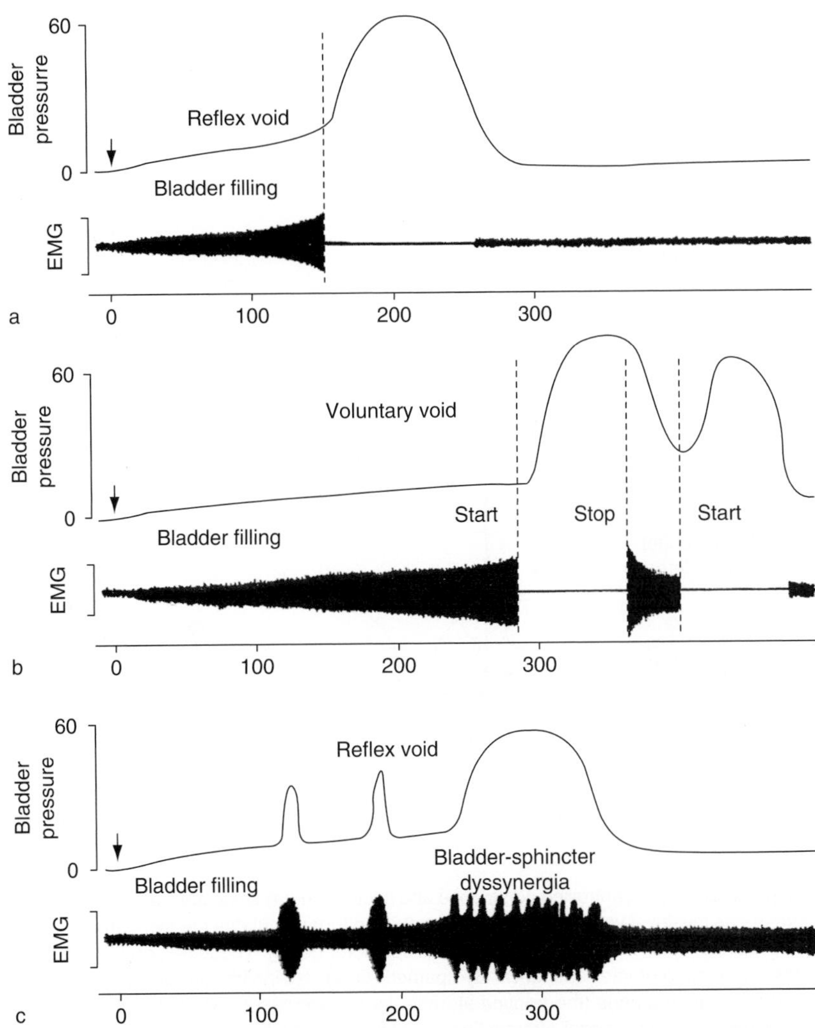

Figure 5 Combined cystometrograms and sphincter electromyograms (EMGs) comparing reflex voiding responses in an infant (a) and in a paraplegic patient (c) with a voluntary voiding response in an adult (b). The abscissa in all records represents bladder volume in milliliters and the ordinates represent bladder pressure in cmH_2O and electrical activity of the EMG recording. On the left side of each trace the arrows indicate the start of a slow infusion of fluid into the bladder (bladder filling). Vertical dashed lines indicate the start of sphincter relaxation, which precedes by a few seconds the bladder contraction in parts (a, b). In part (b), note that a voluntary cessation of voiding (stop) is associated with an initial increase in sphincter EMG followed by a reciprocal relaxation of the bladder. A resumption of voiding is again associated with sphincter relaxation and a delayed increase in bladder pressure. On the other hand, in the paraplegic patient (c) the reciprocal relationship between bladder and sphincter is abolished. During bladder filling, transient uninhibited bladder contractions occur in association with sphincter activity. Further filling leads to more prolonged and simultaneous contractions of the bladder and sphincter (bladder–sphincter dyssynergia). Loss of the reciprocal relationship between bladder and sphincter in paraplegic patients interferes with bladder emptying.

Table 1 Reflexes to the lower urinary tract

Afferent pathway	*Efferent pathway*	*Central pathway*
Urine Storage		
Low-level vesical afferent activity (pelvic nerve)	1. External sphincter contraction (somatic nerves) 2. Internal sphincter contraction (sympathetic nerves) 3. Detrusor inhibition (sympathetic nerves) 4. Ganglionic inhibition (sympathetic nerves) 5. Sacral parasympathetic outflow inactive	Spinal reflexes
Afferent activity from the external urethral sphincter	6. Inhibition of parasympathetic outflow	Spinal reflex
Micturition		
High-level vesical afferent activity (pelvic nerve)	1. Inhibition of external sphincter activity 2. Inhibition of sympathetic outflow 3. Activation of parasympathetic outflow to the bladder	Spinobulbospinal reflexes
	4. Activation of parasympathetic outflow to the urethra	Spinal reflex

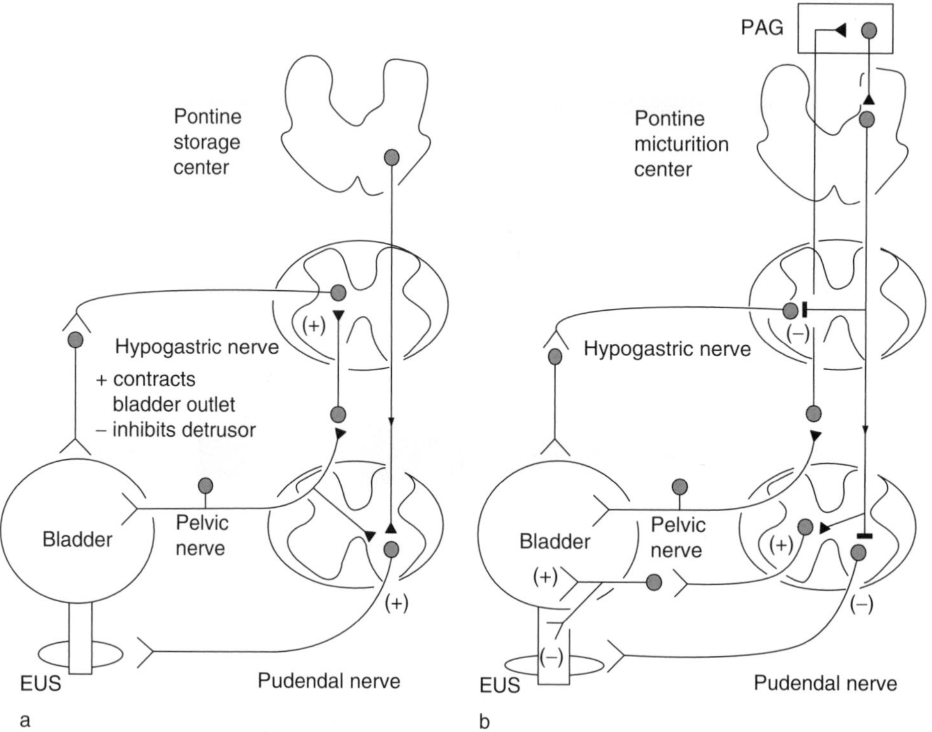

Figure 6 Diagram showing neural circuits controlling continence and micturition. (a) Urine storage reflexes. During the storage of urine, distension of the bladder produces low-level vesical afferent firing, which in turn stimulates the sympathetic outflow to the bladder outlet (base and urethra) and the pudendal outflow to the external urethral sphincter (EUS). These responses occur by spinal reflex pathways and represent guarding reflexes, which promote continence. Sympathetic firing also inhibits detrusor muscle and modulates transmission in bladder ganglia. A region in the rostral pons (the pontine storage center) increases external urethral sphincter activity. (b) Voiding reflexes. During elimination of urine, intense bladder afferent firing activates spinobulbospinal reflex pathways passing through the PMC, which in turn stimulates the parasympathetic outflow to the bladder and urethral sphincter smooth muscle and inhibits the sympathetic and pudendal outflow to the urethral outlet. Ascending afferent input from the spinal cord may pass through relay neurons in the periaqueductal gray (PAG) before reaching the PMC.

and excitatory influences from areas of the brain rostral to the pons (e.g., diencephalon and cerebral cortex) (**Figure** 7). Afferent input from the bladder is thought to reach the PMC after passing through a relay station in the PAG (**Figure 6(b)**).

Suprapontine control of micturition Studies in humans and animals indicate that voluntary control of micturition depends on connections between the frontal cortex, hypothalamus, and other forebrain structures, such as anterior cingulate gyrus, amygdala,

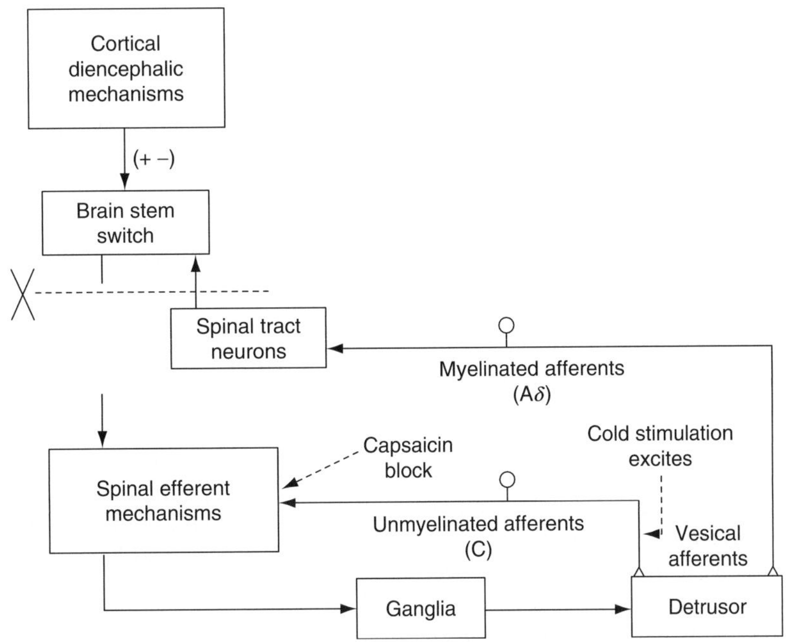

Figure 7 Diagram showing the organization of the parasympathetic excitatory reflex pathway to the detrusor muscle. Scheme is based on electrophysiologic studies in cats. In animals with an intact spinal cord, micturition is initiated by a supraspinal reflex pathway passing through a center in the brain stem. The pathway is triggered by myelinated afferents (A-δ fibers), which are connected to the tension receptors in the bladder wall. Injury to the spinal cord above the sacral segments (X) interrupts the connections between the brain and spinal autonomic centers and initially blocks micturition. However, over a period of several weeks following cord injury, a spinal reflex mechanism emerges, which is triggered by unmyelinated vesical afferents (C, C-fibers); the A-fiber afferent inputs are ineffective. The C-fiber reflex pathway is usually weak or undetectable in animals with an intact nervous system. Stimulation of the C-fiber bladder afferents by instillation of ice water into the bladder (cold stimulation) activates voiding responses in patients with spinal cord injury. Capsaicin (20–30 mg, subcutaneously) blocks the C-fiber reflex in chronic spinal cats, but does not block micturition reflexes in intact cats. Intravesical capsaicin also suppresses detrusor hyperreflexia and cold-evoked reflexes in patients with neurogenic bladder dysfunction.

bed nucleus of the stria terminalis, and septal nuclei, where electrical stimulation elicits excitatory bladder effects. Damage to the cerebral cortex due to tumors, aneurysms, or cerebrovascular disease appears to remove inhibitory control of the PMC, resulting in bladder overactivity.

Human brain imaging studies using single-photon emission computed tomography (SPECT), positron emission tomography (PET), and functional magnetic resonance imaging (fMRI) have examined the areas of the brain involved in the control of micturition. Some studies evaluated the brain areas responsible for the perception of bladder fullness and the sensation of the desire to void during bladder filling, whereas others examined brain activity during micturition or voluntary contractions of the pelvic floor during urine withholding. PET scan studies revealed that two cortical areas (the right dorsolateral prefrontal cortex and the anterior cingulate gyrus) were active (i.e., exhibited increased blood flow) during voiding (**Figure 8**). The hypothalamus, including the preoptic area as well as the pons and the PAG, also showed activity in concert with voluntary micturition. Other PET studies that examined the changes in brain activity

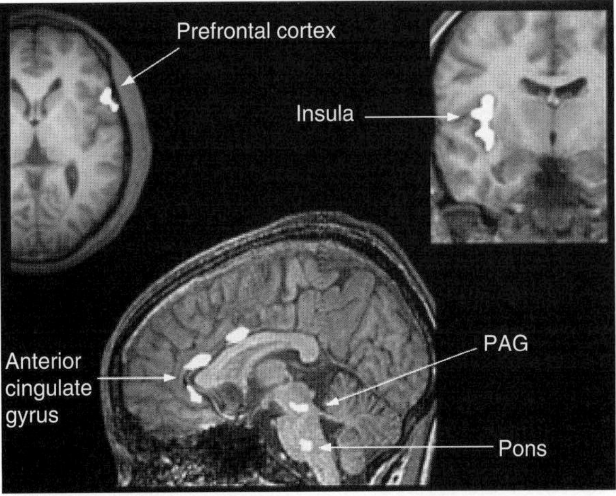

Figure 8 Regions in the human brain identified in PET imaging studies that exhibit differences in activity based on whether the bladder was full or empty (PAG, periaqueductal gray).

during filling of the bladder revealed that increased activity occurred in the PAG, midline pons, mid-cingulate gyrus, and bilaterally in the frontal lobes. The results were consistent with the notion that the

PAG receives information about bladder fullness and then relays this information to other brain areas involved in the control of bladder storage.

Spinal micturition reflex pathway Spinal cord injury rostral to the lumbosacral level eliminates voluntary and supraspinal control of voiding, leading initially to an areflexic bladder and complete urinary retention followed by a slow development of automatic micturition and bladder hyperactivity (**Figure** 5(c)) mediated by spinal reflex pathways. However, voiding is commonly inefficient due to simultaneous contractions of the bladder and urethral sphincter (bladder–sphincter dyssynergia) (**Figure** 5(c)). Electrophysiologic studies in animals have shown that the micturition reflex pathway changes after spinal cord injury. For example, the afferent limb of the micturition reflex in cats with chronic spinal transection above the lumbar level consists of unmyelinated (C-fiber) axons (**Figure** 7), whereas in cats with an intact spinal cord, myelinated (A-δ) afferents activate the micturition reflex (**Figure** 7). In normal cats, capsaicin, a neurotoxin known to disrupt the function of C-fiber afferents, did not block reflex contractions of the bladder or the A-δ-afferent fiber-evoked bladder reflex. However, in cats with chronic spinal injury, capsaicin completely blocked C-fiber-evoked bladder reflexes. The emergence of C-fiber-evoked bladder reflexes seems to be mediated by several mechanisms, including changes in central synaptic connections and alterations in the properties of the peripheral afferent receptors that lead to sensitization of the mechanoinsensitive C-fiber afferents and the unmasking of mechanosensitivity. Evidence for a role of C-fiber bladder afferents in neurogenic bladder overactivity and involuntary voiding in humans has been obtained in studies in which capsaicin or resiniferatoxin, another C-fiber afferent neurotoxin, was administered intravesically to patients with detrusor overactivity due to multiple sclerosis or spinal cord injuries. In these patients the toxins increased bladder capacity and reduced the frequency of incontinence.

Neurotransmitters in Central Micturition Reflex Pathways

Excitatory Neurotransmitters

Excitatory transmission in the central pathways controlling the lower urinary tract involves several types of neurotransmitters, including glutamic acid, neuropeptides, nitric oxide, and adenosine triphosphate (ATP). Pharmacological experiments in rats revealed that glutamic acid is an essential transmitter in the ascending, pontine, and descending limbs of the spinobulbospinal micturition reflex pathway and in spinal reflex pathways controlling the bladder and external urethral sphincter. *N*-Methyl-D-aspartate (NMDA) and non-NMDA glutamatergic synaptic mechanisms appear to interact synergistically to mediate transmission in these pathways.

Inhibitory Neurotransmitters

Several types of inhibitory transmitters, including inhibitory amino acids (γ-aminobutyric acid (GABA) and glycine) and opioid peptides (enkephalins), can suppress the micturition reflex when applied to the central nervous system. Experimental evidence in anesthetized animals indicates that GABA, glycine, and enkephalins exert a tonic inhibitory control in the PMC and regulate bladder capacity. These substances also have inhibitory actions in the spinal cord.

Neurotransmitters with Mixed Excitatory and Inhibitory Actions

Some transmitters (dopamine, 5-hydroxytryptamine, norepinephrine, acetylcholine, and nonopioid peptides, including vasoactive intestinal polypeptide and corticotropin-releasing factor) have both inhibitory and excitatory effects on reflex bladder activity. In some instances the different effects are mediated by different types of receptors. For example, the inhibitory effects of dopamine are mediated by D_1-like (D_1 and D_5) receptor subtypes, and the facilitatory effects are mediated by D_2-like (D_2, D_3, and D_4) receptor subtypes. Loss of forebrain dopaminergic inhibitory mechanisms in patients with idiopathic Parkinson's disease is associated with bladder hyperactivity.

5-Hydroxytryptamine has complex effects on the lower urinary tract; these vary in different species. For example, in the cat, activation of central 5-HT_{1A} receptors inhibits reflex bladder activity; whereas activation of 5-HT_2 receptors enhances urethral sphincter activity. On the other hand, activation of central 5-HT_{1A} receptors in the rat enhances bladder and sphincter reflexes.

Central Neural Control of the Urogenital Organs

The physiologic changes initiated by erotic stimuli can be divided into four distinct phases (excitement, plateau, orgasm, and resolution) that have been designated collectively as the sexual response cycle. Although anatomic differences obviously preclude identical responses in males and females during each phase of the cycle, it is clear that similar vascular responses (skin flush, penile, and clitoral erection), secretory responses (stimulation of the prostate, bulbourethral gland, glands of Littre in the male; Bartholin's and paraurethral glands in the female), and responses of smooth and striated muscles occur in both sexes. The following sections review the central nervous system control of the principal autonomic and somatic responses in the male sexual response cycle (erection, secretion, emission, and ejaculation).

Penile Erection

Penile erection, which is one of the first responses to occur during sexual arousal, is a vascular phenomenon resulting from neurally mediated increase in blood flow to the penile erectile tissue (corpora cavernosa and corpus spongiosum). The erectile tissue consists of large venous sinuses that contain very little blood when the penis is flaccid, but distend considerably when blood flow is increased. Dilation in the arterial supply to the cavernous tissue, coupled with a relaxation of the sinusoidal smooth muscle in the trabecular tissue, is responsible for erection.

Penile erection is primarily an involuntary or reflex phenomenon that can be elicited by a variety of stimuli and by at least two distinct central mechanisms: psychogenic and reflexogenic (**Table 2**). Psychogenic erections are initiated by supraspinal centers in response to auditory, visual, olfactory, tactile, and imaginative stimuli and can be elicited by peripheral neural pathways arising in either the sacral parasympathetic or the thoracolumbar sympathetic nuclei of the spinal cord (**Table 2**). Reflexogenic erections, which are initiated by exteroceptive stimulation of the genital regions, are mediated by a sacral spinal reflex mechanism having an afferent limb in the pudendal nerves and an efferent limb in the sacral parasympathetic nerves. Electrophysiological studies in animals have revealed that the sacral parasympathetic reflex mechanism involved in penile erection is not altered by transection of the spinal cord above the lumbar level, indicating that the reflex circuitry is contained in the spinal cord.

In patients with lower motor neuron lesions involving the sacral spinal cord, reflexogenic erections are abolished but psychogenic erections may still occur via the sympathetic innervation to the penis. In patients with spinal cord lesions above the level of T12, psychogenic erections are abolished but reflexogenic reactions persist. Under normal conditions it is likely that psychogenic and reflexogenic stimuli act synergistically in producing erections. It is also known that psychologic factors, such as guilt and hostility, or endocrine disturbances that influence libido or supraspinal centers, can interfere with erectile reflexes.

Brain imaging studies using fMRI or PET during sexual arousal induced by erotic visual stimuli have identified several brain regions in both males and females that are activated during sexual stimulation. PET studies in males detected activation in three general regions, including limbic–paralimbic areas (anterior cingulate gyrus, orbitofrontal cortex), the striatum (head of the caudate nucleus), and the posterior hypothalamus. Strong signals specifically associated with penile turgidity were observed in the right subinsular region (claustrum, caudate, cingulate gyrus) using fMRI. A PET study also showed increased blood flow in the right prefrontal cortex during orgasm in males, whereas all other cortical areas showed decreases in blood flow.

It has been suggested that the activation of paralimbic areas is correlated with the emotional and motivational states associated with sexual arousal, whereas the activation of the anterior cingulate and hypothalamic regions is related to the affective,

Table 2 Male sexual reflexes

Response	*Afferent nerves*	*Efferent nerves*	*Central pathway*	*Effector organ*
Penile erection Reflexogenic	Pudendal nerve	Sacral parasympathetic	Sacral spinal reflex	Dilation of arterial supply to corpus cavernosum and corpus spongiosum
Psychogenic	Auditory, imaginative, visual, olfactory, tactile	Sacral parasympathetic, lumbar sympathetic	Supraspinal origin	
Glandular secretion	Pudendal nerve	Sacral parasympathetic, lumbar sympathetic	Sacral spinal reflex	Seminal vesicle and prostate
Seminal emission	Pudendal nerve	Lumbar sympathetic	Intersegmental spinal reflex (sacrolumbar)	Contraction of vas deferens, ampulla, seminal vesicles, prostate, and closure of bladder neck
Ejaculation	Pudendal nerve	Somatic efferents in pudendal nerve	Sacral spinal reflex	Rhythmic contractions of bulbocavernosus and ischiocavernosus muscles

autonomic, and endocrine responses. Recent fMRI studies in males have also detected activation in parietal areas known to be involved in attentional processes. The human data correlate well with results from animal experiments which indicate that connections between the limbic system and the hypothalamus (medial preoptic and paraventricular nuclei) are essential for the stimulation of the descending autonomic projections to the spinal cord that trigger psychogenic penile erections.

Glandular Secretion

During the second phase of the sexual response cycle (plateau), activity in parasympathetic pathways stimulates mucus secretion from bulbourethral and Littre's glands and secretion from the seminal vesicles and the prostate gland. Mucus secretion contributes to lubrication of the penis, whereas secretions from the seminal vesicles and prostate provide the bulk of the fluid and chemical factors that contribute to the viability and motility of the spermatozoa. Glandular secretion is thought to be mediated by the parasympathetic system; however, the central reflex mechanisms and the neurotransmitters have not been identified.

Emission–Ejaculation

The third phase of the sexual act (orgasm), which is accompanied by emission and ejaculation of semen, involves the coordination of autonomic and somatic reflex mechanisms at different levels of the lumbosacral spinal cord. During the first step in this process (emission), reflex activity in the thoracolumbar sympathetic outflow elicits rhythmic contractions of the smooth muscle of the seminal vesicles, prostate, ductus deferens, and ampulla, resulting in the ejection of sperm and glandular secretions into the urethra and at the same time a closure of the bladder neck, to prevent the backflow of semen into the bladder. Pharmacologic studies have shown that these responses are mediated by sympathetic neurotransmitters, norepinephrine and ATP, interacting with α-adrenergic receptors and purinergic receptors, respectively, in the peripheral organs. After emission of semen into the proximal urethra, rhythmic contractions of the bulbocavernosus, ischiocavernosus, and paraurethral striated muscles result in ejaculation. The afferent and efferent limbs of the ejaculation reflex are contained in the pudendal nerve (**Table 2**). The sensations accompanying ejaculation constitute the orgasm. Orgasm is not necessarily affected by sympathectomy, provided that the pudendal nerves remain intact. Thus, neither afferent fibers in the sympathetic nerves nor contractions of smooth muscles of the seminal vesicles and ductus deferens are essential for the occurrence of orgasm.

In the spinal cord, ascending fibers carrying sensory information from the sex organs seem to lie primarily within the anterolateral tracts. Thus, bilateral cordotomy for the treatment of chronic pain usually diminishes or completely abolishes orgasm. Cordotomy also severely compromises ejaculatory and erectile mechanisms, although the latter may recover over a period of time. With more extensive damage of the spinal cord in paraplegics, when the site of injury is located rostral to T12, ejaculation occurs in a relatively small percentage of patients in comparison to reflexogenic erections, which are readily elicited. This observation is consistent with the greater complexity of spinal reflex pathways underlying ejaculation and indicates a considerable dependence of these pathways on supraspinal coordinating mechanisms.

Recent studies in rats have identified a population of lumbar spinal neurons located around the central canal in lamina X and medial lamina VII that participate in the ejaculatory reflex. These neurons express neurokinin-1 receptors, receive afferent input from the penis, and project to the thalamus. It was initially hypothesized that these neurons might function primarily in transmitting sensory information from the penis to pleasure centers in the brain. However, it was discovered that these neurons were only activated during ejaculation and not during other components of male sexual behavior. When the neurons were destroyed by intrathecal administration of a toxin (saporin conjugated to a substance P analog) that selectively destroys neurons expressing neurokinin-1 receptors, ejaculatory responses were completely abolished without altering other components of copulatory behavior (penile erection, mounting, intromission). It was concluded that the neurons may be part of a spinal ejaculation generator.

Central Neurotransmitters

Pharmacological studies in animals have implicated many neurotransmitters in the central control of sexual function. The dopaminergic system in the medial preoptic area and the oxytocin and nitric oxide systems in the paraventricular nucleus exert major excitatory effects on erection, while the serotonergic pathways arising in the raphe nuclei seem to be inhibitory. Elsewhere in the brain, noradrenergic, GABAergic, and serotonergic pathways are generally inhibitory, whereas dopaminergic, nitric oxide, and oxytocin pathways facilitate sexual function. Nitric oxide appears to tonically modulate the hypothalamic excitatory circuitry, because intracerebroventricular injections of nitric oxide synthase inhibitors prevent the penile erectile responses in rats induced by dopamine agonists and oxytocin. It is uncertain whether all of these findings in animals are entirely applicable to man. However, recently, apomorphine, a nonselective dopamine receptor agonist that stimulates sexual behavior and penile erection in animals, has been marketed to treat penile erectile dysfunction in patients.

See also: Autonomic and Enteric Nervous System: Apoptosis and Trophic Support During Development; Autonomic Disorders; Autonomic Failure; Autonomic Nervous System: Urogenital Control; Autonomic Nervous System: Neuroanatomy; Autonomic Neuroimmunology; Autonomic Neuroplasticity: Development; Autonomic Nervous System; Sympathetic Nervous System.

Further Reading

Argiolas A and Melis MR (2005) Central control of penile erection: Role of the paraventricular nucleus of the hypothalamus. *Progress in Neurobiology* 76: 1–21.

Coolen LM, Allard J, Truitt WA, et al. (2004) Central regulation of ejaculation. *Physiology & Behavior* 83: 203–215.

de Groat WC (2006) Integrative control of the lower urinary tract: Preclinical perspective. *British Journal of Pharmacology* 147 (supplement 2): S25–S40.

de Groat WC and Yoshimura N (2001) Pharmacology of the lower urinary tract. *Annual Review of Pharmacology and Toxicology* 41: 691–721.

de Groat WC and Yoshimura N (2006) Mechanisms underlying the recovery of lower urinary tract function following spinal cord injury. *Progress in Brain Research* 152: 59–84.

Ferretti A, Caulo M, Del GC, et al. (2005) Dynamics of male sexual arousal: Distinct components of brain activation revealed by fMRI. *NeuroImage* 26: 1086–1096.

Giuliano F and Rampin O (2004) Neural control of erection. *Physiology & Behavior* 83: 189–201.

Kavia RB, Dasgupta R, and Fowler CJ (2005) Functional imaging and the central control of the bladder. *Journal of Comparative Neurology* 493: 27–32.

Marson L and Foley KA (2004) Identification of neural pathways involved in genital reflexes in the female: A combined anterograde and retrograde tracing study. *Neuroscience* 127: 723–736.

Marson L, Platt KB, and McKenna KE (1993) Central nervous system innervation of the penis as revealed by the transneuronal transport of pseudorabies virus. *Neuroscience* 55: 263–280.

Morrison J, Birder L, Craggs M, et al. (2005) Neural control. In: Abrams P, Cardozo L, Khoury S, et al. (eds.) *Incontinence*, pp. 363–422. Jersey: Health Publications Ltd.

Nadelhaft I and Vera PL (1995) Central nervous system neurons infected by pseudorabies virus injected into the rat urinary bladder following unilateral transection of the pelvic nerve. *Journal of Comparative Neurology* 359: 443–456.

Saenz de Tejada I, Cavidad NG, Heaton J, et al. (2001) Anatomy, physiology, and pathophysiology of erectile function. In: Jardin A, Wagner G, Khoury S, et al. (eds.) *Erectile Dysfunction*, pp. 65–102. Jersey: Health Publications Ltd.

Sasaki M (2005) Properties of Barrington's neurones in cats: Units that fire inversely with micturition contraction. *Brain Research* 1033: 41–50.

Truitt WA and Coolen LM (2002) Identification of a potential ejaculation generator in the spinal cord. *Science* 297: 1566–1569.

Autonomic Nervous System: Clinical Testing

C J Mathias, Imperial College London at St. Mary's Hospital, London, UK

Introduction and Basic Principles

The autonomic nervous system, through its major efferent components, the parasympathetic and sympathetic nervous systems, supplies every organ in the body (**Figure 1(a)**). It influences local organ function and integrated systems such as those that control blood pressure and body temperature. The parasympathetic nervous system has a cranial division that innervates the lacrimal glands, iris musculature, salivary glands, heart, and upper gastrointestinal tract; the sacral division supplies the urinary bladder, sexual organs, and lower gastrointestinal tract. The sympathetic nervous system emerges from the thoracic and upper lumbar segments of the spinal cord and innervates every organ in the body. Both systems have pre- and postganglionic segments. Some preganglionic fibers continue through the paravertebral ganglia, forming the greater and lesser splanchnic nerves, with synapses in the celiac, superior mesenteric, inferior mesenteric, and renal ganglia, from where postganglionic unmyelinated fibers emerge; these comprise the enteric nervous system, considered the third component of the autonomic nervous system.

Autonomic activity is controlled and modulated at a number of levels within the neural axis. There are major cerebral centers, especially in the hypothalamus and brain stem, which are influenced by a variety of peripheral sensory inputs, in addition to signals from cortical and other areas of the brain. These central centers and pathways increasingly are being delineated in humans using modern neuroimaging and allied technological advances. Autonomic efferent activity is a composite response to a variety of information derived externally and internally, with integration at different levels of the neuraxis that maintain functioning of the organs and systems in a wide range of environments and stimuli.

Acetylcholine (ACh) is the major transmitter at ganglia acting through the nicotinic receptor and affecting both parasympathetic and sympathetic activity (**Figure 1(b)**). ACh also is the neurotransmitter within the adrenal medulla, which functions as a ganglion, but with an endocrine role. It secretes adrenaline and noradrenaline (NA), which enter the circulation and act on α-adrenoreceptors (in blood vessels) and β-adrenoreceptors (in the heart and lungs). Neurons often contain more than one neurotransmitter and a range of substances, including neuropeptides, that play important roles in influencing autonomic activity and target organ function. The major neurotransmitter at sympathetic postganglionic sites is noradrenaline, formed through a series of metabolic steps from the amino acid tyrosine (**Figure 1(c)**). ACh acts at parasympathetic postganglionic nerve endings and at sympathetic terminals on sweat glands through the muscarinic receptor.

Clinical Features

Localized lesions (**Table 1**), such as Horner's syndrome, cause few symptoms but characteristic signs such as ptosis, a small pupil, and anhidrosis, indicating damage to sympathetic pathways to the face. The underlying cause, however, may be a tumor at the apex of the lung, a lesion within the spinal cord, or dissection of the internal carotid artery. Determining the underlying disorders that contribute to morbidity and mortality therefore is of importance. Generalized disorders (**Table 2**) often are caused by irreversible damage to the autonomic nervous system. There are many manifestations resulting from impairment of different regions and systems (**Table 3**). The symptoms (**Table 4**) and confirmation of postural (orthostatic) hypotension (**Figure 2(a)**) often lead to consideration of autonomic involvement.

In secondary autonomic dysfunction, the clinical manifestations of the primary disorder, such as muscle paralysis in tetraplegia, may predominate. Autonomic features include severe paroxysmal hypertension with headache and facial flushing due to autonomic dysreflexia. In diabetes mellitus, gustatory sweating (hyperhidrosis over the head and neck during food ingestion), impotence, and nocturnal diarrhea occur. Drugs are a frequent cause of autonomic dysfunction (**Table 5**).

Some disorders result in intermittent autonomic impairment. They include neurally mediated syncope, where hypotension and bradycardia contribute to fainting. There often are no discernible abnormalities, autonomic or otherwise, between attacks. Of importance are the history and the association of attacks with provocative factors. In the recently described disorder, the postural tachycardia syndrome (PoTS), palpitations occur during standing and exertion, with an abnormal elevation of heart rate to over 30 beats min^{-1} on postural challenge.

Clinical Investigation of Autonomic Function

The clinical investigative aims are to ascertain the presence and degree of autonomic dysfunction, define the etiological factors, and delineate the site of the lesion. Each is important for diagnosis, prognosis, and management. A variety of tests relating to different systems may need to be performed (**Table 6**). This depends on the presenting problem, the possible underlying disorder, and the strategies planned for treatment and its evaluation. Advances in technology now enable measurements noninvasively and continuously, as exemplified in relation to blood pressure (BP), heart rate (HR), and other cardiovascular variables.

Cardiovascular System

Physiological

Gravitational challenge Postural hypotension is a cardinal manifestation of sympathetic failure. A fall of 20 mmHg systolic or 10 mmHg diastolic blood pressure, or less in the presence of symptoms, warrants further investigation. Measurements are first made with the individual lying flat, with adequate readings to determine the stability of blood pressure before postural change. This is induced either by using a tilt table (to 60°) or by making the individual stand upright. Prolonged tilt-table testing, for up to 45 min, also is used in the investigation of neurally mediated syncope, sometimes in combination with

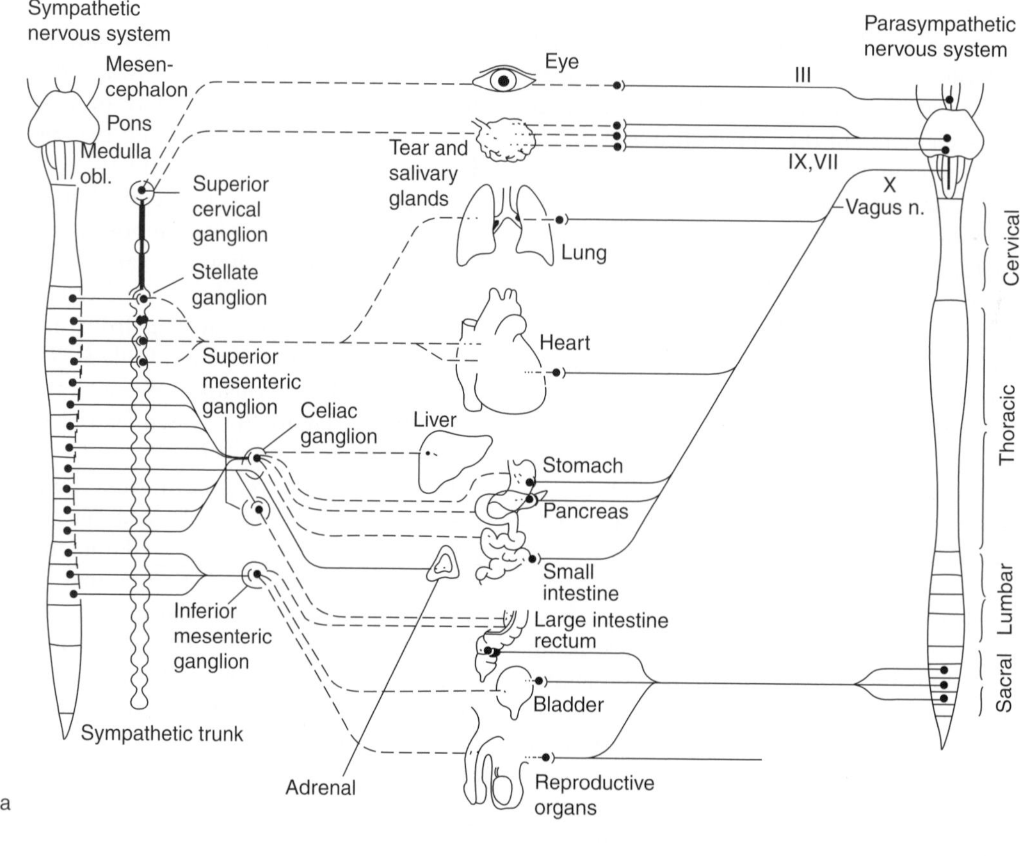

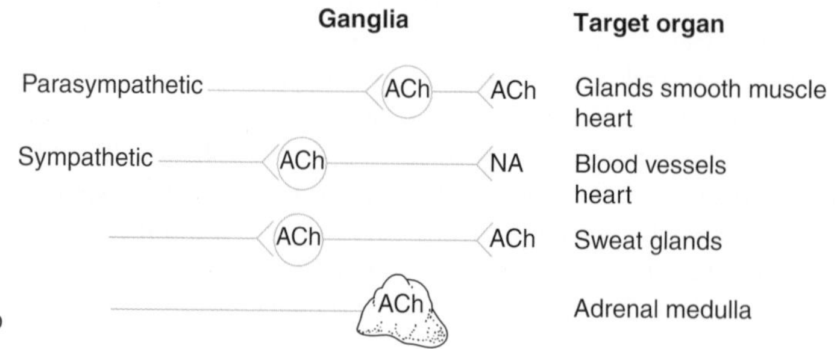

Figure 1 Continued

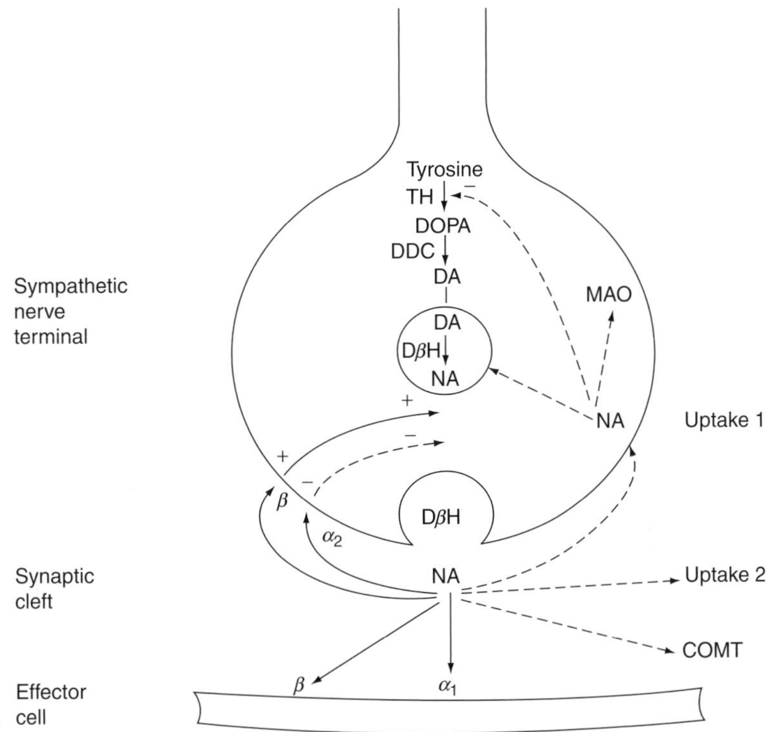

Figure 1 (a) Schema outlining details of the craniosacral parasympathetic and thoracolumbar sympathetic outflows to various target organs. (b) Neurotransmitters subserving the parasympathetic and sympathetic pathways at ganglia and target organs. (c) Steps involved in the formation of noradrenaline (NA) from tyrosine within a sympathetic nerve terminal. NA in granules is released by a process of exocytosis into the synaptic cleft, following which it acts on various α- or β-adrenoreceptors, either pre- or post-synaptically. NA is subject to various processes, which involve uptake 1 into the nerve terminal, following which it is either incorporated into granules, exerts negative feedback on TH, or is metabolized by monoamine oxidase (MAO). Some is taken up into nonneuronal tissues (uptake 2) and some is metabolized by catechol-*O*-methyltransferase (COMT), while the rest spills over into the circulation. ACh, acetylcholine; DA, dopamine; DβH, dopamine β-hydroxylase, DDC, dopa decarboxylase; DOPA, dihydroxyphenylalanine; TH, tyrosine hydroxylase. (a) From Janig W and McLachlan EM (2001) Neurobiology of the autonomic nervous system. In: Mathias CJ and Bannister R (eds.) *Autonomic Failure: A Textbook of Clinical Disorders of the Autonomic Nervous System*, 4th edn., pp. 3–15. Oxford: Oxford University Press. (b) From Mathias CJ (1987) Autonomic dysfunction. *British Journal of Hospital Medicine* 38: 238–243. (c) From Mathias CJ (2000) Disorders of the autonomic nervous system. In: Bradley WG, Daroff RB, Fenichel GM, et al. (eds.) *Neurology in Clinical Practice*, 3rd edn., pp. 2131–3265. Boston: Butterworth-Heinemann.

Table 1 Examples of localized autonomic disorders

Horner's syndrome
Holmes–Adie pupil
Crocodile tears (Bogorad's syndrome)
Gustatory sweating (Frey's syndrome)
Reflex sympathetic dystrophy
Idiopathic palmar, plantar, or axillary hyperhidrosis
Chagas' disease (*Trypanosoma cruzi*)[a]
Surgical procedures[b]
Sympathectomy (regional) causing compensatory hyperhidrosis
Vagotomy and gastric drainage procedures in 'dumping' syndrome
Organ transplantation (heart, lungs)

[a]Listed here because it specifically targets intrinsic cholinergic plexuses in the heart and gut.
[b]Surgery also may cause other localised disorders, such as Frey's syndrome, after parotid surgery.
From Mathias CJ and Bannister R (2001) Investigation of autonomic disorders. In: Mathias CJ and Bannister R (eds.) *Autonomic Failure: A Textbook of Clinical Disorders of the Autonomic Nervous System*, 4th edn., pp. 169–195. Oxford: Oxford University Press.

supraphysiological (negative body pressure) or pharmacological (glyceryl trinitrate) stimuli.

In normal individuals there are only small changes in blood pressure during head-up tilt or standing. In autonomic failure, the fall is often rapid and a return to horizontal restores blood pressure levels (**Figure 2(a)**). There may be considerable variability in the extent of postural hypotension, as many factors in daily life adversely or favorably influence blood pressure in these individuals. Postural hypotension is greater in the morning, after food ingestion, with exercise, in hot weather, and if vasodilator drugs are used (**Table 7**). Water ingestion (500 ml) and various physical maneuvers (such as squatting, leg crossing, and tensing of leg muscles) have the opposite effect and reduce postural hypotension. It is important to exclude nonneurogenic causes that can exacerbate neurogenic postural hypotension (**Table 8**).

There is normally a modest rise in heart rate during postural change. A lack of change in heart rate despite

Table 2 Outline classification of various causes of autonomic failure

Primary
Acute/subacute dysautonomias
Pure cholinergic dysautonomia
Pure pandysautonomia
Pandysautonomia with neurological features
Chronic autonomic failure syndromes
Pure autonomic failure
Multiple system atrophy (Shy–Drager syndrome)
Autonomic failure with Parkinson's disease
Diffuse Lewy body disease
Secondary
Congenital
Nerve growth factor deficiency
Hereditary
Autosomal dominant trait in familial amyloid neuropathy
Autosomal recessive trait
Familial dysautonomia – Riley–Day syndrome
Dopamine β-hydroxylase deficiency
Metabolic diseases
Diabetes mellitus
Chronic renal failure
Chronic liver disease
Alcohol induced
Inflammatory
Guillain–Barré syndrome
Transverse myelitis
Infections
Bacterial – tetanus
Viral – human immunodeficiency virus infection
Prion – fatal familial insomnia
Neoplasia
Brain tumors – especially of third ventricle or posterior fossa
Paraneoplastic, to include adenocarcinomas of lung and pancreas
Trauma
Cervical and high thoracic spinal cord transection
Drugs
By their direct effects
By causing a neuropathy
Neurally mediated syncope
Vasovagal syncope
Carotid sinus hypersensitivity
Situational syncope
Postural tachycardia syndrome

From Mathias CJ (2003) Disorders of the autonomic nervous system. In: Bradley WG, Daroff RB, Fenichel GM, et al. (eds.) *Neurology in Clinical Practice*, 3rd edn., pp. 2403–2420. Boston, MA: Butterworth-Heinemann.

Table 3 Clinical manifestations in chronic primary autonomic failure[a]

Cardiovascular	Postural hypotension
Sweating	Anhidrosis, heat intolerance
Gastrointestinal	Oropharyngeal dysphasia, constipation, occasionally diarrhea
Renal and urinary bladder	Nocturia, frequency, urgency, incontinence, retention
Genital	Erectile and ejaculatory failure in the male
Eye	Anisocoria, Horner's syndrome
Respiratory	Stridor, involuntary inspiratory gasps, apneic episodes
Other neurological deficits	Parkinsonian, cerebellar, and pyramidal features

[a]The respiratory and additional neurological deficits do not occur in pure autonomic failure and usually are part of multiple system atrophy/Shy–Drager syndrome.
From Mathias CJ (1997) Autonomic disorders and their recognition. *New England Journal of Medicine* 10: 721–724.

Table 4 Some of the symptoms resulting from orthostatic hypotension and impaired perfusion of various organs

Cerebral hypoperfusion
Dizziness
Visual disturbances
Blurred – tunnel
Scotoma
Graying out – blacking out
Color defects
Syncope
Impaired cognition
Muscle hypoperfusion
Paracervical and suboccipital ('coathanger') ache
Lower back/buttock ache
Calf claudication
Chest wall and cardiac hypoperfusion
Chest pain
Angina pectoris
Spinal cord hypoperfusion
Lower limb weakness and parasthesiae
Renal hypoperfusion
Oliguria
Nonspecific
Weakness, lethargy, fatigue
Falls

From Mathias CJ, Mallipeddi R, and Bleasdale-Barr K (1999) Symptoms associated with orthostatic hypotension in pure autonomic failure and multiple system atrophy. *Journal of Neurology* 246: 893–898.

a substantial fall in blood pressure usually is indicative of cardiac parasympathetic failure or afferent baroreceptor block. In tetraplegia the heart rate rises in response to the fall in blood pressure, as the sympathetic efferent components of the baroreceptor reflex are interrupted within the cervical spinal cord, while the vagi from the brain stem are intact (**Figure 2(b)**). The heart rate, however, does not usually rise above 110 beats $\min^{-1}$, which is similar to cardiac parasympathetic withdrawal as observed after vagal blockade with atropine. Further elevation probably is dependent on adrenomedullary stimulation and elevation of plasma adrenaline levels, which does not occur in tetraplegics. This adrenal component probably accounts for the greater tachycardia observed in subjects with an intact sympathetic nervous system when blood pressure is lowered substantially, as in hemorrhagic shock.

In neurally mediated syncope (such as vasovagal syncope), autonomic testing indicates normal autonomic function between attacks. Blood pressure often is maintained initially during head-up tilt, but falls during provocation; this may be preceded by a fall in heart rate (**Figure 2(c)**). The former results from withdrawal of sympathetic activity and the latter from vagal overactivity. This differs markedly from the responses to

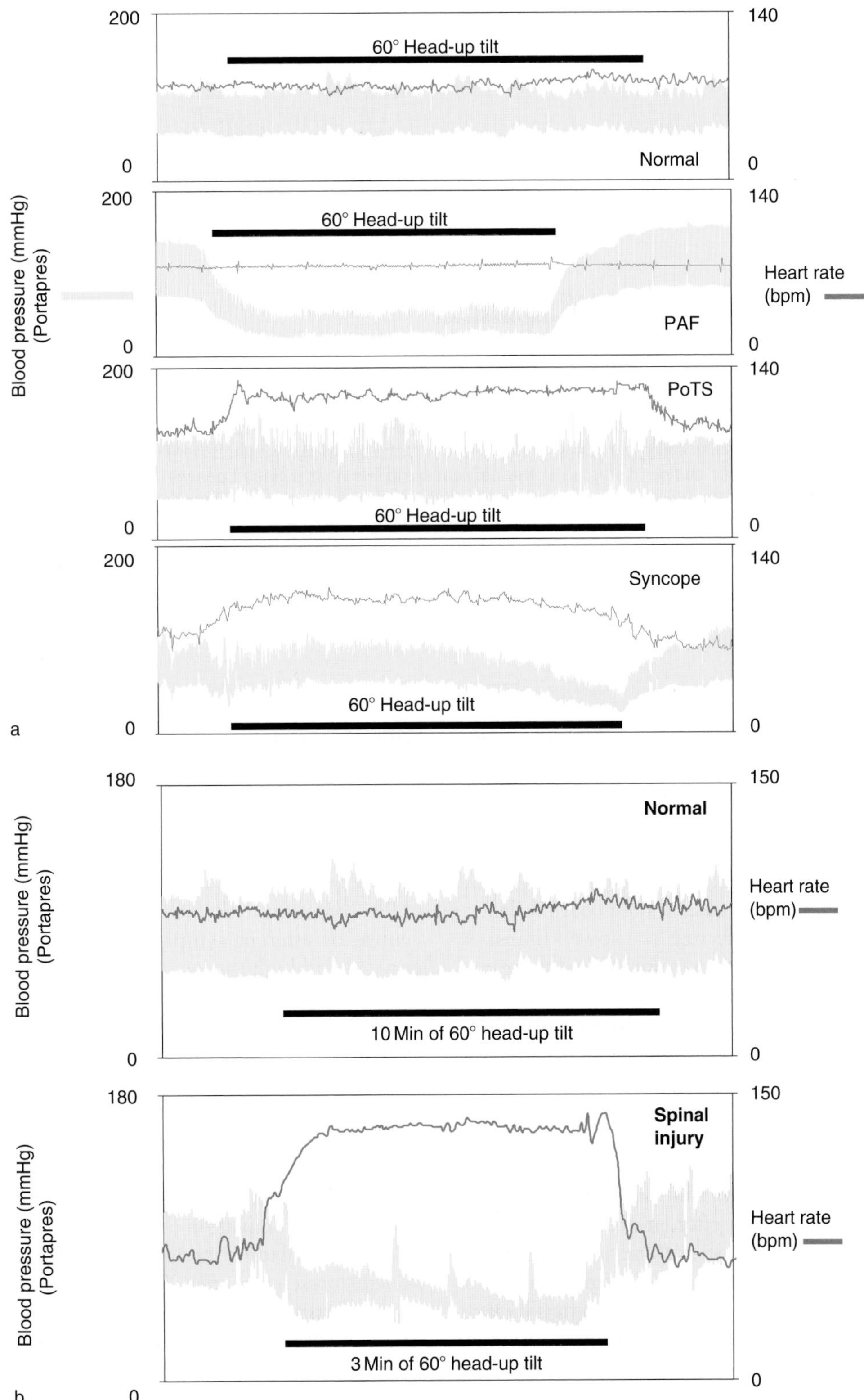

Figure 2 Continued

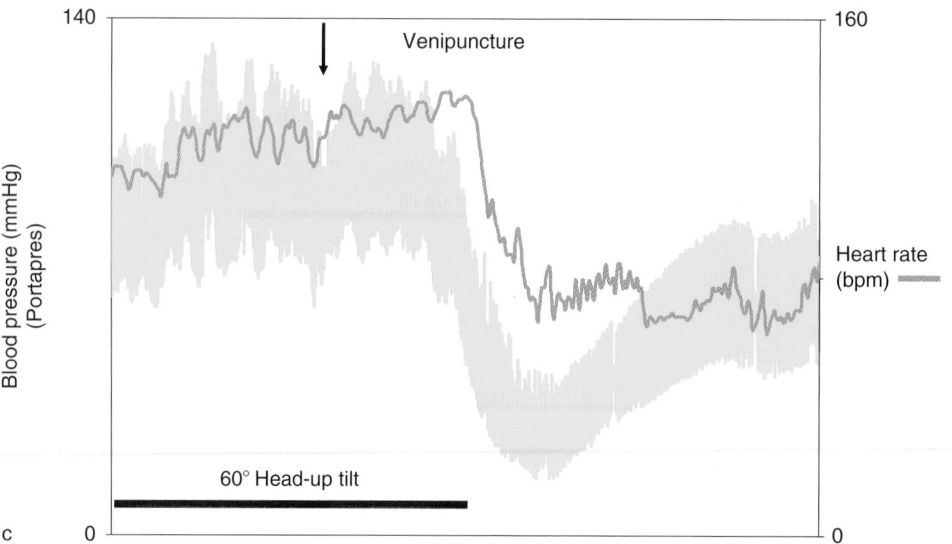

Figure 2 (a) Blood pressure and heart rate (beats per minute, bpm) measured continuously before, during, and after 60° head-up tilt by the Portapres II in a normal person and in individuals with three different autonomic disorders: with pure autonomic failure (PAF), postural tachycardia syndrome (PoTS), and vasovagal syncope. (b) Blood pressure and heart rate measured continuously with the Portapres II in a normal person and in a patient with a high cervical spinal cord lesion. In the spinal injury patient there is a fall in blood pressure because of impairment of the sympathetic outflow disrupted in the cervical spine. Heart rate rises because of withdrawal of vagal activity in response to the rise in pressure. (c) Blood pressure and heart rate with continuous recordings from the Portapres II in a patient with the mixed (cardioinhibitory and vasodepressor) form of vasovagal syncope. (a) From Mathias CJ (2002) To stand on ones' own legs. *Clinical Medicine* 2: 237–245. (b) From Mathias CJ (2006) Orthostatic hypotension and orthostatic intolerance. In: De Groot LJ, Jameson JL, de Kretser D, et al. (eds.) *Endocrinology*, 5th edn., pp. 2613–2632. Philadelphia: Elsevier. (c) From Mathias CJ (2006) Orthostatic hypotension and orthostatic intolerance. In: De Groot LJ, Jameson JL, de Kretser D, et al. (eds.) *Endocrinology*, 5th edn., pp. 2613–2632. Philadelphia: Elsevier.

head-up postural change observed in orthostatic hypotension due to autonomic failure.

In the syndrome PoTS, also called the neuropathic PoTS, the heart rate rises by over 30 beats min^{-1} during standing or head-up tilt, without postural hypotension. There is no evidence of widespread autonomic failure. However, in some cases there is partial denervation, predominantly affecting the lower limbs. A genetic abnormality, linked to a noradrenaline transporter defect has been described in one family. Psychogenic components, including hyperventilation, may contribute. There is an association with mitral valve prolapse, the chronic fatigue syndromes, physical decompensation, and the joint hypermobility syndrome (Ehlers–Danlos III).

Pressor stimuli Pressor stimuli raise blood pressure by increasing sympathetic efferent activity. They act via the periphery (isometric exercise and cutaneous cold (**Figure 3(a)**), or directly by cerebral stimulation (sudden noise and mental arithmetic). Isometric exercise is performed by sustained hand-grip using a dynamometer. In the cold pressor test the hand is immersed in ice for 90 s. In both situations the autonomic responses depend on peripheral stimulation but with a central component. Cerebral stimulation is induced by a sudden noise, such as a starting pistol, or mental arithmetic, utilizing subtraction or addition of serial 7s or 17s (or more complex) tasks. In normal individuals both blood pressure and heart rate increase.

Pressor stimuli do not elevate blood pressure in central or efferent sympathetic lesions. In complete cervical and high thoracic spinal cord lesions, stimuli above the level of the lesion (cutaneous cold and mental arithmetic) do not raise blood pressure, in contrast to stimuli below the lesion, which can elevate blood pressure markedly by inducing spinal cord sympathetic reflexes (**Figure 3(b)**). The heart rate falls in response to stimulation activation of baroreflex afferents and an increase in vagal activity. Stimuli that cause these changes include cutaneous stimuli (such as cold), activation of abdominal or pelvic viscera (by urinary bladder or large bowel distension), and skeletal muscle spasms. The cardiovascular changes to such stimulation are an important component of the syndrome of autonomic dysreflexia. This may be mistaken for a hypertensive crisis as occurs with pheochromocytoma tumors, which secrete excessive amounts of catecholamines. Measurement of plasma

Table 5 Drugs, chemicals, poisons, and toxins causing autonomic dysfunction

Decreasing sympathetic activity
Centrally acting
Clonidine
Methyldopa
Moxonidine
Reserpine
Barbiturates
Anesthetics
Peripherally acting
Sympathetic nerve ending (guanethidine, bethanidine)
α-Adrenoceptor blockade (phenoxybenzamine)
β-Adrenoceptor blockade (propranolol)
Increasing sympathetic activity
Amphetamines
Releasing noradrenaline (tyramine)
Uptake blockers (imipramine)
Monoamine oxidase inhibitors (tranylcypromine)
β-Adrenoceptor stimulants (isoprenaline)
Decreasing parasympathetic activity
Antidepressants (imipramine)
Tranquilizers (phenothiazines)
Antidysrhythmics (disopyramide)
Anticholinergics (atropine, probanthine, benztropine)
Toxins (botulinum)
Increasing parasympathetic activity
Cholinomimetics (carbachol, bethanechol, pilocarpine, mushroom poisoning)
Anticholinesterases
Reversible carbamate inhibitors (pyridostigmine, neostigmine)
Organophosphorous inhibitors (parathion, sarin)
Miscellaneous
Alcohol, thiamine (vitamin B_1) deficiency
Vincristine, perhexiline maleate
Thallium, arsenic, mercury
Mercury poisoning (pink disease)
Ciguatera toxicity
Jellyfish and marine animal venoms
First dose of certain drugs (prazosin, captopril)
Withdrawal of chronically used drugs (clonidine, opiates, alcohol)

From Mathias CJ (2003) Disorders of the autonomic nervous system. In: Bradley WG, Daroff RB, Fenichel GM, et al. (eds.) *Neurology in Clinical Practice*, 3rd edn., pp. 2403–2420. Boston, MA: Butterworth-Heinemann.

noradrenaline and adrenaline levels during these episodes excludes the possibility of catecholamine hypersecretion. In autonomic dysreflexia there is a modest rise in plasma noradrenaline but not adrenaline levels, the former reflecting sympathoneural activation.

Heart rate responses to respiratory change The cardiac vagi rapidly respond to respiratory changes, with a rise in heart rate during inspiration and a fall during expiration (sinus arrhythmia) (**Figure 3(c)**). In normal individuals, deep breathing (6 breaths min^{-1}) and hyperventilation raise heart rate because of vagal withdrawal. This does not occur in autonomic failure affecting cardiac parasympathetic pathways. In diabetes mellitus and alcoholism, where cardiac vagal lesions may occur prior to sympathetic impairment, the heart rate responses to such stimuli often are abnormal.

Valsalva maneuver In the Valsalva maneuver, the blood pressure and heart rate responses to raised intrathoracic pressure provide a measure of the integrity of the baroreflex pathway (**Figure 3(d)**). The individual blows with an open glottis into a sphygmomanometer and maintains a forced expiratory pressure of up to 40 mmHg for 10 s. The rise in intrathoracic pressure impairs venous return, reduces cardiac output, and lowers blood pressure, with a consequent rise in heart rate and blood pressure recovery by increasing sympathetic neural activity. At the end of the maneuver, when intrathoracic pressure returns to normal, there is a blood pressure overshoot because sympathetic stimulation persists and baroreflex activation reduces the heart rate. In sympathetic vasoconstrictor failure the Valsalva maneuver causes blood pressure to fall without stabilization. Following reduction in intrathoracic pressure, there is no blood pressure overshoot and no compensatory bradycardia.

Food challenge Food releases vasodilatatory gut peptides, causing dilatation in the splanchnic circulation, which normally results in an increase in cardiac output, and constriction of skeletal muscle resistance vessels that prevents blood pressure from falling. These compensatory responses to splanchnic vasodilatation do not occur in autonomic failure and may cause postprandial hypotension even in the supine position (**Figure 4(a)**). Measuring the blood pressure during head-up tilt before and 45 min after a balanced 300 ml liquid meal of 530 kcal determines if food unmasks or enhances postural hypotension.

Exercise testing Exercise raises blood pressure in normal individuals. In autonomic failure even modest exercise on a bicycle while supine lowers blood pressure and can, on standing postexercise, aggravate postural hypotension (**Figure 4(b)**). This is due to vasodilatation in exercising muscles, not accompanied by compensatory changes in other vascular beds.

Carotid sinus massage Carotid sinus hypersensitivity is an important cause of falls, especially in the elderly. Carotid sinus massage normally causes minor changes in heart rate with no decrease in blood pressure. In carotid sinus hypersensitivity, bradycardia and hypotension occur especially when massage is performed during head-up tilt (**Figure 4(c)**).

Table 6 Outline of investigations in autonomic failure

Cardiovascular	
Physiological	Head-up tilt (60°)*
	Standing*, Valsalva maneuver*
	Pressor stimuli – isometric exercise,* cutaneous cold,* mental arithmetic*
	Heart rate responses – deep breathing,* hyperventilation,* standing,* head-up tilt*
	Carotid sinus massage
	Liquid meal challenge, modified exercise testing
Biochemical	Plasma noradrenaline – supine and standing, urinary catecholamines and their metabolites
Pharmacological	Noradrenaline – α-adrenoreceptors (vascular)
	Isoprenaline – β-adrenoreceptors (vascular and cardiac)
	Tyramine – pressor and noradrenaline response
	Atropine – parasympathetic cardiac blockade
	Clonidine – noradrenaline suppression and growth hormone stimulation
Sweating	Thermoregulatory sweat test – increase core temperature by 1 °C and assess sweating
	Modified sweat testing for localized abnormalities
	Gustatory sweating
	Sweat gland response – intradermal acetylcholine, quantitative sudomotor axon reflex test (Q-SART)
	Sympathetic skin response
Gastrointestinal	Barium studies, video-cinefluoroscopy, endoscopy, gastric emptying studies
Respiratory	Laryngoscopy, sleep studies to assess apnea and oxygen desaturation
Renal function and urinary	Day and night urine volumes and sodium/potassium excretion
	Urodynamic studies, intravenous urography, ultrasound examination, urethral or anal sphincter electromyography
Sexual	Penile plethysmography
	Intracavernosal papaverine response
Eye	Pupillary function – physiological and pharmacological
	Schirmer's test for lacrimal secretion

Asterisk denotes inclusion of the screening tests used in our London laboratories.
From Mathias CJ and Bannister R (2001) Investigation of autonomic disorders. In: Mathias CJ and Bannister R (eds.) *Autonomic Failure: A Textbook of Clinical Disorders of the Autonomic Nervous System*, 4th edn., pp. 169–195. Oxford: Oxford University Press.

Table 7 Factors influencing orthostatic hypotension

Speed of positional change
Time of day (worse in the morning)
Prolonged recumbency
Warm environment (hot weather, central heating, hot bath)
Raising intrathoracic pressure – micturition, defecation, or coughing
Food and alcohol ingestion
Water ingestion[a]
Physical exertion
Physical maneuvers and positions (bending forward, abdominal compression, leg crossing, squatting, activating calf muscle pump)[b]
Drugs with vasoactive properties (including dopaminergic agents)

[a]This raises blood pressure in autonomic failure.
[b]These maneuvres usually reduce the postural fall in blood pressure, unlike the others.
From Mathias CJ (2003) Autonomic diseases – clinical features and laboratory evaluation. *Journal of Neurology, Neurosurgery & Psychiatry* 74: 31–41.

Table 8 Nonneurogenic causes of postural hypotension

Low intravascular volume	
Blood/plasma loss	Hemorrhage, burns, hemodialysis
Fluid/electrolyte loss	Inadequate intake – anorexia nervosa, vomiting
	Diarrhea – including losses from ileostomy
	Renal/endocrine – salt-losing nephropathy, adrenal insufficiency (Addison's disease), diabetes insipidus, diuretics
Vasodilatation	Drugs – glyceryl trinitrate
	Alcohol
	Heat, pyrexia
	Hyperbradykininism
	Extensive varicose veins
Cardiac impairment	
Myocardial	Myocarditis
Impaired ventricular filling	Atrial myxoma, constrictive pericarditis
Impaired output	Aortic stenosis

From Bannister R and Mathias CJ (2002) Management of postural hypotension. In: Mathias CJ and Bannister R (eds.) *Autonomic Failure: A Textbook of Clinical Disorders of the Autonomic Nervous System,* 4th edn., pp. 342–356. Oxford: Oxford University Press.

Other methods of assessing cardiovascular autonomic and baroreceptor reflex function In addition to autonomic screening and allied tests that are used in clinical autonomic laboratories, there are newer techniques mainly used in clinical research laboratories; these techniques enable continuous noninvasive measurement of cardiac function, peripheral vascular resistance, and blood flow in various regions. Computer and spectral analytical techniques assess spontaneous or induced fluctuations in blood pressure and heart rate, providing a noninvasive measure of the different autonomic systems influencing cardiovascular control.

A variety of techniques determine baroreflex function. Cardiopulmonary baroreceptor afferents can be stimulated by negative pressure to the lower body, induced by placing the individual in a box with an airtight seal, extending up to the midthoracic region, which allows suction to 'unload' the receptors. This normally increases sympathetic neural activity, with a rise in heart rate constriction of resistance vessel, and maintenance of blood pressure. In sympathetic failure, blood pressure rapidly falls during induction of negative pressure.

Carotid sinus afferents may be tested with a cervical collar, whereby elevation or lowering of pressure enables inhibition or stimulation of receptors. Pressor agents (phenylephrine) or vasodilators (glyceryl trinitrate), given either as bolus injections or sublingually, cause transient changes in blood pressure and heart rate, enabling pharmacological assessment of baroreflex sensitivity.

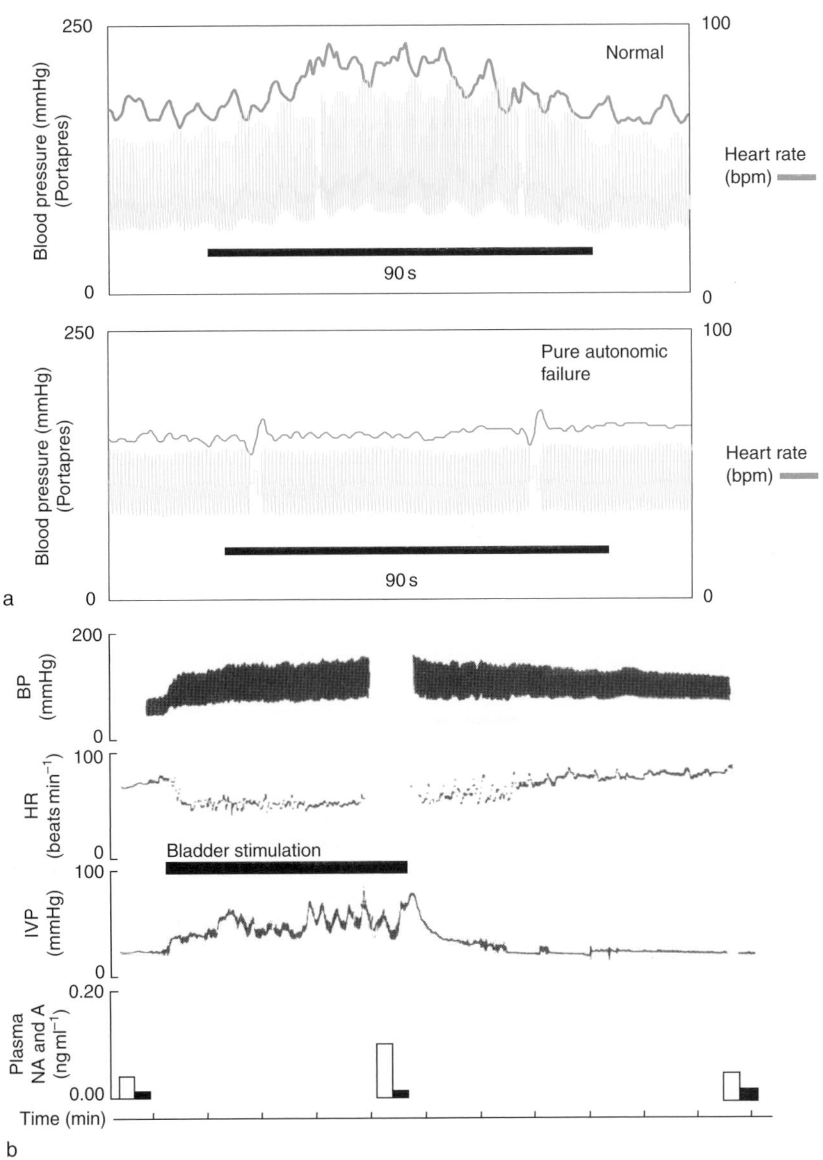

Figure 3 Continued

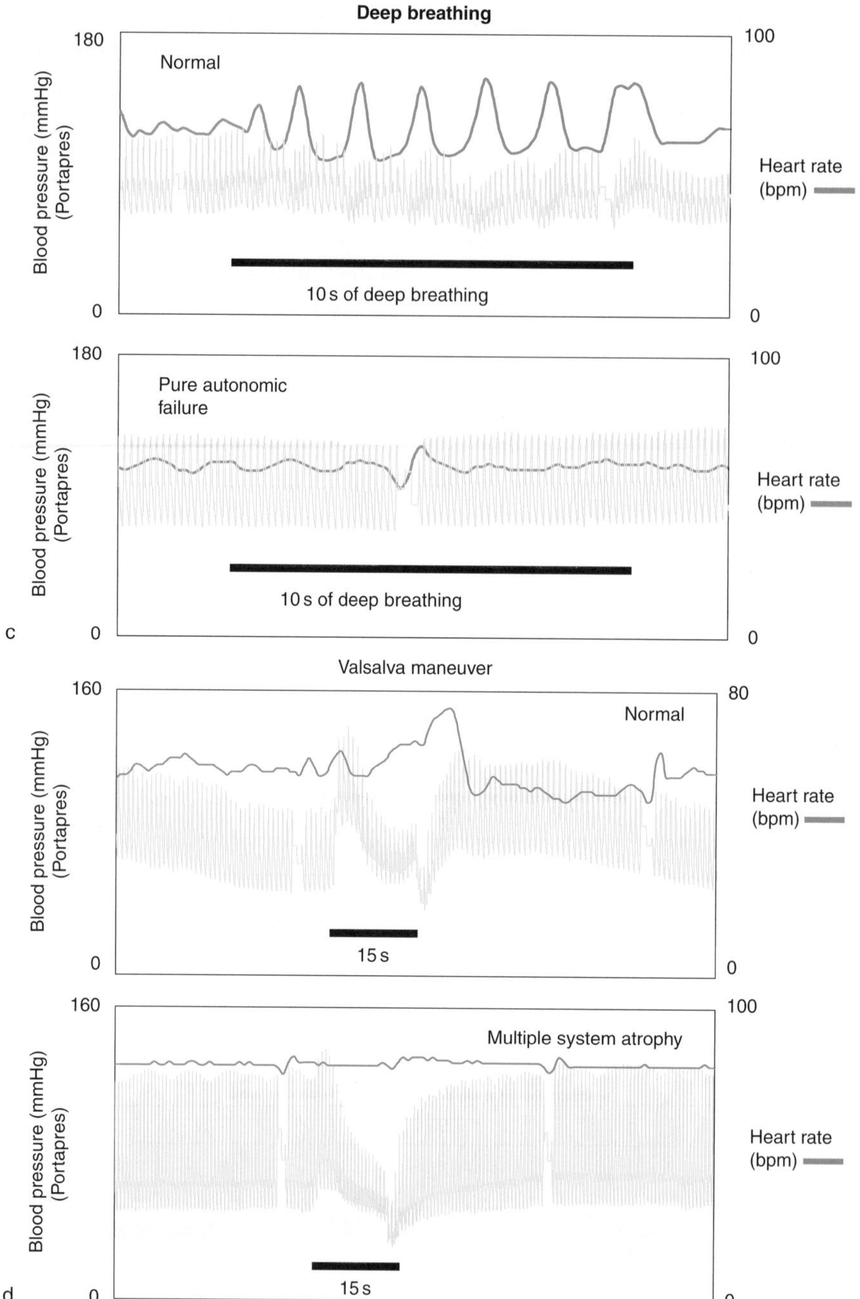

Figure 3 (a) Rise in blood pressure and heart rate (beats per minute, bpm) during cutaneous cold in a normal individual but not in pure autonomic failure. (b) Blood pressure (BP) and heart rate (HR) in a tetraplegic patient before and during bladder stimulation induced by suprapubic percussion of the anterior abdominal wall. This is expected to cause neurogenic bladder contraction. There is a marked rise in BP with a corresponding fall in HR. The breaks in the record indicate where blood was taken for measurement of plasma noradrenaline (NA) and adrenaline (A). There is no change in A (filled rectangles) levels, while NA (open rectangles) levels rise, indicating that the rise in pressure is the result of reflex sympathetic nervous activity occurring via the isolated spinal cord. (c) Rise and fall in heart rate (sinus arrhythmia) in a normal individual and in pure autonomic failure. Normal responses are diminished in autonomic failure affecting the cardiac parasympathetic pathway. (d) Intraarterial recording of blood pressure and heart rate during the Valsalva maneuver in a normal person. There is a fall in blood pressure (with a rise in heart rate), while intrathoracic pressure is elevated. There is a partial recovery in blood pressure with time. Following release of intrathoracic pressure there is a blood pressure overshoot and a corresponding fall in heart rate below basal levels; compare to multiple system atrophy. (a, c, and d) From Mathias CJ and Bannister R (2001) Investigation of autonomic disorders. In: Mathias CJ and Bannister R (eds.) *Autonomic Failure: A Textbook of Clinical Disorders of the Autonomic Nervous System*, 4th edn., pp. 169–195. Oxford: Oxford University Press. (b) From Mathias CJ and Frankel HL (1983) Clinical manifestations of malfunctioning mechanisms in tetraplegic man. *Journal of the Autonomic Nervous System* 7: 303–312.

Pharmacological

The responses to agonist or antagonist drugs provide information on the integrity of autonomic pathways and sensitivity of receptors on target organs. They often predict the response to drug therapy.

Noradrenaline predominantly stimulates α-adrenoreceptors and in sympathetic failure results in an enhanced pressor response. This is of greater magnitude in peripheral lesions, as compared to central lesions. Isoprenaline increases heart rate by stimulation of β_1-adrenorceptors and lowers blood pressure through vasodilatatory β_2-adrenoreceptors, if the latter response is unopposed by sympathetic reflexes, blood pressure falls to a greater extent. The drugs are given intravenously, by bolus injection or continuous infusion. Ideally a dose–response curve to three or more doses should be constructed.

The ability of sympathetic nerve endings to release noradrenaline can be tested by using tyramine, which releases noradrenaline from both granules and cytosol in the nerve terminal. The lack of a pressor response indicates postganglionic denervation and absence of noradrenaline (as in dopamine β-hydroxylase (DBH) deficiency). In incomplete lesions, release of even small amounts of noradrenaline may elicit a response in the presence of pressor supersensitivity.

Cardiac parasympathetic efferent pathways are tested by using the muscarinic blocker atropine. An intravenous bolus of $5\,\mu g\,kg^{-1}$ every 2 min (to a maximum of 4 mg) normally raises heart rate to 110–$120\,beats\,min^{-1}$ because of vagal withdrawal. Vagal efferent lesions result in minimal or no change in heart rate. Atropine testing also helps determine whether vasovagal syncope has a predominantly cardioinhibitory component.

The α_2-adrenoceptor agonist clonidine reduces sympathetic activity predominantly through its central effects. Clonidine lowers levels of plasma noradrenaline if elevated due to increased neural activity such as stress. This suppression does not occur with autonomous secretion, as in pheochromocytoma (**Figure 5**). Another central action of clonidine, mainly through the hypothalamus, is stimulation of growth hormone-releasing hormone that acts on the anterior pituitary to release growth hormone. A rise in serum growth hormone levels occurs in normal individuals and in distal autonomic lesions (such as pure autonomic failure (PAF)), but not in central lesions (such as multiple system atrophy (MSA)) (**Figures 5(a)** and **5(b)**). The clonidine–growth hormone test is used to differentiate these disorders and also in separating MSA from parkinsonian disorders at an early stage when autonomic failure may not be present.

Biochemical

Activation of the sympathetic nervous system, as during head-up tilt, increases synthesis of the neurotransmitter noradrenaline and results in elevation of its plasma levels in normal individuals but not in sympathetic failure. Basal plasma noradrenaline levels are low when there is damage to postganglionic nerve terminals such as PAF, unlike central lesions such as MSA, when they often are within the normal range (**Figure 6**). Basal levels of plasma adrenaline in normal individuals are much lower (around $50\,pg\,ml^{-1}$), and near the detection limit of many assays. Exercise and hypoglycemia cause a greater degree of adrenal stimulation and predominantly raise plasma adrenaline levels. Levels of plasma noradrenaline and/or adrenaline both at rest and during stimulation often are greatly elevated in pheochromocytoma. In DBH deficiency, a condition characterized by postural hypotension due to sympathetic adrenergic failure, present from birth or early childhood, plasma noradrenaline and adrenaline levels are undetectable while plasma dopamine levels are elevated. The enzyme DBH is undetectable in plasma and tissue.

Urinary measurements of catecholamines and their metabolites provide a useful measure of secretion over a period, and are of value in certain situations, as in pheochromocytoma, in which random plasma measurements may miss intermittent secretion. In DBH deficiency, metabolites of noradrenaline and adrenaline (vanillylmandelic acid, normetadrenaline, and metadrenaline) are extremely low, whereas dopamine metabolites (homovanillic acid and 3-methoxytyramine) are normal or elevated.

Noradrenaline spillover techniques, although invasive, provide information on sympathetic activity in the whole body and also regionally in heart, splanchnic, renal, and brain circulations. Radionuclide ($[^{123}I]$ *meta*-iodobenzylguanidine) and positron emission tomography (PET) scanning (using 6-$[^{18}F]$fluorodopamine) allow imaging of sympathetic innervation of cardiac tissue. Some of these research techniques are applied in the clinical setting.

Sudomotor System and Thermoregulation

Sweat glands influence thermoregulation and are innervated by sympathetic cholinergic nerves. In the thermoregulatory sweat test, central temperature is elevated by 1 °C and results in widespread activation of eccrine glands. Sweating is detected by dyes such as quinazarine or Ponso red (which turns from pale pink to bright red on exposure to moisture) (**Figure 7(a)**), or starch iodine (which turns dark brown). In autonomic failure, sweating often is impaired or absent. The sweat glands may be directly tested by applying ACh intradermally or by iontophoresis. This results in sweating via the axon reflex. Sudorometers measure sweat production in localized areas and with appropriate studies help differentiate preganglionic from postganglionic sympathetic lesions.

Local abnormalities of sudomotor function due to denervation and anhidrosis, as in leprosy, can be detected by using quinazarine powder covered over

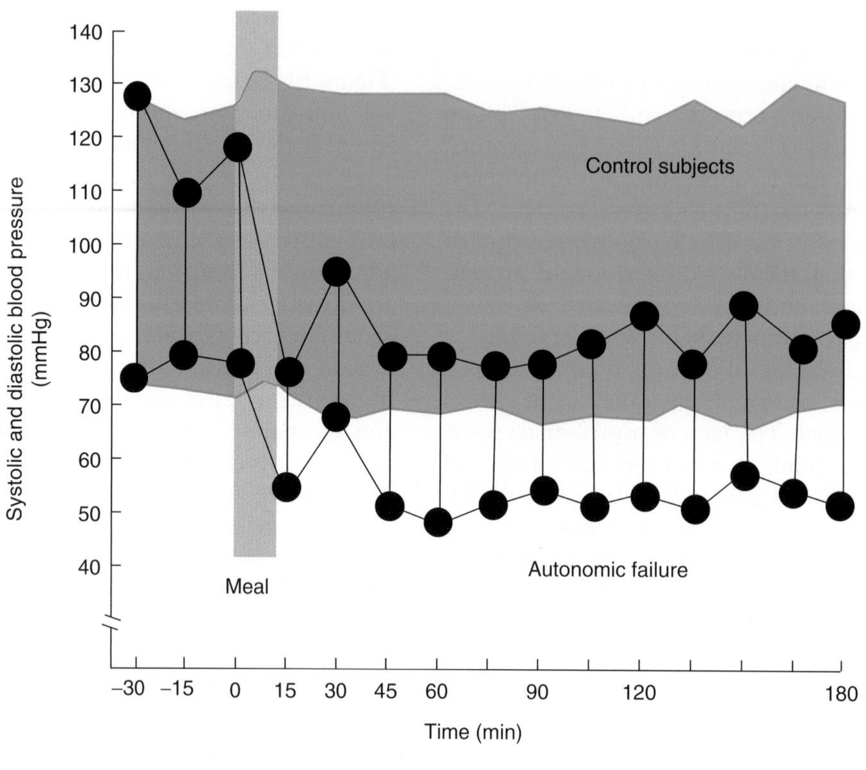

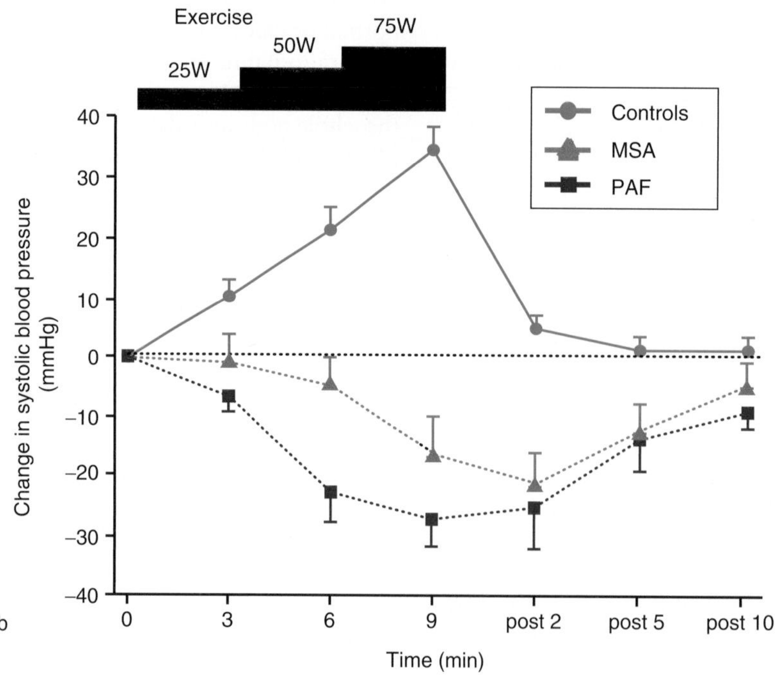

Figure 4 Continued

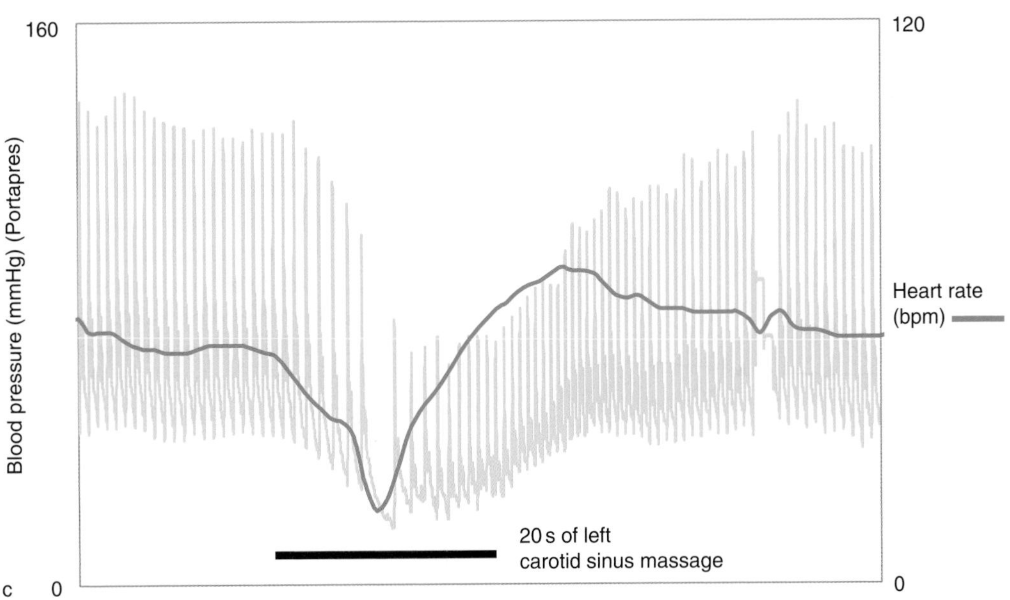

Figure 4 (a) Changes in blood pressure before and after a standard meal in a group of normal persons (controls; green area) and in a patient with autonomic failure (black circles), while in the supine and horizontal position. Bars indicate ± standard error of mean. In the normal persons there is no change in blood pressure after the meal. In the patient with autonomic failure there is a marked fall in blood pressure soon after food ingestion, with levels falling to around 80/50 mmHg and remaining low for 3 h, even in the supine position. (b) Changes in systolic blood pressure during horizontal bicycle exercise at three incremental levels (and on standing, postexercise) in normal persons (controls) and in patients with multiple system atrophy (MSA) and pure autonomic failure (PAF). In both MSA and PAF, unlike controls, there is a fall in blood pressure. (c) Continuous blood pressure and heart rate measured noninvasively (by Portapres) in a patient with falls of unknown etiology. Left carotid sinus massage caused a drop in both heart rate and blood pressure. The findings indicate the mixed (cardioinhibitory and vasodepressor) form of carotid sinus hypersensitivity. (a) From Mathias CJ and Bannister R (2001) Postprandial hypotension in autonomic disorders. In: Mathias CJ and Bannister R (eds.) *Autonomic Failure: A Textbook of Clinical Disorders of the Autonomic Nervous System*, 4th edn., pp. 283–295. Oxford: Oxford University Press. (b) From Smith GDP, Watson LP, Pavitt DV, et al. (1995) Abnormal cardiovascular and catecholamine responses to supine exercise in human subjects with sympathetic dysfunction. *Journal of Physiology (London)* 485, 255–265. (c) From Mathias CJ and Bannister R (2001) Investigation of autonomic disorders. In: Mathias CJ and Bannister R (eds.) *Autonomic Failure: A Textbook of Clinical Disorders of the Autonomic Nervous System*, 4th edn., pp. 169–195. Oxford: Oxford University Press.

with transparent adhesive tape. Gustatory sweating (or Frey's syndrome), which mainly affects the head and neck, may be severe and socially embarrassing and occurs after ingestion of spicy or acidic foods. It occurs in diabetes mellitus or after surgery to the parotid gland. This is because nerve damage is followed by neural regeneration, with aberrant reinnervation of parasympathetic fibers (to the salivary glands with sympathetic cholinergic fibers that supply the face).

Denervation of sweat glands causes patchy anhidrosis and may cause troublesome compensatory hyperhidrosis in innervated areas. This occurs in the Holmes–Adie syndrome associated with progressive anhidrosis (Ross' syndrome) and after percutaneous transthoracic endoscopic sympathectomy performed for palmar, axillary, and facial hyperhidrosis.

Measurements of electrical potentials from electrodes on the foot and hand record sympathetic cholinergic activity to sweat glands (sympathetic skin response (SSR)). Physiological (inspiratory gasps, loud noise, or touch) or electrical (median nerve) stimulations induce the SSR. In distal sympathetic lesions caused by PAF and pure cholinergic dysautonomia (**Figure** 7(**b**)), the SSR is absent. The SSR above and below the level of the lesion may provide a sensitive marker of the integrity of descending sudomotor pathways in spinal cord injuries.

Hypothalamic lesions and sympathetic failure result in impaired thermoregulation. Hyperpyrexia may occur because of the inability to appropriately sweat and vasodilate in the periphery. When exposed to radiant heat, body temperature rises without change in peripheral blood flow or sweating. The reverse, hypothermia, occurs as sympathetic vasoconstrictor failure reduces the capacity to retain heat, especially when muscle paralysis, as in tetraplegics, impairs heat production because of reduced shivering thermogenesis. Hypothermia may be missed unless a low-reading rectal thermometer, or tympanic membrane measurement, is used.

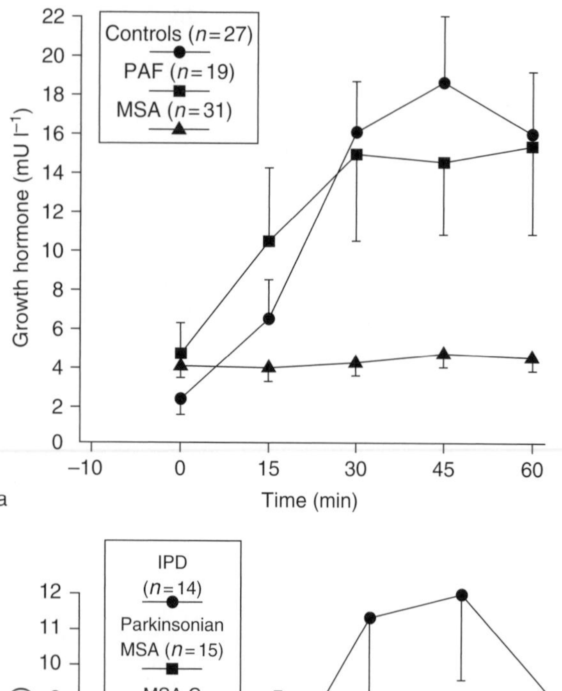

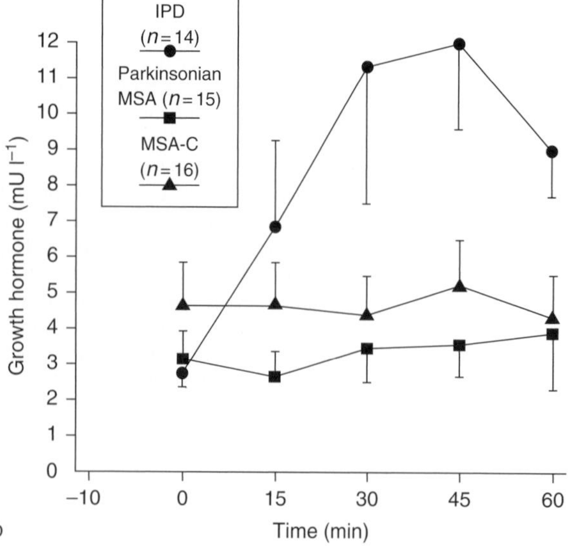

Figure 5 (a) Serum growth hormone concentrations before (0) and at 15 min intervals for 60 min after clonidine (2 μg/kg/min) in normal persons (controls) and in patients with pure autonomic failure (PAF) and multiple system atrophy (MSA). Growth hormone concentrations rise in controls and in patients with PAF with a peripheral lesion; there is no rise in patients with MSA with a central lesion. (b) Lack of serum growth hormone response to clonidine in the two forms of MSA, the cerebellar form (MSA-C) and the parkinsonian form, in contrast to patients with idiopathic Parkinson's disease (IPD) with no autonomic deficit, in whom there is a significant rise in growth hormone levels. (b) From Kimber JR, Watson L, and Mathias CJ (1997) Distinction of idiopathic Parkinson's disease from multiple system atrophy by stimulation of growth hormone release with clonidine. *Lancet* 349: 1877–1881.

Intraneural Recordings of Sympathetic Activity

Skin and muscle sympathetic nerve activity can be recorded using a tungsten microelectrode inserted into a peripheral nerve. Skin sympathetic nerve activity is affected by thermal stimuli, with cold increasing and heat decreasing activity. It is not time-locked to blood pressure changes during the cardiac cycle, which is characteristic of muscle sympathetic nerve activity. In tetraplegia, basal sympathetic discharge is diminished, consistent with low basal plasma noradrenaline levels, while during autonomic dysreflexia there is only a modest and transient increase in sympathetic discharge, despite the disproportionately greater and prolonged elevation in blood pressure. The latter probably is due to decentralization pressor supersensitivity and impaired baroreflex activity. In the Guillain–Barré syndrome, episodes of hypertension and tachycardia are closely correlated with increased sympathetic discharge. In neurally mediated syncope, sympathetic discharge diminishes, causing vasodilatation and hypotension.

The disadvantage of sympathetic microneurography is that it is an invasive technique involving measurements in a restricted region of the body. It is ideal when sympathetic activation is expected, but not when sympathoneural activity is low or absent, as in autonomic failure.

The Eye

Pupillary responses can be tested physiologically or pharmacologically. In Horner's syndrome, infrared television pupillometry indicates reduced pupillary dilatation in the dark. There is a constrictor response to light as occurs normally, but with a redilatation lag in the latter part of the recovery phase. With parasympathetic impairment, as in diabetes mellitus, the speed of the light reflex is reduced and its latency prolonged.

Drugs that stimulate or block autonomic papillary responses can be instilled onto the conjunctiva. Cocaine (4%) blocks the neuronal uptake of noradrenaline and normally causes pupillary dilatation; in sympathetic failure there is a diminished response. Hydroxyamphetamine (0.5%) releases noradrenaline and causes pupillary dilatation in preganglionic but not in postganglionic lesions, thus helping localize the site of lesion. In sympathetic denervation there is dilator supersensitivity of the pupil to 1:1000 adrenaline solution. In parasympathetic denervation (Holmes–Adie pupil) there is constrictor supersensitivity, unlike the normal pupil (**Figure 8**).

Lacrimal secretion measured with a special blotting paper (Schirmer's test) is diminished in parasympathetic failure. The reverse, excessive lacrimation, is found in 'crocodile tears' syndrome (gustolacrimal reflex, or Bogorod's syndrome) during eating.

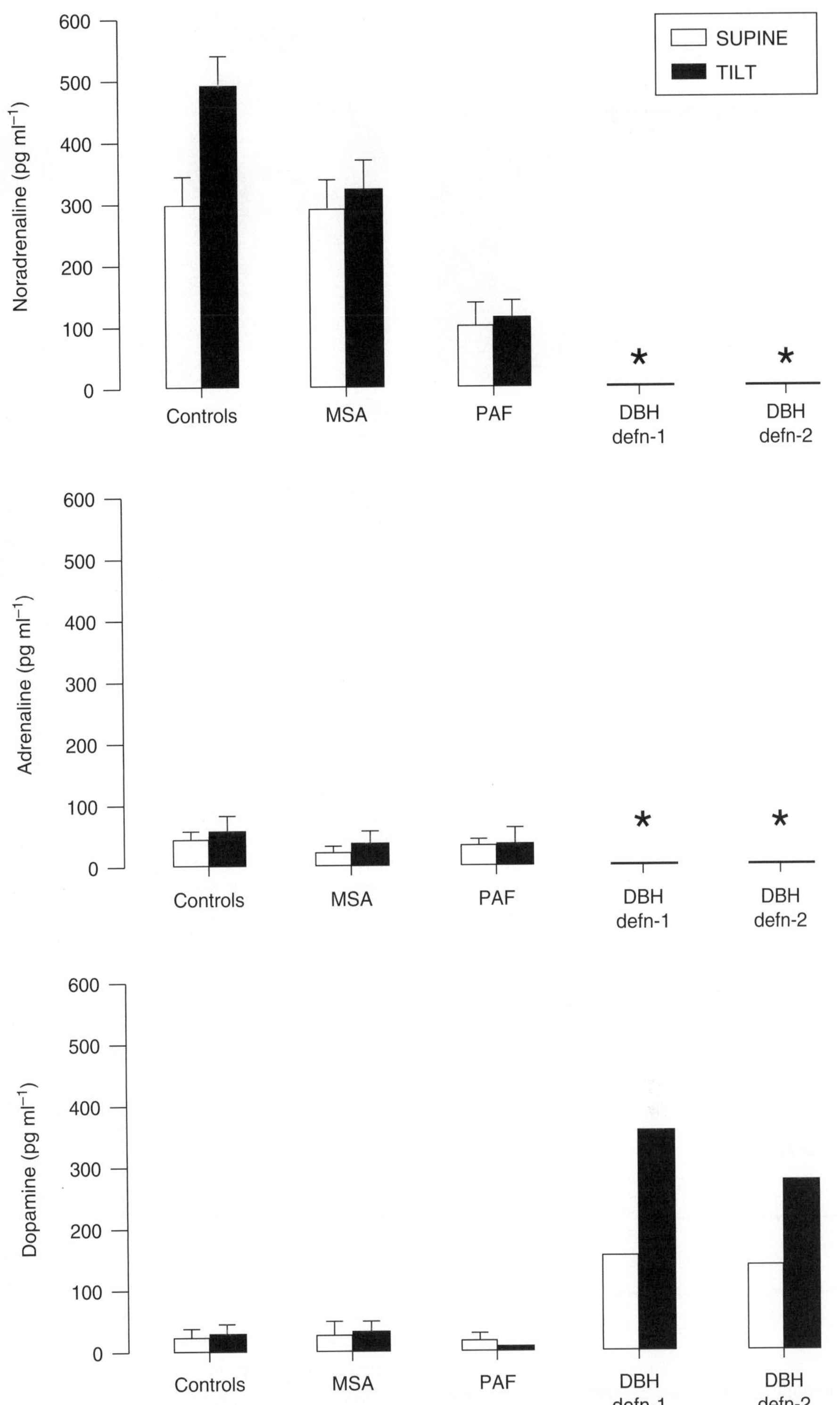

Figure 6 Basal levels of plasma noradrenaline, adrenaline, and dopamine measured by high-pressure liquid chromatography with electrochemical detection, in normal persons (controls), in patients with multiple system atrophy (MSA) and pure autonomic failure (PAF), and in two siblings with a deficiency of dopamine β-hydroxylase (DBH). Levels of noradrenaline are low in PAF and undetectable in DBH deficiency. In the DBH pair, plasma dopamine levels are clearly elevated, indicating a block at this particular stage in the formation of noradrenaline. The asterisks indicate levels of sensitivity, which are $<5\,\mathrm{pg\,ml^{-1}}$ for noradrenaline and adrenaline and $<20\,\mathrm{pg\,ml^{-1}}$ for dopamine. From Mathias CJ, Bannister R, Cortelli P, et al. (1990) Clinical autonomic and therapeutic observations in two siblings with postural hypotension and sympathetic failure due to an inability to synthesize noradrenaline from dopamine because of a deficiency of dopamine beta hydroxylase. *Quarterly Journal of Medicine* 278: 617–633.

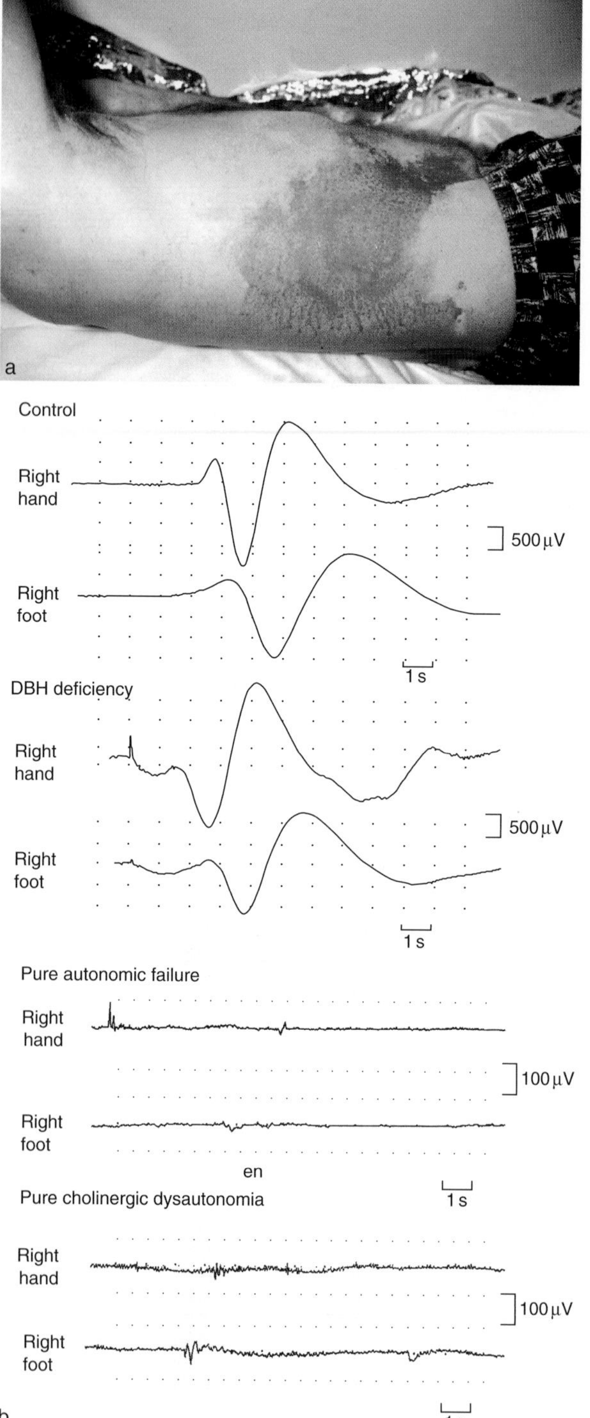

Figure 7 (a) Segmental hyperhidrosis as a presenting feature in a patient in whom there were large areas without sweating. He had the Holmes–Adie syndrome with Ross's variant. (b) The sympathetic skin response (in microvolts) from the right hand and right foot of a normal individual (control) and of a patient with dopamine β-hydroxylase (DBH) deficiency (upper panel). The sympathetic skin response could not be recorded in two patients, one with pure autonomic failure and the other with pure cholinergic dysautonomia (lower panel). (a) From Mathias CJ (1998) Autonomic disorders. In: Bogousslavsky J and Fisher M (eds.) *Textbook of Neurology*, pp. 519–545. Newton, MA: Butterworth-Heinemann. (b) From Magnifico F, Misra VP, Murray NMF, et al. (1998) The sympathetic skin response in peripheral autonomic failure – evaluation in pure autonomic failure, pure cholinergic dysautonomia and dopamine-β-hydroxylase deficiency. *Clinical Autonomic Research* 8:133–138.

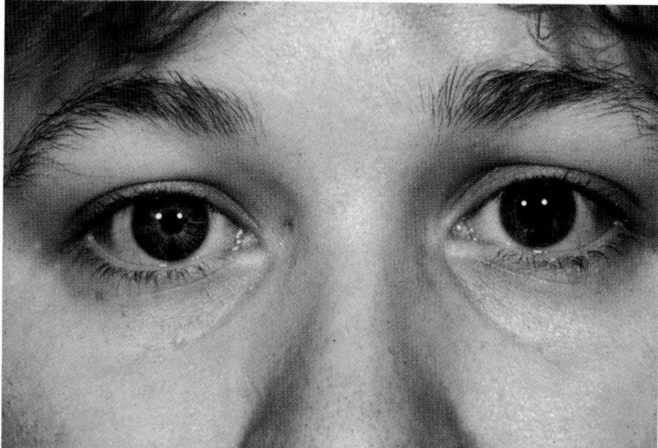

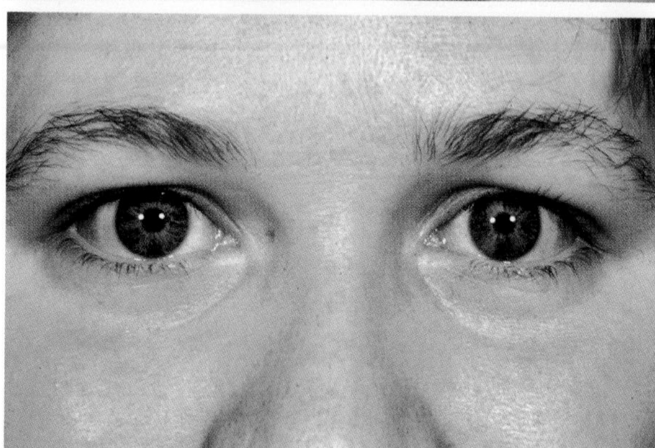

Figure 8 Pupillary dilatation in Holmes–Adie syndrome and its response to pilocarpine. Upper panel: There is a diminished light response, causing a larger left pupil. Lower panel: The response of the left pupil to dilute pilocarpine is greater than that of the normal right pupil. From Mathias CJ (1998) Autonomic disorders. In: Bogousslavsky J and Fisher M (eds.) *Textbook of Neurology*, pp. 519–545. Newton, MA: Butterworth-Heinemann.

Gastrointestinal Tract

Swallowing often is affected in MSA and can be detected by video-cinefluoroscopy. A barium swallow determines esophageal involvement, as in Chagas' disease, a trypanasomal infection that causes dilatation of the lower esophagus. Measurement of esophageal pressure is valuable in achalasia. Upper gastrointestinal endoscopy enables biopsy of tissue, which is diagnostic in systemic amyloidosis. Gastric motility is measured using a technetium-labeled meal and progress of the gamma emitter is tracked by a scintiscan camera. Delayed gastric emptying may complicate amyloidosis and diabetes mellitus.

In diabetes mellitus, diarrhea may be a problem. The reasons include motility abnormalities and bacterial overgrowth in dilated, poorly functioning segments of gut. The response to broad-spectrum antibiotics provides a therapeutic diagnosis. Jejunal aspiration, bacterial culture, and the [^{14}C]glycocholate test may be needed. Gastrointestinal motility and activity can be continuously monitored using indwelling tubes or capsules with telemetric devices.

Large bowel dysfunction and constipation are common in autonomic failure, and a barium enema or colonoscopy may be needed. A rectal biopsy, with Congo red staining, provides definitive evidence in amyloidosis.

Renal Function and Urinary Tract

Nocturnal polyuria is common in autonomic failure and may cause an overnight weight loss of over a kilogram, with a reduction in extracellular fluid volume, which aggravates morning postural hypotension. Day and night measurements of urine volume and sodium excretion provide measurement of diurnal changes. With urinary bladder and sphincter involvement, frequency of micturition and incontinence may occur. An intravenous urogram and/or ultrasound examination of the urinary tract may be needed. Urodynamic measurements provide information as to the nature of bladder dysfunction such as detrusor areflexia or hyperreflexia. Urethral or anal sphincter electromyography is abnormal in MSA but is unaffected in Parkinson's disease.

Sexual Function

In autonomic failure, erectile and ejaculatory sexual dysfunctions are common in the male. Differentiation from psychogenic impotence can be difficult in the

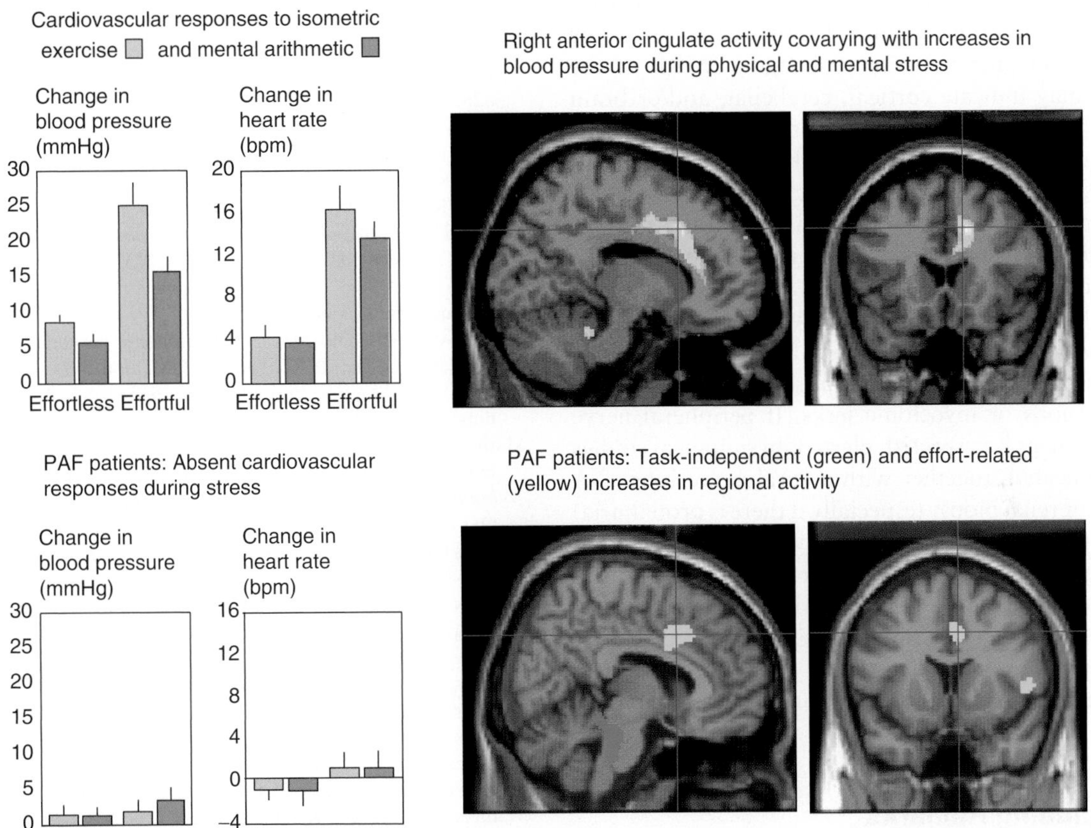

Figure 9 Simultaneous neuroimaging and physiological measurement during cardiovascular autonomic activation in normal persons and in pure autonomic failure (PAF) patients. Changes in blood pressure and heart rate during control (effortless) tasks and effortful isometric exercise and mental arithmetic are shown (top left, normal persons; lower left, PAF patients). Alongside are images showing changes in regional cerebral (mainly right anterior cingulate) activity. Normal persons show increased activity in the right anterior cingulate area and positive covariance with mean arterial blood pressure during isometric exercise and mental arithmetic. In PAF there is an additional increase in activity in the right insula and brain stem (pontine) areas despite no change in blood pressure. The cerebral autonomic centers in PAF are preserved, and the additional areas activated indicate attempts to increase peripheral autonomic activity, which is not possible because of their postganglionic sympathetic lesion. From Mathias CJ (2002) To stand on ones' own legs. *Clinical Medicine* 2: 237–245.

absence of other autonomic features, as both respond well to intracorporeal injections of papaverine. In diabetes mellitus, vascular factors may cause or contribute to impotence.

Respiratory Tract

Nocturnal stridor due to denervation of abductors of the vocal cords may occur in the later stages of MSA. Hypoxia results from laryngeal abductor paresis and/or brain stem dysfunction. The former is detected by laryngoscopy. Blood gas monitoring during sleep is necessary to determine if significant oxygen desaturation warrants procedures such as continuous positive-pressure ventilation or a tracheostomy.

Miscellaneous Investigations

Additional nonautonomic investigations may be needed to determine the etiology of the autonomic disorder and to distinguish between the different groups with primary autonomic failure. Computed tomography (CT) or magnetic resonance imaging (MRI) of the brain may indicate cortical, cerebellar, and/or brain stem atrophy in MSA. Advances in neuroimaging (especially PET and event-related functional MRI), as used in clinical research (**Figure 9**) increasingly are being applied to determine pathophysiological mechanisms, and also aid diagnosis. The electroencephalogram (EEG) is of value when postural hypotension is associated with, or precipitates, seizures and in separating epilepsy from neurally mediated syncope, which results in hypoxia-induced convulsions or myoclonic jerks. If peripheral nerve involvement is suspected, electrophysiological studies are indicated, together with a sural nerve biopsy. A rectal or renal biopsy (especially if there is proteinuria) may confirm the diagnosis of amyloidosis. In amyloidosis, gene mapping to determine mutations is important, especially in familial amyloid polyneuropathy and also in sporadic amyloidosis. Cardiac causes of syncope are common, especially in the elderly, and may necessitate a 24 h ambulatory electrocardiogram (ECG) recording.

Concluding Remarks

Autonomic disorders involve a number of systems and complicate many diseases. Clinical autonomic testing, utilizing a combination of physiological, biochemical, and molecular biological methods along with modern neuroimaging techniques, have advanced our ability to investigate autonomic disorders. Autonomic investigation often provides important information for the clinician and patient, and aids diagnosis, prognosis, and appropriate treatment.

See also: Autonomic Disorders; Autonomic Failure; Autonomic Nervous System: Central Cardiovascular Control; Autonomic Nervous System: Central Urogenital Control; Autonomic Nervous System: Central Thermoregulatory Control; Autonomic Nervous System: Central Control of the Gastrointestinal Tract; Autonomic Nervous System: Central Respiratory Control; Autonomic Dysfunction: Drug-Induced; Autonomic Nervous System; Parasympathetic Nervous System; Sympathetic Nervous System.

Further Reading

Appenzeller O and Oribe E (eds.) (1997) *The Autonomic Nervous System. An Introduction to Basic and Clinical Concepts,* 5th edn. Amsterdam: Elsevier Medical Press.

Cariga P, Catley M, Savic G, et al. (2002) Organisation of the sympathetic skin response in spinal cord injury. *Journal of Neurology, Neurosurgery & Psychiatry* 72: 356–360.

Critchley HD, Corfield DR, Chandler MP, et al. (2000) Cerebral correlates of autonomic cardiovascular arousal: A functional neuroimaging investigation in humans. *Journal of Physiology* 523: 259–270.

Critchley HD, Mathias CJ, and Dolan RJ (2001) Neuroanatomical basis for first- and second-order representations of bodily states. *Nature Neuroscience* 4: 207–212.

Johnson RH, Lambie DG, and Spalding JMK (1984) *Neurocardiology. The Interrelationship between Dysfunction in the Nervous and Cardiovascular Systems.* London: W.B. Saunders.

Low PA (ed.) (1997) *Clinical Autonomic Disorders,* 2nd edn., Philadelphia: Lippincott, Raven.

Mathias CJ (1991) Disorders of the autonomic nervous system. In: Bradley WG, Daroff RB, Fenichel GM, et al. (eds.) *Neurology in Clinical Practice,* 2nd edn., pp. 1661–1685. Boston, MA: Butterworth Publishers.

Mathias CJ (2000) Autonomic dysfunction. In: Grimley-Evans J, Franklin Williams T, Lynn Beattie B, et al. (eds.) *Oxford Textbook of Geriatric Medicine,* 2nd edn., pp. 833–852. Oxford: Oxford University Press.

Mathias CJ (2002) To stand on ones' own legs. *Clinical Medicine* 2: 237–245.

Mathias CJ (2003) Autonomic diseases – clinical features and laboratory evaluation. *Journal of Neurology, Neurosurgery & Psychiatry* 74: 31–41.

Mathias CJ (2003) Autonomic diseases – management. *Journal of Neurology, Neurosurgery & Psychiatry* 74: 42–47.

Mathias CJ and Bannister R (eds.) (2002) *Autonomic Failure: A Textbook of Clinical Disorders of the Autonomic Nervous System,* 4th edn., Oxford: Oxford University Press.

Mathias CJ, Deguchi K, and Schatz I (2001) Observations on recurrent syncope and presyncope in 641 patients. *Lancet* 357: 348–353.

Mathias CJ and Williams AC (1994) The Shy–Drager syndrome (and multiple system atrophy). In: Calne DB (ed.) *Neurodegenerative Diseases*, pp. 742–767. London: Saunders Scientific Publications.

Mathias CJ and Young TM (2004) Water drinking in the management of orthostatic intolerance due to orthostatic hypotension, vasovagal syncope and the postural tachycardia syndrome. *European Journal of Neurology* 11: 613–619.

Autonomic Nervous System: Gastrointestinal Control

J C Bornstein, University of Melbourne, Parkville, VIC, Australia

Introduction

The gastrointestinal system is the most complex organ in the body, outside the brain. Its function is digestion and absorption of nutrients. This requires a wide variety of enabling behaviors, including movement of food via muscular contractions (motility), secretion of water and salt to provide an appropriate medium, and secretion of digestive enzymes. Freed nutrients are separated from waste by absorption through the mucosa. Water and salt are then reabsorbed to maintain whole-body water and electrolyte balance. Blood flow through the intestinal wall is tightly regulated both for these purposes and to ensure effective transport of nutrients. Finally, waste is eliminated via defecation. All these processes are regulated by the autonomic nervous system, although the details vary among regions. This article focuses on the small intestine, with reference to the colon where appropriate.

Neural control of intestinal functions operates at multiple levels (**Figure 1**). The major control is exerted locally by a network of intrinsic neurons, the enteric nervous system (ENS). This is the largest and most complex part of the autonomic nervous system, containing 30 000 000–100 000 000 neurons in humans. The ENS produces complex behaviors in response to physiological stimuli independently of the central nervous system (CNS) and so includes sensory neurons, motor neurons, and integrative neurons (interneurons). Other levels of neural control include coordination of the behavior of different gut regions by reflex pathways running through the CNS and an intermediate-level pathway that passes through prevertebral sympathetic ganglia (intestino-intestinal reflexes).

Intestinal behavior depends critically on whether the animal is fed or fasted. Even the sight or smell of food converts intestinal behavior from a fasted to a fed state. This clearly involves the brain, but both fed and fasted states depend on the ENS for their specific behaviors. In the fed state, intestinal movements fall into three distinct patterns. The most common, segmentation, involves localized constrictions repeated rhythmically at specific sites, where they remain stationary or propagate only short distances. Segmentation mixes chyme with digestive enzymes and with water secreted into the lumen from the blood. Occasionally, segmentation is replaced by a constriction that propagates anally, forcing the content ahead of it (peristalsis). Orally propagating constrictions occur; but are rarer than peristalsis. When food is present, water and salt are secreted into the lumen of the duodenum and jejunum. The water and salt are then reabsorbed in the ileum and colon. Both secretion and absorption require increased blood flow to the intestine.

Once digestion and absorption have been completed, the fasted state appears. This manifests as the interdigestive migrating motor complex (MMC) in the small intestine. MMCs have three distinct phases that appear in the antrum or upper duodenum and propagate slowly along the small intestine to the ileo-colonic junction, taking over an hour to traverse the human gut. Phase I is a period of quiescence with no muscle contractions; this slowly converts to a period of irregular activity (phase II), which then converts to a strong wave of contraction lasting up to 5 min (phase III). In herbivores, phase III contractions interrupt the fed-state behavior and may be important for intestinal propulsion. In the fasted state, phase III contractions clear mucosal debris and accumulated intestinal bacteria. All three phases of interdigestive MMCs depend on enteric neural activity.

The role of the colon is distinctly different: reabsorption of water and storage and excretion of waste. It also regulates small intestinal transit of digesta: colonic distension or bile acids in the colonic lumen cause significant slowing of propulsion along the jejunum via a reflex pathway running through the prevertebral sympathetic ganglia.

Structure of the ENS

Locations of Neurons and Effectors

The basic structure of the gastrointestinal tract (**Figure 2**) is maintained along its length. The intestinal tube is lined by the mucosal epithelium and associated connective tissues and blood vessels. Adjacent to the mucosa is a layer of smooth muscle, the muscularis mucosae. This is virtually nonexistent in small animals, but is substantial in larger species, including humans. The next layer is the submucosa, a layer of connective tissue within which are embedded arterioles, venules, and the cell bodies and processes of neurons grouped into a ganglionated submucous plexus. The major muscle layer, the circular muscle, lies adjacent to the submucosa. This is the thickest muscle layer and its smooth muscle cells are oriented

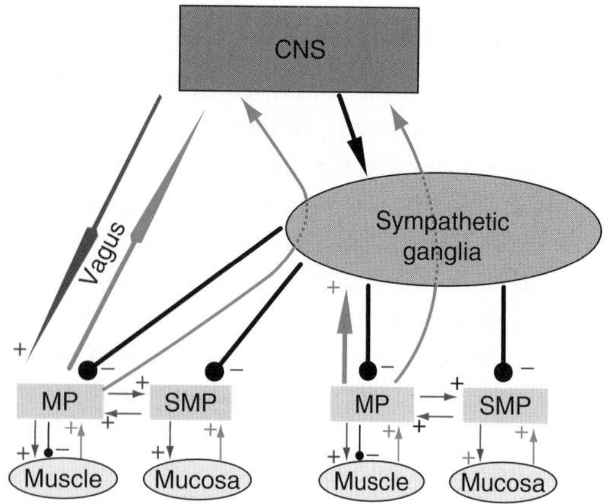

Figure 1 The hierarchy of control of the gastrointestinal tract and its interconnections. Excitatory connections are illustrated by the blue arrows, sensory pathways are indicated by the red arrows, and inhibitory pathways are indicated by the black 'clubs.' The major control of gastrointestinal movements and secretion is exerted within the enteric nervous system. The muscle is directly regulated by the myenteric plexus (MP) and the mucosa is regulated from the submucous plexus (SMP). There are strong excitatory inputs to the SMP from the MP and weaker excitatory connections in the opposite direction. Activity of enteric neurons is modulated by input from the central nervous system (CNS) via the vagus (excitatory) and sympathetic nerves (inhibitory). Enteric ganglia receive sensory input from both muscle and mucosa. Sensory information reaches the CNS via vagal afferents and visceral afferents running with the sympathetic innervation. Intestinofugal neurons, most prominent in the large bowel, send axons to the prevertebral sympathetic ganglia via the sympathetic nerve trunks, activating inhibitory reflexes that regulate activity in the colon and also in the upper small intestine.

parallel to the gut's circumference, so their contractions constrict the lumen. The next layer outward, the myenteric plexus, contains many neurons and their processes in an ordered network of nerve trunks and ganglia. The final muscle layer is the longitudinal muscle, whose smooth muscle cells run parallel to the long axis of the intestine. This is bounded by the serosa, the outer wall of the intestine. The blood supply, with the extrinsic nerves that run with it, enters the intestine via the mesentery and ramifies within the intestinal wall.

The myenteric plexus extends from the upper esophageal sphincter to the anus. The submucous plexus only becomes ganglionated at the pyloric sphincter of the stomach. The submucous plexus can sometimes be subdivided into an inner and outer plexus, each containing different types of neurons.

Effector Targets of Enteric Neurons

The myenteric plexus provides most of the innervation of the muscle coat. In smaller mammals, this comes exclusively from myenteric neurons, but submucosal neurons contribute in larger animals. Smooth muscle cells are not the major cellular target of neurons innervating the circular muscle layer; rather, they contact a different cell type, the interstitial cells of Cajal (ICCs) (black cells in the CM in **Figure 2**). These are organized into a functional network via gap junctions with each other and also form gap junctions with circular muscle cells. Transmitters released from motor neurons apparently alter muscle behavior indirectly via actions on ICCs. This contrasts with the innervation of the longitudinal muscle, which involves direct connections between nerve terminals and muscle cells. A second population of ICCs, adjacent to the myenteric plexus (**Figure 2**), consists of pacemaker cells generating rhythmic depolarizations – slow waves – in the smooth muscle. These depolarizations are small in guinea pigs (1–3 mV), but are much larger in most species (up to 40 mV) and impose an underlying rhythm on contractions evoked by neural activity.

The efferent innervation of the mucosa is largely from submucous neurons (**Figures 1** and **2**), although some myenteric neurons, notably the intrinsic sensory neurons, send axons to the mucosa. Submucous neurons also innervate submucosal arterioles and act as vasodilators. Extrinsic autonomic neurons, whether parasympathetic or sympathetic, typically terminate within myenteric or submucous ganglia, indicating that they modulate the activity of enteric neurons, rather than having a direct effect on the smooth muscle. This raises an interesting dichotomy because sympathetic neurons that modulate intestinal movement or secretion act as interneurons, while neurons in the same ganglia that supply intestinal blood vessels are final motor neurons.

Functional Types of Enteric Neuron

At least 12 functional types of enteric neurons are present in every mammal (**Table 1**). Intrinsic sensory neurons detect the chemistry, texture, and volume of the intestinal contents. Excitatory and inhibitory motor neurons separately innervate the circular and longitudinal muscle layers. There are also orally directed (ascending) interneurons, anally directed (descending) interneurons, two types of secretomotor neurons, and vasodilator neurons. In addition, intestinofugal neurons project to prevertebral sympathetic ganglia as the afferent arm of reflexes coordinating behavior between different intestinal regions. In guinea pigs, the most comprehensively studied species, many of these neurons can be further subdivided on the basis of their neurochemistry and connections (**Table 2**). For example, three classes of anally projecting (descending) interneurons can be discriminated neurochemically in guinea pig ileum: some are

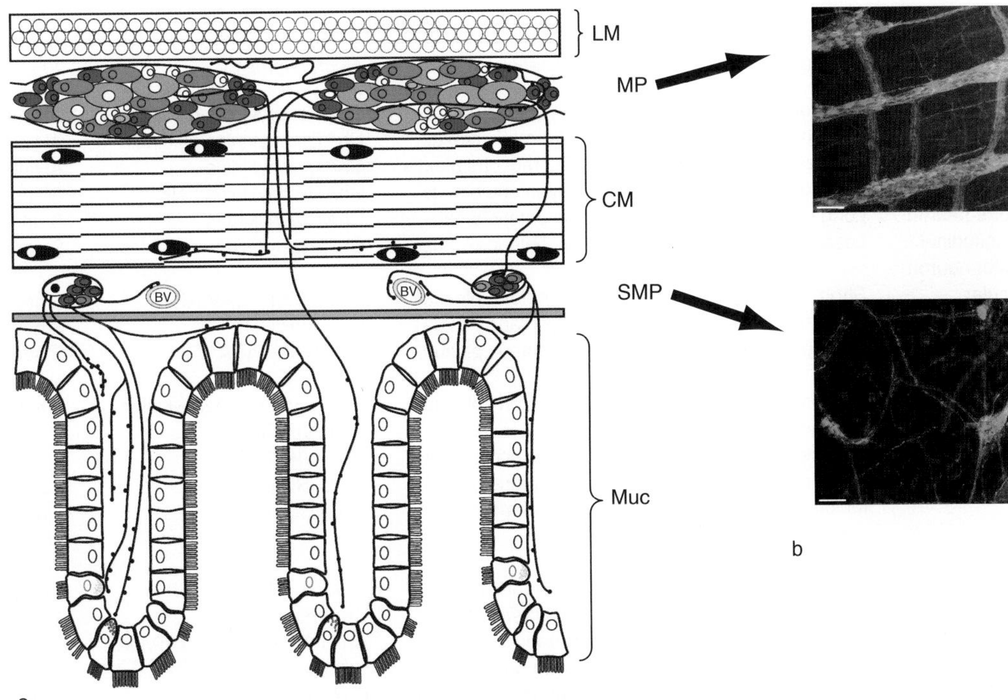

Figure 2 The layers of the intestine and their innervation. (a) Schematic drawing showing the layers of the intestine and where their innervation arises in the intrinsic ganglia. In this case, the projection pattern corresponds to guinea pig small intestine; villi are absent from other parts of the gastrointestinal tract. Neurons are color coded to match the neurons stained immunohistochemically in the micrographs (b), and neuron types not immunoreactive in the micrographs are shown as gray or white. Interstitial cells of Cajal are shown in the CM as interconnected black cells next to the MP and in the deep muscular plexus. Sources of innervation of longitudinal muscle, other ganglia, circular muscle, blood vessels, and mucosa are shown via representative axons. (b) Micrographs showing the MP (top) stained to reveal immunoreactivity for nitric oxide synthase (red) and calretinin (green), and the SMP (bottom) stained to reveal immunoreactivity for neuropeptide Y (red) and vasoactive intestinal peptide (green). Neurons in the MP are grouped into large ganglia (about 100 neurons in guinea pig), while SMP ganglia are much smaller (about 10 neurons). Note the extensive interconnections via interganglionic connectives and the neuropeptide Y reactive innervation of submucosal blood vessels, which arises from sympathetic vasoconstrictor neurons. LM, longitudinal muscle; MP, myenteric plexus; CM, circular muscle; SMP, submucous plexus; Muc, mucosa.

Table 1 Types of neurons that can be deduced from physiological and pharmacological studies to be present in the ENS of all species

Intrinsic sensory neurons
Excitatory motor neurons innervating the circular muscle
Excitatory motor neurons innervating the longitudinal muscle
Inhibitory motor neurons innervating the circular muscle
Inhibitory motor neurons innervating the longitudinal muscle
Motor neurons innervating the muscularis mucosae
Orally directed (ascending) interneurons
Anally directed (descending) interneurons
Cholinergic secretomotor neurons
Noncholinergic secretomotor neurons
Vasodilator neurons
Intestinofugal neurons

immunoreactive for nitric oxide synthase (NOS), vasoactive intestinal peptide (VIP), and other neuropeptides; others are immunoreactive for somatostatin and choline acetyltransferase (ChAT); another group express serotonin (5-hydroxytryptamine; 5-HT) and ChAT. However, neurochemical codes from one location cannot be confidently generalized to other preparations because regional differences occur within the same species (e.g., somatostatin interneurons project orally in guinea pig colon).

Electrophysiological studies in the small intestine and colon of guinea pigs and mice have identified two broad classes of neurons, AH and S neurons (**Figure 3**). Both classes are seen in human and rat tissue, but AH neurons have not been identified in pig tissue. AH neurons have prominent, prolonged, after-hyperpolarizations (AHPs) following their action potentials. They have a common shape (**Figure 3**), whether they are in the myenteric or submucous plexus, with large cell bodies and multiple axons that project circumferentially and ramify in myenteric ganglia, the submucous plexus, and the mucosa. There is strong evidence that they have a sensory function (**Figure 4**).

Table 2 Subtypes of neurons in guinea pig ileum and their distinguishing features

General function	*Subtype*	*Location*	*Neurochemistry*	*Electrophysiology*
Intrinsic sensory neuron	Chemoreceptor	Myenteric plexus	ChAT/tachykinins/calbindin	AH neuron
	Stretch/tension receptor	Myenteric plexus	ChAT/tachykinins/calbindin	AH neuron
	Mucosal mechanoreceptor	Submucous plexus	ChAT/tachykinins	AH neuron
Excitatory circular muscle motor neuron	Short oral projection	Myenteric plexus	ChAT/tachykinins/ENK	S neuron
	Long oral projection	Myenteric plexus	ChAT/tachykinins	S neuron
Excitatory longitudinal muscle motor neuron	Local projections	Myenteric plexus	ChAT/calretinin/tachykinins	S neuron
Inhibitory circular muscle motor neuron	Short anal projection	Myenteric plexus	NOS/VIP/PACAP/NPY	S neuron
	Long anal projection	Myenteric plexus	NOS/VIP/GRP	S neuron
Inhibitory longitudinal muscle motor neuron	Local projection	Myenteric plexus	NOS/VIP/GABA	S neuron
Ascending interneuron	Oral projection	Myenteric plexus	ChAT/calretinin/tachykinins	S neuron
Descending interneuron	Supplies myenteric plexus	Myenteric plexus	NOS/VIP/GRP	S neuron
	Supplies submucous plexus	Myenteric plexus	NOS/VIP/GRP	S neuron
	Supplies myenteric plexus	Myenteric plexus	ChAT/SOM	S neuron
	Inhibitory interneuron supplying submucous plexus	Myenteric plexus	SOM/ChAT	S neuron
	Supplies myenteric plexus	Myenteric plexus	ChAT/5-HT	S neuron
	Supplies submucous plexus	Myenteric plexus	ChAT/5-HT	S neuron
Secretomotor neuron	Cholinergic	Submucous plexus	ChAT/NPY/SOM/CGRP/GAL	S neuron
	Noncholinergic	Submucous plexus	VIP/Dyn/GAL	S neuron
Vasodilator neuron	Cholinergic	Submucous plexus	ChAT/calretinin	S neuron
	Noncholinergic	Submucous plexus	VIP/Dyn/GAL	S neuron
Interplexus interneuron	Noncholinergic	Submucous plexus	VIP/Dyn/GAL	S neuron

ChAT (choline acetyltransferase) is the enzyme that synthesizes acetylcholine; 'tachykinins' refers to substance P and neurokinin-A, which are products of the same gene and are difficult to distinguish immunohistochemically; ENK, enkephalin; PACAP, pituitary adenylyl cyclase-activating peptide; NPY, neuropeptide Y; GRP, gastrin-releasing peptide; GABA, γ-aminobutyric acid; SOM, somatostatin; CGRP, calcitonin gene-related peptide; GAL, galanin; Dyn, dynorphin.

Unlike AH neurons, S neurons have prominent fast excitatory postsynaptic potentials (EPSPs; **Figure 3**). Myenteric S neurons are morphologically diverse, with single axons that project orally or anally, and ramify in the muscle (motor neurons), supply other myenteric ganglia (interneurons) or submucous ganglia (interneurons), or run to the mucosa (secretomotor neurons). Most submucosal S neurons project to the mucosa. Others are vasodilator neurons.

Putting the Neurons in Their Circuits

The functions of neurons can only be fully defined when neurons are placed in physiologically meaningful circuits. Relevant evidence largely comes from analysis of neuronal behavior during simple motility reflexes. Mechanical or chemical stimulation of the intestinal wall elicits polarized reflexes: circular muscle contractions oral to and relaxations anal to the stimulus (ascending excitation and descending inhibition, respectively). Direct recordings from identified neurons during these reflexes, together with pharmacological analyses of transmission in reflex pathways and neurochemical studies of different neuronal subpopulations, have been combined to analyze the enteric neural circuitry. Collectively, the studies reveal the polarized circuit shown in **Figure 5** along with the nature of transmission at most classes of synapses within the circuit. Key features include sensory neurons organized into recurrent excitatory networks around the intestinal circumference and with outputs to interneurons in two polarized feed-forward pathways running orally to excite and anally to inhibit the circular muscle. The sensory neurons have monosynaptic outputs to local excitatory and inhibitory motor neurons, but these also lead to polarized activity because the former run orally and the latter run anally.

The intrinsic sensory neurons are AH neurons and respond to several different stimuli, including muscle tension, increases in muscle length (distension), chemicals in the lumen, and mucosal deformation. The last stimulus is detected by submucosal sensory neurons, but the others are largely the province of myenteric sensory neurons. The myenteric sensory neurons that respond to distension have axons that run anally in addition to their normal circumferential projections. However, most myenteric sensory neurons have only circumferential projections.

The intrinsic sensory neurons transmit to ascending interneurons (S neurons), via acetylcholine (ACh)

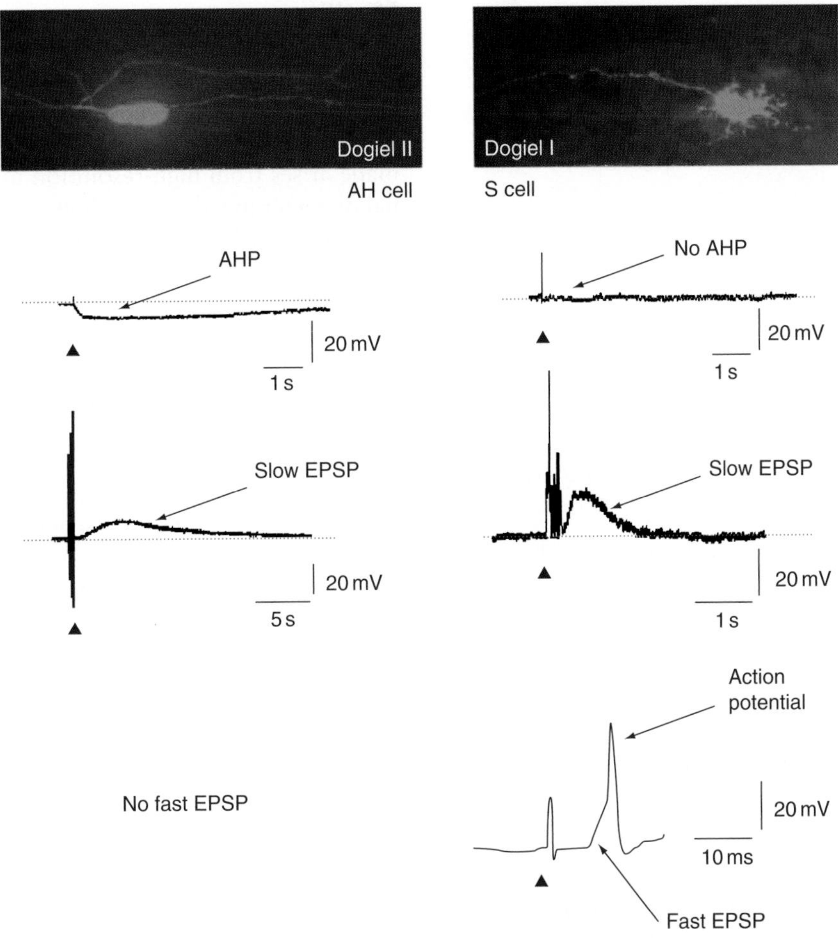

Figure 3 Types of neurons in the enteric nervous system: morphology and electrophysiology. The left side of this figure shows the shape and electrophysiological properties of AH/Dogiel type II neurons. The micrograph shows a neuron of this type injected with biocytin from an intracellular recording electrode and later fixed and stained with streptavidin–Texas red. The neuron is a typical Dogiel type II cell with a large oval cell body, lacking dendrites, but with at least two distinct axons that ramify within the neuron's own ganglion. The traces below show the action potential of one of these neurons, with its characteristic prolonged afterhyperpolarization (AHP) lasting well over 10 s (upper trace) and a slow excitatory postsynaptic potential (slow EPSP) evoked in the same neuron by a train of stimuli delivered to an interganglionic connective (lower trace). Fast EPSPs are not seen in these cells. The right side shows a micrograph of an S neuron, a typical Dogiel type I cell with one axon and short lamellar dendrites. The top trace shows the action potential of such a neuron, which lacks the prolonged AHPs of AH/Dogiel type II neurons. These neurons also exhibit slow EPSPs (middle trace), while they also have prominent fast EPSPs (bottom trace).

acting at nicotinic ACh receptors (nAChRs), ACh acting at muscarinic ACh receptors (mAChR), and tachykinins acting at tachykinin (neurokinin; NK-3) receptors. They transmit to other sensory neurons via NK-3 and/or NK-1 tachykinin receptors. Transmission between ascending interneurons is exclusively via nAChRs. Transmission from interneurons to excitatory motor neurons is via nAChRs, supplemented by tachykinins acting at NK-3 receptors. Transmission in the descending reflex pathways depends on the species and the location. In guinea pig ileum, transmission from intrinsic sensory neurons to descending interneurons and local inhibitory motor neurons is via an, as yet, unidentified transmitter/receptor combination. However, transmission from interneurons to nitric oxide synthase (NOS) interneurons may be mediated by ATP acting at purinergic ($P2Y_1$) receptors, while transmission from these neurons to inhibitory motor neurons is mediated by ATP acting at P2X receptors. No role for the nAChR has been identified in this pathway. By contrast, transmission within the descending inhibitory pathway of both guinea pig and rat colon depends on nAChRs at every functionally defined synapse.

Motor neurons supplying the muscle are S neurons. Their axons run longitudinally within the myenteric plexus before entering the circular muscle, where they branch and run circumferentially. Their projections are polarized according to function. Excitatory motor neurons run orally. Inhibitory motor neurons run

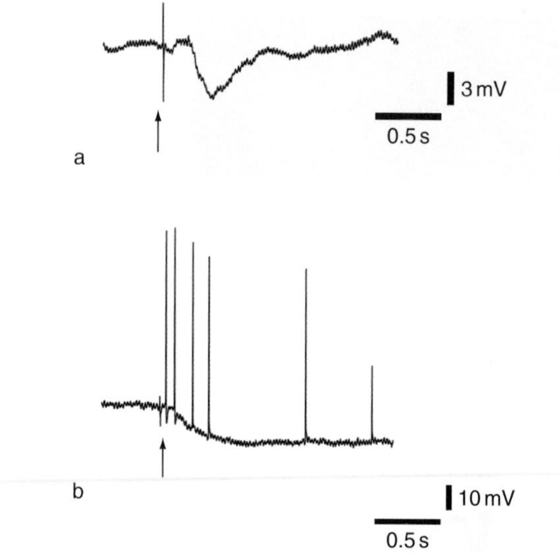

Figure 4 Sensory stimuli can excite both local reflexes and direct firing of action potentials in AH/Dogiel type I neurons. (a) Intracellular recording of the circular muscle response to local application of 10^{-3} mol l^{-1} L-alanine (arrow) to mucosal villi near the recording electrode. A robust and highly repeatable inhibitory junction potential is evoked by the stimulus. (b) The same stimulus evoked a burst of action potentials in an AH/Dogiel type II neuron, the cell body of which lies in the myenteric plexus close to the smooth muscle, in which reflexes of the type shown in (a) can be recorded. This burst of action potentials is characteristic of an intrinsic sensory neuron sensitive to chemical stimuli applied to the mucosa.

anally. Neurons innervating the longitudinal muscle divide close to their cell bodies and ramify locally.

Excitatory transmission to the smooth muscle is largely via mAChRs, but tachykinins play a significant, if lesser, role. Many myenteric neurons contain both ChAT and tachykinins. Some project to the muscle and are almost certainly excitatory motor neurons. Others are ascending interneurons. In guinea pig, all excitatory motor neurons contain and release ACh and most also contain tachykinins, but fewer cholinergic motor neurons may use tachykinins in other species. Inhibitory neuromuscular transmission also involves cotransmission, with at least two transmitters being released from inhibitory motor neurons. Both nitric oxide and ATP have been identified as transmitters from these neurons, although their relative contributions vary between species and locations within species. All inhibitory motor neurons contain NOS, but some interneurons also contain this enzyme. There is no reliable marker for neurons releasing ATP as a transmitter, as ATP is co-stored in vesicles with other neurotransmitters. The NOS inhibitory motor neurons also contain several neuropeptides (e.g., vasoactive intestinal peptide, pituitary adenylyl cyclase-activating peptide) that relax or inhibit the muscle and these may also be cotransmitters from these neurons.

Motility

Segmentation

Segmentation is the most enigmatic of the major behaviors of the intestine. What progress has been made arises from high-resolution mapping of intestinal diameter in video recordings as a function of both time and location. Such maps allow the relationships between the activity of adjacent regions to be determined, and hence whether constrictions propagate and, if so, their direction and speed of propagation. Luminal nutrients, either fatty acids like decanoic and dodecanoic acid or amino acids like L-phenylalanine and L-tryptophan, induce a fed-state-like behavior *in vitro*. This is blocked by tetrodotoxin, which blocks neural activity, and hyoscine, which largely blocks excitatory neuromuscular transmission, and so is dependent on neural activity. In guinea pig intestine, the stationary contractions result from excitatory junction potentials that generate action potentials, while the region outside the contracted area is actively inhibited. Thus, there is a local oscillator circuit that produces periodic local excitation of the circular muscle with an inhibitory surround. The identity of the oscillator is unknown, but it is probably the recurrent network of intrinsic sensory neurons. By contrast, in mice, neural activity enhances or depresses slow-wave-mediated contractions so that two pattern generators, one neural and the other myogenic, interact to produce the final pattern of constriction and dilation. There are no studies of segmentation in human intestine *in vitro*, and *in vivo* studies lack the spatial resolution to determine the relationship between neural and slow-wave activity. Both probably contribute to segmentation.

Peristalsis

Segmentation induced by luminal nutrient is occasionally interrupted by powerful constrictions that propagate rapidly (20 mm s^{-1}) along the intestinal segment, peristalsis and antiperistalsis. Peristalsis is a strong constriction that appears close to the oral end of the segment and then extends anally. Antiperistalsis appears at the anal end and extends orally. Stretching the intestinal wall evokes propagating constrictions similar to those evoked by nutrients, but these always run anally.

Recent data challenge the simple model of peristalsis as arising from ascending excitation and descending inhibition. Importantly, localized stimuli (distension, mucosal distortion) activate anally propagating excitation of circular muscle although the stimuli do not move. This implies that the activity is carried by descending neurons, rather than being due to continuous reactivation of reflexes. This anally

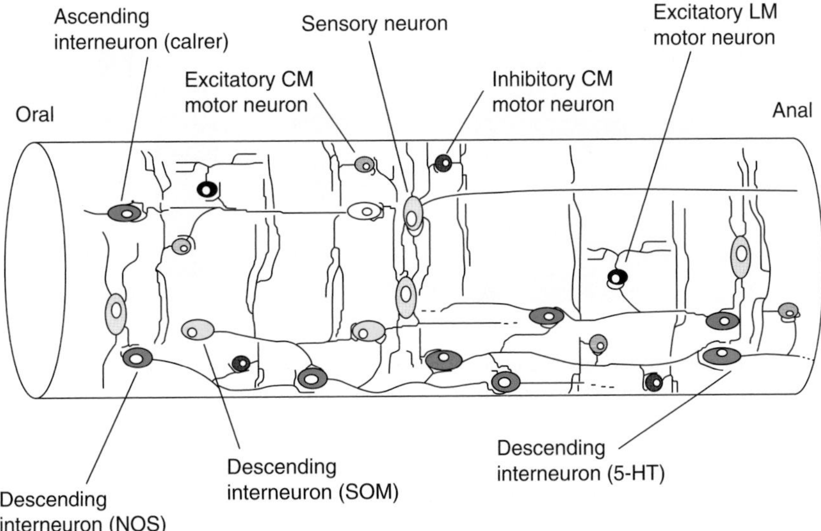

Figure 5 The neural circuitry that can be deduced from physiological, pharmacological, and neuroanatomical studies to be responsible for propulsive motility reflexes in the guinea pig ileum. There are three basic components to the circuit. The intrinsic sensory neurons (pale gray, not subdivided by sensory modality for the sake of some simplicity) project predominantly circumferentially and make synaptic connections with each other as well as having outputs to virtually all other neuron types. Some project anally, but their targets have not been identified. These neurons provide excitatory outputs to ascending interneurons (green), which in turn excite other ascending interneurons, excitatory motor neurons that project orally to supply the circular muscle (CM) (dark gray), and locally projecting longitudinal muscle (LM) motor neurons (excitatory, black). The intrinsic sensory neurons also excite descending interneurons (blue) that contain nitric oxide synthase (NOS) and form the major pathway for descending inhibitory reflexes. The NOS neurons excite inhibitory motor neurons (black) that project anally, ascending interneurons, and two other populations of descending interneurons, those that contain choline acetyltransferase and somatostatin (SOM; yellow) and those that contain choline acetyltransferase and 5-hydroxytryptamine (5-HT; red). The latter contact both excitatory motor neurons and ascending interneurons, but the role and connections of the former are not yet clear. At each point along the pathways, all neurons except the choline acetyltransferase/SOM neurons receive inputs from local intrinsic sensory neurons.

directed pathway probably includes interneurons that contain 5-HT and transmit to excitatory motor neurons via 5-HT_3 receptors, and other interneurons that are excited by ATP acting at P2X receptors. A key finding is that longitudinal muscle is excited simultaneously with the circular muscle, contrary to predictions of earlier models.

Peristalsis in the fed state depends on both the composition and the volume of the digesta at the point of initiation. How these stimuli interact is unknown. However, a neural switch is needed to shift the ratio of segmentation to propulsion to account for changes in intestinal transit resulting from varying luminal content. Despite the myogenic slow waves that are prevalent in many species, peristalsis is predominantly regulated by neural activity. Slow waves impart an underlying rhythm rather than determining the direction or speed of propagation of the constriction.

Interdigestive Motor Complex

Interdigestive migrating motor complexes are difficult to identify *in vitro*, although contraction complexes similar in form and timing to phase III type are seen in isolated mouse colon and small intestine (**Figure 6**). Contraction complexes seen in the jejunum *in vitro* often propagate from anal to oral, so are not identical to the MMCs seen *in vivo*. Nevertheless, the neural circuits regulating these two behaviors must have many neurons in common and the behaviors share several key properties, including very low cycling rates (1 cycle every 2–5 min) and propagation speeds (about 1 mm s^{-1}). The latter are too slow to be accounted for by activity passing along chains of interneurons, because all identified interneurons run longitudinally. By contrast, AH neurons send axons around the intestinal circumference with a slight bias in the anal direction and make excitatory synapses with each other. These transmit via slow EPSPs that can initiate action potentials and can be evoked by physiological stimuli. This arrangement automatically slows longitudinal conduction of activity within the network, so that activity in a recurrent AH neuron network can account for the slow propagation of the contraction complexes. Modeling networks of these neurons indicates that interactions between their AHPs and the slow EPSPs are essential for encoding ongoing stimuli applied to the intestinal wall. Complete suppression of the AHPs causes the network to produce propagating waves of excitation that

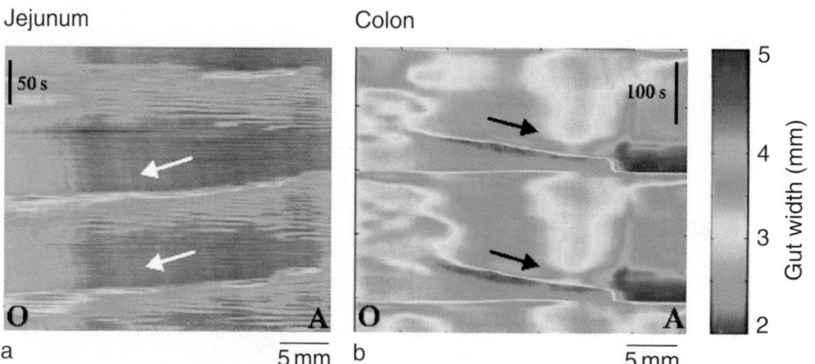

Figure 6 Spatiotemporal maps showing propagating contractile complexes in the mouse jejunum (a) and colon (b). These maps are constructed from video recordings of segments of jejunum and colon *in vitro* as they contract spontaneously. In each frame of the recording, computer software converts the intestine's image into a silhouette, rotates the silhouette until it is horizontal, and then for each horizontal pixel measures the cross section of the silhouette. This value is converted to a color scale, with the minimum diameter represented by dark red and the maximum diameter represented by dark blue. The diameter of the intestine at each point along the segment is then plotted as a multicolored line. The result of this calculation for the next frame is then added below, and so on. In these maps, time increases from top to bottom and the oral end of the segment is on the left. Arrows indicate the direction of apparent propagation of constrictions. Contraction complexes in the jejunum repeat at approximately 2 min intervals and appear to propagate orally. Similar complexes appear in the colon, but propagate anally.

migrate along the intestine at a speed similar to that of MMCs. The activity can propagate in either direction from the point of initiation, but oral to anal propagation is imposed on the network if the point of initiation is at the oral end. Hence, both interdigestive MMCs and the propagating contraction complexes seen *in vitro* may be driven by activity in recurrent AH neuron networks. In this state, AH neurons cannot encode the magnitude of a stimulus and instead act as coordinating interneurons and not sensory neurons. This suggests that there is a physiological switch that converts neurons that detect the composition, consistency, and volume of the digesta into interneurons with no sensory function, and back again.

Secretion of Water and Electrolytes

Like intestinal motility, secretion of water and electrolytes into the intestinal lumen is essential for digestion and is directly controlled by the ENS. Blockade of neural activity leads to net absorption of water and salt. Several neurally mediated secretory responses to physiological stimuli, including distension of the intestinal wall and mechanical deformation of the mucosa, have been identified. Furthermore, the hypersecretion produced by diarrhea-producing toxins, such as cholera toxin, heat-labile *Escherichia coli* enterotoxin, and heat-stable *E. coli* enterotoxin type A, depends critically on enteric neural activity, despite the same toxins directly activating mucosal enterocytes to promote secretion.

The secretomotor neurons are predominantly located within the submucous plexus. There are two classes of submucosal secretomotor neurons: cholinergic secretomotor neurons and noncholinergic secretomotor neurons. The latter contain vasoactive intestinal peptide (VIP), a potent secretagogue, and this is their major neurotransmitter. Their inputs and their potential sources are illustrated in **Figure 7**. While cholinergic secretomotor neurons are under the direct control of circuits arising in the myenteric plexus, the noncholinergic secretomotor neurons have a much more complex role. VIP neurons receive at least seven types of synaptic inputs from several sources (**Table 3**). These neurons receive excitatory input producing fast EPSPs via nAChR from submucosal neurons and from myenteric neurons. Many VIP neurons also exhibit fast EPSPs mediated by ATP acting at P2X receptors, although their source is unknown. In addition, these neurons receive two or more pharmacologically distinct slow EPSPs from myenteric and submucosal neurons. Much of this slow input is mediated by ATP acting at $P2Y_1$ receptors, with the rest being probably mediated by a neuropeptide, possibly VIP. The complexity of their input is underlined by the observation that VIP secretomotor neurons also have strong inhibitory inputs from two distinct sources. Sympathetic axons produce inhibitory synaptic potentials mediated by norepinephrine acting at α_2-adrenoceptors, and this is the only direct input from outside the gut into the secretomotor pathways. A second inhibitory input comes from myenteric neurons. The roles of the diverse pathways converging on these neurons are yet to be clarified.

Intestino-intestinal Reflexes

There are several long pathways by which the behaviors of distant regions of the gut are coordinated.

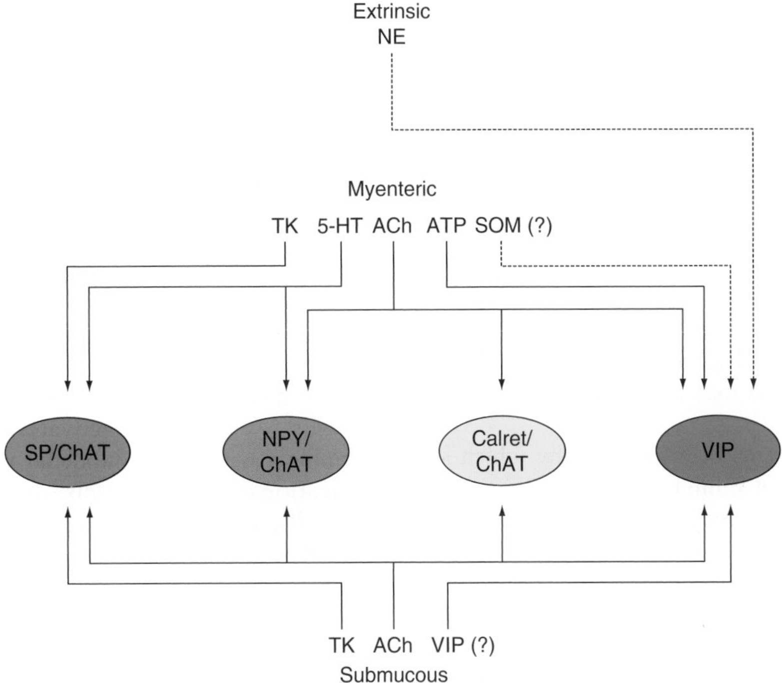

Figure 7 Sources and transmitters of inputs to neurochemically identified neurons in the submucous plexus. Schematic diagram showing the different sources and transmitters of the synaptic inputs to neurochemically identified submucous neurons of the guinea pig ileum. Each discrete class of submucous neurons has a distinct pattern of input, with all receiving excitatory inputs from myenteric and submucous neurons (solid arrows), but only VIP neurons receiving inhibitory inputs (dashed arrows) from both myenteric neurons (probably mediated by somatostatin (SOM)) and the prevertebral ganglia (mediated by norepinephrine (NE)). All classes of neurons appear to receive excitatory input from submucous and myenteric neurons via acetylcholine (ACh) acting at nicotinic receptors. However, the transmitters employed by other excitatory inputs are diverse. Excitatory postsynaptic potentials mediated by tachykinins (TK; notably SP) appear to be confined to the SP/ChAT neurons. Excitatory inputs mediated via serotonin (5-HT) are largely confined to NPY/ChAT neurons, although there may also be an input to SP/ChAT neurons. Excitatory inputs mediated by ATP are largely confined to the VIP neurons. SP/ChAT, substance P/choline acetyltransferase; NPY/ChAT, neuropeptide Y/choline acetyltransferase; Calret/ChAT, calretinin/choline acetyltransferase; VIP, vasoactive intestinal peptide.

Table 3 Sources of inputs to VIP secretomotor neurons

Source	*Synaptic potential*	*Transmitter*	*Receptor*
Myenteric neuron	Fast EPSP	ACh	Nicotinic
Submucous neuron	Fast EPSP	ACh	Nicotinic
Myenteric neuron	Fast EPSP	ATP	P2X
Myenteric neuron	Slow EPSP	ATP	$P2Y_1$
Myenteric neuron	Intermediate EPSP	ATP	$P2Y_1$
Submucous neuron	Slow EPSP	?VIP?	Unknown
Myenteric neuron	IPSP	Somatostatin	Unknown
Sympathetic ganglion cells	IPSP	Norepinephrine	α_2-Adrenoceptor

ACh, acetylcholine; ATP, adenosine 5′-triphosphate; EPSP, excitatory postsynaptic potential; IPSP, inhibitory postsynaptic potential; P2X, $P2Y_1$, purinergic; VIP, vasoactive intestinal peptide.

In general, this coordination takes the form of inhibition of proximal regions by stimuli applied distally. A classic example is the ileal brake, in which fats in the ileum cause marked slowing of transit along the jejunum. This allows more time for the digestion and absorption of other nutrients, thereby increasing the efficiency of the system. Similar reflexes are triggered by bile salt metabolites in the colon or by colonic distension and also slow the arrival of undigested material into regions not involved in digestion or nutrient absorption. Similar reflexes regulate gastric emptying.

A major pathway for these long-distance reflexes is via myenteric neurons whose axons run to, and make synapses in, prevertebral sympathetic ganglia (**Figure 1**). These intestinofugal neurons excite sympathetic neurons within the celiac, superior mesenteric, and inferior mesenteric ganglia, which in turn project to the myenteric plexus of more proximal intestinal regions to inhibit enteric neural circuits regulating motility. Most intestinofugal neurons are interneurons receiving input from local intrinsic sensory neurons and exciting their targets via nAChRs. They also contain various neuropeptides that may act as cotransmitters, producing slow EPSPs in some physiological circumstances. Some intestinofugal neurons may also have sensory functions of their own.

There also appears to be a pathway by which stimuli in the colon can inhibit secretion in the small intestine. Evidence for this is less direct than for the regulation of motility via the intestino-intestinal pathway, but is nevertheless compelling. Celiac neurons that contain somatostatin and norepinephrine have only one identified target in the ileum, the submucous ganglia, and these neurons receive a major portion of the input from intestinofugal neurons. Thus, activation of intestinofugal pathways should inhibit VIP secretomotor neurons in the submucous plexus, as these are the only submucous neurons with inputs from sympathetic axons. This implies that the content of the colon can regulate secretion of water and salt in the small intestine.

Vasomotor Control

Intestinal blood flow is critical for digestion, as well as being a key element of overall blood pressure control. During exercise and other environmental stresses, blood flow to the gut is restricted to allow maintenance of appropriate flows to the brain, heart, and skeletal muscles, for example. This process is regulated by the CNS via vasoconstrictor neurons with cell bodies in the para- and prevertebral sympathetic ganglia, which innervate both mesenteric blood vessels and submucosal arterioles. Vasoconstrictor neurons in prevertebral ganglia do not receive input from intestinofugal neurons, so vasoconstriction is not modulated by direct input from the gut, although pathways that run via the CNS are certainly involved.

On the other hand, there are also enteric vasodilator pathways. Submucosal vasodilator neurons have been identified by stimulation of individual neurons in ganglia and measuring dilation of nearby arterioles. These neurons are excited by cholinergic neurons in distension-activated reflex pathways running from the myenteric plexus, and act upon the blood vessels via mAChRs. How vasodilation, motility, and mucosal secretion are coordinated by the intrinsic pathways will be a fertile ground for research in the near future.

Neural Regulation of Digestive Enzyme Secretion

Secretion of digestive enzymes was thought to be primarily regulated by circulating hormones, notably cholecystokinin (CCK), released by nutrients in the duodenum. However, pancreatic enzyme secretion is also mediated via the autonomic nervous system. Locally released CCK activates vagal afferent nerve terminals within the duodenal wall, which leads to activation of vagal efferents supplying pancreatic acinar cells. This reflex is supplemented by enteric neurons that project to the pancreas to modulate enzyme release.

Summary

All major processes involved in digestion depend on the autonomic nervous system. The intestinal movements and secretion of water and salt depend on the ENS. Coordination of different intestinal regions depends on reflex pathways running through sympathetic ganglia. Coordination of whole-body functions and secretion of digestive enzymes depend on central sympathetic and parasympathetic pathways. Thus, the innervation of the gut reflects its role as 'the way to a man's heart.'

See also: Adenosine Triphosphate (ATP); Autonomic and Enteric Nervous System: Apoptosis and Trophic Support During Development; Autonomic Nervous System: Neuroanatomy; Autonomic Nervous System: Central Control of the Gastrointestinal Tract; Autonomic Neuroeffector Junction; Cotransmission; Enteric Nervous System: Sensory Pathways; Enteric Nervous System: Neural Circuits and Chemical Coding; Enteric Nervous System: Glial Cells and Interstitial Cells of Cajal; Gastrointestinal Signals: Stimulation; Gastrointestinal Signals: Satiety; Gastrointestinal Tract role in Neural Control of Metabolism, Food Intake and Body Weight: A Summary; Muscarinic Receptors: Autonomic Neurons; Nitric Oxide; Peptidergic Receptors; Purinergic Receptors.

Further Reading

Bornstein JC, Furness JB, Kunze WAA, et al. (2002) Enteric reflexes that influence motility. In: Brookes SJH and Costa M (eds.) *Innervation of the Gastrointestinal Tract*, pp. 1–55. London: Taylor & Francis.

Brookes SJH and Costa M (2002) Cellular organisation of the mammalian enteric nervous system. In: Brookes SJH and Costa M (eds.) *Innervation of the Gastrointestinal Tract*, pp. 393–467. London: Taylor & Francis.

Costa M, Brookes SJH, Steele PA, et al. (1996) Neurochemical classification of myenteric neurons in the guinea-pig ileum. *Neuroscience* 75: 949–967.

Furness JB (2006) *The Enteric Nervous System*. Oxford: Blackwell Publishing.

Furness JB, Jones C, Nurgali K, et al. (2004) Intrinsic primary afferent neurons and nerve circuits within the intestine. *Progress in Neurobiology* 72: 143–164.

Galligan JJ (2002) Pharmacology of synaptic transmission in the enteric nervous system. *Current Opinion in Pharmacology* 2: 623–629.

Gwynne RM, Thomas EA, Goh SM, et al. (2004) Segmentation induced by intraluminal fatty acid in isolated guinea-pig duodenum and jejunum. *Journal of Physiology (London)* 556: 557–569.

Owyang C and Logsdon CD (2004) New insights into neurohormonal regulation of pancreatic secretion. *Gastroenterology* 127: 957–969.

Sanders KM, Koh SD, and Ward SM (2006) Interstitial cells of Cajal as pacemakers in the gastrointestinal tract. *Annual Reviews of Physiology* 68: 307–343.

Thomas EA, Sjövall H, and Bornstein JC (2004) Computational model of the migrating motor complex of the small intestine. *American Journal of Physiology – Gastrointestinal and Liver Physiology* 286: G564–G572.

Wood JD (2004) Enteric neuroimmunology and pathophysiology. *Gastroenterology* 127: 635–657.

Autonomic Nervous System: General Overview

N R Keller and D Robertson, Vanderbilt University, Nashville, TN, USA

Introduction

The autonomic nervous system (ANS) is composed of the sympathetic, parasympathetic, and enteric neurons, and the adrenal medulla. It is one of two major divisions of peripheral nerves, the other being the sensorisomatic nervous system. Reciprocally integrated with central autonomic foci, the ANS maintains homeostasis and adaptive responses to physical and mental stress virtually beyond the awareness of the host. The sympathetic system controls fight-or-flight reactions by increasing sympathetic outflow to the heart and other viscera, while the parasympathetic system maintains basal regulation of bodily functions. The enteric nervous system, on the other hand, manages peristalsis and influences glandular secretions. Among the many different aspects of the ANS, this article focuses primarily on neurotransmission within the sympathetic branch (**Figure 1**).

Neurotransmitters were first proposed as chemical stimulants released by nerve terminals early in the twentieth century. Research in the past 10 to 20 years has generated substantial progress in identifying intra- and extracellular signaling molecules, developmental regulatory transcription factors, metabolic pathways, targeted pharmacologic agents, and the influence of genetic mutations in the pathogenesis of autonomic dysfunction. Along with these discoveries, our understanding of the mechanisms underlying various neurotransmitter disorders has also grown. A sampling of genetic, metabolic, and transport errors is highlighted herein.

Synthesis

Dopamine β-Hydroxylase

Dopamine β-hydroxylase deficiency Dopamine β-hydroxylase (DBH) is an oxidoreductase of the copper type II, ascorbate-dependent monooxygenase family, and is responsible for conversion of dopamine (DA) to norepinephrine (NE). DBH deficiency is an autosomal recessive disorder of heterogeneous nature. The human DBH gene contains 12 exons and is situated at the 9q34 locus, immediately adjacent to the blood group locus ABO. All DBH-deficient patients characterized thus far carry one copy of an IVS1 + 2T $\rightarrow$ C mutation in intron 1 that prevents appropriate splicing. *In vitro* analysis, however, has shown that this mutation does not entirely explain undetectable levels of DBH protein in plasma, because some properly spliced DBH message can be detected. Instead, concomitant genetic changes such as −1021C $\rightarrow$ T, which is linked with very low plasma DBH activity, are suspected as inhibitory transcriptional modulators. Thus, NE deficiency may only occur in the context of a specific haplotype, not merely as the result of an individual mutation. Missense mutations of exon 2 (D100E) and of exons 1 and 6 residing *in cis* (V87M and D331N) have also been suggested as pathogenic accomplices.

Both norepinephrine and epinephrine (EPI) are completely absent in plasma, urine, and cerebrospinal fluid (CSF) of humans with DBH deficiency (**Figure 2**), since the enzyme is vital for conversion of DA to NE. Untreated individuals experience a lifetime of severe orthostatic hypotension and mild eyelid ptosis, with some skeletal muscle hypotonia and nasal stuffiness. Perinatal progress is understandably precarious; near-fatal events in the first few days of life include hypotension, hypoglycemia, hypothermia, and seizures. In survivors, delay in eye opening may occur. Because neonates with undetected enzyme insufficiency may succumb to such complications, it is possible that the incidence of DBH deficiency is greater than predicted based on the number of affected adults, although it is still quite rare.

The lack of sympathetic tone severely limits the hemodynamic response to upright posture, with systolic blood pressure (BP) routinely falling well below 100 mmHg with a relatively small increase in heart rate (HR). Presyncopal dizziness is often accompanied by dyspnea, nuchal discomfort, and occasional chest pain. Dizziness is exacerbated by warm environmental temperature. For these reasons, individuals with DBH deficiency are not able to tolerate physical exercise. Hypothermia is seen in infants. Male reproduction is complicated by retrograde ejaculation.

Autonomic testing in DBH deficiency is abnormal for sympathetic function. No increase in BP is elicited in the cold pressor test, isometric handgrip, or phase IV of the Valsalva maneuver. Patients are normal, however, with respect to sweating, lacrimation, salivation, corneal and deep tendon reflexes, sensory function, and the senses of taste and smell. Gastrointestinal function is within normal limits, and bladder function is preserved. Nocturnal polyuria with increased dopamine-mediated natriuresis is a shared feature of the disease.

Sympathetic nerve endings in DBH deficiency appear normal and stain positive for cytosolic

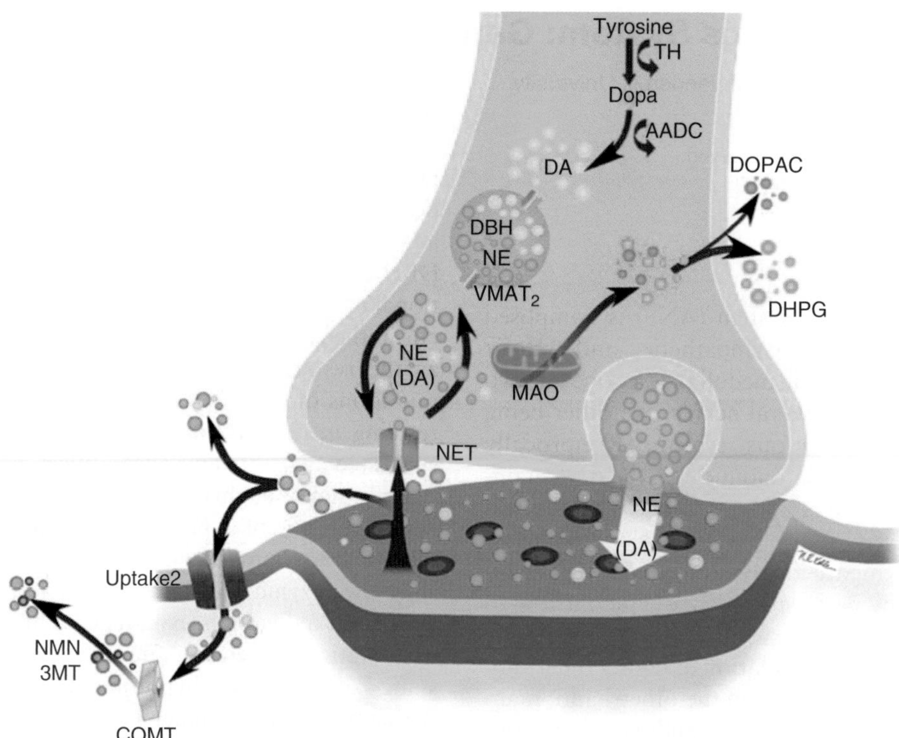

Figure 1 Normal sympathetic neurotransmission. Norepinephrine (NE) is the major neurotransmitter in postganglionic sympathetic and central noradrenergic neurons. Enzymatic conversion of tyrosine to dopamine (DA) takes place in the cytoplasm, whereas transformation of dopamine to NE occurs within the storage vesicles. After release into the synapse, residual catecholamines are pumped back into the presynaptic neuron by the norepinephrine transporter (NET), and returned to storage vesicles by the vesicular monoamine transporter 2 ($VMAT_2$). A smaller portion of the surplus transmitters diffuse into the extracellular space; these transmitters are cleared by the low-affinity, high-capacity 'Uptake 2' transporter, which has recently been renamed the 'extraneuronal monoamine transporter' in humans. Breakdown of cytosolic catecholamines is accomplished in part by monoamine oxidases (MAOs), while extracellular metabolism is conducted by catechol-*O*-methyltransferase (COMT). NMN, normetanephrine; 3MT, 3-methoxytyramine; DBH, dopamine β-hydroxylase; TH, tyrosine hydroxylase; Dopa, dihydroxyphenylalanine; DOPAC, 3,4-dihydroxyphenylacetic acid; AADC, L-aromatic amino acid decarboxylase; DHPG, 3,4-dihydroxyphenylglycol.

tyrosine hydroxylase, the rate-limiting enzyme in DA synthesis. Normally, DA is taken into the vesicles for conversion to NE by DBH. Without dopamine transformation, however, it is released from stimulated noradrenergic nerve terminals in place of NE. DBH is also present in the normal adrenal medulla, thus the effects of enzyme failure are undetectable tissue EPI and NE.

Without NE-mediated negative feedback on tyrosine hydroxylase, cytosolic DA production is amplified. Plasma levels of DA, therefore, are notably elevated; approximately five- to tenfold. Urine DA is more than doubled and CSF levels are also increased. The DA metabolites homovanillic acid (HVA) and 3-methoxytyramine (3MT) are roughly double normal values in urine. Plasma and CSF dihydroxyphenylalanine (DOPA) levels are double to triple normal values. As might be expected, prolactin is low. The NE metabolites 3-methoxy-4-hydroxyphenylglycol (MHPG) and normetanephrine (NM) are below detectable limits in urine.

Patients can be successfully treated with droxidopa (dihydroxyphenylserine, or L-DOPS), which is endogenously converted to NE by L-aromatic amino acid decarboxylase (AADC). Both intraneuronal and extraneuronal conversions take place, since AADC is present in most tissues. Adrenomedullary conversion of DA to NE does not occur, however, and EPI levels remain undetectable. With drug administration, BP increases dramatically and NE concentrations approach the normal range in plasma and urine. Consequently, standing time is greatly enhanced, as is quality of life – so improved, in fact, that in 2005 a 26-year-old patient treated with droxidopa for less than 2 years of her life was able to compete in and finish a 26-mile marathon.

Menkes Disease

Menkes disease is an X-linked fatal neurodegenerative disease caused by impaired copper transport in the gastrointestinal tract, placenta, and the blood–brain barrier, resulting in systemic deficiency of the

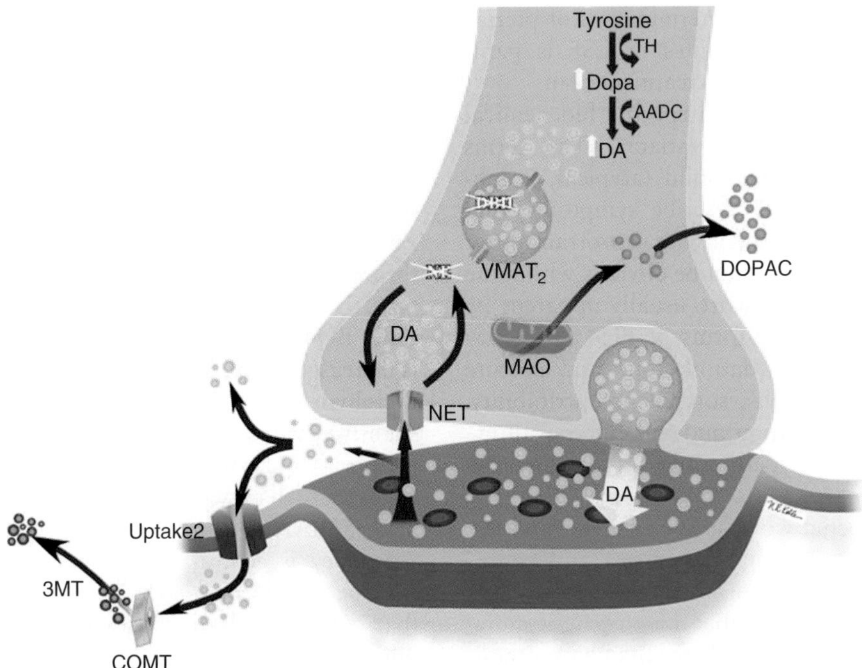

Figure 2 Dopamine β-hydroxylase (DBH) deficiency. Lack of DBH causes the total absence of norepinephrine (NE) in noradrenergic nerve terminals. Sympathetic neurotransmission, therefore, results in the release of dopamine (DA). Because there is no NE-mediated negative feedback on tyrosine hydroxylase (TH), DA production is increased. Reuptake of synaptic DA occurs through the norepinephrine transporter (NET). Enzymatic breakdown yields increased DA metabolites. (Other abbreviations as in **Figure 1**.)

element. Mutations in the *ATP7A* gene are responsible for the defective protein, MNK, a P-type ATPase that belongs to the CPx-type family of transmembrane enzymes necessary for ATP-dependent translocation of metal cations across cell membranes. MNK typically resides in the *trans*-Golgi network, where it delivers copper to local enzymes, including DBH. The ATPase can also be found in cytosolic vesicular compartments and the plasma membrane, where it mediates efflux of excess copper.

Copper is an essential cofactor for DBH. Consequently, plasma catecholamines are abnormally low, although NE levels in the CSF may be normal or elevated. Typically, however, because DBH activity is impaired, plasma and CSF DOPA, DA, and dihydroxyphenylacetic acid (DOPAC) are elevated. Ratios of plasma DOPA to dihydroxyphenylglycol (DHPG), the chief neuronal metabolite of NE, are increased approximately sevenfold, and DOPAC to DHPG ratios are about fivefold greater. CSF ratios follow a similar pattern. In addition, the proportion of HVA to vanillylmandelic acid (HVA:VMA) increases in the urine.

Early detection and treatment of the disease is crucial, since copper deficiency begins during gestation. Affected infants often appear normal until 2 to 3 months of age. Clinical symptoms vary somewhat depending on the exact mutation, but most neonates experience seizures, severe hypotonia, hypothermia, and extreme developmental delay. Connective tissue abnormalities due to impaired lysyl oxidase may appear as hyperelastic skin, hernias, and aortic aneurysms. Copper administration, however, does not correct these developmental defects. The most striking outward symptom may be the appearance of the hair, which is unusually short, sparse, coarse, and twisted, and is often lightly pigmented. It has been compared to steel wool, and the disease was formerly referred to as the 'kinky hair syndrome.' The first indicators of autonomic dysfunction are likely temperature instability and hypoglycemia. Without treatment using parenteral copper-histidine, death typically occurs by 3–4 years of age. Copper-histidine therapy for 3 months ordinarily normalizes serum copper, ceruloplasmin, DA, and NE levels.

Tetrahydrobiopterin

Hyperphenylalaninemia (HPA) and phenylketonuria (PKU) are the result of impaired enzymatic action of phenylalanine-4-hydroxylase (PAH). Underlying the PAH malfunction, however, is an array of direct and indirect mechanisms ultimately linked to tetrahydrobiopterin (BH4) deficiency. BH4 is a cofactor for PAH, tyrosine-3-hydroxylase (TH), and tryptophan-5-hydroxylase (TPH). Impaired hydroxylation of tyrosine and tryptophan results in catecholamine and serotonin deficiencies. Determination of pterins and neurotransmitter metabolites in CSF is paramount in establishing an early treatment plan.

Clinical features, along with specific biochemical criteria, distinguish disease variants. The terms 'severe' (typical, general) and 'mild' (atypical, peripheral, partial) are used to describe symptoms, and indicate the need for treatment with neurotransmitter precursors. Symptoms may not be obvious within the first few weeks of life, but are usually apparent by the fourth month. Severe forms of BH4 deficiency manifest as mental retardation, tone and posture abnormalities, convulsions, somnolence, irritability, dyskinesias, hypersalivation and difficulty swallowing, and intermittent hyperthermia without underlying infection. Neurologic symptoms may follow a diurnal pattern, being worst in the evenings and less bothersome in the mornings after rest and neurotransmitter precursor replenishment.

Cofactor deficiency is attributable to genetic mutations in the enzymes responsible for BH4 biosynthesis or regeneration. Here, focus is limited to faulty metabolic pathways causing catecholamine deficiency. Catechol replacement usually consists of droxidopa, with carbidopa added to inhibit peripheral DA production. Levels of plasma phenylalanine warrant close attention when administering neurotransmitter therapy, since high concentrations of phenylalanine can hinder precursor membrane transport or compete with tyrosine.

The form of HPA encountered most is 6-pyruvoyl-tetrahydropterin synthase (PTPS) deficiency, which also exhibits the largest amount of genetic variability. Over 50% of BH4 deficiencies are PTPS and about 80% of affected individuals have the severe form of the disease. Phenotypic expression ordinarily reflects homozygous, compound heterozygous, and obligate heterozygous enzyme activities. Mutations in the PTS gene have been located in all six exons and generally result in a deficient or inactive enzyme product. An exception is the K129E mutant, which demonstrates heightened enzyme activity in a cell-specific manner *in vitro*, suggesting that posttranslational modification plays a role in inducing the upper limits of enzyme performance.

Analysis of urinary pterins discloses that severe PTPS produces the highest amount of neopterin of the BH4 deficiencies, while levels in the mild form are comparable to the pterin-4α-carbinolamine dehydratase (PCD) deficiency profile. Neurotransmitter metabolites are often normal in the CSF of persons with the mild version, as they are in the rarer intermediate form of PTPS deficiency, whereas urinary pterin concentrations are similar to the severe variety. Milder symptoms can worsen with age, and CSF neurotransmitter levels, although normal soon after birth, decrease in accordance with neurological decline.

Most patients with dihydropteridine reductase (DHPR) deficiency lack enzyme activity entirely, or nearly so. Thus, the majority of patients have the severe phenotype. About half are found in southern Europe. Again, mutations are distributed fairly evenly throughout the gene, and most are of the missense type. Urine biopterin levels are higher than in any other BH4 deficiency, and pterin synthesis in general is increased due to lack of feedback inhibition of GTP cyclohydrolase I (GTPCH). Neopterin is normal or slightly elevated. The CSF neurotransmitter metabolic profile is similar to severe PTPS deficiency, with HVA levels more than twice those of 5-hydroxyindoleacetic acid (5HIAA). An unrelated consequence of DHPR deficiency is central folate depletion, presumed to be due to accumulation of 7,8-dihydrobiopterin that competitively interferes with reduction of 7,8-dihydrofolic acid to the active tetrahydrofolic acid.

Both GTP cyclohydrolase I (GTPCH) deficiency and PCD are relatively rare compared to other BH4 disease varieties, with the former being somewhat variable clinically but in general, more severe. Pterins are virtually undetectable in urine from GTPCH patients since the enzyme participates in early pterin synthesis, while neopterin is elevated in PCD. In CSF, pterins follow a similar trend, although both neopterin and biopterin are increased in PCD. Neurotransmitter metabolites are below normal in both disorders as well.

Metabolism

Monoamine Oxidase

Monoamine oxidase (MAO) is located in the outer mitochondrial membrane where oxidative deamination of neuroactive, vasoactive, and xenobiotic amines occurs (**Figure 3**), producing hydrogen peroxide and aldehydes. MAO-A and MAO-B are encoded by two adjacent gene products on the X-chromosome, with different molecular weights and immunological properties. They share about 70% homology and neurons routinely contain both forms. Norepinephrine and serotonin (5-hydroxytryptamine; 5-HT) are preferential substrates for MAO-A, while dopamine, phenylethylamine, and benzylamine are metabolized by MAO-B. Dopamine, tyramine, and tryptamine are substrates of both isoenzymes.

MAO has been implicated in the pathogenesis of neurodegeneration, including that underlying Parkinson's disease. Active transport of catecholamines into neuronal vesicles is dynamically opposed

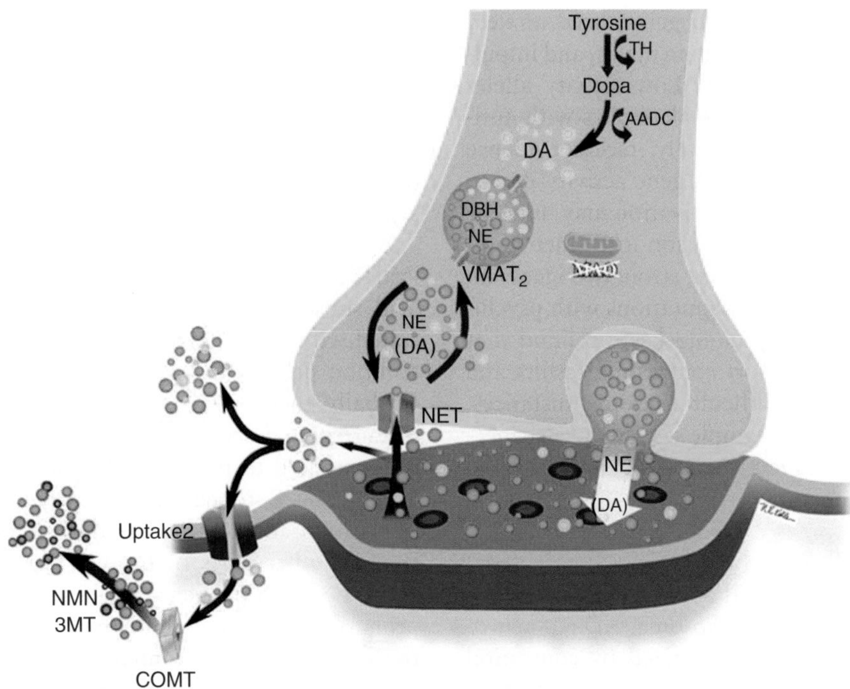

Figure 3 Monoamine oxidase (MAO) deficiency. Norepinephrine (NE) and 5-hydroxytryptamine are metabolized by MAO-A, while dopamine (DA), phenylethylamine, and benzylamine are broken down by MAO-B. Dopamine, tyramine, and tryptamine are substrates of both isoenzymes. Deficiency of only one gene product is reflected by the absence of its corresponding metabolites. Without oxidative deamination of NE by MAO-A in sympathetic nerves, for example, 3,4-dihydroxyphenylglycol will not be formed. Lack of both enzyme subtypes would result in the loss of both 3,4-dihydroxyphenylglycol and 3,4-dihydroxyphenylacetic acid. Metabolites of catechol-*O*-methyltransferase (COMT) will increase. (Other abbreviations as in **Figure 1**.)

by passive leakage into the cytoplasm. Historically, neurotoxicity has been explained by nonenzymatic autooxidation of cytoplasmic catecholamines and the formation of reactive quinones, free radical production, and formation of alkaloids via condensation reactions. The oxidative by-products of MAO, on the other hand, are deaminated catecholaldehyde metabolites. In fact, 3,4-dihydroxyphenylacetaldehyde (DOPAL) and 3,4-dihydroxyphenylglycolaldehyde (DOPEGAL) are formed in such a manner by deamination of DA and NE/EPI, respectively, by MAO and are highly reactive. Prompt aldehyde metabolism to deaminated acids by aldehyde dehydrogenase or to deaminated alcohols by aldose or aldehyde reductase normally neutralizes the deleterious effects. However, any cytoplasmic imbalance that interferes with vesicular uptake, increases vesicular leakage, or inhibits aldehyde metabolism theoretically favors accumulation of neurotoxins and impending cell damage. Centrally, this could be associated with neurodegenerative changes in dopaminergic neurons and parkinsonian-like movement disorders, whereas buildup of toxic NE metabolites in cardiac sympathetic nerves has been postulated to play a role in orthostatic hypotension due to sympathetic denervation of the heart. Unfortunately, we do not have experimental evidence which implicates all these mechanisms in human disease. If such mechanisms are found to be operative, then MAO inhibitors might reduce neurodegeneration caused by overproduction and/or reduced metabolism of cytoplasmic catecholaldehydes.

MAO inhibitors have long been used in the treatment of depression, and the significance of MAO polymorphisms in psychiatric illness has been widely investigated. Deficiency of either isoenzyme alone or the combination of both will yield characteristic neurochemical profiles. One MAO-A mutation found in a Dutch pedigree yields no active enzyme and is associated with mental retardation and impulsive–aggressive behavior in males. Follow-up animal models have also revealed an aggressive phenotype in male mice lacking MAO-A.

In other human studies, significant variations in upstream coding sequences of the MAO-A gene were found across ethnic groups, consisting of 30 bp repeats present in 2, 3, 3.5, 4, or 5 copies. The polymorphism proved to be functional in regulating the MAO-A promoter, and alleles with a higher number of repeat copies (3.5, 4, and 5) demonstrate increased transcriptional activity. High-activity alleles have been associated with panic and generalized anxiety disorders in females, neuroticism in males, and impulsivity and aggression in males. Low-activity alleles have been associated with male alcoholics with antisocial personality disorder. Finally, recent evidence suggests that increased MAO-A gene activity due to incomplete X chromosome inactivation may underlie the greater prevalence of depression in women compared to men. While these data strongly suggest an association of specific genetic mutations with psychiatric disorders, the reader should keep in mind that different groups often report contrasting results and resolution has proved difficult in some instances. Repeat studies and larger sample sizes are clearly necessary before a definitive association can be claimed for any given diagnosis.

MAO-A and-B deficiency (and Norrie disease gene) The MAO genes are situated next to the Norrie disease (ND) gene on the X chromosome. ND is a neurodevelopmental disorder characterized by congenital blindness caused by incomplete retinal vascularization. Approximately one-third of patients also suffer sensorineural deafness. Mental retardation is not uncommon, but the degree of severity is variable. Patients with atypical ND, by contrast, exhibit profound mental retardation, autistic-like behavior, hypogonadism, atonic seizures, altered peripheral autonomic function, and aberrant biogenic amine metabolism. Such severity is ascribed to large deletions encompassing the MAO and ND gene loci together. Combined MAO-AB deficiency results in >93% reductions in plasma DHPG, MHPG, and VMA, along with elevated plasma normetanephrine and metanephrine resulting from increased catechol-O-methyltransferase (COMT) metabolism. Individuals lacking the ND and MAO-B genes, on the other hand, show no signs of mental retardation or disruptive behavior. MAO-B deficiency is characterized by undetectable levels of platelet MAO-B activity and increased urinary phenylethylamine.

Catechol-*O*-Methyltransferase

Most extraneuronal catecholamine metabolism is accomplished by COMT in the liver and kidney. In addition to catecholamines, COMT also metabolizes L-DOPA, DHPG, DOPAC, and catechol estrogens. Conversion of catechols to their O-methylated metabolites requires *S*-adenosyl-L-methionine as a methyl donor. Two isoenzyme forms of COMT exist, one soluble and the other membrane bound. Soluble COMT is present in most tissues and has higher activity levels compared to the membrane-bound form. The main exception is the adrenal medulla, in which membrane-bound COMT is most prevalent. Pheochromocytomas, catecholamine-producing tumors at adrenomedullary and extraadrenal sites, contain vast amounts of membrane-bound COMT and generate high concentrations of O-methylated metabolites, primarily metanephrine and MHPG. Diagnosis is best accomplished by assaying plasma free metanephrines.

A functional polymorphism of the COMT gene (Val158Met) has been identified that causes amino acid substitution. The wild-type Val allele is associated with high enzyme activity, while the Met substitution produces three- to fourfold lower activity. Globally, there is wide allelic variation, with persons of European descent having the greatest heterozygosity compared to those of African or Asian heritage. Specifically, Europeans have nearly equal frequencies of the two alleles, whereas the majority of populations in other parts of the world express the high-activity allele. Furthermore, high-activity COMT is primarily found in nonhuman primates, implying it is the ancestral allele in humans.

A large genetically homogeneous population-based study conducted in Norway recently revealed an association of the Val/Val genotype with systolic BP $\geq$140 mm Hg compared to either Val/Met or Met/Met genotypes. Although the underlying mechanism was not accounted for, the authors postulate that altered dopamine catabolism may be involved. Dopamine-mediated natriuresis is believed to be important in long-term control of BP. Therefore, elevated COMT activity in the kidney would likely increase dopamine breakdown, thus increasing sodium levels and, ultimately, BP.

The COMT polymorphism has been extensively investigated in anxiety disorders. Despite early promising findings, later studies revealed conflicting results and failed to replicate a significant association of this polymorphism with various subtypes of anxiety, including obsessive–compulsive, general anxiety, and panic disorders, or phobias.

A secondary role for COMT is methylation of 2-OH and 4-OH catechol estrogens. Transformation to methoxy estrogens lowers the risk of breast cancer associated with estrogen-mediated DNA adduct formation or oxidative damage arising from catechol estrogen quinone–semiquinone redox cycling. Cancer risk examined by *in vitro* analysis of the COMT Val → Met polymorphism demonstrated that the low-activity Met variant produced two- to threefold lower levels of methoxy estrogens in breast cancer cells compared to the wild-type variant. Therefore, estrogenized women with lower COMT activity may be at increased risk for breast cancer due to elevated tissue concentrations of prooxidative 4-OH catechol estrogens, and lower levels of 2-methoxyestradiol, an antiproliferative metabolite.

Reuptake

Norepinephrine Transporter Dysfunction

Norepinephrine is rapidly recycled by the presynaptic neuron for rerelease. Most NE is removed from the synaptic cleft by active transport via the norepinephrine transporter (NET). The NET is a member of Na^+/Cl^--dependent transporters sharing similar amino acid sequences and pumping mechanisms. Transporters were historically considered to be static proteins intrinsic to the cell surface. Recent evidence, however, has revealed that NET is usually tethered to neuronal vesicles until transmitter release, and that phosphorylation is crucial to cell surface expression. NET is dynamically regulated by both intra- and extracellular signaling molecules, as well as targeted by a host of long-term therapeutic agents and drugs of abuse. Indeed, NET expression is recognized as a means of neuronal adaptation to changes in transmission. Endogenous regulators of NET expression include norepinephrine, acetylcholine, γ-aminobutyric acid (GABA), angiotensin, insulin, natriuretic peptide, nerve growth factor, estrogens, and nitric oxide.

A substitution mutation in the NET gene was discovered in a large family in which affected individuals had orthostatic tachycardia and postural tachycardia syndrome (POTS). The A457P mutation resulted in decreased plasma membrane concentrations of NET in heterozygous individuals. In fact, the altered protein was shown to bind native transporter and reduce its trafficking to the cell surface, making this a dominant-negative mutation. (General NET deficiency is shown in **Figure 4**.) Analysis of 254 additional POTS patients has not yielded similar functional NET mutations to account for diminished NE reuptake. Therefore, POTS remains a dysautonomia of heterogeneous etiology, with NET deficiency a very rare etiology.

POTS is an often debilitating autonomic disorder characterized by excessive tachycardia on standing, elevated plasma NE spillover, and presyncopal/syncopal episodes. Affected individuals are young and primarily women. In addition to syncopal symptoms, common complaints include dizziness, visual changes, head and neck discomfort, poor concentration while

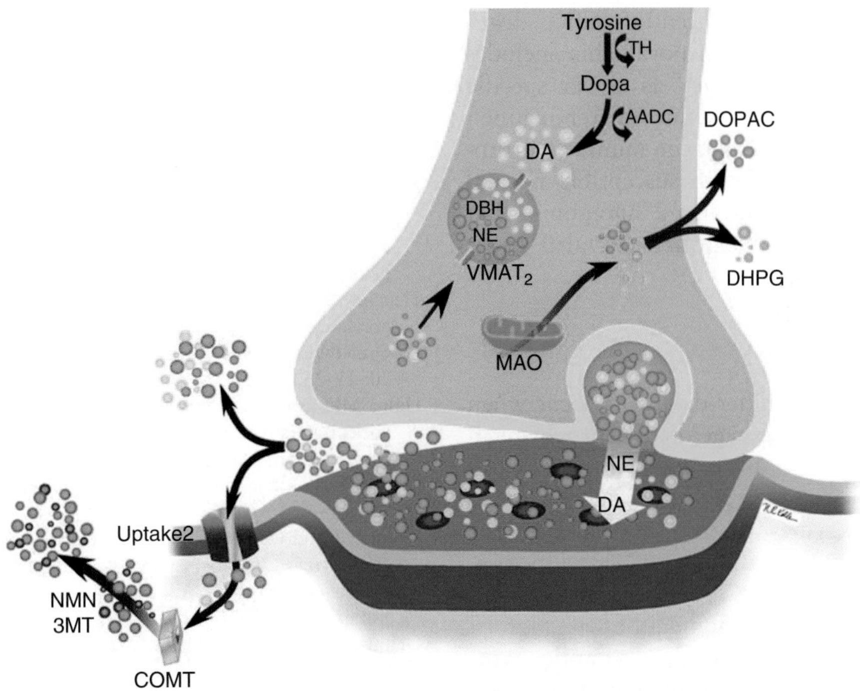

Figure 4 Norepinephrine (NE) transporter deficiency. Residual NE and dopamine (DA) are removed from the synaptic cleft by active transport through the NE transporter. The absence of NE transporter results in an increased concentration of catecholamines in the synapse, prolonging noradrenergic signaling and elevating levels of plasma NE and DA. Because monoamine recycling does not occur, conversion of DA to NE is increased. Continuous neurotransmission can deplete intraneuronal stores of NE, resulting in greater release of precursor and decreased production of 3,4-dihydroxyphenylglycol (DHPG). Metabolites of catechol-*O*-methyltransferase (COMT) are increased. (Other abbreviations as in **Figure 1**.)

standing, fatigue, palpitations, tremor, and anxiety. In familial POTS, levels of plasma NE increase approximately twice as high as in normal individuals on standing. DHPG concentration is decreased supine and upright, thereby lowering the DHPG:NE ratio. This is in contrast to idiopathic POTS, in which NE spillover is similar to that in normal individuals, both supine and standing.

Numerous naturally occurring single-nucleotide polymorphisms (SNPs) have been identified in the NET gene; these SNPs alter protein synthesis or maturation, transport mechanics, antagonist interactions, and kinase-mediated regulation. Many were discovered in cardiovascular phenotypes, and the role of similar functional mutations has been examined in psychiatric profiles. One such case is an A → T polymorphism at −3081 upstream of the transcription initiation site. The T allele was shown to decrease NET promoter function by allowing the binding of transcriptional repressors. A significant association of the mutation was demonstrated with attention-deficit/hyperactivity disorder (ADHD) in a relatively small sample size, but more extensive investigation is needed to confirm these findings.

Although detection of functional NET mutations has thus far been limited, pharmacological NET inhibition is very common. Many antidepressants and medications used in treating attention-deficit disorders block or alter NET transport. This includes first-generation tricyclics as well as newer specific NET inhibitors (e.g., atomoxetine or reboxetine). Pharmacological NET blockade can mimic an orthostatic tachycardia phenotype in susceptible normal subjects. Yohimbine, a central α-2 antagonist that increases synaptic norepinephrine, can also cause orthostatic tachycardia.

Conclusion

In summary, neurotransmitter disorders occur when there is insufficient or exaggerated neurochemical transmission, and can be due to either direct or indirect causes. For example, genetic mutations may affect essential enzymes directly in the synthesis pathway, or indirectly by limiting cofactor function. Multiple mechanisms exist for the clearance and recycling of catecholamines, and gain-of-function or loss-of-function mutations that alter synaptic, cytosolic, or vesicular concentrations often underlie aberrant neurotransmission. Continued investigation of genotype–phenotype associations will provide a better understanding of the pathophysiology of neurotransmitter disorders, allowing new methods for diagnosis and adjustment of treatment options to meet individual needs.

See also: Autonomic Disorders; Autonomic Failure; Autonomic Nervous System: Cardiovascular Control; Autonomic Nervous System: Clinical Testing; Autonomic Nervous System; Cell Culture: Autonomic and Enteric Neurons; Dopamine; Dopamine – CNS Pathways and Neurophysiology; Dopamine Receptors and Antipsychotic Drugs in Health and Disease; Dopamine Neurons: Reward and Uncertainty; Monoamine Transporters: Focus on the Regulation of Serotonin Transporter by Cytokines; Sympathetic Nervous System; Trace Monoamines and Receptors in Mammalian CNS.

Further Reading

Blakely RD, DeFelice LJ, and Galli A (2005) Biogenic amine neurotransmitter transporters: Just when you thought you knew them. *Physiology* 20: 225–231.

Blau N, Thony B, Cotton RRH, et al. (2001) Disorders of tetrahydrobiopterin and related biogenic amines. In: Scriver CR, Beaudet AL, Sly WS, et al. (eds.) *The Metabolic and Molecular Bases of Inherited Disease*, pp. 1725–1776. New York: McGraw-Hill.

Carrel L and Willard HF (2005) X-Inactivation profile reveals extensive variability in X-linked gene expression in females. *Nature* 434: 400–404.

Cox DW and Moore SDP (2002) Copper transporting P-type ATPases and human disease. *Journal of Bioenergetics and Biomembranes* 34: 333–338.

Daniel KG, Harbach RH, Guida WC, et al. (2004) Copper storage diseases: Menkes, Wilson's, and cancer. *Frontiers in Bioscience* 9: 2652–2662.

Eisenhofer G, Kopin IJ, and Goldstein DS (2004) Catecholamine metabolism: A contemporary view with implications for physiology and medicine. *Pharmacological Reviews* 56: 331–349.

Garland EM, Hahn MK, Ketch TP, et al. (2002) Genetic basis of clinical catecholamine disorders. *Annals of the New York Academy of Sciences* 971: 506–514.

Goldstein DS, Eisenhofer G, and Kopin IJ (2003) Sources and significance of plasma levels of catechols and their metabolites in humans. *Journal of Pharmacology and Experimental Therapeutics* 305: 800–811.

Hagen K, Pettersen E, Stovner LJ, et al. (2007) High systolic blood pressure is associated with Val/Val genotype in the catechol-O-methyltransferase gene. *American Journal of Hypertension* 20: 21–26.

Hahn MK, Mazei-Robison MS, and Blakely RD (2005) Single nucleotide polymorphisms in the human norepinephrine transporter gene affect expression, trafficking, antidepressant interaction, and protein kinase C regulation. *Molecular Pharmacology* 68: 457–466.

Keller NR and Robertson D (2006) Familial orthostatic tachycardia. *Current Opinion in Cardiology* 21: 173–179.

Mandela P and Ordway GA (2006) The norepinephrine transporter and its regulation. *Journal of Neurochemistry* 97: 310–333.

Oeltmann T, Carson R, Shannon JR, et al. (2004) Assessment of O-methylated catecholamine levels in plasma and urine for diagnosis of autonomic disorders. *Autonomic Neuroscience* 116: 1–10.

Palmatier MA, Kang AM, and Kidd KK (1999) Global variation in the frequencies of functionally different catechol-O-methyltransferase alleles. *Biological Psychiatry* 46: 557–567.

Robertson D, Biaggioni I, Burnstock G, et al. (eds.) (2004) *Primer on the Autonomic Nervous System*, 2nd edn., San Diego: Elsevier Academic Press.

Robertson D, Goldberg MR, Onrot J, et al. (1986) Isolated failure of autonomic noradrenergic neurotransmission: Evidence for impaired beta-hydroxylation of dopamine. *New England Journal of Medicine* 314: 1494–1497.

Sabol SZ, Hu S, and Hamer D (1998) A functional polymorphism in the monoamine oxidase A gene promoter. *Human Genetics* 103: 273–279.

Shannon JR, Flattem NL, Jordan J, et al. (2000) Orthostatic intolerance and tachycardia associated with norepinephrine-transporter deficiency. *New England Journal of Medicine* 342: 541–549.

Shih JC, Chen K, and Ridd MJ (1999) Monoamine oxidase: From genes to behavior. *Annual Review of Neuroscience* 22: 197–217.

Vincent S, Bieck PR, Garland EM, et al. (2004) Clinical assessment of norepinephrine transporter blockade through biochemical and pharmacological profiles. *Circulation* 109: 3202–3207.

Zahniser NR and Doolen S (2001) Chronic and acute regulation of Na^+/Cl^--dependent neurotransmitter transporters: Drugs, substrates, presynaptic receptors, and signaling systems. *Pharmacology and Therapeutics* 92: 21–55.

Zhu BT (2002) Catechol-O-methyltransferase(COMT)-mediated methylation metabolism of endogenous bioactive catechols and modulation by endobiotics and xenobiotics: Importance in pathophysiologoy and pathogenesis. *Current Drug Metabolism* 3: 321–349.

Relevant Websites

http://www.genetests.org – Medical genetics information resource (funded by the National Institutes of Health).

http://www.bh4.org – Tetrahydrobiopterin deficiencies research (Zürich Research Group).

Autonomic Nervous System: Metabolic Function

N E Straznicky, P J Nestel, and M D Esler, Baker Heart Research Institute, Melbourne, VIC, Australia

Introduction

The autonomic nervous system is a key neuromodulator of cardiovascular, metabolic, and other physiological functions in humans. As such, it provides an important vehicle by which the central nervous system (CNS) maintains homeostasis during periods of acute and chronic changes in the physiological state. Dysfunctions in autonomic regulatory systems may lead to disease. Abdominal obesity, characterized by the accumulation of visceral adipose tissue, is commonly associated with a constellation of risk factors for cardiovascular disease and type 2 diabetes, termed the 'metabolic syndrome.' Evidence shows that sympathetic nervous system (SNS) overactivity plays at least an important, if not pivotal, role in the pathogenesis and complications of the metabolic syndrome. The hyperadrenergic state of the metabolic syndrome may be mechanistically complex because of the involvement of metabolic, central, and reflex factors. A better understanding of the causes and consequences of autonomic dysfunction in the metabolic syndrome may facilitate efforts to prevent and treat this condition. This article provides a brief overview of the anatomical and physiological aspects of the autonomic nervous system as a prelude to considering the evidence showing a role for autonomic dysfunction in the metabolic syndrome.

Organization of the Autonomic Nervous System

There are two main effector or efferent limbs of the autonomic nervous system. (1) Sympathetic innervation is derived from preganglionic fibers whose cell bodies lie in the intermediolateral column of the spinal cord at the level of the thoracic and upper lumbar roots. These nerves synapse with postganglionic sympathetic neurons in the bilateral chain of sympathetic ganglia located close to the spinal cord. Postganglionic sympathetic nerves innervate almost all of the vital organs and tissues of the body by release of neurotransmitters from their varicosities. (2) The parasympathetic nervous system consists of preganglionic fibers that originate in the CNS, namely the midbrain, the medulla oblongata and sacral part of the spinal cord, and their postganglionic connections. The parasympathetic system has terminal ganglia very near or within the organs innervated. The neurotransmitter of all preganglionic autonomic fibers, all postganglionic parasympathetic fibers, and a few postganglionic sympathetic fibers is acetylcholine. Adrenergic fibers comprise the majority of postganglionic sympathetic fibers and noradrenaline is the primary neurotransmitter. The adrenal medulla can be considered as a modified sympathetic ganglion. It is cholinergically innervated and releases catecholamines (primarily epinephrine) directly into the circulation instead of having a postganglionic nerve. The effects of norepinephrine and epinephrine are the result of activation of adrenergic receptors and subtypes thereof (α_1, α_2, β_1, β_2, and β_3). There are two distinct cholinergic receptors, termed nicotinic and muscarinic.

Peripheral autonomic activation usually occurs as part of reflex mechanisms, which involve afferent signals and processing in the CNS. Information on the status of visceral organs is transmitted to the CNS through two main sensory systems: the cranial nerve (parasympathetic) visceral sensory system and the spinal (sympathetic) visceral afferent system. In addition, diverse hormonal signals supply information about the peripheral metabolic state to the brain, for example, insulin from the pancreas and leptin from adipose tissue. The hypothalamus is regarded as the principal CNS site for the integration of autonomic function important in the control of metabolism. These integrated patterns of response involve autonomic, endocrine, and behavioral components (**Table 1**). The lateral hypothalamic area and ventromedial hypothalamic nucleus (VMH) are recognized as the parasympathetic and sympathetic centers respectively, and are functionally important in appetite regulation, glucose, and lipid metabolism. The paraventricular nucleus regulates the outflow of sympathetic and parasympathetic innervation of visceral organs, including the liver, pancreas, and the adrenal glands.

Organs of Metabolism, Their Innervation, and Key Metabolic Processes

The autonomic nervous system regulates metabolic processes through both direct neural effects (**Table 2**) and hormonal effects. The effects of the SNS on glucose and lipid metabolism are mediated by circulating glucagon, epinephrine, and the direct sympathetic innervation of the liver, adipose tissues, and skeletal muscle. The effects of the parasympathetic nervous system on glucose and lipid metabolism are mediated by insulin and the direct parasympathetic innervation of the liver. In general, sympathetic stimulation

Table 1 Hypothalamic areas and function

Area		*Nucleus*	*Function*
Lateral			Sending parasympathetic signals to the body via autonomic nerves; anabolic response: glycogen synthesis, lipogenesis; decreased appetite.
Medial	VMH	Ventromedial hypothalamic nucleus	Sending sympathetic signals to the body via autonomic nerves; catabolic response: gluconeogenesis, lipolysis; increased appetite.
Paraventricular	PVN	Paraventricular hypothalamic nucleus	Receiving information from SCN and ARN; sending information to autonomic nerves and pituitary gland.
	SCN	Suprachiasmatic nucleus	Circadian clock of the brain; receiving afferent input from retina; sending some signals to the PVN.
	ARN	Arcuate nucleus	Sending signals to the PVN and other nuclei in the hypothalamus; sensitive to insulin, leptin and glucose in the blood; playing a key role in energy metabolism.

Adapted from Uyama N, Geerts A, and Reynaert H (2004) Neural connections between the hypothalamus and the liver. *The anatomical Record A* 280: 808–820.

Table 2 Responses of metabolic organs to autonomic nerve impulses

Organ system	*Sympathetic effects*	*Adrenergic receptor type*	*Parasympathetic effects*	*Cholinergic receptor type*
Liver	Glycogenolysis and gluconeogenesis (+++)	α_1, β_2	↑ Glucogenesis ↑ HISS release	M
WAT	↑ Glucose uptake	α_1	↑ Glucose and FFA uptake	
	Lipolysis (+++)	β_1, β_2, β_3		
	Inhibition of lipolysis	α_2	↓ HSL activity ↑ Release of leptin and resistin	
Skeletal muscle	Vasoconstriction	α_1		
	Vasodilation (++)[a] and ↑ glucose uptake	β_2		
	↑ Contractility	β_2		
	Glycogenolysis	β_2		
	↑ HSL activity			
Pancreas				
α cells	↑ Glucagon secretion	β_2		
β cells	↓ Insulin secretion (+++)	α_2		
	↑ Insulin secretion (+)	β_2		
Adrenal medulla	Secretion of epinephrine and norepinephrine			N; M (secondarily)

[a]β_2 vasodilation in skeletal muscle is mediated by epinephrine. Cholinergic receptors, nicotinic (N) and muscarinic (M); HISS, hepatic insulin sensitizing substance; HSL, hormone sensitive lipase; WAT, white adipose tissue.
Responses are designated + to +++ to provide an approximate indication of the importance of sympathetic and parasympathetic nerve activity.
From Flechtner-Mors M, Jonkinson CP, Alt A, Biesalski HK, Adler G, and Ditschuneit HH (2004) Sympathetic regulation of glucose uptake by the alpha1-adrenoreceptor in human obesity. *Obesity Research* 12: 612–620; and Romjin JA and Fliers E (2005) Sympathetic and parasympathetic innervation of adipose tissue: Metabolic implications. *Current Opinion in Clinical Nutrition and Metabolic Care* 8: 440–444.

produces catabolic effects while parasympathetic stimulation produces anabolic effects.

Liver

More than any other organ, the liver contributes to the maintenance of metabolic equilibrium, including glucose and lipid homeostasis. The liver is innervated by the splanchnic sympathetic nerves and vagal parasympathetic nerves. Sympathetic stimulation promotes hepatic glucose production by activating glycogenolysis in fed states and gluconeogenesis in fasted states. Glucose output is increased through rapid activation of the key glycogenolytic enzyme, glycogen phosphorylase, with a concomitant decrease in liver glycogen content. In addition, sympathetic stimulation increases the activity of the gluconeogenic enzyme phosphoenol pyruvate carboxykinase. Parasympathetic stimulation leads to hepatic glycogen synthesis through activation of the enzyme glycogen synthase and an inhibition of glucagon-induced glucose output. Recent research suggests that hepatic parasympathetic nerves may also play a permissive regulatory role in the release of hepatic insulin sensitizing substance, a liver-derived factor that increases glucose deposition in skeletal muscle.

In addition to direct control, autonomic nerves can indirectly influence liver function via the hypothalamus-pancreas and hypothalamus-adrenal axes. Sympathetic outflow from the hypothalamus stimulates the release of catecholamines from the adrenal medulla and the release of glucagon from the pancreatic alpha cells. Parasympathetic outflow stimulates the release of insulin from pancreatic beta cells.

Adipose Tissue

White adipose tissue is the principal tissue for energy storage in humans. When energy needs cannot be met by circulating fuels or stored carbohydrate, lipid is mobilized through the process of lipolysis, a breakdown of triglycerides into glycerol and free fatty acids, to provide fuel to other organs and to deliver substrates to the liver for gluconeogenesis (glycerol) and lipoprotein synthesis (free fatty acids). Adipose tissue lipid metabolism is highly regulated by hormones (insulin and catecholamines) and other factors such as nutritional status (feeding, fasting) and exercise. Sympathetic nerve fibers innervate white adipose tissue and stimulate lipolysis. The catecholamines noradrenaline and epinephrine activate lipolysis via β_1-, β_2-, and β_3-adrenoceptors, initiating a chain of events that activates hormone sensitive lipase, the principal intracellular enzyme responsible for the hydrolysis of triglycerides. A counter-regulatory system causing inhibition of lipolysis is mediated via α_2-adrenoceptors. The lipolytic activity of catecholamines is highest in the visceral fat depot, followed by abdominal subcutaneous fat, and lowest activity occurs in peripheral subcutaneous fat tissue. The SNS innervation of white adipose tissue also regulates fat cell number, as noradrenaline inhibits fat cell proliferation *in vitro*. Insulin is the most potent antilipolytic hormone in adipose tissue. It regulates the anabolic actions of fat cells, by stimulating glucose uptake, free fatty acid uptake (via action of lipoprotein lipase (LPL) on circulating triglycerides), inhibiting lipolysis, reesterifying fatty acids to triglyceride, and possibly by stimulating *de novo* fatty acid synthesis (lipogenesis). The SNS also has stimulatory effects on glucose uptake in adipose tissue, mediated by β- and α_1-adrenoceptors.

LPL is a rate-limiting enzyme, located on the endothelial surface of adipose tissue and skeletal muscle that hydrolyzes the triglyceride component of circulating lipoproteins. Studies in humans indicate that the response of LPL to catecholamines is tissue specific, with no effect or an inhibition in adipose tissue, but stimulation in skeletal muscle.

Recent neuroanatomical studies in rats have demonstrated parasympathetic innervation of white adipose tissue. Parasympathetic input was found to increase adipose tissue insulin sensitivity and modulate endocrine function, namely the secretion of the adipokines leptin and resistin. Interestingly, there is differential innervation of visceral and subcutaneous fat depots by both sympathetic and parasympathetic neurons.

Skeletal Muscle

The complex action of catecholamines is evident in skeletal muscle. Here the hormones have catabolic effects, which lead to mobilization of energy through the breakdown of glycogen (glycogenolysis) and lipids (lipolysis) via β_2-adrenoceptors. Catecholamines also increase muscle contraction and blood flow. β-2 Vasodilation in skeletal muscle is mediated by epinephrine. Skeletal muscle is the principal site of insulinstimulated glucose uptake. There is also evidence of noninsulin-mediated pathways of glucose utilization, activated by increased sympathetic activity. Electrical stimulation of the VMH is accompanied by increased glucose uptake independent of insulin.

Adrenal Medulla

Epinephrine, released from the adrenal medulla, can also directly affect lipid, protein, and carbohydrate metabolism. During physiological (e.g., heavy exercise) and stressful conditions (e.g., hypoglycemia, cold), epinephrine can affect hepatic glucose production by increasing glycogen degradation. It can also stimulate lipolysis of adipose tissue and proteolysis of muscle. Epinephrine affects gluconeogenesis indirectly by increasing the availability of substrates such as glycerol.

Autonomic Regulation of Energy Balance

Energy balance is maintained via a homeostatic system involving both the brain and the periphery. A key component of this system is the hypothalamus. Hormonal signals reflecting the availability and demand for metabolic fuel are relayed via neurons in the arcuate nucleus of the hypothalamus. One neuron group expresses neuropeptide Y and another proopiomelanocortin. Activation of the former leads to increased food intake and decreased energy expenditure, while the opposite effect is elicited by activation of the latter. A large number of circulating hormones and metabolites modulate neuronal function. Under stable conditions, equilibrium exists between orexigenic peptides that stimulate feeding behavior (e.g., ghrelin) and anorectic peptides (e.g., leptin, insulin) that function to restrict excessive body fat accrual by limiting food intake and upregulating SNS activity. The second main input for information relating to energy balance is the brain stem. Vagal afferents that are sensitive to gastrointestinal signals synapse onto and excite neurons in the nucleus tractus solitarus, causing satiation.

Linked to the effects on appetite are thermogenic effects that dissipate heat. Components of human energy expenditure include (1) the obligatory energy expended on basic cellular and physiological functions that require adenosine triphosphate (ATP) (resting metabolic rate); (2) the energy expended in physical activity (exercise and spontaneous physical activity); and (3) adaptive thermogenesis, or that component that is stimulated in response to external stresses such as cold or changes in the amount of food ingested (thermic effect of food). Since the demonstration that sympathetic activity is increased during overfeeding and decreased during starvation, the SNS has been recognized as a pivotal efferent system in the link between diet and thermogenesis. Adaptive thermogenesis requires an uncoupling between substrate oxidation and biological work (phosphorylation). It is mediated in brown adipocytes by a mitochondrial uncoupling protein (UCP-1), which allows protons to leak back across the inner mitochondrial membrane, with a resultant increase in heat production. In small rodents, noradrenaline acts via β-adrenoceptors to activate and increase expression of UCP-1. Although a number of human studies have linked polymorphisms of UCP-1 homologs (UCP-2 and UCP-3) with obesity or low rates of energy expenditure and/or fat oxidation, there is considerable uncertainty about their physiological roles and function in adaptive thermogenesis.

The Metabolic Syndrome

The metabolic syndrome consists of a constellation of factors that raise the risk of cardiovascular disease and type 2 diabetes, namely, visceral obesity, insulin resistance, glucose intolerance, dyslipidemia, hypertension, and a pro-inflammatory state. The global obesity epidemic has been associated with a striking increase in the prevalence of the metabolic syndrome. The age-adjusted prevalence in the US population is estimated at 24% and increases to 44% in adults aged $\geq$60 years. Three competing hypotheses have been advanced to explain the primary etiology of this complex syndrome. In all of these, sympathetic overactivity plays a fundamental role:

1. Insulin resistance, at a cellular or receptor level leads to hyperinsulinemia, sympathetic activation, and hypertension. Obesity, dyslipidemia, and impaired glucose tolerance are components of the syndrome.
2. Obesity, particularly visceral obesity, is the primary problem leading to secondary insulin resistance, sympathetic activation, and attendant features.
3. Increased sympathetic activity of central origin is the primary defect leading to insulin resistance and other metabolic and cardiovascular sequelae.

Evidence for Sympathetic Overactivity in Obesity and the Metabolic Syndrome

Direct intraneural recording of sympathetic nerve traffic by microneurography and the measurement of systemic and regional noradrenaline spillover using tracer methods are currently preferred techniques for the assessment of SNS activity in humans. Microneurography records postganglionic sympathetic nerve action potentials targeted at skeletal muscle vasculature (muscle sympathetic nervous system activity, MSNA). Multifiber recordings of 'bursts' of nerve activity synchronous with the heart beat, and more recently single fiber traces are generated. The use of arterial blood samples in the spillover technique allows an assessment of whole-body SNS activity, while regional venous sampling allows SNS activity to be determined in a variety of organs.

Application of these sensitive newer methods has yielded substantial evidence that human obesity is characterized by increased SNS activity. The magnitude of this increase is influenced by coexistent risk factors and body fat distribution. Furthermore, there is evidence to show that regional SNS activity is heterogeneous in the obese state. In normotensive obese subjects, sympathetic outflow to the kidneys and skeletal muscle vasculature are increased by up to twofold, outflow to skin and the hepatomesenteric circulation are unchanged, while the sympathetic outflow to the heart is reduced, with cardiac noradrenaline spillover being only about 50% of that found in age-matched lean controls. Additional augmentation in MSNA has been demonstrated in obese patients with hypertension or obstructive sleep apnea (OSA). Among the adiposity indexes, abdominal visceral fat is most strongly associated with sympathetic overactivity. MSNA is greater in individuals with central obesity than in those with peripheral obesity and significant positive associations between waist circumference, MSNA, and whole-body noradrenaline spillover have been reported.

Findings in metabolic syndrome subjects mirror those in obesity: higher 24 h urinary noradrenaline excretion, elevated whole-body noradrenaline spillover, and increased MSNA (including single unit discharges), suggesting the occurrence of sympathetic overactivity.

Power spectral analysis of variations in heart rate is based on sophisticated mathematical models to determine the balance between sympathetic and vagal nerve activities in the heart. Low heart rate variability is a risk factor for coronary heart disease. These indirect measures of cardiac autonomic tone show that obesity and the metabolic syndrome are associated with reduced heart rate variability and a relative sympathetic predominance. One criticism of this technique, however, is that it seems to have poor correlation with measurements of noradrenaline spillover from the heart.

Potential Mechanisms Contributing to Sympathetic Activation

Several factors have been implicated in the sympathetic activation that occurs in obesity. Since obesity is a heterogeneous condition, the relative contribution of these factors may vary based upon other modifying biological factors, both genetic and environmental.

Hyperinsulinemia Insulin resistance, defined as a decreased ability of target tissues to increase glucose uptake in response to insulin, is accompanied by compensatory hyperinsulinemia. Animal, human, and epidemiological (e.g., The Normative Ageing Study) studies indicate that sympathetic overactivity and hyperinsulinemia are closely interrelated. According to the Landsberg hypothesis, the insulin resistance of obesity is a mechanism evolved to limit further fuel storage as muscle glycogen and adipose tissue triglyceride, and to increase metabolic rate, thereby restoring energy balance. Thus, hyperinsulinemia-mediated sympathetic nervous activation was viewed as an adaptive response to limit further weight gain. In support of this hypothesis, insulin infusion under euglycemic conditions has been demonstrated to enhance MSNA in healthy subjects and borderline hypertensives. However, there is some doubt as to whether the obese retain their sensitivity to the stimulatory effect of insulin on the SNS. Vollenweider observed a specific impairment of sympathetic neural responsiveness (MSNA) to physiological hyperinsulinemia in obese subjects. Moreover, euglycemic insulin infusion does not activate renal sympathetic nerves in obese insulin-resistant humans and fasting insulin and renal noradrenaline spillover are not correlated. An alternate explanation for the association of hyperinsulinemia and sympathetic activity is that sympathetic activation is the primary etiopathogenetic factor, with insulin resistance being a secondary phenomenon. This notion is discussed below.

Downregulation of β-Adrenergic receptors Several studies have indicated that the response of the SNS to various physiological (under- and over-eating, cold exposure) and pharmacological stimuli is blunted in the obese. Reduced *in vivo* lipolytic response to intraneural electrical stimulation and reduced β_2-adrenoceptor numbers in isolated subcutaneous adipocytes have also been reported in obese compared with lean controls. It is currently unclear whether adrenergic receptor defects lead to SNS overactivity or whether enhanced sympathetic tone downregulates β-adrenergic receptors, which would promote weight gain and obesity. A polymorphism at Arg16Gly of the β_2-adrenoceptor gene has been linked to SNS overactivity. In other studies, the Gly16 allele of Arg16Gly has been associated with blunted vasodilatory responses to β_2-agonist. Thus, it could be speculated that β_2-adrenoceptor polymorphisms might contribute to blunted β_2-adrenoceptor function and increased sympathetic activity.

Baroreflex impairment Impaired baroreflex modulation of heart rate and MSNA at rest, and following vasoactive stimuli, have been consistently demonstrated in obese and metabolic syndrome populations. Potential mechanisms that may influence baroreflex function in this clinical setting include reduced arterial distensibility, altered central integration of afferent vagal nerve traffic, or impaired cardiac muscarinic receptor function. However, since individuals with peripheral and central obesity have a similar degree of baroreflex impairment, the greater sympathetic activation of individuals with a visceral fat depot cannot be explained by this reflex mechanism.

Leptin Leptin is a 167-amino-acid protein hormone produced primarily by adipose tissue that acts on specific receptors in the hypothalamus to decrease appetite and increase energy expenditure. Based on findings in rodents, which showed that leptin administration increases catecholamine turnover in brown adipose tissue, kidneys, and hindlimb vasculature, it has been speculated that the sympathetic activation of obesity may be mediated in part by increased plasma leptin levels. Obesity is known to be associated with circulating hyperleptinemia, reflecting a high fat mass and partial resistance to leptin. It has been postulated that leptin resistance in obesity can be selective, with preservation of sympathetic responsiveness despite resistance to satiety and the metabolic actions of leptin. To date there are no definitive descriptions of the sympathetic actions of leptin in humans. We have previously reported a strong positive correlation between renal noradrenaline spillover and plasma leptin in men of widely differing adiposity. In a subsequent study in overweight and obese metabolic syndrome subjects, plasma leptin was significantly associated with whole-body noradrenaline spillover, but not MSNA (**Figure 1(a)**). These results may reflect differences in regional sympathetic responses to leptin, in concordance with animal studies. Furthermore, animal studies indicate that regional sympathoactivation may be mediated by different neuropeptide pathways: with melanocortins mediating renal sympathoactivation and corticotrophin-releasing hormone mediating brown adipose tissue SNS activation.

Obstructive sleep apnea OSA is common in the obese and has been increasingly implicated in the progression of cardiovascular disease and the increase in blood pressure that results from heightened sympathetic drive. OSA is associated with a marked increase in MSNA not only in the obese but also in lean individuals. It is believed that this increase originates from the stimulation of peripheral chemoreceptors brought about by repetitive nocturnal hypoxemia. Nasal continuous positive airway pressure is the treatment of choice and has been shown to effectively decrease apneic episodes, improve insulin sensitivity, and increase noradrenaline clearance from the circulation, but was without effect on noradrenaline release.

Nonesterified fatty acid Abdominal obesity is linked to increased nonesterified fatty acid (NEFA) concentration and turnover, which are resistant to suppression by insulin. NEFA contributes to the metabolic aspects of the metabolic syndrome by impairing insulin-mediated glucose disposal. Based on epidemiological data that showed an association between NEFA levels and blood pressure (e.g., the Paris Prospective Study), it has been hypothesized that increased delivery of NEFA to the portal circulation could trigger a reflex increase in sympathetic tone. Studies in lean and obese human subjects have demonstrated that acute elevations in NEFA result in decreased baroreceptor sensitivity and an increase in α_1-adrenoceptor-mediated pressor sensitivity. However, in a recent study in healthy adults, whole-body and renal noradrenaline spillover did not change during fatty acid infusion, despite a threefold rise in NEFA.

Angiotensinogen A bidirectional interaction between the SNS and renin–angiotensin system, both of which are activated in the obese state, has been long appreciated. The expression of the angiotensin II precursor angiotensinogen is increased in visceral adipose tissue. Previous work in animals suggests that angiotensin II can trigger peripheral and central sympathoexcitatory effects. Evidence that angiotensin II activates the SNS in humans is less strong. A 3-month treatment with the angiotensin II receptor blocker candesartan in obese hypertensives was accompanied by a 21% reduction in MSNA. This was attributed to the concomitant improvement in insulin sensitivity rather than a direct sympathomodulatory influence, because in lean insulin sensitive

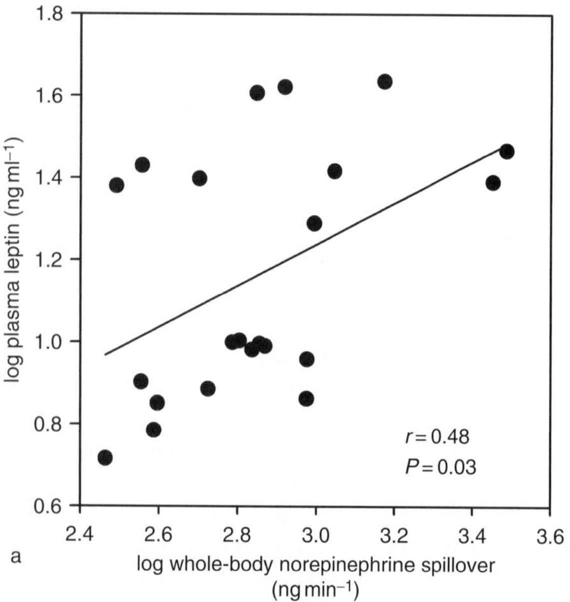

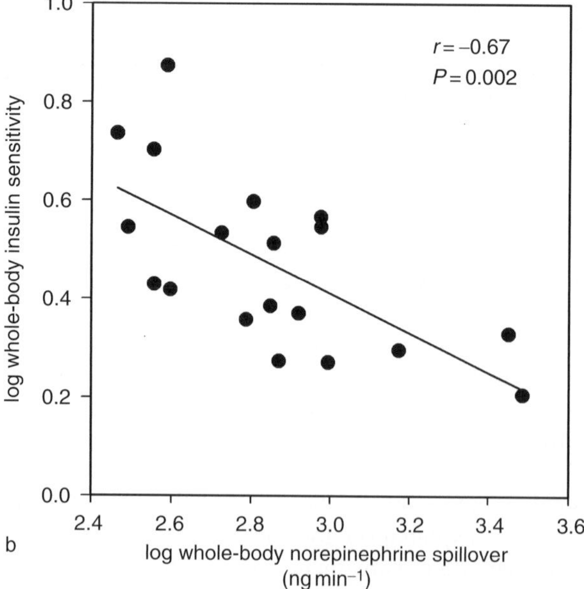

Figure 1 Univariate correlates of whole-body plasma noradrenaline spillover rate with fasting plasma leptin (a) and whole-body insulin sensitivity (b) in obese metabolic syndrome subjects. From Straznicky NE, Lambert EA, Lambert GW, et al. (2005) Effects of dietary weight loss on sympathetic activity and cardiovascular risk factors associated with the metabolic syndrome. *Journal of Clinical Endocrinology and Metabolism* 90: 5998–6005.

hypertensive subjects, angiotensin II receptor blockade did not result in sympathoinhibition.

Metabolic and Cardiovascular Consequences of Sympathetic Overactivity

Insulin resistance/glucose intolerance The notion that sympathetic overactivity precedes the development of insulin resistance is supported by longitudinal data in Japanese men. Masuo et al. reported that elevated plasma noradrenaline levels at baseline predicted future higher blood pressure readings, gain of weight, and higher insulin values in men followed for 10 years. The possible primacy of autonomic dysfunction is also supported by epidemiological data. Healthy participants in The Atherosclerosis Risk in Communities Study with reduced low-frequency power heart rate variability (a marker of decreased parasympathetic input) were at increased risk for developing type 2 diabetes.

Alterations in skeletal muscle blood flow, and hence the rate of delivery of both glucose and insulin, may be one mechanism by which sympathetic activation may modulate insulin sensitivity. Insulin sensitivity is also related to the number of slow-twitch muscle fibers (type 1) and capillary density. Experimental maneuvers that acutely stimulate endogenous noradrenaline release (e.g., lower body negative pressure and bilateral thigh cuff inflation) are accompanied by a concomitant reduction in insulin-mediated glucose uptake in skeletal muscle. These effects have been attributed to a reduction in skeletal muscle blood flow (via α-receptors) and/or a direct metabolic effect of noradrenaline. Conversely, mental stress which increases skeletal muscle blood flow (via β-receptors) improves glucose uptake during euglycemic hyperinsulinemia. Increased sympathetic drive also induces loss of capillary density and facilitates the shunting of glucose and insulin to less metabolically active skeletal muscle beds.

A further mechanism by which chronic sympathetic stimulation may cause insulin resistance is via the promotion of lipolysis. An increase in plasma-free fatty acid concentration can produce insulin resistance of glucose utilization by skeletal muscle and inhibition of the normal suppression by insulin of hepatic glucose production. Together, this could lead to glucose intolerance and insulin resistance. We have previously demonstrated a strong inverse relationship between whole-body plasma noradrenaline spillover rate and insulin sensitivity in obese metabolic syndrome subjects (**Figure 1(b)**).

Dyslipidemia The atherogenic dyslipidemia of metabolic syndrome obesity is characterized by increased liver secretion of very low-density lipoprotein (VLDL), high fasting and postprandial triglyceride levels, low high-density lipoprotein (HDL), and small dense low-density lipoprotein particles. Multiple factors contribute to the observed alterations in lipoprotein metabolism, including excess adiposity, insulin resistance, adipose tissue cytokines, and sympathetic overactivity. A major determinant of increased VLDL secretion in the metabolic syndrome is higher hepatic triglyceride content, derived in part from increased free fatty acid delivery from adipose tissue, and return of triglyceride rich lipoprotein remnants to the liver. Kinetic studies have shown that low HDL concentration is the consequence of increased catabolism, possibly secondary to triglyceride enrichment of the particles, and increased hepatic lipase activity.

Hypertension Elevated arterial blood pressure is common in metabolic syndrome subjects. Contributing mechanisms include (1) increased SNS activity, (2) impaired endothelial function with impaired vasodilation, and (3) increased renal sodium reabsorption, which would increase plasma volume and thus cardiac output.

Altered cytokine production There is some evidence to suggest that the SNS may regulate adipocytokine production. β-Adrenergic agonist induced secretion of interleukin-6 has been demonstrated *in vitro* in murine cultured adipocytes and *in vivo* in humans following isoproterenol infusion. Similarly, epinephrine, noradrenaline, and isoproterenol all induced a concentration-dependent increase in the production of tumor necrosis factor alpha in human mononuclear cells. Conversely, sympathetic antagonism with β-blocker treatment decreases the level of both of these cytokines. Adiponectin is another adipocytokine that is associated negatively with components of the metabolic syndrome, including obesity, insulin resistance, and dyslipidemia. In contrast to the aforementioned cytokines, β-adrenergic agonists reduce adiponectin mRNA in adipose tissue and adiponectin concentration in plasma.

Clinical Management of the Metabolic Syndrome

The underlying risk factors that promote development of the metabolic syndrome are overweight and obesity, physical inactivity, and an atherogenic diet. All current guidelines on the management of the metabolic syndrome emphasize that lifestyle modification (weight loss and physical activity) are first-line therapy. Several large clinical trials such as the Diabetes Prevention Program have found that lifestyle modification is the most efficacious strategy to prevent progression to type 2 diabetes. In this study intense lifestyle changes were accompanied by a 58% risk reduction. Both weight loss and exercise have been documented to exert marked sympathoinhibitory effects and improve metabolic syndrome components. Drug treatment for associated risk factors may be indicated for some patients.

Weight Loss

We have recently reported that moderate weight loss (7% of body weight) is accompanied by a 43% reduction in whole-body noradrenaline spillover and an improvement in all metabolic syndrome components. It is likely that both the decrease in body fat and caloric restriction (negative energy balance) contributed to these changes. It is of note that only subjects who were insulin resistant at baseline observed a reduction in noradrenaline spillover after weight loss, despite the fact that both insulin sensitive and insulin resistant groups lost similar amounts of weight (**Figure 2**). This suggests that insulin resistance is a key driver of sympathetic activity in the metabolic syndrome.

Exercise

Regular aerobic exercise enhances caloric expenditure and is thought to have a synergistic effect on weight loss during a hypocaloric diet in obese individuals. In addition, exercise training provokes beneficial cardiovascular and autonomic adaptations, namely a reduction in blood pressure and sympathetic activity (preferentially involving renal sympathetic outflow) and improvements in insulin sensitivity and lipoprotein profile.

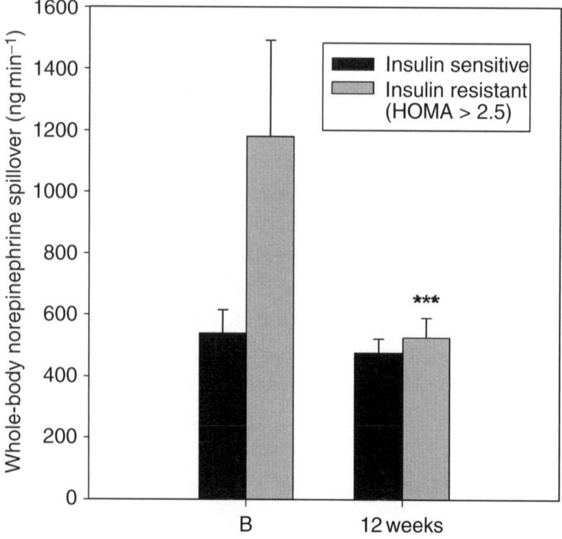

Figure 2 Whole-body noradrenaline spillover rate at baseline (B) and after a 12-week hypocaloric diet, in insulin-resistant and insulin-sensitive obese metabolic syndrome subjects. Insulin-resistant ($n=10$) and insulin-sensitive ($n=9$) subjects were defined as HOMA index ≥ 2.5 and <2.5, respectively. Weight loss was similar in the two groups ($6.4+1.1$ vs. $6.6+1.0$ kg). Data are shown as the mean + SEM. $^{***}P<0.001$ versus baseline values. From Straznicky NE, Lambert EA, Lambert GW, et al. (2005) Effects of dietary weight loss on sympathetic activity and cardiovascular risk factors associated with the metabolic syndrome. *Journal of Clinical Endocrinology and Metabolism* 90: 5998–6005.

Drug Treatment

Modulating sympathetic overactivity is a worthwhile therapeutic goal not only in the management of hypertension itself but also in the cluster of metabolic abnormalities with which it is associated. In choosing antihypertensive drug treatment, potential metabolic side effects need to be considered. Vasoconstricting β-blockers and diuretics can worsen insulin resistance and atherogenic dyslipidemia. β-Blockers also inhibit gluconeogenesis and lipolysis and are typically associated with some weight gain. Antihypertensive drugs for which concomitant beneficial effects on glucose metabolism and SNS activity have been demonstrated include the angiotensin II antagonist candesartan, and imidazoline-I_1-receptors agonists such as moxonidine and rilmenidine, which reduce central sympathetic drive. The latter class of drugs improve insulin-mediated glucose disposal, decrease plasma triglycerides, and do not increase body weight in obese patients with hypertension.

Insulin-sensitizing drugs represent a second-line therapeutic option for metabolic syndrome subjects in whom lifestyle interventions have been unsuccessful. Thiazolidendiones activate nuclear peroxisome proliferator-activated receptor-gamma and improve insulin sensitivity at key metabolic sites (skeletal muscle, adipose tissue, and liver). They also reduce glucose production and have a beneficial effect on the dyslipidemia associated with the metabolic syndrome. Furthermore, antihypertensive and sympathoinhibitory effects have been demonstrated after long-term treatment. Improvement in insulin sensitivity and reduced leptin expression are two potentially contributing mechanisms.

See also: Autonomic Disorders; Autonomic Failure; Autonomic Dysfunction: Drug-Induced; Autonomic Nervous System; Neuropathy: Metabolically-Induced; Sympathetic Nervous System.

Further Reading

Broberger C (2005) Brain regulation of food intake and appetite: Molecules and networks. *Journal of Internal Medicine* 258: 301–327.

Brunton LL, Lazo JS, and Parker K (eds.) (2006) *Goodman and Gilman's Pharmacological Basis of Therapeutics*, 11th edn. New York: McGraw-Hill.

Dulloo AG, Seydoux J, and Jacquet J (2004) Adaptive thermogenesis and uncoupling proteins: A reappraisal of their roles in fat metabolism and energy balance. *Physiology and Behaviour* 83: 587–602.

Eikelis N, Schlaich M, Aggarwal A, Kaye D, and Esler M (2003) Interactions between leptin and the human sympathetic nervous system. *Hypertension* 41: 1072–1079.

Esler M, Straznicky N, Eikelis N, et al. (2006) Mechanisms of sympathetic activation in obesity-related hypertension. *Hypertension* 48: 787–796.

Flechtner-Mors M, Jenkinson CP, Alt A, Biesalski HK, Adler G, and Ditschuneit HH (2004) Sympathetic regulation of glucose uptake by the alpha1-adrenoceptor in human obesity. *Obesity Research* 12: 612–620.

Grundy SM, Hansen B, Smith SC, Cleeman JI, and Kahn RA (2004) Clinical management of the metabolic syndrome. *Circulation* 109: 551–556.

Landsberg L (1986) Diet, obesity and hypertension: An hypothesis involving insulin, the sympathetic nervous system and adaptive thermogenesis. *Quebec Journal of Medicine* 236: 1081–1090.

Masuo K, Katsuya T, Fu Y, et al. (2005) β2- and β3-Adrenergic receptor polymorphisms are related to the onset of weight gain and blood pressure elevation over 5 years. *Circulation* 111: 3429–3434.

Masuo K, Mikami H, Ogihara T, and Tuck ML (1997) Sympathetic nerve hyperactivity precedes hyperinsulinaemia in young, nonobese Japanese population. *American Journal of Hypertension* 10: 77–83.

Nestel P (2003) Metabolic syndrome: Multiple candidate genes, multiple environmental factors – multiple syndromes? *International Journal of Clinical Practice* 134(Supplement): 3–9.

Nonogaki K (2000) New insights into sympathetic regulation of glucose and fat metabolism. *Diabetologia* 43: 533–549.

Puschel GP (2004) Control of hepatocyte Metabolism by sympathetic and parasympathetic hepatic nerves. *The Anatomical Record A* 280: 854–867.

Romjin JA and Fliers E (2005) Sympathetic and parasympathetic innervation of adipose tissue: Metabolic implications. *Current Opinion in Clinical Nutrition and Metabolic Care* 8: 440–444.

Straznicky NE, Lambert EA, Lambert GW, et al. (2005) Effects of dietary weight loss on sympathetic activity and cardiovascular risk factors associated with the metabolic syndrome. *Journal of Clinical Endocrinology and Metabolism* 90: 5998–6005.

Uyama N, Geerts A, and Reynaert H (2004) Neural connections between the hypothalamus and the liver. *The Anatomical Record A* 280: 808–820.

Vollenweider P, Randin D, Tappy L, et al. (1994) Impaired insulin-induced sympathetic neural activation and vasodilation in skeletal muscle in obese humans. *Journal of Clinical Investigation* 93: 2365–2371.

Autonomic Nervous System: Neuroanatomy

G Gabella, University College London, London, UK

Introduction

While the body moves around pursuing its behavior and interacting with the environment by motor and sensory mechanisms, its internal organs, such as the heart, stomach, glands, and bronchi, carry out their activities under the control and monitoring of the autonomic nerves. These constitute the autonomic nervous system (ANS), which is connected to, but relatively independent from, the rest of the nervous system. It has central components, part of the central nervous system (CNS), and peripheral components, spread throughout the body; all its parts, with the exception of solar plexus and enteric nerves, are bilaterally symmetrical. Somatic and visceral functions are distinct from each other but are connected and are part of a whole organism. The activities of viscera are spontaneous (or generated without voluntary intervention, based on reflexes, or 'automatic') and any sensitivity is poorly localized. Yet, these activities and the sensations generated from deep organs, including pain, are integrated with other bodily functions and reach to some extent our awareness. The internal state of the body is monitored and signaled to the brain.

Thus, the ANS controls the motility of the heart, gastrointestinal tract, urinary tract, and reproductive organs; the tone of blood vessels, lymphatic vessels, and airways; movements of the pupil and lens; the activity of the minute muscles of hairs; and the secretion of glands (from the large salivary glands and lachrymal glands to innumerable small glands in the gut, airways, genital organs, and skin) and the secretion of the adrenal medulla. Several of these activities are coordinated with somatic behavior. A major difference with somatic behavior is that visceral functions are not fully initiated and controlled (or dominated) by nerves, as is the case with somatic functions. The heart would continue to beat without nerves (although with a different pattern and a reduced range of adaptations); the intestine would still contract if all nerves were destroyed. Some visceral functions are in fact little influenced by nerves (e.g., the tone of large vessels or the motility of the uterus); other functions, however, are under tight control from autonomic nerves (e.g., ejaculation, motility of urethra and bladder, or movement of the intrinsic muscles of the eye).

General Plan

The central component of the ANS (within the CNS) consists of nuclei in the spinal cord, brain stem, and hypothalamus and nerve pathways connecting them to each other and to somatic nuclei. From the autonomic nuclei of the spinal cord and brain stem emerge nerve fibers that exit the CNS directed to peripheral organs.

The peripheral component consists of ganglia and nerves (ganglionated chains and plexuses) forming an extensive web that spreads from the spinal cord and brain stem to deep inside peripheral organs. Only the sympathetic chain is clearly recognizable by the naked eye; most of the other structures are visible only under a microscope.

The sensory component includes sensory neurons located in dorsal root ganglia and in some cranial nerve ganglia, mixed with somatic sensory neurons; these neurons project simultaneously to the spinal cord (dorsal horn) and to peripheral organs. There are many autonomic sensory neurons in ganglia associated with the vagus and glossopharyngeal nerves.

The peripheral component, however intricate and widespread, consists of nerve pathways leading from motor neurons in the spinal cord and brain stem to contractile and secretory elements (know as effectors) in peripheral organs. Whereas in somatic nerves there is a direct connection between motor neuron and muscle fiber, in autonomic nerves the pathway is interrupted by a ganglion. The autonomic motor neurons issue a fiber that terminates in a ganglion (hence the terms preganglionic neuron and preganglionic fiber). A second set of neurons within a ganglion, the ganglion neurons, which are driven by the preganglionic fibers that reach them, issue nerve fibers, known as postganglionic fibers, that reach the peripheral organ and terminate onto effector cells (muscle cells and secretory cells). Both preganglionic fibers and postganglionic fibers are very thin (0.2–1.0 μm in diameter) and are predominantly nonmyelinated: The conduction velocity is much slower (especially in postganglionic fibers) than in somatic nerve fibers.

Despite their diminutive diameter (and given the physical properties of cell membrane and the size of organelles such as vesicles and microtubules, these fibers are close to the physical limit of miniaturization) and the absence of a myelin sheath, postganglionic fibers can extend from a ganglion near the spine to an arteriole at the end of the foot or in the middle of the scalp.

The two-neuron pathway connecting the brain stem and spinal cord to muscular and secretory effectors is thus relatively slow compared to the somatic motor neurons; by this arrangement, each preganglionic neuron (motor neuron) affects more than one ganglion neuron and each of these affects many effector cells by means of vast spreading of peripheral terminals. The interruption at the level of the ganglion, where the preganglionic fibers form synapses abutting ganglion neurons, adds further delay to transmission, and this matches the relatively slow contractile response of smooth muscle (which also carries large savings in energy cost).

Preganglionic Neurons

In the brain stem, the main bilateral groups of preganglionic neurons are the accessory oculomotor nucleus of Edinger–Westphal which projects to the ciliary ganglion; the superior salivatory nucleus projecting to the pterygopalatine and submandibular ganglia; the inferior salivatory nucleus projecting to the otic ganglion; the dorsal vagal nucleus projecting to ganglia in the trachea, esophagus, and gastrointestinal tract; and part of the nucleus ambiguus projecting to cardiac ganglia.

In the spinal cord, preganglionic neurons are grouped into symmetrical columns at the base of the ventral horn. The largest columns are part of the sympathetic pathways and extend from the last cervical level (C8) through the thoracic levels to the second lumbar level (L2). Other columns, which are part of the parasympathetic pathways, are located in the sacral segments (between S2 and S4).

Preganglionic neurons are all cholinergic and receive two main inputs: one from higher centers, including the micturition center in the pons, the cardiovascular centers in the brain stem, and several autonomic nuclei in the hypothalamus, and the other from the periphery via afferent (sensory) autonomic neurons in dorsal root ganglia. The latter input is either direct (monosynaptic reflex) or mediated by an interneuron.

Preganglionic fibers from preganglionic neurons gather into bundles and exit the cord (or brain stem) within ventral roots. Preganglionic neurons, ganglion neurons, and pre- and postganglionic fibers together form the sympathetic and parasympathetic pathways.

Sympathetic Pathways

The sympathetic pathways extend from the autonomic columns of the thoracolumbar spinal cord to the peripheral organs, including the adrenal gland, heart, and all the blood vessels of the body. They represent the autonomic thoracolumbar outflow and include the sympathetic ganglia.

The axons of preganglionic neurons of the sympathetic pathways (preganglionic fibers) reach the corresponding ventral roots and emerge as separate nerves (called white rami communicants), which in turn reach the paravertebral sympathetic chain (or trunk). The chain is made of ganglia and connecting nerve strands, and it extends bilaterally from the base of the skull to the sacral region in front and to the side of the vertebral bodies. Within the chain, preganglionic fibers spread out either cranially or caudally; with this arrangement, preganglionic fibers, which issue only from the middle part (the thoracic and some of the lumbar part) of the spinal cord, can reach all the ganglia of the chains, which extend for the full length of the spine. The largest ganglia are the superior cervical ganglion, which provides sympathetic innervation to organs of the head including cerebral blood vessels and skin, and the stellate ganglion, the main target of which is the heart.

Some preganglionic fibers reach the adrenal gland. Others travel further away from the spinal cord, form the splanchnic nerves, and reach sympathetic ganglia located in front of the abdominal aorta – the prevertebral ganglia. These ganglia, also known collectively as the celiac or solar plexus, are large and close to each other, and they are densely interconnected. They issue nerves directly to all the organs in the upper part of the abdominal cavity.

All the ganglia previously mentioned, which are an integral part of the sympathetic pathways, contain ganglion neurons that have fibers projecting to the peripheral organs. These are called postganglionic fibers, and they complete the sympathetic pathways.

Parasympathetic Pathways

Some parasympathetic pathways involve the brain stem and cranial ganglia, and some are centered on the sacral region of the spinal cord and the pelvic ganglia. In the brain stem, the main nuclei of the parasympathetic pathways (the cranial parasympathetic outflow) are the Edinger–Westphal nucleus, the superior and inferior salivatory nuclei, the dorsal vagal nucleus, and the nucleus ambiguus. The neurons of these nuclei issue axons and then enter the oculomotor nerve, the facial nerve, the glossopharyngeal nerve, and the vagus nerve, respectively. The oculomotor branch projects to the ciliary ganglion (its neurons innervate the iris and ciliary muscle, thus providing accommodation and pupil constriction). The facial nerve branch projects to the pterygopalatine and submandibular ganglia (their neurons innervate the lachrymal, submandibular, and sublingual glands), and the glossopharyngeal branch projects to the otic ganglion (its neurons innervate the parotid gland). The fibers from the nucleus ambiguus and dorsal vagal nucleus enter the vagus nerve, of which they represent the motor component, and extend a long distance in the neck, thorax, and upper part of the abdomen. The fibers terminate synapsing on neurons in small ganglia close to the tracheal and bronchial muscles and the esophagus and in some enteric ganglia of stomach and intestine. From these minute intramural ganglia, postganglionic fibers emerge that innervate glands and smooth musculature.

In the spinal cord, preganglionic parasympathetic neurons are assembled into columns in the second, third, and fourth sacral segments. Their preganglionic fibers, predominantly cholinergic, project onto pelvic ganglia, whose postganglionic fibers innervate mainly the urogenital organs.

In some organs, typically the heart or the pupil, which receive both sympathetic and parasympathetic fibers, these exert an antagonistic effect. Many organs, however, are controlled predominantly by one or the other of the two pathways.

Autonomic Ganglia

These ganglia range in size from the largest, which in man are a few millimeters wide, to microscopic aggregates of only a few neurons. The largest ones – those of the sympathetic chain, the solar plexus and the pelvic plexus, and the named ganglia of the cranial parasympathetic pathways – have a capsule of connective tissue, an important protection against mechanical stimuli for structures situated close to large pulsating arteries and close to motile organs. The capsule is made of layers of connective tissue and is perforated by nerves entering and exiting the ganglion (also carrying the blood supply). The ganglion is in essence an aggregate of neurons (ganglion neurons). There are large and small neurons (partly related to the length which their axon has to travel to each the target organ and the extent of peripheral spreading). Each ganglion neuron has an axon (which becomes a nerve fiber after it is bundled up and associated with a glial support) that reaches the periphery, the postganglionic fiber. Ganglion neurons have dendrites spreading like trees around their cell body – long and highly branched arborization from large neurons (especially in the sympathetic ganglia of large animal species), short, barely branching arborization from small neurons (especially in small animal species), or even the absence of dendrites in certain ganglion neurons (some sympathetic neurons of the mouse). All ganglion neurons are fully cloaked by glial cells (satellite cells making the neuronal capsule), interrupted only where cell processes leave the cell and where terminal incoming fibers reach the neuron to synapse on it. The satellite cells, connected to each other by gap junction (hence, metabolically coupled), provide a tight barrier around the neuron and isolate it from the extracellular space of the ganglion and from other neurons.

In addition to ganglion neurons, satellite glial cells, and axons in transit and their glial sheath, autonomic ganglia contain fibroblast and other elements of connective tissue and some chromaffin cells. The ganglia are highly vascularized, in contrast to the nerve trunks.

The space between neurons is a busy place; at any one site there are hundreds of nerve fibers carrying impulses in one direction and in the opposite direction preganglionic fibers carrying impulses to the ganglion neurons, postganglionic fibers carrying impulses to the periphery, and many more fibers passing to and from other ganglia. Locally, the input from a preganglionic fiber is distributed to several ganglion neurons (divergence), and each ganglion neuron receives an input from several preganglionic neurons (convergence).

Postganglionic Fibers and Innervation of Viscera

Ganglion neurons of autonomic ganglia all issue an axon (which turns into a nerve fiber), and virtually all these fibers exit the ganglion directed toward a peripheral target organ, along different paths depending on the ganglion of origin. Some form discrete nerves (splanchnic, pelvic, and urinary), whereas others penetrate into somatic nerves; the mixed nerves of autonomic and somatic fibers thus formed reach the periphery, particularly the limbs and all the skin. Upon reaching the target organ, they branch extensively and make contact with muscle elements (mainly smooth muscle cells) and with secretory elements (glands).

Postganglionic fibers exert their effect on muscle and glands by releasing neurotransmitters that stimulate (and sometimes inhibit) contraction and secretion. The neurotransmitters are released from 'terminals' scattered along a substantial part of the terminal portion of the axons. Each ganglion neuron issues one fiber traveling to the periphery, which has many branches within the terminal organ, which in turn have many thousands of terminals. In the bladder muscle, for example, a few thousand ganglion neurons directly innervate many millions of muscle cells.

The relationship between varicosities (nerve endings) and muscle cells (or gland cells) varies from a close contact (a close and specialized junction) in some tissues (e.g., the bladder) to a loose relationship with a wide gap (some blood vessels). Chemical transmission from these endings – the release of the content of a few synaptic vesicles each time an ending is depolarized – completes the activity of the efferent autonomic pathway.

The problem posed by the discrepancy between a small number of postganglionic fibers and a very large number of effector cells is solved by means of various structural arrangements. First, postganglionic fibers branch extensively inside the target organ (but not before). Second, nerve endings (which contain the neurotransmitters packed into axonal vesicles) are found not just as expansions at the anatomical end of each axonal branch: The entire terminal part of each branch (the last 100 μm of its length or more) presents bulbous expansion (or varicosities) that gives a beaded appearance to the axon and functions as nerve endings. Varicosities are expansions 1 or 2 μm wide, regularly spaced in a chain and separated by intervaricose segments that can measure as little as 0.3 μm in diameter. Each expansion is packed with axonal vesicles, and even if it is not the end of an axon, it functions like a nerve ending. The number of functional nerve endings of these so-called varicose axons is therefore 10–100 times greater than the number of anatomical end points of the postganglionic fiber.

Two other mechanisms assist in the uniform activation (or inhibition) of a large number of effector cells. These cells are often coupled to each other by gap junctions that allow the two cells to directly exchange metabolites and ions (which influence the electrical status of a cell). In addition, many effectors display some spontaneous activity (e.g., myogenic contractions of some muscles), and the action of nerves is a modulation and enhancement of some background activity. Lastly, some effectors are able to contract in the absence of nerves (e.g., the muscle of large arteries), and many effectors are affected by chemical agents that do not derive from nerves but may derive from endocrine glands or from secretion at adjacent sites.

Sensory Fibers

The ANS comprises an important sensory component. Its neurons lie in dorsal root ganglia, mixed with the vastly more numerous somatic sensory neurons. Each ganglion neuron issues an axon that splits, sending a branch centrally to the spinal cord and a branch peripherally to a peripheral organ. The peripheral branches travel together with sympathetic and parasympathetic fibers in mixed nerves and are distributed to the same tissues. Some of the central branches terminate onto the autonomic preganglionic neurons, in which case a reflex pathway is formed composed of an afferent and an efferent arm. Others synapse on interneurons, which in turn project to preganglionic neurons or form long pathways to higher centers of the CNS. The latter pathways convey sensations from the viscera, including pain.

Within the spinal cord there is a certain degree of convergence of visceral and somatic afferent nerve fibers, and some incoming fibers may synapse on the same interneuron, despite their different origin. This anatomical feature may explain the phenomenon of referred pain – that is, pain stimuli originating from a peripheral organ that are perceived by the subject as if they were coming from the surface of the body.

Enteric Nerves

Enteric nerves (or the enteric nervous system) are the set of autonomic nerves dedicated to the control of the activity of the stomach and intestine (including the lower part of the esophagus, anal canal, and gallbladder). They are located entirely within the wall of these organs (intramural nerves); they are linked to the sympathetic and parasympathetic outflows but are relatively independent of them. The enteric nerves consist of myriad small aggregates of neurons (enteric ganglia, which are quite different in size and structure from the other autonomic ganglia) connected by strands and forming a mesh or plexus, a relatively thin and flat structure, like a net with knots and meshes of a microscopic texture.

The most extensive and best known of these plexuses is the myenteric, or Auerbach's, plexus, located within the muscle coat, usually between circular and longitudinal muscle layers and extending continuously from the lower part of the esophagus to the stomach, small intestine, colon, rectum, and over the entire circumference of these organs.

The submucosal, or Meissner's, plexus is situated in the submucosa (the layer of dense connective tissue and blood vessels between muscle and mucosa), close to the inner border of the musculature and in proximity to the mucosa (in some animal species there is an additional component of this plexus nearer the mucosa).

There are differences in the pattern of the meshes in the plexuses, size of ganglia, and types of neurons in different parts of the gut, and these possibly reflect the differences in the activities of the different organs. The continuity of the plexuses over the entire length of the gut allows the activity in adjacent segments to be coordinated (even when they are separated by a sphincter); long-distance coordination, however, is achieved through reflexes involving extrinsic neurons (i.e., neurons located outside the gut wall) in the prevertebral ganglia. Still, the continuity of the plexus along the entire gut is compatible with the generation of different patterns of activity, especially motor activity, in different regions.

Invisible to the naked eye, and remarkably autonomous from other parts of the nervous system, the enteric neurons constitute an immense population of cells. Many thousands of neurons are found in every square centimeter of intestinal surface, and the total number in the gut amounts to many millions. Their spatial density is highest in the smallest mammals and decreases in animals of larger body size, in which connections between ganglia become longer. The total number of neurons, however, is highest in the larger animal species.

The ganglia are situated amid other tissues of the gut wall. They contain no connective tissue but consist exclusively of neurons with their processes, glial cells, sometimes other undifferentiated cells, and many nerve fibers in transit. The glia is highly developed, not as individual wrapping around axons and neurons (as in sympathetic ganglia) but as flattened cell bodies and expansions that permeate and fill all the spaces in the ganglion. Axons are usually not individually wrapped by a glial process (but are bundled together in direct contact with each other), and there are no myelinated fibers.

Many enteric neurons have axons and dendrites that may be very long but never leave the plexus, and they connect synaptically with other neurons situated upstream or downstream or elsewhere around the circumference or with neurons situated in deeper or more superficial ganglia. All these should be regarded as interneurons, and they are very abundant. The long axons of other neurons (motor or efferent neurons) exit the plexus and reach the musculature and the glands of the organ.

A small number of fibers exit the ganglia and the gut wall and project to the prevertebral ganglia. Also, some fibers from sympathetic and parasympathetic ganglia reach the intramural plexuses. The sensory fibers originate in the dorsal root ganglia, but there are also afferent fibers entirely intramural, supporting local reflexes.

Enteric neurons are a large population of nerve cells comprising several different categories of neurons. Due to the intricate packing of neurons and supporting glial cells, much research has been performed to identify nerve cell 'types' on the basis of cell appearance, incoming and outgoing connections, location, the presence of chemical features (neuronal markers), the types of transmitter released, its electrical properties, and its metabolic properties. There are descriptions that reveal the remarkable complexity of this nervous tissue.

An intriguing aspect of enteric neurons is that they seem to have highly specific distribution of chemical transmitters and markers (as in the endocrine system) but also have precise synaptic connections (as generally applies in the nervous system).

Central Components of the Autonomic Nervous System

The columns of preganglionic neurons in the spinal cord are the lowest level of a hierarchy of structures in the CNS that preside over the autonomic functions and their coordination. Higher up, in the brain stem, several nuclei and some less distinct groups of neurons constitute the interconnected reticular centers that regulate activities that are entirely or predominantly autonomic, such as cardiac, respiratory, and gastrooesophageal functions. These neurons receive an input from higher centers in the CNS, notably the hypothalamus, and an input from neurons (e.g., in the nucleus of the solitary tract) that receive sensory input from heart, intestine, and airway. The main outputs of the brain stem nuclei are (1) parasympathetic fibers to the heart, bronchi, stomach, etc. in the glossopharyngeal and vagus nerves and (2) sympathetic fibers to the sympathetic preganglionic neurons of the spinal cord. In certain systems, somatic and autonomic functions are tightly linked and integrated to each other (e.g., in respiration or vomiting).

The largest structure in the CNS controlling autonomic functions is the hypothalamus, located between the mesencephalon and the telencephalon, below the thalamus, bordered laterally by the internal capsule and the optic tract. Complex functions such as body temperature (homeostasis), circadian rhythms (the oscillations of certain bodily functions, mainly autonomic, with a 24 h period), feeding, drinking, and reproductive behaviors are regulated essentially from the hypothalamus. The organ exerts the highest and direct control both on the ANS and on the endocrine system; the hypothalamus is linked to the pituitary gland via the infundibular stalk and some of its neurons secrete the hormones oxytocin and vasopressin. Unlike in the rest of the brain, there is no blood–brain barrier in parts of the hypothalamus. In addition to projections to the thalamus, the forebrain, and the pituitary, many hypothalamic neurons project to cardiovascular centers in the brain stem (parabrachial nucleus, nucleus ambiguus, nucleus of the solitary tract, and dorsal motor nucleus of the vagus) and to the preganglionic neurons in the spinal cord.

The hypothalamus is divided, conventionally, into topographic zones that contain distinct nuclei as well as some diffuse groups of neurons. The periventricular zone includes the supraoptic nucleus, the preoptic nucleus (active in temperature control), and the suprachiasmatic nucleus (active in circadian rhythms). Located in the infundibulotuberal zone is the arcuate nucleus, and located in the posterior (or mammillary) zone are the medial mammillary nucleus (that has a major projection to the thalamus) and the tuberomammillary nucleus. The lateral zone contains the lateral hypothalamic nucleus and the lateral tuberal nucleus. In the intermediate (or medial) zone are theparaventricular nucleus, the supraoptic nucleus, the mammillary body, the tuberomammillary nucleus, and the ventromedial nucleus. The ventromedial nucleus contains neurons sensitive to the plasma level of glucose and other metabolites; by integrating sensory inputs from the viscera, this part of the hypothalamus participates in the control of metabolism and food intake and of feeding behaviors in general. The preoptic nucleus contains neurons that are sensitive to small changes in the temperature of the blood circulating through the hypothalamus.

See also: Autonomic and Enteric Nervous System: Apoptosis and Trophic Support During Development; Autonomic Disorders; Autonomic Failure; Autonomic Neuroimmunology; Autonomic Nervous System; Energy Homeostasis: Visceral Control; Parasympathetic Nervous System; Sympathetic Nervous System.

Further Reading

Brading A (1999) *The Autonomic Nervous System and Its Effectors.* Oxford: Blackwell.

Furness J (2005) *The Enteric Nervous System.* Oxford: Blackwell.

Mathias CJ and Bannister R (2001) *Autonomic Failure. A Textbook of Clinical Disorders of the Autonomic Nervous System,* 4th edn. Oxford: Oxford University Press.

Robertson D (2004) *Primer on the Autonomic Nervous System.* San Diego: Elsevier.

Saper CB (1990) Hypothalamus. In: Paxinos G (ed.) *The Human Nervous System,* pp. 389–411. London: Academic Press.

Wilson-Pauwels L, Stewart PA, and Akesson EJ (1997) *Autonomic Nerves, Basic Science, Clinical Aspects, Case Studies.* London: Dekker.

Autonomic Nervous System: Ophthalmic Control

J Pintor, Universidad Complutense de Madrid, Madrid, Spain

Introduction

The eye is the sense organ that permits the detection of light owing to the existence of a sophisticated neuronal array, called the retina, which is responsive to photons. This ocular structure sends the acquired information through the optic nerve to central areas such as the lateral geniculate nucleus and the superior colliculus, on the way to the visual cortex.

The correct functioning of this complex system requires the coordination of several intraocular structures that ultimately permit the perfect focusing of images on the neural retina. Light has to pass through four different media before it reaches the retina: the cornea, aqueous humor, lens, and vitreous humor. Moreover, the composition and structure of some of these media can change due to several physiological mechanisms that are mostly controlled by the nervous system.

Although it is not yet clear whether some pathologies occur as a consequence of ocular nerve malfunction, it is apparent that disorders such as glaucoma may be a result of an imbalance between the formation and drainage of aqueous humor, due to a problem in the ocular innervation.

In this article, the importance of the nervous system controlling the iris and the ciliary body is reviewed.

The Iris and the Ciliary Body Are Located in the Anterior Uvea

The uvea is the part of the eye that is located between the scleral coat and the neuroepithelial cell layer. This region contains two main structures called the iris and the ciliary body (**Figure 1**).

The iris is the colored structure that works like a diaphragm regulating the amount of light passing toward the retina. It is important to control the quantity of light that arrives at the retina otherwise it will be overexposed and accurate vision of the object will be impossible. In brief, the iris works in a similar way to the diaphragm of a photographic camera. The iris has two main muscles, the iris dilator (dilator pupillae) and the iris sphincter (sphincter pupillae). These structures are innervated by the sympathetic and parasympathetic nervous systems, respectively, and work in coordination to control the size of the pupil in order to permit the best optic accuracy under different light conditions.

The ciliary body is located behind the iris. This structure is connected to the ciliary muscle which controls the lens shape in order to properly focus all the images in the retina in a process termed accommodation. Apart from the importance of the ciliary body in accommodation, the main task of this ocular structure is the synthesis of the aqueous humor. This transparent fluid serves as the source of nutrients for the avascular structures within the eye, such as the lens and the cornea. It is also important to remember that an increase in the synthesis of aqueous humor or a reduction in aqueous humor outflow can produce an increase in the intraocular pressure (IOP) leading to the pathology known as glaucoma. The ciliary processes are formed by two main epithelial cells types, pigmented and the nonpigmented ciliary epithelial cells, which are in contact with the aqueous humor. Under these cells vascular and neural plexuses provide nutrients and control the physiology of this ocular area.

Uvea Innervation: General Aspects

Sympathetic, parasympathetic, and sensory divisions of the peripheral nervous system are responsible for the uveal innervation, and they provide the three components regulating the physiology of this ocular area.

The Sympathetic Nerves and Its Neurotransmitters

Uveal sympathetic nerves are derived from the superior cervical ganglion and they innervate both the iris and the ciliary body. The sympathetic innervation originates in the spinal cord. From the segments C_8–T_3, the preganglionic neurons exit the spinal cord, contributing to thoracic nerves I–III, synapsing in the superior cervical ganglia. Postganglionic neurons in these ganglia project to the trigeminal ganglion and cavernous sinus. Some of the axons pass through the trigeminal ganglion and enter the ocular globe via the long posterior ciliary nerves. When in the eye, these axons spread and project to the iris muscle and to blood vessels. Other axons travel along the carotid artery, pass through the ciliary ganglion, and innervate the ciliary body as well as the uveal blood vessels.

The iris contains a dense population of adrenergic fibers, which innervate the iris dilator muscle. In human ciliary processes, the adrenergic nerves are located just beneath the ciliary epithelial cells, while in some animal models it seems that these nerve endings penetrate more and establish contacts with nonpigmented epithelial cells.

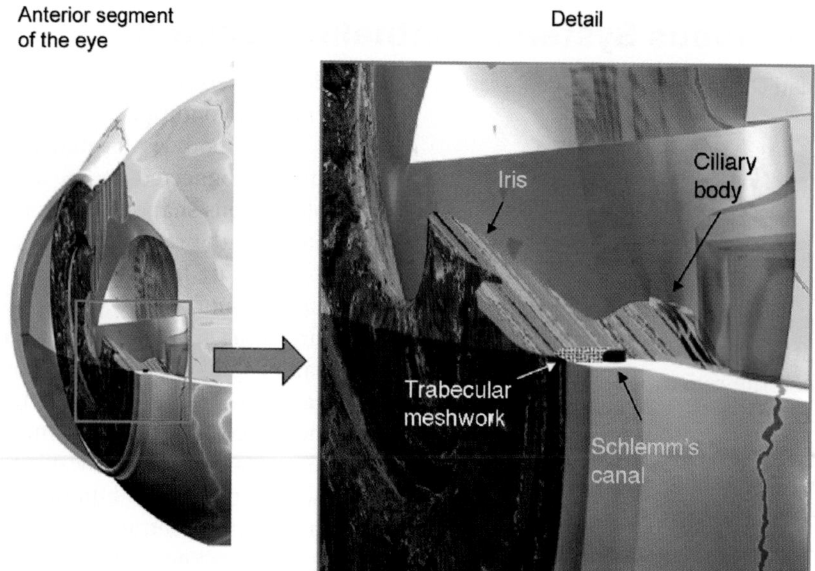

Figure 1 Three-dimensional models of the anterior segment of the human eye. In the selected area the main structures described in this article are depicted.

Noradrenaline (NA) is probably the best representative of sympathetic neurotransmitters in the eye. NA is synthesized in the classical pathway starting from the amino acid tyrosine. With the intermediate steps which produce the substances dihydroxyphenylalanine (DOPA) and dopamine, it is transformed into NA by means of the enzyme dopamine β-hydroxylase.

NA is an important transmitter in the iris since its concentration can be above $20\,\mathrm{nmol\,g^{-1}}$ tissue. This is in clear contrast with the concentration found in the aqueous humor, which is about $0.5\,\mathrm{ng\,ml^{-1}}$. The origin of NA in the aqueous humor seems to be mainly from the iris since iridectomy reduces the concentrations of this transmitter in aqueous humor to below 80%. This does not mean that NA is inconsequential in the ciliary processes; on the contrary when analyzed in detail, NA is important since this transmitter is released after local electrical stimulation. The control of NA release in experimental models has been investigated. These studies demonstrate that α_2-adrenergic receptors, muscarinic, dopamine, and neuropeptide Y (NPY) receptors among others, can all inhibit NA release. In contrast, angiotensin can enhance NA release.

Dopamine is present at low concentrations in the aqueous humor ($0.4\,\mathrm{ng\,ml^{-1}}$) and its origin, in clear contrast to NA, does not come from the iris. It has been suggested that dopamine originates from the ciliary body.

5-Hydroxytryptamine (5-HT), also termed serotonin, is derived from the amino acid tryptophan. Its presence in the aqueous humor is very low although some authors have been able to identify some serotonergic nerve fibers in the ciliary body. The existence of receptors for 5-HT has been demonstrated in areas such as the ciliary body and the iris.

NPY is formed from 36 amino acids, and is amidated at its C-terminal. It is a neurotransmitter which runs in parallel with NA-containing fibers, and may be contained in the same nerve fibers as NA. The iris dilator muscle and the ciliary processes seem to present nerve fibers containing this transmitter.

Co-stored with NPY, opioids have been identified in postganglionic sympathetic nerves that project to the iris, and both sphincter and dilator muscles. To date, little is know about the role of these peptides in the iris, although receptors have been found in this tissue.

Parasympathetic Nerves and Its Neurotransmitters

The parasympathetic innervation that controls the iris and ciliary body comes from the preganglionic nerves located in the Edinger–Westfal nucleus which is located in the midbrain. Axons of these nerves leave the brain through the inferior oculomotor and oculomotor branches and synapse in the ciliary ganglion. The neurons which project to the iris and ciliary body are located in the periphery of the ciliary ganglion. More than 95% of all the fibers leaving this ganglion project to the ciliary body, while the remaining 5% reach the iris sphincter and the blood vessels. Parasympathetic terminals do not have a myelin coat and run in parallel with either the blood vessels or the muscle fibers.

It is clear that the main parasympathetic neurotransmitter is acetylcholine (ACh). Synthesized by the enzyme choline acetyltransferase, this enzyme has been extensively used to localize ocular nerves containing ACh as neurotransmitters. In particular, the presence of this enzyme has been described in the iris sphincter and ciliary muscle. The ACh concentration in the iris of some animals is between 100 and 200 ng mg^{-1} tissue. Nevertheless, one needs to take into account that ACh can be rapidly degraded by acetylcholinesterase (AChE). This enzyme rapidly reduces the local concentrations of Ach, thus explaining the acute effects of AChE inhibitors on miosis and accommodation.

ACh release is strongly controlled by other transmitters by means of prejunctional α_2-adrenoceptors, adenosine A_1, and opioid and nicotinic receptors (amongst others). Vasoactive intestinal polypeptide (VIP), a C-terminal amidated polypeptide containing 28 amino acids, is present in parasympathetic terminals coming from the pterygopalatine ganglion. Since these fibers mainly innervate blood vessels, it is possible to detect them in the iris stroma and ciliary processes. In the ciliary body it has been possible to detect receptors for this peptide, which are positively coupled to adenylate cyclase.

Sensory Innervation and Its Neurotransmitters

Starting in the trigeminal ganglia, the trigeminal neurons extend their axons centrally into the pons and peripherally, via the ophthalmic, maxillary, and mandibular branches to the target organs. The ophthalmic branch divides into frontal, lachrymal, and nasociliary nerves, the latter providing a major branch that travels to the eye with the ciliary nerves. The location of these nerves within the eye demonstrates that in the iris their location is mainly in the sphincter, as well as other ocular areas.

In the ciliary body, sensory nerves are widely distributed, being present in the ciliary processes, blood vessels, ciliary muscle, and scleral spur. The nature of these nerve terminals permits them to be considered as mainly mechanoreceptors and polymodal receptors.

A representative of the neurotransmitters present in sensory terminals is substance P (SP). Nerve fibers that are positive to this transmitter are located in the iris sphincter muscle as well as in the blood vessels of the iris stroma. It has also been possible to find nerves containing SP in the ciliary processes and ciliary muscle. SP is released after the application of certain compounds including capsaicin, bradykinin, or nicotine, which excite the nerve-ending triggering the release of the neurotransmitter. SP release in ocular structures can be modulated by presynaptic transmitters via the activation of α_2-adrenergic and κ-opioid receptors, and its release can be stimulated by means of nicotinic and adenosine receptors.

Calcitonin gene-related peptide is a polypeptide formed by 37 amino acids, which is often co-expressed with SP and therefore both are co-distributed in the iris and ciliary body.

Cholecystokinin (CCK) terminals are not very abundant but they can be found at the iris stroma and are also associated with blood vessels in the ciliary body. The 29-amino-acid polypeptide galanin is present as well, in terminals in the sphincter, dilator, and iris stroma.

Vasopressin (the antidiuretic hormone), a small, cyclized, C-terminal amidated peptide of nine residues, is the transmitter of some nerve fibers which arrive at the iris sphincter muscle. The application of exogenous vasopressin produces contraction in this iris structure.

The Neural Control of the Iris

Although the structure of the iris could be simplified to the iris dilator muscle, iris sphincter muscle, and the stroma, the main part of the control of the physiology of the iris is driven by the harmonization of sympathetic and parasympathetic terminals.

Sympathetic innervation of the iris sphincter muscle is mainly by NA. Neuropharmacological studies reveal that NA, acting through β-adrenoceptors, is able to induce relaxation in the sphincter by means of a mechanism which involves the activation of adenylate cyclase and the corresponding formation of cyclic AMP (cAMP). In some animal species the participation of α-adrenoceptors has also been quoted, but in this case mediating contraction.

The sympathetic regulation of the iris dilator muscle is mainly driven by α_{1B}-adrenoceptors, which contribute to the contraction of this muscle. This role has been demonstrated by means of selective antagonists that clearly abolish transmural contraction of the iris. The mechanism underlying the activation of this receptor seems to be phospholipase C (PLC) activation with the generation of inositol triphosphate (IP_3) and diacylglycerol (DAG). IP_3 releases Ca^{2+} from intracellular stores that together with DAG stimulates protein kinase C (PKC), which in turn phosphorylates myosin light chain (**Figure 2**).

The existence of β-adrenoceptors has been demonstrated in the iris dilator muscle. The effect that is mediated through this receptor is the relaxation of the muscle by means of the cAMP production. This apparent contradictory effect is supposed to exist in order to control the contractility, which is triggered by the α-adrenoceptor.

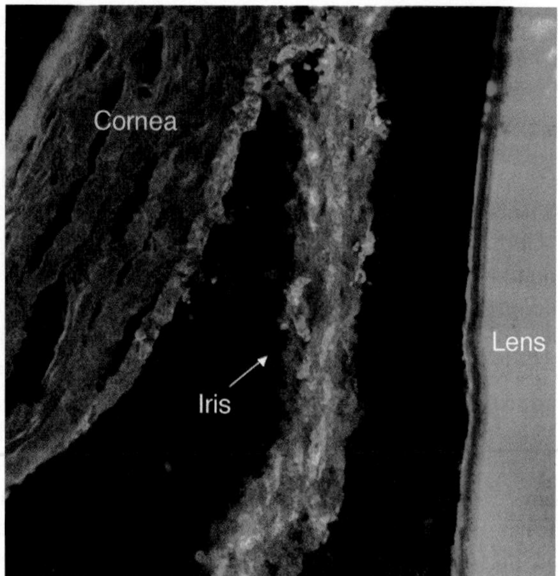

Figure 2 Immunohistochemical section of the iris labeled with an antibody which shows the nerve endings in green .

The neurotransmitter NPY seems to work closely with the α_1-adrenoceptor. On its own, NPY does not modify the physiology of the dilator muscle; nevertheless, it can potentiate the responses of NA acting via the α-adrenoceptor. Considering the mechanism involved in this potentiation it is possible that NPY, in some way, inhibits adenylate cyclase.

Parasympathetic control of the iris sphincter muscle is well known since muscarinic antagonists were able to block the miosis induced by light exposure. Atropine is isolated from the plant *Atropa belladonna*. The word belladonna comes from Italian, meaning beautiful lady; sixteenth-century Italian courtesans used to dilate their pupils by instilling belladonna into their eyes, making them more beautiful according to the perceptions of that time.

Indeed, the sphincter is contracted after treatment with ACh or its analogs. The receptors responsible for these actions are mainly of the muscarinic M3 subtype with fewer M2 receptors. M3 receptors are coupled to PLC and the corresponding generation IP_3 and DAG, while M2 receptors are negatively coupled to adenylate cyclase. Activation of both receptors would induce a more efficient contraction of the iris sphincter muscle.

The muscarinic contraction is a process which develops in two steps. There is one which is dependent on the release of Ca^{2+} from extracellular stores (triggered by IP_3), and a second in which calcium channels are involved. DAG activates PKC, and phosphorylation of calcium channels permits a higher cytosolic Ca^{2+} concentration that, together with PKC, permit a more efficient myosin light-chain phosphorylation.

VIP, which behaves as a co-transmitter in parasympathetic terminals, reverses the contraction triggered by cholinergic agents due to the production of cAMP, which probably inhibits the release of ACh from the parasympathetic nerve terminals.

The parasympathetic regulation of the iris dilator muscle is predominantly relaxant, in this case the effect being mediated by M3 receptors involving the activation of calcium-dependent adenylate cyclase. Also, the presence of M2 receptors has been defined on noradrenergic terminals; therefore, ACh controls the release of NA.

Sensory regulation of the iris sphincter has been identified by the use of selective antagonists such as SP receptor blockers. Capsaicin-induced SP release produces contraction in the iris sphincter muscle, but it has been demonstrated that the effect of SP can vary according to species, and in humans SP does not contract this muscle. These two behaviors are also corroborated by the putative activation of different SP receptors; while in contracting tissues SP receptor seems to be coupled to PLC activation, in humans the SP receptor is coupled to adenylate cyclase.

CCK seems to be more important than SP in humans since this transmitter is able to produce contractions in the human sphincter by a mechanism which involves CCK-A receptors.

The Neural Control of Aqueous Humor Formation

Aqueous humor is a transparent nutritional fluid that supplies nutrients to those intraocular structures that do not have a blood supply, especially the lens and the cornea. This fluid also provides the eye with an internal pressure, the IOP, that is necessary to preserve its shape.

Regulating IOP is a physiological process which depends on the right balance between the production and the drainage of the aqueous humor. Aqueous humor is produced at the ciliary processes while the drainage occurs in the iridocorneal angle via the trabecular meshwork, canal of Schlemm, and epiescleral veins. Equilibrium between production and elimination of the humor is a guarantor of ocular health. In contrast, imbalance brings complications that can lead to pathologies such as glaucoma, defined as an increase in IOP. When this happens it is probable that, to some extent, the elevation in pressure reduces retinal blood supply. If the pressure is strong it can even strangle the optic nerve. In both cases the result is a gradual blindness, starting from the peripheral retina towards the fovea (**Figure 3**).

As can be inferred, the control of the production and drainage of the aqueous humor is a tandem

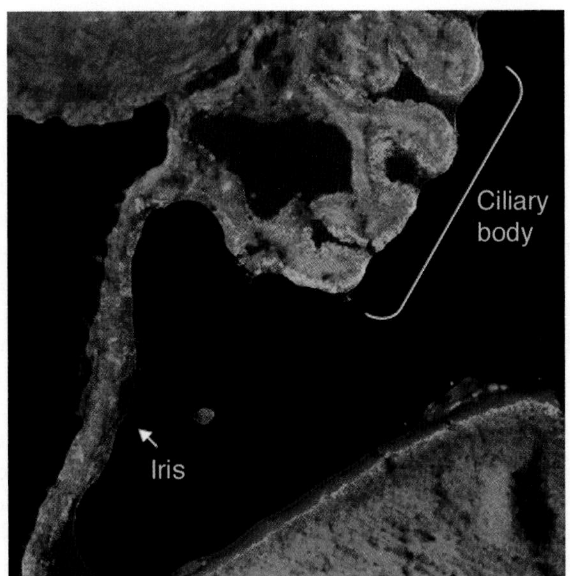

Figure 3 Immunohistochemical section of the iris and ciliary processes. The image shows the ciliary processes with the non-pigmented epithelium in green and the pigmented epithelium and stroma in red.

process that needs to be highly controlled to avoid the unpleasant consequences. The nervous system controls aqueous humor dynamics mainly by means of the sympathetic and parasympathetic pathways.

The Sympathetic Component

Stimulation or denervation are classical methods to study the involvement of the nervous system in the control of IOP. It is necessary to take into consideration that stimulation of the superior cervical ganglion will activate all the sympathetic pathways that arrive at ocular structures, making it difficult sometimes to understand the results obtained. It is, for example, typical that sympathetic denervation produces the release of transmitters from degenerating or degenerated nerve terminals, altering all the intraocular physiology. This may explain some contradictory results described in the literature.

Sympathetic stimulation by, for example, exciting the superior cervical ganglion reduces IOP mainly because it causes a reduction in uveal flow. This effect seems to be controlled by α-adrenoceptors producing vasoconstriction. Paradoxically, high frequencies of stimulation produce the opposite effect: an increase in IOP. It looks as if there is a selective release of neurotransmitters; at low frequencies NA is released while at high frequencies NPY is preferentially released.

Sectioning the superior cervical ganglion produces changes in the circadian variations of aqueous humor production. Also, this ganglionectomy produces an increase in IOP due to catecholamine release from degenerating sympathetic terminals.

Most of the drugs developed to reduce IOP in patients presenting with glaucoma is based on the use of more-or-less selective antagonists for adrenergic receptors, mostly β-adrenergic receptors. The presence of α- and β-receptors in the ciliary processes makes nonselective adrenergic agonists produce an increase in the secretion of the aqueous humor, an effect which is mainly blocked by β-adrenergic antagonists, such as timolol. Due to the circadian involvement of the sympathetic system, the ability of β-antagonists to reduce the production of aqueous humor is effective during the day but not during sleep.

Among the possible candidates responsible for these effects, β_2-adrenergic receptors appear as the best candidates. This receptor is present in the ciliary processes and is positively coupled to adenylate cyclase. Thus, it is cAMP that is the second messenger triggering a series of intracellular changes that , in the end, produce an increase of aqueous humor production and an increase in IOP.

α_2-Adrenergic agonists are also interesting agents to take into account when studying the involvement of the sympathetic nervous system and the control of IOP. It has been shown that clonidine among others can reduce the rate of aqueous humor production reducing concomitantly IOP. Biochemically speaking, α_2-adrenergic receptors are negatively coupled to adenylate cyclase diminishing the levels of cAMP stimulated by the corresponding activation of the β_2-adrenoceptors. There is also a prejunctional effect of α_2-adrenoceptors, diminishing the release of NA from the nerve endings and this may affect all the previously described physiology mediated by α- and β-adrenergic receptors.

Apart from NA, NPY seems to present an important role in the control of IOP. This peptide reduces the blood flow at the ciliary processes. Also, in this ocular structure the presence of receptors for NPY has been described. These receptors are negatively coupled to adenylate cyclase functioning in a similar way to α_2-adrenoceptors.

The Parasympathetic Component

The cholinergic influences on IOP are well known. Indeed, a vast pharmacology for the treatment of glaucoma has been developed, in which cholinomimetics clearly reduce IOP. Cholinergic agonists such as pilocarpine or inhibitors of AChE lower IOP by facilitating the outflow via the trabecular meshwork and Schlemm's canal. Apparently, this facilitation is a consequence of the ciliary muscle contraction and the subsequent pull of the tendon in the scleral spur which opens the corneal–scleral angle. This effect produces a wider spacing between the trabecular sheets, reducing the outflow resistance. The existence of muscarinic receptors in the ciliary processes has been demonstrated. The best candidates are M2 and M4 receptors that mediate adenylate cyclase inhibition. Nevertheless, the results obtained in different experimental animals provide contradictory results on the direct effect of ACh in the ciliary epithelium.

Denervation of some cranial nerves such as facial produces a reduction in the uveal blood flow and consequently a reduction in IOP. The stimulation of this nerve causes just the opposite. These two effects are related with changes in the vasculature (vasodilatation) rather than local effects in the ciliary processes/trabecular meshwork. A transmitter which appears to participate in the vasodilatation is VIP, although intraocular application stimulates the production of the aqueous humor, since this transmitter stimulates adenylate cyclase.

The Purinergic Component of the Eye Autonomic Nervous System

Although purines have been described as co-transmitters of sympathetic and parasympathetic nervous system, little is known about the presence of purines in autonomic nerves of the eye. Evidences nevertheless indicate that they may be stored together with catecholamines in the sympathetic nervous system, and with ACh in the parasympathetic nervous system.

In any case, and before exploring the features of nucleotides in the eye, it is important to note that nucleotides and dinucleotide actions are due to the existence of receptors in cell membranes termed P2 and dinucleotide receptors, respectively. These receptors are divided in ionotropic (P2X and dinucleotide receptors) and metabotropic (P2Y) receptors. All these purinoceptors contribute to the regulation of important processes such as the vascular tone or neurotransmission in the periphery and central nervous system.

In the eye, and in particular in the rabbit iris dilator muscle, the phasic contractions of this muscle present two components, one adrenergic and other nonadrenergic The second component of the iris contraction is mimicked by ATP and the analog α,β-methylene ATP (α,β-meATP). ATP may play a prejunctional action in the iris, modulating the release of sympathetic neurotransmitters. It is not known which receptor is involved in the action of ATP in the iris, but some evidence suggests that it may be a P2Y receptor.

Nucleotides and dinucleotides have interesting actions on aqueous humor dynamics. Indeed, the presence of mononucleotides and dinucleotides in aqueous humor has been described in the relevant literature. This fluid contains measurable amounts of both mono- and dinucleotides. The most relevant adenine mononucleotides found in the whole aqueous humor were AMP, ADP, and ATP, presenting concentration values in the micromolar range. A detailed study has described the concentration of ATP in the immediate area where it is released (the ciliary epithelium), between 4 and 8 μM. Diadenosine polyphosphates have also been discovered, their concentrations being in the nanomolar range.

The effect of mono- and dinucleotides presents a clear pattern of modulation on IOP. Thus, the experiments performed by applying nucleotides to the eye allowed to classify the effect of these molecules under two main groups: on the one hand those elevating IOP and on the other those reducing it. Among the first, the synthetic analogs 2-methylthio ATP and ATP-γ-S presented a clear increase of IOP. This profile fits well with P2Y receptor pharmacology, although the slow destruction of these nucleotides by means of ectonucleotidases cannot be ignored.

On the contrary, β,γ-methylene ATP (β,γ-meATP) and α,β-meATP produced a clear and marked reduction in rabbit IOP. The hypotensive effect produced by these two mononucleotides was blocked by the P2 antagonist PPADS but was unaffected by the adenosine antagonist DPCPX.

β,γ-MeATP and α,β-meATP activate P2 receptors, presumably $P2X_2$ receptors, present in cholinergic terminals that are innervating the ciliary muscle. The activation of this $P2X_2$ receptor would generate an increase in ACh, which contracts the ciliary muscle, pulls the scleral spur, and opens the iridocorneal angle facilitating the drainage of the aqueous humor.

Diadenosine polyphosphates (ApnA) present a similar behavior as that observed for mononucleotides. Of all the tested dinucleotides, only Ap_4A presented a clear reduction in IOP, with the others presenting a hypertensive effect.

Recent results indicate that $P2Y_1$ receptors present in trabecular meshwork cells also modulate IOP. Nucleotides and dinucleotides such as Ap_4A can activate this receptor. The role of this purinergic receptor is to facilitate the drainage of the aqueous humor by changing the shape of these cells and reducing the hydrodynamic resistance of the aqueous humor evacuation.

The Retina

The retina is a very sophisticated neural structure which does not belong to the autonomic nervous system. It is considered as an extension of the brain that is specialized in responding to environmental light by means of a specialized group of neurons termed photoreceptors. The retina presents a clear cellular architecture in which four other neuronal types help to transmit visual information to central visual areas. The organization of the retina presents two cell planes, one vertical consisting of photoreceptors, bipolar, and ganglion cells and another which is horizontal formed by horizontal and amacrine cells. All these cells define two synaptic regions, one formed with the photoreceptors, bipolar, and horizontal cells, and another with bipolar, amacrine, and ganglion cells.

When light stimulates the photoreceptor (cones and rods), a biochemical process occurs so the release of the neurotransmitter glutamate that is released during the darkness stops. Bipolar cells, which contain different types of glutamate receptors, respond to the lack of glutamate by either triggering an action potential and releasing more glutamate or inhibiting it. The presence or absence of glutamate can modify the activity of ganglion cells that projects their axons to central visual areas.

Horizontal cells as well as amacrine cells connect peripheral photoreceptors and peripheral bipolar cells with ganglion cells. These two types of neurons can present different transmitters since in some cases they are excitatory but in others they are inhibitory. Amacrine cells, for instance, can contain ACh or γ-aminobutyric acid (GABA) depending on its function.

Altogether, the retina is an extremely amazing neural array that permits the acquisition of light and its transmission, under the form of an electrical impulse, to arrive to central areas where it will be processed to construct sophisticated images.

The retinal pigmented epithelium (RPE) is part of what is call nonneural retina. This monolayer of cubic epithelial cells is in close relation with the photoreceptors since it is in charge of cone and rod turnover. The RPE is a noninnervated tissue, although some chemical signals can come from photoreceptors or from the choroids. Molecules such as SP and NPY released from the ciliary nerve in the choroids, and with a vasoactive action, may also change the physiology of RPE cells. On the other hand, it has been demonstrated that ATP can also modulate the activity of RPE cells by activating P2Y receptors. The ATP necessary to stimulate these receptors comes from the disc phagocytosis carried out by the RPE cells. So, although from neural origin since photoreceptors are neurons, ATP may not be considered in this case as a neurotransmitter in a classical way. One of the most interesting roles of ATP in the eye occurs in the RPE. During the pathology termed retinal detachment, a variable amount of fluid accumulates between photoreceptors and the RPE. This is potentially dangerous since the detachment of the neural retina drives to blindness. It has been demonstrated that the activation of P2Y receptors by ATP, UTP, and other dinucleotides (as well as NA) can reverse this process permitting the reabsorption of the fluid and therefore permitting the restoration of the right retinal conditions.

Conclusions

It is not easy to join all the experimental data coming from different animal models when describing the role of the nervous system in the iris and ciliary body. Very often, there are differences among species, which need to be taken into account, and that makes it difficult to unify concepts. This article aims to provide a general view of how the nervous system can control relevant processes within the eye, but one should be aware that there are cases in which the results presented here may not fit with the role of the nervous system in certain animal models.

See also: Autonomic Disorders; Autonomic Neuroplasticity and Aging; Diabetic Neuropathy; Retinal Pharmacology: Inner Retinal Layers; Retinal Development: An Overview; Sensory Ganglia; Vision: Light and Dark Adaptation.

Further Reading

Abdel-Latif AA (1989) Calcium-mobilizing receptors, polyphosphoinositides, generation of second messengers and contraction in the mammalian iris smooth muscle: Historical perspectives and current status. *Life Sciences* 45: 757–786.

Bergmanson JPG (1982) Neural control of intraocular pressure. *American Journal of Optometry and Ophysiological Optics* 59: 94–98.

Bognar IT, Wesner MT, and Fuder H (1990) Muscarine receptor types mediating autoinhibition of acetylcholine release and sphincter contraction in the guinea-pig iris. *Naunyn-Schmiedeberg's Archives of Pharmacology* 341: 22–29.

Burnstock G and Sillito AM (2000) *Nervous Control of the Eye.* Amsterdam: Harwood Academic Publishers.

Davson H (1993) The aqueous humour and the intraocular pressure. In: *Physiology of the Eye,* 5th edn., pp. 34–95. New York: Pergamon Press.

Fuder H and Muth U (1993) ATP and endogenous agonists inhibit evoked [^{3}H]-noradrenalin from field stimulated rat iris via A_1 and P2Y like receptors. *Naunyn-Schmiedeberg's Archives of Pharmacology* 348: 352–357.

Gallego R and Belmonte C (1974) Nervous efferent activity in the ciliary nerves related to intraocular pressure changes. *Experimental Eye Research* 19: 331–334.

Jumblatt JE (2000) Innervation and pharmacology of the iris and ciliary body. In: Burnstock G and Sillito AM (eds.) *Nervous Control of the Eye*, pp. 1–40. Amsterdam: Harwood Academic Publishers.

Kaufman PL, Wiedman T, and Robinson JR (1984) Cholinergics. In: Sears ML (ed.) *Handbook of Experimental Pharmacology*, pp. 149–191. Berlin: Springer.

Lütjen-Drecoll J and Rohen E (1989) Morphology of the aqueous outflow pathways in normal and glaucomatous eyes. In: Klein EA (ed.) *The Glaucoma*, pp. 89–123. St. Louis: CV Mosby.

Muramatsu I, Kigoshi S, and Oda Y (1994) Evidence for sympathetic, purinergic transmission in the iris dilator muscle of the rabbit. *Japanese Journal of Pharmacology* 66: 191–193.

Peterson WM, Meggyesy C, Yu K, and Miller SS (1997) Extracellular ATP activates calcium signaling, ion, and fluid transport in retinal pigment epithelium. *Journal of Neuroscience* 17: 2324–2337.

Rohen J (1964) Das auge uns seine hilforgane. In: von Möllendorff W and Bargmann W (eds.) *Handbuch der Mikroskopischen Anatomie des Menshen*, Bd III/2, pp. 189–328. Berlin: Springer.

Autonomic Nervous System: Respiratory Control

B J Undem, The Johns Hopkins University School of Medicine, Baltimore, MD, USA

Introduction

The autonomic nervous system assists in maintaining homeostasis of the internal milieu. In the respiratory system this aids in optimizing the function of gas exchange. Perhaps more important than this, however, is the role the respiratory nervous system plays in host defense. Quietly breathing while reading this article, you are inhaling about 600 ml of air with each breath, which amounts to nearly 10 l of air every minute. The air we breathe contains myriad potential harmful substances, the nature of which depends on the environment in which we are breathing. It is little wonder that a majority of nerves in the respiratory tract are afferent polymodal 'nociceptive' nerves designed to detect potentially noxious chemical perturbations. The afferent nerves communicate with the central nervous system, thereby evoking obvious defensive reflexes such as apnea, sneezing, and coughing. More subtly, these afferent nerves can assist in defending the airspace by initiating autonomic reflexes, resulting in bronchoconstriction, mucus secretion, and alterations in blood flow. The following sections provide a basic overview of the innervation of the respiratory tract with these functions in mind.

Upper Respiratory Tract (Nasal Innervation)

A schematic of the innervation of the nasal respiratory mucosa is depicted in **Figure 1**.

Afferent Innervation of Upper Airways

Sensory nerves innervating the nonolfactory respiratory nasal epithelium are derived from the ophthalmic and maxillial branches of the trigeminal ganglia (the sensory nerves in the olfactory epithelium are not reviewed here). Neurons situated in the trigeminal ganglia project both nociceptive capsaicin-sensitive C-fibers and mechanosensitive A-fibers to the nasal epithelium. Little is known about the specific consequences of stimulation of nasal A-fibers. Stimulation of nasal C-fibers with capsaicin, however, causes sensations of intense burning, and increases the parasympathetic autonomic outflow to the nasal passages and tear ducts.

Autonomic Innervation of Upper Airways

Parasympathetic Preganglionic parasympathetic innervation of the nasal airways originates from the facial nucleus of the brain stem and the superior salivatory nucleus. Preganglionic fibers follow the greater superficial petrosal nerve and the vidian nerve until they synapse on neurons clustered in the sphenopalatine ganglion (also known as the pterygopalatine ganglion). The preganglionic nerves are cholinergic, and upon stimulation cause nicotinic fast excitatory potentials on the postganglionic neurons.

Postganglionic fibers are distributed to the nasal mucosa via the branches of the posterior nasal nerve, where they ultimately innervate serous and mucous glands, arteries, veins, and arteriovenous anastomoses. Postganglionic parasympathetic nerves in the nasal mucosa release acetylcholine, and in some cases nitric oxide and vasoactive intestinal peptide (and related peptides), onto effector cells. The most extensively distributed of all of the muscarinic receptor subtypes in the nose appears to be the M3 receptor, which is responsible for cholinergic glandular secretions.

Nasal secretions are derived from serous/mucous glands and plasma extravasation from blood vessels in the mucosa. Most of the autonomically mediated secretory effects are due to glandular secretion and are inhibited by drugs that block cholinergic muscarinic receptors, indicating a major role of parasympathetic cholinergic nerves. Consistent with this hypothesis, experimental stimulation of the parasympathetic nerves causes glandular secretions rich in mucous glycoproteins, lactoferrin, lysozyme, secretory leukoprotease inhibitor, neutral endopeptidase, and secretory immunoglobulin A (IgA). In experimental animal models, direct stimulation of parasympathetic nasal nerves can also elicit vasodilatation of vessels in the nasal mucosa through both cholinergic and noncholinergic mechanisms. The noncholinergic component (atropine resistant) of the vascular response to parasympathetic nerve stimulation is likely due to either vasoactive intestinal peptide and/or nitric oxide release from the nerves, as both of these transmitters have been localized to parasympathetic nerves in the nose, and both are effective vasodilators. Although immunohistochemical evidence indicates that there are nitric oxide synthase-containing nerves innervating the human nasal mucosa, a nonadrenergic, noncholinergic (NANC) neural control of the nasal blood flow has not yet been explored.

Sympathetic Most of the sympathetic neural output to the human nose originates from cholinergic

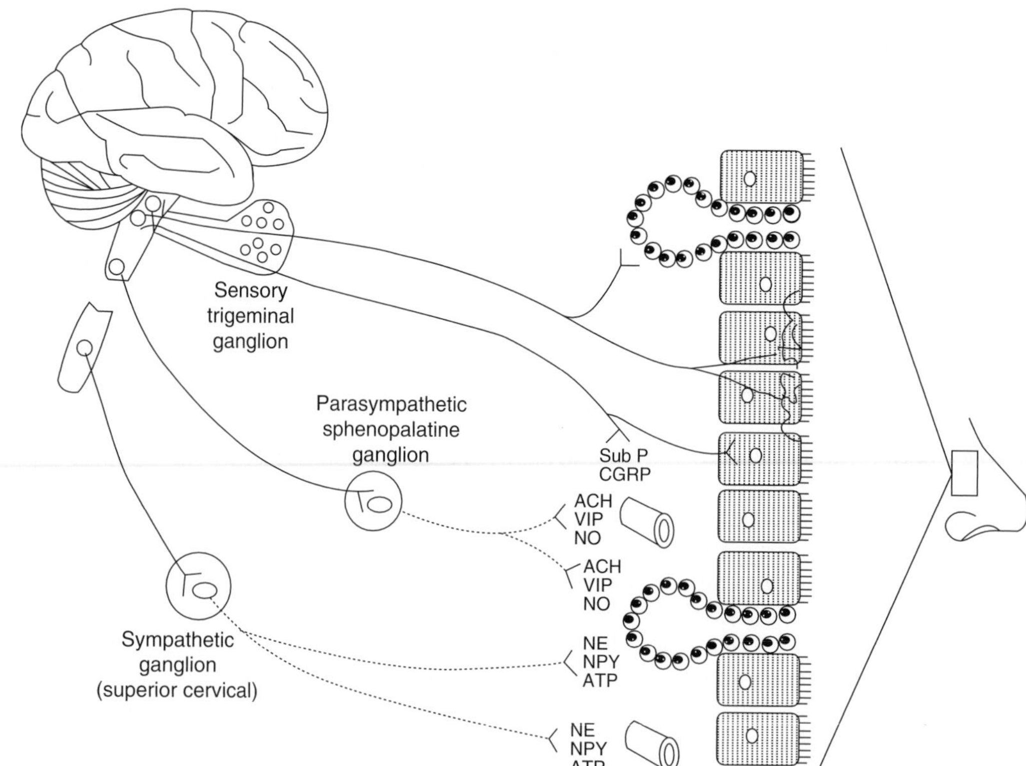

Figure 1 Schematic diagram of the innervation of the nasal respiratory mucosa. The transmitters noted are acetylcholine (ACH), adenosine 5′-triphosphate (ATP), vasoactive intestinal peptide (VIP), nitric oxide (NO), substance P (Sub P), calcitonin gene-related peptide (CGRP), norepinephrine (NE), and neuropeptide Y (NPY). These transmitters and neuropeptides are commonly found in the nasal mucosa, but should not be considered an exclusive list. The cholinergic and the nonadrenergic, noncholinergic (VIP and NO) branches of the parasympathetic system are depicted as arising from the same neurons, but it is also possible that the cholinergic nerves are distinct from the nonadrenergic, noncholinergic nerves, as has been observed in the lower airways.

preganglionic fibers in the thoracolumbar region of the spinal cord; these fibers synapse on neurons in the superior cervical ganglion. The postganglionic fibers form the petrosal nerve, which joins the greater superficial nerve to form the vidian nerve. The vidian nerve therefore contains both the parasympathetic and the sympathetic innervation to the nasal mucosa. The postganglionic sympathetic nerves in the mammalian nose typically contain catecholamines, but may also contain adenosine 5′-triphosphate (ATP) and neuropeptides, most notably neuropeptide Y (NPY).

The major consequence of sympathetic nerve stimulation is vasoconstriction and increases in nasal airway patency. There is also some evidence that sympathetic activity can induce airway glandular secretion through stimulation of serous cells. The effects of sympathetic nerves on nasal blood flow are mediated mainly through adrenoceptors. Stimulation of α1- and α2-adrenoceptors on the smooth muscle of resistance vessels (which control blood flow) and of venous sinusoids (which are responsible for blood pooling, leading to mucosal engorgement and modulation of nasal airway resistance) leads to vasoconstriction and, consequently, to reduced blood flow and reduced blood pooling. Some individuals that take α-adrenoceptor antagonists for treatment of their hypertension experience nasal congestion as a side effect. Conversely, α-adrenoceptor agonists are commonly used in nasal sprays to treat 'stuffy' noses. Activation of β1- and β2-adrenoceptors, also present on the vasculature of the nasal mucosa, can lead to vasodilation of resistance vessels and increased blood flow. The effects of β-receptor stimulation, however, are less pronounced than are those induced by α-receptor stimulation. Both NPY and ATP, transmitters often co-localized with norepinephrine in the nasal mucosa, also cause vasoconstriction and decongestion. These transmitters likely account for any nonadrenergic component of sympathetic nasal vasoconstriction.

Autonomic Reflexes in the Nose

Activation of afferent nerves in the respiratory mucosa can lead to sensations of temperature changes, touch, pain, sneezing, changes in breathing pattern, and changes in autonomic outflow to the nose. Experimental stimulation of nociceptive C-fibers with capsaicin application to one nostril invariably leads to secretions from both the ipsilateral and the contralateral nostrils. The contralateral secretion can be prevented by treatment with muscarinic receptor antagonists, indicating a C-fiber-initiated cholinergic reflex. Activation of nasal afferent nerves can also alter autonomic outflow to the tear ducts and to the lower respiratory tract.

In a majority of human beings, there is a cycling of resistance to nasal airflow between the two nostrils. This so-called nasal cycle is thought to be due to unilateral cyclical changes in parasympathetic and sympathetic drive to the nasal vasculature. The congestion occurs concurrently with a decrease in sympathetic drive, but some evidence also implicates a role for a cyclical increase in parasympathetic drive.

Axon reflexes Generally speaking, stimulation of nasal sensory nerves regulates end-organ activity by increasing or decreasing autonomic reflex tone. However, there is substantial evidence for the existence of local reflex regulation of end-organ activity that occurs independently of centrally mediated autonomic response via sensory axon reflex mechanisms. Subsets of C-fibers innervating the nasal mucosa contain neuropeptides such as substance P, neurokinin A, and calcitonin gene-related peptide in their peripheral (and central) terminals, which can release neuropeptides upon activation. The axon-reflex concept envisages an action potential discharge initiated at a terminal site of the axon reaching a branch in the nerve, where it then antidromically (away from the central nervous system) travels back down a collateral branch of the same nerve, resulting in the local release of bioactive peptides. Release of neuropeptides from sensory nerve endings can lead not only to vasodilatation, but also to increased leukocyte infiltration and increased vascular permeability. These events are collectively referred to as 'neurogenic inflammation.' Stimulation of neuropeptide-containing C-fibers with capsaicin has been reported to evoke neurogenic inflammatory responses in the human nose, and these effects may be amplified when studied in the context of ongoing nasal inflammation (rhinitis).

Lower Respiratory Tract

A schematic of the innervation of the bronchial wall is illustrated in **Figure 2**. The trachea and bronchi are richly innervated by autonomic and sensory nerve fibers. In guinea pigs and rodents a majority of airway nerves are found in the subepthelial nerve plexus, and in a more superficial plexus near the smooth muscle layer. This gets more complex in larger mammals. In the dog trachea, for example, five separate nerve plexi have been identified. Beginning from the serosal surface there is paratracheal plexus, an intramuscular plexus, an outer submucosal plexus, an inner submucosal plexus, and, finally, a mucosal plexus.

Sensory Nerves in the Lower Respiratory Tract

The majority of afferent fibers reach the lower respiratory tract via the vagus nerves. Among the 8000 or so vagal afferent fibers innervating cat lungs, morphological studies reveal that about 5000 of these are C-fibers, with the remaining 3000 being myelinated A-fibers.

The vagal C-fibers innervating the lower respiratory tract have been subdivided into those innervating large airways (bronchial C-fibers), and those innervating more peripheral structures (pulmonary C-fibers). Based on studies in guinea pigs and rodents it appears that these C-fiber subtypes may originate in different vagal ganglia, with the C-fibers innervating the large airways being derived mainly from the jugular or supranodose ganglia, and those found deeper in the tissue arising from nodose ganglia. The nodose and jugular ganglia have different embryological origins, and this may explain why the nodose and jugular C-fibers in the respiratory tract have distinctly different phenotypes.

Most A-fibers in the lungs are relatively low-threshold mechanosensors that can be activated by lung distension. These 'stretch receptors' have been subdivided into rapidly adapting receptors (RARs) and slowly adapting receptor (SARs) based on the action potential accommodation to prolonged suprathreshold lung distension. Another vagal fiber phenotype that conducts action potentials at a velocity faster than C-fibers do, but slower than RAR or SAR fibers do, has been described in the extrapulmonary airways of the guinea pig. These fibers are not stretch receptors, and evoke the cough reflex when stimulated. Myelinated afferent fibers that do not fit into either the RAR or SAR class are also found, along with a subset of C-fibers, innervating neuroepithelial bodies (NEBs) in the airway epithelium. The function of NEB-associated afferent nerves has not yet been worked out.

Although the vast majority of afferent fibers innervating the lower respiratory tract are vagal, sensory neurons situated in the thoracic dorsal root ganglia also project fibers to the lungs. These are thought to be primarily C-fibers, but to date they have not been studied in detail.

Autonomic Innervation of Lower Respiratory Tract

Parasympathetic The parasympathetic branch of the autonomic nervous system is the dominant neural regulator of intra- and extrapulmonary airways. The preganglionic parasympathetic neurons are situated

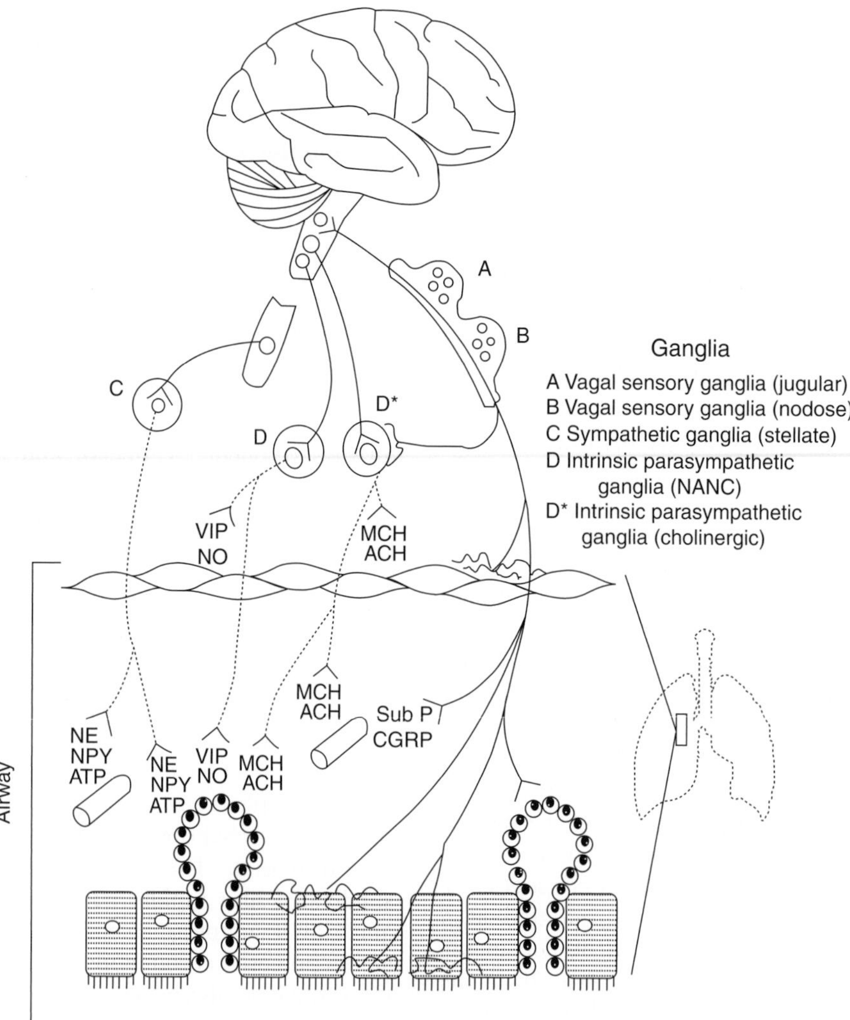

Figure 2 Schematic diagram of the innervation of the bronchi. The transmitters noted are acetylcholine (ACH), adenosine 5′-triphosphate (ATP), vasoactive intestinal peptide (VIP), nitric oxide (NO), substance P (Sub P), calcitonin gene-related peptide (CGRP), norepinephrine (NE), neuropeptide Y (NPY), and methacholine (MCH); the intrinsic parasympathetic ganglia are nonadrenergic, noncholinergic (NANC). The transmitters and neuropeptides depicted are the most commonly studied ones, but should not be considered an exclusive list. Not depicted is the afferent innervation derived from spinal (dorsal root) sensory ganglia. The spinal ganglia are found mainly in the lungs, but where and if the terminate in the airway wall have not been clearly established.

in the brain stem in and around the compact formation of the nucleus ambiguous and the dorsal motor nucleus of the vagus, and in the reticular formation located between these nuclei.

The preganglionic parasympathetic fibers arise at the airway via the vagus nerves, where they synapse with principal ganglia neurons (postganglionic neurons) that are located in small ganglia within or near the airway wall. Much of the preganglionic input is filtered at the parasympathetic ganglia because the excitatory postsynaptic potentials fail to reach action potential threshold. Parasympathetic ganglia are located mainly in the larger airways, but the postganglionic fibers innervate the conducting airways to the bronchioles. As a consequence, the parasympathetic ganglion neurons function to filter, integrate, and distribute parasympathetic activity throughout the branches of the airway tree. The number of neurons within an intrinsic airway parasympathetic ganglion ranges from a few to nearly hundred. Investigators have estimated the total number of intrinsic parasympathetic neurons in the airways of several species, including mouse (about 200), guinea pig (about 250), rat (about 750), ferret (about 4000), and dog (about 15 000).

Neurons in the bronchial parasympathetic ganglia are cholinergic and NANC or both. As in the upper airways, the major parasympathetic NANC neurotransmitters identified thus far in human airways are nitric oxide and vasoactive intestinal peptide, but in some species tachykinins such as substance P have also been localized in intrinsic airway parasympathetic neurons. Several studies have provided evidence that the cholinergic innervation and NANC parasympathetic innervation reflect discrete parasympathetic pathways. In ferrets, the NANC neurons are situated in ganglia that are more superficial than are the ganglia containing mainly cholinergic neurons. In the guinea pig trachea and main bronchi, the parasympathetic NANC nerve fibers are actually derived from neurons situated in ganglia associated with the myenteric plexus of the esophagus (this may explain why the trachea has relatively few intrinsic neurons compared to the rat; see earlier). The preganglionic fibers that innervate the NANC airway-targeted neurons in the guinea pig esophagus are also distinct from those that innervate the cholinergic neurons within the airway. Therefore, in this species at least, separate reflex pathways may exist for parasympathetic cholinergic responses and parasympathetic noncholinergic responses. Circumstantial evidence also supports the hypothesis of separate cholinergic and NANC parasympathetic pathways in cat airways.

The parasympathetic nervous system is the dominant regulator of airway smooth muscle tone in all mammals, including humans. Stimulation of vagal parasympathetic nerves leads to bronchial smooth muscle contraction in all mammals. Invariably, the parasympathetic contractions are the result of cholinergic muscarinic receptor activation. Although bronchial smooth muscle expresses both muscarinic M2 and M3 receptors, the latter are responsible for the smooth muscle contractions. This explains the utility of inhaled antimuscarinic drugs such as ipratropium, and the pseudo-irreversible tiotropium, in the treatment of obstructive lung diseases. There are muscarinic M2 receptors on the cholinergic terminals that serve to inhibit acetylcholine release. Thus, selectively blocking muscarinic M2 receptors in human bronchi can actually enhance nerve-evoked cholinergic contractions.

In many species, including humans, the parasympathetic nervous system also provides relaxant innervation to bronchial smooth muscle. Drugs that block the production or action of nitric oxide largely inhibit the parasympathetically mediated relaxations. As mentioned earlier, at least in guinea pigs, the parasympathetic relaxant innervation is distinct from the cholinergic innervation. In theory, acetylcholine and the noncholinergic transmitters may be co-released from the same nerve terminals. Such co-release, however, has not yet been observed at the level of airway smooth muscle.

Tracheobronchial submucosal glands and goblet cells receive both parasympathetic and sympathetic efferent innervations, though they are dominated by parasympathetic tone. The neurogenic increase in mucus secretion in human airways is largely cholinergic via muscarinic M3 receptor stimulation; however, noncholinergic transmitters, such as vasoactive intestinal peptide, associated with parasympathetic nerves also stimulate mucus secretion. The parasympathetic system may play a role in mucociliary responses. Administration of atropine, a muscarinic receptor antagonist, decreases ciliary beat frequency, suggesting that the role is stimulatory.

Increases in parasympathetic outflow to the airways results in vasodilatation and increases in blood flow. This occurs in both the pulmonary and bronchial (systemic) circulation. In experimental animals, parasympathetic nerve-mediated relaxation of the bronchial vasculature is mediated through both cholinergic and noncholinergic (likely nitric oxide and vasoactive intestinal peptide) mechanisms.

Sympathetic The mammalian respiratory tract receives sympathetic innervation mainly from cell bodies situated in the stellate ganglia. Postganglionic sympathetic nerves typically use norepinephrine as the major neurotransmitter. In some cases NPY and ATP are released along with norepinephrine as co-transmitters. Much of the early work on airway innervation was carried out on dogs. This work led to the conclusion, still reprinted in many textbooks, that the parasympathetic system provides cholinergic contractile innervation to the bronchial smooth muscle, and this is apposed by the sympathetic nervous system, which provides adrenergic relaxant innervation. Indeed, stimulation of sympathetic nerves in the dog leads to β-adrenoceptor-mediated bronchodilation. This is not the case, however, in most species. In guinea pigs, cats, monkeys, and humans, the major, if not only, relaxant innervation to the airway smooth muscle is provided by the parasympathetic nervous system, as already described. Sympathetic, β-adrenoceptor-mediated, bronchial smooth muscle relaxation can occur, however, in all species thus far studied via circulating epinephrine derived from the adrenal medulla.

As in other organs, the sympathetic nervous system innervates the blood vessels in the airways and lungs. The pulmonary and bronchial circulations are under adrenergic control, and perhaps also under sympathetic nonadrenergic (e.g., NPY) control. Studies on intrapulmonary pulmonary arteries have indicated that ATP may also contribute to sympathetic nonadrenergic vasoconstriction. Sympathetic transmitters can constrict and relax vascular smooth muscle, but in the vast majority of studies in experimental animals, the net effect of sympathetic nerve stimulation is vasoconstriction and decreased blood flow.

Autonomic Reflexes in the Lower Airways

The parasympathetic preganglionic neurons are rhythmically active during eupnic breathing and send volleys of action potentials down the preganglionic axons. Intracellular recordings of tracheal ganglion neurons in the cat reveal rhythmic preganglionic volleys in excess of 25 Hz. This baseline activity appears to be due to the effect of pattern generators within the central nervous system, and to reflex activity evoked through the stimulation of certain vagal mechanosensors as a consequence of lung expansion following inspiration. Artificially increasing the respiratory rate increases the baseline parasympathetic drive. This is thought to be due to activation of vagal RAR stretch receptors. The baseline parasympathetic drive to the airways can be essentially terminated by a prolonged deep inspiration, presumably due to activation of vagal SAR stretch receptors. In other words, RARs and SARs have apposing effects on reflex parasympathetic tone.

Both cholinergic and NANC parasympathetic fibers are active during normal respiration in airways, but the net effect is one of cholinergic constriction. Studies using computed tomography indicate that the luminal diameters of small and midsized human bronchi are reduced about 15% by the tonic descending parasympathetic tone. When the vagi are cut (in experimental animals), or if large doses of antimuscarinic drugs are administered, the baseline smooth muscle tone is essentially abolished and the airways maximally dilate.

Inhalation of an irritant or potentially noxious stimulus that is capable of activating nociceptive C-fibers in the airway mucosa will lead to increases in parasympathetic drive and the consequential airway narrowing, mucus secretion, and vasodilation. Several inflammatory mediators are adept at evoking action potential discharge in airway vagal nociceptors, and it is thought that the resulting increase in parasympathetic reflex nerve activity may contribute to the pathophysiology of inflammatory airway diseases (e.g., chronic obstructive pulmonary disease, asthma).

Hypoxic pulmonary vasoconstriction is a major mechanism in the prevention of arterial hypoxemia. There is some evidence that reflex sympathetic outflow to the pulmonary circulation may contribute to this important phenomenon.

Peripheral reflexes A peripheral autonomic reflex in which sensory nerves directly communicate with intrinsic autonomic nerves, independent of the central nervous system, may occur in the lower airways, although this is certainly less prevalent than in the enteric nervous system. Neurokinin-containing C-fiber terminals are found near many bronchial parasympathetic ganglion neurons in several species, including humans. Direct electrophysiological recordings have demonstrated that stimulation of these C-fibers can evoke membrane depolarization and occasionally action potential discharge in the principal ganglion neurons. In guinea pig and human airways, the C-fiber-mediated excitatory postsynaptic potentials are due to stimulation of neurokinin (NK-3) receptors. Consistent with the electrophysiological studies, blockers of NK-3 receptors can inhibit cholinergic bronchoconstriction *in vivo* caused by vagus nerve stimulation (presumably by decreasing the synaptic efficacy within the intrinsic ganglia). There is anatomical evidence of intraganglionic communication within the airways. Specifically, postganglionic parasympathetic NANC fibers have been observed within cholinergic parasympathetic ganglia. The functional significance of this type of intraganglionic integration is not known.

Axon reflexes C-fiber axon reflex-induced neurogenic inflammation (as discussed previously) has been thoroughly investigated in the airways of guinea pigs and rodents. In these species, antidromic stimulation of vagal C-fibers leads to pronounced plasma extravasation and infiltration of leukocytes. The plasma leakage occurs at the postcapillary venules of the bronchial circulation. In guinea pigs, axon reflex responses may also include bronchial smooth muscle contraction via activation of neurokinin NK-1 and NK-2 receptors on the airway smooth muscle.

The effect of axon reflexes in human airways is less obvious. The density of peptidergic C-fiber innervation in the airways of guinea pigs and rats is much greater than that seen in humans. It is therefore likely that the neurogenic inflammatory response that occurs in the lower airways of these species overstates the case seen in humans.

See also: Autonomic Nervous System: Central Respiratory Control; Autonomic Nervous System; Brainstem Respiratory Circuits; Evolution of Vertebrate Respiratory Control; Parasympathetic Nervous System; Respiration; Sympathetic Nervous System.

Further Reading

Brouns I, De Proost I, Pintelon I, et al. (2006) Sensory receptors in the airways: Neurochemical coding of smooth muscle-associated airway receptors and pulmonary neuroepithelial body innervation. *Autonomic Neuroscience* 126–127: 307–319.

Canning BJ (2006) Reflex regulation of airway smooth muscle tone. *Journal of Applied Physiology* 101: 971–985.

Canning BJ and Mazzone SB (2005) Reflexes initiated by activation of the vagal afferent nerves innervating the airways and lungs. In: Undem BJ and Weinreich D (eds.) *Advances in Vagal Afferent Neurobiology*, pp. 403–430. Boca Raton, FL: CRC Press and Taylor & Francis Group.

Carr MJ and Undem BJ (2003) Bronchopulmonary afferent nerves. *Respirology* 8: 291–301.

Coleridge HM and Coleridge JC (1994) Neural regulation of bronchial blood flow. *Respiratory Physiology* 98: 1–13.

Coleridge HM and Coleridge JC (1994) Pulmonary reflexes: Neural mechanisms of pulmonary defense. *Annual Review of Physiology* 56: 69–91.

Ellis JL and Undem BJ (1994) Pharmacology of non-adrenergic, non-cholinergic nerves in airway smooth muscle. *Pulmonary Pharmacology* 7: 205–223.

Kubin L, Alheid GF, Zuperku EJ, et al. (2006) Central pathways of pulmonary and lower airway vagal afferents. *Journal of Applied Physiology* 101: 618–627.

Laitinen LA and Laitinen A (1997) Innervation of the airways. In: Barnes PJ (ed.) *Autonomic Control of the Respiratory System*, pp. 1–21. Amsterdam: Harwood Academic Publishers.

Rogers DF (2002) Pharmacological regulation of the neuronal control of airway mucus secretion. *Current Opinions in Pharmacology* 2: 249–255.

Sarin S, Undem B, Sanico A, et al. (2006) The role of the nervous system in rhinitis. *Journal of Allergy and Clinical Immunology* 118: 999–1016.

Undem BJ and Myers AC (1997) Autonomic ganglia. In: Barnes PJ (ed.) *Autonomic Control of the Respiratory System*, pp. 1–21. Amsterdam: Harwood Academic Publishers.

Widdicombe JG (1991) Neural control of airway vasculature and edema. *American Review of Respiratory Disease* 143: S18–S21.

Wine JJ (2007) Parasympathetic control of airway submucosal glands: Central reflexes and the airway intrinsic nervous system. *Autonomic Neuroscience* 133: 35–54.

Autonomic Nervous System: Urogenital Control

W C de Groat, University of Pittsburgh Medical School, Pittsburgh, PA, USA

Introduction

The urogenital system has two major subdivisions, the urinary tract and the genital tract. The urinary tract, consisting of the kidneys, ureters, urinary bladder, and urethra, is responsible for the production, storage, and elimination of urine. The genital tract, consisting of the penis, clitoris, vagina, testes, ovaries, and uterus, is involved in reproductive activity. Many functions of the urogenital system are controlled by autonomic (parasympathetic and sympathetic) and somatic efferent pathways originating in the lumbosacral spinal cord (**Figure 1**). These pathways regulate the activity of smooth muscle, striated muscle, epithelial cells, and exocrine glands in the urogenital organs via the release of multiple neurotransmitters, including acetylcholine (ACh), norepinephrine, adenosine triphosphate (ATP), nitric oxide (NO), and neuropeptides. This article reviews the innervation of the urogenital organs, the functions controlled by the different neural pathways, and the physiology and pharmacology of neurotransmission.

Lower Urinary Tract

Anatomy

The storage and periodic elimination of urine is controlled by the activity of two functional units in the lower urinary tract, (1) a reservoir (the urinary bladder) and (2) an outlet (consisting of bladder neck, urethra, and striated sphincter muscles). Under normal conditions, the urinary bladder and urethral outlet exhibit a reciprocal activity. During storage, the bladder neck and proximal urethra are closed and the detrusor muscle is quiescent, allowing intravesical pressure to remain low (5–15 cmH_2O) over a wide range of bladder volumes. During voluntary micturition the initial event is a reduction of intraurethral pressure, which reflects a relaxation of the pelvic floor and the paraurethral striated muscles. These changes are followed in a few seconds by a detrusor contraction and a rise in intravesical pressure that is maintained until the bladder empties.

Innervation

The innervation of the lower urinary tract is derived from three sets of peripheral nerves: sacral parasympathetic (pelvic nerves), thoracolumbar sympathetic (hypogastric nerves and sympathetic chain), and sacral somatic nerves (primarily the pudendal nerves) (**Figure 1**).

Parasympathetic pathways The sacral parasympathetic outflow, which in humans originates from the S2 to S4 segments of the spinal cord, provides the major excitatory input to the bladder. Cholinergic preganglionic neurons located in the intermediolateral region of the sacral spinal cord send axons to cholinergic ganglion cells in the pelvic plexus and in the bladder wall (**Figure 2**). Transmission in bladder ganglia is mediated by a nicotinic cholinergic mechanism, which is sensitive to modulation by various transmitter systems, including muscarinic, adrenergic, purinergic, and peptidergic (**Table 1**; **Figure 2**). The ganglion cells in turn excite the bladder smooth muscle. A large proportion of the ganglia and nerves supplying the human lower urinary tract contain acetylcholinesterase (AChE) as well as the vesicular ACh transporter (VAChT), and therefore must be cholinergic. AChE- and VAChT-positive nerves are abundant in all parts of the bladder but are less extensive in the urethra. Neuropeptide Y (NPY) and nitric oxide synthase (NOS) have also been identified in a large percentage (40–95%) of intramural ganglia of the human bladder. Several populations of axonal varicosities have been detected in close proximity to intramural ganglion cells, including (1) substance P-, vasoactive intestinal polypeptide (VIP)-, and calcitonin-gene related peptide-positive axons which are presumably collaterals of extrinsic sensory nerves; (2) tyrosine hydroxylase and NPY axons which are likely to be sympathetic axons; and (3) galanin- and NPY-containing axons; which are thought to be preganglionic nerve terminals.

Parasympathetic neuroeffector transmission in the bladder is mediated by ACh acting on postjunctional muscarinic (M) receptors. Both M_2 and M_3 muscarinic receptor subtypes are expressed in bladder smooth muscle (**Figure 3**); however, studies with subtype-selective muscarinic receptor antagonists and muscarinic receptor knockout mice have revealed that the M_3 subtype is the principal receptor involved in excitatory transmission.

In bladders of various animals, stimulation of parasympathetic nerves also produces a noncholinergic contraction that is resistant to atropine and other muscarinic receptor-blocking agents. ATP (**Table 1**) has been identified as the excitatory transmitter mediating the noncholinergic contraction. ATP excites the bladder smooth muscle by acting on P2X receptors

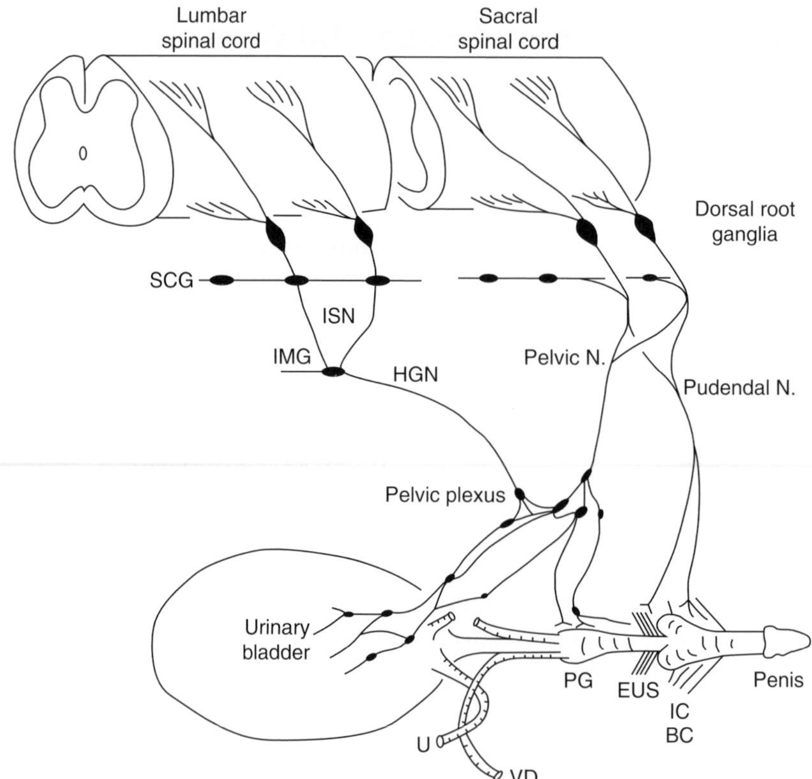

Figure 1 Diagram showing the sympathetic, parasympathetic, and somatic innervation of the urogenital tract of the male cat. Sympathetic preganglionic pathways emerge from the lumbar spinal cord and pass to the sympathetic chain ganglia (SCG) and then via the inferior splanchnic nerves (ISN) to the inferior mesenteric ganglia (IMG). Preganglionic and postganglionic sympathetic axons then travel in the hypogastric nerve (HGN) to the pelvic plexus and the urogenital organs. Parasympathetic preganglionic axons, which originate in the sacral spinal cord, pass in the pelvic nerve to ganglion cells in the pelvic plexus and to distal ganglia in the organs. Sacral somatic pathways are contained in the pudendal nerve, which provides an innervation to the penis, the ischiocavernosus (IC), bulbocavernosus (BC), and external urethral sphincter (EUS) muscles. The pudendal and pelvic nerves also receive postganglionic axons from the caudal sympathetic chain ganglia. These three sets of nerves contain afferent axons from the lumbosacral dorsal root ganglia. Abbreviations: U, ureter; PG, prostate gland; VD, vas deferens.

that are ligand-gated ion channels. Among the seven types of P2X receptors that have been identified in the bladder, the $P2X_1$ subtype is the major subtype expressed in the rat and human bladder smooth muscle. Although purinergic excitatory transmission is not important in the normal human bladder, it has been identified in bladders from patients with pathological conditions such as chronic urethral outlet obstruction or interstitial cystitis.

Smooth muscle contractions are initiated by an increase in intracellular Ca^{2+} concentration that can occur by intracellular release of Ca^{2+} from the sarcoplasmic reticulum or by influx of Ca^{2+} from the extracellular fluid. The former mechanism is an essential step in the cholinergic activation of the detrusor muscle. It has been shown that stimulation of M_3 receptors triggers the formation of inositol trisphosphate (IP_3) and this in turn activates IP_3 receptors on the sarcoplasmic reticulum, which then causes the release of Ca^{2+} (**Figure 3**). On the other hand, activation of P2X purinergic receptors causes the influx of extracellular Ca^{2+} as well as depolarization of the cells, leading to an opening of voltage-gated Ca^{2+} channels. This triggers intracellular Ca^{2+}-induced Ca^{2+} release by activation of ryanodine-sensitive receptors in the sarcoplasmic reticulum. Intracellular Ca^{2+} combines with calmodulin to activate the contractile proteins. Activation of M_2 muscarinic receptors also appears to enhance contractions by suppressing β-adrenergic inhibitory mechanisms by blocking adenylyl cyclase (**Figure 3**) or K^+ channels.

Parasympathetic pathways to the urethra induce relaxation during voiding. In various species, the relaxation is not affected by muscarinic antagonists and therefore is not mediated by ACh. However, inhibitors of NOS block the relaxation *in vivo* during reflex voiding or block the relaxation of urethral smooth muscle strips induced *in vitro* by electrical stimulation of intramural nerves, indicating that NO is the inhibitory transmitter involved in relaxation. NO increases the levels of cyclic guanosine monophosphate (GMP) by stimulating guanylyl cyclase. Cyclic

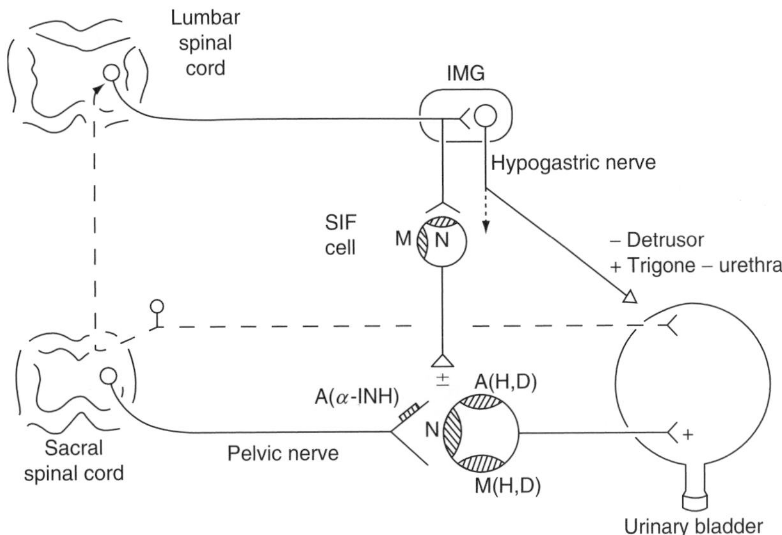

Figure 2 Diagram showing the autonomic innervation of the urinary bladder of the cat and the synaptic mechanisms within bladder ganglia. Nicotinic (N), muscarinic (M), and adrenergic (A) receptors are depicted on a principal ganglion cell and a small intensely fluorescent (SIF) cell. Receptors mediating hyperpolarization (H) and depolarization (D) are also indicated. An α-adrenergic receptor (A, α-INH) mediating presynaptic inhibition is indicated on the preganglionic nerve terminal. Inhibitory and excitatory synaptic mechanisms are designated by – and +, respectively. Postsynaptic adrenergic and muscarinic receptors mediate both hyperpolarizing and depolarizing responses. Sympathetic preganglionic axons make synaptic contact with cells in the inferior mesenteric ganglion (IMG) and send axons through the IMG to make synaptic contact with SIF cells in bladder ganglia. SIF cells have both nicotinic and muscarinic excitatory receptors. Sympathetic efferent pathways can be activated by afferent projections from the urinary bladder.

Table 1 Receptors for putative transmitters in the lower urinary tract

Tissue	*Cholinergic*	*Adrenergic*	*Other*
Bladder body	$+(M_2)$ $+(M_3)$	$-(\beta_2)$ $-(\beta_3)$	+Purinergic ($P2X_1$) –VIP +Substance P (NK-2)
Bladder base	$+(M_2)$ $+(M_3)$	$+(\alpha_1)$	–VIP +Substance P (NK-2) +Purinergic (P2X)
Urothelium	$+(M_2)$ $+(M_3)$	$+\alpha$ $+\beta$	+TRPV1 +TRPM8 +P2X +P2Y +Substance P +Bradykinin (B2)
Urethra	+(M)	$+(\alpha_1)$ $+(\alpha_2)$ $-(\beta)$	+Purinergic (P2X) –VIP –Nitric oxide
Sphincter striated muscle	+(N)		
Adrenergic nerve terminals	$-(M_{2/4})$ $+(M_1)$	$-(\alpha_2)$	–NPY
Cholinergic nerve terminals	$-(M_{2/4})$ $+(M_1)$	$+(\alpha_1)$	–NPY
Afferent nerve terminals			+Purinergic ($P2X_{2/3}$) +TRPV1
Ganglia	+(N) $+(M_1)$	$+(\alpha_1)$ $-(\alpha_2)$ $+(\beta)$	–Enkephalinergic (δ) –Purinergic (P_1) +Purinergic (P2X) +Substance P

VIP, vasoactive intestinal polypeptide; NPY, neuropeptide Y; TRP, transient receptor potential. Letters in parentheses indicate receptor type (M, muscarinic; N, nicotinic; NK-2, neurokinin-A). Plus and minus signs indicate excitatory and inhibitory effects.

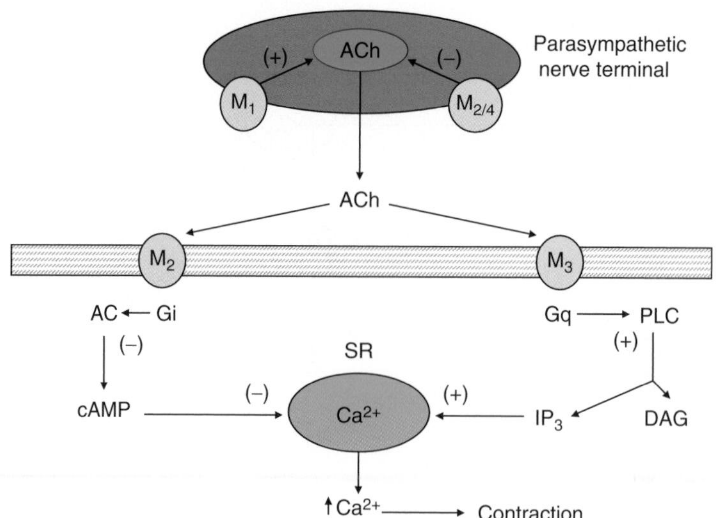

Figure 3 Diagram showing prejunctional modulatory muscarinic receptors on a parasympathetic cholinergic nerve terminal in the urinary bladder. Acetylcholine (ACh) released from the nerve terminals acts on postjunctional M_2 and M_3 muscarinic receptors on the bladder smooth muscle. Activation of M_3 receptors acts via a G-protein (Gq) to stimulate phospholipase C (PLC) and induce formation of diacylglycerol (DAG) and inositol trisphosphate (IP_3); IP_3 in turn triggers the release of Ca^{2+} from the sarcoplasmic reticulum (SR). Increase in Ca^{2+} initiates a muscle contraction. Stimulation of M_2 receptors acts through another G-protein (Gi) to inhibit adenylyl cyclase (AC) and reduce cyclic adenosine monophosphate (cAMP) inhibition of contractile mechanisms.

GMP in turn activates protein kinase G, which produces smooth muscle relaxation by several mechanisms, including activation of K^+ channels and desensitization of the contractile machinery to Ca^{2+}.

Sympathetic pathways Sympathetic preganglionic pathways that arise from the T11 to L2 spinal segments pass to the sympathetic chain ganglia and then to prevertebral ganglia in the superior hypogastric and pelvic plexuses, and also to short adrenergic neurons in the bladder and urethra. Sympathetic postganglionic nerves that release norepinephrine provide an excitatory input to smooth muscle of the urethra and bladder base, an inhibitory input to smooth muscle in the body of the bladder, as well as inhibitory and facilitatory inputs to vesical parasympathetic ganglia (**Figure 2**). Histofluorescence microscopy in animals and humans has shown that the smooth muscle of the bladder base is richly innervated by adrenergic terminals, but the bladder body has a considerably weaker adrenergic innervation. Vesical parasympathetic ganglion cells also receive an extensive adrenergic innervation. Radioligand receptor binding studies show that α-adrenergic receptors are concentrated in the bladder base and proximal urethra, whereas β-adrenergic receptors are most prominent in the bladder body (**Table 1**). These observations are consistent with pharmacological studies showing that sympathetic nerve stimulation or exogenous catecholamines produce β-adrenergic receptor-mediated inhibition of the body and strong α-adrenergic receptor-mediated contractions of the base and urethra and weak contractions of the bladder body. Molecular and contractility studies have shown that β_3-adrenergic receptors elicit inhibition and α_1-adrenergic receptors elicit contractions. The α_{1A}-adrenergic receptor subtype is most prominent in normal bladders, but the α_{1D} subtype is upregulated in bladders from patients with outlet obstruction due to benign prostatic hyperplasia. This finding raises the possibility that enhanced α_1-adrenergic receptor excitatory mechanisms in the bladder body might contribute to irritative lower urinary tract symptoms in these patients with prostate disease.

Activation of β-adrenergic receptors in bladder smooth muscle stimulates adenylyl cyclase and increases cyclic adenosine monophosphate (cAMP), which in turn activates protein kinase A. Protein kinase A is thought to act in part by inducing a hyperpolarization of the cells, either by opening of K^+ channels or by stimulating an electrogenic ion pump. Excitatory responses in the urethra and bladder neck mediated by α_1-adrenergic receptors are attributed to an increased release of Ca^{2+} from intracellular stores.

Neural Modulatory Mechanisms

Presynaptic modulatory mechanisms and synaptic communication between parasympathetic and sympathetic pathways to the bladder have been demonstrated in bladder ganglia and at postganglionic nerve terminals. In the cat, stimulation of sympathetic nerves elicits an initial inhibitory and a delayed facilitatory modulation of cholinergic transmission in parasympathetic bladder ganglia mediated by α_2- and α_1-adrenergic receptors, respectively (**Table 1**;

Figure 2). Transmission in cat bladder ganglia is also modulated by enkephalins released as cotransmitters with ACh from preganglionic nerve terminals. The inhibitory effect of enkephalins can be blocked by the opioid receptor antagonist naloxone and occurs by a presynaptic inhibitory mechanism. Adenosine, presumably derived from ATP released in the ganglia, also exerts an inhibitory action on ganglionic transmission.

Adrenergic and enkephalinergic inhibitory mechanisms in cat bladder ganglia are frequency dependent, being prominent at low frequencies (0.25–0.5 Hz) and markedly reduced at higher frequencies (1–10 Hz) of stimulation. This phenomenon is presumably related to the marked temporal facilitation, which occurs in cat bladder ganglia at frequencies of preganglionic nerve stimulation above 0.5 Hz (**Figure 4**). It has been speculated that cat bladder ganglia function as high-pass filters eliminating parasympathetic input to the bladder when preganglionic activity is low during urine storage, but significantly amplifying the input to the bladder when firing is increased during voiding. Thus the ganglion acts like a gating circuit. Frequency-dependent adrenergic and enkephalinergic inhibitory mechanisms complement this gating function by inhibiting transmission during urine storage, but turning off during voiding to allow complete bladder emptying.

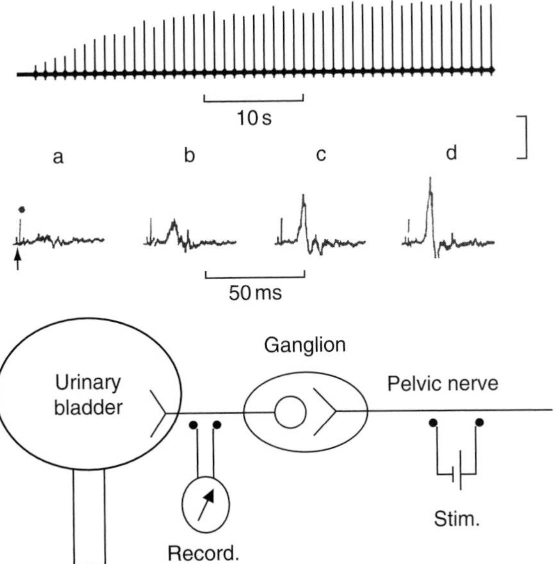

Figure 4 Extracellular recordings on a bladder postganglionic nerve, demonstrating facilitation of bladder ganglionic transmission during repetitive stimulation of preganglionic axons in the pelvic nerve. Top: Action potentials evoked with submaximal (5 V, 1 Hz) stimulation, recorded with a slow time base, showing gradual increase in the amplitude during repetitive preganglionic nerve stimulation. Middle: Sample responses from the preceding action potentials (a, 1st; b, 5th; c, 10th; and d, 20th responses in the series), recorded with a fast time base. Bottom: Diagram showing sites for nerve stimulation and recording.

In the rat, the autonomic pathways to the lower urinary tract are susceptible to modulation at postganglionic sites. For example, NPY, a cotransmitter in cholinergic and adrenergic nerves, acts on postganglionic nerve terminals to suppress the release of both ACh and norepinephrine in the bladder and urethra. The NPY inhibition is also frequency dependent, being most prominent at low frequencies of nerve stimulation. In addition, in the rat urinary bladder, ACh released from parasympathetic nerves can induce a heterosynaptic facilitation and inhibition of norepinephrine release from sympathetic nerves by acting, respectively, on prejunctional M_1 and $M_{2/4}$ muscarinic receptors on the adrenergic terminals. Prejunctional M_1 facilitatory and $M_{2/4}$ inhibitory muscarinic modulatory mechanisms have also been identified on parasympathetic nerve terminals in the rat bladder (**Figure 3**; **Table 1**). These receptors mediate positive and negative cholinergic feedback mechanisms, respectively, that regulate the release of ACh. Inhibitory $M_{2/4}$ mechanisms are dominant at low frequencies of nerve activity and therefore could contribute to urine storage; whereas M_1 facilitatory mechanisms are dominant at high frequencies of nerve stimulation and could contribute to an enhancement of neurally evoked bladder contractions during micturition.

Somatic pathways The external urethral sphincter (EUS), which is composed of striated muscle, receives a somatic innervation via the pudendal nerve from anterior horn cells in the third and fourth sacral segments. Branches of the pudendal nerve and other sacral somatic nerves also carry efferent impulses to muscles of the pelvic floor. Reflex activation of the EUS occurs during bladder filling or during sneezing and coughing, thereby contributing to the maintenance of urinary continence.

Urothelial–Afferent Interactions

The luminal surface of the bladder is covered by a multilayered urothelium, which functions as a highly efficient barrier to movement of water and ionized substances across the bladder wall. Recent studies have also revealed that the urothelium has specialized sensory and signaling properties that allow urothelial cells to respond to their chemical and physical environment and to engage in reciprocal chemical communication with neighboring nerves in the bladder wall or with smooth muscle cells (**Figure 5**). These properties include (1) expression of nicotinic, muscarinic, tachykinin, adrenergic, and capsaicin (TRPV1) receptors; (2) sensitivity to transmitters released from

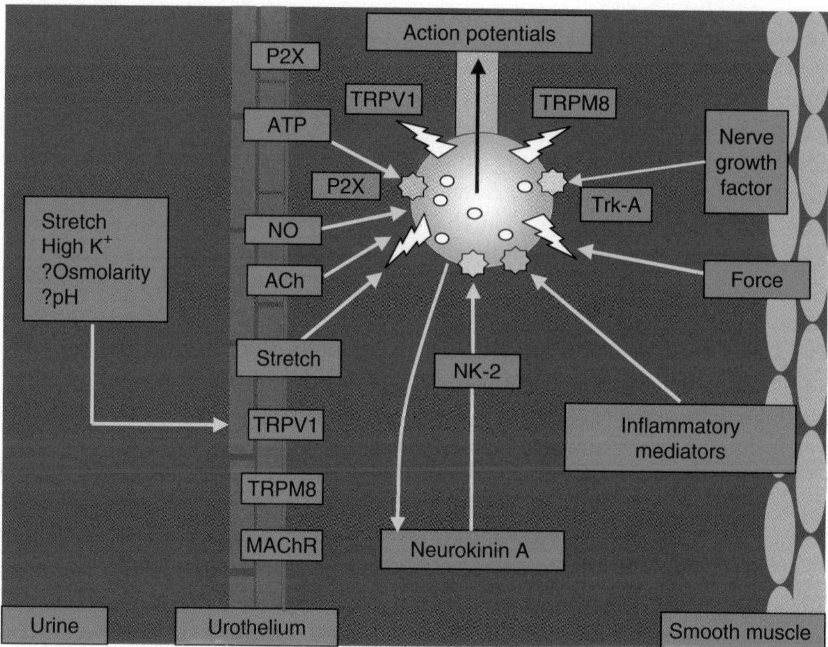

Figure 5 Diagram showing receptors present in the urothelium and in sensory nerve endings in the bladder mucosa, and putative chemical mediators that are released by the urothelium, nerves, or smooth muscle that can modulate the excitability of sensory nerves. Urothelial cells and sensory nerves express common receptors (purinergic, P2X; transient receptor potential vanilloid receptor 1 (sensitive to capsaicin), TRPV1; and transient receptor potential menthol/cold receptor, TRPM8). Distension of the bladder activates stretch receptors and triggers the release of urothelial transmitters, such as adenosine triphosphate (ATP), acetylcholine (ACh), and nitric oxide (NO), that may interact with adjacent nerves. Receptors in afferent nerves or the urothelium (e.g., TRPV1 or P2X) can respond to changes in pH, osmolality, high K^+ concentration, chemicals in the urine, or inflammatory mediators released in the bladder wall. Neuropeptides (neurokinin-A) released from sensory nerves in response to distension or chemical stimulation can act on neurokinin-A (NK-2) autoreceptors to sensitize the mechanosensitive nerve endings. The smooth muscle can generate force, which may influence some mucosal endings. Nerve growth factor released from muscle or urothelium can exert an acute and chronic influence on the excitability of sensory nerves via an action on tyrosine kinase A (Trk-A) receptors.

sensory nerves; (3) close physical association with afferent nerves; (4) ability to release chemical mediators such as ATP, ACh, and NO that can regulate the activity of adjacent nerves and thereby trigger local vascular changes and/or reflex bladder contractions; and (5) release of inhibitory factors that relax smooth muscle.

The role of ATP in urothelial–afferent communication has attracted considerable attention because bladder distension releases ATP from the urothelium and intravesical administration of ATP induces bladder hyperactivity, an effect blocked by administration of P2X purinergic receptor antagonists that suppress the excitatory action of ATP on bladder afferent neurons. Mice in which the $P2X_3$ receptor was knocked out exhibited hypoactive bladder activity and inefficient voiding, suggesting that activation of $P2X_3$ receptors on bladder afferent nerves by ATP released from the urothelium is essential for normal bladder function. It has also been reported that urothelial cells obtained from patients or cats with a chronic painful bladder condition (interstitial cystitis) released significantly larger amounts of ATP in response to mechanical stretching than did urothelial cells from normal patients. This raises the possibility that ATP-mediated signaling between the urothelium and afferent nerves (**Figure** 5) is involved in the triggering of painful bladder sensations.

Transmitter release from the urothelium appears to be mediated by exocytosis associated with vesicle trafficking, similar to the release mechanisms occurring in nerve terminals. Botulinum toxin A blocks release of transmitters from nerves by destroying soluble *N*-ethylmalemide-sensitive factor attachment protein receptors (SNARE proteins; involved in synaptic vesicle trafficking) and also blocks transmitter release from urothelial cells. Based on these observations it has been proposed that the beneficial effects of intravesical injections of botulinum toxin A in the treatment of overactive bladder symptoms may be related in part to the suppression of chemical communication between the urothelium and afferent nerves.

Sex Organs

The physiologic changes initiated by erotic stimuli can be divided into four distinct phases (excitement, plateau, orgasm, and resolution) that have been

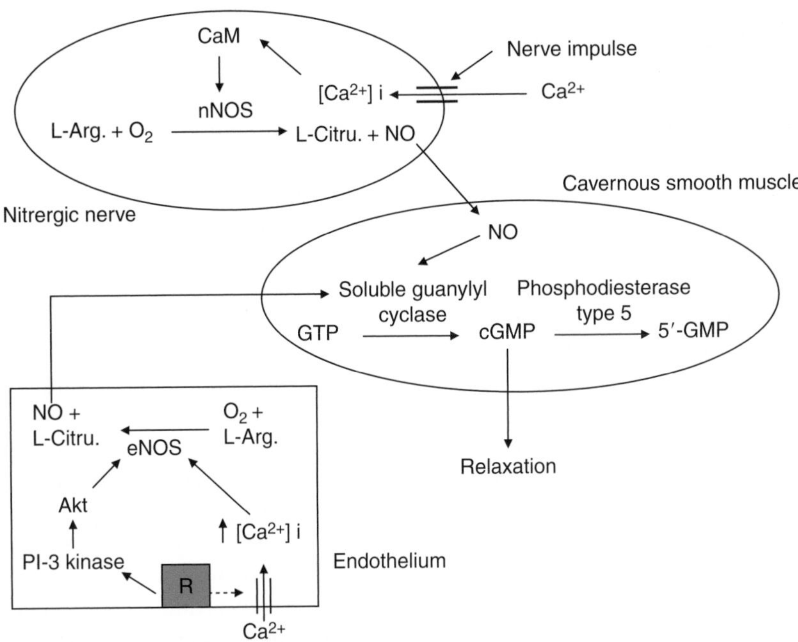

Figure 6 Diagram showing the regulation of cavernosal smooth muscle function by NO liberated from nitrergic nerves and the sinusoidal endothelium. A nerve impulse stimulates Ca^{2+} influx into the nerve terminal followed by synthesis of nitric oxide (NO) by neuronal NO synthase (nNOS). NO released from the nerve terminal activates cyclic guanosine monophosphate (cGMP) formation and relaxation of the cavernous smooth muscle. Changes in blood flow activate shear stress receptors in the endothelium, enhancing the synthesis of NO that elicits further relaxation of the smooth muscle. Nerve stimulation has also been reported to stimulate receptors (R), which activates a signaling mechanism involving phosphatidylinositol 3-kinase (PI-3 kinase) and a serine/threonine protein kinase (Akt), leading to the phosphorylation and activation of endothelial NOS (eNOS) and NO production. Abbreviations: CaM, calmodulin; Arg, arginine; Citru, citrulline.

designated collectively as 'the sexual response cycle.' Although anatomic differences obviously preclude identical responses in males and females during each phase of the cycle, it is clear that similar vascular responses (skin flush, penile and clitoral erection), secretory responses (stimulation of the prostate, bulbourethral gland, glands of Littre in the male; Bartholin's and paraurethral glands in the female), and responses of smooth and striated muscles occur in both sexes. The following sections review the neural control of the peripheral autonomic and somatic pathways involved in male sexual responses.

Innervation

Parasympathetic pathways As described earlier with regard to micturition, three sets of nerves provide an innervation to the urogenital system (**Figure 1**). Parasympathetic preganglionic axons from the sacral spinal cord provide the efferent outflow to erectile tissue in the penis and to the seminal vesicles, prostate, and urethral glands. The sacral pathways have cholinergic ganglionic relay stations in the pelvic plexus and possibly in the effector organs. The postganglionic parasympathetic neurons synthesize and release several transmitters, including NO, ACh, VIP, and ATP. NO is thought to be the major transmitter mediating neurally induced erections (**Figure 6**), whereas ACh appears to be involved in stimulating glandular secretion. The functions of VIP and ATP are uncertain.

Sympathetic pathways The sympathetic innervation of the genital organs, consisting of preganglionic neurons in the thoracolumbar segments of the spinal cord and postganglionic neurons in the paravertebral and prevertebral ganglia (inferior mesenteric and pelvic ganglia), provides an input to the penile erectile tissue as well as to the smooth muscle of ductus deferens, seminal vesicles, urethra, and prostate. Sympathetic postganglionic axons are also carried in the pudendal nerves and in nerves arising in the pelvic plexus. Most sympathetic postganglionic neurons are noradrenergic and release norepinephrine, ATP, and neuropeptides such as NPY. These nerves produce constriction of blood vessels and cause a contraction of the vas deferens and urethra-bladder neck. Some sympathetic nerves arising in the pelvic plexus release NO and presumably induce vasodilation and penile erection.

Somatic pathways The pudendal nerve arising from the S2 to S4 segments of the spinal cord provides an efferent excitatory input to the striated muscles (bulbocavernosus and ischiocavernosus) involved in ejaculation (**Figure 1**). The pudendal nerve also contains the principal afferent pathway from the penis and from the clitoris and vagina in the female. Afferent innervation to the uterine cervix and uterus travel in the pelvic nerves and sympathetic nerves, respectively.

Erection

Penile erection, which is one of the first responses to occur during sexual arousal, is a vascular phenomenon resulting from neurally mediated increase in blood flow to the penile erectile tissue (corpora cavernosa and corpus spongiosum). The erectile tissue consists of large venous sinuses that contain very little blood when the penis is flaccid, but distend considerably when blood flow is increased. Dilation in the arterial supply to the cavernous tissue, coupled with a relaxation of the sinusoidal smooth muscle in the trabecular tissue, is responsible for erection. During the initiation of erection, corpus blood flow increases 7- to 30-fold.

The neural pathways producing erection arise from both the sacral (parasympathetic) and the thoracolumbar (sympathetic) segments of the spinal cord. The postganglionic neurons in these pathways express neuronal NOS (nNOS; **Figure 6**). All smooth muscle in the penis is richly innervated by nerves containing nNOS. The endothelial cells in the penis also express endothelial NOS (eNOS) and can release NO in response to mechanical stimuli (shear stress) associated with changes in blood flow (**Figure 6**). NO directly activates soluble guanylyl cyclase in the penile smooth muscle to increase the formation of cGMP, which in turn activates cGMP-dependent protein kinases (cGK I). Inactivation of cGK I in mice severely reduces reproductive function and markedly reduces the ability of corpus cavernosal tissue to relax in response to neurally, endothelially, or exogenously administered NO. cGK I is thought to act by multiple mechanisms, including suppression of membrane and sarcoplasmic reticulum Ca^{2+} channels and suppression of IP_3-induced intracellular Ca^{2+} release. The effects of NO are terminated by the enzymatic breakdown of cGMP by phosphodiesterase (**Figure 6**). Pharmacologic studies in animals have shown that erections elicited by stimulation of autonomic nerves are reduced by NOS inhibitors and are enhanced by phosphodiesterase (PDE) inhibitors. One type of PDE (PDE5) is highly expressed in penile tissues. Several PDE5 inhibitors are currently used to treat erectile dysfunction.

Endothelial NOS has also been implicated in neurally mediated erections (**Figure 6**). While erections are initiated by NO synthesized in nerves by nNOS, it has been proposed that this mechanism triggers a very transient increase in blood flow and expansion of the penile vasculature and sinusoidal spaces. However, the resulting shear force on the endothelium activates a phosphatidylinositol 3-kinase (PI3K) pathway that in turn stimulates a serine/threonine protein kinase (Akt, protein kinase B), causing direct phosphorylation of eNOS. After phosphorylation, the Ca^{2+} requirement for eNOS activity is reduced, causing increased production of NO, which induces a prolonged penile erection. Intercellular communication via gap junctions is thought to promote the spread of signals throughout the smooth muscle of the penis and amplify the relaxation.

Exogenous ACh acts on postjunctional muscarinic receptors to elicit a contraction or relaxation of *in vitro* preparations of the corpus cavernosum. The latter effect is mediated by M_3 receptors and release of NO from the endothelium. However, exogenous ACh does not produce erection *in vivo*, and atropine, a muscarinic receptor antagonist, does not block erections, indicating that cholinergic mechanisms are not essential for neurally mediated increases in penile blood flow. On the other hand, ACh may act on prejunctional inhibitory receptors on adrenergic nerve terminals to suppress the release of norepinephrine, leading to a decrease in noradrenergic vasoconstriction in the penis and a facilitation of NO-mediated erections.

VIP activates G-protein-coupled receptors to stimulate adenylyl cyclase and increase cAMP, which in turn stimulates cAMP-dependent kinases to suppress contractile mechanisms in vascular smooth muscle and smooth muscle of the trabecular tissue of the penis (**Figure 7**). While it is clear that exogenous VIP can relax human cavernosal tissue strips *in vitro*, it has never been shown that VIP released from nerves is responsible for relaxation of penile smooth muscle *in vitro* or *in vivo*.

The sympathetic adrenergic innervation of the penis that provides an excitatory input to penile blood vessels is thought to be involved primarily in detumescence. Electrical stimulation of sympathetic axons in either the hypogastric or pudendal nerves in various animals produces a substantial reduction in penile blood flow. The effect is blocked by α-adrenergic receptor-blocking agents. In some animals, a partial erection can be elicited by the administration of α-adrenergic blocking agents, suggesting that tonic sympathetic vasoconstrictor mechanisms have an inhibitory influence on erection. Several mechanisms have been implicated in the noradrenergic vasoconstrictor effect, including (1) prejunctional inhibition of parasympathetic nitrergic nerve terminals mediated by α_2-adrenergic receptors (**Figure 7**); (2) activation of postjunctional α_2-adrenergic receptors, which then

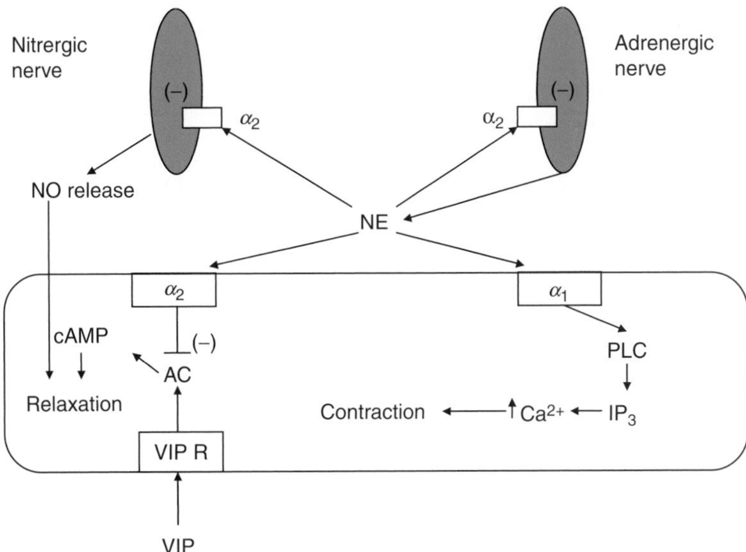

Figure 7 Diagram showing the reciprocal regulation of the contractile tone of penile smooth muscle by norepinephrine (NE) and vasoactive intestinal polypeptide (VIP). NE released from an adrenergic nerve acts on postjunctional α_1-adrenergic receptors in smooth muscle to stimulate the formation of inositol trisphosphate (IP_3) via a phospholipase C (PLC) mechanism. IP_3 triggers the intracellular release of Ca^{2+} that induces a muscle contraction and penile detumescence. NE also can act on α_2-adrenergic receptors located on nerve terminals to suppress transmitter release and on postjunctional receptors to inhibit adenylyl cyclase (AC), and in turn suppress the production of cyclic adenosine monophosphate (cAMP) that induces smooth muscle relaxation and penile erection. VIP acts on receptors (VIP R) to induce relaxation and erection by stimulating AC and increasing cAMP.

inhibits adenylate cyclase and formation of cAMP; (3) activation of postjunctional α_1-adrenergic receptors, which stimulates phospholipase C activity followed by formation of IP_3 and diacylglycerol, leading to release of intracellular Ca^{2+} as well as sensitization of contractile mechanisms to Ca^{2+} (**Figure** 7). Ca^{2+} sensitization has been linked to activation of an intracellular signaling pathway involving Rho kinase, which can be stimulated by G-protein-coupled α-adrenergic receptors. Activation of Rho kinase leads to a change in the phosphorylation state of myosin light-chain kinase, resulting in phosphorylation of myosin and subsequent smooth muscle contraction. Administration of a Rho kinase inhibitor can trigger an increase in intracavernous pressure and an NO-independent erection. Similarly, the expression of a dominant negative construct to downregulate Rho kinase enhances erectile function in rats. Other studies have revealed that NO can relax penile smooth muscle by suppressing Rho kinase activity. These studies have raised the possibility that antagonism of Rho kinase activity may yield new treatments for erectile dysfunction.

Glandular Secretion

During the second phase of the sexual response cycle (plateau), activity in parasympathetic pathways stimulates mucus secretion from bulbourethral and Littre's glands and secretion from the seminal vesicles and the prostate gland. Mucus secretion contributes to lubrication of the penis, whereas secretions from the seminal vesicles and prostate provide the bulk of the fluid and chemical factors that contribute to the viability and motility of the spermatozoa. Glandular secretion is regulated by parasympathetic nerves; however, the transmitters have not been identified. ACh may be involved since cholinomimetic agents stimulate secretion from some glands and AChE-containing nerves have been identified in the prostate and seminal vesicles. However, VIP-containing nerves are also present in these organs.

Emission–Ejaculation

The third phase of the sexual act (orgasm), which is accompanied by emission and ejaculation of semen, involves the coordination of autonomic and somatic reflex mechanisms at different levels of the lumbosacral spinal cord. During the first step in this process (emission), reflex activity in the thoracolumbar sympathetic outflow elicits rhythmic contractions of the smooth muscle of the seminal vesicles, prostate, ductus deferens, and ampulla, resulting in the ejection of sperm and glandular secretions into the urethra and at the same time a closure of the vesical neck to prevent the backflow of semen into the bladder. Pharmacologic studies have shown that these responses are mediated by the sympathetic nerve transmitters, norepinephrine and ATP, interacting with α-adrenergic receptors and purinergic receptors, respectively. Surgical interruption of sympathetic nerves, or the administration of drugs that block α-adrenergic receptors, deplete norepinephrine stores, or block norepinephrine release, blocks emission.

After emission of semen into the proximal urethra, rhythmic contractions of the bulbocavernosus, ischiocavernosus, and the paraurethral striated muscles result in ejaculation. The afferent and efferent limbs of the ejaculation reflex are contained in the pudendal nerve. The sensations accompanying ejaculation constitute the orgasm.

See also: Autonomic Nervous System: Central Urogenital Control; Autonomic Nervous System; Parasympathetic Nervous System; Sympathetic Nervous System.

Further Reading

Andersson KE (2003) Erectile physiological and pathophysiological pathways involved in erectile dysfunction. *Journal of Urology* 170: S6–S13.

Andersson KE and Arner A (2004) Urinary bladder contraction and relaxation: Physiology and pathophysiology. *Physiological Reviews* 84: 935–986.

Beckel JM, Kanai A, Lee SJ, et al. (2006) Expression of functional nicotinic acetylcholine receptors in rat urinary bladder epithelial cells. *American Journal of Physiology – Renal Physiology* 290: F103–F110.

Birder LA, Nakamura Y, Kiss S, et al. (2002) Altered urinary bladder function in mice lacking the vanilloid receptor TRPV1. *Nature Neuroscience* 5: 856–860.

Burnett AL (2001) Novel pharmacological approaches in the treatment of erectile dysfunction. *World Journal of Urology* 19: 57–66.

Burnett AL, Chang AG, Crone JK, et al. (2002) Noncholinergic penile erection in mice lacking the gene for endothelial nitric oxide synthase. *Journal of Andrology* 23: 92–97.

Chancellor MB and Yoshimura N (2002) Physiology and pharmacology of the bladder and urethra. In: Walsh PC, Retik AB, Vaughn ED, et al. (eds.) *Campbell's Urology*, pp. 831–886. Philadelphia: WB Saunders.

Cockayne DA, Hamilton SG, Zhu QM, et al. (2000) Urinary bladder hyporeflexia and reduced pain-related behaviour in P2X3-deficient mice. *Nature* 407: 1011–1015.

de Groat WC (2004) The urothelium in overactive bladder: Passive bystander or active participant? *Urology* 64: 7–11.

de Groat WC and Booth AM (1993) Synaptic transmission in pelvic ganglia. In: Maggi CA (ed.) *The Autonomic Nervous System, Volume 3, Nervous Control of the Urogenital System*, pp. 291–347. London: Harwood Academic Publishers.

Fry CH, Brading AF, Hussain M, et al. (2005) Cell biology. In: Abrams P, Cardozo L, Khoury S, et al. (eds.) *Incontinence*, pp. 363–422. Jersey: Health Publications.

Matsui M, Motomura D, Karasawa H, et al. (2000) Multiple functional defects in peripheral autonomic organs in mice lacking muscarinic acetylcholine receptor gene for the M3 subtype. *Proceedings of the National Academy of Science USA* 97: 9579–9584.

Morrison J, Birder L, Craggs M, et al. (2005) Neural control. In: Abrams P, Cardozo L, Khoury S, et al. (eds.) *Incontinence*, pp. 363–422. Jersey: Health Publications.

Saenz de Tejada I, Cavidad NG, Heaton J, et al. (2001) Anatomy, physiology, and pathophysiology of erectile function. In: Jardin A, Wagner G, Khoury S, et al. (eds.) *Erectile Dysfunction*, pp. 65–102. Jersey: Health Publications.

Toda N, Ayajiki K, and Okamura T (2005) Nitric oxide and penile erectile function. *Pharmacology & Therapeutics* 106: 233–266.

Autonomic Neuroeffector Junction

G Burnstock, Royal Free and University College Medical School, London, UK

Introduction

Within the past 30 years, new discoveries have changed our understanding of the organization of the autonomic nervous system (ANS), including the structure of the autonomic neuroeffector junction and the multiplicity of neurotransmitters which take part in the process of autonomic neuroeffector transmission, as well as cotransmission, neuromodulation, receptor expression, and long-term (trophic) signaling. An outstanding feature of autonomic neurotransmission is the inherent plasticity afforded by its structural and neurochemical organization and the interaction between neural mediators and environmental factors. In this way, autonomic neurotransmission is matched to ongoing changes in demands and can sometimes be compensatory in pathophysiological situations.

Structure of the Autonomic Neuromuscular Junction

The autonomic neuromuscular junction differs in several important respects from the better known skeletal neuromuscular junction; it is not a synapse with the well-defined prejunctional and postjunctional specializations established for the skeletal neuromuscular synapse or ganglionic synapses. A model of the autonomic neuroeffector junction has been proposed on the basis of combined electrophysiologic, histochemical, and electron-microscopical studies. The essential features of this model are that the terminal portions of autonomic nerve fibers are varicose, transmitters being released *en passage* from varicosities during conduction of an impulse, although excitatory and inhibitory junction potentials are probably elicited only at close junctions. Furthermore, the effectors are muscle bundles rather than single smooth muscle cells and are connected by low-resistance pathways (gap junctions) that allow electrotonic spread of activity within the effector bundle. In blood vessels, the nerves are confined to the adventitial side of the media muscle coat, and this geometry appears to facilitate dual control of vascular smooth muscle by perivascular nerves and by endothelial relaxing and contracting factors. Neuroeffector junctions do not have a permanent geometry with postjunctional specializations, but rather the varicosities are continuously moving, and their special relation with muscle cell membranes changes with time, including dispersal and reformation of receptor clusters. For example, varicosity movement is likely to occur in cerebral blood arteries, where there is a continuously increasing density of sympathetic innervation during development and aging and in hypertensive vessels or those that have been stimulated chronically *in vivo*, where there can be an increase in innervation density of up to threefold.

Varicose Terminal Axons

In the vicinity of the effector tissue, axons become varicose, varicosities occurring at 5–10 μm intervals (**Figure 1(a)**), and branches intermingle with other axons to form the autonomic ground plexus, first described by Hillarp in 1946. The extent of the branching and the area of effector tissue affected by individual neurons vary with the tissue. Autonomic axons combined in bundles are enveloped by Schwann cells. Within the effector tissue, they partially lose their Schwann cell envelope, usually leaving the last few varicosities naked.

The density of innervation, in terms of the number of axon profiles per 100 muscle cells in cross-section, also varies considerably in different organs. For example, it is very high in the vas deferens (**Figure 2(a)**), iris, nictitating membrane, and sphincteric parts of the gastrointestinal tract but low in the ureter, uterus, and longitudinal muscle coat of the gastrointestinal tract. In most blood vessels, the varicose nerve plexus is placed at the adventitial border, and fibers rarely penetrate into the medial muscle coat (**Figure 2(b)**).

Junctional Cleft

The width of the junctional cleft varies considerably in different organs. In the vas deferens, nictitating membrane, sphincter pupillae, rat parotid gland, and atrioventricular and sinoatrial nodes in the heart, the smallest neuromuscular distances range from 10–30 nm. The minimum neuromuscular distance varies considerably in different blood vessels. Generally, the greater the vessel diameter, the greater the separation of nerve and muscle. Thus, minimal neuromuscular distances in arterioles and in small arteries and veins are about 50–100 nm, in medium to large arteries the separation is 200–500 nm, whereas in large elastic arteries where the innervation is sparser, the minimum neuromuscular distances are as wide as 1000–2000 nm. Serial sectioning has shown that at close junctions in both visceral and vascular organs, there is fusion of prejunctional and postjunctional

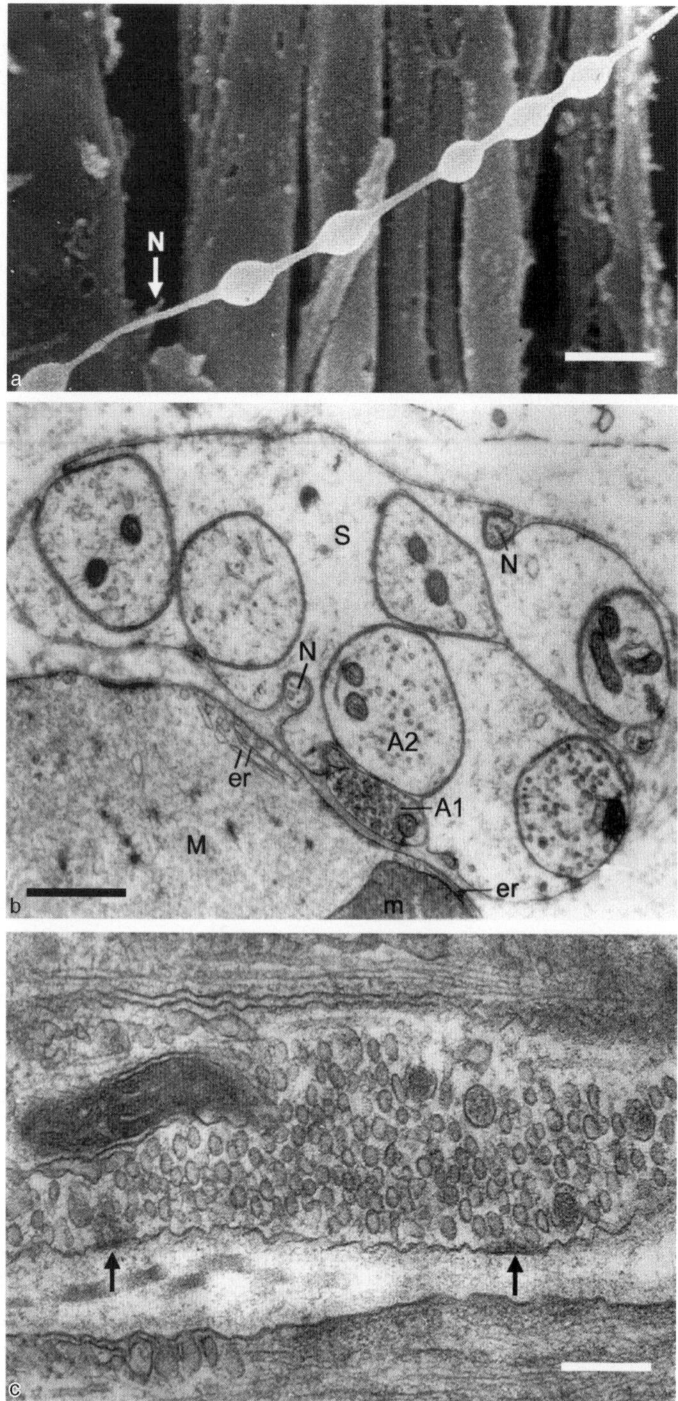

Figure 1 (a) Scanning electron micrograph of a single terminal varicose nerve fiber(N) lying over smooth muscle of small intestine of rat. Intestine was pretreated to remove connective tissue components by digestion with trypsin and hydrolysis with HCl. (b) A medium-size intramuscular bundle of axons within a single Schwann cell (S). There is no perineurial sheath. Some axons, free of Schwann cell processes, contain synaptic vesicles (e.g., A1 and A2). For nerve profile A1, there is close proximity (about 80 nm) to smooth muscle (M) with fusion of nerve and muscle basement membranes. Most of the axons in bundles of this size have few vesicles in the plane of section, but they resemble the vesicle-containing axons of the larger trunks in that they have few large neurofilaments. The small profiles (N), less than 0.25 μm in diameter, are probably intervaricosity regions of terminal axons. m, mitochondria; er, edoplasmic reticulum. (c) Autonomic varicosities with dense prejunctional thickenings and bunching of vesicles, probably representing transmitter release sites (arrows), but there is no postjunctional specialization. Scale bar = 3 μm (a), 1 μm (b), and 0.25 μm (c). (a) Reproduced from *The Airways: Neural Control in Health and Disease*, 1988, 1–22, Autonomic neural control mechanisms: With special reference to the airways, Burnstock G, figure 1, copyright Marcel and Dekker. With kind permission of Springer Science and Buisness Media. (b) Reproduced from *The Journal of Cell Biology*, 1963, vol. 19, pp. 529–550, by copyright permission of The Rockfeller University Press. (c) Courtesy of Phillip R Gordon-Weeks.

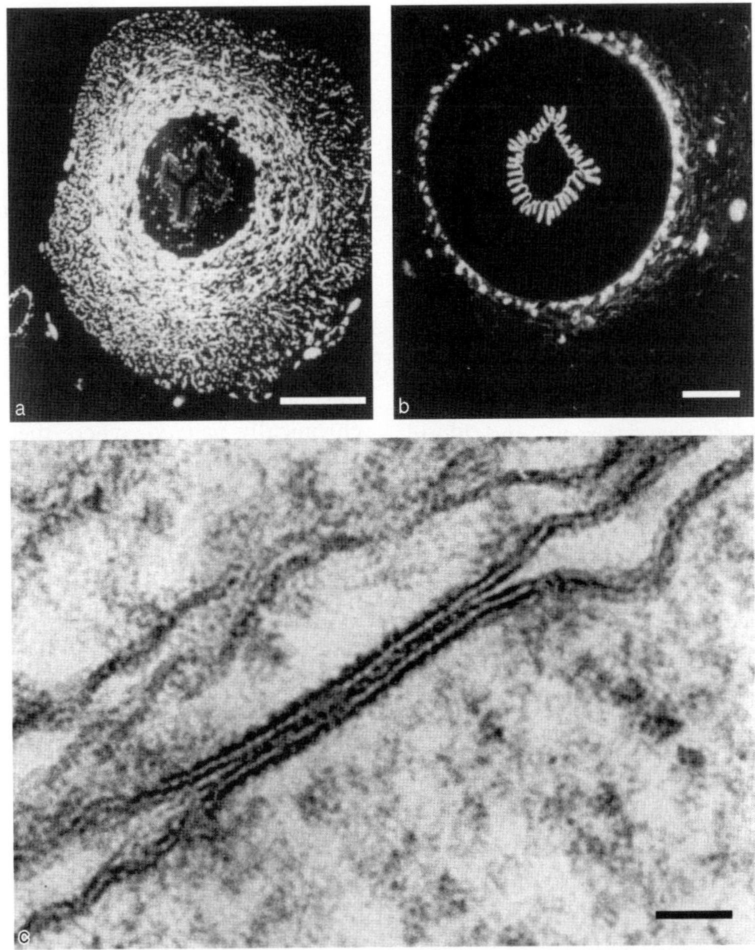

Figure 2 Comparison between the adrenergic innervation of the densely innervated vas deferens of the guinea pig (a) and the rabbit ear artery (b), in which the adrenergic fibers are confined to the adventitial-medial border. The inner elastic membrane shows a nonspecific fluorescence (autofluorescence). (c) A gap junction between two smooth muscle cells grown in tissue culture. Scale bar = 500 μm (a), 50 μm (b), and 50 nm (c). (a, b) Reproduced from *Adrenergic Neurones: Their Organisation, Function and Development in the Peripheral Nervous System*, 1975, Burnstock G and Costa M, plate 9, copyright Chapman and Hall. With kind permission of Springer Science and Buisness Media. (c) Reproduced from *The Journal of Cell Biology*, 1971, Vol. 49, pp. 21–34, by copyright permission of The Rockfeller University Press.

basal lamina (see **Figure 1(b)**). In the longitudinal muscle coat of the gastrointestinal tract, autonomic nerves and smooth muscle are rarely separated by less than 100 nm. However, in the circular muscle coat, close (20 nm) junctions are common, sometimes several axon profiles being closely apposed with single muscle cells.

Prejunctional and Postjunctional Specialization

Although there are many examples of prejunctional thickenings of nerve membranes in varicosities associated with accumulations of small synaptic vesicles, representing sites of transmitter release (see **Figure 1(c)**), there are no convincing demonstrations of postjunctional specializations, such as membrane thickening or folding or indeed absence of micropinocytic vesicles; this is in keeping with the view that even close junctions might be temporary liaisons.

Muscle Effector Bundles and Gap Junctions

The smooth muscle effector is a muscle bundle rather than a single muscle cell – that is, individual muscle cells are connected by low-resistance pathways that allow electrotonic spread of activity within the effector bundle. Sites of electrotonic coupling are represented morphologically by areas of close apposition between the plasma membranes of adjacent muscle cells. High-resolution electron micrographs have shown that the membranes at these sites consist of gap junctions (see **Figure 2(c)**). Gap junctions (or nexuses) vary in size between punctate junctions, which are not easily recognized except in freeze-fracture preparations, and junctional areas more than 1 μm in diameter. The number and arrangement of gap junctions in muscle effector bundles of different sizes in different organs and their relation to density of autonomic innervation have not been fully analyzed. It is interesting that partial denervation has been shown to result in an increase in gap junctions.

Receptor Localization on Smooth Muscle Cells

The distribution of P2X purinoceptors on smooth muscle cells in relation to autonomic nerve varicosities in urinary bladder, vas deferens, and blood vessels has been examined recently by using immunofluorescence and confocal microscopy. Antibodies against the $P2X_1$ receptor, the dominant receptor subtypes found in smooth muscle, and an antibody against the synaptic vesicle proteoglycan SV2 showed clusters of receptors (about $0.9 \times 0.2\,\mu m$ in size) located beneath varicosities. Many more small clusters (about $0.4 \times 0.04\,\mu m$) were present on the whole surface of smooth muscle cells unrelated to varicosities; they may represent pools of receptors that can migrate toward varicosities to form large clusters. In blood vessels, small clusters of P2X receptors are present on cells throughout the medial muscle coat, whereas large clusters are restricted to the muscle cells at the adventitial surface. α_2-Adrenoceptors appear to be located only in extrajunctional regions, so the possibility that noradrenaline (NA) is released from more distant varicosities has been raised. There are hints from studies of receptor-coupled green fluorescent protein chimeras that the receptor clusters are labile, dispersing when a varicosity moves to a new site where clusters reform, perhaps within a 20–30 min timescale.

Model of Autonomic Neuroeffector Junction

A model of the autonomic neuromuscular junction has been proposed on the basis of combined electrophysiologic, histochemical, and electron-microscopical studies described earlier (**Figures 3(a)** and **3(b)**). The essential features of this model are that the terminal portions of autonomic nerve fibers are varicose, transmitter being released *en passage* from varicosities during conduction of an impulse, although excitatory junction potentials (EJPs) and inhibitory junction potentials are probably elicited only at close junctions. Furthermore, the effectors are muscle bundles rather than single smooth muscle cells, which are connected by low-resistance pathways (gap junctions) that allow electrotonic spread of activity within the effector bundle. In blood vessels, the nerves are confined to the adventitial side of the media muscle coat, and this geometry appears to facilitate dual control of vascular smooth muscle by endothelial relaxing and contracting factors and perivascular nerves.

Neuroeffector junctions do not have a permanent geometry with postjunctional specializations, but rather the varicosities are continuously moving, and their special relation with muscle cell membranes changes with time. For example, varicosity movement is likely to occur in cerebral blood arteries, where there is a continuously increasing density of sympathetic innervation during development until old age, and in vessels that have been stimulated chronically *in vivo*, where there can be an increase in innervation density of up to threefold, including an increase in the number of varicosities per unit length of nerve from 10–20 per 100 μm to 30 per 100 mm.

Autonomic effector junctions appear to be suitable not only for neurotransmission but also for neuromodulation. A neuromodulator is defined as any substance that modifies the process of neurotransmission. It may achieve this either by prejunctional action that increases or decreases transmitter release or by postjunctional action that alters the time course or extent of action of the transmitter or by both (**Figure 3(c)**).

Finally, it should be emphasized that if this model of the autonomic effector junction is true, then the earlier emphasis on looking for images of specialized nerve-cell close apposition may not be appropriate. If a varicosity has a passing close relation with a cell and releases transmitter to act on receptors expressed on that cell (e.g., most cells, epithelial cells, or even immune cells), then, in effect, that cell is innervated.

Autonomic Neurotransmission

The Multiplicity of Neurotransmitters in the Autonomic Nervous System

A neurotransmitter is a chemical substance released from nerves on electrical stimulation and which acts on specific receptors on adjacent effector cells to bring about a response, thus acting as a chemical messenger of neural activation. In early studies, acceptance of a substance as a neurotransmitter required satisfaction of the following criteria: (1) the presynaptic neuron synthesizes and stores the transmitter; (2) the transmitter is released in a calcium-dependent manner; (3) there should be a mechanism for terminating the activity of the transmitter, either by enzymatic degradation or by cellular uptake; (4) local exogenous application of the substance should mimic its effects following release due to electrical nerve stimulation; and (5) agents that block or

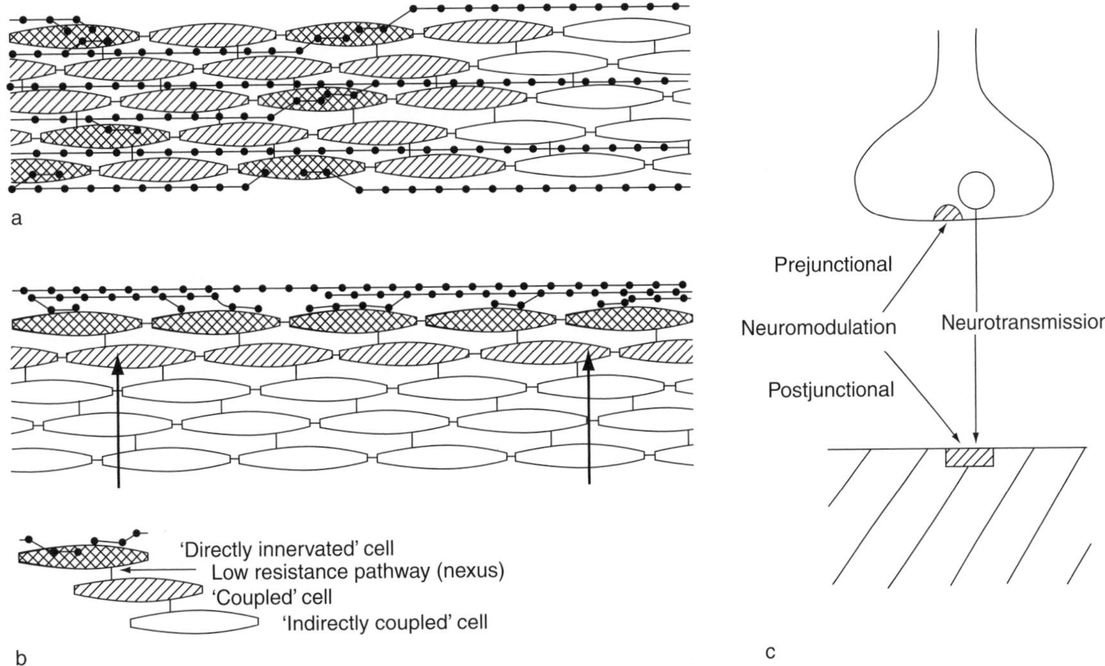

Figure 3 (a) Schematic representation of control of visceral smooth muscle. Directly innervated cells (cross-hatched) are those that are directly activated by neurotransmitter; coupled cells (hatched) are those where junction potentials spread from directly innervated cells, when a sufficient area of the muscle effector bundle is depolarized, a propagated action potential will activate the indirectly coupled cells (white). (b) Schematic representation of control of vascular smooth muscle by nerves (-•-•-) and endothelial factors (arrows). (c) Schematic representation of prejunctional and postjunctional neuromodulation. (a, b) Reproduced from *Adrenergic Neurones: Their Organisation, Function and Development in the peripheral Nervous System*, 1975, Burnstock G and Costa M, figure 18 (a) and (b), copyright Chapman and Hall. With kind permission of Springer Science and Buisness Media. (c) Reproduced from: Royal College of Physicians. *Advanced Medicine 18*. Sanner M (ed). London: Pitman Medical, 1982. Copyright © 1982 Royal College of Physicians. Reproduced by permission.

potentiate the endogenous activity of the transmitter should also affect local exogenous application in the same way.

The classical view of autonomic nervous control as antagonistic actions of NA and acetylcholine (ACh) causing either constriction or relaxation, depending on the tissue, was changed in the early 1960s when clear evidence of a nonadrenergic, noncholinergic (NANC) system was presented. About a decade later, studies of autonomic neurotransmission revealed a multiplicity of neurotransmitters in the ANS. Neurally released substances, including monoamines, amino acids, neuropeptides, adenosine 5′-triphosphate (ATP), and nitric oxide (NO) were identified (see **Table 1**). Since NO does not conform to the constraints of the criteria outlined earlier, although it certainly acts as a rapid chemical messenger in the ANS, a reappraisal of the criteria for defining a neurotransmitter was proposed by Hoyle and Burnstock in 1996, taking into account evidence for nonvesicular, Ca^{2+}-independent release of some classical neurotransmitters and the intracellular site of action of NO. The rapid expansion of the number of proposed autonomic neurotransmitters in recent years, including endothelin, secretoneurin, pituitary adenylate cyclase-activating peptide (PACAP), which is similar in structure to vasoactive intestinal polypeptide (VIP), glutamate, and carbon monoxide, makes it likely that the list is still incomplete.

Cotransmission

The concept of cotransmission was first formulated by Burnstock in 1976, incorporating hints in the earlier literature from both vertebrate and invertebrate systems. It is now well established. Immunohistochemical evidence of coexistence of more than one neurotransmitter should not necessarily be interpreted as evidence of cotransmission, since in order for substances to be termed cotransmitters, it is essential that postjunctional actions to each substance be shown to occur via their own specific receptors. For example, many neuropeptides have slow trophic actions on surrounding tissues, and this may be their primary role, or they may act as neuromodulators. The relative contribution of each transmitter to neurogenic responses is dependent on the parameters of

Table 1 Established and putative neurotransmitters/neuromodulators in the autonomic nervous system

Noradrenaline (NA)
Acetylcholine (ACh)
Adenosine 5′-triphosphate (ATP) and other nucleotides
Nitric oxide (NO)
Carbon monoxide (CO)
5-Hydroxytryptamine (5-HT)
Dopamine (DA)
γ-Aminobutyric acid (GABA)
Glutamate (GLU)
Neuropeptides
Neuropeptide Y (NPY)/pancreatic polypeptide (PP)
Enkephalin (ENK)/endorphin (END)/dynorphin (DYN)
Vasoactive intestinal polypeptide (VIP) and related peptides PHI and PHM
Pituitary adenylate cyclase-activating peptide (PACAP)
Substance P (SP)/neurokinin A (NKA)/neurokinin B (NKB)
Calcitonin gene-related peptide (CGRP)
Somatostatin (SOM)
Galanin (GAL)
Gastrin releasing peptide (GRP)/bombesin (BOM)
Neurotensin (NT)
Cholecystokinin (CCK)/gastrin (GAS)
Angiotensin II (AII)
Adrenocorticotrophic hormone (ACTH)
Secretoneurin
Endothelin (ET)

Reproduced from Burnstock G (2007) Structural and chemical organization of the autonomic nervous system with special reference to nonadrenergic, noncholinergic transmission. In Mathias CJ and Bannister R (eds.) *Autonomic failure. A textbook of clinical disorders of the autonomic nervous system.* Oxford: Oxford University Press, by permission of Oxford University Press.

stimulation. For example, short bursts (1 s) of electrical stimulation of sympathetic nerves at low frequency (2–5 Hz) favor ATP release whereas longer periods of nerve stimulation (30 s or more) favor NA release.

Peptides, purine nucleotides, and NO (identified by localization of nitric oxide synthase (NOS)) are often found together with the classic neurotransmitters, NA and ACh. In fact, the majority of nerve fibers in the ANS, if not all, contain a mixture of different neurotransmitter substances that vary in proportion in different tissues and species and during development and disease. The widespread use of double and triple immunohistochemical labeling techniques has been critical to the demonstration of co-localization of potential cotransmitters within the same nerve fiber and has been invaluable when combined with electron microscopy. Different neurotransmitters within the same varicosity may be localized in the same or separate vesicular populations using post-embedding colloidal gold techniques. In the gastrointestinal tract, many neurons contain multiple transmitters. ATP is a cotransmitter with calcitonin gene-related peptide (CGRP) and substance P (SP) in many sensory–motor nerves and with NO and VIP in enteric NANC inhibitory nerves. Transmitters with seemingly diverse and opposing effector action are sometimes co-localized in the same neuron, but generally they act in the same way and usually synergistically.

The precise combinations of neurotransmitters (and neuromodulators) contained in individual neurons and their projections and central connections, termed their 'chemical coding' by Furness and Costa in 1997, has been defined in studies of the enteric nervous system and peripheral autonomic and sensory ganglia.

Neurotransmission at the sympathetic neuroeffector junctions: Evidence for co-release and roles of NA, ATP, and neuropeptide Y It is now recognized that the main neurotransmitters/neuromodulators in postganglionic sympathetic nerves are NA, ATP, and neuropeptide Y (NPY). These substances are co-released in varying proportions, depending on the tissue and species, and also on the parameters of stimulation. Short bursts at low frequency particularly favor the purinergic component whereas longer periods of nerve stimulation favor the adrenergic component, and NPY release is optimal with high-frequency intermittent bursts of stimulation. A considerable variability in the contribution of a purinergic component to sympathetic neurotransmission has been demonstrated in different blood vessels; for example, rabbit saphenous and mesenteric arteries have a substantial purinergic component, whereas in the rabbit ear artery, the purinergic component is relatively small. In intestinal submucosal arteries, the responses to sympathetic nerve stimulation are mediated solely by ATP, with NA acting as a prejunctional modulator via α_2-adrenoceptors. The initial electrophysiological postjunctional response to sympathetic nerve stimulation is a rapid, transient EJP, which is mediated by ATP. In some vessels, the EJP is followed by a slow depolarization, which is mediated by NA. Postjunctionally, the effects of ATP and NA released as sympathetic cotransmitters are generally synergistic. NA and ATP can depress sympathetic neurotransmission by prejunctional modulation, via α_2-adrenoceptors or predominantly via P1 receptors following extracellular breakdown to adenosine, but also via P2 receptors in some vessels. Prejunctional P2 receptor-mediated increase in NA release has also been reported.

In most tissues, including the vas deferens and many blood vessels, NPY does not act as a genuine neurotransmitter, having little direct postjunctional action, but rather acts as a neuromodulator, often by prejunctional attenuation of NA and ATP release and/or postjunctional potentiation of responses to adrenergic and purinergic components of sympathetic nerve responses. In tissues in which NPY does have a direct vasoconstrictor effect, such as in blood vessels of the spleen and kidney and in coronary and cerebral arteries, the response is characteristically slow in onset and long lasting.

Other substances localized within sympathetic nerves include 5-hydroxytryptamine (5-HT), which is largely taken up by sympathetic nerves and released as a false transmitter. Opioid peptides are also widely distributed in sympathetic neurons where their functional role appears to be related to their prejunctional inhibitory effects on sympathetic neurotransmission.

Neurotransmission at the parasympathetic neuroeffector junctions: The atropine-resistant components of parasympathetic neurotransmission ACh, VIP, ATP, and NO are cotransmitters commonly synthesized in and released from parasympathetic nerves. As with sympathetic cotransmission, the relative functional importance of the cotransmitters in parasympathetic neurotransmission is variable in different tissues and species. For example, NO may be the main mediator of neurogenic vasodilation in cerebral vessels, whereas VIP may be of more importance during neurogenic vasodilation in the pancreas. The coordinated roles of VIP and ACh in parasympathetic neurotransmission were demonstrated in an elegant study of the cat exocrine salivary gland innervation. It showed that VIP and ACh were stored in separate vesicles in the same nerve terminal and were both released on transmural nerve stimulation, but with different stimulation parameters. ACh was released during low-frequency stimulation to increase salivary secretion from acinar cells and to elicit some minor dilatation of blood vessels in the gland. At high stimulation frequencies, VIP was released to produce marked dilatation of the blood vessels in the gland and to act as a neuromodulator postjunctionally on the acinar gland to enhance the actions of ACh and prejunctionally on the nerve varicosities to enhance the release of ACh. ACh was also found to have an inhibitory action on the release of VIP. VIP has since been shown to have a direct vasodilatory action in the submandibular gland in man. PACAP also seems to be present in VIP-containing parasympathetic nerves. NOS is often co-localized with ACh and VIP in parasympathetic nerves innervating blood vessels. Postganglionic nerves from pelvic ganglia containing VIP, ACh, and NOS project to the urethra, colon, and penis. The human bladder body receives a dense parasympathetic innervation comprised predominantly of ACh-containing nerves. In the rodent bladder, ATP is a major cotransmitter in these nerves. However, only a small purinergic component is present in human bladder, except in pathological conditions (discussed later).

Neurotransmission at sensory–motor neuroeffector junctions: The roles of SP, CGRP, and ATP The motor function of sensory nerves, whereby antidromic impulses down collateral fibers result in local release of sensory neurotransmitters, is widespread in autonomic effector systems and forms an important physiological component of autonomic control. To distinguish these nerves from the other subpopulation of afferent fibers that have an entirely sensory role and have terminals containing few vesicles and a predominance of mitochondria, they have been termed 'sensory–motor' nerves.

SP and CGRP are cotransmitters in many unmyelinated, primary afferent nerves. They often coexist in the same large granular vesicles in capsaicin-sensitive nerve terminals. The proportions of coexistence of SP and CGRP vary with species; for example, in the guinea pig, most sensory neurons containing CGRP also contain SP, but in the rat, about 50% of CGRP-containing neurons do not contain SP. In the vasculature, unlike CGRP, SP does not appear to act directly on receptors of the vascular smooth muscle but rather acts via occupation of receptors on endothelial cells lining the lumen to bring about NO release and consequent vasodilation. This action of neurally released SP may be particularly important in the microvasculature, but access of neurally released SP to the endothelium in large vessels is questionable; it is largely released from endothelial cells to act on receptors on endothelial cells to release NO, resulting in vasodilation. ATP is now also established as a cotransmitter with glutamate in small primary sensory nerves mediating mechanical and/or nociceptive signals.

Other neuropeptides and transmitters have been localized in sensory–motor nerves. For example, in the human urinary bladder, VIP, cholecystokinin (CCK), and dynorphin (DYN) are present, together with SP and CGRP, in the afferent projections to the lumbosacral spinal cord. In the guinea pig, dorsal root ganglion neurons containing SP, CGRP, CCK, and DYN project to the epidermis and small dermal blood vessels. NOS has been localized in populations of primary sensory neurons of trigeminal and dorsal root ganglia. Endothelin, a potent vasoconstrictor peptide with mitogenic actions, is also localized in neurons of these sensory ganglia, often co-localized with SP.

There are increasing examples in the literature of cross-talk between sensory–motor, sympathetic, and parasympathetic nerves. In the heart, SP has excitatory effects on cardiac parasympathetic innervation, in contrast to CGRP, which is inhibitory.

Neurotransmission involving intrinsic neurons: Special reference to neurotransmitters localized in nerve cell bodies in the heart, bladder, intestine, and lung Many intrinsic neurons localized within autonomic neuroeffector tissues are part of the postganglionic parasympathetic system, but there are also intrinsic neurons derived from neural crest tissue that is different from that which forms sympathetic and parasympathetic neurons, such as intrinsic neurons abundant in the gut and possibly subpopulations in the heart and airways.

The most extensive system of intrinsic neurons is in the myenteric and submucous plexuses of the gastrointestinal tract. These enteric neurons contain numerous neuroactive substances, of which the majority are involved in neurotransmission or neuromodulation at the ganglion level and/or have a trophic role; only a small percentage are involved in neuromuscular transmission. The chemical coding of enteric neurons has been examined in detail, particularly in the guinea pig. ATP, NO, and VIP mediate NANC inhibitory neurotransmission in the gut in varying proportions depending on the region. ACh and SP are cotransmitters in enteric excitatory neurons.

There are many intrinsic neurons in the heart, particularly in the right atrium. The neurochemical makeup of the intrinsic cardiac ganglia is heterogeneous and includes a variety of neurochemical markers. For example, subpopulations of atrial intrinsic neurons from newborn guinea pigs immunostain for NPY, 5-HT, heme oxygenase-2, and NOS, and these neurons probably also utilize ACh and ATP.

Most airway intrinsic neurons contain choline acetyltransferase, but NOS and VIP are also found in these neurons in humans. Intrinsic ganglia in the human urinary bladder wall contain a number of neuroactive substances (VIP, NOS, NPY, ATP, galanin, and occasionally tyrosine hydroxylase); in the bladder neck, a few intrinsic neurons contain enkephalin and SP. Intramural ganglia containing NPY and VIP have been identified in human urethra.

Autonomic Neuromodulation

Some substances stored and released from nerves do not have direct actions on effector muscle cells but alter the release and/or the actions of other transmitters; these substances are termed neuromodulators. Many other substances (e.g., circulating neurohormones; locally released agents such as prostanoids, bradykinin, histamine, and endothelin; and neurotransmitters from nearby nerves) are also neuromodulators in that they modify the process of neurotransmission. Many substances that are cotransmitters are also neuromodulators. The wide and variable cleft characteristic of autonomic neuroeffector junctions makes them particularly amenable to the mechanisms of neural control mentioned earlier.

Plasticity of the Autonomic Nervous System

There are some examples of altered expression of neurotransmitters/neuromodulators in autonomic nerves during development and aging; following trauma, surgery, and chronic exposure to drugs; and in disease. Neurons possess the genetic potential to produce many neurotransmitters. The particular combination and quantity that result are partly preprogrammed and partly determined by 'trophic' factors and hormones that trigger the expression or suppression of the appropriate genetic machinery. The plasticity of expression of neural substances cocoordinated to environmental cues allows rapid and precise matching of neurotransmission to altered demands. Several neurotransmitters/neuromodulators are themselves trophic molecules, with mitogenic or growth-promoting/-inhibiting properties.

Conclusions

A combination of the variety of neurotransmitters involved in autonomic neurotransmission and the interactions between sympathetic, parasympathetic, and sensory–motor nerves and those arising from intrinsic ganglia, via mechanisms of cotransmission and pre- and postjunctional neuromodulation, indicate the complexity of peripheral autonomic control and the variety of ways by which autonomic dysfunction can occur. Recent advances in the unraveling of these mechanisms, together with molecular identification of specific receptor subtypes and localization and characterization of their expression, and of the long-term effects of dysfunction, will bring advances toward the design of treatment regimes to combat autonomic failure.

See also: Autonomic Neuroplasticity and Aging; Autonomic Neuroplasticity: Development; Autonomic Neuroplasticity and Regeneration; Autonomic Nervous System; Cotransmission; Gap Junction Abnormalities and Disorders of the Nervous System.

Further Reading

Burnstock G (1976) Do some nerve cells release more than one transmitter? *Neuroscience* 1: 239–248.

Burnstock G (1982) Neuromuscular transmitter and trophic factors. In: Samer M (ed.) *Advanced Medicine 18*, pp. 143–148. London: Pitman Medical, Royal College of Physicians.

Burnstock G (1986) Autonomic neuromuscular junctions: Current developments and future directions. *Journal of Anatomy* 146: 1–30.

Burnstock G (1986) The changing face of autonomic neurotransmission. *Acta Physiologica Scandinavica* 126: 67–91.

Burnstock G (1988) Autonomic neural control mechanisms: With special reference to the airways. In: Kaliner MA and Bames PJ (eds.) *The Airways: Neural Control in Health and Disease*, pp. 1–22. New York: Dekker.

Burnstock G (1975) *Adrenergic Neurones: Their Organisation, Function and Development in the Peripheral Nervous System.* London: Chapman and Hall, Elsevier.

Burnstock G (1990) Changes in expression of autonomic nerves in aging and disease. *Journal of the Autonomic Nervous System* 30: S25–S34.

Burnstock G (1992) *The Autonomic Nervous System, Vol. 1: Autonomic Neuroeffector Mechanisms.* Chur, Switzerland: Harwood.

Burnstock G (2004) Cotransmission. *Current Opinions in Pharmacology* 4: 47–52.

Burnstock G and Iwayama T (1971) Fine structural identification of autonomic nerves and their relation to smooth muscle. In: Eränkö O (ed.) *Progress in Brain Research, 34, Histochemistry of Nervous Transmission*, pp. 389–404. Amsterdam: Elsevier.

Burnstock G and Ralevic V (1994) New insights into the local regulation of blood flow by perivascular nerves and endothelium. *British Journal of Plastic Surgery* 47: 527–543.

Campbell GR, Uehara Y, Mark G, et al. (1971) Fine structure of smooth muscle cells grown in tissue culture. *Journal of Cell Biology* 49: 21–34.

Furness JB and Costa M (1987) *The Enteric Nervous System.* Edinburgh: Churchill Livingstone.

Hansen MA, Balcar VJ, Barden JA, et al. (1998) The distribution of single P2X1-receptor clusters on smooth muscle cells in relation to nerve varicosities in the rat urinary bladder. *Journal of Neurocytology* 27: 529–539.

Hillarp NA (1946) Functional organization of the peripheral autonomic innervation. *Acta Anatomica* 2: 1–153.

Hoyle CHV and Burnstock G (1996) Criteria for defining enteric neurotransmitters. In: Gaginella TS (ed.) *Handbook of Methods in Pharmacology*, pp. 123–140. London: CRC Press.

Luff SE (1996) Ultrastructure of sympathetic axons and their structural relationship with vascular smooth muscle. *Anatomy and Embryology (Berlin)* 193: 515–531.

Lundberg JM (1996) Pharmacology of cotransmission in the autonomic nervous system: Integrative aspects on amines, neuropeptides, adenosine triphosphate, amino acids and nitric oxide. *Pharmacological Reviews* 48: 113–178.

Merrillees NCR, Burnstock G, and Holman ME (1963) Correlation of fine structure and physiology of the innervation of smooth muscle in the guinea pig vas deferens. *Journal of Cell Biology* 19: 529–550.

Sandow SL, Whitehouse D, and Hill CE (1998) Specialised sympathetic neuroeffector associations in rat iris arterioles. *Journal of Anatomy* 192: 45–57.

Autonomic Neuroimmunology

H P M van der Kleij, P Forsythe, and J Bienenstock, McMaster University, Hamilton, ON, Canada

Structural Associations

Immune cells occur in close proximity to neural processes at sites where these structures are abundant. Anatomical nerve–immune associations have been reported involving mast cells as well as other bone marrow-derived cells, such as eosinophils and plasma cells in the lamina propria. Further work has demonstrated similar morphologic mast cell–nerve associations in the liver, mesentery, urinary bladder, and skin. In the skin, observations of Egan and co-workers showed that especially unmyelinated axons were associated with mast cells, as well as Langerhans cells (LCs) in primate and murine skin. Many if not most epidermal LCs appeared to be closely associated anatomically with calcitonin gene-related peptide (CGRP)-containing nerves, as determined by confocal laser scanning microscopy. In addition, there is morphological evidence for neural modulation of immune activity *in vivo*. It has been demonstrated clearly that the regions in which lymphocytes reside receive direct sympathetic neural input, and substance P may interact with thymocytes, mast cells, and other cells in the thymus, affecting their development and function.

Nerve–Immune Systems: Interaction

The nervous and immune systems can communicate in a bidirectional fashion as a result of a common set of signal molecules and their receptors. It has been established that the central nervous system (CNS) can regulate immune functioning. For instance, immune responses can be classically conditioned involving a CNS learning paradigm. Accordingly, information can flow not only from the CNS to the immune system but also in the opposite direction. Immune cells can produce neuropeptides such as substance P and β-endorphin. In addition, cells of the immune system possess receptors for a host of hormones and neurotransmitters, including receptors for catecholamines, acetylcholine (ACh), adrenocorticotropic hormone (ACTH), opioid peptides, substance P, CGRP, somatostatin (SOM), and vasoactive intestinal peptide (VIP). These receptors have been shown to respond *in vivo* and/or *in vitro* to the neurotransmitter substances, and their manipulation can alter immune responses. Furthermore, the CNS shares receptors with the immune system for neuropeptides and neurotransmitters as well as for cytokines.

Immune regulation is mediated via a cascade of events requiring cell–cell interactions and subsequent release of their contents, such as cytokines and antibodies, to regulate these responses. Cells of the immune system and their products have been shown to influence peripheral and central neurotransmission, leading to the conceptualization of a bidirectional neuroimmune communication system.

Innate immunity and adaptive (acquired) immunity are the two major immune defense responses of the body. Microbe-associated molecular patterns (MAMPs) derived from bacteria are recognized by pattern recognition receptors such as Toll-like receptors (TLRs). Detection of microorganisms through MAMPs by TLRs is crucial in triggering protective immunity. Studies have shown an essential role for TLRs in the activation of innate and adaptive immunity in animals. Ligand recognition induces a host defense response, which includes the production or regulation of inflammatory cytokines, upregulation of co-stimulatory molecules, and induction of antimicrobial defenses. Dendritic cells (DCs) are key players in the orchestration of immune events. Their function is crucial in both the early innate responses and the subsequent adaptive response. Activation of DCs by TLR ligands is necessary for their maturation and consequent ability to initiate immune responses.

Peripheral neuropeptides can directly attract immature DCs and at the same time they may capture mature DCs at sites of neurogenic inflammation. Low neuropeptide concentrations can improve the motility of DCs to the source of the neuropeptides, primarily sensory nerve fibers, where they may undergo functional and phenotypical maturation, probably due to changes in signal transduction pathways of neuropeptide receptors in immature and mature DCs. These findings provide evidence for a connection between adaptive immunity and the nervous system. The adaptive immune response can be altered by the nervous system releasing neuropeptides at specific local sites of infection or challenge, which may then influence the direction of the T helper cell (Th1/Th2) response. For instance, CGRP can inhibit interferon-γ (IFN-γ) production markedly in a dose-dependent manner; substance P and VIP can suppress interleukin-4 (IL-4) production. Catecholamines may play an important role in regulating adaptive immune responses as Th1 cells express adrenergic receptors. Furthermore, adrenergic receptors on B cells suggest that the sympathetic nervous system may also regulate antibody production in addition to natural killer (NK) cell activity.

Recent reports have identified neural pathways that regulate the peripheral immune response. For instance, two immunomodulatory peptides, VIP and the pituitary adenylate cyclase-activating polypeptide (PACAP), are present in and released from both nerves and immune cells. VIP/PACAP have a general anti-inflammatory effect, both in innate and adaptive immunity. In innate immunity, VIP/PACAP inhibit the production of proinflammatory cytokines and chemokines from macrophages, microglia, and DCs. In adaptive immunity, VIP/PACAP promote Th2-type responses, and reduce the proinflammatory Th1-type responses.

Cytokine-induced illness (behavioral changes) can be seen as a neuroimmune response to activation of innate immunity. The nervous system can regulate immune function and inflammation. The CNS alerts the immune system to environmental changes using the shared neuropeptide, neurotransmitter, and cytokine receptors on immune cells. Neuropeptides, neurohormones, and neurotransmitters of the CNS interact with the immune system, which, in turn, feeds back to the brain, inducing behavioral changes and changes in the immune system. Cytokines produced in the periphery can affect the brain, producing illness and activation of neuroendocrine and autonomic stress-related circuits. Recognition of stimuli such as viruses and bacteria by the immune systems results in transmission of information to the CNS, causing physiological responses, such as fever associated with illness behavior. In the abdomen these responses are a direct result of immune cell-derived proinflammatory cytokines signaling the CNS via the subdiaphragmatic vagus nerve.

Neuropeptides

Neuropeptides are produced primarily in the brain and nervous system, although every tissue in the body can produce and exchange neuropeptides. The immunological activity of neuropeptides is mostly mediated through specific receptors. The presence of specific receptors for substance P, SOM, CGRP, corticotropin-releasing hormone (CRH), melanocortin peptides, VIP, and related peptides has been reported on immune cells. For example, macrophages, DCs, T lymphocytes, and B lymphocytes all have the ability to express neurokinin type 1 (NK-1) receptors. The existence of these specific neuropeptide receptors on immune cells represents the framework for neuropeptides functioning as mediators of neuroimmune interactions.

Substance P, the most important neuropeptide of the tachykinin family, is widely distributed in the central, peripheral, and enteric nervous systems of many species. Substance P functions in the CNS as a neurotransmitter and its NK-1 receptors are localized in distinct areas of the brain important in affecting behavior and the neurochemical response to both psychological and somatic stress. It may also coordinate the response to stress by interacting with the hypothalamic–pituitary–adrenal (HPA) axis and the sympathetic nervous system. Neuronal substance P is stored in vesicles and released from sensory nerves in response to various stimuli, such as leukotrienes, prostaglandins, and histamine. Release of substance P is often accompanied by release of CGRP, since they are often co-localized in autonomic nerves.

The tachykinin family has recently been extended by the discovery of a third tachykinin gene preprotachykinin-C mRNA, encoding a tachykinin called hemokinin-1. An important observation that has been made is the fact that constitutive expression of hemokinin mRNA occurs preferentially in leukocytes and lymphoid tissue, not in the nervous system. Microglial cells, described as resident macrophages in the brain, are a potential source of hemokinin-1. There is the possibility that hemokinin-1 could play a significant role in communication between immune cells in the CNS, and between immune cells and neuronal cells expressing the NK-1 receptor.

Neuropeptides can influence the innate immune response, especially since NK-1 receptors have been found on macrophages and DCs. *In vivo* studies have demonstrated a role for the NK-1 receptor in the initiation of the host response against bacterial and viral pathogens. Upregulation of NK-1 receptor expression by *Salmonella* may significantly increase the substance P-mediated macrophage response, directly killing the bacteria. In addition, substance P can enhance IL-1, IL-6, and tumor necrosis factor (TNF) secretion by macrophages.

CGRP can also be found by itself in some motor and enteric neurons. CGRP inhibits T cell proliferation by inhibiting IL-2 production. In macrophages, it inhibits antigen presentation and other macrophage functions, including phagocytosis. It also stimulates production of a number of cytokines, such as IL-6, IL-10, and TNF. *In vivo* injection of CGRP was shown to inhibit the onset of delayed-type hypersensitivity reactions, supporting the role of CGRP in systemic regulation of immune activation. Endogenous CGRP may generally modulate immune function, presumably, in part, through an inhibitory effect on LC function in the skin. CGRP is capable of inhibiting both the induction and expression of fundamental cellular immune functions. This suggests an interaction between the nervous system and immunological function. These observations imply the likelihood that the CNS may be capable of suppressing the initiation of immune responses, and is also effective in their modulation. Overall, there is substantial experimental evidence demonstrating that CGRP can influence the function and development of inflammatory and immune cells in local microenvironments by specific receptor-mediated mechanisms.

Although classically neuropeptides are released from autonomic and sensory nerves, there is increasing evidence that they may also be synthesized and released from immune cells such as macrophages and eosinophils, particularly in disease. The regulation of synthesis and secretion of these mediators is largely undetermined. Therefore any description of homeostasis, especially under conditions in which immune stimulation or inflammation is present, must by necessity remain incomplete until this knowledge is available.

Neurotrophins

Based on their expression profile, neurotrophins are good candidates for mediating immune–nerve cell interactions. Nerve growth factor (NGF) is the best characterized neurotrophic factor, but the term 'neurotrophin' includes brain-derived neurotrophic factor (BDNF), glial-derived nerve growth factor (GDNF), and neurotrophins 3/5. Neurotrophins influence the proliferation, differentiation, survival, and death of neuronal and nonneuronal cells. However, several nonneuronal functions have also been characterized. For instance, neurotrophic factors are essential for enteric nervous development. Especially GDNF has been identified to control the migration, morphogenesis, and differentiation from precursors to the enteric nervous system. Furthermore, neurotrophins can regulate neuropeptide expression, interact with immunoregulatory cells and epithelial cells, and control motility during inflammation. Neurotrophins also act as an important intermediate in inflammatory pain, contributing to both peripheral and central sensitization.

Neurotrophin receptors are widely expressed in the peripheral and the CNSs, as well as on cells of the immune system. Many immune cells express the high-affinity NGF tyrosine receptor kinase A (TrkA) receptor. This allows NGF to promote the release of inflammatory mediators. Several of these mediators, such as IL-1, IL-4, IL-5, TNF, and IFN, can, in turn, induce the release of NGF. Therefore, NGF seems to be a mediator with functions on both immune cells and nerve cells and is likely an important factor in integrating communication between the nervous and immune systems.

There is increasing evidence that NGF acts on cells of the immune system. NGF promotes differentiation, activation, and cytokine production of mast cells and macrophages, and activates eosinophils. It promotes survival of different cell types, such as mast cells, by their prevention of apoptosis. NGF is a chemoattractant, thereby causing an increase in the number of mast cells as well their degranulation. Furthermore, NGF influences activity of basophils, eosinophils, neutrophils, macrophages, and T cells and is also produced by a wide range of immune cells. Mast cells, T cells, B cells, eosinophils, lymphocytes, and epithelial cells can synthesize NGF.

NGF can have proinflammatory as well as anti-inflammatory effects, dependent on its location and concentration. In mice, nasal treatment with NGF induces airway hyper-responsiveness, measured by electrical field stimulation. NGF-transgenic mice, overexpressing NGF in Clara cells, develop hyper-reactivity in the absence of airway inflammation, suggesting that NGF by itself determines the induction of airway hyper-responsiveness. On the other hand, there is evidence indicating that the increased production of NGF in the CNS during brain diseases, such as multiple sclerosis, can suppress inflammation by switching the immune response to an anti-inflammatory, suppressive mechanism. The fact that NGF may under different circumstances be either proinflammatory or anti-inflammatory is an important example of the complexity of interactions between neuroactive molecules and the immune and inflammatory systems.

Mast Cells–Nerve Interaction

Histological studies reveal an intimate association between mast cells and neurons in both the peripheral and CNSs. Besides the anatomical link, mast cells also form a functional link between the immune and nervous systems. Neuronal mechanisms are involved in mast cell activation and mast cells act as principal transducers of information between peripheral nerves and local inflammatory events.

Direct bidirectional communication between mast cells and superior cervical neurites has been supported, in an *in vitro* co-culture model of murine sympathetic neurons cultured with rat basophilic leukemia (RBL-2H3) cells. Neurites show neurotrophic and neurotropic activity for RBL cells. Intimate synapse-like contact occurs followed by cessation of division, membrane electrophysiologic alterations, maturation of granules in RBL cells, and accumulation of vesicles in neurons close to the point of contact. Activation of nerves with scorpion venom elicited degranulation of RBL cells via substance P, and substance P activation is initiated only at their point of contact. Co-cultures of bone marrow-derived mast cells (BMMCs) and neurites have demonstrated that expression of NK-1 receptors by mast cells lowers the threshold of activation induced by nerve stimulation.

Mast cells and nerves in many tissues are in constant contact with each other and share a number of activating signals, for some of which both cells express receptors (such as vanilloids). Additionally, both mast cells and nerves respond to stimulation by secretion of preformed mediators, many of which are produced by both cells (NGF, neuropeptides, and endothelin-1). Moreover, a large proportion of primary spinal afferent neurons, which contain CGRP and substance P, express the proteinase-activated receptor 2 (PAR-2). Proteases, such as tryptase from degranulated mast cells, have recently been shown to cleave PAR-2 on primary spinal afferent neurons, causing release of substance P, activation of the NK-1 receptor, and amplification of inflammation and thermal and mechanical hyperalgesia. This mechanism of protease-induced neurogenic inflammation may contribute to the proinflammatory effects of mast cells in human disease.

Many studies have shown that mast cell-derived mediators, such as histamine, serotonin, and cytokines, modulate nonadrenergic noncholinergic (NANC) neurotransmission. NANC nerve endings express receptors for histamine (H1 and H3) and serotonin (5-HT_{2A}), and histamine H1 receptor expression at least is upregulated on primary NANC nerves in inflammation. Mast cell mediators such as TNF can sensitize afferent C fibers by lowering their threshold, but can also cause direct release of substance P, neurokinin A, and CGRP from unmyelinated fibers.

In addition to IgE and antigen, cytokines, hormones, and neuropeptides can trigger mast cell secretion. The latter include SOM, neurotensin, PACAP, CGRP, and substance P. CRH, stem cell factor (SCF), and NGF can promote mast cell growth and can also trigger mast cell degranulation. SCF has also been reported to induce mast cells to become responsive to PACAP. Keratinocytes, LCs, fibroblasts, mast cells, and endothelial cells express functional neurokinin receptors, while G-proteins on mast cells can, in addition, be directly activated by substance P. Functional NK-1 receptors are expressed on murine mast cells in the presence of IL-4 and SCF, but are not constitutively expressed by human intestinal mast cells; this changes upon stimulation by IgE receptor cross-linking. This suggests that, *in vivo*, specific tissue conditions such as allergic inflammation may lead to mast cell expression of NK-1 receptors.

Nerve–mast cell interactions are engaged in both homeostatic and pathologic regulation. Synaptic cell adhesion molecule (SynCAM) is involved in most CNS synaptic associations. Furthermore, SynCAM was reported to mediate the attachment of mast cells to neurites by expression in both cell types. This homophilic binding not only results in this immune–nerve association, but also facilitates successful neurotransmission from neurite to mast cell.

Neurogenic Inflammation

Neurogenic inflammation involves a change in function of sensory neurons due to inflammatory mediators, inducing an enhanced release of neuropeptides from the sensory nerve endings. The effects produced by tachykinins and CGRP released from peripheral endings of capsaicin-sensitive primary sensory neurons are collectively referred to as neurogenic inflammation. There is also histological evidence for the presence of cytokines, especially IL-6, within both sensory and autonomic nerves. The release of these cytokines would cause local inflammation and indicates a direct link between neurogenic stimulation and the release of proinflammatory cytokines. Experimental research has shown a role for neurogenic inflammation in disease processes such as arthritis, colitis, bladder inflammation, and asthma.

The postganglionic sympathetic neuron may also play a role in neurogenic inflammation. In the presence of subdiaphragmatic vagotomy, the potency of nociceptor afferents to inhibit the sympathetically dependent bradykinin-induced plasma extravasation was significantly increased. These findings suggest that there is a vagal mechanism affecting neurogenic inflammation. Afferent signals to the brain can generate a reflex, causing an anti-inflammatory response, which is partly mediated by the efferent branch of the vagus nerve. This cholinergic anti-inflammatory pathway is mediated primarily by nicotinic ACh receptors expressed on tissue macrophages and blood monocytes. ACh can interact specifically with macrophage $\alpha 7$ subunits of nicotinic ACh receptors, leading to cellular deactivation and inhibition of cytokine release.

CNS and Stress

Several pathways have been shown to link the brain and the immune system: the autonomic nervous system, via direct neural influences, and the neuroendocrine humoral outflow, via the pituitary. The CNS can and does activate or even inhibit immune and inflammatory events. Stress activation of the HPA axis has such a bimodal effect. CRH activates peripheral mast cells. Corticosterone may inhibit inflammatory events and catecholamines have complex effects. Hypnosis has been shown to inhibit the expression of skin reactivity to allergen in classical tests of allergic hypersensitivity, and it has been shown that classical Pavlovian conditioning could cause mast cell secretion.

In the CNS, mast cells may participate in the regulation of inflammatory responses through interactions with the HPA axis. In the dog, degranulation of CNS mast cells evokes HPA activation in response to histamine release and CRH. Control or peripheral administration of specific antigen results in corticosterone release from the adrenal glands. CRH is also thought to be involved peripherally in tissue responses to stress in the skin, respiratory tract, intestine, and bladder. Many, if not all, of the recorded changes have involved mast cells and neuronal activation, the latter being often mediated by neurotensin and/or substance P.

Central neuropeptides initiate a systemic stress response by activation of neuroendocrinological pathways such as the sympathetic nervous system, the hypothalamic–pituitary axis, and the renin–angiotensin system, with the release of the stress hormones (i.e., catecholamines, corticosteroids, growth hormone, glucagons, and renin). Immobilization stress increases colonic motility and many other physiologic parameters through release of CRH. Both mast cells and nerves are involved, and the effects have been shown to occur through neurotensin and substance P, mediated by cholinergic, adrenergic, and ganglionic pathways. These results provide support for an important link between the nervous and the immune systems in stress-mediated intestinal responses.

Active immune responses represent a source of systemic stress, which impacts the brain and modifies various neuroendocrine and behavioral functions. Therefore, the immune system has been conceived of as a potential contributor to stress-related behavioral abnormalities, such as clinical depression. It is suggested that immunologically induced changes in the brain activate neuropeptides, thereby sustaining an adaptive state of arousal that promotes appropriate behavioral adjustments during infectious illness.

Immune cells such as macrophages are involved in certain types of stress. The cold-water swim test results in the increase of peritoneal substance P and its receptor in peritoneal macrophages. Immune–nerve interaction has been shown to be involved in the skin during stress. Stress can inhibit hair growth *in vivo* and significantly increases the number of hair follicles containing apoptotic cells, activated perifollicular macrophage clusters, and degranulated mast cells, whereas it downregulates the number of intraepithelial lymphocytes. Substance P seems to be a key mediator of stress-induced hair growth inhibition *in vivo*. Increased numbers of substance P-immunoreactive sensory fibers, seen in the dermis of stressed mice, are a result of transient high levels of NGF, suggesting that NGF is a central element in the perifollicular neurogenic inflammation that develops during the murine skin response to stress.

Several inflammatory skin conditions, including atopic dermatitis and psoriasis, are exacerbated by stress. Recent evidence suggests that cross-talk between mast cells, neurons, and keratinocytes might be involved in such exacerbation. CRH and its receptor are present in the skin and their levels are increased following stress. Human mast cells synthesize and secrete CRH in response to IgE receptor cross-linking. Mast cells also express CRH receptors, activation of which leads to the selective release of cytokines and other proinflammatory mediators. CRH receptor antagonists could be used to inhibit stress-induced mast cell activation and provide new therapeutic options for chronic inflammatory conditions exacerbated by stress.

Summary

The brain and nervous system have a bidirectional communication with the immune system. There is a consistent finding of associations between nonmyelinated nerves and lymphocytes, mast cells, eosinophils, and LCs, at which locations bidirectional interaction clearly occurs. Furthermore, it has become apparent that many cells in these different systems possess the same receptors and can make the same molecules. Behavioral changes can be initiated as a result of peripheral inflammation (illness behavior) and fever can be caused by peripheral infections and inflammation as a result of communication with and through the nervous system. The brain routinely receives communication signals from peripheral immune events and can coordinate responses through the HPA axis and the sympathetic, parasympathetic, and NANC nervous systems. Hypoactivity or hyperactivity of these mechanisms could contribute to diseases characterized by either a decreased or an exaggerated inflammatory response. Much remains to be done, but scientific progress promises much greater complexity as well as understanding of these processes. What is clear, however, is that immune mechanisms cannot be ignored in dealing with a functional understanding of the nervous system or even behavior, and the reverse is also true.

See also: Immune System–Neuroendocrine Interactions; Inflammation in Neurodegenerative Disease and Injury;

Neural Repair and Regeneration: Inflammatory Mechanisms and Cytokines; Neuroimmune System: Aging; Neuropeptides in Autonomic Neurons; Neurotrophins: Physiology and Pharmacology; Stress and Parasympathetic Control; Stress, the HPA Axis and Depressive Illness.

Further Reading

Bienenstoc J, Goetzl EJ, and Blennerhassett MG (2003) Autonomic neuroimmunology, In: Burnstock G (ed.) *The Autonomic Nervous System*, vol. 15, pp. 1–392. London: Taylor & Francis.

Black PH (1994) Central nervous system–immune system interactions: Psychoneuroendocrinology of stress and its immune consequences. *Antimicrobial Agents and Chemotherapy* 38: 1–6.

Blalock JE (1984) The immune system as a sensory organ. *Journal of Immunology* 132: 1067–1070.

Chrousos GP and Gold PW (1992) The concepts of stress and stress system disorders. Overview of physical and behavioral homeostasis. *JAMA: The Journal of the American Medical Association* 267: 1244–1252.

Dantzer R (2001) Cytokine-induced sickness behavior: Mechanisms and implications. *Annals of the New York Academy of Sciences* 933: 222–234.

Felten DL, Felten SY, Bellinger DL, et al. (1987) Noradrenergic sympathetic neural interactions with the immune system: Structure and function. *Immunological Reviews* 100: 225–260.

Maier SF and Watkins LR (1998) Cytokines for psychologists: Implications of bidirectional immune-to-brain communication for understanding behavior, mood, and cognition. *Psychological Review* 105: 83–107.

Marshall JS and Waserman S (1995) Mast cells and the nerves – Potential interactions in the context of chronic disease. *Clinical & Experimental Allergy* 25: 102–110.

McDonald DM, Bowden JJ, Baluk P, et al. (1996) Neurogenic inflammation. A model for studying efferent actions of sensory nerves. *Advances in Experimental Medicine and Biology* 410: 453–462.

Purcell WM and Atterwill CK (1995) Mast cells in neuroimmune function: Neurotoxicological and neuropharmacological perspectives. *Neurochemical Research* 20: 521–532.

Steinhoff M, Stander S, Seeliger S, et al. (2003) Modern aspects of cutaneous neurogenic inflammation. *Archives of Dermatology* 139: 1479–1488.

Tracey KJ (2002) The inflammatory reflex. *Nature* 420: 853–859.

van der Kleij HP, Blennerhassett MG, and Bienenstock J (2003) Nerve–mast cell interactions – Partnership in health and disease. In: Bienenstock J, Goetzl EJ, and Blennerhassett MG (eds.) *Autonomic Neuroimmunology*, pp. 139–170. London: Taylor & Francis.

Watkins LR and Maier SF (1999) Implications of immune-to-brain communication for sickness and pain. *Proceedings of the National Academy of Science of the United States of America* 96: 7710–7713.

Williams RM, Bienenstock J, and Stead RH (1995) Mast cells: The neuroimmune connection. *Chemical Immunology* 61: 208–235.

Autonomic Neuroplasticity and Aging

T Cowen, University College London, London, UK

Introduction

Autonomic neurons, in common with neurons from all parts of the nervous system, undergo extensive and continual adaptation in response to intrinsic and extrinsic stimuli. The nature of neural plasticity and the factors that influence it vary throughout life. Broadly, these changes fall into three stages: development, maturation and adulthood, and old age. During development, neurons undergo differentiation toward their adult phenotype. This process includes acquiring appropriate morphology, expression of neurotransmitters and receptors, and patterns of activity and signaling pathways adapted to particular functions and connectivity. Toward the end of this period, usually during the perinatal period in mammals, neurons that have not made appropriate connections with their target tissues (which may be other neurons or effector organs such as muscles and glands) die by programmed cell death. During maturation and adulthood, plasticity takes different forms. The broad outline of the phenotype acquired during development – morphology, neurotransmitter, and other functional characteristics – remains stable. However, neurons retain responsiveness to altered demand. During adulthood and aging, many autonomic neurons retain a remarkable capacity to adapt to changing circumstance. However, a minority of cells become subject to selective vulnerability, undergo atrophy, and may die, probably contributing to the functional losses of old age which include deficits in autonomic control of homeostasis. The molecular mechanisms that underlie autonomic neural plasticity are beginning to be understood. A recurring theme throughout the life span is the interaction between neurons and their target tissues, which has profound effects on many aspects of neural plasticity.

Plasticity in the Developing Autonomic Nervous System

Neuronal Growth

During embryonic development, differentiating neurons of the autonomic nervous system (ANS) migrate to appropriate peripheral locations, such as the pre- and paravertebral autonomic ganglia and the enteric nerve plexuses of the gut. From these locations, the neurons extend dendrites to form synapses with spinal autonomic neurons of the intermediolateral column. They also extend axons toward their peripheral targets which may be smooth muscle effectors in organs such as blood vessels, gut, or iris, or glands such as salivary, pineal, or sweat glands, or glands of the gastrointestinal tract. Guidance cues for migratory neurons and growing neurites are provided by bound molecules of the extracellular matrix (ECM) as well as by diffusible factors. In recent years, it has become increasingly clear that guidance, like other epigenetic processes in the nervous system, is regulated by a combination of positive and negative influences. In the case of neural guidance, bound and diffusible factors may act as chemoattractants or chemorepellents. Chemoattractants include the ECM glycoprotein laminin and the diffusible neurotrophic and other growth factors, while chemorepellents include members of the semaphorin family, netrins, and ECM glycoproteins. Some but not all of these molecules have been investigated in relation to autonomic neurons. The combination of attractive and repulsive signals ensures that growing neurites follow appropriate pathways, including established nerve fascicles, grow toward appropriate targets, and avoid inappropriate ones.

Around the time that neurons approach their target areas, the peripheral tissues start to synthesize a locally specific cocktail of factors, which attract the growing axons and stimulate synaptogenesis. Bound factors such as laminin stimulate growth of autonomic neurites, for example, through contact between receptors on axon terminals and laminin located in glial or smooth muscle basal laminae. Laminin binds to integrin receptors of the tyrosine kinase family, which initiate neuronal growth and survival signals through ERK/MAP-kinases and PI3-kinase. Diffusible growth-promoting factors produced by targets include the neurotrophin family of growth factors, nerve growth factor (NGF), and neurotrophin-3 (NT3) having the most clearly defined functions in the ANS. These molecules generate growth responses in sympathetic and other peripheral neurons by activating the Trk family of tyrosine kinase receptors, of which TrkA and TrkC are the receptors principally responsible for mediating the effects of NGF and NT3, respectively. The contribution of the nonspecific neurotrophin receptor, p75, remains controversial. Some suggest a synergistic role with Trk receptors in promoting growth and survival; others argue for an inhibitory role for p75 in neuronal growth and survival, probably acting in situations where Trk's are absent. A characteristic of these signaling systems is that they often act in combination, with bound and diffusible factors either working synergistically or providing a balance of positive and negative influences.

Locally Specific Phenotype

Target specificity of innervation is achieved by secretion by target tissues of locally specific combinations of factors, which stimulate particular patterns of neuronal growth as well as neuronal synthesis of particular neuropeptides. Growth of large, complex sympathetic neurons (supplying, e.g., the iris where fast physiological reflexes are required) correlates with high levels of synthesis of NGF. In contrast, smaller sympathetic neurons supplying sparse innervation to blood vessels (where much slower, less specific neuromuscular responses are required) are associated with low levels of NGF synthesis in the vessel wall (**Figure 1**). Target-derived factors also ensure the adoption by the neurons of a suitable neurotransmitter phenotype The mechanism underlying establishment of neurotransmitter phenotype has been investigated widely, particularly in the minority of sympathetic nerves of the ANS which utilize acetylcholine as a neurotransmitter. Target-derived molecules are clearly involved in this process; however, their identity remains elusive and may include the cytokines ciliary neurotrophic factor, leukemia inhibitory factor, or cardiotrophin-1 and/or the neurotrophic factor, glial cell-derived neurotrophic factor (GDNF).

Neuronal Survival

The nervous system at large generates more neurons during embryogenesis than are required in maturity.

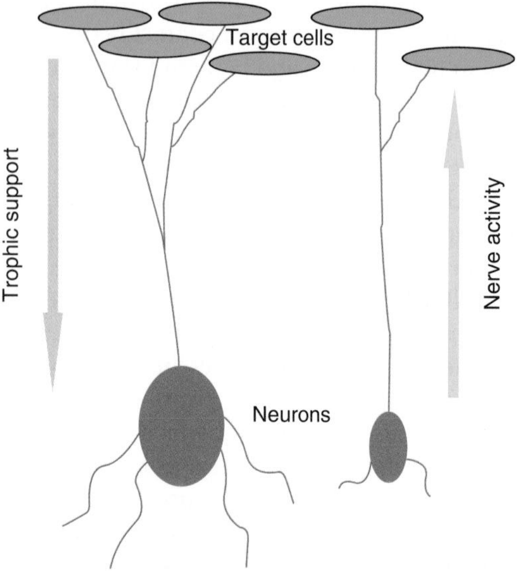

Figure 1 The neurotrophic hypothesis. Target tissues support neuronal growth, survival, and phenotype through their production of neurotrophic and other factors which are taken up by neurons and retrogradely transported to the cell body where they initiate gene transcription. Neurons regulate the supply of these factors through their activity. Targets producing high levels of neurotrophic factors become innervated by large, complex neurons; conversely, targets producing lower levels of these factors become innervated by smaller ones. These different phenotypes are likely to be functionally adaptive. However, they appear to have pleiotropic effects on the survival trajectories of these neurons in old age (see text).

Fine-tuning of the numbers of surviving neurons to the physical size and physiological demands of peripheral organs occurs by ensuring that the cocktail of diffusible growth factors produced by the target tissue provides sufficient but not superfluous support for the innervating neurons. Neurons that reach their target compete for this scarce resource by growing axons bearing receptors which uptake the appropriate factors, generating further growth and survival signals. Those that fail to take up a sufficient supply undergo programmed cell death or apoptosis. This process was extensively investigated in sympathetic neurons of the ANS and only later understood to be important in many areas of the nervous system. This model of developmental regulation of neuron survival has become encapsulated in the 'neurotrophic theory' (**Figure 1**). An interesting recent addition to this concept concerns the emerging evidence that physiological activity of the target, which is at least partly a result of the activity of the innervating neurons, can itself stimulate synthesis or secretion of growth factors. This important observation suggests that neurons have the capacity to regulate their target-derived trophic support.

Similar groups of molecules contribute to the survival as well as to the growth of ANS neurons. Thus, target-derived NGF and NT3 signaling through Trk receptors regulate survival of numbers of sympathetic neurons appropriate to the particular target tissues during development while the survival of gut neurons is regulated partly by laminin, and subsequently by GDNF and NT3. Pruning of neuron numbers takes place at different periods of perinatal life depending on the target area. In general, it starts in the few days before birth and continues for a variable number of weeks after birth. The signaling pathways mediating developmental survival are similar to those driving neurite growth, that is, tyrosine kinase signaling through ERK/MAP-kinase and PI3-kinase. It seems most likely that downregulation of these pathways (i.e., by subthreshold levels of the ligand neurotrophins), particularly of PI3-kinase, is the key signal for neuronal death. The death signal involves a pro-apoptotic shift in the downstream effectors of tyrosine kinase signaling, which include expression of the BCL-2 and caspase families of proteins. Knockout studies in mice indicate that the p75 receptor, signaling through NFκB, is also involved in the developmental regulation of cell death in NGF-dependent autonomic and other neurons.

Plasticity in the Adult and Aging ANS

Neuronal Growth

Adult neurons retain the capacity to respond sensitively to extrinsic growth signals. Body organs continue to grow beyond the period of naturally occurring postnatal neuronal cell death. Existing and relatively stable populations of neurons are therefore required to expand their axonal and dendritic connections by continued growth and synaptogenesis. *In vitro* studies have shown that adult neurons are at least as sensitive as early postnatal neurons in terms of their growth responsiveness to neurotrophins. Indeed, even those neurons that survive into old age retain growth responsiveness to neurotrophins; strong nerve sprouting has been observed *in vivo* in response to treatment with neurotrophic factors. However, there is an age-related reduction in the dose-response curve *in vivo* and *in vitro*.

That plasticity is a physiological feature of adult autonomic neurons is supported by several lines of evidence. Pregnancy and the estrous cycle are accompanied by substantial cyclical changes in the autonomic neurons innervating the uterus. Sympathetic nerve fibers undergo atrophy in late estrus and pregnancy, followed by regeneration. Pregnancy-induced neuronal atrophy may protect the fetus against vasoconstriction caused by homeostatic responses such as fight and flight. Multiple pregnancies may result in the death of some of the innervating sympathetic neurons. Transplantation studies and experimental enlargement of organs demonstrate further that neurons, particularly autonomic neurons, retain an extraordinary ability to expand their field of innervation. Partial obstruction of the bladder can result in a roughly tenfold increase in muscle mass to which the innervating autonomic neurons respond with an increase in size of around 80%. Removal of the obstruction results in reversion of neuron size to control values over a few weeks. Comparable manipulations in the gut and iris – both autonomically innervated tissues – have yielded similar results.

The same neurotrophic factors, which regulate growth of neurons during development, are retained by the adult target tissues, generally at unchanged levels. In the case of adult sympathetic neurons, these factors include NGF and NT3. Of these, NGF produces the substantially stronger growth response, although both factors are present in many of the target tissues of sympathetic neurons. TrkA and TrkC receptors are present on adult sympathetic neurons to mediate the growth responses. The increased target area resulting from the physiological or experimental expansion of organs described above is therefore likely to induce nerve growth by increasing the availability of neurotrophins in the vicinity of existing nerve terminals. The role of the nonspecific p75 receptor remains as controversial in the adult nervous system as it is at earlier stages of development. Some argue that p75 inhibits neuronal growth in adult rodent neurons while others consider that p75 supports the trophic role of NGF by heterodimerization with Trk. These apparently contradictory roles for this receptor may result from different expression ratios of p75 and Trk in different neurons and under different circumstances. Local availability of neurotrophins can affect the expression of both p75 and Trk receptors in the adult nervous system, thus providing a further level at which nerve growth can be regulated.

The signaling pathways involved in adult neuron growth are less well defined than those regulating survival. While inhibition of tyrosine kinase signaling largely inhibits extension of neurites from adult sympathetic neurons *in vitro*, there is less conclusive definition of the downstream pathways involved. This may be because neurite extension involves integrin-as well as neurotrophin-mediated activation of tyrosine kinase signaling, with resulting convergent activation of ERK/MAP-kinase and PI3-kinase.

During aging, some autonomic neurons undergo atrophy (shrinkage of their dendritic and axonal arbors and cell soma), while others continue to grow into old age. This phenomenon has been called 'selective vulnerability' and affects sympathetic neurons projecting to vascular targets and sweat and pineal glands. In contrast, neurons projecting to iris and submandibular gland show no evidence of atrophy. To the extent that the subpopulations of neurons innervating each target tissue exhibit specific patterns of growth or atrophy during aging, selective vulnerability involves interactions between particular target tissues and neurons. However, it has not proved possible to demonstrate that reduced levels of NGF availability from targets causes age-related atrophy of fibers.

Locally Specific Phenotype

Adult neurons retain locally specific morphology and neurochemical phenotype. Thus, subpopulations of adult sympathetic neurons, retrogradely traced from particular target tissues, exhibit different sizes which appear to relate to physiological function in the same way that developing neurons adapt to the physiological function of the end organ. In addition, adult neurons continue to express locally specific patterns of neuropeptides and transmitters. For example, neuropeptide Y expression is highest in superior cervical ganglion (SCG) neurons projecting to iris, lower in those projecting to the middle cerebral artery, and lowest in neurons projecting to the submandibular gland. There is evidence that these patterns of neurotransmitter expression remain plastic in adulthood. Transplantation in adult rats of sweat glands (normally innervated by cholinergic, vasoactive intestinal peptide (VIP)-expressing sympathetic neurons) into contact with catecholaminergic sympathetic neurons induces a shift to a cholinergic, VIP-expressing phenotype in those neurons, which extend processes to innervate the transplanted tissue. There is no obvious loss of this form of plasticity in old age.

Neuronal Survival in Adulthood

While neurons retain growth responsiveness to extrinsic factors in adulthood and even in old age, the factors that determine survival, at least of sympathetic and other (sensory) peripheral neurons, appear to change dramatically during maturation and in aging. Between birth and 12 weeks of age, rodent sympathetic neurons, from being dependent for their survival on an extrinsic source of neurotrophin, become almost completely independent and able to survive in dissociated cell culture for many days and even weeks under serum- and NGF-free conditions (**Figure 2**). The loss of survival dependence on target-derived signals may be seen as an adaptation allowing adult neurons a degree of resistance against an axonal injury that separates the neuron from its target, thus allowing the neuron a period of respite in which to initiate regeneration.

The principal signaling pathway that mediates neuronal survival independence, as this phenomenon is called, is PI3-kinase. Pharmacological and/or genetic inhibition of this pathway at different points leads to the rapid death of cultured adult sympathetic neurons; inhibition of ERK/MAP-kinase is without significant effect. This suggests strongly that while neurons lose their dependence on signaling through Trk receptors to regulate their survival, the downstream PI3-kinase pathway remains a key determinant of neuronal resistance to cell death. It is feasible that autonomic and perhaps other neurons protect their survival in adulthood by cell-autonomous activation of PI3-kinase in a way that resembles a mechanism by which cancer cells avoid apoptosis.

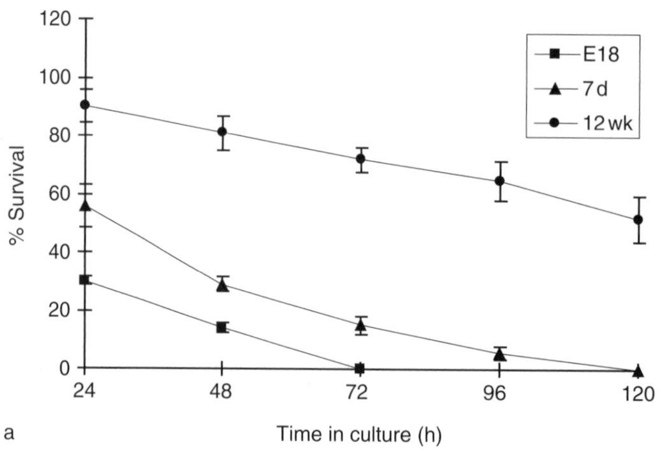

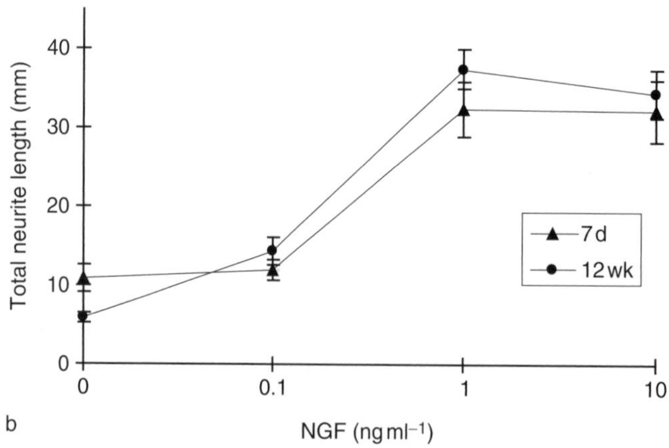

Figure 2 Differential regulation of neuronal growth and survival during maturation. During postnatal maturation, sympathetic neurons lose their survival dependence on NGF. Growth, however, remains NGF dependent. (a) Sympathetic neurons become independent of NGF for their survival between embryonic day 18 (E18) and 12 weeks of age, as demonstrated by a dramatic increase in survival with age in dissociated, pure neuron cell cultures maintained over several days in the absence of serum and NGF. Note that by 12 weeks, approximately 60% of neurons survive for longer than 5 days *in vitro*, whereas no E18 neurons survive longer than 3 days. (b) Postnatal sympathetic neurons require NGF to initiate growth and to extend neurites *in vitro*. Note that neurons fail to grow in the absence of NGF; dose responsiveness is similar in 12-week-old and 7-day-old neurons.

It has recently become clear that production of neuronal progenitors, at least in the central nervous system (CNS), continues throughout life under the influence of a range of epigenetic factors, including physical exercise, brain activity, and injury. It is not clear, however, whether the ANS can continue to generate neuronal precursors in adulthood or whether plasticity mechanisms similar to those seen in development and maturation regulate the differentiation, survival, and integration of progenitor cells into the functioning adult nervous system.

Neuronal Survival in Old Age

Selective vulnerability to neuron cell death is a well-established characteristic of the aging nervous system as a whole, as well as of the ANS. The groups of neurons affected are similar to those which exhibit age-related atrophy (see the earlier section on neuronal growth). Among sympathetic neurons, only certain populations undergo cell death, while many survive into extreme old age. Roughly 50% of vascular-projecting neurons in the SCG are lost in aging rats, while neurons projecting to the iris are less significantly reduced in numbers. Autonomic neurons from the myenteric plexus of the gut are also vulnerable to age-related cell death, with roughly 50% of neurons lost from for small and large intestine of aging rats and guinea pigs (**Figure 3**). There is limited but less secure evidence that similar patterns of myenteric neuron loss occur in humans, at least from the large intestine.

In line with evidence that adult neurons become relatively independent of target-derived trophic support, it is probable that locally selective patterns of neuron death during aging are not the result of

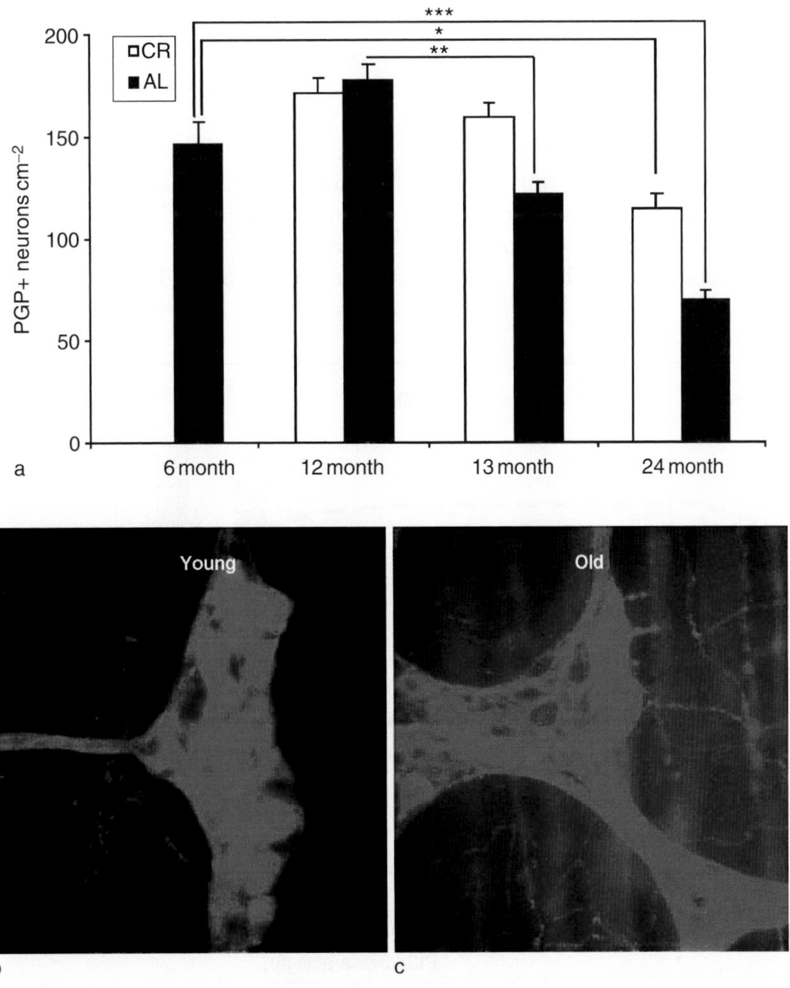

Figure 3 Age- and diet-related loss of myenteric neurons from the rat small intestine. (a) Counts of myenteric neurons using the pan-neuronal marker, PGP 9.5. The histogram shows significant losses of total numbers of neurons, commencing at 12–13 months of age in *ad libitum*-fed (AL) animals, which reaches a peak of 51% at 24 months. There is a much smaller but statistically significant ($p < 0.05$) reduction seen in animals of the same age (24 months) but fed a calorie-restricted (CR) diet, demonstrating substantial protection by CR against age- and diet-related loss of myenteric neurons. Statistically significant differences are indicated by $^{*}p < 0.05$; $^{**}p < 0.01$; $^{***}p < 0.001$. (b, c) Photomicrographs of young (6 months; b) and old (24 months AL; c) PGP-immunostained whole-mount preparations of rat myenteric plexus. Note the large spaces in the old ganglia indicating where neuron cell loss has occurred.

reduced availability of neurotrophins. Studies of different target areas show no evidence of altered levels of NGF or NT3 protein or mRNA in targets of aging sympathetic neurons, mirroring similar observations in targets of vulnerable NGF-dependent neurons in the CNS. The search for causes of age-related neurodegeneration is a high priority in aging research. A number of avenues are being explored including whether expression of the p75 or Trk receptors are selectively altered in subgroups of aging sympathetic neurons vulnerable to age-related cell loss. In the absence of changes in levels of extrinsic factors, these are likely to be cell-autonomously regulated. Subpopulations of neurons that exhibit age-related cell loss project to target areas that throughout life express relatively low levels of neurotrophic factors; conversely neurons that survive successfully into old age tend to project to targets expressing high levels of neurotrophic factors (see the above discussion on locally specific phenotype). These differential levels of neurotrophin synthesis are retained throughout life and are reflected in altered patterns of p75 and Trk receptor expression, which correlate directly with neurotrophin availability. It therefore seems likely that growth factor signaling is set during maturation at levels that are adapted to particular physiological requirements of subgroups of neurons. This maturational setting affects the capacity of neurons to survive into old age.

Oxidative and other forms of free radical damage are heavily implicated in neurodegenerative disease as well as in normal aging. It is now known that neurotrophic factors can influence antioxidant defense in neurons and that free radical levels increase in neurons, such as those of the myenteric plexus, prior to the onset of age-related cell loss. A working hypothesis, then, is that differential, lifelong patterns of neurotrophin signaling are established during early postnatal life, and become more or less fixed in maturity. These differential and adaptive levels of signaling then determine, among other things, neuronal capacity to resist prolonged exposure to free radical damage. The neurons with low levels of signaling succumb first (**Figure 4**).

An alternative way in which neurotrophin signaling might be altered in adulthood and old age concerns the recent discovery that the precursor pro-form of NGF, previously thought to be expressed within target cells prior to its cleavage by enzymes of the proconvertase family, is widely distributed in the adult nervous system, including in nonneural targets of autonomic nerves. ProNGF levels are elevated in the pregnant uterus, in association with the atrophy and degeneration of the innervating sympathetic neurons (see the earlier section on neuronal growth). ProNGF is also the predominant form of NGF present in the Alzheimer's diseased brain and in the spinal cord after injury – two situations known to

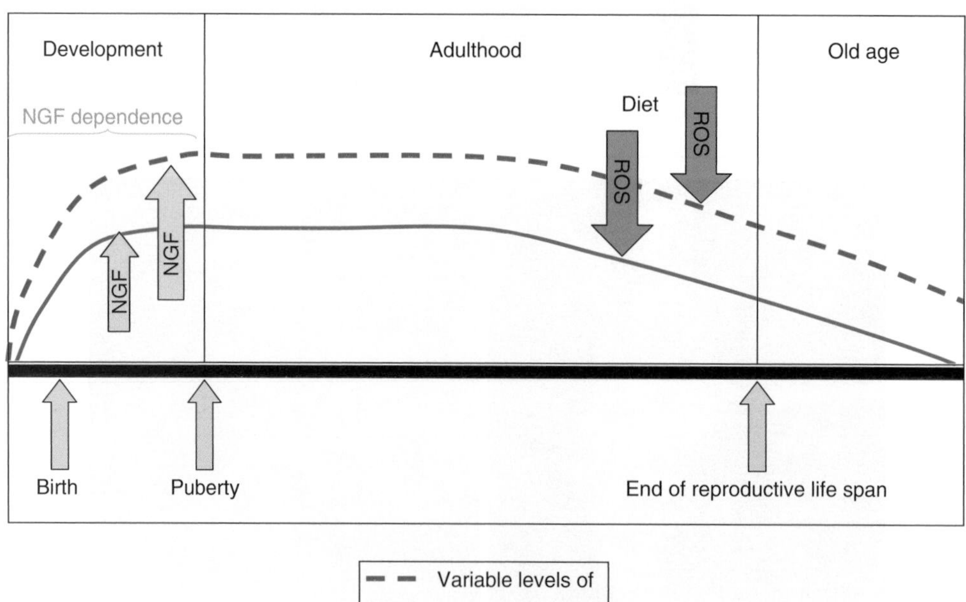

Figure 4 Model illustrating mechanisms regulating lifelong survival of autonomic neurons. During perinatal life, autonomic neurons become increasingly independent of target-derived NGF for their survival as a result of upregulation of intraneuronal PI3-kinase activity. The degree of upregulation is variable (see dashed and solid green lines), locally specific and adaptive, being set initially by the levels of the target-derived NGF (large and small arrows). Later in life (around puberty and after), PI3-kinase activity is maintained independently of NGF availability. Toward the end of the reproductive life span, exposure to reactive oxygen species (ROS) downregulates PI3-kinase activity, increasing vulnerability to age-related neuron cell death. AL feeding (see **Figure 3**) increases exposure to ROS and thereby accentuates this process, while CR provides protection.

be associated with widespread neuron cell death. ProNGF has been shown to induce apoptotic cell death in PC12 cells as well as SCG neurons from young animals. ProNGF signals via the receptor sortilin (first discovered as a member of the Vsp10 family of sorting receptors and later identified as the receptor for neurotensin) acting in combination with p75. At present, there is little evidence from the aging nervous system that proNGF contributes to age-related cell death. However, the sortilin receptor is expressed in the adult (including human) brain; therefore, this possibility cannot be ruled out.

Extrinsic Factors Affecting Neuron Survival

Factors that influence life span have been shown, unsurprisingly, to also affect neuron survival. The classical extender of life span in rodents, caloric restriction (CR), has remarkable effects in rescuing myenteric neurons of rats from age-related cell death. CR also reduces free radical levels in neurons, and appears to achieve this by influencing neurotrophic factor-mediated antioxidant defense. Exercise and the environmental stimuli to which laboratory rodents are exposed (including group vs single animal housing) may also affect health, life span, and, perhaps, cell survival. Further studies are required to quantify these phenomena and to explore the underlying mechanisms.

Conclusions

Autonomic neuronal plasticity is regulated by core mechanisms which are set up during development, modified and consolidated during maturation, and maintained into adulthood and old age. These mechanisms, which are adaptive, may have deleterious pleiotropic consequences for some groups of aging neurons, making them vulnerable to neurodegeneration. Neurodegeneration probably occurs initially as loss of terminal nerve fibers (axons and dendrites) and shrinkage of the cell soma and may progress ultimately to cell death. Interactions between neurons and their target tissues, including the provision of nerve growth and survival promoting factors by targets and the related expression of appropriate receptors in neurons, comprise an intimate and essential interaction at the core of plasticity responses throughout life. Intraneuronal signaling pathways have specific roles in regulating plasticity, including growth and survival responses, and, probably, maintenance of neurotransmitter phenotype and antioxidant defense. Extrinsic factors, including diet and other behaviorally regulated phenomena, have major effects on neural plasticity, probably by acting on intraneuronal signaling pathways.

See also: Autonomic and Enteric Nervous System: Apoptosis and Trophic Support During Development; Autonomic Disorders; Autonomic Failure; Autonomic Nervous System: Neuroanatomy; Autonomic Neuroplasticity: Development; Autonomic Neuroplasticity and Regeneration; Autonomic Nervous System.

Further Reading

Anderson DJ (2000) Genes, lineages and the neural crest: A speculative review. *Philosophical Transactions of the Royal Society of London B: Biological Sciences* 355: 953–964.

Brauer MM, Shockley KP, Chavez R, Richeri A, Cowen T, and Crutcher KA (2000) The role of NGF in pregnancy-induced degeneration and regeneration of sympathetic nerves in the guinea pig uterus. *Journal of the Autonomic Nervous System* 79: 19–27.

Bruno MA and Cuello AC (2006) Activity-dependent release of precursor nerve growth factor, conversion to mature nerve growth factor, and its degradation by a protease cascade. *Proceedings of the National Academy of Sciences of the United States of America* 103: 6735–6740.

Cowen T (2002) Selective vulnerability in adult and ageing mammalian neurons. *Autonomic Neuroscience* 96: 20–24.

Cowen T and Gavazzi I (1998) Plasticity in adult and ageing sympathetic neurons. *Progress in Neurobiology* 54: 249–288.

Dontchev VD and Letourneau PC (2003) Growth cones integrate signaling from multiple guidance cues. *Journal of Histochemistry and Cytochemistry* 51: 435–444.

Fahnestock M, Michalski B, Xu B, and Coughlin MD (2001) The precursor pro-nerve growth factor is the predominant form of nerve growth factor in brain and is increased in Alzheimer's disease. *Molecular and Cellular Neuroscience* 18: 210–220.

Gabella G, Berggren T, and Uvelius B (1992) Hypertrophy and reversal of hypertrophy in rat pelvic ganglion neurons. *Journal of Neurocytology* 21: 649–662.

Gershon MD (1998) V. Genes, lineages, and tissue interactions in the development of the enteric nervous system. *American Journal of Physiology* 275: G869–G873.

Glebova NO and Ginty DD (2005) Growth and survival signals controlling sympathetic nervous system development. *Annual Review of Neuroscience* 28: 191–222.

Miller FD and Kaplan DR (2001) Neurotrophin signalling pathways regulating neuronal apoptosis. *Cellular and Molecular Life Sciences* 58: 1045–1053.

Nykjaer A, Lee R, Teng KK, et al. (2004) Sortilin is essential for proNGF-induced neuronal cell death. *Nature* 427: 843–848.

Orike N, Middleton G, Buchman VL, Cowen T, and Davies AM (2001) Role of PI 3-kinase, Akt and Bcl-2-related proteins in sustaining the survival of neurotrophic factor-independent adult sympathetic neurons. *Journal of Cell Biology* 154: 995–1005.

Thrasivoulou C, Soubeyre V, Ridha H, et al. (2006) Reactive oxygen species, dietary restriction and neurotrophic factors in age-related loss of myenteric neurons. *Aging Cell* 5: 247–257.

Xu XM, Fisher DA, Zhou L, et al. (2000) The transmembrane protein semaphorin 6A repels embryonic sympathetic axons. *Journal of Neuroscience* 20: 2638–2648.

Autonomic Neuroplasticity and Regeneration

P J Kingham and G Terenghi, University of Manchester, Manchester, UK

Introduction

Neuronal plasticity can be defined as the responsiveness of a neuron to its surroundings and how this changes following injury. The normal phenotype and function of a neuron depends on both the integrity of the neuron and also its connection with other neurons and end targets. During development of the nervous system, the terminals of neurons project to the same target and compete for limited quantities of neurotrophic growth factors, which then bind to surface receptors and are internalized and transported to the cell body to provide key signals for survival and differentiation. Disruption of these connections, through injury, initiates a sequence of molecular and cellular responses which act to promote regeneration and if successful reestablish contact with end targets. Therefore, the plastic changes which remodel the autonomic nervous system following injury involve a balance between cell death, due to trophic starvation, and regeneration of surviving neurons which switch from a differentiated to growing phenotype.

Cell Survival and Growth Factors in the Autonomic Nervous System

Nerve growth factor (NGF) is the classic target-derived neurotrophin synthesized by end targets. Sympathetic neurons were the first neurons to be identified as NGF responsive and their targets, vascular and smooth muscle, skin cells such as keratinocytes and melanocytes, and various endocrine tissues produce large quantities of the growth factor. NGF has two known receptors, TrkA and $p75^{NTR}$, which are both expressed by adult sympathetic neurons. TrkA is a transmembrane protein which serves as a receptor tyrosine kinase, and the neurotrophic properties of NGF are largely attributable to its effects on this receptor. $p75^{NTR}$ interaction with TrkA further modulates the responsiveness to NGF. The use of primary cell cultures of sympathetic neurons or the PC12 cell line has led to the identification of numerous intracellular signaling pathways activated by NGF. In brief, binding of NGF to TrkA initiates receptor dimerization leading to phosphorylation of integral tyrosine residues, which results in activation of kinase activity and subsequent phosphorylation of adaptor proteins which link to the mitogen-activated protein (MAP) kinase signaling cascade. Activation of these proteins contributes to altered expression of NGF-responsive genes via the mobilization of transcription factors from the cytoplasm to the nucleus. In addition to the local effect of NGF on the axon terminals, it is retrogradely transported from the end target to neuronal cell bodies within the sympathetic ganglia via postganglionic axons. This occurs through endocytosis of the ligand receptor complex and the formation of signaling endosomes. These vesicles then accumulate in the neuronal soma where high levels of phosphorylated TrkA activate the signaling events necessary for survival.

The importance of NGF signaling for cell survival is illustrated by studies of NGF–/– animals, which show almost complete loss of sympathetic nerves. Lesion to a sympathetic nerve isolates it from the end target source of NGF and leads to a reduction in neuron number. Nevertheless, adult sympathetic neurons do not die acutely after such a challenge but instead undergo a gradual cell death. This resistance to growth factor deprivation suggests that protective mechanisms can be activated following nerve injury. This involves changes both at the lesion site and within the cell bodies. Ultimately, this leads to activation of signaling cascades which act to mimic the effects of NGF and other growth factors which are important for the regulation of autonomic nerve survival.

Early Responses to Nerve Injury

Nerve injury causes damage to axonal membranes and interrupts the retrograde flow of signals from the normal innervation target. Before the ends can be resealed, Ca^{2+} enters the axoplasm and activates calcium-dependent proteases such as calpains, leading to extensive cytoskeletal remodeling. This may underlie new growth cone formation and may direct new intra-axonal protein synthesis, preventing the axons from simply withering away. Both myelinated and unmyelinated (sympathetic postganglionic nerves) axons are surrounded by Schwann cells. In the healthy nerve, Schwann cells appear quiescent and their function is tightly regulated by the axon they ensheath. When this interaction is lost, Schwann cells become highly active. Within hours of nerve injury, Schwann cells produce elevated levels of inflammatory cytokines such as tumor necrosis factor α and interleukin-1 which act to attract infiltrating macrophages, T-cells, and neutrophils. Both resident and recruited macrophages are critical for removal of nerve debris, which might impede the regeneration of axons.

Schwann cells proliferate in large numbers at the distal stump, forming a strand of cells called the band of Büngner, which provides an indispensable pathway for guiding regenerating axons to their end target. Perhaps more importantly, the Schwann cells release a number of neurotrophic factors to compensate for the decline occurring as a result of lost contact with the end target. Following nerve injury, Schwann cells produce increased levels of NGF which remain 10- to 15-fold elevated for at least 2 weeks. High levels of NGF are maintained by interactions between Schwann cells and invading macrophages. NGF may subsequently be taken up by the newly forming growth cones at the proximal stump. It is likely that the growth factor is dispersed diffusely in a gradient fashion around regenerating axons and therefore boosts axonal elongation. Brain-derived neurotrophic factor (BDNF) is another growth factor important for the survival of sympathetic neurons. Quiescent Schwann cells contain low levels of BDNF, which is upregulated following axotomy. In contrast to NGF, the response to denervation is slower and peak levels of BDNF are reached weeks following injury. Ciliary neurotrophic factor (CNTF) is a growth factor important for the survival of parasympathetic ciliary ganglion neurons. Schwann cells in intact nerves contain high levels of CNTF. Following injury, the levels decline and are reestablished only upon reinnervation, suggesting that contact with axons is necessary for the synthesis of CNTF in Schwann cells. Studies exploring the downstream targets of CNTF such as signal transducer and activator of transcription-3 (STAT-3) indicate that this could be an important mediator of early responses to nerve injury. Local release of CNTF by denervated Schwann cells activates the CNTF receptor on nearby axons, leading to phosphorylation of STAT-3 and its retrograde transport to the cell bodies of injured neurons. Therefore, the release of neurotrophic factors by Schwann cells at the injury site can help maintain a viable population of neurons which can switch to a growth phenotype, enabling their axons to regenerate across the lesion and back to the periphery.

Signaling Pathways and Changes in Gene Expression in the Injured Neuron

Chromatolysis, a histological change involving the rough endoplasmic reticulum and nucleus, was first noted in axotomized cell bodies many years ago. It was suggested that this process reflected a change in the types of proteins the neurons were synthesizing. It is now well accepted that nerve injury induces changes in intracellular signaling pathways which result in the induction of transcription factors and gene activation leading to upregulation of growth-associated molecules and a switch from neurotransmission. The recent development of microarray technology has led to the identification of large numbers of neuronal genes regulated by nerve injury. With regard to the autonomic nervous system, these changes have been best studied in the sympathetic, superior cervical ganglion (SCG). This is partly due to the difficulties of injuring selectively parasympathetic axons, given that most of the peripheral neuron cell bodies lie in widely dispersed ganglia closely associated with, or in the end target. Additionally, *in vitro* SCG explants provide an intact ganglion maintaining the association between neurons and nonneuronal cells and allow the subsequent investigation of the role of individual regeneration-induced genes. Indeed, it has been shown that upregulation of leukemia inhibitory factor (LIF) in the satellite cells of the ganglion within 6 h of axotomy mediates the expression of many of the induced genes. Studies are now underway to elucidate the role of the gene products produced during regeneration, a selection of which are discussed below.

Transcription Factors

These proteins play a major role in coupling extracellular signals to altered gene expression in the neuron and contribute to the underlying neuroplasticity of the autonomic nervous system. STATs are transcription factors which are phosphorylated by janus (JAK) kinases in response to cytokine activation of cell surface receptor tyrosine kinases. Activated STATs dimerize and translocate to the nucleus where they activate transcription of cytokine-responsive genes. To date, at least three JAK kinases and six STAT proteins have been shown to participate in this complex signaling pathway. Axotomy of SCG results in the rapid and sustained activation of STAT-1α and STAT-3 proteins in neuronal cell bodies. This activation is dependent on LIF release from satellite cells. The signaling cascade may also in part be mediated by reduced trophic support since NGF and other growth factors have been shown to reduce STAT expression. The STAT pathway may be particularly important for the regulation of neuropeptide expression in the SCG (see below).

Nuclear factor kappa B (NF-κB) is another transcription factor that plays a key role in the regulation of genes involved in nervous system function. NF-κB is a heterodimeric protein, principally composed of a 50 kDa subunit and a 65 kDa subunit. Typically, NF-κB is sequestered in the cytoplasm by the specific inhibitory protein IkB. The phosphorylation and subsequent proteasomal degradation of IkB releases

NF-κB and permits its translocation into the nucleus in which it can bind DNA and regulates transcription. NF-κB is constitutively activated in sympathetic neuron cultures maintained in the presence of NGF and withdrawal of the growth factor downregulates NF-κB activity. This correlates with increased apoptosis and therefore NF-κB might play a role in the survival of sympathetic neurons.

Like NF-κB, activating protein-1 (AP-1) transcription factors are dimeric molecules and comprise two major protein families: the Jun family and the Fos/activating transcription factor (ATF) family. Activation of the transcription factor c-Jun through phosphorylation by the MAP kinase, c-Jun N-terminal kinase (JNK), has been implicated in neuronal stress responses. Neurons contain low basal levels of the c-jun protein and axonal damage results in dramatic increases in expression levels, most likely as a consequence of decreased trophic support. The relative role of c-jun in neuronal cell death and regeneration is still unknown. In sympathetic neuron culture, c-jun is activated following NGF withdrawal and correlates positively with the onset of apoptosis. In contrast, *in vivo* studies of peripheral nerves have shown that deletion of c-jun hinders the expression of genes and proteins associated with axonal regeneration and reduces the speed of target reinnervation. The activity of other transcription factors such as NF-κB and ATF-3 can suppress the c-Jun-dependent transcription of cell death genes. Microarray data from axotomized SCG indicate that all three transcription factors are upregulated, suggesting a complex interaction for the balance of neuronal cell death and regeneration.

Cell Survival Proteins

The expression of a number of genes encoding anti-apoptotic proteins are increased in axotomized SCG. These proteins may contribute to the survival of injured neurons in response to growth factor deprivation. Heat shock proteins (HSPs) are a group of molecules expressed after cellular stress such as nerve injury. It has been suggested that these proteins serve a protective role by stabilizing other protein structures that are sensitive to denaturation and to help them to retain or to restore their native, functional conformations. The production of HSP70 mRNA is increased rapidly following SCG injury, and although protein studies have not been performed, it is likely that the ganglion cells expressing high HSP70 levels show a greater resistance to apoptosis. In mixed peripheral nerves, HSP70 protein remains elevated for days and positively correlates with surviving neurons. One particular stress a neuron encounters following axotomy is an increase in oxidative free radicals. Elevated expression levels of superoxide dismutase following sympathetic axotomy may protect against cell death. Of the pathways leading to apoptosis, mitochondria play a central role releasing a number of proapoptotic molecules resulting in the activation of caspases and fragmentation of DNA. Studies of cultured sympathetic neurons indicate that the translocation of Bax protein, from cytoplasm to mitochondria, is a critical and irreversible step in the cell death pathway following trophic factor deprivation. However, the extent to which the same pathway is used *in vivo* is still not clear. Nevertheless, the observation that Bax inhibitor-1 expression is increased in the SCG following axotomy suggests a protective response.

Neural Transmission Molecules

Noradrenaline (NA) is the principal neurotransmitter of peripheral sympathetic neurons and is stored in vesicles, which are formed in the cell body and then transported into the axon, toward the nerve terminal, by fast axonal transport. Here the neurotransmitter is released upon the arrival of a nerve impulse. Both large (80–100 nm diameter) and small (40–50 nm diameter) vesicles exist, which in addition to NA contain varying types of peptides (see below). ATP is also an important co-transmitter with NA. The phenotype of sympathetic neurons is highly plastic both *in vitro* and *in vivo*. For example, dissociated neurons exposed to medium conditioned by nonneuronal cells can upregulate choline acetyltransferase activity and develop cholinergic characteristics. Likewise, the sympathetic axons innervating rat sweat glands are initially noradrenergic but as the sweat gland matures, NA levels fall and cholinergic properties are acquired. This indicates that sympathetic neurons are highly responsive to extracellular factors. This plasticity is especially apparent in neuropeptide expression levels following nerve injury.

Neuropeptide Y (NPY) is a 36-amino-acid peptide, which is highly concentrated in the large vesicles of sympathetic postganglionic neurons and is co-released with NA. Following axotomy, NPY levels are increased both *in vivo* and in SCG explants. This occurs in response to as yet unidentified factors released from injured nonneuronal cells in the ganglion. In contrast to the high number of adult sympathetic neurons positive for NPY, other neuropeptides such as galanin and vasoactive intestinal polypeptide (VIP) are rarely expressed in healthy neurons. Both are significantly increased in response to axotomy as the result of LIF release in the ganglion. In the case of VIP, it has been shown that LIF promotes the binding of transcription factors to a cyclic AMP responsive element located within the 5′-flanking region of the

VIP gene. Other cytokines such as interleukin-1 can also potentiate the number of VIP-positive sympathetic neurons. In contrast, administration of NGF partially suppresses the increases in galanin and VIP positive neurons following axotomy. This suggests that under physiological conditions, NGF may play a role in controlling the neuropeptide phenotype of sympathetic neurons. Recent microarray data have identified two additional upregulated neuropeptides, cholecystokinin (CCK) and PTHrP, which had not previously been observed in sympathetic neurons. It is tempting to speculate that in addition to their classic transmitter role, neuropeptides are involved in the growth response during sympathetic nerve regeneration. The observation that galanin knockout mice display retarded regeneration of the sciatic nerve indicates this may be the case. Furthermore, VIP increases the neurite extension of sympathetic neuroblasts and the survival of cultured sympathetic neurons deprived of NGF. In addition to affecting neurons, VIP can also act on Schwann cells to increase the expression of extracellular matrix (ECM) molecules necessary for promoting regeneration.

Regulators of Neurite Outgrowth

Although sprouting at the lesion site occurs within hours of injury, it takes several days before axonal outgrowth emerges from the proximal stump. During this time, there is upregulation of many proteins necessary for remodeling the growth cone of regenerating axons. Organized cytoskeletal structures in association with activated intracellular signal transduction pathways determine the shape and motility of growth cones. Many years ago, it was noted that while molecules involved in the synthesis of neurotransmitters such as tyrosine hydroxylase were decreased, important structural proteins including actin and tubulin were increased, indicating an active remodeling process. GAP-43 is a small acidic protein, which is highly associated with outgrowth of neurites during development. Following injury to the adult nerve, GAP-43 is synthesized in the cell body and transported to the regenerating tip where it accumulates at high levels in the growth cones. The N-terminal of GAP-43 binds to the plasma membrane of the growth cone and the C-terminal with the cytoskeletal components of the axon. Calmodulin, an important mediator of signal transduction, is also attached to GAP-43. Phosphorylation of GAP-43 by protein kinase C releases calmodulin and leads to growth cone extension. In contrast, activation of G-proteins at the N-terminal end can lead to inhibition of growth cone extension. Thus, GAP-43 plays an important role in signaling and controls the spreading, branching, and adhesion of the growth cone.

The extension and path finding of growth cones are mediated by substrates such as ECM molecules with which they interact. The basal lamina is the ECM to which regenerating axons attach preferentially. The most potent adhesion molecule for promoting axonal outgrowth in this matrix is laminin, which binds with integrin receptors on the axon surface. An integrin is a heterodimer protein composed of α and β chains, and peripheral nerve injury results in the upregulation of the $\alpha6$, $\alpha7$, and $\beta1$ integrin subunits. Neuronal jun-deficient mice do not show an upregulation of $\alpha7$ integrin after injury, which contributes to the poor regenerative response seen in these animals. In addition, the expression of the fibronectin gene is significantly increased following axotomy of sympathetic neurons, and this may contribute to Schwann cell migration to the regenerating front. Therefore, the increased expression of cell adhesion molecules is likely to lead to activation of intracellular signal transduction pathways and remodeling of the cytoskeleton, facilitating the growth of regenerating axons. This is in part mediated by the nonreceptor tyrosine kinase, pp60c-src, which is also upregulated following nerve injury, and concentrates within the growth cone. Pp60c-src directly links a complex of integrins, vinculin and actin filaments, which can control cytoskeletal transformations. Transient activation of this complex, by phosphorylation events, enables growth cone motility. In addition, phosphorylation of tubulins by other members of the c-src family of tyrosine kinases suppresses the formation of microtubules creating a highly plastic environment in the regenerating axon.

Since regrowth is necessary for functional recovery following axotomy, it is interesting to note that microarray data often produce 'hits' for molecules that are known to regulate neurite outgrowth. For instance, axotomy of the SCG leads to increased expression of the arginase 1 gene, an enzyme involved in polyamine synthesis. Arginine is hydrolyzed to ornithine, which is then converted to the diamine putrescine and then the polyamines spermidine and spermine. Levels of these molecules increase in axotomized ganglia. Pharmacological blockade of polyamine synthesis inhibits the regenerative response, while administration of exogenous polyamines promotes neurite outgrowth and functional recovery following injury to *in vivo* SCG neurons. In contrast, the levels of inhibitors of regeneration such as the proteinase activated receptor or thrombin receptor fall in response to axotomy. Together, these changes lead to the promotion of outgrowth and an enhanced regenerative response.

Conclusions

The aim of this article is to highlight how autonomic nerves respond to nerve injury and produce a regenerative response. The recent use of microarray technology combined with classic protein biochemistry studies have helped to elucidate many of the signaling pathways involved and demonstrate the highly plastic nature of the peripheral nervous system. Activation of mechanisms that compensate for neurotrophic factor deprivation aim to maintain sufficient neuronal numbers which can then switch from a differentiated to growing phenotype and subsequently reinnervate the end target. Further investigation of these endogenous responses to injury should lead to a new generation of therapies to enhance nerve repair.

See also: Apoptosis in Nervous System Injury; Autonomic Neuroplasticity and Aging; Autonomic Neuroplasticity: Development; Axonal Regeneration: Role of Growth and Guidance Cues; Nerve Growth Factor; Peripheral Nerve Regeneration: An Overview; Schwann Cells and Axon Relationship; Transcription Factors in Synaptic Plasticity and Learning and Memory.

Further Reading

Bennet MR, Gibson WG, and Lemon G (2002) Neuronal cell death, nerve growth factor and neurotrophic models: 50 years on. *Autonomic Neuroscience* 95: 1–23.

Boeshore KL, Schreiber RC, Vaccariello SA, et al. (2004) Novel changes in gene expression following axotomy of a sympathetic ganglion: A microarray analysis. *Journal of Neurobiology* 59: 216–235.

Glebova NO and Ginty DD (2005) Growth and survival signals controlling sympathetic nervous system development. *Annual Review of Neuroscience* 28: 191–222.

Hendry IA (1992) Responses of autonomic neurones to target deprivation. In: Hendry IA and Hill CE (eds.) *Development, Regeneration and Plasticity of the Autonomic Nervous System*, pp. 415–462. Chur, Switzerland: Harwood Academic Publishers.

Hou XE and Dahlstrom A (2000) Synaptic vesicle proteins and neuronal plasticity in adrenergic neurons. *Neurochemical Research* 25: 1275–1300.

Ide C (1996) Peripheral nerve regeneration. *Neuroscience Research* 25: 101–121.

Klimaschewski L, Tran TD, Nobiling R, and Heym C (1994) Plasticity of postganglionic sympathetic neurons in the rat superior cervical ganglion after axotomy. *Microscopy Research and Technique* 29: 120–130.

Kury P, Stoll G, and Muller HW (2001) Molecular mechanisms of cellular interactions in peripheral nerve regeneration. *Current Opinion in Neurology* 14: 635–639.

Makwana M and Raivich G (2005) Molecular mechanisms in successful peripheral regeneration. *FEBS Journal* 272: 2628–2638.

Raivich G, Bohatschek M, Da Costa C, et al. (2004) The AP-1 transcription factor c-Jun is required for efficient axonal regeneration. *Neuron* 43: 57–67.

Terenghi G (1999) Peripheral nerve regeneration and neurotrophic factors. *Journal of Anatomy* 194: 1–14.

Zigmond RE and Sun Y (1997) Regulation of neuropeptide expression in sympathetic neurons. Paracrine and retrograde influences. *Annals of the New York Academy of Sciences* 814: 181–197.

Autonomic Neuroplasticity: Development

P G Smith, University of Kansas Medical Center, Kansas City, KS, USA

Introduction

The concept that the nervous system can undergo substantial reorganization has been appreciated for over a century. In 1909, in *Modern Problems in Psychiatry,* Ernesto Lugaro used the term 'plasticity' in the context of the nervous system to convey the idea that chemical interactions responsible for the developmental patterning of the nervous system also provide adaptive mechanisms for behavioral modification and nervous system repair.

While plasticity can occur throughout the nervous system, the autonomic nervous system arguably has been most instructive in revealing the underlying principles. One reason is that the peripheral autonomic nervous system provides a particularly tractable system for studying neuroplasticity. Because its principal neurons are located outside of the central nervous system, it is relatively easy to isolate and remove relatively pure neuronal populations, either for cell culture under regimented conditions or to examine the effects of extirpation or transposition. A second reason is that sympathetic neurons are responsive to the prototypic neurotrophic factor, nerve growth factor (NGF); thus the role of target-derived proteins and autonomic neuronal survival and growth has been recognized and manipulated in this system for over half a century. Moreover, this system has proved not only to be highly complex, with multiple factors controlling development, but also very amenable to investigations using genetic manipulation to elucidate the role of various genes and proteins. Accordingly, the autonomic nervous system has served as a primary model system for advancing our knowledge of cellular and molecular processes regulating nervous system development. Perturbations to these processes have allowed us to more fully appreciate the repertoire of responses that comprise developmental neuroplasticity.

The process of nervous system development is itself inherently plastic. During development, undifferentiated cells undergo variable degrees of migration and cell division, giving rise to pools of precursor cells which provide the anatomical substrate of the nervous system. Precursor cells differentiate, taking on specific properties and becoming responsive to chemical cues in their environment, and eventually elaborate axons which find their way to appropriate targets. Neuron numbers are further refined through apoptosis in response to limitations in available target-derived survival factors, thus matching neuronal numbers to target mass and cell type. Target innervation density is then established through the actions of both propulsive and repulsive signals, and retrogradely transported proteins contribute to final determination of transmitter phenotype. In the perinatal period, hormonal factors play increasingly important roles in refining nerve–target relationships. Perturbations to any of these dynamic processes can lead to significant alterations in nervous system properties, which are appropriately defined as 'developmental neuroplasticity.'

Developmental autonomic neuroplasticity is defined here broadly to include pre- and postnatal developmental processes associated with sympathetic and parasympathetic nervous system formation and those perturbations that affect structure, phenotype, and function. These include (1) molecular signaling events regulating autonomic neuronal ontogeny and the impact of their disruption through genetic manipulation and transpositioning or removing individual structures, (2) the establishment and maintenance of autonomic projections to peripheral targets, and (3) establishment, maintenance, and alterations of neurochemical phenotype during development or in response to injury.

Autonomic Gangliogenesis

A defining event in the ontogeny of the autonomic nervous system is the formation of motor ganglia. These consist of aggregates of sympathetic and parasympathetic postganglionic neurons, located in the periphery, which receive excitatory cholinergic synaptic input from preganglionic neurons located in the central nervous system. Autonomic postganglionic neurons derive from the neural crest, a transient structure that originates from cells migrating rostrocaudally from the neural tube shortly after closure. Neural crest precursor cells give rise to a variety of cell types, including melanocytes, peripheral glia, cartilage, smooth muscle, adrenomedullary chromaffin cells, and sensory and autonomic neurons. Those destined to become autonomic ganglion neurons of the peripheral nervous system migrate ventrally through the rostral somitic mesoderm to populate sympathetic and parasympathetic ganglia. Any perturbation that disrupts normal ganglion formation will severely impact autonomic nervous system organization and function.

The generation of autonomic neuronal precursors is complex, involving a variety of transcription factors and signaling proteins (**Figure 1**, **Table 1**). A pool of

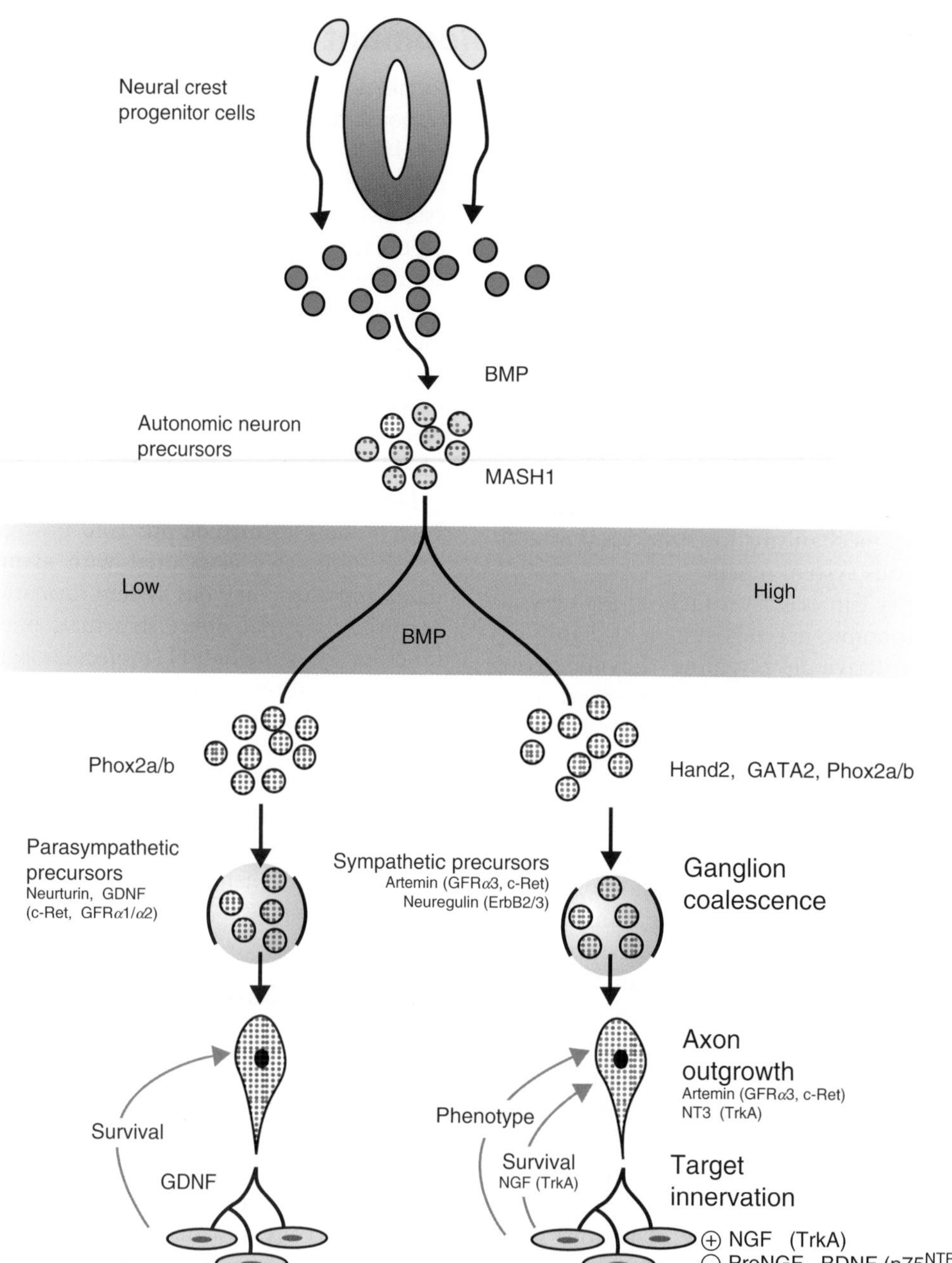

Figure 1 Schematic diagram indicating processes involved in the ontogeny of the autonomic nervous system, and the roles of some regulatory proteins. Neural crest gives rise to progenitor cells which differentiate into mammalian achaete–scute homolog-1 (MASH1)-expressing autonomic neuronal precursors under the influence of members of the bone morphogenetic protein (BMP) family. Neuronal identity appears to be determined within a BMP gradient, with sympathetic neurons differentiating within high concentrations and parasympathetic neurons differentiating within low concentrations. Parasympathetic neurons require expression of members of the Phox2 family of transcription factors, which are also required for sympathetic neurons, in addition to Hand2 and Gata2. Coalescence of precursors into ganglia requires artemin acting on the GFRα3/c-Ret receptor complex and neuregulin acting on ErbB receptors for sympathetic ganglia, and glial cell-derived neurotrophic factor (GDNF) and neurturin acting on GFRα1 or GFRα2 complexed with c-Ret. Sympathetic axon outgrowth requires both artemin and neurotrophin-3 (NT3), which acts on the TrkA receptor, whereas parasympathetic axon outgrowth may be dependent on GDNF or related proteins. Target-derived factors regulate target innervation: in sympathetic neurons, nerve growth factor (NGF) acting on TrkA can promote neuronal survival and increase innervation density, whereas proNGF and brain-derived neurotrophic factor (BDNF) acting selectively on the $p75^{NTR}$ diminish target innervation density. GDNF regulates innervation density of some parasympathetic targets. Target-derived factors can also influence neuronal neurochemical phenotype.

Table 1 Factors regulating autonomic nervous system development, and effects of mutations of regulatory genes

Factor	*Role in development*	*Effect of deletion*
MASH1, CASH1	Transcription factors specifying autonomic neuron identity	Autonomic neurons fail to form
Bone morphogenetic protein	Regulates formation of autonomic precursors, and differentiation of sympathetic and parasympathetic neurons	Sympathetic and parasympathetic neurons fail to form
Phox2a/b	Transcription factors required for catecholaminergic phenotype	Sympathetic and parasympathetic neurons fail to develop
Hand2, Gata2	Transcription factors specifying sympathetic fate	Absence of sympathetic neurons
Neuregulin, ErbB2/3	Promote sympathetic precursor migration	Inappropriate sympathetic trunk ganglion formation
Hepatocyte growth factor	Neuroblast survival and differentiation	Reduction in sympathetic ganglion neurons
Semaphorin 3A	Neural crest cell migration	Abnormal sympathetic ganglion formation
Neurotrophin-3	Sympathetic axon outgrowth	Impaired axon extension, fewer surviving sympathetic neurons due to inaccessibility to target-derived trophic factors; impaired cholinergic sympathetic phenotype
Glial cell-derived neurotrophic factor, c-Ret, GFRα1/2	Parasympathetic neuron survival, neuroblast migration; axon guidance	Reduced numbers of parasympathetic postganglionic neurons, impaired axon outgrowth
Nerve growth factor	Target-derived protein regulating sympathetic neuronal survival in early postnatal development, and target innervation density	Augmented cell death of sympathetic neurons; absent or diminished sympathetic innervation of some targets
Artermin, c-Ret, GFRα3	Sympathetic neuroblast migration; rostral sympathetic ganglion formation; sympathetic axon outgrowth	Misplaced superior cervical ganglion; impaired sympathetic axon outgrowth
Neurturin	Maintenance of parasympathetic target innervation	Reduced parasympathetic innervation

autonomic neuron progenitor cells differentiates from neural crest precursors under the influence of local expression of bone morphogenetic proteins (BMPs). Differentiation of both sympathetic and parasympathetic precursor cells is dependent upon their expression of the mouse or chick achaete–scute homolog transcription factors, MASH1 or CASH1, respectively, and animals lacking these proteins fail to form autonomic neurons. The transcription factors Phox2a and Phox2b are also involved in expression of catecholaminergic properties, which are characteristic of both sympathetic and parasympathetic neuronal precursors. Phox transcription factors are required for early survival of both sympathetic and parasympathetic autonomic neuron precursors. Sympathetic neuronal precursors are further dependent upon the Hand2 and Gata2 transcription factors.

Autonomic neuronal precursors migrate ventrally toward their ultimate destination, guided by local signaling proteins which, for truncal sympathetic neurons, include neuregulin and Sema3A, acting on the ErbB and neuropilin 1 receptors, respectively. Artermin, acting on the c-Ret and GFRα3 receptors, appears to play an important role in sympathetic ganglion coalescence and in providing chemotactic signals responsible for rostral migration of superior cervical ganglion precursor neurons. The factors regulating parasympathetic neuron migration are less understood, but evidence indicates that in some cranial parasympathetic ganglia, precursors are dependent initially upon glial cell-derived neurotrophic factor (GDNF) via the c-Ret and GFRα1 receptors for ganglion cell proliferation and migration, and subsequently on neurturin and the GFRα2 receptor for neuronal survival and ganglion development.

During normal development, different regions of the neural crest give rise to specific structures. In chick–quail chimeras, for example, rostral parasympathetic ganglion neurons derive from mesencephalic neural crest, whereas sympathetic ganglion neurons derive from the truncal neural crest extending below somite 4. However, premigratory precursor cells apparently are not committed to specific fates. Thus, neural crest cells from the vagal region that would normally give rise to enteric ganglion cells, when transplanted caudally, give rise to apparently normal sympathetic ganglion neurons and adrenomedullary cells. This suggests that neural crest derivatives at this stage remain multipotent and substantially plastic, and differentiation into mature neural crest derivatives is determined largely by peri- and postmigratory environments.

While most migrating neural crest cells become irrevocably committed during the process of migration and gangliogenesis, it is now clear that a small population of precursor cells remains undifferentiated and may give rise to nascent neurons and glia throughout development and maturity. These presumptive stem cells may serve as a source of cells that can contribute to continued plasticity in the developing and adult organism.

Plasticity of Autonomic Neuroeffector Pathways

Axon Outgrowth

Following migration and coalescence of ganglion cells, connectivity between the central nervous system and peripheral targets must be established. Autonomic neurons elaborate axons in the prenatal and early postnatal periods. Preganglionic axons traverse the anterior region of the somitic mesoderm. Somitic mesoderm appears to play a critical role in preganglionic axon guidance, and removal or transpositioning of somites leads to major alterations in axon trajectory. Preganglionic axons project to the region of their intended ganglion, even if the ganglion has been removed. Once having entered the ganglion, preganglionic axons synapse with the postganglionic neurons; while some synapses are present in the prenatal period, the first postnatal week is characterized by dramatic synaptogenesis and establishment of a functional pathway, at least in the sympathetic inferior cervical ganglion of the rat. Ganglion neurons are initially polyinnervated by convergent inputs from multiple preganglionic neurons, and normal reductions through axon pruning taking place postnatally.

Autonomic postganglionic neurons elaborate axons shortly after exiting the cell cycle, and postganglionic axons first reach their intended targets in the late postnatal period in rodents; substantial increases in numbers of target projecting neurons occur through the first 2–3 postnatal weeks. In the case of sympathetic axonal outgrowth, this appears to be regulated predominantly by two locally produced proteins. Artemin, a member of the GDNF family of trophic factors, is produced by vascular smooth muscle cells in early development, and sympathetic axons at this developmental stage express the artemin receptors c-Ret and GFRα3. Neurotrophin 3 is also produced locally and activates sympathetic TrkA receptors, and is required for normal postganglionic axon outgrowth. The association of these growth factors with vascular smooth muscle is believed to account for the fact that sympathetic pathways associate closely with blood vessels as they course to their target destinations. Parasympathetic axon outgrowth is dependent upon GDNF and its receptors c-Ret and GFRα1. Disruptions of these growth factors or their receptors have serious consequences; without sufficient axon outgrowth, not only are targets hypoinnervated, but neurons are also in turn deprived of required target-derived survival factors and hence undergo abnormally extensive cell death, leading to fewer ganglion neurons. In some cases, diminished postganglionic axon outgrowth affects the mature neurochemical phenotype because of deprivation of target-derived proteins that influence neurotransmitter properties.

One feature of the postganglionic sympathetic system is a dramatic ability to reorganize in response to injury in the postnatal period. For example, innervation of the orbit normally derives exclusively from the ipsilateral superior cervical ganglion, and damage to this pathway in the adult results in sustained denervation and impairment, known clinically as Horner's syndrome. In this syndrome, the pupil is abnormally constricted (loss of excitatory sympathetic innervation to the pupil dilator muscle) and the upper eyelid fails to retract normally (ptosis, due to loss of excitatory sympathetic innervation to the superior tarsal smooth muscle which elevates the eyelid). However, in the neonatal rat, removal of one superior cervical ganglion results in sprouting of spared intracranial vascular sympathetic axons from the contralateral intact ganglion, which take atypical pathways to reach the orbital targets. Despite having significant phenotypic differences from normal resident innervation, such as expression of neuropeptide Y, as is typical of vascular sympathetic neurons but not normal resident tarsal muscle innervation, this aberrant pathway largely restores functional control of the target (**Figure 2**). The ability to establish contralateral orbital innervation is lost completely during the first postnatal month, largely through differentiation of smooth muscle-like myofibroblasts that define their trajectory to the orbit; this coincides with the time course reported for the loss of artemin expression during development in similar cell types, implicating a developmental reduction in artemin expression as a cause of the postnatal diminution in pathway plasticity.

Neuronal Survival: The Neurotrophic Hypothesis

Once the terminal axon reaches the target, a critical process takes place in which the relationship between target and neuron is established and refined, thus ensuring that the neuroeffector pathway is effective and appropriate. This process relies on target-derived proteins that can modify both neuronal survival and terminal axon growth.

The prototypical neurotrophin NGF has long been associated with sympathetic postganglionic axon sprouting *in vitro*, so it is not surprising that this protein would be considered to be the primary candidate for controlling sympathetic innervation. An abundance of

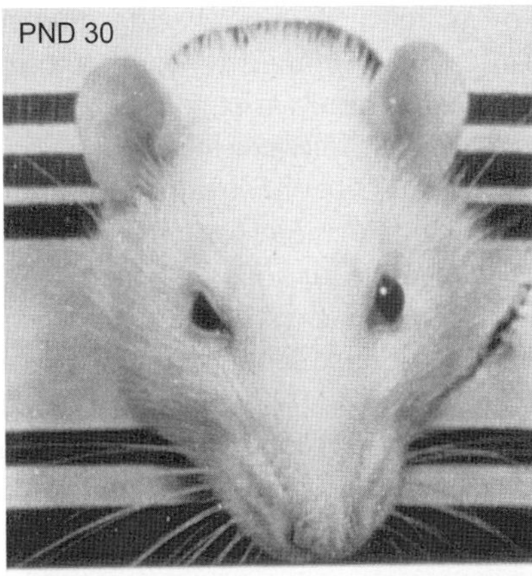

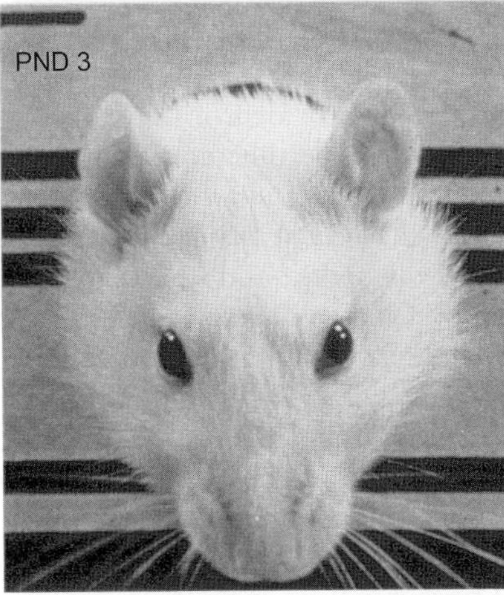

Figure 2 Plasticity in postganglionic pathway formation in the rat. The right superior cervical ganglion was removed from the rat in the top panel on postnatal day (PND) 30, whereas the same ganglion was removed from the rat in the bottom panel on PND 3. The photographs are of the awake rats at 3 months of age. Following PND 30 ganglionectomy, contraction of the superior tarsal muscle in the right eyelid is impaired, leading to ptosis and reduced palpebral fissure width. In the rat receiving ganglionectomy on PND 3, there is minimal evidence of impaired contraction; recovery is due to the formation of an atypical pathway deriving from intact contralateral neurons that normally project to cranial blood vessels to the denervated orbital targets.

information supports the idea that NGF (and many other trophic proteins in different contexts) is required for neuronal survival. Sympathetic neurons are produced in overabundance during development, and a relatively large proportion (in some cases >50%) undergoes developmental cell death in the perinatal period, leading to the ganglion cell numbers ultimately observed in the adult. Those neurons that do survive are believed to be the ones that have obtained access to adequate amounts of a target-derived survival protein (trophic factor, from the Greek term *trophos*, meaning 'to nourish'). However, trophic proteins are believed to be produced in limited quantities. In those sympathetic axons that reach the target first and have access to the greatest amounts of NGF, which binds to the NGF-selective receptor TrkA and is transported retrogradely to the cell body as receptor–ligand complex, programmed cell death (apoptosis) is prevented. Axons that cannot compete for adequate neurotrophic factor because they are misguided or arrive after target innervation has already achieved optimal density ultimately are eliminated through programmed cell death. The principle that an oversupply of developing neurons competes for target-limited trophic factor, thus matching neuronal numbers to target needs, represents the essence of the 'neurotrophic hypothesis.' While less is known about the trophic requirements of parasympathetic neurons, members of the GDNF family of ligands appear to serve similar roles. Modulating the amount of trophic factor available can induce considerable plasticity in the numbers of postganglionic neurons; addition of exogenous NGF or upregulation of its synthesis in peripheral tissues, such as epidermis, results in increased numbers of surviving sympathetic ganglionic neurons. Similarly, depletion of endogenous NGF during development, inactivation of the NGF gene, or null mutations of the TrkA gene encoding the NGF receptor all result in marked reductions in numbers of sympathetic ganglion neurons. A comparable situation has been shown to be the case for cranial parasympathetic neurons, the survival of which is reliant first upon GDNF and subsequently upon neurturin, acting, respectively, on GFRα1 and GFRα2 complexed with the coreceptor c-Ret.

Regulation of Target Innervation

The establishment of effective neuroeffector transmission is dependent upon the presence of appropriate numbers of terminal autonomic axons within the target tissue. Visceral targets show a wide range in the numbers of sympathetic and parasympathetic fibers present per unit tissue mass (innervation density). In light of NGF's well-documented role, not only in inducing sympathetic neuronal survival but also in eliciting robust outgrowth *in vitro*, it is not surprising that this neurotrophin would be a logical candidate for regulating sympathetic innervation density in peripheral targets. Indeed, early studies of several different targets revealed a positive correlation between levels of NGF mRNA or protein and the relative densities of sympathetic innervation. In keeping with this idea, mice with null mutations for NGF or TrkA fail to develop normal target innervation. Conversely, conditions favoring increased target levels of NGF can lead to increased target innervation. For example, spontaneously hypertensive rats display abnormally elevated levels of NGF in the mesenteric vasculature, and these vessels are also hyperinnervated by sympathetic axons. Further, in keeping with the central tenet of the neurotrophic hypothesis, that neurons compete for limited amounts of trophic factors, neonatal destruction by capsaicin of sensory nociceptor nerves (which express TrkA and compete for target NGF) results in sympathetic axon hyperinnervation; this presumably occurs because reduced competition makes more NGF available to sympathetic axons. Accordingly, alterations in target NGF levels can result in significant plasticity in target innervation density.

While there is little doubt that NGF plays an important role in sympathetic target innervation, recent lines of investigation show that the relationship is far more complex. As indicated earlier, NGF is required for neuronal survival; therefore, loss of innervation in the absence of NGF or TrkA could be due solely to cell death rather than to an absence of local growth-promoting properties. Indeed, using mouse mutants lacking genes for both NGF and the apoptotic factor Bax (thus circumventing cell death that would normally occur in the absence of NGF), Ginty and colleagues found that the dependency on NGF for development of target innervation was highly variable, with innervation of some targets being drastically reduced (e.g., salivary glands, heart) while others were unaffected (trachea). This supports the idea that growth-promoting factors other than NGF contribute to target sympathetic innervation.

A second complicating factor is that, in addition to propulsive effects of proteins such as NGF and GDNF, autonomic target innervation appears to be controlled by proteins that are repulsive in nature. For example, sympathetic neurons are responsive to semaphorins and can be repulsed from targets expressing high levels of these proteins. Moreover, neurotrophins show surprising complexity. In addition to the TrkA receptor mediating sympathetic neuron survival and axon outgrowth, the pan-neurotrophin receptor $p75^{NTR}$ can facilitate ligand binding to TrkA, but can also mediate repulsive cues. Brain-derived neurotrophic factor (BDNF), a neurotrophin closely related to NGF, shows little affinity for TrkA but binds avidly to $p75^{NTR}$. When $p75^{NTR}$ is selectively activated, this inhibits sympathetic axonal outgrowth and contributes to sympathetic axon pruning (and in some cases can cause cell death). Importantly, in addition to BDNF, the precursor form of NGF, proNGF, shows relatively little affinity for TrkA but strongly binds $p75^{NTR}$, and, like BDNF, can be repulsive and pro-apoptotic to sympathetic neurons. Thus the extent to which NGF regulates target innervation density will depend upon which form predominates.

The balance between TrkA and $p75^{NTR}$ activation has important implications for developmental plasticity of sympathetic innervation. Some targets, such as those of the female reproductive tract, undergo considerable innervation plasticity in the postnatal period. For example, uterine innervation changes through puberty and beyond, under the influence of the gonadal steroid hormone estrogen; in fact, changes in estrogen lead to reductions in myometrial sympathetic innervation at the estrus stage of each estrous cycle. Available evidence shows that estrogen increases uterine expression of BDNF, and that fluctuations in this hormone underlie physiological axon pruning that takes place during puberty and throughout the reproductive cycle.

Neurochemical Plasticity

Neural crest progenitor neurons give rise to a richly diversified range of peripheral autonomic neurons. Fully differentiated parasympathetic neurons typically express a cholinergic phenotype, which includes choline acetyltransferase and vesicular acetylcholine transporter (VAChT), a nitrergic phenotype (neuronal nitric oxide synthase), and the neuropeptide vasoactive intestinal polypeptide (VIP). Most mature sympathetic neurons are noradrenergic, expressing the proteins tyrosine hydroxylase (TH), dopamine β-hydroxylase (DBH), vesicular monoamine transporter-2 (VMAT2), and the norepinephrine transporter; a significant proportion of noradrenergic sympathetic nerves, mainly those innervating blood vessels, also contain neuropeptide Y. About 5% of mature sympathetic neurons, which project to eccrine sweat glands and periosteum, are cholinergic. Neural crest progenitor cells also give rise to autonomic ganglionic small intensely fluorescent (SIF) cells, which are dopaminergic (express TH but not DBH), and adrenomedullary chromaffin cells (synthesize epinephrine and contain phenylethanolamine *N*-methyltransferase (PNMT), in addition to TH and DBH).

The differentiation of parasympathetic and sympathetic neurons appears to be regulated by way of a BMP gradient; progenitors most proximate to BMP-producing tissues (e.g., dorsal aorta) acquire a noradrenergic phenotype, while those exposed to lower concentrations become cholinergic parasympathetic neurons. Differentiation of both types of neurons requires MASH1 and Phox2a/b transcription factors, and noradrenergic neurons additionally require the Hand2 and Gata2 transcription factors. Differentiation of neural crest progenitors into adrenomedullary chromaffin cells is, in part, under control of the local environment; corticosteroid derived from the surrounding adrenal cortex is required for PNMT expression. Local factors have also been implicated in phenotype acquisition and maintenance in SIF cells, which are located proximate to blood vessels and may require endothelial factors for differentiation.

The common origins of both sympathetic and parasympathetic neurons underlie sustained similarities throughout development and presage responses observed with certain perturbations. Experimental observations in neural crest-derived precursor cells reveal that the biochemical machinery necessary for acetylcholine synthesis is present well before commitment to final cell fate. Similarly, catecholaminergic properties have also been described as a feature of autonomic neuron progenitors prior to expression of other pan-neuronal properties.

Transmitter Phenotype 'Switching'

While genetic programming and instructive signals during precursor migration and gangliogenesis appear to play a major role in determining autonomic neuronal phenotype, target-derived proteins are also important in determining or refining mature neuronal phenotype. The best studied example of this is the transition from an adrenergic to a cholinergic phenotype in some sympathetic neurons. Early studies showed that neonatal sympathetic neurons cultured under certain conditions (e.g., in culture medium conditioned by cardiac myocytes) lose their noradrenergic phenotype, and this is replaced over time by cholinergic properties. In an extensive series of experiments, Story Landis and colleagues extended these findings *in vivo* by showing that sweat glands, which are innervated by cholinergic sympathetic axons in the adult, are actually innervated by axons with a catecholaminergic phenotype at postnatal day 4; however, these catecholaminergic properties (dense core vesicles, catecholamine histofluorescence, and TH immunoreactivity) are downregulated over the next 10 days and are replaced by cholinergic properties (VIP immunoreactivity and cholinesterase activity). An impressive body of evidence has been accrued to support the idea that these neurons undergo a 'phenotype shift' from catecholaminergic to cholinergic. Evidence in favor of this hypothesis includes several findings: (1) destruction of catecholaminergic fibers by neonatal administration of selective neurotoxins 6-hydroxydopamine, guanethidine, or antisera to NGF prior to the shift prevents sweat gland cholinergic innervation, even though cholinergic nerves should be unaffected by these treatments, (2) transplantation of the sweat glands to ectopic sites where sympathetic nerves normally do not show a cholinergic phenotype (trunk hairy skin, ocular anterior chamber co-cultures with the superior cervical ganglion) induces cholinergic properties, (3) replacement of sweat glands with a target that normally is not innervated by cholinergic sympathetic axons (salivary gland) prevents cholinergic properties from appearing in the resident innervation, and (4) mouse mutants that fail to develop sweat glands also fail to develop cholinergic fibers to the footpad. Further studies have shown that the shift is facilitated by an interactive loop with the target, in which noradrenergic transmission accelerates acquisition of cholinergic properties, and that a similar phenotypic shift occurs to sympathetic innervation to the periosteum; periosteal cells induce a VIP and VAChT phenotype in resident innervation and also when transplanted to ectopic sites. The factors responsible for this shift have not been definitively identified, but members of the gp130 cytokine family, which includes leukemia inhibitory factor and cardiotropin, promote cholinergic properties in cultured neurons and are strong candidates.

Recent studies imply that the establishment of cholinergic traits in sympathetic neurons is more complex. Using antibodies with high sensitivity to VAChT, it appears that cholinergic properties are demonstrable in subpopulations of sympathetic neurons prior to target innervation, and that sweat gland-innervating fibers can express VAChT at an age supposedly preceding the phenotype shift. Thus, the extent to which cholinergic properties are conferred *de novo* by the target or are instead an intrinsic property of sympathetic neurons undergoing normal development *in vivo* remains uncertain. Nonetheless, it is clear that some targets produce factors that are capable of inducing neurochemical phenotype plasticity in developing sympathetic neurons.

Another factor that can influence neurochemical phenotype is injury. Normally, VIP-immunoreactive neurons are only rarely encountered in the superior cervical ganglion. Following axotomy, however, VIP is dramatically upregulated *in situ* or under culture conditions. This implies that the factors normally responsible for providing a stable neurochemical phenotype can be affected by injury.

See also: Autonomic Nervous System: Neuroanatomy; Autonomic Neuroplasticity and Aging; Autonomic Neuroplasticity and Regeneration; Autonomic Nervous System Development; Nerve Growth Factor; Neural Crest.

Further Reading

Ernsberger U (2001) The development of postganglionic sympathetic neurons: Coordinating neuronal differentiation and diversification. *Autonomic Neuroscience* 94: 1–13.

Francis NJ and Landis SC (1999) Cellular and molecular determinants of sympathetic neuron development. *Annual Review of Neuroscience* 22: 541–566.

Glebova NO and Ginty DD (2005) Growth and survival signals controlling sympathetic nervous system development. *Annual Review of Neuroscience* 28: 191–222.

Honma Y, Araki T, Gianino S, et al. (2002) Artemin is a vascular-derived neurotropic factor for developing sympathetic neurons. *Neuron* 35: 267–282.

Howard MJ (2005) Mechanisms and perspectives on differentiation of autonomic neurons. *Developmental Biology* 277: 271–286.

Huber K (2006) The sympathoadrenal cell lineage: Specification, diversification, and new perspectives. *Developmental Biology* 298: 335–343.

Hyatt-Sachs H, Schreiber RC, Bennett TA, et al. (1993) Phenotypic plasticity in adult sympathetic ganglia *in vivo*: Effects of deafferentation and axotomy on the expression of vasoactive intestinal peptide. *Journal of Neuroscience* 13: 1642–1653.

Keast JR (2006) Plasticity of pelvic autonomic ganglia and urogenital innervation. *International Review of Cytology* 248: 141–208.

LeDourin NM, Creuzet S, Couly G, et al. (2004) Neural crest cell plasticity and its limits. *Development* 131: 4637–4650.

Mione MC, Cavanagh JF, Lincoln J, et al. (1990) Long-term chemical sympathectomy leads to an increase of neuropeptide Y immunoreactivity in cerebrovascular nerves and iris of the developing rat. *Neuroscience* 34: 369–378.

Morrison SJ, White PM, Zock C, et al. (1999) Prospective identification, isolation by flow cytometry, and in vivo self-renewal of multipotent mammalian neural crest stem cells. *Cell* 96: 737–749.

Smith PG, Fan Q, and Zhang R (1999) Divergence of smooth muscle target and sympathetic pathway cell phenotypes in the orbit of the developing rat. *Journal of Comparative Neurology* 408: 352–364.

Young HM, Anderson RB, and Anderson CR (2004) Guidance cues involved in the development of the peripheral autonomic nervous system. *Autonomic Neuroscience* 112: 1–14.

Autophagy and Neuronal Death

P G H Clarke, University of Lausanne, Lausanne, Switzerland

Autophagy and Its Role in the Mediation of Autophagic Cell Death

Autophagy

Although this article is primarily concerned with neurons, and the implication of autophagy in their death, it will be necessary to draw general principles from studies on other cell types, because autophagy is a general phenomenon occurring in virtually all types of cell, and the most convincing molecular analyses of its role in cell death have been done in nonneuronal cell lines.

Autophagy is the mechanism by which cells degrade parts of their own cytoplasm using the lysosomal machinery. There are several types of autophagy, including microautophagy, the direct capture of tiny portions of cytosol by invagination of lysosomal membranes; chaperone-mediated autophagy, a specific mechanism for degrading cytosolic proteins containing a particular pentapeptide consensus motif; pexophagy, the specific autophagocytosis of peroxisomes; and macroautophagy, which involves the engulfment of sizeable regions of cytoplasm, including organelles, in double-membrane vesicles called autophagosomes. Macroautophagy is the best-studied type of autophagy, and the only type that has been studied in detail in relation to cell death. This article will therefore deal primarily with macroautophagy.

Macroautophagy (**Figure 1**) is initiated by the formation of autophagosomes (about 400–800 nm in diameter) from cup-shaped double-membranous structures called isolation membranes or phagophores, which engulf cytosolic components, including organelles. The isolation membrane then closes to form the autophagosome. The origin of the isolation membrane is still a matter of debate. There is evidence that it may arise from various sources including smooth endoplasmic reticulum and the *trans*-Golgi network, but recent research on yeast indicates that a major source of its membrane is an independent punctate structure called the 'pre-autophagosomal structure.' The autophagosome fuses with a lysosome to form an autolysosome, where the enclosed material is broken down. The term 'autophagic vacuole' includes both autophagosomes and autolysosomes.

Autophagy (including macroautophagy) is involved in the normal turnover of cell contents and is enhanced by cellular stresses, against which it provides protection, for example, by replenishing the pool of free amino acids in the case of amino acid depletion, or by eliminating damaged proteins. Also, by reducing the size of stressed cells, autophagy reduces their metabolic burden. Thus, in many situations, autophagy promotes the health and survival of cells.

Autophagic Cell Death: Origins of the Concept

Despite the life-promoting roles of autophagy, macroautophagy has also been associated with cell death, and the term 'autophagic (or type 2) cell death' is used as a morphological classification for dying cells containing numerous autolysosomes. This occurs frequently during embryonic development and is probably the most common type of cell death in insect metamorphosis, but (macro)autophagic features are also associated with many cases of pathological cell death including heart failure, excitotoxicity, and neurodegenerative diseases.

Historically, the development of electron microscopy permitted the discovery of (macro)autophagy in the early 1960s, and this was soon followed by numerous ultrastructural studies from the mid-1960s onward, showing an abundance of autolysosomes in dying cells in many situations, including most cases of metamorphosis. Nevertheless, even as late as the 1990s, only a few authors considered that the autophagy was instrumental in the cell death.

The reasons for this reluctance were multiple. One was that autophagy had from the moment of its discovery been understood to play physiological roles in healthy cells, for example, the provision of breakdown products for reuse and the elimination of abnormal proteins, and several authors interpreted its presence in the dying neurons to reflect an unsuccessful survival-promoting mechanism for eliminating damaged regions of cytoplasm. Many other authors were influenced by the (then) widely accepted 'suicide bag hypothesis' of De Duve, discoverer of the lysosome, according to which cell death is achieved by the release of hydrolases from the lysosomes; the status of this hypothesis is still controversial. Then, as the suicide bag hypothesis gradually fell out of favor in the 1970s and 1980s, the simultaneous rise in popularity of a somewhat rigid dichotomy according to which all cell death had to be apoptosis or necrosis did not encourage openness to alternative mechanisms of cell death. Indeed, proponents of the apoptosis–necrosis dichotomy maintained that

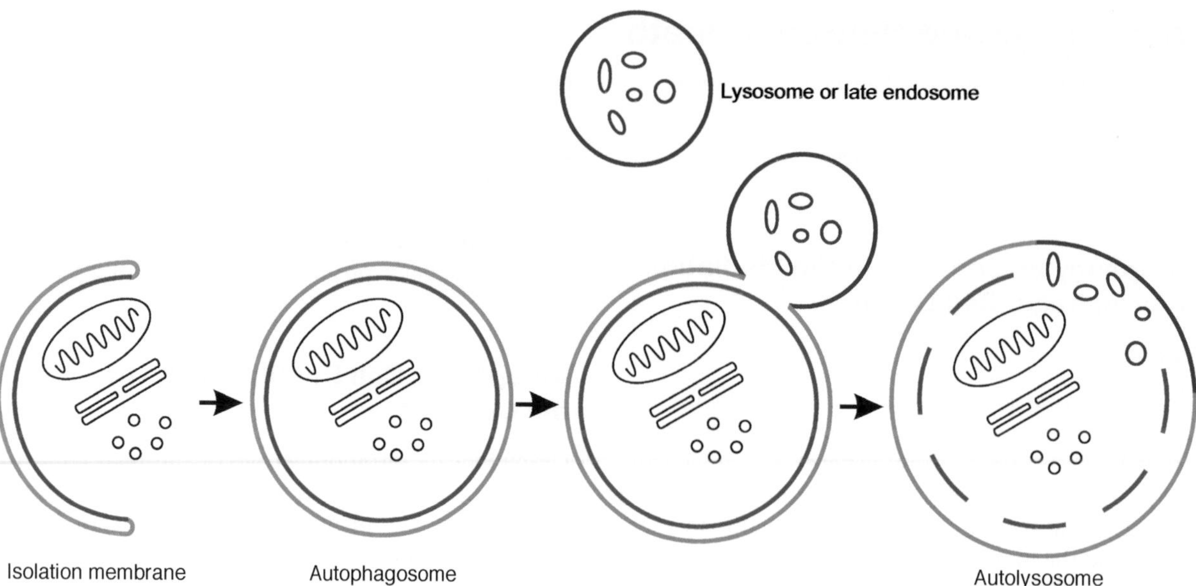

Figure 1 Macroautophagy. An isolation membrane (formed from smooth endoplasmic reticulum, the *trans*-Golgi network, or the 'pre-autophagosomal structure') enwraps cytosolic components and closes to form the two-membraned autophagosome, which then fuses with a lysosome or late endosome to give an autolysosome, in which the engulfed contents and the inner autophagosomal membrane (shown in green) are degraded.

autophagic dying cells were in fact undergoing apoptosis and that the autolysosomes were either a protective reaction or an irrelevant epiphenomenon. And, finally, it has to be admitted that a death-mediating role for the autophagy had not been proved, and in several cases very strong autophagy can occur without neuronal death.

The idea of autophagy-mediated cell death was, however, supported in the 1980s by experiments on neuronal death in the target-deprived isthmo-optic nucleus in chick embryos. (The isthmo-optic nucleus is the source of efferents from the brain to the retina in birds.) This neuronal death was characterized by abundant autolysosomes that ultimately filled most of the cytoplasm, and also by the loss of DNA from the nucleus to neighboring lysosomes. The fact that a cell's own DNA was being degraded by autophagy went against the view that the autophagy was a survival-promoting reaction to cellular stress.

Prevention of Autophagic Cell Death by Pharmacological Inhibitors of Autophagy

Nevertheless, a death-promoting role for autophagy gained only limited acceptance until it could be proved that inhibiting it prevented cell death. Initial evidence for this was provided in the 1990s by the death-preventing effects of 3-methyladenine (3-MA), an inhibitor of the formation of autophagic vacuoles that has been described as 'specific' but only in the limited sense that it does not alter the overall level of protein synthesis. Sandvig and van Beurs first showed, in 1992, that cell death, in this case toxin induced, could be prevented by 10 mM 3-MA. Subsequently, similar doses of 3-MA were shown to prevent (partially or completely) or delay cell death with autophagic characteristics in many situations including sympathetic neurons deprived of nerve growth factor, telencephalic neurons exposed to chloroquine, and cerebellar granule neurons deprived of serum and potassium. In all cases, the dying cells were shown to contain numerous autophagic vacuoles, and their rescue by 3-MA was accompanied by a reduction in their content of autophagic vacuoles. The suppression by 3-MA of autophagy is probably due to its inhibition of class III phosphatidylinositol 3-kinase (PI3-K), but it was uncertain whether this is also the basis of its protection against autophagic cell death, because its pharmacological profile is poorly characterized and it probably affects other enzymes. It was therefore important to test whether better-characterized inhibitors of PI3-K (LY294002 and wortmannin) could have similar protective effects. In many situations, these inhibitors are proapoptotic, because they inhibit the powerfully protective class I PI3-K pathway, so a protective effect due to inhibition of class III PI3-K (and hence autophagy) can easily be masked; but in serum-deprived PC12 cells, LY294002, wortmannin, and 3-MA have all been shown to be protective, apparently through the blockade of autophagy.

Prevention of Autophagic Cell Death by Interference with Autophagy Genes

However, even the better-characterized PI3-K inhibitors affect other cellular processes as well as autophagy, and definitive evidence for the death-mediating role of autophagy was provided only recently, by studies involving RNA interference of specific autophagy genes.

Our understanding of the control mechanisms of autophagy (particularly macroautophagy) depends to a great extent on intensive studies on autophagy in yeast, where about 30 genes controlling the initiation and execution of autophagy have been identified during the last decade. Until recently, these were grouped into three main gene families (apg, aut, and cvt), according to the genetic screens in which they were detected, but the functional distinctions between these families do not appear to be very clear-cut, and in the current terminology all the genes are grouped into the single 'atg' (autophagy gene) family (atg1–atg29 at the time of writing, but new members continue to be discovered). A detailed description of how these genes control autophagy would be beyond the scope of this chapter, but it is highly relevant to our present concerns that many of the yeast genes have vertebrate (including mammalian) homologs, and that certain of them, including atg5, atg6 (beclin 1), and atg7, are essential for the formation of autophagosomes.

This fact was used in two key papers in 2004 in which macroautophagy was blocked by RNA interference of atg5, atg6 (beclin 1), and atg7 (as well as by pharmacological inhibitors of autophagy) in cell lines whose apoptotic machinery had been deactivated genetically or pharmacologically. In both papers, pure autophagic cell death occurred and both the autophagy and the accompanying cell death were prevented by the RNA interference (and by the pharmacological inhibitors). Although a role for the autophagy genes in processes other than autophagy cannot be entirely ruled out, the fact that silencing each of the three genes prevented the autophagic cell death is strong evidence that the (macro)autophagy is not merely an epiphenomenon, or a defensive reaction, but is actually involved in mediating the cell death.

The Importance of Autophagy-Mediated Cell Death in Relation to Apoptosis

Autophagic cell death, as judged morphologically, seems to be the commonest type of cell death in physiological situations of massive cell death leading to the destruction of a tissue, as in many cases of metamorphosis and in some radical cases of mammalian embryonic tissue remodeling, whereas apoptosis appears to be the usual mechanism where sporadic dying cells occur in a tissue destined to survive. Thus, if autophagy could be assumed to mediate cell death in all cases of morphologically identified autophagic cell death, one could conclude that the autophagic death mechanism was of almost equal importance to the apoptotic mechanism.

Unfortunately, this is currently uncertain. While the reliability of 3-MA in protecting against many different cases of autophagic cell death does suggest that the autophagic death mechanism is of widespread importance, the more specific studies with RNA interference (or antisense) are still few in number, and situations have been reported in which massive autophagy can occur in cells without them dying. Moreover, there is evidence that a lysosomal, presumably autophagic, mechanism can initiate caspase activation and apoptosis. This is clearly different from autophagic cell death, which in many cases has been shown to be caspase independent, but does mean that morphological evidence for autophagy cannot be taken as proof of autophagy-mediated cell death. Thus, although the existence of an autophagic death mechanism is now difficult to deny, its generality and importance are still matters of debate.

Indeed, it has recently been argued that autophagy may mediate cell death only in very artificial situations where apoptosis has been deactivated (as in the two RNA interference studies mentioned above). Even if this were true, it would not detract from the importance of autophagic cell death in many pathological situations, where apoptosis may indeed have been deactivated either genetically (as in cancers that have become resistant to apoptosis-promoting chemotherapy) or pharmacologically (as in anticaspase neuroprotective protocols). But it has recently been shown that downregulation of atg5 by antisense technology protected against interferon-γ-induced autophagic cell death in HeLa cells whose apoptotic machinery had not been inhibited. Moreover, pharmacological blockade of autophagy by inhibition of PI3-kinase actually enhances the apoptotic machinery by increasing caspase-3 activation, but it can still prevent or delay cell death.

Thus, the autophagic death mechanism can be effective without the artificial deactivation of apoptosis, but its generality and importance are still not entirely clear.

Autophagy in Neuronal Death and Neurodegeneration

Autophagy in Neuronal Death

Although our mechanistic understanding of autophagic cell death has come largely from studies of

Table 1 Reports of neuroprotection by 3-MA

Situation	*Autophagic morphology?*	*Protection by 3-MA?*	*Caspase-dependent?*	*Reference*
Cerebellar granule neurons deprived of serum and potassium	Yes	Partial protection	Partially	Kaasik et al. (2005)
Cerebellar granule neurons in low potassium	Yes	Yes	Yes	Canu et al. (2005)
Telencephalic neuronal cultures exposed to chloroquine	Yes	Yes	No	Zaidi et al. (2001)
Sympathetic neurons treated with cytosine arabinoside	Yes	Cell death delayed	No	Xue et al. (1999)
Sympathetic neurons deprived of NGF	Yes	Cell death delayed	No	Xue et al. (1999)

nonneuronal cells, there is considerable morphological evidence for autophagic 'neuronal' death in all the main situations where neurons die: in natural development, in various pathological situations, and in experimental models, as is discussed below. In addition, there are a few studies showing the prevention of autophagic neuronal death by 3-MA (see **Table 1**).

Autophagic Neuronal Death during Development

Reports of autophagic neuronal death occurring naturally during development are relatively few, and most concerned anuran metamorphosis, including the death of the Rohon–Beard neurons, a transient population of sensory neurons that undergoes 100% cell death. In mammals, one is able to find only one relevant report; it concerned autophagic neuronal death in the developing cerebral cortex. This paucity of reports suggests that autophagic cell death plays only a relatively minor role in naturally occurring neuronal death in mammals (and other higher vertebrates). This fits with the generalization made above, that autophagic cell death occurs most commonly in physiological situations of massive cell death leading to the destruction of a tissue. However, caution is required, because in many studies isolated autophagic dying cells may have been mistaken for phagocytes, which they resemble morphologically and in their expression of autophagic markers.

Failure in competition for retrograde neurotrophic support is believed to be a major cause of naturally occurring neuronal death, and numerous studies of neuronal death in development have involved axotomy or other means of depriving neurons of retrograde support. In some cases, the resulting neuronal death was autophagic, but in many others it was clearly not. The reasons for the differences are unclear, but one factor may be the developmental stage. This was first indicated by an elegant study by Decker in 1978 on motor neuronal death in larval frogs. He found that very early axotomy caused a pyknotic (apoptosis-like) morphology, whereas very late axotomy caused classic chromatolysis. But axotomy at an intermediate stage caused the 'genesis of numerous secondary lysosomes in degenerating cells' – in other words, cell death with an autophagic morphology.

Studies on the isthmo-optic nucleus of chick embryos showed an age dependence that was similar to the above but not quite so clear cut. Early deprivation of retrograde support by blocking axonal transport in the isthmo-optic axons led to isthmo-optic neuronal death with a mixed morphology that was both pyknotic and autophagic, whereas later transport blockade caused a purer form of autophagic cell death with only minimal pyknosis (**Figure 2**). This neuronal death was also characterized by strong endocytic activity, a phenomenon that has since been observed in several subsequent studies of stressed, but not necessarily dying, neurons. Isthmo-optic neuron death could also be provoked by de-afferentation, but this caused no signs of autophagy, and when combined with blockade of retrograde support it decreased the autophagic characteristics of the dying neurons.

Neuronal Autophagy in Acute Neurological Conditions

The neuronal cell death in virtually all acute neurological conditions (e.g., stroke, traumatic brain injury, and neonatal asphyxia) shares a common mechanism: excitotoxicity, excessive depolarization that is usually due to the excessive activation of glutamate receptors, especially the *N*-methyl-D-aspartate (NMDA) subtype. Excitotoxic neuronal death is generally considered to be necrosis (most commonly) or apoptosis or a combination of the two, and, until recently, the presence of enhanced autophagy in these conditions was largely ignored. However, over the last few years, morphological evidence for intense autophagy and an increase in the autophagosomal marker LC3-II have been reported in several experimental models of cerebral hypoxia–ischemia, and an increase in the autophagy gene beclin 1 has been reported in a model of traumatic brain injury. NMDA receptor activation has likewise

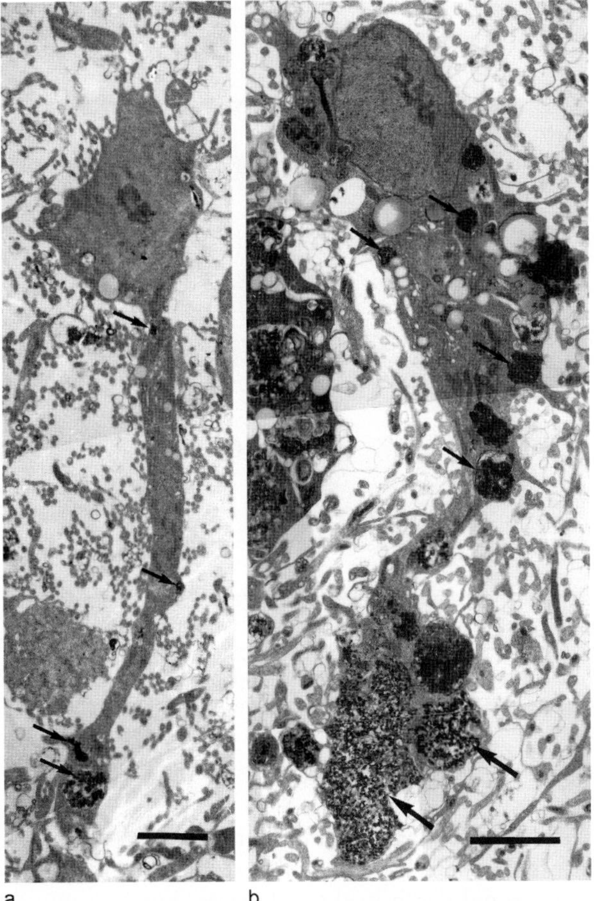

Figure 2 Electron micrographs of two autophagic dying neurons in the isthmo-optic nucleus of 14-day-old chick embryos. The death of both neurons was provoked by an injection of colchicine in their axonal target territory, which blocked axoplasmic transport, depriving them of retrograde support. (a) Early stage of neuronal death. Several vacuoles are found in the cytoplasm of the soma and main dendrite. The nucleoplasm and cytoplasm are somewhat denser than in a healthy neuron. (b) Advanced stage of neuronal death. The vacuoles are larger and more numerous, many containing membranous whorls. The arrows indicate vacuoles labeled with intravascularly injected horseradish peroxidase (in this and several other cases autophagic dying neurons have been shown to be strongly endocytic). Scale bar = 2 μm. From Hornung JP, Koppel H, and Clarke PGH (1989) Endocytosis and autophagy in dying neurons: An ultrastructural study in chick embryos. *Journal of Comparative Neurology* 283: 425–437.

been shown to induce autophagic neuronal death, in organotypic hippocampal cultures. This neuronal death was also characterized by strong endocytosis of exogenous horseradish peroxidase (**Figure 3**). However, it is currently unknown whether the autophagy in acute neurological conditions and excitotoxicity mediates cell death.

Autophagy in Neurodegenerative Diseases

In contrast to acute neurological conditions, neurodegenerative diseases involve progressive neuronal degeneration over periods of many months or years. Changes in the endosomal–lysosomal system, including increased macroautophagy, have been reported in virtually all neurodegenerative diseases including Alzheimer's, Huntington's, and Parkinson's diseases, prion diseases, and amyotrophic lateral sclerosis. The causes and roles of the increased macroautophagy are difficult to establish in human diseases, but additional information from experimental models provides some preliminary hypotheses. From models of Alzheimer's, Huntington's, and Parkinson's diseases, there is evidence that the macroautophagy may in many cases be involved in clearing protein aggregates from affected neurons, and hence be protective, but may also lead to autophagic neuronal death.

In Huntington's disease, the autophagy seems to be primarily protective. This disease involves massive neuronal death in the striatum as a result of the presence of an expanded polyglutamine repeat in the Huntington gene product. The dying neurons have a strongly autophagic morphology, and the autophagy appears to be a defense mechanism because the experimental enhancement of autophagy in fly and mouse models of Huntington's disease reduces the accumulation of polyglutamines as well as the neuronal death, whereas inhibition of autophagy has the opposite effect on both.

In Parkinson's disease, the situation is more ambiguous. The best-known neuropathological characteristics of this disease are the degeneration of dopaminergic neurons of the substantia nigra, and the presence of cytoplasmic inclusions called Lewy bodies in these neurons before they die. Lewy bodies contain ubiquitinated aggregates of α-synuclein and other proteins. There are reports that this neuronal death can have an autophagic morphology. Some cases of early-onset Parkinson's disease involve a mutation in the α-synuclein gene. In cultured PC12 cells, overexpression of mutant but not wild-type α-synuclein causes an impairment in the ubiquitin–proteasome system and the presence of ubiquitinated protein aggregates, an accumulation of autophagic vacuoles, and increased nonapoptotic autophagic cell death. Thus, although the increased autophagy may be an attempt to protect the cells by clearing the protein aggregates, it may also be involved in mediating the neuronal death.

Alzheimer's disease is characterized by the presence of (extracellular) β-amyloid plaques and filamentous tangles, primarily in the hippocampus and cerebral cortex. Both are currently believed to be involved in the degenerative changes in these brain regions. Pronounced macroautophagy has been demonstrated in the affected neurons, and β-amyloid has been shown to be generated by the proteolytic cleavage of β-amyloid precursor protein. In a mouse model of the

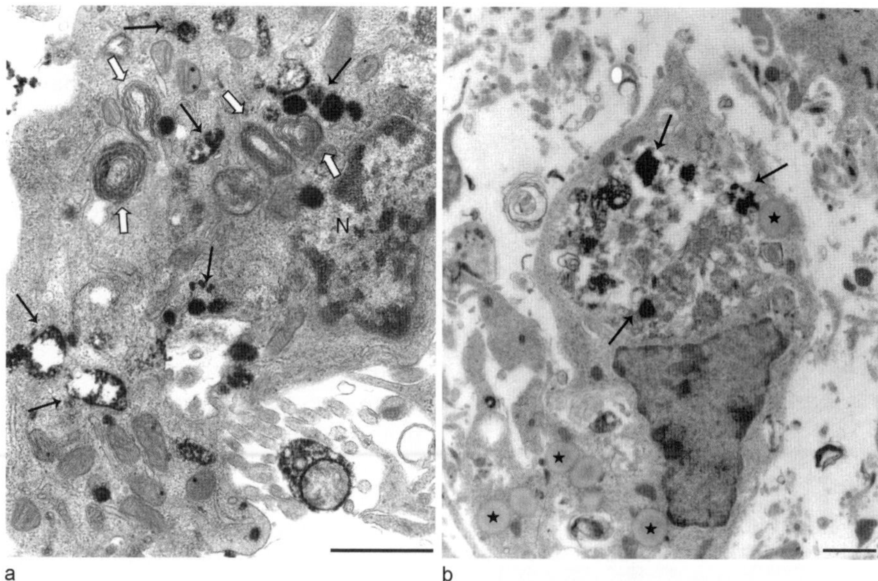

Figure 3 Autophagic degenerating neurons in the CA1 region of organotypic hippocampal cultures after 2 h (a) or 8 h (b) of exposure to 100 μM NMDA. (a) Thick white arrows, membranous whorls (autolysosomes); black arrows, endosomes labeled with horseradish peroxidase; N, nucleus. (b) Clumps of peroxide labeling within a large vacuole; stars, putative unlabeled endosomes. Scale bar = 1 μM. Reproduced from Borsello T, Croquelois K, Hornung JP, and Clarke PGH (2003) *N*-methyl-D-aspartate-triggered neuronal death in organotypic hippocampal cultures is endocytic, autophagic and mediated by the c-Jun N-terminal kinase pathway. *European Journal of Neuroscience* 18: 473–485, with permission from Blackwell Publishing.

disease, a similar neuronal macroautophagy occurs, and this happens rather early, before the extracellular β-amyloid deposits, but the maturation of autophagosomes to autolysosomes appears to be impaired. At later stages, there is a further accumulation of autophagosomes, and these are rich in β-amyloid. Inducing or inhibiting macroautophagy elicits parallel changes in macroautophagy and β-amyloid production, suggesting that in this case the macrophagy may contribute to the disease process, but not necessarily through autophagic cell death.

Neuronal Autophagy in Lysosomal Storage Diseases

Lysosomal storage diseases are caused by mutations in the genes encoding various lysosomal hydrolases, leading to the accumulation (or 'storage') of partially digested substances in lysosomes. Different lysosomal storage diseases cause degenerative and other changes in different organs of the body, including in some cases the brain (e.g., in Tay–Sachs disease and Niemann–Pick C disease). Whereas most neurodegenerative diseases involve increased lysosomal digestion, lysosomal storage diseases are caused by a 'decrease' in one particular component of lysosomal digestion, but this can lead to complex changes in many different cellular signaling pathways. Since the genetic mutation directly affects the lysosomal system, autophagic digestion must presumably be affected. There have been few studies of autophagy in neuronal death in these diseases, but in a mouse model of Niemann–Pick C disease there was massive degeneration of cerebellar Purkinje cells, which had features consistent with autophagic cell death.

See also: Alzheimer's Disease: Neurodegeneration; Amyotrophic Lateral Sclerosis (ALS): Disease Mechanisms; Epilepsy: Neuronal Death; Huntington's Disease: Neurodegeneration; Intracellular Calcium and Neuronal Death; Neuroprotection: Pharmacological Approaches; Parkinson's Disease: Alpha-Synuclein and Neurodegeneration; Programmed Cell Death.

Further Reading

Borsello T, Croquelois K, Hornung JP, and Clarke PGH (2003) *N*-methyl-D-aspartate-triggered neuronal death in organotypic hippocampal cultures is endocytic, autophagic and mediated by the c-Jun N-terminal kinase pathway. *European Journal of Neuroscience* 18: 473–485.

Bursch W and Ellinger A (2005) Autophagy – a basic mechanism and a potential role for neurodegeneration. *Folia Neuropathologica* 43: 297–310.

Canu N, Tufi R, Serafino AL, Amadoro G, Ciotti MT, and Calissano P (2005) Role of the autophagic-lysosomal system on low potassium-induced apoptosis in cultured cerebellar granule cells. *Journal of Neurochemistry* 92(5): 1228–1242.

Clarke PGH (1990) Developmental cell death: Morphological diversity and multiple mechanisms. *Anatomy and Embryology* 181: 195–213.

Hornung JP, Koppel H, and Clarke PGH (1989) Endocytosis and autophagy in dying neurons: An ultrastructural study in chick embryos. *Journal of Comparative Neurology* 283: 425–437.

Kaasik A, Rikk T, Piirsoo A, Zharkovsky T, and Zharkovsky A (2005) Up-regulation of lysosomal cathepsin L and autophagy during neuronal death induced by reduced serum and potassium. *European Journal of Neuroscience* 22(5): 1023–1031.

Larsen KE and Sulzer D (2002) Autophagy in neurons: A review. *Histology and Histopathology* 17: 897–908.

Nixon RA, Wegiel J, Kumar A, et al. (2005) Extensive involvement of autophagy in Alzheimer disease: An immuno-electron microscopy study. *Journal of Neuropathology and Experimental Neurology* 64: 113–122.

Shimizu S, Kanaseki T, Mizushima N, et al. (2004) Role of Bcl-2 family proteins in a non-apoptotic programmed cell death dependent on autophagy genes. *Nature Cell Biology* 6: 1221–1228.

Stefanis L, Larsen KE, Rideout HJ, Sulzer D, and Greene LA (2001) Expression of A53T mutant but not wild-type alpha-synuclein in PC12 cells induces alterations of the ubiquitin-dependent degradation system, loss of dopamine release, and autophagic cell death. *Journal of Neuroscience* 21: 9549–9560.

Xue L, Fletcher GC, and Tolkovsky AM (1999) Autophagy is activated by apoptotic signalling in sympathetic neurons: An alternative mechanism of death execution. *Molecular and Cellular Neuroscience* 14(3): 180–198.

Yu L, Alva A, Su H, et al. (2004) Regulation of an ATG7-beclin 1 program of autophagic cell death by caspase-8. *Science* 304: 1500–1502.

Yu WH, Cuervo AM, Kumar A, et al. (2005) Macroautophagy – a novel beta-amyloid peptide-generating pathway activated in Alzheimer's disease. *Journal of Cell Biology* 171: 87–98.

Zaidi AU, McDonough JS, Klocke BJ, et al. (2001) Chloroquine-induced neuronal cell death is p53 and Bcl-2 family-dependent but caspase-independent. *Journal of Neuropathology and Experimental Neurology* 60(10): 937–945.

Aversive Emotions: Genetic Mechanisms of Serotonin

S E Ahmari, M D Alter, and R Hen, Columbia University, New York, NY, USA

Introduction

Anxiety and fear responses are necessary components of adaptive behavior. Fear of predators and normal vigilant behavior are essential for survival in the wild. It is when these responses are uncontrollable that they stop being protective, instead becoming maladaptive and pathological, ultimately impairing an individual's ability to function in the world.

Though a complete understanding of the neurobiological mechanisms underlying normal and pathological fear and anxiety remains a future goal and motivator for research, a role for serotonin, also known as 5-hydroxytryptamine (5-HT), in these emotions has been firmly established. Both clinical evidence, ranging from empirical data (serotonin reuptake inhibitors have been found effective in treating a wide range of anxiety disorders; see **Table 1**) to human genetic studies, and animal experiments have supported a role for this neurotransmitter in both normal anxiety and pathological disease states. For example, there is evidence from human linkage analyses implicating monoamine oxidase A (MAO-A; a molecule which plays a key role in the degradation of serotonin) in the pathophysiology of panic disorder, and also suggesting a role for the serotonin transporter and two serotonin receptors (5-HT1B and 5-HT2A) in obsessive–compulsive disorder (OCD). It is likewise clear that neurotransmitters other than serotonin play a role in the development of anxiety and fear responses. However, though the story is ultimately likely to be very complex, with particular neurotransmitters and receptors being important at different times in development and with gene–environment interactions and epigenetics playing important roles in the ultimate determination of behavioral phenotypes, focusing on the role of serotonin will serve as a starting point for discussion. One way of shedding light on the role of this neurotransmitter in fear and anxiety is through examination of the combination of the effects of serotonin-related genetic factors and early life stress on development and the impact of these factors on the development of anxiety disorders in later life.

Relationship between Fear and Anxiety

Though there is a close relationship between fear and anxiety, and animal models of fear conditioning have allowed us to gain important insights into the neural circuits underlying anxiety, it is apparent that there is a distinction between these responses despite substantial overlap in involved brain regions. Past models have differentiated fear as being a response to real threats and anxiety as being an abnormal response to a perceived threat in the absence of immediate danger to the organism. Alternatively, one can understand fear as being stimulus specific (akin to 'phobia' in humans), whereas anxiety is more generalized. Despite this distinction, it seems apparent that there is substantial overlap in the neural circuitry of fear and anxiety. For example, rodent studies of fear conditioning have clearly demonstrated the involvement of cortical and thalamic projections to the lateral nucleus of the amygdala, where substantial intra-amygdala processing occurs. Projections then arise from the central nucleus of the amygdala to areas of the brain stem, hypothalamus, and hippocampus, leading to fear responses. However, there is evidence that different amygdala output tracts may differentially subserve fear versus anxiety responses. For example, by using bright light exposure and corticotropin-releasing hormone-enhanced startle paradigms to model anxiety in comparison to stimulus-specific fear conditioning, Davis and colleagues demonstrated differential involvement of the bed nucleus of the stria terminalis in anxiety versus the central nucleus of the amygdala in fear, despite very similar output projections to the hypothalamus and brain stem. Human neuroimaging studies of perception of fearful faces (trait anxiety) and fear conditioning have clearly demonstrated involvement of the amygdala, hippocampus, and thalamus; similarly, neuroimaging studies of anxiety disorders have also implicated the amygdala (post-traumatic stress disorder (PTSD), phobia, OCD, panic disorder), prefrontal cortex (PTSD, generalized anxiety disorder (GAD), OCD, phobia), hippocampus (PTSD, phobia, panic disorder), and striatum (OCD, phobia). The remainder of this article will focus on the relationship between serotonin and the development of anxiety responses in both animals and humans, with the understanding that there is likely a close relationship between anxiety and fear responses, particularly in disorders such as PTSD.

Early Life Stress and Mental Health

A multitude of research studies associate early life stress with the development of mental health problems. Morbidity and mortality from a wide array of medical and mental illnesses are increased in abused children. Even less severe forms of early-life

Table 1 Pharmacological dissection of anxiety disorders

Disorder	*BDZ*	*Buspirone*	*IMI*	*CMI*	*MAOI*	*SSRI*
GAD	+	+	+	+	0	+
Social phobia	+	(+)	0	(+)	+	+
OCD	0	(+)	(+)	+++	(+)	+++
Panic disorder	+	0	+	+++	+	+

GAD, generalized anxiety disorder; OCD, obsessive–compulsive disorder; BDZ, benzodiazepines, compounds with activity at the γ-aminobutyric acid (GABA) receptor; buspirone, partial 5-hydroxytryptamine (5-HT) 1A agonist (partial agonist at a particular serotonin receptor); IMI, imipramine, a tricyclic antidepressant without activity at serotonin receptors; CMI, clomipramine, a tricyclic antidepressant with significant activity at serotonin receptors; MAOI, monoamine oxidase (MAO) inhibitors, compounds which inhibit MAO and therefore inhibit the breakdown of serotonin; SSRI, selective serotonin reuptake inhibitors, which prevent reuptake of serotonin into neurons, leading to increased concentrations in the synaptic cleft and enhanced postsynaptic activation of serotonin receptors; +, effective anxiolysis; (+), small or discrepant effects; +++, extensive anxiolysis; 0, no effect. Adapted from McNaughton N and Corr PJ (2004) A two-dimensional neuropsychology of defense: Fear/anxiety and defensive distance. *Neuroscience and Biobehavioral Reviews* 28(3): 285–305, with permission from Elsevier.

stress have been associated with an increase in adverse outcomes. For example, children brought up in families characterized by a lack of parental warmth or either over- or underregulation of children's behavior are more likely to have a host of medical or psychiatric problems. One could speculate that such children might start to have aberrations in the regulation of normal anxiety and fear responses early in development, which ultimately puts them at greater risk for development of disordered mood in later life.

Still, not all children exposed to these stressful environments have adverse outcomes. A particularly interesting line of research has demonstrated that at least a portion of these effects is mediated by genetic variability of the serotonin system. In a landmark study, Caspi and colleagues demonstrated that boys with a hypofunctioning polymorphism in the MAO-A gene are at increased risk for adolescent and adult antisocial behavior if they were abused during childhood. Since MAO-A is one of the primary enzymes responsible for the breakdown of serotonin in the central nervous system, this finding demonstrates a potential interaction between serotonin genetics and adolescent and adult psychopathology. Similarly, in a separate study in the same prospective cohort but utilizing both boys and girls, individuals homozygous (s/s) or heterozygous (l/s) for the short allele of a hypofunctioning polymorphism in the serotonin transporter (SERT) gene were found to have a markedly increased risk for depression if subjected to child abuse, but individuals homozygous for the normal functioning long allele (l/l) had no increase in depression when exposed to abuse. Many studies have since examined the relationship between SERT and MAO polymorphisms and various forms of psychopathology, demonstrating associations in OCD, panic disorder, specific and social phobia, GAD, and anxious personality traits.

When examining the multiple studies demonstrating the detrimental effects of early life stress and genetic variance on mental health, the logical next question is, how are these detrimental effects being mediated? An attractive hypothesis is that early life represents a particularly sensitive period for the development of neural circuits and that disruption of the normal processes of establishing these circuits makes a person more vulnerable to the development of psychopathology.

Sensitive Periods of Development: A Role for Serotonin?

A sensitive period is a general term that applies whenever experience has a particularly strong and lasting effect over a limited time (**Figure 1**). A classic example is that of filial imprinting, which occurs shortly after birth as a young animal learns to recognize and bond with its parent. An important characteristic of sensitive periods is that the effects of experience are not readily amenable to change after the sensitive period has occurred. For example, once an animal has undergone imprinting, the attachment to the parent will be strong and persistent; moreover, attachment to an alternative parental figure will be much more difficult to achieve. This is thought to occur because, though sensitive periods are marked by behaviors such as attachment to a parent or speaking a language, they are actually a property of underlying neural circuits. The properties of the neural circuits are shaped by experience during the sensitive period, and once formed, they may be modulated but do not easily change their basic structure. Furthermore, because behaviors, complex behaviors like language or anxiety responses in particular, represent the coordination of many neural circuits, the true effect of the persistence of circuit-level changes occurring during a sensitive period will tend to be underestimated.

Sensitive periods are of particular interest to researchers and clinicians because they represent periods in development during which certain capacities are readily shaped or altered by experience and therefore may serve as key points during which interventions may have a lasting impact on the development of normal versus abnormal brain circuits.

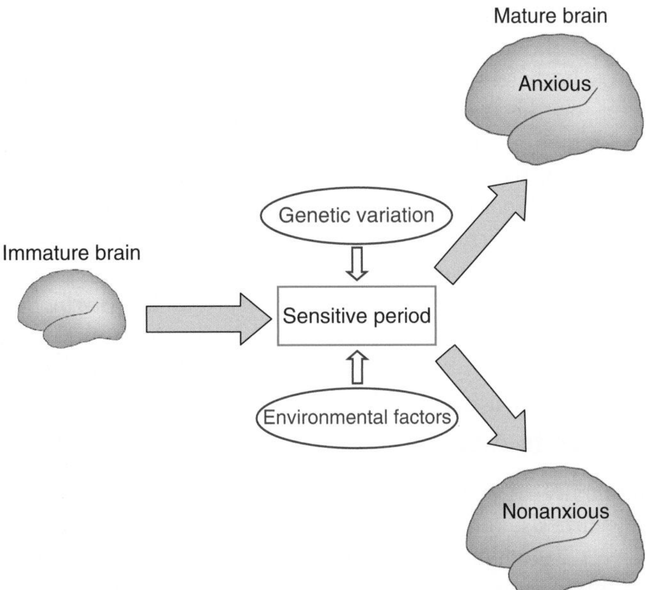

Figure 1 Sensitive period. The immature brain goes through a sensitive period of development during which it is particularly sensitive to environmental and genetic factors. Changes that occur during this period are persistent and are reflected as differences in adult behavior.

Research on the serotonin system underscores an important role of serotonin in the development of anxiety. Many regions innervated by the serotonin system, including the amygdala, hippocampus, anterior and posterior cingulate, and prefrontal cortex, have been suggested to play a role in anxiety and anxiety-like behavior. Studying the genetics of the serotonin system has helped to outline circuits important in anxiety and to further suggest that serotonin affects these circuits during sensitive periods of their development. This work has repeatedly pointed to early life as a sensitive period of development and the amygdala and hippocampus as regions of importance. Together, these lines of research are lending insight into how a molecule traditionally believed to be involved only in synaptic transmission can have multiple roles in shaping brain development and complex behavioral responses.

Serotonin Has Multiple Roles

Initially isolated from serum in 1949 and detected in the mammalian brain in 1953, serotonin is traditionally viewed as a modulatory neurotransmitter. Serotonergic neurons cluster in the brain stem, primarily in the dorsal and median raphe nuclei, and send projections throughout the brain and spinal cord. Dorsal raphe neurons preferentially send projections to the basal ganglia (striatum, substantia nigra); in addition, they typically have small fusiform boutons, diffusely branching fibers, and likely more release of extrasynaptic serotonin. In contrast, neurons in the median raphe nuclei mainly project to components of the limbic system (hippocampus, amygdala) and have large, spherical varicosities with extensive repeated synapses. Notably, there is significant overlap in projection areas from these two sets of nuclei, and single serotonergic neurons have been found to innervate multiple regions in a synaptically connected circuit.

Through its diverse projections, the serotonin system plays a role in modulating sleep, appetite, memory/cognitive function, impulsivity, sexual behavior, motor function, and limbic/affective responsiveness. In addition to this traditional understanding of serotonin as a neuromodulator, increasing evidence demonstrates that neurotransmitters, including serotonin, have important roles in development. Serotonin in particular has been closely studied in this regard. Pharmacologic studies have demonstrated that serotonin can modulate a number of developmental events, including cell division, neuronal migration, cell differentiation, and synaptogenesis.

At first glance, it appears unlikely that one neurotransmitter could have such diverse modulatory functions and also play a role in the regulation of complex neural circuit formation. However, the serotonin system seems ideally suited for this position as it is itself quite complex and diverse; it therefore appears perfectly capable of carrying out multiple tasks. This complexity is clearly demonstrated by the multitude of receptors through which serotonin exerts its effects. Fifteen distinct genes encode for serotonin receptors; these are further classified into five families on the basis of their second messenger systems. Splice variants and posttranslational processing expand the repertoire of viable receptor products to 30. In addition, receptors are located throughout the body and brain, and their expression is both temporally and spatially dynamic. Depending on the receptor type, its location, and the developmental stage of the organism, stimulation of serotonin receptors can be excitatory or inhibitory. Thus, combinations of variables can lead to vastly different downstream effects from this single neurotransmitter, which may be particularly important in the development of anxiety disorders, including abnormal fear responses.

Disruption of Serotonin Homeostasis during Development Leads to Anxiety

Adding further to the complexity of the serotonin receptor system, serotonin synthesis, reuptake, and degradation are further levels at which serotonin function is regulated. Disruption at each of these dynamically regulated levels of serotonin synthesis has been demonstrated to have long-standing effects on anxiety behavior. These effects appear to be mediated at least in part through changes that occur during development.

Excessive extracellular serotonin during development is associated with increased anxiety. Deletion of the MAO-A or SERT gene in mice leads to an accumulation of extracellular serotonin in the brain. MAO-A is the principal enzyme responsible for degradation of monoamines in early postnatal life. MAO-B compensates for the deletion later in life, but mice lacking MAO-A have a ninefold increase in brain serotonin levels during the first postnatal week. SERT is another molecule responsible for the removal of serotonin from the synaptic cleft. In the absence of SERT, serotonin continues to be released but is unable to be removed and accumulates in the synaptic cleft, leading to significant alterations in receptor binding and downstream effects. Knockout mice with deletions of either MAO-A or SERT genes have marked alterations in behavior characterized by increased anxiety and aggressiveness.

Despite this clear anxiety behavioral phenotype, putative anxiety circuits have not been carefully examined in these mice. Notably, SERT and MAO-A knockout mice have been found to have defects in the formation of brain circuits involved in sensory processing. It is interesting that both these mutations lead to abnormal development of thalamocortical axons important in the formation of barrel receptive fields and retinal axons important in visual processing. This abnormal development has been shown to be mediated by the serotonin 5-HT1B receptor. Double knockouts of MAO-A and 5-HT1B or triple knockouts of MAO-A, SERT, and 5-HT1B resulted in normal axonal development. It is tempting to speculate that excessive serotonin might mediate abnormal development of anxiety circuits in a manner similar to the serotonin-dependent disruption of thalamocortical and retinal circuits in these mice. Disruption of anxiety circuits during a developmental sensitive period would help to explain the persistence of anxiety-like behavior in these mice throughout their life span.

Serotonin levels which are abnormally low can also lead to the development of abnormal behaviors. For example, Pet-1 is a transcription factor that regulates transcription of several genes important in serotonergic function. In mice, knockout of Pet-1 leads to an 80% reduction in brain serotonin levels. It is interesting that although the levels of serotonin in these mice move in directions opposite those in the MAO-A and SERT knockout mice, at a behavioral level Pet-1 knockout mice resemble MAO-A and SERT knockout mice, with increased anxiety and aggressive behavior.

The recent identification of a neuronal-specific form of tryptophan hydroxylase provides another example of the effects of reduced serotonin in the developing brain. Functionally significant polymorphisms in the gene that encodes for this neuron-specific form, TPH2, have been found in humans and mice. In mice, the hypofunctioning allele is found in the BALB strain, which is known for its significantly increased baseline levels of anxiety. However, this allele is not found in the less anxious strain, C57. There is good evidence that these behavioral effects of decreased serotonin may be mediated through effects on developmental processes. For example, in rodents, pharmacologic depletion of serotonin in the mother or embryo results in alterations in neurogenesis, neuronal migration, and dendritic maturation.

Taken together, the evidence from mouse models shows that disruption of serotonergic homeostasis in either direction can have clear effects on anxiety behavior. There is also good evidence that serotonergic dysregulation can have developmental effects on circuit formation, neurogenesis, neuronal migration, and dendritic maturation. Putting these observations together lends credence to the idea that serotonin's effects on development, at least in part, underlie serotonin's effects on behavior.

Anxiety Phenotype Develops in Early Life

In the above models, it is not clear when particular gene deletions are mediating their effects. For example, germline deletions of MAO-A and SERT result in changes in serotonin levels throughout the life of an animal, thus making it difficult to determine when these changes might be most influential. To address this issue of timing, researchers have used fluoxetine, a selective inhibitor of serotonin transporter activity, to mimic the effects of disrupting the SERT gene during a restricted time. In one particular study, newborn mice were given fluoxetine for only the first 3 weeks of life. In this way, serotonin levels were altered only during a limited time of postnatal development. Nonetheless, throughout the life span, these mice were behaviorally indistinguishable from the SERT knockout mice with respect to exploration in novel environments, one measure of anxiety behavior.

Mice deficient in the 5-HT1A receptor have been demonstrated to have increased anxiety-related behavior. In an elegant study to address the timing of this effect, Gross and colleagues showed that genetic replacement of the serotonin 5-HT1A receptor during only the first 3 weeks of life was able to completely rescue normal anxiety behavior in mice deficient in 5-HT1A. On the other hand, replacement of the 5-HT1A receptor throughout life with the exception of the first 3 weeks did not restore normal behavior. Mice deficient in 5-HT1A receptor during only the first 3 weeks of life continued to display the characteristic 5-HT1A knockout anxiety phenotype that is similar to the anxiety phenotype of MAO-A and SERT knockout mice. There is also substantial evidence that experience has potent effects during this same period. Rats exposed to high levels of maternal care during the first 3 weeks of life have decreased levels of anxiety relative to rats that receive low levels of maternal care. Similarly, lengthy maternal separation during the same period results in increased anxiety-related behavior.

In humans and other primates, this developmentally sensitive period appears to be somewhat longer, with research suggesting a time course beginning during the third trimester of pregnancy and extending to periadolescence. Nonetheless, as discussed previously, early life appears to be a particularly potent period for the effects of experience and genetic variation.

Early Life: What Is Happening at the Circuit Level?

It is important to consider what is occurring in the brain during the developmentally sensitive period of early life. Numerous studies have demonstrated that different brain regions develop and mature at different rates and different times. In a general sense, phylogenetically primitive regions such as the hippocampus or amygdala develop earlier than the phylogenetically more advanced regions such as the human frontal cortex. It is therefore important to consider not only timing but location in order to understand the effects of experience on development.

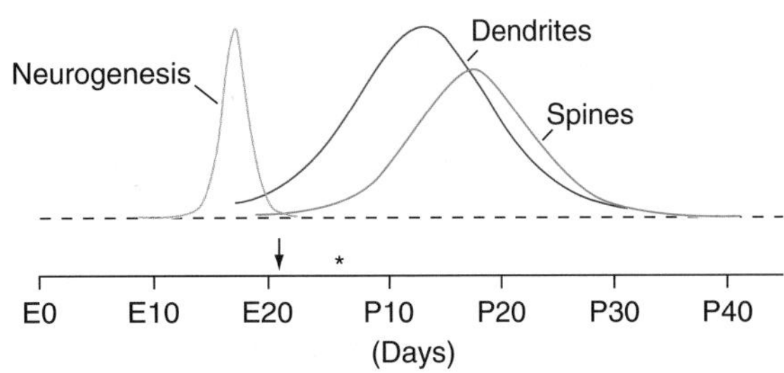

Figure 2 Development of rat hippocampus. Development of the rat hippocampus proceeds in a stepwise fashion in which late prenatal neurogenesis is followed by early postnatal dendritic outgrowth and spine development. These postnatal changes are coincident with the sensitive period for the development of an anxiety phenotype. E, embryonic; P, postnatal; *, birth. Adapted from Ben-Ari Y (2001) Developing networks play a similar melody. *Trends in Neurosciences* 24(6): 353–360, with permission from Elsevier.

The rat hippocampus has been particularly well studied in regard to its developmental time course, and as discussed previously, the hippocampus has also been well demonstrated to be involved in anxiety and anxiety-related behavior. Within the first 3 weeks, there is a dramatic rise and fall of dendritic growth and dendritic spine formation in the rat hippocampus that is fundamental to the formation of synapses and circuits (**Figure 2**). During this same time, other events occur that are likewise important for the development of brain circuits. In the rat hippocampus, there is a dramatic increase in the spontaneous synchronous firing of neurons that is a hallmark of developing neuronal networks. The timing of this spontaneous network activity directly precedes the emergence of hippocampal-dependent functions such as context-dependent learning.

In concordance with the broadened period of vulnerability, in humans these processes are proposed to occur over an extended period beginning earlier, in the third trimester of fetal development, and continuing later, throughout much of childhood. In summary, the period during which animals are most vulnerable to experience and genetic variability is the same period during which brain circuits important in anxiety are developing.

Anxiety Circuits Disrupted?

We have seen that early life represents a period of increased vulnerability to the effects of both experience and genetic variation. Furthermore, it is clear that this period is an important time in brain development and circuit formation. But what about the final piece of the hypothesis? Do changes in serotonin function during early postnatal life affect circuit formation in the anxiety system? The developmentally sensitive period for the establishment of an anxiety phenotype that persists into adulthood is highly suggestive of an effect at the circuit level.

Examination of 5-HT1A knockout mice, which display more anxiety-related behaviors than do wild-type mice, lends support to this hypothesis. Lesion studies in these mice have specifically implicated the hippocampal circuit in anxiety-related behaviors. In humans, a functional polymorphism in the 5-HT1A gene promoter is associated with increased anxiety-related personality traits, as well as an increased incidence of a subtype of panic disorder characterized by the additional presence of agoraphobia.

Also in support of the hypothesis that changes in serotonin system gene expression can modulate anxiety and fear circuits, Hariri and colleagues have shown that in humans, a polymorphism in the SERT gene is associated with particularly robust changes in the sensitivity of the amygdala to fearful or threatening stimuli. Individuals with the hypo-functioning SERT allele (s/s) had a fivefold increase in amygdala activation in response to fearful stimuli. Another group of researchers also demonstrated this increased amygdala activation and further suggested changes in neural circuitry by demonstrating that individuals with at least one copy of the short allele had increased coupling of amygdala and prefrontal cortex activity. Because of the likely relation between fear and anxiety, this change in fear responsiveness may underlie an increased risk for anxiety disorders in these individuals. It is important to note that these differences in regional brain activation were much more readily apparent than were overt behavioral differences in these individuals, underscoring the point that a change in neural circuit function does not necessarily result in a behavioral phenotype in the case of an animal model, or in an anxiety disorder in the case of humans.

Sensitive Period Revisited: Gene–Environment Interactions

Current evidence suggests that anxiety circuits appear to be subject to a sensitive period of development. Gene knockout studies and phenotypic characterization of individuals with functional polymorphisms have highlighted several genes of the serotonin system that affect the developmental course of circuits involved in anxiety. Evidence for this is both direct, as with functional neuroimaging of individuals with SERT polymorphisms, and indirect, as evidenced by the lifelong persistence of anxiety phenotypes based on genetic variation and early experience.

The existence of a sensitive period in the development of neural circuits involved in anxiety becomes particularly interesting when the myriad influences are examined for their combined effects. These effects can be antagonistic, as seen when the 5-HT1B receptor is removed in the MAO-A and SERT knockout mice. Removal of this receptor blocks the detrimental effects of increased serotonin on the formation of thalamocortical and retinal circuits. The combined effects can also be additive, exemplified by an exacerbation of the anxiety phenotype in SERT knockout mice when BDNF, another molecule important in circuit formation, is also removed.

It is important to note that these combined effects are not limited to gene–gene interactions. As discussed at length above, the normal functioning (l/l) allele of the SERT polymorphism appears to be protective against many of the behavioral consequences of early life stress in humans and in primates, whereas those that are homozygous (s/s) or heterozygous (l/s) for the hypo-functioning short allele have less resilience in the face of early life adverse events.

Conclusions, Future Directions, and Resilience

Understanding that psychopathology, particularly anxiety responses, can represent developmentally mediated changes in neural circuits is not an end but rather a beginning. It affords an opportunity to more carefully dissect not only the factors involved in psychopathology, but also the resilience factors that protect individuals with circuit-level dysfunction from developing illness.

Understanding brain function and pathology as an intermingling of individual circuits, each subject to its own developmental trajectory, also serves as a framework to probe for deeper understanding of the process of brain development. Defining circuits for a particular research focus opens the door for a careful dissection of genetic and environmental influences on neural function that is not available when using the gross outcome measures of behavior or disease.

Furthermore, it is possible that dysfunction in neural circuit formation may represent a more fundamental dysfunction of molecular circuits. This certainly may be the case in the 5-HT1A knockout mouse, in which the 5-HT1A receptor represents a critical regulator of negative feedback on serotonin function. Understanding dysfunction of molecular circuits would not only enhance understanding of the fundamentals of brain function but may also provide a yet more sensitive phenotypic marker of risk of psychopathology and a potential marker for treatment responsiveness.

Ultimately, however, attempts must be made to relate neural changes to behaviors or cognitive functions. Having defined changes in molecular or neural circuits, one then has a carefully defined subset of individuals to probe for subtle differences in higher-order functions. It is through this iterative process that progress in understanding brain function and behavioral pathology can be made.

See also: Aggression: Hormonal Basis; Aggression: Neurochemical and Molecular Mechanisms; Amygdala: Contributions to Fear; Anxiety Disorders; Aversive Emotions: Molecular Basis of Unconditioned Fear; Obsessive–Compulsive Disorder; Serotonin (5-Hydroxytryptamine; 5-HT): Neurotransmission and Neuromodulation; Serotonin (5-Hydroxytryptamine; 5-HT): CNS Pathways and Neurophysiology; Serotonin (5-Hydroxytryptamine; 5-HT): Receptors.

Further Reading

Ansorge MS, Zhou M, Lira A, Hen R, and Gingrich JA (2004) Early-life blockade of the 5-HT transporter alters emotional behavior in adult mice. *Science* 306(5697): 879–881.

Ben-Ari Y (2001) Developing networks play a similar melody. *Trends in Neurosciences* 24(6): 353–360.

Caspi A, McClay J, Moffitt TT, et al. (2002) Role of genotype in the cycle of violence in maltreated children. *Science* 297(5582): 851–854.

Caspi A, Sugden K, Moffitt TT, et al. (2003) Influence of life stress on depression: Moderation by a polymorphism in the 5-HTT gene. *Science* 301(5631): 386–389.

Charney DS (2003) Neuroanatomical circuits modulating fear and anxiety behaviors. *Acta Psychiatrica Scandinavica. Supplementum* (417): 38–50.

Gaspar P, Cases O, and Maroteaux L (2003) The developmental role of serotonin: News from mouse molecular genetics. *Nature Reviews Neuroscience* 4(12): 1002–1012.

Gingrich JA, Ansorge MS, Merker R, Weisstaub N, and Zhou M (2003) New lessons from knockout mice: The role of serotonin during development and its possible contribution to the origins of neuropsychiatric disorders. *CNS Spectrums* 8(8): 572–577.

Gingrich JA and Hen R (2001) Dissecting the role of the serotonin system in neuropsychiatric disorders using knockout mice. *Psychopharmacology* 155(1): 1–10.

Gordon JA and Hen R (2004) The serotonergic system and anxiety. *Neuromolecular Medicine* 5(1): 27–40.

Gross C and Hen R (2004) The developmental origins of anxiety. *Nature Reviews Neuroscience* 5(7): 545–552.

Gross C, Zhuang X, Stark K, et al. (2002) Serotonin1A receptor acts during development to establish normal anxiety-like behaviour in the adult. *Nature* 416(6879): 396–400.

Horn G (2004) Pathways of the past: The imprint of memory. *Nature Reviews Neuroscience* 5(2): 108–120.

Knudsen EI (2004) Sensitive periods in the development of the brain and behavior. *Journal of Cognitive Neuroscience* 16(8): 1412–1425.

Levitt P, Harvey JA, Friedman E, Simansky K, and Murphy EH (1997) New evidence for neurotransmitter influences on brain development. *Trends in Neuroscience* 20(6): 269–274.

McNaughton N and Corr PJ (2004) A two-dimensional neuropsychology of defense: Fear/anxiety and defensive distance. *Neuroscience and Biobehavioral Reviews* 28(3): 285–305.

Nemeroff CB (2004) Neurobiological consequences of childhood trauma. *Journal of Clinical Psychiatry* 65(supplement 1): 18–28.

Noda M, Higashida H, Aoki S, and Wada K (2004) Multiple signal transduction pathways mediated by 5-HT receptors. *Molecular Neurobiology* 29(1): 31–39.

Phelps EA and LeDoux JE (2005) Contributions of the amygdala to emotion processing: From animal models to human behavior: Review. *Neuron* 48: 175–187.

Ressler KJ and Nemeroff CB (2000) Role of serotonergic and noradrenergic systems in the pathophysiology of depression and anxiety disorders. *Depression and Anxiety* 12(supplement 1): 2–19.

Whitaker-Azmitia PM (2001) Serotonin and brain development: Role in human developmental diseases. *Brain Research Bulletin* 56(5): 479–485.

Zhuang X, Gross C, Santarelli L, Compan V, Trillat AC, and Hen R (1999) Altered emotional states in knockout mice lacking 5-HT1A or 5-HT1B receptors. *Neuropsychopharmacology* 21(supplement 2): 52S–60S.

Aversive Emotions: Molecular Basis of Unconditioned Fear

J B Rosen, University of Delaware, Newark, DE, USA

Interest in the scientific study of emotions in psychology, biology, and neuroscience has waxed and waned for many years. In the latter part of the nineteenth century, Darwin brought emotion to the forefront of biology with a functionally and biologically grounded analysis of emotional expression. Darwin considered that emotion and the expressions of emotional behavior developed because of their importance in intra- and interspecies communication that promotes survival. Many emotions were considered innate and developed through evolutionary and functional mechanisms, just as body morphology evolved through environmental pressures. However, the scientific study of emotions has always been hampered by the subjectivity of emotional experience and therefore was not as prominent a field of study in psychology and neuroscience as other fields of behaviorism and cognition were throughout much of the twentieth century. In the last few decades, the study of the neurobiology and molecular biology of emotions, particular aversive emotions such as fear, has recently become tractable in both animals and humans because of the ability to measure behavioral and neural emotional responses, and it consequently has become a major field of neuroscientific study in the twenty-first century.

Great strides in our knowledge of the neuroscience of fear stems from the use of simple conditioning paradigms that measure fear with species-specific defensive responses to elucidate a fundamental and detailed understanding of the neuroanatomy of fear that can then be followed by the study of the molecular biology within these neuroanatomical circuits of fear. Combining a number of learning procedures, particularly Pavlovian fear conditioning paradigms in rodents that produce robust learning and memory of fear, with neuroscience techniques has been very successful in developing an understanding of neurobehavioral systems in the brain that are crucial for learning, memory, and the expression of the emotion fear. On the heels of the study of learned fear, in recent years the neuroscientific study of unconditioned fear has begun and is now accumulating data that suggest the neural and molecular mechanisms of unconditioned fear may have similarities to, but also important differences from, those of conditioned fear.

Conditioned and Unconditioned Fear

Fear conditioning paradigms in animals have a long history in psychological research and theory. Pavlov not only paired a bell with the arrival of food to induce conditioned or learned appetitive responses to the bell, but also paired bells and other neutral stimuli to unconditioned aversive events, such as shocks, to induce aversive responses to these formerly neutral, but now conditioned, stimuli. Thus, just as stimuli contingently paired with food become predictive signals for the arrival of food, stimuli contingently paired with shock develop into predictive signals for the arrival of the shock. Pavlovian fear conditioning therefore is an adaptively important process that helps organisms survive and is one of the most rapid types of learning in many animal species, with significant learning in a single trial and typically reaching asymptote within a handful of trials. Although Pavlovian conditioning by pairing lights, tones, contexts, and shocks is a fairly simple paradigm to perform, the associative process is not merely a transfer of the ability to elicit behavioral responses from one stimulus to another but is, instead, cognitive in nature and involves animals making predictions, representations, and relations between events, situations, and circumstances. Pavlovian learning and memory processes have immense biological significance by increasing the survivability of animals in a world where the predictive value of stimuli changes.

Because Pavlovian fear conditioning is learning about predictive relationships between conditioned and unconditioned stimuli, and not about learning new behavioral responses, innate, functionally adaptive species-specific defensive responses are typically displayed during fear and experimentally are used as measures of conditioned fear. These include autonomic changes in heart rate and blood pressure, and increases in behavioral responses such as freezing and reflexes such as startle. All these measures are considered preparatory responses to the anticipated aversive stimulus that are induced by presentation of the conditioned stimulus and are, thus, thought to be indicative of a state of fear.

Whereas learning to be afraid of stimuli that we encounter is adaptive and has great significance for survival, the ability to be innately fearful of some stimuli has similar significance and may be important for encounters and circumstances that organisms typically do not survive (and therefore cannot learn from). It is suggested that in humans fear of heights, the dark, and small animals may be examples of unconditioned fears. Unconditioned fear may also be a basis for some anxiety disorders, particularly some specific phobias. However, it is very difficult, if not impossible, to determine whether these fears are innate or are due to learning. However, this may be resolved in laboratory animals that have been bred for generations in a laboratory environment with no contact with putative unconditioned fear stimuli. Although there are a number of different types of unconditioned fear stimuli or situations that have been studied, such as neophobia and brightly lit environments, this article focuses on fear of predators and predator odors. It has been know since 1913 that laboratory rats display unconditioned fear responses to cats, and unconditioned fear to predators and predator odors have been studied most extensively.

The use of ecologically and biologically based stimuli, such as exposure to predators and odors from predators, allows for the controlled, scientific study of unconditioned fear. Use of these innate fear-inducing stimuli exploits the evolutionarily shaped unconditioned fear of predators and predator odors. The physiological and behavioral responses to these predators and predator odors are similar to those evoked by conditioned fear stimuli (e.g., cardiovascular changes, freezing, startle, and vigilance), indicating that the same species-specific defensive responses are used during unconditioned fear states. What distinguishes unconditioned from conditioned fear is that it is unnecessary to initially induce unconditioned fear with a painful stimulus and, significantly, that unconditioned fear appears without learning or prior exposure to the unconditioned stimulus. Also, fear responses in rats to repeated exposure to live predators (e.g., cat) and some predator odors (e.g., ferret fur odor or trimethylthiazoline, a synthetic fox odor) do not habituate. This does not mean that learning cannot occur with exposure to predators and predator odors. Robust conditioned flavor avoidance is supported by synthetic fox odor. Learning of cue and contextual fear can be supported by exposure to a live cat and some predator odors (e.g., cat fur), but this is difficult to demonstrate with other predator odors (e.g., synthetic fox odor, ferret fur odor, and cat feces), except in specific types of environments. However, sensitization can be produced by exposure to these predators and predator odors that may last for at least 3 weeks. These learning effects of predators and predator odors (both nonassociative sensitization and associative conditioning) may recruit different neural circuits from those necessary for unconditioned fear responses.

Neuroanatomical Circuits of Unconditioned Fear

Before discussing the molecular biology of unconditioned fear, I delineate the neural circuits where molecular changes are to be found. If we were to envision the neurocircuitry of unconditioned fear, it would probably be similar to that subserving conditioned fear. Pavlovian fear conditioning in rodents has been very important for delineating specific neuroanatomical circuits of fear conditioning (**Figure 1**). Numerous studies have demonstrated that the lateral, basal, and central nuclei of the amygdala are all necessary for fear conditioning. The lateral nucleus is particularly important as a site of stimulus integration because it receives information from association cortices and from sensory and pain-relay nuclei of the thalamus. The basal nucleus also integrates afferents from the lateral nucleus and other brain regions. The central nucleus receives input from the lateral and basal nuclei and is the major output pathway to many subcortical regions that mediate numerous fear-related responses and behaviors.

Whereas the neuroanatomy of conditioned fear is fairly well established, the neuroanatomy of unconditioned fear is just beginning to be dissected. Early studies of exposure to a cat or cat fur supported the view that conditioned and unconditioned neural circuits are the same by showing that fear behavior diminished in rats with large lesions or inactivation of the amygdala. However, more recent studies with smaller lesions of discrete nuclei of the amygdala found that the lateral, basal, and central nuclei of the amygdala are not necessary for the display of

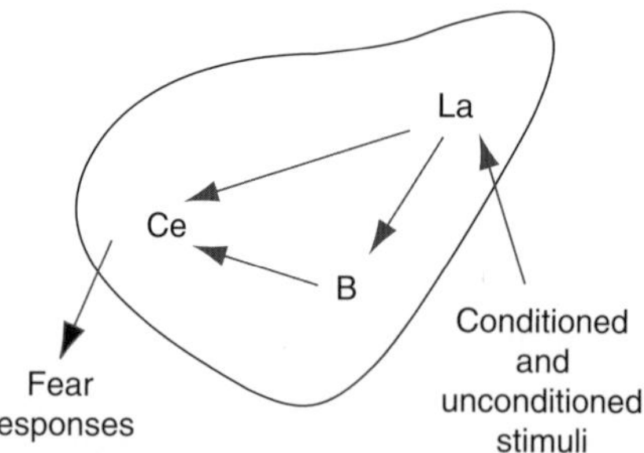

Figure 1 An amygdala fear circuit for fear conditioning. Input from thalamus and cortex with information about conditioned and unconditioned stimuli enters the lateral nucleus of the amygdala (La). The lateral nucleus projects to the basal (B) and central (Ce) nuclei of the amygdala. Output from the Ce nucleus projects to midbrain and brain stem areas to generate different kinds of conditioned fear behavior. This circuit is hypothesized to be important for fear conditioning, but not for unconditioned predator fear. However, it is recruited for associative and nonassociative fear induced by unconditioned predator fear.

unconditioned fear to various predator odors. A different neural fear circuit running through the medial hypothalamus may underlie innate defensive behavior to predators and predator odors (**Figure 2**). The core of this medial hypothalamic defense circuit includes the anterior hypothalamic nucleus (AHN), dorsomedial part of the ventromedial hypothalamic nucleus (VMHdm), and the dorsal premammillary nucleus (PMd). These nuclei are highly interactive, where the AHN has strong reciprocal connections with the VMHdm and PMd, and the VMHdm sends efferents to the PMd. This system also lies as an interface between sensory inputs and motor outputs, similar to the way that the amygdala does. The circuit is thought to play a major role in the somatomotor integration and initiation of aversively motivated behavior. For example, rats with PMd lesions display significantly reduced unconditioned freezing and escape during exposure to a live cat or cat fur odor. Also, lesion or inactivation of major inputs to this system (the medial nucleus of the amygdala or bed nucleus of the stria terminalis) also reduces freezing to predator odors.

The AHN and VMHdm have substantial olfactory afferents from main and accessory olfactory systems via the medial and accessory basal nuclei of the amygdala and the bed nucleus of the stria terminalis. VMHdm afferents from posterior intralaminal thalamus supply auditory inputs, paralleling thalamic auditory afferents of the lateral nucleus of the amygdala. Indirect input from the lateral nucleus of the amygdala by way of hippocampus and ventral subiculum provides processed visual information to the AHN. The VMHdm also receives substantial visceral and nociceptive information from the parabrachial nucleus in the brain stem. Efferents from the hypothalamic nuclei to the amygdala, septum, bed nucleus of the stria terminalis, and medial prefrontal cortex are able to influence processing in numerous regions of the telencephalon. Finally, the three core hypothalamic nuclei of the circuit each have substantial projections to the periaqueductal gray and other brain stem areas to generate and control a variety of fearlike behaviors.

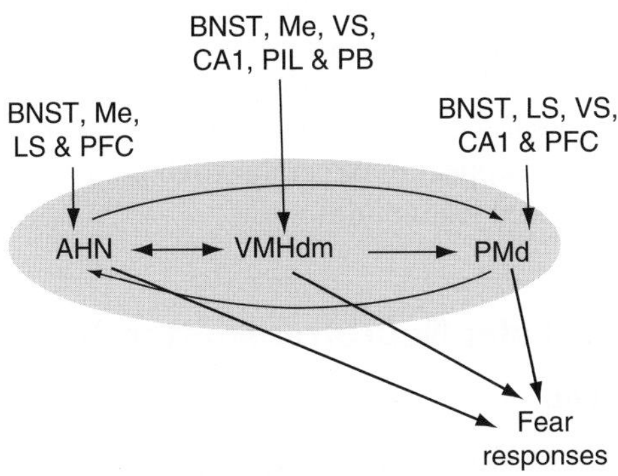

Figure 2 A medial hypothalamic fear circuit for unconditioned predator and predator odor fear. The anterior hypothalamic nucleus (AHN), the dorsomedial ventromedial hypothalamic nucleus (VMHdm), and the dorsal premammillary nucleus (PMd) make up the core of this circuit. Input into this system comes for numerous areas. Inputs from the bed nucleus of the stria terminalis (BNST), medial nucleus of the amygdala (Me), lateral septum (LS), ventral subiculum (VS), CA1 region of hippocampus (CA1), and several regions of the medial prefrontal cortex (PFC), posterior intralaminar nucleus of thalamus (PIL), and parabrachial nucleus (PB) are shown, but other inputs also exist. All core nuclei of the medial hypothalamic circuit project to brain stem regions (particularly strong to the periaqueductal gray) that control and generate fear responses.

Although the fear circuits for conditioned and unconditioned fear may be different, continued research will surely find a functional overlap between these circuits. Neuroanatomical studies demonstrate that the lateral nucleus of the amygdala receives projections from the VMHdm. The prefrontal cortex has been shown to project to both amygdala nuclei and the AHN and PMd. It has been known for many years that electrical and chemical stimulation of the ventromedial hypothalamus elicits fear behavior and that this is modulated by stimulation of the amygdala. Furthermore, lesions of the bed nucleus of the stria terminalis disrupt contextually fear-conditioned freezing and lesions of the medial nucleus of the amygdala interfere with fear-potentiated startle. Probably there is much interaction between fear conditioning and unconditioned fear circuits during both associative and nonassociative learning to unconditioned predator stimuli that will be elucidated in the coming years.

Molecular Biology of Unconditioned Fear

The study of the molecular and cellular mechanisms of unconditioned fear to predators and predator odors is in its infancy and sketchy at best, but it will eventually include the expression of molecules at all levels of neural transmission, signaling transduction, and gene expression. To data, the data demonstrate altered expression and participation of molecules at the levels of gene expression, neurotransmitters, and receptors following the exposure of rodents to these unconditioned fear stimuli.

Inducible Transcription Factors (Immediate-Early Genes)

The neural circuitry of unconditioned fear to predators and predator odors suggested by brain-lesion techniques is corroborated by measures of immediate-early genes that are highly responsive to environmental stimulation. Changes in the expression of immediate-early gene mRNA or protein levels, such as c-fos and egr-1, not only can be used as markers for transient neural activity, but, because these genes are inducible transcription factors, they provide a pattern of gene expression in these brain regions that may regulate some of the long-lasting changes in behavior (e.g., nonassociative sensitization, and associative conditioning) induced by exposure to predator stimuli.

Exposing rats to cats induces c-fos mRNA in regions of the medial hypothalamic defensive circuit, including the ventromedial hypothalamus, dorsal premammillary nucleus, bed nucleus of the stria terminalis, and lateral septum. Significantly, c-fos in these regions does not increase with restraint or hypoxia, other common stress-inducing manipulations. Furthermore, c-fos in the medial nucleus of the amygdala is increased with both cat exposure and restraint stress. Other studies of fear conditioning also demonstrate that c-fos increased in the medial nucleus of the amygdala, indicating that the activation of the medial nucleus of the amygdala is not specific for unconditioned fear to predators or predator odors but is activated during many types of fear or stress.

Studies investigating exposure to predator odors (cat and ferret fur and synthetic fox odor) induce similar patterns of c-fos mRNA and Fos protein expression as seen with exposure to a live cat, but more brain regions have been examined. Thus, these predator odors are found to induce c-fos mRNA or Fos protein in the main olfactory bulbs, paraventricular nucleus of the hypothalamus, locus coeruleus, cuniform nucleus, and all divisions of the periaqueductal gray (the brain region important for the fear behaviors freezing and escape). Interestingly, the subnuclei of the amygdala (lateral and basal nuclei), regions critical for all types of fear conditioning (i.e., auditory, visual, olfactory, and contextual fear conditioning induced by shock), do not display increased c-fos expression following live predator or predator odors. In addition, the expression of another immediate-early gene, egr-1, also is not induced in the lateral and basal nuclei of the amygdala of rats exposed to a live cat. These data begin to delineate and corroborate lesion studies showing differences in brain circuits for conditioned and unconditioned fear.

The expression of these inducible transcription factor genes not only provides patterns of neural activation during behavior, but also suggests that these changes may alter long-term responses to predators and predator odors and, in the wild, may facilitate the survival of the animal. To date, few studies have examined the role of inducible transcription factors in unconditioned fear. However, these are some data demonstrating that the reduction of translation of egr-1 in the lateral nucleus of the amygdala interferes with fear conditioning but not with unconditioned fear to synthetic fox odor. Further research will elucidate the importance of inducible transcription factors in unconditioned and conditioned fear.

Phosphorylated Cyclic AMP Responsive Element-Binding Protein

Expression of other inducible transcription factors has not been investigated. However, there are a couple of studies examined the phosphorylation of cyclic AMP responsive element-binding protein (CREB) induced by cat exposure. CREB is a constitutive factor that on phosphorylation (pCREB) becomes an active transcription factor that has been shown to be involved in many forms of learning and memory, including fear conditioning within the amygdala. pCREB had been found to increase in the lateral, basal, central, and medial nuclei of the amygdala, as well as in the periaqueductal gray of rats following exposure to a live cat. This is interesting because live-cat exposure induces a long-lasting sensitization that can be demonstrated in a number of anxiety tests. Of further interest is that pCREB does not increase in the ventromedial hypothalamus, a nucleus of the medial hypothalamic defense circuit. Taken together with the findings that the inducible transcription factors, c-fos and egr-1, are not induced in the lateral and basal nuclei of the amygdala by a live cat, cat odor, ferret odor, or synthetic fox odor but are induced in the ventromedial hypothalamus, the data suggest that unconditioned predator fear and predator-induced sensitization engage different neural circuits. Circuits for predator-induced sensitization appear to be similar to those necessary for shock-induced fear conditioning.

Extracellular Neurotransmitter Signaling

Dopamine

The effect of unconditioned predator fear on the dopamine system has been studied more than other neurotransmitter systems. Dopamine release and turnover in the prefrontal cortex and amygdala (subnuclei of the amygdala were not specified), which are known to increase in response to stress, also increase with exposure to synthetic fox odor. Dopamine neurons in the ventral tegmental area that project to the prefrontal cortex and amygdala are activated by synthetic fox odor, whereas dopamine neurons in the substantia nigra do not. Unconditioned fear induced by synthetic fox odor also interferes in a non-match-to-sample short-term working memory task that is dependent on prefrontal cortex processing. Dopamine in other regions, such as hypothalamus, striatum, and nucleus accumbens, has also been shown to increase with exposure to predator odors but not as consistently as the increase in the prefrontal cortex. Pharmacologically, dopamine D1 receptor agonist infused in the prefrontal cortex of mice increases fear and anxious behavior to predator and predator odor exposure. These data suggest that unconditioned predator and predator odor exposure, similar to other fear- and stress-inducing stimuli, enhance dopamine neurotransmission in the prefrontal cortex and interfere with working memory tasks that are dependent on the prefrontal cortex.

Serotonin

In general, pharmacological serotonin 5-HT1_A receptor agonism and blockade of serotonin reuptake reduce fear and anxious behavioral responses in rats and mice to predators and predator odors. In mice, there appears to be differential effects of 5-HT1_A receptor agonists to reduce predator risk-assessment behavior, and 5-HT2_A receptor antagonism to reduce flight or escape behavior. Mutant mice with a knockout of the serotonin transporter were shown to have increased vulnerability to predator-induced sensitization, as shown by the increase anxiety-like behavior in the elevated-plus maze and a light–dark box test after exposure to a cat. The increased vulnerability to unconditioned fear in these mice may be similar to the enhanced vulnerability to stress in humans with the short allele of the serotonin transporter.

Norepinephrine

Norepinephrine plays a role in vigilance and conditioned fear, and therefore it should also modulate unconditioned fear. Norepinephrine was shown to increase in the hippocampus in mice exposed to a cat and in the bed nucleus of the stria terminalis of rats exposed to synthetic fox odor. Complementary to increases in norepinephrine, the blockade of norepinephrine neurotransmission in the bed nucleus of the stria terminalis with a local infusion of clonidine decreased unconditioned fear-related freezing responses to synthetic fox odor. Furthermore, in a drug schedule of intermittent methylphenidate treatment for 3 weeks that sensitized dopamine and norepinephrine systems, rats given yohimbine, which facilitates norepineprhine neurotransmission, displayed increased avoidance of a predator odor. Together, these studies confirm that norepinephrine neurotransmission is important for unconditioned fear to predators and predator odors.

γ-Aminobutyric Acid

γ-Aminobutyric acid (GABA) levels have been shown to increase in the amygdala and nucleus accumbens with exposure to a predator or predator odor. These increases are probably compensatory because the facilitation of $GABA_A$ neurotransmission with the administration of systemic benzodiazepine agonist reduces defensive threat/attack and risk-assessment behaviors to cat and cat odors in mice. Unconditioned freezing in rats to synthetic fox odor is blocked by local infusion of muscimol, a $GABA_A$ agonist, into the bed nucleus of the stria terminalis or medial nucleus of the amygdala. Interestingly, muscimol infusion into the basal nucleus of the amygdala of rats only delays freezing in response to synthetic fox odor for a few minutes, but freezing reaches levels-seen in vehicle-treated rats. Furthermore, in a study investigating fear behavior of sheep in response to a predator (a dog), an infusion of bicuculline, a $GABA_A$ antagonist into the sheep's central nucleus of the amygdala had only a marginal effect on predator-elicited fear, but sensitized predator fear when measured 2 days after the initial exposure to the dog. These data suggest that the basal and central nuclei of the amygdala play only a minor modulatory role in unconditioned fear to predators, but may have a more substantial role in long-lasting fear sensitization induced by predator fear.

Glutamate

Both competitive and noncompetitive antagonism of N-methyl-D-aspartate (NMDA) receptors has effects on fear induced by exposure to a predator and to predator odor. In rat pups, CGP 43487 given at postnatal day 5 reduces freezing elicited by exposure to an adult male rat at postnatal days 13 and 20 but not at postnatal day 30. In adult rats, CPP and AP7 do not affect rat defensive behavior to cat exposure, but they significantly interfere with long-lasting enhancement of fear on a number of anxiety tests. This is reminiscent of the role of NMDA receptors in long-term potentiation, in which NMDA receptors are involved in long-term synaptic plasticity by not short-term potentiation. The role of NMDA glutamate receptors in the amygdala appears to be involved in the long-lasting fear sensitization induced by predator exposure, possibly by strengthened synaptic transmission in amygdala–periaqueductal gray circuits. Glutamate has also been shown in increase in the rat nucleus accumbens in response to synthetic fox odor. Other brain regions have not been examined yet.

Corticotropin-Releasing Hormone and Corticosterone

Corticotropin-releasing hormone (CRH) and corticosterone are part of the hypothalamic–pituitary–adrenal axis. This system is important for the regulation of glucose metabolism to provide fuel for diverse behavioral and physiological demands, including fear. Blood plasma corticosterone levels are increased by exposure to predator and predator odors. Exogenous CRH induces fear and anxiety-like behavior. Both CRH and corticosterone are involved in the consolidation of fear conditioning, and the amygdala is a major locus of action. Chronic high levels of corticosterone enhance memory of learned and unconditioned fear, and they also induce CRH expression in the central nucleus of the amygdala and the bed nucleus of the stria terminalis. Corticosterone enhancement of unconditioned startle appears to be through CRH in the bed nucleus of the stria terminalis but not the amygdala. However, exposure to predator and predator odor increases CRH release in the amygdala. In sheep exposed to a dog (a fear stimulus in sheep), CRH release in the amygdala occurs in two waves. The first wave lasts for only a few minutes and is independent of corticosterone, whereas the second wave of CRH release is corticosterone dependent. Interestingly, repeated exposure to the dog potentiates fear to the dog and enhances only the corticosterone-dependent CRH release in the amygdala. Furthermore, infusion of a CRH antagonist into the amygdala had little effect on unconditioned fear behavior to the dog, but interfered with the potentiation of fear to subsequent dog exposure. These data suggest that CRH in the amygdala is not critical for unconditioned predator fear but that corticosterone-regulated CRH in the amygdala is important for nonassociative sensitization to predator stimuli. Whether corticosterone-induced enhancement of CRH in the bed nucleus of the stria terminalis is involved in sensitization to predators, as it is in sensitization to cocaine, awaits further research.

Cholesystokinin

Cholesystokinin (CCK) is known to induce anxiety-like feelings in humans. In rats, it appears to play a role in unconditioned predator fear and sensitization induced by predator exposure. CCK is increased in the basal nucleus of the amygdala in mice exposed to synthetic fox odor compared to mice not exposed to this odor, but only after a light–dark box anxiety test. Blockade of CCK-2 receptors, but not CCK-1 receptors, decreases fearfulness in rats during exposure to a cat and decreases cat-exposure-induced sensitization demonstrated in the elevated plus maze. An interesting difference in response to cat exposure has been found in PVG hooded and Sprague-Dawley rats, in which PVG hooded rats freeze more during exposure to a cat than do Sprague-Dawley rats. PVG hooded rats also have a greater abundance of CCK-2 receptors than and several genetic variations in the code for CCK-2 receptors different from those of Sprague-Dawley rats. The importance of these differences for unconditioned fear is not known at present, but it is interesting to note that Sprague-Dawley and Long-Evans rats display robust freezing to synthetic fox odor, whereas Wistar rats do not. There are also mouse strain differences in response to predator odors. Whether these are related to CCK or other neurotransmitter systems is not known.

Conclusion

Investigation of the molecular basis of unconditioned fear, particularly fear of predators and predator odors, has seen substantial progress in the last few years. As shown by lesion and expression of immediate early genes, a neural circuit for unconditioned predator fear through the medial hypothalamus is emerging that may be distinct from an amygdala circuit for fear conditioning. However, there appears to be an overlap of these circuits and recruitment of the amygdala fear-conditioning circuit for associative and nonassociative conditioning to predator stimuli. Although inducible transcription factors seem to be expressed in the unconditioned fear circuit during exposure to predator stimuli, nonasssociative sensitization of fear induced by predator exposure appears to induce pCREB, a phosphorylated constitutive transcription factor known to be involved in fear conditioning and other types of learning, in amygdala nuclei of the fear-conditioning circuit. A number of neurotransmitter systems also seem to be involved in unconditioned fear and sensitization induced by predator exposure. The list of neurotransmitter systems involved will certainly grow. Investigation of signal transduction pathways has not started, but will probably contribute to our understanding of unconditioned fear in the coming years. Continued research with ethologically based fear stimuli should provide more neuroanatomical and molecular detail of unconditioned fear and should have a substantial influence on our understanding of normal fear and related anxiety disorders.

See also: Amygdala: Contributions to Fear; Aversive Emotions: Genetic Mechanisms of Serotonin; Emotion: Neuroimaging; Emotional Control of the Autonomic Nervous System; Fear Conditioning and Synaptic Plasticity; Genetics of Human Anxiety and Its Disorders.

Further Reading

Adamec RE, Blundell J, and Burton P (2006) Relationship of the predatory attack experience to neural plasticity, pCREB expression and neuroendocrine response. *Neuroscience and Biobehavioral Reviews* 30: 356–375.

Adamec RE, Burton P, Blundell J, Murphy DL, and Holmes A (2006) Vulnerability to mild predator stress in serotonin transporter knockout mice. *Behavioural Brain Research* 170: 126–140.

Blanchard DC, Canteras NS, Markham CM, Pentkowski NS, and Blanchard RJ (2005) Lesions of structures showing FOS expression to cat presentation: Effects on responsivity to a cat, cat odor, and nonpredator threat. *Neuroscience and Biobehavioral Reviews* 29: 1243–1253.

Blanchard RJ, Blanchard DC, and Hori K (1989) An ethoexperimental approach to the study of defense. In: Blanchard RJ, Brain PF, Blanchard DC, and Parmigiani S (eds.) *Ethoexperimental Approaches to the Study of Behavior,* vol. 48, pp. 114–136. Dordecht: Kluwer Academic.

Blundell J, Adamec R, and Burton P (2005) Role of NMDA receptors in the syndrome of behavioral changes produced by predator stress. *Physiology & Behavior* 86: 233–243.

Canteras NS (2002) The medial hypothalamic defensive system: Hodological organization and functional implications. *Pharmacology, Biochemistry & Behavior* 71: 481–491.

Day HE, Masini CV, and Campeau S (2004) The pattern of brain c-fos mRNA induced by a component of fox odor, 2,5-dihydro-2,4,5-trimethylthiazoline (TMT), in rats, suggests both systemic and processive stress characteristics. *Brain Research* 1025: 139–151.

Dielenberg RA, Hunt GE, and McGregor IS (2001) 'When a rat smells a cat': Distribution of Fos immunoreactivty in rat brain following exposure to predatory odor. *Neuroscience* 104: 1085–1097.

Dielenberg RA and McGregor IS (2001) Defensive behavior in rats towards predatory odors: A review. *Neuroscience and Biobehavioral Reviews* 25: 597–609.

Fendt M, Endres T, Lowry CA, Apfelbach R, and McGregor IS (2005) TMT-induced autonomic and behavioral changes and the neural basis of its processing. *Neuroscience and Biobehavioral Reviews* 29: 1145–1156.

Masini CV, Sauer S, and Campeau S (2005) Ferret odor as a processive stress model in rats: Neurochemical, behavioral, and endocrine evidence. *Behavioral Neuroscience* 119: 280–292.

Morrow BA, Elsworth JD, and Roth RH (2002) Fear-like biochemical and behavioral responses in rats to the predator odor, TMT, are dependent on the exposure environment. *Synapse* 46: 11–18.

Morrow BA, Roth RH, and Elsworth JD (2000) TMT, a predator odor, elevates mesoprefrontal dopamine metabolic activity and disrupts short-term working memory in the rat. *Brain Research Bulletin* 52: 519–523.

Rosen JB (2004) The neurobiology of conditioned and unconditioned fear: A neurobehavioral system analysis of the amygdala. *Behavioral and Cognitive Neuroscience Reviews* 3: 23–41.

Takahashi LK, Nakashima BR, Hong H, and Watanabe K (2005) The smell of danger: A behavioral and neural analysis of predator odor-induced fear. *Neuroscience and Biobehavioral Reviews* 29: 1157–1167.

Walker DL, Toufexis DJ, and Davis M (2003) Role of the bed nucleus of the stria terminalis versus the amygdala in fear, stress, and anxiety. *European Journal of Pharmacology* 463: 199–216.

Awareness: Functional Imaging

G Rees, University College London, London, UK

Introduction

This article describes attempts to identify experimentally the neural correlates of the contents of conscious experience in humans. In general usage, the term 'conscious' can be used in both a transitive and intransitive sense. For example, it is possible to be conscious (as opposed to being unconscious, asleep, or in some other state). But it is also possible, when awake, to be conscious of a particular thing (as opposed to not being conscious of that thing, or conscious of some other thing at that particular moment). This distinction means that any experimental account of the neural correlates of consciousness must at a minimum specify the neural structures and processes associated with each sense of the word conscious.

The Waking State

The contents of consciousness are accessible and can be reported by individuals who are in the waking state. Changes in the level of consciousness, such as those induced artificially by anesthesia or pathologically by brain damage leading to coma, lead to individuals' being rendered unconscious and unable to report the contents of their consciousness. Sensory processing is altered, leading to an absence of conscious representation. Neural structures mediating level of consciousness are therefore surmised to facilitate the representation of conscious contents. Understanding how sensory processing is altered during changes in the level of consciousness can therefore help understand how the contents of consciousness are represented in the human brain.

Sensory processing is profoundly altered by changes in level of consciousness resulting from different causes. Typically, responses can still be obtained from the primary sensory cortices, but responses of association and the higher cortices are reduced or absent. For example, primary auditory cortex activation can still be elicited during sleep, during anesthesia, or after brain damage leading to coma. Similarly, less dramatic modifications of level of consciousness associated with meditation modify activity only in areas of the posterior sensory cortex, apart from primary visual cortex.

Anesthesia represents perhaps the most common deliberate manipulation of the level of consciousness, and it is crucial for surgical practice. Although there is increasing knowledge about the molecular and cellular targets of anesthetic agents, their effects on sensory processing in the human brain at a systems level remain poorly understood. Inhalational agents modulate brain stem nuclei and (by inference) their ascending modulatory efferent pathways. For example, isoflurane and halothane cause relative reductions in regional cerebral glucose metabolism in midbrain reticular formation, and propofol-induced depression of thalamic and reticular blood flow are closely linked. Similarly, halothane and enflurane or isoflurane depress spontaneous single-unit firing in the reticular formation of cat. However, brain stem responses to sensory stimulation (e.g., auditory evoked potentials) appear intact.

In the auditory cortex, most investigations of the effects of anesthesia on sensory processing have focused on auditory stimulation and are consistent with the general position outlined previously. In contrast, very little is known about processing in the other sensory cortices under anesthesia. For the visual system, a dose-dependent reduction in activation of the primary visual cortex is observed with thiopental, but that study did not measure the depth of anesthesia so we cannot correlate such findings with the level of consciousness. Subanesthetic isoflurane affects task-induced activation in the parieto-frontal (but not visual) cortices during performance of a visual search task. Visual evoked potentials can still be obtained during anesthesia, although somewhat unreliably in the operative environment, indicating some preservation of cortical processing. Similarly, little is known about brain areas involved in tactile processing in the unconscious human, despite the obvious clinical requirement to eliminate somatosensory sensation during surgical procedures. The clinical utility of monitoring somatosensory evoked potentials indicates that cortical processing can persist, although presumably without awareness. Two neuroimaging studies have examined localized cortical responses to somatosensory stimuli. Both showed diminished responses in the primary somatosensory cortex to electrical or vibrotactile stimulation but showed preserved responses in either the thalamus or secondary somatosensory cortex. This suggests impaired thalamocortical information transfer during anesthesia, but it is somewhat inconsistent with observations from other techniques. For example, intraoperative intrinsic signals from human somatosensory cortices to vibrotactile stimuli delivered during anesthesia are topographically specific to the stimulated digit. The reasons for this inconsistency are unclear, but it may suggest that some somatosensory cortical responsiveness is retained during anesthesia.

Taken together, these studies suggest that the waking state (compared to the anesthetized state) is associated with significant modifications in the responsiveness of sensory cortices to stimulation, particularly affecting secondary and association cortices. One unifying model for the actions of anesthetic agents is that the interruption of the dynamic connectivity between the thalamus and cortex represents a common mechanism underlying the observed changes in sensory processing and unconsciousness. Because changes in thalamocortical connectivity can also be observed in persistent vegetative state and resolve on recovery, it is conceivable that a single underlying mechanism underlies all changes in sensory processing associated with modification of the level of consciousness.

Rarely, awareness during anesthesia is reported subsequent to the administration of an anesthetic. The neural substrates of such experience remain poorly understood due to its rarity. A single case has recently been reported of an individual in a minimally conscious state who could not report her experiences overtly; however, when the patient was asked to imagine one of two different experiences, brain imaging revealed different patterns of brain activation. One possible explanation of these findings is that the individual was able to formulate an intention to follow the instructions given by the experimenter, implying concomitant awareness. How this can arise in an apparently comatose patient unable to report by any overt means and its implications for understanding the mechanisms of coma and changes in level of consciousness more generally remain to be explored.

Conscious Access: Neural Correlates of Perceptual Awareness

When people are awake, a characteristic feature of awareness is that its contents can be reported. We have subjective conscious access to the contents of our consciousness. But it is equally apparent that much of our mental life remains unconscious and inaccessible to report. Experimental enquiry must therefore distinguish what is special concerning the neural activity specifically associated with the contents of conscious experience, in contrast to other activity that merely reflects the unconscious correlates of perception or behavior. The principal methodological approach of the work reviewed next is thus to contrast the neural activity evoked by conscious versus unconscious processing, using a number of experimental paradigms.

Spontaneous Variation in the Contents of Consciousness

Bistable Stimulation

When a physical stimulus allows two different perceptual interpretations, awareness typically fluctuates spontaneously between the two possible perceptual interpretation. For example, binocular rivalry occurs when two different images are presented, one to each eye separately. Conscious experience fluctuates spontaneously, each monocular view being in turn visible for a few seconds while the other is suppressed. Because these fluctuations in awareness occur without any change in the physical stimulus, binocular rivalry (and other forms of bistable perception) form a useful model system with which to distinguish the neural correlates of these changes in the contents of consciousness from brain activity related solely to physical stimulus characteristics.

Investigations of the neural correlates of bistable stimulus perception identified two distinct patterns of findings. First, brain activity time-locked to perceptual transitions can be identified in the prefrontal and parietal cortices for both binocular rivalry and a wide variety of bistable stimuli. Moreover, the involvement of these cortical areas is independent of any requirement to make overt reports of the perceptual transitions. Similar areas are also active during spontaneous changes in the contents of consciousness in very different paradigms, during the emergence of a figure when the subject views either fragmented figures or stereo pop-out stimuli. Activation in these areas of the parietal and prefrontal cortices is associated with selective attention. More recently, activation of different parietal and prefrontal loci, thought to be involved in working memory, have been associated with the maintenance of a stable bistable percept. Although these studies do not identify whether these different psychological processes are specifically associated with such prefrontal and parietal activation, they offer a rich line of future investigation.

A second major pattern of findings associated with studies of bistable perception is that fluctuations in brain activity specifically associated with the particular contents of consciousness can be identified in the ventral visual cortex. Strikingly, such fluctuations can be identified at all levels of the human geniculostriate pathway, from the lateral geniculate nucleus (LGN) to V1 to category-specific areas such as the fusiform face area. Indeed, the local pattern of activity in cortically evoked responses carries sufficient information that automated algorithms can be used to blindly decode and accurately predict the fluctuations in perceptual experience that the observer reports. Significantly, response fluctuations that are further along the ventral visual pathway are generally larger than those observed at earlier stages, and at later stages in processing they are equal in magnitude to responses evoked to physical (nonrivalrous) stimulus alteration. This pattern of findings has been interpreted as suggesting that, at anatomically early stages of the ventral visual pathway, both seen and perceptually suppressed images are physically represented, but that competition between the two is increasingly resolved at higher stages of the visual system. Such findings are broadly consistent with event-related potential (ERP) measurements in humans and single-unit recordings in monkeys.

The involvement of both ventral visual cortical structures (representing the current contents of consciousness) and the frontoparietal structures (whose activity is either time-locked to perceptual transitions or associated with perceptual maintenance) suggests the involvement of a distributed network in perceptual rivalry. Consistent with such a hypothesis, large-scale changes in intra- and interhemispheric neuronal synchronization can be observed during rivalrous fluctuations. One model that might account for these data is that distributed object representations in the ventral visual pathway compete for perceptual dominance, biased by top-down signals reflecting working memory and attention from the frontal and parietal cortices. In the context of bistability, a conscious percept is thus represented by both an activated representation of the perceptual content and activity in the frontal and parietal cortices. Such a hypothesis is consistent with emerging theoretical models of consciousness that posit the involvement of both sensory representations and a global workspace, the latter tentatively identified with the frontal and parietal cortices.

Near-Threshold Stimuli

A detection threshold can be defined for any type of sensory stimulus, for example, by presenting it in temporal proximity to a mask or by physically degrading it in some fashion. At detection threshold, only a proportion of stimuli presented to observers will be detected, whereas the others fail to reach awareness. This spontaneous trial-to-trial variability in awareness of stimulation can be used to determine the neural correlates of stimulus awareness without confounding by changes in physical stimulus characteristics. Alternatively, brain responses to different types of fully masked and invisible stimuli can be used to determine neural correlates of unconscious processing.

Fully invisible stimuli can nevertheless evoke modest but reliable activation throughout the visual cortex. For example, activity specific to the orientation of a simple visual grating can be decoded from the primary visual cortex activity even when the stimulus is completely invisible to the observer. Similarly, when a simple target such as an achromatic disc is briefly flashed, robust activity is elicited in the corresponding retinotopic location of V1 (plus V2 and V3), even when the target is rendered completely invisible by a surrounding mask. Invisible images of tools evoke both activity in the primary visual cortex and category-specific activity in visually responsive dorsal stream structures. Invisible motion can activate the V5/middle temporal (MT) visual motion area. At higher levels of the visual system, activation selective for the presence (although undetected and invisible to the observer) of a word or for invisible faces and objects can also be observed. When an object or word is not noticed during the attentional blink, unconscious activation of the ventral visual pathway corresponding to the object identity or semantic processing can be observed. Unconscious sensory activation by complex visual stimuli is not confined to the cortical visual pathways but extends to subcortical structures. For example, the amygdala can be activated by fearful-face stimuli that have been rendered invisible through masking, in response to the emotional content of invisible words or during the suppression in binocular rivalry.

Complementing these studies in normal observers, unconscious activation of the visual cortex can also be identified in patients with visual extinction resulting from damage to the right parietal cortex. Areas of both the primary and extrastriate visual cortex that are activated by a seen left-visual-field stimulus are also activated by an unseen and extinguished left-visual-field stimulus. The unconscious processing of an extinguished face stimulus extends even to the face-selective cortex in the fusiform face area. Taken together, these findings indicate that, even without awareness, extensive processing of visual stimuli can take place in the normal visual cortex, including the generation of category-specific responses.

Invisible stimuli evoke activity associated with their mere presence and/or identity; it is equally clear that, when stimuli become visible, this is associated with an increase in activity in the corresponding category-specific areas of the visual cortex. For example, simple detection of a low-contrast grating stimulus is associated with retinotopically specific enhancement of activity in the primary visual cortex. Visually presented objects can be made difficult to identify by degrading them, and in such circumstances occipitotemporal activity shows a close correlation with recognition performance. At higher levels of the ventral visual cortex, successful detection of a face stimulus during the attentional blink is associated with significantly higher activity in the fusiform face area. Such successful detection of visually presented stimuli in the attentional blink is associated with event-related potentials that occur relatively late in stimulus processing (after several hundred milliseconds), suggesting that such activation may be associated with a late wave of activation that spreads through a distributed network of the cortical association areas. A similar pattern of enhanced activation in category-specific areas of the visual system is seen in other paradigms comparing conscious and unconscious perceptions. For example, the introduction of a flicker between successively presented visual images makes it particularly difficult to detect large-scale changes between the images. When such a change is successfully perceived, however, enhanced activity in regions of the visual cortex with category specificity for the detected change can be observed that may precede conscious detection of the change.

Successful conscious detection of stimuli is associated not just with changes in activation of the ventral visual cortex but also with changes in coupling between the early retinotopic cortex and higher visual areas. When the interval between a target and a nonoverlapping metacontrast mask is varied over a few hundred milliseconds, observers report that visibility of the target first declines before rising again, in a U-shaped function. Strikingly, the degree of coupling (or effective connectivity) between the early retinotopic cortex and higher visual areas shows a pattern of changes with target–mask stimulus onset asynchrony (SOA) that closely parallel this behavioral function, suggesting that effective connectivity may play a role in determining visibility.

As for bistable stimuli, conscious detection of stimuli close to the threshold involves not just the sensory cortices but areas of the frontal and parietal cortices. For example, successful detection of a change is associated with parietofrontal activation that may reflect the successful deployment of attention. At the detection threshold, differences in parietal and prefrontal activity emerge following electrical activity associated with conscious perception at approximately 100 ms. Parietal cortical activation is also associated with conscious recognition of visual verbal stimuli. Successful identification of an object evokes event-related negativity and is associated with occipital responses and modulation of the parieto-occipital alpha rhythm. Finally, conscious detection of flicker is associated with the activation of the frontal and parietal cortices, indicating that these areas are important not just for awareness of spatially extended objects but also for conscious detection of temporally distinct visual events.

Hallucinations

A hallucination is a false sensory perception in the absence of an external stimulus, and it is often associated with drug use, sleep deprivation, and neurological or psychiatric disease. During visual hallucinations, brain activation can be observed in object-selective regions of the ventral visual pathway whose selectivity corresponds to the content of the hallucination. During auditory hallucinations in schizophrenia, the subcortical nuclei and limbic structures and the primary auditory cortex are activated. Finally, hallucination of a supernumerary limb after stroke is associated with activation of the medial prefrontal cortex in the supplementary motor area. Hallucinations are therefore associated with the activation of either the sensory or motor cortex whose modality and neuronal specificities corresponds closely to that of the hallucinatory perceptual content.

Modifying the Contents of Consciousness

Changes in the contents of consciousness can be driven by spontaneous fluctuations. They can also be systematically altered by either top-down signals (e.g., directing attention toward or away from some aspect of a stimulus) or by altering the spatial or temporal context in which a stimulus is presented.

Attention

When subjects are engaged in a demanding task, irrelevant but highly salient stimuli outside the immediate focus of attention can go entirely unnoticed. This is known as inattentional blindness, and it suggests that consciousness may depend on attention. Consistent with this hypothesis, brain activity evoked by irrelevant sensory stimulation in the ventral occipital and temporal cortices is reduced when attention is withdrawn. Moreover, when inattentional blindness results for unattended words, then brain activity no longer differentiates between such meaningful words and random letters. This suggests that attention is required both for brain activity associated with the higher processing of sensory stimuli and for their subsequent representation as the contents of consciousness. However, although attention may be necessary for conscious representation, it is not sufficient. Varying the availability of attentional resources changes the activity evoked in the primary visual cortex by even completely invisible stimuli. Thus, attention can act on invisible stimuli and need not change their visibility as a consequence.

Illusions

An illusion occurs when real external stimuli are misperceived and represented in consciousness in an incorrect fashion. The content of the illusory perception typically depends on the spatial and temporal context in which it occurs. For example, sensory aftereffects are illusory sensory perceptions in the absence of sensory stimulation that occur following an extended period of adaptation to a sensory stimulus. Aftereffects contingent on prior adaptation to color or motion activate V4 or V5/MT, respectively, and the time course of such activation reflects phenomenal experience. Perception of illusory or implied motion in a static visual stimulus results in the activation of V5/MT, and the perception of motion at unstimulated regions of the visual field on the path of apparent motion is associated with retinotopically specific activation of V1 mediated by feedback from V5/MT. Perception of illusory contours activates areas of the early retinotopic extrastriate cortex. Similarly, when a moving grating is divided by a large gap, observers report seeing a moving phantom in the gap and there is enhanced activity in the locations in the early retinotopic visual cortex corresponding to the illusory percept. The primary visual cortex represents the perceived angular size of an object rather than its actual size on the retina. Finally, when a uniformly illuminated surface is placed eccentrically on a dynamic textured background, after a few seconds, it is perceived to disappear and be replaced by the background texture. This texture filling-in is associated with the reduction of activity in the primary visual cortex associated with the representation of the now-invisible surface.

Common to all these experimental paradigms are changes in phenomenal experience without corresponding changes in physical stimulus. Perceptual illusions can also be used to manipulate feelings of ownership of a rubber hand. When participants feel that the hand is theirs, then activity in the premotor cortex reflects this conscious perception of ownership. Thus, both perceptual and motor illusions lead to a specific modification of brain activity. Altered brain activity is observed in areas of the brain known to contain neurons whose stimulus specificities encompass the attribute represented in consciousness.

Imagery

Conscious perception can be created by the act of imagination. In patients with implanted electrodes for presurgical epilepsy mapping, single neurons in the human medial temporal lobe that fire selectively when particular visual stimuli are presented are also activated when the individual imagines the same stimuli. Similarly, neuronal populations elsewhere in the ventral visual pathway with stimulus specificity for faces or places are activated during imagery of these categories of object. These findings are consistent with the notion that activity in functionally specialized neuronal populations in sensory cortices is associated with the representation of an object or image in consciousness.

Summary and Outstanding Problems

Findings from a wide variety of neuroimaging experiments are remarkably consistent in reaching the following conclusions. First, sensory stimuli that remain unconscious can nevertheless undergo extensive processing in the sensory cortices. Second, compared to stimuli that remain unconscious, consciously perceived stimuli elicit greater levels of activity in functionally specialized areas of the sensory cortex corresponding to the attributes represented in consciousness. Finally, consciously perceived stimuli also elicit activity in specific areas of the prefrontal and parietal cortices that may be associated with higher-level cognitive processes such as attention and working memory.

This article has focused exclusively on studies of awareness employing a single method, functional neuroimaging. In practice, findings from a single technique are never considered in isolation and identifying experimentally the neural correlates of the contents of conscious experience in humans depends on considering converging evidence from multiple methodologies. It is therefore appropriate to consider some of the important limitations of neuroimaging methodologies in order to understand the potential contribution of these data in the context of findings from other experimental techniques.

Causality

All the work discussed in this article has attempted to correlate changes in brain activity with changes in the contents of consciousness. Such correlational studies cannot determine whether such neural activity plays a causal role in determining the contents of consciousness. In particular, they cannot show whether the areas that have been identified in association with particular contents of consciousness are either necessary or sufficient for such conscious experiences to occur. The identification of necessary and sufficient brain activity requires explicit experimental manipulation of that activity. If manipulation of brain activity changes consciousness, then a causal role for that brain activity can be inferred. Manipulations of brain activity can be performed with direct electrical stimulation, using transcranial magnetic stimulation and as a consequence of brain lesions (although in humans this depends on the presence of pathological conditions and is not under direct experimental control!).

Report

The studies discussed in this article seek to relate neural activity to visual awareness, and in doing so they must distinguish the neural correlates of such activity from the merely unconscious concomitants, such as unconscious perception or behavior. Typically, keeping unconscious aspects of perception or behavior constant can allow this distinction to be made. For example, many of the studies reviewed here employed physically identical stimuli that elicited different conscious percepts. However, with very few exceptions, such studies also required that participants report whether they were aware of the stimulus. Because different behaviors are required to make different reports of awareness, it is possible that this might confound the interpretation of such studies. For example, when masked words are consciously identified and then named (as opposed to when they are not consciously identified and not named), then activity is observed in the frontoparietal cortex. However, such activity may reflect the (unconscious) processes associated with generating verbal output rather than the neural correlates of awareness. Although the differences in behavior elicited by the various reports of awareness are typically much more subtle, and, for example, frontal and parietal activsergenity associated with awareness can be observed in the absence of report or when performance confounds are controlled, further study of the unconscious correlates of behavior associated with report is indicated.

Conclusion

The most parsimonious account of currently available data is that the current contents of visual consciousness consist of a representation in the primary visual cortex and ventral visual pathway corresponding to the attributes represented in consciousness, together with activity in specific parietal (and perhaps) prefrontal structures. Future work must seek to more closely delineate the qualitative and quantitative differences between sensory representations of stimuli that either reach consciousness or remain unconscious, determine whether the frontal and parietal cortices play a causal role in representing the contents of consciousness, and more closely define the relationship between particular higher cognitive processes involved in awareness and corresponding patterns of brain activation.

See also: Binocular Rivalry; Cognition: An Overview of Neuroimaging Techniques; Consciousness: Neural Basis of Conscious Experience; Consciousness: Neurophysiology and Visual Awareness in; Consciousness: Philosophy; Consciousness: Theories and Models; Neuroimaging; The AIM Model of Dreaming, Sleeping, and Waking Consciousness.

Further Reading

Alkire MT, Haier RJ, and Fallon JH (2000) Toward a unified theory of narcosis: Brain imaging evidence for a thalamocortical switch as the neurophysiologic basis of anesthetic-induced unconsciousness. *Consciousness and Cognition* 9(3): 370–386.

Antognini JF, Buonocore MH, Disbrow EA, and Carstens E (1997) Isoflurane anesthesia blunts cerebral responses to noxious and innocuous stimuli: A fMRI study. *Life Sciences* 61(24): PL 349–354.

Baars BJ (2005) Global workspace theory of consciousness: Toward a cognitive neuroscience of human experience. *Progress in Brain Research* 150: 45–53.

Bahrami B, Lavie N, and Rees G (2007) Attentional load modulates responses of human primary visual cortex to invisible stimuli. *Current Biology* 17(6): 509–513.

Barnes J, Howard RJ, Senior C, et al. (1999) The functional anatomy of the mccollough contingent colour after-effect. *NeuroReport* 10(1): 195–199.

Beck DM, Rees G, Frith CD, and Lavie N (2001) Neural correlates of change detection and change blindness. *Nature Neuroscience* 4(6): 645–650.

Blake R and Logothetis NK (2002) Visual competition. *Nature Reviews Neuroscience* 3(1): 13–21.

Bonhomme V, Fiset P, Meuret P, et al. (2001) Propofol anesthesia and cerebral blood flow changes elicited by vibrotactile stimulation: A positron emission tomography study. *Journal of Neurophysiology* 85(3): 1299–1308.

Cannestra AF, Black KL, Martin NA, et al. (1998) Topographical and temporal specificity of human intraoperative optical intrinsic signals. *NeuroReport* 9(11): 2557–2563.

Carmel D, Lavie N, and Rees G (2006) Conscious awareness of flicker in humans involves frontal and parietal cortex. *Current Biology* 16(9): 907–911.

Dehaene S, Kerszberg M, and Changeux JP (1998) A neuronal model of a global workspace in effortful cognitive tasks. *Proceedings of the National Academy of Sciences of the United States of America* 95(24): 14529–14534.

Dehaene S, Naccache L, Cohen L, et al. (2001) Cerebral mechanisms of word masking and unconscious repetition priming. *Nature Neuroscience* 4(7): 752–758.

Dierks T, Linden DE, Jandl M, et al. (1999) Activation of Heschl's gyrus during auditory hallucinations. *Neuron* 22(3): 615–621.

Dueck MH, Petzke F, Gerbershagen HJ, et al. (2005) Propofol attenuates responses of the auditory cortex to acoustic stimulation in a dose-dependent manner: A fMRI study. *Acta Anaesthesiologia Scandinavica* 49(6): 784–791.

Ehrsson HH, Spence C, and Passingham RE (2004) That's my hand!: Activity in premotor cortex reflects feeling of ownership of a limb. *Science* 305(5685): 875–877.

Fang F and He S (2005) Cortical responses to invisible objects in the human dorsal and ventral pathways. *Nature Neuroscience* 8(10): 1380–1385.

Ffytche DH, Howard RJ, Brammer MJ, David A, Woodruff P, and Williams S (1998) The anatomy of conscious vision: An fMRI study of visual hallucinations. *Nature Neuroscience* 1(8): 738–742.

Fiset P, Paus T, Daloze T, et al. (1999) Brain mechanisms of propofol-induced loss of consciousness in humans: A positron emission tomographic study. *Journal of Neuroscience* 19(13): 5506–5513.

Frith C, Perry R, and Lumer E (1999) The neural correlates of conscious experience: An experimental framework. *Trends in Cognitive Sciences* 3(3): 105–114.

Grill-Spector K, Kushnir T, Hendler T, and Malach R (2000) The dynamics of object-selective activation correlate with recognition performance in humans. *Nature Neuroscience* 3(8): 837–843.

Hadjikhani N, Liu AK, Dale AM, Cavanagh P, and Tootell RB (1998) Retinotopy and color sensitivity in human visual cortical area v8. *Nature Neuroscience* 1(3): 235–241.

Haynes JD, Deichmann R, and Rees G (2005) Eye-specific effects of binocular rivalry in the human lateral geniculate nucleus. *Nature* 438(7067): 496–499.

Haynes JD, Driver J, and Rees G (2005) Visibility reflects dynamic changes of effective connectivity between v1 and fusiform cortex. *Neuron* 46(5): 811–821.

Haynes JD and Rees G (2005) Predicting the orientation of invisible stimuli from activity in human primary visual cortex. *Nature Neuroscience* 8(5): 686–691.

Haynes JD and Rees G (2005) Predicting the stream of consciousness from activity in human visual cortex. *Current Biology* 15(14): 1301–1307.

He S, Cohen ER, and Hu X (1998) Close correlation between activity in brain area mt/v5 and the perception of a visual motion aftereffect. *Current Biology* 8(22): 1215–1218.

Heinke W, Fiebach CJ, Schwarzbauer C, Meyer M, Olthoff D, and Alter K (2004) Sequential effects of propofol on functional brain activation induced by auditory language processing: An event-related functional magnetic resonance imaging study. *British Journal of Anaesthesia* 92(5): 641–650.

Heinke W and Schwarzbauer C (2001) Subanesthetic isoflurane affects task-induced brain activation in a highly specific manner: A functional magnetic resonance imaging study. *Anesthesiology* 94(6): 973–981.

Hirsch J, De LaPaz RL, Relkin NR, et al. (1995) Illusory contours activate specific regions in human visual cortex: Evidence from functional magnetic resonance imaging. *Proceedings of the National Academy of Sciences of the United States of America* 92(14): 6469–6473.

Huettel SA, Guzeldere G, and McCarthy G (2001) Dissociating the neural mechanisms of visual attention in change detection using functional MRI. *Journal of Cognitive Neuroscience* 13(7): 1006–1018.

Kerssens C, Hamann S, Peltier S, Hu XP, Byas-Smith MG, and Sebel PS (2005) Attenuated brain response to auditory word stimulation with sevoflurane: A functional magnetic resonance imaging study in humans. *Anesthesiology* 103(1): 11–19.

Kleinschmidt A, Buchel C, Zeki S, and Frackowiak RS (1998) Human brain activity during spontaneously reversing perception of ambiguous figures. *Proceedings: Biological Sciences* 265 (1413): 2427–2433.

Kourtzi Z and Kanwisher N (2000) Activation in human mt/mst by static images with implied motion. *Journal of Cognitive Neuroscience* 12(1): 48–55.

Kreiman G, Koch C, and Fried I (2000) Imagery neurons in the human brain. *Nature* 408(6810): 357–361.

Kumar A, Bhattacharya A, and Makhija N (2000) Evoked potential monitoring in anaesthesia and analgesia. *Anaesthesia* 55(3): 225–241.

Lau HC and Passingham RE (2006) Relative blindsight in normal observers and the neural correlate of visual consciousness. *Proceedings of the National Academy of Sciences of the United States of America* 103(49): 18763–18768.

Laureys S, Faymonville ME, Degueldre C, et al. (2000) Auditory processing in the vegetative state. *Brain* 123(part 8): 1589–1601.

Laureys S, Faymonville ME, Luxen A, Lamy M, Franck G, and Maquet P (2000) Restoration of thalamocortical connectivity after recovery from persistent vegetative state. *Lancet* 355 (9217): 1790–1791.

Lee SH and Blake R (2002) V1 activity is reduced during binocular rivalry. *Journal of Vision* 2(9): 618–626.

Lee SH, Blake R, and Heeger DJ (2005) Traveling waves of activity in primary visual cortex during binocular rivalry. *Nature Neuroscience* 8(1): 22–23.

Leopold DA and Logothetis NK (1996) Activity changes in early visual cortex reflect monkeys' percepts during binocular rivalry. *Nature* 379(6565): 549–553.

Lou HC, Kjaer TW, Friberg L, Wildschiodtz G, Holm S, and Nowak M (1999) A 15o-h2o pet study of meditation and the resting state of normal consciousness. *Human Brain Mapping* 7(2): 98–105.

Luck SJ, Vogel EK, and Shapiro KL (1996) Word meanings can be accessed but not reported during the attentional blink. *Nature* 383(6601): 616–618.

Lumer ED, Friston KJ, and Rees G (1998) Neural correlates of perceptual rivalry in the human brain. *Science* 280(5371): 1930–1934.

Lumer ED and Rees G (1999) Covariation of activity in visual and prefrontal cortex associated with subjective visual perception. *Proceedings of the National Academy of Sciences of the United States of America* 96(4): 1669–1673.

Marois R, Yi DJ, and Chun MM (2004) The neural fate of consciously perceived and missed events in the attentional blink. *Neuron* 41(3): 465–472.

Martin E, Thiel T, Joeri P, et al. (2000) Effect of pentobarbital on visual processing in man. *Human Brain Mapping* 10(3): 132–139.

McGonigle DJ, Hanninen R, Salenius S, Hari R, Frackowiak RS, and Frith CD (2002) Whose arm is it anyway? An fMRI case study of supernumerary phantom limb. *Brain* 125(Pt. 6): 1265–1274.

Mendola JD, Conner IP, Sharma S, Bahekar A, and Lemieux S (2006) fMRI measures of perceptual filling-in in the human visual cortex. *Journal of Cognitive Neuroscience* 18(3): 363–375.

Mendola JD, Dale AM, Fischl B, Liu AK, and Tootell RB (1999) The representation of illusory and real contours in human cortical visual areas revealed by functional magnetic resonance imaging. *Journal of Neuroscience* 19(19): 8560–8572.

Meng M, Remus DA, and Tong F (2005) Filling-in of visual phantoms in the human brain. *Nature Neuroscience* 8(9): 1248–1254.

Morris JS, Ohman A, and Dolan RJ (1999) A subcortical pathway to the right amygdala mediating "unseen" fear. *Proceedings of the National Academy of Sciences of the United States of America* 96(4): 1680–1685.

Moutoussis K and Zeki S (2002) The relationship between cortical activation and perception investigated with invisible stimuli. *Proceedings of the National Academy of Sciences of the United States of America* 99(14): 9527–9532.

Moutoussis K and Zeki S (2006) Seeing invisible motion: A human fMRI study. *Current Biology* 16(6): 574–579.

Muckli L, Kohler A, Kriegeskorte N, and Singer W (2005) Primary visual cortex activity along the apparent-motion trace reflects illusory perception. *PLoS Biology* 3(8): e265 [online].

Murray SO, Boyaci H, and Kersten D (2006) The representation of perceived angular size in human primary visual cortex. *Nature Neuroscience* 9(3): 429–434.

Myles PS, Leslie K, McNeil J, Forbes A, and Chan MT (2004) Bispectral index monitoring to prevent awareness during anaesthesia: The b-aware randomised controlled trial. *Lancet* 363 (9423): 1757–1763.

Naccache L, Gaillard R, Adam C, et al. (2005) A direct intracranial record of emotions evoked by subliminal words. *Proceedings of the National Academy of Sciences of the United States of America* 102(21): 7713–7717.

Nachev P and Husain M (2007) Comment on "detecting awareness in the vegetative state." *Science* 315(5816): 1221 [author reply 1221].

Niedeggen M, Wichmann P, and Stoerig P (2001) Change blindness and time to consciousness. *European Journal of Neuroscience* 14(10): 1719–1726.

O'Craven KM and Kanwisher N (2000) Mental imagery of faces and places activates corresponding stiimulus-specific brain regions. *Journal of Cognitive Neuroscience* 12(6): 1013–1023.

Ogawa T, Shingu K, Shibata M, Osawa M, and Mori K (1992) The divergent actions of volatile anaesthetics on background neuronal activity and reactive capability in the central nervous system in cats. *Canadian Journal of Anaesthetics* 39(8): 862–872.

Owen AM, Coleman MR, Boly M, Davis MH, Laureys S, and Pickard JD (2006) Detecting awareness in the vegetative state. *Science* 313(5792): 1402.

Pasley BN, Mayes LC, and Schultz RT (2004) Subcortical discrimination of unperceived objects during binocular rivalry. *Neuron* 42(1): 163–172.

Pessoa L and Ungerleider LG (2004) Neural correlates of change detection and change blindness in a working memory task. *Cerebral Cortex* 14(5): 511–520.

Pins D and Ffytche D (2003) The neural correlates of conscious vision. *Cerebral Cortex* 13(5): 461–474.

Plourde G, Belin P, Chartrand D, et al. (2006) Cortical processing of complex auditory stimuli during alterations of consciousness with the general anesthetic propofol. *Anesthesiology* 104(3): 448–457.

Polonsky A, Blake R, Braun J, and Heeger DJ (2000) Neuronal activity in human primary visual cortex correlates with perception during binocular rivalry. *Nature Neuroscience* 3(11): 1153–1159.

Portas CM, Krakow K, Allen P, Josephs O, Armony JL, and Frith CD (2000) Auditory processing across the sleep-wake cycle: Simultaneous eeg and fMRI monitoring in humans. *Neuron* 28(3): 991–999.

Portas CM, Strange BA, Friston KJ, Dolan RJ, and Frith CD (2000) How does the brain sustain a visual percept? *Proceedings: Biological Sciences* 267(1446): 845–850.

Ress D and Heeger DJ (2003) Neuronal correlates of perception in early visual cortex. *Nature Neuroscience* 6(4): 414–420.

Rees G (2001) Seeing is not perceiving. *Nature Neuroscience* 4(7): 678–680.

Rees G, Frith CD, and Lavie N (1997) Modulating irrelevant motion perception by varying attentional load in an unrelated task. *Science* 278(5343): 1616–1619.

Rees G and Lavie N (2001) What can functional imaging reveal about the role of attention in visual awareness? *Neuropsychologia* 39(12): 1343–1353.

Rees G, Russell C, Frith CD, and Driver J (1999) Inattentional blindness versus inattentional amnesia for fixated but ignored words. *Science* 286(5449): 2504–2507.

Rees G, Wojciulik E, Clarke K, Husain M, Frith C, and Driver J (2000) Unconscious activation of visual cortex in the damaged right hemisphere of a parietal patient with extinction. *Brain* 123 (Pt. 8): 1624–1633.

Rees G, Wojciulik E, Clarke K, Husain M, Frith C, and Driver J (2002) Neural correlates of conscious and unconscious vision in parietal extinction. *Neurocase* 8(5): 387–393.

Ritzl A, Marshall JC, Weiss PH, et al. (2003) Functional anatomy and differential time courses of neural processing for explicit inferred, and illusory contours: An event-related fMRI study. *Neuroimage* 19(4): 1567–1577.

Rodriguez E, George N, Lachaux JP, Martinerie J, Renault B, and Varela FJ (1999) Perception's shadow: Long-distance synchronization of human brain activity. *Nature* 397(6718): 430–433.

Rudolph U and Antkowiak B (2004) Molecular and neuronal substrates for general anaesthetics. *Nature Reviews Neuroscience* 5(9): 709–720.

Sakai K, Watanabe E, Onodera Y, et al. (1995) Functional mapping of the human colour centre with echo-planar magnetic resonance imaging. *Proceedings: Biological Sciences* 261(1360): 89–98.

Senior C, Barnes J, Giampietro V, et al. (2000) The functional neuroanatomy of implicit-motion perception or representational momentum. *Current Biology* 10(1): 16–22.

Silbersweig DA, Stern E, Frith C, et al. (1995) A functional neuroanatomy of hallucinations in schizophrenia. *Nature* 378(6553): 176–179.

Sterzer P, Haynes JD, and Rees G (2006) Primary visual cortex activation on the path of apparent motion is mediated by feedback from hmt+/v5. *Neuroimage* 32(3): 1308–1316.

Sterzer P, Russ MO, Preibisch C, and Kleinschmidt A (2002) Neural correlates of spontaneous direction reversals in ambiguous apparent visual motion. *Neuroimage* 15(4): 908–916.

Tong F and Engel SA (2001) Interocular rivalry revealed in the human cortical blind-spot representation. *Nature* 411(6834): 195–199.

Tong F, Nakayama K, Vaughan JT, and Kanwisher N (1998) Binocular rivalry and visual awareness in human extrastriate cortex. *Neuron* 21(4): 753–759.

Tononi G, Srinivasan R, Russell DP, and Edelman GM (1998) Investigating neural correlates of conscious perception by frequency-tagged neuromagnetic responses. *Proceedings of the National Academy of Sciences of the United States of America* 95(6): 3198–3203.

Tootell RB, Reppas JB, Dale AM, et al. (1995) Visual motion aftereffect in human cortical area mt revealed by functional magnetic resonance imaging. *Nature* 375(6527): 139–141.

Vuilleumier P, Sagiv N, Hazeltine E, et al. (2001) Neural fate of seen and unseen faces in visuospatial neglect: A combined event-related functional MRI and event-related potential study. *Proceedings of the National Academy of Sciences of the United States of America* 98(6): 3495–3500.

Weil RS, Kilner JM, Haynes JD, and Rees G (2007) Neural correlates of perceptual filling-in of an artificial scotoma in humans. *Proceedings of the National Academy of Sciences of the United States of America* 104(12): 5211–5216.

Wiedemayer H, Fauser B, Armbruster W, Gasser T, and Stolke D (2003) Visual evoked potentials for intraoperative neurophysiologic monitoring using total intravenous anesthesia. *Journal of Neurosurgical Anesthesiology* 15(1): 19–24.

Wunderlich K, Schneider KA, and Kastner S (2005) Neural correlates of binocular rivalry in the human lateral geniculate nucleus. *Nature Neuroscience* 8(11): 1595–1602.

Yi DJ, Woodman GF, Widders D, Marois R, and Chun MM (2004) Neural fate of ignored stimuli: Dissociable effects of perceptual and working memory load. *Nature Neuroscience* 7(9): 992–996.

Zeki S, Watson JD, and Frackowiak RS (1993) Going beyond the information given: The relation of illusory visual motion to brain activity. *Proceedings: Biological Sciences* 252(1335): 215–222.

Axon Guidance by Glia

A Faissner, Ruhr-University, Bochum, Germany

Evidence for Conducive Functions of Astroglia

Astroglial Cells Constitute Axonal Growth Pathways *In Vivo*

Astrocyte-derived basal laminae Astrocyte surfaces were originally recognized as a favorable substrate for axon outgrowth. Müller cells, for example, produce a specialized structure toward the inner surface of the retina, the glial endfeet. Retinal ganglion cell axons elongate along the endfeet-derived basal lamina that contains characteristic extracellular matrix (ECM) components such as laminin-1, collagen IV, heparan sulfate proteoglycan (HSPG), and nidogen. The response of axonal growth cones to laminin-1 is regulated by integrins, heterodimeric receptors of the ECM.

Astrocyte monolayers Astrocyte monolayers constitute an excellent growth substrate for axons *in vitro*. For example, type I protoplasmic astrocytes with extended surfaces prepared from embryonic or perinatal central nervous system (CNS) tissues are more efficient than those obtained from later postnatal stages. Primary astrocytes are more efficient soon after plating and tend to lose beneficial properties with time. Subtypes may exist that are less supportive for axon extension. The neurite growth-promoting properties are also regulated by the spatial arrangement of astrocytes.

Astrocytes in three dimensions It has been observed that astrocytes that still support axon growth when used as monolayers may have inhibitory properties when assembled in tissuelike structures. For example, dorsal root ganglion axons confronting astrocytes lodged in three dimensions in a cellulose acetate tube will grow readily into tubes filled with embryonic but not with aged or matured astrocytes. The model in this regard mimics the dorsal root entry zone, a reputed stop area for centripetal axons in the adult. In this context, the astrocyte–meningeal cell interface may construct a growth barrier.

Multiple Control Mechanisms Regulate the Astrocyte-Dependent Establishment of Cortical Connections via the Corpus Callosum

The Corpus Callosum and the Glial Sling

Anatomy of the corpus callosum and the blueprint hypothesis The corpus callosum connects the left with the right hemisphere in the developing nervous system. In the mouse, it consists of approximately 3 million myelinated fibers that link corresponding regions of the cortices. More that 50 human genetic aberrations have been described that lead to some form of dysgenesis of the corpus callosum. Two populations of midline glia, the indusium griseum and the glial wedge, are necessary for the adequate guidance of axons in this process. The emerging glial structures construct a transient bridge of astrocytes that connects the left with the right hemisphere of the developing telencephalon. This glial bridge, also called the glial sling, supports the reciprocal growth of cortical axons, and the experimental interruption of the sling leads to the formation of acallosal mice. In this situation, the cortical connecting axons role up on either side of the cerebral midline and form the bundles of Probst, longitudinal fascicles of misdirected axons. Growth promotion of cortical axons can be restored by the implantation of nitrocellulose filters that are covered with embryonic astrocytes-derived membranes. The blueprint hypothesis of axon growth states that channels walled by astrocyte surfaces might provide a mechanical growth and guidance substrate for growth cones. Molecular specializations of both growth cone and astrocyte surfaces are presumably implicated in the regulation of these interactions.

Molecular bases of neuron–glia interactions The mechanistic concept of axon guidance has been modified and geared toward an interpretation which emphasizes molecular signals in the growth environment, the readout by specific growth cone-based receptors, and integration of these influences by signal transduction cascades which eventually modulate growth cone movements (**Table 1**). Thus, a conduit function of astroglia based on the chemorepellent slit-2 has been proposed as an additional guidance principle in the corpus callosum. The molecular analysis of the signaling system involved in corpus callosum formation has progressed and additional gene families have been identified. Thus, the wnt5 protein is expressed by cells of the glial wedge and plays a role in guidance of connecting axons on the contralateral side of the developing corpus callosum formation. Wnt5a-dependent guidance in this context is mediated by the protein tyrosine kinase Ryk that is required in the axon.

Cell Adhesion Molecules Play a Pivotal Role in Neural Cell Interactions

General Definition of Cell Adhesion Molecules

Cell adhesion molecules (CAMs) were first identified during the study of sponge aggregation and

Table 1 Families of adhesion molecules involved in axon guidance

Adhesion molecules	*Features*	*Functions*
Ig superfamily (IgSf)	At least one immunoglobulin (Ig-like) domain, often in combination with fibronectin type III domains	Ca-independent adhesion, either in the homophilic or in the heterophilic mode
L1CAM, NCAM	First IgSf members found in the nervous system	Axon growth and guidance, synapse formation, and plasticity; human L1CAM mutated in the human MASA syndrome that is associated with mental retardation
TAG-1	GPI-linked member of the IgSf	Axon fasciculation
Cadherin superfamily	Characterized by at least one cadherin domain	Primarily homophilic, Ca^{2+}-dependent adhesion; roles in sorting out of cells, synapse formation
N-cadherin	Classical cadherin	Neuron–neuron and neuron–astrocyte adhesion; elimination leads to premature death and cardiovascular malformations
Laminin-1	Founding member of the laminin gene family that comprises more than ten members; laminins are heteromultimers formed by α, β, and γ chains; several genes for each subunit have been distinguished	Excellent axon growth promoting substrate and, for this reason, obligatory cell substrate constituent in many cell culture protocols; as a component of basal lamina, laminin-1 promotes axon growth in the peripheral nervous system; the outgrowth promotion is read out via integrins
Integrins	Large family of heterodimeric receptors formed by α and β subunits; many genes for either type of subunit have been described	Integrins are receptors of extracellular matrix constituents; a subgroup interacts with the sequence RGD in target molecules; more than 20 Itg receptors have been identified
Netrin-1, -2	Homologies to the γ_1 subunit of laminin-1	Netrins are released by midline glia and build up a chemotactic gradient that attracts axons from commissural neurons

subsequently discovered in mammals. One distinguishes Ca^{2+}-dependent and Ca^{2+}-independent adhesion mechanisms that are served by separate gene families, the immunoglobulin (Ig) and the cadherin gene superfamilies. CAMs interact either in the homophilic mode with identical members of a gene family or in the heterophilic mode with distinct partners that may or may not belong to the same gene family.

The Cadherins and Ca^{2+}-Dependent Adhesion

N-cadherin was the first classical cadherin discovered in the CNS and it mediates neuron–neuron and neuron–astrocyte interactions. Classical cadherins contain five cadherin repeat motifs, calcium binding sites, and a transmembrane domain that result in an overall molecular mass of approximately 100 kDa. Subgroups of cadherin-related neuronal receptors and of protocadherins as characterized by the cadherin domain have been discovered, many of which are expressed in the CNS. There, expression patterns correlate with neuroanatomical networks, and current concepts propose a functional role in wiring and synaptogenesis within these systems. This interpretation is consistent with the fact that classical cadherins function in a homophilic, calcium-dependent manner. The constitutive cadherin domain motif has also been found in so-called unconventional cadherin-like molecules that play a role in epithelial cell junction formation or in *Drosophila* tumors (e.g., the gene *fat*). The cytoplasmic domain of classical cadherins interacts with specialized linker proteins called catenins. These are required for the functional activity of cadherins, and the component β-catenin is involved in signal transduction to the cell nucleus and is an important mediator of the wnt signaling pathway, which links cadherins to differentiation processes and to cancer pathology.

Immunoglobulin Superfamily

The Ig superfamily constitutes the other dominating family of CAMs in the nervous system. Its characteristic feature is the Ig domain that consists of 90–100 amino acids arranged in seven antiparallel β-pleated sheets that fold into a globular structure. In many molecules, the Ig loop is combined with one or several fibronectin type III (FNIII) domains, followed by a transmembrane domain. Some IgCAMs are connected to the membrane by a glycosylphosphatidylinositol (GPI) anchor that confers augmented mobility within the membrane plane. GPI-linked proteins are thought to be associated with transmembrane glycoproteins that convey signal transduction processes – for example, the neurexin family member Caspr/paranodin in the case of the IgCAM contactin. On the mechanistic level, Ig superfamily members mediate calcium-independent adhesion processes that may be either homophilic or heterophilic. Several neuronal adhesion molecules with pronounced expression on axonal surfaces, such as L1CAM or TAG-1/axonin-1 CAM, have been grouped as AxCAMs, highlighting their prominent role in axon fasciculation.

IgCAM Interactions Launch Signal Transduction

Functional participation of IgCAMs in neurite outgrowth requires the activation of downstream signaling mechanisms, including the variation of intracellular calcium in the growth cone. The transduction mechanisms elicited by IgCAMs converge with those launched via N-cadherin and necessitate the basic fibroblast growth factor (bFGF) receptor that *cis* interacts with these CAMs within the membrane. Selected isoforms of NCAM are expressed by astrocytes *in vitro* and mediate neuron–astrocyte adhesion. With regard to heterophilic interactions, the small isoform of the receptor protein tyrosine phosphatase (RPTP)-β/ζ is expressed by astrocytes, a transmembrane protein that interacts with the neuronal adhesion molecule contactin, and other CAMs of the Ig superfamily. These examples illustrate the functional involvement of IgCAMs in conducive neuron–astrocyte interactions. Mutations of the human L1CAM lead to mental retardation, hypotrophy of the corticospinal tract, and hydrocephalus, emphasizing the importance of the Ig superfamily for development of CNS structures.

Integrins as Extracellular Matrix Receptors

The integrins constitute the third prominent gene family of membrane-based CAMs. These consist of α and β subunits that assemble to heterodimers and mediate the interactions of neural cells with ECM components. In some cases, interactions between integrins and Ig superfamily members have been reported, and the activation of integrins involves signal transduction pathways with small GTP-binding proteins such as *ras*.

Evidence for Segregating Functions of Glia

Gene Families That Mediate Axon Growth Inhibition Counterbalance Adhesion Molecule Gene Families

Growth cone collapse The investigation of neural cell interactions resulted in the identification of the Ig and cadherin superfamilies and of growth- and motility-promoting ECM constituents. However, in the 1980s it became clear that in addition to growth promotion, growth inhibitory molecules also contribute to the regulation of cell migration and growth cone movement (**Table 2**). The phenomenon of growth cone collapse had been described in the context of the interaction of sympathetic with retinal neurites, or when growth cones were confronted with myelin fractions or membrane preparations obtained from inhibitory territories. These interactions invariably resulted in collapse and retraction of the growth cone that remained paralyzed for 30–60 min *in vitro* prior to resuming growth and exploratory behavior. It was quickly realized that this inhibitory effect might equally affect guidance and inhibition of regeneration.

Eph kinases and ephrins The systematic investigation of the visual projection led to the identification of RAG (retinal axon guidance molecule), a GPI-linked protein with a gradient-like distribution in the tectum that proved homologous to a gene family subsequently renamed ephrins. One distinguishes the GPI-linked ephrins A, which interact with complementary EphA-type tyrosine kinases, from the transmembrane ephrins B, which recognize the EphB-type tyrosine kinases. Both groups contain a large number of both ephrin- and Eph-type kinase genes, and a certain degree of freedom in the mutual combinations has been noted. A remarkable trait of the Eph kinases and the corresponding ligands is that the molecules are expressed in reciprocal gradients, which is similar to the concept that Sperry developed in his chemoaffinity hypothesis that posited a chemical cell surface expressed code of positional information. Manipulation of expression levels of Eph genes resulted in a graded repositioning of axonal projections in the visual system. These studies provided impressive support both for the Sperry hypothesis and for the notion that the Eph–ephrin signaling system is an important contributor to the code. Whether the members of the pairs are expressed in neuronal and glial lineages, respectively, remains to be established.

Semaphorins and neuropilins Another gene family mediating growth cone collapse is the semaphorins, which comprise an increasing number of members and have been described in *Drosophila*, chicken, mouse, and human. A constitutive structural feature of these proteins is the sema domain of 500 amino acids that is shared by all family members. Semaphorin-dependent growth cone collapse is mediated by the plexin receptors, which can be categorized into subtypes A–D. In some cases, the plexins interact in the membrane in *cis* with the neuropilins NP1 or NP2 that act as coreceptors in these cases. The semaphorin IIIa dimer, for example, interacts with plexin A1 and neuropilin 1 to launch growth cone collapse using a signal transduction pathway that

Table 2 Gene families involved in repulsion and inhibition

Inhibitory molecules	*Features*	*Functions*
Semaphorins	Characterized by the semaphorin domain; seven structural classes reported in vertebrates and invertebrates	Sema3 molecules inhibit axon growth from several classes of neurons.
Plexins and neuropilins	Transmembrane semaphorin receptors of the growth cone	Detect and mediate Sema-dependent axon growth and guidance.
Ephrins	Family of genes that is divided into the GPI-linked ephrin A and the transmembrane located ephrin B molecules	Ephrins are the long sought-after ligands of the Eph kinases, originally described as orphan receptors.
Eph kinases	Tyrosine kinases located in the membranes of growth cones; one distinguishes Eph-A and Eph-B kinases; Eph kinases activated by interaction with complementary ephrin-A or -B ligands	In the CNS, activation of Eph kinases by ephrins in certain (but not all) combinations leads to growth cone collapse. Eph kinases and ephrins construct complementary gradients in the nervous system that encode positional information, for example, in the visual system.
Nogo proteins	Nogo proteins A–C have been described; they belong to the inhibitory components of myelin that are involved in the prevention of regeneration (in addition to MAG and OMgp)	Nogo proteins activate the Nogo receptor NgR that interacts with the low-affinity NGF receptor p75. Activation of these receptors results in growth cone collapse.
Chondroitinsulfate proteoglycans (CSPGs)	Composed of a core glycoprotein and at least one chondroitin sulfate carbohydrate chain; CSPGs include the lectican family, NG2, and phosphacan	CSPGs are found enriched in glial boundaries during development and in glial scars of the lesioned CNS. They are associated with axon growth inhibition. Axon growth stimulatory chondroitin sulfates have also been described.
Tenascin-C and tenascin-R	Composed of egf-type repeats and fibronectin-type 3 modules and represent the best-characterized glycoproteins of the neural extracellular matrix; tenascins assemble to multimers and are distributed in discrete patterns	Tenascins are inhibitory in certain situations and are capable of forming boundaries *in vivo* and *in vitro*. On the other hand, stimulation of axonal growth has also been reported. These molecules are thus best suited to mediate the ambivalent influences of astrocytes on axon growth and guidance.

involves small GTP-binding proteins, such as Rho and Rac1. The contributions of these inhibitory molecules in the context of neuron interactions with astrocytes remain to be worked out in detail.

Nogo and myelin-based inhibitors By comparison, more is known about the myelin sheath-derived inducers of growth cone collapse that are believed to underlie myelin-dependent inhibition of regeneration. These include the Nogo glycoproteins, with Nogo-A as the principal myelin-based inhibitor of growth, a member of the larger Reticulin gene family. Nogo-A contains a region called Nogo-66, a looplike structure which induces growth cone collapse by interacting with the complementary Nogo receptor NgR. NgR is GPI anchored to the growth cone membrane and part of a receptor complex that also contains the low-affinity NGF receptor p75. Interestingly, two other myelin components inhibitory to axon growth have been detected – the Ig superfamily member MAG (myelin-associated glycoprotein) and OMGP (oligodendrocyte–myelin glycoprotein). Both are also able to activate the Nogo receptor complex, which suggests a common downstream pathway of myelin-dependent inhibition. This pathway involves the downstream activation of RhoA family-type proteins that play a major role in growth cone collapse (**Table 3**).

Glial Boundaries between Developmental Compartments of the Central Nervous System

Rhombomeres of the hindbrain The brain stem is subdivided into rhombomeres, and the paraxial mesoderm is substructured into somites. The rhombomeres in the hindbrain emerge beginning with the six somite (6s) stage and can be identified as a series of eight swellings along the neuraxis which are separated by grooves at stage 16s. The axons of motor neurons located in pairs of rhombomeres are destined to innervate specific branchial arches and leave the even but not the odd-numbered rhombomeres at defined exit points. The motor nuclei of the cranial nerves and their motor tracts are confined to subsets of rhombomeres along the rostrocaudal axis. Cells of neighboring compartments do not mingle, and intercellular communication by gap junctions is limited to cells within a given compartment. The apparent segmentation of the hindbrain correlates with

Table 3 Gene families involved in choice decisions at the midline

Floor plate-based signals	*Growth cone-based receptors*	*Functions and comments*
Netrin-1, -2 (laminin gene SF; unc-6 in the nematode)	Dcc ('deleted in colorectal cancer', IgSF)	Netrin attracts commissural neurons toward the floor plate. The gene products unc-5 and unc-6 were found to control circumferential axon growth in the nematode.
Sonic hedgehog (Shh)	Patched and smoothened	Shh is a morphogen with fate-instructing functions in early development (e.g., induces the floor plate). It serves as chemoattractant at this stage.
Slit-1, -2, -3 (extracellular matrix components)	Robo-1, -2, -3 (IgSF members)	Slit builds a chemorepellent gradient to drive axons away from the floor plate after crossing. When the robo receptor is mutated, the axons circle incessantly back and forth across the midline ('roundabout' mutation in *Drosophila*). There are several *slit* and *robo* genes in mammals.
Sema3a	Neuropilin and plexin-A, -B, and -C receptors	Acts as chemorepellent that prevents return to the floor plate after crossing.
Ephrin B2	EphB kinases	The Eph–ephrin connection mediates membrane-based reciprocal avoidance behaviors in certain combinations.
NrCAM (IgSF member)	Tag-1/axonin-1 (IgSF member)	The NrCAM–Tag-1 interactions are required to enable the growth cone to cross the midline.

distinct expression patterns of transcription factors of the homeobox gene family that code positional identities in *Drosophila* and vertebrates. The limits of expression of these genes in the mouse spinal cord progress in a caudorostral direction and are strictly coherent with boundaries of rhombomeric pairs in ascending order. Also, other genes are expressed in register with rhombomere boundaries, such as various members of the Eph tyrosine kinase and complementary ephrin ligand gene families and zinc finger transcriptional activators such as *Krox-20*. The Hox code can be altered by treatment with retinoic acid, which caudalizes the spinal cord, including cranial nerve nuclei. These findings concur to qualify the rhombomeres as true compartments with homeotic identity.

The prosomeric model of the prosencephalon Analogous patterns of transcription factors at more rostral positions of the neuraxis have prompted the view that compartments may also exist in diencephalic or telencephalic structures. A detailed analysis of the distribution of numerous transcription factors, such as Otx2, Emx1, Emx2, and Gbx2, in the developing rostral CNS has founded the view that one compartment yields the mesencephalon and six prosomeres preconfigure the diencephalon and the telencephalon with its cortical hemispheres. A more detailed assessment of compartments of the CNS is beyond the scope of this article and is available elsewhere. Interestingly, early axonal pathways often follow the boundaries between transcription factor expression territories. This is also true for interrhombomeric boundaries, where the extension of axonal fiber pathways has been documented.

Specialized cells in the compartment interface Closer inspection of the cellular populations in boundary regions revealed a reduced interkinetic nuclear migration of neuroepithelial cells and an increased intercellular space. IgCAMs such as NCAM that are expressed in rhombomeres may mediate intercellular adhesion in this context. The boundary cells exhibit an unusual fan-shaped array and abut with their endfeet on both the pial and the ventricular surfaces. In zebra fish, a so-called glial curtain has been proposed to separate individual rhombomeres. It has been envisaged that the specialized boundary cells construct a privileged pathway for outgrowing axons. An instructive influence of glial cordones on axon pathways is supported by transplantation experiments in which fiber tracts follow ectopic boundaries. The mechanism and molecular bases of hypothesized boundary functions for axon guidance are unknown, and differential adhesion or local inhibition concepts are being discussed.

Midline Glia as a Signaling Center for Growth Cone Guidance Decisions

Genetic control of the midline in *Drosophila* In *Drosophila*, the midline of the ladderlike nervous system involves the so-called midline glia, which is controlled by a battery of genes. Upstream, *single-minded* regulates the fate determination of midline cells in general. Glial cells emerge under the regulatory influence of *spitz*, a homolog of transforming growth factor-α, and the transcription factor *pointed*, which intervenes in the expression of *gcm* (glial cell missing), an additional gene required for the generation of glial cells in *Drosophila*. Homologs of *gcm* have been described, but they seem to play a different role in vertebrates. Elimination of the midline glia prevents the separation of commissural fibers and eliminates the longitudinal connections of the characteristic ladderlike nervous system. The commissural fibers of the system begin to extend between three pairs of midline glia cells and the pair of MP1 neurons. Separation of the commissural fiber systems involves migration of the midline glia in the correct direction. A genetic screen has led to the identification of a number of additional genes which control this migratory behavior, one of which, called *klötzchen*, seems to implicate the spectrin cytoskeleton of midline glia. In all cases, deficits in midline glia migration lead to errors in the separation of the connecting commissural fiber systems.

The midline glia in the optic nerve chiasm A comparable class of boundaries associated with glial cell types has been described in the midline of the developing nervous systems of vertebrates. Thus, glial cells separate the left and right axonal projection systems at the optic chiasm. Advancing axons are attracted by netrin, interact with this glial population, and are directed either to the ipsilateral or to the contralateral cortex, as required in the context of binocular vision. Several candidate molecules have been considered to be involved in the directional growth cone choice. Among these are the glycoprotein L1, the hyaluronate receptor CD44, and chondroitin sulfate proteoglycan(s), as visualized by the expression of chondroitin sulfate epitopes in boundary glia. Chondroitin sulfate proteoglycans (CSPGs) are thought to indicate inhibitory territories in the developing nervous system that are also secured by ephrins.

The floor plate of the vertebrate neural tube Analogous situations are also observed in the floor and roof plates of the developing spinal cord (**Figure 1**). In the latter case, the dorsal midline cells express keratan sulfate and chondroitin sulfate epitopes linked to proteoglycan core proteins. Thereby, the CSPGs outline a boundary region that is not traversed by the commissural axons. When chondroitin/keratan sulfate-expressing proteoglycans are exposed as patterned substrates alternating with the glycoprotein laminin-1, various cultured neurons and their processes avoid the proteoglycan-rich regions and grow out on the laminin-1 substrates.

In the ventral half of the spinal cord, commissural axons of the dorsal horn are attracted toward the floor plate that releases the chemoattractant netrin-1. Netrin-1 binds to the growth cone-based receptor deleted in colorectal cancer (*dcc*) of the Ig superfamily. This mechanism is highly conserved because it has already evolved in the nematode *Caenorhabditis elegans*, in which unc-5 guides circumferential axons via the unc-6 receptor – homologs of netrin-1 and dcc, respectively. Sonic hedgehog (*Shh*) represents a second chemoattractant that acts in a similar direction. When the axons reach the midline, the commissural fibers interact with the floor plate via an adhesive mechanism that involves the Ig superfamily members NrCAM as expressed in the midline and TAG-1/axonin-1 as heterophilic ligand in the growth cone membrane. The GPI-linked axonin-1/TAG-1 protein is downregulated after the crossing has occurred and L1CAM appears on the longitudinally oriented fibers. Concerted interactions of these Ig superfamily members thus seem to be required to regulate the crossing step. Subsequently, the axons lose responsiveness to the attractant netrin-1 and develop

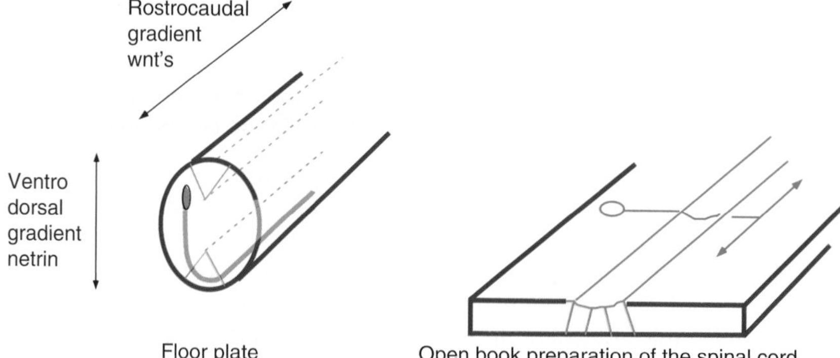

Figure 1 Axon guidance at the midline of the developing spinal cord. The figure shows the developing spinal cord of the rat at approximately embryonic day 13. At this stage, axons emanating from the commissural neurons migrate toward the ventral midline. There, axons contact the floor plate, midline glia (outlined in green) that defines an axon growth choice point. The growth cone migration and its behavior at the midline can be monitored *in vitro* with the so-called 'open book' preparation (right). Many genes are involved in growth cone guidance at the midline, some of which are listed in **Table 3**.

a sensitivity to the chemorepulsive signaling protein slit. This glycoprotein is released by midline glia and induces growth cone repulsion mediated by the Ig superfamily receptor roundabout (*robo*). *Robo* as well as *slit* were originally discovered in *Drosophila*, in which axons fail to stay on the ipsilateral side after crossing the midline, leading to the characteristic picture of incessantly circling axons. Several *slit* and *robo* genes have since been identified in mammals. A further inhibitory system is provided by semaphorins of the sema3 class and the complementary plexin and neuropilin receptor complexes NP1 and NP2, which mediate growth cone collapse. In addition, evidence has been presented that Eph kinase–ephrin interactions are also involved in the midline choice decision of the growing corticospinal projection. Thus, ephrin B3 is a constituent of the midline and prevents recrossing of these axons when they enter the gray matter after having migrated to the contralateral side of the spinal cord. Subsequently, the wnt proteins are involved in stimulating neurite outgrowth and regulating anterior–posterior guidance of commissural axons. These proteins utilize at least three signal transduction pathways and are implicated in many aspects of axon growth and interneuronal wiring (**Table 3**).

The Ambivalent Properties of Astrocytes Are Reflected by Extracellular Matrix Components That Embody Both Axon Inhibitory and Stimulatory Properties

Glycoproteins of the Extracellular Matrix

The pericellular space is organized by a superstructure of interacting glycoproteins and proteoglycans of the ECM. Astrocytes *in vitro* produce many of the ECM glycoproteins originally described in other tissues, namely fibronectin, laminin-1, vitronectin, thrombospondin, and tenascin-C. Laminin-1 is a functional component of astroglial endfeet in limiting membranes, for example, in the developing retina, and forms an excellent growth substrate for axon extension of many neuronal cell types. The chemodiffusible chemoattractants netrin-1 and netrin-2, which guide outgrowing commissural axons toward the floor plate of the midline in the spinal cord, are structurally related genes. Fibronectin has been found in association with blood vessels, a structure in which astrocytes contribute to the formation of the blood–brain barrier which isolates the CNS from the bloodstream.

The Tenascin Gene Family

Tenascin-C is transiently expressed by immature astrocytes in the developing CNS in which its distribution follows functional neuroanatomical subdivisions. For example, in the barrel field it delineates the emerging barrel field structure in layer IV. The glycoproteins of the tenascin gene family are characterized by structural motifs common to tenascin-C (Tnc), tenascin-R (Tnr), tenascin-X (Tnx), tenascin-Y (Tny), and tenascin-W (Tnw). A cysteine-rich N-terminus is followed by a series of egf-type repeats, fibronectin type III modules, and, finally, homologies to fibrinogen-β and -γ at the C-terminus. Different from this organization, a tenascin-like pair-rule gene in *Drosophila* contains the characteristic egf-type repeats but is devoid of other structural elements. The egf-type repeats of tenascins display a characteristic arrangement of cysteines that has also been found in the ECM protein reelin and is distinct from the one in the egf-type repeat modules of Notch or the laminin genes. The N-terminus links monomers in Tnr to trimers and in Tnc to hexamers under nonreducing conditions. As viewed with the electron microscope, the hexamer appears as a typical six-armed structure that has been designated hexabrachion and seems to be conserved during evolution. Two isoforms which are distinguished by one FNIII motif have been described in Tnr, a gene which is expressed in oligodendrocytes at later stages of development.

Multiple Isoforms of Tnc

Tnc possesses an alternative splice site between the fifth and the sixth FNIII module, where as many as six and nine additional FNIII repeats can be inserted in mouse and human Tnc, respectively. Up to 30 alternatively spliced variants in this region of Tnc have been revealed, approximately 50% of the theoretically possible 64 isoforms assuming a binary combinatorial code. In the human, theoretically 512 or 2^9 splice variants are conceivable on the ground of 9 potential alternatively spliced modules. Therefore, Tnc seems suited to specify pericellular microenvironments or to distinguish glial sublineages. Tnc is associated with numerous pathological conditions, including glial tumors.

Tnc Is a Multimodular and Multifunctional ECM Component

The characteristic hallmark of Tnc is its antiadhesive property for many cell types. In situations that expose a Tnc-rich environment alternating with laminin-1, the glycoprotein deflects growth cones and neuronal cell bodies, consistent with its boundary-like distribution in several CNS territories. On the other hand, homogeneous substrates containing Tnc promote neurite outgrowth of most neuronal cell types studied to date. Neurite outgrowth promotion could be mapped to the distal splice site that surrounds

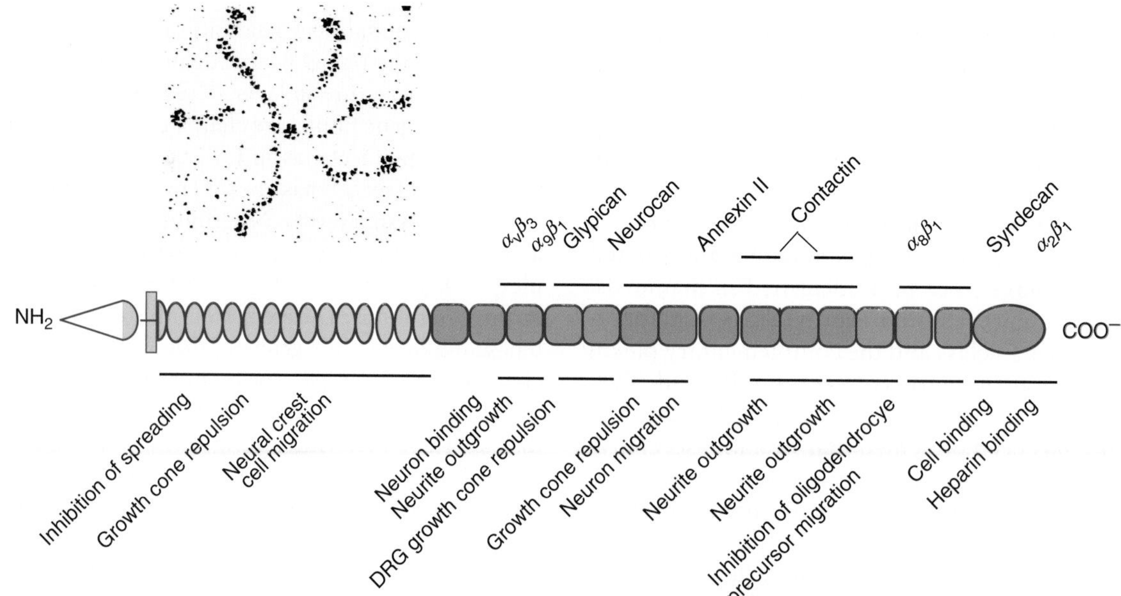

Figure 2 Structure–function relationship of tenascin-C. The glycoprotein of the extracellular matrix (ECM) tenascin-C is characterized by a modular structure with fibronectin type III domains (boxes), some of which are alternatively spliced (pink boxes). On the N-terminal end, the sequence comprises egf-type repeats (yellow elipses). Many functions have been ascribed to tenascin-C that are mediated by distinct receptors and/or ligands. The corresponding binding sites are underlined and complementary receptors are indicated. $\alpha_n\beta_m$ pairs of letters indicate specific integrin heterodimers.

cassette D. A site interfering with the motility of oligodendrocyte precursors was located to the FNIII module pair TNfn78, whereas the antiadhesive qualities could be mapped to other domains (**Figure 2**). Several Tnc receptors have been described, including the Ig superfamily member contactin, which mediates Tnc-dependent stimulation of neurite outgrowth, and the integrins $\alpha_v\beta_3$, $\alpha_7\beta_1$, $\alpha_8\beta_1$, and $\alpha_9\beta_1$. In the ECM, Tnc interacts with the proteoglycans phosphacan and neurocan. A number of studies on the Tnc −/− mutants suggest altered responses to stress, behavioral modifications, and deficits in the stem cell compartments of several organs, including the brain, during development, in the adult, and in response to lesion. These observations and the strong association of Tnc with human pathology in CNS cancer, hippocampal sclerosis, and various types of lesion warrant further studies of this versatile multifunctional glycoprotein. The antiadhesive glycoproteins of the CNS also comprise the thrombospondins, which have been found to mediate an astrocyte-derived signal for synaptic maturation.

Proteoglycans of the Extracellular Matrix

Key features of proteoglycans and heparan sulfate proteoglycans Proteoglycans represent the second class of ECM components expressed in the CNS. These components are defined as glycoproteins that carry at least one covalently linked glycosaminoglycan chain. One can distinguish HSPGs, which are mainly membrane bound, from CSPGs and keratan sulfate proteoglycans (KSPGs) of the nervous system, which are preferentially recovered from saline detergent-free extracts.

Two important HSPG subfamilies are the membrane-based syndecans and the GPI-linked glypicans. Central roles of the HSPGs may reside in the support of signaling processes in the context of tissue development. Thus, bFGF binds to a specific motif in heparan sulfate carbohydrate chains that expose the factor to its (the FGF receptor is a membrane bound tyrosine kinase) receptor. Similar mechanisms have been proposed for wnt signaling that plays important roles in axon growth and synapse formation.

CSPGs and KSPGs of the central nervous system With regard to CSPGs, the members of the lectican family – brevican, neurocan, versican, and aggrecan – have been identified in the CNS. These CSPGs possess a binding site for hyaluronic acid, a lectin-type sequence, and further distinct structural motifs. Versican is expressed by mature oligodendrocytes, whereas neurocan and aggrecan have been detected in neurons. The core glycoproteins may carry the HNK-1 epitope, a carbohydrate structure also expressed by neural recognition molecules, or other N-linked carbohydrates, for example, of the Lewis X-type, which are recognized by specific monoclonal antibodies. Further CSPGs that have been described in the CNS with the help of specific monoclonal antibodies include CAT 301 and NG2, a marker of oligodendrocytes. CAT 301 upregulation during early embryonic development of the spinal cord depends on activation of the NMDA receptor, suggesting a role in plasticity. The possible roles of CSPGs in synaptic reorganization have been highlighted by the finding that injection of chondroitinase ABC, which degrades the particular glycosaminoglycans of CSPGs, restores plasticity of the adult visual cortex.

Inhibition of axon outgrowth by CSPGs On the functional level, many studies support the view that CSPGs inhibit axon outgrowth in an otherwise supportive environment. Thus, DRG axons extend profusely on a laminin-1-coated substrate but strictly avoid territories that have been replenished with CSPG preparations. Comparable findings have been obtained with defined CSPGs such as neurocan and versican and a variety of cell types, including neural crest, which led to the conclusion that CSPGs play a central role in axon growth inhibition and guidance. These observations motivated the analysis of CSPGs in the context of regeneration inhibition. Numerous studies concur that CSPGs are strongly upregulated by reactive astrocytes in a broad spectrum of lesion paradigms. Therefore, these components are considered an important inhibitory compartment that plays a pivotal role in the lack of regeneration of the CNS. This interpretation has gained support by the observation that injection of chondroitinase ABC into the lesioned spinal cord enhances plasticity and reactive sprouting and hence entails some improvement of the afferent sensory function. Several possibilities are conceivable to explain the mechanistic aspects of the inhibitory properties of CSPGs, and it is plausible that these components bind growth cone collapse, inducing molecules such as semaphorins to particular structural sequences in the glycosaminoglycan chains. On the other hand, CSPGs have been found associated with axon growth and regeneration in the peripheral nerve, which emphasizes that overall matrix composition and the lineage and age of the neurons involved need to be considered.

Phosphacan and Receptor Tyrosine Phosphatases of the Central Nervous System

The CSPG phosphacan/DSD-1-PG DSD-1-PG/phosphacan is one of the more abundant soluble CSPGs in postnatal mouse brain and is homologous to the secreted proteoglycan phosphacan from rat tissues. The GAGs of phosphacan are composed of chondroitin sulfate (CS)-A and CS-C motifs, a keratan sulfate chain, and the DSD-1-epitope. This unique structure was discovered with the MAb 473HD, requires the sulfation of the carbohydrate backbone, and contains CS-D dimers and dermatansulfate. The DSD-1-epitope displays neurite outgrowth promoting properties, which possibly involves its capacity to bind pleitrophin. Phosphacan is a splice variant and corresponds to the complete extracellular region of the largest isoform of the transmembrane receptor protein tyrosine phosphatase-β (RPTP-ζ/β). RPTP-ζ/β proteins occur as large or short receptors which possess a transmembrane domain and two cytoplasmic tyrosine phosphatase modules. The additional phosphacan short isoform (PSI) that corresponds to the N-terminal sequence has been described in the mouse, and several isoform variants have been found in *Xenopus*.

Receptor protein tyrosine phosphatases in neuron–glia interactions The different isoforms of RPTP-ζ/β are developmentally regulated, and astrocytes from various parts of the CNS express the short RPTP-ζ/β receptor. RPTP-$\zeta/\beta_{\text{long}}$ is expressed in the ventricular zone of the developing and the subventricular zone of the adult CNS and also by oligodendrocyte precursors. Neuronal expression has also been observed which may partially be due to PSI that is strongly expressed by cortical neurons. The spatiotemporal expression patterns of RPTP-ζ/β isoforms during development, maintenance, and pathology of the CNS have been correlated with cell–cell signaling, cellular proliferation, migration, differentiation, axon outgrowth, synaptogenesis, synaptic function, and tissue regeneration. Based on the prominent glial expression of phosphacan and RPTP-ζ/β receptors, the proteins have been considered as possible mediators of neuron–glia interactions. In the adult CNS, phosphacan occurs in the perineuronal nets of parvalbumin-expressing neurons, surrounding axon terminals and glial endfeet but not the synaptic clefts. It has been hypothesized that CSPGs associate with hyaluronic acid in these structures to build a neural ECM comparable to that in connective tissue. CSPGs in the perineuronal ECM may constitute an important element in limiting synaptic plasticity. On the functional plane, phosphacan interacts with the Ig superfamily members contactin/F3/F11, axonin-1/TAG-1, NrCAM, and NgCAM and hence might intervene in both homophilic and heterophilic interactions. Therefore, it is plausible to assume that IgSF constituents represent neuronal ligands of RPTP-ζ/β receptors expressed in glial membranes and serve as molecular mediators at the interface between these two cellular lineages. Interestingly, both RPTP-ζ/β

and the IgCAMs are linked to signal transduction pathways and hence will act back on the expressing cells in the context of reciprocal signaling mechanisms. In particular, the phosphotyrosine–phosphatase modules of RPTP-ζ/β may antagonize tyrosine kinase-based activities. ECM ligands of phosphacan include tenascin-C and tenascin-R. The integration into ECM superstructures might explain to some extent why the elimination of the phosphacan gene does not yield serious developmental deficits in mice.

See also: Axon Guidance: Building Pathways with Molecular Cues in Vertebrate Sensory Systems; Axonal Regeneration: Role of Growth and Guidance Cues; Axonal Regeneration: Role of the Extracellular Matrix and the Glial Scar; Axonal Pathfinding: Extracellular Matrix Role; Cadherins and Synapse Organization; Cell Adhesion Molecules at Synapses; Glia and Stroke; Glial Responses to Injury; Glial Responses to Virus Infection.

Further Reading

Bandtlow CE and Zimmermann DR (2000) Proteoglycans in the developing brain: New conceptual insights for old proteins. *Physiological Reviews* 80: 1267–1290.

Brembeck FH, Rosario M, and Birchmeier W (2006) Balancing cell adhesion and Wnt signaling, the key role of beta-catenin. *Current Opinion in Genetics & Development* 16: 51–59.

Carulli D, Laabs T, Geller HM, and Fawcett JW (2005) Chondroitin sulfate proteoglycans in neural development and regeneration. *Current Opinion in Neurobiology* 15: 116–120.

Ciani L and Salinas PC (2005) WNTs in the vertebrate nervous system: From patterning to neuronal connectivity. *Nature Reviews Neuroscience* 6: 351–362.

Doetsch F (2003) The glial identity of neural stem cells. *Nature Neuroscience* 6: 1127–1134.

Faissner A, Heck N, Dobbertin A, and Garwood J (2006) DSD-1-Proteoglycan/Phosphacan and receptor protein tyrosine phosphatase-beta isoforms during development and regeneration of neural tisssues. *Advances in Experimental Medicine and Biology* 557: 25–53.

Gotz M and Huttner WB (2005) The cell biology of neurogenesis. *Nature Reviews Molecular Cell Biology* 6: 777–788.

Granderath S and Klambt C (1999) Glia development in the embryonic CNS of *Drosophila. Current Opinion in Neurobiology* 9: 531–536.

Joester A and Faissner A (2001) The structure and function of tenascins in the nervous system. *Matrix Biology* 20: 13–22.

Johnson KG, Ghose A, Epstein E, Lincecum J, O'Connor MB, and Van Vactor D (2004) Axonal heparan sulfate proteoglycans regulate the distribution and efficiency of the repellent slit during midline axon guidance. *Current Biology* 14: 499–504.

Kiecker C and Lumsden A (2005) Compartments and their boundaries in vertebrate brain development. *Nature Reviews Neuroscience* 6: 553–564.

Kleene R and Schachner M (2004) Glycans and neural cell interactions. *Nature Reviews Neuroscience* 5: 195–208.

Kullander K and Klein R (2002) Mechanisms and functions of Eph and ephrin signalling. *Nature Reviews Molecular Cell Biology* 3: 475–486.

Lemke G (2001) Glial control of neuronal development. *Annual Review of Neuroscience* 24: 87–105.

Lemke G and Reber M (2005) Retinotectal mapping: New insights from molecular genetics. *Annual Review of Cell and Developmental Biology* 21: 551–580.

Levine JM, Reynolds R, and Fawcett JW (2001) The oligodendrocyte precursor cell in health and disease. *Trends in Neurosciences* 24: 39–47.

Liu BP and Strittmatter SM (2001) Semaphorin-mediated axonal guidance via Rho-related G proteins. *Current Opinion in Cell Biology* 13: 619–626.

Orend G (2005) Potential oncogenic action of tenascin-C in tumorigenesis. *International Journal of Biochemistry & Cell Biology* 37: 1066–1083.

Rhodes KE and Fawcett JW (2004) Chondroitin sulphate proteoglycans: Preventing plasticity or protecting the CNS? *Journal of Anatomy* 204: 33–48.

Sandvig A, Berry M, Barrett LB, Butt A, and Logan A (2004) Myelin-, reactive glia-, and scar-derived CNS axon growth inhibitors: Expression, receptor signaling, and correlation with axon regeneration. *Glia* 46: 225–251.

Yamaguchi Y (2001) Heparan sulfate proteoglycans in the nervous system: Their diverse roles in neurogenesis, axon guidance, and synaptogenesis. *Seminars in Cell and Developmental Biology* 12: 99–106.

Axon Guidance: Building Pathways with Molecular Cues in Vertebrate Sensory Systems

F Wang, Duke University Medical Center, Durham, NC, USA

Introduction

The functions and signaling mechanisms of many important axon guidance molecules may be described individually, but here we attempt to take a more systematic view to describe how various chemotropic guidance cues and adhesion molecules work together to help build precise connections in sensory systems, from the periphery to the brain. We focus on the cell surface/secreted molecules that regulate important pathfinding events in the vertebrates, especially in the mammalian sensory systems. Among the sensory systems, the pathfinding mechanisms are best understood in the visual and the olfactory systems, and to a lesser extent in the somatosensory system. The auditory and the gustatory systems, however, have not been very well characterized.

Building the Visual System

In vertebrate animals, visual stimuli are transmitted by retinal ganglion cells (RGCs) from the retina to subcortical relay stations and eventually to the primary visual cortex. The underlying pathway is formed during development by axons navigating through complex environments, involving decisions at multiple steps, in a remarkably stereotypical and precise manner.

Pathfinding and Topographic Mapping by RGCs

RGC axons project from the retina to the superior colliculus (SC) in mammals, or to the optic tectum (OT) in amphibians and avians, before reaching the lateral geniculate nucleus (LGN) of the thalamus. Significant progress has been made in revealing how RGC axons are guided to make several important choices: where to exit the retina, where to form the optic nerve, to cross or not to cross at the optic chiasm, and how to map the visual field topographically onto the SC and the LGN.

Finding and exiting the optic disk The first step in building the visual pathway is for the RGC axons to exit the retina and project into the optic nerve. RGC axons travel in radial routes inwardly toward the fovea of the retina, where the optic disk is located. Studies in zebra fish have revealed that RGCs express the chemokine receptor CXCR4, while the optic disk expresses the ligand stromal cell-derived factor-1 (SDF-1), and that SDF-1 attracts RGCs toward the optic disk (**Figure 1(a)**). In addition, cell adhesion molecules and bone morphogenetic protein (BMP) receptor type I, chondroitin sulfate proteoglycan (CSPG), and Sonic hedgehog (SHH) molecules have also been implicated in directing this process. Once the RGC axons reach the optic disk, they require netrin-1 to project into the optic nerve head (**Figure 1(b)**). In netrin-1 and DCC (deleted in colorectal cancer) mutant mouse embryos, many RGC axons fail to exit the eye and instead they project aberrantly into other regions of the retina, resulting in the hypoplasia of the optic nerve.

Forming the optic nerve and growing toward the optic chiasm After the RGC axons exit the eye, their growth along the optic nerve is guided in part by Slit proteins (Slit1 and Slit2). Slit1 and Slit2 are repellent molecules for RGC axons and are expressed in overlapping and complementary domains surrounding the optic nerve and adjacent to the optic chiasm (**Figures 1(b)** and **1(c)**). In this way, Slit1 and Slit2 help establish repulsive barriers for a narrow corridor that channels retinal axons toward the optic chiasm. In the absence of both Slit1 and Slit2, retinal axons are defasciculated and project ectopically into the preoptic area. Slits also help to prevent RGC axons from growing back toward the other eye after they reach the chiasm. In zebra fish, mutations in the Slit receptor gene *Robo2* lead to random projections of RGC axons and complete disruption of optic nerve formation. The responses and sensitivities of RGC axons to Slit proteins appear to be regulated by heparan sulfotransferases.

Crossing or not crossing at the optic chiasm At the optic chiasm, RGC axons have to decide whether to stay on the ipsilateral side or to cross the midline to the contralateral side. The percentage of RGC axons crossing at the optic chiasm differs among different species. Fish and birds have no binocular vision and all of their axons cross the midline. In higher mammals, RGCs located in the nasal retina project their axons to the contralateral side, while axons from RGCs of the temporal retina stay ipsilaterally. In mammalian species with less binocular vision, such as mice, only a small subpopulation of RGCs in the ventrotemporal (VT) region of the retina will project their axons ipsilaterally. Studies in *Xenopus* and mice

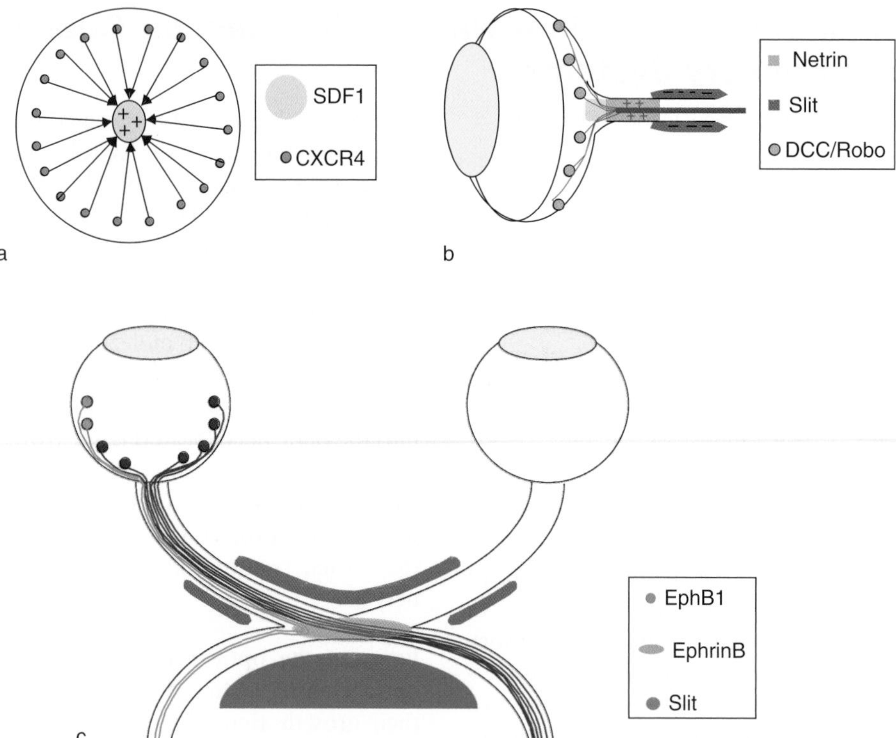

Figure 1 Molecules guide retinal ganglion cells at the optic disk, optic nerve, and optic chiasm. (a) Expressed stromal cell-derived factor-1 (SDF-1) attracts the chemokine (CXCR4)-expressing retinal ganglion cells toward the optic disk. (b) Netrin-1 attracts retinal ganglion cells into the optic nerve. Slit proteins prevent retinal ganglion cell axons from straying away. (c) Slits channel retinal ganglion cell axons toward the optic chiasm. EphrinB prevents EphB1-expressing retinal ganglion cells from crossing the optic chiasm.

have demonstrated that the repulsive ligand ephrinB is expressed at the optic chiasm. EphrinB prevents the crossing of ipsilaterally projecting RGC axons which express the receptor EphB1 (**Figure 1(c)**). In EphB1 null mice, ipsilateral projections are dramatically reduced and almost all RGC axons cross the midline. The expression of EphB1 in these noncrossing RGCs is likely to be regulated by the zinc-finger transcription factor Zic2 and the Lim-homeodomain factor Islet2.

Topographic mapping in the optic tectum/superior colliculus Leaving the optic chiasm, RGC axons project to several subcortical regions. The most prominent midbrain target is the optic tectum of fish, amphibians, and avians, or the SC of mammals. In the OT/SC, RGC axons form an ordered map of the visual field, the positions on the retina being topographically mapped onto the OT/SC along two orthogonally oriented axes: the temporal–nasal (T–N) axis of the retina maps along the anterior–posterior (A–P) axis of the OT/SC (**Figure 2**, upper panel), whereas the dorsal–ventral (D–V) axis of the retina maps along the lateral–medial (L–M) axis of the OT/SC (**Figure 2**, lower panel). The T–N to A–P mapping is primarily governed by the repulsive guidance of the EphA–ephrinA interactions. The EphA expressions in RGCs have a high-to-low gradient along the T–N axis, whereas the ephrinA expressions in the OT/SC show a low-to-high gradient along the A–P axis. These two countergradients together control either the sites of RGC axon termination in fish and amphibians, or the locations of interstitial collateral sprouting in avians and mammals (**Figure 2**, upper panel). The D–V to L–M mapping is achieved by functions of two sets of guidance cues, ephrinB and Wnt. EphrinB is expressed in a high-to-low gradient along the M–L axis in the OT/SC, while EphB is expressed in a low-to-high gradient along the V–D axis in the retina. High-level EphB signaling results in repulsion, while low-level signaling leads to attraction. EphB–ephrinB signaling results in a net attraction of all RGC axons to the medial part of the OT/SC. This is balanced by Wnt protein signaling. Wnt3 causes repulsion of RGC axons through the Ryk receptor, but attraction is through the Frizzled (Fzl) receptor. Wnt3 is expressed in a high-to-low gradient along the M–L axis in the OT/SC. Ryk is expressed in a low-to-high gradient along the V–D axis in the retina, while Fzl expression is equal along the V–D axis. The net result is that Wnt3 causes RGC axons to project toward the lateral part of the OT/SC. The combination of ephrinB and Wnt signaling is required for topographic mapping of RGC axons from the retina to the OT/SC (**Figure 2**, lower panel). This projection can also be refined by

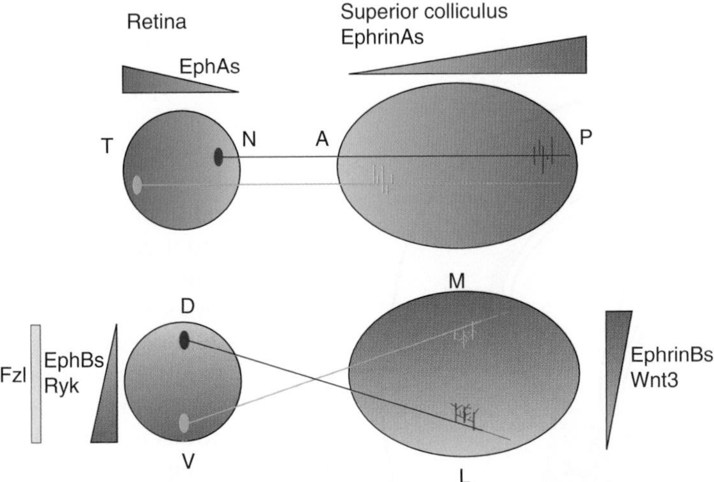

Figure 2 Topographic mapping of retinal ganglion cell projection in the superior colliculus, showing the temporal–nasal (T, N), the anterior–posterior (A, P), the dorsal–ventral (D, V), and the lateral–medial (L, M) axes. Schematic representations of graded receptors (EphA, EphB, Fzl, Ryk) and ligands (Wnt3 protein, ephrinA, ephrinB) that guide retinocollicular topographic mapping.

spontaneous correlated neural activity during a brief critical period.

Topographic mapping and segregation into eye-specific layers in LGN The lateral geniculate nucleus of the thalamus is the other main target of RGCs and is the relay station for visual input to the cortex. RGC axons again form ordered projections that map the visual field onto the LGN. Molecular mechanisms similar to those that govern the retinotopic map in the OT/SC also regulate the retinogeniculate projections. Gradients of ephrins in the LGN and countergradients of Ephs in RGCs regulate the topographic mapping in the LGN. Moreover, in mammals, input from the two eyes segregates into stereotyped eye-specific layers in the LGN. The precise pattern of segregation requires the function of EphrinAs, whereas the segregation of the eye input *per se* depends on neural activity.

From LGN to the Visual Cortex

The next connection in the visual pathway in mammals is for LGN neurons to relay the visual input that they receive from RGCs to the visual cortex. Compared with the pathfinding of RGCs, our understanding of the molecular mechanisms that guide the LGN neurons projections is quite limited. Here we summarize the current findings in this field, focusing on the guidance cues that are supported by *in vivo* evidence from mutant mice.

Getting to the cortex via thalamocortical projections LGN neurons follow precise pathways to relay information from the thalamus to the visual cortex. Their axons first project ventrally through the ventral thalamus. As they approach the dorsal border of the hypothalamus, they make a sharp turn and extend dorsolaterally through a narrow corridor, called the internal capsule (IC), into the ventral telencephalon. All thalamocortical axons (TCAs) likewise follow this path. Slit1 and Slit2 present in the hypothalamus prevent the TCAs from entering into the hypothalamus and also push TCAs away from the midline and toward the IC. Projecting into the IC requires recently identified 'corridor cells,' which express the membrane-bound form of neuregulin (Nrg1-CRD) that helps to attract and/or permit the growth of thalamic axons into the IC (**Figure 3**). Further extension of these axons to the cortex depends on the secreted form of neuregulin, containing an immunoglobulin-like (Ig) domain (Nrg1-Ig), which is expressed in the pallium and acts as a long-range chemoattractant (**Figure 3**). TCAs express the receptor Erb4 for both isoforms of neuregulin. Thalamocortical projection is severely disrupted in either the neuregulin or the Erb4 knockout mice in manners consistent with the roles of corridor cells and neuregulin in attracting TCAs. In addition, netrin-1 restricts the width of the internal capsule, and the semaphorin Sema6A is involved in the guidance of some TCAs.

Innervating and mapping in the visual cortex The molecular guidance cues that direct either the LGN axons toward the visual cortex or the subsequent axon branch projections into layer IV of the cortex are unknown at present. In terms of cellular events, axons initially accumulate and wait below the developing visual cortex in a zone called the subplate, which contains the first postmitotic neurons of the cerebral cortex, the subplate neurons. Subplate

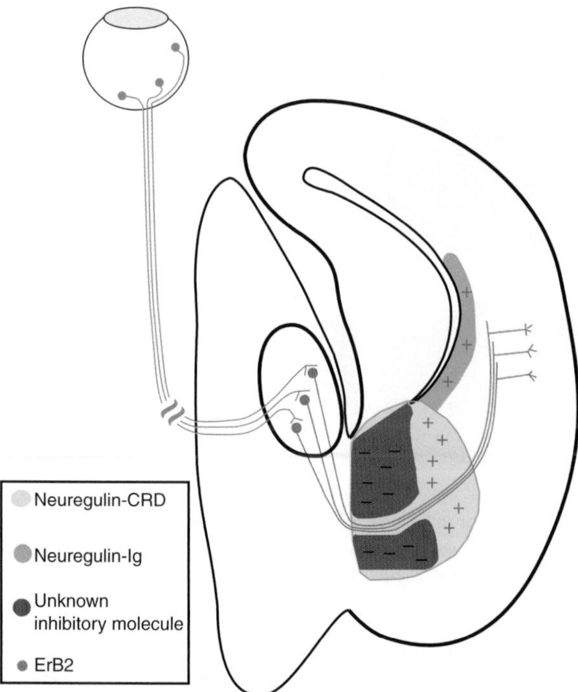

Figure 3 Thalamocortical axon projections are guided in part by neuregulin isoforms. Thalamocortical axons which express ErB2, the receptor for neuregulin, project through a narrow corridor toward cortex. The corridor is made attractive by the presence of a membrane-bound form of neuregulin (Nrg-CRD). In addition, a secreted form of neuregulin containing an immunoglobulin-like (Ig) domain (Nrg-Ig) acts as a long-range attractant for these axons.

neurons have been shown to play important roles in controlling where LGN axons wait and where they project, but the identities of the guidance molecules expressed by subplate neurons are unknown. LGN axons also form topographic maps of the visual field in the visual cortex. Here again, interactions between gradients of EphA/ephrinA direct the formation of the map, but they are not the only factors. Moreover, activity-dependent processes are likely to play a major role in shaping the visual connectivity in the cortex.

Building the Olfactory System

Odorants are detected by olfactory sensory neurons (OSNs) residing in the nose of mammals. OSNs project their axons to the olfactory bulb (OB), the first relay station in the olfactory pathway, and form a sensory map that encodes the quality of odorous chemicals. Mitral (and tufted) cells in the olfactory bulb project to multiple olfactory centers via the lateral olfactory tract (LOT). In the following sections, we review receptors and cues that have been shown to play important roles in guiding the OSN projection and the formation of LOT.

Formation of an Olfactory Sensory Map in the Olfactory Bulb

In the nasal epithelium of mammals, each OSN expresses only one odorant receptor (OR) from a family of about 1000 genes that all encode seven-transmembrane proteins. All neurons expressing the same OR, although randomly distributed in a broad circumferential zone along the dorsoventral axis in the epithelium, project convergently to a pair of glomeruli, each located at a stereotyped position on one side of the olfactory bulb. In this way, the identity of a chemical group (or groups) that is recognized by a given OR is mapped onto two anatomically conserved locations in the bulb (**Figure 4**). How do neurons expressing a specific OR find their target with such precision?

ORs play important roles in OSN targeting. Since the choice to express a given OR is tightly linked to the choice of the axonal projection site in the olfactory bulb, a model in which ORs dictate OSN axon targeting was proposed and supported by several lines of evidence. mRNAs of ORs are found in the OSN axons. OR proteins are concentrated not only in dendrites (to detect odorants) but also on axon termini in the glomeruli. Swapping either the entire coding sequence or the DNA sequence for a few amino acids of one OR with those of another resulted in altered OSN axonal convergence points in the bulb.

How might ORs influence where OSNs form the glomeruli? Two recent studies have shed a light on this puzzle. The first study strongly suggests that ORs signal through G-proteins to regulate intracellular cyclic adenosine monophosphate (cAMP) levels and that different cAMP concentrations determine the differential expressions of axon guidance molecules along the A–P axis. All OSNs express the olfactory-specific G-protein, G_{olf}, throughout their life span. In addition, they express the generic α subunit of G-proteins, G_s, at younger stages. ORs are coupled to these stimulatory G-proteins through a conserved DRY (Asp-Arg-Tyr) motif in their cytoplasmic loop after the third transmembrane domain. Mutating this G-protein binding site on a given OR, I7, caused the mutant ($I7_{DRY}$) OSNs to project diffusely in a broad domain in the olfactory bulb and failure to form glomeruli. This lack of axonal convergence can be partially reverted by expressing either a constitutively active G_s or a constitutively active cAMP response element-binding protein (CREB) in $I7_{DRY}$ neurons. Interestingly, co-expressing a dominant-negative protein kinase A (PKA) with the wild-type I7 results in these neurons projecting to a drastically anteriorly shifted position. In contrast, putting either a constitutively active G_s or a constitutively active PKA into

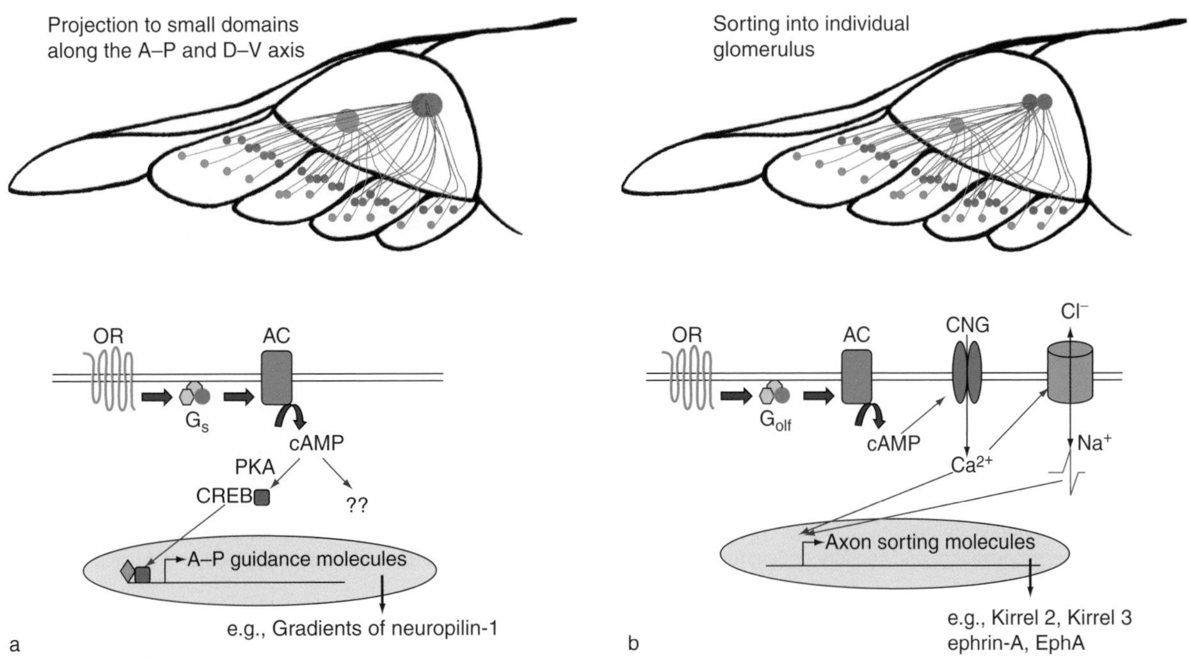

Figure 4 Formation of the olfactory sensory map is guided by odorant receptors (ORs). (a) ORs activate G-protein (G_s)signaling/cAMP signaling to guide the projection of olfactory sensory neurons along the anterior–posterior (A–P) axis and dorsal–ventral (D–V) axis to a small domain in the olfactory bulb. (b) Odorant receptors activate olfactory-specific G-protein (G_{olf}) signaling to induce calcium entry and neuronal depolarizations. These neuronal activities in turn regulate the expression of axon sorting molecules that help sort olfactory sensory axons into distinct glomeruli. AC, adenylate cyclase; CNG, cyclic nucleotide gated channel; cAMP, cyclic adenosine monophosphate; CREB, cAMP response element-binding protein; PKA, protein kinase A.

I7-OSNs, causes them to converge at a much more posterior location in the olfactory bulb. These results led the authors to propose a model in which ORs instruct OSN targeting indirectly via OR-derived cAMP signals (most likely by activating G_s) and that different amounts of cAMP regulate the expression of A–P axon guidance molecules, probably through activating CREB (**Figure 4(a)**). Indeed, the authors have found that expression of the neuropilin-1 receptor positively correlates with the cAMP level, and that neuropilin-1 is expressed in a low-to-high A–P gradient in the glomeruli layer of the olfactory bulb. However, this model also implies that each OR can determine a unique level of cAMP, a hypothesis that remains to be tested. Moreover, since cAMP has been shown to have profound effects on growth cone turning responses, it is conceivable that OR-derived cAMP also control OSN growth cones navigation locally.

Another study used transgenic mice in which the majority of OSNs express one particular OR. This has led to the discovery that ORs also control the expression of the homophilic adhesive molecules (such as the Ig domain containing Kirrel2/Kirrel3 proteins) and the repulsive molecules (such as ephrinA5/EphA5) in an OR-specific and activity-dependent manner (**Figure 4(b)**). Different ORs, through OR-evoked activity, determine the 'on' or 'off' expression of certain cell adhesion or repulsion molecules, such that different classes of OSNs have distinct combinatorial adhesive/repulsive codes. Once OSNs project to a narrow domain in the olfactory bulb, the adhesive/repulsive molecules then control the sorting of olfactory axons expressing different ORs into different glomeruli within that domain, by both homophilic and mutual repulsive interactions. At the present, it is not known how ORs control the expression of these sorting molecules through OR-dependent neuronal activity. Neural activity may be spontaneous and random, as spontaneous activity is known to be required for the formation of the olfactory map. Spontaneous and odor-evoked activities are also required for the maintenance of the olfactory sensory map.

From the Olfactory Bulb to Higher Olfactory Centers

Mitral and tufted (M/T) cells are the projection neurons in the olfactory bulb; they send axons through the lateral olfactory tract onto several structures of the olfactory cortex. M/T axons are known to wait in the LOT (for about 2 days in the mouse) before sending collateral branches to the olfactory cortex. M/T cells with dendrites that receive input from one particular glomerulus project collaterals to multiple cortical areas in a highly distributed and complex, but not random, manner. Because the organizational

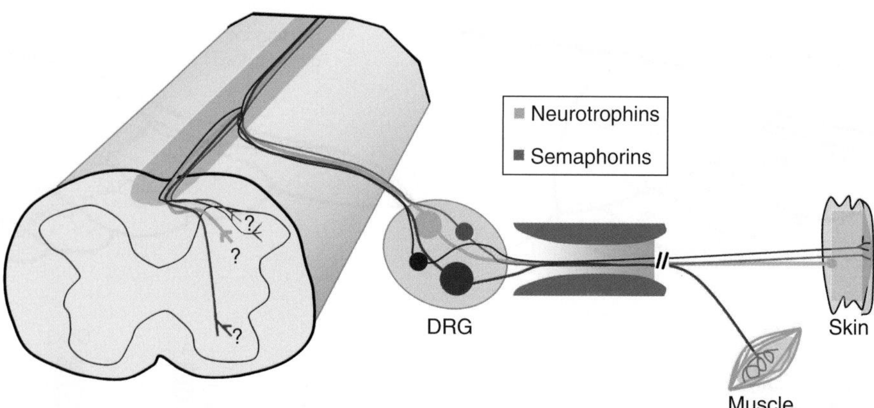

Figure 5 Peripheral projections of somatosensory neurons are guided in part by neurotrophins and semaphorins. Schematic representation of axonal projections of dorsal root ganglion (DRG) neurons. The molecules that guide the central axons into different layers of the spinal cord are unknown and are depicted as '?' in the drawing. Two classes of molecules, neurotrophins (including nerve growth factors, brain-derived neurotrophic factor, neurotrophins 3, 4, and 5, and glial-derived neuronal factors) and semaphorins, direct the peripheral axon outgrowth of sensory neurons by acting as chemoattractants and surround-repellents, respectively. Neurotrophins also promote the innervation, axon arborization, and sensory ending formation of DRG neurons in the target organs such as skin and muscle.

principle of the M/T projections in olfactory cortices is unclear, studies of this pathfinding process have largely focused on the formation of the LOT.

M/T cell projection to and along the LOT requires both short-range and long-range guidance cues. The short-ranged cues are believed to be at the surface of specialized cells called LOT cells. These cues help guide the M/T axons into the LOT and promote their elongation, although their molecular identities are unknown. M/T axonal projections in the LOT are also guided by long-range chemorepulsive molecules expressed in the septum, such that the axons avoid innervating septum regions *in vivo*. Slit1 and Slit2 together account for the repulsive ligands expressed in septum that repel M/T axons expressing Robo receptors. In Slit1 and Slit2 double-mutant embryos, M/T projection is completely disorganized, with many axons entering the septum region.

Building the Somatosensory System

The vertebrate somatic sensory system transmits to the brain information about physical stimuli that the body experiences. The stimuli can be painful, thermal, mechanical, or proprioceptive. Each of these modalities is detected by distinct types of primary sensory neurons and is processed by different central pathways. Although it is generally true that the body surface is topographically mapped onto subcortical processing centers and ultimately to the somatosensory cortex, surprisingly little is known about the molecular mechanisms that control the establishment of the somatosensory map and the neural circuit for each distinct modality.

The cell bodies of somatosensory neurons reside in a series of ganglia located outside the spinal cord (dorsal root ganglia) or the hindbrain (trigeminal ganglia), with the exception of the facial proprioceptive neurons, the cell bodies of which reside in the trigeminal mesencephalic nuclei inside the brain stem. Each sensory neuron grows two major branches stemming from a unipolar axon: a peripheral axon that innervates a specific body target, such as skin, viscera, or muscles, and a central axon that projects into the spinal cord or brain stem and forms specific synapses (**Figure 5**). Therefore, each neuron has two pathfinding tasks, one to the periphery and the other to the central nervous system (CNS), and the two processes must be coordinated such that the body is faithfully mapped onto the brain. The molecular mechanisms that guide the central axons of sensory neurons are largely unknown at present, but significant progress has been made in understanding the guidance of peripheral somatic sensory axons.

Peripheral Growth and Projection of Somatic Sensory Neurons Are Guided by Neurotrophins and Semaphorins

The growth of peripheral sensory axons follows stereotyped trajectories; this process is partly guided by neurotrophins (NTs) and semaphorins (**Figure 5**). NTs (including nerve growth factor, brain-derived neurotrophic factor, NT3, NT4, NT5, and glial-derived neuronal factor) are expressed and secreted by peripheral target tissues. They exert many biological effects on sensory neurons, acting not only as survival and differentiation factors, but also as chemoattractive factors for sensory axons. Both *in vitro* and *in vivo* experiments have demonstrated that the chemotropic effects of NTs (for promoting axon elongation and target innervation) can be separated from their chemotrophic (survival) effects. For example, nerve growth factor-soaked beads, when embedded *in vivo* in embryos, can attract sensory axons growing toward them. The peripheral axons grow in fascicles, forming nerve bundles. This fasciculation is regulated by class III secreted semaphorins, which are expressed in a pattern surrounding the axon pathways, and sensory neurons express receptors that recognize these repellents (i.e., neuropilins 1 and 2 and plexins A3 and A4). Mice deficient in any of these receptors or semaphorin3A have severely defasciculated sensory axons. Besides NTs and semaphorins, little is known about the peripheral target selection and innervation process of sensory axons.

Summary

In summary, we have briefly reviewed the axon guidance molecules that work in concert to orchestrate the pathfinding processes that lead to the formation of visual, olfactory, and somatosensory systems. A general scheme is emerging, in that sensory neurons form topographic maps in their projections to the brain, such that sensory information is transformed into distinct spatial activities for the brain to process. These maps are guided first roughly by molecular gradients, and then by refinement with neuronal activity-dependent axon sorting. En route to their targets, neurons are guided to project through stereotyped pathways. These pathways are laid down by a combination of attractive and repulsive cues that usually result in a narrow channel for axons to navigate.

See also: Axon Guidance: Morphogens as Chemoattractants and Chemorepellants; Axon Guidance: Guidance Cues and Guidepost Cells; Axon Guidance by Glia; Axonal Regeneration: Role of Growth and Guidance Cues; Fovea: Primate; Olfactory Cortex: Comparative Anatomy; Olfactory Neuron Patterning and Specification; Olfactory Coding; Olfactory Bulb Anatomy; Olfactory Bulb Physiology; Olfactory Bulb Mapping; Optic Nerve, Optic Chiasm and Optic Tracts; Optic Tectum: Development and Plasticity; Retina: An Overview; Retinal Ganglion Cells: Receptive Fields; Retinal Development: An Overview; Retinal Development: Cell Type Specification; Somatosensory Plasticity; Visual Cortex: Mapping of Functional Architecture Using Optical Imaging.

Further Reading

Deiner MS, Kennedy TE, Fazeli A, et al. (1997) Netrin-1 and DCC mediate axon guidance locally at the optic disc: Loss of function leads to optic nerve hypoplasia. *Neuron* 19: 575–589.

Imai T, Suzuki M, and Sakano H (2006) Odorant receptor-derived cAMP signals direct axonal targeting. *Science* 314: 657–661.

Li HS, Chen JH, Wu W, et al. (1999) Vertebrate Slit, a secreted ligand for the transmembrane protein Roundabout, is a repellent for olfactory bulb axons. *Cell* 96: 807–818.

Li Q, Shirabe K, Thisse C, et al. (2005) Chemokine signaling guides axons within the retina in zebrafish. *Journal of Neuroscience* 25: 1711–1717.

Lopez-Bendito G, Cautinat A, Sanchez JA, et al. (2006) Tangential neuronal migration controls axon guidance: A role for neuregulin-1 in thalamocortical axon navigation. *Cell* 125: 127–142.

McLaughlin T and O'Leary DD (2005) Molecular gradients and development of retinotopic maps. *Annual Review of Neuroscience* 28: 327–355.

Nguyen-Ba-Charvet KT, Plump AS, Tessier-Lavigne M, et al. (2002) Slit1 and Slit2 proteins control the development of the lateral olfactory tract. *Journal of Neuroscience* 22: 5473–5480.

Patel TD, Jackman A, Rice FL, et al. (2000) Development of sensory neurons in the absence of NGF/TrkA signaling *in vivo*. *Neuron* 25: 345–357.

Pfeiffenberger C, Yamada J, and Feldheim DA (2006) Ephrin-As and patterned retinal activity act together in the development of topographic maps in the primary visual system. *Journal of Neuroscience* 26: 12873–12884.

Plump AS, Erskine L, Sabatier C, et al. (2002) Slit1 and Slit2 cooperate to prevent premature midline crossing of retinal axons in the mouse visual system. *Neuron* 33: 219–232.

Schmitt AM, Shi J, Wolf AM, et al. (2006) Wnt-Ryk signalling mediates medial-lateral retinotectal topographic mapping. *Nature* 439: 31–37.

Serizawa S, Miyamichi K, Takeuchi H, et al. (2006) A neuronal identity code for the odorant receptor-specific and activity-dependent axon sorting. *Cell* 127: 1057–1069.

Wang F, Nemes A, Mendelsohn M, et al. (1998) Odorant receptors govern the formation of a precise topographic map. *Cell* 93: 47–60.

Williams SE, Mason CA, and Herrera E (2004) The optic chiasm as a midline choice point. *Current Opinion in Neurobiology* 14: 51–60.

Yaron A, Huang PH, Cheng HJ, et al. (2005) Differential requirement for plexin-A3 and-A4 in mediating responses of sensory and sympathetic neurons to distinct class 3 semaphorins. *Neuron* 45: 513–523.

Axon Guidance: Guidance Cues and Guidepost Cells

L Ma, University of Southern California, Los Angeles, CA, USA
K Shen, Stanford University, Stanford, CA, USA

Guidepost cells, also called landmark cells, were originally used to refer to specific cells in the developing insect limb that help peripheral nerve cells find their central targets. As vividly depicted by their name, these cells provide local cues to guide pioneer neurons to navigate through an unfamiliar territory in embryonic tissues. In combination with other cellular mechanisms, such as selective adhesion, they ensure the proper development of stereotypic neural connections that underlie complex behavior.

Although they have been historically studied in insects, guidepost cells have been demonstrated in many organisms as well as in both peripheral and central nerve systems. In fact, they now take a broader meaning and more appropriately refer to the intermediate targets in the course of neural development. In addition to guiding axons, they provide local guidance cues to control many other processes during nerve development. This article reviews the historical studies in insect cells, their role in midline crossing, and their emerging function in synapse formation.

Historical Studies of Guidepost Cells in Grasshoppers

The concept of 'guidepost cells' was first suggested by Australian biologist Michael Bate in his seminal study of peripheral pioneer neurons in grasshopper *Locusta migratoria*. In insects, peripheral sensory neurons are produced by a small number of epidermal cells at the body surface. Most of these cells invade their central targets by hopping along the preexisting nerve fibers that are established by pioneer neurons early in development. So how do pioneer neurons connect to the central nervous system (CNS) in the first place? To understand this, Bate followed them in the developing antenna and limbs by both light and electron microscopy. He discovered that these neurons, also with cell bodies born at the edge of each appendage, have to extend their axons through the entire length of the antenna or the limb to reach the CNS. Interestingly, through the course of their journey, not all axons follow a linear path, the shortest route. Instead, some of them, especially those in the limb, take an indirect route and change growth directions several times before reaching the central target. Based on this observation, Bate concluded that the connection path is determined by the interaction of the pioneer axons and the extrinsic cues in the limb. He suggested that these cues are provided by a group of cells, 'the signpost,' that "occur consistently at intervals along the developing appendage and seem to provide a series of stepping stones between the tip of the limb and its base."

The use of 'guidepost cells' or 'landmark cells' was adopted in subsequent studies of peripheral pioneer neurons in grasshopper limb buds, in which the simple nerve connection and the easy access for experimental manipulation provided an attractive system to further test this idea. Using newly developed staining techniques, many more cells have been found that can potentially serve as guidepost functions in the insect limb. They have some characteristic features, which were later used as major criteria to define guidepost cells in other systems: (1) they are located along the route of pioneer axonal path; (2) they are contacted by the growth cone of pioneer axons; (3) they are separated from each other but within reach of growth cone filopodia; (4) they are distinct from neighboring cells because they can be labeled by specific antisera on the surface, and they do not line up to form a continuous path for pioneer growth cones to follow; and (5) they form a special connection with axons because dye can pass from pioneer neurons to these cells. These physical and morphological attributes lead to a 'guidepost hypothesis,' which suggests that the placement of a series of distinctive cells is used to guide pioneer neurons along their trajectory from a distance.

The hypothesis was later tested in a cell ablation experiment, in which the putative guidepost cells that can be recognized immunologically were selectively killed with high-intensity light. The removal of guidepost cells in the developing grasshopper limb buds caused the pioneer neurons to wander away from their normal axonal pathway, and often resulted in the formation of multiple branches. Therefore, cells in the embryonic tissues are present to guide pioneer neurons during early development.

Guidepost Cells and Guidance Cues in Axon Guidance

Following the initial studies in the grasshopper, guidepost cells have also been described in several other invertebrates, including leech, moth, and fruit fly. Although the use of the term has been limited to these simple systems, the cellular studies of axonal pathfinding in grasshoppers have provided two important concepts that have simplified our understanding of how complex neural pathways are connected. First, they have revealed that the long and sometimes irregular axonal trajectories can be broken down into short segments, and at each segment axonal growth is regulated by local cues at choice points. These choice points contain a cluster of individual cells as the guidepost cells in insects or often a group of functionally specialized cells, which are referred to as intermediate targets in vertebrates. Second, they have shown that complex neural networks are established in two phases, by pioneer neurons that enter the axon-free environment in early development and by later born neurons that face a more complex environment filled with intertwined nerve fibers from early projecting neurons. Whereas the pioneer neurons are guided by local guidance cues, those later developing axons can follow preexisting nerve fibers by selective fasciculation, a cellular mechanism that has also been well studied in grasshoppers.

How do guidepost cells regulate the growth of axons? What are the guidance cues they provide to change the growth direction of developing axons? Ramon y Cajal proposed a century ago that diffusible chemicals in the embryonic environment could guide axonal growth over a long range, just like the chemotaxis of single motile cells. They do so by attracting growth cones, the motile sensors at the leading edge of a neuron that have exquisite sensitivity to detect chemical cues in the embryonic tissues. *In vitro* studies of cultured neurons initially demonstrated the existence of factors secreted by intermediate targets of axons. In the past 15 years, many molecules have been identified by genetic screening and biochemical purification to serve as guidance cues. They include several families of highly conserved extracellular molecules – Netrins, Semaphorins, Slits, and Ephrins – as well as classic neurotrophins and morphogens.

The biochemical and biophysical properties of guidance cues separate them into two classes, diffusible and contact mediated, although their distinction has become less clear in recent studies. Diffusible cues are secreted by cells from a long distance and can either attract axons toward the target or repel axons away from it. Contact-mediated cues are often associated with extracellular matrix or cell surface and regulate the adhesiveness of the growth environment, so they either provide a permissive surface on which axons grow or create an unfavorable region that axons tend to avoid. As shown by embryological, tissue culture, and genetic studies, these molecules are present in developing nervous system and work together to control the growth and guidance of axons during their long journey to their eventual synaptic partners. The following sections review recent studies of midline guidance in both invertebrates and vertebrates and use it as an example to demonstrate now a simple change in axon growth direction is regulated by guidance cues at a specific choice point.

Guidepost Cells at the Midline

The midline in the animal CNS is important for establishing neural pathways used for bilateral communication. In the insect nerve cord or the vertebrate spinal cord, a group of neurons called commissures extend their axons across the midline. Once crossed, the axons turn rostrally, join the longitudinal nerve tracks, and eventually synapse on the contralateral side of the brain.

Cells in the ventral midline have been found to serve as intermediate targets to guide the commissural axons to cross the midline. In flies, there are several midline glial cells that are important for controlling the cross because mutations affecting their formation greatly reduce the fidelity of the growth of commissures crossing the midline. In the vertebrate spinal cord, the main action occurs at the floor plate, a region that contains several layers of neuroepithelial cells with distinct morphological and immunological features at the ventral midline. They are derived from the initial neural fold, but their properties are induced by a molecule, Sonic hedgehog, secreted from the notochord that lies ventrally below them. In mouse mutants of Danforth's short tail or Gli2 (a zinc-finger transcription factor) in which the floor plate is missing, commissural axons grow abnormally when they reach the ventral midline. The same has been documented in zebra fish, in which mutations affecting floor plates also result in the misguidance of spinal commissural axons. Therefore, the ventral midline cells provide a unique system for understanding axon guidance at a choice point.

Attracting to the Midline

The initial characterization in grasshoppers demonstrated that a single filopodium of the sensory growth cone makes a direct contact with guidepost cells and leads the axon in the correct direction. The distance between each pair of guidepost stops is less than 100 μm, which might be sufficient for filopodia to search in space and find the guidepost cells by differential adhesion. However, for commissural neurons, especially those from vertebrates, their targets are hundreds of micrometers away. How do they know where to find the intermediate target?

The ventral midline cells appear to secrete diffusible molecules that attract commissural growth cones to cross midline. The presence of diffusible factors was first demonstrated in the study of vertebrate commissural neurons in culture. When the floor plate from chick or rodent embryos was cultured adjacent to the dorsal spinal cord explant in three-dimensional collagen gels, commissural axons in the explant grew out toward the floor plate. Subsequent biochemical purification identified a family of secreted protein Netrins that can stimulate commissural axon outgrowth and attract them toward COS cell aggregates that produce the protein. Netrins are secreted by the floor plate cells and bind DCC, a cell surface receptor that is exclusively expressed on commissural neuron axons. Genetic analyses in mice, fruit flies, as well as *Caenorhabditis elegans* have established that this ligand and receptor pair is required for midline attraction because mutations in these two genes lead to the failure of commissural axons growing toward the midline. However, interestingly, in the mouse netrin-1 mutant spinal cord, some axons still reach the floor plate. This is due to the presence of another molecule, Sonic hedgehog, also secreted from the floor plate.

How do soluble factors attract axons over a long distance? Target-derived factors have been thought to form a diffusible gradient that guides the growth cone, a theory proposed by Ramon y Cajal after the discovery of growth cones. Recent studies of netrin expression in chicks demonstrated that proteins secreted in the spinal cord do form a dorsal–ventral gradient, with the highest amount concentrated at the floor plate. In addition, growth cones in culture can respond to netrin gradients and turn toward the source that provides the protein. Therefore, chemoattraction provides a simple mechanism to explain the guidance of embryonic axons toward their targets.

Repelling Away from the Midline

After commissural axons are attracted to the midline, most of them do not stay there but instead leave the midline and project to the contralateral side of the spinal cord. A remarkably conservative and reproducible feature is that they cross the midline only once. How is this achieved?

The guidepost cells that provide attraction at the ventral midline also secrete another family of extracellular molecule, Slits. These molecules bind and activate cell surface receptor Robos on the commissural axons and repel them away from the midline after crossing. This was first demonstrated in a genetic screen in flies, in which a mutation in Slit caused all the commissural neurons to collapse at the midline. Interestingly however, the initial study of the Robo receptor revealed a very different phenotype. With only Robo eliminated, commissural axons freely cross and recross the midline and wrap around it to give the roundabout phenotype that is very different from Slit. This is because there are multiple Robo isoforms in flies. When another receptor Robo2 is also deleted in flies, the same Slit phenotype is observed, suggesting that both Robo1 and Robo2 can signal commissural axons to leave the midline, whereas only Robo1 is needed to prevent recrossing. Therefore, slits provide a negative guidance cue to commissural axons and directly activate Robo receptors to drive them away from the midline, thus preventing axon from recrossing. The same molecular mechanism is also used in vertebrates. When all three Slit genes are deleted from the mouse spinal cord, a considerable amount of commissurals are stalled near the floor plate.

In addition to preventing commissural axons from recrossing, the Slit proteins secreted by the midline cells serve as a guide to determine how far each interneuron axon should extend and which fascicle to join. Three Robo 1 homologs are differentially expressed on axonal fascicles, with Robo only on the medial fascicle, Robo and Robo3 on the intermediate ones, and all three (Robo, Robo2, and Robo3) on the most lateral one. The amount of Robo expressed on their surface provides a combinatorial code to determine the position of longitudinal fascicles. When the receptor level is perturbed by overexpression or knockdown, the fascicle positions along the lateral axis either shift away or move closer to the midline accordingly.

Slits are expressed in the midline at the same time as the attractive factor netrins. How do the commissural axons avoid being repelled from the midline before crossing? In flies, this is accomplished by an intracellular protein Comm, which appears to interact with Robo receptors and keeps them away from the growth cone membrane before crossing. In post-crossing axons, Comm expression is downregulated and thereby surface Robos are increased to respond to the repulsive signal from slits. Interestingly, no Comm homolog has been found in vertebrates, but a third Robo-like receptor, Robo3/Rig1, which is expressed also only on the precrossing axons, appears to serve the same function. In the Robo3/Rig1 mouse mutant, commissural neurons reach the ventral side of the spinal cord but stay away from the floor plate, mimicking the defect initially found for the Comm mutation in flies.

Adhesion at the Midline

The initial study of pioneer neuron guidance in grasshoppers suggests that cell adhesion may play an important role in controlling growth direction by guidepost cells. Studies in the past 20 years have identified many cell adhesion molecules that form homophilic or heterophilic dimers on the cell surface. In the vertebrate floor plate, several immunoglobulin superfamily cell adhesion molecules – axonin-I, NrCAM, and NgCAM – have been shown to be involved in proper control of the behavior of commissural axons at the floor plate. If their functions are perturbed by antibodies in chick embryos, the axons reach the floor plate but fail to cross the midline and turn to the ipsilateral side of the spinal cord.

Another cell surface molecule, Fasciclin II (FasII), provides a permissive substrate for the later born neurons to join the fascicle and extend along the longitudinal axis after crossing. In fly FasII mutants, these bundles do not form tightly, even though they can turn and move away from the midline, as repelled by slits.

Guidepost Function in Synapse Formation

The concept of guidepost cells has also been used in a different developmental context. During the development of neural circuits, axons are guided to different regions of the nervous system, where they contact their synaptic partners. Synaptic target selection and synapse formation are also critical steps to achieve the precise assembly of neuronal circuits. Several studies have shown that distinct populations of guidepost cells are important at the level of synaptogenesis. The following sections summarize the discoveries of these studies.

Caja-Retzius Cells and Certain GABAergic Interneurons as Guidepost in Hippocampus

In hippocampus, the cell bodies of pyramidal neurons are located in the basal region and send out dendrites toward the pial surface. Distinct subcellular domains of pyramidal dendrites are innervated by two populations of afferents. Distal portions of pyramidal dendrites receive input from the ipsilateral entorhinal afferents, whereas the proximal dendrites form synapses with the commissural fibers. This layer structure is established during development with the help of two populations of guidepost cells. Caja-Retzius cells and a set of GABAergic interneurons are early developing neurons found in the two afferent layers. They synapse with the entorhinal afferents and the commissural axons, respectively. These synapses are transient in nature since some of the Caja-Retzius cells and GABAergic interneurons die later. The disappearance of transient synapses is accompanied by synaptogenesis of afferents with the pyramidal dendrites as the synapses are transferred from the guidepost cells to mature synaptic targets.

The significance of the guidepost cells in the development of hippocampal circuit was demonstrated by cell lesion experiments. When the Caja-Retzius cells are ablated, the laminar innervation of the entorhinal axons is impaired. It is interesting to note that the ingrowth of entorhinal axons into the hippocampus precedes the extension of pyramidal dendrites. It is conceivable that the early existence of the Caja-Retzius cells is important to hold the presynaptic terminal in place before the true postsynaptic target arrives. Since in the absence of the Caja-Retzius cells laminar projection of the entorhinal axons is impaired, this suggests that early synaptogenesis might be important for the stabilization of axon arbors. Indeed, several *in vivo* time-lapse studies have shown that synapse formation in the CNS is very dynamic, with constant synapse formation and disassembly, and branch addition and retraction. The presence of synapses on axon branches increases the stability of the branches. Therefore, the laminar distribution of the guidepost cells provides a scaffold for the presynaptic terminals on the afferent fibers, which consequently stabilizes the axonal arborization and achieves laminar innervation.

Subplate Neurons as Guidepost in the Maturation of Visual Cortical Circuit

In the mature visual system of vertebrate animals, thalamic inputs directly innervate primary visual cortex layer 4 neurons. Functional organization of the visual cortex, such as ocular dominance columns and orientation columns, emerges through specific synaptic circuit formation. Interestingly, thalamic inputs first form synapses with another population of neurons, the subplate neurons, before connecting to the layer 4 neurons. At this early time, subplate axons innervate the layer 4 neurons and relay the information from thalamus to the cortex. Later, during the critical period of cortical activity-dependent plasticity, the subplate neurons gradually die through programmed cell death. In the meantime, adult circuit forms in which thalamic inputs directly synapse onto layer 4 neurons.

Ablation of subplate neurons results in the failure of segregation of the thalamic inputs into ocular dominance columns and the formation of orientation columns. This strongly suggests that subplate neurons are essential for synaptic remodeling and maturation of neural circuit. In support of this notion, one study showed that subplate ablation prevents the upregulation of $GABA_A$ receptor expression and perturb the maturation of inhibitory circuits in the layer 4 neurons. Collectively, subplate neurons act as a relay station at early stages of cortical development and are indispensable for patterning mature synaptic circuits.

Vulval Epithelial Cells as Guidepost in the Development of the Egg-Laying Synaptic Circuit

In *C. elegans*, egg-laying behavior is controlled by a pair of motor neuron HSNs. The cell bodies of HSNs are situated just posterior to the vulva, with its axon guided ventrally and then anteriorly. The HSN axons defasciculate dorsally from the ventral nerve cords near the vulval opening and innervate the vulval muscles by forming a cluster of neuromuscular junctions onto the muscle arms. The HSN axons also form synapses onto the VC motor neurons in this region, which in turn innervate muscles. Genetic and developmental analysis revealed that the surrounding vulval epithelial cells play an important guidepost role in the formation of this egg-laying neural circuit. In the absence of these epithelial cells, HSNs form ectopic synapses, which are located more anteriorly than normal. Molecularly, an immunoglobulin superfamily protein, SYG-2, was found to perform the guidepost function. SYG-2 is expressed transiently by the guidepost epithelial cells at early stages of HSN synapse formation. SYG-2 directly binds its receptor on HSN, another immunoglobulin superfamily protein, SYG-1, and clusters SYG-1 at the segment of HSN axon near the vulva. SYG-1 induces accumulation of synaptic vesicles/presynaptic active zone components and directs the location and target selection of HSN presynaptic specialization. In the absence of functional SYG-1 or SYG-2, HSN has a reduced number of synapses formed at its normal location. Instead, the majority of synapses are formed onto adjacent body wall muscles, inappropriate synaptic targets, at anterior ectopic locations. Developmentally, the axons of HSN reach the synaptic region prior to the outgrowth of the postsynaptic VC dendrites. Presynaptic terminals can be observed in this segment of the HSN axons before the VC dendrites reach the same region. The transient synapses of HSN are likely to form directly onto the guidepost cells before being 'handed over' to the late maturing postsynaptic targets. The guidepost vulval epithelial cells have several important functions in the development of the egg-laying organ. These cells attract the migrating sex myoblast, which gives rise to the vulval muscles. They also stimulate the branching of the VC motor neurons. Therefore, guidepost cells not only spatially and temporally control the maturation of the egg-laying neurons and muscles but also regulate the assembly of the neural circuits at the synapse formation level.

The three types of synaptic guidepost cells mentioned previously are probably examples of similar cell types that have not been discovered. Guidepost cells seem to be particularly important in synapse formation where there is temporal discrepancy between axonal and dendritic development. It is conceivable that guidepost cells stabilize axons by forming transient synapses, which disappear upon the arrival of true postsynaptic dendrites.

Conclusion

The function of guidepost cells in axon guidance is well established. They are frequently the sources of axon guidance molecules that attract or repel axon growth cones. The emerging roles of guidepost cells in synapse formation and neural circuit maturation reveal previously unknown complexity during synaptic circuit assembly.

See also: Axon Guidance: Morphogens as Chemoattractants and Chemorepellants; Axon Guidance: Building Pathways with Molecular Cues in Vertebrate Sensory Systems; Axonal Regeneration: Role of Growth and Guidance Cues; Axonal Pathfinding: Netrins; Growth Cones.

Further Reading

Bate CM (1976) Pioneerneurones in an insect embryo. *Nature* 260: 54–56.

Bentley D and Caudy M (1983) Pioneer axons lose directed growth after selective killing of guidepost cells. *Nature* 304: 62–65.

Dickson BJ (2002) Molecular mechanisms of axon guidance. *Science* 298: 1959–1964.

Ghosh A and Shatz CJ (1992) Involvement of subplate neurons in the formation of ocular dominance columns. *Science* 255 (5050): 1441–1443.

Ho RK and Goodman CS (1982) Peripheral pathways are pioneered by an array of central and peripheral neurones in grasshopper embryos. *Nature* 297: 404–406.

Kanold PO, Kara P, Reid RC, and Shatz CJ (2003) Role of subplate neurons in functional maturation of visual cortical columns. *Science* 301(5632): 521–525.

Kanold PO and Shatz CJ (2006) Subplate neurons regulate maturation of cortical inhibition and outcome of ocular dominance plasticity. *Neuron* 51(5): 627–638.

Kidd T, Bland KS, and Goodman CS (1999) Slit is the midline repellent for the robo receptor in *Drosophila*. *Cell* 96: 785–794.

Kidd T, Brose K, Mitchell KJ, et al. (1998) Roundabout controls axon crossing of the CNS midline and defines a novel subfamily of evolutionarily conserved guidance receptors. *Cell* 92: 205–215.

Long H, Sabatier C, Ma L, et al. (2004) Conserved roles for Slit and Robo proteins in midline commissural axon guidance. *Neuron* 42: 213–223.

Palka J, Whitlock KE, and Murray MA (1992) Guidepost cells. *Current Opinion in Neurobiology* 2: 48–54.

Sabatier C, Plump AS, Ma L, et al. (2004) The divergent Robo family protein rig-1/Robo3 is a negative regulator of slit responsiveness required for midline crossing by commissural axons. *Cell* 117: 157–169.

Sanes JR and Yamagata M (1999) Formation of lamina-specific synaptic connections. *Current Opinion in Neurobiology* 9(1): 79–87.

Serafini T, Colamarino SA, Leonardo ED, et al. (1996) Netrin-1 is required for commissural axon guidance in the developing vertebrate nervous system. *Cell* 87: 1001–1014.

Serafini T, Kennedy TE, Galko MJ, Mirzayan C, Jessell TM, and Tessier-Lavigne M (1994) The netrins define a family of axon outgrowth-promoting proteins homologous to *C. elegans* UNC-6. *Cell* 78: 409–424.

Shen K and Bargmann CI (2003) The immunoglobulin superfamily protein SYG-1 determines the location of specific synapses in *C. elegans*. *Cell* 112(5): 619–630.

Shen K, Fetter RD, and Bargmann CI (2004) Synaptic specificity is generated by the synaptic guidepost protein SYG-2 and its receptor, SYG-1. *Cell* 116(6): 869–881. [Erratum in *Cell* 117 (4): 553, 2004].

Super H, Martinez A, Del Rio JA, and Soriano E (1998) Involvement of distinct pioneer neurons in the formation of layer-specific connections in the hippocampus. *Journal of Neuroscience* 18(12): 4616–4626.

Tessier-Lavigne M and Goodman CS (1996) The molecular biology of axon guidance. *Science* 274: 1123–1133.

Tessier-Lavigne M, Placzek M, Lumsden AG, Dodd J, and Jessell TM (1988) Chemotropic guidance of developing axons in the mammalian central nervous system. *Nature* 336: 775–778.

Axon Guidance: Morphogens as Chemoattractants and Chemorepellants

J B Thomas and S Yoshikawa, Salk Institute for Biological Studies, San Diego, CA, USA

Introduction

The growth cones of developing neurons are guided to their targets by attractive and repulsive cues in the extracellular environment. Receptors on the growth cones recognize these cues and transduce signals that ultimately lead to changes in the direction of growth. Several families of molecules acting as guidance cues have been identified, including Netrins, Slits, Ephrins, and Semaphorins. In addition to these families, secreted signaling molecules from families of classical morphogens, known for their roles in controlling cell fates in a concentration-dependent manner, can act as axon guidance molecules. Sonic hedgehog (Shh) of the hedgehog family, bone morphogenetic protein 7 (BMP7) of the transforming growth factor-beta (TGF-β) family, and members of the Wnt family have all been found to function in the guidance of specific classes of neurons.

For each of the three morphogen families, the receptors and canonical signaling pathways through which they control cell fate are well studied and involve transcriptional regulation as their output. However, some members of these families can also signal through noncanonical pathways and control processes such as cell movements and cell polarity. The roles of these families in axon guidance, plus the finding that they likely act directly on the growth cone, argue for the activation of noncanonical signaling pathways that lead to cytoskeletal rearrangement.

Shh as a Chemoattractant for Commissural Neurons

Commissural neurons residing in the dorsal region of the spinal cord adjacent to the roof plate project axons through the floor plate at the ventral midline (**Figure 1(a)**). These commissural axons initially extend away from the roof plate and take a ventral and circumferential pathway through the dorsal spinal cord. Midway to the ventral midline, the axons change course and project ventrally and medially to the floor plate. After crossing the floor plate, they then make an abrupt turn and project anteriorly on the contralateral side of the spinal cord.

Commissural axons are guided to the ventral midline by Netrin-1, a chemoattractant secreted by cells of the floor plate and the adjacent periventricular zone. In *Netrin-1*$^{-/-}$ mutant mice, as well as in mice mutant for the Netrin receptor deleted in colorectal cancer (DCC), many commissural axons project abnormally, often failing to enter the ventral region of the spinal cord. However, some commissural axons in these mutants do manage to reach the floor plate, suggesting the existence of one or more additional factors in the floor plate. Indeed, the floor plate of *Netrin-1*$^{-/-}$ mice is still effective in reorienting commissural axons when juxtaposed to spinal cord explants, indicating the presence of another chemoattractant.

Shh was implicated as the additional factor since it is secreted by the floor plate and is known to have long-range effects within the spinal cord. In addition, Shh can cause rapid changes in the growth cone behavior of cultured retinal ganglion cells. An initial indication that Shh might be the missing floor plate factor was garnered by demonstrating the ability of Shh-expressing COS cells to reorient commissural axons in spinal cord explants. This reorienting effect of Shh is mediated through the Shh signaling component Smoothened (Smo), since the Smo inhibitor cyclopamine abolishes the effect.

Assessing the axon guidance role of Shh *in vivo* is more challenging because of its role in early patterning of the spinal cord. Shh acts as a morphogen to generate distinct neuronal cell types within the ventral spinal cord in a concentration-dependent manner. Although Shh is not directly required for generation of the dorsally located commissural neurons, these neurons do project their axons ventrally through regions that are patterned by Shh. This rules out the analysis of mice mutant for Shh, Smo, or the Shh receptor patched (Ptc) since other guidance cues in the ventral spinal cord of these mutants might be altered. However, using the Cre/loxP recombinase system and the *Wnt-1* promoter to drive expression of Cre recombinase, Smo can be selectively removed in commissural neurons developing within an otherwise normally patterned spinal cord. In these conditionally mutant mice, commissural axons project abnormally into the ventral spinal cord, sometimes failing to make the medioventral turn toward the floor plate and instead projecting along the edge of the spinal cord. Thus, Shh appears to be acting as a chemoattractant for commissural neurons. Ultimately, the commissural axons in these conditional Smo mutants do reach the ventral midline and form

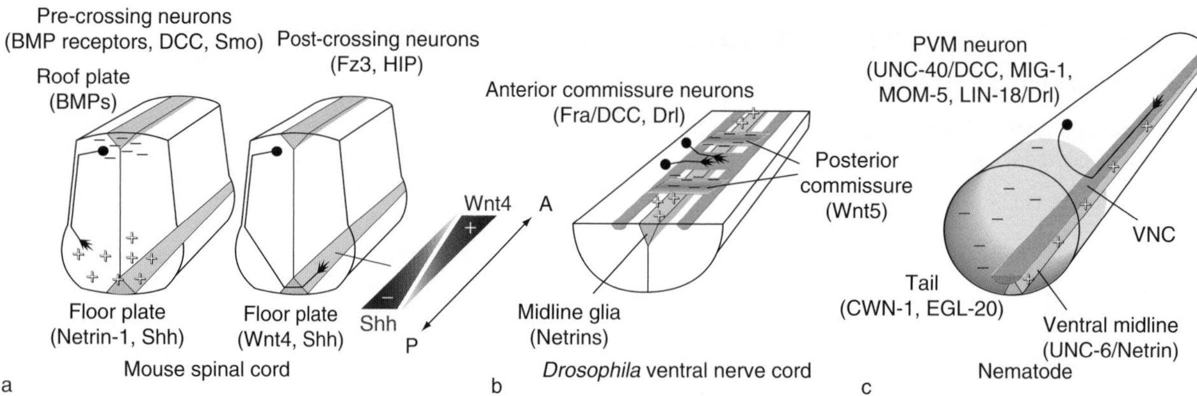

Figure 1 Morphogens as axon guidance molecules. Schematic of the axon projections of neurons in the developing mammalian spinal cord (a) and the ventral nerve cords (VNC) of *Drosophila* (b) and nematode (c). The floor plate (green) secretes Netrin-1 (+), which acts as a chemoattractant for commissural axons expressing the Netrin receptor DCC. The floor plate also secretes Shh, which functions in concert with Netrin-1 to attract commissural axons expressing the Shh signaling component Smo. The roof plate (orange) secretes BMP7 (−), which repels the initial growth cones of commissural neurons away from the dorsal midline, probably as a heterodimer with GDF7. The specific BMP receptors in commissural neurons mediating the repulsive response are not known. After crossing the floor plate, growth cones of commissural neurons turn anteriorly in response to an increasing posterior-to-anterior (P-to-A) gradient of Wnt4, plus a decreasing gradient of Shh, secreted by floor plate cells. Fz3 and HIP are thought to mediate the responses to Wnt4 and Shh, respectively. In *Drosophila*, the major axon tracts (gray) consist of the bilaterally symmetric longitudinal connectives running in the anterior/posterior axis and, connecting the two sides in each segment, an anterior and posterior commissure. Like the vertebrate floor plate, the midline glia (green) secrete Netrins (+) which attract commissural axons expressing the DCC homolog Fra. Commissural axons that project through the anterior commissure, all of which express the Drl receptor, are repelled from the posterior commissures of their segment of origin and the adjacent anterior segment by Wnt5 (−). In the nematode, migrating cells that end up at the ventral midline (green) secrete the Netrin homolog UNC-6 (+), as does the VNC (gray). Axons expressing UNC-40, the nematode DCC homolog, are attracted to the VNC. Wnt proteins CWN-1 and EGL-20 (−) are expressed by cells in the tail and act to repel PVM and other anteriorly projecting neurons expressing the Fz receptors MIG-1 and MOM-5. The Drl/Ryk homolog, LIN-18, acts redundantly in Wnt-mediated repulsion.

a normal-looking commissure, arguing that Netrin-1, in the absence of Shh, is sufficient to attract commissural axons to the floor plate. If Shh is the only additional guidance factor secreted by the floor plate, the expectation is that *Smo* mutant commissural neurons should not be able to reach the floor plate in a *Netrin-1*$^{-/-}$ mutant background.

Although a loss-of-function phenotype is the gold standard for demonstrating that a gene is required for a particular process, a phenotype does not necessarily reveal the underlying mechanism. For example, rather than acting directly on the growth cones of commissural neurons, Shh could be activating a retrograde signal within the neurons that regulates the expression of genes encoding other guidance receptors. The definitive test for a guidance molecule's direct action on growth cones is the *in vitro* growth cone turning assay. Here, the spatial and temporal application of a factor can be precisely controlled by pulsing it through a pipette, creating a gradient of purified protein to which isolated cultured neurons can respond. Importantly, in this context Shh has the ability to rapidly attract growth cones and this attraction is abolished by cyclopamine. Thus, Shh, signaling through Smo, can indeed act as a chemoattractant.

Shh as a Chemorepellant for Postcrossing Commissural Neurons

After crossing the midline, commissural axons turn anteriorly along the contralateral side of the floor plate. RNA interference (RNAi) experiments in chick embryos have implicated Shh in this guidance event. When Shh levels are reduced by injection of Shh dsRNA, commissural axons tend to either stall at the floor plate exit point or project posteriorly. A decreasing posterior-to-anterior gradient of Shh suggests that in this case Shh is acting as a chemorepellant, a notion supported by its ability to repel postcrossing commissural axons in spinal cord explants. Importantly, this repulsion is not mediated by Smo and Ptc, since neither is expressed by commissural neurons by the time they have crossed the midline. Furthermore, cyclopamine has no effect on turning behavior. Instead of Ptc, another Shh receptor, hedgehog interacting protein (HIP), has been implicated. Lowering HIP function by RNAi knockdown results in the same turning defects observed when Shh is downregulated.

A model has thus emerged that depending on the receptor employed, Shh can act as either a chemoattractant or a chemorepellant. However, in contrast to Shh chemoattraction, it is not known whether HIP-mediated repulsion by Shh is a result of direct action on the growth cone rather than the transcriptional control of other guidance receptors, such as those involved in Wnt-mediated attraction of post-crossing commissural axons.

BMP7 as a Chemorepellant for Commissural Neurons

The early phase of commissural axon growth within the dorsal spinal cord is unaltered in either *Netrin-1*$^{-/-}$ mutants or mice lacking a floor plate, suggesting the existence of additional non-floor plate-derived factors in their guidance. The proximity of commissural neurons to the roof plate and the stereotyped projections of their axons away from the dorsal midline suggests a role of the roof plate in repulsion (**Figure 1**(a)). This idea is supported by experiments showing that when juxtaposed to a spinal cord explant, the roof plate can reorient commissural axons away from the side facing the roof plate. Furthermore, the reorienting activity can be mimicked by COS cells expressing BMP7, one of the BMPs secreted by the roof plate.

BMP7 functions most efficiently in the axon reorienting assay by forming heterodimers with GDF7, another BMP family member expressed by roof plate cells but which on its own has no axon reorienting activity. In *BMP7*$^{-/-}$ mutant mice, there is a significant increase over wild type in the number of commissural axons that initially project medially and dorsally instead of laterally and ventrally. Similar numbers are seen in *GDF7*$^{-/-}$ mutants. However, in each case, these defects are corrected such that later in development the projection pattern of commissural neurons resembles that of wild type. Thus, in addition to BMP7 and GDF7, other cues must be operating–repulsive ones from the roof plate and/or attractive ones from more lateral positions in the dorsal spinal cord. The former possibility is bolstered by the finding that roof plates from *BMP7*$^{-/-}$ *GDF7*$^{-/-}$ double mutants, although compromised in their ability to redirect commissural axons within spinal cord explants, still retain some residual axon-reorienting activity.

Similar to the patterning role of Shh in the ventral spinal cord, BMP family members function early in development to generate specific dorsal cell types. In fact, GDF7 is required to generate a specific subclass of commissural neurons, raising the question of whether the observed guidance effects of BMP7 are direct or indirect. Although the guidance role of BMP7 has not been cleanly separated from its inductive role, as was done for Shh using the conditional *Smo* mutation, several lines of evidence argue against an indirect mechanism involving changes in cell fate. First, *BMP7*$^{-/-}$ mutants show no obvious defects in dorsal spinal cord patterning, presumably due to redundancy with one or more of the other roof plate BMPs. Second, commissural neurons in spinal cord explants from *BMP7*$^{-/-}$ mutants are able to reorient to the same extent as those from wild type in response to a wild-type roof plate, suggesting that these neurons retain at least some of their wild-type properties, including the expression of an appropriate BMP receptor complex capable of transducing the BMP7 signal. Finally, although *in vitro* growth cone turning assays to test for repulsion have not been carried out, BMP7 alone, as well as BMP7:GDF7 heterodimers, is capable of causing rapid growth cone collapse of commissural neurons *in vitro*. Collectively, these studies advance the case for a direct role on the growth cone in axon guidance.

Wnt Proteins as Chemorepellants

Wnt5 in *Drosophila*

In addition to specifying cell fates, Wnt proteins play diverse signaling roles in the developing nervous system. These include presynaptic axon remodeling during vertebrate synaptogenesis, the maturation of the *Drosophila* neuromuscular junction, and the polarity of neurons along the anterior–posterior axis in the nematode *Caenorhabditis elegans*. As first shown in *Drosophila*, Wnt family members have also turned out to function as axon guidance molecules.

As in vertebrates, the large number of commissural neurons in the *Drosophila* embryonic ventral nerve cord (the fly counterpart to the spinal cord) project their axons across the midline to the contralateral side (**Figure 1**(b)). Analogous to the vertebrate floor plate, specialized cells at the midline, the midline glia, divide the two halves of the ventral nerve cord and play a critical role in axon guidance. Like the floor plate, midline glia secrete Netrins that act as chemoattractants for commissural axons expressing Frazzled (Fra), the fly homolog of the Netrin receptor DCC.

Once attracted to the midline, commissural axons do not cross randomly. Instead, in each segment they reproducibly choose one of two distinct tracts that connect the two sides, either the anterior or the posterior commissure. This choice of whether to project anteriorly or posteriorly into the appropriate commissure is controlled by one of the seven fly members of the Wnt family, Wnt5, and a receptor to which it binds, derailed (Drl). The Drl receptor is expressed on the growth cones and axons of all neurons that project through the anterior commissure. In *drl* mutants, many of the anterior commissure axons abnormally project through the posterior commissure; conversely, when misexpressed by posterior commissure neurons, Drl switches their axonal projections to the anterior commissure. Drl is therefore not only required by anterior commissure neurons for their guidance but also sufficient to dictate the choice of the anterior commissure for crossing axons.

In *wnt5* mutants, as in *drl* mutants, anterior commissure axons project abnormally through the posterior commissure. Since Wnt5 is expressed by cells associated with the posterior commissure (and the Drl receptor by anterior commissure neurons), it is functioning as a chemorepellant to keep Drl-expressing axons from entering the posterior commissure. Such a repulsive activity was illustrated *in vivo* by misexpressing Wnt5 in the midline glia. In these embryos, anterior commissure axons are prevented from crossing the midline in a Drl-dependent manner, overriding Netrin-mediated attraction.

Although Drl binds Wnt5 and is required for Wnt5 function in guidance, it is not yet clear whether Drl transduces the Wnt5 signal within growth cones or whether it acts as a coreceptor with a member of the other family of Wnt receptors, the Frizzled (Fz) proteins. Fz receptors are good candidates for transducing such a signal since they have been shown not only to mediate canonical Wnt signaling resulting in transcriptional regulation but also to mediate noncanonical signaling involved in controlling cell movements. In support of the coreceptor hypothesis, the mammalian homolog of Drl, Ryk, can bind to the cysteine-rich domain of Fz proteins when the two are co-expressed in tissue culture cells, suggesting the ability of Fz and Ryk to form a complex. Regardless of which receptor(s) transduces the signal, it is clear that Wnt5 controls the guidance of anterior commissure axons and that the Drl receptor is essential for Wnt5 signaling.

However, does Wnt5 act directly or indirectly in these guidance events? The difficulty in purifying active Wnt proteins has hindered efforts to test Wnt5 for repellant activity in an *in vitro* growth cone turning assay. However, a key observation *in vivo* does suggest a direct role in guidance: Drl can dictate commissure choice in a Wnt5-dependent manner when misexpressed on growth cones shortly before they make the choice. This makes it less likely that a retrograde signal, followed by a round of transcription and translation, is required.

Mammalian Wnts

Wnt proteins, acting through the Ryk receptor, also control axon guidance events in the developing mammalian nervous system. Similar to Drl expression on anterior commissure axons in *Drosophila*, Ryk is expressed on axons that project across the midline within the corpus callosum. In *Ryk*$^{-/-}$ mutant mice, callosal axons reach and cross the midline but often fail to project away from it on the contralateral side. The Wnt ligand in this case appears to be Wnt5a, which together with Wnt5b is most closely related to *Drosophila* Wnt5. Wnt5a binds to Ryk and is expressed by midline glia known to be required for proper guidance of the callosal axons. The spatial and temporal expression pattern of Wnt5a suggests that it is acting as a chemorepellant to guide Ryk-expressing axons away from the midline, a notion supported by the finding that Wnt5a can repel cortical axons in explants from wild-type, but not *Ryk*$^{-/-}$, brains.

Like callosal axons, axons of corticospinal neurons also express Ryk. These axons project posteriorly from the brain through a dorsal region of the spinal cord that exhibits an increasing posterior-to-anterior gradient of a number of Wnts, including Wnt1 and Wnt5a. Anti-Ryk antibodies block both the posterior growth of corticospinal axons *in vivo* and the repellant activity of Wnts on cortical axons in cultured explants, providing good evidence that Ryk is required for posterior guidance by interpreting the Wnt gradient as a chemorepellant one. The striking similarities between mammalian Wnt-mediated corticospinal axon guidance and *Drosophila* Wnt5-mediated commissure choice argue for a deep-rooted conservation of function: both fly Wnt5 and mammalian Wnts are acting as chemorepellants that signal via the Drl/Ryk receptor to control guidance in the anterior–posterior axis.

Wnts in the Nematode

In the nematode *C. elegans*, the conserved cue Netrin (UNC-6 in nematode) guides axons to the ventral nerve cord at the midline (**Figure 1(c)**). Once they enter the ventral nerve cord, growth cones choose to project either anteriorly or posteriorly. For example, the axon of the posterior ventral microtubule cell (PVM) enters the nerve cord and projects anteriorly toward the head. Two Wnts, CWN-1 and EGL-20, are expressed in the posterior end of the embryo and act redundantly as chemorepellants to guide the PVM axon, as well as others, anteriorly. In *cwn-1:egl-20* double mutants, the PVM axon often projects posteriorly and this guidance defect is enhanced by misexpression of EGL-20 anteriorly. In contrast to *Drosophila* and mouse, Wnt repulsion of the PVM axon is mediated primarily through two Fz receptors, with LIN-18, the nematode homolog of Drl/Ryk, acting redundantly, presumably through a parallel pathway.

Wnt Proteins as Chemoattractants

After crossing the floor plate of the spinal cord, commissural axons turn anteriorly toward the brain (**Figure 1a**). Observations of cultured spinal cord explants provided the first indication that this guidance event might be controlled by a chemoattractant. Near the posterior cut end of spinal cord explants, commissural axons project anteriorly in a normal fashion, but near the anterior cut end, they either stall or project randomly along the anterior–posterior axis. This suggests that an increasing posterior-to-anterior gradient of a diffusible chemoattractant, which slowly escapes from the cut ends of explants, guides commissural axons anteriorly. A candidate molecule approach led to the Wnt family, particularly Wnt4, whose mRNA levels exhibit an increasing posterior-to-anterior gradient in the floor plate. Wnt4-expressing COS cells juxtaposed to the anterior cut end of explants can redirect postcrossing axons anteriorly, in effect rescuing the projection defects near the cut end. Further support for this guidance event being mediated by a Wnt protein comes from the analysis of mice deficient for Fz3, one of the Wnt receptors expressed by commissural neurons. In these mice, commissural growth cones emerge from the floor plate and tend to project randomly along the anterior–posterior axis of the spinal cord.

Similar to Wnt function in *Drosophila* and nematode, Wnt4 appears to control anterior–posterior guidance. However, in contrast to the repellant activity of Wnts in the guidance of *Drosophila* commissural axons, the nematode PVM axon, and of mammalian callosal and corticospinal axons, Wnt4 is acting as a chemoattractant in the guidance of postcrossing commissural axons. Notably, this guidance event does not involve Ryk since commissural neurons do not express the receptor. Instead, it seems that a member of the Fz family of Wnt receptors, acting in the absence of Ryk, mediates the guidance. A model similar to that for Shh has been proposed in which Wnts can act as either chemoattractants or chemorepellants depending on whether a Fz or Ryk receptor is involved. However, the relative roles of these receptors in Wnt-mediated guidance in nematode suggest that the situation may turn out to be more complicated.

Conclusion

Hedgehog, TGF-β, and Wnt family members not only have roles in cell fate specification and early embryonic patterning but also act as axon guidance molecules. For Shh and BMPs, morphogen gradients initially used to specify cell types in the spinal cord are reused later in development to attract or repel axons. There is evidence that Shh, BMP7, and Wnt proteins act directly on the growth cone rather than through canonical signaling pathways to the nucleus. The current challenge, and one which is faced generally in studies of guidance molecules, is to identify the signaling pathways engaged by the receptors and to understand how these signals are integrated within the growth cone to achieve changes in direction of growth.

See also: Axon Guidance: Building Pathways with Molecular Cues in Vertebrate Sensory Systems; Axon Guidance: Guidance Cues and Guidepost Cells; Axonal Regeneration: Role of Growth and Guidance Cues; Axonal Pathfinding: Netrins; Axonal Pathfinding: Guidance Activities of Sonic Hedgehog (Shh); Axonal Pathfinding: Extracellular Matrix Role; Growth Cones; Morphogens: History.

Further Reading

Bourikas D, Pekarik V, Baeriswyl T, et al. (2005) Sonic hedgehog guides commissural axons along the longitudinal axis of the spinal cord. *Nature Neuroscience* 8: 297–304.

Butler SJ and Dodd J (2003) A role for BMP heterodimers in roof plate-mediated repulsion of commissural axons. *Neuron* 38: 389–401.

Charron F, Stein E, Jeong J, McMahon AP, and Tessier-Lavigne M (2003) The morphogen sonic hedgehog is an axonal chemoattractant that collaborates with netrin-1 in midline axon guidance. *Cell* 113: 11–23.

Dickson BJ (2002) Molecular mechanisms of axon guidance. *Science* 298: 1959–1964.

Fradkin LG, Garriga G, Salinas PC, Thomas JB, Yu X, and Zou Y (2005) Wnt signaling in neural circuit development. *Journal of Neuroscience* 25: 10376–10378.

Keeble TR, Halford MM, Seaman C, et al. (2006) The Wnt receptor Ryk is required for Wnt5a-mediated axon guidance on the contralateral side of the corpus callosum. *Journal of Neuroscience* 26: 5840–5848.

Liu Y, Shi J, Lu CC, et al. (2005) Ryk-mediated Wnt repulsion regulates posterior-directed growth of corticospinal tract. *Nature Neuroscience* 8: 1151–1159.

Lu W, Yamamoto V, Ortega B, and Baltimore D (2004) Mammalian Ryk is a Wnt coreceptor required for stimulation of neurite outgrowth. *Cell* 119: 97–108.

Lyuksyutova AI, Lu CC, Milanesio N, et al. (2003) Anterior–posterior guidance of commissural axons by Wnt-frizzled signaling. *Science* 302: 1984–1988.

Pan CL, Howell JE, Clark SG, et al. (2006) Multiple Wnts and frizzled receptors regulate anteriorly directed cell and growth cone migrations in *Caenorhabditis elegans*. *Developmental Cell* 10: 367–377.

Schnorrer F and Dickson BJ (2004) Axon guidance: Morphogens show the way. *Current Biology* 14: R19–R21.

Yoshikawa S, McKinnon RD, Kokel M, and Thomas JB (2003) Wnt-mediated axon guidance via the *Drosophila* Derailed receptor. *Nature* 422: 583–588.

Axonal and Dendritic Identity and Structure: Control of

C G Dotti and A Gärtner, University of Leuven and Flanders Institute of Biotechnology, Leuven, Belgium
F Calderon de Anda, Picower Institute for Learning and Memory, MIT, Boston, USA

Introduction

In its simplest biological application, the term polarity is used to describe the existence of an asymmetric process, whether this refers to the shape of an organism, an organ or a single cell; directed movement; the distribution of molecules and organelles; and intracellular vectorial flows, centrifugal and centripetal. The relevance of understanding the cellular and molecular basis of polarity lies in the fact that most basic biological functions are a consequence of asymmetry. For instance, during development, the formation of the main body axes, that is, rostral–caudal and ventral–dorsal, is the consequence of the establishment of polarized domains. Polarized activities also control the proper three-dimensional organization of organs and tissues and the coordinated orientation of cells in two dimensions in certain tissues, a process called planar polarity. At the single-cell level, to cite only a few examples, polarity is important for asymmetric cell division, which generates cellular diversity; for the migratory movement of cells in response to external stimuli; for the differential uptake and release of ligands; and for the reception and propagation of electrical signals through dendrites and axons in neurons.

A polarized neuron is characterized by a very particular molecular and supramolecular organization of axons and dendrites, that is, a differential distribution of both cytoplasmic and membrane components. Most simply seen, the capacity of axons to transmit signals over long distances is due to the segregation of voltage-gated sodium channels and synaptic vesicles to this domain. Dendrites, on the other hand, are capable of receiving incoming information owing to the presence of neurotransmitter receptors. At the cytoplasmic level, the microtubule binding protein MAP2 is specifically distributed in dendrites, and axons are enriched in tau. Moreover, microtubules in axons are always oriented with their plus end toward the periphery, whereas in dendrites they exhibit a mixed polarity. Neurons, however, establish a clearly polarized phenotype very early in development, before they have acquired the capacity to transmit and receive electrical signals.

Neuronal polarization becomes visible shortly after the last mitosis, through the sprouting of a number of extensions which, due to the lack of a clear axonal or dendritic identity, are called 'minor neurites.' In the typical case of neurons generated in the ventricular or subventricular zones, these initial extensions constitute the cell's migratory force for reaching its final position in the cortex. It is of great importance to understand how neurons establish their polarity. Defects in this early process will have severe consequences in the way the brain is organized into different layers, nuclei, and areas, leading to brain dysfunction. Moreover, failures in the definition of axons and dendrites at later stages will lead to defective wiring of the brain. This article describes the morphological events and mechanisms leading to the regulated generation of the primordial neurites and how they differentiate into precursors of functional axons and dendrites. The anatomical changes occurring during the development of polarity have been well described for a variety of neurons *in situ*. However, the molecular, biochemical, and cellular biological knowledge on how a neurite initially forms from the round cell body and how this neurite later acquires axonal or dendritic characteristics mainly relies on studies performed in hippocampal neurons taken from rat embryos, which spontaneously polarize under *in vitro* conditions. After a short description of the morphological phases of neuronal polarization *in situ*, this article describes how neuronal polarization is understood as a result of the large number of studies performed in hippocampal neurons in culture.

From Round to Polar: Morphological Steps

Developmental Stages *In Situ*

In situ, cortical pyramidal neurons are generated in the proliferative layer of the dorsal telencephalon and follow mainly a radial path toward the pial surface, where they occupy the most superficial layer of the developing cortical plate and populate the cortex in an inside first, outside last sequence, generating the six-layered cortex. In the mouse neocortex, neurogenesis is initiated at embryonic day (E) 10, peaks at E15, and finishes around birth. Initially, ventricular zone neuroepithelial cells give rise, through symmetric proliferative divisions, to neurons and radial glia cells (E10–E12). While the newly generated neurons begin to differentiate, radial glia cells continue to divide either symmetrically, into two neurons, or asymmetrically, into one neuron and one radial glia cell. In addition, new neurons arise from symmetric neurogenic divisions of basal precursors in the subventricular zone. Irrespective of the place and mode by which they are generated, neurons start to migrate toward upper cortical layers through a four-phase process. In phase 1, ventricular zone-derived neurons acquire a bipolar morphology, with one neurite facing toward the ventricle and one neurite toward the opposite, pial surface and migrating to the subventricular zone. In phase 2, neurons pause for approximately 24 h, forming more neurites, making the cells appear multipolar. In phase 3, a subpopulation of neurons extend a process toward the ventricle. Finally, in phase 4, cells migrate, after again acquiring a bipolar shape with a pia-directed leading process, to their final destination, where neurons attain their final shape and polarity. Thus, neuronal polarization begins at the place where neurons are generated, through the sprouting of initial asymmetric neurites. In addition, *in vivo* morphological progression studies indicate that the first 'role' that polarization plays is to support neuronal migration – quite different from the electrical transmission role of polarity in the mature neuron.

Developmental Stages *In Vitro*

Although the initial development of neurons is described well *in situ* on a morphological basis, most data describing functional and molecular aspects underlying the establishment of polarity are derived from a more easily accessible *in vitro* model system, that is, isolated rat hippocampal neurons developing in low-density culture. An early study on the polarization of these neurons demonstrated that polarization occurs in a most stereotyped and reproducible manner: Shortly after dissociation from the brain, embryonic hippocampal neurons are round (named stage 1 of development). Within the next 24 h, the round cells sprout four to six morphologically and ultrastructurally identical neurites (called stage 2). Finally, one of these neurites elongates more rapidly than the others, becoming the cell's axon. This is, according to this study, the first stage of polarity and represents developmental stage 3. A most important implication of that model and subsequent work was that polarity is established at the transition between stages 2 and 3 (see **Figure 1(a)**). A different perspective has arisen from more recent work showing that the neurite that appears first from the cell body of the new neuron later becomes the axon. The data from this recent work are consistent with the view that the immediate postmitotic neuron corresponds to stage 0, which represents a round neuron with a polarized cytoplasm marking the site of axonal outgrowth. The stage after the neuron grows the first sprout from that site is named stage 1. Stage-2 and stage-3 neurons in the new model correspond to stages 2 and 3 in the earlier classification (**Figures 1(a)** and **1(b)**). The data from the newer work suggest that the establishment of polarity occurs immediately after mitosis, at the transition between stage 0 and stage 1 (see **Figure 1(b)**). **Figure 2(a)** illustrates the progression of polarity in actual neurons. The important difference in these two concepts of when polarity is first established is that they imply different mechanistic principles. Thus, from the earlier model, the random choice of a neurite for an axonal destiny implies that polarized growth is a stochastic process, in which the cross-talk between one neurite and the

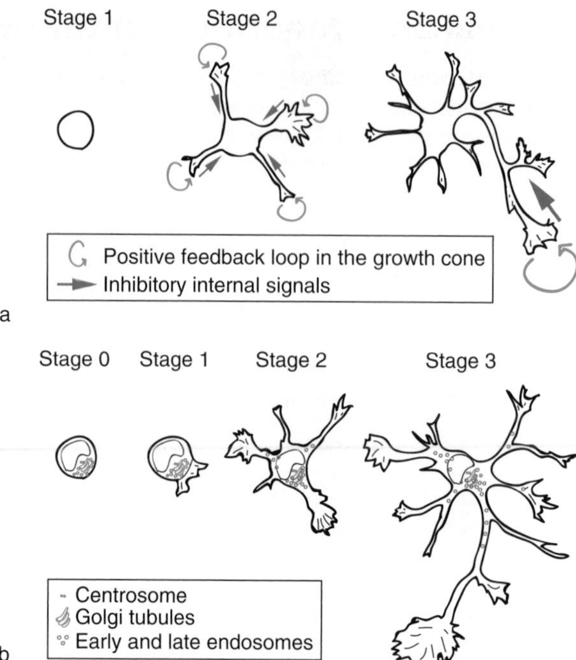

Figure 1 Establishment of polarity in dissociated hippocampal neurons. (a) This model proposes that neuronal polarization starts at the stage 2–3 transition. Positive and negative feedback regulation leads to the disruption of stage-2 symmetry. Red circles with arrows represent molecular signaling cascades that form positive feedback loops in the growth cone and promote neurite extension. Blue arrows represent a negative signal that is generated at each growth cone, propagates throughout the cell, and modulates the positive feedback loops in other neurites. The growth cone with highest positive feedback (produced probably by the environment) generates a strong negative signal, precluding other neurites from growing; meanwhile the neurite with the highest positive feedback grows as an axon. (b) The second model proposes an intrinsic polarization that is established at the immediate postmitotic stage and maintained through differentiation. Modified from Dotti CG, Sullivan CA, and Banker GA (1988) The establishment of polarity by hippocampal neurons in culture. *Journal of Neuroscience* 8: 1454–1468. From De Anda FC, Pollarolo G, Da Silva JS, Camoletto PG, Feiguin F, and Dotti CG (2005) Centrosome localization determines neuronal polarity. *Nature* 436: 704–708.

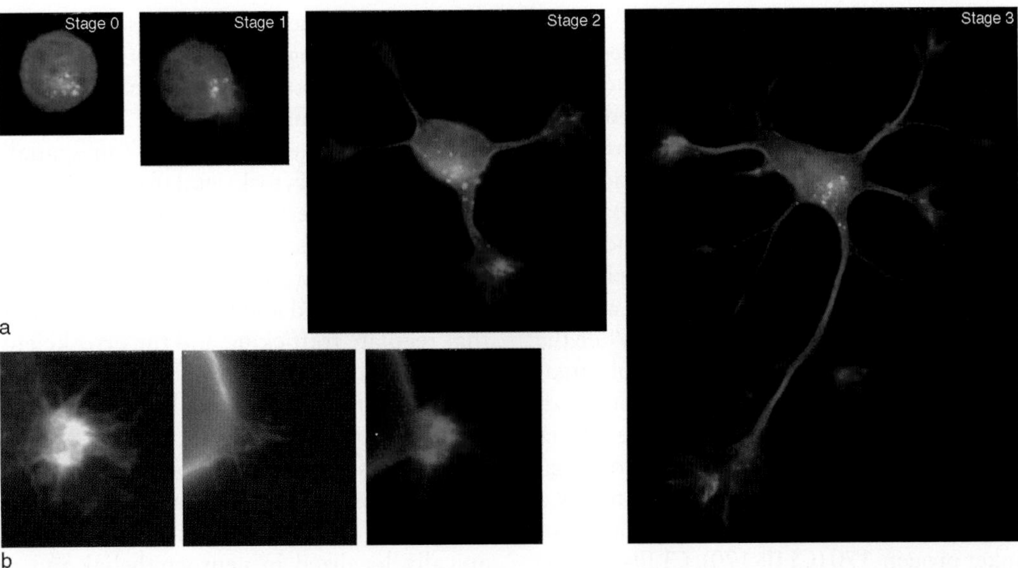

Figure 2 (a) Intracellular polarization through different stages of neuronal differentiation as proposed in the model in **Figure1(b)**. Neurons were fixed after different times in culture and stained for tubulin (blue), F-actin (red), and endosomes (green). (b) A neuronal sprout containing unstable F-actin with microtubules entering the sprout. Left panel, actin (red), middle panel, tubulin (blue), and right panel, merging of tubulin (blue) and actin (red) signals.

most appropriate environmental cue decides final outcome (i.e., axonal outgrowth). On the contrary, the temporal hierarchy described by the second model, suggesting that the first neurite has greater likelihood of becoming an axon, does not permit environmental influence, implying that the neurite that later grows into an axon is intrinsically predisposed. As will be shown, a number of early and recent observations indicate that the models overlap and that the establishment of neuronal polarity is, in the end, the con sequence of an initial intrinsically mediated predisposition later confirmed through extrinsic signaling.

From Round to Polar: Creating a Positional 'Lead'

Irrespective of the primary cause determining which neurite later becomes axon, the fact is that before this occurs, neurites must form from the round cell body of an immediately postmitotic neuron. This we consider as a most critical stage of neuronal polarization as it is the precondition for the later development of neurites into either axons or dendrites.

The sphere is the most economical cell architecture. Thus, the occurrence of the earliest morphological change leading to the appearance of the first neurite sprout is necessarily the product of energy-consuming activities that, on the one hand, disrupt the natural sphere-sustaining activities and, on the other hand, activate expansive forces. The site where sphere-sustaining forces are disrupted and consequently where neurite-inducing forces act must be tightly oriented in space. As described in the *in situ* polarization process, the place where the first neurite emerges is a most critical event as it determines the fidelity of migration. Therefore, the following section describes the most conspicuous intracellular constituents behind the disruption of the postmitotic neuron sphere as well as those involved in generating the primordial neurite and the way in which these activities are restricted in space.

Disrupting the neuronal sphere will require, among many other activities, changes in the dynamics of microtubules and actin filaments. In the quiescent, nonmotile stage, actin filaments are organized in a tight meshwork that surrounds the entire cell, directly underneath the plasma membrane. The action of a spatially restricted signal, whether a centrifugal action from the inside of the cell (e.g., microtubules or transported membrane) or a centripetal force from an environmental signal acting on the plasma membrane, will affect the stability of the tight actin meshwork, producing a localized change in this structure. Commonly, this change is the destabilization of the tight actin structure, leading to the formation of an actin cortex 'pore' allowing for the penetration of microtubules and thus increasing their chances to directly contact the plasma membrane at a spatially restricted place (**Figure 2(b)**). In fact, when the tight actin cortical organization is broken, highly motile actin-containing structures appear on the surface of cells: filopodia and lamellipodia, rich in highly dynamic actin and microtubules. In a neuron these structures precede the formation of a neurite. Neurites therefore arise as a consequence of the destabilization of the actin cortex, the polymerization and stabilization of microtubules, and the directed membrane delivery via these microtubule tracks from the center of the cell. Microtubule polymerization can work in addition as one of the centrifugal forces inducing a localized change in the cortical actin, to trigger the escape of microtubules to the periphery to make the neurite. Microtubules polymerize in a polarized fashion from the minus to the plus end. While the minus end is either embedded in the microtubule organizing center (MTOC) or protected by minus end-binding proteins, the plus end dynamically explores the cytoplasm, and its dynamics and localization (e.g., its connection with the cell cortex) are regulated by a number of plus end-binding proteins, such as cytoplasmic linker protein-170 (CLIP-170), CLIP-115, lissencephaly 1 (Lis1), and end-binding protein 1. These proteins, which collectively are called plus end tracking proteins, are key players during neuronal migration, and their dysfunction leads to migration defects and thus severe brain malformations. It is interesting that *Lis1* haploinsufficiency results in reduced filamentous actin at the leading edge of migrating neurons. To summarize, changes in the dynamics of the actin and microtubule skeleton are major events leading to the disruption of the sphere and in turn to neurite production. However, attention should be drawn to many other events, such as localized endocytosis, local translation of messenger RNAs, calcium release, and mitochondria trafficking, without which the microtubule–actin tandem would not be able to act as a mechanical force.

How does the neuron position the microtubule–actin sphere-breaking/neurite-generating activities so that a neurite is made at the right place? This is a most critical issue as *in situ* the primary neurites must arise from a precise pole in order to more efficiently interact with the cues required to guide the neuron's migration. The precise place where microtubule–actin-based activities occur can be theoretically determined by two mechanisms. The first most economically profits from the asymmetric organization of the cytoplasm of a recently generated neuron. In fact, one immediate consequence of cytokinesis is the occurrence of an asymmetrically positioned centrosome, the organelle from which microtubules are nucleated, in the daughter cells (**Figure 3(a)**). The asymmetric positioning of this organelle adds to the separation 'force,' which will result in stronger contact of microtubules with the nearby actin cortex and thus the plasma membrane (**Figure 3(b)**). Since an asymmetrically positioned centrosome is thought to play a role in other polar processes, such as the orientation of the migratory axis of motile cells, it can certainly be important in controlling the emergence of neurites in neurons.

The second mechanism that can spatially restrict the activity of cytoskeletal forces depends on the localization of certain molecules to a restricted place on the membrane. In parallel to studies on bud site selection on yeast, where the membrane of the future bud site is marked and recruits a cascade of molecules that regulate trafficking and the cytoskeleton, such a hypothetical membrane mark could also be a determining feature in neuronal polarity (**Figure 3(c)**). Good candidate molecules to exert such a role are the proteins that regulate polarity during asymmetric cell division, such as the partitioning-defective proteins (Pars). Indeed, Par3 and Par6 are described as apically localized in neuroepithelial and radial glia cells. Their sorting to one of the two cells during division provides that cell with precursor properties, while the other cell becomes a neuron. Therefore, it is quite possible that small membrane patches containing Par3/6 are sufficient to bestow asymmetry to the nascent neuron, irrespective of where the centrosome is positioned. In this scenario, one can envision that these or similar molecules are responsible for 'weakness' on the cortical actin or the plasma membrane, in turn allowing microtubules to enter and support neurite expression. Whether an asymmetric, centrosome-mastered cytoplasm creates a major positional lead for polarity or whether molecular 'polarizers,' working independently or in tandem with centrosome-emerging forces establish initial asymmetry should be tested (**Figures 3(b)–3(d)**).

Irrespective of their degree of participation or hierarchical positioning in the polarity cascade in the postmitotic neuron (i.e., whether a molecular polarizer determines the positioning of the centrosome or vice versa), both these mechanisms support the idea of a mitotically determined process, strongly indicating that the decision of polar growth axis is intrinsically controlled.

From Round to Polar: Making a Neurite Become an Axon

Once sphere-breaking, neurite-inducing forces have been positioned in the sphere, a morphological deformation will result that eventually will become a neurite. Later, from the first or another neurite (depending on the model, **Figure 1(a)** or **1(b)**), the axon will form.

The growth of all neurites, independent of whether they become the axon or a dendrite, is dependent on the directed activity of the microtubule–actin

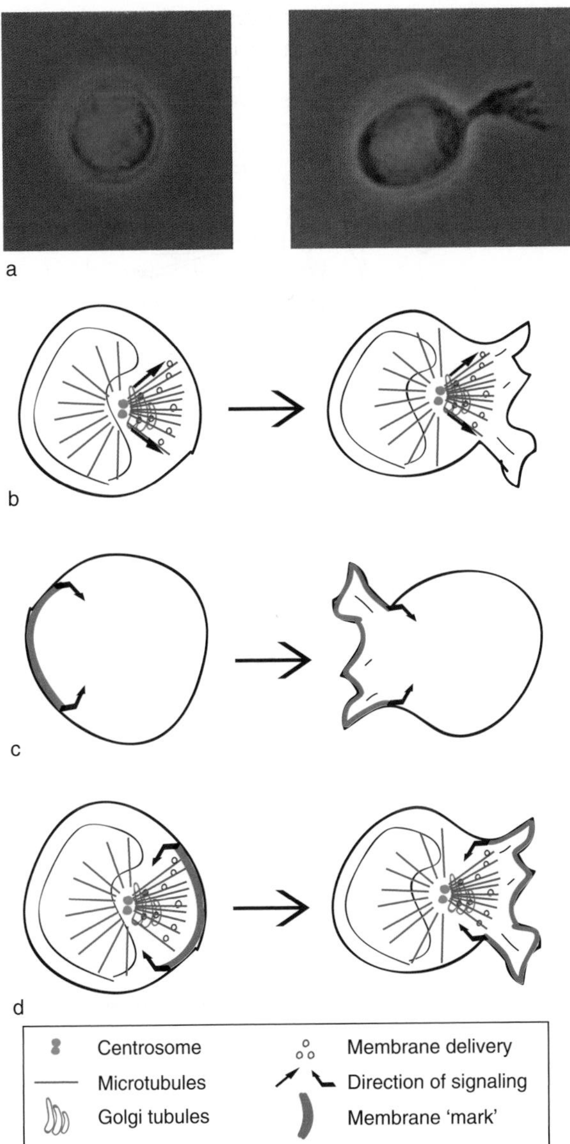

Figure 3 Ways neuronal asymmetry might be produced and might induce sprout formation. (a) The intracellular polarization of neurons coincides with the first sprout. Neurons were fixed after 1–3 h, and the centrosome was labeled in red and the nucleus in blue. (b) Forces originating from the asymmetric centrosome provide directionality and could regulate where the sphere is disrupted. (c) Disruption of symmetry might be explained by a membrane 'mark' that directs neuronal sprouting. (d) Another possibility is that inherited membrane asymmetries signal the inside of the cell to produce and maintain a cytoplasmic polarization.

cytoskeleton and selective membrane transport. The number of neurites will depend on the intrinsic program of the neuron: monopolar, bipolar, or multipolar. In fact, microtubules irradiate from the centrosome/MTOC in all directions of the sphere and are transported into all neurites. Actin is also similarly responsible for the presence of filopodia and lamellipodia in all neurites, and all neurites contain membrane carriers and mitochondria. Obviously, suppression of microtubule polymerization or membrane trafficking or stabilization of actin dynamics results in the lack of neurite outgrowth. The important issue, however, is which mechanisms neurons use to select only one of these neurites for fast growth, so that it later becomes the axon. One possible mechanism is through the simple 'passive' accumulation of 'neurite-making forces' in the place where the initial breakage was made. In this scenario, the neurite formed first will eventually become the axon because it is the place where these forces acted for the longest period and, consequently, where the growth threshold signal is surpassed first. Since the decision on the place on the sphere where the first neurite forms is defined by the positioning of the centrosome, which in turn is a consequence of the plane of cleavage of the last mitosis, this hierarchical, 'quantitative' mechanism implies, necessarily, that the decision of axon destiny is an intrinsic, mitotically predisposed process (**Figure 4(a)**).

The alternative mechanism is a 'qualitative' one, mediated by the 'sorting' to a single neurite, not necessarily the first, of molecules (called 'polarizers') with the specific capacity to affect microtubule–actin–membrane dynamics. This model of axon selection is illustrated in **Figure 4(b)**. In line with such a predicted mechanism, it has been reported that certain molecules, such as the ganglioside converting enzyme PMGS or the small guanosine triphosphatases (GTPases) Rap1B and Cdc42, are preferentially transported to one neurite before it becomes the axon. Clearly, the polarization of such molecules to the tip (growth cone) of a single neurite can induce a special rate of growth (axonal) by selectively affecting the stability of the actin filaments or microtubules or by increasing membrane fusion or, most likely, by affecting all them in line or in parallel. Consistently, both PMGS and Rap1B signal to the pleiotropic PI3K, whose phosphorylating activity impinges on members of the actin and microtubule regulating proteins. Other molecules whose polarization before axonal growth can induce this growth are the Rho GTPases, a family of proteins with a very well elucidated role in the control of actin cytoskeleton dynamics. Studies of neurite elongation have shown that in general, RhoA inhibits and Cdc42 and Rac1 promote neurite elongation. RhoA activation induces stress fibers in starved fibroblast and leads to a more rigid actin cytoskeleton, whereas Cdc42 induces filopodia and Rac, membrane ruffles. If these actions apply to neurons, one would predict that Rho activity is low and Cdc42 and Rac1 activity are high in the growing

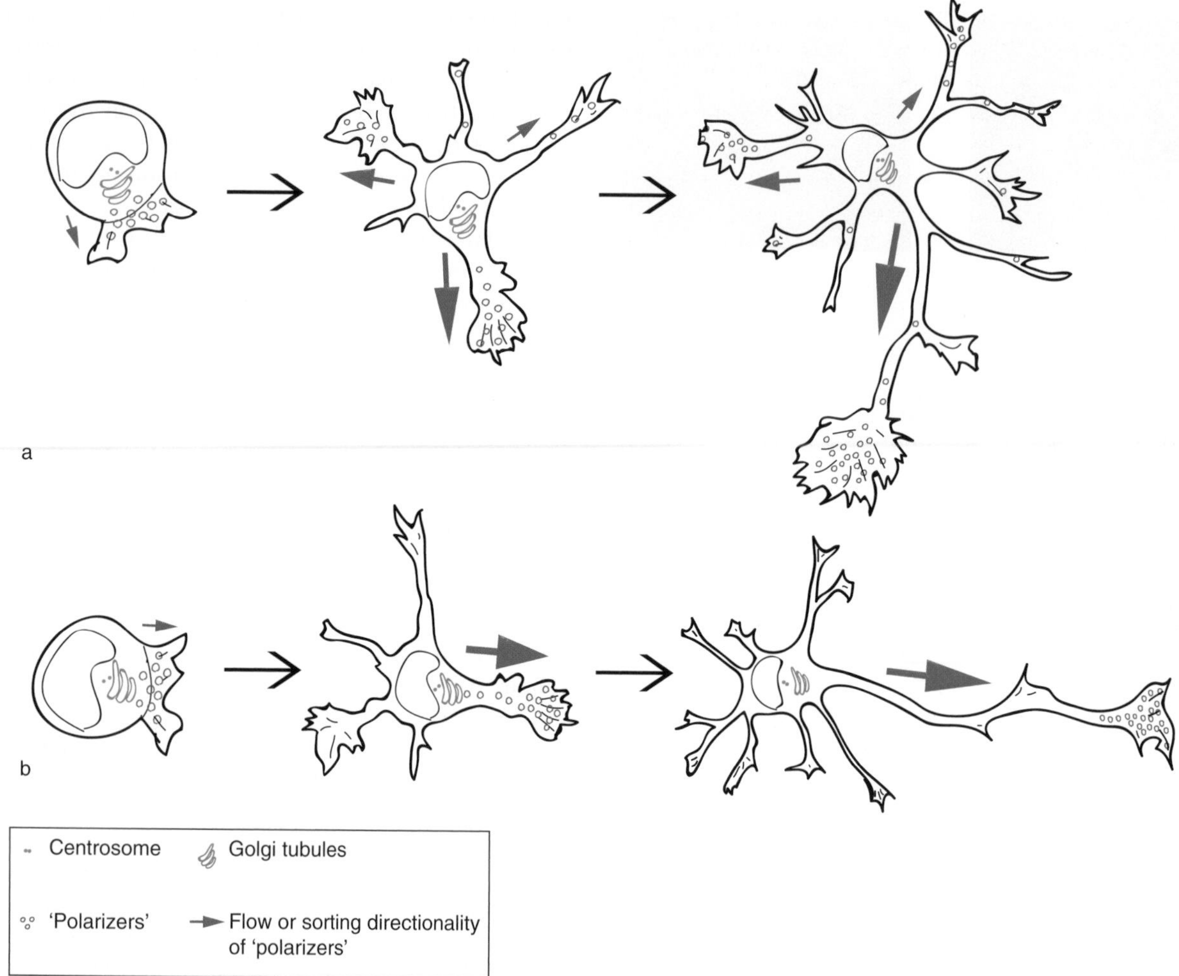

Figure 4 Sorting or cytoplasmic flow could select one neurite for sustained growth. (a) A quantitative mechanism would imply the delivery of 'polarizers' to all neurites, but with a preference for the one that began to grow first. (b) A second 'quantitative' model suggests a sorting process selectively sorting to the axon.

axon. In line with this, inhibition of RhoA has been reported to lead to the outgrowth of multiple axons, and the inhibition of Rac or Cdc42 inhibits axonal outgrowth. As mentioned already, Cdc42 has been found to be accumulated before axonal outgrowth in a single growth cone. From the latter mechanism arises a most basic question: what determines the sorting of the polarizers? One possibility is that it is the consequence of the time–hierarchy quantitative process described in the first place. This would not only reconcile quantitative and qualitative mechanisms but would also stress that the establishment of neuronal polarity is absolutely intrinsically defined. In this scenario, the extracellular environment would play a role in sustaining and guiding the already formed axon. The observation that the activity of PMGS increases locally on the addition of growth factors, an event also observed with the insulin-like growth factor receptor, is additional evidence of the importance of extracellular stimuli in growth sustenance. Moreover, there is abundant literature on the role of different types of extracellular ligands on the induction of axonal or dendritic growth.

A valid conclusion from all these findings is that a quantitative and centrifugal intracellular mechanism confers preferential, but not exclusive, growth capacity on a single neurite, with the extracellular environment playing a critical role in confirming such a predisposition. Thus, neuronal polarity is the result of the synchronous and sequential effect of intrinsic and extrinsic inputs.

Sustaining Polarized Growth

The sudden growth of the 'selected' neurite is a most important aspect of neuronal polarization, signaling to the cell that the normal one-axon, several-dendrites phenotype is finally established. A corollary is that axonal fate is active whereas dendritic fate is passive (lack of sudden growth destines the remaining neurites of the cell to become dendrites). Obviously, final dendritic architecture requires more processes than a growth rate different from that of axons, and therefore dendritic destiny will be an active process as well. Science is only beginning to understand the mechanisms that lead to dendritic determination once the axon is formed, but the mechanisms responsible for rapid growth of the newly formed axon are fairly well understood. One important, well-described mechanism required for rapid growth is the destabilization of the actin skeleton in the growth cone of the selected neurite. Thus, neurons treated with the actin depolymerizing drug cytochalasin D extend, not one, but several rapidly growing neurites with axonlike characteristics. More important, local administration of this drug to any growth cone provides that selected neurite with axonal growth capacity. Another critical component behind rapid growth is the stability of microtubules. Indeed, the axon contains a higher proportion of acetylated, and thus more stable, tubulin, and a dephosphorylated version of microtubule binding protein tau, with a differential capacity to regulate microtubule dynamics depending on its phosphorylation state, is present in a distal–proximal gradient in axons.

The process of induction of rapid growth and of its maintenance requires the contribution of certain sustaining molecules, probably the same as the aforementioned 'polarizers.' Thus, for instance, the Par proteins described earlier are essential for axonal growth. The role of Par in axonal growth is mediated via kinases, such as the atypical protein kinase C ζ and PI3K, with a profound effect on cytoskeleton dynamics. Recently, it has been shown that the inhibition of the serine–threonine kinase GSK-3 leads to the outgrowth of supernumerous axons and thus is involved in regulating polarity. GSK-3 is interesting since among its many substrates are several microtubule-binding proteins, such as APC, tau, and Map1B. Its activity could therefore control polarization via the regulation of microtubule stability, microtubule bundling, and the dynamics at microtubule ends. In fact, phosphorylation of Crmp-2, a microtubule-binding protein whose ectopic overexpression leads to the outgrowth of extra axons, by GSK-3 reduces its microtubule-binding activity. On the other hand, GSK-3-mediated phosphorylation of APC changes the capacity of APC to bind and stabilize microtubules, influencing neuronal polarity and axon outgrowth.

Thus, any molecule with a critical role in the modulation of one or more of the critical determinants of the neurite-producing forces, whether microtubules, actin, membrane fusion/retrieval, or other events leading to large structural changes, will, to a greater or lesser degree, have an effect on neuronal polarity.

Conclusions and Summary

- *In situ* excitatory pyramidal neurons polarize in the ventricular and subventricular zone immediately after the last neurogenic division by sprouting two opposite neurites, which support migration.
- *In vitro* hippocampal neurons polarize immediately after mitosis: The neurite grown first eventually becomes the axon. Whether the first neurite *in situ* likewise becomes the axon still needs to be confirmed.
- The position from which the first neurite sprouts is marked by the presence of the centrosome and membrane organelles. It remains to be established firmly whether such spatial correlation alone determines the formation of the first sprout or whether a preexisting neurite-forming signal is needed.
- The first neurite grows as the axon because it is in the spot receiving growth information for the longest amount of time. It remains to be established whether the sorting of specific molecules primes such polarized delivery.
- A second neurite can become the axon if the first is lost. It remains to be established whether the selection of the new axon follows the same principles that govern the selection of the first.
- Dendrites begin to differentiate only after axonal destiny is conferred on one neurite.

See also: Axon Guidance: Building Pathways with Molecular Cues in Vertebrate Sensory Systems; Axonal Pathfinding; Dendrite Development, Synapse Formation and Elimination; Glutamate Regulation of Dendritic Spine Form and Function; Interstitial Axon Branching/Collateral Elimination; Neurogenesis in the Intact Adult Brain; Neurogenesis and Neural Precursors, Progenitors, and Stem Cells in the Adult Brain; Vesicular Sorting to Axons and Dendrites.

Further Reading

Bradke F and Dotti CG (2000) Changes in membrane trafficking and actin dynamics during axon formation in cultured hippocampal neurons. *Microscopy Research and Technique* 48: 3–11.

Bradke F and Dotti CG (2000) Establishment of neuronal polarity: Lessons from cultured hippocampal neurons. *Current Opinion in Neurobiology* 10: 574–581.

Craig AM and Banker G (1994) Neuronal polarity. *Annual Reviews in Neuroscience* 17: 267–310.

Da Silva JS and Dotti CG (2002) Breaking the neuronal sphere: Regulation of the actin cytoskeleton in neuritogenesis. *Nature Reviews of Neuroscience* 3: 694–704.

De Anda FC, Pollarolo G, Da Silva JS, Camoletto PG, Feiguin F, and Dotti CG (2005) Centrosome localization determines neuronal polarity. *Nature* 436: 704–708.

Dent EW and Gertler FB (2003) Cytoskeletal dynamics and transport in growth cone motility and axon guidance. *Neuron* 40: 209–227.

Dotti CG and Banker GA (1987) Experimentally induced alteration in the polarity of developing neurons. *Nature* 330: 254–256.

Dotti CG, Sullivan CA, and Banker GA (1988) The establishment of polarity by hippocampal neurons in culture. *Journal of Neuroscience* 8: 1454–1468.

Goslin K and Banker G (1989) Experimental observations on the development of polarity by hippocampal neurons in culture. *Journal of Cell Biology* 108: 1507–1516.

Götz M and Huttner WB (2005) The cell biology of neurogenesis. *Nature Reviews of Molecular and Cell Biology* 6: 777–788.

Horton AC and Ehlers MD (2003) Neuronal polarity and trafficking. *Neuron* 40: 277–295.

Kriegstein AR and Noctor SC (2004) Patterns of neuronal migration in the embryonic cortex. *Trends in Neuroscience* 27: 392–399.

Noctor SC, Martinez-Cerdeno V, Ivic L, and Kriegstein AR (2004) Cortical neurons arise in symmetric and asymmetric division zones and migrate through specific phases. *Nature Neuroscience* 7: 136–144.

Wiggin GR, Fawcett JP, and Pawson T (2005) Polarity proteins in axon specification and synaptogenesis. *Developmental Cell* 8: 803–816.

Axonal and Dendritic Transport by Dyneins and Kinesins in Neurons

L S B Goldstein, University of California San Diego School of Medicine, La Jolla, CA, USA

Size Problems in Neurons

Neurons are usually the largest and most highly polarized cells in multicellular organisms. In addition to the cell body, neurons have a dendritic compartment and an axonal compartment. Focusing on the axon, some simple size calculations reveal just how substantial the burden of their size is. Consider a 1 m motor neuron in a human adult. The 1 m axon can have a diameter ranging from a few tenths of a micrometer up to several micrometers as a consequence of variability in growth in diameter. The diameter of the cell body is on the order of 20–30 μm. Thus, the axon can be as much as or more than a thousand times the volume of the cell body. In addition to extraordinary volume, neurons must move materials such as synaptic proteins and vesicles synthesized in the cell body an enormous distance to reach the synapse where they have essential functions. Since diffusion is inadequate to support distant synapses and large axons and dendrites supplied with appropriate materials, neurons use a supply system composed of microtubule tracks and two types of microtubule-dependent molecular motor proteins, called kinesins and dyneins. These supply systems physically move the required large amounts of material from the cell body to the synapse and also keep the axon and dendrite supplied with ion channels, cytoskeletal proteins, membrane proteins, organelles, and many other components needed for axonal function and viability. In addition, although neurons use electrical signaling in the form of the action potential to communicate with targets and to drive electrical networks, a great many growth factors are moved to and then secreted at axonal termini to signal forward to targets. Simultaneously, targets release growth factors, which must complex with receptors and then be moved back to cell bodies to send chemical signals. Collectively these movements define axonal and dendritic transport. Axonal transport is better understood and consists of fast and slow components. In the case of movement from cell body to synapse, axonal transport is called anterograde, whereas from synapse to cell body, it is called retrograde. Dendrites have comparable burdens but also have a more complicated organization of cytoskeletal elements and a more highly branched structure with shorter processes than the axon. As a result, less is known about transport in dendrites although most work suggests that similar systems and principles are at work.

Microtubule Organization

Microtubules have structural, kinetic, and biochemical polarity (**Figure 1**). 'Minus' ends in nonpolarized cells are anchored at the centrosome whereas 'plus' ends radiate out toward the cell periphery. In axons, minus ends point toward the cell body, and plus ends point toward the synapse. Dendrites are more complex, with microtubules generally having a mixed polarity orientation in proximal dendrites but having plus end distal organization in the distal dendrites that have been studied. As discussed below, molecular motor proteins have defined directions of movement with respect to microtubules. Thus, a motor protein that walks from minus to plus along a microtubule is likely to play a role in anterograde transport in axons whereas its role in dendrites may be more complex. Similarly, motors that move from plus to minus along the microtubule will have retrograde functions in the axon and, again, more complex roles in the dendrites. In fact, because of the complex organization of microtubules in dendrites, principles relating direction of movements and microtubule organization to functions in the dendrites are at present poorly understood.

Assays for Axonal Transport

Many types of assays have been used to define the characteristics of axonal transport. One type of assay relies on radioactively labeling proteins during synthesis in the cell body and then measuring the rate at which different types of radioactive proteins appear at different distances along the axon (**Figure 2**). This type of assay led to the discovery that axons have both fast transport, which consists of proteins associated with vesicles, and slow axonal transport, which consists largely of soluble proteins and cytoskeletal proteins such as tubulin, actin, and neurofilaments. A second type of assay relies on physically interrupting the movement of materials along the axon by crushing a nerve containing many axons or by tying a thread around the nerve (ligation) to block physical movement (**Figure 3**). This assay has been used most commonly with sciatic nerve axons in which cell bodies are clustered together in the vicinity of the spinal cord and axonal termini are distributed in the periphery of the animal. Thus, ligating a sciatic nerve blocks both anterograde transport and retrograde transport and

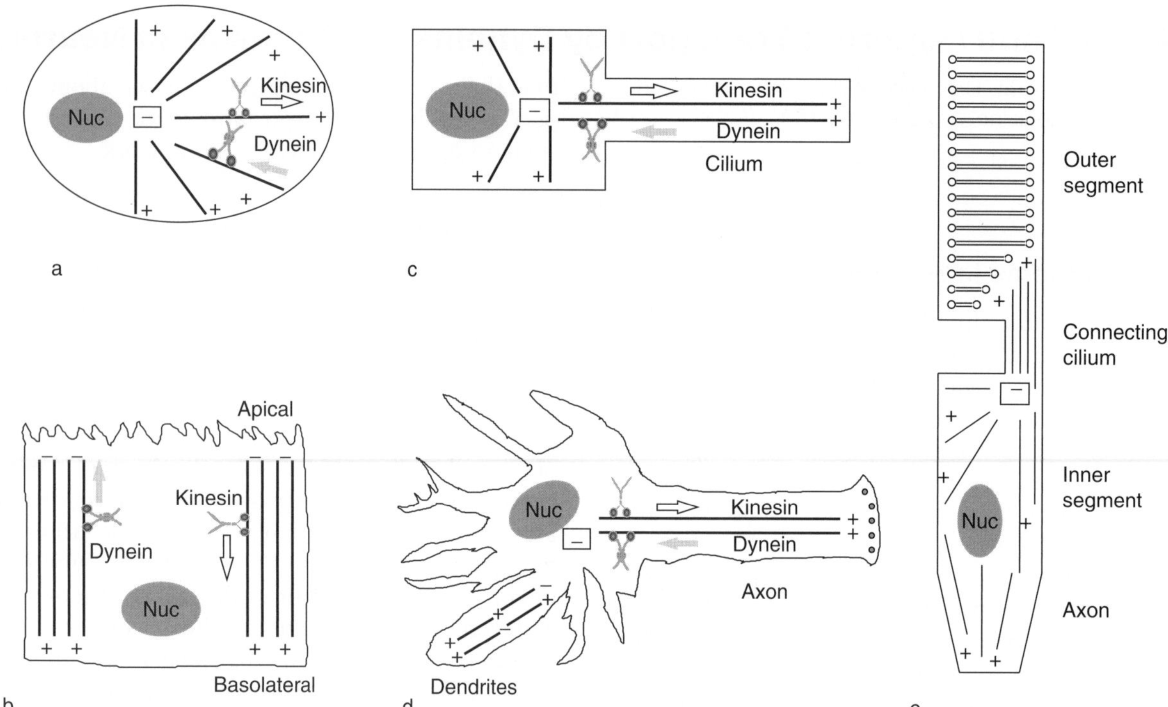

Figure 1 Schematic diagrams of microtubule organization in a variety of cell types. (a) Nonpolarized cell, (b) polarized cell, (c) ciliated cell, (d) neuron, (e) rod photoreceptor. + and −, polarity of microtubules; Nuc, nucleus. Likely directions of motor-mediated movements are indicated (arrows). Reproduced with permission from Goldstein LS and Yang Z (2000) Microtubule-based transport systems in neurons: The roles of kinesins and dyneins. *Annual Review of Neuroscience* 23: 39–71.

leads to accumulation over time of anterograde materials on the proximal side (nearest the spinal cord) and retrograde materials on the distal side (nearest the periphery) of the ligature. Accumulated materials can be assessed either by biochemical means, such as electrophoresis and Western blotting, or by microscopic methods, such as immunofluorescence for specific proteins and organelles. These types of assay have not only led to the realization that different materials are moved at different rates in axons; they have also revealed that some materials that move from cell body to axonal terminus, such as the anterograde kinesin molecular motors, do not return to the cell body and are thus likely to be degraded at the terminus. These assays can be used to characterize both normal behavior and behavior in a variety of mutations that disrupt axonal transport or cause disease by interfering with proper axonal functions. Finally, another important type of assay of axonal transport relies on direct visualization of moving vesicles and organelles in the light microscope in live cells; these methods often use specific labeling of proteins or organelles with fluorescent molecules.

Kinesins

Kinesin molecular motor proteins were originally discovered in the giant axon of the squid. Although not recognized at the time, this new type of molecular motor protein, although biochemically distinct from myosins and dyneins, is the founding member of a diverse but related collection of molecular motor proteins that were deployed for many different functions in eukaryotic cells. The founding member of this family, originally called conventional kinesin, and now called kinesin-1, is a heterotetrameric protein composed of two kinesin heavy chains (KHCs) and two kinesin light chains (KLCs) (**Figure 4**). The motor activity of kinesin-1 resides in KHC in an N-terminal region of approximately 350 amino acids. This protein domain has been shown to be capable of generating adenosine 5′-triphosphate (ATP)-dependent movements along microtubules *in vitro*, has been crystallized and solved structurally, and has proven to be conserved and shared among all kinesin-related motor proteins. The nonmotor portion of KHC has two structural domains. One structural domain is an alpha-helical coiled coil that mediates KHC dimerization. The other region is of unknown structure and complexes with the KLCs to form the so-called tail or cargo binding domain of kinesin-1. KLC has a short coiled coil domain at its N-terminus that may mediate the dimerization and/or association with KHC, and then has a region of six relatively well-conserved tetratricopeptide repeat (TPR) units, whose structure has been solved in other proteins and which are predicted

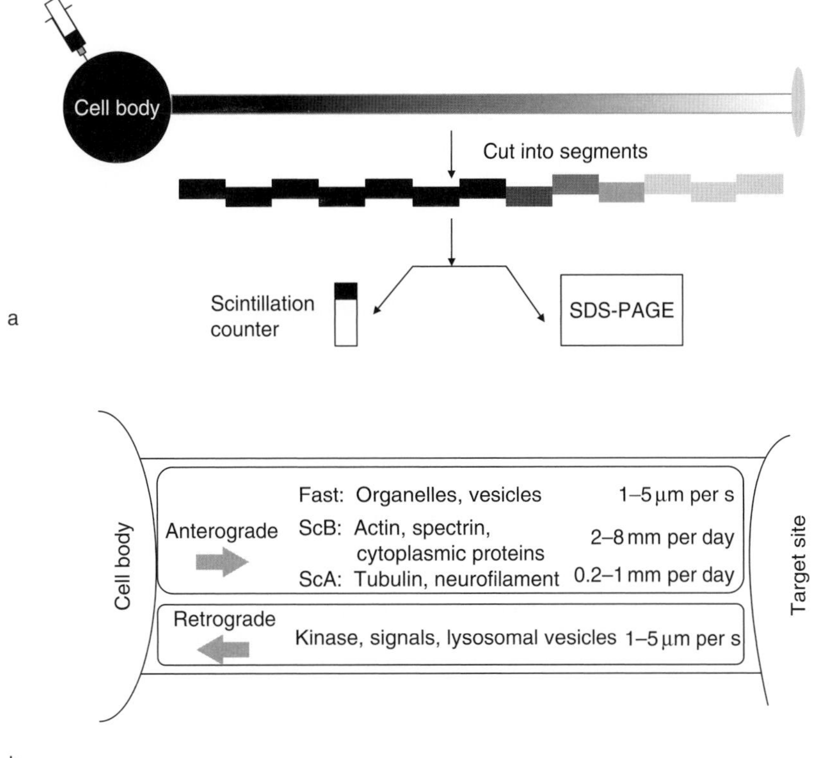

Figure 2 Typical axonal transport assays and pathways. (a) Pulse-labeling experiments are typically used for measuring axonal transport rates and character. Radiolabeled protein precursors are injected into a region containing the cell bodies of particular neurons (e.g., dorsal root ganglion). After different times, the nerves are harvested and cut into segments. The segments can be used to evaluate rates of transport using scintillation counting, sodium dodecyl sulfate polyacrylamide gel electrophoresis (SDS-PAGE), and autoradiography, or other analytical methods. (b) Diagram of anterograde and retrograde transport pathways indicating potential cargoes and transport rates. Reproduced with permission from Goldstein LS and Yang Z (2000) Microtubule-based transport systems in neurons: The roles of kinesins and dyneins. *Annual Review of Neuroscience* 23: 39–71.

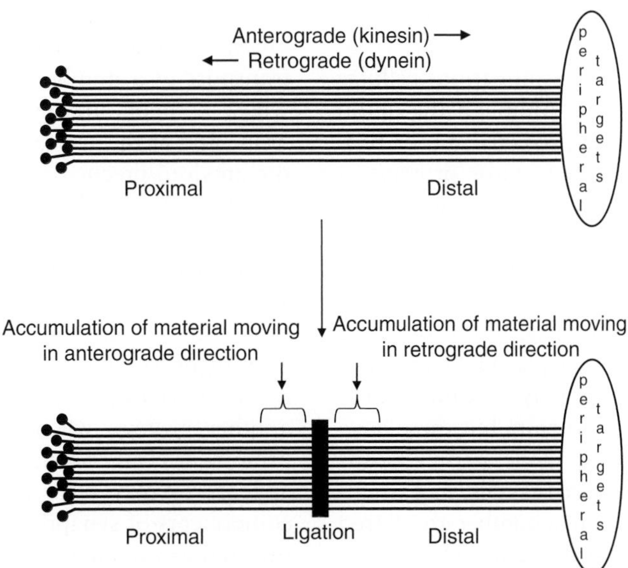

Figure 3 Schematic diagram of sciatic nerve ligation assay. Cell bodies clustered near the spinal cord define the proximal end of the nerve and are the source of materials moving in the anterograde direction. Axonal contacts with peripheral targets at the distal end of the nerve are the major source of material moving in the retrograde direction. Ligation (tying off) of the nerve results in accumulation of materials at the site of ligation.

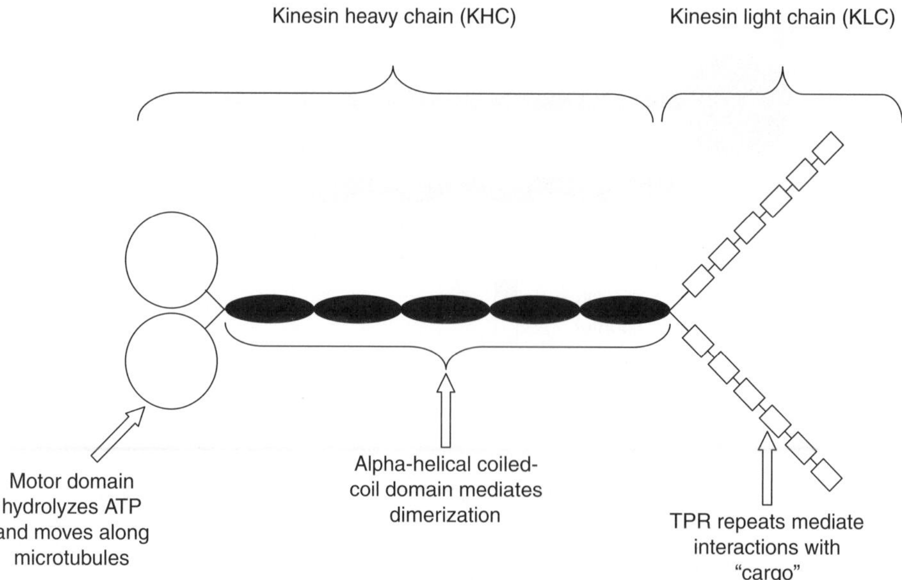

Figure 4 Schematic diagram of kinesin-1 structure. Kinesin-1 is a heterotetramer consisting of two kinesin heavy chain (KHC) subunits and two kinesin light chain (KLC) subunits. KHC contains the motor domain, the dimerization domain, and a short terminal domain that interacts with the KLC subunits. Cargo binding may be mediated by KHC and KLC together or individually. KLC also plays roles in regulating KHC motor domain activity. ATP, adenosine 5′-triphosphate; TPR, tetratricopeptide repeat.

to form a triple alpha-helical groove composed of three TPR units. Thus, KLC is predicted to have two of these triple alpha-helical grooves that may be important in cargo binding and kinesin regulation.

The sequence of the KHC motor domain led to the discovery that eukaryotic genomes encode many different kinesin motor proteins. These proteins are all predicted to share a common motor domain but to have a variety of structurally diverse 'tail' domains attached that may harness force-generating activities by kinesins to a variety of cargoes in various cellular types and functions (**Figure** 5). Although the initial expectation was that all these predicted motor proteins would be plus end-directed motor proteins, a surprising degree of diversity in behavior was discovered. Some of these kinesin motor proteins mediate minus end-directed movements, and some may not be motor proteins at all and may mediate the rapid depolymerization of microtubules. Both of these types of proteins nonetheless couple ATP hydrolysis to their activities. Some kinesins may have other functional activities in the motor domain, but these have not yet been clearly elucidated. While many kinesins appear to play roles in nonneuronal cells, in particular in the process of cell division, a number are shared between neurons and nonneuronal cells. Many kinesins are specific only to neurons, where they may be involved in axonal or dendritic transport processes. For example, in mammals, on the order of 10–15 of the known kinesins among the 50 or so predicted to be encoded in the genome are specialized for cell division. In *Drosophila*, because of the smaller genome, proportionally fewer are used in mitosis and in neurons. A key question is which motor proteins move which types of vesicles and organelles in each type of neuron.

In neurons, although there is a large diversity of different kinesin motor proteins and cargoes, some simple principles have begun to emerge. For example, kinesin-1 appears to play roles in the anterograde transport of a number of different types of vesicles whose contents are needed along the axonal membrane and at the synapse, such as vesicles containing the amyloid precursor protein (APP), whose processing may cause Alzheimer's disease. Kinesin-1 also appears to play a role in the transport of neurofilaments, which are required for radial growth of axons, and in the movement of mitochondria. It is interesting to note that recent work suggests that mitochondrial movement by kinesin-1 may be mediated by a special form of kinesin-1 in which KLC is replaced by another protein, whose name is Milton. A second prominent class of kinesins, referred to as kinesin-3, appears to play roles in the movement of protein components of synaptic vesicles and may also play a role in the movement of mitochondria. Other kinesins may move mRNA particles, components of postsynaptic complexes and postsynaptic receptors, as well as other organelles, including peroxisomes, elements of endoplasmic reticulum, and the Golgi apparatus.

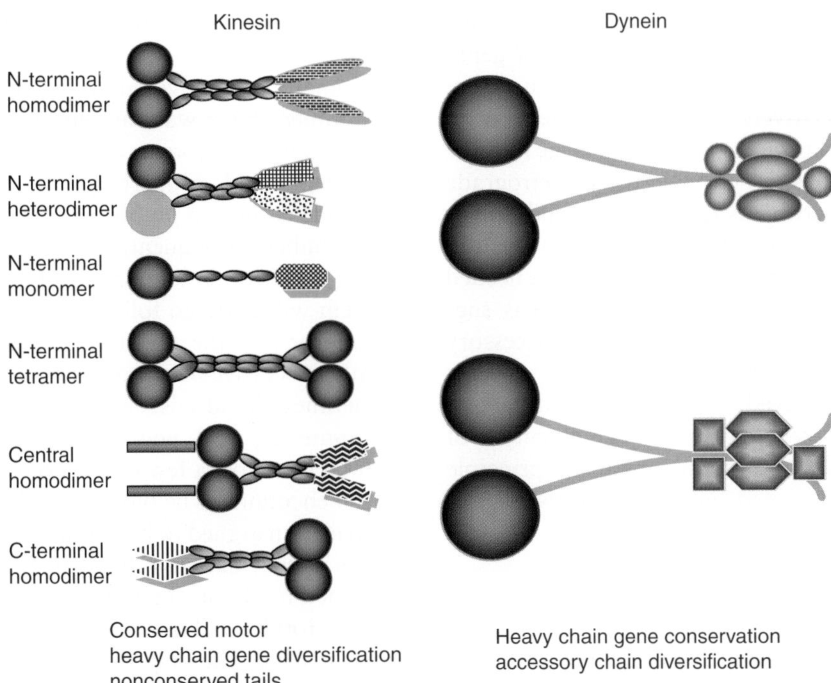

Figure 5 Principles of kinesin and dynein diversification. Kinesins appear to have evolved by gene duplication and divergence such that a conserved motor (shaded ovals) has become harnessed to a wide array of diverse tail domains. The location of the motor domain in the heavy chain polypeptide allows kinesin motors to be divided into several groups: an N-terminal group, a central group, and a C-terminal group. In addition, N-terminal kinesins have been found as homodimers, heterodimers, homotetramers, and monomers. Cytoplasmic dyneins appear to have conserved the bulk of their principal motor polypeptide and have generated far fewer divergent genes. Accessory chain diversification may be the principal mechanism for functional specialization. Reproduced with permission from Goldstein LS and Yang Z (2000) Microtubule-based transport systems in neurons: The roles of kinesins and dyneins. *Annual Review of Neuroscience* 23: 39–71.

An important and unresolved question is whether the movements of nonmembrane particles and proteins are mediated by direct interactions with the motor proteins or by binding to vesicle or organelle membranes and hitchhiking.

Another type of kinesin used for movements in some types of neurons is called kinesin-2. This kinesin is generally a heterotrimer of two nonidentical motor subunits associated with a third nonmotor domain protein, which may form part of the tail. This class of motor protein has been found to be used primarily for movement of materials in cilia and is thus necessary for the construction of cilia and perhaps for signaling mediated by cilia. In neurons in particular, this class of kinesin appears to be used most prominently in sensory neurons of various types, which often use modified cilia as their sensory endings within which receptors for odors or chemical molecules are bound and signals are transmitted. Vertebrate photoreceptors have a modified cilium as their light-sensing domain, the so-called outer segment, and this modified cilium also appears to require the use of this kinesin-2 type of motor protein for the movement of photoreceptive components such as opsin.

Dyneins

Dyneins provide a radically different example of the way in which motor proteins may be utilized to generate force and movement within neurons. In fact, most of the common principles learned from studying kinesins or myosins, which appear to have a shared evolutionary ancestor, appear not to be true for dyneins. A typical cytoplasmic dynein protein complex such as that used in neurons is many times the size of a native kinesin protein complex. The dynein complex generally consists of two copies of a dynein heavy chain and then several intermediate, light intermediate, and light chains, which together with the dynein heavy chains form the dynein complex and define its functions (**Figure 5**). The motor region of dynein is present in the heavy chains and is structurally and mechanistically different from kinesins and myosins, as it appears to use a mechanism most similar to that found in AAA ATPases. Cargo binding and regulatory activities appear to be conferred by the other subunits of dynein. Unlike kinesins and myosins, cytoplasmic dynein heavy chains are not highly duplicated and diversified in eukaryotic genomes.

Mammals have at most two or three cytoplasmic dynein heavy chain genes and a few each of genes encoding the other subunits. It has been argued that this small and relatively nondiverse collection of dynein heavy chain proteins carry out a large number of different functions in neurons, such as retrograde transport of compartments derived from endosomes and retrograde transport of components involved in signaling from the synapse to the nucleus. The needed diversity of dynein function and regulation is suggested to be provided by diversity in the accessory subunits. An unanswered question is why dyneins and kinesins have evolved so differently in eukaryotic organisms. It is conceivable that this divergent set of structural principles reflects an unknown principle about the regulatory needs in complex cells such as neurons.

Motor Regulation

Logic and experimental evidence suggests that molecular motor proteins must be regulated so that they do not move unless they are associated with an appropriate cargo. This issue is best understood for kinesin-1, which has been the subject of many studies of regulation and attachment. Early biochemical studies of kinesin-1 revealed that even though the protein could generate movement along microtubules *in vitro*, its ATPase activity measured in solution was much lower than could account for the physical parameters of movement. The answer was revealed in a series of studies that removed the tail domain from conventional kinesin in a series of recombinant constructs. These studies found that the kinesin motor domain with its tail domains attached has a very low ATPase activity while the kinesin motor domain with the tail domain removed has an ATPase activity that is high enough to account for the rates of movement seen *in vitro* and *in vivo*. In addition, structural studies revealed that kinesin-1 could exist in either a folded conformation, in which the tail and the head appeared to be associated, or in an extended conformation, in which the tail and the motor were separated; the latter conformation looks most appropriate for generating active movement. Although all the details have not yet been elucidated, considerable evidence now suggests that some combination of coupling kinesin-1 to an appropriate cargo combined with the action of soluble proteins leads to activation of the motor activity, perhaps by unfolding the motor. There is some evidence that phosphorylation mediated by signaling kinases may also be an important part of this regulation, although the details of how this is done are not yet clear. An intriguing possibility is that the signaling kinases might regulate motor protein association with appropriate cargoes in the cell.

Numerous biophysical studies have revealed that like other track-walking enzymes such as polymerases, motor proteins exhibit variation in a potentially regulatable property referred to as processivity. A simple definition of this property is that it refers to the number of sequential reaction cycles or steps along a microtubule or actin filament that a motor protein will undergo following a single encounter of the motor with the appropriate track. For example, a highly processive motor protein will take many steps, sometimes hundreds, along a microtubule after encountering it. A poorly processive motor protein will take at most a few or even only one step after an initial encounter. The biophysical basis of this behavior has been argued to be a result of the fraction of the motor's enzymatic/mechanochemical cycle that is occupied by the strongly bound state of the motor to the microtubule. Thus, a poorly processive motor is one for which the bulk of the reaction cycle is occupied by either a weak or a detached state of association of the motor to the microtubule whereas a highly processive motor is characterized by a large fraction of the cycle's being occupied by the strongly bound state. One consequence is that in solution, a single molecule of a poorly processive motor cannot walk very far, if at all, along the microtubule, whereas a highly processive motor can walk for great distances. Another consequence is that as the number of highly processive motors increases on moving particles, velocity is not expected to increase although the distance traveled is expected to increase. For poorly processive motors, both velocity and distance traveled are expected to increase as motor number increases. High processivity is a property that one might expect of a motor whose natural function is to move small vesicles over long distances. Indeed kinesin-1 appears to have as one of its functions the movement of small vesicles over long distances, and most work suggests that only a very few kinesin motors are found on most vesicles. Another consequence of this biophysical behavior is that processivity itself might represent a regulatory event for motor proteins. For example, as discussed below, kinesin-3 may be an example of clustering of single molecules of a poorly processive motor leading to highly processive behavior because one motor will remain bound while the others are detached. Cytoplasmic dynein in neurons also appears to exhibit this property, but the mediator of enhanced dynein processivity appears to be a protein complex called dynactin. Dynactin has been suggested to link dynein to some cargoes such as vesicles and organelles and also to enhance the processivity of the dynein motor by providing a static microtubule binding function that tethers the dynein vesicle complex to a microtubule when the motor protein is in the detached state. Dynactin may thus be required for processive movement of vesicles that have small numbers of dynein molecules on their surfaces.

Another possible regulatory mechanism has emerged from studies of kinesin-3, which as a soluble native protein appears to be monomeric and has been generally found to be a poorly processive motor protein. This motor protein has a lipid-binding domain in its tail, and there is intriguing evidence that induced clustering of lipids can bring multiple kinesin-3 subunits together on the surface of a vesicle. This clustering may allow the kinesin-3 proteins to oligomerize and initiate processive movement through the action of multiple motor domains that become clustered and coordinated.

A final biophysical behavior that has been observed and that might be related to processivity and the ability of one or more molecules to generate movement is that some motor proteins, such as kinesin-1, track along a single protofilament of a microtubule in a highly linear movement process. Motor proteins such as dynein, however, appear to wander over the microtubule surface as they generate net force and movement. Whether this is a biophysical curiosity or is in some way related to the differing functions of these two motor protein types remains to be answered.

Mechanisms of Motor Protein Attachment to Vesicles and Organelles

The mechanisms by which dynein and kinesin motor proteins attach to vesicles, organelles, and other cargoes in neurons remain poorly understood. A number of candidate proteins have been found by use of yeast two-hybrid assays and other biochemical measures of protein interactions. In many of these cases, however, it is unclear whether the interactions that have been identified are required for attachment of motor proteins to cellular cargoes or whether these interactions reflect transient regulatory interactions that may occur on the surface of the cargo. In only a very few cases has strong evidence been reported for a role of an identified protein or lipid to link a motor protein to a cargo. For example, for kinesin-3 there is strong evidence that lipid associations can in some circumstances be sufficient to link this motor protein to a vesicle. How important the lipid interaction is for normal kinesin-3 attachment, and whether other protein factors might be involved, is unclear. For kinesin-1, a number of proteins have been suggested to be important for attachment, including APP, which is implicated in causing Alzheimer's disease. Evidence for this function is strong but not yet definitive, and there is some evidence that these interactions may also, or alternatively, play a role in kinesin-1 regulation. Another set of candidates for kinesin-1 attachment proteins is the c-Jun N-terminal kinase (JNK) scaffolding proteins, which may play roles in damage signaling in axons and dendrites and for kinesin-1 attachment to vesicles. There is some evidence that JNK scaffolding proteins are part of the mechanism by which kinesin-1 may be attached to APP vesicles. For dyneins, there is widespread acceptance that dynactin is important for attachment of dynein to vesicles and organelles, but definitive evidence for this point is lacking, and there is some evidence that is inconsistent with this view. At this time, considerably more experimental evidence is needed to unravel the relationship between attachment of motors to cargoes and their regulatory control. Novel methods may be needed to detect and demonstrate the functions of these interactions.

Conclusion and Perspectives

Although much has been learned about the general features of the systems that underlie movement in axons and dendrites of neurons, it is clear that many important details are not yet known. In particular, how motor proteins are attached to cargoes and regulated is particularly poorly understood. Understanding these issues is not only important for understanding the special functions of neurons in networks and systems; an emerging theme is that insults to these systems may be important in the development of a wide range of neurodegenerative disorders such as Alzheimer's disease and Huntington's disease.

See also: Axonal Transport Disorders; Axonal mRNA Transport and Functions; Axonal Transport and Alzheimer's Disease; Axonal Transport and Neurodegenerative Diseases; Axonal Transport Tracers; Axonal and Dendritic Identity and Structure: Control of; Microtubule Associated Proteins in Neurons; Mitochondrial Organization and Transport in Neurons; Perceptual Learning and Sensory Plasticity; Slow Axonal Transport.

Further Reading

Blasius TL, Cai D, Jih GT, et al. (2007) Two binding partners cooperate to activate the molecular motor Kinesin-1. *Journal of Cell Biology* 176: 11–17.

Cai D, Hoppe AD, Swanson JA, and Verhey KJ (2007) Kinesin-1 structural organization and conformational changes revealed by FRET stoichiometry in live cells. *Journal of Cell Biology* 176: 51–63.

Goldstein LS (2003) Do disorders of movement cause movement disorders and dementia? *Neuron* 40: 415–425.

Goldstein LS and Yang Z (2000) Microtubule-based transport systems in neurons: The roles of kinesins and dyneins. *Annual Review of Neuroscience* 23: 39–71.

Hackney DD (2007) Biochemistry: Processive motor movement. *Science* 316: 58–59.

Hirokawa N and Takemura R (2005) Molecular motors and mechanisms of directional transport in neurons. *Nature Reviews Neuroscience* 6: 201–214.

Lawrence CJ, Dawe RK, Christie KR, et al. (2004) A standardized kinesin nomenclature. *Journal of Cell Biology* 167: 19–22.

Miki H, Okada Y, and Hirokawa N (2005) Analysis of the kinesin superfamily: Insights into structure and function. *Trends in Cell Biology* 15: 467–476.

Nakata T and Hirokawa N (2007) Neuronal polarity and the kinesin superfamily proteins. *Science's STKE: Signal Transduction Knowledge Environment* 372: pe6.

Pfister KK, Fisher EM, Gibbons IR, et al. (2005) Cytoplasmic dynein nomenclature. *Journal of Cell Biology* 171: 411–413.

Schroer TA (2004) Dynactin. *Annual Review of Cell and Developmental Biology* 20: 759–779.

Stokin GB and Goldstein LS (2006) Axonal transport and Alzheimer's disease. *Annual Review of Biochemistry* 75: 607–627.

Vale RD (2003) The molecular motor toolbox for intracellular transport. *Cell* 112: 467–480.

Relevant Websites

http://www.proweb.org – Kinesin Home Page.

http://www.dynein.org – The Dynein Homepage, Faculty of Biological Sciences, University of Leeds, UK.

Axonal Injury in Demyelinating Disease and CNS Injury

G A Criste, G J Kidd, and B D Trapp, Cleveland Clinic Foundation, Cleveland, OH, USA

Introduction

Multiple sclerosis (MS) is an inflammatory neurodegenerative disease of the central nervous system (CNS). MS is the leading cause of nontraumatic neurological disability in young adults in the United States and Europe, and approximately 2.5 million people worldwide suffer from this disease. The majority of MS patients initially exhibit a relapsing-remitting (RR) disease course (RR-MS) characterized by alternating episodes of neurological disability and recovery. Patients eventually convert to a secondary-progressive disease course (SP-MS) characterized by persistent and increasing neurological disability. Classically, MS has been characterized as a disease stemming from focal immune-mediated demyelination, which causes disability through block of conduction along demyelinated axons. More recently, however, the fundamental role of axonal injury in promoting irreversible disability in MS has been recognized and efforts to reduce or prevent axon loss have become a major focus of MS research.

Evidence that axonal injury is an early and persistent event in the progression of MS pathology comes from several sources. Histopathological studies have documented axonal loss in MS lesions (**Figure 1**), and neuroimaging approaches, including magnetic resonance imaging (MRI) and MR spectroscopy (MRS), have indicated that axon loss is pivotal and dynamic during the evolution of MS pathology. Proposed mechanisms of axonal injury in MS, however, remain diverse and speculative. Although generally considered to be sequelae of inflammatory demyelination, the limited success of immunotherapy in treating progressive disability in MS patients has diverted attention to neuroprotective therapies. The finding that diffused, irreversible axonal transection occurs early in the course of this complex disease has underscored the need to develop axonal neuroprotective treatments.

Axonal Pathology in MS Lesions

Histological evidence of axonal pathology was reported in the early literature on MS. These reports include descriptions of axonal swellings, axonal transection, and Wallerian degeneration, as well as discussions regarding the functional consequences of such pathology. However, the extent of axonal loss was controversial. In their classical works on MS, both Charcot and Marburg described MS pathology largely in terms of demyelination and reactive gliosis, and emphasized the relative sparing of axons in the lesions. In 1936, Putnam reported a 50% loss of axons in MS lesions from 11 patients. That same year, however, Greenfield and King reported normal axon densities in more than 90% of MS lesions from 13 patients that they studied. The differences between these studies were suggested to result from more sensitive axon staining in the latter.

More recent studies using contemporary technology such as MRI, MRS, and confocal microscopy have demonstrated that axonal transection begins at disease onset and that cumulative axonal loss provides the pathological substrate for the progressive disability that most long-term MS patients experience. Moreover, postmortem studies have shown that several histopathological abnormalities, including axonal loss, can be detected in the normal-appearing white matter (NAWM) and cortical gray matter of patients with MS, suggesting a more diffuse pathology than previously thought.

Axonal Damage Occurs during the Early Stage of MS

One important clue that axonal function was disrupted in MS lesions came from studies of amyloid precursor protein (APP). APP is transported by fast axonal transport and is normally present in axons at low levels that are not detectable by immunohistochemistry. APP accumulation in axons located in active MS lesions and at the border of chronic active MS lesions suggested axonal impairment in MS lesions. In particular, some APP immunoreactive structures exhibited the morphology of terminal axonal swellings, suggesting that axonal transection had occurred. The number of APP-labeled axonal swellings correlated with the degree of inflammation in the lesions. Similar swellings were detected with neurofilament immunostaining. Using confocal microscopy and computer-based three-dimensional reconstruction, axonal ovoids were identified as terminal ends of transected axons (**Figure 2(a)**). In one study quantifying axonal transections in 11 patients with disease duration ranging from 2 weeks to 27 years, extensive axonal destruction was demonstrated in cerebral white matter MS lesions. Active lesions contained over 11 000 terminal ends per cubic millimeter (**Figure 2(b)**), the edge of chronic active lesions contained over 3000 terminal ends per cubic millimeter, and the core of chronic

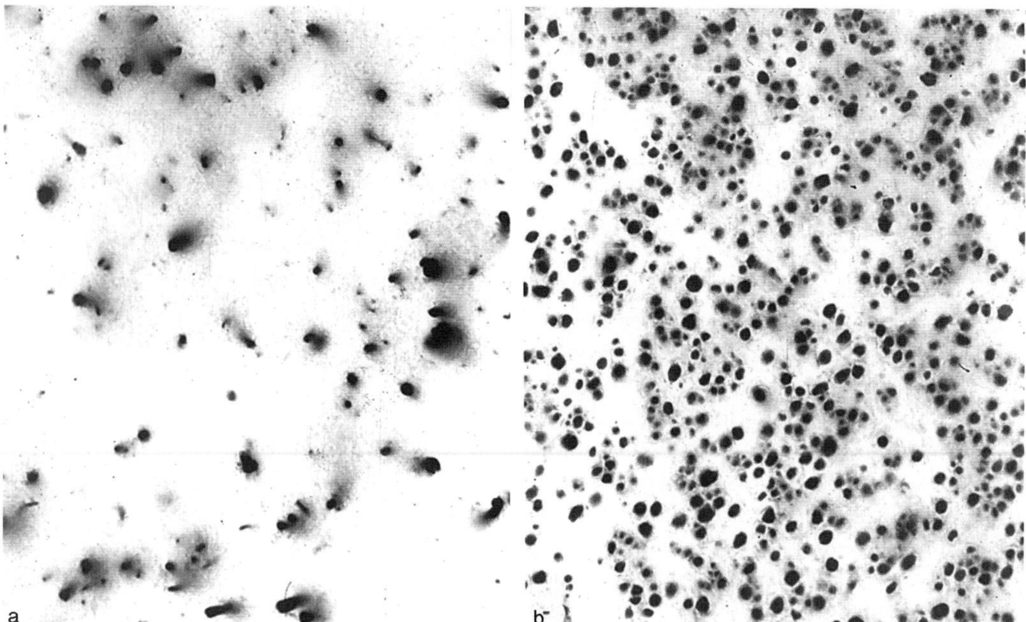

Figure 1 Axonal loss in MS patients. Axonal density is significantly decreased in a demyelinated area of MS spinal cord (a) compared to control cord (b). Reproduced from Bjartmar C, Yin X, and Trapp BD (1999) Axonal pathology in myelin disorders. *Journal of Neurocytology* 28(4–5): 383–395, with kind permission from Springer Science and Business Media.

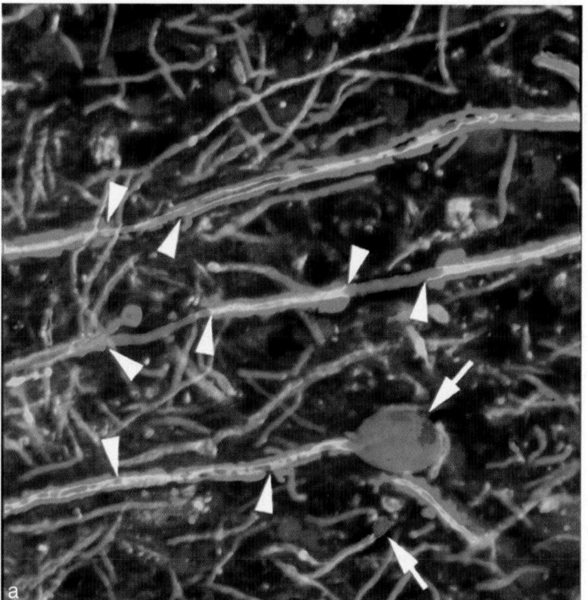

Activity of lesion	*Transected axons/mm³ (mean + sem)*
Active	11 236 + 2775*
Chronic active	
Edge	3138 + 688*
Core	875 + 246*
Nonlesion white matter	15 + 3*
Control white matter	0.7 + 0.7

b

Figure 2 Axonal transection during inflammatory demyelination. Axonal transection occurs during inflammatory demyelination and induces formation of terminal axonal ovoids (a, arrows). When quantified (b), transected axons are abundant in MS lesions and appear to correlate with inflammatory activity of the lesion (* = statistically significant at $p < 0.01$). Reproduced with permission from Trapp BD, Peterson J, Ransohoff RM, Rudick R, Mork S, and Bo L (1998) Axonal transection in the lesions of multiple sclerosis. *New England Journal of Medicine* 338: 278–285, Copyright © 1998 Massachusetts Medical Society. All rights reserved.

active lesions contained on average 875 terminal ends per cubic millimeter. In contrast, less than one transected axon was found per cubic millimeter in control white matter.

Together, these data demonstrate a positive correlation between axonal transection and degree of inflammation in cerebral white matter MS lesions undergoing demyelination. The presence of axonal ovoids in patients with short disease duration clearly demonstrated that axonal transection begins at an early stage of MS.

Axonal Loss Also Occurs in NAWM

It is well established that axons, once severed, will undergo relatively rapid Wallerian degeneration distal to the site of transection. Unlike axons, CNS myelin can persist for a long time after transection of the proximal axon. Remaining myelin sheaths may appear as empty tubes or as degenerating ovoids by microscopy. Despite this histological pathology, the white matter may appear grossly normal and appear unremarkable by conventional neuroimaging techniques.

Immunohistochemical evidence suggestive of Wallerian degeneration, such as discontinuous staining of axonal neurofilaments and presence of terminal axonal ovoids, has been demonstrated in NAWM from MS brains. Reductions in axonal density by 19–42% in areas without plaque have been reported in the lateral corticospinal tract of MS patients with lower limb weakness. By comparing axonal density in lesions and in NAWM from the cervical spinal cords of SP-S patients, one study showed how extensive axon loss can be. The average reduction in axonal density in lesions of lateral and posterior columns was 61%. In NAWM, however, the average decrease in axonal density was as much as 57%. By taking into account both decreased axonal density and reductions in tissue volume, another study showed that total axonal loss in the corpus callosum of MS patients averaged 53%; disease duration was between 5 and 34 years. The measured reduction in axonal density was only 34% in the same material. These findings emphasize the need to consider both tissue volume and axonal density to properly assess the degree of total axonal loss. These studies also indicate that white matter may appear normal judged by immunohistochemistry for myelin, or on MRI scans, but may still exhibit a considerable axonal dropout, especially in chronic patients with long disease duration.

Wallerian degeneration in NAWM has also been described by immunohistochemistry in an MS patient with short disease duration. The patient succumbed to a fatal brain stem lesion just after a 9-month history of RR-MS with few permanent neurological signs. Demyelinated lesions were not found in the spinal cord postmortem. However, the ventral column of the spinal cord, which contains axons projecting from the brain stem lesion, exhibited a 20% axonal loss. Microscopy revealed myelin ovoids and signs of myelin degradation by activated microglia, characteristic of Wallerian degeneration. Since much of the myelin remains, these can represent 'invisible lesions' as far as MRI and immunostaining for myelin are concerned.

Neuronal Pathology Also Occurs in MS Cortex

In addition to the more commonly described white matter locations, MS lesions can also involve the gray matter. The histopathological features and the clinical significance of such lesions are not completely understood. Gray matter lesions are difficult to detect macroscopically and histologically. On conventional T2-weighted MRI images, MS lesions in the cerebral cortex are much less obvious than white matter lesions. For these reasons, the frequency of cortical lesions has often been underestimated. Recent studies have shown that the use of gadolinium enhancement resulted in an increased detection of cortical lesions on MRI scans by 140%. A total of 26% of these enhancing lesions arose within or adjacent to the cerebral cortex. These results suggested that conventional MRI under-reports the presence of cortical lesions, when compared to neuropathological analysis.

In postmortem MS cortex, neurite transection is common, and TUNEL-positive neurons were detected in ~50% of chronic lesions, indicating that apoptotic loss of neurons can be significant. Interestingly, compared to white matter lesions, inflammation is substantially reduced in these lesions. Gray matter lesions contained fewer inflammatory cells, no perivascular cuffs, and consisted mainly of reactive microglia. T cells have been proposed to take a central role in the pathogenesis of MS. While T lymphocytes are found in MS white matter lesions, fewer are detected in cortical lesions that extended through both white and gray matter. In intracortical demyelinated lesions, T cell numbers were not different to T lymphocyte density in nondemyelinated cerebral cortex within the same tissue block.

It has been hypothesized that injury to neurons in cortical or subcortical lesion may provide the biological correlate to the cognitive dysfunction many MS patients experience. In fact, executive and cognitive functional deficits arise in 40–70% of these patients. Increased knowledge regarding mechanisms of neuronal damage in cortical MS lesions will contribute to the understanding of the functional significance of such lesions.

Mechanisms of Axonal Injury in MS

The pathophysiology of axonal injury in MS is poorly understood. It seems likely that different mechanisms may contribute to axonal degeneration at different stages of the disease. Elucidating the cellular and molecular mechanisms that promote axonal loss in MS is important to the development of future neuroprotective therapies.

Genetic Susceptibility to Axonal Injury

Current evidence indicates that interactions between multiple genes influence the outcome of MS in individual patients. One line of evidence suggesting a genetic influence on axonal damage in inflammatory demyelination comes from studies of Theiler's murine encephalomyelitis virus (TMEV) disease, a viral-induced model of inflammatory CNS demyelination. Infected animals with susceptible genetic backgrounds develop neurological impairment and pathological changes comparable to those in MS. For example, infected SJL/J mice develop chronic demyelination, neurological deficits, and extensive loss of axons in the spinal cord, while C57BL/6x129 mice are resistant to TMEV-induced disease. When the major histocompatibility complex (MHC) class I protein is absent from C57BL/6x129 mice, they develop similar demyelination to SJL/J mice after infection, but do not go on to develop functional disability. In contrast, C57BL/6x129 mice lacking MHC class II developed a chronic inflammation with neurological deficits and axonal degeneration in spinal cord white. These results indicate that MHC molecules are involved in susceptibility to demyelination and axonal damage and highlight the possible role that genetic differences can play on the development of pathology during inflammatory demyelination.

Genes encoding trophic factors that are involved in neuroprotection and repair are at least as important as the immune-related genes that are thought to promote pathology. In this context, ciliary neurotrophic factor (CNTF) is an interesting candidate to promote resistance to axonal injury in MS. One study described a correlation between CNTF null mutations and early onset of MS symptoms. This suggests that axonal loss, which is the basis of disability, may be accelerated in these individuals through lack of CNTF-mediated trophic support of neurons and oligodendrocytes. Although a second study reported no correlation between CNTF genotype and disease, the initial results accord well with observations of experimental allergic encephalomyelitis (EAE) in CNTF knockout mice. After induction of EAE with myelin oligodendrocyte glycoprotein (MOG), CNTF$^{-/-}$ mice showed a significantly earlier disease onset and a delayed recovery from relapses.

Axonal Injury and Inflammation

Current knowledge suggests that MS is a primary inflammatory demyelinating disease of the CNS, and that disease activity reflects CNS inflammation, although much of the disease may not be detected clinically. For example, most RR-MS patients exhibit progressive brain atrophy and persistent inflammation, identified as gadolinium-enhancing lesions by MRI, regardless of their clinical symptoms. Since axon pathology, including transection of axons, correlates with the degree of inflammation in MS lesions, early axonal transection may reflect vulnerability of the demyelinated axons to inflammation. Indeed, the inflammatory microenvironment contains a variety of substances that could potentially injure axons, including proteolytic enzymes, cytokines, oxidative products, and free radicals produced by activated immune and glial cells. The ability of cytotoxic CD8^{+} T-cells to mediate axonal transection has been described in MS tissue, EAE mice, and *in vitro*. Treatment of EAE mice with the AMPA/kainate glutamate receptor antagonist NBQX resulted in increased oligodendrocyte survival and reduced axonal damage.This suggests that excitotoxicity mediated by glutamate is involved in tissue damage in acute lesions. Directly or indirectly, inflammation may also affect energy metabolism of axons (discussed below). Inflammatory intermediates such as nitric oxide (NO) can act directly on mitochondria, while local inflammatory edema may interfere with the blood supply and potentially induce an ischemic mechanism of axonal degeneration.

Regardless of the molecular mechanisms involved, inflammation is a major factor behind accumulating axonal pathology at early stages of MS. For that reason, aggressive anti-inflammatory treatment during RR-MS may have indirect but long-term effects in preventing axonal injury in addition to important direct effects on inflammation.

Loss of Neuroprotective Effects of Myelin

The past decade has seen a deeper understanding of the intricate interdependence between the myelin-forming cells and its associated neuron. For example, axonal neuregulin has been found to control the proliferation and particularly the survival of oligodendrocytes and Schwann cells, ensuring a good match between the axon surface area requiring myelination and the number of surviving myelinating cells. Myelin-forming cells in turn have a profound influence on axonal morphology and physiology, affecting axonal diameter, neurofilament phosphorylation, cytoskeletal organization, and axonal transport rates. It follows therefore that disease processes that directly affect myelin-forming cells might also influence the underlying axon as well.

Peripheral nervous system (PNS) dysmyelinating diseases (in which myelin forms abnormally) are clearly associated with axonal changes. Charcot–Marie–Tooth neuropathy type 1 (CMT1) is a genetically heterogeneous group of chronic dysmyelinating peripheral neuropathies. Mutations affecting the myelin genes peripheral myelin protein 22 (PMP22), protein zero (P_0), and connexin-32 account for most CMT1 cases. While the electrophysiological consequences of dysmyelination account for some of the neurological symptoms in CMT1, the main contributor to clinical progression is axon loss, as determined by measurements of nerve conduction amplitudes and motor unit numbers. As in MS, this may be related to abnormal glial–axonal interactions. Thus, axonal degeneration may be a final common outcome in a wide variety of myelin-related diseases.

In the CNS, myelin protein defects also result in axonal pathology, even though they may cause little or no myelin abnormality. Myelin-associated glycoprotein (MAG), a member of the immunoglobulin gene superfamily, is a low-abundance myelin protein enriched in the innermost myelin membrane that lies adjacent to the axon. In MAG-deficient mice, postnatal myelination progresses as in wild-type animals. From the age of 5 weeks, however, progressive axonal atrophy including reductions in axonal caliber, neurofilament spacing, neurofilament phosphorylation, and increased Wallerian degeneration is observed. MAG inhibits neurite outgrowth and causes growth cone collapse *in vitro*, suggesting that it can modulate the axonal cytoskeleton. The findings indicate that MAG has direct or indirect long-term modulating effects on the cytoskeleton via axonal kinases or phosphatases.

The X-linked myelin proteolipid protein (PLP/DM20) is the major structural protein of compact CNS myelin. Mutations, deletions, or duplications involving the PLP gene cause Pelizaeus Merzbacher disease (PMD) and spastic paraplegia of varying severity in humans. In mice, PLP mutations result in premature oligodendrocyte death and dysmyelination. Many of these phenotypes, however, are 'gain-of-function' effects due to toxicity of misfolded proteins encoded by the mutated genes. In contrast, PLP-null mice remain competent to myelinate CNS axons of all calibers and to assemble compacted myelin sheaths. Ultrastructurally, however, the electron-dense 'intraperiod' lines in myelin remain condensed, correlating with its reduced physical stability. From the age of 6 weeks, PLP-deficient mice exhibit focal axonal swellings containing dense bodies and mitochondria in CNS regions containing mainly small-diameter axons. The accumulation and distribution of organelles and neurofilaments in axonal swellings indicate impairment of retrograde axonal transport. A length-dependent axonal degeneration in the absence of demyelination and inflammation in patients with PLP-null mutations has also been reported. Replacement of PLP with P_0 in PLP-null mice resulted in normal-appearing CNS compact myelin with PNS-like membrane ultrastructure. Although initially indistinguishable neurologically from wild-type or PLP-null mice, the P_0-CNS mice exhibited accelerated rates of axonal damage and rapid progression of neurological deficits and morbidity. These results suggest that PLP gene expression is required for axonal maintenance independent of its roles in myelin formation.

Unlike other myelin proteins, CNP is not crucial to myelin assembly but is essential for axonal maintenance. CNP is distributed throughout the cell soma and in noncompacted regions of myelin. In mice lacking CNP, myelin assembly was not visibly affected. However, at about 4 months of age, the mice developed motor deficits that progress with age. Histological analysis revealed late-onset axonal pathology characterized by abnormal axonal swellings and degeneration of many axons.

These data from myelin protein defects support dual roles of the oligodendrocytes, first, in the formation of the myelin sheath, and second, in the maintenance of the underlying axon. Understanding the signaling pathways and effector molecules, such as kinases and phosphatases, may lead to new therapeutics that promote axonal preservation in demyelinating and dysmyelinating diseases.

Mitochondrial Involvement in Axonal Injury

Recent evidence suggests a hypoxia-like metabolic injury as a pathogenetic component of axonal injury in MS. Although this model was largely derived from studies of white matter injury in models of ischemia and neurotrauma, recent observations suggest that such a mechanism operates in inflammatory brain lesions in MS as well. In this model, ischemia leads to ATP depletion. The resulting energy crisis impairs the function of ATP-dependent ion channels (e.g., Na^+–K^+ ATPase, Na^+–Ca^{2+} ATPase) leading to an elevation in intracellular Na^+ concentration. Accumulation of axoplasmic Na^+ through non-inactivating Na^+ channels, together with membrane depolarization, promotes reversal of Na^+–Ca^{2+} exchange and axonal Ca^{2+} overload. Ultimately, the pathological increase in intracellular Ca^{2+} drives Ca^{2+}-dependent enzymes to damage the axon.

Mitochondrial damage could occur through several mechanisms. Astrocytes, activated microglia, and macrophages in the CNS release substantial amount of NO in MS lesions. One mechanism of the toxic action of NO is the impairment of mitochondrial function, and NO-treated white matter has been shown to undergo ATP depletion and irreversible damage. Mitochondrial dysfunction has also been implicated by microarray analysis of postmortem MS motor cortex. Decreases in nuclear-encoded mitochondrial genes from four of the five respiratory chain complexes have been reported, which raises the interesting possibility that inherent defects in these organelles may arise in MS and further compromise energy production capacity.

Microvascular pathology also contributes a major role in the hypoxic MS pathology, starving mitochondria of oxygen and reducing their capacity to produce ATP. Edema within inflammatory lesions leads to focal disturbance of microcirculation with subsequent ischemia. Such a mechanism is likely to have greatest impact in CNS locations where room for tissue expansion is limited, such as the spinal cord and the optic nerves. Inflammatory damage to the vessel walls can also lead to activation of the clotting cascade resulting in local microvascular thrombosis and ischemic injury to the axons similar to a stroke.

Axonal compensation after demyelination may place increased demands on axonal mitochondria. Demyelinated axons revert to continuous rather than saltatory conduction by express Na^+ channels along the length of the axolemma. However, propagation of action potential under this circumstance involves much greater ion movements and hence ATP consumption. Recent reports on the beneficial aspects of Na channel blockers such as flecainide, phenytoin, and tetrodotoxin in attenuating axonal pathology in animal models of MS may reflect reduced energy demands on the demyelinated axons. Na^+ channel blockers can also reduce neuroinflammation through actions on the microglia and macrophages in EAE and MS. Thus, in addition to direct neuroprotective effect, Na channel blockers may reduce axonal degeneration through anti-inflammatory effects.

Strategies for Axonal Protection

Therapeutic strategies aimed at preventing axonal and neuronal damage are crucial in heading off permanent disability in MS patients. Most neuroprotective strategies are based on the emerging models presented above, which have provided novel perspectives on providing axonal protection.

Anti-Inflammatory Strategies

The correlation between axonal damage and magnitude of inflammation suggests that the axons may be innocent bystanders in the surrounding inflammatory milieu during active demyelination. The clinical observation that the topographic pattern of irreversible, progressive neurological deficits in MS depends on the localization of the previous inflammatory attacks seems to favor this interpretation. Anti-inflammatory agents are most effective during the RR phase when inflammation dominates the clinical picture. Since many axons are lost during this phase, early treatment with these agents is likely to be highly beneficial.

Remyelination Strategies

Myelin contributes both as a physical barrier between the axon and surrounding milieu and by providing trophic support, and so strategies that aid in remyelination can confer axonal protection. Remyelination is common in MS but not sufficiently robust to promote full and sustained recovery. Strategies based on enhancing production of endogenous myelinating cells or supplementing them by transplantation of progenitors are being explored at present.

Interruption of the Secondary Injury Cascade in Axons

In the hypoxia-like models of axonal injury, different cellular insults result in impairment in energy production that leads to a reversal of the Na^+–Ca^{2+} exchanger. The resulting surge in Ca^{2+} levels in the axoplasm drives enzymatic processes that lead to cellular destruction. Based on these pathogenetic mechanisms, treatment with Na channel blockers and Na^+–Ca^{2+} channel blockers may prevent this Ca^+-driven autolysis of neuron. In EAE models and *in vitro* studies, Na channel blockers like phenytoin and flecainide may have succeeded in providing axonal protection through this mechanism. Bepridil, an inhibitor of Na^+–Ca^{2+} exchange has been shown to protect axons from injury caused by NO *in vitro*. These drugs are worth looking into as some of them have already been in the market for decades. Once proven effective in MS, we have an immediate addition to our arsenal against MS that has a relatively well-established safety profile.

Conclusion

Axonal loss is the major cause of irreversible disability in MS. Axonal damage, including complete transection of the axon, begins early in MS and correlates with inflammatory activity. Extensive reductions in axon number may also impact nonlesioned white and gray matter. Understanding the mechanisms underlying axonal degeneration and how these mechanisms operate at different stages of disease may ultimately produce novel therapeutic interventions that enhance axonal survival in MS brains and delay progression of neurological disability in MS patients.

See also: Axonal Regeneration: Role of Growth and Guidance Cues; Axonal Injury: Neuronal Responses; Demyelinating Diseases; Neuropathy: Metabolically-Induced; Premenstrual Dysphoric Disorder and Postpartum Major Depression: Chronobiology; Wallerian Degeneration.

Further Reading

Arnold DI (2005) Changes observed in multiple sclerosis using magnetic resonance imaging reflect a focal pathology distributed along axonal pathways. *Journal of Neurology* 252(supplement 5): v25–v29.

Bjartmar C, Kidd G, Mork S, Rudick R, and Trapp BD (2000) Neurological disability correlates with spinal cord axonal loss and reduce *N*-acetyl aspartate in chronic multiple sclerosis patients. *Annals of Neurology* 48: 893–901.

Bjartmar C and Trapp BD (2001) Axonal and neuronal degeneration in multiple sclerosis: Mechanisms and functional consequences. *Current Opinion in Neurology* 14: 271–278.

Bjartmar C and Trapp BD (2003) Axonal degeneration and progressive neurologic disability in multiple sclerosis. *Neurotoxicity Research* 5: 157–164.

Bjartmar C, Wujek JR, and Trapp BD (2003) Axonal loss in the pathology of MS: Consequences for understanding the progressive phase of the disease. *Journal of Neurological Sciences* 206: 165–171.

Bjartmar C, Yin X, and Trapp BD (1999) Axonal pathology in myelin disorders. *Journal of Neurocytology* 28(4–5): 383–395.

Bruck W (2005) The pathology of multiple sclerosis is the result of focal inflammatory demyelination with axonal damage. *Journal of Neurology* 252(supplement 5): v3–v9.

Evangelou N, Konz D, Esiri MM, Smith S, Palace J, and Matthews PM (2000) Regional axonal loss in the corpus callosum correlates with cerebral white matter lesion volume and distribution in multiple sclerosis. *Brain* 123(Pt. 9): 1845–1849.

Kornek B and Lassmann H (1999) Axonal pathology in multiple sclerosis. A historical note. *Brain Pathology* 9: 651–656.

Peterson JW, Kidd GJ, and Trapp BD (2005) Axonal degeneration in multiple sclerosis: The histopathological evidence. In: Waxman S (ed.) *Multiple Sclerosis as a Neuronal Disease*, pp. 165–184. New York: Elsevier.

Stys PK (2004) Axonal degeneration in multiple sclerosis: Is it time for neuroprotective strategies? *Annals of Neurology* 55: 601–603.

Trapp BD, Peterson J, Ransohoff RM, Rudick R, Mork S, and Bo L (1998) Axonal transection in the lesions of multiple sclerosis. *New England Journal of Medicine* 338: 278–285.

Relevant Website

http://www.nationalmssociety.org – The National Multiple Sclerosis Society.

Axonal Injury: Neuronal Responses

G Raivich, University College London, London, UK

Introduction

Injury to neurons results in a host of many different structural, electrophysiological, and molecular responses. Some of these are associated with and may play an important role in the mounting of a successful response and the ensuing recovery of function. Others reflect the enhanced likelihood of cell death, either within hours or days of the injury, particularly in neonatal animals, or during a delayed form after a brief or lengthy regenerative phase, and/or the gradual withering away and atrophy, processes that can be addressed with neurotrophic therapy.

Axonal injury and the ensuing neuronal response can also induce a strong glial and immune reaction. These reactions can have a positive or a negative impact on neuronal regeneration. For example, cell death and the ensuing inflammation can impart a strong stimulus for the generation of new neurons. The nonneuronal responses will also provide the cellular basis for the immune surveillance of the injured brain, a key component of their overall function. The aim of this article is to present a concise overview of specific molecular, structural, and electrophysiological changes occurring in the injured neuron itself.

Early Axonal Reaction

The Initiation

The physiological and molecular signals that act as sensors of axonal injury and induce the regenerative program are still a subject of debate. Generally, three major changes appear to provide separate signals, even though they may act in a complementary and/or synergistic way:

1. Ionic flux, injury-induced discharges. There is a rapid entry of extracellular ions such as calcium and sodium through the transiently opened axonal plasma membrane, before it is resealed, resulting in depolarization and a sequence of injury-induced action potentials. The calcium influx and bystander activation of intra-axonal proteases and cytoskeleton remodeling underlie growth cone formation and may also be involved in changing intra-axonal protein synthesis.

Early changes in the cell body of axotomized neurons precede the arrival of molecules traveling by rapid axonal transport. They include the rapid elevation of calcium and cAMP, ribosomal RNA processing, and the increase of gap junction proteins such as connexin-43 in neighboring astrocytes. However, inhibition of antidromic electrical discharges does not interfere significantly with the somewhat later induction in transcription factors, glucose utilization, or the onset of regeneration.

2. The tip of the injured axon is briefly exposed to the intracellular content of neighboring axons and Schwann cells. These substances include the ciliary neurotrophic factor (CNTF), neurotrophin-3 (NT3), and fibroblast growth factors (FGFs), which are released following injury, deposited locally, and can then act on the proximal tips of the injured axons to induce a specific response. At a later time, the axons are also exposed to the extracellular environment of the inflamed neural tissue, containing a host of different proinflammatory cytokines, such as interleukin-1 and interleukin-6.

Particularly instructive is the role of CNTF and the nuclear transcription factor, signal transducer and activator of transcription-3 (STAT3), the downstream target of activated CNTF receptor. Phosphorylation and activation of STAT3 in the proximal part of injured axons clearly precede that in axotomized neurons, pointing to the retrograde transport of phosphorylated STAT3 as one of the injury-mediating signals. Neuron-specific deletion of STAT3 leads to enhanced posttraumatic cell death, suggesting that this signal is involved in promoting survival of lesioned neurons. Moreover, deletion of CNTF, a neurotrophin that is abundantly present in myelinated Schwann cells but not in or around the cell bodies of axotomized motor neurons, causes a significant delay in the appearance of the phosphorylated STAT3 and its nuclear translocation in neuronal cell bodies.

3. Block of retrograde transport. Last but not least, axonal injury also interferes with the retrograde flow of trophic signals from the normal innervation target. The data on nerve growth factor (NGF), the model neurotrophin that is retrogradely transported from periphery to neuronal cell body, are particularly instructive. In fact, analysis of its retrograde transport in the injured sciatic nerve shows a very drastic decrease immediately following axotomy. However, this initial, very strong decrease is only transient, for approximately 48 h. Nevertheless, there is a much longer, still threefold reduction that continues until the onset of regeneration. Moreover, NGF deprivation by neutralizing the endogenous activity with specific antibodies also induces axotomy-like changes

even in the intact, NGF-sensitive sensory and sympathetic neurons, particularly in the expression of transcription factors such as c-jun and a series of neuropeptides including galanin, vasoactive intestinal peptide (VIP), SP, CGRP, and NPY.

Retrograde Transfer of Injury Signal

The second, intermediate phase involves the posttraumatic generation and activation of axoplasmic proteins with the nuclear localization signal (NLS) sequences. These NLS proteins function as positive and retrogradely transported injury signals The same regulation also applies to importins, also known as karyopherins, an intermediary class of proteins that permit NLS proteins to become attached to the retrogradely transporting dynein motors and allow them to transfer injury signal(s) to the nucleus of the injured neurons. There are several ways to verify that these are indeed injury signals. For example, interference with this importin/dynein-mediated transport of NLS proteins strongly reduces the neurite outgrowth from cultured adult neurons from previously unperturbed sensory ganglia. It also blocks the much more vigorous outgrowth from cultured neurons that were conditioned by peripheral injury.

The Cell Body Response – Morphology

In the injured neurons, the axonal disconnection is followed by a complex of structural, molecular, as well as electrophysiological changes such as the long-term hyperexcitability or LTH. The structural changes are exemplified by the appearance of growth cones at the proximal tip of the lesioned axons, the swelling of the neuronal cell body associated with a strong increase in cellular metabolism, and protein synthesis.

Chromatolysis

The injured and swollen neuronal cell bodies show augmentation and regional dispersion of areas of packed rough endoplasmic reticulum (rER), the so-called Nissl bodies (NBs) (aka Tigroid substance, Nissl shoals), in neuronal cytoplasm, shown in **Figure 1**. Neuronal nuclei leave their typical cell body midpoint and move to a more eccentric position. On the electron

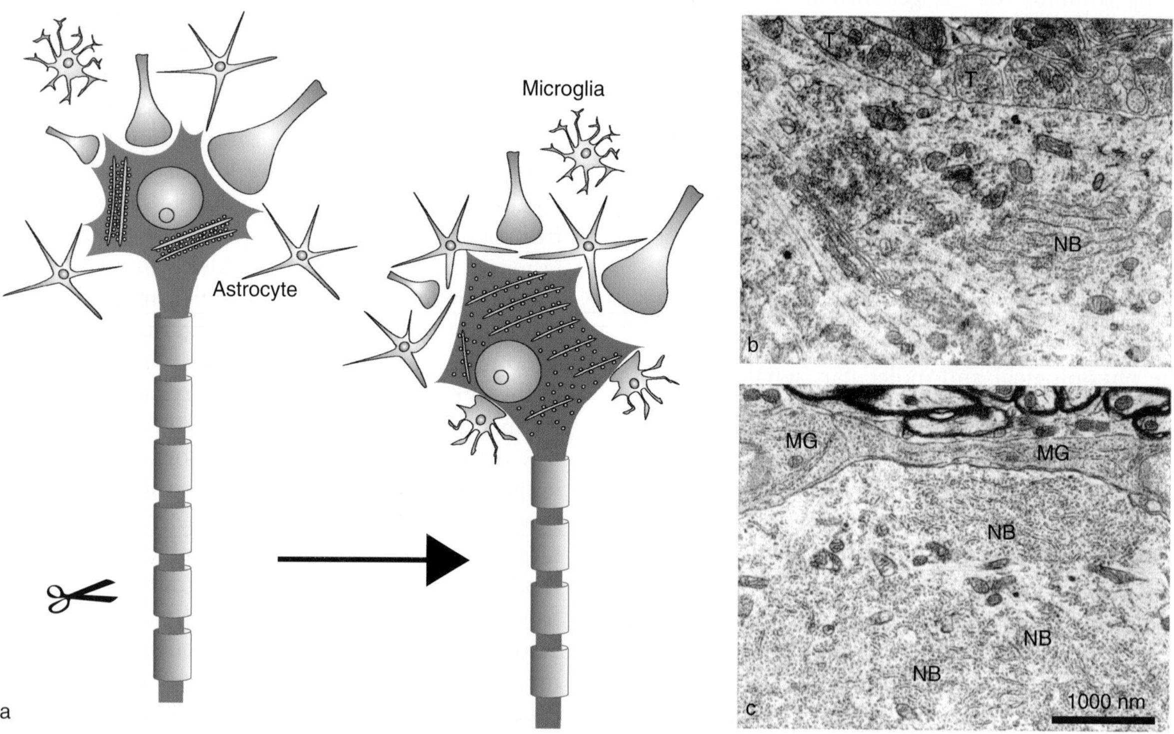

Figure 1 (a–c) Cellular and ultrastructural changes following axonal injury. (a) Schematic summary, (b, c) electron microscopy of a normal (b) and an axotomized rat facial motor neuron, 10 days after lesion (c) at higher magnification. (a) Axotomy causes a detachment of neurite terminals and filling in of empty cell surfaces with CNS glia. Inside the neuron, there is a rearrangement of rough endoplasmitic reticulum cisternae or Nissl bodies (NBs). (b, c) These NB cisternae are normally arranged in parallel (b), but take a more random and curvilinear appearance after axotomy (c). Note the enormous increase in ribosomes and enlargement of NB area. The neuronal cell surface is covered by microglial cells (MG), the neurite terminals (T) are detached. (b, c) Reproduced from Kreutzberg GW and Raivich G Neurobiology of degeneration and regeneration. In: May M and Schaitkin BM (eds.) *The Facial Nerve*, 2nd edn. New York: Thieme; 2000, pp. 67–80. Reprinted by permission of Thieme.

microscopic level, such neurons are seen to have greatly increased numbers of cellular organelles (**Figures 1(b)** and **1(c)**). Ribosomes, either in free form or bound to the endoplasmic reticulum, are especially abundant. The organization of the rER undergoes characteristic changes. The cisternae are no longer arranged in parallel. Due to the augmentation and spreading of the rER, single NBs can no longer be seen; the cells also stain more intensively with Nissl dyes under light microscopy. The contour of the cell body's surface, which is normally smooth, becomes convex and wavy, reflecting the cellular hypertrophy and instability of the plasmalemma.

These overall changes, some of which are easily visible at the light microscopical level using basophilic dyes (eccentric nuclei, increased staining, dissolution of NBs, eccentric nuclei), were termed chromatolysis or chromatolytic reaction by Franz Nissl at the end of the nineteenth century; the term that is still used as a synonym of cellular and molecular changes in the lesioned neurons or retrograde reaction following axonal injury.

Synaptic Stripping

Structurally, these early changes are followed by retraction of neurite endings on the cell bodies of axotomized neurons. This phenomenon, called synaptic stripping, leads to a partial de-afferentation of the injured neurons and may persist for a very long time after initial injury and interfere with long-term recovery (**Figure 1(c)**). The emptied spaces are gradually taken over by the neighboring reactive astrocytes, which can wrap around injured neurons with multilammellar stacks of thin astrocyte processes.

Central Sprouting

In addition to the generation of new axons at the injury site, supernumerary axonal sprouts can also develop more proximally, at the nodes of Ranvier, a response first noted by Santiago Ramón y Cajal, and occasionally even at the level of the injured neuronal cell body. These latter structures have been termed dendraxons because they exhibit axonal properties and markers but frequently, though not always, grow out of the reorganized dendritic structures.

Peripheral axotomy can also induce the sprouting of central sensory processes of the affected dorsal root ganglia (DRG) neurons in the spinal cord as well as the appearance of perineuronal neurite baskets in the DRGs themselves. Together with the long-term hyperexcitability of the injured sensory neurons, both of these latter processes have been implicated to contribute to the posttraumatic neuropathic pain.

The Cell Body Response – Molecular Changes

The prompt arrival of signals for cellular injury and stress is followed by the induction of factors involved in the regulation of transcription and translation; later on, also by adhesion molecules, growth-associated proteins, and structural components required for axonal elongation. Moreover, the introduction of the different differential cDNA display and gene chip array technologies has led to a dramatic increase in the number of identified genes regulated in injured and regenerating neurons.

Translation and Signal Transduction

One of the earliest biochemical changes following axonal injury is the transient upregulation of polyamine-producing enzymes ornithine decarboxylase and transglutaminase. This has been linked to an increase in mRNA metabolism, and protein synthesis resulting in polyamines such as putrescine, spermine, and spermidine shows a strong enhancing effect on neurite outgrowth both *in vitro* and *in vivo*. The induction of polyamine-producing enzymes is followed by the upregulation of the classical, intracellular stress and growth signaling pathways – the phosphatidylinositol-3 kinase (PI3K) and the mitogen-associated protein kinase (MAPK) cascades. Both pathways are involved in stimulating neurite outgrowth. Expression of active protein kinase B, also known as Akt, a downstream target of PI3K, promotes axonal regeneration as well as neuronal survival after neonatal axonal injury. Expression of active ras, the upstream inducer of ERK1 (MAPK1) and ERK2 (MAPK2), promotes adult survival. Interference with ERK1 and ERK2 also retards axonal outgrowth.

Transcription Factors

Activation, phosphorylation, and nuclear localization of transcription factors appear as the primary signals mediating changes in gene expression in the injured neuron. The majority of transcription factors reported in the context of injury – c-Jun, JunD, ATF3, P311, Sox11, c-EBPβ, and STAT3 – go up. However, some are actually going down, for example, Islet-1, Fra-2, and ATF2. Although all of these transcriptional changes are probably involved in fine-tuning the posttraumatic and regenerative response, the data on their importance using deletion studies are now available for c-Jun, c-EBPβ, and STAT3. In the case of c-EBPβ, deletion of this transcription factor interferes with the injury-induced expression of the molecules important for the coordination of the membrane cytoskletal interface, tubulinα 1, and neuromodulin/GAP43. Deletion of neuronal STAT3 reduces the posttraumatic survival of axotomized neurons, to the extent observed following the deletion of CNTF and LIF, pointing to STAT3 as the transcription factor that mediates these effects.

However, particularly pronounced effects are observed for neuronal deletion of c-Jun, briefly summarized in **Figure 2**. The absence of neuronal c-Jun (**Figure 2(a)**) shows the gene deletion strategy, **Figure 2(b)** the

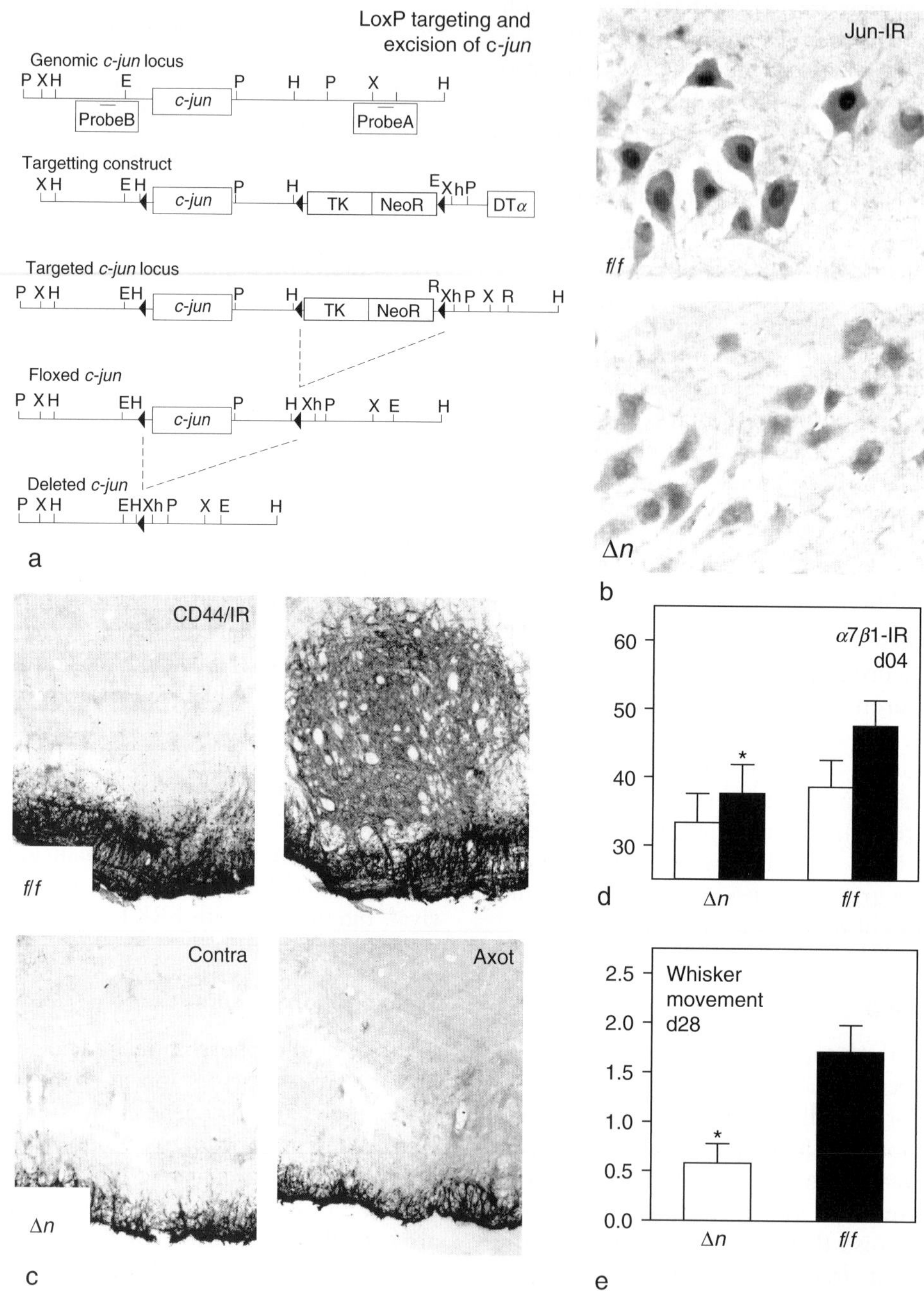

Figure 2 Deletion of the *c-jun* gene in neurons interferes with the neuronal response to injury following transection of the facial nerve. (a) A summary of the three-step gene deletion strategy: (1) by introducing the loxP sites (◄) around the *c-jun* gene and the thymidine kinase (TK) and neomycine resistance *neoR* cassette, (2) removing the floxed *TK* and *neoR* cassette, and (3) deletion of the floxed *c-jun* gene through *cre* recominase under the control of *nestin* promoter active in central neuroepithelial cells. The strategy restricts *c-jun* deletion to neural cells, to avoid embryonic lethality due to global *c-jun* gene deficiency. (b) Absence of c-Jun immunoreactivity (Jun-IR) in neurons with gene deletion (Δ*n*), but not in control animals with two copies of floxed *c-jun* gene (*f/f*), which behave as wild-type controls (not shown). (c–e) Absence of neuronal c-jun blocks the axotomy-induced appearance of neuronal hyaluronic acid receptor CD44 (c), the upregulation of the laminin receptor α7β1 integrin (d) and interferes with normal motor recovery (e). Axotomized side: axot in (c), filled bar in (d); contralateral side: contra in (c), empty bar in (d). Reproduced from Raivich G, Bohatschek M, DaCosta C, et al. (2004) The AP-1 transcription factor c-Jun is required for efficient axonal regeneration. *Neuron* 43: 57–67.

disappearance of nuclear c-Jun immunoreactivity – interferes with the induction of most proteins expressed after injury. There is a complete block in the expression of the adhesion molecule CD44 (**Figure 2(c)**), which functions as a receptor for hyaluronic acid. There is also a strong decrease in the laminin receptor $\alpha7\beta1$ integrin (**Figure 2(d)**) and neuropeptide galanin that are associated with axonal regeneration, a significant reduction in the speed of reinnervation and functional recovery (**Figure 2(e)**), and a complete inhibition of central axonal sprouting. In addition, absence of neuronal c-Jun inhibits the recruitment of lymphocytes and activation of brain microglia, showing the importance of neuronal factors for the chemoattraction of neighboring nonneuronal cells. As *in vitro*, deletion of c-Jun interferes with posttraumatic neuronal cell death. However, it also causes severe neuronal atrophy, showing that saving neurons from cell death does not automatically translate into the maintenance of a well-nourished and functional state.

Metabolic Changes

The normal neuronal response to injury is associated with a strong increase in energy metabolism, in the uptake of glucose, and in the enzyme activity of the hexose monophophate shunt (glucose-6-phosphate dehydrogenase and 6-phosphogluconate dehydrogenase). This may be related to increased production of pentoses needed for the increased synthesis of ribonucleotides. Axotomized neurons show a massive induction in the receptors for transferrin, a carrier protein for the trivalent iron ion (Fe^{3+}) and an associated uptake of iron ions into the affected facial motor nucleus. There is also an increase in the neuronal uptake of fatty acids, which is probably involved in the provision of additional constituents to the newly formed axonal membranes.

Neurotransmission and Neurochemical Differentiation

As a general rule, proteins associated with neurotransmission are decreased after injury. This includes the acetylcholine-synthesizing enzyme cholinacetyltransferase and the degrading enzyme acetylcholinesterase in the cholinergic sympathetic and cranial and spinal motor neurons, enzymes involved in catecholamine synthesis (noradrenaline, dopamine), in adrenergic sympathetic and dopaminergic nigrostriatal neurons, or substance P and CGRP in the peripheral sensory neurons responsible for the pain modality. This downregulation is frequently accompanied by reduction in the level and activation of neurotrophin receptors that may be responsible for the neurochemical shift toward a more immature and embryonic phenotype. In sensory, sympathetic, and motor neurons, this shift is also accompanied by an upregulation of other neuropeptides, such as galanin and VIP, which are more typical of their embryonic phenotype.

Typically, the synthesis of neurofilament proteins is also reduced. However, there is strong upregulation of peripherin and the tubulin subunits T-$\alpha1$ and T-$\beta3$. Tubulins are major components of the microtubules which are assembled from α- and β-subunits. These microtubules are prominent axonal organelles that function as a carrier for kinesin, a molecular motor that drives axonal transport. Changes in cytoskeletal proteins thus appear important in promoting axonal regeneration. However, they may also lead to the overt morphological alterations at the level of the injured neuronal cell body.

Cell Adhesion Molecules and Cell Surface–Cytoskeletal Interactions

Increased expression or *de novo* appearance of adhesion molecules on the cell surface of the injured neuron appears to be one of the primary motors in changing the growth status of the lesioned neuron and permitting it to extend along, as well as sprout additional fibers in the injured peripheral nerve. Axotomized neurons upregulate a variety of integrin subunits that serve as receptors for extracellular matrix and cell surface molecules. They increase in receptors for the hyaluronic acid and galactoside residues such as CD44 and galectin-1 and express cell multimodal surface molecules involved in homophilic binding such as ninjurin and gicerin/CD146. The latter molecule is also a heterophilic receptor for the neurite outgrowth factor, a member of the laminin family. Another cell surface molecule strongly upregulated after axotomy is FLTR3, which contains numerous leucin-rich repeats and a fibronectin III-type extracellular domain and can form complexes with receptors for the FGF and promote FGF signaling.

Although the common assumption is that most of these molecules are involved in regeneration proper, some may also contribute to aberrant sprouting. For example, deletion of $\alpha7$ integrin subunit is associated with a strong upregulation of its corresponding $\beta1$counterpart, but also with a significant increase in central axonal sprouting. This may imply compensatory mechanisms involving other closely related α subunits. Central axonal sprouting on injury-induced neuronal cell surface molecules such as L1 and the close homolog of L1 (CHL1) relies on $\beta1$-integrin function and may be involved in the effects observed in $\alpha7$-deficient mice.

Successful axonal regeneration is accompanied by the appearance of numerous, functionally diverse families of molecules that regulate the surface–cytoskeletal interaction. The best known so far is the GMC family that includes cytoplasmic proteins GAP43, MARCKS, and CAP23. These molecules codistribute with the phosphoinositol-4,5-diphosphate, PI(4,5)P2, at the semicrystalline, raft regions of the cell membrane. These GMC molecules are functionally exchangeable, bind exclusively to acidic phospholipids such as PI-(4,5)-P2, calcium/calmodulin, protein kinase C, and actin filaments, modify the raft recruitment of signaling molecules such as src, regulate actin cytoskeleton polymerization, organization, and disassembly, and have a formative role in the appearance of filopodia and microspikes, in addition to the process of neurite outgrowth.

Posttraumatic Neuronal Death and Cell Death Signals

Several cell surface molecules have been shown to mediate neuronal cell death following neonatal and adult axonal injury, including fas, tumor necrosis factor receptors (TNFRs) 1 and 2, and FAS. Some of these receptors carry a cytoplasmic death domain that exerts a pro-apoptotic signal through FADD; expression of dominant negative form of FADD has been shown to block the axotomy-induced cell death signal due to fas and TNFR. Recent evidence also points to glutamate excitotoxicity.

Associated downstream cytoplasmic cell death signals play a key role in promoting neuronal cell death in neonatal and immature animals that are sensitive to axonal injury, for example, noxA. Some heat shock proteins, for example, hsp27, are neuroprotective. Deletion of bax or inhibition of caspase 3 or the whole family of caspases has also been shown to prevent cell death, but with an enhanced and more persistent effect for broad caspase inhibitors than caspase 3 alone. In both cases, bax and caspases appear to act downstream of the phosphorylation of the transcription factor c-Jun and the decrease in neuronal metabolism, suggesting a sequence of events beginning with phospho-Jun, leading to atrophy, activation of bax, and ending with the initiation of the caspase cascade, resulting in neuronal cell death.

See also: Axonal Regeneration: Role of Growth and Guidance Cues; Axonal Regeneration: Role of the Extracellular Matrix and the Glial Scar; Axonal Injury in Demyelinating Disease and CNS Injury; Neurogenesis in the Intact Adult Brain; Neurogenesis and Neural Precursors, Progenitors, and Stem Cells in the Adult Brain; Neuronal Plasticity after Cortical Damage; Nogo-A: Its Role in Axon Regeneration; Peripheral Nerve Regeneration: An Overview; Sensory Re-education.

Further Reading

Benn SC and Woolf CJ (2004) Adult neuron survival strategies – Slamming on the brakes. *Nature Reviews Neuroscience* 5: 686–700.

Herdegen T and Leah JD (1998) Inducible and constitutive transcription factors in the mammalian nervous system: Control of gene expression by Jun, Fos and Krox, and CREB/ATF proteins. *Brain Research Reviews* 28: 370–490.

Kiryu-Seo S, Hirayama T, Kato R, and Kiyama H (2005) Noxa is a critical mediator of p53-dependent motor neuron death after nerve injury in adult mouse. *Journal of Neuroscience* 25: 1442–1447.

Kreutzberg GW and Raivich G (2000) Neurobiology of degeneration and regeneration. In: May M and Schaitkin BM (eds.) *The Facial Nerve*, 2nd edn., pp. 67–80. New York: Thieme.

Lieberman AR (1971) The axon reaction: A review of the principal features of perikaryal responses to axon injury. *International Review of Neurobiology* 14: 49–124.

Makwana M and Raivich G (2005) Molecular mechanisms in successful peripheral regeneration. *FEBS Journal* 272: 2628–2638.

Moran LB and Graeber MB (2004) The facial nerve axotomy model. *Brain Research Reviews* 44: 154–178.

Nissl F (1890/1892) Über die Veränderungen der Ganglienzellen am Facialiskern des Kaninchens nach Ausreissung der Nerven. *Versammlungen des Südwestdeutschen Psychiatervereins in Karlsruhe* 1890: 22 *Allgemeine Zeitschrift für Psychiatrie (Berlin)* 48: 197–198.

Perlson E, Hanz S, Medzihradszky KF, Burlingame AL, and Fainzilber M (2004) From snails to sciatic nerve: Retrograde injury signaling from axon to soma in lesioned neurons. *Journal of Neurobiology* 58: 287–294.

Raivich G, Bohatschek M, DaCosta C, *et al.* (2004) The AP-1 transcription factor c-Jun is required for efficient axonal regeneration. *Neuron* 43: 57–67.

Ramón y Cajal S (1928) *Degeneration and Regeneration of the Nervous System.* Oxford: Oxford University Press.

Sun W and Oppenheim RW (2003) Response of motoneurons to neonatal sciatic nerve axotomy in Bax-knockout mice. *Molecular and Cellular Neuroscience* 24: 875–886.

Sung YJ and Ambron RT (2004) Pathways that elicit long-term changes in gene expression in nociceptive neurons following nerve injury: Contributions to neuropathic pain. *Neurological Research* 26: 195–203.

Wiese S, Beck M, Karch C, and Sendtner M (2004) Signalling mechanisms for survival of lesioned motoneurons. *Acta Neurochirurgica Supplement* 89: 21–35.

Woolf CJ (2004) Dissecting out mechanisms responsible for peripheral neuropathic pain: Implications for diagnosis and therapy. *Life Sciences* 74: 2605–2610.

Zigmond RE, Hyatt-Sachs H, Mohney RP, *et al.* (1996) Changes in neuropeptide phenotype after axotomy of adult peripheral neurons and the role of leukemia inhibitory factor. *Perspectives on Developmental Neurobiology* 4: 75–90.

Axonal mRNA Transport and Functions

F P G van Horck and C E Holt, University of Cambridge, Cambridge, UK

Introduction

Neurons are highly polarized cells consisting of a cell body, multiple dendrites, and a single axon. During development, axons and their highly dynamic structures at the tip, the growth cones, navigate through the complex environment of the embryo to reach their synaptic targets, which are often long distances from their cell bodies. Along the pathway, growth cones encounter and respond to molecular guidance cues that are present in gradients or localized to discrete choice points. Once they arrive at their destination, axons must select the appropriate neurons with which to form synapses. The specialized functions of the axon and growth cone require a subset of proteins different from those needed in the cell body or in the dendritic compartment. Neuronal function therefore relies on an asymmetric distribution of protein and protein complexes, both during development and throughout the life of mature neurons.

Asymmetric protein distribution can be achieved by two mechanisms: axonal transport of proteins and local protein synthesis. It was long thought that the cell body was the exclusive site of protein synthesis, with newly synthesized proteins being transported into the neuronal processes by selective axonal and dendritic transport mechanisms. This view was altered by the discovery of local protein synthesis in dendrites, beginning with the detection of polyribosomes in postsynaptic structures and followed by the identification of dendritically targeted messenger RNAs (mRNAs). Local translation in dendrites is now known to play an important role in synaptic plasticity and long-term memory storage. The concept of local protein synthesis in axons has been more controversial and has been accepted only since the early 2000s. Although electron microscopy studies in the 1970s of growing axons *in vivo* revealed the presence of polyribosomes in axons, the failure to unambiguously detect components of the protein translation machinery in biochemical axonal preparations hampered further progress. Moreover, *in situ* hybridization studies revealed the presence of mRNAs in young hippocampal neurons but not in mature axons. This transient appearance of mRNAs in developing axons was thought to reflect the fact that transport mechanisms that mediate RNA sorting in mature neurons had not yet developed. However, the process of axonal translation gained more support when individual mRNAs, such as β-actin, were detected in both invertebrate and vertebrate developing axons. Evidence that these mRNAs are translated in axons was initially provided from studies in invertebrates, in which mRNAs injected into isolated axons were shown to be translated into functional proteins. Studies in invertebrates have been viewed with caution, however, since invertebrate axons have both receptive and transmissive capacities and therefore combine the functions of both dendrites and axons. Subsequent work has shown that vertebrate axons can also actively synthesize proteins. Indeed, vertebrate axons have been suggested to represent about 5% of total neuronal protein synthesis. There is, however, a relative paucity in our understanding of the molecular mechanism underlying axonal mRNA transport and translation as the field of local translation in neurons has been dominated by studies in dendrites. This article focuses on the mRNAs that have been identified in axons, the mechanism of axonal mRNA transport, and the way translation of axonal mRNAs controls critical functions in neuronal behavior.

Which mRNAs Are Transported into the Axon?

To date, the main group of mRNAs identified in axons and growth cones are those encoding cytoskeletal proteins, such as β-actin, or regulators of cytoskeletal dynamics, such as actin depolymerizing factor (ADF)/cofilin, profilin, β-thymosin, microtubule-associated protein (MAP)1b, and tau. However, mRNAs for proteins involved in cell signaling, such as RhoA and transmembrane receptors such as EphB2 have also been found, albeit to a lesser extent. Most axonal mRNAs have been identified by candidate approaches, such as by using fluorescent *in situ* hybridization on axonal cultures *in vitro*. Other approaches involve the generation of pure axonal preparations and subsequent biochemical analysis of mRNAs associated with ribosomal proteins or molecular analysis of the mRNA content. The latter approach has also led to the identification of mRNAs in regenerating adult axons. In this approach dorsal root ganglia (DRG) neurons are damaged in the animal and then allowed a period of recovery in order to allow induction of the regeneration program (see the section titled 'Axonal mRNA transport in adult axons') prior to neuronal isolation. Purified axons of these so-called injury-conditioned neurons contain, besides cytoskeletal mRNAs, those encoding heat shock proteins, resident endoplasmic reticulum (ER) proteins, proteins associated with neurodegenerative diseases, antioxidant proteins, and metabolic proteins. This implies that axons can contain a wide range of mRNAs. The repertoire of axonal mRNAs is likely to vary depending on neuronal cell type, function, and developmental stage.

Mechanism of Axonal mRNA Transport

Targeting Elements

As has been found in nonneuronal cells, mRNA localization in axons depends on *cis*-acting sequences, called zipcodes. Zipcodes are essential for the transport of the mRNAs, and removal or mutations to them impair the targeting of the transcript. Zipcodes are usually present in the 3′ untranslated region (UTR) of the mRNA and are occasionally found in the 5′UTR or even in the coding region. Zipcodes can be short segments with defined nucleotide sequences but can also be secondary or tertiary structures such as stem loops. A well-studied example in axons is β-actin mRNA. Localization of the β-actin mRNA into growth cones is dependent on the 3′UTR. Mutation analysis has defined a 54-nucleotide sequence within the 3′UTR that is necessary and sufficient for targeting β-actin mRNA into the growth cone of chick forebrain axons. β-actin mRNA localization can be inhibited by antisense oligonucleotides directed against the 54-nucleotide sequence as well as by mutations that disrupt the secondary structure. Similarly, the mRNA for the MAP tau is targeted to the axon by an AU-rich sequence in the 3′UTR. Fusion of the tau 3′UTR axonal localization signal to a gene normally not present in axons results in axonal localization of its mRNA. In most cases, the mRNA localization signals are more complex and lack a simple consensus. Together with the fact that axonal localization signals are not highly conserved among species, it is not yet possible to predict from the primary sequence whether an mRNA is targeted into the axon.

RNA-Binding Proteins

The targeting elements in mRNAs are recognized by RNA-binding proteins (RNBPs), which are essential for mRNA transport. To date only a limited number of zipcode-binding RNBPs have been identified in axons. These include zipcode-binding proteins (ZBPs)-1 and 2, HuD, cytoplasmic polyadenylation element binding protein (CPEB), survival of motor neuron protein (SMN), and fragile X mental retardation protein (FMRP); all of which are proteins with well-established roles in mRNA localization in oocytes, embryos, and fibroblasts. RNA binding is modulated by RNA-binding domains such as RNA recognition motif, heterogeneous nuclear ribonucleoprotein K homology domains, and Arg-Gly-Gly boxes. RNBPs often harbor multiple RNA-binding modules and have the potential to interact with multiple mRNAs, as well as with other proteins involved in transport or translation. Together they form large protein complexes that are actively transported into the axon (**Figure 1**).

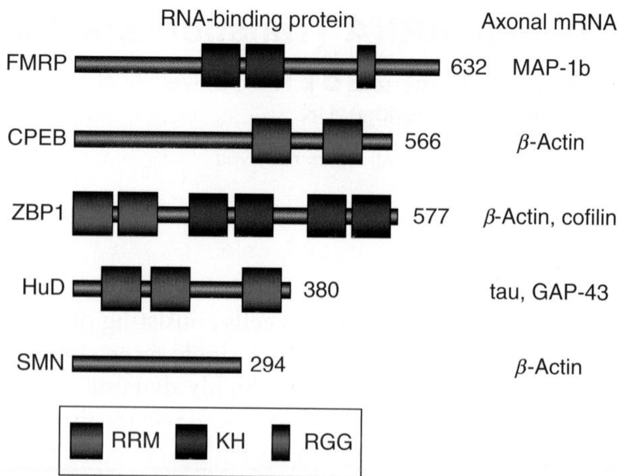

Figure 1 Examples of RNA-binding proteins (RNBPs) and messenger RNA (mRNA) complexes identified in axons and growth cones. mRNAs are depicted that have been shown to interact either directly or indirectly with the indicated RNBP in axons. RNBPs interact with the mRNAs via conserved RNA-binding modules, such as RNA recognition motif (RRM), heterogeneous nuclear ribonucleoprotein K homology (KH), and Arg-Gly-Gly (RGG) boxes. Survival of motor neuron protein (SMN) lacks any of these recognizable RNA-binding domains. Numbers refer to amino acids. FMRP, fragile X mental retardation protein; CPEB, cytoplasmic polyadenylation element binding protein; ZBP1, zipcode-binding protein 1; HuD, a zipcode-binding RNBP; MAP, microtubule-associated protein; GAP, growth-associated protein.

Active Transport

There are several mechanisms known to localize mRNA to specific subcellular regions, including mRNA diffusion and anchoring, local degradation, and active transport along cytoskeletal filaments. Of these, active transport of mRNAs has been implicated as the major localization mechanism in neurons. A consensus view of this process has come primarily from studies on dendrites, as well as from *Xenopus* and *Drosophila* early development systems. Findings on mRNA localization in axons are limited but in agreement with the current model of active transport, which encapsulates the following multistep pathway: RNA granule formation, cytoskeleton transport and release, anchoring, and translation.

RNA granule formation As soon as mRNAs are generated, they are subject to multiple dynamic interactions with RNBPs involved in mRNA processing and nuclear export. Once in the cytoplasm, some proteins dissociate from the mRNAs whereas others remain attached. For example, both ZBP1 and 2 associate with the β-actin mRNA zipcode in the nucleus. However, while ZBP1 remains bound to β-actin mRNA in the cytoplasm, ZBP2 is a shuttling RNBP that is thought to be released for return to the nucleus. Next, proteins that are required for cytoplasmic transport, translation, degradation, and anchoring associate with the mRNA, resulting in the formation of large particles, called ribonucleoprotein (RNP) complexes or RNA granules. RNA granules can be identified by phase light microscopy, in which they appear as large, clustered aggregates. Biochemical characterization of these granules from neuronal extracts has revealed the presence of many localized mRNAs, RNBPs, ribosomal proteins, and proteins involved in the regulation of translation, as well as motor proteins that connect RNA granules to the cytoskeletal transport network. The assembly of transport particles, containing multiple mRNAs and both nuclear and cytoplasmic RNBPs, is therefore an important step in mRNA localization.

Movement along the cytoskeleton Active transport is characterized by directed movement of RNA granules at a speed that is too fast to be explained by cytoplasmic flow. Live cell imaging of mRNA or RNBP movement shows that active transport can reach velocities of 0.1–2 $\mu m\ s^{-1}$. How is such rapid transport achieved? RNA granules contain motor proteins that connect the RNP complexes with the cytoskeleton. In an analogous fashion to vesicle transport, motor proteins mediate RNP transport along cytoskeletal filaments in an adenosine triphosphate-dependent manner. In axons, the shaft consists of actin filaments and bundles of stabilized microtubules, which extend, together with some dynamic microtubules, into the growth cone central domain and are therefore an ideal track for transport into the axon.

Active transport via the cytoskeleton is mediated by specific motor proteins. Kinesin and dynein are microtubule-based motor proteins, whereas myosin is required for transport along actin filaments. In general, anterograde movement along plus end microtubules (i.e., away from the cell body, into the axon) is mediated by kinesin, whereas retrograde transport along minus end microtubules (i.e., toward the cell body, out of the axon) might involve dynein. However, the kinesin motor family comprises many family members with different expression patterns and cargo specificities, and some family members have also been shown to mediate retrograde transport. Kinesin-2 is involved in anterograde transport and is expressed in axons. One of the microtubule-based motor subunits of kinesin-2, KIF3A, interacts with HuD in HuD-tau mRNA-containing granules in axons. Moreover, inhibition of KIF3A expression abolishes tau mRNA transport into the axon, resulting in decreased levels of axonal tau protein. Since tau is important for microtubule stabilization, the decrease in expression levels causes neurite retraction. Other RNBPs have been shown to interact with several different kinesin family members, albeit not specifically in axons or growth cones. In *Drosophila* neuronal cells, FMRP granule transport depends on both kinesin-1 and cytoplasmic dynein. Knockdown of either kinesin-1 or cytoplasmic dynein is sufficient to impede both anterograde- and retrograde-directed FMRP transport, indicating that the two transport mechanisms are coupled.

The identity of the motor proteins suggests that axonal mRNAs are transported along microtubules. Indeed, treatment of neurons with inhibitors of microtubule polymerization, such as nocodazole or colchicine, reduces transport of ZBP1, SMN, and FMRP granules into the axon, showing that these granules are predominantly transported along microtubules. It should be noted that granule movement is not solely unidirectional. Detailed analysis of enhanced green fluorescent protein (eGFP)-tagged RNBP dynamics shows a variety of movements, uni- and bidirectional, as well as oscillatory movements with no net displacement. The cytoskeleton of the peripheral domain of growth cones is comprised mainly of actin filaments, which are either arranged in a network forming the lamellipodia or in f-actin bundles forming the filopodia. RNBPs such as ZBP1, SMN, and FMRP are detected in developing growth cones, suggesting that within the growth cone, transport can also take place along actin filaments. Indeed, the actin depolymerizing drug cytochalasin D blocks transport of eGFP–SMN, but it mainly inhibits short-distance-directed movements and oscillatory movements. Thus axonal mRNA transport seems to involve long-distance-directed movement of RNA granules along microtubules in the axon shaft as well as short-distance-directed movements along actin filaments in growth cones.

mRNA release and anchoring To restrict protein synthesis to a specific subcellular domain, mRNAs must not be translated until they reach their destination. This implies that a tight coupling between mRNA transport and repression of protein translation exists, and RNBPs are thought to play a key role in this regulation. ZBP1, for example, is not only essential for β-actin mRNA localization; it also represses the translation of β-actin mRNA during transport in neuroblastoma cells. Once an mRNA

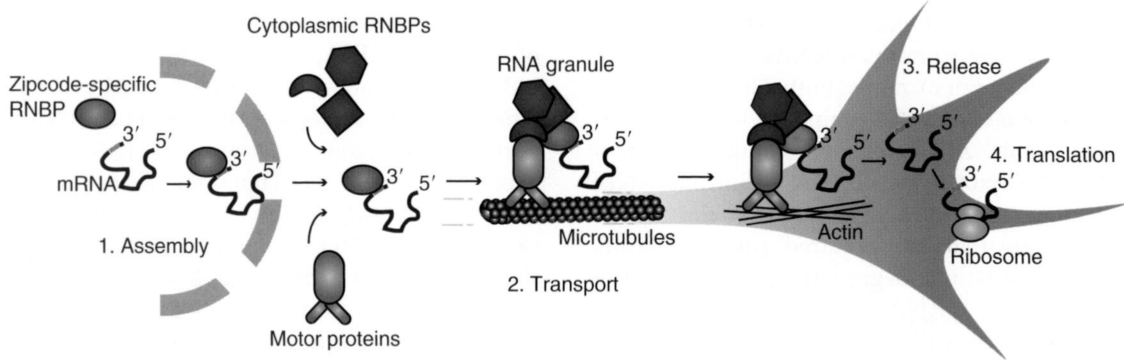

Figure 2 Model for axonal messenger RNA (mRNA) transport. Axonal mRNA localization is a multistep process, involving ribonucleoprotein (RNP) assembly in nucleus and cytoplasm, followed by active transport along microtubules and actin filaments. Once arrived at the growth cone, transported mRNAs are released and anchored until needed for translation. RNBP, RNA-binding protein.

granule arrives at its site of action, the mRNA has to be released and anchored in the correct place until needed for translation. In the case of ZBP1, this has been shown to involve a phosphorylation-dependent mechanism. The nonreceptor kinase Src phosphorylates ZBP1, which reduces the ability of ZBP1 to bind to β-actin mRNA and thereby releases the translational repression. It is interesting that this regulatory step seems to take place in the growth cones, suggesting that external cues might induce this process leading to local β-actin translation. The molecular mechanism of mRNA anchoring in neurons is not well understood. One of the components of the translation machinery, elongation factor 1a, has been implicated in anchoring β-actin mRNA to the actin filaments, where polysomes are also known to reside. Furthermore, FMRP is associated with translating polyribosomes in neuronal cells, where it is thought to regulate translation (**Figure 2**).

mRNA translation mRNA translation in axons is thought to involve the cap-dependent translation machinery. Cap-dependent mRNAs are abundant in growth cones, and molecular cues shown to stimulate translation in growth cones (see below) induce the rapid phosphorylation of the translation initiation factor eIF4E binding protein 1 (4EBP) and the translation initiation factor eIF4E. Hypophosphorylated 4EBP blocks translation initiation by sequestering the rate-limiting eIF4E. Hyperphosphorylation causes 4EBP to release eIF4E and allows translation initiation to proceed. In turn, eIF4E is phosphorylated by the kinase target of rapamycin (TOR, which is inhibited by rapamycin). Although rapamycin blocks translation in retinal growth cones, translation is not completely inhibited, indicating that cap-independent translation may also be involved. Which other components of the translation initiation machinery are required for axonal mRNA translation and how the process is spatiotemporally regulated are currently unknown.

Functions of Axonal mRNA Transport

Axonal mRNA transport serves to spatially restrict protein synthesis to the axon and growth cone. How is this ability to locally transport mRNAs important functionally for the axon? Axons are continuously exposed to molecular cues, which regulate their outgrowth and pathfinding. *In vitro* assays on cultured neurons have revealed that molecular cues are able to regulate mRNA transport. For example, in chick forebrain neurons, the neurotrophic factor Neurotrophin-3 (NT-3) increases β-actin mRNA localization in the growth cone within a short timescale (10 min). This increase is blocked when, prior to NT-3 stimulation, the axons are incubated with antisense oligonucleotides that interfere with the binding of ZBP1 to β-actin mRNA, indicating that NT-3 stimulates axonal mRNA transport rather than local mRNA stability. Indeed, detailed analysis of ZBP1 granule movement shows that NT-3 increases the rapid ZBP1 transport into the growth cone. Furthermore, cue-induced stimulation of mRNA transport seems to be selective since, for example, in axons from regenerating adult DRG neurons, brain-derived neurotrophic factor and nerve growth factor selectively induce the transport of a subset of mRNAs, including β-actin, vimentin, and peripherin. The precise functional significance of this finding is unknown but is presumed to reflect proteins that are required for axonal responses in both developing and adult axons.

Axonal mRNA Transport in Developing Axons

Developing axons are characterized by their ability to grow, navigate, and innervate their target region. Each of these processes requires cytoskeletal rearrangements. Significantly, mRNAs encoding cytoskeletal proteins and cytoskeletal-associated proteins have been isolated from axons of developing neurons from a wide range of organisms. Given the importance of cytoskeletal dynamics in developing neurons, this suggests multiple roles for local translation in regulating axon and growth cone behavior, such as axon growth and pathfinding.

Axon growth Addition of protein synthesis inhibitors to cultured *Xenopus* retinal or spinal axons does not affect their rate of extension over the short term (1 h), suggesting that local translation of axonal mRNAs is not critical for short-term axon outgrowth. In contrast, disruption of the interaction between ZBP1 and *β*-actin mRNA by antisense oligonucleotides has been shown to impair the persistent forward movement of growth cones. This apparent discrepancy could reflect a difference in the requirement for translation and mRNA transport. Several *in vitro* studies indicate that mRNA transport is important for long-term axonal growth. Expression of a mutant ZBP1, which cannot release the *β*-actin mRNA, reduces neurite outgrowth in hippocampal neurons. Similarly, manipulation of HuD levels in rat cortical neurons results in significant changes in neurite outgrowth. More-direct evidence that axonal transport is important for growth comes from studies on SMN. Transport of SMN from the nucleus into the cytoplasm requires exon 7. Neurons transfected with an exon 7 deletion mutant (SMNΔ7) have shorter processes, both of the minor neurites and the incipient axon. It is important to note that this axon outgrowth defect can be rescued by redirecting SMNΔ7 into neurites by an axonal targeting sequence from growth-associated protein 43, indicating that SMN-mediated mRNA transport is critical for axon outgrowth. Further complexity is apparent from the ability of some translated mRNAs to inhibit axon outgrowth. For example, mRNA of *β*-thymosin, a negative regulator of actin polymerization, is accumulated in the distal neurites of *Lymnaea* neurons, and reducing the *β*-thymosin mRNA levels in isolated neurites increases the rate of neurite outgrowth. Thus, locally synthesized proteins can both positively and negatively regulate axonal growth.

Axon guidance As axons grow, their growth cones encounter a wide variety of signals that influence their navigation. Growth cones react to local attractive and repulsive cues in their environment by extending and retracting filopodia (fingerlike protrusions) and lamellipodia (flattened, veillike extensions), which determines the net direction of migration. Turning is caused by extension of filopodia on one side and withdrawal or collapse of filopodia on the other. Using two different chemotropic assays, the behavior of *Xenopus* retinal ganglion cell (RGC) axons *in vitro* has provided an insight into the role of local mRNA translation in growth cone guidance. In turning assays, growth cones grow toward a gradient of an attractant and turn away from a gradient of repellent over the course of 30–60 min. In collapse assays, bath application of a repellent causes complete withdrawal of growth cone filopodia and lamellipodia within 10 min. Embryonic *Xenopus* RGC growth cones are attracted to a source of netrin-1, but this attractive response is blocked by protein synthesis inhibitors. Moreover, protein synthesis inhibitors block the repulsive turning as well as the collapse response of RGCs toward the repellents Slit-2 and Semaphorin 3A (Sema3A). It is important to note that this inhibition of chemotropic responses by protein synthesis inhibitors also takes place in growth cones that have been severed from their cell bodies, showing that the translation occurs locally, in the growing axon. Indeed, netrin-1, Slit-2, and Sema3A are able to induce rapid (within 5–10 min) protein translation in the growth cone as measured by phosphorylation of eIF4E and eIF4EBP and the incorporation of labeled amino acids. It is striking that netrin-1, Slit-2, and Sema3A elicit local translation by activating different intracellular pathways involving mitogen-activated protein kinases (MAPKs). For example, netrin-1 and Slit-2, but not Sema3A, induce translation that depends on p38 MAPK. The netrin-1, Slit, and Sema3A pathways are dependent on mammalian TOR, indicating that cue-induced translation is cap-dependent.

These findings have provided evidence that growth cone turning and collapse depend on local protein synthesis, but which mRNAs are translated in these processes? Since growth cone turning involves remodeling of the microtubule and actin cytoskeleton, translation of mRNAs encoding cytoskeletal regulators are good candidates. Consistent with this hypothesis, Slit-2 has been shown to induce a protein synthesis-dependent increase in ADF/cofilin in *Xenopus* RGC growth cones. Moreover, Sema3A-mediated growth cone collapse in DRG neurons is dependent on the local translation of the small GTPase RhoA. RhoA is a key regulator of the actin cytoskeleton, and its mRNA is targeted to the growth cone by a specific sequence located in its 3′UTR. The local translation of RhoA is both sufficient and necessary for Sema3A-mediated growth cone collapse, suggesting that basal levels of RhoA are not sufficient to mediate the Sema3A responses. Thus, the local translation of cytoskeletal proteins provides the developing axon and growth cone with a mechanism to rapidly modulate the cytoskeleton in response to extracellular cues (**Figure 3**).

Local mRNA translation in growth cones also plays a role in growth cone adaptation, a process whereby the growth cone desensitizes and resensitizes to a guidance cue. Adaptation is thought to be an integral part of the steering mechanism, enabling growth cones to adjust their sensitivity in a way that allows them to navigate in gradients of axon guidance cues. In *Xenopus* spinal neurons, attractive turning responses toward netrin-1 or brain-derived neurotrophic factor are attenuated if either of the chemoattractants is present in the bath. After 30 min, the desensitization disappears, and the growth cones regain their normal turning behavior. Protein synthesis inhibitors block this resensitization process, indicating that locally synthesized proteins play a role in regaining responsiveness. Collapse assays in *Xenopus* retinal growth cones show that desensitization and resensitization can even occur at a much shorter timescale (2–5 min) in response to repellents. Resensitization in response to Sema3A and netrin-1 under repulsive conditions is ligand-specific and requires new protein synthesis. Growth cone resensitization to Sema3A and netrin-1 correlates with a protein synthesis-dependent increase of their respective receptors, neuropilin and deleted in colorectal cancer, at the plasma membrane, indicating that resensitization might involve, at least partly, the translation of guidance cue receptors. In this way, growth cones can adjust their sensitivity as a function of exposure to a specific ligand.

Although functional studies on axonal mRNA transport are mostly conducted *in vitro*, local protein synthesis has also been implicated in axon navigation *in vivo*. When growing over long distances,

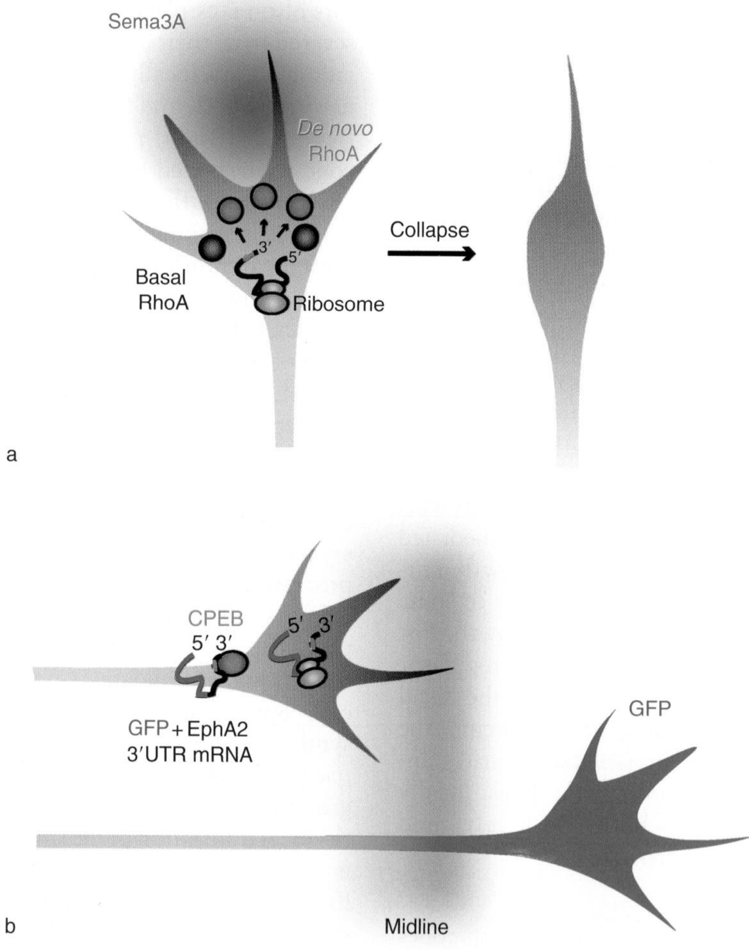

Figure 3 Diagram showing the role of axonal messenger RNA (mRNA) transport and local translation in growth cone behavior in developing axons. (a) Growth cones exposed to Sema3A induce the translation of RhoA, a small guanosine triphosphatase. The newly synthesized RhoA is required and sufficient to mediate Sema3A-induced growth cone collapse. (b) Local translation at the midline intermediate target. Translation of a reporter construct encompassing green fluorescent protein (GFP) and the 3′ untranslated region (3′UTR) of the EphA2 receptor is upregulated as growth cones enter the midline and reaches high levels as they emerge from the midline. Translation of the reporter construct is dependent on the cytoplasmic polyadenylation element sequence in the 3′UTR, suggesting that cytoplasmic polyadenylation element binding protein (CPEB) may be involved in mRNA localization or translational control.

axons use intermediate targets to which they are directed by local attractive and repulsive cues. After reaching the intermediate target, the axons must continue their journey and change responsiveness to the cues that guide them to the next target. One way to change responsiveness at an intermediate target is by changing the repertoire of guidance receptors at the plasma membrane. Studies in chick embryonic commissural neurons have suggested that this can be achieved by local translation of a receptor. Spinal commissural axons are initially attracted toward the midline floor plate, but after crossing the midline, axons lose responsiveness to the midline attractants and gain responsiveness to a new set of guidance cues. When these commissural axons are transfected with a construct encompassing the 3′UTR of the EphA2 receptor and GFP, protein expression is detected in the cell bodies but not in the growth cones. However, the protein is expressed in the growth cones of axons that have crossed the midline, suggesting that translation of the reporter construct is selectively activated after midline crossing. Translation is dependent on a cytoplasmic polyadenylation element binding element in the 3′UTR, indicating that CPEB might target the EphA2 mRNA to the growth cone.

Thus, local protein synthesis may play a crucial role in axon guidance. It has been known for decades that axons that are separated from their cell body retain their ability to grow and even make accurate guidance decisions. The ability of axons to locally translate proteins may explain this observation. By locally synthesizing proteins, the growth cone can autonomously respond to its environment. Different guidance cues along the pathway might stimulate the transport and the translation of a specific subset of mRNA needed for immediate growth cone turning, adaptation, and responses at intermediate targets. Moreover, protein synthesis may play a role presynaptically in the establishment of arbors and synapses, and studies in invertebrates indeed suggest a role for local translation in regulating synaptic connections in response to activation. It should be noted that not all cues elicit protein synthesis in the growth cones, indicating that local translation is not a ubiquitous mechanism in cue-directed growth.

Axonal mRNA Transport in Adult Axons

In contrast to developing axons, neuronal function of mature axons is not dependent on major cytoskeletal rearrangements like those involved in growth cone steering but rather on maintaining the integrity of the cytoskeleton, except when an axon is disconnected from its targets on injury. Neuronal regeneration is dependent on many factors, including severity of the injury and cell type. Peripheral nervous system neurons have regenerative capacity, whereas neurons of the central nervous system do not. Successful regeneration requires the tip of the cut axon to be remodeled and reformed into a growth cone. Experiments in cultured adult sensory and retinal axons have demonstrated that the formation of a growth cone, after axotomy, depends on local axonal protein synthesis (and degradation). It is important to note that the intrinsic regenerative capacity of the different types of neurons correlates with the levels of intra-axonal protein synthetic machinery: sensory neurons with high regenerative ability contain high levels of ribosomal proteins and activated translational initiation factors, whereas these proteins are absent in retinal axons, which have a low regenerative ability. Similar to developing axons, growth cone reformation after injury depends on activation of MAP kinases. A mechanism for the role of translation and MAP kinases has been proposed in the peripheral nervous system. Nerve injury leads to the activation of MAP kinases in the axon as well as the translation of the intermediate filament protein vimentin and the nuclear import protein β-importin. Vimentin then undergoes proteolytic cleavage, and these vimentin fragments bind to the activated MAP kinases to couple them to the importin–dynein complex for transport to the cell body. This is thought to provide a mechanism to retrogradely transport an injury signal to the nucleus to facilitate the regenerative response. Maintenance of regenerative growth, after the initiation of growth cone formation, might be regulated by extracellular stimuli. A proteomic approach has identified multiple axonally synthesized proteins in injury-conditioned DRG axons, but the growth-promoting neurotrophins selectively increase the levels of some cytoskeletal mRNAs into the axon, presumably to retain growth cone motility. Thus, axonal mRNA translation is important for multiple steps during regeneration, including growth cone formation, generation of an injury signal, and maintenance of axonal growth (**Figure 4**).

Axonal mRNA Transport and Neurological Diseases

Although *in vivo* data on the function of axonal mRNA transport and translation are still lacking, the importance of these processes becomes evident when the function of key regulators such as RNBPs is impaired. For example, mutations in the FMR1 gene on the X chromosome lead to the fragile X syndrome, which is the most common cause of inherited mental retardation. Mutations in the FMR1 gene lead to transcriptional silencing of the gene and a consequent lack of FMRP and FMRP-transported mRNAs. Furthermore, HuD is associated with

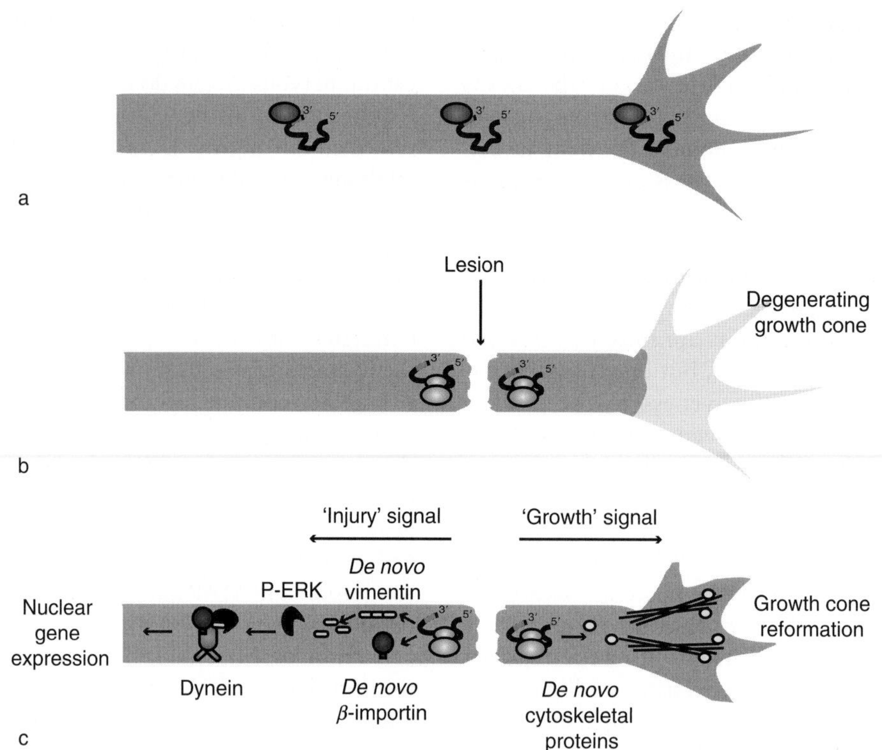

Figure 4 Local translation in injured mature axons. (a) Uninjured axons transport mRNA along the axon, presumably for cytoskeletal maintenance. (b) Once an axon is lesioned, the growth cone undergoes degeneration and loss of connections with its targets. At the site of the lesion, protein translation is induced. (c) Newly synthesized proteins include cytoskeletal proteins such as β-actin for growth cone reformation and of vimentin and β-importin for the generation of an injury signal. Vimentin protein is proteolytically cleaved, which generates fragments that link phospho-extracellular signal-regulated kinase (P-ERK) to β-importin to the dynein motor protein. This complex is retrogradely transported along microtubules into the nucleus to regulate gene expression to facilitate the regeneration process. Adapted from Willis DE and Twiss JL (2006) The evolving roles of axonally synthesized proteins in regeneration. *Current Opinion in Neurobiology* 16: 111–118.

paraneoplastic encephalomyelitis, a neurological disorder caused by an autoimmune reaction against HuD. Moreover, low levels of SMN protein result in spinal muscular atrophy, which is an autosomal recessive disorder and the leading hereditary cause of infant mortality and is characterized by a loss of α motor neurons in the spinal cord. In most cases, the causes of these neurological diseases have not been directly attributed to specific defects in axonal mRNA transport. However, knockdown of SMN in zebra fish has been shown to result in motor axon-specific outgrowth and pathfinding defects, emphasizing the importance of regulated mRNA transport for neural development. Future studies on the identification of RNP complexes and their functions might explain the cellular consequences of RNA misregulation that underlie distinct types of neurological diseases.

Conclusion

The concept of local translation at growth cones has gained increasing acceptance during the past 5 years, and it is now clear that mRNAs are transported into the axon and growth cone and can be translated in response to environmental cues. Axonal mRNA transport serves to spatially restrict protein synthesis to the axon and growth cone, providing axons with the capacity to autonomously regulate their structure and function during axonal growth, navigation, and regeneration. A general model about the molecular mechanism of neuronal mRNA transport and translation has come from studies in dendrites, and although this mechanism seems to be highly conserved, many aspects of axonal mRNA function remain completely undefined. For instance, characterization of the identity of RNA granules and determination of the ways environmental cues regulate mRNA localization within the milieu of the elongating axon are incompletely understood. One recently proposed hypothesis is that RNBPs transport mRNAs as a set that encodes proteins with a common biological function. Some dendritic RNBPs, for example, transport a coherent subset of mRNAs to specifically mediate synaptic plasticity. It will be fascinating to determine whether axonal RNA granules function in an analogous fashion to regulate specific axon and growth cone function. Elucidating how axonal mRNA transport and translation regulate growth and pathfinding *in vivo* should provide key insights into the ways dysfunction in these processes can lead to neurological diseases.

See also: Actin Cytoskeleton in Growth Cones, Nerve Terminals, and Dendritic Spines; Axon Guidance: Guidance Cues and Guidepost Cells; Axonal Regeneration: Role of Growth and Guidance Cues; Axonal Transport Disorders; Axonal and Dendritic Transport by Dyneins and Kinesins in Neurons; Dendritic RNA Transport: Dynamic Spatio-Temporal Control of Neuronal Gene Expression; Growth Cones; Myosin Transport and Neuronal Function; RNA Binding Protein Methods.

Further Reading

Bassel GJ and Kelic S (2004) Binding proteins for mRNA localization and local translation, and their dysfunction in genetic neurological disease. *Current Opinion in Neurobiology* 14: 574–581.

Brittis PA, Lu Q, and Flanagan JG (2002) Axonal protein synthesis provides a mechanism for localized regulation at an intermediate target. *Cell* 110: 223–235.

Campbell DS and Holt C (2001) Chemotropic responses of retinal growth cones mediated by rapid local protein synthesis and degradation. *Neuron* 32: 1013–1026.

Martin KC (2004) Local protein synthesis during axon guidance and synaptic plasticity. *Current Opinion in Neurobiology* 14: 305–310.

Ming GL, Wong ST, Henley J, et al. (2002) Adaptation in the chemotactic guidance of nerve growth cones. *Nature* 417: 411–418.

Piper M and Holt C (2004) RNA translation in axons. *Annual Reviews in Cell and Developmental Biology* 20: 505–523.

Piper M, Salih S, Weinl C, Holt CE, and Harris WA (2005) Endocytosis-dependent desensitization and protein synthesis-dependent resensitization in retinal growth cone adaptation. *Nature Neuroscience* 8: 179–186.

St. Johnston D (2005) Moving messages: The intracellular localization of mRNAs. *Nature Reviews in Molecular Cell Biology* 6: 363–375.

Willis DE and Twiss JL (2006) The evolving roles of axonally synthesized proteins in regeneration. *Current Opinion in Neurobiology* 16: 111–118.

Wu KY, Hengst U, Cox LJ, et al. (2005) Local translation of RhoA regulates growth cone collapse. *Nature* 436: 1020–1024.

Axonal Pathfinding

D Mortimer and G J Goodhill, The University of Queensland, St. Lucia, QLD, Australia

Introduction

Building a brain requires the precise formation of connections between vast numbers of neurons, often separated by significant distances. Axon pathfinding is thus a crucial process in the development of a functioning nervous system. A better understanding of the principles and mechanisms underlying axon pathfinding has both clinical and broader practical importance. From a clinical perspective, our ability to treat or prevent some neurological defects will be improved by a better understanding of how axon guidance can fail. Furthermore, the regeneration of damaged nerves requires axons to reconnect to their appropriate targets; hence, understanding axon guidance will be necessary for the development of therapeutic techniques. In a broader context, our understanding of axon pathfinding ties into our understanding of the nervous system as a whole: what principles underlie its formation and function? can we harness those principles in order to improve our own engineering processes, such as in the construction of self-wiring computers?

Mathematical and computational models are very useful tools for understanding the constraints on nervous system development. Ultimately, such constraints are quantitative and set by the physics of the system; hence, they must be modeled mathematically to yield the best predictive power and generate strong hypotheses. Models may also prove valuable in the development of therapies. For instance, how can axons be made to grow toward a specific target area? What additional information is needed by axons which are failing to develop correctly? Sufficiently detailed theoretical models have the potential to guide experimental research in axon pathfinding through simulations done *in silico*.

Experimental Data

Guidance Cues

Axons grow along their correct trajectories by following a molecular map consisting of spatiotemporal patterns of guidance cue molecules. Four main families of molecules have been identified based on their guidance abilities – the netrins, the Slits, the semaphorins, and the ephrins – consisting of approximately 100 distinct molecules altogether (although several other classes of molecules also provide guidance information for axons, including the neurotrophins and some classical morphogens). The netrins were first identified as attractive guidance cues which direct contralaterally projecting neurons toward the midline. The transmembrane proteins DCC and Unc-5 have been shown to act as receptors for netrin-1. Both netrins and their receptors are highly conserved between species. Netrin-1 is known to have a bifunctional role, typically attracting growth cones expressing only the DCC receptor but repelling growth cones expressing both DCC and Unc-5. The Slits and their receptors, the roundabout family (the Robos), were first identified as repellents preventing contralaterally projecting neurons from recrossing the midline. Subsequently, they were also shown to stimulate axon outgrowth and branching. The semaphorins appear to act primarily as short-range cues which repel axons from particular regions or, by forming the walls of corridors, hem axons into a preferred path. However, they have also been reported to act as long-range chemoattractants. Semaphorins are classified by their structure into eight groups. They signal through multimeric receptor complexes, with the precise structure of a complex determining its specificity for a semaphorin subgroup. Semaphorin receptor molecules include the neuropilins, the plexins, and the cell adhesion molecule L1. The ephrins are substrate-bound molecules best known for their role in the formation of topographic maps in the central nervous system. For example, the graded expression of the Eph tyrosine kinases – the ephrin receptors – in the retina combined with graded expression of the ephrins in the tectum aid in the formation of an ordered topographic mapping between the two structures. Similar strategies appear to orchestrate the formation of other topographic maps. Although ephrin/Eph signaling induces axon repulsion in these examples, under other contexts the ephrins can also act to attract axons.

The Growth Cone

Growing axons are tipped by special sensorimotor structures known as growth cones. These probe their local environment and, depending on the signals they detect, direct axon outgrowth, turning, branching, and pruning. Growth cones exhibit complex morphology, as illustrated in **Figure 1**. They are conconceptually divided into three sections: an actin-rich peripheral region, a transitional region, and a central region containing organelles and microtubules. Fingerlike protuberances extending from the edge of the growth cone known as filopodia are supported by bundles of filamentous actin (F-actin). These appear to act as sensory devices, extending the effective

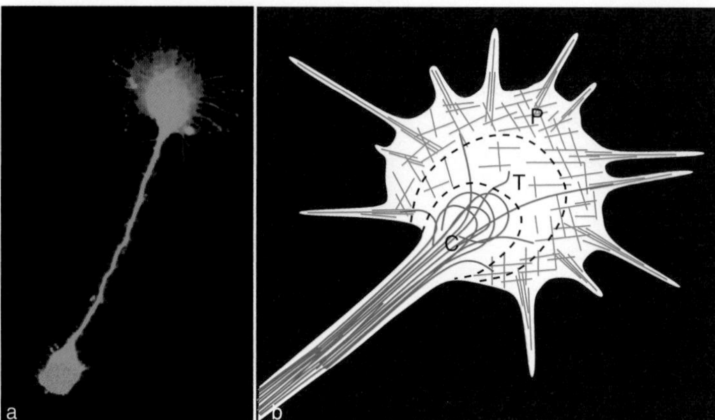

Figure 1 (a) Rat superior cervical ganglion neuron *in vitro*. The neuron was grown on a substrate of laminin for 48 h and then fixed and stained for β-tubulin. The cell body is at the bottom left, and the growth cone is at the top right. (b) Anatomy of the growth cone. This figure illustrates the division of the growth cone into three domains: transitional (T), peripheral (P), and central (C). The peripheral domain is rich in actin, which exhibits two forms of organization: loosely linked networks giving rise to lamellipodial structures and tight bundles supporting filopodia. The central domain contains organelles and microtubules which extend from the axon shaft into the growth cone. Rat SCG neuron image courtesy of Z Pujic.

sensing range of the growth cone. Structures based on more chaotic meshworks of F-actin are known as lamellipodia and have also been implicated in growth cone movement.

Growth cones undergo constant morphological change thought to be driven by remodeling of the actin network controlled by actin-binding proteins such as myosin, ARP2/3, and WASP. Growth cone motility appears to be driven by an 'actin treadmill.' Unpolymerized G-actin diffuses outward into the peripheral region and preferentially polymerizes near the cell membrane. The entire network of polymerized actin is drawn toward the central region by myosin and undergoes depolymerization in the transitional region. This constant cycle of polymerization, retrograde flow, and depolymerization is thought to generate traction when coupled to a permissive substrate through adhesion molecules.

Microtubules also play a significant, although unclear, role in growth cone motility and axon guidance. Interactions between microtubules and actin in the transitional region appear to have a strong influence on axon outgrowth and guidance.

Guidance cues influence growth cone behavior, and subsequent axon outgrowth, through cytoskeletal effectors activated or inhibited by cascades of intracellular second messengers triggered by receptor binding. Several molecules have been implicated as playing roles in this process, particularly calcium and the cyclic nucleotides cAMP and cGMP. These have been the focus of much attention, with the finding that, in some circumstances, changing the relative concentrations of these molecules within the growth cone can switch the effect of several guidance cues from attraction to repulsion or vice versa. Further intriguing findings have demonstrated that protein synthesis occurring locally within the growth cone is necessary for correct growth cone behavior.

Theoretical and Computational Models of Axon Pathfinding

Axon Extension and Branching

Experimental evidence suggests that tubulin molecules are synthesized only in the soma and then assembled into microtubules predominantly in the growth cone. This implies that axon outgrowth is limited by the rate at which microtubules can be transported to regions of active extension. A number of theoretical models have explored this idea, including various effects such as diffusive and active transport of tubulin monomers, competition between neurites for tubulin, viscoelastic stretching of axon segments, calcium-induced microtubule depolymerization, and varying intrinsic rates of tubulin polymerization and depolymerization within different growth cones. These models have been successively refined, ultimately incorporating compartment-based modeling with dynamic compartment allocation. Most strikingly, this modeling program has demonstrated that small variations in polymerization and depolymerization rates in the growth cones of different neurites can lead to sharp changes in elongation rate, including growth cone pausing and neurite retraction.

Growth cone behavior and axon extension are thought to be mediated by partially independent but related processes. This has led to modeling work focused on characterizing the interaction between the two. The majority of this work has been phenomenological, using sophisticated mathematical machinery to better describe experimental data. Work in the mid-1980s showed that in some circumstances, axon elongation can be regarded as a one-dimensional random walk, with parameters varying with neuronal type, substrate, and chemical environment. In the mid-1990s, this was extended by performing correlation analysis which showed that axon outgrowth dynamics exhibit significant anticorrelation. Subsequent analysis showed significant correlation between the dynamics of microtubule polymerization in the central region and growth cone advance.

Growth Cone Morphology

The complex and dynamic nature of growth cone morphology further complicates our understanding of the contribution of growth cones to axon pathfinding. By statistically analyzing time-lapse images of growth cones undergoing dynamic changes in morphology, researchers have developed probabilistic rules specifying the likelihood of filopodial initiation and retraction, and also the spatial distribution of filopodia in terms of a limited number of parameters. In this model, growth cone morphology is described by the instantaneous length and angle of each filopodium, and the dynamics of the filopodia are characterized by the following parameters: the rate at which filopodia extend, the rate of retraction, the average rate at which new filopodia are initiated (modeled as a Poisson process), and the shape parameters for a gamma distribution that gives the time over which a filopodium extends before retracting. The model also specifies a simple conditionally random rule for where a filopodium initiates, and it assumes that filopodia extend radially from the center of the growth cone. Computer simulation then gives qualitatively realistic morphologies, which also satisfy quantitative constraints such as the correct average number of filopodia. By mapping the effects of external cues onto the parameters of the model, one can hope to gain some intuition as to how those cues might operate.

Other researchers have directly modeled the processes underlying actin dynamics in order to understand filopodial formation, stability, and behavior. This analysis indicates that the maximum length of a filopodium is determined by the number of bundled actin filaments in its core. For less than approximately 10 bundled actin filaments, the strain exerted on the bundle by the membrane is sufficient to cause buckling for even very short filopodia. As the number of included filaments increases, it becomes less likely that a filopodium will buckle; however, more G-actin is required for the structure to continue extending. Thus, when the number of filaments is too large, a filopodium is also unable to extend. The best trade-off between stability and G-actin depletion is achieved with approximately 30 actin filaments. The model predicts average filopodia lengths between 1 and 10 μm, which are in agreement with experimental data. Similar work has examined the mechanisms behind lamellipodial structures.

Axon Turning and Guidance

Axons are thought to be guided by external cues through two processes: gradient-based guidance, in which the growth cone attempts to climb or descend a concentration gradient (chemoattraction or chemorepulsion), and contact-mediated guidance, in which the growth cone interacts with small regions of highly concentrated guidance cues and modulates its behavior accordingly. For gradient-based guidance, the growth cone must detect a potentially shallow gradient in the presence of noise, whereas for contact-mediated guidance the growth cone is essentially involved in a search process.

Gradient guidance Single cell chemotaxis – the attraction or repulsion of organisms such as bacteria, leukocytes, or slime molds by chemical gradients – has received a large amount of theoretical attention. Much of this can be directly applied to the case of the growth cone. Of particular relevance is a seminal contribution by Berg and Purcell, who argued that gradient detection by any small sensing device is fundamentally limited by statistical fluctuations, both due to variations in local ligand concentration and due to the inherent stochasticity of receptor binding. Growth cones are believed to sense and respond to gradients by comparing receptor binding across their spatial extent: the side of a growth cone exposed to the highest concentration of ligand will, on average, also display the largest amount of receptor binding. If growth cones do use such a spatial-sensing strategy, then in order for a growth cone to detect and reliably respond to a chemical gradient, the noise due to fluctuations in receptor binding cannot be much larger than the difference in receptor binding across its spatial extent. By modeling the physics of receptor–ligand interaction, one can estimate the limitations growth cones face when responding to chemical gradients. If the root mean squared error in a concentration measurement is given by σ_C, then the error associated with taking the difference between two such measurements is $\sigma_{\Delta C} = \sqrt{2}\sigma_C$. This gives an order of magnitude lower bound on the difference in concentration, ΔC, that the growth cone can detect: $\Delta C_{min} \approx \sigma_{\Delta C}$. Using a simple model of receptor binding mechanics, it has been shown that for the timescale on which growth cone behavior usually occurs (~100 s), growth cones can detect gradients with a steepness of between 1% and 10%, depending on whether the guidance cue is diffusing freely or is substrate bound. However, over much longer periods (several days), experimental work has demonstrated that growth cones can respond to gradients of 0.2% or less across their spatial extent. Furthermore, at sufficiently high and low concentrations, almost all or almost no receptors are bound: this leads to further reductions in sensitivity. Experimental and theoretical work has confirmed that the highest sensitivity is achieved when approximately half of the receptors are bound, which occurs when the concentration is equal to the dissociation constant for binding.

Aside from these general constraints on gradient detection, models have also been developed which directly simulate the behavior of growth cones and axons in the presence of guidance cue gradients. Several models have focused on biochemical networks which are putatively responsible for growth cone motility and guidance. Such models are difficult to construct, partly due to the complexity of growth cone biochemistry and partly due to the lack of experimental data on important quantities such as reaction rate constants, concentrations, and interactions between molecular species. One model has focused on the Rho-GTPase signaling network, which is known to play an important role in actin-driven cell motility. Due to incomplete experimental data, the investigators took a qualitative approach, simulating the behavior of several plausible interaction networks and kinetic constants and using the results of these simulations to form hypotheses about the underlying mechanisms of growth cone motility. They found that the Rho-GTPase network undergoes a sharp transition in its dynamics when a threshold concentration of a particular signaling molecule is reached. The authors linked these two dynamic behaviors to different modes of growth cone motility, developing a model which could reproduce some experimentally observed phenomena.

Other models have placed less emphasis on specific biochemical mechanisms and have focused instead on the potential role of filopodia in axon guidance or on more general signal-processing strategies that a growth cone may implement, such as temporally or spatially averaging receptor inputs in order to reduce noise. Spatial averaging involves pooling the inputs from multiple receptors, whereas temporal averaging combines information from different time points. An interesting conclusion from this study is that spatial averaging provides the most benefit when the average is taken over approximately one-third of the growth cone's spatial extent. This optimum averaging range occurs because although spatial averaging reduces noise in a local concentration measurement, it also reduces the spatial resolution of the measurement. Because gradient detection requires concentration measurements to be made at multiple locations, the advantages gained in noise reduction are offset by the loss in resolution. Assuming growth cones use such a strategy, this has implications for the intracellular signaling network, suggesting that second messenger molecules implementing the spatial averaging process must diffuse at a rate much slower than expected for cytosolic compounds. One possibility is that spatial averaging is achieved through membrane-bound molecules.

In addition to detecting a chemical gradient, the growth cone must also amplify the possibly extremely shallow gradient in receptor binding in order to achieve a definite motile response. Understanding the mechanisms underlying this amplification has been a general focus for experimental and modeling work on microbial chemotaxis. In one influential model, amplification is achieved by coupling the external signal to a pattern formation system involving local activation and long-range inhibition. The system begins in a spatially symmetric, but unstable, steady state. Symmetry is broken by the external signal, which pushes the system into a stable, asymmetric state that reflects the direction in which the symmetry was broken. A difficulty with this approach is that the system then becomes stuck, unable to respond to new inputs such as a change in the external signal. Several additional mechanisms have been proposed to work around this, each postulating a second process which serves to reset the system to its original, unstable state. Further generalizations of this class of models suggest mechanisms for the formation of filopodia and generate testable predictions for the spatiotemporal distribution of such structures.

Contact-mediated guidance Axons are also guided by cues which are more tightly localized in space, referred to as short-range or contact-mediated cues. For example, filopodial contact with single cells expressing appropriate cues can entirely redirect an axon's trajectory. For this kind of guidance, noise is less of an issue because the signal is essentially binary: either the growth cone contacts the cue or it does not. In this situation, an appropriate theoretical framework is that of stochastic search. The question of how filopodial dynamics of the growth cone affects its ability to locate and respond to highly localized guidance cues has been addressed with the aid of the models describing growth cone morphology in terms of filopodial dynamics. The efficacy with which a growth cone is able to locate a guidance cue has been mapped against the parameters defining the dynamics, suggesting some behaviors one might expect to observe depending on the geometry of the guidance cue distribution. This work suggests that filopodial dynamics are set, and possibly modulated, in order to increase a growth cone's ability to detect and respond to relevant cues. In a more abstract approach, growth cone movements were described by a combination of stochastic (e.g., deflection by random adhesion to the substrate) and deterministic (e.g., a tendency to move in the direction of past axon extension) motions. The growth cone was found to more effectively respond to short-range guidance signals when the two processes contributed equally. This prompted the authors to propose that growth cones modulate the relative influence of stochastic and deterministic movements depending on the importance of short-range cues at different stages in development – a suggestion consistent with experimental observations of growth cone behavior.

Axon–Axon Interactions

The models described so far consider the guidance of single neurons in isolation. However, the development of the nervous system involves the correct guidance of many axons simultaneously, and it is well established that axons use one another as additional sources of information during development. One of the earliest computer models of axon guidance attempted to explain the characteristic 'sheetlike' pattern of axon outgrowth observed in the formation of the ventral commissure of the spinal cord. Using a descriptive model of individual axon behavior, including several experimentally observed features of ventral commissural axon growth – a tendency for straight growth, for initial outgrowth to be directed ventrally, and for growing axons to extend preferentially over the surrounding matrix and not other axons – this work attempted to distinguish the most important features of individual axon behavior for the formation of axon sheets. From computer simulations, the authors concluded that initially polarized outgrowth, a suitably high density of neurons, preferential adhesivity for extension over the substrate rather than other axons, and a tendency for straight growth were sufficient to generate the observed patterns. Another model examined the possibility that growth cones secrete diffusible guidance cues in order to attract or repel one another to create or break up axon bundles.

Topographic Map Formation

A specific example of axon guidance that has been well studied theoretically is the formation of the topographic map between the retina and optic tectum/superior colliculus. In 1963, Roger Sperry first proposed that such maps could arise because gradients of molecular labels in the retina are matched to gradients of labels in the tectum. The subsequent discovery of gradients of Eph receptors and their ligands, the ephrins, in the retina and tectum confirmed this prediction. However, a large number of experiments investigating how such matching might work in detail have suggested that several other constraints are also important. Since the 1970s, numerous theoretical models of such map formation have been proposed in order to gain insight into this complexity. Some of the simplest propose sorting mechanisms, whereby an initial random map is refined by comparing the retinal origin of axons terminating at neighboring sites in the tectum. Others have hypothesized that tectal labels are at least partly induced or modified by transport of retinal labels into the tectum. Several models have highlighted the importance of competition in map formation, both between axons for tectal target space and between tectal targets for axons. Another important theme has been cooperative effects between axons, somewhat similar to the axon–axon interactions discussed previously. Increasing data on the precise role of Eph/ephrins in map formation have provided new challenges for such models, many of which are yet to be addressed.

Guidance Cue Patterning

A further area of active research aims to understand how guidance cue patterns are generated in the first place, and how effectively particular patterns can guide axons. The modeling of gradient systems in developing organisms has a long history, and gradients are thought to be a primary means for generating spatial ordering. In general, molecules expressed as gradients in order to provide spatial information are known as morphogens, and several classical morphogens have been shown to also guide axons.

A number of models have been proposed to explain how gradients of appropriate shape and stability could be set up. The simplest model assumes that the molecule of interest is diffusing away from a continuous source through a homogeneous medium. More complex models recognize the inhomogeneous nature of the medium, degradation of molecules, binding of molecules to cells, endocytosis, and active transport processes. A further complication which arises when attempting to generalize results from one experimental model to another is that of scaling: gradients form on a typical length scale, and different species have embryos of different sizes at developmental stages when axon wiring is forming. Hence, a system which works in one embryo may not work in another.

Additional constraints are placed on the formation of gradients useful for axon guidance. The ability of a growth cone to detect and respond to a gradient varies with background concentration and gradient slope. The minimum gradient that a growth cone can detect over a particular background concentration specifies limitations on gradient-based guidance. It allows the construction of optimal gradients, in which the gradient slope is always equal to the minimal detectable gradient for the growth cone. Following this line of argument, coupled with estimates of parameters central to the model, it can be shown that the maximum distance over which growth cones can be guided by an optimal gradient is on the order of 1 cm.

Future Directions

This article provided an overview of the kinds of models which have been applied to axon guidance and how these have helped us to understand axon pathfinding. However, research of this kind is still at an early stage. Important questions remain to be answered and are the focus of active research. For instance, how sensitive can growth cones be to gradients of guidance cues? How close do they come to achieving fundamental sensitivity limits? What are the actual mechanisms they use to detect gradients? How do developing neurites integrate information from multiple guidance cues? What searching strategies do growth cones use to locate local guidance cues? What roles do axon branching and pruning play in axon guidance? How do microtubules and the F-actin cytoskeleton interact to support axon outgrowth and steering? How much of a role do axon–axon interactions play in the formation of the nervous system, and when are they important?

In addition to fresh modeling approaches, answering these questions will require significant experimental advances. A wealth of experimental data is available on axon guidance, but most studies have been aimed at identifying guidance cue molecules or intracellular molecules mediating or eliciting particular behaviors. Although such data are obviously crucial, in order to generate sufficiently constrained models, data of a more quantitative nature are needed. Recognizing this need, experimental techniques for producing well-controlled and flexible patterns of guidance cues have been developed. Ultimately, these techniques should allow us to develop better constrained models and, using them, obtain additional power to tease apart the mechanisms and principles underlying axon guidance.

Finally, the discovery that new neurons are constantly being born in adult brains opens up another area for exploration. These nascent neurons must somehow find their way to their appropriate niches and extend axons to make functional connections. Modeling axon guidance thus has a central role to play in understanding both the initial development and the normal functioning of the nervous system.

See also: Axon Guidance: Morphogens as Chemoattractants and Chemorepellants; Axon Guidance: Building Pathways with Molecular Cues in Vertebrate Sensory Systems; Axon Guidance: Guidance Cues and Guidepost Cells; Axonal Regeneration: Role of Growth and Guidance Cues; Cognitive Neuroscience: An Overview; Computational Methods; Growth Cones; Neuroplasticity: Computational Approaches.

Further Reading

Berg HC and Purcell EM (1977) Physics of chemoreception. *Biophysical Journal* 20: 193–219.

Buettner HM (1996) Analysis of cell-target encounter by random filopodial projections. *American Institute of Chemical Engineering Journal* 42: 1127–1138.

Dickson BJ (2002) Molecular mechanisms of axon guidance. *Science* 298: 1959–1964.

Goodhill GJ (1998) Mathematical guidance for axons. *Trends in Neurosciences* 21: 226–231.

Goodhill GJ, Gu M, and Urbach JS (2004) Predicting axonal response to molecular gradients with a computational model of filopodial dynamics. *Neural Computation* 16: 2221–2243.

Goodhill GJ and Xu J (2005) The development of retinotectal maps: A review of models based on molecular gradients. *Network: Computation in Neural Systems* 16: 5–34.

Gordon-Weeks PR (2000) *Neuronal Growth Cones*. Cambridge, UK: Cambridge University Press.

Hely TA and Willshaw DJ (1998) Short-term interactions between microtubules and actin filaments underlie long-term behaviour in neuronal growth cones. *Proceedings of the Royal Society of London, Series B* 265: 1801–1807.

Katz MJ, George EB, and Gilbert LJ (1984) Axonal elongation as a stochastic walk. *Cell Motility* 4: 351–370.

Krottje JK and van Ooyen A (2007) A mathematical framework for modeling axon guidance. *Bulletin of Mathematical Biology* 69: 3–31.

Maskery SM and Shinbrot T (2005) Deterministic and stochastic elements of axonal guidance. *Annual Review of Biomedical Engineering* 7: 187–221.

Meinhardt H (1999) Orientation of chemotactic cells and growth cones: Models and mechanisms. *Journal of Cell Science* 112: 2867–2874.

Mortimer D, Fothergill T, Pujic Z, Richards LJ and Goodhill GJ (2008) Growth cone chemotaxis. *Trends in Neurosciences* 31: 90–98.

Sakumura Y, Tsukada Y, Yamamoto N, and Ishii S (2005) A molecular model for axon guidance based on cross talk between Rho GTPases. *Biophysical Journal* 89: 812–822.

van Veen MP and van Pelt J (1994) Neuritic growth rate described by modeling microtubule dynamics. *Bulletin of Mathematical Biology* 56: 249–273.

Willshaw DJ and von der Malsburg C (1979) A marker induction mechanism for the establishment of ordered neural mappings: Its application to the retinotectal problem. *Philosophical Transactions of the Royal Society of London, Series B* 287: 203–243.

Axonal Pathfinding: Extracellular Matrix Role

P Letourneau, University of Minnesota, Minneapolis, MN, USA

Introduction

Normal behavior and other neural activities depend on the correct wiring of neural circuits during development. A critical step in forming neural circuits is the growth of axons from nerve cell bodies to the sites where synaptic connections are made. In spanning between neural somata and their synaptic targets, growing axons forge pathways that become the axonal tracts and peripheral nerves of the mature nervous system. The routes that axons take to reach their targets are determined by motile activities at their tips, called growth cones. Growth cones extend fine protrusions that adhere to nearby cells and surfaces. These adhesive contacts provide a toehold from which further protrusions are made. As growth cones crawl forward, they choose a path by detecting and responding to the spatial and temporal distributions of extracellular guidance molecules encountered in their local environments. Five major families of extracellular molecules – netrins, neurotrophins, semaphorins, slits, and ephrins – provide positive and negative cues that orient the migration of growth cones to their targets. These guidance molecules bind receptor proteins on growth cones and initiate cytoplasmic signals that regulate the motility and adhesive contacts that determine the advance, retreat, turning, branching, and stopping of growth cones. This article describes molecules that play a key role in axonal pathfinding by mediating the adhesive interactions necessary for growth cone migration.

Mechanism of Growth Cone Migration

Cytoskeletal Dynamics

Growth cone migration and axonal elongation involve the cytoskeletal components, microtubules and actin filaments. Axon elongation requires the advance and polymerization of microtubules, which are bundled in the axon but which spread apart in the growth cone, where individual microtubules dynamically probe forward to the front of a growth cone via polymerization and movement involving microtubule motor molecules (**Figure 1**). Axonal growth occurs where the main microtubule bundle and associated organelles advance in the growth cone, as determined by the positions and stabilization of these forward 'pioneering' microtubules. These dynamic microtubules project forward into an actin filament network that fills flattened dynamic projections, called lamellipodia, and fingerlike filopodia. This extensive filament system is continually remodeled, as actin filaments initiate and polymerize at the front margin and are then moved back to be fragmented and depolymerized, recycling subunits to the front. Multiple actin-binding proteins regulate this dynamic organization of actin filaments.

Growth cone migration is driven by forces produced within this actin filament domain. Actin polymerization creates protrusive forces that expand lamellipodia and elongate the tips of filopodia. Myosin motor molecules bind actin filaments and generate mechanical forces that move cargo bound to the myosin tail domains or pull on actin filaments to create tensions. Myosin II motor activity pulls actin filaments rearward, where they are depolymerized. Tensions generated by myosin II activity in the actin-rich leading margin can either direct or halt microtubule advance, depending on the situation. Myosin II-generated tensions produce the exploratory movements of lamellipodia and filopodia, whereas excessive levels of tension may sweep microtubules back in a contracting actin network that can collapse a growth cone. It is in the context of these dynamic cytoskeletal activities that adhesive interactions are critical to growth cone migration (**Figure 2**).

Adhesive Contacts of Growth Cones

Growth cone plasma membranes contain adhesion receptors that bind noncovalently to adhesion molecules on other cells or surfaces. Lamellipodia and filopodia initiate adhesive interactions as they explore their environment, and if these bonds persist, receptors cluster to form discrete adhesive contacts, which include intracellular adhesion complexes. Adhesion complexes remain in place or shift rearward as a growth cone advances. These adhesive complexes play two roles in growth cone migration. First, they include proteins that anchor actin filaments at adhesive sites. These links constitute a 'clutch' that stops the retrograde movement of actin filaments and permits the advance of microtubules and axonal organelles (**Figure 2**). Without stabilization provided by adhesive contacts growth cone migration fails, and tensions within the axonal cytoskeleton cause axonal retraction. Second, these complexes include proteins of signaling cascades, protein kinases, protein phosphatases, and Rho GTPases, which act on proteins that regulate the organization of actin filaments and microtubules. Thus, adhesive contacts provide points

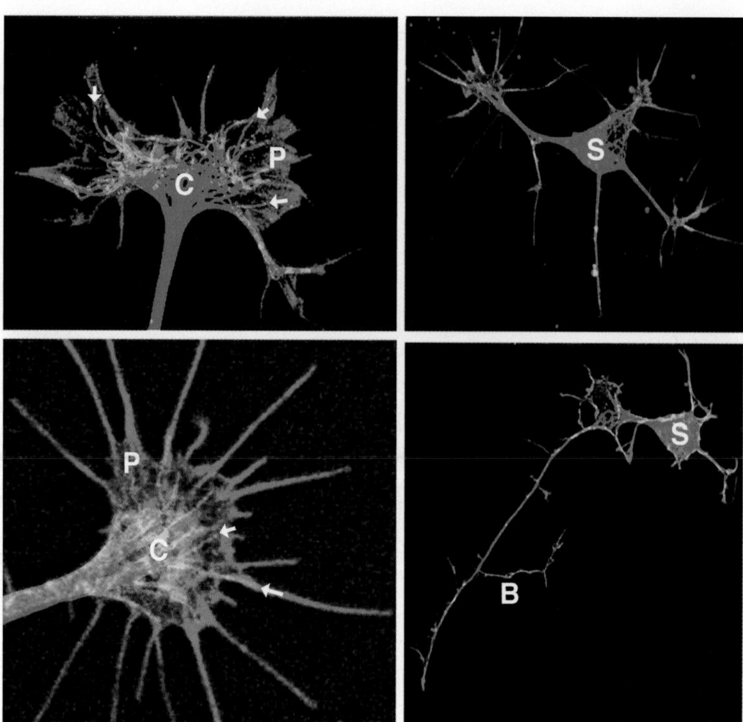

Figure 1 The distribution of microtubules and actin filaments in developing neurons and in axonal growth cones. Microtubules (green) are densely packed in the neuronal cell bodies (S) and are bundled in the axons and branches. Actin filaments are arrayed in filament networks and bundles in the peripheral domains (P) of the growth cones and along the shafts of the axons, where small areas of actin filament dynamics may give rise to collateral branches (B). In a growth cone, the microtubules from the central bundle of the central domain (C) splay apart and individual microtubules extend into the P domain and into filopodia (arrows).

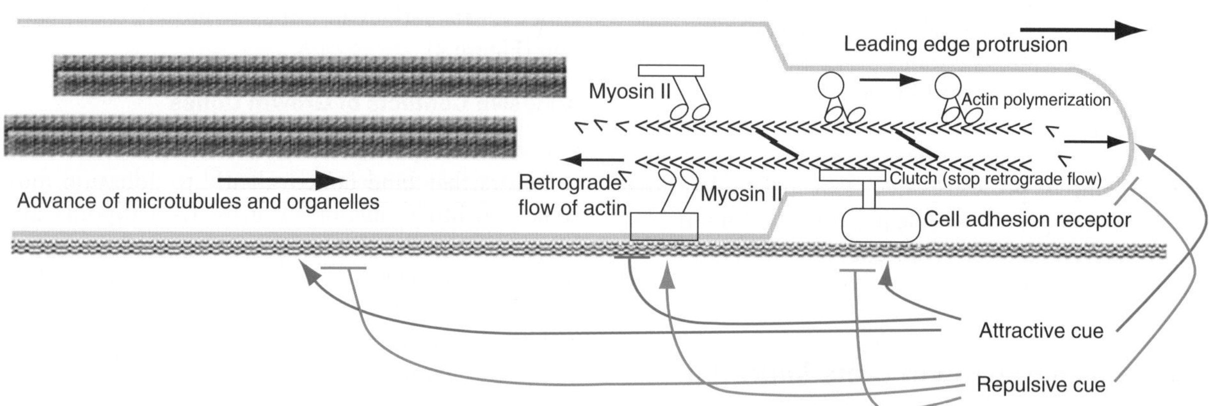

Figure 2 A model of the mechanism of growth cone migration. Actin polymerization pushes the leading margin of the growth cone forward. Forces generated by myosin II pull actin filaments backwards, where filaments are disassembled. When growth cone receptors make adhesive contacts with a surface, a 'clutch' links the adhesive contact to actin filaments of the leading edge, and the retrograde flow of actin filaments stops. This permits the advance of microtubules and organelles and promotes axonal elongation. Intracellular signaling generated by attractive and repulsive axonal guidance cues interacts with the molecular mechanisms of actin polymerization, myosin II force generation, adhesive contacts, and microtubule advance to regulate the paths of growth cone migration.

of stability that are essential to growth cone migration, and they are signaling centers from which regulatory activities promote growth cone motility.

The genetic regulation that determines neuronal phenotype also directs expression of receptors for adhesive ligands and guidance cues by the growth cones of neurons of a particular type. Extracellular positive and negative axonal guidance cues, whether surface bound or soluble, signal through their receptors to modulate an interacting set of pathways that regulate cytoskeletal and membrane dynamics. Thus, growth cone behaviors reflect a complex integration of signaling events triggered at multiple receptors for guidance cues and adhesion molecules. By locally regulating the interplay of adhesive contacts and cytoskeletal dynamics within a growth cone, guidance cues determine the pathways of axonal growth (**Figure 2**).

Three major types of adhesive interactions promote growth cone navigation. Growth cones migrate within extracellular spaces that contain a complex mixture of glycoproteins, organized into an extracellular matrix (ECM) of fibers, protein aggregates, and basal laminae, which are discrete ECM layers at tissue interfaces. One major adhesive interaction of growth cones involves binding of integrin receptors to adhesive ECM proteins, especially the laminins. Two other major adhesive interactions involve growth cone contacts with cells or other axons along their pathways. These interactions involve two groups of adhesive molecules, the cadherins and the immunoglobulin superfamily of cell adhesion molecules (IgCAMs). Cadherins are expressed on all tissue types, including neurons and axons. Cadherin adhesions involve homophilic binding between like cadherin molecules on two interacting cells. Weaker heterophilic interactions between different cadherins can also occur. IgCAMs are also expressed on all tissues, including neurons. Adhesive interactions of IgCAMs can involve homophilic interactions, similar to cadherins, but also heterophilic interactions in which an IgCAM on a growth cone binds a different IgCAM on adjacent cells. Even heterophilic interactions of IgCAMs with non-IgCAMs occur.

Integrin adhesion receptors Integrin receptors are heterodimers of alpha and beta subunits. More than 20 integrin heterodimers have been identified in humans. The binding specificity for ECM components depends on the particular combination of alpha and beta subunits in a heterodimer. The 12 integrin dimers that are expressed in the mammalian nervous system include receptors for collagens, laminin-1 and laminin-5, fibronectin, tenascin, thrombospondin, vitronectin, and VCAM-1. A growth cone can express multiple integrins, allowing interactions with multiple ECM molecules. The cytoplasmic domains of integrins lack enzymatic activities, but when integrins bind adhesive ligands, conformational shifts in the cytoplasmic domains trigger formation of focal contacts that involve integrin clustering and creation of docking sites for proteins that initiate signaling and links to the cytoskeleton. When lamellipodia and filopodia of growth cones bind laminin-1, proteins that localize to the contact sites include paxillin, talin, vinculin, zyxin, and focal adhesion kinase (FAK). Vinculin and talin link actin filaments to the adhesive contacts, providing a clutch for growth cone migration (**Figure 3**). The presence of the adapter protein paxillin and activation of FAK initiates further protein interactions and signaling by Src family kinases, MAP kinases, and Rho GTPases. Activation of Rac1 and Cdc42 GTPases promotes actin polymerization by regulating actin-binding proteins and actin filament dynamics. Thus, when integrins on growth cones bind laminin-1, growth cone migration is stimulated by increased actin filament polymerization to protrude the leading margin and by the establishment of adhesions to stabilize these protrusions and promote the advance of microtubules.

Cadherins and IgCAMs Stimulation of N-cadherin and IgCAMs, such as NCAM and L1, by ligand binding between cells leads to activation of the FGF receptor tyrosine kinase, which triggers signals involving PLC-gamma, DAG lipase, cytoplasmic [Ca^{2+}] elevation, and activation of MAPK. IgCAMs also signal via Src kinases to activate Rac1, PI3K, and MAPK. Cadherin signaling is also reported to activate Rac1. Thus, several pathways activated by cadherins and IgCAMs promote actin filament polymerization. Adhesive binding of cadherins and IgCAMs provides anchorage for actin filaments, creating the clutch necessary for growth cone migration. The cytoplasmic tails of many cadherins, such as N-cadherin, bind catenins, which bind actin filaments and link N-cadherin adhesive sites to the actin cytoskeleton in growth cones (**Figure 4**). The cytoplasmic domain of L1 binds the cytoskeletal linker ankyrin, but L1–ankyrin interactions are involved in stable adhesive junctions, such as at nodes of Ranvier, and not in growth cone migration. Members of the ezrin–moesin–radixin (ERM) proteins mediate actin filament binding to membranes, and interactions of L1 (and other IgCAMs) with ERM proteins may serve as a clutch in growth cone protrusions that bind via L1 or other IgCAMs.

These adhesion receptors can be regulated in ways that are important to growth cone pathfinding. The expression levels of integrin receptors on growth cones are increased when laminin levels are low or when ECM proteins, such as proteoglycans, which interfere with laminin–integrin binding, are present. These responses would maintain growth cone adhesion and migration in environments when access to laminin is reduced. L1 is endocytosed from central regions and recycled to the leading margin of growth cones, increasing availability of L1 for adhesive contacts of lamellipodia and filopodia. The functions of adhesion receptors can also be modulated from the cytoplasm in an 'inside-out' manner or via *cis* interactions with other components of the plasma membrane. An important manner in which guidance

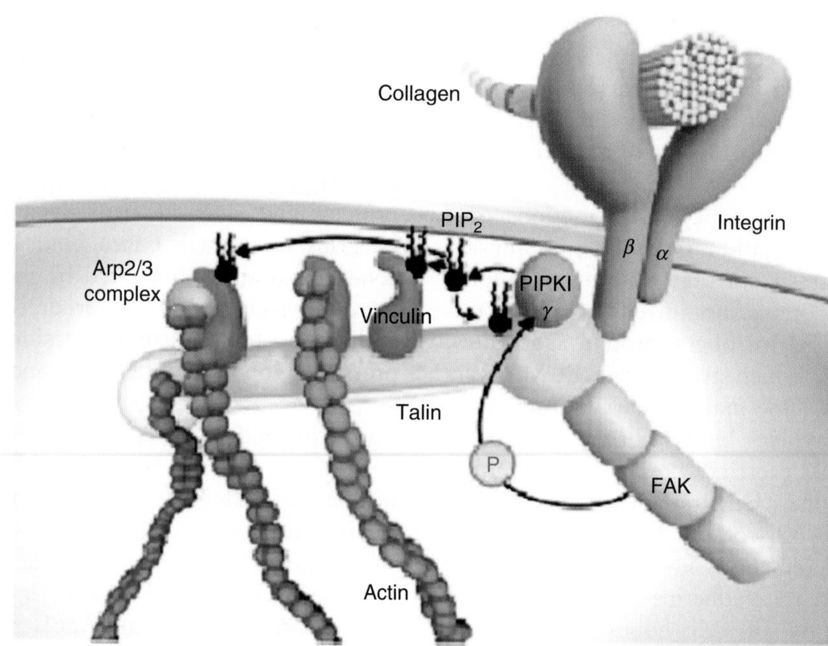

Figure 3 A model of integrin binding to ECM molecules and the formation of intracellular adhesive complexes. An alpha–beta integrin heterodimer is shown bound to a collagen fibril, and the intracellular adhesion complex is pictured, showing the proteins vinculin and talin, which are involved in linkage to actin filaments, and FAK kinase, which initiates signaling cascades. The Arp2/3 complex nucleates actin filament assembly. Reproduced from Brakebusch C and Fässler R (2003) The integrin–actin connection, an eternal love affair. *EMBO Journal* 22: 2324–2333, with permission from Nature Publishing Group.

molecules exert their positive and negative effects on growth cone pathfinding is by modulating the functions of adhesive receptors (**Figure 2**). For example, the negative cue semaphorin 3A may inhibit growth cone migration by blocking integrin-mediated cell adhesion. In addition, adhesion mediated by N-cadherin is inhibited by the negative guidance cue Slit protein via its receptor, Robo. Thus, the negative or repulsive effects of semaphorin 3A and Slit on growth cone pathfinding can involve these inhibitory effects on growth cone adhesion. On the other hand, the attractant netrin signals to activate the kinase FAK, which promotes integrin-mediated adhesion, suggesting that positive guidance cues activate adhesive interactions of growth cones.

Adhesion Molecules and Growth Cone Pathfinding

What are the roles of these adhesion molecules in the pathfinding behaviors of growth cones? Major pathways that are followed by many growing axons offer multiple adhesive ligands for growth cone migration, such as laminins, fibronectin and collagens in the ECM, and cadherins and IgCAMs on adjacent cells and axons. These multiple options for adhesion may provide redundancy, ensuring growth cones form sufficient adhesive contacts for effective migration. The first growth cones that 'pioneer' a pathway have limited options for binding to ECM or cell surface adhesion molecules on adjacent cells, whereas growth cones that enter an established pathway can track along previously extended axons by binding to cadherins and IgCAMs expressed on the surfaces of axonal shafts. Several *in vivo* examples of pathfinding roles of adhesion molecules are described in the following sections.

Laminins

Laminins are large adhesive glycoproteins (MW 1 000 000 Da) that consist of heterotrimers of alpha, beta, and gamma chains. Ten laminin chains are known, forming 11 known heterotrimers with widely varied expression throughout different tissues. The laminins present several domains that mediate laminin binding to several cell surface receptors and to other ECM components. The most common laminins are typically present in basal laminae, an ultrastructural ECM layer associated with epithelia, muscle cells, Schwann cells, and glia. Laminin-1, which has been studied the most, promotes axonal growth *in vitro* from virtually every type of neuron, indicating that laminins have broad roles in promoting growth cone migration. Examples of growth cone migration along basal laminae include growth cones of Rohon–Beard

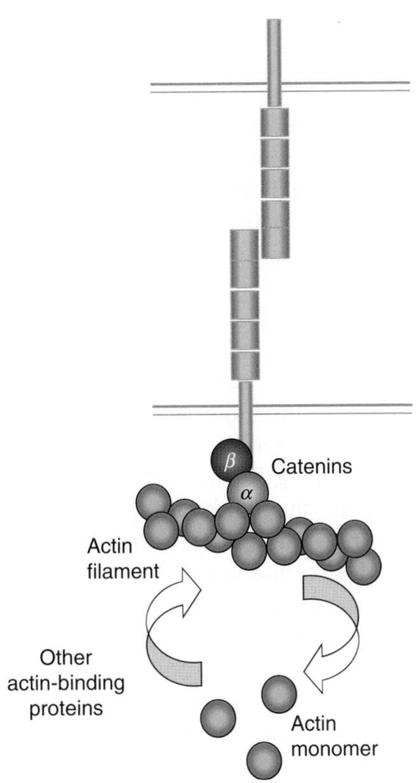

Figure 4 A model of homophilic adhesion between cadherin adhesion molecules and the intracellular binding to actin filaments. Cadherin molecules bind between cells and become linked to actin filaments by way of interactions with alpha and beta catenins. From Weis WI and Nelson WJ (2006) Re-solving the cadherin–catenin–actin conundrum. *Journal of Biological Chemistry* 281: 35593–35597.

neurons in *Xenopus*, growth cones of retinal ganglion cells in the retina and optic nerve, and pioneer axons in the grasshopper limb bud. However, in addition to basal laminae, laminin is transiently expressed in the loose cellular environments of developing tissues, including the nervous system, on cell surfaces and associated with sparse ECM fibers. The growth cones that pioneer pathways, such as the corticofugal pathway of the neocortex or the medial longitudinal fasciculus from the brain into the spinal cord, migrate within loose extracellular spaces in the wall of the immature central nervous system (CNS), where the cells are labeled in a punctate manner by laminin antibodies. The expression of laminin on these cells is transient, and eventually laminin immunoreactivity is restricted to the basal lamina at the outer boundary of the CNS wall. In the developing peripheral nervous system (PNS), laminin is expressed in basal laminae and at early stages in the mesenchyme through which motor and sensory axons extend. Schwann cells express abundant laminin, forming the basal laminae that enclose axon–Schwann cell units. This punctate cellular expression of laminin diminishes during development, although laminin remains present in basal laminae.

In view of the wide distribution of laminins and the ability of laminin-1 to promote robust axonal growth from many neuronal types, it is thought that laminins function permissively, providing adhesion that is required for growth cone migration, but not in an instructive manner to influence pathfinding decisions. Laminins and other ECM molecules may broadly promote growth cone migration along a pathway, whose boundaries are defined not by the absence of adhesive ECM molecules but, rather, by the expression in adjacent tissues of negative guidance cues, such as slits or semaphorins. This 'surround repulsion' occurs in both developing CNS and PNS. Several mutational studies have reported specific errors in pathfinding when a laminin is absent or blocked. Laminin function is essential for growth cone turning in the grasshopper limb bud, and zebra fish with mutations in the laminin-alpha-1 chain exhibit multiple axon guidance defects throughout the CNS, but not in every location. These results suggest that laminin-mediated adhesion is essential for growth cone navigation in at least some instances.

Fibronectin

Fibronectin is a large adhesive glycoprotein (MW 250 000 Da) that is widely distributed in the ECM, including within ganglia and the endoneurium of the PNS. Like the laminins, the fibronectin molecule contains multiple domains that mediate binding to other ECM components and to multiple cellular receptors, including several integrin heterodimers. During development of the PNS and CNS, fibronectin is present in a punctate distribution in loose cellular spaces of immature nervous tissue, and eventually fibronectin expression diminishes as development ends, especially in the CNS. In tissue culture studies, fibronectin promotes axonal growth, but not as vigorously as does laminin. In addition, axonal growth by PNS neurons on fibronectin surfaces exceeds the responses of CNS neurons, probably because PNS neurons express higher levels of fibronectin receptors than CNS neurons. Evidence is lacking for a requirement for fibronectin in growth cone pathfinding.

Integrins

Because neurons express multiple integrin subunits and because many ECM components, such as collagen, laminin, or fibronectin, can bind more than one integrin heterodimer, the essential roles of particular integrins in growth cone pathfinding are not clearly defined. Mouse knockouts of α_1 and α_6 integrins, which are laminin-1 receptors, do not reveal clear defects, however, injections of anti-β_1 integrin, part of several neuronal receptors for ECM proteins, into *Xenopus* embryos disrupts retinal axonal pathfinding. Similarly, conditional knockout of β_1 integrin in sensory neurons results in deficits in innervation of skin, where sensory axons extend through the dermal ECM and along the epidermal basal lamina. The $\alpha_4\beta_1$ integrin heterodimer is specifically implicated in the growth and arborization of sympathetic axons within cardiac muscle. In *Drosophila*, mutations in the integrins α-PS1 and-PS2 lead to pathfinding errors.

Cadherins

Cadherins are characterized as single-pass transmembrane proteins that contain an ectodomain of five cadherin repeats and a conserved cytoplasmic tail. Binding of calcium ion stabilizes an extended rodlike structure of the ectodomain, which is necessary for optimal adhesion by alignment of cadherin molecules on apposing cells. There are at least 100 cadherins, and most are expressed in the developing vertebrate brain on immature cells, neurons, and glia. Their functions are numerous, including cell sorting, boundary formation, target recognition, synaptogenesis, and synapse function. Regarding axonal pathfinding, the widely expressed N-cadherin stimulates *in vitro* axonal growth from a variety of CNS and PNS neurons. *In vivo* studies involving antibody injection or genetic mutation also implicate N-cadherin in axon growth and fasciculation. These results indicate that cadherins promote growth cone migration along axons in highly populated common pathways, but it is unclear whether cadherins play a role in the pathfinding of early pioneer growth cones. In some cases, a common pathway may be shared by several classes of elongating axons, which are distinguished by the expression of different cadherins. For example, the tectofugal projections of chickens express four different cadherins among different axon fascicles. These cadherins may mediate specific pathfinding, as the formation of homophilic adhesions of growth cones to axons expressing the same cadherin directs growth cones along specific axon fascicles toward their targets. Forced expression of specific cadherins causes growth cones to abnormally follow fascicles that express the same cadherin. Finally, growth cones often share expression of specific cadherins with neurons in their particular target. Thus, cadherins also have roles in target recognition and subsequent synaptogenesis.

L1 and NCAM IgCAMs

Proteins that contain an immunoglobulin (Ig)-like domain constitute the Ig superfamily, which makes up more than 2% of human genes, constituting the largest gene family. The neuronal Ig superfamily includes a large number of molecules, which have functions in axonal pathfinding not only as cell adhesion molecules but also as axonal guidance cues and as receptors of guidance cues. This discussion is restricted to two members of this large family, L1 and NCAM. The IgCAM L1 is widely expressed on axons in the developing CNS and PNS, and tissue culture studies show that substrates coated with L1 promote homophilic adhesion and axonal growth from many neuronal types. Spontaneous human mutations in the L1 gene and mouse L1 knockout studies both indicate important roles for L1 in brain development and function. Multiple anatomical and functional deficits result from human and mouse L1 mutations, including a failure of corticospinal axons to decussate in the hindbrain pyramids. Crossing defects were not found in other tracts or were not so extensive. L1 also interacts in *cis* with receptors for other guidance cues, including the semaphorin 3A receptor, suggesting that the defect in pyramidal decussation observed in L1 mutants could be due to disrupted pathfinding responses to semaphorin 3A and other guidance cues, as well as to reduced growth cone tracking along axons.

Another prominent neuronal IgCAM is NCAM, the first neuronal IgCAM identified. NCAM is widely expressed on immature neurons and glia and also on other embryonic cells, such as myoblasts. In tissue culture studies, NCAM mediates neuronal adhesion and axon growth. In addition to homophilic adhesive interactions, NCAM also forms heterophilic adhesive interactions. Antibodies against NCAM can induce axon defasciculation *in vitro* and *in vivo*. Several isoforms of NCAM are expressed, and in some situations NCAM carries a carbohydrate polysialic acid (PSA) moiety that reduces NCAM adhesion. In NCAM-deficient mice defects in fasciculation of hippocampal axons were observed, but in general only minor defects in development or behavior were observed. Perhaps, in the absence of NCAM, other cell adhesion molecules serve the same functions.

Adhesion Molecules and Axonal Regeneration

When the pathfinding phase of circuit construction ends, as growth cones reach their targets, the expression of neural cell adhesion molecules and their receptors is downregulated. However, injury or damage to nervous tissues can disconnect neural circuits, and axons must regenerate in order to reconnect neurons. When axons are injured in the PNS, axon regeneration is often robust, leading to varying degrees of functional recovery. Schwann cells, which ensheath all PNS axons, stimulate axonal regeneration by upregulating their expression of growth factors, and laminins, fibronectin, and cadherin, as substrates for growth cones. Axon regeneration in the PNS is also promoted by increased expression of integrin receptors by regenerating neurons. In the CNS of adult mammals, regeneration of injured axons is poor, and recent research has focused on inhibitory components of myelin and glial scars that block growth cone adhesion and trigger signals that inhibit growth cone motility. In lower vertebrates, CNS regeneration is often successful, and this involves the upregulation of expression of adhesive ligands, such as L1 and cadherins, as demonstrated in regenerating zebra fish optic nerves and spinal cords.

Several strategies for improving axonal regeneration in mammalian model systems, and eventually humans, emphasize measures to improve the adhesive environment for growth cone migration. When stem cells that express L1 are transplanted into a mammalian CNS lesion, increased regeneration of corticospinal axons occurs. Purkinje cells transfected to express L1 and GAP43 show enhanced axonal regeneration. *In vitro* regeneration of axons by adult neurons on laminin and fibronectin is improved by transfection of neurons to express increased levels of the appropriate α integrin chains. Finally, many natural and synthetic bridges have been designed that include adhesion molecules to promote axonal regeneration across lesion sites. These studies demonstrate that strategies to increase the adhesive interactions of regenerating growth cones can stimulate axonal regeneration after injuries in adults. Probably, improved axonal regeneration in adults will also require an increase in the intrinsic ability of adult neurons to sprout and grow axons. This may involve upregulation of genes for adhesion receptors, for other guidance cue receptors, and for proteins that drive the dynamic cytoskeletal functions of immature neurons.

Summary

Growth cone adhesion is integral to the mechanism of growth cone migration and pathfinding. Adhesive interactions of growth cones provide stability for lamellipodial and filopodial protrusions of growth cones and also act as signaling centers that regulate actin and microtubule dynamics and organization in a migrating growth cone. The adhesive interactions of growth cones are also a target of guidance cues that determine where growth cones turn, branch, and stop migrating. Migrating growth cones make three kinds of adhesive contacts with ECM and with other cells. These contacts involve integrin receptors, which recognize laminin and other ECM components; cadherins, which form homophilic adhesions; and IgCAMs, which can form homophilic and heterophilic adhesive interactions. Major pathways of growth cone migration during development contain one or, perhaps more typically, multiple adhesive ligands available to growth cones. The navigational decisions of growth cone pathfinding are based on local differences in adhesive stability for growth cone protrusions and in dynamic protrusive activity, as based on adhesive signaling and the integration of signaling triggered from other guidance cues.

See also: Axon Guidance: Morphogens as Chemoattractants and Chemorepellants; Axon Guidance: Building Pathways with Molecular Cues in Vertebrate Sensory Systems; Axon Guidance: Guidance Cues and Guidepost Cells; Axonal Pathfinding: Netrins; Axonal Pathfinding: Guidance Activities of Sonic Hedgehog (Shh); Growth Cones; Semaphorins.

Further Reading

Brakebusch C and Fässler R (2003) The integrin–actin connection, an eternal love affair. *EMBO Journal* 22: 2324–2333.

Clegg DO, Wingerd KL, Hikita ST, and Tolhurst EC (2003) Integrins in the development, function and dysfunction of the nervous system. *Frontiers in Bioscience* 8: d723–d750.

Colognato H, French-Constant C, and Feltri ML (2005) Human diseases reveal novel roles for neural laminins. *Transactions in Neuroscience* 28: 480–486.

Dent EW and Gertler FB (2003) Cytoskeletal dynamics and transport in growth cone motility and axon guidance. *Neuron* 40: 209–227.

Gordon-Weeks PR (2000) *Neuronal Growth Cones.* Cambridge, UK: Cambridge University Press.

Hortsch M (2003) Neural cell adhesion molecules – brain glue and much more! *Frontiers in Bioscience* 8: d357–d359.

Huber AB, Kolodkin AL, Ginty DD, and Cloutier JF (2003) Signaling at the growth cone: Ligand–receptor complexes and the control of axon growth and guidance. *Annual Review of Neuroscience* 28: 509–563.

Hynes RO (2002) Integrins. Bidirectional, allosteric signaling machines. *Cell* 110: 673–687.

Kamiguchi H (2003) The mechanism of axon growth: What we have learned from the cell adhesion molecule L1. *Molecular Neurobiology* 28: 219–228.

Kiryusho D, Berezin V, and Bock E (2004) Regulators of neurite outgrowth: Role of cell adhesion molecules. *Annals of the New York Academy of Sciences* 1014: 140–154.

Redies C, Treubert-Zimmermann U, and Luo J (2003) Cadherins as regulators for the emergence of neural nets from embryonic divisions. *Journal of Physiology (Paris)* 97: 5–15.

Sakisaka T and Takai Y (2005) Cell adhesion molecules in the CNS. *Journal of Cell Science* 118: 5407–5410.

Suter DM and Forscher P (2000) Substrate–cytoskeletal coupling as a mechanism for regulation of growth cone motility and guidance. *Journal of Neurobiology* 44: 97–113.

Thiery JP (2003) Cell adhesion in development: A complex signaling network. *Current Opinion in Genetics & Development* 13: 365–371.

Weis WI and Nelson WJ (2006) Re-solving the cadherin–catenin–actin conundrum. *Journal of Biological Chemistry* 281: 35593–35597.

Wen Z and Zheng JQ (2006) Directional guidance of nerve growth cones. *Current Opinion in Neurobiology* 16: 52–58.

Axonal Pathfinding: Guidance Activities of Sonic Hedgehog (Shh)

F Charron, Institut de Recherches Cliniques de Montréal (IRCM) and University of Montreal, Montreal, QC, Canada
M Tessier-Lavigne, Genentech, Inc., South San Francisco, CA, USA

Introduction

Over the past decade, genetic, biochemical, and molecular approaches have led to the identification of four major conserved families of guidance cues with prominent developmental effects: the netrins, slits, semaphorins, and ephrins. More recently, members from three other families of secreted signaling molecules have been shown to act as guidance cues: the wingless/Wnt, the hedgehog (Hh), and the decapentaplegic/bone morphogenetic protein/transforming growth factor-β (Dpp/BMP/TGF-β) families. In addition to their axon guidance properties, these molecules share a common characteristic of having been previously identified as morphogens controlling cell fate and tissue patterning. This discovery has opened the door to the study of an entirely new set of axon guidance cues.

This article focuses on the role of the morphogen Shh in axon guidance. After briefly introducing the role and the signaling pathway of Shh in patterning the neural tube, we discuss the emerging evidence that Shh is reused later in development to guide axons, both in the developing spinal cord and the retina. We conclude by discussing the implications for Shh signaling in axon guidance.

The Hedgehog Family in Cell Fate Specification and Tissue Patterning

Morphogens are signaling molecules produced in a restricted region of a tissue; they provide positional information by diffusing from their source to form long-range concentration gradients. A cell's program of differentiation in response to a morphogen is dictated by its position within the gradient and thus on its distance from the morphogen source. Two criteria have gained acceptance as the evidence needed to qualify a secreted signaling protein as a morphogen: it must have a concentration-dependent effect on its target cells and it must exert a direct action at a distance. In vertebrates, Sonic hedgehog (Shh) has been shown to fulfill these criteria and function as a morphogen to specify cell fate in the developing neural tube.

Hedgehog proteins are found in insects and vertebrates, but not nematodes. They are encoded by a single hedgehog gene in flies, and three in mammals: *Shh*, Indian hedgehog (*Ihh*), and Desert hedgehog (*Dhh*). In vertebrate embryos, one of the first steps in the development of the nervous system is the specification of the diverse neural cell fates of the neural tube. Shh is secreted by the notochord and by floor plate cells at the ventral midline of the neural tube, and functions as a graded signal for the generation of distinct classes of ventral neurons along the dorsoventral axis of the neural tube (**Figure 1(a)**). In agreement with its role as a morphogen, Shh is able to induce a range of ventral spinal cord cell fates in a concentration-dependent manner and has been shown to exert a direct action at a distance to specify neural tube cell fate.

Much evidence indicates that the cell fate specification and tissue patterning activities of Hhs are mediated by members of the Ci/Gli transcription factor family, but the signaling mechanisms that lead to activation of these transcription factors are not fully elucidated. Genetic and biochemical experiments have shown that Hhs activate signaling by binding to their receptor Patched (Ptc), which leads to the relief of Ptc-mediated inhibition of Smoothened (Smo), also a transmembrane protein, which can then activate downstream signaling (**Figure 2**). Smo associates directly with a Ci-containing complex, which contains the atypical kinesin Costal-2 (Cos2), the protein kinase Fused (Fu), and the Suppressor of Fused (Su(fu)). This complex constitutively suppresses pathway activity. Activation of Hh signaling reverses this regulatory effect and allows Ci to activate transcription of Hh target genes, thus specifying cell fate.

Roles of Shh in Axon Guidance

Shh Is a Chemoattractant for Commissural Axons

During spinal cord development, commissural neurons, which differentiate in the dorsal neural tube, send axons that project toward and subsequently across the floor plate, forming axon commissures (**Figure 1(b)**). These axons project toward the midline in part because they are attracted by netrin 1, a long-range chemoattractant secreted by the floor plate. In mice mutant for *netrin 1* or the netrin 1 receptor gene, *Dcc* (*deleted in colorectal cancer*), many commissural axon trajectories are foreshortened, fail to invade the ventral spinal cord, and are misguided. However, some of them do reach the midline, indicating that other guidance cues cooperate with netrin 1 to guide these axons. Further analysis of *netrin 1* knockout

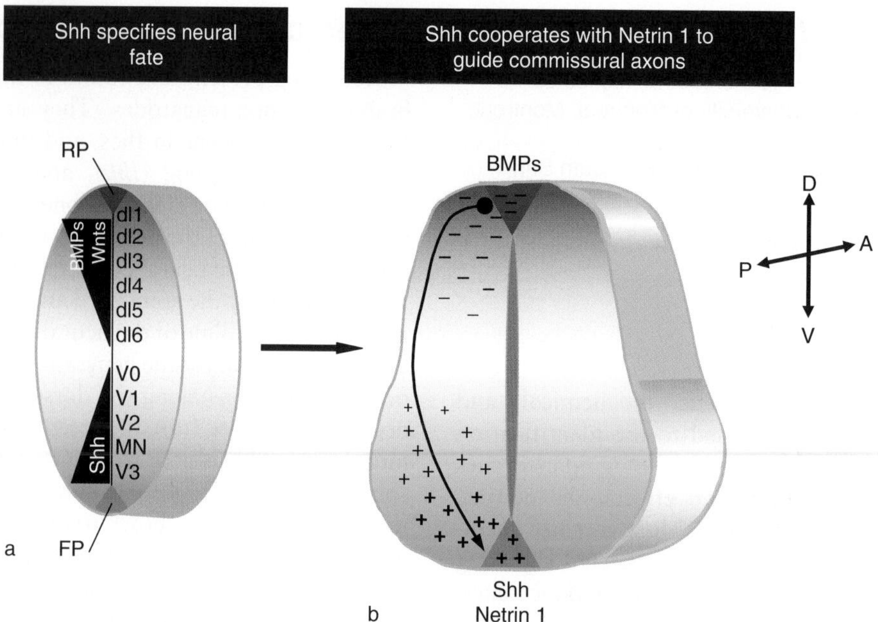

Figure 1 Neuronal cell fate specification and guidance of precrossing commissural axons by Sonic hedgehog (Shh) and netrin 1; cross-section representations of the developing neural tube. Shh is first used to pattern neural progenitors in the spinal cord and then appears to be reused as a guidance cue for commissural axons. (a) In the early neural tube, Shh – together with bone morphogenetic protein (BMP) and Wnt – protein concentration gradients act to specify neural cell fate in the ventral (and dorsal, for BMP and Wnt) spinal cord. (b) Later, the axons of differentiated commissural neurons are attracted to the ventral midline by the combined chemoattractant effects of netrin 1 and Shh. BMPs also contribute to commissural axon guidance by repelling axons from the dorsal midline. V0–V3, ventral interneuron subpopulations; dI1–dI6, dorsal interneuron subpopulations; MN, motor neurons; RP, roof plate; FP, floor plate; D, dorsal; V, ventral; P, posterior; A, anterior. Adapted from Charron F and Tessier-Lavigne M (2005) Novel brain wiring functions for classical morphogens: A role as graded positional cues in axon guidance. *Development* 132: 2251–2262.

mice suggests that the floor plate might actually express an additional diffusible attractant(s) for commissural axons.

Given its expression by the floor plate and its long-range effects in the spinal cord, Shh was a candidate for a midline-derived axonal guidance cue. Shh was indeed shown to function as an axonal chemoattractant that can mimic the netrin-1-independent chemoattractant activity of the floor plate in *in vitro* assays. The chemoattractant activity of Shh, like the chemoattractant activity of floor plate derived from *netrin 1* mutants, can be blocked by cyclopamine, which blocks the actions of Shh in cell fate determination by inhibiting the Shh signaling mediator Smo. This shows that Smo is required for Shh-mediated axon attraction and, importantly, that the netrin-1-independent chemoattractant activity of the floor plate also requires Hh signaling. Since Shh is the only Hh family member expressed in the spinal cord at this stage, these results suggest that Shh functions as a floor plate-derived chemoattractant for commissural axons.

While the reorienting effect of Shh could be due to a direct chemoattractant effect, an alternative explanation was suggested by the fact that Shh is a potent morphogen. Since in these assays commissural axon turning occurs within the spinal cord tissue explant, it seemed possible that Shh was not acting directly on the axons but rather was repatterning and altering the expression of guidance cues by cells within the explant, which then secondarily guided the axons to the Shh source. Arguing against this possibility was the finding that the spinal cord explants used to assess chemoattractant activity are at a developmental stage at which they have apparently lost the competence to be repatterned by Shh, as assessed using a battery of markers of dorsoventral patterning.

A direct action of Shh in attracting the axons was supported further by two sets of experiments. First, Shh was shown to attract the growth cones of isolated *Xenopus* spinal axons in dispersed cell culture in a cyclopamine-dependent manner, proving that Shh, acting via Smo, can function as a chemoattractant, at least for these *Xenopus* axons. A second way of providing evidence that Shh can act directly on commissural axons to guide them relied on blocking Shh signaling selectively in commissural neurons without blocking it in the terrain through which their axons course. This was achieved by conditional inactivation of a floxed allele of *Smo* using the Cre recombinase expressed under the control of the Wnt1 promoter, which drives expression in the dorsal spinal cord (as well as in neural crest progenitors). When Cre, driven by this promoter, was used to delete a floxed *Smo* allele in the dorsal spinal cord, commissural axon trajectories were defective in the ventral spinal cord, where Cre is not expressed (**Table 1**). This result strongly implies that the axonal misrouting is not due to repatterning of the ventral spinal cord, and must instead reflect a guidance defect arising from loss of Smo function in commissural neurons. Taken together, these results suggest that Shh functions to guide commissural axons both *in vitro* and *in vivo* by acting directly as a chemoattractant on these axons through an Smo-dependent signaling mechanism.

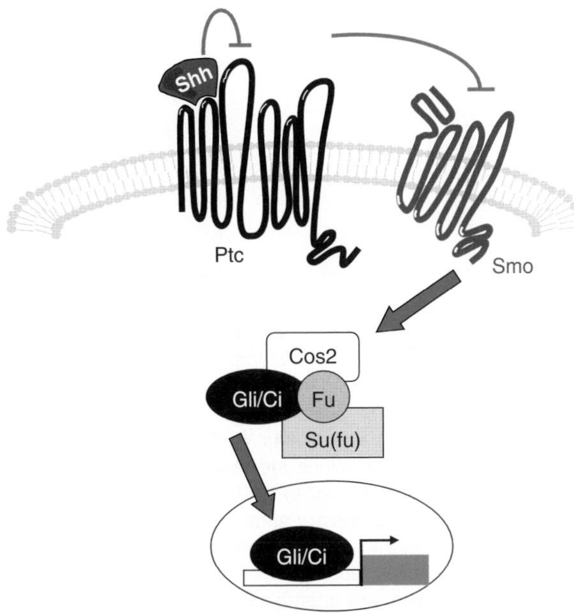

Figure 2 The Sonic hedgehog (Shh) signaling pathway. Genetic and biochemical experiments have shown that hedgehogs activate signaling by binding to their receptor Patched (Ptc; a 12-pass transmembrane protein), which leads to the relief of Ptc-mediated inhibition of Smoothened (Smo), a 7-pass transmembrane protein, which can then activate downstream signaling. Smo associates directly with a Gli/Ci-containing complex which contains the atypical kinesin Costal-2 (Cos2), the protein kinase Fused (Fu), and the Suppressor of Fused (Su(fu)). This complex constitutively suppresses pathway activity by leading to the proteolytic cleavage of Ci, which acts as a transcriptional repressor. Activation of hedgehog signaling reverses this regulatory effect and leads to the production of full-length Ci, which activates transcription of hedgehog target genes.

Boc Is a Receptor for Shh in the Guidance of Commissural Axons to the Floor Plate

Although Shh acts through Smo to attract commissural axons to the floor plate, Smo does not bind Shh, and the binding receptors acting with Smo to mediate the effect of Shh in axon guidance remain elusive. Cdon (cell adhesion molecule-related/downregulated by oncogenes) and Boc (biregional Cdon binding protein) have been implicated in enhancing muscle differentiation. They are type I transmembrane proteins consisting of four to five immunoglobulin (Ig) and two to three fibronectin type III (FNIII) repeats in the extracellular domain. This domain architecture is highly related to that of axon guidance receptors of the Robo and Dcc families. Both Cdon and Boc share a high degree of homology in their extracellular domains and are expressed during early stages of central nervous system (CNS) development. Interestingly, mice with homozygous mutations in Cdon display a microform of holoprosencephaly (HPE), a developmental defect of the forebrain and midface caused by a failure to delineate the midline. In both humans and mice, many forms of HPE are caused by disruptions in the Shh signaling pathway, suggesting that Cdon and Boc might regulate Shh signaling, a possibility supported by the finding that a Cdon/Boc homolog in *Drosophila*, iHog, is required for Hh signaling. Taken together, these observations suggested that Cdon and/or Boc might function in Shh-mediated axon guidance in vertebrates.

The ability of Boc and Cdon to bind Shh was tested and they were found to bind specifically and directly to Shh, suggesting that they could function as Shh receptors. These results are consistent with recent reports demonstrating that Cdon, Boc, and iHog bind and

Table 1 *In vivo* experiments supporting a role for Shh in axon guidance

Gene	*Species*	*Experiment*	*Phenotype*	*Reference*
Smo	Mouse	Conditional inactivation of *Smo* in commissural neurons	Commissural axons project abnormally and invade the motor columns	Charron et al. (2003)
Boc	Mouse	*Boc* inactivation	Commissural axons project abnormally and invade the motor columns	Okada et al. (2006)
Shh	Chick	Ectopic expression of Shh in the optic chiasm	Retinal axons are prevented from crossing the chiasm	Trousse et al. (2001)
Shh	Chick	Silencing of *Shh* by RNA interference	Commissural axons stall at the contralateral floor plate border, with some axons randomly turning caudally or rostrally	Bourikas et al. (2005)

signal through the Hh pathway. To explore their involvement in commissural axon guidance, the expression of Cdon and Boc was examined in the developing mouse spinal cord. Boc is expressed by differentiating commissural neurons and the Boc protein, but not Cdon, is expressed by growing commissural axons. To assess the functions of these receptors in commissural axon guidance, Boc and Cdon mutant mice were generated and sections of spinal cord from Boc and Cdon mutant embryos were analyzed. Although commissural axon projections appeared normal in Cdon mutant animals, abnormal projections of commissural axons were observed in Boc mutants: the axons were highly dispersed and invaded the ventral spinal cord, with ectopic projections extending over the motor columns. Analysis of various neural tube markers indicated that neuronal patterning is occurring normally in these animals. This phenotype is similar to that observed in mice following conditional removal of the Shh signaling mediator Smo in commissural neurons (using a Wnt1 promoter to drive Cre expression; see earlier) and suggests that Boc acts in the same pathway as Smo to guide commissural axons in response to Shh. Finally, using an *in vitro* commissural axon turning assay to test the role of Boc in Shh-mediated axon turning, it was shown that RNA interference (RNAi)-mediated knockdown of Boc impaired the ability of commissural axons to turn toward an ectopic source of Shh *in vitro*. These results indicates that Boc is required for commissural axons to respond to the chemoattractive effect of Shh. Collectively, these data suggest that Boc plays an essential role as a receptor for Shh in commissural axon guidance.

Consistent with a role for Boc in axon pathfinding, an earlier study showed defects in forebrain axon guidance in zebra fish treated with morpholinos directed against Boc. However, based on indirect evidence, these results were interpreted to reflect a role for Boc as a repulsive ligand. In the mammalian spinal cord, the aforementioned results argue that Boc functions as a receptor for Shh-mediated attraction; whether it can also function as a repulsive ligand in other contexts in mammals remains to be explored. Additionally, it will be of interest to determine whether Boc and Cdon also play a role in the guidance of other axonal tracts in the developing mammal and fish embryos.

Shh Guides Commissural Axons along the Longitudinal Axis of the Spinal Cord

After commissural axons have reached and crossed the floor plate, they make a sharp anterior turn toward the brain (**Figure 3**). The molecules involved in the dorsoventral projection of commissural axons to and at the floor plate have been well described, but it is only recently that cues controlling anteroposterior guidance have been identified. Remarkably, the guidance of commissural axons to the floor plate is not, apparently, the only effect of Shh on commissural axons: recent evidence suggests that Shh also guides postcrossing commissural axons in the rostral direction along the longitudinal axis of the spinal cord. Using a subtractive hybridization approach to identify guidance cues responsible for the rostral turn of postcrossing commissural axons in chick embryos, Bourikas and colleagues identified differentially expressed candidates whose function they investigated by RNAi-mediated *in ovo* gene silencing. Unexpectedly, one of their candidates turned out to be Shh. In agreement with these results, silencing of the *Shh* gene by a different RNAi construct or injection of a hybridoma producing a function-blocking Shh antibody led to axon stalling at the contralateral floor plate border, with some axons

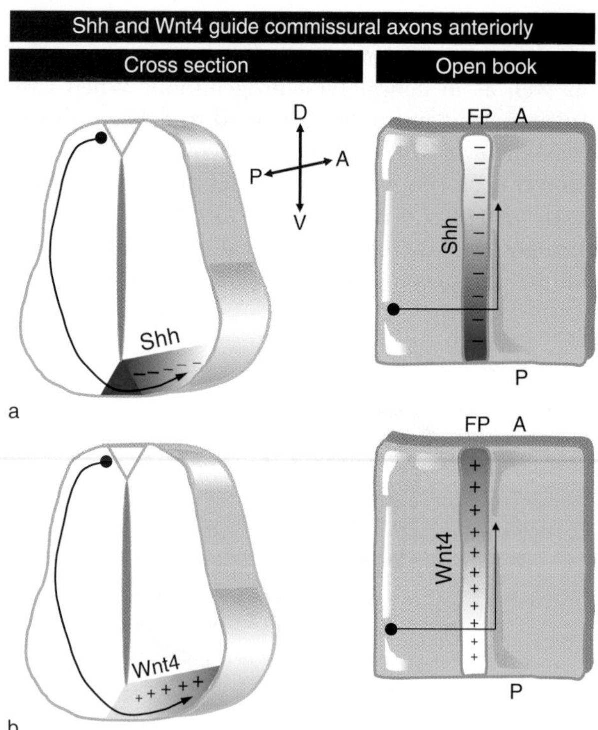

Figure 3 Guidance of postcrossing commissural axons by Sonic hedgehog (Shh) and Wnt4. (a) After crossing the floor plate, chick commissural axons are repelled from the posterior pole by an increasing anterior to posterior Shh gradient. (b) In rodents, commissural axons are attracted to the anterior pole by an increasing posterior-to-anterior Wnt4 gradient. Left panels in a and b are cross-section representations of the developing spinal cord, and right panels are open-book representations. FP, floor plate; D, dorsal; V, ventral; P, posterior; A, anterior. Adapted from Charron F and Tessier-Lavigne M (2005) Novel brain wiring functions for classical morphogens: A role as graded positional cues in axon guidance. *Development* 132: 2251–2262.

turning caudally or rostrally, apparently in a random manner. Importantly, marker analysis revealed that the patterning of the spinal cord was not apparently affected by these manipulations, suggesting that these experiments were performed after neural cell fate specification by Shh has occurred. Finally, postcrossing commissural axons were shown to avoid ectopic Shh *in vivo*. Together, these results provide strong evidence that Shh is essential for the normal guidance of commissural axons along the longitudinal axis of the spinal cord, at least in chick embryos.

An Shh gradient could guide commissural axons along the longitudinal axis directly, or could alternatively be acting only indirectly by controlling a graded distribution of a distinct guidance cue. Two lines of evidence, however, were provided for a direct role of Shh. The first came from an investigation of the receptor mechanism for this guidance. Interestingly, neither cyclopamine nor Smo RNAi interfered with the rostral turn of commissural axons along the longitudinal axis, suggesting that Smo might not be involved in this process. Instead, RNAi-mediated silencing of *Hip1*, a gene encoding an Shh-binding membrane protein transiently expressed in commissural neurons at the time when they cross the floor plate, as well as in the periventricular region, resulted in the same postcrossing phenotype as seen with Shh RNAi. These results, which contrast with the essential role of Smo in Shh-mediated attraction of commissural axons to the floor plate, suggest that Hip1 might be involved in transducing an Shh guidance signal in postcrossing commissural neurons. The relatively restricted expression of Hip1 mRNA to commissural neurons would be consistent with a direct action of Shh on these axons. A second line of evidence that supports a direct role for Shh was obtained by *in vitro* experiments, which showed that postcrossing commissural axons from spinal cord explants can be repelled by Shh beads *in vitro*. Taken together, these results suggest a model in which Shh could be functioning directly through Hip1 as a chemorepellent for postcrossing commissural axons.

Prior to the finding that Shh controls the anteroposterior guidance of commissural axons in chicks, Wnt4 was reported to play a role in this process in rodents. Using a novel *in vitro* assay, evidence was obtained that the activity responsible for the anterior guidance of postcrossing commissural axons in rodents is an increasing posterior to anterior gradient of a diffusible attractant. Although it is not yet known whether Wnt4 guides postcrossing commissural axons in chicks and whether Shh guides postcrossing commissural axons in rodents, it is nonetheless interesting to note that, if this is the case, the complementary Wnt4 and Shh gradients might act cooperatively in the rostral guidance of commissural axons. Additionally, since Boc plays a role in the guidance of precrossing commissural axons, it will be interesting to determine whether it also plays a role in the guidance of postcrossing commissural axons.

Shh Is a Negative Regulator of Retinal Ganglion Cell Axon Growth

Retinal ganglion cell (RGC) axons growing toward the diencephalic ventral midline are faced with the decision to project either contralaterally or ipsilaterally in response to guidance cues at the optic chiasm (**Figure 4**). Homozygous inactivation of the mouse *Pax2* gene alters the development of the optic chiasm, and RGC axons never cross the midline in these mice. Interestingly, whereas in wild-type mice Shh expression is downregulated in the chiasm as RGC axons are migrating toward this region, Shh expression is ectopically maintained along the ventral midline in $Pax2^{-/-}$ mice. These observations raise the possibility that the continuous expression of Shh at the ventral midline might contribute to preventing RGC axon crossing. In agreement with this idea, Trousse and colleagues found that ectopic expression of Shh in the midline region interferes with RGC axon growth and prevents them

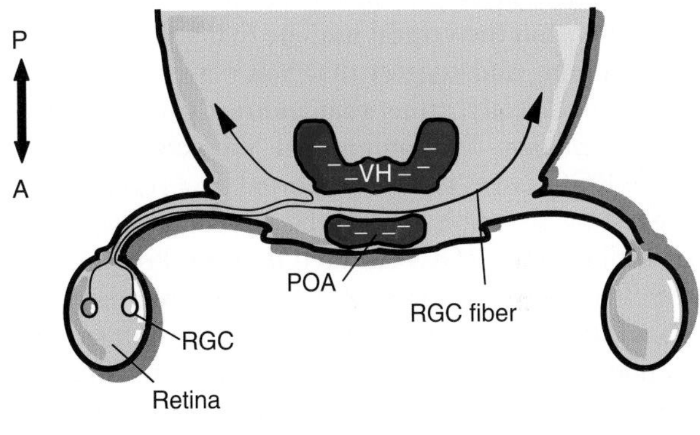

Figure 4 Sonic hedgehog (Shh) expression at the chiasm border defines a barrier within the ventral midline implicated in guiding retinal ganglion cell (RGC) axons. RGC axons growing toward the diencephalic ventral midline are faced with the decision to project either contralaterally or ipsilaterally in response to guidance cues at the optic chiasm. The Shh expression domain is shown in blue. Shh can inhibit retinal axons *in vitro*, suggesting that *in vivo* it may be acting on the axons directly rather than by altering the expression of distinct guidance cues in the chiasm, although conclusive evidence for this guidance function *in vivo* remains to be obtained. A, anterior; P, posterior; POA, preoptic area; VH, ventral hypothalamus. Adapted from Charron F and Tessier-Lavigne M (2005) Novel brain wiring functions for classical morphogens: A role as graded positional cues in axon guidance. *Development* 132: 2251–2262.

from crossing the midline (**Figure 4**). Consistent with the idea that Shh might be acting directly on RGC axons, it was shown that these manipulations do not affect patterning and neural differentiation in the eye. Further experiments will be required to determine whether the chiasm region is repatterned in these experiments, but *in vitro* experiments support the idea that Shh acts directly to control RGC axon migration: addition of exogenous recombinant Shh to retinal explants decreases the number and length of growing axons, without interfering with the rate of proliferation and differentiation of cells in the explant, and time-lapse analysis shows that addition of Shh to retinal explants rapidly causes growth cone arrest and subsequent retraction of RGC axons. Since the response of the growth cone to many extracellular guidance cues appears to be modulated and in some cases perhaps even mediated by intracellular cyclic nucleotide levels (cyclic adenosine monophosphate and cyclic guanosine monophosphate, cAMP and cGMP), the possibility was explored that the effect of Shh on retinal axons *in vitro* might be due to a change in cAMP levels. In agreement with this, addition of Shh to retinal growth cones was shown to decrease intracellular levels of cAMP, a finding consistent with the observation that lowering cAMP levels favors growth inhibition.

Taken together, these results provide evidence that Shh expression at the chiasm border helps define a barrier within the ventral midline that serves to guide RGC axons, and suggest that Shh may be acting on the axons directly, rather than indirectly by repatterning the chiasm. The finding that Shh acts directly to guide RGC axons is also supported by a recent study suggesting that Shh acts directly and rapidly on the growth cone of RGCs cultured *in vitro*. Proving that the effect *in vivo* is direct will, however, require additional studies, such as identifying the mechanism that mediates retinal growth cone responses to Shh and showing that cell-autonomous inhibition of this signaling pathway in the neurons results in guidance defects *in vivo*.

Molecular Mechanism underlying Shh-Mediated Axon Guidance

The aforementioned studies, summarized from several laboratories, have provided evidence that Shh plays a role in the guidance of at least two types of axons: commissural and RGC axons. In the case of precrossing commissural axons, Shh acts as a chemoattractant and its effect is mediated by Boc, a recently identified Shh receptor related to Dcc and Robo family members. In addition, the chemoattractant effect of Shh requires the canonical Shh signaling molecule Smo. After commissural axons have crossed the floor plate, evidence indicates that Shh then functions as a chemorepellent to direct their anterior migration. This later effect of Shh appears to require the Shh binding protein Hip1 and might occur independently of Smo activity. Finally, in the case of RGCs, Shh acts as a negative regulator of axonal growth, at least *in vitro*, but the receptor and the signaling molecules involved in this effect remain elusive.

Although the chemoattractant and chemorepellent effects of Shh might appear to be inconsistent with one another, many other guidance cues have been shown to be bifunctional and exert opposite effects, depending on the context. For example, extrinsic factors can convert netrin attraction to repulsion by modulating cyclic nucleotide levels. Thus, the opposite effects of Shh on pre- and postcrossing commissural and retinal axons might be due to an intrinsic or extrinsic factor that modulates cyclic nucleotide levels. Alternatively, as the molecular mechanisms underlying the effects of Shh on commissural and retinal axons are poorly understood, it is also possible that these pre- and postcrossing effects are mediated by distinct signaling pathways that result in opposite guidance effects – a possibility that also has a precedent in the case of netrins, which can attract axons by activating Dcc family receptors and repel them by activating UNC5 family receptors.

In all of the studies summarized here, Shh was shown to act rapidly (in an hour or less) to affect growth cone morphology. Although these results appear inconsistent with the model that the Shh axon guidance effect function through the transcriptional signaling pathway to the nucleus, this needs to be formally proved, as none of the studies described here has addressed this issue directly. Nonetheless, even if a transcriptional response is found to be required for their guidance effects, additional local signaling would still be required to be elicited in the growth cone in order to generate a polarized response leading to growth cone turning in a specific direction. Indeed, a purely transcriptional response consisting of a retrograde signal to the nucleus, followed by an anterograde signal back to the growth cone, cannot account for the polarized turning effect of a guidance cue. Further studies aimed at understanding the molecular mechanisms underlying growth cone turning by Shh will be necessary to identify the molecules linking Shh signaling to localized growth cone effects.

In this regard, despite many efforts, the canonical Hh signaling pathway is only beginning to be understood, and many intermediate signaling molecules remain to be identified and characterized. Thus, it is possible that the signaling proteins eliciting the growth cone effects are simply components of the signaling pathway required for cell fate specification that are awaiting identification and further characterization to uncover their function in axon guidance.

Alternatively, Shh might be acting through entirely different signaling pathways in axon guidance and cell fate specification, including the use of a different receptor. For example, although Smo and Boc are required for Shh-mediated commissural axon guidance to the floor plate, it is not known whether Ptc, the canonical Shh-binding component of the Shh receptor, is involved. This finding contrasts with chick postcrossing commissural axon guidance, in which Smo does not appear to be required for the rostral turn away from the Shh gradient. Additional experiments on commissural and retinal axons will be required to determine the receptor components and intracellular signaling molecules mediating the effects of Shh on axon guidance.

Conclusion

The discovery that the morphogen Shh can be reused to guide axons has generated considerable excitement in the field, and it will be interesting to see to what extent its signaling components are conserved between the morphogenic and guidance responses. The characterization of the Shh chemotropic signaling pathway will open new avenues to study how guidance signals regulate the motility and steering of the growth cones, and will help elucidate the mechanisms directing the complex wiring of the nervous system.

See also: Axon Guidance: Morphogens as Chemoattractants and Chemorepellants; Axon Guidance: Building Pathways with Molecular Cues in Vertebrate Sensory Systems; Axon Guidance: Guidance Cues and Guidepost Cells; Axonal Pathfinding: Netrins; Growth Cones; Sonic Hedgehog and Neural Patterning; Wnt Pathway and Neural Patterning.

Further Reading

Bourikas D, Pekarik V, Baeriswyl T, et al. (2005) Sonic hedgehog guides commissural axons along the longitudinal axis of the spinal cord. *Nature Neuroscience* 8: 297–304.

Charron F, Stein E, Jeong J, et al. (2003) The morphogen sonic hedgehog is an axonal chemoattractant that collaborates with netrin-1 in midline axon guidance. *Cell* 113: 11–23.

Charron F and Tessier-Lavigne M (2005) Novel brain wiring functions for classical morphogens: A role as graded positional cues in axon guidance. *Development* 132: 2251–2262.

Connor RM, Allen CL, Devine CA, et al. (2005) BOC, brother of CDO, is a dorsoventral axon-guidance molecule in the embryonic vertebrate brain. *Journal of Comparative Neurology* 485: 32–42.

Dickson BJ (2002) Molecular mechanisms of axon guidance. *Science* 298: 1959–1964.

Fuccillo M, Joyner AL, and Fishell G (2006) Morphogen to mitogen: The multiple roles of hedgehog signalling in vertebrate neural development. *Nature Reviews Neuroscience* 7: 772–783.

Ingham PW and McMahon AP (2001) Hedgehog signaling in animal development: Paradigms and principles. *Genes and Development* 15: 3059–3087.

Kolpak A, Zhang J, and Bao ZZ (2005) Sonic hedgehog has a dual effect on the growth of retinal ganglion axons depending on its concentration. *Journal of Neuroscience* 25: 3432–3441.

Lyuksyutova AI, Lu CC, Milanesio N, et al. (2003) Anterior–posterior guidance of commissural axons by Wnt-frizzled signaling. *Science* 302: 1984–1988.

Okada A, Charron F, Morin S, et al. (2006) Boc is a receptor for sonic hedgehog in the guidance of commissural axons. *Nature* 444: 369–373.

Tessier-Lavigne M and Goodman CS (1996) The molecular biology of axon guidance. *Science* 274: 1123–1133.

Trousse F, Marti E, Gruss P, et al. (2001) Control of retinal ganglion cell axon growth: A new role for Sonic hedgehog. *Development* 128: 3927–3936.

Varjosalo M and Taipale J (2007) Hedgehog signaling. *Journal of Cell Science* 120: 3–6.

Axonal Pathfinding: Netrins

S W Moore and T E Kennedy, McGill University, Montreal, QC, Canada
M Tessier-Lavigne, Genentech, Inc., South San Francisco, CA, USA

Introduction

Directing growing axons to their targets is an essential step toward establishing appropriate connections in the nervous system. The growth cone, located at the motile tip of an axon, senses cues within its environment to guide extending axons. Extracellular guidance cues can attract or repel axons. They may be transmembrane, glycosylphosphatidylinositol (GPI)-linked, or secreted cues. Secreted cues have the unique capacity to function at a distance by diffusing away from their source. The netrin family contains both secreted members (netrin-1–netrin-4) and GPI-linked members (netrin-G1 and netrin-G2).

The discovery of the netrin family can be traced back to observations made in the late nineteenth century by the Spanish neuroscientist Santiago Ramón y Cajal. He was the first to propose that gradients of diffusible cues might guide axons. One location where he thought this may occur is in the developing spinal cord where commissural axons extend ventrally toward the floor plate (**Figure 1(a)**). Specifically, he proposed that a gradient of a diffusible cue, emanating from the floor plate at the ventral midline, would function to attract growing commissural axons. In the 1980s, co-culture studies using explants of embryonic neural tube provided experimental evidence for the presence of guidance cues secreted by the floor plate. Initial experiments demonstrated that axon bundles would extend from an explant containing the cell bodies of commissural neurons, through a collagen matrix toward a floor plate explant cultured at a distance (**Figure 1(b)**). Subsequent experiments placed an ectopic floor plate perpendicular to the normal trajectory of commissural axons, resulting in the axons being redirected within the neural epithelium toward the ectopic floor plate (**Figure 1(c)**). An activity that promoted commissural axon outgrowth was then identified in lysates of embryonic chick brains. Two related proteins were purified, their corresponding cDNAs cloned, and recombinant protein shown to mimic the ability of the floor plate to attract commissural axons. Sequence analysis identified homology to UNC-6, a protein required for the circumferential guidance of cells and axons in the roundworm *Caenorhabditis elegans*. These proteins were named netrins, based on the Sanskrit word *netr*, meaning 'one who guides.'

Netrin Family Members

Four secreted netrins (netrin-1–netrin-4) have been identified in vertebrates, along with two membrane-anchored forms, netrin-G1 and netrin-G2. Thus far, orthologs of netrin-2 have only been identified in birds and fish. Mammals express secreted netrin-1, netrin-3, and netrin-4. The netrin family can be divided into three subfamilies: netrins 1–3 (sometimes called 'classical netrins'), netrin-4, and netrin-G1 and netrin-G2. Based on sequence, netrin-4 and the netrin-Gs are more similar to laminins than to netrins 1–3, which exhibit a high degree of similarity to each other (**Figure 2(b)**). Another compelling argument for this subdivision is based on evolutionary conservation. Orthologs of netrins 1–3 have been detected in all bilaterally symmetrical animals studied so far, while orthologs of netrin-4 and netrin-Gs have only been found in vertebrate species (**Figure 2(c)**). Functional differences also support this distinction. Different receptors mediate the function of netrins 1–3 and netrin-Gs. Furthermore, netrins 1–3 similarly elicit axon outgrowth from embryonic spinal commissural neurons, while netrin-Gs do not.

Netrin Structure

Despite these differences, all netrins are classified into a single family on the basis of their size, approximately 600 amino acids, the presence of two characteristic conserved N-terminal domains, domains V and VI, and a more variable C-terminal domain, domain C. Domains V and VI in netrins are homologous to domains V and VI found at the N-terminal ends of the extracellular matrix protein laminin (**Figure 2(a)**). Laminins are large secreted heterotrimers made up of α, β, and γ subunits. Domains V and VI of netrin-4 and netrin-Gs are most similar to β subunits of laminin, while those of netrins 1–3 are more similar to the γ subunits (**Figure 2(b)**).

Domain VI, at the N-terminal end of netrins and laminins, is composed of approximately 300 amino acids. In laminins, this domain interacts with heparin, cell surface receptors, and extracellular matrix (ECM) proteins, and is required for calcium-dependent multimerization that generates a larger matrix of laminin molecules. Genetic studies in *C. elegans* have demonstrated that the highly conserved sequence SXDXGXS/TW within domain VI of the netrin UNC-6 is required for both axon attraction and repulsion. The middle

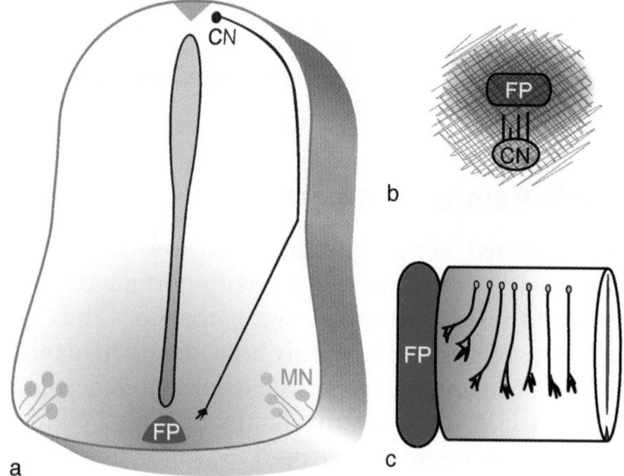

Figure 1 (a) The floor plate (FP) attracts commissural neuron (CN) axons, in part, by releasing netrin-1 (MN, motor neuron). (b) Commissural neurons will extend axon bundles from an explant of dorsal embryonic spinal cord through a collagen matrix when either floor plate or cells expressing netrin are placed within 200–300 μm. (c) An ectopic floor plate or a cell line expressing netrin-1 will attract commissural neuron axons when placed perpendicular to their trajectory.

netrin domain, domain V, is approximately 150 amino acids long and contains three cysteine-rich epidermal growth factor (EGF) repeat subdomains named V-1, V-2, and V-3. Mutation of the third EGF repeat (V-3) disrupts chemoattraction in *C. elegans*, whereas repulsion is lost following mutation of either V-2 or V-3 domain.

The C-terminal domain C of secreted netrins (netrins 1–4) exhibits relatively limited sequence similarity to domains found in the complement C3, C4, and C5 protein family (i.e., CC3, CC4, and CC5), secreted frizzled-related proteins (sFRPs), type I C-proteinase enhancer proteins (PCOLCEs), and tissue inhibitors of metalloproteinases (TIMPs). The majority of netrin protein in the embryonic or mature central nervous system (CNS) is not freely soluble. Structure–function analyses of domain C suggest that it contributes to binding netrins to cell surfaces or extracellular matrix. Domain C includes many basic amino acids that may bind negatively charged sugars associated with proteoglycans, such as heparan sulfate proteoglycans and chondroitin sulfate proteoglycans. Presentation of

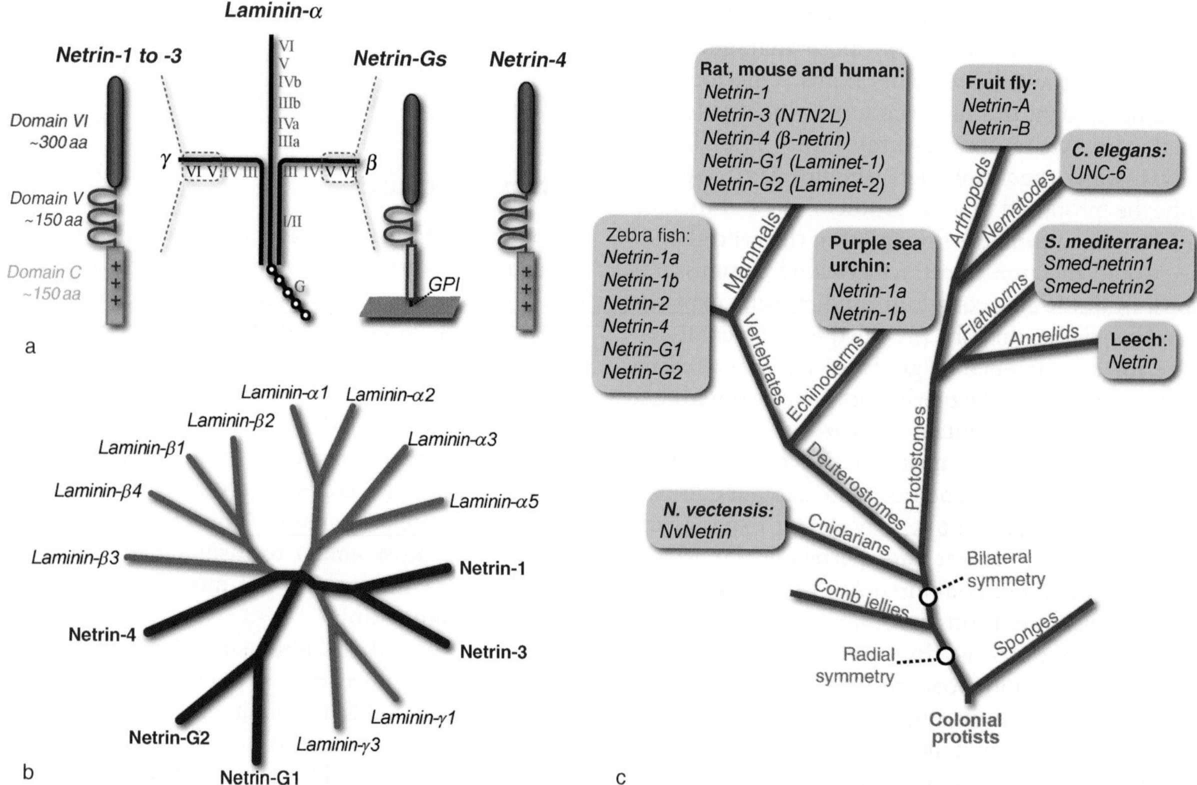

Figure 2 (a) Netrins contain N-terminal domains V and VI related to corresponding N-terminal domains of laminins. Domain V is composed of cysteine-rich epidermal growth factor repeats. Domain C in secreted netrins contains many positively charged, basic residues (GPI, glycophosphatidylinositol). (b) Phylogenetic tree based on the sequences of domains VI and V in human netrins and laminins. (c) Evolutionary tree diagram highlighting the presence of netrin homologs in a wide variety of bilaterally symmetrical organisms.

netrins bound to surfaces may be a common mode of action in the netrin family; although the domain C is not conserved in the netrin-Gs, a C-terminal GPI link anchors them to cell surfaces.

Netrin Expression and Function during Development

Orthologs of netrin-1 have a firmly established, evolutionarily conserved role as secreted cues that direct axon guidance relative to the midline of developing bilaterally symmetric nervous systems. In fact, a netrin-1 ortholog has been identified in the stellate sea anemone *Nematostella vectensis* – an organism thought to exhibit some of the earliest hallmarks of bilateral symmetry (**Figures 2(c)** and **3(g)**). Multiple lines of evidence support the conclusion that netrin-1 attracts commissural axons toward and repels subsets of motor neuron axons away from the midline. Expression of a netrin-1 ortholog at the midline early in neural development is highly conserved (**Figure 3**). Ectopic expression of netrin-1 is sufficient to alter axon extension. Furthermore, netrin-1 ejected from a micropipette functions as a chemoattractant or chemorepellent, depending on the axons stimulated. Genetic disruption of the netrin UNC-6 in *C. elegans* disrupts circumferential axon extension toward and away from the ventral midline. Loss of netrin-1 function in mice generates major disruptions in multiple axon commissures, including the spinal ventral commissure, the corpus callosum, and the hippocampal commissure.

Netrin-1 also influences axon guidance away from the midline. It is required for retinal ganglion cell axons to exit the retina, for dopaminergic axon guidance in the ventral midbrain, and for thalamocortical projections. Netrin-1 also directs neural precursor cell migration, attracting cells that will become inferior olivary, pontine, luteinizing hormone-releasing hormone (LHRH), antidiuretic hormone (ADH), and oxytocin neurons, and repelling striatal neuronal precursors, cerebellar granule cells, spinal accessory neurons, and oligodendrocyte precursor cells. Although netrin-3 can mimic the ability of netrin-1 to attract spinal commissural axons and repel trochlear motor neuron axons *in vitro,* its expression in the spinal cord begins after the initial commissural axons have pioneered the path to the floor plate. Netrin-3 may, however, influence guidance of dorsal root ganglia axons to peripheral targets in the developing peripheral nervous system (PNS).

Very little is known regarding the role of netrin-4 in the development of the nervous system. It is widely expressed in the developing olfactory bulb, retina, and dorsal root ganglia, as well as by cerebellar granule, hippocampal, and cortical neurons. A relatively low level of netrin-4 expression has been detected adjacent to floor plate cells in the developing spinal cord, but only after the first commissural axons have crossed the midline. As discussed later, netrin-Gs do not appear to have a major role in the outgrowth and guidance of axons. They are expressed primarily by neurons, with very limited expression outside the nervous system. Netrin-G1 is expressed in the dorsal thalamus, olfactory bulb, and inferior colliculus, whereas netrin-G2 is expressed in the cerebral cortex.

Netrin-Induced Signal Transduction

Research investigating the signaling mechanisms elicited by netrins has focused on netrin-1, and there is little known about signaling by other netrins. Here we review current insight into netrin-1 signal transduction during axon guidance. The DCC ('deleted in colorectal cancer') family and the UNC5 homolog family are well-established receptors for netrin-1. In vertebrate species, these include DCC and neogenin, as well as UNC5A–UNCD. Both classes of receptors are single-pass transmembrane proteins that are also members of the large immunoglobulin superfamily.

Chemoattractant Signaling in Response to Netrin-1

DCC is required for axon chemoattraction to netrin-1. Although initially identified in vertebrates as a potential tumor suppressor, the importance of DCC in axon guidance was realized shortly after the discovery of netrin-1 as a result of mutational studies done in the roundworm *C. elegans*. While mutation of the *netrin* homolog *unc-6* affects both ventrally and dorsally directed neurons, mutations of the *C. elegans* homolog of *dcc*, *unc-40*, primarily disrupt ventrally directed axons. It was quickly realized that this phenotype was consistent with a potential role for UNC-40/DCC during axon attraction to netrins/UNC-6. Loss of DCC function in mice was then shown to cause neurodevelopmental defects very similar to loss of netrin-1 function, including disruption of the spinal ventral commissure, corpus callosum, and hippocampal commissure.

Structurally, the extracellular domain of DCC contains six fibronectin type 3 (FN3) repeats and four immunoglobulin (Ig) repeats. Netrin-1 appears to bind to the FN3 repeats of DCC, though reports conflict as to the exact domain involved. The DCC intracellular domain has no known catalytic activity; rather, it contains several putative phosphorylation

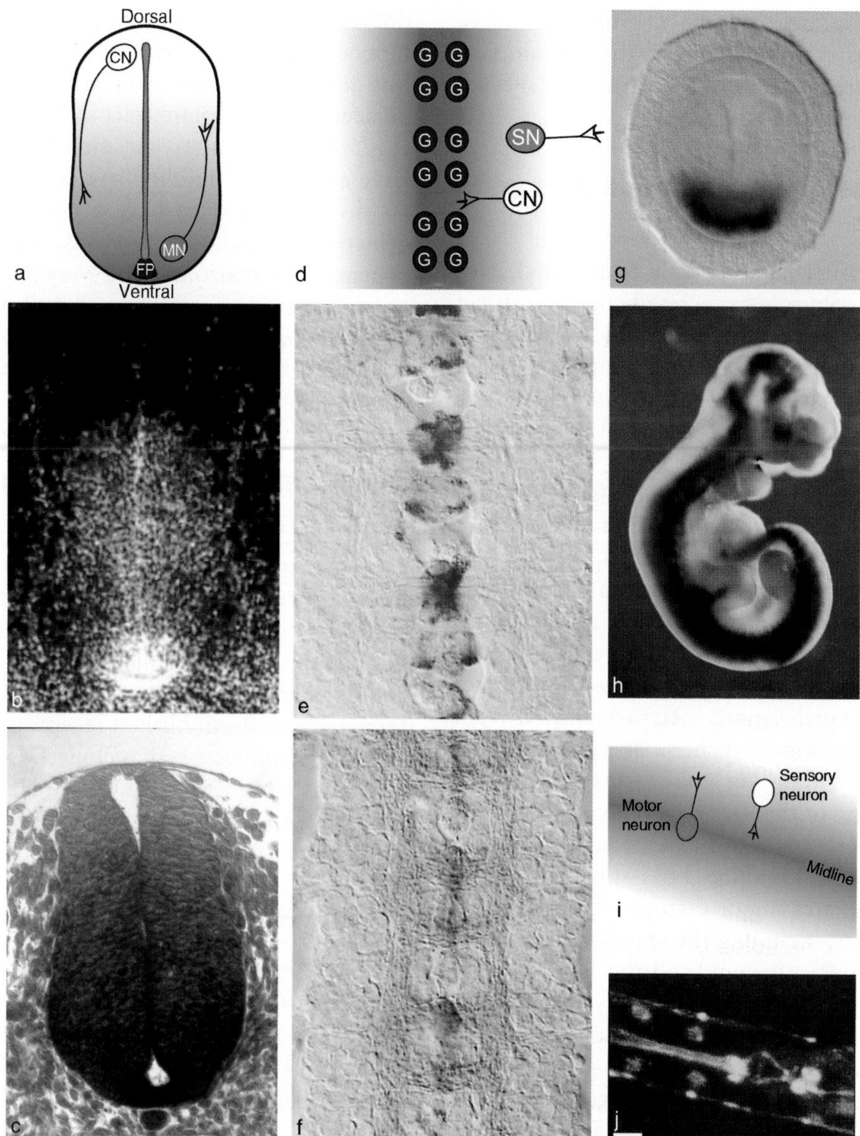

Figure 3 Midline expression of netrin homologs in a variety of organisms. The diagrams in panels (a), (d), and (i) illustrate simplified models of the distributions of netrin protein at the midline in the developing mouse spinal cord, fruit fly and *C. elegans* nematode worm, respectively. In each case, sensory neurons extend toward, while motor neurons extend away from, the source of netrin protein at the midline. *In situ* hybridization illustrates floor plate cells expressing netrin-1 in the embryonic day 9.5 mouse spinal cord (b). Whole-mount staining for expression of a β-galactosidase reporter gene shows netrin-1 expression in an entire E12.5-day-old mouse embryo, illustrating netrin-1 expression along the full rostrocaudal extent of the spinal cord, the developing brain, and the peripheral nervous system (h). *In situ* hybridization reveals netrin expression in the fruit fly (e) and the stellate sea anemme *N. vectensis* (g) during neural development. The distribution of netrin protein in the embryonic day 9.5 mouse spinal cord (c), as well as fruit fly (f) and *C. elegans* (j) embryos are also shown. CN, commissural neuron; MN, motoneuron; SN, segmental nerve. (b) Reprinted from Serafini T, Colamarino SA, Leonardo ED, et al. (1996) Netrin-1 is required for commissural axon guidance in the developing vertebrate nervous system. *Cell* 87: 1001–1014, with permission. (e, f) Reprinted from Harris R, Sabatelli LM, and Seeger MA (1996) Guidance cues at the *Drosophila* CNS midline: Identification and characterization of two *Drosophila* netrin/UNC-6 homologs. *Neuron* 17: 217–228, with permission. (g) Reprinted from Matus DQ, Pang K, Marlow H, et al. (2006) Deep evolutionary roots for bilaterality in the metazoa. *Proceedings of the National Academy of Sciences of the United States of America* 103: 11195–11200, with permission. (j) Reprinted from Wadsworth WG, Bhatt H, and Hedgecock EM (1996) Neuroglia and pioneer neurons express UNC-6 to provide global and local netrin cues for guiding migrations in *C. elegans*. *Neuron* 16: 35–46, with permission.

and binding sites for intracellular proteins. Based on particularly strong evolutionary conservation, three domains (P1, P2, and P3) within the intracellular domain of DCC have been described. The P2 domain is rich in proline residues, containing four PXXP putative SH3 domain-binding motifs, while the P3 domain contains several highly conserved possible phosphorylation sites.

Numerous studies have now described signal transduction events implicated in the response to netrin-1, but our understanding of the mechanisms underlying how a growth cone responds to netrin-1 as a chemoattractant remains fragmentary. Netrin-1-induced multimerization of DCC via its P3 intracellular domain is thought to be essential for chemoattraction. Netrin-1 binding DCC activates the Rho GTPases Rac1 and Cdc42, key intracellular coordinators of cytoskeletal and adhesive interactions. Furthermore, intracellular proteins reported to associate with the intracellular domain of DCC include the adapter protein Nck1, the tyrosine kinases Fak and Fyn, the actin-binding proteins Ena/Vasp and N-Wasp, and the serine/threonine kinase Pak, a downstream effector of Rac1 and Cdc42 activation. Netrin-1 binding DCC also triggers generation of phosphoinositides and their breakdown by phospholipase C into inositol 1,4,5-trisphosphate (IP_3) and diacylglycerol, leading to calcium release from intracellular stores and activation of protein kinase C. **Figure 4(a)** outlines a speculative model of the signal transduction events that occur during chemoattraction to netrin-1.

Chemorepellent Signaling in Response to Netrin-1

Axonal repulsion in response to netrin-1 requires a member of the UNC5 protein family. Four family members are present in mammals: UNC5A, UNC5B, UNC5C, and UNC5D. In *C. elegans*, in contrast to mutation of the *dcc* homolog *unc-40* that affects ventrally extending axons, mutation of *unc-5* predominantly causes defects in dorsally directed axons (i.e., those extending away from the source of UNC-6 netrin). Misexpression of *unc-5* is sufficient to redirect them dorsally, away from the ventral source of UNC-6. The extracellular domain of UNC5 homologs contains two membrane-proximal thrombospondin repeats and distally two immunoglobulin repeats. Netrin-1 binds to both immunoglobulin repeats. The intracellular domain of UNC5 contains three identified conserved domains: a ZU5 domain, a DCC-binding (DB) domain, and a death domain (DD). Although the specific function of the ZU5 domain is not known, a homologous domain is present in the scaffolding protein zona occludens-1 found at tight junctions, and deletion of the UNC5 ZU5 domain in *Drosophila* disrupts UNC5-mediated chemorepulsion.

Interestingly, depending on the distance of the growth cone from the source of netrin, different signaling mechanisms appear to be engaged. Long-range repulsion to netrin requires expression of both UNC5 and DCC, whereas short-range repulsion requires UNC5, but DCC is not essential. One hypothesis for this difference is that DCC and UNC5 together may form a more sensitive netrin receptor complex that is able to respond to the lower concentrations of netrin present at greater distances. At long range, the cytoplasmic domains of UNC5 and DCC associate directly. At short range, genetic studies in *C. elegans* have stressed the importance of an association between UNC5 cytoplasmic ZU5 and DD domains. Several proteins that interact with the UNC5 intracellular domain have been identified. These include the tyrosine kinase Src1, the tyrosine phosphatase Shp2, the F-actin anticapping protein Mena, the structural protein ankyrin, and the adapter protein Max1. Netrin-1-mediated growth cone repulsion triggers tyrosine phosphorylation of UNC5's intracellular domains at multiple sites. A speculative model outlining intracellular events that occur during short- and long-range netrin-mediated growth cone repulsion is illustrated in **Figure 4(b)**.

Regulating Growth Cone Response to Netrin-1

Single growth cones have been shown to have the capacity to rapidly switch between responding to netrin as an attractant or a repellent. The mechanisms that control this shift in responsiveness have been the topic of intense investigation.

Altering UNC5 or DCC expression is one mechanism that influences how a growth cone responds to netrin. For example, the homeobox transcription factor 'even-skipped' promotes *unc5* expression in the fruit fly *Drosophila melanogaster*. Local protein synthesis within the growth cone also appears to contribute to the attraction of axons to netrin-1, though the exact mechanism of action is not clear. DCC expression can be downregulated through proteolytic degradation by either extracellular metalloproteinases or intracellularly, being ubiquitinated by an interaction with Siah-1, a RING domain-containing protein. The intracellular domains of UNC5 proteins are substrates for proteolysis by caspases.

The concentration of cyclic nucleotides within an axonal growth cone has a profound influence on how it responds to guidance cues. In particular, the intracellular concentration of cyclic adenosine monophosphate (cAMP), which activates protein kinase A (PKA), can regulate the response to netrin-1. High levels of cAMP are associated with attractant responses to netrin-1, while growth cones with low concentrations are repelled. Although cAMP and PKA activation may regulate intracellular signal transduction, a specific mechanism regulating the direction of the response made by a growth cone has not been identified. It has been shown that activating PKA leads to the recruitment of DCC from an intracellular pool of vesicles to the growth cone plasma membrane, thereby enhancing chemoattraction to netrin-1. In addition, activating protein kinase C (PKC) leads to endocytosis of UNC5 proteins inducing neurons to

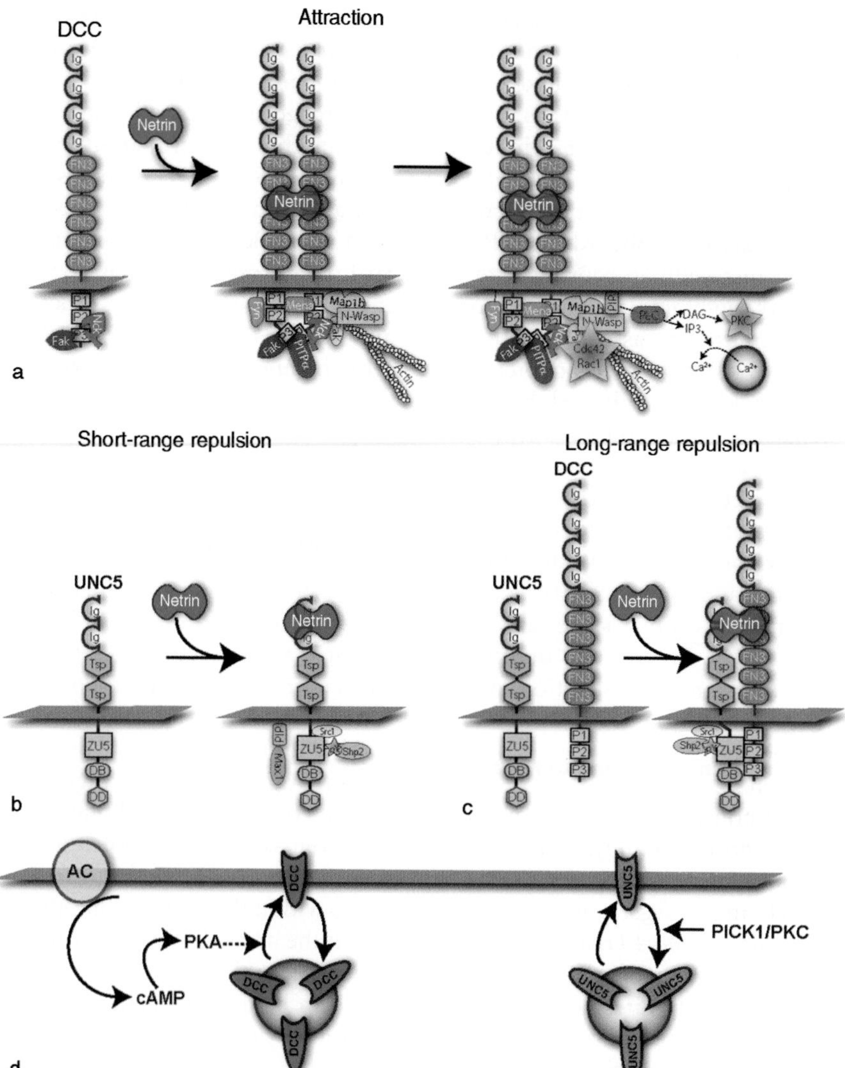

Figure 4 Speculative models outlining chemoattractive, chemorepellent, and modulatory signaling in response to netrin-1. (a) Chemoattraction to netrin-1 can be divided into three conceptually different stages: in the absence of netrin-1, during the initial response to netrin-1, and during cytoskeletal remodeling triggered by netrin-1. The adapter protein Nck1 and the tyrosine kinase Fak associate with the intracellular domain of the DCC ('deleted in colorectal cancer') protein in the absence of netrin-1. Upon netrin-1 binding, DCC multimerizes through association of its P3 domains. Phosphatidylinositol transfer protein-α (PITPα) can also bind the P3 domain of DCC and promote the generation of phosphoinositides (PIPs) by phosphatidylinositol 3-kinases (PI3Ks). PIPS can then be hydrolyzed by phospholipase C (PLC) into inositol 1,4,5-trisphosphate (IP$_3$) and diacylglycerol (DAG), leading to Ca^{2+} release from intracellular stores and activation of protein kinase C (PKC). Netrin-1 binding to DCC leads to phosphorylation of the intracellular domain of DCC and association of proteins such as Fak, Fyn, and Pak. Intracellular Ca^{2+} increases can lead to activation of the Rho GTPases Cdc42 and Rac, and cause remodeling of the cytoskeleton through proteins such as the Wiskott–Aldrich syndrome protein (N-Wasp), Ena/Vasp, and Map1b. (b) Short- and long-range repellent signaling to netrin-1. In the absence of netrin-1, Mena and ankyrin link UNC5 to the cytoskeleton. Upon binding netrin-1, UNC5 is tyrosine-phosphorylated, independently of DCC, by tyrosine kinases such as Src1. Netrin-1 induces recruitment of the tyrosine phosphatase Shp2 to a phosphorylated tyrosine residue in the ZU5 domain. PIPs have been proposed to regulate interaction of Max-1 with UNC5. (c) Long-range repulsion to netrin-1 requires association between the intracellular domains of UNC5 and DCC. (d) Both DCC and UNC5 can traffic between intracellular vesicular pools and the cell surface. Protein kinase A (PKA) can recruit DCC to the plasma membrane. PKC induces endocytosis of UNC5 from the cell surface. AC, adenylate cyclase; cAMP, cyclic adenosine monophosphate; PICK1, protein interacting with C kinase-1.

switch their response to netrin-1 from repulsion to attraction. These findings suggest that neuromodulatory factors that regulate PKA or PKC may influence axon outgrowth by altering which netrin receptors are presented by the growth cone.

Other Potential Netrin Receptors

Receptors for secreted netrins, other than DCC, neogenin, and UNC5 proteins, have been suggested. The G-protein-coupled adenosine A2B receptor has been proposed as a receptor for netrin-1. Through an influence on intracellular cAMP concentration, it is possible that A2B may influence the response to netrin-1; however, contrary to an early report, A2B is neither expressed by nor required for spinal commissural axon guidance. Integrins, best known for their role as receptors for extracellular matrix components, have also been implicated as netrin receptors. Specifically, an adhesive interaction between netrin-1 and integrins $\alpha6\beta4$ and $\alpha3\beta1$ has been suggested to contribute during development of the pancreas; however, evidence of this *in vivo* remains to be demonstrated. Interestingly, the region of netrin reported to interact with these integrins is not in domains V and VI that are homologous to laminins, which are ligands for integrins. Rather, $\alpha6\beta3$ and $\alpha3\beta1$ integrins bind to a highly charged sequence of basic amino acids at the C-terminus of netrin-1.

Netrin Expression in the Mature Nervous System

Roles for netrins beyond directing axon and cell migration during development are beginning to be identified. Netrins and their known receptors are expressed in the adult nervous system: netrin-1 is expressed by neurons, Schwann cells in the PNS, and oligodendrocytes in the CNS. Subcellular fractionation of mature CNS white matter has determined that netrin-1 protein is enriched at the interface – known as periaxonal myelin – between axons and oligodendrocytes, suggesting that a function for netrin-1 in the mature CNS may be to regulate axon oligodendroglial interactions. Interestingly, as the mammalian spinal cord matures, DCC expression falls while UNC5 homolog expression increases. This suggests that UNC5 repellent signaling may be the dominant mode of responsiveness to netrin in the adult spinal cord. Although the functional significance of netrin-1 expression in the adult CNS remains unknown, an intriguing hypothesis is that netrins contribute to maintaining appropriate connections in the intact CNS by restraining inappropriate axonal sprouting.

The expression of netrin-1 by myelinating oligodendrocytes raises the possibility that it might function as a myelin-associated inhibitor of axon growth following injury. Netrin-1 does not appear to be a major component of an injury-induced glial scar in the mature spinal cord, but essentially normal netrin-1 expression persists on either side of the injury site. Although an influence of netrin-1 on axon regeneration in the adult CNS has not been demonstrated directly, these findings suggest that netrin-1 may be a component of CNS myelin that inhibits axon regeneration by neurons expressing UNC5 following injury. Such a role may explain, in part, why increasing cAMP in neurons promotes the ability of axons to grow in the adult mammalian CNS, as increasing cAMP recruits DCC to plasma membranes of growth cones and converts netrin-mediated repulsion to attraction. Interestingly, studies carried out in lamprey, a primitive vertebrate with the capacity for substantial axon regeneration following spinal lesion, correlated poor axonal regeneration following lesion with neuronal expression of UNC5 protein.

Netrin-3, netrin-4, netrin-G1, and netrin-G2 are also expressed in the adult brain. In humans, mutation of the netrin-G1 gene is a rare cause of the childhood neurodevelopmental disorder known as Rett syndrome. This disease is characterized by normal early development followed by loss of purposeful use of the hands, distinctive hand movements, slowed brain and head growth, gait abnormalities, seizures, and mental retardation. In mice, netrin-G1 deficiency does not lead to any obvious changes in neural circuitry, but does lead to altered synaptic responses and defects in sensorimotor gating behavior. Similarly, the netrin-G2 receptor NGL-2 influences the formation of glutamanergic synapses through an interaction with the postsynaptic scaffold protein PSD-95. Together, these findings indicate that netrin-G proteins have a role in the maturation, refinement, and maintenance of synapses, rather than in the guidance of axons.

Netrin Outside of the Nervous System

Netrins are expressed in many tissues. Netrin-1 is expressed in the heart, tongue, lung, inner ear, intestine, mammary gland, and pancreas, netrin-3 is expressed in the bowel, pancreas, and muscle, and netrin-4 is expressed in the intestine, pancreas, spleen, vascular networks, kidney, ovaries, and lung. Functional roles for netrins have been demonstrated in several developing tissues. Netrin-1 has been implicated in vascular patterning, although some disagreement remains regarding whether it functions principally as a repellent, acting via UNC5B, or an attractant, acting via undefined receptors, or both (in different vascular beds). Both netrin-1 and netrin-4 shape the developing lung through an influence on branching of the epithelial endoderm. In the developing mammary gland, netrin-1 is expressed by a layer of luminal cells, and an interaction with neogenin is required for proper organization of the terminal end buds.

Conclusion

Homologs of netrin-1 are evolutionarily conserved guidance cues that function as chemoattractants and chemorepellents that direct cell and axon migration in the developing nervous system. The receptor DCC is essential for chemoattraction to netrin-1, and UNC5 proteins are required for chemorepulsion. The netrin-G subfamily contributes to the maturation, refinement, and maintenance of synapses. In addition to these roles in the CNS, netrins also influence the development of a variety of other tissues.

See also: Axon Guidance: Building Pathways with Molecular Cues in Vertebrate Sensory Systems; Axon Guidance: Guidance Cues and Guidepost Cells; Axonal Pathfinding: Extracellular Matrix Role; Cetacean Brains; Growth Cones; Rett Syndrome.

Further Reading

Barallobre MJ, Pascual M, Del Rio JA, et al. (2005) The Netrin family of guidance factors: Emphasis on Netrin-1 signalling. *Brain Research – Brain Research Reviews* 49: 22–47.

Dickson BJ (2002) Molecular mechanisms of axon guidance. *Science* 298: 1959–1964.

Fazeli A, Dickinson SL, Hermiston ML, et al. (1997) Phenotype of mice lacking functional Deleted in colorectal cancer (Dcc) gene. *Nature* 386: 796–804.

Harris R, Sabatelli LM, and Seeger MA (1996) Guidance cues at the *Drosophila* CNS midline: Identification and characterization of two *Drosophila* netrin/UNC-6 homologs. *Neuron* 17: 217–228.

Hedgecock EM, Culotti JG, and Hall DH (1990) The *unc-5*, *unc-6*, and *unc-40* genes guide circumferential migrations of pioneer axons and mesodermal cells on the epidermis in *C. elegans*. *Neuron* 4: 61–85.

Hinck L (2004) The versatile roles of 'axon guidance' cues in tissue morphogenesis. *Developmental Cell* 7: 783–793.

Huber AB, Kolodkin AL, Ginty DD, et al. (2003) Signaling at the growth cone: Ligand–receptor complexes and the control of axon growth and guidance. *Annual Review of Neuroscience* 26: 509–563.

Ishii N, Wadsworth WG, Stern BD, et al. (1992) UNC-6, a laminin-related protein, guides cell and pioneer axon migrations in *C. elegans*. *Neuron* 9: 873–881.

Keleman K and Dickson BJ (2001) Short- and long-range repulsion by the *Drosophila* Unc5 netrin receptor. *Neuron* 32: 605–617.

Kennedy TE, Serafini T, de la Torre JR, et al. (1994) Netrins are diffusible chemotropic factors for commissural axons in the embryonic spinal cord. *Cell* 78: 425–435.

Kennedy TE, Wang H, Marshall W, et al. (2006) Axon guidance by diffusible chemoattractants: A gradient of netrin protein in the developing spinal cord. *Journal of Neuroscience* 26(34): 8866–8874.

Matus DQ, Pang K, Marlow H, et al. (2006) Deep evolutionary roots for bilaterality in the metazoa. *Proceeding of the National Academy of Sciences of the United States of America* 103: 11195–11200.

Moore SW, Kennedy TE, and Tessier-Lavigne M (2007) Netrins and their receptors. In: Bagnard D (ed.) *Axon Growth and Guidance*. Austin, TX: Landes Biosciences.

Ramón y Cajal S (1899) *Texture of the Nervous System of Man and the Vertebrates*. New York: Springer.

Serafini T, Colamarino SA, Leonardo ED, et al. (1996) Netrin-1 is required for commissural axon guidance in the developing vertebrate nervous system. *Cell* 87: 1001–1014.

Serafini T, Kennedy TE, Galko MJ, et al. (1994) The netrins define a family of axon outgrowth-promoting proteins homologous to *C. elegans* UNC-6. *Cell* 78: 409–424.

Tessier-Lavigne M, Placzek M, Lumsden AG, et al. (1988) Chemotropic guidance of developing axons in the mammalian central nervous system. *Nature* 336: 775–778.

Wadsworth WG, Bhatt H, and Hedgecock EM (1996) Neuroglia and pioneer neurons express UNC-6 to provide global and local netrin cues for guiding migrations in *C. elegans*. *Neuron* 16: 35–46.

Axonal Regeneration: Role of Growth and Guidance Cues

M Bähr and P Lingor, Georg-August-University Göttingen, Göttingen, Germany

Introduction

Compared to other organ systems, the central nervous system (CNS) in the adult human has one of the lowest regenerative capacities following pathological insults, and within the phylogenetic tree the mammalian nervous system is the least capable to renew itself. Besides traumatic nerve injury, a great number of neurodegenerative diseases exhibit marked axonal pathologies, which often precede the histologically more impressive neuron cell death and possibly represent an early stage in disease progression. A functional restoration in a structure as complex as the nervous system therefore requires not only the preservation of cell bodies, but also the maintenance and/or reconstruction of axonal connections to their physiological target regions.

Increased life expectancy has resulted in a continuous rise in the incidence and prevalence of neurodegenerative diseases. The understanding of basic cellular and molecular mechanisms underlying the lack of CNS regeneration is thus not only of interest for the life science community but also of major socioeconomic relevance. Clearly, a deeper insight into mechanisms of axonal de- and regeneration is indispensable for the identification of future treatment strategies in the field of clinical neurosciences.

In this article, developmental mechanisms of axon guidance which regulate axonal pathfinding during embryogenesis are discussed. Interestingly, numerous molecules with developmental roles persist throughout adulthood and may take new functions in the lesioned nervous system, inhibiting or fostering axonal regeneration.

Developmental Guidance Cues: Orchestrating a Symphony of a Thousand

During development, a multitude of signals are required in order to shape naive cells to form adult tissue. Molecular gradients of the so-called morphogens have been identified to regulate cell fate by the induction of transcriptional changes, which, for example, results in cell-type specification according to a cell's position in the gradient. Cellular processes, such as axons, similarly react to molecular gradients of guidance molecules, which act to direct motile structures toward a target area. Four major groups of largely conserved developmental guidance molecules are known to date: ephrins, semaphorins, netrins, and slits. Recent evidence, however, suggests that morphogens are able to route growth cones in a similar way, in addition to their classical function in the commitment of cell fate. The versatile calcium ion influences growth cone development in a precise spatiotemporal manner, including the mediation of pathway cross-talk.

Classical Morphogens with Novel Old Functions

Morphogens are signaling molecules that during development have a locally confined expression in a specific source region from where a concentration gradient is established. The exact mechanism of gradient formation is still a matter of debate, and evidence supports diffusive mechanisms as well as repeated cycles of endo- and exocytosis. Cells positioned in morphogen gradients react to concentration differences by modification of their differentiation program inducing cell fate specification. In addition to this classical morphogen role, evidence is growing for morphogens as axon guidance molecules (**Figure 1**).

Sonic hedgehog (Shh) is a member of the hedgehog protein family of which two other members are known in mammals. It is produced in the notochord and in the ventral midline of the CNS – the floor plate cells, where it is involved in shaping the dorsoventral axis of the neural tube. Signaling through its receptor patched (Ptc), Shh induces the release of the G-protein-associated mediator smoothened (Smo), which eventually activates zinc finger transcription factors of the Ci/GLI family.

As a midline-derived molecule, Shh has been shown to function as a chemoattractant for commissural neurons, guiding their axons toward the midline. This could be inhibited by the inactivation of the Shh signaling mediator Smo, suggesting a direct action of Shh on the axons rather than the induction of secondary guidance cues. In contrast to commissural axons of the spinal cord, retinal ganglion cells (RGCs) are prevented from crossing the midline by continuous expression of Shh. Here, Shh signaling has been shown to induce an inhibition of cyclic AMP (cAMP) production in the cell, which is known to result in growth inhibition. The opposing effects of the same guidance cue on axonal growth of different cell types remain puzzling. It is likely to assume different signaling pathways resulting in diametrical effects or an interplay with other guidance cues present at the same time. A context-specific effect of Shh, for example, dependent on intracellular cAMP levels, is also proposed.

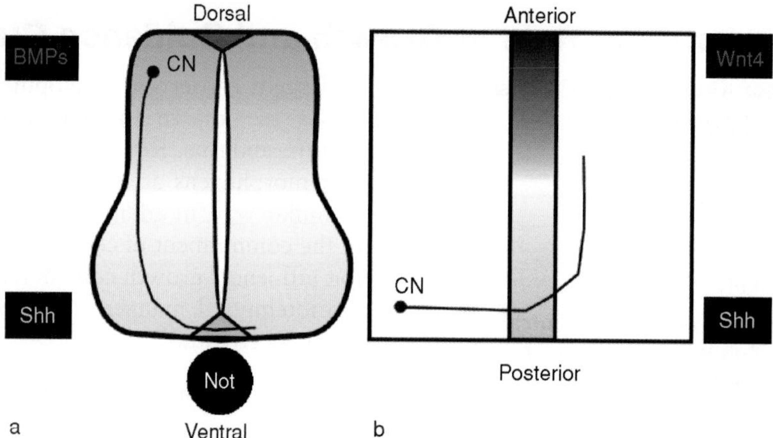

Figure 1 Morphogens involved in commissural axon guidance during embryonic development. Gradients of bone morphogenetic proteins (BMPs) and sonic hedgehog (Shh) shape the dorsoventral axis, while gradients of Shh and Wnt4 regulate the posteroanterior guidance. Cross-section of developing spinal cord (a) and 'open-book' conformation (b). CN, commissural neuron.

Similar to the regulation of the fate of ventral neurons by Shh, members of the bone morphogenetic protein (BMP) family regulate the fate of dorsal interneurons and commissural neurons. BMPs as well as the related growth and differentiation factors (GDFs) belong to the large superfamily of transforming growth factors-β (TGF-β) and signal through type I and type II TGF-β receptors. Through receptor-regulated SMADs and co-SMADs, the signal is transduced to the nucleus, where transcription is activated. BMPs have been shown to equally act as guidance cues for commissural axons after their initial role as morphogens. Heterodimers of GDF7 and BMP7 guide axon growth of commissural neurons, which could be shown by defective outgrowth of commissural axons in knockout mice lacking GDF7, BMP7, or both. A role for BMPs in the development of RGC axons has been suggested by studies in which a deletion of the BMP receptor Ib resulted in misrouting of ventral RGC axons, enabling them to enter the optic nerve head.

The Wingless/Wnt family of morphogens is expressed by roof-plate cells and participates in the developmental specification of dorsal interneurons. Binding to the Frizzled (Fz) receptors is a common denominator for members of the Wnt family, but signaling pathways that are employed downstream of Fz can be diverse: Wnt binding to the Fz receptor and its coreceptor LRP5/6 eventually results in the inhibition of glycogen synthase kinase 3β (GSK3β). In the absence of Wnt signaling, GSK3β phosphorylates β-catenin, which is then ubiquitinated and targeted for degradation by the proteasome. In this so-called canonical pathway, Wnt signaling thus induces stabilization of β-catenin. Stabilized β-catenin induces transcriptional activation of Wnt target genes via association with the lymphoid enhancer factor (Lef/Tcf). Additionally, GSK3β phosphorylates microtubule-associated proteins (MAPs), and Wnt signaling therefore may directly alter the cytoskeletal structure. Other Wnt signaling pathways involve c-Jun-N-terminal kinase (JNK) activation (the so-called planar cell polarity pathway), directing cytoskeletal organization and coordinated cell polarization as well as the release of intracellular calcium and consequent calcineurin activation (Wnt/Ca^{2+} pathway).

Similar to other morphogens, Wnt family members can act as guidance cues for developing neurons. For example, Wnt5 has been shown to repel anterior commissural axons in *Drosophila*. In mouse embryos, Wnt4 acts as a chemoattractant for postcrossing commissural axons. Again, molecules from the same family show opposing effects on axon outgrowth, which may be explained by utilization of different signaling pathways or, as in the case of Wnt5 signaling in *Drosophila*, even the employment of the unconventional receptor Derailed (Drl).

Neurotrophins

As early as in 1939 limb bud transplantation experiments performed by V Hamburger in chick embryos suggested the existence of a target-derived soluble factor that guides the developing axons of sensory neurons. The protein, now known as nerve growth factor (NGF), was later identified by S. Cohen and R. Levi-Montalcini and turned out to serve as growth and survival factor for sympathetic neurons. Besides NGF, the family of neurotrophins – named after their survival and growth-promoting effects on neurons – today comprises three other structurally related proteins in mammals: brain-derived neurotrophic factor (BDNF), neurotrophin 3, and neurotrophin 4/5. All neurotrophins share a common signaling mechanism involving binding to receptor tyrosine kinases (trk) and the so-called low-affinity neurotrophin receptor p75NTR.

Upon neurotrophin binding, the appropriate trk receptor dimerizes and induces the activation of multiple signaling pathways, of which protein kinase A (PKA), phospholipase C-γ (PLC-γ), and phosphatidylinositol-3-kinase (PI3K) are the most important ones. Neurite elongation and axon guidance mediated by neurotrophins are likely to be transduced via the actin depolymerizing factor (ADF)/cofilin pathway and consequent f-actin rearrangement (**Figure 2**).

NGF is able to precisely control neurite outgrowth via local action on distinct cellular segments. Targets of sympathetic and sensory neurons secrete NGF and thus a chemoattractant function of the molecule was suggested. Later, NGF was shown to foster sensory axon elongation and arborization. In culture, NGF induces outgrowth of dorsal root ganglion (DRG) neurons.

Very similar to NGF, other members of the neurotrophin family have effects on neuronal growth, survival, and axon guidance. BDNF, for example, is able to promote the elongation of RGC axons *in vivo*. Motor neurons show an expression of all neurotrophin receptors during development, and this coincides with the expression of all neurotrophins in their muscular target tissue, promoting the hypothesis of neurotrophins as target-derived guidance cues for developing motor neurons. Interestingly, neurotrophins may act in a chameleon-like manner, changing their chemoattractant properties into a repulsive character, depending on the preconditioning of the neuron: NT-3 acted as a chemoattractant for DRG neurons, except when the culture was pretreated with NT-4/5. Preconditioning of DRG cultures with NT-3 rendered them unsusceptible toward NGF-mediated attraction. Furthermore, growth cones of DRG neurons pretreated with NGF were collapsed by local application of BDNF. The activation and cross-talk of different intracellular signaling pathways as well as a regulation of receptor expression in preconditioned cultures may be discussed as a reason for the differential effects mentioned.

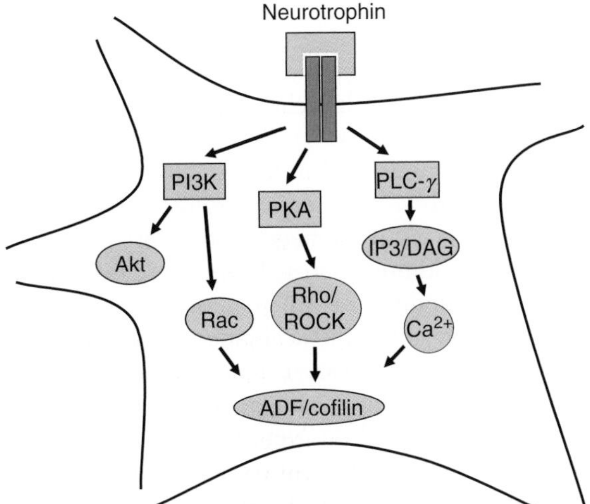

Figure 2 Signaling pathways of neurotrophins involved in axon guidance. Neurotrophin binding to the trk receptor results in activation of the phosphatidylinositol-3-kinase (PI3K), protein kinase A (PKA) or phospholipase C (PLC). Signaling through intermediate molecules results in decreased phosphorylation and thus activation of actin depolymerizing factor (ADF)/cofilin inducing neurite growth.

Besides the neurotrophin family, many other growth factors may act as guidance cues, of which only two examples are given: glial cell line-derived neurotrophic factor (GDNF), one of the most potent survival factors for dopaminergic neurons, has chemoattractive functions on peripheral neurons via activation of cyclin-dependent kinase 5 (Cdk5). Studies in rat explant cultures identified hepatocyte growth factor/scatter factor (HGF/SF) as a mesenchyme-derived chemoattractant for developing motor neurons. In addition to its function as survival factor, HGF/SF thus was able to promote axon growth into the limb bud after binding to the c-Met trk.

Ephrins

A model system for the study of axonal guidance cues in the developmental period is the retino-tectal projection. Axons of RGCs follow a precise topographical projection pattern innervating their target areas in the tectum: RGCs in the dorso-ventral axis terminate along the medio-lateral axis of the tectum, while the temporo-nasal distribution in the retina finds its tectal representation in the antero-posterior axis. The involvement of a chemical gradient derived from the target tissue which regulates the targeted outgrowth of RGC axons was suggested by stripe explant experiments. RGC axons originating from the temporal half of the retina would preferentially grow into stripes derived from the rostral part of the tectum, while stripes derived from the caudal colliculus induced growth cone collapse. A similar preference was observed for RGC axons from the dorsal retinal part, which preferred the lateral tectum stripes and vice versa. When the stripes were treated with phosphatidylinositol-specific phospholipase C (PI-PLC) this effect was no longer observed, suggesting that the guidance molecule is membrane bound via a GPI anchor. The guidance molecules responsible for this part of retinal pathfinding are now known as ephrins: two of them, ephrinA5 and ephrinA2, play a pivotal role in the establishment of the topographical map in the retinotectal projection. Studies with knockout mice for either ephrinA5 or ephrinA2 showed a

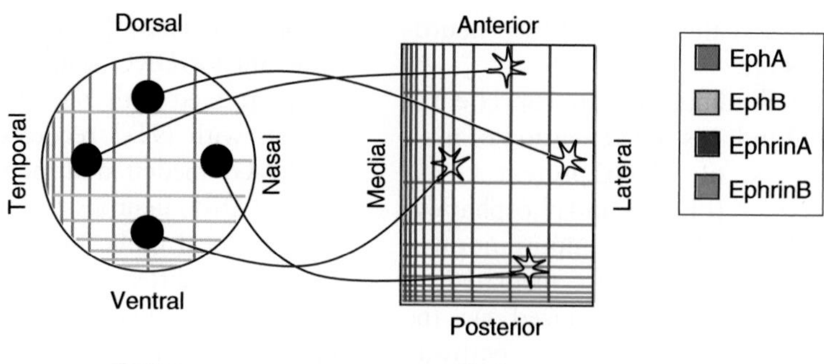

Figure 3 Establishment of retinotectal projections by gradients of Eph receptors and ephrins. Temporonasal gradients of EphA and dorsoventral gradients of EphB define RGCs projecting to the tectum. Anteroposterior gradients of ephrinB and mediolateral gradients of ephrinA result in a retinotopical termination of RGC axons in the tectum.

disorganization of the retinotectal projection, while the double knockout shows an even more prounounced disturbance of axonal wiring. Two families of ephrins are known to date: the GPI-anchored ephrinAs and the transmembrane domain-linked ephrinBs. Trks, termed EphA and EphB, respectively, are responsible for the mediation of the Ephrin signal. It is now confirmed that ephrinA/EphA gradients establish the antero-posterior tectal architecture, while ephrinB/EphB gradients determine the mediolateral projections (**Figure 3**).

Semaphorins

The large family of semaphorins comprises proteins that are either secreted or membrane-bound and that were initially identified for their ability to collapse growth cones of cultured neurons. Up to now, more than 20 semaphorins have been identified and classified into eight subclasses. In invertebrates, subclasses 1 and 2 are expressed, while vertebrates express subclasses 3–7. Nonneurotropic DNA viruses encode for V class semaphorins. All semaphorins are characterized by the so-called sema domain, which is located in the N-terminal region. Semaphorins of classes 3 and 4 homodimerize, which appears to be required for their function.

Neuropilins were the first receptors identified to bind class 3 semaphorins, and they establish target selectivity by formation of multimeric complexes. As is now known, neuropilins bind only to the secreted class 3 semaphorins. However, because of the shortness of their cytoplasmic domain, neuropilins alone are not capable of transducing the signal and therefore form complexes with another type of semaphorin receptors – the plexins. Four subfamilies of plexins have been identified in mammals, termed plexin-A through plexin-D. In contrast to neuropilins, plexins do not need to heteromerize for signal transduction. In their monomeric form plexins represent the receptors for membrane-bound semaphorins of classes 1, 4, 5, 6, and 7.

Additionally, the transmembrane cell adhesion molecule L1 has also been shown to form a complex with neuropilin-1 and act as a semaphorin receptor, modulating the effects of Sema3A: in its soluble form, L1 is able to block the Sema3A-induced growth cone collapse, while L1-deficient mice do not respond to Sema3A-mediated repulsion (**Figure 4**).

Netrins

In the search for factors responsible for the pathfinding of commissural axons in the spinal cord, the biochemical purification revealed two largely homologous proteins – Netrin-1 and -2. To date, Netrins 1–4, which are secreted, and Netrin-G, which is GPI-anchored, are known to exist in vertebrates. Mammalian netrins show homologies to the *Caenorhabditis elegans* UNC-6 protein, which in turn is related to laminin. Two protein families, DCC (deleted in colon cancer) and UNC-5, give rise to netrin receptors.

Netrins seem to have specialized in modeling bilateral symmetry via either chemoattraction (via DCC) or repulsion (via UNC-5). Netrin-1 stimulates outgrowth of commissural axons at different levels in the CNS, and *netrin-1*-deficient mice show severe defects in the establishment of the forebrain commissures. In addition, Netrin-1 is required for developmental pathfinding of RGC axons as they enter the optic nerve, and genetic ablation of Netrin-1 leads to optic nerve hypoplasia. Thalamocortical and corticospinal as well as hippocampal and cerebellar projections equally depend on Netrin-1 signaling for their proper development. One of the intriguing netrin features is its ability to change from attractant to repellent depending on developmental state and intracellular cAMP levels.

Slits

Similar to the midline-forming activity of netrins, the secreted and chemorepellent slits are involved in the

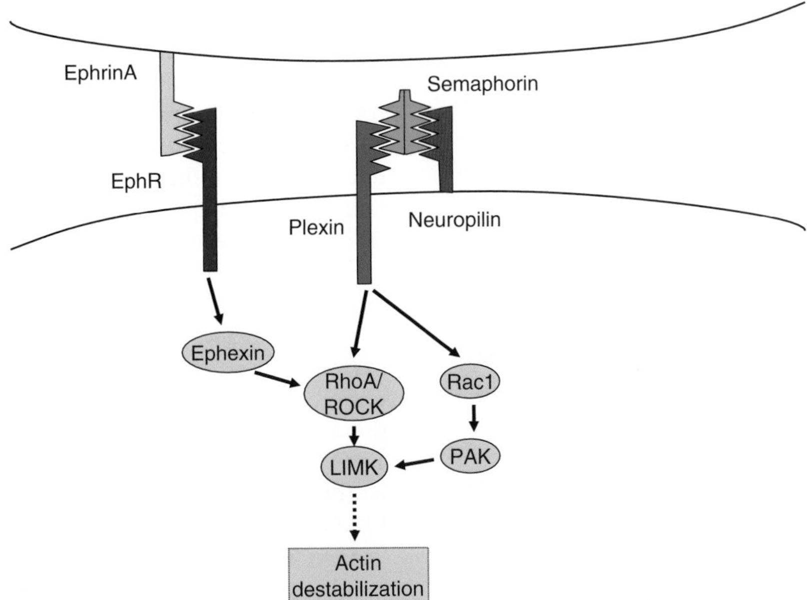

Figure 4 Guidance cue signaling activating growth inhibitory pathways. Membrane-bound ephrins bind to Eph receptors while soluble semaphorins bind to neuropilin/plexin receptors. Both receptor types can elicit growth inhibitory signaling via activation of the RhoA/ROCK cascade, finally inducing actin destabilization and growth inhibition. PAK, p21-activated kinase.

formation of symmetrical structures, such as the optic chiasm. Once bound to axonal Roundabout (Robo) family receptors, slits drive axons away from the midline. Commissural axons initially express low levels of Robo but upregulate the receptor after they successfully crossed to the contralateral side, which prevents them from recrossing the midline. Slits 1–3 have been identified in vertebrates so far, which are able to bind to all three known receptors: Robo 1, Robo 2, and Rig-1.

Left to say, that most of the molecules described above have functions outside the nervous system. Vasculogenesis represents one of the other major processes, where semaphorins, ephrins, netrins, and slits exert their function as guidance cue in development and pathology.

Calcium

The divalent calcium ion is a multifunctional player with roles as a second messenger in the transduction of intracellular signals and as mediator of the action potential. Intracellular cytoplasmic calcium levels ($[Ca^{2+}]_i$) seem to regulate growth cone motility. It has been shown that increased $[Ca^{2+}]_i$ decreases growth cone motility and a reduction of $[Ca^{2+}]_i$ promotes it. However, continuous elevation of $[Ca^{2+}]_i$ can result in adaptation of growth cone behavior, suggesting that targets of Ca^{2+} signaling are downregulated or adjust their sensitivity to the stimulus.

Growth cone motility is not only regulated by baseline cytoplasmic calcium levels, but largely responds to local fluctuations of $[Ca^{2+}]_i$, the so-called calcium transients. Such temporally and locally confined $[Ca^{2+}]_i$ shifts can be generated by calcium flux through ion channels and/or calcium release from intracellular stores. Numerous neurotransmitters are known to regulate calcium channels and thus calcium influx, which in turn can lead to repulsion or attraction of growth cones. For example, the inhibitory neurotransmitter γ-aminobutyric acid (GABA) can act as an attracting stimulus for growth cones, when binding to $GABA_A$ receptors, while exerting repulsive properties, when acting through the $GABA_B$ receptor. The precise spatial and temporal regulation of calcium levels within the growth cone appears to be crucial for the regulation of filopodial dynamics. A proposed binary response to calcium transients, however, greatly oversimplifies the complexity of the growth cone responses. For example, filopodial protrusion can be stimulated by short local Ca^{2+} elevations, but the repeated induction of calcium transients results in the opposite effect.

Downstream targets of calcium include proteins regulating actin dynamics, such as myosin family members, fodrin, α-actinin, gelsolin, and ADF/cofilin. Intracellular calcium sensor proteins, such as calmodulin (CaM), bind calcium and regulate diverse signaling cascades via Ca^{2+}/CaM-dependent phosphatases and kinases. One of the most prominent family members is the Ca^{2+}/CaM-dependent protein kinase II (CaMKII), which shows isoform-specific calcium sensitivity and controls developmental neurite extension.

Guidance Cues in the Adult and Lesioned Nervous System – Play It Again, Sema!

For a long time, the function of axonal guidance cues seemed limited to the developmental period, where neuron–target connection have yet to be established. After successful target innervation, the role of guidance cues seemed obsolete. There is, however, a growing body of evidence that molecules acting as guidance cues and that initially have been attributed classical developmental roles persist throughout adulthood. Even more importantly, experimental data suggest their involvement in the pathophysiology of the lesioned CNS. Next to commonly known inhibitory molecules derived from CNS myelin, the persistence or reexpression of repulsive guidance cues seems to be at least partially responsible for the insufficient regenerative response of the CNS. On the other hand, molecules with neurotrophic properties during embryogenesis may help to overcome regenerative inhibition and support the survival and outgrowth of lesioned neurons.

Myelin-Based Inhibitors of Axon Growth

In contrast to developing axons, neuronal processes in the adult CNS are ensheathed by oligodendrocytes allowing for fast, saltatory conduction of the action potential. Oligodendrocyte-derived myelin has been established as one of the most important inhibitors of axonal growth, while Schwann cells derived from peripheral nerves showed to be permissive toward regenerating axons.

Myelin-based inhibitors of axon growth, that is, Nogo, oligodendrocyte-myelin glycoprotein (OMgp), and myelin-associated glycoprotein (MAG), were initially proposed to account for most of the myelin-derived inhibitory activity. If this were the case, knockouts for one or several of the known myelin-based inhibitors should have markedly improved regenerative capacities in CNS lesion models. However, mutant mice lacking one or all three Nogo isoforms (Nogo-A, B, and C) showed only moderate to no increase in regeneration after spinal cord injury. A similar failure to increase regeneration was observed in MAG knockout mice.

The Nogo receptor (NgR1) transduces the Nogo signal by binding of the 66-amino-acid extracellular domain of Nogo. For signaling, NgR trimerizes with the p75NTR, a versatile receptor initially identified as the low-affinity neurotrophin receptor, and the recently identified LINGO-1, a leucine-rich repeat transmembrane protein. Besides Nogo, MAG and OMgp signal through the trimeric NgR1/p75NTR/LINGO-1 receptor complex. Similar to Nogo knockouts, NgR-deficient mice show a persistent axonal growth inhibition by myelin *in vitro*, while axons from p75NTR-deficient mice were less inhibited, suggesting a more influential role for p75NTR in the regenerative response. Nevertheless, regeneration of corticospinal tract axons after spinal hemisection was not improved either in NgR- or p75NTR-deficient mice, questioning the major role of Nogo/NgR in the mediation of regenerative inhibition. Several hypotheses have been brought forward to explain the persistent lack of regeneration in mice deficient for MAG, Nogo, p75NTR, or NgR. Myelin-based inhibition may employ pathways different from NgR/p75NTR-receptor signaling to induce inhibition. For example, the tumor necrosis factor receptor family member TROY, which is highly expressed in the adult cortex, is able to form a trimeric complex with NgR and LINGO-1 and thus substitute for p75NTR (**Figure 5**).

Next to the 'classical' myelin-based inhibitors, a number of other outgrowth-inhibiting molecules are known to be expressed in the adult CNS. Chondroitin sulfate proteoglycans (CSPGs), such as versican and brevican, are present on differentiated oligodendrocytes and participate in the inhibition of the regenerative response.

Neurocan and phosphacan are two nervous system-specific CSPGs that show high-affinity binding to the extracellular matrix protein Tenascin-C and other cell adhesion molecules, such as Ng-CAM/L1, N-CAM, and TAG-1/axonin-1. Tenascin-R is yet another extracellular matrix protein, which is expressed on oligodendrocytes of adult mice and is upregulated after lesion injury *in vivo*, strongly suggesting an inhibitory role in axonal regeneration.

Developmental Guidance Cues in the Adult

In spite of the overwhelming presence of myelin-derived regeneration inhibitors in the adult, it was equally tempting to suggest that molecules known for their repulsive function as guidance cues during developmental axonal pathfinding additionally act as inhibitory substrates in adulthood.

As outlined previously, ephrins are involved in developmental patterning and axonal guidance and exert their action after binding to transmembrane trks, the Eph receptors. *In situ* hybridization studies in mice revealed that ephrins and Eph receptors are also widely expressed in the adult CNS, although the expression is attenuated compared to the embryonic tissue. Especially, the EphA4 receptor shows a high expression pattern throughout the CNS, while ephrinBs and EphB receptors are highly expressed in regions of known plasticity, such as the olfactory bulb, the hippocampus, and the cerebellum.

A graded expression pattern of ephrinAs in the superior colliculus similar to that during development,

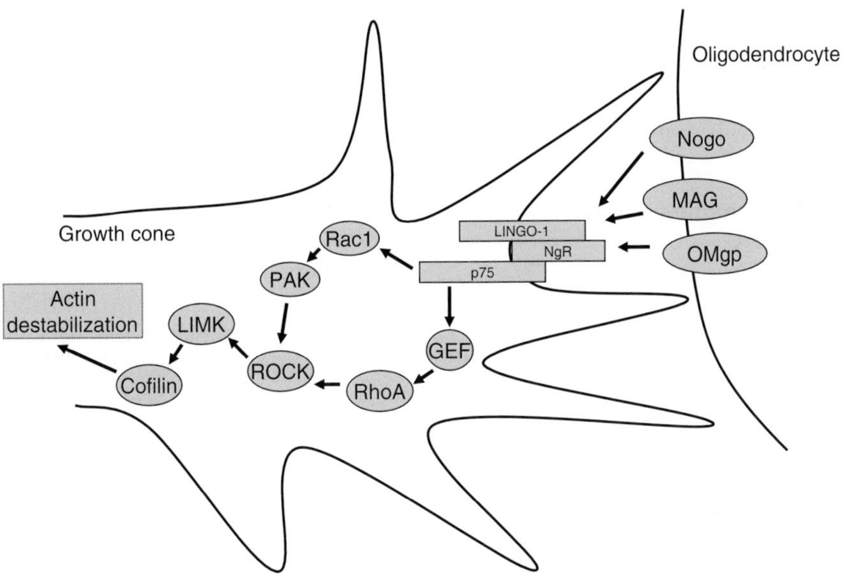

Figure 5 Inhibitory pathways triggered by myelin-based growth inhibitors. Nogo, myelin-associated glycoprotein (MAG), and oligodendrocyte-myelin glycoprotein (OMgp) signal through the trimeric NgR/LINGO-1/p75 receptor. RhoA activation via guanine nucleotide exchange factors (GEFs) induces activation of Rho-associated protein kinase (ROCK) and LIM kinase (LIMK). LIMK phosphorylates cofilin, inducing actin destabilization and growth inhibition.

was observed in the adult mouse before and after differentiation by axotomy. This suggests a requirement for topographic guidance information for functional regeneration of the retinocollicular projection.

Spinal cord hemisection induced an upregulation of EphA4 in astrocytes at the lesion site of wild-type mice. Knockout mice lacking EphA4 were shown to better regenerate corticospinal and rubrospinal axons after spinal cord hemisection than their wild-type littermates. In this paradigm, the regeneration response was improved even though the 'classical' inhibitory signaling cascade via NgR1/p75 was not affected. Myelinating oligodendrocytes in the adult spinal cord further show an expression of ephrinB3 and postnatal cortical neurons express the EphA4 receptor showing sensitivity toward ephrinB3. These data strongly suggest a persistent inhibitory function for developmental repellent molecules.

In analogy to ephrins, a permanent role for semaphorins in the adult CNS has been recently described. Oligodendocytes and astrocytes of postnatal rat optic nerves show expression of multiple semaphorin family members, and RGCs express the neuropilin-1 semaphorin receptor. Of all semaphorins tested, Sema5A had the most pronounced inhibitory effect on growth cone collapse of RGCs. When RGCs were seeded on P8 optic nerve explants, outgrowth was significantly increased after treatment with a Sema5A neutralizing antibody, suggesting an inhibitory role for Sema5A in the regeneration of adult RGC axons. Sema4D/CD100 is expressed on oligodendrocytes of adult mice and acts as a strong inhibitor for postnatal sensory and cerebellar granule cell axons. After spinal cord lesion, Sema4D expression was transiently highly upregulated in oligodendrocytes at the periphery of the lesion. Class 3 semaphorins show an upregulation in the scar tissue of spinal cord-lesioned rats, and receptors for semaphorins are expressed on lesioned cortico- and rubrospinal tract axons.

The involvement of netrins and slits in regenerative paradigms of the adult CNS is less clear. Recently, however, netrin-1 has been shown to promote outgrowth of dopaminergic neuron axons via its DCC receptor, while slit-2 repelled dopaminergic axon growth via its Robo receptor *in vitro*. Whether axonal regeneration of dopaminergic neurons *in vivo* is equally regulated by netrins and slits remains to be determined.

One common denominator of inhibitory molecules seems to be the signaling via the Rho/ROCK pathway. Similar to NgR1/p75/LINGO-1, EphA4 activates the small GTPase RhoA, which in turn leads to an activation of the Rho kinase (ROCK). Rho kinase activates LIM kinase-1 (LIMK1), which phosphorylates cofilin at its serine 3 residue and thus inactivates cofilin. As a potent regulator of actin filament dynamics, the inactivation of cofilin results in actin polymerization and reduced axonal growth.

The role of classical morphogens for axonal regeneration in the lesioned system has not been studied to a great extent yet. However, there is experimental evidence to suggest neuroprotective effects. Supranigral administration of Shh provided an increased survival of dopaminergic neurons after lesion with the selective dopaminergic neurotoxin

1-methyl-4-phenyl-1,2,3,6-tetrahydropyridine (MPTP) in common marmosets and improved their locomotor activity. Adenovirally administered Shh has also been shown to protect motor neurons after axotomy of the facial nerve in adult rats.

Neurotrophic Factors Promoting Axonal Regeneration in the Adult

In contrast to the presence of inhibitory molecules, trophic support derived from neurotrophic factors seems rather limited after CNS lesions. In the mechanically lesioned spinal cord of adult rats, for example, only a marginal upregulation of BDNF and GDNF and its appropriate receptors was observed. Schwann cells of corresponding nerve roots, however, showed a much more pronounced increase in NGF and GDNF mRNA levels, indicating the better regenerative capacity of the peripheral nervous system.

The administration of neurotrophic factors in a therapeutical approach thus seems promising. In the visual system, regeneration of RGC axons can be studied in a crush model of the optic nerve, where the integrity of the nerve is conserved and the transected axons face a nonpermissive environment for regeneration. Several growth factors have been shown to promote not only the survival of RGCs after crush, but equally improve the regrowth of RGC axons into the optic nerve.

Ciliary neurotrophic factor (CNTF) has been shown to promote regeneration of RGC into peripheral nerve grafts, and axotomized RGCs express the appropriate receptors CNTF receptor α and leukemia inhibitory factor receptor (LIFR). The application of CNTF and a Nogo-neutralizing antibody was even synergistic toward promotion of the regenerative response. GDNF, a member of the TGF-β superfamily and potent survival factor for dopaminergic neurons, has been tested for proregenerative properties in spinal cord injury. Mini-guidance channels filled with GDNF and placed into a hemisection gap lesion promoted the regrowth of spinal cord axons into the implant and increased the number of propriospinal neurons.

The combination of growth factors has been the subject of several studies to promote axonal regeneration. For example, fibroblast growth factor-2 (FGF-2), Neurotrophin-3 (NT-3), and BDNF had a synergistic effect on regeneration of RGC axons following crush axotomy, being more effective than treatment with either growth factor alone. BDNF and aFGF promote regeneration of spiral ganglion cell axons into the lesioned organ of Corti. BDNF and NT-3 enhanced propriospinal axonal regeneration in the adult rat spinal cord. The effects of NT-3 and BDNF on regeneration have been shown to be at least partly mediated by enhanced polymerization of f-actin in growth cones.

Intrinsic Regenerative Capacity

The proregenerative effect of neurotrophic factors decreases with the age of neurons. Several factors have been brought forward to be responsible for age-dependent susceptibility to neurotrophic factors and decreased intrinsic regenerative capacity. Differential downregulation of neurotrophin receptors on adult neurons and the increased signaling of inhibitory pathways are likely to contribute to this. An age-dependent reaction of the nervous system toward lesion and a better restoration ability of the peripheral nervous system has been well known from clinical experience. The marked difference between regeneration responses of central and peripheral neurons in the adult CNS may also be explained by differential regulation of transcriptional programs upon lesioning. Pseudo-unipolar DRG neurons, for instance, show a pronounced regenerative response in their peripheral branch, while the ascending central branch is unable to regenerate after injury. Expression of the transcription factor c-Jun, in DRG neurons, is markedly upregulated after lesion of the peripheral branch, while central lesions induce only weak c-Jun upregulation. This correlates with the restricted regeneration potential of the central branch. However, placing a dorsal root into the hemisection site, promoting regeneration into the transplant, again induced marked c-Jun expression in DRG neurons. A very similar effect was shown on the upregulation of growth-associated protein 43 (GAP-43) expression, which is increased equally after lesion of the peripheral branch and almost nonexistent after central lesion. The upregulation of c-Jun and GAP-43 in paradigms of regeneration suggested that these proteins are involved in shaping the cell-intrinsic answer to successful regeneration. Therefore, an overexpression, for example, by viral vectors, has been proposed. Several studies, however, showed that simple overexpression of GAP-43 may not be sufficient to induce regeneration. In the current picture, GAP-43 and c-Jun are standing at the end of a transcriptional program involving the expression of a whole plethora of other so-called regeneration-associated genes (RAGs). However, c-Jun may also act as a two-sided sword in regeneration and death signaling. It has been shown that c-Jun is upregulated after optic nerve axotomy. Moreover, proximal axotomy resulted in stronger c-Jun upregulation and in fewer surviving RGCs than distal optic nerve transection. The arguments for c-Jun as an important apoptosis mediator and potential therapeutical target are supported by the fact that siRNA-mediated downregulation of c-Jun may indeed protect RGCs from axotomy-induced cell death. Thus, the reactivation of growth-associated, developmentally regulated programs is a necessary prerequisite for axonal regeneration. However, it also renders the adult neuron more susceptible to apoptotic cell death.

In striking contrast to the mammalian nervous system, the CNS of amphibians and fish shows a vigorous regeneration reaction following traumatic lesion. Goldfish, for example, readily regrow their axons after optic nerve transection to functionally reinnervate the tectum and even restore vision. Fish RGC do not degenerate after axotomy, but show a pronounced hypertrophy with dense nucleoli, suggesting an activation of protein and RNA synthesis. Similar to mammals, trophic factors, like NGF, stimulate the outgrowth of explanted fish RGC *in vitro*. Proteins secreted by the sheathing cells of the optic nerve (axogenesis factor-1 and -2) have also been found to promote regeneration. Myelin-based inhibitors of axon growth are also present in the fish optic nerve. However, their composition seems to be different from mammalian myelin: RGC explants from fish regenerate readily on fish myelin, but fail to regenerate on rat myelin. Finally, lesion-induced intrinsic transcription programs in fish and amphibians may lead to expression of different target genes than in mammals. For example, the cell adhesion molecule Contactin 1 (Cntn1) is duplicated in fish and activated after optic nerve lesion, suggesting a regeneration-promoting role similar to L1, N-CAM, or TAG-1. The retinol-binding protein purpurin was shown to be upregulated following axotomy and promoted neurite outgrowth of retinal explants. In summary, differences in the intrinsic regenerative capacity by activation of specific transcriptional programs in fish and amphibians may be at least as important for successful regeneration as a more permissive environment.

Preconditioning lesions to the peripheral branch of DRG axons induced an increased regenerative response in the central branch of the DRG in the dorsal column. This was accompanied by an elevation of intracellular cAMP and inhibition of PKA. Increased regeneration in response to application of neurotrophic factors has also been shown to be mediated by elevated levels of cAMP. As could be demonstrated in cerebellar neurons, priming with BDNF or GDNF incurred resistance against MAG-induced outgrowth inhibition via cAMP elevation and activation of PKA. This involves inhibition of phosphodiesterase by activation of the extracellular-signal regulated kinase (ERK). Direct injection of a cell-permeable cAMP analog (dibutyryl-cAMP; db-cAMP) into the dorsal column of a lesioned spinal cord fostered regeneration. Elevated cAMP levels may modulate the response to lesion via activation of the cAMP response element binding protein (CREB) transcription factor. However, CREB activation alone may be insufficient, as was implicated in a study combining neurotrophin treatment and cAMP in a model of spinal cord injury (**Figure 6**).

Taken together, axonal regeneration in the adult mammalian CNS is restrained by a multitude of intrinsic and extrinsic factors: an inhibitory environment consisting of repulsive guidance cues and myelin-derived growth inhibitors paired with lack of

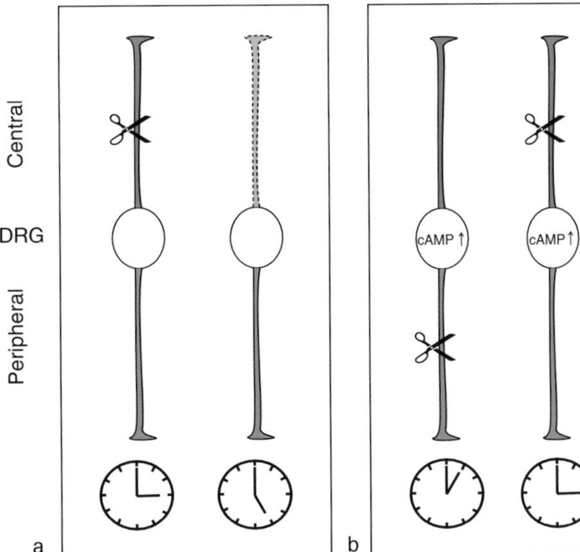

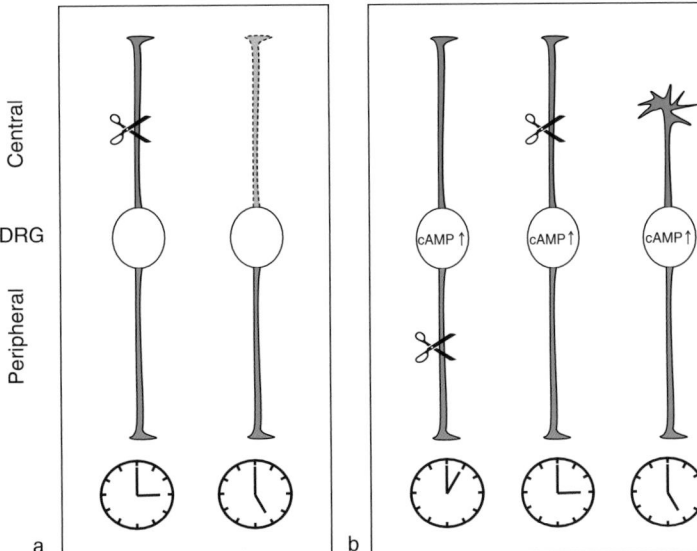

Figure 6 Preconditioning may change intrinsic growth capacity via induction of growth-associated genes. Traumatic lesion to the central branch of the dorsal root ganglion (DRG) results in its degeneration. Preconditioning lesions to the peripheral branch of the DRG and subsequent lesion of the central branch induces a regenerative response in the central branch. This is accompanied by an elevation of intracellular cyclic AMP (cAMP) levels.

trophic support and pro-regenerative transcriptional programs result in insufficient functional renewal capacity. Strategies to overcome regenerative failure will therefore most likely be based on a combination of approaches, targeting several pathways simultaneously.

See also: Axon Guidance: Building Pathways with Molecular Cues in Vertebrate Sensory Systems; Axonal Injury: Neuronal Responses; Axonal Regeneration: Role of the Extracellular Matrix and the Glial Scar; Nogo-A: Its Role in Axon Regeneration; Peripheral Nerve Regeneration: An Overview.

Further Reading

Charron F and Tessier-Lavigne M (2005) Novel brain wiring functions for classical morphogens: A role as graded positional cues in axon guidance. *Development* 132: 2251–2262.

Coleman MP and Perry VH (2002) Axon pathology in neurological disease: A neglected therapeutic target. *Trends in Neurosciences* 25: 532–537.

Gillespie LN (2003) Regulation of axonal growth and guidance by the neurotrophin family of neurotrophic factors. *Clinical and Experimental Pharmacology and Physiology* 30: 724–733.

Gomez TM and Zheng JQ (2006) The molecular basis for calcium-dependent axon pathfinding. *Nature Reviews Neuroscience* 7: 115–125.

Herdegen T, Skene P, and Bähr M (1997) The c-Jun transcription factor-bipotential mediator of neuronal death, survival and regeneration. *Trends in Neurosciences* 20: 227–231.

Klein R (2004) Eph/ephrin signaling in morphogenesis, neural development and plasticity. *Current Opinion in Cell Biology* 16: 580–589.

Luo L (2000) Rho GTPases in neuronal morphogenesis. *Nature Reviews Neuroscience* 1: 173–180.

McKerracher L and Winton MJ (2002) Nogo on the go. *Neuron* 36: 345–348.

Properzi F and Fawcett JW (2004) Proteoglycans and brain repair. *News in Physiological Sciences* 19: 33–38.

Sandvig A, Berry M, Barrett LB, Butt A, and Logan A (2004) Myelin-, reactive glia-, and scar-derived CNS axon growth inhibitors: Expression, receptor signaling, and correlation with axon regeneration. *Glia* 46: 225–251.

Axonal Regeneration: Role of the Extracellular Matrix and the Glial Scar

J Silver, K P Horn, S A Busch and A L Yonkof,
Case Western Reserve University School of Medicine,
Cleveland, OH, USA

Introduction to the Problem of Regeneration Failure

Unlike their counterparts in the peripheral nervous system (PNS), axons of the central nervous system (CNS) fail to regenerate following injury. With the exception of olfactory sensory neurons within the olfactory bulb and hypothalamo-neurohypophysial axons within the hypothalamus, regeneration of axonal tracts is not observed following disruption of the CNS in adult mammals. The regenerative success of these two pathways has been attributed to the specialized support cells present in these regions: olfactory ensheathing cells in the olfactory bulb and tanycytes in the mediobasal hypothalamus. Interestingly, numerous studies have been conducted using these two cell types, along with Schwann cells, the growth-promoting glia of the PNS, as bridging cells in other regions of the CNS with some success. Typically, adult CNS axons are, at best, capable of abortive sprouting, which results in the failure of these neurons to re-innervate their targets. In fact, the vast majority of functional recovery reported following CNS lesions in experimental models is thought to be due to plasticity of spared or uninjured processes and not to new growth of damaged fibers. Functional recovery of adult CNS neurons is limited because tissue damage following injury initiates a cascade of events that lead to structural and molecular changes within the lesion that are inhibitory to regeneration. There are numerous obstacles for regenerating axons to overcome, but our focus in this article is on the role of the extracellular matrix (ECM) in the glial scar.

The Glial Scar

What Is the Glial Scar?

The glial scar is a complex milieu consisting mainly of reactive astrocytes and secreted ECM molecules. In more severe injuries in which the blood–brain barrier (BBB) is broken down and the meninges are compromised, connective tissue elements, such as fibroblasts, as well as invading immunological cells, such as macrophages, may mix with astrocytes in the lesion (**Figure 1**). Immediately following damage to the CNS, astrocytes begin to undergo a process known as reactive gliosis, in which they become extremely hypertrophic, dramatically upregulate expression of intermediate filament proteins such as glial fibrillary acidic protein (GFAP) and vimentin, and increase secretion of ECM components. Although the response is termed gliosis, implying an increase in cell division, proliferation of astrocytes is actually rather limited and confined to the penumbra surrounding the lesion core. Recent evidence has shown that depletion of the proliferating population of astrocytes following injury prevents reestablishment of the BBB and results in increased lesion size and an exacerbated inflammatory response. Therefore, despite its devastating effect on axonal regeneration, the glial scar plays an important role in secluding the damaged site and protecting the surrounding uninjured tissue from secondary injury.

Development of the glial scar begins within hours of injury, subsequent to the introduction of non-CNS molecules and cell types into the parenchyma of the brain or spinal cord via disruption of the BBB. The BBB remains permeable to blood and serum components for up to 2 weeks, continuously bathing the CNS with numerous exogenous proteins and factors. There is a correlation between the areas of the most significant BBB breakdown, highest infiltration of activated macrophages, and maximal glial scaring. As the lesion matures, the enlarged and entangled astrocytic processes and secreted ECM components form a tenacious and rubbery scar.

Inhibitory Nature of the Glial Scar

Historically, the association of dystrophic axonal endings with dense reactive astrocytic processes within a CNS lesion led to the hypothesis that the glial scar presents a physical barrier to regenerating fibers. The distinct morphology of struggling growth cones was first described by Ramón y Cajal, and for many years these growth cones were thought to be dead endbulbs or retraction clubs and were considered sterile and incapable of growth. Recent evidence suggests that dystrophic endings retain their ability to regenerate and are capable of returning to an active state. Injured spinal cord neurons can grow into a peripheral nerve graft 4 weeks after injury, despite having stalled in the lesion for an extensive period of time. Shrunken cell bodies of injured rubrospinal tract neurons (**Figure 1**, E) increase in size following brain-derived neurotrophic factor (BDNF) stimulation 1 year post lesion and can extend growth-associated protein-43 (GAP-43)-positive axons into a peripheral nerve graft.

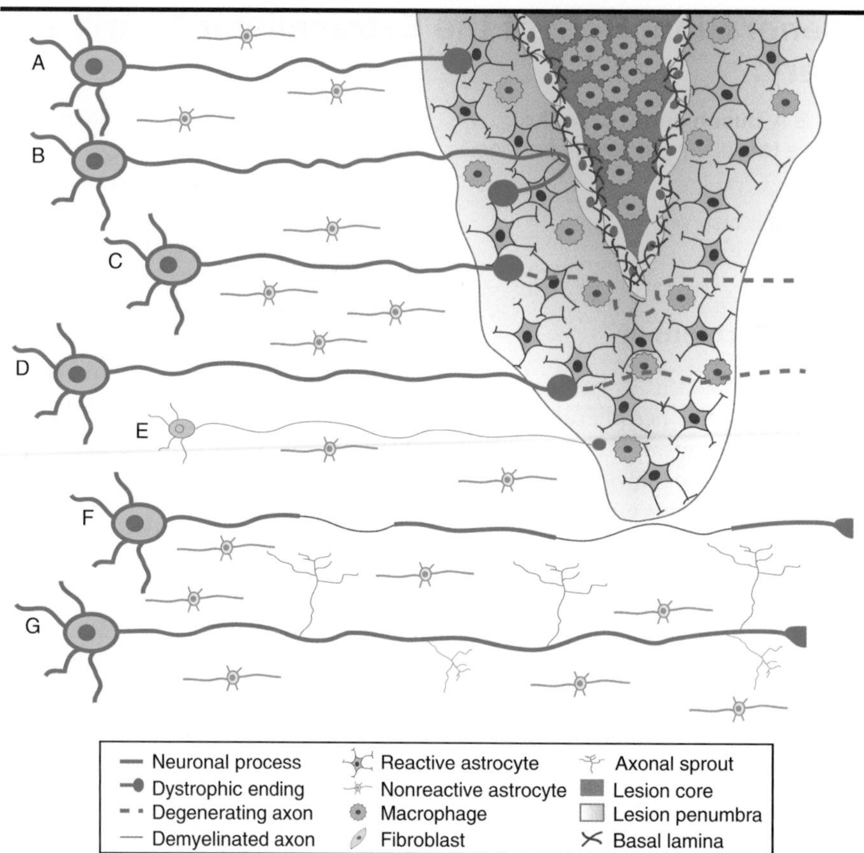

Figure 1 Cellular and molecular elements of a typical penetrating spinal cord lesion. The lesion core is devoid of neurons and astrocytes and is populated by phagocytic macrophages that can cause cavitation. Fibroblasts deposit basal lamina and form a barrier between the lesion core and lesion penumbra. Macrophages and reactive astrocytes occupy the lesion penumbra, where reactive astrocytes secrete inhibitory CSPGs in an increasing gradient toward the lesion core. Other important factors in the lesion penumbra include semaphorins, slits, and Eph/ephrins. A and B, Neurons have been axotomized from the lesion and fail to regenerate, resulting in a dystrophic club ending. B, Dystrophic neuron turns away from the intensely inhibitory lesion core and stops growing in a region of high levels of inhibitory proteoglycan. C, D, Neurons undergo Wallerian degeneration (dotted line) as a result of damage. C, Neuron has been deformed by the lesion and subsequently degenerates because of structural damage and inflammation. D, Axon dies only as a result of secondary damage due to inflammation. E, 'Ghost' neuron atrophies as a result of axotomy. The axon forms a dystrophic endbulb, and the cell body shrinks. The neuron is dormant but may be awakened by growth factors. F, Undamaged axon becomes demyelinated (purple line) both near the lesion and in regions farther away from the lesion. G, Undamaged axon sprouts to re-innervate the spinal cord. Note a contusive injury would be similar to a penetrating injury; however, it would lack the fibroblast infiltrates with their associated ECM components (basal lamina).

Even at extended time periods post lesion, the supposedly sterile endings in the corticospinal tract have been shown to be capable of sprouting.

Axonal Response to the Glial Scar

While initial studies indicated that axons retract significantly from the lesioned area after axotomy, improved tracing techniques have revealed that dystrophic axons actually persist in the vicinity of the lesion (**Figure 1**, A and B). This implies that even though they are stalled and do not elaborate synapses, an intrinsic mechanism is in place that allows them to maintain their viability and stability. It is possible to generate dystrophic endings on adult sensory neurons *in vitro* by culturing them on inverse gradients of the growth-promoting ECM substrate laminin and the potently inhibitory chondroitin sulfate proteoglycan (CSPG) aggrecan (see below; **Figure 2**). Time-lapse imaging of these dystrophic endings using this model first demonstrated that these endbulbs are, in reality, quite dynamic. The endbulbs actively rearrange their cytoskeletons, rapidly recycle their plasma membranes, and alter the location of their repertoire of integrin receptors, which bind to ECM substrates and are critical for adhesion and migration. Further work has shown that alteration of integrin expression and membrane turnover also occur *in vivo* following spinal cord injury.

It is essential to distinguish between the states of dystrophy and collapse when describing axonal responses to injury. Following direct contact with myelin components and mature oligodendrocytes, growth cones of mature neurons become shrunken,

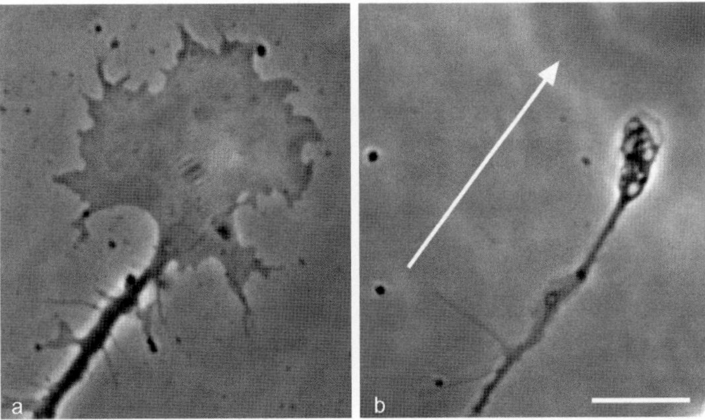

Figure 2 Normal and dystrophic growth cones. (a) Normal growth cone of adult DRG on laminin. The growth cone uses filopodia and lamellopodia to sample the environment for cues. (b) Dystrophic growth cone on a decreasing gradient of laminin plus an increasing gradient of inhibitory proteoglycan (arrow). The growth cone forms a bulbous, dystrophic ending when it encounters inhibitory CSPG, as is characteristic in the CNS lesion environment. Scale bar = 10 μm.

retracting their filopodia and lamellipodia, and enter a temporary state of quiescence. This process is known as collapse and has been well characterized *in vitro*. After some time, growth cones will begin to extend once more only to collapse again upon renewed contact with the collapsing agent. While collapse also results in stalled growth, it is very different from dystrophy, which is a highly dynamic state. Collapse is also prevalent in development and occurs frequently when pathfinding axons encounter transitions between ECM substrates. Observations of the inhibitory nature of myelin *in vitro* led to its implication as another barrier to CNS regeneration, and numerous studies of its major inhibitory element, the Nogo receptor, have used antibodies to block its binding of myelin components to enhance growth of regenerating fibers *in vivo*. While many of these studies have had some success in generating limited functional improvement, recent evidence has suggested that this is due, at least in part, to an increase in plasticity of uninjured fibers.

Immature Astrocytes Promote Regeneration while Mature Astrocytes Make Scars and Block Growth

Embryonic and early postnatal animals have a much higher capacity for regeneration following damage to the CNS than adults. While the neurons undoubtedly differ in their intrinsic growth potential (see below), there are also differences between embryonic and adult astrocyte responses to injury. When hippocampal or retinal ganglion cell (RGC) neurons are cultured on astrocyte explants from lesioned animals, neurite outgrowth is inhibited by reactive mature astrocytes and actually promoted by reactive immature astrocytes. In fact, mature reactive astrocytes can even inhibit the enhanced growth potential of embryonic neurons. Remarkably, when immature astrocytes are transplanted into the adult brain, they can alter the response of endogenous astrocytes and repress formation of the glial scar.

What Are the Molecular Triggers of Glial Scarring?

While the factor or combination of factors responsible for stimulating glial scar formation has not yet been elucidated, there are several putative candidates. Expression of both transforming growth factor beta (TGF-β) 1 and 2 is upregulated in the CNS following injury, and TGF-β2 is known to radically increase production of proteoglycans by astrocytes. The serine protease thrombin has also been implicated as a potential trigger of reactive gliosis. Thrombin, an integral element of the blood coagulation cascade, has additionally been shown to be involved in the chemotaxis and adhesion of monocytes, macrophages, and neutrophils, as well as the regulation of neurite outgrowth, induction of astrocyte proliferation, and activation of microglia in the CNS. Inflammatory cytokines appear to possibly contribute to glial scar formation as well. Members of the interleukin family may also be involved as interleukin 1, which is released by mononuclear phagocytes, has been shown to generate an inflammatory response in astrocytes. Application of interferon gamma has been shown to increase proliferation of astrocytes *in vitro* and treatment of the injured brain with it causes scar expansion. Basic fibroblast growth factor also increases proliferation of astrocytes *in vitro* and its expression increases in the CNS after injury. However, it is important to note that astrocyte proliferation is only a small part in the formation of the glial scar. It is thought that the presence of macrophages is in large part responsible for stimulating astrocytes to take on a reactive state and to construct the glial scar. When Zymosan A, a potent activator of macrophages, is injected into the CNS, extensive migration of astrocytes away from the inflammatory site leads to cavitation, severe upregulation of proteoglycans, and substantial glial scarring. *In vitro* work has further shown that astrocyte migration away from areas of inflammation can cause axonal stretching and secondary axotomy through neuronal–glial interactions. Therefore, factors derived from blood and/or serum coupled with macrophage invasion combine to contribute to the formation of the glial scar and secondary lesion phenomena, including the demyelination of nearby otherwise healthy axons.

Proteoglycans

Proteoglycans Are ECM Molecules

There are four classes of proteoglycan: CSPG, dermatan sulfate proteoglycan, heparan sulfate proteoglycan, and keratan sulfate proteoglycan. Proteoglycans are components of the ECM that are involved in adhesion, growth, receptor binding, migration, barrier formation, and interactions with other ECM molecules. Proteoglycans consist of a protein core to which glycosaminoglycan (GAG) side chains are bound by a tetrasaccharide link (**Figure 3**). These GAG chains are composed of repeating disaccharide units formed from two alternating monosaccharides, usually either *N*-acetylglucosamine or *N*-acetylgalactosamine followed by uronic acid. GAG chains can vary dramatically in length, and as many as 100 can be attached to the core protein. The number of GAG chains along with their sulfation pattern and the core protein to which they are bound yield a tremendous amount of variability within and between classes.

CSPGs are inhibitory ECM molecules CSPGs include aggrecan, brevican, neurocan, NG2, phosphacan (also classified as a keratan sulfate proteoglycan), and veriscan. CSPGs are present throughout development and are crucial for axon guidance in the

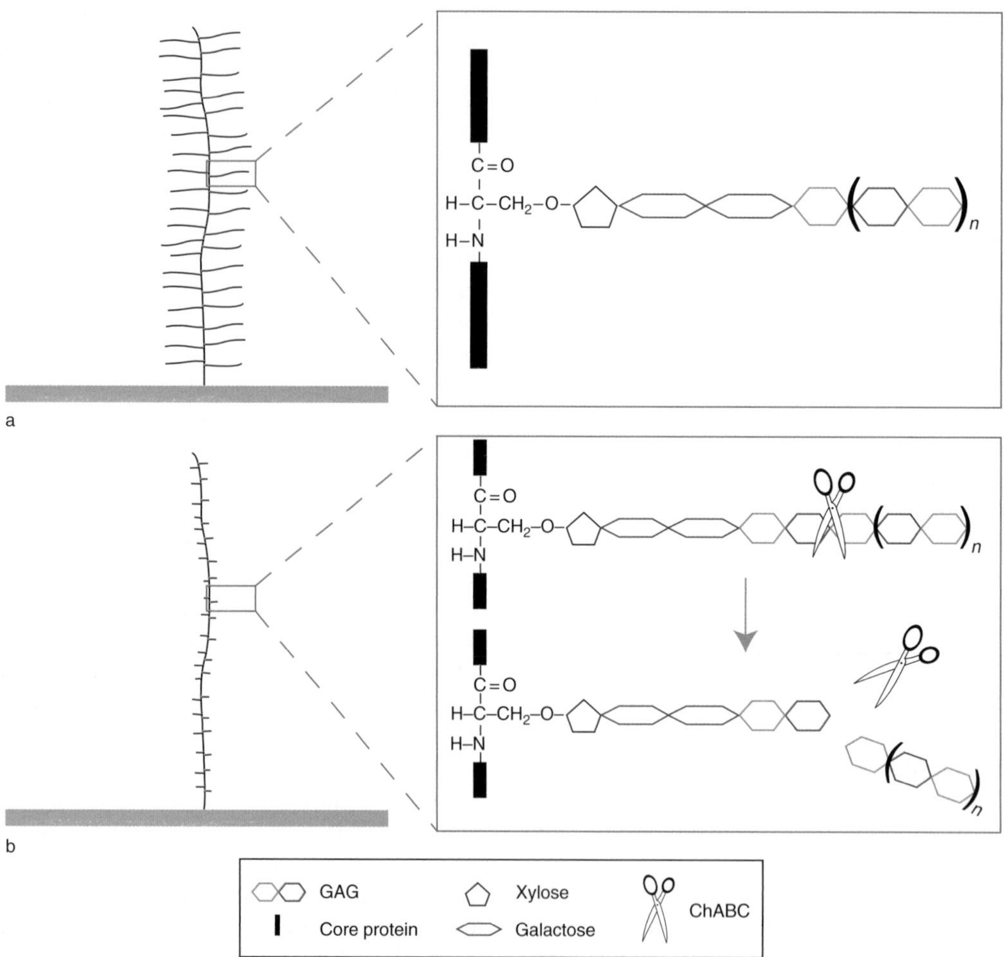

Figure 3 CSPGs and the action of chondroitinase ABC. CSPGs are often anchored to cell membranes, hyaluronan, and other ECM components (gray bar). (a) CSPGs are composed of a core protein linked to GAG side chains by a tetrasaccharide linker of xylose, glucose, and uronic acid. GAGs are disaccharides of *N*-acetylglucosamine or *N*-acetylgalactosamine and uronic acid. These disaccharides repeat to form a long chain. (b) Chondroitinase ABC (ChABC) is a bacterial enzyme that cleaves GAG chains, resulting in a stub of carbohydrates. This abolishes the majority of inhibition from CSPGs.

roof plate, the midline of the rhombencephalon and mesencephalon, the dorsal root entry zone (DREZ), the optic tract and chiasm, and the retina. An elegant study utilized the enzyme chondroitinase ABC (ChABC) from the bacterium *Proteus vulgaris* to digest the GAG chains of CSPGs in the developing retina. The absence of the CSPG GAG chains resulted in pathfinding errors of RGC neurons whose axons grew aberrantly into regions of the retina from which they are normally excluded. The expression of CSPGs changes after the critical period as the animal matures and the axons have reached their targets.

More on the inhibitory nature of CSPGs The physical nature of the glial scar barrier takes several weeks to months to fully develop, which cannot explain the inhibition to axonal growth observed acutely following injury. Interestingly, CSPG expression is robustly upregulated by mature astrocytes in the glial scar within 24 h of CNS injury in adult mammals and maintained for several months. This evidence, along with their inhibitory role in embryonic pathfinding in specific regions, implicated CSPGs as a barrier to regeneration in the CNS. In the early 1990s, it was shown that CSPGs are potently inhibitory to adult neurite outgrowth both *in vitro* and *in vivo* and play a significant role in CNS regeneration failure. CSPGs are not only potent inhibitors of axonal growth when encountered alone, but they can also block the growth-promoting effects of substrates such as laminin, fibronectin, and L1. *In vitro* adult neurites avoid stripes of laminin mixed with aggrecan when given the choice between this combination of substrates and stripes containing laminin alone. When RGCs are cultured on scar explants that are removed from the adult (but not the neonatal) brain, they can only extend long neurites if the wound tissue is pretreated with ChABC, indicating that CSPGs are potent inhibitors of neurite outgrowth *in vivo* as well.

Different neuronal populations experience varying degrees of growth retardation when exposed to CSPGs *in vitro*. RGCs are able to extend further up a stepwise gradient of aggrecan on laminin than either forebrain neurons or embryonic dorsal root ganglion cells (DRGs), although growth was significantly stalled at the transitions between steps. Immature neurons seem to be able to grow on CSPGs by upregulating integrin receptor expression until a threshold concentration of the inhibitory substrate is reached, and they can grow no further. This threshold is different for different populations of neurons. As mentioned earlier, adult DRGs grow up a smooth gradient of CSPG until the potently inhibitory rim is reached, where they stall and form dystrophic endings (**Figure 2**). In development, some neurons are capable of growing through proteoglycan-rich areas. For example, embryonic hippocampal neurons and thalamocortical fibers growing through the subplate are capable of extending neurites on oversulfated CSPGs. *In vitro*, when neurons are plated on low concentrations of CSPGs mixed with laminin, neurites fasciculate tightly as they would in the subplate *in vivo*. In both embryonic and adult systems *in vitro* and *in vivo*, the balance of proteoglycans and other ECM molecules is critical to the determination of growth cone behavior as well as bundling and branching patterns of axons.

Neuronal populations have different thresholds for proteoglycan inhibition when presented in a gradient *in vivo*, but all populations will eventually become dystrophic within a spinal cord lesion (**Figure 1**). Proteoglycan concentration is highest at the lesion epicenter and decreases in a gradient throughout the penumbra. As axons enter the lesion, they receive more inhibitory cues, and form increasing numbers of dystrophic endings as they approach the core of the developing glial scar. Axons of motor neurons are not able to grow into the lesion at all, but instead sprout in the undamaged region adjacent to the lesion. Serotonergic neurons are capable of growing to the lesion edge, and sensory axons can grow further into, but not through the lesion. Amazingly, adult sensory neurons have the ability to regenerate robustly through intact and degenerating white matter when gently microtransplanted (without creating a scar) well distal to a large lesion in the spinal cord (i.e., beyond the zone of BBB disruption). However, regeneration stops when the rapidly growing axons reach the lesion environment, indicating the potently inhibitory nature of the glial scar.

Chondroitinase ABC

ChABC digests inhibitory GAG chains on proteoglycans The use of ChABC is currently the most effective method of removing the inhibition of CSPGs (**Figure 3**). A single injection of ChABC into the normal uninjured brain causes a widespread reduction in the concentration of CSPGs within the ECM for at least 4 weeks. It is important to note that ChABC digestion only cleaves the GAG side chains of CSPGs and that some carbohydrate residues remain attached to the core protein. These residues can be labeled with specialized antibodies to demonstrate digestion of CSPGs in a ChABC-specific manner. While the core proteins of CSPGs are somewhat inhibitory to neurite outgrowth, the vast majority of the inhibitory nature of CSPGs lies within their GAG side chains, which is evident from the work by various laboratories showing that removal of CSPGs from the glial scar enhances regeneration and can provide some functional recovery. Administration of ChABC following a nigrostriatal pathway lesion promoted regeneration of dopaminergic neurons to the point of re-innervating their target area in the striatum. In animals receiving a bilateral dorsal column injury, intrathecal delivery of ChABC induced growth of ascending sensory and descending motor axons into and possibly slightly past the lesion, which resulted in improved function of both proprioception and locomotion. ChABC treatment has also resulted in increased regeneration of Clarke's nucleus neurons following a hemisection lesion. Similar CSPG digestion and improved regeneration have also been observed in cats receiving hemisection and rats receiving contusion injuries. This gives hope that-continuing efforts to increase the efficacy of this technique could potentially lead to treatments for spinal cord injury in primates and hopefully in humans. Typically, ChABC is administered to the CNS through a pump or by direct injection. ChABC then remains active for 3 days *in vivo*. An alternative strategy might be to genetically engineer CNS cells to express ChABC, allowing for controlled, continuous delivery of the enzyme.

The perineuronal net and the regulation of synaptic plasticity Proteoglycans in the gray matter of the brain and spinal cord may encapsulate synapses with a lattice-like accumulation of ECM known as the perineuronal net (PNN) and prevent neuritic sprouting and synaptic plasticity. PNNs consist mostly of CSPGs produced as the animal passes through critical periods of CNS maturation. As the animal matures, PNN production increases and the potential for plasticity decreases as the synaptic connections are firmly established. Application of ChABC to the uninjured cerebellum in rats induces sprouting of unmyelinated Purkinje cell axon terminals. This CSPG-mediated inhibition of sprouting prevents changes in functional connectivity, which could lead to formation of aberrant synaptic connections. However, it may be advantageous to restore plasticity following injury. For instance, when the adult visual cortex is treated with ChABC, plasticity can be recovered. This has important applications in restoration of vision in adult strabismic animals that have been monocularly deprived since weaning and are therefore cortically blind in one eye. ChABC-induced plasticity could also be important during rehabilitation after stroke and in treating epilepsy.

Other Inhibitory ECM Molecules

In addition to CSPGs, there are many other molecules known to be upregulated following CNS injury. It is well known that axon repellants such as semaphorins, slits, and Eph/ephrins play critical roles in embryonic development, but their roles as inhibitors of regeneration in the mature CNS are just beginning to be understood.

Semaphorins The expression of semaphorin 3, a chemorepellant to growing axons during development, is upregulated by invading fibroblasts following stabbing types of injuries to the spinal cord, cortex, and lateral olfactory tract. Semaphorin 3's high-affinity receptor, neuropilin 1, is upregulated in neurons attempting to regenerate into the lesion. Regenerating axons are excluded from areas near the lesion core that contain semaphorin 3, indicating that semaphorins may be another inhibitory cue present in fibroblast-rich lesions of the CNS.

Eph/ephrins The Eph receptor and its ligand, ephrin, function in cell migration, axon guidance, and tissue patterning during development and are upregulated following CNS injury. Often, the interaction of the Eph receptor and ephrin ligand results in cell repulsion. This finding suggests that Eph/ephrin signaling might be inhibitory to axon regrowth at the lesion site, or may be important in cell migration and tissue repair. Recent work has shown a correlation between segregation of reactive astrocytes and meningeal fibroblasts and ephrinB2 and EphB2 expression, respectively, following spinal cord injury in the adult rat. EphrinB2-positive astrocytes and infiltrating meningeal fibroblasts expressing EphB2 undergo a period of intermingling, which is followed by strict segregation during formation of the glial/mesenchymal scar. EphrinB3 has been characterized as a myelin-based inhibitor of axon outgrowth and is present on myelinating oligodendrocytes. Additionally, adult corticospinal neurons express EphA4, the receptor for ephrinB3, suggesting that these axons may be inhibited by Eph/ephrin signaling in the lesion environment.

Slits Slit proteins also play important roles in cell migration and axon guidance during development and have been implicated in regeneration failure because they are upregulated after injury. The slit proteins are ligands of glycipan-1 and the Robo receptor. Cortical astrocytes express both slit and glycipan-1 mRNA following cortical injury. Robo-1 mRNA is also present on macrophages and fibroblasts following injury, which further implies that slit-mediated repulsion could be playing a role in the inhibitory environment of the lesion. While ChABC treatment removes much of the inhibition in the glial scar due to CSPGs, additional strategies must be developed to overcome inhibition resulting from other molecules present in the lesion environment.

Adult Neurons Lack a Robust, Intrinsic Regrowth Potential

Using ChABC and other treatments to remove the extrinsic inhibitory cues of the glial scar may not be sufficient to allow long-distance regeneration because adult neurons also lack a strong intrinsic ability to regenerate following a CNS lesion. Multiple strategies have been used *in vitro* and *in vivo* to successfully amplify the growth potential of adult neurons: application of neurotrophic factors, elevation of intracellular cyclic adenosine monophosphate (cAMP) levels, and utilization of a conditioning lesion.

Neurotrophins/Growth Factors Can Increase Axonal Outgrowth and Overcome the Glial Scar

Treatment of transected dorsal column axons with nerve growth factor (NGF) or neurotrophin-3 (NT-3) increased their growth into and out of a peripheral nerve bridge as well as through the glial scar and back into host tissue. Administration of NGF into a dorsal column crush lesion resulted in increased sprouting into the lesion. Using either intrathecal or adenoviral application of NT-3 or NGF to the injured promoted regeneration of DRG axons through the DREZ and into the spinal cord, restoring nociceptive function. It is critical to understand, however, that treatment with neurotrophins can cause regenerating axons to overcome inhibition by CSPGs and other components of the glial scar, whereas ChABC treatment simply removes the source of inhibition. The increase in the capacity for outgrowth of adult axons observed following neurotrophin application is most likely due to increased expression of growth-enhancing genes in the damaged neuron.

Conditioning Lesions/cAMP Can Also Increase Axonal Outgrowth

Conditioning lesions also enhance the capability of adult neurons to regenerate and this is due at least in part to an elevated cAMP concentration within the cell body. This effect can be mimicked in adult sensory neurons by injecting cAMP into the DRG. cAMP is known to act through protein kinase A (PKA) to influence cellular sprouting and outgrowth, and it may be possible to take advantage of PKA-mediated pathways in order to overcome inhibition by the glial scar and myelin. Inhibition of the Rho GTPase increases neurite outgrowth on proteoglycan- and myelin-containing substrates, and the inhibitory activity of CSPGs within the glial scar is mediated though the Rho/ROCK signaling pathway.

Inflammation: Good or Bad?

Inflammation, although intimately linked with scar formation, may also have a positive effect on regeneration by stimulating an intrinsic growth response in neurons. Using Zymosan A (a purified yeast cell wall extract) to activate macrophages outside of the CNS compartment via intravitreal injection promotes regrowth of RGCs into and slightly past an optic nerve crush. Injection of Zymosan A into the DRG *in vivo* enhances the ability of those neurons to extend neurites on a smooth gradient of aggrecan and laminin, as well as through the DREZ and into the spinal cord following a crush injury to the dorsal root. One mediator of the Zymosan A effect is mannose, a monosaccharide. Surprisingly, treatment of neurons with mannose alone can increase their capacity to regenerate, which may be due to an increase in the overall energy reserves of the neurons. Another effector of increased axon growth, a protein secreted by macrophages, has recently been found. Thus, the combination of mannose, macrophage-derived factor, and cAMP should act synergistically to stimulate axon growth.

Combination Treatments Stimulate Regeneration Best

Recent work has combined stimulation of the intrinsic ability of adult neurons to regenerate using neurotrophins or growth factors with glial scar remodeling to decrease the inhibitory nature of the CSPGs within the glial scar. Retinal lesions, which result in denervation of the superior collicus, cause some minimal sprouting of uninjured RGC axons; however, this sprouting is not robust enough to provide functional recovery. Combining ChABC treatment of the superior colliculus with BDNF stimulation of the RGC cell bodies promoted an increase in sprouting greater than either treatment alone. Some of these sprouting fibers displayed synaptic markers indicating a real potential for some functional recovery. Perhaps the therapeutic value of ChABC treatment will be an increase in functional recovery through an enhanced sprouting response.

Conclusion

The physical and chemical barrier of the glial scar is a formidable opponent for axons attempting to regenerate. Reactive astrocytes, macrophages, fibroblasts, and a host of ECM components including inhibitory CSPGs all play a part in restricting CNS axon regeneration. Strategies to promote regeneration have had partial success, mostly resulting in sprouting of undamaged axons and limited regrowth of damaged axons. In order to overcome the inhibitory environment of the glial scar, a multipartite strategy will most likely be required that (1) removes extrinsic inhibitory factors present within the scar, (2) provides a growth-promoting bridge across or around the scar, and (3) stimulates the neurons to increase their intrinsic capacity for growth, which might allow for long-distance regeneration and functional recovery.

See also: Axonal Regeneration: Role of Growth and Guidance Cues; Axonal Pathfinding: Extracellular Matrix Role; Extracellular Matrix Molecules: Synaptic Plasticity and Learning; Glial Responses to Injury; Glial Cells: Microglia During Normal Brain Aging; Glial Cells: Astrocytes and Oligodendrocytes During Normal Brain Aging; Nogo-A: Its Role in Axon Regeneration; Peripheral Nerve Regeneration: An Overview.

Further Reading

Bareyre FM, Kerschensteiner M, Raineteau O, et al. (2004) The injured spinal cord spontaneously forms a new intraspinal circuit in adult rats. *Nature Neuroscience* 7(3): 269–277.

Berardi N, Pizzorusso T, and Maffei L (2004) Extracellular matrix and visual cortical plasticity: Freeing the synapse. *Neuron* 44(6): 905–908.

Bray GM, Villegas-Perez MP, Vidal-Sanz M, and Aguayo AJ (1987) The use of peripheral nerve grafts to enhance neuronal survival, promote growth and permit terminal reconnections in the central nervous system of adult rats. *Journal of Experimental Biology* 132: 5–19.

Bush SA and Silver J (2007) The role of extracellular matrix in CNS regeneration. *Current Opinion in Neurobiology* 17(1): 120–127.

Corvetti L and Rossi F (2005) Degradation of chondroitin sulfate proteoglycans induces sprouting of intact Purkinje axons in the cerebellum of the adult rat. *Journal of Neuroscience* 25(31): 7150–7158.

Ellezam B, Dubreuil C, Winton M, et al. (2002) Inactivation of intracellular Rho to stimulate axon growth and regeneration. *Progress in Brain Research* 137: 371–380.

Filbin MT (2003) Myelin-associated inhibitors of axonal regeneration in the adult mammalian CNS. *Nature Reviews Neuroscience* 4(12): 703–713.

Fu SY and Gordon T (1997) The cellular and molecular basis of peripheral nerve regeneration. *Molecular Neurobiology* 14 (1–2): 67–116.

Goldshmit Y, Galea MP, Wise G, Bartlett PF, and Turnley AM (2004) Axonal regeneration and lack of astrocytic gliosis in EphA4-deficient mice. *Journal of Neuroscience* 24(45): 10064–10073.

Jone LL, Oudega M, Bunge MB, and Tuszynski MH (2001) Neurotrophic factors, cellular bridges, and gene therapy for spinal cord injury. *Journal of Physiology* 533(Pt. 1): 83–89.

Ramón y Cajal S (1928) *Degeneration and Regeneration of the Nervous System*, May RM (trans.). London: Oxford University Press.

Reier PJ and Houle JD (1988) The glial scar: Its bearing on axonal elongation and transplantation approaches to CNS repair. *Advances in Neurology* 47: 87–138.

Rhodes KE and Fawcett JW (2004) Chondroitin sulphate proteoglycans: Preventing plasticity or protecting the CNS? *Journal of Anatomy* 204: 33–48.

Sandvig A, Berry M, Barrett LB, Butt A, and Logan A (2004) Myelin-, reactive glia-, and scar-derived CNS axon growth inhibitors: Expression, receptor signaling, and correlation with axon regeneration. *Glia* 46(3): 225–251.

Schwab JM, Failli V, and Chédotal A (2005) Injury-related dynamic myelin/oligodendrocytes axon-outgrowth inhibition in the central nervous system. *Lancet* 365: 2055–2057.

Silver J and Miller JH (2004) Regeneration beyond the glial scar. *Nature Reviews Neuroscience* 5(2): 146–156.

Steinmetz MP, Horn KP, Tom VJ, et al. (2005) Chronic enhancement of the intrinsic growth capacity of sensory neurons combined with the degradation of inhibitory proteoglycans allows functional regeneration of sensory axons through the dorsal root entry zone in the mammalian spinal cord. *Journal of Neuroscience* 25(35): 8066–8076.

Tom VJ, Steinmetz MP, Miller JH, Doller CM, and Silver J (2004) Studies on the development and behavior of the dystrophic growth cone, the hallmark of regeneration failure, in an *in vitro* model of the glial scar and after spinal cord injury. *Journal of Neuroscience* 24(29): 6531–6539.

Willson CA, Miranda JD, Foster RD, Onifer SM, and Wittemore SR (2003) Transection of the adult rat spinal cord upregulates EphB3 receptor and ligand expression. *Cell Transplantation* 12(3): 279–290.

Yin Y, Cui Q, Li Y, et al. (2003) Macrophage-derived factors stimulate optic nerve regeneration. *Journal of Neuroscience* 23(6): 2284–2293.

Axonal Transport and ALS

E L F Holzbaur, University of Pennsylvania School of Medicine, Philadelphia, PA, USA

Introduction

Motor neuron diseases involve the selective degeneration and death of motor neurons, leading to muscle atrophy. While some forms of motor neuron disease show an early onset consistent with developmental defects, the most prevalent forms are later in onset and are characterized by inexorable and fatal neurodegeneration.

The cellular and molecular mechanisms leading to the specific degeneration and death of motor neurons have been extensively examined. While many hypotheses have been proposed, there is a particular focus on the unique size and shape of motor neurons, which are huge cells with extended processes. Due to their morphology, motor neurons are uniquely dependent on active intracellular transport. Therefore, specific defects in transport, particularly along axons that may extend up to a meter in length, may lead to degeneration and cell death. Here, we focus on the hypothesis that defects in axonal transport may be a critical component of the neuronal degeneration observed in motor neuron diseases such as amyotrophic lateral sclerosis.

Amyotrophic Lateral Sclerosis

Amyotrophic lateral sclerosis (ALS), also known as Lou Gehrig's disease, has a typical age of onset of 50–60 years, usually leading to death within 5 years due to widespread loss of motor neurons and subsequent muscle atrophy. Currently, there are no available therapies that significantly affect the course of the disease.

The worldwide prevalence of ALS is ~5 in 100 000. Most of these cases (~90%) are sporadic, with no known etiology. Proposed triggers for the disease include environmental influences, such as viral infection or exposure to toxins, or immune dysfunction. However, there are not yet sufficient epidemiological data to support any specific cause. Significantly, the course of disease in sporadic cases is indistinguishable from that observed in the majority of familial cases, consistent with a common pathway of degeneration.

Familial ALS

About 10% of ALS cases are familial, and the identification of the genetic defects involved has provided significant insights into the pathophysiology of the disease, while also raising many further questions. The most common gene linked to ALS encodes Cu/Zn superoxide dismutase 1 (SOD1); *SOD1* is a ubiquitously expressed gene involved in the conversion of superoxide. More than 100 different mutations within the *SOD1* sequence have been identified in patients with familial ALS. These mutations are not limited to the active site of the enzyme; instead they are found throughout much of the coding sequence in *SOD1*. Some but not all of the mutations inactivate the enzyme. As loss of *SOD1* expression does not cause disease in knockout mice, most of the known mutations in *SOD1* are thought to act via a toxic gain-of-function mechanism involving alterations in enzyme folding, stability, or aggregation rather than changes in enzyme activity.

Alternative genetic defects have also been identified in patients with either typical or variant forms of familial ALS (**Table 1**). The affected proteins include alsin, the product of the *ALS2* gene, senataxin, the product of the *SETX* ('amyotrophic lateral sclerosis 4′) gene, and vesicle-associated membrane protein (VAMP)-associated protein B (VAPB), the product of the *VAPB* ('amyotrophic lateral sclerosis 8′) gene. Two of these proteins, alsin and VAPB, are implicated in vesicle trafficking pathways in the cell. A less severe form of inherited motor neuron disease is due to a mutation in the *DCTN1* gene encoding the $p150^{Glued}$ subunit of dynactin, which is required for transport within the neuron. While neurofilaments may play a role in the pathogenesis of ALS, mutations in neurofilament genes are not a significant cause of inherited ALS.

Animal Models of ALS

The identification of mutations in the *SOD1* gene in familial ALS patients led quickly to the development of transgenic animal models for the disease. Analysis of knockout mice demonstrated that loss of *SOD1* gene expression is not sufficient to induce disease. Instead, overexpression of mutant human *SOD1* results in a mouse model with degenerative disease that mirrors the human disease in several aspects.

The first, and most widely studied model involves the overexpression of the human *SOD1* gene with a G93A mutation. This model has a fully penetrant phenotype with disease onset 3–4 months after birth; affected animals die within 4–5 months. Pathological analysis indicates that motor neurons are primarily affected, and the degeneration and death of motor neurons leads to muscle atrophy. A number of other *SOD1* mutations identified in patients with familial

Table 1 ALS-linked genes involved in transport and trafficking

Protein	*Gene*	*Position*	*Onset*	*Clinical presentation*
SOD1	*SOD1* (*ALS1*)	21q22	Adult	Amyotrophic lateral sclerosis
Alsin	*ALS2*	2q33	Juvenile	Spastic paraparesis; slowly progressive
VAPB	*VAPB* (*ALS8*)	20q13	Adult	Amyotrophic lateral sclerosis
p150^{Glued}	*DCTN1*	2p13	Adult	Vocal cord paralysis; slowly progressive weakness of hands, limbs

ALS have also been expressed in mouse models; these models vary in age of onset and severity of disease but show a common phenotype of motor neuron disease. Rat models have also been developed, and display pathology very similar to that seen in the mouse.

Mouse models of other genetic defects linked to ALS have also been developed. For example, knockout of the alsin (*ALS2*) gene associated with juvenile ALS have been generated, which demonstrate only subtle neuropathological defects. At the cellular level, defects in endosomal trafficking are apparent, but these defects are not sufficient to induce significant motor neuron loss or muscle atrophy. Mouse models with mutations or targeted defects in dynein or dynactin function are described in the following sections.

Mechanisms of Cell Death

Initial studies of both mouse models and patient autopsy material revealed two important observations: axonal degeneration of motor neurons and formation of protein aggregates. Based on these observations, and the relative specificity of the disease for motor neurons, a hypothesis was developed that axonal transport was blocked, potentially by the formation of aggregates of mislocalized or mutant protein, and that this in turn leads to the degeneration of the axon, followed by cell death.

Axonal transport The large size and extended processes of motor neurons make them particularly dependent on axonal transport (**Figure 1**). Diffusion alone is not sufficient to maintain cells of this size. Long-distance transport within the cell is dependent on microtubules, which form a polarized network extending out from the microtubule-organizing center (MTOC), positioned near the nucleus within the cell body. Transport outward from the cell body to the cell periphery is driven by members of the kinesin superfamily of molecular motors (**Figure 1**). This anterograde transport is required to supply the distal axon and neuromuscular junction with newly synthesized proteins and lipids, including the replenishment of synaptic vesicle components. Multiple kinesins drive anterograde transport, with some functional specification of motor for cargo, but also some redundancy in the system.

Retrograde transport, from cell periphery to cell center, is driven by the molecular motor, cytoplasmic dynein, and its activator, dynactin (**Figure 1**). In contrast to the extensive nature of the kinesin superfamily, retrograde transport is driven by a single major motor, cytoplasmic dynein. Retrograde transport is required to return damaged and misfolded proteins to the cell center for efficient degradation. Retrograde transport is also critical in signaling pathways, such as neurotrophic factor signaling and injury response pathways. Both the motor domain of dynein and several of the subunits of dynactin are encoded by single genes. The resulting lack of redundancy suggests that mutations in genes encoding subunits of either dynein or dynactin would have a deleterious effect on the neuron, and this prediction is consistent with observations to date.

Active transport is essential for a cell that can extend up to a meter, and thus axonal transport may be an 'Achilles heel' for the motor neuron. Therefore, it is a reasonable hypothesis that defects in axonal transport, either caused directly by mutations in motor proteins or indirectly due, for example, to the accumulation of toxic protein aggregates which might impede efficient transport, could be sufficient to induce distal axonal degeneration. Further, the key role of retrograde transport in pathways such as neurotrophic factor signaling suggests that inhibition of transport may lead more directly to cell death through the perturbation of signal transduction pathways and consequently the activation of apoptosis.

Alternative models for motor neuron death While defects in transport processes provide a plausible model for motor neuron degeneration, and are well supported by data from both genetic and cellular studies, this aspect of motor neuron biology may represent just one piece of the puzzle. Alternative models to explain motor neuron-specific cell death in ALS have also been proposed, including the formation of toxic protein aggregates, excitotoxicity, the destabilization of neuromuscular junctions, oxidative damage due to either mutant *SOD1* or dysfunctional

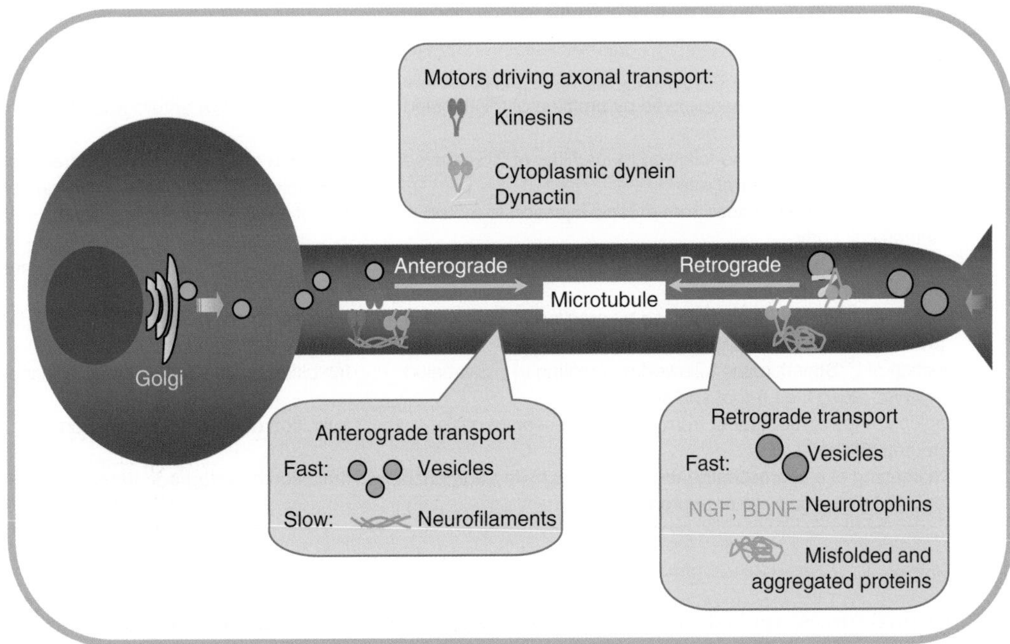

Figure 1 Axonal transport in the motor neuron is driven by microtubule motor proteins. Kinesins drive fast anterograde transport of vesicles carrying newly synthesized proteins and lipids from the Golgi outward along the axon toward the neuromuscular junction. Cytoplasmic dynein and its activator dynactin drive retrograde transport of vesicles, neurotrophins (NGF, nerve growth factor; BDNF, brain-derived neurotrophic factor), and misfolded and aggregated proteins from the cell periphery back to the cell body. Slow anterograde transport of cytoskeletal polymers such as neurofilaments involves the combined activities of both kinesin and cytoplasmic dynein.

mitochondria, and inflammatory processes sparked by activated microglia.

It is critical to note that these pathological processes are not mutually exclusive. It is actually most likely that some initial trigger, such as an accumulation of toxic protein aggregates or the loss of adequate trophic factor support, can lead to activation of cellular stress pathways, which in turn can lead to the triggering of caspase activity, leading to cell death. There is also likely to be a parallel activation of an inflammatory response, which in turn may lead to more widespread death of motor neurons. This model is consistent with recent work demonstrating that while motor neurons are the primary site of disease onset, non-cell-autonomous mechanisms contribute to disease progression.

Evidence for Defective Transport in Mouse Models of ALS

Neurofilament aggregation One of the first indications that defects in axonal transport might be linked to ALS came from the observation of neurofilament aggregates along the axons of motor neurons from affected patients; similar aggregates are observed in animal models. To test for a link between neurofilament aggregation and defects in axonal transport, transgenic mice were generated to overexpress the neurofilament subunit NF-H. Analysis of these mice demonstrated an inhibition in the transport of newly synthesized neurofilaments from cell body outward to the distal axon (**Table 2**). The transport of other cytoskeletal proteins, such as tubulin and actin, was also found to be inhibited.

Slow axonal transport Neurofilaments and other cytoskeletal polymers are transported down the axon at a rate of 0.2–8 mm day^{-1}, in a process known as 'slow' axonal transport. This transport is orders of magnitude slower than the transport of vesicular cargos in 'fast' axonal transport, at rates of ~200–400 mm day^{-1}. For a long time, the mechanisms driving the slow anterograde transport of neurofilaments were unclear. However, recent progress has shown that slow transport along the axon involves the same microtubule motor proteins that drive fast transport. The key difference is that for cargo such as neurofilaments that undergo slow transport, rapid excursions in either the anterograde or retrograde direction are punctuated by prolonged pauses. Both the frequent changes in direction and the extended pauses contribute to an overall net slow outward movement.

The reasons why cytoskeletal polymers are transported in such an apparently inefficient mechanism remain to be determined, but one possibility is that

Table 2 Defective axonal transport in mouse models of motor neuron disease

Model	*Experimental approach*	*Observed defect*
$Tg^{NF\text{-}H}$	Injection of [^{35}S]methionine followed by profiling of ^{35}S-labeled proteins along the sciatic nerve	Inhibition of slow anterograde transport
$SOD1^{G93A}$	Injection of [^{35}S]methionine followed by profiling of ^{35}S-labeled proteins along the ventral root	Inhibition of slow anterograde transport, selective inhibition of fast anterograde transport
	Tracking of retrograde tracer from muscle injection to motor neuron cell body	Inhibition of retrograde transport
	Electron microscopy of axon hillock and initial segment of anterior horn cells	Aberrant accumulation of neurofilaments and mitochondria
$SOD1^{G37R}$	Injection of [^{35}S]methionine followed by profiling of ^{35}S-labeled proteins along the L5 root and sciatic nerve	Inhibition of slow anterograde transport
$SOD1^{G85R}$	Injection of [^{35}S]methionine followed by profiling of ^{35}S-labeled proteins along the L5 root and sciatic nerve	Inhibition of slow anterograde transport
$Tg^{dynamitin}$	Tracking of retrograde tracer from muscle injection to motor neuron cell body	Inhibition of retrograde transport
Loa	Live imaging of a fluorescently labeled tetanus toxin fragment in cultured embryonic motor neurons	Inhibition of retrograde transport

the oppositely oriented forces, applied simultaneously by plus and minus end-directed motors, result in a continuous tension on the cargo. In the case of neurofilaments, this tension may keep the long, flexible filaments stretched out along microtubules, and therefore may counteract aggregation. Support for this hypothesis comes from the observation that knockdown of dynein expression or inhibition of dynein/dynactin function leads to the distal accumulation of neurofilaments along neuronal processes.

Axonal transport defects in mSOD1 mice – the slowing of slow transport Direct assays for anterograde transport have been performed in mouse models expressing mutant SOD1 (**Table 2**). Injection of [^{35}S] methionine into the spinal cord of either $SOD1^{G93A}$ or $SOD1^{G37R}$ mice, followed by the profiling of ^{35}S-labeled proteins along the axon, allows the analysis of transport rates of newly synthesized proteins in comparison to those observed in wild-type control mice. In these studies, fast anterograde transport was not generally inhibited, although there may have been some preferential inhibition of specific cargos. However, inhibition of slow axonal transport was observed, including delays in the transport of both neurofilaments and tubulin. This inhibition of transport occurred in both models prior to disease onset.

Defects in retrograde transport in mSOD1 mice Retrograde transport has also been investigated in mice expressing mutant SOD1. In studies in which a retrograde neurotracer was injected into muscle and transport to the cell bodies of motor neurons was assayed, significant inhibition of transport was observed in transgenic mice expressing $SOD1^{G93A}$ as compared to wild-type littermate controls (**Table 2**). The observed inhibition was at a very early point in the course of disease; inhibition was significant by 50 days after birth, well prior to the onset of clinical disease in this model. This observation suggests that disruption of retrograde transport is a very early event in the course of motor neuron degeneration and death.

Studies on axonal transport in embryonic motor neurons cultured from the $SOD1^{G93A}$ mouse also show a significant slowing of retrograde transport (**Table 2**). Live cell analysis of the retrograde motility of a fluorescently labeled fragment of tetanus toxin shows a shift in the speed profile, indicative of an increased frequency of pauses and oscillatory movements, as compared to the more efficient transport observed in motor neurons from wild-type mice.

The only other marker that shows such an early defect in the $SOD1^{G93A}$ model is the disruption of the Golgi apparatus, which has been observed by 30 days after birth. Disruption of the Golgi is also observed in patient tissue. This is potentially of interest, since in addition to its role in retrograde axonal transport, cytoplasmic dynein drives endoplasmic reticulum (ER)-to-Golgi trafficking in the cell body and is required to maintain an intact Golgi apparatus near the MTOC of the cell. Inhibition of dynein leads to the disruption of the Golgi apparatus.

Thus, both the observed slowing of retrograde transport in the $SOD1^{G93A}$ model and the disruption of the Golgi observed in ALS patients and in a mouse model are consistent with a functional disruption in the activity of cytoplasmic dynein within the neuron. The inhibition of slow anterograde axonal transport observed in mice expressing mutant *SOD1* may also be consistent with a defect in dynein function, as dynein interacts directly with neurofilaments as they move along the axon.

Yet to be determined is the mechanism by which the expression of mutant *SOD1* can disrupt dynein function. One possibility is that the disruption is indirect, due to the activation of stress response pathways by the accumulation of misfolded SOD1 protein. Alternatively, the mechanism may be more direct. Dynein co-localizes with SOD1 aggregates in neurons expressing mutant *SOD1*; these aggregates may deplete the pool of active motors along the axon. Further work will be required to address this question.

Defective Retrograde Transport Is Sufficient to Cause Motor Neuron Degeneration

The dynamitin transgenic mouse In order to test the hypothesis that disruption of retrograde axonal transport is sufficient to induce motor neuron degeneration, a transgenic mouse was developed with a targeted inhibition of dynein/dynactin function in motor neurons. The dynactin subunit dynamitin was expressed at high levels in motor neurons; overexpression of this subunit is sufficient to disrupt the integrity of the dynactin complex and thus effectively disrupts dynein-mediated transport. The resulting $Tg^{dynamitin}$ line exhibits late-onset, slowly progressive degeneration of motor neurons, leading to preferential loss of large-caliber motor neurons and subsequent muscle atrophy.

Loa and Cra1 mutant mice Further evidence linking defects in axonal transport to the development of motor neuron disease comes from the analysis of two lines of *N*-ethyl-*N*-nitrosourea (ENU)-mutated mice, Loa and Cra1. Both lines were found to have point mutations in the gene encoding the heavy chain of cytoplasmic dynein, which includes the motor domain of the enzyme. While neither of the point mutations maps to the motor domain of dynein, both are in domains involved in subunit interactions within the dynein holoenzyme. Therefore, the two mutations are not predicted to abrogate dynein function, but are more likely to make the motor either less stable or less effective in driving motility along the axon.

Both Loa and Cra1 heterozygous mice display phenotypes remarkably similar to the phenotype of the $Tg^{dynamitin}$ mouse. Both lines exhibit late-onset, slowly progressive degeneration of motor neurons. Importantly, while the mutations are ubiquitously expressed, the degeneration is limited to motor neurons in heterozygous mice. Consistent with an impairment but not a block in dynein function, neither mutation is lethal when heterozygous. When homozygous, both mutations are lethal within a day of birth, which may be due to an impaired ability of the pups to suckle resulting from abnormal development of facial motor neurons. However, there are no gross perturbations of the nervous system, suggesting that the mutant dynein, while functionally impaired, is sufficient to drive normal development. In contrast, a complete knockout of dynein heavy chain is lethal early in embryogenesis in the mouse.

The effects of the Loa mutation on intracellular transport have been investigated. Transport of a retrograde tracer, a fragment of tetanus toxin, was not significantly affected in embryonic motor neurons cultured from heterozygous mice, suggesting that either the impairment is subtle, or that the wild type gene is haplosufficient over this time scale. In neurons cultured from Loa/Loa mice, there was a decrease in the extent of high-speed transport of the retrograde tracer, and a marked increase in the frequency of pauses. These data indicate a significant defect in the efficiency of retrograde transport (**Table 2**).

A somewhat unexpected but potentially important result has come from studies examining the effects of crossing either the Loa or Cra1 mutant mice to transgenic mice overexpressing mutant *SOD1* (the $SOD1^{G93A}$ line). Rather than showing an additive or even synergistic effect on motor neuron cell death, mice expressing both the $SOD1^{G93A}$ mutant and either the Loa or Cra1 mutation showed an extension of life span (by 28% and 12%, respectively). Disease onset was also delayed. The mechanism behind this result is not yet understood. One interesting possibility is that the impairment of dynein function induced by either the Loa or Cra1 mutation delays the transmittal of a signal indicative of cellular stress, thus slowing the onset of cell death pathways.

Alternatively, it has been proposed that the expression of mutant dynein may rescue the axonal transport defect caused by the expression of mutant *SOD1*, perhaps by righting the balance between anterograde and retrograde transport and therefore restoring axonal homeostasis. Numerous studies have suggested that anterograde and retrograde transport along the axon are intimately linked, as mutations or inhibitory reagents that should selectively affect transport in a single direction instead have a bidirectional inhibitory effect. Mechanisms of motor coordination may involve either direct links between motors or between motors and adapter proteins such as dynactin, or more indirect links through coordinated regulation of motor protein activity.

Mutant Dynactin in Motor Neuron Disease

Clear demonstration of a link between the axonal transport machinery and motor neuron disease in humans came from the identification of a mutation in the gene encoding the $p150^{Glued}$ subunit of dynactin as the cause of an inherited form of motor neuron disease.

Patients with a G59S mutation in the *DCTN1* gene display an autosomal dominant, slowly progressive form of inherited motor neuron disease. Onset in early adulthood is marked by vocal fold paralysis, leading to breathing difficulties, as well as progressive weakness of the muscles in the face and hands, with distal limb atrophy developing later. Unlike typical ALS, this mutation is not uniformly fatal.

Dynactin is a required activator for most dynein functions in the cell. Dynactin enhances the processivity of dynein, meaning that the motor complex takes longer runs along the microtubule per binding encounter. Dynactin also links dynein to some intracellular cargos. The p150^{Glued} mutation identified in affected patients is a G59S substitution in a critical domain of the polypeptide, known as the CAP-Gly domain, which binds directly to microtubules. The mutation affects the ability of the polypeptide to bind to microtubules, but also increases the propensity of the polypeptide to misfold and to aggregate. Thus, at the cellular level, the G59S mutation both inhibits dynein function and results in the formation of intracellular protein aggregates. Together, these detrimental effects lead to the specific death of motor neurons. Pathological protein aggregates labeled with antibodies to both dynein and dynactin have been observed in motor neurons from affected patients, consistent with this hypothesis.

More recently, four additional mutations have been identified in the coding region of the *DCTN1* gene in patients with ALS. In all cases reported to date, there are insufficient genetic data to determine if these mutations are causative, contributory, or merely allelic variants. Further analysis will be required to determine how significant dynactin mutations are to the development of full-blown ALS.

Why Are Motor Neurons So Susceptible to Defects in Axonal Transport?

Consistently, studies on mouse models of motor neuron disease demonstrate defects in retrograde transport as well as enhanced protein aggregation. Defects in anterograde motors can also lead to neurodegenerative disease. For example, one form of Charcot–Marie–Tooth disease is caused by a point mutation in a kinesin motor protein. However, motor neurons appear to be uniquely susceptible to defects in retrograde axonal transport.

Three different mechanisms may explain these observations. The first is that the health of a motor neuron depends on a continual interaction with its environment. For example, a continuous supply of neurotrophic factors may be required. Neurotrophins, secreted by target tissues, bind to receptors on the motor neuron and are then transported from the synapse back to the cell body to activate signal transduction pathways and modulate gene expression patterns. Inhibition of retrograde transport would block this continuous signaling and thus could deleteriously affect the cell. None of the models studied to date show a complete block in retrograde transport, but instead show a partial inhibition. Slowed or inefficient neurotrophin signaling over time may lead to the late-onset and progressive cell death seen in ALS patients, as well as to patients expressing mutant dynactin.

Alternatively, cell death may result from the defective clearance of misfolded or aggregated proteins. Protein aggregation is a common feature observed in both patients and model systems. Further, dynein has been implicated as a key player in the clearance of misfolded proteins through autophagy. Mutations in dynein/dynactin or alterations in the function of the complex may lead to accumulation of toxic protein aggregates along the axon, which would further impede transport. This could represent a generalized defect in clearance of misfolded proteins or a more specialized defect – for example, in the clearance of mutant SOD1 protein or of aggregated neurofilaments. This mechanism has the potential to amplify the initial defect; dynein co-localizes with aggregates of mutant SOD1 protein and this association may deplete the pool of active motor protein. Dynein also interacts with neurofilaments; loss or inhibition of dynein leads to neurofilament aggregation, a common pathological marker of motor neuron disease.

Finally, alterations in axonal transport may be secondary to the real defect. Inhibition of dynein function may lead to more critical insults to the neuron, such as the disruption of dynein-mediated trafficking (e.g., from the ER to the Golgi). Motor neurons may be uniquely sensitive to trafficking defects, given their large size and metabolic load. Or, the key defect could be in the dynein-driven motility of intracellular organelles, such as the mitochondria. Mislocalization of mitochondria may lead to mitochondrial dysfunction, which in turn may be sufficient to initiate cell death pathways. In support of this possibility, mislocalization of mitochondria has been observed in motor neurons of patients with ALS.

Defects in Transport, Trafficking, and Cytoskeletal Proteins Linked to Neurodegenerative Disease

Recent progress in identifying the genetic causes of inherited neurodegenerative diseases has turned up some intriguing results. As already noted, the identification of mutations in motor proteins has provided strong support for the hypothesis that active axonal transport is critical to maintain the health of neurons. Further, mutations in other proteins (such as Huntingtin or amyloid precursor protein) may also affect axonal transport either directly or indirectly. Together, these data suggest that neurons are absolutely dependent on axonal transport, and are uniquely vulnerable to defects in the motors and other cytoskeletal proteins involved in this transport. A better understanding of the underlying cell biology of motor neurons will therefore be required to fully understand the development of pathology in patients with ALS, and to design rational approaches to therapeutic interventions.

See also: Amyotrophic Lateral Sclerosis (ALS): Disease Mechanisms; Amyotrophic Lateral Sclerosis (ALS); Animal Models of Motor and Sensory Neuron Disease; Axonal Injury: Neuronal Responses; Axonal Transport Disorders; Axonal and Dendritic Transport by Dyneins and Kinesins in Neurons; Axonal Transport and Neurodegenerative Diseases; Demyelinating Diseases; Microtubule Associated Proteins in Neurons; Microtubules: Organization and Function in Neurons; Slow Axonal Transport; Vesicular Sorting to Axons and Dendrites.

Further Reading

Boillee S, Vande Velde C, and Cleveland DW (2006) ALS: A disease of motor neurons and their nonneuronal neighbors. *Neuron* 52: 39–59.

Bruijn LI, Miller TM, and Cleveland DW (2004) Unraveling the mechanisms involved in motor neuron degeneration in ALS. *Annual Review of Neuroscience* 27: 723–749.

Collard JF, Cote F, and Julien JP (1995) Defective axonal transport in a transgenic mouse model of amyotrophic lateral sclerosis. *Nature* 375: 61–64.

Hafezparast M, Klocke R, Ruhrberg C, et al. (2003) Mutations in dynein link motor neuron degeneration to defects in retrograde transport. *Science* 300: 808–812.

He Y, Francis F, Myers KA, et al. (2005) Role of cytoplasmic dynein in the axonal transport of microtubules and neurofilaments. *Journal of Cell Biology* 168: 697–703.

Kieran D, Hafezparast M, Bohnert S, et al. (2005) A mutation in dynein rescues axonal transport defects and extends the life span of ALS mice. *Journal of Cell Biology* 169: 561–567.

LaMonte BH, Wallace KE, Holloway BA, et al. (2002) Disruption of dynein/dynactin inhibits axonal transport in motor neurons causing late-onset progressive degeneration. *Neuron* 34: 715–727.

Levy JR, Sumner CJ, Caviston JP, et al. (2006) A motor neuron disease-associated mutation in $p150^{Glued}$ perturbs dynactin function and induces protein aggregation. *Journal of Cell Biology* 172: 733–745.

Ligon LA, LaMonte BH, Wallace KE, et al. (2005) Mutant superoxide dismutase disrupts cytoplasmic dynein in motor neurons. *NeuroReport* 16: 533–536.

Munch C, Sedlmeier R, Meyer T, et al. (2004) Point mutations of the p150 subunit of dynactin (DCTN1) gene in ALS. *Neurology* 63: 724–726.

Pasinelli P and Brown RH (2006) Molecular biology of amyotrophic lateral sclerosis: Insights from genetics. *Nature Reviews Neuroscience* 7: 710–723.

Puls I, Jonnakuty C, LaMonte BH, et al. (2003) Mutant dynactin in motor neuron disease. *Nature Genetics* 33: 455–456.

Puls I, Oh SJ, Sumner CJ, et al. (2005) Distal spinal and bulbar muscular atrophy caused by dynactin mutation. *Annals of Neurology* 57: 687–694.

Sasaki S, Warita H, Abe K, et al. (2005) Impairment of axonal transport in the axon hillock and the initial segment of anterior horn neurons in transgenic mice with a G93A mutant SOD1 gene. *Acta Neuropathology* 110: 48–56.

Williamson TL and Cleveland DW (1999) Slowing of axonal transport is a very early event in the toxicity of ALS-linked SOD1 mutants to motor neurons. *Nature Neuroscience* 2: 50–56.

Zhang B, Tu P, Abtahian F, et al. (1997) Neurofilaments and orthograde transport are reduced in ventral root axons of transgenic mice that express human SOD1 with a G93A mutation. *Journal of Cell Biology* 139: 1307–1315.

Axonal Transport and Alzheimer's Disease

L S B Goldstein, University of California at San Diego, La Jolla, CA, USA

Basic Features of Alzheimer's Disease

The initiating symptoms or presenting symptoms of Alzheimer's disease (AD) are well known. Problems with memory, language, and cognition are all characteristic features of the behavioral symptoms that signal the initiation of this terrible neurodegenerative disorder. Most AD occurs relatively late in life, generally after 65 years of age. It afflicts 10% of people over the age of 65 and 50% of people over the age of 85. The time from diagnosis to fatality is highly variable, ranging anywhere from a few years to over a decade or more. The tragedy of AD is that not only are there substantial deficits in cognition accompanied by neurodegenerative changes in the brain, there are presently no effective drugs and no effective treatments, and the disease progresses inexorably in its course to end stage, in which a persistent vegetative state is common.

The pathological hallmarks in the brains of people who have had AD are well recognized in postmortem autopsy material. These features include amyloid beta (Aβ) plaques and neurofibrillary tangles. Amyloid plaques consist of aggregated Aβ peptides derived by proteolytic processing of the amyloid precursor protein (**Figure 1**). Amyloid plaques are generally associated with dystrophic neurites, which can be abnormal axons or dendrites that are characterized by substantial swelling and distension, and accumulations of vesicles and organelles. Dystrophic neurites surround and invade the plaques, but are also in regions of diseased brains that lack amyloid plaques. Neurofibrillary tangles are aggregates of a protein called tau that normally binds to and stabilizes microtubules, and which, following hyperphosphorylation and conformational change, forms characteristic paired helical filaments. Although amyloid plaques and neurofibrillary tangles are the hallmark diagnostic pathological features, their correlation with cognitive characteristics of AD is not consistent and is controversial. Many researchers have argued that loss of synapses in key areas of the diseased brain is the earliest pathological feature of AD that is most highly correlated with cognitive defects. Neuronal death is an invariant though relatively late feature of AD, as is loss of axons and white matter. The loss of axons, white matter, and synapses could be a consequence of early changes in axonal transport in AD.

Although most AD is late onset, there are well-recognized hereditary forms (familial AD, or FAD) that can begin as early as the fourth decade of life and are highly virulent in their course. Mutations in the gene encoding amyloid precursor protein (APP) and mutations in the genes encoding presenilin (PS) proteins are the major recognized causes of FAD. Duplications of the APP gene can also give rise to mid- to late-onset AD. Two other major risk factors are known. One major genetic risk factor is the allelic state in the gene (*APOE*) encoding apolipoprotein E (ApoE), which has a role in cholesterol trafficking. While the $\varepsilon 2$ variant of the *APOE* gene is protective, the $\varepsilon 4$ variant predisposes to late-onset AD. Another major risk factor is traumatic brain injury. How these risk factors lead to AD is far from clear, and indeed, what molecular defects and pathways lead to any form of AD remain controversial.

The dominant hypothesis proposed to account for the molecular and cellular changes found in AD is the amyloid cascade hypothesis, which proposes that AD is caused by toxic Aβ_{40} or Aβ_{42} peptide fragments liberated by proteolysis of APP (**Figure 2**). In this hypothesis, the hereditary FAD mutations are proposed to generate excess toxic forms of Aβ, and apolipoprotein E variants are proposed to affect the rates of turnover of these peptides. Most important, the amyloid cascade hypothesis and its major variants indicate that the toxic Aβ peptides initiate a biochemical and cellular cascade of pathology leading to synaptic loss, neurofibrillary tangles, and ultimately neuronal death. While there is considerable evidence for the amyloid cascade hypothesis, principally in the human APP FAD mutations, questions remain, and there are a number of alternative hypotheses that have been proposed over the years. Another class of hypothesis suggests that early deficits in axonal transport either cause, or contribute substantively to, the development of AD.

Axonal Transport Is Disrupted in AD

There has been considerable evidence for some time that axonal transport is disrupted at late stages of AD. For example, although basal fore-brain cholinergic neurons exhibit phenotypes that suggest that they lack retrograde neurotrophic input from nerve growth factor, diseased brains appear to retain high levels of

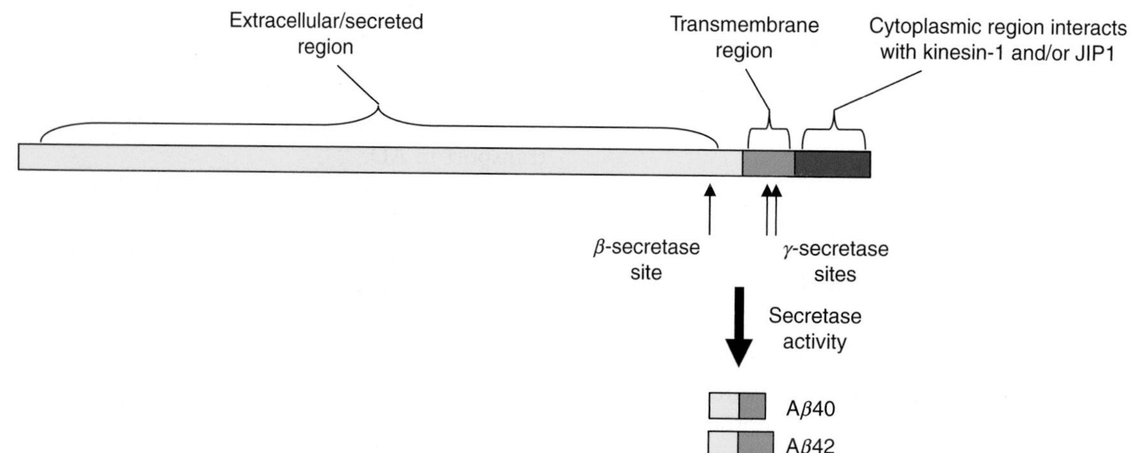

Figure 1 Schematic diagram of amyloid precursor protein. The cytoplasmic region is thought to interact with the axonal transport machinery either directly via binding to the kinesin light chain subunit of kinesin-1, or indirectly via binding to the Jun N-terminal kinase-interacting protein 1, JIP1. Sites for proteolytic activity by β- and γ-secretases are shown, which together liberate amyloid beta (Aβ) peptides that are either 40 or 42 amino acids in length.

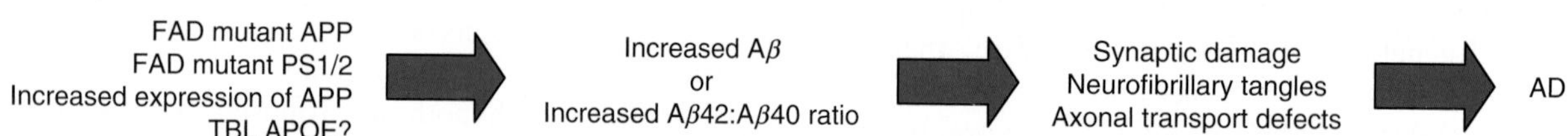

Figure 2 A standard view of the amyloid cascade hypothesis in which the various mutations, genetic variants, and environmental insults induce a series of molecular events caused by increasing levels of toxic human amyloid beta (Aβ_{40} or Aβ_{42}), culminating in Alzheimer's disease (AD). FAD, familial Alzheimer's disease; APP, amyloid precursor protein; PS1/2, presenilins 1 and 2; TBI, traumatic brain injury; APOE, apolipoprotein E.

nerve growth factor in the cortex where basal forebrain cholinergic neurons ordinarily connect to cells that secrete nerve growth factor (**Figure 3**). The levels of cholinergic enzymes in the cortex appear also to be reduced in AD, even though basal forebrain cholinergic neurons appear to express normal levels of cholinergic markers even after the onset of AD.Finally, there are reports that axonal transport is defective in neurons taken from brains of AD patients postmortem, and there are long-standing observations of destruction of the microtubule cytoskeleton in axons in diseased neurons. These disruptions have for some time been thought to be related to the hyperphosphorylation and aggregation of tau protein into neurofibrillary tangles, although what events are causative and what events are consequences of each other remain unclear.

The observation that axonal transport is defective at late stages of AD has led a number for workers over the years to propose that axonal transport defects might be early, possibly causative, events in AD. However, evidence for early defects in axonal transport in AD was lacking until recently. Recent work has provided evidence for early changes in axonal transport in postmortem AD brains and in animal models of AD in which APP is overexpressed. One striking example is the finding of dystrophic axons early in AD and in AD models prior to the formation of amyloid plaques, and thus not associated with, or in the vicinity of, amyloid plaques. These dystrophic axons exhibit signs of axonal transport failure, including massive accumulations of vesicles and organelles in axons that appear to be distended by these accumulations. This type of phenotype is characteristic of mutations that disrupt the axonal transport machinery directly in *Drosophila* and in the mouse. In this context, there is some evidence that deficits in axonal transport can induce formation of, or enhance accumulation of, Aβ. A final line of evidence that raises the possibility that axonal transport deficits are relevant to the initiation of AD is the body of evidence that suggests that there are age-dependent changes and reductions in microtubules in axons. Observed changes include accumulation of APP and other transported material in axons, increases in axonal dystrophies, and some direct evidence that transport itself is reduced in aging neurons relative to younger neurons.

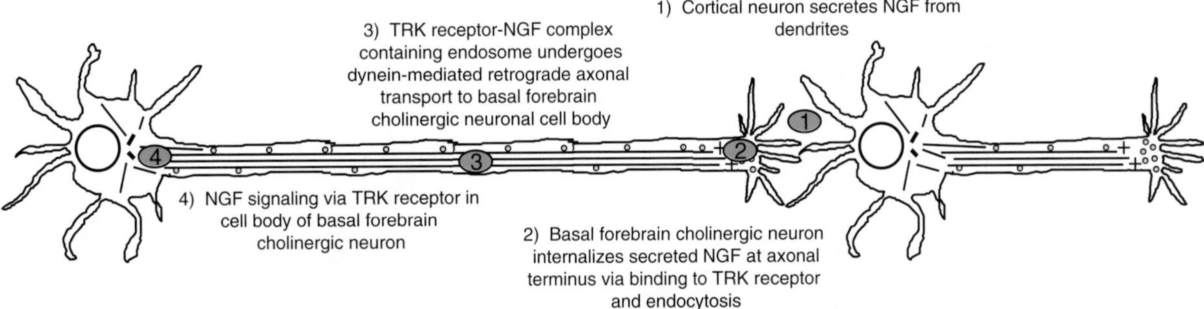

Figure 3 Pathway of nerve growth factor (NGF) signaling from cortical neurons to target cholinergic neurons in the basal forebrain. This transport and signaling pathway appears to be defective in Alzheimer's disease. Secretion of NGF by cortical neurons (1) is followed by internalization at presynaptic endings of basal forebrain cholinergic neurons (2). Retrograde transport of vesicles containing NGF and activated TrkA in postendosomal carriers (3) culminates in arrival at the basal forebrain cholinergic neuron cell body and signaling (4).

Roles and Effects of AD Risk Factors and Genes in Axonal Transport

Several reports suggest that APP and its proteolytic products may play normal roles in, and/or be able to interfere with, axonal transport. For example, deletion of the APP gene in *Drosophila* and in the mouse appears to generate mild defects in axonal transport. Overexpression of APP and its relatives dramatically interferes with axonal transport in *Drosophila* and in the mouse, and even simple duplication of the APP gene in the mouse appears to significantly inhibit some forms of axonal transport – in particular, retrograde transport of the nerve growth factor (NGF)–neurotrophin receptor signaling complex to the basal forebrain. Although this could be the result of, and often has been interpreted as such, an excess of toxic human Aβ, strong evidence in *Drosophila* and in the mouse suggests that even if human Aβ is toxic to axonal transport, other factors, perhaps the C-terminus of the APP protein, may be bigger contributors to these defects.

The mechanism of APP-induced defects in axonal transport remains unknown. One suggestion, based on biochemical evidence from a number of groups, is that the cytoplasmic tail of APP can form either a direct or indirect interaction with the kinesin light chain subunits of conventional kinesin, kinesin-1. If indirect, this interaction may occur via formation of a ternary complex with the Jun kinase scaffolding protein called JIP1 (c-Jun N-terminal kinase-interacting protein 1). There is also evidence from the squid and *Drosophila* that the C-terminus of APP interacts with the axonal transport machinery and is required for interference with axonal transport in overexpression situations. This C-terminus is also required to generate movements of APP in *Drosophila* neurons and in squid axoplasm. These findings are consistent with the proposal that overexpression of APP poisons axonal transport by competing for either JIP1 or kinesin light chain, which are needed for axonal transport of other materials (**Figure 4**). Although several groups have provided evidence in favor of this hypothesis, other groups do not agree. Strong additional evidence in support of this idea comes from the observation that reductions in kinesin-1 subunits in *Drosophila* or in the mouse enhance transport defects caused by overexpression of APP.

There is considerable evidence that FAD mutations and deletion mutations in the gene encoding presenilin 1 may also interfere with axonal transport. One obvious mechanism that is consistent with the amyloid cascade hypothesis is that presenilins might cause axonal transport defects by enhancing or altering Aβ production in neurons, which is then toxic to axonal transport. However, the presenilin mutations that cause familial forms of AD in humans can poison axonal transport of APP, and perhaps other proteins, in the mouse in the absence of human APP. These findings, combined with considerable evidence that kinesin-1 is required for APP transport in axons and with evidence that presenilin and β-secretase and related proteins may be in a common axonal vesicle with APP suggest that these processes may be tied together in a way that does not rely on APP processing to generate Aβ. For example, one possible mechanism of presenilin influence on the axonal transport machinery may be via control of kinesin light chain phosphorylation by glycogen synthase-3β (GSK-3β) or other kinases. A related and surprising finding is that overexpression of β-secretase, which is another protease that may generate Aβ from APP, can generate defects in axonal transport of APP, and may shift the cellular sites of Aβ production and secretion. These ideas have been hotly debated in the literature; further evidence is required to establish or refute these proposals.

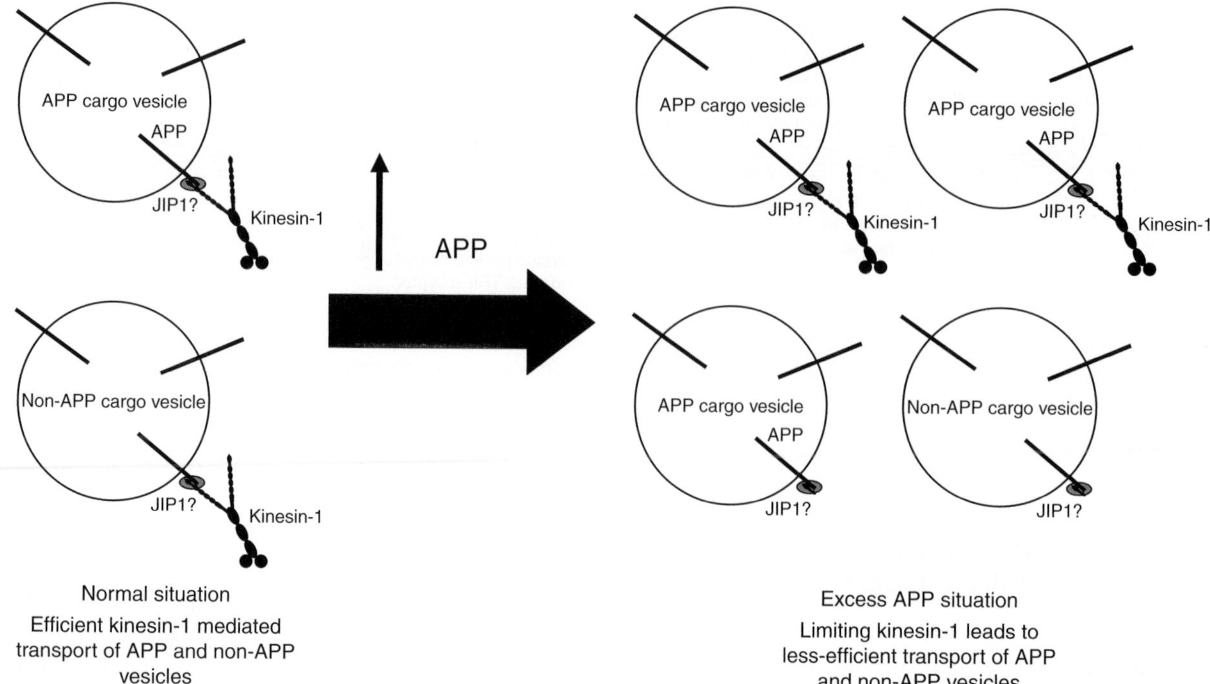

Figure 4 A possible mechanism by which excess amyloid precursor protein (APP) poisons axonal transport. In the normal situation, there is enough membrane-bound kinesin-1 to move APP and non-APP vesicular and organellar cargoes. In the presence of excess APP, competition for limiting kinesin-1 may occur so that some vesicles (intermittently) lack kinesin-1 and can stall in the axon, leading to axonal blockages and transport deficits. Although this schematic shows kinesin-1 as the limiting protein, other proteins or adapters, such as Jun N-terminal kinase-interacting protein 1 (JIP1), could be the vulnerable limiting component.

Another possible link of axonal transport to the pathogenesis of AD is via tau protein. This protein, which aggregates into neurofibrillary tangles, appears to play a normal role in the control of vesicle transport along microtubules, perhaps most prominently for vesicles transported by kinesin-1. There is considerable evidence that tau protein, which binds to microtubules, may directly regulate the movement of vesicles by kinesin-1 along microtubules. This function, in addition to the potential role of tau in initiating the polymerization of microtubules or in enhancing the stability of microtubules, may play important roles in the effects of tau protein and its mutants or of inappropriate phosphorylation variants on the axonal transport machinery. There is also evidence that overexpression of tau protein and its disease-causing mutations poison axonal transport directly.

Another important risk factor for AD is traumatic brain injury. There is evidence that injury to the brain induces so-called diffuse axonal injury, among other consequences. Diffuse axonal injury is of particular interest because it is effectively injury to the axons, which has been reported to result in large accumulations of APP and Aβ peptides at injury sites in axons. The possibility that axonal injuries cause biochemical changes in the axonal transport machinery is likely, given recent evidence that injury to axons induces activation of the damage-signaling kinase, Jun kinase (JNK). JNK is known to phosphorylate tau protein, and has been implicated in some aspects of axonal transport regulation. Thus traumatic brain injury may exert its influence on axonal transport defects via JNK activation and phosphorylation of targets such as tau and perhaps the kinesin heavy chain subunit of kinesin-1.

Another possible risk factor for AD is cholesterol. It is known that genotypic variants at the apolipoprotein E gene in humans have a large influence on the probability of developing late-onset AD. The $\varepsilon 4$ variant confers the highest risk and the $\varepsilon 2$ variant is protective. There is accumulating evidence that cholesterol and phospholipids might have important regulatory effects, not only on vesicular traffic, but also on kinesin-mediated axonal transport. One important kinesin family that is known to be essential for the transport of synaptic vesicle proteins is kinesin-3. This type of kinesin motor can bind to phospholipids in membrane bilayers, and has been reported to be activated by lipid clustering in vesicular membranes via oligomerization of the motor protein. Since cholesterol may have an impact on the ability of phospholipids to cluster, such a mechanism might affect transport directly. Another possible linkage of cholesterol to AD comes from the pediatric disease called Neimann–Pick

C type 1 (NPC1), which is a disorder of cholesterol transport and metabolism caused by defects in the *NPC1* gene. Such mutations appear to interfere with cholesterol trafficking in nonneuronal cells, possibly via the endocytic pathway. It is less clear what NPC1 mutations do in neurons, but it is clear that the mutations induce significant neurodegeneration in patients who have them, and ultimately lead to death. The striking clue is that these mutations appear to generate excess Aβ peptides and to induce the aberrantly phosphorylated forms of tau protein that lead to the formation of neurofibrillary tangles. Further work will be necessary to establish the precise molecular link between cholesterol trafficking and AD. It is intriguing, however, that defects in movement of cholesterol and related components in axons may be an important part of the problem.

Finally, Aβ peptide has been reported by a number of groups to induce phenotypes in cultured neurons that mimic those seen in neurons bearing axonal transport mutations. One group has reported that this may be a result of the effects of Aβ peptides on the state of microtubules; this influence appears to be mediated by extracellular Aβ peptides. Further work is necessary to clarify this association.

Role of Axonal Transport Failure in AD

The key issue is whether axonal transport defects are an early or late event in AD, and whether they are a consequence of Aβ toxicity, or a stimulus for Aβ generation or turnover, or both. Although further work is clearly required to resolve these issues, a testable type of hypothesis, presented in **Figure 5**, encapsulates some key features. Key elements are that axonal transport defects can induce synaptic defects directly, activate kinases that may induce tau hyperphosphorylation, and instigate or enhance Aβ

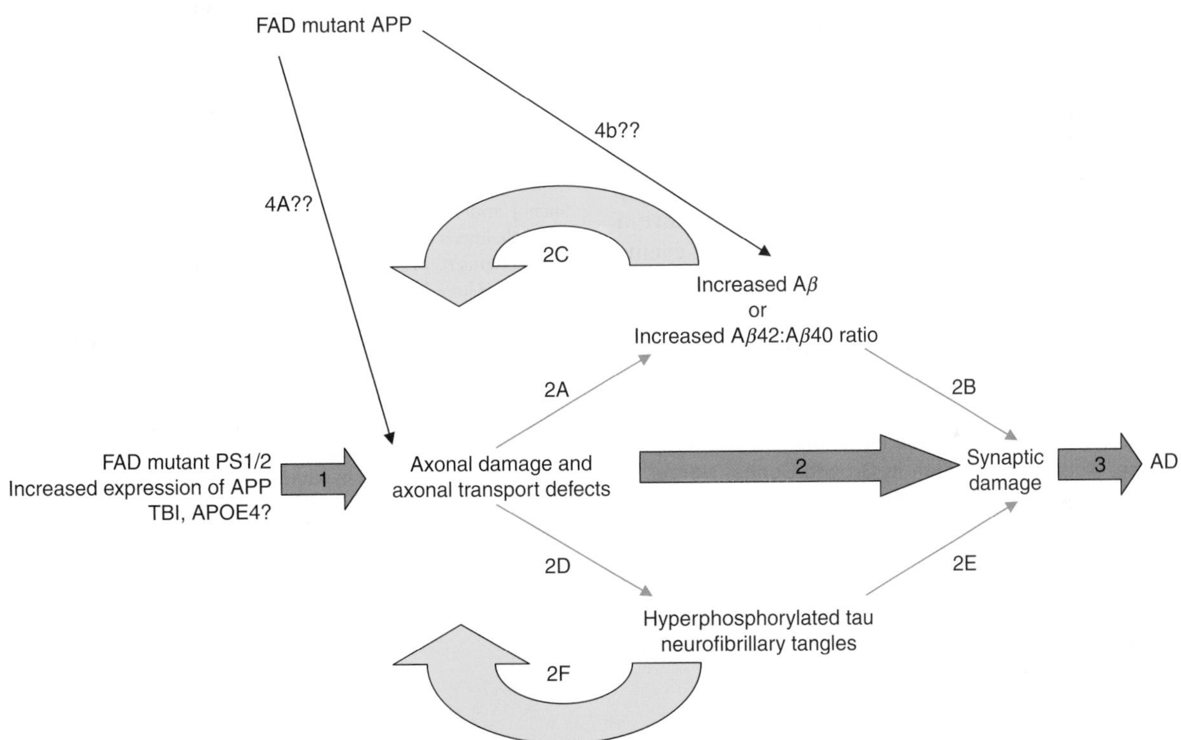

Figure 5 One possible alternative to the amyloid cascade hypothesis that incorporates axonal transport deficits into early stages of Alzheimer's disease (AD) causation or progression. (1) In familial Alzheimer's disease (FAD), mutant presenilins (PS1/2), excess amyloid precursor protein (APP), traumatic brain injury (TBI), and apolipoprotein E ε4 (APOE4) are suggested to initiate neuronal damage by interfering with axonal transport and/or causing axonal damage. (2) Axonal damage and axonal transport deficits may cause synaptic damage directly, or indirectly by the indicated alternative pathways. Axonal damage might lead to increased amyloid beta or increased ratio of the two types (Aβ_{40}, Aβ_{42}) (2A), which may cause synaptic damage directly (2B), or may lead to further axonal damage and transport defects in an autocatalytic loop (2C). Axonal damage may also lead to activation of Jun N-terminal kinase, which could initiate hyperphosphorylation of tau protein and formation of neurofibrillary tangles (2D), which may cause synaptic damage (2E) and cause further defects in axonal transport and axonal damage (2F) in an autocatalytic loop. Synaptic damage caused by all of these insults may be the primary neuronal defect causing cognitive symptoms in AD (3). How FAD mutant APP might initiate these proposed events is unknown, but could happen either by directly interfering with axonal transport or causing axonal damage (4A) or by increasing Aβ formation (4B).

formation, which may have its own toxic consequences. Unique aspects of this class of hypothesis include an autocatalytic spiral of events and multiple entry points, including mutation and environmental insults. In at least one form of hereditary AD, it can be argued strongly that this hypothesis may have validity in view of evidence that moderate increases in APP expression can induce significant axonal transport deficits in the absence of human Aβ. Although these ideas focus on axons, it is likely that comparable phenomena and behaviors exist in dendrites, for which less information is currently available.

Conclusion and Perspectives

The fundamental problem in evaluating whether defects in axonal transport can induce or significantly accelerate early stages of AD derives from the nature of the experimental systems. Only humans develop true AD, but axonal transport and biochemical defects induced by agents and mutations that can cause AD can only be rigorously studied at present in animal models of the disease. Further resolution of these issues will require noninvasive imaging suitable for studying axonal transport in living human patients, and the development of true human neuronal models of disease, perhaps using human embryonic or other types of stem cells. Finally, it is likely that truly effective therapies for AD will require a detailed and mechanistic understanding of the events leading from causative insults to the biochemistry of neuronal function and viability; axonal transport appears to be an important process in disease progression, and its role needs to be rigorously evaluated.

See also: Acetylcholinesterase Inhibitors and Alzheimer's Disease; Actin Cytoskeleton in Growth Cones, Nerve Terminals, and Dendritic Spines; Aging of the Brain and Alzheimer's Disease; Animal Models of Alzheimer's Disease; Axonal and Dendritic Transport by Dyneins and Kinesins in Neurons; Axonal Transport and Neurodegenerative Diseases; Dementia; Microtubule Associated Proteins in Neurons; Microtubules: Organization and Function in Neurons; Slow Axonal Transport; Stroke: Neonate vs. Adult.

Further Reading

Goldstein LS (2003) Do disorders of movement cause movement disorders and dementia? *Neuron* 40: 415–425.

King ME, Kan HM, Baas PW, et al. (2006) Tau-dependent microtubule disassembly initiated by prefibrillar beta-amyloid. *Journal of Cell Biology* 175: 541–546.

Lee EB, Zhang B, Liu K, et al. (2005) BACE overexpression alters the subcellular processing of APP and inhibits Abeta deposition *in vivo*. *Journal of Cell Biology* 168: 291–302.

Muresan Z and Muresan V (2005) c-Jun NH_2-terminal kinase-interacting protein-3 facilitates phosphorylation and controls localization of amyloid-beta precursor protein. *Journal of Neuroscience* 25: 3741–3751.

Pigino G, Morfini G, Pelsman A, et al. (2003) Alzheimer's presenilin 1 mutations impair kinesin-based axonal transport. *Journal of Neuroscience* 23: 4499–4508.

Satpute-Krishnan P, DeGiorgis JA, Conley MP, et al. (2006) A peptide zipcode sufficient for anterograde transport within amyloid precursor protein. *Proceedings of the National Academy of Sciences of the United States of America* 103: 16532–16537.

Shen J and Kelleher RJ, III (2007) The presenilin hypothesis of Alzheimer's disease: Evidence for a loss-of-function pathogenic mechanism. *Proceedings of the National Academy of Sciences of the United States of America* 104: 403–409.

Stokin GB and Goldstein LS (2006) Axonal transport and Alzheimer's disease. *Annual Review of Biochemistry* 75: 607–627.

Stokin GB, Lillo C, Falzone TL, et al. (2005) Axonopathy and transport deficits early in the pathogenesis of Alzheimer's disease. *Science* 307: 1282–1288.

Terry RD (1996) The pathogenesis of Alzheimer disease: An alternative to the amyloid hypothesis. *Journal of Neuropathology & Experimental Neurology* 55: 1023–1025.

Axonal Transport and Huntington's Disease

F Saudou and S Humbert, Institut Curie and CNRS UMR 146, Orsay, France

Introduction

Huntington's disease (HD) is a dominantly inherited neurodegenerative disease caused by abnormal polyglutamine (polyQ) expansion in a specific protein. HD is characterized by uncontrolled movements (chorea), personality changes, and dementia. Patients typically die 10–20 years after the appearance of the first clinical symptoms. In HD, the specific dysfunction and death of neurons in the striatum and cerebral cortex has been reported. The mutated protein is huntingtin, a large ubiquitous protein, the function of which is not fully understood. However, there is evidence that huntingtin is an antiapoptotic protein. Indeed, it is necessary for development since knockout mice die soon after gastrulation and conditional knockout mice, in which huntingtin is selectively turned off in brain and testis during adulthood, show a progressive degenerative neuronal phenotype and sterility. Moreover, overexpression of normal huntingtin protects striatal cells from a variety of apoptotic stimuli. Huntingtin becomes toxic when it contains an abnormal polyQ expansion. PolyQ-huntingtin induces the formation of neuritic and intranuclear inclusions, neuronal dysfunction, and, finally, neuronal death. The precise mechanisms underlying these phenomena are not well understood, and it is not known how increased neuronal death in the brain relates to huntingtin function and dysfunction.

Evidence suggests that huntingtin is involved in intracellular transport and that axonal transport is impaired in HD. Herein, we review studies demonstrating that huntingtin belongs to a protein complex that controls intracellular transport and that the alteration of this complex may participate in the pathogenesis of HD.

Intracellular Transport and Neurons

Neurons are highly polarized cells composed of a cell body and an extensive network of cell processes: dendrites, and a single axon. The axonal extension can reach up to 1 m in length. Most protein synthesis is restricted to the cell body. Therefore, active transport is required to supply the axon with newly synthesized materials. Conversely, external signals have to be transmitted to the cell body from the cell's extremities, where the signaling cascade is activated. Here also, active transport of receptor–ligand complexes is required for a quick and appropriate response. Due to the length of their processes, transport efficiency is of particular importance in neurons.

To allow efficient communication between cell bodies and axon termini, molecular motor proteins continuously shuttle vesicles and organelles along the microtubules and actin filaments that make up the cellular cytoskeleton. This transport process mostly involves microtubules, which are polarized structures with 'plus' (cell extremity) and 'minus' (cell body) ends. Molecular motors are considered to be unidirectional: Dynein complexes are connected to retrograde transport (from plus to minus end), whereas kinesins are connected to anterograde transport (from cell body to plasma membrane).

These two classes of motor complexes are required for efficient transport in neurons. Disruption of these complexes has dramatic consequences on neurons and leads to severe diseases.

Huntingtin and Intracellular Dynamics

There is growing evidence to suggest that huntingtin plays a role in intracellular dynamics. Cytoplasmic huntingtin co-localizes with microtubules and interacts with β-tubulin. Huntingtin is also found in neurites co-localizing with vesicles. In addition to this specific distribution, huntingtin interacts with many proteins that are implicated in intracellular trafficking. The best-described example is the case of huntingtin-associated protein 1 (HAP1), which is discussed later. Huntingtin also interacts with huntingtin-interacting protein 1 (HIP1), which binds to the clathrin light chain and regulates clathrin assembly, linking huntingtin to the control of endocytosis. A role of huntingtin in the control of endocytosis is further supported by the observation that huntingtin bound to the HAP40 protein acts as an effector of the small guanosine triphosphatase Rab5. In the pathological situation, a reduction of endosomal motility and endocytic activity in HD fibroblasts and mutant cells is observed. Finally, huntingtin associates with PACSIN1, a protein that is part of the endocytic machinery, which could act as a negative regulator.

In addition to regulating endocytosis at the plasma membrane, huntingtin may also regulate secretion and trafficking of proteins or organelles from the Golgi region. Indeed, HIP14, another huntingtin interacting protein that contains an ankyrin domain, is found at the Golgi and at vesicles. This protein is a palmitoyl transferase involved in the sorting of proteins from the Golgi region. Also, huntingtin binds optineurin, a protein associated with Rab8 and myosin VI. This complex is found at the Golgi and in vesicles. The depletion of optineurin leads to the disruption of the Golgi ribbon structure and to the reduction in the transport of VSV-G from the Golgi complex to the plasma membrane. Finally, two other proteins that are dysregulated in HD participate in the sorting of organelles from the Golgi complex. The DNAJ-containing HSJ1b protein, possibly by regulating clathrin uncoating, promotes the processing of vesicles that contain the brain-derived neurotrophic factor (BDNF) from the Golgi to the cytoplasm. HSJ1b protein colocalizes with the transglutaminase 2 (TGase 2) enzyme that has a negative role in BDNF sorting from the Golgi. Interestingly, TGase 2 is inhibited by cystamine and by its reduced form, cysteamine, resulting in an increase in the sorting of BDNF-containing vesicles from the Golgi and in the release of this trophic factor from the cell. Treating HD mice with cysteamine results in an increase in BDNF release in the brain and in the reduction of the loss of striatal neurons in HD mutant mice. That a compound such as cysteamine, which is a Food and Drug Administration-approved drug, is neuroprotective by enhancing the intracellular processing of vesicles suggests that strategies aiming at restoring the defects in intracellular trafficking of organelles are of therapeutic interest.

Huntingtin Associates with the Molecular Motor Complex

The identification of HAP1 and its characterization with respect to huntingtin led to the hypothesis that both proteins may have an important function in the axonal transport machinery. Indeed, huntingtin and HAP1 are transported both anterogradely and retrogradely in rat sciatic nerves. Both proteins are associated with synaptic vesicles and localize with microtubules in neurites. The discovery in 1997 that HAP1 associates with the p150^{Glued} subunit of dynactin further supported this possibility. Since then, a growing body of evidence has suggested that huntingtin has a major function in the control of axonal transport and that this function requires the HAP1 protein. Indeed, HAP1 binds not only to the p150^{Glued} subunit of dynactin but also to the kinesin light chain 2, a subunit of the kinesin-1 complex. In agreement, depletion of HAP1 reduces kinesin-dependent transport of amyloid precursor protein vesicles. In addition to the indirect association of huntingtin to dynactin through HAP1, huntingtin also interacts with the dynein light chain of the dynein complex. Finally, huntingtin and HAP1 comigrate in the same fractions after sucrose gradient fractionation, further supporting the idea that huntingtin belongs to the molecular motor complex dynein/dynactin/kinesin-1.

Huntingtin Stimulates Axonal Transport

In addition to the previously discussed biochemical evidence, videomicroscopy approaches in cells and in *Drosophila* have further established a functional role of both HAP1 and huntingtin in axonal transport. In *Drosophila*, reduction in huntingtin protein level by the RNAi approach results in axonal transport defects. Furthermore, huntingtin directly promotes intracellular transport of vesicles along microtubules in neuroblastoma and striatal cells. Indeed, using BDNF as a marker of intracellular trafficking in cells, Gauthier and collaborators showed that expression of huntingtin enhances the velocity of BDNF-containing vesicles while reducing the percentage of time spent pausing. In support of a positive role of huntingtin in stimulating axonal transport of vesicles that contain BDNF, downregulation of huntingtin by RNAi approaches resulted in a decrease in the velocity of moving vesicles and an increase in the percentage of time spent pausing. BDNF was chosen because it is an important factor in HD. It is produced in the cortex and is transported to the striatum, the major site of degeneration in HD, where it supports neuronal differentiation and survival. It inhibits polyQ–huntingtin-induced neuronal death and its level is abnormally low in HD patients. BDNF is synthesized from the large precursor protein pre-pro-BDNF that is proteolytically processed and moves through the Golgi apparatus to the trans-Golgi network, where it is packaged into vesicles. BDNF-containing vesicles are then transported along microtubules (MTs) to the plasma membrane and subsequently released through the regulated secretory pathway. BDNF-containing vesicles are immunopositive for the classical markers of secretion and their activity-dependent release requires an intact MT network because it is blocked by nocodazole, an MT depolymerizing agent.

The huntingtin-dependent transport of BDNF vesicles is bidirectional because huntingtin stimulates both anterograde and retrograde transport in axons. It requires the HAP1 protein. Indeed, short N-terminal fragments of huntingtin that do not contain the HAP1 interacting region are unable to stimulate intracellular transport. Also, BDNF transport is reduced after downregulation of HAP1 protein and, in these conditions, huntingtin is unable to enhance BDNF trafficking. The HAP1-dependent transport is not restricted to BDNF vesicles since the movement of amyloid precursor protein vesicles is also reduced by decreasing HAP1 levels. Since it is known that the movement of amyloid precursor protein vesicles is kinesin dependent, one can infer an important role of huntingtin and HAP1 in both anterograde and retrograde transport.

Although these studies revealed a role for huntingtin in axonal transport, the extend to which axonal transport depends on huntingtin remains unknown. Huntingtin stimulates the dynamic of BDNF-containing vesicles; however, whether other types of vesicles are regulated by huntingtin remains to be established. The observations that transport of amyloid precursor protein vesicles depends on HAP1 and that downregulation of huntingtin alters the general axonal transport in *Drosophila* strongly suggest that huntingtin regulates the transport of other small vesicles.

Milton, a *Drosophila* ortholog of HAP1, participates in the axonal transport of mitochondria, thus raising the possibility that huntingtin and HAP1 could also regulate the transport of mitochondria. However, the velocity of mitochondria is not regulated by huntingtin, and whereas HAP1 overexpression leads to the redistribution of BDNF vesicles in cells, it has no effect on mitochondria. Conversely, Milton, although known to redistribute mitochondria in cells, has no effect on BDNF vesicles. Therefore, HAP1 and Milton show specificity in the type of cargoes they are transporting.

HD Pathological Situation and Axonal Transport Defects

Several studies indicate an alteration of axonal transport in the polyQ situation. However, depending on the stage of the disease, different mechanisms may be involved.

In striatal cells and in neurons, whereas normal full-length huntingtin stimulates BDNF transport, full-length huntingtin with an abnormally expanded polyQ expansion has lost this stimulatory ability. Also, a reduction in transport of BDNF is observed in striatal cells derived from knockin mice in which a CAG expansion has been inserted in the endogenous mouse huntingtin gene. These mutant striatal cells express full-length huntingtin at endogenous levels. Similarly, polyQ-containing polypeptides encompassing amino acids 1–548 of huntingtin inhibit fast axonal transport in isolated axoplams without the formation of detectable morphological aggregates. Thus, polyQ expansion of huntingtin results in a loss of huntingtin function in transport. This phenotype is mediated by HAP1, which forms a protein complex by binding to the $p150^{Glued}$ subunit of dynactin and simultaneously binding to a domain located in the N-terminal region of huntingtin. The formation of this complex subsequently stimulates microtubule-based transport. In contrast, when huntingtin contains an abnormal polyQ expansion, it interacts more strongly with HAP1 and $p150^{Glued}$, leading to the detachment of the molecular motors from the microtubules and thus to less efficient transport of BDNF vesicles.

The presence of an abnormal polyQ expansion in huntingtin leads to a reduced anterograde and retrograde transport of BDNF-containing vesicles. In cortical neurons that deliver BDNF to striatal neurons, the reduced transport of BDNF in the pathological situation has a direct consequence on the ability of these cells to release BDNF, rendering striatal neurons particularly vulnerable. In agreement, cortical ablation of BDNF in mice induces striatal dendrite deficits followed by neuronal loss. Once released, BDNF binds to its receptor TrkB; the BDNF–TrkB complex is endocytosed in striatal neurons and transported retrogradely to the nucleus, where it activates downstream signaling pathways. The consequences of axonal transport defects in these neurons remain to be established.

A critical step in HD pathogenesis is the cleavage of polyQ–huntingtin into N-terminal fragments containing the polyQ expansion. These fragments translocate into the nucleus and induce the formation of aggregates. In the nucleus, polyQ–huntingtin fragments cause neuronal death by a gain-of-function mechanism, leading to the dysregulation of transcriptional activity. In addition, N-terminal huntingtin fragments and their aggregates accumulate in axonal processes and terminals. What are the consequences of these aggregates on axonal transport? In *Drosophila*, N-terminal huntingtin polypeptide fragments containing the polyQ expansion accumulate in axonal inclusions and cause axonal trafficking defects. This subsequently induces neuronal death and organismal death. Furthermore, mutant huntingtin aggregates alter vesicular and mitochondrial transport in mammalian neurons. These aggregation-induced defects in transport may be due to a physical blockage of vesicles but may also involve the sequestration by mutant huntingtin aggregates of motor proteins (particularly $p150^{Glued}$ and kinesin heavy chain) from other cargoes and pathways. Interestingly, this aggregation-induced alteration of general transport appears to be independent of the protein context since it has also been reported to occur in other polyQ disorders.

In summary, in early stages of the disease, the role of huntingtin in transport is lost following the alteration of the huntingtin (soluble form)/HAP1 interaction. In later stages, polyQ–huntingtin forms neuritic aggregates that contribute to a trafficking defect by a gain-of-function mechanism.

Disruption of Axonal Transport and Neurodegenerative Disorders

Transport defects have also been described in other polyQ neurodegenerative disorders. The polyQ-containing androgen receptor, the mutation of which is responsible for spinobulbar muscular atrophy, inhibits fast axonal transport. Non-polyQ neurodegenerative diseases have also been linked to transport failure. In Alzheimer's disease (AD), mutation of the amyloid precursor protein, responsible for familial forms of AD, disrupts axonal transport. Similarly, overexpression of tau, a neuronal microtubule-associated protein that accumulates in AD, inhibits organelle trafficking. Several different neurodegenerative disorders are caused by mutation within motor proteins. For instance, Charcot–Marie–Tooth disease type 2 is caused by a loss-of-function mutation in the kinesin KIF1B, and hereditary spastic paraplegia type 10 is caused by a missense mutation in KIF5A. Finally, a mutation in the p150^{Glued} subunit of dynactin has been identified in a family with an autosomal dominant form of lower motor neuron disease. Furthermore, inhibition of dynein-mediated axonal transport causes neurodegeneration in mouse models.

Conclusion

Slowing of transport might be a general phenomenon in neurodegenerative diseases. It can be either causative or contributory, depending on the disease. Neurons are highly susceptible to dysregulation of transport. Indeed, because transport in neurons occurs over very long distances, a decrease in intracellular trafficking has dramatic consequences on their viability. The development of accurate models of intracellular transport in neurons should therefore help researchers understand the molecular mechanisms underlying these diseases.

Axonal transport is a promising target for designing and testing treatments to slow or block neurodegeneration. In the case of HD, compounds that enhance transport or rescue huntingtin dysfunction might be of therapeutic interest since huntingtin directly controls transport of the pro-survival factor BDNF. This is of utmost importance because no treatment currently exists for this devastating disorder.

See also: Axonal Transport Disorders; Axonal and Dendritic Transport by Dyneins and Kinesins in Neurons; Axonal Transport and Neurodegenerative Diseases; Cell Replacement Therapy for Huntington's Disease; Huntington's Disease; Huntington's Disease: Neurodegeneration; Triplicate Repeats: Huntington's disease.

Further Reading

Altar CA, Cai N, Bliven T, et al. (1997) Anterograde transport of brain-derived neurotrophic factor and its role in the brain. *Nature* 389: 856–860.

Borrell-Pages M, Canals JM, Cordelieres FP, et al. (2006) Cystamine and cysteamine increase brain levels of BDNF in Huntington disease via HSJ1b and transglutaminase. *Journal of Clinical Investigation* 116: 1410–1424.

Cattaneo E, Zuccato C, and Tartari M (2005) Normal huntingtin function: An alternative approach to Huntington's disease. *Nature Reviews Neuroscience* 6: 919–930.

Chang DT, Rintoul GL, Pandipati S, and Reynolds IJ (2006) Mutant huntingtin aggregates impair mitochondrial movement and trafficking in cortical neurons. *Neurobiology of Disease* 22: 388–400.

Gauthier LR, Charrin BC, Borrell-Pages M, et al. (2004) Huntingtin controls neurotrophic support and survival of neurons by enhancing BDNF vesicular transport along microtubules. *Cell* 118: 127–138.

Gunawardena S, Her LS, Brusch RG, et al. (2003) Disruption of axonal transport by loss of huntingtin or expression of pathogenic polyQ proteins in *Drosophila*. *Neuron* 40: 25–40.

Hafezparast M, Klocke R, Ruhrberg C, et al. (2003) Mutations in dynein link motor neuron degeneration to defects in retrograde transport. *Science* 300: 808–812.

Heerssen HM, Pazyra MF, and Segal RA (2004) Dynein motors transport activated Trks to promote survival of target-dependent neurons. *Nature Neuroscience* 7: 596–604.

Hirokawa N and Takemura R (2005) Molecular motors and mechanisms of directional transport in neurons. *Nature Reviews Neuroscience* 6: 201–214.

LaMonte BH, Wallace KE, Holloway BA, et al. (2002) Disruption of dynein/dynactin inhibits axonal transport in motor neurons causing late-onset progressive degeneration. *Neuron* 34: 715–727.

Lee WC, Yoshihara M, and Littleton JT (2004) Cytoplasmic aggregates trap polyglutamine-containing proteins and block axonal transport in a *Drosophila* model of Huntington's disease. *Proceedings of the National Academy of Sciences of the United States of America* 101: 3224–3229.

Li SH, Gutekunst CA, Hersch SM, and Li XJ (1998) Interaction of huntingtin-associated protein with dynactin P150Glued. *Journal of Neuroscience* 18: 1261–1269.

Li XJ and Li SH (2005) HAP1 and intracellular trafficking. *Trends in Pharmacological Sciences* 26: 1–3.

Pal A, Severin F, Lommer B, Shevchenko A, and Zerial M (2006) Huntingtin–HAP40 complex is a novel Rab5 effector that regulates early endosome motility and is upregulated in Huntington's disease. *Journal of Cell Biology* 172: 605–618.

Piccioni F, Pinton P, Simeoni S, et al. (2002) Androgen receptor with elongated polyglutamine tract forms aggregates that alter axonal trafficking and mitochondrial distribution in motor neuronal processes. *FASEB Journal* 16: 1418–1420.

Sahlender DA, Roberts RC, Arden SD, et al. (2005) Optineurin links myosin VI to the Golgi complex and is involved in Golgi organization and exocytosis. *Journal of Cell Biology* 169: 285–295.

Saudou F, Finkbeiner S, Devys D, and Greenberg ME (1998) Huntingtin acts in the nucleus to induce apoptosis but death does not correlate with the formation of intranuclear inclusions. *Cell* 95: 55–66.

Szebenyi G, Morfini GA, Babcock A, et al. (2003) Neuropathogenic forms of huntingtin and androgen receptor inhibit fast axonal transport. *Neuron* 40: 41–52.

Trushina E, Dyer RB, Badger JD 2nd, et al. (2004) Mutant huntingtin impairs axonal trafficking in mammalian neurons *in vivo* and *in vitro*. *Molecular and Cellular Biology* 24: 8195–8209.

Yanai A, Huang K, Kang R, et al. (2006) Palmitoylation of huntingtin by HIP14 is essential for its trafficking and function. *Nature Neuroscience* 9: 824–831.

Axonal Transport and Neurodegenerative Diseases

S Roy, V M-Y Lee, and J Q Trojanowski, University of Pennsylvania School of Medicine, Philadelphia, PA, USA

Introduction

The primary functions of the neuron are to receive, process, and transmit information. A typical neuron has multiple dendrites and a single elongated process, the axon. The dendrites and the cell bodies play a role in the collection and processing of information, and the axon is responsible for the transmission of information to other neurons via synapses. Whereas dendrites are usually in close proximity to the neuronal cell bodies and have some capacity for protein synthesis, axons can extend up to tremendous lengths, sometimes several feet, and are generally devoid of the protein synthetic machinery. As a result, axonal proteins are synthesized in the cell bodies and subsequently transported into the axons and synapses. This process is called axonal transport. Axonal transport is essential for the survival of axons and synapses, and it occurs throughout the life of the neuron.

The unique architecture of the axon combined with its dependence on the remotely located cell body for growth and maintenance makes it especially vulnerable to a variety of insults. Not surprisingly, axonal and synaptic defects are characteristic features of most neurodegenerative diseases, and it was long suspected that defects in axonal transport could play a major role in these diseases. Although correlative evidence linking axonal transport defects and neurodegenerative diseases was well known, recent experimental evidence from several laboratories suggests that many neurodegenerative diseases may be a direct consequence of altered or defective axonal transport. In this article, we first highlight the basic principles of the mechanisms of axonal transport. Then, we review some of the evidence that links axonal transport to neurodegenerative diseases, including pathogenic mutations in human genes that encode proteins involved in axonal transport and studies of animal models of disease showing impairments of axonal transport.

Basic Mechanisms of Axonal Transport

Although many specific details of axonal transport mechanisms are not fully understood, the basic principles are well established. In general, to move any cargo, two components are required – motors to move the cargo and rails to transport it. In addition, motor proteins often require adaptor/linker proteins to attach the motors to the cargo. For the rails, the neurons use the structural network of the cell, the cytoskeleton. Most long-range transport is dependent on the microtubules, but actin filaments also play a role in short-range movements. Microtubules in axons are polarized (unlike dendrites), allowing a directional bias in the movement of the motors and their cargo. For the motors, the neurons have an abundance of small molecular machines, known as kinesins and dyneins, moving the cargo mainly anterogradely (away from the cell body to the axon tip) or retrogradely (toward the cell body), respectively (**Figure 1**). The anterograde transport consists of components required for maintaining synapses, cytoskeletal proteins required for maintaining axonal structure and function, as well as a variety of other proteins required for neuronal function, including mitochondria, various chaperones, and glycolytic enzymes. The retrograde transport consists mainly of endosomal/lysosomal organelles that carry degraded proteins back to the cell bodies for degradation and also neurotrophic factors required for neuronal survival. In reality, however, most transport is bidirectional, with a bias either in the anterograde or in the retrograde direction.

Our understanding of the mechanisms of transport has increased dramatically during the past few decades due to advances in the development of model systems to dissect basic transport mechanisms and advances in biochemistry and genetics that have led to the identification and characterization of the molecular motors and their cellular functions. Model systems developed to visualize the transport process revealed that the bulk of proteins are transported anterogradely in two broad, discrete groups. Whereas some of the proteins are transported rapidly, at rates of 100–400 mm day^{-1} (1–5 $\mu m\, s^{-1}$), quickly reaching up to the tip of the axon, others move at rates that are several orders of magnitude slower, at approximately 0.2–5 mm day^{-1} (0.0002–0.05 $\mu m\, s^{-1}$). These two components are called fast axonal transport and slow axonal transport, respectively. Whereas fast transport mainly comprises vesicular cargo, including synaptic proteins, ion channels, and other components, slow axonal transport is primarily composed of transported cytoskeletal proteins, mainly microtubules, neurofilaments, and actin, along with many additional cytosolic proteins involved in neuronal homeostasis. Visualization of slow transport of the cytoskeleton revealed that despite moving with overall distinct dynamics, both slow and fast transport use similar

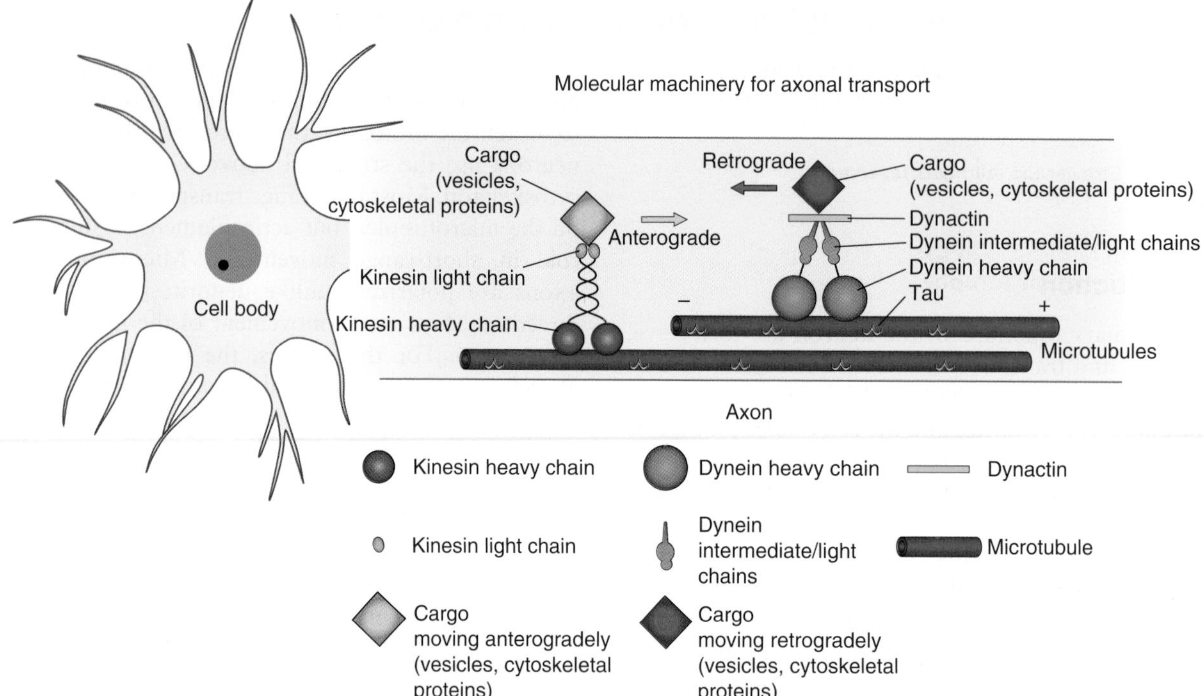

Figure 1 Basic principles of axonal transport. Cargo is transported anterogradely and retrogradely by the molecular motors kinesin and dynein, respectively. A variety of additional adaptor/linker proteins may also help in binding kinesin to the cargo (not shown). Dynactin, a multiprotein complex, links dynein to its cargo and is thought to increase the efficiency of the motor. Long-range movement occurs along cytoskeletal 'tracks,' mainly microtubules.

underlying mechanisms. These studies showed that slow cargo moves rapidly like fast cargo, but unlike fast cargo, it pauses for prolonged times during transit. Thus, the overall movement is infrequent and intermittent, leading to a slow overall rate of the transported cytoskeletal population. It has also been shown that slow and fast transport share similar molecular motors.

Progress in biochemistry and genetics has led to the identification of approximately 45 members of the kinesin superfamily (Kifs), grouped into 14 subfamilies, and two members of the cytoplasmic dynein family. Besides this large array of motor proteins, there are also a number of linker/adaptor proteins responsible for binding cargo to the motor proteins that have been identified. This heterogeneity is thought to play a major role in the recognition of specific motors to their cargo. Many excellent reviews on the subject are available.

Axonal Transport and Neurodegenerative Diseases

The following developments have dramatically highlighted the role of axonal transport disruptions in human neurodegenerative diseases: (1) the discovery of motor protein mutations in human neurodegenerative diseases, (2) axonal transport defects in animal and *in vitro* cellular models of neurodegenerative diseases, and (3) newly discovered roles in axonal transport regulation for known pathogenic proteins involved in neurodegenerative diseases. With a focus on specific diseases, we discuss some of the recent finings linking axonal transport to neurodegenerative diseases.

Kinesin Mutation in Hereditary Spastic Paraplegia

Hereditary spastic paraplegias (HSPs), also known as familial spastic paraplegias, represent an autosomal dominant inherited group of neurodegenerative diseases. Patients present in their thirties or forties with symptoms in the lower limbs, with gradual proximal spread of symptoms. Neuropathologically, a distal axonopathy is seen, with severe degeneration and gliosis of the distal corticospinal tracts and relative sparing of the tracts in the brain stem and proximal cord. Due to the peculiar distal-to-proximal 'dying back' axonopathy observed in this disease, it had been hypothesized that dysfunctions in axonal transport leading to selective damage of the distant portions of the axons may be responsible for the pathogenesis of HSPs.

Indeed, a missense mutation in one of the genes encoding a major kinesin protein (the gene for kinesin heavy chain Kif 5a) was found in a family with HSP. The same mutation was found in all affected members of the family, as well as in some presymptomatic members. This mutation occurs within a functional motor domain of the kinesin protein and a homologous mutation in yeast has been found to decouple kinesin binding to microtubules, highlighting the functional role of the kinesin mutation in the pathology of HSPs.

Dynactin Mutations in Distal Spinal and Bulbar Muscular Atrophy

Dynactin is a large protein complex linked to the retrograde motor dynein, and it is thought to link the motor to its cargo and/or increase the processivity of the motor. Animal models disrupting the dynein/dynactin complex develop late-onset motor neuron degeneration, and missense mutations in a dynein subunit cause progressive motor neuron degeneration in mice. Due to the central role of the dynein/dynactin complex in axonal transport, and based on data from the animal studies, it was hypothesized that mutations in the dynein/dynactin complex could play a role in neurodegeneration. Indeed, mutations in the gene encoding a subunit ($p150^{Glued}$) of the dynactin complex have been reported in a family with the neurodegenerative disease distal spinal and bulbar muscle atrophy (SBMA). In familial SBMA cases, the disease is transmitted in an autosomal dominant fashion and is manifested as is a primary lower motor-type neuropathy with patients presenting in their thirties, often with breathing difficulty due to vocal fold paralysis, which later leads to weakness in the face and distal extremities. Neuron loss as well as inclusions containing dynein and dynactin were also seen in autopsy studies from one patient. The mutation reduces the binding affinity of dynactin to microtubules and also causes subtle defects in dynein function.

Kinesin Mutations in Charcot–Marie–Tooth Disease

Charcot–Marie–Tooth (CMT) disease comprises a heterogeneous group of inherited peripheral neuropathies characterized by motor and sensory deficits, often presenting in young adults as tingling, numbness, and loss of deep reflexes. The progression of the disease varies among individuals, with symptoms ranging from mild neuropathy to complete disability. Two basic forms can be recognized, with primary demyelinating (CMT1) or axonal (CMT2) types of degeneration predominating. Various genes have been implicated in CMT syndromes, including several genes known to play a role in myelination (PMP22 and MPZ) and genes for gap junction proteins (Connexin 32). However, in a remarkable series of studies, it was shown that mutations in a kinesin subunit protein (Kif 1B beta) can lead to an axonal type of CMT (CMT type 2A).

While studying animal models of kinesin knockout mice, it was found that heterozygous knockout mice for one of the kinesins, Kif 1B, developed progressive muscle weakness with normal motor nerve conduction velocities, symptoms resembling axonal type CMT, CMT type 2. Incidentally, the gene for CMT type 2A had been mapped to the same interval as the gene for Kif 1B, and genomic analysis of pedigrees with CMT type 2A revealed that these patients had a mutation in the Kif 1B gene. It was further shown that the mutant motor protein may not tightly bind to microtubules, thus suggesting a loss of function of the Kif 1B protein in patients with CMT type 2A.

Axonal Transport Defects in Alzheimer's Disease and Other Tauopathies

The story of the role of axonal transport in Alzheimer's disease (AD) is a rapidly developing one. Neuropathologically, the two hallmarks of AD are deposits of fibrillar Aβ into diffuse and neuritic plaques in the extracellular space, and filamentous accumulations of tau proteins as neurofibrillary tangles and neuropil threads within neurons and their processes. Amyloidogenic Aβ peptides are generated by proteolytic processing of the Aβ precursor proteins (APPs), conveyed by fast axonal transport. Several enzymes are involved in the proteolytic processing of APPs to Aβ, including the gamma- and beta-secretases (presenilins and BACE, respectively). Human mutations of both APPs and presenilins are seen in familial AD cases, highlighting the critical roles of these two proteins in AD. On the other hand, tau is a microtubule binding protein thought to stabilize microtubules *in vivo*.

The accumulation of tau in neuronal cell bodies and axons, as well as axonal swellings seen in AD, prompted the notion that axonal transport failure was important in the pathogenesis of this disease. Recent observations indicate that defects in axonal transport can occur long before the onset of severe symptoms in animal models of AD as well as in patients with AD. Interestingly, global reduction of axonal transport by reduction of kinesin levels can also exacerbate AD-type pathology in mouse models, further highlighting the role of diminished axonal transport in the pathogenesis of this disease.

The previous observations suggest that axonal transport defects play a role in the pathogenesis of AD. In addition, many key proteins directly involved in AD are also thought to play roles in the regulation of axonal transport. Studies of several different model systems have suggested that APP may act as a receptor for kinesin, thus proposing a direct link between a pathogenic protein and the axonal transport machinery. However, other studies have been unable to confirm these results. It has also been proposed that APPs may be transported in a vesicular complex containing presenilins and BACE, the gamma- and beta-secretase enzymes, and altered processing of APPs within this complex can lead to exacerbation of AD. These findings suggest that misregulation of APPs, either directly from known APP mutations (as in familial AD) or indirectly via proteins associating/interacting with APPs, can lead to disruption of fast axonal transport in general, thus leading to axonal depletion of critical components and neurodegeneration.

Another interesting line of evidence highlighting the role of presenilins in the regulation of fast axonal transport in AD derives from studies of presenilin mutant mice. As mentioned previously, presenilins are proteins responsible for regulated proteolysis of APPs, and mutations in presenilins are seen in most cases of early familial AD. Several studies indicate that presenilins interact with glycogen synthase kinase-3β (GSK-3β). GSK-3β is a kinase with many different roles, including the phosphorylation of kinesin light chains, and it has been shown that GSK-3β-mediated phosphorylation of kinesin light chains led to detachment of the kinesin motor from the cargo, thus preventing further transport of cargo. By using transgenic presenilin mutant mice, it was also shown that mutant or absent presenilin increased GSK-3β levels, thereby phosphorylating kinesin light chains, detaching the kinesins from their cargo, and impairing axonal transport.

Roles in axonal transport regulation have also been assigned to tau. Because tau binds to microtubules and is thought to stabilize them, it was proposed that dysfunctions in tau can destabilize microtubules and lead to a failure of axonal transport. Indeed, human tau mutations impair the ability of tau to bind to and stabilize microtubules, and tau overexpression in cellular models can lead to defects in axonal transport. In addition, the mechanisms causing the accumulation of tau in AD are also beginning to be understood. Defective axonal transport of disease-associated mutant tau has also been demonstrated in mouse models, providing experimental evidence for a mechanism of tau accumulation.

A final line of evidence suggesting a role for axonal transport defects in AD comes from studies of ApoE4, a gene whose allelic state is associated with an increased risk for AD. Mice expressing human ApoE4 exhibit defects in axonal transport, and the receptor for ApoE4, ApoER2, binds to JIP1/2, a protein that appears to mediate the binding of APPs to kinesin I. Thus, it can be postulated that overexpression of ApoE4 protein can lead to misregulation of JIP1/2-mediated binding of kinesin to APPs, leading to defects in fast axonal transport.

Stabilization of microtubules is also being explored as a therapeutic option in AD. The proof of principle for this concept was demonstrated in a study that showed that a microtubule-stabilizing drug (paclitaxel) could ameliorate the neurodegenerative phenotype in transgenic tau mice by offsetting the loss of tau function by stabilizing the microtubules and correcting the fast axonal transport defects in these mice.

Axonal Transport and Huntington's Disease

Neuropathologically, Huntington's disease is characterized by atrophy and degeneration of striatal neurons, with aggregates of pathological polyglutamine containing the protein huntingtin. Huntingtin is a predominantly cytoplasmic protein that associates with vesicles and moves in the fast axonal transport component. Although it has been known for several years that polyglutamine repeats within the huntingtin protein cause a gain of deleterious function leading to neurodegeneration, the exact pathogenic role of the repeats is unclear. By infusing huntingtin-harboring pathological polyglutamine repeats into a model for studies of fast axonal transport, it was shown that fast axonal transport was specifically inhibited by pathologically expanded polyglutamine repeats (but not normal proteins), along with inhibition of neurite extension in cultured cells. Furthermore, disruption of the *Drosophila* huntingtin gene also caused axonal transport defects. These findings lend support to a model in which aggregates of polyglutamine repeats disrupt fast axonal transport. Whether the disruption of axonal transport is a direct effect of the polyglutamine repeats or a secondary phenomenon remains to be established.

Axonal Transport and Amyotrophic Lateral Sclerosis

The histopathologic observation of prominent neurofilament-rich inclusions in the axons of spinal motor neurons of patients with amyotrophic lateral sclerosis (ALS) led to the hypothesis that disrupted axonal transport of proteins may play a role in the pathogenesis of the disease. However, the first direct evidence that axonal transport is disrupted in ALS awaited the development of transgenic mouse models of familial ALS based on expression of mutant superoxide dismutase (SOD-1) protein mutant mice that replicate several key aspects of ALS. Studies of these SOD-1 transgenic mouse models of ALS showed that the transport of neurofilament proteins was retarded in these animals, even before the mice were symptomatic, thereby implicating impaired axonal transport as an early deficit in ALS. Mutations in the p150 subunit of dynactin have also been reported in ALS patients.

Axonal Transport and Synucleinopathies

Synucleinopathies, also known as α-synucleinopathies, are a group of neurodegenerative disorders in which the primary pathology is the intracytoplasmic accumulation of α-synuclein primarily in neurons and, in some cases, glial cells. These disorders include Parkinson's disease, dementia with Lewy bodies, the Lewy body variant of AD, multiple system atrophy, and neurodegeneration with brain iron accumulation. In familial forms of Parkinson's disease, autosomal dominant missense mutations in genes encoding for α-synuclein are seen, suggesting a role of α-synuclein in the pathogenesis of these disorders. α-Synuclein is a highly conserved protein belonging to a multigene family that includes β-synuclein and γ-synuclein. α-Synuclein is strongly expressed in neurons, highly enriched in presynaptic terminals, and transported predominantly in the slow component. Axonal transport abnormalities of α-synuclein have been proposed in synucleinopathies, based on the observation that axonal α-synuclein pathology is pronounced in the disease and also on experimental evidence suggesting that α-synuclein may play a role in transport of presynaptic vesicles. Age-related retardation in the normal transport of α-synuclein was also seen in a study. Collectively, these studies propose a model in which age-related retardation of α-synuclein transport leads to accumulations of the protein over time, predisposing to the α-synuclein pathology in axons. Although these findings are interesting, many questions remain unanswered. The physiological role of synuclein is far from clear, and much work needs to be done to identify the role of synuclein in neurodegenerative disorders.

Conclusions and Future Directions for Research

A growing body of evidence implicates axonal transport defects in the etiology and pathogenesis of neurodegenerative diseases. Although motor protein defects in neurodegenerative diseases are direct evidence for this, it is likely that many other disease proteins are directly or indirectly linked to the complicated machinery of axonal transport. Thus, studies on the molecular mechanisms of transport are necessary to facilitate our understanding of the role of axonal transport in these diseases. Since pathogenic proteins in AD have roles in the regulation of axonal transport, and these proteins have been extensively studied, the time is perhaps ripe to uncover the links between AD and impaired axonal transport. With greater understanding of the other neurodegenerative diseases, it is likely that many more links to axonal transport will be uncovered. Another exciting but largely neglected avenue of research is drug discovery efforts to counteract axonal transport impairments as therapeutic interventions for neurodegenerative diseases. Indeed, if axonal transport defects are shown to be part of a common mechanism of disease in many neurodegenerative disorders, the discovery of drugs that modulate axonal transport could result in the development of important therapeutic interventions for the treatment of these disorders.

See also: Axonal Transport Disorders; Axonal and Dendritic Transport by Dyneins and Kinesins in Neurons; Axonal Transport and ALS; Axonal Transport and Alzheimer's Disease; Axonal Transport and Huntington's Disease; Hereditary Spastic Paraplegia; Slow Axonal Transport.

Further Reading

Brown A (2003) Axonal transport of membranous and nonmembranous cargoes: A unified perspective. *Journal of Cell Biology* 160: 817–821.

Chevalier-Larsen E and Holzbaur EL (2006) Axonal transport and neurodegenerative disease. *Biochimica et Biophysica Acta* 1762: 1094–1108.

Duncan JE and Goldstein LS (2006) The genetics of axonal transport and axonal transport disorders. *PLoS Genetics* 2: e124.

Hirokawa N and Takemura R (2005) Molecular motors and mechanisms of directional transport in neurons. *Nature Reviews Neuroscience* 6: 201–214.

Munch C, Rosenbohm A, Sperfeld AD, et al. (2005) Heterozygous R1101K mutation of the DCTN1 gene in a family with ALS and FTD. *Annals of Neurology* 58: 777–780.

Munch C, Sedlmeier R, Meyer T, et al. (2004) Point mutations of the p150 subunit of dynactin (DCTN1) gene in ALS. *Neurology* 63: 724–726.

Puls I, Jonnakuty C, LaMonte BH, et al. (2003) Mutant dynactin in motor neuron disease. *Nature Genetics* 33: 455–456.

Puls I, Oh SJ, Sumner CJ, et al. (2005) Distal spinal and bulbar muscular atrophy caused by dynactin mutation. *Annals of Neurology* 57: 687–694.

Reid E, Kloos M, Ashley-Koch A, et al. (2002) A kinesin heavy chain (KIF5A) mutation in hereditary spastic paraplegia (SPG10). *American Journal of Human Genetics* 71: 1189–1194.

Roy S, Zhang B, Lee VM, and Trojanowski JQ (2005) Axonal transport defects: A common theme in neurodegenerative diseases. *Acta Neuropathologica (Berlin)* 109: 5–13.

Trojanowski JQ, Smith AB, Huryn D, and Lee VM (2005) Microtubule-stabilising drugs for therapy of Alzheimer's disease and other neurodegenerative disorders with axonal transport impairments. *Expert Opinion on Pharmacotherapy* 6: 683–686.

Zhao C, Takita J, Tanaka Y, et al. (2001) Charcot–Marie–Tooth disease type 2A caused by mutation in a microtubule motor KIF1Bbeta. *Cell* 105: 587–597.

Axonal Transport Disorders

P N Hoffman, P C Wong, J W Griffin, and D L Price, The Johns Hopkins University, Baltimore, MD, USA

Fast (Anterograde and Retrograde) Transport and Slow Transport

Membranous organelles, such as vesicles, are transported at rapid velocities (i.e., up to 400 mm day^{-1}) in both the anterograde (i.e., away from the cell body) and retrograde (i.e., toward the cell body) directions. For example, proteins critical for synthesis of transmitters and their uptake, vesicle trafficking, transmitter release and recycling, etc. are, for the most part, rapidly delivered to terminals where they are essential for events involved in synaptic transmission. Some protein moves bidirectionally. Trophic factor receptors (e.g., the Trk receptors) are transported as inactive (nonphosphorylated) proteins in an anterograde direction to terminals, whereas active Trk receptors (phosphorylated) and their ligands (nerve growth factor and brain-derived neurotrophic factor) are carried retrograde to cell bodies where they influence transcription and, ultimately, the biology and survival of neurons. Moreover, transported proteins can be involved in disease in different ways. APP, the amyloid precursor protein, is a type 1 transmembrane protein that is rapidly transported anterograde along β- and γ-secretases in the central nervous system and is the source, when cleaved by these secretases, of the amyloid β (Aβ) peptide. Neurotoxic Aβ peptides are released at synaptic terminals where they accumulate as Aβ oligomers (and amyloid) which interfere with synaptic communication. When mutated, dynactin, a motor protein involved in retrograde transport, is the cause of amyotrophic lateral sclerosis (ALS), a motor neuron disease (**Table 1**). Moreover, tetanus toxin and some viruses (poliovirus and rabies virus) enter terminals and are carried by retrograde transport to cell bodies where their activities cause diseases such as tetanus, polio, and rabies.

The Cytoskeleton and Motor Proteins

Tubulin and microtubule-associated protein, including tau, as well as specific motor complex proteins (kinesin, dynein, dynactin, etc.), are transported within axons and provide the pathways and mechanisms for the bidirectional movement of membranous organelles and other constituents (e.g., mitochondria) in axons. Microtubules, which provide the rails for much of transport, have a uniform polarity, with their plus ends directed toward axonal terminals. Similarly, dynein, dynactin, and other components of retrograde motors play critical roles in the transport of components from terminals toward cell bodies. Kinesins, dynein, dynactin, and other proteins play key roles in the movement of membranous organelles toward either the plus end (anterograde direction) or minus end (retrograde direction) of axonal microtubules. Anterograde transport relies on kinesins, which are members of a family of highly conserved motor proteins that are critical for intracellular trafficking and transport. In neurons, kinesin is the principal motor for anterograde transport, with kinesin-associated proteins mediating binding of cargo.

The cytoskeletal proteins (including actin, tubulin, and tau) move slowly down axons. Tubulin, actin, clathrin, dynein, dynactin, and glycolytic enzymes are members of slow component b (SCb) and move at a rate of 2–8 mm day^{-1}. Some tubulins and the neurofilament (NF) proteins, NF-L, -M, and -H, which play distinct roles in filament assemblies, appear to be slowly transported in the anterograde direction at velocities of 1–5 mm day^{-1}. Evidence is consistent with the idea that some of these proteins, including the NF proteins, move in the slow component because their movement, although sometimes rapid, is asynchronous and infrequent. Thus, the NF proteins pause, go, and pause – a phenomenon leading to an overall slow rate of transport. Kinesin and dynein and its partners may participate in forward and return movements of these proteins, respectively.

Axonal Injury

Axonal transection (acute axotomy) of peripheral nerves is the simplest model used to investigate the effects of injury on transport processes. Both fast anterograde and retrograde transport are interrupted, and rapidly moving membranous organelles (and bidirectionally moving mitochondria) accumulate at the cut ends of the proximal and distal nerve stumps. Subsequently, the distal stump degenerates (Wallerian degeneration (Wlds)), and regenerating sprouts extend from the proximal stump at the rate of 3–5 mm day^{-1}. Axonal regeneration requires the presence of cytoskeletal proteins and membranous elements supplied by 'slow' and 'fast' anterograde transport, respectively. After axonal transection, changes occur in the expression and transport of a variety of proteins, including those of the cytoskeleton. Elevated expression of actin and tubulin is associated with the increased delivery of these proteins to regenerating axons. At the same time, levels of NF proteins are reduced, resulting in

Table 1 Select genetic forms of ALS

Subtype	*Gene locus*	*Onset*	*Inheritance*
ALS1	*SOD1*/21q21	Adult	Autosomal dominant
ALS2	*ALS2*/2q33	Juvenile	Autosomal recessive
ALS3	18q21	Adult	Autosomal dominant
ALS4	*SNTX*/9q34	Juvenile	Autosomal dominant
ALS5	15q15	Juvenile	Autosomal recessive
ALS6	16	Adult	Autosomal dominant
ALS	*DCTN1*/2q13	Adult	Autosomal dominant
ALS8	20q13 *VAPD*	Adult	Autosomal dominant

a decrease in the amounts of NF proteins transported in the proximal stump. Because NF polymers play a major role, along with neuregulin signaling in determing axonal caliber in myelinated nerve fibers, the decrease in transport of NF proteins is associated with reductions in axonal caliber that originate in proximal axons (near the cell body) and propagate distally at rates comparable to the velocity of NF transport. This phenomenon is referred to as somatofugal axonal atrophy. After axonal injury, alterations in levels of the membrane-associated proteins transported in the fast component are more subtle with the notable exception of increased expression of GAP-43, a membrane-associated phosphoprotein involved in the guidance of growth cones. Alterations in gene expression, levels of specific proteins, and axonal transport usually return to normal after regenerating axons have reinnervated targets.

Disorders Associated with Abnormalities of Axonal Transport

Impairments of axonal transport have profound consequences for the structure and functions of nerve cells. Alterations in axonal transport can be classified by the time course (acute, subacute, or chronic), the character of the pathology (accumulations of membranous or cytoskeletal organelles, swelling or shrinkage of axons), the locations of abnormalities within axons (proximal vs. distal, as exemplified by proximal NF axonal swellings vs. distal dying back axonopathies), the regions of the nervous systems affected by disease (central nervous system or peripheral nervous system (PNS)), and types of neuron (motor, sensory, or other) vulnerable to disease.

Axonal degeneration occurs in a variety of disorders of motor and sensory neurons, including genetic disease (Charcot–Marie–Tooth disease, mutant *SOD1* and *DCTN1*-linked fALS, giant axonal neuropathy linked to mutations in the *GAN* gene, etc.), toxic neuropathies (*n*-hexane, vincristine, Taxol, and cisplatin), demyelinating diseases (Guillain–Barré syndrome), and neurodegenerative diseases such as ALS and Alzheimer's disease. Neuroaxonal dystrophy, a prototype of membranous neuropathy, is characterized by the presence of membranous organelles accumulating in the distal regions of axons. Similar pathology occurs in experimental disorders (vitamin E deficiency and intoxication with *p*-bromophenylacetylurea) and in a variety of human illnesses, including those that occur in childhood (infantile neuroaxonal dystrophy) and in aging (dystrophy of axons in the dorsal columns). It has been speculated that the accumulation of rapidly transported membranous materials may occur because of failure of local utilization, excessive anterograde delivery, or failure of recycling/removal (turnaround) by retrograde transport.

Several experimental and clinical axonopathies are associated with the pathology of the cytoskeleton, particularly the accumulation of NFs. Experimental exposure to β,β'-iminodipropionitrile (IDPN) is a prototype for this type of disorder. This toxin selectively blocks transport of NF, perhaps by altering interactions of NFs with microtubules or with anterograde motors. The transport of other cytoskeletal proteins, including actin and tubulin, is relatively unaffected, and fast anterograde and retrograde transport appear to be relatively normal. After systemic intoxication with IDPN, NF synthesis is not altered, and newly synthesized NF proteins continue to enter the axon normally. Because transport of these proteins along axons is markedly impaired, NFs accumulate and distend proximal axons (with secondary local demyelination). NF axonal pathology occurs in other experimental models, including aluminum intoxication; the overexpression of wild-type NF transgenes (altering the stoichiometry of NF protein); the expression of a mutant NF transgene which encodes a protein that fails to interact properly for assembly; and progressive motor neuronopathy, a disease in mice homozygous for mutations in *Tbce* (tubulin-specific chaperone E). Hereditary canine spinal muscular atrophy, an inherited motor neuron disorder in Brittany spaniels, shows severe accumulation of NF in proximal axons, closely resembling lesions seen in IDPN models and in ALS. Disorganization of the NF network occurs in humans with giant axonal neuropathy linked to mutations in *GAN*, which encodes gigaxonin, a member of the cytoskeletal BTB/Kelch family. Dynein and dynactin (*DCTN1*) play roles in retrograde axonal transport, and alterations in the biology of these proteins are associated with disease. For example, slowly progressive autosomal dominant ALS has been linked to a G59S mutation in the $P150^{glued}$ subunit of *DCTN1*. The dynactin complex, which includes P150 and dynamitin, provides a linker between cargo, microtubules, and dynein. In the index family, the mutation occurs in a motif in the *P150*^{glued} subunit that mediates binding to microtubules. Modeling studies suggest that this mutation impacts on protein structure to create steric hindrance and distortion of the folding of the microtubule-binding domain. Consistent with this concept is the observation that the mutant protein binds less well to microtubules. Members of our group at Johns Hopkins have created a mutant *P150*^{glued} model of fALS with a robust motor neuron disease phenotype.

Axonal abnormalities are well documented in ALS (Lou Gehrig's disease). NF accumulates in the neuronal cell bodies and in proximal axons; the latter lesions resemble those produced by IDPN exposure and appear to reflect the impaired transfer of newly synthesized NF proteins from the cell body to the axon. Eventually, in ALS, axons degenerate and neurons die. Importantly, mutant *SOD1* fALS mice exhibit weakness and muscle atrophy associated with pathological features of human SOD1-linked ALS. These mice exhibit evidence of impaired axonal transport and develop proximal axonal swellings and more distal Wallerian degeneration followed by death of motor neurons. In a number of disorders, lesions occur more distally (NF swellings and accumulation of membranous organelles) and are often associated with distal axonal degeneration. These diseases are termed 'dying back' neuropathies because the distal axons/terminals exhibit conspicuous and early abnormalities distally and extend retrograde toward the cell bodies. Recent evidence suggests that similar processes occur in disease of the CNS.

Axon Regeneration and Protection

Intriguingly, in both experimental and human disease, regeneration and recovery can occur. The classical example of this process is regeneration after axotomy in the PNS. Usually, the distal axons of the PNS show Wallerian degeneration after transection, and eventually they regenerate successfully. Significantly, degeneration of these axons can be prevented genetically by the presence of an unusual gene. Significantly, mice (C57BL/Wlds) expressing a mutant gene (*Wlds*) exhibit slow Wallerian degeneration of axons distal to transections. The protective gene encodes an N-terminal fragment of ubiquitination factor E4B (Ube4b) fused to a nicotinamide mononucleotide adenyltransferase. The mechanism of this semidominant protection is not fully understood, and it has been suggested to be related to altered ubiquitination, changed purine nucleotide metabolism, or sirtuin-like activity. This unique genetic model provides an opportunity to discover new pathways whereby axons may be protected from degeneration.

See also: Amyotrophic Lateral Sclerosis (ALS); Axonal Injury: Neuronal Responses; Axonal and Dendritic Transport by Dyneins and Kinesins in Neurons; Axonal mRNA Transport and Functions; Axonal Transport and ALS; Axonal Transport and Alzheimer's Disease; Axonal Transport and Huntington's Disease; Axonal Transport and Neurodegenerative Diseases; Axonal Transport Tracers; Microtubule Associated Proteins in Neurons; Slow Axonal Transport.

Further Reading

Bomont P, Cavalier L, Blondeau F, et al. (2000) The gene encoding gigaxonin, a new member of the cytoskeletal BTB/kelch repeat family, is mutated in giant axonal neuropathy. *Nature Genetics* 26(3): 370–374.

Borchelt DR, Wong PC, Becher MW, et al. (1998) Axonal transport of mutant superoxide dismutase 1 and focal axonal abnormalities in the proximal axons of transgenic mice. *Neurobiology of Disease* 5(1): 27–35.

Bruijn LI, Miller TM, and Cleveland DW (2004) Unraveling the mechanisms involved in motor neuron degeneration in ALS. *Annual Review of Neuroscience* 27: 723–749.

Carmeliet P and Tessier-Lavigne M (2005) Common mechanisms of nerve and blood vessel wiring. *Nature* 436(7048): 193–200.

Coleman MP and Perry VH (2002) Axon pathology in neurological disease: A neglected therapeutic target. *Trends in Neuroscience* 25(10): 532–537.

Cork LC, Griffin JW, Munnell JF, Lorenz MD, Adams RJ, and Price DL (1979) Hereditary canine spinal muscular atrophy. *Journal of Neuropathology & Experimental Neurology* 38: 209–221.

Ehlers MD, Kaplan DR, Price DL, and Koliatsos VE (1995) NGF-stimulated retrograde transport of trkA in the mammalian nervous system. *Journal of Cell Biology* 130(1): 149–156.

Gold BG, Griffin JW, and Price DL (1985) Slow axonal transport in acrylamide neuropathy: Different abnormalities produced by single-dose and continuous administration. *Journal of Neuroscience* 5: 1755–1768.

Griffin JW, Hoffman PN, Clark AW, Carroll PT, and Price DL (1978) Slow axonal transport of neurofilament proteins: Impairment by β,β'-iminodipropionitrile administration. *Science* 202: 633–635.

Hadano S, Hand CK, Osuga H, et al. (2001) A gene encoding a putative GTPase regulator is mutated in familial amyotrophic lateral sclerosis 2. *Nature Genetics* 29: 166–173.

Hafezparast M, Klocke R, Ruhrberg C, et al. (2003) Mutations in dynein link motor neuron degeneration to defects in retrograde transport. *Science* 300(5620): 808–812.

Hanz S, Perlson E, Willis D, et al. (2003) Axoplasmic importins enable retrograde injury signaling in lesioned nerve. *Neuron* 40(6): 1095–1104.

Hirokawa N and Takemura R (2005) Molecular motors and mechanisms of directional transport in neurons. *Nature Reviews Neuroscience* 6(3): 201–214.

Hoffman PN, Cleveland DW, Griffin JW, Landes PW, Cowan NJ, and Price DL (1987) Neurofilament gene expression: A major determinant of axonal caliber. *Proceedings of the National Academy of Sciences of the United States of America* 84: 3472–3476.

Hoffman PN, Griffin JW, and Price DL (1984) Control of axonal caliber by neurofilament transport. *Journal of Cell Biology* 99: 705–714.

Holzbaur ELF (2004) Motor neurons rely on motor proteins. *Trends in Cell Biology* 14(5): 233–240.

Koliatsos VE, Clatterbuck RE, Winslow JW, Cayouette MH, and Price DL (1993) Evidence that brain-derived neurotrophic factor is a trophic factor for motor neurons *in vivo*. *Neuron* 10(3): 359–367.

Lazarov O, Morfini GA, Lee EB, et al. (2005) Axonal transport, amyloid precursor protein, kinesin-1, and the processing apparatus: Revisited. *Journal of Neuroscience* 25(9): 2386–2395.

Lee MK, Marszalek JR, and Clevel and DW (1994) A mutant neurofilament subunit causes massive, selective motor neuron death: Implications for the pathogenesis of human motor neuron disease. *Neuron* 13: 975–988.

Mack TG, Reiner M, Beirowski B, et al. (2001) Wallerian degeneration of injured axons and synapses is delayed by a Ube4b/Nmnat chimeric gene. *Nature Neuroscience* 4(12): 1199–1206.

Martin N, Jaubert J, Gounon P, et al. (2002) A missense mutation in tbce causes progressive motor neuronopathy in mice. *Nature Genetics* 32(3): 443–447.

Price DL, Ackerley S, Martin LJ, Koliatsos VE, and Wong PC (2005) Motor neuron diseases. In: Brady ST, Siegel GJ, Albers RW and Price DL (eds.) *Basic Neurochemistry*, pp. 731–743. Burlington, MA: Elsevier.

Price DL, Griffin J, Young A, Peck K, and Stocks A (1975) Tetanus toxin: Direct evidence for retrograde intraaxonal transport. *Science* 188: 945–947.

Puls I, Jonnakuty C, LaMonte BH, et al. (2003) Mutant dynactin in motor neuron disease. *Nature Genetics* 33(4): 455–456.

Rowland LP and Shneider NA (2001) Amyotrophic lateral sclerosis. *New England Journal of Medicine* 344(22): 1688–1700.

Schroer TA (2004) Dynactin. *Annual Review of Cell and Developmental Biology* 20: 759–779.

Vallee RB, Williams JC, Varma D, and Barnhart LE (2004) Dynein: An ancient motor protein involved in multiple modes of transport. *Journal of Neurobiology* 58(2): 189–200.

Verpoorten N, DeJonghe P, and Timmerman V (2006) Disease mechanisms in hereditary sensory and autonomic neuropathies. *Neurobiology of Disease* 21(2): 247–255.

Williamson TL and Cleveland DW (1999) Slowing of axonal transport is a very early event in the toxicity of ALS-linked SOD1 mutants to motor neurons. *Nature Neuroscience* 2: 50–56.

Wong PC, Cai H, Borchelt DR, and Price DL (2002) Genetically engineered mouse models of neurodegenerative diseases. *Nature Neuroscience* 5(7): 633–639.

Wong PC, Pardo CA, Borchelt DR, et al. (1995) An adverse property of a familial ALS-linked SOD1 mutation causes motor neuron disease characterized by vacuolar degeneration of mitochondria. *Neuron* 14: 1105–1116.

Axonal Transport Tracers

L G Bilsland and G Schiavo, Cancer Research UK London Research Institute, London, UK

Introduction

The formation of highly ordered and specialized neural networks during development is critical for the assembly of a functional nervous system. To gain a better understanding of these higher brain structures, it is necessary to correlate functional information with the intricate anatomy of specific neural circuitry, a process that becomes essential to evaluate changes in neural networks in circumstances of disease and injury. The discovery that exogenous proteins can be retrogradely transported in peripheral nerves initiated the development of neuroanatomical mapping techniques based on axonal transport tracing. The anatomical tracers now available can be classified according to their direction of axonal transport. Anterograde tracers are taken up by the cell soma or dendrites and transported down the axon to the nerve terminals. In contrast, retrograde tracers are taken up at the nerve terminals and transported toward the soma. Transport of tracers is predominantly in either an anterograde or a retrograde direction, although on closer examination it appears that most tracers are transported bidirectionally to some extent. In addition, several probes, termed transsynaptic tracers, are transferred from the original neuron to synaptically connected neurons. Neuroanatomical studies based on the detection of anterograde, retrograde, and transsynaptic tracers can provide a detailed knowledge of cellular projections and consequently morphology. Furthermore they allow investigation of neuronal connectivity that can divulge information on the organization and potential regulation of neural networks.

Early tracing studies were based on the induction of axonal degeneration. Following a brain lesion, axon degeneration was traced back to the cell soma by means of Golgi silver impregnation techniques. In the 1970s, advances in anterograde tracing included the injection of radiolabeled amino acids into specific brain regions. The anterograde transport of radiolabeled peptides and proteins was then monitored by autoradiographic techniques. This technique is still in use, although it has largely been superseded by tracers that require simpler detection methods and have improved anatomical resolution.

Anterograde and Retrograde Transport Tracers

Horseradish Peroxidase

The plant enzyme horseradish peroxidase (HRP) was one of the first retrograde tracers used. Application of HRP is generally by pressure or iontophoretic injection into the region of interest, where it is taken up by either damaged or intact axons and is endocytosed at nerve terminals via a fluid-phase mechanism. Histochemical detection of HRP is based on the production of a permanent electron-dense precipitate after HRP reacts with hydrogen peroxide and diaminobenzidine tetrahydrochloride (DAB) or tetramethylbenzidine. HRP labeling is, however, restricted to the cell soma and proximal dendrites. Conjugation of HRP with wheat germ agglutinin (WGA) significantly increases its ability to label axon terminals and allows transsynaptic transfer.

Fluorescent Dyes

The most frequently used axonal tracers are a range of fluorescent, hydrophilic dyes which include fast blue, rhodamine isothiocyanate, lucifer yellow, propidium iodide, and diamidino yellow. These dyes can be injected into the region of interest where they are taken up by fluid-phase endocytosis, or alternatively they can be microinjected directly into the cell soma. Several dyes can be used simultaneously to trace multiple cellular projections. Generally these dyes are bidirectionally transported, although some, such as fast blue and diamidino yellow, are predominantly transported in the retrograde direction. The majority of these probes are, however, not transsynaptically transferred. They are sensitive tracers, providing good resolution of cellular morphology. However, as with all fluorescence-based techniques, photobleaching can be a problem as well as saturating signals, which might obscure fine morphological details.

Fluorogold

Developed in 1986, the fluorescent retrograde tracer fluorogold has largely superseded the use of fluorescent dyes. Fluorogold can be applied by either pressure or iontophoretic injection into the region of interest, where it is taken up by damaged neurons and at nerve terminals by fluid-phase endocytosis. The advantage of this tracer lies in the availability of specific antibodies to fluorogold. Immunostaining can then be converted to a permanent electron-dense precipitate through the use of a biotinylated secondary antibody followed by processing with an avidin–HRP conjugate and then reaction with DAB. Fluorogold provides sensitive staining of the axonal and somatodendritic compartments but does not undergo transcytosis.

Carbocyanine Dyes

This highly lipophilic family of dyes, which includes DiI (1,1′-dioctadecyl-3,3,3′,3′-tetramethylindocarbocyanine perchlorate) and DiO (3,3′-dioctadecyloxacarbocyanine perchlorate), are fluorescent membrane tracers. Carbocyanine dyes can be applied via pressure injection, placement of dye crystals in the region of interest, or biolistic methods and can also be used postfixative. These lipophilic dyes diffuse laterally (2 mm $month^{-1}$) through the phospholipid bilayers of the cell membrane, giving extensive labeling of cellular processes and hence detailed morphological information. Transsynaptic transfer of these dyes is not a common occurrence unless two cell membranes are very closely associated. *In vivo*, membrane fragments labeled with carbocyanine dyes can also become endocytosed into vesicles, which are then transported bidirectionally, but not transsynaptically. Thus *in vivo*, the detailed membrane labeling typical of carbocyanine dyes is replaced with a granular cytoplasmic labeling that prevents good resolution of cellular processes.

Biocytin

Extensive labeling of neuronal arborization can also be achieved with biocytin, a peptide derivative of biotin. Applied either by iontophoretic injection or as a pellet, biocytin is actively internalized and rapidly transported, predominantly in the anterograde direction, although with large injection volumes, retrograde transport is also seen. The high affinity of biocytin for avidin enables detection either with avidin–HRP followed by DAB staining or through the use of fluorescent avidin conjugates. However, biocytin is rapidly catabolized and therefore is only suitable for short-term tracing.

Fluorescent Microbeads

Introduced in the 1980s, fluorescent microbeads consist of either polystyrene or latex beads that are labeled with dyes of different colors, permitting double labeling experiments. Pressure injection of the microbeads in the brain results in a highly restricted region of uptake, the mechanism of which has not been defined, although it is likely to be due to local phagocytosis. Uptake is followed by retrograde transport of the beads. The microspheres produce stable fluorescent labeling. However, the signal is restricted to the cell soma; thus, information on cellular projections is very limited.

Dextran Amines

Fluorescent or biotinylated dextran amines were first investigated as transport tracers in the late 1980s and early 1990s. Dextrans can be applied either by pressure or iontophoretic injection or via application of dye crystals to the region of interest, where they undergo fluid-phase endocytosis. Transport then occurs in either an anterograde or retrograde direction depending on the molecular weight of the dextran. Dextrans of greater than 10 kDa are predominantly anterogradely transported, whereas dextrans of molecular weight of 3 kDa are transported retrogradely on injection in an acidic vehicle. This differentiation based on molecular weight represents a major advantage over other bidirectional tracers, allowing clearer and unequivocal results. Following standard aldehyde fixation, these hydrophilic polysaccharides are easily detected either by fluorescence or by processing of biotinylated dextran amines with an avidin–HRP conjugate followed by DAB staining to generate an electron-dense product. Dextrans are highly sensitive, nontoxic tracers that provide detailed labeling of fine cellular arborization.

Cholera Toxin B Subunit

Cholera toxin (CT) is a bacterial protein toxin produced by *Vibrio cholerae*, which binds to cellular membranes with high affinity. CT is composed of an active subunit (A), which induces toxicity by continuously activating stimulatory G-proteins (Gs), thereby increasing cyclic adenine monophosphate levels, and a homopentameric binding subunit (B). The B subunit (CTB) binds to the monosialoganglioside GM1 that is clustered in membrane microdomains on the plasma membrane of different cell types, including neurons. Once bound at the nerve terminal, CT becomes internalized in endocytic vesicles and then undergoes retrograde transport (at a rate of 102 mm day^{-1}). Recombinant CTB maintains the specific binding of full-length CT and is completely atoxic, thus constituting an ideal transport probe. CTB was first used as a retrograde tracer, but recently it has been found to also be a very sensitive anterograde transport marker that gives fine definition of neuronal processes and nerve terminals. Application of CTB subunit is either by iontophoresis or pressure injection into the region of interest. Cellular labeling can then be detected with either a fluorescent or a biotinylated secondary antibody or by the use of fluorescently conjugated CTB.

The examples of anterograde and retrograde transport tracers discussed above are well established and widely used. However, their transport, and consequently their labeling, is restricted to the neurons where their uptake occurs, which greatly limits their ability to map neuronal circuitry. Transsynaptic tracers circumvent this limitation by spreading from the initial neurons to synaptically connected second- and third-order neurons, thus providing information on neuronal connectivity.

Transsynaptic Tracers

Plant Lectins

Plant lectins are carbohydrate-binding proteins that have high binding affinity for specific sugar components of glycoproteins and glycolipids on cell membranes. One of the first applications of lectins as neuronal tracers was WGA conjugated to HRP, which significantly enhanced the tracing sensitivity of free HRP. However, *Phaseolus vulgaris* leucoagglutinin, WGA, and barley lectin (BL) are also commonly used as tracers. Application of lectins is by iontophoresis or pressure injection into the region of interest. Following binding at the plasma membrane, lectins are actively internalized and transported predominantly in an anterograde direction, although retrograde transport is also observed. Lectins label neuronal arborization and the cell soma and are also transsynaptically transferred, thus providing information on neuronal connectivity. Although specific antilectin antibodies exist, their direct conjugation with fluorescent dyes, biotin, or HRP represents a better alternative and permits their direct detection by fluorescence microscopy, immunohistochemistry, or enzymatic methods, as discussed previously.

Unfortunately, several disadvantages are associated with lectin-mediated neuronal tracing. Following injection into the brain, lectins cannot be targeted to a specific neuronal subtype, and nonspecific labeling can occur. Furthermore, lectins can be released from neurons and taken up by neighboring glial cells. The bidirectional transport of lectins can also make interpretation of neuronal circuitry data very difficult. A further disadvantage is that WGA administration can induce a severe immune response in the host and marked inflammation at the injection site. However, the use of transgenic technology in combination with plant lectin tracers can minimize some of these issues, as discussed in the section titled 'Genetic targeting with specific neuronal promoters.'

Tetanus Toxin

Tetanus toxin (TeNT) is a bacterial protein toxin produced by *Clostridium tetani*. After entering the bloodstream, TeNT binds with high affinity to a receptor complex at the neuromuscular junction. Neurospecific binding of TeNT involves the interaction with polysialogangliosides of the G1b series, such as GT1b and GD1b (**Figure 1 (a)**), and glycosylphosphatidylinositol-anchored proteins, including Thy-1. TeNT internalization requires the integrity of membrane microdomains, and it is preceded by the lateral sorting of TeNT from the lipid component of its receptor complex. A specialized clathrin-mediated endocytic pathway delivers TeNT into stationary early endosomes that mature into fast retrogradely transported carriers in a mechanism dependent on the small GTPases Rab5 and Rab7. Following its transcytosis into inhibitory interneurons, full-length TeNT blocks neurotransmitter release by cleaving vesicle-associated membrane protein (VAMP)/synaptobrevin, a SNARE protein localized on synaptic vesicles. This induces a spastic paralysis, which is often fatal. *In vivo*, TeNT retrograde transport and trancytosis occur not only in motor neurons/inhibitory interneurons but also in other neuronal networks, such as the pre- and postganglionic neurons innervating the superior cervical ganglion and the avian paravertebral ganglion.

TeNT consists of two associated fragments, a 50 kDa light (L) chain and a 100 kDa heavy (H) chain. The L chain has a zinc endopeptidase activity responsible for the cleavage of VAMP/synaptobrevin. The H chain consists of two domains, the N-terminal domain (H_N), which is involved in the pH-dependent membrane penetration and translocation of the catalytic domain, and the C-terminal portion (H_C) responsible for neurospecific binding. A recombinant H_C fragment (TeNT H_C) retains the high affinity for neuronal membranes and retrograde transport capacity of the native toxin but lacks toxicity. Consequently, TeNT H_C provides a useful biological tool for the investigation of both retrograde axonal transport and neuronal connectivity. Generation of hybrid proteins consisting of TeNT H_C linked to reporters such as green fluorescent protein (GFP), lacZ, or HRP permits the tracing of axonal transport.

The ability of the H_C fragment of TeNT to be endocytosed and transported in a manner similar to that of the native toxin has allowed the development of a real-time transport assay that permits the quantitative analysis of the retrograde transport of fluorescent TeNT H_C in purified primary motor neurons in culture. The speed distribution of TeNT H_C carriers *in vitro* overlaps with that seen *in vivo* (0.8–3.6 $\mu m\, s^{-1}$). The axonal transport of TeNT H_C occurs primarily by a cytoplasmic dynein-mediated process along microtubules, although there is evidence to suggest that fast retrograde transport of TeNT H_C also involves F-actin and myosin motors,

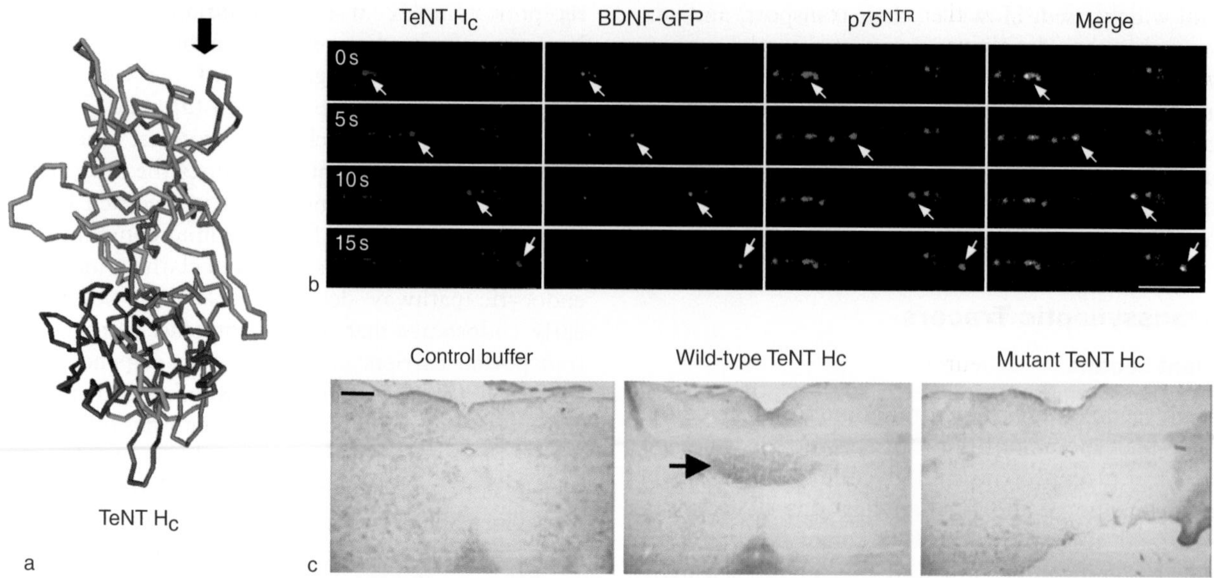

Figure 1 The binding fragment of tetanus toxin is an axonal retrograde marker *in vitro* and *in vivo.* (a) Structure of the TeNT H_C fragment. The black arrow indicates the position of one of the two ganglioside binding pockets, which are necessary for highly specific binding to the neuronal surface and trafficking. For further details, see Lalli G, Bohnert S, Deinhardt K, Verastegui C, and Schiavo G (2003) The journey of tetanus and botulinum neurotoxins in neurons. *Trends in Microbiology* 11(9): 431–437. (b) Axonal retrograde transport of fluorescently tagged. Binding fragment of tetanus toxin (TeNT H_C) in differentiated spinal cord motor neurons in culture. This probe (in blue) is cotransported in axonal carriers with brain-derived nerve growth factor fused with green fluorescent protein (BDNF-GFP) and with an antibody specific for the neurotrophin receptor $p75^{NTR}$. The motor neuron cell body is located out of view to the right. Arrows point out a TeNT H_C carrier positive for the neurotrophin and $p75^{NTR}$. (c) Retrograde labeling of the hypoglossal nucleus after intramuscular injection in the tongue of TeNT H_C. Mice were injected with control buffer, wild-type TeNT H_C, and a TeNT H_C mutant in which the ganglioside binding region highlighted in purple in (a) has been deleted. Only the specimen treated with the wild-type protein shows a strong staining in the region of the hypoglossal nucleus (arrow), which indicates that TeNT H_C undergoes retrograde transport *in vivo.* Scale bar = 5 μm (b), 167 μm (c). (b) Adapted from Deinhardt K, Salinas S, Verastegui C, et al. (2006) Rab 5 and Rab control endocytic sorting along the axonal retrograde transport pathway. *Neuron* 52: 293–305; Sinha K, Box M, Lalli G, et al. (2000) Analysis of mutants of tetanus toxin H_C fragment: Ganglioside binding, cell binding and netrograde axonal transport properties. *Molecular Neurobiology* 37: 1041–1051.

in particular myosin Va. Significantly, this assay has also identified that TeNT H_C is retrogradely transported in vesicles that also contain nerve growth factor, brain derived neurotrophic factor, and their receptors, $p75^{NTR}$ and TrkB (**Figure 1(b)**). Therefore, it appears that TeNT H_C exploits an essential physiological route of transport, which validates this assay as a tool to investigate endogenous retrograde transport in both health and disease. With this *in vitro* assay, a significant deficit in retrograde transport, which may contribute to disease pathogenesis, has been shown in embryonic motor neurons isolated from animal models of motor neuron disease, including the 'legs at odd angles' mouse, which has a mutation in dynein heavy chain, and the mutant $SOD1^{G93A}$ mouse.

Several studies have taken advantage of the specific retrograde transport and transsynaptic transfer of TeNT H_C for the investigation of neuronal circuitry. Early studies based on the detection of TeNT H_C labeled with colloidal gold by electron microscopy established the validity of the method. More recently, injection of either lacZ-TeNT H_C protein or complementary DNA into the tongue of mice produced strong β-galactosidase staining in the hypoglossal nucleus after retrograde transport in the hypoglossal motor neuron axons (**Figure 1(c)**). Furthermore, β-galactosidase staining is evident in the lateral reticular formation and several midbrain projections, such as the mesencephalic trigeminal nucleus, which are second- and third-order neurons of this network. Similarly, to investigate the regulation of locomotor activity, GFP–TeNT H_C was injected into the medial gastrocnemius muscle. Following internalization at the neuromuscular junction, GFP–TeNT H_C was retrogradely transported to the cell soma of motor neurons innervating the medial gastrocnemius muscle. Specific transsynaptic transfer allowed identification of the premotor neurons responsible for controlling the activity of gastrocnemius motor neurons. This tracing technique is not restricted to use in mammalian systems and has been shown to be effective in other organisms.

However, interpretation of circuits and connectivity revealed by the use of TeNT H_C as a neuronal tracer must take into account the dependence of TeNT H_C uptake on synaptic activity. Hence, the tracer will be internalized rapidly at the most active synapses and more slowly at less active synapses via synaptic vesicle recycling. It is possible, therefore, that labeling of weak synaptically connected second-order neurons may coincide with the timing of labeling of strong synaptically connected third-order neurons. Furthermore, the extent of labeling is dependent on the quantity and quality of TeNT H_C initially administered.

Neurotropic Viruses

The application of neurotropic viruses as neuronal tracers exploits their infectious nature and ability to replicate within the target cell, which provides a self-amplifying signal. Furthermore, their ability to cross synapses to synaptically connected cells permits visualization of higher-order neurons and the mapping of neuronal circuits. A variety of viruses, including herpes simplex virus types 1 and 2 (HSV-1 and HSV-2), rabies and pseudorabies virus type 1, adenovirus, and adeno-associated virus, have been used for tracing motor, visual, and auditory neuronal networks *in vivo*. After injection into the brain, neurotropic viruses bind to specific receptors, are endocytosed via a variety of different pathways, and are then transported to the nucleus, where the viral genome is replicated. Viruses can be manipulated to express a capsid protein fused with a GFP variant, or else a GFP variant is translated on viral replication, thus allowing infected cells to be identified easily (**Figure 2**). The majority of neurotropic viruses are retrograde tracers. However, adeno-associated viruses and strain H129 of HSV 1 are predominantly anterograde tracers.

Viral neuronal tracing, however, has several limitations. As a consequence of their virulence, the infection and replication of certain viruses (e.g., herpes virus) can induce neuronal degeneration. Following replication, certain viruses can also spread to local glial cells or nonsynaptically connected neurons, thus generating nonspecific labeling. The extent of neuronal tracing is dependent on the viral strain, the dose, and the postinoculation time. However, use of a virulent strain to increase the tracing capacity may exacerbate neuronal death or increase nonspecific viral labeling. This drawback is partially avoided in case of viruses characterized by a long incubation period before the onset of symptoms (e.g., rabies). Alternatively, attenuated viruses display a limited spread to nonneuronal cells and a longer neuronal survival time postinoculation, thus extending the time available for transsynaptic transfer. It is important to note that certain viruses bind to receptors that are specific to neuronal membranes. For example, canine adenoviruses (CAVs) bind to the neuronal

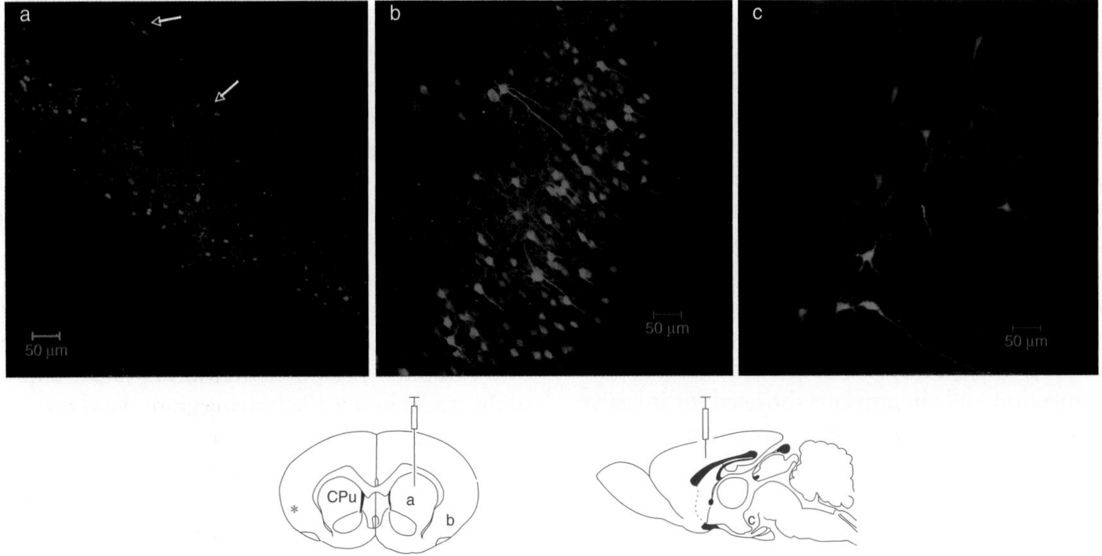

Figure 2 Retrograde transport of adenoviruses in the rat central nervous system. Both a canine adenovirus serotype 2 (CAV-2) vector expressing green fluorescent protein (CAVGFP) and a human adenovirus serotype 5 (HAd5) expressing DsRed2 (AdRFP) were stereotactically coinjected into the rat striatum (10^9 particles for each vector). Rat brains were cut rostrocaudally into 150 µm sections and were screened simultaneously for GFP and RFP expression. Each confocal image represents an approximate depth of 130 µm. At the site of injection (a), there was a dense concentration of AdRFP-transduced striatal astrocytes (~70%) and neurons (~25%) as well as some neurons transduced by both vector (yellow; white arrows). Only CAVGFP underwent significant retrograde transport from the striatum to afferent neurons in the cortex (b) and substantia nigra (c). Schematic inserts show midsagittal and frontal cross sections of the rat brain and the location of the injection site relative to the location of the transduced cells. Red asterisk represents contralateral cortex. CPU, caudate putamen. Adapted from Soudais C, Laplaec-Builhe C, Kissa K, Kremer EJ (2001) Preferential transduction of neurons by canine adenovirus vectors and their efficient retrograde transport *in vivo*. *FASEB Journal* 15: 2283–2285.

human coxsackievirus and adenovirus receptor. Consequently, CAV infection is restricted to neuronal cells in rat spinal cord cultures *in vitro*, in rat striatum *in vivo*, and in human brain biopsies *ex vivo* (**Figure 2**). Similarly, the rabies virus does not infect glial cells. In contrast, human adenovirus serotype 5 predominantly infects glial cells, and strains of HSV-1 and pseudorabies virus can also infect glial cells. The use of viral tracing technology therefore requires careful consideration and understanding of the benefits and limitations of each viral strain.

Genetic Tracing Technology

The development of multiple anterograde and retrograde tracers, as listed above, has provided extensive information on neuronal projections. Furthermore, the advent of transsynaptic labeling has enabled significant advances in neuroanatomical mapping. In spite of their flexibility, these tools also have severe limitations. One of the major drawbacks is their rather cumbersome targeting: The application of such tracers is normally by stereotactic injection or by administration of dye crystals into a chosen brain region. In addition, these methods do not allow the labeling of a specific neuronal population and often give rise to nonreproducible results between animals.

Recent advances in genetic technology have led to the development of novel methods for the mapping of neural networks. Specific labeling of a neuronal subset can be achieved via expression of a transgene encoding a transsynaptic tracer fused to a marker molecule, under the control of a neuronal-specific promoter. Continuous expression of the transgene enhances cellular detection and may promote a greater synaptic transfer, increasing the sensitivity of neuronal mapping. Unambiguous conclusions can therefore be made regarding the connectivity of specific neuronal subsets after monitoring axonal transport of the tracer. Genetic mapping will also prevent the nonspecific labeling that occurs after tracer injection and circumvents the need for invasive brain surgery that may damage the neural circuit under investigation. Furthermore, this technology can be used to investigate the development of neuronal circuits *in utero* and may provide functional insights since transsynaptic transfer of tracers will occur only in active circuits.

Genetic Targeting with Specific Neuronal Promoters

Transgenic expression of WGA or BL under the control of the rat olfactory marker protein gene promoter, which is expressed almost exclusively by olfactory sensory neurons in olfactory epithelium (OE) and vomeronasal epithelium (VNO), has provided reproducible mapping of olfactory circuitry. Lectin immunostaining was evident in the axons, dendrites, and cell soma of OE and VNO neurons, the first-order neurons. Furthermore, transsynaptic transfer permitted the labeling of neurons of the main and accessory olfactory bulbs, which represent the terminal fields of OE and VNO neurons (second-order), and of granule cells of the olfactory cortex (third-order), which are innervated by neurons from the main and accessory olfactory bulbs. However, immunostaining was not restricted to these areas but was also present in several neuromodulatory regions of the olfactory cortex, which indicated that retrograde transport had also occurred. BL expression has also been used to map sensory input in the olfactory cortex. Co-expression of BL with two distinct odorant receptor genes generated a transsynaptic tracer that could be monitored in localized areas of the OE, through the mitral and tufted relay neurons in the olfactory bulbs, to specific clusters of olfactory cortical neurons.

Similarly, following use of an L7 (Pcp2) promoter, WGA messenger RNA was expressed in cerebellar Purkinje cells. WGA was also detected in the synaptically connected deep cerebellar nuclei (second-order) and the red nucleus and thalamic ventrolateral nucleus, the third-order targets of cerebellar efferent axons. Furthermore, insight into neuronal connectivity in the *Drosophila* visual system was revealed by expression of WGA under the GAL4/UAS promoter, which drives expression in the R1–R6 photoreceptors in the retina. WGA immunoreactivity was present in the retina (first-order neurons), in monopolar cells (second-order targets of the photoreceptor cells), and also in the medulla, the target of the monopolar cells (third-order).

Transgenic mice expressing the GFP–TeNT H_C fusion protein have also been used to investigate neuronal circuitry. GFP–TeNT H_C under the control of a calbindin promoter, which drove the expression of the tracer in specific brain regions, has been studied. The transgenes also contained an internal ribosome entry sequence to allow co-expression of the lacZ reporter. This enabled differentiation of the neurons at the origin of the circuit, which expressed both β-galactosidase and GFP–TeNT H_C, and those receiving the protein as a consequence of transcytosis, which were only GFP–TeNT H_C-positive. Beta-galactosidase was detected in Purkinje cells, whereas GFP–TeNT H_C was also present in cerebellar granule cells, basket cells, and Golgi cells (second-order cells) due to the retrograde transport of GFP–TeNT H_C. Expression of a TeNT H_C–GFP fusion protein under the control of the orexin promoter has also been used to investigate the neuronal regulation of orexin-expressing neurons in the lateral hypothalamic area involved in states of sleep and wakefulness. The use of TeNT H_C as a tracer has distinct advantages over lectins since it is only transported retrogradely, whereas lectins are transported bidirectionally.

Potentially viral vectors can also be genetically targeted to specific neuronal populations via insertion of a neuron-type specific promoter into the coding region of the viral glycoprotein gene. Further manipulation via insertion of additional genes, such as β-galactosidase or WGA, into the glycoprotein-coding region will induce selective protein expression. Indeed, in the olfactory system, intranasal application of an adenovirus encoding a WGA transgene under the chromogranin A promoter induced a pattern of lectin expression in the OE neurons (first-order), the main olfactory bulbs (second-order), and the main terminal fields of the olfactory cortex (third-order) similar to that obtained with transgenic expression of lectins under the olfactory marker protein promoter.

Genetic Recombination Strategies

The Cre/loxP recombination system has also been used to direct expression of transsynaptic tracers in selected neuronal populations. Breeding mice that express WGA only after excision of a loxP-flanked lacZ gene with mice expressing Cre recombinase under the control of the L7 (Pcp2) promoter generated offspring expressing WGA selectively in Purkinje cells. Anterograde transport analysis of WGA revealed connections with neurons of the deep cerebellar and vestibular nuclei (second-order) and the red nucleus of the midbrain and the thalamic ventrolateral nucleus (third-order).

This technology has also been applied in combination with viral tracing methods. A pseudorabies virus has been generated that requires Cre mediated recombination for its replication. Subsequent viral infection of transgenic mice that expressed Cre recombinase only in neuropeptide Y-expressing neurons consequently induced viral replication only in this neuronal population. Tracing of the subsequent viral spread has revealed the anatomical location of synaptic inputs regulating these neurons. More recently, a viral transcomplementation method was also employed to restrict viral expression to selected neuronal populations. A pseudotyped, GFP-expressing rabies virus, unable to spread transsynaptically, was used to infect brain slices in which relatively isolated neuronal populations had been transfected with both viral receptors and the viral glycoprotein gene required for transsynaptic spread. Following replication in transfected cells, the virus could spread transsynaptically, in a retrograde direction only, to monosynaptically connected neurons, identified by enhanced GFP expression, but no further due to the lack of glycoprotein in those neurons. Potentially, this technology may be used in combination with single-neuron electroporation to allow the electrophysiological characterization of synaptic contacts of selected neurons.

The tetracycline transactivator system provides an alternative strategy for the genetic regulation of neuronal mapping. In one example, mice transgenic for a TeNT H_C–alkaline phosphatase (AP) fusion protein under the tetracycline operator promoter were bred with mice expressing the tetracycline-controlled transactivator under the control of a cyclic phosphodiesterase splice variant, PDE102A, promoter. Expression of the fusion protein in the offspring was restricted to striatal spiny neurons, although retrograde transfer of TeNT H_C–AP revealed second-order connections with neurons of the ventral tegmental area/substantia nigra pars compacta, the dorsal raphe nucleus, and the mesopontine tegmentum.

Recombination systems can also provide greater control over the onset of transgenic tracer expression. As described above transgenic mice expressing WGA only after Cre recombinase-mediated excision of a lacZ gene have been developed. Microinjection of an adeno-associated virus expressing Cre recombinase and GFP, as a marker of viral infection, into the cerebellar cortex or visual cortex induced expression of the WGA protein in the infected areas and allowed its transsynaptic transfer to connected neurons.

Concluding Remarks

Over recent years the use of transsynaptic tracers has led to significant developments in the anatomical mapping of neural circuitry. Ongoing developments in genetic technology have further facilitated the mapping of functional interactions within selected neuronal populations, allowing their in-depth functional analysis, such as by electrophysiological recordings. The ability to identify specific connections within a given neural network *in utero* will now permit the onset of functionality in specific circuits to be determined. Ultimately, correlation of anatomical circuitry with neuronal functions will lead to a greater understanding of higher brain functions.

See also: Axonal mRNA Transport and Functions; Axonal Transport and ALS; Axonal Transport and Alzheimer's Disease; Axonal Transport and Huntington's Disease; Axonal Transport and Neurodegenerative Diseases; Botulinum and Tetanus Toxins; Retrograde Transsynaptic Influences; Retrograde Neurotrophic Signaling.

Further Reading

Bohnert S, Deinhardt KD, Salinas S, and Schiavo G (2006) Uptake and transport of clostridium neurotoxins. In: Alouf JE, and Popoff MR (eds.) *Bacterial Protein Toxins,* 3rd edn., pp. 390–407. London: Elsevier.

Braz JM, Rico B, and Basbaum AI (2002) Transneuronal tracing of diverse CNS circuits by Cre-mediated induction of wheat germ agglutinin in transgenic mice. *Proceedings of the National Academy of Sciences of the United States of America* 99: 15148–15153.

Callahan CA, Yoshikawa S, and Thomas JB (1998) Tracing axons. *Current Opinion in Neurobiology* 8: 582–586.

DeFalco J, Tomishima M, Liu H, et al. (2001) Virus-assisted mapping of neural inputs to a feeding center in the hypothalamus. *Science* 291: 2608–2613.

Deinhardt K, Salinas S, Verastegui C, et al. (2006) Rab 5 and Rab control endocytic sorting along the axonal retrograde transport pathway. *Neuron* 52: 293–305.

Enquist LW and Card JP (2003) Recent advances in the use of neurotropic viruses for circuit analysis. *Current Opinion in Neurobiology* 13: 603–606.

Kobbert C, Apps R, and Bechmann I (2000) Current concepts in neuroanatomical tracing. *Progress in Neurobiology* 62: 327–351.

Kremer EJ (2004) CAR chasing: Canine adenovirus vectors – all bite and no bark? *Journal of Gene Medicine* 6: S139–S151.

Lalli G, Bohnert S, Deinhardt K, Verastegui C, and Schiavo G (2003) The journey of tetanus and botulinum neurotoxins in neurons. *Trends in Microbiology* 11: 431–437.

Reiner A, Veenman CL, Medina L, et al. (2000) Pathway tracing using biotinylated dextran amines. *Journal of Neuroscience Methods* 103: 23–37.

Roux S, Colasante C, Saint cloment C, et al. (2005) Internalisation of a GFP-tetanus toxin c-terminal fragment fusion protein at mature mouse neuromuscular junctions. *Molecular and Cellular Neuroscience* 30: 572–582.

Sano H, Nagai Y, and Yokoi M (2007) Inducible expression of retrograde transsynaptic genetic tracer in mice. *Genesis* 45: 123–128.

Sinha K, Box M, Lalli G, et al. (2000) Analysis of mutants of tetanus toxin H_C fragments: Ganglioside binding, cell binding and netrograde axonal transport properties. *Molecular Neurobiology* 37: 1041–1051.

Soudais C, Laplace-Builhe C, Kissa K, Kremer EJ (2001) Preferential transduction of neurons by canine adenovirus vectors and their efficient retrograde transport *in vivo. FASEB Journal* 15: 2283–2285.

Ugolini G (1995) Specificity of rabies virus as a transneuronal tracer of motor networks: Transfer from hypoglossal motoneurons to connected second-order and higher order central nervous system cell groups. *Journal of Comparative Neurology* 356: 457–480.

Vercelli A, Repici M, Garbossa D, and Grimaldi A (2000) Recent techniques for tracing pathways in the central nervous system of developing and adult animals. *Brain Research Bulletins* 51: 11–28.

Wickersham IR, Lyon DC, Barnard RJO, et al. (2007) Monosynaptic restriction of transsynaptic tracing from single, genetically targeted neurons. *Neuron* 53: 639–647.

Yoshihara Y (2002) Visualising selective neural pathways with WGA transgene: Combination of neuroanatomy with gene technology. *Neuroscience Research* 44: 133–140.